Welding
Principles and Practices

Welding
Principles and Practices

Sixth Edition

Edward R. Bohnart

WELDING: PRINCIPLES AND PRACTICES, SIXTH EDITION

Published by McGraw Hill LLC, 1325 Avenue of the Americas, New York, NY 10019. Copyright ©2024 by McGraw Hill LLC. All rights reserved. Printed in the United States of America. Previous editions ©2018, 2012, and 2005. No part of this publication may be reproduced or distributed in any form or by any means, or stored in a database or retrieval system, without the prior written consent of McGraw Hill LLC, including, but not limited to, in any network or other electronic storage or transmission, or broadcast for distance learning.

Some ancillaries, including electronic and print components, may not be available to customers outside the United States.

This book is printed on acid-free paper.

1 2 3 4 5 6 7 8 9 LWI 28 27 26 25 24 23

ISBN 978-1-266-73337-6 (bound edition)
MHID 1-266-73337-X (bound edition)
ISBN 978-1-265-86873-4 (loose-leaf edition)
MHID 1-265-86873-5 (loose-leaf edition)

Portfolio Manager: *Beth Bettcher*
Product Developer: *Erin Kamm*
Marketing Manager: *Lisa Granger*
Content Project Managers: *Maria McGreal, Rachael Hillebrand*
Buyer: *Rachel Hirschfeld*
Content Licensing Specialist: *Gina Oberbroeckling*
Cover Image: *Glow Images*
Compositor: *Aptara®, Inc.*

All credits appearing on page or at the end of the book are considered to be an extension of the copyright page.

Library of Congress Cataloging-in-Publication Data

Names: Bohnart, Edward R., author.
Title: Welding: principles and practices / Edward R. Bohnart.
Description: Six edition. | New York, NY : McGraw Hill LLC, [2024] |
  Includes index.
Identifiers: LCCN 2022024592 (print) | LCCN 2022024593 (ebook) | ISBN
  9781266733376 | ISBN 9781265868734 (spiral bound) | ISBN 9781265867294
  (ebook) | ISBN 9781265861926 (ebook other)
Subjects: LCSH: Welding.
Classification: LCC TS227 .S22 2024 (print) | LCC TS227 (ebook) | DDC
  671.5/2—dc23/eng/20220809
LC record available at https://lccn.loc.gov/2022024592
LC ebook record available at https://lccn.loc.gov/2022024593

The Internet addresses listed in the text were accurate at the time of publication. The inclusion of a website does not indicate an endorsement by the authors or McGraw Hill LLC, and McGraw Hill LLC does not guarantee the accuracy of the information presented at these sites.

mheducation.com/highered

Contents

Preface xi
Acknowledgments xiv

UNIT 1

Introduction to Welding and Oxyfuel 1

Chapter 1 History of Welding 2

Overview 2
The History of Metalworking 3
Welding as an Occupation 7
Industrial Welding Applications 7
Review 11

Chapter 2 Industrial Welding 13

Fabrication 14
Maintenance and Repair 14
Industries 14
Review 32

Chapter 3 Steel and Other Metals 34

History of Steel 35
Raw Materials for the Making of Steel 36
The Smelting of Iron 46
Steelmaking Processes 49
Metalworking Processes 65
Metal Internal Structures 70
Physical Properties of Metals 72
Effects of Common Elements on Steel 77
Types of Steel 82
SAE/AISI Steel Numbering System 87
ASTM Numbering System 92
Unified Numbering Designation 92
Types of Cast Iron 92
Aluminum-Making in the World 95

Titanium-Making in the United States 97
Unique Metals 99
Effects of Welding on Metal 99
Review 109

Chapter 4 Basic Joints and Welds 112

Types of Joints 113
The Four Weld Types 113
Weld Size and Strength 114
Weld Positions 120
Strength of Welds 124
Common Weld and Weld-Related Discontinuities 125
Review 134

Chapter 5 Gas Welding 137

Oxyacetylene Welding 137
Gases 139
General Cylinder Handling, Storage, and Operation Safety Concerns 143
Welding Equipment 148
Supporting Equipment 160
Safety Equipment 161
General Safety Operating Procedures 163
Review 164

Chapter 6 Flame Cutting Principles 166

Oxyacetylene and Other Fuel Gas Cutting 166
Oxygen Lance Cutting 175
Review 176

| Chapter 7 | Flame Cutting Practice: Jobs 7-J1–J3 | 178 |

Review of Flame Cutting Principles 179
Cutting Different Metals 181
Cutting Technique 182
Surface Appearance of High Quality Flame Cuts . . . 186
Arc Cutting. 187
Practice Jobs. 187
Job 7-J1   Straight Line and Bevel Cutting 193
Job 7-J2   Laying Out and Cutting Odd Shapes 197
Job 7-J3   Cutting Cast Iron 198
Review 201

| Chapter 8 | Gas Welding Practice: Jobs 8-J1–J38 | 203 |

Sound Weld Characteristics. 204
The Oxyacetylene Welding Flame. 205
Setting Up the Equipment 207
Flame Adjustment 209
Closing Down the Equipment 211
Safety 212
Practice Jobs. 212
Low Carbon Steel Plate 212

Heavy Steel Plate and Pipe 223
Review 229

| Chapter 9 | Braze Welding and Advanced Gas Welding Practice: Jobs 9-J39–J49 | 232 |

Braze Welding 232
Welding Cast Iron 236
Welding of Aluminum 238
Welding Other Metals with the
Oxyacetylene Process. 242
Hard Facing (Surfacing). 244
Review 247

| Chapter 10 | Soldering and Brazing Principles and Practice: Jobs 10-J50–J51 | 250 |

Soldering and Brazing Copper Tubing 250
Soldering 251
Practice Jobs: Soldering. 257
Torch Brazing (TB) 260
Practice Jobs: Brazing 269
Review 274

UNIT 2

Shielded Metal Arc Welding   277

| Chapter 11 | Shielded Metal Arc Welding Principles | 278 |

Process Capability 279
Operating Principles 279
Welding Power Sources 279
Machines for Shielded Metal Arc Welding 283
Multiple-Operator Systems 289
Power Supply Ratings 290
Cables and Fasteners 290
Electrode Holders. 293
Other Electric Arc Processes 294
Personal Safety Equipment 294
Review 298

| Chapter 12 | Shielded Metal Arc Welding Electrodes | 300 |

Introduction 300
Shielded Metal Arc Welding Electrodes 301

Functions of Electrode Coverings 302
Composition of Electrode Coverings. 302
Identifying Electrodes 305
Electrode Selection 307
Specific Electrode Classifications 315
Packing and Protection of Electrodes 325
Review 328

| Chapter 13 | Shielded Metal Arc Welding Practice: Jobs 13-J1–J25 (Plate) | 330 |

Introduction 330
Approach to the Job 331
Learning Welding Skills 333
Practice Jobs. 338
Job 13-J1   Striking the Arc and Short
Stringer Beading 339
Job 13-J2   Stringer Beading. 341

Job 13-J3	Weaved Beading	343
Job 13-J4	Stringer Beading	345
Job 13-J5	Weaved Beading	347
Job 13-J6	Welding an Edge	349
Job 13-J7	Welding an Edge Joint	350
Job 13-J8	Welding a Lap Joint	352
Job 13-J9	Welding a Lap Joint	354
Job 13-J10	Stringer Beading	356
Job 13-J11	Stringer Beading	357
Job 13-J12	Welding a Lap Joint	359
Job 13-J13	Welding a Lap Joint	361
Job 13-J14	Welding a T-Joint	363
Job 13-J15	Welding a T-Joint	365
Job 13-J16	Welding a T-Joint	367
Job 13-J17	Welding a T-Joint	369
Job 13-J18	Stringer Beading	370
Job 13-J19	Weaved Beading	372
Job 13-J20	Weaved Beading	374
Job 13-J21	Welding a Lap Joint	375
Job 13-J22	Welding a T-Joint	378
Job 13-J23	Welding a T-Joint	380
Job 13-J24	Welding a T-Joint	382
Job 13-J25	Welding a T-Joint	383
Review		385

Chapter 14 Shielded Metal Arc Welding Practice: Jobs 14-J26–J42 (Plate) 388

Introduction		388
Practice Jobs		389
Job 14-J26	Stringer Beading	389
Job 14-J27	Weave Beads	391
Job 14-J28	Welding a Single-V Butt Joint (Backing Bar Construction)	393
Job 14-J29	Welding a T-Joint	395
Job 14-J30	Welding a Single-V Butt Joint (Backing Bar Construction)	397
Job 14-J31	Welding a Square Butt Joint	399
Job 14-J32	Welding an Outside Corner Joint	400
Job 14-J33	Welding a Single-V Butt Joint	403
Job 14-J34	Welding a T-Joint	406
Job 14-J35	Welding a T-Joint	408
Job 14-J36	Welding a T-Joint	410
Job 14-J37	Welding a T-Joint	412
Job 14-J38	Welding a Single-V Butt Joint (Backing Bar Construction)	414
Job 14-J39	Welding a Square Butt Joint	416
Job 14-J40	Welding an Outside Corner Joint	418
Job 14-J41	Welding a T-Joint	420
Job 14-J42	Welding a Single-V Butt Joint (Backing Bar Construction)	422
Review		424

Chapter 15 Shielded Metal Arc Welding Practice: Jobs 15-J43–J55 (Plate) 428

Introduction		428
Practice Jobs		429
Job 15-J43	Welding a Single-V Butt Joint	429
Job 15-J44	Welding a T-Joint	431
Job 15-J45	Welding a T-Joint	434
Job 15-J46	Welding a Lap Joint	435
Job 15-J47	Welding a Lap Joint	437
Job 15-J48	Welding a Single-V Butt Joint	439
Job 15-J49	Welding a T-Joint	441
Job 15-J50	Welding a T-Joint	443
Job 15-J51	Welding a Single-V Butt Joint	444
Job 15-J52	Welding a Coupling to a Flat Plate	447
Job 15-J53	Welding a Coupling to a Flat Plate	449
Job 15-J54	Welding a Single-V Butt Joint (Backing Bar Construction)	450
Job 15-J55	Welding a Single-V Butt Joint	452
Tests		454
Review		465

Chapter 16 Pipe Welding and Shielded Metal Arc Welding Practice: Jobs 16-J1–J17 (Pipe) 469

Introduction	469
Shielded Metal Arc Welding of Pipe	476
Joint Design	477
Codes and Standards	483
Practice Jobs	498
Tools for Pipe Fabrication	511
Review	516

UNIT 3

Arc Cutting and Gas Tungsten Arc Welding — 521

Chapter 17 Arc Cutting Principles and Arc Cutting Practice: Jobs 17-J1–J7 — 522

Arc Cutting. — 522
Learning Arc Cutting Skills. — 535
Practice Jobs. — 535
Job 17-J1 Square Cutting with PAC — 538
Job 17-J2 Bevel Cutting with PAC — 538
Job 17-J3 Gouging with PAC — 538
Job 17-J4 Hole Piercing — 539
Job 17-J5 Shape Cutting with PAC. — 539
Job 17-J6 Gouging with CAC-A. — 543
Job 17-J7 Weld Removal with CAC-A — 544
Review — 546

Chapter 18 Gas Tungsten Arc and Plasma Arc Welding Principles — 549

Gas Shielded Arc Welding Processes — 549
Gas Tungsten Arc Welding — 554
TIG Hot Wire Welding — 580
Review — 590

Chapter 19 Gas Tungsten Arc Welding Practice: Jobs 19-J1–J19 (Plate) — 593

Gas Tungsten Arc Welding of Various Metals — 593
Joint Design and Practices — 604
Setting Up the Equipment — 607
Safe Practices. — 608
Arc Starting — 609
Welding Technique — 611
Practice Jobs. — 613
Review — 628

Chapter 20 Gas Tungsten Arc Welding Practice: Jobs 20-J1–J17 (Pipe) — 630

Joint Design — 631
Porosity in Gas Tungsten Arc Welds — 635
Practice Jobs. — 635
Review — 650

UNIT 4

Gas Metal Arc, Flux Cored Arc, and Submerged Arc Welding — 655

Chapter 21 Gas Metal Arc and Flux Cored Arc Welding Principles — 656

Overview — 656
GMAW/FCAW Welding Equipment — 665
Summary — 702
Review — 706

Chapter 22 Gas Metal Arc Welding Practice with Solid and Metal Core Wire: Jobs 22-J1–J23 (Plate) — 708

Operating Variables That Affect Weld Formation — 708
Weld Defects — 717
Safe Practices. — 720

Care and Use of Equipment — 722
Welding Technique — 726
Process and Equipment Problems — 727
Practice Jobs. — 727
Gas Metal Arc Welding of Other Metals — 745
Review — 749

Chapter 23 Flux Cored Arc Welding Practice (Plate), Submerged Arc Welding, and Related Processes: FCAW-G Jobs 23-J1–J11, FCAW-S Jobs 23-J1–J12; SAW Job 23-J1 — 751

Flux Cored Wire Welding — 751
Flux Cored Arc Welding—Gas Shielded Practice Jobs. — 763

Flux Cored Arc Welding—Self-Shielded 768
Flux Cored Arc Welding—Self-Shielded
Practice Jobs.................................. 769
Automatic or Mechanized
Welding Applications........................ 771
Submerged Arc Welding Semiautomatic
Practice Job.................................. 780
Choice of Welding Process 784
Review 785

Chapter 24 Gas Metal Arc Welding Practice: Jobs 24-J1–J15 (Pipe) 790

Industrial Applications of GMAW
Pipe Welding 790
Use of Equipment and Supplies............... 792
Welding Operations 794
Practice Jobs................................. 795
Review 811

UNIT 5

High Energy Beams, Automation, Robotics, and Weld Shop Management 815

Chapter 25 High Energy Beams and Related Welding and Cutting Process Principles 816

Introduction 816
High Energy Beam Processes 817
Summary 828
Review 830

Chapter 26 General Equipment for Welding Shops 832

Screens and Booths 832
Work-Holding Devices....................... 833
Preheating and Annealing Equipment 845
Sandblasting Equipment 846
Spot Welder 846
Hydraulic Tools 847
Power Squaring Shears 850
Small Hand Tools............................ 851
Portable Power Tools........................ 853
Machine Tools 860
Review 868

Chapter 27 Automatic and Robotic Arc Welding Equipment 870

Arc Control Devices......................... 872
Arc Monitoring 875
Controller Communication 881
Robotic Arc Welding Systems................ 882
Robot Ratings................................ 885
Robot Programming......................... 886
Training, Qualification, and Certification 887
Conclusion 889
Review 889

Chapter 28 Joint Design, Testing, and Inspection 891

Joint Design 892
Code Welding................................ 898
Nondestructive Testing (NDT)............... 902
Destructive Testing 920
Visual Inspection 936
Summary 943
Review 947

Chapter 29 Reading Shop Drawings 949

Introduction 949
Standard Drawing Techniques................ 952
Types of Views.............................. 962
Review 967

Chapter 30 Welding Symbols 978

Fillet Welds and Symbol 980
Weld-All-Around Symbol 983
Groove Welds................................ 983
Edge Welds 986
Combination Symbols 986
Contour Symbol.............................. 986
Applications of Welding Symbols............ 987
Review 988

Chapter 31 Welding and Bonding of Plastics	**994**

Know Your Plastics 995
Characteristics of Plastics 998
Welding as a Method of Joining
Plastics 999
Inspection and Testing 1010
Practice Jobs 1014
Tack Welding 1015
Review 1024

Chapter 32 Safety	**1026**

Safety Practices: Electric Welding Processes 1027
Safety Practices: Oxyacetylene Welding
and Cutting 1043
Review 1052

Appendices

A: Conversion Tables 1058
B: Illustrated Guide to Welding Terminology 1071
C: Welding Abbreviation List 1086
D: Major Agencies Issuing Codes,
    Specifications, and Associations 1088
E: Sources of Welding Information 1090
F: Metric Conversion Information
    for the Welding Industry 1091
Glossary 1093
Index 1109

Preface

Welding: Principles and Practices, 6e, is both a revision and an expansion of the *Theory and Practice of Arc Welding,* which was first published in 1943. The previous editions have enjoyed success during the years as a major text used in the training of welders by industry and the schools.

This book is designed to be used as the principal text for welding training in career schools, community technical college systems, technical junior colleges, engineering schools, and secondary technical schools. It is also suitable for on-the-job training and apprenticeship programs. It can serve as a supplementary text for classes in building construction, metalworking, and industrial technology programs.

Welding: Principles and Practices, 6e, provides a course of instruction in welding, other joining processes, and cutting that will enable students to begin with the most elementary work and progressively study and practice each process until they are skilled. Both principles and practice are presented so that the student can combine the "why" and the "how" for complete understanding.

The chapters have been arranged into sections to facilitate training programs with reduced contact time segments. Each section maintains the twofold approach of Welding Principles, in which students are introduced to fundamentals that will enable them to understand what is taking place in the application of the various processes, and Welding Practices, where they learn the necessary hands-on skills.

Welding: Principles and Practices, 6e, presents the fundamental theory of the practice in gas, arc, gas-shielded and self-shielded processes, welding, brazing, soldering, and plastic welding processes. The various applications of these processes are covered such as manual, semiautomatic, mechanized, automatic, and robotic methods. Current industrial practices are cited with use of various national welding codes and standards. The content is based on the American Welding Society along with other leading welding authorities.

Welding is an art, technology, and engineering science. It requires the skillful manipulation of the weld pool, a thorough knowledge of welding processes, and the characteristics of the type of material being used. Students can be assured of success if they are willing to spend the time required in actual practice work and the study of the principles presented in this text until they thoroughly understand their significance. Faithful adherence to this course of study will enable them to master the current industrial material joining and cutting processes thoroughly.

The Sixth Edition of *Welding: Principles and Practices* Includes:

Connect

The sixth edition of *Welding* is now available online with Connect, McGraw Hill Education's integrated assignment and assessment platform. Connect also offers SmartBook for the new edition, which is the first adaptive reading experience proven to improve grades and help students study more effectively. All of the title's website and ancillary content is also available through Connect, including:

- A Student Workbook with additional questions and practice to promote students' learning.
- An Instructor's Manual that includes answers to the Review questions and Student Workbook questions.
- Lecture Slides for instructor use in class.

Updated Content

Every chapter complies with the most current AWS Standards. The terminology is current so students know the most recent terms to use when they begin to practice. Additional information on many different topics including, safety, lead welding, arc wandering, gas metal arc braze welding, and more are also included in the text.

Instructors
Student Success Starts with You

Tools to enhance your unique voice

Want to build your own course? No problem. Prefer to use an OLC-aligned, prebuilt course? Easy. Want to make changes throughout the semester? Sure. And you'll save time with Connect's auto-grading, too.

65%
Less Time Grading

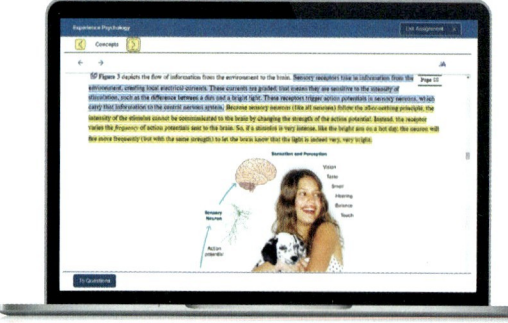

Laptop: Getty Images; Woman/dog: George Doyle/Getty Images

A unique path for each student

In Connect, instructors can assign an adaptive reading experience with SmartBook® 2.0. Rooted in advanced learning science principles, SmartBook 2.0 delivers each student a personalized experience, focusing students on their learning gaps, ensuring that the time they spend studying is time well-spent.
mheducation.com/highered/connect/smartbook

Affordable solutions, added value

Make technology work for you with LMS integration for single sign-on access, mobile access to the digital textbook, and reports to quickly show you how each of your students is doing. And with our Inclusive Access program, you can provide all these tools at the lowest available market price to your students. Ask your McGraw Hill representative for more information.

Solutions for your challenges

A product isn't a solution. Real solutions are affordable, reliable, and come with training and ongoing support when you need it and how you want it. Visit **supportateverystep.com** for videos and resources both you and your students can use throughout the term.

Students
Get Learning that Fits You

Effective tools for efficient studying

Connect is designed to help you be more productive with simple, flexible, intuitive tools that maximize your study time and meet your individual learning needs. Get learning that works for you with Connect.

Study anytime, anywhere

Download the free ReadAnywhere® app and access your online eBook, SmartBook® 2.0, or Adaptive Learning Assignments when it's convenient, even if you're offline. And since the app automatically syncs with your Connect account, all of your work is available every time you open it. Find out more at **mheducation.com/readanywhere**

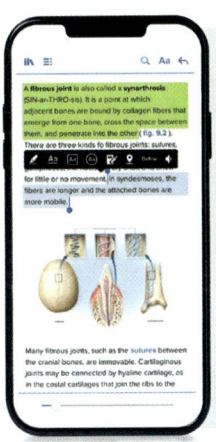

"I really liked this app—it made it easy to study when you don't have your textbook in front of you."

- Jordan Cunningham, Eastern Washington University

iPhone: Getty Images

Everything you need in one place

Your Connect course has everything you need—whether reading your digital eBook or completing assignments for class, Connect makes it easy to get your work done.

Learning for everyone

McGraw Hill works directly with Accessibility Services Departments and faculty to meet the learning needs of all students. Please contact your Accessibility Services Office and ask them to email accessibility@mheducation.com, or visit **mheducation.com/about/accessibility** for more information.

Acknowledgments

Throughout the process of revising *Welding: Principles and Practices,* 6e, many individuals and organizations contributed their thoughts, counsel, and expertise to the project.

I would also like to express thanks to the instructors who reviewed this textbook, thereby ensuring that it is clear, focused, accurate, and up-to-date.

Reviewers

Karsten Illg
College of Lake County

Thomas Looker
Edison State

Joshua O'Neal
Barstow Community College

Pete Stracener
South Plains College

Joel Ziegler
Northland Community and Technical College

Technical Editors

Richard Bremen
Barstow Community College

Troy Miller
Central Community College

Finally, I would like to thank all of the individuals and corporations that aided in the extensive photo research program necessary for this edition. Because of your help, *Welding: Principles and Practices,* 6e, contains hundreds of new and updated color photos and art pieces.

- ACF Industries
- Agfa Corporation
- Allegheny Ludlum Corporation
- American Welding Society
- Ansul/Tyco Fire Protection Products
- Arch Machines
- Arcos Corporation
- Atlas Welding Accessories
- Baldor Electric Company
- BHP Billiton
- Binzel-Abicor
- Black & Decker, Inc.
- Bluco Corporation
- Boeing
- BUG-O
- Bunting Magnetic Company
- Caterpillar, Inc.
- Circlesafe Aerosol/Circle Systems, Inc.
- Clausing Industrial, Inc.
- CM Industries, Inc.
- Combustion Engineering Company
- Computer Weld Tech., Inc.
- Contour Sales Corporation
- Crane Company
- CRC-Evans Pipeline International, Inc.
- D. L. Ricci Corp
- Dakota Creek Industries
- De-Sta Company
- DoAll Company
- Donaldson Company
- Drader Manufacturing
- Dreis and Krump Manufacturing Company
- Dukane
- E.H. Wachs Company
- Editorial Image, LLC
- Elderfield and Hall, Inc.
- Electro-Technic Products, Inc.
- Empire Abrasive Equipment Company
- Enerpac, Inc.
- Enrique Vega
- ESAB Welding and Cutting Products
- Fibre-Metal Products Company
- Foerster Instruments

- Fox Valley Technical College
- Fronius International GmbH
- G.A.L. Gage Company
- Gasflux Company
- General Electric Company
- General Welding & Equipment Company
- Gentec
- Gullco
- Haney Technical Center
- Heritage Building Systems
- Hobart Brothers Company
- Hornell, Inc. Speedglas
- Hossfeld Manufacturing Company
- Howden Buffalo, Inc.
- Hypertherm, Inc.
- IMPACT Engineering
- Industrial Plastics Fabrication
- Interlaken Technology Corp
- ITW Jetline—Cyclomatic
- Jackson Products Company
- Jackson Safety, Inc.
- John E. White III
- Kaiser Aluminum & Chemical Corporation
- Kamweld Products Company
- Kromer Cap Company, Inc.
- Laramy Products Company, Inc.
- Lenco dba NLC, Inc.
- Lincoln Electric Company
- Lockheed Martin Aeronautics
- MAG IAS, LLC
- Magna Flux Corp
- Magnatech Limited Partnership
- Malcom
- Manitowoc Company, Inc.
- Manufactured Housing Institute
- Mathey Dearman
- McGraw-Edison
- Metal Fabricating Institute
- Micro Photonics, Inc.
- Miller Electric Mfg. Company
- Milwaukee Electric Tool Corporation
- Mine Safety Appliances Co.
- Mitutoyo
- Modern Engineering Company
- Motoman, Inc.
- NASA
- National Welding Equipment Company
- Navy Joining Center
- NES Rentals
- Newage Testing Instruments, Inc.
- Nooter Corporation
- North American Manufacturing Company
- Northeast Wisconsin Technical College
- NovaTech
- Pandjiris
- Phoenix International
- Pipefitters Union, St. Louis, MO
- Piping Systems, Inc.
- Plumbers and Pipefitters Union, Alton, IL
- Praxair, Inc.
- Prior Scientific
- Rexarc
- Robvon Backing Ring Company
- Rogers Manufacturing Inc.
- Schuler AG
- Seelye, Inc.
- Sellstrom
- Servo-Robot Corporation
- Shaw Pipeline Services
- Sheet Metal and Air Conditioning Contractors' National Association
- Smith Equipment
- South Bend Lathe Co.
- St. Louis Car. Company
- Stanley G. Flagg & Company
- Sypris Technologies, Inc.—Tube Turns Division
- Team Industries, Inc.
- TEC Torch Company
- The Welding Encyclopedia
- Thermacote Welco
- Thermadyne Industries, Inc.
- Tim Anderson
- Tony DeMarco
- Torit Donaldson Company
- TransCanada Pipelines Ltd.
- UA Local 400
- United Association
- United States Steel Corporation
- Uvex Safety
- Wegener
- Welding Engineering Company, Inc.
- Wells Manufacturing Company
- Widder Corporation
- Wilson Industries, Inc.
- Wilson Products
- Wisconsin Wire Works
- Woodard/CC Industries
- Wyatt Industries
- Zephyr Manufacturing Company

Union Recognition

Recognition is due the United Association of Plumbers and Pipe Fitters National as well as the locals in Kaukauna, Wisconsin, St. Louis, Missouri, and Alton, Illinois; the International Association of Bridge, Structural, Ornamental, and Reinforcing Iron Workers; and the Sheet Metal Workers International Association. Their focus on skill training for the workforce in quality, productivity, and safety ensures that the practices presented in the text are current.

About the Author

Edward R. Bohnart is the principal of Welding Education and Consulting located in Wisconsin. He launched his consulting business after a successful career with Miller Electric Manufacturing Company, where he directed training operations. He is a graduate of the Nebraska Vocational Technical College in welding and metallurgy and has studied at both the University of Nebraska at Omaha and the University of Omaha. Previous certifications include AWS-SCWI, CWE, CWS, CWSR, CRAW-T, and AWS Certified Welder.

Bohnart was selected in the 2011 Class of Counselors of the American Welding Society, and he is also an AWS Distinguished Member and national past President. He remains active with the SkillsUSA organization and is past chair of the AWS Skills Competition Committee, which conducted the USOpen Weld Trials to select the TeamUSA welder for the WorldSkills Competition. He was the United States of America's Welding Technical Expert for the WorldSkills Competition from 1989 to 2009. Bohnart chaired the AWS C5 Committee on Arc Welding and Cutting Processes and remains on the committee as an advisor.

The American Welding Society has recognized Ed Bohnart with the National Meritorious Award, George E. Willis Award, and Plummer Memorial Educational Lecture Award. The Wisconsin State Superintendents Technology Education Advisory Committee has acknowledged him with the Technology Literacy Award. The state of Nebraska Community College System has appointed him Alumnus of the Year, and the Youth Development Foundation of SkillsUSA has honored Bohnart with the SkillsUSA Torch Carrier Award.

Ed has been active on the Edison Welding Institute Board of Trustees and on the American Institute of Steel Constructions Industry Round Table. He has served on the industrial advisory boards for Arizona State University, The University of Wisconsin–Stout, Fox Valley Technical Colleges, and the Haney Technical Center industrial advisory boards.

He has lectured at a number of major institutions, such as the Massachusetts Institute of Technology, Colorado School of Mines, Texas A&M, Arizona State University, and the Paton Institute of Welding, Kiev, Ukraine.

UNIT 1

Introduction to Welding and Oxyfuel

Chapter 1
History of Welding

Chapter 2
Industrial Welding

Chapter 3
Steel and Other Metals

Chapter 4
Basic Joints and Welds

Chapter 5
Gas Welding

Chapter 6
Flame Cutting Principles

Chapter 7
Flame Cutting Practice: Jobs 7-J1–J3

Chapter 8
Gas Welding Practice: Jobs 8-J1–J38

Chapter 9
Braze Welding and Advanced Gas Welding Practice: Jobs 9-J39–J49

Chapter 10
Soldering and Brazing Principles and Practice: Jobs 10-J50–J51

1
History of Welding

Overview

You are about to begin the learning process of preparing yourself for a position in one of the fastest growing industries in the world of work—the *welding industry*.

Welding is the joining together of two pieces of metal by heating to a temperature high enough to cause softening or melting, with or without the application of pressure, and with or without the use of filler metal. Any filler metal used has either a melting point approximately the same as the metals being joined or a melting point that is below these metals but above 800 degrees Fahrenheit (°F).

New methods, new applications, and new systems have continued to develop over the last few decades. Continuing research makes welding a dynamic leader in industrial processes. The industry has made tremendous progress in a short period of time. Furthermore, it has made a major contribution toward raising the standard of living of the American people. By simplifying and speeding up industrial processes and making it possible to develop environmentally sound industries like wind and solar power, hybrid power vehicles, plants to produce organic fuels, and continued development in nuclear, fossil fuels along with continued space exploration and utilization, it has increased the world's supply of goods, Fig. 1.1.

Chapter Objectives

After completing this chapter, you will be able to:

1-1 Explain the history of metalworking and welding.
1-2 Explain the development of modern welding.
1-3 Give details of the mission of welding in the industrial process.
1-4 Describe the diverse welding processes.
1-5 List the various welding occupations.
1-6 Define welder qualifications and characteristics.
1-7 Express the duties and responsibilities of a welder.
1-8 Recognize welder safety and working conditions.
1-9 Identify trade associations and what responsibility they have in the welding industry.
1-10 Establish goals to keep you up-to-date in the field.

Fig. 1-1 Use of natural energy sources (green energy) such as solar, wind turbines, and bio-fuels like ethanol are getting a tremendous amount of interest in the way of research, development, and real applications. As they continue to develop, other issues will need to be dealt with, such as ROI. Welding plays a very important role in the manufacture of these green energy sources. (top) Stocktrek Images/Getty Images; (middle) Bear Dancer Studios/Mark Dierker; (bottom) Mark A. Dierker/McGraw Hill

Welding is usually the best method to use when fastening metal. If you want to build something made of metal, you can fasten the parts by using screws or rivets, bending the parts, or even gluing the parts. However, a quality, long-lasting, attractive, safe product is best fabricated by using one of the many types of prevailing welding processes.

The History of Metalworking

Metalworking began when early people found that they could shape rocks by chipping them with other rocks. The first metal to be worked was probably pure copper because it is a soft, ductile metal that was widely available. **Ductile** means easily hammered, bent, or drawn into a new shape or form. Excavations in Egypt and in what is now the United States indicate the use of copper as early as 4000 B.C. and before 2000 B.C., respectively. More than 4,000 years ago, copper mines on the peninsula of Sinai and the island of Cyprus were worked. Welding began more than 3,000 years ago when hot or cold metals were hammered to obtain a forge weld. Forged metals, bronze and iron, are mentioned in the Old Testament.

Archaeologists have determined that bronze was developed sometime between 3000 and 2000 B.C. Iron became known to Europe about 1000 B.C., several thousand years after the use of copper. About 1300 B.C. the Philistines had four iron furnaces and a factory for producing swords, chisels, daggers, and spearheads. The Egyptians began to make iron tools and weapons during the period of 900 to 850 B.C. After 800 B.C. iron replaced bronze as the metal used in the manufacture of utensils, armor, and other practical applications. A welded iron headrest for Tutankhamen (King Tut) was crafted around 1350 B.C.

The famous Damascus swords and daggers were made in Syria about 1300 B.C. These were sought after because of their strength and toughness. Their keen edge was likely capable of severing heavy iron spears or cutting the most delicate fabric floating in the air. The swords were made by forge-welding iron bars of different degrees of hardness, drawing them down, and repeating the process many times.

The working of metals—copper, bronze, silver, gold, and iron—followed one another in the great ancient civilizations. By the time of the Roman Empire, the use of iron was common in Europe, the Near East, and the Far East. The Chinese developed the ability to make steel from wrought iron in A.D. 589. The Belgians were responsible for most of the progress made in Europe, due to the high degree of metalworking skill developed by their workers. By the eighth century, the Japanese manufactured steel by repeated welding and forging and controlled the amount of carbon in steel by the use of fluxes. They produced the famous Samurai sword with a blade of excellent quality and superior handiwork.

The blast furnace was developed for melting iron about the years A.D. 1000 to 1200. One such furnace was in the Province of Catalonia in Spain. The fourteenth and fifteenth centuries saw great improvements in the design of blast furnaces. The first cast iron cannon was produced in the early 1600s.

About the middle of the eighteenth century, a series of inventions in England revolutionized the methods of industry and brought on what later came to be known as the Industrial Revolution. Our present factory system of mass production was introduced. An American, Eli Whitney, developed the idea of interchanging parts in the manufacture of arms. By the beginning of the nineteenth century, the working of iron with the use of dies and molds became commonplace. Early in the twentieth century, Henry Ford was involved in developing the assembly line method for manufacturing automobiles.

Early Developments in Welding

At the beginning of the nineteenth century, Edmund Davy discovered **acetylene**, a gas that was later used in oxyacetylene welding, heating, and cutting. The electric arc was first discovered by Sir Humphry Davy in 1801 while he was conducting experiments in electricity. He was concerned primarily with the possibilities of the use of the arc for illumination. By 1809, he had demonstrated that it was possible to maintain a high voltage arc for varying periods of time. By the middle of the nineteenth century, workable electrical-generating devices were invented and developed on a practical basis. These inventions were the forerunner of the present arc welding process.

The first documented instance of fusion welding was done by Auguste de Meritens in 1881. He welded lead battery plates together with a carbon electrode. Two of his pupils, N. Benardos and S. Olszewski, saw the possibilities of this discovery and experimented with the arc powered by batteries that were charged from high voltage dynamos. After four years of work, they were issued a British patent for a welding process using carbon electrodes and an electric power source. Applications of the process included the fusion welding of metals, the cutting of metals, and the punching of holes in metal. Although they experimented with solid and hollow carbon rods filled with powdered metals, the solid electrodes proved more successful. Repair welding was the primary goal of the inventors.

Bare metal electrode welding was introduced in 1888 by N. G. Slavianoff, a Russian. His discovery was first recognized in Western Europe in 1892. C. L. Coffin was one of the pioneers of the welding industry in the United States. In 1889, he received a patent on the equipment and process for flash-butt welding. In 1890, he received additional patents for spot-welding equipment. In 1892, working without knowledge of Slavianoff's work, he received a patent for the bare metal electrode *arc welding* process. By the turn of the century, welding was a common method of repair. At this time, welding was given added impetus by the development of the first commercial *oxyacetylene welding* torch by two Frenchmen, Foresche and Picard. Bare electrode welding became the prevailing electric arc welding method used in the United States until about 1920.

Bare metal electrode welding was handicapped because the welds produced by these electrodes were not as strong as the metal being welded and the welding arc was very unstable. In 1907, Kjellberg, a Swedish engineer, received a patent covering the electrode-coating process. The coating was thin and acted only as a stabilizer of the arc rather than as a purifier of the weld metal. It produced welds that were little better than those made with bare electrodes. In 1912, Kjellberg received another patent for an electrode with a heavier coating made of asbestos with a binder of sodium silicate. See Fig. 1-2. Benardos patented a process in 1908 that has come into popular use in the past few decades. This is the **electroslag** process of welding thick plates in one pass.

Welding technology and its industrial application progressed rather slowly until World War I. Prior to that time, it was used chiefly as a means of maintenance and repair. The demands of the war for an increased flow of goods called for improved methods of fabrication.

ABOUT WELDING

Shipbuilding
Through 1945, some 5,171 vessels of all types were constructed to American Bureau of Shipping standards during the Maritime Commission wartime shipbuilding program. At this time in shipbuilding history, welding was replacing riveting as the main method of assembly.

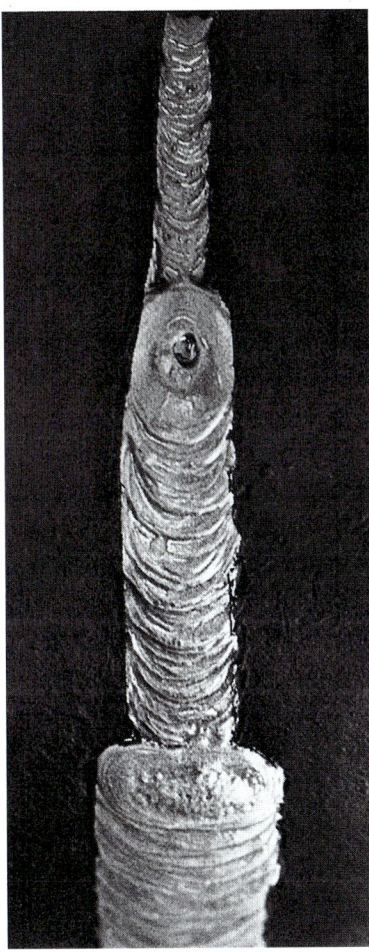

Fig. 1-2 The ability to make multipass welds such as this one, on plate and pipe, led to the growth of the industry. Welds are sound and have uniform appearance. *Edward R. Bohnart*

At the end of World War I, welding was widely accepted. Research on coated electrodes through the 1920s resulted in electrode coatings and improved core wire. This significant development was the main reason for the rapid advancement of the stick welding process. This term has now been superseded by the term *shielded metal arc welding* (SMAW). The development of X-raying goods made it possible to examine the internal soundness of welded joints which indicated a need for improved methods of fabrication.

The Development of Modern Welding

During the postwar period, the design of welding machines changed very little. Since welding was first done with direct current (d.c.) from battery banks, it was only natural that as welding machines were developed, they would be d.c. machines. In the late 1920s and during the 1930s, considerable research was carried on with alternating current (a.c.) for welding. The use of a.c. welding machines increased through the early 1930s. One of the first high frequency, stabilized a.c. industrial welding machines was introduced in 1936 by the Miller Electric Manufacturing Company. The a.c. welding machines have since become popular because of the high rate of metal deposition and the absence of arc blow.

World War II spurred the development of inert gas welding, thus making it possible to produce welds of high purity and critical application. A patent was issued in 1930 to Hobart and Devers for the use of the electric arc within an inert gas atmosphere. The process was not well received by industry because of the high cost of argon and helium and the lack of suitable torch equipment.

> **SHOP TALK**
>
> **Beams**
> Beams used in bridges must be welded on both sides. In automated systems, a second station can handle the reverse side, or a turnover station is used to get the beam back to be sent through a second time.

Russell Merideth, an engineer for the Northrop Aircraft Company, was faced with the task of finding an improved means of welding aluminum and magnesium in the inert atmosphere. Because of a high burnoff rate, the magnesium procedure was replaced by a tungsten electrode, and a patent was issued in 1942. Later in 1942, the Linde Company obtained a license to develop the *gas tungsten arc welding* (GTAW) [or *tungsten inert gas* (TIG)] *process*, also known as HELIARC, used today, Fig. 1-3. The company perfected a water-cooled torch capable of high amperage.

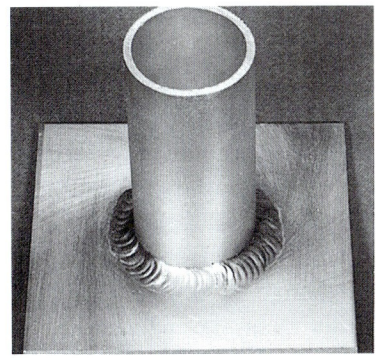

Fig. 1-3 An aluminum weld made using the TIG process. The welding of aluminum is no longer a problem and can be done with the same ease as that of steel. *Edward R. Bohnart*

GTAW welding was first done with rotating d.c. welding machines. Later, a.c. units with built-in high frequency were developed. In about 1950, selenium rectifier type d.c. welding machines came into use, and a.c.-d.c. rectifier welding machines with built-in frequency for GTAW welding became available in the 1950s. Since that time, the Miller Electric Manufacturing Company has developed the Miller controlled-wave a.c. welder for critical welds on aircraft and missiles. Now, many manufacturers of welding machines produce square-wave a.c. machines.

The use of aluminum and magnesium increased at a rapid rate as a result of (1) the development of GTAW welding, and (2) the desirable characteristics of reduced weight and resistance to corrosion. As the size of weldments increased, thicker materials were employed in their construction. It was found that for aluminum thicknesses above 1/4 inch, GTAW welding required preheating. Since this was costly and highly impractical for large weldments, a number of welding equipment manufacturers engaged in the search for another welding process.

In 1948, the U.S. patent office issued a patent for the **gas metal arc welding (GMAW)** process. The GMAW term superseded the earlier terms of *metal inert gas* (*MIG*) and *metal active gas* (MAG).

The GMAW process concentrates high heat at a focal point, producing deep penetration, a narrow bead width, a small heat-affected zone, and faster welding speeds resulting in less warpage and distortion of the welded joint and minimum postweld cleaning. The use of GMAW has increased very rapidly; it is now used in virtually all industries. A GMAW or similar process is responsible for over 70 percent of welds being performed today. In the early 1950s, the gas shielded flux cored arc welding (FCAW) process was developed, Fig. 1-4. It was referred to as "dual shield" as it had a flux but also required external gas shielding. Late in the 1950s, self-shielded flux cored wires were introduced. And in the early 1970s, all position flux cored wires became available. Metal cored wires came along shortly after this. The solid wire, metal cored wire, and flux cored wire use nearly the same equipment; however, since flux cored wires produce a slag that covers the entire weld, it is considered a separate process.

During the 1980s and continuing today, rapid changes are evolving in the welding industry as engineers devise more advanced filler metal formulas to improve arc performance and weld quality on even the most exotic of materials. Even though our history is vague in the areas of welding and filler metal development, it has shown that advancements are inevitable and will continue, such as exotic multiple gas mixes, state-of-the-art electrodes, on-board computers, hybrid processes, and robotic welding. Some processes were developed for limited applications

Fig. 1-4 Two different sized production fillet welds on steel made with the flux cored arc welding process. Edward R. Bohnart

and are used to fill a particular need. Other methods are evolving that may significantly change the way welds will be made in the future.

The following processes involve the use of the electric arc:

- Arc spot welding
- Atomic-hydrogen welding
- Electrogas
- Plasma arc welding
- Stud welding
- Submerged arc welding
- Underwater arc welding

Other specialized processes include:

- Cold welding
- Electron beam welding
- Explosive welding
- Forge welding
- Friction welding
- Friction stir welding
- Laser welding
- Oxyhydrogen welding
- Thermit welding
- Ultrasonic welding
- Welding of plastics

Today, there are over 90 welding processes in use. The demands of industry in the future will force new and improved developments in machines, gases, torches, electrodes, procedures, and technology. The shipbuilding, space, and nuclear industries conduct constant research for new metals, which in turn spurs research in welding. For example, the ability to join metals with nonmetallic materials is the subject of much effort. As industry expands and improves its technology, new welding processes will play an indispensable part in progress.

Currently, five welding associations provide guidance and standards related to the welding industry.

- American National Standards Institute (ANSI)
- American Petroleum Institute (API)
- American Society of Mechanical Engineers (ASME)
- American Welding Society (AWS)
- American Bureau of Shipping (ABS)

Welding as an Occupation

A student needs to learn all phases of the trade. Welding, reading drawings, math, and computer knowledge will secure a successful career. Many qualified welders are certified by the AWS, ASME, and API. The tests are difficult and require many hours of practice.

Because welders hold key positions in the major industries, they are important to the economic welfare of our country. Without welding, the metal industry would be seriously restricted; many of the scientific feats of the past and the future would be impossible. As long as there are metal products, welders will be needed to fabricate and repair them.

Fig. 1-5 Welding is generally considered a nontraditional occupation for women. However, it can be a very lucrative and in-demand skill for those women choosing this career path. A procedure is being used setting up a plasma arc gouging operation.
Andersen Ross/The Image Bank/Getty Images

JOB TIP

Job Hunting
Looking for a job is a job! When you begin, make a list of what you plan to do in the next week. Assess what kind of job you want. As you complete items on your list, you not only will be closer to your goal, but you also will be in control of the job-hunting process and will be less stressed.

Welding is gender friendly, Fig. 1-5. Thousands of women are employed throughout the industry. Many women find the work highly satisfying and are paid well at a rate equivalent to that of men.

Welding is done in nearly every country in the world. You may wish to work in the oil fields of the Near East or in our own country. You may wish to work in some jungle area of South America or Africa, constructing buildings, power plants, pipelines, or bridges. Our many military installations throughout the world offer jobs for civilian workers. Employment opportunities for welders are plentiful in all parts of the United States.

Keep in mind that the field of welding can offer you prestige and security. It can offer you a future of continuous employment with steady advancement at wages that are equal to other skilled trades and are better than average. It can offer you employment in practically any industry you choose and travel to all parts of the world. It is an expanding industry, and your chances for advancement are excellent. Welders have the opportunity to participate in many phases of industrial processes, thus giving them the broad knowledge of the field necessary for advancement to supervisory or technical positions.

Industrial Welding Applications

Welding is not a simple operation. The more than 90 different welding processes are divided into three major types: arc, gas, and resistance welding. A number of other types, such as induction, forge, thermit, flow welding, and brazing are used to a somewhat lesser extent.

Resistance welding includes spot welding, seam welding, flash welding, projection welding, and other similar processes that are performed on machines. These welding areas are not the subject of this text. Because of the specialized nature of the machines, operators are usually taught on the job. They are semiskilled workers who do not need

Fig. 1-6 Welding in the vertical position. Miller Electric Mfg. Co.

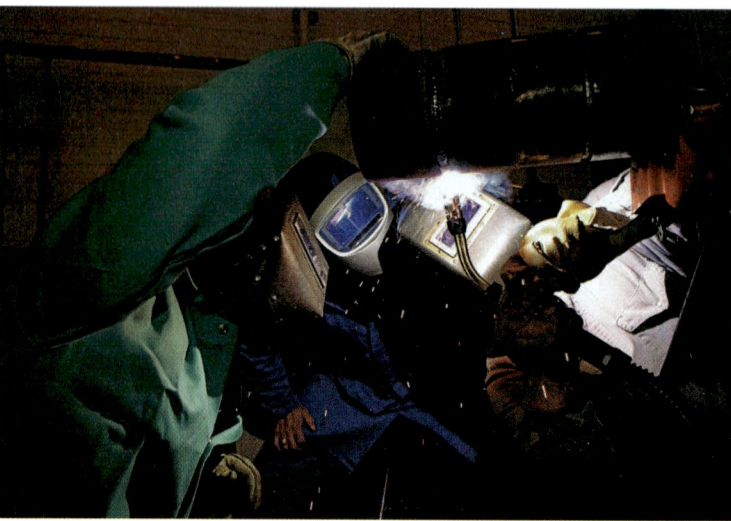

Fig. 1-7 Instructor observing students practicing for a 5G position pipe weld test. The welder is working out of the overhead position on the pipe and getting into the vertical position. The progression of the weld is uphill. The flux cored arc welding process is being used and is being applied in a semiautomatic fashion. Miller Electric Mfg. Co.

specific hands-on welding skills. The arc and gas welding processes will be extensively covered later in this text.

In a sense, welders are both artists and scientists. Arc and gas welders have almost complete control of the process. Much of their work demands manipulative skill and independent judgment that can be gained only through training and a wide variety of job experience. They must know the properties of the metals they weld; which weld process to use; and how to plan, measure, and fabricate their work. They must use visualization skills and be precise, logical, and able to use their heads as well as their hands. Most welders are expected to be able to weld in the vertical and overhead positions, Figs. 1-6 and 1-7, as well as in the flat and horizontal positions.

Gas welders may specialize in oxyacetylene or GTAW processes. Some welders are skilled in all the processes. You should acquire competence in shielded metal arc SMAW, GTAW, and GMAW processes for both plates and pipes.

Qualifications and Personal Characteristics

The standards are high in welding. In doing work in which lives may depend on the quality of the welding—high-rise buildings, bridges, tanks and pressure vessels of all kinds, aircraft, spacecraft, and pipelines—welders must be certified for their ability to do the work, and their

Fig. 1-8 Using a method of weld inspection known as magnetic-particle testing in pipe fabrication. This non-destructive method followed by radiograph and/or ultrasonic testing assures weld soundness for critical pipe welds. Mark A. Dierker/McGraw Hill

work is inspected, Figs. 1-8 and 1-9. Welders are required to pass periodic qualification tests established by various code authorities, insurance companies, the military, and other governmental inspection agencies. Certifications are issued according to the kind and gauge of metal and the specific welding process, technique, or procedure used. Some welders hold several different certifications simultaneously.

Fig. 1-9 Workers using a crane to lift a cask filled with highly radioactive fuel bundles at a Hanford, Washington, nuclear facility. The construction of this type of vessel relies heavily upon welding. U.S. Department of Energy, The Monroe Evening News/AP Images

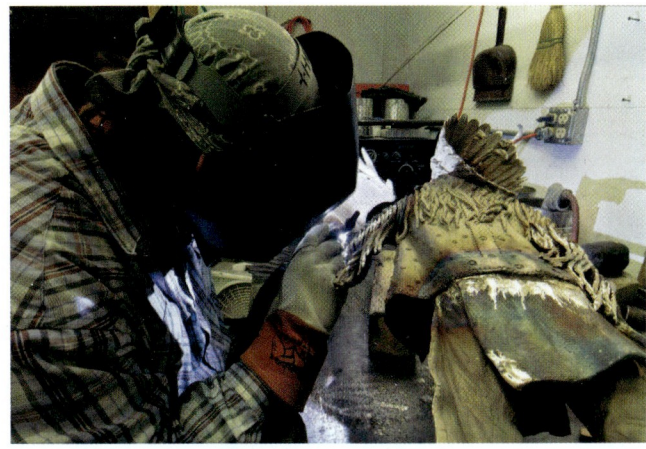

Fig. 1-10 A large amount of art metalwork is done with welding processes. Leon Werdinger/Alamy Stock Photo

The welder must perform certain basic tasks and possess certain technical information in order to perform the welding operation. In making a gas weld, the welder attaches the proper tip to the torch and adjusts the welding regulators for the proper volume and pressure of the gases. The welder must also regulate the flame according to the needs of the job.

For electric arc welding, the welder must be able to regulate the welding machine for the proper welding current and select the proper electrode size and type, as well as the right shielding gas.

Welding requires a steady hand. The welder must hold the torch or electrode at the proper angle, a uniform distance from the work, and move it along the line of weld at a uniform speed.

During the welding process, the welder should use visualization skills to form a mental picture of how the weld will be created. Although much of the work is single pass, welds made on heavy material often require a number of passes side by side and in layers according to the specified weld procedure.

Welders must also be able to cut metals with the oxyacetylene cutting torch and with the various cutting procedures involving the plasma arc cutting machine. Flame cutting is often the only practical method for cutting parts or repairing steel plate and pipe. **Plasma arc cutting** is used to cut all types of metals. Proper use of an electric or pneumatic grinder will save many hours in the welding process.

The master welder is a master craftsperson, Fig. 1-10. Such a person is able to weld all the steels and their alloys, as well as nickel, aluminum, tantalum, titanium, zirconium, and their alloys and claddings. From heavy pressure vessels requiring 4-inch plate to the delicate welding of silver and gold, the welds are of the highest quality and can be depended upon to meet the requirements of the job.

The following welding occupations require a high school education:

- Welding operator
- Welder fitter
- Combination welder
- Master welder
- Welding supervisor
- Welding analyst
- Inspector
- Welding supervisor
- Welding superintendent
- Equipment sales
- Sales demonstrator
- Sales troubleshooter
- Welding instructor
- Robotics welder operator
- Job or fabrication shop owner

> **SHOP TALK**
>
> **Medical Alert**
> The technology of medical heart pacemakers continues to change. Some pacemakers are less likely to be prone to interference by electromagnetic fields. People who weld and have pacemakers are safer if there are other people nearby to help if they have problems. Waiting 10 seconds between each weld may be a good strategy for those with pacemakers.

Fig. 1-11 Welders in the construction industries are called upon to weld in many unusual positions. Here a welder is making an attachment to a building beam in the overhead position. The shielded metal arc welding process is being used and is being applied in the typical manual fashion. Note the safety gear and fall protection devices. Mofles/iStock/Getty Images

Certain welding occupations also require a college education:

- Welding engineer (metallurgical)
- Welding development engineer
- Welding research engineer
- Welding engineer
- Technical editor
- Welding professor
- Certified welding inspector (AWS/CWI)
- Corporation executive
- Owner of welding business
- Sales engineer

Many people in the welding occupations listed entered the industry as welders and were able to improve their positions by attending evening classes at a university or community college.

Safety and Working Conditions

Welders work on many kinds of jobs in almost any environment. They may do light or heavy welding, indoors or outdoors, in spacious surroundings or cramped quarters. Often they work in awkward positions in boiler shops, shipyards, tanks, and piping systems. The work may be extremely noisy (hearing protection will be necessary), and welders may have to work on scaffolds high off the ground (necessitating the use of a safety harness), Fig. 1-11. On some jobs there may be considerable lifting, tugging, and pushing as equipment and materials are placed in position.

A large number of unsafe situations must be of concern to the welder who is conscious of the need to work in a safe environment. Very often accidents are caused as a result of some small, relatively unimportant condition. Extremely dangerous hazards usually get the attention of the welder and are, therefore, rarely a cause of accidents.

Job hazards may include fire danger, burns, "sunburn" from electric arcs, noxious fumes from materials vaporized at high temperatures, eyestrain, welders flash, and electric shock. These hazards can be minimized or eliminated by the use of the proper protective clothing and safety shoes, welding hood, face shields, goggles, respiratory equipment, and adequate ventilation. When performing jobs, welders always take precautionary measures for their own safety and the safety of others in the area.

You are encouraged to study the various safety practices and regulations presented in this text. Safety precautions related to specific processes are presented in the principle chapters (Chapters 1–6, 10, 11, 12, 18, 21, 25–32). Safe welding technique and the safe use of equipment are given in the practice chapters (Chapters 7–10, 13–17, 19, 20, 22–24). Before you begin to practice welding, you should read Chapter 32, Safety, which summarizes the safety measures described elsewhere and presents the precautions to be followed both in the school shop and in industry.

There are several ways of helping to secure your place in this fast-paced field. These methods can assist you in staying current with the most recent changes in technology and help you network with other professionals.

1. Read trade journals, service manuals, textbooks, and trade catalogs.
2. Join associations such as the American Welding Society.
3. Research topics on the Internet.
4. Trade tips with your peers.

ABOUT WELDING

Welding Processes

The welding process using electron beams was first developed in the 1950s by the French Atomic Energy Commission, by J. A. Stohr. During this same time, the Russians were perfecting a method of solid-state joining called friction welding. In the United States, General Motors started using an electroslag welding process.

CHAPTER 1 REVIEW

Multiple Choice

Choose the letter of the correct answer.

1. When did humans learn the art of welding? (Obj. 1-1)
   a. Early 1990s
   b. Around the birth of Christ
   c. Between 3000 and 2000 B.C.
   d. Welding started between World Wars I and II

2. Name four metals that were used by early metalworkers. (Obj. 1-1)
   a. Copper, bronze, silver, gold
   b. Zinc, pewter, aluminum, lead
   c. Silver, mercury, vanadium, gold
   d. Cast iron, steel, brass, tin

3. Which metal was probably the first to be worked by early metalworkers? (Obj. 1-1)
   a. Pewter
   b. Gold
   c. Copper
   d. Tin

4. When was fusion welding, as we know it, first developed? (Obj. 1-1)
   a. In 1888 by N. G. Slavianoff
   b. In 1892 by C. L. Coffin
   c. In 1881 by Auguste de Meritens
   d. In 1930 by Hobart & Devers

5. Electric arc welding using an electrode was developed around what period? (Obj. 1-1)
   a. 1880–1900
   b. 1930–1942
   c. 1750–1765
   d. 1950–1965

6. In what country was a patent first issued for electric arc welding? (Obj. 1-1)
   a. France
   b. China
   c. Russia
   d. United States

7. What invention gave the electric arc welding process its greatest boost? (Obj. 1-1)
   a. Covered electrodes
   b. Oxyacetylene gas mixture
   c. Workable electric generating devices
   d. Both a and c

8. Oxyacetylene welding was developed around what period? (Obj. 1-1)
   a. 1720–1740
   b. 1890–1900
   c. 1930–1942
   d. 1950–1965

9. Using American Welding Society Standards, name four popular welding processes in use today. (Obj. 1-2)
   a. SMAW, GTAW, GMAW, ESW
   b. MCAW, CAW, EBW, OHW, LBW, ARTW
   c. SSW, ROW, FLB, AAW
   d. GLUEW, STKW, GASW, MIGW

10. When was a patent issued for the GTAW process? (Obj. 1-2)
    a. 1936
    b. 1942
    c. 1948
    d. 1965

11. When was a patent issued for the GMAW process? (Obj. 1-2)
    a. 1936
    b. 1942
    c. 1948
    d. 1965

12. What is welding? (Obj. 1-3)
    a. Hammering two pieces of metal together until they become one
    b. Using rivets or screws to attach metal
    c. Bending and shaping metal
    d. Joining together two pieces of metal by heating to a temperature high enough to cause softening or melting, with or without the application of pressure and with or without the use of filler metal

13. Welding is _____ and there are many jobs available for both men and women. (Obj. 1-3)
    a. Gender friendly
    b. Nonskilled
    c. Easy learning
    d. Filler type

14. In electric arc welding, which of the following does the welder *not* have to regulate? (Obj. 1-4)
    a. Cruise control
    b. Welding current
    c. Electrode
    d. Shielding gas

15. Even with the proper equipment, which of the following would be very difficult to weld? (Obj. 1-4)
    a. Aluminum
    b. Magnesium

c. Stainless steel
d. All of these

16. Which of the following lists *classifications* of welding occupations? (Obj. 1-5)
    a. Combination welder, welder fitter, welding inspector, welding instructor, welding engineer
    b. Junk yard welder, wanna-be welder, art welder, stick welder
    c. Inside welder, outside welder, underwater welder, upside-down welder
    d. Flat welder, vertical welder, horizontal welder, overhead welder

17. Which of the following is *not* a welding occupation? (Obj. 1-5)
    a. Certified welding inspector
    b. Pilot technician
    c. Instructor
    d. Technical editor

18. Welders are required to pass periodic qualification tests established by various _____. (Obj. 1-6)
    a. Code authorities
    b. Training agencies
    c. Insurance companies
    d. Both a and c

19. If you are a competent welder, you need to know _____ skills. (Obj. 1-7)
    a. Drawing
    b. Math
    c. Layout skills
    d. All of these

20. Job hazards can be minimized or eliminated by the use of _____. (Obj. 1-8)
    a. Protective clothing
    b. Face shields
    c. Adequate ventilation
    d. All of these

21. Which of the following is *not* a trade association related to the welding industry? (Obj. 1-9)
    a. AWS
    b. ASME
    c. AUPS
    d. API

22. Establishing goals such as _____ will help secure your achievement as a skilled welder. (Obj. 1-10)
    a. Join a professional organization
    b. Nonstop classroom work
    c. Friendly equipment
    d. Trade vocation

Review Questions

Write the answers in your own words.

23. Is welding a recent industrial process? Explain. (Obj. 1-1)
24. Name four metals that were used by early metalworkers. Which metal was the first to be worked? (Obj. 1-1)
25. When did the manufacture of steel begin? (Obj. 1-1)
26. In what country was the patent for electric arc welding first issued? (Obj. 1-2)
27. What invention gave electric arc welding its greatest boost? (Obj. 1-2)
28. Name four important welding processes. (Obj. 1-4)
29. Is it true that industry uses MIG/MAG welding only for special applications because of its instability? Explain. (Obj. 1-4)
30. Name 10 occupational classifications in the welding industry. (Obj. 1-5)
31. Name three welding occupations that require a college degree. (Obj. 1-5)
32. Which welding process contributed most to aluminum welding? (Obj. 1-10)

INTERNET ACTIVITIES

Internet Activity A

Search on the Web for books about welding at England's Cambridge International Science Publishing. Make a list of books that sounds interesting to you.

Internet Activity B

Using the Internet, search for safety and health guidelines for welding and write a report on your findings. You may want to try the American Welding Society's Web site: www.aws.org.

Design credits: Cinema and movie icon: Denis Meshkov/123RF; and Illustration of Welding icons: Mr. Nachai Sorasee/123RF

2

Industrial Welding

Chapter Objectives

After completing this chapter, you will be able to:

2-1 Name the two major functions welding has in industry.

2-2 Name several industries that have found welding to be an advantage.

2-3 Explain why welding plays an important part in manufacturing.

2-4 Discuss how companies save thousands of dollars by using welding for maintenance and repair.

2-5 Explain why welding replaced riveting in the fabrication of pressure vessels.

It may be said that welding has two major functions in industry: (1) as a means of fabrication and (2) for maintenance and repair. It would be difficult to find a single industry that does not use welding in either of these classifications. It is common for a great many industries to use the process in both capacities.

The following industries have found welding to be an advantage:

- Aircraft
- Automotive
- Bridge building
- Construction equipment
- Farm equipment and appliances
- Furnaces and heating equipment
- Guided missiles and spacecraft
- Jigs and fixtures
- Machine tools
- Military equipment
- Mining equipment
- Oil drilling and refining equipment
- Ornamental iron work
- Piping
- Quality food service equipment
- Railroad equipment

- Residential, commercial, and industrial construction
- Sheet metal
- Steel mill equipment
- Tanks and boilers
- Tools and dies
- Watercraft

Fabrication

Welding fabrication has grown rapidly because of its speed and economy. Welding plays an important part in manufacturing processes for the following reasons:

- *Greater design flexibility and lower design costs* Welded design affords an easy, quick means of meeting the functional requirements. This freedom results in lower costs and improvements in the service life of the product.

- *An elimination of patterns* Welded designs are built directly from standard steel shapes. Since **patterns** are not required, this saves the cost of pattern drawings, pattern making, storage, and repairing.

- *Lower cost of material* Rolled steel is a stronger, stiffer, more uniform material than castings. Therefore, fewer pounds are required to do an equivalent job. Also, since rolled steel costs one-quarter to one-half as much as a casting, the cost of material is cut by as much as three-quarters by using welded steel fabrication. By replacing riveting, welded fabrication saves more than 35 percent. Welded steel construction eliminates connecting members such as gusset plates, simplifies drafting, and cuts material costs and weight from 15 to 25 percent. Welding permits the use of simple **jigs** and **fixtures** for speeding layouts and fabrication.

- *Fewer worker-hours of production* Roundabout casting procedures are eliminated, machining operations are minimized, and worker-hours are not wasted because of defective castings that must be rejected. Special machines can be produced with only slight modifications of the standard design, saving time and money. With welding and today's highly efficient fabrication methods, production is straightforward and faster.

- *Absorption of fixed charges* Instead of purchasing parts from an outside concern, the company may make them in its own shop. The additional work applied to existing production facilities and supervision absorbs overhead costs, thereby resulting in more efficient production. This extra work means more jobs for the welder.

- *Minimized inventory and obsolescence charges* Inventory for welding is approximately 10 percent of that required for casting. The standard steel parts used for welding may be purchased on short notice from any steel mill or jobber. New design developments do not make stocked material obsolete.

> **JOB TIP**
>
> **Occupations**
> Employment can be found in many types of businesses that supply project equipment:
> 1. Stamping
> 2. Finishing
> 3. Forming
> 4. Fabricating
> 5. Tooling
> 6. Assembly

Maintenance and Repair

Hundreds of companies save thousands of dollars by using welding for maintenance and repair as follows:

- *The addition of new metal* Hard-facing worn parts usually produce a part that is more serviceable than the original at a substantial saving over replacement cost.

- *Repair and replacement of broken parts* Immediate repair by welding forestalls costly interruptions in production and saves expensive replacements, Fig. 2-1.

- *Special needs* Production equipment, shop fixtures, and structures of many kinds can be adapted to meet particular production requirements.

Industries

Aircraft

Wood, fabric, and wire were the principal materials to be found in the first airplanes. With the change to metal construction, the industry became one of the foremost users of the welding process in all its forms. Aircraft welding was first tried in 1911 and used in warcraft production by the Germans in World War I. Some few years after the war, the tubular steel fuselage was developed, and production lines were set up.

Today welding is universally accepted by the aircraft, missile, and rocket industries. The development of supersonic aircraft and missiles, involving increased stresses, higher temperatures, and high speeds, have presented fabrication problems that only welding can meet.

Fig. 2-1 A welding instructor demonstrates bronze repair of a damaged gear tooth on a sintered manufactured gear blank. Mark A. Dierker/McGraw Hill

- Brackets
- Control columns, quadrants, and levers
- Armor plate assemblies
- Gun mounts
- Gas and liquid tanks
- Fuselage wing and tail assemblies
- Engine cowlings
- Wheel boots
- Ammunition boxes and chutes

Fig. 2-3 Welded jigs and fixtures are used with various welding processes. In this case they are being used for a friction stir weld on an orange peel section of an end cap for a large aluminum fuel storage vessel. NASA

Fig. 2-2 A 5th generation F-35 Lightning II Joint Strike Stealth Fighter, capable of Mach speed with vertical takeoff and landing capability. Currently considered the world's most advanced fighter plane. U.S. Air Force

Fig. 2-4 Welding is an essential tool in the construction of the flight and nonflight hardware for our deep space exploration as to Mars. NASA

Welding processes employed in the aerospace industries, Figs. 2-2, 2-3, 2-4, and 2-5, include all types of **fusion** and **solid-state welding (SSW), resistance welding (RW), brazing,** and **soldering.** Aircraft welders performing manual welding operations are required to have **qualification certification.**

Welding is used in the fabrication of the following aircraft units:

- Pulley brackets
- Structural fitting, mufflers, and exhaust manifolds
- Axle and landing gear parts and assemblies
- Struts and fuselages
- Engine bearers, mounts, and parts

Industrial Welding **Chapter 2** 15

Fig. 2-5 An automatic friction stir welding machine weld being observed on an aerospace component. NASA

- Mounting brackets for radios and other equipment
- Oil coolers and heaters
- Nose cones and rocket shells
- Space capsules
- Space shuttles

Automotive

Welding processes for the manufacture of passenger cars were first introduced during World War II. Since that time, the automobile industry has employed welding on a large scale.

Welding is the method of fabrication for the whole automobile. It is the joining process used to build the body, frame, structural brackets, much of the running gear, and parts of the engine. Welding also is a necessary process in the service and repair of automotive equipment.

Welding is often used to construct fire-fighting equipment, Figs. 2-6 and 2-7. Fire trucks have welded tanks, bodies, frames, aerial supports, and a number of brackets.

Welding is used extensively in military automotive construction and in the construction of all types of vehicles, Fig. 2-8. Many people are not familiar with the many types of military automotive equipment since much of this equipment is of a secret nature. However, this is not the case with freight carriers. Most people recognize the many types of trucks that can be seen on the highways. You may have wondered about the great variety of designs, sizes, and shapes. These would not be possible but for the flexibility of the welding process and the use of standard sheets and bars. Individual designs and needs may be handled on a mass-production basis.

Many dollars have been saved by passenger car owners and truck operators by the application of welding as a method of repair. Alert mechanics who were quick to realize the utility of the process have applied it to automotive repairs of all types: engine heads, engine blocks, oil pans, cracked and broken frames, engine and body brackets, and body and fender repairs.

Construction Machinery

The highway to Alaska was built in nine months. This 1,600-mile highway, through what had been considered impassable terrain, would have been impossible without imagination, daring, and proper equipment. Construction crews encountered many serious problems. Huge tonnage

SHOP TALK

Classic Cars

Restoring a classic car may require separating rusty panels and using arc spot or plug welding. To do this, drill out each spot weld and hammer the panels out to the correct shape. Then clean them and make them ready for reassembly. Finally, refill the holes with appropriate metal from an electrode.

Fig. 2-6 These life- and property-saving vehicles must perform in all types of conditions. Welding plays an essential role in their manufacture. Carbon steel, stainless steel, and aluminum are all typically used in fire-fighting vehicles. Mark A. Dierker/McGraw Hill

Fig. 2-7 Note the two water cannons being used to fight this fire. One is also equipped with a basket the fire fighters can use to rescue people. Light-weight aluminum is extensively used and welded to accommodate this type of service. Mark A. Dierker/McGraw Hill

Fig. 2-8 Robotic welders assemble a vehicle chassis. Glow Images

had to be hauled over frozen tundra, mud, and gumbo. Replacement and repair centers were located hundreds of miles away from some construction sites. Both the climate and the job required equipment that could take heavy punishment.

Equipment manufacturers met the challenge with welded equipment: pullers, pushers, scrapers, diggers, rollers, haulers, and graders. This equipment had to meet several requirements. Irregularly shaped parts and movable members had to be very strong, yet light, so that economical motive power could be employed. Relatively low first cost was important, but more important was the need for strength, rigidity, and light weight.

The design of an earth-moving unit, fabricated entirely by the welded method from mill-run steel plates and shapes, reduced the weight of the total earth-moving machine from 15 to 20 percent over the conventional method of manufacturing. The welded joint, which fuses the edges of the parts, was substituted for heavy reinforcing sections involved in the other common methods of joining parts. The greater strength and rigidity of rolled steel, combined with the fact that a welded joint unifies the parts to produce a rigid and permanent unit, the joints of which are at least as strong as the section joined, paved the way for more powerful tractors and for larger earth-moving units. A further reduction of weight is now affected by the use of high tensile, low alloy steels. The additional savings of 15 to 25 percent may be added to the load-carrying capacity of the machines. Welded, rolled-steel construction also lends itself to repair and maintenance. Breakdowns can be repaired in the field by the welding process.

America continues its vast road building and repair program over the entire country and is also clearing vast land areas for many types of construction projects. These jobs would be impossible to accomplish without the tremendous advances that have been made in earth-moving equipment. Today thousands of yards of earth can be moved in a fraction of the time and for a fraction of the cost formerly required with nonwelded equipment.

Some idea of the tremendous job that is being done can be gained from the following comparisons. One giant off-highway truck is powered by a 3,400-horsepower diesel engine. When fully loaded, it holds 290 cubic yards of earth and weighs 390 tons. The design operating weight is 1,230,000 pounds, Fig. 2-9. The body and box-beam frame are constructed of high tensile strength steel, completely fabricated by welding. Assuming that one wheelbarrow will hold 150 pounds of dirt, it would take 5,200 loads to remove the same quantity of earth. Of course, there is no way to compare the hauling distance these trucks can cover in a unit of time compared with the wheelbarrow. Figure 2-10 shows a scraper bowl that can move 44 cubic yards of dirt when fully loaded. If the same amount of earth were moved in the wheelbarrow cited previously, it would take 789 loads to finish the job. Figure 2-11 shows a large crane capable of moving 5.5 million pounds in one lift. It is designed to lift large pressure vessels into place. It can also position wind turbine generating towers and other large structures.

Fig. 2-9 An off-highway truck used for moving earth. Note its size in comparison with the worker. Martin Barraud/Getty Images

Fig. 2-11 This type of large-capacity crawler crane can lift up to 5.5 million pounds in a single lift. This design employs an innovative lift-enhancing mechanism, which eliminates the need for a counterweight wagon. This feature, called the VPC (variable position counterweight), never touches the ground and extends or retracts as needed by the crane's lift. The Manitowoc Company, Inc.

Fig. 2-10 A scraper bowl that can move up to 44 cubic yards of earth. iStock/Getty Images Plus

For video that shows the type of shovel that is used to fill this type of truck, please visit www.mhhe.com/welding.

ABOUT WELDING

Robotic Welding Cells

Some industries, such as automaking, are now using integrated, preengineered robotic welding cells. Each cell is the size of a pallet. The cell has a programmable robot arm with a welding torch. Cells can accomplish GMAW, GTAW, and plasma and laser-cutting methods. They have safety features such as hard perimeter fencing, an arc glare shield, and light curtains.

Household Equipment

The application of welding to household equipment plays an important part in today's economy. Welded fabrication is popular in those industries engaged in the manufacture of tubular metal furniture, kitchen and sink cabinets, sinks, furnaces, ranges, refrigerators, and various kinds of ornamental ironwork. Much of this work is done with RW, GTAW, GMAW, and brazing processes.

Welded fabrication permits the use of stainless steels, aluminum, and magnesium—materials that can provide light weight, strength and rigidity, and long life. Welding is adaptable to mass-production methods and saves material. In addition, it permits flexibility of design and contributes to the pleasing appearance of a product, Fig. 2-12.

Fig. 2-12 An example of all-welded outdoor furniture. The welding processes used in fabrication include RW, SMAW, GMAW, GTAW, and brazing. This example is welded aluminum.
Tanu Stock Photo/iStock/Getty Images

Jigs and Fixtures

Welding is a valuable aid to tooling in meeting requirements for mass production, and it affords outstanding economies. For these reasons welded jigs and fixtures are used universally in the aerospace industry, Fig. 2-3, page 15. A standard differentiation of jigs from fixtures is that a jig is what mounts onto a workpiece, while a fixture has the workpiece placed on it, into it, or next to it. The blue structures in the figure are fixtures and the yellow structures are jigs, with the aluminum orange peel structures in between.

Among the advantages of welded steel jigs and fixtures are maximum strength and accuracy for close tolerances. Cost saving as high as 75 percent, time saving as much as 85 percent, and weight saving as much as 50 percent can be achieved. Welding permits a wide range of application, simplifies design, and minimizes machining. Modifications to meet design changes can be easily made.

Machine Tools

For the manufacture of machine tools, steel has certain advantages over cast iron:

- *Steel is two to three times stiffer*

- *Steel has about four times the resistance to fatigue* Cast iron has a fatigue resistance of about 7,500 pounds per square inch (p.s.i.), whereas that of steel is 28,000 to 32,000 p.s.i.

- *Steel costs one-quarter to one-half as much* Rolled steel costs approximately 40 percent as much per pound as cast iron and 25 percent as much as cast steel.

- *Steel is three to six times stronger in tension* A test of two equal-sized bars showed that the cast iron bar broke at 26,420 p.s.i., but the mild steel bar withstood 61,800 p.s.i.

- *Steel can withstand heavy impacts* In a test, one blow of a 9-pound sledge shattered a cast iron part. Twenty blows of the sledge merely bent the duplicate part, which was built of steel.

- *Steel is uniform and dependable* Its homogeneous structure, devoid of blowholes and uneven strains, makes possible more economical and more structurally sound designs.

- *Steel can be welded without losing desirable physical properties* A weld in steel, made by the shielded metal arc process, has physical properties equal to or better than those of cold-rolled steel, Table 2-1.

Today some manufacturers make and feature full lines of broaching, drilling, boring, and grinding machines with welded members, Fig. 2-13. Presses, brakes,

Table 2-1 Physical Properties of a Steel Weldment

Tensile strength	65,000–85,000 p.s.i.
Ductility	20–30% elongation/2 in.
Fatigue resistance	25,000–32,000 p.s.i.
Impact resistance	50–80 ft-lb (Izod)

Fig. 2-13 Several rail cars in various stages of completion. Note the size of this fabrication area. Air carbon arc gouging is being used, as well as many of the typical arc welding processes like GMAW and FCAW. Alberto Incrocci/Riser/Getty Images

Fig. 2-14 A large three-spindle, vertical-bridge, numerically controlled profiler. MAG IAS, LLC

and numerous types of handling machinery are now **weldments.** Other manufacturers are using welding for many of the accessories, pipelines, chip pans, and subassemblies.

The flexibility of welding in this work is exemplified by a milling machine bed made in one piece and constructed without any manufacturing difficulty in lengths varying from 10 to 25 feet. Broaching machines, which are used for a great variety of accurate machining of plane and curved surfaces, depend on welded construction. The manufacture of machine tools requires the use of bars, shapers, and heavy plates up to 6 inches in thickness. Bed sections for large planers and profilers may be as long as 98 feet, Figs. 2-14 and 2-15.

A horizontal cylinder block broach, among the largest (almost 35 feet long) equipment of its kind made, has construction features that would be almost impossible without welding. For example, a rough bed for a certain broach would have necessitated a casting of over 40,000 pounds. Heat-treated alloy parts, mild carbon plate, sheet steel, iron and steel castings, and forgings were used where suitable. Because of welded fabrication, the bed weighed only 26,000 pounds. Each member was made of material of the correct mechanical type, and the whole was a rigidly dependable unit.

Nuclear Power

Nuclear power depends upon the generation of large, concentrated quantities of heat energy, rapid heat removal, a highly radioactive environment, and changes in the properties of radioactive materials. The heart of the process is a reactor pressure vessel used to contain the nuclear reaction, Fig. 2-16.

Fig. 2-15 A mechanic grinding welds on the bed of a profiler. MAG IAS, LLC

ABOUT WELDING

Aerowave

The nuclear facility at General Electric uses aluminum chambers from an Aero Vac, a fabrication shop. To avoid problems with tungsten spitting, the shop uses the asymmetric technology of an Aerowave, an a.c. TIG machine from Miller Electric. The Aerowave has a low primary current draw. The technology provides a fast travel speed and the ability to weld thick metals at a given amperage. The operator can independently adjust the current in each a.c. half cycle from 1 to 375 amperes (A). The duration of the electrode negative portion of the cycle can be changed from 30 to 90 percent. The frequency can be adjusted from 40 to 400 hertz (Hz). An operator can fine-tune the penetration depth and width ratios of the weld bead. Has been updated to the Dynasty with similar features.

In addition to the reactor vessel, the system includes a heat exchanger or steam generator and associated equipment such as piping, valves, and pumps, as well as purification, storage, and waste disposal equipment. Welding is necessary for the fabrication of all

Fig. 2-16 This large reactor for a nuclear power plant is designed to withstand high pressure and high temperatures. Wall thicknesses are typically over several inches thick and welding plays a critical role in their manufacture. Note the large lifting eyes. Large crawler cranes can lift this type mega ton load into position in one lift. Frank Hormann/AP Images

Fig. 2-17 A 'Y' branch shop fabricated undergoing NDE to assure weld quality through soundness testing. GMAW root pass followed by FCAW fill and final beads. Piping Systems, Inc.

Fig. 2-18 A high pressure flange SAW to a heavy wall pipe of 20 inches in diameter and 3-1/2 inches thick. GMAW was used for the root bead and extensive visual, ultrasonic, and radiography are used to assure weld soundness. Mark A. Dierker/McGraw Hill

these units. The production of nuclear energy would not be possible but for the highly developed processes of today.

Piping

High pressure pipeline work, with its **headers** and other fittings, is a vast field in which welding has proved itself. The number of ferrous and nonferrous alloys used as piping materials is increasing. Industry requires better materials to meet the high heat and high pressure operating conditions of power plants, nuclear plants, oil refineries, chemical and petrochemical plants, and many other manufacturing plants where steam, air, gas, or liquids are used. Pressure of over 1,000 p.s.i. and temperatures ranging from −200 to +1,200° F are not uncommon in high pressure pipelines, Figs. 2-17 and 2-18. Marine lines and generator stations have installations operating at 1,250 p.s.i. with 950° F at turbine throttles. Demands for equipment in the steel mills, oil refineries, and other industries in which such lines operate emphasize reductions in size and weight and streamlining the appearance of piping as well as the flow. The lines are becoming increasingly complex: recirculation units, boosters, headers, and miscellaneous accessories and fittings are introduced into the lines, making them take on the appearance of complex electrical lines. Small pipe is connected with large pipe; T's, bends, return valves, and other fittings are introduced into the lines, Fig. 2-19.

The design of fittings for welded pipe is flexible and simple. Many fittings required by mechanically connected systems can be eliminated. The absence of projections inside the pipe produces less resistance to flow, Fig. 2-20. Because welded piping systems have permanently tight connections of greater strength and rigidity, maintenance costs are reduced. Other advantages of welded fabrication include a more pleasing appearance and easier, cheaper application of insulation.

With the development of welded fittings, the pipe fabricator realized the possibility of easily making any conceivable combination of sizes and shapes. Practically all overland pipeline is welded, Fig. 2-21.

Fig. 2-19 A pipe header being fabricated in the shop. The roller allows the header to be position in the most advantageous position for weld quality and productivity, with a number of branches coming off. *Piping Systems, Inc.*

Railroad units fabricated by welding include streamlined diesel and electric locomotives (Figs. 2-22 and 2-23), passenger cars, subways, freight cars, tank cars, refrigerator cars, and many other special types. Most new track is of the continuous welded type, and battered rail ends of old track are built up with the process.

Weight is an important consideration in the design of freight cars. In an entire trainload, a freight engine may be called upon to haul a string of up to 100 freight cars whose deadweight alone amounts to 2,400 to 3,400 tons. Welding and alloy steel construction reduced the weight of 50-ton boxcars from a light weight (empty) of 48,000 to 36,000 pounds, a reduction of approximately 25 percent. These cars are also able to carry a 25 percent greater load than their former 50-ton capacity.

Large cars are 50 feet in length and have a capacity of up to 100 tons and an empty weight of approximately 61,500 pounds. This is an increase of 100 percent in load-carrying capacity and only 35 percent in empty weight. Cars constructed for automobile shipping service are 89 feet in length.

Another development in freight car construction is the super-sized aluminum gondola. The car is a third longer and almost twice as high as the ordinary gondola.

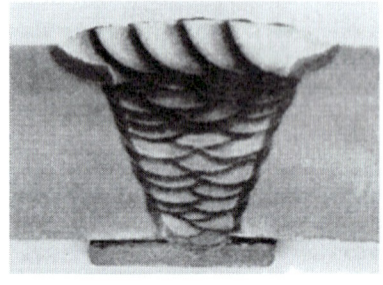

Fig. 2-20 An etched cross section of a multipass weld in chromemoly pipe. Note the sequence of weld beads. *Nooter Corp.*

Piping is used for the transportation of crude petroleum and its derivatives, gas and gasoline, in all parts of the country. Overland welded-pipe installations are both efficient and economical. Successive lengths of pipe are put together so cheaply in the field that total construction costs are materially reduced. These lines can be welded to the older lines without difficulty.

No better example of the extreme reliability and speedy construction available in the welding of pipeline can be cited than the oil line running from Texas to Illinois. This is almost 2,400 miles of pipe joined by welding.

Railroad Equipment

Welding is the principal method of joining materials used by the railroad industry. The railroads first made use of the process as a maintenance tool, and it has been extended to the building of all rolling stock. It is also used extensively in the construction and repair of equipment on the right-of-way.

Fig. 2-21 Cross-country pipeline being made on location. Note the rugged terrain. Follow the pipe down this hill. It is heading to the bottom across the valley and up the far hill. The equipment is lowering the pipe into the trench. The pipe cannot be easily rotated so both the 5G and 6G welding positions would be commonly used. The 6G welds are often referred to as "Arkansas Bell Hole." *Historic England Archive/Heritage Images/Alamy Stock Photo*

Fig. 2-22 Welding frames of railway cars. Note the safety gear and fresh air welding helmet. Bruce Forster/Stone/Getty Images

The principal welding processes are shielded metal arc, gas metal arc, and gas tungsten arc. A considerable amount of gas and arc cutting is also used in the fabrication of the plate.

In addition to the weight saving and the greater payloads, the tight joints that welding makes possible have virtually eliminated infiltration of dust and moisture into the cars, thus reducing damage claims against railroads. This can be effected only through welded fabrication, for no matter how carefully riveted joints are made, the constant jolting to which freight cars are subjected loosens the joints and opens seams that permit dust and moisture infiltration. Welded freight cars have a life of 20 to 40 years with very little maintenance required.

By employing low alloy steel and welded fabrication, car builders have been able to reduce the weight of a refrigerator car by as much as 6½ tons. This leakproof construction, made possible by welding and the corrosion-resistant properties of the low alloy steel that make up the car body, make it possible to keep the car in active service because the damage done to riveted refrigerator cars by leaking brine solutions is prevented.

Tank cars, too, have benefited through the use of corrosion-resistant steels and the permanently tight joints. The old type of construction caused many leakage problems. Often liquids are carried that require preheating to make them flow easily when the cars are being emptied. The alternate heating and cooling to which the riveted joints were subjected pulled them loose. Many of the tank cars being built today are 61 feet long and have a capacity of 30,300 gallons.

Another means of freight transportation in the United States is the highway trailer carried by railroad flat cars, commonly called piggyback service. Automobile carriers are also another means of freight transportation.

As many as twelve standard cars can be hauled on one 89-foot flat car equipped with triple decks. These flat cars weigh 56,300 pounds and can transport 130,000 pounds.

Hopper cars were redesigned with a saving in weight and the development of interior smoothness that permits free discharge of the load. With riveted construction, it was difficult to clean these cars when unloading because coal, carbon black, sulfur, sand, and cement adhered to the rivet heads and laps.

Fig. 2-23 These rack gears being MAG welded are made from forged and heat treated 8630 material. It is being welded to the beam section which is A514, 100K yield material. Four to six welders were engaged in preheating and welding. It was preheated and welded from both sides via fixturing and head and tail stock positioners to reduce distortion. This rack is used to move a 3.5 million ton counterweight during lifts. The Manitowoc Company, Inc.

The car is 90 feet long and weighs 96,400 pounds: some 60 pounds per foot less than ordinary steel gondolas of similar design. The underframe is steel, and the extensive use of aluminum saves 11,000 pounds in weight.

Fig. 2-24 A crew of 23 arc welders supply their own light for the photographer as they apply 100 feet of weld in a matter of minutes to the underframe of a Center Flow dry bulk commodities railroad car, with a capacity of 5,250 cubic feet. ACF Industries

Fig. 2-25 The end view of a covered hopper railcar. Note its clean smooth appearance. Ease of cleaning and painting are other reasons for making welding a good choice in fabricating products like this. Stayorgo/iStockphoto/Getty Images

The increase in the demand for covered hopper cars to transport many powdered and granular food products, as well as chemicals, has resulted in the Center Flow car of modern welded design, Fig. 2-24. These cars are 50 or more feet in length and have capacities of from 2,900 to 5,250 cubic feet and loading capacities up to 125 tons.

Many hopper cars are now made of aluminum. A steel hopper car weighs 72,500 pounds, but one made of aluminum weighs only 56,400 pounds. This makes it possible to carry heavier loads and thus reduce the cost of moving freight. Steel is still needed, however, for highly stressed parts like the underframe.

Hopper cars have two, three, or four compartments. Each compartment has a loading hatch and hopper outlet, providing for up to four different kinds of materials in the load. A four-compartment hopper car requires 4,000 feet of weld. Welding is the largest and most important production operation, Fig. 2-25. One-piece side plates are 112 inches by 49 feet by $5/16$ inch thick. Steel members are welded with the submerged arc process, and the aluminum is welded with the gas metal arc process. A considerable amount of gas tungsten and shielded metal arc welding is also used.

A standard steel passenger coach has a weight of about 160,000 pounds. Experimentation with welded construction for this type of coach produced a coach weight of 98,000 pounds. In addition, the modern welded coaches are stronger, safer, and considerably more comfortable than former types.

Not only has the rolling stock of the railroads been improved through the use of welding, but the rails they roll on are also undergoing great change. The old 90- to 100-pound rail sections of 39-foot lengths are giving way on several roads to continuous rail of $1/4$-mile lengths. This so-called continuous rail is not actually continuous but consists of a number of lengths welded together at the joints.

Track maintenance workers are continually faced with the problem of batter in rail ends that occurs at the joints. Battered rail ends cause a jolting when the train passes over them, which in turn means discomfort for passengers, shifting freight load, and wear on rolling stock. Battered rail ends can be eliminated by one of two methods: either replace the rail or build up the battered end. By employing welding and building up battered ends to a common level somewhere near the rail body level, both rails may remain in service. However, as old rails are replaced by continuous welded rail, service life is longer, and joint maintenance is decreased.

Inside the railroad shop, welding is likewise doing the heavy lifting. In locomotive maintenance, it is used on pipe and tubing repair, frames, cylinders, hub liners, floors, housings, tender tanks, and for streamlining shrouds. On freight cars, posts and braces, bolster and center plate connections, and other underframe and superstructure parts are repaired, strengthened, or straightened. Passenger car bolsters and cracked truck sides are repaired without difficulty. Vestibule and baggage-car side doors and inside trim are weld-repaired as standard practice.

Shipbuilding

The Naval Limitation Treaties of the 1920s and 1930s were the impetus behind the research program that led to a new conception of welding in ship construction. Under these treaties the various nations agreed to limit not only the number of capital vessels built, but also their weight. The Navy's reaction, therefore, was to build the most highly effective ships possible by any method within the limitations of the treaties. A capital ship must be light in weight and highly maneuverable, but it must have adequate defensive armor plate, gun power, and strength. It must be built to take as well as to give punishment.

That welded ships can take it is borne out by the story of the USS *Kearney,* which limped into port on October 18, 1941. This fighting ship, blasted amid-ships by a torpedo, came home under its own power, putting the stamp of approval on a type of construction in which our Navy had been a leader for years. It is highly improbable that any other than a welded type of ship could have reached home, and it was impossible that any other could have rejoined its command, as did the *Kearney,* a few months later. Since World War II, a large number of similar occurrences involving military and nonmilitary ships have been recorded.

Military watercraft fabricated by welding include aircraft carriers, battleships, destroyers, cruisers, and atomic-powered submarines.

The standard specifications for Navy welding work, which cover all welding done for the Bureau of Ships, are concerned with a variety of structures, such as watertight and oiltight longitudinals, bulkheads, tanks, turret assemblies, rudder crossheads, pressure vessels, and pipelines. Air, steam, oil, and water lines in various systems are all of homogeneous welded construction, Fig. 2-26.

Fig. 2-26 The main steam system piping for the engine room of a nuclear submarine. Crane Co.

Some idea of the immensity of these units may be gathered from the fact that gun turrets of a 35,000-ton battleship are built from welded materials ranging from one-half to several inches thick. The units weigh 250 tons each. The sternposts weigh 70 tons; and the rudders, 40 tons. The welded rudder of the carrier USS *Lexington* weighed 129 tons and was a 12½-foot thick (not 12½-inch) fabrication.

It is now possible to construct submarine hulls with a seam efficiency of 100 percent, as against the 70 percent efficiency of riveted hulls. Caulking is unnecessary because the hull is permanently leakproof. Hull production time is reduced by approximately 25 percent, and the total weight of the hull is reduced by about 15 percent because of the use of butt joints and groove welded plate. The smooth lines of the welded plate make hulls more streamlined and, therefore, faster and more maneuverable. They foul less quickly because of their smooth lines and can stay away from bases longer.

SHOP TALK

Repairing Welds
The very first step to repairing a weld is knowing the base material. From there, you can figure out the matching electrode and the correct preheat and interpass temperatures.

Fig. 2-27 This Tri-Hull 319-foot long high speed ferry can carry 760 passengers and 200 vehicles at speeds of 40 knots or 46 mph. With 38,000 HP from four diesel engines. Made from lightweight aluminum, the boat was fabricated using the pulsed gas metal arc welding process. Miguel Medina/AFP/Getty Images

Fig. 2-28 Welder at work in shipyard. Note the overhead position and leather safety protection. Kevin Fleming/Corbis/VCG/Getty Images

The hulls and power plants of nuclear submarines are also constructed of all-welded alloy steel plate instead of castings. Reductions in weight and size are accomplished along with improved structural strength. Between these savings and the weight reductions possible with welded piping and accessories, the modern submarine is made into a fabrication of far greater potential use. Hull strength for longer underwater runs, resistance to depth bombs, and deeper dives; increased power plant efficiency; and an overall decrease in weight per horsepower make the submarine an outstanding example of a unit welded for its purpose.

Ships differ widely in type and conditions of service. They range from river barges to large cargo and passenger vessels, Fig. 2-27. The adoption of the construction methods used in building ocean-going "Liberty ships" during World War II has reduced construction time from keel laying to launching by more than 20 percent.

Prefabrication, preassembly, and welding are the reasons for the dramatic reduction in building time, Fig. 2-28. Parts and substructures are shaped in advance. Accessories, pipelines, and necessary preassemblies are constructed in many cases far away from the scene of the actual building of the ship's hull. After completion, they are transported to the site and then installed as units into the vessel.

A completely riveted freighter would require in its construction thousands of rivets, averaging about 1 pound each. From a labor and timesaving standpoint, there is a reduction of 20 to 25 percent in deadweight that can be used largely for cargo carrying. In many ships today, there are only 200 rivets. A welded ship uses approximately 18 percent less steel than one that is riveted. In other words, in every six 10,000-ton vessels built, enough steel is saved to build another ship. Today's cargo ships weigh from 10 to 15 percent less than their 1918 counterparts, despite the fact that their deadweight capacity is 2,000 tons greater. Smoothness of hull construction has materially increased the speed of the vessels and reduced hull maintenance costs by 25 percent.

Oil tankers are of such size that only the welding process with its great saving in weight and strength makes construction possible. A fairly recently constructed tanker is more than 100 feet longer than one of the world's largest passenger liners, the *Queen Elizabeth II,* which is 1,031 feet long and 119 feet wide. The deck would dwarf a football field. The tanker is 105 feet high. It is powered by an 18,720-horsepower diesel engine, with a second one in reserve. It is designed to carry 276,000 tons of cargo and costs $20 million. Tankers now on the drawing boards will have a capacity of 600,000 tons.

Structural Steel Construction

The welding process has been applied to the construction of hydroelectric units, power generation units, bridges, commercial buildings, and private dwellings. The construction of such superpowered projects as the Bonneville, Grand Coulee, Hoover Dam, and the Tennessee Valley Authority projects called for entirely new methods of construction for water turbine parts and water power machinery.

Welded construction was first used extensively for the 74,000-horsepower hydraulic turbines in the Bonneville project on the Columbia River near Portland, Oregon.

The units alone of this power project involved 80,000 feet of gas cutting, 118,000 linear feet of welding, 286,000 pounds of electrodes, and 6,450,000 pounds of rolled plate steel. The Bonneville project demonstrated that the following advantages were realized by the use of welded members:

- A large number of patterns could be eliminated.
- Parts were ready more quickly for machining.
- Because of the use of steel plate, there was the practical assurance that machine work would not expose defects with resultant replacement and delay. This was important because of the necessity for quick delivery.
- Weights could be figured accurately, allowing close estimates for material costs.
- The amount of metal allowed for machining was reduced, simultaneously saving the time necessary for machining.
- Exact scale models could be made and tested under the same conditions as large units.
- Composite construction could be used. This type of construction involves the welding together of plate steel and castings or forgings, a combination of mild steel and alloy steel, or a combination of two alloy steels.
- Welding was also responsible for the usual saving in weight, together with greater strength, and improved quality, efficiency, and flexibility of design.

Bridges Bridges are constructed wholly or in part by the welding process. For over 50 years, steel bridges, both highway and railroad, have been constructed by this means, and the number of welded-steel bridges is increasing, Fig. 2-29.

Typical of the weight reduction possible in bridge construction is a saving of 42½ tons in a bascule span of a highway bridge built in Florida. One hundred tons were eliminated in the counterweights. Fixed and expansion bridge shoes had welded rolled-steel slabs for strength, reliability, and economy.

Savings in typical steel bridges, resulting from welded construction, range up to 20 percent. If these savings were extended to the long-range road building program that has been initiated by the federal government, enough steel could be saved to build a highway girder bridge approximately 800 miles long. Cost comparisons of actual rivet construction and welded construction have demonstrated that there is a 5.5:1 advantage in cost for welding construction.

Although cost and weight are important considerations, the strength of welded steel tips the scale in its favor. A welded-butt joint is the best type of joint. It has

Fig. 2-29 A welder performing FCA welding on a bridge. Keep in mind all the welded joints and thermal cutting that would typically go into the fabrication and construction of a bridge. David S. Moyer/McGraw Hill

the greatest strength and the most uniform stress distribution. The flow of stress in a riveted joint, however, is not uniform; it has a number of stress concentrations at various points. Just the punching of a hole in a plate for the rivet causes high stress concentrations when the plate is loaded.

Most rivets are driven hot. A hot rivet always shrinks upon cooling after being driven. This means that all rivets tend to shrink lengthwise, thus producing locked-up tensile stress in the rivet body, even without an external load. It also means that the rivet shrinks transversely so that it never quite fills the hole. The holes must be reamed so that the rivet is not deformed by holes that do not line up. This operation adds extra cost to the job.

The foundation pilings of many bridges have cutting edges made of welded steel plate. Tower caissons are made in sections and, because they are watertight, are floated to the site and filled with concrete. All-welded bridge floors are fairly common. Reinforcing girders, crossbeams, and other members have been constructed with a saving of as much as 50 percent in both weight and time.

Industrial and Commercial Buildings All types of buildings are welded during construction. Welding

> **JOB TIP**
>
> **Career in Welding**
> A career in welding offers numerous opportunities to advance in the industry. As you gain skill, you can continue to succeed. With thorough experience in the field, many welders develop an interest in other related jobs, such as
>
> 1. Shop supervisor
> 2. Maintenance engineer
> 3. Robot operator
> 4. Robot technician
> 5. Degreed welding engineer
> 6. Teacher
> 7. Shop owner
> 8. Instructor to industries

has become a major method of making joints in structures. The fact that there are no holes needed for rivets is an advantage in the design of trusses and plate girders. Flange angles are not needed in plate girders, and single plates can be used for stiffeners instead of angles. Rigid frame structures are possible, permitting the bent-rib type of roof construction that gives maximum headroom, no diagonal cross-bracing members, and no shadow lines from truss members, Fig. 2-30. In multiple-story buildings, the rigid frame permits shallow beam depths that allow lower story heights.

Welding reduces construction and maintenance costs due to smooth lines of construction, decreased weight of moving elements such as cranes, and ease of making alterations and new additions. First cost is materially less because of a saving in weight of materials, which may be as much as 10 to 30 percent. Many building units can be fabricated in the shop under controlled conditions, thus reducing expensive on-site work. Interiors are open and unrestricted; there are no columns in the way.

Excavation is speeded up by the use of digging equipment with abrasion-resisting teeth, made economically possible by welding. Piling sections and reinforcing steel are flame-cut and welded. Welding replaces riveting in the shop fabrication and field erection of columns, beams, and girder sections. Flame-cutting is used to prepare gussets and perform field trimming operations. Incidentally, most of the construction equipment used on the job (such as cranes, bulldozers, and concrete mixers) is welded.

After the structural steel framework of the building is complete, continued use of welding also speeds up the mechanical installations. Pipelines and electrical conduits are welded into continuous lengths. Air ducts and smoke risers are fabricated to the required shapes by welding and cutting. Welded electrical junction and panel boxes are secured to the columns and beams by welding. Transformers, switchboards, furnaces, ventilating equipment, tanks, grating, railing, and window sashes are partially or completely prefabricated. Once located, their installation and connections are made with the aid of welding. Changes or additions to the building or its equipment are greatly aided by this method.

The construction industry has long felt the need to solve the problems of creating housing for a mass market. Some architects have turned to a steel-fabricated welded structure as a solution. Such prefabricated housing has the following advantages:

- The construction method uses factory-produced materials of many kinds that are standard, readily available, and accurate.

Fig. 2-30 Industrial building interiors take on an entirely new appearance. Arc welded rigid frames replace conventional truss sawtooth framing. Note the absence of columns and the improved headroom. The Lincoln Electric Co.

Fig. 2-31 Steel home construction saves owners thousands in upkeep, insurance, and energy costs. Heritage Building Systems

- A large part of the construction can be shop-fabricated under controlled conditions and mass produced, thus requiring less site labor.
- Site erection is fast, thus providing for an overall reduction in cost. Steel home construction, Fig. 2-31, is also enjoying increased popularity.
- Construction materials weigh less, are stronger, and lend themselves to acoustical treatment more easily than standard materials.
- Prefabricated modules provide flexibility of design and floor plan arrangements.
- A higher factor of earthquake, flood, and wind resistance is possible.

Tank and Pressure Vessel Construction

The growth of cities and towns has increased both the number and the size of tanks needed for the storage of water, oil, natural gas, and propane. The increase in the number of automobiles, trucks, and aircraft has increased the need for storage facilities for petroleum products. In addition, our space and missile programs have created the need for the storage of oxygen, nitrogen, and hydrogen in large quantities. The fertilizer industry requires volume storage facilities for ammonia. The basic materials for many industries, supplying such diverse products as tires, fabrics, soap, and food products, are stored in pressure vessels. Tanks and vessels of all types have become one of the principal applications of welding.

Welding replaced riveting in the fabrication of pressure vessels approximately 65 years ago, Figs. 2-32 and 2-33. This improved the service performance of a pressure vessel through the elimination of two common areas of service failure in riveted vessels: leakage and corrosion around rivets.

The construction and maintenance costs of both welded tanks and pressure vessels are also reduced. Less material is used in the construction of a welded vessel. A riveted joint develops a strength equal to only 80 percent of the tank plate, whereas a welded joint develops a strength 20 to 30 percent greater than the plate. It is, therefore, possible to reduce the plate thickness and still obtain the same design strength by welding. Some of the heavier pressure vessels, 3 to 5 inches in thickness, cannot be fabricated in any other way because it is impossible to rivet plates of this thickness with any degree of success. In addition, there is further saving because it is unnecessary to punch the plates and caulk the seams of a welded joint. Maintenance costs of welded tanks are practically negligible, and the joints are permanently tight, Figs. 2-34 and 2-35.

One of the leading pressure vessel manufacturers points to the following seven factors in support of welded construction.

- Elimination of thickness limit of about 2¾ inches for successful riveting, and elimination of leakage at high pressure
- Elimination of thickness limit for forge and hammer welding, which was about 2 inches

Fig. 2-32 Riveted construction formerly used in constructing pressure vessels. Each rivet was a point of breakdown. Compare with today's all-welded vessel shown in Fig. 2-33. Edward R. Bohnart

Industrial Welding **Chapter 2** 29

Fig. 2-33 This steam generator plant has a capacity of 127,000 pounds and contains more than 9 miles of tubing. The plant produces steam from controlled nuclear fission. Nooter Corp.

- Elimination of caustic embrittlement in riveted boiler drums
- Economy in weight through higher joint efficiency and elimination of butt-straps and rivets
- A reduction in size to meet the same service requirements
- Greater flexibility of design, permitting uniform, or at least gradual, stress distribution
- Elimination of all fabricating stresses in the completed vessel by heat treatment

To these achievements of welding in the fabrication of pressure vessels might be added increased speed of fabrication, Fig. 2-36, reduction of corrosion for longer life, and smooth interiors of chemical and food vessels for sanitation, Fig. 2-37. By eliminating the size

Fig. 2-34 A water tank constructed of plate 1½ inches thick, which is 240 feet in diameter and has a capacity of 11 million gallons of water. Nooter Corp.

Fig. 2-36 An oil refinery sphere being constructed in the field indicates the mobility and flexibility of the welding process. Nooter Corp.

Fig. 2-35 World's largest titanium tower—10 feet in diameter. A considerable amount of gas metal arc welding is used on this type of work. Edward R. Bohnart

Fig. 2-37 Automatic gas-shielded metal arc welding of brewery vessels. Hobart Brothers Co.

limit on pressure vessels, welding made a direct contribution to our productive capacity and technology.

Miscellaneous Applications

A few miscellaneous applications are illustrated in Figs. 2-38 through 2-42 so that the student may appreciate the flexibility of the welding process.

Fig. 2-40 Gear-reducing unit. All parts were flame-cut, and unit includes all types of joints and welds. General Electric Company

Fig. 2-38 All-welded fabricated gear. The parts of the gear were flame-cut. The Lincoln Electric Co.

Fig. 2-41 All of the fabrications shown in this chapter used gas and arc cutting as a fabricating tool. Shown here is a multiple-torch application, burning natural gas and oxygen, which is cutting out parts that will later become part of a weldment. Praxair, Inc.

Fig. 2-39 Constructing an underground intercontinental ballistic missile base. Welding and cutting are used extensively. The Lincoln Electric Co.

Fig. 2-42 Turbine blades being inspected and adjusted. These types of devices travel at very high velocities and at extreme temperatures. Welding plays an important role in the fabrication of the turbine blades and housings. Issues with dissimilar metals and superalloys must be considered. Howden Buffalo Inc.

CHAPTER 2 REVIEW

Multiple Choice

Choose the letter of the correct answer.

1. The two major functions of welding in industry are _____. (Obj. 2-1)
   a. Tool and die
   b. Fabrication; maintenance and repair
   c. Stocks and trades
   d. Steel and aluminum

2. Which of the following industries have found welding to be an advantage? (Obj. 2-2)
   a. Aircraft
   b. Piping
   c. Railroad equipment
   d. All of these

3. The manufacturer of which of the construction machinery has *not* met the challenge with welded equipment? (Obj. 2-2)
   a. Pullers
   b. Scrapers
   c. Rollers
   d. Electrodes

4. Welded household equipment fabrication permits the use of _____. (Obj. 2-2)
   a. Stainless steel
   b. Aluminum
   c. Magnesium
   d. All of these

5. Using jigs and fixtures results in cost saving of _____ percent to industry. (Obj. 2-2)
   a. 75
   b. 35
   c. 50
   d. 85

6. For the manufacture of machine tools, what advantage(s) does steel have over cast iron? (Obj. 2-2)
   a. Steel is two to three times stiffer
   b. Steel has four times the resistance to fatigue
   c. Steel is three to six times stronger in tension
   d. All of these

7. Advantages of welded fabrication include _____. (Obj. 2-2)
   a. A pleasing appearance
   b. A workable product
   c. A ridged product
   d. Elimination of porosity

8. It is now possible to construct submarine hulls with a seam efficiency percentage of _____. (Obj. 2-2)
   a. 30
   b. 100
   c. 80
   d. 90

9. For over _____ years, steel bridges, both highway and railroad, have been of welded construction. (Obj. 2-2)
   a. 30
   b. 50
   c. 75
   d. 100

10. The growth of cities and towns has increased both the number and the size of welded tanks needed for _____. (Obj. 2-2)
    a. Water storage
    b. Oil
    c. Gas
    d. All of these

11. Welding fabrication has grown rapidly because of _____. (Obj. 2-3)
    a. Design and flexibility
    b. Low cost
    c. Special production needs
    d. Speed and economy

12. Which of the following were *not* principal materials to be found on the first airplanes? (Obj. 2-3)
    a. Wood
    b. Fabric
    c. Wire
    d. None of these

13. Aircraft welding was first tried and used in warcraft production in _____. (Obj. 2-3)
    a. 1903
    b. 1911
    c. 1927
    d. 1932

14. What country first introduced warcraft production? (Obj. 2-3)
    a. United States
    b. Great Britain
    c. Germany
    d. France

15. Welding processes for the manufacture of passenger cars were first introduced during _____. (Obj 2-3)
    a. World War I
    b. The Great Depression
    c. World War II
    d. None of these
16. Railroad cars have a capacity up to _____ tons. (Obj. 2-3)
    a. 50
    b. 70
    c. 100
    d. 120
17. Navy standard specifications for welding work, which cover all welding done for the Bureau of Ships, are concerned with which of the following structures? (Obj. 2-3)
    a. Bulkheads
    b. Pipelines
    c. Rudder crossheads
    d. All of these
18. The cost of rolled steel over a casting is _____. (Obj. 2-4)
    a. ¼
    b. ½
    c. ¼ to ½
    d. ⅓ to ¾
19. By replacing riveting in shipbuilding, welding uses _____ percent less steel. (Obj. 2-5)
    a. 18
    b. 25
    c. 50
    d. 65
20. Welding replaced riveting in the fabrication of pressure vessels about _____ years ago. (Obj. 2-5)
    a. 40
    b. 50
    c. 55
    d. 70

Review Questions

Write the answers in your own words.

21. List the advantages of welding as a means of fabrication. (Obj. 2-1)
22. List the advantages of welding when used for maintenance and repair. (Obj. 2-1)
23. Steel has several advantages for construction. Name them. (Obj. 2-2)
24. What are some of the advantages of welding in pressure and overland piping? (Obj. 2-2)
25. List some of the advantages of welded construction when applied to pressure vessels. (Obj. 2-2)
26. List at least five advantages that can be gained in the application of welding to building construction. (Obj. 2-2)
27. Do bridges commonly have all-welded construction? (Obj. 2-2)
28. What features of welded construction make it resistant to earthquakes, floods, and high winds? (Obj. 2-2)
29. Is welding limited in its application to piping and pressure vessels because the process is not dependable at high pressures and temperatures? Explain your answer. (Obj. 2-3)
30. List some of the types of watercraft that are fabricated by welding. (Obj. 2-3)
31. How has welding in bridge construction progressed in recent years? (Obj. 2-3)
32. List at least 10 products used by the military that are manufactured wholly or in part by welding. (Obj. 2-3)
33. How is welding used by the railroads? (Obj. 2-4)
34. Identify three weaknesses of rivet construction. (Obj. 2-5)
35. Can a tank with a wall thickness of over 3 inches be riveted? Can it be welded? (Obj. 2-5)

INTERNET ACTIVITIES

Internet Activity A

Suppose you wanted to find some information on steel home construction. How would you find it on the Internet? What search engine would you use? What would be your key word(s)?

Internet Activity B

Using your favorite search engine, use "welding" as your key word. Choose a topic of interest to you from the results of the search. Then write a brief report about it. Share it with other people in your class.

Design credits: Cinema and movie icon: Denis Meshkov/123RF; and Illustration of Welding icons: Mr. Nachai Sorasee/123RF

Steel and Other Metals

Chapter Objectives

After completing this chapter, you will be able to:

3-1 Describe the steelmaking process.

3-2 List the metalworking processes used to shape and improve the characteristics of solidified steel.

3-3 State the proper use of each of the heat-treating processes for steel.

3-4 Describe the internal structures of metals.

3-5 Describe the physical and mechanical properties of metals.

3-6 Name the various alloying elements used in steelmaking and their effects.

3-7 List the various types of ferrous metals and their applications.

3-8 List the various types of nonferrous metals and their applications.

3-9 Describe the various systems used to designate metals.

3-10 Explain the heating and cooling effects on a weldment brought about by welding and how these effects can be controlled.

The welding process joins metals, plastics, and glass without the use of mechanical fastening devices. In this study, we are primarily concerned with its application to metals. Metals are separated into two major groups: ferrous metals and nonferrous metals.

Ferrous metals are those metals that have a high iron content. They include the many types of steel and its alloys, cast iron, and wrought iron.

Nonferrous metals are those metals that are almost free of iron. The nonferrous group includes such common metals as copper, lead, zinc, titanium, aluminum, nickel, tungsten, manganese, brass, and bronze. The precious metals (gold, platinum, and silver) and radioactive metals such as uranium and radium are also nonferrous.

Steel is a combination of iron and carbon. Iron is a pure chemical element. Oxides of iron are found in nature, and iron ore is abundant throughout the world. Because iron is not strong enough and hard enough to be used in structural members, it must be combined with carbon to produce the characteristics necessary for steel forms. Up to a certain point, the more carbon steel contains, the stronger and harder the steel will be but will have less ductility and a more crack-sensitive microstructure. Although it is possible to weld nearly all of the ferrous and nonferrous metals and alloys, this chapter will be concerned principally with steels in the low and medium carbon ranges. These are the steels that the student will be primarily concerned with in the practice of welding. It has been estimated that nearly 80 percent of all weldments are fabricated from steel and that 85 percent of the total amount of steel welded is in the mild (low carbon) steel classification.

History of Steel

The ancient Assyrians are credited with the first recorded use of iron about 3700 B.C. Since the use of iron in making weapons gave them an advantage over other nations, they became the most powerful nation of their time. From about 1350 B.C. to A.D. 1300, all of the iron tools and weapons were produced directly from iron ore.

Low carbon iron was first produced in relatively flat hearth furnaces. Gradually the furnaces were increased in height, and the charge was introduced through the top. These shaft furnaces produced molten high carbon iron. Shaft furnaces were used in Europe after A.D. 1350. The modern blast furnace is a shaft furnace.

Accurate information about the first process for making steel is not available. Tools with hardened points and edges have been found that date back to 1000 to 500 B.C. Early writers mention steel razors, surgical instruments, files, chisels, and stone-cutting tools several hundred centuries before the Christian era.

Prior to the Bessemer process of making steel, only two methods were used. The cementation process increased the carbon content of wrought iron by heating it in contact with hot carbon in the absence of air. The crucible process consisted of melting wrought iron in crucibles to which carbon had been added. Both of these processes were known and used by the ancients.

During the Middle Ages both the cementation and crucible processes were lost to civilization. The cementation process was revived in Belgium about the year A.D. 1600, while the crucible process was rediscovered in England in 1742. The crucible process eventually came to be used chiefly for making special steels. The cementation process was highly developed and was also used extensively in England during the eighteenth and nineteenth centuries. This process is still used to a limited extent. The crucible process has been replaced by the various electric furnace processes for making special alloy steels and carbon tool steels.

Steelmaking in the United States

The history of the iron and steel industry in North America extends back over 300 years, beginning with a successful ironworks in Saugus, Massachusetts, about 20 miles northeast of Boston. It operated from 1646 to 1670. Through the support of the American Iron and Steel Institute, this site has been restored and is open to the public.

Very little steel was manufactured in America during the early days. The first patent was issued in Connecticut in 1728. A succession of events spurred the growth of the steel industry:

- New uses for iron
- The discovery of large iron ore deposits in northern Michigan
- The development of the Bessemer and open hearth processes
- The Civil War and America's explosive industrial growth following the war
- The expansion of the railroads
- World Wars I and II

Currently the largest steel producer in the world is China at 500.5 metric tons. The European Union is second with 198.0 metric tons, followed by Japan with 118.7 metric tons, and the United States with 91.4 metric tons. Other major producers are Russia at 68.5, India at 55.2, South Korea at 53.6, Germany at 45.8, Ukraine at 37.1, Brazil at 33.7, Italy at 30.6, Turkey at 26.8, Taiwan at 19.9, France at 17.9, Spain at 18.6, and Mexico at 17.2.

Annual steel production in the United States, as indicated, is just over 90 metric tons. Steelmaking facilities have changed greatly over the last few decades. Where there used to be nearly 250 blast furnaces, there are now only 36 blast furnaces for the production of iron and no open hearth furnaces being used. The principal reason for this reduction is the increased use of recycled steel. With more than 1,220 furnaces worldwide it is possible to meet the demand for steel. Nearly 40 percent of all industrial jobs in the United States involve the making of steel or the use of steel.

The perfection of the welding process as a means of joining metals has speeded up and expanded the use of steel. The adaptability of steel to manufacturing processes and its ability to join with many other metals to give a wide variety of alloys have also contributed to its widespread use.

With the continued development of GTAW, GMAW, flux cored arc welding (FCAW), laser beam cutting (LBC), and plasma arc cutting (PAC), the welding and cutting of aluminum, stainless steel, titanium, and other alloys have become routine production applications.

In this chapter you will study the important characteristics of iron, steel, and other metals so that you will have a basic understanding of the nature of metals and the various results of the welding process.

Raw Materials for the Making of Steel

Huge quantities of raw materials are needed to produce the vast amount of steel needed by humankind. The United States is well-supplied with the basic resources such as iron ore, limestone, and coal. The principal supplies of manganese, tin, nickel, and chromium necessary to the making of steel and its alloys are found in other countries.

As stated earlier, iron occurs in nature in the form of compounds with oxygen. In order to obtain the iron, the oxygen is removed in a blast furnace by contact with carbon. Coke, a coal product, is the usual source of carbon. In this process the iron becomes contaminated with some of the carbon. This extra carbon is in turn removed in the steelmaking processes by using controlled amounts of oxygen. The resulting product is formed into various shapes by rolling or other processes. Steel may also be heat treated.

Iron Ore

Iron is a metallic element that is the most abundant and most useful of all metals. It occurs in the free state only in limited quantities in basalts and in meteorites. Combined with oxygen and other elements in the form of an ore mixed with rocks, clay, and sand, iron is found in many parts of the world. About 5 percent of the Earth's crust is composed of iron compounds. For economic reasons it is mined only in those locations that have very large deposits.

In the United States, nearly all the ore is mined in northern Minnesota near Lake Superior. This ore is principally taconite. Taconite has a low natural percentage of iron, but the iron percentage can be enhanced by grinding the taconite into powder, separating out the silica, and reconstituting it into pellets with an iron content of about 65 percent. The pellets can be fluxed or acid prepared. The fluxed pellets have limestone in them; the acid pellets do not. Fluxed pellets also have a slightly lower level of silica. Extensive deposits of iron ore are located in Brazil, which is noted as one of the best sources. The purest iron ore comes from Sweden and is responsible for the high quality of Swedish steel.

The following iron ores are listed in the order of their iron content:

- **Magnetite** (Fe_3O_4) is a brownish ore that contains about 65 to 70 percent iron. This is the richest iron ore and the least common.

> **JOB TIP**
>
> **Working with Others**
> Do you like working with people? Your career success—and that of the business—depends on how well you work with others as a team, and whether or not you are skilled at dealing with customers.

- **Hematite** (Fe_2O_3), known as red iron, contains about 70 percent iron. It is widely mined in the United States.
- **Limonite** ($2Fe_2O_3 \cdot H_2O$) contains from 52 to 66 percent iron.
- **Siderite** ($FeCO_3$) contains about 48 percent iron.
- **Taconite** is a low grade ore that contains from 22 to 40 percent iron and a large amount of silica. It is green in color.
- **Jasper** is an iron-bearing rock. The ore is predominantly magnetite or hematite.

Iron ore is mined by the underground and open pit methods. The method chosen primarily depends on the depth of the ore body below the surface and the character of the rock surrounding the ore body.

In underground mining, a vertical shaft is sunk in the rock next to the ore body. Tunnels are drilled from this shaft and blasted horizontally into the ore body at a number of levels. In open pit mining, the mineral is lying relatively near the surface. The earth and rock covering the ore body are first removed. Blast holes are then drilled, and explosives shatter the ore to permit easy digging. The loosened ore is hauled out of the pit by truck, train, or conveyor belt.

For many years in the past, the great bulk of iron ore had only to be mined and shipped directly to the blast furnace. This is no longer the case. With the great drain on the ore bodies due to the rapidly expanding steel industry, the supply of high grade ore suitable for direct shipment was seriously depleted by the end of World War II. To solve this problem, steel producers and mining companies started upgrading ore quality by crushing, screening, and washing the ore in order to obtain a more suitable feed for the blast furnace. Figures 3-1 and 3-2, pages 38–40, detail the complex treatment that low grade ore receives before it is ready for the blast furnace.

Oxygen

Oxygen is the most abundant element on earth. Almost half the weight of the land, 21 percent by weight of the air,

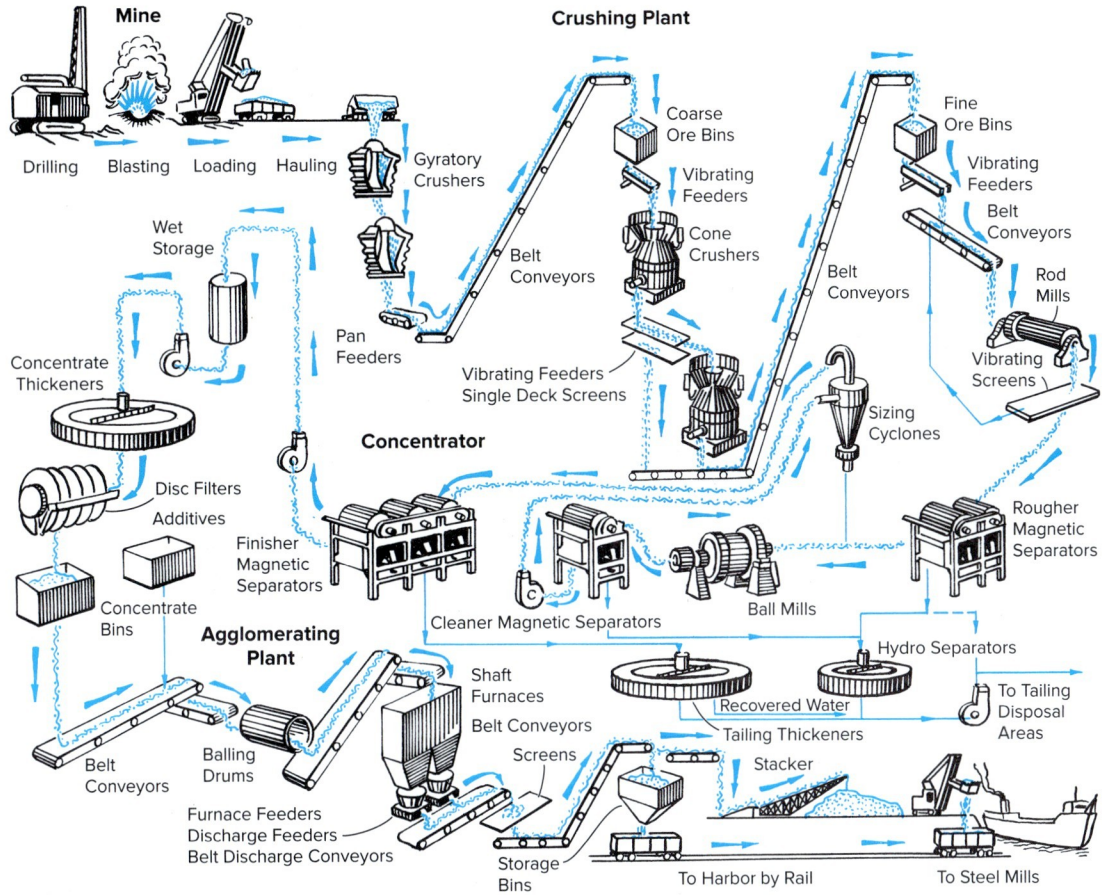

Fig. 3-1 This flowchart shows the complex treatment that taconite ore receives before it is sent to the blast furnaces.

and nearly 90 percent by weight of the sea consists of oxygen. Most of the oxygen for commercial purposes is made through the electrolysis of water and the liquefaction and subsequent distillation of air.

The steel industry is a major consumer of oxygen. The gas goes into most of the standard steel mill processes from the blast furnace to the finished product. Oxygen is used in the steelmaking process to purify the material. When directed onto the surface of molten iron, it oxidizes the carbon, silicon, manganese, and other undesirable elements. It also speeds up the steelmaking process by supporting the combustion of other fuels. These oxyfuel flames provide much higher temperatures than fuels burned in air. Figure 3-3, pages 41–42, illustrates the processing of oxygen for industrial use.

Fuels

Heat is indispensable in the manufacture of iron and steel. It is also essential in making steel mill products. To supply the heat required, the steel industry depends on three major natural fuels—coal, oil, and natural gas. Coal is the most important of these fuels.

Coal Coal supplies more than 80 percent of the iron and steel industry's total heat and energy requirements. More than 100 million tons have been consumed by this industry in one year. This is enough to supply more than 15 million homes with their average yearly supply of fuel for heating. A large part of the coal is used in making coke for use in the blast furnace. About 1,300 pounds of coke are used for each ton of pig iron produced.

Not all types of bituminous coal can be used to make coke. Coke must be free from dust, the right size to permit rapid combustion, strong enough to carry the charge in the blasting furnace, and as free as possible from sulfur. Coal of coking quality is mined in 24 states; however, West Virginia, Pennsylvania, Kentucky, and Alabama supply nearly 90 percent of the coal used in the steel industry.

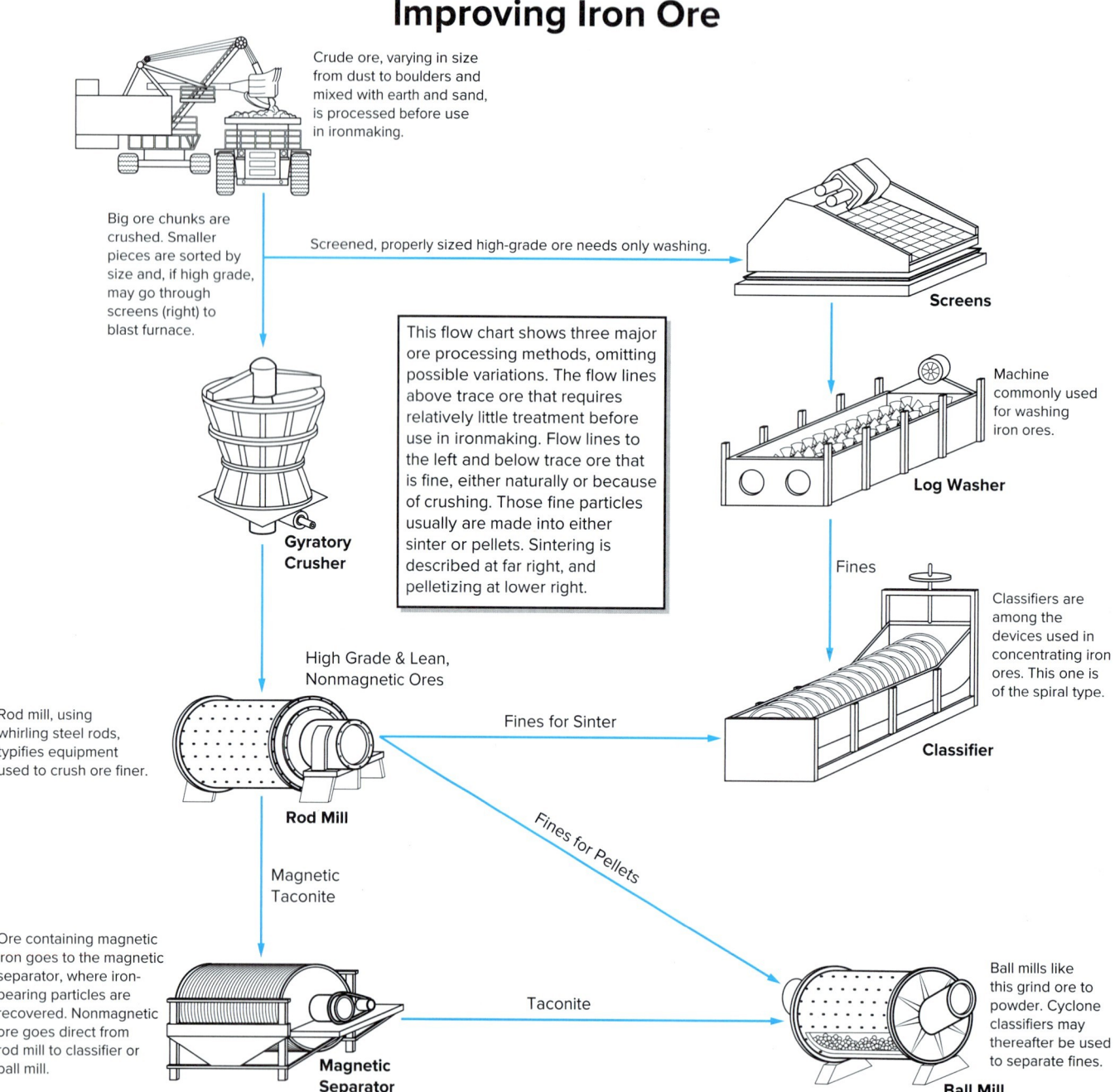

Fig. 3-2 Ore processing methods. Adapted from American Iron & Steel Institute

Oil Oil is used extensively by the industry both as a fuel and as a lubricant for machinery and products. The heaviest grades of oil are most commonly consumed in steel plants. About 70 percent of the fuel oil used by the steel industry is consumed in melting iron. More than 20 percent of the industry's fuel oil is burned in heating and annealing furnaces where steel products are given special heat treatments. The remainder is used in a wide variety of applications.

Natural Gas Natural gas is burned in reheating furnaces and in other places where clean heat is necessary. It contains almost no objectionable constituents, leaves no wastes or residues, and has a flame temperature as high as 3,700°F.

Natural gas contains more heating value than all other gases employed: it delivers 1,000 British thermal units (Btu) per cubic foot as compared with about 500 Btu for coke oven gas, 300 Btu for blue water gas, and 85 Btu for

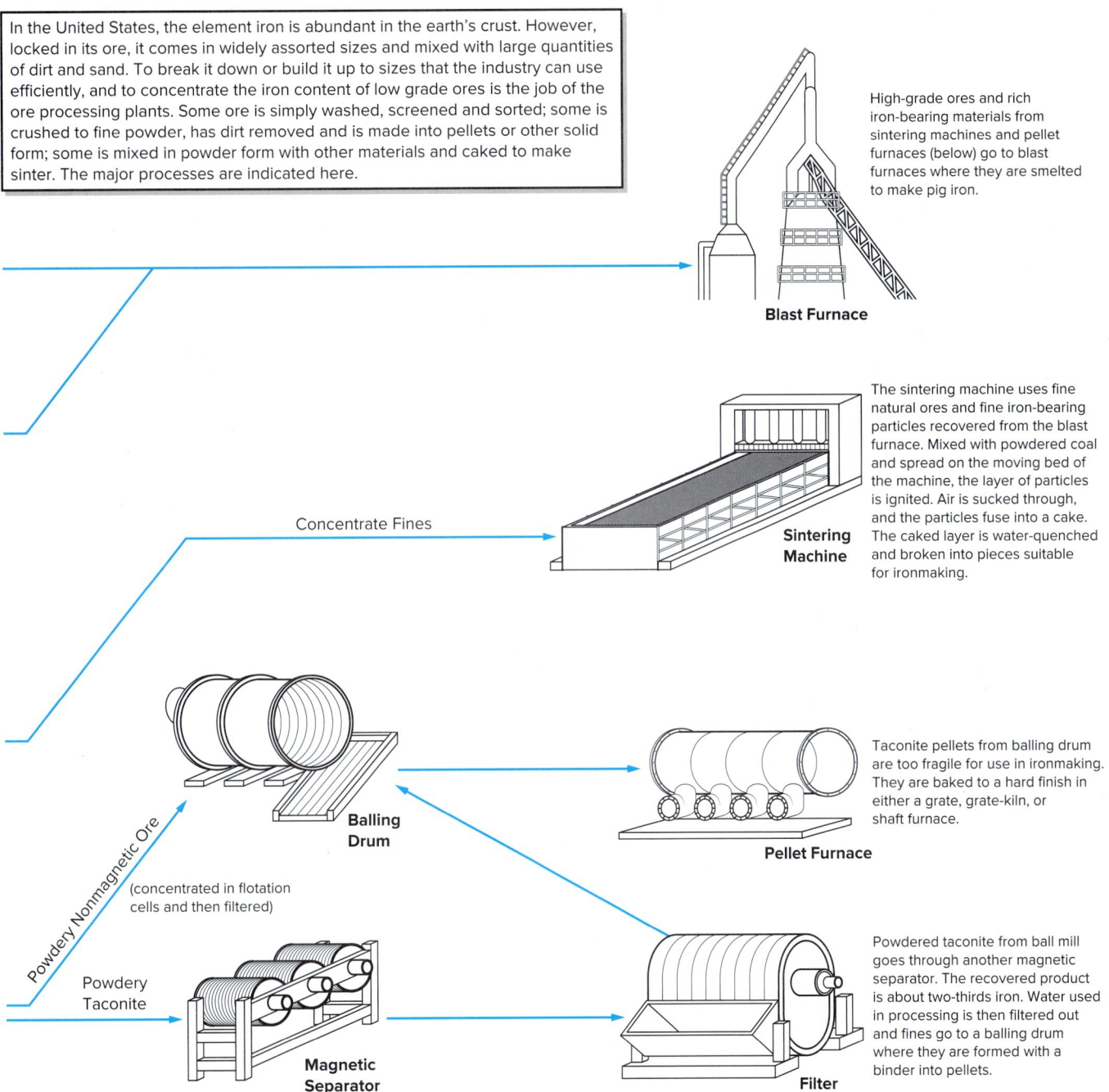

Fig. 3-2 Ore processing methods. *(Continued)*

blast furnace gas. At peak capacity the industry consumes over 400 billion cubic feet of natural gas per year. About 50 percent of this amount is consumed in heat-treating and annealing furnaces.

Coke The heat required for smelting iron in blast furnaces is obtained from the burning of coke. **Coke** may be defined as the solid residue obtained when coal is heated to a high temperature in the absence of air. This causes the gases and other impurities to be released. Coke is a hard, brittle substance consisting chiefly of carbon, together with small amounts of hydrogen, oxygen, nitrogen, sulfur, and phosphorus. In recent years, it has found some use as a smokeless domestic fuel.

The Manufacture of Coke Prior to 1840, charcoal was the only fuel used in the United States for iron smelting. In 1855, anthracite coal became the leading blast furnace

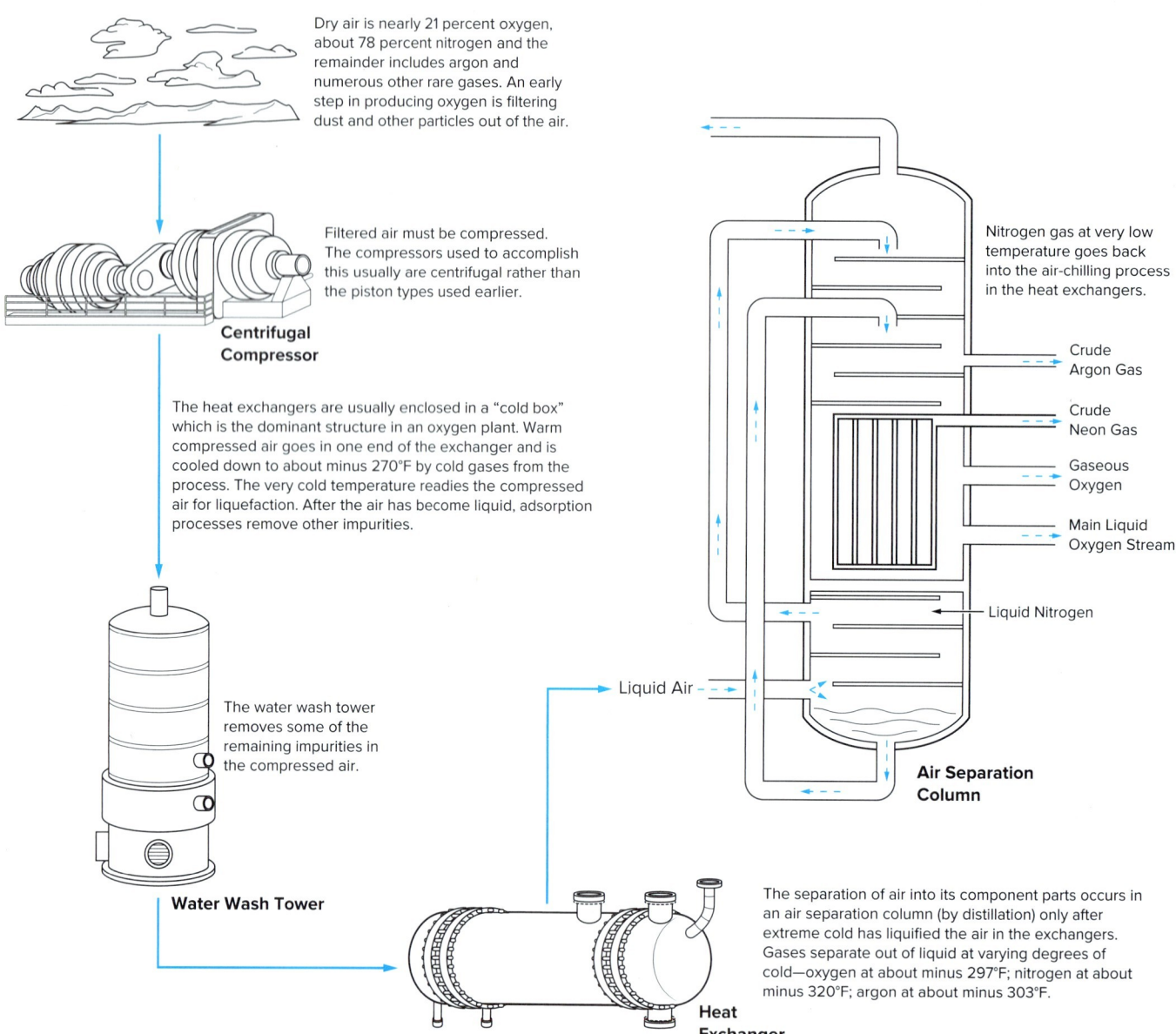

Fig. 3-2 Ore processing methods. *(Concluded)*

fuel because it was readily available, and charcoal was becoming more difficult to obtain. Another natural fuel, raw bituminous coal, was first burned in 1780 following the opening of the Pittsburgh coal seam. It was discovered in 1835 that by coking this coal, a product more suited to the needs of the blast furnace could be produced. In 1875, coke succeeded anthracite as the major blast furnace fuel.

By 1919, the coal chemical process of producing coke was developed, and coke became the leading fuel of the steel industry. The process, in addition to recovering the chemicals in coal, makes possible the production of stronger coke from a greater variety of coals. Figure 3-4, page 43, illustrates the mining of coal and its manufacture into coke. A coke oven is shown in Fig. 3-5, page 44. The volatile products that pass out of the ovens are piped to the chemical plant where they are treated to yield gas, tar, ammonia liquor, ammonium sulfate, and light oil. Further refinement of the light oil produces benzene, toluene, and other chemicals.

From these, basic chemicals are produced such varied products as aviation gasoline, nylon, printing inks, pharmaceuticals, perfumes, dyes, TNT, sulfa drugs, vitamins, soaps, and artificial flavors.

Coke production in the United States exceeds 64 million tons per year of which more than 92 percent is consumed as blast furnace fuel.

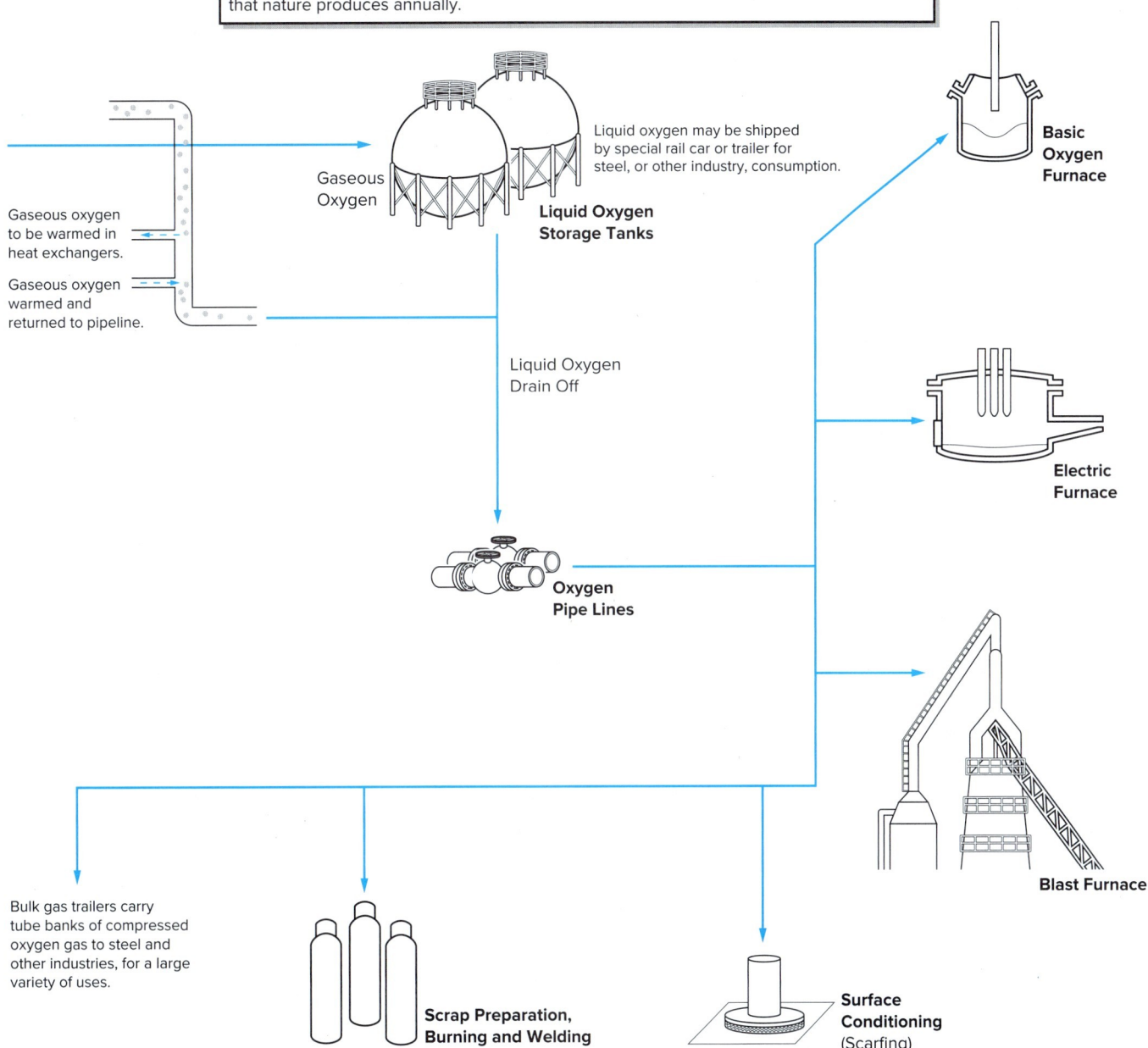

Fig. 3-3 Oxygen for steelmaking. Adapted from American Iron & Steel Institute

Steel Scrap

The earliest methods of making steel could not make use of scrap. Today basic oxygen furnaces (BOFs), which include early blast furnaces and electric furnaces, are very capable of using scrap, and nearly 66 percent of the steel currently used is recycled. Steel is generally made using a continuous caster that produces slabs, billets, or blooms. A BOF may take up to 80 percent liquid metal directly

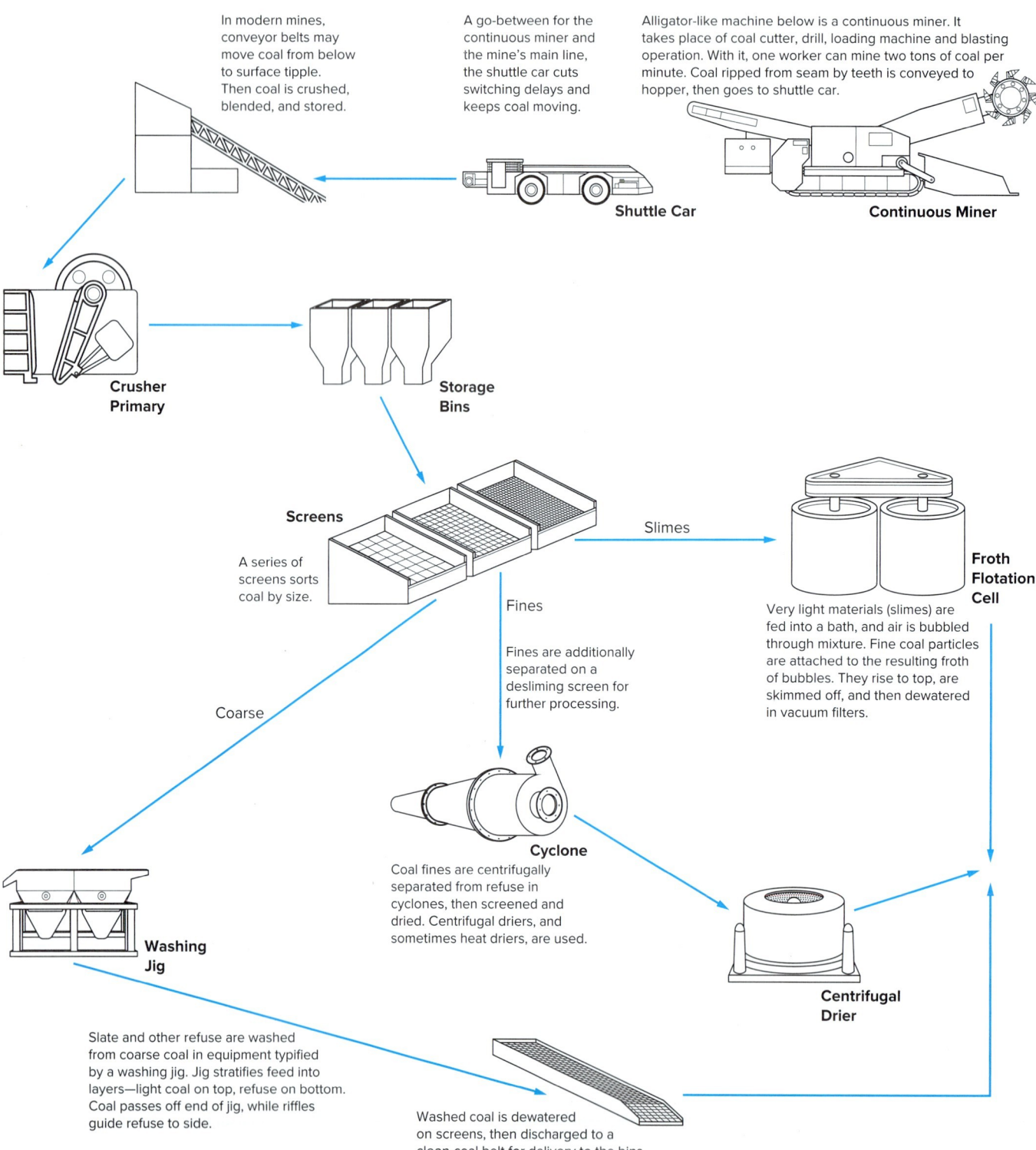

Fig. 3-3 Oxygen for steelmaking. *(Concluded)*

from the blast furnace and then have up to 20 percent scrap added. An integrated producer using this method can better control and produce higher grades of steel than a steelmaker who simply melts scrap. Since steel has no memory, what once was a juice can may become part of your car this year and in years to come be part of a bridge. Blast furnaces do not use scrap except in the form of sinter (i.e., in powdered form).

Steel mills recycle any of their own products that is not usable, and they also recycle items such as packing cases.

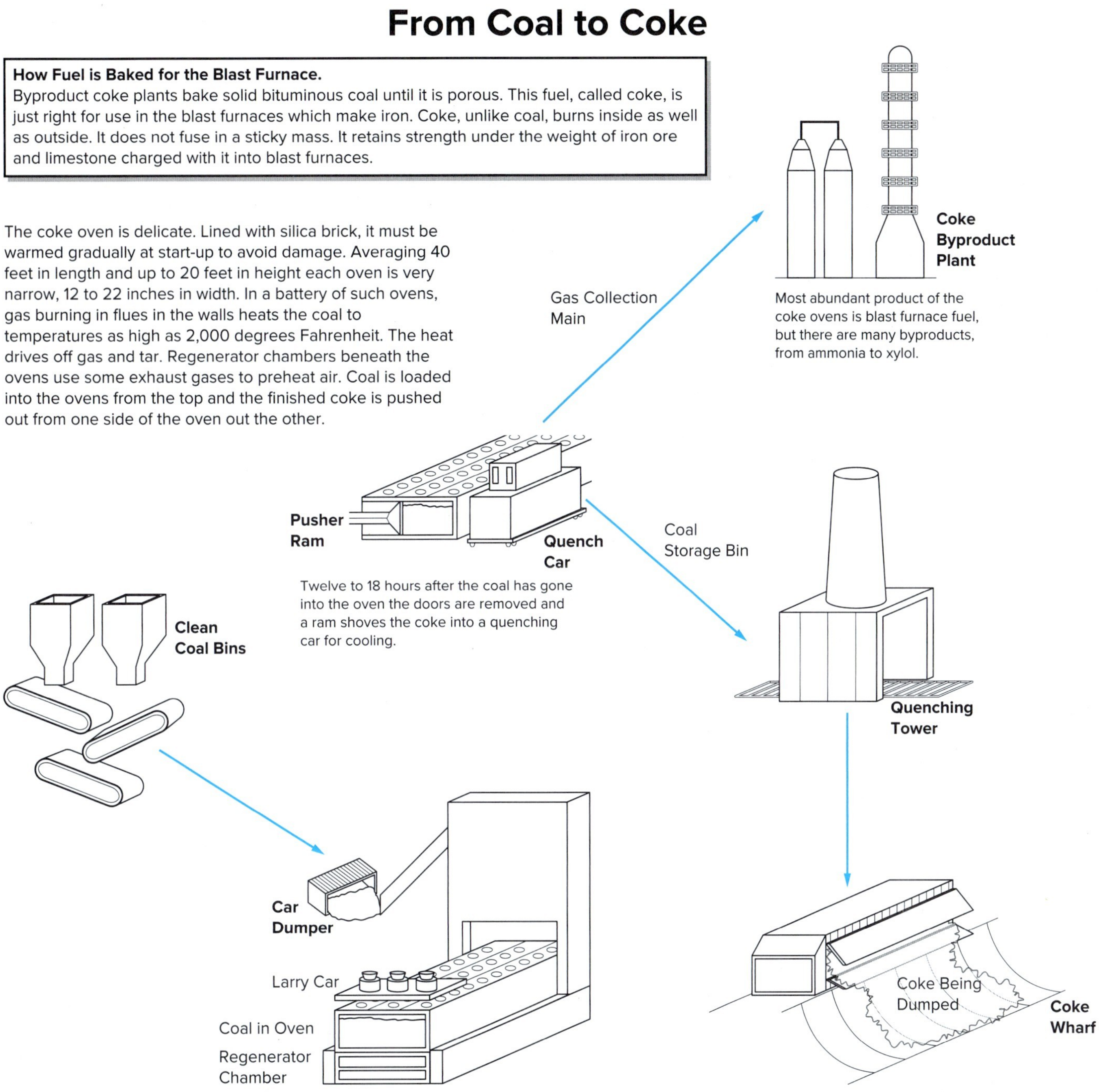

From Coal to Coke

How Fuel is Baked for the Blast Furnace.
Byproduct coke plants bake solid bituminous coal until it is porous. This fuel, called coke, is just right for use in the blast furnaces which make iron. Coke, unlike coal, burns inside as well as outside. It does not fuse in a sticky mass. It retains strength under the weight of iron ore and limestone charged with it into blast furnaces.

The coke oven is delicate. Lined with silica brick, it must be warmed gradually at start-up to avoid damage. Averaging 40 feet in length and up to 20 feet in height each oven is very narrow, 12 to 22 inches in width. In a battery of such ovens, gas burning in flues in the walls heats the coal to temperatures as high as 2,000 degrees Fahrenheit. The heat drives off gas and tar. Regenerator chambers beneath the ovens use some exhaust gases to preheat air. Coal is loaded into the ovens from the top and the finished coke is pushed out from one side of the oven out the other.

Gas Collection Main

Coke Byproduct Plant
Most abundant product of the coke ovens is blast furnace fuel, but there are many byproducts, from ammonia to xylol.

Pusher Ram

Quench Car
Twelve to 18 hours after the coal has gone into the oven the doors are removed and a ram shoves the coke into a quenching car for cooling.

Coal Storage Bin

Clean Coal Bins

Quenching Tower

Car Dumper

Larry Car

Coal in Oven

Regenerator Chamber

Coke Being Dumped

Coke Wharf

Fig. 3-4 Producing coke. Adapted from American Iron & Steel Institute

However, one of the best sources of scrap steel is from old automobiles. Scrap steel has become such a valuable commodity that the American metal market actually tracks the price of certain grades of scrap daily.

Limestone

Limestone is used as a flux in the blast furnace. It is a sedimentary rock commonly found all over the world. There are large deposits in many parts of the United States, especially in the Appalachian Mountains, the Rocky Mountains, and the Mississippi River Valley.

Limestone consists largely of *calcium carbonate* in varying degrees of purity. Common chalk is a form of pure limestone. The color of the limestone changes with the presence of different types of impurities. It is white when pure and may also be found as gray, yellow, or black due to such impurities as iron

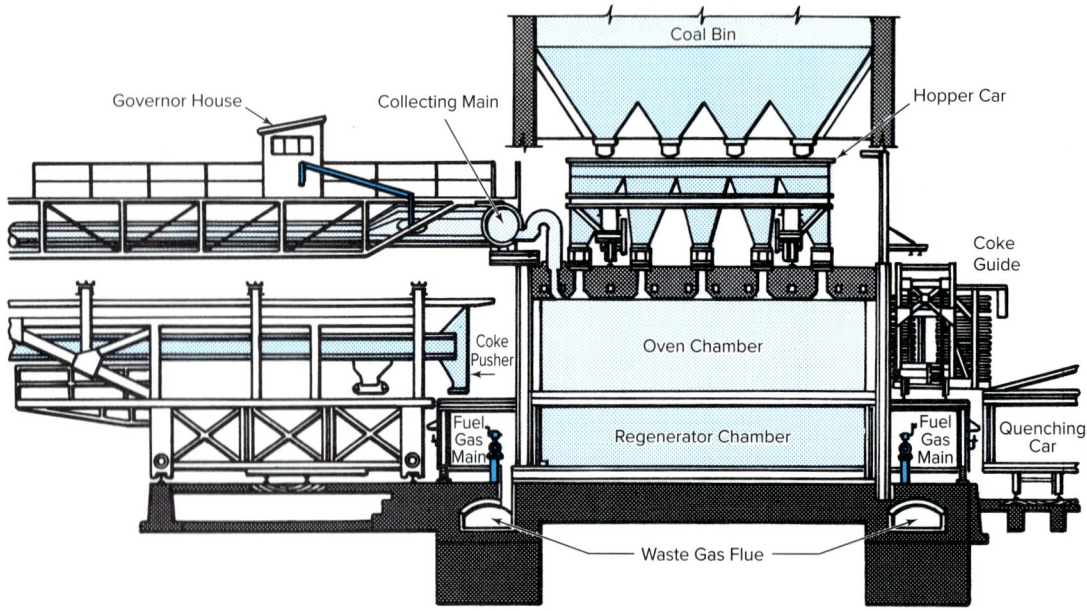

Fig. 3-5 A schematic diagram of a coal-chemical coke oven. Coal falls from bins into a hopper car, which runs on top of many narrow ovens, dropping in coal. Heat, in the absence of air, drives gases from the coal to make coke. The collected gases are valuable byproducts for chemicals.

oxide and organic matter. The properties of the rock change if certain compounds are present: silica makes it harder, clay softer, and magnesium carbonate turns it to dolomite, which is pinkish in color. Limestone may contain many fossils and loosely cemented fragments of shells.

Limestone is one of the chief fluxes used in steelmaking to separate the impurities from the iron ore. Many of the impurities associated with iron ores are of a highly **refractory** nature; that is, they are difficult to melt. If they remained unfused, they would retard the smelting operation and interfere with the separation of metal and the impurities. The primary function of limestone is to make these substances more easily fusible. Figure 3-6 shows the steps taken to process limestone.

Refractory Materials

Refractory materials may be defined as nonmetallic materials that can tolerate severe or destructive service conditions at high temperatures. They must withstand chemical attack, molten metal and slag erosion, thermal shock, physical impact, catalytic actions, pressure under load in soaking heat, and other rigorous abuse. Melting or softening temperatures of most refractory materials range from 2,600°F for light duty fireclay to 5,000°F for brick made from magnesia in its purest commercial form.

Refractory materials have an almost unlimited number of applications in the steel industry. Among the most important are linings for blast furnaces, steelmaking furnaces, soaking pits, reheating furnaces, heat-treating furnaces, ladles, and submarine cars.

Refractory materials are produced from quartzite, fireclay, alumina (aluminum oxide), magnesia (magnesium oxide), iron oxide, natural and artificial graphites, and various types of coal, coke, and tar. The raw materials are crushed, ground, and screened to proper sizes for use in making bricks and other forms of linings. They are

SHOP TALK

Slag and Stubends

Two environmental concerns in the industry are the recycling of welding slag and the stubends of electrodes. Manufacturers of welding consumables can reuse the slag. Unfortunately, the cost is too high for the return cost of the residual material. If there are no forbidden substances, the least expensive option at this point is to transfer the material to a dump.

The Purifying Stone

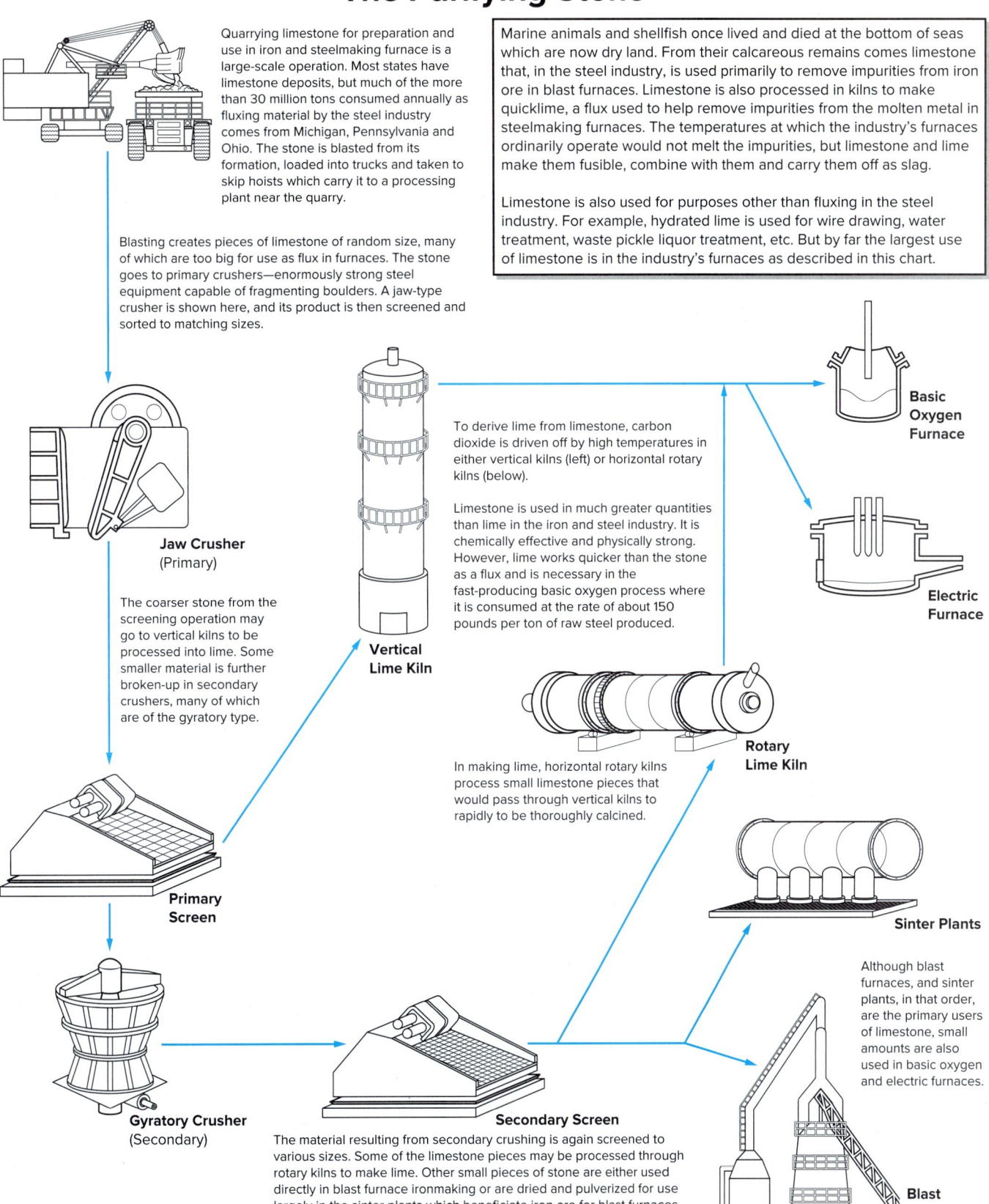

Quarrying limestone for preparation and use in iron and steelmaking furnace is a large-scale operation. Most states have limestone deposits, but much of the more than 30 million tons consumed annually as fluxing material by the steel industry comes from Michigan, Pennsylvania and Ohio. The stone is blasted from its formation, loaded into trucks and taken to skip hoists which carry it to a processing plant near the quarry.

Marine animals and shellfish once lived and died at the bottom of seas which are now dry land. From their calcareous remains comes limestone that, in the steel industry, is used primarily to remove impurities from iron ore in blast furnaces. Limestone is also processed in kilns to make quicklime, a flux used to help remove impurities from the molten metal in steelmaking furnaces. The temperatures at which the industry's furnaces ordinarily operate would not melt the impurities, but limestone and lime make them fusible, combine with them and carry them off as slag.

Limestone is also used for purposes other than fluxing in the steel industry. For example, hydrated lime is used for wire drawing, water treatment, waste pickle liquor treatment, etc. But by far the largest use of limestone is in the industry's furnaces as described in this chart.

Blasting creates pieces of limestone of random size, many of which are too big for use as flux in furnaces. The stone goes to primary crushers—enormously strong steel equipment capable of fragmenting boulders. A jaw-type crusher is shown here, and its product is then screened and sorted to matching sizes.

Jaw Crusher (Primary)

The coarser stone from the screening operation may go to vertical kilns to be processed into lime. Some smaller material is further broken-up in secondary crushers, many of which are of the gyratory type.

To derive lime from limestone, carbon dioxide is driven off by high temperatures in either vertical kilns (left) or horizontal rotary kilns (below).

Limestone is used in much greater quantities than lime in the iron and steel industry. It is chemically effective and physically strong. However, lime works quicker than the stone as a flux and is necessary in the fast-producing basic oxygen process where it is consumed at the rate of about 150 pounds per ton of raw steel produced.

Vertical Lime Kiln

Basic Oxygen Furnace

Electric Furnace

Primary Screen

In making lime, horizontal rotary kilns process small limestone pieces that would pass through vertical kilns to rapidly to be thoroughly calcined.

Rotary Lime Kiln

Sinter Plants

Gyratory Crusher (Secondary)

Secondary Screen

The material resulting from secondary crushing is again screened to various sizes. Some of the limestone pieces may be processed through rotary kilns to make lime. Other small pieces of stone are either used directly in blast furnace ironmaking or are dried and pulverized for use largely in the sinter plants which beneficiate iron ore for blast furnaces.

Although blast furnaces, and sinter plants, in that order, are the primary users of limestone, small amounts are also used in basic oxygen and electric furnaces.

Blast Furnace

Fig. 3-6 Limestone. Adapted from American Iron & Steel Institute

combined with certain binders, and the prepared batches are fed to the forming machines. The most common methods for forming refractory bricks are power pressing, extrusion, and hand molding. Most refractory bricks are fired in kilns at high temperature to give them permanent strength.

Iron Blast Furnace Slag

Slag is the residue produced from the interaction of the molten limestone and the impurities of the iron. It contains the oxides of calcium, silicon, aluminum, and magnesium, small amounts of iron oxide, and sulfur. Slag may be processed for use in the manufacture of cement and concrete blocks, road materials, insulating roofing material, and soil conditioner.

Carbon

Carbon is a nonmetallic element that can form a great variety of compounds with other elements. Compounds containing carbon are called organic compounds.

In union with oxygen, carbon forms carbon monoxide and carbon dioxide. When carbon combines with a metal, it may form compounds such as calcium carbide and iron carbide.

Three forms of pure carbon exist. The diamond is the hard crystalline form, and graphite is the soft form. Carbon black is the amorphous form.

In addition to being important as an ingredient of steel, carbon is used for industrial diamonds and abrasives and arc carbons of all kinds. As graphite, it forms a base for lubricants and is used as a lining for blast furnaces.

Fig. 3-7 This blast furnace will produce over 1,800 tons of pig iron daily. The furnace stack and other accessories, fabricated by welding, contain over 2,400 tons of steel plate and structurals. Nooter Corp.

The Smelting of Iron

The Blast Furnace

The first step in the conversion of iron ore into steel takes place in the blast furnace, Figs. 3-7 and 3-8. In this towering cylindrical structure, iron is freed from most of the impurities associated with it in the ore.

The furnace is charged with iron ore, limestone, and coke. A blast of preheated air burns the coke, producing heat to melt the iron, which falls to the bottom of the furnace. The molten limestone combines with most of the impurities in the ore to form a slag that separates from the liquid iron because it is lighter and floats. The liquid iron

Fig. 3-8 In a blast furnace operation, new charge enters from the top while liquid iron and slag are drawn away below. Thomas Saupe/Getty Images

Blast Furnace Ironmaking

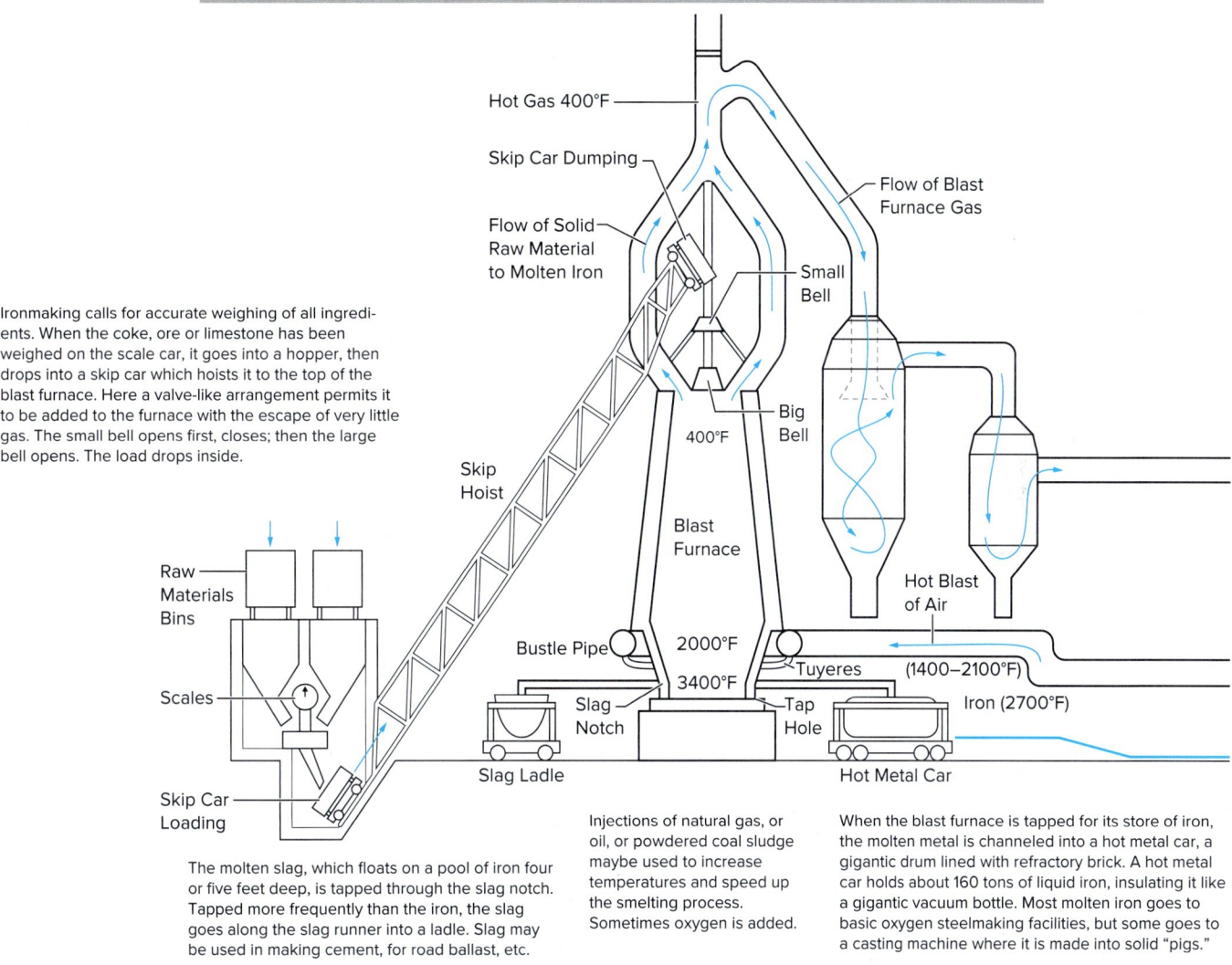

Fig. 3-9 The blast furnace process. Adapted from American Iron & Steel Institute *(Continued)*

and the liquid slag are removed periodically from the bottom of the furnace. This is a continuous process; as a new charge is introduced at the top, the liquid iron and slag are removed at the bottom. The progress of the charge through the furnace from the time it enters the top until it becomes iron is gradual; five to eight hours are required. The process is illustrated in Figs. 3-9 and 3-10, page 49.

The liquid iron is poured into molds to form what is known as *pigs* of iron. **Pig iron** is hard and brittle. It contains considerable amounts of dissolved carbon, manganese, silicon, phosphorus, and sulfur. Steelmaking is the process of removing impurities from pig iron and then adding certain elements in predetermined amounts to arrive at the properties desired in the finished metal. While several of the elements added are the same as those removed, the proportions differ.

Nearly all of the pig iron produced in blast furnaces remains in the molten state and is loaded directly

Hot air is indispensable in a blast furnace. As much as four and one-half tons of it may be needed to make one ton of pig iron. It is blown in at the bottom of the furnace and roars up through the charge of iron ore, coke, and limestone that has been dumped in from the top. Fanned by the air, the coke burns. Its gases reduce the ore to metallic iron by removing oxygen from it while the limestone causes the earthy matter of the ore to flow. Freed, the heavy metal settles to the bottom. From there, 4,000 to 10,000 tons of pig iron are drawn off per day.

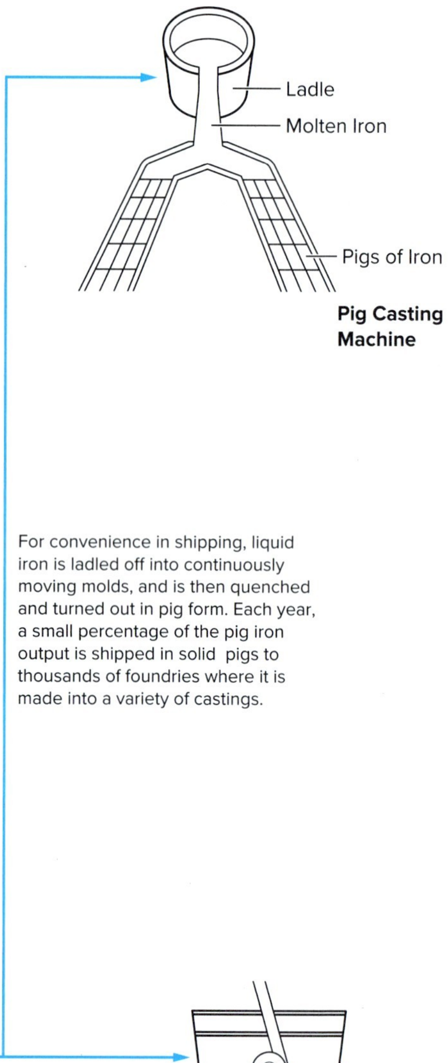

Pig Casting Machine

Ladle
Molten Iron
Pigs of Iron

For convenience in shipping, liquid iron is ladled off into continuously moving molds, and is then quenched and turned out in pig form. Each year, a small percentage of the pig iron output is shipped in solid pigs to thousands of foundries where it is made into a variety of castings.

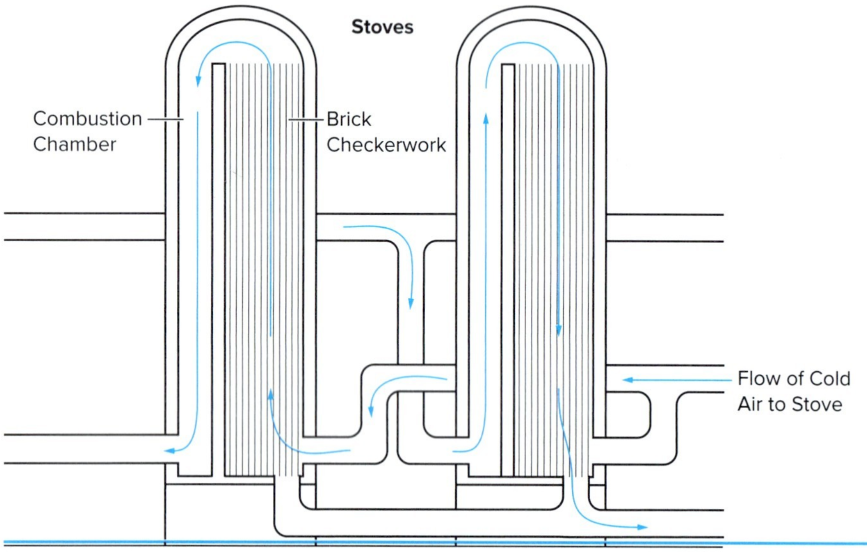

Stoves

Combustion Chamber
Brick Checkerwork
Flow of Cold Air to Stove

Air for the blast furnace is heated in huge stoves. At least two stoves are needed for each blast furnace. One stove heats while the other blows hot air into the bustle pipe and through tuyeres to the bottom of the furnace. In a combustion chamber in the stove being heated, cleaned exhaust gases from the blast furnace are mixed with air and burned to raise the temperature of refractory brick.

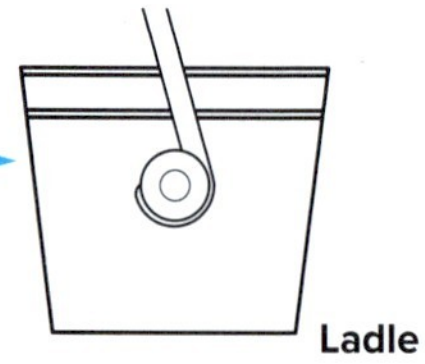

Ladle

A ladle full of molten iron joins limestone, scrap steel and alloying materials in a basic oxygen furnace to form a special heat of steel meeting rigid specifications.

Fig. 3-9 The blast furnace process. *(Concluded)*

into steelmaking furnaces. A small amount is solidified and transported to iron foundries that remelt it. Then the iron is cast into a wide variety of products ranging from toys to cylinder heads for automobile engines.

A modern blast furnace may be as much as 250 feet in height and 28 feet in diameter. The furnace shaft is lined with refractory materials, and this lining is water cooled to withstand high temperatures. Flame temperatures as high as 3,500°F and gas temperatures of 700°F are generated. As much as 10 to 12 million gallons of water per day may be used to cool a furnace. A furnace may operate for several years before relining is necessary.

The number of blast furnaces in the United States has declined over the past 30 years, but the total annual pig iron production has increased greatly. Enlarged furnaces, refined and controlled raw materials, and much higher blast temperatures are responsible for increased

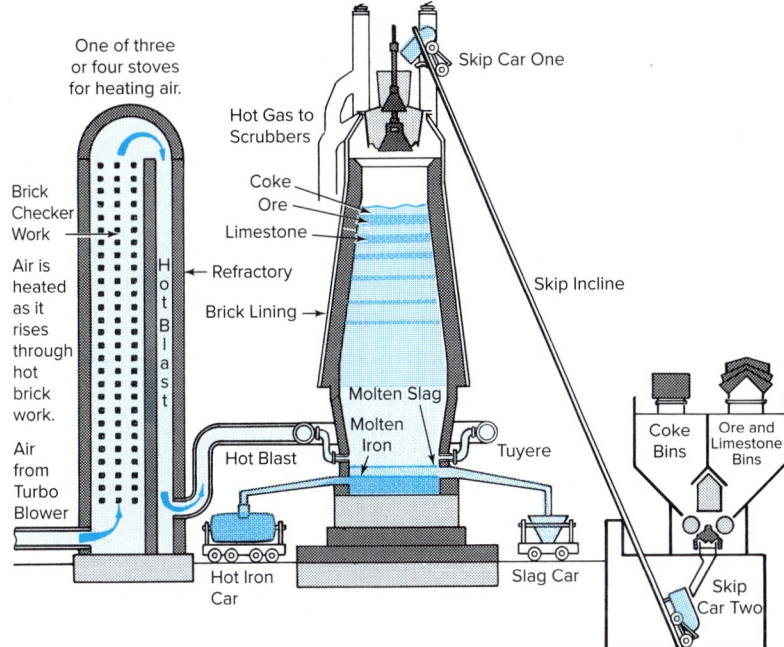

Fig. 3-10 Schematic diagram of a blast furnace, hot blast stove, and skiploader. Ore, limestone, and coke are fed in at the top of the furnace. Preheated air, delivered at the bottom, burns the coke and generates gases and heat required to separate iron from the ore. Source: American Iron & Steel Institute

production. The number of furnaces probably will continue to decrease as the production rate for leading furnaces exceeds 3,000 net tons per day.

Steelmaking Processes

You have read that steel was used in a primitive form for several thousand years. However, this early steel was not strong nor did it have the variety of properties necessary for extensive use. It was produced by the cementation and the crucible processes. In recent times, two major developments have made it possible to produce large quantities of steel with a variety of properties at a competitive cost.

The first of these developments was the Bessemer furnace invented in 1856 in both Europe and the United States. The second was the open hearth furnace which was invented 12 years later in the United States. Figure 3-11, pages 50–51, shows the modern steelmaking process from raw materials to finished product.

For video on steelmaking operations, please visit www.mhhe.com/welding.

Cementation Process

Cementation is the oldest method of steelmaking. It consists of heating wrought iron with carbon in a vacuum. This increases the carbon content of surfaces and edges which can then be hardened by heating and quenching. The metal is not molten during steelmaking. Hence impurities are not removed from the iron, and only the surface of the metal is affected. It is probable that most of the steel of ancient times was produced in this way.

A later improvement of this process was the stacking of alternate layers of soft, carbon-free iron with iron containing carbon. The layers were then heated so that the pieces could be worked. The layers of soft and hard metal strengthened the internal structure of the steel. Much of this steelmaking was centered in Syria during the Middle Ages, and the steels became known as the famous Damascus steels, used widely for swords and spears of the highest quality.

The steel made by this process was further improved by the crucible process that came into use in the eighteenth century.

Crucible Process

The crucible process was revived in England during the early 1740s. Steel produced by the cementation process was melted in a clay crucible to remove the impurities. While fluid, the slag was skimmed off the top. Then the metal was poured into a mold where it solidified into a mass that could be worked into the desired shape. In the United States graphite crucibles, with a capacity of about 100 pounds of metal, were used in a gas-fired furnace. This process produced a steel of uniform quality that was free of slag and dust.

Electric Furnace Processes

Electric furnaces are of two types: (1) the electric arc type and (2) the induction furnace. The first *electric arc furnace* had a capacity of 4 tons. It was put into operation in France by the French metallurgist Paul Heroult in 1899 and introduced into the United States in 1904. The modern furnace, Fig. 3-12, page 52, has a charge of 80 to 100 tons. A few furnaces hold a charge of 200 tons and produce more than 800 tons of steel in 24 hours. These large furnaces are made possible by the increase of electric power capacity, the production

Steel and Other Metals   Chapter 3   49

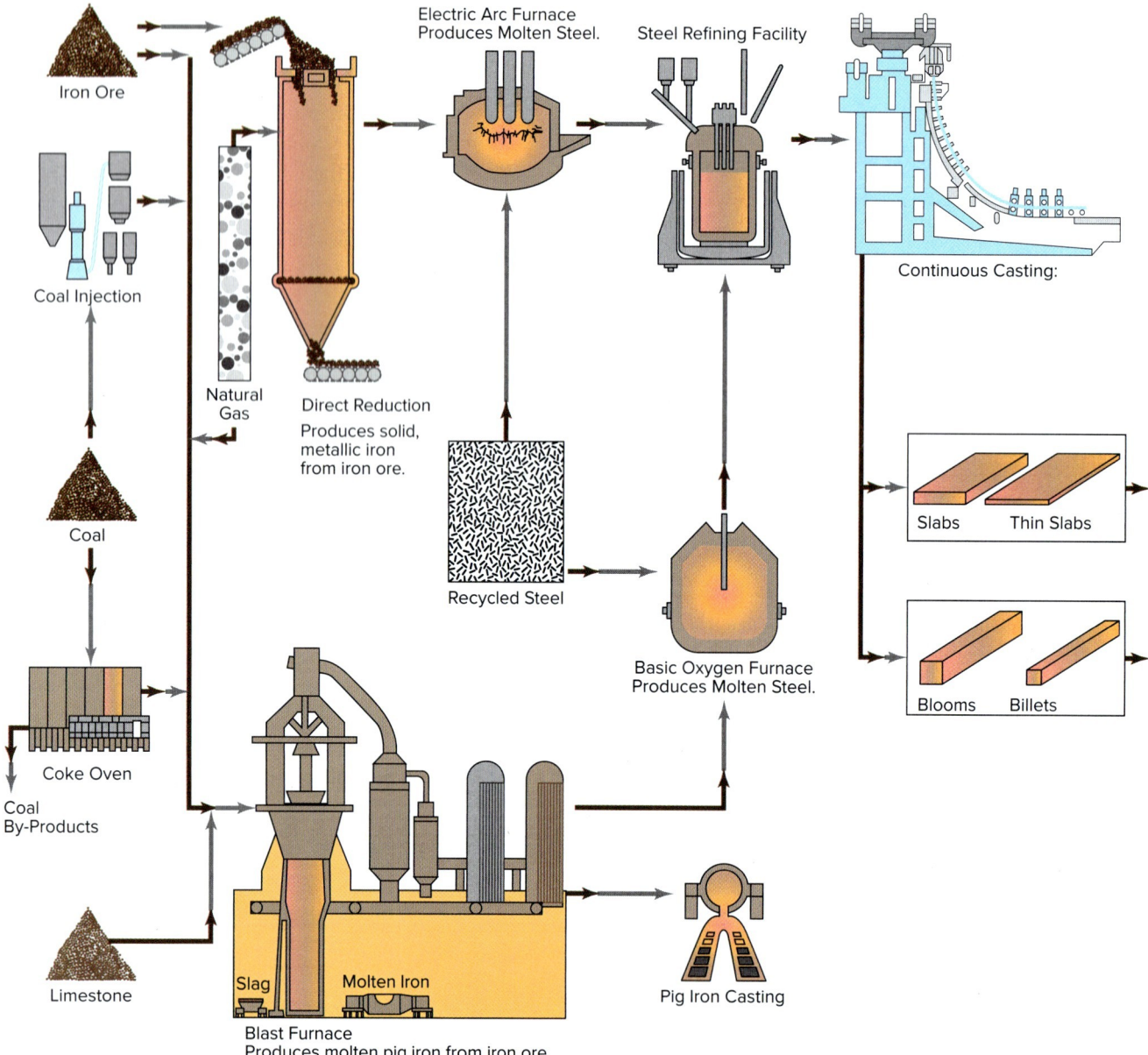

A FLOWLINE ON STEELMAKING

This is a simplified road map through the complex world of steelmaking. Each stop along the routes from raw materials to mill products contained in this chart can itself be charted. From this overall view, one major point emerges: Many operations—involving much equipment and workers—are required to produce civilization's principal and least expensive metal.

The raw materials of steelmaking must be brought together, often from hundreds of miles away, and smelted in a blast furnace to produce most of the iron that goes into steelmaking furnaces. Air and oxygen are among the most important raw materials in iron and steelmaking.

Continuous casting machines solidify steel into billets, blooms, and slabs. The metal is usually formed first at high temperature, after which it may be cold-formed into additional products.

Fig. 3-11 A "road map" of raw materials to mill products. Source: American Iron & Steel Institute

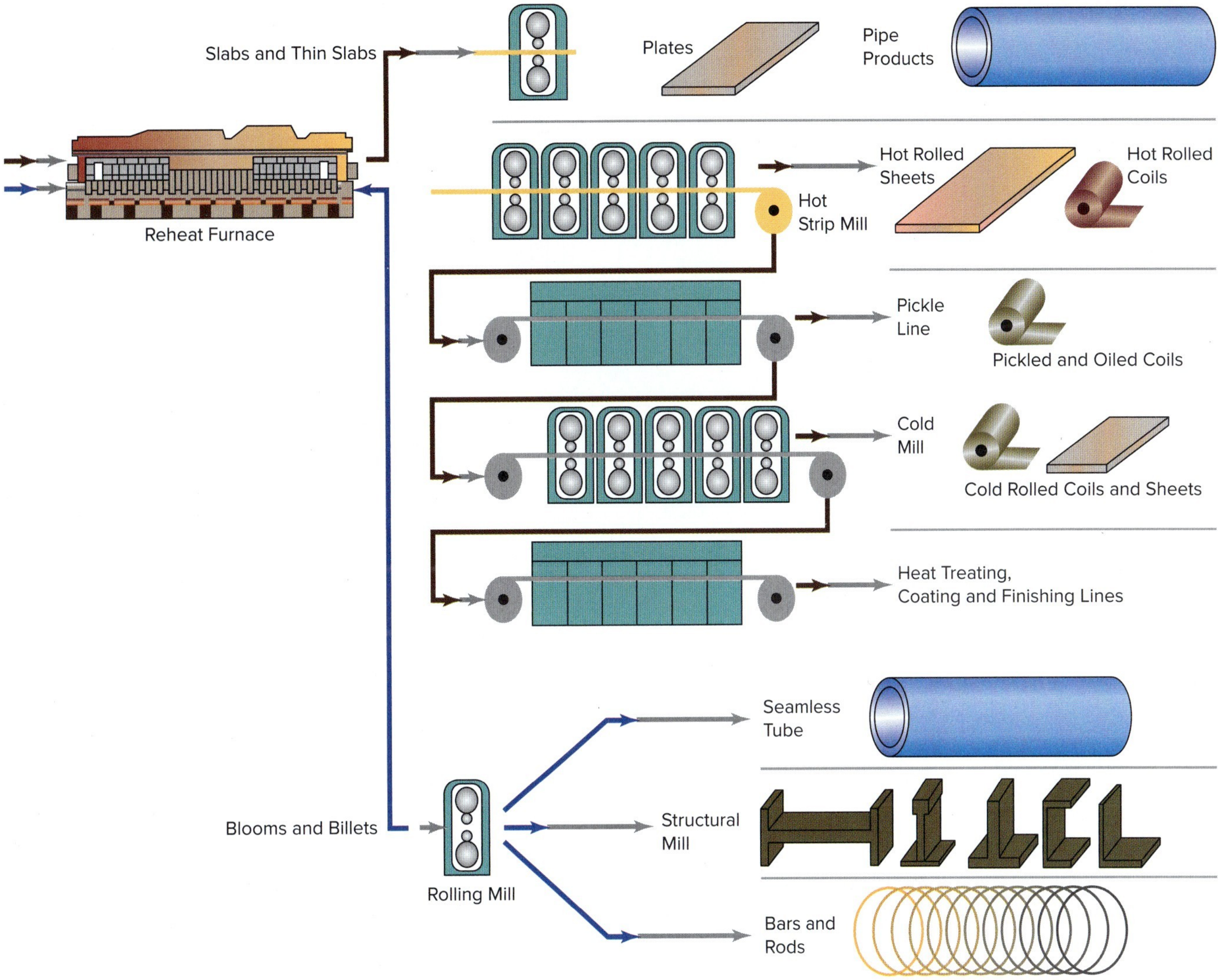

Fig. 3-11 A "road map" of raw materials to mill products. *(Concluded)*

of large graphite (carbon) electrodes, the development of improved refractory materials for linings, and better furnace design.

Electricity is used solely for the production of heat and does not impart any properties to the steel. These furnaces have three electrodes ranging from 4 to 24 inches in diameter. They produce a direct arc with three-phase power and are supplied with electric current through a transformer. Newer furnaces have electrical capacity between 900–1,000 kVA per ton of steel being processes. This could amount to 42,000 amperes, thus a very high energy arc. Keep this in mind next time you are welding at a few hundred amperes. The electrodes enter the furnace through the roof. The roof is removable and can be swung aside to charge the furnace.

The charge consists almost entirely of scrap with small amounts of burned lime and mill scale. The furnaces are circular and can be tilted to tip the molten steel into a ladle, Fig. 3-13, page 53. They may be lined with either basic (magnesite, dolomite) or acid (silica brick) refractory materials, Fig. 3-14, page 53.

For video showing an electric furnace in operation, please visit www.mhhe.com/welding.

Before World War II, practically all alloy, stainless, and tool steels were produced in electric furnaces. Today, however, ordinary steels may also be produced in those

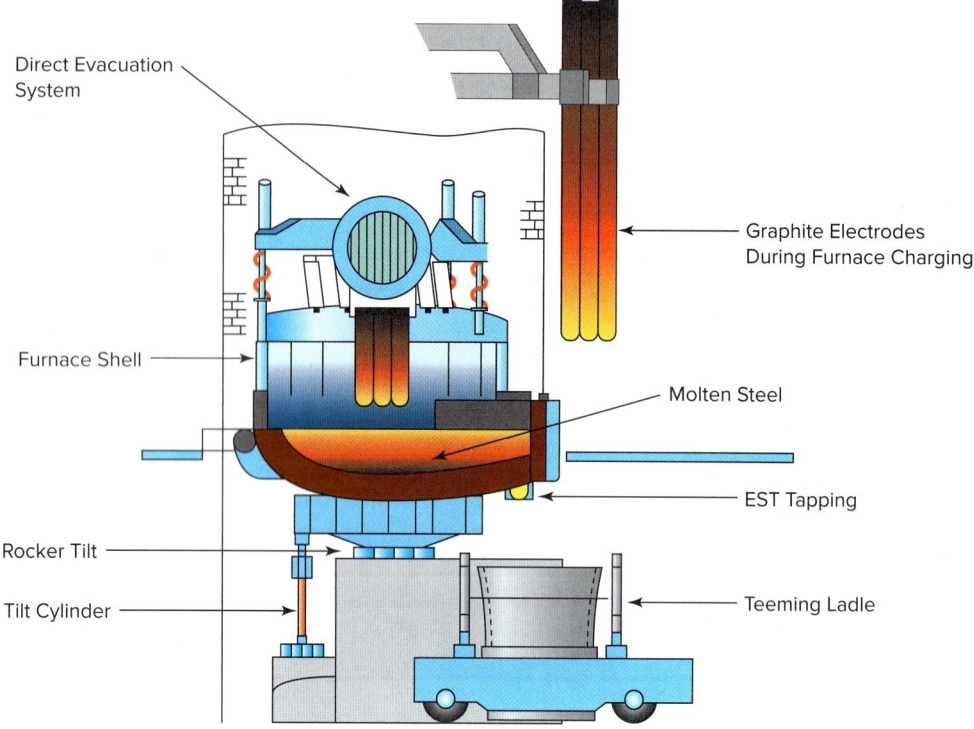

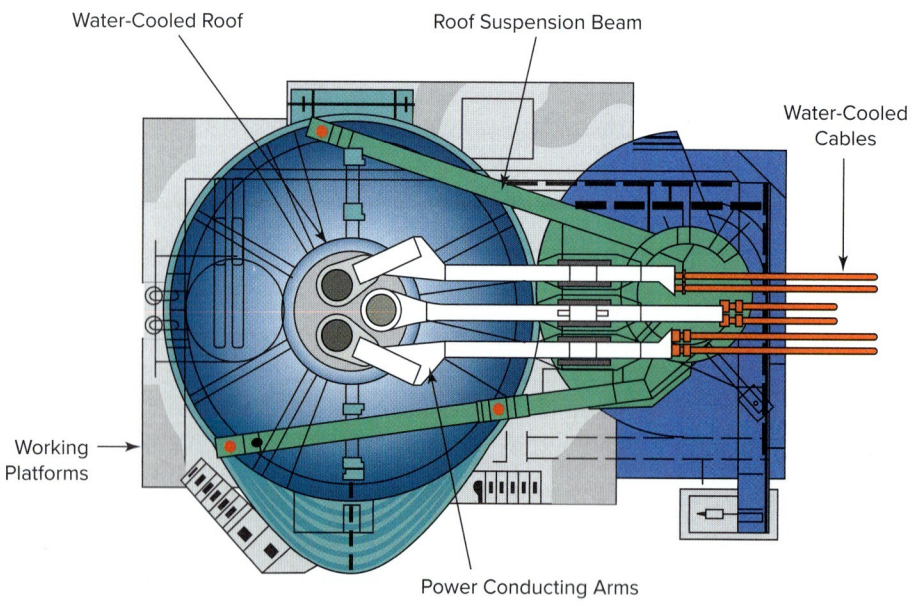

Fig. 3-12 The electric furnace process. Source: American Iron & Steel Institute

areas where there are large supplies of scrap and favorable electric power rates.

The *electric induction furnace,* Fig. 3-15, page 54, is essentially a transformer with the molten metal acting as the core. It consists of a crucible, usually made of magnesia, surrounded by a layer of tamped-in magnesia refractory. Around this is a coil made of copper tubing, forming the primary winding that is connected to the current source. The coil is encased in a heavy box with a silica brick bottom lining. A lip is built into the box to

Fig. 3-13 Making a "pour" from an electric furnace. Note the large electrodes through which the electric current flows to provide the arcing that produces the heat to melt the metal. United States Steel Corporation

allow the metal contents to run out as the furnace is tilted forward.

The charge consists of scrap of the approximate composition desired plus necessary ferro-alloys to give final chemical composition within specifications. Scrap may be any size that will fit into the furnace. A 1,000-pound charge can be melted down in 45 minutes. After melting is complete, the metal is further heated to the tapping temperature in about 15 minutes. During this time small additions of alloys or deoxidizers are added. When the proper temperature is obtained, the furnace is tilted and liquid metal runs out over the lip into a ladle or directly into a mold.

Oxygen Process

The oxygen process, also known as the *Linz-Donawitz* process, was first established in Linz, Austria, in 1952, and in Donawitz, Austria, a short time later. The process was first used in the United States in 1954.

The Linz-Donawitz process is a method of pig iron and scrap conversion whereby oxygen is injected downward over a bath of metal. A fairly large amount of hot metal is necessary to start the

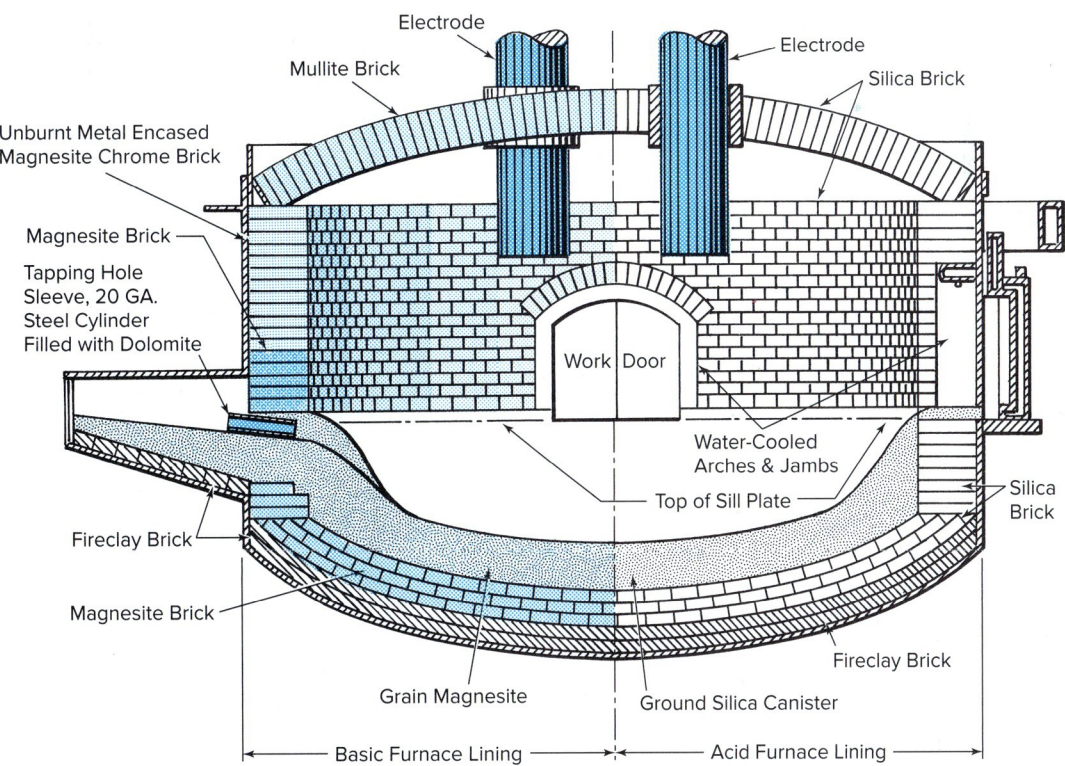

Fig. 3-14 The electric arc furnace produces heat through arcing action from electrodes to metal. Electrodes move down as metal melts.

Steel and Other Metals   Chapter 3   53

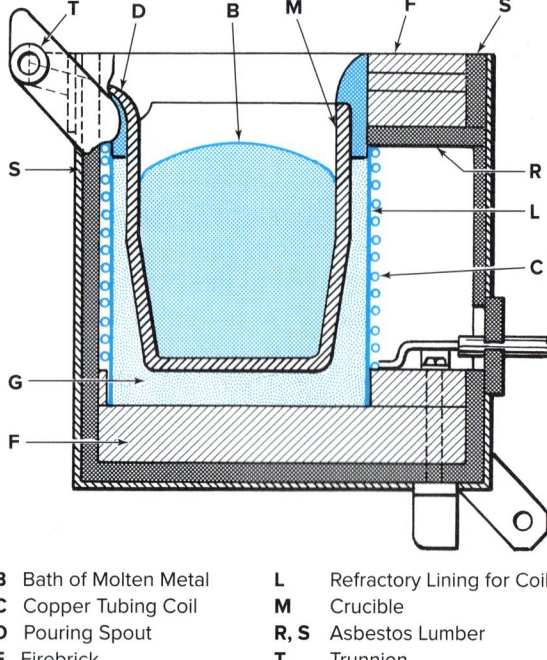

B Bath of Molten Metal	L Refractory Lining for Coil
C Copper Tubing Coil	M Crucible
D Pouring Spout	R, S Asbestos Lumber
F Firebrick	T Trunnion
G Powdered Refractory	

Fig. 3-15 Cross section of an electric induction furnace. Heat is generated by means of transformer action, where the bath of molted metal (B) acts as the core of the secondary winding; water-cooled copper coil (C) carries the primary electric current.

Fig. 3-16 Charging hot metal into a 150-ton basic oxygen furnace. H. Mark Weidman Photography/Alamy Stock Photo

oxidizing reaction so that the scrap content is limited to about 30 percent of the charge. A pear-shaped vessel is charged with molten pig iron and scrap while the vessel is in a tilted position, Fig. 3-16. Then the vessel is turned upright. Fluxes are added, and high purity oxygen is directed over the surface of the molten metal bath by the insertion of a water-cooled lance into the vessel mouth, Fig. 3-17.

The chemical reaction of the oxygen and fluxes refines the pig iron and scrap into steel. The temperature reaches 3,000°F, and the refining continues for 20 to 25 minutes.

When the refining is complete, the lance is withdrawn. The furnace is tilted, and the steel is tapped through a hole in the side near the top. The slag is also removed, and the furnace is ready for another charge. The complete process is shown in Figs. 3-18 and 3-19, page 56.

The main advantage claimed for the process is that it takes only 45 minutes to complete. Heats as large as 300 tons are made. Steels of any carbon content can be produced. While alloy and stainless steels have been made by the oxygen process, the holding time in the vessel to

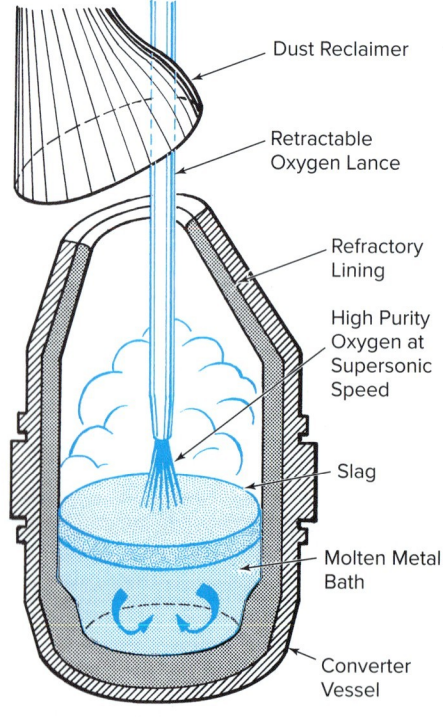

Fig. 3-17 Basic oxygen steelmaking furnace. After scrap and hot metal are charged into the furnace, the dust cap is put on, and oxygen is blown through the lance to the surface of the molten metal in order to burn out impurities.

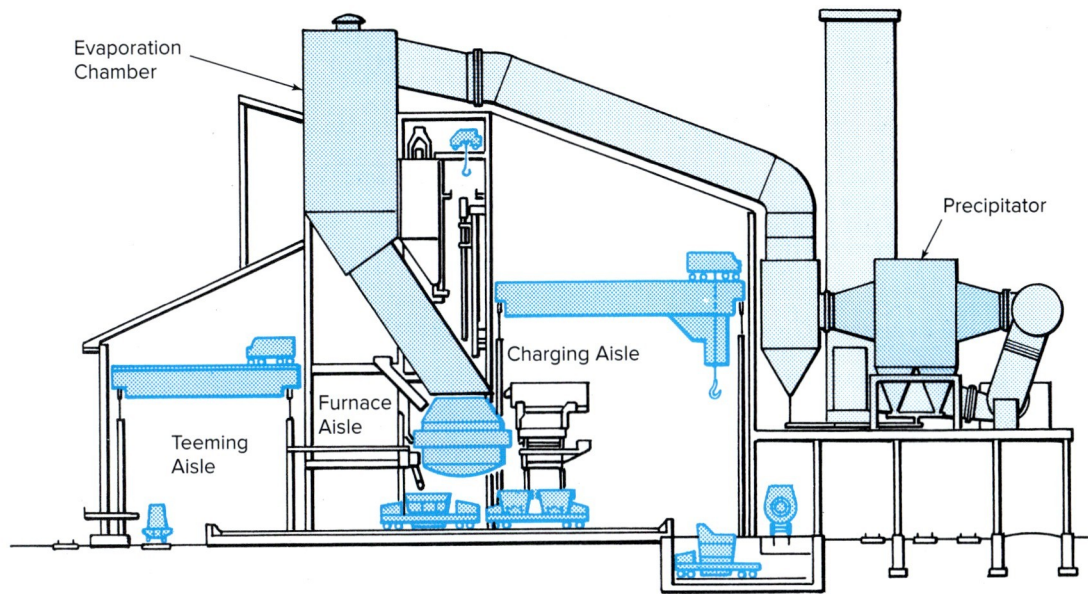

Fig. 3-18 Cross section of a basic oxygen steel plant. The furnace (converter vessel), nearly 18 feet in diameter and 27 feet high, is just left of center. The charging box at the right of the converter is 25 feet above floor level. The entire steelmaking cycle takes about 45 minutes from tap to tap.

obtain the desired chemical composition largely eliminates the short time cycle advantage and, in general, only carbon steels are produced.

Vacuum Furnaces and Degassing Equipment

The melting of steel and other alloys in a vacuum reduces the gases in the metal and produces metal with a minimum of impurities. The gases formed in a vacuum furnace are pulled out of the metal by vacuum pumps. Figure 3-20 (pages 57–58) illustrates the various vacuum melters and degassers. There are two general types of furnaces used for vacuum melting. The two processes are called vacuum induction melting and consumable electrode vacuum arc melting.

Vacuum Induction Melting Vacuum induction melting was first used in the 1940s. The charge is melted in a conventional induction furnace contained within an airtight, water-cooled steel chamber, Fig. 3-21, page 59. The furnace resembles induction furnaces used for air-melt processes. Advantages of the vacuum induction process include freedom from air contamination, close control of heat, and fewer air inclusions.

Consumable Electrode Vacuum Arc Melting Consumable electrode melting is a refining process for steel prepared by other methods. Steel electrodes of a predetermined composition are remelted by an electric arc in an air-tight, water-cooled crucible. The principle of operation is similar to arc welding. (Refer to Chapter 12.)

The furnace consists of a water-cooled copper crucible, a vacuum system for removing air from the crucible during melting, and a d.c. power source for producing the arc, Fig. 3-22, page 59. The electrode is attached to an electrode holder that feeds the electrode during the remelting operation to maintain the arc. The copper crucible is enclosed by a water jacket that provides the means of controlling ingot solidification.

In general, both of these processes produce high quality steel and steel alloys. The equipment has the following advantages:

- Production of alloys too expensive to manufacture by air-melt processes
- Use of reactive elements
- Decreased amounts of hydrogen, oxygen, and nitrogen in the finished product
- Improved mechanical properties
- Close heat control
- Better hot and cold workability

Vacuum Degassing The vacuum degassing of molten steel is a refining operation. Its purpose is to reduce

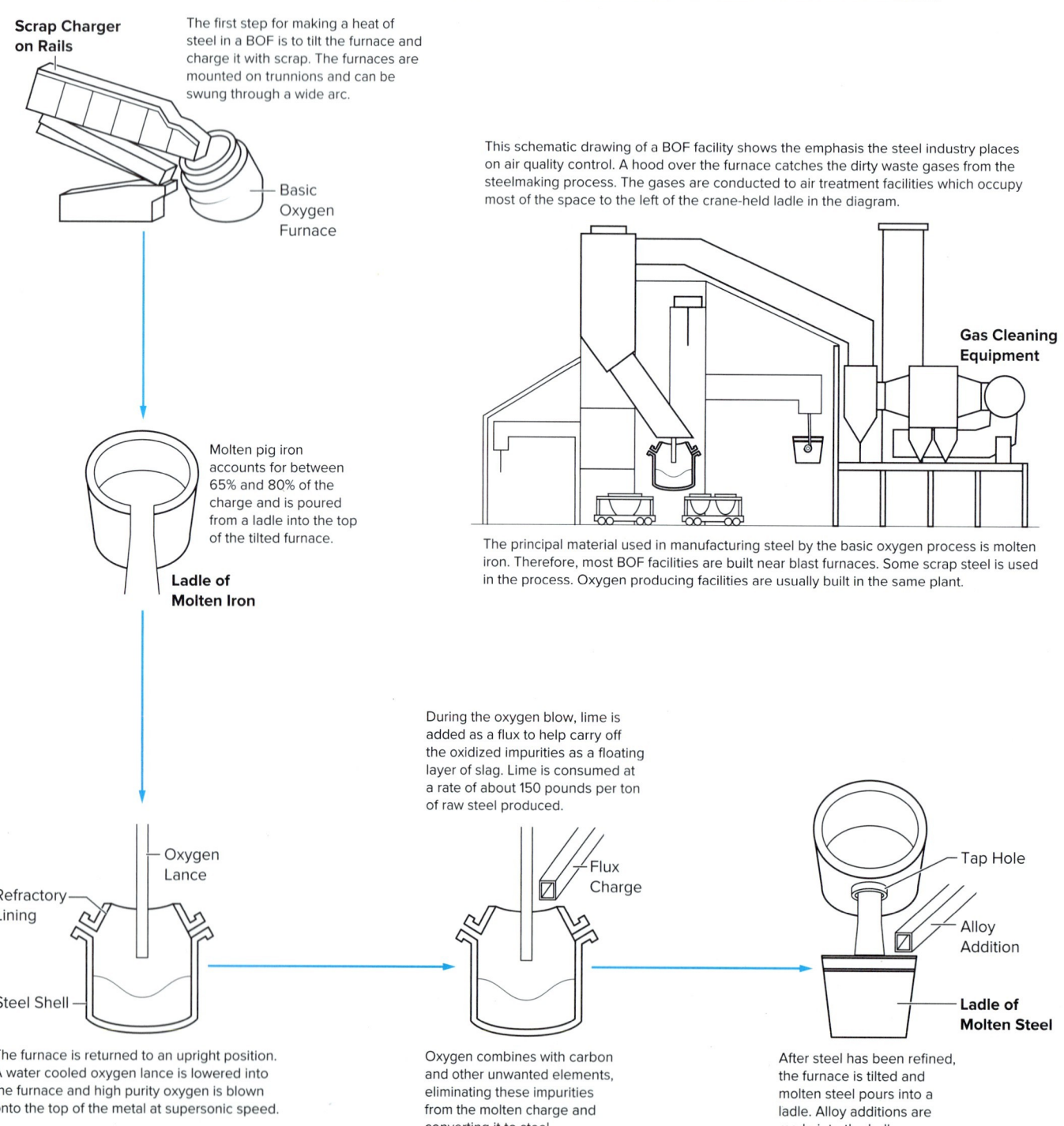

Fig. 3-19 The basic oxygen process. Adapted from American Iron & Steel Institute

Vacuum Processing of Steel

Steels for special applications are often processed in a vacuum to give them properties not otherwise obtainable. The primary purpose of vacuum processing is to remove such gases as oxygen, nitrogen, and hydrogen from molten metal to make higher-purity steel.
Many grades of steel are degassed by processes similar to those shown on this page. Even greater purity and uniformity of steel chemistry than is available by degassing is obtained by subjecting the metal to vacuum melting processes like those shown on the facing page.

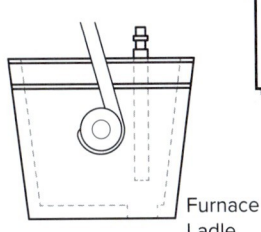
Furnace Ladle

The Vacuum Degassers

In vacuum stream degassing (left), a ladle of molten steel from a conventional furnace is taken to a vacuum chamber. An ingot mold is shown within the chamber. Larger chambers designed to contain ladles are also used. The conventionally melted steel goes into a pony ladle and from there into the chamber. The stream of steel is broken up into droplets when it is exposed to vacuum within the chamber. During the droplet phase, undesirable gases escape from the steel and are drawn off before the metal solidifies in the mold.

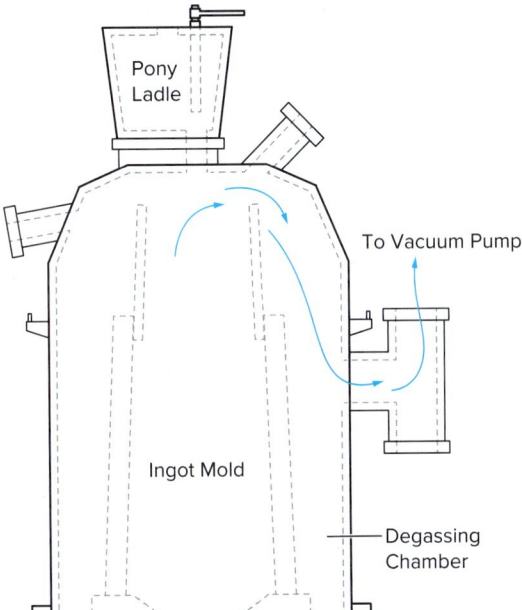

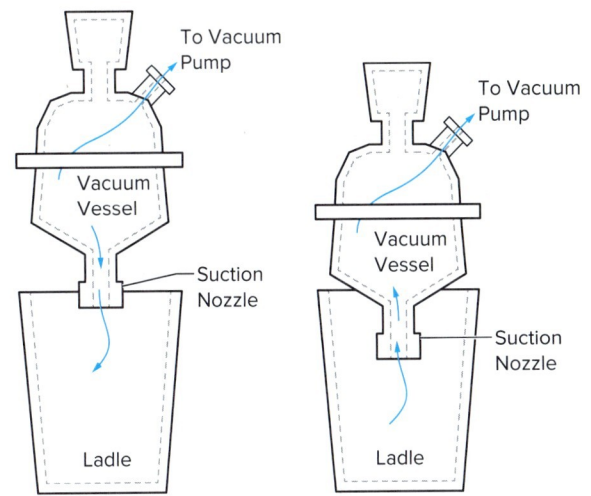

Ladle degassing facilities (right) of several kinds are in current use. In the left-hand facility, molten steel is forced by atmospheric pressure into the heated vacuum chamber. Gases are removed in this pressure chamber, which is then raised so that the molten steel returns by gravity into the ladle. Since not all of the steel enters the vacuum chamber at one time, this process is repeated until essentially all the steel in the ladle has been processed.

Fig. 3-20 Vacuum degassing and melting. Source: American Iron & Steel Institute *(Continued)*

the amounts of hydrogen, oxygen, and nitrogen in steel. The process is carried out after the molten metal is removed (tapped) from the furnace and before it is poured into ingots and castings. It is based on the principle that the solubility of a gas in liquid steel decreases as pressure decreases. There are three processes used nowadays.

Stream Degassing Steel is poured into a tank from which the air has been already removed. After degassing, it is collected in an ingot mold or ladle, Fig. 3-23, page 60.

Ladle Degassing A ladle of molten steel is placed in a tank and then air is removed from the tank, thus exposing the metal to the vacuum, Fig. 3-24, page 60. This method has the advantage of being able to process smaller amounts of steel than stream degassing.

Vacuum Lifter Degassing A vacuum is created in a chamber suspended above a ladle of steel. The metal is forced upward into the vacuum chamber through nozzles by means of atmospheric pressure, Fig. 3-25, page 61.

The Vacuum Melters

Vacuum melting by either of the two processes shown on this page has helped make possible steels for many advances in space flight, nuclear science, electronics, and industry. A third process called electroslag remelting is coming into increasing favor; it is an extension, in some ways, of the consumable-electrode method described below.

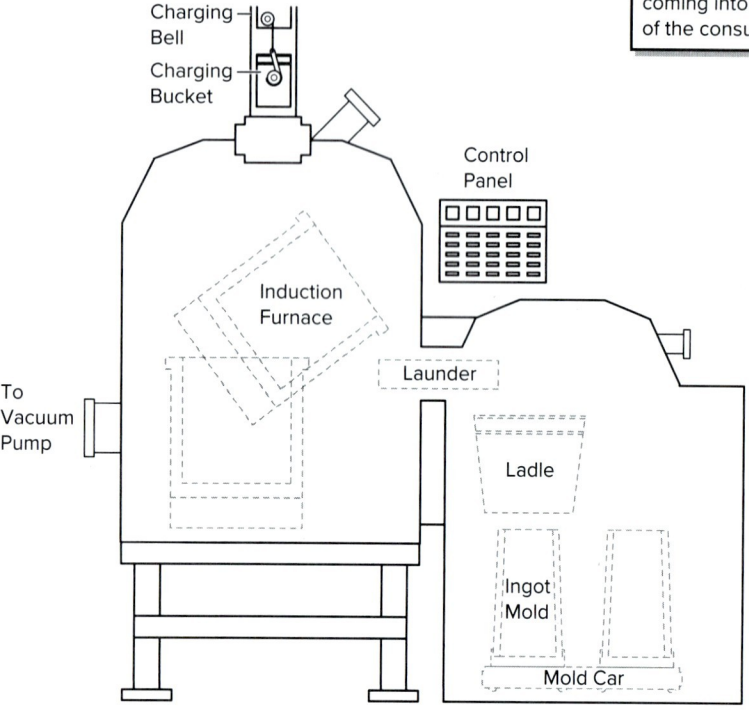

The vacuum induction process above melts and refines steel in a furnace surrounded by an electrical coil. A secondary current induced in the steel provides melting heat. The entire furnace is in a vacuum. Scrap or molten steel is charged to the furnace, from which most of the atmosphere has been evacuated. In the type of vacuum induction facility illustrated, after the gases are eliminated, the furnace tilts and pours newly refined steel into the trough (launder) which conveys it into a holding ladle from which it can be cast into separate ingot molds. All of these operations are remotely controlled within three separate vacuum chambers sealed off from each other.

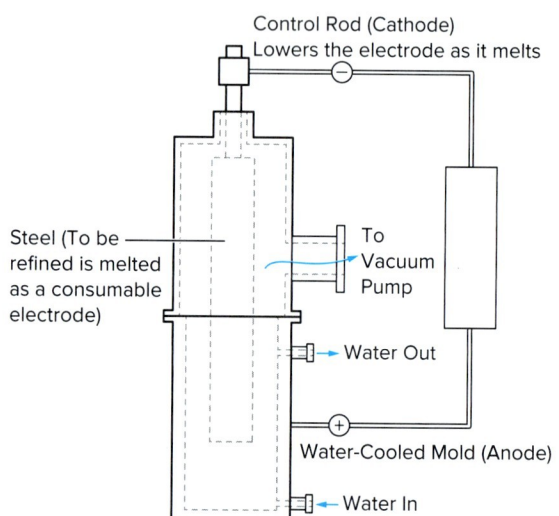

A vacuum arc process, called the "consumable electrode" process, remelts steels produced by other methods. Its purpose is to improve the purity and uniformity of the metal. The solid steel performs like a gigantic electrode in arc welding with the heat of the electric arc melting the end of the steel electrode. The gaseous impurities are drawn off by the vacuum in the chamber as the molten steel drops into the water-cooled mold below. The remelted product is almost free of center porosity after it solidifies. Inclusions are minimized.

Fig. 3-20 Vacuum degassing and melting. *(Concluded)*

The following benefits are generally derived from the degassing operation:

- The reduction of hydrogen eliminates flaking of the steel.
- The reduction of oxygen promotes internal cleanliness. Oxygen reduction, however, is not as low as that achieved in vacuum-melted steels.
- Nitrogen content is reduced slightly.
- The transverse ductility (flexibility across the grain of the metal) of most degassed forged products is nearly double that of air-cast steel.

Continuous Casting of Steel

Continuous casting is the process by which molten steel is solidified into a semifinished billet, bloom, or slab for subsequent finishing. Prior to the use of continuous casting in the 1950s, steel was poured into stationary molds to form ingots. Since that time, continuous casting has taken over this operation to achieve improved yield, quality, productivity, and cost efficiency. Figure 3-26, page 61, shows some examples of continuous caster configurations.

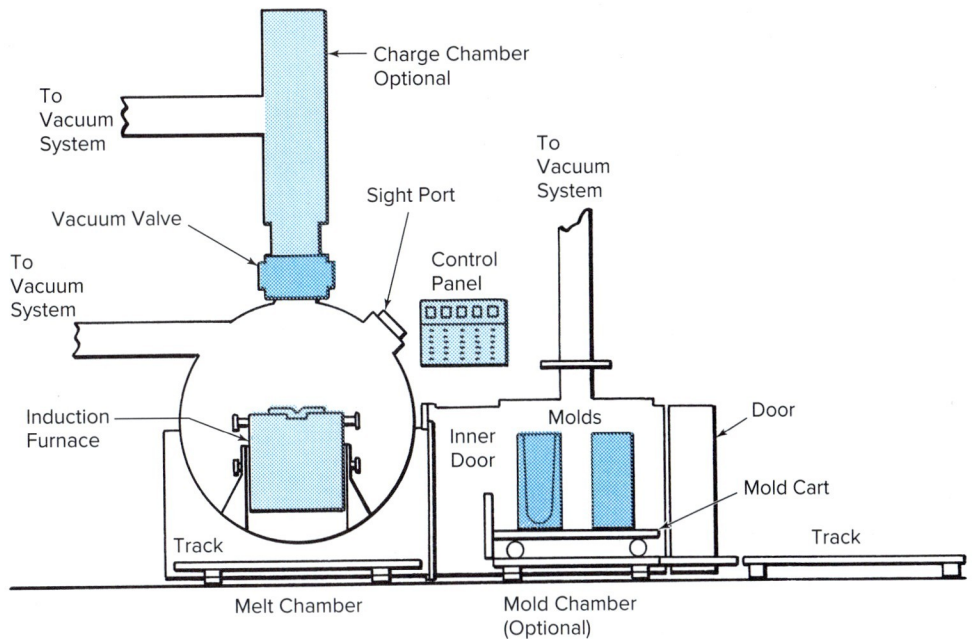

Fig. 3-21 Cross section of a typical vacuum induction furnace inside a vacuum chamber.

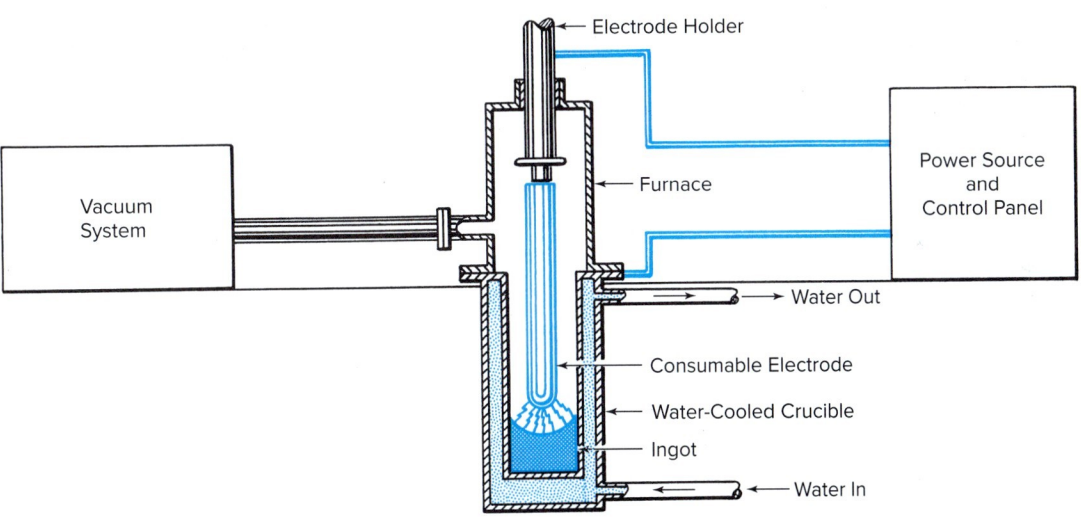

Fig. 3-22 Schematic drawing of a consumable electrode remelting furnace. Direct current produces an arc that melts a single electrode. Circulating water cools the ingot mold.

Steel from the BOF or the electric furnace is tapped into a ladle and taken to the continuous casting machine. The tundish is located under the ladle that has been raised onto the turret, which will rotate the ladle into the proper pour position. Refer to Fig. 3-26, which covers the flow through the continuous caster.

Depending on the product end use, various shapes are cast. The trend is to have the melting, casting, and rolling processes linked so that the casting shape substantially conforms to the finished product. The near-net-shape cast section is most commonly used for beams and flat-rolled products and greatly improves operation efficiency. The complete operation from liquid metal to finished rolling can be achieved within two hours.

How Continuous Casting Works To begin the casting operation, the mold bottom must be plugged with a steel dummy bar that seals it. The bar is held in

Steel and Other Metals **Chapter 3** 59

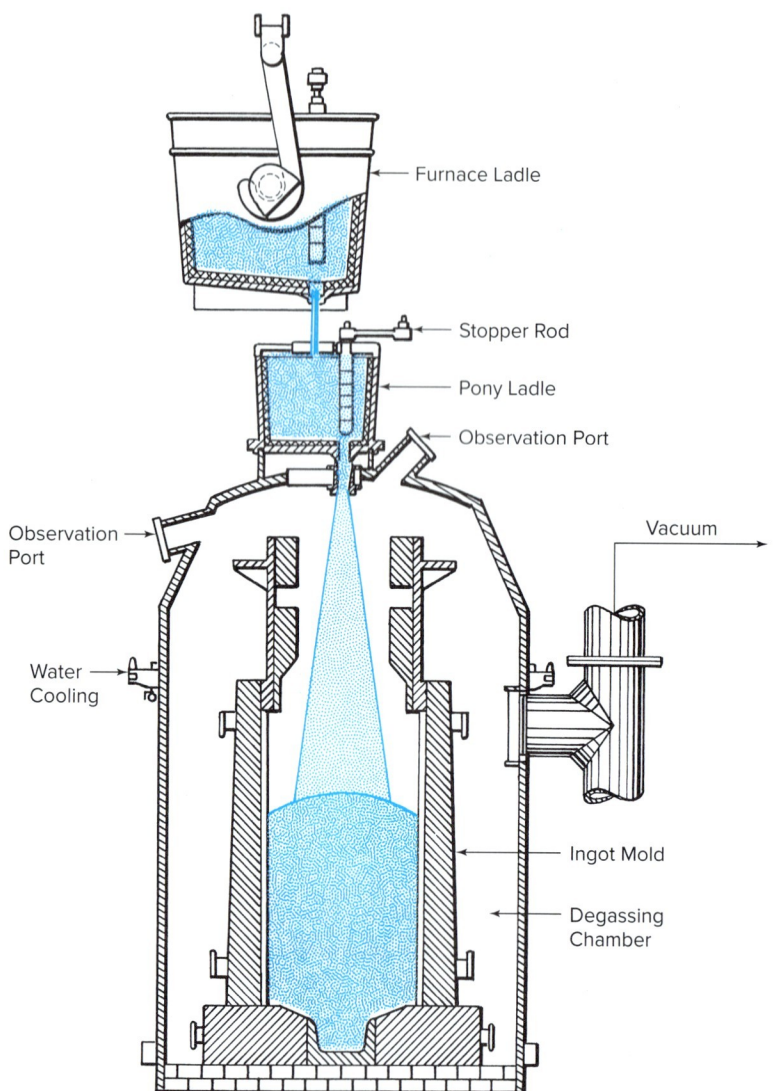

Fig. 3-23 Cross section of a vacuum degassing unit shows principal components. Molten steel at the top pours into a pony ladle that measures steel into the vacuum unit, permitting the escape of hydrogen and other gases.

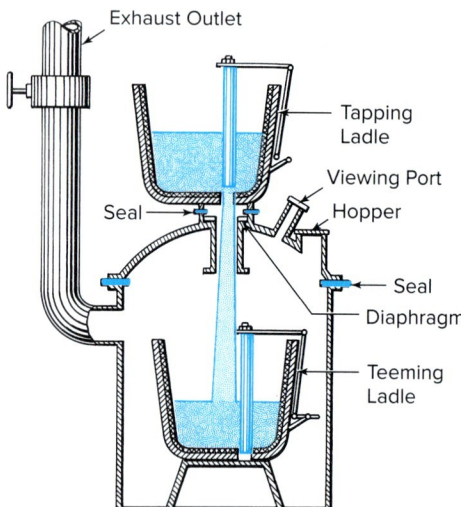

Fig. 3-24 The ladle degassing process substitutes a ladle for the ingot mold used in the stream degassing process.

place hydraulically by the straightener withdrawal units (see Fig. 3-26, item 6). The liquid metal is prevented from flowing into the mold via this bar. The steel poured into the mold is partially solidified, producing a steel strand with a solid outer shell and a liquid core. Once the solid steel shell is about 0.4 to 0.8 inch in the primary cooling area, the straightener withdrawal units withdraw the partially solidified strand out of the mold along with the dummy bar. Liquid steel is continuously poured into the mold to replenish the withdrawn steel at an equal rate. The withdrawal rate is dependent upon the cross section, grade, and quality of steel being produced, and may vary between 12 and 300 inches per minute. The time required for this casting operation is typically 1.0 to 1.5 hours per heat to avoid excessive ladle heat losses. (A *heat* is one pouring of a specified amount of molten metal.)

When exiting the mold, a roller containment section is entered by the strand (see Fig. 3-26, items 4 and 5) in which water or a combination of water and air is sprayed onto the strand, solidifying it. This area preserves cast shape integrity and product quality. Extended roller containments are used for larger cross sections. When the strand has solidified and passed through the straightener withdrawal units, the dummy bar is removed. Following the straightener, the strand is cut into the following as-cast products: slabs, blooms, billets, rounds, or beam blanks, depending on machine design.

Typically billets will have cast section sizes up to about 7 inches square. Bloom section sizes will range from approximately 7 inches square to about 15 inches by 23 inches. Round castings will be produced in diameters of anywhere from 5 to 20 inches. Slab castings will range in thickness from 2 to 16 inches and can be over 100 inches wide. Beam blanks take the shape of dog bones and are subsequently rolled into I beams. The aspect ratio is the width-to-thickness ratio and is used to determine the dividing line between blooms and slabs. A product with a 2.5:1 aspect ratio or greater is considered to be a slab.

The casting process comprises the following sections:

- A tundish, located above the mold, to feed liquid steel to the mold at a regulated rate
- A primary cooling zone or water-cooled copper mold through which the steel is fed from the tundish to generate a solidified outer shell sufficiently strong to maintain the strand shape as it passes into the secondary cooling zone
- A secondary cooling zone in association with a containment section positioned below the mold through which the still mostly liquid strand passes and is sprayed with water, or both water and air, to further solidify the strand
- An unbending and straightening section for all except straight vertical casters
- A severing unit (cutting torch or mechanical shears) to cut the solidified strand into pieces for removal and further processing

Liquid Steel Transfer There are two steps involved in transferring liquid steel from the ladle to the molds. Initially the steel must be transferred from the ladle to the tundish. Next the steel is transferred from the tundish to the molds. The tundish-to-mold steel flow is regulated by orifice control devices of various designs: slide gates, stopper rods, or metering nozzles. Metering nozzles are controlled by the tundish steel level adjustment.

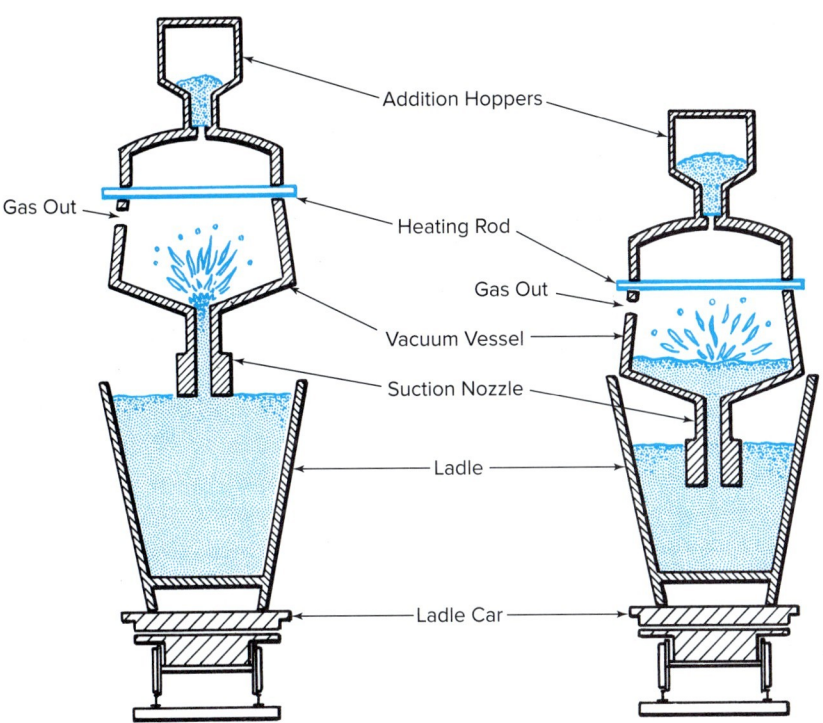

Fig. 3-25 Vacuum lifter degassing works on the principle of atmospheric pressure pushing steel upward into a newly created vacuum. After the steel is exposed to the vacuum for the proper time, it is returned to the lower ladle.

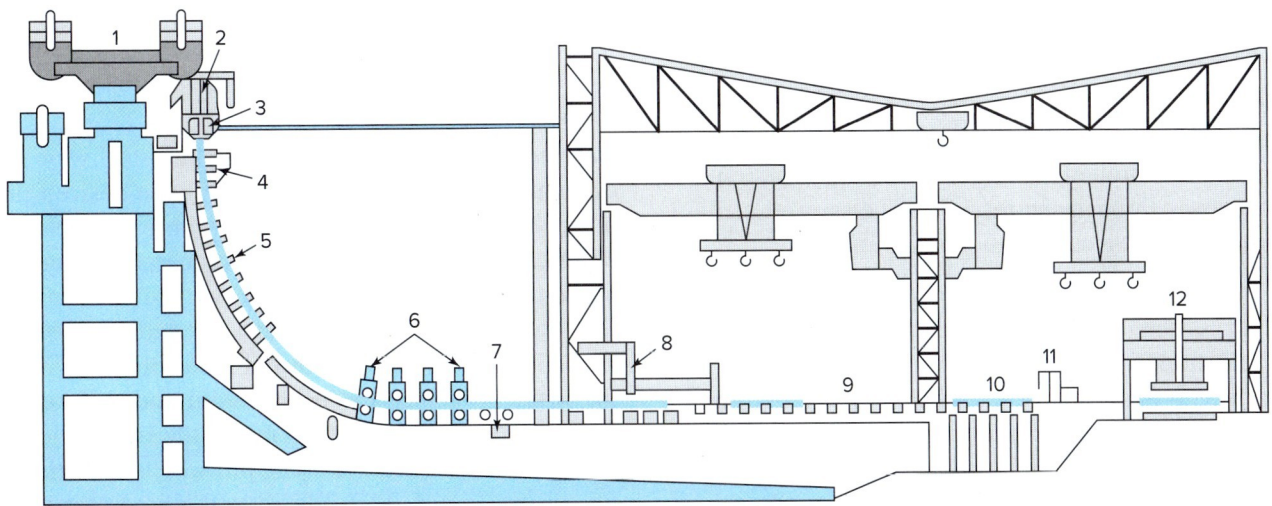

Fig. 3-26 Examples of continuous casters. Liquid steel flows (1) out of the ladle, (2) into the tundish, and then into (3) a water-cooled copper mold. Solidification begins in the mold, and continues through (4) the first zone, and (5) the strand guide. In this configuration, the strand is (6) straightened, (8) torch-cut, and then (12) discharged for intermediate storage or hot-charged for finished rolling. Source: American Iron & Steel Institute

Tundish Overview The typical shape of the tundish is rectangular; however, delta and T shapes are also used. Nozzles are located along the bottom of the tundish to distribute liquid steel to the various molds. The tundish also serves several other key functions:

- Enhances oxide inclusion separation
- Provides a continuous flow of liquid steel to the mold during ladle exchanges
- Maintains a steady metal height above the nozzles to the molds, thereby keeping steel flow constant and hence casting speed constant
- Provides more stable stream patterns to the mold(s)

Mold The purpose of the mold is to allow the establishment of a solid shell sufficient in strength to contain its liquid core upon entry into the secondary spray cooling zone. Key product elements are shape, shell thickness, uniform shell temperature distribution, defect-free internal and surface quality with minimal porosity, and few nonmetallic inclusions.

The mold is an open-ended box structure containing a water-cooled inner lining fabricated from a high purity copper alloy. Mold heat transfer is critical though quite complex, generally requiring computer modeling to aid in proper design and operating practices.

Mold oscillation is necessary to minimize friction and sticking of the solidifying shell and to avoid shell tearing and liquid steel breakouts that can wreak havoc on equipment and cause machine downtime to handle clean up and repairs. Oscillation is achieved either hydraulically or via motor-driven cams or levers that support and oscillate the mold. Friction is further reduced between the shell and mold through the use of mold lubricants such as oils or powdered fluxes.

Secondary Cooling Typically, the secondary cooling system comprises a series of zones, each responsible for a segment of controlled cooling of the solidifying strand as it progresses through the machine. The sprayed medium is either water or a combination of air and water.

Three basic forms of heat transfer occur in this region:

- **Radiation,** or the transfer of heat energy from the surface to the atmosphere, is the predominant form of heat transfer in the upper regions of the secondary cooling chamber.
- **Conduction,** or the transfer of heat energy through direct contact of the material's solid structure, is the process used for the transfer of heat from the product through the shell and through the thickness of the rolls by their direct contact.
- **Convection,** or heat transfer by moving airflow, is the heat transfer mechanism that occurs by quickly moving sprayed water droplets or mist from the spray nozzles. The droplets penetrate the steam layer next to the steel surface, and the water then evaporates, cooling the surface.

The purpose of the spray chamber is to:

- Enhance and control the rate of solidification and allow full solidification in this region
- Regulate strand temperature via spray-water intensity adjustment
- Control the machine containment cooling

Once the strand has been cooled to a sufficient level, it is straightened. Next the strand is transferred on roller tables to a cutoff machine, which cuts the product into ordered lengths. Sectioning can be achieved via either torches or mechanical shears. Then, depending on the shape or grade, the cast section will either be placed in intermediate storage, hot-charged for finished rolling, or sold as a semifinished product. Prior to hot rolling, the product will enter a reheat furnace and have its thermal conditions adjusted to achieve optimum metallurgical properties and dimensional tolerances.

The continuous casting process has evolved from a batch process into a sophisticated continuous process. This transformation has occurred through understanding principles of mechanical design, heat transfer, steel metallurgical properties, and stress-strain relationships to produce a product with excellent shape and quality. Currently, the process has been optimized through careful use of electromechanical sensors, computer control, and production planning to provide a highly automated system design.

Solidification

Casting and Soaking Ingots If molten steel were to be cast into molds having the shape of the desired product, we would always be dealing with cast steel in our structures. Since *cast steel* is generally inferior to *wrought steel* (metal that is to be worked mechanically), the molten steel is poured into ingot molds or continuance casting and allowed to cool until solidified. To give the inside a chance to become solid and still keep the outside from cooling off too much, the ingot is lowered into a furnace called a *soaking pit,* which heats the steel for rolling. Figure 3-27 illustrates the processes used in solidifying steel.

Deoxidation In most steelmaking processes, the primary reaction involved is the combination of carbon and

The First Solid Forms of Steel

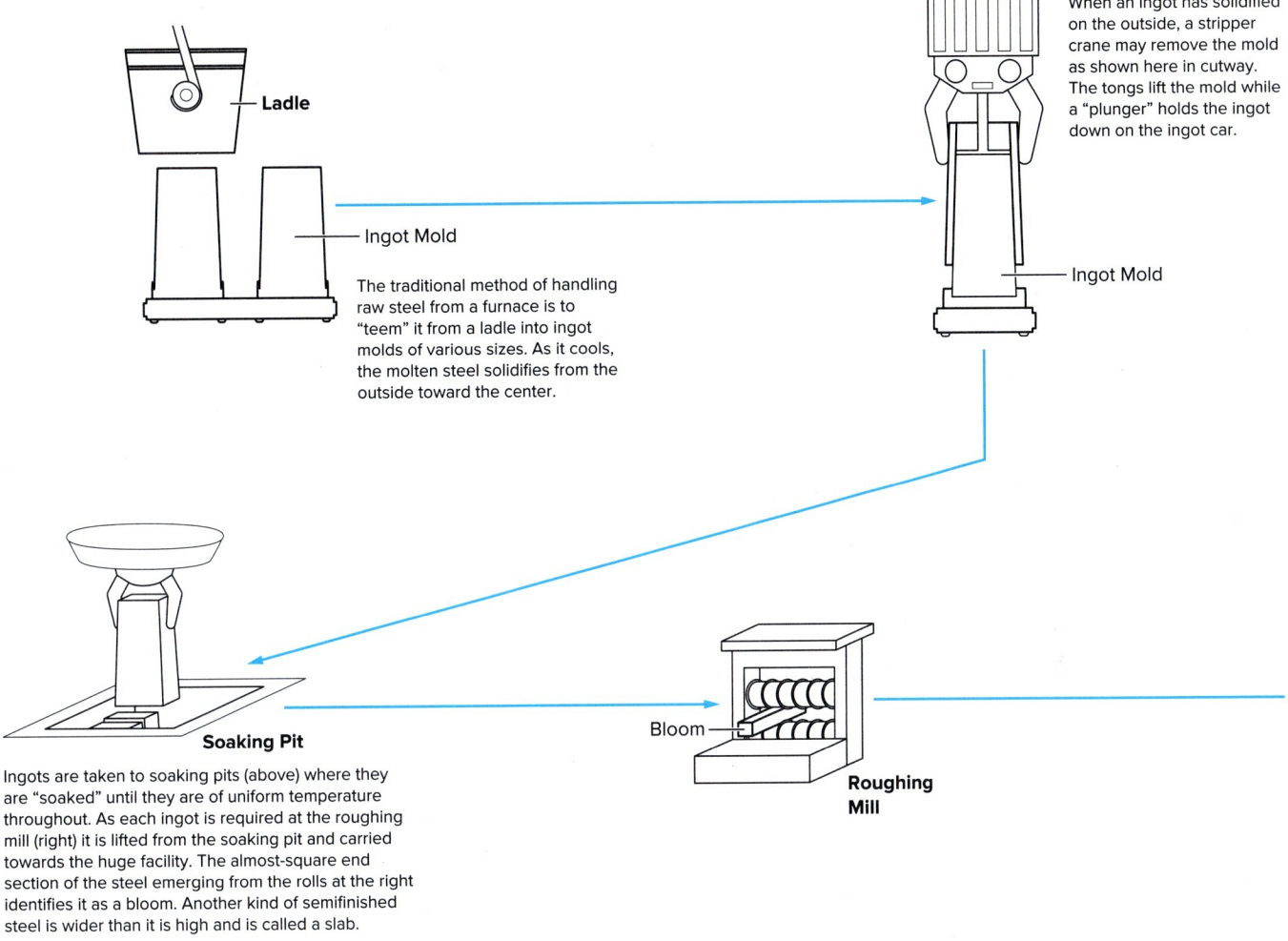

Fig. 3-27 Producing solid steel. Adapted from American Iron & Steel Institute *(Continued)*

oxygen to form a gas. Proper control of the amount of gas evolved during solidification determines the type of steel. If no gas is evolved, the steel is termed *killed* because it lies quietly in the molds. Increasing degrees of gas evolution result in killed, semikilled, capped, and rimmed steel.

Killed Steels Because killed steels are strongly deoxidized, they are characterized by a relatively high degree of uniformity in composition and properties. This uniformity of killed steel renders it most suitable for applications involving such operations as forging, piercing, carburizing, and heat treatment.

Semikilled Steels Semikilled steels are intermediate in deoxidation between killed and rimmed grades. Consequently, there is a greater possibility that the carbon will be unevenly distributed than in killed steels, but the composition is more uniform than in rimmed steels. Semikilled steels are used where neither the cold-forming and surface characteristics of rimmed steel nor the greater uniformity of killed steels are essential requirements.

Capped Steels The duration of the deoxidation process is curtailed for capped steels so that they have a thin low carbon rim. The remainder of the cross section, however, approaches the degree of uniformity typical of semikilled steels. This combination of properties has resulted in a great increase in the use of capped steels over rimmed steels in recent years.

Rimmed Steels Rimmed steels have the surface and cold-forming characteristics of capped steels. They are only slightly deoxidized so that a brisk evolution of gas occurs as the metal begins to solidify. The low

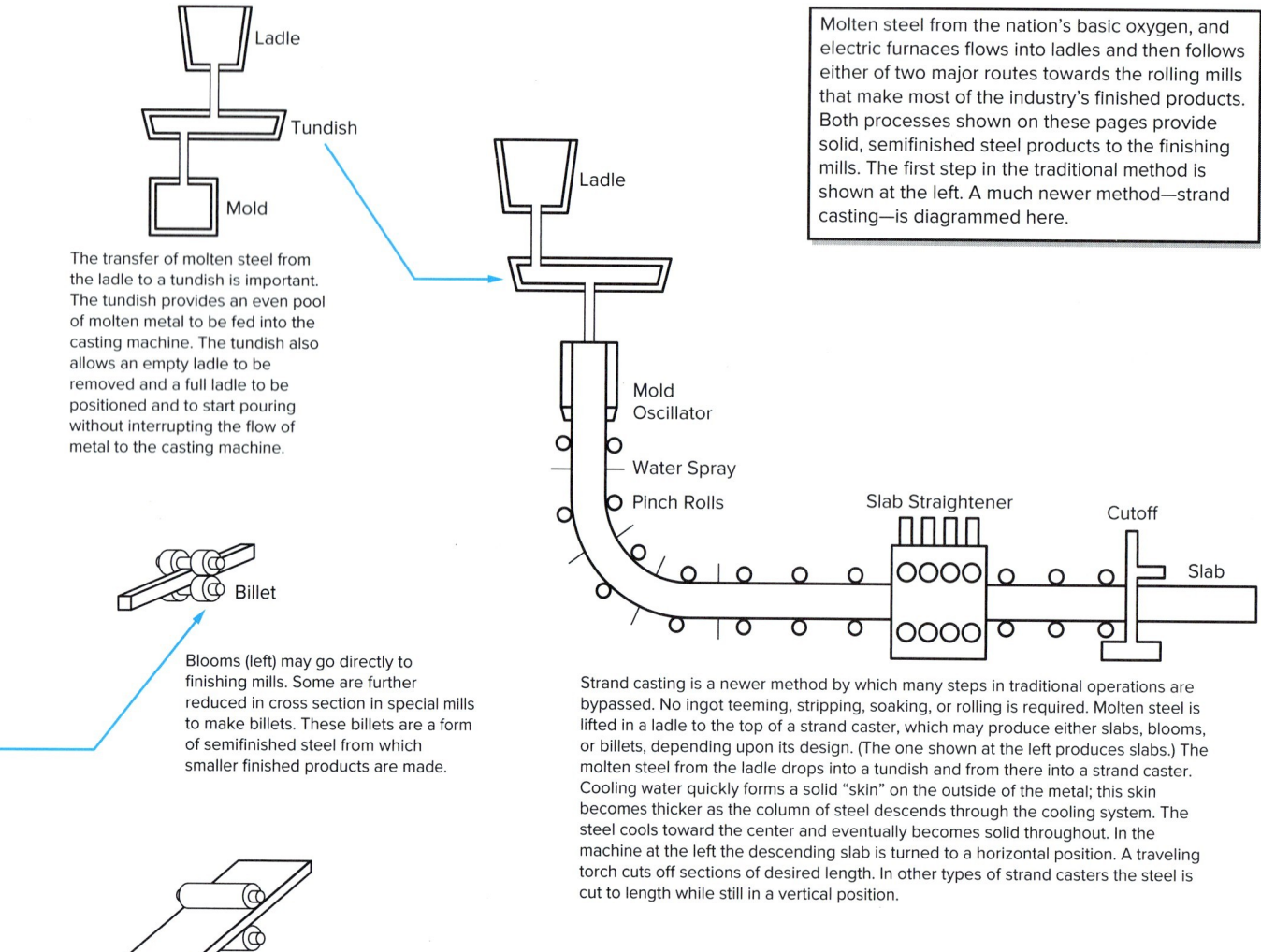

Fig. 3-27 Producing solid steel. *(Concluded)*

carbon surface layer of rimmed steel is very ductile. Rolling rimmed steel produces a very sound surface. Consequently, rimmed grades are adaptable to applications involving cold forming and when the surface is of prime importance.

Environmental Progress in the Steel Industry

As in most industries, there is a concern for our environment. The steel industry has had environmental expenditures amounting to more than $8 billion over the past 25 years. In a typical year, over 15 percent of the steel industry's capital spending is for environmental facilities. Costs to operate and maintain environmental facilities amount to $10 to $20 per ton of steel produced, which makes it a significant portion of the budget.

The amount of energy required to produce a ton of steel decreased by almost 45 percent from 1975 to 1998 as a result of technological improvements and energy conservation measures. Much of this was brought about by more accurate and efficient microprocessor controls.

This environmental and steelmaking efficiency had an impact on jobs in the steel industry. In the late sixties, 600,000 people were employed in the steel industry, while currently there are fewer than 200,000. With two-thirds less labor the nation is producing more steel. It used to take 12 labor hours to produce a ton of steel in the open hearth furnace, while currently it only takes 45 labor minutes in a basic oxygen furnace coupled with a continuous casting operation. The increased use of the electric arc furnace and minimills (which use scrap as the basic component) has greatly improved efficiency and made for more environmentally friendly production. It is also interesting to note that 50 percent of the steel types—chemistries, coatings, etc.—produced today did not even exist a decade ago. These stronger and more chemically resistant steel products have allowed automobile manufacturers to improve fuel efficiency.

The air quality in many North American steelmaking cities has greatly improved since 1970 because the discharge of air and water pollutants has been reduced by over 90 percent.

To achieve this, innovative environmental ideas such as the Bubble Concept were pioneered by the steel industry. (The Bubble Concept is a regulatory concept that provides flexibility and cost-effective solutions by establishing an emissions limit for an entire plant instead of on a process-by-process basis.)

The steel industry has made great strides in terms of recycling. Most hazardous wastes once generated by the steel industry are now being recycled for recovery for beneficial reuse, and over 95 percent of the water used for steel production and processing is recycled. Steel's recycling rate of 66 percent is far higher than that of any other material, capturing more than twice as much tonnage as all other materials combined. Each year steel recycling saves enough energy to electrically power more than 18 million households or to meet the needs of Los Angeles for more than 8 years.

The steel industry has worked cooperatively with federal environmental agencies, environmental groups, and others in such efforts as the coke oven emission regulatory negotiations, the Canadian Industry Program for Energy Conservation, and the U.S. Environmental Protection Agency's (EPA's) Common Sense Initiative to shape reasonable and cost-effective regulations acceptable to all stakeholders.

Metalworking Processes

After steel has been cast into ingot molds and solidified, it may be put through one or more of several metalworking processes to shape it and to further improve its characteristics. *Forging* and *rolling* serve two fundamental purposes. They serve the purely mechanical purpose of getting the steel into the desired shape, and they improve the mechanical properties by destroying the cast structure. This breaking up of the cast structure, also called "orienting the grain," is important chiefly in that it makes the steel stronger, more ductile, and gives it a greater shock resistance.

Forging

The method of reducing metal to the desired shape is known as forging. It is usually done with a steam hammer. The piece is turned and worked in a manner similar to the process used by the blacksmith when hand forging.

Considerable forging is done today with hydraulic presses instead of with hammers. The press can take cooler ingots and can work to closer dimensions.

Another forging process is **drop forging,** in which a piece of roughly shaped metal is placed between die-shaped faces of the exact form of the finished piece. The metal is forced to take this form by drawing the dies together. Many automobile parts are made in this way.

Rolling

Steel is nearly always rolled hot except for finishing passes on sheet. After rolling, ingots are known as blooms, billets, or slabs, depending on their size and shape.

- A **bloom** is square or oblong with a minimum cross-sectional area of 36 square inches.
- A **billet** is also square or oblong, but it is considerably smaller than a bloom. A bloom may sometimes be preheated and rolled into billets.
- A **slab** is oblong. It varies in thickness from 2 to 6 inches and in width from 5 to 6 feet.

Steel may be also rolled into bars of a wide variety of shapes such as angles, rounds, squares, round-cornered squares, hexagons, and flats as well as pipe and tubing.

ABOUT WELDING

SkillsUSA Competitions

SkillsUSA competitions are a way that students can show their potential. Winners of the state levels can go on to the National Competition. For many students of welding, the competitions advance their skills and give them more information about the trade. Opening up doors and increasing confidence are the main goals of this national organization, which also offers scholarships. Locate it on the Web at www.skillsusa.org.

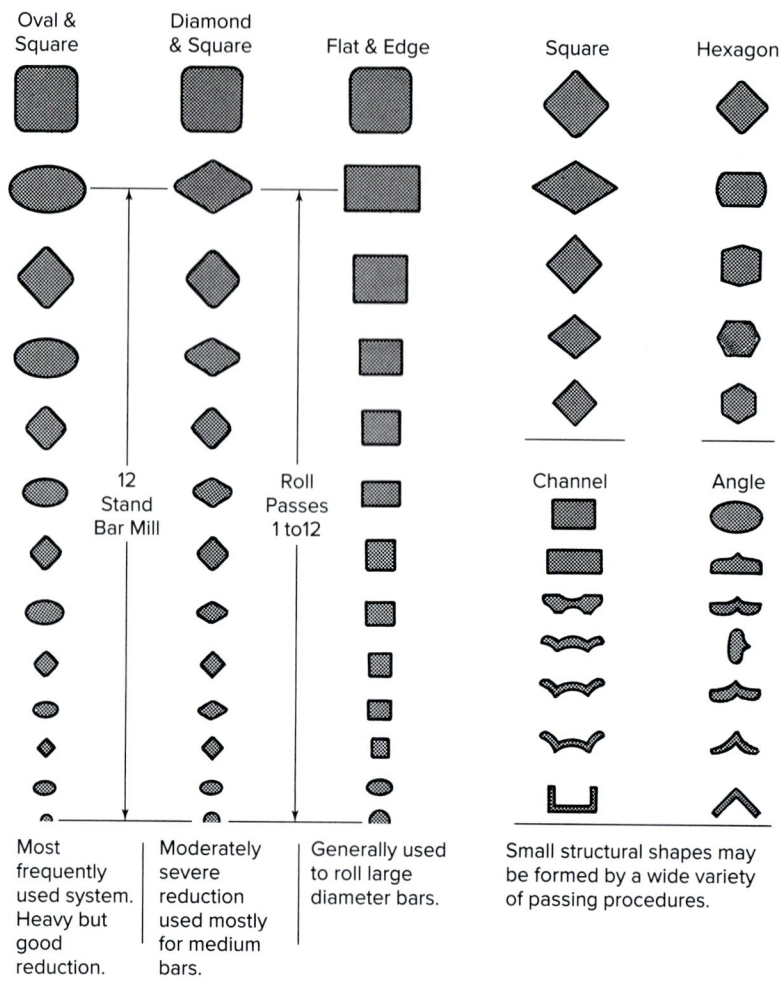

Fig. 3-28 Bar mill roll passes. Each vertical line of roll passes indicates the steps in rolling the bar or section shown at the bottom.

Figure 3-28 illustrates the various shapes produced by hot rolling.

About one-half of the rolled steel products made in the United States are flat rolled. These include such items as plates, sheet, and strip. Plates are usually thicker and heavier than strip and sheet. Figure 3-29 summarizes the processes for rolling steel.

Flat-rolled steel is divided into two major categories: hot rolled and cold rolled. Hot-rolled steel is usually finished at temperatures between 900 and 2,400°F. Untreated flat steel that is hot rolled is known as **black iron.** Cold-rolled products are reduced to their final thickness by rolling at room temperature. The surface finish is smooth and bright. If the sheets are coated with zinc, they are known as **galvanized** sheets; if they are coated with tin, they are known as **tin plate.**

Terne plate is sheet coated with an alloy of lead and tin.

Tubular steel products are classified according to two principal methods of manufacture: the welded and seamless methods. Welded tubing and pipe are made by **flash welding** steel strip. In this process, the metal pieces are heated until the contacting surfaces are in a plastic (semisolid) state and then forced together quickly under pressure. Seamless tubing or pipe is made from billets by two processes known as piercing and cupping. In **piercing,** a heated steel bar is pierced by a mandrel and rolled to the desired diameter and wall thickness, Fig. 3-30, page 68. In the **cupping process,** heated plate is formed around cup-shaped dies.

Steel may also be shaped into wire, bars, forgings, extrusions, rails, and structured shapes, Fig. 3-31, page 69. These are the basic steel shapes with which the welder fabricator works.

In the rolling operation, the grains are oriented in the direction of rolling. Just like a piece of wood, steel has more strength with the grain, less strength across the grain, and even less through the grain. Figure 3-32, page 69, describes this situation. Most rolled metals have this "anisotropy" property. To describe the direction effect of this grain orientation, the letters X, Y, and Z are used to identify the direction. This is why when taking a welder test using test plates the grain orientation must be known and the welds are made perpendicular to the grain or X direction.

Drawing

Drawing is the operation of reducing the cross section and increasing the length of a metal bar or wire by drawing it through a series of conical, tapering holes in a die plate. Each hole is a little smaller than the preceding one. Shapes varying in size from the finest wire to sections having a cross-sectional area of several square inches are drawn.

Extrusion

Some metals lend themselves to forming by pressing through an opening, rather than by drawing or rolling. Brass rod is usually formed in this way. By the extrusion process, perfectly round rods are obtainable. The metal to be extruded is placed in a closed chamber fitted with an opening at one end and a piston at the other

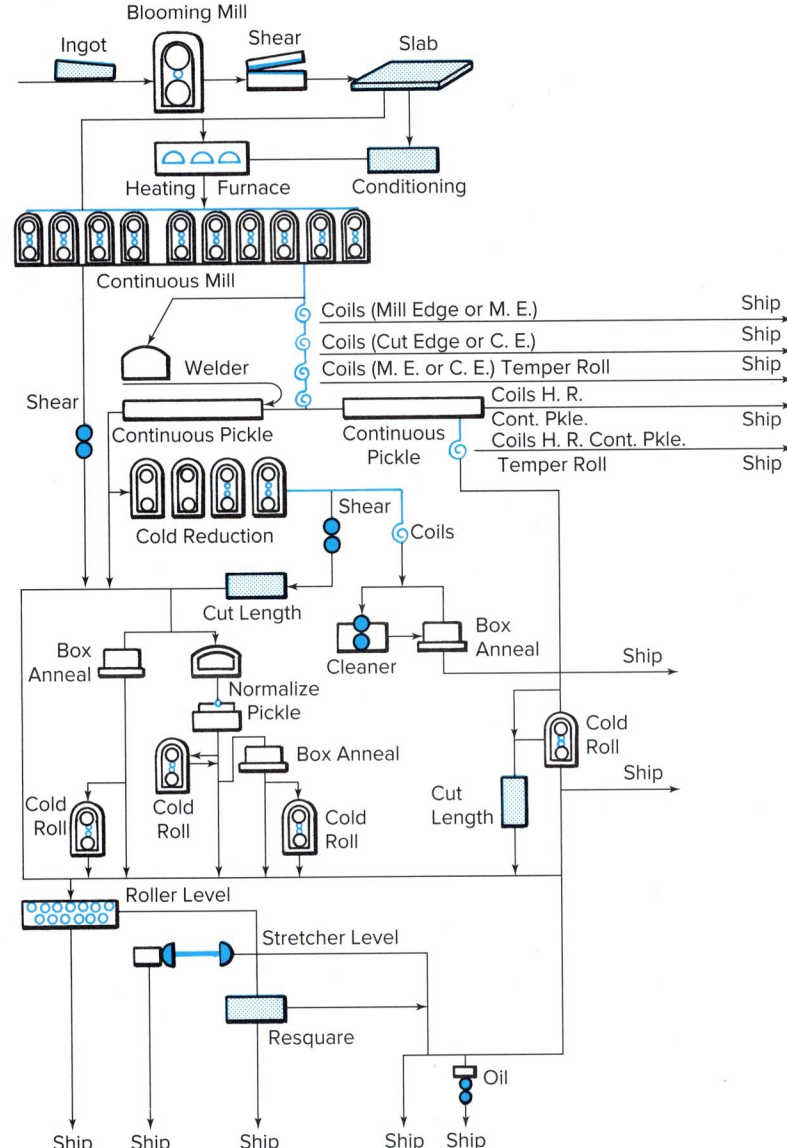

Fig. 3-29 Schematic diagram of a series of steel mill processes for the production of hot- and cold-rolled sheet and strip steel.

machining. Through various processes of heat treatment we can make a metal easier to machine, draw, or form by making it softer, or we can increase the hardness so that it will have wear resistance.

The important variables in any heat treatment process are (1) carbon content, (2) temperature of heating, (3) time allowed for cooling, and (4) the cooling medium (water, oil, or air).

Hardening Hardening is a process in which steel is heated above its critical point and then cooled rapidly. The critical point is the point at which the carbon, which is the chief hardening agent, changes the structure of the steel. This produces a hardness that is superior to that of the steel before heating and cooling. Only medium, high, and very high carbon steel can be treated in this way. The 24-ton vessel shown in Fig. 3-34, page 69, is ready for quenching. It has just been heated to 1,950°F and will be dunked for immersion cooling. The furnace uses low-sulfur gas for heating and can reach a temperature of 2,300°F.

Case Hardening Case hardening is a process that gives steel a hard, wear-resistant surface while leaving the interior soft and tough. The process chosen may be cyaniding, carburizing, nitriding, flame hardening, hard surfacing by welding, or metal spraying. Plain carbon steels and alloy steels are often case hardened.

Cyaniding Cyaniding is a method of surface-hardening low carbon steels. Carbon and

end and is forced out through the opening by hydraulic pressure.

Cold Working

Cold working is the shaping of metals by working at ordinary temperatures. They may be hammered, rolled, or drawn.

Heat Treatment

Heat treatment, Fig. 3-33, page 69, is the process of heating and cooling a metal for the purpose of improving its structural or physical properties. Very often this is done to remove stresses caused by welding, casting, or heavy

> **SHOP TALK**
>
> **People in Welding**
>
> It's interesting to hear from people about their careers in the welding field. Nick Peterson is a manager for Miller Electric, a manufacturer of welding machines and equipment. In an interview with *American Welder*, Nick spoke about his career motivation: "As a welder, I loved seeing my work after it was complete, taking pride in building something, in a job well done. The best part about being a manager is the diversity. One day I'm in a mine, the next an aerospace facility. There's always a new challenge."

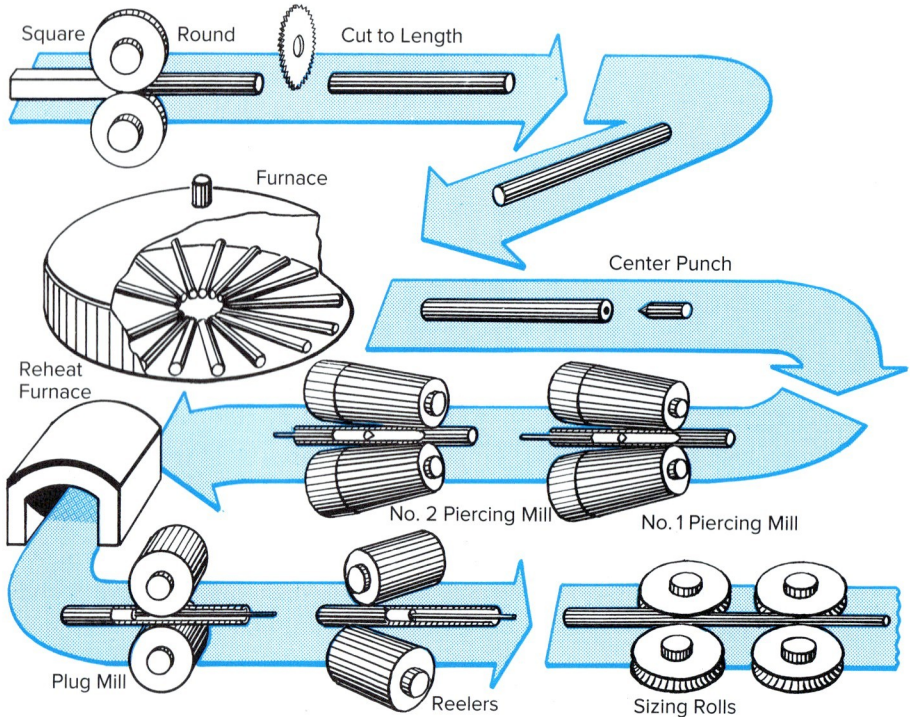

Fig. 3-30 Path of a solid steel round on its route toward becoming a tube. Included are heating, piercing, rolling, and sizing operations.

nitrogen are absorbed in the outer layer of the steel to a depth of 0.003 to 0.020 inch.

Cyaniding can be done in either liquid or gas form. Liquid cyaniding involves heating the parts in a bath of cyanide salts at a temperature of 1,550 to 1,600°F. The steel is held at this temperature for up to two hours, depending upon the depth of hardening desired. Then it is quenched in brine, water, or oil. Gas cyaniding involves case-hardening low carbon steels in a gas carburizing atmosphere that contains ammonia. The steel is heated to a temperature of 1,700°F and quenched in oil. These processes form a hard, but very thin, surface over the steel, beneath which the rest of the metal is still in a relatively soft condition.

Carburizing Carburizing is a process whereby low carbon steel is made to absorb carbon in its outer surface so that it can be hardened. The depth to which the carbon will penetrate depends upon the time a heat is held, the temperature that is reached, and the carburizing compound used.

Carburizing may be done with carbonaceous solids (solid substances containing carbon), cyanidizing liquids, or hydrocarbon gases. The carburizing process selected depends on the nature of the job, the depth of hardening desired, and the type of heat-treatment equipment available.

In gas welding, the surface of a weld may be hardened by the use of a carbonizing flame while welding or heating.

Nitriding Nitriding is a case-hardening process that is used only with a group of low alloy steels. These steels contain elements such as vanadium, chromium, or aluminum that will combine with nitrogen to form nitrides. The nitrides act as a super hard skin on the surface of the steel.

The parts are heated in a nitrogenous atmosphere, usually ammonia gas, to a temperature of 900 to 1,000°F. Nitrogen is slowly absorbed from the ammonia gas. Because of the low temperature and the fact that quenching is unnecessary, there is little distortion or warpage.

Flame Hardening Flame hardening is the most recent of the hardening processes. It permits localized treatment with complete control.

The steel must contain enough carbon for hardening to take place. The article is heat treated and drawn. Then the surface to be hardened is exposed to a multiple-tipped oxyacetylene flame that heats it quickly to a high temperature. It is cooled quickly by water. The depth of hardness can be controlled by the temperature of the water.

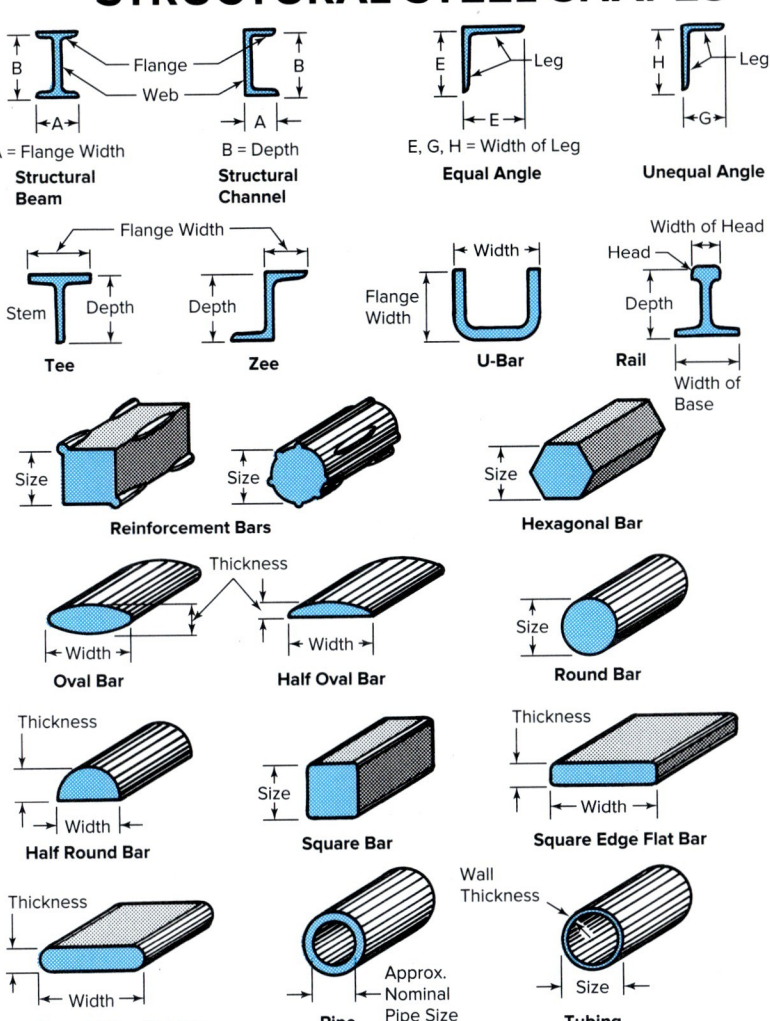

Fig. 3-31 Commercial structural steel shapes.

Hardening the surfaces of gear teeth is an example of flame hardening. Flame hardening can also be used on certain types of cast iron. Flame hardening has the advantage in that it can be used on parts that are too bulky to put into a furnace.

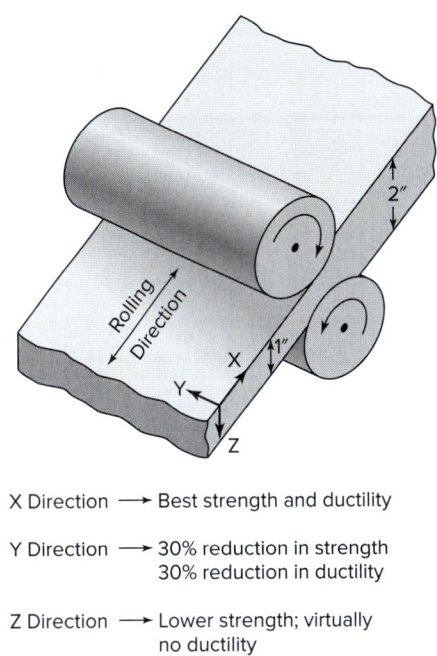

X Direction → Best strength and ductility

Y Direction → 30% reduction in strength
30% reduction in ductility

Z Direction → Lower strength; virtually no ductility

Fig. 3-32 Rolling directions. From *Welding Inspection Technology*, 4/e, slide set; 2000

Fig. 3-33 All-welded pressure vessel being removed from a heat-treating furnace. Nooter Corp.

Fig. 3-34 Vessel ready for quenching, or quick cooling. The metal changes from white-hot to black in less than three minutes. Nooter Corp.

Steel and Other Metals Chapter 3

> **JOB TIP**
>
> **In the Know**
> Over half of the products manufactured in the United States are impacted by welding. The competitive student comes to this marketplace with some knowledge of lasers, robotics, fusion welding, welding design, materials science, solid-state welding, metallurgy, computer modeling, and nondestructive evaluation.

Annealing Annealing includes several different treatments. The effects of annealing are:

- To remove stresses
- To induce softness for better machining properties
- To alter ductility, toughness, or electrical, magnetic, or other physical properties
- To refine the crystalline structure
- To produce a definite microstructure

The changes that take place in a metal depend on the annealing temperature, the rate of cooling, and the carbon content.

When it is desired to produce maximum softness and grain refinement in previously hardened steel, the steel is heated slightly above the critical range and cooled slowly. Either the metal is allowed to cool in the furnace, which is gradually cooled, or it is buried in lime or some other insulating material.

Another form of annealing is stress-relief annealing. It is usually applied only to low carbon steels. The purpose here is to relieve the stress caused by working of the steel, such as in welding. The material is heated to a point just below the critical range and allowed to cool normally.

It is important to note here that the difference between hardening and softening of steels is due to the rate of cooling. Fast cooling hardens, and slow cooling softens. Both tempering and annealing reduce the hardness of a material.

Tempering Tempering is a process wherein the hardness of a steel is reduced after heat treatment. It is also used to relieve the stresses and strains caused by quenching. This is usually done by heating the hardened steel to some predetermined temperature between room temperature and the critical temperature, holding it at that temperature for a length of time, and cooling it in air or water. The reduction of hardness depends upon the following three factors: (1) the tempering temperature, (2) the amount of time the steel is held at this temperature, and (3) the carbon content of the steel. Generally, as the temperature and time are increased, the hardness will be decreased. The higher the carbon content at a given temperature and time, the higher the resulting hardness.

Normalizing The purpose of normalizing is to improve the grain structure of a metal and return it to normal by removing stresses after welding, casting, or forging. These stresses are caused by uneven cooling following these operations.

Normalizing is done by heating the steel to a temperature similar to that used for annealing and then cooling it in still air. Normalizing requires a faster rate of cooling than that employed for annealing, and it results in harder, stronger metal than that which is obtained by annealing.

Metal Internal Structures

Metallurgy is the science that deals with the internal structure of metals. In welding metallurgy, we are concerned about the various changes that take place in the metals when they are cut or joined with thermal processes such as welding or thermal cutting. Especially problematic are mechanical property changes.

In order to understand metallurgical properties of metals, it is important to have an understanding of atomic structure and the various states of matter. The four states of matter are solids, liquids, gases, and plasmas. These four states of matter must be dealt with each time a piece of metal is welded or thermally cut.

The atomic arrangements that make up these four states of matter are so small they cannot be seen, even with the most powerful microscopes. Inside these extremely small atoms there are subatomic particles, including electrons (which carry a negative charge) and protons (which carry a positive charge). The attracting and repelling forces of these particles affect the properties of the material.

For example, a solid such as steel has an atomic structure such that when a process attempts to force the atoms closer together, a strong repulsive action counteracts the compressive forces. If, on the other hand, a process attempts to pull the atoms further apart, a strong attractive action counteracts the tensile forces. The atoms try to maintain a home position even though they are constantly in a state of vibration.

 vs

Fig. 3-35 Solid versus liquid. *American Welding Society*

As heat energy such as from a welding arc is put on to a solid such as steel, the atomic movement becomes more active. As the temperature rises, the atomic structure expands. If the temperature continues to rise above the melting temperature of the steel, the atoms are able to move freely and the solid becomes a liquid. If the temperature is increased still further, the vaporization temperature will be reached and the liquid will turn into a gas. If the gas is superheated, it will ionize and become a plasma. Gas plasma is simply a gas that has become an electrical conductor. This form of plasma occurs in the welding arc and thus this name is applied to such processes as plasma arc cutting and plasma arc welding.

A graphic example of the transition from liquid to solid or solid to liquid is shown in Fig. 3-35, which depicts an example of a solid railroad rail and a pour of liquid metal, which eventually was formed into the rail.

Solid metals take on a three-dimensional crystalline structure because the atoms align themselves into orderly layers, lines, and rows. Looking at the broken surface of a metal or weld, this crystalline structure is quite evident. The metal has not been crystallized because it is old or has been overheated but because all metals are crystalline in nature.

The most common phases, or crystalline structures, of metals are body-centered cubic (BCC), face-centered cubic (FCC), body-centered tetragonal (BCT), and hexagonal close-packed (HCP). These crystalline structures can be represented in Table 3-1. The table

Table 3-1 Metals and Their Phases (Crystalline Structures)

Structure Name	Description
BCC	**Body-centered cubic:** a cube with an atom at each of the eight corners and a single atom at the center of the cell. **Example Metals:** carbon steels, iron, chromium, molybdenum, and tungsten.
FCC	**Face-centered cubic:** describes a cube with one atom at the center of each of the six faces. **Example Metals:** carbon steel and iron heated above its transformation temperature, aluminum, nickel, silver, copper, and austenitic stainless steels.
BCT	A **body-centered tetragonal:** unit cell has one axis elongating to form the shape of a rectangle, with an atom in the center. **Example Metals:** alloy steels and higher carbon, when rapidly cooled, form martensite, a very hard, crack-susceptible phase.
HCP	In a **hexagonal close-packed** structure, two hexagons (six-sided shapes) form the top and bottom of a prism with an atom located at the center and at each point of the hexagon. A triangle is located midpoint between the top and bottom prism, with an atom at each point of the triangle. **Example Metals:** magnesium, cadmium, and zinc.

Adapted from American Welding Society, *Welding Inspection Technology*, 4th ed., pp. 8–7, Figure 8.7, 2000

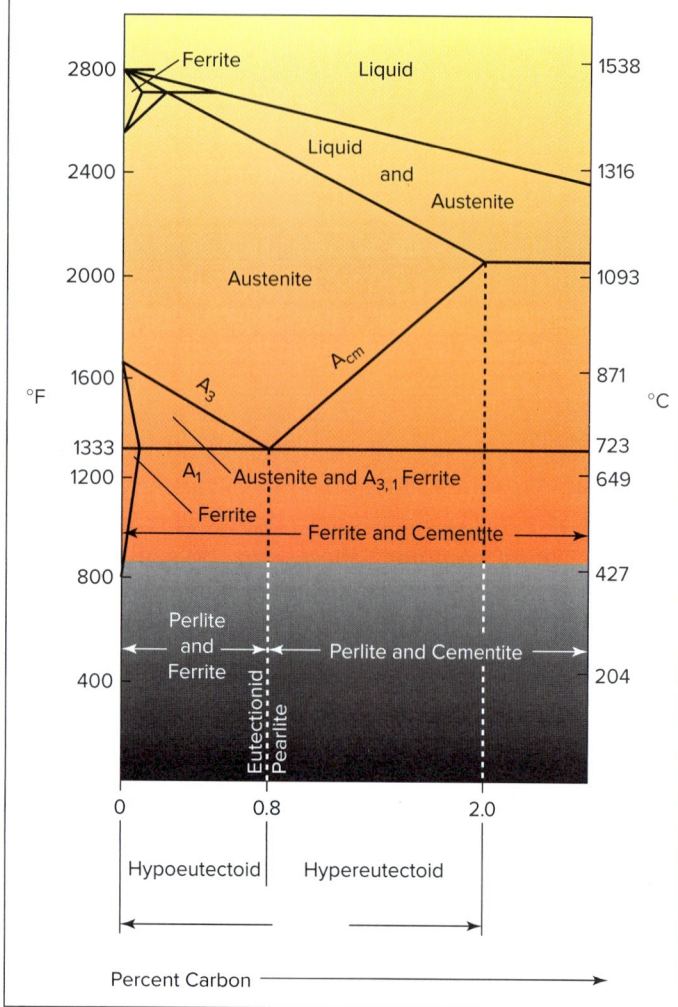

Fig. 3-36 Iron carbon phase. From *Welding Inspection Technology*, 4/e, slide set; 2000

describes these crystal structures and the metals that they impact. Figures 3-36 and 3-37 illustrate an iron carbon phase diagram and the five circles that show the various phases steel goes through as it is heated and cooled.

As molten metal (liquid) solidifies into its solid crystalline structure, it starts at the interface between the molten weld metal and the cooler unmelted heat-affected zone. These clusters of atoms form grains and grain boundaries, as seen in Fig. 3-38, page 74. Note the imaginary mold denoted by the dashed line. This is very similar in principle to the molten steel in a ladle being poured into an ingot mold as in the steelmaking process. A weld is considered a cast structure because of this similarity to the making of steel.

The grain (crystal) size will have an effect on the mechanical properties of the metal.

Fine-Grained Metals Have

- Good tensile strength
- Good ductility
- Good low temperature properties

Coarse-Grained Metals Have

- Slightly lower strength
- Slightly less ductility
- Good high temperature properties

Welding has a marked effect on grain size depending on such factors as heat input, cooling rate (preheat), long or short arc, slow or fast travel speed, welding on the high or low end of the parameter ranges, and the process selected.

Another method of affecting mechanical properties is alloying. This changes the orderly rows, lines, and layers of the three-dimensional crystalline structure the pure metal would take. Small atoms such as those of carbon, nitrogen, and hydrogen can occupy spaces between the larger atoms in a material structure. This is known as interstitial alloying. Larger atoms such as those of copper and nickel will replace other atoms in a material structure. This is known as substitutional alloying. The additions of these types of alloying elements create irregularities in the orderly arrangement of the atoms in these structures. Figure 3-39, page 74, shows a representation of this effect. Note that the presence of an alloying element exerts various degrees of atomic attraction and repulsion. This distorts or strains the grain structure, which tends to increase the internal energy of the metal and results in improved mechanical properties. See the colored insert.

The elements used for alloying will be discussed later in this chapter in the section titled Effects of Common Elements on Steel.

Physical Properties of Metals

It is very important for the welder to be familiar with the physical properties of metals and the terms and measurements used to describe them. For convenience, the definitions of common properties have been divided into three general classifications: those related to the absorption and transmission of energy, the internal structure of the metal, and resistance to stress.

Properties Related to Energy

Melting Point The melting point is the temperature at which a substance passes from a solid to a liquid condition. For water this is 32°F. Steel has a melting point around 2,700°F, depending upon the carbon range. The

The Five Circles

This is a broad simplification, but consider yourself looking through a microscope lens as the highly polished and etched steel is heated and cooled at various rates. If you can grasp what is taking place in these five circles, it will aid you greatly in understanding more complex issues related to the heating and cooling of steel.

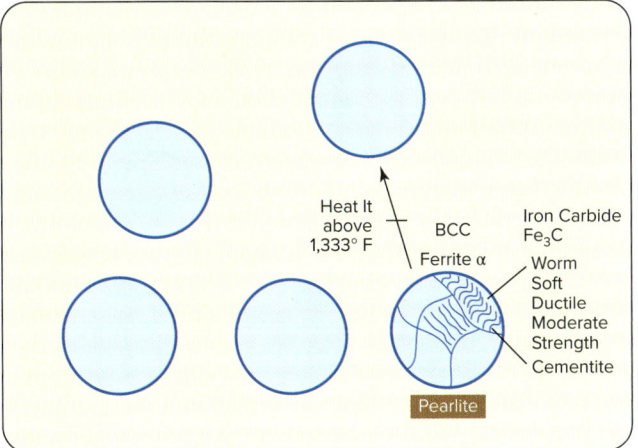

Cell 1: Steel at room temperature, which is in the process of being heated. This structure is pearlite (BCC) and is made up of ferrite which is alpha iron and cementite. It is also referred to as iron carbide Fe₃C. It can be observed as having a worm-like appearance. Pearlite is a soft, ductile form of steel with moderate strength (tensile and yield).

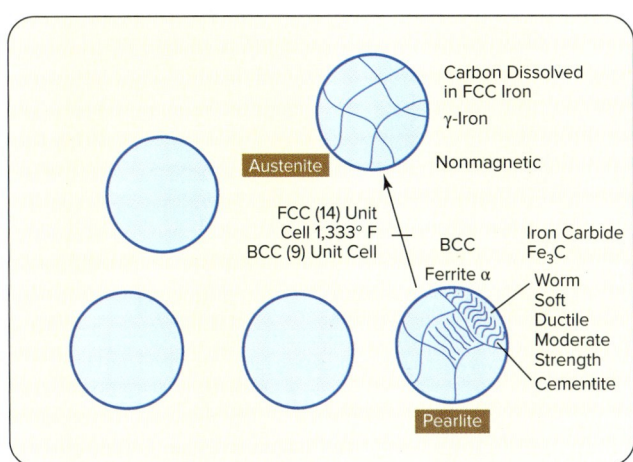

Cell 2: Steel has a transformation temperature that is dependent upon its carbon content. In this exercise we will be using 1,333° F as a fixed temperature. BCC steel has its atoms arranged into a 9-unit cell configuration and is magnetic. Above 1,333° F it is still solid, but the atoms realign into FCC, a 14-unit cell configuration. (Note Table 3-1.) In this phase the solid steel can absorb large amounts of carbon and is referred to as austenite. In austenite, carbon is soluble up to 2% by weight, whereas in ferrite carbon is soluble up to only 0.02% by weight. As the carbon dissolves into the FCC iron, also called gamma iron, it becomes nonmagnetic.

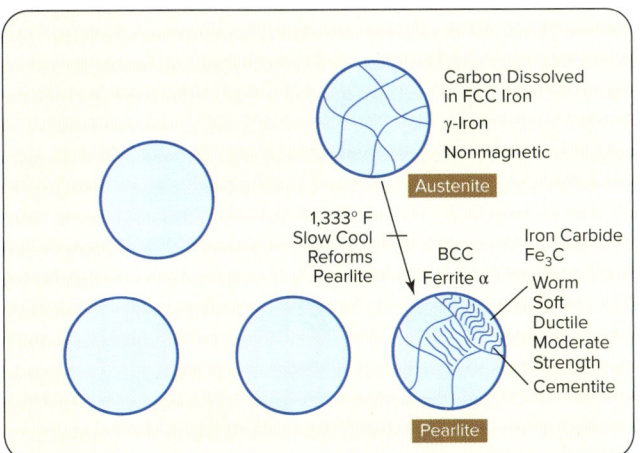

Cell 3: If cooled very slowly the austenite will all return to pearlite. The slow cooling would be like leaving the steel in a furnace, then turning off the furnace and over a day or so it cools to room temperature. This slow cooling (annealing) allows the carbon time to come out of solution and reform the soft, ductile pearlite structure.

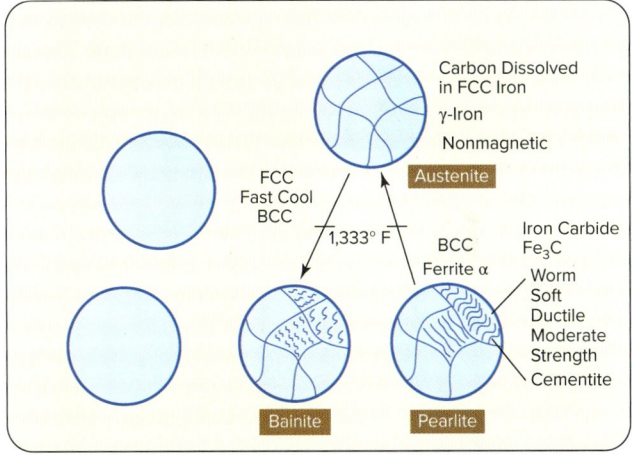

Cell 4: This cell shows a somewhat faster cooling rate. Instead of leaving the steel in the furnace, it is removed from the furnace and allowed to cool in still air. This faster cooling rate (normalizing) will form a bainite structure. It can be observed as not having the worm-like appearance but more like a slug, shorter and broader. Bainite is harder, stronger, and less ductile than pearlite, thus has higher tensile and yield strength.

Fig. 3-37 Transformation of steel upon heating and cooling at various rates. Adapted from Kenneth W. Coryell *(Continued)*

higher the carbon content is, the lower will be the melting point. The higher the melting point, the greater the amount of heat needed to melt a given volume of metal. The temperature of the heat source in welding must be above the melting point of the material being welded. For example, the temperature of a flame produced by the burning of acetylene with air is not as high as the temperature of the flame produced by the burning of acetylene with oxygen. Thus, it does not have the ability to melt the same materials that the oxyacetylene flame has.

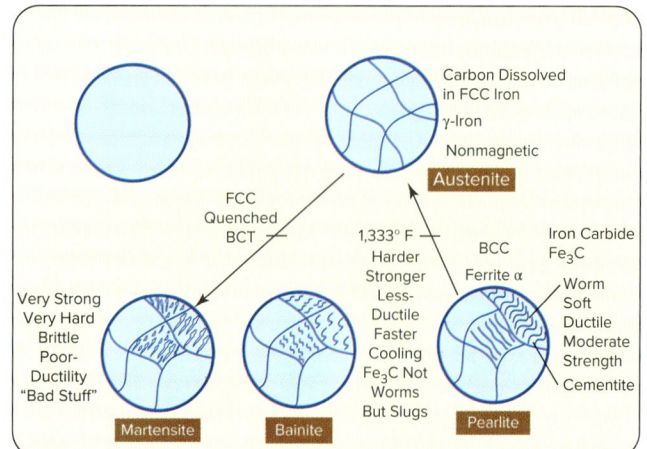
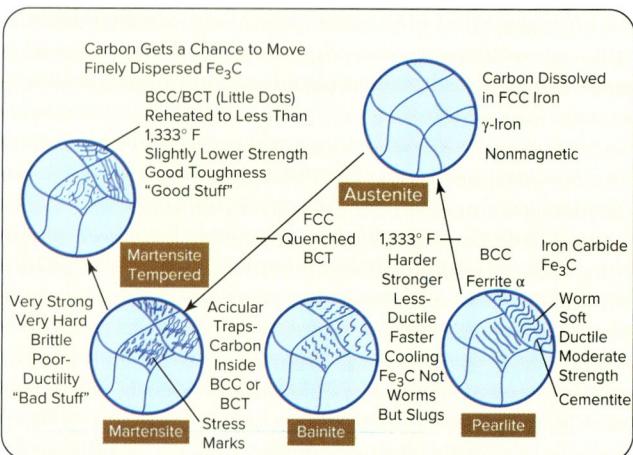

Cell 5: This cell shows what a very fast cooling rate would do to the steel. It would be like taking the steel directly out of the furnace and quenching it in brine water, plain water, or oil. This would not give the carbon time to diffuse out of the austenite and would form an acicular (needle-like) structure called martensite (BCT). The trapped carbon will make the steel very strong (high tensile and yield), but at a great sacrifice to ductility. This is "bad stuff" as it will be very hard and brittle and prone to cracking. Martensite must be dealt with—the case is not will it crack, but when will it crack.

Cell 6: To keep some of the good characteristics of martensite (such as strength) but bring back some of the ductility, the steel can be tempered. This is done by heating it below the transformation temperature and then cooling slowly. This gives the carbon a chance to move and finely disperse the Fe_3C. It is a combination of BCC/BCT. The little dots you can see are much like the dowel rods used in wood working to give more strength at a joint or, in this case, the grain boundaries. This will form "good stuff" as it has good strength and toughness.

Generally, cooling rate is most critical when steel is heated above the transformation temperature and is much less critical if heated below the transformation temperature. It is not possible to transform steel between martensite, bainite, and pearlite without first taking it through the austenitic phase.

Fig. 3-37 Transformation of steel upon heating and cooling at various rates. *(Concluded)*

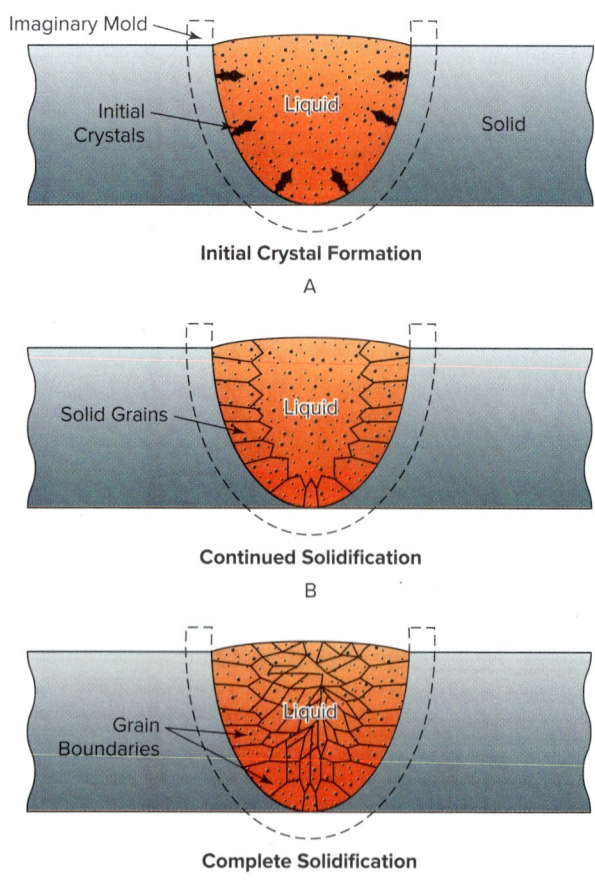

Fig. 3-38 Solidification of molten weld metal. Adapted from American Welding Society, *Welding Inspection Technology,* 5th ed., pp. 8–5, Figure 8-4, 2000

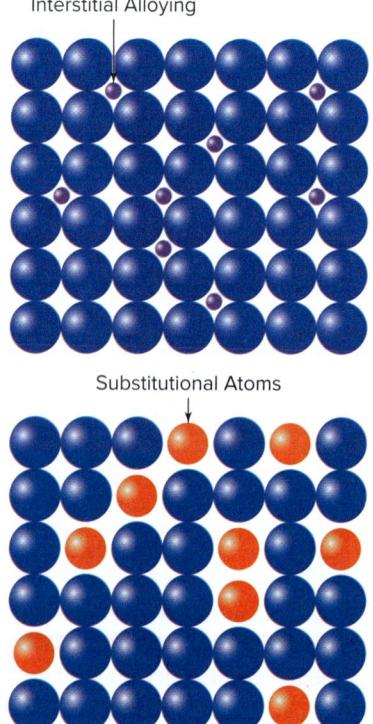

Fig. 3-39 (A) Smaller atoms, such as carbon, nitrogen, and hydrogen, tend to occupy sites between the atoms that form the grain structure of the base metal. This is known as interstitial alloying. (B) Alloying elements with atoms close to the size of those of the base metal tend to occupy substitutional sites. That is, they replace one of the base metal atoms in the grain structure. This is known as substitutional alloying. Adapted from American Welding Society, *Welding Inspection Technology,* 4th ed., pp. 8–7, Figure 8.7, 2000

Weldability Weldability is the capacity of a metal substance to form a strong bond of adherence while under pressure or during solidification from a liquid state.

Fusibility Fusibility is the ease with which a metal may be melted. In general, soft metals are easily fusible, whereas harder metals melt at higher temperatures. For example, tin, lead, and zinc are more easily fused than iron, chromium, and molybdenum.

Volatility Volatility is the ease with which a substance may be vaporized. A metal that has a low melting point is more volatile than a metal with a high melting point. Volatility is measured by the degree of temperature at which a metal boils under atmospheric pressure.

Electrical Conductivity The electrical conductivity of a substance is the ability of the substance to conduct electrical current.

Electrical Resistance The opposition to electric current as it flows through a wire is termed the resistance of the wire. Electrical resistance is measured by a unit called the **ohm.** Lead has 10 times the resistance of copper. This means that lead wire would have to be 10 times as large as the copper wire to carry the same amount of current without loss. A poor conductor heats up to a greater extent than a good conductor when the same amount of current is passed through each.

Thermal Conductivity The thermal conductivity of a substance is the ability of the substance to carry heat. The heat that travels to both sides of the groove face during the welding of a bevel butt joint is the proof that metals conduct heat. The heat is rapidly conducted away from the groove face in a good thermal conductor, but slowly in a poor one. Copper is a good conductor, and iron is a poor conductor. This accounts for the fact that copper requires more heat for welding than iron, although its melting point (1,981°F) is lower than the melting point of iron (2,750°F).

Coefficient of Thermal Expansion The coefficient of thermal expansion is the amount of expansion a metal undergoes when it is heated and the amount of contraction that occurs when it is cooled. The increase in the length of a bar 1 inch long when its temperature is raised 1°C is called the **linear coefficient of thermal expansion.** The higher the coefficient, the greater the amount of expansion and, therefore, the greater the contraction upon cooling. Expansion and contraction will be discussed in more detail under Effects of Welding on Metal, pages 99–109.

Hot Shortness Hot shortness is brittleness in metal when hot. This characteristic should be kept in mind in the handling of hot metals and in jig construction and clamping.

Overheating A metal is said to be overheated when the temperature exceeds its critical range, that is, it is heated to such a degree that its properties are impaired. In some instances, it is possible to destroy the original properties of the metal through heat treatment. If the metal does not respond to further heat treatment, it is considered to be burned and cannot meet the requirements of a heavy load. In arc welding, excess welding current or too slow a travel speed may cause overheating in the weld deposit.

Properties Related to Internal Structure

Specific Gravity Specific gravity is a unit of measurement based on the weight of a volume of material compared with an equal volume of water. Aluminum has a specific gravity of 2.70; thus, it is almost 2¾ times heavier than water. When two molten metals are mixed together, the metal with the lower specific gravity will be forced to the top, and the metal with the higher specific gravity will sink to the bottom.

Density A metal is said to be dense when it is compact and does not contain such discontinuities as slag, inclusions, and porosity. Density is expressed as the quantity per unit volume. The density of low carbon steel, for example, is 0.283 pound per cubic inch. The density of aluminum, a much lighter metal, is only 0.096 pound per cubic inch.

Porosity Porosity is the opposite of density. Some materials are porous by their very nature and allow liquids under pressure to leak through them. Materials that are porous have an internal structure that lacks compactness or have other discontinuities that leave voids in the metal.

Properties Related to Stress Resistance

An important physical property of a metal is the ability of that material to perform under certain types of stress. Stresses to which metal fabrications are subjected during both welding and service include the following:

- *Compression:* squeezing
- *Shear:* strain on a lap joint pulled in opposite directions

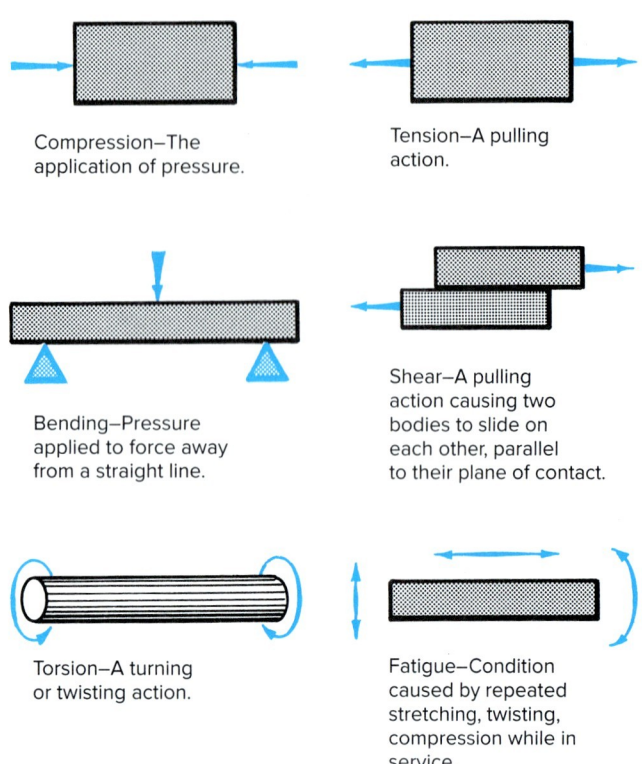

Fig. 3-40 Types of stresses on loads imposed on weldments.

- *Bending:* deflection as a result of a compressive force
- *Tension:* pulling in opposite directions
- *Fatigue:* result of repeated cycles of forces applied and released in all directions
- *Torsion:* twisting force in opposite directions

These typical stresses are illustrated in Fig. 3-40.

Plasticity The ability of a material to deform without breaking is its plasticity. Strength combined with plasticity is the most important combination of properties a metal can have. Metals having these properties can be used in structural fabrications. For example, if a member of a bridge structure becomes overloaded, the property of plasticity allows the overloaded member to flow so that the load becomes redistributed to other parts of the bridge structure.

Strength Strength is the ability of a material to resist deformation. It is usually expressed as the **ultimate tensile strength** in pounds per square inch. The ultimate tensile strength of a material is its resistance to breaking. Cast iron has an approximate tensile strength of 15,000 p.s.i. One type of stainless steel, on the other hand, has reached a strength of 400,000 p.s.i.

Toughness Although there is no direct method of measuring the toughness of materials accurately, a material may be assumed to be tough if it has high tensile strength and the ability to deform permanently without breaking. Toughness may be thought of as the opposite of brittleness since a tough metal gives warning of failure through deformation whereas a brittle material breaks without any warning. Copper and iron are tough materials.

Impact Resistance Impact resistance may be defined as the ability of a material to withstand a maximum load applied suddenly. The impact resistance of a material is often taken as an indication of its toughness.

Brittleness Brittle materials fail without any such warning as deformation, elongation, or a change of shape. It may be said that a brittle material lacks plasticity and toughness. A piece of chalk is very brittle.

Hardness The ability of one material to penetrate another material without fracture of either is known as hardness. The greater the hardness, the greater the resistance to marking or deformation. Hardness is usually measured by pressing a hardened steel ball into the material. In the Brinell hardness test the diameter of the impression is measured, and in the Rockwell hardness test the depth of the impression is measured. A hard material is also a strong material, but it is not very ductile. The opposite of hardness is softness.

Malleability The ability a material possesses to deform permanently under compression without breaking or fracturing is known as the malleability of the metal. Metals that possess this characteristic can be rolled or hammered into thinner forms. Metals must have malleability in order to be forged.

Elastic Limit Loading a material will cause it to change its shape. The ability of the material to return to its original shape after the load has been removed is known as **elasticity.** The **elastic limit** is the greatest load that may be applied after which the material will return to its original condition. Once the elastic limit of a material has been reached, it no longer behaves elastically. It will now behave in a plastic manner and permanent deformation occurs. For practical purposes, the elastic limit is required in designing because it is usually more important to know what load will deform a structure than what load will cause a fracture or break.

Modulus of Elasticity Some materials require higher stresses to stretch than others do. In other words, some materials are stiffer than others. To compare the stiffness of one metal with that of another, we must determine what is known as the **modulus of elasticity** for each of them. The modulus of elasticity is the ratio of the stress to the strain. It is a measure of relative stiffness. If the modulus is high, the material is more likely to resist movement or distortion. A material that stretches easily has a low modulus.

Yield Point When a sample of low or medium carbon steel is subjected to a tension test, a curious thing happens. As the load on the test specimen is increased slowly, a point is found at which a definite increase in the length of the specimen occurs with no increase in the load. The load at this point, expressed as pounds per square inch, is called the **yield point** of the material. Nonferrous metals and types of steel other than low and medium carbon steels do not have a yield point.

Resilience Resilience (springiness) is the energy stored in a material under strain within its elastic limit that causes it to resume its original shape when the load is removed. Resilience is a property of all spring steels.

Ductility Ductility is the ability of a material to be permanently deformed (stretched) by loading and yet resist fracture. When this happens, both elongation and reduction in area take place in the material. The amount of stretching is expressed as **percent of elongation.** Metals with high ductility may be stretched, formed, or drawn without tearing or cracking. Gold, silver, copper, and iron are metals with good ductility. A ductile metal is not necessarily a soft metal.

Fatigue Failure Failure of metals under repeated or alternating stresses is known as fatigue failure. When a metal is broken in a tensile machine, it is found that a certain load is required to break it. The same material, however, will fail when a much smaller load has been applied and removed many times. A spring, for example, may fail after it has been in service for months even though its loading has not been changed or increased. In designing parts subjected to varying stresses, the fatigue limit of a material is more important than its tensile strength or elastic limit. The **fatigue limit** is that load, usually expressed in pounds per square inch, which may be applied for an indefinite number of cycles without causing failure.

Cyclic loading is another way of referring to fatigue testing. If a load can be applied tens of millions of times without causing failure, it is assumed that this load or a lesser load can be applied indefinitely without failure. This level of loading is called the **endurance limit of the material** and is the maximum load that can be applied at which no failure will occur, no matter how many cycles the load is applied.

Resistance to Corrosion The ability of metals to resist atmospheric corrosion and corrosion by liquids or gases is often very important. Corrosion is the gradual wearing away or disintegration of a material by a chemical process. The action of oxygen on steel to form rust is a form of slow corrosion. Corrosion may be measured by (1) determining the loss in strength of tensile samples, (2) determining loss in weight of materials that dissolve in the corroding medium, or (3) determining gain in weight when a heavy coating of rust is formed.

Table 3-2 (pp. 78–79) lists types of metals and their physical and chemical properties. Study this table carefully in order to acquire a basic understanding of the differences in metals and the part these differences play in choosing a metal for a particular job.

Effects of Common Elements on Steel

Nonmetals

Carbon Pure carbon is found in its native state both as diamond, a very hard material, and as graphite, a soft material. Carbon is a part of coal, petroleum, asphalt, and limestone. Commercially it can be obtained as lampblack, charcoal, and coke.

As indicated previously, the amount of carbon present in steel determines its hardness and has serious implications for welding. Increased carbon content increases the tensile strength of steel but reduces its ductility and weldability. If the carbon content is above 0.25 percent, sudden cooling from the welding temperature may produce a brittle area next to the weld. The weld itself may be hard and brittle if an excess of carbon is picked up from the steel being welded. If other alloying elements are added to promote high tensile strength, good weld qualities can be retained. In general, an effort is made to use steels of low or medium carbon content.

Boron Boron is a nonmetallic element that is plentiful and occurs in nature in combination with other elements, as in borax. Pure boron is a gray, extremely hard solid with a melting point in excess of 400°F. It increases the hardenability of steel, that is, the depth to which the steel will harden when quenched. Boron's

Table 3-2 Metals and Their Properties

Metal	Melting Point (°F)	Linear Expansion per 10 ft Length per 100°F Rise in Temperature in Inches	Heat Conductivity (British Thermal Units per Hour per Square Foot per Inch of Thickness per Degree Fahrenheit)	Density (lb/in.³)	Shrinkage Allowance in Castings (in./ft)	Brinell Hardness Hard	Brinell Hardness Soft	Approximate Tensile Strength (p.s.i.)	Approximate Analysis of Chemical Composition (%)	Remarks
Allegheny metal	2,640	0.115		0.283			140	90,000[1]–120,000		[1]Annealed
Aluminum—cast—8% copper	1,175				0.1875			20,000	Aluminum 92; copper 8	
Aluminum—pure	1,218	0.148	1,393	0.096		40	23	12,000–28,000	Aluminum	
Aluminum—5% silicon	1,117	0.146		0.093				18,000	Aluminum 95; silicon 5	
Ambrac—A	2,100	0.109		0.310				50,000–130,000	Copper 75; nickel 20; zinc 5	
Antimony	1,166	0.075		0.245				1,000	Antimony	
Bismuth	520								Bismuth	
Boron	3,992			0.094					Boron	
Brass—commercial high	1,660	0.115	756	0.306	0.1875			46,000	Copper 66; zinc 34	
Bronze—tobin	1,625	0.119		0.304				54,000	Copper 60; zinc 39; tin 1	
Bronze—muntz metal	1,625								Copper 60; zinc 40	
Bronze—manganese	1,598	0.119		0.302			95	60,000	Copper 96; tin 3.75; phosphorus 0.25	
Bronze—phosphor	1,922	0.119		0.321		160	100	45,000		
Cadmium	610								Cadmium	
Carbon	6,332[2]								Carbon	[2]Greater than
Chromium	2,740			0.235					Chromium	
Cobalt	2,700	0.082		0.312				34,400	Cobalt	
Copper—deoxidized	1,981	0.118		0.307	0.25			32,000–55,000	Copper 99.99	
Copper—electrolyte	1,981	0.106	2,640	0.322		107	42	20,000–70,000	Copper 99.93; oxygen 0.07	
Duriron	2,310	0.104		0.253	0.1875				Silicon 14.5; carbon 0.85; manganese 0.35; iron (base)	Acid resisting iron
Everdur	1,866	0.113	871	0.306	0.1875	180	80	52,000–100,000	Copper 94.8–96; silicon 3–4; manganese, 1–1.2	Wt. cast 0.294 (lb/in³)
Gold	1,945	0.094	2,046	0.697				14,000	Gold	
Iron—cast	2,300	0.067	338	0.260	0.125		193	15,000		
Iron—malleable	2,300			0.268	0.125			53,000		
Iron—pure	2,786	0.078	467	0.283			84	38,500	Iron	
Iron—wrought	2,900	0.078	419	0.278			90	48,000	Iron; slag	
Lead—pure	620	0.181	240	0.411	0.312		6	1,780	Lead	
Lead—chemical	620	0.193		0.410	0.312			1,780	Lead 99.92; copper 0.08	
Manganese	2,246			0.268					Manganese	
Molybdenum	4,532		100	0.309				42,500–154,000	Molybdenum	

Material								Composition	Notes
Monel metal	2,480	0.093		174	0.318	190	015	79,000–109,000[3]	Nickel 67; copper 28; iron; manganese; silicon; carbon; sulfur	[3]Sheets from soft to full hard
Nichrome	2,460	0.091			0.295			100,000	Nickel 60; iron 24; chromium 16; carbon 0.1	
Nickel	2,646	0.083		413	0.319		112	61,000–109,000[4]	Nickel	[4]Sheets from soft to full hard
Nickel silver—18%	1,955	0.122			0.309	158	77	58,000–95,000	Copper 65; nickel 18; zinc 17	
Platinum	3,218								Platinum	
Silicon	2,588						65		Silicon	Welding rods
Silphos	1,300							33,000[5]	Silver 15; copper 80; phosphorus[5]	[5]Joint
Silver—pure	1,762	0.126		2,919	0.380		65	36,000	Silver	[6]1% carbon
Steel—hard (0.40–0.70% carbon)	2,500	0.076		312[6]	0.283	590	240	75,000		
Steel—low carbon (less than 0.15%)	2,700	0.076			0.283		138	50,000		
Steel—medium (0.15–0.40% carbon)	2,600	0.076			0.283		180	60,000		
Steel—manganese	2,450					255		70,000–100,000	Carbon 1.0–1.45; manganese 12–15; silicon 0.10–0.20; iron	
Steel—amsco nickel manganese	2,450								Carbon 1.3; manganese 14; nickel 5; silicon 0.35; iron	Welding rods
Steel—nickel—3½%	2,600				0.284			80,000	Nickel 3.25–3.75; carbon 0.15–0.25; manganese 0.50–0.80	
Steel—cast	2,600		0.25		0.282		144	58,000		
Stainless steel—18-8	2,550	0.073			0.279		140	89,000–100,000	Chromium 18; nickel 8; carbon 0.16; iron (base)	
Stainless steel—18-8 low carbon	2,640				0.280		140	80,000[7]	Chromium 18; nickel 8; carbon 0.07; iron (base)	[7]Annealed
Tin	450	0.139	0.083	450	0.263		15	5,000	Tin	[8]Moh's scale
Tungsten	6,152			1,381	0.678	9[8]	442		Tungsten	[9]Cannot be melted with torch
Tungsten carbide	9									
Vanadium	3,182				0.199				Vanadium	[10]Av. hard rolled
Zinc—cast	786	0.169			0.248		45	9,000	Zinc	
Zinc—rolled	786	0.169	0.312	770	0.258			31,000[10]	Zinc	

SHOP TALK

In the Future

Future opportunities in welding may come from

1. The use of mesh as a reinforcement option
2. The replacement of old concrete bridges with fabricated steel

---

effectiveness is limited to sections whose size and shape permit liquid quenching.

Boron also intensifies the hardenability characteristics of other elements present in the steel. It is very effective when used with low carbon steels. Its effect, however, is reduced as the carbon content of the steel increases.

Silicon Silicon is the main substance in sand and sandstone. It forms about one-fourth of the Earth's crust. This element is added mainly as a deoxidizing agent to produce soundness during the steelmaking process. A large amount of silicon may increase the tensile strength. If the carbon content is also high, however, the addition of silicon increases the tendency to cracking. Large amounts are alloyed with steel to produce certain magnetic qualities for electrical and magnetic applications.

Phosphorus Phosphorus is usually present in iron ore. Small amounts improve the machinability of both low and high carbon steel. It is, however, an impurity as far as welding is concerned, and the content in steel should be kept as low as possible. Over 0.04 percent phosphorus makes welds brittle and increases the tendency to cracking.

Sulfur Sulfur is considered to be a harmful impurity in steel because it makes steel brittle and causes cracking at high temperatures. Steel picks up some sulfur from coke used in the blast furnace. Most of the sulfur present in the blast furnace is fluxed out by the lime in the furnace.

The sulfur content in steel should be kept below 0.05 percent. Sulfur increases the tendency of the weld deposit to crack when cooling and may also cause extreme porosity if the weld penetration is deep. Sulfur does, however, improve the machinability of steel. Steels containing sulfur may be readily welded with low hydrogen electrodes.

Selenium This element is used interchangeably with sulfur in some stainless steels to promote machinability.

Metals

Manganese Manganese is a very hard, grayish-white metal with a reddish luster. In its pure state it is so hard that it can scratch glass. It was first used to color glass during glassmaking. Today it is one of the most useful metals for alloying steel. The addition of manganese increases both tensile strength and hardness. The alloy is a steel that can be readily heat treated. Special care must be exercised in welding since manganese steels have a tendency to porosity and cracking.

High manganese steels are very resistant to abrasion. In amounts up to 15 percent, manganese produces very hard, wear-resistant steels that cannot be cut or drilled in the ordinary way. They must be machined with carbide-tipped tools. Because of their high resistance to abrasion, manganese steels are used in such equipment as rock crushers, grinding mills, and power shovel scoops.

Molybdenum Molybdenum is a silvery white metal that increases the toughness of steel. Since it also promotes tensile strength in steels that are subject to high temperatures, it is an alloying element in pipe where high pressure and high temperature are common. Molybdenum also increases the corrosion resistance of stainless steels.

Molybdenum steels may be readily welded if the carbon content is low. Preheating is required for welding if the carbon content is above 0.15 percent. It can be hardened by quenching in oil or air rather than water.

Chromium Chromium is a hard, brittle, grayish-white metal that is highly resistant to corrosion. It is the principal element in the straight chromium and nickel-chromium stainless steels. The addition of chromium to low alloy steels increases the tensile strength, hardness, and resistance to corrosion and oxidation. Ductility is decreased. This is true at both high and low temperatures. Chromium is used as an alloying element in chrome steel and as plating metal for steel parts such as auto bumpers and door handles. The depth of hardness is increased by quenching chromium in oil.

Steels that contain chromium are easily welded if the carbon content is low. However, the presence of a high percentage of carbon increases hardness. Thus preheating and sometimes postheating are required to prevent brittle weld deposits and fusion zones.

Nickel Nickel is a hard, silvery white element. It is used extensively for plating purposes and as an alloying element in steel. In combination with chromium, nickel is an important alloy in stainless steels. Nickel increases the strength, toughness, and corrosion resistance of steel. Nickel-chromium steels are readily welded. Heavy sections should be preheated.

Niobium The use of niobium in steel has been largely confined to stainless steels in which it combines with carbon and improves the corrosion resistance. More recently it has been added to carbon steel as a means of developing higher tensile strength.

Cobalt Cobalt is a tough, lustrous, silvery white metal. It is usually found in nature with iron and nickel. Cobalt is used as an alloying metal in high speed steel and special alloys when high strength and hardness must be maintained at high temperatures. It is also added to some permanent magnet steels in amounts of 17 to 36 percent. Cobalt is being used increasingly in the aerospace industry.

Copper Copper is a soft, ductile, malleable metal that melts at 1,984°F. It has an expansion rate 1½ times greater than that of steel, and its thermal conductivity is 10 times greater. Only silver is a better conductor of heat and electricity. Copper is also highly corrosion resistant.

Copper is added to steel to improve its resistance to atmospheric corrosion. In the small amounts used (0.10 to 0.40%), copper has no significant effect on physical properties. Its other effects are undesirable, particularly the tendency to promote hot shortness (brittleness when hot), thereby lowering surface quality.

Copper is used for roofing, plumbing, electrical work, and in the manufacture of such alloys as brass, bronze, and German silver. When used in silver and gold jewelry, it increases hardness.

Brass is the most common class of copper alloys. Zinc is the alloying element in brass. Bronzes are produced when other alloying elements such as zinc, tin, silicon, aluminum, phosphorus, and beryllium are added to copper.

Aluminum Aluminum is never found in nature in its pure state. It is derived chiefly from bauxite, an aluminum hydroxide. It is one of the lightest metals: its weight is about one-third that of iron. It is a good conductor of heat and electricity and is highly resistant to atmospheric corrosion. Aluminum is ductile and malleable. It can be easily cast, extruded, forged, rolled, drawn, and machined. Aluminum can be joined by welding, brazing, soldering, adhesive bonding, and mechanical fastening. Aluminum melts at about 1,200°F.

Aluminum is used in both carbon and alloy steels. When used in the making of alloy steels, it has several important functions. Because it combines easily with

> **JOB TIP**
>
> **The Welding Workplace**
> In the past, some young people did not want to go into welding because of this image: dirty, dangerous, smoky, oily, and hot! But now a career in this skilled trade is quite different. Shops have air cleaners and fume exhausters. Tools are easier to handle and often computerized. The welding workplace has vastly changed over the last 20 years and has growth opportunities for young people who are motivated and hardworking. And it is certain that manufacturing needs such individuals in order to expand and prosper in the global marketplace.

oxygen, it is a reliable deoxidizer and purifier. It also produces fine austenitic grain size. When aluminum is present in amounts of approximately 1 percent, it promotes high hardness in steel. Inert gas welding of aluminum is very successful because the gas protects the weld from oxide formation.

Aluminum finds many commercial uses, especially in aircraft, trucks, trains, and in the construction industry. Types of aluminum and aluminum alloys are discussed on page 96.

Titanium and Zirconium These metals are sometimes added in small amounts to certain high strength, low alloy steels to deoxidize the metal, control fine grain size, and improve physical properties.

Lead Lead sulfide is the most important lead ore. Lead is a soft, malleable, heavy metal. It has a very low melting point: approximately 620°F. Lead has little tensile strength. It is highly resistant to corrosion.

Additions of lead to carbon and alloy steels improve machinability without significantly affecting physical properties. Increases of 20 to 40 percent in machinability ratings are normal, and in many instances even greater improvement is realized. The economic factors of each job must be studied. For leaded steels to be economical, the machining must involve considerable removal of metal, and the machine tools must be able to take advantage of the increased cutting speeds.

To date, leaded carbon steels have been used mainly for stock that is to be free machined. Lead is added to a base composition with high phosphorus, carbon, sulfur, and nitrogen content to obtain the optimum in machinability.

Lead is used extensively in the plumbing industry, on cable coverings, and in batteries. It is also used in making such alloys as solder, bearing metals, and terne plate.

Tungsten Tungsten is a steel-gray metal that is more than twice as heavy as iron. It has a melting point above 6,000°F.

Tungsten improves the hardness, wear resistance, and tensile strength of steel. In amounts from 17 to 20 percent and in combination with chromium and molybdenum, it produces a steel that retains its hardness at high temperature. Tungsten is a common element in high speed and hot-worked steels and in hard-surfacing welding rods that are used for building up surfaces that are subject to wear.

Vanadium Vanadium increases the toughness of steel and gives it the ability to take heavy shocks without breaking. Vanadium also has a high resistance to metal fatigue and high impact resistance, thus making steel containing vanadium excellent for springs, gears, and shafts. When heat treated, a vanadium steel has a fine-grain structure. Vanadium steel may require preheat for welding.

Types of Steel

Carbon Steels

Steel may be defined as refined pig iron or an alloy of iron and carbon. Besides iron, steel is made up of carbon, silicon, sulfur, phosphorus, and manganese. Carbon is the most important alloying ingredient in steel. An increase of as little as 0.1 percent carbon can materially change all the properties of steel.

The carbon content of the steel has a direct effect on the physical properties of steel. Increases in the carbon content reduce the melting point of the steel. It becomes harder, has a higher tensile strength, and is more resistant to wear. Harder steel has the tendency to crack if welded, and it is more difficult to machine. As carbon increases, steel also loses some of its ductility and grows more brittle. The addition of carbon makes it possible to heat-treat steel. High and medium carbon steels usually cost more than low carbon steel.

The physical properties of steel are so dependent upon its carbon content and its final heat treatment that types of steels range from the very soft steels, such as those used in the manufacture of wire and nails, to the tool steels, which can be hardened and made into cutting tools to cut the softer steels and other metals. Carbon steels are usually divided into low carbon steels, medium carbon steels, high carbon steels, and tool steels.

Low Carbon Steels Those steels whose carbon content does not exceed 0.30 percent and may be as low as 0.05 percent are low carbon steels. They are also referred to as *mild steels*. These steels may be quenched very rapidly in water or brine and do not harden to any great extent. General-purpose steels of 0.08 to 0.29 percent carbon content are found in this classification, and they are available in standard shapes from steel warehouses. Machine steel (0.08–0.29% carbon) and cold-rolled steel (0.08–0.29% carbon) are the most common low carbon steels. Low carbon steels are produced in greater quantities than all other steels combined, and they make up the largest part of welded fabrication. Thus they are the type of steel that will be stressed in this text.

The weldability of low carbon steels is excellent. They go into most of the structures fabricated by welding such as bridges, ships, tanks, pipes, buildings, railroad cars, and automobiles.

Medium Carbon Steels Medium carbon steels are those steels that have a carbon content ranging from 0.30 to 0.59 percent. They are considerably stronger than low carbon steels and have higher heat-treat qualities. Some hardening can take place when the steel is heated and quenched. They should be welded with shielded metal arc low hydrogen electrodes and other low hydrogen processes. More care must be taken in the welding operation, and best results are obtained if the steel is preheated before welding and normalized after welding. This ensures maximum tensile strength and ductility.

Medium carbon steels are used in many of the same structures indicated for mild steel, except that these structures are subject to greater stress and higher load demands.

High Carbon Steels Steels whose carbon content ranges from 0.60 to 0.99 percent are known as high carbon steels. While the ultra high carbon steels contain carbon between 1.0 to 2.0 percent, they are more difficult to weld than low or medium carbon steels. They can be heat treated for maximum hardness and wear resistance. Preheating and heat treatment after welding eliminate hardness and brittleness at the fusion zone. These steels are used in springs, punches, dies, tools, military tanks, and structural steel.

Alloy Steels

Steel is classified as alloy steel when the content of alloying elements exceeds certain limits. The amounts of alloying elements lie within a specified range for commercial alloy steels. These elements are added to obtain a desired effect in the finished product as described on pages 88 to 91. Alloy steels are readily welded by welding processes such as MIG/MAG and TIG.

High Strength, Low Alloy Steels The high strength, low alloy steels make up a group of steels with chemical compositions specially developed to give higher physical property values and materially greater corrosion resistance than are obtainable from the carbon steel group. These steels contain, in addition to carbon and manganese, other alloying elements that are added to obtain greater strength, toughness, and hardening qualities.

High strength, low alloy steel is generally used when savings in weight are important. Its greater strength and corrosion resistance require less reinforcement and, therefore, fewer structural members than fabrications made with carbon steel. Its better durability is also an advantage in these applications. Among the steels in this classification are oil-hardening steel, air-hardening steel, and high speed steel.

High strength, low alloy steel is readily adaptable to fabrication by shearing, plasma cutting, laser cutting, water jet cutting, punching, forming, riveting, welding without quenching, and tempering heat treatment by the fabricator.

Stainless and Heat-Resisting Steels As the name implies, stainless and heat-resisting steels possess unusual resistance to corrosion at both normal and elevated temperatures. This superior corrosion resistance is accomplished by the addition of chromium to iron. The corrosion resistance of the stainless steels generally increases with increasing chromium content. It appears that when chromium is present, a thin layer of chromium oxide is bonded to the surface, and this oxide prevents any further *oxidation* (ordinary rusting, which is the most common kind of corrosion). Eleven and five-tenths percent chromium is generally accepted as the dividing line between low alloy steel and stainless steel. Although other elements such as copper, aluminum, silicon, nickel, and molybdenum also increase the corrosion resistance of steel, they are limited in their usefulness.

Some stainless steels have practically an indefinite life even without cleaning. Stainless steels are also resistant to corrosion at elevated temperatures which are the result of oxidation, carburization, and sulfidation (deterioration of the surface caused by the action of oxygen, carbon, and sulfur, respectively). Users of stainless steel have experienced some difficulty with pitting. This usually occurs when the material is exposed to chlorides, or at points where the steel is in contact with other materials, such as leather, glass, and grease. Pitting can be materially reduced by treating the area with strong oxidizing agents such as some chromates or phosphates. The addition of molybdenum to austenitic nickel-chromium steels also helps control pitting.

The uses for stainless steels are many, and there are many varieties to choose from. Stainless steels have the following advantages:

- They resist corrosion and the effects of high temperatures.
- They maintain the purity of materials in contact with them.
- They permit greater cleanliness than other types of steel.
- Stainless-steel fabrications usually cost little to maintain.
- Low strength-to-weight ratios are possible both at room and elevated temperatures.
- They are tough at low temperatures.
- They have high weldability.
- They are highly pleasing in appearance and require a minimum of finishing.

In general, stainless steels are produced in either the electric arc or the induction furnace. The largest tonnages by far are melted in electric arc furnaces.

Stainless and heat-resisting steels are commonly produced in finished forms such as plates, sheets, strip, bars, structural shapes, wire, tubing, semifinished castings, and forgings. These steels fall into five general classifications according to their characteristics and alloy content:

1. Five percent chromium, hardenable (martensitic) 500 series
2. Twelve percent chromium, hardenable (martensitic) 400 series
3. Seventeen percent chromium, nonhardenable (ferritic) 400 series
4. Chromium-nickel (austenitic) 300 series
5. Chromium-nickel-manganese (austenitic) 200 series

Series 400 and 500 (Martensitic) Steels in these two groups are primarily heat resisting and retain a large part of their properties at temperatures up to 1,100°F. They are somewhat more resistant to corrosion than alloy steels, but they are not considered true stainless steels.

These steels contain carbon, chromium, and sometimes nickel in such proportions that they will undergo hardening and annealing. Chromium content in this group ranges from 11.5 to 18 percent; and carbon from 0.15 to 1.20 percent, Table 3-3 (p. 84).

Table 3-3 Typical Compositions of Martensitic Stainless Steels

AISI Type	Composition (%)[1]		Other[2]
	Carbon	Chromium	
403	0.15	11.5–13.0	0.5 silicon
410	0.15	11.5–13.5	—
414	0.15	11.5–13.5	1.25–2.5 nickel
416	0.15	12.0–14.0	1.25 manganese, 0.15 sulfur (min.), 0.060 phosphorus, 0.60 molybdenum (opt.)
416Se	0.15	12.0–14.0	1.25 manganese, 0.060 phosphorus, 0.15 selenium (min.)
420	0.15 (min.)	12.0–14.0	—
431	0.20	15.0–17.0	1.25–2.5 nickel
440A	0.60–0.75	16.0–18.0	0.75 molybdenum
440B	0.75–0.95	16.0–18.0	0.75 molybdenum
440C	0.95–1.20	16.0–18.0	0.75 molybdenum

[1]Single values denote maximum percentage unless otherwise noted.

[2]Unless otherwise noted, other elements of all alloys listed include maximum contents of 1.0% manganese, 1.0% silicon, 0.040% phosphorus, and 0.030% sulfur. The balance is iron.

Because of their lower chromium content, steels in the martensitic groups do not offer quite as much corrosion resistance as types in the ferritic and austenitic groups. They are satisfactory for mildly corrosive conditions. They are suitable for applications requiring high strength, hardness, and resistance to abrasion and wet or dry erosion. Thus they are suitable for coal-handling equipment, steam and gas turbine parts, bearings, and cutlery.

These steels are satisfactory for both hot and cold working. They are air hardening and must be cooled slowly or annealed after forging or welding to prevent cracking.

Series 400 (Ferritic) The chromium content of this group ranges from 11.5 to 27 percent, and the carbon content is low, generally under 0.20 percent, Table 3-4. There is no nickel. Ferritic stainless steels cannot be hardened by heat treatment although hardness may be increased by cold working. Suitable hot or cold working, followed by annealing, is the only means of refining the grain and improving ductility.

Table 3-4 Typical Compositions of Ferritic Stainless Steels

AISI Type	Composition (%)[1]			Other[2]
	Carbon	Chromium	Manganese	
405	0.08	11.5–14.5	1.0	0.1–0.3 aluminum
430	0.12	14.0–18.0	1.0	—
430F	0.12	14.0–18.0	1.25	0.060 phosphorus, 0.15 sulfur (min.), 0.60 molybdenum (opt.)
430FSe	0.12	14.0–18.0	1.25	0.060 phosphorus, 0.060 sulfur, 0.15 selenium (min.)
442	0.20	18.0–23.0	1.0	—
446	0.20	23.0–27.0	1.5	0.25 nitrogen

[1]Single values denote maximum percentage unless otherwise noted.

[2]Unless otherwise noted, other elements of all alloys listed include maximum contents of 1.0% silicon, 0.40% phosphorus, and 0.030% sulfur. The balance is iron.

Stainless steels in the ferritic group have a low coefficient of thermal expansion and good resistance to corrosion. They are adaptable to high temperatures. Since their ductility is fair, they can be fabricated by the usual methods such as forming, bending, spinning, and light drawing. Welding is possible, but the welds have low toughness and ductility, which can be improved somewhat by heat treatment. These steels may be buffed to a high finish resembling chromium plate.

For the most part, ferritic stainless steels are used for automotive trim, applications involving nitric acid, high temperature service requiring resistance to scaling, and uses that call for low thermal expansion.

Series 200 and 300 (Austenitic) The chromium content of the austenitic group ranges from 16 to 26 percent, the nickel from 3.5 to 22 percent, and the carbon from 0.15 to 0.08 percent, Table 3-5. These steels are more

Table 3-5 Typical Compositions of Austenitic Stainless Steels

AISI Type	Composition (%)[1]			
	Carbon	Chromium	Nickel	Other[2]
201	0.15	16.0–18.0	3.5–5.5	0.25 nitrogen, 5.5–7.5 manganese, 0.060 phosphorus
202	0.15	17.0–19.0	4.0–6.0	0.25 nitrogen, 7.5–10.0 manganese, 0.060 phosphorus
301	0.15	16.0–18.0	6.0–8.0	
302	0.15	17.0–19.0	8.0–10.0	
302B	0.15	17.0–19.0	8.0–10.0	2.0–3.0 silicon
303	0.15	17.0–19.0	8.0–10.0	0.20 phosphorus, 0.15 sulfur (min.), 0.60 molybdenum (opt.)
303Se	0.15	17.0–19.0	8.0–10.0	0.20 phosphorus, 0.06 sulfur, 0.15 selenium (min.)
304	0.08	18.0–20.0	8.0–12.0	
304L	0.03	18.0–20.0	8.0–12.0	
305	0.12	17.0–19.0	10.0–13.0	
308	0.08	19.0–21.0	10.0–12.0	
309	0.20	22.0–24.0	12.0–15.0	
309S	0.08	22.0–24.0	12.0–15.0	
310	0.25	24.0–26.0	19.0–22.0	1.5 silicon
310S	0.08	24.0–26.0	19.0–22.0	1.5 silicon
314	0.25	23.0–26.0	19.0–22.0	1.5–3.0 silicon
316	0.08	16.0–18.0	10.0–14.0	2.0–3.0 molybdenum
316L	0.03	16.0–18.0	10.0–14.0	2.0–3.0 molybdenum
317	0.08	18.0–20.0	11.0–15.0	3.0–4.0 molybdenum
321	0.08	17.0–19.0	9.0–12.0	Titanium (5 × % carbon min.)
347	0.08	17.0–19.0	9.0–13.0	Niobium + tantalum (10 × % carbon min.)
348	0.08	17.0–19.0	9.0–13.0	Niobium + tantalum (10 × % carbon min., but 0.10 tantalum max.), 0.20 cobalt

[1]Single values denote maximum percentage unless otherwise noted.
[2]Unless otherwise noted, other elements of all alloys listed include maximum contents of 2.0% manganese, 1.0% silicon, 0.045% phosphorus, and 0.030% sulfur. The balance is iron.

numerous and more often used than steels of the 400 series. They differ widely from the chromium alloys due principally to their stable structure at low temperatures. They offer a low yield point with high ultimate tensile strength at room temperatures, a combination that makes for ductility. They are not hardenable by heat treatment, but they harden when cold worked to a degree varying with each type.

Austenitic stainless steels provide the maximum resistance to corrosion, and they are well suited to standard fabrication. They have the ductility required for severe deep drawing and forming. They are easily welded. By controlling the chromium-nickel ratio and degree of cold reduction, a material with high tensile strength is produced that is especially suitable for lightweight welded structures.

At high temperatures, the chromium-nickel types have good oxidation resistance and high rupture and creep-strength values. They are very satisfactory for high temperature equipment because of their relatively high coefficient of thermal expansion.

The chromium content of the duplex group ranges from 18.0 to 29.0 percent, the nickel from 2.5 to 8.5 percent, and the carbon from 0.03 to 0.08 percent, Table 3-6.

These steels are characterized by a low carbon, BCC ferrite, FCC austenite microstructure. Interest in these alloys over the 300-series austenitic stainless-steel alloys is due to their resistance to stress corrosion cracking, crevice corrosion, general corrosion, and pitting. From a strength standpoint, they have yield strengths that are twice that of the 300-series alloys, so they are used where thinner sections and weight reduction is desirable. These duplex stainless-steel (DSS) alloys have the advantages of both the ferritic and austenitic stainless steels, but also some of the disadvantages. Normally postweld heat treatment (PWHT) is not necessary or recommended. The DSS alloys have weldability characteristics better than those of ferritic stainless steels but worse than those of austenitic steels. Good mechanical and acceptable corrosion resistance is available from these alloys in the as-welded condition for most applications. It is essential to follow a qualified welding procedure to control the cooling rate. Very rapid cooling rates are to be avoided. This can best be accomplished by controlling the heat input. The welding procedure must contain minimum and maximum values of all parameters controlling heat input as well as specified interpass and preheat control.

Table 3-6 Chemical Compositions of Typical Duplex Stainless Steels

Alloy	UNS Number	Composition[1,2,3]					
		C	Cr	Ni	Mo	N	Other Elements
329	S32900	0.08	23.0–28.0	2.5–5.0	1.0–2.0	—	
44LN	S31200	0.030	24.0–26.0	5.5–6.5	1.2–2.0	0.14–0.20	
DP3	S31260	0.030	24.0–26.0	5.5–7.5	2.5–3.5	0.10–0.30	0.20–0.80 Cu; 0.10–0.50 W
2205	S31803	0.030	21.0–23.0	4.5–6.5	2.5–3.5	0.08–0.20	
2304	S32304	0.030	21.5–24.5	3.0–5.5	0.05–0.6	0.05–0.20	
255	S32550	0.04	24.0–27.0	4.5–6.5	2.9–3.9	0.10–0.25	1.5–2.5 Cu
2507	S32750	0.030	24.0–26.0	6.0–8.0	3.0–5.0	0.24–0.32	
Z100[4]	S32760	0.030	24.0–26.0	6.0–8.0	3.0–4.0	0.2–0.3	0.5–1.0 Cu; 0.5–1.0 W
3RE60	S31500	0.030	18.0–19.0	4.25–5.25	2.5–3.0	—	
U50[4]	S32404	0.04	20.5–22.5	5.5–8.5	2.0–3.0	0.20	1.0–2.0 Cu
7MoPLUS	S32950	0.03	26.0–29.0	3.5–5.2	1.0–2.5	0.15–0.35	
DP3W	S39274	0.03	24.0–26.0	6.0–8.0	2.5–3.5	0.24–0.32	0.2–0.8 Cu; 1.5–2.5 W

[1]Single values are maximum percentages.
[2]2.5 Mn max.
[3]0.70–1.0 Si max.
[4]Z100—Zeron 100; U50—Uranus 50.

Source: American Welding Society, *Welding Handbook*, Vol. 4, 8th ed., p. 310.

Tool Steels Tool steels are either carbon or alloy steels capable of being hardened and tempered. They are produced primarily for machine tools that cut and shape articles used in all types of manufacturing operations. Tool steels vary in chemical composition depending upon the end use. They range from plain carbon types with no appreciable alloying elements to high-speed cutting types containing as much as 45 percent of alloying elements.

There are many different types of tool steel including high speed, hot work, cold work, shock-resisting, mold, special-purpose, and water-hardening tool steels. They have a carbon range from 0.80 to 1.50 percent carbon and may also contain molybdenum, tungsten, and chromium.

Tool steels are usually melted in electric furnaces, in comparatively small batches, to meet special requirements. They are produced in the form of hot- and cold-finished bars, special shapes, forgings, hollow bar, wire, drill rod, plate, sheets, strip, tool bits, and castings.

Tool steels may be used for certain hand tools or mechanical fixtures for cutting, shaping, forming, and blanking materials at normal or elevated temperatures. They are also used for other applications when wear resistance is important.

Tool steels are rarely welded and must be preheated to do so. After-treatment is also necessary. Tool steel is most often welded to resurface cutting tools and dies. Special hard-surfacing electrodes are required for this work, depending upon the type of deposit required. (See Chapter 12, pp. 330–333.)

Carbon Equivalency The importance of carbon as an alloy has been demonstrated. It has the most pronounced effect on the ease with which a metal will harden upon cooling from elevated temperatures. The amount of carbon present in a particular alloy is very important. The higher the carbon content, the higher the hardness of the steel.

While carbon is very important, other alloys will also promote hardenability. So the carbon equivalency of these alloys must be understood. There are a variety of formulas that will aid in calculating the carbon equivalency (CE). The following formula is one example and is intended for use with carbon and alloy steels that contain more than 0.5 percent carbon, 1.5 percent manganese, 3.5 percent nickel, 1 percent chromium, 1 percent copper, and 0.5 percent molybdenum.

$$CE = \%Carbon + \frac{\%Mn}{6} + \frac{\%Ni}{15} + \frac{\%Cr}{5} + \frac{\%Cu}{1} + \frac{\%Mo}{4}$$

Once the carbon equivalency has been determined, a better understanding of the proper preheat and interpass requirements are known and applied to welding techniques and methods. It must be understood that with increased hardenability the possibility of cracking also increases.

SAE/AISI Steel Numbering System

The various types of steels are identified by a numbering system developed by the Society of Automotive Engineers (SAE) and the American Iron and Steel Institute (AISI). It is based on a chemical analysis of the steel. This numbering system makes it possible to use numerals on shop drawings that indicate the type of steel to be used in fabrication.

In the case of the simple alloy steels, the second digit generally indicates the approximate percentage of the predominant alloying element in the steel. Usually the last two or three digits indicate the average carbon content in *points,* or hundredths of 1 percent. Thus the digit *2* in *2340* identifies a nickel steel. The digit *3* denotes approximately 3 percent nickel (3.25–3.75), and *40* indicates 0.40 percent carbon (0.35–0.45). The digit *7* in *71360* indicates a tungsten steel of about 13 percent tungsten (12–15) and 0.60 percent carbon (0.50–0.70).

The first number designations for the various types of SAE/AISI steels are given in Table 3-7. The specific classification numbers and the alloy amounts they denote are given in Tables 3-8 through 3-18 (pp. 88–90).

Consult Table 3-19 (p. 91) which gives the mechanical properties of various ferrous metals. Note that in the case of steel, the tensile strength and hardness increases, and the ductility decreases as the carbon content increases.

Table 3-7 First digit of SAE/AISI Numbering System

The first digit is for the major alloying element:

1 — Carbon
2 — Nickel
3 — Nickel-chromium
4 — Molybdenum
5 — Chromium
6 — Chromium-vanadium
7 — Tungsten
8 — Nickel-chromium-molybdenum
9 — Silicon-manganese

Table 3-8 Carbon Steels

SAE No.	Carbon Range (%)	Manganese Range (%)	Phosphorus, Max. (%)	Sulfur, Max. (%)
1010	0.05–0.15	0.30–0.60	0.045	0.055
1015	0.10–0.20	0.30–0.60	0.045	0.055
X1015	0.10–0.20	0.70–1.00	0.045	0.055
1020	0.15–0.25	0.30–0.60	0.045	0.055
X1020	0.15–0.25	0.70–1.00	0.045	0.055
1025	0.20–0.30	0.30–0.60	0.045	0.055
X1025	0.20–0.30	0.70–1.00	0.045	0.055
1030	0.25–0.35	0.60–0.90	0.045	0.055
1035	0.30–0.40	0.60–0.90	0.045	0.055
1040	0.35–0.45	0.60–0.90	0.045	0.055
1045	0.40–0.50	0.60–0.90	0.045	0.055
1050	0.45–0.55	0.60–0.90	0.045	0.055
1055	0.50–0.60	0.60–0.90	0.040	0.055
1060	0.55–0.70	0.60–0.90	0.040	0.055
1065	0.60–0.75	0.60–0.90	0.040	0.055
X1065	0.60–0.75	0.90–1.20	0.040	0.055
1070	0.65–0.80	0.60–0.90	0.040	0.055
1075	0.70–0.85	0.60–0.90	0.040	0.055
1080	0.75–0.90	0.60–0.90	0.040	0.055
1085	0.80–0.95	0.60–0.90	0.040	0.055
1090	0.85–1.00	0.60–0.90	0.040	0.055
1095	0.90–1.05	0.25–0.50	0.040	0.055

Table 3-9 Free-Cutting Steels

SAE No.	Carbon Range (%)	Manganese Range (%)	Phosphorus Range (%)	Sulfur Range (%)
1112	0.08–0.16	0.60–0.90	0.09–0.13	0.10–0.20
X1112	0.08–0.16	0.60–0.90	0.09–0.13	0.20–0.30
1115	0.10–0.20	0.70–1.00	0.045 max.	0.075–0.15
X1314	0.10–0.20	1.00–1.30	0.045 max.	0.075–0.15
X1315	0.10–0.20	1.30–1.60	0.045 max.	0.075–0.15
X1330	0.25–0.35	1.35–1.65	0.045 max.	0.075–0.15
X1335	0.30–0.40	1.35–1.65	0.045 max.	0.075–0.15
X1340	0.35–0.45	1.35–1.65	0.045 max.	0.075–0.15

Table 3-10 Manganese Steels

SAE[1] No.	Carbon Range (%)	Manganese Range (%)	Phosphorus, Max. (%)	Sulfur, Max. (%)
T1330	0.25–0.35	1.60–1.90	0.040	0.050
T1335	0.30–0.40	1.60–1.90	0.040	0.050
T1340	0.35–0.45	1.60–1.90	0.040	0.050
T1350	0.45–0.55	1.60–1.90	0.040	0.050

[1]The silicon range of all SAE basic alloy steels is 0.15–0.30%. For electric alloy steels, the silicon content is 0.15% minimum.

Table 3-11 Nickel Steels

SAE[1] No.	Carbon Range (%)	Manganese Range (%)	Phosphorus, Max. (%)	Sulfur, Max. (%)	Nickel Range (%)
2315	0.10–0.20	0.30–0.60	0.040	0.050	3.25–3.75
2330	0.25–0.35	0.50–0.80	0.040	0.050	3.25–3.75
2340	0.35–0.45	0.60–0.90	0.040	0.050	3.25–3.75
2345	0.40–0.50	0.60–0.90	0.040	0.050	3.25–3.75
2515	0.10–0.20	0.30–0.60	0.040	0.050	4.75–5.25

[1]The silicon range of all SAE basic alloy steels is 0.15–0.30%. For electric alloy steels, the silicon content is 0.15% minimum.

Table 3-12 Nickel-Chromium Steels

SAE[1] No.	Carbon Range (%)	Manganese Range (%)	Phosphorus, Max. (%)	Sulfur, Max. (%)	Nickel Range (%)	Chromium Range (%)
3115	0.10–0.20	0.30–0.60	0.040	0.050	1.00–1.50	0.45–0.75
3120	0.15–0.25	0.30–0.60	0.040	0.050	1.00–1.50	0.45–0.75
3130	0.25–0.35	0.50–0.80	0.040	0.050	1.00–1.50	0.45–0.75
3135	0.30–0.40	0.50–0.80	0.040	0.050	1.00–1.50	0.45–0.75
3140	0.35–0.45	0.60–0.90	0.040	0.050	1.00–1.50	0.45–0.75
X3140	0.35–0.45	0.60–0.90	0.040	0.050	1.00–1.50	0.60–0.90
3145	0.40–0.50	0.60–0.90	0.040	0.050	1.00–1.50	0.45–0.75
3150	0.45–0.55	0.60–0.90	0.040	0.050	1.00–1.50	0.45–0.75
3215	0.10–0.20	0.30–0.60	0.040	0.050	1.50–2.00	0.90–1.25
3220	0.15–0.25	0.30–0.60	0.040	0.050	1.50–2.00	0.90–1.25
3240	0.35–0.45	0.30–0.60	0.040	0.050	1.50–2.00	0.90–1.25
3245	0.40–0.50	0.30–0.60	0.040	0.050	1.50–2.00	0.90–1.25
3250	0.45–0.55	0.30–0.60	0.040	0.050	1.50–2.00	0.90–1.25
3312	0.17 max.	0.30–0.60	0.040	0.050	3.25–3.75	1.25–1.75
3415	0.10–0.20	0.30–0.60	0.040	0.050	2.75–3.25	0.60–0.95

[1]The silicon range of all SAE basic alloy steels is 0.15–0.30%. For electric alloy steels, the silicon content is 0.15% minimum.

Table 3-13 Molybdenum Steels

SAE[1] No.	Carbon Range (%)	Manganese Range (%)	Phosphorus, Max. (%)	Sulfur, Max. (%)	Chromium Range (%)	Nickel Range (%)	Molybdenum Range (%)
X4130	0.25–0.35	0.40–0.60	0.040	0.050	0.80–1.10	—	0.15–0.25
4140	0.35–0.45	0.60–0.90	0.040	0.050	0.80–1.10	—	0.15–0.25
4150	0.45–0.55	0.60–0.90	0.040	0.050	0.80–1.10	—	0.15–0.25
4320	0.15–0.25	0.40–0.70	0.040	0.050	0.30–0.60	1.65–2.00	0.20–0.30
X4340	0.35–0.45	0.50–0.80	0.040	0.050	0.50–0.80	1.65–2.00	0.20–0.30
4615	0.10–0.20	0.40–0.70	0.040	0.050	—	1.65–2.00	0.20–0.30
4620	0.15–0.25	0.40–0.70	0.040	0.050	—	1.65–2.00	0.20–0.30
4640	0.35–0.45	0.50–0.80	0.040	0.050	—	1.65–2.00	0.20–0.30
4815	0.10–0.20	0.40–0.60	0.040	0.050	—	3.25–3.75	0.20–0.30
4820	0.15–0.25	0.40–0.60	0.040	0.050	—	3.25–3.75	0.20–0.30

[1]The silicon range of all SAE basic alloy steels is 0.15–0.30%. For electric alloy steels, the silicon content is 0.15% minimum.

Table 3-14 Chromium Steels

SAE[1] No.	Carbon Range (%)	Manganese Range (%)	Phosphorus, Max. (%)	Sulfur, Max. (%)	Chromium Range (%)
5120	0.15–0.25	0.30–0.60	0.040	0.050	0.60–0.90
5140	0.35–0.45	0.60–0.90	0.040	0.050	0.80–1.10
5150	0.45–0.55	0.60–0.90	0.040	0.050	0.80–1.10
52100	0.95–1.10	0.20–0.50	0.030	0.035	1.20–1.50

[1]The silicon range of all SAE basic alloy steels is 0.15–0.30%. For electric alloy steels, the silicon content is 0.15% minimum.

Table 3-15 Corrosion- and Heat-Resisting Alloys

SAE[1] No.	Carbon, Max. (%)	Manganese, Max. (%)	Silicon, Max. (%)	Phosphorus, Max. (%)	Sulfur, Max. (%)	Chromium Range (%)	Nickel Range (%)
30905	0.08	0.20–0.70	0.75	0.030	0.030	17.00–20.00	8.00–10.00
30915	0.09–0.20	0.20–0.70	0.75	0.030	0.030	17.00–20.00	8.00–10.00
51210	0.12	0.60	0.50	0.030	0.030	11.50–13.00	
X51410	0.12	0.60	0.50	0.030	0.15–0.50	13.00–15.00	
51335	0.25–0.40	0.60	0.50	0.030	0.030	12.00–14.00	
51510	0.12	0.60	0.50	0.030	0.030	14.00–16.00	
51710	0.12	0.60	0.50	0.030	0.030	16.00–18.00	

[1]The silicon range of all SAE basic alloy steels is 0.15–0.30%. For electric alloy steels, the silicon content is 0.15% minimum.

Table 3-16 Chromium-Vanadium Steels

SAE[1] No.	Carbon Range (%)	Manganese Range (%)	Phosphorus, Max. (%)	Sulfur, Max. (%)	Chromium Range (%)	Vanadium Min. (%)	Vanadium Desired (%)
6135	0.30–0.40	0.60–0.90	0.040	0.050	0.80–1.10	0.15	0.18
6150	0.45–0.55	0.60–0.90	0.040	0.050	0.80–1.10	1.15	0.18
6195	0.90–1.05	0.20–0.45	0.030	0.035	0.80–1.10	0.15	0.18

[1]The silicon range of all SAE basic alloy steels is 0.15–0.30%. For electric alloy steels, the silicon content is 0.15% minimum.

Table 3-17 Tungsten Steels

SAE[1] No.	Carbon Range (%)	Manganese, Max. (%)	Phosphorus, Max. (%)	Sulfur, Max. (%)	Chromium Range (%)	Tungsten Range (%)
71360	0.50–0.70	0.30	0.035	0.040	3.00–4.00	12.00–15.00
71660	0.50–0.70	0.30	0.035	0.040	3.00–4.00	15.00–18.00
7260	0.50–0.70	0.30	0.035	0.040	0.50–1.00	1.50–2.00

[1]The silicon range of all SAE basic alloy steels is 0.15–0.30%. For electric alloy steels, the silicon content is 0.15% minimum.

Table 3-18 Silicon-Manganese Steels

SAE[1] No.	Carbon Range (%)	Manganese Range (%)	Phosphorus, Max. (%)	Sulfur, Max. (%)	Silicon Range (%)
9255	0.50–0.60	0.60–0.90	0.040	0.050	1.80–2.20
9260	0.55–0.65	0.60–0.90	0.040	0.050	1.80–2.20

[1]The silicon range of all SAE basic alloy steels is 0.15–0.30%. For electric alloy steels, the silicon content is 0.15% minimum.

Table 3-19 Mechanical Properties and Chemical Composition of Various Ferrous Metals

Material	Specification	Chemical Analysis			Mechanical Properties			
		Carbon (%)		Others (%)	Yield Strength	Tensile Strength (lb/in²)	Elongation (% per 2 in)	Brinell Hardness
Cast iron, gray, grade 20	ASTM A48-56	3.00–4.00			—	20,000	—	163
gray grade 30	ASTM A48-56	3.00–4.00			—	30,000	—	180
nickel		2.00–3.50	Nickel 0.25–0.50		—	40,000	—	310
chrome-nickel		2.00–3.50	Nickel 1.00–3.00	Chromium 0.50–1.00	—	53,000	—	510
white		2.00–4.00	Silicon 0.80–1.50		—	46,000	—	420
malleable	ASTM A47-52	1.75–2.30	Silicon 0.85–1.20		35,000	53,000	18	140
Iron, wrought, plates	ASTM A42-55	0.08	Silicon 0.15	Slag 1.20	26,000	46,000	35	105
Iron, wrought, forgings	ASTM A75-55	0.01–0.05	Iron 99.45–99.80		25,000	44,000	30	100
Steel, cast, low carbon		0.11	Manganese 0.60	Silicon 0.40	35,000	60,000	22	120
Steel, cast, medium carbon		0.25	Manganese 0.68	Silicon 0.32	44,000	72,000	18	140
Steel, cast, high carbon		0.50			40,000	80,000	17	182
Steel, rolled, carbon	SAE 1010	0.05–0.15			28,000	56,000	35	110
Steel, rolled, carbon	SAE 1015	0.15–0.25			30,000	60,000	26	120
Steel, rolled, carbon	SAE 1025	0.20–0.30			33,000	67,000	25	135
Steel, rolled, carbon	SAE 1035	0.30–0.40			52,000	87,000	24	175
Steel, rolled, carbon	SAE 1045	0.40–0.50			58,000	97,000	22	200
Steel, rolled, carbon	SAE 1050	0.45–0.55			60,000	102,000	20	207
Steel, rolled, carbon	SAE 1095	0.90–1.05			100,000	150,000	15	300
Steel, rolled, nickel	SAE 2315	0.10–0.20	Nickel 3.25–3.75		90,000	125,000	21	230
Steel, rolled, nickel–chromium	SAE 3240	0.35–0.45	Nickel 1.50–2.00	Chromium 0.90–1.25	113,000	136,000	21	280
Steel, rolled, molybdenum	SAE 4130	0.25–0.35	Chromium 0.50–0.80	Molybdenum 0.15–0.25	115,000	139,000	18	280
Steel, rolled, chromium	SAE 5140	0.35–0.45	Chromium 0.80–1.10		128,000	150,000	19	300
Steel, rolled, chromium–vanadium	SAE 6130	0.25–0.35	Chromium 0.80–1.10	Vanadium 0.15–0.18	125,000	150,000	18	310
Steel, rolled, silicon–manganese	SAE 9260	0.55–0.65	Manganese 0.60–0.90	Silicon 1.80–2.20	180,000	200,000	12	390

Table 3-20 Uses for Steel by Carbon Content

Carbon Class	Carbon Range (%)	Typical Uses
Low/mild	0.05–0.15	Chain, nails, pipe, rivets, screws, sheets for pressing and stamping, wire
	0.16–0.29	Bars, plates, structural shapes
Medium	0.30–0.59	Axles, connecting rods, shafting
High	0.60–0.99	Crankshafts, scraper blades
		Automobile springs, anvils, bandsaws, drop hammer dies
		Chisels, punches, sand tools
		Knives, shear blades, springs
Ultra high	1.0–2.0	Milling cutters, dies, taps
		Lathe tools, woodworking tools
		Files, reamers
		Dies for wire drawing
		Metal-cutting saws
		Over 2.0% carbon by weight is referred to as-cast iron. Carbon steels can successfully undergo heat treatment with a carbon content in the range of 0.30 to 1.70% by weight.

(Carbon is indicated by the last two digits in the specification number.) Note too that in the case of cast irons, the alloying element is the important factor that increases tensile strength in the metal.

Refer to Table 3-20 to become familiar with the various uses of the different grades of steel.

ASTM Numbering System

The ASTM numbering system was developed by the American Society of Testing Materials (ASTM). It is based on the form such as sheet, plate, pipe, tube, forging, casting, and so on. Chemical properties may or may not be specified, but mechanical properties and the intended applications are. If the chemical properties are specified, they will be shown as SAE or AISI designation.

For example, ASTM A210 covers seamless medium-carbon steel boiler and superheater tubes, whereas ASTM A514 is high-strength quenched and tempered steel, with basic tensile and yield strength in the 100,000 p.s.i. range. It is typically used for structural steel, while A517 is used for pressure vessels. You need to be familiar not only with the technical designations of steel but all the trade names as well. A514 is often referred to as a T1 steel and the name is now owned by International Steel Group. Some additional ASTM specifications are ASTM A830 covers AISI/SAE carbon steel plate, while ASTM A576 covers AISI/SAE carbon steel bars. The ASTM standards for iron and steel products are available in eight volumes and cover over 8,000 pages. Some typical ASTM designated carbon steels used in construction, pressure vessels, and piping are listed in Table 3-21.

Unified Numbering Designation

The Unified Numbering System (UNS) has been developed by ASTM and SAE and several other technical societies, trade associations, and U.S. government agencies, Table 3-22 (p. 94).

A UNS number, which is a designation of chemical composition and not a specification, is assigned to each chemical composition of a metallic alloy. The UNS designation of an alloy consists of a letter and five numerals. The letters indicate the broad class of alloys; the numerals define specific alloys within that class. Existing designation systems, such as the SAE/AISI system for steels, have been incorporated into UNS designations. For more information on the UNS designations refer to SAE J1086 and ASTM E527.

Types of Cast Iron

Cast iron is an iron-based material containing 91 to 94 percent iron and such other elements as carbon (2.0 to 4.0%), silicon (0.4 to 2.8%), manganese (0.25 to 1.25%), sulfur (0.2% maximum), and phosphorus (0.6% maximum), Table 3-23 (p. 94). Alloyed cast irons are produced

Table 3-21 Application, Mechanical Properties, and Chemical Compositions of Typical ASTM Carbon Steels

Application	ASTM Standard	Type or Grade	Typical Composition Limits (%)[1] C	Mn	Si	Tensile Strength k.s.i.	MPa	Min. Yield Strength k.s.i.	MPa
\multicolumn{10}{c}{Structural Steels}									
Welded buildings, bridges, and general structural purposes	A36	—	0.29	0.80–1.20	0.15–0.40	58–80	440–552	36	248
Welded buildings and general purposes	A529	—	0.27	1.20	—	60–85	414–586	42	290
General-purpose sheet and strip	A570	30, 33, 36,	0.25	0.90	—	49–55	338–379	30	207
		40, 45, 50	0.25	1.35	—	60–65	414–448	45	310
General-purpose plate (improved toughness)	A573	58	0.23	0.60–0.90	0.10–0.35	58–71	440–489	32	221
		65	0.26	0.85–1.20	0.15–0.40	65–77	448–531	35	241
		70	0.28	0.85–1.20	0.15–0.40	70–90	483–621	42	290
\multicolumn{10}{c}{Pressure Vessel Steels}									
Plate, low and intermediate tensile strength	A285	A	0.17	0.90	—	45–65	310–448	24	165
		B	0.22	0.90	—	50–70	345–483	27	186
		C	0.28	0.90	—	55–75	379–517	30	207
Plate, manganese-silicon	A299	—	0.30	0.90–1.40	0.15–0.40	75–95	517–655	40	276
Plate, low temperature applications	A442	55	0.24	0.60–0.90	0.15–0.40	55–75	379–517	30	207
		60	0.27	0.60–0.90	0.15–0.40	60–80	414–552	32	221
Plate, intermediate and high temperature service	A515	55	0.28	0.90	0.15–0.40	55–75	379–517	30	207
		60	0.31	0.90	0.15–0.40	60–80	414–552	32	221
		65	0.33	0.90	0.15–0.40	65–85	448–586	35	241
		70	0.35	1.20	0.15–0.40	70–90	483–621	38	262
Plate, moderate and low temperature service	A516	55	0.26	0.60–1.20	0.15–0.40	55–75	379–517	30	207
		60	0.27	0.85–1.20	0.15–0.40	60–80	414–552	32	221
		65	0.29	0.85–1.20	0.15–0.40	65–85	448–586	35	241
		70	0.31	0.85–1.20	0.15–0.40	70–90	483–621	38	262
Plate, carbon-manganese-silicon heat-treated	A537	1[3]	0.24	0.70–1.60	0.15–0.50	65–90	448–621	45	310
		2[4]	0.24	0.70–1.60	0.15–0.50	75–100	517–689	55	379
\multicolumn{10}{c}{Piping and Tubing}									
Welded and seamless pipe, black and galvanized	A53	A	0.25	0.95–1.20	—	48 min.	331	30	207
		B	0.30	0.95–1.20	—	60 min.	414	35	241
Seamless pipe for high temperature service	A106	A	0.25	0.27–0.93	0.10 min.	48 min.	331	30	207
		B	0.30	0.29–1.06	0.10 min.	60 min.	414	35	241
		C	0.35	0.29–1.06	0.10 min.	70 min.	483	40	276
Structural tubing	A501	—	0.26	—	—	58 min.	400	36	248
\multicolumn{10}{c}{Cast Steels}									
General use	A27	60–30	0.30	0.60	0.80	60 min.	414	30	207
Valves and fittings for high temperature service	A216	WCA	0.25	0.70	0.60	60–85	207–586	30	207
		WCB	0.30	1.00	0.60	70–95	483–655	36	248
		WCC	0.25	1.20	0.60	70–95	483–655	40	276
Valves and fittings for low temperature service	A352	LCA[4,5]	0.25	0.70	0.60	60–85	414–586	30	207
		LCB[4,5]	0.30	1.00	0.60	65–90	448–621	35	241
		LCC[4,5]	0.25	1.20	0.60	70–95	483–655	40	276

[1] Single values are maximum unless otherwise noted.
[2] k.s.i = kilopounds per square inch.
[3] Normalized condition.
[4] Quenched and tempered condition.
[5] Normalized and tempered condition

Source: American Welding Society, *Welding Handbook*, Vol. 4, 8th ed., p. 12.

Table 3-22 Unified Numbering Systems

Primary Series of UNS Numbers

UNS Series	Metal Group
Axxxxx	Aluminum and aluminum alloys
Cxxxxx	Copper and copper alloys
Dxxxxx	Steels—designated by mechanical property
Exxxxx	Rare earth and rare earth-like alloys
Fxxxxx	Cast irons
Gxxxxx	AISI and SAE carbon and alloy steels
Hxxxxx	AISI H-steels
Jxxxxx	Cast steels
Kxxxxx	Miscellaneous steels and ferrous alloys
Lxxxxx	Low melting metals and alloys
Mxxxxx	Miscellaneous nonferrous metals and alloys
Nxxxxx	Nickel and nickel alloys
Pxxxxx	Precious metals and alloys
Rxxxxx	Reactive and refractory metals and alloys
Sxxxxx	Stainless steels, valve steels, Superalloys
Txxxxx	Tool steels
Wxxxxx	Welding filler metals
Zxxxxx	Zinc and zinc alloys

Examples of UNS Designations

UNS series	Traditional
A03190	AA 319.0 (Aluminum alloy casting)
A92024	AA 2024 (Wrought aluminum alloy)
C26200	CDA 262 (Cartridge brass)
G12144	AISI 12L14 (Leaded alloy steel)
G41300	AISI 4130 (Alloy steel)
K93600	Invar (36% nickel alloy steel)
L13700	Alloy Sn 70 (Tin-lead solder)

Examples of UNS Designations

UNS series	Traditional
N06007	Nickel-chromium alloy (Hastelloy G)
N06625	Alloy 625 (Nickel-chromium-molybdenum-columbium alloy)
R58210	Alloy 21 (Titanium alloy)
S30452	AISI 304N (Stainless steel, high nitrogen)
S32550	Ferralium 255 (Duplex stainless steel)
T30108	AISI A-8 (Tool steel)
W30710	AWS E307 (Stainless steel electrode)
Z33520	Alloy AG40A (Zinc alloy)

A Typical Entry from Metals and Alloys in the Unified Numbering System

Unified Number	Description	Chemical Composition	Cross-Reference Specifications
N10003	Ni-Mo Alloy. Solid solution strengthened (Hastelloy N)	Al 0.50 max B 0.010 max C 0.04-0.08 n Co 0.20 max Cr 6.0–80 Cu 0.35 max Fe 5.0 max Mn 1.00 max Mo 15.0-18.0 Ni rem P 0.015 max S 0.020 max Si 1.00 max V 0.50 max W 0.50 max	AMS 5607; 5771 ASME SB434; SFA5.14 (ERNiMo-2) ASTM B366; B573 AWS A5.14 (ERNiMO-2)

Source: ASTM International.

Table 3-23 Compositions of Cast Irons (Percentage of Constituents)

	Iron	Total Carbon (%)	Silicon (%)	Sulfur (%)	Phosphorus (%)	Manganese (%)
Gray iron	Balance	2.0–4.0	1.0 min.	0.2	0.6	1.0 max.
Malleable iron	Balance	2.0–3.0	0.9–1.8	0.2 max.	0.2 max.	0.25–1.25
Nodular iron	Balance	3.2–4.1	1.8–2.8	0.03 max.	0.1 max.	0.80 max.
White iron	Balance	2.5–4.0	0.4–1.6	0.15	0.4	0.3–0.8

by adding chromium, copper, nickel, and molybdenum. One of the differences between cast iron and steel is in the amount of carbon present. Most cast irons contain 2.5 to 3.5 percent carbon.

Cast iron cannot be formed by forging, rolling, drawing, bending, or spinning because of its low ductility and lack of malleability. Gray cast iron produces castings that have low ductility and low tensile strength. The material fractures readily when subjected to bending or pulling stresses, successive shocks, or sudden temperature changes.

The gray cast iron does have excellent compressive strength. While the name "cast iron" refers to a wide variety of materials, the four major classes of cast iron generally used today are gray iron, white iron, nodular iron, and malleable iron. Gray cast iron and malleable iron are the most common types used commercially.

Gray Iron

Gray cast iron may be fusion welded or braze welded without difficulty if preheating before welding and cooling after welding are controlled. It is low in ductility and has moderate tensile strength and high compression strength. Corrosion resistance and tensile strength can be improved by adding nickel, copper, and chromium as alloying materials. Gray cast iron has high machinability.

White Iron

White cast iron is produced through a process of rapid cooling which causes the carbon to combine with the iron. There is no free carbon as in gray cast iron. This causes white cast iron to be hard, brittle, and very difficult to machine except with special cutting tools. It is so difficult to weld that it is considered unweldable. White cast iron is not generally used for castings. It is the first step in the making of malleable iron. White iron has a fine grain structure and a silvery white appearance when fractured.

Malleable Iron

Malleable iron forms when white cast iron has been heat treated by a long annealing process that changes the combined iron and carbon structure into iron and free carbon. The tensile strength, impact strength, ductility, and toughness are higher than that of gray or white cast iron. The material may be bent and formed to a certain degree. In many respects, the mechanical properties approach those of low carbon steel. Fusion welding destroys the properties of malleable iron in the weld area. Because of fast cooling, the area reverts back to chilled cast iron and must be heat treated. Braze welding is recommended because of the relatively lower temperature (1,500°F) used. If a piece of malleable iron is broken, the fracture will show a white rim and dark center.

Nodular Iron

Nodular iron is also referred to as **ductile iron.** Amounts of magnesium and/or cerium are added to the iron when it is produced. Without these alloys, the graphite (free carbon) produces a notch effect which lowers the tensile strength, toughness, and ductility of the iron. The alloys change the shape of the graphite particles from flakes to spheroids and so reduce the notch effect. The silicon content of nodular iron is higher than that in other irons. Nodular iron approaches the tensile strength and ductility of steel. It has excellent machinability, shock resistance, thermal shock resistance, wear resistance, and rigidity.

Nodular iron is readily fusion welded with a filler rod containing nickel. Both preheating and postheating are necessary, and the weldment must be cooled slowly.

Aluminum-Making in the World

The world's largest producers of aluminum are China at 5,896,000 tons, Russia at 4,102,000 tons, United States at 3,493,000 tons, Canada at 3,117,000 tons, and Australia at 1,945,000 tons.

The refining of bauxite ore is the fundamental production process of reducing alumina to aluminum by means of electricity. This has remained essentially unchanged. However, improved materials, operating practices, and computerization have made the process more efficient. In the aluminum-smelting operations, work is being done on improved electrode materials as well as more efficient combustion burners for the purification of the molten aluminum. This work has led to the fact that in the last 50 years, the average amount of electricity needed to make a pound of aluminum has been reduced from 12 to about 7 kilowatt-hours. The primary products produced and their industrial applications are:

- Sheet (cans, construction materials, and automotive parts)
- Plate (aircraft and space fuel tanks)
- Foil (household aluminum foil, building insulation, and automotive parts)

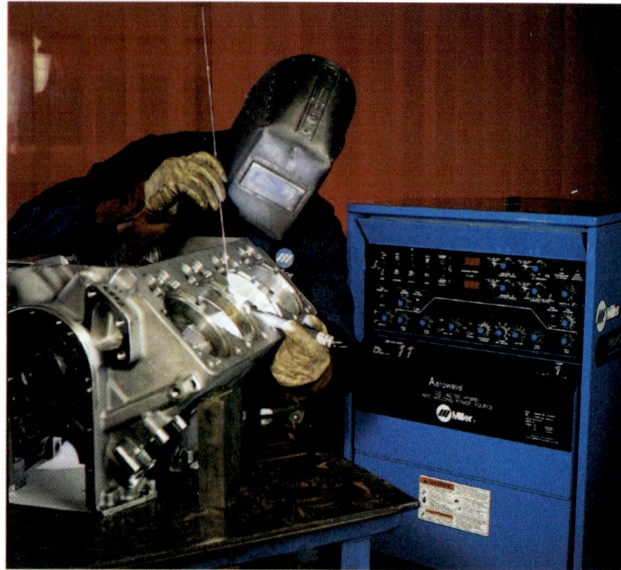

Fig. 3-41 Use of aluminum for the transportation industry for an engine block; previously made of cast iron. This block is undergoing some GTAW repair. Miller Electric Mfg. Co.

- Rod, bar, and wire (electrical transmission lines and the nonrust staples in tea bags)
- Extrusions (storm windows, bridge structures, and automotive parts)

Because of its strength, light weight, and durability, aluminum has become extremely popular in the transportation industry. Figure 3-41 shows an aluminum engine block; previously, these were made of cast iron. Aluminum can provide a weight savings of more than 50 percent compared to an equivalent steel structure.

Environmental Progress in the Aluminum Industry in the United States

Pollution prevention initiatives have been recognized for superior environmental performances due to reduction of waste. Focus has been on reducing air emissions, water discharges, and solid waste. The primary plants are equipped to capture pollutants and recycle raw materials, and industry furnaces are continually reducing the amount of chlorine gas used.

The recycling of aluminum is very important because of the environmental, as well as the economic, impact on the product. The amount of aluminum that has been recycled in the last decade has doubled. Recycling saves almost 95 percent of the energy needed to extract aluminum from its original bauxite ore. Because of the nearly 10,000 recycling centers around the nation and environmental concerns nearly two-thirds of the aluminum beverage cans produced are recycled.

Table 3-24 Designations for Aluminum Alloy Groups

Major Alloying Element	Designation[1]
99.0% min. aluminum and over	1XXX
Copper	2XXX
Manganese	3XXX
Silicon	4XXX
Magnesium	5XXX
Magnesium and silicon	6XXX
Zinc	7XXX
Other element	8XXX
Unused series	9XXX

[1]Aluminum Association designations.
Source: American Welding Society

Types of Aluminum

A four-digit numbering system is used to identify pure aluminum and wrought aluminum alloys, Table 3-24. The first digit indicates the major alloying group. For example, IXXX identifies an aluminum that is at least 99.00 percent pure; 2XXX is an aluminum with copper as the major alloying element.

The three categories of aluminum that find the most welding applications are commercially pure aluminum, wrought aluminum alloys, and aluminum casting alloys.

- ***Commercially pure wrought aluminum*** (1100) is 99 percent pure aluminum with just a little iron and silicon added. It is easily welded, and the welds have strengths equal to the material being welded.

- ***Wrought aluminum-manganese alloy*** (3003) contains about 1.2 percent manganese and a minimum of 97 percent aluminum. It is stronger than the 1100 type and is less ductile. This reduces its work ability. It can be welded without difficulty, and the welds are strong.

- ***Aluminum-silicon-magnesium-chromium alloy*** (6151) has silicon and magnesium as its main alloys. The welds are not as strong as the material being welded, but weld strength can be improved by heat treatment.

- ***Aluminum-magnesium-chromium alloy*** (5052) is strong and highly resistant to corrosion. Good ductility permits the material to be worked. Cold working will produce hardness.

- ***Aluminum-magnesium-silicon alloy*** (6053) is readily welded and can be heat treated.

Titanium-Making in the United States

Titanium is a versatile metal because of its light weight, physical properties, and mechanical properties. Titanium mineral concentrates are mainly produced from heavy-mineral sands containing ilmenite and/or rutile and also titaniferous slags made by the smelting of ilmenite with carbon. These mineral sands and titaniferous slags are processed by pigment manufacturers in the United States. However, the United States has become increasingly dependent on importing titanium mineral concentrates.

Titanium dioxide, as ilmenite or leuxocene ores, is typically associated with iron. This material can be mined from rutile beach sand in one of its purest forms. In the manufacture of titanium dioxide pigment, the principal raw materials are rutile and ilmenite. It is estimated that approximately one-third of the world supply of titanium dioxide pigment is found in the United States. In fact the United States exports up to 362,000 tons in a typical year.

Titanium sponge is produced in a retort by the vapor phase reduction of titanium tetrachloride with magnesium (the Kroll process) or sodium (the Hunter process) metal. (Titanium sponge got its name because of its spongelike appearance at this point of the processing.) The Kroll process is the most common method used worldwide for sponge production.

Titanium ingot is produced by the melting of sponge, scrap, or a combination of both. Alloying elements such as vanadium and aluminum are added to produce the typical properties required in the finished product. Russia and the United States produce the bulk of the world supply of ingots. The vacuum arc remelt (VAR) process is used to refine the material. A vacuum melter is shown in Fig. 3-20, pages 57–58. In this figure a steel electrode is shown, though when refining titanium, a titanium electrode would be used.

Titanium mill products are formed by rolling, forging, drawing, or extruding slabs and ingots in such products such as billet, bar, rod, wire, plate, sheet, pipe, strip, and tube. Titanium can also be cast into a variety of products.

Scrap and waste are produced at each step of the production process as well as the fabrication process. Titanium scrap is a large source of feedstock material with the growth in the cold hearth melting capacity. Over two-thirds of the titanium scrap consumed is of new material that has never been placed into service.

Titanium has many properties that make its replacement with other materials very difficult. This is especially true in aerospace and defense industry applications. It has great impact properties and durability with excellent mechanical strength (comparable to mild steel). It has a modulus of elasticity half that of stainless steel, which means it is much more flexible. If used for structures, it can take much more loading; for example, from an earthquake or other periods of violent movement. It is very lightweight—about 60 percent the density of steel, half that of copper, and 1.7 times that of aluminum. Titanium's coefficient of thermal expansion is half that of stainless steel and copper and one-third that of aluminum. It is virtually equal to that of glass and concrete, making titanium highly compatible with these materials. Thus, thermal stress on titanium is very low. It is immune to environmental attack. Figure 3-42 is a comparison of titanium to some other metals.

Since titanium is so corrosion resistant, it is inert to human body fluids, making it a natural material to use for implants such as hip and knee joint replacements. Titanium actually allows bone growth to adhere to the implant. Titanium plate and mesh can support broken bones and are commonly used for reconstructive surgery applications. Figure 3-43, page 98, shows some of the medical uses of titanium. Table 3-25 (p. 98) lists a few of the common types of titanium available and their mechanical properties.

The largest single demand for titanium is in the commercial aerospace industry where it is used for both engines and airframes. Other major applications are in the following industrial sectors: chemical processing, pulp and paper equipment, power generation, oil and gas exploration and processing, heat exchangers,

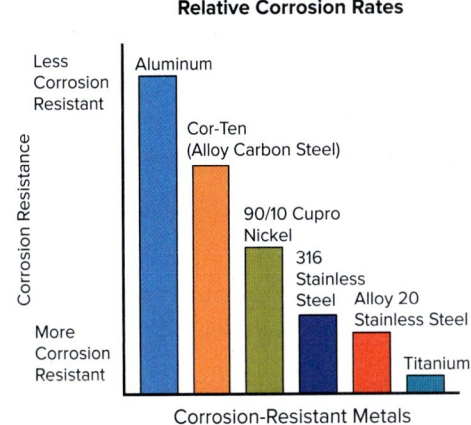

Fig. 3-42 Titanium compared to other fairly corrosion-resistant metals.

pollution control equipment, desalination plants, and metal finishing. Some emerging areas in consumer products are in bicycles, tennis rackets, golf clubs, and other sport equipment, and automotive and motor cycle components such as springs, valves, and connecting rods. Other consumer uses include wheelchairs, medical implants, watches, cameras, eyeglass frames, writing pens, and jewelry. Table 3-26 lists a few of the common types of titanium available and their chemical compositions.

Titanium is nontoxic and does not require serious limitation on its use due to health hazards. However, it is pyrophoric, which means it can produce its own heat when in the presence of oxidizing elements such as oxygen. In large pieces such as ingots, tube, pipe, bar, plate, or sheet, heating presents no problems with excessive burning or oxidation. But in small pieces with a lot of surface contact area to the air (of which a major component is oxygen), it can ignite and burn at extremely high temperatures. These small pieces may occur in the form of machining or grinding chips. Large amounts of chips or other finely divided titanium should be avoided for this reason. These fine particles should be stored in nonflammable containers and in isolated areas. An effective storage method is to submerse the particles in water with a thin layer of oil on top. If a fire should occur, properly trained personnel should use dry sand, powdered graphite, or commercially available Metal-X* to extinguish the titanium. Other flammable material in the area can be extinguished with large quantities of water.

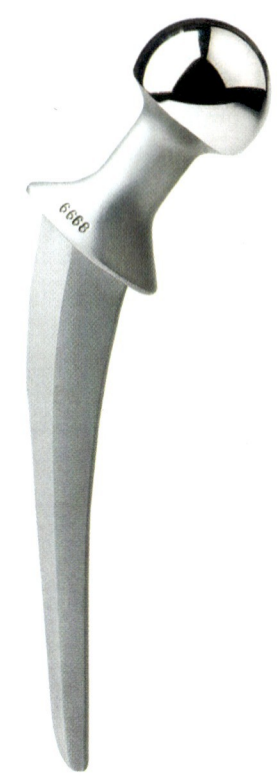

Fig. 3-43 Medical use of titanium for hip implants.
Comstock/Alamy Stock Photo

---

*Metal-X is a registered trademark for powder produced by Ansul Manufacture in Marinette, Wisconsin.

Table 3-25 Mechanical Properties of Various Titanium Alloys

ASTM Specification		Tensile Strength (k.s.i.)	Elongation (%)	Fatigue Endurance Limits (% of TS)
Gr. 1		50	35	50
Gr. 2		70	28	50
Gr. 3		85	25	50
Gr. 4		100	23	50
Gr. 5	Sheet	150	12	55–60
	Rod	140	15	
	Fastener stock	175	14	
Gr. 9		95	15	50
Gr. 12		87	22	50
Gr. 23		143	16	55–60
Gr. 32		125	15	None specified

Table 3-26 Titanium Chemical Composition Limits*

ASTM No.	Nitrogen	Carbon	Hydrogen	Oxygen	Iron	Max. Others Each	Max. Others Total	Min. Ti Content	Mo	Ni
1	0.03	0.10	0.015	0.18	0.20	0.10	0.40	99.175	NA	NA
2	0.03	0.10	0.015	0.25	0.30	0.05	0.30	98.885	NA	NA
3	0.05	0.10	0.015	0.35	0.30	0.05	0.30	98.885	NA	NA
4	0.05	0.10	0.015	0.40	0.50	0.05	0.30	98.635	NA	NA
12	0.03	0.08	0.015	0.25	0.30	0.10	0.40	Remainder	0.2–0.4	0.6–0.9
									Al	V
9	0.02	0.10	0.015	0.15	0.25	0.10	0.40	Remainder	2.5–3.5	2.0–3.0

*Values given are in percentages.

NA = not applicable.

Table 3-27 Unique Metal Products

Product	Application	Market
Zircaloys (alloyed)	Navel nuclear propulsion systems	Defense
	Civil nuclear electrical power	Energy
Zircadynes (unalloyed)	Chemical processing	Corrosion control
Niobium (commercial purity)	Chemical processing	Corrosion control
	Jewelry	Consumer
Niobium-hafnium-tungsten	Aircraft and aerospace engine plants	Aerospace
Niobium-titanium	Aircraft and aerospace fasteners	Aerospace
Niobium-titanium-tin-vanadium	Superconductors	Energy
		Medical
Titanium (commercial purity)	Chemical and seawater piping systems	Corrosion control
Titanium (high purity)	Sputtering targets	Electronics
Titanium-aluminum-vanadium	Military and civil aircraft, recreation	Air transport, defense, recreation
Titanium-niobium-zirconium	Hip implants	Medical
Titanium-nickel	Eyeglass frames	Personal
Titanium-niobium	Autoclaves, piping, valves	Mining
Hafnium (commercial purity)	Naval nuclear propulsion systems	Defense
	Civil nuclear electrical power	Energy
	Nickel-base superalloy additive	Building material, air transport
	Plasma arc cutting electrodes	Metalworking
Vanadium-chromium-titanium	Nuclear fusion reactor construction	Energy
Nickel-zirconium-vanadium-titanium	Automotive, consumer electronics, batteries	Transport, consumer
Tantalum-tungsten	Naval missile systems	Defense

Unique Metals

Many other metals may be encountered in the metalworking industry. Table 3-27 lists various metals, their applications, and market areas where they may be encountered.

Effects of Welding on Metal

Expansion and Contraction

All materials when loaded or stressed will deform, shrink, or stretch. Stress may also be caused by the welding process. During welding or cutting, the heat of the gas flame or electric arc causes the metal to expand. This causes a strain on the article being welded. The metal expands, but it is not free to move because of other welds, tacking, jigs, and the design of the weldment. The metal contracts when it cools. This combination of expansion due to heat, contraction due to cooling, and the conditions of restraint present in every weldment, causes stresses to build up in the weldment during fabrication.

Distortion and Stress in Welding

There are two major aspects of contraction in all types of welding, namely, *distortion* and *stress*. **Distortion,** also called *shrinkage,* usually means the overall motion of parts being welded from the position occupied before welding to that occupied after welding. **Stress,** on the other hand, is a force that will cause distortion later unless it is relieved. For example, the *distortion* of a weld made between two parts, held everywhere with absolute rigidity during welding, is probably zero, but the *stresses* due to shrinkage may be great, Fig. 3-44.

Temporary distortion and stress occur while welding is in progress. Stresses that remain after the welded members have cooled to a normal temperature are those that affect the weldment. These are referred to as **residual stresses.** These stresses must be relieved or they will cause cracking or fracture in the weld and/or the plate. The shrinkage of a completely stress-relieved weld between two parts free to move during welding may be large, whereas the residual stresses may be small. The V-groove butt joint in Fig. 3-44 may be high in stresses, and yet there is no distortion, whereas that in Fig. 3-45 shows marked distortion and may contain very little stress.

Stress and distortion are affected by the physical properties of the base metal, the welding process, and the welder's technique. The remainder of this chapter will be devoted to the causes and control of stress and distortion. The student should refer to Chapter 4, Basic Joints and Welds, as necessary.

Physical Properties of Metal and Distortion Distortion is the result of heating and cooling and involves stiffness and yielding. Heat changes the physical properties of metals, and these changes have a direct effect on distortion. When the temperature of a metal increases during welding, the physical properties change as follows:

- The yield point lowers.
- The modulus of elasticity decreases.
- The coefficient of thermal expansion increases.
- The thermal conductivity decreases.
- The specific heat increases.

The differing physical properties of the various metals also affect the amount of distortion and stress that can be expected.

Yield Point The yield point of steel is the point at which it will stretch and elongate under load even though the load is not increased. The higher the yield point of the weld and the base metal next to the weld, the greater the amount of residual stress that can act to distort the assembly. The lower the yield point, the less likely or severe the residual stress is.

Coefficient of Thermal Expansion The coefficient of thermal expansion is the amount of expansion a metal undergoes when heated and the amount of contraction that occurs when it is cooled. If the metals we weld did not change in length when they were welded, there would be no distortion of the part being welded and no shrinkage. A high coefficient tends to increase the shrinkage of the weld metal and the base metal next to the weld, thus increasing the possibility of distortion in the weldment. If a metal also has a high thermal conductivity, thus allowing the spread of heat over a larger area, the problems of distortion are greater.

Thermal Conductivity Thermal conductivity is a measure of the flow of heat through a metal. A metal with low thermal conductivity retards the flow of heat from the weld, thus causing a concentration of heat at the weld area. This increases the shrinkage of the weld and the plate next to it.

Modulus of Elasticity The modulus of elasticity is a measure of the relative stiffness of a metal. If the modulus is high, the material is more likely to resist movement and distortion.

Fig. 3-44 Single V-groove butt joint that shows no distortion but may be high in stresses. Edward R. Bohnart

Fig. 3-45 V-groove butt joint shows evidence of extreme distortion, but it may contain no shrinkage stresses. Edward R. Bohnart

Causes and Control of Distortion Distortion is one of the serious problems that the welder must contend with. Very often the ability to control distortion in a weldment is the difference between a satisfactory and an unsatisfactory job. A great deal of welding engineering and welding experience has been devoted to this subject. The great advance in the welding industry is a tribute to the success of these studies.

The student welder is urged to learn as much as possible about the control of distortion. During your welding practice you are urged to experiment with the various welding methods and techniques that will be presented in this text. On the job you should be capable of making welds free of defects that have good physical properties. Welders should not attempt welds that they feel they may not be able to do well enough to meet requirements.

The Types of Distortion

Lengthwise Shrinkage (Longitudinal Contraction) If a weld is deposited lengthwise along a strip of steel that is not clamped or held in any way, the strip will bow upward at both ends as the weld cools, Fig. 3-46. This is due to the contraction of the weld reinforcement above the plate surface. Weld beads that are small, or that have deep penetration and are flat, do not cause as much deformation as those that are convex.

If a welding procedure can be followed that will keep the heat on both sides of the plate nearly the same, very little distortion will occur. By depositing weld beads on the opposite side of the strip, it can be brought back to its original form. Excess weld deposit is to be avoided since it adds nothing to the strength of the joint and increases the cost of welding.

Keeping the welds balanced about the neutral axis of the joint is key in minimizing distortion. The neutral axis is the center of gravity of the joint. This can best be described as looking at a joint and visualizing a hole being drilled through the neutral axis. If a solid rod were inserted through this hole, the joint would balance about this axis on the rod. Then, this becomes the center of gravity. See Fig. 3-47 for examples of the neutral axis of a T-joint and a butt joint. Keeping the welds as close to the neutral axis as possible or balancing the weld sequences about the neutral axis will help eliminate distortion.

Crosswise Shrinkage (Transverse Contraction) If two plates are being butt welded and are free to move during welding, they will be drawn together at the opposite end due to the contraction of the weld metal upon cooling, Fig. 3-48. This is known as transverse contraction. Transverse contraction can be controlled. If the seam to be welded is short, it may be tack-welded at the opposite end, Fig. 3-49, page 102. If the seam is long, it may have to be tacked in several places, Fig. 3-50, page 102. The frequency and size of the tack welds depend upon the thickness of plate, the type of material, and the type of edge preparation. They are usually twice as long as the thickness of plate and spaced at intervals of 8 to 12 inches. The tacks also prevent the plates from buckling out of plane to each other.

Long seams can also be controlled by the use of clamping devices and wedges. The wedge is advanced along the seam ahead of the weld during the weld operation. Clamping devices help keep the plates on the same plane.

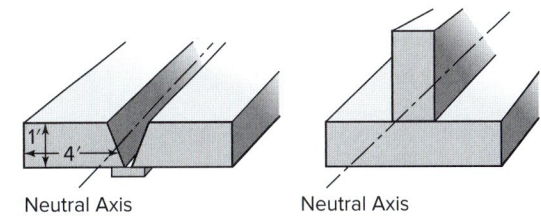

Fig. 3-47 Neutral axis of T- and butt joints.

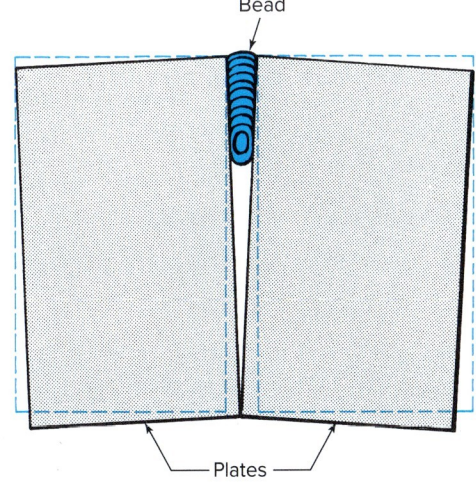

Fig. 3-46 Dotted lines indicate position before welding. Solid lines indicate position after welding.

Fig. 3-48 Position of plates before welding is indicated by dotted lines. Solid lines show position after welding.

Prespacing is another means of controlling expansion and contraction. The amount of spacing depends upon a number of variables such as length of seam, type of material, thickness of plate, and speed of welding. The welder will learn from experience the proper spacing for the job at hand.

Warping (Contraction of Weld Deposit) In welding beveled edges of thick plates such as single V-butt joints and U-groove joints, the plates will be pulled out of line with each other, Fig. 3-51. This is so because the opening at the top of the groove is greater than at the bottom, resulting in more weld deposit at the top and hence more contraction because most of the weld is above the neutral axis, as explained in Fig. 3-47, page 101.

The greater the number of passes, the greater will be this warping. Warping can be counteracted by setting up plates before welding so that they bow in the opposite direction, Fig. 3-52. This is not always possible, however, and the use of various clamping devices may be necessary.

Clamping or highly restraining joints causes a lot of internal stress. If the joint members are allowed to move, the distortion is evident. If they are restrained, there is much more residual stress built up. If the residual stress exceeds the yield strength of the metal, it will cause the metal to deform. If the residual stress exceeds the ultimate tensile strength of the material, it will cause the material to crack. In many cases the stresses applied from improper joint design, welding techniques, and heating cycles cause more forces acting on the joint than the loads it was designed to carry in service.

Angular Distortion Fillet welds contain both longitudinal and transverse stresses, Fig. 3-53. When a fillet weld

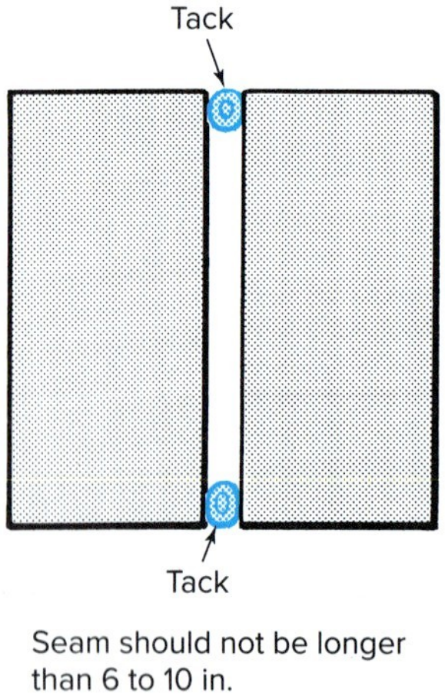

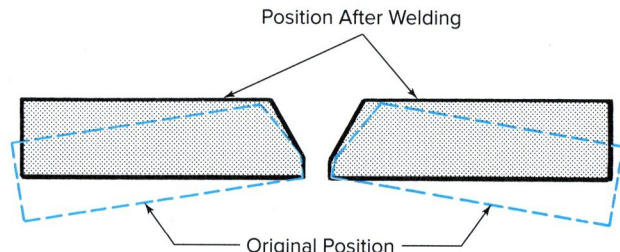

Fig. 3-52 Preset plates. Dotted lines indicate original position. Position after welding is indicated by solid lines.

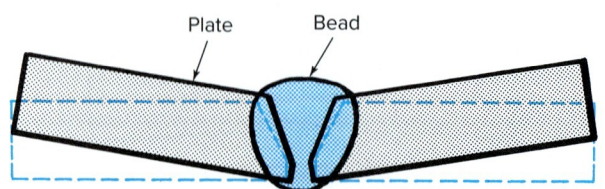

Fig. 3-49 Short seam, tack welded.

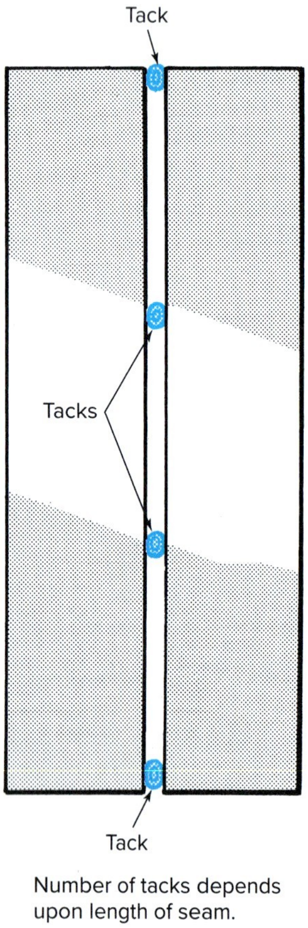

Fig. 3-50 Long seam, tack welded.

Fig. 3-51 Warpage after welding. Dotted lines indicate original position. Position after welding is indicated by solid lines.

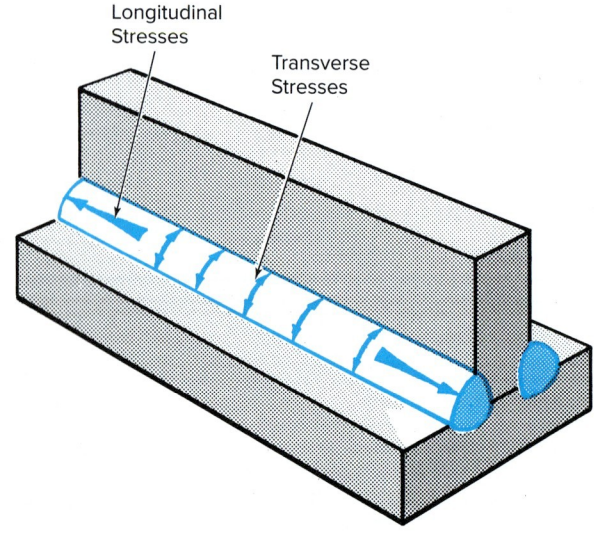

Fig. 3-53 Fillet weld metal shrinkage causes both longitudinal and transverse stresses.

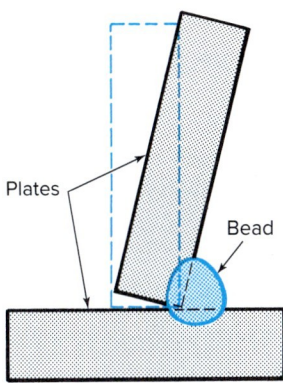

Fig. 3-54 How T-joints may be affected by fillet welding. Dotted lines indicate original position. Position after welding is indicated by solid lines.

is used in a T-joint, it will pull the vertical member of the joint toward the side that is welded. It will also bow somewhat because of the longitudinal stresses. Here again, for a given type of material and type of groove, the greater the size of the weld, the greater the number of passes, and the slower the speed of welding, the greater the amount of distortion. Figure 3-54 illustrates this angular distortion.

Effect on Butt Joints and Groove Welds The following factors affect the shrinkage *perpendicular* to the weld:

- Cross-sectional area of weld for a given thickness of plate: the larger the cross section, the greater the shrinkage.
- The angle of the bevel is not nearly as important as the free distance spacing between roots (root opening) and type of groove in causing distortion perpendicular to the weld.
- Total heat input: the greater the localized total heat input, the greater the amount of distortion.
- *Rate of heating:* other factors being equal, a greater rate of heat input results in less distortion.
- Weld wandering like the *backstep* procedures lessen the amount of distortion. (The backstep procedure will be explained on page 105.)
- *Peening,* properly used, is effective in reducing the amount of distortion, but it is not recommended except under very careful control and supervision. Peening will be discussed under Control of Residual Stress, page 107.

Effects of Angular Distortion The following factors affect angular distortion:

- The angular distortion of V-joints free to move increases with the number of layers.
- Butt weld angular distortion is greatest in V-grooves, next in U-grooves, less in double-V and double-U grooves, and least in square grooves.
- Angular distortion may be controlled by peening every fill pass layer to a suitable extent. Root and cover passes are generally not peened.
- Angular distortion may be practically eliminated in double-V and double-U groove welds by welding alternately on both sides in multilayer welding about the neutral axis.
- The time of welding and size of electrode have an important bearing on angular distortion.
- Rate of heating: other factors being equal, a greater rate of heat input results in less angular distortion.

Effect on Fillet Welds Distortion affects fillet welds in the following ways.

- Shrinkage increases with the size of the weld and decreases as the rate of heat input increases.
- If the weld is intermittent, shrinkage is proportional to the length of the weld.
- Shrinkage may be decreased materially by choosing suitable sequences and procedures of welding and peening.
- Transverse shrinkage is less for a lap joint than for a V-butt joint. Angular distortion is reduced by preheating and by suitably arranging the sequence of welding and staggering.

Prevention of Distortion
Before Welding

Design Joints should be such that they require a minimum amount of filler metal, and they should be arranged so that they balance each other, avoiding localized areas of extensive shrinkage. For example, when welding heavy materials the double-V butt joint should be used instead of the single-V joint whenever possible.

Selection of Process and Equipment Achieving higher welding speeds through the use of powdered iron manual electrodes or semiautomatic and full automatic submerged arc welding processes reduces the amount of base metal that is affected by the heat of the arc, thereby decreasing the amount of distortion.

Prebending Shrinkage forces can be put to work by prebending the parts to be welded, Fig. 3-55, page 104. The plates are bent in a direction opposite to the side being welded. The shrinkage of the weld metal is restrained during welding by the clamps. Clamping reduces warping and is more effective when the welded members are allowed to cool in the clamps. However, clamping does not entirely eliminate warping. When the clamps are removed after cooling, the plates spring back so that they are pulled into alignment.

Spacing of Parts Another method is to space parts out of position before welding, Fig. 3-56. Experience is necessary to determine the exact amount to allow for a given job. The arms will be pulled back to the proper spacing by the shrinkage forces of the welding. Figure 3-57 shows the vertical plate of a T-joint out of alignment. This is done before welding. When the completed weld shrinks, it will pull the vertical plate into the correct position. Before welding, the welder should make sure that all joints are fitted properly and do not have too great a gap. If, however, bad fitup does occur, joint spacers such as strip, bar, or ring can be inserted in the joint root to serve as a backing. Care must be exercised so that proper procedures are followed when using joint spacers.

Jigs and Fixtures Jigs and fixtures prevent warping by holding the weldment in a fixed position to reduce movement. They are widely used in production welding since, in addition to reducing warping, they permit positioning of the weldment and its parts, thus materially increasing the speed of welding.

Strong Backs Strong backs are temporary stiffeners for the purpose of increasing the resistance to distortion. They are removed after the welding is completed and cooled. This method is used in shipbuilding and for other large structures.

Distortion Control During Welding

Distortion may be reduced by using a sequence of welding known as **wandering** that provides for making welds at different points of the weldment. The shrinkage set up by one weld is counteracted by the shrinkage set up by another. This is accomplished by employing either the method shown in Fig. 3-58, called *chain intermittent fillet welds,* or that shown in Fig. 3-59, called *staggered intermittent fillet welds.* If the job requires a continuous weld, the unwelded spaces may be welded in the same order. Less longitudinal distortion is achieved with the staggered than with the chain intermittent fillet weld.

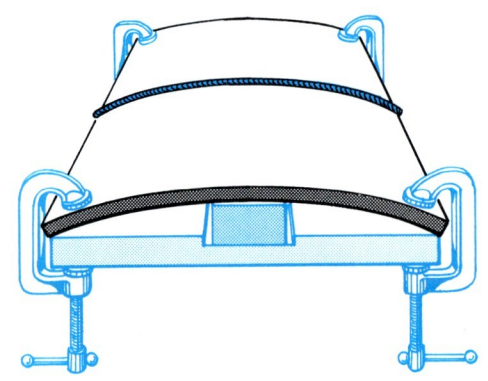

Fig. 3-55 Parts are prebent and restrained.

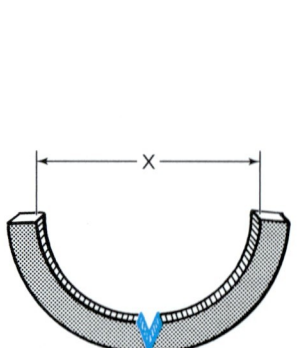

Fig. 3-56 Parts are spread before welding to reduce distortion.

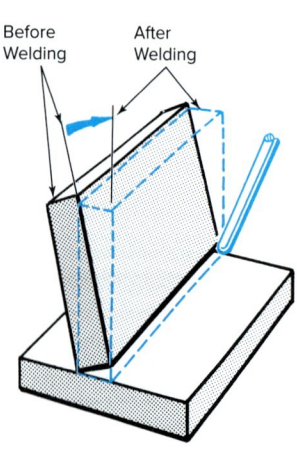

Fig. 3-57 The plate is set up out-of-square away from the weld. The weld shrinkage will bring it back to square.

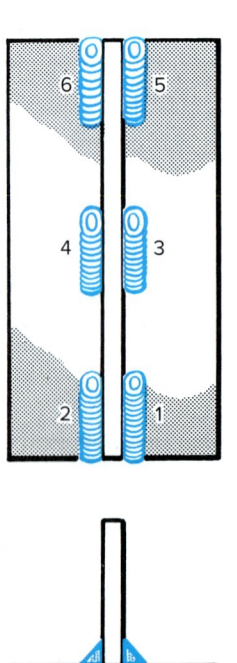
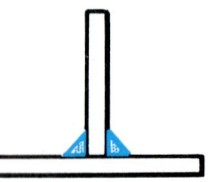

Fig. 3-58 Chain intermittent fillet welds.

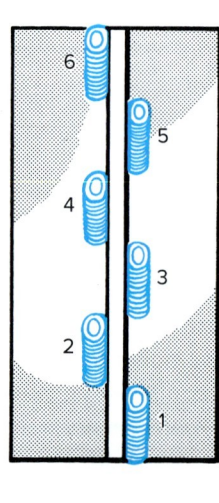
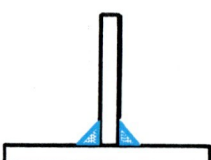

Fig. 3-59 Staggered intermittent fillet welds.

If it is found advisable to distribute the stresses or to help prevent their accumulation, the *backstep* method of welding may be employed. This method consists of breaking up the welds in short sections and depends upon welding in the proper direction. The general progression of welding is from left to right, but each bead is deposited from right to left, Fig. 3-60. Backstep welding reduces locked-up stresses and warping.

Distortion may also be held at a minimum through the use of the **skip-stop, backstep method,** a combination of skip and backstep welding. This is done in the sequence shown in Fig. 3-61. The direction of welding is the same as that employed in the backstep method except that the short welds are not made in a continuous sequence. One is made at the beginning of the joint, a section is skipped, and the second weld is applied near the center. Then a third weld is applied after further spacing. After the end of the seam has been reached, a return is made to the beginning, and the unwelded sections are completed.

In a **balanced welding sequence** an equal number of welders weld on opposite sides of a structure at the same time, thus introducing balanced stresses. The wandering techniques given previously and balanced welding all contribute to the simultaneous completion of welded connections in large fabrications. Thus distortion caused by restraint and reduction in local heating is reduced.

Correction of Distortion After Welding

If warping has occurred in a structure, the following corrective measures may be used.

- *Shrinkage* Shrinkage consists of alternate heating and cooling, frequently accompanied by hammering or mechanical working, thus shrinking excess material in a wrinkle or buckle.

- *Shrink welding* Shrink welding is a variation of shrinkage in which the heat is applied by running beads of weld metal on the convex side of a buckled area. On cooling, the combined shrinkage of the heated base metal and the added weld metal remove the distortion. The beads of weld metal may then be ground off if a smooth surface is desired.

- *Added stiffening* Added stiffening is a technique that can be used only on plate. It consists of pulling the plate into line with strong backs and welding additional stiffeners to the plate to make it retain its plane. A benefit is also derived from the shrinkage in the connecting welds.

Summary of Distortion Control

Following is a summary of the basic means that can be applied in the control of distortion.

- *Metal expansion* Some metals expand more than others. A metal with a high coefficient of expansion distorts more than one with a lower coefficient. For example, stainless steel has a high coefficient, and care must be taken to keep distortion to a minimum.

- *Distortion effects* The kind of welding process used has an influence on distortion. Gas welding produces more distortion than shielded metal arc welding and the forms of automatic and semiautomatic welding. Use higher deposition rate processes. Use higher speed welding methods—metal cored electrodes and mechanized welding. Use welding methods that give deeper penetration and thus reduce the amount of weld

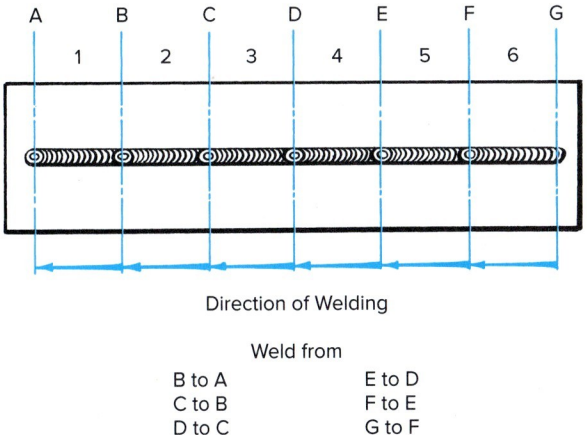

Fig. 3-60 Example of procedure and sequence of welding by the backstep method.

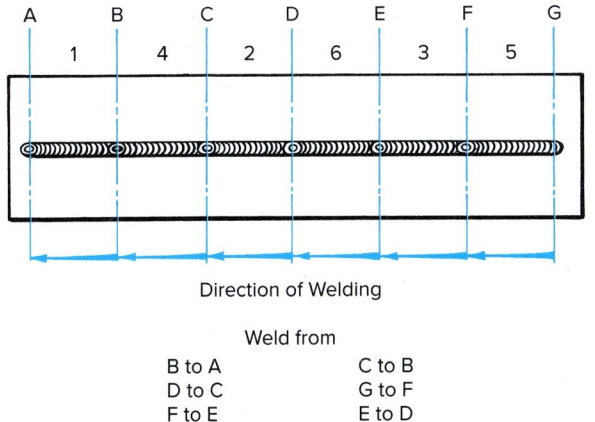

Fig. 3-61 Example of procedure and sequence of welding by the skip-stop, backstep method.

metal needed for the same strength and amount of heat input.

- **Use of welding positioners** Use welding positioners to achieve maximum amount of flat position welding, thus allowing the use of larger diameter electrodes or welding procedures with higher deposition rates and faster welding speeds.

- **Balanced forces** One shrinkage force can be balanced with another by prebending and presetting in a direction opposite to the movement caused by weld shrinkage. Shrinkage will pull the material back into alignment.

- **Forcible restraints** Effective control can be achieved by restraining the parts forcibly through the use of clamps, fixtures, and tack welds. Use strong backs and tack welds to maintain fitup and alignment. Control fitup at every point. The welder must be careful not to overrestrain the parts. This increases the stresses during welding and the tendency for cracking. Weld toward the unrestrained part of the member.

- **Clamping parts during fabrication** During fabrication the parts can be clamped or welded to a heavy fixture that can be stress relieved with the weldment, thus ensuring dimensional tolerance and stability.

- **Heat distribution** Distribute the welding heat as evenly as possible through a planned welding sequence and planned weld positions. Welders should start welding at points on a structure where the parts are very restrained or fixed and move on to points having a greater relative freedom of movement. On assemblies, joints that will have the greatest amount of shrinkage should be welded first and then the joints with less expected shrinkage last. Complex structures should be welded with as little restraint as possible. Sequence subassemblies and final assemblies so that the welds being made continually balance each other around the neutral axis of the section.

- **Increase speed with heat** As the heat input is increased, the welding speed should be increased.

- **General rule about warping** All other things being equal, a decrease in speed and an increase in the number of passes increase warping.

- **Welding from both sides** Welding from both sides reduces distortion, Fig. 3-62. Welding from both sides at the same time all but eliminates distortion.

- **Welding direction** The welding direction should be away from the point of restraint and toward the point of maximum freedom.

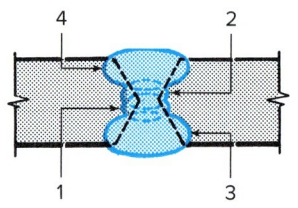

Fig. 3-62 Distortion is reduced by welding from both sides.

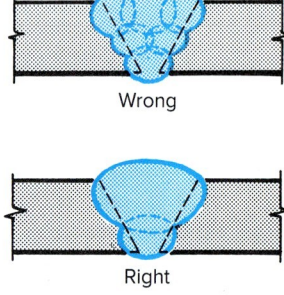

Fig. 3-63 Minimum number of passes.

- **Wandering sequences** The employment of wandering sequences of welding, such as skip welding and backstep welding, prevents a local buildup of heat and thus reduces shrinkage.

- **End fixing** To make sure that welds will not fail at the end and carry on into the joint, the welds on the ends of members should be fixed or welded around the end. This is referred to as *boxing* and is used when a fillet weld is wrapped around the corner of a member as a continuation of the principal weld.

- **Avoid overwelding** Too much welding increases distortion. Excessive weld size and too many weld passes cause additional heat input, Fig. 3-63. A stringer bead produces less distortion than a weave bead, and a single pass is better than several passes. Use the smallest leg size permissible when fillet welding. Use fewer weld passes.

- **Reduce weld metal** Excessive widths of groove welds add nothing to strength, but they increase weld shrinkage as well as welding costs. The root opening, included angle, and reinforcement should be kept to a minimum, Fig. 3-64. Select joints that require little weld metal. For example, choose a double-V groove butt

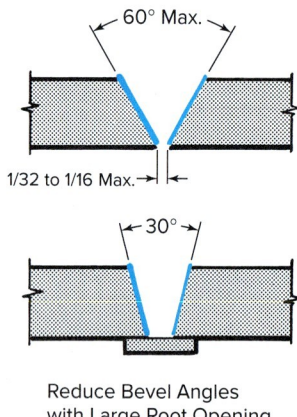

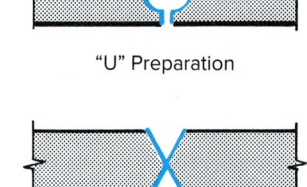

Fig. 3-64 Correct edge preparation and good fitup.

Fig. 3-65 The construction industry uses a tremendous amount of welding. This welder is working on some internal structure of a tunnel. Dmitry Kalinovsky/Shutterstock

joint instead of a single-V groove butt joint. Weld those joints that cause the most contraction first.

- **Fix tack welds first** Weak welds or cracked tack welds should be chipped or melted out before proceeding with the weld.

- **Peening** Peening the weld is effective. Too much peening, however, causes a loss of ductility and impact properties.

Control of Residual Stress

A welded structure develops many internal stresses that may lead to cracking or fracture. Under normal conditions, these stresses are not a threat to the structure. However, for certain kinds of code welding requirements, and on those structures where there is chance of cracking, stress relief is necessary, Fig. 3-65. Both hot and cold processes are used.

Preheating It is often necessary to control or reduce the rate of expansion and contraction in a structure during the welding operation. This is done by preheating the entire structure before welding and maintaining the heat during welding. Considerable care must be taken to make sure that preheat is uniform throughout the structure. If one part of the structure is heated to a higher temperature than another, internal stresses will be set up, thus offsetting the advantages sought through preheating. After the weld is completed, the structure must be allowed to cool slowly. Preheating may be done in a furnace with an oxyfuel flame or high frequency induction heating. Preheating is also used to slow the cooling rate when working on materials with a sensitive microstructure. The carbon equivalence, thickness, amount of hydrogen in the welding process and the restrained the joint is being welded contribute to the preheating temperature.

Postheating The most common method of stress relieving is postheating. This kind of heat treatment must be done in a furnace capable of uniform heating under temperature control. The method of heating must not be injurious to the metal being treated. The work must be supported so that distortion of any part of the weldment is prevented. The rise in temperature must be gradual, and all parts of the weldment must be heated at a uniform rate. Mild steel is usually heated to about 1,100 to 1,200°F. Other steels may require a higher temperature, depending upon the yield characteristics of the metal. Some alloy steels are brought up to a temperature of 1,600°F or higher.

When the weldment reaches the maximum temperature, it is permitted to soak. The length of time depends upon the thickness of plate and the plasticizing rate of the steel of which it is made. The rate is usually 1 hour per inch of thickness. The weldment should be permitted to soak long enough to ensure relief to the thickest part.

The reduction of temperature must be gradual and at a rate that will ensure approximately uniform

Table 3-28 Suggested Preheat Temperatures

Carbon Equivalent (%)	Temperature (°F)
Up to 0.45	Optional
0.45–0.60	200–400°F
Above 0.60	400–700°F

temperatures throughout all parts. Structures of different thicknesses sometimes may require as long as 48 hours to cool.

The temperature at which the work may be withdrawn from the furnace depends upon the varying thicknesses and rigidity of the members. This may be as low as 200°F or as high as 600°F. It is important that the air surrounding the furnace be quiet enough to ensure uniform temperature. Table 3-28 gives some suggested preheat temperatures based on the carbon equivalency of the steel.

It is not always possible to heat the entire structure, as, for example with pipelines. In such cases it may be possible to relieve stress by heating only one portion of the structure at a time. It is important that the member be able to expand and contract at will. Otherwise, additional stresses will be introduced into the structure that may be greater than the original stresses being treated.

Full Annealing Annealing is superior to all other methods, but it is very difficult to handle. Work that is fully annealed must be heated up to 1,600 to 1,650°F. This causes the formation of a very heavy scale and there is danger of collapse on some types of weldments.

Cold Peening In cold peening, the bead is hammered to stretch it and counteract shrinkage due to cooling. Cold peening of the weld metal causes plastic flow, thereby relieving the restraint that causes the residual stress. In effect the toes of the weld that were in tension have now been placed in compression. This greatly reduces the tendency for cracking. Effective peening requires considerable judgment. Peening is identical to cold working the steel and, if overdone, will cause it to lose ductility and become work hardened. Overpeening may result in cracks or the introduction of new residual stresses.

The weld should be peened at a temperature low enough to permit the hand to be placed upon it. The root and face layers of the weld should not be peened. A pneumatic chisel with a blunt, rounded edge is used. Hand peening cannot be controlled properly.

Vibratory Stress Relieving This technique uses a low frequency, high amplitude vibration to reduce the residual stress levels to the point where they cannot cause distortion or other problems. A vibration generator is clamped to the workpiece or attached to the tooling fixture. The vibration level is then adjusted to create the desired amplitude, and sine waves pass through the parts, relaxing the microstructure. It usually takes between 15 and 30 minutes for a full treatment. The size and weight of the workpiece determines the time required.

Cryogenic Stress Relieving Cryogenic stress relieving takes various structures at a very slow rate down from room temperature to –300°F by exposing them to liquid nitrogen vapors. This is done at roughly 1°F per minute. Once a structure reaches its proper temperature, it is allowed to soak at a holding temperature for 24 to 36 hours. The molecules in the structure get closer together as the temperature drops. They are at a much lower energy level and a better bond forms. If a fracture zone exists due to a weak molecular bond or no bond, this is improved and these voids are filled in. As with steel if these voids are filled in, the structure can withstand more force. At the end of this holding period, the structure is slowly warmed back up to room temperature. This is again done very slowly at a rate of approximately 1°F per minute. This process helps align the molecules so there is less stress. A conventional heat treatment can be done prior to the cryogenic process.

Mechanical Loading In mechanical loading, the base metal is stressed just at the point of yielding by application of internal pressure to a pressure vessel: This procedure works well with a simple weldment. On vessels where the plate is not of uniform thickness or where there are reinforcements, however, not all the metal can be stretched to the same extent. It is important that only a very small yielding takes place. Otherwise, there is danger of strain hardening or embrittling the steel. Hydraulic pressure, rather than air pressure, is used because of the dangers in connection with air pressure if the vessel should rupture.

Reducing Stress through Welding Technique While it is impossible to completely control the residual stresses due to the welding operation without preheating or postheating, the amount of stress can be minimized if the following welding procedures are employed:

- The product should be designed to incorporate the types of joints having the lowest residual stress.

- The degree of residual stress caused by the welding process should be considered when choosing a process.
- Plan assembly welding sequences that permit the movement of component parts during welding, which do not increase joint fixity, and in which joints of maximum fixity are welded first.
- Avoid highly localized and intersecting weld areas.
- Use electrodes that have an elongation of at least 20 percent in 2 inches.
- Peening is an effective method of reducing stresses and partly correcting distortion and warping. Root and face layers and layers more than 1/8 inch thick should not be peened.

CHAPTER 3 REVIEW

Multiple Choice

Choose the letter of the correct answer.

1. Metals are separated into which of the two major groups? (Obj. 3-1)
   a. Alloys and castings
   b. Slabs and plates
   c. Billets and blooms
   d. Ferrous and nonferrous
2. The first recorded use of iron dates back to what century? (Obj. 3-1)
   a. 3700 B.C.
   b. 1350 B.C.
   c. 1000 B.C.
   D. A.D. 1600
3. Steel is made up of what two elements? (Obj. 3-1)
   a. Coke and limestone
   b. Granite and oxygen
   c. Carbon and iron
   d. None of these
4. What are the basic raw materials that go into the steelmaking process? (Obj. 3-1)
   a. Heat, oxygen, and slag
   b. Limestone, coke, and iron ore
   c. Scrap, rutile, and cellulous
   d. All of these
5. Which of the following is a steelmaking process? (Obj. 3-1)
   a. Electric
   b. Basic oxygen
   c. Open hearth
   d. Both a and b
6. Once the steel has been produced it must be processed into usable material by which method? (Obj. 3-2)
   a. Bending and twisting
   b. Forging
   c. Rolling
   d. Both b and c
7. A steel structure that is 2 to 6 inches thick and 5 to 6 feet wide is referred to as a _____. (Obj. 3-2)
   a. Bloom
   b. Billet
   c. Slab
   d. None of these
8. The operation of reducing the cross section and increasing the length of a metal bar or wire by pulling it through a series of conical, tapering holes in a die plate is _____. (Obj. 3-2)
   a. Excursion
   b. Extrusion
   c. Drawing
   d. Both a and b
9. In which grain orientation direction does a rolled steel section have the least strength? (Obj. 3-2)
   a. X
   b. Y
   c. Z
   d. E
10. Which of the following is an important variable in any heat-treating process? (Obj. 3-3)
    a. Carbon content
    b. Temperature of heating
    c. Time allowed for cooling
    d. All of these
11. Which of the following is a state of matter? (Obj. 3-4)
    a. Solids
    b. Liquids
    c. Gases
    d. All of these
12. _____ is the capacity of a metallic substance to form a strong bond of adherence under pressure or when solidifying from a liquid state. (Obj. 3-5)
    a. Weldability
    b. Fusibility
    c. Volatility
    d. Electrical conductivity

13. _____ is the ease with which a metal may be vaporized. (Obj. 3-5)
    a. Weldability
    b. Fusibility
    c. Volatility
    d. Thermal conductivity
14. The _____ of a substance is the ability of the substance to conduct electric current. (Obj. 3-5)
    a. Electrical conductivity
    b. Thermal conductivity
    c. Melting point
    d. Catalyst
15. The _____ is the amount of expansion a metal undergoes when it is heated and the amount of contraction that occurs when it is cooled. (Obj. 3-5)
    a. Expansion coefficient
    b. Thermal conductivity
    c. Coefficient of thermal expansion
    d. Vaporization
16. _____ is the weight of a volume of material compared with an equal volume of water. (Obj. 3-5)
    a. Mass
    b. Density
    c. Specific gravity
    d. Both a and c
17. Porosity means _____. (Obj. 3-5)
    a. How dense a material is
    b. How well a material will let a fluid pass through it
    c. How tall the material is
    d. How much the material weighs compared to an equal volume of water
18. Loading of a material will cause it to lose its form. The ability of the material to return to its original shape after the load has been removed is known as its _____. (Obj. 3-5)
    a. Plasticity
    b. Impact property
    c. Elastic limit
    d. None of these
19. The endurance level of a material undergoing a fatigue test is _____. (Obj. 3-5)
    a. The maximum load level at which no fatigue cracking will occur no matter how many times it is cycled
    b. The time in service before a fatigue crack can be expected
    c. 120 percent of the tension stress failure point
    d. Calculated after the failure of the material
20. Which of the alloying elements are used in the making of steel? (Obj. 3-6)
    a. Krypton, neon, and lanthone
    b. Carbon, silicon, and manganese
    c. Chromium, nickel, and niobium
    d. Both b and c
21. A medium carbon steel has a carbon percent range of _____. (Obj. 3-7)
    a. 0.08% to 0.30%
    b. 0.30% to 0.60%
    c. 0.60% to 1.70%
    d. Both a and b
22. The refining of bauxite ore is the fundamental production process for making which nonferrous metal? (Obj. 3-8)
    a. Copper
    b. Titanium
    c. Gold
    d. Aluminum
23. Which ASTM number is used to identify a titanium alloy made up of 98.885 percent titanium, 0.05 percent nitrogen, 0.10 percent carbon, 0.015 percent hydrogen, and 0.35 percent oxygen? (Obj. 3-9)
    a. 1
    b. 2
    c. 3
    d. All of these
24. When a material is heated, it expands, and when it cools, it contracts. This is true for all the materials listed except _____. (Obj. 3-10)
    a. Steel
    b. Aluminum
    c. Titanium
    d. Water
25. Distortion in weldments can occur in which directions? (Obj. 3-10)
    a. Longitudinal
    b. Transverse
    c. Diagonal
    d. Both a and b
26. Which of the following methods can be used for controlling expansion and contraction forces? (Obj. 3-10)
    a. Sequencing the welds
    b. Joint design
    c. Presetting the joint
    d. All of these
27. How can the residual stresses of welding be controlled? (Obj. 3-10)
    a. Preheat
    b. Peening

c. Vibratory or cryogenically
   d. All of these
28. A method used to put the toes of a weld in compression instead of tension to reduce distortion and prevent cracking is called _____. (Obj. 3-10)
   a. Channeling
   b. Slugging
   c. Peening
   d. None of these

Review Questions

Write the answers in your own words.

29. List five events that happened in this country to spur the growth of steelmaking. (Obj. 3-1)
30. Describe at least five steps in the continuous casting process. (Obj. 3-1)
31. Name at least four steelmaking processes. (Obj. 3-1)
32. What can be done to control warping? (Obj. 3-2)
33. Name the five types of carbon steels and give the range of carbon content for each. (Obj. 3-4)
34. List 10 applications for the use of titanium. (Obj. 3-8)
35. Name three ways to avoid overwelding. (Obj. 3-10)
36. Name three factors that are important in the choice of weld joints. (Obj. 3-10)
37. If a weld is deposited lengthwise along a strip of steel, what effect will it have upon the piece of steel when it cools? (Obj. 3-10)
38. What are the two major aspects of contraction in all types of welding? Explain. (Obj. 3-10)

INTERNET ACTIVITIES

Internet Activity A

Using your favorite search engine, look up the history of steelmaking. Make a timeline showing when improvement or changes took place in the steelmaking industry.

Internet Activity B

Find a Web site about a steel manufacturer. See what is new today in steelmaking. Write a report about your findings.

Design credits: Cinema and movie icon: Denis Meshkov/123RF; and Illustration of Welding icons: Mr. Nachai Sorasee/123RF

Basic Joints and Welds

4

The fabrication of welded structures is a highly competitive business. In order to survive, the shop doing this type of work must take advantage of every economy possible. Thus it adopts the latest improvements and designs, makes full use of materials, and eliminates unnecessary operations. The selection of the wrong type of weld joint may result not only in a great loss of time and money but may also contribute to the breakdown of the weldment in use, thus specifically damaging the reputation of the manufacturer and, in general, contributing to a distrust in welding. Only personnel who have practical and technical training along these lines can possibly hope for satisfactory results.

Although weld joint design and selection are the responsibility of the engineering department, the welder still should be concerned about weld joint design and welding procedures. Recognition of the requirements for a particular type of weld will lead to work of higher quality and accuracy. It is the welder's responsibility to understand fundamental joint design and welding procedures. A welder who has a practical understanding of the values of weld joint design and the characteristics of different types of welds can provide valuable assistance to the supervisory personnel and the engineering department. Best results are obtained where this kind of cooperation is found.

Chapter Objectives

After completing this chapter, you will be able to:

4-1 Describe the five basic joints and the welds applied to each.

4-2 Measure fillet and groove weld sizes.

4-3 Determine the position of welding for groove and fillet welds on plate and pipe.

4-4 List the factors that will affect the strength of a welded joint.

4-5 Describe the difference between a weld discontinuity and a weld defect.

4-6 Describe visual inspection and its limitations and advantages.

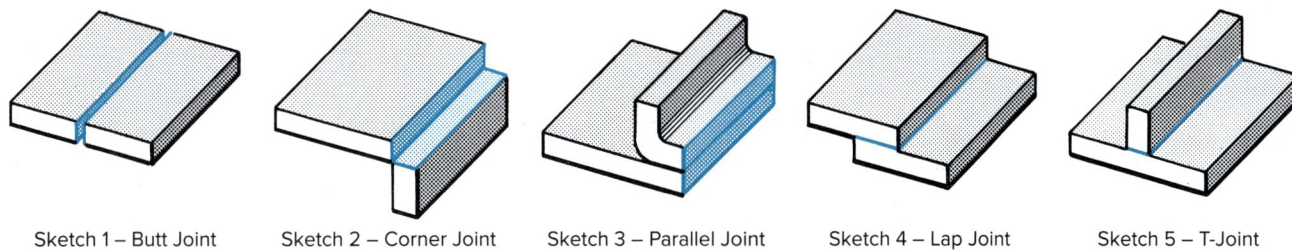

Fig. 4-1 Basic types of joints. An important aspect of the joint is where the members approach closest to each other. This is called the joint root and may be a point, a line, or an area.

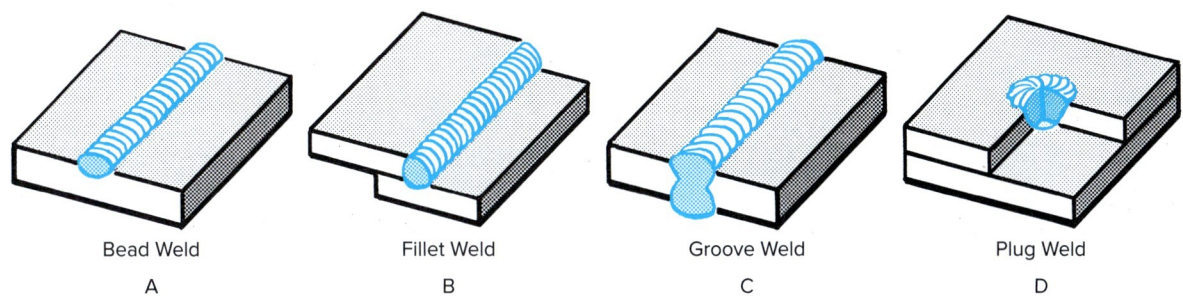

Fig. 4-2 Types of welds.

Types of Joints

There are only five basic types of joints: the butt joint, the corner joint, the parallel joint, the lap joint, and the T-joint (Fig. 4-1). In Chapter 28, the most common joints will be described in terms of their use, advantages and disadvantages, joint preparation, and economy.

The Four Weld Types

It is important to understand the four basic weld types. They are the bead (surface) weld, fillet weld, groove weld, and plug (slot) weld. These four are depicted in Fig. 4-2.

- **Bead welds** These welds, also called surface welds, are single-pass deposits of weld metal, as illustrated in Fig. 4-2A. Bead welds are used to build up a pad of metal and to replace metal on worn surfaces.

- **Fillet welds** These welds consist of one or more beads deposited in a right angle formed by two plates, as shown in Fig. 4-2B. They are used for lap joints and T-joints. Fillet welds take a triangular cross section due to the location they are placed in the weld joint. The weld symbol used for fillet welds takes the same triangular shape as the weld, so it is easily recognized. Open corner joints are also welded with fillet welds. It is important to understand the terminology applied to a fillet weld. The various parts of a fillet weld are shown in Fig. 4-3. An important aspect of a weld is its profile. The face of the weld can be convex, concave, or flat. This aspect is important when determining the size of the fillet weld, which will be covered later in this chapter.

- **Groove welds** These welds consist of one or more beads deposited in a groove, such as shown in Fig. 4-2C. Groove welds are used for butt joints. The butt joint can be left unprepared with square edges, or it can be prepared with a bevel or a J-groove. If both members are beveled or J-grooved when they are brought together, they take the shape of a V or U and that is how these grooves are referred to on a butt joint—namely, as a V-groove or a

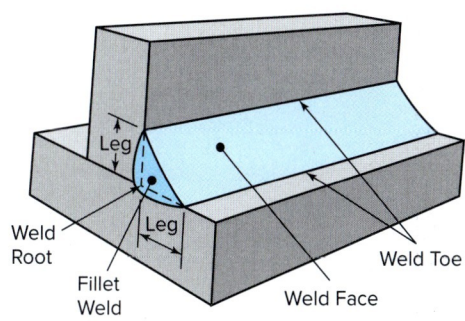

Fig. 4-3 Face, toe, root, and leg of a fillet weld on a T-joint.
Source: *Welding Inspection Technology,* 5th ed., pp. 4–24, Figure 4.21.

U-groove butt joint. This weld is applicable on both plate and pipe. Figure 4-4 shows the various parts of a groove weld.

- **Plug welds** These welds, which are similar to slot welds, are used for filling slotted or circular holes in lap joints, as shown in Fig. 4-2D, page 113. If the hole, or slot, is large, a fillet weld may be made around the faying surface of the joint. Figure 4-5 shows slot welds, plug welds, and an example of a hole with a fillet weld in it. The plug weld may or may not completely fill the joint as shown. The hole or slot may be open at one end.

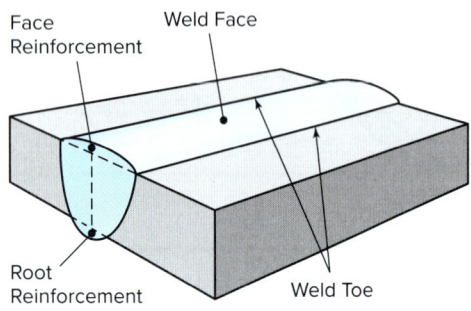

Fig. 4-4 Weld, toe, root, and face reinforcement of a square groove butt joint. Source: *Welding Inspection Technology,* 5th ed., pp. 4–23, Figure 4.20, 2008.

Weld Size and Strength

Weld Size

When the design engineer determines the load-carrying capacity of the welded joint, they will specify this on the drawing. This is done through the use of welding symbols that will be covered in Chapter 30.

Groove Welds A *groove weld* is measured and sized by its depth of penetration/fusion into the joint. This size does not include reinforcement of the face or root of the weld. Groove welds are generally referred to as partial joint penetration (PJP) welds or complete joint penetration (CJP) welds. If a groove weld symbol has no size reference, then it should be considered to be a CJP weld. If the design engineer wanted to have a PJP weld, then this would be designated on the welding symbol. Figure 4-6 shows a CJP groove preparation on the end of a pipe. Figure 4-7 illustrates the CJP V-groove butt joint. Figure 4-8 shows the fusion terms that apply to groove welds. Weld interface is the line separating the weld from the heat-affected zone (HAZ). Note again that the reinforcement on the face or root does not count as part of the weld size.

Seal Welds On many structures, the strength of the joint may be derived from its riveted construction. However, to

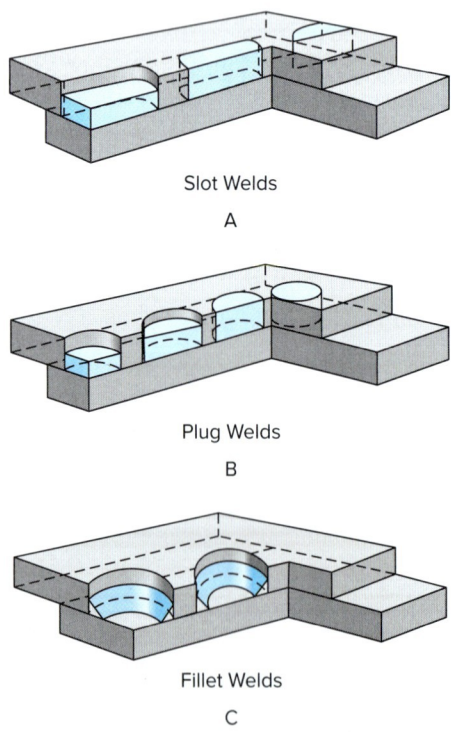

Fig. 4-5 (A) Slot welds, (B) plug welds, and (C) fillet welds in a hole. Source: AWS A3.0-2010 STANDARD WELDING TERMS AND DEFINITIONS, P. 31 FIGURE B.15 (D), (E) AND (F).

Fig. 4-6 For code quality work on CJP groove welds the appropriate end preparation must be provided. As indicated on this machine cut pipe nipple practice pipe. A typical bevel angle of 37½° is used to produce a 75° groove angle for butt joints. The root face will need to be applied. Mark A. Dierker/McGraw Hill

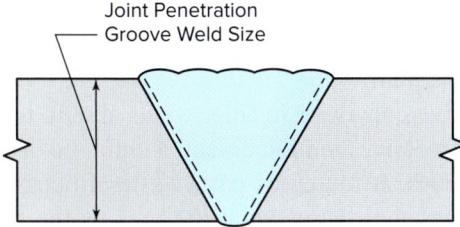

Fig. 4-7 Complete joint penetration groove welds where the maximum load-carrying capacity is required for the joint.

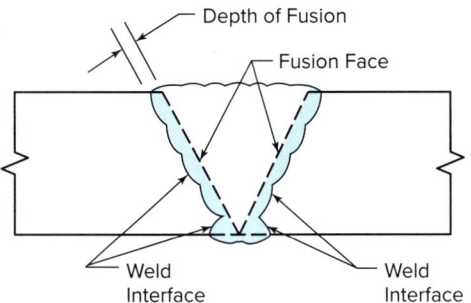

Fig. 4-8 V-groove with backing weld. Dotted line denotes the original groove face that is now fused into the weld.

excessive reinforcement above the allowable limit is a waste of time and weld metal, but it also decreases the working strength of the joint because of a concentration of stresses at the toe of the weld. The steep entrance angle greatly reduces the endurance limit under fatigue loading. It is obvious, on the other hand, that a lack of reinforcement or insufficient penetration into the joint will decrease the size of the weld. Proper reinforcement should not exceed ⅛ inch. Figures 4-10 and 4-11, page 116, depict, respectively, the measurement and reinforcement of CJP groove welds.

The width of a groove weld should not be more than ¼ inch greater than the width of the groove face. This allows for a maximum amount of fusion beyond the groove face of ⅛ inch on each side of the joint. Metal deposited beyond the groove face is a waste of time and filler metal. It also adds to the overall heat input and increases resultant residual stresses. The excess deposit adds cost to the joint and decreases its strength.

CJP welds are usually designed to possess the maximum physical characteristics of the base metal. Those CJP welds that meet code requirements, such as the butt joint in piping, must have better physical properties than those used

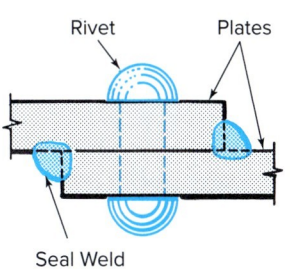

Fig. 4-9 Riveted and seal-welded lap joint.

make sure that the joints will not leak or allow moisture into the joint, continuous welds are run the entire length of the joints' seal. These welds are called seal welds. They are usually single-pass welds deposited along the root of the joint, Fig. 4-9. They must be sound, but they are not expected to carry a heavy load.

Many welders believe that high reinforcement increases the strength of the welded joint. This is not true. Not only

SHOP TALK

AWS Student Membership

The American Welding Society, formed in 1919, is a worldwide group with over 50,000 members. Students can join at a discount and receive a journal about welding. Being with AWS is a great way to meet professionals and learn about jobs. Locate AWS on the Web. It is a useful group, providing publications and conferences about the welding craft. Being top-notch in your field means keeping up with the newest methods, and belonging to AWS is a fun way to do that.

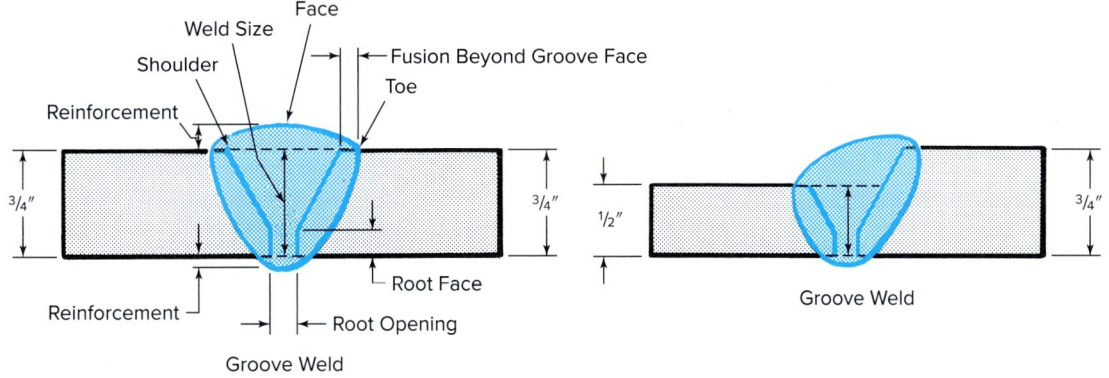

Fig. 4-10 Measurement of groove welds.

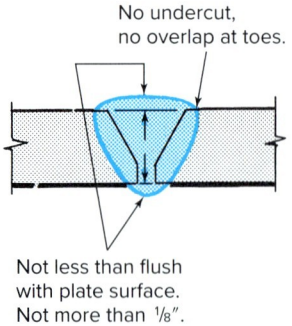

Fig. 4-11 Reinforcement for groove welds.

in the fabrication of noncode production components. The idea is to have a weld that is fit for its intended purpose. The minimum size called for on the welding symbol must be made for these groove welds to fit their intended purpose.

Figure 4-12 depicts a partial joint penetration groove weld. In this case the design engineer has specified that this will give sufficient strength to carry the load intended. If a PJP weld is what was intended, this would be specified on the welding symbol and is acceptable. If, however, a CJP weld was intended and only a PJP weld was made, then this is an improper weld and will need to be repaired or replaced. This would then be referred to as an *incomplete joint penetration weld,* which would be considered a defective weld.

Fillet Welds The most common weld used in industry is the *fillet weld*. Fillet welds can be as strong as or stronger than the base metal if the weld is the correct size and the proper welding techniques are used. When discussing the size of fillet welds, the weld contour must first be determined. **Contour** is the shape of the face of the weld. Figure 4-13 shows a cross-sectional profile of the three types of fillet weld contours: flat, convex, and concave.

When discussing fillet weld size, familiarity with the various parts of a weld is required. Figure 4-14 shows a convex fillet weld and the associated terms. The size of a **convex fillet weld** is generally considered to be the length of the leg referenced. On **convex** and **flat contour fillet welds** the size and leg are the same, and this is what the design engineer specifies on the weld symbol.

The convex fillet weld, in contrast to the concave and flat fillet welds, has less of a tendency to crack as a result of shrinkage setup cooling. The welder will also find it easier to keep from undercutting at the toes when making this type of weld.

Excessive convexity in fillet welds should be avoided just as excessive reinforcement should be avoided in groove welds. It increases cost, wastes filler metal, and concentrates more stresses at the toes of the weld. If a fillet weld is specified to be convex, it should only have a slight amount of convexity. This is generally based on the width of the weld face. Table 4-1 lists the acceptable amounts of convexity.

For concave fillet welding, the size and leg are two different dimensions. The leg is the dimension from the weld

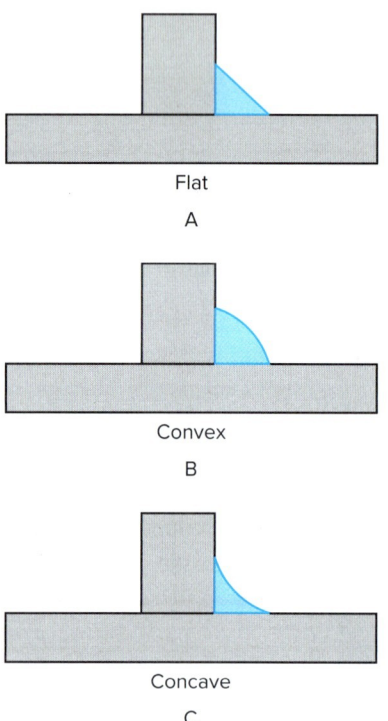

Fig. 4-13 Fillet face contours.

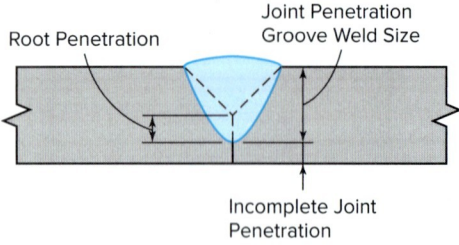

Fig. 4-12 Partial joint penetration V-groove weld butt joint. It would be considered incomplete joint penetration only if a CJP groove was called for. Source: *Welding Inspection Technology,* 5th ed., pp. 4–24, Figure 4.23.

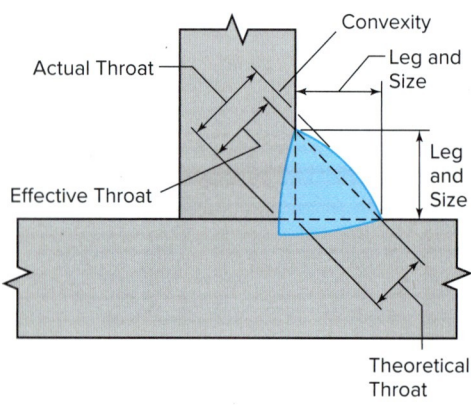

Fig. 4-14 Convex fillet weld. From *Welding Inspection Technology,* 5th ed., pp. 4–25, Figure 4.26 top.

Table 4-1 Maximum Convexity Allowable on Fillet Welds

Width of Weld Face of Total Joint or Individual Weld Bead (in.)	Maximum Convexity (in.)
≤5/16	1/16
>5/16	1/8
≥1	3/16

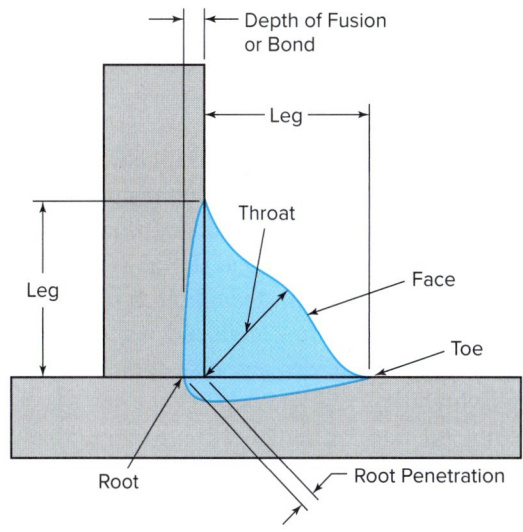

Fig. 4-16 Ideal fillet weld shape.

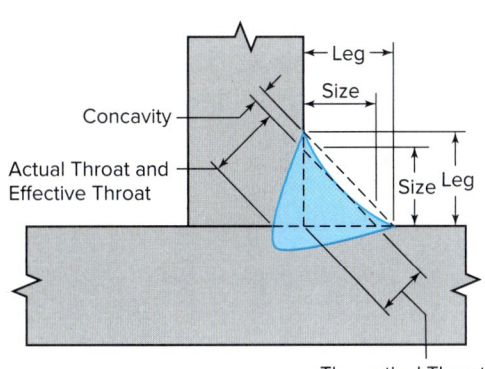

Fig. 4-15 Concave fillet weld. From *Welding Inspection Technology*, 5th ed., pp. 4–25, Figure 4.26 bottom.

toe to the start of the joint root. The actual size of a convex fillet weld, as shown in Fig. 4-15, is measured as the largest right triangle that can be inscribed within the weld profile. A special fillet weld gauge is used to measure concave fillet welds. If the weld is flat, either the concave or convex fillet weld gauge can be used.

The concave fillet weld, as compared to the flat or convex fillet welds, has a gradual change in contour at the toe. Stress concentrations are improved over the other types, which generally give this weld contour a better endurance limit under fatigue loading. Figure 4-16 shows a very flat entrance angle at the toes with a slight amount of convexity.

All three types of fillet profiles—concave, flat, or convex—are widely used. The position of welding, process, type of consumables (gas, electrode), type of joint, and job requirements are some of the factors that determine the type of fillet contour to be specified.

Fillet welds can also be measured in a slightly more complex way—by determining throat size. Three different throat sizes may be referred to when discussing the size of fillet welds, as seen in Figs. 4-14 and 4-15.

Design engineers sometimes refer to the **theoretical throat** of a weld. As Figs. 4-14 and 4-15 show, the theoretical throat extends from the point where the two base metal members join (the beginning of the joint root), to the top of the weld, minus any convexity on the convex fillet weld and concavity on the concave fillet weld, to the top of the largest right triangle that can be inscribed in the weld. The theoretical measurement looks at the weld as if it were an actual right triangle. The penetration is not figured into the theoretical throat size.

The **effective throat** of a fillet weld is measured from the depth of the joint root penetration. This is an important consideration as the penetration is now considered part of this dimension. However, no credit is given for the convexity. Some people consider convexity as reinforcement and thus indicates more strength. The exception is a fillet weld where too much convexity is detrimental to the overall joint strength. Excess convexity increases stresses at the weld toes and can lead to cracking. On convex and concave fillet welds the effective throat is measured to the top of the largest right triangle that can be drawn in the weld. This measurement can be used to indicate the size of the weld. The outward appearance of the weld may look too small, but if penetration can be ensured, then the weld will be of sufficient strength.

JOB TIP

Welding Connects our World

From the technology in our pockets to the rockets that take our astronauts into space. None of it would be possible without the welding industry. The American Welding Society Youtube page highlights and introduces the trade with many exciting videos.

Learn more and start your welding journey at https://www.youtube.com/videoaws

The **actual throat** of a fillet weld is the same as the effective throat on a concave fillet weld. But as can be seen in Fig. 4-14, page 116, there is a difference. This throat dimension can also be used to indicate size and strength. If anything other than the theoretical throat is used to size a fillet weld, the welding procedure would have to be carefully written and an in-process inspection would be required to ensure that the joint is being properly penetrated. The overall reduction in fillet weld size, increased speed of welding, reduced heat input, and reduction of internal stresses and distortion may make the effort worthwhile.

The general rule for the fillet weld size is that the leg should be the same size as the thickness of the metals. If ¼-inch thick plate is being welded, a ¼-inch leg fillet is needed to properly join the members. Consider again the ¼-inch thick plate. Imagine ½-inch legs on the fillet. This would result in what is termed overwelding. This weld is not just twice as large as required, but its volume is three times that required. This wastes weld metal, the welder's time, causes more nonuniform heating which results in more distortion, and may even weaken the structure because of residual stress. Figure 4-17 shows correct and incorrect fillet welds.

A weld or weld joint is no stronger than its weakest point. Even though the weld in Fig. 4-17B would appear to be much stronger, it will not support more stress than the weld in Fig. 4-17A. It may even support less stress due to the additional residual stresses built up in the joint that is overwelded.

When metals of different thickness are to be joined, such as welding a ¼-inch thick plate onto a ½-inch thick plate in the form of a T-joint, the rule for fillet weld size is that the size of the fillet weld leg should equal the thickness of the metal being welded. Since there are two different thicknesses, one method is to make an unequal leg fillet weld. Figure 4-18 shows correct

Fig. 4-17 Correct and incorrect fillets.

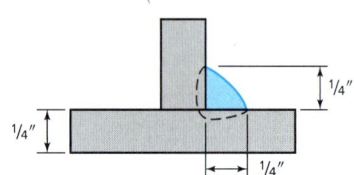

Correctly made fillet weld.
Leg appropriate for thickness of plate.

A

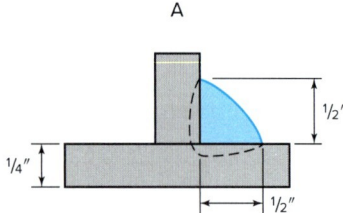

Overwelded (base metal will break at toes of weld). Legs of weld too large for thickness of plate.

B

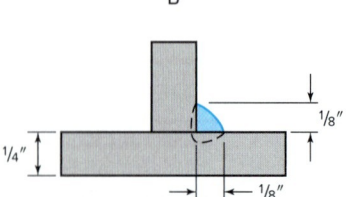

Underwelded (weld may break through throat of weld). Need larger legs on fillet.

C

Fig. 4-18 Unequal leg fillet and equal leg fillet.

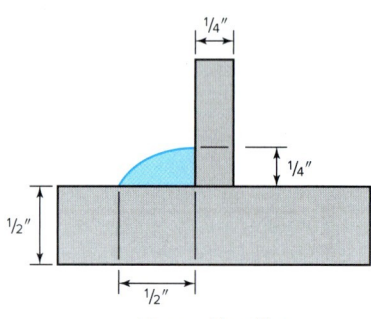

Unequal leg fillet

A

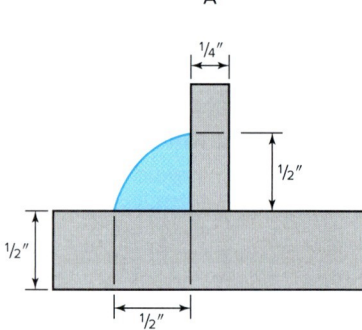

Equal leg ½" fillet (wasted weld metal, time, and extra heat input). Weakest point will be at the toe of the weld on the ¼" plate.

B

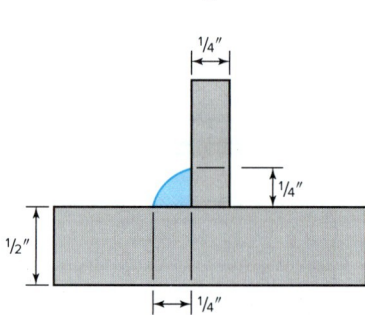

Equal leg ¼" fillet (less time, less weld metal, less heat input equals better weld) just as strong as welds A and B.

C

and incorrect examples. The correct, *unequal leg fillet* weld has a ¼-inch weld leg on the ¼-inch plate and a ½-inch weld leg on the ½-inch plate. This is considered by some to be the best way to handle this weldment. However, consider the results of making the weld with an *equal leg fillet*. There would then be two choices: a ½-inch fillet or a ¼-inch fillet. In this instance, the ¼-inch fillet would be the more practical, since a weldment is no stronger than its weakest point. The extra welds in the ½-inch fillet will also require more time, electrode wire, and put more heat into the metal, causing more residual stress.

Weld Length

Fillet and groove welds are usually made along the full length of the joint. In some cases, the full strength of a fillet welded joint can be achieved by only welding a portion of the joint. The effective length of a fillet weld is measured as the overall length of the full-sized fillet weld. The start and stop of the weld must be allowed for in the length measurement. Good welding techniques make it possible to have excellent starts and weld craters that are filled to the weld's full cross section. The weld starts and stops are not square, so an allowance must be made when measuring the length to account for the start and stop radius.

If a specific weld length is specified, it will be shown on the drawing. In some cases the fillet weld will be made at intermittent intervals. The space between the welds is determined by the center-to-center distance of the welds, which is called the **pitch.** If intermittent fillet welds are called for, then the welding symbol will indicate their length and pitch.

The weld area and stress are easily calculated. Multiplying the weld length with the weld size equals the weld area:

$$\text{area} = \text{weld length} \times \text{weld size}$$

It is important to understand that this will determine how much stress the joint can take. The design engineer is aware of the base material properties and the loads it will see in service and applies the following formula.

$$\text{stress} = \frac{\text{load}}{\text{weld area}}$$

Safety margins are built in to ensure the weld is able to withstand the load. The designer applies the weld size and length to the drawing via the welding symbol. Weld efficiency can be lost due to overwelding, so it is important to follow the specifications on the drawing and not overweld.

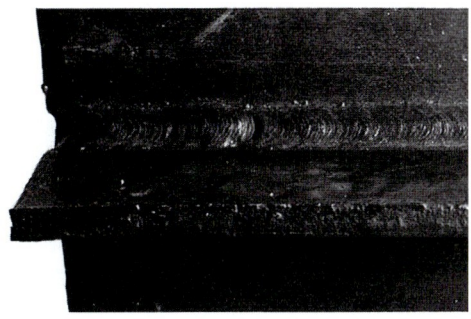

Fig. 4-19 A T-joint welded with a continuous single-pass fillet weld. Edward R. Bohnart

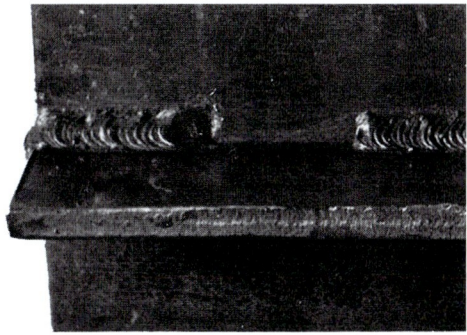

Fig. 4-20 A T-joint with intermittent welds of equal length and equally spaced. Edward R. Bohnart

Continuous Welds Continuous welds, Fig. 4-19, extend across the entire length of the joint from one end to the other. On structures that are to develop maximum strength and tightness, it is necessary to weld all of the seams completely.

Intermittent Welds Intermittent welds, illustrated in Fig. 4-20, are a series of short welds spaced at intervals. They cannot be used where maximum strength is required or where it is necessary that the work be watertight or airtight. However, on work that is not critical, the cost of welding can be considerably reduced by the use of intermittent welds.

The frequency, length, and size of the welds depend upon the thickness of the plates, the type of joint, the method of welding, and the service requirements of the job. Intermittent welds are usually employed in lap and T-joints. They are rarely, if ever, used for groove welds.

Tack Welds Welded fabrications are often made up of many parts. In the process of assembly before welding, some means is necessary to join the parts to the whole. This is done by a series of short welds spaced at intervals called **tack welds,** Fig. 4-21, page 120.

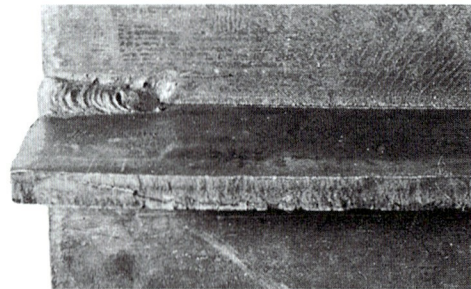

Fig. 4-21 Tack welds. Edward R. Bohnart

Some welders do not attach enough importance to the tack welding procedure and the remelting of tack welds in the major welding operation. There are many instances when a welder has failed an important test by being careless in tack welding. Tack welds must be strong. Not only must they be able to hold the part in the position in which it is to be welded, but they must also be able to resist the stress exerted on them when expansion and contraction occur during welding. Cold working, which is often necessary, imposes a severe load on the tack welds.

The number and size of the tack welds depend upon the thickness of the plate, the length of the seam, the amount of cold working to be done, and the nature of the welding operation. Tack welds must have good fusion and good root penetration. They should be flat and smooth—not convex and lumpy. It is advisable to use more heat for tack welding than for the major welding operation.

Stringer Bead A **stringer bead** is a weld made by moving the weld pool along the intended path in a straight line. With certain welding processes and electrodes a forward, backward, or whipping motion may be applicable. A stringer bead is welded along the line of travel with little or no side-to-side or weaving motion. Because of the faster travel speeds, stringer beads have very fast cooling rates that can impact the grain structure and also affect the distortion level. Figure 4-22A represents a stringer bead motion.

Weave Bead A **weave bead** is a weld made by moving the weld pool along the intended path but with a side-to-side oscillation, Fig. 4-22B. This is generally done to increase the weld size. Most codes or specifications will limit the width of a weave bead. The reduced travel speed will increase the heat input and slow the cooling rate. This will impact the grain structure and affect the distortion level. Controlling the maximum weave width will also help eliminate slag inclusions and incomplete fusion type discontinuities.

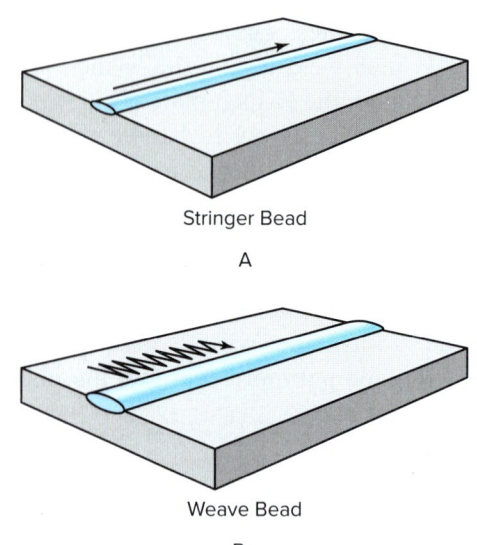

Fig. 4-22 Stringer bead and weave bead identification. From *Welding Inspection Technology*, 5th ed., pp. 4–27, Figure 4.31.

> **ABOUT WELDING**
>
> **Hephaestus**
> Hephaestus is the Greek god of fire and blacksmithing. So in the late 1880s, Victorian researcher Nikolai Benardos honored his arc welding methods with the name *electrohephaestus*. The Roman counterpart to Hephaestus is the god Vulcan.

Weld Positions

The four basic positions for welding are flat, horizontal, vertical, and overhead. They are also designated with a number system to aid in brevity in oral or written communication. They are defined as follows:

- *Flat position (1)* The **flat position (number 1)** is the welding position used to weld from the upper side of the joint at a point where the weld axis is approximately horizontal and the weld face lies in an approximately horizontal plane.

- *Horizontal position (2)* The **horizontal position (number 2)** is the fillet welding position in which the weld is on the upper side of an approximately horizontal surface and against an approximately vertical surface. For groove welds it is the position in which the weld face lies in an approximately vertical plane and the weld axis at the point of welding is approximately

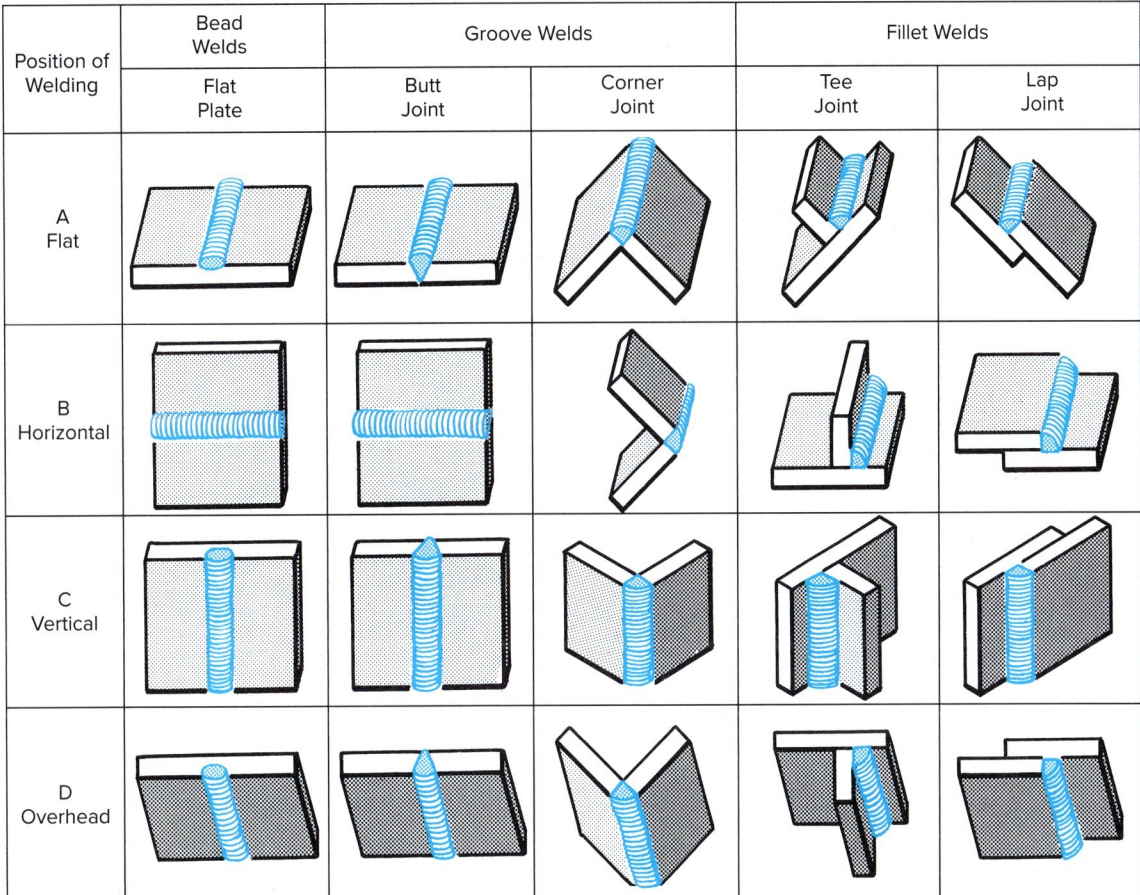

Fig. 4-23 Positions of welding.

horizontal, while the plate remains in an approximate vertical orientation.

- *Vertical position (3)* The **vertical position (number 3)** is the welding position in which the weld axis at the point of welding is approximately vertical, and the weld face lies in an approximately vertical plane. Welding travel may be up or down. When travel is up, the welding end of the electrode or torch is pointed upward at an angle, ahead of the weld. When the travel is down, the end of the electrode or torch is pointed up and at an angle to the weld pool. The travel direction up or down is an essential variable in most codes and so the welder must follow what is stated and proven in the welding procedure.

- *Overhead position (4)* The **overhead position (number 4)** is the welding position in which welding is performed from the underside of the joint. The overhead position is the reverse of the flat position.

Figure 4-23 shows the different positions of welding. Figure 4-24, page 122, shows examples of welds and welding positions.

When discussing groove welds, a "G" is used and a number is assigned to signify the welding position (see Fig. 4-23).

Plate Weld Designations are:

1G   flat position groove weld
2G   horizontal position groove weld
3G   vertical position groove weld
4G   overhead position groove weld

Pipe Weld Designations are:

1G   flat position groove weld, pipe axis is horizontal, and the pipe is rotated
2G   horizontal groove weld, pipe axis is vertical
5G   multiple-position (overhead, vertical, and flat) groove weld, pipe axis is horizontal, and the pipe is not rotated
6G and 6GR   multiple-position groove weld, pipe axis is 45° from horizontal, and the pipe is not rotated. The R designates a restricting ring.

Figure 4-25, page 122, represents a graphic view of these groove weld positions on plate and pipe.

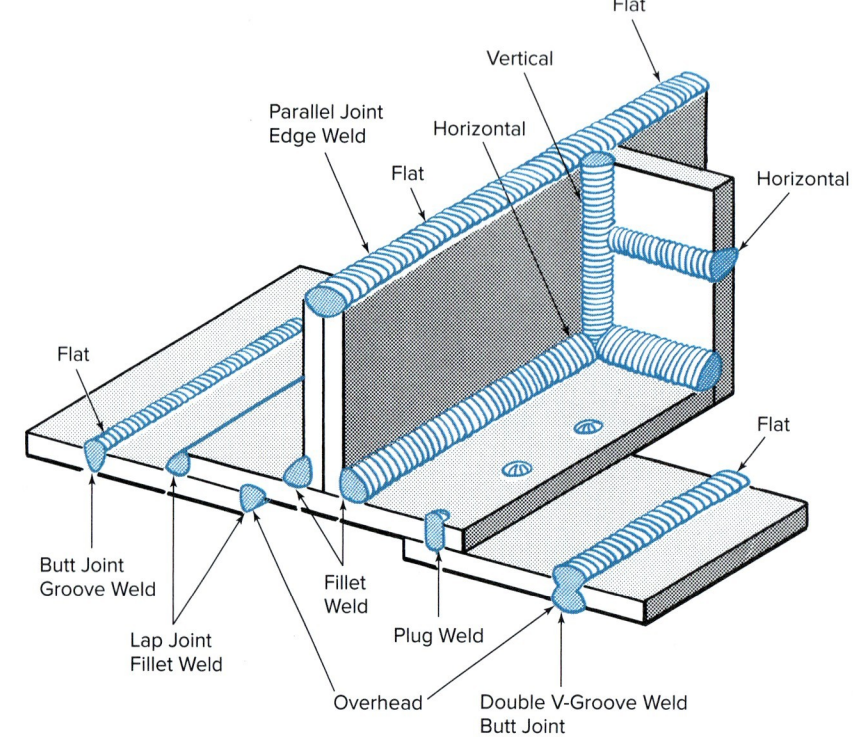

Fig. 4-24 Examples of welds and positions of welding.

When discussing fillet welds, the letter *F* is used and a number is assigned to signify the welding position, Fig. 4-26.

Plate Positions are Designated as:

1F flat position fillet weld
2F horizontal position fillet weld
3F vertical position fillet weld
4F overhead position fillet weld

Pipe Positions are Designated as:

1F flat position fillet weld, pipe axis is 45° from the horizontal, and the pipe is rotated
2F horizontal fillet weld, pipe axis is vertical
2FR horizontal fillet weld, pipe axis is horizontal, and the pipe is rotated
4F overhead fillet weld, pipe axis is vertical
5F multiple-position (overhead, vertical and horizontal) fillet weld, pipe axis is horizontal, and the pipe is not rotated
6F multiple-position fillet weld, pipe axis is 45° from horizontal, and the pipe is not rotated

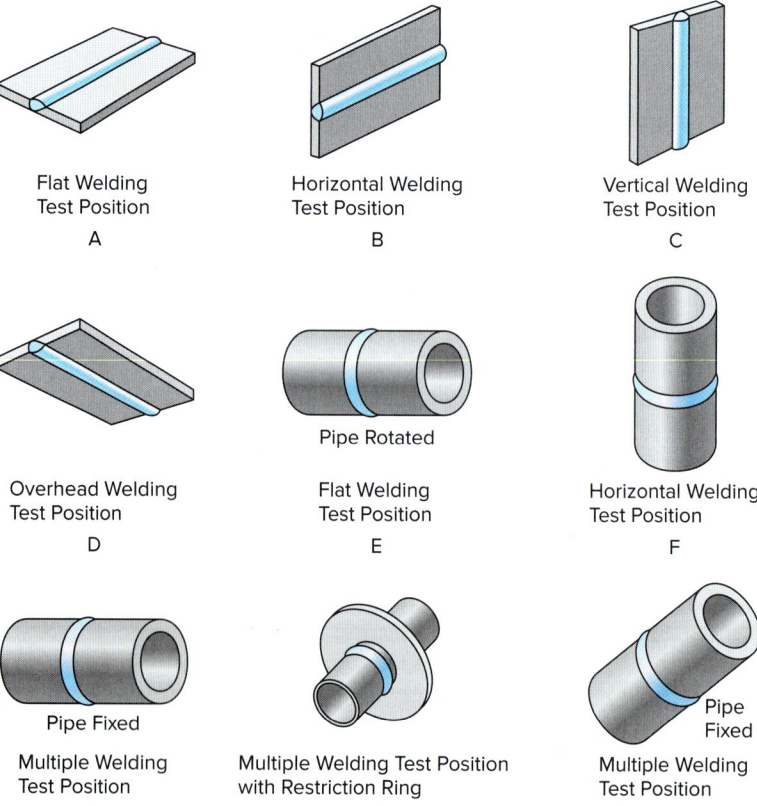

Fig. 4-25 Groove weld positions. From AWS 3.0:2010 STANDARD TERMS AND DEFINITIONS, P. 85, FIGURE B.17 (A), (B), (C), (D), AND P. 88, FIGURE B.19 (A), (B), (C), (D), (E) ARE COMBINED.

122 Chapter 4 Basic Joints and Welds

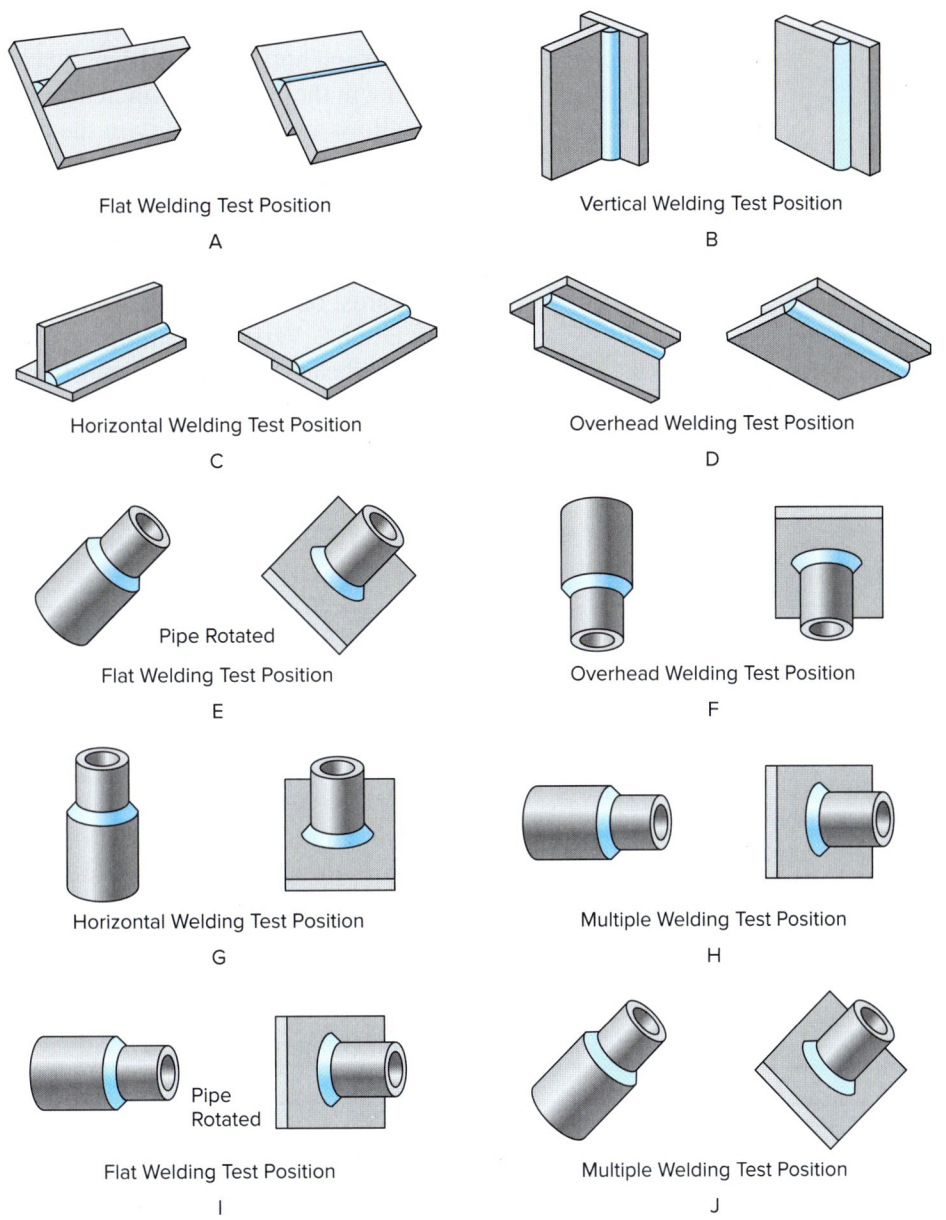

Fig. 4-26 Fillet weld positions. From AWS 3.0:2010 STANDARD TERMS AND DEFINITIONS, P. 85, FIGURE B.17 (A), (B), (C), (D), AND PG.88, FIGURE B.19 (A), (B), (C), (D), (E) ARE COMBINED.

Figure 4-26 represents a graphic view of these fillet weld positions on plate and pipe.

The positions just described are welder test positions. The weld test plates or pipes should be positioned as close to those illustrated as possible. However, in production welding it is not always possible to have the joint axis and weld face rotations lined up vertically, horizontally, or at 45°. Figures 4-27 to 4-29, pages 124–126, show how to calculate the welding position for production welds that are not easily determined. In order to use these diagrams, you must visualize the axis of the weld and the rotation of the weld face and apply it to the tabulated information.

ABOUT WELDING

Feed Speed

How fast (and how much) filler metal goes into a weld is called the wire *feed speed* and is measured in inches per minute or millimeters per second. The higher the speed, generally, the higher the amperage.

Basic Joints and Welds **Chapter 4** 123

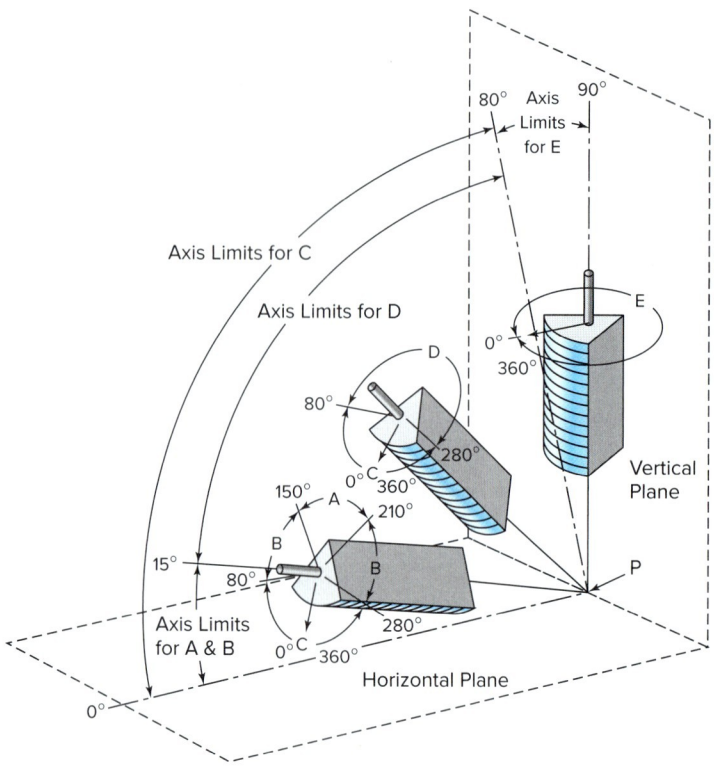

Tabulation of Positions of Groove Welds

Position	Diagram Reference	Inclination of Axis	Rotation of Face
Flat	A	0–15°	150–210°
Horizontal	B	0–15°	80–150° 210–280°
Overhead	C	0–80°	0–80° 280–360°
Vertical	D	15–80°	80–280°
	E	80–90°	0–360°

Notes:
1. The horizontal reference plane is always taken to lie below the weld under consideration.
2. The inclination of the weld axis is measured from the horizontal reference plane toward the vertical reference plane.
3. The angle of rotation of the weld face is determined by a line perpendicular to the weld face at its center which passes through the weld axis. The reference position (0°) of rotation of the weld face invariably points in the direction opposite to that in which the weld axis angle increases. When looking at point P, the angle of rotation of the weld face is measured in a clockwise direction from the reference position (0°).

Fig. 4-27 Production welding position diagram for groove welds in plate. From AWS 3.0:2010 STANDARD TERMS AND DEFINITIONS, P. 82, FIGURE B.16A

The welding position is an essential variable for the welder. If the welder is attempting to weld in a position that they are not qualified for, it will cause the code work being done to be rejected.

The trend in most shops is toward welding in the flat and horizontal positions wherever possible. Welding in these positions increases the speed of welding, allows more flexibility in the choice of process, and ensures work of better appearance and quality. Vertical and overhead welding find their widest application in those industries where the fabrications are large and permanent. Such conditions exist in shipyards, on construction projects, and in piping installations. Vertical welding is done more often than overhead welding, and most welders find it a less difficult position to weld in. However, welders must be able to weld in all positions. Inability to do so limits their possibilities of advancement to a higher job classification and prevents them from taking advantage of all the job opportunities they may encounter.

Strength of Welds

In general, welded joints are as strong as, or stronger than, the base metal being welded. It is not always necessary that this be so. Good welding design specifies welds that require the minimum amount of weld metal that is adequate for the job at hand. Weld metal costs a good deal more than base metal and requires labor costs for its application.

The strength of a welded joint depends upon the following factors:

- Strength of the weld metal
- Type of joint preparation
- Type of weld
- Location of the joint in relation to the parts joined
- Load conditions to which the weld will be subjected
- Welding process and procedure
- Heat treatment
- Skill of the welder

Approximately ¼ inch should be added to the designed length of fillet welds for starting and stopping the arc. The craters in the welds should be filled.

The location of the welds in relation to the parts joined, in many cases, has an effect on the strength of the welded joint. Repeated tests reveal that, when other factors are equal, welds having their linear dimension transverse (at right angles) to the lines of stress are approximately 10 to 15 percent stronger per average than by a single weld or welds close together. In Fig. 4-31, page 126, a single weld at A is not as effective as welds at both A and B in resistance to the turning effect. Two small welds at A and B are much more effective than a large single weld at A or B only. If possible, welded joints should be designed so that bending or prying action is minimized.

In some designs, it may be desirable to take into account the stress distribution through the welds in a joint. Any abrupt change in surface (for example, a notch or saw cut in a square bar under tension) causes stress concentration and increases the possibility of fracture. As an illustration of this principle, the weld shown in Fig. 4-32, page 126, would have considerably more concentration of stress than that in Fig. 4-33, page 126. The weld shown in Fig. 4-34, page 126, allows a minimum of stress concentration and improved service. Under many load conditions, the stress is greater at the ends of the weld than in the middle. Therefore, it is advisable in such cases to box the bead around the joint as indicated in Fig. 4-35, page 126. When this is done, far greater resistance to a tearing action on the weld is obtained. The length of this boxing (end return) should be a minimum of twice the size of the weld specified. If flexibility is required in this joint, the boxing should not exceed four times the size of the weld specified.

Common Weld and Weld-Related Discontinuities

General Considerations

A *weld discontinuity* is any interruption in normal flow of the structure of a weldment. The interruption can be found in the physical, mechanical, or metallurgical characteristics of the material or weldment. If the discontinuity exceeds the acceptance criteria being used, it becomes a defect. All metals, heat-affected zones, and welds have discontinuities. The HAZ is the base metal next to the weld that did not melt but was hot enough to change its mechanical properties or its microstructure properties. As all metals are crystalline structures, the interruptions at each of the grain boundaries reflect an interruption of the normal flow of the material. But size, location, extent, and other factors must be applied to see if the product is fit for a purpose. When a defect is indicated, it means that the defect exceeds the acceptable limits of the code or specification being applied.

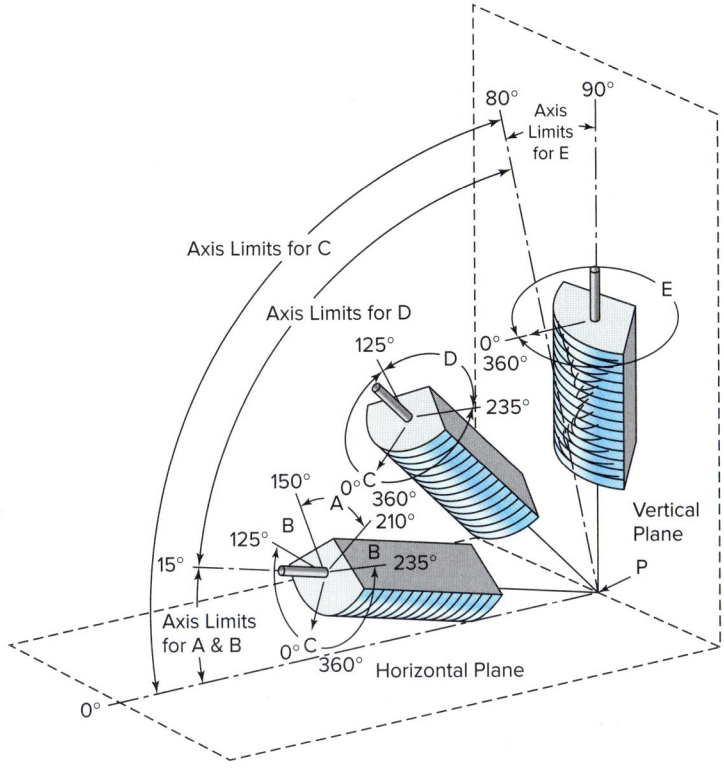

Tabulations of Positions of Fillet Welds

Position	Diagram Reference	Inclination of Axis	Rotation of Face
Flat	A	0–15°	150–210°
Horizontal	B	0–15°	125–150°
			210–235°
Overhead	C	0–80°	0–125°
			235–360°
Vertical	D	15–80°	125–235°
	E	80–90°	0–360°

Notes:
1. The horizontal reference plane is always taken to lie below the weld under consideration.
2. The inclination of the weld axis is measured from the horizontal reference plane toward the vertical reference plane.
3. The angle of rotation of the weld face is determined by a line perpendicular to the weld face at its center which passes through the weld axis. The reference position (0°) of rotation of the weld face invariably points in the direction opposite to that in which the weld axis angle increases. When looking at point P, the angle of rotation of the weld face is measured in a clockwise direction from the reference position (0°).

Fig. 4-28 Production welding position diagram for fillet welds in plate.
From AWS 3.0:2010 STANDARD TERMS AND DEFINITIONS, P. 83, FIGURE B.16B.

unit length than welds that have their linear dimension parallel to the lines of stress, Fig. 4-30, page 126.

Resistance to a turning effect of one member at a joint is best obtained by welds that are well separated, rather

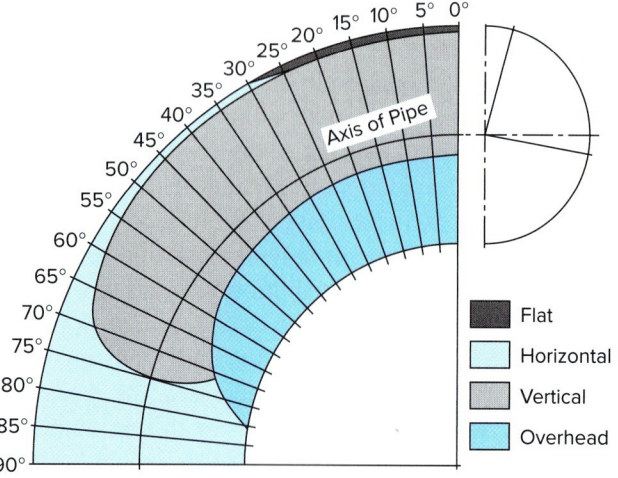

Fig. 4-29 Production welding position diagram for groove welds in pipe. Positions for circumferential groove welds indicated by shaded areas for pipe with axis varying from horizontal (0°) to vertical (90°). From AWS 3.0:2010 STANDARD TERMS AND DEFINITIONS, P. 83, FIGURE B.16B.

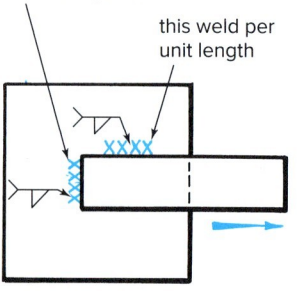

Fig. 4-30 Transverse welds are stronger than welds parallel to lines of stress.

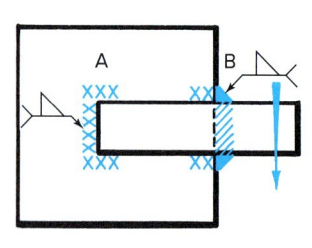

Fig. 4-31 Example of proper placement of welds to resist turning effect of one member of the joint.

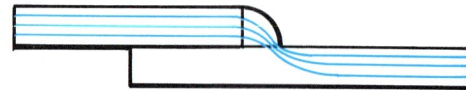

Fig. 4-32 A lap weld having poor distribution of stress through the weld. Excessive convexity.

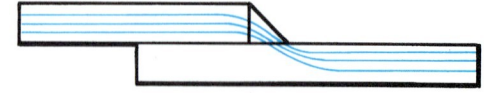

Fig. 4-33 A lap weld having a more even distribution of stress through the weld than that shown in Fig. 4-32.

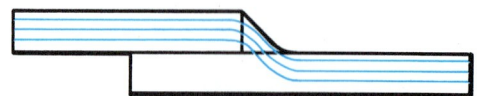

Fig. 4-34 A lap weld in which there is a uniform transfer of stress through the weld.

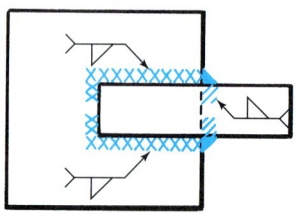

Fig. 4-35 Example of weld boxing around the corners to obtain resistance to tearing action on welds when subjected to eccentric loads.

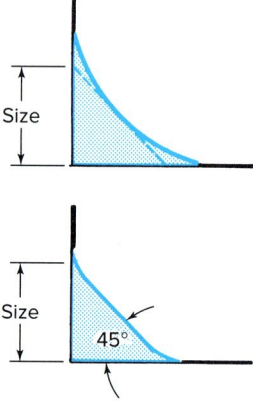

Fig. 4-36 Desirable fillet weld profiles.

> **JOB TIP** 
>
> **Calling About a Job**
> When calling a business about a job, start by briefly saying who you are and that you are looking for employment with the company. Give a couple of qualifications you have for the job you want. Ask if you can provide your resume, and when it would be good to talk with someone further. If you write this all down on an index card, then the call won't be as hard to make.

Fillet Weld Profiles

Figure 4-36 shows flat and concave fillet weld profiles that are considered desirable. Figure 4-37 illustrates a slightly convex profile that is also acceptable. Thus, we are again reminded that the welder should try to avoid excess convexity. Convex fillet welds are acceptable, providing the convexity is within the limits indicated by Table 4-1, page 117.

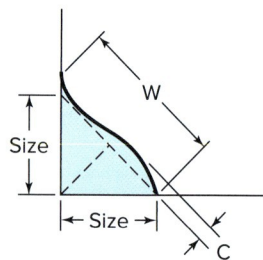

Convexity, C, shall not exceed 0.15 ± 0.03 inches

Fig. 4-37 Acceptable fillet weld profile.

Figure 4-38 shows profiles of weld defects that result from poor welding technique.

Fillet Weld with Insufficient Throat Reduction of the effective throat, Fig. 4-38A, materially reduces the size of the weld. This abrupt change in the face concentrates stress at the center. The smaller size of the weld and the stress concentration weaken the weld and invite

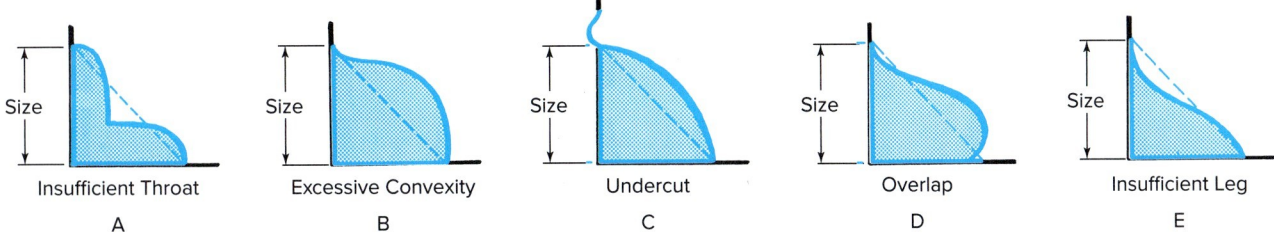

Fig. 4-38 Defective fillet weld profiles.

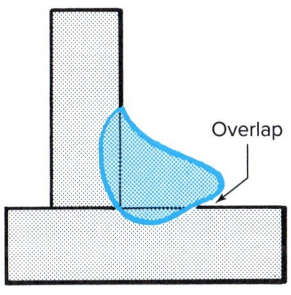

Fig. 4-39 *Overlap* is an overflow of weld metal beyond the toe of the weld.

joint failure. This defect is usually caused by too fast travel and excessive welding current.

Fillet Weld with Excessive Convexity The weld metal in this type of defect, Fig. 4-38B, may contain a great deal of slag and porosity. There may also be poor fusion at the root of the weld and poor fusion of the weld metal to plate surfaces, Figs. 4-39 and 4-40. Stress concentrates at the toe of the weld. This weld defect is usually caused by low welding current and a slow rate of travel.

Fillet Weld with Excess Undercut Figure 4-38C shows the melting away of the base metal next to the weld toe. This cutting away of one of the plate surfaces at the edge of the weld is known as undercut. Excess undercut like this one decreases the thickness of the plate at that point. Any material reduction in plate thickness leads to plate weakness. The situation invites joint failure because the designed load of the joint is based on the original plate thickness. The possibilities of failure at this point are increased when undercutting occurs at the toe of the weld, a point where there is high stress concentration. This weld defect is usually caused by excessive arc length, incorrect electrode angle, incorrect electrode location, fast travel, and excessive welding current.

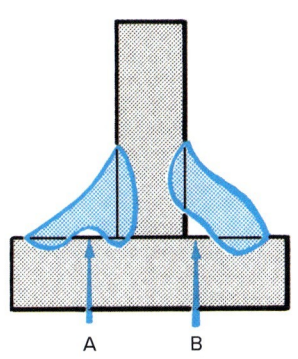

Fig. 4-40 At problem area A, there is incomplete fusion in the fillet welds. At problem area B, the weld has bridged the joint root and is an incomplete fusion.

Fillet Weld with Overlap Overlap, shown in Figs. 4-38D and 4-39, is protrusion of the weld metal beyond the weld toe and base metal. It can be likened to applying a wad of chewing gum to a surface. When load is applied to the gum, it will peel from the surface. The welded joint will act in the same way under load, and the result will be weld failure. It is obvious that overlap must be avoided if we are to prevent a peeling off of the weld metal when load is applied. Failure of the joint is certain when overlap is located in the weld. This is a serious defect and should be avoided. It may be caused by low welding current, slow travel, or improper electrode manipulation.

Fillet Weld with Insufficient Leg A reduction in leg length, Fig. 4-38E, means a reduction in the size of the fillet weld. If the demands of a joint require a fillet of a certain size, any reduction of that size results in a weld that does not possess the physical properties needed for safe operation. Failure is sure to result. This defect is usually caused by improper electrode angle and faulty electrode manipulation. In addition, these faults in welding technique may be accompanied by too fast travel.

Fillet Weld with Incomplete Fusion This defect is usually found at the root of the weld and at the plate surfaces (fusion face), Fig. 4-40. Incomplete fusion is usually caused by welding with the current too low, an improper speed of travel, and/or improper electrode manipulation. When these conditions exist during welding, the deposited weld metal may have slag inclusions and porosity (gas entrapment).

Fillet Weld with Various Other Discontinuities Figure 4-41, page 128, represents many of the possible defects that can be encountered in the base material or the weld bead.

Porosity Porosity is cavity-type discontinuities (referred to as *pores*) formed by gas entrapment during solidification. The discontinuities are spherical and may be elongated. Contamination of the filler metal or base metal or

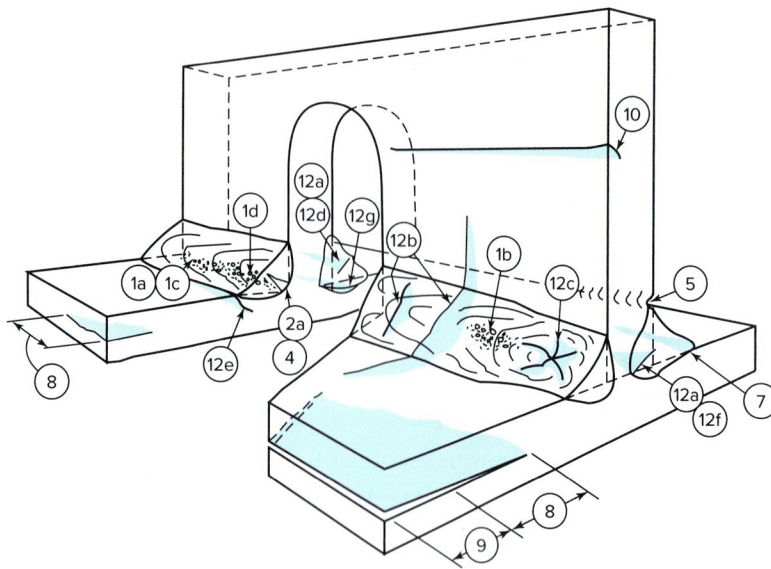

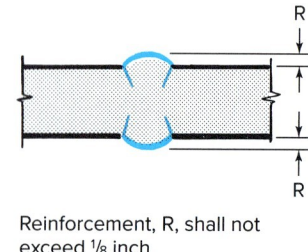

Reinforcement, R, shall not exceed ⅛ inch

Fig. 4-43 Acceptable butt weld profile.

Fig. 4-41 Discontinuities in a single-pass double fillet weld on a T-joint. 1a, 1c. Uniformly scattered and piping porosity. 1b. Cluster porosity. 1d. Aligned porosity. 2a. Slag inclusion. 4. Incomplete fusion. 5. Undercut. 7. Overlap. 8. Lamination. 9. Delamination. 10. Seam and lap. 12a. Longitudinal crack. 12b. Transverse crack. 12c. Crater crack. 12d. Throat crack. 12e. Toe crack. 12f. Root crack. 12g. Underbead and heat-affected zone (HAZ) cracks.

Groove Weld Profiles

Figure 4-43 shows an acceptable groove weld profile. It should be noted that the recommended reinforcement does not extend more than ⅛ inch above the surface of the plate. Figure 4-44 shows defective butt weld profiles.

Groove Weld with Insufficient Size A decrease in size, Fig. 4-44A, reduces the size of the butt weld. The thickness of the weld is less than the thickness of the plate, and the weld will not be as strong as the plate. Failure of the weld under maximum load is certain. This defect is usually caused by a combination of high welding current and travel that is too fast. Although penetration at the root of the weld may be complete and fusion to the plate surfaces may be excellent, these desirable characteristics cannot overcome insufficient weld size.

Groove Weld with Excessive Convexity This is the opposite of a concave profile, Fig. 4-44B. It may be less strong than the weld with insufficient size, due to concentration of stress in the weld. Comparative strength, of course, depends upon the degree of convexity of one weld and the size insufficiency of the other. Excessive convexity may be caused by travel that is too slow or low welding current. Even though complete penetration and good

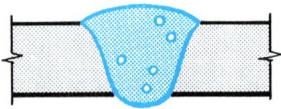

Fig. 4-42 Porosity.

improper gas shielding will generally lead to porosity. Figure 4-42 is an example of subsurface porosity in a groove weld. Generally porosity is not considered to be as severe a concern as cracks or incomplete fusion. The rounded shape the discontinuities take does not concentrate stress, as would a crack or fusion-type defect. Some general guidelines for porosity are found in Table 4-2. This is for structural steel type welding requirements; other applications may differ significantly.

Table 4-2 Acceptable Porosity Limits Guideline

Type of Weld and Location	Diameter (in.)	Sum of Diameters of Individual Porosity Pores (in.)	Length of Weld (in.)
Groove—transverse to tensile loading	No visible piping porosity allowed	N/A	N/A
Groove—fillet	>¹⁄₃₂ ≤ ³⁄₈	³⁄₈	1
Groove—fillet	≥³⁄₈	¾	12
Fillet—CJP groove	≤³⁄₃₂ piping porosity	Single pore	4

128 Chapter 4 Basic Joints and Welds

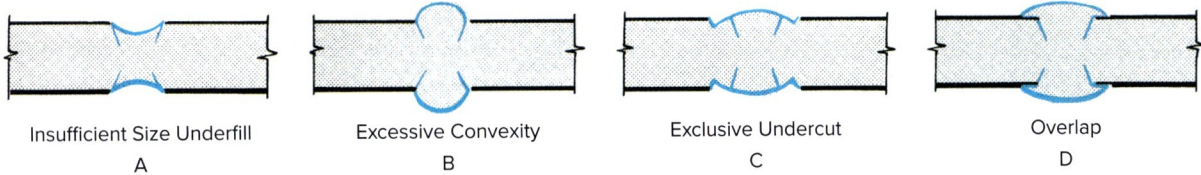

| Insufficient Size Underfill | Excessive Convexity | Exclusive Undercut | Overlap |
| A | B | C | D |

Fig. 4-44 Defective butt weld profiles.

fusion may exist, these desirable characteristics cannot overcome the loss of strength due to extreme convexity. There is also the possibility of porosity and slag inclusion in the weld. The defect wastes material and time, thus increasing costs. Very poor appearance will also result.

Groove Weld with Undercut As with the fillet weld, a cutting away of the plate surface at the toe of the weld results in a reduction of actual plate thickness, Fig. 4-44C. The reduction in plate surface, together with the concentration of stress at the toe due to the sharp corner, may cause failure of the welded joint at this point.

Undercutting may be acceptable if its depth and length do not exceed the acceptance requirement of the code or specification being applied. Table 4-3 gives an example of allowable undercut. Because the undercut has a radius and is not a sharp notch, the stress concentrations are not as high as once believed. Undercut is a discontinuity to be avoided, but it does not need to be repaired unless it exceeds the acceptance criteria. It is usually caused by high welding current, travel that is too fast, or improper electrode manipulation.

Groove Weld with Overlap Overlap (Fig. 4-44D) results from poor fusion. It is basically an incomplete fusion at the toe of the weld. Most codes or specifications will not allow any amount of lack of fusion. Overlap is usually caused by low welding current, slow rate of travel, or improper electrode manipulation. A weld with excessive convexity and overlap usually contains a certain amount of porosity and poor fusion. Figures 4-42 through

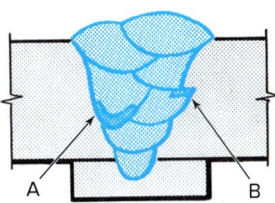

Fig. 4-45 Slag inclusions, between passes at A, and at undercut at B.

Fig. 4-46 Incomplete fusion from oxide or dross of center of joint, especially in aluminum.

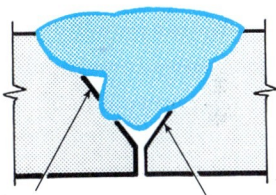

Fig. 4-47 Incomplete fusion and incomplete penetration in a groove weld.

4-47 illustrate the defects that may be found alone or in combination.

Groove Weld with Various Other Discontinuities Figure 4-48, page 130, shows many of the possible defects that can be encountered in the base material or the weld bead.

Other Discontinuities Found on Groove and Fillet Welds

Cracks A weld crack is a fracture-type discontinuity that has a sharp tip and a length much greater than its width

Table 4-3 Acceptable Guideline Undercut Limits

Material Thickness (in.)	Depth of Undercut (in.)	Length of Undercut
<1	≤1/32	Unlimited
<1	1/32–1/16[1]	2 in. in any 12 in. of weld
≥1	≤1/16	Unlimited

[1] Not including depths equal to 1/16 in.

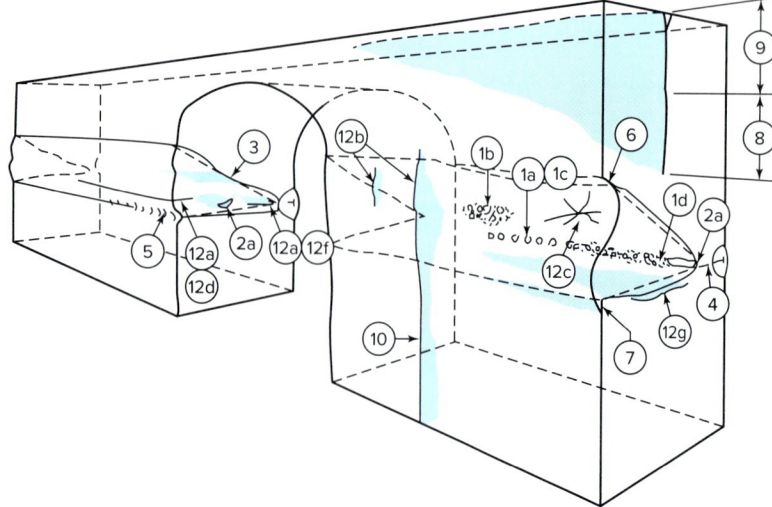

Fig. 4-48 Single-bevel groove weld in a butt joint. 1a and 1c. Uniformly scattered and piping porosity. 1b. Cluster porosity. 1d. Aligned porosity. 2a. Slag inclusion. 3. Incomplete fusion. 4. Incomplete joint penetration. 5. Undercut. 6. Underfill. 7. Overlap. 8. Lamination. 9. Delamination. 10. Seam and lap. 12a. Longitudinal crack. 12b. Transverse crack. 12c. Crater crack. 12d. Throat crack. 12f. Root crack. 12g. Underbead and heat-affected zone (HAZ) cracks.

or opening. In most codes or specifications, cracks of any length, location, or orientation are not allowed. In this case all cracks would be defects and must be repaired. Because they have a sharp tip, cracks are considered a stress riser. They can also propagate rapidly across the joint or weldment.

Cracks can generally be classified as either hot cracks or cold cracks. Insufficient ductility at high temperatures will cause hot cracks. These cracks move between grains in the weld metal or at the weld interface. If cracks occur once the weld metal has solidified, then they are considered cold cracks. The weld metal, heat-affected zone, or base metal can be affected by cold cracks. Cold cracks occur because of improper welding procedure or techniques or the welding service condition. Figure 4-49 is an example of a crack starting in the crater area of a weld. The crater was not properly filled to the full cross section of the weld. A small crack formed in the crater due to the shrinkage forces, and the crack propagated out of the crater all the way around the joint.

Hydrogen Cracking Hydrogen cracking is also referred to as delayed cracking. In certain situations, inspection will be delayed for up to several days to let this type of crack manifest itself. Cracking of this nature is brought about by one of the following four factors:

- Presence of hydrogen
- Hard grain structures
- Amount of restraint in the joint
- Low temperature operation of weldment

Hydrogen in the form of moisture can come from many sources. The coating on the SMAW electrode, the flux in the core of a FCAW electrode, oxides on the metal, lubricants, contamination on the plate or the filler metal, and even the moisture in the air. For particular materials, such as hard grain structure type steel, this problem is very pronounced. Some examples of materials susceptible to hydrogen cracking are the carbon-manganese and low alloy steels. Only the hard grain structures are sensitive to

> **SHOP TALK**
>
> **Saving Time and Money**
> Fillet welding can be done about four to seven times faster than butt welding. This results in time and money savings. Welders with lower skills can produce consistent and high quality fillet welds, while this is not true of butt welding. Fillet welds can be verified visually. This also results in time and money savings. A butt weld currently can be verified by a radiograph which is costly. The fillet welds help deliver projects on time and under budget.

Fig. 4-49 Longitudinal crack propagating from crater crack. *American Welding Society*

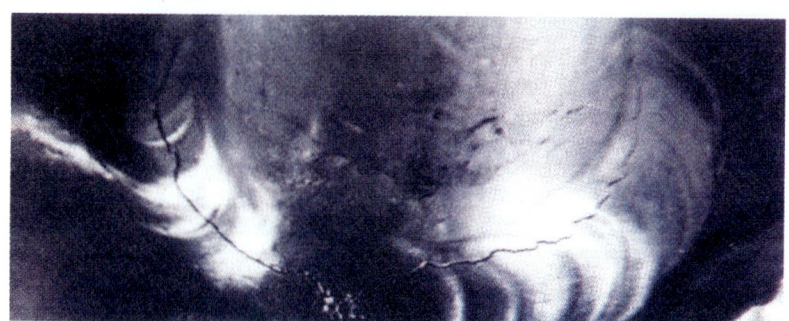

this type of cracking. Low heat input with its fast cooling rate will create more problems as the hydrogen cannot escape from the weld area and is trapped. Also if the joint is highly restrained, it will have a greater tendency for this type of cracking. If the service condition of the product will be at low temperature, this type of cracking will also be more prevalent. Since the service condition of the weldments, along with the amount of restraint applied to the joint and the type of metal used for fabrication can be hard to control, the best method to eliminate delayed cracking or hydrogen cracking is to eliminate or reduce all levels of hydrogen to acceptable levels. The hydrogen can be in the form of contaminants in or on the metal, the electrodes, or shielding gas. Any source of moisture can create hydrogen-induced cracking. It is also good practice to use proper welding procedures that will help control the cooling rate. Use of preheat and interpass temperature and postweld heat treatment may be required as well. This type of cracking is usually found in the HAZ. These cracks may not open to the surface initially, so they are sometimes called **underbead cracks.** This makes them difficult to locate and is why final inspection may be delayed to allow the crack to propagate to the surface.

Incomplete Fusion Incomplete fusion is a weld discontinuity that occurs when the weld metal is in contact with other weld metal, the joint groove, or root face but does not fuse with it. In most codes and specifications incomplete fusion is not allowed no matter the size, length, location, or orientation to the applied load. It is always considered a defect and must be repaired. Improper welding techniques, joint preparation, or joint design will result in incomplete fusion. Lack of welding heat or access of the arc (or other heat source) to the required fusion area will result in incomplete fusion. The joint and prior welds must be properly cleaned of oxides. Incomplete fusion can occur even if the heat and access are available if the welds are being made over dense, tightly adhering oxides. Figures 4-39 and 4-40, page 127, and Figs. 4-46 and 4-47, page 129, show incomplete fusion at various locations.

Incomplete Joint Penetration Incomplete joint penetration occurs when the weld metal does not extend all the way into the root of the joint. If the weld metal penetrates the root but does not fuse, then it is referred to as incomplete fusion. Incomplete joint penetration may be caused by not dissolving surface oxides or impurities, but most generally it is due to not applying sufficient heat and arc force to penetrate the root of the joint. The areas above the joint root will reach melting temperatures first. If improper techniques, process, joint geometry, or consumables are used, then the molten metal will bridge over, insulating the root from the arc force. The welder needs to pay very close attention to the shape of the weld pool when making the root pass to ensure that a notch is not formed, which would indicate that the root has been bridged. Great care must be taken to ensure complete penetration. This discontinuity is generally not acceptable in most codes or specifications. No amount of incomplete penetration is generally allowed, so it is considered a defect that must be repaired. It is undesirable because at the root of the joint, which may be subject to tension or bending forces, the weld size is not as large as required and a failure will result. Even if the structure does not have bending or tension forces at the root, the shrinkage forces of the weld cooling may lead to cracks. These cracks may propagate from the root out into the base metal or out through subsequent weld passes. It is important to not provide a crack-initiating site by getting complete joint penetration. Figures 4-40 and 4-47 show incomplete joint penetration as well as incomplete fusion. Figures 4-14 and 4-15, pages 116–117, show excellent penetration on a fillet weld. You will note that penetration is measured by how far the weld penetrates into the joint—not the base metal. The line indicating the depth of the effective throat is also an indication of the amount of root penetration. It can be seen that the weld actually penetrated deeper into the bottom plate; however, this adds nothing to joint strength and is not considered part of the joint penetration. Figure 4-14 shows the root penetration on a fillet welded T-joint.

Inclusions Inclusions are entrapped solid materials that are not intended to be in the weld joint. Examples of materials found in inclusions are slag, flux, oxides, and tungsten. Virtually anything solid found in the joint that should not be there would be called an inclusion. Inclusions can be located in between weld passes, and/or between a weld pass and the joint groove or root face. Figure 4-45, page 129, shows an example of slag inclusions between passes and trapped in a section of undercut located along the groove face. If the inclusion is open to the surface, it should be repaired, as there is no fusion at that location. However, if there are subsurface inclusions, they are not as critical. If subsurface inclusions are of a specific size and separated by sufficient distance, then they may be considered to be discontinuities and allowed to remain in the weldment. If subsurface inclusions exceed the acceptance requirement of the code or specification being used, then they will need to be repaired. Generally, inclusions do not pose as severe a problem as porosity. This is due to the fact that being solid, inclusions can transmit a certain amount of load. They generally do not have sharp edges, so in some situations fairly large inclusions are allowable.

For example, on a statically loaded nontubular steel structure, a weld size of ½ inch would allow an inclusion of up to approximately 5/16 inch as long as there is at least 1-inch clearance between the edges of this type of inclusion. Inclusions should be avoided if at all possible. They can be caused by a number of factors, but can best be controlled by following the welding procedure, proper weld bead location, and proper welding techniques. Cleaning between passes will greatly reduce the possibility of creating inclusions.

Underfill This condition exists when the weld face or root surface extends below the surface of the material being welded. It results from poor welder observation and technique. Some amount of underfill is usually allowable depending upon the acceptance criteria of the code or specification being used. In most cases, underfill will provide better fatigue properties than overwelding. Figure 4-44A, page 129, depicts the underfill discontinuity.

Discontinuities, Defects, and Visual Inspection

Welding students often ask questions such as:

Which defect contributes to joint failure?
Which ones may pass inspections?
Are there jobs on which weld defects are permitted?

Knowing the construction and use of the weldment helps the welder answer these questions. In general, the welder's goal should be to avoid all defects; they all contribute to weld and joint failure. Minor allowances may be permitted if the work is not critical. Little tolerance is permitted in critical or code work because high strength is necessary due to load conditions of heat, pressure, or stress.

The criticality of a discontinuity is one way of assessing the importance of classifying it as a defect. The actual repair of a discontinuity may create more problems and increase residual stresses than if the defect were left alone. Is the discontinuity linear (cracklike) or nonlinear (spherical, porosity, or tungsten inclusion)? Is the end condition sharp (cracklike), or does it have a radius (undercut)? Does the discontinuity break the surface or is it subsurface? Higher stresses will be localized on surface opening discontinuities. Is the discontinuity longitudinal to the load or transverse to the load? Will the loading be fatigue or impact or simply a static load? What is the ambient temperature of the weldments? The engineer will take all these issues into consideration when determining the design of the weldments, materials, welding procedures, the acceptance criteria, and which code or specification to follow.

If the welder expects to be recognized as a professional, they will always strive to do work that is sound and of good appearance. Inability to do so with the high quality equipment available suggests carelessness on the part of the welder. It will surely lessen the regard that the shop supervisor and other workers have for the welder.

Because the welder is watching every inch of weld as it is being made, they are in the best position to do the visual inspection. Visual inspection needs to be done before welding with the proper joint fitup and selection of materials. It should be done during the welding after tacking, root pass, fill passes, and cap passes. Visual inspection should also be done after welding is completed to determine weld dimensions, overall part dimensional accuracy, and whenever postweld heat treatment is required. The American Welding Society has a certification program that recognizes those people who have the work experience and have passed a multiple-part test covering welding fundamentals, code interpretation, and practical inspection techniques. Each portion of this three-part test has a maximum 2-hour time limit. However, weld quality cannot be "inspected in." It must be built in. Weld quality is built in by each person who is involved with the project. As a professional welder, you should be properly trained for the work you are doing. If it is code work, you will have to take a certification test to demonstrate the required welding skills. Some companies will conduct knowledge tests that may include your ability to recognize weld discontinuities or defects. A professional welder should have available the welding procedure for the work being done, and in many cases the code requires the procedure to be displayed and the welder educated on its use. Since welders will be required to determine if the acceptance criteria have been met, they will require tools to measure the weld size and joint geometry. Figure 4-50 shows some tools usable for welding inspection. Kits such as the one shown in Fig. 4-50 are available from sources such as the American Welding Society. For a welder doing fillet welds perhaps all that is required is the appropriate fillet weld gauge. Figure 4-51 shows the proper use of fillet gauges.

Handheld Scanner Technology Includes the Palm™ Organizer Visual inspection methods and tools have not changed much over the last 50 years. However, there is a dramatic change occurring in the equipment that is now available to assist the person inspecting welds. Generically speaking, these devices are known as **handheld weld scanners.** By simply aiming the scanner at a joint preparation or finished weld and activating it, the operator quickly obtains a multitude of measurements and validates their geometry against preset thresholds.

The handheld scanner then supplies data via visual displays, computer-saved records, and strip chart printouts. Typical features measured on the prepared joint include

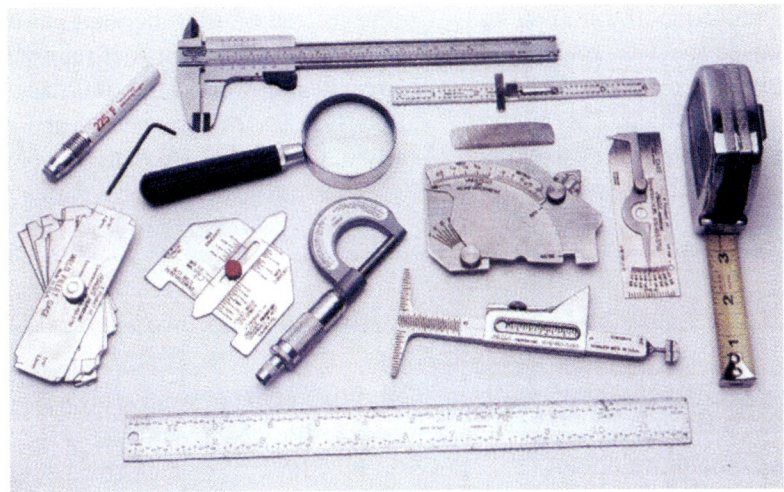

Fig. 4-50 Visual inspection tools. American Welding Society

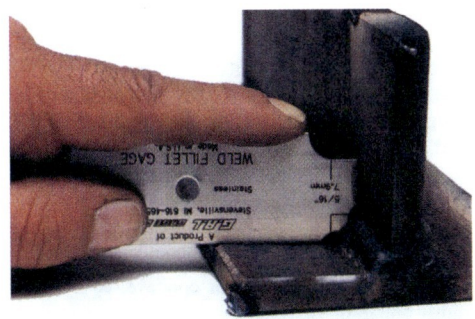 

Fig. 4-51 Measuring fillet weld size. On the left, a concave fillet weld; on the right, a convex fillet weld. American Welding Society

root opening, groove angle, material thickness, and root face size. (See Fig. 4-52 for an example of a typical display.) For the finished weld, features such as leg length, weld length, skips, toe angle, concavity/convexity, and undercut can be measured directly. In addition, for both preweld joints and finished welds, other information can be calculated including actual joint volume to be filled, throat size (theoretical), off seam amount, and percent over welding. These handheld devices can determine whether a weld is good or bad as well as or better than conventional gauges. However, their real value is that they can turn the produced data into useful information that can improve the effectiveness of welding operations.

Visual inspection (VI) by the welder can be a very effective tool in controlling overall weld quality. Although VI is limited to the visible surface of the welds, it is understood that the external surfaces of weldments see the highest stresses in service. Discontinuities opening to the surface are considered to be critical to the overall weldment fitness for the intended purpose. VI is a cost-effective

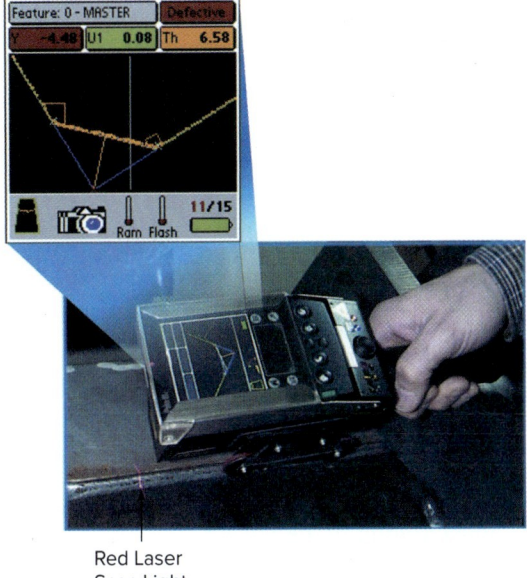

Red Laser
Scan Light

Fig. 4-52 Portable handheld scanner for weld inspection. Servo-Robot Corp.

Basic Joints and Welds **Chapter 4** 133

inspection method, and when performed by trained and qualified people, VI can uncover the vast majority of those defects that would otherwise be discovered later by more expensive nondestructive test methods. This method allows for the discovery and repair of defects as they may occur instead of after the joint has been completed, when there will be the expense of removing and then replacing a great deal of weld metal. It is easy to see the importance and efficiency of discovering and correcting a defect as early in the welding operation as possible.

CHAPTER 4 REVIEW

Multiple Choice

Choose the letter of the correct answer.

1. Which of the following is a type of weld joint? (Obj. 4-1)
   a. Butt
   b. Lap
   c. Corner, edge
   d. All of these

2. Which of the following is a type of weld? (Obj. 4-1)
   a. Fillet
   b. Groove
   c. Bead or surface
   d. All of these

3. The purpose of a seal weld is _____. (Obj. 4-1)
   a. To prevent moisture or other fluids from entering or exiting the joint
   b. To provide rigidity
   c. To provide strength
   d. To be a pivot

4. Which of the following is a weld profile? (Obj. 4-2)
   a. Convex
   b. Flat
   c. Isotope
   d. Both a and b

5. The size of a weld determines its _____. (Obj. 4-2)
   a. Load-carrying capacity
   b. Surface appearance
   c. Defects and inclusions
   d. Upward or downward position

6. A CJP weld is a _____. (Obj. 4-2)
   a. Can just penetrate weld
   b. Complete joules preparation weld
   c. Complete joint penetration weld
   d. None of these

7. A PJP weld is a _____. (Obj. 4-2)
   a. Present joint process weld
   b. Procedure just produced weld
   c. Partial just provided weld
   d. Partial joint penetration weld

8. Which type of weld is stronger? (Obj. 4-2)
   a. PJP
   b. CJP
   c. Both a and b have equal strength
   d. This does not have anything to do with strength

9. If the weld face of a fillet weld is 5/16 inch, what is the maximum convexity allowed? (Obj. 4-2)
   a. 1/16 inch
   b. 1/8 inch
   c. 3/16 inch
   d. Both a and c

10. A 3G weld is what type and position? (Obj. 4-3)
    a. Flat position groove
    b. Vertical position groove weld
    c. Horizontal position groove weld
    d. More information is needed to determine the type and position

11. A 5G position weld is made on _____. (Obj. 4-3)
    a. Sheet metal
    b. I-beams
    c. H-beams
    d. Pipe

12. All welding should be done in this position if possible. (Obj. 4-3)
    a. 1
    b. 2
    c. 3
    d. None of these

13. The strength of a welded joint depends upon which of the following? (Obj. 4-4)
    a. Strength of the weld metal
    b. Type of joint preparation and type of weld
    c. Type of load condition and location of joint in relation to load
    d. All of these

14. On a boxing weld where flexibility is required, the length of the boxing weld should be how long compared to the weld size? (Obj. 4-4)
    a. Twice
    b. Four times
    c. Five times
    d. None of these
15. An imperfection in a weld that exceeds the acceptance criteria is called a _____. (Obj. 4-5)
    a. Discontinuity
    b. Defect
    c. Both a and b
    d. None of these
16. Undercut is allowable if _____. (Obj. 4-5)
    a. It is not too deep or too long in relation to the thickness and size of the weldments
    b. It is twice the size of the weldment
    c. The joint is a crucial one
    d. Undercut is never allowable
17. On a groove weld the buildup is referred to as _____, and on a fillet weld the buildup is referred to as _____. (Obj. 4-5)
    a. Convexity; reinforcement
    b. Reinforcement; convexity
    c. Convexity; convexity
    d. Reinforcement; reinforcement
18. In some cases porosity is allowable if the cavities are not large in diameter and if _____. (Obj. 4-5)
    a. They don't exceed a combined dimension in a certain length of weld
    b. It is a groove weld with transverse loading
    c. It is a groove weld with tensile loading
    d. Both b and c
19. Who is in the best position to observe every inch of weld being deposited? (Obj. 4-6)
    a. Welder
    b. Supervisor
    c. AWS certified welding inspector
    d. Boss
20. A critical discontinuity would _____. (Obj. 4-6)
    a. Be linear (cracklike)
    b. Have a sharp end condition
    c. Be in a fatigue and/or impact loading condition
    d. All of these

Review Questions

Write the answers in your own words.

21. Write a brief definition of (a) continuous welds, (b) intermittent welds, and (c) seal welds. (Obj. 4-1)
22. Name the four fundamental types of welds. (Obj. 4-1)
23. How are groove welds measured? Fillet welds? (Obj. 4-2)
24. A fillet weld may take on three face characteristics. Name them and describe each type. (Obj. 4-2)
25. What are the factors that determine whether a joint is to have open or closed roots? (Obj. 4-3)
26. Draw a simple sketch of a groove weld in a single-bevel butt joint and name all of its parts. (Obj. 4-3)
27. Draw a simple sketch of a fillet weld in a T-joint and list the names of all its parts. (Obj. 4-3)
28. Name three defects that may be found in fillet welds. (Obj. 4-5)
29. Name three defects that may be found in groove welds. (Obj. 4-5)
30. List four reasons why visual inspection is a very effective inspection tool. (Obj. 4-6)
31. Match the appropriate callout letter on the drawing (Fig. 4-53) with the most appropriate weld terminology number. (Obj. 4-2)

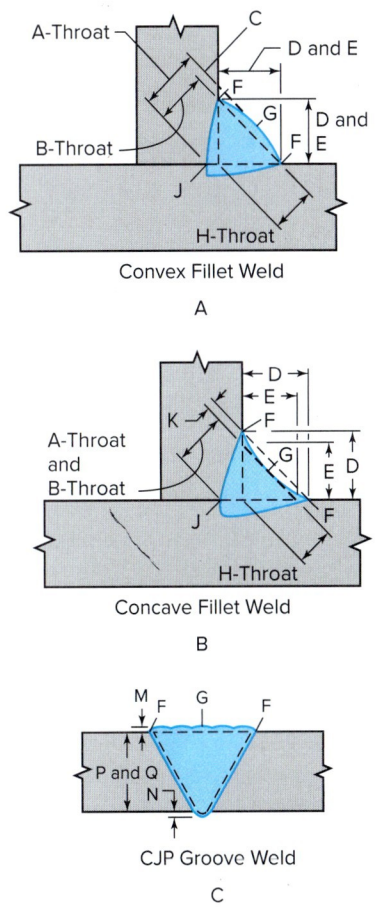

Fig. 4-53 For Question 31.

Letter Callouts

A. _____ H. _____
B. _____ J. _____
C. _____ K. _____
D. _____ M. _____
E. _____ N. _____
F. _____ P. _____ or _____
G. _____ Q. _____ or _____

Weld Terminology Number

1. Root
3. Face reinforcement
2. Root reinforcement
4. Toe
5. Face
6. Leg
7. Size
8. Convexity
9. Reinforcement
10. Concavity
11. Actual
12. Effective
13. Theoretical
14. Elbow
15. Joint penetration
16. Groove weld size

INTERNET ACTIVITIES

Internet Activity A

Edge is the publication (now online) of the Gas and Welding Distributor Association (GAWDA). This group provides many member services. Find its Web site to name the services GAWDA provides.

Internet Activity B

Use a favorite search engine to find out what is new in fillet welding.

Design credits: Cinema and movie icon: Denis Meshkov/123RF; and Illustration of Welding icons: Mr. Nachai Sorasee/123RF

5

Gas Welding

Chapter Objectives

After completing this chapter, you will be able to:

5-1 Describe the history of oxyacetylene welding (OAW).
5-2 List and describe the properties and distribution systems for the gases used for OAW.
5-3 Explain the safety issues of OAW.
5-4 List the equipment used for the OAW process.

Oxyacetylene Welding

Oxyacetylene welding (OAW), Fig. 5-1, page 138, is a way to join metal by heating the surfaces to be joined to the melting point with a gas flame, fusing the molten metal into a homogeneous mass, and then letting it solidify into a single unit. The flame at the cone reaches temperatures as high as 5,800 to 6,300°F. It is produced by burning acetylene in an oxygen-rich gas atmosphere mixed in the proper proportions in a welding torch. A filler rod may or may not be used to intermix with the molten pool of the metal being welded.

During the first part of this century, oxyacetylene welding became the major welding process both for fabrication and construction and for maintenance and repair. It had wide application because it can be used to weld practically all of the major metals. Today, however, we find that its use is limited for industrial production purposes. It is slower than the other welding processes, and many of the prime metals such as aluminum, titanium, and stainless steel can be welded more easily with other processes. The oxyacetylene process is still used for performing such operations as brazing, soldering, and metalizing; welding metals with low melting points; and general maintenance and repair work. Welding on pipe with small diameters is still being done with the oxyacetylene process, Fig. 5-2, page 138.

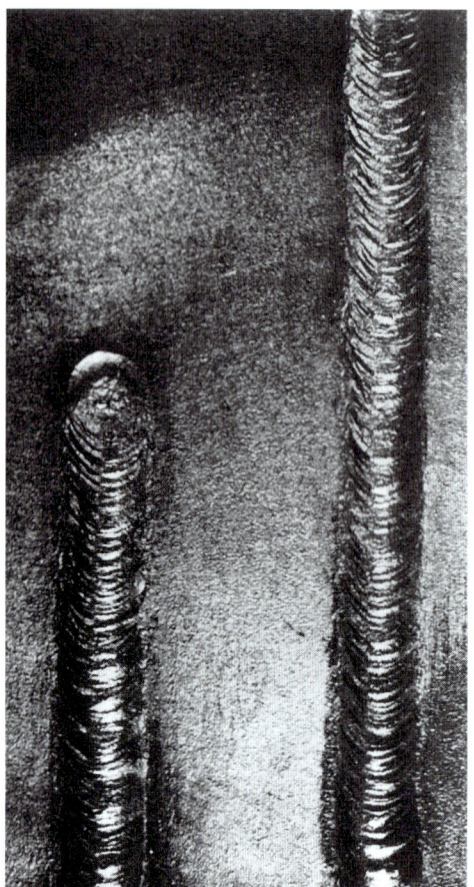

Fig. 5-1 Welds made with the oxyacetylene process on steel plate. They were welded in the up position. Gas welding produces sound welds of good appearance. Welds can be applied in all positions. Edward R. Bohnart

Fig. 5-2 Oxyacetylene welding a small diameter pipe on a construction site. This is a radiant heating installation in the floor. Edward R. Bohnart

While the oxyacetylene process is not used as much as it once was, it has a wide enough application to make it a necessary skill. It is an excellent means through which the student welder can observe the effect of heat and the flow of molten metal. It also develops the coordination of both hands that is the basis of good technique for all welding processes. In industry some jobs may require the ability to weld with both the oxyacetylene process and the electric arc process.

The History of Oxyacetylene Welding

The oxyacetylene process had its beginning many centuries ago. The early Egyptians, Greeks, and Romans used an alcohol or oil flame to fuse metals.

In the nineteenth century various gases were tested in experimental welding. They were used in the laboratory and in working with precious metals. In 1847, Robert Hare of Philadelphia fused platinum with an oxyhydrogen flame. In 1880, the production of oxygen and hydrogen through the electrolysis of water made possible the distribution of these gases in cylinders under pressure. Experiments were also done with oxygen-coal gas and air-hydrogen flames in the late 1800s.

A number of discoveries led to the development of the oxyacetylene process:

- In 1836, Edmund Davey discovered acetylene gas.
- In 1862, acetylene gas was produced from calcium carbide.
- In 1895, Thomas L. Willson began to produce calcium carbide commercially. It was first used for residential lighting.
- In 1895, Henry LeChatelier, a French chemist, announced his discovery that the combustion of acetylene with oxygen produced a flame hotter than any other gas flame.
- In 1900, Edmond Fouche, a Frenchman, invented a high pressure acetylene torch. He later designed a low pressure torch that worked on the injector principle.

ABOUT WELDING

Thomas Willson

Thomas Willson had the foresight to look at other uses for acetylene gas, and in 1903, after mixing it with oxygen, created the oxyacetylene torch. It could reach temperatures of 6,000°F. This was hot enough for welding or to use as a metal-cutting device. The torch was quickly adopted by the automotive and shipbuilding industries and revolutionized both.

- In 1906, Eugene Bourbonville brought the first welding torch to this country. The process was first used for maintenance and repair.
- During World War I, oxyacetylene welding came into its own as a production tool.

Gases

The oxyacetylene welding process makes use of two principal gases: oxygen and acetylene. However, a number of other fuel gases can be used for cutting and heating. These include propane, natural gas, and a gas called by the trade name of Mapp® gas. At one time, hydrogen was also used extensively. Study the relative gas temperatures given in Table 5-1. Figure 5-3 shows the type of cylinders for the different gases, their sizes, and capacities.

Oxygen

Oxygen is the gaseous chemical element in the air that is necessary for life. It is the most abundant chemical element in the crust of the Earth. It has no color, odor, or taste. It does not burn, but by combining with other

Table 5-1 Various Fuel Gas Efficiencies

Fuel Gas	Btus (ft^3)	Combined Intensity (usable heat) Btu (s/ft^2 of flame cone area)	Flame Temp.[1]	Oxygen per ft^3 of Fuel[1]	Approx. Normal Velocity (ft/s)
Acetylene	1,433	12,700	5,420°F	1.04	17.7
Mapp®[2]	2,381	5,540	5,301°F	2.4	7.9
Propane	2,309	5,500	5,190°F	4.00	11.9
Natural gas[3] (Mpls./St. Paul)	918	5,600	5,000°F	1.50	15.2
Hydrogen	275	7,500	4,600°F	0.25	36

[1]All the figures above are based on neutral flame conditions. Slightly higher flame temperatures can be attained for most of these gases by burning them in an oxidizing flame. Oxidizing flames require higher oxygen consumption. If the often quoted 6,300°F temperature for acetylene were used, then figures for the other four gases would be respectively higher.
[2]Mapp® gas data supplied by Dow Chemical Co.
[3]Natural gas data vary with exact composition in different geographic areas. City gas is as low as 4,400°F.

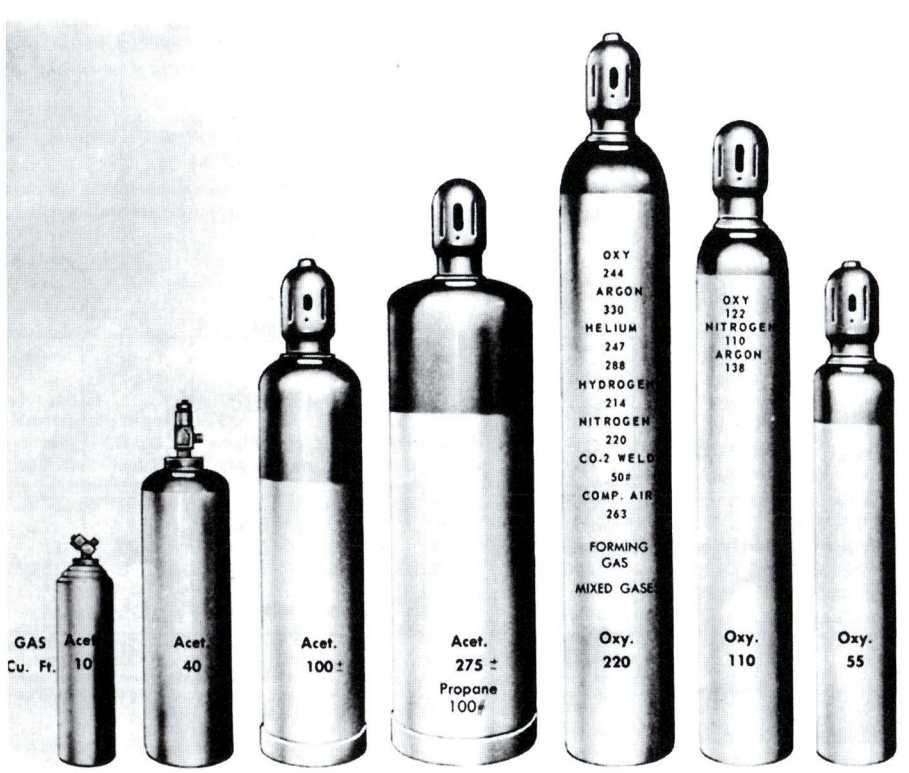

Fig. 5-3 Gas cylinder sizes and capacities. Thermadyne Industries, Inc.

elements, it supports their combustion. Substances that burn in air do so much more vigorously in pure oxygen. Other substances, such as iron, that do not burn in air will do so in oxygen. It is this property that makes it useful in cutting iron and steel.

Oxygen Production and Distribution

There are two commercial processes used in the production of *oxygen*. One of these is the separation of air into oxygen and nitrogen by liquefying the air. The other method is the separation of water into oxygen and hydrogen by the *electrolysis* of water, or in other words, by passing an electric current through it. By far, the greatest part of the oxygen used commercially today is manufactured by the liquefaction process.

Oxygen is distributed in steel or aluminum cylinders, Fig. 5-4. Steel cylinders are made from a single plate of high grade steel that has been heat treated to develop maximum strength and hardness. Internal construction is shown in Fig. 5-5. Because of the high pressure involved, these cylinders undergo rigid testing and inspection. Oxygen is also distributed in bulk tanks that are made according to the same specifications.

The aluminum cylinders are typically used for medical gases and are generally of the smaller size. Typically they are made of the high strength 6061-T6 series aluminum alloy.

Valve Mechanisms Each cylinder has a valve that must be opened to release oxygen. Some manufacturers use a double-seated valve that is perfectly tight when completely open or closed, Fig. 5-6. Valves that are not fully opened or closed cause oxygen to leak through the stem and result in a serious waste of materials. Other manufacturers use a type of valve that requires only a turn or two to open. The valve is protected from damage by an iron cap that screws on the neck ring of the cylinder. This cap should always be in place except when the cylinder is in use.

Cylinders are charged with oxygen at a pressure of about 2,200 p.s.i. at 70°F. An increase in the temperature of the gas causes it to expand

Fig. 5-5 Sectional view of an oxygen cylinder. Oxygen cylinders are seamless, drawn-steel vessels having a malleable iron neck ring shrunk on at the top and a cylinder valve screwed into the neck.

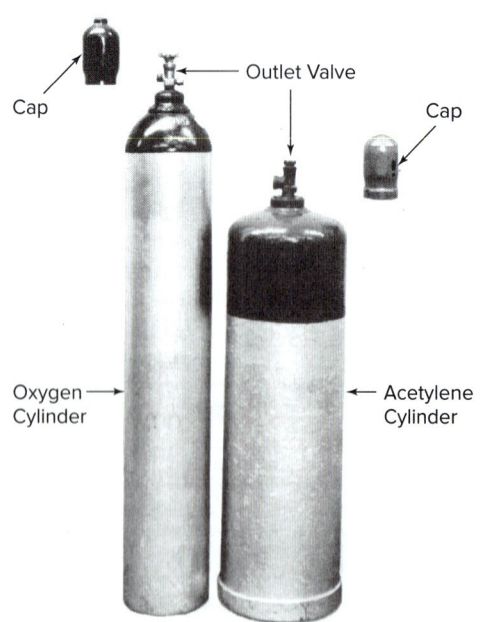

Fig. 5-4 Cylinders of oxygen and acetylene. Thermadyne Industries, Inc.

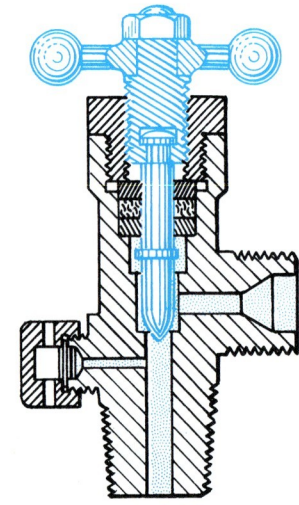

Fig. 5-6 This type of oxygen cylinder valve is double seated to prevent leakage when open. It should always be opened all the way. The valve is constructed to operate efficiently under high pressure.

and increase the pressure within the tank. A decrease in temperature causes the gas to contract and reduce the pressure within the tank. To prevent excess pressure, every cylinder valve has a safety device to blow off the oxygen long before there is any danger of explosion. Nevertheless, cylinders should not be stored where they might become overheated. If this occurs, the safety device will burst, and the oxygen will be lost.

Capacity of Cylinder There are three cylinder sizes generally used for welding and cutting. Gas suppliers fill their tanks with varying amounts of gas. The large size contains 220 to 244 cubic feet of oxygen. It weighs about 148 to 152 pounds when full and 130 to 133 pounds when empty. The middle size contains 110 to 122 cubic feet of oxygen. It weighs 89 to 101 pounds when full and 79 to 93 pounds when empty. A small size has 55 to 80 cubic feet of gas and weighs about 67 pounds when full and 60 pounds when empty. All of the oxygen in the cylinder may be used.

Safety in Handling It is important to remember that pure oxygen under pressure is an active substance. It will cause oily and greasy materials to burst into flame with almost explosive violence. The following safety precautions should be strictly observed:

- Take special care to keep oil and grease away from oxygen. Never store oxygen cylinders near oil, grease, or other combustibles.
- When using the cylinders, do not place them where oil might drop on them from overhead bearings or machines.
- Never use oxygen in pneumatic tools or to start internal combustion engines.
- Never use oxygen to blow out pipe or hose lines, dust clothes, or to create head pressure in a tank of any kind.
- Do not store oxygen cylinders near an acetylene generator, carbide, acetylene, or other fuel-gas cylinders.
- Do not use the cylinder as a roller or lift it by the cap.
- Keep cylinders away from the welding operation and close the cylinder valve when work is completed.
- Keep cylinders away from any electrical contact.

Acetylene

Acetylene is the most widely used of all the fuel gases, both for welding and for cutting. It is generated as the result of the chemical reaction that takes place when calcium carbide comes in contact with water. Laboratory tests have shown oxyacetylene flame temperatures up to approximately 6,300°F. Thus it has a very rapid rate of preheating. Acetylene has a peculiar odor. If lighted, it

> **JOB TIP**
>
> **Listening**
> Listening is as easy as breathing—right? Actually, no; it takes an active mind to listen on the job. You must stop thinking about what you are going to say next. Focus solely on understanding. After the person finishes, repeat in your own words, what you have understood.

burns with a smoky flame and gives off a great deal of carbon as soot.

An effective welding fuel gas must possess the following characteristics:

- High flame temperature
- High rate of flame propagation
- Adequate heat content
- Minimum chemical reaction of the flame with base and filler metal

Acetylene most closely matches all these requirements and is used for welding purposes. For heating and cutting, other fuel gases can be used.

Acetylene Production and Distribution

Commercial acetylene is made from calcium carbide, which is commonly referred to as *carbide*. **Carbide** is a gray, stonelike substance that is the product of smelting coke and lime in an electric furnace. It is distributed in standard steel drums containing 100 pounds of carbide for use in acetylene generators. Several sizes are available: lump (3½ × 2 inches); egg, nut, quarter, or pea (¼ × ½ inch); rice and 14 ND (0.055 × 0.0173 inch).

Carbide is also available in a special form known as *carbic-processed carbide,* which is carbide compressed into briquets. The cakes are used in acetylene generators. They are 4 × 3 inches and weigh about 2½ pounds. They are packed in drums containing 40 cakes each.

Like oxygen, acetylene is distributed in cylinders. (See Fig. 5-4.) These cylinders are constructed differently from oxygen cylinders because of the fact that free acetylene should not be stored at a pressure above 15 p.s.i. After much study, the problem of combining safety with capacity was solved by packing the cylinders with a porous material saturated with **acetone,** a liquid chemical having the property of dissolving or absorbing many times its own volume of acetylene. The cylinder itself is a strong steel container. It is packed completely full. In such cylinders, the acetylene is perfectly safe, but it is still considered to

Fig. 5-7 Three basic types of acetylene cylinders and their internal construction.

be unstable and must be handled with care. Figure 5-7 shows three basic types of acetylene cylinders and their internal construction.

Valve Mechanisms Acetylene is drawn off through a valve that in some cylinders is located in a recessed top; and in others, on a convex top. This valve does not have to stand the high pressure that the oxygen valve is subjected to and is, therefore, much simpler in construction. It needs to be opened only about 1½ turns. This is done so that the cylinder can be turned off quickly if a fire starts in any part of the welding apparatus. Safety fuse plugs are also provided.

Capacity of Cylinder A full cylinder of acetylene has a pressure of about 225 p.s.i. Two sizes are generally used for welding and cutting. The large size contains about 300 cubic feet of acetylene and weighs about 232 pounds when full and 214 pounds when empty. The small size contains about 100 cubic feet of acetylene and weighs about 91 pounds when full and 85 pounds when empty. Two special sizes containing 10 cubic feet and 40 cubic feet are also available.

Not all the acetylene in the cylinder can be used. The maximum practical use of the gas is reached when the oxyacetylene flame begins to lengthen and loses much of its heat. The acetylene pressure regulator reading will be about 35 p.s.i. This varies with temperature.

Safety in Handling Remember that acetylene will burn, and like any other combustible gas, it will form an explosive mixture with air. The following precautions should be observed:

- Do not leave acetylene cylinders on their sides. Store and use them with the valve end up.
- Store the cylinders in a well-protected, ventilated, dry location. They should not be near highly combustible material such as oil or excelsior, or stoves, radiators, furnaces, and other sources of heat. Keep the valve cap on when the cylinder is not in use.
- If the outlet valve becomes clogged with ice, thaw it with warm (not boiling) water applied only to the valve. The fusible safety plugs with which all cylinders are provided melt at the boiling point of water. Never use a flame for thawing the valve.
- Handle acetylene cylinders carefully. Rough handling, knocks, and falls may damage the cylinder, valve, or fuse plugs and cause leakage.
- If acetylene leaks around the valve spindle when the valve on the cylinder is opened, close the spindle, move the cylinder to an outside area, and advise the supplier immediately. Only trained personnel should service and repair cylinder valves.
- Never tamper with the fuse plugs.

Propane Gas

Propane is a *hydrocarbon* present in petroleum and natural gas. It is used primarily for oxyfuel heating, cutting, soldering, and brazing. It is sold and transported in steel cylinders containing 20 to 100 pounds of the liquefied gas. It can also be supplied by tank car and bulk delivery. The oxypropane flame temperature is approximately 5,190°F. Because this temperature is less than that of oxyacetylene, it takes longer to bring the steel to the melting point. Propane is used extensively for soldering and alloy brazing.

Mapp® Gas

Mapp® gas is a liquefied acetylene compound that is a fuel gas for oxyfuel heating and cutting. It has a strong smell that is an aid in discovering leaks. When mixed with oxygen, the flame has a temperature of 5,301°F. This temperature is higher than that of the oxypropane flame, but not as high as that for the oxyacetylene flame. Although the heating and cutting may be somewhat slower due to the lower temperature, users indicate that overall expenses are lower due to reduced handling costs and lower gas costs. The use of this gas as a fuel gas for heating and cutting is growing.

Mapp® Gas Distribution Mapp® gas is distributed in bulk or in steel cylinders similar in appearance to acetylene cylinders. They have a shutoff valve similar to that on acetylene cylinders.

Capacity of Cylinder Although Mapp® gas is an acetylene product, it is liquefied and stabilized so that it can be used at pressures as high as 375 p.s.i. at 170°F as compared with 15 p.s.i. for acetylene. The explosive limits of Mapp® gas are lower than those of acetylene. Since the gas can be stored in its free state and at high pressures, a 120-pound cylinder contains as much gas as five 240-pound cylinders of acetylene. This reduces handling cost for the user.

Safety in Handling Mapp® gas, like all other flammable gases, forms an explosive mixture with the air. The same general precautions that have been recommended in the handling of acetylene cylinders should be observed in the handling of these cylinders. It is the safest of industrial fuels. The explosive limits of Mapp® gas vapor in air and oxygen are much narrower than acetylene and about the same as propane and natural gas. You can smell it at concentrations as low as 0.01 percent and find leaks that cannot be detected with other gases.

General Cylinder Handling, Storage, and Operation Safety Concerns

Information on general cylinder safety concerns is available from many sources such as OSHA Regulations (Standards–1910 Subpart Q Welding, Cutting, and Brazing CFR–1910.253). All oxyacetylene welders should understand and follow these regulations:

1. All portable cylinders used for the storage and shipment of compressed gases shall be constructed and maintained in accordance with the regulations of the U.S. Department of Transportation, 49 CFR Parts 171–180.
2. Compressed gas cylinders shall be legibly marked, for the purpose of identifying the gas content, with either the chemical or the trade name of the gas. Such marking shall be by means of stenciling, stamping, or labeling and shall not be readily removable. Whenever practical, the marking shall be located on the shoulder of the cylinder. This method conforms to the American National Standard Method for Marking Portable Compressed Gas Containers to Identify the Material Contained, ANSI Z48.1-1954.
3. Compressed gas cylinders shall be equipped with connections complying with the American National Standard Compressed Gas Cylinder Valve Outlet and Inlet Connections, ANSI B57.1-1965.
4. All cylinders with a water weight capacity of over 30 pounds (13.6 kilograms) shall be equipped with means of connecting a valve protection cap or with a collar or recess to protect the valve.
5. Cylinders shall be kept away from radiators and other sources of heat.
6. Inside of buildings, cylinders shall be stored in a well-protected, well-ventilated, dry location, at least 20 feet from highly combustible materials such as oil or excelsior. Cylinders should be stored in clearly identified assigned places away from elevators, stairs, or gangways. Assigned storage spaces shall be located where cylinders will not be knocked over or damaged by passing or falling objects, or subject to tampering by unauthorized persons. Cylinders shall not be kept in unventilated enclosures such as lockers and cupboards.
7. The valves on empty cylinders shall be closed.
8. For cylinders designed to accept a valve protection cap, the cap shall always be in place, hand-tight, except when cylinders are in use or connected for use.
9. For fuel-gas cylinder storage inside a building, cylinders, except those in actual use or attached ready for use, shall be limited to a total gas capacity of 2,000 cubic feet or 300 pounds of liquefied petroleum gas.
10. Oxygen cylinders shall not be stored near highly combustible material, especially oil and grease, or near acetylene or other fuel-gas cylinders, or near any other substance likely to cause or accelerate fire.
11. Oxygen cylinders in storage shall be separated from fuel-gas cylinders or combustible materials (especially oil or grease), a minimum distance of 20 feet or by a noncombustible barrier at least 5 feet high having a fire-resistance rating of at least one-half hour.

SHOP TALK

Citigroup Center
After 20 years, it was revealed that a skyscraper in New York, Citigroup Center, almost toppled over due to a structural weakness. The building had been designed to have welded joints, but bolted joints were used instead. When the mistake was noticed, welders were flown in to weld all the joints. During these repairs, a hurricane headed toward them. Fortunately, the hurricane altered course and the welders were able to complete the task.

12. Cylinders, cylinder valves, couplings, regulators, hoses, and apparatus shall be kept free from oily or greasy substances. Oxygen cylinders or apparatus shall not be handled with oily hands or gloves. A jet of oxygen must never be permitted to strike an oily surface, greasy clothes, or enter a fuel oil or other storage tank.
13. A cradle, boat, or suitable platform shall be used when transporting cylinders by a crane or derrick. Slings or electric magnets shall not be used for this purpose. Valve-protection caps shall always be in place on cylinders designed to accept a cap.
14. Cylinders shall not be dropped, roughly handled, or struck or permitted to strike each other violently. Rough handling, knocks, or falls are liable to damage the cylinder, the valve, or the safety devices and cause leakage.
15. Valve-protection caps shall not be used for lifting cylinders from one vertical position to another. Bars shall not be used under valves or valve-protection caps to pry cylinders loose when frozen to the ground or otherwise fixed; the use of warm (not boiling) water is recommended. Valve-protection caps are designed to protect cylinder valves from damage.
16. Unless cylinders are secured on a special truck, regulators shall be removed and valve-protection caps, when provided for, shall be put in place before cylinders are moved.
17. Cylinders not having fixed hand wheels shall have keys, handles, or nonadjustable wrenches on valve stems while these cylinders are in service. In multiple-cylinder installations, only one key or handle is required for each manifold.
18. Cylinder valves shall be closed before cylinders are moved.
19. Cylinder valves shall be closed when work is finished.
20. Cylinders shall be kept far enough away from the actual welding or cutting operation so that sparks, hot dross, or flame will not reach them, or fire-resistant shields shall be provided.
21. Cylinders shall not be placed where they might become part of an electric circuit. Contacts with third rails, trolley wires, or the like shall be avoided. Cylinders shall be kept away from radiators, piping systems, and layout tables that may be used for grounding electric circuits, such as for arc welding machines. Any practice such as the tapping of an electrode against a cylinder to strike an arc shall be prohibited.
22. Cylinders shall never be used as rollers or supports, whether full or empty.
23. The numbers and markings stamped into cylinders shall not be tampered with.
24. No person, other than the gas supplier, shall attempt to mix gases in a cylinder. No one, except the owner of the cylinder or person authorized by the owner, shall refill a cylinder.
25. No one shall tamper with safety devices on cylinders or valves.
26. A hammer or wrench shall not be used to open cylinder valves. If valves cannot be opened by hand, the supplier shall be notified.
27. Cylinder valves shall not be tampered with nor should any attempt be made to repair them. If trouble is experienced, the supplier should be sent a report promptly, indicating the character of the trouble and the cylinder's serial number. Supplier's instructions as to its disposition shall be followed.
28. Fuel-gas cylinders shall be placed with valve end up whenever they are in use. Liquefied gases shall be stored and shipped with the valve end up.
29. Before connecting a regulator to a cylinder valve, the valve shall be opened slightly and closed immediately. The valve shall be opened while standing to one side of the outlet; never in front of it. Never crack a fuel-gas cylinder valve near other welding work or near sparks, flame, or other possible sources of ignition.
30. Before a regulator is removed from a cylinder valve, the cylinder valve shall be closed and the gas released from the regulator.
31. Nothing that may damage the safety device or interfere with the quick closing of the valve shall be placed on top of an acetylene cylinder when it is in use.
32. If cylinder valves or fittings are found to have leaks that cannot be stopped by closing the valve or fitting, the cylinders shall be plainly tagged and taken outdoors away from sources of ignition. A sign should be placed near the leaky cylinders warning unauthorized personnel not to approach them. The supplier should be promptly notified, and the supplier's instructions for the return of the cylinders followed.
33. Safety devices shall not be tampered with.
34. Fuel gas shall never be used from cylinders through torches or other devices equipped with shutoff valves without reducing the pressure through a suitable regulator attached to the cylinder valve or manifold.
35. The cylinder valve shall always be opened slowly.
36. An acetylene cylinder valve shall not be opened more than one and one-half turns of the spindle, and preferably no more than three-fourths of a turn.

Manifold Distribution

Where it is necessary to supply a number of work stations and conserve space, both oxygen and the fuel gas may be supplied by a *manifold system,* Figs. 5-8, 5-9, and 5-10.

The acetylene manifolds must be equipped with a flash arrestor, Figs. 5-11 and 5-12, page 146, to prevent flashback through the manifold into the cylinders. In addition, each cylinder is connected to the manifold by means of an

Fig. 5-8 Note torch and filler metal angles and location on this practice plate. Gases are being supplied via a manifold system as shown in Fig. 5-9. Bench top surface being protected with fire brick.
Location: Northeast Wisconsin Technical College Mark A. Dierker/McGraw Hill

Fig. 5-9 This manifold system is being used to supply oxygen and acetylene to the booths on both sided of it.
Location: Northeast Wisconsin Technical College Mark A. Dierker/McGraw Hill

Fig. 5-10 A typical station outlet for both oxygen and acetylene pipelines from the source. Stations for the attachment of oxygen and acetylene regulators are placed at convenient points around the shop or plant where welding or cutting is to be done.

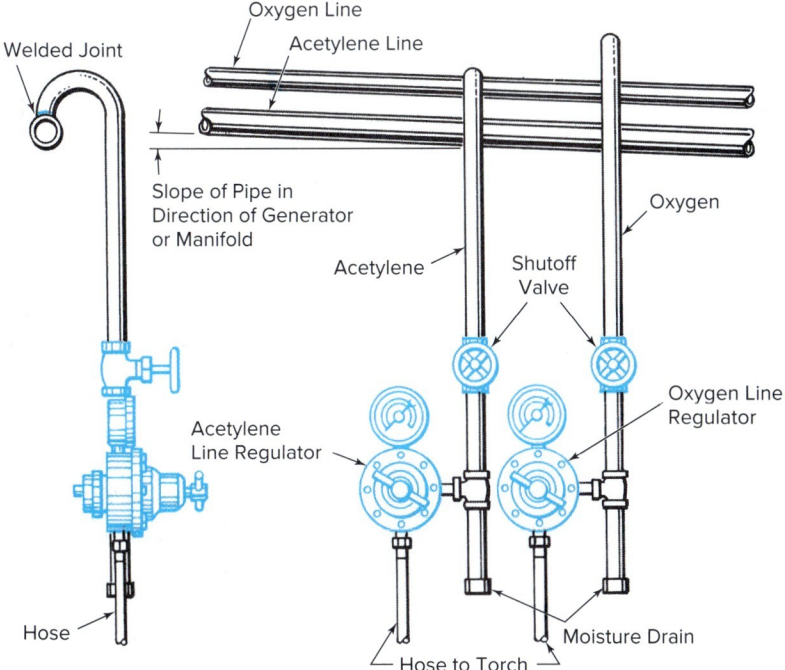

Gas Welding Chapter 5 145

Fig. 5-11 An acetylene gas manifold system. Thermadyne Industries, Inc.

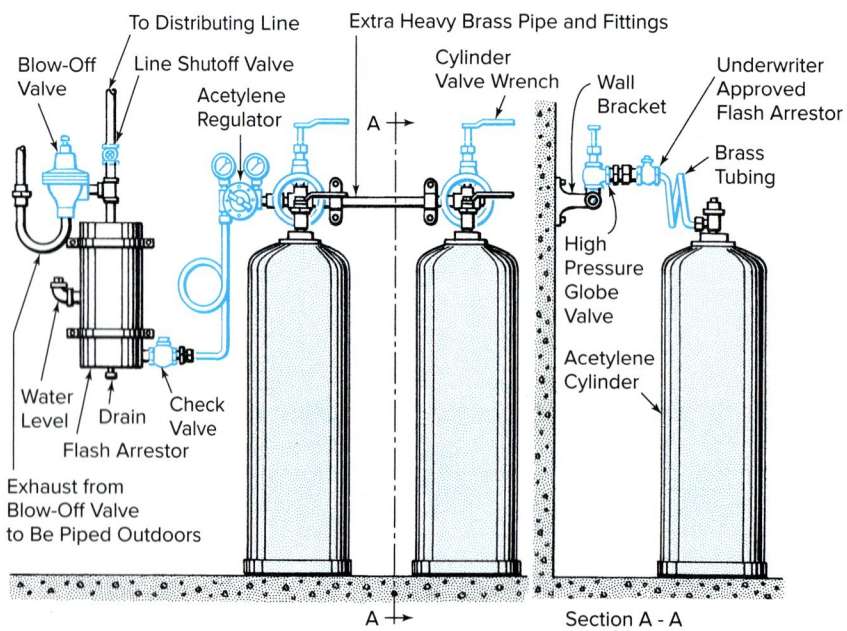

Fig. 5-12 Acetylene cylinder manifolds are constructed of extra heavy pipe and fittings to conform to the regulations of the National Board of Fire Underwriters.

Fig. 5-13 Oxygen manifold system. Note master regulator and flexible copper tubing leading from each tank. Thermadyne Industries, Inc.

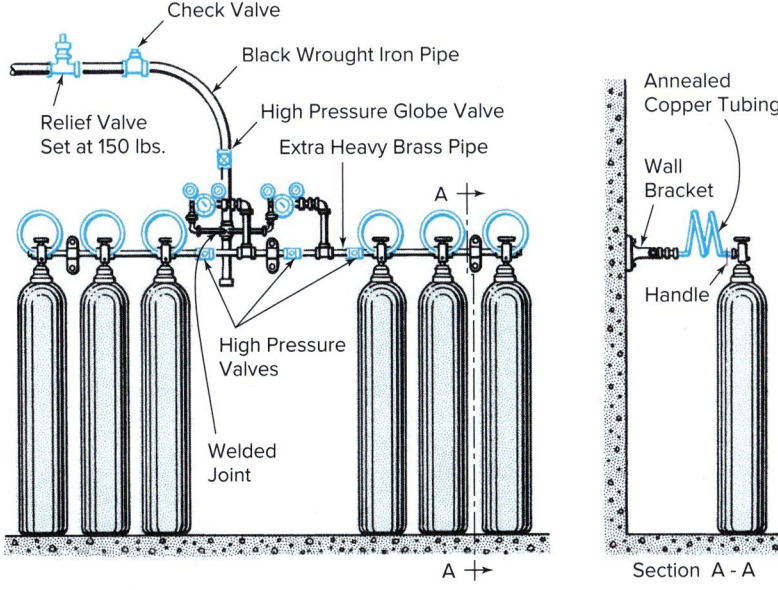

Fig. 5-14 An oxygen manifold. It is often convenient to place all of the oxygen cylinders at one point where they may be readily handled and to connect them together to an oxygen manifold as illustrated.

individual pigtail flash arrestor and backcheck valves. There is the full cylinder pressure of 2,000 p.s.i. in the oxygen manifold pipes. The manifold regulator reduces the pressure to 50 or 75 pounds in the line that goes to the various station outlets in the shop. Figures 5-13 and 5-14 present an oxygen manifold installation.

Both types of manifolds have a pressure regulator for the purpose of reducing and controlling the pressure of the gas to the work station. The work station is also equipped with an acetylene and oxygen regulator for further pressure control at the welding or cutting torch, as shown in Fig. 5-10, page 145.

Manifold piping systems for acetylene or acetylenic compounds should be steel or wrought iron. Copper can be used for oxygen piping systems. Manifold systems range from very simplistic to highly sophisticated automatic systems for virtually any gas control requirement. Where high capacity is required, *liquid vessel* systems can be used. Figure 5-15, page 148, shows a sophisticated design, which automatically switches between the primary and reserve supply with no interruption of service.

Acetylene Generators

Large users of acetylene generate their own gas in an acetylene generator, Fig. 5-16, page 148.

The carbide-to-water generators are used to produce acetylene gas for welding and cutting. Small amounts of calcium carbide are fed into relatively large amounts of water. The water absorbs the heat given off by the chemical reaction, and the gas is cooled and purified by bubbling through the water. Generators of this type can produce up to 6,000 cubic feet of acetylene per hour.

There are two classes of generators: low pressure, in which the acetylene pressure is less than 1 p.s.i., and medium pressure, in which the acetylene pressure is 1 to 15 p.s.i. You will recall that free acetylene cannot be used at a pressure higher than 15 p.s.i.

Acetylene generators are not used by individual companies or schools but are used by large gas suppliers. The acetylene product is then provided to the consumer in cylinders. This has been brought about due to the various regulations that must deal with the byproducts from an environmental aspect. Additional concerns come from the

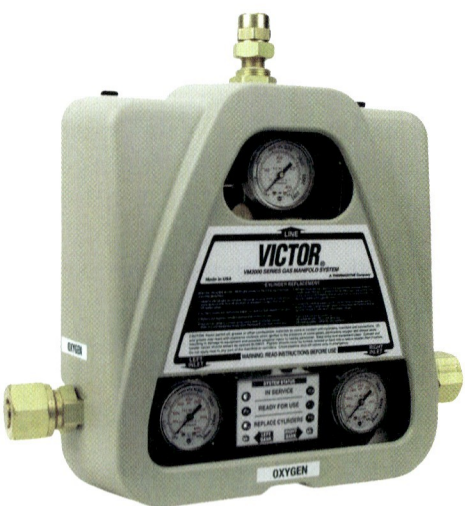

Fig. 5-15 An automated manifold system utilizing the larger capacity liquid storage cylinders. Victor Equipment Company/Thermadyne Industries, Inc.

Fig. 5-16 A large acetylene generator for commercial supply. It can use carbide sizes from lump to 14 ND. Rexarc

- Acetylene welding hose
- Hose couplings
- Single-purpose cutting torch or welding torch
- Cylinders and cart for portability
- Flash arrestor and check valves (protective equipment)
- Flint lighter to ignite torch

Approved protective equipment shall be installed in an OAW system to prevent:

- Backflow of oxygen into the fuel-gas supply system
- Passage of a flashback into the fuel-gas supply system
- Excessive back pressure of oxygen in the fuel-gas supply system

The two functions of the protective equipment may be combined in one device or provided by separate devices. The *protective equipment* should be located in the main supply line, at the head of each branch line in a manifold system, or at each location where fuel gas is withdrawn. The *backflow protection* should also be provided to prevent fuel gases from flowing into the oxygen system. *Flashback protection* should be provided to prevent flame from passing into the fuel-gas system or oxygen gas system.

Much of this equipment is illustrated in Fig. 5-17A and B. Note that the welding torch comes with a wide assortment of tips that provide a choice in the volume of heat desired. When welding or flame cutting, the welder must wear a pair of protective goggles to prevent harm

> **ABOUT WELDING**
>
> **Shrinkage**
>
> Factors causing contraction or shrinkage during welding:
>
> 1. Heat input
> 2. Mass of the structure
> 3. Ambient temperature
> 4. Cooling rate
> 5. External and internal restraints

requirement for operating licenses, inspections, and use of trained operators.

Welding Equipment

Oxyacetylene welding requires the following equipment:

- Oxygen regulator
- Acetylene regulator
- Oxygen welding hose

> **JOB TIP**
>
> **The Internet**
> You can go job hunting on the Web.

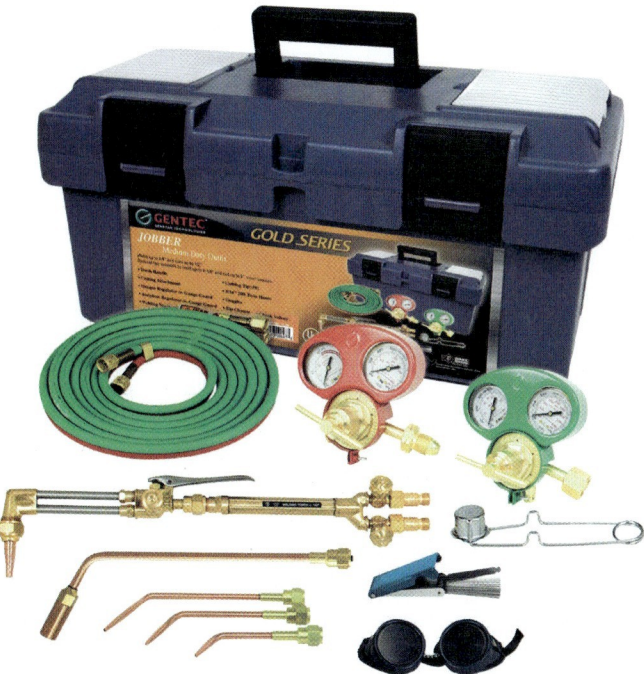

Fig. 5-17A A travel kit for a medium duty torch for cutting, heating, welding, brazing, and soldering. Note the cutting tip, welding tips, heating tip, and ancillary equipment such as storage case. Gentec

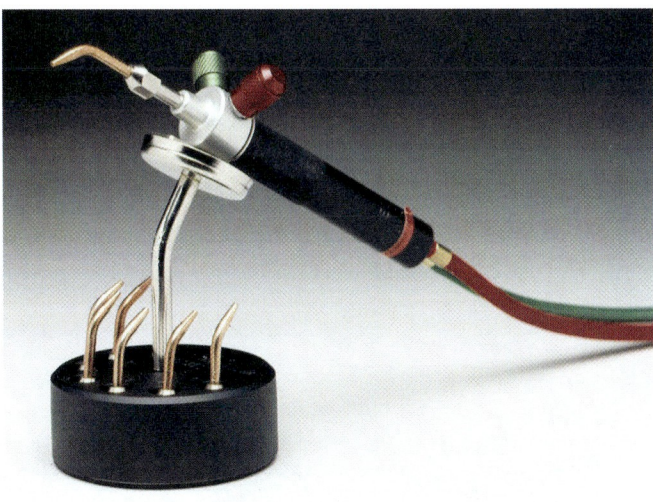

Fig. 5-17B Light duty torch for welding, principally for jewelry and other fine work. Note the various tips and magnetic holder. Smith Equipment

to the eyes from hot particles of metal, sparks, and glare. A flint lighter is also required to light the torch. The equipment is set up for welding in Fig. 5-18, page 150. Portable tank outfits are mounted on a truck similar to that shown in Fig. 5-19, page 150.

Pressure Regulators

The pressure at which oxygen and acetylene is compressed into cylinders is much too high for direct use in welding and cutting. Some means must be provided to reduce the high internal cylinder pressure—about 2,200 p.s.i. in the case of oxygen—to the relatively low pressure of 0 to 45 p.s.i. required for welding and cutting. The flame must also be steady and uniform. This can be accomplished only if the gas pressures do not fluctuate. Pressure regulators carry out both of these all-important functions. They reduce the high cylinder pressure to that used for welding, and they can maintain that pressure without variation during the welding operation.

Design Figure 5-20, page 150, illustrates regulator design. They have a union nipple (A) for attachment to the cylinders and an outlet connection (B) for the hose leading to the torch. Two pressure gauges are mounted on the body of the regulator: one (C) shows the pressure in the cylinder, and the other (D) shows the pressure of the gas being supplied to the torch. The pressure gauge that shows the pressure in the cylinder is also useful in indicating to the welder the amount of gas remaining in the cylinder.

The working pressure is adjusted by means of a hand screw (E). When this screw is turned to the left, or counterclockwise, the valve mechanism inside the regulator is shut off, and gas cannot pass through the regulator to the torch. Turning the pressure-adjusting screw to the right, or clockwise, presses it against the regulator mechanism. The valve opens, and gas passes through the regulator to the torch at the pressure shown on the working pressure gauge. Any pressure can be set up by turning the handle until the desired pressure is indicated. Figure 5-21, page 151, illustrates the internal mechanism of a typical single stage gas regulator.

Safety Precautions The following precautions regarding the use of regulators should be observed:

- Inspect all nuts and connections before use to detect faulty seats that may cause leakage of gas.
- Keep clean and free from oil and grease.
- Before attaching the regulator, crack the cylinder valve to blow out contaminants.
- Before opening the valve of a cylinder (always open cylinder valves slowly) to which a regulator has been attached, be sure that the pressure-adjusting screw has been completely released by turning it to the left. Failure to do this allows the full cylinder pressure to hit the engaged mechanism of the regulator. Not only will the regulator be damaged, but it may also burst and injure the welder.

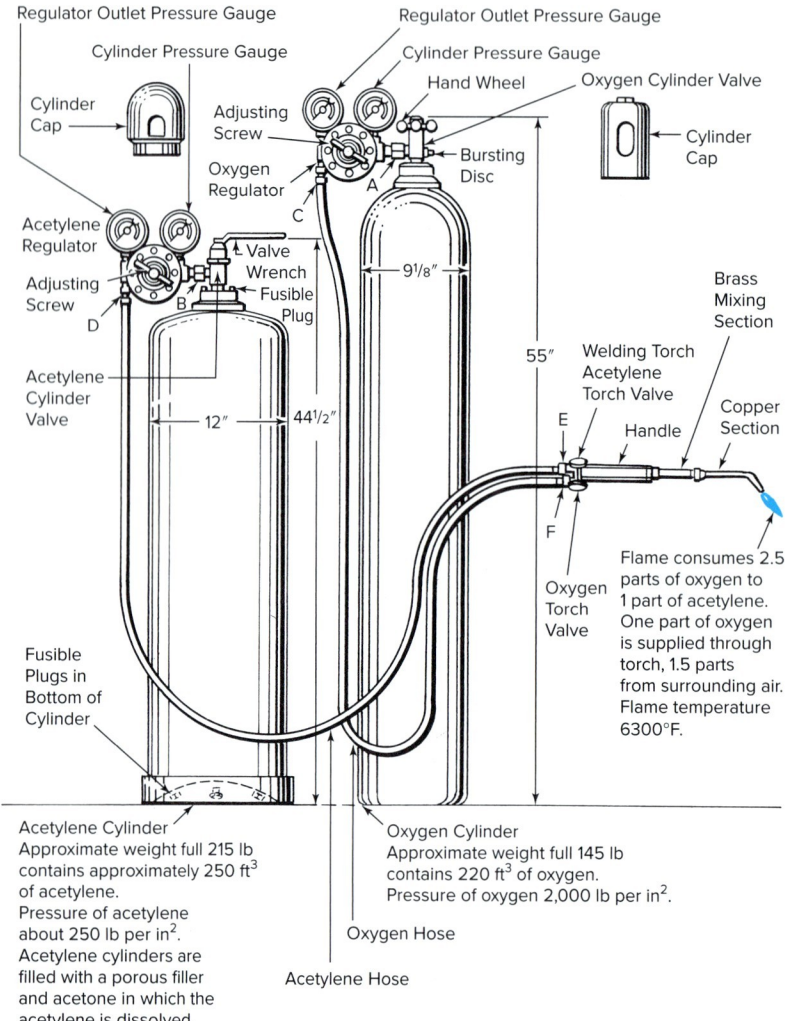

Fig. 5-18 Diagram showing cylinders, regulators, hose, and welding torch properly connected and ready to operate. Torch and/or regulators need to be equipped with approved flash arrestors and check valves.

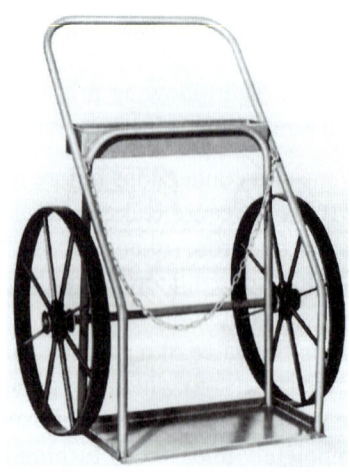

Fig. 5-19 Steel handtruck used to carry oxygen and acetylene tank and welding equipment. A portable outfit offers wide flexibility in the use of gas welding in both shop and field. *Thermadyne Industries, Inc.*

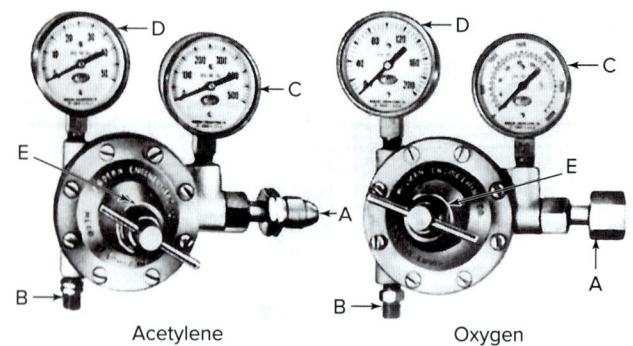

Fig. 5-20 Oxygen and acetylene welding regulators for use with individual tanks. *Thermadyne Industries, Inc.*

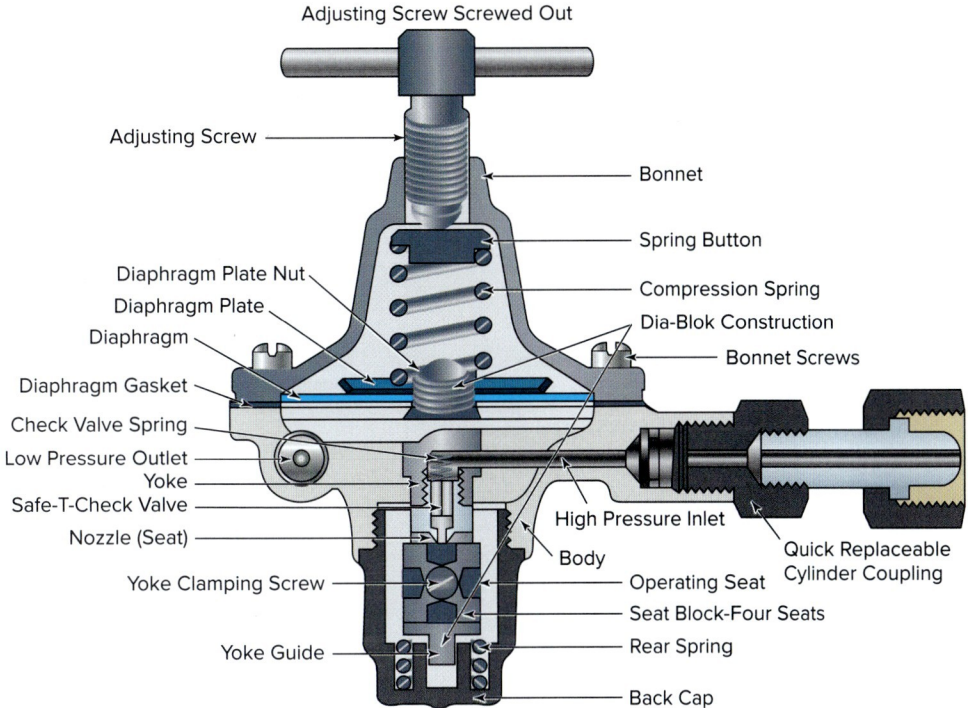

Fig. 5-21 Internal construction of single-stage cylinder regulator.
Adapted from Thermadyne Holdings Corporation

- Never attempt to connect an acetylene regulator to a cylinder containing oxygen or vice versa. Do not force connections that do not fit, and be sure that all connections are tight.
- Use regulators only for the gas and pressures for which they are intended.
- Have regulators repaired only by skilled mechanics who have been properly trained and qualified.

For video on regulator burnout, please visit www.mhhe.com/welding.

Regulator Construction To better understand the workings of the internal mechanism of a gas regulator, study Fig. 5-22. Let us consider an oxygen regulator as an example. The oxygen under high pressure and coming directly from the oxygen cylinder—enters the regulator at the left at 2,200 p.s.i., and it must leave the regulator at the desired pressure for welding, perhaps 10 p.s.i.

Before oxygen is allowed to enter at the left, tension on the spring (S) puts pressure on the flexible diaphragm, deflecting it to the left, which in turn causes it to contact the valve stem, opening the valve (V). If the oxygen tank is opened and the gas is permitted to flow into the

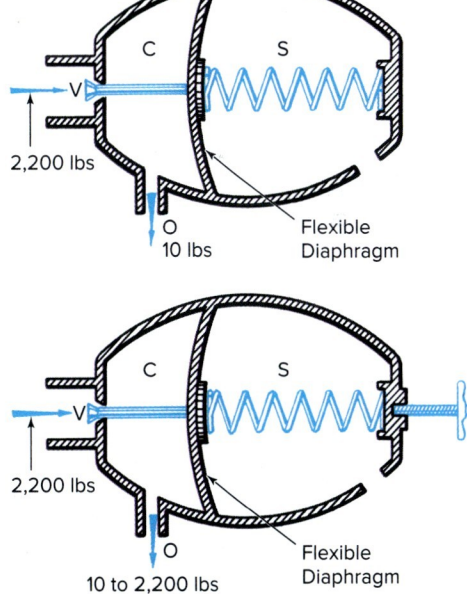

Fig. 5-22 Connecting the outlet shown in the top sketch to the inlet of the lower demonstrates the principle of two-stage regulators.

chamber (C), the full force of 2,200 p.s.i. pressure is exerted against the diaphragm, causing it to move to the right. Thus, the pressure against the valve stem is removed so that the valve (V) closes.

Gas Welding **Chapter 5** 151

When, because of the requirements of the welding or cutting flame, oxygen has been withdrawn from the chamber (C), the pressure falls below a certain point. Therefore, the tension spring (S) becomes the greater force and deflects the diaphragm to the left, reopening the valve to permit more high pressure oxygen to enter the chamber. Bear in mind that the force that opens the valve is provided by the tension of the spring, while the force that closes the valve is provided by the high gas pressure from the tank. When these two forces are balanced, a constant flow of oxygen to the torch results. When the tension of the spring is properly adjusted by means of the regulator-adjusting screw (Figs. 5-20 and 5-21, pages 150–151), the constant pressure desired is maintained in the chamber (C). Thus, a constant pressure can be withdrawn from the chamber, providing an even flow of oxygen to the torch.

Regulators are designated as *single-stage* and *two-stage regulators*. The operation of the single-stage regulator has just been explained. In the two-stage regulator, the pressure reduction is accomplished in two stages. In the first stage, the spring tension has been set so that the pressure in the high pressure chamber is a fixed amount. For example, it may be set at 150 p.s.i. The gas then passes into a second reducing chamber that has a screw adjustment similar to that in the single-stage regulator. This adjustment makes it possible to obtain any desired pressure. The principle of the two-stage regulator is shown in Fig. 5-22, page 151, in which the upper sketch represents the first stage; and the lower, the second. Figure 5-23 illustrates three typical regulator designs.

For the most part, single-stage regulators are used with manifold systems. In these systems, the 2,200 p.s.i. pressure is reduced at the manifold before it enters the piping system through a heavy duty regulator on the manifold. The relatively low manifold pressure is further lowered at the work station to the required pressure by the single-stage regulator.

Regulator springs are made of a good grade of spring steel. The diaphragm may be made of brass, sheet spring steel, stainless steel, or rubber.

When a number of stations are serviced by a line gas system, individual oxygen and acetylene regulators are required at each station. These regulators are smaller than cylinder regulators because they are not subject to high cylinder pressure. Figure 5-24 shows the difference in the internal construction of a tank and a line regulator.

Although the mechanical details of regulator construction vary among different manufacturers, the fundamental operating principles are the same for all oxyacetylene regulators used for welding and cutting.

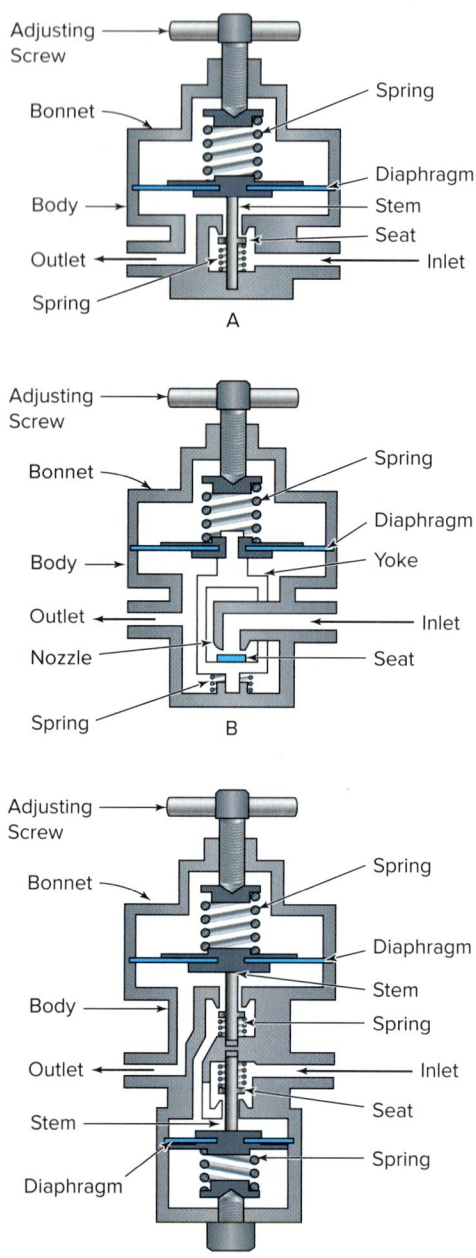

Fig. 5-23 Typical gas regulators: (A) single-stage stem type, (B) single-stage nozzle type, and (C) two-stage nozzle type. Adapted from Thermadyne Holdings Corporation

Welding Torches

The **welding torch** is an apparatus for mixing oxygen and acetylene in the proportions necessary to carry on the welding operation. It also provides a handle so that the welder can hold and direct the flame while welding. The handle has two inlet gas connections: one for oxygen and the other for acetylene. Each inlet has a valve that controls the volume of oxygen or acetylene passing through. By means of these valves, the desired proportions of oxygen and acetylene are allowed to flow through the torch where

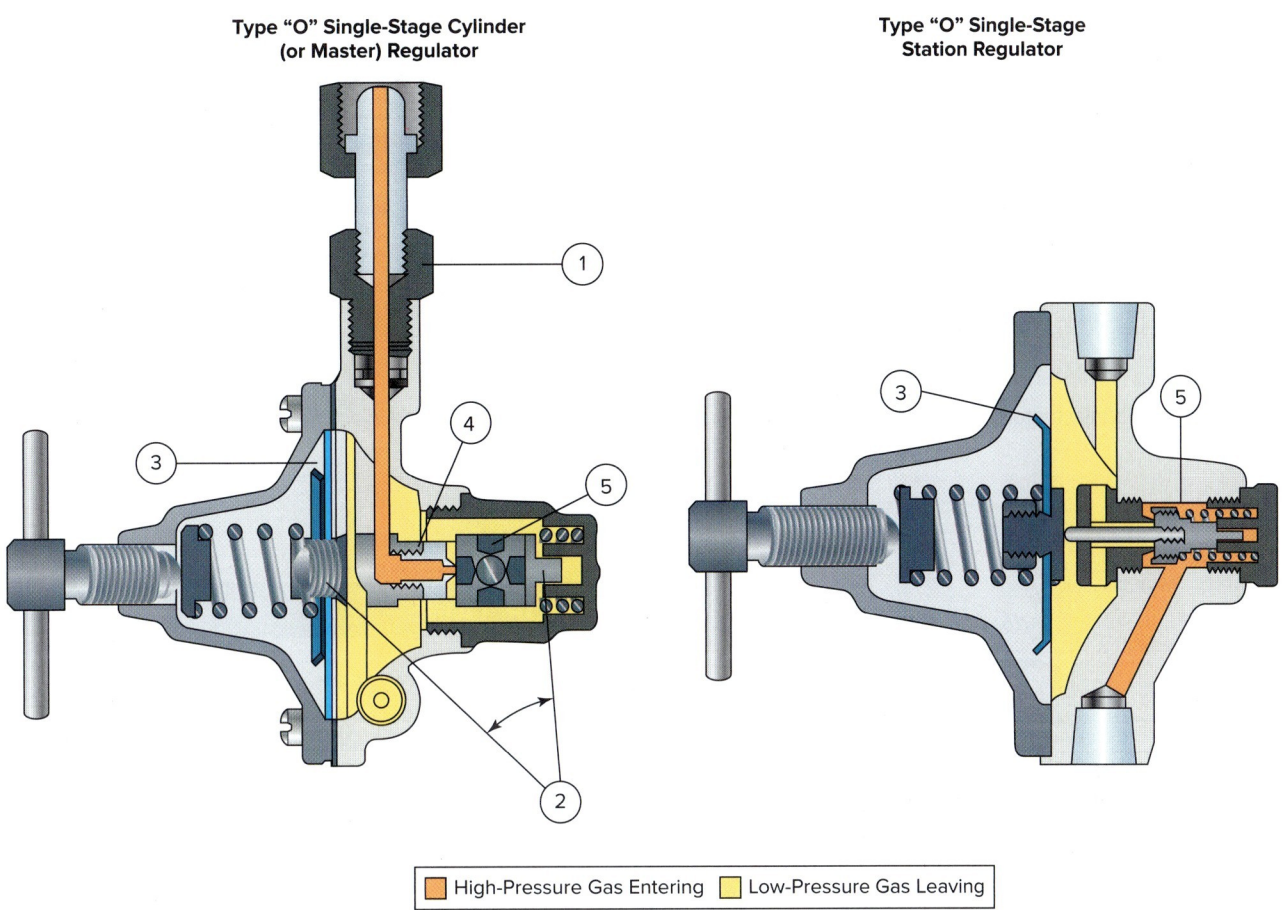

Fig. 5-24 Internal construction of tank and line regulator. There are two major types of regulators. First, the *single-stage cylinder* type is the most used and is shown in the top drawing. (When gases are manifolded, a manifold version of this type is used, called a *master*.) The second kind of regulator, shown underneath the first type, is the *single-stage station* regulator. It is used to reduce the outlet pressure of the master regulator to the pressures commonly used for welding and cutting at the individual stations.

1. Quick replacement cylinder coupling—used on cylinder type regulators only for added strength.
2. Dia-Blok construction—diaphragm and seat are positively connected by means of a yoke, so that both the seat and diaphragm move at the same time—providing long seat life and minimum pressure fluctuation.
3. Diaphragm—stainless-steel diaphragms are used in single-stage cylinder regulators and master regulators, whereas reinforced rubber diaphragms are used in the lower pressure station regulators.
4. Safe-T-Check valve—located in the nozzle of both single-stage cylinder and master regulators. This valve will automatically close if the seat of the regulator is off the nozzle when full cylinder pressure enters. It protects against seat failure.
5. Regulator seat—a multiseat block is used on cylinder type single-stage regulators, which can be rotated for seat change. A single seat is used on the lower pressure single-stage station regulator.

Adapted from Thermadyne Holdings Corporation

they are thoroughly mixed before issuing from the torch tip. Each torch can be supplied with a wide range of welding tip sizes so that a large number of flame types and sizes can be set up for the various thicknesses of metal to be welded.

Two types of oxyacetylene welding torches are in common use: the *injector* and the *equal* (balanced-pressure) types. Figure 5-25, page 154, shows an internal view of the injector torch. The acetylene is carried through the torch and tip at low pressure by the suction force of the higher oxygen pressure passing through the small orifice of the injector nozzle, Fig. 5-26, page 154. The mixing head and injector are usually made as an integral part of the tip, and they are designed to correspond to the various tip sizes. The oxygen pressure used with this type of torch is considerably higher than that used with the equal-pressure torch.

In the equal-pressure (balanced-pressure) torch, Fig. 5-27, page 155, both gases are delivered through the torch to the tip at essentially equal pressures. The mixer

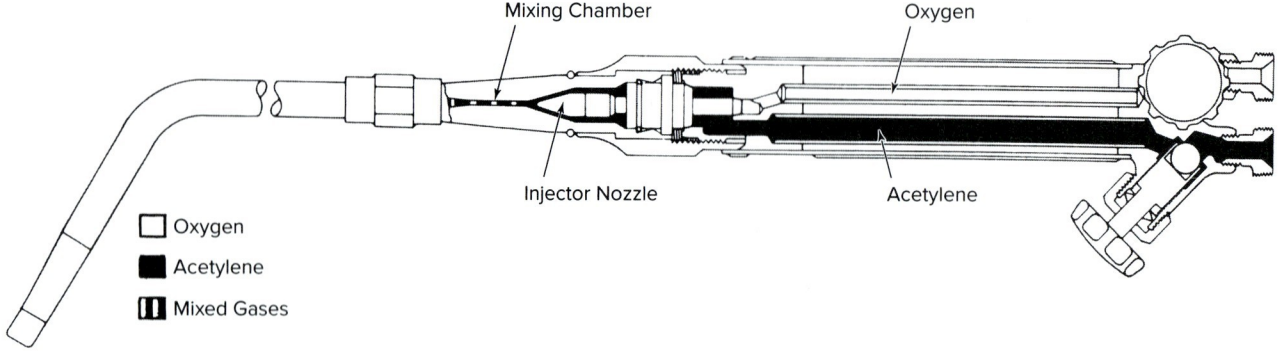

Fig. 5-25 This injector torch is designed for both low and medium pressure operations. The injector is shown in Fig. 5-26.

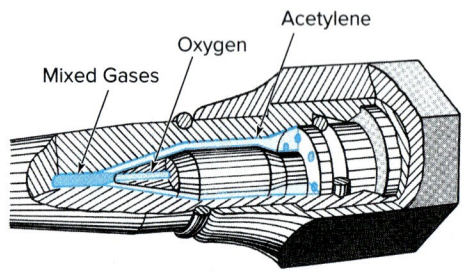

Fig. 5-26 As oxygen issues at relatively high velocity from the tip of the injector, it draws the proper amount of acetylene into the stream. Oxygen and acetylene are thoroughly mixed before issuing from the torch tip.

or mixing head is usually a separate replaceable unit in the body of the torch into which a variety of tips may be fitted. See Fig. 5-28, page 156, for a comparison of the equal-pressure and injector types of mixing chambers. A standard mixer can provide for the variety of tip sizes.

Light gauge sheet metal welding and aircraft welding are usually done with a torch equipped with a smaller handle than the standard torch. This type of welding requires a delicate touch, and the heavier standard torch would be clumsy. An internal view of the balanced-pressure torch and the cutting attachment for light welding is shown in Fig. 5-29, page 157. Note the high pressure oxygen valve (poppet).

For lightweight portable type work, see Fig. 5-30, page 157. One great advantage of the OAW process is that it requires no outside source of power. It is a self-contained system.

Torch Tips Oxyacetylene welding requires a variety of flame sizes. For this reason a series of interchangeable heads or tips of different sizes and types are available. The size of the tip is governed by the diameter of the opening at the end of it. The tip size is marked on the side of the tip. The most common system consists of numbers that may range from 000 to 15. In this system the larger the number, the larger the hole in the tip and the greater the volume of heat that is provided. An increased volume of gas is also required for the flame. Table 5-2 (p. 156) gives information concerning one manufacturer's balanced-pressure torch and tip sizes.

You will recall that tips for injector torches are provided with individual mixers so that the mixer and tip are one unit. Tips for medium- or equal-pressure torches do not have the mixer as part of the tip. Mixers are part of the torch and serve a number of tip sizes. Also available are universal mixers designed to give proper gas mixtures for a full range of tip sizes.

Most welding tips are made of pure drawn copper because of the ability of this material to dissipate heat rapidly. Since copper is soft and tips are subject to considerable wear, both at the tip and threaded ends, tips must be handled carefully. The following precautions should be observed:

- Do not remove a tip with pliers. Pliers make heavy gouge marks on the tip. Manufacturers provide a

> **SHOP TALK**
>
> **Transistorized Advantages**
> Transistorized welding machines have several advantages as advanced welding power sources.
> 1. Stable arc.
> 2. Current, and its shape in pulse mode, can be precisely adjusted.
> 3. Stable against disturbances like voltage fluctuation.
> 4. Arc current and voltage feedback can be controlled from outside (and automated) sensors.

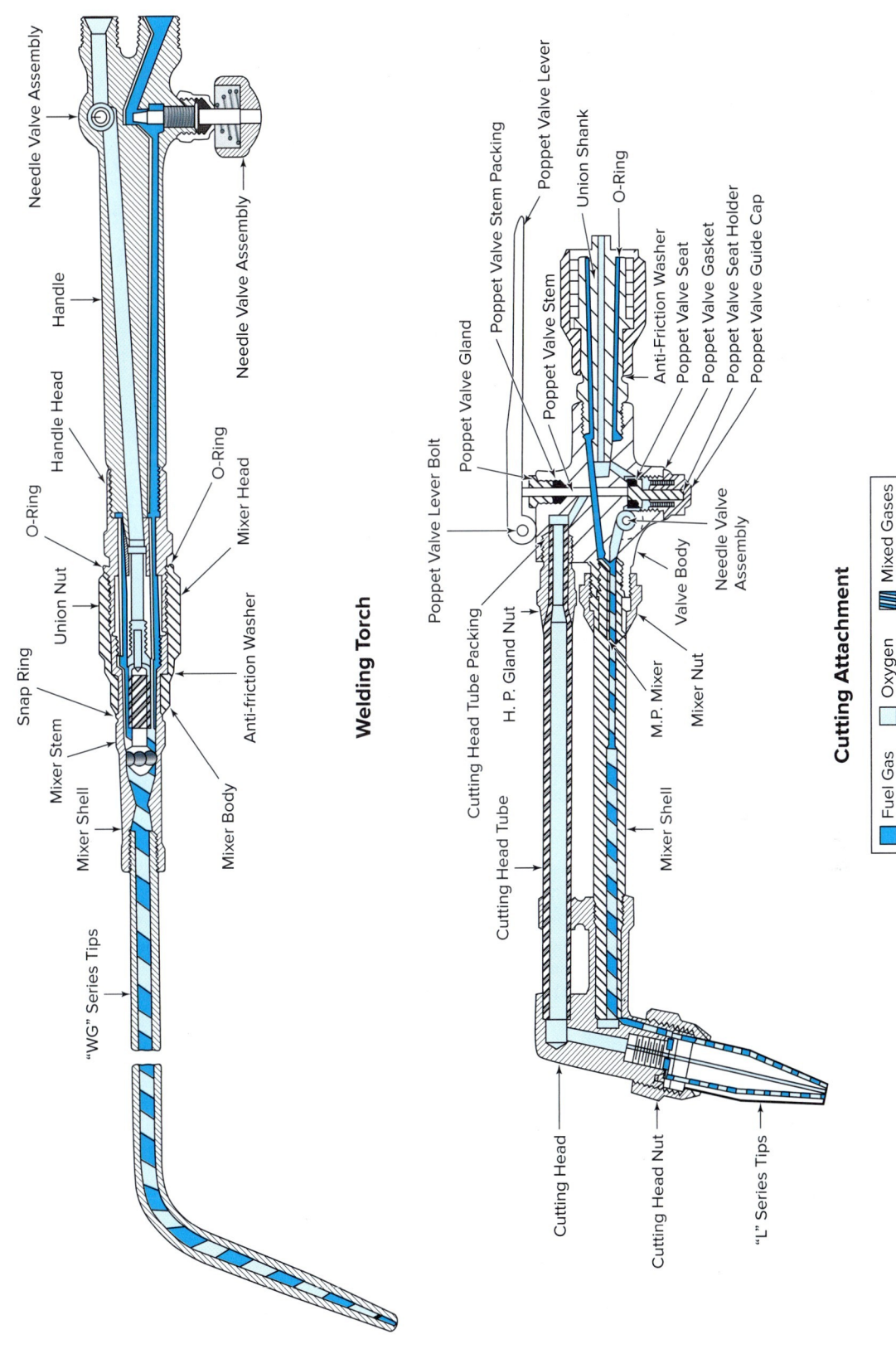

Fig. 5-27 Internal view of balanced-pressure welding torch and cutting attachment. Adapted from Thermadyne Holdings Corporation

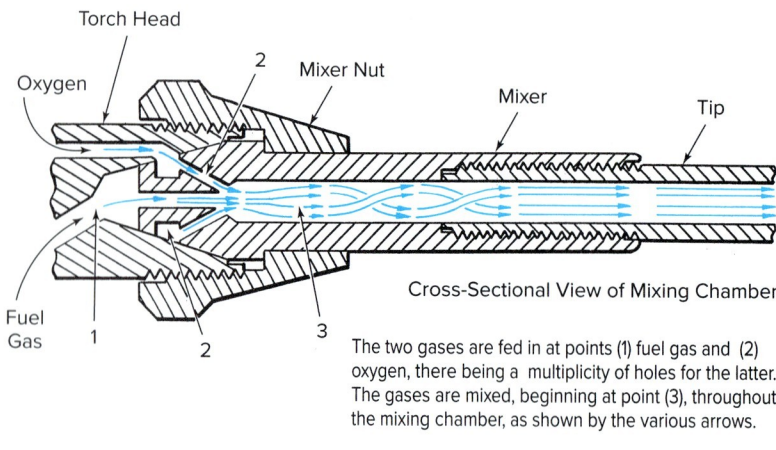

The two gases are fed in at points (1) fuel gas and (2) oxygen, there being a multiplicity of holes for the latter. The gases are mixed, beginning at point (3), throughout the mixing chamber, as shown by the various arrows.

A

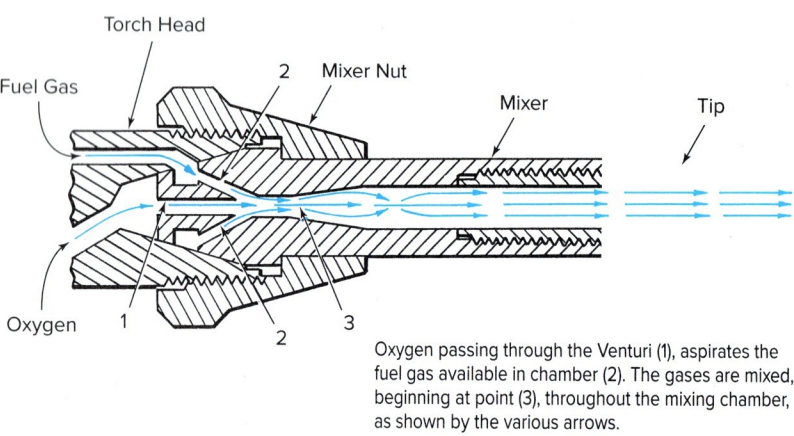

Oxygen passing through the Venturi (1), aspirates the fuel gas available in chamber (2). The gases are mixed, beginning at point (3), throughout the mixing chamber, as shown by the various arrows.

B

Fig. 5-28 Basic elements of welding torch mixers: (A) positive (balanced-pressure) type, (B) injector type.

Table 5-2 Gas Pressures for Different-Sized Welding Tips

Tip No.	Thickness of Metal (in.)	Acetylene Pressure (lb)	Oxygen Pressure (lb)	Oxygen Consumption per Hour	Lineal Ft Welded per Hour
1	1/32	1/2	1/2	7.80	30
2	1/16	1	1	7.90	25
3	3/32	1	1 1/2	8.10	20
4	1/8	1	2	9.75	15
5	3/16	1 1/2	2 1/2	16.80	9
6	5/16	2	2 1/2	26.40	6
7	3/8	3	5	39.35	5
8	1/2	5	8	51.15	4
9	5/8	8	14	69.10	3
10	3/4 & up	10	18	80.00	2

Note: Oxygen consumption per hour was measured with the Hydrex Flow Indicator, with maximum size flame. Information is intended for estimating purposes only and should amply cover adverse conditions.

Source: Modern Engineering Co.

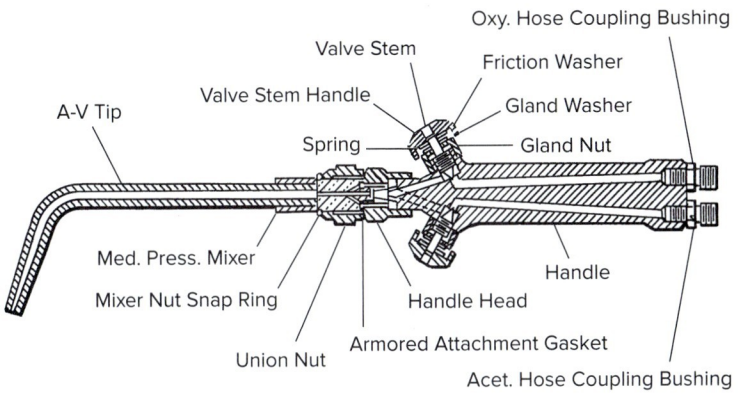

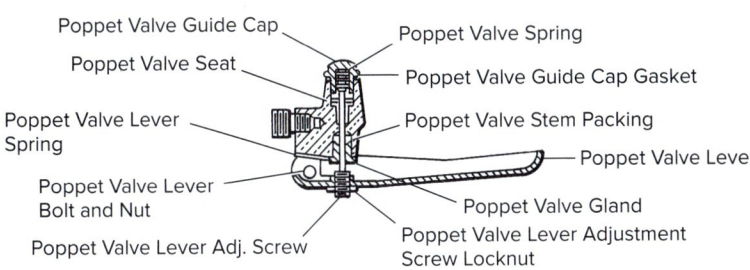

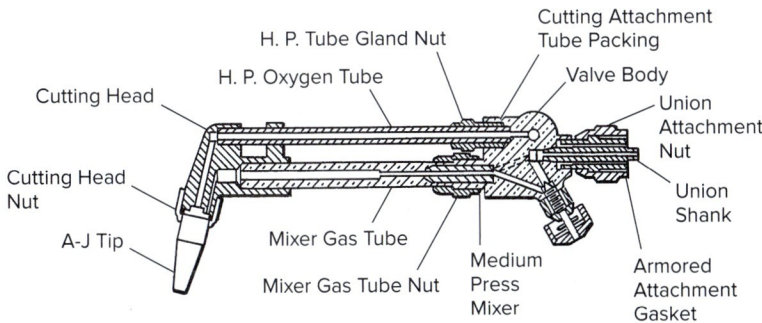

Fig. 5-29 Internal view of the balanced-pressure torch used for light welding. A detachable cutting head is also shown.

Fig. 5-30 Portable oxyfuel setup. Victor Equipment Company/Thermadyne Industries, Inc.

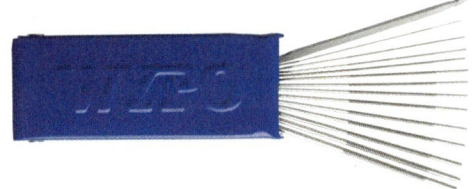

Fig. 5-31 Tip cleaners used to clean out the orifices of welding and cutting tips. David A. Tietz/Editorial Image, LLC

wrench for their tip design that should be used at all times.

- Never insert or remove a tip while the tip tube is hot. Allow the tip and the tip tube to cool first.
- Keep the orifice at the end of the tip clean at all times. During welding, weld spatter, scale, and molten metal may partially close the orifice and cause the welding flame to be very uneven. The tip will also erode unevenly from the heat of the flame. The orifice should be cleaned often with tip cleaners, Fig. 5-31. Do not scratch the tip end on the firebrick or the metal you are welding. Some welders like to use a wood block for cleaning. The block removes contamination on the outside of the tip end, but it does not remove the particles on the inside of the orifice.
- Do not use the tip as a hammer. This is the quickest way to destroy a tip.
- Protect the seat of the tip. If the tip is nicked through dropping or other rough treatment, it will leak at the joint and be dangerous or impossible to use.

Oxygen and Acetylene Hose

The *hose* used for welding and cutting, Fig. 5-32, page 158, is especially manufactured for the purpose. It must be strong enough to resist internal wear, flexible enough so that it does not interfere with the welder's movement, and able to withstand a great deal of abrasive wear on the job. The proper designation hose must be used such as RMACGA Grade T hose for fuel gas (including acetylene) to prevent hose failure. Grade R and RM are for acetylene only.

Hose is usually made of three layers of construction. The inner lining is composed of a high grade of gum rubber. This is surrounded by layers of rubber-impregnated fabric. The outside cover is made of a colored vulcanized rubber that is plain or ribbed to

Gas Welding **Chapter 5** 157

Fig. 5-32 Gas welding hose with fittings attached. It is a ³⁄₁₆-inch diameter. The diameter is determined by the volume of gas required and the length of the hose. Red is the fuel gas hose while green is the oxygen hose. *David A. Tietz/Editorial Image, LLC*

Fig. 5-33 Standard oxygen and acetylene hose connections. Oxygen: right-hand thread; acetylene: left-hand thread. Note groove indicating LH thread on all fuel gas fittings. *Thermadyne Industries, Inc.*

provide maximum life. The oxygen hose is green, and the word *oxygen* is sometimes molded on the hose. The acetylene hose is red, and the word *acetylene* is sometimes molded on the hose.

A *black hose* is used for inert gas and air. Different-sized hoses may be required for different types of welding and cutting operations, depending upon the amount of gas that is required, the length of hose used, and the pressures that are needed. Hoses can be obtained in sizes of ³⁄₁₆, ¼, ⅜, and ½ inch. The ³⁄₁₆ I.D. (inside diameter) hose is very flexible and light; it is used for light welding, such as aircraft welding, and as a whip for pipe welding. The ½-inch I.D. is usually used for heavy cutting. Hose may be single or double. The double hose is actually two pieces joined by a web that prevents tangling the hose.

Hose Connections Only hose connections made for the purpose should be used for connecting the hose to the regulator and the welding and cutting torch. A standard hose connection, Fig. 5-33, consists of a nipple that is inserted in the end of the hose and a nut that attaches the nipple to the torch or regulator. So, there is no danger of attaching the wrong hose to the wrong regulator or torch connection, the oxygen coupling has a right-hand thread, and the acetylene coupling has a left-hand thread. The hose connection nuts are marked *STD. OXY* for the oxygen and *STD. ACET* for the acetylene. In addition, the acetylene nuts have a groove cut around their center to indicate a left-hand thread. Clamps or ferrules connect the hose tightly to the nipple to ensure a leakproof connection.

Care of Hose The welder must use and care for the hose correctly in the interest of both economy and safety. Hose is subject to a great deal of wear even under normal conditions; and if it is abused, the wear is considerable. The welder is urged to take the following precautions:

- Always use a hose to carry only one kind of gas. A combustible mixture may result if it is used for first one gas and then another.
- Test the hose for leaks frequently by immersing the hose at normal working pressure under water. A leaking hose is a serious hazard and a waste of gas. If there is a leak in the connection with the proper nipple and nut, cut off the hose a few inches back and remake the connection. Leaks in other locations should be repaired by cutting off the bad section and inserting a hose coupling as a splice.
- Clamp all of the hose connections or fasten them securely so that they will withstand a pressure of at least 300 p.s.i. without leakage.
- Hose showing leaks, burns, worn places, or other defects are unfit for service and must be repaired or replaced.
- Do not attempt to repair a hose with tape. Tape has a tendency to break down the hose material, and it is not a permanent repair.
- Handle the hose carefully when welding. Avoid dragging it on a greasy floor. The hose should not be allowed to come in contact with flame or hot metal. The hose should be protected from falling articles, from vehicles running over it, and from being stepped on. It should not be kinked sharply.
- At the end of the day or at the end of the job, roll up the hose and hang it where it will be out of the way. Spring-loaded hose reels are available.

Lighters

The welding torch should be lighted with a *friction spark lighter,* Fig. 5-34. The flints for friction lighters can be

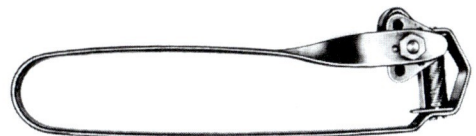

Fig. 5-34 Triple-flint torch friction lighter.
Thermadyne Industries, Inc.

easily replaced at small cost when worn out. Matches should never be used because the welder's hand has to be too close to the torch tip and may be burned when the gases ignite. If the welder carries matches in their pockets, there is also the danger that they will ignite during welding and cause severe burns.

Filler Rod

While a great deal of oxyacetylene welding or autogenous welding is done by merely fusing the metal edges together, most gas welding is done with the addition of a filler rod. The filler rod provides the additional metal necessary to form a larger weld bead. Filler rods are available for the welding of mild steel, cast iron, stainless steel, various brazing alloys, and aluminum. The usual rod length is 36 inches, and the diameters are $\frac{1}{16}$, $\frac{3}{32}$, $\frac{1}{8}$, $\frac{5}{32}$, $\frac{3}{16}$, $\frac{1}{4}$, $\frac{5}{16}$, and $\frac{1}{8}$ inch. Welding rods are available in bundles of 50 or 100 pounds net weight and in boxes of 10, 50, 100, and 300 pounds net weight.

Steel rods are copper coated to keep them from rusting. The copper-coated rods are triple deoxidized GTAW filler, usually an ER70S-2, ER70S-3, or ER70S-6 class. RG-65, RG-60, and RG-45 are typically uncoated. Some types of aluminum rods are flux coated to improve their working characteristics. Both steel and aluminum rods are 28 inches in length.

The American Welding Society has set up the following AWS classification numbers for steel gas-welding rods: RG-65, RG-60, and RG-45. The letter R indicates a welding rod and the G indicates that it is used with gas welding. The numbers designate the approximate tensile strength of the weld metal produced in thousand pounds per square inch. For example, 45 designates a rod with a tensile strength of approximately 45,000 p.s.i.

Rod Characteristics *Gas welding rods* produce welds of varying tensile strengths depending upon the nature of the base metal. Welds made on alloy steels will produce weld composition between that of the base metal and the filler metal. Gas welding rods can be used in all positions, limited only by the skill of the welder.

Class RG-65 RG-65 gas welding rods are of low alloy steel composition and may be used to weld sheet, plate, tubes, and pipes of carbon and low alloy steels. They produce welds in the range of 65,000 to 75,000 p.s.i.

Class RG-60 RG-60 gas welding rods are of low alloy composition and may be used to weld carbon steel pipes for power plants, process piping, and other severe service conditions. They produce welds in the range of 50,000 to 65,000 p.s.i. This type of welding rod is used extensively and is considered a general-purpose welding rod. It produces highly satisfactory welds in such materials as carbon steels, low alloy steels, and wrought iron.

Class RG-45 RG-45 gas welding rods are general-purpose welding rods of low carbon steel composition. They may be used to weld mild steels and wrought iron, and they produce welds in the range of 40,000 to 50,000 p.s.i.

Fluxes

A **flux** is a cleaning agent used to dissolve oxides, release trapped gases and slag, and cleanse metal surfaces for welding, soldering, and brazing.

The oxides of all commercial metals, except steel, have higher melting points than the metals themselves and do not flow away readily. The function of the flux is to combine with oxides to form a fusible slag having a melting point lower than the metal so that it flows away from the weld area. Since there is a wide variation in the chemical characteristics and the melting points of the different oxides, there is no one flux that is satisfactory for all metals.

The melting point of a flux must be lower than that of either the metal or the oxides formed so that it will be liquid. The ideal flux has exactly the right fluidity at the welding temperature so that it can blanket the molten metal to protect it from atmospheric oxidation. It must remain close to the weld area instead of flowing all over the job.

Fluxes are available as dry powders, pastes, thick solutions, and coatings on filler rod. Powdered fluxes are packed in tin cans or glass jars. When not in use, they should be stored in a closed container because they lose their welding properties if exposed too long to the atmosphere.

Fluxes differ in their composition and the way in which they work according to the metals with which they are to be used. In cast iron welding, slag forms on the surface of the pool, and the flux breaks the slag up. In welding aluminum, the flux combats the tendency for the heavy slag to mix with the melted aluminum and weaken the weld. Flux will be explained in more detail in connection with the welding of those metals that require the use of a flux.

Other Gas Welding Processes

Oxyhydrogen Welding Oxyhydrogen welding (OHW) is a form of gas welding that was once used extensively. Today it has only a limited use.

The oxyhydrogen flame is produced by burning two volumes of hydrogen (H_2) with one volume of oxygen (O_2). The result is a flame with a temperature of approximately 4,100°F. The flame is almost invisible, making it somewhat difficult to adjust the welding torch.

The equipment necessary is very similar to that used for oxyacetylene welding. The same torches, mixers, tips, and hose are used for both. There is some difference in the regulators. A standard oxygen regulator is used on the oxygen cylinder. A regulator specifically designed for use with hydrogen must be used with the hydrogen cylinder.

Because of the relatively low flame temperature, the process is used principally in the welding of metals having low melting points, such as aluminum, magnesium, and lead. It is also used to a limited extent in the welding of very light gauge steel and brazing operations.

The oxyhydrogen flame is still used extensively in the welding of lead. The lower flame temperature is ideal for lead because it has a low melting point. Oxyhydrogen is used for welding thicknesses of lead up to ¼ or ⅛ inch. For greater thicknesses, the oxyacetylene flame is generally used because of the greater heat input required. Another advantage of oxyhydrogen welding is that there is no deposit of carbon, which accelerates corrosion in welded assemblies.

Fuel Gases Standard oxyacetylene welding equipment can also be used with propane, butane, city gas, and natural gas. It is necessary to have a special fuel gas regulator for use with these gases. Suitable heating and cutting tips are available in a variety of sizes. City gas and natural gas are supplied by pipelines, while propane and butane are stored in cylinders or delivered in liquid form to storage tanks on the user's property.

Because of the oxidizing nature of the flame and the relatively low flame temperature, these gases are not suitable for welding ferrous materials. They are used extensively for both manual and mechanized brazing and soldering operations. The plumbing, refrigeration, and electrical trades use propane in small cylinders for many heating and soldering applications. The torches are designed to be used with air as the combustion-supporting gas.

Air-Acetylene Welding (AAW) Fuel gas burned with air has a lower flame temperature than that obtained when the same gas is burned with oxygen. You will recall that oxygen mixed with a fuel gas produces the hottest flame temperature. Air contains approximately ⅘ nitrogen by volume, which is neither a fuel gas nor a supporter of combustion. Thus, acetylene burned with air produces lower flame temperatures than the other gas combinations. The total heat content is also lowered.

Torches for use with air-acetylene are generally designed to draw in the proper quantity of air from the atmosphere to provide combustion. The acetylene flows through the torch at a supply pressure of 2 to 15 p.s.i. and serves to suck in the air. For light work, the acetylene is usually supplied from a small cylinder that is easily transportable.

The air-acetylene flame is used for welding lead up to approximately ¼ inch in thickness. The greatest application is in the plumbing and electrical industry, where it is used extensively for soldering and brazing copper tubing with sweat-type joints.

Supporting Equipment

The welding shop should be equipped with a great deal of equipment that is needed for the preparation of the work and is necessary to the welding process:

- A welding table with either a cast iron top, slotted to permit the use of hold-down clamps, or a firebrick top, Fig. 5-35.
- C-clamps, carpenter clamps, various other types of clamps, straightedges, metal blocks, V-blocks, and a steel square for holding and lining up parts.
- Grinders, air-chisels, files, and hand chisels for beveling plate.

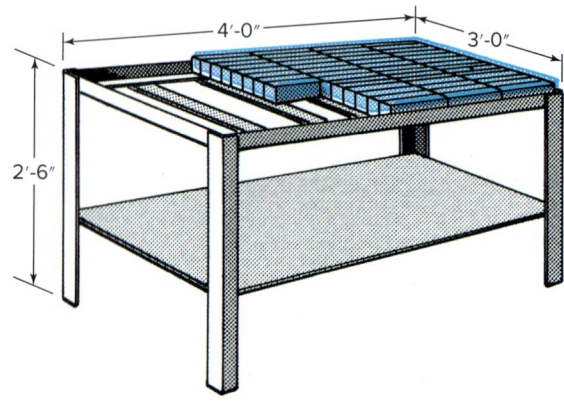

Fig. 5-35 Welding table: angle iron construction with welded joints and firebrick top.

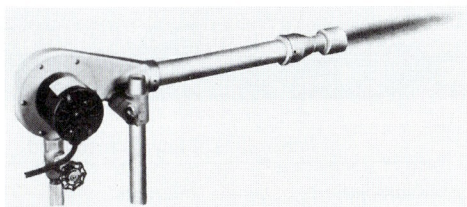

Fig. 5-36 A typical gas-electric blowtorch used for preheating. North American Manufacturing Co.

> **ABOUT WELDING**
>
> **Sensing Devices**
> Automatic sensing devices can control joint alignment, so the welder does not have to constantly adjust the equipment controls.

- A cutting torch, gas or electric, for beveling or for repair work.
- Carbon, in the form of rods, plates, or paste. It is highly fire-resistant and is useful to protect surfaces and holes, to back up welds, to control and shape the flow of metal, and to support and align broken parts.
- Preheating equipment, Fig. 5-36, and a number of materials to provide for slow cooling, such as fiber cement, hydrated lime, and other commercially available materials.
- A wire power brush for cleaning scale and slag.
- Some jobs may need no finishing of any kind, while others may require filing, grinding, drilling, or even considerable machining. The welder should know the use of the part and the type of finishing that is going to be done before the job is welded. With this information, a good welder can keep the amount of excess metal deposited close to the finish requirements, thus keeping the amount of postwelding work to a minimum.

Safety Equipment

Welding and Cutting Goggles

Welders must wear specially designed goggles to protect their eyes from infrared and ultraviolet rays, flying sparks, and particles of hot metal, Fig. 5-37. The eyes must also be protected from the heat coming from the job. Just any kind of colored lens does not give the necessary protection.

Filter lenses are made of special optical glass of various diameters and tinted green or brown. These lenses not only filter out the harmful rays, but they also minimize the effect of glare to permit the welders to see their work clearly.

The shade of the lens must be selected carefully. Lenses can be obtained in light, medium, and dark shades, Table 5-3 (p. 162). Lenses that are too dark cause eye strain because welders are not able to see their work clearly. Lenses that are too light cause the eyes to suffer the effects of light and heat.

The outer lens is a clear glass or plastic of optical quality and $3/64$ to $1/16$ inch thick. This cover lens protects the filter lens from spatter. When the outer lens becomes pitted and reduces vision, it should be replaced. Treated lenses that have longer life than the untreated type can be purchased.

Goggle frames are usually made of a tough, heat-resistant material similar to Bakelite. They should be light in weight and fit the face so that they are comfortable. Although they should provide adequate ventilation, it is important that they do not leak light.

It is foolhardy to attempt to save a few cents by purchasing an inferior grade of goggles. Purchase only the best. Never use oxyacetylene welding and cutting goggles for welding with the electric arc.

Some welders prefer to use eye shields, Fig. 5-38, page 162. These shields also have the clear cover lens and the colored filter lens. The lenses are the same size as those used in the arc welding helmets. Eye shields provide a wide range of vision and can be used over eyeglasses. They may be fitted with a headband and work on a swivel.

It is also possible to obtain eyeglass-type frames that have the correct lenses. Their use is not recommended for welding or cutting because they do not give any protection from the sides. Thus the eyes are exposed to injurious rays, glare, sparks, and hot particles of metal. They may

Fig. 5-37 Deep cup oxyacetylene welding goggles. This type can be worn over eyeglasses. Thermadyne Industries, Inc.

Table 5-3 The Proper Shade of Welding Lens to Use for Different Types of Welding and Cutting and the Percentage of Rays Transmitted

Shade	Recommended Uses	Percentage of Rays Transmitted		
		Noninjurious Visible Rays	Injurious Infrared	Injurious Ultraviolet
2	Reflected glare and low temperature furnace work	28.0%	0.87%	1.075%
3	Light brazing and lead burning	16.0	0.43	1.035
4	Acetylene burning and brazing	6.5	None	0.097
5	Light acetylene welding and cutting	2.0	None	0.046
6	Standard shade for acetylene gas welding	0.8	None	None
8	Heavy acetylene welding, electric arc cutting and welding up to 75 amperes (A)	0.25	None	None
10	Electric arc cutting and welding between 75 and 250 A	0.014	None	None
12	Electric arc cutting and welding above 250 A	0.002	None	None
14	Carbon arc cutting and welding	0.0003	None	None

Fig. 5-38 This multi-purpose face shield provides uninterrupted protection for grinding, brazing, welding, and cutting operations. Sellstrom Manufacturing Company

be used by inspectors or onlookers since they are usually a safe distance away from the work.

Protective Clothing and Gloves

Welders must always remember that they are working with fire and that they should avoid any clothing that is highly flammable. Sparks, molten bits of metal, and hot scale are hazards. Under no circumstances should a sweater be worn. The body should be protected by an apron, a shop coat, or coveralls that resist fire. The head and hair should be protected by a welder's cap. For overhead welding and cutting, ear protection is desirable. Avoid wearing low-cut shoes and clothing with cuffs and open pockets.

All welders wear the same kinds of clothing on the job: jeans (without cuffs), heavy shirts, high-top shoes, and the necessary protective clothing.

The heat coming from the welding job may be very intense. There will also be a shower of sparks, hot material to handle, and a hot welding torch to hold. This makes it necessary for welders to protect their hands with gloves. Gloves should be made of nonflammable material. However, for light welding jobs it is the common practice to wear an ordinary canvas glove with a cuff. Gloves must be kept free from grease and oil because of the danger involved in contact with oxygen.

Oxyfuel cutting and heating torches are subject to three types of phenomena: backfires, sustained backfires, and flashbacks. It is important to understand these terms and be able to identify these reactions if they occur. Chapters 7 and 8 cover this topic in more detail.

When there is danger of sharp or heavy falling objects and when the working space is confined, hard hats or head protectors should be worn.

Flash Guard Check Valves

Check valves prevent the reverse flow of mixed gases in torch hoses or regulators. One valve is required for the acetylene hose and one for the oxygen hose. They may be attached to the torch or to the regulators.

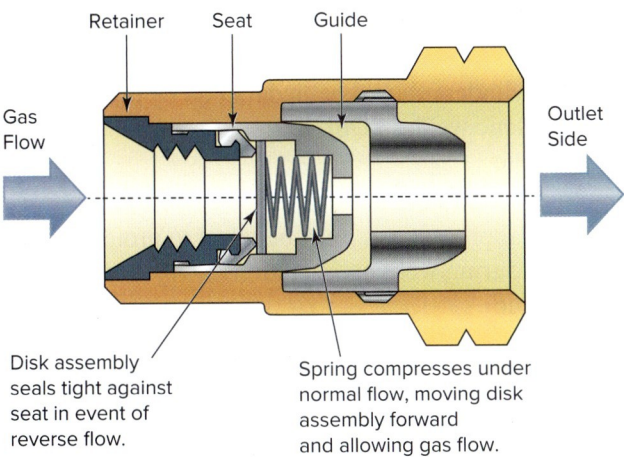

Fig. 5-39 Check valve for an oxygen-fuel system.
Adapted from Thermadyne Holdings Corporation

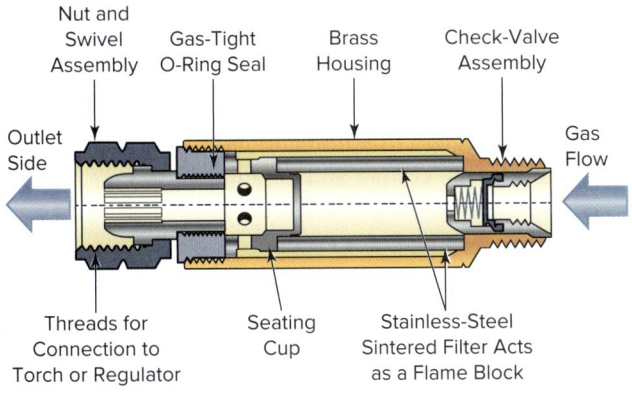

Fig. 5-40 Flashback arrestor for an oxygen-fuel system.
Adapted from Thermadyne Holdings Corporation

Some models fit internally into the torch. The check valve permits forward flow of gas and closes when gas begins to flow in a reverse direction. The forward flow causes a disk to move overcoming a slight spring tension. However, a reverse flow would cause the disk to seat tightly against a seal, preventing any backflow. Figure 5-39 shows the internal working of the check valve. This reverse flow can be caused by a blocked torch tip, excess gas or oxygen pressure, lack of pressure, or unsafe start-up or shut-down procedures.

Flash arrestors are generally made of a sintered metal alloy. The sintered stainless-steel filter prevents flame from moving upstream of the arrestor. Figure 5-40 is the internal working of a flashback arrestor. Neither of these devices protects the torch or tip. The best way to ensure that accidents don't happen is for the welder to follow all safety operating procedures. However, even safety equipment, if used improperly, can create hazardous conditions, which increase the risk of accidents and/or injury.

General Safety Operating Procedures

Protective equipment, hose, and regulators:

1. Equipment shall be installed and used only in the service for which it is approved and as recommended by the manufacturer.
2. Approved protective equipment shall be installed in fuel-gas piping.
3. Hose for oxyfuel gas service shall comply with the Specification for Rubber Welding Hose, 1958, Compressed Gas Association and Rubber Manufacturers Association.
4. When parallel lengths of oxygen and acetylene hose are taped together for convenience and to prevent tangling, not more than 4 out of 12 inches shall be covered by tape.
5. Hose connections shall comply with the Standard Hose Connection Specifications, 1957, Compressed Gas Association.
6. Hose connections shall be clamped or otherwise securely fastened in a manner that will withstand, without leakage, twice the pressure to which they are normally subjected in service, but in no case less than a pressure of 300 p.s.i. Oil-free air or an oil-free inert gas shall be used for the test.
7. Hose showing leaks, burns, worn places, or other defects rendering it unfit for service shall be repaired or replaced.
8. Pressure-reducing regulators shall be used only for the gas and pressures for which they are intended. The regulator inlet connections shall comply with Regulator Connection Standards, 1958, Compressed Gas Association.
9. When regulators or parts of regulators, including gauges, need repair, the work shall be performed by skilled mechanics who have been properly instructed.
10. Gauges on oxygen regulators shall be marked *Use No Oil*.
11. Union nuts and connections on regulators shall be inspected before use to detect faulty seats that may cause leakage of gas when the regulators are attached to the cylinder valves.

More information about equipment can be found in Chapter 26.

CHAPTER 5 REVIEW

Multiple Choice

Choose the letter of the correct answer.

1. What type of fuel did Egyptians, Greeks, and Romans use centuries ago for fusing metal? (Obj. 5-1)
   a. Acetylene
   b. Hydrogen
   c. Alcohol or oil
   d. Natural gas

2. The hottest flame gas known is oxyacetylene, which has a maximum temperature of _____ °F. (Obj. 5-2)
   a. 1,550
   b. 3,280
   c. 6,300
   d. 9,260

3. Oxyacetylene welding can only be used to weld basic metals such as steel and cast iron. (Obj. 5-2)
   a. True
   b. False

4. Name the gas that should be used for welding purposes. (Obj. 5-2)
   a. Acetylene
   b. Mapp®
   c. Propane
   d. Natural gas

5. How is acetylene produced? (Obj. 5-2)
   a. Chemical reaction of calcium carbide and water
   b. Chemical reaction in an acetylene generator
   c. From a chemical that is a byproduct of smelting coke and water
   d. All of these

6. How is oxygen distributed? (Obj. 5-2)
   a. Gaseous form in cylinders
   b. Liquid form in cylinders
   c. Portable oxygen generators extracting from tap water
   d. Both a and b

7. Acetylene is distributed in cylinders, which are hollow vessels that can safely contain the gas at any pressure. (Obj. 5-2)
   a. True
   b. False

8. Cylinders containing liquids must be used in the upright position. (Obj. 5-2)
   a. True
   b. False

9. Safety procedures for handling, storing, and using gas cylinders should be understood by every oxyacetylene welder. (Obj. 5-3)
   a. True
   b. False

10. Special precautions must be taken to protect the body during welding operations. (Obj. 5-3)
    a. True
    b. False

11. The OSHA regulation covering oxygen-fuel gas welding and cutting is _____. (Obj. 5-3)
    a. 2000.007
    b. 1947.619
    c. 1910.253
    d. 1776.059

12. The device used to lower the cylinder pressure to a safe working pressure is the _____. (Obj. 5-3)
    a. Cylinder valve
    b. Regulator
    c. Torch
    d. None of these

13. What device must be installed in the oxyacetylene welding system to eliminate the flame from going back into the system and creating a hazard? (Obj. 5-4)
    a. Flash arrestor
    b. Check valve
    c. Gas hose
    d. Both a and b

14. Regular sun glasses are adequate to wear for OAW. (Obj. 5-4)
    a. True
    b. False

15. The two gauges on a regulator indicate _____. (Obj. 5-4)
    a. Atmospheric pressure
    b. Cylinder pressure
    c. Working pressure
    d. Both b and c

16. Which regulator is more effective in maintaining a consistent flame? (Obj. 5-4)
    a. Single stage
    b. Dual stage

17. The regulator connection to the cylinder valve should be in compliance with _____. (Obj. 5-4)
    a. OSHA section 1107
    b. Compressed Gas Association Standard, 1958

c. NEMA section 1
   d. AWS 3.0
18. Fuel-gas fittings have left-hand threads so they cannot be interchanged with an oxygen fitting. How can you generally tell by looking at a fitting if it has LH threads? (Obj. 5-4)
   a. It is color coded.
   b. It is made of steel.
   c. It has a groove machined into it.
   d. All of these.

Review Questions

Write the answers in your own words.

19. Name four gases used in various forms of gas welding. (Obj. 5-2)
20. What is the approximate temperature of the oxyacetylene flame? (Obj. 5-2)
21. List four precautions in the use and care of welding tips. (Obj. 5-3)
22. Is it absolutely necessary that the eyes of the welder be protected while performing the welding operation? Explain. (Obj. 5-3)
23. What kinds of protective clothing are worn for oxyacetylene welding? (Obj. 5-3)
24. List the three main functions of a torch. (Obj. 5-4)
25. What does the RG-60 filler material classification designate? (Obj. 5-4)
26. Explain the difference in the design of the balanced pressure torch and the injector torch. (Obj. 5-4)
27. Explain the function of the pressure regulators. (Obj. 5-4)
28. Explain the function of the gas economizer. (Obj. 5-4)

INTERNET ACTIVITIES

Internet Activity A

Use the American Welding Society web site. (Remember, AWS is an organization.) Click on the links to see what might be of interest to you and write a report.

Internet Activity B

Look up oxyacetylene torch tips on the Internet. Choose three of the torch tips and describe the kind of jobs for which they are used.

Design credits: Cinema and movie icon: Denis Meshkov/123RF; and Illustration of Welding icons: Mr. Nachai Sorasee/123RF

6

Flame Cutting Principles

In the majority of fabricating shops and in the field, the welder must be able to do manual flame cutting (Fig. 6-1). The oxyfuel gas cutting (OFC) torch has become a universal tool. OFC is widely used for straight-line shape cutting. It is also used as a means of scrapping obsolete metal structures. OFC devices can be used to fabricate metal structures (Fig. 6-2).

Cutting processes have made it possible to fabricate structures requiring heavy thicknesses of metal from rolled steel. Formerly, these structures had to be cast. The combination of cutting and welding processes created an industry devoted to the fabrication of heavy machinery and equipment from rolled steel. Oxyfuel gas cutting increases the speed of fabrication and eliminates many costly joining, shaping, and finishing operations.

Oxyacetylene and Other Fuel Gas Cutting

Oxyacetylene cutting (OFC-A) is limited to the cutting of ferrous materials (materials containing iron). Stainless steel, manganese steels, and nonferrous materials are not readily cut because they do not oxidize rapidly. Many of these materials may be cut by the arc or water jet process. Most ferrous materials have an affinity for oxygen. Even under normal conditions, the oxygen in the

Chapter Objectives

After completing this chapter, you will be able to:

- **6-1** Describe oxyfuel cutting principles.
- **6-2** Identify oxyfuel cutting equipment.
- **6-3** Explain the safety issues of oxyfuel cutting.
- **6-4** Describe various support equipment for oxyfuel cutting.
- **6-5** Identify various cutting techniques.
- **6-6** Describe the oxygen lance cutting process.

Fig. 6-1 A construction worker using an OFC torch to scrap an obsolete riveted structure. Iowefoto/Alamy Stock Photo

Fig. 6-2 Cutting bar stock which is part of the structural steel on a building construction job. Johner Images/Getty Images

air attacks these materials to form an iron oxide that we recognize as rust. Thus, the rusting process is a slow form of oxygen cutting. At elevated temperatures, the oxidation process is increased. Oxygen cutting requires that the part to be cut be raised to a temperature of 1,500 to 1,600°F (usually a cherry red). A stream of pure oxygen is directed onto the hot metal causing it to burn rapidly.

Steel burns in pure oxygen after having reached its kindling temperature just as paper burns in air. The main difference is that burning paper gives off carbon dioxide and water vapor. These products of combustion are gaseous and pass off into the air. When steel burns, however, it gives off iron oxide which is solid at room temperature. Its melting point is below the melting point of steel. The heat generated by the burning iron and the oxyacetylene flame is high enough to melt the iron oxide so that it runs off as molten dross, exposing more iron to the oxygen jet. Thus, the jet can be moved along to produce a clean cut.

Cutting is really a process of very rapid rusting. It is thus considered a chemical cutting process, unlike the arc and water jet cutting that uses a mechanical method of materials removal.

The oxyfuel gas cutting process makes use of several other fuel (flammable) gases in addition to acetylene. These fuel gases include propane, natural gas, propylene, and Mapp gas. Refer to Table 6-1 for a review of the characteristics of these gases and their distribution.

Equipment

Oxyfuel gas cutting requires the following equipment:

- Single-purpose cutting torch, Figs. 6-3 and 6-4, or a welding torch to which an adaptable cutting head has been attached
    - Flint lighter
    - Oxygen regulator
    - Acetylene regulator
    - Oxygen welding hose with couplings attached to each end
    - Acetylene welding hose with couplings attached to each end
    - Flame arresters and check valves

> **ABOUT WELDING**
>
> **ASME Code**
> Welding and brazing qualifications are a crucial part of the ASME code now called the Boiler and Pressure Vessel Code. This code, which began in 1915, explains what an operator's qualifications must be and proper procedures. It also provides requirements for new construction of pressure-related items so that you will be safer when you use them.

> **JOB TIP**
>
> **Experience Required**
> The American Welding Society has a certificate that shows prospective employers a certain skill level that has been achieved. It can be a ticket to that first interview.

Table 6-1 Properties of Common Fuel Gases

	Acetylene	Propane	Propylene	Methylacetylene Propadiene (MPS)	Natural Gas
Chemical formula	C_2H_2	C_8H_8	C_3H_6	C_3H_4 (Methylacetylene, propadiene)	CH_4 (Methane)
Neutral flame temperature, °F	5,600	4,580	5,200	5,200	4,600
Primary flame heat emission, Btu/ft^3	507	255	433	517	11
Secondary flame heat emission, Btu/ft^3	963	2,243	1,938	1,889	989
Total heat value (after vaporization), Btu/ft^3	1,470	2,498	2,371	2,406	1,000
Total heat value (after vaporization), Btu/lb	21,500	21,800	21,100	21,100	23,900
Total oxygen required (neutral flame), vol. O_2/vol. fuel	2.5	5.0	4.5	4.0	2.0
Oxygen supplied through torch (neutral flame), vol. O_2/vol. fuel	1.1	3.5	2.6	2.5	1.5
ft^3 oxygen/lb fuel (60°F)	16.0	30.3	23.0	22.1	35.4
Maximum allowable regulator pressure, p.s.i.	15	150	150	150	Line
Explosive limits in air, %	2.5–8.0	2.3–9.5	2.0–10	3.4–10.8	5.3–14
Volume-to-weight ratio, ft^3/lb (60°F)	14.6	8.66	8.9	8.85	23.6
Specific gravity of gas (60°F) Air = 1	0.906	1.52	1.48	1.48	0.62

Modified from the American Welding Society, *Welding Handbook*, Vol. 2, 8th ed., p. 454, Table 14.1.

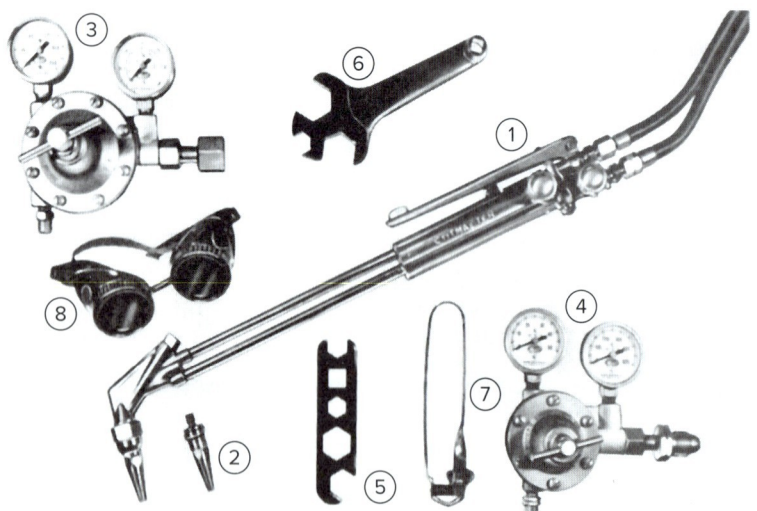

Fig. 6-3 Equipment for oxyacetylene and other fuel gas cutting. 1. Cutting torch. 2. Cutting tip. 3. Oxygen regulator. 4. Acetylene regulator. 5. Torch wrench. 6. Tank and regulator wrench. 7. Lighter. 8. Goggles. Thermadyne Industries, Inc.

Portable tank outfits are mounted on a truck similar to that shown in Fig. 5-21, page 151. Review pages 148–160 in Chapter 5 concerning the regulators, hose, and other equipment.

Cutting Torch The **cutting torch**, Fig. 6-4A, premixes oxygen and acetylene or other fuel gases in the proportions necessary for cutting. The oxygen furnished to the preheating flame is regulated by a preheat valve on the side of the handle. A high pressure oxygen valve operated by a lever controls the oxygen.

On some torches, the preheating oxygen and acetylene mix in the cutting tip. The cutting torch, Fig. 6-4B, tip mixes oxygen and acetylene of other fuel gases in the proportions necessary for cutting.

The cutting head may be at a 45 or 90° angle to the body of the torch. Torches with straight heads may also be obtained.

Figure 6-5 illustrates the internal construction of a standard cutting torch.

On some jobs, such as pipe welding and repair work, cutting is a small part of the welding operation. For such jobs, the adaptable cutting attachment is used for cutting. In a few seconds, it can be attached to the welding torch without disconnecting the hose from the torch. The construction and operation of the cutting attachment is similar to the type of cutting torch in which the preheating gases mix in the handle.

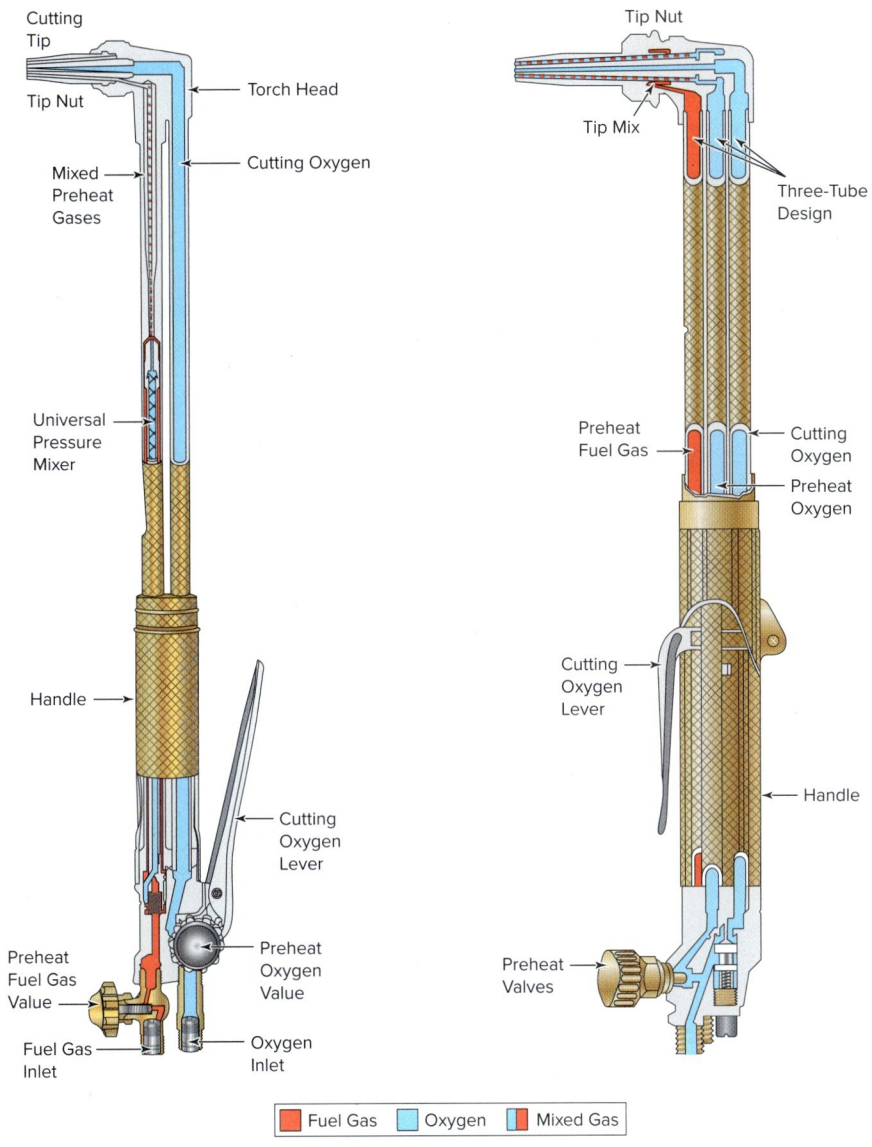

Fig. 6-4A Typical premixing-type cutting torch. Adapted from *Welding Handbook,* Vol. 2, 8th ed., p. 451, Figure 14.2

Fig. 6-4B Typical tip mixing-type cutting torch. Adapted from *Welding Handbook,* Vol. 2, 8th ed., p. 451, Figure 14.2

Cutting tips are designated as standard or high speed. The *standard tip* has a straight bore cutting oxygen port and is typically used with oxygen pressures in the 30 to 60 p.s.i. range, while the *high speed tips* have a diverging cutting oxygen port that flares out toward the opening. This flaring out allows much higher oxygen pressure (60–100 p.s.i.), while maintaining a uniform oxygen jet at supersonic velocities. These high speed tips are typically used only for machine cutting and will yield an increase in travel speed of 20 percent over standard tips, Fig. 6-8 (p. 172). The cutting oxygen orifice size is not usually affected by the type of fuel gas being used. However, the preheat orifices need to be of the appropriate design for the type of fuel gas being used. The various fuel gases require different volumes of oxygen and fuel as shown in Table 6-2. Tips used for acetylene are usually one piece, while the other fuel cases may be one- or two-piece tips. See Fig. 6-8B. Acetylene tips are flat on the flame end. Tips for methylacetylene-propadiene (MPS) have a flat surface on the flame end. Most propylene tips have a slight recess, and natural gas and propane tips usually have a deeper recess or cupped end.

Tips may be obtained for flame machining, gouging, scarfing, and rivet cutting, Fig. 6-9 (p. 173).

Figure 6-6 (p. 171) illustrates the internal construction of a standard adaptable cutting attachment.

Cutting Tips The **cutting tip**, Fig. 6-7 (p. 171), has a central hole through which the high pressure oxygen flows. Around this center hole are a number of preheating flame holes. Cutting tips may be obtained in various shapes and sizes. The thicker the metal that is to be cut, the larger the size of the center hole must be. Table 6-2 (p. 172) lists different-sized cutting tips and the gas pressures appropriate for them.

Lighters The cutting torch should be lighted with a friction lighter, Fig. 6-3. The flints of friction lighters can be easily replaced at small cost when worn out. Matches should never be used because the thermal cutter's hand has to be too close to the torch tip and may be burned when the gases ignite. There is also the danger that the supply of matches that the thermal cutter may be carrying in a pocket will ignite and cause severe burns.

Goggles The thermal cutter must wear protective goggles to prevent harm to their eyes from sparks, hot particles of metal, and glare. Suitable goggles and lenses are discussed in Chapter 5 (pp. 161–162).

Flame Cutting Principles **Chapter 6** 169

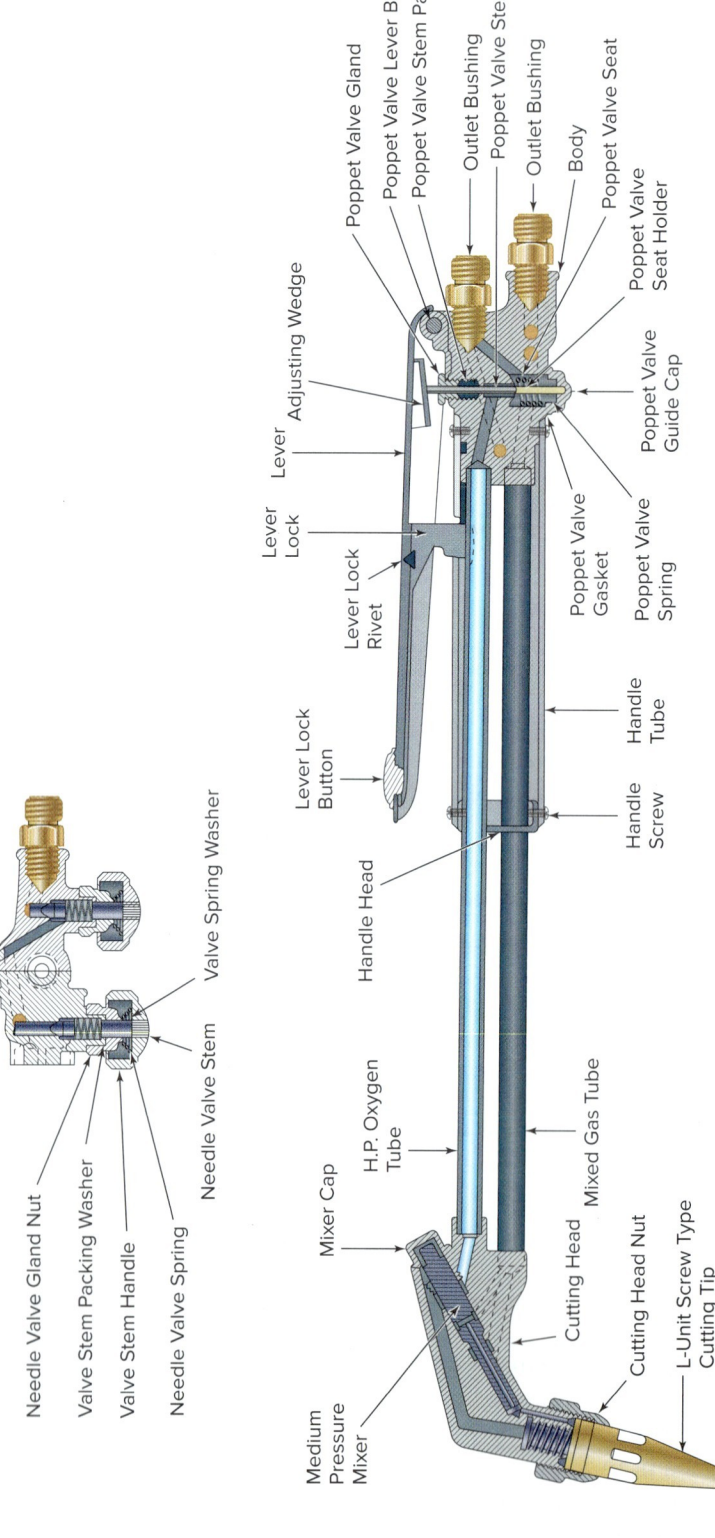

Fig. 6-5 Internal construction of a standard oxyfuel gas cutting torch. Adapted from Thermadyne Holdings Corporation

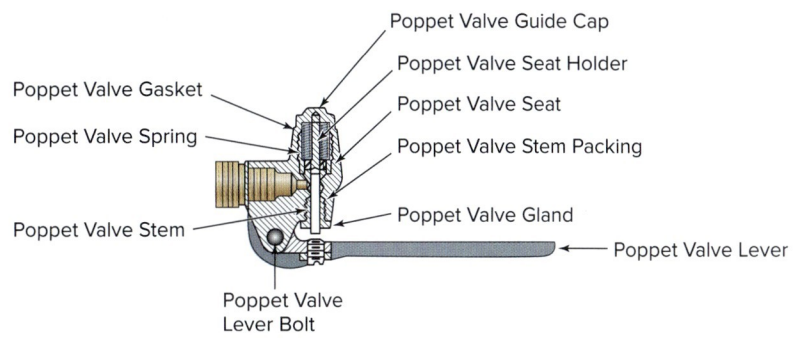

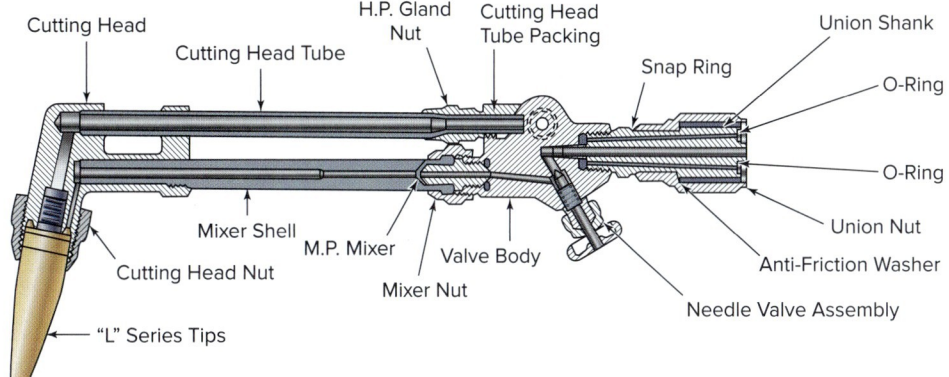

Fig. 6-6 Internal construction of a standard oxyfuel gas adaptable cutting attachment. Adapted from Thermadyne Holdings Corporation

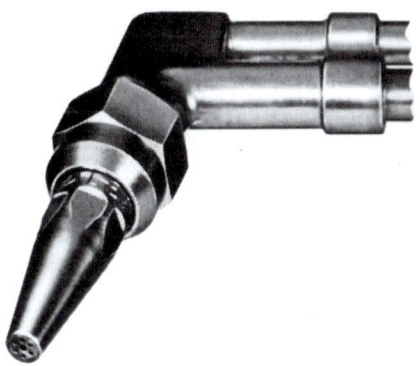

Fig. 6-7 Standard cutting tip attached to cutting torch head. Thermadyne Industries, Inc.

SHOP TALK

Safety
Today's consumers assume that you will be working in the most environmentally friendly way possible. Corporate responsibility means taking precautions to prevent injury to the Earth and its inhabitants.

Gloves The heat coming from a cutting job may be very intense. There may also be a shower of sparks and hot material which makes it necessary for the thermal cutter to use gloves to protect the hands. For the best protection, gloves should be of nonburnable material. It is common practice, however, to wear an ordinary canvas glove with a cuff, which can be purchased at very small cost.

Gloves should be kept free from grease and oil because of the danger involved in contact with oxygen.

Oxyfuel Gas Cutting Machines

A very large part of the cutting done today is performed by oxyfuel gas cutting machines, Fig. 6-10. These machines have a device to hold the cutting torch and guide it along the work at a uniform rate of speed. It is possible to produce work of higher quality and at a greater speed than with the hand cutting torch. Machines may be used for cutting straight lines, bevels, circles, and other cuts of varied shape.

Small portable cutting machines are used with only one torch. Large permanent installations can make use of several cutting torches to make a number of similar shapes at the same time. A multiple-torch cutting machine and its automatic controls are shown in Figs. 6-11 and 6-12.

Maximum productive capacity is achieved through the use of stationary cutting machines developed for production cutting of regular and irregular shapes of practically

Flame Cutting Principles **Chapter 6** 171

Table 6-2 Cutting Tip Size, Speed, Pressure, and Gas Flow Rates for Various Steel Thicknesses

Thickness of Steel (in.)	Diameter of Cutting Orifice (in.)	Cutting Speed (in./min)	Cutting Oxygen (Approx. Pressure, p.s.i.)	Acetylene (Approx. Pressure, p.s.i.)	MPS	Natural Gas	Propane
1/8	0.020–0.040	16–32	15–45 (10)	3–9 (4)	2–10	9–25	3–10
1/4	0.030–0.060	16–26	30–55 (15)	3–9 (4)	4–10	9–25	5–12
3/8	0.030–0.060	15–24	40–70 (20)	6–12 (4)	40–10	10–25	5–15
1/2	0.040–0.060	12–23	55–85 (25)	6–12 (4)	6–10	15–30	5–15
3/4	0.045–0.060	12–21	100–150 (30)	7–14 (5)	8–15	15–30	6–18
1	0.045–0.060	9–18	110–160 (40)	7–14 (5)	8–15	18–35	6–18
1½	0.060–0.080	6–14	110–175 (50)	8–16 (5)	8–15	18–35	8–20
2	0.060–0.080	6–13	130–190 (60)	8–16 (5)	8–20	20–40	8–20
3	0.065–0.085	4–11	190–300 (70)	9–20 (6)	8–20	20–40	9–22
4	0.080–0.090	4–10	240–360 (80)	9–20 (6)	10–20	20–40	9–24
5	0.080–0.095	4–8	270–360 (90)	10–25 (6)	10–20	25–50	10–25
6	0.095–0.105	3–7	260–500 (100)	10–25 (7)	20–40	25–50	10–30
8	0.095–0.110	3–5	460–620 (130)	15–30 (7)	20–40	30–55	15–32
10	0.095–0.110	2–4	580–700 (150)	15–35 (8)	30–60	35–70	15–35
12	0.110–0.130	2–4	720–850 (170)	20–40 (9)	30–60	45–95	20–40

Notes:
1. Preheat oxygen consumptions: Preheat oxygen for acetylene = 1.1 to 1.25 × acetylene flow rate in cubic feet per hour; preheat oxygen for natural gas = 1.5 to 2.5 × natural gas flow rate in cubic feet per hour; preheat oxygen for propane = 3.5 to 5 × propane flow rate in cubic feet per hour.
2. Higher gas flows and lower speeds are generally associated with manual cutting, whereas lower gas flows and higher speeds apply to machine cutting. When cutting heavily scaled or rusted plate, use high gas flow and low speeds. Maximum indicated speeds apply to straight-line cutting; for intricate shape cutting and best quality, lower speeds will be required.

Data extracted from the American Welding Society, *American Welding Handbook*, Vol. 2, 8th ed., p. 464, Table 14.2.

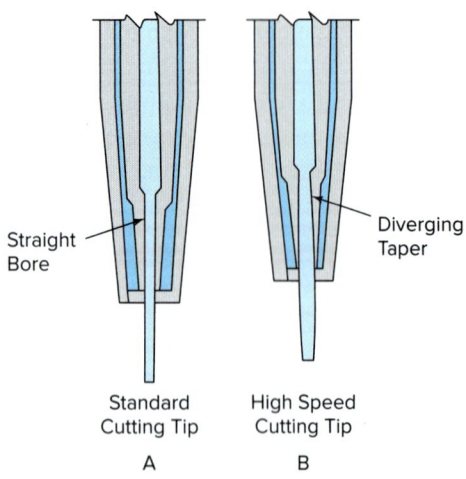

Fig. 6-8 Oxyfuel gas cutting tips. American Welding Society, *Welding Handbook*, Vol. 2, 8th ed., p. 457, Figure 14.5.

any design. These machines can be particularly adapted to operations in which the same pattern or design is to be cut repeatedly, Fig. 6-13. A number of cutting torches are mounted on the machine so that a number of parts of the same shape can be cut simultaneously. These machines can be used for straight line or circle cutting. They can be guided by hand or a template.

Cutting machines may be guided by various types of tracing devices. One type follows a pattern line of tracer ink and electrically controls the movement of the torch by means of a servomechanism.

Some units make use of a tracer roller, which is magnetized and kept in contact with a steel pattern, Fig. 6-14. The tracer follows the exact outer contour of the pattern and causes the cutting tools to produce a cut in exactly the same shape. Figure 6-15 shows a pattern tracer in use.

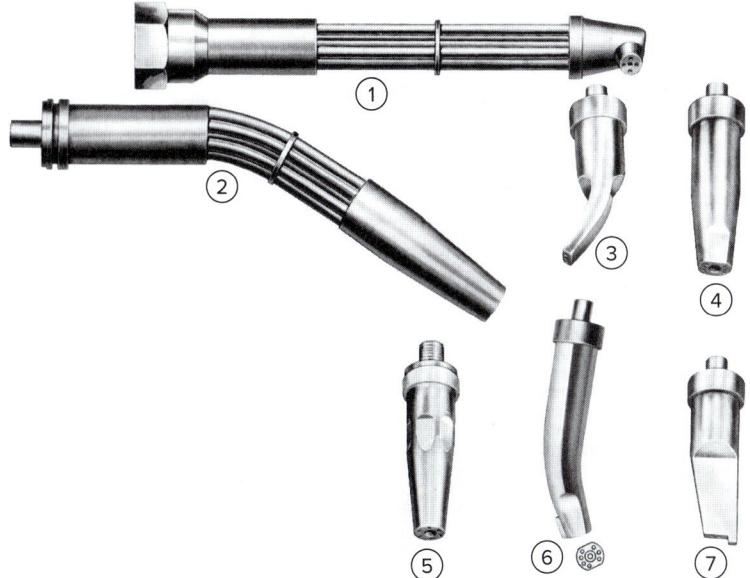

Fig. 6-9 Special-purpose cutting tips: 1. Close quarters cutting attachment: for use when cutting tubes, pipe, reinforcing bars, and other cuts where space does not permit the use of standard tips. For oxyacetylene only. 2. Long cutting tip: for special purposes and all gases. Available in any length in straight or desired angle. 3. Rivet-cutting tip: for slicing rivet heads off flat. For oxyacetylene only. 4. Straight gouging tips: for removing narrow strips of surface metal from steel plates, forgings, castings, and weldments. For oxyacetylene only. 5. Rivet-burning tips: for removing rivet heads in such a manner so as not to damage the plate. For oxyacetylene only. 6. Bent gouging tip: a bent gouging nozzle with a stainless-steel skid for longer wear. For oxyacetylene only. 7. Sheet metal cutting tip: for making clean cuts in light gauge sheet metal. Stainless-steel skid is used as a guide. For oxyacetylene only.
Thermadyne Industries, Inc.

JOB TIP

Specializing

Expect specialized growth in:

1. Developing areas (such as Mexico and Asia) needing infrastructure
2. Welding automation and laser beam
3. GTAW for precision work with new metals
4. GMAW with mixed gas shielding
5. Sheet metal

Stack Cutting

In addition to cuts made through a single thickness of material, oxygen cuts are made through several thicknesses at the same time, Fig. 6-16. This is a machine cutting process known as **stack cutting.** The plates in the stack must be clean and flat and have their edges in alignment where the cut is started. The plates must be in tight contact so that there is a minimum of air space between them. It is usually necessary to clamp them together.

Stack cutting is particularly suitable for cutting thin sheets. A sheet $\frac{1}{8}$-inch thick or less warps, and the edge is

Fig. 6-10 This "ULTRA-LINE" oxyfuel gas cutting machine is used for straight line and bevel cutting. Its speed range is 3–100 in./min and it uses knurled drive wheels for positive traction. It is light weight (aluminum one-piece housing 22 lbs), portable with 6-foot track sections. ESAB

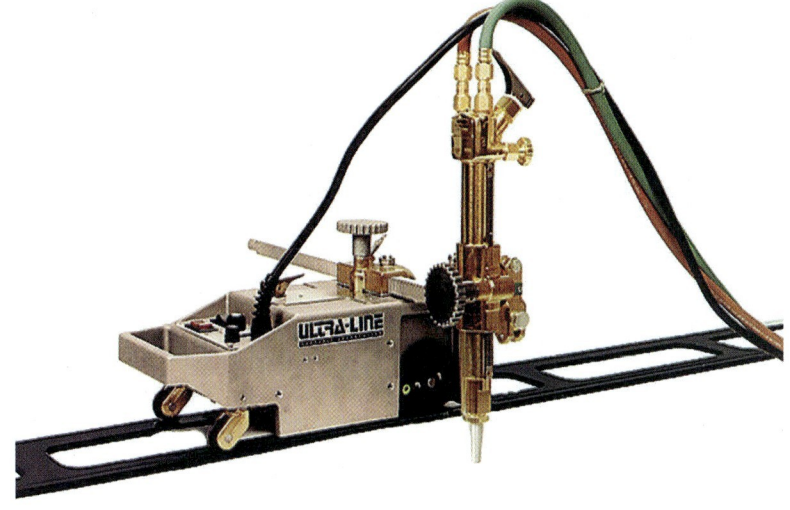

Flame Cutting Principles   Chapter 6

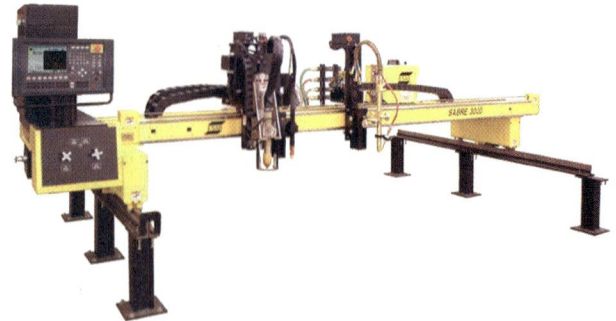

Fig. 6-11 Multiple-torch shape-cutting low profile gantry cutting system for cutting squares, rectangles, circles, and other shapes without templates. It utilizes CNC (Computer Numerical Control) processing and has a speed range of 2–400 in./min. Note the six oxyfuel cutting torches and the two other thermal cutting torches. See Chapter 17 for additional information on arc cutting. ESAB

> **SHOP TALK**
>
> **Blowtorch**
> An alternative use for the newly developed blowtorch was discovered in 1890 by a bank robber in England who used one to get into a vault.

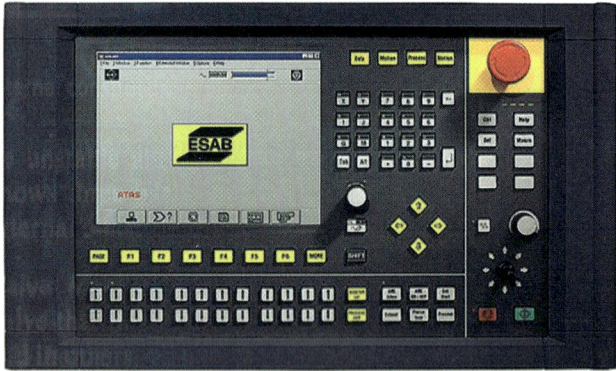

Fig. 6-12 The CNC digital control unit for the cutting machine shown in Fig. 6-11. ESAB

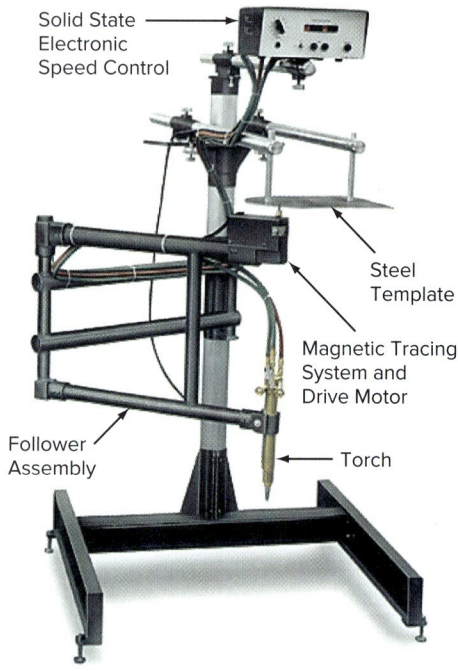

Fig. 6-14 A magnetic tracer machine with a template follower in the torch arm of the Ultra-Graph cutting machine. ESAB

Beam Cutter

The beam cutter is a portable structural fabricating tool, Fig. 6-17. From one rail setting the operator can trim, bevel, and cope beams, channels, and angles. The beam rail is positioned across the flanges. Two permanent magnets lock and square the rail in position. Variable speed power units are used on both the horizontal and vertical drives. A squaring gauge enables the operator to adjust the tip quickly from bevel to straight trim cuts.

This all-position flame-cutting machine weighs only 60 pounds. It is easily moved and set in position by one operator, so, setup time is minimal. The beam cutter provides clean, accurate cuts in a minimum amount of time. It has become a very important tool for use in the construction industry.

Fig. 6-13 Multiple-gas cutting by machine control. Metal Fabricating Institute

rough with dross if it is cut singly. If stack cut, the edges are straight and smooth and free from dross. Stack cutting may be used on plates up to ½-inch thick.

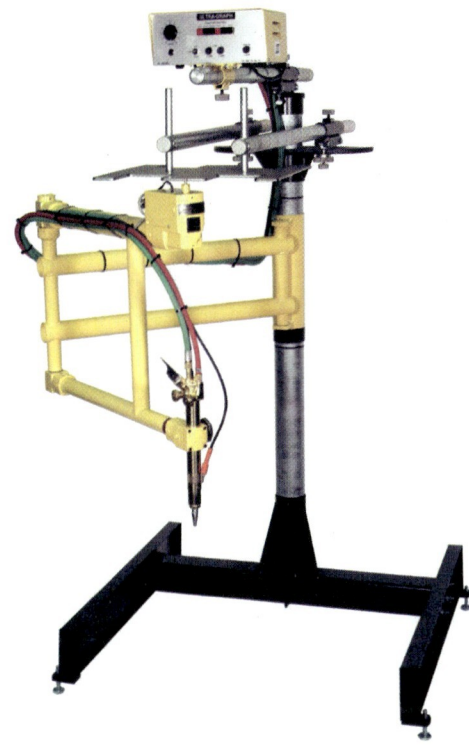

Fig. 6-15 A cutting machine with a pattern tracer. ESAB

Fig. 6-16 An oxyfuel stack cutting operation. Note the plates are being held tightly together with welds. American Welding Society, *Welding Science and Technology,* Vol 1 of Welding Handbook, 9th ed., p. 468, Figure 14.13.

Oxygen Lance Cutting

Oxygen lance cutting (OLC) is a method of cutting heavy sections of steel that would be very difficult by any other means. The lance is merely a length of black iron pipe fitted with a valve at one end to which an oxygen hose is connected.

Oxygen pressure of 75 to 100 p.s.i. is used. The pipe size may vary from $\frac{1}{4}$ to $\frac{3}{8}$ inch.

In order to start the cut, it is necessary to preheat the cutting end of the pipe (lance) to a cherry red with an oxyfuel cutting or welding torch. Once it is cherry red, the oxygen flow is started. The steel pipe burns in a self-sustaining, exothermic reaction, and the heating torch is

Fig. 6-17 Beam cutter used in heavy steel construction. BUG-O

removed. When the burning end of the lance is brought close to the workpiece, the work is melted by the heat of the flame. Figure 6-18 is a schematic of the OLC operation. See Fig. 6-19 for its use on cutting cast steel.

The lance is slowly consumed during the operation and must be replaced from time to time.

The oxygen lance is useful for piercing holes in heavy thicknesses of steel, cutting off large risers in the foundry, and opening holes in steelmaking equipment that has become plugged with solidified metal.

ABOUT WELDING

Silver Brazing

Silver brazing is effective for manufacturing or maintenance, even when joining unlike metals. This method uses a minimum of alloy, which keeps costs down.

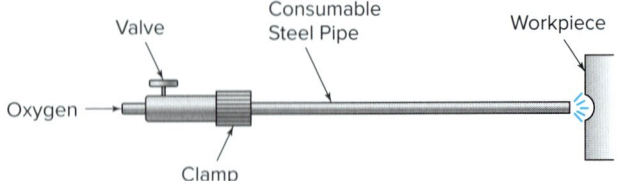

Fig. 6-18 Schematic view of oxygen lance cutting.
American Welding Society, *Welding Handbook,* Vol. 2, 8th ed., p. 479, Figure 14.25.

Fig. 6-19 An oxygen lance being used to cut cast steel.
American Welding Society, *Welding Handbook,* 8th ed. Vol. 2, p, 479, Figure 14.26.

CHAPTER 6 REVIEW

Multiple Choice

Choose the letter of the correct answer.

1. Which of the following metals can be readily cut with the OFC process? (Obj. 6-1)
   a. Stainless steel
   b. Aluminum
   c. Low carbon steel
   d. Cast iron
2. OFC _____. (Obj. 6-1)
   a. Is a mechanical cutting process
   b. Is a chemical cutting process
   c. Relies upon rapid rusting of the metal
   d. Both b and c
3. Which of the following pieces of equipment is *not* required for OFC? (Obj. 6-2)
   a. Cutting torch with flame arresters and check valves
   b. Matches for lighting the torch
   c. Oxygen and fuel gas regulator
   d. Oxygen and fuel gas hoses
4. The size of the cutting tip is determined by the _____. (Obj. 6-2)
   a. Plate thickness to be cut
   b. Skill of the thermal cutter
   c. Type of material to be cut
   d. Type of fuel gas to be used
5. Goggles must be worn during OFC to protect the thermal cutter from _____. (Obj. 6-3)
   a. Sparks
   b. Hot metal
   c. Glare
   d. All of these
6. Gloves must be _____. (Obj. 6-3)
   a. Worn to pick up red-hot metal
   b. Worn to keep hands warm on cold days
   c. Kept free from grease and oil
   d. All of these
7. Automatic cutting machines can be used to cut _____. (Obj. 6-4)
   a. Straight lines
   b. Bevels
   c. Circles
   d. All of these
8. Stack cutting is never done on sheet metal ⅛-inch thick or thinner. (Obj. 6-5)
   a. True
   b. False
9. OLC is done using oxygen pressures of _____. (Obj. 6-6)
   a. 10–25 p.s.i.
   b. 50–100 p.s.i.
   c. 75–100 p.s.i.
   d. 2,200 p.s.i. directly out of the cylinder
10. OLC uses a black iron pipe to deliver the oxygen to the cut area. (Obj. 6-6)
    a. True
    b. False

Review Questions

Write the answers in your own words.

11. Explain the chemical reaction that takes place in the OFC of steel. (Obj. 6-1)
12. List at least six metals OFC will not effectively cut. (Obj. 6-1)

13. List all the equipment required for OFC. (Obj. 6-2)
14. Write down how the OFC-A torch works. (Obj. 6-2)
15. Sketch and label the end of the OFC-A cutting tip and describe the purpose of the multiports. (Obj. 6-2)
16. Describe how the OFC torch should be properly lighted. (Obj. 6-3)
17. OFC machines provide for what advantages in cutting? (Obj. 6-4)
18. Describe two methods to use to guide an OFC machine. (Obj. 6-4)
19. In stack cutting, what conditions must exist for this technique to work? (Obj. 6-5)
20. Describe how the material is brought to kindling temperature with the OLC process. (Obj. 6-6)

INTERNET ACTIVITIES

Internet Activity A

Use the Internet to find photos or artwork of flame cutting. Describe what is being pictured in each photo or piece of artwork. Explain what flame cutting can be used for. Place your photos and/or artwork in a binder to share with your class.

Internet Activity B

Research water jet cutting on the Internet. Name the advantages of water jet cutting.

Design credits: Cinema and movie icon: Denis Meshkov/123RF; and Illustration of Welding icons: Mr. Nachai Sorasee/123RF

Flame Cutting Practice:

Jobs 7-J1–J3

7

To be a successful skilled craftsperson, a welder must be able to perform manual flame cutting (oxyfuel cutting), Fig. 7-1. A number of flame cutting procedures are commonly used in the preparation and treatment of metal in a wide variety of industries. The process is used in such different industries as multiple-piece production in manufacturing, shipbuilding, steel production, construction, scrapping and salvage, and for rescue work by police and fire departments.

It is important that welding students learn the art of flame cutting before they are introduced to the various electric welding processes. Much of the material used for electric welding is heavier than that used for gas welding practice, and edge preparation becomes a problem. In those schools that do not have stocks of prepared material available or mechanized-cutting equipment to prepare plates, the students must be able to cut so that they can prepare their own material in the shortest possible time with the least waste.

In many cases, however, the training program may be such that students do not need any knowledge of cutting. The purpose of the training may be to train welders for certain production jobs that require limited skill. If this is true, the instructor may ask you to begin with the arc welding course.

Chapter Objectives

After completing this chapter, you will be able to:

7-1 Describe flame cutting principles.
7-2 Identify OFC techniques.
7-3 Perform visual inspection of oxyfuel cuts.
7-4 Describe OFC equipment set up and take down.
7-5 Perform troubleshooting of cut quality.
7-6 Perform straight line and curve square cuts as well as bevel cuts on a variety of shapes.

Fig. 7-1 These young welders are being observed by their instructor as to the proper procedure for cutting a bevel on a flat bar. A track type cutting machine is being used. Note the PPE being used. Hill Street Studios/Blend Images/Getty Images

Since most oxyfuel cutting (OFC) that a welder will be called upon to do in industry will be with straight carbon steel, the emphasis in this chapter will be on steel. It is important, however, for the skilled welder to know about the wider applications of OFC and to understand basic principles and techniques.

Review of Flame Cutting Principles

The cutting of metals by the oxyacetylene, oxyhydrogen, and oxyfuel gas processes is based on the fact that all metals oxidize to a greater or lesser degree, depending upon the physical conditions around them. Wrought iron and steel oxidize quite rapidly, even under ordinary atmospheric conditions. When oxidation occurs in the air around us, it is called rusting and can readily be recognized by the oxide that forms on the metal. Rusting is, of course, a slow process, but it illustrates the tendency of ferrous metals to combine with oxygen.

After iron or steel is heated to a red heat and cooled, it is covered with a thick scale. These metals oxidize more rapidly when hot than when cold. If the temperature of the steel is raised even higher to a white heat and a stream of pure oxygen is directed against the white hot spot, it burns rapidly. This can be demonstrated in the shop by taking a thin piece of steel wire and heating it to a red heat and then submerging it in a vessel containing oxygen. The heated end will immediately burst into flame and burn vigorously until the wire is burned away or the oxygen is consumed.

The reaction between oxygen and iron causes a considerable amount of heat. It forms a molten oxide (dross) that flows or is blown away, exposing more metal to the action of the oxygen. In order to keep this action going, it is necessary to supply heat to the point of cutting from an external source.

The function of the cutting torch, Fig. 7-2, is to provide a flame to heat the metal to a red heat, to maintain that heat, and to direct a stream of oxygen on the heated point of cutting. Always keep in mind that the high-pressure oxygen jet does the cutting. Practically all trouble in cutting is caused by the cutting tip becoming burred or obstructed by small particles of molten dross adhering to it. An unobstructed, cylindrical jet of oxygen always produces a smooth cut. Any obstruction, either in the bore or at the end of the tip, retards the speed of the cut and produces a rough cut. With the torch on and the pressure at the proper setting, a clean cutting tip should sound like ripping paper when the torch handle is pressed down.

Vertical Cutting Progressions Position the material to be cut in a vertical position and then work from the bottom to the top. Use gravity to help remove the molten material.

Flame Cutting Effects on Steel

For many years after the development of the oxyacetylene cutting process, steel fabricators were reluctant to use the process in the preparation of metals for welding. They felt that harmful metallurgical changes took place on the surface of the cut, then surface cracking occurred and

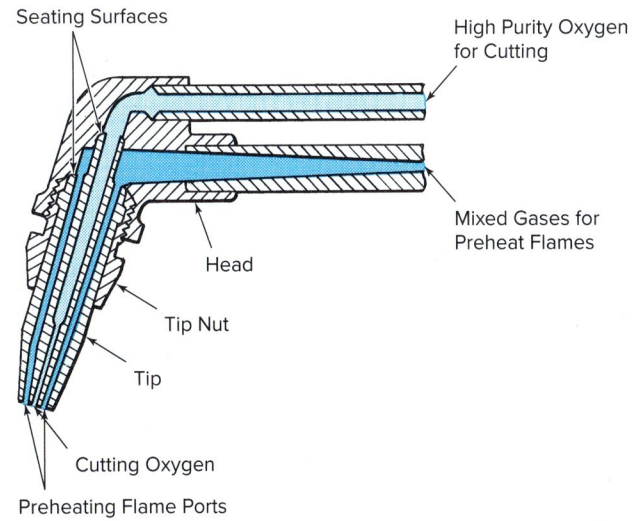

Fig. 7-2 An oxyfuel cutting torch is designed to supply mixed gases for the heating flames and a stream of high purity oxygen to do the actual cutting.

locked-up stresses were present. Engineering investigation has proved that the oxyacetylene-cut edge is superior to that of some mechanical procedures which, because of violent contact, damage the plate edge.

Care must be exercised in flame cutting steels with high carbon content or alloy content that have air-hardening properties. When cut, these steels have a tendency for hardened zones and cracking along the edge of the cut. They can be successfully flame cut by preheating, postheating, or both. The cut should be allowed to cool slowly.

When applied to large areas, the cutting procedure produces a certain amount of expansion and contraction that may cause distortion if it is not controlled. This problem is not as serious as in welding, and the usual techniques used to control distortion when welding may be applied to the cutting procedure. See Chapter 3, pages 100 to 103, for ways to control distortion.

Comparison of Fuel Gases

The function of the fuel gas is to feed the preheat flames that maintain the temperature of the materials being cut so that the cutting process can continue without interruption. For many years, acetylene gas was the only gas used for flame cutting. Today a number of different gases are being used, such as *acetylene, propane, Mapp® gas, natural gas,* and *propylene*. While acetylene is still the workhorse of the industry, each of the other gases has its particular advantage, depending upon the application. For example, hydrogen was once used exclusively for underwater cutting because it can be safely compressed to high pressures and regulated to overcome the pressure exerted by the water at salvage operation depths. Natural gas, which can be similarly compressed, is used now for underwater cutting.

The choice of the type of fuel gas to use depends upon the following factors:

- Cost of the gas
- Availability of the gas
- Flame temperature needed
- Length of the cutting procedure
- Length of cut required
- Speed of cutting

When used as a fuel gas, the flame temperature of acetylene is the highest, followed by propylene, Mapp® gas, propane, and natural gas in that order. All of the gases give satisfactory cuts when used under the proper conditions. For best results, it is essential that the cutting technique, torches, and tips employed be designed for the particular fuel gas used.

Flame Temperature Of the five principal gases used today, acetylene gives the hottest flame—350°F hotter than its nearest rival. Acetylene has a flame temperature of 5,650°F; Mapp® gas and propylene, 5,300°F; and propane and natural gas, 4,800°F. Acetylene burns much faster than the other gases and much closer to the tip of the torch. Therefore, it can be adjusted to a very intense, concentrated flame that is exactly what is needed for cutting. The oxyacetylene flame is ready to cut in less than half the time of its nearest competitor and five times as fast as the slowest.

Heat Concentration In a comparison of the heat produced by the various gases, measured in British thermal units (BTU's), Mapp® gas and propane have 2,400; acetylene, 1,450; and natural gas, 1,000. Although acetylene is third in BTU's, it concentrates the most BTU's in the smallest area of any gas because of the way it burns. In the oxyacetylene flame, BTU's are created mainly in the inner cone. With the other gases, BTU's are largely dispersed through the outer flame envelope, Fig. 7-3. Thus, acetylene concentrates heat at the point of cutting.

Cost Acetylene is among the higher priced fuel gases. Its cost is offset, however, by the reduction in the use of oxygen in the cutting process. Acetylene requires less oxygen per cubic foot of fuel than any of the other four fuel gases. The following amounts of oxygen are needed for 1 unit of fuel gas: acetylene 1.5 : 1; natural gas 2 : 1; Mapp® gas 3.4 : 1; and propane 4.7 : 1.

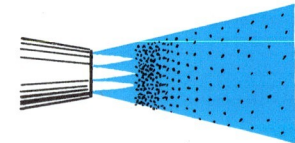

Acetylene

BTU's Concentrated

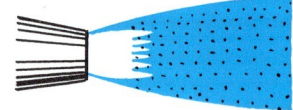

Other Gases

BTU's Dispersed

Fig. 7-3 The BTU concentration of the oxyfuel flame.

> **JOB TIP**
>
> **Education**
> Make the most of your education. Use this textbook as a tool to gain understanding. Each chapter begins with a list of the objectives. Scan the chapter before reading to note the major headings. After reading the chapter, use the review questions to check your understanding. Go back to reread the major headings; could you explain to someone what each section was about? When you do your reading assignments, you will feel prepared for class.

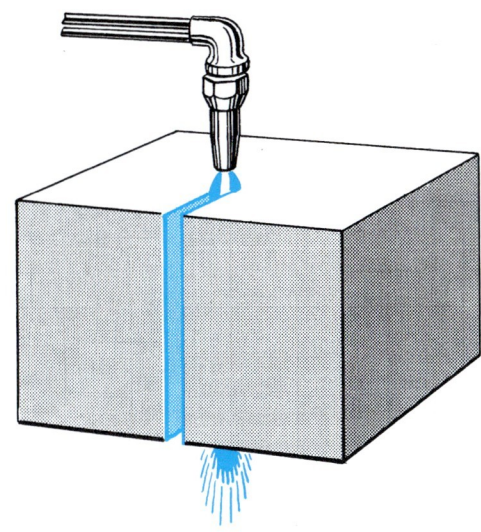

Fig. 7-4 A bushy flame heats the entire depth of thick plate.

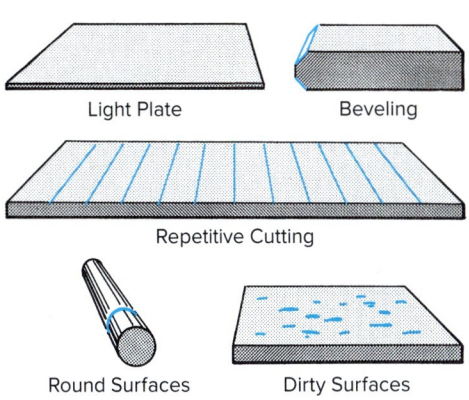

Fig. 7-6 Where acetylene works best.

Special Application Although acetylene has many advantages, no one gas can do everything best. There are a number of applications in which propane, Mapp® gas, propylene, and natural gas are better choices than acetylene.

For cutting plate of 6 inches or heavier, propane, propylene, or Mapp® gas are better choices. Because a deep area must be heated, the dispersed BTU's, the *bushy* flame, and the high total BTU content of these gases do a good job in heavy plate cutting, Fig. 7-4.

Natural gas is widely used in steel mills for removing surface defects because it is the most economical fuel gas. Most mills use fuel gas with air.

Acetylene is ideal for cutting steel plate. In the ¼-inch to 1-inch range, acetylene preheats up to 50 percent faster and allows cutting 25 percent faster than its closest competitor. In repetitive cutting, acetylene is far superior because this kind of cutting requires frequent preheats and starts.

Bevel cutting and cutting on rounded surfaces are natural applications for acetylene. The flame is held at an angle to the surface so that the heat tends to bounce off the work at the opposite angle, Fig. 7-5. In order to get started, a concentrated flame is needed. On surfaces that are covered with rust, scale, grease, or paint, the concentrated acetylene flame cuts through the barriers. Figure 7-6 presents the applications for which the acetylene flame works best.

Cutting Different Metals

Oxyfuel cutting finds its widest use in the cutting of carbon and low alloy steels. The normal oxygen cutting methods must be varied for metals with high alloy content such as titanium, cast iron, and stainless steel. You may think that this is peculiar since all metals oxidize. Some metals, however, do not form an oxide that melts at a lower temperature than the base metal. This high temperature oxide protects the metal surface and prevents further oxidation. Generally, as the amounts of alloying materials, including carbon, increase, the oxidation rate decreases. This is one reason why cast iron, which has a high carbon content, is more difficult to cut than steel.

There are a number of methods that overcome to some degree the effects of carbon and alloying materials that interfere with the cutting process.

Preheating It has been pointed out that high heat increases the rate of oxidation. As a consequence, many materials that are difficult to cut can be cut more easily if they are heated to a temperature approaching their melting point.

Waster Plate A low carbon steel plate may be clamped along the line of cut. The melting of this plate causes a great amount of iron oxide to be formed, making it possible for the cut to continue without interruption. A welding bead of low carbon steel may also be deposited along the line of cut to serve the same purpose.

Wire Feed Low carbon steel wire may be fed into the preheat flames to produce rapid oxidation. The heat that is generated by the burning of the wire rapidly brings the

Fig. 7-5 A concentrated flame is needed for bevel cutting.

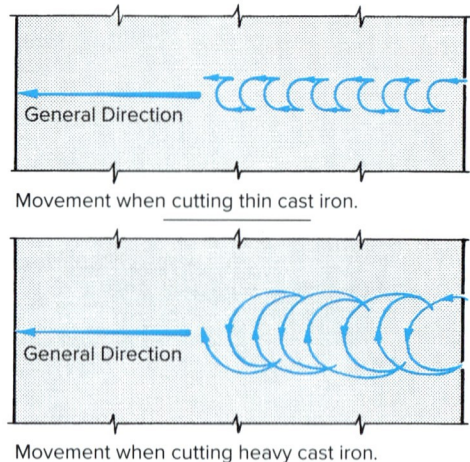

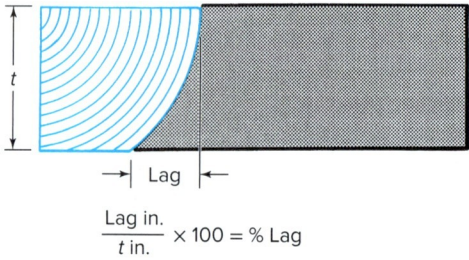

$$\frac{\text{Lag in.}}{t \text{ in.}} \times 100 = \% \text{ Lag}$$

Fig. 7-8 Lag is the amount by which the bottom of the cut lags behind the top. Lag is usually expressed as a percent of the plate thickness. If this were a 1-inch-thick plate and the amount of lag were ½ inch, there would be a 50 percent lag.

Fig. 7-7 Typical cutting torch manipulation for cutting thin cast iron (top) and heavy cast iron (bottom).

surface of the plate to ignition temperature. This method can be used to obtain rapid starts in heavy plate. It can also be used for cutting cast iron. Short lengths of wire can be fed manually, or special equipment can be attached to the cutting torch for continuous iron wire, or powder feed.

Oscillatory Motion When the torch is moved from side to side, Fig. 7-7, more material is heated to the ignition temperature so that additional material is oxidized and blown out of the kerf. This is one method used in the cutting of cast iron, and it produces a rough cut.

Difficult-to-cut materials may also be cut using other methods such as oxygen arc cutting, flux cutting, and powder cutting. These methods are explained in Chapter 6.

Cutting Technique

Welders must manipulate the torch when cutting so that they create the proper kerf and drag for the material being cut.

Kerf

Kerf is the gap created as the material is removed by cutting. Control of the kerf is important to the accuracy of the cut and the squareness of the face of the cut. The width of the kerf is determined by the size of the cutting tip, speed of cutting, oxygen pressure, and torch movement. Since oxygen pressure is directly affected by the thickness of the material being cut, the width of the kerf increases as the thickness of the material increases. A rough cut also increases the width of the kerf.

Drag

When steel is cut, lines form on the face of the work that are caused by the flow of the high pressure oxygen. When the torch is held in a vertical position and the cutting conditions are correct, this line is vertical from top to bottom, Fig. 7-18 (1), page 186. This condition is referred to as **zero drag.** If the speed of cutting is increased or if the oxygen pressure is not set high enough for the thickness of the material being cut, the drag lines at the bottom of the kerf lag behind the top of the kerf. The amount of lag can be expressed as a percentage of the plate thickness, Fig. 7-8. In reverse drag, the drag lines at the top of the kerf lag behind those at the bottom. Reverse drag may result from too much oxygen or a speed of travel that is too slow. Both drag and reverse drag must be avoided because they may cause the cut to be lost.

Straight Line Cutting Cutting along a straight line that has been laid out on the material to be cut may be done with a great degree of accuracy, depending upon the skill of the welder. Very often a straightedge or angle iron is clamped along the line of cut to act as a guide. See Table 7-3, page 190, for hand cutting data. The torch is held perpendicular to the plate with the holes positioned as shown in Fig. 7-9. Mechanized cutting is somewhat more accurate than manual cutting.

Bevel Cutting Bevel cutting is one of the common operations used in beveling the edges of plate and pipe for welding. The torch tip is held sideways at the angle desired with the holes positioned as shown in Fig. 7-9. Cuts may be made by either hand or machine in straight or irregular lines.

In order to do the jobs that are outlined for advanced gas welding, the shielded metal arc, gas tungsten arc, and gas metal arc welding of plate and pipe, it will be necessary to prepare bevel joints.

For plate beveling, the student will use the flame cutting machine provided in the welding shop, Fig. 7-10. This type of machine is often referred to as a "track burner" on the shop floor. A commercial machine, Fig. 7-11, may

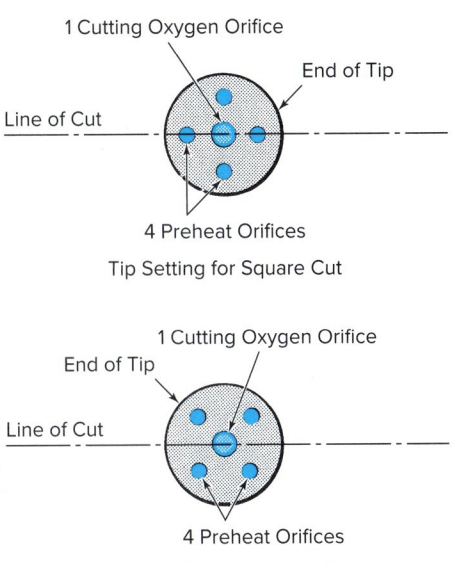

Fig. 7-9 For square and bevel cutting, the tip should be set so that the preheat orifices are positioned as shown. Cutting tips with more than four preheat orifices should be positioned in the same way.

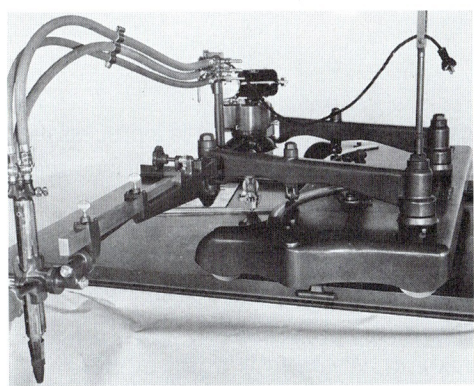

Fig. 7-10 A standard flame cutting machine for plate shape-cutting and beveling. A machine of this type has extreme flexibility. *Edward R. Bohnart*

Fig. 7-11 Two mechanized cutting torches are mounted in a fixed position over the pipe that is being rotated. This is set up in a high production high quality pipe spool fabrication shop. To cut pipe to length with the appropriate beveled ends. Location: Piping System's Inc. Mark A. Dierker/McGraw Hill

be available for the beveling of pipe. A pipe-beveling machine may also be constructed in the school shop. The machine shown in Fig. 7-12 was built by students. It is manually powered. The torch may be purchased from any welding equipment distributor. Study Table 7-1 for machine cutting data. The operator of this mechanized equipment must be properly trained and monitor the cut being made.

Flame Piercing Piercing is the process of forming a hole in a metal part. This may be for the purpose of producing a bolt hole or starting a cut in from the edge of a plate.

Fig. 7-12 A welding student is beveling pipe with a shop-built beveling machine. Location: Northeast Wisconsin Technical College Mark A. Dierker/McGraw Hill

Flame Cutting Practice: Jobs 7-J1–J3 **Chapter 7** 183

Table 7-1 Machine Flame Cutting Data for Clean Mild Steel (Not Preheated)

Thickness of Steel (in.)	Diameter of Cutting Orifice (in.)	Oxygen Pressure (p.s.i.)	Cutting Speed (in./min)	Gas Consumption Oxygen (ft³/h)	Gas Consumption Acetylene (ft³/h)
1/8	0.0200–0.0400	15–30	22–32	17–55	5–9
1/4	0.0310–0.0595	11–35	20–28	36–93	6–11
3/8	0.0310–0.0595	17–40	19–26	46–115	6–12
1/2	0.0310–0.0595	20–55	17–24	63–125	8–13
3/4	0.0380–0.0595	24–50	15–22	117–159	12–15
1	0.0465–0.0595	28–55	14–19	130–174	13–16
1 1/2	0.0670–0.0810	22–55	12–15	185–240	14–18
2	0.0670–0.0810	22–60	10–14	185–260	16–20
3	0.0810–0.0860	30–50	8–11	207–332	16–23
4	0.0810–0.0860	40–60	6.5–9	293–384	21–26
5	0.0810–0.0860	50–65	5.5–7.5	347–411	23–29
6	0.0980–0.0995	45–65	4.5–6.5	400–490	26–32
8	0.0980–0.0995	60–90	3.7–4.9	505–625	31–39
10	0.0995–0.1100	70–90	2.9–4.0	610–750	37–45
12	0.1100–0.1200	69–105	2.4–3.5	720–880	42–52

Note: In this table, acetylene pressures have been omitted because they are mainly a function of equipment design and are not directly related to the thickness of the material to be cut. It is suggested that for acetylene pressure settings, the charts of the equipment manufacturer be consulted.

Source: American Welding Society

Piercing takes more time than edge starting to bring the metal to the temperature required for cutting. The time required can be reduced by center punching or chisel nicking the plate to raise a small burr of metal that will tend to catch the flame and reach the cutting temperature much quicker. A spot is heated to a bright red, the torch is raised slowly to about ½ inch above the plate, and a limited amount of oxygen is turned on. As soon as the plate is pierced, the cut can continue in a regular manner. Care must be taken so that the dross that is formed on the surface of the plate is not blown back into the torch orifice. Figure 7-13 shows the following sequence of operations:

1. The cutting torch is held still until a spot on the surface just begins to melt.
2. The cutting torch is raised until the end of the tip is about ½ inch above the plate.
3. Meanwhile the cutting oxygen lever is slowly depressed, and the cutting torch moved slightly to one side at the same time to start a small spiral motion. As the cutting action starts, the dross will be blown out the opposite side of the pool.
4. When the cut has pierced all the way through the plate, the cutting torch is lowered and the cut is continued with a spiral motion until the desired diameter of hole has been obtained.

Flame Scarfing Flame scarfing is a process used mainly in the steel mills to remove cracks, surface seams, scabs, breaks, decarburized surfaces, and other defects on the surfaces of unfinished steel shapes. The operation is done before the steel reaches the final finishing departments.

Flame Washing Washing is a procedure similar to that of scarfing. It is used in the steel mills to remove unwanted metal such as fins on castings and to blend in riser pads and sand washouts in castings. It is also used to remove rivets without damage to the riveted parts. A skilled thermal cutter can remove a rivet from a hole, a threaded bolt from a threaded hole, or a piece of pipe from a threaded flange without destroying the hole or the threads, Figs. 7-14 through 7-16.

Flame Gouging Flame gouging provides a means for quickly and accurately removing a narrow strip of surface metal from steel plate, forgings, and castings without penetrating through the entire thickness of metal. The process is generally used for the removal of weld defects, tack welds, and metal from the back side of weld seams. When used for removing defects, the procedure is called **spot gouging.** Care must be taken so that only a small area of the weld is affected. Edge preparation of plate for welding

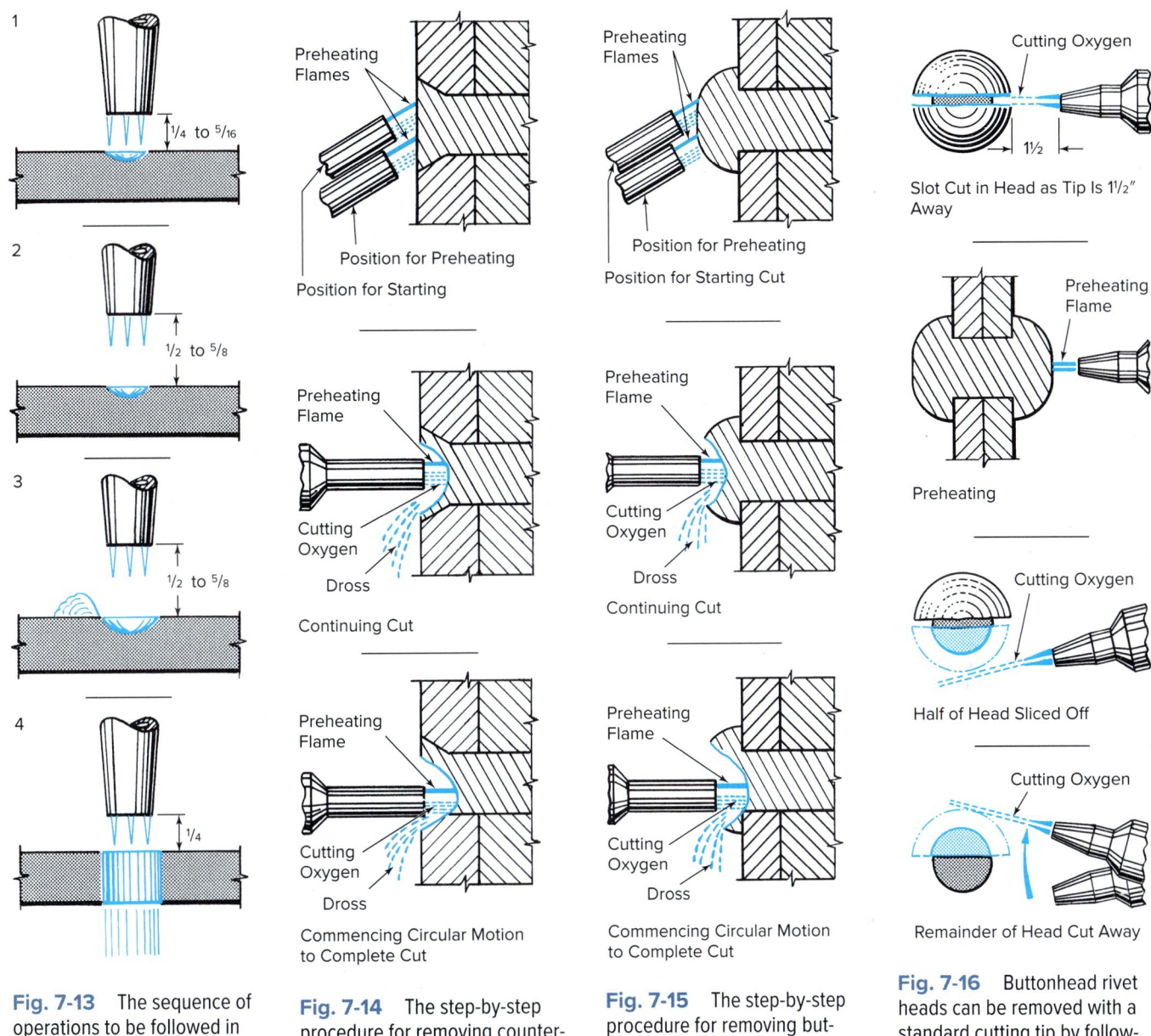

Fig. 7-13 The sequence of operations to be followed in piercing a hole in steel plate.

Fig. 7-14 The step-by-step procedure for removing countersunk rivet heads with a special rivet-cutting tip.

Fig. 7-15 The step-by-step procedure for removing buttonhead rivets with a special rivet-cutting tip.

Fig. 7-16 Buttonhead rivet heads can be removed with a standard cutting tip by following this step-by-step procedure.

is also frequently done by flame gouging. Figure 7-17 shows the following sequence of operations:

1. To start a gouge at the edge of a piece of plate, the plate edge is preheated to a bright red.
2. As the tip angle is reduced to the operating angle, the cutting oxygen is turned on to make a gradual starting cut.
3. and 4. When the desired depth of groove is reached, the gouging tip is moved along the line of cut.

There are two methods of preparing a U-type butt joint for welding by flame gouging. In the first method, the plates are tacked together in the form of a square butt joint. Then they are welded on one side, and the opposite side is gouged out to the root of the weld to form a U-groove. In the second method, the plates are tacked together in the form of a square butt joint. The U-groove is cut to a depth within ⅛ inch of the thickness of the plate, and the plate is welded with as much penetration as possible through to the back side. The reverse side is then welded with an electrode having deep penetration. U-grooves may also be formed with the arc cutting processes. Special tips are generally used for gouging.

Surface Appearance of High Quality Flame Cuts

Since one of the principal fields for oxyfuel flame cutting is in the preparation of plate and pipe for welded fabrications, it is important that the cuts be accurate, smooth, and free of dross that will contaminate the weld or cause the welder to spend a great deal of time in grinding or chipping.

The shoulder at the top edge of the plate should be sharp and square. The face of the cut should be smooth and not deeply grooved, fluted, or ragged. If it is square cut, the face should be square with the plate surface. Bevel cuts should be approximately the degree specified. The dross on the back side should be at a minimum and should be easily removed.

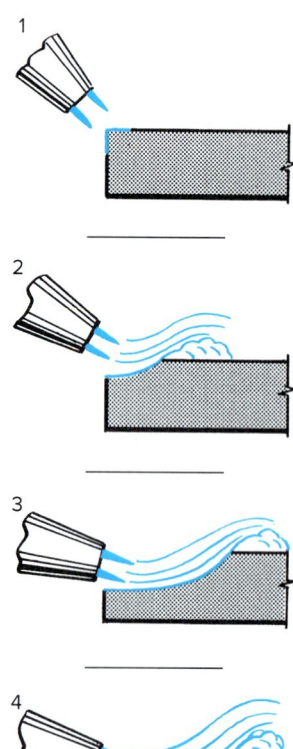

Fig. 7-17 The sequence of operations to be followed in flame gouging a piece of plate.

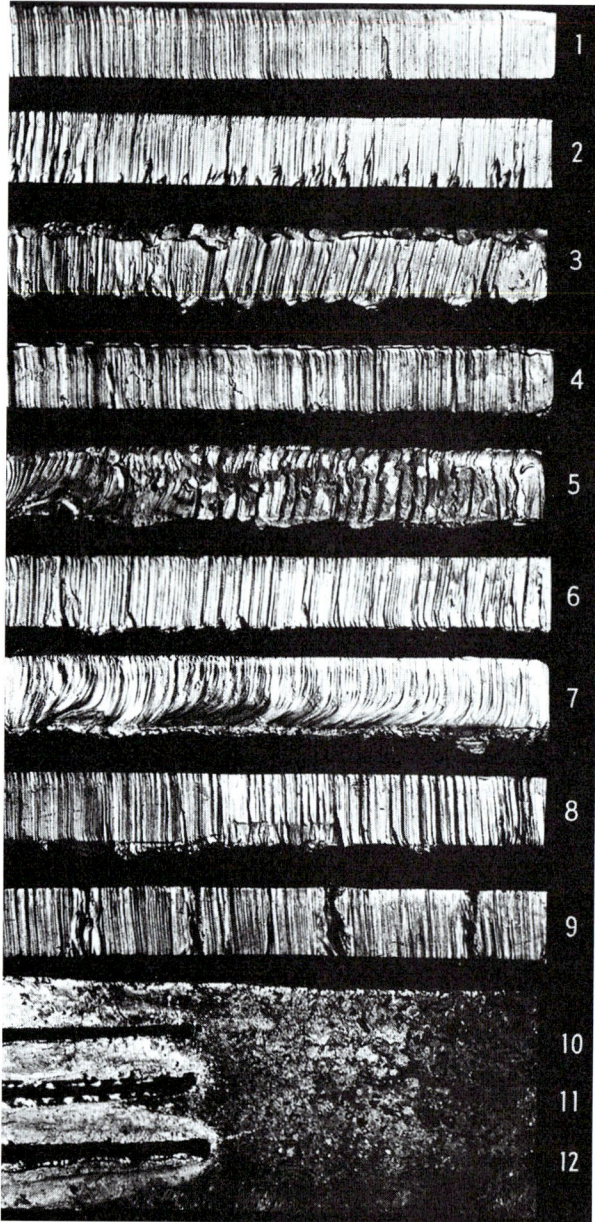

Fig. 7-18 These photographs show good cuts and the common faults that may occur in hand cutting. Quality terms are discussed on page 197. Praxair, Inc.

Figure 7-18 shows high quality cuts and a variety of defects in the following order:

1. This is *a correctly made cut* in 1-inch plate. The edge is square, and the draglines are vertical and not too pronounced. There is little dross along the bottom edge.
2. *Preheat flames were too small* for this cut so that the cutting speed was too slow, causing bad gouging at the bottom.
3. *Preheat flames were too long.* The top surface has melted over, the cut edge is irregular, and there is too much dross.
4. *Oxygen pressure was too low.* The top edge has melted over because of too slow of a cutting speed.
5. *Oxygen pressure was too high and the tip size too small* so that the entire control of the cut has been lost.
6. *Cutting speed was too slow* so that irregularities of the draglines are pronounced.
7. *Cutting speed was too high* so that there is a pronounced break and lag to the dragline, the cut is irregular, and there is too much dross on the bottom side.

ABOUT WELDING

Welding Curtains

To protect passersby from incidental exposure to nonionizing radiation, transparent welding curtains are used. Remember that these curtains are not to be used as welding filter plates. Instead, use a helmet with the proper shade of filter.

8. *Torch travel was unsteady* so that the cut is wavy and irregular.
9. *Cut was lost* and not carefully restarted. Bad gouges were caused at the restarting point.
10. *Correct procedure* was used in making this cut.
11. *Too much preheat* was used and the tip was held too close to the plate so that bad melting at the top edge occurred.
12. *Too little preheat* was used and the flames were held too far from the plate so that the heat spread opened up the kerf at the top. The kerf is too wide at the top and tapers in.

Conditions Affecting Quality of Cut

The following is a list of important conditions that should be met in order to make accurate and smooth cuts. Study these recommendations thoroughly so that you will be able to follow them during your flame cutting practice:

- *Uniformity of oxygen pressure regulation* This is important especially when cutting heavy plate thicknesses. Make sure that the oxygen and fuel regulators are working properly and do not creep or fluctuate.

- *High oxygen purity* If the oxygen contains impurities, the chemical reaction that takes place in the cutting operation will produce poor quality cuts. This is not a serious problem today because the manufacturers of the gas take great care in its production.

- *Proper oxygen operating pressure* High oxygen operating pressures usually do not improve the quality of the cut, and they are a serious waste of oxygen.

- *Proper tip size* If the wrong tip size is being used, cutting is sure to be unsatisfactory. Too large a tip wastes oxygen and fuel. Too small a tip results in poor cutting and wastes time.

- *Smoothness of bore and proper cutting orifice* The bore may become out of round or nicked by improper cleaning methods. Always use a tip drill of the proper size. Do not use a tip drill excessively. Do not use makeshift reamers for this purpose, and do not scrape the tip on the metal being cut. Remember that the tip is made of soft copper, and it will be damaged by the much harder steel.

- *Cleanliness of preheat orifices* If the preheat orifices become dirty or clogged with oxide, the preheat flames will be distorted and the jet of high pressure oxygen will not be directed squarely along the line of cut. This produces ragged and uneven cuts.

- *Uniformity of steel being cut* Any variation in the thickness of the material, its shape, or the plate's surface condition alters the cutting procedure.

- **Low enough speeds for light and moderate thicknesses** A common error is to proceed at excessive speeds, which results in a cut of poor quality.

- **A gradual increase in speed** when starting the cut in very heavy thickness.

- **High enough speeds for heavier thicknesses** If the speed is not high enough, the lower portion of the cut will be ragged. To increase the speed, it is necessary to increase the tip size.

- **Uniformity of torch movement** The thermal cutter should make every effort to develop a smooth, even torch movement. Uneven movements make rough, ragged cuts.

- **Cutting thin material** Angle the torch toward the direction of the cut to prevent the material from wanting to fuse itself back together.

Arc Cutting

In the large welding fabricating shops, there is a growing use of the various arc cutting processes in the fabrication of heavy weldments of various kinds. Consult Chapter 6 for a review of these processes and practice the processes available in your school welding shop. You are strongly urged to become as skilled as possible in their use.

Practice Jobs

Instructions for Completing Practice Jobs

Jobs 7-J0* (p. 187), 7-J1 (p. 193), and 7-J2 (p. 197) are necessary for you to practice in order to be able to cut low carbon steel. Refer to Table 7-2. Before you begin the jobs, study the provided information and drawing that accompanies them. The title block gives the size and type of stock. A pictorial view of the job is shown to help you interpret the side and top views. These jobs also provide experience in layout. Before making the cuts, check your final layout with the drawing for accuracy. Job 7-J3 (p. 198) is an optional activity that your instructor may assign. A drawing has not been given for this job. Any casting of a suitable grade of cast iron may be used for practice.

Job 7-J0 Setting Up Equipment and Closing Down Equipment

Secure the following equipment:

- Cylinder of fuel
- Cylinder of oxygen
- Oxygen regulator
- Fuel regulator

---

*To distinguish job numbers from illustration numbers, the letter *J* is used. Job 7-J0 is thus Job #0, Chapter 7.

Table 7-2 Job Outline: OFC Practice

Job Number	Type of Job	Material[1] Type	Material[1] Size (in.)	Gas Pressure (p.s.i.)[2] Acet.	Gas Pressure (p.s.i.)[2] Oxy.	Diameter of Cutting Orifice (in.)	Text Reference
7-J0	Setting up equipment and closing down equipment	OFC system	NA	5	20	0.030–0.060	213
7-J1	Straight line and bevel cutting	CS	1/4 × 7 × 8	5	20	0.030–0.060	218
7-J2	Laying out and cutting odd shapes	CS	1/4 × 8-1/2 × 11	5	20	0.030–0.060	222
7-J3	Cutting cast iron	Gray CI	Available	5	25–60	0.030–0.060	224

[1]CS = carbon steel, CI = cast iron.

[2]For specifics on your equipment, check your tip size and use the manufacturer's recommended pressure settings.

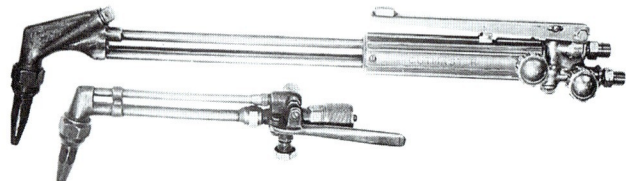

Fig. 7-19 Standard oxyfuel cutting torch and adaptable cutting head. Edward R. Bohnart

- Oxygen hose
- Fuel hose
- Cutting torch with tip, Fig. 7-19
- Cutting goggles
- Fuel tank wrench
- Torch wrench
- Friction lighter

Before setting up the equipment, review pages 1050 to 1052 in Chapter 32.

The equipment should be set up in the following manner:

1. During storage some dirt and dust will collect in the cylinder valve outlet. To make sure that this dirt and dust will not be carried into the regulator, clear the cylinder valves by blowing them out. Crack the valves only slightly, Fig. 7-20. Open the fuel cylinder away from open flame.
2. Attach the oxygen regulator, Fig. 7-21, and the fuel regulator, Fig. 7-22, to the respective cylinders.
    Note that the oxygen regulator has a right-hand thread and that the fuel regulator has a left-hand thread. Make sure that regulator nipples are in line with the valves of the tank so that they may seat properly.

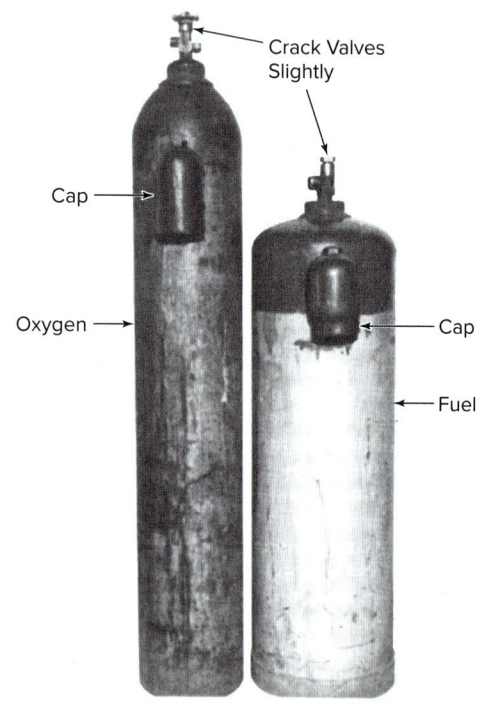

Fig. 7-20 Oxygen and fuel cylinders. Note the difference in the valves on each cylinder. The oxygen valve has a male connection, and the fuel valve has a female connection. Edward R. Bohnart

3. Attach the welding hose to the regulators, Fig. 7-23. The oxygen hose is green and has a right-hand thread. The fuel hose is red and has a left-hand thread. Make sure that the nipples of the hose are in line with the connections of the regulator to ensure correct seating. Tighten nuts securely and make sure that the connections do not leak. Use only approved leak testing solution.

Fig. 7-21 Attaching an oxygen regulator to an oxygen cylinder. One gauge is for the high pressure in the oxygen cylinder while the other is for the working pressure in the torch. Location: Northeast Wisconsin Technical College Mark A. Dierker/McGraw Hill

Fig. 7-22 Attaching a fuel regulator to a fuel cylinder. Note the gauge on the left is for the low pressure fuel gas for the torch. In this case acetylene is being used and red lines at 50 p.s.i. The gauge on the right is for the high pressure in the fuel gas cylinder. Location: Northeast Wisconsin Technical College Mark A. Dierker/McGraw Hill

Fig. 7-23 Attaching the oxygen hose to the oxygen regulator. Location: Northeast Wisconsin Technical College Mark A. Dierker/McGraw Hill

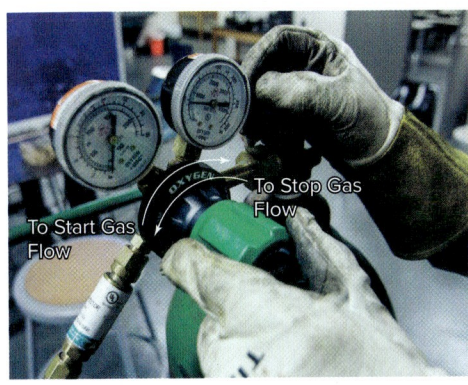

Fig. 7-24 Regulator-adjusting screw is turned to the right to cause the flow of gas through the regulator and hose to the torch. Location: Northeast Wisconsin Technical College Mark A. Dierker/McGraw Hill

> **SHOP TALK**
>
> **Weld Repairs**
> Before making a repair weld, you can examine the amount of cracking by magnetic particle testing or liquid dye penetrant testing.

4. It is important that no grit or dirt be allowed to remain in the regulators or the hose. This can be prevented by blowing out regulator and hose. After having made sure that the regulator-adjusting screw is fully released, open the cylinder valves slowly, so as not to damage the regulator. Adjust the regulators by turning the adjusting screws to the right, Fig. 7-24, to permit the gases to flow through both hoses. Turn the adjusting screws to the left again after enough gas has been allowed to flow to clear regulators and hose.
5. When connecting the hose to the torch, Fig. 7-25, connect the green oxygen hose to the torch outlet marked *oxygen* and the red fuel hose with the left hand threaded nut with the groove machined around it to the torch outlet marked *fuel*. Make sure that nipples of the hose connections are in line with the torch outlets for proper seating. Be certain that connections do not leak.
6. To adjust the working pressure, Table 7-3, open the torch valves and set the pressure on the regulator by turning the adjusting screw to the right, Fig. 7-24. Close the torch valves when pressure has been set.
7. To light the torch, open the fuel preheat valve and light the gas with a friction lighter, Fig. 7-26. The fuel will burn with a smoky yellow flame, Fig. 7-27,

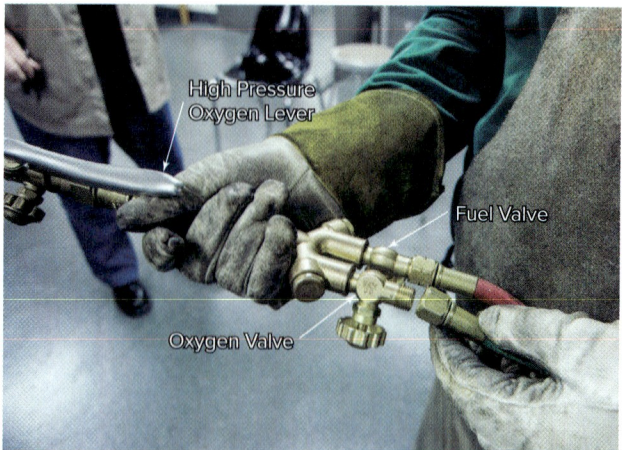

Fig. 7-25 Connecting the oxygen hose to the oxygen outlet on the cutting torch. The fuel hose is already connected.
Location: Northeast Wisconsin Technical College Mark A. Dierker/McGraw Hill

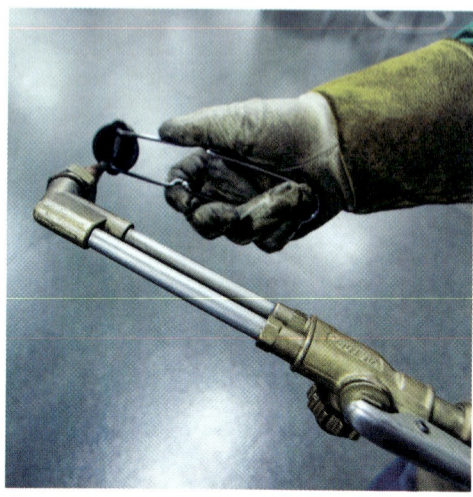

Fig. 7-26 Lighting the cutting torch with a friction lighter.
Location: Northeast Wisconsin Technical College Mark A. Dierker/McGraw Hill

and give off fine black soot. When the fine black soot disappears from the flame, stop adding fuel gas.

8. Open the oxygen preheat valve slowly. The flame will gradually change in color from yellow to blue, Fig. 7-28. It will show the characteristics of the excess fuel flame, also called the **carburizing** or *reducing flame*. There will be three distinct parts to the flame: a brilliant but feathery edged inner cone surrounded by a secondary cone, and a bluish outer envelope forming a third zone. As you adjust

Table 7-3 Hand Flame Cutting Data for Mild Steel (Not Preheated)

Thickness of Steel (in.)	Diameter of Cutting Orifice (in.)	Oxygen Pressure (p.s.i.)	Cutting Speed (in./min)	Gas Consumption	
				Oxygen (ft³/h)	Acetylene (ft³/h)
1/8	0.0200–0.0400	15–30	20–30	18–55	6–9
1/4	0.0310–0.0595	11–20	16–26	37–93	7–11
3/8	0.0310–0.0595	17–30	15–24	47–115	7–12
1/2	0.0400–0.0595	20–31	12–22	66–125	10–13
3/4	0.0465–0.0595	24–35	12–20	117–143	12–15
1	0.0465–0.0595	28–40	9–18	130–160	13–16
1 1/2	0.0595–0.0810	30–45	6–12	150–225	15–20
2	0.0670–0.0810	22–50	6–13	185–231	16–20
3	0.0670–0.0810	33–55	4–10	207–290	16–23
4	0.0810–0.0860	42–60	4–8	235–388	20–26
5	0.0810–0.0860	49–70	3.5–6.4	281–437	20–29
6	0.0980–0.0995	36–80	3.0–5.4	400–567	25–32
8	0.0995–0.1100	57–77	2.6–4.2	505–625	30–39
10	0.0995–0.1100	66–96	1.9–3.2	610–750	36–46
12	0.1100–0.1200	58–86	1.4–2.6	720–905	42–55

Note: Acetylene pressures have been omitted from this table, since they depend mainly on torch and tip design rather than on external factors such as thickness of the material. It is therefore recommended that for specific fuel pressure settings particularly, and also for oxygen pressure values, reference be made to the equipment manufacturer's charts for the apparatus being used.

Source: American Welding Society

Fig. 7-27 Atmospheric burning of acetylene. Open acetylene valve until smoking of the flame disappears. David A. Tietz/Editorial Image, LLC

Fig. 7-28 Carburizing cutting flame. It is an excess acetylene flame. More oxygen is required for preheat flames. David A. Tietz/Editorial Image, LLC

Fig. 7-29 Neutral cutting flame without excess oxygen or acetylene. At proper adjustment, the temperature of the flame is approximately 6,300°F. David A. Tietz/Editorial Image, LLC

Fig. 7-30 Oxidizing cutting flame. It is an excess oxygen flame used only for rough work. It reduces preheat time. David A. Tietz/Editorial Image, LLC

the oxygen valve, you will notice that the secondary cone gets smaller and smaller until it finally disappears. Just at this point of complete disappearance the neutral flame is formed, Fig. 7-29. If you continue to adjust the oxygen valve, an **oxidizing** (excess oxygen) flame will be formed. The entire flame will decrease in size, and the inner cone will become much less sharply defined. The sound of the flame becomes a shrill hissing noise, Fig. 7-30.

9. Practice adjusting the flame several times until you are completely familiar with the neutral flame.

The fastest preheat time is achieved with an oxidizing flame because it is hotter and more concentrated than the carburizing flame and because it speeds up the oxidizing processes, Fig. 7-31. If the cut surface is to be used for welding, the neutral flame is recommended. The cut surface will also have a cleaner appearance with the use of the neutral flame.

10. Check the preheat flame by pressing the high pressure lever shown in Fig. 7-25 and observe any change in the flame, Fig. 7-32. Excess fuel feather should not appear.

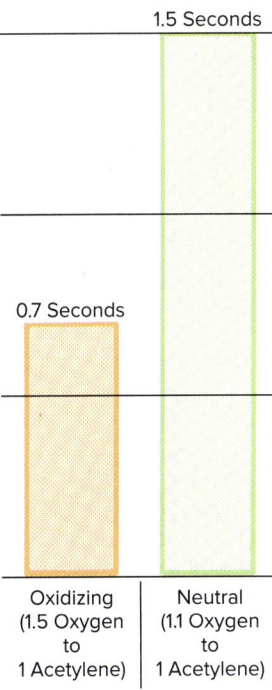

Fig. 7-31 Comparative preheat time for two flame types.

JOB TIP

Credibility

Others will like working with you the more they find you have credibility. Here are ways to increase your credibility on the job:

1. Competence: get the training you need to become qualified and knowledgeable.
2. Trustworthiness: treat others fairly, with respect.
3. Dynamism: be active, confident, and hard-working.
4. Coorientation: see how your company's goals for success fit with your own.

Fig. 7-32 Neutral flame with the cutting lever depressed. The cutting jet is straight and smooth. David A. Tietz/Editorial Image, LLC

Flame Cutting Practice: Jobs 7-J1–J3   **Chapter 7**   **191**

Fig. 7-33 Position while cutting over an appropriate area. Location: Northeast Wisconsin Technical College Mark A. Dierker/McGraw Hill

11. Hold the torch as indicated in Fig. 7-33 for both preheating and cutting. Figures 7-34 and 7-35 illustrate two of the several positions the welder may have to use when cutting.

Fig. 7-34 Plates can be cut at the point of fabrication and need not be carried to another machine for cutting and back again for fabrication. Monty Rakusen/Cultura/Getty Images

Fig. 7-35 Making a manual bevel cut on a pipe. As might be required in the field. It would be impossible to find a shape that could not readily be cut with the handheld cutting torch. Location: UA Local 400 Mark A. Dierker/McGraw Hill

Attaching the Adaptable Cutting Head to the Oxyfuel Torch The adaptable cutting head is attached in the following manner:

1. Detach the welding tip and neck from the welding torch handle by unscrewing the shell nut, Fig. 7-36.

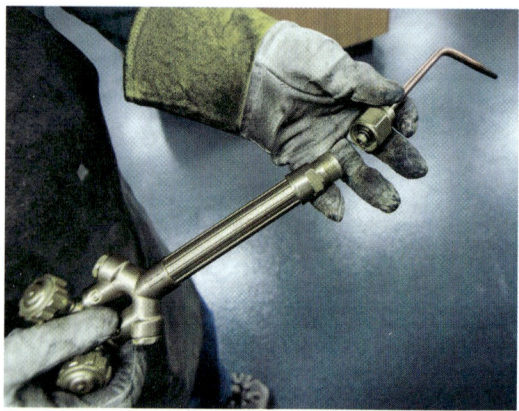

Fig. 7-36 Detaching the welding head and tip from a standard oxyfuel torch body. Location: Northeast Wisconsin Technical College Mark A. Dierker/McGraw Hill

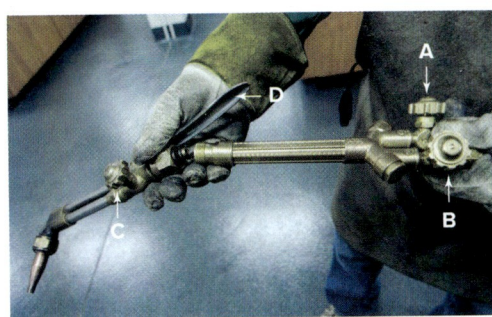

Fig. 7-37 Attaching an adaptable cutting head to the welding torch body. The welding torch may now be used as a cutting tool.
Location: Northeast Wisconsin Technical College Mark A. Dierker/McGraw Hill

2. Attach the adaptable cutting head to the welding torch handle by screwing the shell nut to the handle, Fig. 7-37.
3. Light the torch by opening oxygen valve A, Fig. 7-37, on the welding torch handle wide to permit the passage of oxygen to oxygen preheat valve C on the adaptable cutting head. Open fuel valve B, Fig. 7-37, on the welding torch handle and light the gas with a friction lighter.
4. Open the oxygen preheat valve C on the adaptable cutting head until the proper flame is obtained.
5. Check the preheat flame by pressing the high pressure lever D, Fig. 7-37, on the adaptable cutting head and observe the effect upon the flame. You are now ready to begin cutting.

Closing Down Equipment

When the cutting operation is finished, the equipment may be taken down in the following manner:

1. To extinguish the flame, close the oxygen valve on the torch. Next close the fuel valve on the torch. This is just opposite as to what past thinking was. If welders/burners are in the habit of turning off the oxygen first and then the fuel and an emergency situation is encountered like a sustained backfire or flashback into the torch, they will deal with it instinctively the correct way. Sustained backfire or flashback is a sustained burning of the flame back inside the torch, usually at the mixer, but under certain situations can burn upstream past the mixer. They are recognized by a hissing or squealing sound and/or a smoky, sharp-pointed flame. The torch should be shut down immediately. Severe damage to the torch, as well as an increased risk of fire, would result if the valves were not turned off immediately in the correct order. Sustained backfires and flashbacks are normally contained inside the torch with the quick action of the welder in shutting off the valves. This prevents the flame from burning through to the outside of the torch. Sustained backfire and burnback are serious in that if it is not recognized quickly and the valves turned off the flame can burn through the torch body and spew fire and molten metal.
2. Close the fuel and oxygen cylinder valves.
3. Open the fuel and the oxygen valves on the torch to permit the trapped gas to pass out of the regulators and the hose. Then, release the regulator-adjusting screws by turning to the left.
4. Disconnect the hose from the torch.
5. Disconnect the hose from the regulator.
6. Remove the regulators from the cylinders and put protective caps on the cylinders.

Job 7-J1 Straight Line and Bevel Cutting

Objective

To make a straight line and bevel cuts on a flat plate with the oxyfuel hand cutting torch.

General Job Information

Almost all cutting that the thermal cutter does on the job, whether it be a straight line or shape cut, is made with either a square edge or a beveled edge. On many of the jobs in manufacturing plants, it is not necessary for the welders to be able to cut. It is essential, however, that they are able to cut in fabricating plants and on outside construction jobs. Cutting is a quick and economical method of shaping a plate for fabrication. Refer to Fig. 7-38 for the many shapes that can be flame cut by hand or machine.

For video of OFC being made, please visit www.mhhe.com/welding.

Cutting Technique

It is assumed that you have your equipment set up, have selected the proper tip size, and have practiced flame adjustment. Study Table 7-3, page 190, which lists gas pressures. In the actual cutting, the left hand is used to steady the torch, the elbow or forearm is rested on a convenient

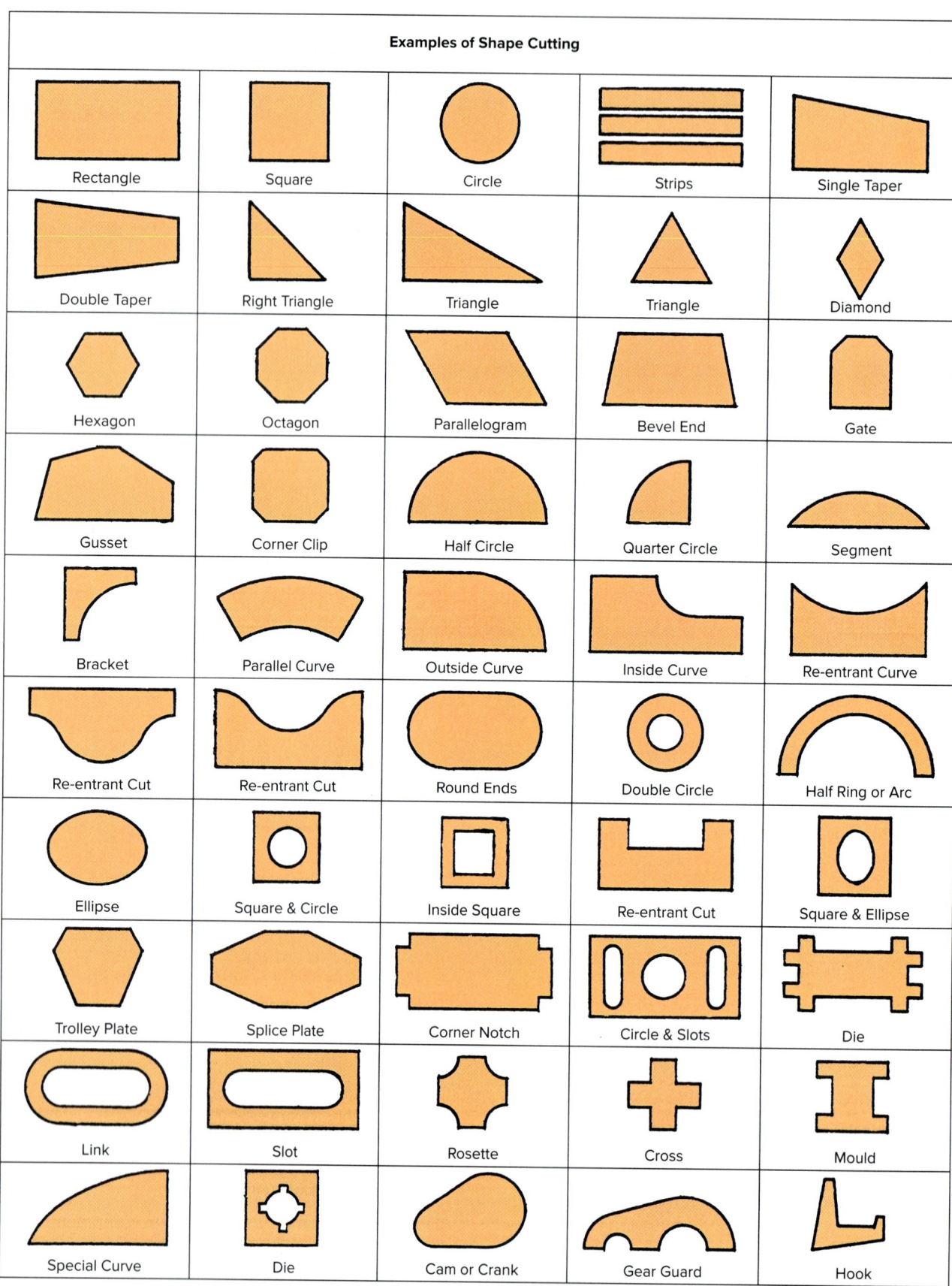

Fig. 7-38 Examples of shape cutting.

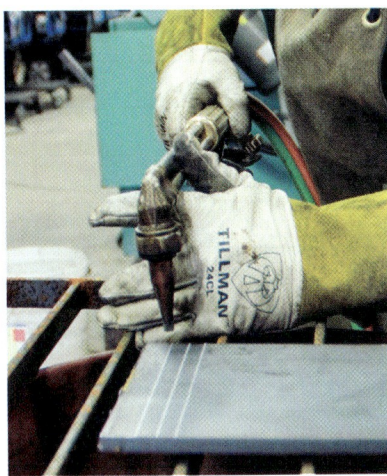

Fig. 7-39 Position of the cutting torch when making square cuts in the flat position. Location: Northeast Wisconsin Technical College
Mark A. Dierker/McGraw Hill

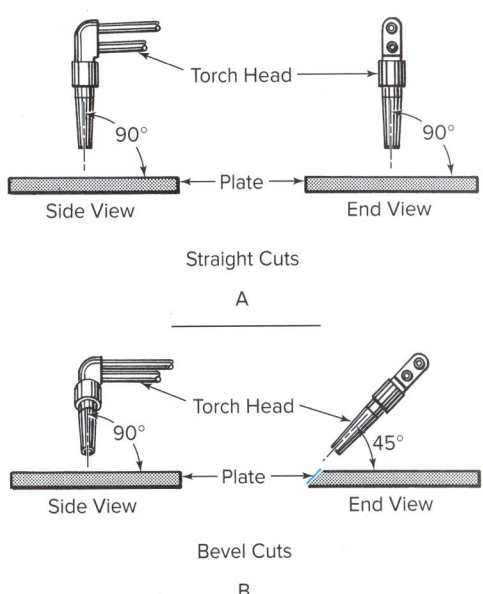

Fig. 7-40 Position of the cutting head when making (A) straight and (B) bevel cuts in the flat position.

support, and the right hand is used to move the torch along the line of cut, Fig. 7-39. The thumb on the right hand is used to operate the oxygen cutting lever. The tips of the preheating flame cones are held about 1/16 inch above the surface of the material.

Hold the torch steady. When a spot of metal at the edge of the plate has been heated to a cherry red, press the lever controlling the oxygen cutting jet and start the cutting action. Move the torch slowly but steadily along the line of cut. If the torch head wavers, a wide cut will result. Besides producing a poor cut, this is a waste of time and material.

For square cuts, the cutting head must be held perpendicular, Fig. 7-40A. For bevel cuts, the cutting head should be tilted at an angle corresponding to the angle of bevel desired, Fig. 7-40B. Figure 7-41A shows the torch positioned for bevel cuts as instructed in Fig. 7-41B.

If the cut has been started properly, a shower of sparks will fall from the underside of the plate. This indicates that the cutting action is penetrating through the work. The forward movement should be just fast enough so that the cut continues to penetrate the full plate thickness.

If the torch is moved forward too rapidly, the cutting jet will fail to go clear through the plate, the molten pool will be lost, and the cutting will stop. If this happens, close the cutting valve immediately and preheat the joint where the cut stopped until it is a bright red. Open the cutting valve to restart the cut.

If, on the other hand, the torch is not moved forward rapidly enough, the preheating flames will tend to melt the edges of the cut. This may produce a ragged edge or fuse the metal together again.

Watch the cutting action closely. Make sure that the dross flows freely from the kerf and does not gather at the

Fig. 7-41A Position of the cutting torch when making bevel cuts in the flat position. Location: Northeast Wisconsin Technical College
Mark A. Dierker/McGraw Hill

bottom of the groove, thus hindering the cut and producing rough edges.

The speed of travel is determined by the cutting action and varies with the thickness of metal and the tip size being used. Do not cut beyond your comfort range. It is better to reposition yourself and start anew.

Operations

1. Obtain plate (1/4 × 7 × 8 inches).
2. Obtain a square head and scale, center punch, and scribe from the toolroom.

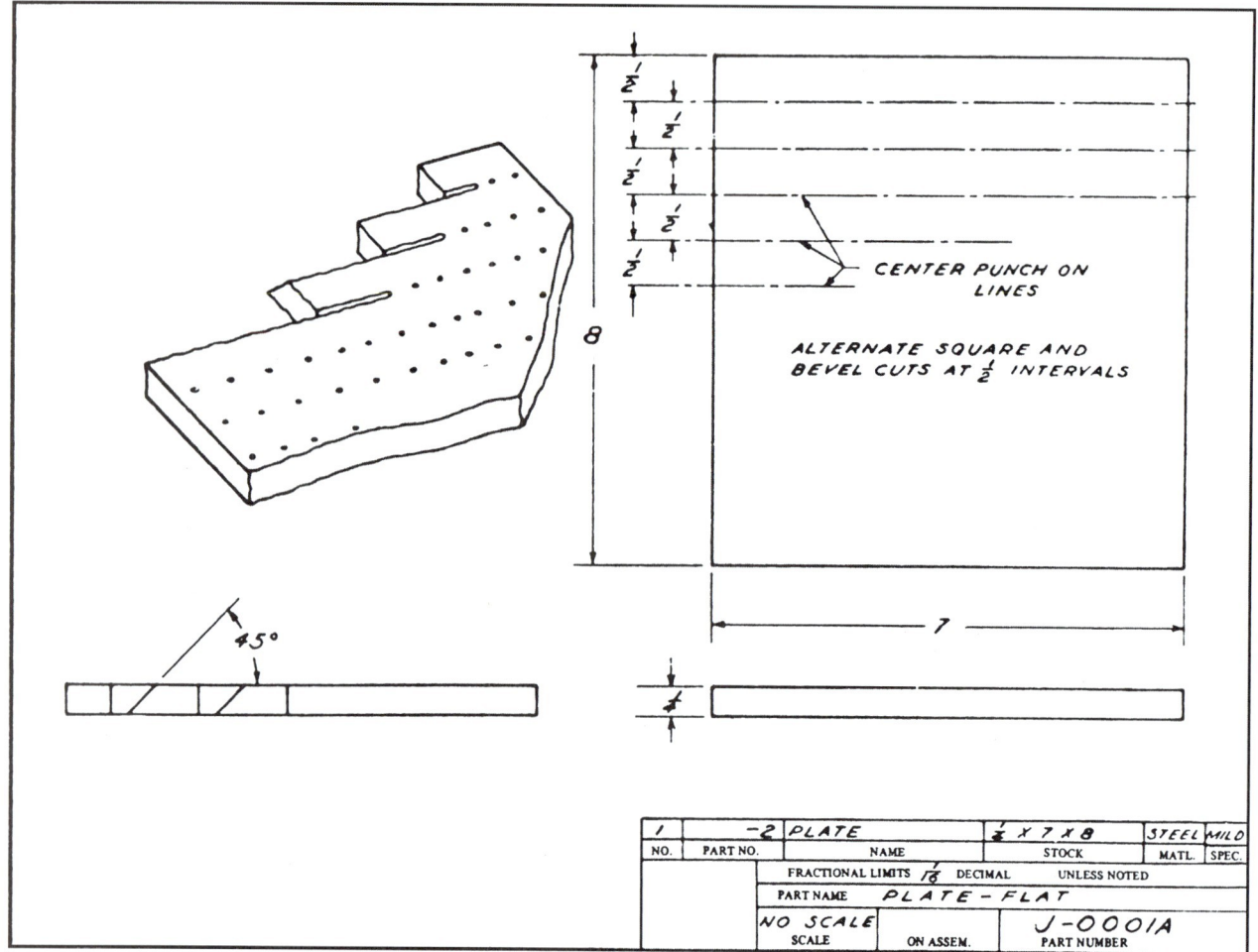

Fig. 7-41B Drawing of part; Job 7-J1.

3. Lay out parallel lines as shown on the job drawing, Fig. 7-41B.
4. Mark the lines with a center punch.
5. Set up the oxyfuel cutting equipment. (Refer to pp. 187–193.)
6. Adjust the regulators for 5 pounds fuel pressure and 20 pounds oxygen pressure. Check for specifics on your equipment. Adjust the regulator pressure according to the tip size and the manufacturer's suggested pressure settings.
7. Light the torch, adjust it to the proper flame, and proceed to cut as shown on the job drawing, Fig. 7-41B.
8. Chip the dross from the cut and inspect it.
9. Practice until you can produce even, clean cuts consistently.
10. Take down the oxyfuel cutting equipment.

Inspection

The top edges of the cut should be sharp and in a straight line—not ragged. The bottom edge will have some oxide adhering to it, but it should not be excessive. The face of the cuts should be square and smooth. Every effort should be made to remove gouging on the surface of the cut. If a clean cut has been made, the piece will fall when the cut is finished. There will be no tendency for the plate to weld together at the bottom of the cut. Compare your cut with Fig. 7-18 (1).

ABOUT WELDING

TWI

Welding measurement tools are available at Britain's TWI (The Welding Institute), which has a Web site and a magazine called *Connect*. The tools available include a weld gauge and a linear angular measurement gauge. The group also offers online technology files, which are detailed research essays about materials-joining technology.

Terms Used to Describe Oxygen-Cut Surface Quality

- *Flatness (F)* The distance between the two closest parallel planes between which all points of the cut surface lie.
- *Angularity (A)* The deviation in degrees at any point of the cut surface from the specified angle.
- *Draglines (D)* Lines that appear on the oxygen-cut surface. Their contours and directions do not affect the quality of the cut surface.
- *Roughness (R)* Roughness consists of recurring peaks and valleys in the oxygen-cut surface. This can be determined by samples of acceptable quality or by comparison to the AWS C4.1 Surface Roughness Guide for Oxygen Cutting. See Fig. 7-42 for an example of AWS C4.1, which is a plastic replica available from the American Welding Society and represents a national standard.
- *Top edge rounding (T)* Melting of the top edge of an oxygen-cut surface.
- *Notch (N)* Gouges in an oxygen-cut surface significantly deeper than the overall surface roughness.
- *Dross (DRS)* Deposit resulting from the oxygen cutting process that adheres to the base metal or cut surface.

Refer back to Fig. 7-18, page 186, for photos showing good cuts and common faults.

Disposal
Cut strips should be discarded in the waste bin.

Job 7-J2 Laying Out and Cutting Odd Shapes

Objectives
1. To lay out simple shapes on flat plate.
2. To cut out shapes on flat plate, using the oxyfuel hand cutting torch.

General Job Information
You will do many jobs for which you must be able to lay out squares, parallel lines, and circles. You must also be prepared to cut in other than a straight line. Forms will take on many odd shapes and if you develop in school the flexibility to follow shaped lines, you will have little difficulty on the job. Study the examples in Fig. 7-38, page 194, carefully.

Cutting Technique
Hold the torch head in the position shown in Fig. 7-43. The cutting technique is basically the same as in Job 7-J1

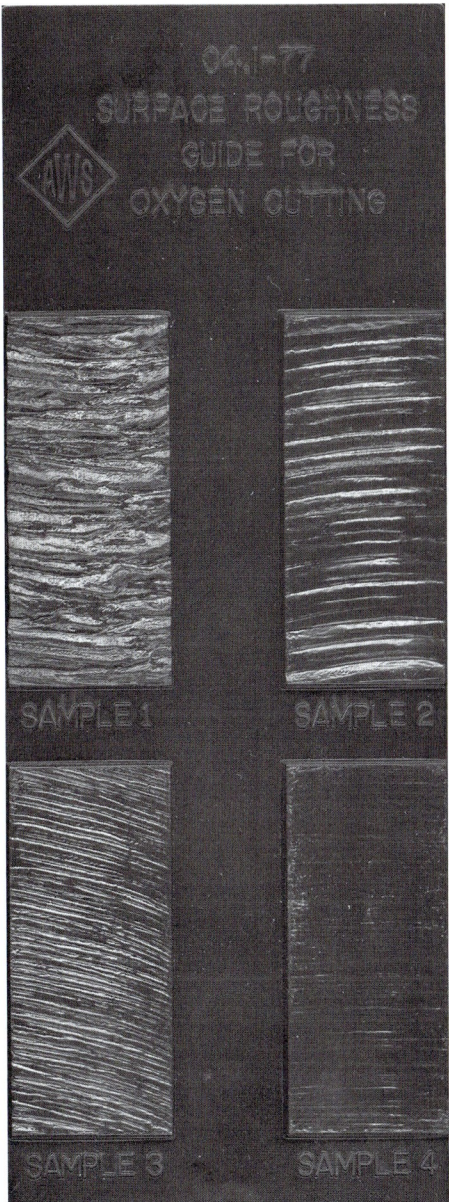

Fig. 7-42 Surface roughness guide for oxygen cutting. For materials up to and including 4 in. thick, Sample 3 should be used, and for materials over 4 in. and up to 8 in. thick, Sample 2 should be used. American Welding Society. C4.6M:2006 (ISO 9013:2002 IDT), Figure C.8

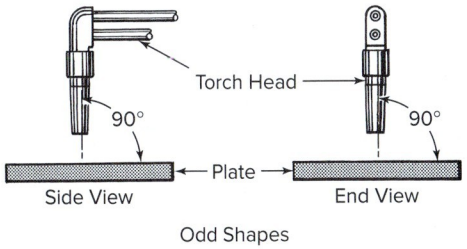

Fig. 7-43 Position of the cutting torch head when making square cuts.

except that starts will have to be made in the plate away from the edge. This requires more time to bring the plate up to the kindling temperature.

After the spot has been brought up to a cherry red, raise the torch about ½ inch above the plate and turn on the cutting oxygen slowly. When the plate has been pierced through, lower the torch to the normal cutting height and complete the cut. Here again cut edges may be square or beveled, depending upon the nature of the cut. When cuts must be precise, drill a hole along the edge of the line and start the cut in the hole.

Operations

1. Obtain a flat plate (¼ × 8½ × 11 inches).
2. Obtain a square head and scale, dividers, scribe, and center punch from the toolroom.
3. Lay out the shapes as shown on the job drawing, Fig. 7-44B.
4. Mark the lines with a center punch.
5. Set up the oxyfuel cutting equipment.
6. Adjust the regulators for 5 pounds fuel pressure and 20 pounds oxygen pressure. Check for specifics on your equipment. Adjust the regulator pressure according to the tip size and the manufacturer's suggested pressure settings.
7. Light the torch, adjust it to the proper flame, and proceed to cut as shown in the job drawing.
8. Chip the dross from the cut and inspect it.
9. Practice these cuts until you can produce even, clean cuts consistently. Compare them with the cuts in Fig. 7-18 (1), page 186, and Fig. 7-44A.
10. Take down the oxyfuel cutting equipment.

Inspection

The cut edge of both the pieces cut from the plate and the plate itself should be as clean as possible. The top edges should be sharp. They should show no evidence of raggedness. Inspect the bottom of the cuts for excessive dross. The faces of cut should be square, smooth, and free of gouging. If the cuts have been made cleanly, the pieces will fall from the plate when entirely cut through.

Disposal

Return the plate to the scrap bin. It may be used for Job 13-J1, Striking the Arc and Short Stringer Beading, in Chapter 13.

Job 7-J3 Cutting Cast Iron

Objective

To make cuts in cast iron using a carburizing flame and an oscillatory motion of the torch.

General Job Information

Most industrial welders do not have occasion to cut cast iron with oxyfuel flame. If skill in this technique is necessary in your area, you should complete the job after you have had some experience with the hand and machine cutting of low carbon steel.

Cutting cast iron with the oxyfuel flame is somewhat more difficult than the cutting of steel. You will recall that the oxidation process slows down as the amount of carbon and other alloy materials is increased. Cast iron, of course, is relatively high in carbon compared with low carbon steel.

Cutting Technique

The technique for cutting cast iron is quite different from that for cutting steel. The preheat flame must be larger for cast iron and should be adjusted for an excess of fuel. The oxygen pressure required is from 25 to 200 percent greater than that for cutting steel of the same thickness. The torch tip is moved back and forth constantly across the line of cut.

The quality of the cast iron also determines the ease with which it can be cut. Good quality gray iron, such as that usually used for machinery parts, is less difficult to cut than the furnace iron which is for cheap castings.

When cutting a poor grade of cast iron, the dross does not flow freely. The fluidity of the dross can be increased by mixing it with a flux. Feeding a steel rod into the kerf is a form of flux cutting. This may be done at the start of the cut and may have to be continued for the length of the cut, Fig. 7-45.

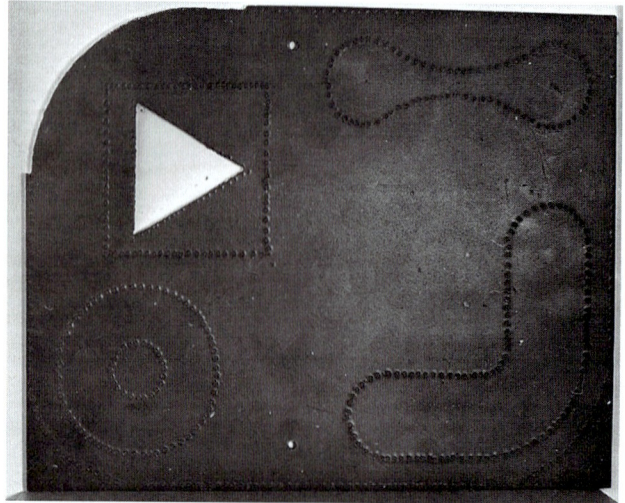

Fig. 7-44A Practice plate for shape cutting. Note the method of laying out guidelines and center punching. *Edward R. Bohnart*

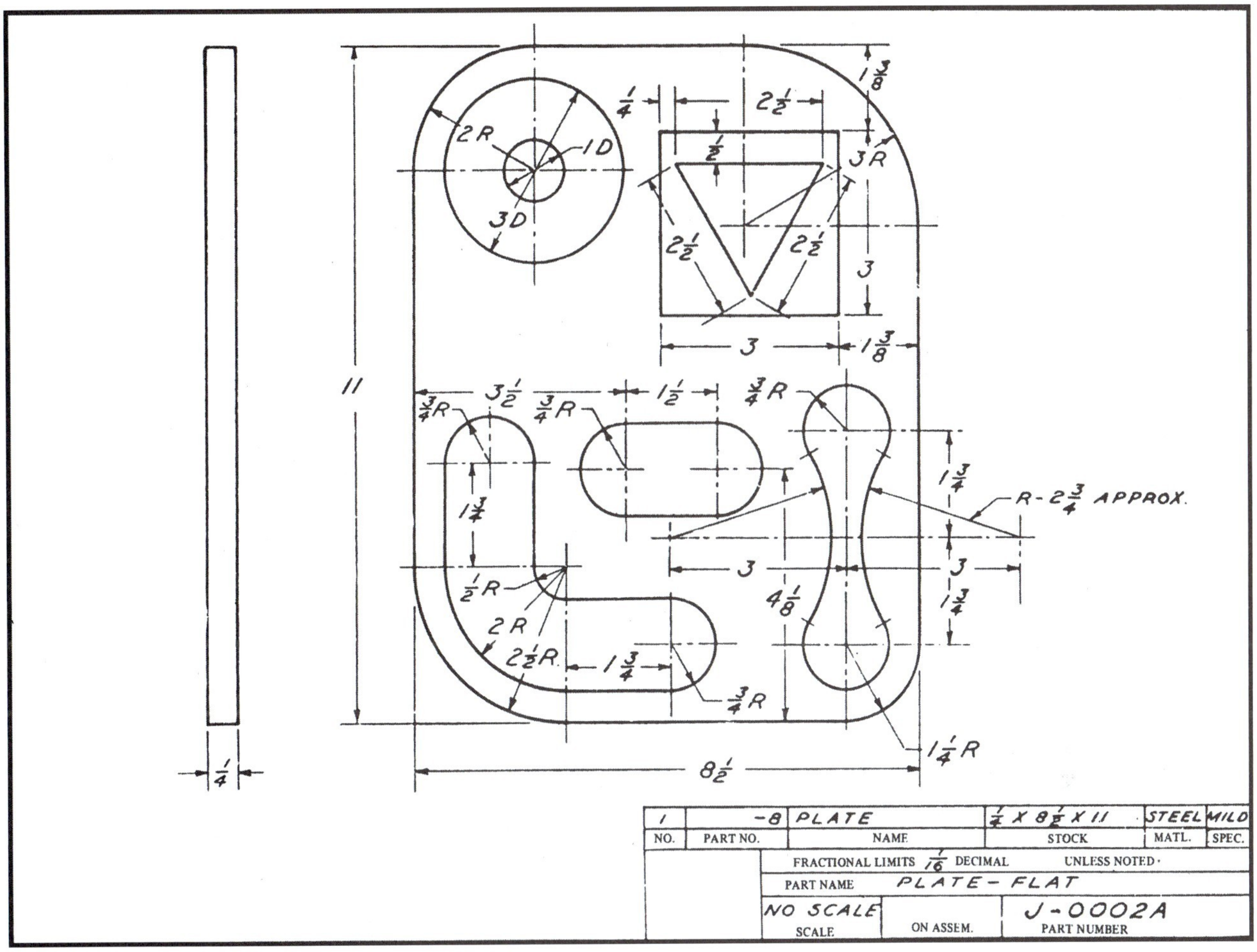

Fig. 7-44B Drawing of part; Job 7-J2.

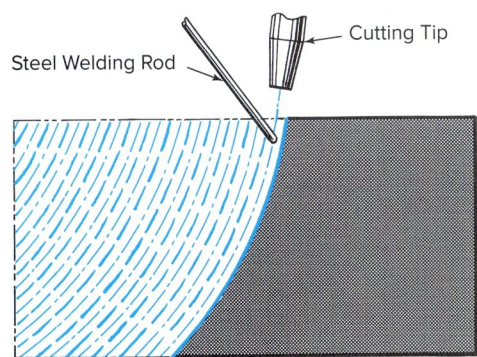

Fig. 7-45 The use of a steel welding rod for starting a cut on cast iron is an example of flux cutting. This method is used frequently on low-grade castings, as well as on those that are somewhat oxidation-resistant.

Operations

1. Select a piece of high quality gray cast iron of whatever thickness is available.

2. Adjust the excess fuel preheat flames with the cutting oxygen valve open.
3. Close the cutting oxygen valve and begin preheating along the front edge of the casting from top to bottom, Fig. 7-46. This makes starting and continuing the cut easier.
4. When the line of the cut is heated, position the torch tip at about a 45° angle with the inner cones of the flame about ⅛ to ¼ inch above the surface of the casting, Fig. 7-47.
5. Heat a semicircular area about ¾ inch in diameter until it is molten, Fig. 7-46. Open the cutting oxygen valve to start the cut. Use an oscillating (back and forth) motion throughout the cut, Fig. 7-47C. When the cut extends through the bottom of the casting, change the tip angle to approximately 75°, Fig. 7-47B. Be sure to keep the metal hot so that the cut is not lost. If the cut is lost, heat up one side and part of the end of the cut again just as if starting a new cut. When cutting low-grade castings, the entire cut may

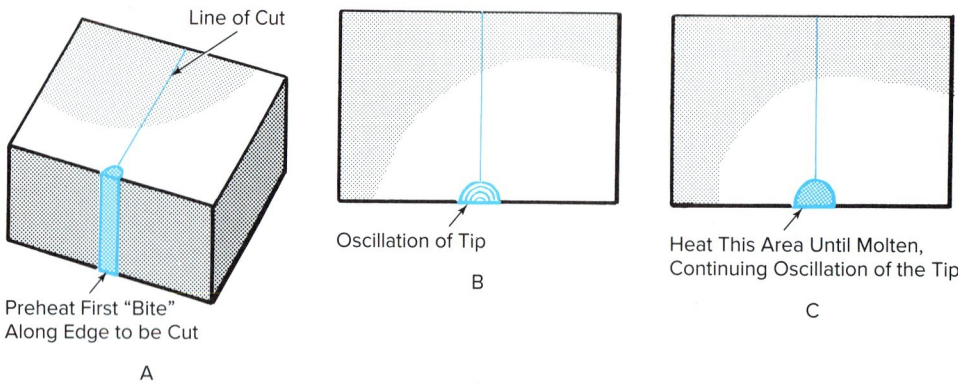

Fig. 7-46 The first step in cast iron cutting is to preheat the entire length of the edge to be cut. The tip is then directed at the starting point at the top and the side-to-side motion begun.

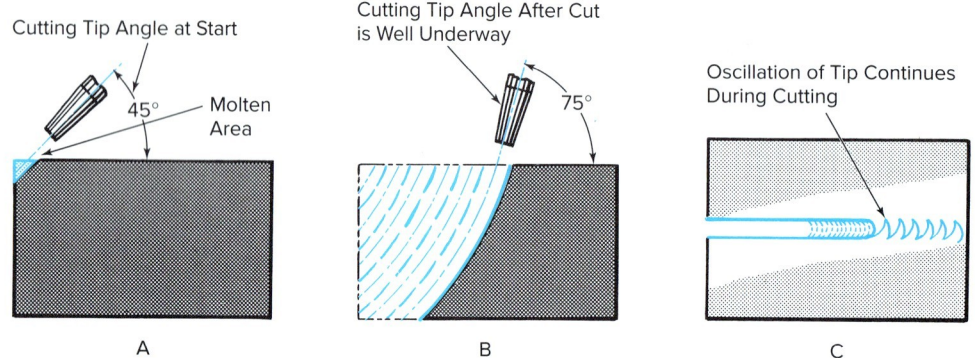

Fig. 7-47 At the start of cutting, the tip makes an angle of about 45° with the top surface. As the cutting oxygen lever is gradually depressed so that the cutting will start, the side-to-side motion of the tip is continued.

SHOP TALK

Inverter Power Sources
New welding inverter power sources
1. Have good arc control
2. Are portable
3. Are reliable
4. Allow tungsten inert gas (TIG), MIG, MAG, stick, and flux cored wire welding
5. Sport recessed controls
6. Have storage for gloves and helmets

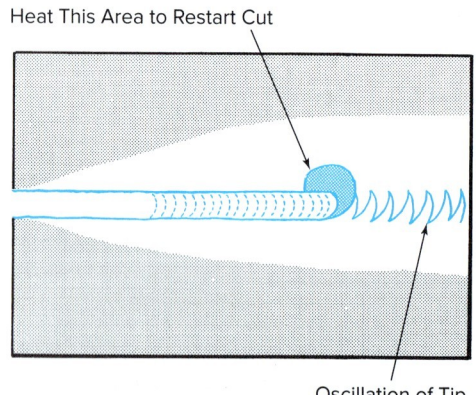

Fig. 7-48 Restarting the cut.

have to be done as a series of definite steps in this way, Fig. 7-48.
6. There is always considerable lag when cutting cast iron even if the tip is brought up to the vertical position. When the cutting has reached the far edge of the plate, carry it down the far side at about the same angle as the angle of lag until the cut is completed, Fig. 7-49.
7. Practice these cuts until you can produce even, clean cuts consistently.
8. Take down the oxyfuel cutting equipment.

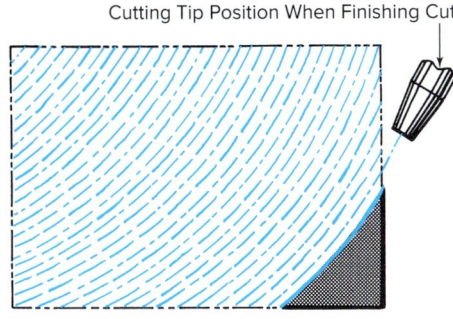

Fig. 7-49 Finishing the cut.

Inspection

Inspect the surface of the cut and compare it with Fig. 7-18 (1), page 186. The face of the cut will not be as smooth as that possible with steel plate. It should, however, be in a fairly straight line and square from top to bottom.

Disposal

Cut strips should be discarded in the waste bin.

CHAPTER 7 REVIEW

Multiple Choice

Choose the letter of the correct answer.

1. Which of the following gases can be used for OFC? (Obj. 7-1)
   a. Acetylene
   b. Mapp
   c. Propylene
   d. All of these

2. The choice of the type of fuel gas used should be based on cost of the gas, availability of gas, flame temperature, length of cut, and speed of cutting. (Obj. 7-1)
   a. True
   b. False

3. The kerf describes _____. (Obj. 7-1)
   a. Dross left along the bottom of the cut
   b. A saw-toothed surface left by the cut
   c. The material removed by the cutting oxygen
   d. Both a and b

4. Why does raising a small burr on the surface of the steel to be pierced reduce the time required to make a pierce? (Obj. 7-1)
   a. Lowers the kindling temperature
   b. Increases the flame temperature
   c. Presents a surface that catches the flame
   d. Doesn't help at all

5. Propylene gives the highest temperature flame. (Obj. 7-1)
   a. True
   b. False

6. Acetylene has the highest BTU rating. (Obj. 7-1)
   a. True
   b. False

7. The lines that are formed on the face of the cut surface are referred to as _____ lines. (Obj. 7-2)
   a. Kerf
   b. Drag
   c. Straight
   d. None of these

8. Which of the following is a technique or procedure used for the treatment of metal with the oxyfuel process? (Obj. 7-2)
   a. Straight line and bevel cutting
   b. Scarfing
   c. Gouging
   d. All of these

9. Which of the following conditions affects the quality of the cut? (Obj. 7-3)
   a. Uniform oxygen pressure, high oxygen purity, and proper oxygen cutting pressure
   b. Proper tip size, uniformity of steel being cut, and uniformity of torch movement
   c. Smoothness of bore of cutting orifice and cleanliness of preheat orifices
   d. All of these

10. The function of the high pressure lever on the gas cutting torch is to _____. (Obj. 7-4)
    a. Start the cutting operation
    b. Preheat the metal
    c. Blow dust off the bench
    d. Replenish the oxygen in the air being burned with the preheat flames

11. The neutral flame is the hottest preheat flame. (Obj. 7-4)
    a. True
    b. False

12. A special welding torch is necessary in order to use the adaptable cutting head. (Obj. 7-4)
    a. True
    b. False
13. When turning off the cutting torch, the fuel valve is closed first. (Obj. 7-4)
    a. True
    b. False
14. A special tip is required for gouging. (Obj. 7-4)
    a. True
    b. False
15. When cutting cast iron, the oxygen pressure is less than for steel of the same thickness. (Obj. 7-4)
    a. True
    b. False
16. If the torch is moved forward too rapidly, the top edge will be melted and will produce a ragged edge and will allow the metal to fuse together again. (Obj. 7-5)
    a. True
    b. False
17. It is not important in current manufacturing operations for a welder to be able to use the OFC process. (Obj. 7-6)
    a. True
    b. False
18. The preheat flames of the cutting torch should be in contact with the metal surface while the cut is in progress. (Obj. 7-6)
    a. True
    b. False

Review Questions

Write the answers in your own words.

19. Explain the reaction that takes place in the cutting of ferrous metals with the oxyfuel cutting process. (Obj. 7-1)
20. Name four factors that determine the type of fuel gas to use. (Obj. 7-1)
21. Why must care be taken when flame cutting steels with high carbon content or alloy content that have air-hardening properties? What can prevent these problems? (Obj. 7-1)
22. Identify the applications in which propane, Mapp gas, and natural gas are better choices than acetylene. (Obj. 7-1)
23. Identify the OFC techniques. (Obj. 7-2)
24. Describe how the plate and pipe should look after flame cutting. (Obj. 7-3)
25. Name the OFC equipment needed for a job. (Obj. 7-4)
26. Explain why high oxygen operating pressure is not necessary. (Obj. 7-5)
27. Is it possible to weld directly on an edge that has been flame cut? Explain. (Obj. 7-5)
28. Explain how your hands are used in the actual cutting process. (Obj. 7-6)

INTERNET ACTIVITIES

Internet Activity A

Suppose you wanted to find some information on flame cutting fuel gases. How would you find it on the Internet? What would be your key word(s)? Name two Web sites that you found.

Internet Activity B

Use the Internet to find information on flame cutting torches. Create a presentation using photos or drawings of the torches. You may want to include photos or art of the cuts the torches make.

Design credits: Cinema and movie icon: Denis Meshkov/123RF; and Illustration of Welding icons: Mr. Nachai Sorasee/123RF

8

Gas Welding Practice:
Jobs 8-J1–J38

Chapter Objectives

After completing this chapter, you will be able to:

8-1 Demonstrate safe and practical setting up and closing down of the oxyacetylene equipment.

8-2 Adjust the equipment and perform tip maintenance to produce the best flame operation.

8-3 Demonstrate ability to manipulate the torch and filler rod to make sound welds on sheet metal and pipe.

As a new student of welding, these practice jobs in oxyacetylene welding are your first experience in actually performing a welding operation. The practice jobs in this text have been carefully selected, according to the recommendations of a number of skilled welders, supervisors of welding, and teachers of welding. Thus, the sequence represents many years of experience in the welding industry. Table 8-4, on page 231, provides a job outline for Chapter 8. We recommend that students complete the jobs in the order shown because they are organized on the basis of learning difficulty in an orderly sequence. If you complete the assigned reading and welding practice, you will acquire enough skill and knowledge to qualify for a job in industry in any one of the many branches of welding.

It is desirable for students to begin practice with the oxyacetylene welding process because it offers many learning advantages. The student is able to develop that coordination of mind, eye, and hands that is the basis for all welding practice. The action of a molten pool of metal can be clearly observed. The significance of fusion and penetration is better understood than when the beginner practices with the electric arc processes. The ability to control the size, direction, and appearance of the molten pool is quickly acquired.

Your undivided attention as a welder to your work is required at all times. You must not only see that you are fusing the weld metal properly to the base metal, but you must also observe the welding flame itself to see that it is not burning an excess of oxygen or acetylene.

Review Chapter 5 before beginning practice so that you have a thorough knowledge of the equipment used in the process. Before you begin welding, you must also learn the following information in this chapter:

- How to set up the equipment in a safe and practical manner
- How to adjust the equipment for the best operation
- How to light and adjust the torch for the proper flame
- How to manipulate the torch for sound welding

It is also important that you recognize the basic characteristics of a sound weld of good appearance.

Sound Weld Characteristics

Fusion

Fusion is the complete blending of the two edges of the base metal being joined or the blending of the base metal and the filler metal being added during welding.

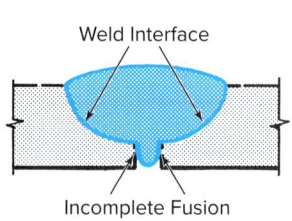

Fig. 8-1 Incomplete fusion.

When there is a lack of fusion, we have the condition referred to as incomplete fusion, Fig. 8-1.

Incomplete fusion (IF) may be caused by adding molten metal to solid metal or by lack of fluidity in the molten weld pool. IF may also be caused by improperly beveling the pieces to be welded, by improper inclination of the torch, by improper use of the filler rod, or by faulty adjustment and manipulation of the welding flame.

> **ABOUT WELDING**
>
> **Spatter**
>
> *Spatter* is the metal pieces thrown out from the welding arc, which are not part of the weld.

Beginning welding students frequently do not prepare the plates properly for welding. The beveling is not deep enough to extend entirely through the section to be welded, or it is not wide enough.

A common fault of beginning welders is that they add filler rod to the surfaces of the material being joined before they are in the proper condition for fusion to take place. Sometimes one surface is in fusion, but the other is not. When a lack of fusion takes place on only one side, the overall strength of the weld is lessened just as if a lack of fusion were present on both sides.

In some cases, the surfaces of the metal are brought to a state of fusion too soon so that oxide has a chance to form on the edges of the weld. When the filler rod is added, IF occurs because a film of oxide separates the surface and the added filler material.

Often through faulty torch manipulation, some of the molten weld metal is forced ahead. The surface of this metal is not in the proper state of fusion, and IF will occur.

Penetration

The term **penetration** refers to the depth to which the base metal is melted and fused in the root of the joint. *Fusion* is the essential characteristic of a good weld. It is possible to have penetration through the joint and at the same time have a condition of incomplete fusion, Fig. 8-2.

A sound weld must penetrate through to the root of the joint. The cross section of weld metal should be as thick as the material being welded.

In their desire to complete the weld as soon as possible, beginning welders have the tendency to hasten over the most important part of the work, which is to penetrate to the root of the weld. Incomplete penetration reduces the thickness of the metal at the weld and provides a line of weakness that will break down when the joint is under stress and strain.

The principal cause of incomplete penetration is improper joint preparation and alignment. The material may not be beveled in the proper manner, the root face may be too thick, or the root opening may be too much or too little. Improper heat, poor welding technique, or using a filler rod that is too large may also cause incomplete penetration.

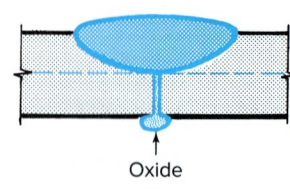

Fig. 8-2 Poor penetration.

Weld Reinforcement

A groove weld is said to be *reinforced* when the weld metal is built up above the surface of the metal being welded. Beginning welding students may build up weld metal above the surface of the base metal because they feel that the more weld metal that is added to the bead, the stronger the weld. Metal above the surface of the plate adds little strength to the joint. In fact, too much reinforcement reduces the strength of the welded joint by introducing stresses at the center and edges of the weld bead. If the weld is sound from the root of the joint to the surface of the joint, there is no need to add a great amount of face or root reinforcement.

Some reinforcement is desirable to offset possible pockets and other discontinuities on the surface of the weld. It should not be more than $1/16$ inch above the surface of the plate. There should be an absence of undercutting where the weld bead joins the plate. The edge of the weld bead should flow into the plate surface with an appearance of perfect fusion.

Satisfactory Weld Appearance

Careful visual inspection of each bead that student welders complete is necessary before going on to the next bead. They must be familiar with the characteristics of a good weld in order to set the standard of performance that they will try to attain. At this time, we will limit our concern with visual inspection. It is assumed that the student is familiar with Chapter 28 on inspection and testing.

- The weld should be of consistent width throughout. The two edges should form straight parallel lines.
- The face of the fillet weld should be slightly convex with a groove weld reinforcement of not more than $1/16$ inch above the plate surface. The convexity reinforcement should be even along the entire length of the weld. It should not be high in one place and low in another.
- The face of the weld should have fine, evenly spaced ripples. It should be free of excessive spatter, scale, and pitting.
- The edges of the weld should be free of undercut or overlap. The edges should be fused into the plate surface so that they do not have a distinct line of demarcation. The weld should appear to have been molded into the surface.
- Starts and stops should blend together so that it is difficult to determine where they have taken place.

> **JOB TIP**
>
> **Communication**
> Perhaps you have heard the saying, "Some people couldn't get along with a welding robot!" It's clear that working with people is important on the job. Effective working relationships are those based on communication that is direct, honest, respectful, and sincere.

- The crater at the end of the weld should be filled and show no porosity or holes. There is always a tendency to undercut at the end of the weld because of the high heat that has built up. Proper torch and filler rod manipulation prevents undercutting.

  If the joint is a butt joint, check the back side for complete penetration through the root of the joint. A slight bead should form on the back side.
- The root penetration and fusion of lap and T-joints can be checked by putting pressure on the upper plate until it is bent double. If the weld has not penetrated through the root, the plate will crack open at the joint as it is being bent. If it breaks, observe the extent of the penetration and fusion at the root. It will probably be incomplete in fusion and penetration.

The Oxyacetylene Welding Flame

Before setting up the welding equipment and adjusting the flame, the student should have a good understanding of the various types of oxyacetylene flames and their welding applications. A knowledge of the chemistry of the flame is also important.

The tool of the oxyacetylene welding process is not the welding torch, but the flame it produces. The sole purpose of the various items of equipment is to enable the welder to produce an oxyacetylene flame suited for the work at hand. The flame must be of the proper size, shape, and chemical type to operate with maximum efficiency.

There are three chemical types of oxyacetylene flames, depending upon the ratio of the amounts of oxygen and acetylene supplied through the welding torch. These flames are the *neutral flame,* the *excess acetylene flame,* and the *excess oxygen flame.* Their applications are given in Tables 8-1 and 8-2.

The Neutral Flame

The name **neutral flame** is derived from the fact that there is no chemical effect of the flame on the molten weld metal during welding. The metal is clean and clear and flows easily.

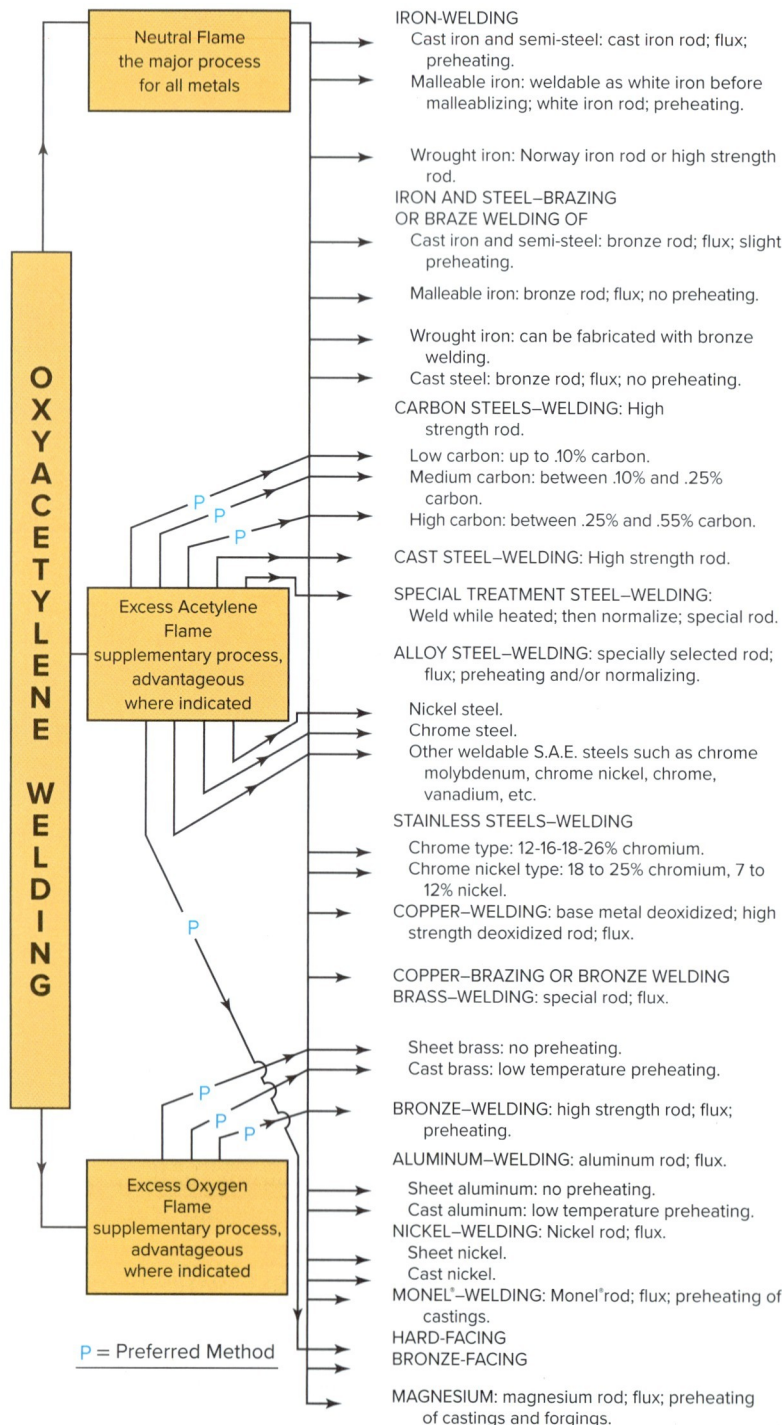

Table 8-1 Various Types of Flame That Are Recommended for Different Metals and Alloys

Source: The Welding Encyclopedia. Modified from the Oxy-Acetylene Committee of the International Acetylene Association

The neutral flame, Fig. 8-3, is obtained by burning an approximately one-to-one mixture of acetylene and oxygen. The pale blue core of the flame, Fig. 8-4, is known as the **inner cone**. The oxygen required for the combustion of the carbon monoxide and hydrogen in the outer envelope of the flame is supplied from the surrounding air, Fig. 8-5. The temperature at the inner cone ranges from 5,800 to 6,300°F.

The neutral flame is used for the majority of oxyacetylene welding and cutting operations. It serves as a basic point of reference for making other flame adjustments.

The Excess Acetylene Flame

The *excess acetylene flame* has an excess of acetylene gas in its mixture. The flame has three zones: the *inner cone,* the *excess acetylene feather,* and the *outer envelope,* Fig. 8-6. This **acetylene feather** contains white hot carbon particles, some of which are introduced into the weld pool during welding. For this reason, the flame adjustment is also referred to as a *carburizing flame*. It is also known as a *reducing flame* because it tends to remove the oxygen from iron oxides when welding steel.

The excess acetylene flame is used for many gas welding applications, high test pipe welding, and for certain surfacing applications. Its effect on steel is to cause the weld pool to boil and be very cloudy. Welds are hard and brittle. The temperature of this flame ranges between 5,500 and 5,700°F, depending upon the oxygen content. The temperature increases as the amount of oxygen increases.

The Excess Oxygen Flame

This type of flame adjustment has an excess of oxygen in the mixture. The flame has only two zones like the *neutral flame,* but the *inner cone* is shorter and may be sharper. It is nicked on the sides and has a purplish tinge. See Figs. 8-3 and 8-7. At high temperatures, the excess oxygen in this flame mixture combines readily with many metals to form oxides that are hard, brittle, and of low strength, thus seriously reducing weld quality.

A slightly oxidizing flame is used in braze welding and bronze surfacing. A more strongly oxidizing flame is used in fusion welding certain brasses and bronzes.

Table 8-2 Gas Welding Data for the Welding of Ferrous Metals

Base Metal	Flame Adjustment	Flux	Welding Rod
Steel, cast	Neutral	No	Steel
Steel pipe	Neutral	No	Steel
Steel plate	Neutral	No	Steel
Steel sheet	Neutral	No	Steel
	Slightly oxidizing	Yes	Bronze
High carbon steel	Reducing	No	Steel
Manganese steel	Slightly oxidizing	No	Base metal composition
Cromansil steel	Neutral	No	Steel
Wrought iron	Neutral	No	Steel
Galvanized iron	Neutral	No	Steel
	Slightly oxidizing	Yes	Bronze
Cast iron, gray	Neutral	Yes	Cast iron
	Slightly oxidizing	Yes	Bronze
Cast iron, malleable	Slightly oxidizing	Yes	Bronze
Chromium nickel	Slightly oxidizing	Yes	Bronze
Chromium nickel steel castings	Neutral	Yes	Base metal composition 25-12 chromium nickel
Chromium nickel (18-8) and (25-12)	Neutral	Yes	Columbium stainless steel or base metal composition
Chromium steel	Neutral	Yes	Columbium stainless steel or base metal composition
Chromium iron	Neutral	Yes	Columbium stainless steel or base metal composition

On steel, it causes excessive foaming and sparking of the molten weld pool. Welds have very poor strength and ductility. The temperature of this flame ranges between 6,000 and 6,300°F, depending upon the oxygen content. Beyond a certain point, an increase in the oxygen content reduces the flame temperature.

Setting Up the Equipment

If you are a student in a welding school, the oxyacetylene welding shop is probably equipped with a manifold system of welding stations. They have the regulators and welding torch already set up and attached, ready for welding. It is essential, however, that you know how to set up a portable welding outfit for welding. This type of equipment is found on construction jobs, in repair shops, and wherever maintenance welding is performed.

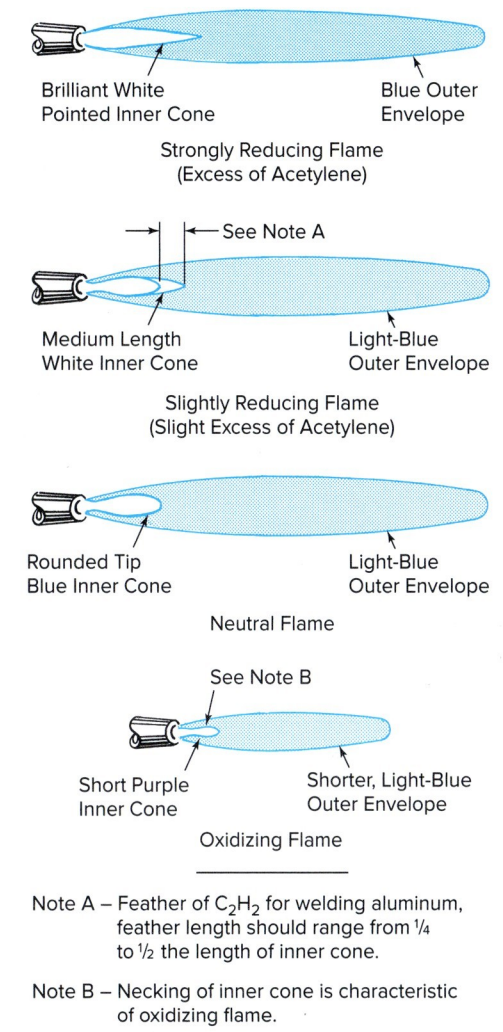

Fig. 8-3 Typical oxyacetylene flames.

Fig. 8-4 Neutral welding flame does not have an excess of oxygen or acetylene. It is used for welding steel and cast iron. Its temperature is approximately 6,300°F. David A. Tietz/Editorial Image, LLC

Gas Welding Practice: Jobs 8-J1–J38 **Chapter 8** 207

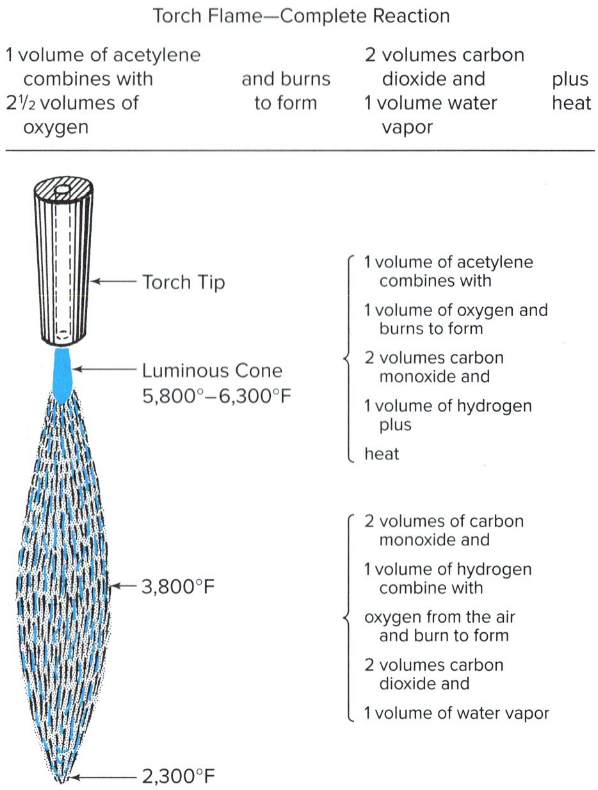

Fig. 8-5 The products of combustion of the oxyacetylene flame.

Torch Flame—Complete Reaction

1 volume of acetylene combines with 2½ volumes of oxygen and burns to form 2 volumes carbon dioxide and 1 volume water vapor plus heat

- Torch Tip
- Luminous Cone 5,800°–6,300°F: 1 volume of acetylene combines with 1 volume of oxygen and burns to form 2 volumes carbon monoxide and 1 volume of hydrogen plus heat
- 3,800°F: 2 volumes of carbon monoxide and 1 volume of hydrogen combine with oxygen from the air and burn to form 2 volumes carbon dioxide and 1 volume of water vapor
- 2,300°F

Fig. 8-6 Carburizing welding flame commonly used for welding white metal, stainless steel, hard facing, some forms of pipe welding, and soldering and brazing. David A. Tietz/Editorial Image, LLC

Fig. 8-7 Oxidizing welding flame commonly used for brazing with bronze rods. David A. Tietz/Editorial Image, LLC

Review pages 187 to 193 in Chapter 7 before setting up the welding equipment. The equipment necessary to perform oxyacetylene welding includes the following:

- Oxygen cylinder
- Acetylene cylinder
- Oxygen regulator
- Acetylene regulator
- Welding torch (also called a *blowpipe*)
- Oxygen hose
- Acetylene hose
- Wrenches
- Friction lighter
- Filler rod
- Gloves and goggles

Use the following procedure to set up the equipment:

1. Set up the cylinders. Secure them on a portable hand truck or fasten them to a wall so that they cannot fall over. Remove the protector caps from the cylinders.
2. Dirt and dust collect in the cylinder valve outlets during storage. To make sure that dirt and dust will not be carried into the regulators, clear the cylinder valves by blowing them out. Crack the valves only slightly. Open the acetylene cylinder outdoors and away from open flame.
3. Attach the oxygen regulator, Fig. 7-21, page 189, and the acetylene regulator, Fig. 7-22, page 189, to the tanks. Always make certain your hands are free of oil and grease prior to handling the regulators as a small amount inside the oxygen regulator connections could create a fire or explosion. Observe that the oxygen regulator has a right-hand thread and that the acetylene regulator has a left-hand thread. Make sure that the regulator nipples are in line with the valves of the tanks so that they may seat properly. Be careful not to cross the threads in the connecting nuts.
4. Attach the welding hoses to the flashback arrestors and tighten them to the regulators. These arrestors are designed to prevent a *flashback* (small fire) from entering the regulators (Fig. 7-23, page 189). The oxygen hose is green or black and has a right-hand thread. The acetylene hose is red and has a left-hand thread. Make sure that the nipples of the hose are in line with the connections on the regulator to ensure correct seating. Make sure that the threads in the nuts are not crossed. Tighten the nuts securely with a wrench and make sure the connections do not leak.
5. It is important that no grit or dirt be allowed to remain in the regulators or the hoses. This can be

prevented by blowing them out. After having made sure that the regulator-adjusting screw is fully released so that the regulator will not be damaged, turn on the tanks. Be sure that you use the proper acetylene tank wrench. Adjust the regulator by turning the adjusting screw to the right (Fig. 7-24, p. 189) to permit the gases to flow through. (After turning off the tanks, turn the regulator-adjusting screw to the left again after enough gas has flowed to clear the regulator and hose.)

6. Connect the green or black oxygen hose to the torch inlet marked *oxygen* and the red acetylene hose to the torch inlet marked *acetylene,* Fig. 7-25, page 190. The torch body should have reverse flow check valves to prevent backflow of gases. The oxygen and acetylene hoses are then connected to the check valves that are connected to the torch body. Make sure that the nipples of the hose connections are in line with the torch inlet for proper seating. Make sure that the threads in the nuts are not crossed. Tighten the nuts securely with a wrench and make sure the connections do not leak.
7. Open the valve of the oxygen and acetylene tanks. Open acetylene cylinders no more than one-half turn, in case a quick emergency shutdown is necessary. Be sure you have the proper acetylene tank wrench.
8. Select the proper welding tip for the job at hand and adjust the oxygen and acetylene regulators for the proper working pressure, Table 8-3. To adjust the working pressure, open the torch valves and set the correct pressure on the regulators by turning the adjusting screw to the right, Fig. 7-24, page 189. Close the torch valves when the pressure has been set. You are now ready to light the torch and adjust the flame.

Flame Adjustment

Proper flame adjustment for the job at hand is essential. If the flame is not correct, the properties of the weld metal will be affected, and a sound weld of maximum strength and durability will be impossible. The *oxidizing flame* causes oxides to form in the weld metal, and an *excess acetylene* introduces carbon particles into the weld. Both of these conditions are detrimental to good welding under certain conditions. The most common fault is the presence of too much oxygen in the welding flame. It is not easily detected because of its similarity to the neutral flame. Too much acetylene is readily seen by the presence of the slight feather.

A flame may also be too harsh. This is caused by setting the regulator at too high gas pressure. The gas flows in greater quantity and with greater force than it should. A harsh flame disturbs the weld pool, causing the metal to spatter around it. High pressure is often the major cause of poor fusion. It is impossible to make a weld of good appearance with a high pressure flame. The flame is noisy, and the inner cone is sharp.

Table 8-3 Tip Size for Steel Plates of Different Thickness

Tip Size	[1]Diameter of Hole (in.)	Thickness of Plate	Average Length of Flame (in.)	Approximate Pressure of Regulator		Approximate No. of Cubic Feet of Gas Used per Hour		Diameter of Rod (in.)
				Acetylene	Oxygen	Oxygen	Acetylene	
1	0.037	22–16 ga.	3/16	1	1	4.0	4.0	1/16
2	0.042	1/16–1/8 in.	1/4	2	2	5.0	5.0	1/16–1/8
3	0.055	1/8–3/16 in.	5/16	3	3	8.0	8.0	1/8
4	0.063	3/16–5/16 in.	3/8	4	4	12.0	12.0	3/16
5	0.076	5/16–7/16 in.	7/16	5	5	19.0	19.0	3/16
6	0.086	7/16–5/8 in.	1/2	6	6	23.0	23.0	1/4
7	0.098	1/2–3/4 in.	1/2	7	7	35.0	35.0	1/4
8	0.1065	5/8–1 in.	9/16	8	8	48.0	48.0	1/4
9	0.116	1 in. or over	5/8	9	9	57.0	57.0	1/4
10	0.140	Heavy duty	3/4	10	10	95.0	95.0	1/4
11	0.147	Heavy duty	7/8	10	10	100.0	100.0	1/4
12	0.149	Heavy duty	7/8	10	10	110.0	110.0	1/4

[1]There is no standardization in tip sizes, so this table gives approximations only. The pressures are correct for the hole size indicated.
Source: Modern Engineering Co.

If the gas pressure is correct for the tip size, the flame is quiet and soft. The molten weld pool is not disturbed. Spatter and sparks are at a minimum. Good fusion results, and the weld pool may be advanced along the weld joint with little risk of incomplete fusion.

The condition of the welding tip also helps in securing a quiet, even flame. The hole in the tip end must be absolutely round, and the internal passage of the tip must be free of dirt and metal particles. An out-of-round hole or foreign matter in the tip changes the direction of gas flow, causing a harsh flame with an imperfect inner cone.

Procedure for Lighting the Torch:

1. Keeping the tip facing downward, open the acetylene valve and light the gas with the friction lighter. The acetylene will burn with a smoky yellow flame and will give off quantities of fine black soot, Fig. 8-8.
2. Open the oxygen valve slowly. The flame will gradually change in color from yellow to blue and will show the characteristics of the excess acetylene flame. There will be three distinct parts to the flame: (1) a brilliant but feathery edged inner cone, (2) a secondary cone, and (3) a bluish outer envelope of flame, Fig. 8-6.
3. If you continue turning the oxygen valve, you will notice that the secondary cone gets smaller and smaller until it finally disappears completely. Just at this point of complete disappearance, the neutral flame is formed, Fig. 8-4. If you continue to turn the oxygen valve, the flame will go past the neutral stage, and an excess oxygen (oxidizing) flame will be formed. The entire flame will decrease in size. The inner cone will become shorter and much bluer in color. It is usually more pointed than the neutral flame, Fig. 8-7.
4. Practice adjusting the flame several times until you are familiar with all three flames and can adjust the neutral flame with precision.

Fig. 8-8 Atmospheric burning of acetylene. Open the acetylene valve until the smoke disappears. David A. Tietz/Editorial Image, LLC

Backfire

Improper operation of the welding torch or defective equipment may cause the flame to go out with a loud snap or pop and the flame goes out. Reignition may take place if sufficient heat is present. This is not normally a safety concern and, in fact, many manufacturers induce backfires during design and production tests to ensure flame integrity of torches and tips. This is called **backfire.** However, a sustained backfire—that is, a sustained burning of the flame back inside the torch, usually at the mixer—under certain situations can burn upstream past the mixer. Sustained backfires can be recognized by a hissing or squealing sound and/or a smoky, sharp-pointed flame. The torch should be shut down immediately, first by shutting off the oxygen valve and then the fuel gas valve. Severe damage to the torch, as well as an increased risk of fire, would result if the valves are not turned off immediately. Sustained backfires are normally contained inside the torch with the quick action of the welder in shutting off the valves. This prevents the flame from burning through to the outside of the torch. The sustained backfire is serious in that if it is not recognized quickly and the valves turned off, the flame can burn through the torch body and spew fire and molten metal. The torch should be turned off at the valves. Check the equipment and the job to see which of the following causes may be at fault:

- ***Operating the torch at lower pressure than that required for the tip size used.*** The gases are flowing too slowly through the tip, and they are burning faster than the speed at which they flow out of the tip. Increasing the gas pressure will correct this. Always use a smaller tip with higher flow rate than a large tip with reduced flow rate. Make certain the safety devices such as check valves and flashback arrestors are of the proper size for the pressure and flow requirements.
- ***Touching the tip to the work.*** This smothers the flame and does not allow for combustion to take place. The welder should always keep the inner cone of the flame from touching the work.
- ***The tip may become overheated.*** This can result from overuse, from welding in a corner, or from being too close to the weld. The tip must be cooled before going ahead with the work. The job conditions should also be corrected.
- ***A loose tip or head.*** A loose tip may be the result of overheating, which causes expansion of the threaded joint. The obvious correction is to tighten the tip. The screw threads on the tip may also be crossed.

- ***The inside of the tip may have carbon deposits or small metal particles inside the hole.*** These particles become hot and cause preignition of the gases. Clean the tip out with standard tip cleaners.
- ***The seat of the tip may have dirt on it, or it may have become nicked through careless handling.*** Dirt and nicks prevent a positive seat at the joint where the tip is screwed into the stem. Gas escapes at this point, causing preignition. The dirt should be cleaned out. If the seat is nicked, it may be possible to reseat it with a seating tool. A new tip may have to be used.

Flashback

A **flashback** occurs when the flame burns back inside the torch and causes a shrill hissing or squealing. The oxygen valve should be closed first, thus eliminating the oxygen to the fire (flashback). *Never* shut off the fuel first as this gives the flashback the opportunity to seek other fuel such as soot or other flammable material and continue to burn. The torch should be allowed to cool. Before relighting, blow oxygen through the torch tip for a few seconds to clear out any soot that may have formed in the passage.

To prevent the flashback from traveling upstream into the hose, regulator, and gas supply systems, flashback arrestors are used. Most flashback arrestors used are designed with sintered stainless-steel elements. The remainder uses designs of coiled tubing to quench the heat associated with a flashback flame. These elements quench the flame front as it moves back upstream toward the torch and/or regulator. Flashback arrestors are reliable in performing the intended function if the device is used according to the manufacturer's instructions. The dry (sintered metal alloy) **flash arrestor** was first developed to fit onto the regulator outlet. The design has both a check valve and a sintered stainless-steel filter or tube. The flashback arrestor would permit forward passage (and also reverse flow) of gas but would extinguish a flame, thus preventing a flame from moving upstream of the arrestor. This type of arrestor offers almost total protection against a flame entering the gas source, and yet it allows good downstream flow of gases. A large flashback arrestor uses a large sintered filter to allow high gas flow through it. While small arrestors may provide better balance when mounted on the torch, they may actually reduce flow rate to the point where they may contribute to torch flashback. Dry torch flashback arrestors are available that will allow flows great enough to handle all tips and high volume heating nozzles, even if attached to units that also contain check valves. Regulator and torch arrestors both may be used in the same oxyacetylene system.

Neither check valves nor flashback arrestors will protect the torch or tip. As an oxyacetylene welder, you must recognize the situation and promptly shut the system down and take corrective action. Repeated flashbacks indicate that there is a serious defect in the equipment or that it is being used incorrectly. The orifice, barrel, or mixing chamber may be clogged. Oxides may have formed, flow rates may not be sufficient, or gas pressure may be incorrect. Never use gas pressures higher than those recommended by the manufacturer of the equipment you are using. Use flashback arrestors and check valves that do not overly restrict the flow. Always follow safe operating procedures. For more information contact the Compressed Gas Association, 1235 Jefferson Davis Highway, Suite 501, Arlington, VA 22202-3269, and ask for document E-2 on check valves and document TB-3 on flashback arrestors.

Closing Down the Equipment

When the welding practice is finished for the day, take down the equipment. If you are welding with a portable outfit on the job, use the following method:

1. Turn off the flame by first closing the oxygen valve on the torch. Then close the acetylene valve on the torch. This will extinguish the flame.
2. Close the oxygen and acetylene cylinder valves.
3. Open the acetylene and the oxygen valves on the torch to permit the trapped gases to pass out of the regulators and the hoses.
4. Release the regulator-adjusting screws by turning them to the left.
5. Disconnect the hoses from the torch.
6. Disconnect the hoses from the regulators.
7. Remove the regulators from the cylinders and replace the protective caps on the cylinders.

If you are working on a line system in the school shop, use the following procedure:

1. Turn off the flame as stated in step 1.
2. Close the oxygen and acetylene valves on the line system.
3. Open the acetylene and the oxygen valves on the torch to permit the trapped gases to pass out of the regulators and the hoses.
4. Release the regulator-adjusting screws by turning them to the left.
5. Close the torch valves and hang the torch on the gas economizer.

ABOUT WELDING

Lockout

Lockout means a device that prevents a mechanism such as a switch or valve from being turned on. On the device is a warning tag, with the date and name of the person who closed down the mechanism. This method helps prevent injury on the job. Two persons working together should have their own locks and tags.

Safety

At this point it would be well to review a few important safety tips for gas welding and cutting:

- Wear goggles with the right filter lenses.
- Wear gauntlet gloves of heat-resistant leather; keep them away from oil and grease.
- Don't wear oily or greasy clothes. Oil or grease, plus oxygen, will burn. Don't blow off your clothes or the work with oxygen; it's wasteful and dangerous.
- Woolen clothes are better than cotton. They burn less easily and protect better against heat.
- Wear fire-resistant apron, sleeves, and leggings when you do heavy cutting. Keep cuffs rolled down and pockets closed. Don't wear low-cut shoes.
- Oxygen should never be substituted for compressed air.
- Do not use matches for lighting the torch.
- When welding or cutting material containing, or coated with, lead, zinc, aluminum, cadmium, or beryllium, wear an air-supply respirator. Fluoride-type fluxes give off poisonous fumes. Avoid heavy smoke.
- Wear a respirator when you weld galvanized iron, brass, or bronze in confined areas. Even if ventilation is good, the respirator is a good idea.
- Don't let your clothing become saturated with oxygen or air-rich in oxygen; you're chancing burns if you do.
- Don't work with equipment that you suspect is defective. Have leaking or damaged equipment repaired by a qualified person.
- Never use oil on, or around, oxygen regulators, cylinder connections, or torches. Keep your hands clean.
- Blow water or dust out of new welding hose with compressed air or oxygen—never acetylene.
- Never use oily compounds on hose connections.
- Never interchange hose connections. Standard oxygen hose is green; fittings have right-hand threads. Standard hose for acetylene is red; fittings have left-hand threads and a grooved nut. Don't force connections.
- Unnecessarily long hoses tend to kink and leak.
- Store hose on reels. Inspect hose every week for cracks, wear, and burns.
- Test hose for leaks at normal working pressure by submerging it in water. Test connections for leaks with soapy water.
- Do not hang the torch and hose on the regulators and cylinders.
- Cut off parts of hose burned by flashback and remake connections. Flashback weakens the inner wall.
- Don't repair hose with friction tape.

Practice Jobs

Instructions for Completing Practice Jobs

Your instructor will assign appropriate practice in gas welding from the jobs listed in the Job Outline in Table 8-4. Before you begin a job, study the specifications given in the Job Outline. Then turn to the pages indicated in the Page column for that job and study the welding technique described. For example, find the specifications for Job 8-J8. According to the outline for this job, you are to practice beading on a $\frac{1}{8}$-inch mild steel plate. You are to use a type RG45 mild steel filler rod with $\frac{3}{32}$-inch diameter. The weld is to be done in the overhead position. Information on the welding technique proper for this job is given on page 219.

Low Carbon Steel Plate

It is common practice to limit the application of oxyacetylene welding of low carbon steel to steel plate with a maximum thickness of about 11 gauge ($\frac{1}{8}$ inch). The electric arc welding processes such as flux cored arc, shielded metal arc, gas metal arc, and gas tungsten arc are used for heavier plate. The oxyacetylene process is still used a good deal in maintenance and repair, for welding cast iron, braze-welding, brazing, and soft soldering. It is also used for the welding of small diameter steel pipe.

It was pointed out earlier that the skills learned in the practice of oxyacetylene welding are the basis for learning techniques in the other welding processes. These gas welding practice jobs provide practice in the coordinated use of both hands. They also provide the means for observing the flow of molten metal and the appearance of fusion and penetration taking place. You are urged to develop the highest possible skill in torch manipulation and visual inspection.

Carrying a Pool Without Filler Rod

The first steps in learning to weld are to get the *feel* of the torch and to observe the action of the flame on steel and the flow of molten metal. Probably the best way to accomplish these important steps is to practice carrying a pool (autogenous welding) without melting-through.

Autogenous Welding, Flat Position: Job 8-J1 An important factor in carrying a pool of molten metal across the plate is the speed of travel of the flame. The weld should move forward only when the pool of molten metal is formed. If travel is too fast, the bead will be narrow, and you will lose the molten pool. Incomplete fusion may also result. If travel is too slow, the bead will become too wide, and it will melt through the plate. A pool that has been moved at the proper rate will produce a weld with closely formed ripples and be of uniform width.

1. Check the Job Outline for plate thickness. Follow the manufacturer's recommendations for tip size and study Table 8-3.
2. Adjust the torch for a neutral flame, Fig. 8-4.
3. Grasp the torch like a pencil, Fig. 8-9. If it is more comfortable for you, you may grip it like a fishing pole or hammer. The hose should be slack so that the torch is in balance. Hold the torch so that the tip is at an angle of 45 to 60° with the plate.
4. Hold the torch in one spot so that the inner cone of the flame is about 1/8 inch above the surface of the plate until a molten pool of metal is formed that is about 3/16 to 3/8 inch in diameter.
5. Carry the molten pool across the plate, forming a bead of even width and even ripples. Be sure the weld pool is fluid so that IF does not result. Develop the technique illustrated in Figs. 8-10 and 8-11. After completing the weld, draw the flame from the end of the weld pool slowly to prevent hydrogen from contaminating the weld pool. Not doing so will result in a small hole (crater) at the end of the pool.
6. Examine the other side of the plate for even penetration. There must be no evidence of burn-through. Also note the effect of expansion and contraction on the plate.
7. Practice the job until you can run beads that are passed by the instructor. Acceptable beads are shown in Fig. 8-12A and D.

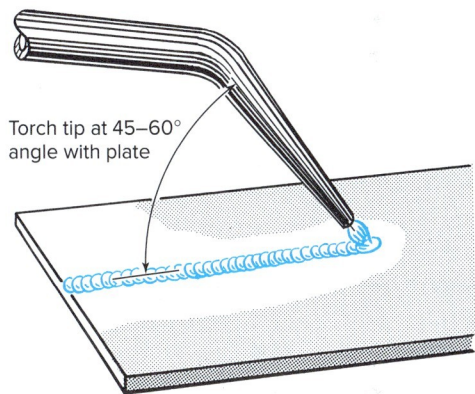

Fig. 8-10 Proper tip angle when welding in the flat position.

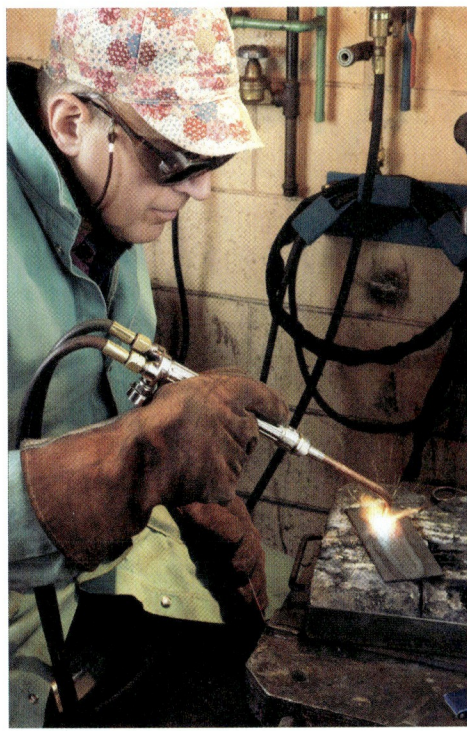

Fig. 8-9 Ripple welding. A light welding torch can be held like a pencil. A larger torch is held like a hammer. Prographics

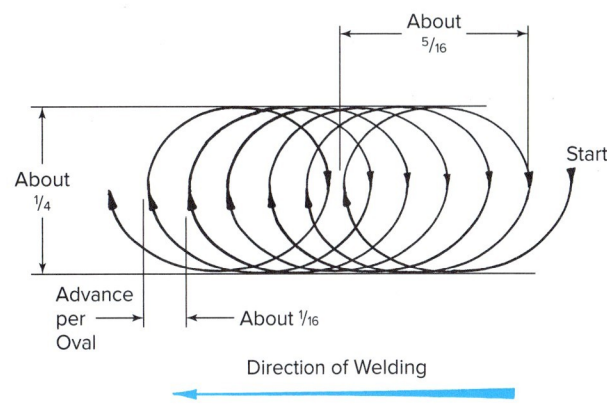

Fig. 8-11 The recommended torch motion when carrying a pool on flat plate.

Gas Welding Practice: Jobs 8-J1–J38   **Chapter 8**   213

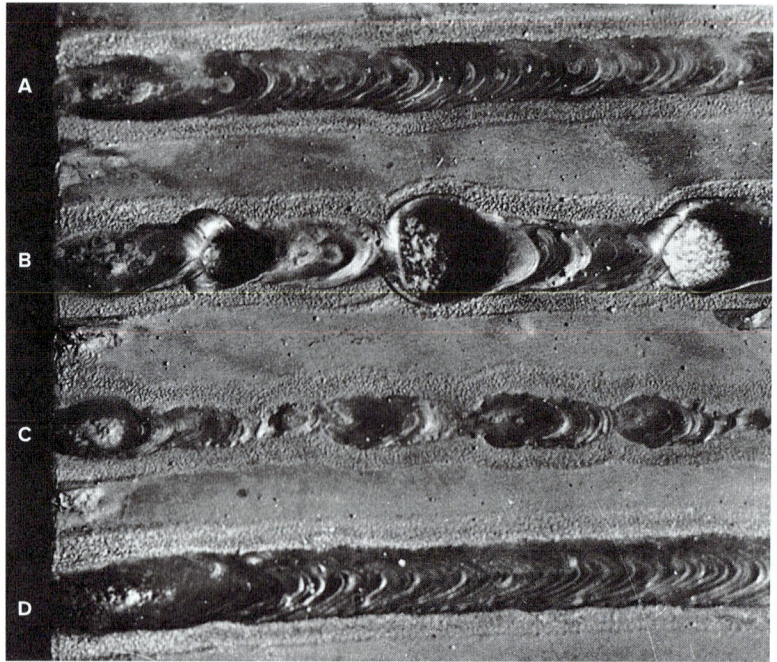

Fig. 8-12 Correct welding technique will result in strips that look like A and D. Excess heat will cause holes to be burned into the sheet as in B. Insufficient heat and uneven torch movement will result in a strip like C. Praxair, Inc.

8. Experiment with carburizing and oxidizing flame adjustments. Notice the effect of each flame on the weld pool. These flames are shown in Figs. 8-6 and 8-7.

Autogenous Welding, Vertical Position: Job 8-J2 This job should be practiced with flame adjustment of different types and sizes. It is only through firsthand experience and observation that you will learn the amount of heat required to obtain a uniform, smooth weld.

1. Set up the plate in a vertical position. The direction of welding is from the bottom to the top.
2. Adjust the torch for a neutral flame.
3. Grip the torch like a hammer and point it in the direction of travel. As in welding in the flat position, the tip should be at an angle of 45 to 60° with the plate.
4. Form a molten pool about 3/16 to 3/8 inch in diameter as in welding in the flat position.
5. After forming the pool, begin to travel up the plate. Use a short, side-to-side motion across the weld area and move up the plate with a semicircular motion. This technique provides equal heat distribution and uniform control of the molten metal.
6. When you find that the weld is getting too hot, pull the torch away so that the flame does not play directly on the weld pool. Begin welding again after the weld has cooled somewhat. If the pool gets too hot, it will burn through or the weld pool will become so fluid that it will run down the face of the plate.
7. Practice this job until you can consistently produce beads that are uniform and have smooth ripples.

Making a Bead Weld with Filler Rod

Although edge and corner joints may be welded without filler rod, autogenous filler rod is necessary for most joints to build up a weld bead to the required strength. The independent but coordinated motion of both hands is necessary to weld with a filler rod. The welding torch is held in the right hand, and the filler rod is held in the left hand. *Right hand* is used to describe the predominate hand, and *left hand* is used to describe the secondary hand. All the subsequent examples should be considered in this fashion. It is obvious that if you are left-handed, your primary hand will be your left and your secondary hand will be your right. If you are ambidextrous, this will not pose any problem. If you want to be the most versatile welder, you may want to do limited practice welding with both hands. On occasion it may be difficult to get in the most advantageous position, and you may have to make a weld using the opposite hand. Each hand has a specific function in the welding operation.

Beading, Flat Position: Job 8-J3

1. Check the Job Outline for the thickness of the plate, and size and type of filler rod. The rod size to use for a particular job usually depends upon the thickness of the plate. A general rule to follow is to use a rod diameter equal to the thickness of the metal being welded. Follow the manufacturer's recommendations for tip size.
2. Adjust the torch for a neutral flame.
3. Hold the torch so that the tip is at an angle of 45 to 60° with the plate. Hold the filler rod in your left hand. Slant it away from the torch tip at an angle of about 45°, Figs. 8-13 and 8-14.
4. Heat an area about 1/2 inch in diameter to the melting temperature. Hold the end of the filler rod near the flame so that it will be heated while the plate is being heated.
5. When the molten pool has been formed, touch it with the tip of the filler rod. Always dip the rod in

Fig. 8-13 Torch and rod being held as typical for right-handed welder. Programphics

Fig. 8-14 Torch and filler rod position when welding in the flat position. The torch may also be held like a hammer.
Location: Northeast Wisconsin Technical College Mark A. Dierker/McGraw Hill

SHOP TALK

Droopers

Droopers are not the new fashion trend. Rather, the term refers to constant curve welding machines that have a negative volt–amp curve. In these machines, amperage is only slightly varied while the voltage changes with different arc lengths.

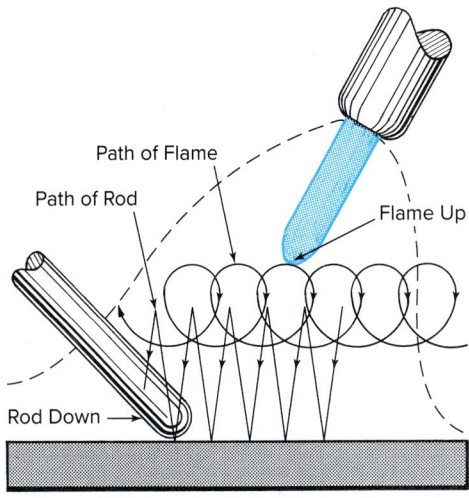

Fig. 8-15 Torch and filler rod movement when making a bead weld in the flat position.

the center of the molten weld pool. Do not permit filler metal to drip from above the pool through the air. Oxygen in the air will attack the molten drop of filler metal and oxidize it.

6. Move the torch flame straight ahead. Rotate the torch to form overlapping ovals, Fig. 8-15. Add filler rod as needed as the molten pool moves forward. Raise and lower the rod, and add just enough filler metal to supply the metal necessary to build up the weld to the desired width and height. When the welding rod is not in the molten pool, hold the end in the cooler outer envelope of the flame. If the filler rod is of the proper size, it should not melt in this outer envelope. You must keep it in the outer envelope to prevent it from oxidizing in the unprotected air outside this envelope. Learning proper filler metal-handling techniques at this stage of your learning will assist you greatly when you begin practicing the welding of more exotic materials like nickel, stainless steel, and titanium with the gas tungsten arc welding process. Remember that the cone of the flame should not touch the molten weld pool. It should be about $1/8$ inch above the surface of the weld pool. In the beginning, you will have trouble with the filler rod sticking. Either the weld pool is not hot enough, or the rod has been applied to a cold area outside of the molten pool. Release the rod by placing the flame directly on it.

7. Practice with different flame adjustments. Also observe the effect of various speeds of travel. If the molten pool is moved too slowly, it will grow so large and fluid that it burns through, Fig. 8-16B.

If it is moved too rapidly, the filler rod will be deposited on cold metal, resulting in incomplete fusion, Fig. 8-16C. The bead will also be very narrow. Make sure that the flame protects the molten weld pool so that it will not be contaminated by the surrounding air.

8. Continue practicing until the beads conform to the desirable characteristics shown in Figs. 8-16A and D and 8-17 and are acceptable to your instructor.

Beading, Vertical Position: Job 8-J4 Applying a bead weld in the vertical position is somewhat more difficult than welding in the flat position. After you have mastered the vertical position, however, you may find that it is your favorite position of welding. In this position, you will learn how to control molten metal that has a tendency to run with the torch flame and filler rod. You must time the application of heat, in the molten pool the movement of the filler rod, and the movement upward to obtain a uniform weld bead.

1. Check the Job Outline for plate thickness and size and type of filler rod. Follow the manufacturer's recommendations for tip size.
2. Adjust the torch for a neutral flame.
3. Place the plate in the vertical position. Grip the torch like a hammer and point it upward in the direction of travel at about 45° angle.
4. Heat an area about ½ inch in diameter at the bottom of the plate to the melting point. Place the tip of the filler rod near the torch flame so that it will become preheated while the plate is being heated.
5. When a molten pool has been formed, apply the tip of the filler rod to the forward edge of the molten pool of metal. Move the torch flame and filler rod away from and toward each other so that they cross at the center of the weld deposit, Figs. 8-18 and 8-19. The distance of cross movement should not exceed three times the diameter of the filler rod. A slight circular motion of the torch flame may be easier for you than a straight side-to-side movement. The rod can be used straight or bent as desired. A bend of about 45 to 90° in the filler rod a few inches from the tip will enable you to keep your hand away from the heat and also move the rod freely. In the beginning, you may find the bent rod difficult to manipulate and may prefer to practice with the straight rod.
6. Weld evenly up the plate, manipulating the torch as instructed in step 5. Do not allow the pool to become too hot. If the molten metal becomes highly fluid, or if there is an excess of metal in the weld pool, it will spill out of the weld pool. If the weld pool seems to be getting too hot, raise the flame slightly so that it will not play directly upon the pool and give it a chance to chill slightly. The flame can also be directed momentarily on the filler rod, but be careful not to burn it. You must not bring the rod to a white heat or molten condition unless you are ready to

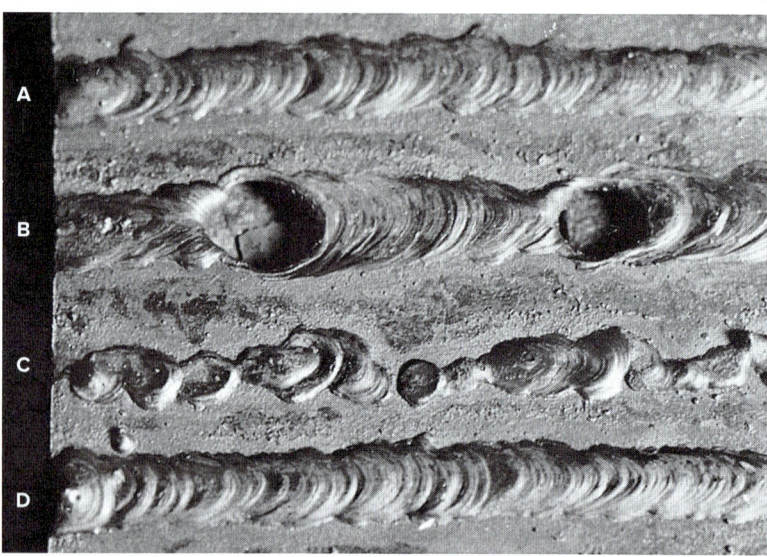

Fig. 8-16 Welds A and D are satisfactory for beginners. Weld B shows overheating, and weld C indicates insufficient heating, little fusion, and improper melting of the filler rod. Praxair, Inc.

Fig. 8-17 Weld beads in ⅛-inch mild steel plate welded in the flat position. A no. 3 tip and a 3⁄32-inch filler rod were used. The flame was neutral. Edward R. Bohnart

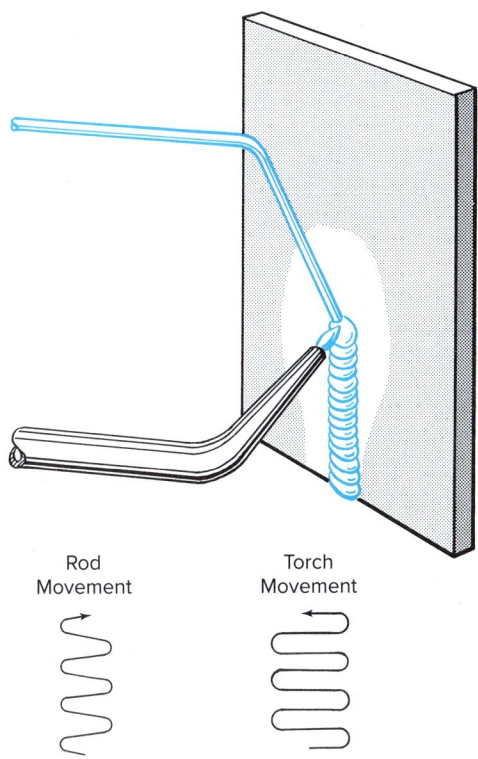

Fig. 8-18 Manipulation of the torch and rod for beading in the vertical position.

Fig. 8-19 Welding in the vertical position. Location: Northeast Wisconsin Technical College Mark A. Dierker/McGraw Hill

add rod to the weld pool. Add only enough rod to form the weld size desired.

7. Practice this job with flames of different types and sizes in order to observe their effects on the molten pool of weld metal. Practice until the beads pass inspection. Satisfactory beads are shown in Fig. 8-20.

Autogenous Welding, Horizontal Position: Job 8-J5 Some trainees find welding in the horizontal position the most difficult of all positions to master. Effort to apply any

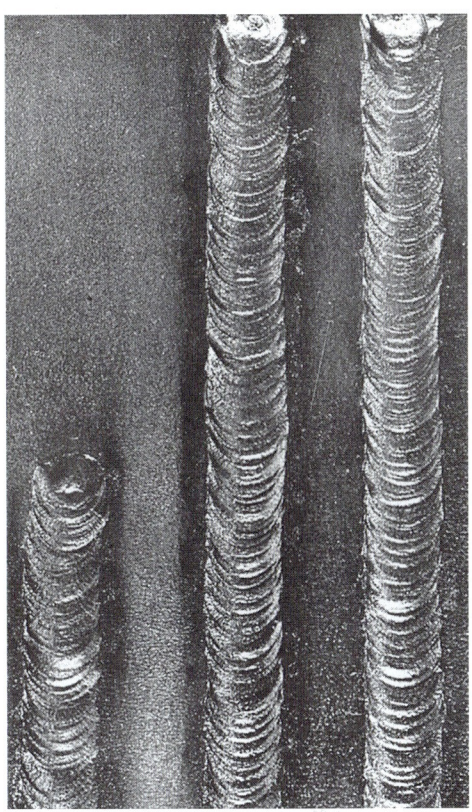

Fig. 8-20 Weld beads in 1/8-inch mild steel plate welded in the vertical position. A no. 3 tip and a 3/32-inch high-test filler rod were used. The flame was neutral. Edward R. Bohnart

material (such as plaster, paint, adhesive, weld metal, etc.) to a vertical surface requires careful manipulation of the material to keep it of uniform thickness. The tendency of the material is to sag toward the bottom. Close observation of the molten weld pool and careful torch movement will overcome this difficulty and make it possible to produce welds that have somewhat the same appearance as welds made in the flat, vertical, or overhead position.

1. Check the Job Outline for plate thickness. Select the proper tip size.
2. Adjust the torch for a neutral flame.
3. Place the plate in a vertical position. Welding should be from right to left or from left to right if you are left-handed. Hold the torch as shown in Fig. 8-21.
4. Heat the plate until a molten pool is formed. Carry the pool forward with a semicircular torch motion. There will be a tendency for the pool to sag on the lower edge. This can be prevented by directing the flame upward and forward. It is important that the weld pool does not become too liquid because this will cause the pool to sag at the lower edge.
5. Experiment with different flame types and sizes.

Fig. 8-22 Weld beads in ⅛-inch mild steel plate welded in the horizontal position. A no. 3 tip and a ³⁄₃₂-inch mild steel filler rod were used. The flame was neutral. Edward R. Bohnart

Fig. 8-21 Welding in the horizontal position on steel plate. Note torch and filler rod angles. Prographics

6. Practice this job until you can produce weld beads that are of uniform width and have smooth ripples.

Beading, Horizontal Position: Job 8-J6

1. Check the Job Outline for plate thickness and type and size of filler rod. Follow the manufacturer's recommendations for tip size.
2. Adjust the torch for a neutral flame.
3. Place the plate in a vertical position. Welding progresses from right to left. Hold the torch and rod as shown in Fig. 8-21.
4. After the usual molten pool has been formed, carry the pool forward with a semicircular torch motion. As the weld bead moves along, the bead will sag and build up on the lower edge. To correct this, direct the flame upward and in the direction of welding while adding filler rod to the upper edge of the weld pool.
5. Experiment as before with different flame types and sizes. Practice until you can do satisfactory work. Good-quality beads are shown in Fig. 8-22.

Autogenous Welding, Overhead Position: Job 8-J7 Most welding students will find welding in the overhead position no more difficult than welding in the vertical and horizontal positions. It will take some time to get used to the position in which your arms and body are placed. You may find it very tiresome to hold your torch arm above your head. Grip your torch lightly and be as relaxed as possible. If the torch is held too tightly, you will not have free movement of the torch and you will tire rapidly.

1. Check the Job Outline for plate thickness. Select the proper tip size.
2. Adjust the torch for a neutral flame.
3. Place the plate in the overhead position. Welding progresses from right to left. Hold the torch as shown in Fig. 8-23.
4. Form the weld pool as before. The side-to-side torch movement will determine the width of the **ripples.** The weld ripple is part of the weld bead. Look at Fig. 8-20 on page 217. What you see are the welding beads. However, the small little freeze lines in the beads are called *ripples,* as if you were throwing a rock in a pool and seeing the ripples. The weld pool will not fall as long as it is not permitted to become too fluid and too large. The torch should also be directed in the direction of welding at an angle of about 45°.

Fig. 8-23 Welding in the overhead position on steel plate. A small torch and steel filler rod are used. Location: Northeast Wisconsin Technical College Mark A. Dierker/McGraw Hill

5. Experiment with various flame types and sizes. Observe the effect that different flames have on the molten pool of metal.
6. Check the ripple appearance carefully. It should be smooth and of even width the entire length of the ripple weld.

Beading, Overhead Position: Job 8-J8 In doing this job you will be required to hold both arms above your head. Both hands will need to be coordinated in their movement overhead and this will probably be a little difficult for a short time. You will also experience some difficulty getting used to the position in which your arms and body are placed. Again you are cautioned to grip the torch and welding rod lightly and to be as relaxed as possible. If you are tense, you will not be able to respond readily to the demands of the weld bead.

1. Check the Job Outline for plate thickness and type and size of filler rod. Follow the manufacturer's recommendations for tip size.
2. Adjust the torch for a neutral flame.
3. Place the plates in the overhead position. The torch and the filler rod have the same relation to each other as in other positions except that they are overhead, Fig. 8-23.
4. Start the weld pool as before. Control the filler rod and the size of the pool. The weld pool will not fall as long as it is not allowed to get too large. The weld pool must be kept shallow and under control. Do not permit it to become highly fluid. If the pool is kept fairly small and not allowed to form a drop, you will have little difficulty. Add filler rod sparingly. Too much filler rod will cause the weld metal to drip. Make sure that the hot filler rod and the weld pool are protected by the torch flame at all times to prevent oxidation.
5. Continue practice until the beads have the characteristics shown in Fig. 8-24 and are acceptable to your instructor.

Fig. 8-24 Weld beads in ⅛-inch mild steel plate welded in the overhead position. A no. 3 tip and a ³⁄₃₂-inch, high-test filler were used. The flame was neutral.
Edward R. Bohnart

JOB TIP

Networking
Networking is one of your best sources for a job. Tell everyone—from friends, to relatives, to strangers you meet—what kind of job you are looking for. Join a professional group to gain more knowledge and to meet others who work in your field. Follow industry news to learn about job leads.

Welding Edge and Corner Joints: Jobs 8-J9–J16

Edge joints are sometimes referred to as **flange joints.** The flange is formed by turning over the edge of the sheet to a height equal to the thickness of the plate. Flange, edge, and corner joints are used a great deal on sheet metal less than 16 gauge in thickness. These joints may be welded without filler rod by fusing the two plates together. Filler rod should be used, however, when maximum strength is desired.

1. Check the Job Outline for plate thickness and type and size of filler rod. Follow the manufacturer's recommendations for tip size.
2. Set up and tack the plates about every 2 inches.
3. Adjust the torch for a neutral flame.
4. Welding will be in the flat position. The positions of the torch and filler rod are similar to those used for welding flat plate in the flat position. Figure 8-25 shows the torch without the filler rod. Note that the flame should be directed at the inside edges of the plate.
5. Form the molten pool and move it along the seam. Make sure that the edge surfaces are molten and that you are melting through to the root of the weld. The corner joint should show penetration through to the back side. Do not add too much filler rod to the weld. These joints do not require much reinforcement on the bead. Move the torch with a slight crosswise or circular motion over the surface of the joint. If the movement is too wide and too much filler rod is added, the weld metal will spill over the sides of the joint. Add the filler rod to the center or forward edge of the molten weld pool. Make sure that the pool and filler rod are protected at all times with the torch flame to prevent the formation of oxides and nitrides.

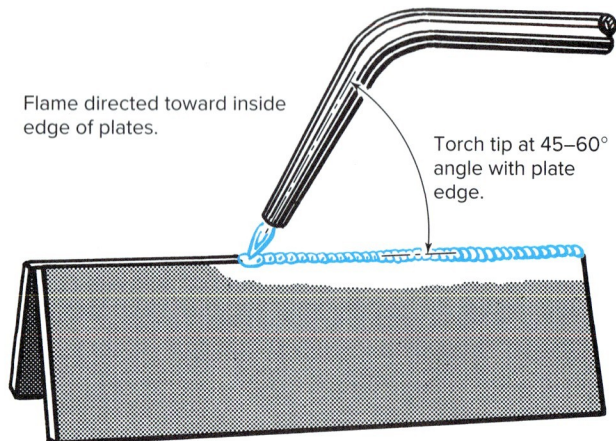

Fig. 8-25 Position of the torch when making an edge weld in the flat position.

1. Check the Job Outline for plate thickness and type and size of filler rod. Follow the manufacturer's recommendations for tip size.
2. Adjust the torch for a neutral flame.
3. Space the plates so that penetration can be secured through the root of the joint and expansion can take place during welding. The plates will draw together if not spaced. They should be spaced about 1/16 inch apart at the starting end and about 1/8 inch apart at the other end, Fig. 8-28.
4. Tack the weld at each end and in the center. Be sure that the edge of each plate is molten before you add filler rod to bridge the gap.
5. Play the flame on the first tack weld and remelt a small pool on its surface. This pool should be increased in size by adding a limited amount of filler rod. Be sure that the pool is molten before you add the filler rod.
6. Move the molten pool along the open seam from right to left with a small circular motion. As the flame moves from side to side, it should play directly first on the edge of one sheet and then on the edge of the other so that they are brought to the state of fusion at the same time. Preheat the rod as you move along. The weld pool must not be too fluid, or you will burn a hole in the plate, and excessive weld metal will hang

6. Continue practice until the welds have the characteristics shown in Figs. 8-26 and 8-27 and are acceptable to your instructor.
7. Test the welds by bending the pieces flat so that the back side of the weld is under tension. After the sheet has been bent over as far as you can hammer, place the joint in a vise and squeeze the sheets on each side of the weld down on top of each other. Inspect the underside of the weld bead carefully. The weld should not crack or break.
8. After you are able to make welds of good appearance and high strength in the flat position, practice in the other positions.
9. Weld a test joint in each of the positions being practiced. Weld one edge only and pry the plates apart to check penetration and fusion at the inside corners of both plates and along the edge surface.

Welding a Square Butt Joint: Jobs 8-J17 and J18

Square butt joints require no edge preparation. For this reason, the joint is seldom used for material more than 12 gauge in thickness. In order to secure 100 percent penetration in thicker stock, welding must be done from both sides.

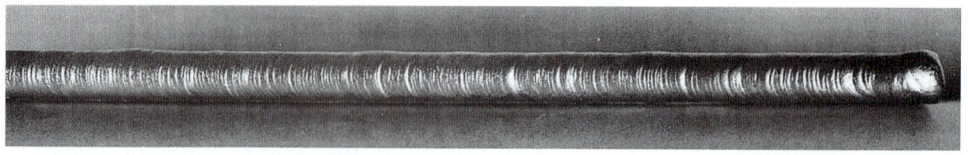

Fig. 8-26 An edge joint welded in the flat position. Note the even, smooth weld. It is made in 1/8-inch steel plate. A no. 3 tip and a 3/32-inch filler rod were used. The flame was neutral.
Edward R. Bohnart

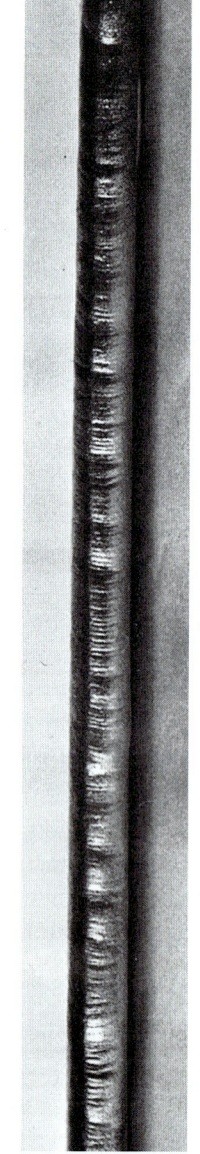

Fig. 8-27 An edge joint welded in the vertical-up position. It is made in 1/8-inch steel plate. A no. 3 tip and a 3/32-inch filler rod were used. The flame was neutral.
Edward R. Bohnart

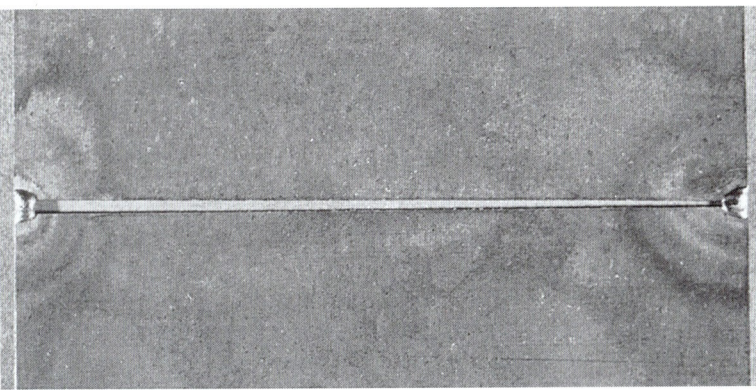

Fig. 8-28 Proper spacing of the plates for a butt weld. Praxair, Inc.

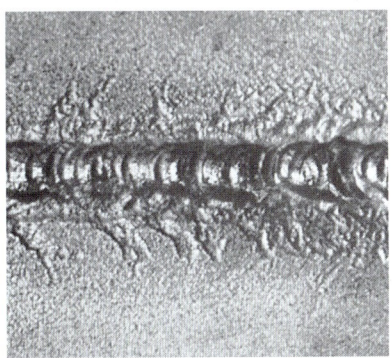

Fig. 8-29 Complete joint penetration groove weld, showing root penetration through to the back side of a butt joint is essential for welds of maximum strength. Edward R. Bohnart

Fig. 8-30 Hammering welded square butt joint in vise positioned so that root of the weld is exposed to the most stress. Prographics

underneath. Keep the following important points in mind as you weld:
- The joint crack should be the center of the molten weld bead.
- The plate edges must be completely fused, and penetration must extend through to the root.
- Weld root and face reinforcement should not be higher than $1/16$ inch. It should be of uniform height and width and have smooth, even ripples.

7. Continue practice until the welds conform to the characteristics described in step 6 and are acceptable to your instructor. A good weld will show a slight bead on the back side, which indicates that complete penetration and fusion have taken place, Fig. 8-29. You should not be able to see the edges of the sheet metal on the back side.

8. To test your welds, clamp the sheet in a vise so that the weld is parallel to the jaws of the vise and about $1/8$ inch above the jaws. The back side of the weld should be toward you. Then hammer the top sheet away from you so that the root of the weld is in tension. After the sheet has been bent over as far as you can hammer it, place the joint in the vise and squeeze the sheets on each side of the weld until they are back to back, Fig. 8-30. The weld should not crack or break. If it cracks or breaks, you will probably notice that the plate edge was not fused and that it is in its original square-cut condition. This indicates that the plate edges were not brought to a molten condition and the filler rod was shoved through a cold seam.

9. After you are able to make welds of good appearance and good quality in the flat position, practice in the other positions. Note the vertical butt weld shown in Fig. 8-31.

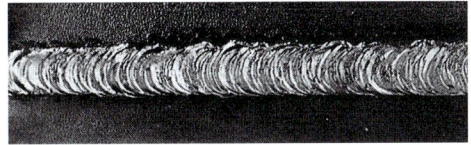

Fig. 8-31 A vertical butt joint, groove weld in $1/8$-inch plate. A no. 3 tip and a $3/32$-inch, high-test filler rod were used. The flame was neutral. Edward R. Bohnart

Welding a Lap Joint: Jobs 8-J19–J22

The lap joint is made with a fillet weld. It is one of the most difficult to weld with the oxyacetylene process because it requires careful attention to the distribution of heat on all surfaces to be welded. The top sheet may be so hot that it burns away while the bottom sheet may not be hot enough. This adds to the difficulty of obtaining fusion at the root of the joint. If the joint is welded from one side only, the weld is under shearing stress, and the joint cannot reach maximum strength. Welding from both sides strengthens the joint, but this is not always possible on the job.

1. Check the Job Outline for plate thickness and type and size of filler rod. Follow the manufacturer's recommendations for tip size.
2. Adjust the torch for a neutral flame.
3. Set up the plates in the flat position with one plate overlapping the other.
4. Tack weld all four corners of the two plates.
5. Weld from right to left. Position the torch and filler rod as shown in Fig. 8-32. You will find that you will have to vary the position of both the torch and the filler rod depending upon the condition of the upper plate edge and the molten pool. Move the torch with a slight semicircular motion. The edge of the upper plate will have a tendency to melt before the root of the joint and the surface of the lower plate have had a chance to melt. This can be overcome by directing the flame to the lower plate and also by protecting the upper plate with the welding rod as shown in Fig. 8-32. The rod absorbs some heat to prevent excessive melting of the top plate. Torch movement must be uniform and not too wide, or the completed weld will be too wide, and the appearance will be rough. The weld face should be slightly convex.
6. Continue practice until the welds have the characteristics shown in Fig. 8-33 and are acceptable to your instructor.
7. To test your weld, place it in a vise, Fig. 8-34, and pry the two sheets apart. Bend the upper sheet backward until it forms a T with the bottom plate. Inspect the root of the weld for fusion and complete penetration. The weld should be thoroughly fused to the bottom plate and not pull away.
8. After you are able to make welds of good appearance and of good quality in the flat position, practice in the other positions.

Welding a T-Joint: Jobs 8-J23–J25

A T-joint also requires fillet welds. In welding a T-joint from both sides, you will have to apply more heat than for the other joints and manipulate the filler rod more. It is a good joint to practice in preparation for the welding of small diameter pipe with the oxyacetylene process.

1. Check the Job Outline for plate thickness and type and size of filler rod. Follow the manufacturer's recommendations for tip size.
2. Adjust the torch for a neutral flame.

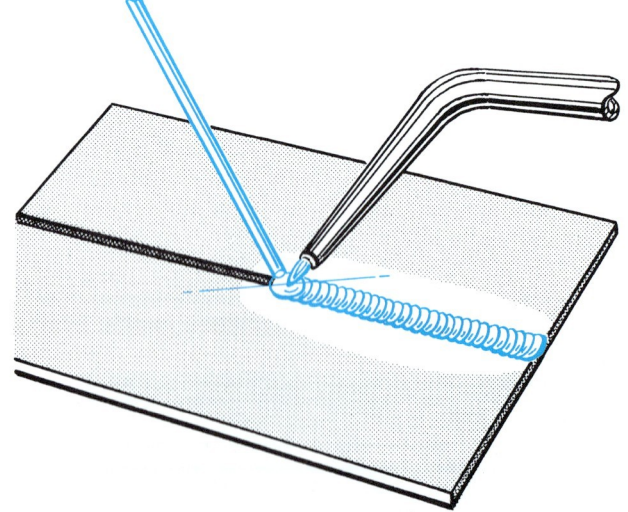

Fig. 8-32 Positions of the torch and filler rod when making a lap joint, fillet weld in flat position. The upper plate edge is protected by the filler rod.

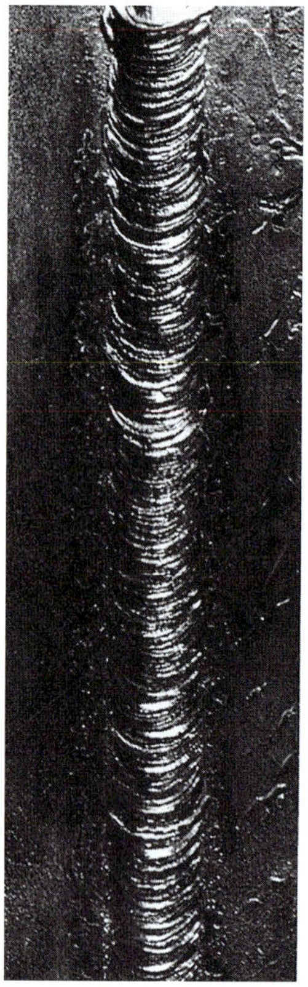

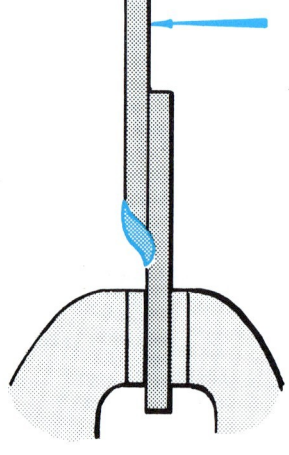

Fig. 8-33 A lap joint welded in the vertical-up position in 1/8-inch steel plate. A no. 3 tip and 3/32-inch mild steel filler rod were used. The flame was neutral.
Edward R. Bohnart

Fig. 8-34 To test a fillet weld on a lap joint, pry the sheet apart and inspect the fusion and penetration.

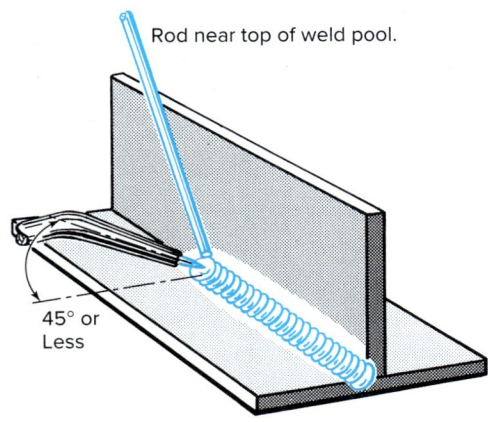

Fig. 8-35 Positions of the torch and filler rod when welding a T-joint in the horizontal position.

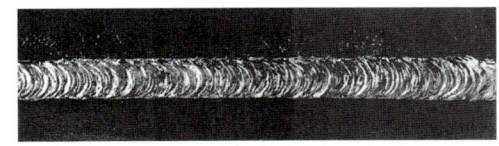

Fig. 8-36 A fillet weld in a T-joint welded in the horizontal position. Stock is 1/8-inch plate. A no. 4 tip and 1/8-inch mild steel filler rod were used. The flame was neutral. *Edward R. Bohnart*

3. Set up the plates and tack weld them at each end of the joint.
4. Place the joint on the table so that welding will be in the horizontal position.
5. The torch and filler rod should be held in about the same position as that used when welding a lap joint, Fig. 8-35.
6. Weld from right to left. Pay careful attention to flame control. There is a tendency for the vertical plate to melt before the flat plate. This causes undercutting along the edge of the weld on the vertical plate. Direct more heat toward the bottom plate and toward the corner of the joint in order to secure fusion at the root. Be careful that you do not form a pocket in the corner and thus overheat the tip. Move the torch back and forth slightly or in a semicircular movement as required for proper heat distribution. Feed the filler rod just above center at the forward edge of the molten pool so that it comes between the flame and the vertical plate. Thus the filler rod absorbs some heat and prevents undercut in the vertical plate. The weld bead should be equally divided between the two plates and have a slightly convex face.
7. Clean the scale produced by the first weld from the back side of the joint. This can be done by moving the flame rapidly back and forth along the joint so that the oxide scale expands and pops from the surface.
8. Weld the joint from both sides. You will find the second side somewhat harder than the first side because of the weld on the other side. There is more material to bring to the welding temperature. The welding technique is different because the bottom plate has a tendency to melt before the root of the weld and the vertical plate. You must direct more heat toward the root of the weld and the vertical plate than on the flat plate. Move the weld pool slowly along the line of weld and pay careful attention to the state of fusion at the root of the joint. Bridging from plate to plate is a common fault because root penetration is difficult.
9. Continue practice until the welds have the characteristics shown in Fig. 8-36 and are acceptable to your instructor.
10. Make up a test joint welded only from one side. Bend the vertical plate toward the face of the weld. Examine the root of the weld for penetration and fusion. The weld should be thoroughly fused to the flat plate and should not peel away. Inspect the edge of the vertical plate. It should show that fusion has taken place. The raw, square-cut edges should not be intact.
11. After you are able to make welds of good appearance and of good quality in the flat position, practice in the other positions. A T-joint welded in the vertical-up position is shown in Fig. 8-37.

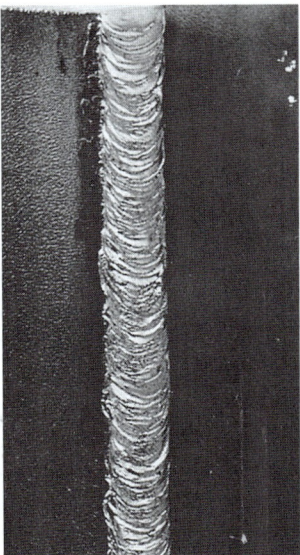

Fig. 8-37 A fillet weld in a T-joint welded in the vertical position travel up. Stock is 1/8-inch plate. A no. 4 tip and 1/8-inch mild steel filler rod were used. The flame was neutral. *Edward R. Bohnart*

Heavy Steel Plate and Pipe

Heavy Plate Welding Practice: Jobs 8-J26–J33

These jobs will give you the opportunity to practice oxyacetylene welding on heavier steel. You will gain experience in handling a greater amount of heat and larger pools of molten metal.

Practice Jobs 8-J26–J33 on heavy plate with high test welding rod before actual practice on pipe. Practice in heavy plate welding will reduce the number of hours in pipe practice necessary.

All of the welding techniques learned while completing the gas welding jobs on 1/8-inch plate apply to heavier plate. Your principal concern in welding heavy plate will be that the plate surface is melted and that the weld pool is fluid so that fusion, not IF, can take place. In making a V-butt weld, make sure that the plate edges at the bottom of the V are in a molten state when the filler rod is applied. There must be fusion through to the back side. Make sure that the tip is the proper size so that the flame is large enough to provide enough heat to melt the plate. A flame that is too large for the tip size will be harsh and cause the completed weld to be rough. Proper regulator pressure is also important to ensure a soft flame.

Backhand Welding of Plate

Up to this point, all of your welds were made with the *forehand method of welding*. The direction of travel was from right to left, and the flame and filler rod were in front of the completed weld bead. Remember if you are left-handed, this would have been reversed as well as the backhand method. The *backhand method* is another technique that has certain advantages over the forehand method, especially when welding heavy plate and pipe.

In the welding of plate by the backhand method, the welding progresses from left to right. The flame is directed backward at the partially completed weld. The welding rod is held between the partially completed weld and the flame, Figs. 8-38 and 8-39. This permits the flame to be directed at all times on the surfaces of the material to be welded and ahead of the weld pool with little motion of the torch and filler rod. In welding beveled pipe or plate, a narrower angle may be used for forehand welding. This speeds up the welding operation. The surface appearance of a backhand weld is not as smooth as that of a forehand weld. The ripples are coarse, and there are fewer of them per unit of weld length. The penetration bead on the back side at the root of the weld should look about the same as in forehand welding. Welders usually

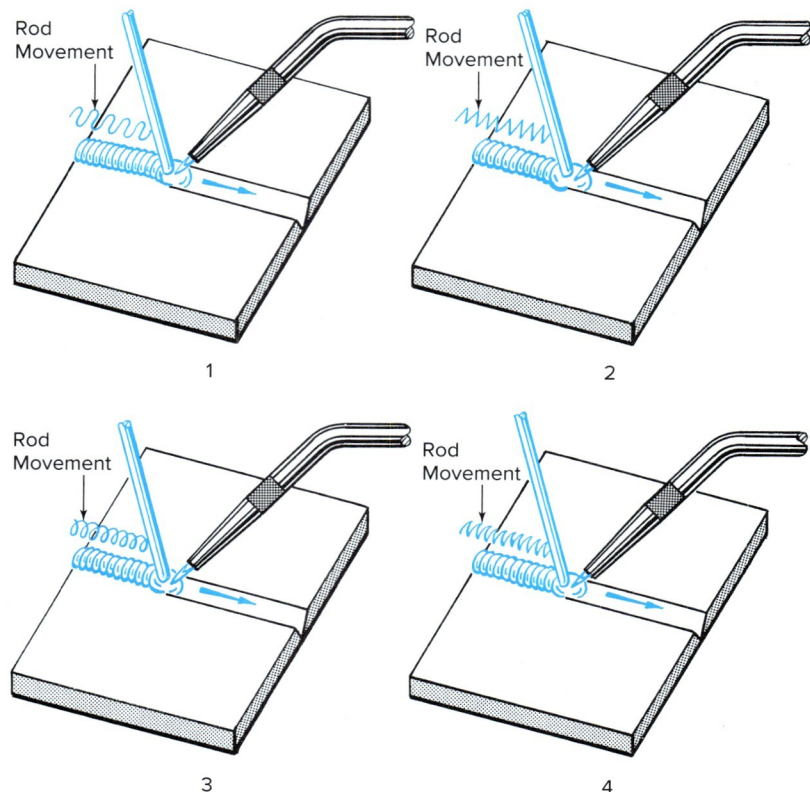

Fig. 8-38 The backhand method of welding from left to right. The flame precedes the welding rod, but the weld advances from left to right instead of from right to left. Note rod manipulation.

Fig. 8-39 Welding a butt joint in the overhead position, using the backhand welding technique. Location: Northeast Wisconsin Technical College Mark A. Dierker/McGraw Hill

find it easier to get complete penetration through the back side of a V-butt joint with the backhand method.

Pipe Welding Practice: Jobs 8-J34–J38

Pipe welding was first done with the oxyacetylene process. This was the only process used for a number of years both in construction and overland pipeline. For the

SHOP TALK

Flexible Automation
The term refers to robots used for complex shapes and applications. In these welding paths, torch-angle manipulation is needed. In contrast, *fixed automation* means an electronic welding system for circles or straight paths.

most part, pipe now is welded with the shielded metal arc, gas metal arc, flux cored arc, or gas tungsten arc processes. You will have an opportunity to practice with these other processes in the later chapters of this text. Oxyacetylene welding is now generally confined to small diameter piping and in field installations. Oxyacetylene welding of small pipe up to 2 inches in diameter is faster than with the electric arc. Welding may be done in all positions with a single or multipass procedure and either the forehand or backhand technique. The choice of procedure and welding technique varies with the type of joint, the diameter and wall thickness of the pipe, the position of welding, and the size of the filler rod. The number of weld passes necessary for a particular pipe joint should be determined by the rule of one weld pass for each $\frac{1}{8}$-inch thickness of pipe wall over $\frac{3}{16}$ inch.

Study Fig. 8-40, which presents standard pipe joints.

On each pipe fabrication job, the welding procedure to be followed in order to comply with the appropriate codes for pressure piping, boiler, and pressure vessel welding includes the following:

- Welding process to be used
- The nature of the pipe material
- The type of filler metal to be used in welding
- The position of welding
- The preparation of the material, whether by machining, flame cutting, chipping or grinding, or a combination of these
- Use or nonuse of backup rings—the welder must qualify for each.
- Size of the welding tip and oxygen and acetylene pressure
- The nature of the flame
- Method of welding, whether forehand or backhand
- The number of layers of welding for the various pipe thicknesses
- The cleaning process between each layer
- Treatment of weld defects
- Heat treatment or stress relieving if required

Prepare the pipe for welding according to the following procedure:

1. Check the Job Outline, Table 8-4 (p. 231), for size and type of pipe and filler rod.
2. Before making a butt weld on pipe that is over $\frac{1}{8}$-inch thick, bevel the ends of each pipe at an angle of 30 to $37\frac{1}{2}°$ to form a 60 to 75° included angle. Space the pipe at the root at a distance of $\frac{3}{32}$ to $\frac{1}{8}$ inch, Fig. 8-41. Pipe that has a wall thickness less than $\frac{1}{8}$ inch may be successfully welded without beveling by spacing the pipe ends apart a distance equal to the wall thickness and fusing the pipe ends ahead of the weld deposit. This is a similiar technique as that practiced in welding square butt welds.
3. Adjust the torch for a neutral flame. In some pipe welding situations, a stronger weld may be run with a slight carburizing flame.

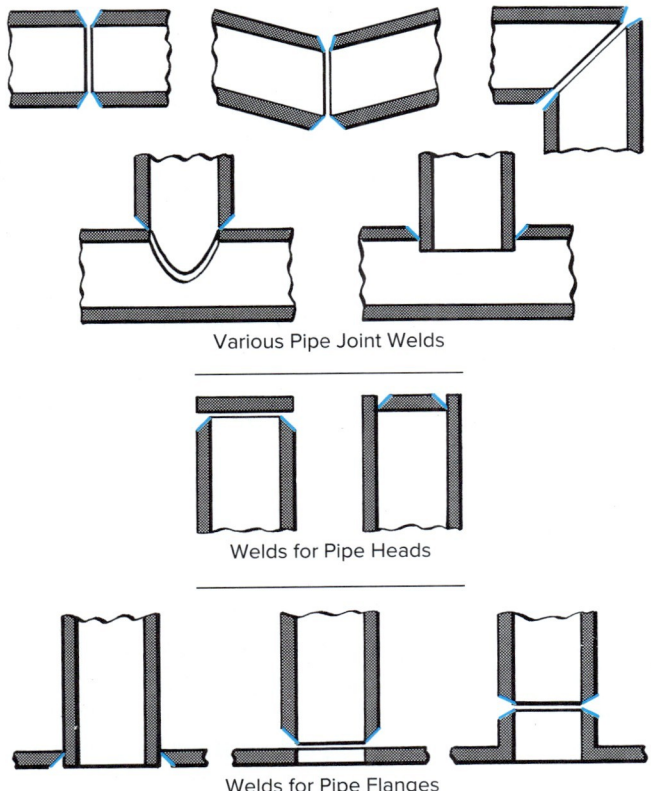

Fig. 8-40 Pipe joints.

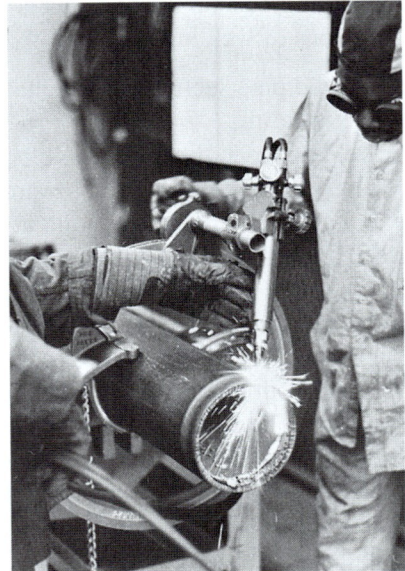

Fig. 8-41 Students beveling the edge of a pipe with a standard beveling machine. Edward R. Bohnart

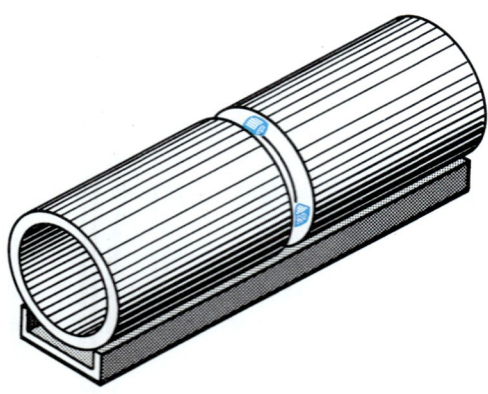

Fig. 8-42 Tack welding a beveled-pipe joint.

4. Place two sections of beveled pipe on a supporting jig like that shown in Fig. 8-42. Space the beveled edges at least 1/8 inch all the way around.
5. Make a tack weld on the top of the pipe. Make sure that the tack is fused to the surface of each pipe bevel and deep into the root of the joint.
6. When the tack weld has cooled to a black heat, roll the pipe around and make another tack weld on the opposite side. Before making this tack, make sure that the spacing is equal all the way around the pipe.
7. Make tack welds at the quarter points as instructed in steps 5 and 6 so that the joint is held by four equally spaced tack welds.

Roll Positions: Job 8-J34 Even though you are able to gas weld in all positions as a result of the practice you have had up to this point, it is recommended that your first pipe practice be a roll weld. Welding around the circumference of a section of stationary pipe requires torch and rod angles that are changing constantly, and this is not like any of the other positions of welding.

1. Place the tacked pipe sections in the channel iron jig used for tack welding.
2. Start the weld between the 2 and 3 o'clock positions. Hold the torch so that it is pointed toward the top of the pipe and along the line of the joint at about a 45° angle. This is somewhat similar to the position used in the vertical position of welding on plate.
3. Begin the weld by heating an area slightly wider than the width of the bevel. Add the filler rod when the groove faces of both pipe ends has been brought to the molten state. The weld should be built up to the size desired before moving on. Move the torch in a short, semicircular movement, and the rod with a short, crosswise motion. The torch motion and the rod motion should cross at the center of the joint. A tack weld should be thoroughly fused with the advancing weld pool. Make sure that fusion is obtained at the root of the joint and on the groove face. Penetration must be through to the back side, and a small bead should be formed on the inside of the pipe.
4. Carry the weld up on the pipe to a point slightly below the 12 o'clock position at top center. The weld is being done in the vertical up and flat positions. Give the pipe a one-quarter turn toward you and begin a new weld at the end of the previous weld.
5. Repeat the procedure until the entire pipe is welded. The pipe is welded in quarter sections. Rotate each quarter section toward you as it is completed, Fig. 8-43. Make sure that fusion is obtained at the root of the weld and on the groove surface. The weld must penetrate the back side, and a small bead should be formed on the inside of the pipe.
6. Continue practice until the welds have the appearance shown in Fig. 8-44 and the characteristics described in step 8, page 221.
7. Make up a test weld and remove coupons as shown in Figs. 16-29 and 16-30, page 485. Subject them to face- and root-bend tests as described under Root-, Face-, and Side-Bend Soundness Tests, pages 486–487. Figure 8-45 shows the typical appearance of larger pipe specimens after testing.

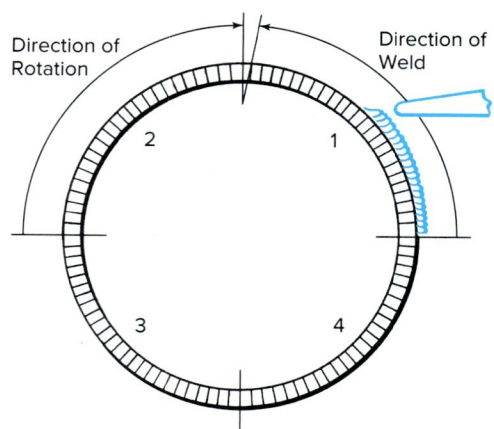

Fig. 8-43 Follow this sequence for roll pipe welding with a gas torch.

Horizontal Fixed Position (Forehand Technique): Job 8-J35

1. Tack the pipe joint in the same manner as directed for roll welding. Place it in the horizontal fixed position.
2. Start welding at the bottom of the pipe between two tack welds at the 5 or 7 o'clock position, depending upon the side of the pipe you are welding from. Use the welding technique described in step 3 for the roll positions, page 226. Welding is in the overhead position. The pool must not become too fluid, and just enough filler rod must be added to form the bead size desired.

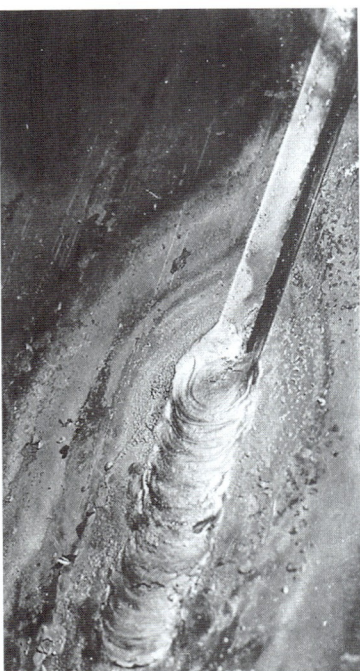

Fig. 8-44 A butt joint, V-groove weld in pipe. Note the complete fusion and penetration at the leading point of the weld. Note also the size in relation to the width of the joint. *Praxair, Inc.*

Fig. 8-45 A good weld in low carbon steel will bend at least 90° without cracking or fracturing. These specimens have been bent double with no sign of failure. *Edward R. Bohnart*

3. As the weld progresses upward, change the technique to that learned in making vertical welds. The weld pool must not be too fluid, and control of the pool must be maintained through proper torch and rod manipulation. The force and pressure of the torch flame is used to keep the weld pool in place. When you find that the weld is getting too hot, pull the torch away so that the flame does not play directly on the weld pool. Begin welding again after the weld has cooled somewhat. Careful use of the filler rod is necessary.

4. As you approach the top of the joint, the welding technique will merge into that used for welding in the flat position. Travel will be a little faster, and you must be careful not to burn through.

5. After you have reached a point a little past center on the top, return to the bottom of the pipe. Concentrate the flame on the end of the previously completed weld and melt the front surface so that good fusion and penetration is secured at this point. The new weld bead must be thoroughly fused with the previously laid bead.

6. Carry the welding pool up to the top of the pipe. Use the same welding technique employed on the first side. Make sure that all tack welds are melted and become a part of the weld pool.

7. When you are about ¼ inch away from the end of the completed weld at the top of the pipe, concentrate the flame on the face of the completed weld and at the bottom of the V in order to establish a weld pool that will join the two welds.

8. Inspect the weld. The face of the weld should be only about ⅛ inch wider on each side than the width of the bevel shoulders. Weld face reinforcement should be about 3/32 inch high. It should be of uniform width and height without high or low areas. Ripples should be smooth and uniform. It should be very difficult to find the spot at the top and bottom where the welding started and ended. The edges of the weld should be of even width and parallel all around the pipe with no undercut. The edges should flow into the pipe surfaces as if molded.

9. Continue practice until the welds have the appearance shown in Fig. 8-44 on page 227, the characteristics described in step 8, and are acceptable to your instructor.

10. Make a test weld and cut it up into face and root test coupons. Use the testing procedure described in step 7 in the procedure for welding pipe in the roll position, page 226.

Backhand Welding of Pipe

It has already been pointed out that the backhand method of welding for heavy plate provides the advantage of increased speed of welding along with a decrease in the amount of filler rod and gases consumed. Practice with this process increases one's skill in the use of the gas welding torch. Backhand pipe welding is based on the principle that hot steel will absorb carbon, which lowers the melting point. Thus, less heat is required for fusion so that welding speed increases.

The flame is adjusted for excess acetylene or carburizing. The acetylene feather is about 1½ times the length of the inner cone. When the steel is heated, it picks up carbon from the excess acetylene feather, thus lowering the melting point of the base metal at the surface. This action takes place to a slight depth only because carbon cannot penetrate very deeply into the steel within the time required to complete the weld.

The lowering of the melting point means that a thin layer of metal at the surface is ready for fusion sooner than it would be if a neutral flame were being used. The result is that thorough fusion between the base metal and weld metal is obtained without deep melting. During the actual welding, the carbon that was absorbed by the surface of the base metal becomes distributed through the weld deposit. Larger flames and welding rods can be used. The V-angle is smaller, and the root of the weld can be fused readily without serious burn-through.

Vertical Welds in Pipe in the Horizontal Fixed Position: (Backhand Technique) Job 8-J37

1. Check the Job Outline for pipe size and type and size of filler rod. Follow the manufacturer's recommendations for tip size.
2. Adjust the torch for a neutral or carburizing flame, depending on the method of welding you wish to practice and the type of filler rod used.
3. Tack weld two pieces of beveled pipe and set them up in the horizontal fixed position.
4. Start the weld at the top of the joint at about the 12 o'clock position and carry it down one side to the bottom. Do not start the weld at a tack weld. Hold the torch and filler rod as shown in Fig. 8-46. Note the backhand position of the torch. The method of depositing the bead is shown in Fig. 8-47. Advance the filler rod with a circular motion, and at the same time move the flame back and forth in the direction of travel. Time these movements so that the rod and flame part and then come together like an accordion. When you reach the 3 o'clock position, you will be welding in the vertical position. The pressure of the burning gases, the

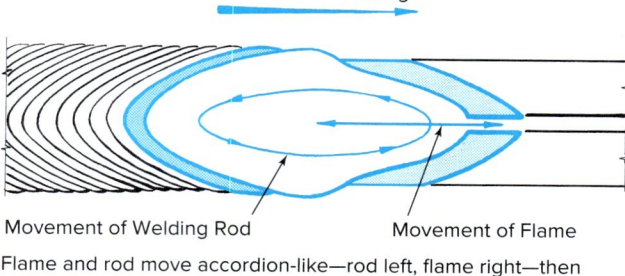

Fig. 8-47 This detail of Fig. 8-46 demonstrates the motion of flame and rod in backhand welding.

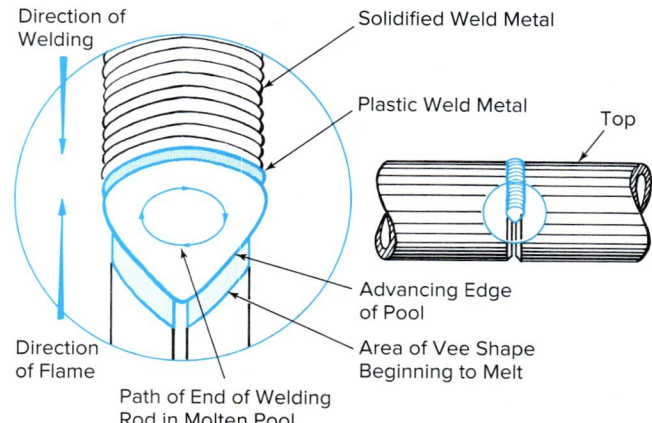

Fig. 8-48 In making a position weld (5G) in pipe, part of the weld is done in the vertical position. In the backhand technique, welding in the vertical position can be done from top to bottom. The path of the end of the welding rod in the molten pool is changed to keep the pool spread out. The pool itself is kept small.

flame, and careful rod manipulation will keep the weld pool from running out, Fig. 8-48. As you approach the 6 o'clock position, you will be welding in the overhead position. The most important factor in pool control in this position is to prevent the formation of a drop of metal. The metal will not fall until a drop is formed. The cohesiveness of the weld pool will also assist in keeping the molten metal from sagging and dripping.

5. Restart the weld at the top and carry it down the opposite side to meet the first weld at the bottom of the joint. Repeat the welding technique practiced on the second side. The changes in torch position for control of the pool are summarized in Fig. 8-49.
6. Continue practice until the welds have the desirable weld appearance shown in Fig. 8-45, the characteristics described in step 8, page 227, and are acceptable to your instructor.
7. Weld a test joint in the same position and following the same welding procedure used for this practice

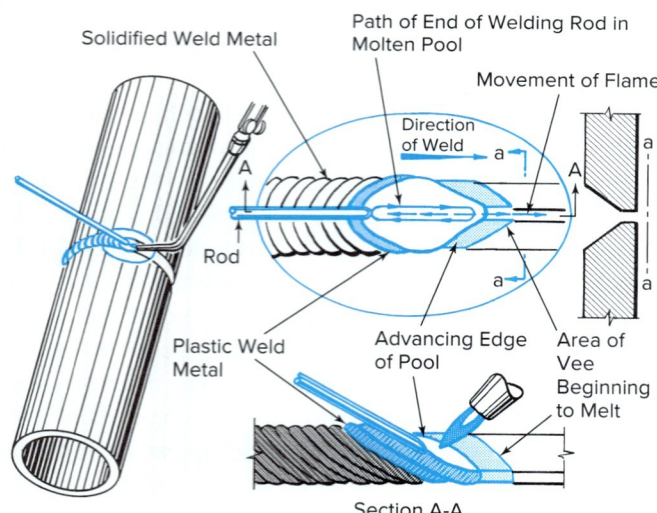

Section A-A shows average position of rod and flame with respect to the pool.

Fig. 8-46 The essential elements in the backhand method for pipe welding.

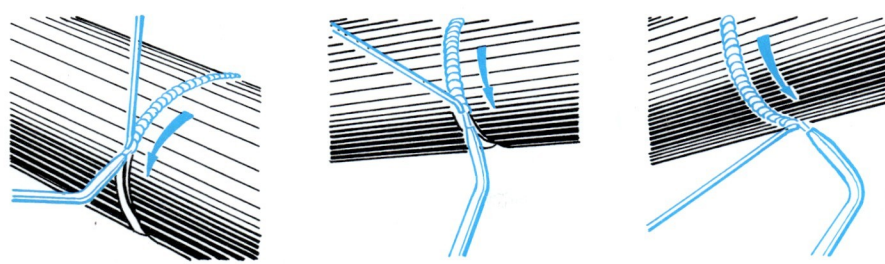

Fig. 8-49 The angle of the torch flame is changed as a position pipe (5G) weld is made so that as much advantage as possible can be taken of the pressure of the burning gases in controlling the pool.

job. Take standard face- and root-bend coupons from the test joint and subject them to bending tests.

Horizontal Welds in Plate and Pipe in the Vertical Fixed Position (Backhand Technique): Jobs 8-J36 and 8-J38 Horizontal welding is the most difficult of all gas welding positions and becoming proficient will take a great deal of practice. In order to save time and cut the cost of materials, you are asked to practice 8-J36 on plate. After you have mastered welding on plate, you will find that welding pipe in this position is relatively easy.

1. Check the Job Outline for plate or pipe size and type and size of filler rod. Follow the manufacturer's recommendations for tip size.
2. Adjust the torch for a neutral or carburizing flame, depending on the method of welding you wish to practice.
3. Tack weld two pieces of beveled plate or pipe and set them up in the vertical fixed position.
4. In making this weld, travel from left to right. The flame is ahead of the filler rod and points back at the completed weld. Apply heat to the lower side of the V. The weld has a tendency to build up on the lower panel of the joint, and a shelf must be provided on the lower bevel to support the pool. The backhand motion of the torch and filler rod causes the molten weld metal to assume a diagonal shape. The completed weld has coarse diagonal ripples that are not close together, Fig. 8-50. The support that is supplied by the built-up metal on the lower bevel makes it possible to keep the pool somewhat larger than in a vertical weld and fairly molten. This permits rapid welding. Manipulate the filler rod more than in forehand welding and use a circular motion to keep the welding pool spread out rather thin and to prevent too much metal sagging at the lower side of the weld. In multilayer welding, starting and stopping points of beads are staggered alternately with successive beads.
5. Continue practice until the welds conform to the desirable weld appearance shown in Fig. 8-45, the characteristics described in step 8, page 227, and are acceptable to your instructor. Make sure that you have good penetration and fusion on the inside of the pipe.
6. Weld a test joint in the same position and following the same welding procedure that you used for this practice job. Take standard face- and root-bend coupons from the test joint and subject the coupons to bending tests.

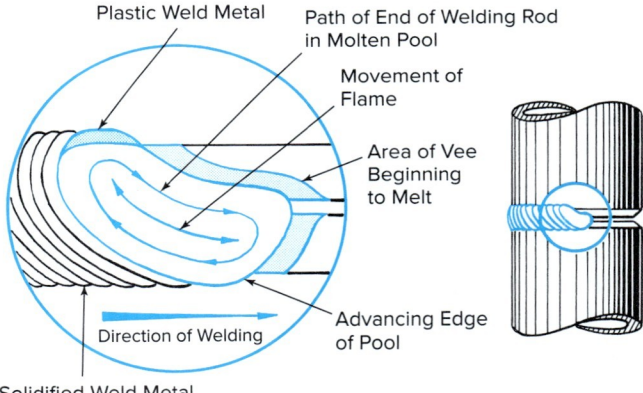

Fig. 8-50 In making a horizontal weld in pipe axis in the vertical fixed position (2G), the paths of the torch and the welding rod tip are altered to assist in the control of the pool.

CHAPTER 8 REVIEW

Multiple Choice

Choose the letter of the correct answer.

1. When shutting off the gas welding torch, which valve should be shut off first? (Obj. 8-1)
   a. Oxygen cylinder valve
   b. Acetylene cylinder valve
   c. Oxygen torch valve
   d. None of these

2. If the regulator adjusting screw becomes hard to operate, which lubricant should be used? (Obj. 8-1)
   a. Oil
   b. Grease
   c. No lubricants should be used around any oxyacetylene fittings or parts
   d. Either a or b

3. Which of the following is a part of the oxyacetylene flame? (Obj. 8-2)
   a. Outer envelope
   b. Inner cone
   c. White inner cone (acetylene feather)
   d. All of these
4. What is another name for the excess acetylene flame? (Obj. 8-2)
   a. Oxidizing
   b. Carburizing
   c. Reducing
   d. Both b and c
5. Braze welding and bronze surfacing are usually performed with what type of flame? (Obj. 8-2)
   a. Oxidizing
   b. Carburizing
   c. Neutral
   d. Both b and c
6. Which is the tool for the oxyacetylene welder? (Obj. 8-2)
   a. Torch
   b. Flame
7. What is the preferred tool for lighting the gas welding torch? (Obj. 8-2)
   a. Matches
   b. Cigarette lighter
   c. Flint spark lighter
   d. None of these
8. What is a flashback? (Obj. 8-2)
   a. Violent explosion out of the tip
   b. Flame burning back inside the torch
   c. Shrill hissing sound or squealing sound
   d. Both b and c
9. Which of the following is an AWS classification number for a gas welding filler rod? (Obj. 8-3)
   a. E-71-T1
   b. RG-45
   c. RG-60
   d. Both b and c
10. In tacking up a pipe joint, make sure there is no root opening by having the pipe butted tight together. (Obj. 8-3)
    a. True
    b. False

Review Questions

Write the answers in your own words.

11. Name the equipment that is needed to perform oxyacetylene welding. (Obj. 8-1)
12. Explain how you would close down a line system for welding. (Obj. 8-1)
13. Name the three stages of the oxyacetylene flame. (Obj. 8-2)
14. Name at least four causes of torch backfire. (Obj. 8-3)
15. What are the important factors in control of the pool in the vertical position? (Obj. 8-3)
16. What is meant by a *roll position weld*? (Obj. 8-3)

Define each word. (Obj. 8-2)

17. Fusion
18. Penetration
19. Reinforcement
20. Undercut

INTERNET ACTIVITIES

Internet Activity A

Search the Internet to see what kinds of courses are offered in gas welding practice. List two courses you could take, after you complete this one, to further your knowledge of gas welding. Describe the course.

Internet Activity B

Choose a topic from this chapter that you would like to know more about and research it on the Internet. Either write a report or make a drawing showing what information you found.

Table 8-4 Job Outline: Gas Welding Practice: Jobs 8-J1–J38

Job Order[1]	Text No.	Joint	Operation	Material Type	Thickness	Filler Rod Type	Size	Welding Position[2]	Text Reference
1st	8-J1	Flat plate	Autogenous welding	Mild steel	1/8	None		1	213
2nd	8-J2	Flat plate	Autogenous welding	Mild steel	1/8	None		3	214
3rd	8-J3	Flat plate	Beading	Mild steel	1/8	Mild steel RG45	3/32	1	214
4th	8-J9	Edge joint	Flange welding	Mild steel	1/8	Mild steel RG45	3/32	1	216
5th	8-J10	Corner joint	Groove welding	Mild steel	1/8	Mild steel RG45	3/32	1	217
6th	8-J4	Flat plate	Beading	Mild steel	1/8	Mild steel RG45	3/32	3	218
7th	8-J11	Edge joint	Flange welding	Mild steel	1/8	Mild steel RG45	3/32	3	218
8th	8-J12	Corner joint	Groove welding	Mild steel	1/8	Mild steel RG45	3/32	3	219
9th	8-J5	Flat plate	Autogenous welding	Mild steel	1/8	None		2	219
10th	8-J6	Flat plate	Beading	Mild steel	1/8	Mild steel RG45	3/32	2	219
11th	8-J13	Edge joint	Flange welding	Mild steel	1/8	Mild steel RG45	3/32	2	219
12th	8-J14	Corner joint	Groove welding	Mild steel	1/8	Mild steel RG45	3/32	2	219
13th	8-J7	Flat plate	Autogenous welding	Mild steel	1/8	None	3/32	4	219
14th	8-J8	Flat plate	Beading	Mild steel	1/8	Mild steel RG45	3/32	4	220
15th	8-J15	Edge joint	Flange welding	Mild steel	1/8	Mild steel RG45	3/32	4	220
16th	8-J16	Corner joint	Groove welding	Mild steel	1/8	Mild steel RG45	3/32	4	220
17th	8-J19	Lap joint	Fillet welding	Mild steel	1/8	Mild steel RG45	3/32	1	220
18th	8-J20	Lap joint	Fillet welding	Mild steel	1/8	Mild steel RG45	3/32	3	220
19th	8-J17	Square butt joint	Groove welding	Mild steel	1/8	Mild steel RG45	3/32	1	222
20th	8-J18	Square butt joint	Groove welding	Mild steel	1/8	Mild steel RG45	3/32	3	222
21st	8-J21	Lap joint	Fillet welding	Mild steel	1/8	Mild steel RG45	3/32	2	222
22nd	8-J23	T-joint	Fillet welding	Mild steel	1/8	Mild steel RG45	3/32	2	222
23rd	8-J22	Lap joint	Fillet welding	Mild steel	1/8	Mild steel RG45	3/32	4	222
24th	8-J24	T-joint	Fillet welding	Mild steel	1/8	Mild steel RG45	3/32	3	222
25th	8-J25	T-joint	Fillet welding	Mild steel	1/8	Mild steel RG45	3/32	4	222
26th	8-J26	Flat plate	Beading—backhand	Mild steel	3/16	Steel, high test RG-60	1/8	1	223
27th	8-J27	Flat plate	Beading	Mild steel	3/16	Steel, high test RG-60	1/8	3	223
28th	8-J28	Lap joint	Fillet welding—backhand	Mild steel	3/16	Steel, high test RG-60	1/8	1	223
29th	8-J29	Flat plate	Beading—backhand	Mild steel	3/16	Steel, high test RG-60	1/8	2	223
30th	8-J30	Lap joint	Fillet welding—backhand	Mild steel	3/16	Steel, high test RG-60	1/8	2	223
31st	8-J31	Lap joint	Fillet welding	Mild steel	3/16	Steel, high test RG-60	1/8	3	223
32nd	8-J32	Beveled butt joint	Groove welding—backhand	Mild steel	3/16	Steel, high test RG-60	1/8	1	223
33rd	8-J33	Beveled butt joint	Groove welding	Mild steel	3/16	Steel, high test RG-60	1/8	3	224
34th	8-J34	Beveled butt joint	Groove welding	2- to 4-in. pipe	Standard	Steel, high test RG-60	1/8 or 5/32	1G-roll	226
35th	8-J35	Beveled butt joint	Groove welding—forehand	2- to 4-in. pipe	Standard	Steel, high test RG-60	1/8 or 5/32	5G	226
36th	8-J37	Beveled butt joint	Groove welding—backhand	2- to 4-in. pipe	Standard	Steel, high test RG-60	5/32 or 3/16	3G	226
37th	8-J36	Beveled butt joint	Groove welding—backhand	Mild steel plate	3/16	Steel, high test RG-60	1/8 or 5/32	5G	228
38th	8-J38	Beveled butt joint	Groove welding—backhand	2- to 4-in. pipe	Standard	Steel, high test RG-60	1/8 or 5/32	2G	229

[1] It is recommended that the student do the jobs in this order. In the text, the jobs are grouped according to the type of operation to avoid repetition.

[2] 1 = flat, 2 = horizontal, 3 = vertical, 4 = overhead, 5 = multiple positions pipe.

Design credits: Cinema and movie icon: Denis Meshkov/123RF; and Illustration of Welding icons: Mr. Nachai Sorasee/123RF

Braze Welding and Advanced Gas Welding Practice:

Jobs 9-J39–J49

Chapter Objectives

After completing this chapter, you will be able to:

9-1 Define, describe, and demonstrate braze welding.

9-2 Describe and demonstrate gas welding on various ferrous and nonferrous metals.

9-3 Demonstrate and describe hard facing.

Most industrial gas welding is on the ferrous and nonferrous metals and alloys covered in this chapter. The practice jobs have been carefully selected on the basis of the practices being followed in industry. Table 9-3, on page 249, provides a Job Outline for this chapter. It is recommended that the students complete the jobs in the order shown. As a beginning welder, you may not have an opportunity on the job to weld all of these materials until you gain more experience. Mastery of the practice jobs listed in the Job Outline, however, will provide you with the variety of gas welding skills that is the mark of the skilled combination welder. Because this unit continues the discussion of gas welding practice begun in Chapter 8, the practice jobs are numbered consecutively. Thus, the first job in this chapter will be 9-J39.

Braze Welding

Braze welding is a form of torch brazing (TB). As AWS defines it,

> A welding process that uses a filler metal with a liquidus above 840°F and below the solidus of the base metal; the base metal is not melted. Unlike brazing, in braze welding the filler metal is not distributed in the joint by capillary action. (American Welding Society, Standard Welding Terms and Definitions: ANSI/AWS 3.0)

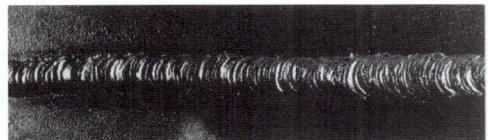

Fig. 9-1 A braze-welded butt joint in ¼-inch steel plate. The weld was made with a no. 4 tip and a ⁵⁄₃₂-inch bronze rod in the flat position. Edward R. Bohnart

In braze welding (BW), one makes edge, groove, fillet, plug, or slot welds by using a nonferrous filler metal. Intentional capillary action is not a factor in distribution of the brazing filler metal. Joint designs for braze welding are the same as for oxyacetylene welding. Some differential chemical attack and galvanic corrosion may occur in braze welding.

In **torch brazing,** a bronze filler rod supplies the weld metal, and the oxyacetylene flame furnishes the heat.

The joint design for braze welding is the same as for fusion welds. The welding technique is the same, except that the base metal is not melted. It is, in fact, raised only to the tinning temperature. A **bond** (molecular union) is formed between the bronze welding rod and the prepared surface of the work. The welds compare favorably with fusion welds, Fig. 9-1.

Heavy materials have to be beveled, and the weld must be made with several passes to ensure a weld throughout the total thickness of the metal. Each pass should be cleaned before applying the next layer of weld metal.

Industrial Uses

Braze welding is particularly adaptable to the joining and repairing of such metals as cast iron, malleable iron, copper, brass, and dissimilar metals. It is also used for the building up of worn surfaces, Fig. 9-2, a process that is referred to as **bronze surfacing.** Parts that are going to be subjected to high stress or high temperature should not be braze welded. Bronze loses its strength when heated to a temperature above 500°F. The process also cannot be used when a color match is desired.

Braze welding has the following commercial advantages:

- The speed of welding is increased because less heat is required than for fusion welding.
- The welding of cast iron is greatly simplified because the base metal is not melted. Preheating is reduced or eliminated.
- Since the work does not have to be brought up to a high temperature, expansion and contraction are reduced.
- The bronze weld yields as the work cools until the temperature reaches about 500°F. This reduces locked-up stresses.

Fig. 9-2 Building up a gear tooth with bronze in a series of layers, each of which is about ¹⁄₁₆-inch thick. A flux-coated brazing rod is being used. Location: Northeast Wisconsin Technical College Mark A. Dierker/McGraw Hill

- Malleable cast iron can be welded only by braze welding, since the higher heat of fusion welding destroys the malleability.
- The coating on galvanized iron is less affected by the low temperatures of braze welding. This in turn minimizes the distortion in sheet metal fabrication.
- Braze welding is a good method for joining dissimilar metals such as cast iron and steel.

Bronze Filler Rod

The melting temperature of bronze filler rod is approximately 1,600°F. This is considerably below the melting point of such metals as carbon steel, low alloy steel, and cast iron. The filler rods are copper alloys. They usually contain a little less than 60 percent copper and 40 percent zinc. These metals provide high tensile strength and high ductility. Small quantities of tin, iron, manganese, and silicon are also added. These metals produce a rod that is free flowing and has a deoxidizing effect on the weld. They decrease the tendency to fume and increase the hardness of the weld metal for greater wear resistance. Special filler rods are also available for use as a bearing bronze.

Flux

Braze welding depends on the fact that molten metal with low surface tension flows easily and evenly over the surface of a heated and chemically clean base metal, Fig. 9-3. The surface of the base metal contains oxides

Fig. 9-3 The brazing action. After a spot at the edge of the plate has been heated to a visible red, some bronze from the fluxed rod is melted on the hot area. The bronze filler metal spreads and runs evenly over the hot surface. This practice is known as *tinning*.
Location: Northeast Wisconsin Technical College Mark A. Dierker/McGraw Hill

that can be removed only through the use of a **flux.** Since brazing is a bonding process, the surface of the base material must be free of contamination of any kind.

Two methods are in general use for applying this flux. In the first method, the end of the filler rod is heated and dipped into a supply of powdered flux. This causes the flux to adhere to the rod. It flows over the work surface as the rod is melted. Note the flux on the rod in Fig. 9-3. In the second method, liquid flux is added to the stream of acetylene flowing to the torch through a special unit designed for that purpose, Figs. 9-4 and 9-5. The flux becomes an integral part of the welding flame. This second method applies the flux evenly and increases the speed of welding. This process is used by those manufacturing industries that engage in mass production, Fig. 9-6.

Fig. 9-4 Gasfluxer introduces flux into the flame for bronze and silver brazing. It uses a highly volatile liquid flux and can be used with any fuel gas. The Gasflux Co.

For video of mechanized torch brazing with flux injected into flame, please visit www.mhhe.com/welding.

Braze Welding Practice: Jobs 9-J39–J44

Practice in braze welding includes running beads on plate in the flat and vertical positions, lap and T-joints, and single-bevel butt joints. The metals should include steel, gray cast iron, and copper or brass if available.

1. No edge preparation is necessary if beads are being run on flat plate. Generally, joint design for braze welds is the same as for fusion welds in base metals of like thickness. Bevel the edges of

SHOP TALK

Starting Up

It is common for resistance welding operators to start prior to having stabilized the electrode force. They start before stabilization in order to speed up production. Even so, electrodes must get up to at least 95 percent of the chosen weld force setting before a weld should be started.

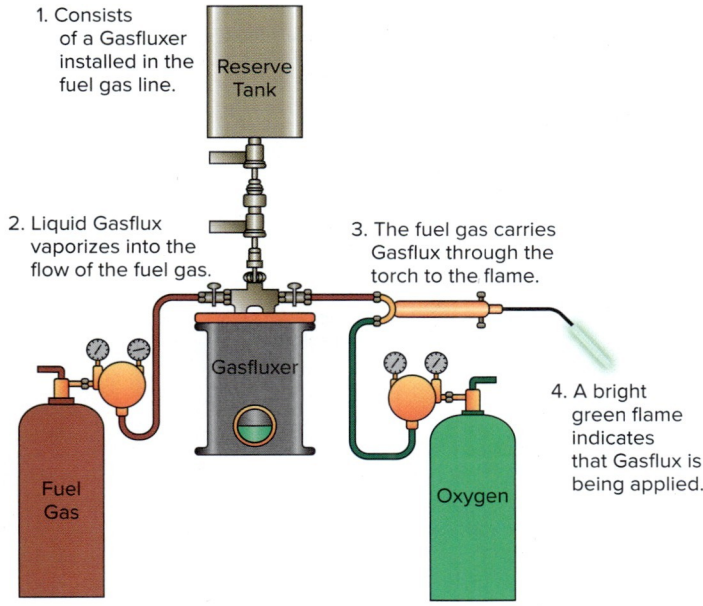

Fig. 9-5 How the liquid fluxer is connected to the torch and gas supply. *The Gasflux Co.*

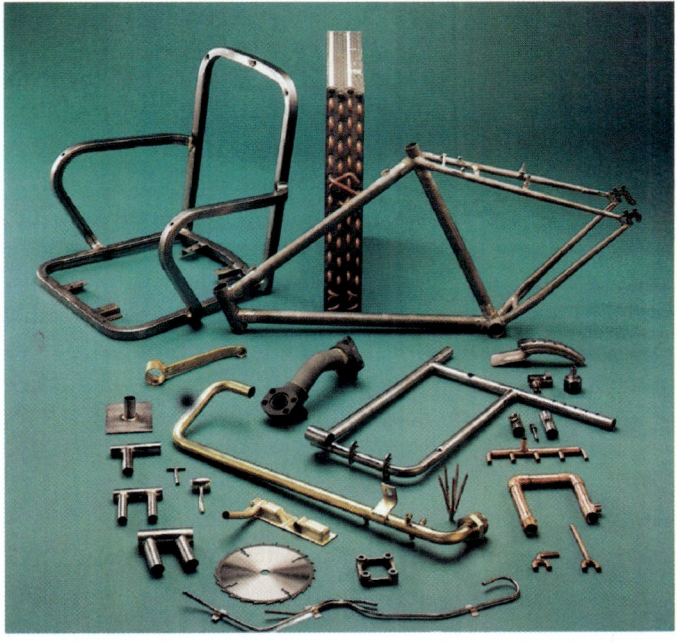

Fig. 9-6 Some typical brazing and soldering applications. *The Gasflux Co.*

metal thicker than 1/8 inch to provide a groove for welding.
2. Clean the surface of the metal with a wire brush to remove any foreign substances.
3. Position the work so that it runs slightly uphill. This prevents the molten bronze from flowing ahead to plate surfaces that are not hot enough.

4. Adjust the torch for a slightly oxidizing flame.
5. Heat the filler rod and dip it into the flux. Heat the base metal to just the right temperature. The proper temperature is indicated when the base metal begins to glow. At this point, melt a small amount of the rod and let it spread over the joint. This is referred to as **tinning.** If filler rod is applied before it is hot enough, the molten bronze will not flow over the surface. It will form drops that will not adhere to the material being welded. If the base metal is too hot, the molten bronze will boil and form little balls on the surface of the material being welded. Bronze weld metal that is too hot will also burn and give off a white smoke. This is the zinc that is burning with the oxygen in the air. It forms zinc oxide. The bronze will flow readily over the surface when the work has reached the correct temperature.
6. Carry the weld forward with a slightly circular movement of the torch like that used in fusion welding while adding filler rod to the weld pool. As the weld progresses, keep the end of the filler rod well-fluxed. Keep in mind that proper heating of the base metal is critical.
7. If more than one layer of beads is necessary, as in the welding of a second pass in a groove, make sure that you obtain through-fusion between the new bronze weld and the previously deposited bronze bead, and a good bond on the bevel and plate surfaces. Be careful that there are no inclusions of slag or oxide and other contaminants.
8. Continue to practice until you can make braze welds that are of uniform width and height. They should be smooth, with fine ripples, and free of pits and other porosity. They should be brightly colored. The edges should flow into the plate without overlap or other signs of lack of a bond. A deposit of white residue on the weld indicates overheating and burning.
9. Test the welds by making butt, lap, or T-joints, and testing them as you did your fusion welds in steel. Examine them for evenness and depth of bond. Find out whether the bronze has flowed to the root of the joint. Good braze welds often tear out cast iron when tested.

Powder Brazing

Powder brazing is a form of brazing that is used in mass production industries. A specially designed torch, Fig. 9-7, or a small hopper attachment unit that can be attached to a standard welding torch, provides the powder. The mixer matches the powder flow range to the gas flow

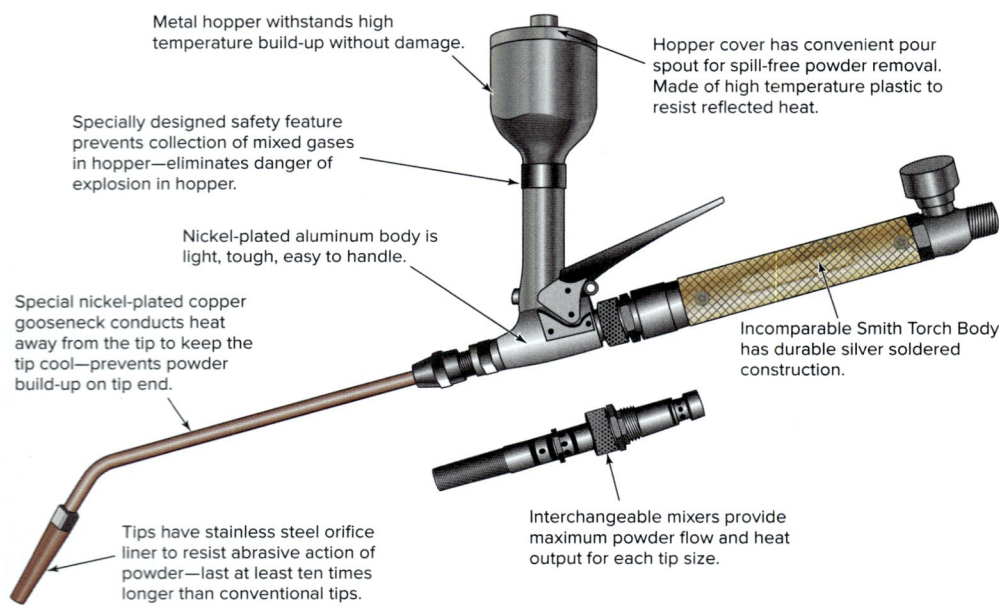

Fig. 9-7 Powder brazing torch. Adapted from Smith Equipment

range of the tip to provide maximum performance with each tip size.

The powder brazing process can be used as a means of hard surfacing, brazing, and buildup welding. It produces high quality work at high speeds. A wide selection of overlay powders is available with a large variety of hardness ratings, machineability, and resistance to heat and corrosion. Brazing alloys are also available for use in many applications.

Welding Cast Iron

Characteristics of Cast Iron

You will recall that cast irons are classified as gray iron, white iron, nodular iron, and malleable iron. Both gray cast iron and malleable iron are used commercially. We will be concerned here with the fusion welding of gray cast iron. Malleable iron cannot be fusion welded with the oxyacetylene process and must be braze welded.

Gray cast iron is an alloy of iron, carbon, and silicon. The carbon in gray cast iron may be present in two forms: in a carbon and iron solution and as a free carbon in the form of graphite. When it is broken, the fractured surface has a gray look due to the presence of the graphite particles. Gray cast iron is easy to machine because graphite is a fine lubricant. The presence of graphite also causes gray cast iron to have low ductility and tensile strength.

Welding Applications

Gray cast iron may be braze welded or fusion welded. For the most part, cast iron is welded in maintenance and repair work. Welding is seldom used as a fabricating process. Braze welding is preferred since it can be applied at a low temperature, and the bronze weld is highly ductile. Fusion welding is used when the color of the base metal must be retained and when the welded part is to be subjected to service temperatures over 500°F. Table 9-1 summarizes the various cast iron welding procedures. Before practice welding, review Chapter 3, pages 92–95, for additional information concerning the nature of cast iron.

Preheating

Control of expansion and contraction is very important in cast iron welding. The bulk and shape of the casting and

> **JOB TIP**
>
> **Interview**
> Preparing for an interview starts by listing your qualifications in a resume. Once you have that information down on paper, you'll feel ready to answer the interview questions. Decide what you have to offer that business. You'll know best what to offer by having some knowledge of the company.

Table 9-1 Summary of Cast Iron Welding Procedures

Cast Iron Type	Procedure	Treatment	Properties
Gray iron	Weld with cast iron	Preheat and cool slowly	Same as original
Gray iron	Braze weld	Preheat and cool slowly	Weld better; heat-affected zone as good as original
Gray iron	Braze weld	No preheat	Weld better; base metal hardened
Gray iron	Weld with steel	Preheat if at all possible	Weld better; base metal may be too hard to machine; if not preheated, needs to be welded intermittently to avoid cracking
Gray iron	Weld with steel around studs in joint	No preheat	Joint as strong as original
Gray iron	Weld with nickel	Preheat preferred	Joint as strong as original; thin hardened zone; machinable
Malleable iron	Weld with cast iron	Preheat, and postheat to repeat malleablizing treatment	Good weld, but slow and costly
Malleable iron	Weld with bronze	Preheat	As strong, but heat-affected zone not as ductile as original
White cast iron	Welding not recommended		
Nodular iron	Weld with nickel	Preheat preferred; postheat preferred	Joint strong and ductile, but some loss of original properties; machinable; all qualities lower in absence of preheat and/or postheat

Source: The James F. Lincoln Arc Welding Foundation

whether or not light sections join heavy sections affect preheating and welding technique. When fusion welding cast iron, all parts of the casting must be able to expand equally to prevent cracking and locked-in stress in the job. If the torch is applied directly to the cold casting and the joint to be welded is raised to the melting point, the expansion of the heated metal will cause a break or crack in the relatively cold casting surrounding the weld. If it does not break, severe internal stresses are locked in that may later cause a failure under service.

Small castings can be preheated with the oxyacetylene flame during the welding operation if the entire casting is heated evenly. Large castings may have to be preheated in a firebrick furnace built around the casting. Heating is usually done with gas- or oil-fired burners. The furnace is covered with a heat-resistant material to retain the heat and keep out cold drafts. The casting is welded through a large hole in the heat-resistant material. When the weld is completed, the casting is again raised to an even heat all over. Then it is buried in heat-resistant material and allowed to cool very slowly.

Filler Rod

Fusion welding requires the use of a good grade of cast iron filler rod that matches the material being welded. The rod must contain enough silicon to replace the silicon that tends to burn out during welding. Silicon assists in the flow of the molten metal during welding and retards oxidation that may lead to the formation of slag inclusions and blowholes. Good cast iron welding rods contain 3 to 4 percent silicon.

Cast iron filler rods are supplied in diameters of $\frac{3}{16}$, $\frac{1}{4}$, $\frac{3}{8}$, and $\frac{1}{2}$ inch and in lengths of 12 to 18 inches. They carry the AWS-ASTM classification of RCI, RCI-A, and RCI-B. They may be round, square, or hexagonal in shape. If the welder needs a longer or heavier rod, they can weld two or more rods together.

Flux

The problem in welding cast iron, as in welding other metals, is to prevent oxide from forming and, when it does, to remove it from the weld. The flux dissolves the oxide, floats off other impurities such as sand, scale, and dirt, and increases the fluidity of the molten metal.

The student welder must learn to apply flux properly. Too much flux can cause as much trouble as too little. Excessive flux becomes entrapped in the molten metal and causes blowholes and porosity. Also, the molten iron will combine with certain elements in the flux if it is applied in excess. You will learn by experience the right amount to use. The amount that adheres to the hot end of the welding rod when it is dipped in the flux is usually enough. Do not throw additional quantities into the weld as you are welding.

Cast Iron Welding Practice: Jobs 9-J45 and J46

The same general welding procedure may be used for beading and groove welding. The following steps are for groove welding:

1. Prepare the metal as in other welding. If the material is more than ⅛-inch thick, bevel the edges, leaving about ⅛-inch thickness at the root face. Clean all dirt, rust, and scale from the surface. It is assumed that you are practicing with small pieces of casting that have been beveled.
2. Select the proper size tip and adjust the torch for a neutral flame.
3. Play the torch flame over the entire work until the entire joint has been preheated. Then, starting at the right edge, direct the flame at the bottom of the groove until the metal there has been melted.
4. Heat the bottom of the groove and the side walls until they are molten and the metal flows to the bottom of the groove. If the metal gets too hot and the molten pool tends to run away, raise the flame slightly. Hold the torch at a 90° angle to the work, using a motion like that for welding steel plate. Keep the sides and the bottom in a molten condition. If the torch is held at an angle, the flame will blow the molten metal ahead of the weld, and incomplete fusion will result. Make sure that the surfaces of the groove are fused ahead of the weld pool so that the molten metal is not forced ahead to colder plate surfaces to produce incomplete fusion.
5. Filler rod that has been fluxed should be added to the weld pool to fill up the groove. Never hold the filler rod above the weld and melt it drop by drop into the molten pool.
6. Gas bubbles or impurities can be floated to the surface of the weld by the addition of flux and the use of the flame. Skim these impurities off the surface of the weld with the filler rod. Impurities left in the weld are defects that weaken the joint.
7. After the weld has been completed, heat the entire casting to the same temperature throughout and allow it to cool very slowly.
8. Continue to practice until you can make good quality welds that are satisfactory to the instructor. Be particularly concerned about fusion along the edges of the weld, a smooth surface without holes and depressions, and good penetration and fusion on the back side.
9. Test one of your completed welds. Use the same procedure used in testing butt joints with groove welds in sheet metal. Place the specimen in a vise with the weld above the vise jaws. Break it in pieces with a hammer. A sound weld will cause the break to take place in the casting. You may also wish to break or saw through the weld to examine the weld metal. The weld metal should be sound and have no slag inclusions or blowholes.

> **ABOUT WELDING**
>
> **Aluminum**
>
> *Hot* and *fast* are the key words for aluminum welding. Hot—because aluminum has high thermal conductivity. Fast—to avoid, on thin aluminum, excessive burn-through.

Welding of Aluminum

Aluminum is one of the metals that has become important to industry. It is light-weight (about one-third as heavy as steel), yet it has a high strength-to-weight ratio. It is a good conductor of electricity and has high resistance to corrosion.

Aluminum can be welded by gas welding and many other welding processes. It is comparatively easy to fabricate. Much of the aluminum used in industry is welded with the gas metal arc and gas tungsten arc processes. It is highly desirable, however, for welding students to practice welding aluminum with the oxyacetylene process so that they can better judge the flow of metal and heat ranges.

Characteristics of Aluminum

The three categories of aluminum that have the most welded applications are commercially *pure aluminum, wrought aluminum alloys,* and *aluminum casting alloys.*

Pure aluminum melts at 1,220°F, whereas weldable commercial aluminum alloys have a melting range of 900 to 1,220°F. Compare these temperatures with steel, which melts at about 2,800°F, and copper, which melts at about 1,980°F. Aluminum does not change color during heating, and when the melting temperature is reached, it collapses suddenly. This characteristic is called **hot-shortness.** Because aluminum is hot-short, it requires support when hot.

Aluminum is a good **thermal conductor.** It conducts heat about three times faster than iron, and requires higher heat input than that used in welding steel.

Aluminum expands during heating. For this reason, aluminum welds have a tendency to crack because of the shrinkage that takes place in the weld metal when cooling. Too much restraint of the parts during cooling may also result in weld cracking. The speed of welding is also important. Welding at a slow rate of speed causes more heat input into the part being welded. This increases the rate of expansion and contraction.

The aluminum weld pool oxidizes very rapidly. It forms an oxide with a melting point of 3,700°F, which must be removed either chemically, with a flux, or mechanically, with a paddle.

Before practice welding, review Chapter 3, pages 95–96, for additional information concerning the nature of aluminum.

Filler Rod

The selection of the proper filler rod for aluminum welding depends on the composition of the base metal. The AWS-ASTM classification R1100 is recommended for welding commercially pure aluminum sheet (1100); and R4043, for other types of aluminum. Castings require special rods or consider aluminum brazing.

Aluminum filler rods are obtainable in sizes of $\frac{1}{16}$, $\frac{3}{32}$, $\frac{1}{8}$, $\frac{5}{32}$, $\frac{3}{16}$, and $\frac{1}{4}$ inch in diameter. Standard lengths are 36 inches.

As a rule, the diameter of the filler rod should equal the thickness of the metal being welded. A rod that is too large melts too slowly. Thus, it retards the fluid action of the weld pool, and may cause a lack of fusion. On the other hand, a rod that is too small melts too fast and may be burned. There is also not enough filler metal for the weld pool. This overheats the weld and may cause burn-through.

Flux

The welding of aluminum requires the use of *flux* to remove the aluminum oxides that are on the surface and those that are formed during welding. The flux-oxide compound forms a slag, which is expelled from the weld pool by the action of the flame and the molten metal.

A number of commercial fluxes are available. Those in powdered form, which are mixed with alcohol or water to make a paste, are the most practical. Mixing should be done in a glass, ceramic, aluminum, or stainless-steel container. Containers of steel, copper, or brass may contaminate the flux and should not be used.

When welding sheet aluminum, a brush is normally used to spread the flux on the welding area and to coat the welding rod. When welding cast aluminum, the flux is applied to the end of the filler rod by dipping its heated tip into the flux. During welding, the flux melts and forms a protective coating over the molten metal and the heated end of the filler rod. The flux is melted in advance of the weld pool, and thus cleans the base metal before it is welded. The flux that is applied to the base metal should be heated slowly before welding to ensure that excess moisture is driven off. This is done to prevent spattering and porosity.

Some welders prefer filler rods that are prefluxed. When such rods are used, it is not always necessary to apply flux to the base metal. Better results are generally obtained, however, if flux is applied to the work. The period between fluxing and welding should not exceed 45 minutes.

Flame Adjustment

Acetylene is the most popular gas to use with oxygen for welding aluminum because of its high heat and wide availability. Hydrogen is also used with oxygen for the welding of aluminum. Because of its lower heat, it is used chiefly in the welding of the thinnest gauges of aluminum sheet.

In order to produce clean, sound welds with maximum welding speed, a neutral flame is ideal, though a slightly reducing flame (carburizing) is highly satisfactory. You will recall that aluminum oxidizes readily. The reason for the excess acetylene flame is to make sure that the gas mixture does not stray to the excess oxygen side. An oxidizing flame causes a bead that balls up. The weld has poor fusion, poor penetration, and much porosity.

Aluminum may also be welded with the oxyhydrogen flame. The same torch and type of tip are used. Because of the lower temperature of the oxyhydrogen flame, a larger tip size is required. You may have some difficulty adjusting the flame. It is very similar, however, to the adjustment of the oxyacetylene flame, Fig. 9-8.

The neutral oxyhydrogen flame is a white to pale violet color, fairly large, and rather indistinct. The small, well-defined cone in the center has a bluish tinge.

The reducing flame is larger and ragged. Its color varies from pale blue to reddish-violet, depending upon the amount of excess hydrogen. There is no well-defined cone at the center. This flame is usually used for welding.

The oxidizing flame has a very short, blue inner cone. The flame is thinner than the reducing flame and ranges in color from white to transparent. Because these differences are very hard to see, a flowmeter may be the best means for ensuring the desired hydrogen-oxygen ratio.

A soft flame is essential in all oxy-gas welding of aluminum. A strong or noisy flame causes turbulence and lowers weld quality. The proper pressure adjustment at the regulators and precise torch adjustment are a must.

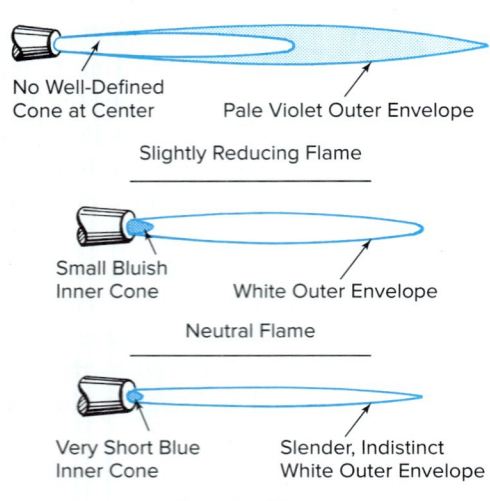

Fig. 9-8 A neutral flame (center) tending toward a slightly reducing flame (top) is recommended for the oxyhydrogen welding of aluminum.

Aluminum Welding Technique

Edge Preparation As with all other metals, the type of edge preparation depends upon the thickness of the metal being welded. On aluminum butt joints up to a thickness of 1/16 inch, no edge preparation is necessary. A flange-type joint may also be used. Material from 1/16 inch to 5/32 inch can be welded in the form of a square butt joint, but the plate edges should be notched. For aluminum plate 3/16 to 7/16-inch thick, the edge preparation is like that for a single-V butt joint with a 90° V. For thicknesses of 1/2 inch and over, the double-V butt joint with a 90° V on each side is used if both sides are accessible. Table 9-2 lists the rod and tip sizes and gas pressures for welding various thicknesses of aluminum.

Cleaning Since grease, oil, and dirt cause weld porosity and interfere with welding, they should be removed from the welding surfaces. Commercial degreasing agents are available for this purpose. Wire brushing may also be used to remove heavy oxide films from the plate surfaces. Removal of oxides permits more effective fluxing action.

Preheating Like cast iron, cast aluminum requires careful preheating before welding and slow cooling after the weld is completed. Preheating is always recommended for gas welding aluminum when the mass of the base metal is such that heat is conducted away from the joint area faster than it can be supplied to produce fusion. A large difference in temperature between the surrounding metal and the weld area increases welding difficulties. The preheat temperature must stay below 700°F. Preheat temperature-indicating compounds can be used to advantage. These are available in both crayon and liquid form. They are accurate and are made in almost any temperature range.

Table 9-2 Rod Sizes, Tip Sizes, and Gas Pressures for the Welding of Aluminum

Aluminum Thickness (in.)	Filler Rod Diameter (in.)	Oxyacetylene Welding			Oxyhydrogen Welding		
		Orifice Diameter in Tip of Torch (in.)	Oxygen Pressure (p.s.i.)	Acetylene Pressure (p.s.i.)	Orifice Diameter in Tip of Torch (in.)	Oxygen Pressure (p.s.i.)	Hydrogen Pressure (p.s.i.)
0.020	3/22	0.025	1	1	0.035	1	1
0.032	3/32	0.035	1	1	0.045	1	1
0.050	3/32	0.045	2	2	0.065	2	1
0.064	3/32	0.045	2	2	0.065	2	1
0.080	1/8	0.055	3	3	0.075	2	1
1/8	1/8	0.065	4	4	0.095	3	2
3/16	5/32	0.065	5	5	0.095	3	2
1/4	3/16	0.075	5	4	0.105	4	2
5/16	3/16	0.085	5	5	0.115	4	2
3/8	3/16	0.095	6	6	0.125	5	3
1/2	1/4	0.100	7	7	0.140	8	6
5/8	1/4	0.105	7	7	0.150	8	6

Source: Kaiser Aluminum & Chemical Corp.

Jigs and Fixtures Aluminum and many of its alloys are weak when hot. Aluminum parts, especially thin stock, should be supported adequately. *Jigs* and *fixtures* maintain alignment and reduce buckling and distortion. It is especially important for the edges of the joint to be spaced and aligned correctly. Tack welding before final welding also aids in maintaining alignment and minimizing distortion. Tack welds should be placed carefully. They must be small enough not to interfere with welding, and they must have good penetration and fusion.

Welding Positions Aluminum welding with gas can be done in all positions. *Overhead welding*, however, is difficult and should be attempted on the job only by highly experienced welders. The *flat position* is preferable and should be used whenever possible.

The *horizontal welding* technique is similar to that for flat welding. Special manipulation of the rod and torch is necessary, however, to offset the tendency of molten metal to sag and build up the lower edge of the weld bead. Butt joints are not practical in the horizontal position.

Vertical welding is usually performed on materials ¼ inch or thicker. It is almost impossible to keep thinner materials from burning through, and the weld pool becomes too fluid and spills over. Torch and welding rod angles are about the same as for all other vertical welding.

Postweld Cleaning Thorough cleaning after welding aluminum is very important. Flux residues on gas-welded sections corrode aluminum if moisture is present. Small sections may be cleaned by a 10- to 15-minute immersion in a cold 10 percent sulfuric acid bath, or a 5- to 10-minute immersion in a 5 percent sulfuric acid bath held at 150°F. Acid cleaning should always be followed by a hot or cold water rinse. Steam cleaning also may be used to remove flux residue, particularly on parts that are too large to be immersed. Brushing may also be necessary to remove adhering flux particles.

JOB TIP

Finding Work

Finding work when you are just starting out can be easier if you take advantage of the career centers at vocational schools and colleges, and at job fairs sponsored by government and nonprofit groups. Some centers also provide listings of businesses that offer job shadowing, apprenticeships, or internships, which often lead to full employment.

Sheet Aluminum Welding Practice: Jobs 9-J47–J49

1. Prepare, clean, and tack the joint. It is recommended that you practice on aluminum sheet ranging from 14 gauge to about ³⁄₁₆-inch thick. Select the proper tip size for the thickness of the metal being welded. Select the filler rod according to the type of aluminum being welded.
2. Apply flux to the filler rod and the work.
3. The welding technique used to weld aluminum is not much different than that used in the welding of steel. Remember that aluminum does not change color when heated, it burns through readily, it forms oxides rapidly, and it has a high rate of expansion and contraction. Start welding about 1½ inches from the edge of the sheet and travel to the opposite edge. Heat the entire plate or joint evenly.
4. After a small pool has been formed, point the flame in the direction of welding at an angle of about 30°, Fig. 9-9. This position heats the plate ahead of you, reduces the problem of burning through, and increases the welding speed. Do not let the cone of the flame touch the weld pool. A distance of ⅛ to ¼ inch is advised. Make sure the weld pool is in a molten state before adding filler rod. Stir the filler rod in the weld pool. This action causes the oxides to rise to the surface of the weld pool and reduces porosity. Side-to-side movement depends upon the thickness of the plate and the size of the weld bead desired. Side-to-side movement should not be used for thin sheets. On heavier sheets, the amount of movement depends upon the size of the weld bead desired.

As the end of the joint is approached, flatten out the angle of the torch until the flame is reaching mainly the welding rod. This is particularly recommended for the thinner gauges of aluminum.

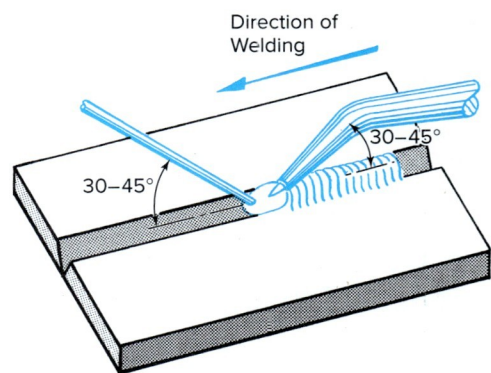

Fig. 9-9 Typical angles for torch and filler rod range from 30 to 45° in the welding of aluminum.

Fig. 9-10 A corner weld in aluminum sheet. The ripples should be smooth, and the weld should show no overhang.
Edward R. Bohnart

5. Return to the beginning of the short weld and restart the weld. Weld toward the other edge of the plate.
6. Clean the weld and base metal by brushing with hot water.
7. Continue to practice these welds on various thicknesses of aluminum until you can make welds of satisfactory appearance that are acceptable to your instructor. Note the weld characteristics shown in Fig. 9-10. Be particularly concerned about fusion along the edges of the weld; a smooth surface without pits, holes, or depressions; and good penetration on the back side.

   The beads should not be too wide, and their edges should be parallel.
8. Test a few of these welds as you did for steel by bending the plates back. The weld should not fracture or peel off the surface of the metal.

SHOP TALK

Squeeze Time

Squeeze time means the time between when the electrode is told to advance the part and when most of the force of the electrode has been reached. It is how fast the electrode clamps down on the workpiece during resistance welding operations.

Welding Other Metals with the Oxyacetylene Process

Magnesium Alloys

Magnesium is used extensively by the aircraft industry. It is two-thirds as heavy as aluminum and less than one-quarter as heavy as steel. It has a melting point of 1,202°F. It burns with a brilliant light and is used for flares and photographic flashes. When magnesium is used in a fabrication that is to be welded, it is usually alloyed with aluminum. Magnesium may be welded by the gas process, but gas metal arc and gas tungsten arc are widely used today.

Magnesium has a high coefficient of expansion that increases as the temperature increases. Distortion and internal stresses are conditions that must be provided for. Sheet metal should be welded with one pass.

Pure magnesium has little strength. Its strength is increased when alloyed with aluminum, manganese, or zinc. It has a high resistance to corrosion.

The following factors must be taken into consideration in the welding of magnesium alloys with the oxyacetylene flame:

- Joint design is similar to that for a corresponding thickness of aluminum. Thicknesses up to 1/8 inch do not need to be beveled. Thicknesses above 1/8 inch should have a 45° bevel.
- Lap and fillet welds are not recommended because of the possibility of flux entrapment.
- Magnesium alloys have low strength at a temperature just below the melting point. The joints must be well supported so that they will not collapse.
- The filler rod must be of the same composition as the base metal.
- A flux is used to prevent oxidation. It is a powder, and must be mixed with water to form a paste. The filler rod and both sides of the base metal should be coated with flux. Use flux sparingly, since too much may cause flux inclusions in the weld and make final cleaning much more difficult.
- A neutral flame is recommended for welding. The flame must never enter the oxidizing stage.
- Oil, grease, and dirt near the weld should be removed by a degreasing solvent that does not contain fluorides. The oxide film on the surface of the base metal should be removed by wire brushing or by filing. Preheating is not recommended.
- Welding technique is similar to that used in the welding of aluminum. The inner cone of the flame should be just above the weld pool. The tip should be held at an angle of 30 to 45°, depending on the thickness of the material being welded. Thin materials require a smaller angle than heavier materials do. The heat should be concentrated in order to prevent buckling and cracking.
- When the weld is completed, the flux should be removed by brushing with hot water. Welded parts should be treated with a solution of 0.5 percent sodium dichromate. Small parts are boiled in the

solution for two hours. Large parts are heated to about 150°F and brushed with the hot solution. The parts should be rinsed with clear water and dried with a hot air blast. This treatment gives the material a yellowish appearance.

Lead

Environmental and health concerns mandate proper use of lead base metal and lead filler metal while welding. Care should be taken in their application, particularly with respect to fume inhalation and ingestion.

Pure lead is a heavy, soft metal with a dull gray surface appearance. When cut, the cross section has a bright metallic luster. Lead is highly malleable and highly ductile. It has very little tensile or compression strength. Antimony may be added to increase the strength of pure lead. Lead is a poor conductor of electricity.

Lead is used in making pipe and containers for corrosive liquids. It is an element in many useful alloys including solder, antifriction, and antifriction metals. Lead is also widely used in oil refineries, chemical plants, paper mills, and the storage battery industry.

Lead welding is commonly referred to as *lead burning*. This term is incorrect, however, because the lead is not burned. The welding process produces true fusion of the base metal. Lead has a low melting point, about 600°F. It can be welded with oxyacetylene, oxy-propane, oxy-natural gas, or oxyhydrogen. A special small torch, Fig. 9-11, with small tip sizes, is used for welding lead with oxyacetylene.

Other tools that would be of use for the lead welder are:

- Welding torch tips ranging in drill size from 78 to 68. The smaller tips are for use with 6-pound lead (lead sheet is sized by pounds per square foot; this would be 1/10-inch thick) and lighter, and the larger tips for heavier lead. Depending upon conditions and the welder's experience, torch tips larger than 68 drill size are sometimes used. Light fingertip torches with valves easily accessible are preferred by the welder.
- Mallets and dressers, of materials that will not dent or mar the lead, for dressing the sections of lead to be joined so that they are properly aligned for welding.
- Wooden turn pins for expanding one end of a length of pipe so it can easily be inserted into another section of pipe to form a cup or lap joint.
- Shave hooks, scrapers, and wire brushes for cleaning or removing oxide from the lead in the areas that are to become part of the joint.

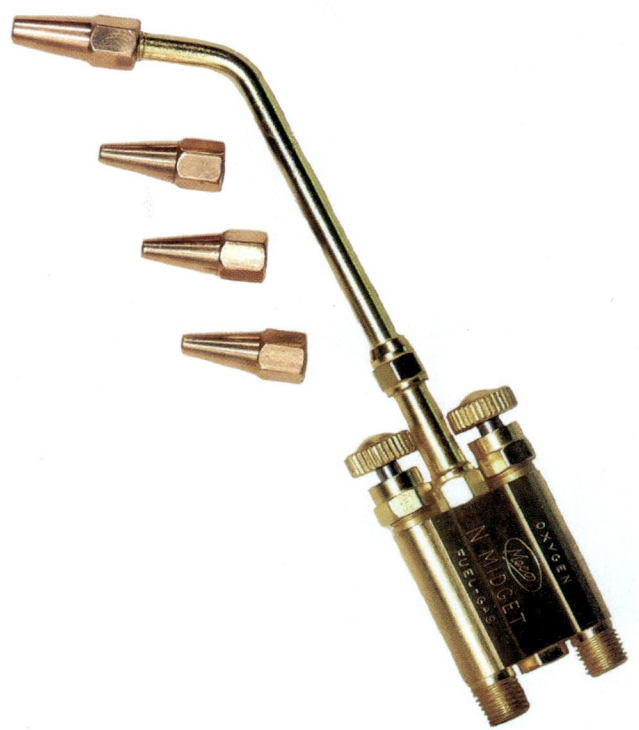

Fig. 9-11 Midget welding and brazing torch for lead welding.
TM Technologies

- Various tools for uniformly cutting the lead.
- Metal molds for use on vertical butt seams.

The following factors must be taken into consideration in the welding of lead with the oxyacetylene flame:

- Lead is suitable for all of the standard welded joints, Fig. 9-12. Edge preparation is easy, since the metal is soft. The main problem is to make sure that the surface to be welded is clean. This is done with a special scraping tool. Welding may be done in all positions. Overhead welding is most difficult. If required, use lap joint and no filler metal. The pressure of the flame and molecular tension will hold the molten pool in place. The vertical position is accomplished with mostly lap joints. If butt joints are to be welded, molds must be used. It is more of a casting operation than welding, starting at the bottom of the joint and moving upward.
- Because of the soft nature of the material, the work must be well supported to prevent collapse and severe distortion.
- Edge and flange joints may be welded without the addition of filler rod. When filler rod is necessary, it can be made by cutting lead sheet into strips. Special molds are also available for casting rod. The *V* of an angle iron can be used as a mold.

Braze Welding and Advanced Gas Welding Practice: Jobs 9-J39–J49   Chapter 9   243

Fig. 9-12 Standard joint design in lead welding. It is possible to weld in all positions. Tony DeMarco

- No flux is required.
- Flame adjustment should be slightly excess acetylene.
- The torch technique is somewhat different from that used with any other metal. Lead has a low melting point and will quickly fall through or develop holes. When filler rod is added to the weld pool, it is essential that both reach the state of fusion at the same time. Care must also be taken that the weld pool does not get too large and too fluid. Some welders withdraw the torch when the pool becomes too fluid and let it solidify. Then they form a new pool that overlaps the first, and again withdraw the torch at the proper time. This bead may be likened to a series of overlapping spot welds. Other lead welders have developed such skill that they have complete control of the molten metal, and they carry a continuous bead with a torch technique somewhat like that used for welding aluminum.
- Lack of fusion is not a problem in lead welding. The chief difficulties are excess fusion and burn-through.

Hard Facing (Surfacing)

Hard facing is the process of welding on worn metal surfaces a coating, edge, or point of metal that is highly capable of resisting abrasion, corrosion, erosion, high temperature, or impact. The process is also called *metal surfacing.* A surface that is worn away can be restored to its original state, or additional qualities can be given to the material. The part to be surfaced may be prepared as shown in Fig. 9-13.

Hard facing has the following advantages:

- Hard-faced surfaces will last 2 to 40 times as long as low carbon steel, depending on the type of surfacing alloy and the service required.
- The service life of a part is greatly increased. Production schedules are maintained due to fewer replacements and more continuous operation of equipment.
- Surfacing materials need be applied only to those surfaces that will be subjected to severe conditions. Cheaper material may be used to make the part being surfaced.
- Parts can be rebuilt again and again without being replaced.
- Special cutting edges can be formed for special situations.

ABOUT WELDING

Welding in Space

In the late 1960s, Russian cosmonauts did welding in space by the electron beam method. The United States tried the technique in space five years later.

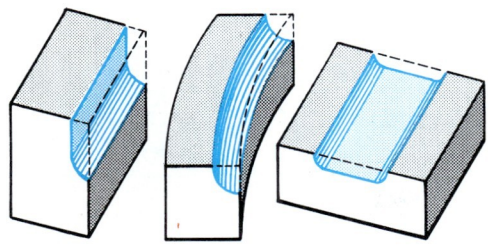

Machined recesses for hard-facing alloy. Dotted lines indicate finish–ground dimensions.

Fig. 9-13 A sectional view showing the method of machining steel edges and corners before hard facing with Haynes-Stellite alloy rod.

Welding Processes

A great deal of hard facing is done with the oxyacetylene process, Fig. 9-14. This is the oldest process used. Hard facing is also done with the gas tungsten arc, gas metal arc, shielded metal arc, flux cored arc, atomic hydrogen, metal spraying, and plasma arc welding processes.

Oxyacetylene may be used for most applications. When a flawless surface is required, the gas tungsten arc process is highly desirable. The shielded metal arc process is used when a heavy deposit of metal is necessary. When heavy sections are to be covered with a light coating of surfacing material, the atomic hydrogen process is preferred. Metal spraying can be used for a thin deposit on almost all materials. The plasma arc process will bond almost any material to any other. Ceramics may even be bonded to metals.

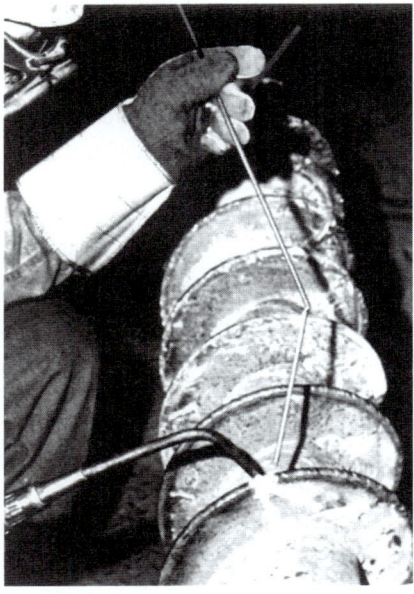

Fig. 9-14 Welder uses oxyacetylene torch and Haynes-Stellite no. 6, a hard-surfacing rod, to hard surface edges of a screw conveyor for handling crushed limestone. Praxair, Inc.

Surfacing Materials

There are many types of hard-surfacing rods to be used with the oxyacetylene process. Nearly all of the surfacing alloys have a base of iron, nickel, cobalt, or copper.

- The *cobalt-base rods* are composed of cobalt-chromium-tungsten alloys. This material has high resistance to oxidation, corrosion, abrasion, and heat. The rods have a high degree of hardness at service temperatures of 1,500°F. They recover full hardness at room temperature. There are a number of rod types in this category to meet specific conditions.
- *Nickel-base rods* are recommended for applications involving metal-to-metal wear, temperatures to 1,000°F, and corrosion. They recover full hardness at room temperature after being heated. Several types are available.
- *Iron-base rods* are designed to resist abrasion and impact in varying degrees, depending upon the type used. Many of these rods have a high chromium content. They are available as cast rods and tubing.
- *Copper-base alloys* are employed chiefly to resist corrosion and metal-to-metal wear. They are excellent materials for bearing surfaces. They have poor abrasion resistance.
- *Tungsten carbide rods* are made in two forms: a tube type and a composite type. The tube type is a steel tube containing crushed tungsten carbide particles in a choice of grain sizes. When deposited, the tungsten carbide grains are held in a matrix of tough steel. The tube rod provides excellent resistance to severe abrasion and good resistance to impact. The composite type is a cast rod containing controlled grain sizes of crushed tungsten carbide. One is a nickel-base alloy with excellent corrosion resistance and severe abrasion resistance. Another tungsten carbide composite rod is a cobalt-base alloy with excellent high temperature hardness and good corrosion resistance. A third rod has large, hard, sharp tungsten particles in a matrix of bronze. This rod has good corrosion and wear resistance.
- A wide variety of spray hard-surfacing powders are available for all types of service. Some are nickel-base alloys that resist abrasion and corrosion. Others have a cobalt base and resist abrasion, high temperature, erosion, and corrosion. Still others have a tungsten carbide and nickel base and resist severe abrasion. These powders are applied with standard oxyacetylene spraying equipment.

Application of Hard-Facing Materials to Steel

Hard-surfacing materials may be applied to ordinary steels, high carbon and alloy steels, cast iron, and malleable iron.

1. Thoroughly clean the surface of the base metal by grinding, machining, filing, or wire brushing. Round all sharp corners.
2. Preheat the parts to be hard faced. Small parts can be preheated with the welding torch. Large parts will have to be preheated in a furnace. The parts should be heated to a very dull red, about 800°F. Be careful not to heat the parts too much, or oxidation in the form of scale will form on the surface of the part to be surfaced.
3. Select a tip size that allows slow, careful deposition and good heat control without overheating the base metal.
4. Adjust the torch to an excess acetylene flame. This flame adjustment causes the deposited metal to spread freely. Too little acetylene will cause the material to foam or bubble. On the other hand, too much acetylene will leave a heavy black carbon deposit. The inner cone of the flame should not be more than 1/8 inch from the surface of the base metal. The excess acetylene flame prepares the steel surface by melting an extremely thin surface layer. This gives the steel the watery, glazed appearance known as *sweating*, which is necessary to the hard-surfacing application.
5. Direct the flame at an angle of 30 to 60° to the base metal. Heat a small area until sweating appears, Fig. 9-15. This indicates that the base metal is ready for the application of the hard-surfacing material.
6. Withdraw the flame and bring the end of the hard-surfacing rod between the inner cone of the flame and the hot base metal. The tip of the inner cone should just about touch the rod, and the rod should lightly touch the sweating area, Fig. 9-16. The end of the rod will melt. The backhand welding technique may also be used, Fig. 9-17. If the base metal is not hot enough, the surfacing material will not spread uniformly. If the conditions are correct, the deposited metal will spread and flow as bronze does in braze welding.
7. The pressure of the flame should be used to move the pool along. Do not stir the weld pool with the rod. Avoid melting too much base metal, because it dilutes the hard-surfacing material, thus reducing its service characteristics. Pinholes may be present in the deposited metal. These may be caused by particles of rust or scale that have not floated to the surface with the flame. Pinholes are also caused when there is not enough acetylene in the flame, when the hard-surfacing rod is applied before the base metal has reached a sweating surface heat, and when the welding flame is removed suddenly from the pool of molten metal.
8. Cool the part slowly. This may be done by furnace cooling or covering it with powdered lime, powdered heat-resistant material, warm dry sand, wood ashes, or a heat-resistant blanket.
9. After the part has cooled, it may be necessary to remove the high spots or form the part by grinding. Do not use a speed that is slower than 2,800 nor more than 4,200 surface feet per minute.

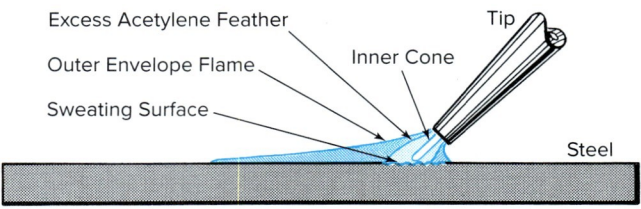

Fig. 9-15 The proper angle at which to hold the torch and the position of the flame for producing sweating.

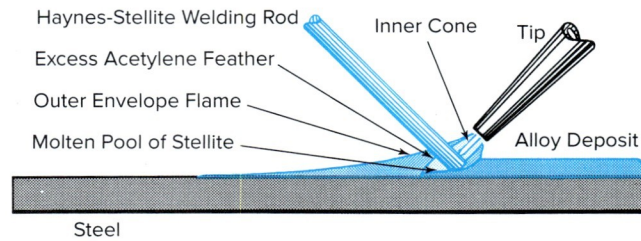

Fig. 9-16 The approximate relationship of torch tip, hard-facing rod, pool, and base metal during hard facing with the forehand technique.

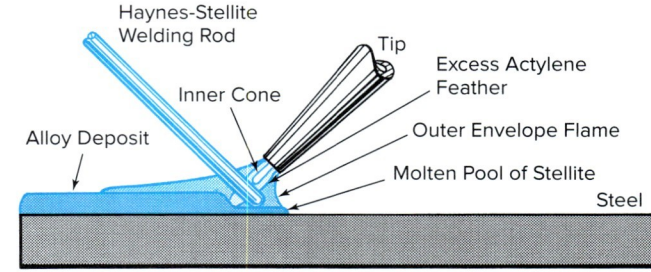

Fig. 9-17 The backhand technique for hard facing, as in welding, has the hard-facing rod between the flame and the completed deposit. The approximate relationship of torch tip, hard-facing rod, deposit, pool, and base metal is shown.

10. Practice hard facing until you can produce deposits that have the hardness claimed for the rod. Use a hardness-testing machine.

Special Hard-Surfacing Techniques

Cast Iron Cast iron does not sweat like steel, and less acetylene should be used in the flame. A cast iron flux is used. The hard-surfacing material is applied in a thin layer, and then additional layers are deposited to build up the desired thickness. Do not melt the base metal too deeply.

Alloy Steels High manganese steels, silicon steels, and some forms of stainless steels are difficult to hard face because of their tendency to crack and reluctance to sweat. These materials require special care in preheating and postcooling.

Because of the high rate of expansion of stainless steel, it must be heated evenly and cooled slowly to prevent uneven internal stresses.

Copper Copper is relatively difficult to surface. Brass and other alloys with low melting points cannot be hard faced satisfactorily.

(-Tool Steel) Hard surfacing high speed steel is not satisfactory because many times cracks will be formed in the steel below the coating. If it is attempted, the steel must be fully annealed. Heat must be kept even over the entire part. When the application is completed, the entire part must be brought to an even red heat and permitted to cool very slowly.

CHAPTER 9 REVIEW

Multiple Choice

Choose the letter of the correct answer.

1. Because the weld pool is highly fluid, braze welding cannot be done in the vertical position. (Obj. 9-1)
   a. True
   b. False
2. What is the primary function of the welding flux? (Obj. 9-1)
   a. To remove surface oxides from base metal
   b. To remove heavy rust
   c. To remove oil
   d. To remove grease
3. Braze welding is carried out with an excess acetylene flame. (Obj. 9-1)
   a. True
   b. False
4. Malleable cast iron can be welded only with a high test steel filler rod. (Obj. 9-2)
   a. True
   b. False
5. Gray cast iron can only be braze welded. (Obj. 9-2)
   a. True
   b. False
6. Why is cast iron preheated? (Obj. 9-2)
   a. To lower its melting temperature
   b. To raise its melting temperature
   c. To control expansion and contraction
   d. To eliminate expansion and contraction
7. What quality does hot-shortness describe when welding aluminum? (Obj. 9-2)
   a. Soft and ductile at its melting temperature
   b. Does not change color as it is heated
   c. Brittle, cracks and collapses at its melting temperature
   d. Improved impact strength at elevated temperature
8. Aluminum melts at approximately 2,800°F. (Obj. 9-2)
   a. True
   b. False
9. Aluminum oxide has a melting temperature about how many times greater than that of aluminum? (Obj. 9-2)
   a. 1
   b. 2
   c. 3
   d. 4
10. Flux is required when using OFW to weld aluminum. (Obj. 9-2)
    a. True
    b. False
11. An oxidizing flame should not be used for welding aluminum. (Obj. 9-2)
    a. True
    b. False

12. Aluminum is welded exclusively with the oxyacetylene process. (Obj. 9-2)
    a. True
    b. False
13. Fusion welding of aluminum with the oxyacetylene process is not widely used. (Obj. 9-2)
    a. True
    b. False
14. Magnesium welds often contain locked-up stresses and distortion. (Obj. 9-2)
    a. True
    b. False
15. What is magnesium usually alloyed with? (Obj. 9-2)
    a. Aluminum
    b. Manganese
    c. Zinc
    d. All of the above
16. Magnesium cannot be welded because it burns. (Obj. 9-2)
    a. True
    b. False
17. What is the approximate melting point of lead? (Obj. 9-2)
    a. 150°F
    b. 300°F
    c. 600°F
    d. 900°F
18. What is another name for hard facing? (Obj. 9-3)
    a. Wearing facing
    b. Scratch surfacing
    c. Metal surfacing
    d. None of these
19. When a flawless surface is required with hard facing, what welding process is the best choice? (Obj. 9-3)
    a. Gas tungsten arc
    b. Plasma arc
    c. Shielded metal arc
    d. Gas metal arc
20. Which metals are difficult to hard surface? (Obj. 9-3)
    a. Ordinary steels and alloy steels
    b. Cast iron and high carbon steels
    c. Alloy steels and malleable iron
    d. Brass and high speed steel

Review Questions

Write the answers in your own words.

21. What is braze welding? (Obj. 9-1)
22. Name some advantages of powder brazing. (Obj. 9-1)
23. Describe two methods used for applying flux. (Obj. 9-1)
24. What is the reason for preheating a casting prior to welding? (Obj. 9-2)
25. What problems might the student welder run into when applying flux? Explain. (Obj. 9-2)
26. Explain why you should clean aluminum before and after welding it. (Obj. 9-2)
27. Name three categories of aluminum. (Obj. 9-2)
28. Define hard facing. (Obj. 9-3)
29. List five advantages of hard facing. (Obj. 9-3)
30. Name all types of hard-surfacing rods. (Obj. 9-3)

INTERNET ACTIVITIES

Internet Activity A

Use the Internet to describe some of the environmental and health hazards involved with working with lead. Prepare a one-page report.

Internet Activity B

Locate a Web site about braze welding. Make a drawing of the backhand method to braze weld cast iron. Then make two other drawings to show what tinning and penetration would look like for that weld. Share and discuss your drawings with the class.

Table 9-3 Job Outline: Advanced Gas Welding and Braze Welding Practice: Jobs 9-J39–J49

Job No.	Joint	Operation	Material Type	Material Thickness	Filler Rod Type	Filler Rod Size	Welding Position[1]	Text Reference
9-J39	Flat plate	Beading (braze welding)	Mild steel	1/8	Bronze RCuZn-C	1/8	1	241
9-J40	Lap joint	Fillet (braze welding)	Mild steel	1/8	Bronze RCuZn-C	1/8	1	241
9-J41	T-joint	Fillet (braze welding)	Mild steel	1/8	Bronze RCuZn-C	1/8	2	241
9-J42	Beveled-butt joint	Groove (braze welding)	Mild steel	3/16	Bronze RCuZn-C	1/8	1	241
9-J43	Flat casting	Beading (braze welding)	Cast iron	3/16	Bronze RCuZn-C	1/8	1	241
9-J44	Beveled-butt joint	Groove (braze welding)	Cast iron	3/16 to 1/4	Bronze RCuZn-C	1/8	1	241
9-J45	Casting	Beading (fusion)	Cast iron	3/16 to 1/4	Cast iron RCI	3/16 or 1/4	1	245
9-J46	Beveled-butt joint	Groove welding (fusion)	Cast iron	1/4 to 1/2	Cast iron RCI	3/16 or 1/4	1	245
9-J47	Flat plate	Beading	Sheet aluminum	1/8	Aluminum R1100 or R4043	3/32 or 1/8	1	248
9-J48	Outside corner joint	Groove welding	Sheet aluminum	1/8	Aluminum R1100 or R4043	1/8	1	248
9-J49	Square butt joint	Groove welding	Sheet aluminum	1/8	Aluminum R1100 or R4043	1/8	1	248

Note: It is recommended that students complete the jobs in the order shown. At this point continue welding practice with those other metals described in the chapter that are available in the school shop. Hard facing should also be practiced.

It is very important that you become proficient in gas cutting. (See Chapter 7.) Practice should include straight line, shape, and bevel cutting with both the hand cutting torch and the machine cutting torch. This is a skill that is also necessary for the electric arc welder.

[1] 1 = flat, 2 = horizontal, 3 = vertical, 4 = overhead, 5 = multiple positions pipe.

Design credits: Cinema and movie icon: Denis Meshkov/123RF; and Illustration of Welding icons: Mr. Nachai Sorasee/123RF

10

Soldering and Brazing Principles and Practice:

Jobs 10-J50–J51

Soldering and Brazing Copper Tubing

This chapter deals mainly with the soldering and brazing of copper tubing. Table 10-6, on page 276, provides a Job Outline for this chapter. It is recommended that the students complete the jobs in the order shown.

Copper was one of the first metals used. People made tools for handicraft and agriculture, weapons for hunting and war, and decorative and household articles from copper. Pieces of copper pipe, buried for centuries, have been found in excellent condition. This is a testimonial to copper's durability and resistance to corrosion.

Today, copper in the form of pipes and tubing is used in the installation of plumbing and heating, Fig. 10-1. It is light, strong, and corrosion resistant, and is available in hard and soft tempers. Tubing comes in a wide variety of diameters and wall thicknesses, with clean, efficient fittings to serve every purpose. Joints are made simply and effectively by soldering and brazing.

Copper tubing and pipe are widely used in shipbuilding, oil refineries, chemical plants, and, in general, by those industries in which corrosion and scaling are problems. Applications include saltwater lines, oil lines, refrigeration systems, vacuum lines, chemical lines, air lines, and low pressure steam lines.

Chapter Objectives

After completing this chapter, you will be able to:

10-1 Define and perform soldering.

10-2 Define and perform brazing.

10-3 Demonstrate ability to troubleshoot soldered and brazed joints.

Fig. 10-1 Soldering a copper tube with a butane torch.
Programatics

Definitions of Soldering and Brazing

The following definitions for soldering and brazing are adapted from those established by the American Welding Society as stated in their *Welding Handbook, Vol. 2: Welding Processes,* 8th ed.

Soldering is defined as a group of joining processes that produce coalescence of materials by heating them to the soldering temperature and by using a filler metal (solder) having a liquidus not exceeding 840°F and below the solidus of the base metals. Solder differs from brazing in that brazing filler metals have a liquidus above 840°F. The solder is distributed between closely fitted faying surfaces of the joint by *capillary action*. **Capillary action** is the flow of a liquid when it is drawn into a small space between closely fitted (faying) surfaces.

Brazing is a process that joins materials by heating them in the presence of a filler metal having a liquidus above 840°F but below the solidus of the base metal. Heat may be provided by a variety of processes. The filler metal distributes itself between the closely fitted surfaces of the joint by capillary action, Fig. 10-2. *Brazing* differs from *braze welding* in that braze welding is done by melting and depositing the filler metal directly in groove and fillet joints exactly at the points where it is to be used. Capillary action is not a factor in the distribution of the brazing filler metal during braze welding. For additional information on braze welding, see Chapter 9.

Soldering must meet all of the three following criteria:

- The parts must be joined without melting the base metals.
- The filler metal must have a liquidus temperature below 840°F.
- The filler metal must wet the base metal surface and be drawn into or held in the joint by capillary action.

Other terms describing soldering and brazing are defined in the Glossary. You are urged to look up any words you do not know.

Soldering

Filler Metals

Selection of the proper materials is an important preliminary step in the soldering of copper tube joints. Selection depends on the metals to be joined, the expected service, operational temperatures, and the expansion, contraction, and vibration that will be experienced during service.

Under ordinary circumstances, a properly made joint will be stronger than the tube itself for stresses of short duration. Because solder is somewhat plastic, however, it may give when stress is maintained at high temperatures over long periods of time. The stress that causes failure under these conditions is less than what would produce a break with short-time loads. This condition is known as *creep,* and the creep strength of various types of solder varies widely. Different elements serve different roles in the solder alloy:

Silver (Ag) provides mechanical strength, but has worse ductility than lead. In absence of lead, it improves resistance to fatigue from thermal cycles.

Copper (Cu) lowers the melting point, improves resistance to thermal cycle fatigue, and improves wetting properties of the molten solder. It also slows down the rate of separation in the component parts such as copper from the board and part leads in the liquid solder.

Bismuth (Bi) significantly lowers the melting point and improves wetting action. In presence of sufficient lead and tin, bismuth forms crystals of $Sn_{16}Pb_{32}Bi_{52}$ with melting point of only 203°F, which diffuses along the grain boundaries and may cause a joint failure at relatively low temperatures. A high-power part pretinned with an alloy of lead can, therefore, de-solder under load when soldered with a bismuth-containing solder.

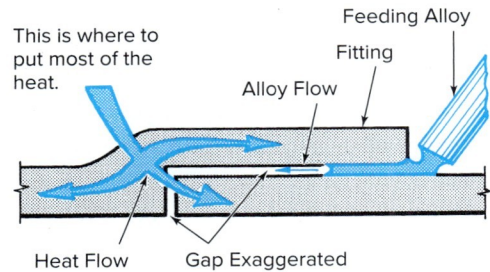

Fig. 10-2 Capillary action pulls and distributes the brazing alloy into the gap where it wets both metal surfaces.

Table 10-1 Tin/Lead Solder

	Composition				Specification	
Solder	Tin	Lead	Antimony	Solidus (°F)	Liquidus (°F)	ASTM B32–86
40–60	40	60	—	360	460	Sn40a
60–40	60	40	—	360	375	Sn60
50–50	50	50	—	360	420	Sn50
95–5	95	—	5	452	464	Sb5

Note: These types of solder make up the largest portion of solders used. They are used on copper, most copper alloys, lead, high nickel alloys, and steel.

Caution: Lead-bearing solders are not to be used in potable water systems.

Tin (Sn) is commonly used because it is the cheapest material and has appropriate surface activity to form intermetallic bonds with a wide range of surfaces—the alternative element for this use would be indium, which is rarer and more expensive.

Indium (In) lowers the melting point and improves ductility. In presence of lead, it forms a ternary compound (three different elements) that undergoes phase change at 237°F.

Zinc (Zn) lowers the melting point and is low-cost. However, it is highly susceptible to corrosion and oxidation in air; therefore, zinc-containing alloys are unsuitable for some purposes, e.g., wave soldering, and zinc-containing solder pastes have shorter shelf life than zinc-free.

Antimony (Sb) is added to increase strength without affecting wetting action.

Tables 10-1 and 10-2 list the composition and their specifications. The **pasty range** is the difference between the melting and solid temperatures. Solder is semisolid in the pasty range and is referred to as the "working range."

Tin-Lead Solders *Tin-lead* and *tin-lead-other alloy solders* are the most widely used solders, and they are suitable for joining most metals. A good grade of 50–50 tin-lead solder is generally used for all service at room temperatures and for low pressure steam (up to 15 p.s.i.), as well as for moderate pressures with temperatures up to 250°F. This solder is classified as 50A. It is molten at 420°F and solid at 360°F. The pasty range of 60° will produce a much wetter, "flatter" bead than 60–40. The 60–40 solder has a working range of 17°. This solder is used for electronic or copper foil work. The liquid temperature and narrow pasty range make it easy to form and maintain consistent high, rounded, beaded seams on copper. It is able to maintain a smooth finish bead due to its low melting point and is easily reworked.

Table 10-2 Melting Ranges of Solders Containing Other Metals

Solder Alloy	Melting Point (°C)	Melting Point (°F)
5Sn-95Pb	307	585
0.5Sn-92.5Pb-2.5Ag	280	536
Sn/5Sb	243	469
100Sn	232	450
99.3Sn-0.7Cu	227	440
96.5Sn-3.5Ag	221	430
52In-48Sn	118	244
Sn/3.0Ag/0.5Cu	219	426
Sn/3.8Ag/1.0Cu	217	423
Sn/3.5Ag/1.0Cu/3Bi	213	415
50In-50Pb	209	402
45Sn-55Pb	204	400
55Sn-45Pb	193	379
60Sn-40Pb	186	368
63Sn-37Pb	183	361
62Sn-36Pb-2Ag	179	354
97In-3Ag	143	289
Sn/57Bi	139	282

Note: Notice the amounts of tin, silver, copper, indium, and bismuth.
Source: From *Welding Handbook, 9/e*

The 40–60 solder has a much wider pasty range of 100°. The wider range allows for more dressing of the solder as in pipe or tube work.

When the tin content is increased, the flow of the solder and its wetting characteristics are increased. Because the pasty range is wide, however, care must be taken to keep the joint from being moved when it is cooling.

Tin-Antimony-Lead Solders *Antimony* is added for higher strength. The gain in higher tensile strength and creep strength is offset, however, because these solders are more difficult to work with than tin-lead solders. They have poorer flow and capillarity characteristics. The tin-antimony-lead solders may be used for joints in operating temperatures around 300°F. They are not recommended for aluminum, zinc-coated steels, or other alloys having a zinc base.

Tin-Antimony Solders If higher strengths than those provided by tin-lead solders and tin-lead-antimony solders are required, a 95–5 tin-antimony solder should be used for temperatures up to 250°F. It has a melting point of 464°F and is completely solid at 452°F. Its very narrow pasty range (12°F) makes it difficult to use in the vertical position. The absence of lead, a toxic substance, makes it highly desirable for food handling equipment.

Tin-Zinc Solders These solders are used for joining aluminum. Their melting point ranges from 390 to 708°F, and they are solid at 390°F. As the zinc content increases, the melting temperature increases. The type containing 91 percent tin and 9 percent zinc both melts and solidifies at 390°F. It wets aluminum readily, flows easily, and possesses a high resistance to corrosion with aluminum.

Cadmium-Silver Solders Improper use of solders containing cadmium may cause health hazards. Therefore, care should be taken in their application, particularly with respect to fume inhalation. The most common solder in this classification is 95 percent cadmium and 5 percent silver. It has a melting temperature of 740°F and is solid at 640°F. When it is used for butt joints in copper tube, the joint has a tensile strength of 2,600 p.s.i. at temperatures up to 425°F.

Cadmium-Zinc Solders The cadmium-zinc solders are used to join aluminum with joints of wide clearance and provide a strong, corrosive-resistant joint. They have a melting range of 509 to 750°F and solidify at 509°F. The solder containing 90 percent zinc has a wide pasty range of 241°F.

Zinc-Aluminum Solders The 95 percent zinc and 5 percent aluminum solder is in common use. It is a high temperature solder that melts at 720°F. Because of the high zinc content, it also has high resistance to corrosion.

Paste Solders These solders are composed of finely granulated solder, generally 50–50 lead-tin, which is in suspension in a paste flux. The flux paste makes cleaning the copper unnecessary. Care must be taken in making a vertical joint, because the solder and flux have a tendency to run down the tube.

Forms of Solders The solder in general use for soldering copper tubing is commercially available in solid wire form. The wire comes in diameters of 0.010 to 0.30 inch on spools weighing 1, 5, 10, 20, 25, and 50 pounds. Flux may be incorporated with the solder in single or multiple hollows or in external parallel grooves.

Other forms available for special applications include pig, slabs, cakes, bars, paste, tape, ingots, creams, ribbon, preforms, powder, foil, and sheet. An unlimited range of sizes and shapes may be preformed to meet special requirements.

Fluxes

A **soldering flux** is a liquid, solid, or gaseous material that, when heated, improves the wetting of metals with solder. The flux does not clean the base metal. If the base metal has been cleaned, however, the flux removes the tarnish films and oxides from both the metal and solder. When applied to a properly cleaned surface, flux performs the following functions:

- Protects the surface from oxidation during heating
- Permits easy displacement by the filler metal so that it flows into the joint
- Floats out the remaining oxides ahead of the molten solder
- Increases the wetting action of the molten solder by lowering its surface tension

Fluxes are classified based on their ability to remove metal tarnishes (activity). Fluxes may be classified into three groups: *inorganic fluxes* (most active—highly corrosive), *organic fluxes* (moderately active—intermediate), and *rosin fluxes* (least active—noncorrosive), Table 10-3. The type of flux to use depends on the metal being soldered, the oxidation rate of the metal, and the resistance of the oxide to removal. Such metals as aluminum, stainless and high alloy steels, and aluminum bronzes form a hard oxide film when exposed to air. They require a highly active and corrosive flux. A milder flux can be used with copper because of its slow rate of oxidation and

---

SHOP TALK

Skills Needed

The welding student who wants to work in a manufacturing environment should demonstrate math and reading skills. The ability to work with others, good eye–hand coordination, and manual dexterity are also required.

Table 10-3 Metal Solderability Chart and Flux Selector Guide

Metals	Solderability	Rosin Fluxes			Organic Fluxes (Water Soluble)	Inorganic Fluxes (Water Soluble)	Special Flux and/or Solder
		Non-activated	Mildly Activated	Activated			
Platinum, gold, copper, silver, cadmium plate, tin (hot dipped), tin plate, solder plate	Easy to solder	Suitable	Suitable	Suitable	Suitable	Not recommended for electrical soldering	
Lead, nickel plate, brass, bronze, rhodium, beryllium copper	Less easy to solder	Not suitable	Not suitable	Not suitable	Suitable	Suitable	
Galvanized iron, tin-nickel, nickel-iron, low carbon steel	Difficult to solder	Not suitable	Not suitable	Not suitable	Suitable	Suitable	
Chromium, nickel-chromium, nickel-copper, stainless steel	Very difficult to solder	Not suitable	Not suitable	Not suitable	Not suitable	Suitable	
Aluminum, aluminum-bronze	Most difficult to solder	Not suitable	Not suitable	Not suitable	Not suitable	—	Suitable
Beryllium, titanium	Not solderable						

Source: American Welding Society, *Welding Handbook*, vol. 2, 8th ed., p. 435, Table 13.13.

the ease of removal of the oxide. Good soldering practice requires the selection of the mildest flux that will perform satisfactorily in a specific application.

Highly Corrosive Fluxes The highly corrosive fluxes consist of such inorganic acids and salts as zinc chloride, ammonium chloride, sodium chloride, potassium chloride, hydrochloric acid, and hydrofluoric acid. They are available as liquids, pastes, and dry salts. Corrosive fluxes are recommended for those metals requiring a rapid and highly active fluxing action. These fluxes leave a chemically active residue after soldering and will cause severe corrosion at the joint if not removed.

Intermediate Fluxes Intermediate fluxes are weaker than the inorganic salt types. They consist mainly of such mild organic acids and bases as citric acid, lactic acid, and benzoic acid. They are very active at soldering temperatures, but this activity is short-lived, since they are also highly volatile at soldering temperatures. These fluxes are useful for quick soldering operations. The residue does not remain active after the joint has been soldered, and it can be removed readily with materials requiring a mild flux.

Noncorrosive Fluxes The electrical industry is a large user of a noncorrosive flux composed of water and white resin dissolved in an organic solvent such as abietic acid or benzoic acid. The residue from these fluxes does not cause corrosion. Noncorrosive fluxes are effective on copper, brass, bronze, nickel, and silver.

Paste Fluxes Many types of paste flux, ranging from noncorrosive to corrosive, are available. A paste flux can be localized at the joint and will not spread to other parts of the work where it would be harmful. The body of the flux is composed of petroleum jelly, tallow, lanolin and glycerin, or other moisture-retaining substances.

Joint Design

Although our emphasis in this chapter concerns the soldering of pipe joints, it is well to consider briefly other joints that are soldered in industry. Figure 10-3 presents typical soldered joint designs, and should be studied carefully. Generally, the joint design depends on the service requirements of the assembly. Other factors include the heating method, assembly requirements before soldering, the number of items to be soldered, and the method of applying the solder. If the service conditions are severe, the design should be such that the strength of the joint is equal to or greater than the load-carrying capacity of the weakest member of the assembly. The joint must be accessible, because the solder is normally face-fed into the joint. For high production parts, solder in the form of wire, shims, strip, preformed, powder, a precoat, or solder flux-paste may be preplaced.

Clearance between the parts being joined should be such that the solder can be drawn into the space between them by capillary action, but not so large that the solder cannot fill the gap. Capillary attraction cannot function well if the clearance is greater than 0.010 inch. A clearance

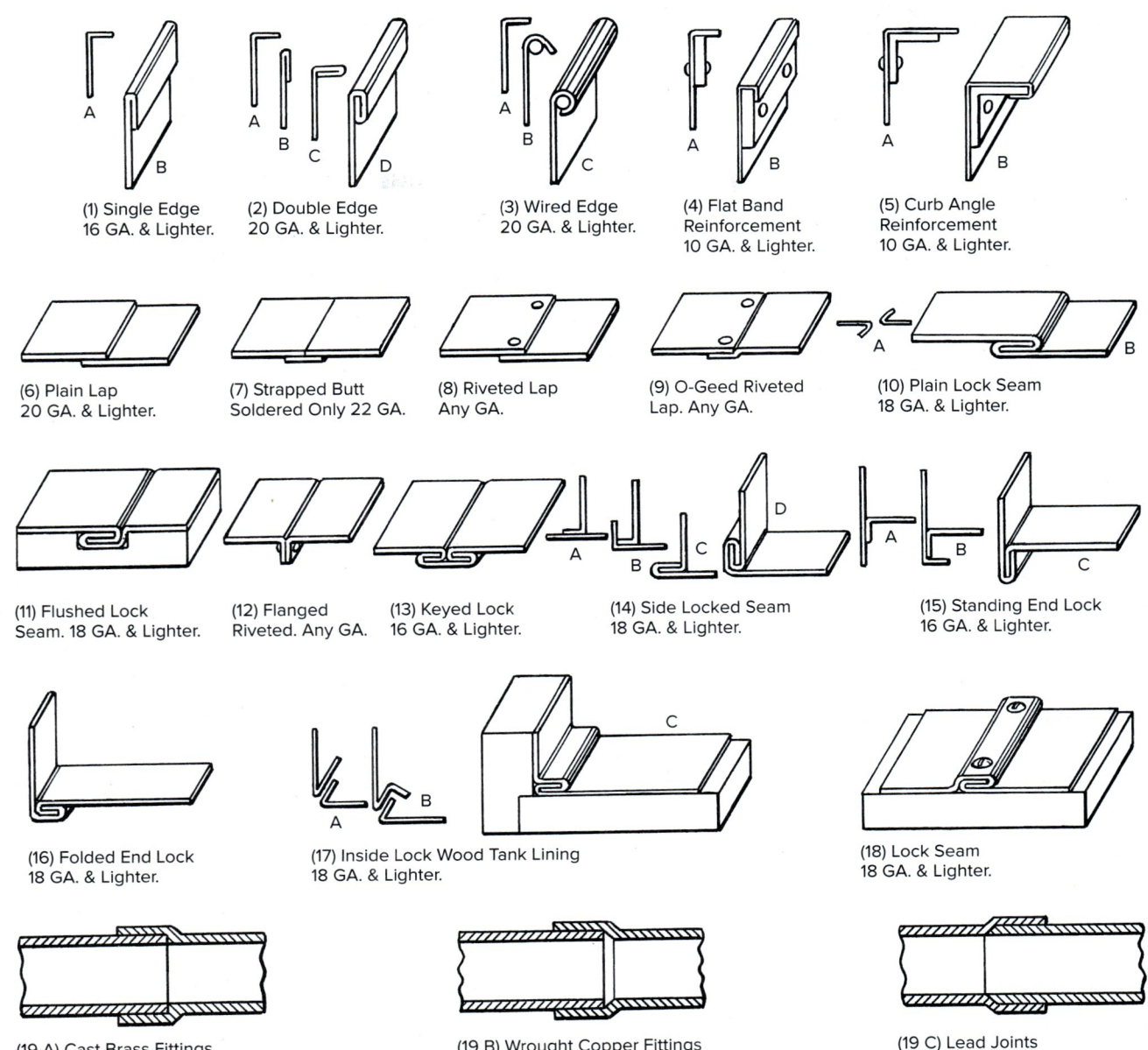

Fig. 10-3 Typical solder joint designs.

range of 0.003 to 0.005 inch is recommended. A joint's tensile strength is reduced as the clearance increases beyond the recommended amount.

Heating Methods

Obviously, heat is necessary to carry out the soldering application. The solder must melt while the surface is heated to permit the molten solder to flow over the surface. Heat may be applied in one of several ways, depending upon the application. Methods include soldering irons, dip soldering, induction heating, resistance heating, oven heating, spray gun heating, and flame heating. In our soldering practice, we will be concerned with flame heating.

ABOUT WELDING

Production Welder

Your duties may range from basic tasks like cutting, soldering, brazing, and welding different metal components to advanced activities like MIG/MAG, and TIG welding using aluminum or stainless steel.

The type of the torch to be used depends upon the size, mass, and design of the assembly. Time is also an important factor. Fast soldering requires a high temperature flame and large tip size. Slower techniques require a low temperature flame and small tip size. Fuel gas that burns with oxygen will also burn with air. The highest flame temperatures are reached with acetylene; and lower temperatures, with propane, butane, propylene, natural gas, and manufactured gas, in the order named. Care must be taken to avoid a sooty flame, since the carbon deposited on the base metal prevents the solder from flowing. In general, the oxyacetylene welding torch, the air-acetylene torch (Fig. 10-4), or the propane torch is used depending upon the amount of heat necessary. A small handheld pressure tank with an attached stem is available for small jobs.

Preparation for Soldering

This practice course will deal with the soldering of copper tubing, since this is the type of work a skilled welder will probably do in industry. Other forms of soldering are usually special applications that are not done by welders.

Joint Preparation The material covered here applies to the preparation of copper tubing for both soldering and brazing.

In order to make the assembly, the copper tubing is cut to various lengths and soldered to copper fittings. The end of the tube to be soldered should be square and free from burrs. The outer surface of the end of the tube should be round and within 0.001 to 0.002 inch of the specified diameter for a distance of 1 inch.

Use a hacksaw with a straightedge ring jig or a square-end sawing vise to cut ends off, Fig. 10-5. A bandsaw equipped for making perfectly square cuts will also do a good job. A pipe cutter, Fig. 10-6, may be used, but you must be careful not to put so much pressure on the cutter that it deforms the tube, Fig. 10-7. A pipe cutter also makes reaming necessary. Regardless of the method of

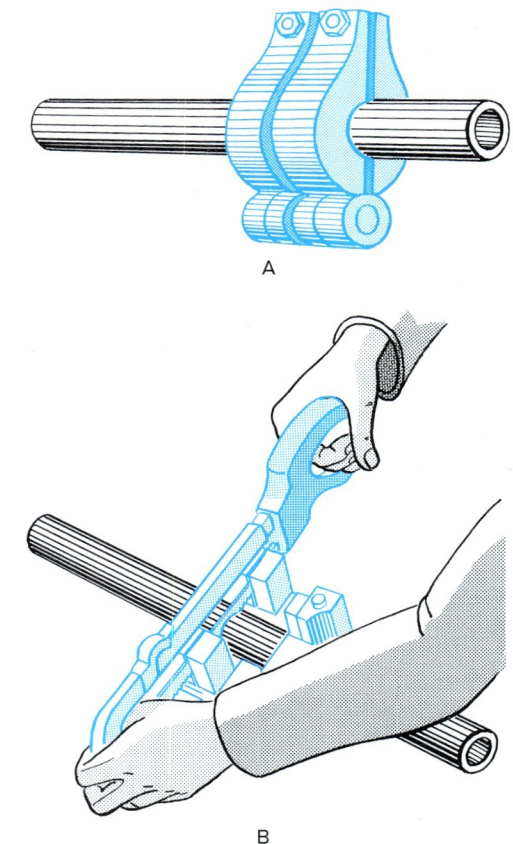

Fig. 10-5 (A) How to use a straightedge ring jig to saw off a pipe with a square end. (B) A square-end sawing vise holds the pipe and guides your hacksaw for a square cut.

cutting, all burrs on the outside and inside of the tubing should be removed. A hand file can be used for this purpose. If the end of the tube is out of round, a plug sizing can be used to round it off, Fig. 10-8.

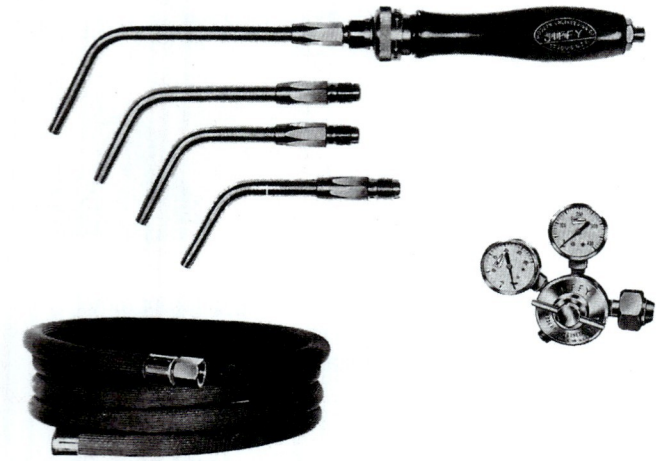

Fig. 10-4 The air-acetylene torch and accessories used for soldering and light brazing. Note the various tip sizes. Thermadyne Industries, Inc.

Fig. 10-6 Bench chain vise holding copper tubing to be square cut with a pipe cutter. Location: UA Local 400 Edward R. Bohnart

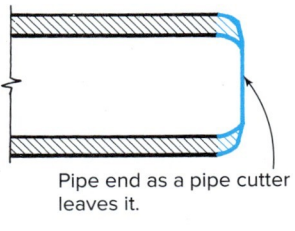

Pipe end as a pipe cutter leaves it.

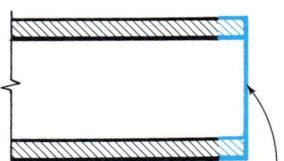

Pipe end as is wanted for soldering or brazing.

Fig. 10-7 The top sketch shows, with some exaggeration, why using a pipe cutter makes extra work in preparing ends for soldering or brazing such as removing the burr.

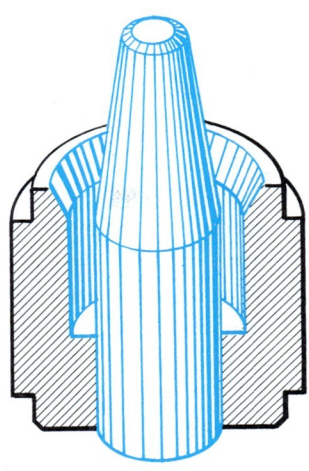

Fig. 10-8 For tubing such as copper type B tubing that has walls made to close tolerances, a plug-type tool is very effective in getting the outside diameter to size, true, and round.

Fig. 10-10 Remove the burr on the inside of the pipe with the deburring tool shown below. Then clean with the wire brushes.
Location: UA Local 400 Edward R. Bohnart

Precleaning and Surface Preparation Care in cleaning the surface of the material to be soldered is essential. A dirty surface impairs the wetting and alloying action because it prevents the solder from flowing as a thin film. All foreign materials such as oil, paint, pencil markings, lubricants, general atmospheric dirt, and oxide films must be removed before soldering. The strength and adherence of the solder is a function of the surface contact area of the solder to the base metal. Contact may be improved by roughening the surface of the base metal.

Two methods of surface cleaning are employed: mechanical and chemical. The mechanical method is more widely used in soldering and brazing of tube or pipe.

Mechanical Cleaning Several methods of cleaning may be used, depending on the nature of the job and the availability of the equipment. These methods include *grit-* or *shot-blasting, mechanical sanding* or *grinding, filing, hand sanding* (Fig. 10-9), *cleaning with stainless-steel wool, wire brushing* (Fig. 10-10), and *scraping* with a knife or shave hook. *Sandcloth* is the most widely used method of mechanical cleaning for copper, brass, and the softer metals.

The end of the tube should be cleaned for a distance only slightly more than that required for the full insertion of the tube into the cup of the fitting. The cup of the fitting should also be cleaned. Cleaning beyond these areas wastes filler metal and may permit the solder to flow beyond the desired areas. Excessive cleaning may reduce the outer diameter of the tubing, and thus cause clearance problems. If the tubing is very dirty, it may require both mechanical and chemical cleaning.

Chemical Cleaning Chemical cleaning is usually done for production operations. Either *solvent* or *alkaline degreasing* is recommended. The *vapor condensation* type solvents probably leave the least residual film on the surface. Because the equipment required for this process is expensive, dipping the tube ends into a liquid solvent or in detergent solutions are suitable alternate procedures.

Acid cleaning, also called **pickling,** removes rust, scale, oxides, and sulfides. The inorganic acids—hydrochloric, sulfuric, phosphoric, nitric, and hydrofluoric—are used singly or mixed. Hydrochloric and sulfuric acid are the most frequently used. The tubing should be thoroughly washed in hot water after pickling and dried as quickly as possible.

Practice Jobs: Soldering

Instructions for Completing Practice Job 10-J50

The following procedure will serve as a guide in soldering practice. Use copper tubing with diameters of ½ to 1 inch and copper fittings to match the tube diameters. Select a good grade of 50–50 and/or 95–5 solder. Heat may be applied by

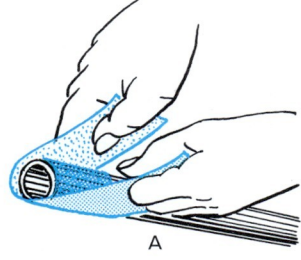 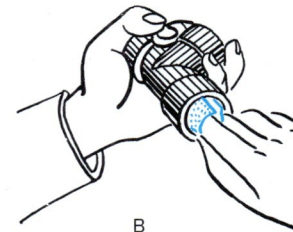

Fig. 10-9 (A) Use abrasives to get scale and dirt off the end of the pipe. (B) Clean out the cup and chamfer of each fitting outlet with abrasives.

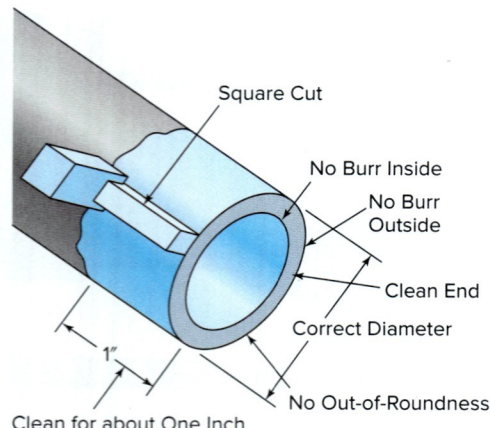

Fig. 10-11 This sketch is a checklist of the requirements for a pipe that has been cleaned and sized for soldering.

a gas-air torch or a blowtorch; the gas-air torch is preferred. Practice in all positions.

1. Cut the tubing into 12-inch lengths that can be recut for each new joint. (Review the information on cutting.)
2. Remove the burrs and straighten up the ends of the tubing. Clean the surface thoroughly. (Review the information previously given for these operations.) Figure 10-11 illustrates properly prepared tubing.
3. Flux the tube and fitting surfaces as soon as possible after cleaning. The preferred flux is one that is mildly corrosive and contains zinc and ammonium chlorides in a petrolatum base. Because the chemicals have a tendency to settle from long standing, stir the paste thoroughly when you open a new can. Use solder flux brushes to apply the flux evenly, Fig. 10-12.

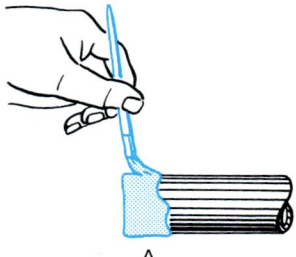
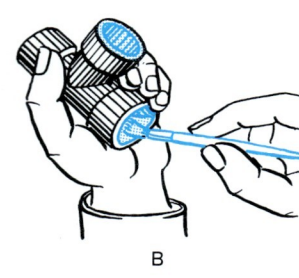
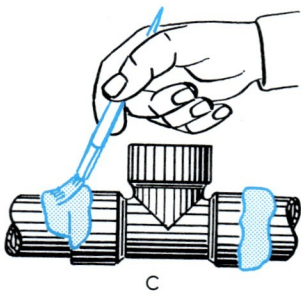

Fig. 10-12 Brush flux on the end of the pipe as soon as you have cleaned it. Flux all the cups on a fitting as soon as they are clean. Brush more flux around the joint after you have fitted the two parts together.

4. Assemble the joint by inserting the tube into the fitting. Make sure that the tube is hard up against the stop of the socket. A small twist helps to spread the flux over the two surfaces. The joint is now ready for soldering. The joint may be held in position in a vise. Frequently, a large number of joints are cleaned, fluxed, and assembled before soldering. Do not let the assembling get more than two or three hours ahead of soldering, and never leave prepared joints unfinished overnight.
5. Adjust the flame as shown in Fig. 10-13. Play the flame on the fitting. Keep it moving so as to heat as large an area as possible, Fig 10-14. Do not point the flame into the socket. When the metal is hot enough, move the flame away.
6. When the joint is the correct temperature, touch the end of the solder wire to the joint. If the joint has been made properly, a ring of solder will be observed almost instantly all the way around the joint. Opinions differ as to whether a fillet is desirable. Never apply the flame directly on the solder. It should melt on contact with the surface of the base metal and be drawn into the joint by the natural force of capillary attraction, regardless of whether the solder is being fed upward, downward, or sideways. If the solder does not melt, remove the solder and apply more heat. Then apply the solder again. Avoid overheating, which may burn the flux and destroy its effectiveness. If the flux has been burned, the solder will not enter the joint, and the joint must

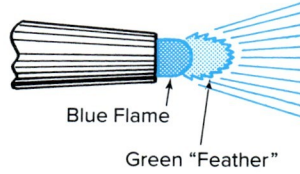

Fig. 10-13 Add enough acetylene to the mixture to produce a slight green "feather" on the blue cone. This is a reducing flame.

Fig. 10-14 (A) Heat the pipe near the fitting all around its circumference. (B) After the pipe is warmed up, shift the heat to the fitting. Keep the flame pointed toward the pipe.

be opened, recleaned, and refluxed. Overheating of cast fittings may also cause the fittings to crack.

7. While the joint is still hot, remove surplus solder and flux with a rag or brush, Fig. 10-15. This improves the appearance of the assembly and removes any chance of continued corrosive action by the surplus flux.
8. Allow the joint to cool naturally for some time before applying water, particularly if cast fittings are used. Too rapid cooling has been known to crack cast fittings.
9. Practice these joints in all positions until you can make joints with relative ease and have mastered the control of heat and the flow of solder. You are now ready for a check test.

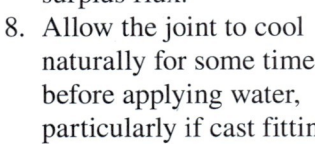

Fig. 10-15 Use water and a stiff brush to remove excess flux after completing a joint.

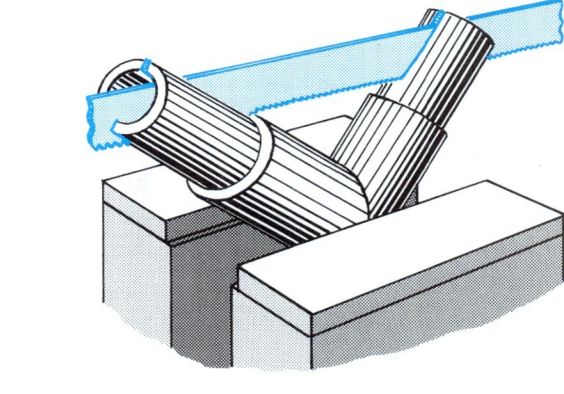

Fig. 10-16 Cut the fitting and tube along the center line.

Check Tests

You will perform two kinds of tests to check the soundness of your soldered joints: surface inspection and a water pressure test.

Surface Inspection:
1. Select a 1-inch T-connection and solder a length of tubing about 12 inches long into each opening of the connection. Following the procedure as outlined, complete the solder joints. Solder the three cups of the fitting in different positions: horizontal, vertical, and overhead.
2. After the joints have cooled and the excess flux has been cleaned off, cut the tubing at each end. Leave about 1 inch of tubing sticking out. Cut the three joints lengthwise along the center line with a hacksaw, Fig. 10-16.
3. Place the half tube and fitting in a vise, with the tube end down and the face of the fitting cup flush against the jaws of the vise. Tighten the vise until the tube end is flattened. Pull the pipe away from the fitting, Fig. 10-17.
4. Inspect the soldered surfaces of the tubing and the fitting. A perfect joint will have the entire cup area of the fitting completely covered with solder, which will have a grayish appearance. Defects include unsoldered areas and flux inclusions.

- *Unsoldered areas:* Fittings are designed to allow an ample safety factor in the joint and permit a bare spot or two if they do not cause leaks. Bare spots running circumferentially may not leak, but they weaken the joint. A bare spot of the same size running lengthwise may leak. Unsoldered areas may be the result of improper

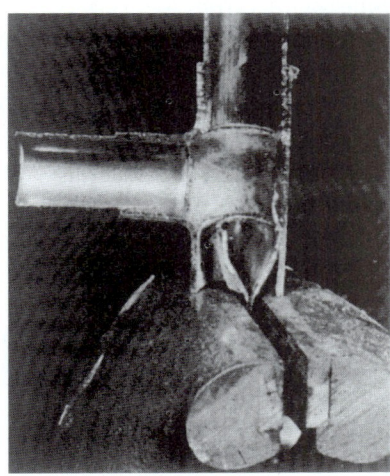

Fig. 10-17 Separate the tube from the fitting. American Welding Society

fluxing and heating or improper cleaning, if the surface does not have the glassy appearance of dried flux. If the unsoldered area is covered with flux, it is caused by a flux inclusion.
- *Flux inclusions:* A flux inclusion indicates that the flux had no chance to flow ahead of the solder. It may be caused by feeding the solder into the joint improperly. On small tubing, the solder should be fed at one point. Shiny areas indicate that although the metal is tinned on both surfaces, there is a flux inclusion between them. This is as serious as though there were no solder at all on these areas. The cause may be that the fitting was too loose on the tube.

Water Pressure Test:
1. Make up a closed-line assembly composed of joints made in all positions. Solder the connections with both the silver alloys and the copper-phosphorus alloys.

2. Braze a male-to-female fitting into the line so that you can introduce water under pressure into the tubing. Although the line can be tested with air pressure, the water-pressure test detects more leaks. If flux has sealed any pinholes on the inside of the assembly, water dissolves it and leaks through.
3. Pump water into the assembly at line pressure and observe for leaks. This may require from several hours to a day under test pressure.

Torch Brazing (TB)

Brazing is one of the oldest joining processes. It was first used to join ornamental gold fabrications with gold-silver and gold-copper-silver alloys as filler metals. At the beginning of the Iron Age, copper-zinc alloys, called **spelter,** were developed for joining iron and steel. These alloys are strong and easily melted. They have a vigorous wetting action on clean, fluxed ferrous metals. The early silver solders survive today as silver-base brazing alloys and are of great importance. Silver alloys are used extensively in brazing joints in copper tubing.

> **JOB TIP**
>
> **Turning a Job into a Career**
> In any interview, you're going to be asked why you want to work there. Of course you need the money, but that isn't your answer. Figure out what you like about that company. What does the company do that interests you? What does the industry mean to you? Let the company know that you understand what the job requires and that you're able to do it.

> **ABOUT WELDING**
>
> **Sumerians**
> As early as 3000 B.C., Sumerians were using hard soldering to make swords.

The essential differences between brazing and soldering are the much higher melting temperatures of the brazing filler metals and the special fluxes used for brazing. Brazing makes a joint stronger than soft solder, but the higher temperature necessary to melt the brazing filler metal anneals the copper tube in the heat-affected zone.

Strong, leaktight brazed connections for copper water tube may be made with brazing alloys melting at temperatures between 1,100 and 1,500°F. These are sometimes referred to as *hard solders,* a term not universally accepted.

The highest temperature at which a brazing material is completely solid is the **solidus temperature**. At the **liquidus temperature,** the brazing material is completely melted. This is the minimum temperature at which brazing will take place. The difference between the solidus and liquidus temperatures is known as the **melting range.** The melting range may be important in the selection of the brazing material, particularly as an indication of the rapidity with which the alloy will *freeze* after brazing.

Industrial Applications

The brazing process is used in joining copper and other metals. The process has the following advantages:

- Brazed joints are stronger than threaded joints because the pipe or tube is not notched or mutilated. The joints are as strong as the fittings themselves, Fig. 10-18.
- Vibration does not loosen brazed joints. If a system is damaged, the brazed joint will hold together longer than the threaded joint.
- Brazed joints do not leak. A sound joint will stay pressure- or vacuum-tight throughout its service life.
- Corrosion resistance is one of the main requirements of the kinds of piping and fittings commonly assembled by brazing. When copper, brass, or copper-nickel alloy is used to combat rust and deterioration, the joining material must resist corrosion too. The silver alloy filler metal used for brazing is generally as resistant to attack as are these metals themselves.
- Streamlined design, which brazing makes possible, means that there will be less pressure drop, reduction of dead weight, less clogging, and reduced tendency to pit or erode piping near the fittings.
- Accurate assemblies can be made by brazing. Pipe or tubes can be cut to exact dimensions, because no guesswork allowance for threading is necessary, Fig. 10-19. The angle

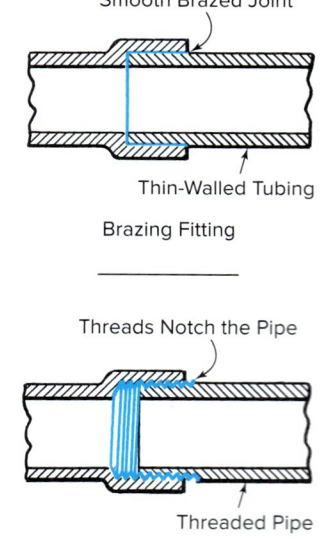

Fig. 10-18 Brazed joints have all the strength of pipe because they have no notches to cause weak spots.

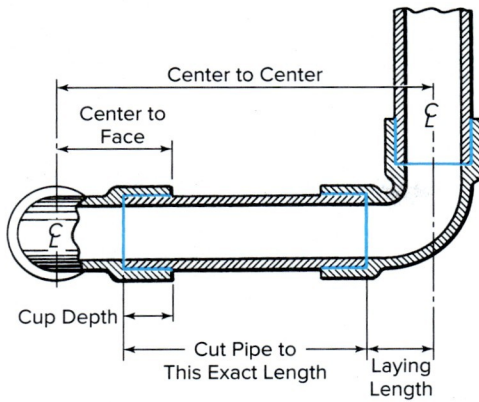

Fig. 10-19 You can work to accurate measurements by brazing because the cut ends of every piece of pipe are seated against precision-machined shoulders in the fitting.

of a fitting on a pipe can be preset. There is never a need for overtightening or slacking off in order to line up parts.
- Temporary or emergency piping can be assembled rapidly by brazing.
- Brazed piping can be taken apart, and all the pieces can be reused.

Filler Metals

Four factors should be considered when selecting a brazing filler metal:

1. Compatibility with base metal and joint design
2. Service requirements for the brazed assembly
3. Brazing temperature required
4. Method of heating

The American Welding Society lists the following classifications of brazing filler metals:

- BAlSi—aluminum
- BAg—silver base
- BAu—gold base
- BCu—copper
- BCu-P—copper phosphorus
- RBCuZn—copper zinc
- BMg—magnesium base
- BNi—nickel base

The composition and melting ranges of filler metals in these classifications are given in Table 10-4.

Aluminum-Silicon Filler Metals (BAlSi) These are used exclusively for brazing aluminum. They require flux. Type 2 is used as a cladding and applied with dip and furnace brazing. Type 3 is a general-purpose metal for dip and furnace application. Type 4 is used for torch brazing and dip and furnace brazing. It is highly corrosion resistant. Type 5 is used for dip and furnace brazing at temperatures lower than type 2.

For video of mechanized torch brazing, please visit www.mhhe.com/welding.

Copper-Phosphorus Filler Metals (BCuP) We will be concerned with these types of filler metals in the practice course. They are used primarily for joining copper and copper alloys, but they may also be used for joining other nonferrous metals. With copper, these types of filler metals are self-fluxing, but fluxes are recommended for other metals and copper alloys. Type 1 is used for preplacing in joints and is suited for resistance and furnace brazing. Types 2, 3, and 4 are all highly fluid filler metals that are suited for close clearance. Type 5 is used for joints where the clearance is less.

Gold Filler Metals (BAu) Gold alloys are used to join parts in electron tube assemblies and for missile components. They are suitable for induction, furnace, and resistance brazing. They require a flux.

Copper (BCu) and Copper-Zinc (RBCuZn) Filler Metals These filler metals are used for joining both ferrous and nonferrous metals with borax–boric acid flux. Since copper and copper-zinc alloys are extremely fluid, they require close fits. Overheating will cause volatilization of the zinc. These filler metals should not be used to join copper alloys or stainless steels because of interior corrosion resistance. This group is used for joining ferrous metals, nickel, and copper-nickel alloys.

Magnesium Filler Metals (BMg) Magnesium alloys are used for joining magnesium with the torch, dip, and furnace brazing processes.

Nickel Filler Metals (BNi) These materials are used when extreme heat and corrosion resistance are required. Typical applications include jet and rocket engines, food and chemical processing equipment, automobiles, cryogenic and vacuum equipment, and nuclear reactor components. Nickel alloys are very strong and may have high or low ductility, depending on the brazing method. The filler metal is supplied as a powder, paste, or sheet, or it is formed with binder materials into wire and strip.

Type 1 is highly corrosive and cannot be used with thin sheets. Type 2 has the lowest melting point and is the least corrosive of the group. Type 3 is a chromium-free alloy

Table 10-4 Summary of Brazing Filler Metals

AWS Classification	Nominal Composition (%)					Brazing Range (°F)	Uses
	Ag	Cu	Al	Ni	Other		
BAlSi-2	—	—	92.5	—	Si, 7.5	1,110–1,150	For joining aluminum alloys and cast alloys. All of these filler metals are suitable for furnace and dip brazing. BAlSi-3 and BAlSi-5 are suitable for torch brazing.
BAlSi-3	—	4	86	—	Si, 10	1,060–1,120	
BAlSi-5	—	—	90	—	Si, 10	1,090–1,120	
BAlSi-6	—	—	90	—	Si, 7.5; Mg, 2.5	1,125–1,150	Vacuum brazing filler metals. Magnesium is present as an O_2 getter.
BAlSi-8	—	—	86.5	—	Si, 12; Mg, 1.5	1,080–1,120	
BAlSi-10	—	—	86.5	—	Si, 11; Mg, 2.5	1,080–1,120	
BAlSi-11	—	—	88.4	—	Si, 10; Mg, 1.5; Bi, 0.1	1,090–1,120	
BCuP-1	—	95	—	—	P, 5	1,450–1,700	For joining copper and its alloys with some limited use on silver, tungsten, and molybdenum. Not for use on ferrous or nickel-base alloys. Are used for cupro-nickels, but caution should be exercised when nickel content is greater than 30 percent. Suitable for all brazing processes. Lap joints are recommended but butt joints may be used.
BCuP-3	5	89	—	—	P, 6	1,300–1,500	
BCuP-5	15	80	—	—	P, 5	1,300–1,500	
BCuP-7	5	88	—	—	P, 6.8	1,300–1,500	

AWS Classification	Nominal Composition (%)						Brazing Range (°F)	Uses
	Ag	Cu	Zn	Al	Ni	Other		
BAg-1	45	15	16	—	—	Cd, 24	1,145–1,400	For joining most ferrous and nonferrous metals except aluminum and magnesium. These filler materials have good brazing properties and are suitable for placement in the joint or for manual feeding into the joint. All methods of heating may be used. Lap joints are generally used; however, butt joints may be used.
BAg-2	35	26	21	—	—	Cd, 18	1,295–1,550	
BAg-4	40	30	28	—	2	—	1,435–1,650	
BAg-6	50	34	16	—	—	—	1,425–1,600	
BAg-8	72	28	—	—	—	—	1,435–1,650	
BAg-13	54	40	5	—	1	—	1,575–1,775	
BAg-18	60	30	—	—	—	Sn, 10	1,325–1,550	
BAg-20	30	38	32	—	—	—	1,410–1,600	
BAg-22	49	16	23	—	4.5	Mn, 7.5	1,290–1,525	
BAg-24	50	20	28	—	2	—	1,305–1,550	
BAg-26	25	38	33	—	2	Mn, 2	1,475–1,600	
BAg-28	40	30	28	—	—	Sn, 2	1,310–1,550	

AWS Classification	Nominal Composition (%)						Brazing Range (°F)	Uses
	Ni	Cu	Cr	B	Si	Other		
BCu-1	—	100	—	—	—	—	2,000–2,100	For joining various ferrous and nonferrous metals. They can also be used with various brazing processes. Avoid overheating the Cu-Zn alloys. Lap and butt joints are commonly used.
BCu-2	—	86.5	—	—	—	O, 13.5	2,000–2,100	
RBCuZn-A	—	59	—	—	—	Zn, 41	1,670–1,750	
RBCuZn-C	—	58	—	—	0.1	Zn, 40; Fe, 0.7; Mn, 0.3 Sn, 1	1,670–1,750	
RBCuZn-D	10	48	—	—	0.2	Zn, 42	1,720–1,800	
BCuZn-E	—	50	—	—	—	Zn, 50	1,610–1,725	
BCuZn-F	—	50	—	—	—	Zn, 46.5; Sn, 3.5	1,580–1,700	
BCuZn-G	—	70	—	—	—	Zn, 30	1,750–1,850	
BCuZn-H	—	80	—	—	—	Zn, 20	1,830–1,950	

(Continued)

Table 10-4 (Concluded)

AWS Classification	Nominal Composition (%)						Brazing Range (°F)	Uses
	Ni	Cu	Cr	B	Si	Other		
BAu-1	—	63	—	—	—	Au, 37	1,860–2,000	For brazing of iron, nickel, and cobalt-base metals where resistance to oxidation or corrosion is required. Low rate of interaction with base metal facilitates use on thin base metals. Used with induction, furnace, or resistance heating in a reducing atmosphere or in a vacuum and with no flux. For other applications, a borax–boric acid flux is used.
BAu-2	—	20.5	—	—	—	Au, 79.5	1,635–1,850	
BAu-4	18.5	—	—	—	—	Au, 81.5	1,740–1,840	
BAu-6	22	—	—	—	—	Au, 70 Pd, 8	1,915–2,050	
BCo-1	17	—	—	—	8	Cr, 19 W, 4 B, 0.8 C, 0.4 Co, 59	2,100–2,250	Generally used for high temperature properties and compatibility with cobalt-base metals.

Source: Engineers Edge, LLC

with a narrow melting range and free-flowing characteristics. Type 4 is similar to type 3. It is used for joints containing large gaps and for forming large ductile fillets. Type 5 resists oxidation up to 2,000°F. It is used for high strength joints needed in elevated temperature service, as in nuclear reactor components. Type 6 is a free-flowing filler metal. It produces a minimum amount of corrosion with most nickel and iron-base metals. Type 7 makes a strong leak-proof joint at relatively low temperatures. It is used for thin-walled tube assemblies. Ductility increases with time.

Silver Filler Metals (BAg) Silver and copper-phosphorus filler metals are the materials we are primarily concerned with in this practice course. Silver alloys are used for joining virtually all ferrous and nonferrous metals, with the exception of aluminum, magnesium, and several other metals with low melting points. They are generally free-flowing. Best results are obtained when the clearance between the tube or pipe and the bore of the fitting is held between 0.002 and 0.005 inch. A flux is required.

Types 1 and 1a are general-purpose metals, which are free-flowing and have low melting points. Type 2 is suitable for general purposes at higher temperatures. Type 3 is used for brazing carbide tool tips to shanks and for corrosion-resistant joints in stainless steel. Type 4 is also for carbide tip brazing, but at higher temperatures. Types 5 and 6 are general-purpose metals for higher brazing temperatures. Type 7 is a cadmium-free filler metal with a low melting point used for furnace brazing. Type 8 is a silver-copper

> **SHOP TALK**
>
> **Knowledge Required**
> To be a prepared welder, you need to have an understanding of metallurgy, the AWS standards, and equipment repair.

eutectic used for vacuum sealing parts. It is free-flowing, but does not wet well on ferrous metals. Type 8a resembles type 8, but the addition of lithium makes it self-fluxing on ferrous metals and alloys in a dry, protective atmosphere. Type 13 is a filler metal with a high melting point that is used in aircraft and aircraft engine construction. Type 18 does not contain cadmium and zinc. It is made of sterling silver, and lithium is added to promote self-fluxing. It has a low melting point. This alloy is used for brazing stainless steel for ultra-high-speed aircraft.

Specialized Filler Metals There are also available a number of special-purpose alloy filler metals based on such uncommon metals as gold, platinum, and palladium. They are used, for the most part, in brazing vacuum sealed components. Soldering and brazing of aircraft engines, nuclear systems, electronic equipment, pressure vessels, biomedical components, and aerospace vehicles have continued to push the development of new filler metals.

Forms of Filler Metals Most of the filler metals are ductile so that they can be rolled or drawn to wire or strip in various standard forms. They may also be supplied as powders. Nickel filler metals are available only as powders, although the powders can be bonded into wire or strip with plastic binder materials.

Fluxes

You will recall the importance of fluxing in the soldering process. It carries the same, or greater, responsibility in the brazing process. The importance of using the right flux becomes clear when we consider all the things the flux is called upon to do. It has to stay on the tube without blowing or washing away while being heated. The flux prevents oxidation from spoiling the clean metal surfaces when they are being heated for brazing. As the brazing alloy flows in, the flux has to flow out of the joint without leaving impurities or inclusions. Finally, it should be possible to clean all flux off the parts easily after the joint has been made. Recommended types can be washed off with water. Fluxes are available in the following forms:

- Paste or liquid
- Powder
- Solid coating preapplied on the brazing filler metal
- Vapor

Entirely apart from the protection it gives the joint, flux also acts as a temperature indicator. Were it not for this feature, it would be difficult to get the base metal hot enough for brazing without overheating it. When fluxed parts are warmed up, the water in the paste boils off at 212°F. Further heating, to about 600 to 700°F, makes the flux begin to work (bubble). At 800°F, the flux begins to melt. At 1,100°F, the flux is a clear, waterlike fluid, and the bright metal brazing surface can be seen beneath it. At 1,150 to 1,300°F, the brazing temperature has been reached. If the parts reach 1,600°F, the flux will lose its protective qualities. See Fig. 10-20.

Brazing fluxes include the following materials:

- **Borates** Sodium, potassium, and lithium borate compounds are used in high temperature fluxes whose melting range is 1,400°F. Their oxide-dissolving characteristics are good.
- **Fused borax** This material is used as an active flux at high temperatures.
- **Fluoborates** These are compounds of fluorine, boron, and active metals such as sodium and potassium. They have better flow properties and oxide removal properties than the borates. Protection against oxidation is of short duration.
- **Fluorides** Fluorides containing sodium, potassium, lithium, and other elements form active fluxes and react readily with even the most stable oxides at elevated temperatures. They work well in dissolving the refractory metal oxides, and they assist in brazing with silver filler metals. Fluorides increase the capillary flow of brazing filler metals.
- **Chlorides** Chlorides are similar in their fluxing characteristics to fluorides, although they are less effective.
- **Boric acid** Boric acid is a commonly used base for brazing fluxes and is used principally as a cleaning agent. It assists in removing the flux residue from the metal surface after brazing.
- **Alkalies** These are hydroxides of sodium and potassium. They elevate the temperature at which fluxes are effective. Alkalies have the ability to absorb moisture from the air, which limits their usefulness.
- **Wetting agents** Chemical wetting agents are commonly used in paste and liquid fluxes to improve contact between the flux and the metal interfaces.
- **Water** Water is present in all fluxes. Hard water cannot be used effectively; if no other water is available, alcohol should be used. Study Table 10-5 for the correct use of commercially available brazing fluxes.

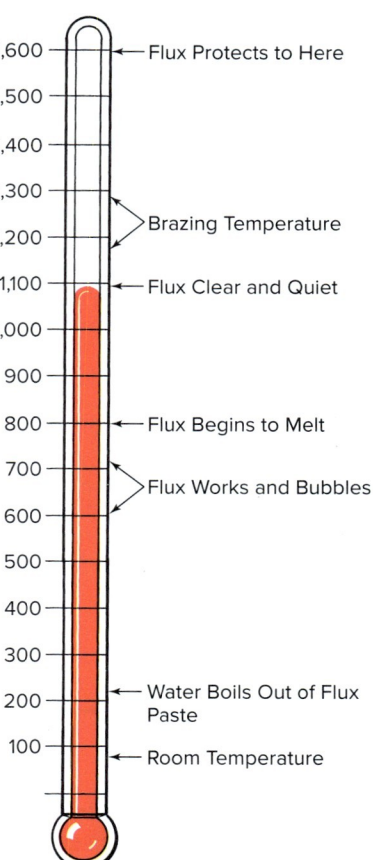

Fig. 10-20   Here is how flux behaves as the temperature rises.

Table 10-5 Information on Applications for Brazing Fluxes[1]

AWS Brazing Flux Type No.	Metal Combinations for which Various Fluxes Are Suitable		Effective Temperature Range of Flux (°F)	Major Constituents of Flux	Physical Form	Methods of Application[2]
	Base Metals	Filler Metals				
1	Aluminum and aluminum alloys	Aluminum-silicon (BAlSi)	700–1,190	Fluorides; chlorides	Powder	1, 2, 3, 4
2	Magnesium alloys	Magnesium (BMg)	900–1,200	Fluorides; chlorides	Powder	3, 4
3A	Copper and copper-base alloys (except those with aluminum); iron-base alloys; cast iron; carbon and alloy steel; nickel and nickel-base alloys; stainless steels; precious metals (gold, silver, palladium, etc.)[3]	Copper-phosphorus (BCuP) Silver (BAg)	1,050–1,600	Boric acid, borates, fluorides, fluoborate wetting agent	Powder Paste Liquid	1, 2, 3
3B	Copper and copper-base alloys (except those with aluminum); iron-base alloys; cast iron; carbon and alloy steel; nickel and nickel-base alloys; stainless steels; precious metals (gold, silver, palladium, etc.)	Copper (BCu) Copper-phosphorus (BCuP) Silver (BAg) Gold (BAu) Copper-zinc (RBCuZn) Nickel (BNi)	1,350–2,100	Boric acid Borates Fluorides Fluoborate Wetting agent	Powder Paste Liquid	1, 2, 3
4	Aluminum-bronze; aluminum-brass[4]	Silver (BAg); copper-zinc (RBCuZn); copper-phosphorus (BCuP)	1,050–1,600	Borates Fluorides Chlorides	Powder Paste	1, 2, 3
5	Copper and copper-base alloys (except those with aluminum); nickel and nickel-base alloys; stainless steels; carbon and alloy steels; cast iron and miscellaneous iron-base alloys; precious metals (except gold and silver)	Copper (BCu); copper-phosphorus (BCuP) Silver (BAg 8–19) Gold (BAu); copper-zinc (RBCuZn) Nickel (BNi)	1,400–2,200	Borax Boric acid Borates	Powder Paste Liquid	1, 2, 3

[1]This table provides a guide for classification of most of the proprietary fluxes available commercially. For additional data, consult AWS specification for brazing filler metal A5.8 ASTM B260; consult also AWS Brazing Manual, 1991 ed.
[2]1-sprinkle dry powder on joint; 2-dip heated filler metal rod in powder or paste; 3-mix to paste consistency with water, alcohol, monochlorobenzene, etc.; 4-molten flux bath.
[3]Some Type 3A fluxes are specifically recommended for base metals listed under Type 4.
[4]In some cases Type 1 flux may be used on base metals listed under Type 4.

Joint Design

As in our study of the soldering process, our main concern in brazing practice will be joints in copper tubing and piping. It is well, however, to consider briefly the other types of joints that are brazed in industry.

The design of a joint to be brazed depends upon a number of factors. The most important are:

- **Composition of base and filler metals** Members to be brazed may be of similar or dissimilar materials.
- **Types of joints** In selecting the type of joint for brazing, the following conditions must be given consideration: the brazing process to be used, the fabrication or manufacturing techniques required before brazing, the quantity of production, the method of applying filler metal, and service requirements such as pressure, temperature, corrosion, and tightness. The two basic types of brazed joints are the butt joint and the lap joint. Lap joints have the highest efficiency, but they have the disadvantage of increasing the thickness of the joint. The strength of butt joints is usually less than that of lap joints. Figure 10-21 illustrates the various types of brazed joints.
- **Service requirements** Your attention is again called to the importance of the service a part is expected to provide. The use to which an assembly is subject is the determining factor. The characteristics of the material to be carried in the system affect

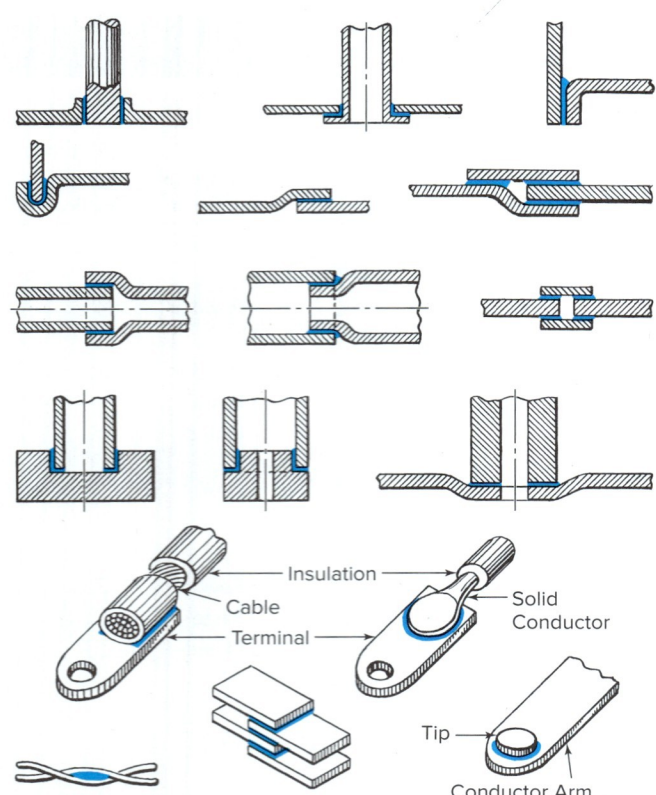

Fig. 10-21 Typical joints for brazing.

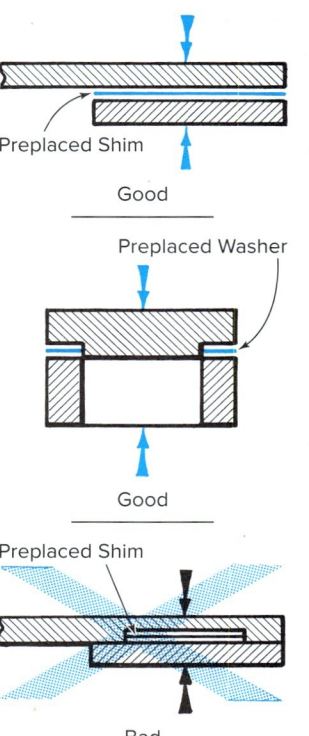

Fig. 10-22 Preplacement of brazing filler metal in shim form. The bad shim is locked in so that there is no room for expansion or for gases to escape.

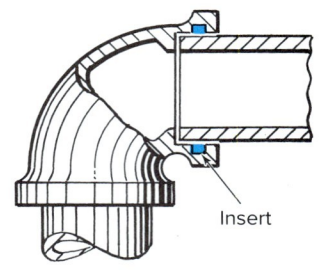

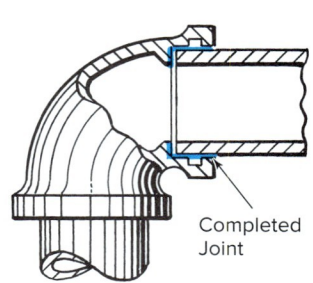

Fig. 10-24 Preinserted silver-brazing alloy in a groove flows both ways through the cup to make a joint.

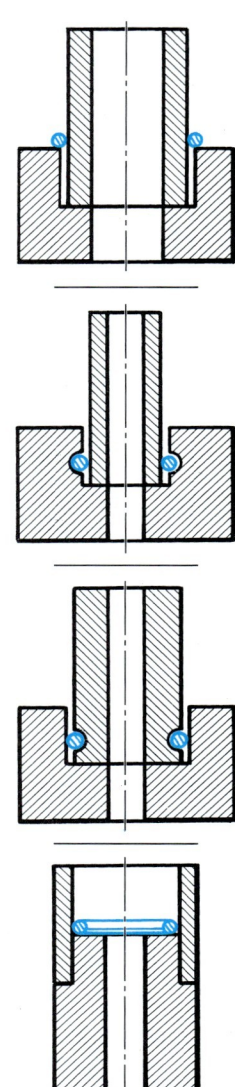

Fig. 10-23 Methods of preplacing brazing filler metal in wire form.

joint design. Is it liquid, solid, or gas? What are its temperature and pressure ranges? Is it corrosive? The joint must maintain the properties in the base metal and filler metal that the material demands. These properties include tensile strength and resistance to impact, fatigue, and extremes of temperature and pressure.

- **Stress distribution** Joints should be designed to avoid stress concentrations at the brazed area that may cause tearing.
- **Placement of brazing filler metal** Before designing a brazed joint, it is necessary to select the brazing process and the manner in which the filler metal will be applied to the joint. In most torch-brazed joints, the filler metal is simply *face-fed*. Mass factory production, however, may require the use of automatic equipment for the preplacement of brazing filler metal. See Figs. 10-22, 10-23, and 10-24.
- **Electric conductivity** In brazing an electrical joint, consideration must be given to the resistance set up by the brazed joint. In general, brazing filler materials have a lower electrical conductivity than copper. One approach is to use a shorter lap, thus reducing the bulk of the joint.
- **Pressure tightness** Wherever possible, the lap joint should be used in the fabrication of pressure-tight assemblies. In making these joints, the entire surface to be joined must have uniform coverage. There must be no channels or bare spots through which leakage can occur. It is also important in brazing a closed assembly that it be vented in some way to

provide an outlet for the air or gases enclosed. If gases are not allowed to escape during brazing, they will create pressure on the filler metal flowing through the joint and retard capillary attraction.

Heating Methods

Several methods of heating are used to produce brazed joints. In selecting the method to be used, consideration should be given to the heat requirements of the joint and the materials being brazed, accessibility to the joint, production quotas, compactness and lightness of design, and the mass of the component.

Torch Brazing (TB) Four different kinds of torches are used for the brazing process, depending on the fuel-gas mixtures. Mixtures include:

- Air-gas
- Air-acetylene
- Oxyacetylene
- Oxyhydrogen
- City gas
- Natural gas
- Propane
- Propylene
- Butane
- Other oxyfuel gases

Air-gas torches provide the lowest flame temperatures and the least heat. Both air-gas and air-acetylene torches can be used to advantage on small parts and thin sections.

Torches that employ oxygen with city gas, natural gas, propane, propylene, or butane provide a higher flame temperature. Like air-gas torches, they are suitable for small components, lower heating speeds, and certain brazing alloys.

Oxyhydrogen torches are often used for brazing aluminum and other nonferrous alloys. The temperature they produce is higher than those of the torches previously considered and lower than that of the oxyacetylene torch. The danger of overheating is reduced. Excess hydrogen provides the joint with additional cleaning and protection during brazing.

Oxyacetylene torches provide the widest range of heat control and the highest temperatures of all the torches considered. They may be used in a variety of situations and with most filler materials. Because of the high heat possible, extreme skill must be exercised to avoid local over-heating. It is an advantage to have the torch constantly moving over the work.

Torch heating is limited to those brazing filler metals that may be used with flux or are self-fluxing. This includes the aluminum-silicon, silver, copper-phosphorus, and copper-zinc classifications.

Furnace Brazing (FB) Furnace brazing is widely used when (1) the parts to be brazed can be preassembled or jigged to hold them in position, (2) the brazing filler metals can be preformed and preplaced, and (3) exacting atmosphere control is necessary. The method of heating varies according to the application and relative cost of the fuel. Some furnaces are heated by gas or oil, but the majority are heated electrically. The parts to be brazed are usually assembled and placed in trays for loading into the furnace. The flame does not make direct contact with the parts being brazed. Loading and unloading may be either manual or automatic.

Induction Brazing (IB) Induction heating is used on parts that are self-jigging or that can be fixtured in such a manner that effective heating will not be reduced by the fixture. Parts to be brazed act as a short-circuited resistance unit in the electric circuit and are heated as a result. Most of the heat generated by this method is relatively near the surface. The interior is heated by thermal conduction from the hot surface.

> **ABOUT WELDING**
>
> **Being Effective**
> In today's welding environment, you will need to read drawings, perform basic shop math, read measurement tools, and communicate using up-to-date welding terminology.

Resistance Brazing (RB) This process is used when small areas are to be brazed and the material is high in electrical conductivity. The heat is provided by the resistance of the parts to the flow of high current, supplied by a transformer that is brought to the brazing area, through conductors made of carbon, molybdenum, tungsten, or steel. A typical application is the brazing of conductors into the commutator slots in large electric motors or generators.

Dip Brazing (DB) Two methods are in common use:

- Molten metal dip brazing
- Molten chemical (flux) dip brazing

The molten metal dip process is limited to the brazing of small assemblies such as wire connections and metal strips. The filler metal is melted in a graphite crucible, which is externally heated. A cover of flux is maintained over the molten filler metal. Clean parts are immersed into the molten metal. Care must be taken that the mass to be brazed does not lower the temperature of the molten metal below that necessary for brazing.

The molten chemical dip process is also limited to small assemblies that can be dipped. The flux is heated in a metal or a ceramic container to a fluid condition. Heat may be applied externally or by means of resistance heating of the flux itself. Parts should be cleaned, assembled, and preferably held in jigs before dipping. Brazing filler metal is preplaced as rings, washers, slugs, or cladding on the base metal. After dipping, the flux is drained off while the parts are hot. Any flux remaining must be removed by water or chemical means after cooling.

Infrared Brazing (IRB) This process is used mostly for brazing of small parts. The heat source is a high intensity quartz lamp that produces a radiant heat. These lamps can produce up to 5,000 watts of radiant energy.

Diffusion Brazing (DFB) The demands of the nuclear and aerospace industries have forced the development of many special processes. These processes have been given the general term **diffusion brazing.** They employ a filler metal that diffuses into the base metal under a specific set of conditions of time, temperature, and pressure. The joint brazed by diffusion bonding has a higher melting point than those joined by the normal brazing process. Thus, diffusion bonding permits the high service temperatures so necessary to these types of assemblies.

Brazeable Metals

As we pointed out earlier in the chapter, our purpose here is to provide practice in the soldering and brazing of copper tubing and pipe. These are the kinds of jobs that you will be likely to encounter as a welder. The other forms of brazing discussed in this chapter are production processes performed by semiskilled factory workers. It will add to your confidence and worth as a welder, however, to have some knowledge of the reaction of other metals when brazed.

Aluminum and Aluminum Alloys Aluminum and most of its alloys can be brazed. Aluminum alloys with high magnesium content are difficult to braze because of poor fluxing and wetting. Aluminum filler metals usually contain silicon and belong to the BAlSi group. A flux is usually required.

Magnesium and Magnesium Alloys Most forms can be brazed with filler metals of the BMg group. Torch and furnace brazing have limited applications, but dip brazing can be used for all magnesium alloys. Corrosion resistance depends upon the thoroughness of flux removal.

Copper The two types of copper for industrial use are oxygen-bearing copper and oxygen-free copper. The oxygen-bearing coppers contain a small percentage of oxygen in the form of cuprous oxide. At temperatures above 1,680°F, the oxide is active and reduces the ductility of the brazing materials. The tensile strength is not affected by heating.

The oxygen-free coppers do not contain copper oxide and are not subject to oxygen migration or hydrogen embrittlement during brazing. Filler metals of the BAg and BCuP groupings are used.

Low Carbon and Low Alloy Steels These steels can be brazed without difficulty. All of the processes may be used. The BAg, BCu, and BCuZn groups are the best suited filler metals. A flux is necessary.

Stainless Steels This category covers a wide range of base metals. Most of the BAg, BCu, and BCuZn filler metals may be used. The BAg grades that contain nickel are generally best for corrosion resistance. Filler metals containing phosphorus should be avoided on highly stressed parts because brittle nickel and iron phosphides may be formed at the joint interface. The BNi filler materials should be used for all applications above 800°F to obtain maximum corrosion resistance. A flux is necessary.

High Carbon and High Speed Tool Steels The brazing of high carbon steel is best accomplished prior to or at the same time as the hardening operation. The hardening temperatures for carbon steels range from 1,400 to 1,500°F. Therefore, filler metals having brazing temperatures above 1,500°F should be used after hardening. When brazing and hardening are done at the same time, filler metals having a solidus at or below the hardening temperature should be used. A flux is necessary.

Nickel and High Nickel Alloys These metals may be brazed by the standard processes. They are subject to embrittlement when mixed with sulfur and metals with low melting points such as lead, bismuth, and antimony. Nickel and its alloys are subject to stress corrosion cracking during brazing and should, therefore, be annealed before brazing. When high corrosion resistance is desired, the BAg groups are used. The BNi filler metals offer the greatest corrosion and oxidation resistance and elevated-temperature strength. The BCu materials may also be used.

Cast Iron The brazing of cast iron is somewhat difficult. It requires thorough cleaning with electrochemical flame, grit-blasting, or chemical methods. When the percentage of silicon and graphitic carbon in cast iron is relatively low, the brazing alloys wet without difficulty. Where the percentage is high, wetting is difficult. The high temperature filler metals such as BCu may be used. When silver brazing alloys with low melting points are used, they reduce the oxidation effect and make wetting easier. Cast iron with high carbon content has a low melting point, and care must be taken to use as low a temperature as possible to avoid melting the surface area.

Other Metals Aircraft and missile development have brought into use a number of metals and materials that lend themselves to brazing. These include metals such as tungsten, tantalum, molybdenum, niobium, beryllium, titanium, and zirconium. Most of these can be brazed with the silver-based filler metals as well as other filler metals to meet special conditions.

In addition to the reactive and refractory metals listed previously, brazing may also be applied to certain types of ceramics and to join graphite to graphite and graphite to metals. Another expanding application is the brazing of dissimilar metals. There appears to be no limit to the uses of the various brazing processes.

Practice Jobs: Brazing

Brazing materials suitable for joining copper tubing may be divided into two classes: the silver alloys and the copper-phosphorous alloys. The two classes have fairly wide differences in their melting and flowing characteristics. The welder should consider these characteristics as well as the time required to make a joint when selecting filler metal. For joining copper tube with copper or bronze capillary fittings, any brazing alloy in the BAg and BCuP classifications provides needed strength and tightness.

Instructions for Completing Practice Job 10-J51

The following general procedure should be followed for brazing copper tubing and pipe. Brazing should be done in each position until joints of acceptable quality are produced.

1. Select copper tubing with diameters ranging from ½ to 1½ inches and copper fittings to match the tube diameters. Cut the tubing into 12-inch lengths that can be recut for each new joint. Review the information concerning cutting methods on page 263.
2. Refer to pages 263 to 265 for the proper way to measure, cut, and clean the tube ends. Tube ends and sockets must be thoroughly cleaned and free of burrs.
3. Choose and apply a flux in accordance with the recommendations of the manufacturer of the brazing alloy. Apply the flux with a brush to the cleaned area of the tube end and the fitting socket. Do not use too much flux, and do not apply flux to areas outside the brazing area. Be especially careful not to get flux inside the tube. Apply flux immediately after cleaning, even though the parts are not going to be brazed immediately. The flux prevents oxidation, which gradually affects clean metal surfaces even at ordinary room temperatures. If the fluxed surfaces dry out, add some fresh flux when you are ready to braze. Use flux that has the consistency of honey. Add water if it gets stiff. A great deal of brazing trouble can be prevented by keeping the flux just right—not too thick and not too watery. Never use a dirty brush for applying flux.
4. Assemble the joint by inserting the tube into the socket of the fitting, hard against the stop. Support the two parts so that they are lined up true and square. The joint may be positioned with the aid of a vise, V-blocks, clamps, or other devices. Avoid massive metal supports or clamps with large areas in contact near the joint, because they will tend to suck the heat out of the pipe and the fitting.

   The strength of a brazed copper tube joint depends to a large extent on maintaining the proper clearance between the outside of the tube and the socket of the fitting. Check the clearance gap carefully. Figure 10-25 indicates that as the clearance increases, the tensile strength of the brazed joint is reduced. Because copper tubing and braze-type fittings are accurately made for each other, the tolerances permitted for each ensure that the capillary space will be kept within the limits necessary for a joint of satisfactory strength.
5. Brush additional flux at the joint around the chamfer of the fitting. A small twist of the tube and fitting helps to spread the flux over the two surfaces. The joint is now ready for brazing.

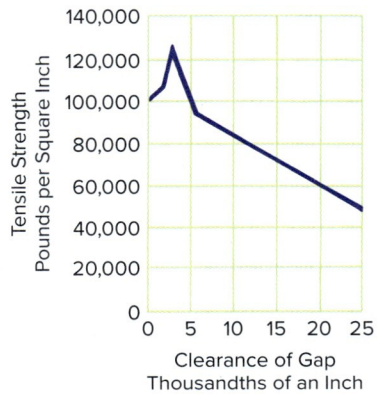

Fig. 10-25 Tensile tests with brazing alloy having a strength of 42,000 p.s.i. show that a small clearance gap makes a joint with as much as three times the strength of the alloy itself.

6. Use an oxyacetylene torch for brazing, and adjust the flame to slightly excess acetylene. Propane and other gases are sometimes used on small assemblies.

7. Heat the tube, beginning at about 1 inch from the edge of the fitting. Sweep the flame around the tube in short strokes, up and down, at right angles to the run of the tube. It is very important that the flame be in constant motion to avoid burning through the tube. If the flame is permitted to blow on the tube, it may also wash the flux away. Heating the tube first makes it expand. This causes the tube to press against the cup, and some heat gets carried through to warm up the fitting. Generally, the flux may be used as a guide to how long to heat the tube. Continue heating after the flux starts to bubble and until the flux becomes quiet and transparent, like clear water. This indicates that the tube has reached the brazing temperature.

8. Switch the flame to the fitting at the base of the cup. Heat uniformly by sweeping the flame from the fitting to the tube until the flux on the fitting stops bubbling. Avoid overheating the fittings.

9. When the flux appears liquid and transparent on both the tube and the fitting, start sweeping the flame back and forth along the axis of the joint to maintain heat on the parts to be joined, especially toward the base of the cup of the fitting. The flame must be kept moving to avoid burning the tube or fitting.

10. It is helpful to brush some flux on the brazing rod and to play the flame on the rod briefly to warm it up. Apply the brazing rod at a point where the tube enters the socket of the fitting. Because the temperature of the joint is hot enough to melt the brazing alloy, keep the flame away from the rod as it is fed into the joint. Keep both the fitting and the tube heated by moving the flame back and forth from one to the other as the filler alloy is drawn into the joint, Fig. 10-26. When the proper temperature is reached, the alloy will flow readily into the space between the tube's outer wall and the fitting socket, drawn in by the natural force of capillary attraction,

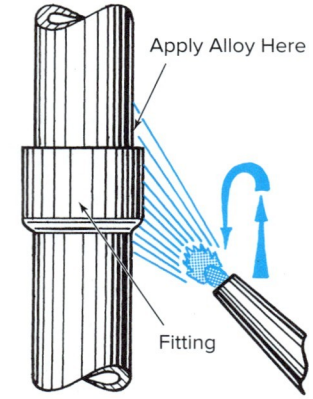

Fig. 10-26 Pull the alloy in with a brushing motion of the torch. Concentrate a good deal of the heat on the base of the cup.

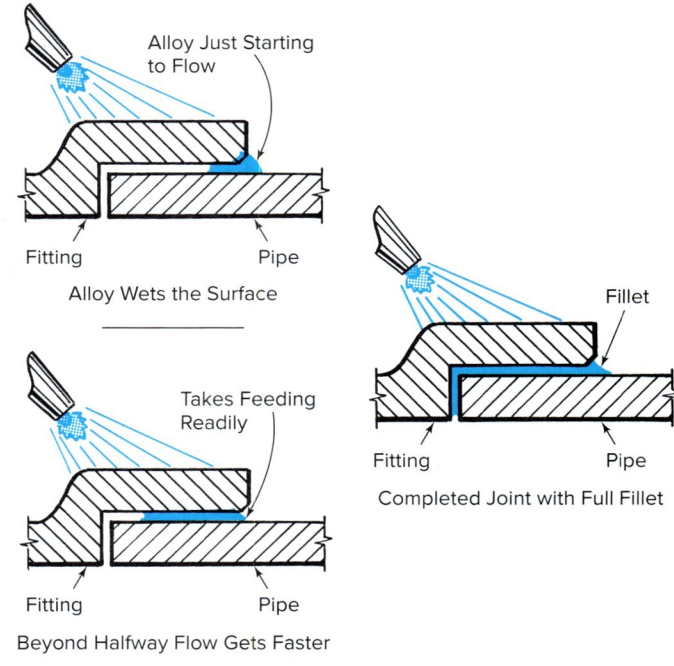

Fig. 10-27 Wet metal surfaces cause brazing alloy to flow into the gap.

Fig. 10-27. When the joint is filled, a continuous fillet of brazing alloy will be visible completely around the joint. Stop feeding as soon as the joint is filled. If the alloy fails to flow or has a tendency to ball up, it indicates either oxidation of the metal surfaces, or insufficient heat on the parts to be joined. If the work starts to oxidize during heating, there is not enough flux or the flux is too thin. If the brazing alloy flows over the outside of either member of the joint, one member is overheated, the other member is underheated, or both members are at the wrong temperature.

11. After the brazing alloy has set, clean off the remaining flux with a wet brush or cloth. Wrought fittings may be chilled quickly. It is advisable, however, to allow cast fittings to cool naturally to some extent before cleaning. All fluxes must be removed before inspection or testing.

12. Practice these joints in all positions until you have mastered the control of heat and the flow of the brazing material. You are now ready for a check test.

13. Use the same tests outlined for testing a soldering joint, pages 266 to 267. All the conditions and results are similar. Carry out both the surface inspection and the water pressure test.

Brazing Pipe

For tube or pipe 1 inch or larger in diameter, it is difficult to bring the whole joint up to the proper temperature at the same time. A double-tip torch maintains the proper temperature over the larger areas. A mild preheating of the whole assembly, both pipe and fitting, is recommended for the larger sizes. The heating then can proceed as outlined in the steps given.

If difficulty is encountered in getting the entire joint up to the desired temperature at the same time, a portion of the joint is heated to the brazing temperature and the alloy applied, Fig. 10-28. The joint is divided into sectors, and each is given individual treatment. The size of the pipe and fitting determine how much of the fitting cup circumference can be heated and brazed successfully. At the proper brazing temperature, the alloy is fed into the joint, and then the torch is moved to the next sector. The process is repeated, overlapping the previous sector. This procedure is continued until the joint is complete all around.

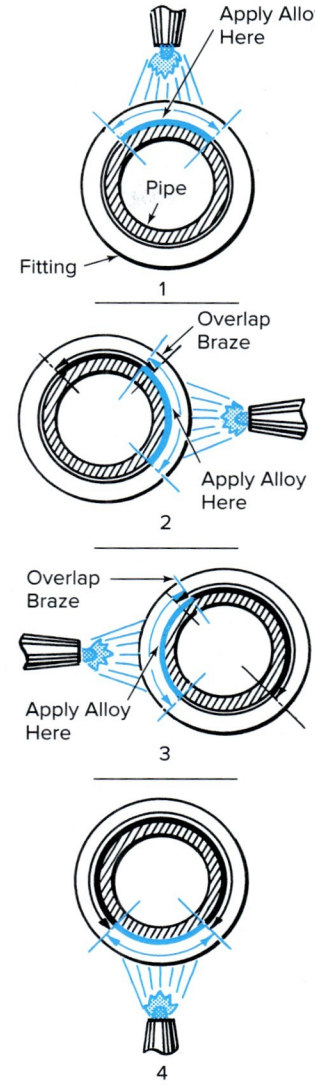

Fig. 10-28 Work in overlapping sectors around the pipe until you have completed the fillet.

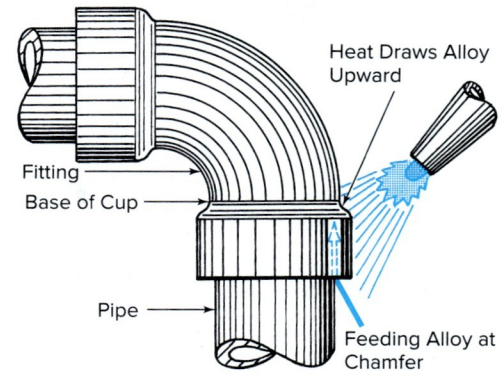

Fig. 10-29 Concentrate heat at the base of the cup to draw the alloy upward.

This may cause the alloy to run down the tube. If the alloy runs, take the heat away and allow the alloy to set. Then reheat the band of the fitting to draw up the alloy. Filler metal is added in the manner indicated in Fig. 10-29.

Brazed joints can be made in close quarters where screwed piping or flanges would be difficult or even impossible to handle. The torch can be formed so that you can reach hard-to-braze joints. Bend the extension on the torch or the tip, if it is a gooseneck type, to direct the flame where you want it, Fig. 10-30.

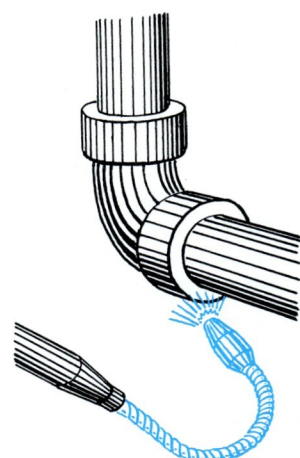

Fig. 10-30 Bend the flexible extension on your torch tip to put the flame where you want it.

Find the most comfortable place to stand or sit where you can keep the torch in a generally horizontal position. It may be helpful to rig a polished metal mirror that will enable you to observe the flux on the far side of the joint. The torch flame will supply illumination. It may also be necessary to protect all the surfaces around the joint. Wet rags, sheet metal, or heat-resistant sheeting may be used as a protective shield.

Horizontal and Vertical Joints

Joints in the horizontal and vertical positions can be brazed with the same ease as those in the flat position. This is possible since the filler material is drawn into the joint by capillary attraction and not the action of gravity. The major problem is overheating. If the surface of the bare metal is too hot, the brazing alloy will run out of the joint.

When making horizontal joints, it is preferable to start applying the filler metal at the top, then the two sides, and finally the bottom. Make sure that the filler alloy overlaps. On vertical joints, it does not matter where the start is made. If the opening of the socket of the fitting is pointed down, care should be taken to avoid overheating the tube.

Fittings That Can Be Brazed

Pipe fittings and valves of all types are commercially available for joining copper tubing and pipe by brazing. The following fittings require special techniques:

- *Couplings, Ls, Ts, and crosses.* These are available in a variety of sizes and types. The cups must be cleaned thoroughly and well-fluxed.

- *Fittings with both brazed and threaded ends.* Do not braze next to a screwed joint. Heat damages the compound that seals the threads, and the fitting may leak. Make up the brazed end first and then the threaded end, Fig. 10-31.

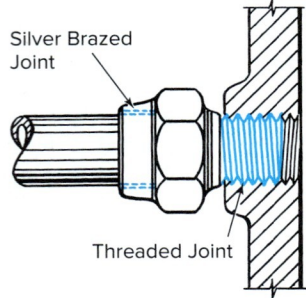

Fig. 10-31 Always make up the silver-brazed joint first, because heat will damage pipe-thread compound. See that necessary unions are provided for disassembly and servicing.

- *Unions.* Protect the ground sealing surfaces of unions with a generous supply of flux, Fig. 10-32. This will keep them from tarnishing and keep the surface from being damaged when heat is applied to make the joints. Do not play the torch flame directly on the ground surfaces. If the union is assembled during brazing, run the nut up only by hand.
- *Flanges.* Heat should be applied to the hub of the flange, Fig. 10-33. Heat warps a seating surface. Large flanges may need preheating from another source in order to make sure the flange is evenly heated all over. The entire assembly must be cooled slowly.
- *Return bends.* These must be free at one end while the brazing operation is carried on in order to provide for expansion.
- *Valves.* Do not remove the valve bonnet. It helps to stiffen the valve during brazing. The valve should be opened wide and then backed off just a little so that it will not be jammed in the open position. Wrap a wet cloth around the bonnet to protect it.

Each cup should be cleaned, fluxed, and brazed like any other fitting. A light coat of flux on the valve seat

JOB TIP

High Energy Beams and Computers in Welding

Interesting areas of future research in industrial laser processing will include:

1. Using low-power carbon monoxide (CO) lasers to focus and direct a pulsed electrical arc
2. Designing highly compact neodymium: yttrium-aluminum-garnet (Nd:YAG) focus heads to repair materials without disassembly
3. Creating computer simulations for process refinement
4. Micromachining and microjoining

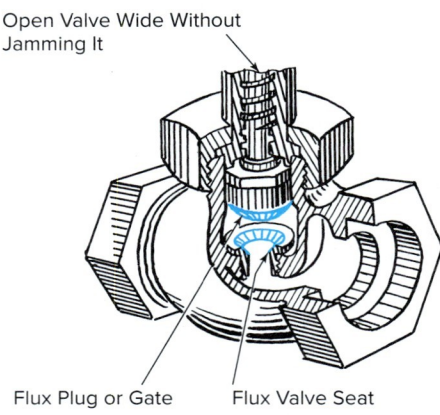

Fig. 10-34 Open a valve wide and then close it part of a turn. Flux the sealing surfaces to prevent damage while you are heating the valve.

and on the plug will protect these surfaces, Fig. 10-34. Do not direct the flame toward the body of the valve. Keep it directed near the ends and toward the pipe. Most of the heat should be at the base of the cup.

Aids to Good Brazing

Many of the points set out here have been mentioned elsewhere in the chapter. They are of such importance, however, that they are condensed and repeated for you to review.

Fluxing Fluxing is a very important operation. Oxidation forms surface deposits; flux reduces this action and soaks up the few oxides that may form.

Do not expect flux to clean a dirty metal surface. The surface must be cleaned by chemical or mechanical means. The flux will protect after the surface is cleaned.

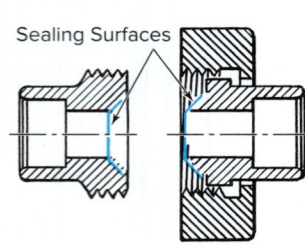

Fig. 10-32 Flux the sealing surfaces of a union to prevent them from tarnishing in the heat required during brazing.

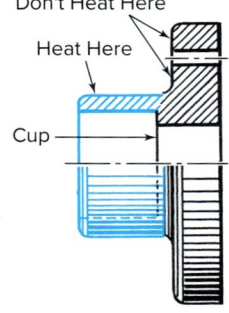

Fig. 10-33 When brazing a flange, heat the base of the cup—not the outer part of the fitting.

High pressure joints that take longer to braze than standard connections, and metals like copper, nickel, and steel, which have a greater tendency to oxidize than other metals, require a thicker coating of flux. One way to apply more flux is to keep the mixture thick.

Watch the flux closely during brazing. The clear, still, watery look it gets at about 1,100°F tells you that the metal is almost hot enough to melt a silver brazing alloy. Overheating destroys the effectiveness of the flux.

Support Keep a piping assembly well-supported so that there are no strains on it while you work. Allow for the expansion that results from heating the joint.

Poor Capillary Action If the alloy runs down the pipe, the joint may be too hot, the surfaces may not be clean or sufficiently fluxed, or the joint clearance may be too great, Fig. 10-35. Correction is obvious.

Timing Timing—knowing when to flux, when to heat, and when to braze—is extremely important. Brazing is a speedy operation. In fact, one of the worst things you can do is to take too long.

The time to put on flux is right after a surface has been cleaned. If the flux has dried out, put on more moist flux.

When you begin the heating operation, be careful not to heat the workpiece beyond the brazing temperature. It is important to know when to stop heating. When the flux shows that the metal is hot enough, start feeding the filler alloy. Work as rapidly as the alloy can be flowed in. Take the torch away as soon as the fillet is formed. Nothing is gained by more heating.

Fig. 10-36 Note all the various pipe sizes, elbows, tees, and transition bushing from one pipe size to another. This will be visually, leak inspected and sized to the according print.
Location: UA Local 400 Edward R. Bohnart

Two points, then, are most important: apply flux immediately after cleaning, and do not heat the assembly any longer than necessary.

Testing To make certain the joints soldered or brazed are "fit for purpose." That is, do not leak or break apart. To verify quality a sample test configuration such as shown in Fig. 10-36 may be used.

Supplies Sample joints soldered or brazed will need to be practiced to develop the skill to produce consistent acceptable results. These supplies can be expensive so adequate secured storage is required as shown in Fig. 10-37.

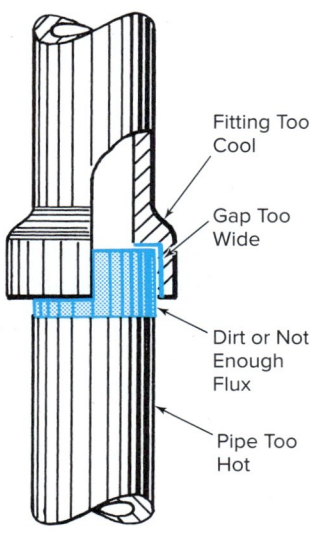

Fig. 10-35 Some of the things that will make silver brazing alloy flow the wrong way.

Aftercare Be sure to remove the flux after you have finished brazing. This may be done while the joint is hot, or you can wait until it cools. Scrub the joints with water. The inside of piping systems should be flushed with water to take away the flux inside.

Soldering and Brazing Principles and Practice: Jobs 10-J50–J51 **Chapter 10** 273

Fig. 10-37 Controlled storage for copper soldering and brazing supplies. Location: UA Local 400 Edward R. Bohnart

Disassembly It is often necessary to take brazed piping apart. First, brush flux around the area of the fillet at the edge of the cup. Put the pipe in a vise and heat the joint as you would to braze it. It takes about as much time and heat to take a joint apart as it did to braze it. When it is up to brazing heat, pull the tubing out of the fitting. If the alloy is melted, an easy pull does the job.

Pipe and fittings can be used again after they are taken apart. Wipe the molten alloy off the pipe and out of the cup before it sets. After the parts have been cleaned, they can be used again like new material.

CHAPTER 10 REVIEW

Multiple Choice

Choose the letter of the correct answer.

1. Copper is used in which of the following industrial applications? (Obj. 10-1)
   a. Saltwater, oil, vacuum, chemical, and air lines
   b. Refrigeration systems
   c. Low pressure steam lines
   d. All of these

2. What is the melting temperature variation that separates soldering from brazing? (Objs. 10-1 and 10-2)
   a. 440°F
   b. 540°F
   c. 740°F
   d. 840°F

3. What is capillary action? (Obj. 10-1)
   a. Repulsion of liquids out of large gaps
   b. Attraction of liquids into large gaps
   c. Repulsion of liquids out of small gaps between faying surfaces
   d. Attraction of liquids into small gaps between faying surfaces

4. One criterion for soldering is that _____. (Obj. 10-1)
   a. Both joined surfaces should melt
   b. The filler metal should be drawn into or held in the joint by capillary action
   c. The filler metal should flow like a liquid at room temperature
   d. The joint is strong at high temperature

5. Which of the following is not a type of solder? (Obj. 10-1)
   a. Tin-lead
   b. Cadmium-silver
   c. Silver-niobium
   d. Tin-zinc

6. Which of the following is not a property of flux for soldering and brazing? (Objs. 10-1 and 10-2)
   a. Liquid
   b. Solid
   c. Gaseous
   d. Cleans oils and grease

7. What is the clearance range required for capillary action to take place with soldering and brazing? (Objs. 10-1 and 10-2)
   a. 0.0003 to 0.0005 inch
   b. 0.003 to 0.005 inch
   c. 0.03 to 0.05 inch
   d. 0.3 to 0.5 inch

8. What type of torch is used for soldering? (Obj. 10-1)
   a. Oxyacetylene welding torch
   b. Air-acetylene torch
   c. Propane torch
   d. All of these

9. Which of the following lists a defect that may be encountered in a soldered joint and explains the cause of each? (Obj. 10-1)
   a. Unsoldered area: improper fluxing, improper heating, improper cleaning, improper fit up
   b. Flux inclusions: improper feeding of flux into joint, improper fit up
   c. Burning a hole through the material
   d. Both a and b

10. Brazing is one of the most recent of joining processes. (Obj. 10-2)
    a. True
    b. False
11. What are the melting temperature ranges of brazing alloys? (Obj. 10-2)
    a. 110 to 150°F
    b. 1,100 to 1,500°F
    c. 1.1 to 1.5°kF
    d. Both b and c
12. Which of the following is not an advantage of brazing copper tubing? (Obj. 10-2)
    a. Stronger than threaded joint
    b. Vibration does not loosen
    c. Cannot be taken apart
    d. Corrosion resistant
13. Which of the following is not a type of braze filler metal? (Obj. 10-2)
    a. BCu—copper
    b. BAlSi—aluminum
    c. BTi—titanium
    d. BAu—gold
14. Which of the following are used as brazing filler metals? (Obj. 10-2)
    a. Gold base
    b. Aluminum
    c. Copper phosphorus
    d. All of these
15. Which of the following are not fluxes for brazing? (Obj. 10-2)
    a. Acrylics
    b. Fluorides
    c. Chlorides
    d. Hydroxides of sodium and potassium
16. Which of the following is not a fuel gas for brazing? (Obj. 10-2)
    a. Acetylene
    b. Hydrogen
    c. Helium
    d. Natural gas
17. Which of the following is an appropriate brazing method? (Obj. 10-2)
    a. FB—furnace brazing
    b. IB—induction brazing
    c. DB—dip brazing, IRB—infrared brazing
    d. All of these
18. Resistance brazing has which of the following properties? (Obj. 10-2)
    a. The heat is provided by resistance to the flow of electricity through the parts to be brazed.
    b. It is useful for large areas and parts resistant to the flow of electricity.
    c. Flux is provided by a molten metal dip in a graphite crucible.
    d. The heat is produced by a lamp with 5,000 watts of radiant energy.
19. What is an important step to follow when brazing? (Obj. 10-2)
    a. Apply flux to the inside of the tube.
    b. Make sure flux has the consistency of water.
    c. Use massive metal supports with large areas of contact to suck the heat out of the fitting.
    d. Heat the tube starting at about 1 inch from the edge of the fitting, sweeping the flame around the tube in short strokes.
20. Poor fluxing can result in _____. (Obj. 10-3)
    a. Oxidation and surface deposits
    b. Leaking high pressure joints
    c. Nickel and copper oxidation
    d. All of these

Review Questions

Write the answers in your own words.

21. Describe the steps required to prepare a joint for soldering or brazing. (Objs. 10-1 and 10-2)
22. Name the functions that a soldering flux performs. (Obj. 10-1)
23. Describe a soldering flux. (Obj. 10-1)
24. In soldering, how may heat be applied? (Obj. 10-1)
25. What is the essential difference between soldering and brazing? (Objs. 10-1 and 10-2)
26. Explain the difference between solidus temperature and liquidus temperature. What is that difference called? (Obj. 10-2)
27. What are the most important factors in the design of a joint to be brazed? (Obj. 10-2)
28. When would you use resistance brazing? (Obj. 10-2)
29. What kind of fittings require special techniques? (Obj. 10-2)
30. Name some good aids to brazing. (Obj. 10-3)

INTERNET ACTIVITIES

Internet Activity A

Find out how to desolder by looking on the Internet. Then, explain to another classmate how to desolder.

Internet Activity B

Using your favorite search engine, find what artistic functions brazing can be used for.

Internet Activity C

Use your favorite search engine to find out what *babbit* solder is.

Table 10-6 Job Outline: Soldering and Brazing Practice: Jobs 10-J50 and J51

Job No.	Joint	Operation TB	Material Type	Diam. (in.)	Filler Rod/Flux Type	Size	Welding Position[1]	Text Reference
10-J50	Pipe fitting, socket	Solder copper tubing	Copper	½ to 1	50–50 or 95–5, appropriate flux	Available	2, 5, and 6	265
10-J51	Pipe fitting, socket	Braze copper pipe and tubing	Copper	½ to 1½	BCuP or BAg, appropriate flux	Available	2, 5, and 6	277

Note: It is recommended that students complete the jobs in the order shown.

[1] 2 = horizontal, 5 and 6 = multipositions.

Design credits: Cinema and movie icon: Denis Meshkov/123RF; and Illustration of Welding icons: Mr. Nachai Sorasee/123RF

UNIT 2

Shielded Metal Arc Welding

Chapter 11
Shielded Metal Arc Welding Principles

Chapter 12
Shielded Metal Arc Welding Electrodes

Chapter 13
Shielded Metal Arc Welding Practice: Jobs 13-J1–J25 (Plate)

Chapter 14
Shielded Metal Arc Welding Practice: Jobs 14-J26–J42 (Plate)

Chapter 15
Shielded Metal Arc Welding Practice: Jobs 15-J43–J55 (Plate)

Chapter 16
Pipe Welding and Shielded Metal Arc Welding Practice: Jobs 16-J1–J17 (Pipe)

11

Shielded Metal Arc Welding Principles

Shielded metal arc welding (SMAW) is manual arc welding in which the heat for welding is generated by an electric arc established between a flux-covered consumable metal rod called the **electrode** and the work. For this reason, the process is also called **stick electrode welding.** The combustion and decomposition of the electrode creates a gaseous shield that protects the electrode tip, weld puddle, arc, and the highly heated work from atmospheric contamination, Fig. 11-1. Additional shielding is provided for the molten metal in the weld pool by a covering of molten slag (**flux**).

> For video showing the fundamentals of SMAW, please visit www.mhhe.com/welding.

The weld metal is supplied by the metal core of the consumable electrode, and the physical properties vary with the electrode classification. Certain electrodes also have metal powder mixed with the electrode coverings. The shielding and filler metal largely control the mechanical, chemical, metallurgical, and electrical characteristics of the weld. Characteristics of electrodes for shielded metal arc welding are presented in Chapter 12.

Chapter Objectives

After completing this chapter, you will be able to:

11-1 List the percentage of usage of SMAW in the industry.

11-2 Name the components that make up the schematic representation of the shielded metal arc.

11-3 Know the maximum arc temperature of an SMAW electrode.

11-4 List the four constant current welding machines.

11-5 List the common type and uses of constant current welding machines.

11-6 Name the power supply ratings.

11-7 Name the characteristics of the four basic types of welding machines.

11-8 Choose the correct cable size based on the application.

11-9 List the welder's safety equipment.

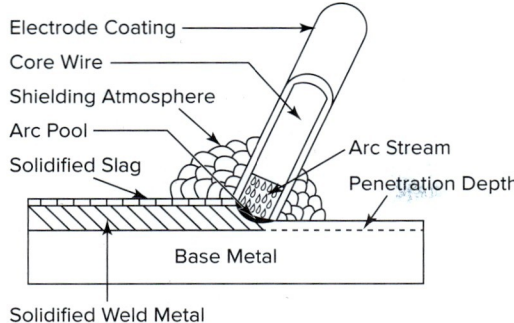

Fig. 11-1 Schematic representation of the shielded metal arc. Adapted from AWS Standard Terms and Definitions, AWS3.0:2010, page 12

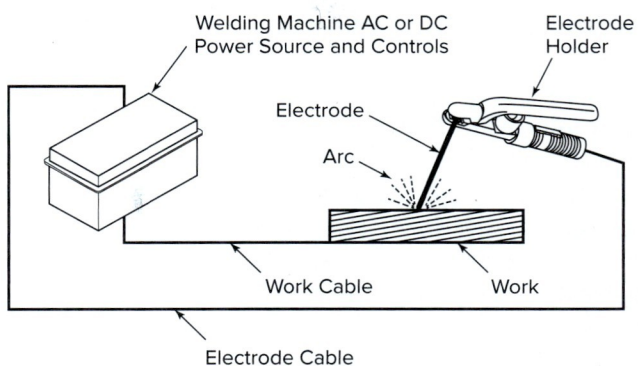

Fig. 11-2 Elements of a typical electric circuit for shielded metal arc welding. Adapted from Welding Handbook, 9/e

Process Capability

Shielded metal arc welding is one of the most widely used of the various electric arc welding processes. Rather than replacing shielded metal arc welding, the popularity of the other processes is causing the total amount of welding performed to increase.

According to a recent market study of equipment and process usage among the industrial market segments, SMAW represents 42 percent, GMAW/FCAW 34 percent, GTAW 13 percent, SAW 9 percent, and others 2 percent of the total welding. However, when looking at the dominant markets using SMAW, 56 percent of construction welding is SMAW and 61 percent of maintenance and repair is SMAW.

Equipment for shielded metal arc welding is less complex, more portable, and less costly than that for other arc welding processes. This type of welding can be done indoors and outdoors, in any location, and in any position.

Shielded metal arc electrodes are available to match the properties and strength of most base metals. Metals most easily welded by the shielded metal arc process are carbon and low alloy steels, stainless steels, and heat-resisting steels. Cast iron and the high strength and hardenable types of steel can also be welded providing the proper preheating and postheating procedures are employed. Copper and nickel alloys can be welded by this process although gas metal arc (GMAW, also known as MIG/MAG) welding and gas tungsten arc (GTAW, also known as TIG) welding are preferred.

The shielded process is not used for welding the softer metals such as zinc, lead, and tin, which have low melting and boiling temperatures.

Processes that employ continuously fed electrode wires (gas metal arc) or no filler metal (gas tungsten arc) have some advantage in deposition efficiency. After each pass, the slag formed by the shielded metal arc electrode must be removed. Deslagging causes some loss in deposition efficiency.

Operating Principles

Shielded metal arc welding equipment sets up an electric circuit that includes the welding machine, the work, electric cables, the electrode holder and electrodes, and a work clamp, Fig. 11-2. The heat of the electric arc brings the work to be welded and the consumable electrode to a molten state. The heat of the arc is intense. Temperatures as high as 9,000°F (5,000°C) have been measured at its center.

Welding begins when an electric arc is started by striking the work with the electrode. The heat of the arc melts the electrode and the surface of the base metal next to the arc. Tiny globules of molten metal form on the tip of the electrode and are transferred by the arc into the molten pool on the work surface. In flat welding the transfer is induced by the force of gravity, molecular attraction, and surface tension. When the welding is in the vertical and overhead positions, electromagnetic, along with molecular attraction and surface tension are the forces that cause the metal transfer. Because of the high temperature of the arc, melting takes place almost instantaneously as the arc is applied to the weld. After the weld is started, the arc is moved along the work, melting and fusing the metal as it progresses.

Welding Power Sources

The successful application of the welding process requires proper tools and equipment. Welders will find it impossible to do good work if they have inferior equipment, if they lack equipment, or if they choose the wrong type of equipment for a particular job. The welding industry has spent many years in experimentation and practical application, as well as millions of dollars in developing the equipment necessary to carry on welding processes.

Each welding process reaches its maximum efficiency when used with the power source designed for it. Each type of power source has certain fundamental electrical differences that make it best suited for a particular process.

A welding machine must be able to meet changing arc load and environmental conditions instantly. It must deliver the exact amount of electric current at precisely the right time to the welding arc in order to control arc behavior and make welds of consistent quality.

The arc welding process requires enough electric current (measured in *amperes*) to produce melting of the work and the metal electrode and the proper voltage (measured in *volts*) to maintain the arc. Depending on their size and type, electrodes require 17 to 45 volts and approximately 100 to 500 amperes. The current can be alternating or direct, but it must be provided through a source that can be controlled to meet the many conditions encountered on the job.

Welding machines are available in a wide variety of types and sizes to suit the demands of different welding processes, operations, and types of work. There are a number of different makes, each having its own particular design. The control panel of each model is different in many respects. The selection of a particular machine may also depend upon cost, portability, and personal preference.

Welding power sources are also known as *power supplies* and *welding machines*. All machines may be classed by (1) *output slope*, whether constant current or constant voltage, and (2) *power source type*, such as transformer, transformer-rectifier, inverter, or generator.

Types of Output Slope

Welding machines have two basic types of output slope—*constant current*, which is also referred to as *variable voltage*, and *constant voltage*, which is also referred to as *constant potential*.

Output slope, sometimes called the *volt–ampere characteristic* or *curve*, is the relationship between the output voltage and output current (amperage) of the machine as the current or welding workload is increased or decreased, Fig. 11-3.

Output slope largely determines how much the welding current will change for a given change in load voltage. Thus, it permits the welding machine to control the welding heat and maintain a stable arc.

The output slope of the machine indicates the type and amount of electric current it is designed to produce.

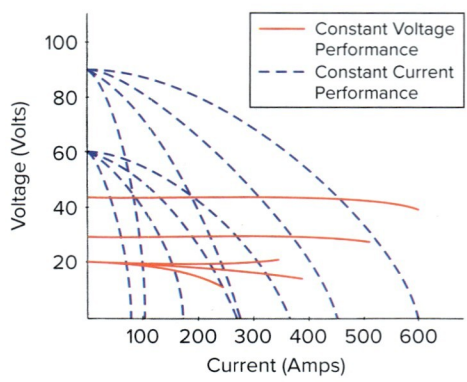

Fig. 11-3 Typical output slopes for constant current and constant voltage power sources. Source: From Welding Handbook, 9/e

Each arc welding process has a characteristic output slope:

- SMAW and GTAW require a steep output slope from a constant current welding machine. Constant current is necessary to control the stability of the arc properly.
- GMAW and FCAW require a relatively flat output slope from a constant voltage power source to produce a stable arc.
- Submerged arc welding is adaptable to either slope, depending on the application and extra control equipment.

Some d.c. welding machines, especially the inverter type, may combine the two basic types of output slope in a single unit. By turning a selector switch on the machine, a steep slope (constant current) or a flat slope (constant voltage) can be produced.

Types of Power Source

Constant current welding machines are also classified by the manner in which they produce welding current. There are four general types:

- ***Engine-driven generators*** These are powered by a gas or diesel combustion engine. These machines can also be found with an a.c. or d.c. electric motor as their power source; however, they are no longer being manufactured and are rarely found in industry today. Both may produce d.c., a.c., or a.c.-d.c. welding current.
- ***Transformer-rectifiers*** These use a basic electrical transformer to step down the a.c. line power voltage to a.c. welding voltage. The welding voltage is in turn passed through a rectifier to convert the a.c. output to d.c. welding current. Transformer-rectifiers may be either d.c. or a.c.-d.c. machines.

- *A.C. transformers* These use an electrical transformer to step down the a.c. line power voltage to a.c. welding voltage.
- *Inverters* This type of power source increases the frequency of the incoming primary power, thus providing for a smaller-size machine and improved electrical characteristics for welding, such as faster response time and more control for pulse welding. Inverters may be either constant current, constant voltage, or both. They can produce a.c. or d.c. welding current.

It is important to select the right power source for each job, Fig. 11-4. A study of the job will usually indicate whether alternating or direct current should be used. This will determine the selection of either an a.c. or d.c. power source or a combination a.c.-d.c. power source. For shielded metal arc welding and gas tungsten arc welding, the performance characteristics of any of these four power sources must be the constant current rather than the constant voltage type. The constant voltage machine is preferred for semiautomatic and automatic arc welding processes such as gas metal arc.

Constant Current Characteristics

Constant current welding machines are primarily used for shielded metal arc welding and gas tungsten arc welding because current remains fairly constant regardless of changes in arc length. They are also called *drooping voltage,* variable voltage, or *droopers* because the load voltage decreases, or droops, as the welding current increases.

These machines can also be adapted to spray arc MIG/MAG welding (d.c.), flux cored arc, or submerged arc welding (a.c. or d.c.) with a special arc voltage sensing control and variable speed wire feeder. This control senses the changes in arc voltage and corresponding arc length and varies the wire feed speed to maintain a constant arc length voltage. Constant current welding machines, however, are not as satisfactory for spray arc MIG/MAG welding as constant voltage machines because of the poorer arc starting and more complicated equipment.

Output Slope Constant current welding machines have a steep output slope and are available in both d.c. and a.c. welding current. The steeper the slope of the volt–ampere curve within the welding range, the smaller the current change for a given change in arc voltage, Fig. 11-5. Some jobs require a steep volt-ampere curve (*A*) while other jobs require the use of a less steep volt-ampere curve (*B*). The constant current power source enables the welder to control welding current in a specific range by simply changing the length of the arc as welding progresses.

Open Circuit and Arc Voltage **Open circuit voltage** is the voltage generated by the welding machine when no welding is being done. The machine is running idle. **Arc voltage** is the voltage generated between the molten tip of the electrode and the surface of the molten weld pool during welding. It is also referred to as arc length. **Load voltage** is the voltage at the output terminals of the welding machine when an arc is going. Load voltage is the combination of the arc voltage plus the voltage drop in the welding circuit. Open circuit voltage generally runs between 50 and 100 volts. Arc voltage runs between 18 and 36 volts. The two open circuit voltages expressed in Fig. 11-5 are approximately 90 volts (*A*) and slightly below 50 volts (*B*). The arc voltages range from 32 (long arc) to 22 volts (short arc).

Open circuit voltage drops to arc voltage when the arc is struck, and the welding load is on the machine. Arc voltage is determined by the arc length held by the welder and the type of electrode used. When the arc is lengthened, the arc voltage increases and the current decreases. If the arc is shortened, the arc voltage decreases and the current increases. Note that in Fig. 11-5, there is a difference of 40 amperes between the long arc and the short arc. Just how much change takes place depends on the open circuit voltage setting. Although the total current range between the long arc and the short arc in Fig. 11-5 is 40 amperes, the range that the two open circuit voltages have in common is only 15 amperes. The minimum open circuit voltage (*B*) produces a much wider current range than the maximum circuit voltage (*A*).

Open circuit voltage on constant current machines is higher (about 80 volts) than on most constant voltage machines (about 50 volts). Arc voltage depends on the physical arc length at the point of welding and is controlled by the welder in shielded metal arc welding and gas tungsten arc welding. Arc voltage is much lower than open circuit voltage: normally about 20 to 30 volts for shielded metal arc welding, 10 to 15 volts for gas tungsten arc, and 25 to 35 volts for submerged arc.

There is no voltage adjustment dial on constant current welding machines. However, open circuit voltage can be adjusted with the fine current adjustment dial on welding machines having dual control.

In shielded metal arc welding and gas tungsten arc welding, the welder cannot hold the arc length truly constant. Rather wide changes in arc length and, therefore, a voltage change, will occur. With a constant current machine, relatively small changes in current (amperage) result from changes in arc length. Thus the welding heat and burnoff rate of the electrode are influenced very little, and the welder is able to maintain good control of the weld pool and the welding operation.

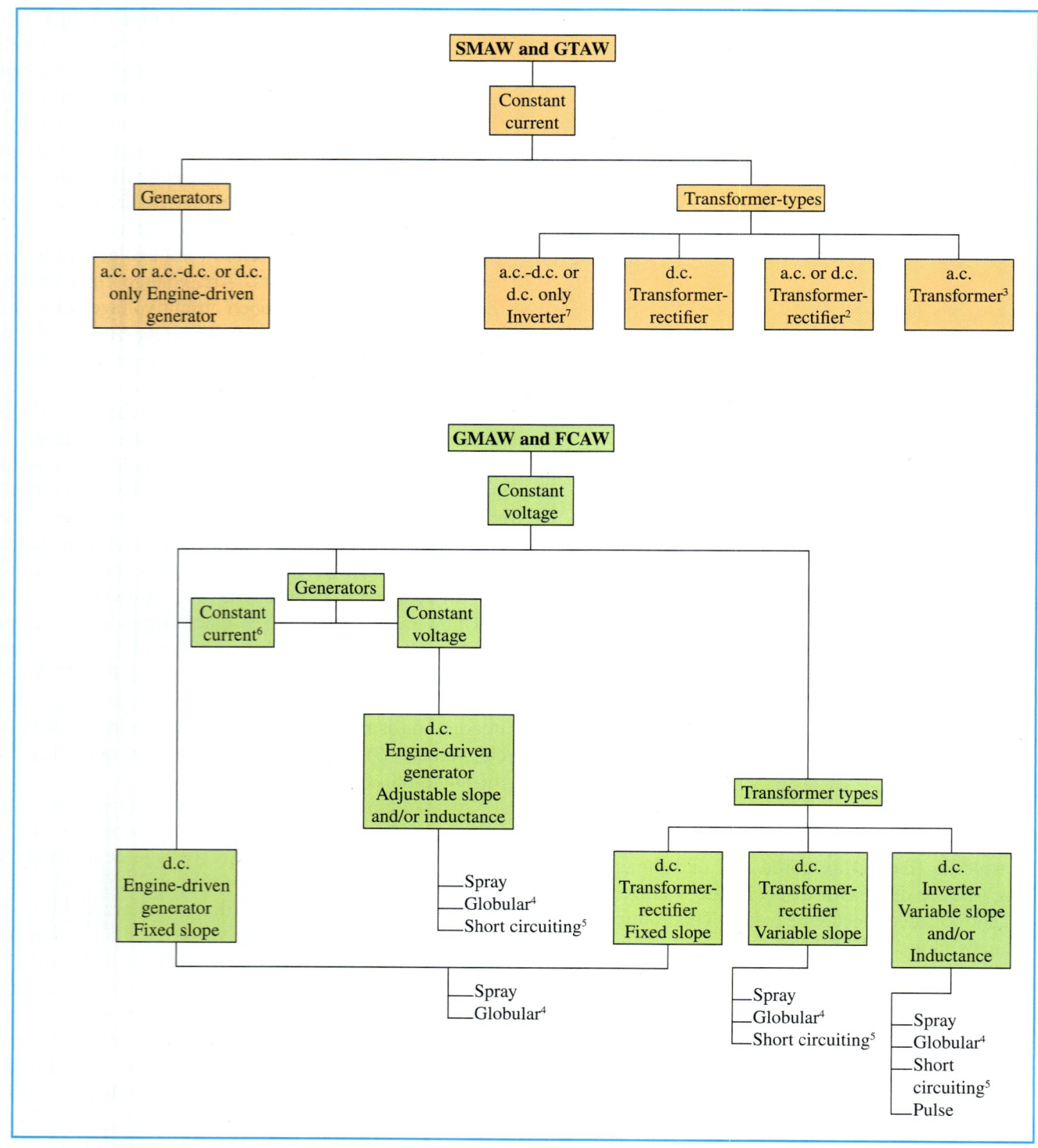

Fig. 11-4 Common types and uses of arc welding machines.[1] Source: Republic Steel Co.

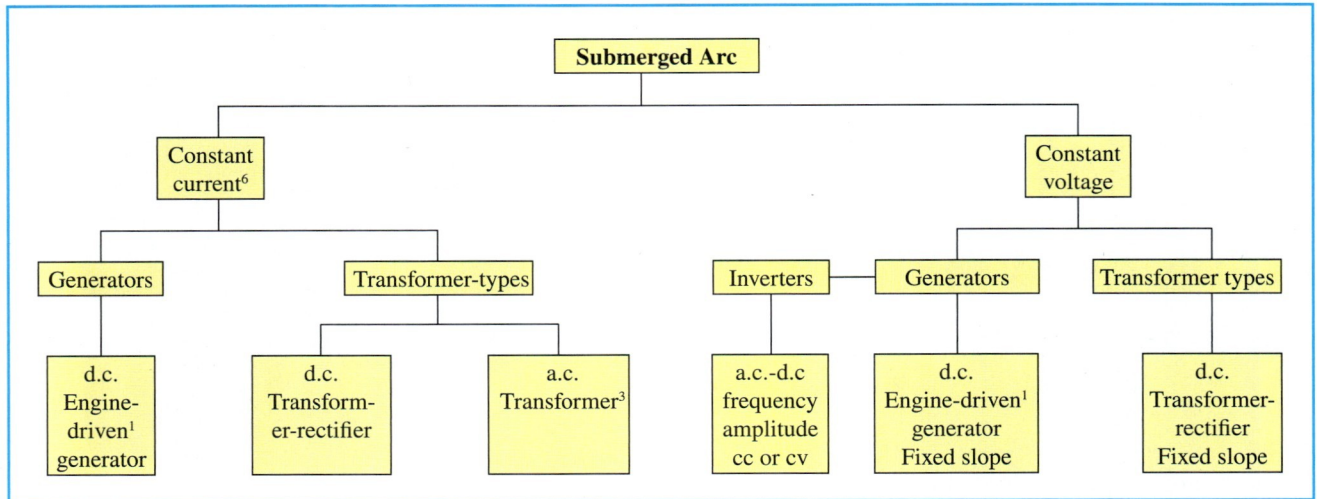

Fig. 11-4 (Concluded)

[1]This figure shows the most common uses of welding machines for processes noted. Other combinations are possible.

[2]The a.c. can be used only on single-phase line power. The d.c. is most commonly used on three-phase power, except for most a.c.-d.c. units.

[3]Single phase only.

[4]Mild steel carbon dioxide welding—not normally used for stainless.

[5]Inductance usually added. Some shorting-arc machines have a fixed slope set at a steeper angle with adjustment made by varying the inductance.

[6]Used with voltage-sensing control and variable speed wire-fed unit.

[7]Inverter—a.c. or d.c. may be used.

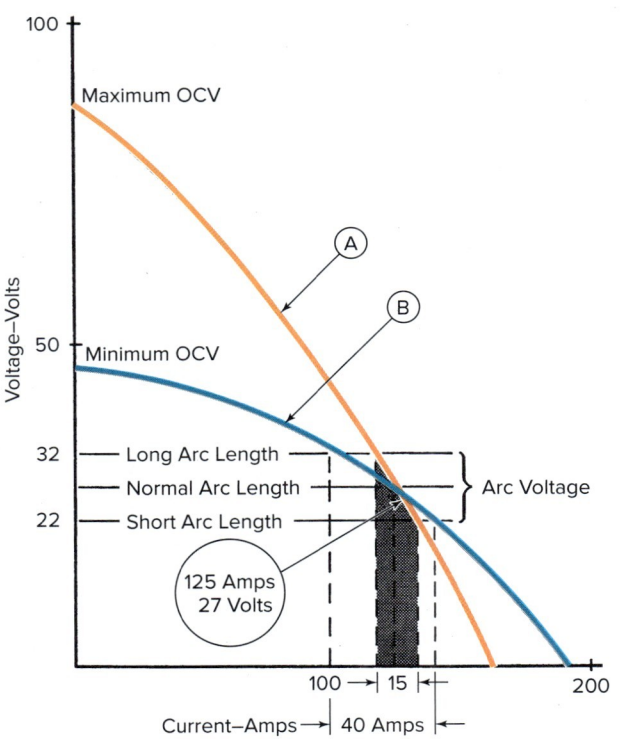

Fig. 11-5 Two possible output slopes for a constant current welding power source. The steep slope (A) gives minimum current variation. The flatter slope (B) indicates the variation in current by changing the length of the arc. Note that changing the open circuit voltage (OCV) changes the slope of the curve.

Machines for Shielded Metal Arc Welding

Engine-Driven Generator Welding Machines

Generators are classified by the type of engine that drives the generator. The common generator welding machine consists of a hydraulic or internal combustion engine, a d.c. generator, and an exciter connected to it. The engine drives the generator that provides the current for welding.

These types of engine-driven generators are very popular in the field where electric power supply is not readily available. They are typically driven by a gasoline or diesel internal combustion engine. Generators used in the field are illustrated in Figs. 11-6, 11-7, and 11-8.

The engine-driven generators have a very wide use due chiefly to their desirable characteristics for many arc welding operations.

- They have a forceful or soft penetrating arc.
- They are versatile; they can be used to weld all metals that are weldable by the arc process.
- They are flexible, with the proper electrode, processes and gases can be used for many applications.
- They can also function as generators to power lights or other tools like plasma arc cutting equipment.

Fig. 11-6 Engine-driven d.c. generator for SMAW. Note remote control and engine gauges. The Lincoln Electric Co.

Fig. 11-8 A versatile engine-driven generator (alternator) capable of SMAW, GTAW, GMAW, FCAW, and PAC, while providing power for a grinder for weld-joint preparation. Miller Electric Mfg. Co.

Fig. 11-7 Engine-driven generator (alternator) for SMAW. Note a.c.-d.c. polarity, wire, stick switch, and 240V/120V auxillary power receptacles are located behind protective doors. Miller Electric Mfg. Co.

- They can also have on board air compressors for doing air carbon arc gouging.
- They are durable and have a long machine life where portability and limited electrical power are the key requirements.

Sizes Machine sizes are determined on the basis of amperage (current output). They range from 100-ampere rated machines for home use to more than 750-ampere rated machines for use with automatic submerged arc welding equipment.

In the manual welding operation, the machine is not required to provide current continuously. The machine is idle part of the time while the welder changes electrodes, adjusts the heat setting, sets up the work, cleans the welds between passes, and shifts welding positions. For machines rated at 200 amperes or more, the output rating is based on a 60 percent duty cycle. This means that the machine can deliver its rated load output for six out of every ten minutes. Machines rated at 100 amperes or less are usually rated on the basis of a 50 percent duty cycle. In many cases, these are portable machines that are not used on a continuous basis. Fully automatic power supply units are usually rated at a 100 percent duty cycle.

A welding machine will deliver a higher current value than its rating indicates. Thus, a welder rated at 200 amperes will deliver from 200 to 250 amperes for welding. The actual output is always somewhat higher than the rated output. However, a machine should not be used at or beyond its maximum capacity over an extended period. A larger machine is indicated. These machines provide a steady supply of current over a wide range of welding voltage.

Maintenance The engine-driven generator machines are very durable and give long service if properly maintained. Repair costs may be somewhat higher than for other types of welding machines because many moving parts provide the opportunity for wear and malfunction. The control rheostat should be inspected, cleaned frequently, and replaced when necessary. Brushes need frequent inspection for wear and must be replaced when worn down. Brushes should ride free in the brush holders.

The manufacturer's owner's manual should be followed for all maintenance items.

Controls of the Engine-Driven Generator

Start and Stop Controls These controls are for the purpose of starting and stopping the engine that drives the generator of the welding machine.

Polarity Switch Electrode negative and electrode positive are used in d.c. welding. When welding with **d.c. electrode negative (DCEN)**, the electrode is connected to the negative terminal of the power source, and the work is connected to the positive terminal. When welding with **d.c. electrode positive (DCEP)**, the electrode is connected to the positive terminal of the power source, and the work is connected to the negative terminal.

The polarity switch (if provided) changes to either electrode positive or electrode negative. This controls the direction of current flow. In the electron theory, it is said to flow from negative to positive.

Volt–Ampere Meters Meters sometimes serve a dual purpose. On some machines, they indicate polarity as well as current values in volts and amperes. Some machines have individual meters for volts and amperes. Others have single meters that indicate both the volt and ampere readings. A switch or button must be engaged to obtain individual readings on a combination volt-ampere meter. In order to obtain a true indication of the volt and ampere values being used, the welder must have some other person check the meters while they are welding. Digital meters such as these will hold the display for 3 seconds after the arc is broken.

Current Controls Students will better understand the function of the current controls if they keep in mind that the ampere rating can be compared with the amount of water flowing through a pipe. Amperage is the quantity of current and determines the amount of heat produced at a weld. Voltage is like pressure behind the water in the pipe. It is a measure of the force of the current. Voltage determines the ability to strike an arc and maintain its consistency. If voltage is too high, the arc is too harsh and may produce arc blow. On the other hand, if it is too low, it is very difficult to maintain the arc.

Dual control welding generators provide a wide range of welding current selection and make it possible to provide precise heat values. Good arc control is possible because the output slope of the welding machine can be readily adjusted. A wheel or lever permits the welder to set the current at a definite rate.

There are two types of dual control generators: tapped-step current control and continuously variable current control.

Dual Tapped-Current Control On generators with fixed current steps, the coarse adjustment dial selects the current range, called *steps* or *taps*. It is impossible to secure a current value between two steps by setting the dial between them. The dial must be set squarely on the desired step to prevent burning out the control device and taps.

The fine adjustment dial trims the current between the steps. Whether the fine adjustment dial is to be set high or low depends on the type and size of the electrode, the thickness of the metal, whether a soft or digging arc is required, the arc starting, any restricting characteristics, and the position of welding.

Dual Control In constant current or constant voltage generators with dual control, the **coarse adjustment** switch adjusts current. (Some manufacturers refer to the coarse adjustment dial as the *job selector*, mode switch, or *electrode selector*.)

The fine dial adjusts the current (amperage) or the voltage. The operator adjusts the output slope/inductance-dig for a given current setting by manipulating both the coarse adjustment switch and dig/inductance control dial.

A wheel or knob on both the amperage and the voltage setting devices gives the welder continuous control of both the amperage and voltage ratings. This separate, complete control of both voltage and amperage makes it possible to set up any volt-ampere combination within the output range of the particular machine being used—from minimum voltage at minimum current to maximum voltage at maximum current. Thus, a welding arc having various characteristics may be secured.

Remote Control Welding machines may be obtained that are equipped with remote-control, current-control units. The welding machine may be installed in any remote location, and the welder may adjust the current without leaving the welding work site. This is timesaving on work where it is necessary for the welder to leave the fabrication to readjust current.

Air Filters Wear in arc welding machines is costly since it necessitates not only the cost of replacement parts and labor, but also the loss of production due to nonuse of the machine. Bearing and engine wear is critical and may be reduced through the use of air filters. The filter is replaced or cleaned regularly with high pressure air, a commercial solvent, or steam.

D.C. Transformer-Rectifier Welding Machines

Transformer-rectifier welding machines, Figs. 11-9 and 11-10, have many designs and purposes. Flexibility is one reason for their wide acceptance by industry. These machines can deliver either DCEN or DCEP. They may be used for stick electrode welding, gas tungsten arc welding, submerged arc welding, multiple-operator systems, and stud welding.

All transformer-rectifier machines have two basic parts: (1) a transformer for producing and regulating the alternating current that enters the machine and (2) a rectifier that converts the alternating current to direct current. A third important part is a ventilating fan. The fan keeps the rectifier from overheating and thus shortening its life.

The simple design of the transformer-rectifier improves arc stability and makes it easy to hold a short arc. The arc itself is soft and steady. Continuous current control is available over the range of the machine being used. Transformer-rectifiers have no major rotating parts so they consume little power while idle and operate quietly.

A.C.-D.C. Transformer-Rectifier Welding Machines

An a.c.-d.c. welding machine is used for stick electrode welding and gas tungsten arc welding in which alternating current is required for most nonferrous metals and direct current is required for stainless steel. The a.c.-d.c. welders have the versatility of the d.c. welder and the special advantages of the a.c. welder in a single machine (Figs. 11-11 and 11-12).

Fig. 11-9 Industrial 3 phase SMAW welding machine. Note controls and volt-amp metering. The Lincoln Electric Co.

Fig. 11-11 A.C.-D.C. SMAW machine. It is electrically controlled and has optional remote control capability. Miller Electric Mfg. Co.

Fig. 11-10 Industrial 3 phase transformer rectifier. Note louvers on front of case to allow cooling air flow. Miller Electric Mfg. Co.

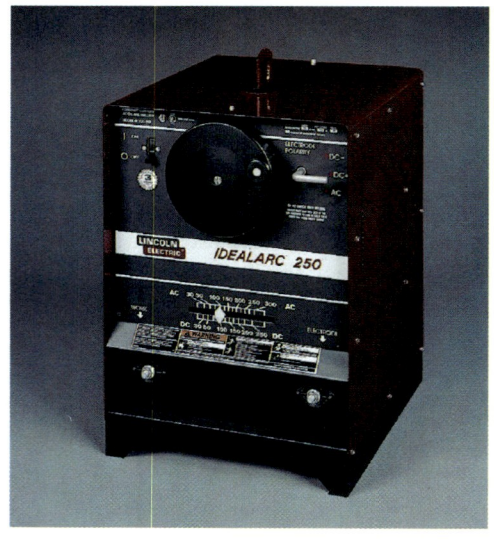

Fig. 11-12 A.C.-D.C. SMAW machine. It is mechanically controlled. Note the large handle wheel on the front panel. The Lincoln Electric Co.

The a.c.-d.c. machines permit the welder to select either alternating or direct current and electrode negative or electrode positive. They are essentially a.c. transformer-rectifiers. A switch permits the welder to use only the transformer part of the machine for a.c. welding. By flipping a switch or turning a dial, output current is directed through the rectifier. This converts it to d.c. welding. The rectifier circuit is similar to all other transformer-rectifier machines.

High frequency arc-starting devices, water and gas flow controls, balance controls for a.c. operation, and other welding controls such as remote control are often built into the machine. Controls for two models are identified in Figs. 11-13 and 11-14. Figure 11-15 shows one type of a.c.-d.c. welder in use.

A.C. Transformer Welding Machines

The transformer type is the most popular a.c. welding machine, Figs. 11-16 and 11-17. The function of the transformer is to step down the high voltage of the input current (for example, 120, 220, or 440 volts) to the high amperage, low voltage current required for welding.

The characteristics of a.c. current are such that there is a reversal of current each $\frac{1}{120}$ of a second. This constant reversal of current keeps the effect of the magnetic field at a minimum, thus reducing the tendency to arc blow. Arc blow results in excessive spatter and interferes with metal transfer.

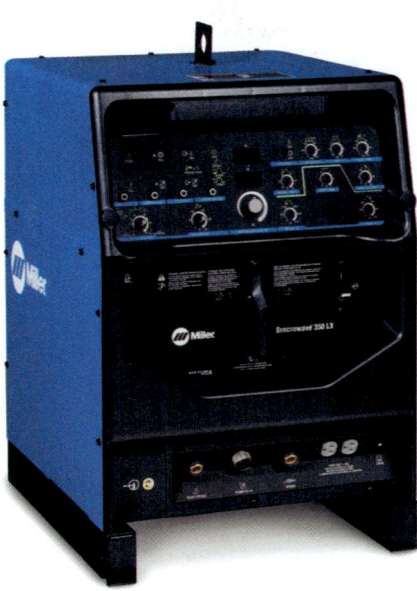

Fig. 11-14 A 300 amp a.c.-d.c. GTAW/SMAW machine. This machine has square wave output which is advantageous for TIG and stick welding. Miller Electric Mfg. Co.

Fig. 11-15 A Lincoln a.c.-d.c. transformer-rectifier welding machine. In this model, which has a 300-ampere capacity, a handwheel permits continuous control of current. The a.c.-d.c. current selector is at the top right. A direct-reading current indicator is at the bottom of the machine. The Lincoln Electric Co.

Fig. 11-13 A portable SMAW/GTAW welding machine. Note cart, storage, foot control TIG Torch, spare torch parts, work clamp, and flow meter. The Lincoln Electric Co.

While the arc may be somewhat more difficult to start than one produced by direct current, the absence of arc blow and higher voltage makes the arc easy to hold once it is obtained. This condition also permits the use of larger electrodes, resulting in faster speeds on heavy materials.

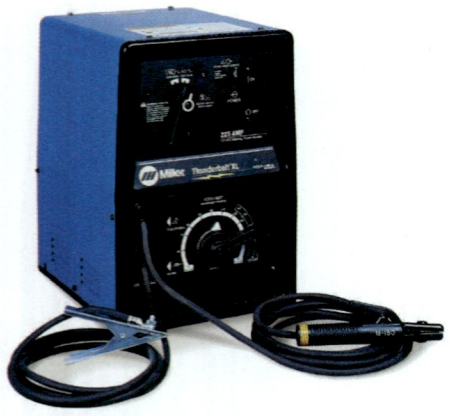

Fig. 11-16 A simple SMAW welding machine with mechanical control of the welding output. This hand lever permits continuous control of current. *Miller Electric Mfg. Co.*

Fig. 11-18 A 450 amp d.c. inverter power source. This machine is equipped with a microprocessor to control the various process output capability. *Miller Electric Mfg. Co.*

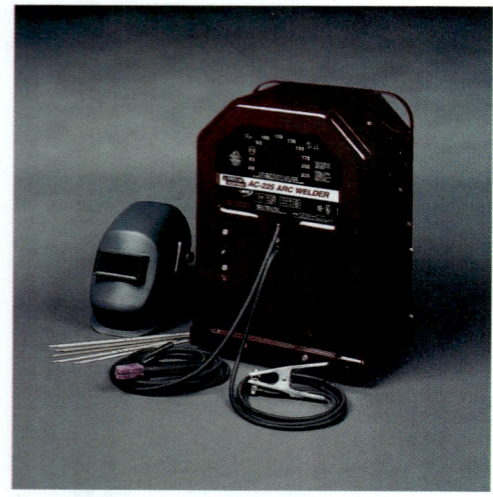

Fig. 11-17 A simple SMAW welding machine with mechanical control of the welding output. Note the switch type stepped current control. *The Lincoln Electric Co.*

Fig. 11-19 A 275 amp d.c. inverter power source. One person portable for use in construction applications. *The Lincoln Electric Co.*

Other advantages of a.c. power sources include lower cost, decreased power consumption, high overall electrical efficiency, noiseless operation, and reduced maintenance.

Best welding performance can be obtained in the flat and horizontal positions on groove and fillet welds with certain types of heavily coated shielded arc electrodes. An a.c. power source is especially suited for heavy work.

D.C. and A.C.-D.C. Inverter Welding Machines

The inverter welding machines are lightweight. Because of this, the inverter is portable and versatile. Inverters may be either constant current, constant voltage, or both constant current and constant voltage (c.c.-c.v.). This means that one machine can perform several different processes, Figs. 11-18 and 11-19.

Cost Comparisons: Arc Power Sources

There are three main areas of cost in considering the type of machine to purchase: the cost of purchasing the equipment, operating efficiency, and maintenance. A recent check indicated all types cost nearly the same.

It is in the second area—operating efficiency—that real cost advantages appear. In a comparative study, transformer-rectifiers operated at 64 to 72 percent efficiency and inverters operate at 85 percent efficiency.

Thus, the cost of operating a transformer-rectifier machine is higher than the more efficient inverter.

The third area for cost comparison is maintenance. Maintenance on engine-driven generator machines requires replacing brushes, cutting the commutator, lubricating bearings, and engine wear and tear. In a transformer-rectifier type and inverter, on the other hand, there are no moving parts except for the cooling fan. For practical purposes, the transformer-rectifier and inverter machine has no maintenance problem because it has no rotating parts to wear. However, the inverter power sources rely heavily upon the use of printed circuit boards. Like most highly electronic products a failure in a PCB can be costly.

Consult Table 11-1 for a comparison of the overall characteristics of the three basic types of welding machines.

Multiple-Operator Systems

On construction jobs, in steel mills, and in shipyards, it is often necessary for a large number of welders to work within a limited area. If single-operator machines are used for each welder, the space becomes too crowded for efficient work and may be highly dangerous. A multiple-operator power unit, Fig. 11-20, can be installed away from the work site and be connected to control panels located close to the welding operator. A large number of welding stations may be supplied from each unit. When using direct current, all welders must weld with the same polarity.

The welding current may be supplied by static-rectifiers, or transformers. Most installations are direct current. Power sources may range in size from 600 to 2,500 amperes. A multiple-operator installation costs less, saves space and cable, lowers operating costs, and requires less maintenance.

Fig. 11-20 A multiple operator power source capable of providing eight areas from one primary connection. Miller Electric Mfg. Co.

Table 11-1 Characteristics of Three Basic Types of Welding Machines (300-ampere capacity)

Characteristics	a.c.-d.c. Transformer-Rectifier	a.c. Transformer	d.c. Inverter
Weight (lb.)	427	573	76
Floor area (in^2)	440	576	300
Volume (ft^3)	12.8	16.7	2.95
Efficiency (rated load)	84.70	64.70	85
No-load input (W)	576	1,060	
Safety	Fair	Good	Good
Electrode choice	Least	Good	Good
Electrode cost	Premium	Standard	Standard
Current change with input voltage	Large	Most	None
Current change with warm up	Least	Low	None
Arc blow	Low	Moderate	Moderate
Noise	Low	Low	Low
Maximum output (%)	140	138	150
Welder life	Best	Good	Good
Weld quality	Good	Good	Excellent
Maintenance cost	Least	Medium	High

Power Supply Ratings

Welding power sources are rated by standards that have been set by The National Electrical Manufacturers Association (NEMA) and the Occupational Safety and Health Administration (OSHA). These standards provide guidelines for the manufacture and performance of power sources.

Welding power sources are rated by current output, open circuit voltage, duty cycle, efficiency of output, and power factor.

- **Current Output** Welding machines are rated on the basis of current output in amperes. Amperage may range from 200 amperes or less for light or medium work to over 2,000 amperes for submerged arc welding.

- **Open Circuit Voltage** The maximum allowable open circuit voltage of machines used for manual welding is 80 volts for a.c. or a.c.-d.c. machines, while d.c. machines with very smooth output (less than 2 percent ripple) can be 100 volts. Some constant current machines used for machine (automatic) welding are rated up to 125 open circuit voltage. Constant voltage types are normally rated from 15 to 50 open circuit voltage.

- **Duty Cycle** The duty cycle is the percentage of any given 10-minute period that a machine can operate at its rated current without overheating or breaking down.

  A machine that can be used at its rated amperage on a continuous basis is rated as 100 percent duty cycle. Continuous, automatic machine welding requires this type of machine.

  A machine that can be used at its capacity 6 out of every 10 minutes without damage is rated as 60 percent duty cycle. This type of machine is satisfactory for heavy SMAW and GTAW.

  Machines used for welding in small shops or on farms for light repair usually have a 20 percent duty cycle.

- **Efficiency** Efficiency is the relationship of secondary power output to primary power input and is indicated in percent. It is determined by losses through the machine when actually welding at rated current and voltage.

  Most transformer-rectifiers average about 70 percent and inverters 85 percent.

- **Power Factor** The power factor is a measure of how effectively the welding machine makes use of a.c. primary line power. The primary power used is divided by the amount total drawn and is usually expressed in percent.

  Three-phase d.c. transformer-rectifiers have a power factor of about 75 percent as compared with single-phase a.c. power units with a power factor of about 55 percent. If low power factor is an issue, welding machines can be purchased with power factor correction.

Cables and Fasteners

Power Cable

For welding machines connected to an electric power line, conductors of ample capacity and adequately insulated for the voltage transmit the power. It is necessary to ground the frame of the welding machine so that a break in the insulation of the cable or the machine does not cause an electrical shock. The machine is often grounded by a portable cable with an extra conductor that is fastened to the machine frame on one end and a solid ground on the other. Since the voltage on the power side of the machine is much greater than on the output side of the machine, it is important that portable power cable be adequately insulated with tough abrasion-resisting insulation so that it will stand up under the rough usage ordinarily encountered when it is dragged around welding shops.

Electrode and Work Cable

To complete the electric circuit between the welding machine and the work, two cables of adequate current-carrying capacity are required. An *electrode cable*, also called a welding cable, is attached to the electrode holder. A *work cable* is attached to the work. Rubber-covered multistrand copper cable, Fig. 11-21, especially designed for welding, is generally used.

Cable, especially electrode cable, must have high flexibility so that it will not interfere with and tire the welder when working. Vertical and overhead welding, by the very nature of the position that the welder must maintain, imposes a severe strain. An electrode cable with poor flexibility increases this difficulty and encourages fatigue that, in turn, makes it difficult or impossible for the welder to maintain speed and good quality of work.

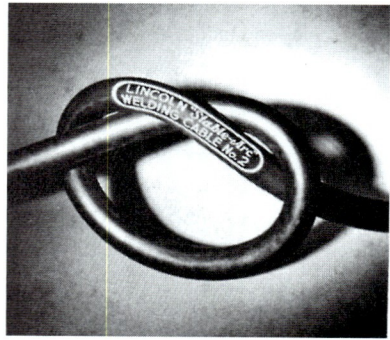

Fig. 11-21 Welding cable with thousands of fine wires has greater flexibility. A specially finished, durable paper wrapper allows the conductor to slip easily when the cable is bent.
The Lincoln Electric Co.

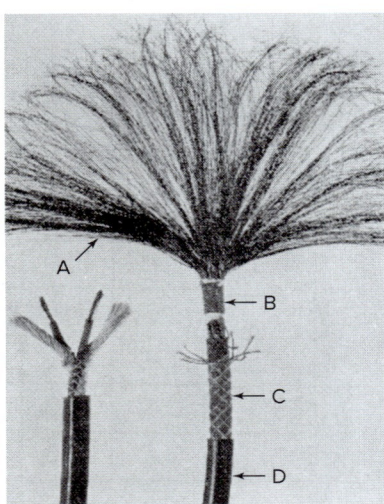

Fig. 11-22 Internal construction of welding cable. Edward R. Bohnart

High flexibility is the result of the core construction. The cable is woven of thousands of very fine, almost hairlike strands of copper wire. The greater the number of strands in a given-size cable, the more flexible it is. Figure 11-22 shows the inner construction of rubber-covered copper cable. Note the following components:

A. Thousands of hairlike copper wires are stranded in a special manner for extra flexibility.
B. Paper wrapping around stranded wires allows the conductor to slip within the rubber covering when the cable is bent sharply.
C. Extra strength is obtained by the open-braided reinforcement of extra cotton cords embedded in the covering.
D. The special composition and curing of the heavy rubber covering makes it tough, durable, elastic, and waterproof.

It is not necessary for work cable to have the flexibility of the electrode cable since it is not handled by the welder in the act of welding. Nevertheless, the same type of cable is often used for the work cable.

The amperage of the welding machine and its distance from the work are important considerations when selecting the size of welding cable, Table 11-2. The greater the amperage and the greater the distance between the work and the welding machine, the larger the cable size must be. The larger the cable number, the smaller the cable size is. Resistance increases as the diameter of the cable decreases. Therefore, if the cable is too small, it will overheat and affect the welding operation. Resistance also increases as the length of the cable increases. The welding machine should be set up as close to the work as possible. Welding cables that are too long may cause a voltage drop so great that it has serious effects on the welding current and arc.

Table 11-2 Choosing the Right Size Welding Cable

Using the right size welding cable on a power source is vitally important in obtaining sound welds. For example, a 100-ft lead of 4/0 cable at 500 A will have a voltage drop of 4 V. At a load voltage of 40 V, the power loss would amount to 10%. The current setting should be raised 10% to compensate for this loss. The following table gives the recommended sizes of cable to be used as approved by the American Welding Society.

Welder Cap. (A)	Total Length of Cable Being Used						
	50	75	100	125	150	175	200
100	2	2	2	2	1	1/0	1/0
150	2	2	1	1/0	2/0	2/0	3/0
200	2	1	1/0	2/0	3/0	4/0	4/0
250	2	1/0	2/0	3/0	4/0		
300	1	2/0	3/0	4/0			
350	1/0	3/0	4/0				
400	1/0	3/0	4/0				
450	2/0	3/0					
500	3/0	4/0					

Source: From *Welding Handbook*, 9/e.

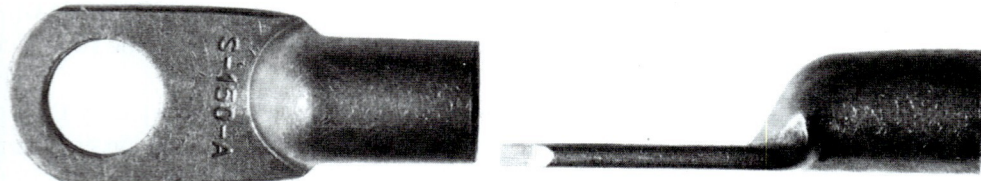

Fig. 11-23 Electrode and work cable lugs. Edward R. Bohnart

Cable Lugs and Work Clamps

Suitable lugs, Fig. 11-23, are required on both the electrode cable and the work cable to connect them to the welding machine. These lugs should be soldered or fastened mechanically to the cables. The connections that attach the cable lugs to the terminal posts on the welding machine must be tight. Loose connections cause the lugs to become hot, melting the solder that holds the cables to the lugs. Then the connector posts and cable lugs will burn and interfere with the welding current.

The other end of the electrode cable is connected to the electrode holder. The other end of the work cable should have a means of connecting it to the work. If this connection is not secure, arcing across the connection will burn up the connection and may cause unsatisfactory welding conditions.

If the work is of such a nature that it can be done on a welding bench or in a permanent fixture, the work cable is usually bolted to the bench or fixture. When the welder must work on a variety of structures in different parts of the shop, many types of work clamping devices may be used, including a copper hook, a heavy metal weight, a C-clamp, and specialized work clamps, Figs. 11-24 and 11-25.

Figure 11-26 shows a type of rotating work clamp. This clamp stops the twisting and turning of welding cable attached to welding fixtures and positioners where the work rotates. The clamp is welded to the part thus permitting the work clamp to be attached in seconds to almost any shape. This clamp is generally used in fabricating tanks and pressure vessels, and on weld positioners.

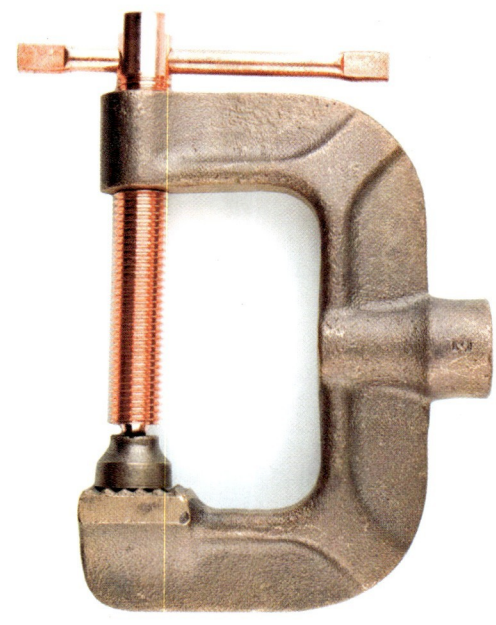

Fig. 11-25 A "C" clamp type work clamp. David A. Tietz/Editorial Image, LLC

Fig. 11-24 A spring loaded type work clamp. David A. Tietz/Editorial Image, LLC

Fig. 11-26 A rotary type work connection. Adapted from Lenco dba NLC, Inc.

Fig. 11-27 Insulated quick connectors. David Moyer/McGraw Hill

When the work is of such nature that different lengths of cable leads are necessary, cables may be made up in specific lengths and equipped with cable connectors so that they can be quickly and easily attached to make up any desired length. The connector shown in Fig. 11-27 has a cam-type action that ensures a positive stop and lock and cannot come loose or accidentally fall apart.

Electrode Holders

Metal Electrode Holders

A metal *electrode holder* is the device used for holding the electrode mechanically. It conveys electric current from the welding cable to the electrode, and it has an insulated handle to protect the welder's hand from heat.

The jaws of the holder should be designed so that they may grip the electrode firmly at any desired angle. They should be made of a metal that has high electrical conductivity and possesses the ability to withstand high temperatures, Figs. 11-28 and 11-29. On many holders, the jaws can be replaced with new ones when they become so badly burned that they do not grip securely. The holder should be light in weight, well-balanced, and have a comfortable grip. Although it should be easy to change the electrode, the holder must be sturdy enough to withstand rough usage. The current-carrying parts must be large enough to prevent overheating, which causes the handle to become too hot for the welder to hold. For the same reason, the size of the holder must be in line with the size of the welding machine; that is to say, a larger electrode holder is required for a 400-ampere welding machine than for a 200-ampere machine. Most holders are fully insulated and may be laid anywhere on the work without danger of a short circuit. This is especially convenient for work in close quarters.

Figure 11-30 shows an electrode holder of relatively new design. It holds electrodes burned to a very short stub. Its twist-type locking device permits electrode-gripping power in excess of 2,000 pounds. There is never danger of dislodging the electrode when attempting to break through slag to start the arc. The electrode position is always known. There is positive contact between the holder and electrode, thus reducing both heating up of the holder and electrode waste. The handle is fully insulated so that it stays cool even with high duty cycles.

Another holder is the angle-head, screw-clamp, fully insulated holder, shown in Fig. 11-31. It is available in sizes of 400 and 600 amperes and takes electrodes from $\frac{1}{16}$ through $\frac{5}{16}$ inch in diameter. Such holders are efficient and reduce costs since electrode stub loss is at a

Fig. 11-28 A 300 amp electrode holder. This type is spring loaded to grip the electrode. David Moyer/McGraw Hill

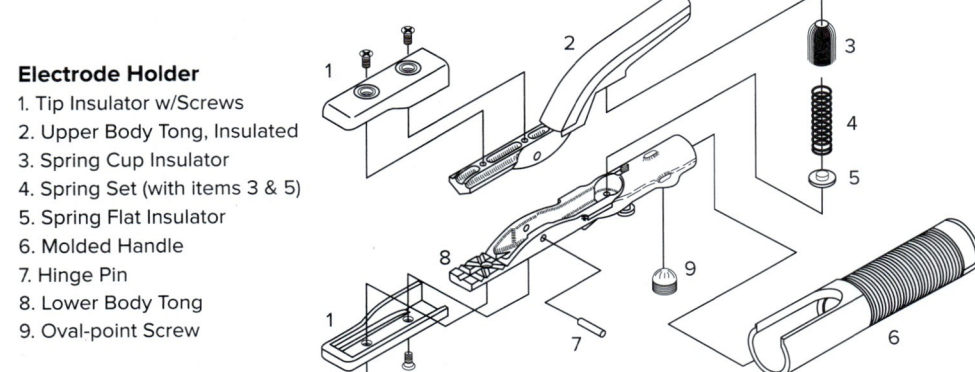

Electrode Holder
1. Tip Insulator w/Screws
2. Upper Body Tong, Insulated
3. Spring Cup Insulator
4. Spring Set (with items 3 & 5)
5. Spring Flat Insulator
6. Molded Handle
7. Hinge Pin
8. Lower Body Tong
9. Oval-point Screw

Fig. 11-29 Exploded view of a spring type electrode holder. Adapted from Lenco dba NLC, Inc.

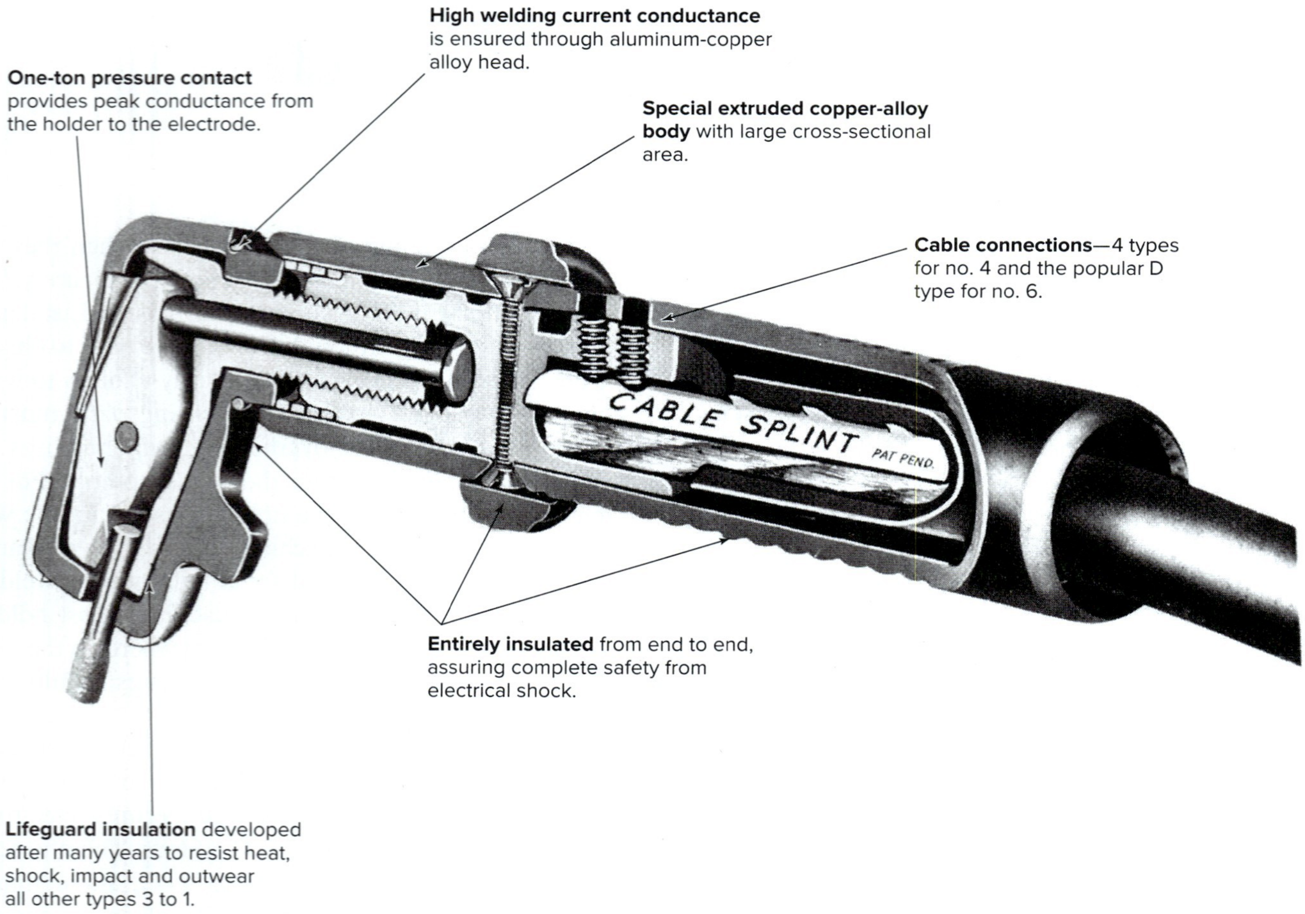

One-ton pressure contact provides peak conductance from the holder to the electrode.

High welding current conductance is ensured through aluminum-copper alloy head.

Special extruded copper-alloy body with large cross-sectional area.

Cable connections—4 types for no. 4 and the popular D type for no. 6.

Entirely insulated from end to end, assuring complete safety from electrical shock.

Lifeguard insulation developed after many years to resist heat, shock, impact and outwear all other types 3 to 1.

Fig. 11-30 Short-stub electrode holder with twist-type locking device. 2010 Bernard Welding Equipment Co.

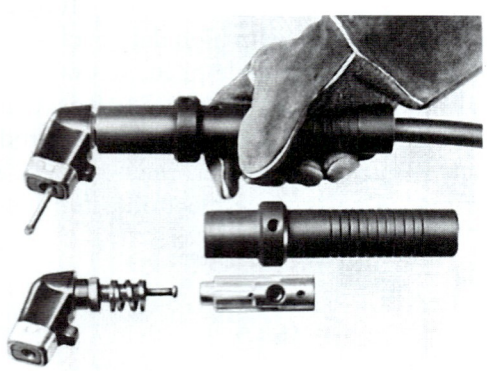

Fig. 11-31 This type of electrode holder has an angled head, and it is shorter and lighter than other holders of comparable capacity. Jackson Products

minimum and holder maintenance is low. The screw-on head can be replaced to make a "new" holder.

The holder is usually attached to the welding cable by means of solderless connections. It is important that good contact be maintained to prevent overheating.

Other Electric Arc Processes

The electric arc generates the heat necessary for several major welding processes. Chapter 18 will present the principles of gas tungsten arc welding (GTAW), and Chapter 21 will present the principles of gas metal arc welding (GMAW). Arc cutting, which utilizes much of the same equipment as arc welding, is presented in Chapter 6.

Personal Safety Equipment

Hand and Head Shields

The brilliant light caused by an electric arc contains two kinds of invisible rays that injure the eyes and skin unless protection is provided. One of these invisible rays is known as *ultraviolet* and the other as *infrared*. Repeated exposure, either directly or indirectly, results in painful but not permanent injury to the eyes. Welders speak of this as "hot sand in the eyes." The rays may also produce a severe case of sunburn and sometimes infection. The rays affect the eyes at any distance within 50 feet; and the skin, at any distance within 20 feet.

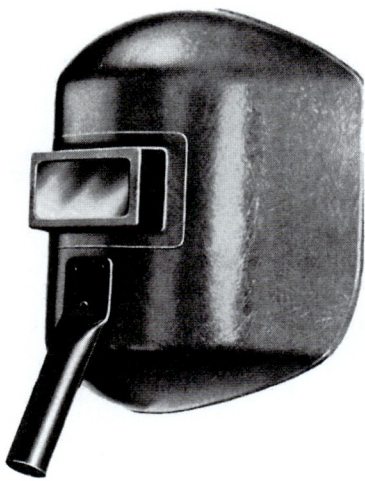

Fig. 11-32 The hand shield usually used by foremen and inspectors. Fibre-Metal Products Co.

Shields protect the welder not only against harmful light rays, but also against the hot globules of metal that the welding operation gives off, especially in the vertical and overhead positions. The hand shield, Fig. 11-32, has a handle so that the person using it may hold the shield in front of their face. Welding inspectors and supervisors use this type of shield. It is not suitable for welders since they can work with only one hand while using it. It is impossible to manipulate the electrode and perform other necessary operations at the same time with the other hand.

The head shield, Figs. 11-33 and 11-34, also called a hood or helmet, is worn like a helmet. It is attached to an adjustable headband, Fig. 11-35, which allows it to be moved up or down as the wearer desires. A helmet and safety cap combination, Fig. 11-36, offers practical,

Fig. 11-34 Note the vertical position of the large viewing window. Jackson Products

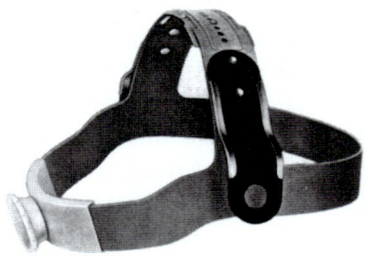

Fig. 11-35 An adjustable free-floating headband with adjustable crown. Fibre-Metal Products Co.

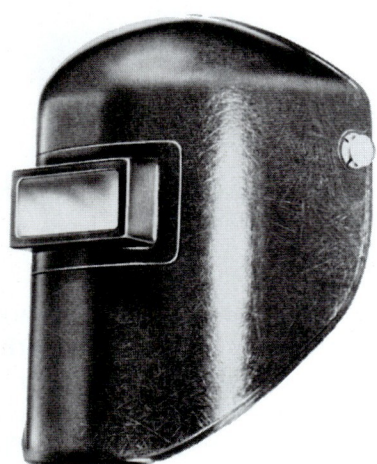

Fig. 11-33 The standard head shield or hood. Fibre-Metal Products Co.

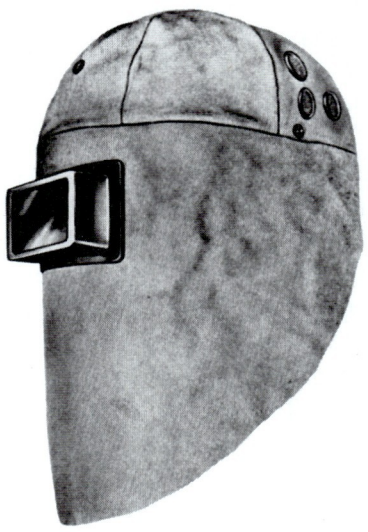

Fig. 11-36 Chrome leather helmets are ideal for those hard-to-get-into areas. Fibre-Metal Products Co.

Shielded Metal Arc Welding Principles Chapter 11 295

dependable protection when there is danger of falling or flying objects. Both hands are free to grasp the electrode holder and carry out accompanying operations. Partial protection is provided for the top of the head, but the operator must also wear a leather or nonflammable cap for adequate protection. This cap should be smooth and have no pockets or turned-up edges that will hold hot globules of metal. Most caps have a small bill that can be positioned to give additional protection to the ear that is most exposed to weld splatter and slag especially when welding out of position.

Both the hand shield and the head shield are constructed of a heat-resisting, pressed-fiber insulating material. The shields are fully molded at the top and bottom to protect the head and neck from metal particles, fumes, and dangerous light rays. They are usually black to reduce reflection and have a window frame for holding the protective lens which permits the welder to look at the arc safely.

The size of the protective lens is 2 × 4½ inches or 4½ × 5¼ inches. It is colored so that it can screen out ultraviolet rays, infrared rays, and most of the visible rays from the electric arc. A variety of shades of color may be obtained. The density of the color chosen depends on the brilliance of the arc, which varies with the size of the electrode and the volume of current. The shade for welding with metallic electrodes with current values up to 300 amperes is shade no. 10. Shade no. 12 is recommended for current values beyond 300 amperes and for shielded gas arc welding.

Good quality lenses are guaranteed to absorb 99.5 percent or more of the infrared rays and 99.75 percent or more of the ultraviolet rays present. The lenses have absorbed 100 percent of these rays as reported in tests conducted by the U.S. Bureau of Standards. The purchase of cheap filter lenses is to be discouraged.

The side of the protective glass exposed to the weld pool is protected by a clear polycarbonate plastic cover lens. This is to protect the more costly filter lens from molten metal spatter and breakage.

When the clear plastic cover becomes dirty from smoke and fumes and pitted from weld spatter, it should be replaced. Welding with clouded glass impairs vision and causes eye strain. Many welders prefer to protect both sides of the colored lens with a clear plastic cover.

Another type of head shield is shown in Fig. 11-37. This differs from the others only in that it is equipped with a flip front. The holder, which contains the protective and clear cover lenses, flips up at a touch of the finger. A clear plastic cover in the stationary section protects the eyes from hot scale when inspecting hot welds and from flying particles of slag and steel during cleaning. This type is especially useful when working in close quarters where it is difficult to raise the helmet.

Fig. 11-37 The flip-front welding helmet permits the welder to inspect and brush the weld without lifting the hood. Edward R. Bohnart

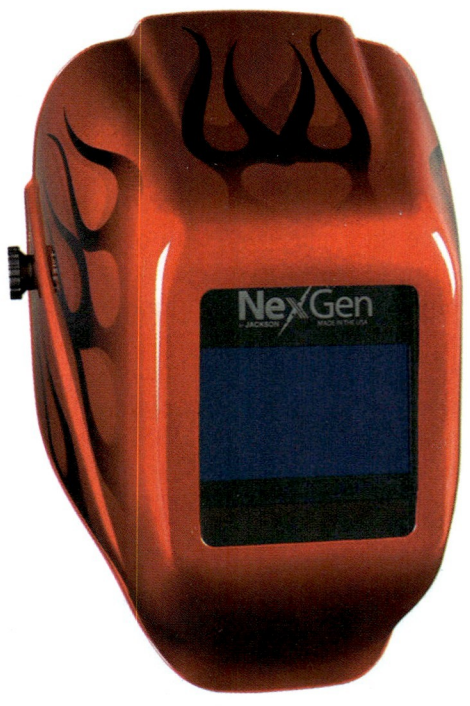

Fig. 11-38 An electronic filter helmet improves productivity and reduces neck strain. Jackson Products

The helmet in Fig. 11-38 is an autodarkening electronic filter helmet. It is useful when working in close quarters, doing high production work, or where inadvertent arc strikes must be avoided. These helmets can have a single shade or variable shades. The filter can switch from light to dark in less than 1/10,000 (100 millionths) of a second. They are battery powered and may have a solar battery booster.

Safety Glasses

Whenever you are in the work area, safety glasses should be worn. The practice of wearing them, Fig. 11-39, behind the hood is an important safety precaution. This is especially true in shops where there are a number of welders working close to each other. It is almost impossible under these conditions to prevent severe arc flash reaching the eyes without safety glasses with side shields. Safety glasses can absorb more than 99.9 percent of harmful ultraviolet rays that can cause arc burn. The lenses are usually made of a high impact polycarbonate material and must meet ANSI Z87.1 standards. Welding with tinted or dark lenses should be discouraged because it causes eyestrain due to compounding the shading of the lens. Safety glasses also protect the welder's eyes while they are inspecting the recently completed weld, chipping slag, wire brushing, and grinding.

Goggles should also be worn by helpers, foremen, inspectors, and others who work with welders. The goggles should be light in weight, well ventilated, and comfortable. In order to protect the eyes from side glare, they should be provided with tented side shields, and the lenses should have a tint.

Protective Clothing

Gloves are necessary to protect the hands. Gloves are made of leather or some other type of fire-resistant material. Leather capes, sleeves, shoulder garments with a detachable bib, and aprons protect the clothing and body of the welder from harm, Fig. 11-40. They are really necessary when the welder is called upon to do vertical and overhead work. If much of the work is to be done in a sitting position, the welder should wear overalls or a split-type apron, because full aprons form a lap for hot

Fig. 11-40 An apron helps protect the welder from sparks and heat. Northeast Wisconsin Technical College Mark A. Dierker/McGraw Hill

particles. Leather overalls may be worn with leather jackets. High-top shoes must always be worn when welding. The feet and legs can be further protected by the use of leggings and spats. Burns on the feet are quite painful, become infected easily, and are slow to heal. Rolled sleeves and turned up trouser cuffs provide lodging places for hot metal and should be avoided.

The clothing of choice for welders should be 100 percent cotton or wool. Cotton denim shirt and pants are popular attire for today's welders as shown by a student welder in Fig. 11-41. The clothing should be thick enough to prevent injurious ultraviolet rays from penetrating to the skin. Shirts should be long sleeve and button at the cuff. Shirttails are

Fig. 11-39 Wear your safety glasses even underneath your welding helmet. Miller Electric Mfg. Co.

Fig. 11-41 Thick clothing prevents injurious ultraviolet rays from penetrating to the skin. Miller Electric Mfg. Co.

tucked into the welder's pants and shirts should button to the neck. The pants should be long enough so they cover the top of the leather boots and they should not have cuffs. However, at times it becomes necessary to wear further protective clothing either for welding processes and/or positions. When this is the case, the welder has many options to choose from to protect the skin from sparks and hot globules of molten metal thrown out by the arc during the welding operation. Figure 11-42 shows a student welder who is well protected from sparks, rays, and heat. However, the welder can be the best judge of the protective clothing needed for the job.

Ear Protection

Because of the noise and possibility of hot weld spatter or slag entering the ear canal, this area must be protected. Either full ear muffs that cover the entire ear or simple ear

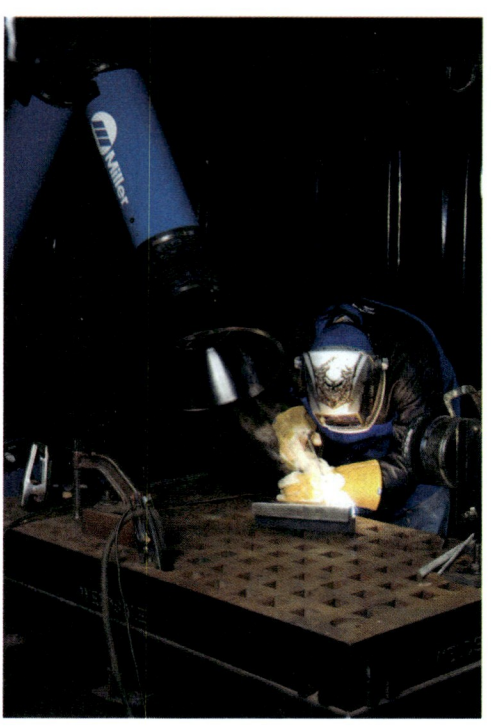

Fig. 11-43 A well-protected student welder. Note the platen table, clamps, and the position of the fume extractor for optimum protection. Miller Electric Mfg. Co.

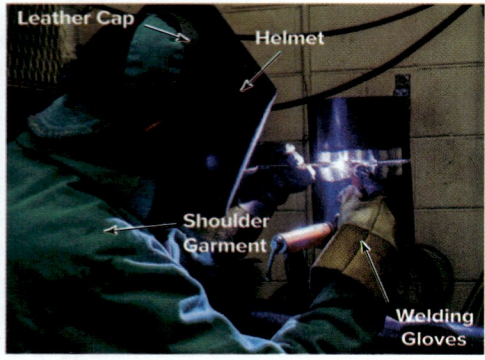

Fig. 11-42 A well-protected student welder. Note the gloves, helmet, skull cap, safety glasses, and fire-resistant jacket. Northeast Wisconsin Technical College. Mark A. Dierker/McGraw Hill

plugs can be used. A welder's cap with a bill on it can be used to deflect hot material away from the ear area.

Fume Protection

Always use proper ventilation. When welding keep your head out of the fume plume. If this is not possible, use a fume extractor as in Fig. 11-43, or a respirator.

CHAPTER 11 REVIEW

Multiple Choice

Choose the letter of the correct answer.

1. Shielded and metal arc electrodes are available to match the _____ and _____ of most metals. (Obj. 11-1)
   a. Ductility and fatigue
   b. Density and plasticity
   c. Properties and strength
   d. Stress and hardness

2. Metals most easily welded by the SMAW process are _____. (Obj. 11-2)
   a. Carbon and low alloyed steels
   b. Stainless steels
   c. Heat-resisting steels
   d. All of these

3. The SMAW process is used for welding _____. (Obj. 11-3)
   a. Steel
   b. Tin
   c. Zinc
   d. All of these

4. SMAW power sources are known as _____. (Obj. 11-4)
   a. Constant voltage
   b. Constant potential
   c. Constant current
   d. Constant output

5. A transformer-rectifier can supply _____. (Obj. 11-5)
   a. a.c. power
   b. DCEP power
   c. DCEN power
   d. All of these

6. Arc voltage is when _____. (Obj. 11-6)
   a. The voltage is being generated between the electrode and the work
   b. The machine is on but idle; no welding is being done
   c. The machine is off
   d. The machine is converting to volts

7. The duty cycle is the percentage of any given _____ minutes that a machine can operate at its rated current without overheating or breaking down. (Obj. 11-6)
   a. 10
   b. 20
   c. 30
   d. 40

8. A long welding arc _____. (Obj. 11-7)
   a. Decreases amperage
   b. Increases amperage
   c. Increases voltage
   d. Both a and c

9. The size of an electrode holder must be in line with the size of _____. (Obj. 11-8)
   a. A welding machine
   b. An electrode
   c. A cable
   d. All of these

10. The clothing a welder wears should be of _____ material. (Obj. 11-9)
    a. Cotton
    b. Wool
    c. Nylon
    d. Both a and b

Review Questions

Write the answers in your own words.

11. Name the four basic types of arc welding machines. Which type is used most? (Objs. 11-1 and 11-4)
12. What is the difference between an engine-driven generator welding machine and a transformer-rectifier welding machine? Give the general use of each. (Obj. 11-2)
13. Do a.c. welding machines have more wearing parts than d.c. machines? Explain. (Obj. 11-4)
14. Are d.c. welding machines capable of welding faster than a.c. machines? Explain. (Objs. 11-5 and 11-7)
15. How is the output of an arc welding machine indicated? (Obj. 11-6)
16. List three advantages of a.c. arc welding. (Obj. 11-7)
17. Explain the use of multiple-operator systems. (Obj. 11-7)
18. How is the current delivered to the electrode and the work? Explain. (Obj. 11-8)
19. Compare the characteristics of engine-driven generator, transformer, and transformer-rectifier welding machines and inverters. (Obj. 11-9)
20. What device is used to hold the electrode? Give a general description of this device. (Obj. 11-9)

INTERNET ACTIVITIES

Internet Activity A

Use a search engine to find a company that sells videos about welding. If it has photos of various frames of the video, you can see, in color, what the welding process looks like. Make a color drawing of the process you saw and describe it.

Internet Activity B

Look on the Internet to see if you can locate helmets. Now look for a helmet that can be used for closed-in spaces that have little or no ventilation. Report on what you found.

Design credits: Cinema and movie icon: Denis Meshkov/123RF; and Illustration of Welding icons: Mr. Nachai Sorasee/123RF

12

Shielded Metal Arc Welding Electrodes

Introduction

Shielded metal arc welding (SMAW), often referred to by welders in the field as *stick electrode welding*, is one of the most widely used welding processes in maintenance, fabrication in the field, and repair of metals. This popularity has resulted from the development of flux-covered electrodes capable of making welds having physical properties that are equal or superior to the material being welded.

Leading manufacturers of welding equipment and supplies are constantly developing new welding electrodes and improving existing ones. Two other groups have had a major part in research and development: the American Welding Society and the American Society for Testing and Materials. These societies are continually improving the specifications and methods of classifying electrodes and filler rods so that the welder can readily select the best electrode for a particular job. Electrodes are designed to meet the needs of each welding application.

It is not the purpose of this chapter to discuss in detail all the different types of electrodes that are used in the welding process. At this time, we are concerned chiefly with those steel electrodes that will be used for the practice jobs in Unit 2 of this text. Such electrodes comprise about 80 percent of the total number used by industry. Many welders who have spent years as journeymen have never had occasion to use anything but low carbon steel electrodes.

Chapter Objectives

After completing this chapter, you will be able to:

12-1 List the major functions of the SMAW electrode coating.

12-2 Describe the composition of an electrode covering.

12-3 Determine the maximum arc length of an SMAW electrode.

12-4 List the basic systems for identifying steel electrodes.

12-5 Explain the electrode selection process.

12-6 List the AWS electrode classification system.

12-7 List the operating characteristics of the fast fill, fast follow, and fast freeze electrodes.

12-8 Describe the characteristics of the low hydrogen electrode.

12-9 Describe the characteristics of the iron powder electrode.

12-10 List the reasons for keeping mineral coated electrodes dry.

It is, of course, your objective to be able to deposit welds that have the most desirable physical and chemical properties, soundness, and appearance that the electrodes are capable of giving. Study Fig. 12-1 carefully. Learn to recognize good and bad welds and to understand the factors involved in good and bad welding. After you are thoroughly familiar with the characteristics of the electrodes discussed in this chapter, you should consult the catalogs of electrode suppliers and the materials available from the American Welding Society for information on other types of electrodes, such as those for high tensile steel, alloy steels, nonferrous materials, and surfacing materials.

Shielded Metal Arc Welding Electrodes

The general definition of a shielded metal arc welding **covered electrode**, as given in the American National Standard and the American Welding Society's "Standard Welding Terms and Definitions," A3.0 is as follows:

A composite filler metal electrode consisting of a core of a bare electrode or metal cored electrode to which a covering sufficient to provide a slag layer on the weld metal has been applied. The covering may contain materials providing such functions as shielding from the atmosphere, deoxidation, and arc stabilization, and can serve as a source of metallic additions to the weld.

Introduction to Covered Electrodes

The type of covering influences the degree of penetration of the arc and the crater depth. The proper electrode selection, therefore, makes it possible to obtain sound welds in close-fitting joints and to avoid burning through poorly fitted joints. Since the covering influences the extent of penetration, it affects the extent of recrystallization and annealing of previously deposited layers. This characteristic improves the internal (radiographic) quality of the weld. The low electrical conductivity of the covering permits the use of electrodes in narrow grooves. The covering also reduces weld spatter.

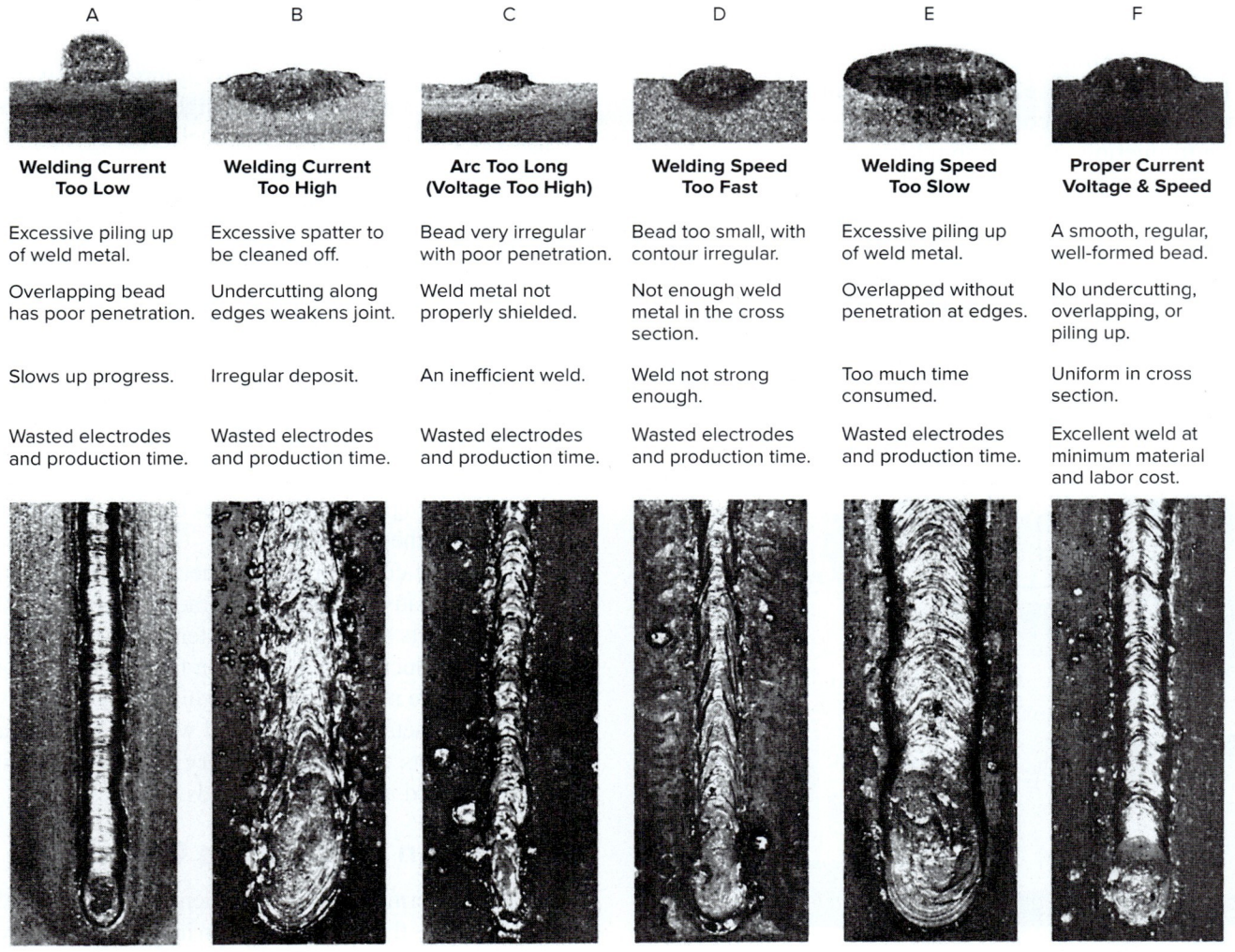

Fig. 12-1 Plan and elevation views of welds made with shielded arc electrodes under various conditions. Hobart Brothers Co.

Functions of Electrode Coverings

Protective Gaseous Atmosphere and Slag Covering

The covering materials on the electrode provide an automatic cleansing and deoxidizing action in the molten weld pool. By supplying a protective gaseous atmosphere and blanket of molten slag for the weld metal, the covering excludes harmful oxygen and nitrogen, Fig. 12-2. The extent of gaseous and slag protection depends on the type of covering. In addition to protection, slag performs the following functions:

- Acts as a scavenger in removing oxides and impurities
- Slows down the freezing rate of molten metal
- Slows down the cooling rate of solidified weld metal
- Controls the shape and appearance of the deposit
- Affects operating characteristics (DCEP, alternating current, etc.)

Alteration or Restoration of Base Metal

To a large extent, the covering controls the composition of the weld metal, either by maintaining the original composition of the core wire of the electrode or through the introduction of additional elements. In this way, alloying elements are added to the weld metal or lost elements are restored. For example, molybdenum and vanadium may be added to the covering because these alloys produced better physical properties in the weld metal than those possessed by the core wire. On the other hand, manganese is usually included in the covering to maintain the same manganese content in the weld metal that was present in the core wire. The coating may also be balanced to adjust the carbon and silicon content of the weld deposit.

The addition of large amounts of iron powder to the coating of an electrode increases the speed of welding and improves the weld appearance. In the intense heat of the arc, the iron powder is converted to steel and contributes metal to the weld deposit.

Low hydrogen electrodes greatly improve the welding of high carbon and alloy steels, high sulfur steels, and phosphorus-bearing steels. Such steels tend to be porous and crack under the bead. The reduction of the hydrogen content in the weld eliminates these harmful characteristics.

Control of Arc Characteristics

The covering makes starting the arc easier as you begin the weld. The covering not only helps you to maintain a steady arc throughout the operation, but it also serves as an insulator for the core wire of the electrode. Coverings allow greater variation in arc length. Better arc control permits the use of higher currents and larger electrodes.

As a rule, when welding with an SMAW electrode, the maximum arc length is never greater than the diameter of the bare end of the electrode.

The covering concentrates the heat of the arc on the work and thus causes an increased melting rate. The flux contains elements that ionize at the temperature of the arc and form a conducting path between the poles during the periods when the metal arc is extinguished by the transfer of globules of metal. The decreased welding time usually more than offsets the time necessary to remove the slag from the shielded arc electrode welds.

Composition of Electrode Coverings

The type of covering affects the arc length and the welding voltage, as well as the welding position in which the electrode can be used. Its composition is very important. The covering should have a melting point lower than that of both the core

> **JOB TIP**
>
> **Automation**
> An area with potential is automation for such uses as:
> 1. Sensing and feedback control for welding process automation.
> 2. Welding process control system design, vision-based.
> 3. Through-arc process sensing and control.
> 4. Welding robotics and automation.

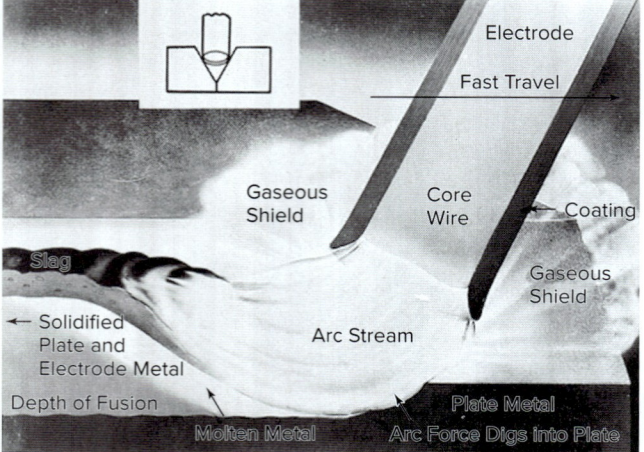

Fig. 12-2 Cross section of arc action, gaseous shielding, and metal flow when welding with covered electrode. Lincoln Electric

wire and the base metal. The slag must have a lower density than the solidifying weld metal in order to be expelled quickly and thoroughly. The slag must also be able to solidify quickly when the electrode is used for overhead and vertical welding. These are all essential functions that ensure the formation of sound welds having the physical and mechanical properties required for the job. The coating is primarily responsible for the differences between electrodes; the core wires for carbon steel electrodes are usually made of the same type of steel. However, some alloys are transferred more effectively through the core wire than through the coating.

Materials for Electrode Coverings

Materials used in the coverings of shielded arc electrodes may be classified according to their purposes:

- Fluxes
- Deoxidizers
- Slagging ingredients
- Alloying ingredients
- Gas reducers
- Binders
- Arc stabilizers
- Shielding gas

Sodium and potassium silicates are universally used as binders; some organic gums also have a limited use for this purpose. Ferro-alloys and pure metals serve as deoxidizers and alloying ingredients. The alkaline earth metals are the best arc stabilizers. Wood flour, wood pulp, refined cellulose, cotton linters, starch, sugar, and other organic materials provide a shield of reducing gases. Fluxes and slagging ingredients include silica, alumina, clay, iron ore, rutile, limestone, magnesite, mica, and many other minerals, as well as some synthetic materials, such as potassium titanate and titanium dioxide.

Polarity Interchangeability

The composition of the covering determines the best polarity of the electrode used for d.c. applications. Some coverings operate more efficiently with electrode negative (DCEN). Other coverings make the electrode more efficient with electrode positive (DCEP). Both types have advantages that make them preferable for certain applications. Coverings have also been developed that operate equally well on either polarity. These types are also efficient in a.c. welding.

For video of the polarity of SMAW, please visit www.mhhe.com/welding.

Tables 12-1 through 12-3 demonstrate the important functions of the covering and its direct effect on the chemical and physical properties of the weld. Study carefully the properties of the various types of electrodes described on the following pages and in Table 12-4. It will also be beneficial at this time to become familiar with the current values given in Table 12-5.

Table 12-1 Physical Properties of Welds and Covered Electrodes

	Covered Electrode Weld Metal	Base Metal
Ultimate strength, p.s.i.	60–75,000	55–70,000
Yield point, p.s.i.	45–60,000	30–32,000
Elongation, % in 2 in.	20.0–40.0	30.0–40.0
Elongation-free bend, %	35.0–60.0	
Reduction of area, %	35.0–65.0	60.0–70.0
Density, g/cm	7.80–7.85	7.85
Endurance limit, p.s.i.	26–30,000	26–30,000
Impact (Izod), ft/lb.	40–70	50–80

Table 12-2 Comparison of Core Wire Composition: Filler Metal and Covered Electrodes

Element (Percentage)	Filler Metal Core Wire	Shielded Arc Weld Metal
Carbon	0.10–0.15	0.08–0.15
Manganese	0.40–0.60	0.30–0.50
Silicon	0.025 max.	0.05–0.30
Sulfur	0.04 max.	0.035 max.
Phosphorus	0.04 max.	0.035 max.
Oxygen	0.06 max.	0.04–0.10
Nitrogen	0.006 max.	0.01–0.03

Table 12-3 Weld Metal Properties Afforded by Types of Covered Electrodes

	Minimum Weld Metal Properties		
Property No.	Tensile Strength (p.s.i.)	Yield Point (p.s.i.)	Percentage of Elongation
E60XX	62,000	50,000	17–25
E70XX	70,000	57,000	22
E80XX	80,000	67,000	19
E90XX	90,000	77,000	14–17
E100XX	100,000	87,000	13–16
E120XX	120,000	107,000	14

Table 12-4 Welding Characteristics of Mild Steel Electrodes

	Type of Coating	Position of Welding	Type of Current[1]	Penetration	Rate of Deposition	Appearance of Bead	Spatter	Slag Removal	Minimum Tensile Strength (p.s.i.)	Yield Point (p.s.i.)	Minimum Elongation in 2 in. (%)
E6010	High cellulose sodium	All positions	DCEP	Deep	Average rate	Rippled and flat	Moderate	Moderately easy	62,000	50,000	22
E6011	High cellulose potassium	All positions	DCEP, a.c.	Deep	Average rate	Rippled and flat	Moderate	Moderately easy	62,000	50,000	22
E6012	High titania sodium	All positions	DCEN, a.c.	Medium	Good rate	Smooth and convex	Slight	Easy	67,000	55,000	17
E6013	High titania potassium	All positions	DCEP, DCEN, a.c.	Mild	Good rate	Smooth and flat to convex	Slight	Easy	67,000	55,000	17
E7014	Iron powder titania	All positions	DCEP, DCEN, a.c.	Medium	High rate	Smooth and flat to convex	Slight	Easy	70,000	60,000	17
E7015	Low hydrogen sodium	All positions	DCEP	Mild to medium	Good rate	Smooth and convex	Slight	Moderately easy	70,000	60,000	22
E7016	Low hydrogen potassium	All positions	DCEP, a.c.	Mild to medium	Good rate	Smooth and convex	Slight	Very easy	70,000	60,000	22
E6020	High iron oxide	Flat hor. fillets	Flat: d.c., a.c. hor. fillets: DCEN, a.c.	Deep	High rate	Smooth and flat to concave	Slight	Very easy	62,000	50,000	25
E7024	Iron powder titania	Flat hor. fillets	DCEN, DCEP, a.c.	Mild	Very high rate	Smooth and slightly convex	Slight	Easy	72,000	60,000	17
E6027	Iron powder iron oxide	Flat hor. fillets	Flat: d.c., a.c. hor. fillets: DCEN, a.c.	Medium	Very high rate	Flat to concave	Slight	Easy	62,000	50,000	25
E7018	Iron powder low hydrogen	All positions	DCEP, a.c.	Mild	High rate	Smooth and flat to convex	Slight	Very easy	72,000	60,000	22
E7028	Iron powder low hydrogen	Flat hor. fillets	DCEP, a.c.	Mild	Very high rate	Smooth and slightly convex	Slight	Very easy	72,000	60,000	22
E7048	Iron powder low hydrogen	All positions vertical down	DCEP, a.c.	Mild	High rate	Smooth and slightly convex	Slight	Easy	72,000	58,000	22

[1]DCEP means direct current, electrode positive (reverse polarity).
DCEN means direct current, electrode negative (straight polarity).

Source: National Cylinder Gas Division, Chemetron Corp.

Table 12-5 Current Range for Mild and Low Alloy Steel Electrodes

Electrode Diameter (in.)	Current Range (A) — Electrode Type									
	E6010, E6011	E6012	E6013	E6020	E6027	E7014	E7015, E7016	E7018	E7024, E7028	E7048
1/16	—	20–40	20–40							
5/64	—	25–60	25–60							
3/32	40–80	35–85	45–90	—	—	80–125	65–110	70–100	100–145*	
1/8	75–125	80–140	80–130	100–150	125–185	110–160	100–150	115–165	140–190	80–140
5/32	110–170	110–190	105–180	130–190	160–240	150–210	140–200	150–220	180–250	150–220
3/16	140–215	140–240	150–230	175–250	210–300	200–275	180–255	200–275	230–305	210–290
7/32	170–250	200–320	210–300	225–310	250–350	260–340	240–320	260–340	275–365	
1/4	210–320	250–400	250–350	275–375	300–420	330–415	300–390	315–400	335–430	
5/16	275–425	300–500	320–430	340–450	375–475	390–500	375–475	375–470	400–525*	

*These values do not apply to the E7028 classification.

Fig. 12-3A Location of electrode classification number for covered end-grip welding electrodes.

Identifying Electrodes

The identifying system for covered arc welding electrodes requires that the electrode classification number be imprinted or stamped on the electrode covering, within 2½ inches of the grip and of the electrode. (See Fig. 12-3A and B.)

AWS Classification of Carbon Steel Electrodes

The many types of electrodes that are available to weld carbon and low alloy steels are classified in the *AWS Filler Specifications A5.1* and *A5.5*. These booklets can be purchased from the American Welding Society.

Other agencies concerned with electrode approvals, specifications, and classifications include:

- American Society of Mechanical Engineers (ASME)
- American Bureau of Shipping (Bureau of Ships)
- U.S. Department of Defense (for the Army, Navy, and Air Force)
- U.S. Coast Guard
- Canadian Welding Bureau, Division of Canadian Standards Association
- Municipal, county, and state organizations responsible for welding standards

The various electrodes are classified with a system of numbers and organized on the basis of the chemical composition and mechanical properties of the deposited undiluted weld metal, type of covering, welding

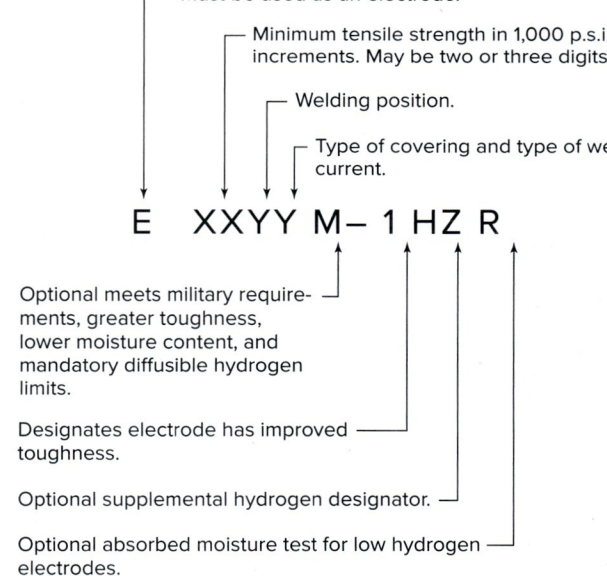

Fig. 12-3B SMAW electrode classification designators.

position, and type of current required for best results. Compare the various characteristics of carbon steel electrodes in Tables 12-3 and 12-4, pages 303 to 304.

The numbering system provides a series of four- or five-digit numbers prefixed with the letter *E*. The *E* indicates electrode for electric welding. The first two digits (three digits, for a five-digit number) multiplied by one thousand express the minimum tensile strength in thousands of pounds per square inch. For example, *60* in E6010 electrodes means 60,000 p.s.i.; *70* in E7010 electrodes means 70,000 p.s.i.; and *100* in E10010 electrodes means 100,000 p.s.i. The next-to-last digit indicates the position of welding. Thus, the *1* in E6010 means that welding can be done in all positions—flat, horizontal, vertical, and overhead. The *2* in E7020 means that the electrode should be used for flat and horizontal fillet welding. The last digit indicates the type of current and the type of covering on the electrode. Further interpretation of the classification numbers is presented in Tables 12-6 and 12-7 and Fig. 12-3B.

> **SHOP TALK** 
>
> **Contact Lenses**
>
> If you wear contact lenses, keep a spare pair, or glasses, at work in case you lose a lens. Research indicates wearing contact lenses poses no problems for welders in most normal situations. Wear contact lenses in industrial environments, in combination with appropriate industrial safety eyewear. Safety and medical personnel should not discriminate against an employee who can achieve vision correction by use of contact lenses. Let others know you wear contacts. All first-aid personnel should learn how to remove contact lenses.

Table 12-6 AWS Electrode Classification System

Digit	Significance	Example
1st two or 1st three	Min. tensile strength (stress relieved)	E-60XX = 60,000 p.s.i. (min.)
		E-110XX = 110,000 p.s.i. (min.)
2nd to last	Welding position	E-XX1X = all positions
		E-XX2X = horizontal and flat
		E-XX3X = flat
		E-XX4X = vertical down
Last	Power supply, type of slag, type of arc, amount of penetration, presence of iron powder in coating	See Table 12-7

Note: Prefix "E" (to left of a 4- or 5-digit number) signifies arc welding "electrode."

Table 12-7 Interpretation of the Last Digit in the AWS Electrode Classification System

F-No Classification	Current	Arc	Penetration	Covering & Slag	Iron Powder (%)
F-3 EXX10	DCEP	Digging	Deep	Cellulose-sodium	0–10
F-3 EXXX1	AC & DCEP	Digging	Deep	Cellulose-potassium	0
F-2 EXXX2	AC & DCEN	Medium	Medium	Rutile-sodium	0–10
F-2 EXXX3	AC & DC	Light	Light	Rutile-potassium	0–10
F-2 EXXX4	AC & DC	Light	Light	Rutile-iron powder	25–40
F-4 EXXX5	DCEP	Medium	Medium	Low hydrogen-sodium	0
F-4 EXXX6	AC or DCEP	Medium	Medium	Low hydrogen-potassium	0
F-4 EXXX8	AC or DCEP	Medium	Medium	Low hydrogen-iron powder	25–45
F-I EXX20	AC or DC	Medium	Medium	Iron oxide-sodium	0
F-I EXX24	AC or DC	Light	Light	Rutile-iron powder	50
F-I EXX27	AC or DC	Medium	Medium	Iron oxide-iron powder	50
F-I EXX28	AC or DCEP	Medium	Medium	Low hydrogen-iron powder	50

Note: Iron powder percentage is based on weight of the covering.

Electrode Selection

The selection of the proper electrode for a given welding job is one of the most important decisions that the welder faces. The nature of the deposited weld metal and its suitability as a joining material for the pieces being welded depend, of course, on the selection of the proper type of electrode. The continued use of the shielded metal arc welding process is due in great measure to the high quality of the electrodes available.

Electrodes may be generally grouped according to:

- Operating characteristics
- Type of covering
- Characteristics of deposited metal

Size is also an important consideration.

It is important to understand that many electrode classifications contain the same basic core wire. The differences in operating characteristics and in the physical and mechanical nature of the deposited weld metal are mostly determined by the materials in the covering.

You are urged to study carefully the characteristics of the basic welding electrodes described in this section. A thorough knowledge will assist you not only in acquiring the technical knowledge necessary to choose the proper electrode for a given job, but also in mastering the manipulative techniques of the welding operation.

As we have seen, there is a large number of electrodes that can be selected. In order to reduce the number of tests a welder has to take to become qualified (certified), many codes group electrodes based on their ease of welding, rather than on mechanical properties. One such grouping is from the AWS D1.1 Structural Welding Code—Steel. (See Table 12-7.) If any of the electrodes in the more difficult F4 grouping is used on the welder qualification test, a welder who passes the test will be qualified on all the electrodes in this group, as well as on all lesser group numbers and the electrodes they represent. Students must become knowledgeable about welding with as many electrodes as are practical for their job market. However, many companies use this grouping system to eliminate the need for their welders to qualify for welding procedure specifications (WPSs) unless absolutely necessary.

Size of Electrodes

Selecting the proper size of electrode to use on a given job is as important as selecting the right classification of electrode. The following points should be taken into consideration:

- ***Joint design*** A fillet weld can be welded with a larger electrode than an open groove weld.
- ***Material thickness*** Obviously a larger size electrode can be used as the thickness of material increases.
- ***Thickness of weld layers*** The thickness of the material to be welded and the position of welding are factors. More weld metal can be deposited in the flat and horizontal positions than in the vertical and overhead positions.
- ***Welding position*** A larger size electrode can be used in the flat and horizontal positions than in the vertical and overhead positions.
- ***Amount of current*** The higher the current value used for welding, the larger the electrode size.
- ***Skill of the welder*** Some welders become so expert that they can handle large size electrodes in the vertical and overhead positions.

All classes of covered electrodes are designed for multiple-pass welding. The size varies with the types of joints and welding position.

- The first pass for pipe welding and other bevel butt joints should be welded with $\frac{1}{8}$-inch or $\frac{5}{32}$-inch electrodes. This is necessary in order to obtain good fusion at the root and to avoid excessive melt-through. The remaining passes may be welded with $\frac{5}{32}$-inch or $\frac{3}{16}$-inch electrodes in all positions, and $\frac{3}{16}$ inch or larger in the flat position.
- For flat position welding of beveled-butt joints that have a backing bar, a $\frac{3}{16}$-inch electrode can be used for the first pass, and an electrode $\frac{7}{32}$ inch or larger for the remainder.
- For fillet welds in the flat position and other deep groove, flat position joints, $\frac{3}{16}$-inch, $\frac{7}{32}$-inch, or $\frac{1}{4}$-inch electrodes may be used. Extra heavy plates can be welded with larger electrodes.
- Out-of-position fillet and groove welding is usually done with $\frac{5}{32}$-inch electrodes. For certain jobs, $\frac{3}{16}$-inch electrodes may be used.
- The sizes of low hydrogen electrodes generally used for vertical and overhead welding are $\frac{1}{8}$ inch and $\frac{5}{32}$ inch. Electrodes for flat and horizontal welding may be $\frac{3}{16}$ inch or larger.

Job Requirements

The requirements of the job are the basis for the proper selection of electrodes. Look the job over carefully to determine just what the electrode must do. Study the variable factors given in Table 12-8. Following are a few conditions to check:

- Skill of the welder
- Code requirements (if any)
- Properties of base metal

Table 12-8 Relative Ratings of Factors Affecting the Preliminary Selection of Electrodes

Variable Factors (a)	E6010	E6011	E6012	E6012X	E6013	E7014	E7016	E7018	E6020	E7024	E6027
1 Groove butt welds, flat (>¼ in.)	4	5	3	2	8	8	7	9	10	9	10
2 Groove butt welds, all positions (>¼ in.)	10	9	5	4	8	8	7	6	(b)	(b)	(b)
3 Fillet welds, flat or horizontal	2	3	8	7	7	7	5	7	10	10	7
4 Fillet welds, all positions	10	9	6	4	7	7	8	6	(b)	(b)	(b)
5 Current	DCEP	a.c. DCEP	a.c. DCEN	a.c. DCEN	a.c. d.c.	d.c. d.c.	a.c. DCEP	a.c. d.c.	a.c. d.c.	a.c. d.c.	a.c. d.c.
6 Thin material (<¼ in.)	5	7	8	10	9	8	2	2	(b)	7	(b)
7 Heavy plate or highly restrained joint	8	8	6	(b)	8	8	10	9	8	7	8
8 High sulfur or off-analysis steel	(b)	(b)	5	4	3	3	10	9	(b)	5	(b)
9 Deposition rate	5	5	7	7	7	9	5	8	9	10	10
10 Depth of penetration	10	9	6	5	5	5	7	7	8	4	8
11 Appearance, undercutting	6	6	8	7	9	9	7	10	9	10	10
12 Soundness	6	6	3	3	5	5	10	8	9	8	9
13 Ductility	6	7	4	3	5	5	10	10	10	5	10
14 Low-temperature impact strength	8	8	4	4	5	5	10	10	8	9	9
15 Low spatter loss	1	2	6	6	7	7	6	8	9	10	10
16 Poor fitup	6	7	10	10	8	8	4	4	(b)	8	(b)
17 Welder appeal	7	6	8	8	9	10	6	8	9	10	10
18 Slag removal	10	8	6	6	8	8	4	7	8	8	8

(a) Rating (for same size electrodes) is on a comparative basis for electrodes listed in this table: 10 is the highest value. Ratings may change with a change in size.
(b) Not recommended.

- Position of the joint
- Type and preparation of joint
- Heat-treating requirements
- Environmental job conditions
- Expansion and contraction problems
- Amount of weld required
- Tightness of fitup
- Type of welding current available
- Thickness and shape of base metal
- Specifications and service conditions
- Demands of production and cost considerations

Operating Characteristics of Electrodes

The nature of the materials that go into the covering of an electrode usually determine not only the physical and mechanical properties of the weld deposit, but also the operating characteristics of the electrode. Different electrodes require different welding techniques. Thus, electrodes may be grouped according to their operating characteristics and the requirements of the joints to be welded as *fast fill*, *fast follow*, and *fast freeze*.

Fast Fill The fast-fill electrode deposits weld metal rapidly. It is the opposite of the fast-freeze electrode. The fast-fill group includes the heavy-coated, iron powder electrodes that are widely used for fillet and deep groove deposition. The fast-fill electrode is especially designed for fast, flat position welding. It has high metal deposition and permits easy slag removal. There is little undercutting. It burns with a soft arc and has shallow penetration that causes little mixing of the base metal and weld metal. Bead appearance is very smooth. It has a flat to slightly convex face, and there is little spatter, Fig. 12-4H and I.

Certain electrodes have been developed for out-of-position welding that have faster freezing characteristics. An example of this type of electrode is the EXX14. The EXX24 and EXX27 electrodes are generally used for flat fillets and groove welding, Fig. 12-4.

Fast Follow This group is also known as *fill-freeze* electrodes. They have characteristics that in some degree combine both fast-freeze and fast-fill requirements. In making lap welds or light gauge sheet metal welds, little additional weld metal is needed to form the weld. The most economical way to make the joint is to move rapidly. Because it is necessary to make the crater follow the arc as rapidly as possible, the electrode is called fast follow. It burns with a moderately forceful arc and has medium penetration. This, together with the lower current and lower heat input, reduces the problem of excessive melt-through.

These electrodes have complete slag coverage, and beads are formed with distinct, even ripples, Fig. 12-4D and E. Many production shops use them as general-purpose electrodes, and they are also widely used for repair work. Although they may be used for all-position work, the fast-freeze electrodes are preferable. Many shops that engage in light gauge sheet steel fabrication use the fast-follow electrode for vertical position, travel-down welding. Examples of these electrodes are the EXX12 for a.c. and DCEN welding and the EXX13 for a.c., DCEN, and DCEP welding.

Fast Freeze Fast-freeze electrodes have the ability to deposit weld metal that solidifies, or freezes, rapidly. This is important when there is some chance of slag or weld metal spilling out of the joint and when welding in the vertical or overhead positions.

These electrodes have a snappy, deep-penetrating arc. They have little slag and produce flat beads, Fig. 12-4B and C. With few exceptions, they produce X-ray quality weld deposits and are used for pipe and pressure vessel code work. They are widely used for all-position welding, in both fabrication and repair.

Combination Types Some joints require the characteristics of both fast-fill and fast-freeze electrodes. When fast-freeze is required, the best electrodes are the EXX10, DCEP, EXX11 a.c., and DCEP types. An electrode with a share of both characteristics is the all-position iron powder electrode EXX14. These electrodes do not have as much fast fill as an EXX24, nor do they have the degree of fast freeze of the EXX10. Rather, they are a compromise between the two types.

Low Hydrogen Low hydrogen electrodes are those that have coverings containing practically no hydrogen. They produce welds that are free from underbead and microcracking and have exceptional ductility, Fig. 12-4F and K. The electrodes eliminate porosity in sulfur-bearing steels and ensure X-ray quality deposits. Because they reduce preheat requirements, their chief use is in the welding of hard-to-weld steels and high-tensile alloy steels. Examples of these types of electrodes are the EXX18 and EXX28 classifications.

Iron Powder Iron powders are added to the covering of many types of electrodes. In the intense arc heat, the iron powder is converted to steel, thus contributing additional metal to the weld deposit. When iron powder has been added to the electrode covering in relatively large amounts (30 percent or more), the speed of welding is appreciably increased, the arc is stabilized, spatter is reduced, and slag removal is improved, Fig. 12-5. The weld appearance is very smooth, Fig. 12-4, welds H and I. Examples of these types of electrodes are the EXX14, EXX18, EXX24, EXX27, and EXX28.

Fig. 12-4 Comparative appearance of weld beads made with different types of electrodes. Edward R. Bohnart

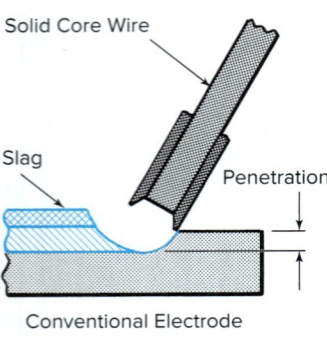

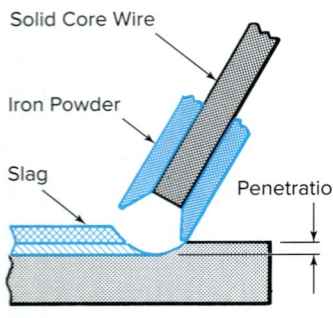

Fig. 12-5 A thicker coating of iron powder on an electrode creates a crucible effect at the tip of the electrode, making more efficient use of arc energy. They are sometimes referred to as drag rods because the coating can be dragged on the workpiece.

Type of Base Metal

The nature of the material to be welded is of prime importance. Satisfactory welds cannot be made if the weld metal deposited does not have the same physical and chemical qualities as the material being welded.

If the material specifications are not known, simple tests such as the spark test, torch test, chip test, magnetic test, color test, fracture test, and sound test may be performed. The information in Tables 12-9 through 12-12 will help you recognize the various metals according to their characteristics. Such information can, however, identify only broad categories of material. The welder may be able to tell, for example, that the material is steel rather than cast iron, but there are so many types of alloy and stainless steels that it is necessary to know the correct analysis of the steel as designated by the manufacturer. For example, if the steel is high in sulfur or carbon, or if it contains certain alloys, the E7015, E7016, E7018, or E7028 electrode should be chosen. These electrodes reduce the tendency toward underbead cracking that is characteristic of such steels. They produce welds of high tensile strength and ductility without stress relieving, and they reduce or eliminate the need for preheat.

Nature of Welding Current

Welding machines produce two types of welding current, alternating current and direct current. Direct current can be either electrode negative (DCEN) or electrode positive (DCEP). The nature of the current available also influences the selection of the electrode. For example, if only a.c. equipment is available, E6010 and E7015 electrodes are eliminated from consideration, since they are designed to operate on direct current. If only d.c. equipment is available, current is not a limiting factor. Although designed for use primarily with alternating current, E6011, E6013, E7014, E7016, E7018, and E7028 electrodes perform adequately with direct current.

Thickness and Shape of Material to Be Welded

Whether the material is of heavy gauge or light gauge, partially determines electrode size. *As a general rule, never use an electrode having a diameter that is larger than the thickness of the material being welded.*

For light gauge sheet metal work ($3/32$ inch and thinner), the E6013 electrode is usually the choice. The E6013 electrode was designed for this type of work, and it has the least penetration of any electrode in the E60XX series.

Joint Design and Fitup

There are many kinds of welded joints, and each type has particular requirements for welding. Welding fillets differ a great deal from groove welding butt joints. Butt joints may be either square or deep grooved. The fitup may have large gaps, or it may be too tight.

In cases in which poor fitup is unavoidable, the E6013 electrode should be chosen for use with the d.c. equipment and with a.c. equipment. This electrode bridges gaps very well because of the globular transfer of metal through the arc stream.

ABOUT WELDING

Electric Arc Lighting

The notion of electric arc lighting began in London with the experiments of Sir Humphrey Davy in 1801. He showed, using a battery, that an arc could be made to form between two carbon electrodes. Thirty-five years later, his cousin Edmund Davy described the properties of acetylene (which, in another 34 years, a German, Friedrich Wohler, figured out how to produce from calcium carbide).

Table 12-9 Temperature Data for Metals and Alloys

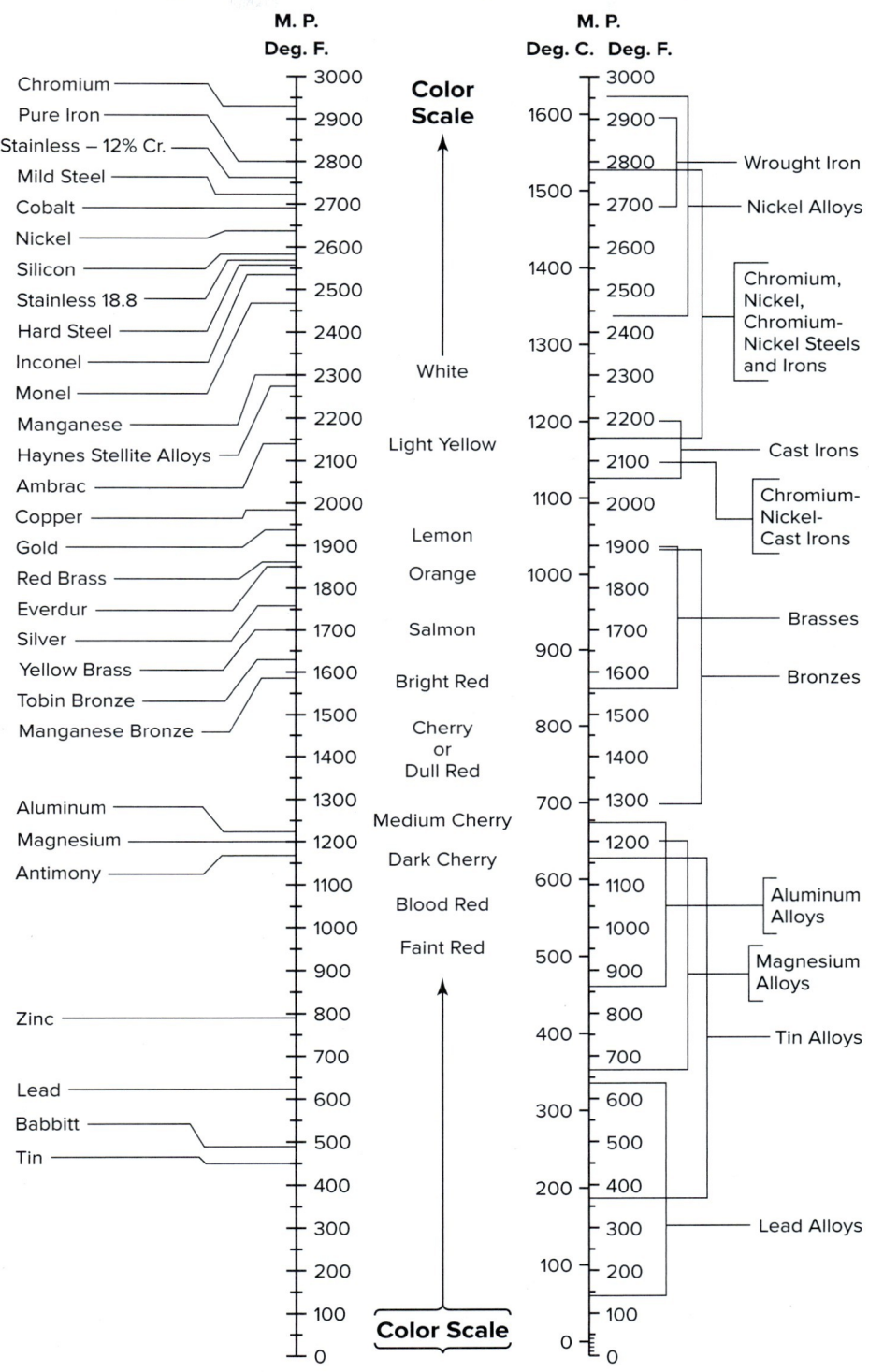

Table 12-10 Identification of Metals by Appearance

	Alloy Steel	Copper	Brass and Bronze	Aluminum and Alloys	Monel	Nickel	Lead
Fracture	Medium gray	Red color	Red to yellow	White	Light gray	Almost white	White; crystalline
Unfinished surface	Dark gray; relatively rough; rolling or forging lines may be noticeable	Various degrees of reddish brown to green due to oxides; smooth	Various shades of green, brown, or yellow due to oxides; smooth	Evidence of mold or rolls; very light gray	Smooth; dark gray	Smooth; dark gray	Smooth; velvety; white to gray
Newly machined	Very smooth; bright gray	Bright copper red color dulls with time	Red through to whitish yellow; very smooth	Smooth; very white	Very smooth; light gray	Very smooth; white	Very smooth; white

	White Cast Iron	Gray Cast Iron	Malleable Iron	Wrought Iron	Low Carbon Steel and Cast Steel	High Carbon Steel
Fracture	Very fine silvery white, silky crystalline formation	Dark gray	Dark gray	Bright gray	Bright gray	Very light gray
Unfinished surface	Evidence of sand mold; dull gray	Evidence of sand mold; very dull gray	Evidence of sand mold; dull gray	Light gray; smooth	Dark gray; forging marks may be noticeable; cast—evidences of mold	Dark gray; rolling or forging lines may be noticeable
Newly machined	Rarely machined	Fairly smooth; light gray	Smooth surface; light gray	Very smooth surface; light gray	Very smooth; bright gray	Very smooth; bright gray

Source: American Welding Society

Welding Position

The position of welding is a very important consideration in the choice of an electrode. Certain types of electrodes can be used only in the flat position; others perform equally well in all positions. The type of position also has an influence on costs. Welding is most economical in the flat position, then horizontal, and then vertical; the overhead position is the least economical. As you develop skill in welding, you will understand the limitations that welding in the vertical and overhead positions places on the choice of an electrode.

The size of the electrode to be used is strongly influenced by the position of welding. V-butt joints in the vertical and overhead positions are usually welded with a small diameter electrode, in order to obtain complete penetration at the root of the weld. In multilayer welding, the other passes can be made with large electrodes. Welding in the vertical and overhead positions should never be attempted with an electrode larger than $3/16$ inch in diameter.

For production work, the largest electrode size that can be handled should be used. This permits higher welding current values, thus increasing the speed of welding. Higher deposition rates are also achieved. The larger the diameter of the electrode, the greater the quantity of weld deposited in a unit of time. The cost of labor is also reduced because fewer stops are necessary to change electrodes.

If welding must be done in an overhead, vertical, or horizontal position, electrodes of the EXX20, EXX24, EXX27, and EXX28 classifications cannot be used, leaving the choice to be made among the remaining electrodes in the series. The EXX15, EXX16, and EXX18 electrodes, though classified for all positions, are more difficult to handle in the vertical and overhead positions.

In general, the welder will find that electrodes in the EXX12, EXX13, EXX20, EXX24, EXX27, and EXX28 classifications are easiest to handle in the horizontal and flat positions. Vertical and overhead welding is easiest with the EXX10 and EXX11.

Table 12-11 Identification of Metals by Flame Test (Footnotes are listed below Table 12-12)

	Alloy[1] Steel	Copper	Brass and Bronze	Aluminum and Alloys[2]	Monel	Nickel	Lead[3]
Speed of melting (from cold state)		Slow	Moderate to fast	Faster than steel	Slower than steel	Slower than steel	Very fast
Color change while heating		May turn black and then red; copper color may become more intense	Becomes noticeably red before melting	No apparent change in color	Becomes red before melting	Becomes red before melting	No apparent change
Appearance of slag		So little slag that it is hardly noticeable	Various quantities of white fumes, though bronze may not have any	Stiff black scum	Gray scum; considerable amounts	Gray scum; less slag than Monel	Dull gray coating
Action of slag		Quiet	Appears as fumes	Quiet	Quiet; hard to break	Quiet; hard to break	Quiet
Appearance of molten pool		Has mirror-like surface directly under flame	Liquid	Same color as unheated metal; very fluid under slag	Fluid under slag	Fluid under slag film	White and fluid under slag
Action of molten pool under blowpipe flame		Tendency to bubble; pool solidifies slowly and may sink slightly	Like drops of water; with oxidizing flame will bubble	Quiet	Quiet	Quiet	Quiet; may boil if too hot

	White Cast Iron[4]	Gray Cast Iron	Malleable Iron[5]	Wrought Iron	Low Carbon Steel and Cast Steel	High Carbon Steel
Speed of melting (from cold state)	Moderate	Moderate	Moderate	Fast	Fast	Fast
Color change while heating	Becomes dull red before melting	Becomes dull red before melting	Becomes red before melting	Becomes bright red before melting	Becomes bright red before melting	Becomes bright red before melting
Appearance of slag	A medium film develops	A thick film develops	A medium film develops	Oily or greasy appearance with white lines	Similar to molten metal	Similar to molten metal
Action of slag	Quiet; tough, but can be broken up	Quiet; tough, but possible to break it up	Quiet; tough, but can be broken	Quiet; easily broken up	Quiet	Quiet
Appearance of molten pool	Fluid and watery; reddish white	Fluid and watery; reddish white	Fluid and watery; straw color	Liquid; straw color	Liquid; straw color	Lighter than low carbon steel; has a cellular appearance
Action of molten puddle under blowpipe flame	Quiet; no sparks; depression under flame disappears when flame is removed	Quiet; no sparks; depression under flame disappears when flame is removed	Boils and leaves blowholes; surface metal sparks; interior does not	Does not get viscous; generally quiet; may be slight tendency to spark	Molten metal sparks	Sparks more freely than low carbon steel

Table 12-12 Identification of Metals by Chips

	Alloy[1] Steel	Copper	Brass and Bronze	Aluminum and Alloys[2]	Monel	Nickel	Lead[3]
Appearance of chip		Smooth chips; saw edges where cut	Smooth chips; saw edges where cut	Smooth chips; saw edges where cut	Smooth edges	Smooth edges	Any shaped chip can be secured because of softness
Size of chip		Can be continuous if desired	Can be continuous if desired	Can be continuous if desired	Can be continuous if desired	Can be continuous if desired	Can be continuous if desired
Facility of chipping		Very easily cut	Easily cut; more brittle than copper	Very easily cut	Chips easily	Chips easily	Chips so easily it can be cut with penknife

	White Cast Iron[4]	Gray Cast Iron	Malleable Iron[5]	Wrought Iron	Low Carbon Steel and Cast Steel	High Carbon Steel
Appearance of chip	Small broken fragments	Small partially broken chips, but possible to chip a fairly smooth groove	Chips do not break short as in cast iron	Smooth edges where cut	Smooth edges where cut	Fine grain fracture; edges lighter in color than low carbon steel
Size of chip		1/8 in.	1/4–3/8 in.	Can be continuous if desired	Can be continuous if desired	Can be continuous if desired
Facility of chipping	Brittleness prevents chipping a path with smooth sides	Not easy to chip because chips break off from base metal	Very tough, therefore harder to chip than cast iron	Soft and easily cut or chipped	Easily cut or chipped	Metal is usually very hard, but can be chipped

[1] Alloy steels vary so much in composition and consequently in results of tests that experience is the best solution to identification problems.
[2] Because of white or light color and extremely light weight, aluminum is usually easily distinguishable from all other metals; aluminum alloys are usually harder and slightly darker in color than pure aluminum.
[3] Weight, softness, and great ductility are distinguishing characteristics of lead.
[4] Very seldom used commercially.
[5] Malleable iron should always be braze welded.

Conditions of Use

The service requirements are of utmost importance. The type of structure and the stress that it will encounter in use must be considered. Tensile strength, ductility, and fatigue resistance are important weld characteristics that help to determine choice of electrode. Note the variation in weld metal properties among the electrodes compared in Table 12-3, page 303.

Engineering Specifications

All code requirements and engineering specifications must be noted carefully in any determination of the correct electrode to be used. The type of electrode to be used is specified in the code requirements.

Production Efficiency

Several electrode classifications have high deposition characteristics. Compare the rates of deposition given in Table 12-4, page 304. Full advantage cannot always be made of these characteristics because of the nature of the material, the type of joint, and the position of the work. For example, the electrodes classified as having a very high rate of deposition in Table 12-4 can be used only for flat and horizontal fillet welds.

The principal factor in the cost of a welding job is the speed with which the welding can be done. Electrode cost is small by comparison. E7024, E7028, and E6027 electrodes permit the highest rate of deposition; E6020, E6013, E6011, and E6010 follow in the order given. Type of steel, not speed, should govern the choice of the E7015 or E7018. The E7028 is similar to the E7018, but it has a much thicker coating that contains a higher percentage of iron powder; thus, its deposition rate is much higher.

Welding speed is increased by using electrodes with large diameters, particularly in flat and horizontal position welding. The E7024, E7028, and E6020 classifications, followed by the E6027, give the highest increases in welding speeds as the diameters of the electrodes are increased.

Job Conditions

Is the material clean, rusty, painted, or greasy? What is the type of surface treatment required for the finished job? Is the completed job to be stress relieved or heat treated? Are the welds in a prominent location so that weld appearance is important? Only a welder with a thorough knowledge of electrode characteristics can answer these questions by choosing the best electrode for the job. It is important to study manufacturer's specifications.

Summary of Factors Affecting the Selection of Electrodes

The foregoing are just a few of the considerations that make it necessary for you to become highly familiar with the nature of the different electrode classifications.

Selection of the proper electrode size and type for a given welding job requires a thorough knowledge of electrodes coupled with common sense. Careful study of the physical, chemical, and working characteristics of electrodes and practice in their use will enable you to make the proper selection without difficulty.

In summary, the following factors are the most important to consider:

- Type of joint and position of welding
- Type of welding current
- Properties of the base metal
- Thickness of the base metal
- Depth of penetration desired
- Weld appearance desired
- Whether the work is required to meet code specifications
- Tensile strength, ductility, and impact strength required of the weld deposit
- Design and fitup of the joint to be welded
- Nature of slag removal

Specific Electrode Classifications

Carbon Steel Electrodes

The carbon steel electrodes for welding low and medium carbon steels carry AWS classification numbers E6010, -11, -12, -13, -20, -27 or E7014, -15, -16, -18, -24, -28, and -48.

Heavily Covered Mild Steel, Shielded Arc Electrodes

E6010: All-Position, DCEP (Fast-Freeze Type) This electrode is the best adapted of the shielded arc types for vertical and overhead welding. It is, therefore, the most widely used electrode for the welding of steel structures that cannot be readily positioned and which require considerable multiple-pass welding in the vertical and overhead positions. Although the majority of applications are on mild steel, E6010 electrodes may be used to advantage on galvanized plate and on some low alloy steels. In welding steel, the forceful arc bites through the galvanizing and the light slag to reduce bubbling and prevent porosity. Typical applications include shipbuilding, structures such as buildings and bridges, storage tanks, pipe welding, tanks, and pressure vessels.

The quality of the weld metal is of a high order (Fig. 12-4B, p. 318), and the specifications for this classification are correspondingly rigid. The essential operating characteristics of the electrode are:

- Strong and penetrating arc, enabling penetration beyond the root of the butt or fillet joint.
- Quickly solidifying weld metal, enabling the deposition of welds without excessive convexity and undercutting.

> **JOB TIP**
>
> **An Interview**
>
> In an interview for the Hobart Institute, welding manager Anthony Morgan explained what he looks for in hiring:
>
> 1. People who view their work with a sense of pride.
> 2. People who want to push their limits to see how much can be learned and accomplished.
> 3. A strong work ethic.

- Low quantity of slag with low melting and low density characteristics, so that it does not become entrapped nor interfere with oscillating and weaving techniques.
- Adequate gaseous atmosphere to protect molten metal during welding. Electrodes of this type are usable only with DCEN.

The E6010 electrode is commonly classified as the *cellulosic* type. The electrode coating contains considerable quantities of cellulose, either in a treated form or as wood flour or other natural forms. During welding, the cellulose is changed to carbon dioxide and water vapor, forming the gaseous envelope that excludes the harmful oxygen and nitrogen in the air. The water vapor from minerals containing water and the vapor retained by the binder are also liberated during welding.

This type of cellulose electrode needs a certain amount of moisture present in its coating and should not be stored in dry rod ovens. If improperly stored, operating characteristics will be adversely affected.

The slag-forming materials of the E6010 covering include titanium dioxide and either magnesium or aluminum silicates. Ferromanganese is used as a **deoxidizer**, or **degasifier**, as it is often called. Since there is usually no increase in manganese in the weld deposit over that of the core wire, the manganese enters the slag as an oxide. The common binder for the coating materials is sodium silicate solution, which also is a slag-forming material. The core wire is low carbon rimmed steel. It usually contains 0.10 to 0.15 percent carbon, 0.40 to 0.60 percent manganese, a maximum of 0.40 percent sulfur and phosphorus, and a maximum of 0.025 percent silicon.

E6011: All-Position, Alternating Current and DCEP (Fast-Freeze Type) The operating characteristics, mechanical properties, and welding applications of the E6011 resemble those of the E6010, but the E6011 requires alternating current. Although it may also be used with DCEP, it loses many of its beneficial characteristics with this polarity.

The penetration, arc action, slag, and fillet-weld appearance are similar to those obtained with the E6010 type, Fig. 12-4C, page 309. The weld deposit is free from porosity, holes, and pits. The slag can be removed readily. Fillet and bead contours are flat rather than convex. E6011 electrodes may be used in all-position welding.

The E6011 coverings are classified as the high cellulose potassium type. Small quantities of calcium and potassium are present in addition to the other ingredients usually found in the E6010-type coverings. The core wire is identical to that used for E6010 electrodes.

E6013: All-Position, Alternating Current and DCEN or DCEP (Fill-Freeze Type) Slag removal is easy, and the arc can be established and maintained readily. This is especially true of electrodes with small diameters ($1/16$, $5/64$, and $3/32$ inch). Consequently, it permits satisfactory operation with lower open-circuit voltage. Originally, this electrode was designed specifically for light gauge sheet metal work and for vertical welding from the top down.

The covering of the E6013 contains rutile, siliceous materials, cellulose, ferromanganese, potassium, and liquid silicate binders. An important difference is that easily ionized materials are incorporated in the covering. This feature permits the establishment and maintenance of an arc with alternating current at low welding currents and low open-circuit voltages.

The molten metal is slightly more fluid than that of the E6010 electrode, but not to the extent that the E6013 cannot be used in all-position welding. The molten metal and slag characteristics control the shape of the weld deposit, Fig. 12-4E, page 309. The arc action tends to be quiet, and the bead surface is smoother, with a finer ripple. These electrodes are suitable for making fillet welds and groove welds with a flat or slightly convex appearance.

E7014: All-Position, Alternating Current and DCEN or DCEP (Fast-Fill Type) The covering of this electrode is similar to that of the E6013, but the addition of iron powder makes it much thicker. This thicker coating forms a crucible that allows this electrode to be dragged and the correct arc length is maintained. The deposition rate is somewhat higher.

The E7014 is suitable for welding mild steel in all positions. The weld beads have a smooth surface with fine ripples, and the slag is easily removed. The fillet welds made with the E7014 are flat to slightly convex. It is a good electrode for production welding on plate of medium thickness.

Low Hydrogen Electrodes Low hydrogen electrodes are a result of research during World War II. The object of this research was to find an electrode for welding armor plate that would require the use of less strategic alloys. Today, these electrodes are no longer considered emergency tools. They are used because they have superior mechanical properties and because many are custom-made to match the heat-treating properties of alloy steels.

The name stems from the fact that the coatings are free of minerals that contain hydrogen. The lack of hydrogen is an important characteristic because hydrogen causes underbead cracking in high carbon and alloy steels. By eliminating hydrogen, underbead cracking is prevented and difficult steels can be welded with little or no preheat. These electrodes also produce porosity-free welds in high sulfur steels and eliminate hot-shortness in phosphorus-bearing steels. The addition of iron powder in the coating increases the deposition (melting) rate.

SMAW electrodes that have classifications ending in 5, 6, or 8 are considered low hydrogen electrodes. These electrodes are low in hydrogen-bearing compounds, so only traces of hydrogen and moisture are present in the arc atmosphere. The core contains 0.08 to 0.13 percent carbon, 0.40 to 0.60 percent manganese, and a maximum of 0.04 percent sulfur and phosphorus. A typical analysis of the deposit from a low hydrogen electrode is 0.08 percent carbon, 0.56 percent manganese, and 0.25 percent silicon.

A low hydrogen electrode has a core of mild steel or low alloy steel. The mineral covering consists of alkaline earth carbonates, fluorides, silicate binders, and ferroalloys. This covering produces the desired weld metal analysis and mechanical properties. E7018 and E7028 have iron powder in their covering. During welding, the covering forms a carbon dioxide shield around the arc. On the job, these electrodes must not be exposed to humid air because of their tendency to absorb a considerable amount of moisture.

A wide range of weld properties is possible by adding a number of alloying elements to the covering. Such additions may include carbon, manganese, chromium, nickel, molybdenum, and vanadium.

The arc is not harsh and is moderately penetrating. The slag is heavy, friable (easily crumbled), and easily removed. A short arc must be used in welding. A long arc permits **hydrogen pickup** (an increase in moisture), which causes porosity and slag inclusions in the weld deposit. High quality, radiographically sound welds can be made with proper welding techniques. See Fig. 12-4F, J, and K.

Although low hydrogen electrode sizes up to and including $5/32$ inch may be used in all positions, they are not truly all-position welding electrodes such as those in the E6010 classification. The larger diameters are useful for horizontal fillet welds in the horizontal and flat positions.

The mechanical properties produced by these electrodes are far superior to those furnished by conventional electrodes, such as the E6010 and E6013. Tensile strengths of 120,000 p.s.i. and better, with high ductility, are obtainable in the as-welded condition. With heat treatment, tensile strength can be increased to as much as 300,000 p.s.i.

Following is a description of each electrode in this class.

E7015: All-Position, DCEP (Low Hydrogen) The coating of this electrode is high in calcium compounds and low in hydrogen, carbon, manganese, sulfur, and phosphorus. It contains a trace of silicon. It is referred to as the low hydrogen, sodium type because sodium silicate is used as a binder for the covering.

The arc is moderately penetrating. The slag is heavy, friable, and easily removed. The deposited metal is flat and may be somewhat convex. Welding with a short arc is essential for low hydrogen, high quality deposits. Welding in all positions is possible with sizes up to $5/32$ inch. Larger electrodes can be used in the horizontal and flat positions.

These electrodes are recommended for the welding of alloy steels, high carbon steels, high sulfur steels, malleable iron, sulfur-bearing steels, steels that are to be enameled, spring steels, and for the mild steel side of clad plates. Very often pre- and postheating may be eliminated by the use of this electrode.

E7016: All-Position, Alternating Current and DCEP (Low Hydrogen) The E7016 electrode has all the characteristics of the E7015 type. An added advantage is that it may be used with either alternating or direct welding current. The core wire and covering are similar to the E7015, except for the addition of potassium silicate or other potassium salts. The potassium compounds make these electrodes suitable for alternating current. A typical weld produced by this electrode is shown in Fig. 12-4J.

E7018: All-Position, Alternating Current and DCEP (Low Hydrogen, Iron Powder) The coating of this electrode contains a high percentage of iron powder, from 25 to 40 percent, in combination with low hydrogen ingredients. The covering of the E7018 is similar to, but thicker than, the covering of the E7015 and E7016 electrodes. The slag is heavy, friable, and easily removed. The deposited metal is flat, Fig. 12-4K, and its appearance is better than welds made with the E7015. The deposit may be slightly convex in a fillet or groove weld.

Welding may be done in all positions with electrode sizes up to $5/32$ inch. Larger diameters are used for fillet and groove welds in the horizontal and flat positions. A short arc must be held at all times, and special care taken to keep the covering in contact with the molten pool when welding in the vertical-up position. A long arc will cause porosity in the weld deposit. The deposition rate of E7018 is somewhat higher than that of the E7015.

The strength of the deposited weld metal can be improved through the addition of certain alloys to the coverings, rather than by changing the composition of the core wire. Adding alloying elements to the coating is more economical and can be better controlled. These electrodes are available in the E8018 through E12018 classifications (tensile strength 80,000 p.s.i. through 120,000 p.s.i.).

Usually the applications of the E7018 electrodes require specific mechanical and chemical properties so that the weld metal will meet the requirements of the base metal, adjust to stress-relieving and heat treatment, withstand extreme loading and fatigue, and resist cracking.

E7028: Horizontal and Flat Positions, Alternating Current and DCEP (Low Hydrogen, Iron Powder) These electrodes are similar to the E7018 electrodes, with some differences. The E7028 is suitable only for horizontal and flat position welding, whereas the E7018 can be used in all positions. The coating of the E7028 contains a higher percentage of iron powder (50 percent) than the E7018 and, as a result, is thicker and heavier. The deposition rate of the E7028 is higher than the rate of the E7018.

The penetration is not deep, and the weld appearance is flat to concave, with a smooth, fine ripple, Fig. 12.4F. The slag coating is heavy and easily removed. The E7028 has the characteristics of the fast-fill type of electrode, Fig. 12-6, page 318.

E7048: Flat, Horizontal, Overhead, and Vertical Down Positions, Alternating Current and DCEP (Low Hydrogen, Iron Powder) These electrodes are similar to the E7018 electrodes.

SHOP TALK

Electrical Burns
An electrical burn is similar to a thermal burn. If there is an accident, turn off the power. Then, use a nonconductive material to pull the person away from the live wires. If they are not breathing, perform CPR. Use ice or clean, cold compresses on the burn. Obtain medical attention immediately.

Fig. 12-6 Deep groove welds have an excellent appearance and wash-in, easy slag removal. E7028 iron powder electrodes produce a smooth, clean, flat cover pass. *Edward R. Bohnart*

However, they have been especially formulated for vertical down welding techniques. These techniques are popular with this process on cross-country pipelines. Because of the increasingly stronger pipe being specified for line pipe, the cellulose-type electrodes do not possess the required physical or mechanical properties. This has brought about the development of the E7048 electrode.

Other Covered Electrodes

E7024: Horizontal Fillet and Flat Position, Alternating Current and DCEN or DCEP (Iron Powder, Titania) The E7024 classification indicates a covering with a high percentage of iron powder, usually 50 percent of the weight of the covered electrode, in combination with fluxing ingredients similar to those commonly found in the E6013 electrodes. The E7024 electrodes may be used for those applications that usually require E6013.

E7024 electrodes are also referred to as *contact electrodes*, since the electrode coating may rest on the surface of the joint to be welded. During actual welding, the electrode *drags*, resulting in an effective shielding of the weld pool from the atmosphere. Many welders prefer to hold a short, free arc. In addition to melting the core wire and the base metal, the arc melts iron powder in the electrode coating to provide greater deposition of metal per ampere, Fig. 12-5, page 310. Hence, it makes possible greater welding speed. It has been determined that one-third of the weld metal deposited comes from the covering.

In comparing the E7024 electrodes with conventional electrodes, contact welding has the following advantages:

- Less weld spatter and, therefore, higher deposition efficiencies even at high welding current
- Lower nitrogen content within the weld metal
- Sounder metal with less tendency for such defects as cracking, porosity, and slag inclusions
- Welds that are practically self-cleaning and have greater freedom from spatter
- Smoother weld appearance, approaching the bead quality obtained by automatic welding

E7024 electrodes are well suited for fillet welds in mild steel. The welds are slightly convex in profile. They have a very smooth surface and an extremely fine ripple that approaches the appearance of machine-made welds, Figs. 12-7 and 12-8. The electrode is characterized by a smooth, quiet arc, very low spatter, and low penetration. It can be used at high lineal speed. It operates with alternating current, DCEN, or DCEP, although alternating current is preferable.

Alloy Steel Electrodes

Because of the high strength requirements of many industrial fabrications, the use of high strength alloy steels has increased a great deal. The ability to fabricate by welding

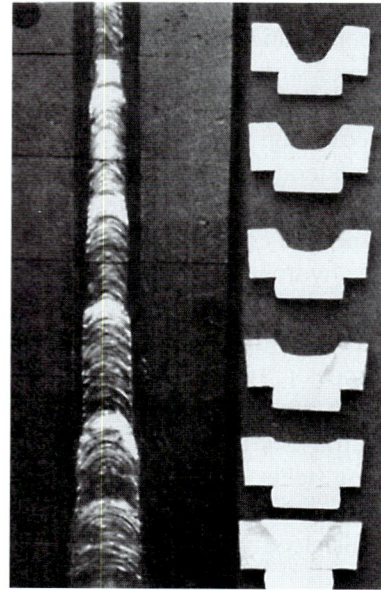

Fig. 12-7 A multipass groove weld, bevel butt joint in ¾-inch plate welded with an iron powder electrode (E7024). *Lincoln Electric*

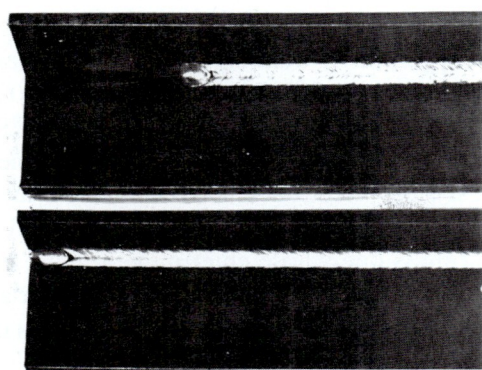

Fig. 12-8 Iron powder electrodes provide high welding speeds with good appearance. The bottom joint was made with a single iron powder electrode, and the top was welded with a single E6013 electrode. The iron powder electrode was 36 percent faster and gave the weld a better appearance.
Edward R. Bohnart

is a major factor in this increase. A shielded arc electrode capable of producing weld deposits with a tensile strength exceeding 100,000 p.s.i. has been developed. The core wire of this electrode is composed of alloy steel instead of low carbon steel. The electrode covering has the lime-ferritic nature typical of the low hydrogen types, and it may contain powdered iron.

Operating characteristics of electrodes for welding alloy steel resemble those of the low hydrogen electrodes in the E7015, -16, and -18 classifications. They are available, however, in tensile-strength classifications of 80XX, 90XX, 100XX, 110XX, and 120XX.

You will recall that the first two or three digits of the electrode identification number indicate tensile strength. Thus, the E11018 electrode is rated at a tensile strength of 110,000 p.s.i., whereas the E7018 electrode has a tensile strength of only 70,000 p.s.i. Both electrodes weld in all positions, use alternating current or DCEP, and have 30 percent iron powder in their covering. They have medium arc force and penetration, and their slag is heavy, friable, and easily removed.

The Stainless-Steel Electrodes Stainless steel is the popular term for the chromium and chromium-nickel steels. It is a tough, strong material that is highly resistant to corrosion, high temperatures, oxidation, and scaling. There is a large variety of stainless steels and electrodes to weld them. Both the base metals and the electrodes are expensive, and should be handled with care.

Metallurgically, the stainless steels are classified as martensitic, ferritic, precipitation-hardening, duplex, and austenitic. In order to understand the various uses for these steels, it is important that you have an understanding of these terms.

- **Martensitic stainless steel** is an air-hardening steel containing chromium as its principal alloying element in amounts ranging from 4 to 12 percent. It is normally hard and brittle. Martensitic stainless steel requires both preheating and postheating for welding.
- **Ferritic stainless steel** is a magnetic, straight-chromium steel containing 14 to 26 percent chromium. It is normally soft and ductile, but it may become brittle when welded. Preheating and postheating are necessary for successful welding.
- **Precipitation-hardening (PH) stainless steel** has the ability to develop high strength with reasonably simple heat treatment. The precipitation hardening is promoted by one or more alloying elements, such as copper, titanium, niobium, and aluminum. The PH stainless steels do not require preheat. SMAW electrodes are not available for all the PH stainless-steel types. Thus, high joint strengths may not be available if a matching electrode is not available. Where high-strength welds are not required, standard austenitic stainless-steel electrodes such as E308 or E309 can be used.
- **Duplex stainless steel (DSS)** is characterized by a low carbon, body-centered-cubic ferrite, face-centered-cubic austenite microstructure. Duplex types resist stress corrosion cracking and pitting, and have yield strengths twice that of the 300 series alloys. A typical DSS alloy of 2304 would contain 0.030 percent carbon, 21.5 to 24.5 percent chromium, 3.0 to 5.5 percent nickel, 0.05 to 0.6 percent molybdenum, and 0.05 to 0.20 percent nitrogen. These alloys are easier to weld than the ferrite types, but are more difficult to weld than the austenitic types. Postweld heat treatment (PWHT) is normally not necessary or recommended. SMAW electrodes are readily available for the common DSS grades.
- **Austenitic stainless steel** contains both chromium and nickel. The nickel content usually ranges from 3.5 to 22 percent, and the chromium content from 16 to 26 percent. It is strong, ductile, and resistant to impact. Austenitic stainless steel is nonmagnetic when annealed but slightly magnetic when cold worked. Heat treatment is not necessary during welding.

Selection of the proper type of electrode for stainless steel welding is critical. A different welding method is required for each type. The resultant weld must have the required tensile strength, ductility, and corrosion resistance equal to the base metal. There are also the problems of matching the color of the base metal and producing a smooth bead that will require a minimum amount of grinding.

Table 12-13 Identification of Stainless-Steel Electrodes

Austenitic or Chromium Nickel Types AISI 300 Series

AISI Designation	Popular Trade Names	AWS Electrode Recommended
302	18/8 or 19/9	E308-15, E308-16
[1]303		
304		
308		
309	25/12	E309-15, E309-16, [2]E309Cb-15 E310-15, E310-16
310	25/20	[2]E310Cb-15, [1]E310Cb-16, [2]E310Mo-15, [1]E310Mo-16
312	29/9	E312-15, E312-16, E316-15, E316-16
316	18/12Mo	[2]E316Cb-15, E316Cb-16
317	18/12Mo	E317-15, E317-16
318	18/12Mo Cb	
330	15/35	E330-15, E330-16
347	18/8Cb	E347-15, E347-16

Straight-Chromium Types

AISI 400 Series		
410	12 chromium	E410-15
430	16 chromium	E430-15
442	18 chromium	E442-15
446	28 chromium	E446-15
AISI 500 Series		
502	5 chromium	E502-15, 502-18 505-18
505	9 chromium	E502-16, 505-18

[1]AISI 303 has additional sulfur added to improve machinability. Use lime type E308-15 when welding.
[2]Niobium and/or molybdenum are added to these coatings to aid in prevention of carbide precipitation or to improve strength at elevated temperatures (decrease creep rate), respectively.

The identification numbers for stainless-steel electrode classifications are somewhat different from those used in the AWS system for carbon steel electrodes. They are based on the American Iron and Steel Institute (AISI) classifications of metal alloys, Table 12-13. The first digits in the identification number refer to the AISI metal classification number instead of the tensile strength. Thus, the electrodes in the E308XX through E309XX series are suitable for austenitic stainless steels. (See Chapter 3, pp. 87 and 92, for more information on the AISI classification system.)

The last two digits of the electrode identification number refer to the position of welding and operating characteristics as given in the AWS classification system. E309-15, for example, may be used in all positions with DCEP. It is a low hydrogen electrode with a medium arc and penetration.

Stainless-steel electrodes are designated as having *lime, lime-titania,* or *titania* coverings. In general, a lime-type covering is one whose chief mineral ingredients include limestone and fluorspar. It contains minor amounts of titanium dioxide (up to 8 percent). Coverings containing more than 20 percent titanium dioxide are usually considered as the titania type; those containing between 8 and 20 percent are considered to be the lime-titania type.

The lime-type covering is designed for DCEP only. These electrodes produce welds that have a convex face and are desirable for root passes in which the full throat section prevents cracking. They may be used in the vertical and overhead positions. The slag completely covers the weld, provides a rapid wetting action, and produces welds with a minimum amount of spatter. The coating also produces a flux that drives the impurities from the weld, thereby ensuring a weld that is free of porosity and has the mechanical and corrosion properties expected. These electrodes are in the E3XX-15 classification.

The titania-type covering is designed for either alternating current or DCEP. Electrodes with such coverings are preferred to the lime-type electrodes because of the smooth arc action, fine bead appearance, and very easy slag removal. They produce slightly concave welds that require a minimum of cleaning, grinding, and polishing time. These electrodes are in the E3XX-16 classification. They are also available for DCEP only. Electrodes with the lime-titania covering may be used with DCEP or DCEN only or with alternating current, DCEP, or DCEN. They are all-position electrodes. They weld the straight chromium (ferritic) and chromium-molybdenum stainless steels and, to some extent, the chromium-nickel (austenitic) stainless steels.

Lime is used for the electrode covering because it tends to eliminate hydrogen, which causes underbead cracking. Since chromium has an affinity for carbon and materials high in carbon, carbon compounds such as alkaline earth carbonatis, which is used to eliminate hydrogen in low hydrogen electrodes, cannot be put in the covering for stainless-steel electrodes. Manganese and silicon are included to reduce oxidation. Titanium promotes arc stability, produces an easily removable slag, and prevents carbon precipitation. Niobium also prevents carbon precipitation.

Hard-Facing Electrodes

Hard-facing is the deposition of an alloy material on a metal part by one of several welding processes to form a protective surface. This can also be done by metal spraying. Depending on the alloy added, the surface resists abrasion, impact, heat, corrosion, or a combination of these factors.

- **Abrasion** is the wear on surfaces that are subjected to continuous grinding, rubbing, or gouging action. The forces move parallel to the surface of the component.
- **Impact** causes metal to be lost or deformed as a result of chipping, upsetting, cracking, or crushing forces. The force is a striking action that is perpendicular to the absorbing members.
- **Corrosion** is the destruction of a surface from atmospheric chemical contamination and from oxidation or scaling at elevated temperatures.

Hard-facing may be applied to new parts to improve their resistance to wear during service or to worn parts to restore them to serviceable condition. The wear-resistant material is applied only to those surfaces of a component where maximum wear takes place.

Alloys for hard-facing are usually available as bare cast or tubular rod, covered solid or tubular electrodes, or solid wire and powder. Electrodes for metal arc welding have been classified on the basis of the following service conditions:

1. Resistance to severe impact
2. Resistance to severe abrasion
3. Resistance to corrosion and abrasion at high temperature
4. Resistance to severe abrasion with moderate impact
5. Resistance to abrasion with moderate-to-heavy impact

In most cases, only a single type of electrode is needed. However, there are some conditions that require two types of electrodes. For example, when severe abrasion is encountered in combination with severe impact, a Class 1 electrode should be used for buildup metal. A second type of electrode is used to deposit a material that has high abrasion characteristics. The first material cushions impact loads and supports the hard deposit that resists abrasion.

There are many different kinds of hard-facing materials. Generally, they have a base of iron, nickel, copper, or cobalt. Alloying elements may be carbon, chromium, molybdenum, tungsten, silicon, manganese, nitrogen, vanadium, and titanium. The alloying elements form hard carbides that contribute to the properties of the hard-facing metals. A high percentage of tungsten or chromium with a high carbon content forms a carbide that is harder than quartz. Materials with high chromium content have high resistance to oxidation and scaling. Nickel, cobalt, and chromium have high corrosion resistance.

Hard-facing electrodes are classified on the basis of the type of service they perform and are divided between ferrous and nonferrous base alloys, Table 12-14.

Ferrous Base Alloys

Austenitic Steel Electrodes There are two major types of austenitic steel electrodes: those containing a high percentage of manganese and those containing chromium, nickel, and iron. The latter are known as stainless steels. Electrodes made of austenitic manganese steel provide metal-to-metal wear resistance coupled with impact, and surface protection or replacement of worn areas where abrasion is associated with severe impact. They have been widely used to resurface battered-down railway trackwork. Surfacing of power shovel dippers is another common application.

The stainless-steel types of hard-facing electrodes are used for corrosion-resistance overlays and for joining or buildup purposes. Certain types are good heat-resistant alloys and serve for surface protection against oxidation up to 2,000°F.

Martensitic Steel Electrodes The carbon in martensitic steels is the major determinant of their characteristics. These steels are inexpensive and tough. The alloys with little carbon are tougher and more crack-resistant than the types with greater amounts of carbon. When used as electrodes, they can be built up to form thick, crack-free deposits of high strength and some ductility. Abrasion resistance is moderate, and increases with carbon content and hardness. The deposits with little hardness may be machinable with tools, whereas grinding is advisable for those with higher hardness. These electrodes are used for building up the surfaces of shafts, rockers, and other machined surfaces.

Iron Electrodes Iron-base alloys with high carbon content are called **irons** because they have the characteristics of cast iron and are used for facing heavy cast iron machinery parts. They have a moderate-to-high alloy content of chromium, molybdenum, or nickel. Irons resist abrasion

ABOUT WELDING

ELECTROSLAG

Prior to 1977, highway bridges used electroslag welding (ESW). When cracks were discovered during bridge inspections in Pennsylvania, the government found that this method could no longer be used. However, recent developments now allow ESW to be used in certain applications.

Table 12-14 Properties of Hard-Facing Electrodes

Abrasion	Friction	Impact	Hardness	Ductility	Machinability	Corrosion	Hard-Facing Guide
1	2	5	1	6	6	1	Carbide type—powder form
1	1	5	1	5	6	1	Carbide type—high abrasion resistance
3	2	5	1	5	6	2	Carbide type—good abrasion, moderate toughness
4	3	4	1	5	5 4*	5	Semiaustenitic type—high carbon, chromium alloy
5	4	1	6	2	4	6	Austenitic type—11 to 14% manganese
5	5	1	6	1	4	1	Austenitic type—18-8 and 25-20 stainless
5	2	5	1	6	6 4*	4	Martensitic type—high speed tool steel
2	2	5	1	6	6 4*	4	Martensitic type—5% chrome tool steel
5	4	3	2	4	5 3*	5	Martensitic type—low carbon chrome, manganese
5	5	3	3	4	4 3*	5	Martensitic type—medium carbon, chrome, manganese
6	6	2	6	3	3 1*	6	Ferritic type—low carbon with 5% molybdenum
6	6	1	6	1	1	6	Ferritic type—conventional low carbon electrode

Scale 1 to 6
1. Excellent
2. Very good
3. Good
4. Fair
5. Poor
6. Very poor
*Annealed condition.

better than the austenitic and martensitic steels, up to the point at which they lack toughness to withstand the associated impact. Hard-facing with iron-base electrodes is usually limited to one or two layer overlays. Cracking often results from the deformation of a soft base under the harder overlay. Proper support of the tough iron-base alloys is important.

Nonferrous Base Alloys

Cobalt-Base Surfacing Metals (Rods and Electrodes) Cobalt-base alloys usually contain 26 to 33 percent chromium, 3 to 14 percent tungsten, and 0.7 to 3.0 percent carbon. Three grades are available in which hardness, abrasion resistance, and crack sensitivity increase as the carbon and tungsten increase. They have high resistance to oxidation, corrosion, and heat. Metals surfaced with cobalt-base alloys stand up under some types of service at temperatures as high as 1,800°F. Thus, they are often used in the manufacture of exhaust valves for internal combustion engines.

Composite Tungsten Carbide Materials Tungsten carbide hard-facing material is supplied in the form of mild steel tubes filled with crushed and sized granules of cast tungsten carbide, usually in proportions of 60 percent carbide and 40 percent tungsten, by weight. The carbide is very hard, tough, and abrasion resistant. Deposits containing large, undissolved amounts of tungsten carbide have more resistance to all types of abrasion than any other welded overlay.

As the heat of welding melts the steel tube, the molten metal dissolves some of the tungsten carbide to form a matrix of high tungsten steel or iron. This serves as an anchor and support for the undissolved granules of carbide. The highest abrasion resistance is achieved with oxyacetylene welding. Some of the carbide is dissolved by shielded metal arc welding, thus decreasing abrasion resistance.

Copper-Base Surfacing Metals (Rods and Electrodes) Various alloys of copper with aluminum, silicon, tin, and zinc are used for corrosion resistance, as well as wear applications. Alloys with aluminum contain 9 to 15 percent aluminum and up to 5 percent iron. They are the hardest of the copper-surfacing alloys, achieving a reading of 380 on the Brinell scale. For this reason, copper-base surfacing alloys are used extensively to minimize metal-to-metal wear. Alloys with tin and zinc are also used for bearing surfaces.

Nickel-Base Surfacing Metals (Rods and Electrodes) Nickel-base surfacing metals contain a relatively high degree of chromium and lesser amounts of carbon, boron, silicon, and iron. Their hardness and abrasion resistance increase with the carbon, boron, silicon, and iron content. In general, nickel-base alloys are hard and have satisfactory resistance to abrasion, oxidation, corrosion, and heat. Hot strength and resistance to high-stress abrasion, however, is somewhat lower than for the cobalt-base group.

Aluminum Electrodes

Aluminum is the most widely fabricated metal after steel. This popularity is due primarily to such factors as its wide availability, strength, light weight, good workability, and pleasing appearance. The increase in many applications is due principally, however, to the development of dependable, high speed welding processes for joining the majority of aluminum alloys. More than two dozen major welding processes are used; gas metal arc and gas tungsten arc are used more than the others. A limited amount of shielded metal arc welding is used. Aluminum can be welded throughout a thickness range of 0.00015 inch in foil to 6 inches in plate.

The high strength liquid and airtight joints now produced at high speeds in welding aluminum have extended the use of this lightweight material for freight cars of all types, tank cars, hopper cars, ships, barges, trucks, piping, and many other industrial applications.

The major alloying elements used in aluminum electrodes are magnesium (with zinc and with or without manganese) and silicon (with or without copper). See Table 12-15, page 324.

The two types of covered electrodes generally available are the E1100 and 4043 alloys. Alloy E1100 is commercially pure aluminum (99 percent aluminum) giving a weld deposit with a minimum tensile strength of 12,000 p.s.i. Alloy 4043 contains approximately 95 percent aluminum and 5 percent silicon. It has an ultimate tensile strength of 30,000 p.s.i. The 4043 (AL-2) electrode is a general-purpose wire. It gives better fluidity during welding than the 1100 (AL-43) electrode.

In applications where corrosive factors are important, an electrode should be selected that has a composition as close to that of the base metal as practical.

The presence of moisture in the electrode coating is a major cause of a porous weld structure. Since it is essential that the covering be completely dry, it is advisable to bake all *doubtful* electrodes and those taken from previously opened packages at 350 to 400°F for an hour before welding. After baking, they should be stored in a heated cabinet until used.

Ductility and cracking, known as **hot-shortness**, are problems in the welding of aluminum. These conditions are influenced by the degree of dilution between base metal and filler metal. Joints that require a great amount

Table 12-15 Composition of Weld Metal; Aluminum and Aluminum Alloy Welding Rods and Bare Electrodes

AWS Classification	Silicon (%)	Iron (%)	Copper (%)	Manganese (%)	Magnesium (%)	Chromium (%)	Nickel (%)	Zinc (%)	Titanium (%)	Other Elements (%) Each	Other Elements (%) Total	Aluminum (%)
ER1100	—	—	0.05–0.20	0.05	—	—	—	0.10	—	0.05	0.15	99.00 min.
ER1260	—	—	0.04	0.01	—	—	—	—	—	0.03	—	99.60 min.
ER2319	0.20	0.30	5.8–6.8	0.20–0.40	0.02	—	—	0.10	0.10–0.20	0.05	0.15	Remainder
ER4145	9.3–10.7	0.8	3.3–4.7	0.15	0.15	0.15	—	0.20	—	0.05	0.15	Remainder
ER4043	4.5–6.0	0.8	0.30	0.05	0.05	—	—	0.10	0.20	0.05	0.15	Remainder
ER4047	11.0–13.0	0.8	0.30	0.15	0.10	—	—	0.20	—	0.05	0.15	Remainder
ER5039	0.10	0.40	0.03	0.30–0.50	3.3–4.3	0.10–0.20	—	2.4–3.2	0.10	0.05	0.10	Remainder
ER5554	—	—	0.10	0.50–1.0	2.4–3.0	0.05–0.20	—	0.25	0.05–0.20	0.05	0.15	Remainder
ER5654	—	—	0.05	0.01	3.1–3.9	0.15–0.35	—	0.20	0.05–0.15	0.05	0.15	Remainder
ER5356	—	—	0.10	0.05–0.20	4.5–5.5	0.05–0.20	—	0.10	0.06–0.20	0.05	0.15	Remainder
ER5556	—	—	0.10	0.50–1.0	4.7–5.5	0.05–0.20	—	0.25	0.05–0.20	0.05	0.15	Remainder
ER5183	0.40	0.40	0.10	0.50–1.0	4.3–5.2	0.05–0.25	—	0.25	0.15	0.05	0.15	Remainder
R-C4A	1.5	1.0	4.0–5.0	0.35	0.03	—	—	0.35	0.25	0.05	0.15	Remainder
R-CN42A	0.7	1.0	3.5–4.5	0.35	1.2–1.8	0.25	1.7–2.3	0.35	0.25	0.05	0.15	Remainder
R-SC51A	4.5–5.5	0.8	1.0–1.5	0.50	0.40–0.60	0.25	—	0.35	0.25	0.05	0.15	Remainder
R-SG70A	6.5–7.5	0.6	0.25	0.35	0.20–0.40	—	—	0.35	0.25	0.05	0.15	Remainder

of filler metal should be avoided. Cracking is generally caused by the low strength of some weld metal compositions at elevated temperatures. This can be overcome through the use of one of the higher magnesium-content aluminum alloys, since they give the weld higher strength during solidification. The type of joint and the welding technique also influence cracking.

Specialized Electrodes

There is a large variety of specialized electrodes to meet the conditions presented by such metals as nickel and high nickel alloys, copper and copper alloys, magnesium and magnesium alloys, and titanium and titanium alloys.

High nickel electrodes have been developed for the welding of gray iron castings, ductile iron, malleable iron, and other iron-base metals. Special high nickel alloy, filler-metal compositions are capable of welding dissimilar metal combinations. A number of electrodes that contain 50 percent or more nickel are used in the welding of nickel and its alloys. These electrodes include the following combinations of alloys:

- Nickel-copper alloys
- Monel-nickel-copper alloys
- Age-hardenable Monel-nickel-copper alloys
- Age-hardenable Inconel
- Inconel-nickel-chromium-iron alloys
- High nickel alloy filler metal

Copper electrodes contain tin and silicon in addition to copper. These copper alloys are classified with the prefix E plus the chemical abbreviations of the metals they contain. For example, a copper-aluminum electrode would be designated ECuAl (E = electrode; Cu = copper; Al = aluminum).

Copper-silicon alloys (ECuSi) are often referred to as **silicon bronzes.** The core wire contains about 3 percent silicon and may contain small percentages of manganese and tin. Copper-silicon electrodes weld copper-silicon metal and copper-zinc (brass).

Copper-tin alloys (ECuSn-A) are usually referred to as **phosphor bronzes.** Phosphor bronze A contains about 8 percent tin. Both copper and tin are deoxidized with phosphorus. Phosphor bronze electrodes weld copper, bronze, brass, and cast iron. They are also used for overlaying steel. These electrodes require preheat, especially on heavy sections. They are used with DCEP.

The copper-nickel (ECuNi) electrodes contain 70 percent copper and 30 percent nickel. They are used for the standard copper-nickel alloys and must not be preheated or allowed to overheat. They are used with DCEP.

Copper-aluminum (ECuAl) electrodes are of two types: the copper-aluminum type used with gas welding and the copper-aluminum-iron type for shielded metal arc welding. They weld aluminum bronzes, manganese bronzes, some nickel alloys, many ferrous metals and alloys, and combinations of dissimilar metals. They are used with DCEP.

Aluminum bronze (ECuAl-A2) is a copper alloy electrode containing 65 to 90 percent aluminum and 0.5 to 5.0 percent iron. This electrode produces a deposit with higher tensile strength, yield strength, hardness, and lower ductility than a pure copper-aluminum electrode. It is used for joining aluminum-bronze plate and sheet, repairing castings, and for joining nonferrous metals such as silicon bronze to steel. It is used with DCEP.

Nickel-aluminum bronze (ECuNiAl) and the manganese-nickel-aluminum-iron (ECuMuNiAl) alloys are used to weld similar base metals for ship propellers and ship fittings. They are used with DCEP.

The selection of the proper type of electrode for a particular welding job is vital to the successful completion of the job. A welder will never be in the position of knowing all there is to know about the various types of electrodes. You are urged to consult the various welding suppliers, catalogs, and the excellent material available through the American Welding Society in order to develop a comprehensive understanding of the differences between and uses of electrodes.

See Table 12-16 for a list of AWS filler metal specifications for the SMAW process.

Packing and Protection of Electrodes

Standard Sizes and Lengths

The standard sizes and lengths of end-gripping electrodes are shown in Table 12-17. In all cases, standard size refers to the diameter of the core wire exclusive of the coating.

Table 12-16 AWS Filler Metal Specifications for SMAW Electrodes

Material	Specification Number
Carbon steel	A5.1
Low alloy steel	A5.5
Stainless steel	A5.4
Cast iron	A5.15
Nickel alloys	A5.11
Aluminum alloys	A5.3
Copper alloys	A5.6
Surfacing alloys	A5.13

Table 12-17 Sizes and Lengths of End-Gripping Electrodes

Standard Size (Diameter of Core Wire, in.)	Standard Length (Length of Core Wire, in.)
1/16	9
5/64	9, 12
3/32	12
1/8	14
5/32	14
3/16	14
7/32	14, 18
1/4	18
5/16	18
3/8	18

In 18- and 36-inch lengths, center gripping of the electrode is standard. End gripping is standard for all other lengths.

Packing

Electrodes are always suitably packed, wrapped, boxed, or crated to protect against damage during shipment or storage as follows:

- Bundles of 50 pounds net weight
- Boxes of 25 or 50 pounds net weight
- Coils, reels, or spools of 200 pounds or less

Marking

All bundles, boxes, coils, and reels usually contain the following information:

- Classification
- Manufacturer's name and trade designation
- Standard size and length (weight instead of length in case of reels and coils)
- Guarantee

The welder should always give careful attention to the manufacturer's recommendations concerning heat settings, type and preparation of joint, base metal, welding technique, welding position, and nature of the welding current. Any references to moisture control are also important.

Moisture Control

A perfectly dry electrode is the first requirement for a perfect welding job when the job requires low hydrogen electrodes or other moisture-prone electrodes. If the electrode is not dry, the welder cannot be sure of a sound weld. Welding with moist electrodes leads to increased arc voltage, spatter loss, undercutting, and poor slag removal. The weld deposit may suffer from porosity, underbead cracking, and rough appearance. When present, these conditions cause rejects and expensive reworking and, if undetected, contribute to product failure.

All mineral-covered electrodes are *thirsty*. The minute they are unpacked, they start absorbing moisture—too much moisture for a sound weld. Outside of a laboratory it is impossible to tell when an electrode has absorbed enough moisture from the air to be unsafe. Electrodes, therefore, require antimoisture protection. This is especially true when field welding is being done. A number of building and construction codes contain the specific provision that some kind of electrode holding and conditioning equipment is provided on the job site.

Electrode Ovens Electrode manufacturers recommend oven storage at specified holding temperatures to preserve and maintain factory baked-in quality. Oven protection is not only recommended, but mandatory for the storage of low hydrogen and hard-facing electrodes and others made from special alloys such as the following:

- Iron powder
- Stainless steel
- Aluminum
- Inconel®
- Monel®
- Brass
- Bronze

The covering of low hydrogen electrodes, for example, is reduced to less than 0.2 percent moisture at manufacture, and the electrodes are packaged in moisture-proof containers. Within 2 hours at 80 percent humidity, the electrodes may contain up to 13 times the allowable moisture content for U.S. Government specifications. Within 24 hours, they may contain up to 26 times the 0.2 percent allowed.

Electrode drying ovens have capacities varying from 12 to 1,000 pounds, and temperature controls to 1,000°F. The smaller ovens, Fig. 12-9, are portable, making them convenient for shop or field welding. The larger ovens, Figs. 12-10 and 12-11, provide for central storage and baking for the entire shop.

The increased use of the submerged arc and flux cored arc welding processes has caused a problem in regard to flux moisture. Flux that is unprotected will pick up moisture, which results in welds with hydrogen inclusions. Figure 12-12 shows a portable type of holding oven that can be used to protect flux or flux cored wire.

Fig. 12-9 The electric dry-rod electrode oven on the job. Phoenix International

Fig. 12-11 An electrically heated shop electrode oven. This oven has the capacity to hold 350 pounds of 18-inch electrodes. It is equipped with a thermostat control with a temperature range from 100°F to 550°F. Phoenix International

Fig. 12-10 An electrically heated shop electrode oven. This oven has the capacity to hold 400 pounds of 18-inch electrodes, and it can reach a maximum temperature of 800°F with digital readout. It is also equipped with a 150 CFM recirculating system. Phoenix International

Fig. 12-12 This portable flux holding oven may also be used for flux cored wire. Phoenix International

Many welders protect their electrodes with a small leather electrode carrier that is strapped to the welder's waist. These carriers offer limited protection for moisture, and they are not satisfactory for the electrodes listed previously. The oven shown in Fig. 12-9 is portable and can be plugged into auxiliary current in the field. Such ovens also offer good protection after the current has been turned off.

Table 12-18 covers some typical storage and conditioning recommendations. Always consult the electrode manufacturer for specific requirements.

Table 12-18 Storage and Drying Conditions of SMAW Electrodes

AWS Classification	Storage Conditions After Removal from Manufacturer's Packaging		Drying Conditions (Because of Variations in Covering Compositions, Consult the Manufacturer's Exact Drying Conditions)
	Ambient Air	Holding Ovens	
EXX10-X EXX11-X	Ambient temperature	100–120°F	Not recommended
EXX13-X E7020-X E7027-X	60–100°F 50% max. relative humidity	100–120°F	250–300°F 1 hour at temperature
EXX15-X EXX16-X EXX18M(1) EXX18-X	Not recommended. Some of these electrodes may carry an R designator for moisture resistant, but this does not imply they can be stored in ambient air.	250–300°F	500–800°F Depending on the code being used, the electrodes may be put through this redrying or baking conditioning only once, or the coating binders may be adversely affected.

CHAPTER 12 REVIEW

Multiple Choice

Choose the letter of the correct answer.

1. What is a major function of the coating of an SMAW electrode? (Obj. 12-1)
   a. Acts as a scavenger, removing oxides and impurities
   b. Influences the lack of penetration
   c. Liquefies the base metal
   d. None of these
2. The popularity of SMAW and its electrodes is a result of the development of _____. (Obj. 12-2)
   a. Flux cored electrodes
   b. Bare wire electrodes
   c. Arc electrodes
   d. Both a and b
3. What determines the maximum arc length of an SMAW electrode? (Obj. 12-3)
   a. The end of the electrode
   b. The coated end of the electrode
   c. The base metal
   d. Both a and b
4. For the electrode E6011, the 60 signifies _____. (Obj. 12-4)
   a. 60 times
   b. 60 p.s.i.
   c. 60,000 p.s.i.
   d. 60,000 pounds
5. The welder should never use an electrode that is _____ the thickness of the base metal. (Obj. 12-5)
   a. Less than
   b. Equal to or less than
   c. Greater than
   d. Larger than
6. What two groups have had a major part in the research and development of welding electrodes? (Obj. 12-6)
   a. The American Welding Society
   b. The American Society for Testing and Materials
   c. The American Petroleum Institute
   d. Both a and b
7. Which of the following electrodes is considered a fast-freeze electrode? (Obj. 12-7)
   a. E7018
   b. E7024

    c. E6010
    d. None of these
8. Which of the following is *not* a low hydrogen electrode? (Obj. 12-8)
    a. E7015
    b. E7014
    c. E7016
    d. E7018
9. The system for identifying steel electrodes in the welding industry is _____. (Objs. 12-6 and 12-9)
    a. The color identification system
    b. The electrode classification number system
    c. The set-type system
    d. Both a and b
10. A properly dried low hydrogen electrode is the first requirement for _____. (Obj. 12-10)
    a. A perfect welding job
    b. A porosity free weld
    c. An undercut free weld
    d. None of these

Review Questions

Write the answers in your own words.

11. What is the function of the coating on a shielded metal arc electrode? (Obj. 12-1)
12. Give a definition of a shielded metal arc welding electrode. (Obj. 12-2)
13. In general terms, describe tensile strength and elongation in welds made with shielded metal arc electrodes. How do these properties compare with those of the base metal? (Obj. 12-2)
14. List the standard sizes of electrodes. (Obj. 12-5)
15. List the important considerations in selecting the type of electrode to use. (Obj. 12-5)
16. List the service conditions for hard-facing electrodes. (Obj. 12-5)
17. Give the AWS classification numbers for the following electrode characteristics: (Obj. 12-6)
    a. DCEP
    b. A.C.
    c. Iron powder
    d. Low hydrogen
18. Explain the functions of fast-fill, fast-follow, and fast-freeze electrodes. (Obj. 12-7)
19. What is the purpose of heavy slag-type, free-flowing electrodes? Name some of the other types of electrodes used for shielded metal arc welding. (Objs. 12-7 to 12-9)
20. How are low hydrogen electrodes best protected from moisture in the field? (Obj. 12-10)

INTERNET ACTIVITIES

Internet Activity A

Use the Internet to find ESAB's Material Safety Data Sheet for a.c. and d.c. stainless-steel covered electrodes. What are the fire and explosion hazards? What do the reactivity data state, and what are the health hazards?

Internet Activity B

Use the same Web site that you used in Activity A. Name the free publications you can get from ESAB for electric and gas welding safety. Find their address and send for a free copy.

Design credits: Cinema and movie icon: Denis Meshkov/123RF; and Illustration of Welding icons: Mr. Nachai Sorasee/123RF

13

Shielded Metal Arc Welding Practice:

Jobs 13-J1–J25 (Plate)

Chapter Objectives

After completing this chapter, you will be able to:

- **13-1** Describe uses for the **SMAW** process.
- **13-2** Explain the approach and operating characteristics of welding with the **SMAW** process.
- **13-3** Demonstrate the ability to troubleshoot the **SMAW** equipment and process.
- **13-4** Demonstrate the ability to weld with the **SMAW** process on various joints, with various welds, in various positions.

Introduction

Practice in shielded metal arc welding (SMAW), also called stick electrode welding, is the basis for learning the other types of welding, Fig. 13-1. SMAW accounts for less than 50 percent of the total amount of welding being done, while gas metal arc welding (MIG/MAG) and flux cored arc welding (FCAW) make up over 50 percent of the welding being done. Submerged arc welding (SAW) follows with less than 10 percent in popularity.

Stick welding has great versatility. It is a dependable process that welds any steel in any position. It is readily available at modest cost. It will probably be used as long as any type of welding is being done and especially in maintenance and repair work.

The practice jobs in this and the next two chapters are the basis for the mastery of welding. The jobs are numbered consecutively and include all of the basic joints and welds in all positions and provide experience with all of the commonly used electrodes. As you practice, you will also learn how to set the welding machine for all kinds of welding situations. A welder must be able to set the welding machine to meet the varying conditions of the job in order to produce satisfactory welds.

The jobs in this chapter provide practice on a number of fundamental joints and welds. Welded joints include the edge, lap, and T-joint. You will learn how to run stringer

Fig. 13-1 A young welder who has completed training is welding a steel fabrication on the job with the shielded metal arc welding process. ZEFF Travel Photo/Alamy Stock Photo

Fig. 13-2 A student welder making a fillet weld on a lap joint is seated at a workbench in order to be as relaxed and comfortable as possible. Note the protective gear being worn. Location: Northeast Wisconsin Technical College Mark A. Dierker/McGraw Hill

and weaved beads and make fillet welds in the flat, horizontal, and vertical positions. You will experience the different characteristics and heat requirements of various types of electrodes with electrode negative, electrode positive, and alternating current. You will also begin to read shop drawings and become familiar with welding symbols. The satisfactory completion of these jobs will provide a sound basis for the more advanced jobs that require more skill to master.

Approach to the Job

Prior to starting any welding job, the welder should make sure that the lenses in the welding hood are clean, of the proper shade, and the hood is properly adjusted.

Be relaxed. This is necessary if you are to avoid fatigue. When fatigue sets in, it is almost impossible to do good work. Grip the electrode holder lightly. If the electrode holder is held too tightly, your arm will tire, making it difficult to produce a sound weld. The electrode cable should be draped around the shoulder or across the lap to take the weight of it off the holder. When sitting or standing, assume the most comfortable position possible, Fig. 13-2. No part of the body should be under the slightest strain. You should try to position yourself so that the entire length of the welding rod may be welded without obstruction and without putting your arms in a bind. It is equally important that you are at all times protected by use of the proper protective equipment.

Review of Operating Characteristics

What follows are descriptions of those basic arc welding techniques that the welding student must be able to apply in order to improve with practice. Discuss these points with your instructor to make sure that you understand them.

Polarity In a d.c. circuit, the current always flows in one direction. The electron theory has the current flow from negative to positive. One line of the circuit is the positive side, and the other line is the negative side. When welding is being performed with **electrode positive (DCEP)**, the electrode lead becomes the positive (+) side of the circuit and the work lead becomes the negative (−) side of the circuit, Fig. 13-3. When welding is being performed with **electrode negative (DCEN)**, the electrode lead becomes the negative (−) side of the circuit and the work lead becomes the positive (+) side of the circuit, Fig. 13-4, page 332. Electrode positive will produce the deepest penetrating weld, while at a given amperage (heat) electrode negative will produce the shallowest penetration but greatest deposition rate.

The type of electrode determines the kind of polarity necessary for operation. In a.c. welding, there is no choice

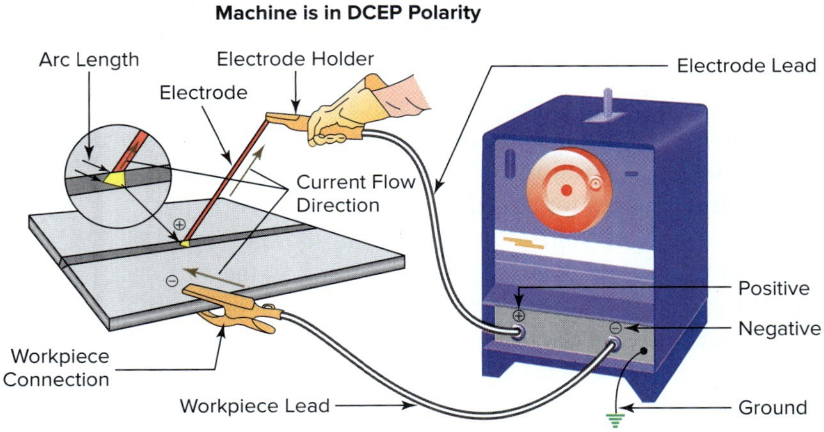

Fig. 13-3 Direction of current when welding is being done with electrode positive polarity.

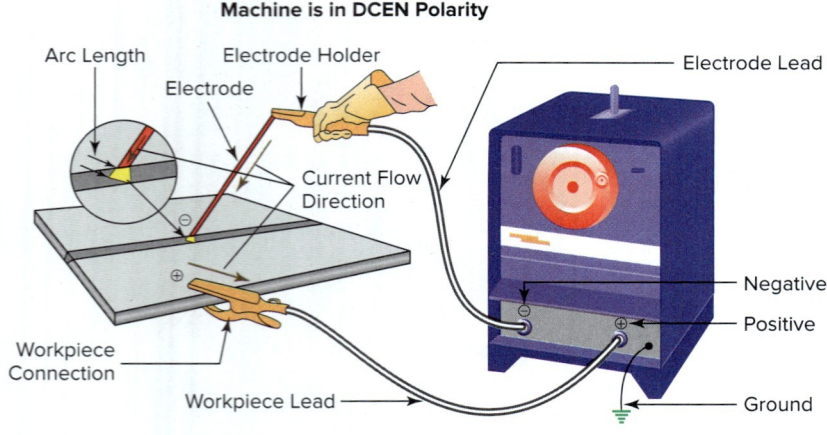

Fig. 13-4 Direction of current when welding is being done with electrode negative polarity.

of polarity since the circuit is alternately positive and negative, first on one side and then on the other.

Arc Length and Arc Voltage A fairly short arc is necessary for SMAW. A short arc permits the heat to be concentrated on the work, and it is more stable since it reduces the effect of arc blow. The arc, consisting of vapors from the coating, surrounds the electrode metal, the arc, and weld pool and prevents air from destroying the weld metal.

With a long arc a great deal of the heat is lost into the surrounding area, preventing good penetration and fusion. The arc is very unstable, the effect of arc blow is increased, and the arc has a tendency to blow out. The greater arc length permits air to reach the molten globule of metal as it passes from the electrode to the weld pool itself. Thus the weld metal absorbs oxygen and nitrogen, both of which are harmful. Weld metal deposited with a long arc has low strength, poor ductility, and high porosity. The weld shows poor fusion and excessive spatter.

The length of arc gap that the welder should maintain depends on the type and size of electrode, the position of welding, and the amount of current.

A shorter arc is maintained for vertical, horizontal, and overhead welding than for flat welding. The normal arc voltage in the flat position for $\frac{1}{8}$- to $\frac{1}{4}$-inch electrodes is 22 to 40 volts. Voltages for the other positions are 2 to 5 volts less. Since the arc voltage increases with the arc length, a shorter arc gap is necessary when welding in the vertical, horizontal, and overhead positions. A 20-volt arc with lightly coated electrodes is approximately $\frac{1}{8}$ inch long, and a 40-volt covered electrode arc is approximately $\frac{1}{4}$ inch long. The type of coating on the electrode also has some effect on the arc length.

Good welders can tell if they are welding with the proper arc length by the appearance of the metal transfer and the weld pool and by listening to the sound of the arc. With a long arc they can see the globule of molten metal flutter from side to side, throwing off small particles of metal in all directions. The proper arc gap transfers the metal in a steady stream that is not easily apparent. A short arc makes a sharp crackling sound like grease frying in a pan. A long arc is identified by a hissing sound and explosions at steady intervals.

Arc Blow The welding student will find arc blow a disturbing feature of arc welding. **Arc blow** is the swerving of an electric arc from its normal path because of magnetic forces. It may be likened to a number of magnets all pulling at a ball of steel. The ball will waver between the different magnets and will finally be attracted toward the magnet with the strongest pull.

Arc blow is especially noticeable when welding with large diameter electrodes. Excessive arc blow may cause lack of fusion, porosity, and excessive spatter.

Arc blow may be controlled somewhat by moving the work clamp to another location and by changing the position of the electrode.

Current Values It is impossible to give a table that will indicate the exact amount of current needed to burn a certain size of electrode for every welding condition. The best that can be done is to offer a basic heat table, the values of which can be increased or decreased for the different conditions. The following have a bearing on the amount of current:

- *Type and size of electrode:* A $\frac{3}{16}$-inch electrode requires higher heat than a $\frac{1}{8}$-inch electrode. An electrode-negative polarity electrode requires more heat than a positive polarity electrode.
- *Material to be welded:* Materials that are good conductors require more heat than materials that are poor conductors.

SHOP TALK

Save the Earth

Companies can lessen their impact on the planet by:

1. Using an environmental management system.
2. Assessing the life cycle of each product.
3. Limiting the use of natural resources and focusing on renewable ones.
4. Proper use and disposal of hazardous substances.
5. Minimizing energy consumption.

- *Size and shape of the work:* A piece of material 20 inches square and 1 inch thick requires more heat for welding than one that is 10 inches square and ⅛ inch thick.
- *Type of joint:* The first pass in a T-joint can be welded with a higher current than the first pass of an open, single-V, groove butt joint.
- *Position of the weld:* Welding in the flat and horizontal positions can be done with a higher current than welding in the vertical and overhead positions.
- *Obstructions around the weld site:* A weld that must be made behind obstructions requires a change in heat.
- *Kind and condition of equipment:* No two machines can be set in exactly the same manner. Even two machines of the same size and make differ somewhat.
- *Speed required:* Speed has a direct bearing on the current setting.
- *Skill of the welder:* Some welders are able to use higher heats than others on the same type of work.

Table 13-1 is a simplified guide to approximate current settings. Note that the decimal equivalent of the fractional sizes also gives some indication of the heat setting. Thus 5/32 inch is 0.156, and the corresponding current range is ±150. Table 13-2 lists both voltage and current settings according to electrode classification and size.

Electrode Angle The angle of the electrode to the work affects the quality of the weld. The proper angle prevents undercut and slag inclusions and makes it easier to deposit the filler metal in the weld. Because it influences surface tension and the force of gravity on the molten metal, the correct angle promotes uniformity of fusion and weld contour in vertical and overhead welding.

Learning Welding Skills

Study Table 13-3 (p. 334), the Welder's Troubleshooter, thoroughly before you begin welding practice. Look up your weld defects in the table during practice and correct them as instructed before beginning the next job.

Table 13-1 How Heat Ranges Coincide with the Decimal Equivalent of the Fractional Electrode Size

Diameter of Electrode (in.)	Decimal Equivalent	Current Range (A)
1/16	0.062	±60
3/32	0.093	±75
1/8	0.125	±100
5/32	0.156	±150
3/16	0.187	±200
1/4	0.250	±250
5/16	0.312	±300
3/8	0.375	±400

Table 13-2 Current Variations with Covered Electrode Classifications and Sizes

Electrode Dia.	E6010 E6011	E6013	E7016	E7018	E7028	E7024
1/16		20–40 A[1] 17–20 V[2]	No size	No size	No size	No size
3/32	30–80 A 22–24 V	30–80 A 17–21 V	60–100 A 17–22 V	70–120 A 17–21 V	No size	100–145 A 20–24 V
1/8	80–120 A 24–26 V	70–120 A 18–22 V	80–120 A 18–22 V	100–150 A 18–22 V	140–190 A 21–25 V	140–190 A 21–25 V
5/32	120–160 A 24–26 V	120–170 A 18–22 V	140–190 A 20–24 V	120–200 A 20–24 V	180–250 A 22–26 V	180–250 A 22–26 V
3/16	140–220 A 26–30 V	140–240 A 20–24 V	170–250 A 21–25 V	200–275 A 21–25 V	230–305 A 23–27 V	230–305 A 23–27 V
7/32	170–250 A 26–30 V	210–300 A 21–25 V	240–320 A 23–27 V	260–340 A 22–26 V	275–365 A 23–28 V	275–365 A 27–28 V
1/4	200–300 A 28–32 V	200–350 A 22–26 V	300–400 A 24–28 V	300–400 A 23–27 V	335–430 A 24–29 V	335–430 A 24–29 V
5/16	250–450 A 28–32 V	250–450 A 23–27 V	375–475 A 24–28 V	375–470 A 23–28 V	400–525 A 24–30 V	400–525 A 24–30 V

[1] A indicates current in amperes.
[2] V indicates voltage, arc volts.

Welding with Shielded Metal Arc Electrodes

Molten metal combines with the oxygen and nitrogen in the surrounding air to form oxides and nitrides. Thus, covered electrodes are required to get appropriate weld strength, Table 13-4 (p. 337).

All industries use shielded arc (covered) electrodes for metal arc welding because they reduce the formation of oxides and nitrides by protecting the molten weld pool. Protection is accomplished in the following manner.

- The covering is consumed at a lower rate than the metallic core wire. Thus, it projects over the core wire so that it affords mechanical protection to the wire and the arc and controls the direction of the arc, Fig. 13-5A, page 337.
- A protective shield of gases is formed as the coating burns, and the atmosphere is thus excluded, Fig. 13-5B.
- Slag is formed over the deposited weld metal, Fig. 13-5C. It protects the deposit from the atmosphere while it is cooling.

Table 13-3 The Welder's Troubleshooter

Problem	Cause	Solution
Arc blow	• Magnetic field causing a d.c. arc to blow away from the point at which it is directed. Arc blow is particularly noticeable at ends of joints and in corners.	• Proper location of the work connection on the work. Placing the work connection in the direction the arc blows from the point of welding is often helpful. • Separating the work connection in two or more parts is helpful. • Weld toward the direction the arc blows. • Hold a short arc. • Arc blow is not present with a.c. welding.
Brittle joints	• Air-hardening base metal. • Improper welding procedure. • Unsatisfactory electrode.	• When welding medium carbon steel or certain alloy steels, the fusion zone may be hard as a result of rapid cooling. Preheat at 300–500°F before welding. • Multiple-layer welds will tend to anneal hard zones. • Heating at 1,100–1,200°F after welding will generally reduce hard areas formed during welding. • Austenitic electrodes will often work on special steels, but the fusion zone will generally contain an alloy that is hard.
Brittle welds	• Unsatisfactory electrode. • Excessive welding current causing coarse grained and burnt metal. • High carbon or alloy base metal that has not been taken into consideration.	• Covered electrodes must be used if ductile welds are required. Low hydrogen electrodes give the best ductility. • Do not use welding current that is too high. It may cause coarse grain structure and oxidized deposits. • A single-pass weld may be more brittle than a multiple-layer weld because it has not been refined by successive layers of weld metal. • Welds may absorb alloy elements from the base metal and become hard. Do not weld a steel unless the analysis and characteristics are known.
Corrosion	• Type of electrode used. • Improper weld deposit for corrosive media. • Metallurgical effect of welding. • Improper cleaning of weld.	• Covered electrodes produce welds that are more resistant to corrosion than the base metal. • Do not expect more from the weld than you do from the base metal. On stainless steels use electrodes that are equal to or better than the base metal. • When welding austenitic stainless steel, be sure the analysis of the steel and welding procedure is correct so that the groove welding does not cause carbide precipitations. Condition can be corrected. Anneal at 1,900–2,100°F and quench. • Certain metals, such as aluminum, require careful cleaning of all slag to prevent corrosion.

Table 13-3 (Continued)

Problem	Cause	Solution
Cracked welds	• Joint too rigid. • Welds too small for size of parts joined. • Improper welding procedure. • Poor welds. • Improper preparation of joints.	• Design the structure and develop a welding procedure to eliminate rigid joints. • Do not use too small a weld between heavy plates. Increase the size of welds. • Do not make welds in string beads. Make weld full size in short section 8 to 10 inches long. Use low hydrogen electrodes. • Plan welding sequence to leave ends free to move as long as possible. • Ensure welds are sound and the fusion is good. • Preheating parts to be welded is helpful. • Prepare joints with a uniform free space of the proper width. In some cases a free space is essential. In other cases a shrink or press fit may be required.
Distortion	• Shrinkage of deposited metal pulls parts together and changes relative positions. • Nonuniform heating of parts during welding causes them to distort before welding is finished. Final welding of parts in distorted position prevents the maintenance of proper dimensions. • Improper welding sequence.	• Clamp or tack parts properly to resist shrinkage. • Separate or preform parts to compensate for shrinkage of welds. • Distribute welding to prevent excessive local heating. Preheating is desirable in some heavy structures. • Removal of rolling or forming strains before welding is sometimes helpful. • Study structure and develop a definite sequence of welding.
Incomplete penetration	• Improper preparation of joint. • Too large an electrode. • Too low welding current. • Too fast a welding speed.	• Be sure to allow the proper free space at the bottom of a weld. • Do not expect too much penetration from an electrode. • Use small diameter electrodes in a narrow welding groove. • Use enough welding current to obtain proper penetration. Do not weld too rapidly. • Use a backup bar if possible. • Chip or cut out the back of the joint and deposit a bead of weld metal at this point.
Poor fusion	• Improper diameter of electrode. • Improper welding current. • Improper welding technique. • Improper preparation of joint.	• When welding in narrow Vs, use an electrode small enough to reach the bottom. • Use sufficient welding current to deposit the metal and penetrate into the plates. Heavier plates require more current for a given electrode than light plates. • Be sure the weave is wide enough to thoroughly melt the sides of a joint. • The deposited metal should sweat on the plates and not curl away from it.
Poor weld appearance	• Poor welding technique, improper current setting, or electrode manipulation. • Characteristic of electrode used. • Welding in improper position for which electrode is designed. • Improper joint preparation.	• Use the proper welding technique for the electrode. • Use an electrode designed for the type of weld and the position in which the weld is to be made. • Do not make fillet welds with downhand-electrodes unless the parts are positioned. • Do not use too high welding currents. • Use a uniform weave or rate of travel at all times. • Prepare all joints properly.

Table 13-3 (Concluded)

Problem	Cause	Solution
Porous welds	• Characteristic of some electrodes. • Improper welding procedure. • Not sufficient pooling time to allow entrapped gas to escape. • Poor quality base metal.	• Some electrodes produce sounder welds than others. Be sure the proper electrodes are used. If they are low hydrogen electrodes, they must be dry. • Pooling keeps the weld metal molten longer and often ensures sounder welds. • A weld made of a series of stringer beads is apt to contain minute pinholes. Weaving will often eliminate this trouble. • Do not use welding currents that are too high. • In some cases the base metal may be at fault. Check it for segregations and impurities.
Spatter	• Characteristic of some electrodes. • Excessive welding current for the type or diameter of electrode used. • Some coated electrodes produce large spatter.	• Select proper type of electrode. • Do not use too much welding current. On power sources equipped with arc control set for softer arc. • Paint parts next to weld with antispatter compound. This prevents spatter from welding to parts, and they can be removed easily.
Undercut	• Too high welding current. • Improper manipulation of electrode. • Attempting to weld in a position for which the electrode is not designed.	• Use a moderate welding current and do not try to travel too rapidly. • Do not use too large an electrode. If the pool of molten metal becomes too large, undercut may result. • Avoid too much weaving. • A uniform weave will aid greatly in preventing undercut in groove welds. • If an electrode is held too near the vertical plane when making a horizontal fillet weld, undercut will occur on the vertical plate.
Warping (thin plates)	• Shrinkage of deposited weld metal. • Excessive local heating at the joint. • Improper preparation of joint. • Improper clamping of parts. • Improper welding procedure.	• Select electrode with high welding speed and moderate penetrating properties. • Weld rapidly to prevent overheating the plates adjacent to the weld. • Do not have wide spaces between the parts to be welded. • Clamp parts next to the joint properly. Use backup strip to cool parts rapidly. • Use backstep or skip welding sequence. • Hammer joint edges thinner than rest of plate before welding. This elongates edges, and the weld shrinkage causes them to pull back to the original shape.
Welding stresses	• Joints too rigid. • Improper welding procedure. • Stress occurs in all welds, especially in heavy parts.	• Slight movement of parts during welding will reduce welding stresses. • Make weld in as few passes as practical. • Peen weld metal as appropriate. • Heat finished product at 1,100–1,200°F for one hour per inch of thickness. • Develop welding procedure that permits all parts to be free to move as long as possible.

Adapted from Westinghouse Electric Co.

Table 13-4 Shielded Arc Properties of Weld Metal Deposited with Covered Electrodes

Tensile strength	65,000–77,000 p.s.i.
Ductility or elongation in 2"	20–26%
Impact	30–70 ft-lb (Izod)
Fatigue	28,000–32,000 p.s.i.
Density	7.84–7.86

Resistance to corrosion is better than mild steel.
It should be noted that these characteristics permit the weld metal to be flanged or bent cold with success.
As deposited (not stress relieved).

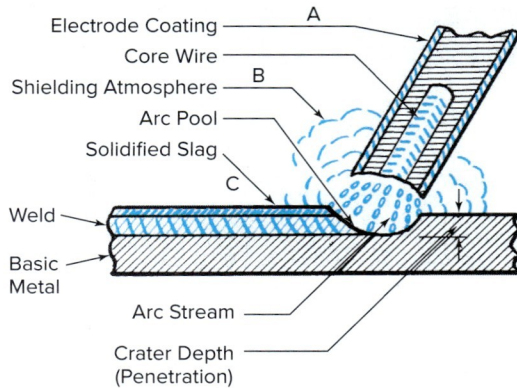

Fig. 13-5 Action of the shielded arc electrode.

Shielded arc welding metal electrodes produce weld metal characteristics that are superior to the base metal.

In this course, we will be concerned with those types of shielded arc welding metal electrodes that are widely used in industry and covered electrodes in the following AWS-ASTM classifications:

1. *E6010:* covered electrode; DCEP only; all position
2. *E6011:* covered electrode; DCEP or a.c.; all position
3. *E6013:* covered electrode; DCEP or a.c.; all position
4. *E7028:* iron powder, low hydrogen; DCEP or a.c.; flat and horizontal position
5. *E7016:* low hydrogen; DCEP or a.c.; all position
6. *E7018:* iron powder, low hydrogen; DCEP or a.c.; all position

All electrodes in other tensile strength classifications produce welds resembling those in the 60XX and 70XX classifications shown in Fig. 13-6. Study Table 13-5 (p. 338)

Fig. 13-6 Comparative appearance of weld beads made with various electrodes. Edward R. Bohnart

Table 13-5 Weld Characteristics Obtained in Mild Rolled Steel with Covered Electrodes

Operating Variables[1]	Resulting Weld Characteristics			
	Arc Sound	Penetration and Fusion	Electrode Burnoff	Appearance of Bead
Normal amps, normal volts, normal speed (A)	Sputtering hiss plus irregular energetic crackling sounds	Fairly deep and well-defined	Normal appearance. Coating burns evenly	Smooth, well-defined bead and no under-cutting. Excellent fusion—no overlap
Low amps, normal volts, normal speed (B)	Very irregular Sputtering Few crackling sounds	Poor: not very deep nor defined	Not greatly different from above	Bead lies on top of plate. Note there is not the overlap produced by bare rod; excessive pilling
High amps, normal volts, normal speed (C)	Rather regular explosive sounds	Deep, long crater	Coating is consumed at irregular high rate—watch carefully	Broad, rather thin bead; good fusion; excessive spatter and undercutting; irregular deposit
Low volts, normal speed, normal amps (D)	Hiss plus steady sputter	Small	Coating too close to crater. Touches molten metal and results in porosity	Bead lies on top of plate but not so pronounced as for low amps. Somewhat broader
High volts, normal speed, normal amps (E)	Very soft sound plus hiss and few crackles	Wide and rather deep	Note drops at end of electrode. Flutter and then drop into crater	Wide, spattered, irregular bead; pilling
Low speed, normal amps, normal volts (F)	Normal	Crater normal	Normal	Wide bead—large overlap base metal and bead heated for considerable area; pilling
High speed, normal amps, normal volts (G)	Normal	Small, rather well-defined crater	Normal	Small bead—undercut. The reduction in bead size and amount of undercutting depend on ratio of speed and amps

[1] See Fig. 13-7 for welds A through G.

carefully. It lists operating variables and weld characteristics that will be helpful as you begin practice with covered electrodes.

You will observe fusion of the weld metal and base metal so that you will gain a good understanding of its action and importance. The arc requires that the welder hold an almost perfect arc gap for a weld of good appearance and quality. Practice will help you develop a steady hand and make it easier for you to make good welds.

Study Fig. 13-7 to learn what operating and weld characteristics you should strive for and what you should avoid when welding.

Practice Jobs

Instructions for Completing Practice Jobs

Your instructor will assign appropriate practice in arc welding from the jobs listed in the Job Outline, Table 13-6, page 387. Before you begin a job, study the specifications given in the Job Outline. The Text Reference column lists the pages

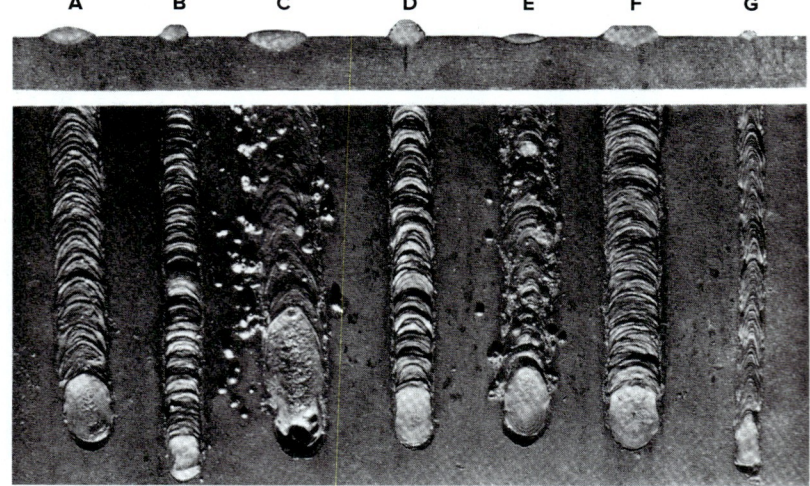

Fig. 13-7 Plan and elevation views of welds made with electrodes under various operating conditions. (A) Current, voltage, and speed normal. (B) Current too low. (C) Current too high. (D) Voltage too low, short arc. (E) Voltage too high, long arc. (F) Speed of travel too low. (G) Speed of travel too high. Edward R. Bohnart

in the chapter on which the job drawing and job description appear. Pay particular attention to the drawing that accompanies the job description. The title block gives the size and type of stock, the number, type and size of electrodes, and the tolerances required. A pictorial view of the job is shown to help you interpret the top, front, and left side views. Refer to Chapter 30, Welding Symbols, as necessary to interpret the welding symbols shown on the drawings.

As an example of how to find the information necessary to complete a job, turn to the Job Outline and find Job 13-J11, Stringer Beading. The Job Outline states that the weld is to be made on flat plate, the direction of travel is down, and the position of welding is vertical. The welding current may be either direct current (electrode positive polarity) or alternating current. Use the E6010 electrode with direct current, electrode positive, and the E6011 electrode with alternating current. The job drawing and job description may be found on pages 345 to 347. Turn to page 346 and study the job drawing. The title block states that the stock is $3/16 \times 7 \times 8$ mild steel plate, the fractional limits are $1/8$-inch, and $5/32$-inch electrodes are required. The front view indicates $1/16$-inch reinforcement, $1/16$-inch penetration, and beads that are $5/16$ inch wide. The top view shows the direction of travel and instructs you to make beads on both sides of the plate. The welding symbol shown in the pictorial view tells you that these are surface beads and directs you to read note A below the view for additional information.

Complete the jobs in the order assigned by your instructor. At certain points in your practice, you will test a sample weld for soundness. Refer to Job 13-J14, page 363, for an example of such a check test. As you progress, you will be asked to pass performance tests for welder qualification. After completing Job 14-J28, for example, you will take Test 1 in the flat position. Depending on local requirements, your instructor may also ask you to pass Test 3 in the flat position. At the conclusion of the course in shielded metal arc welding (plate), you will be required to take the complete series of welder qualification tests. These tests are described on pages 920–934.

Job 13-J1 Striking the Arc and Short Stringer Beading

Objective
To strike the arc and deposit short stringer beads on flat steel plate in the flat position with E6013 electrodes.

General Job Information
Striking the arc is a basic action throughout the entire welding operation. It will occur each time the welding operation is started.

Fig. 13-8 Position of the electrode and electrode holder when striking the arc. Location: Northeast Wisconsin Technical College
Mark A. Dierker/McGraw Hill

Welding Technique
The arc is established by lightly touching the plate with the electrode and then withdrawing it, Fig. 13-8. Current flows immediately, and the arc will be maintained if the gap is not too great. The heat of the arc causes the plate, electrode coating, and the electrode to melt, making it possible to weld. If the electrode is not withdrawn fast enough, it will stick to the plate and may be freed by a quick twist of the wrist.

Striking the arc may be accomplished by a straight up-and-down motion (Fig. 13-9A), or by a scratching motion

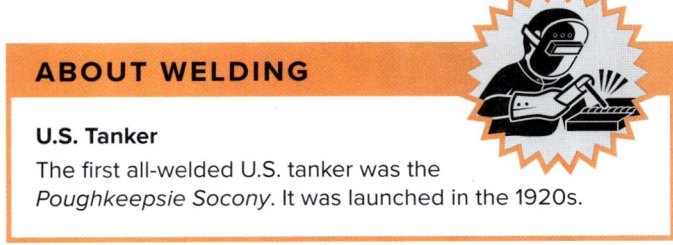

ABOUT WELDING

U.S. Tanker
The first all-welded U.S. tanker was the *Poughkeepsie Socony*. It was launched in the 1920s.

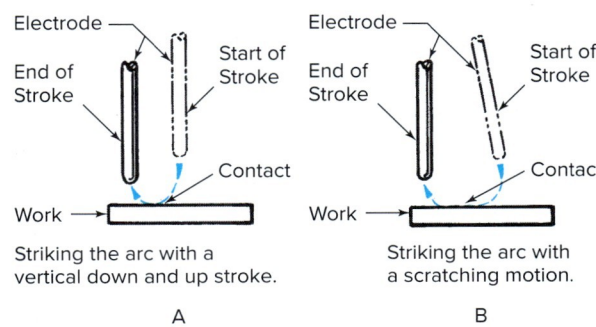

Fig. 13-9 Two methods of striking the arc.

(Fig. 13-9B), page 339. Use the method that causes you the least trouble.

After you have mastered striking the arc, practice depositing sort stringer beads. Hold the electrode at the starting point for a short time to allow fusion to occur and the bead to form. Advance the electrode at a uniform rate of travel, feeding steadily downward as the electrode is melted into the pool. Consume the electrode until it reaches the last classification numbers. This will reduce waste and help your instructor control costs.

It is important that the correct arc gap be maintained. A sharp crackling sound, an even transfer of the molten ball of metal across the arc gap, and lack of spatter are indications of correct arc gap. An arc that is held too short is very erratic. It will stick to the plate and go out. A long arc is recognized by a steady hiss, like escaping steam. Penetration is poor, and overlap occurs along the sides of the bead. There is a considerable amount of spatter.

In order to get a good "hot start" and prevent arc strikes, start the arc ⅜ inch ahead of where you want to start the weld. With a rapid motion and correct arc length, move to the starting point of the weld. This will preheat the starting point

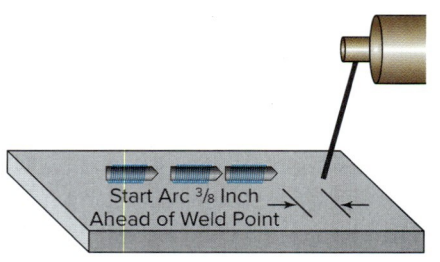

Fig. 13-10 Use this hot-start method with the rapid motion and correct arc gap to eliminate cold starts, poor fusion, and poor weld profile at the start.

and help stabilize the arc, which will provide good fusion at the start of the weld and in the arc strike area, Fig. 13-10.

Operations

1. Obtain plate; check the job drawing, Fig. 13-11, for size.
2. Obtain a square and scale, dividers, scribe, and center punch from the toolroom.
3. Lay out parallel lines as shown in the job drawing.
4. Mark the lines with a center punch. They should look like those in Fig. 13-8, page 339.

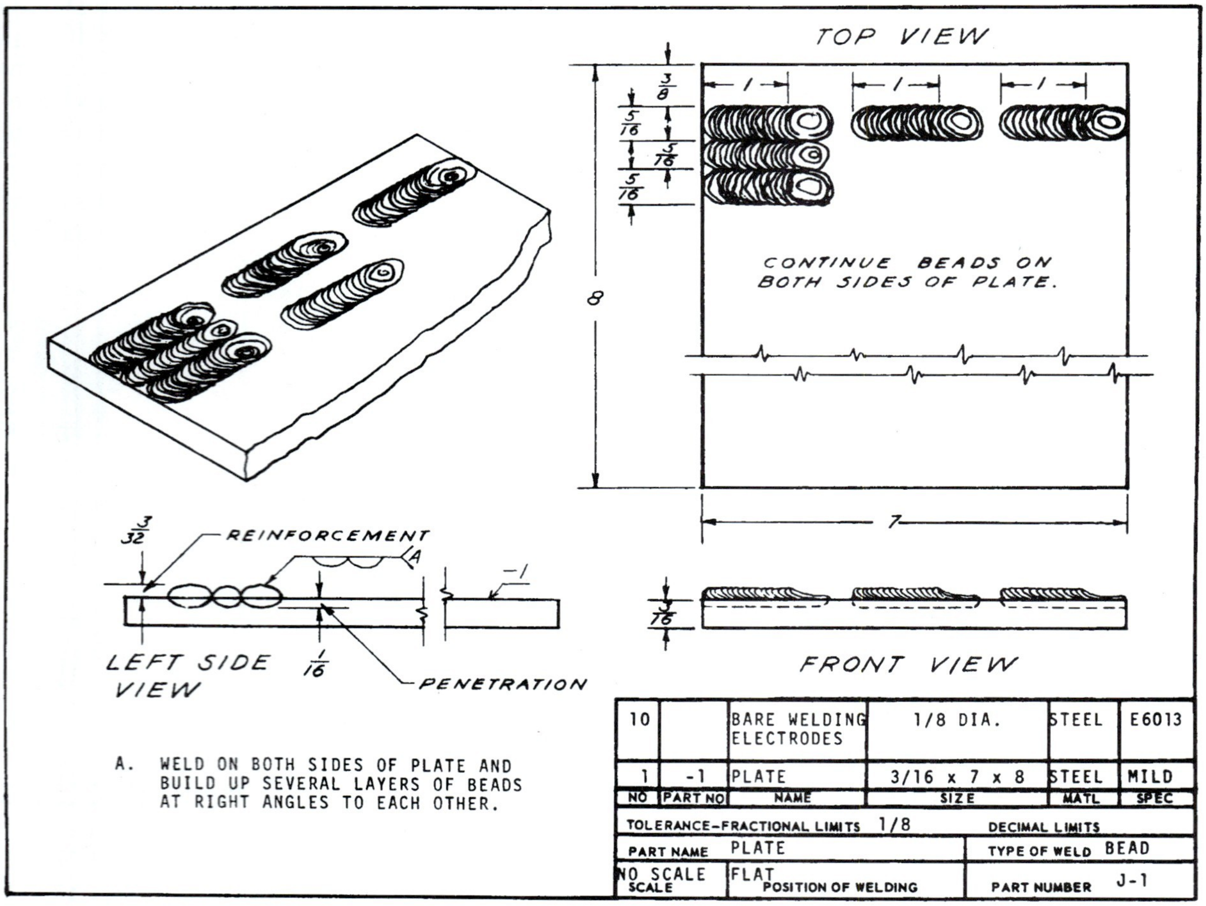

Fig. 13-11 Job drawing J1.

340   Chapter 13   Shielded Metal Arc Welding Practice: Jobs 13-J1–J25 (Plate)

5. Obtain electrodes of each quantity, type, and size specified in the job drawing.
6. Set power source for 70 to 120 amperes. Set a d.c. power source for DCEN.
7. Lay the plate in the flat position on the welding table. Make sure it is well-grounded.
8. Practice striking the arc and make beads as instructed in the job drawing. Hold the electrode in the position shown in Fig. 13-8. Strike the arc by touching the plate with either of the motions shown in Fig. 13-9.
9. Chip the slag and brush the beads and inspect. Refer to Inspection, below.
10. Practice this job until you can strike an arc freely, without sticking, and where desired.

Inspection

Compare the beads with Fig. 13-12, and check them for the following weld characteristics:

Width and height: Uniform
Appearance: Smooth with even ripples; free of voids and high spots
Size: Refer to the job drawing.
Face of beads: Slightly convex
Edges of beads: Good fusion, no overlap, no undercut
Beginnings: Full size
Penetration and fusion: To plate surfaces
Surrounding plate surfaces: Free of spatter and arc strikes

Disposal

Discard completed plates in the waste bin. Plates must be filled with beads on both sides before disposal.

Job 13-J2   Stringer Beading

Objective

To deposit stringer beads on flat plate in the flat position with DCEN and/or a.c. shielded metal arc electrodes (AWS E6013).

General Job Information

It has been found beneficial to start practice with medium-sized electrodes ($\frac{1}{8}$- and $\frac{5}{32}$-inch diameters) although the tendency in industry today is toward the use of larger-sized electrodes ($\frac{3}{16}$- and $\frac{1}{4}$-inch diameters). You should practice with $\frac{3}{16}$-inch diameter electrodes after you have mastered the use of the $\frac{5}{32}$-inch size.

This job requires E6013 electrodes. The power source may be either DCEN or alternating current. Study the section in Chapter 12 concerning this electrode. It is classified as an all-position electrode, but is used for the most part in the flat and horizontal positions. Vertical-down welding is also a major use.

E6013 electrodes used with alternating current require lower welding currents. They are particularly suited to welding thin metals. Their arc is soft, and penetration is very light. The bead is smooth, and the ripple is fine. E6013 electrodes produce a flat fillet weld. They are suitable for groove welds because of the flat bead shape and easily removed slag.

The principal differences between alternating current and direct current welding are (1) the establishing of the arc and (2) the length of the arc maintained when welding. The a.c. arc is established by scratching or dragging the electrode over the work. The constant reversal of alternating current causes starting to be somewhat difficult and makes it necessary to hold a longer arc than when welding with a direct current. The longer arc makes it more difficult to weld in the vertical and overhead positions.

One major advantage of welding with alternating current is the absence of arc blow. This permits the use of larger electrodes and more current so that heavy steels can be welded faster. The same basic techniques used in welding with direct current may also be used in welding with alternating current. A.C. welds have good penetration, are of high quality, and are similar in appearance to those produced with direct current.

Welding Technique

Pay close attention to your *current setting*. Hold the electrode at a 90° angle to the plate. This is called the work angle, Fig. 13-9A. Then tilt the electrode 10 to 20° in the direction of travel. This is called the **travel angle**, Fig. 13-9B.

Too short an arc gap allows the coating to contact the molten weld pool and causes slag inclusions. If the arc is too long, the metal is transferred in large irregular drops and is exposed to the surrounding air. This produces rough welds, excessive spatter, and oxidation. The sound of the arc and the action of the weld pool indicate the correct arc gap.

When starting the bead, strike a long arc and hold it on one spot long enough to heat the base metal. Gradually close the arc gap until a little pool of molten metal of the proper size is formed.

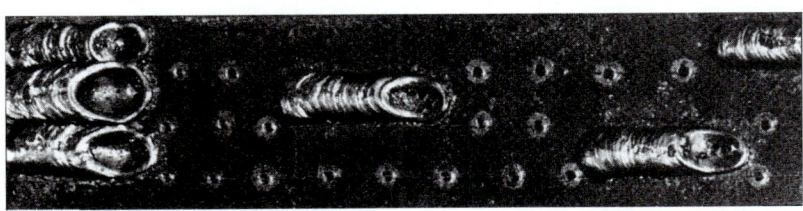

Fig. 13-12 Method of marking parallel lines and typical appearance of short stringer beads welded in the flat position with coated electrodes. Edward R. Bohnart

JOB TIP

Areas of Growth

1. Infrastructure repair
2. Transportation, including marine and aerospace
3. Automotive use of aluminum
4. Processes such as friction stir, resistance, plasma arc, and capacitor discharge

The pool should be fused well into the base metal before moving forward with the bead. A slight back-and-forth motion along the line of weld may be used. Most welders, however, prefer to advance along the line of weld without this motion. They pace their travel by the formation of the bead and feed the electrode downward with a slow steady movement to maintain a constant arc gap.

When the electrode is consumed or for any reason the arc is broken, a special procedure must be used to ensure a good start with good fusion and appearance. Remove the slag, and wire brush the crater area. Restart the arc at the forward edge of the crater. Bring it back across the crater to the edge of the already deposited metal, and then forward again in the direction of welding, Fig. 13-9, page 339. If you do not bring the electrode back far enough, there will be a depression between the starting and stopping points. If you bring it back too far, the weld metal will pile up in a large lump.

You will notice that a crater will form at the end of each weld bead and at each stopping point. These craters are a helpful indication of penetration. They should be filled to the full cross section of the weld to prevent crater cracks and for craters to meet size and strength requirements. The crater can be filled with a number of techniques.

1. Move the arc to the back of the weld pool. Pause the arc a short time over the crater while maintaining the proper arc length. Then quickly break the arc.
2. Use a slight circular motion to fill the crater.
3. Reverse the direction (back-stepping) and leave the crater on top of the previously deposited weld metal. Since this weld has some convexity, the cross section should be appropriate.

Never use a long arc to fill the crater. This is a bad technique, and when using the low hydrogen electrodes, long arcing will cause porosity.

Weld on both sides of the plate. When the plate is completely filled, deposit additional beads at right angles to the beads already on the plate. Be sure to practice traveling in all directions: left to right, right to left, and away from yourself.

To execute good stop and restarts, review Fig. 13-13.

Operations

1. Obtain plate; check the job drawing, Fig. 13-14, for size.
2. Obtain a square head and scale, dividers, scribe, and center punch from the toolroom.
3. Lay out parallel lines as shown on the job drawing.
4. Mark the lines with a center punch. They should look like those in Fig. 13-15.
5. Obtain electrodes of each quantity, type, and size specified in the job drawing.
6. Set a d.c. power source for electrode negative at 110 to 190 amperes, or an a.c. power source at 120 to 210 amperes.
7. Lay the plate in the flat position on the welding table. Make sure that the plates are well-connected to the work connection.
8. Make beads on $5/8$-inch center lines as shown on the job drawing. Hold the electrode at a 90° angle to the plate and then tilt it 10 to 20° in the direction of welding, Fig. 13-16, page 344.
9. Chip the slag from the beads, brush, and inspect. Refer to Inspection.
10. Make additional beads between the already deposited beads as shown on the job drawing.

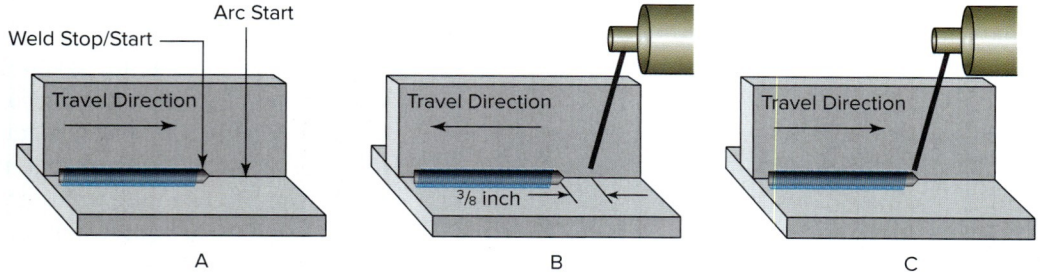

Fig. 13-13 Stops and restarts. (A) Do not fill crater at his point. (B) As you have been doing, start the arc $3/8$ inch ahead of crater and move rapidly to the crater. (C) Dwell long enough to get proper fill and begin welding in normal fashion. This is being shown on a T-joint with a fillet weld. However, this technique will work with beads on plate as well.

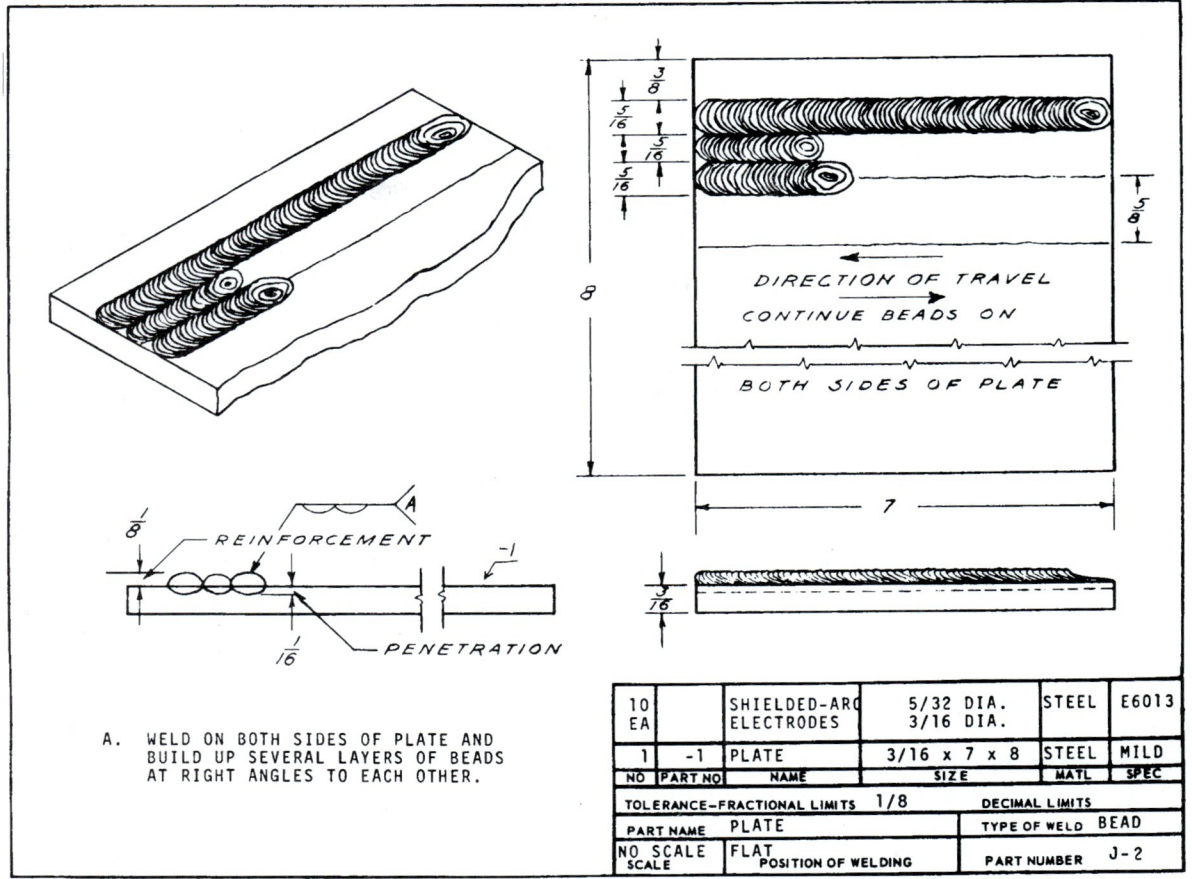

Fig. 13-14 Job drawing J2.

11. Chip the slag from the beads, brush, and inspect. Refer to Inspection.
12. Practice these beads until you can produce uniform beads consistently with both types of current.

Inspection

Compare the beads with Fig. 13-15 and check them for the following characteristics:

Width and height: Uniform
Appearance: Smooth with close ripples; free of voids and high spots. Restarts should be difficult to locate.
Size: Refer to the job drawing.
Face of beads: Slightly convex
Edges of beads: Good fusion, no overlap, no undercut
Starts and stops: Free of depressions and high spots
Beginnings and endings: Full size, craters filled
Penetration and fusion: To plate surfaces and adjacent beads
Surrounding plate surfaces: Free of spatter
Slag formation: Full coverage, easily removable

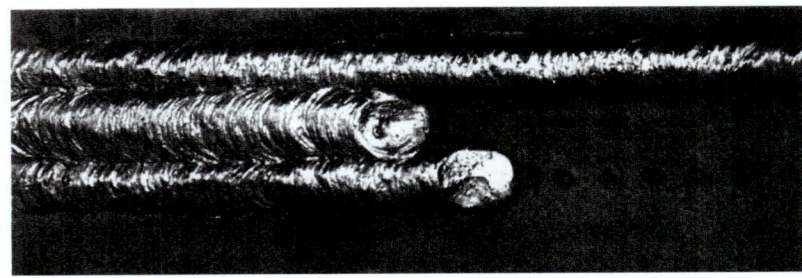

Fig. 13-15 Method of marking parallel lines and typical appearance of stringer beads welded in the flat position with DCEN or a.c. shielded arc electrodes.
Edward R. Bohnart

Disposal

Discard completed plates in the waste bin. Plates must be filled with beads on both sides.

Job 13-J3   Weaved Beading

Objective

To deposit weaved beads on flat steel plate in the flat position with a DCEN and/or a.c. shielded metal arc electrode (AWS E6013).

Shielded Metal Arc Welding Practice: Jobs 13-J1–J25 (Plate)   Chapter 13   343

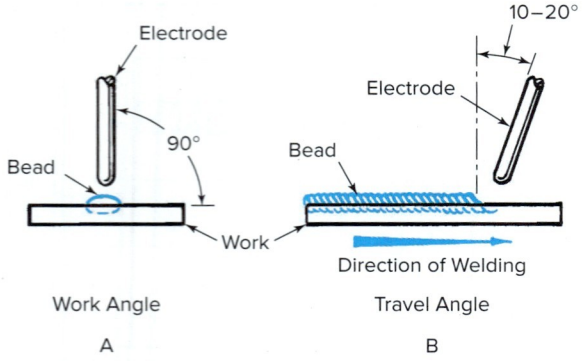

Fig. 13-16 Electrode position for stringer beads in the flat position.

General Job Information

It is often necessary to deposit beads wider than stringer beads. In order to do this, the electrode must be weaved. Weaved beads are usually necessary in the welding of deep-groove joints and multipass fillet welds. They are called weave beads.

Welding Technique

Pay close attention to your current setting. The electrode position should be approximately that shown in Fig. 13-17. Hold the electrode at a 90° angle to the plate, Fig. 13-17A. Then tilt it 10 to 20° in the direction of welding, Fig. 13-17B. Move the electrode back and forth, advancing it no more than 1/8 inch with each weave, Fig. 13-17C. Hesitate at the sides of the weld to prevent undercutting along the edges of the weld with the arc.

It is not good practice to weave beads wider than three times the diameter of the electrode. The weld deposit must be molten until the desired shape is formed. Advancing the arc too far ahead and then delaying the return to the molten pool allows the pool to cool. This causes trapped slag and poor appearance. You will not have difficulty with undercutting with this electrode, but it is well to hesitate an instant at the sides. Electrode motion should be smooth and rhythmic along the entire weld.

Starts and stops are made and craters are filled as practiced in previous jobs.

Weld on both sides of the plate. When the plate is completely filled, deposit additional beads at right angles to the beads already on the plate. Practice traveling in all directions.

Operations

1. Obtain plate; check the job drawing, Fig. 13-18, for size.
2. Obtain a square head and scale, dividers, scribe, and center punch from the toolroom.
3. Lay out parallel lines as shown on the job drawing.
4. Mark the lines with a center punch. They should look like those in Fig. 13-19.
5. Obtain electrodes of each quantity, type, and size specified.
6. Set a d.c. power source for electrode negative at 110 to 190 amperes, or an a.c. power source at 120 to 210 amperes.
7. Lay the plate in the flat position on the welding table. Make sure that the plates are connected to the work connection.
8. Make weaved beads between parallel lines as shown on the job drawing. Hold the electrode as shown in Fig. 13-17.
9. Chip the slag from the beads, brush, and inspect. Refer to Inspection.
10. Make stringer beads between the weaved beads already deposited.
11. Chip the slag from the beads, brush, and inspect. Refer to Inspection.
12. Practice weaved beads and stringer beads until you can produce uniform beads consistently with both types of current.

Inspection

Compare the beads with Figs. 13-19 and 13-20 and check them for the following weld characteristics:

Width and height: Uniform
Appearance: Smooth with close ripples; free of voids and high spots. Restarts should be difficult to locate.
Size: Refer to the job drawing.
Face of beads: Slightly convex
Edges of beads: Good fusion, no overlap, no undercut
Starts and stops: Free of depressions and high spots
Beginnings and endings: Full size, craters filled
Penetration and fusion: To plate surfaces and adjacent beads
Surrounding plate surfaces: Free of spatter
Slag formation: Full coverage, easily removable

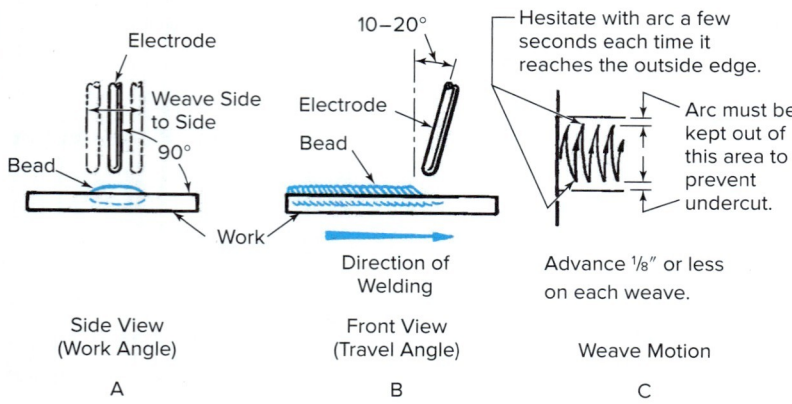

Fig. 13-17 Electrode position for weave beads.

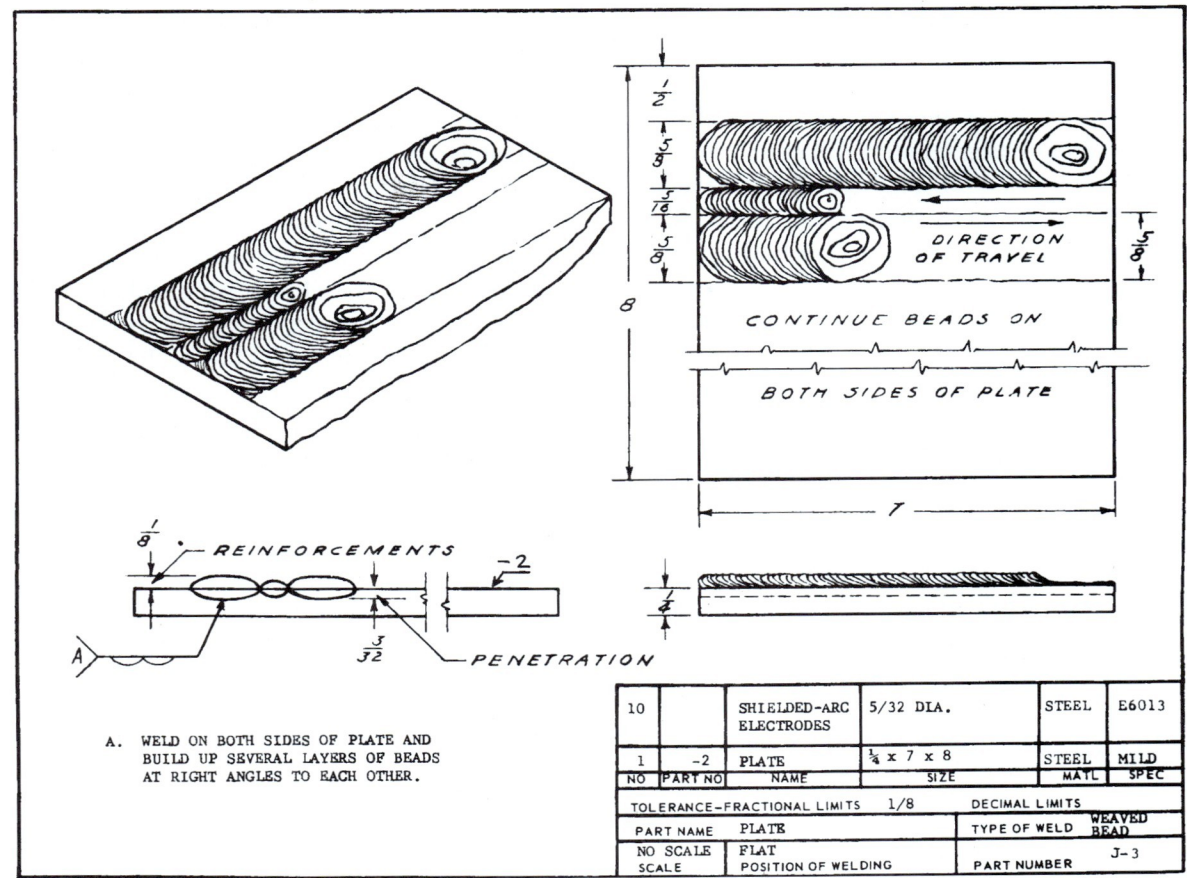

Fig. 13-18 Job drawing J3.

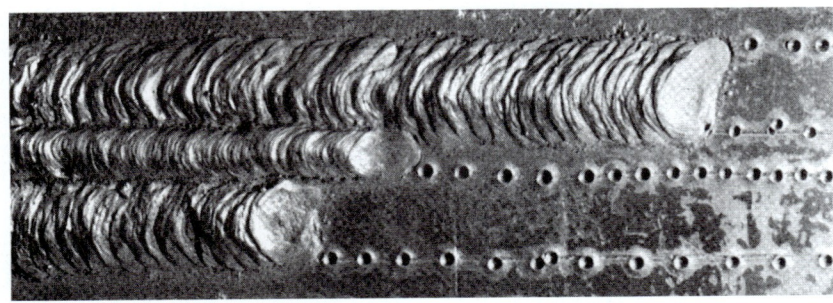

Fig. 13-19 Method of marking parallel lines and typical appearance of weaved beads welded in the flat position with DCEN shielded arc electrodes. *Edward R. Bohnart*

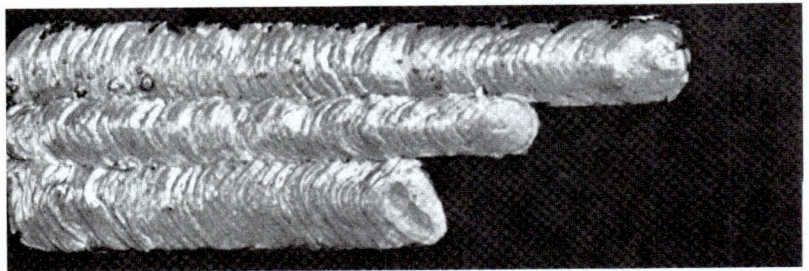

Fig. 13-20 Typical appearance of weave beads welded in the flat position with the E6013 a.c. type of shielded arc electrode. *Edward R. Bohnart*

Disposal

Discard completed plates in the waste bin. Plates must be filled with beads on both sides.

Job 13-J4 Stringer Beading

Objective

To deposit stringer beads on flat steel plate in the flat position with DCEP and/or a.c. shielded metal arc electrodes (AWS E6010–E6011).

General Job Information

This job is basically the same as Job 13-J2, but different types of electrodes are used, resulting in different arc characteristics and deposition.

This job requires either a DCEP or a.c. power source and both E6010 and E6011 electrodes. Study the section in Chapter 12 concerning these electrodes.

The electrodes produce welds of high quality in all positions. Their spray-type arc produces deep penetration, so precise electrode manipulation is necessary in order to control undercut and weld spatter. The thickness of the covering is held to a minimum. The weld pool wets and spreads well, but it sets up very quickly. Thus, these electrodes have fine out-of-position welding characteristics. Fillet and groove welds are flat in profile and have a rather coarse, unevenly spaced ripple. The current and voltage ranges are similar for both electrodes. The coating is slightly heavier on the E6011. The ductility, tensile strength, and yield strength of the deposited weld metal are higher than can be obtained with an E6010 electrode. Both of these electrodes are used whenever welding of high quality is required, such as in shipyards, field construction, pressure vessel fabrication, piping, and other code welding applications.

Welding Technique

Pay close attention to your heat setting. It is better to start with the heat setting a little too hot and have to adjust it down rather than start cold and have to adjust it up. The electrode position should be approximately that shown in Fig. 13-16, page 344. The arc gap for the E6010 electrode (electrode positive) may be less than for the E6013 (electrode negative) electrode. Do not permit the coating to contact the molten pool. Observe the weld metal transfer closely and listen to the sound of the arc for correct arc gap.

Start the arc and proceed with the bead in much the same manner as practiced in Job 13-J3. A slight forward and backward motion along the line of weld may be used. These electrodes will have a tendency to undercut the plate along the edges of the weld. To prevent undercutting, do not move the arc past the forward end of the crater until enough metal has been deposited to advance a weld of the required size and all undercut has been filled.

Starts and stops and filling craters are performed as practiced in previous jobs.

Weld on both sides of the plate. When the plate is completely filled, deposit additional beads at right angles to the beads already on the plates. Practice traveling in all directions. You should also practice with the 3/16-inch diameter electrodes after you have mastered the 5/32-inch size.

Operations

1. Obtain plate; check the job drawing, Fig. 13-21, for size.

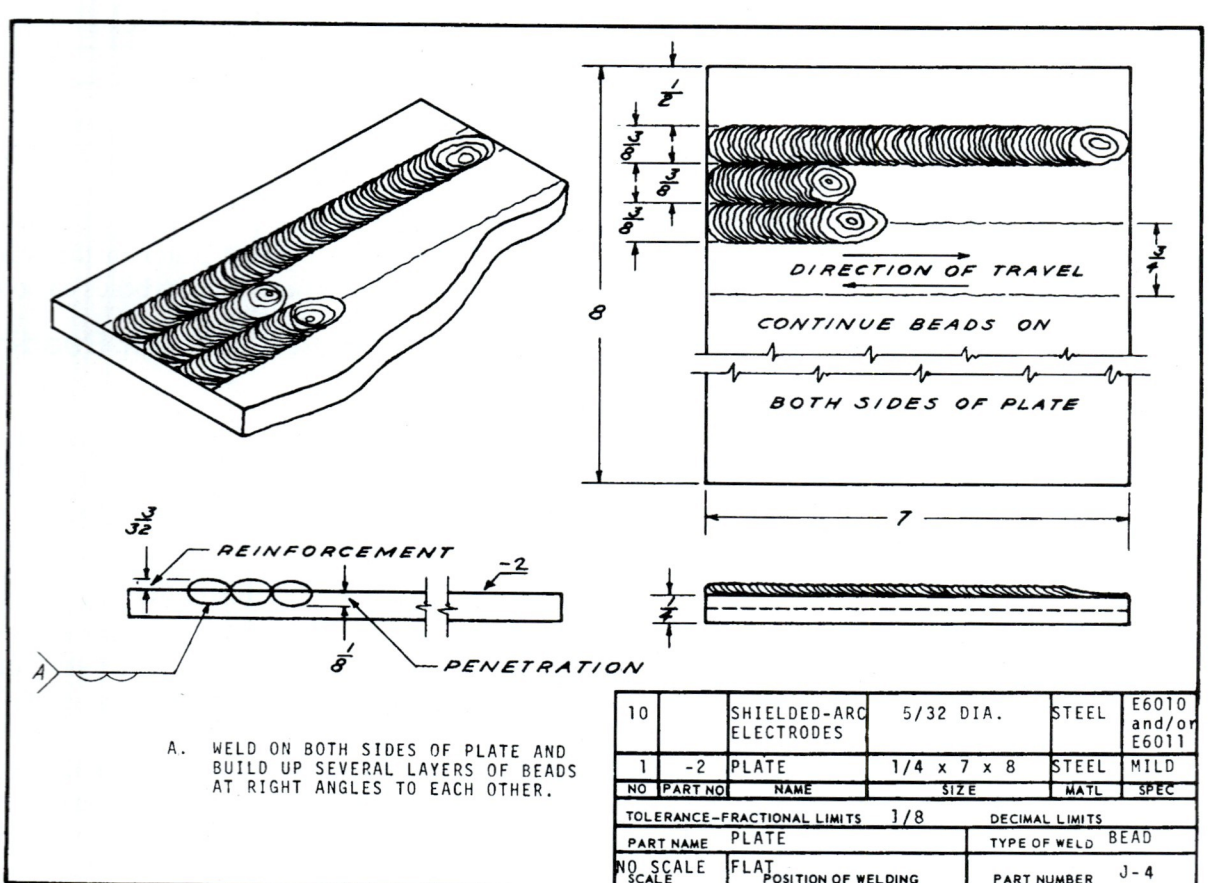

Fig. 13-21 Job drawing J4.

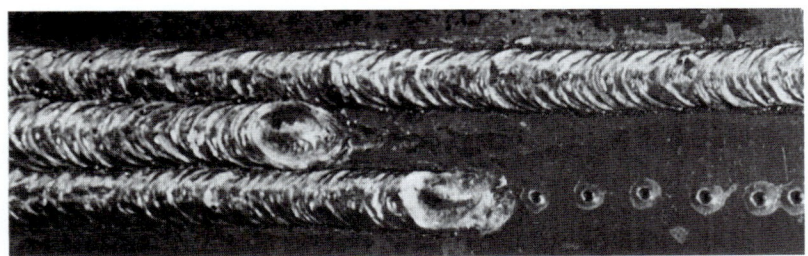

Fig. 13-22 Method of marking parallel lines and typical appearance of stringer beads welded in the flat position with DCEP or a.c. shielded arc electrodes.
Edward R. Bohnart

2. Obtain a square head and scale, dividers, scribe, and center punch from the toolroom.
3. Lay out parallel lines as shown on the job drawing.
4. Mark the lines with a center punch. They should look like those in Fig. 13-22.
5. Obtain electrodes of each quantity, type, and size specified in the job drawing.
6. Set the power source for 110 to 170 amperes. Set a d.c. power source for reverse polarity.
7. Lay the plate in the flat position on the welding table. Make sure it is connected with the work connection.
8. Make stringer beads on ¾-inch center lines as shown on the job drawing. Hold the electrode as shown in Fig. 13-16.
9. Chip the slag from the beads, brush, and inspect. Refer to Inspection.
10. Make additional beads between the beads already deposited as shown on the job drawing.
11. Chip the slag from the beads, brush, and inspect. Refer to Inspection below.
12. Practice these beads until you can produce uniform beads consistently with both the E6010 and E6011 electrodes.

Inspection

Compare the beads with Fig. 13-22 and check for the following weld characteristics:

Width and height: Uniform
Appearance: Smooth with close ripples; free of voids and high spots. Restarts should be difficult to locate.
Size: Refer to the job drawing.
Face of beads: Slightly convex
Edges of beads: Good fusion, no overlap, no undercut
Starts and stops: Free of depressions and high spots
Beginnings and endings: Full size, craters filled
Penetration and fusion: To plate surfaces and adjacent beads
Surrounding plate surfaces: Free of spatter
Slag formation: Full coverage, easily removable

Disposal

Discard completed plates in the waste bin. Plates must be filled with beads on both sides.

Job 13-J5 Weaved Beading

Objective

To deposit weaved beads on flat steel plate in the flat position with DCEP and/or a.c. shielded metal arc electrodes (AWS E6010–E6011).

General Job Information

This job is basically the same as Job 20-J3, except that different types of electrodes are used, resulting in different arc and deposition characteristics. In industry you will do more weaved beading with electrode positive electrodes than with electrode negative electrodes.

Welding Technique

Pay close attention to the current setting. Study the electrode position and weaving technique shown in Fig. 13-17, page 344. The method of depositing beads is the same as for electrode negative electrodes except that undercutting is a problem. This can be overcome only by longer hesitation at the sides of the weld and by keeping the movement of the electrode within the confines of the bead width. Bead reinforcement will not be as high as with electrode negative electrodes, and penetration will be deeper. There will be less slag present. It will be less fluid and will solidify more quickly.

Starts and stops are made and craters are filled as practiced in previous jobs.

Weld on both sides of the plate. When the plate is completely filled, deposit additional beads at right angles to the beads already on the plate. Practice traveling in all directions.

Operations

1. Obtain plate; check the job drawing, Fig. 13-23, page 348, for size.
2. Obtain a square head and scale, dividers, scribe, and center punch from the toolroom.
3. Lay out parallel lines as shown on the job drawing.
4. Mark the lines with a center punch. They should look like those shown in Fig. 13-24, page 348.
5. Obtain electrodes of each quantity, type, and size specified in the job drawing.
6. Set a d.c. power source for electrode positive at 110 to 170 amperes or an a.c. power source at 110 to 190 amperes.

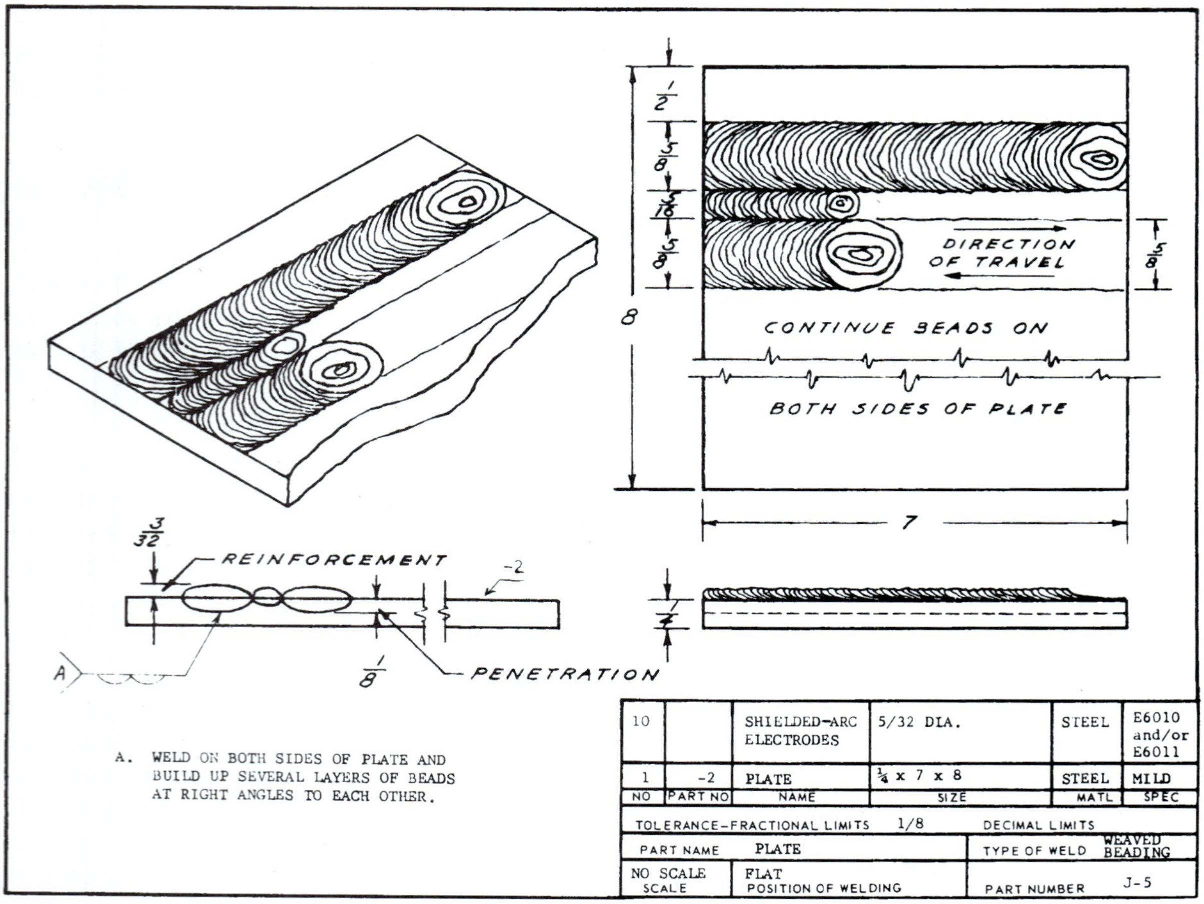

Fig. 13-23 Job drawing J5.

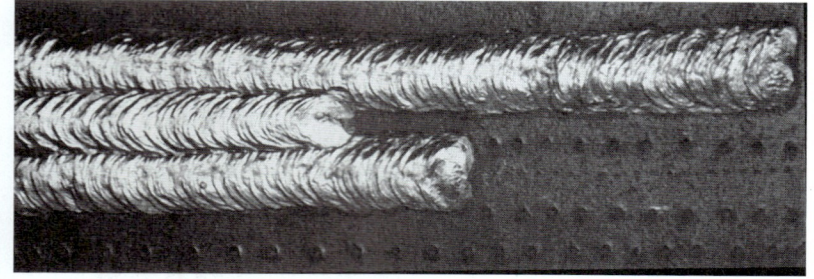

Fig. 13-24 Method of marking parallel lines and typical appearance of weave beads welded in the flat position with DCEP shielded arc electrodes.
Edward R. Bohnart

7. Lay the plate in the flat position on the welding table. Make sure it is connected to the work connection.
8. Make weaved beads between parallel lines as shown on the job drawing. Hold the electrode as shown in Fig. 13-17, page 344.
9. Chip the slag from the beads, brush, and inspect. Refer to Inspection.
10. Make stringer beads between the weaved beads already deposited.
11. Chip the slag from the beads, brush, and inspect. Refer to Inspection.
12. Practice these beads until you can produce uniform beads consistently with both types of electrodes.

Inspection

Compare the beads with Figs. 13-24 and 13-25 and check them for the following characteristics:

Width and height: Uniform
Appearance: Smooth with close ripples; free of voids and high spots. Restarts should be difficult to locate.
Size: Refer to the job drawing.
Face of beads: Slightly convex
Edges of beads: Good fusion, no overlap, no undercut
Starts and stops: Free of depressions and high spots
Beginnings and endings: Full size, craters filled
Penetration and fusion: To plate surfaces and adjacent beads
Surrounding plate surfaces: Free of spatter
Slag formation: Full coverage, easily removable

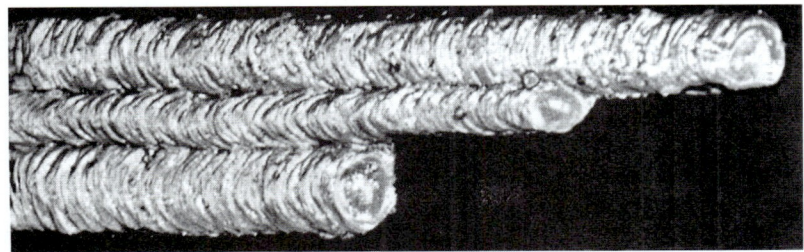

Fig. 13-25 Typical appearance of weave beads welded in the flat position with the E6011 a.c. type of shielded arc electrode. *Edward R. Bohnart*

3. Set a d.c. power source for electrode negative at 110 to 190 amperes or an a.c. power source at 120 to 210 amperes.
4. Set up the plates and tack weld them at each corner.
5. Place the joint in the flat position on the welding table. Make sure it is connected to the work connection.
6. Make welds as shown on job drawing. Hold electrode as shown in Figs. 13-25 and 13-27.

Disposal

Discard completed plates in the waste bin. Plates must be filled with beads on both sides before disposal.

Job 13-J6 Welding an Edge

Objective

To edge weld a parallel joint in the flat position with DCEN and/or a.c. shielded metal arc electrodes (AWS E6013).

General Job Information

The parallel joint is used extensively on vessels and tanks that are not subjected to high pressures. The joint will not stand very much load when subjected to tension and bending.

It is an economical Edge Weld to set up. The amount of electrode needed for welding is small since some of the base metal is melted off and combines with the electrode deposit to form the weld. It is an easy joint to weld.

Welding Technique

The joint can be welded over a wide range of current settings. Hold the electrode at a 90° angle to the plates to be welded, Fig. 13-26A. Then tilt it 10 to 20° in the direction of travel, Fig. 13-26B. Note the positions of the electrode and electrode holder in Fig. 13-27. It is good practice to weave slightly so that there is complete coverage of the joint area. It is important, however, not to weave so wide that the metal hangs over the sides of the plate. Be very careful to fuse both edges and to secure good penetration along the center of the joint.

Starts and stops are made and craters are filled as practiced in previous jobs.

Weld both edges of the joint. Practice traveling in all directions.

Operations

1. Obtain plates; check the drawing for quantity and size.
2. Obtain electrodes of each quantity, type, and size specified in the job drawing, Fig. 13-28, page 350.

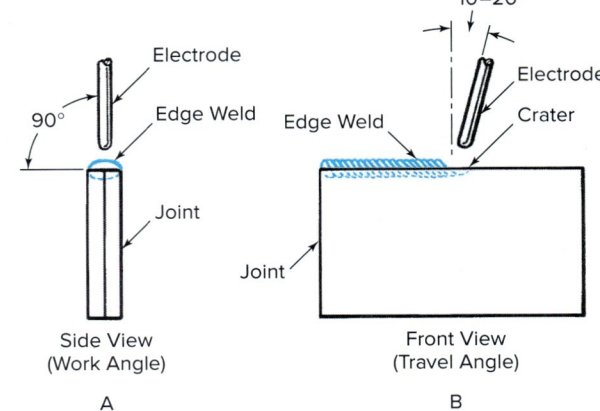

Fig. 13-26 Electrode position when welding a Parallel Joint.

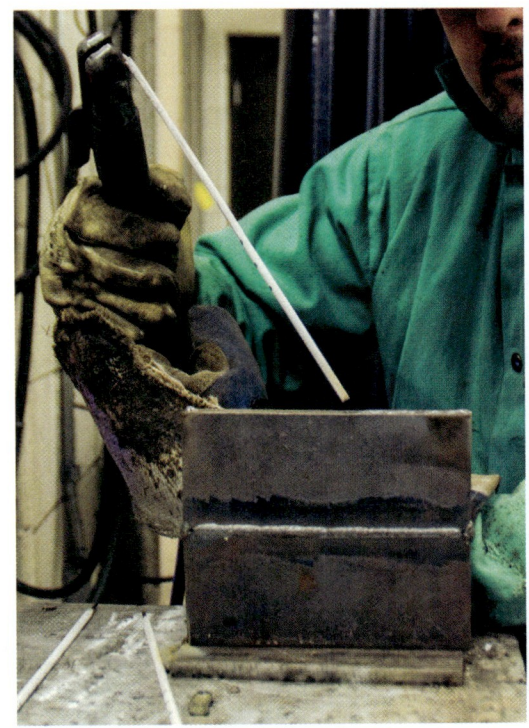

Fig. 13-27 Welding an Edge Weld. The electrode and electrode holder are positioned as instructed in Fig. 13–26. *Location: Northeast Wisconsin Technical College Mark A. Dierker/McGraw Hill*

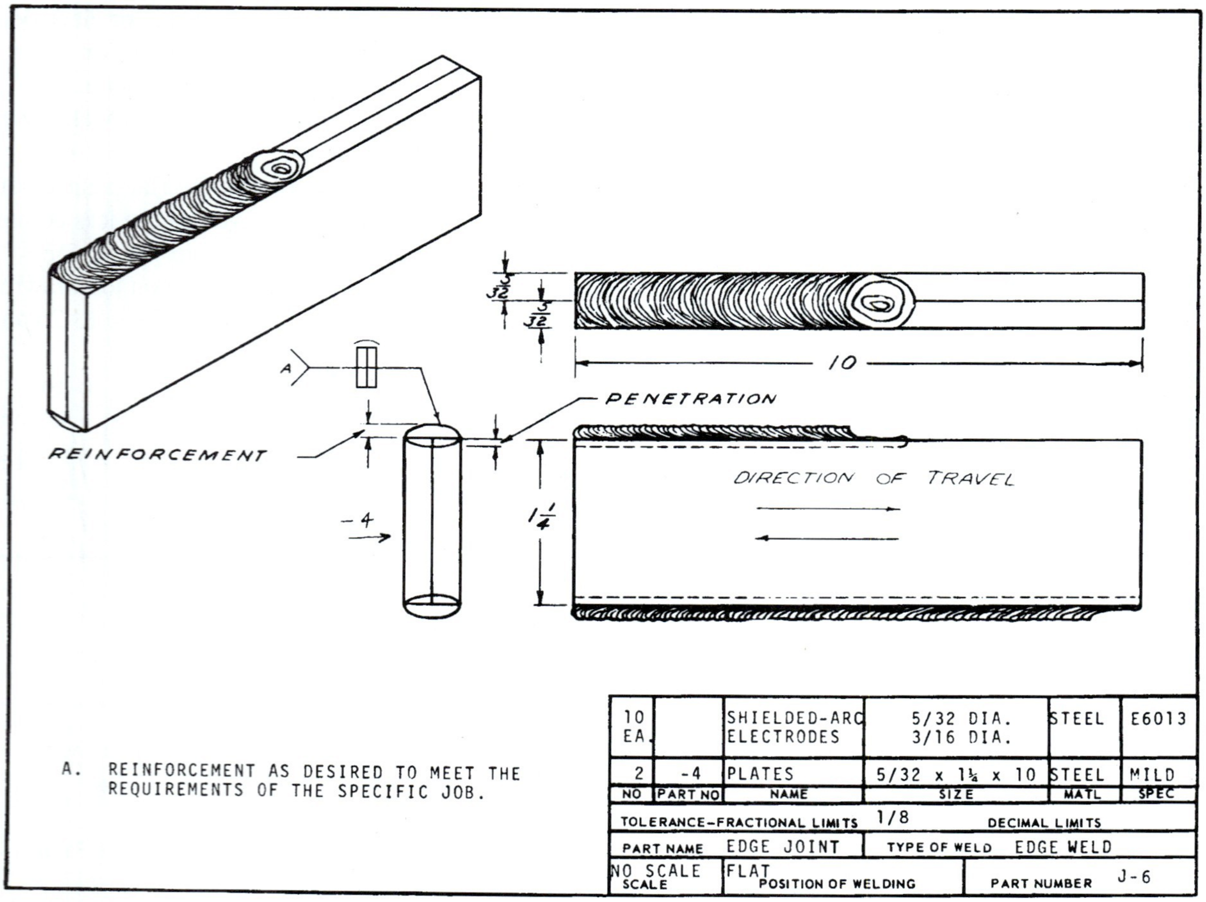

Fig. 13-28 Parallel Joint and Edge Weld for Job drawing J6.

7. Chip the slag from the welds, brush, and inspect. Refer to Inspection.
8. Practice these welds until you can produce good welds consistently with 5/32-inch diameter electrodes.

Inspection

Compare the welds with Fig. 13-29 and check them for the following characteristics:

Width and height: Uniform
Appearance: Smooth with close ripples; free of voids and high spots. Restarts should be difficult to locate.
Size: Refer to the job drawing.
Face of welds: Slightly convex
Edges of welds: Good fusion, no overlap, no undercut, no weld metal hanging over the edges of the joint
Starts and stops: Free of depressions and high spots
Beginnings and endings: Full size, craters filled
Penetration and fusion: To plate surfaces
Surrounding plate surfaces: Free of spatter
Slag formation: Full coverage, easily removable

Disposal

Run stringer beads on the unused sides of the joints or put the completed joints in the scrap bin so that they are available for stringer bead practice.

Job 13-J7 Welding a Parallel Joint

Objective

To weld an Edge Weld in the flat position with DCEP and/or a.c. shielded metal arc electrodes (AWS E6010–E6011).

Fig. 13-29 Typical appearance of a Parallel Joint and Edge Weld, welded in the flat position with DCEN or a.c. shielded arc electrodes. *Edward R. Bohnart*

General Job Information

This joint is the same as that welded in Job 13-J6 except that a different type of electrode is used.

Welding Technique

The Parallel Joint can be welded over a wide range of current settings. The electrode position should be approximately that shown in Fig. 13-26, page 349. Use the same basic welding technique practiced in Job 13-J6. The arc gap will be less and the current adjustment will be slightly lower than with electrode negative. Welds will not have very high buildup, and slag will be less.

Practice starts and stops carefully. Weld in all directions.

Operations

1. Obtain plates. Check the job drawing, Fig. 13-30, for quantity and size.
2. Obtain electrodes of each quantity, type, and size specified in the job drawing.
3. Set a d.c. power source for electrode positive at 110 to 170 amperes or an a.c. power source at 110 to 190 amperes.
4. Set up the plates and tack weld them at each corner.
5. Place the joint in the flat position on the welding table. Make sure it is connected to the work connection.
6. Make welds as shown on the job drawing. Hold the electrode as shown in Fig. 13-26, page 349.
7. Chip the slag from the welds, brush, and inspect. Refer to Inspection below.
8. Practice these welds until you can produce good welds consistently with both the E6010 and E6011 electrodes and with $5/32$-inch diameter electrodes.

Inspection

Compare the welds with Fig. 13-31, page 352, and check them for the following characteristics:

Width and height: Uniform

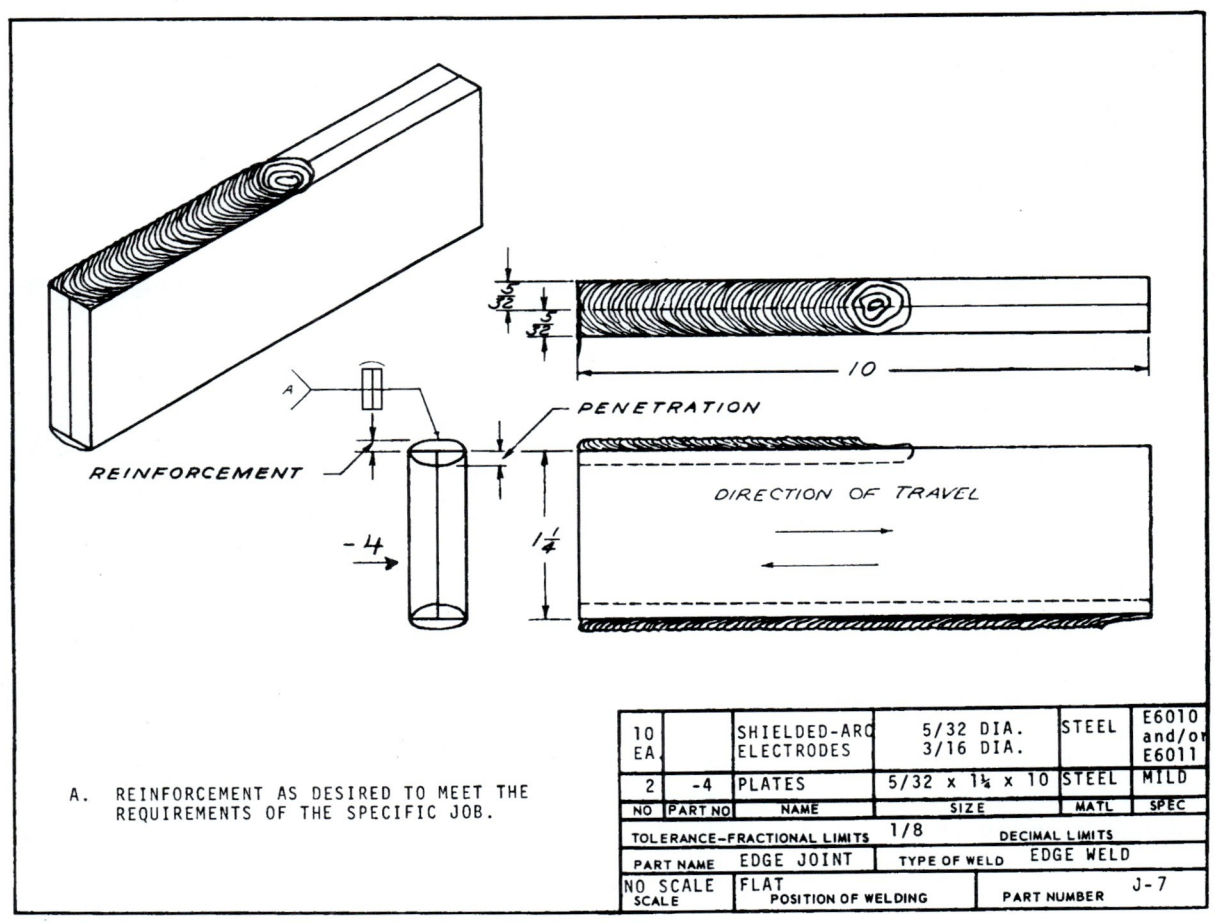

Fig. 13-30 Parallel Joint and Edge Weld for Job drawing J7.

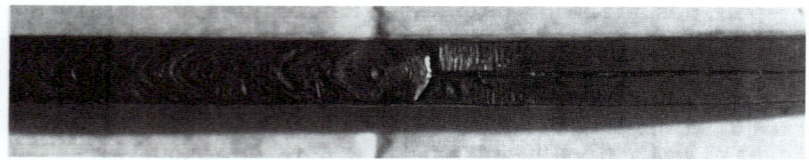

Fig. 13-31 Typical appearance of a Parallel Joint and Edge Weld, welded in the flat position with DCEP or a.c. shielded arc electrodes. Edward R. Bohnart

Appearance: Smooth with close ripples; free of voids and high spots. Restarts should be difficult to locate.
Size: Refer to the job drawing.
Face of welds: Slightly convex
Edges of welds: Good fusion, no overlap, no undercut, no weld metal hanging over the edges of the joint
Starts and stops: Free of depressions and high spots
Beginnings and endings: Full size, craters filled
Penetration and fusion: To plate surfaces
Surrounding plate surfaces: Free of spatter
Slag formation: Full coverage, easily removable

Disposal

Run stringer beads on the unused sides of the joints or put the completed joints in the scrap bin so that they are available for stringer bead practice.

Job 13-J8 Welding a Lap Joint

Objective

To weld a lap joint in the horizontal position by means of a single-pass fillet weld with DCEN and/or a.c. shielded metal arc electrodes (AWS E6013).

General Job Information

The lap joint is used extensively in industry for tank, structural, and shipyard construction. It is an economical joint because it needs very little joint preparation and fitup. The lap joint is strongest when it is double-lap welded on both sides. This job and the jobs that follow specify a single pass. For heavier plate, several passes must be made with both stringer and weaved beads.

Welding Technique

The adjustment of current should not be too high. The electrode position should be approximately that shown in Figs. 13-32 and 13-33. A fairly close arc gap is necessary, and it should be directed toward the root of the joint and toward the flat plate surface. Use a slight back-and-forth motion as needed along the line of weld. This preheats the joint ahead of the weld, results in proper convexity, and keeps the slag washed back over the deposited metal. The strength of a fillet weld depends upon the throat size.

Complete penetration must be secured at the root of the joint, and good fusion must be obtained with the two plate surfaces. The top edge of the top plate has a tendency to burn away. This can be prevented by making sure that the bead formation is full and that the arc is not played along the top surface. Undercutting is not a problem with the E6013 electrode.

There should be a straight line of fusion with the top and bottom plates and a smooth transition along the edge of the weld where it enters the plate. A rate of travel that is too slow causes an excess deposit of weld metal. When it rolls over on the bottom plate, it changes the contour of the weld abruptly and increases the possibility of poor fusion to the plate surface. Excess weld metal

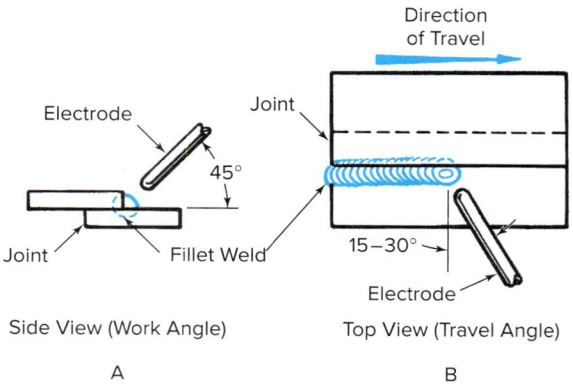

Fig. 13-32 Electrode position for lap joints in the horizontal position.

Fig. 13-33 Welding lap joints in the flat position. The electrode and electrode holder are positioned as instructed in Fig. 13-32. Location: Northeast Wisconsin Technical College Mark A. Dierker/McGraw Hill

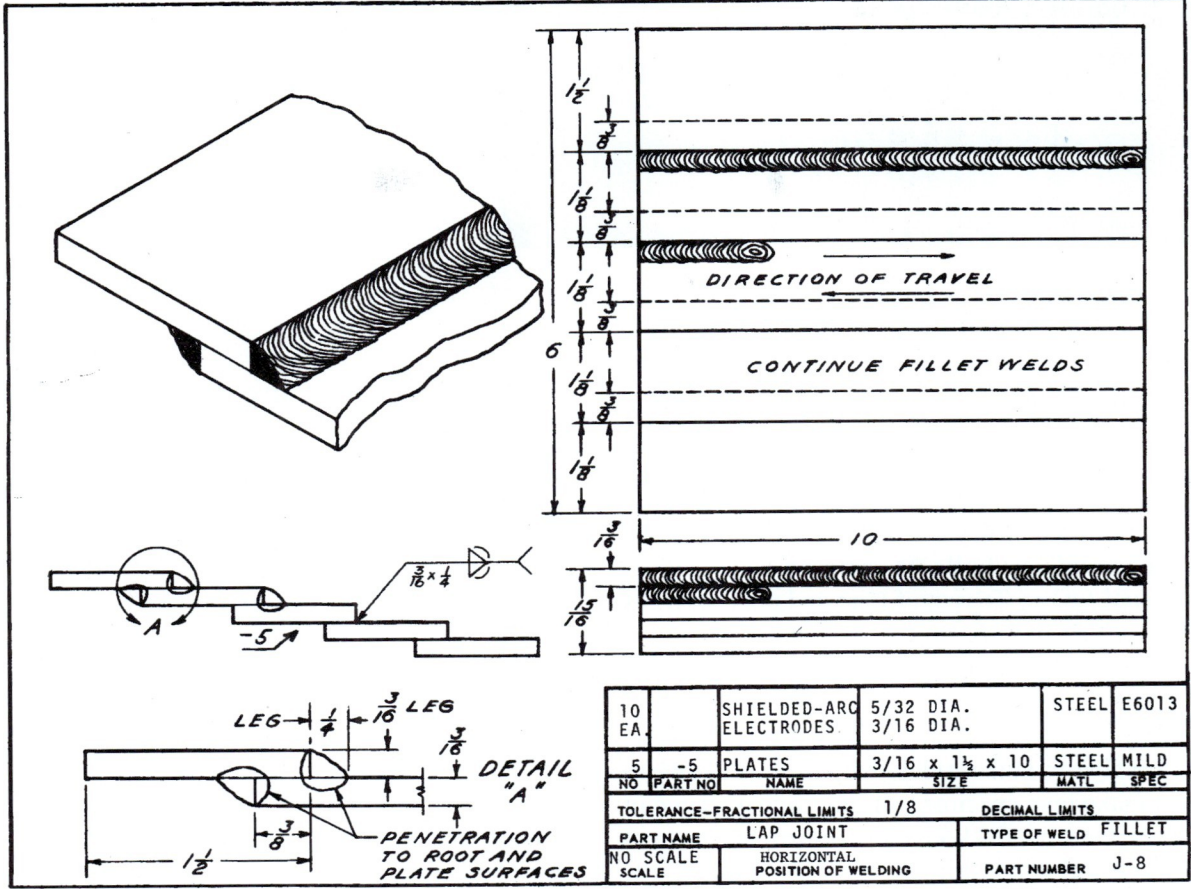

Fig. 13-34 Job drawing J8

also weakens the joint by causing stress concentration at this point.

Practice stops and starts. Travel in all directions.

Operations

1. Obtain plates; check the job drawing, Fig. 13-34, for the correct quantity and size.
2. Obtain electrodes of each quantity, type, and size specified in the job drawing.
3. Set a d.c. power source for electrode negative at 110 to 190 amperes or an a.c. power source at 120 to 210 amperes.
4. Set up the plates and tack weld them at each corner.
5. Place the joint in the flat position on the welding table. Make sure that the plates are connected with the work connection.
6. Make welds as shown on the job drawing. Hold the electrode as shown in Figs. 13-32 and 13-33.
7. Chip the slag from the welds, brush, and inspect. Refer to Inspection.
8. Practice these welds until you can produce good welds consistently with both types of current and with $5/32$-inch diameter electrodes.

Inspection

Compare the welds with Fig. 13-35, page 354, and check them for the following characteristics:

Width and height: Uniform
Appearance: Smooth with close ripples; free of voids and high spots. Restarts should be difficult to locate.

SHOP TALK

Using Antifreeze
Welders working on aluminum commonly use water-cooled, compact torches. Antifreeze added to the torch's water recirculator can keep it from overheating or freezing. The antifreeze needs to be deionized and not contain any leak preventors. Contact the manufacturer of the coolant system for recommendations.

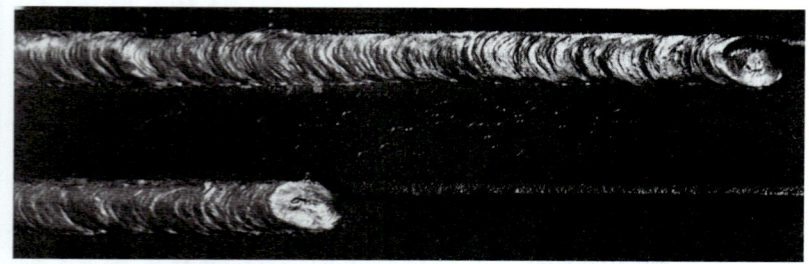

Fig. 13-35 Typical appearance of a fillet weld in a lap joint welded in the horizontal position with DCEN or a.c. shielded arc electrodes. Edward R. Bohnart

Size: Refer to the job drawing. Check with a fillet weld gauge.
Face of welds: Slightly convex
Edges of welds: Good fusion, no overlap, no undercut
Starts and stops: Free of depressions and high spots
Beginnings and endings: Full size, craters filled
Penetration and fusion: To root of joint and plate surfaces
Surrounding plate surfaces: Free of spatter
Slag formation: Full coverage, easily removable

Check Test

After you are able to make welds that are satisfactory in appearance, make up a specimen similar to that shown in Fig. 13-36. Use the plate thickness, welding procedure, and electrode size specified for this job. Weld on one side only.

Break the joint as shown in Fig. 13-37. Examine the surface of the fracture for soundness. The weld must not be porous and must show complete fusion and penetration at the root and to both plate surfaces. The joint should break through the throat of the weld, and the weld should not peel away from the plate.

Disposal

Put completed joints in the scrap bin so that they will be available for further use. Unwelded plate surfaces can be used for beading.

Job 13-J9   Welding a Lap Joint

Objective

To weld a lap joint in the horizontal position by means of a single-pass fillet weld with DCEP and/or a.c. shielded metal arc electrodes (AWS E6010–E6011).

General Job Information

This joint is the same as that practiced in Job 13-J8 except that a different type of electrode is used.

Welding Technique

Current adjustment must not be too high. The electrode position should be approximately that shown in Figs. 13-32 and 13-33, page 352. The welding technique is basically the same as that used for the previous job. You should, however, hold a shorter arc. Be careful to avoid undercut both on the flat plate surface and along the top edge of the upper plate.

Remember that the arc movement must be within the limits of the desired weld size.

Practice stops and starts. Travel in all directions.

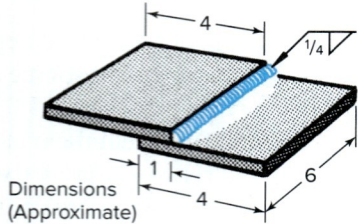

Fig. 13-36 Joint for fillet weld test specimen.

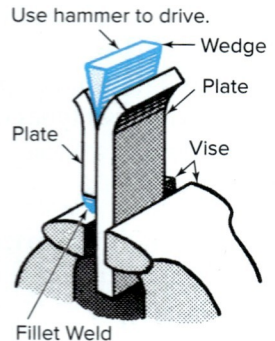

Fig. 13-37 Wedge test for fillet welds on lap joints.

Operations

1. Obtain plates; check the job drawing, Fig. 13-38, for size.
2. Obtain electrodes of each quantity, type, and size specified in the job drawing.
3. Set a d.c. power source for electrode positive at 110 to 170 amperes or an a.c. power source at 110 to 190 amperes.
4. Set up the plates and tack weld them at each corner.
5. Place the joint in the flat position on the welding table. Make sure it is connected to the work connection.
6. Make welds as shown on the job drawing. Hold the electrode as shown in Figs. 13-32 and 13-33.
7. Chip the slag from the welds, brush, and inspect. Refer to Inspection below.

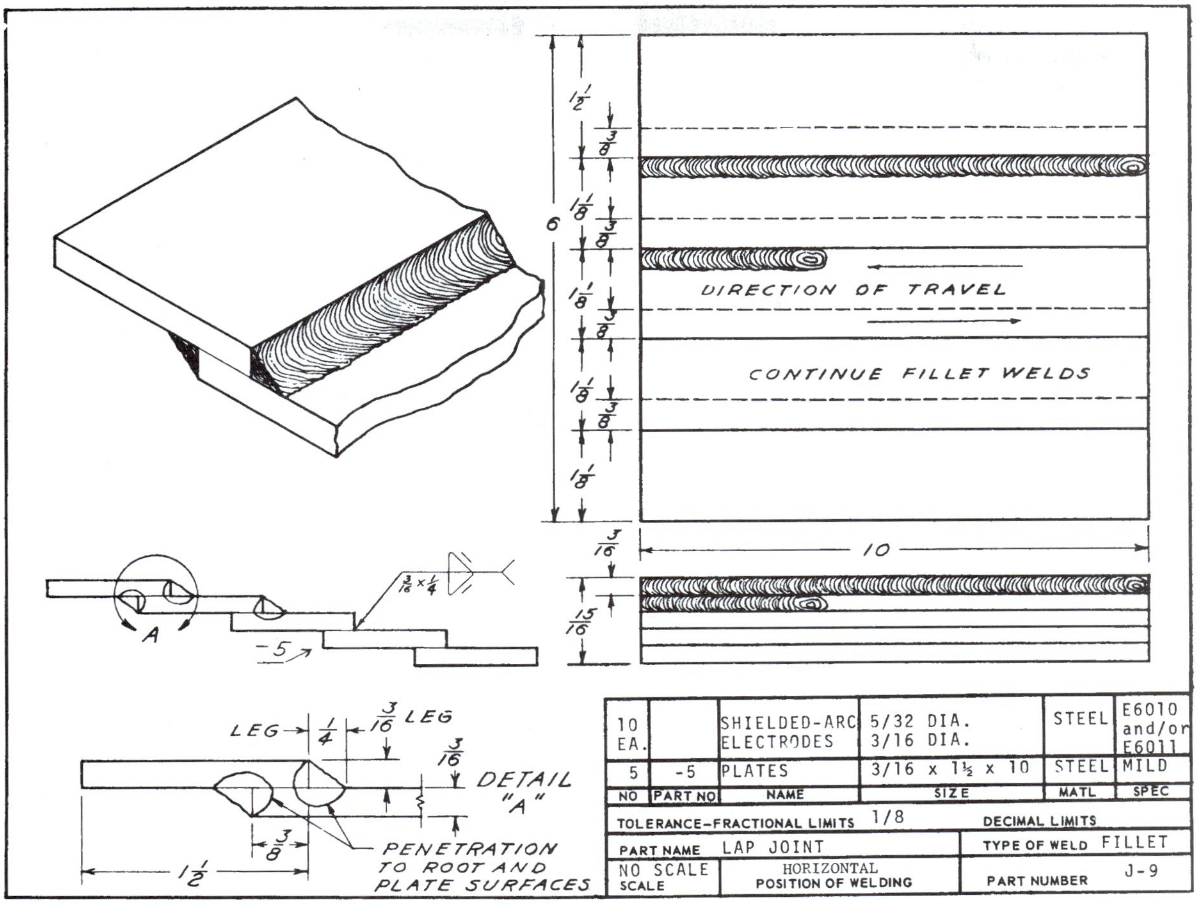

Fig. 13-38 Job drawing J9.

8. Practice these welds until you can produce good welds consistently with both types of electrodes and with 5/32-inch diameter electrodes.

Inspection

Compare the welds with Fig. 13-39 and check them for the following characteristics:

Width and height: Uniform
Appearance: Smooth with close ripples; free of voids and high spots. Restarts should be difficult to locate.
Size: Refer to the job drawing. Check with a fillet weld gauge.
Face of weld: Flat
Edges of weld: Good fusion, no overlap, no undercut
Starts and stops: Free of depressions and high spots
Beginnings and endings: Full size, craters filled
Penetration and fusion: To root of joint and to plate surfaces
Surrounding plate surfaces: Free of spatter
Slag formation: Full coverage, easily removed

Check Test

After you are able to make welds that are satisfactory in appearance, make up a specimen similar to that shown in Fig. 13-36. Use the plate's thickness, welding technique, and electrode type and size used in this job. Weld on one side only. Break the joint as shown in Fig. 13-37. Examine the surfaces for soundness. The weld must not be porous, and it must show complete fusion and penetration at the root and to the plate surfaces. The joint should break through the throat of the weld.

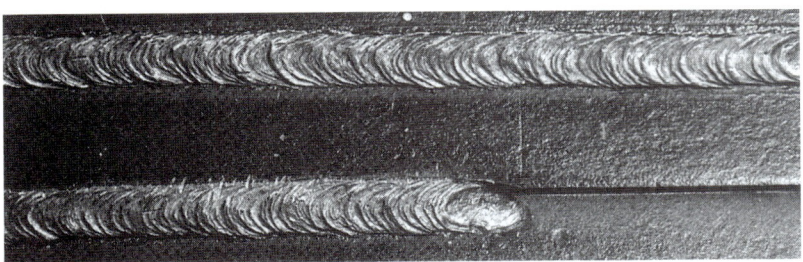

Fig. 13-39 Typical appearance of a fillet weld in a lap joint welded in the horizontal position with DCEP or a.c. shielded arc electrodes. *Edward R. Bohnart*

Disposal

Put completed joints in the scrap bin so that they will be available for further use. Unwelded plate surfaces can be used for beading.

Job 13-J10 Stringer Beading

Objective

To deposit horizontal stringer beads on flat steel plate positioned in the vertical position with DCEP and/or a.c. shielded metal arc electrodes (AWS E6010–E6011).

General Job Information

Horizontal groove welds and fillet welds are welded with stringer beads. The finishing layer is most frequently made in the same manner practiced on this job.

Welding Technique

Current adjustment must not be too high. The electrode position should be approximately that shown in Figs. 13-40 and 13-41. Hold the electrode at a 15° angle downward from the horizontal, Fig. 13-40A, and 15° forward in the direction of the travel, Fig. 13-40B. The electrode is manipulated in a roughly circular motion known as **oscillation.** The actual positions of the electrode and holder are shown in Fig. 13-41.

This job provides practice in running a straight weld bead without the guidance of a line to follow. It also provides practice in fusing a weld bead to the plate and to another weld bead.

Practice laying beads across the plate in the horizontal position. Begin at the bottom of the plate with the first bead. Remove the slag thoroughly after each pass and deposit another bead along the side. It should be fused into the preceding bead for one-third to one-half of its width.

You will find it necessary to use a slightly lower welding current and to hold a shorter arc than when welding in

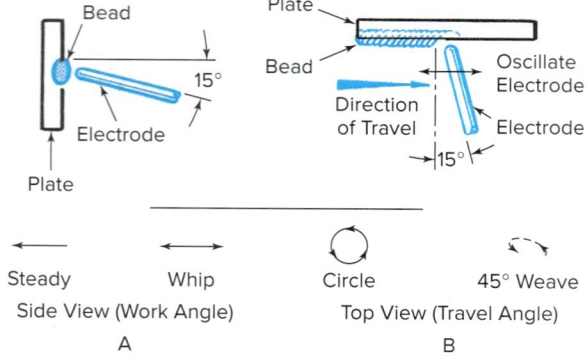

Fig. 13-41 Welding beads in the horizontal position. The electrode and electrode holder are positioned as instructed in Fig. 13-40. Location: Northeast Wisconsin Technical College
Mark A. Dierker/McGraw Hill

the flat position. Watch the weld carefully so that it does not sag on the bottom edge. Sagging can be prevented by moving the electrode back and forth. Be sure to get good penetration into the plate as well as good fusion between beads. Clean the slag from each bead before applying the next bead. Each bead should lap over the preceding bead to the extent of 30 to 50 percent. To prevent undercut, keep electrode movement within the bead size.

Practice starts and stops. Practice travel from left to right and from right to left. Weld on both sides of the plate and lay several layers on each side.

Operations

1. Obtain plate; check the job drawing, Fig. 13-42, for size.
2. Obtain electrodes in each quantity, type, and size specified in the job drawing.
3. Stand the plate in the vertical position on the welding bench. Make sure that it is connected to the work connection.
4. Set a d.c. power source for electrode positive at 110 to 160 amperes or an a.c. power source at 110 to 180 amperes.
5. Make stringer beads as shown on the job drawing. Hold the electrode as shown in Figs. 13-40 and 13-41.
6. Chip the slag from the beads, brush, and inspect. Refer to Inspection.
7. Practice these beads until you can produce uniform beads consistently with both types of electrodes and with $5/32$-inch diameter electrodes.

Fig. 13-40 Electrode position when welding stringer beads in the horizontal position.

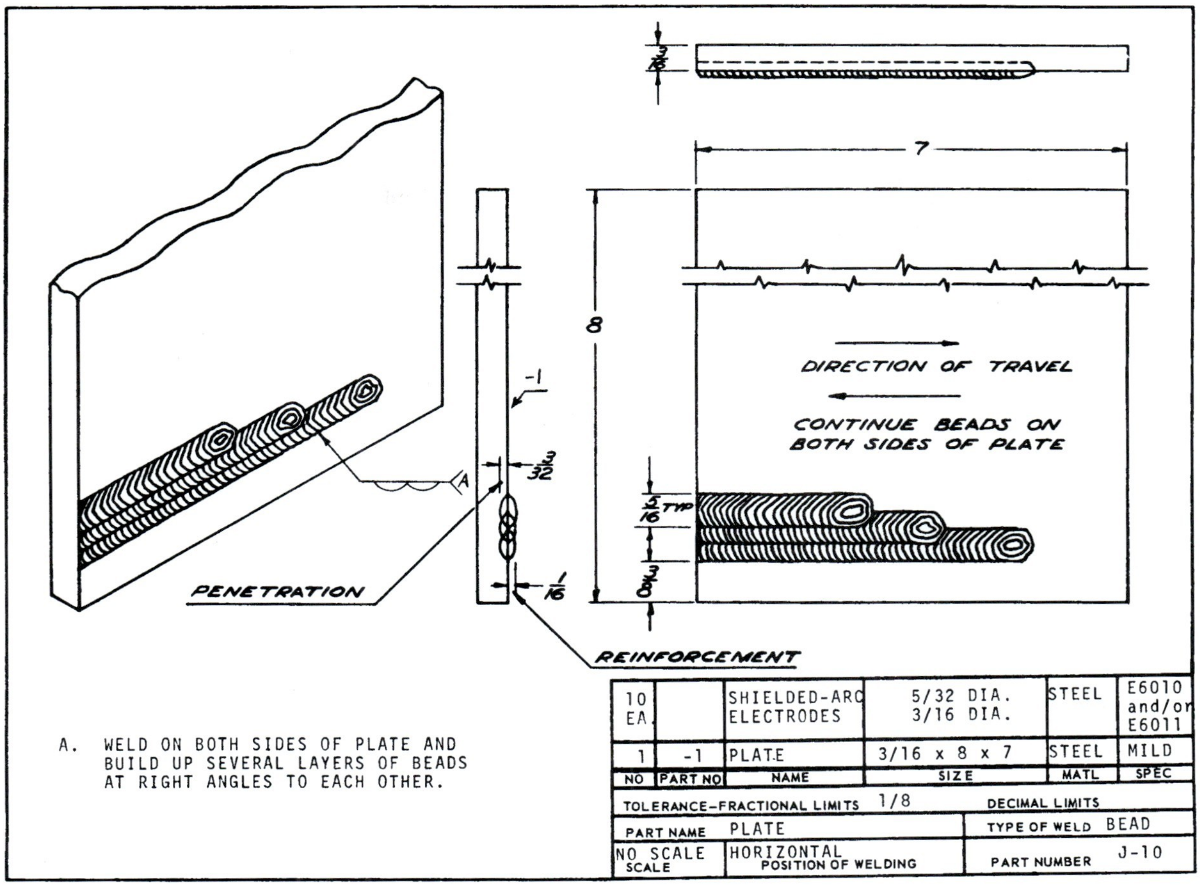

Fig. 13-42 Job drawing J10.

Inspection

Compare the beads with Fig. 13-43 and check them for the following characteristics:

Width and height: Uniform
Appearance: Smooth with close ripples; free of voids and high spots. Restarts should be difficult to locate.
Size: Refer to the job drawing.
Face of beads: Slightly convex
Edges of beads: Good fusion, no overlap, no undercut
Starts and stops: Free of depressions and high spots
Beginning and endings: Full size, craters filled
Penetration and fusion: To plate surfaces and adjacent beads
Surrounding plate surfaces: Free of spatter
Slag formation: Full coverage, easily removable

Disposal

Discard completed plates in the waste bin. Plates must be filled on both sides.

Job 13-J11 Stringer Beading

Objective

To deposit stringer beads on flat steel plate; vertical position, travel down; DCEP and/or a.c. shielded metal arc electrodes (AWS E6010–E6011).

General Job Information

Welding downhill on a vertical surface is frequently done in industry, especially on

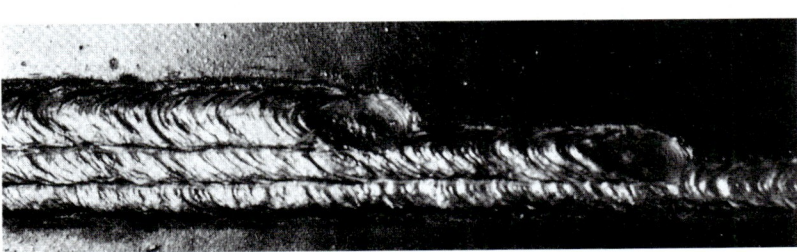

Fig. 13-43 Typical appearance of stringer beads welded in the horizontal position with DCEP or a.c. shielded arc electrodes. *Edward R. Bohnart*

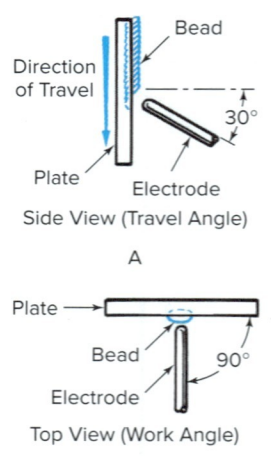

Fig. 13-44 Electrode position when making stringer beads, vertical position, travel down.

noncritical work. This method of welding is often used on some cross country pipeline and on light gauge sheet metal. On flat surfaces penetration is not too good, and there is danger of slag inclusions.

Welding Technique

Pay particular attention to current adjustment. The electrode position should be approximately that shown in Fig. 13-44. Tilt the electrode 30° downward in the direction of travel, Fig. 13-44A, and at a 90° angle with the plate surface, Fig. 13-44B. A very short arc gap must be maintained so that the slag will not run ahead of the weld deposit. Weaving is not generally used, and the speed of travel is fast. This results in beads that are narrow and do not have very high reinforcement. Undercutting is not a problem.

Practice stops and starts. Weld on both sides of the plate.

Operations

1. Obtain plate; check the job drawing, Fig. 13-45, for size.
2. Obtain a square head and scale, dividers, scribe, and center punch from the toolroom.
3. Lay out parallel lines as shown on the job drawing.
4. Mark lines with the center punch.
5. Obtain electrodes of each quantity, type, and size specified in the job drawing.
6. Stand the plate in the vertical position on the welding table. Make sure that it is connected to the work connection.
7. Set a d.c. power source for electrode positive at 110 to 170 amperes or an a.c. power source at 110 to 190 amperes.

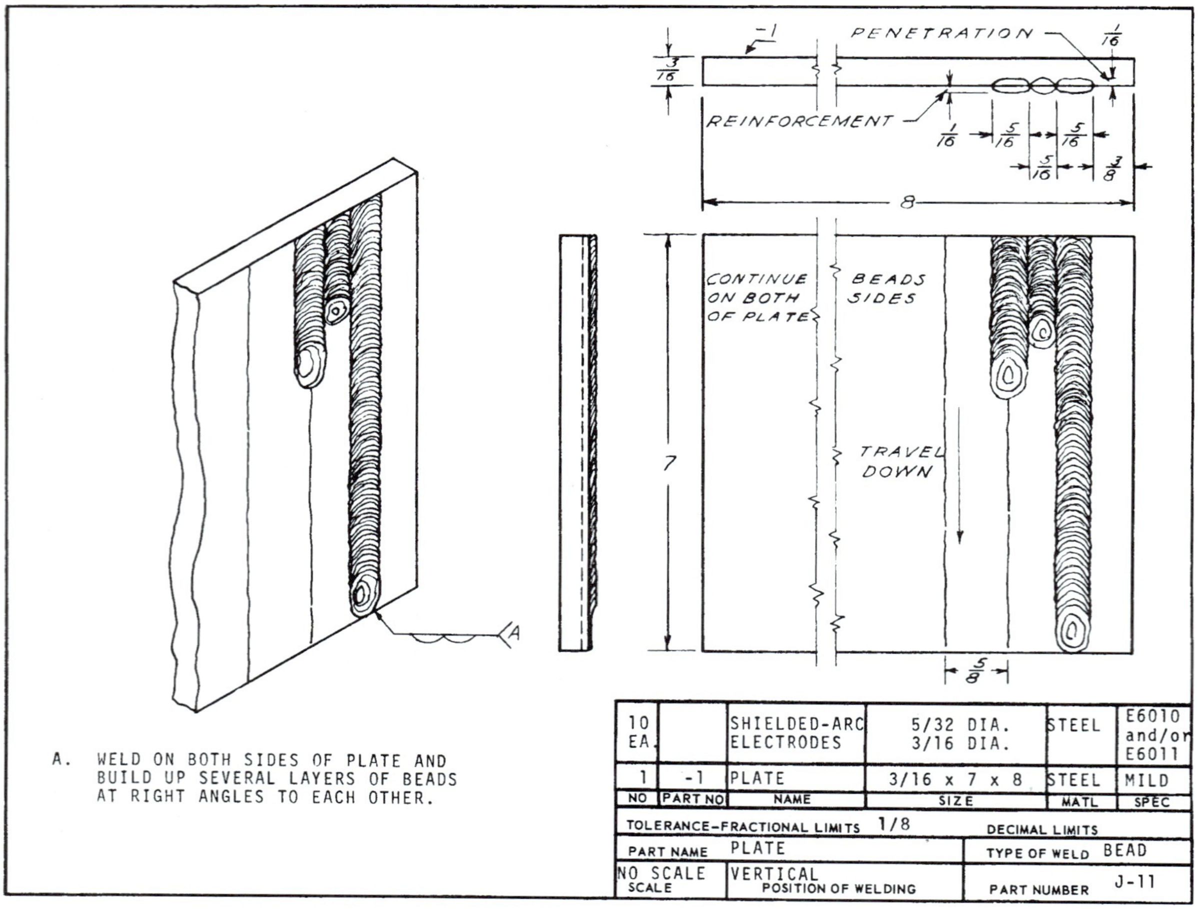

Fig. 13-45 Job drawing J11.

8. Make stringer beads on 5/8-inch center lines as shown on the job drawing. Hold the electrode as shown in Fig. 13-44.
9. Chip the slag from the beads, brush, and inspect. Refer to Inspection.
10. Make additional stringer beads between already deposited beads as shown on the job drawing.
11. Chip the slag from the beads, brush, and inspect. Refer to Inspection, below.
12. Practice these beads until you can produce uniform beads consistently with both types of electrodes and with 5/32-inch diameter electrodes.

Inspection

Compare the beads with Fig. 13-46 and check them for the following weld characteristics:

Width and height: Uniform
Appearance: Smooth with close ripples; free of voids and high spots. Restarts should be difficult to locate.
Size: Refer to the job drawing.
Face of weld: Flat with very little reinforcement
Edges of weld: Good fusion, no overlap, no undercut
Starts and stops: Free of depressions and high spots
Beginnings and endings: Full size, craters filled
Penetration and fusion: To plate surfaces and adjacent beads
Surrounding plate surfaces: Free of spatter
Stag formation: Full coverage, easily removable

Disposal

Discard completed plates in the waste bin. Plates must be filled on both sides.

Job 13-J12 Welding a Lap Joint

Objective

To weld a lap joint in the vertical position by means of a single-pass fillet weld, travel down, with DCEN and/or a.c. shielded metal arc electrodes (AWS E6013).

General Job Information

You will find that lap joints are often welded downhill on noncritical work. This is especially true on lighter plate construction such as sheet metal weldments. This joint should not be used when maximum strength is required. It is usually made with straight polarity electrodes in order to take advantage of the greater convexity and absence of undercut that are characteristics of these electrodes.

Welding Technique

Current adjustment should be high. Tilt the electrode 30° downward in the direction of travel, Fig. 13-47A. Hold it at a 45° angle to the plate surface as shown in Fig. 13-47B. The correct position for an actual weld is shown in Fig. 13-48, page 360. Hold a close arc gap and

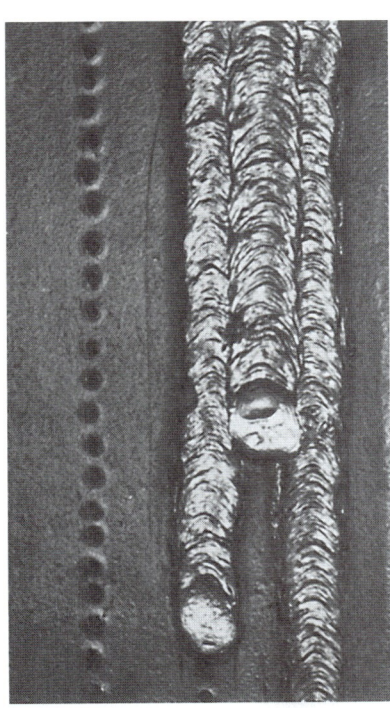

Fig. 13-46 Typical appearance of stringer beads on flat plate welded in the vertical position, travel down, with DCEP or a.c. shielded arc electrodes. *Edward R. Bohnart*

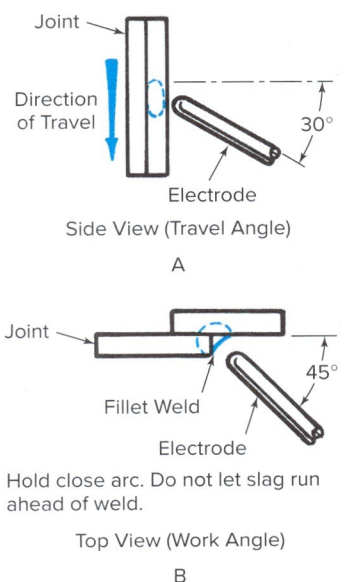

Fig. 13-47 Electrode position when welding lap joints in the vertical position, travel down.

Fig. 13-48 Welding lap joints in the vertical position, travel down. The electrode and electrode holder are positioned as instructed in Fig. 13-47. Location: Northeast Wisconsin Technical College Mark A. Dierker/McGraw Hill

proceed at a uniform rate of travel. Do not weave the electrode. A slight oscillating (circular) motion may be used. Slag must be kept back on weld deposit and not allowed to run ahead of the weld. Good fusion and penetration are difficult and should be watched very closely. Fusion to the plate surfaces and penetration at the root of the joint are essential. Undercut will not be a problem.

Practice starts and stops. Remember to clean the slag from the weld before making new starts.

Operations

1. Obtain plates; check the job drawing, Fig. 13-49, for quantity and size.
2. Obtain electrodes of each quantity, type, and size specified in the job drawing.
3. Set a d.c. power source for electrode negative at 110 to 190 amperes or an a.c. power source at 120 to 200 amperes.
4. Set up the plates and tack weld them at each corner.
5. Stand the plates in the vertical position on the welding table. Make sure that they are connected to the work connection.
6. Make welds as shown on the job drawing. Hold the electrode as shown in Figs. 13-47, page 359, and 13-48.
7. Chip the slag from the welds, brush, and inspect. Refer to Inspection below.

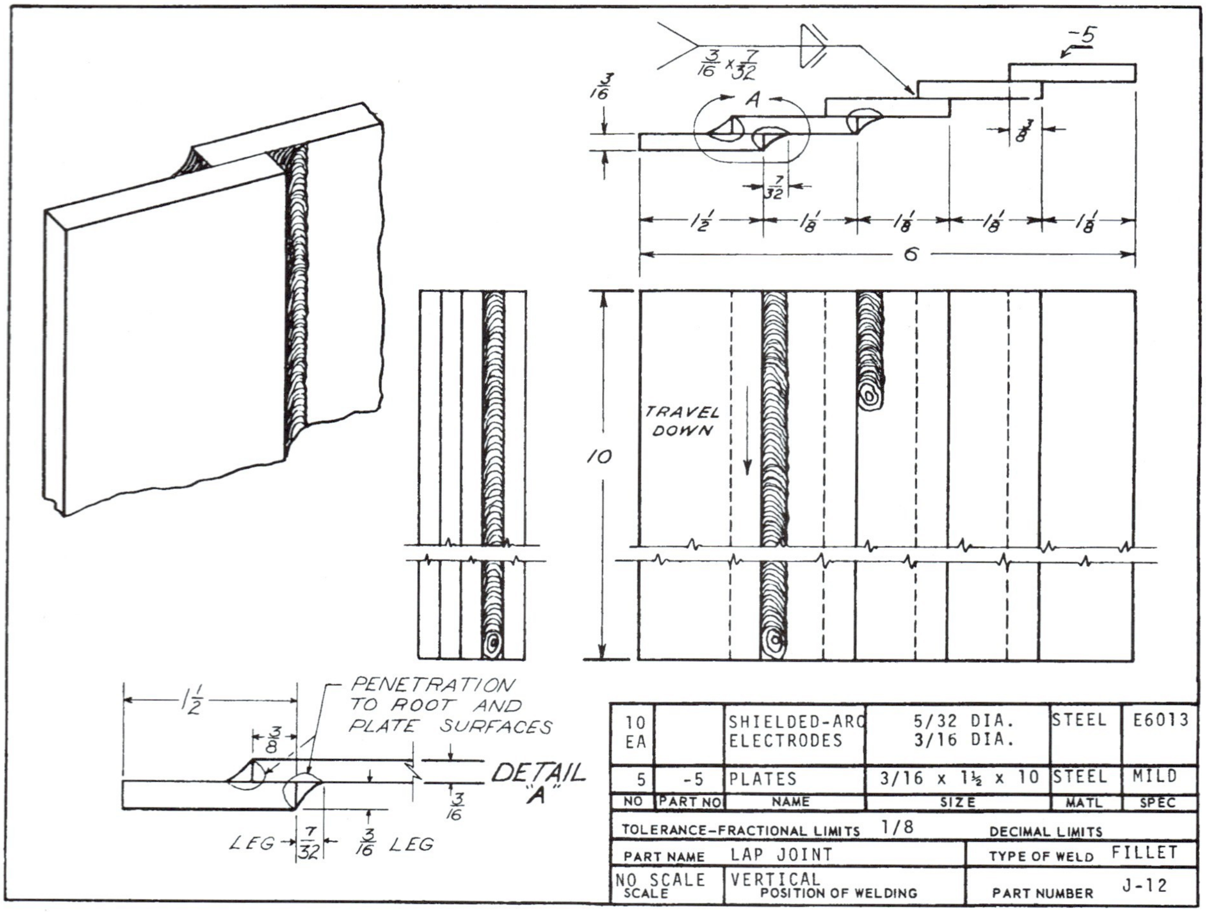

Fig. 13-49 Job drawing J12.

360   Chapter 13   Shielded Metal Arc Welding Practice: Jobs 13-J1–J25 (Plate)

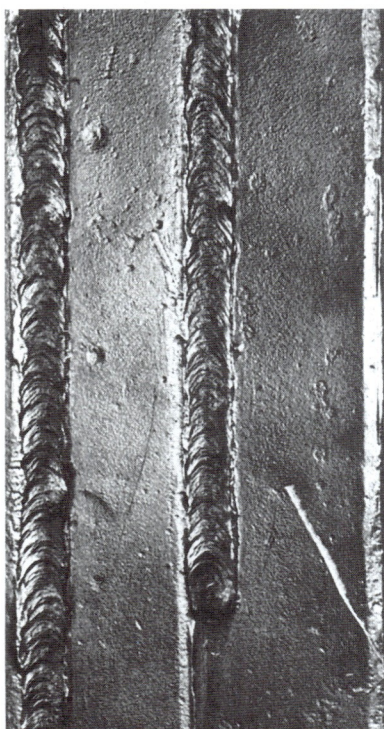

Fig. 13-50 Typical appearance of a fillet weld in a lap joint welded in the vertical position, travel down, with DCEN or a.c. shielded arc electrodes. *Edward R. Bohnart*

8. Practice these welds until you can produce good welds consistently with both types of current and with 5/32-inch diameter electrodes.

Inspection
Compare the welds with Fig. 13-50 and check them for the following characteristics:

Width and height: Uniform
Appearance: Very smooth with fine ripples; free of voids and high spots. Restarts should be difficult to locate.
Size: Refer to the job drawing.
Face of weld: Slightly concave
Edges of weld: Good fusion, no overlap, no undercut
Starts and stops: Free of depressions and high spots
Beginnings and endings: Full size, craters filled
Penetration and fusion: To the root of the joint and both plate surfaces
Surrounding plate surfaces: Free of spatter
Slag formation: Full coverage, easily removable

Disposal
Put completed joints in the scrap bin so that they will be available for further use. Unwelded plate surfaces can be used for beading.

Job 13-J13   Welding a Lap Joint

Objective
To weld a lap joint in the horizontal position by means of a single-pass fillet weld with DCEP and/or a.c. shielded metal arc electrodes (AWS E6013).

General Job Information
This position is similar to that used in Jobs 13-J11 and 12. The horizontal position is frequently necessary for tank and structural work.

Welding Technique
Current adjustment must not be too high. Electrode position should be approximately that shown in Fig. 13-51. Direct the electrode at the root of the weld and at the vertical plate surface. Tilt the electrode 30° up from the horizontal position, Fig. 13-51A, and 30° toward the direction of travel, Fig. 13-51B. A slight back-and-forth motion may be used with very little side movement. The molten pool must not become too hot since it makes the weld spill off. Be sure to manipulate the electrode so that the plate edge is not burned away and so that the arc does not come in contact with the surface of the vertical plate beyond the weld lines, thus causing undercut.

Practice starts and stops. Travel in all directions.

Operations
1. Obtain plates; check the job drawing, Fig. 13-52, page 362, for quantity and size.
2. Obtain electrodes of each quantity, type, and size specified in the job drawing.
3. Set a d.c. power source for electrode negative at 110 to 190 amperes or an a.c. power source at 115 to 200 amperes.

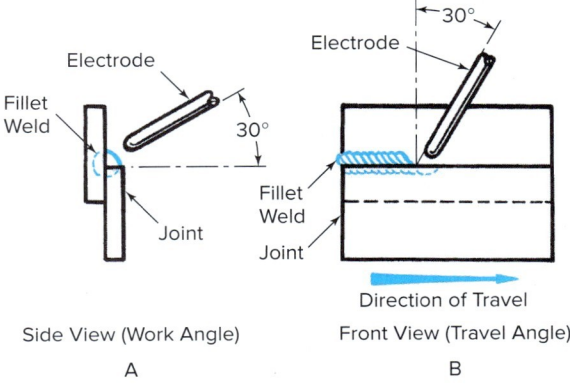

Fig. 13-51 Electrode position when welding lap joints in the horizontal position.

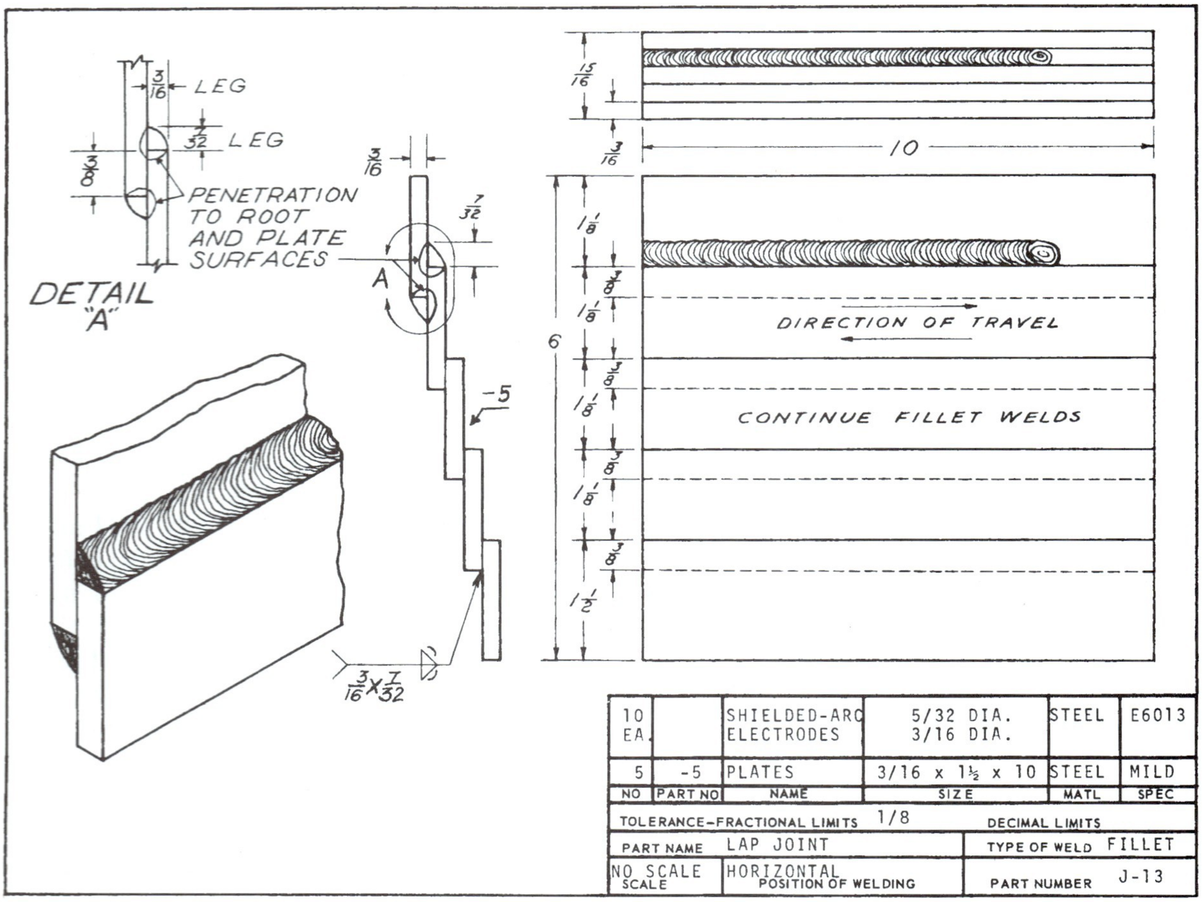

Fig. 13-52 Job drawing J13.

4. Set up the plates and tack weld them at each corner.
5. Stand the plates in the vertical position on the welding table. Make sure that they are connected to the work connection.
6. Make welds as shown on the job drawing. Hold the electrode as shown in Fig. 13-51, page 361.
7. Chip the slag from the welds, brush, and inspect. Refer to Inspection below.
8. Practice these welds until you can produce good welds consistently with both types of current and with 5/32-inch diameter electrodes.

Inspection

Compare the welds with Fig. 13-53 and check them for the following characteristics:

Width and height: Uniform
Appearance: Smooth with close ripples; free of voids. Restarts should be difficult to locate.

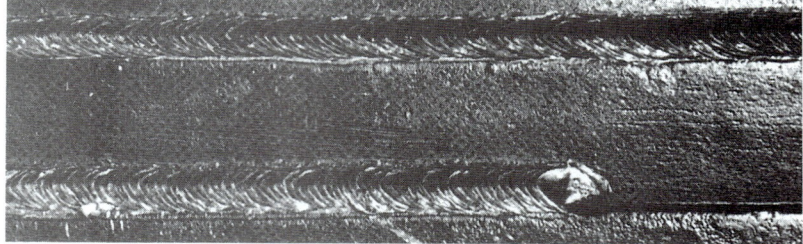

Fig. 13-53 Typical appearance of a fillet weld in a lap joint welded in the horizontal position with DCEN or a.c. shielded arc electrodes. *Edward R. Bohnart*

Size: Refer to the job. Check with a fillet weld gauge.
Face of weld: Slightly convex
Edges of weld: Good fusion, no overlap or hangdowns, no undercut
Starts and stops: Free of depressions and high spots
Beginnings and endings: Full size, craters filled
Penetration and fusion: To root of joint and plate surfaces

Surrounding plate surfaces: Free of spatter
Slag formation: Full coverage, easily removable

Check Test
Perform a check test, following the procedures outlined in Job 13-J9.

Disposal
Put completed joints in the scrap bin so that they will be available for further use. Unwelded plate surfaces can be used for beading.

Job 13-J14 Welding a T-Joint

Objective
To weld a T-joint in the flat position by means of single-pass fillet weld with DCEN and/or a.c. shielded metal arc electrodes (AWS E6013).

General Job Information
The T-joint requires a fillet weld. It is widely used in all forms of welded fabrication, Fig. 13-54. It is a very effective joint when welded from both sides or when complete penetration is secured from one side. Welding is done with all types of electrodes in all positions and can be single or multiple pass.

Welding Technique
Current adjustment should be high. Place the joint on the welding table as shown in Fig. 13-55A and hold the electrode at a 90° angle to the table. Tilt it 5 to 20° in the direction of travel, Fig. 13-55B. The actual position for welding is illustrated in Fig. 13-56, page 364.

When you start to weld, direct the electrode over the edge of the plate. After the plate edge has been preheated, deposit a bead of the desired width and height. Then advance the electrode along the line of weld with a slight back-and-forth motion. This helps to preheat the root of the weld and prevents the slag from running ahead of the weld. Direct the electrode precisely at the root of the weld. Never allow the arc to contact the plate surface outside the width of the weld bead formation. Keep a short arc to obtain good fusion at the root of the joint. Never deposit too much metal in

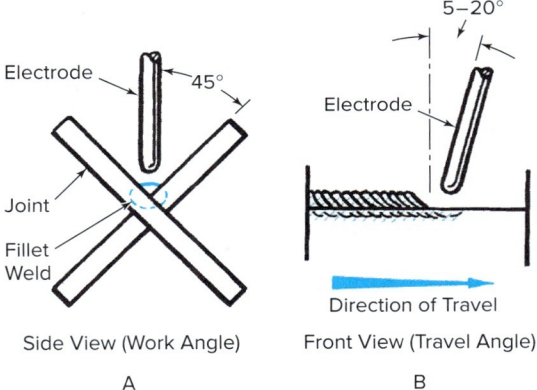

Fig. 13-55 Electrode position when welding a T-joint with a single-pass fillet weld in the flat position.

Fig. 13-54 A woman welding truss braces in industry. This job is similar to the practice jobs on T-joints. *Edward R. Bohnart*

JOB TIP

Keep Learning

In an interview in *American Welder*, career-welder Ernest Levert discusses how he developed a way to electron beam weld live missiles for Lockheed.

To those entering the field, he says, "Hard work does pay off. You can't stop learning. Once you do, you put a cap on your potential."

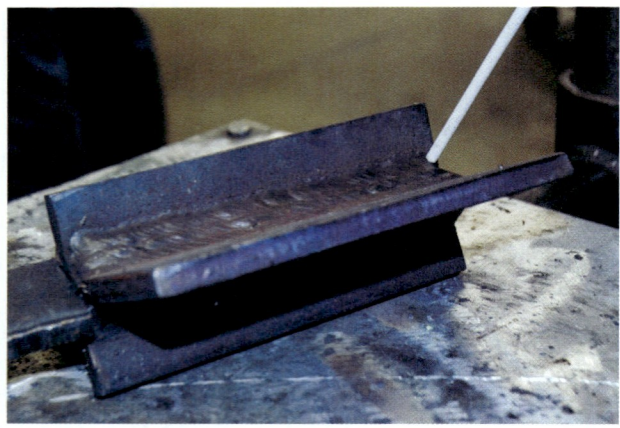

Fig. 13-56 Welding a T-joint in the flat position. The electrode and electrode holder are positioned as instructed in Fig. 13-55. Location: Northeast Wisconsin Technical College Mark A. Dierker/ McGraw Hill

one pass to avoid porosity, slag inclusion, incomplete fusion, and excess convexity.

Practice many starts and stops and be sure to fill the craters at the end of the weld. Travel in all directions.

Operations

1. Obtain plates; check the job drawing, Fig. 13-57, for quantity and size.
2. Obtain electrodes of each quantity, type, and size specified in the job drawing.
3. Set a d.c. power source for electrode negative at 160 to 220 amperes or an a.c. power source at 180 to 230 amperes.
4. Set up the plates and tack weld them at each corner.
5. Place the joint on the welding table in the position shown in Fig. 13-56.
6. Make welds in the flat position as shown in the job drawing. Hold the electrode as shown in Figs. 13-55 and 13-56.
7. Chip the slag from the welds, brush, and inspect. Refer to Inspection.
8. Practice these welds until you can produce good welds consistently with both types of current.

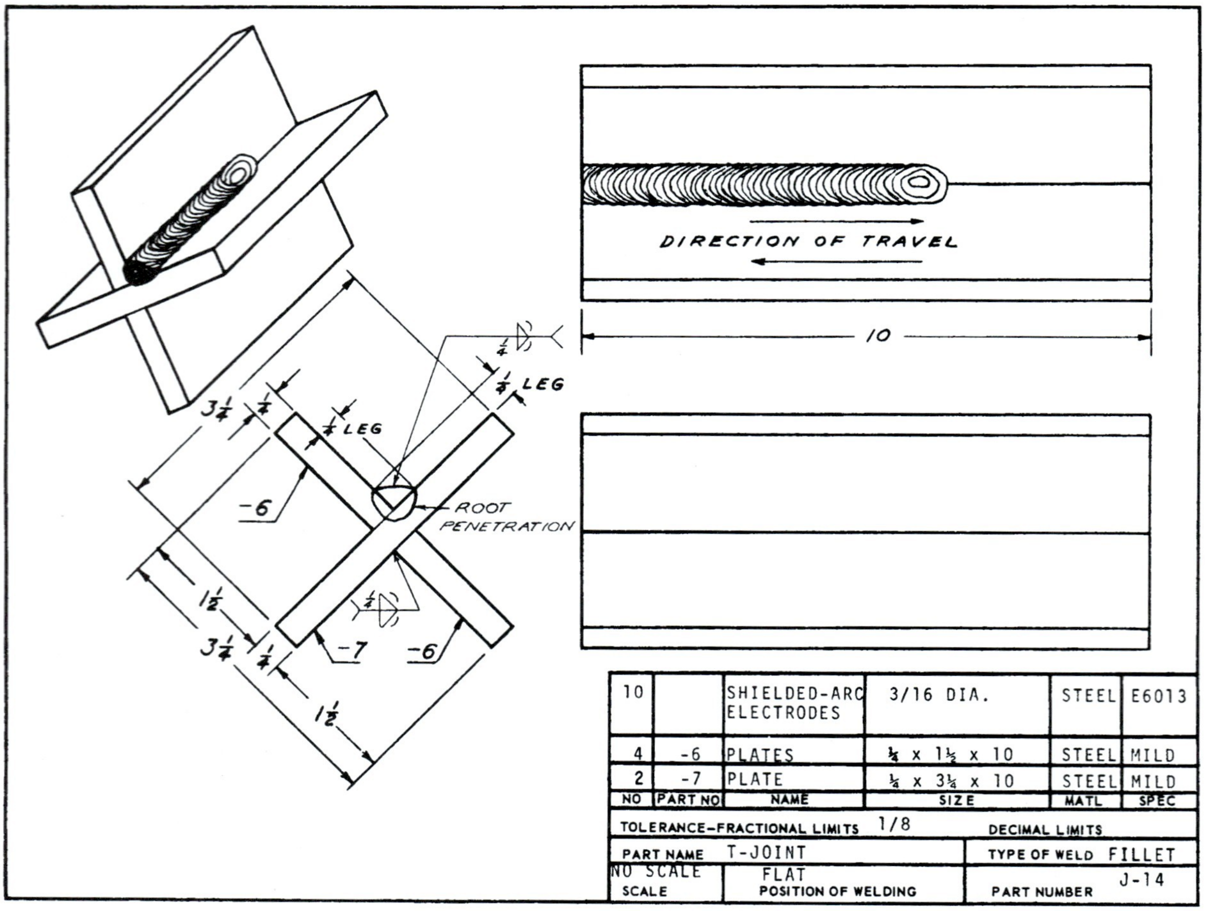

NO	PART NO	NAME	SIZE	MATL	SPEC
10		SHIELDED-ARC ELECTRODES	3/16 DIA.	STEEL	E6013
4	-6	PLATES	¼ x 1½ x 10	STEEL	MILD
2	-7	PLATE	¼ x 3¼ x 10	STEEL	MILD

TOLERANCE-FRACTIONAL LIMITS 1/8		DECIMAL LIMITS	
PART NAME	T-JOINT	TYPE OF WELD	FILLET
SCALE: NO SCALE	POSITION OF WELDING: FLAT	PART NUMBER	J-14

Fig. 13-57 Job drawing J14.

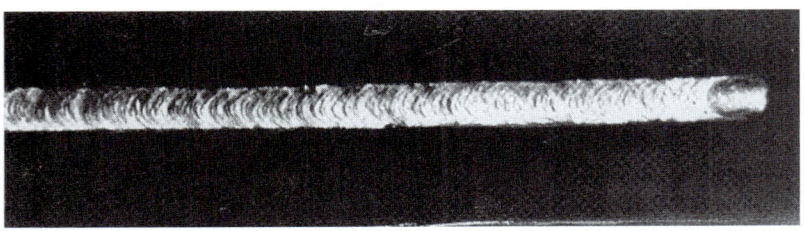

Fig. 13-58 Typical appearance of a single-pass fillet weld in a T-joint welded in the flat position with DCEN or a.c. shielded arc electrodes. *Edward R. Bohnart*

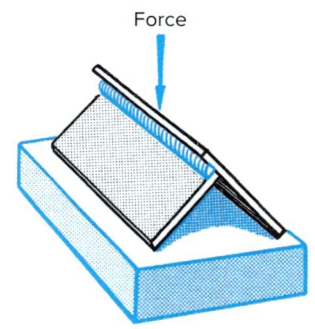

Fig. 13-60 Method of breaking check-test specimen.

Inspection

Compare the welds with Fig. 13-58 and check them for the following characteristics:

Width and height: Uniform
Appearance: Smooth with close ripples; free of voids and high spots. Restarts should be difficult to locate.
Size: Refer to the job drawing. Check with a fillet weld gauge.
Face of welds: Flat to very slightly convex
Edges of welds: Good fusion, no overlap, no undercut
Starts and stops: Free of depressions and high spots
Beginnings and endings: Full size, craters filled
Penetration and fusion: To root of joint and plate surfaces
Surrounding plate surfaces: Free of spatter
Slag formation: Full coverage, easily removable

Check Test

After you are able to make welds that are satisfactory in appearance, make up a specimen like the one shown in Fig. 13-59. Use the plate thickness, welding technique, and electrode type and size specified for this job. Weld on one side only. Break the finished weld as shown in Fig. 13-60. Examine the surfaces for soundness. The weld must not be porous and must show complete fusion and penetration at the root and to the plate surfaces. The joint should break evenly through the throat of the weld.

Disposal

Save the completed single-pass joints for your next job, which will be multipass fillet welding.

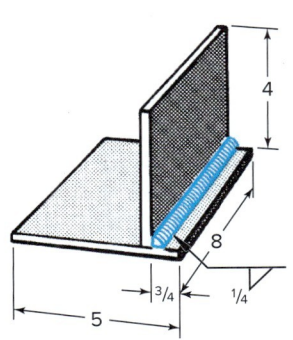

Dimensions Approximate

Fig. 13-59 Joint for fillet weld test specimen.

Job 13-J15 Welding a T-Joint

Objective

To weld a T-joint in the flat position by means of multipass fillet welds, weaved beading technique, with DCEN and/or a.c. shielded metal arc electrodes (AWS E6013).

General Job Information

The welding of T-joints in the flat position often requires more than one pass. Single-pass fillet welds should not be more than 1/16 to 1/8 inch larger than the size of the electrode being used. On multipass fillet welds, not more than 1/8 inch thickness of weld metal should be deposited on each layer. A T-joint is a convenient joint on which to practice weaved beading in grooves since the T-joint in the flat position forms a groove that is similar to the groove formed by beveled-butt welds in heavy plate.

You should understand that a T-joint made of 1/4-inch plate does not require three passes to develop maximum strength. The additional passes are only for practice purposes and to make full and economical use of materials.

Welding Technique

Current adjustment should be selected with care. Electrode position should be approximately that shown in Fig. 13-61, page 366. Hold the electrode at a 90° angle to the table, Fig. 13-61A, and tilt it 5 to 20° in the direction of travel, Fig. 13-61B. It is assumed that you are using the practice joints welded for Job 13-J14. The second and final pass must be weaved from side to side to ensure complete fusion and size, Fig. 13-61A. Bear in mind that the weave should not be as wide as the width of the desired weld. It should be only as wide as the underneath pass. Hesitate slightly (maybe half a second) on each side, and move the electrode smoothly and quickly across the face of the underneath weld. Remember that you cross the middle of the weld twice as often as you touch either side. To

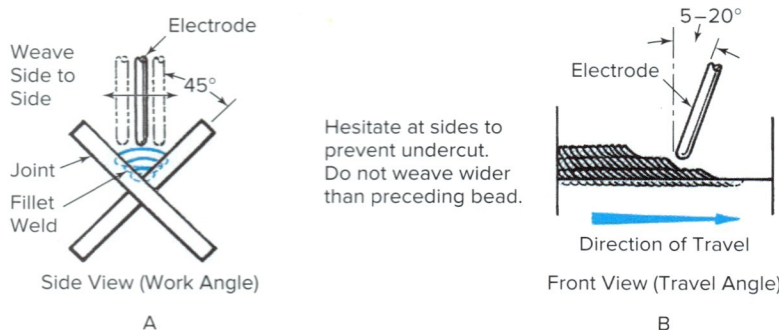

Fig. 13-61 Electrode position when welding a T-joint with multipass fillet welds in the flat position.

produce an even deposit across the weld, you must have a fast crossover speed. Otherwise, you will pile up weld metal in the middle.

It is very important to remove slag thoroughly between each pass. Failure to do this will result in slag inclusions.

Practice many starts and stops and be sure to fill the craters at the end of the weld. Travel in all directions.

It is advisable to have several joints available so that one joint may be cooling while the other is being worked on. Joints must not become too overheated.

Operations

1. Select good quality joints welded in Job 13-J14.
2. Obtain electrodes of each quantity, type, and size specified in the job drawing, Fig. 13-62.
3. Set a d.c. power source for electrode negative at 150 to 240 amperes or an a.c. power source at 165 to 260 amperes.
4. Place the joint in the flat position on the welding table. Make sure that the plates are connected to the work connection.

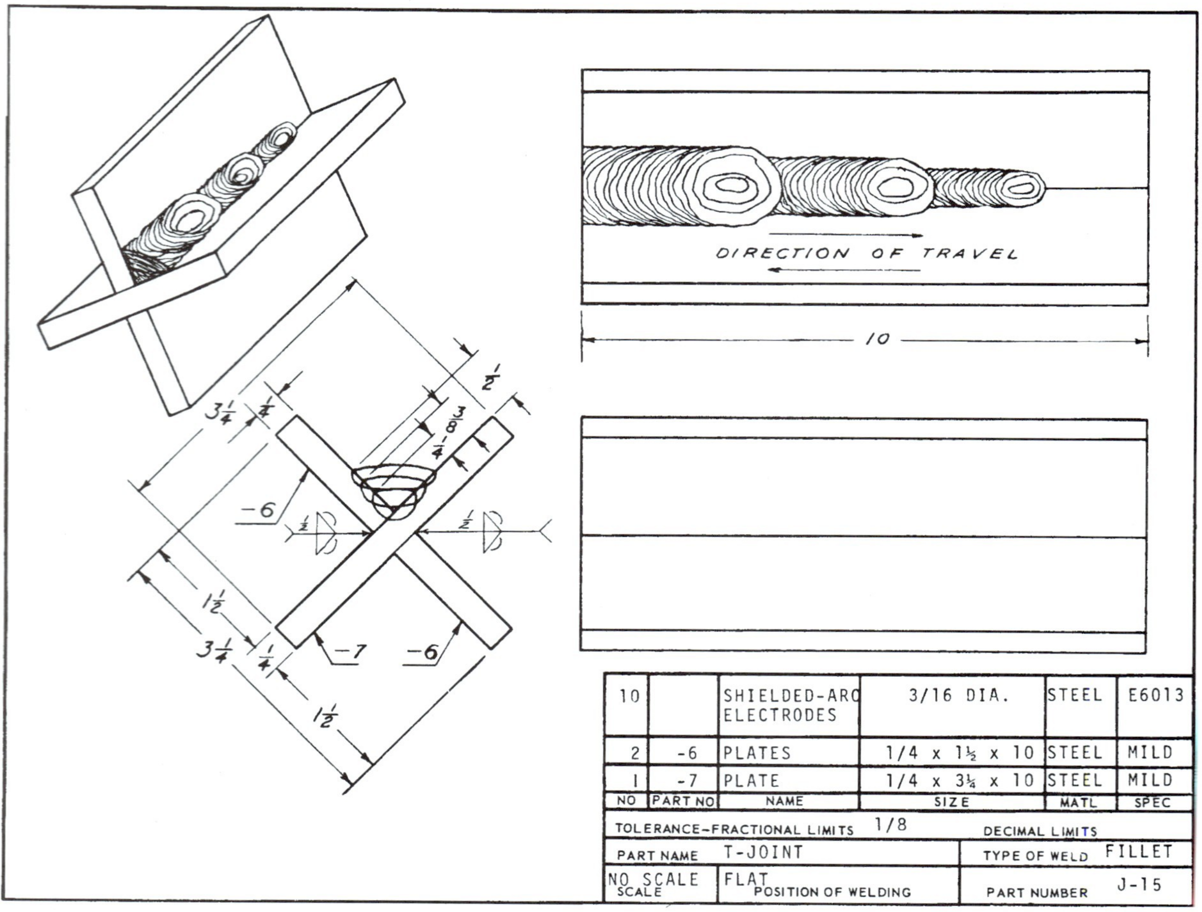

10		SHIELDED-ARC ELECTRODES	3/16 DIA.	STEEL	E6013
2	-6	PLATES	1/4 x 1½ x 10	STEEL	MILD
1	-7	PLATE	1/4 x 3½ x 10	STEEL	MILD
NO	PART NO	NAME	SIZE	MATL	SPEC

TOLERANCE-FRACTIONAL LIMITS 1/8	DECIMAL LIMITS	
PART NAME T-JOINT	TYPE OF WELD	FILLET
NO SCALE FLAT POSITION OF WELDING	PART NUMBER	J-15

Fig. 13-62 Job drawing J15.

5. Make the second pass as shown on the job drawing. Hold the electrode as shown in Fig. 13-61.
6. Chip the slag from the weld, brush, and inspect. Refer to Inspection.
7. Make the third pass as shown on the job drawing.
8. Chip the slag from the weld, brush, and inspect. Refer to Inspection.
9. Practice these welds until you can produce good welds consistently with both types of electrodes.

Inspection

Compare each pass with Figs. 13-63 and 13-64 and check it for the following characteristics:

Width and height: Uniform
Appearance: Smooth with close ripples; free of voids and high spots. Restarts should be difficult to locate.
Size: Refer to the job drawing. Check with a fillet weld gauge.
Face of Welds: Flat to slightly convex
Edges of welds: Good fusion, no overlap, no undercut
Starts and stops: Free of depressions and high spots
Beginnings and endings: Full size, craters filled
Penetration and fusion: To root of joint, preceding passes, and plate surfaces
Surrounding plate surfaces: Free of spatter
Slag formation: Full coverage, easily removable

Disposal

Put completed joints in the scrap bin so that they will be available for further use. Square edges can be butted together to make square butt joints, and unwelded plate surfaces can be used for beading.

Job 13-J16 Welding a T-Joint

Objective

To weld a T-joint in the flat position by means of a single-pass fillet weld, with DCEP and/or a.c. shielded metal arc electrodes (AWS E6010–E6011).

General Job Information

This job is similar to Job 13-J14, except that welding is done with electrode positive electrodes.

Welding Technique

Current adjustment should be high. Electrode position should be approximately that shown in Fig. 13-55, page 363. Use the welding technique practiced in 13-J14. The arc gap will be slightly less, and you will have to be more careful to avoid undercut than when using electrode negative electrodes. Slag will not be as heavy, but it must be removed thoroughly.

Practice starts and stops and be sure to fill the craters at the end of the weld. Travel in all directions.

Operations

1. Obtain plates; check the job drawing, Fig. 13-65, page 368, for quantity and size.
2. Obtain electrodes of each quantity, type, and size specified in the job drawing.
3. Set the power source for 140 to 215 amperes. Set a d.c. power source for electrode positive.
4. Set up the plates and tack weld them at each corner.
5. Place the joint in the flat position on the welding table. Make sure that the plates are connected to the work connection.
6. Make the weld as shown on the job drawing. Hold the electrode as shown in Fig. 13-55.
7. Chip the slag from the weld, brush, and inspect. Refer to Inspection.
8. Practice these welds until you can produce good welds consistently with both types of electrodes.

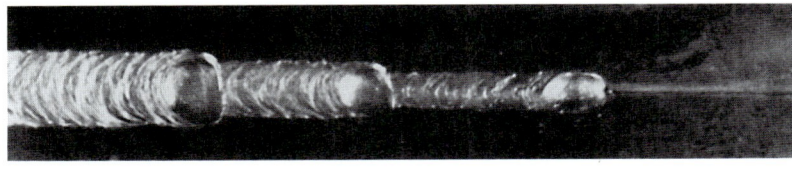

Fig. 13-63 Typical appearance of a multipass fillet weld in a T-joint welded in the flat position with DCEN shielded arc electrodes. *Edward R. Bohnart*

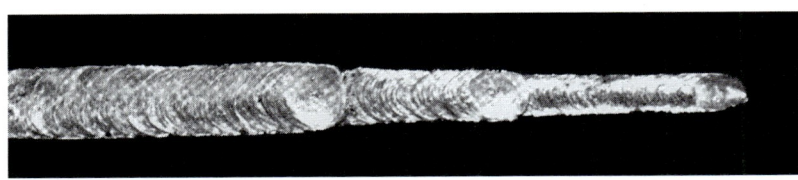

Fig. 13-64 Typical appearance of a multipass fillet weld in a T-joint welded in the flat position with a.c. type (E6013) shielded arc electrodes. *Edward R. Bohnart*

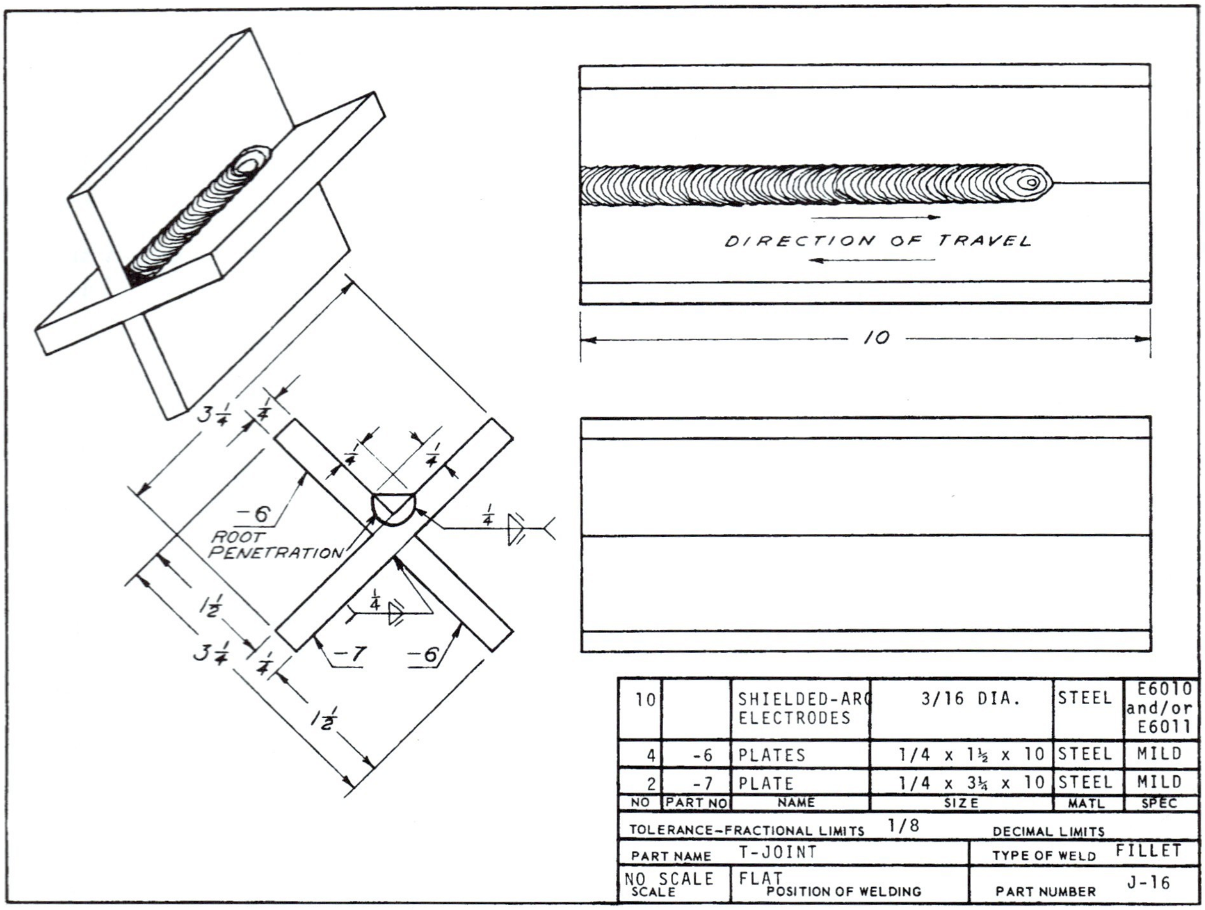

NO	PART NO	NAME	SIZE	MATL	SPEC
10		SHIELDED-ARC ELECTRODES	3/16 DIA.	STEEL	E6010 and/or E6011
4	-6	PLATES	1/4 x 1½ x 10	STEEL	MILD
2	-7	PLATE	1/4 x 3¼ x 10	STEEL	MILD
TOLERANCE–FRACTIONAL LIMITS	1/8	DECIMAL LIMITS			
PART NAME	T-JOINT	TYPE OF WELD	FILLET		
SCALE	NO SCALE	POSITION OF WELDING	FLAT	PART NUMBER	J-16

Fig. 13-65 Job drawing J16.

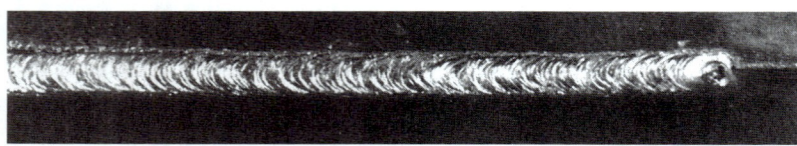

Fig. 13-66 Typical appearance of a single-pass fillet weld in a T-joint welded in the flat position with DCEP or a.c. shielded arc electrodes. Edward R. Bohnart

Inspection

Compare the welds with Fig. 13-66 and check them for the following characteristics:

Width and height: Uniform
Appearance: Smooth with close ripples; free of voids and high spots. Restarts should be difficult to locate.
Size: Refer to the job drawing. Check with a fillet weld gauge.
Face of welds: Flat
Edges of welds: Good fusion, no overlap, no undercut
Starts and stops: Free of depressions and high spots
Beginnings and endings: Full size, craters filled
Penetration and fusion: To root of joint and plate surfaces
Surrounding plate surfaces: Free of spatter
Slag formation: Full coverage, easily removable

Check Test

After you are able to make welds that are satisfactory in appearance, make up a test specimen like that made in Job 13-J14. Use the same plate thickness, welding technique, and electrode type and size as in this job. Weld on one side only.

Break the completed weld and examine the surfaces for soundness. The weld must not be porous. It must show complete fusion and penetration at the root and to the plate surfaces. The joint should break evenly through the throat of the weld.

Disposal

Save the completed joints for your next job, which will be multipass fillet welding.

Job 13-J17 Welding a T-Joint

Objective

To weld a T-joint in the flat position by means of multipass fillet welds, weaved beading technique, with DCEP and/or a.c. shielded metal arc electrodes (AWS E6010–E6011).

General Information

This job is similar to Job 13-J15, except that electrode positive electrodes are used for welding. The greater amount of multipass welding is done with electrode positive. This is especially true of code welding.

Welding Technique

Current adjustment should be selected with care. The electrode position should be approximately that shown in Fig. 13-61, page 366. Use the welding technique outlined in Job 13-J15. Arc gap should be slightly less. To prevent undercut, hesitate longer at the sides of the weld than when welding with electrode negative and control the width of your weave carefully.

Clean slag thoroughly between each pass. Practice starts and stops and be sure to fill the craters at the end of the weld. Travel in all directions.

It is advisable to have several joints available so that one joint may be cooling while the other is being worked on. Joints must not become overheated.

Operations

1. Select good quality joints welded in Job 13-J16.
2. Obtain electrodes of each quantity, type, and size specified in the job drawing, Fig. 13-67.
3. Set the power source for 140 to 215 amperes. Set a d.c. power source for electrode positive.
4. Place the joint in the flat position on the welding table. Make sure it is connected to the work connection.
5. Make a second pass as shown on the job drawing.

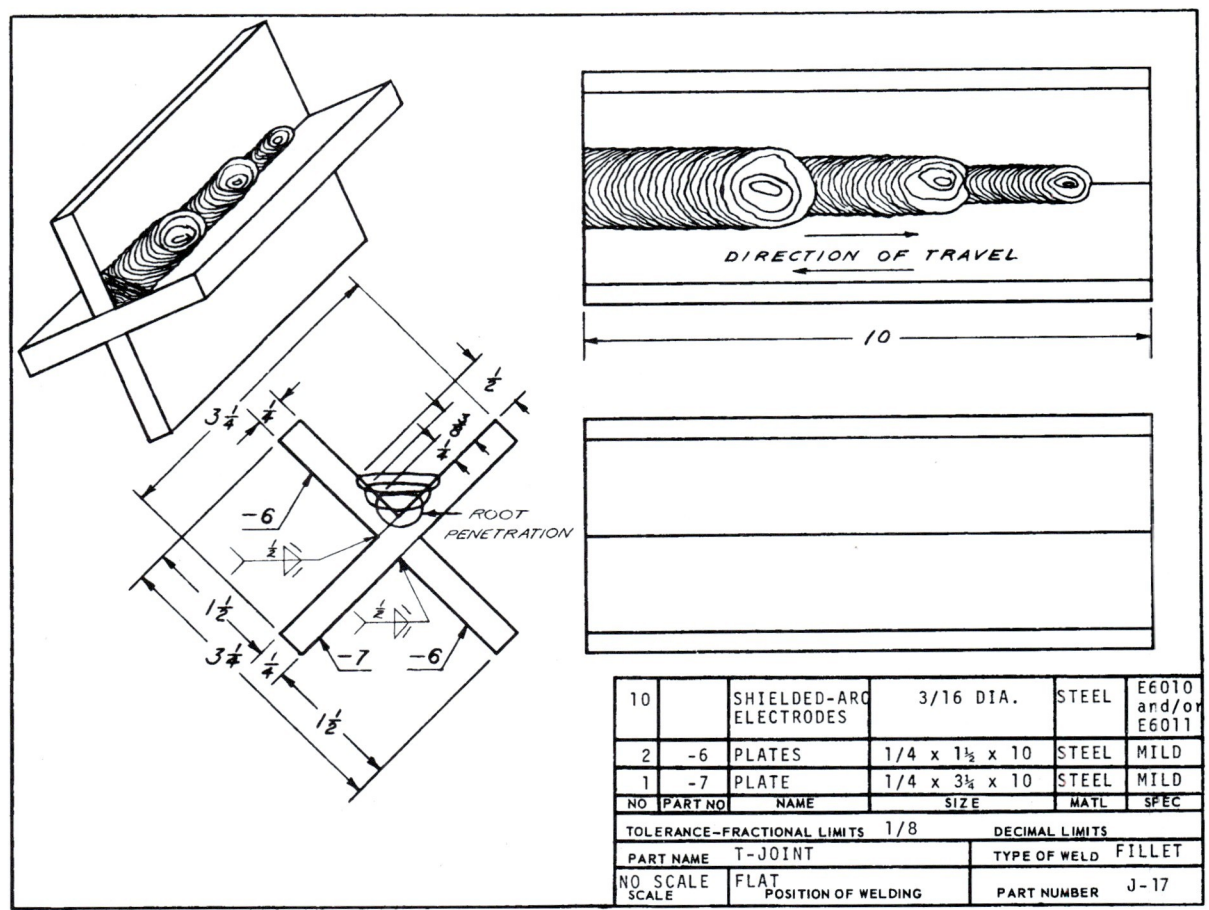

Fig. 13-67 Job drawing J17.

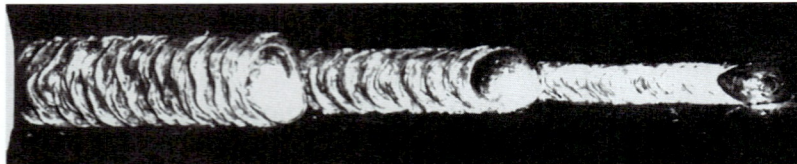

Fig. 13-68 Typical appearance of a multipass fillet weld in a T-joint welded in the flat position with DCEP shielded arc electrodes. *Edward R. Bohnart*

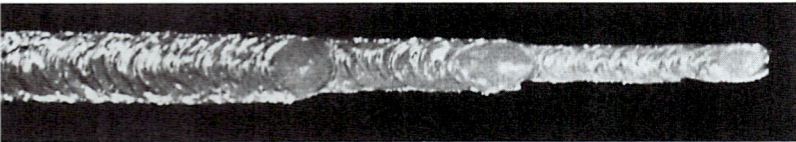

Fig. 13-69 Typical appearance of a multipass fillet weld in a T-joint welded in the flat position with a.c. type (E6011) shielded arc electrodes. *Edward R. Bohnart*

6. Chip the slag from the weld, brush, and inspect. Refer to Inspection.
7. Make a third pass as shown on the job drawing.
8. Chip slag from the weld, brush, and inspect. Refer to Inspection.
9. Practice these welds until you can produce good welds consistently with both types of electrodes.

Inspection

Compare each pass with Figs. 13-68 and 13-69 and check it for these weld characteristics:

Width and height: Uniform
Appearance: Smooth with close ripples; free of voids and high spots. Restarts should be difficult to locate.
Size: Refer to the job drawing. Check with a fillet weld gauge.
Face of welds: Flat
Edges of welds: Good fusion, no overlap, no undercut
Starts and stops: Free of depressions and high spots
Beginnings and endings: Full size, craters filled
Penetration and fusion: To the root of the joint, preceding passes, and plate surfaces
Surrounding plate surfaces: Free of spatter
Slag formation: Full coverage, easily removable

Disposal

Put completed joints in the scrap bin so that they will be available for further use. Square edges can be butted together to make square butt joints, and unwelded plate surfaces can be used for beading.

Job 13-J18 Stringer Beading

Objective

To deposit stringer beads on flat steel plate in the vertical position, travel up, with DCEP and/or a.c. shielded metal arc electrodes (AWS E6010–E6011).

General Job Information

For the most part, vertical welding is done by traveling up, especially on critical work. Pressure piping, shipbuilding, pressure vessels, and structural steel are some of the fields of welding using this procedure. The technique presented here is like that required for a lap joint and the first pass on multipass fillet welds.

Welding Technique

Current adjustment must not be too high. Tilt the electrode downward about 5 to 15°, Fig. 13-70A, and hold it at an angle of 90° to the plate, Fig. 13-70B. Basic welding motions used in vertical welding (travel up) are shown in Fig. 13-71. Figure 13-72 illustrates the actual welding position. The arc gap should be short when depositing weld metal. Lengthen the gap on the upward stroke, and do not break the arc. This type of motion permits the deposited weld metal to solidify so that a shelf is formed on which additional metal is deposited. Some welders prefer to move the electrode back and forth slightly as it is advanced so that it stays in the molten pool. If you use this method, you will find it difficult to keep the weld from building up too high and spilling off. The up-and-down rocking movement shown in Fig. 13-70A produces flatter welds, and there is less danger from slag inclusions, Fig. 13-70A.

Practice starts and stops. Weld on both sides of the plate.

Operations

1. Obtain plate; check the job drawing, Fig. 13-73, for size.
2. Obtain a square head and scale, dividers, scribe, and center punch from the toolroom.

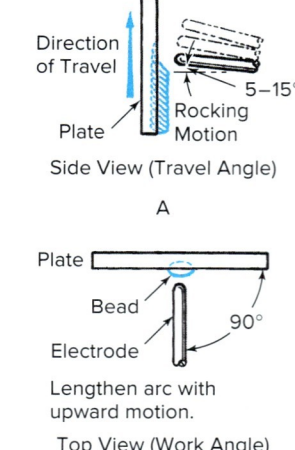

Fig. 13-70 Electrode position for making stringer beads in the vertical position, travel up.

3. Lay out parallel lines as shown on the job drawing.
4. Mark the lines with a center punch.
5. Obtain electrodes of each quantity, size, and type specified in the job drawing.
6. Stand the plate in the vertical position on the welding table. Make sure that it is connected to the work connection.

Fig. 13-72 Welding beads in the vertical position, travel up. The electrode and electrode holder are positioned as instructed in Fig. 13-70. Location: Northeast Wisconsin Technical College
Mark A. Dierker/McGraw Hill

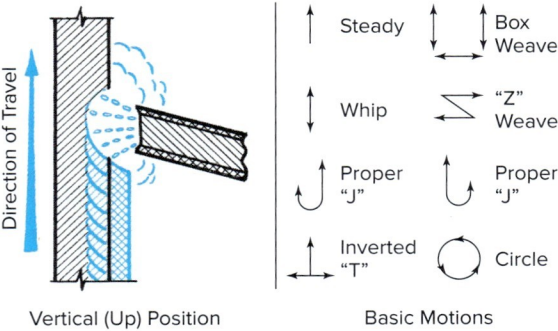

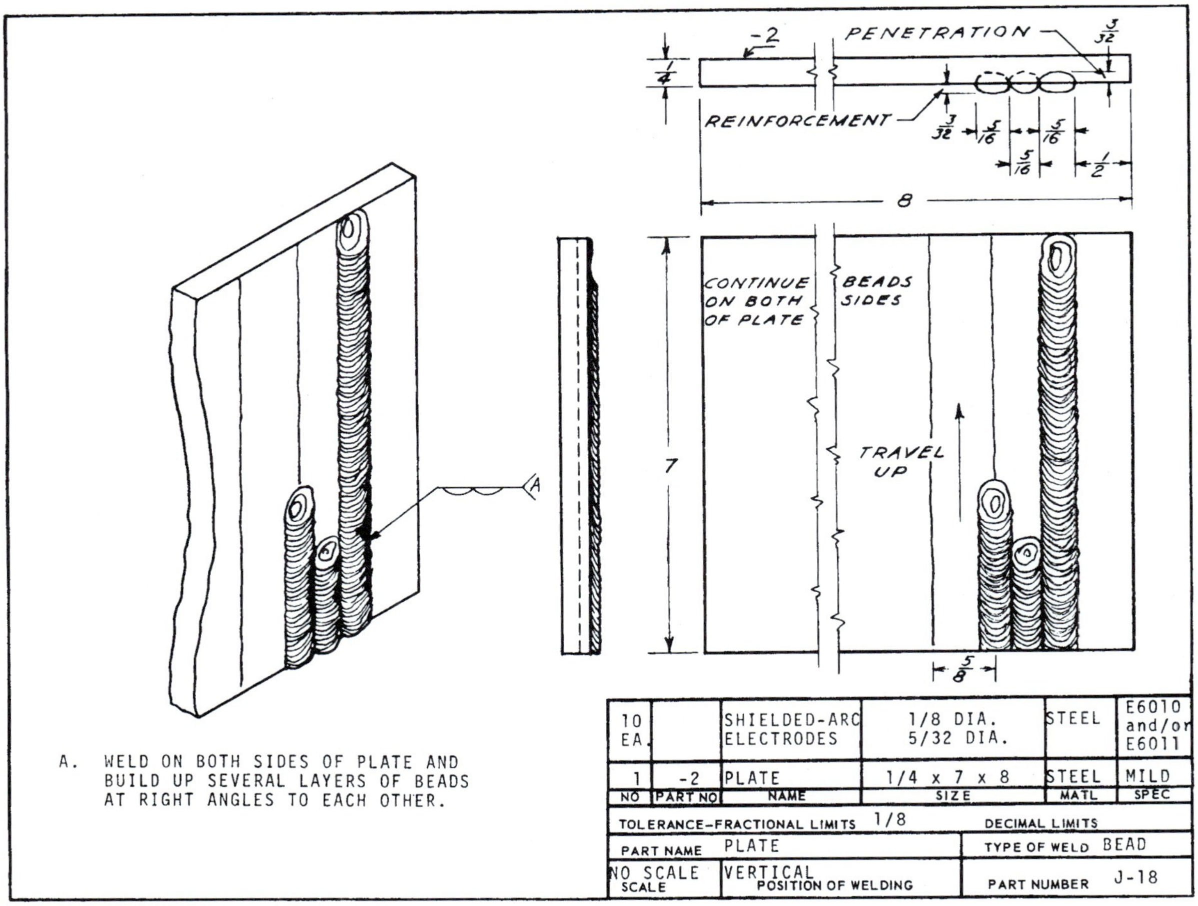

Fig. 13-71 Basic motions that can be employed in welding in the vertical position, travel up.

Fig. 13-73 Job drawing J18.

Shielded Metal Arc Welding Practice: Jobs 13-J1–J25 (Plate) **Chapter 13** 371

7. Set the power source for 75 to 125 amperes. Set a d.c. power source for electrode positive.
8. Make stringer beads on 5⁄8-inch center lines. Manipulate the electrode as instructed in Figs. 13-70 through 13-72, pages 370–371.
9. Chip slag from the beads, brush, and inspect. Refer to Inspection.
10. Make additional stringer beads between the beads already deposited as shown on the job drawing.
11. Chip the slag from the beads, brush, and inspect. Refer to Inspection, below.
12. Practice these beads until you can produce good beads consistently with both types of electrodes and with 1⁄8-inch diameter electrodes.

Inspection

Compare the beads with Fig. 13-74 and check them for the following weld characteristics:

Width and height: Uniform
Appearance: Smooth with close ripples; free of voids and high spots. Restarts should be difficult to locate.
Size: Refer to the job drawing.
Face of weld: Convex with slight reinforcement
Edges of weld: Good fusion, no overlap, no undercut
Starts and stops: Free of depressions and high spots
Beginnings and endings: Full size, craters filled
Penetration and fusion: To plate surfaces and adjacent beads
Surrounding plate surfaces: Free of spatter
Slag formation: Full coverage, easily removable

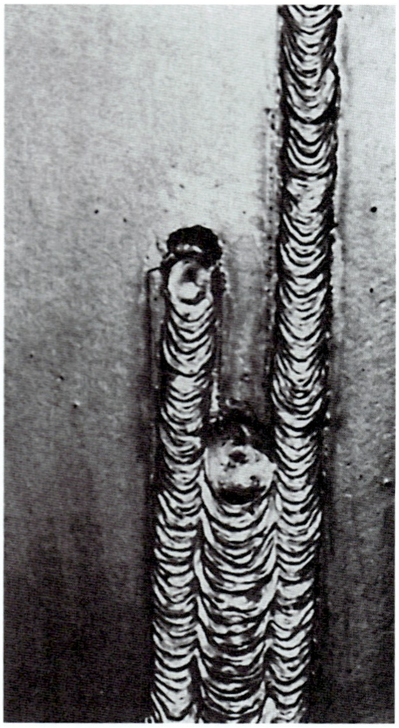

Disposal

Discard completed plates in the waste bin. Plates must be filled on both sides.

Job 13-J19 Weaved Beading

Objective

To deposit weaved beads on flat plate in the vertical position, travel up, with DCEP and/or a.c. shielded metal arc electrodes (AWS E6010–E6011).

General Job Information

When welding in the vertical position and when it is necessary to weave the bead, the direction of travel is usually up. This technique is used on pressure piping and pressure vessels, and in shipbuilding and structural welding. DCEP electrodes are generally selected. The travel-up method is employed in the welding of multipass groove and fillet welds in the vertical position.

Welding Technique

Current adjustment must not be too high, but the arc must be hot enough to ensure good fusion. Tilt the electrode downward about 10° from the horizontal position, Fig. 13-75A, and hold it at a 90° angle to the plate, Fig. 13-75B. A shelf is formed at the bottom of the plate that is the desired width and height. Weave the electrode across the face of the weld, hesitating at the sides of the weld to eliminate undercut, Fig. 13-75C. You may advance the weld by

Fig. 13-74 Typical appearance of stringer beads on flat plate welded in the vertical position, travel up, with DCEP or a.c. shielded arc electrodes. Edward R. Bohnart

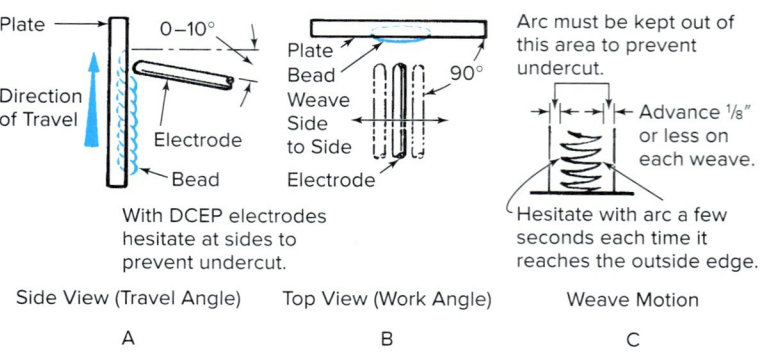

Fig. 13-75 Electrode position when making weave beads in the vertical position, travel up.

keeping the electrode in the pool, but be careful that the molten metal does not become so fluid that it spills out. When this occurs, run the electrode up on either side of the weld. Lengthen the arc, but do not break it. Do not keep the arc out of the crater any longer than necessary since this allows the crater to cool and causes excessive spatter ahead of the weld. Remember that *undercut* is the gouging out of the base metal by the arc over a wider surface than is being covered by weld deposit. *Do not weave the electrode beyond the desired width of the bead,* Fig. 13-75C.

Practice starts and stops. Weld on both sides of plate.

Operations

1. Obtain plate; check the job drawing, Fig. 13-76, for quantity and size.
2. Obtain a square head and scale, dividers, scribe, and center punch from the toolroom.
3. Lay out parallel lines as shown on the job drawing.
4. Mark lines with the center punch.
5. Obtain electrodes of each quantity, type, and size specified in the job drawing.
6. Stand the plate in the vertical position on the welding table. Make sure that it is connected to the work connection.
7. Set the welding power source for 75 to 125 amperes. Set a d.c. power source for electrode positive.
8. Make weaved beads between the parallel lines that are ¾ inch apart as shown on the job drawing. Manipulate the electrode as instructed in Fig. 13-75.
9. Chip the slag from the beads, brush, and inspect. Refer to Inspection.
10. Make stringer beads between the weaved beads already deposited. Manipulate the electrode as instructed in Figs. 13-70 and 13-71, pages 370–371.
11. Chip the slag from the beads, brush, and inspect. Refer to Inspection.

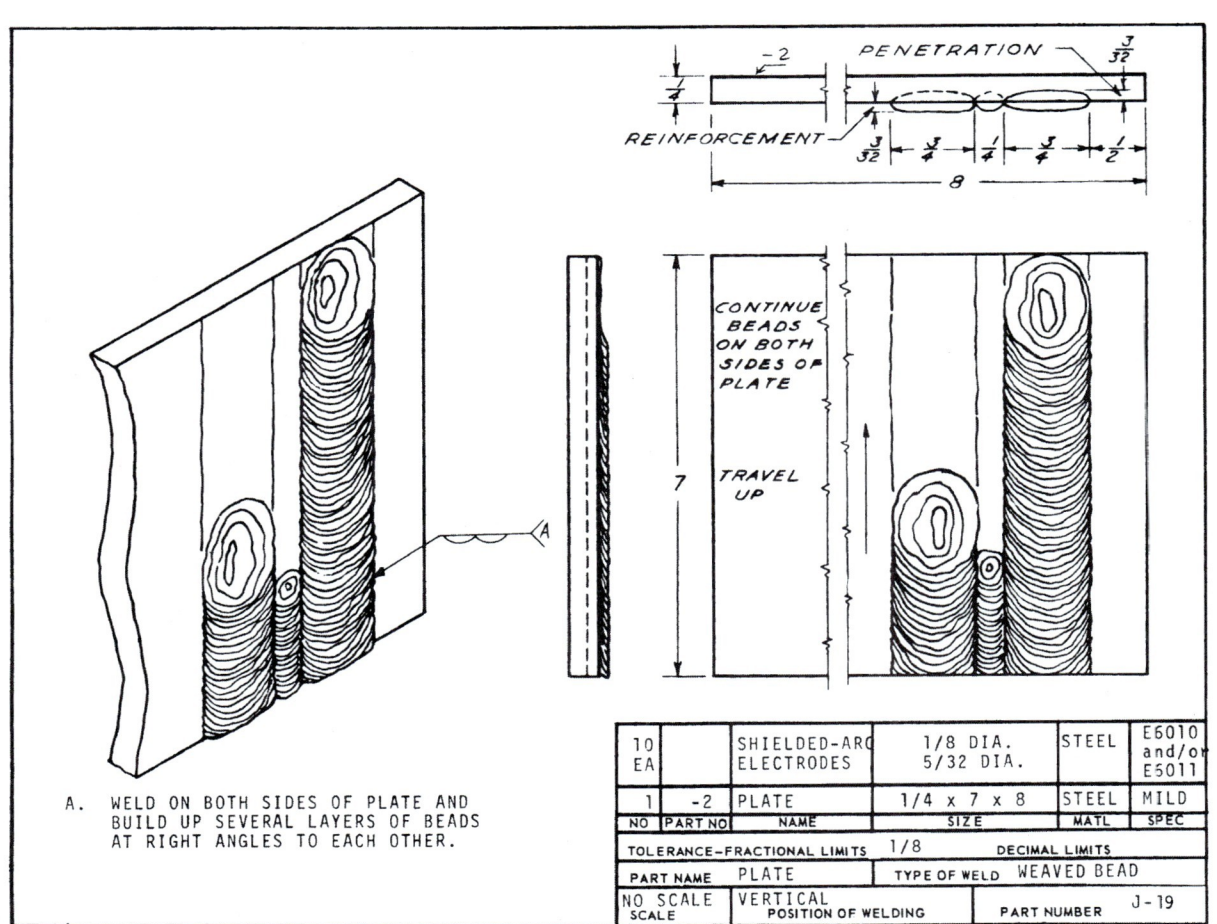

Fig. 13-76 Job drawing J19.

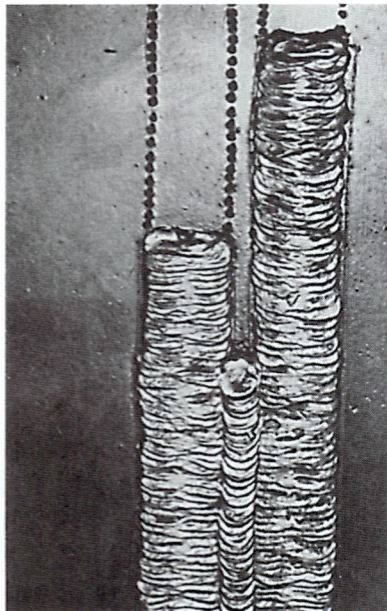

Fig. 13-77 Typical appearance of weave and stringer beads on flat plate welded in the vertical position, travel up, with DCEP or a.c. shielded arc electrodes. *Edward R. Bohnart*

12. Practice these beads until you can produce good beads consistently with both types of electrodes and with ⅛-inch diameter electrodes.

Inspection

Compare the beads with Fig. 13-77 and check them for the following weld characteristics:

Width and height: Uniform
Appearance: Smooth with close ripples; free of voids and high spots. Restarts should be difficult to locate.
Size: Refer to the job drawing.
Face of weld: Convex with slight reinforcement
Edges of weld: Good fusion, no overlap, no undercut
Starts and stops: Free of depressions and high spots
Beginnings and endings: Full size, craters filled
Penetration and fusion: To plate surface and adjacent beads

Surrounding plate surfaces: Free of spatter
Slag formation: Full coverage, easily removable

Disposal

Discard completed plates in the waste bin. Plates must be filled on both sides.

Job 13-J20   Weaved Beading

Objective

To deposit weaved beads on flat plate in the vertical position, travel up, with DCEN and/or a.c. shielded metal arc electrodes (AWS E6013).

General Job Information

This job is like Job 13-J19, except that electrode negative electrodes are specified for welding. Although the use of electrode negative electrodes for vertical work is not very extensive, there is enough of it to justify your spending some time on the application.

Welding Technique

Current adjustment will be slightly higher than for electrode positive electrodes. The electrode position should be approximately that shown in Fig. 13-72, page 371. Because of the higher buildup and the greater amount of slag, the speed of travel should be faster than with electrode positive electrodes. Undercutting is not a problem so that it is unnecessary to hesitate as long at the sides of the weld. Stringer beads should be deposited as instructed in Job 13-J18.

Practice starts and stops. Weld on both sides of the plate.

Operations

1. Obtain plate; check the job drawing, Fig. 13-78, for size.
2. Obtain a square head and scale, dividers, scribe, and center punch from the toolroom.
3. Lay out parallel lines as shown on the job drawing.
4. Mark the lines with a center punch.
5. Obtain electrodes of each quantity, type, and size specified in the job drawing.
6. Stand the plate in the vertical position on the welding table. Make sure that it is connected to the work connection.
7. Set power source for 80 to 140 amperes. Set a d.c. power source for electrode negative.
8. Make weaved beads between the parallel lines that are ¾ inch apart as shown on the job drawing.

ABOUT WELDING

Welding Machine

The first variable d.c. welding machine was manufactured by an Ohio company, Lincoln Electric, in 1907. You can locate Lincoln Electric today on the Web. The Web site has an online technology topic library.

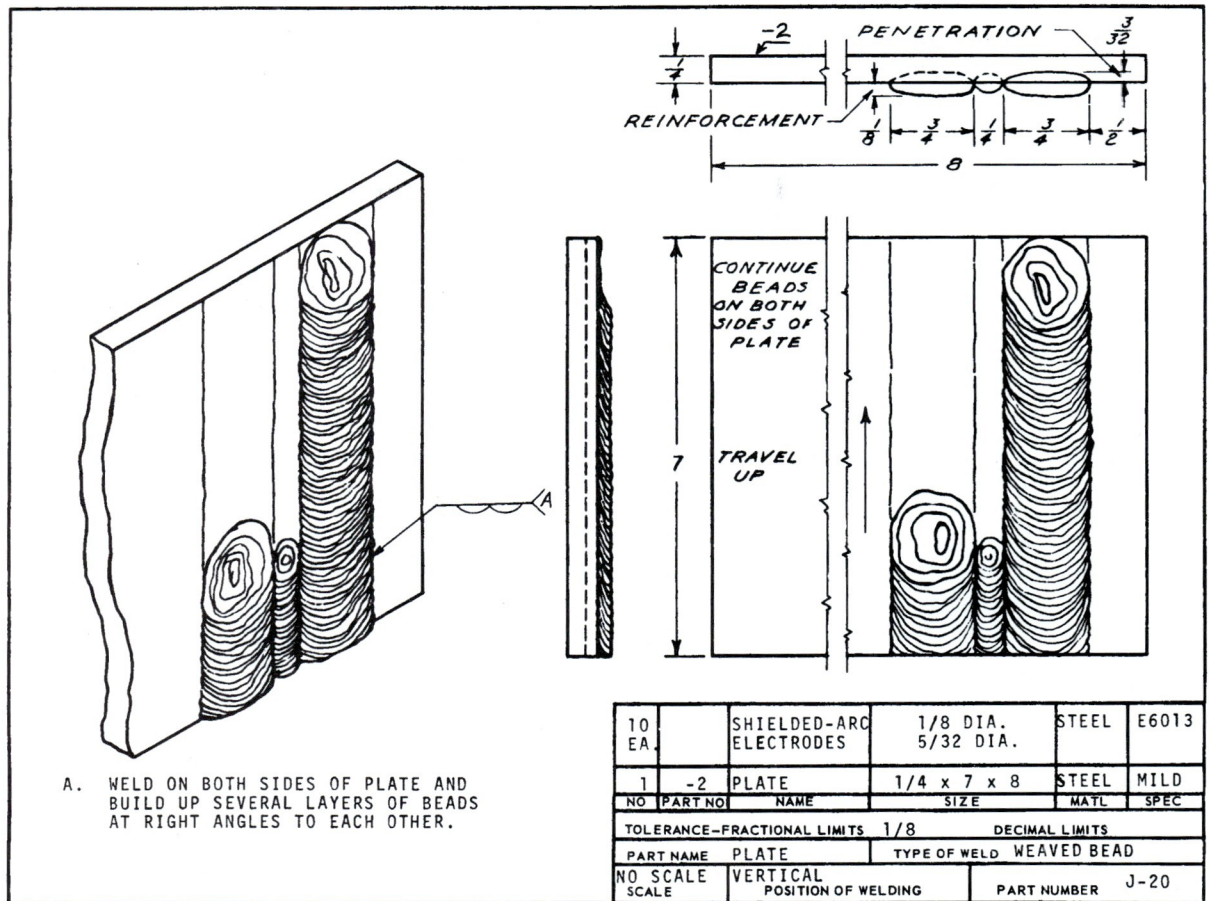

Fig. 13-78 Job drawing J20.

Manipulate the electrode as instructed in Fig. 13-71, page 371, and Fig. 13-75, page 372.
9. Chip the slag from the beads, brush, and inspect. Refer to Inspection.
10. Make stringer beads between the weaved beads already deposited. Manipulate the electrode as instructed in Figs. 13-70 and 13-71, pages 370–371.
11. Chip the slag from the beads, brush, and inspect. Refer to Fig. 13-79 and Inspection.
12. Practice these beads until you can produce good beads consistently with the E6013 electrode, with both types of current, and with ⅛-inch diameter electrodes.

Inspection

Compare the beads with Fig. 13-79, page 376, and check them for the following weld characteristics:

Width and height: Uniform
Appearance: Smooth with close ripples; free of voids and high spots. Restarts should be difficult to locate.
Size: Refer to the job drawing.
Face of weld: Convex with slight reinforcement
Edges of weld: Good fusion, no overlap, no undercut
Starts and stops: Free of depressions and high spots
Beginnings and endings: Full size, craters filled
Penetration and fusion: To plate surfaces and adjacent beads
Surrounding plate surfaces: Free of spatter
Slag formation: Full coverage, easily removable

Disposal

Discard completed plates in the waste bin. Plates must be filled on both sides.

Job 13-J21   Welding a Lap Joint

Objective

To weld a lap joint in the vertical position by means of a single-pass fillet weld, travel up, with DCEP and/or a.c. shielded metal arc electrodes (AWS E6010–E6011).

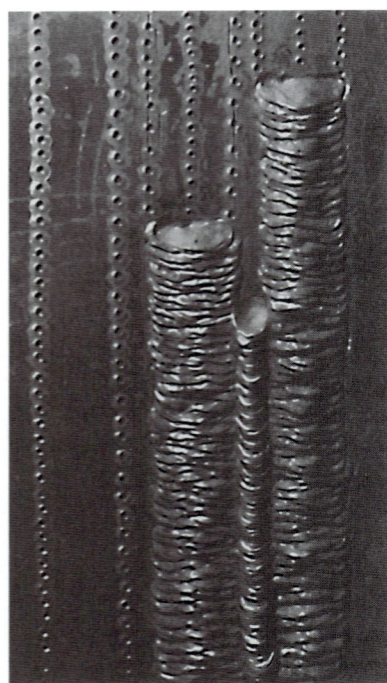

Fig. 13-79 Typical appearance of weave and stringer beads on flat plate welded in the vertical position, travel up, with DCEN or a.c. shielded arc electrodes. *Edward R. Bohnart*

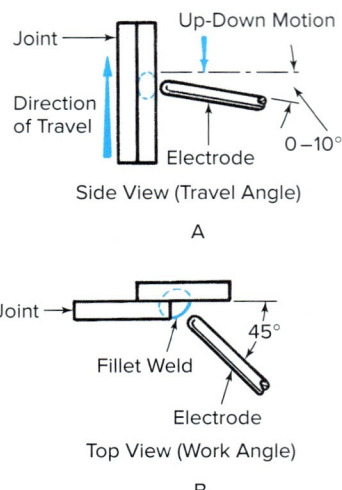

Fig. 13-80 Electrode position when welding lap joints in the vertical position, travel up.

General Job Information

On critical work, the direction of travel for lap joints in the vertical position is usually up. This is true for the shipbuilding, pressure vessel, and structural steel industries. The technique is used for single-pass lap welds and for the first pass on multipass lap welds. When welding heavier plate, the lap joint is often welded with the multipass stringer bead procedure. The first bead should be put into the root with good penetration. The second bead should fuse thoroughly into three-fourths of the first bead and extend ¼ inch out on the bottom plate. The third bead should fuse with the first two and extend to the edge of the top plate.

The lap joint may also be welded with the weaved fillet weld procedure for the last one or two passes, depending upon the thickness of the material.

Welding Technique

Current adjustment must not be too high. Hold the electrode 10° downward from the horizontal position, Fig. 13-80A, and at a 45° angle to the plate surface shown in Fig. 13-80B. Form a shelf at the bottom of the joint that is the required size of the weld. Manipulate the electrode with a rocking back-and-forth motion. Deposit metal with a short arc gap and lengthen the arc gap on the upward stroke. Do not break the arc on the upward stroke. When the electrode is in the molten pool, you may use a slight circular motion to form a full bead. Keep the electrode motion within the confines of the weld width so that the edge of the upper plate is not burned away and so that undercut does not occur on the surface of the lower plate.

Operations

1. Obtain plates; check the job drawing, Fig. 13-81, for quantity and size.
2. Obtain electrodes of each quantity, type, and size specified in the job drawing.
3. Set a d.c. power source for electrode positive at 120 to 170 amperes or an a.c. power source at 130 to 190 amperes.
4. Set up the plates and tack weld them at each corner.
5. Stand the plates in the vertical position on the welding table. Make sure that they are connected to the work connection.
6. Make welds as shown on the job drawing. Manipulate the electrode as instructed in Fig. 13-80.
7. Chip the slag from the welds, brush, and inspect. Refer to Inspection.
8. Practice these welds until you can produce good welds consistently with both types of electrodes.

Inspection

Compare the weld with Fig. 13-82 and check it for the following characteristics:

Width and height: Uniform
Appearance: Smooth with close ripples; free of voids and high spots. Restarts should be difficult to locate.

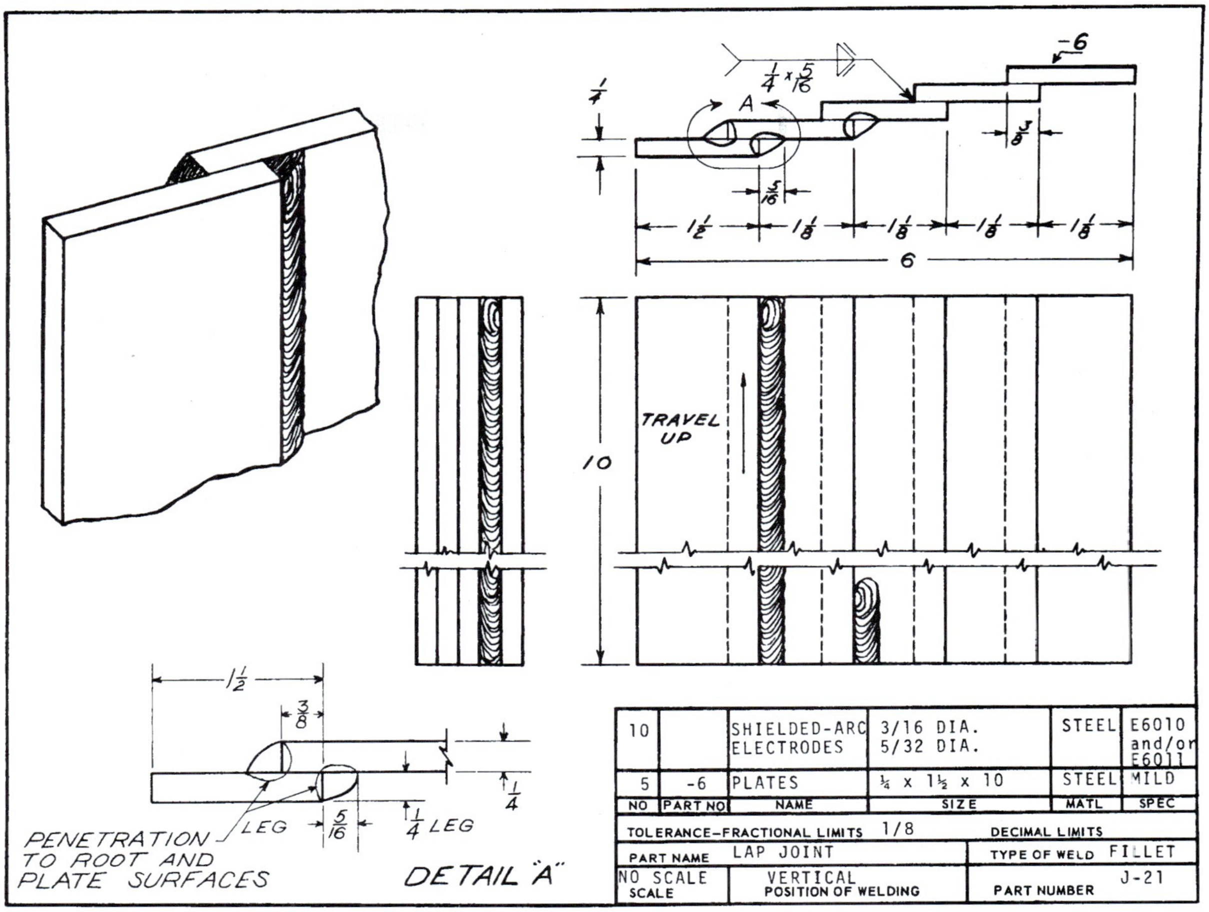

Fig. 13-81 Job drawing J21.

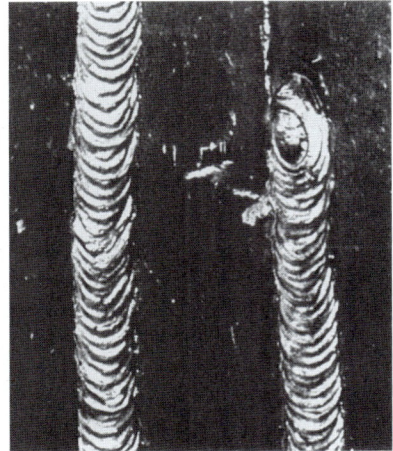

Fig. 13-82 Typical appearance of a fillet weld in a lap joint welded in the vertical position, travel up, with DCEP or a.c. shielded arc electrodes. Edward R. Bohnart

Size: Refer to the job drawing. Check with a fillet weld gauge.
Face of weld: Slightly convex and full
Edges of weld: Good fusion, no overlap, no undercut
Starts and stops: Free of depressions and high spots
Beginnings and endings: Full size, craters filled
Penetration and fusion: To the root of the joint and both plate surfaces
Surrounding plate surfaces: Free of spatter
Slag formation: Full coverage, easily removable

Check Test

After you are able to make welds that are satisfactory in appearance, make up a specimen like that made in Job 13-J11, Fig. 13-36, page 354. Use the thickness of plate, welding technique, and electrode types and size specified for this job. Weld on one side only. Break the joint as shown in Fig. 13-37, page 354. Inspect the weld for lack of penetration and fusion at the root of the joint and to the plate surfaces. The weld metal should be sound and show no evidence of porosity, slag inclusions, or gas pockets. The weld should break evenly through the throat.

Disposal

Put completed joints in the scrap bin so that they will be available for further use. Unwelded plate surfaces can be used for beading.

Job 13-J22 Welding a T-Joint

Objective

To weld a T-joint in the horizontal position by means of a single-pass fillet weld, with DCEN and/or a.c. shielded metal arc electrodes (AWS E6013).

General Job Information

Horizontal fillet welding is a large part of the work that the operator will do in the field. Welding procedures may include single-pass and multipass techniques with all types of electrodes. The final size of a fillet weld determines whether it is to be single pass or multipass. Multipass procedures may require either the weaved or stringer bead welding techniques. It is particularly important for you to become proficient in the next few jobs.

Welding Technique

Current should be high enough to maintain a fluid pool. Direct the electrode at the root of the weld and toward the vertical plate, Fig. 13-83A, and tilt it 30° in the direction of travel, Fig. 13-83B. The position for an actual weld is shown in Fig. 13-84. Advance the electrode at a uniform rate of travel. You may keep it in the weld pool for the entire length of the weld. Some welders, however, prefer to use a slight back-and-forth motion, which preheats the root ahead of the weld, aids the formation of the weld

Fig. 13-84 Welding a T-joint in the horizontal position. The electrode and electrode holder are positioned as instructed in Fig. 13-83. Location: Northeast Wisconsin Technical College
Mark A. Dierker/McGraw Hill

deposit against the vertical plate, and helps to keep the slag from running ahead of the weld.

Undercut is not a serious problem with electrode negative. If the arc is too long or if the angle of the electrode is incorrect, however, undercutting may result. Direct the arc about equally against both legs of the fillet weld with just enough direction against the vertical plate to permit the weld metal to build up on it. Control the speed to obtain good weld contour and a full throat section of the fillet weld. Hold the arc in the crater long enough to build up the molten weld metal to the desired height.

Practice starts and stops. Travel in all directions.

Operations

1. Obtain plates; check the job drawing, Fig. 13-85, for quantity and size.
2. Obtain electrodes of each quantity, type, and size specified in the job drawing.
3. Set a d.c. power source for electrode negative at 110 to 190 amperes or an a.c. power source at 120 to 210 amperes.
4. Set up the plates and tack weld them at each corner.
5. Place the joint in the horizontal position on the welding table. Make sure that it is connected to the work connection.
6. Make welds as shown on the job drawing. Manipulate the electrode as in Fig. 13-83.
7. Chip the slag from the welds, brush, and inspect. Refer to Inspection.
8. Practice these welds until you can produce good welds consistently with both types of current and with $5/32$-inch diameter electrodes.

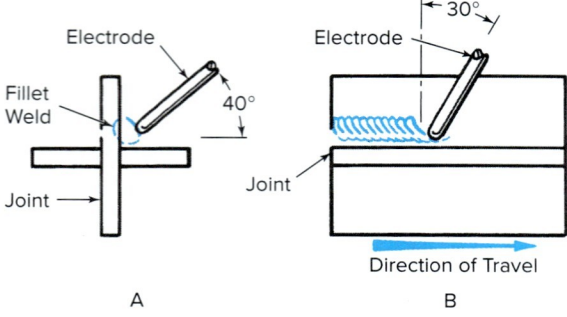

Fig. 13-83 Electrode position when welding a T-joint in the horizontal position.

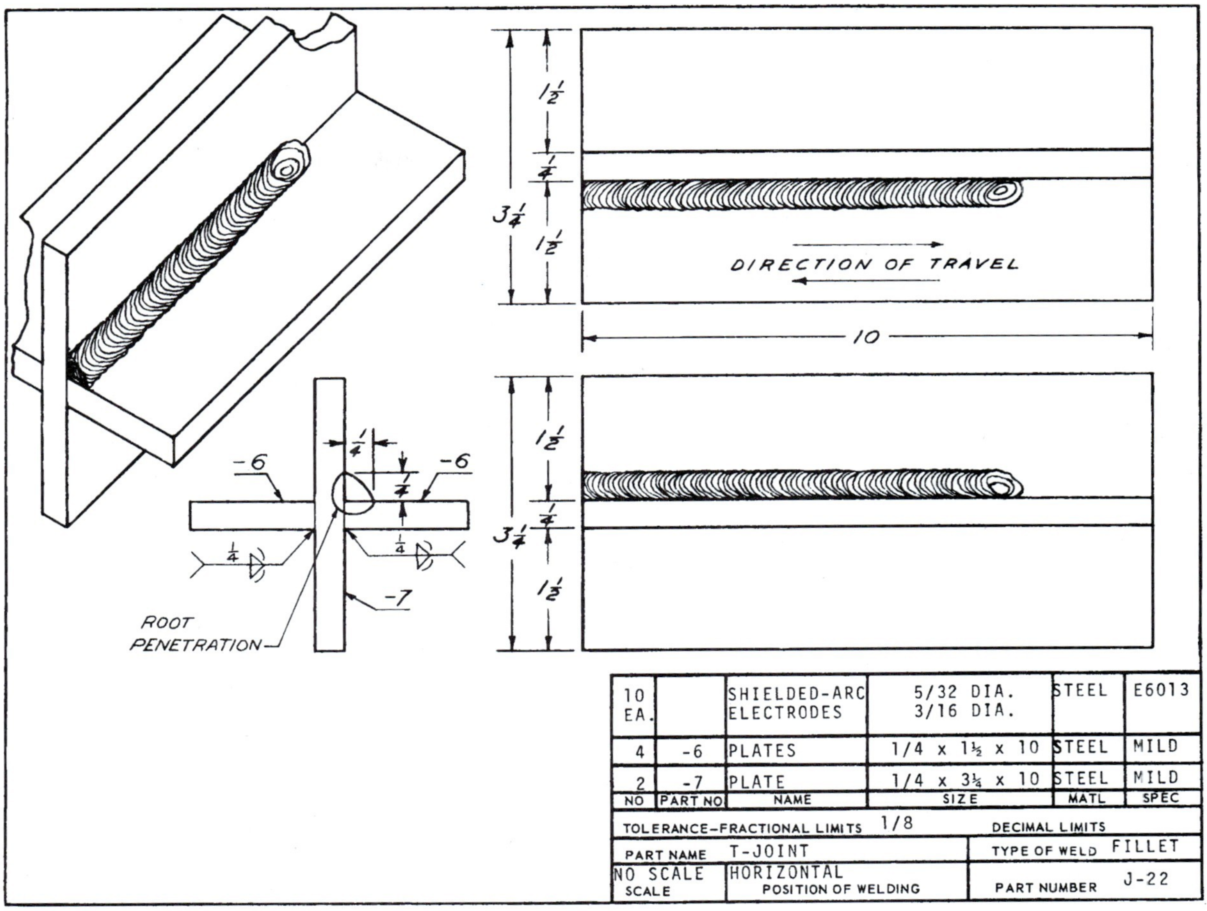

Fig. 13-85 Job drawing J22.

Inspection

Compare the welds with Fig. 13-86 and check them for the following characteristics:

Width and height: Uniform
Appearance: Smooth with close ripples; free of voids. Restarts should be difficult to locate.
Size: Refer to the job drawing. Check with a fillet weld gauge.
Face of weld: Slightly convex
Edges of weld: Good fusion, no overlap, no undercut
Starts and stops: Free of depressions and high spots
Beginnings and endings: Full size, craters filled
Penetration and fusion: To root of joint and plate surfaces
Surrounding plate surfaces: Free of spatter
Slag formation: Full coverage, easily removable

Check Test

After you are able to make welds that are satisfactory in appearance, make up a specimen similar to that made in Job 13-J14, Fig. 13-58, page 365. Use the plate thickness, welding technique, and electrode type and size specified for this job. Weld on one side only. Break the finished weld as shown in Fig. 13-60, page 365, and examine the surfaces for soundness. The weld must not be porous and must show complete fusion and penetration at the root of the joint and to the plate surfaces. The joint should break evenly through the throat of the weld.

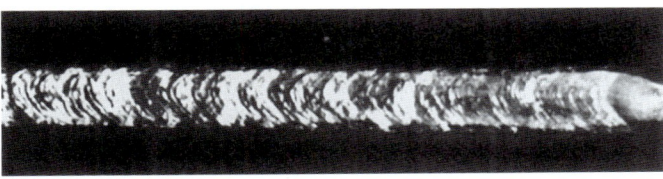

Fig. 13-86 Typical appearance of a single-pass fillet weld in a T-joint welded in the horizontal position with DCEN or a.c. shielded arc electrodes. Edward R. Bohnart

Disposal

Save the completed joints for your next job, which will be multipass fillet welding.

Job 13-J23 Welding a T-Joint

Objectives

1. To weld a T-joint in the horizontal position by means of multipass fillet welds using the stringer bead technique with DCEN and/or a.c. shielded metal arc electrodes (AWS E6013).
2. To weld a cover pass in the horizontal position by means of a weave bead with DCEP and/or a.c. shielded metal arc electrodes (AWS E6010–E6011).

General Job Information

Large fillets are often welded with multipass stringer beads. On many jobs, the final layer is made up of stringer beads. On other jobs, a weave bead is applied as the final bead. This is especially true in piping and pressure vessel work.

For practical purposes ¼-inch plate would not require more than one pass. The additional passes specified in this job are for practice purposes and the full and economical use of material.

Welding Technique

Current adjustment should be similar to that used for the first pass. Electrode position varies with the sequence of passes. The electrode is held at a 50° angle for the second pass, a 30° angle for the third pass, a 50° for the fourth pass, and a 40° angle for the fifth and sixth passes, Fig. 13-87. It is also tilted at a 30° angle in the direction of travel. Electrode movement is the same as for the preceding job. Beads should be so proportioned that the size of the final weld is accurate and to specification.

Be very critical of the angle of the electrode and the arc length to make sure that undercut does not develop in the vertical plate. The second pass should lay about half on the flat plate and half on the first bead. For the third pass, lay down another bead that covers the other half of the first pass and also fuses with the vertical plate and the second pass, Fig. 13-87. Follow this same procedure in making the next layer of three passes, one at the bottom, one in the middle, and one at the top as shown in Fig. 13-87.

The technique of weaving is difficult. Current adjustment must not be too low. A smaller electrode is used because it is hard to handle a large amount of weld metal with this technique. An out-of-position electrode is also necessary.

The electrode position should be approximately that shown in Fig. 13-88. The electrode is tilted at a 35° angle to the flat plate, Fig. 13-88A, and in a 30° angle toward the direction of travel, Fig. 13-88B. Weld metal is deposited only on the downward stroke. The upward stroke should be fast and the arc should be lengthened, but it should not go out. The top and bottom edges of the weld should be your guide for the width of the electrode movement. Split the weld edge with your electrode. Hesitate an instant at the top and bottom edges of the weld to eliminate undercut.

Bear in mind that all multipass welding requires thorough cleaning between passes. Practice starts and stops. Travel in all directions.

Operations

1. Select good quality joints welded in Job 13-J2.
2. Obtain electrodes of each quantity, type, and size specified in the job drawing, Fig. 13-89.
3. Set a d.c. power source for electrode negative at 100 to 190 amperes or an a.c. power source at 125 to 210 amperes.
4. Place the joint in the horizontal position on the welding table. Make sure that it is connected to the work connection.
5. Make passes two through six with 5/32-inch electrodes as shown on the job drawing. Hold the electrode as instructed in Fig. 13-87 for each pass.
6. Chip the slag from the weld, brush, and inspect between each pass. Refer to Inspection.
7. Set a d.c. power source for electrode positive at 80 to 140 amperes or an a.c. power source at 80 to 160 amperes.

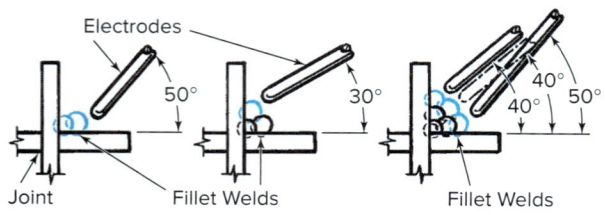

Fig. 13-87 Electrode position when welding a T-joint with multipass fillet welds using the stringer bead technique.

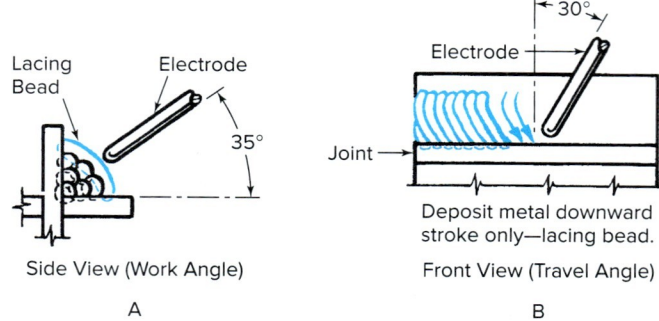

Fig. 13-88 Electrode position when welding the final pass on a T-joint using the lacing bead technique.

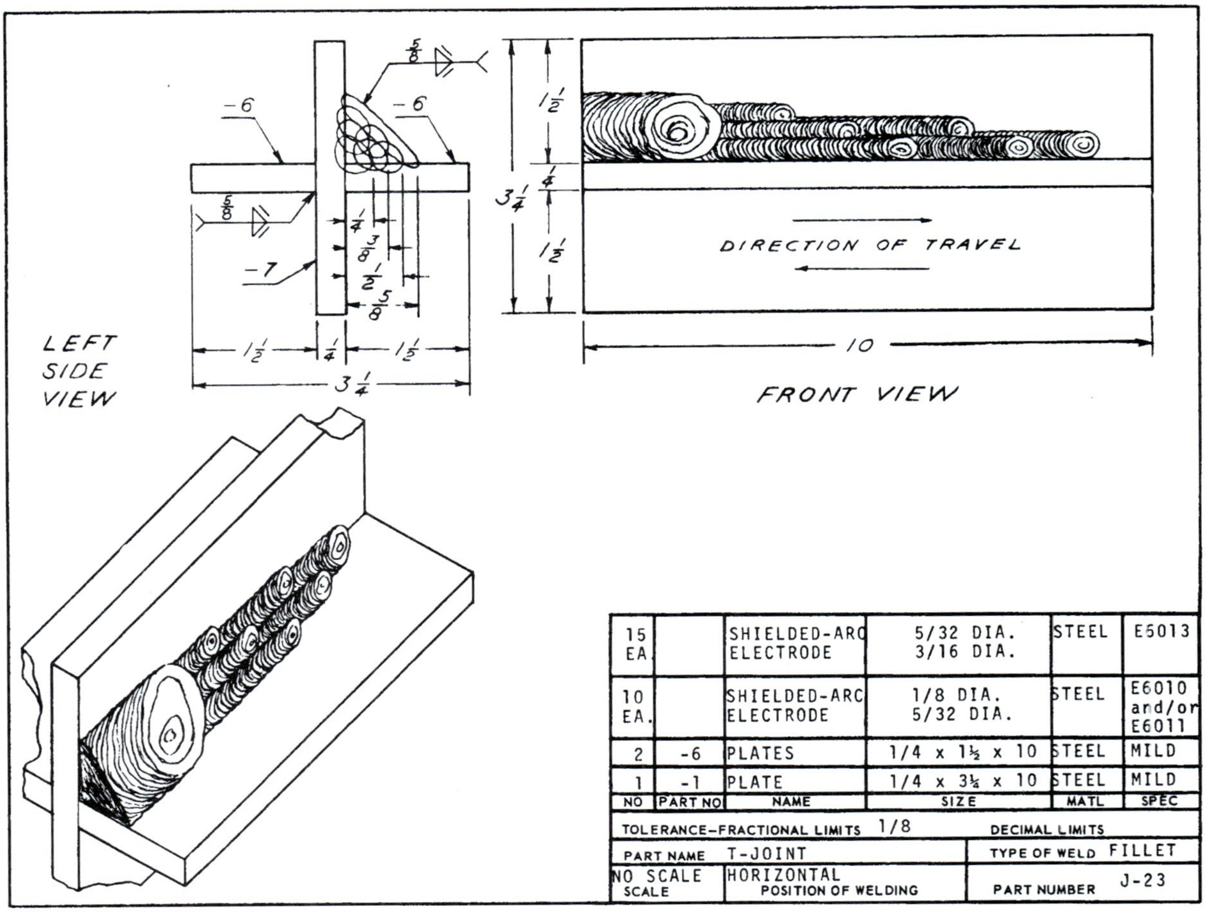

Fig. 13-89 Job drawing J23.

8. Make the lacing bead with ⅛-inch electrodes as shown on the job drawing. Manipulate the electrode as instructed in Fig. 13-88.
9. Chip the slag from the weld, brush, and inspect. Refer to Inspection below.
10. Practice these welds until you can produce good welds consistently with the E6013 electrode in 5/32-inch diameters for the stringer beads and the E6010 and E6011 electrode in ⅛-inch diameters for the weaving pass.

Inspection

Compare passes two through six with Fig. 13-90 and the lacing pass with Fig. 13-91. Check each pass for the following weld characteristics:

Width and height: Uniform
Appearance: Smooth with close ripples; free of voids. Restarts should be difficult to locate.

Fig. 13-90 Typical appearance of a multipass fillet weld done with the stringer bead technique in a T-joint welded in the horizontal position with DCEN or a.c. shielded arc electrodes. *Edward R. Bohnart*

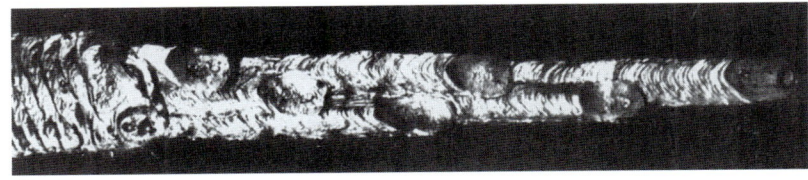

Fig. 13-91 Typical appearance of a multipass fillet weld done with the weaving bead technique for the cover pass. *Edward R. Bohnart*

Size: Refer to the job drawing. Check with a fillet weld gauge.
Face of weld: Flat
Edges of weld: Good fusion, no overlap, no undercut
Starts and stops: Free of depressions and high spots
Beginnings and endings: Full size, craters filled
Penetration and fusion: To root of joint, adjacent beads, and plate surfaces
Surrounding plate surfaces: Free of spatter
Slag formation: Full coverage, easily removable

Disposal

Put completed jobs in the scrap bin so that they can be used again later. Plate edges can be used for square butt welding, and unwelded plate surfaces can be used for beading.

Job 13-J24   Welding a T-Joint

Objective

To weld a T-joint in the horizontal position by means of a single-pass fillet weld with DCEP and/or a.c. shielded arc electrodes (AWS E6010–E6011).

General Job Information

This is the same type of joint practiced in Job 13-J22 except that electrode positive is used. Electrode positive is not the best choice for single-pass fillet welds, but since this is the practice on many jobs, you should master the technique. Beads have a flat face, and there is the tendency toward undercutting.

Welding Technique

Current adjustment should be somewhat lower than for electrode negative electrodes. More care should be given to electrode position. Hold the electrode at a 35° angle to the flat plate, Fig. 13-92A, and at an angle of 40° in the direction of travel, Fig. 13-92B. The welding technique is similar to that used for electrode negative except that the arc gap should be closer and more care must be exercised to prevent undercut.

If the arc is too long or if the angle of the electrode is incorrect, undercutting will result. Direct the arc at the root of the weld and about equally against both legs of the fillet with just enough direction against the vertical plate to permit the weld metal to build up on it. Control the speed of welding to obtain good weld contour and the full throat section of the fillet. Hold the arc in the crater long enough to build up the molten metal to the desired height and travel at uniform speed.

Practice starts and stops. Travel in all directions.

Operations

1. Obtain plates; check the job drawing, Fig. 13-93, for quantity and size.
2. Obtain electrodes of the quantity, type, and size specified in the job drawing.
3. Set a d.c. power source for electrode positive at 110 to 170 amperes or an a.c. power source at 115 to 190 amperes.
4. Set up the plates and tack weld them at each corner.
5. Place the joint in the horizontal position on the welding table. Make sure that it is connected to the work connection.
6. Make welds as shown on the job drawing. Manipulate the electrode as instructed in Fig. 13-92.
7. Chip the slag from the welds, brush, and inspect. Refer to Inspection below.
8. Practice these welds until you can produce good welds consistently with both types of electrodes.

Inspection

Compare the welds with Fig. 13-94 and check them for the following characteristics:

Width and height: Uniform
Appearance: Smooth with close ripples; free of voids. Restarts should be difficult to locate.
Size: Refer to the job drawing. Check with a fillet weld gauge.
Face of weld: Flat
Edges of weld: Good fusion, no overlap, no undercut
Starts and stops: Free of depressions and high spots
Beginnings and endings: Full size, craters filled
Penetration and fusion: To the root of the joint and plate surfaces
Surrounding plate surfaces: Free of spatter
Slag formation: Full coverage, easily removable

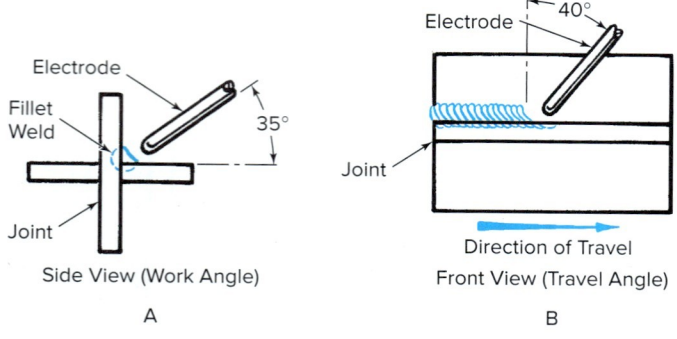

Fig. 13-92 Electrode position when welding a T-joint with a single-pass fillet weld in the horizontal position.

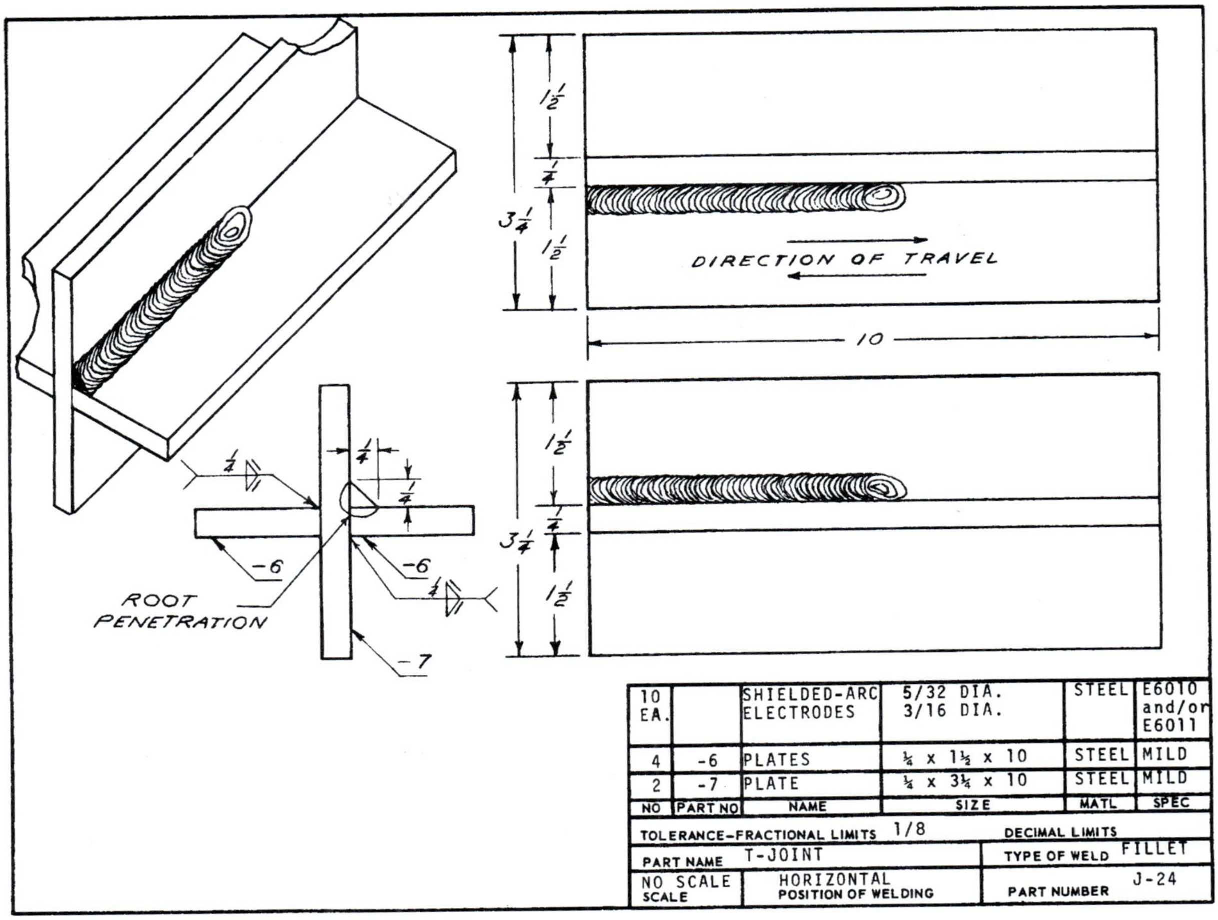

Fig. 13-93 Job drawing J24.

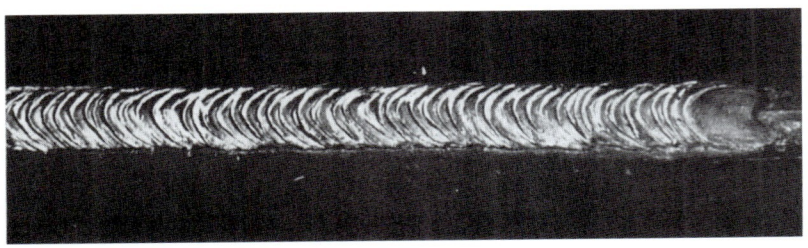

Fig. 13-94 Typical appearance of a single-pass fillet weld in a T-joint welded in the horizontal position with DCEP or a.c. shielded arc electrodes. *Edward R. Bohnart*

Disposal
Save the completed joints for your next job, which will be multipass fillet welding.

Check Test
After you are able to make welds that are satisfactory in appearance and are approved by your instructor, make up a test specimen like that shown in Fig. 13-59, page 365. Use the plate thickness, welding technique, and electrode type and size specified for this job. Weld on one side only.

Break the completed weld as shown in Fig. 13-60, page 365, and examine the surfaces for soundness. They must not be porous and must show fusion and penetration at the root of the weld and to the plate surfaces. The weld should break through the throat. It should not peel off the plate surfaces.

Job 13-J25 Welding a T-Joint

Objectives

1. To weld a T-joint in the horizontal position by means of multipass fillet welds and the stringer bead technique with DCEP and/or a.c. shielded metal arc electrodes (AWS E6010–E6011).
2. To weld an overlay in the horizontal position by means of a weave bead with DCEP and/or a.c. shielded metal arc electrodes (AWS E6010–E6011).

General Job Information

This job is similar to Job 13-J23 except that a different type of electrode is used.

Welding Technique

Current adjustment should be somewhat lower than for electrode negative electrodes. More care should be given to electrode position. The angle of the electrodes is slightly different with electrode positive. It is held at a 45° angle to the flat plate to deposit the second bead, Fig. 13-95A. The angle is changed to 35° for the third bead, Fig. 13-95B. The fourth bead is deposited at a 45° angle, and the angle is changed to 35° for the fifth and sixth beads, Fig. 13-95C. The electrode is also tilted at a 30° angle in the direction of travel. The welding technique for both stringer and lacing beads is similar to that practiced in Job 13-J23. Refer to Fig. 13-88, page 380, for deposition of the weave bead. Undercut is a problem and can be remedied in the usual manner. Review the welding technique described in Job 13-J23.

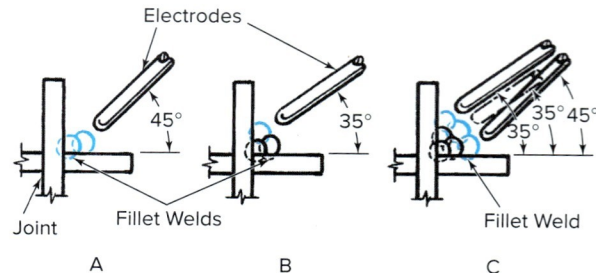

Fig. 13-95 Electrode position when welding a T-joint with the stringer bead technique.

Clean all passes thoroughly. Practice starts and stops. Travel in all directions.

Operations

1. Use joints welded for Job 13-J24.
2. Obtain electrodes of each quantity, type, and size specified in the job drawing, Fig. 13-96.

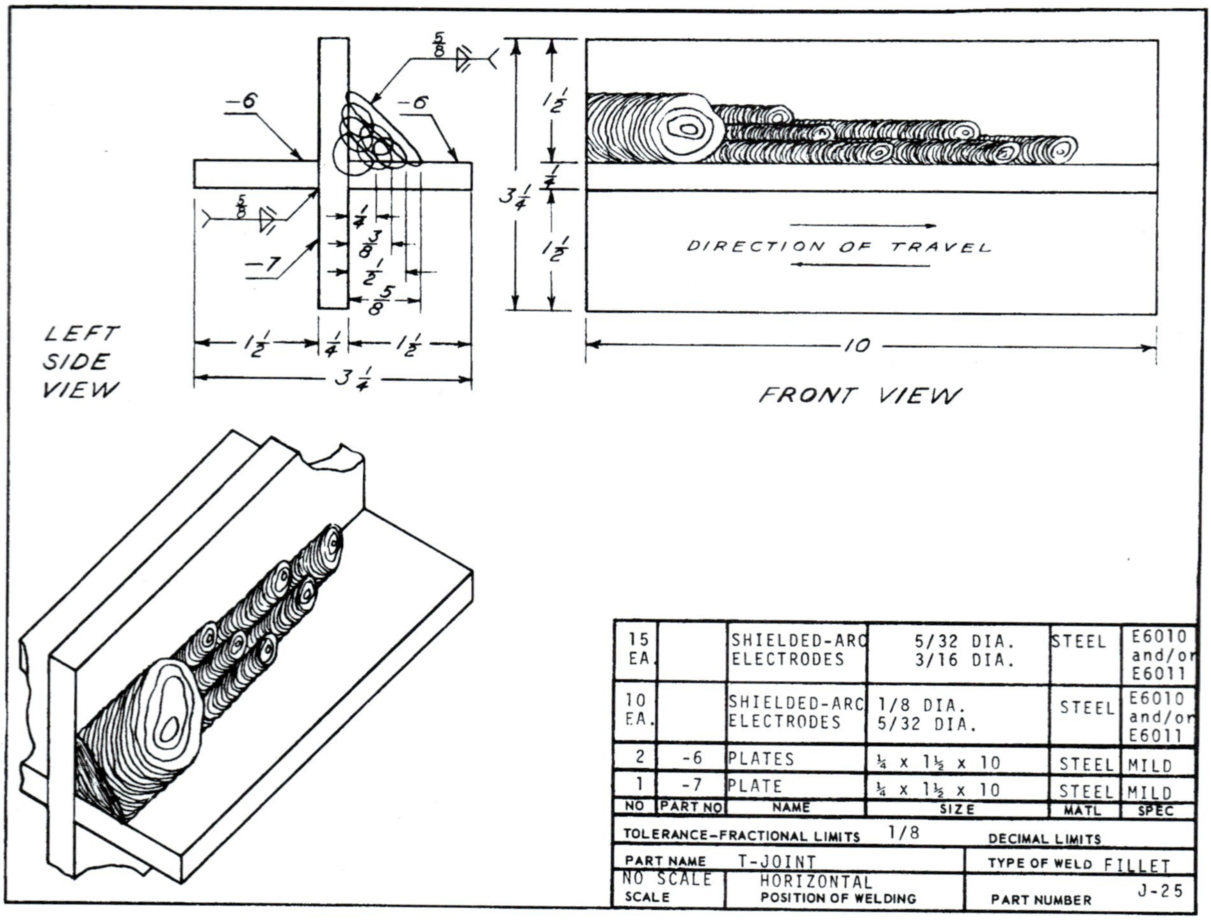

Fig. 13-96 Job drawing J25.

3. Set a d.c. power source for electrode positive at 110 to 170 amperes or an a.c. power source at 115 to 190 amperes.
4. Place the joint in the horizontal position on the welding table. Make sure that it is connected to the work connection.
5. Make passes two through six with 5/32-inch electrodes as shown on the job drawing. Hold the electrode as instructed in Fig. 13-95 for each pass.
6. Chip the slag from the weld, brush, and inspect between each pass. Refer to Inspection.
7. Set a d.c. power source for electrode positive at 75 to 125 amperes or set an a.c. power source at about the same heat value.
8. Make the weaving bead with 1/8-inch electrodes as shown on the job drawing. Manipulate the electrode as shown in Fig. 13-88, page 380.
9. Chip the slag from the weld, brush, and inspect. Refer to Inspection.
10. Practice these welds until you can produce good welds consistently.

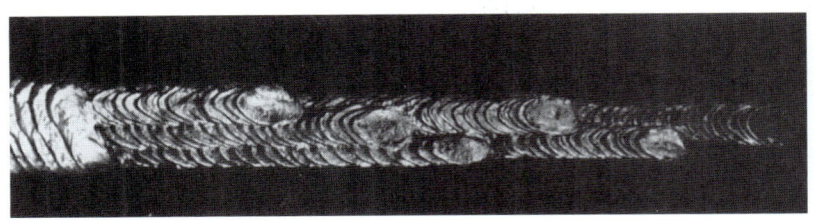

Fig. 13-97 Typical appearance of a multipass fillet weld done with the stringer bead technique and a weaving overlay in a T-joint welded in the horizontal position with DCEP or a.c. shielded arc electrodes. Edward R. Bohnart

Inspection

Compare each pass with Fig. 13-97 and check it for the following weld characteristics:

Width and height: Uniform
Appearance: Smooth with close ripples; free of voids. Restarts should be difficult to locate.
Size: Refer to the job drawing. Check with a fillet weld gauge.
Face of weld: Flat
Edges of weld: Good fusion, no overlap, no undercut
Starts and stops: Free of depressions and high spots
Beginnings and endings: Full size, craters filled
Penetration and fusion: To the root of the joint, adjacent beads, and plate surfaces
Surrounding plate surfaces: Free of spatter
Slag formation: Full coverage, easily removable

Disposal

Put completed joints in the scrap bin so that they can be used again later. Plate edges can be used for square butt welding, and unwelded plate surfaces can be used for beading.

CHAPTER 13 REVIEW

Multiple Choice

Choose the letter of the correct answer.

1. SMAW is a useful process when doing maintenance and repair welding because it can be used on a variety of steel and other metals. It is also portable and can be used in tight access. (Obj. 13-1)
   a. True
   b. False
2. DCEP has which of the following characteristics? (Obj. 13-2)
   a. Deep penetrating weld
   b. Shallow penetrating weld
   c. Eliminates arc blow
   d. Gives the greatest deposition
3. Which of the following is correct? (Obj. 13-2)
   a. The longer the arc, the lower the voltage.
   b. Arc length has no effect on voltage.
   c. The longer the arc, the higher the voltage.
   d. None of these.
4. If typical voltages for SMAW are in the 20- to 40-volt range, what would be a good voltage for vertical and horizontal welding? (Obj. 13-2)
   a. On the high end of the range near 40 volts
   b. Near the middle of the range around 30 volts
   c. At the low end of the voltage range or 20 volts
   d. All of these
5. Arc blow can be controlled by _____. (Obj. 13-2)
   a. Moving the work clamp to another location
   b. Changing the position of the electrode

c. Increasing the current
d. Both a and b
6. The tilt of the electrode in the direction of travel is called _____. (Obj. 13-2)
   a. Work angle
   b. Electrode angle
   c. Travel angle
   d. Transverse angle
7. Distortion can best be controlled by _____. (Obj. 13-3)
   a. Clamping and tacking parts properly
   b. Performing parts to compensate for shrinkage of welds
   c. Sequencing welds to distribute heat and/or preheating heavy structures
   d. All of these
8. Weld spatter is best controlled by _____. (Obj. 13-3)
   a. Using the proper type and size of electrode with the proper current
   b. Holding as long an arc as possible
   c. Laying the electrode angle down as low as possible
   d. Traveling as fast as possible to spread the spatter around
9. Which produces the wider bead? (Obj. 13-4)
   a. Stringer
   b. Weave
   c. Beading
   d. Fillet
10. Some typical ductility or percent of elongation of SMAW covered electrodes would be_____. (Obj. 13-4)
    a. 80 to 94
    b. 40 to 67
    c. 20 to 26
    d. 5 to 16

Review Questions

Write the answers in your own words.

11. List some of the advantages of the Parallel Joint. (Obj. 13-1)
12. The last layer of a horizontal weld is sometimes welded in what manner? Explain the technique. (Obj. 13-2)
13. List six items that affect the amount of current (heat) that is required for the SMAW process. (Obj. 13-2)
14. What are the effects of too long an arc gap on the weld? (Obj. 13-2)
15. What are the characteristics of good weld beads? (Obj. 13-2)
16. Is there any difference in the deposit of an electrode negative and an electrode positive electrode? Explain. (Obj. 13-2)
17. Undercut can best be controlled by doing what five things? (Obj. 13-3)
18. Explain the technique for eliminating the crater at the end of a bead. (Obj. 13-3)
19. What are the results of heat adjustment that is too hot? Too cold? (Obj. 13-3)
20. Explain the technique for running a fillet weld in a lap joint in the horizontal position. (Obj. 13-4)

INTERNET ACTIVITIES

Internet Activity A

Using your favorite search engine, see if you can make a list of the chemicals used to make E6010 and E6013 electrodes. Remember some of the leading electrode manufacturers are Lincoln Electric, Hobart, and ESAB.

Internet Activity B

Using your favorite search engine, investigate issues dealing with welding fumes. Make sure your sources are reputable and make a report to your instructor.

Design credits: Cinema and movie icon: Denis Meshkov/123RF; and Illustration of Welding icons: Mr. Nachai Sorasee/123RF

Table 13-6 Job Outline: Shielded Metal Arc Welding Practice: Jobs 13-J1–28 (Plate)

Job No.	Joint	Type of Weld	Position	DCEP	DCEN	A.C.	AWS Specif. No. d.c.	AWS Specif. No. a.c.	Text Reference
13-J1	Flat plate	Striking the arc—short beading	Flat 1		X		E6013	E6013	349
13-J2	Flat plate	Beading—stringer	Flat 1		X	X	E6013	E6013	351
13-J3	Flat plate	Beading—weaved	Flat 1		X	X	E6013	E6013	354
13-J4	Flat plate	Beading—stringer	Flat 1	X		X	E6010	E6011	356
13-J5	Flat plate	Beading—weaved	Flat 1	X		X	E6010	E6011	358
13-J6	Parallel Joint	Edge	Flat 1		X	X	E6013	E6013	359
13-J7	Parallel Joint	Edge	Flat 1	X		X	E6010	E6011	361
13-J8	Lap joint (plates flat)	Fillet—single pass	Hor. 2F		X	X	E6013	E6013	363
13-J9	Lap joint (plates flat)	Fillet—single pass	Hor. 2F	X		X	E6010	E6011	365
13-J10	Flat plate	Beading—stringer	Hor. 2	X		X	E6010	E6011	366
13-J11	Flat plate	Beading—stringer—travel down	Ver. 3	X		X	E6010	E6011	368
13-J12	Lap joint	Fillet—single pass—travel down	Ver. 3F		X	X	E6013	E6013	370
13-J13	Lap joint (plates vert.)	Fillet—single pass	Hor. 2F		X	X	E6013	E6013	372
13-J14	T-joint	Fillet—single pass	Flat 1F		X	X	E6013	E6013	374
13-J15	T-joint	Fillet—weaved multipass	Flat 1F		X	X	E6013	E6013	376
13-J16	T-joint	Fillet—single pass	Flat 1F	X		X	E6010	E6011	378
13-J17	T-joint	Fillet—weaved multipass	Flat 1F	X		X	E6010	E6011	380
13-J18	Flat plate	Beading-stringer—travel up	Ver. 3	X		X	E6010	E6011	381
13-J19	Flat plate	Beading—weaved—travel up	Ver. 3	X		X	E6010	E6011	383
13-J20	Flat plate	Beading—weaved—travel up	Ver. 3		X	X	E6013	E6013	385
13-J21	Lap joint	Fillet—single pass-travel up	Ver. 3F	X		X	E6010	E6011	386
13-J22	T-joint	Fillet—single pass	Hor. 2F		X	X	E6013	E6013	389
13-J23	T-joint	Fillet—stringer—multipass and weave pass	Hor. 2F		X	X	E6013	E6013	391
13-J24	T-joint	Fillet—single pass	Hor. 2F	X		X	E6010	E6011	393
13-J25	T-joint	Fillet—stringer—multipass and weave pass	Hor. 2F	X		X	E6010	E6011	394

14

Shielded Metal Arc Welding Practice:

Jobs 14-J26–J42* (Plate)

Introduction

This second section of the course in shielded metal arc welding (SMAW) provides practice in a number of weld joints that are more difficult to weld and require greater skill than those experiences that you have had up to this point. You will also have the opportunity to practice with a series of new electrodes of the low hydrogen and low hydrogen with iron powder types.

This chapter will also provide considerable practice in welding various types of basic joints in the flat, horizontal, and vertical positions. You will have your first experience welding in the overhead position. See Table 14-1 on page 427 for a list of all jobs outlined in this chapter. For economic and quality reasons in the shop every effort is made to position the work so that welding can be in the flat or horizontal positions, but it is not always possible to position all of the welding in this way. A portion of the work will always need to be welded in the vertical and overhead positions, Fig. 14-1. A greater amount of out-of-position welding is encountered in field work. It is not possible to position a building or all of the joints on a pipeline. When you have achieved a high degree of skill in welding in the vertical

Chapter Objectives

After completing this chapter, you will be able to:

14-1 Demonstrate ability to make overhead position welds with cellulose electrodes.

14-2 Demonstrate ability to make groove welds with and without backing in the flat, horizontal, and vertical (travel up and down) positions.

14-3 Demonstrate ability to make fillet welds with stringer and weave bead techniques in the horizontal and vertical positions with cellulose and low hydrogen electrodes.

14-4 Demonstrate ability to make outside corner welds in the vertical position with cellulose electrodes.

---

*The jobs in Chapters 13 to 15 are numbered consecutively because these units constitute a course in shielded metal arc welding.

Fig. 14-1 Making a weld in the vertical (3) position. Note the protective clothing. Location: Northeast Wisconsin Technical College Mark A. Dierker/McGraw Hill

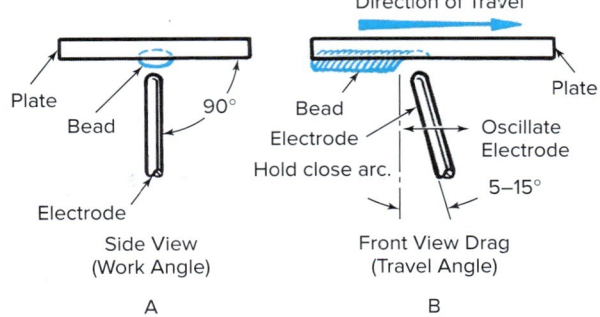

Fig. 14-2 Electrode position when making stringer beads in the overhead (4) position.

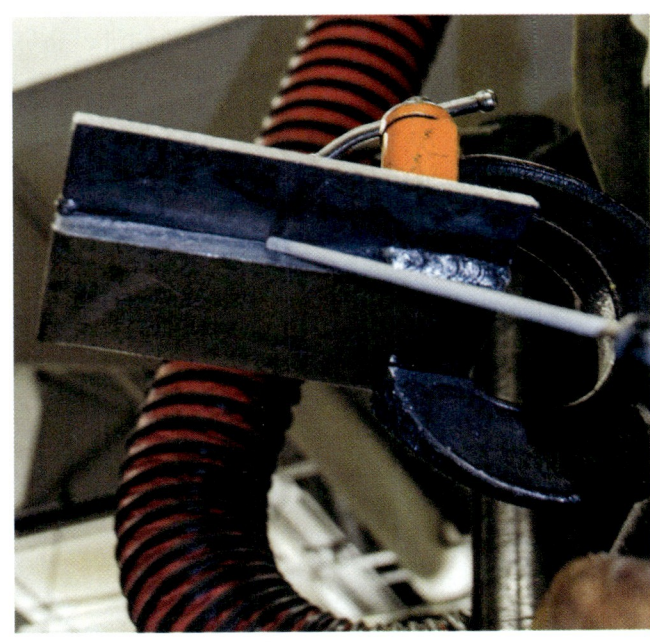

Fig. 14-3 Position of the electrode and electrode holder when welding in the overhead (4) position. This weld will be made from left to right with the proper drag angle. Location: Northeast Wisconsin Technical College Mark A. Dierker/McGraw Hill

and overhead positions, you are well on your way to becoming a skilled welder able to perform the tasks required by industry.

Practice Jobs

Job 14-J26   Stringer Beading

Objective
To deposit stringer beads on flat steel plate in the overhead position with DCEP and/or a.c. shielded metal arc electrodes (AWS E6010–E6011).

General Job Information
Although the tendency today in shops is to eliminate as much overhead welding as possible by positioning the work, you must be able to weld in this position. Much of the work in the piping, shipbuilding, and structural fields is overhead. Overhead welding will seem quite difficult at first, mainly because of the position that you must assume. In order to minimize fatigue, you should be as comfortable as possible. If you are standing, drape the heavy welding cable over your shoulders to reduce the weight on the arm you are welding with. Never wrap the weld cables around your body.

Welding Technique
Current adjustment should be about the same as for vertical welding. Hold the electrode at a 90° work angle to the plate, Fig. 14-2A, and at a drag travel angle of 5 to 15°, Fig. 14-2B. Move the electrode back and forth in an oscillating manner. Figure 14-3 shows the actual position for welding. You should be in such a position that you can see the metal being deposited behind the arc if the direction of travel is away from you. It is often necessary, however, that the direction of travel be toward the welder. This is usually true of pipe welding.

Hold a close arc gap, and do not let the pool become too molten. A slight oscillating motion back and forth along the line of weld preheats the base metal ahead of the weld and keeps the slag washed back on the weld. If the weld pool gets too hot, move forward faster. If it becomes too cold, slow down your rate of travel. The molten weld metal is held up by molecular attraction and surface tension until each drop solidifies. Study the basic motions shown in Fig. 14-4, page 390.

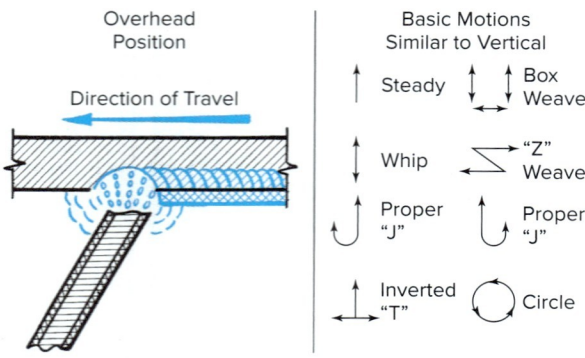

Fig. 14-4 Basic motions when welding in the overhead (4) position.

Practice starts and stops. Travel should be in all directions. Weld on both sides of the plate.

Operations

1. Obtain plate; check the job drawing, Fig. 14-5, for size.
2. Obtain a square head and scale, dividers, scribe, and center punch from the toolroom.
3. Lay out parallel lines as shown on the job drawing.
4. Mark the lines with a center punch.
5. Obtain electrodes of each quantity, type, and size specified in the job drawing.
6. Fasten the plate in the overhead position. Use an overhead welding jig, Fig. 14-6.
7. Set the power source at 110 to 170 amperes. Set a d.c. power source for electrode positive.
8. Make stringer beads on 5/8-inch center lines as shown on the job drawing. Manipulate the electrode as instructed in Figs. 14-2 through 14-4.
9. Chip the slag from the beads, brush, and inspect. Refer to Inspection.
10. Make additional stringer beads between the beads already deposited as shown on the job drawing.
11. Chip the slag from the beads, brush, and inspect. Refer to Inspection.
12. Practice these beads until you can produce good beads consistently with both types of electrodes and with 5/32-inch diameter electrodes.

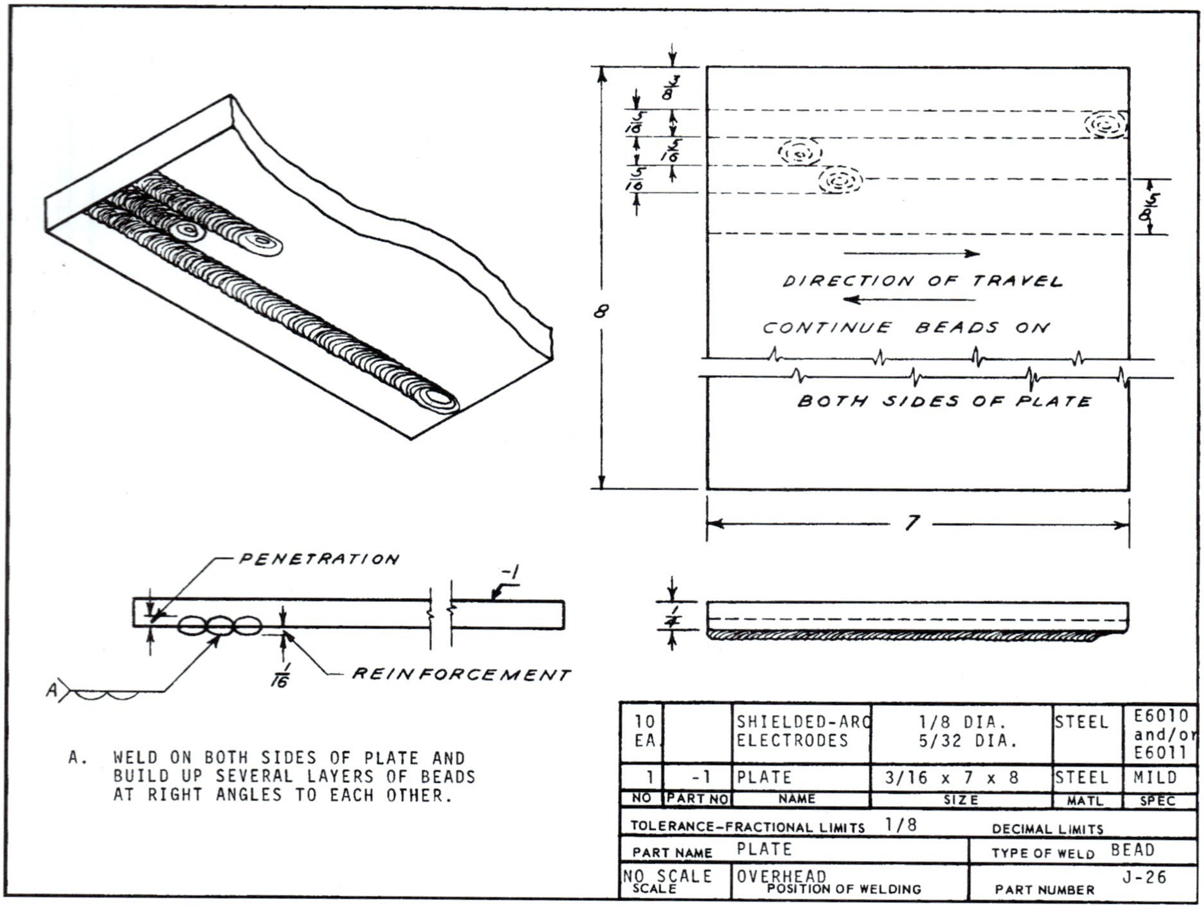

Fig. 14-5 Job drawing J26.

Fig. 14-6 Overhead welding jig. *Edward R. Bohnart*

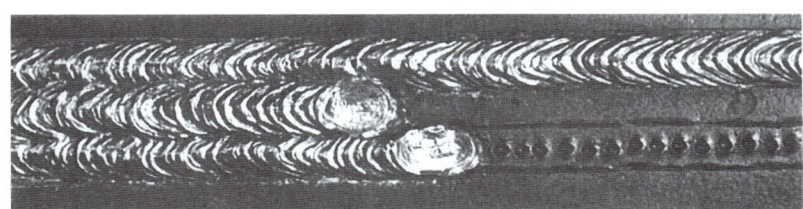

Fig. 14-7 Typical appearance of stringer beads welded in the overhead (4) position with DCEP or a.c. shielded metal arc coated electrodes. *Edward R. Bohnart*

Inspection

Compare the beads with Fig. 14-7 and check them for the following weld characteristics:

Width and height: Uniform
Appearance: Smooth with close ripples; free of voids and high spots. Restarts should be difficult to locate.
Size: Refer to the job drawing.
Face of beads: Slightly convex
Edges of beads: Good fusion, no overlap, no undercut
Starts and stops: Free of depressions and high spots
Beginnings and endings: Full size, craters filled
Penetration and fusion: To plate surface and adjacent beads
Surrounding plate surfaces: Free of spatter
Slag formation: Full coverage, easily removable

Disposal

Discard completed plates in the waste bin. Plates must be filled on both sides.

Job 14-J27 Weave Beads

Objective

To deposit weave beads on flat steel plate in the overhead position with DCEP and/or a.c. shielded metal arc electrodes (AWS E6010–E6011).

General Job Information

On many jobs it is necessary to make weave beads in the overhead position. This is more difficult than stringer beads and has to be practiced diligently. The technique is commonly used in pipe welding.

Welding Technique

Current adjustment must not be too high. Hold the electrode at a 90° work angle to the plate, Fig. 14-8A, and at a travel angle no more than 10° in the direction of travel, Fig. 14-8B. Weave the electrode from side to side and hesitate at the sides to prevent undercut.

It is essential that the arc gap be close and that the gap be uniform for the entire width of the bead. You may advance the weld by keeping the electrode in the pool, but you must prevent the molten metal from becoming so fluid that it sags. If sagging occurs, advance the electrode along either side of the weld in the box weave shown in Fig. 14-4, page 390. Lengthen the arc but do not break it when you correct the sag. Lengthening the arc with a conventional power source will reduce amperage. This in turn reduces the melt rate of the electrode. However, with solid-state and inverter-type power sources this may not be the case. They are equipped with a closed loop feedback electronic circuitry to maintain the amperage that has been set. With these types of power sources do not lengthen the arc, as this will increase voltage with no drop off in amperage so the total heat increases and will cause the possibility for more sag. The technique to use with these types of power sources is to maintain the shorter arc length and simply move out of the weld pool to allow it to cool down and control the sag in this manner. Figure 14-9, page 392, shows this relationship on

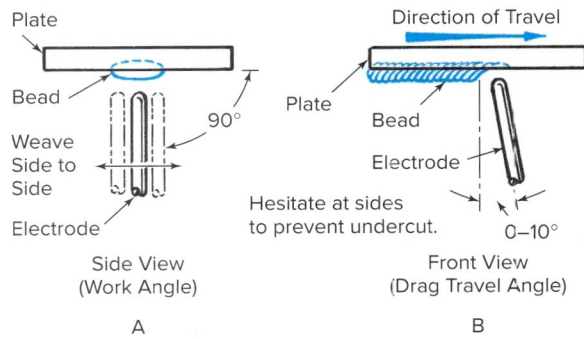

Fig. 14-8 Electrode position when making weave beads in the overhead (4) position.

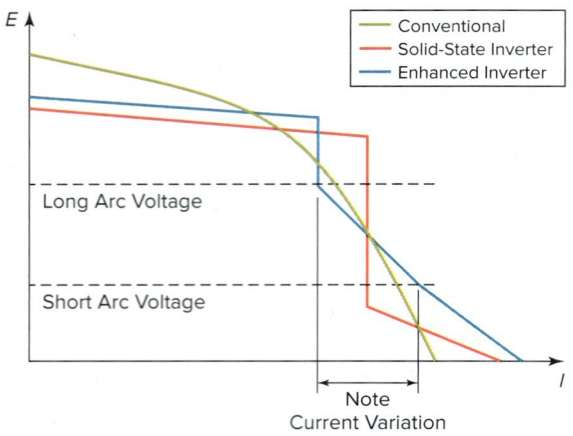

Fig. 14-9 Conventional, solid-state inverter, and enhanced inverter power source volt-amp curves.

various power source volt-amp curves. Compare this to the volt-amp curve of the power source you are using; this will help you understand what type of manipulative technique to use. This long arc or **whipping** technique is very usable on cellulose-type electrodes (E6010–E6011); however, it should never be used with the low hydrogen-type electrodes (E7016–E7018). Do not keep the arc out of the crater any longer than necessary. Keeping it out too long will cause excess spatter ahead of the weld. Move the electrode across the face of the weld rapidly so that you do not overheat the metal deposited in the middle of the weld and cause it to sag. Making beads wider than ¾ inch is not good welding practice.

Practice starts and stops. Travel in all directions. Fill both sides of the plate.

Operations

1. Obtain plate; check the job drawing, Fig. 14-10, for size.
2. Obtain a square head and scale, dividers, scribe, and center punch from the toolroom.
3. Lay out parallel lines as shown on the job drawing.
4. Mark the lines with a center punch.
5. Obtain electrodes of each quantity, type, and size specified in the drawing.

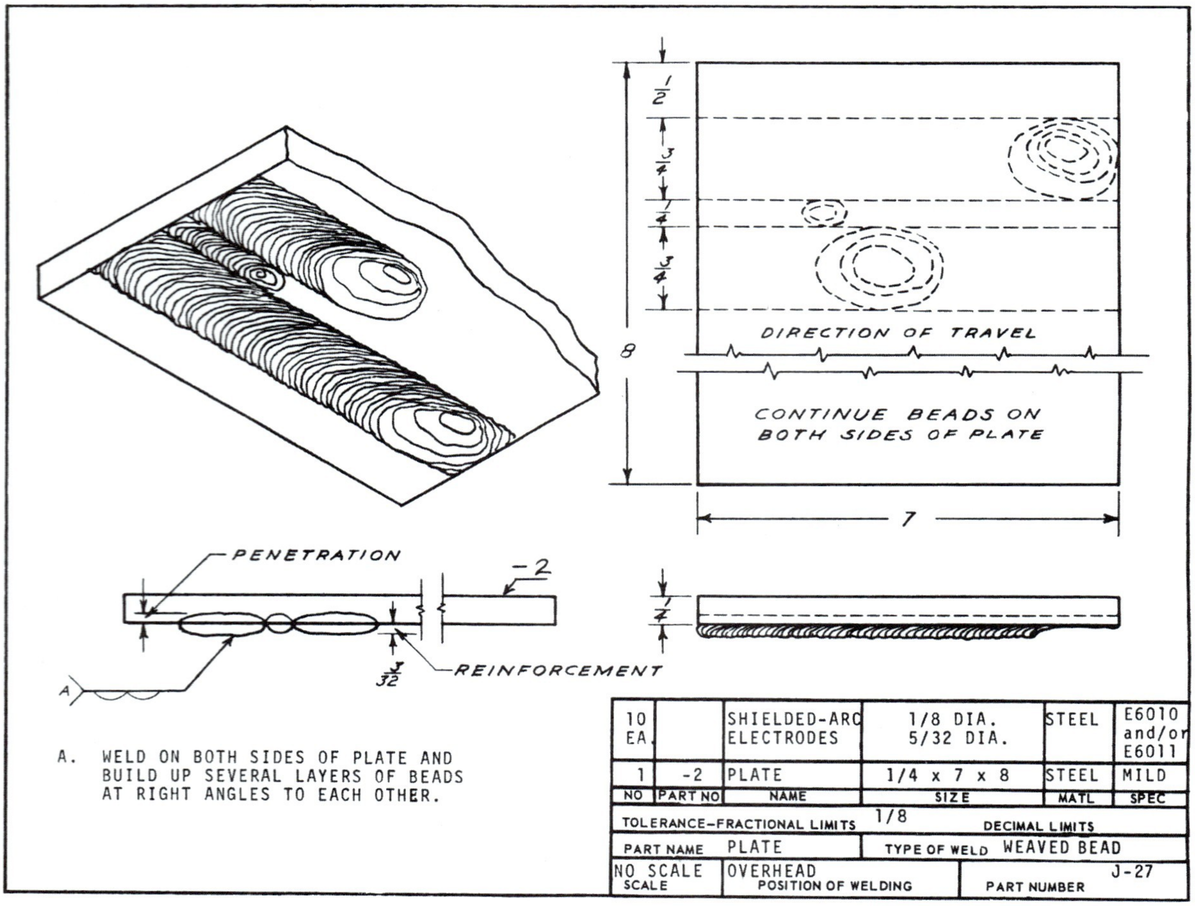

Fig. 14-10 Job drawing J27.

6. Fasten the plate in the overhead position. Use an overhead welding jig.
7. Set the power source at 75 to 125 amperes. Set a d.c. power source for electrode positive.
8. Make weave beads between the ¾-inch parallel lines. Manipulate the electrode as instructed in Fig. 14-8, page 391.
9. Chip the slag from the beads, brush, and inspect. Refer to Inspection.
10. Make stringer beads between the weave beads already deposited. Current should be set a little higher (110- to 170-ampere range). Manipulate the electrode as in Figs. 14-2 through 14-4, pages 389–390.
11. Chip the slag from the beads, brush, and inspect. Refer to Inspection.
12. Practice these beads until you can produce good beads consistently with both types of electrodes and with 5⁄32-inch diameter electrodes.

Inspection

Compare the beads with Fig. 14-11 and check them for the following weld characteristics:

Width and height: Uniform
Appearance: Smooth with close ripples; free of voids and high spots. Restarts should be difficult to locate.
Size: Refer to the job drawing, Fig. 14-10.
Face of beads: Slightly convex
Edges of beads: Good fusion, no overlap, no undercut
Starts and stops: Free of depressions and high spots
Beginnings and endings: Full size, craters filled
Penetration and fusion: To plate surfaces and adjacent beads
Surrounding plate surfaces: Free of spatter
Slag formation: Full coverage, easily removable

Disposal

Discard completed plates in the waste bin. Plates must be filled on both sides.

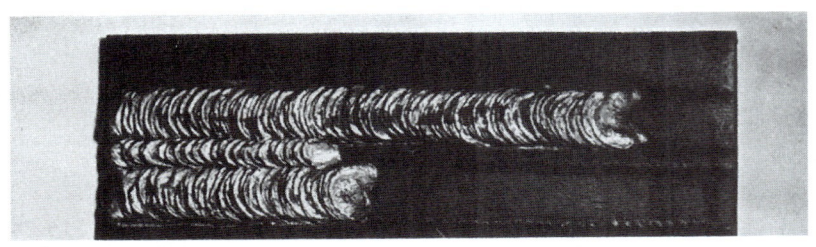

Fig. 14-11 Typical appearance of weave beads welded in the overhead (4) position with DCEP or a.c. shielded metal arc coated electrodes. *Edward R. Bohnart*

Job 14-J28 Welding a Single-V Butt Joint (Backing Bar Construction)

Objective

To weld a single-V butt joint assembled with a backing bar in the flat position by means of multipass groove welds with stringer and weave bead technique with DCEP and/or a.c. shielded metal arc electrodes (AWS E6010–E6011).

General Job Information

This form of joint construction is used quite frequently in piping, pressure vessel, and shipyard construction. The backing material creates the effect of welding a joint with a chill ring. The single-V butt joint with backing bar is identical to the test specified by code authorities such as the Bureau of Shipping, Department of the Navy, and by the American Welding Society. The root opening varies with different welding shops. If it is as large as ⅜ inch, two or three root passes are needed to weld the joint. For economic reasons, it should be kept as narrow as possible but yet attain proper weld quality.

Welding Technique

First Pass Current adjustment can be high. Hold the electrode at a 90° work angle to the joint, Fig. 14-12A, page 394, and at a 10 to 20° drag travel angle, Fig. 14-12B. Use the familiar technique for running stringer beads with a very slight weave. Secure good fusion with the backing bar and the root and groove faces of the bevel plates. Keep the face of the weld as flat as possible.

Second and Third Passes Current adjustment can be high. Electrode position should be approximately that shown in Fig. 14-12. Weave wider beads than for the first pass. Do not advance the electrode rapidly. This will cause coarse ripples and even voids. Keep the side-to-side motion within the limits of the finished bead width to avoid undercutting. Close arc gap control is important. Remove the slag thoroughly so that good fusion may be obtained with preceding passes and with the groove faces of the beveled plate. Use the edges of the plate as a guide for determining the width of the last pass.

Practice starts and stops. Travel in all directions.

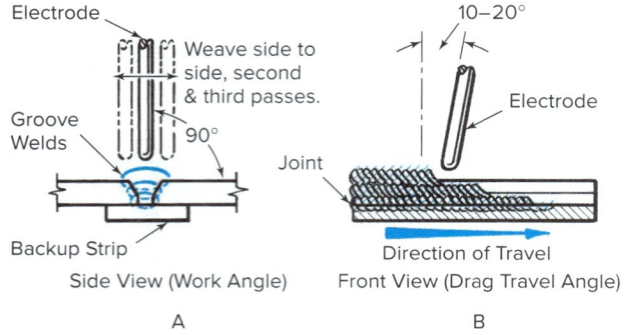

Fig. 14-12 Electrode position when welding a single V-groove butt joint in the flat (1G) position.

Operations

1. Obtain plates; check the job drawing, Fig. 14-13, for quantity and size.
2. Obtain electrodes of each quantity, type, and size specified in the job drawing.
3. Set a d.c. power source for electrode positive at 110 to 170 amperes or an a.c. power source at 115 to 190 amperes.
4. Set up the plates and tack weld them as shown on the job drawing and Fig. 14-14. Make sure the backing bar is tight against the back side of the plate.
5. Place the joint in the flat position on the welding table. Make sure it is connected to the work connection.
6. Make the first pass with a 5/32-inch electrode. Manipulate the electrode as instructed in Fig. 14-12.

> **JOB TIP**
>
> **Centerline**
> Success in business is all about innovation. Take, for example, Centerline, a designer and maker of equipment for resistance welding. It began in Ontario in 1957, founded by Don Beneteau and Fred Wigle. No longer a two-person shop, it now has four manufacturing facilities and employs 350 people worldwide. Its success is due to foreseeing what industry will be needing.

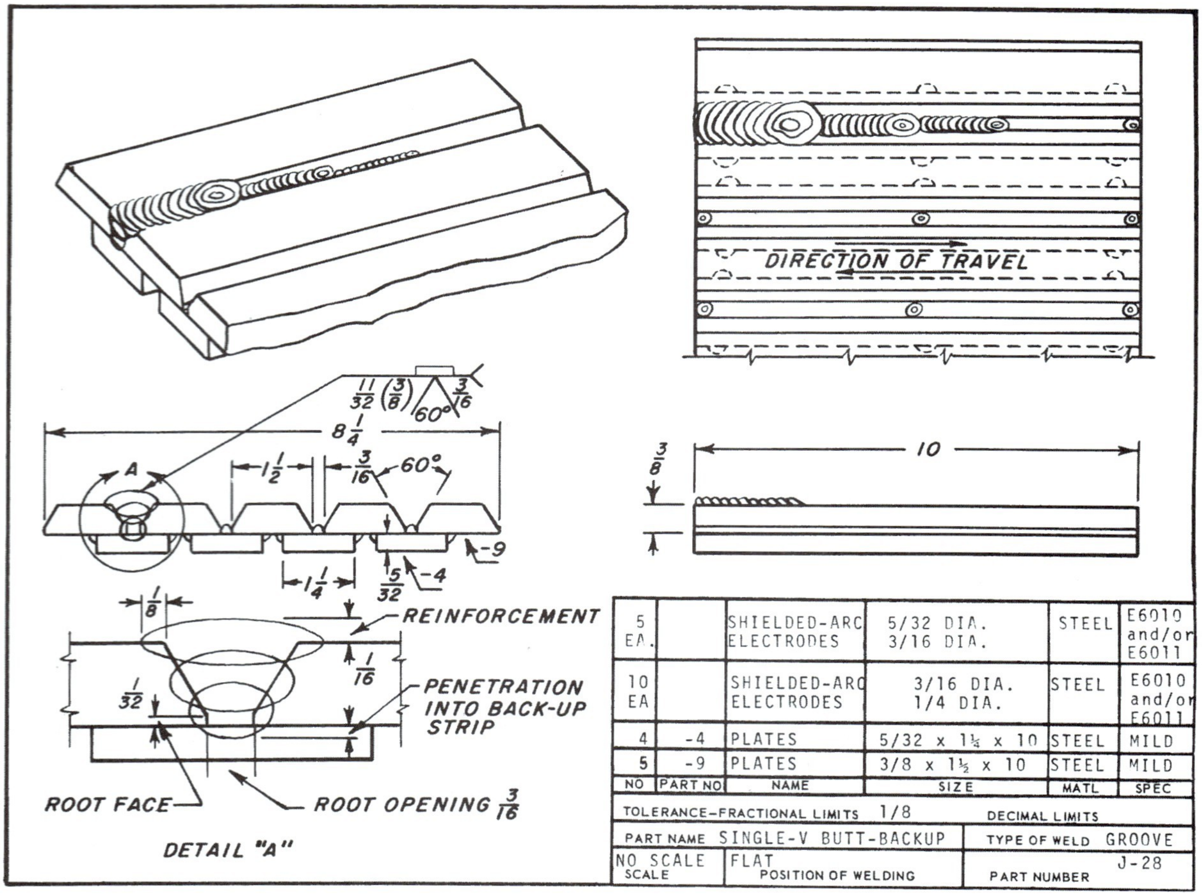

Fig. 14-13 Job drawing J28.

394 Chapter 14 Shielded Metal Arc Welding Practice: Jobs 14-J26–J42 (Plate)

Fig. 14-14 Backing bar applied to the back side of a V-groove butt joint. *Edward R. Bohnart*

7. Chip the slag from the welds, brush, and inspect. Refer to Inspection.
8. Readjust the power source for 140 to 215 amperes.
9. Make the second and third passes with a ³⁄₁₆-inch electrode.
10. Chip the slag from the welds, brush, and inspect.
11. Practice these welds until you can produce good welds consistently with both types of electrodes and with ³⁄₁₆-inch electrodes for the first pass and ¼-inch electrodes for the others.

Inspection

Compare the weld with Figs. 14-14 and 14-15 and check each pass for the following weld characteristics:

Width and height: Uniform
Appearance: Smooth with close ripples; free of voids and high spots. Restarts should be difficult to locate.
Size: Refer to the job drawing, Fig. 14-13; check with butt weld gauge.
Face of welds: First two passes flat; last pass slightly convex
Edges of welds: Good fusion, no overlap, no undercut

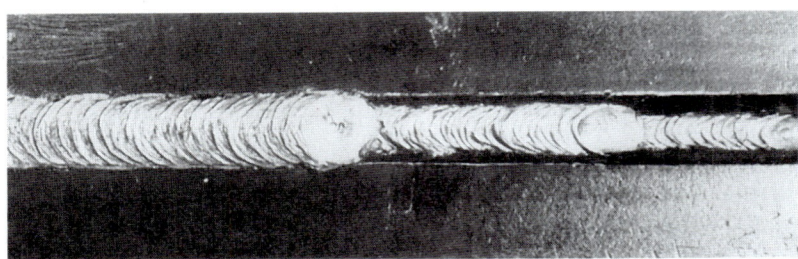

Fig. 14-15 Joint construction and appearance of a typical multipass weld in a V-groove butt joint with a backing bar welded in the flat (1G) position with DCEP or a.c. shielded metal arc coated electrodes. *Edward R. Bohnart*

Starts and stops: Free of depressions and high spots
Beginnings and endings: Full size, craters filled
Penetration and fusion: To the backing bar, preceding passes, and plate surfaces
Surrounding plate surfaces: Free of spatter
Slag formation: Full coverage, easily removable

Disposal

Put completed joints in the scrap bin so that they will be available for further use. The plate can be cut and beveled between welds for further butt joints, and unwelded plate surfaces can be used for beading.

Qualifying Test

At this point you should be able to pass Test 1 in the flat position. (See Chapter 15, p. 455.) This is typical of the test recommended by the American Welding Society.

It will also be desirable for you to pass Test 3 (fillet weld) in the flat position. (See Chapter 15, p. 457.)

Job 14-J29   Welding a T-Joint

Objective

To weld a T-joint in the horizontal (2F) position by means of multipass fillet welds, weave bead technique, with DCEP and/or a.c. shielded metal arc electrodes (AWS E6010–E6011).

General Job Information

It is often the practice to make large fillet welds in the horizontal position with few passes and the weave bead technique. This is especially true in shipyard and construction work. This technique can deposit more weld metal in a single pass than other methods.

Welding Technique

Current adjustment should be high. Hold the electrode at a 40° work angle to the flat plate, Fig. 14-16A, page 396, and at a 30° drag travel angle, Fig. 14-16B. The first pass should be run in the same manner as any other single-pass fillet weld. Keep the face of the weld as flat as possible. Observe the formation of the weld pool; if it has a notch as shown in Fig. 14-17, page 396, penetration is not being achieved. Use proper electrode manipulation, amperage,

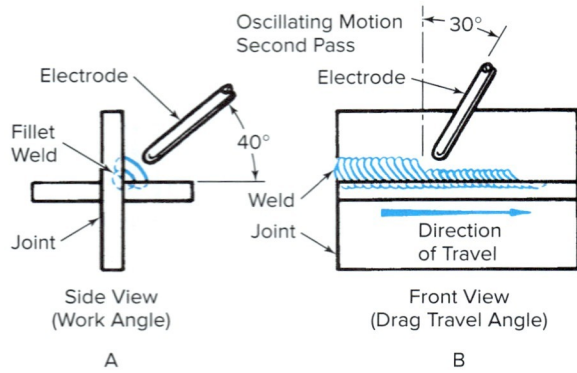

Fig. 14-16 Electrode position when welding a T-joint in the horizontal (2F) position.

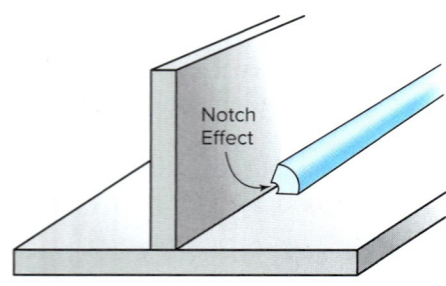

Fig. 14-17 Notch effect, which indicates penetration (fusion) to the root of the joint is not being achieved.

and travel speed to ensure penetration to the root of the joint. If the weld pool displays a completely formed radius on the leading edge, penetration is being achieved. Make sure there is no undercut along the toes of the welds, especially on the vertical joint member. Check for overlap along the toes of the weld, especially on the horizontal joint member.

The second pass is weaved over the first pass. Direct the electrode toward the vertical plate so that the weld metal is forced up on its surface. To prevent undercut, keep the electrode within the limits of the desired weld width. Use a very short arc and pause at the top of the bead against the vertical plate. Then bring the arc toward the bottom plate at a normal rate of travel and at a 30° angle. Return the arc to the top of the fillet weld with a semicircular, counterclockwise motion. Secure good fusion with the preceding pass and the plate surfaces.

Practice starts and stops. Travel in all directions.

Operations

1. Obtain plates; check the job drawing, Fig. 14-18, for quantity and size.
2. Obtain electrodes of the type and size specified in the job drawing.
3. Set a d.c. power source for electrode positive at 140 to 215 amperes or an a.c. power source at 140 to 225 amperes.
4. Set up the plates as shown on the job drawing and tack weld them at each corner.
5. Place the joint in the horizontal position on the welding table. Make sure that it is connected to the work connection.
6. Make the first pass as shown on the job drawing. Manipulate the electrode as instructed in Fig. 14-16.
7. Chip the slag from the weld, brush, and inspect. Refer to Inspection.
8. Make the second pass as shown on the job drawing. Be careful with your current setting and electrode movement.
9. Chip the slag from the weld, brush, and inspect. Refer to Inspection.
10. Practice these welds until you can produce good welds consistently with both types of electrodes.

Inspection

Compare each pass with Fig. 14-19 and check it for the following weld characteristics:

Width and height: Uniform
Appearance: Smooth with close ripples, free of voids. Restarts should be difficult to locate.
Size: Refer to the job drawing, Fig. 14-18. Check with a fillet weld gauge.
Face of weld: Flat to slightly convex
Edges of weld: Good fusion, no overlap, no undercut
Starts and stops: Free of depressions and high spots
Beginnings and endings: Full size, craters filled
Penetration and fusion: To the root of the joint and plate surfaces
Surrounding plate surfaces: Free of spatter
Slag formation: Full coverage, easily removable

Disposal

Put completed joints in the scrap bin so that they will be available for further use. Square edges can be butted together for square butt joints, and unwelded surfaces can be used for beading.

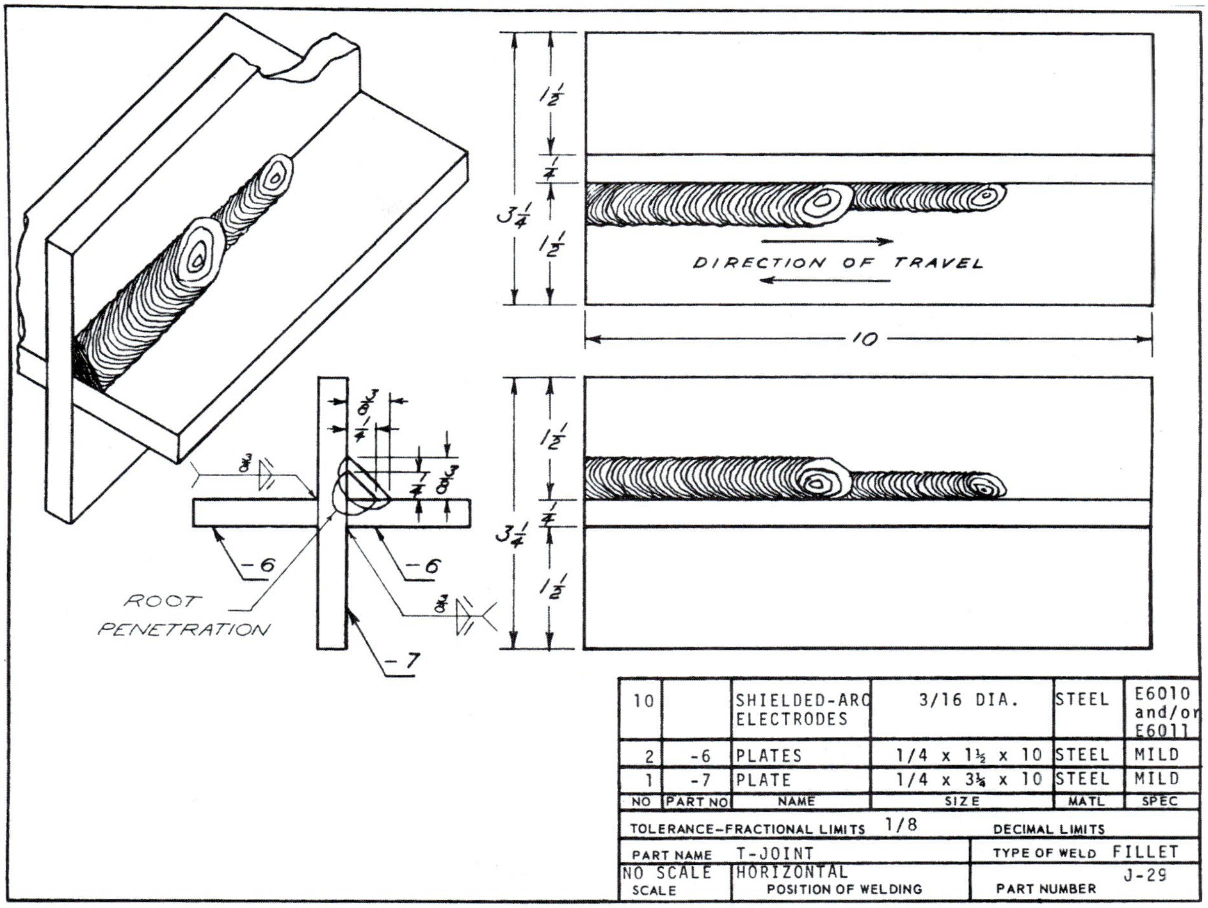

Fig. 14-18 Job drawing J29.

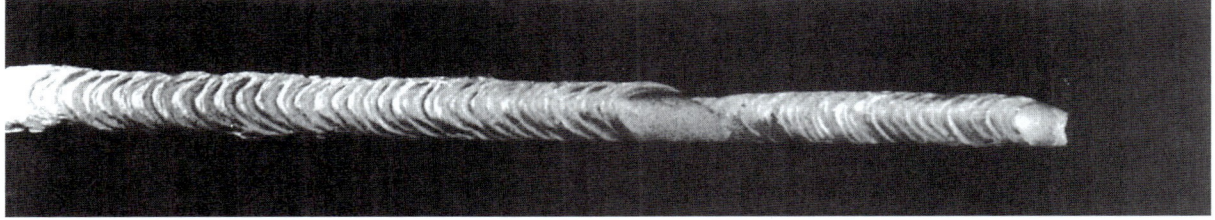

Fig. 14-19 Typical appearance of multipass fillet welds in a T-joint welded in the horizontal (2F) position, modified weave technique, with DCEP or a.c. shielded metal arc coated electrodes. Edward R. Bohnart

Job 14-J30 Welding a Single-V Butt Joint (Backing Bar Construction)

Objective

To weld a single-V groove butt joint (backing bar construction) in the horizontal (2G) position by means of multipass groove welds, stringer bead technique, with DCEP and/or a.c. shielded metal arc coated electrodes (AWS E6010–E6011).

General Job Information

This joint design and position will be encountered in pipe, tank, and shipyard welding. The single-V groove butt joint is identical to the joint that is required by various code authorities such as the Bureau of Shipping, Department of the Navy, and recommended by the American Welding Society. The root opening varies with different welding shops. If it is as large as ⅜ inch, two or three root passes are needed to weld the joint.

Shielded Metal Arc Welding Practice: Jobs 14-J26–J42 (Plate) **Chapter 14** 397

For economic reasons, keep the root opening as narrow as possible.

Welding Technique

First Pass Current adjustment must be high. Hold the electrode at a 90° work angle to the joint, Fig. 14-20A. Use the stringer bead technique. Make sure that you are obtaining fusion with the backing bar and root faces of the plates.

All Other Passes Current adjustment can be higher than for the first pass. Hold the electrode at a 30° work angle to the lower plate for the second pass (Fig. 14-20B) and a 45° angle for the third pass (Fig. 14-20C). Stringer bead technique may be used. It is important that all passes be fused to the preceding passes and plate surfaces. Avoid sagging and undercut.

Four cover passes are needed to fill up the groove completely. They must be at least flush with the plate and should have a slight reinforcement. Hold the electrode at an angle of 5 to 15° from the horizontal for the fourth, fifth, sixth, and seventh passes, Fig. 14-20D. Make sure that the fourth and seventh passes do not extend more than 1/16 inch beyond the edge of the beveled plate. In this way, the edge of the beveled plate is your guide for the final width of the cover passes.

For all passes, tilt the electrode 5 to 15° travel angle or with a drag.

Practice starts and stops. Travel from left to right and from right to left. Very thorough cleaning between passes is essential.

Operations

1. Obtain plates; check the job drawing, Fig. 14-21, for quantity and size.
2. Obtain electrodes of each quantity, type, and size specified in the job drawing.
3. Set a d.c. power source for electrode positive at 110 to 170 amperes or an a.c. power source at 110 to 190 amperes.
4. Set up the plates and tack weld them as shown on the job drawing.
5. Place the joint in the horizontal position on the welding table. Make sure that it is connected with the work connection.
6. Make the first, second, and third passes with 5/32-inch electrodes as shown on the job drawing. Manipulate the electrode as instructed in Fig. 14-20A, B, and C.
7. Chip the slag from the welds, brush, and inspect between each pass. Refer to Inspection.
8. Increase the current setting to 140 to 215 amperes.
9. Make the fourth, fifth, sixth, and seventh passes with 3/16-inch electrodes as shown on the job drawing. Manipulate the electrode as instructed in Fig. 14-20D and E.
10. Chip the slag from the welds, brush, and inspect between each pass. Refer to Inspection, below.
11. Practice these welds until you can produce good welds consistently with both types of electrodes. Practice with 3/16-inch electrodes for all passes.

Inspection

Compare each pass with Fig. 14-22 and check it for the following weld characteristics:

Width and height: Uniform

Appearance: Smooth with close ripples; free of voids and high spots. The lap over of the beads should be well proportioned. Restarts should be difficult to locate.

Size of weld: Refer to the job drawing, Fig. 14-21. Check convexity with a butt weld gauge.

Face of weld: Some reinforcement

Edges of weld: Good fusion, no overlap, no undercut

Starts and stops: Free of depressions and high spots

Beginnings and endings: Full size, craters filled

Penetration and fusion: To the backing bar, preceding passes, and plate surfaces

Surrounding plate surfaces: Free of spatter

Slag formation: Full coverage, easily removable

Disposal

Put completed joints in the scrap bin so that they will be available for further use.

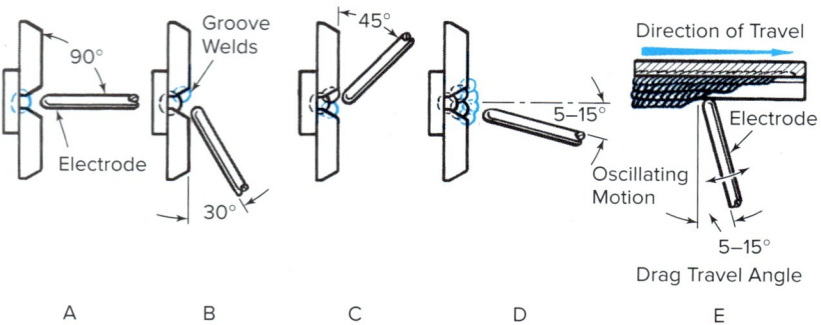

Fig. 14-20 Electrode position when welding a V-groove butt joint in the horizontal (2G) position.

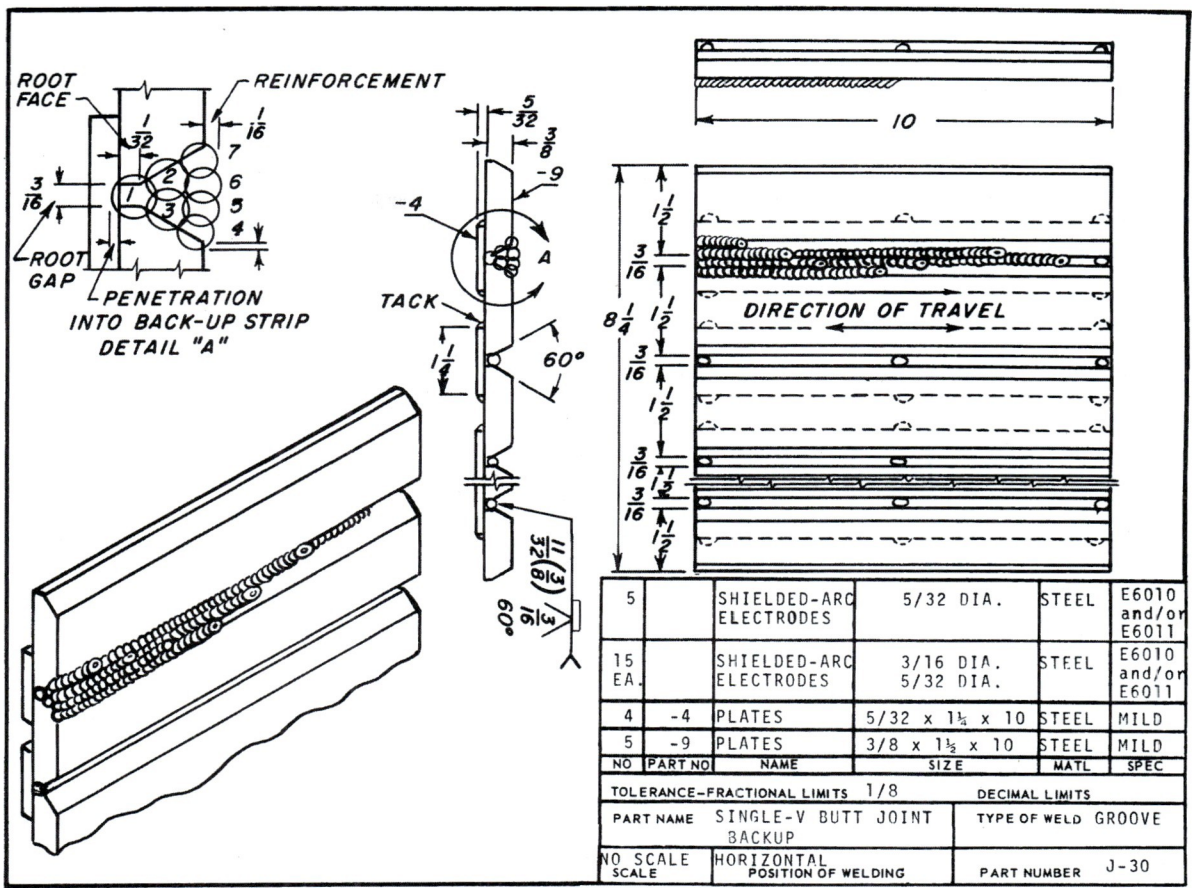

Fig. 14-21 Job drawing J30.

Fig. 14-22 Joint construction and typical appearance of a multipass single V-groove weld in a butt joint welded in the horizontal (2G) position with DCEP or a.c. shielded metal arc coated electrodes. *Edward R. Bohnart*

The plates can be cut and beveled between welds for further butt welding, and unwelded plate surfaces can be used for beading.

Qualifying Test

At this point, you should be able to pass Test 1 in the horizontal position. (See Chapter 15, p. 455.) This is the test recommended by the American Welding Society.

It will also be desirable for you to pass Test 3 (fillet weld) in the horizontal position. (See Chapter 15, p. 457.)

Job 14-J31 Welding a Square Butt Joint

Objective

To weld a square butt joint in the flat (1G) position from both sides of the plate with DCEP and/or a.c. shielded metal arc electrodes (AWS E6010–E6011).

General Job Information

The square butt joint is widely used in industry on ordinary work. When welded from both sides on thicknesses of metal not exceeding ¼ inch, it is a highly efficient joint. The usual practice, however, is to weld the joint from one side only. If this procedure is followed, the strength of the joint varies with the depth of penetration that, in turn, depends on the size of electrode, the amount of current, the amount of gap when setting up the plates, travel speed, and the thickness of plates. When welding from one side, complete penetration without gapping is doubtful on plates heavier than 3/16 inch. These are referred to as *open root* type weld procedures and are the most difficult to perform.

Welding Technique

The heat setting should be high. Hold the electrode at a 90° work angle to the plates, Fig. 14-23A, and tilt it 10 to 20° travel angle in the direction of travel, Fig. 14-23B. Move the electrode back and forth along the line of weld. This preheats the metal ahead of the weld and minimizes the tendency to burn through. It also forces the slag back over the top of the weld, thus lessening the danger of forming slag inclusions. It is important that the rate of travel and the arc gap be steady and uniform. If travel is too fast, undercut will result from the insufficient buildup of weld metals. If travel is too slow, the pool will become too hot and burn through. A long arc gap causes poor appearance, poor penetration, excess spatter, and weld metal with poor physical characteristics. The arc must not be held so short, however, that the slag touches the molten pool.

Practice starts and stops. Weld travel should be in all directions.

Operations

1. Obtain plates; check the job drawing, Fig. 14-24, for quantity and size.
2. Obtain electrodes of each quantity, type, and size specified in the job drawing.
3. Set a d.c. power source for electrode positive at 160 to 190 amperes or an a.c. power source at 160 to 210 amperes.
4. Set up the plates and tack weld them as shown on the job drawing and Fig. 14-25.
5. Place the joint in the flat position on the welding table. Make sure the plates are connected with the work connection.
6. Make the first pass as shown on the job drawing. Manipulate the electrode as shown in Fig. 14-23.
7. Chip the slag from the welds, brush, and inspect. Refer to Inspection.
8. Clean the back side of the first pass before welding the second pass.
9. Make the second pass as shown on the job drawing. Current should be higher.
10. Chip the slag from the welds, brush, and inspect. Refer to Inspection.
11. Practice these welds until you can produce good welds consistently with both types of electrodes.

Inspection

Compare the welds with Fig. 14-25 and check them for the following characteristics:

Width and height: Uniform
Appearance: Smooth with close ripples; free of voids and high spots. Restarts should be difficult to locate.
Size: Refer to the job drawing, Fig. 14-24. Check with a butt weld gauge.
Face of welds: Slightly convex
Edges of welds: Good fusion, no overlap, no undercut
Starts and stops: Free of depressions and high spots
Beginnings and endings: Full size, craters filled
Penetration and fusion: To the root of the weld and plate surfaces
Surrounding plate surfaces: Free of spatter
Slag formation: Full coverage, easily removable

Disposal

Put completed plates in the scrap bin. They can be used for beading practice.

Job 14-J32 Welding an Outside Corner Joint

Objective

To weld an outside corner joint in the flat position by means of multipass groove welds, weave bead technique, with DCEP and/or a.c. shielded metal arc electrodes (AWS E6010–E6011).

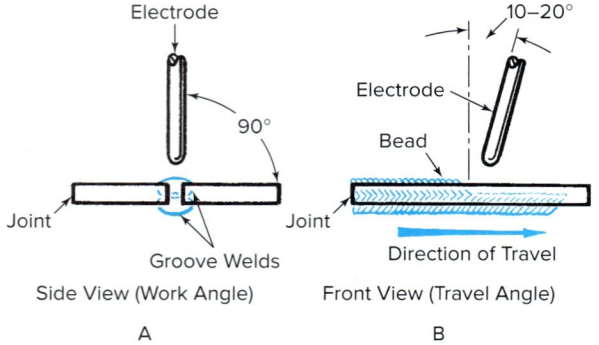

Fig. 14-23 Electrode position when welding a square groove butt joint in the flat (1G) position.

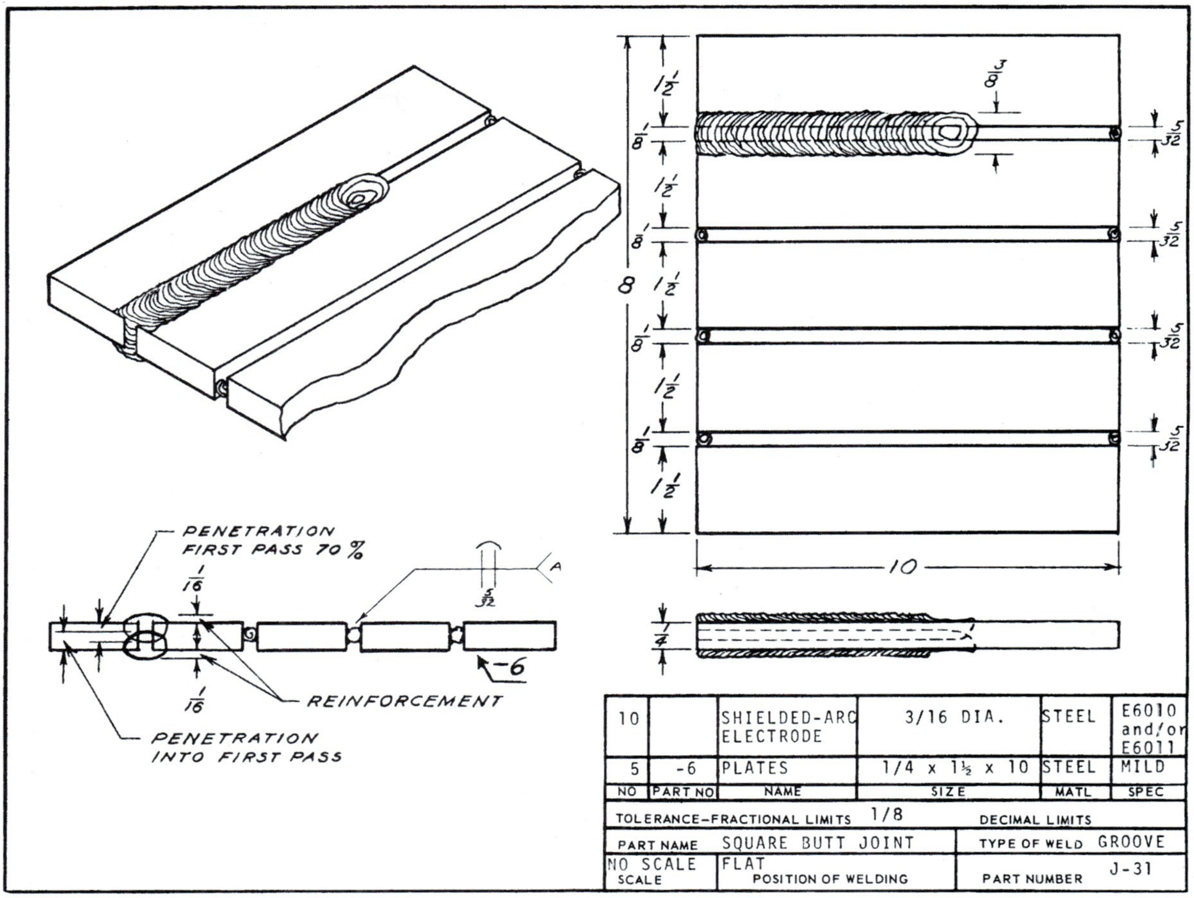

Fig. 14-24 Job drawing J31.

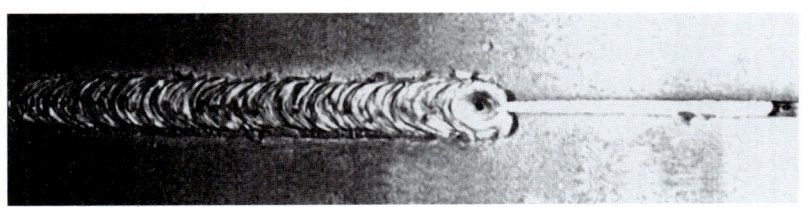

Fig. 14-25 Joint construction and typical (4G) appearance of a square groove weld in a butt joint welded in the flat (1G) position with DCEP or a.c. shielded metal arc coated electrodes. *Edward R. Bohnart*

General Job Information

The outside corner joint is not welded as frequently as butt, lap, and T-joints. It is a good joint to practice on since it can be prepared quickly. Welding conditions are similar to those for V-groove butt joints. To develop maximum strength, full penetration must be secured through the back side, as well as a full radius contour. The addition of a fillet weld in the inside corner adds greatly to efficiency of the corner joint. Joint preparation is inexpensive, but electrode costs are high for heavy plate.

Welding Technique

First Pass Current adjustment should not be too high. Hold the electrode at a 90° work angle to the line of weld, Fig. 14-26A, page 402, and tilt it 10 to 20° travel angle in the direction of welding, Fig. 14-26B. Use the stringer bead welding technique to deposit the first pass in the bottom of the groove. This weld must penetrate through to the back side and have good fusion to both plates, Fig. 14-27, page 402. A close arc is essential. The correct combination of welding current, electrode position, and speed of travel will produce penetration through to the back side. To achieve complete penetration, form a little hole at the leading edge of the weld pool right under the tip of the electrode. Learn how to keep this hole throughout the entire welding operation. Do not let it get so large that you lose control and burn through the joint. The presence of this hole during welding is your assurance that you are melting through to the back side of the groove. This technique is referred to as **keyhole welding.**

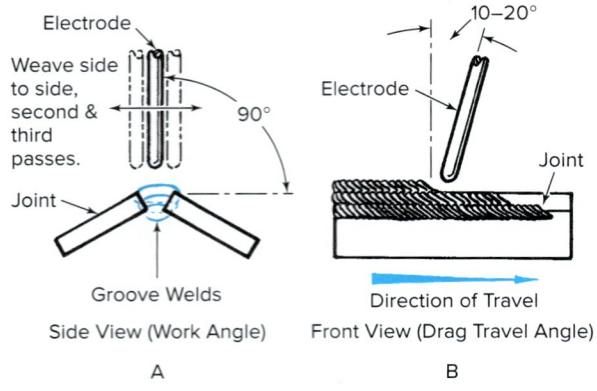

Fig. 14-26 Electrode position when welding a single V-groove on an outside corner joint in the flat (1G) position.

Operations

1. Obtain plates; check the job drawing, Fig. 14-28, for the correct quantity and size.
2. Obtain electrodes of each quantity, type, and size specified in the job drawing.
3. Set the power source for 75 to 125 amperes for the first pass. Set a d.c. power source for electrode positive.
4. Set up the plates and tack weld them as shown on the job drawing and in Fig. 14-29.
5. Place the joint in the flat position on the welding table. Make sure it is connected with the work connection.
6. Make the first pass with a ⅛-inch electrode as shown on the job drawing. Manipulate the electrode as instructed in Fig. 14-26.
7. Chip the slag from the welds, brush, and inspect. Refer to Inspection.
8. Readjust the power source for 110 to 170 amperes.
9. Make the second and third passes with a 5/32-inch electrode as shown on the job drawing. Manipulate the electrode as instructed in Fig. 14-26.
10. Chip the slag from the welds, brush, and inspect. Refer to Inspection.
11. Practice these welds until you can produce good welds consistently with both types of electrodes. Also practice with 5/32-inch diameter electrodes for the first pass, and 3/16-inch diameter electrodes for the other passes.

If you do burn through the joint, move the electrode in a whiplike motion to control the size of the weld pool and the rate at which weld metal is added. If the pool is too hot and burning through, whip ahead along one of the plate edges about ½ inch and whip back into the pool to deposit more metal. Alternate this whiplike motion along opposite plate edges.

If the bottom edges of the groove do not seem to be heating and melting enough and if you are having difficulty maintaining the hole, keep the electrode directly in the bottom of the V-groove, slow the forward movement, and maintain a circular motion.

Second and Third Passes Current adjustment can be hot. Electrode position should be approximately that shown in Fig. 14-26. The weave bead technique may be used. The third pass requires a wider weave than the second. Do not weave past the edges of the plate and hesitate at the sides to eliminate burning away of the plate edges. Make certain that you are securing good fusion with the preceding passes and with both plate surfaces. Remove the slag thoroughly between each pass. Practice starts and stops. Travel in all directions.

Inspection

Compare the back side of the first bead with Fig. 14-27 and all passes with Fig. 14-29. Check them for the following weld characteristics:

Width and height: Uniform
Appearance: Smooth with close ripples; free of voids and high spots. Restarts should be difficult to locate.
Size: Refer to the job drawing, Fig. 14-28.
Face of welds: First two passes flat; last pass convex

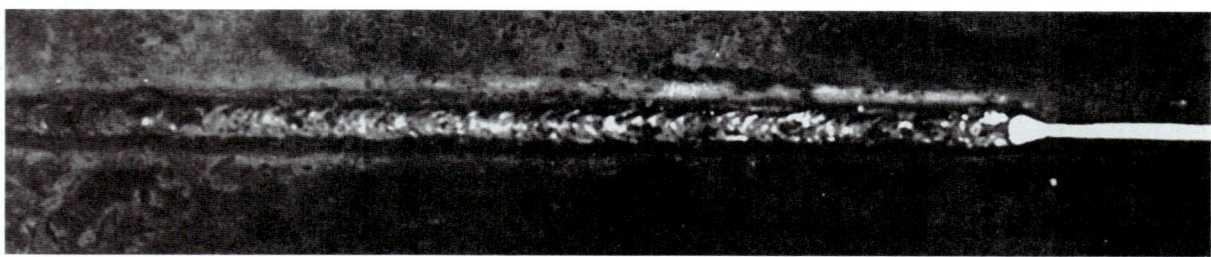

Fig. 14-27 Penetration required through the back side of a corner joint. Note the penetration hole, *keyhole,* at the leading edge of the weld bead. Edward R. Bohnart

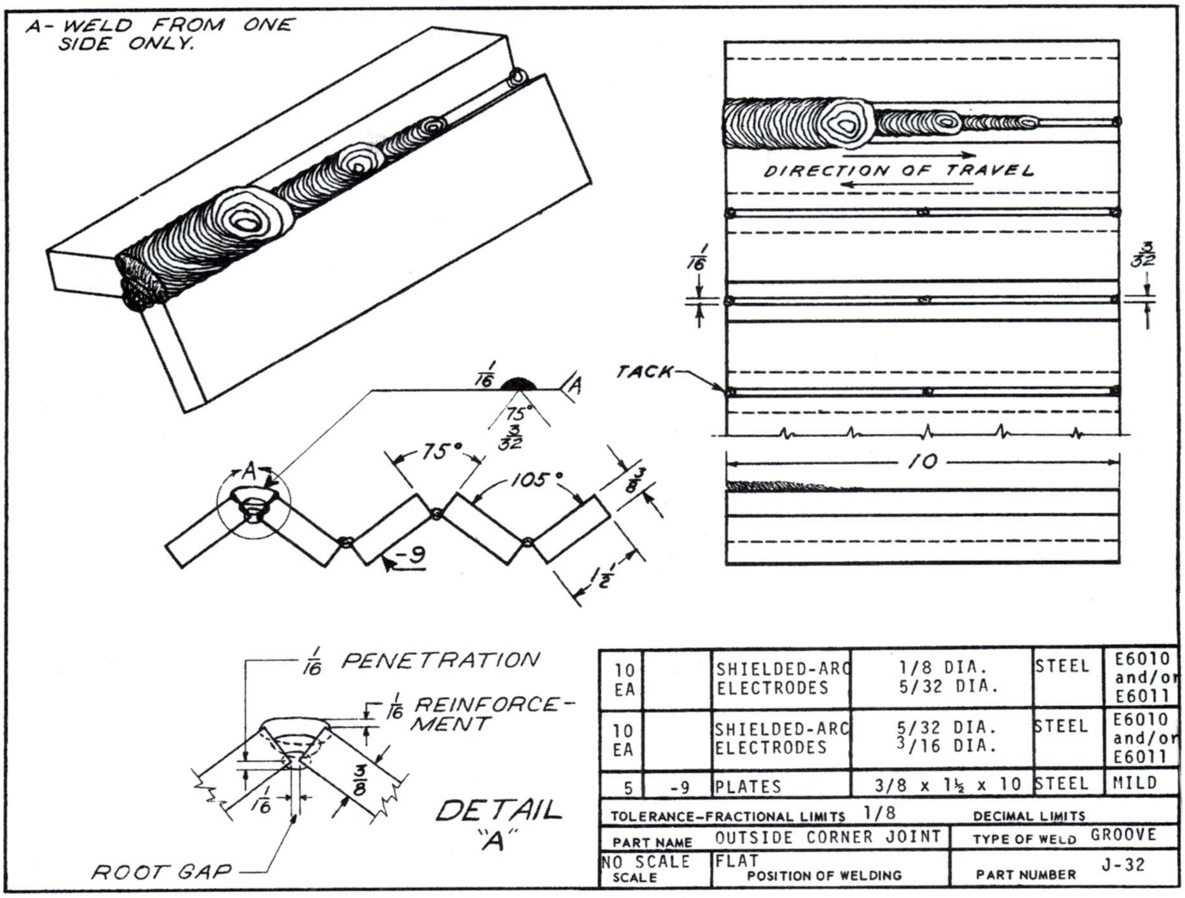

Fig. 14-28 Job drawing J32.

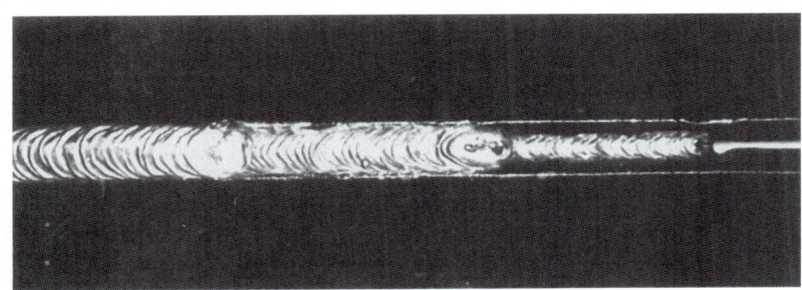

Fig. 14-29 Joint construction and typical appearance of a multipass single V-groove weld in an outside corner joint welded in the flat (1G) position with DCEP or a.c. shielded metal arc coated electrodes. *Edward R. Bohnart*

Edges of welds: Good fusion, no overlap, no undercut; no excess metal at edges of joint
Starts and stops: Free of depressions and high spots
Beginnings and endings: Full size, craters filled
Back side: Complete and uniform penetration (See Fig. 14-27.)
Penetration and fusion: Through the back side, preceding passes, and plate surfaces

Surrounding plate surfaces: Free of spatter
Slag formation: Full coverage, easily removable

Disposal

Put completed joints in the scrap bin so that they will be available for further use. Inside corners can be used for practicing fillet welds, and unwelded plate surfaces can be used for beading.

Job 14-J33 Welding a Single-V Butt Joint

Objective

To weld a single-V butt joint in the flat (1G) position by the means of multipass groove welds, weave bead technique, with DCEP and/or a.c. shielded metal arc electrodes (AWS E6010–E6011).

General Job Information

The single-V butt joint is used in pipe welding and critical plate welding. The joint structure and welding procedure are similar to those in the ASME piping and plate tests. This is an *open root* type procedure.

Welding Technique

First Pass Current adjustment must not be too high. Hold the electrode at a 90° work angle to the plates, Fig. 14-30A, and tilt it 10 to 20° drag travel angle, Fig. 14-30B. Hesitate at the sides of the weld to prevent undercut. Hold a close arc. Use a back-and-forth motion and keep the electrode within the space provided by the root gap. The arc must not be broken on the forward motion. Do not let drops of metal be deposited ahead of the weld because they may interfere with the progress of the weld. You will recall the need to keep a small hole at the leading edge of the weld pool. This technique is referred to as keyhole welding. A small bead must be formed on the reverse side of the groove, Fig. 14-31. The face of the weld must be kept as flat as possible.

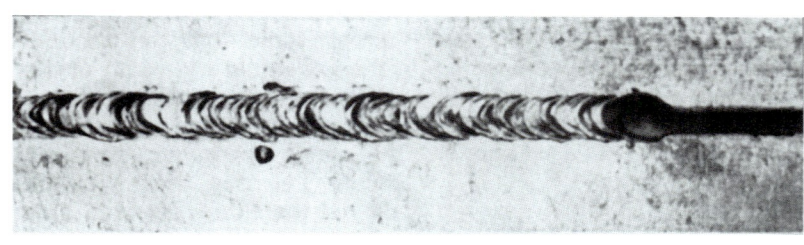

Fig. 14-31 The back side (root) of a single V-groove butt joint showing penetration and fusion. Note the penetration hole, *keyhole*, at the leading edge of the weld. Edward R. Bohnart

Second Pass Use the weave bead technique. A uniform weave and rate of travel is important to obtain a weld of uniform quality and appearance. Avoid undercutting. Obtain fusion with the underneath pass and the groove face of the plate. The face of the weld must be flat. Remove the slag thoroughly.

Third Pass Use the same type of weave as for the second pass, but widen the path of the electrode. Use the edges of the beveled plate as a guide for bead width. The finished weld should be slightly convex. The proper V-groove weld profile is shown in Fig. 14-32.

Practice starts and stops on all passes. Travel in all directions.

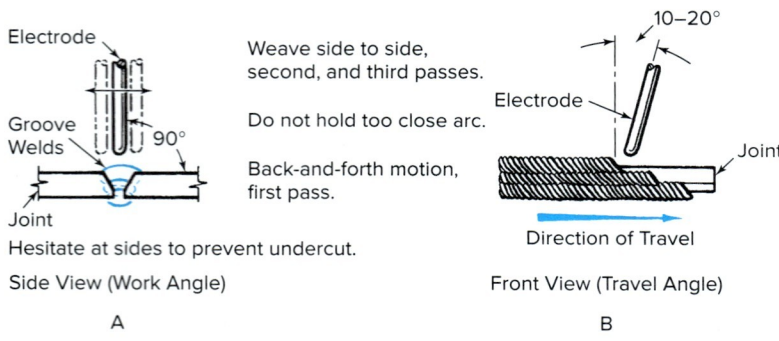

Fig. 14-30 Electrode position when welding single V-groove butt joint in the flat (1G) position.

Operations

1. Obtain plates; check the job drawing, Fig. 14-33, for quantity and size.
2. Obtain electrodes of each quantity, type, and size specified in the job drawing.
3. Set the power source at 75 to 125 amperes. Set a d.c. power source for electrode positive.
4. Set up the plates and tack weld them as shown on the job drawing and Fig. 14-34.

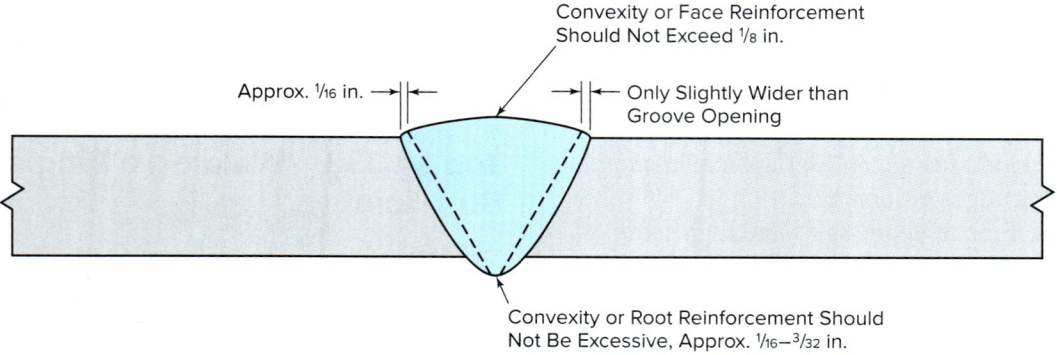

Fig. 14-32 A proper V-groove weld profile. Do not overweld the joint. Just fill the groove as noted.

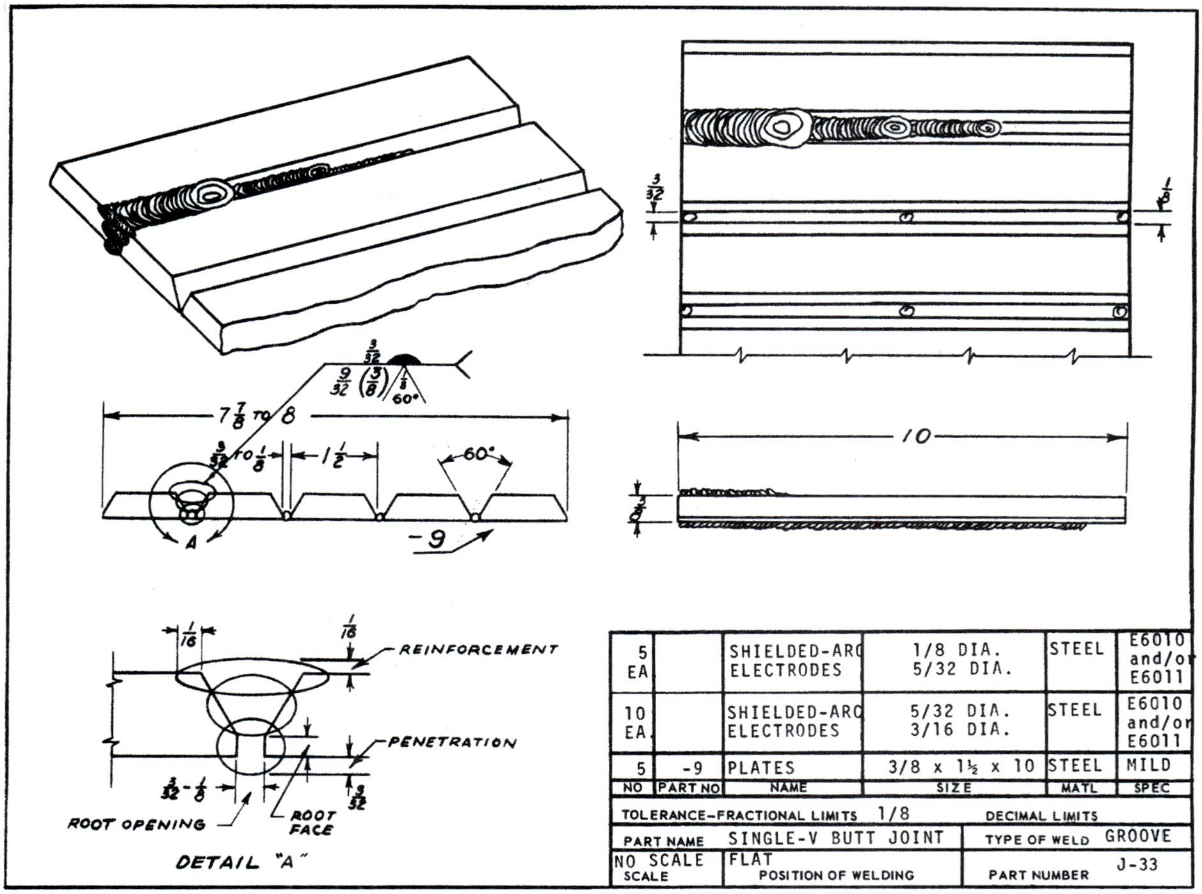

Fig. 14-33 Job drawing J33.

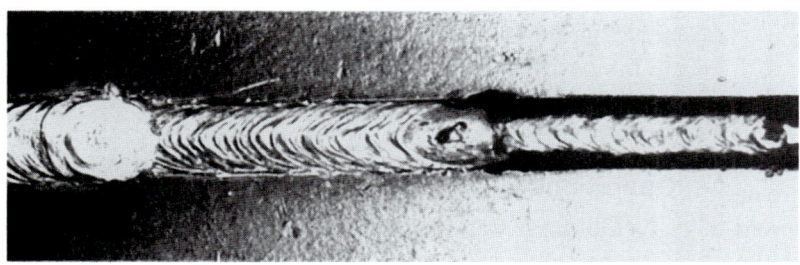

Fig. 14-34 Joint construction and typical appearance of a multipass single V-groove weld in a butt joint welded in the flat position with DCEP or a.c. shielded metal arc electrodes. *Edward R. Bohnart*

5. Place the joint in the flat position on the welding table. Make sure it is connected with the work connection.
6. Make the first pass with ⅛-inch diameter electrodes as shown on the job drawing. Manipulate the electrode as shown in Fig. 14-30.
7. Chip the slag from the weld, brush, and inspect. Refer to Inspection.
8. Increase the power supply setting to 110 to 170 amperes.
9. Make the second and third passes with a 5/32-inch electrode as shown on the job drawing.
10. Chip the slag from the welds, brush, and inspect. Refer to Inspection.
11. Practice these welds until you can produce good welds consistently with both types of electrodes. You are also asked to practice this job using 5/32-inch diameter electrodes for the first pass and 3/16-inch diameter electrodes for the second and third passes.

Inspection

Compare the back side of the first bead with Fig. 14-31 and all passes with Fig. 14-34. Check them for the following weld characteristics:

Width and height: Uniform
Appearance: Smooth with close ripples; free of voids and high spots. Restarts should be difficult to locate.

Size: Refer to the job drawing, Fig. 14-33, page 405; check with a butt weld gauge.
Face of welds: First two passes flat; last pass slightly convex
Edges of welds: Good fusion, no overlap, no undercut
Starts and stops: Free of depressions and high spots
Beginnings and endings: Full size, craters filled
Penetration and fusion: Through the back side and all plate surfaces, Fig. 14-31, page 404
Surrounding plate surfaces: Free of spatter
Slag formation: Full coverage, easily removable

Disposal

Return completed joints to the scrap bin so that they will be available for further use. The plate can be beveled between welds for further butt welding, and unwelded plate surfaces can be used for beading.

Qualifying Test

At this point, you should be able to pass Test 2 in the flat position. (See Chapter 15, p. 456.) This test is similar to the ASME piping and plate tests.

Job 14-J34   Welding a T-Joint

Objective

To weld a T-joint in the horizontal (2F) position by means of multipass fillet welds, stringer bead technique, with low hydrogen, shielded metal arc electrodes (AWS E7016).

General Job Information

All low hydrogen electrodes produce deposits that are practically free of hydrogen. This reduces underbead and microcracking on low alloy steels and thick weldments. Underbead cracks usually occur in base metal just below the weld metal and are caused by hydrogen absorption from the arc atmosphere. Low hydrogen electrodes produce sound welds on troublesome steels such as the high sulfur, high carbon, and low alloy grades. They are recommended for welds that are to be porcelain enameled.

The E7016 electrode produces welds of highest quality and may be used for practically all code work. The welds are high in tensile strength and ductility. Their excellent physical qualities reduce the tendency for cracking at the root of the weld, which is caused by shrinkage stresses. This in turn reduces the need for preheat and stress relief requirements so that the job is done quickly and at lower cost.

Welding Technique

Use alternating current or direct current (electrode positive). Some manufacturers recommend DCEP whenever possible with electrodes having diameters of $5/32$ inch and smaller and a.c. with larger sizes. Although the E7016 is an all-position electrode, sizes larger than $1/8$ inch or $5/32$ inch are difficult to use for vertical and overhead welding.

The weld metal freezes rapidly even though the slag stays relatively fluid. The arc is quiet with medium penetration and little spatter. The slag is moderately heavy: It produces good weld metal protection, excellent bead appearance, and easy cleaning. The beads are slightly convex and have distinct ripples.

The currents used with these electrodes generally are higher than those recommended for the cellulose electrodes of the same diameter. As short an arc as possible should be used in all positions for best results. Stringer beads or small weave passes are preferred to wide weave passes. A long arc and whipping will cause porosity and trapped slag. Use lower currents with direct current than with alternating current. Point the electrode directly into the joint and tip the holder forward slightly in the direction of travel. See Fig. 13-92, page 382. Govern travel speed by the desired bead size.

Current adjustment should be somewhat higher for the first pass than for the other stringer passes. The electrode position will vary with the sequence of passes, Fig. 13-85, page 379. The beads should be so proportioned that the size of the final welds is accurate. Recall the lacing technique, and refer to Job 13-J23, for details. Be sure to clean all the passes thoroughly. Practice starts and stops. Travel in all directions.

Operations

1. Obtain plates; check the job drawing, Fig. 14-35, for the correct quantity and size.
2. Obtain electrodes of each quantity, type, and size specified in the job drawing.
3. Set a d.c. power source for electrode positive at 140 to 200 amperes. If an a.c. power source is used, set it somewhat higher than for d.c.
4. Set up the plates as shown on the job drawing and tack weld them at each corner.
5. Place the joint in the horizontal position on the welding table. Make sure that it is connected with the work connection.
6. Make all welds (first to sixth pass inclusive) with $5/32$-inch electrodes as shown on the job drawing. Manipulate the electrode as instructed in Fig. 14-36.

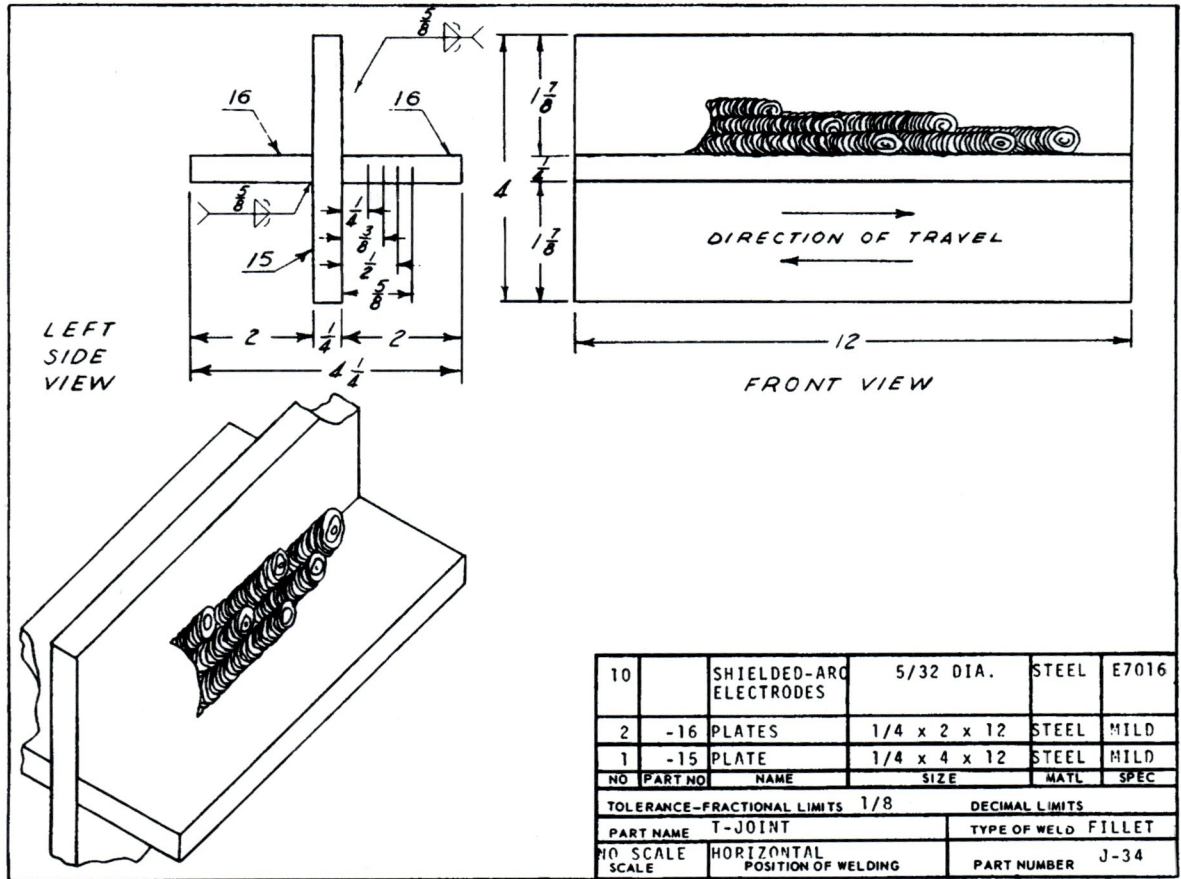

Fig. 14-35 Job drawing J34.

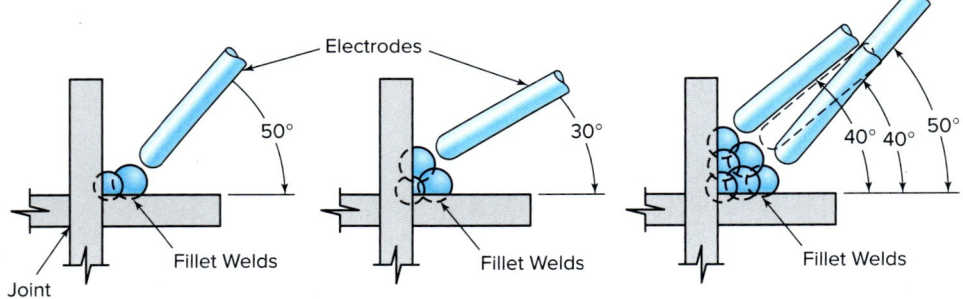

Fig. 14-36 Electrode position when welding a T-joint with a multipass fillet weld using the stringer bead technique.

7. Chip the slag from the weld, brush, and inspect between each pass. Refer to Inspection.
8. Set a d.c. power source for electrode positive at 80 to 140 amperes. If an a.c. power source is used, set it at a somewhat higher amperage.
9. Chip the slag from the welds, brush, and inspect. Refer to Inspection.
10. Practice these welds until you can produce good welds consistently.

Inspection

Compare each pass with Fig. 14-37, page 408, and check for the following weld characteristics:

Width and height: Uniform
Appearance: Very smooth with fine ripples; free of voids and high spots. Restarts should be difficult to locate.
Size: Refer to the job drawing, Fig. 14-35. Check with a fillet weld gauge.

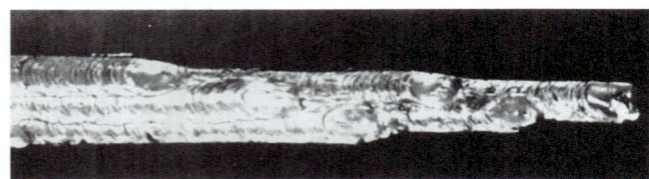

Fig. 14-37 Typical appearance of a multipass fillet weld in a T-joint welded in the horizontal position with E7016 low hydrogen electrodes for the stringer passes and DCEP or a.c. shielded metal arc coated electrodes. *Edward R. Bohnart*

Face of weld: Flat to slightly convex
Edges of weld: Good fusion, no overlap, no undercut
Starts and stops: Free of depressions and high spots
Beginnings and endings: Full size, craters filled
Penetration and fusion: To the root of the weld and plate surfaces
Surrounding plate surfaces: Free of spatter
Slag formation: Full coverage, easily removable

Disposal

Put completed joints in the scrap bin so that they will be available for further use. Plate edges can be beveled for beveled-butt welding, and unwelded plate surfaces can be used for beading.

Job 14-J35 Welding a T-Joint

Objective

To weld a T-joint in the vertical (3F) position by means of a single-pass fillet weld, travel up, with DCEP and/or a.c. shielded metal arc electrodes (AWS E6010–E6011).

General Job Information

A T-joint in the vertical position is frequently used in the fabricating and structural steel industries. Welding is usually up, but welding down is sometimes allowed. Whether single or multiple passes are specified depends upon the use of the joint and the thickness of the plate.

Welding Technique

Current adjustment should be high enough to ensure good fusion and penetration to the root of the joint and to the plate surfaces. Hold the electrode at a 45° work angle to the plates, Fig. 14-38A, and tilt it no more than 10° push travel angle, Fig. 14-38B. Figure 14-39 shows the actual welding position. Use the familiar up-and-down motion recommended for the vertical position. It is necessary to leave the crater at short intervals to prevent the metal from becoming overheated. Overheating makes it spill off and run down. Hold a close arc gap when depositing metal. The arc gap can be lengthened on the upward motion, but it must not be broken. This long arc technique is referred to as whipping the electrode. This technique is used to help control the deep penetration characteristics of the cellulose E6010–E6011 electrodes. With some of the newer power sources with closed loop feedback to maintain output amperage, this long arc technique does not decrease

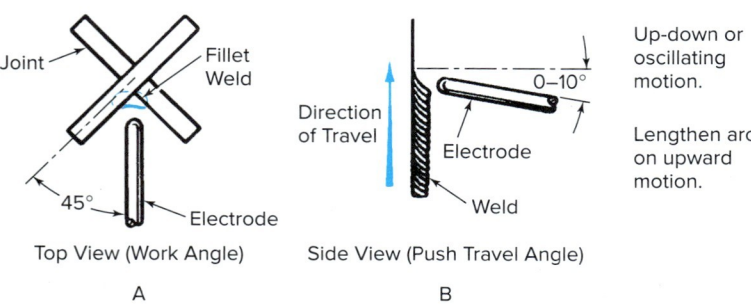

Fig. 14-38 Electrode position when welding a single pass fillet weld in a T-joint in the vertical position with cellulose-type shielded metal arc coated electrodes, travel up.

Fig. 14-39 Electrode and electrode holder positioned as instructed in Fig. 14-38.
Prographics/Don Peach

amperage. With the older power source design or the newer inverter power sources designed for this type of electrode, the amperage will drop off as the arc is lengthened. You will need to consult with your instructor (review the volt-amp curve) for the proper technique to use for the welding machine you are using. If your machine does not let the welding amperage drop off, you will need to maintain a short arc length and just use an upward motion in and out of the weld pool to control the penetration, undercut, and weld profile. Keep the electrode in the weld pool long enough to ensure fusion and penetration, to form the required weld contour, and to prevent undercut.

Practice starts and stops.

Operations

1. Obtain plates; check the job drawing, Fig. 14-40, for quantity and size.
2. Obtain electrodes of each quantity, type, and size specified in the job drawing.
3. Set a d.c. power source for electrode positive at 110 to 170 amperes or an a.c. power source at 110 to 180 amperes.
4. Set up the plates as shown on the job drawing and tack weld them at each corner.
5. Stand the joint in the vertical position on the welding table. Make sure the joint is connected to the work connection.
6. Make welds as shown on the job drawing. Manipulate the electrode as instructed in Fig. 14-38.
7. Chip the slag from the welds, brush, and inspect. Refer to Inspection below.
8. Practice these welds until you can produce good welds consistently with both types of electrodes, if possible, and with $3/16$-inch diameter electrodes.

Inspection

Compare the weld with Fig. 14-41, page 410, and check it for the following weld characteristics:

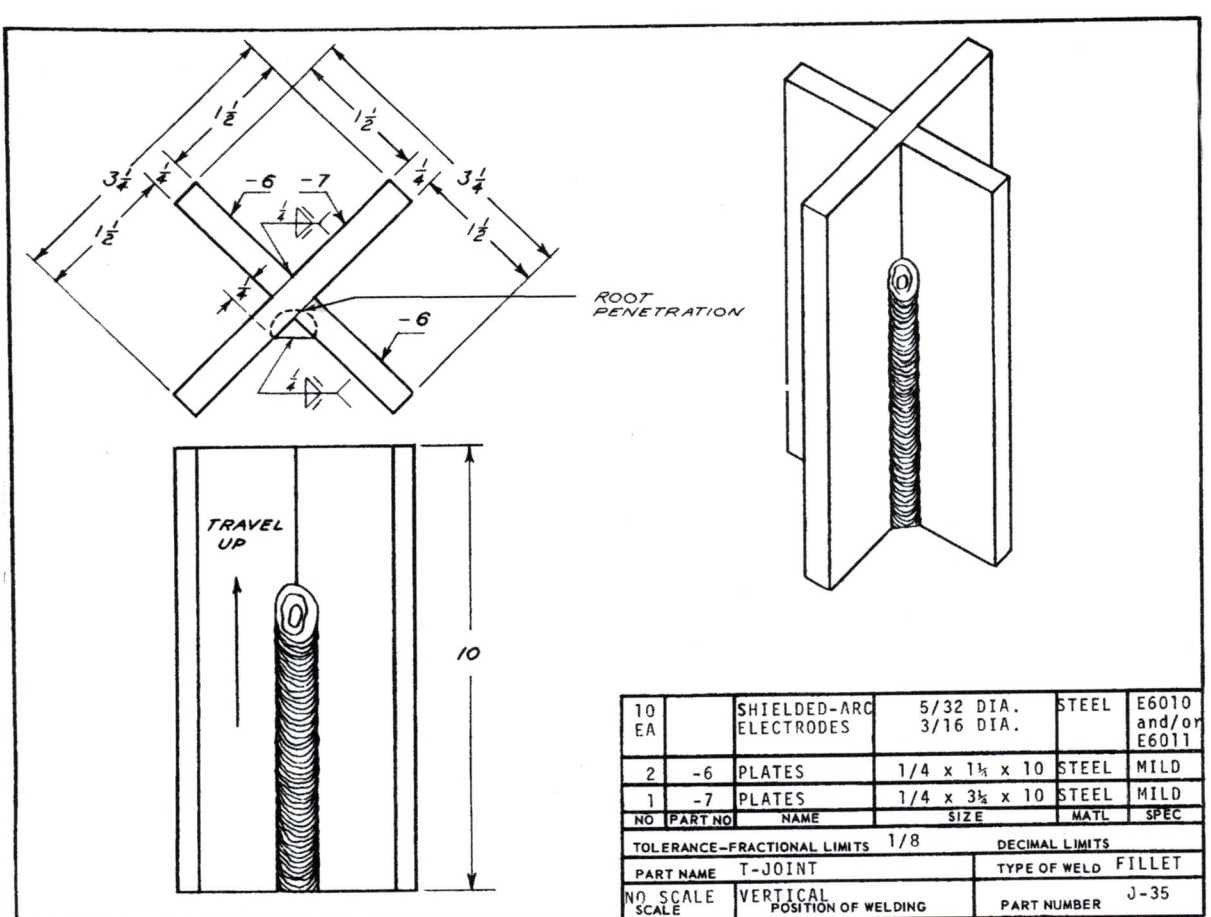

Fig. 14-40   Job drawing J35.

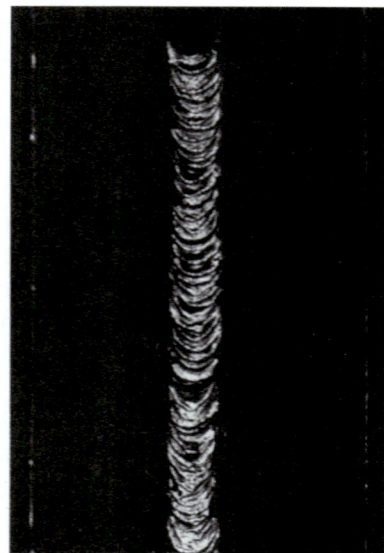

Fig. 14-41 Typical appearance of a single-pass fillet weld in a T-joint welded in the vertical (3F) position, travel up, with DCEP or a.c. shielded metal arc cellulose coated electrodes. *Edward R. Bohnart*

Width and height: Uniform
Appearance: Smooth with close ripples; free of voids and high spots. Restarts should be difficult to locate.
Size: Refer to job drawing, Fig. 14-40, page 409. Check with a fillet weld gauge.
Face of weld: Flat
Edges of weld: Good fusion, no overlap, no undercut
Starts and stops: Free of depressions and high spots
Beginnings and endings: Full size, craters filled
Penetration: To the root of the joint and both plate surfaces
Surrounding plate surfaces: Free of spatter
Slag formation: Full coverage, easily removable

Check Test

After you are able to make welds that are satisfactory in appearance, make up a test specimen similar to that made as a check test in Job 13-J14, Figs. 13-57 and 13-58, pages 364–365. Use the same plate thickness, welding technique, and electrode type and size used in this job. Weld on one side only.

Break the completed weld and examine the surfaces for soundness. The weld must not have porosity, must show good fusion, and penetration to the root of the joint and to the plate surfaces. The joint should break evenly through the throat of the weld.

Disposal

Save the completed joints for your next job which will be multipass fillet welding in the vertical position.

Job 14-J36   Welding a T-Joint

Objective

To weld a T-joint in the vertical position by means of multipass fillet welds, weave bead technique, travel up, with DCEP and/or a.c. shielded metal arc electrodes (AWS E6010–E6011 cellulose electrodes).

General Job Information

For practical purposes, a joint in ¼-inch plate would not require three passes. Three passes are required in this job to permit full use of the plate and to develop the ability to weld weave beads in the vertical position. Multipass fillet welds in the vertical position are usually welded up with electrode positive electrodes. Electrode negative electrodes are sometimes used.

Welding Technique

Second Pass Use the joint from the previous job. Current adjustment should be high enough to ensure good fusion and penetration. Hold the electrode at a 45° work angle to the plate, Fig. 14-42A, and tilt it no more than 10° travel angle down from the horizontal position, Fig. 14-42B. Figure 14-39, page 408, shows the actual position. You will find that it is possible to keep the electrode in the weld pool and advance it by weaving from side to side. A smooth, even motion across the face of the weld will form a flat face, and hesitation at the sides of the weld will prevent undercut. Hold a short arc gap but do not touch the molten metal with the electrode.

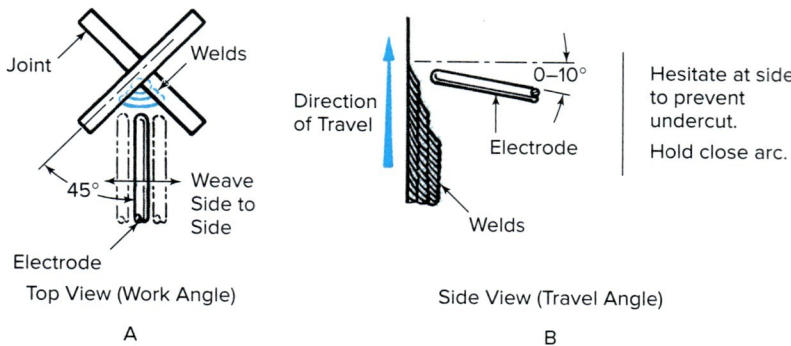

Fig. 14-42 Electrode position when fillet welding a T-joint in the vertical position with weave bead, travel up.

In welding both this second pass and the third pass, the upward movement when you reach the bead edges controls the appearance and thickness of the bead being deposited. A large upward movement will result in wider spaced ripples and a thinner buildup. A small upward movement will produce more closely spaced ripples and a thicker buildup.

Third Pass Current adjustment will have to be higher for the larger electrode. The electrode position is the same as for previous passes. When using larger electrodes and higher heats, run the electrode up quickly and away from the pool at each side. Do not break the arc on this movement and lengthen it on the upward stroke. This long arcing technique is referred to as whipping the electrode. This technique is used to help control the deep penetration characteristics of the cellulose E6010–E6011 electrodes. With some of the newer power sources with closed loop feedback to maintain output amperage, this long arc technique does not decrease amperage. With the older power source design or the newer inverter power sources designed for this type of electrode, the amperage will drop off as the arc is lengthened. You will need to consult with your instructor (review the volt-amp curve) for the proper technique to use for the welding machine you are using. If your machine does not let the welding amperage drop off, you will need to maintain a short arc length and just use an upward motion in and out of the weld pool to control the penetration and weld profile. When the weld pool has cooled enough, bring the electrode back to the crater and deposit additional metal. Keep the width of the weave within the limits of the final bead width.

Practice starts and stops.

Operations

1. Use the joints welded in Job 14-J38, Fig. 14-40, page 409.
2. Obtain electrodes of each quantity, type, and size specified in the job drawing, Fig. 14-43.
3. Set the power source at 110 to 170 amperes. Set a d.c. power source for electrode positive.
4. Stand the joint in the vertical position on the welding table. Make sure that it is connected with the work connection.

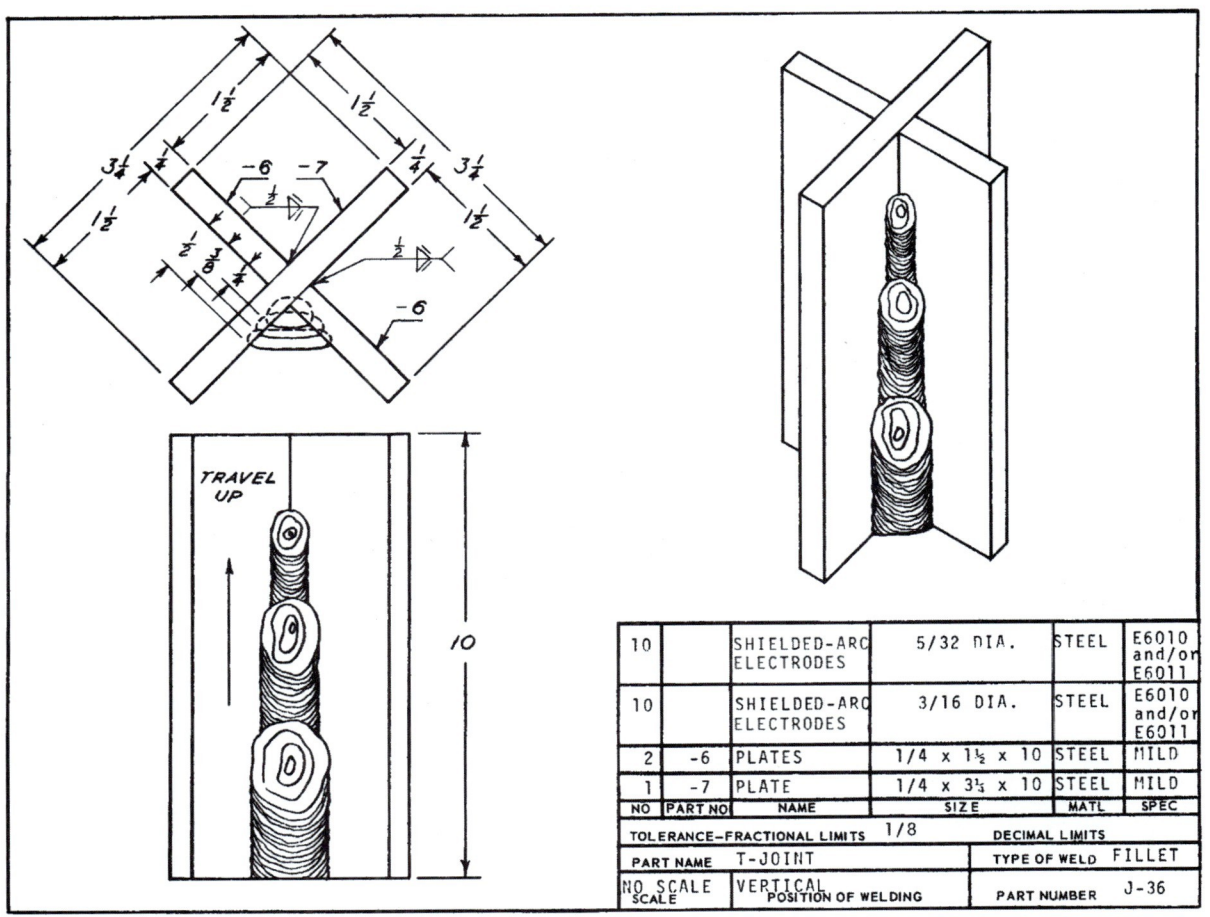

Fig. 14-43 Job drawing J36.

5. Make the second pass with a ⁵⁄₃₂-inch electrode as shown on the job drawing. Manipulate the electrode as instructed in Fig. 14-42, page 410.
6. Chip the slag from the weld, brush, and inspect. Refer to Inspection.
7. Increase the power source setting to 140 to 215 amperes.
8. Make the third pass with ³⁄₁₆-inch electrodes as shown on the job drawing. If you find it difficult with ³⁄₁₆-inch electrodes, practice this pass with ⁵⁄₃₂-inch electrodes at 110 to 170 amperes.
9. Chip the slag from the weld, brush, and inspect. Refer to Inspection.
10. Practice these welds until you can produce good welds consistently with both types of electrodes.

Inspection

Compare each pass with Fig. 14-44 and check it for the following weld characteristics:

Width and height: Uniform
Appearance: Smooth with close ripples; free of voids and high spots. Restarts should be difficult to locate.
Size: Refer to the job drawing, Fig. 14-43, page 411. Check with a fillet weld gauge.

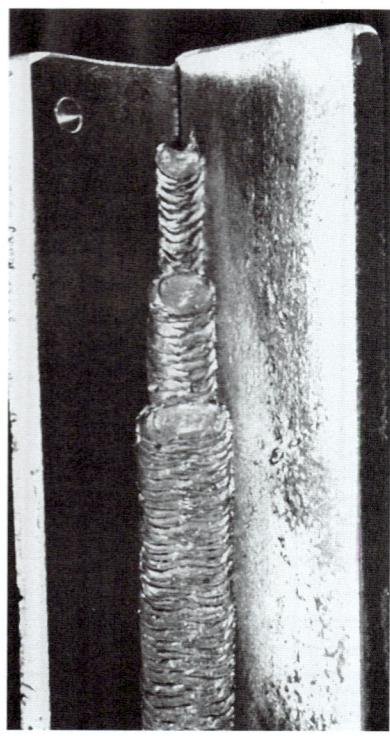

Fig. 14-44 Typical appearance of a multipass fillet weld in a T-joint welded in the vertical position, travel up, with DCEP or a.c. shielded metal arc cellulose coated electrodes. *Edward R. Bohnart*

Face of weld: All passes flat
Edges of weld: Good fusion, no overlap, no undercut
Starts and stops: Free of depressions and high spots
Beginnings and endings: Full size, craters filled
Penetration and fusion: To preceding passes and plate surfaces
Surrounding plate surfaces: Free of spatter
Slag formation: Full coverage, easily removable

Disposal

Put completed joints in the scrap bin so that they will be available for further use. Plate edges can be used for square butt welding, and unwelded plate surfaces can be used for beading.

Job 14-J37   Welding a T-Joint

Objective

To weld a T-joint in the vertical position by means of multipass fillet welds, weave bead technique, travel up, with low hydrogen, iron powder, shielded metal arc electrodes (AWS E7018).

General Job Information

The addition of 25 to 40 percent iron powder to a low hydrogen electrode covering has increased its usage for mild steel and the low alloy, high sulfur, and high carbon grades. Iron powder also increases the deposition rate and gives better restrike characteristics than the E7016 type. The arc action is smoother and more stable. The deposit wets more readily so that undercutting is prevented. The bead is smoother and slag removal is easier than with most E7016 electrodes. All of the desirable low hydrogen characteristics are retained in this electrode.

The coatings containing iron powder are thicker than regular low hydrogen coatings.

Welding Technique

Use a.c. or DCEP. You will find that the arc is smooth and quiet with shallow penetration and little spatter. Since the coating of these electrodes is heavier than normal, vertical and overhead welding is usually limited to smaller diameters. Currents used are somewhat higher than for the E6010s of corresponding size. A short arc must be maintained at all times.

Vertical-up welding is done with either a ⅛- or ⁵⁄₃₂-inch electrode. Do not use a whip technique or take the electrode out of the molten pool. Use a small triangular weave. Build up a small shelf of weld metal and deposit layer upon layer of metal with the weave as the weld progresses up the joint. Point the electrode directly into the joint and slightly upward to permit the arc force to assist in controlling the pool. See Fig. 14-42, page 410. Travel slow enough to maintain the shelf without causing the molten metal to spill off. Use currents in the lower portion of the quoted range. Keep the width of the weave within the confines of the bead width. You have learned to make sure that you obtain good fusion at the root of the joint with the first pass and that all other passes are thoroughly fused with the underneath pass and the side walls of the joint.

Be sure to clean all passes thoroughly. Practice starts and stops.

Operations

1. Obtain plates; check the job drawing, Fig. 14-45, for quantity and size.
2. Obtain electrodes in each quantity, type, and size specified in the job drawing.
3. Set a d.c. power source for electrode positive at 150 to 220 amperes. An a.c. power source may be used at a somewhat higher heat.
4. Set up two sets of plates as shown in the job drawing and tack weld them at each corner.
5. Stand the joint in the vertical position on the welding table. Make sure it is connected with the work connection.
6. Make all passes with 5/32-inch diameter electrodes as shown in the job drawing. Manipulate the electrode as instructed in Fig. 14-42. Weld both joints. (This is to keep the joint from getting too hot.)
7. Chip the slag from the welds, brush, and inspect between each pass. Refer to Inspection.
8. Practice these welds until you can produce good welds consistently.

Inspection

Compare each pass with Fig. 14-46, page 414, and check it for the following weld characteristics:

Width and height: Uniform

Appearance: Very smooth with fine ripples; free of voids and high spots. Restarts should be difficult to locate.

Sizes: Refer to the job drawing, Fig. 14-45. Check with a fillet weld gauge.

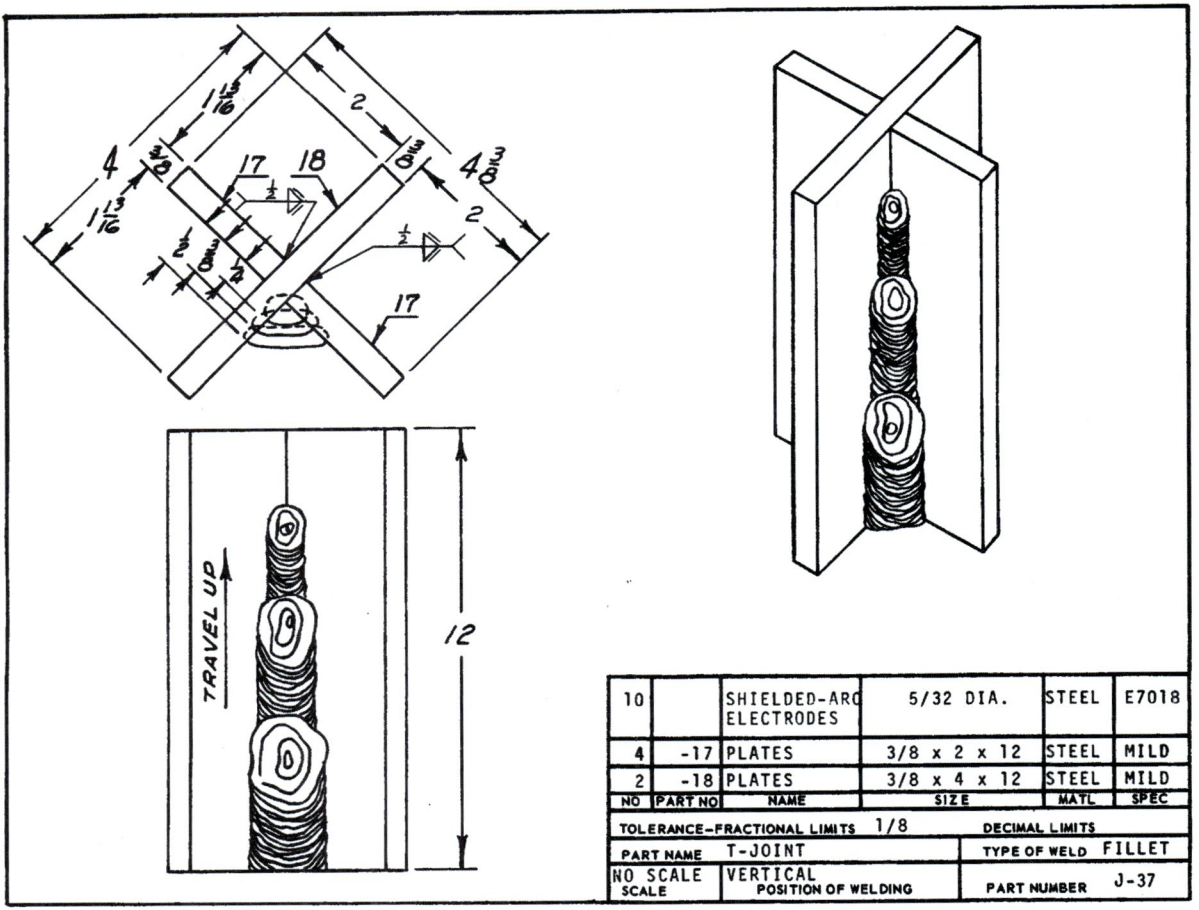

Fig. 14-45 Job drawing J37.

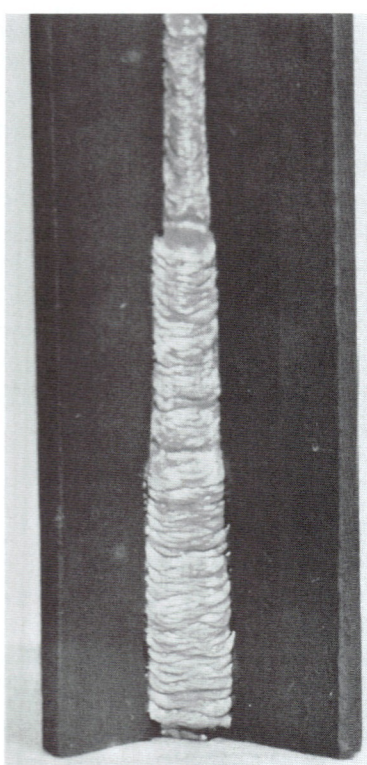

Fig. 14-46 Typical appearance of multipass fillet weld in a T-joint welded in the vertical (3G) position, travel up, with low hydrogen, iron powder, shielded metal arc coated electrodes (DCEP or a.c.).
Edward R. Bohnart

Face of weld: Flat to slightly convex
Edges of weld: Good fusion, no overlap, no undercut
Starts and stops: Free of depressions and high spots
Beginnings and endings: Full size, craters filled
Penetration and fusion: To the root of the weld and plate surfaces
Surrounding plate surfaces: Free of spatter
Slag formation: Full coverage, easily removable

Disposal

Put completed joints in the scrap bin so that they can be used again. Plate edges can be beveled for beveled-butt welding, and unwelded plate surfaces can be used for beading.

Job 14-J38 Welding a Single-V Butt Joint (Backing Bar Construction)

Objective

To weld a single-V butt joint (backing bar construction) in the vertical position by means of multipass groove welds, weave bead technique, travel up, with DCEP and/or a.c. shielded metal arc electrodes (AWS E6010–E6011).

General Job Information

The single-V butt joint with a backing bar is found quite frequently in piping, pressure vessel, and construction. This type of joint is often required by code authorities. It is also recommended by the American Welding Society. The backing material creates the effect of welding a joint with a chill ring. In some situations a nonmetal backing material is used. It is generally ceramic to withstand the welding temperatures and is not fused during welding. The root opening varies with different welding shops. If it is as large as ⅜ inch, two or three root passes are necessary because of the greater separation between the plates. Another welding procedure is to use stringer beads.

Welding Technique

First Pass Current adjustment can be high. Tilt the electrode no more than 10° push travel angle, Fig. 14-47A, and hold it at a 90° work angle to the surface of the plates, Fig. 14-47B. Use an up-and-down motion for running stringer beads in the vertical position. The weld deposit must be fused into the backing bar and the root surfaces of the two plates. Keep the face of the weld flat. If the joint is to be set up with a wider root gap, two or three root passes should be made. Be sure that all passes are fused with each other.

Second Pass The current setting must not be too high. Check Fig. 14-47 for the electrode position. Use the technique for running weave beads in the vertical position. Hesitate at the sides to prevent undercut. Use the edges of the first bead as a guide for the width of the new pass. Keep the face of the weld flat. Reinforcement should not be so high that it covers the sharp edges of the beveled plate since these edges may be used as a guide for the last pass.

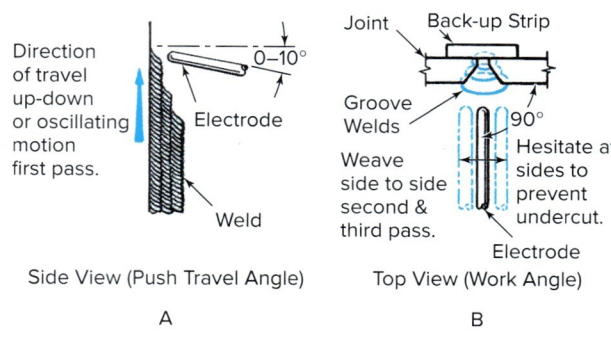

Fig. 14-47 Electrode position when welding a single V-groove weld on a butt joint in the vertical (3G) position, travel up.

Third Pass Current adjustment must be high enough to ensure good fusion and penetration together with smooth appearance. The electrode position is the same as for the other passes. Use the technique for running weave beads in the vertical position. Do not weave beyond the sharp edges of the plate bevel. Hesitate long enough at the sides to prevent undercut. Travel must be fast enough to prevent high convexity. The proper V-groove weld face profile is shown in Fig. 14-32, page 404.

Practice starts and stops. Clean slag thoroughly between each pass.

Operations

1. Obtain plates; check the job drawing, Fig. 14-48, for quantity and size.
2. Obtain electrodes of each quantity, type, and size specified in the job drawing.
3. Set the power source at 110 to 170 amperes. Set a d.c. power source for electrode positive.
4. Set up the plates as shown on the job drawing. Tack weld them at each corner and at the center of the back side.
5. Stand the joints in the vertical position on the welding table. Make sure that they are connected with the work connection.
6. Make welds as shown on the job drawing. Manipulate the electrode as instructed in Fig. 14-47.
7. Chip the slag from the welds, brush, and inspect. Refer to Inspection.
8. Practice these welds until you can produce good welds consistently with both types of electrodes and with 3/16-inch diameter electrodes.

Inspection

Compare each pass with Fig. 14-49, page 416, and check it for the following weld characteristics:

Width and height: Uniform
Appearance: Smooth with close ripples; free of voids and high spots. Restarts should be difficult to locate.
Size: Refer to the job drawing, Fig. 14-48. Check the last pass with a butt weld gauge.
Face of weld: First two passes: flat; last pass: slight reinforcement

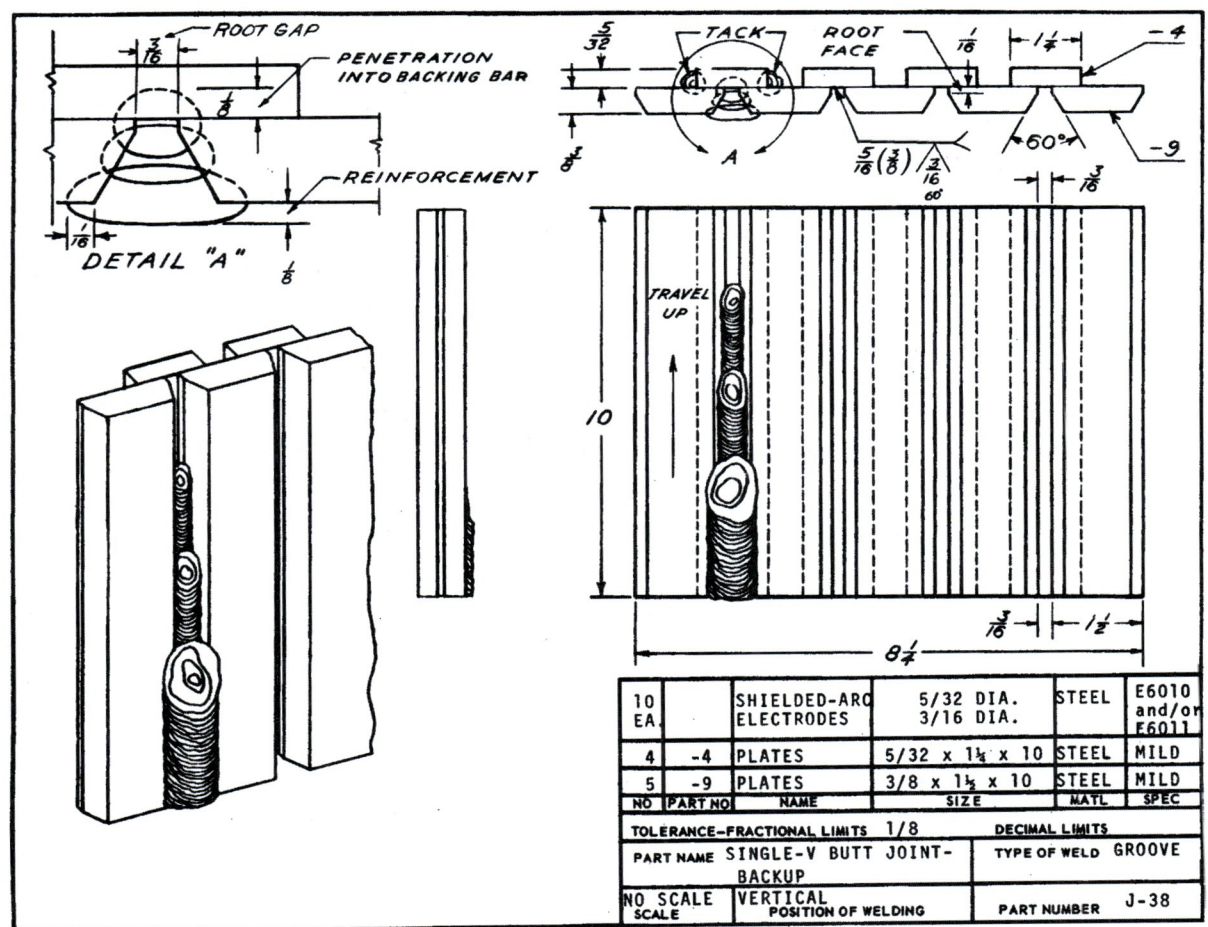

Fig. 14-48 Job drawing J38.

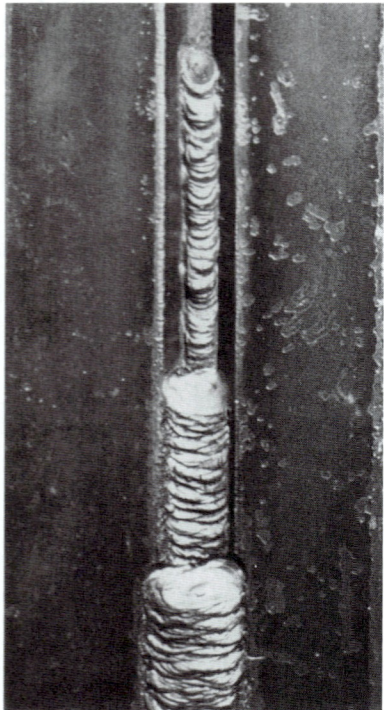

Fig. 14-49 Typical appearance of a multipass groove weld in a single V-groove weld on a butt joint with a backing bar, welded in the vertical position, travel up, with DCEP or a.c. shielded metal arc cellulose coated electrodes. *Edward R. Bohnart*

Edges of weld: Good fusion, no overlap, no undercut
Starts and stops: The weld should be free of depressions and high spots
Beginnings and endings: Full size, craters filled
Penetration and fusion: To the backing bar, preceding passes, and plate surface
Surrounding plate surfaces: Free of spatter
Slag formation: Full coverage, easily removable

> **JOB TIP**
>
> **Customers**
> You and the supervisor share a mutual interest. Ask questions; listen for factual information. What bothers the supervisor most? Accept responsibility and convey the complaint to the proper person in your company. Let the supervisor know what you can and will do (not what you can't do). Being positive will help keep you from taking it personally.

Disposal

Put completed joints in the scrap bin so that they will be available for further use. The plate can be beveled between welds for further butt welding, and unwelded plate surfaces can be used for beading.

Qualifying Test

At this point you should be able to pass Test 1 in the vertical position. (See Chapter 15, p. 455.) This is to AWS recommendations. It will also be desirable for you to pass Test 3 (fillet weld) in the vertical position.

Job 14-J39 Welding a Square Butt Joint

Objective

To weld a square butt joint in the vertical position by means of multipass groove welds, stringer and weave bead techniques, travel up, with DCEP and/or a.c. shielded metal arc electrodes (AWS E6010–E6011 cellulose electrodes).

General Job Information

The purpose of this job is to develop the ability of depositing the first pass in the root of an open butt joint. This job should be practiced with great care. It is the same procedure for making the first pass in a butt weld in pipe. This type of joint is assigned since it lends itself readily to practice and it can be prepared quickly and economically.

Welding Technique

First Pass The current setting must not be too high, and yet the arc must be hot enough to penetrate the back side of the plate thoroughly and to melt away the root faces of the plates. Tilt the electrode a 10° travel angle down from the horizontal position, Fig. 14-50A, and hold it at a 90° work angle to the surface of the plates, Fig. 14-50B. Stringer bead technique may be used. Be sure to maintain the small hole (keyhole) at the leading edge of the weld

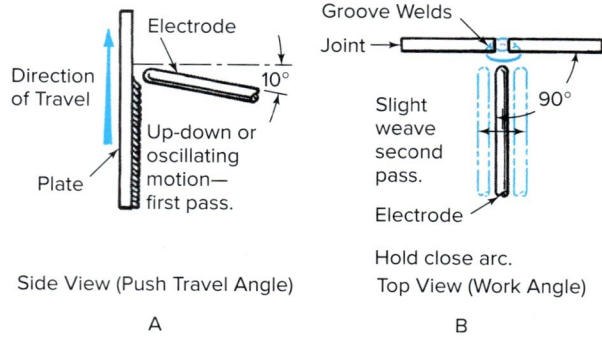

Fig. 14-50 Electrode position when welding a square groove butt joint in the vertical (3G) position, travel up.

pool. A small bead should be formed on the back side of the weld, Fig. 14-51.

Second Pass Current adjustment and electrode position are the same as for first pass. Do not weave beads that are too wide. Adjust the speed of travel to avoid excessive convexity and undercut.

Practice starts and stops, especially on the first pass.

Operations

1. Obtain plates; check the job drawing, Fig. 14-52, for quantity and size.
2. Obtain electrodes of each quantity, type, and size specified in the job drawing.
3. Set the power source for 90 to 120 amperes. Set a d.c. power source for electrode positive.
4. Set up the plates and tack weld as shown on the job drawing.
5. Stand the plates in the vertical position on the welding table. Make sure that they are connected to the work connection.

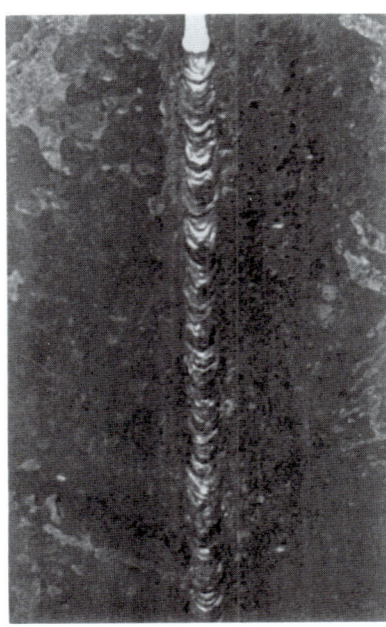

Fig. 14-51 Penetration and fusion through the root side of a square butt joint welded in the vertical position, travel up.
Edward R. Bohnart

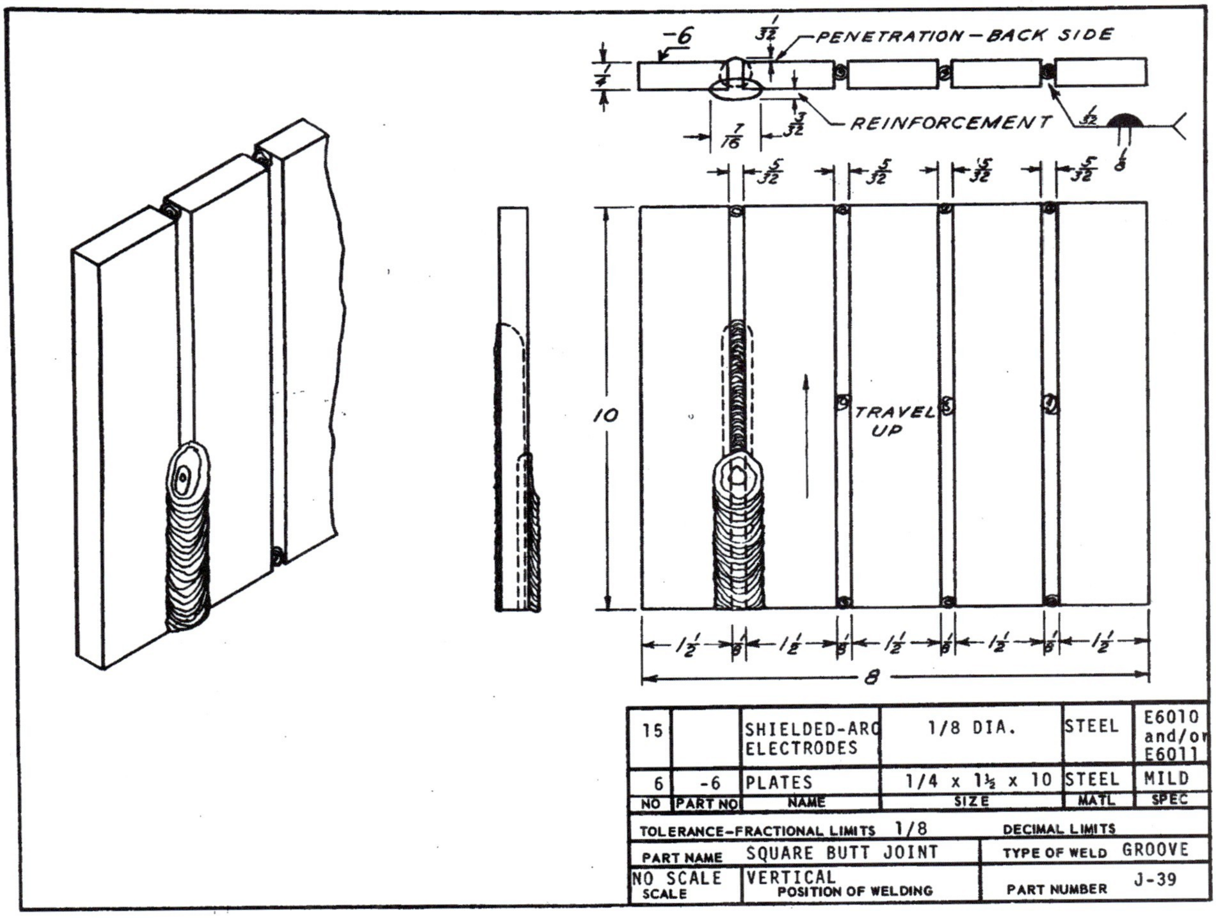

Fig. 14-52 Job drawing J39.

Shielded Metal Arc Welding Practice: Jobs 14-J26–J42 (Plate) **Chapter 14** 417

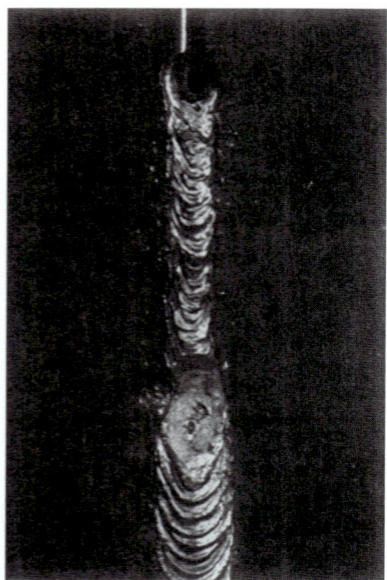

Fig. 14-53 Typical appearance of a multipass single square groove weld in a butt joint welded in the vertical position, travel up, with DCEP or a.c. shielded metal arc cellular coated electrodes. *Edward R. Bohnart*

6. Make the first pass as shown on the job drawing. Manipulate the electrode as instructed in Fig. 14-50, page 416.
7. Chip the slag from the weld, brush, and inspect. Refer to Inspection.
8. Make the second pass as shown on the job drawing.
9. Chip the slag from the weld, brush, and inspect. Refer to Inspection.
10. Practice these welds until you can produce good welds consistently with both types of electrodes.

Inspection

Compare the back side of the first bead with Fig. 14-51, page 417, and both passes with Fig. 14-53. Check each pass for the following weld characteristics:

Width and height: Uniform
Appearance: Smooth with close ripples; free of voids and high spots. Restarts should be difficult to locate.
Size: Refer to the job drawing, Fig. 14-52, page 417. Check the last pass with a butt weld gauge.
Face of weld: Very slight reinforcement
Edges of weld: Good fusion, no overlap, no undercut
Starts and stops: Free of depressions and high spots
Beginnings and endings: Full size, craters filled
Back side: Complete and uniform penetration
Penetration and fusion: Through the back side and plate surfaces. See Fig. 14-51.
Surrounding plate surfaces: Free of spatter
Slag formation: Full coverage, easily removable

Disposal

Put completed joints in the scrap bin so that they will be available for further use. Joints can be cut between the welds for further butt welding, and unwelded plate surfaces can be used for beading.

Job 14-J40   Welding an Outside Corner Joint

Objective

To weld an outside corner joint in the vertical position by means of multipass groove welds, weave bead technique, travel up, with DCEP and/or a.c. shielded metal arc electrodes (AWS E6010–E6011 cellulose electrodes).

General Job Information

The outside corner joint is included in this text chiefly because of its all-around utility. It offers practice in depositing root passes and in multipass weave technique, and it requires little edge preparation since it has a square plate edge.

Welding Technique

First Pass The current adjustment must not be too high. Tilt the electrode no more than 10° travel angle down from the horizontal position, Fig. 14-54A, and hold it at a 90° work angle to the intersection of the plates, Fig. 14-54B. Use the standard root pass technique.

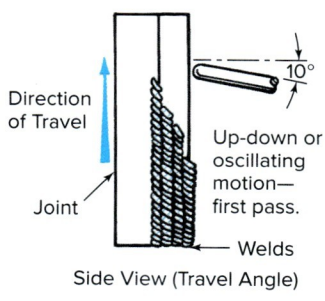

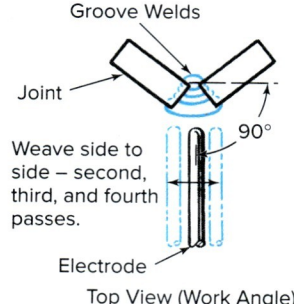

Fig. 14-54 Electrode position when welding a single V-groove in a corner joint in the vertical (3G) position, travel up.

Second and Third Passes The current adjustment can be higher than for the first pass. Check the electrode position with Fig. 14-54. Use the weave bead technique. Secure good fusion with the preceding passes and the walls of the plate. Do not undercut. The face of welds should be kept flat.

Fourth Pass The current adjustment and electrode position are approximately those used for previous passes. Use the weave bead technique. Let the edges of the plate be your guide for bead width. The face of the weld should be slightly convex to provide full reinforcement.

Practice starts and stops.

Operations

1. Obtain plates; check the job drawing, Fig. 14-55, for quantity and size.
2. Obtain electrodes of each quantity, type, and size specified in the job drawing.
3. Set the power source for 75 to 125 amperes for the first pass. Set a d.c. power source for electrode positive.
4. Set up the plates and tack weld them as shown on the job drawing.
5. Stand the joints in the vertical position on the welding table. Make sure that they are connected with the work connection.
6. Make the first pass with a 1/8-inch electrode as shown on the job drawing. Manipulate the electrode as instructed in Fig. 14-54.
7. Chip the slag from the weld, brush, and inspect. Refer to Inspection.
8. Readjust the power source for 110 to 170 amperes.
9. Make the second, third, and fourth passes with 5/32-inch electrodes as shown on the job drawing.
10. Chip the slag from the welds, brush, and inspect between each pass. Refer to Inspection.
11. Practice these welds until you can produce good welds consistently with both types of electrodes.

Inspection

Compare the back side of the first bead with Fig. 14-56, page 420, and all passes with Fig. 14-57, page 420. Check each pass for the following weld characteristics:

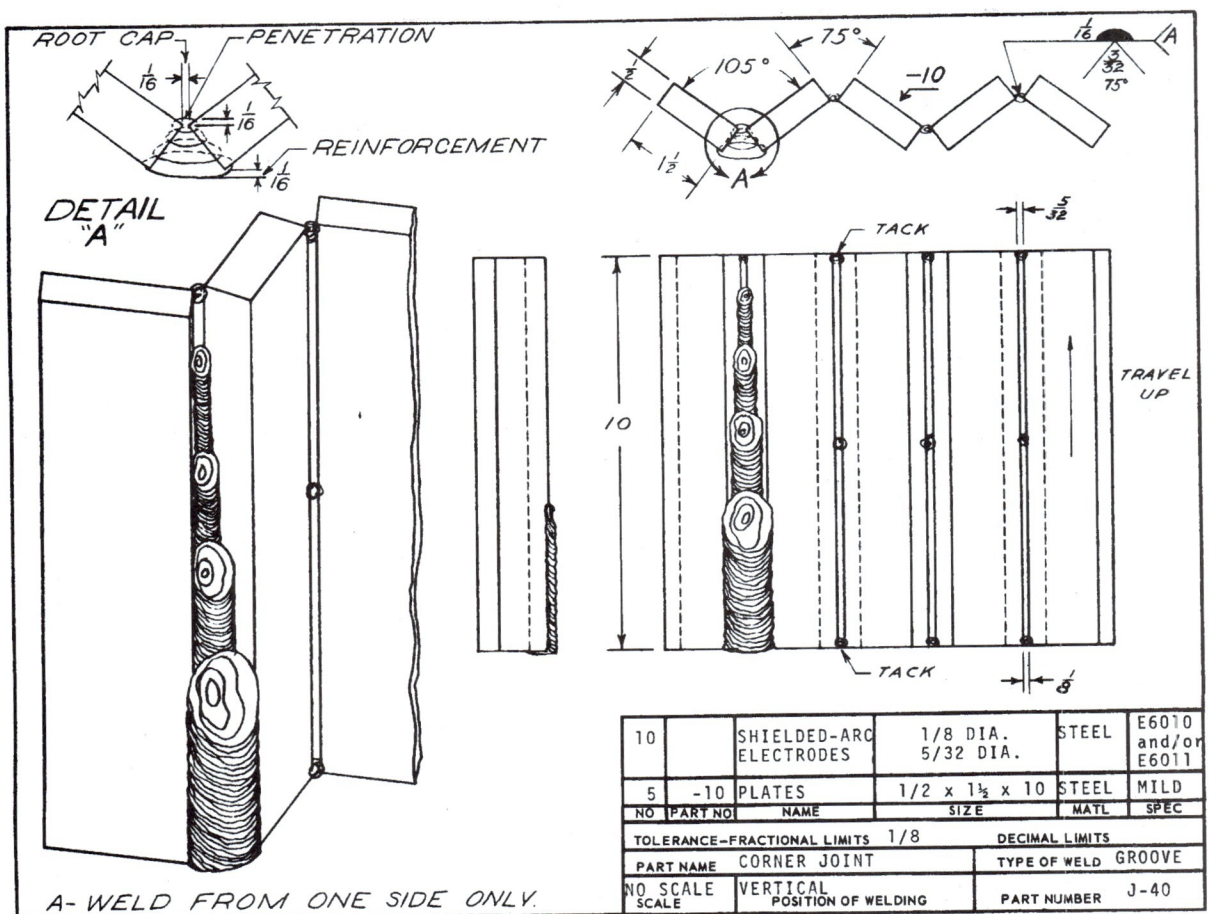

Fig. 14-55 Job drawing J40.

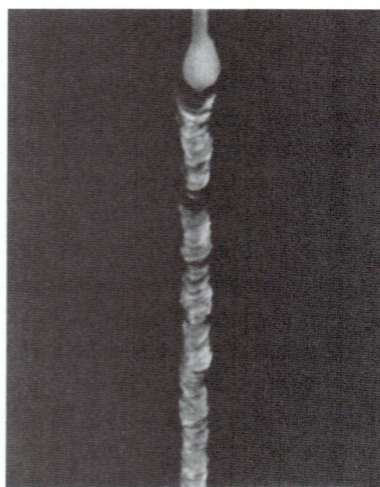

Fig. 14-56 Penetration and fusion through the root side of the outside corner joint shown in Fig. 14-57. *Edward R. Bohnart*

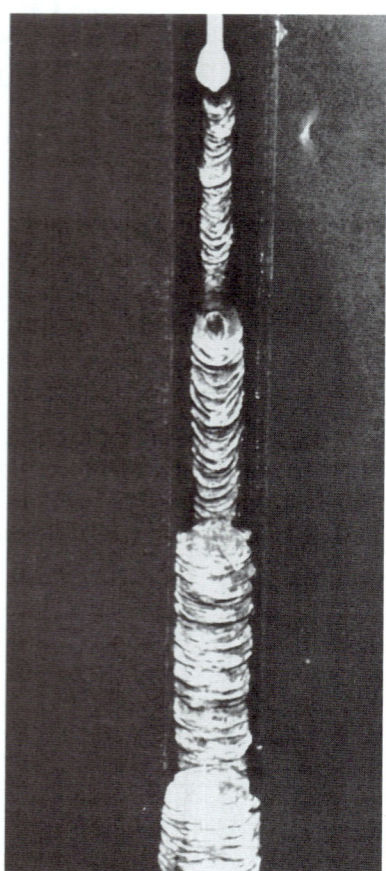

Fig. 14-57 Typical appearance of a multipass single V-groove weld in an outside corner joint welded in the vertical position, travel up, with DCEP or a.c. shielded metal arc, low hydrogen, iron powder, coated electrodes. Note the "keyhole" at the forward edge of the root pass. *Edward R. Bohnart*

Width and height: Uniform
Appearance: Smooth with close ripples; free of voids and high spots. Restarts should be difficult to locate.
Size: Refer to the job drawing, Fig. 14-55, page 419.
Face of weld: First three passes: flat; last pass: slightly convex
Edges of weld: Good fusion, no overlap, no undercut
Starts and stops: Free of depressions and high spots
Beginnings and endings: Full size, craters filled
Back side: Complete and uniform penetration
Penetration and fusion: Through the back side and to all plate surfaces. See Fig. 14-56.
Surrounding plate surfaces: Free of spatter
Slag formation: Full coverage, easily removable

Disposal

Put completed joints in the scrap bin so that they will be available for further use. Fillet welds can be run on the reverse side, or the flat unwelded surfaces can be used for beading.

Job 14-J41 Welding a T-Joint

Objective

To weld a T-joint in the horizontal position by means of multipass fillet welds, weave bead technique, with low hydrogen, iron powder, shielded metal arc electrodes (AWS E7018).

General Job Information

The purpose of this job is to practice making large size fillet welds with low hydrogen, iron powder electrodes. Review all of the material concerning this type of electrode in Chapter 12 and Job 14-J40. Note that the weld symbol in the job drawing carries a diagonal line (/) to indicate a flat face and 12 for the length of the weld.

Welding Technique

First Pass Current adjustment should be somewhat higher than for flat welding. Hold the electrode at a 40° work angle to the flat plate, Fig. 14-58A, and tilt it 30° drag travel angle, Fig. 14-58B. As with other low hydrogen electrodes, a short arc should be maintained at all times. Keep the electrode in the pool so that the slag does not run ahead of the weld. The E7018 electrode can be used with the coating in contact with the work. This is known as the **drag technique**. Use very little weave movement for the first pass. The face of the weld should be kept as flat as possible. Make sure that penetration and fusion is secured at the root of the joint and that there is no undercutting along the side walls.

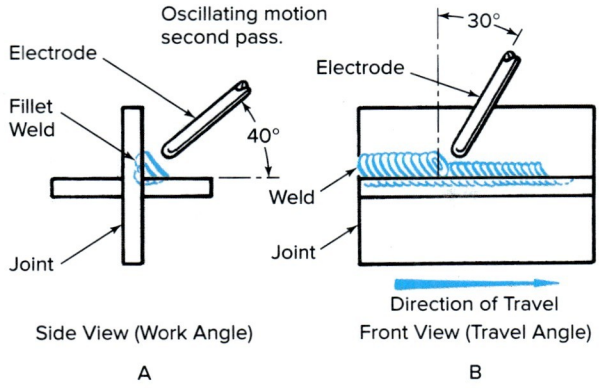

Fig. 14-58 Electrode position when fillet welding a T-joint in the horizontal (2F) position.

Second Pass Deposit the second pass over the first pass with a slight oscillating motion, Fig. 14-58B. Direct the electrode toward the vertical plate in order to force the weld metal up on its surface. Keep a short arc and restrict electrode movement within the limits of the desired weld width to prevent undercut. The circular motion should be counterclockwise. Secure good fusion with the underneath pass and the plate surfaces.

Practice starts and stops. Travel in all directions. These electrodes pick up moisture; be sure to keep them dry.

Operations

1. Obtain plates; check the job drawing, Fig. 14-59, for quantity and size.
2. Obtain electrodes of each quantity, type, and size specified in the job drawing.
3. Set a d.c. power source for electrode positive at 150 to 220 amperes. An a.c. power source may be used at a somewhat higher heat.
4. Set up two sets of plates as shown on the job drawing and tack weld them at each corner.
5. Place the joints in the horizontal position on the welding table. Make sure that they are connected with the work connection.
6. Make the first pass with $5/32$-inch diameter electrodes as shown on the job drawing. Manipulate the electrodes as instructed in Fig. 14-58.

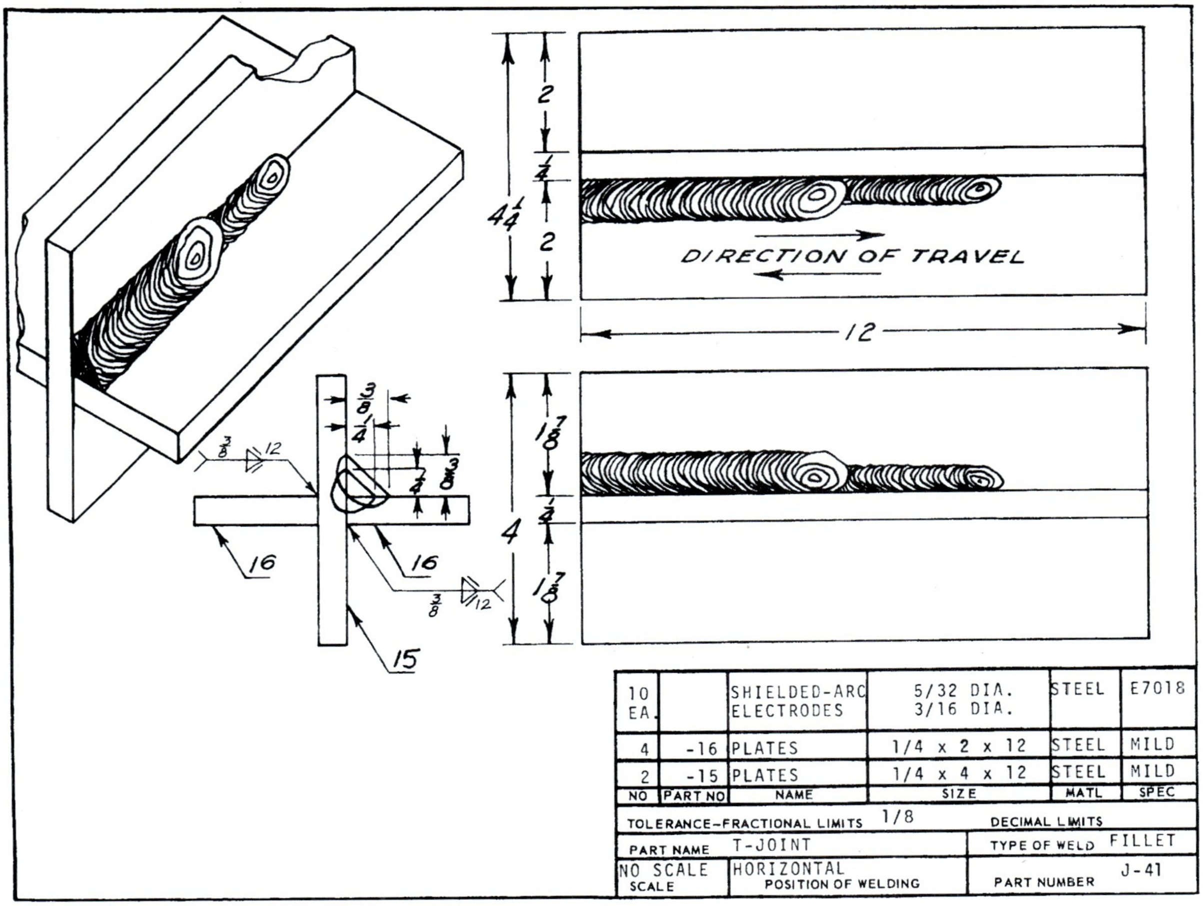

Fig. 14-59 Job drawing J41.

Weld on both joints. (This is to keep the joint from getting too hot.)
7. Chip the slag from the welds, brush, and inspect. Refer to Inspection.
8. Readjust the d.c. power source for 200 to 280 amperes. Set an a.c. power source somewhat higher.
9. Make the second pass with 3/16-inch diameter electrodes as shown on the job drawing. Alternate your welding between both joints.
10. Chip the slag from the welds, brush, and inspect. Refer to Inspection.
11. Practice these welds until you can produce good welds consistently. Change your procedure by using 3/16-inch diameter electrodes for the first pass. You may also wish to weld a third pass with the 3/16-inch size.

Inspection

Compare both passes with Fig. 14-60 and check them for the following weld characteristics:

Width and height: Uniform
Appearance: Very smooth with fine ripples; free of voids and high spots. Restarts should be difficult to locate.
Size: Refer to the job drawing, Fig. 14-59, page 421. Check with a fillet weld gauge.
Face of weld: Flat to slightly convex
Edges of weld: Good fusion, no overlap, no undercut
Starts and stops: Free of depressions and high spots
Beginnings and endings: Full size, craters filled
Penetration and fusion: To the root of the weld and plate surfaces
Surrounding plate surfaces: Free of spatter
Slag formation: Full coverage, easily removable

Disposal

Put completed joints in the scrap bin so that they will be available for further use. Plate edges can be beveled for beveled-butt welding, and unwelded plate surfaces can be used for beading.

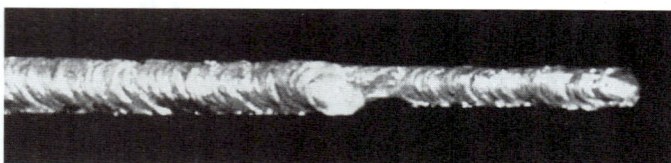

Fig. 14-60 Typical appearance of a multipass fillet weld in a T-joint welded in the horizontal (2F) position with shielded metal arc, low hydrogen, iron powder, coated electrodes (DCEP or a.c.).
Edward R. Bohnart

Job 14-J42 Welding a Single-V Butt Joint (Backing Bar Construction)

Objective

To weld a single-V butt joint with backing bar construction in the vertical position by means of multipass groove welds, the weave bead technique, travel down, with DCEP and/or a.c. shielded metal arc electrodes (AWS E6010–E6011 cellulose electrode).

General Job Information

The single-V butt joint with backing bar, welded by the downhill method, is used extensively in pipeline, tank, and shipyard work. It is suitable for both plate and pipe. The backing bar in plate is comparable to the backing ring for butt joints in pipe. Generally, the joint is used with pipe 12 inches or larger. The downhill welding technique is faster than the uphill technique.

> **ABOUT WELDING**
>
> **Confined Space**
> Avoid working in a confined space in which there is no easy way from the outside to turn off power, gas, and other supplies used inside (in addition to controls inside).

The skill learned in the practice of this joint will assist you in learning to weld pipe, which is presented in Chapter 16.

Welding Technique

Downhill welding is used in the interest of high production. In the piping field, two welders usually work on a joint; each welder takes one side of the joint. They are able to weld as many as five 12-inch pipe joints in an hour. Many welders in the field use 3/16-inch electrodes for all passes. For our initial practice, the first bead will be done with 5/32-inch electrodes.

The current adjustment should be somewhat higher than for vertical upwelding. Tilt the electrode 15 to 20° using a drag travel angle, Fig. 14-61B, and hold it at a 90° work angle to the surface of the plates, Fig. 14-61C.

Root Pass The first pass should be run with a fairly close arc, without weaving, and with a higher than normal rate of travel. The weld pool and the slag always have a tendency to run ahead of the arc. When this happens, it can be overcome by using a shorter arc, increasing the electrode angle, and increasing the rate of travel. Usually one or a combination of all three must be employed. There is no problem with undercutting along the sides of the weld. Make sure that fusion

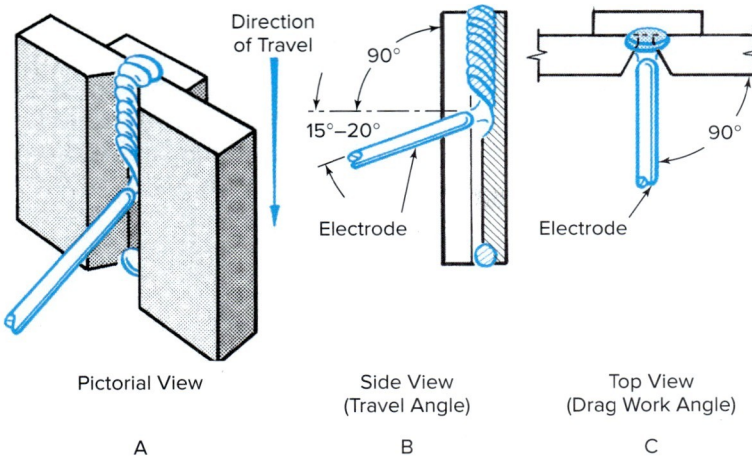

Fig. 14-61 Electrode position when welding a single-V groove butt joint with backing bar in the vertical (3G) position, travel down.

and penetration are secured at the root of the weld with the backing rod and the root faces of the beveled plates. Keep the face of the weld as flat as possible. There may be some trouble with holes or voids in the face of the weld. They are usually caused by slag piling up ahead of the arc. Watch arc length, electrode angle, and speed of travel.

Filler and Cover Passes The second, third, and fourth passes should be run like the first pass. For each pass, the side-to-side motion should be a little wider than for the preceding pass. Watch carefully for holes and voids in the face of the weld. There is also a tendency for the face of the weld to become concave. This can be avoided by pausing for a shorter time at the edges of the weld in making the weave motion. Because undercutting is not a problem, the pause at the sides of the weld should be short.

Make sure that fusion and penetration are secured between weld passes and into the beveled walls of the plate. The face of each completed pass should be as flat as possible. The final finish pass should be slightly convex to ensure that it is full and not below the surface of the plate. In making the final pass, do not weave beyond the sharp edges of the plate at each side.

Practice varying the heat setting, the angle of the electrode, the length of the arc, and the speed of travel. Starts and stops should also be practiced.

Operations

1. Obtain plates; check the job drawing, Fig. 14-62, for quantity and size.

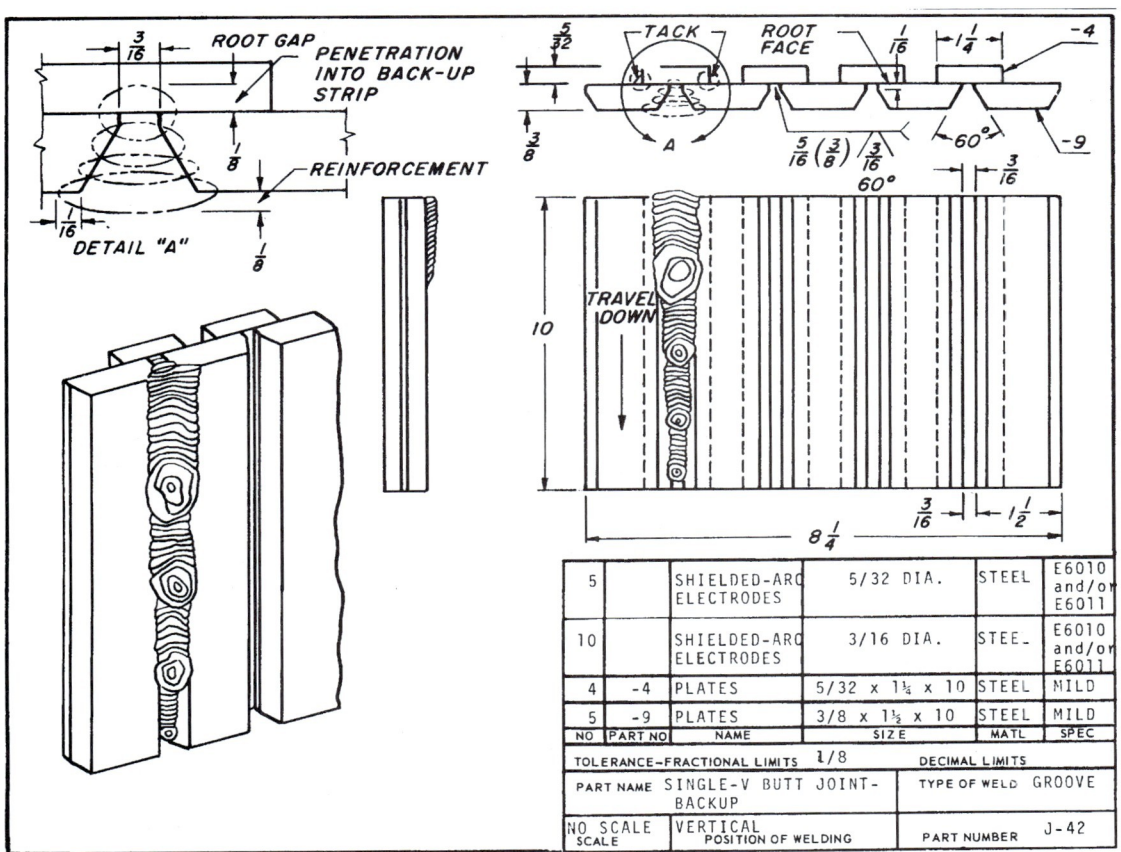

Fig. 14-62 Job drawing J42.

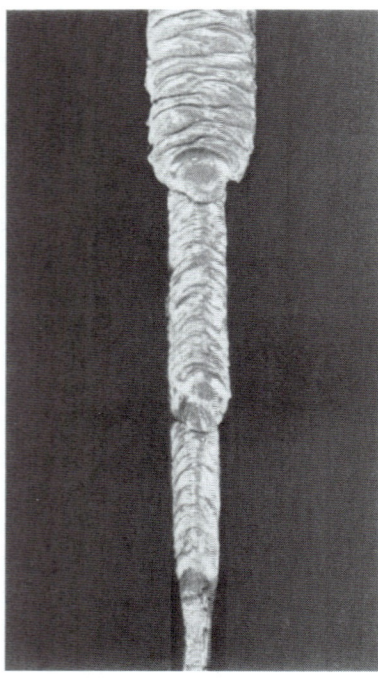

Fig. 14-63 Typical appearance of a multipass single V-groove weld on a butt joint with a backing bar in the vertical (3G) position, travel down, with DCEP or a.c. shielded metal arc cellulose coated electrodes. *Edward R. Bohnart*

2. Obtain electrodes of each quantity, type, and size specified in the job drawing.
3. Set the power source for 125 to 200 amperes. Set a d.c. power source for electrode positive.
4. Set up the plates as shown on the job drawing and tack weld them at each corner.
5. Stand the joint in the vertical position on the welding table. Make sure that it is connected with the work connection.
6. Make the first pass with 5/32-inch electrodes as shown on the job drawing. Manipulate the electrode as instructed in Fig. 14-61, page 423.
7. Chip the slag from the weld, brush, and inspect. Refer to Inspection.
8. Readjust the power source to 130 to 225 amperes.
9. Make the second, third, and fourth passes with 3/16-inch electrodes as shown on the job drawing.
10. Chip the slag from the welds, brush, and inspect. Refer to Inspection.
11. Practice these welds until you can produce good welds consistently. Change your procedure by using 3/16-inch electrodes for the first pass.

Inspection

Compare all passes with Fig. 14-63 and check them for the following weld characteristics:

Width and height: Uniform
Appearance: Very smooth with fine ripples; free of voids and high spots. Restarts should be difficult to locate.
Size: Refer to the job drawing, Fig. 14-62, page 423. Check with a fillet weld gauge.
Face of weld: Flat to slightly convex
Edges of weld: Good fusion, no overlap, no undercut
Starts and stops: Free of depressions and high spots
Beginnings and endings: Full size, craters filled
Penetration and fusion: To the root of the weld and plate surfaces
Surrounding plate surfaces: Free of spatter
Slag formation: Full coverage, easily removable

Disposal

Put completed joints in the scrap bin so that they will be available for further use. Plate edges can be beveled for beveled-butt welding, and unwelded plate surfaces can be used for beading.

CHAPTER 14 REVIEW

Multiple Choice

Choose the letter of the correct answer.

1. Wrapping the weld cables around your body should not be done because _____. (Obj. 14-1)
   a. You might short out the welding machine
   b. The weld cables may get pulled on and cause you to be injured
   c. They get hot and will burn you
   d. Coiling the cables around your body will reduce output

2. Overhead welds are made at an amperage _____. (Obj. 14-1)
   a. Lower than that of vertical welds
   b. Higher than that of vertical welds
   c. Similar to that of vertical welds
   d. Much the same as that for the flat position

3. The forces acting to hold the molten weld pool to the plate are referred to as _____. (Obj. 14-1)
   a. Molecular attraction and surface tension
   b. Magnetic repulsion and attraction
   c. Positive ion repulsion and attraction
   d. Arc plasma repulsion and attraction

4. When welding stringer beads overhead what should the drag travel angle be? (Obj. 14-1)
   a. 0–10°
   b. 10–20°
   c. 15–25°
   d. 5–15°
5. When welding single V-groove butt joints with a backing bar, the root opening should be kept as narrow as possible to _____. (Obj. 14-1)
   a. Reduce the number of weld beads for economic reasons
   b. Reduce heat input into the joint
   c. Reduce distortion forces
   d. All of these
6. The weld face should have no more than what fraction of an inch of reinforcement? (Obj. 14-2)
   a. ¼
   b. ³⁄₁₆
   c. ⅛
   d. ¹⁄₁₆
7. The total width of the final passes should only be what fraction of an inch wider than the widest opening of the V-groove? (Obj. 14-2)
   a. ⅛
   b. ¼
   c. ½
   d. ⅝
8. While filling the V-groove, the weld faces should _____. (Obj. 14-2)
   a. Be kept convex
   b. Be kept ropey
   c. Be kept flat
   d. Have as much reinforcement as possible
9. When welding vertical down with a E6010 or E6011 electrode, which technique should be used to keep the slag and weld pool from running ahead of the arc? (Obj. 14-2)
   a. Use a shorter arc
   b. Increase electrode angle
   c. Increase travel speed
   d. All of these
10. When tacking the V-groove butt joint and backing bar together, how should the backing bar fit to the back of the joint? (Obj. 14-2)
    a. Loose
    b. Gapping
    c. Tight
    d. Doesn't matter how it fits because it will be filled in with weld metal
11. When making 2G butt joints with stringer beads, the work angle should be 90° from the vertical for all passes. (Obj. 14-2)
    a. True
    b. False
12. When making square groove butt joints and welding from only one side, what is the maximum thickness in inches that should be attempted to get complete penetration with the SMAW process? (Obj. 14-2)
    a. ³⁄₁₆
    b. ¼
    c. ⁵⁄₁₆
    d. ⅜
13. Moving the electrode back and forth along the line of the weld on an open root butt joint is done to _____. (Obj. 14-2)
    a. Preheat the metal ahead of the weld
    b. Minimize the tendency to burn through
    c. Force the slag back over the top of the weld
    d. All of these
14. Keeping a small hole in the center of the weld pool on the root pass of an open root type joint is called what technique? (Obj. 14-2)
    a. Notch effect
    b. Burn through
    c. Keyhole
    d. All of these
15. When making a vertical-up groove weld, the travel angle should be _____. (Obj. 14-2)
    a. 0–10° drag
    b. 0–10° push
    c. 10–25° drag
    d. 10–25° push
16. When making vertical welds, which method is considered faster? (Obj. 14-2)
    a. Stringer beads
    b. Weave beads
    c. Uphill technique
    d. Downhill technique
17. On a 2F T-joint weld why is a 40° work angle used to favor the vertical plate? (Obj. 14-3)
    a. To get better root penetration
    b. To make the weld wider
    c. To force the weld metal up on this surface
    d. To reduce spatter
18. To prevent undercut along the top toe of a weave bead you should _____. (Obj. 14-3)
    a. Keep the electrode weave within the limits of the desired weld width
    b. Use a very short arc

c. Pause at the top of the bead (toe) against the vertical plate
   d. All of these
19. Which type of electrode is used with a *whipping the arc* technique? (Obj. 14-5)
   a. E6010, E6018
   b. E7016, E7018
   c. Cellulose electrodes
   d. Both a and c
20. Which electrode is considered low hydrogen with iron powder in its coating? (Obj. 14-5)
   a. E6010
   b. E6011
   c. E7016
   d. E7018

Review Questions

Write the answers in your own words.

21. Explain why every effort should be made to make welds in the flat and horizontal position and describe a situation when this is not possible. (Obj. 14-1)
22. Describe the welding techniques used for welding in the overhead position with cellulose-type electrodes. (Obj. 14-1)
23. Describe the purpose of the *whipping the electrode* technique and how the type of power source being used will affect this technique. (Obj. 14-2)
24. When making an open root groove weld with a cellulose-type electrode, name the technique and explain how the root penetration is controlled. (Obj. 14-2)
25. Draw a sketch of the proper weld profile of an open root V-groove weld. Measure the dimension of the width beyond the groove, face, and root reinforcement. (Obj. 14-2)
26. Name the advantages of using low hydrogen electrodes. (Obj. 14-3)
27. What effect do low hydrogen electrodes have on strength and the cost of welding? (Obj. 14-3)
28. Describe the advantage of the E7018 electrode over the E7016 low hydrogen electrode. (Obj. 14-3)
29. Since whipping the electrode is not allowed with the low hydrogen electrodes, explain how a vertical-up weld is accomplished with an E7018 electrode versus an E6010 electrode. Also describe the requirement for using more or less current with the low hydrogen electrode. (Obj. 14-3)
30. List the visual inspection criteria for the outside corner weld. (Obj. 14-4)

INTERNET ACTIVITIES

Internet Activity A

Describe how you would use the Internet to look for a job. Find a job that you would be interested in and explain why.

Internet Activity B

What's new in inspections? Keep current with inspection trends. Find the AWS Web site to see what is new in inspections. Remember that AWS uses these letters in its Web site address and that AWS is an organization.

Design credits: Cinema and movie icon: Denis Meshkov/123RF; and Illustration of Welding icons: Mr. Nachai Sorasee/123RF

Table 14-1 Job Outline: Shielded Metal Arc Welding Practice: Jobs 14-J26–J42 (Plate)

Job No.	Joint	Type of Weld	Position	Type of Electrode DCEP	DCEN	a.c.	AWS Specif. No. d.c.	AWS Specif. No. and/or	AWS Specif. No. a.c.	Text Reference
14-J26	Flat plate	Beading—stringer	Over. (4)	X		X	E6010		E6011	400
14-J27	Flat plate	Beading—weaved	Over. (4)	X		X	E6010		E6011	402
14-J28	Single-V butt joint backing bar	Groove welding—weaved—multipass	Flat (1G)	X		X	E6010		E6011	404
14-J29	T-joint	Fillet welding—weaved—multipass	Hor. (2F)	X		X	E6010		E6011	406
14-J30	Single-V butt joint backing bar	Groove welding—stringer—multipass	Hor. (2G)	X		X	E6010		E6011	408
14-J31	Square butt joint	Groove welding—multipass	Flat (1G)	X		X	E6010		E6011	410
14-J32	Outside corner joint	Groove welding—weaved—multipass	Flat (1G)	X		X	E6010		E6011	411
14-J33	Single-V butt joint	Groove welding—weaved—multipass	Flat (1G)	X		X	E6010		E6011	414
14-J34	T-joint	Fillet welding—stringer—multipass	Hor. (2F)	X		X	E7016		E7016	417
14-J35	T-joint	Fillet welding—single pass—travel up	Ver. (3F)	X		X	E6010		E6011	419
14-J36	T-joint	Fillet welding—weaved—multipass—travel up	Ver. (3F)	X		X	E6010		E6011	421
14-J37	T-joint	Fillet welding—weaved—multipass—travel up	Ver. (3F)	X		X	E7018		E7018	423
14-J38	Single-V butt joint backing bar	Groove welding—weaved—multipass—travel up	Ver. (3G)	X		X	E6010		E6011	425
14-J39	Square butt joint	Groove welding—multipass—travel up	Ver. (3G)	X		X	E6010		E6011	427
14-J40	Outside corner joint	Groove welding—weaved—multipass—travel up	Ver. (3G)	X		X	E6010		E6011	429
14-J41	T-joint	Fillet welding—weaved—multipass	Hor. (2F)	X		X	E7018		E7018	431
14-J42	Single-V butt joint backing bar	Groove welding—weaved—multipass—travel down	Ver. (3G)	X		X	E6010		E6011	433

NA = not applicable.

15

Shielded Metal Arc Welding Practice:

Jobs 15-J43–J55 (Plate)

Introduction

This third chapter in shielded metal arc welding (SMAW) practice may be considered as advanced welding. It provides practice in joints that are basic to code welding required for pipe and building construction. (As stated previously, the practice jobs are numbered consecutively through the three units.) Successful completion of these practice jobs is the final step in development of skill as a welder. A welder who is able to satisfactorily weld an open-root butt joint with complete penetration through to the back side in the vertical and overhead positions has achieved the manipulative skill that provides the basis for all types of welding application.

Much of this chapter is doing overhead welding. Use all the normal safety precautions and protective clothing to protect from spatter, slag, and hot metal. Earplugs and the bill on your welders cap should be used to deflect hot metal away from the ears.

The chapter also provides considerable experience in the use of both the hand and machine oxyacetylene cutting torches in the preparation of practice joints and test specimens. In addition, it adds to the student's experience in the use of various hand and machine tools. See the Job Outline, Table 15-5, page 468, for a list of all the practice jobs covered in this chapter.

Chapter Objectives

After completing this chapter, you will be able to:

15-1 Describe techniques for making V-groove welds on butt joints with and without backing in various positions with various electrodes and current.

15-2 Describe techniques for making fillet welds on lap and T-joints in various positions on plate and pipe with various electrodes and current.

15-3 Explain various test methods and acceptance criteria.

15-4 Produce weld specimens that can pass groove, fillet, and pressure tests with the SMAW process on plate.

The abilities to weld, to use the various cutting processes, to read job drawing and sketches, to do simple layout, and to be able to use the basic metalworking tools will have been acquired through the completion of these practice jobs. If student welders are able to perform these skills satisfactorily, they can feel confident that they have the necessary skills to satisfy the demands of industry for those jobs in which shielded metal arc welding is performed, Fig. 15-1.

Practice Jobs

Job 15-J43 Welding a Single-V Butt Joint

Objective

To weld a single-V butt joint in the vertical position by means of multipass groove welds, weave bead technique, travel up, with DCEP and/or a.c. shielded metal arc electrodes (AWS E6010–E6011 and E7016–E7018).

General Job Information

The single-V butt joint in the vertical position will be encountered in pipe and critical plate welding. The joint design and welding procedure are similar to those required in the ASME plate and pipe tests.

Welding Technique

First Pass The current setting should not be too low. Hold the electrode at a 90° angle to the plate surfaces (work angle) and no more than 10° down from the horizontal (travel angle), Fig. 15-2A, page 430.

The travel angle can be either a push or drag. The push angle is when the arc is being pushed along by the electrode angle, and the drag angle is when the arc is being dragged along by the electrode angle. When doing vertical welding, you should utilize the push travel angle. This angle uses the arc force to help overcome the effects of gravity.

Use the root pass technique in which an up-and-down motion is employed. On the upward motion do not allow the electrode to travel much more than 2 inches and do not extinguish the arc.

Make sure that the small hole is present at the point of welding during the welding operation. This small hole is referred to as a *keyhole*. This is your guarantee that you are securing penetration through to the back side. Complete penetration is necessary for maximum strength. It is also necessary when making a test weld. Keep the face of the weld as flat as possible.

The up-and-down motion of an electrode is referred to as the *whipping technique*. The upward motion lengthens the arc gap, which raises the voltage. With older welding machines, this long arc high voltage causes the amperage (current-heat) to drop. This cools the weld and reduces the melt rate (deposition rate) of the electrode. The forward motion allows the welder to control the keyhole size and thus the penetration.

Fig. 15-1 Pipe welders working in unison to shielded metal arc weld a 5G position pipe butt joint with progression in the downhill direction for the root pass. Multi-Operator equipment is being used to supply the welding arc and auxiliary power for power cleaning and grinding tools. Miller Electric Mfg. Co.

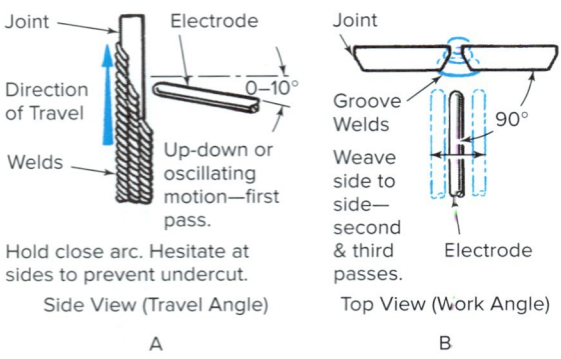

Fig. 15-2 Electrode position when welding a butt joint in the vertical position (3G).

With some of the more recent advancements in welding power sources, this technique will not work as well. These welding machines use closed loop feedback circuits, which maintain the current once it has been set. So when you perform long arcing with this type of welding machine, the voltage does go up but the amperage level is maintained. Instead of cooling the weld and reducing the melt rate, the opposite happens. Many welders describe it as "the arc is getting hotter." Check with your instructor to determine which type of power source you are using. If it is a closed loop feedback power source with a vertical volt-amp curve, you should not long arc the electrode. Instead merely move the arc out of the weld pool (keyhole) area and let the weld pool cool down (dry up). Then at the appropriate time bring the arc back into the keyhole and continue the weld. By watching the size of the keyhole produced and by moving the electrode forward and backward, the penetration level can be controlled. Do not use the up-and-down motion with this type of welding machine. This whipping motion technique should never be used on low hydrogen electrodes, those that end with a classification number of 5, 6, or 8. These electrodes should never be long arced or porosity will result.

Second and Third Passes Current adjustment must be increased. Hold the electrode at a 90° angle to the plate surfaces and weave it from side to side, Fig. 15-2B. A uniform weave and rate of travel is important to obtain a weld of uniform quality and appearance. Movement across the face of the weld must be fast to prevent excessive convexity in the center of the weld. Maintain a close arc gap at all times. Keep electrode movement within the limits of the bead width and hesitate at the sides of the weld to avoid undercut.

Practice starts and stops on all passes.

Operations

1. Obtain plates; check the job drawing, Fig. 15-3, for quantity and size.
2. Obtain electrodes of each quantity, type, and size specified in the job drawing.
3. Set the power for 75 to 125 amperes. Set a d.c. power source for electrode positive.
4. Set up the plates and tack weld them as shown on the job drawing. Grind and feather the tack welds.
5. Place the joint in the vertical position on the welding table. Make sure that it is connected with the work connection.
6. Make the first pass with 1/8-inch electrodes as shown on the job drawing. Grind the crater prior to restarting.
7. Chip the slag from the weld, brush, and inspect. Refer to Inspection, below.
8. Increase the current to 130 to 160 amperes.
9. Make the second and third passes with 5/32-inch electrodes as shown on the job drawing.
10. Chip the slag from the weld, brush, and inspect. Refer to Inspection, below.
11. Practice these welds until you can produce good welds consistently with E6010 or E6011 electrodes.

Once these welds are acceptable to your instructor, make a vertical-up V-groove butt joint using E6010 or E6011 for the root pass and E7016 or E7018 for all subsequent passes. Use a steady short arc gap with the low hydrogen electrodes; do not use the whipping motion with these electrodes. Use 3/32- or 1/8-inch diameter electrodes at 60 to 120 amperes and 80 to 150 amperes, respectively. Perform the same inspection and qualification testing that was done on the previous joints.

Inspection

Compare the back side of the first pass with Fig. 15-4 and all passes with Fig. 15-5, page 432. Check them for the following weld characteristics:

Width and height: Uniform
Appearance: Smooth with close ripples; free of voids and high spots. Restarts should be difficult to locate.
Size: Refer to the job drawing. Check with a butt weld gauge.
Face of weld: First two passes, flat; last pass, slight convexity
Edges of weld: Good fusion, no overlap, no undercut
Starts and stops: Free of depressions and high spots

NO	PART NO	NAME	SIZE	MATL	SPEC
10		SHIELDED-ARC ELECTRODES	3/32 DIA.	STEEL	E7016-E7018
10		SHIELDED-ARC ELECTRODES	1/8 DIA.	STEEL	E6010 and / or E6011 for Root Pass; E7016 or E7018 for All Other Passes
20		SHIELDED-ARC ELECTRODES	5/32 DIA.	STEEL	E6010 and/or E6011
5	-9	PLATES	3/8 x 1½ x 10	STEEL	MILD

TOLERANCE–FRACTIONAL LIMITS 1/8	DECIMAL LIMITS
PART NAME SINGLE BEVEL BUTT JOINT	TYPE OF WELD GROOVE
SCALE NO SCALE; POSITION OF WELDING VERTICAL	PART NUMBER J-43

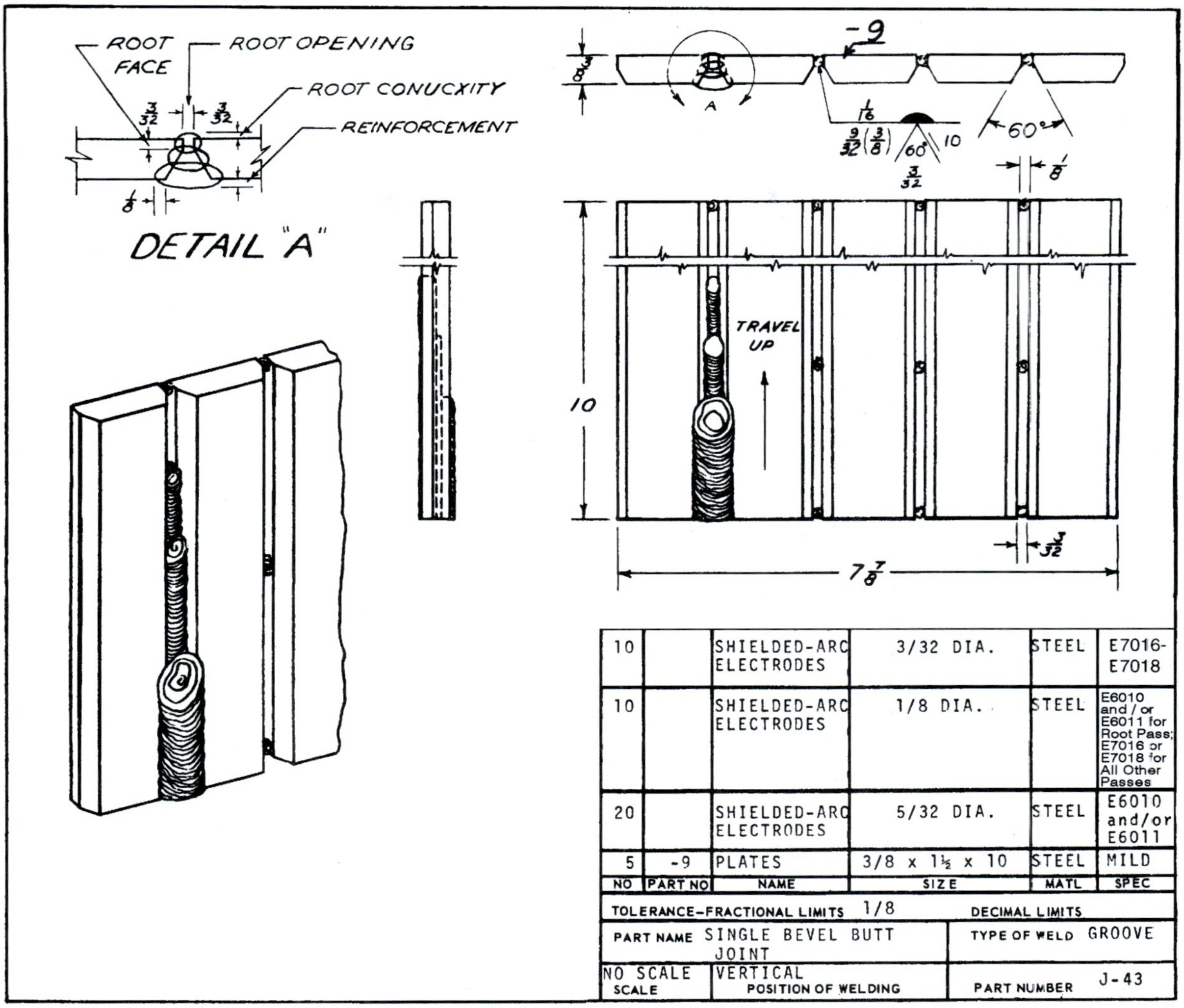

Fig. 15-3 Job drawing J43.

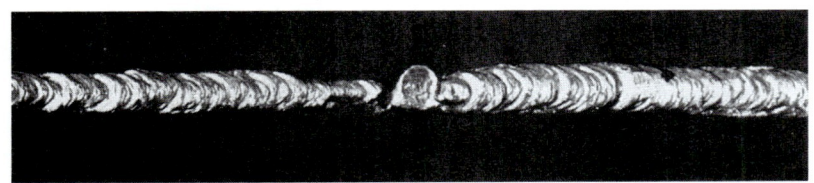

Fig. 15-4 Penetration and fusion through the back side of a V-groove butt joint. Note the spot in the center of the bead where there is a break in the penetration. This can be caused by failure to burn through a tack or a poor tie-in on a restart. Edward R. Bohnart

Beginnings and endings: Full size, craters filled
Back side: Complete and uniform penetration
Penetration and fusion: Through the back side, preceding passes, and all plate surfaces. See Fig. 15-4.
Surrounding plate surfaces: Free of spatter
Slag formation: Full coverage, easily removable

Disposal

Put completed joints in the scrap bin so that they will be available for further use. Joints can be cut and beveled between welds for further butt welding, and unwelded plate surfaces can be used for beading.

Qualifying Test

At this point you should be able to pass a 2G test in the vertical position. (See Test 2 on p. 456.) This test is similar to the test for pipe or plate developed by the ASME.

Job 15-J44 Welding a T-Joint

Objective

To weld a T-joint in the overhead position by means of a single-pass fillet weld, stringer bead technique, with

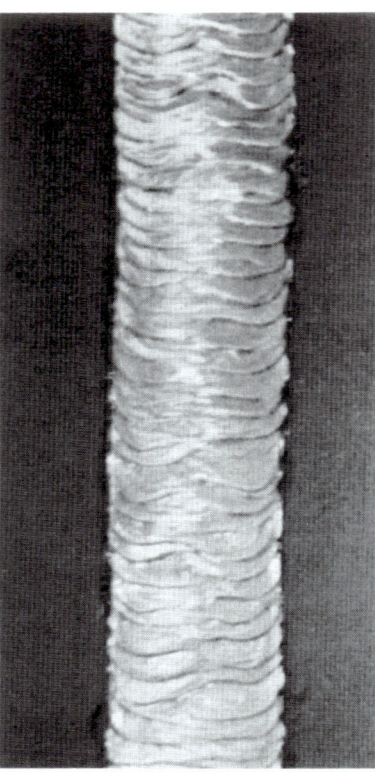

A B

Fig. 15-5 (A) Typical appearance of a multipass groove weld in a V-groove butt joint without backing, welded in the vertical position (3G), travel up, with DCEP or a.c. shielded arc electrodes. (B) Closeup of the weld bead. *Edward R. Bohnart*

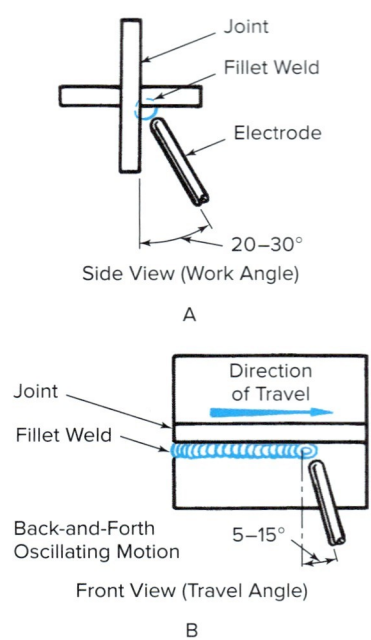

Fig. 15-6 Electrode position when welding a T-joint in the overhead position (4F).

DCEP and/or a.c. shielded metal arc electrodes (AWS E6010–E6011).

General Job Information

The T-joint welded in the overhead position is used extensively in shipyard and structural work.

Welding Technique

Current adjustment should be high. Hold the electrode at a 20 to 30° angle to the vertical plate, Fig. 15-6A, and tilt it no more than 15° in the direction of travel, Fig. 15-6B. Figure 15-7 shows the actual welding position. Use a back-and-forth motion along the line of weld. Hold a close arc when depositing metal. The arc must not be extinguished on the forward motion. Pay close attention to fusion and penetration at the root of the joint and to the plate surfaces. Slag must not be permitted to run ahead of the metal being deposited. Avoid overheating the weld pool to prevent excessive convexity and overlap. While undercut is a problem, it can be overcome in the usual manner. Watch your speed of travel; the bead must be permitted to build up to its full size.

Practice starts and stops. Travel from left to right and from right to left.

Operations

1. Obtain plates; check the job drawing, Fig. 15-8, for quantity and size.
2. Obtain electrodes of the quantity, type, and size specified in the job drawing.

Fig. 15-7 Electrode and electrode holder positioned as instructed in Fig. 15-6. Note how the joint is held by the jig. *Edward R. Bohnart*

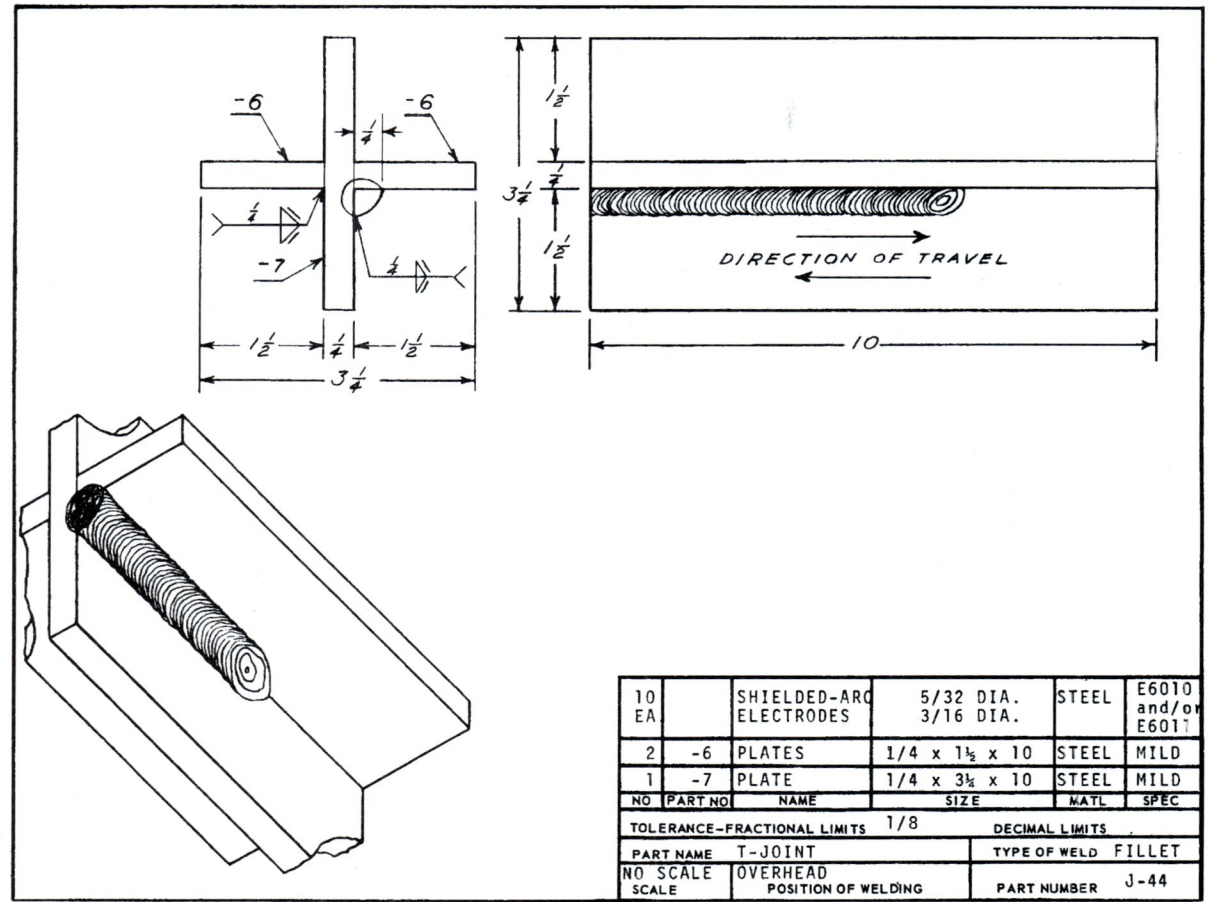

Fig. 15-8 Job drawing J44.

3. Set the power source for 120 to 170 amperes. Set a d.c. power source for electrode positive.
4. Set up the plates as shown on the job drawing and tack weld them at each corner.
5. Fasten the joint in the overhead welding jig. See Fig. 15-7.
6. Make the weld as shown on the job drawing. Manipulate the electrode as instructed in Fig. 15-6.
7. Chip the slag from the weld, brush, and inspect. Refer to Inspection.
8. Practice these welds until you can produce good welds consistently with both types of electrodes.

Inspection

Compare the weld with Fig. 15-9, page 434, and check it for the following characteristics:

Width and height: Uniform
Appearance: Smooth with close ripples; free of voids and high spots. Restarts should be difficult to locate.
Size: Refer to the job drawing. Check with a fillet weld gauge.
Face of weld: Flat
Edges of weld: Good fusion, no overlap, no undercut
Starts and stops: Free of depressions and high spots
Beginnings and endings: Full size, craters filled
Penetration and fusion: To the root of the joint and both plate surfaces
Surrounding plate surfaces: Free of spatter
Slag formation: Full coverage, easily removable

Disposal

Save the completed joints for your next job, which will be multipass fillet welding in the overhead position.

Check Test

After you are able to make welds that are satisfactory in appearance, make up a test specimen similar to that made in Job 13-J14. (Refer to Figs. 13-57 and 13-58, pp. 364–365.) Use the plate thickness, welding technique, and electrode type and size specified for this job. Weld on one side only.

Break the completed weld and examine the surfaces for soundness. The weld must not be porous. It must show complete fusion and penetration at the root of the joint and to the plate surfaces. The joint should break evenly through the throat of the weld.

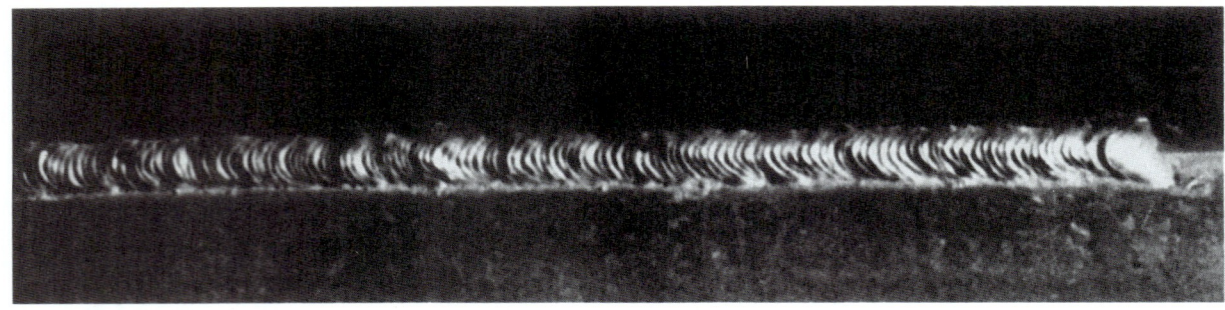

Fig. 15-9 Typical appearance of a single-pass fillet weld in a T-joint welded in the overhead position with DCEP or a.c. shielded arc electrodes. *Edward R. Bohnart*

Job 15-J45 Welding a T-Joint

Objective
To weld a T-joint in the overhead position by means of multipass fillet welds, stringer bead technique, with DCEP and/or a.c. shielded metal arc electrodes (AWS E6010–E6011).

General Job Information
When more than one pass is necessary for fillet welds in the overhead position, it is usually the practice to apply stringer beads. It is to be understood that the thickness of plate used in this job would not require a ½-inch fillet weld. The plate is merely being used as much as possible.

Welding Technique
Current adjustment can be high. The angle of the electrode to the vertical plate changes according to the pass being deposited, Fig. 15-10. It is 20 to 30° for the second pass, 35 to 45° for the third and fourth passes, 45° for the fifth pass, and 20 to 30° for the sixth pass. The electrode is tilted no more than 15° in the direction of travel for all passes. Use the stringer bead technique and deposit each pass in the sequence shown on the job drawing. You will have difficulty in obtaining the proper proportions, but practice will overcome it. Each bead must be fused with the preceding bead and the plate surface. The face of each layer should be as flat as possible.

Practice starts and stops. Travel from left to right and from right to left. Be sure to remove slag from each bead before starting the next bead.

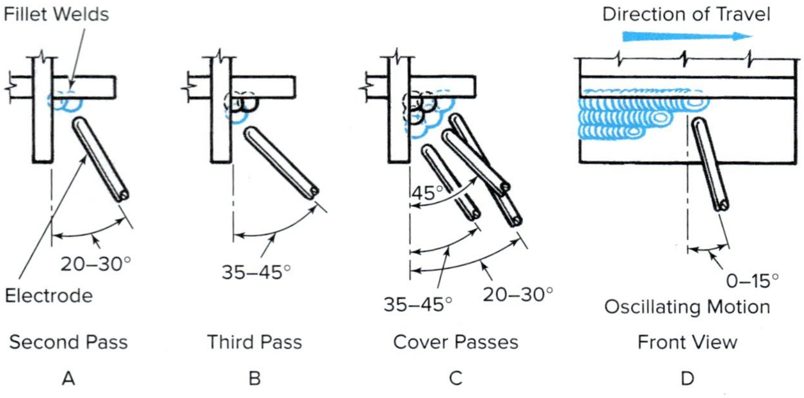

Operations
1. Use the joints welded in Job 15-J44.
2. Obtain electrodes of the quantity, type, and size specified in the job drawing, Fig. 15-11.
3. Set the power source for 120 to 160 amperes. Set a d.c. power source for electrode positive.
4. Fasten the joint in the overhead position with an overhead welding jig.
5. Make welds as shown on the job drawing. Manipulate the electrode as shown in Fig. 15-10.
6. Chip the slag from the welds, brush, and inspect between each pass. Refer to Inspection.
7. Practice these welds until you can produce good welds consistently with both types of electrodes.

Inspection
Compare each pass with Fig. 15-12 and check it for the following weld characteristics:

Width and height: Uniform
Appearance: Smooth with close ripples; free of voids and high spots. Restarts should be difficult to locate.
Size: Refer to the job drawing. Check with a fillet weld gauge.
Face of weld: Flat
Edges of weld: Good fusion, no overlap, no undercut
Starts and stops: Free of depressions and high spots
Beginnings and endings: Full size, craters filled

Fig. 15-10 Electrode position when welding a T-joint in the overhead position (4F) using the stringer bead technique.

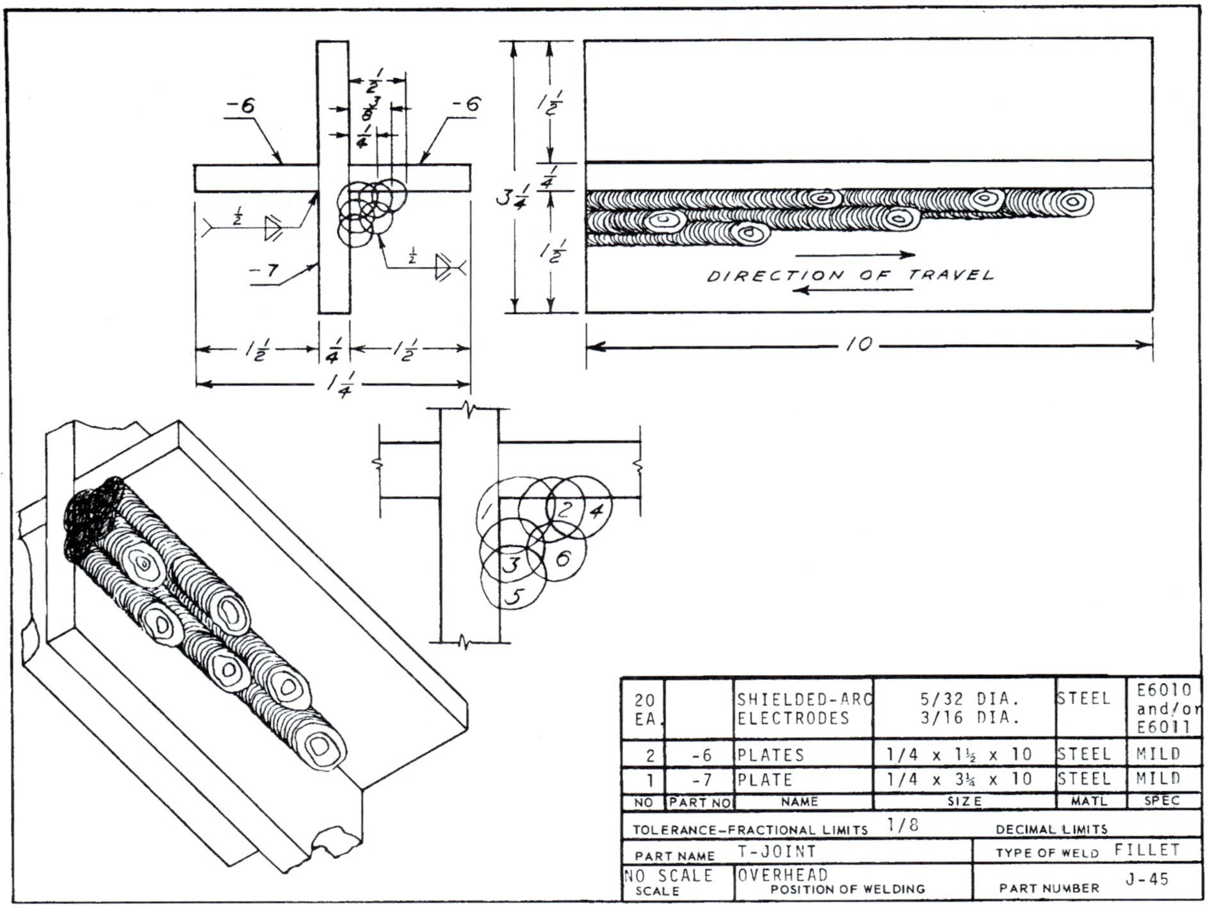

Fig. 15-11 Job drawing J45.

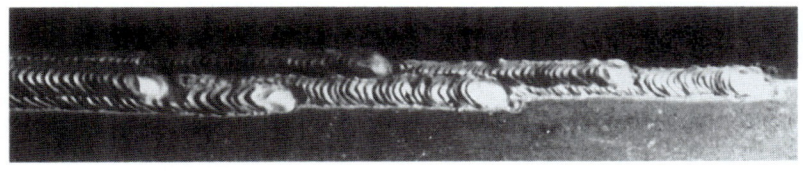

Fig. 15-12 Typical appearance of a multipass fillet weld in a T-joint welded in the overhead position (stringer bead technique) with DCEP or a.c. shielded arc electrodes. Edward R. Bohnart

Penetration and fusion: To preceding beads and both plate surfaces
Surrounding plate surfaces: Free of spatter
Slag formation: Full coverage, easily removable

Disposal

Put completed joints in the scrap bin so that they will be available for further use. Plate edges can be used for square butt welding, and unwelded surfaces can be used for beading.

Job 15-J46   Welding a Lap Joint

Objective

To weld a lap joint in the overhead position by means of a single-pass fillet weld with DCEP and/or a.c. shielded metal arc electrodes (AWS E6010–E6011).

General Job Information

The overhead lap joint is found very frequently in tank, structural, and shipyard work. Because of the size and nature of these objects, it is impractical to position them. The lap joint can be welded with three stringer passes instead of a single weaved bead.

Welding Technique

Current adjustment must not be too high. Hold the electrode at a 30° angle to the lower plate, Fig. 15-13A, page 436, and tilt it 20° in the direction of travel, Fig. 15-13B. Figure 15-14, page 436, shows the actual position for welding. A back-and-forth whipping motion

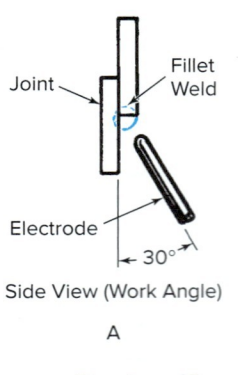

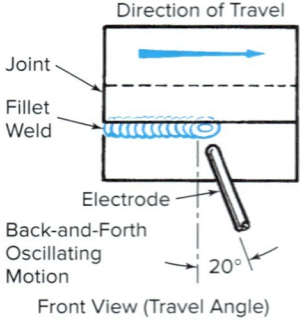

Fig. 15-13 Electrode position when welding a lap joint in the overhead position.

may be used. Do not extinguish the arc on the forward motion. The whiplike action helps to prepare the root of the joint for the weld deposit by preheating it, allows the weld pool to cool off so that excessive convexity and spilling are prevented, and washes the slag back over the weld deposit. Do not allow the arc to contact the light edge of the outer plate. At no time should the arc touch the sections of the plate surface that are outside the final bead width.

Practice starts and stops. Travel from left to right and from right to left.

Fig. 15-14 Electrode and electrode holder positioned as instructed in Fig. 15-13. Note how the joints are held by the jig. *Mark A. Dierker/McGraw Hill*

Operations

1. Obtain plates; check the job drawing, Fig. 15-15, for quantity and size.
2. Obtain electrodes of the quantity, type, and size specified in the job drawing.
3. Set the power source for 140 to 215 amperes. Set a d.c. power source for electrode positive.
4. Set up the plates and tack weld them as shown on the job drawing.
5. Fasten the joints in the overhead position with an overhead welding jig. See Fig. 15-14.
6. Make welds as shown on the job drawing. Manipulate the electrode as shown in Fig. 15-13.
7. Chip the slag from the welds, brush, and inspect. Refer to Inspection.
8. Practice these welds until you can produce good welds consistently with both types of electrodes and by using the stringer bead technique with $\frac{5}{32}$-inch electrodes.

Inspection

Compare the welds with Fig. 15-16 and check them for the following weld characteristics:

Width and height: Uniform
Appearance: Smooth with close ripples; free of voids and high spots
Size: Refer to the job drawing. Check with a fillet weld gauge.
Face of weld: Flat
Edges of weld: Good fusion, no overlap, no undercut. The edge of the upper plate must not be burned away.
Starts and stops: Free of depressions and high spots
Beginnings and endings: Full size, craters filled
Penetration and fusion: To the root of the joint and both plate surfaces
Surrounding plate surfaces: Free of spatter
Slag formation: Full coverage, easily removable

Disposal

Put completed joints in the scrap bin so that they will be available for further use. Unwelded plate surfaces may be used for beading.

Check Test

Make up test specimens and carry out the usual lap joint (fillet-weld) test.

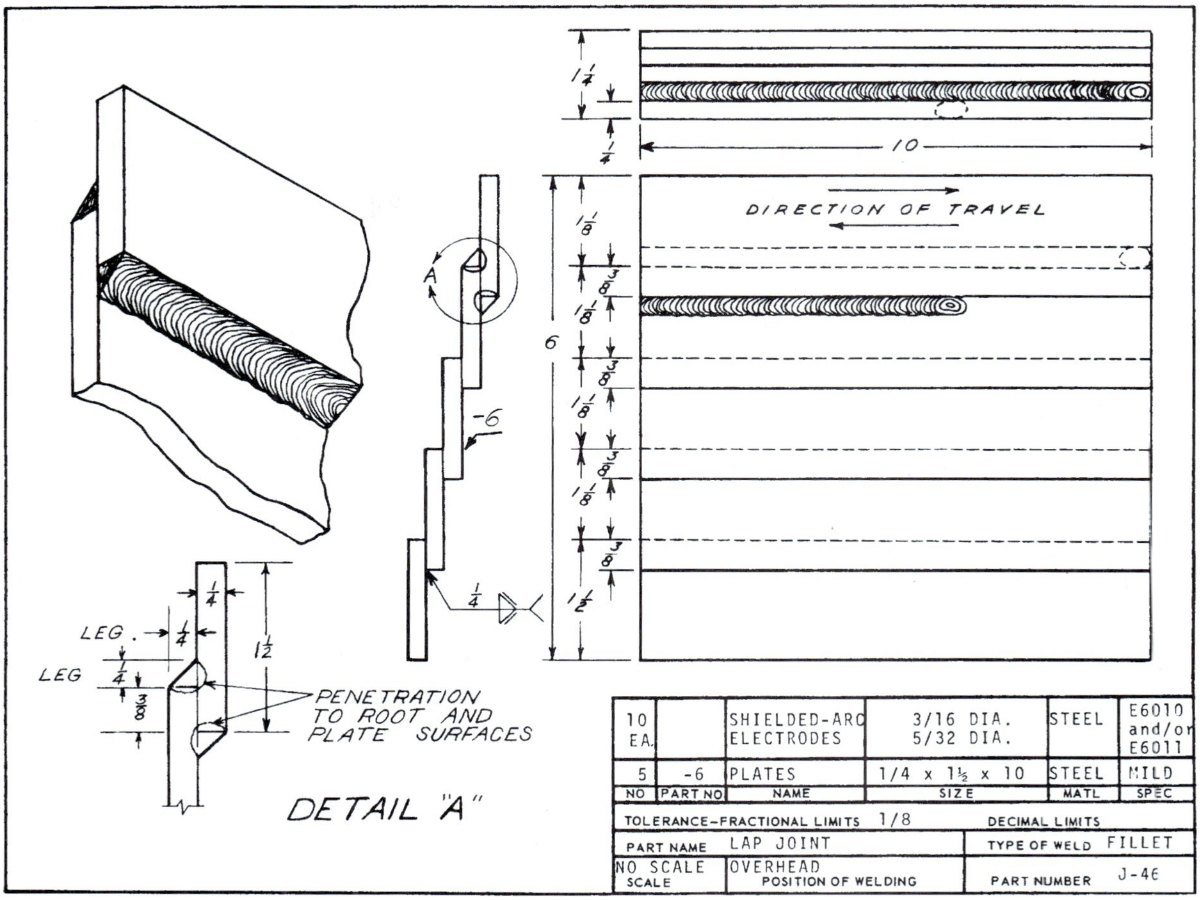

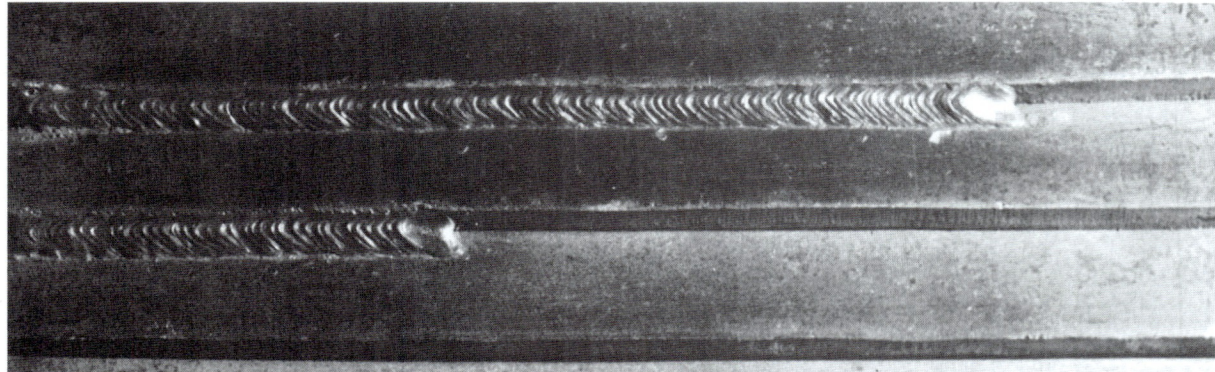

Fig. 15-15 Job drawing J46.

Fig. 15-16 Typical appearance of a single-pass fillet weld in a lap joint welded in the overhead position with DCEP or a.c. shielded arc electrodes. Edward R. Bohnart

Job 15-J47 Welding a Lap Joint

Objective
To weld a lap joint in the overhead position by means of a single-pass fillet weld with DCEP and/or a.c. shielded metal arc electrodes (AWS E7016–E7018).

General Job Information
This job is similar to Job 15-J46 except that low hydrogen electrodes are used.

Welding Technique
Current adjustment may be somewhat higher than for electrode positive. See Fig. 15-13 for the correct electrode

position. Use the welding technique that you practiced in the previous job. The slag will be somewhat more difficult to control than with reverse polarity. The bead will have a tendency to become convex and to run down the vertical plate surface. Undercut is not a problem in this job.

Practice starts and stops. Travel from left to right and from right to left.

Operations

1. Obtain plates; check the job drawing, Fig. 15-17, for quantity and size.
2. Obtain electrodes of the quantity, type, and size specified in the job drawing.
3. Set the power source for 140 to 230 amperes. Set a d.c. power source for electrode positive.
4. Set up the plates and tack weld them as shown on the job drawing.
5. Fasten the joints in the overhead position with an overhead welding jig.
6. Make welds as shown on the job drawing. Manipulate the electrode as instructed in Fig. 15-13, page 436.
7. Chip the slag from the welds, brush, and inspect. Refer to Inspection.
8. Practice these welds until you can produce good welds.

Inspection

Compare the welds with Fig. 15-18 and check them for the following characteristics:

Width and height: Uniform
Appearance: Smooth with close ripples; free of voids and high spots. Restarts should be difficult to locate.
Size: Refer to the job drawing. Check with a fillet weld gauge.
Face of weld: Slightly convex
Edges of weld: Good fusion, no overlap, no undercut. The edge of the upper plate must not be burned away.
Starts and stops: Free of depressions and high spots
Beginnings and endings: Full size, craters filled
Penetration and fusion: To the root of the joint and both plate surfaces
Surrounding plate surfaces: Free of spatter
Slag formation: Full coverage, easily removable

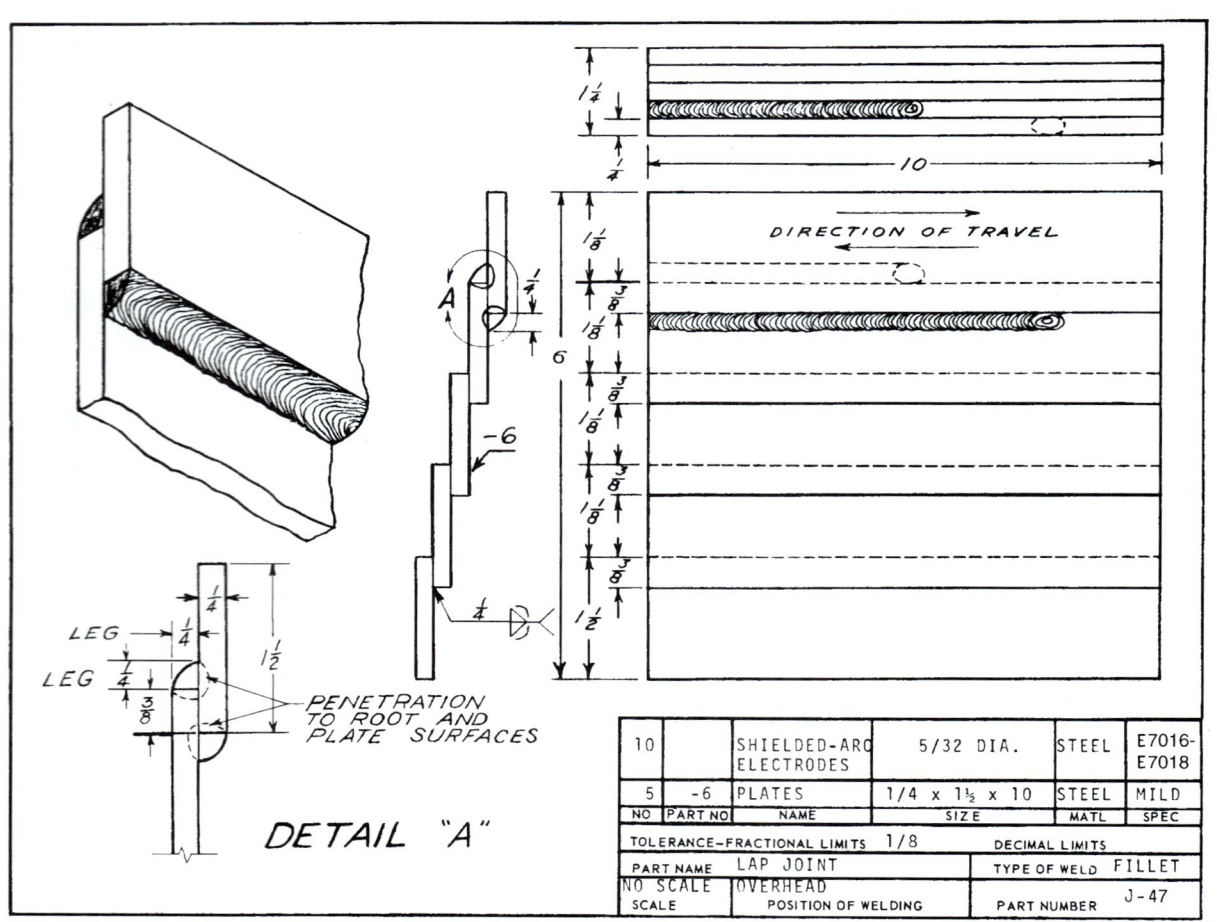

Fig. 15-17 Job drawing J47.

Fig. 15-18 Typical appearance of a single-pass fillet weld in a lap joint welded in the overhead position with DCEP or a.c. shielded arc electrodes. *Edward R. Bohnart*

Disposal

Put completed joints in the scrap bin so that they will be available for further use. Unwelded plate surfaces can be used for beading.

Check Test

Make up a test specimen and carry out the usual lap joint (fillet weld) test.

Job 15-J48 Welding a Single-V Butt Joint

Objective

To weld a single-V butt joint in the vertical position by means of multipass groove welds, weave bead technique, travel down, with DCEP and/or a.c. shielded metal arc electrodes (AWS E6010–E6011).

General Job Information

The single-V butt joint is welded by the downhill method in pipeline work, tank, and shipyard welding. When used in pipe, it is usually limited to sizes under 14 inches in diameter. While the downhill technique is somewhat difficult to master, it has the advantage of speed. When butt joints in pipe are welded by the downhill method without a backing ring, the process is somewhat slower than when a backing ring is used. The cost of the backing ring, however, and the time spent installing it erases the advantage of speed when applied to pipe 10 inches and under.

The skill acquired in the practice of this joint will assist you in learning to weld pipe, which is presented in Chapter 16.

Welding Technique

Current adjustment should be a little higher than that used for a joint with a backing strip and much higher than that used for uphill welding. Electrode angle, arc length, and speed of travel are very important. Tilt the electrode 15 to 20° in the direction of travel, Fig. 15-19A, and hold it at a 90° angle to the surface of the plates, Fig. 15-19B.

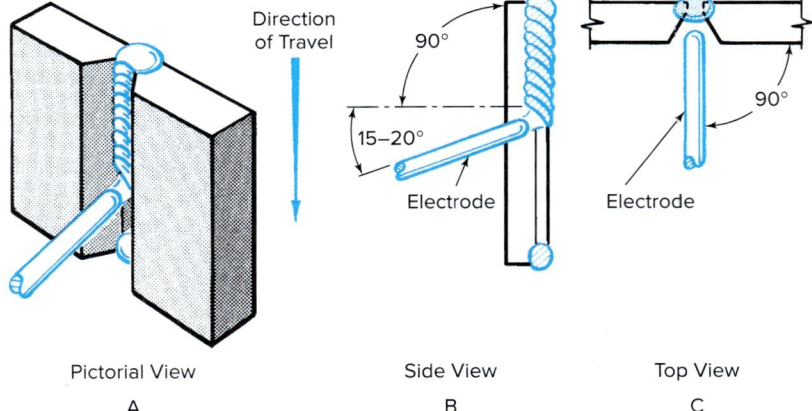

Fig. 15-19 Electrode position when welding a V-groove butt joint without backing the vertical position, travel down (3G).

First Pass The first pass must be run with a close arc. Some welders like to drag the coating in contact with the work and proceed downward at a rapid rate. Very little side-to-side motion is necessary unless there is a tendency to burn through and cause an enlarged hole. The back side of the joint should show a weld of uniform width and height without holes or excessive burn-through. The face side of the weld should be flat with good fusion into the side walls of the plate. There will be no real problem with undercut. If you are having difficulty with this pass, practice with different current settings. Some welders can handle a wider gap than others can.

Second and Third Passes This joint should be welded in three passes. The second and third passes should be run with a larger electrode. Weave the electrode slightly and increase the side-to-side motion with each pass. Watch carefully for holes and voids in the face of the weld. There is also a tendency for the face of the weld to become concave. This can be avoided by pausing for a shorter time at the edges of the weld when making the weave motion. Since there is no problem with undercutting, it is not necessary to pause at the side of the weld to any extent.

Make sure that fusion and penetration are secured between beads and into the groove face. The face of each completed bead should be as flat as possible. The final bead should have a slight contour in order to make sure that it is full and not below the surface of the plate. When making the final pass, do not weave beyond the sharp edges of the plate at each side.

Review Job 14-J42 for additional information on downhill welding. Practice varying the heat setting, electrode angle, arc length, and speed of travel. Starts and stops should also be practiced.

Operations

1. Obtain plates; check the job drawing, Fig. 15-20, for quantity and size.
2. Obtain electrodes of each quantity, type, and size specified in the job drawing.
3. Set the power source for 125 to 200 amperes. Set a d.c. power source for electrode positive.
4. Set up the plates and tack weld them as shown on the job drawing.
5. Stand the joint in the vertical position on the welding table. Make sure that it is connected with the work connection.
6. Make the first pass with 5/32-inch electrodes as shown on the job drawing. Manipulate the electrode as instructed in Fig. 15-19, page 439.
7. Chip the slag from the weld, brush, and inspect. Refer to Inspection.
8. Readjust the power source to 130 to 225 amperes.
9. Make the second and third passes with 3/16-inch electrodes as shown on the job drawing.
10. Chip the slag from the welds, brush, and inspect. Refer to Inspection.
11. Practice these welds until you can produce good welds consistently.

Inspection

Compare each pass with Fig. 15-21 and check it for the following weld characteristics:

Width and height: Uniform
Appearance: Very smooth with fine ripples; free of voids and high spots. Restarts should be difficult to locate.

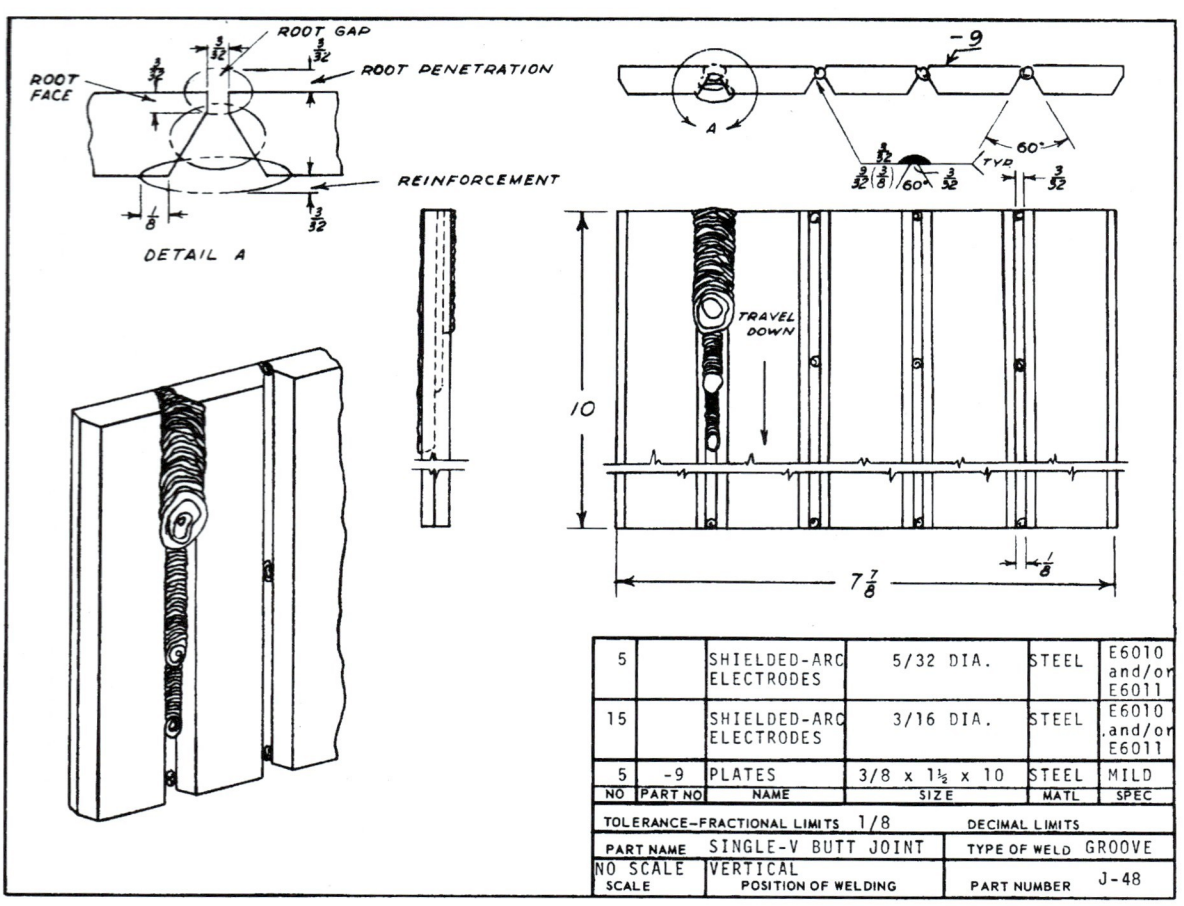

Fig. 15-20 Job drawing J48.

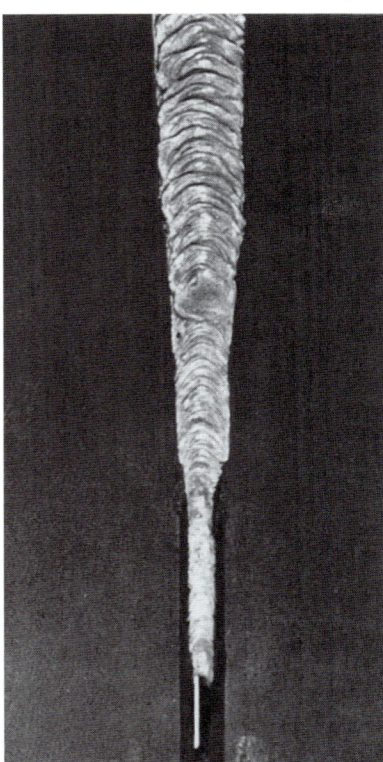

Fig. 15-21 Typical appearance of a multipass groove weld in a V-groove butt joint without a backing strip welded in the vertical position, travel down, with DCEP or a.c. shielded arc electrodes.
Edward R. Bohnart

Size: Refer to the job drawing. Check with a butt weld gauge.
Face of weld: Flat to slightly convex
Edges of weld: Good fusion, no overlap, no undercut
Starts and stops: Free of depressions and high spots
Beginnings and endings: Full size, craters filled
Penetration and fusion: To the root of the weld and plate surfaces
Surrounding plate surfaces: Free of spatter
Slag formation: Full coverage, easily removable

Disposal

Put completed joints in the scrap bin so that they will be available for further use. The plate edges can be beveled for bevel-butt welding, and unwelded plate surfaces can be used for beading.

Qualifying Test

At this point, you should make up test plates. Cut face and root bevel test coupons from the plates and test them. (See Test 2, p. 456.) This test is similar to the API test requirements.

Job 15-J49 Welding a T-Joint

Objective

To weld a T-joint in the overhead position by means of multipass fillet welds, weave-bead technique, with DCEP and/or a.c. shielded metal arc electrodes (AWS E6010–E6011).

General Job Information

The welder is frequently asked to make large fillets in the overhead position with a minimum number of passes and large electrodes. This is especially true of shipyard and structural welding.

Welding Technique

First Pass Current setting can be high. Hold the electrode at a 30° angle to the vertical plate, Fig. 15-22A, and tilt it 5 to 15° in the direction of travel, Fig. 15-22B. Use the welding technique learned in Job 15-J44.

Second Pass The current is increased to burn the larger electrode. The electrode position is the same as for the first pass. Weave the electrode slightly with a whiplike oscillating motion. Remember that because the electrode is larger, more metal is being deposited in the crater so there is more slag present. Keep the electrode motion within the limits of the weld size to prevent undercut.

Note: If it is desirable to get more use out of the plate, a third layer may be applied. This should consist of stringer beads deposited according to the procedures practiced in making beads 4, 5, and 6 in Job 15-J45.

Operations

1. Obtain plates; check the job drawing, Fig. 15-23, page 442, for the correct quantity and size.
2. Obtain electrodes of each quantity, type, and size specified in the job drawing.
3. Set the power source for 120 to 170 amperes. Set a d.c. power source for electrode positive.

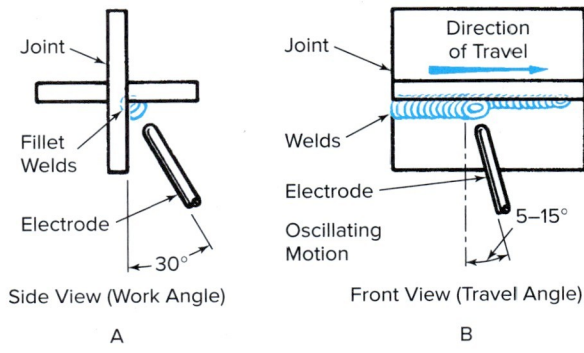

Fig. 15-22 Electrode position when welding a T-joint in the overhead position (4F).

NO	PART NO	NAME	SIZE	MATL	SPEC
10		SHIELDED-ARC ELECTRODES	5/32 DIA.	STEEL	E6010 and/or E6011
10		SHIELDED-ARC ELECTRODES	3/16 DIA.	STEEL	E6010 and/or E6011
2	-6	PLATES	1/4 x 1½ x 10	STEEL	MILD
1	-7	PLATE	1/4 x 3¼ x 10	STEEL	MILD

TOLERANCE–FRACTIONAL LIMITS 1/8	DECIMAL LIMITS
PART NAME T-JOINT	TYPE OF WELD FILLET
NO SCALE SCALE / OVERHEAD POSITION OF WELDING	PART NUMBER J-49

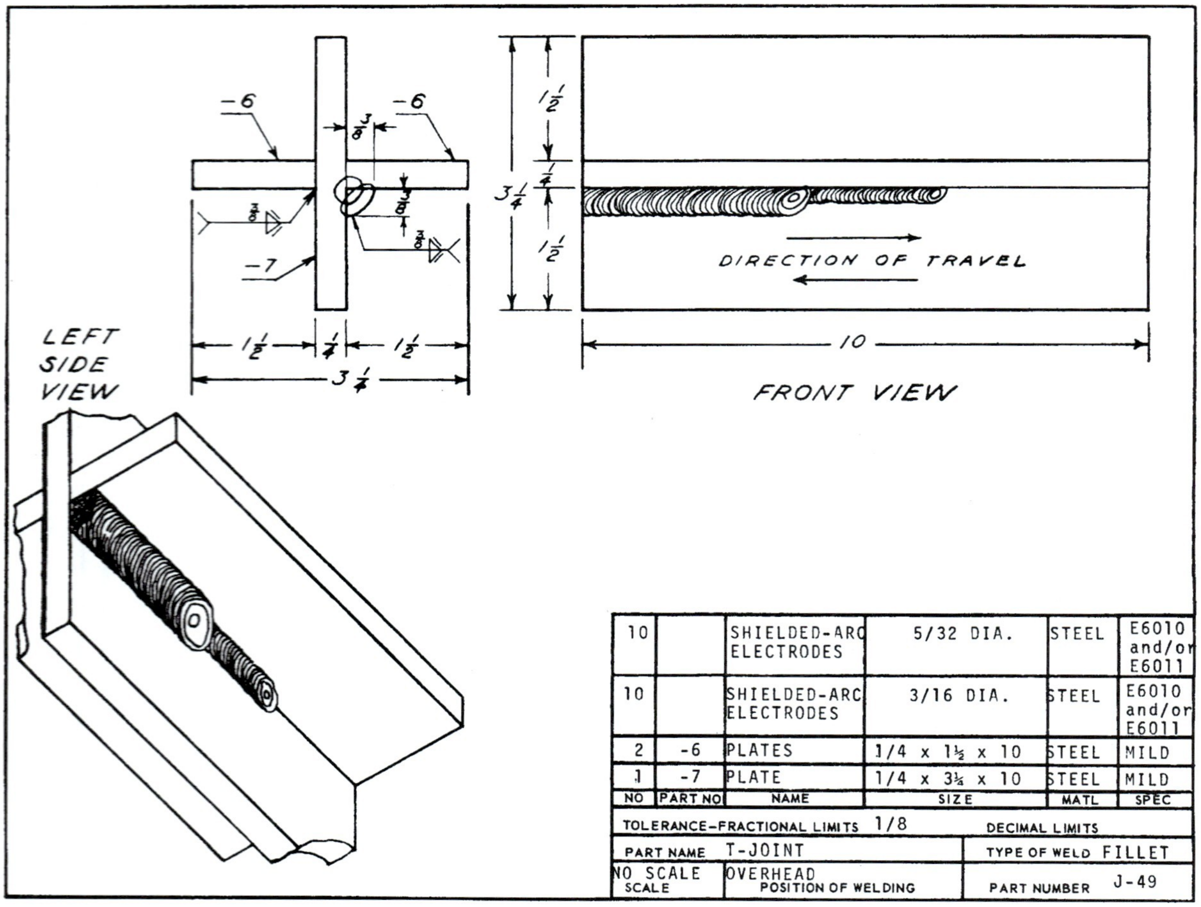

Fig. 15-23 Job drawing J49.

4. Set up the plates as shown on the job drawing and tack weld them at each corner.
5. Fasten the joint in the overhead position with an overhead welding jig.
6. Make the first pass with 5⁄32-inch electrodes as shown on the job drawing. Manipulate the electrode as instructed in Fig. 15-22, page 441.
7. Chip the slag from the weld, brush, and inspect. Refer to Inspection.
8. Increase the power source setting to 140 to 225 amperes.
9. Make the second pass with 3⁄16-inch electrodes as shown on the job drawing.
10. Chip the slag from welds, brush, and inspect. Refer to Inspection.
11. Practice these welds until you can produce good welds consistently with both types of electrodes.

Inspection

Compare each pass with Fig. 15-24 and check it for the following characteristics:

Width and height: Uniform

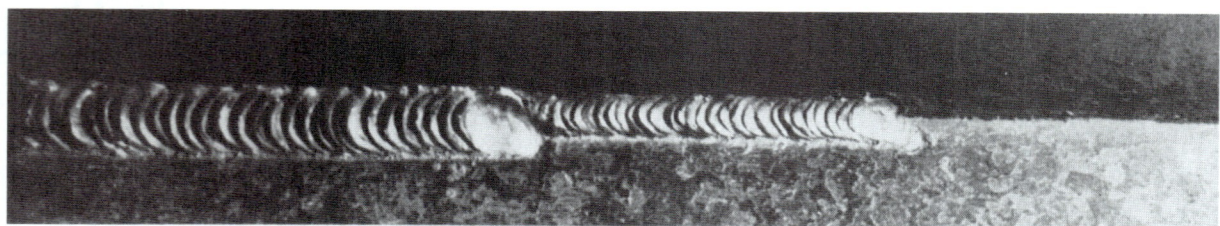

Fig. 15-24 Typical appearance of a multipass fillet weld in a T-joint welded in the overhead position (4F) (weave bead technique) with DCEP or a.c. shielded arc electrodes. Edward R. Bohnart

Appearance: Smooth with close ripples; free of voids and high spots. Restarts should be difficult to locate.
Size: Refer to the job drawing. Check with a fillet weld gauge.
Face of weld: Flat
Edges of weld: Good fusion, no overlap, no undercut
Starts and stops: Free of depressions and high spots
Beginnings and endings: Full size, craters filled
Penetration and fusion: To the root of the joint and both plate surfaces
Surrounding plate surfaces: Free of spatter
Slag formation: Full coverage, easily removable

Disposal

Put completed joints in the scrap bin so that they will be available for further use. The plate edges can be used for square butt welding, and unwelded surfaces can be used for beading.

Job 15-J50 Welding a T-Joint

Objective

To weld a T-joint in the overhead position by means of multipass fillet welds, weave-bead technique, with DCEP and/or a.c. shielded metal arc electrodes (AWS E6010–E6011).

General Job Information

This joint is used extensively in marine construction, especially barge construction, and in heavy fabrication. It differs from the previous job only in that the size of the electrode is larger, the size of the weld is larger, and more weld metal is carried in each pass.

Welding Technique

First Pass Current adjustment can be high. See Fig. 15-22, page 441, for the correct electrode position. Maintain a close arc gap. Weave the electrode by moving it forward with a rapid whipping motion and at the same time lengthen the arc slightly. Do not break the arc and return it to the crater when the crater has solidified. This prevents the molten pool from becoming too hot and spilling off, preheats the root of the joint ahead of the deposit, and forces the slag over the deposit. Because better control of weld metal and arc is possible, undercut, overlap, and convexity are reduced, and the appearance is uniform.

Second Pass The same technique is used for the second pass as for the first pass, but more metal is deposited. You will need practice to increase the deposition. Practice starts and stops. Travel in all directions.

Note: If it is desirable to get more use out of the plate, a third layer can be applied. This should consist of stringer beads deposited according to the procedures practiced in making beads 4, 5, and 6 in Job 15-J45.

Operations

1. Obtain plates; check the job drawing, Fig. 15-25, page 444, for quantity and size.
2. Obtain electrodes of the quantity, type, and size specified in the job drawing.
3. Set the power source for 140 to 215 amperes. Set a d.c. power source for electrode positive.
4. Set up the plates as shown on the job drawing and tack weld them.
5. Fasten the joint in the overhead position with an overhead welding jig.
6. Make the first pass as shown on the job drawing. Manipulate the electrode as instructed in Fig. 15-22, page 441.
7. Chip the slag from the weld, brush, and inspect. Refer to Inspection.
8. Make the second pass as shown on the job drawing.
9. Chip the slag from the weld, brush, and inspect. Refer to Inspection.
10. Practice these welds until you can produce good welds consistently with both types of electrodes.

Inspection

Compare both passes with Fig. 15-26, page 444, and check them for the following characteristics:

Width and height: Uniform
Appearance: Smooth with close ripples; free of voids and high spots. Restarts should be difficult to locate.
Size: Refer to the job drawing. Check with a fillet weld gauge.
Face of weld: Flat
Edges of weld: Good fusion, no overlap, no undercut
Starts and stops: Free of depressions and high spots
Beginnings and endings: Full size, craters filled
Penetration and fusion: To the root of the joint and both plate surfaces
Surrounding plate surfaces: Free of spatter
Slag formation: Full coverage, easily removable

Disposal

Put the completed joints in the scrap bin so that they will be available for further use. The plate edges can be used for square butt welding, and the unwelded surfaces can be used for beading.

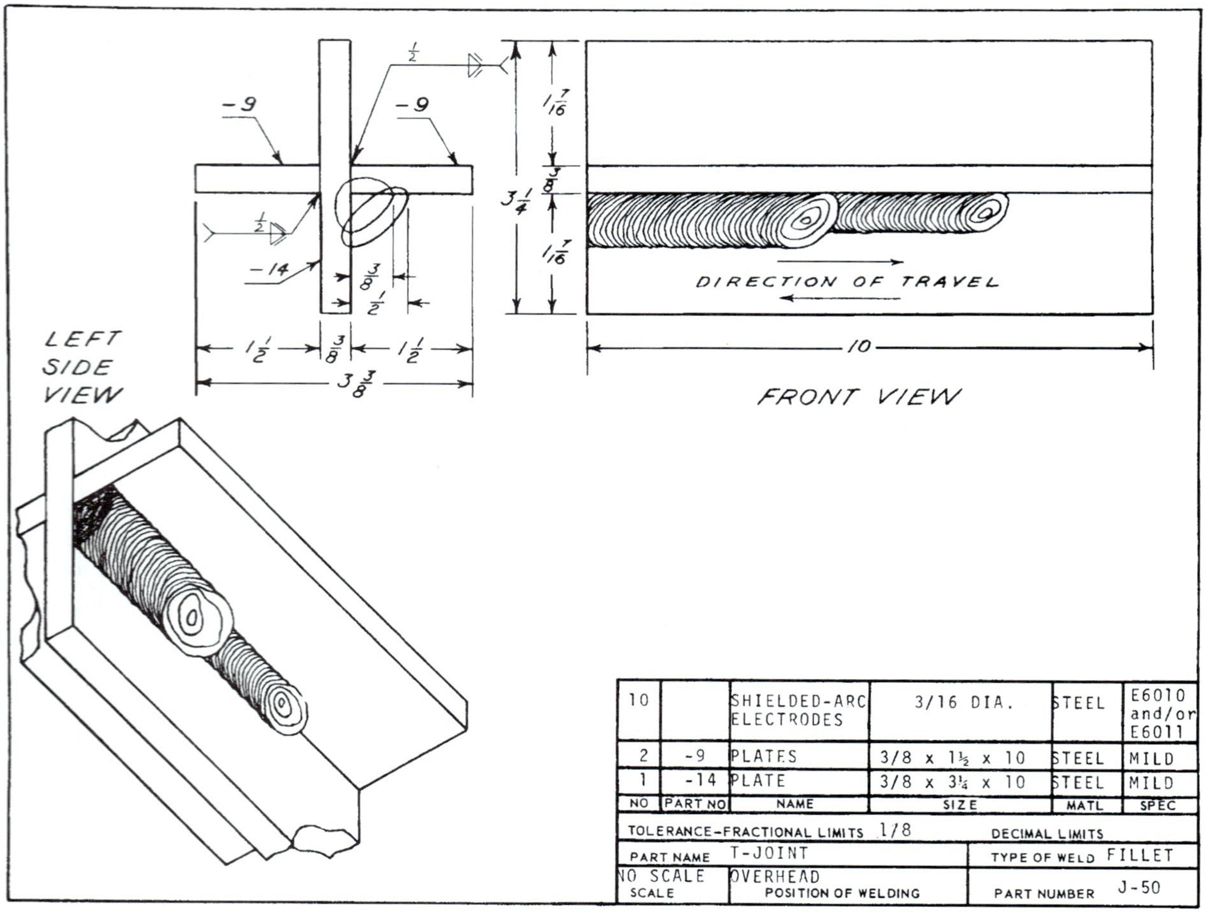

Fig. 15-25 Job drawing J50.

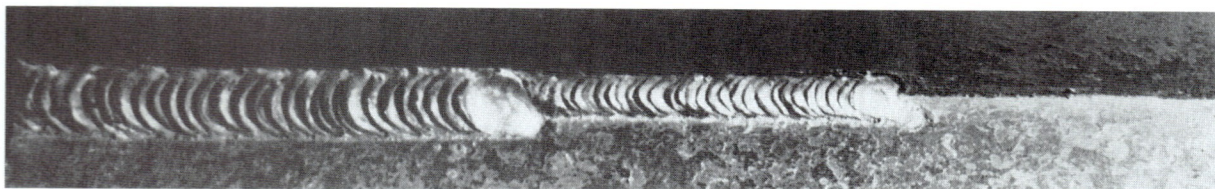

Fig. 15-26 Typical appearance of a multipass fillet weld in a T-joint welded with a slight whiplike motion in the overhead position (4F) with large DCEP or a.c. shielded arc electrodes. *Edward R. Bohnart*

Check Test

After you are able to make welds that are satisfactory in appearance, make up a specimen similar to that made in previous jobs. Use the plate thickness, welding technique, and electrode type and size specified for this job. Weld the first pass only and on one side.

Break the finished weld and examine the surfaces for soundness. The weld must not be porous. It must show good fusion and penetration at the root and to the plate surfaces. The joint should break evenly through the throat of the weld.

Job 15-J51 Welding a Single-V Butt Joint

Objective

To weld a single-V butt joint in the horizontal position (2G) by means of multipass groove welds, stringer bead technique, with DCEP and/or a.c. shielded metal arc electrodes (AWS E6010–E6011 and E7016–E7018).

General Job Information

The beveled-butt joint in the horizontal position is used in pipe and critical plate welding. The joint and the welding

444 Chapter 15 Shielded Metal Arc Welding Practice: Jobs 15-J43–J55 (Plate)

procedure are similar to those required for ASME plate and pipe tests. Some companies demand that the finish bead be weaved. However, most companies prefer stringer beading.

Welding Technique

The angle of the electrode changes according to the pass, Fig. 15-27. The angle is 90° to the top plate for the first pass, 30° to the bottom plate for the second pass, and 45° to the top plate for the third pass. For the cover passes, the electrode should be held at an angle of 5 to 15° to the bottom plate. The electrode should lean from 5 to 15° in the direction of travel for all passes.

First Pass The current control must not be too high. Hold a short arc gap so that weld metal is forced into the gap at the root of the joint. A back-and-forth motion may be used. Do not break the arc on the forward motion. This movement preheats the metal ahead of the deposit and allows the deposited metal to cool to prevent sagging on the bottom plate.

Complete penetration must be secured through the back side, Fig. 15-28. Keep in mind that the correct combination of welding current, electrode position, and speed of travel produces penetration through to the back side. Penetration is complete if there is a little hole at the leading edge of the weld crater right under the tip of the electrode. Learn how to form and keep this hole throughout the welding operation without letting it get too large and causing you to lose control and burn through the plate. The presence of this hole during welding is your assurance that you are melting through to the back side of the groove. This is called keyhole welding.

All Other Passes Current control can be much higher. Electrode position should be approximately that shown in Fig. 15-27B through E. The stringer bead technique may be used. It is important that each pass be fused to the preceding passes and to the plate surfaces. Avoid sagging and undercut. Thorough cleaning is essential. Travel from left to right and from right to left.

Operations

1. Obtain plates; check the job drawing, Fig. 15-29, page 446, for quantity and size.
2. Obtain electrodes of each quantity, type, and size specified in the job drawing.
3. Set the power source for 80 to 130 amperes. Set a d.c. power source for electrode positive.
4. Set up the plates and tack weld them as shown on the job drawing.
5. Place the joints in the horizontal position on the welding table. Make sure that it is connected with the work connection.
6. Make the first, second, and third passes with ⅛-inch electrodes as shown on the job drawing. Manipulate the electrode as shown in Fig. 15-27.
7. Chip the slag from the welds, brush, and inspect between each pass. Refer to Inspection.
8. Increase the power source setting to 112 to 170 amperes.
9. Make the fourth, fifth, sixth, and seventh passes with 5/32-inch electrodes as shown on the job drawing.
10. Chip the slag from the welds, brush, and inspect between each pass. Refer to Inspection.
11. Practice these welds until you can produce good welds consistently.

If you are burning through the plate, you may use a whiplike or circular motion to control the size of the weld pool and the rate at which weld metal is added. If the pool is

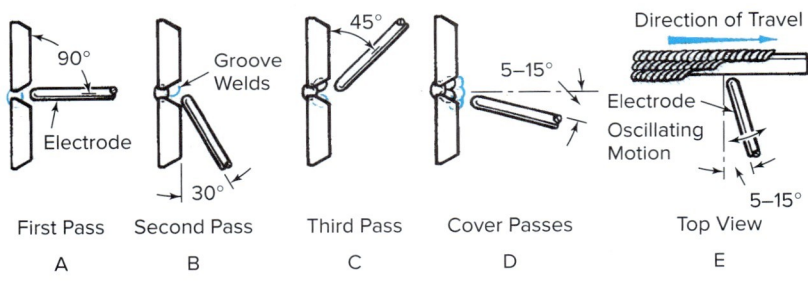

Fig. 15-27 Electrode position when welding a V-groove butt joint in the horizontal position (2G).

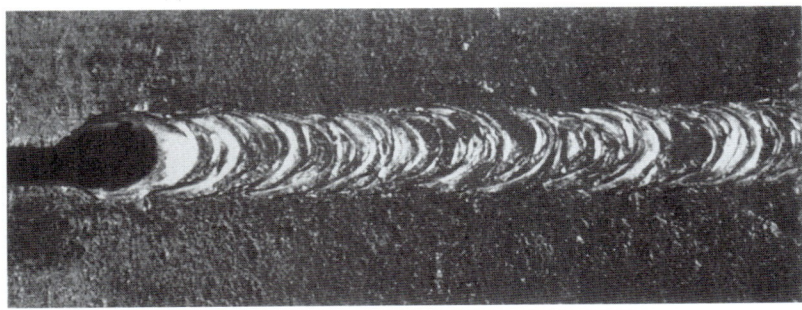

Fig. 15-28 Penetration and fusion through the back side of a V-groove butt joint. Note the hole at the leading edge of the weld. This is referred to as a *keyhole*. Edward R. Bohnart

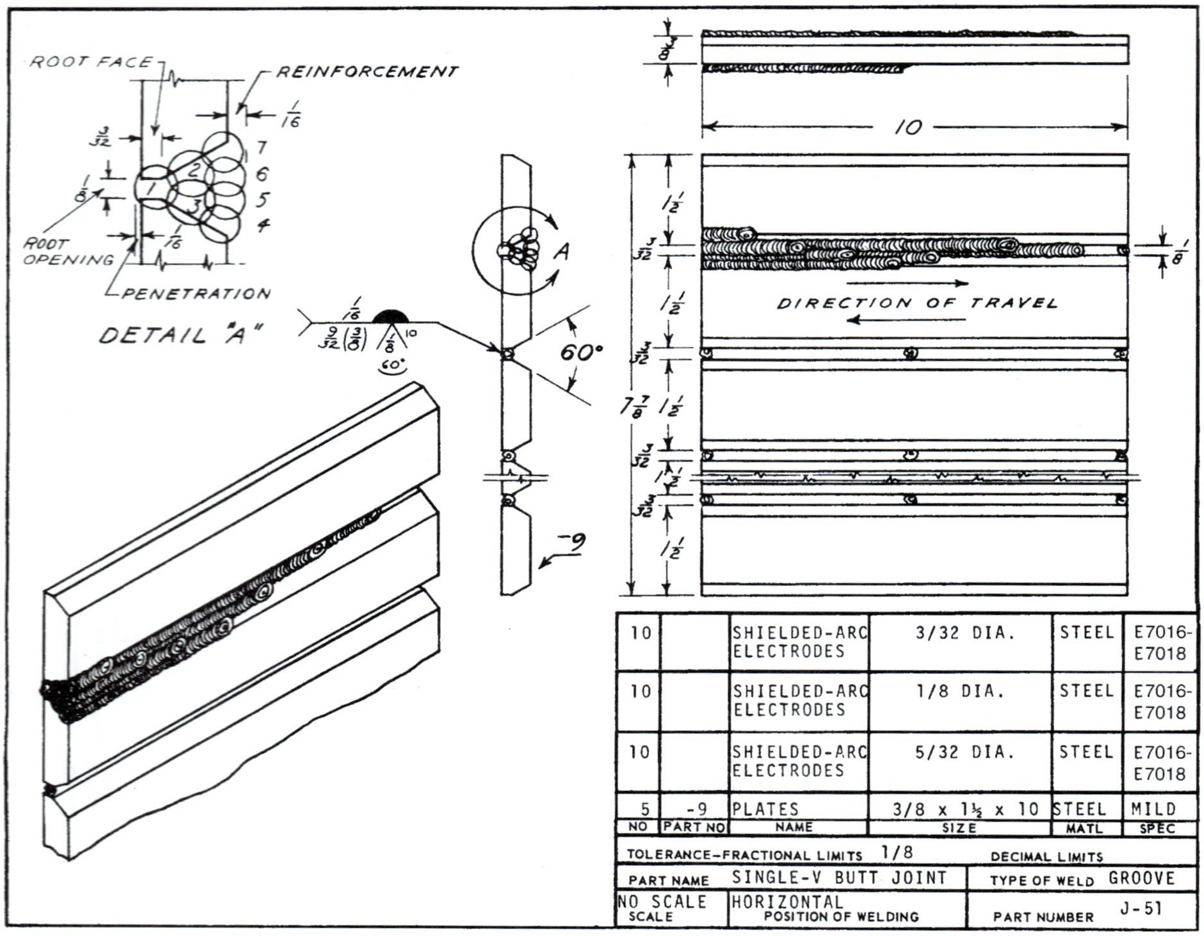

Fig. 15-29 Job drawing J51.

too hot and burning through, whip ahead along one of the plate edges about ½ inch and whip back into the pool to deposit more metal. Alternate this whipping motion along opposite plate edges for each forward movement.

If the root faces of the plates do not seem to be heating and melting enough and you are having difficulty maintaining the hole, keep the electrode directly in the bottom of the V, slow the forward movement, and maintain a circular motion.

Once these welds are acceptable to your instructor, make a horizontal V-groove butt joint using E6010 or E6011 for the root pass and E7016 or E7018 for all subsequent passes. Use a steady short arc gap with the low hydrogen electrodes; do not use the whipping motion with these electrodes. Use $3/32$ or $1/8$-inch diameter electrodes at 60 to 120 amperes and 80 to 150 amperes, respectively. Perform the same inspection and qualification testing that was done on the previous joints.

Inspection

Compare the back side of the first pass with Fig. 15-28, page 445, and all passes with Fig. 15-30. Check all passes for the following weld characteristics:

Width and height: All passes uniform

Appearance: Smooth with close ripples; free of voids and high spots. The lapover of the beads should

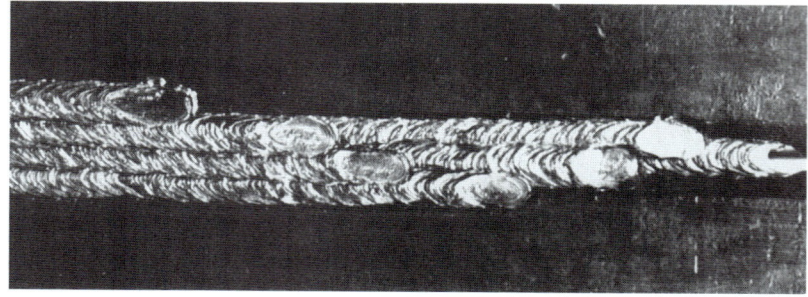

Fig. 15-30 Typical appearance of a multipass groove weld in a V-groove butt joint without a backing strip welded in the horizontal position (2G) with DCEP or a.c. shielded arc electrodes. Edward R. Bohnart

446 Chapter 15 Shielded Metal Arc Welding Practice: Jobs 15-J43–J55 (Plate)

be well-proportioned. Restarts should be difficult to locate.

Size: Refer to the job drawing. Check convexity with a butt weld gauge.

Face of weld: Some reinforcement

Edges of weld: Good fusion, no overlap, no undercut

Starts and stops: Free of depressions and high spots

Beginnings and endings: Full size, craters filled

Back side: Complete and uniform penetration

Penetration and fusion: Through the back side and to all plate surfaces.

Surrounding plate surfaces: Free of spatter

Slag formation: Full coverage, easily removable

Disposal

Put completed joints in the scrap bin so that they will be available for further use. The plate can be cut and beveled between welds for further butt welding, and unwelded plate surfaces can be used for beading.

Qualifying Test

At this point you should be able to pass Test 2 in the horizontal position. (See p. 456.) This test is similar to the ASME test in pipe or plate.

Job 15-J52 Welding a Coupling to a Flat Plate

Objective

To weld pipe or coupling to a flat plate by means of a single-pass horizontal fillet weld with DCEP and/or a.c. shielded metal arc electrode (AWS E7016–E7018.)

General Job Information

This job is similar to those found in tank work in which couplings and fittings are welded into the shell, heads, and bottom. It is also included in this chapter to develop flexibility in positioning the electrode and to give experience in welding in close quarters.

Welding Technique

The current adjustment should be rather high. Hold the electrode at a 45° angle to the plate, Fig. 15-31A, and tilt it 10 to 20° in the direction of travel, Fig. 15-31B. The actual position for welding is shown in Fig. 15-32. The welding technique is not different than that for running stringer beads and single-pass fillet welds in the horizontal position. The position of the electrode is constantly changing in respect to the welder as it travels around the outside of the coupling, however, and space is limited.

Practice starts and stops. Travel in all directions.

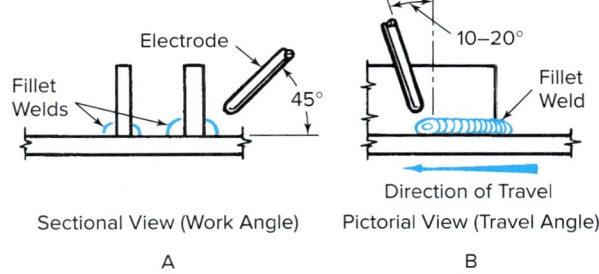

Fig. 15-31 Electrode position when welding couplings with electrode negative.

Fig. 15-32 Electrode and electrode holder positioned as instructed in Fig. 15-31. Note the typical appearance of the fillet welds in couplings welded in the horizontal position (2F) with DCEP or a.c. shielded arc electrodes. Tim Anderson

Operations

1. Obtain plate and pieces of pipe; check the job drawing, Fig. 15-33, page 448, for quantity and size.
2. Obtain electrodes of each quantity, type, and size specified in the job drawing.
3. Set the power source for 90 to 150 amperes. Set a d.c. power source for electrode positive.
4. Place the plate in the flat position on the welding table. Make sure the plate is connected to the work connection.
5. Tack weld the small pipe to the plate.
6. Make inside and outside fillet welds with ⅛-inch electrodes as shown on the job drawing. Manipulate the electrode as instructed in Fig. 15-31.
7. Chip the slag from the welds, brush, and inspect. Refer to Inspection.
8. Increase the power source setting to 140 to 200 amperes.

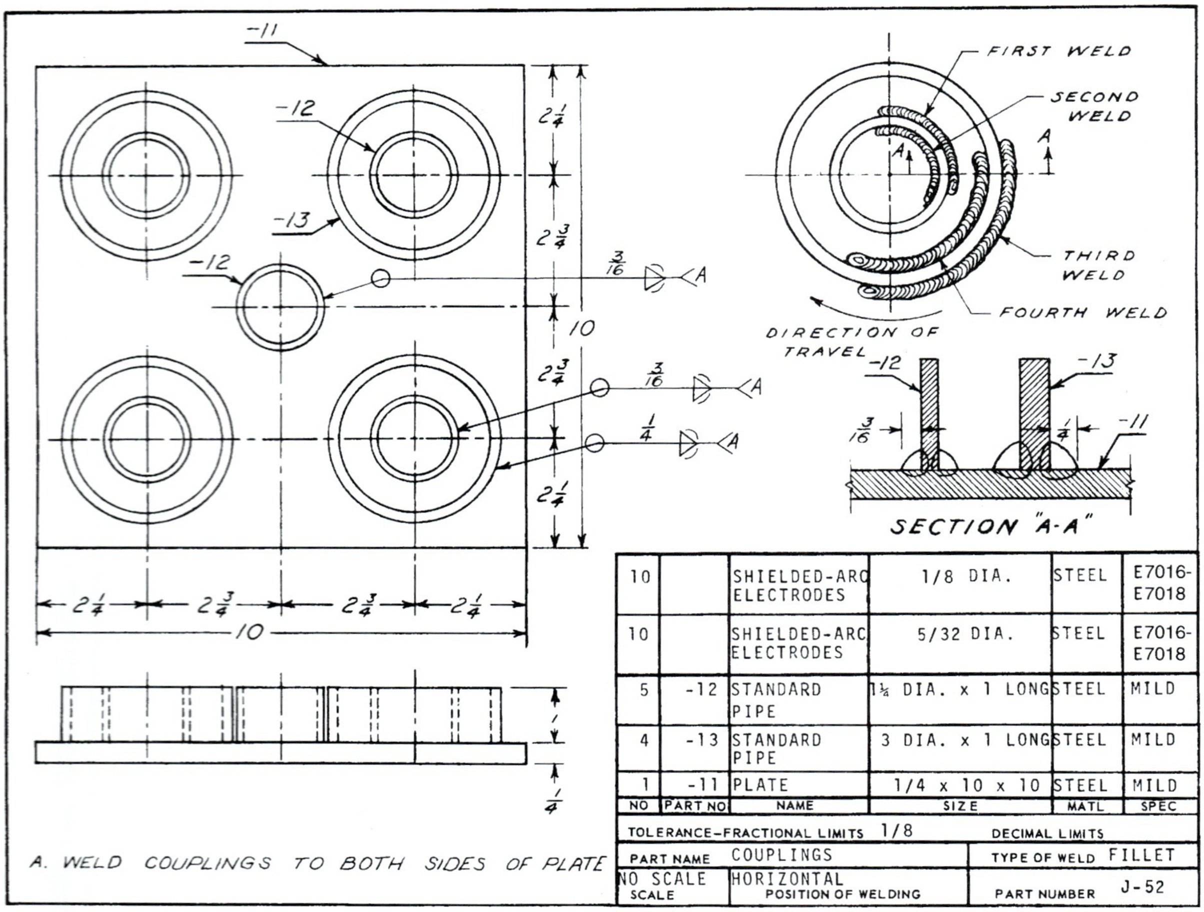

Fig. 15-33 Job drawing J52.

9. Tack weld the large pipe to the plate.
10. Make inside and outside fillet welds with 5/32-inch electrodes as shown on the job drawing.
11. Chip the slag from the welds, brush, and inspect. Refer to Inspection, below.
12. Practice these welds until you can produce good welds consistently.

Inspection

Compare the welds with Fig. 15-34 and check them for the following characteristics:

Width and height: Uniform
Appearance: Smooth with close ripples; free of voids. Restarts should be difficult to locate.
Size: Refer to the job drawing. Check with a fillet weld gauge.
Face of weld: Slightly convex
Edges of weld: Good fusion, no overlap, no undercut
Starts and stops: Free of depressions and high spots

Beginnings and endings: Full size, craters filled. Pay particular attention to points where the end of the weld overlaps the starting point.

Fig. 15-34 Typical appearance of fillet welds on couplings welded in the horizontal position (2F) with DCEP or a.c. shielded arc electrodes. Tim Anderson

Penetration and fusion: To the root of the joint and plate and pipe surfaces
Surrounding plate surfaces: Free of spatter
Slag formation: Full coverage, easily removable

Disposal

Put completed plate in the scrap bin so that it will be available for further use. The unwelded plate surfaces between the couplings can be used for beading.

Job 15-J53 Welding a Coupling to a Flat Plate

Objective

To weld pipe or coupling to a flat plate by means of a single-pass horizontal fillet weld with DCEP and/or a.c. shielded metal arc electrode (AWS E6010–E6011).

General Job Information

This job is identical to Job 15-J52 except that cellulose electrodes are used for welding.

Welding Technique

The current adjustment cannot be as high as with low hydrogen electrodes. More care must be taken with these electrodes. Undercut is a possibility, and arc blow may be a problem. Note in Fig. 15-35 that the electrode is held slightly closer to the plate. The angle is 40° instead of 45° to prevent undercut. Also note that the face of the weld is flat instead of convex. Review the material concerning arc blow, Table 13-3, page 334. Use a welding technique like that used in the previous job, but hold a closer arc gap.

Practice starts and stops. Travel in all directions.

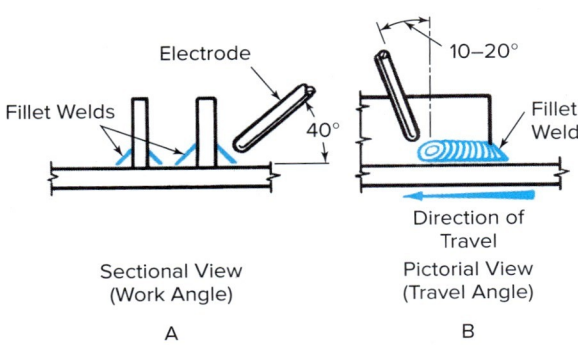

Fig. 15-35 Electrode position when welding couplings with DCEP.

Operations

1. Obtain plate and pieces of pipe; check the job drawing, Fig. 15-36, page 450, for quantity and size.
2. Obtain electrodes in each quantity, type, and size specified in the job drawing.
3. Set the power source for 110 to 170 amperes. Set a d.c. power source for electrode positive.
4. Place the plate in the flat position on the welding table. Make sure the plate is connected with the work connection.
5. Tack weld the small pipe to the plate.
6. Make inside and outside fillet welds with $5/32$-inch electrodes as shown on the job drawing. Manipulate the electrode as instructed in Fig. 15-31, page 447.
7. Chip the slag from the welds, brush, and inspect. Refer to Inspection, below.
8. Increase the power source setting from 140 to 240 amperes.
9. Tack weld the large pipe to the plate.
10. Make inside and outside fillet welds with $3/16$-inch electrodes as shown on the job drawing.
11. Chip the slag from the welds, brush, and inspect. Refer to Inspection.
12. Practice these welds until you can produce good welds consistently with both types of electrodes.

Inspection

Check them for the following characteristics:

Width and height: Uniform
Appearance: Smooth with close ripples; free of voids. Restarts should be difficult to locate.
Size: Refer to the job drawing. Check with a fillet weld gauge.
Face of weld: Flat
Edges of weld: Good fusion, no overlap, no undercut
Starts and stops: Free of depressions and high spots
Beginnings and endings: Full size, craters filled. Pay particular attention to points where the end of the weld overlaps the starting point.
Penetration and fusion: To the root of the joints and plate and pipe surfaces
Surrounding plate surfaces: Free of spatter
Slag formation: Full coverage, easily removable

Disposal

Put completed plate in the scrap bin so that it will be available for further use. The unwelded plate surfaces between the couplings can be used for beading.

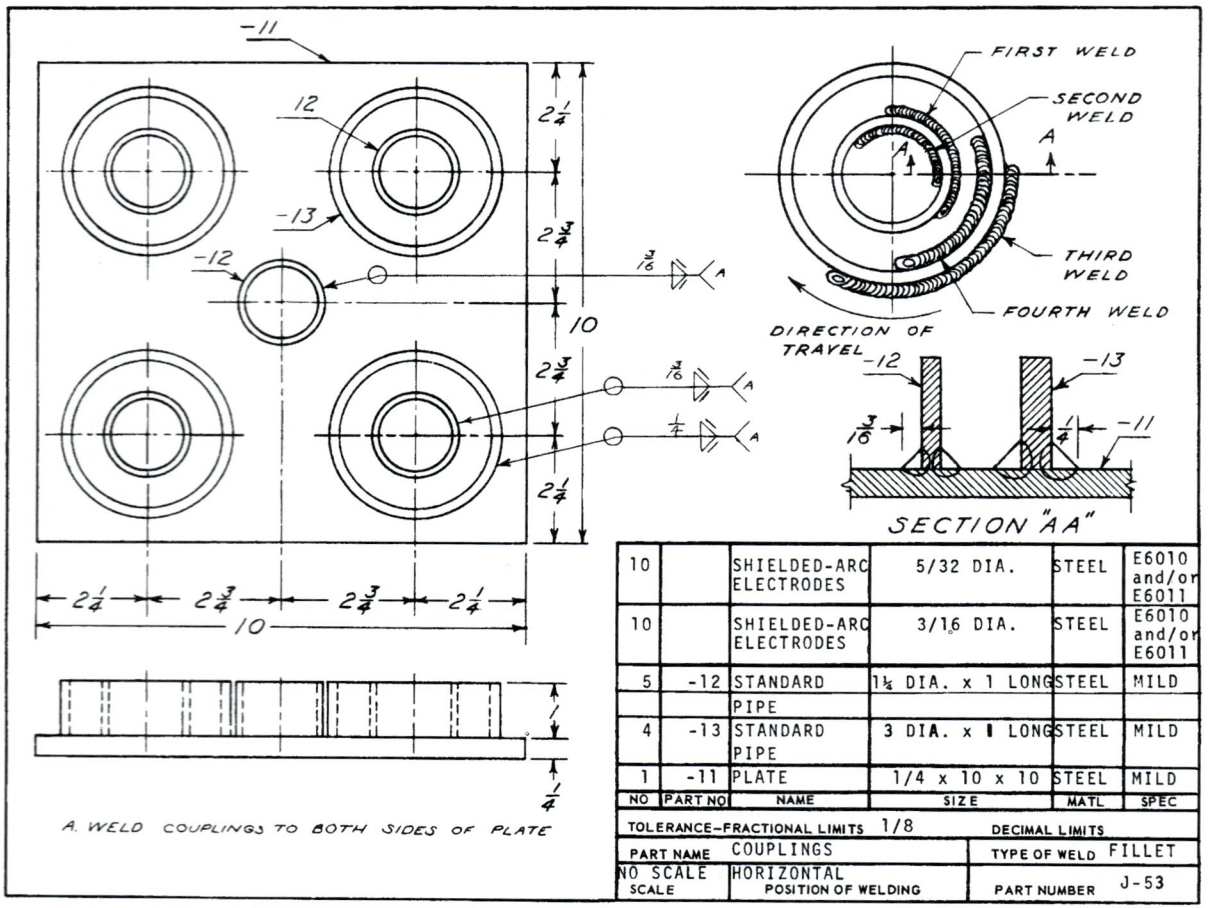

Fig. 15-36 Job drawing J53.

Job 15-J54 Welding a Single-V Butt Joint (Backing Bar Construction)

Objective

To weld a single-V butt joint assembled with a backing bar in the overhead position by means of multipass groove welds, stringer bead technique, with DCEP and/or a.c. shielded metal arc electrodes (AWS E6010–E6011 and/or E7016–E7018).

General Job Information

The beveled-butt joint in plate with a backing bar is similar to joints in pipe and tanks welded with a backing ring or chill ring. This job is identical to the joint and position required by various code authorities and recommended by the American Welding Society. Often the test joint is set up with a ⅜-inch root opening. Two or three root passes are necessary when welding a joint with this gap. More passes are necessary to complete the joint because of greater separation between the plates.

Welding Technique

Tilt the electrode 5 to 10° in the direction of travel, Fig. 15-37A. Hold it at 90° to the surface of the plates for the first pass, Fig. 15-37B; 20 to 35° to the surface of each plate for the second and third passes, Fig. 15-37C and D; and 10 to 20° to the surface of each plate for the fourth and fifth passes, Fig. 15-37E. Like the first pass, the last pass is made at a 90° angle to the surface of the plates, Fig. 15-37F.

First Pass Current adjustment should be high. A short arc is essential to ensure proper metal transfer. A whipping motion should be employed. Make sure that you are obtaining fusion to the backing bar and the root faces of the two plates. Keep the weld as flat as possible.

Second Through Sixth Passes Current adjustment should be high. Use the stringer bead technique and decrease the whipping motion. If the metal becomes too hot, lengthen the arc and whip the electrode forward until the crater has cooled. Fusion must be secured with the

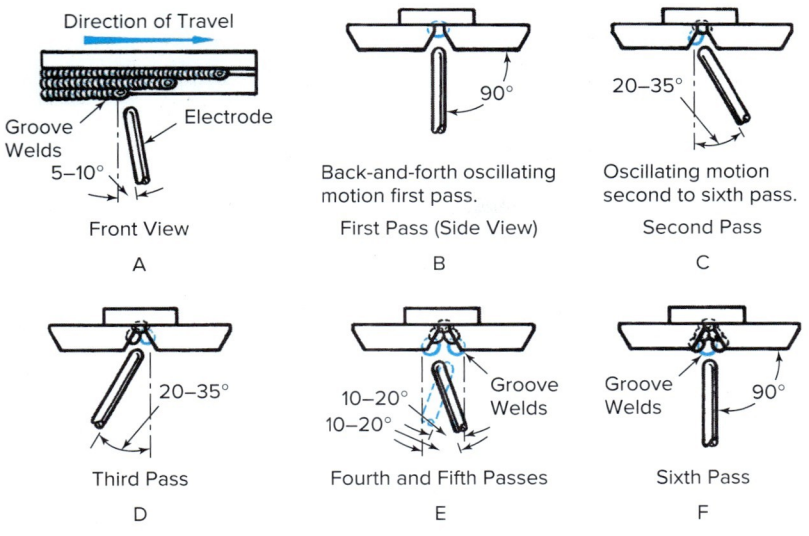

preceding passes and the walls of the groove. Be sure to remove the slag between each pass. Avoid undercut for all passes.

Practice starts and stops. Travel in all directions.

Operations

1. Obtain plates; check the job drawing, Fig. 15-38, for quantity and size.
2. Obtain electrodes of each quantity, type, and size specified in the drawing.
3. Set the power source for 110 to 170 amperes. Set a d.c. power source for electrode positive.
4. Set up the plates and tack weld them as shown on the job drawing.
5. Fasten the joint in the overhead position with an overhead welding jig.

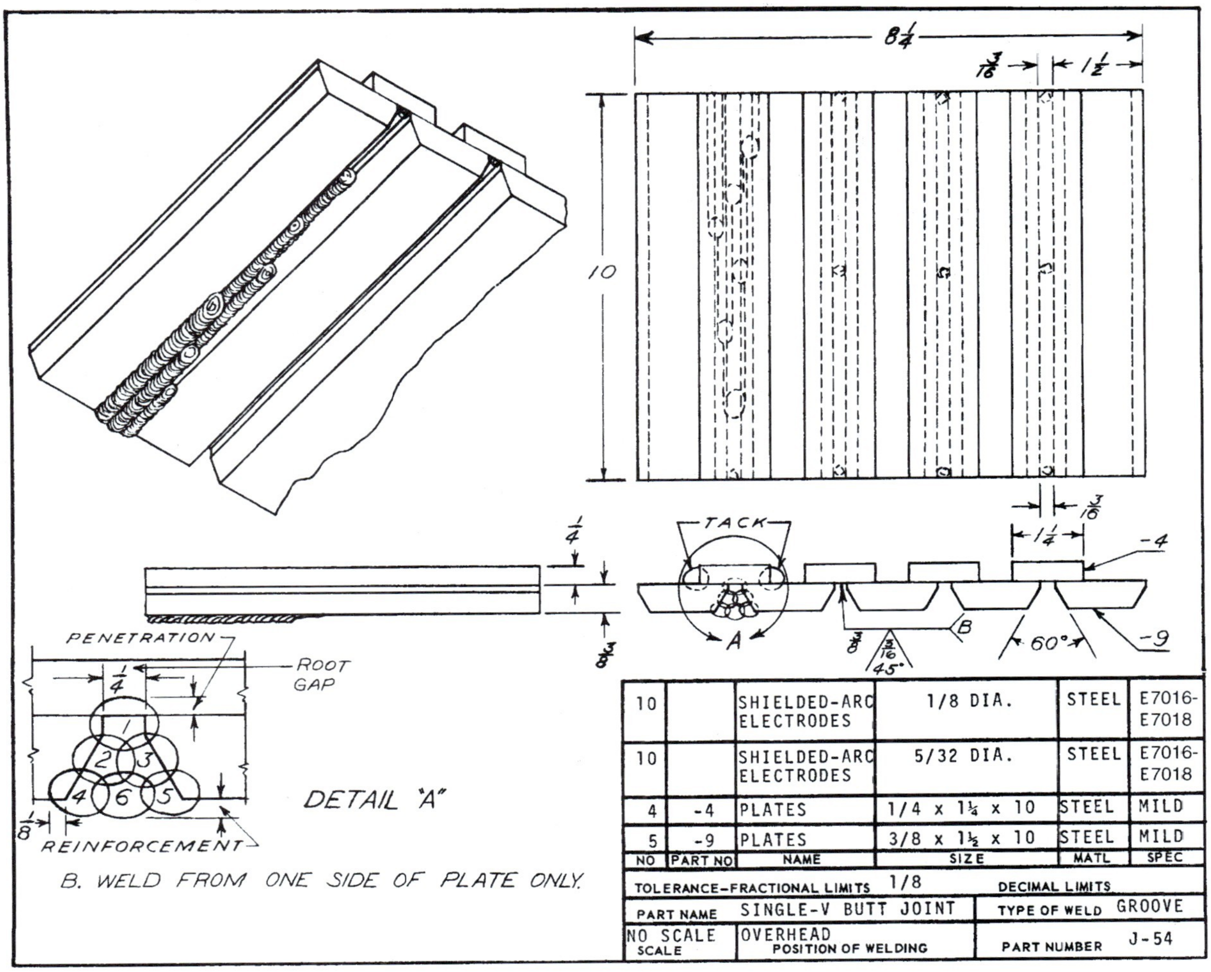

Fig. 15-37 Electrode position when welding a V-groove butt joint in the overhead position (4G).

Fig. 15-38 Job drawing J54.

6. Make all passes with 5/32-inch electrodes as shown on the job drawing. Manipulate the electrode as instructed in Fig. 15-37, page 451.
7. Chip the slag from the welds, brush, and inspect between each pass. Refer to Inspection.
8. Practice these welds until you can produce good welds consistently.

Once these welds are acceptable to your instructor, make an overhead V-groove butt joint with backing using E7016 or E7018 for all passes. Use a steady short arc gap with the low hydrogen electrodes; do not use the whipping motion with these electrodes. Use 3/32- or 1/8-inch diameter electrodes at 60 to 120 amperes and 80 to 150 amperes, respectively. Perform the same inspection and qualification testing that was done on the previous joints.

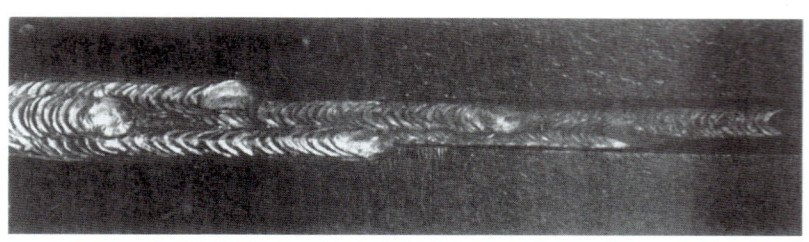

Fig. 15-39 Typical appearance of a multipass groove weld in a V-groove butt joint with a backing strip welded in the overhead position with DCEP or a.c. shielded arc electrodes. *Edward R. Bohnart*

Inspection

Compare each pass with Fig. 15-39 and check it for the following weld characteristics:

Width and height: Uniform
Appearance: Smooth with close ripples; free of voids and high spots. Restarts should be difficult to locate.
Size: Refer to the job drawing. Check the last pass with a butt weld gauge.
Face of weld: Slight reinforcement
Edges of weld: Good fusion, no overlap, no undercut
Starts and stops: Free of depressions and high spots
Beginnings and endings: Full size, craters filled
Penetration and fusion: To the backing bar, preceding passes, and plate surfaces
Surrounding plate surfaces: Free of spatter

Disposal

Put completed joints in the scrap bin so that they will be available for further use. The plate can be cut and beveled between the welds for further butt welding, and unwelded plate surfaces can be used for beading.

Qualifying Test

At this point you should be able to pass a 4G test in the overhead position. (See Test 1, p. 455.) This is identical to Test 1 of the Department of the Navy Requirements and similar to AWS recommendations.

It will also be desirable for you to take a 4F test (fillet weld in the overhead position) (Test 3, p. 457).

Job 15-J55 Welding a Single-V Butt Joint

Objective

To weld a single-V butt joint in the overhead position by means of multipass groove welds, weave bead technique, with DCEP and/or a.c. shielded metal arc electrodes (AWS E6010–E6011 and/or E7016–E7018).

General Job Information

The single-V butt joint in the overhead position will be encountered in pipe and critical plate welding. Both joint construction and welding procedure are similar to those in the ASME test.

Welding Technique

First Pass Current adjustment should be high enough to ensure penetration through the back side. Hold the electrode as shown in Fig. 15-40A and use the root pass technique practiced in the preceding job. A

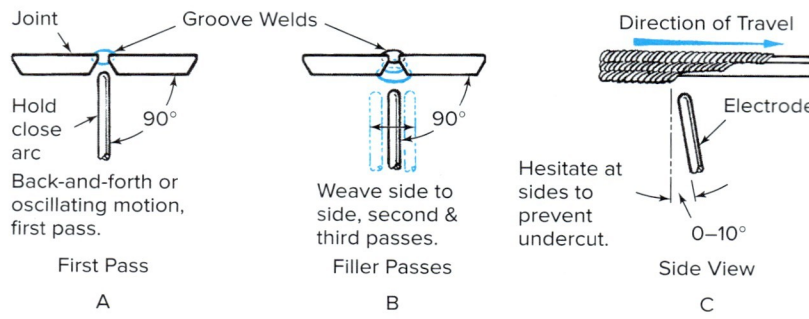

Fig. 15-40 Electrode position when welding a V-groove butt joint in the overhead position (4G).

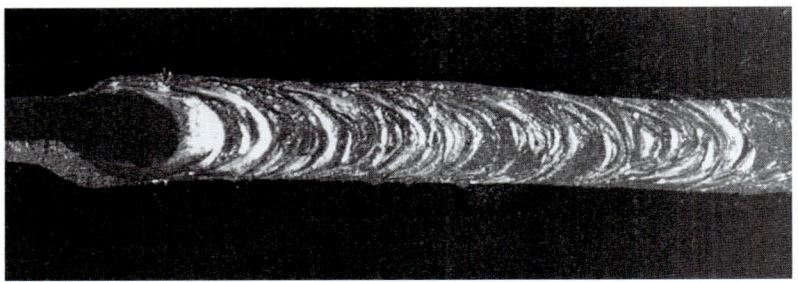

Fig. 15-41 Penetration and fusion through the back side of a V-groove butt joint without a backing strip. *Edward R. Bohnart*

small uniform bead should be formed on the back side (Fig. 15-41), and the face of the weld should be flat. Keep a keyhole at the forward edge of the weld pool as you proceed with the weld pool. If the keyhole gets too large, move forward along the bevel until the pool freezes. Then go back to the keyhole and proceed with the weld as before.

Second and Third Passes Current adjustment can be high. Hold the electrode as shown in Fig. 15-40B and use weave bead technique. Move the electrode across the face of the weld with a rapid crossover movement so that you do not deposit too much metal in the center of the bead, causing it to sag. The side-to-side movement should be kept within the limits of the required weld size. Hesitate at the sides of the weld to prevent undercut. Remember that undercutting is caused by gouging out the plate with the arc and failing to fill it up with deposited weld metal.

Practice starts and stops on all passes.

Operations

1. Obtain plates; check the job drawing, Fig. 15-42, for quantity and size.
2. Obtain electrodes of each quantity, type, and size specified in the job drawing.

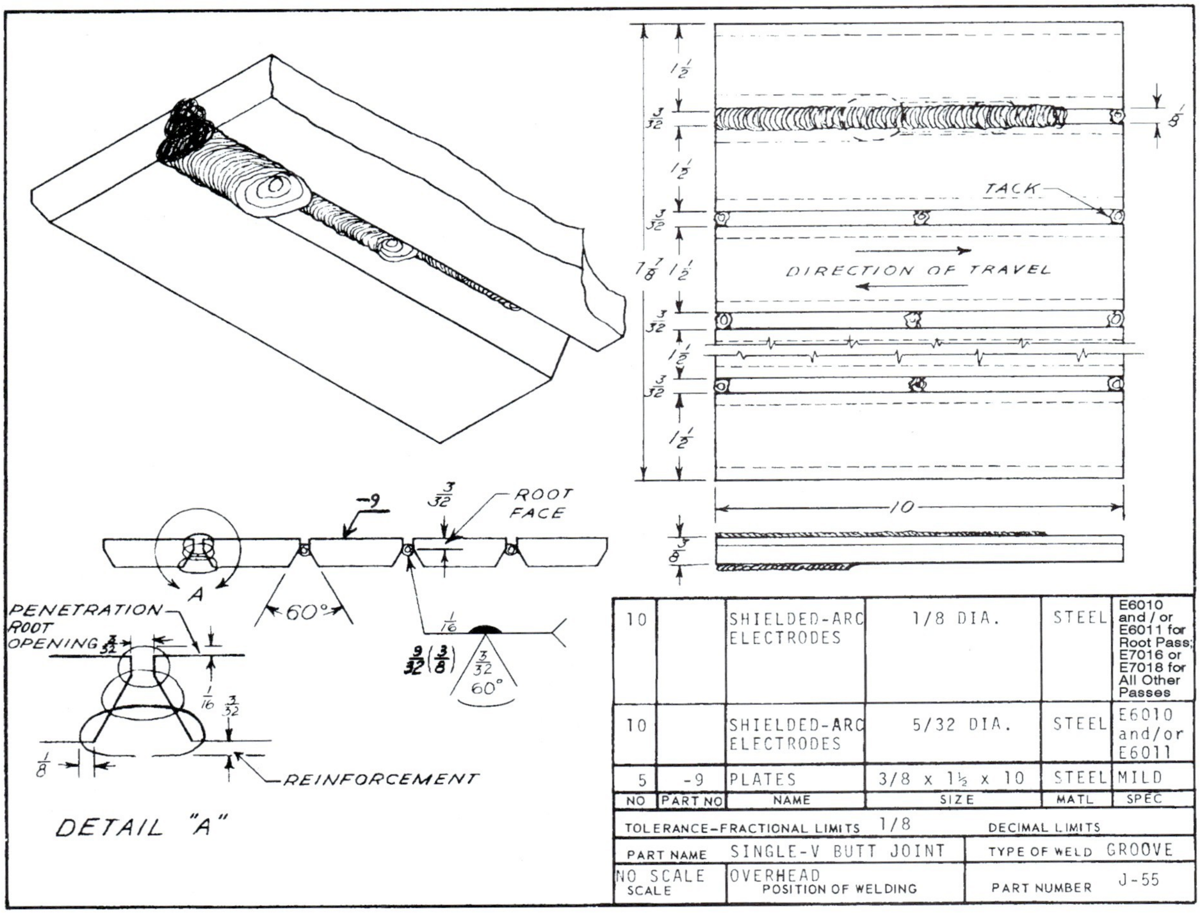

Fig. 15-42 Job drawing J55.

3. Set the power source for 75 to 125 amperes. Set a d.c. power source for electrode positive.
4. Set up the plates and tack weld them as shown on the job drawing.
5. Fasten the joints in the overhead position with an overhead welding jig.
6. Make the first and second passes with ⅛-inch electrodes as shown on the job drawing.
7. Chip the slag from the welds, brush, and inspect. Refer to Inspection.
8. Increase the setting of the power source to 110 to 170 amperes.
9. Make the third pass with ⁵⁄₃₂-inch electrodes as shown on the job drawing.
10. Chip the slag from the weld, brush, and inspect. Refer to Inspection.
11. Practice these welds until you can produce good welds consistently.

Once these welds are acceptable to your instructor, make an overhead V-groove butt joint using E6010 or E6011 for the root pass and E7016 or E7018 for all subsequent passes. Use a steady short arc gap with the low hydrogen electrodes; do not use the whipping motion with these electrodes. Use ³⁄₃₂- or ⅛-inch diameter electrodes at 60 to 120 amperes and 80 to 150 amperes, respectively. Perform the same inspection and qualification testing that was done on the previous joints.

Inspection

Compare the back side of the first pass with Fig. 15-41, page 453, and all passes with Fig. 15-43. Check each pass for the following weld characteristics:

Width and height: Uniform
Appearance: Smooth with close ripples; free of voids and high spots. Restarts should be difficult to locate.
Size: Refer to the job drawing. Check with a butt weld gauge.
Face of weld: First two passes flat; last pass slightly convex
Edges of weld: Good fusion, no overlap, no undercut
Starts and stops: Free of depressions and high spots
Beginnings and endings: Full size, craters filled
Back side: Complete and uniform penetration
Penetration and fusion: Through the back side, preceding passes, and all plate surfaces. See Fig. 15-41.
Surrounding plate surfaces: Free of spatter
Slag formation: Full coverage, easily removable

Disposal

Put completed joints in the scrap bin so that they will be available for further use. Joints can be cut and beveled between welds for further butt welding, and unwelded surfaces can be used for beading.

Qualifying Test

At this point you should be able to pass a 4G test in the overhead position. (See Test 2, p. 456.) This test is similar to the ASME test for pipe and plate.

You should also take Test 4, page 458, which is a simulated pressure vessel.

Tests

General Information

Each test in this section will be conducted after the student has completed the job in which it is assigned and at the conclusion of the course. No student should be considered as having completed the course who has not fulfilled the requirements of Tests 1, 3, and 4. Only students of exceptional ability are expected to meet the requirements of Test 2.

Types of Tests

The tests that follow are especially devised to determine the student's ability to produce sound and pressure-tight welds. These tests are standard tests used by various code authorities to qualify welders for welding of high quality and performance. They are also similar to those recommended by the American Welding Society.

The fillet weld soundness test is prescribed for the testing of fillet welds. For the testing of groove welds, the tests require the

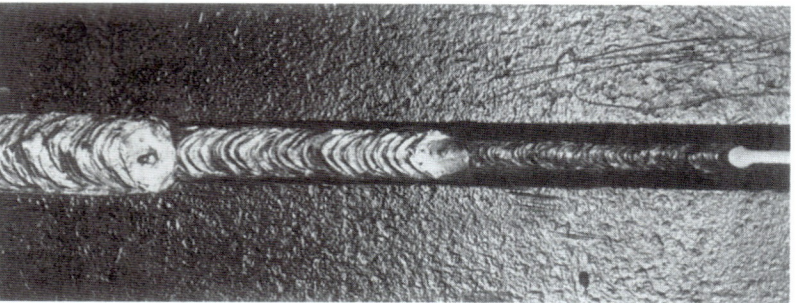

Fig. 15-43 Typical appearance of a multipass groove weld in a V-groove butt joint welded without a backing strip in the overhead position (4G) (weave bead technique) with DCEP or a.c. shielded arc electrodes. Edward R. Bohnart

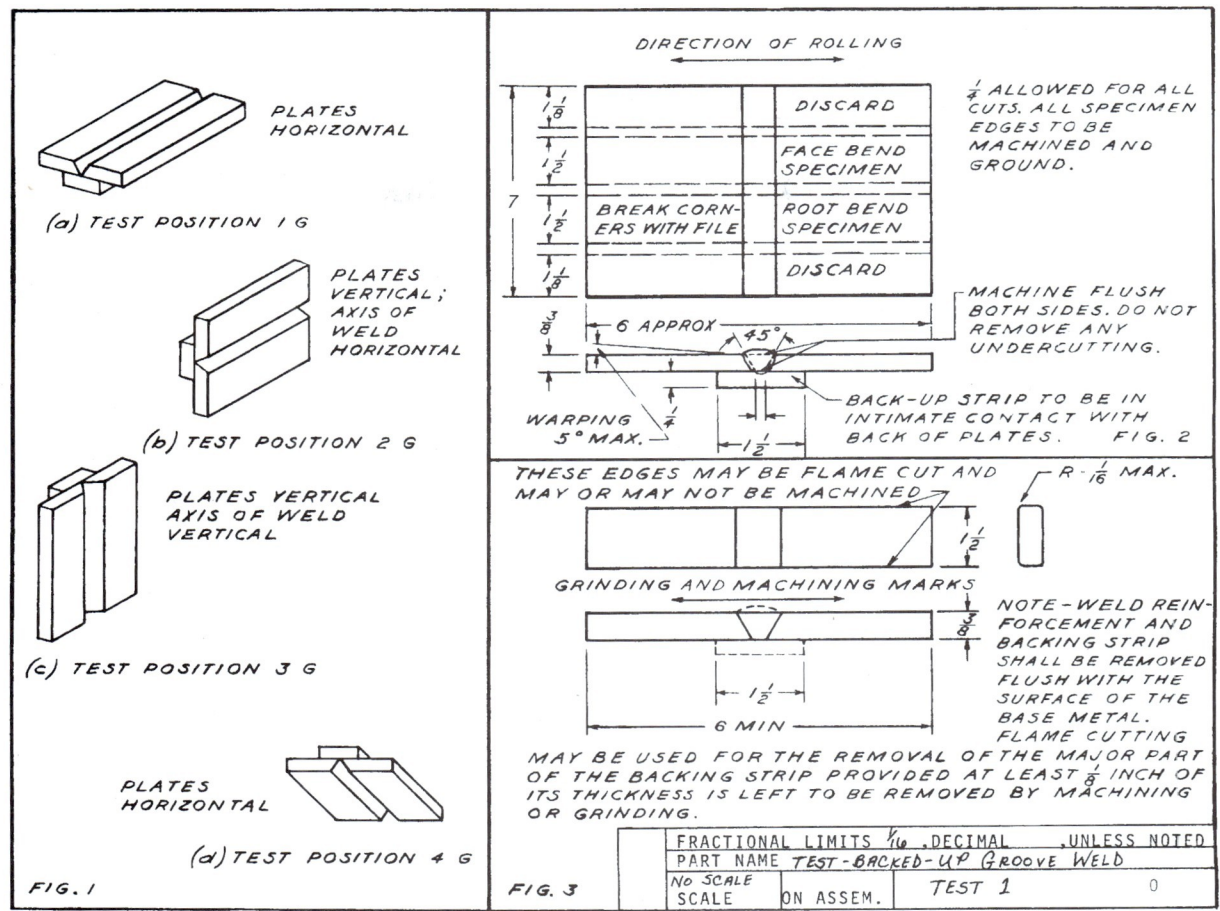

Fig. 15-44 Test 1, Figs. 1 to 3.

welding of beveled-butt joints with and without a backing plate. Specimens are taken from the test joint and are subjected to the standard face- and root-bend test procedures.

In order to determine the student's ability to produce welds that are pressure tight and do not leak, the student will fabricate a special small vessel completely by welding. It will be subjected to 2,500 p.s.i. of internal pressure. All of the types of electrodes used in the course are used in making the testing unit.

Base Metal and Its Preparation

The base metal is mild steel plate. For each type of joint, the length of the weld and the dimensions of the base metal are such as to provide enough material for the test specimens required.

- *Groove welds with backing bar:* The preparation of the base metal for welding is for a single-V groove butt joint meeting the requirements of Fig. 15-44, Test 1, Fig. 2.
- *Groove welds without a backing bar:* The preparation of the base metal for welding is for a single-V groove butt joint meeting the requirements of Fig. 15-45, Test 2, Fig. 2 (p. 456).
- *Fillet welds:* The preparation of the base metal for welding is shown in Fig. 15-46, Test 3, Fig. 2 (p. 457).
- *Pressure testing:* The preparation of the base metal for welding is shown in Fig. 15-47, Test 4 (p. 458).

Position of Test Welds

The positions for the tests are those illustrated in Test 1, Fig. 1; Test 2, Fig. 1; Test 3, Fig. 1; and Test 4. If students pass the tests in the more difficult positions, they need not be tested in the less difficult positions. The

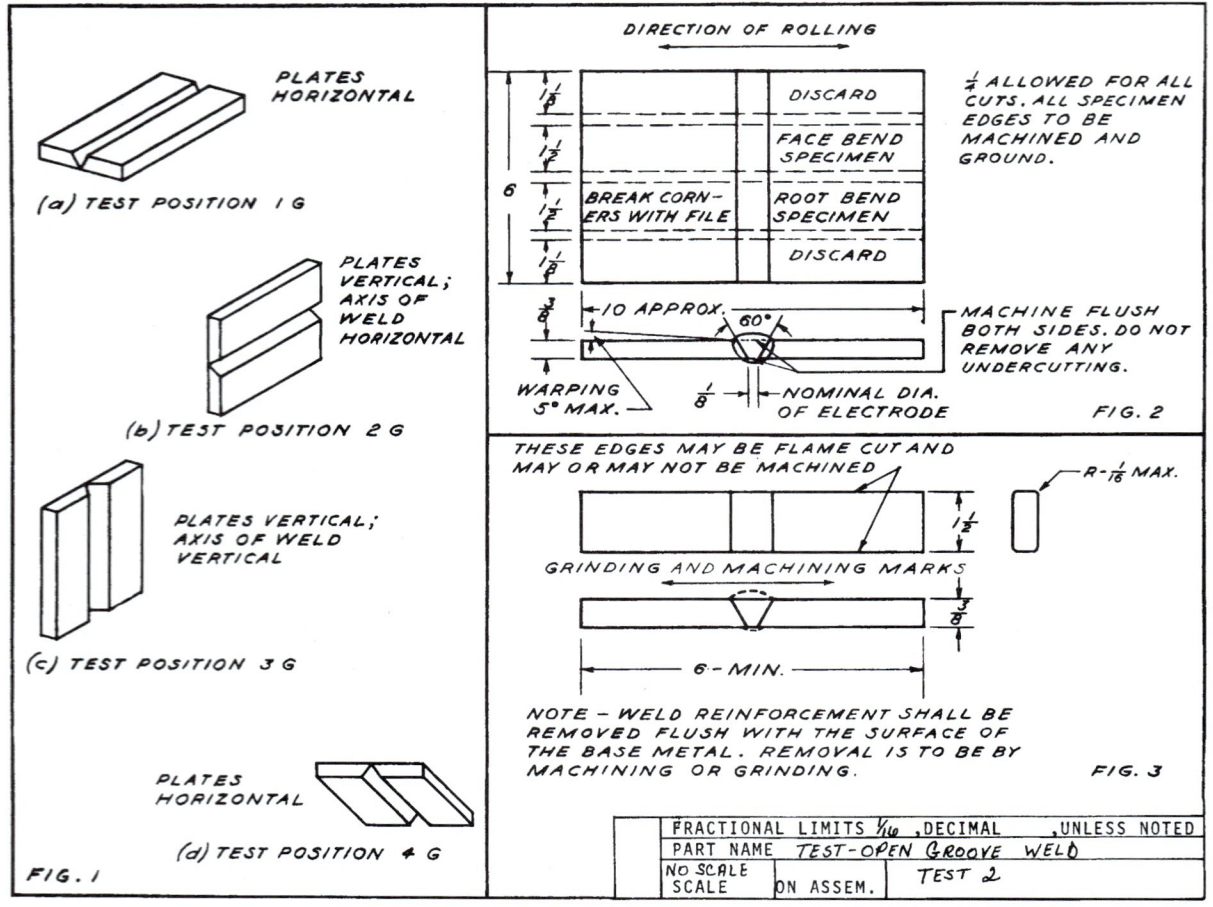

Fig. 15-45 Test 2, Figs. 1 to 3.

student should be considered as having passed such tests in accordance with the following provisions:

- *Groove welds:* If students pass the tests in the vertical and overhead positions, they need not be tested in the flat or horizontal position. Passing groove weld tests will also qualify for fillet welds.
- *Fillet welds:* If students pass the tests in the vertical or overhead position, they need not be tested in either the flat position or the horizontal position. Fillet weld tests only qualify for fillet welds.
- *Pressure welds:* The vessel must be welded as specified in Table 15-1.

Type of Electrodes Used for Testing

In order to reduce the number of tests required for production welders, many of the code authorities have grouped electrodes. For example, the AWS used F numbers to group electrodes. Table 15-2 covers some of the typical electrodes used.

If you pass a test with any electrode in a group, you are also qualified to use any other electrode in that group. The AWS also allows you to weld with electrodes in any lower group F-number. This is due to the understanding that the easiest to run electrodes are in the F1 group and the most difficult to run electrodes are shown in the F4 group. This provision allows for a further reduction of the number of production weld test plates required. Because of the drastic difference in manipulative techniques between the F4 and F3 electrodes, student welders should pass test plates using both types of electrodes.

Number of Test Welds Required

- *Groove welds:* One test weld as shown in Test 1, Fig. 2, page 455, and/or Test 2, Fig. 2, is made for each position in which the student is to be tested.

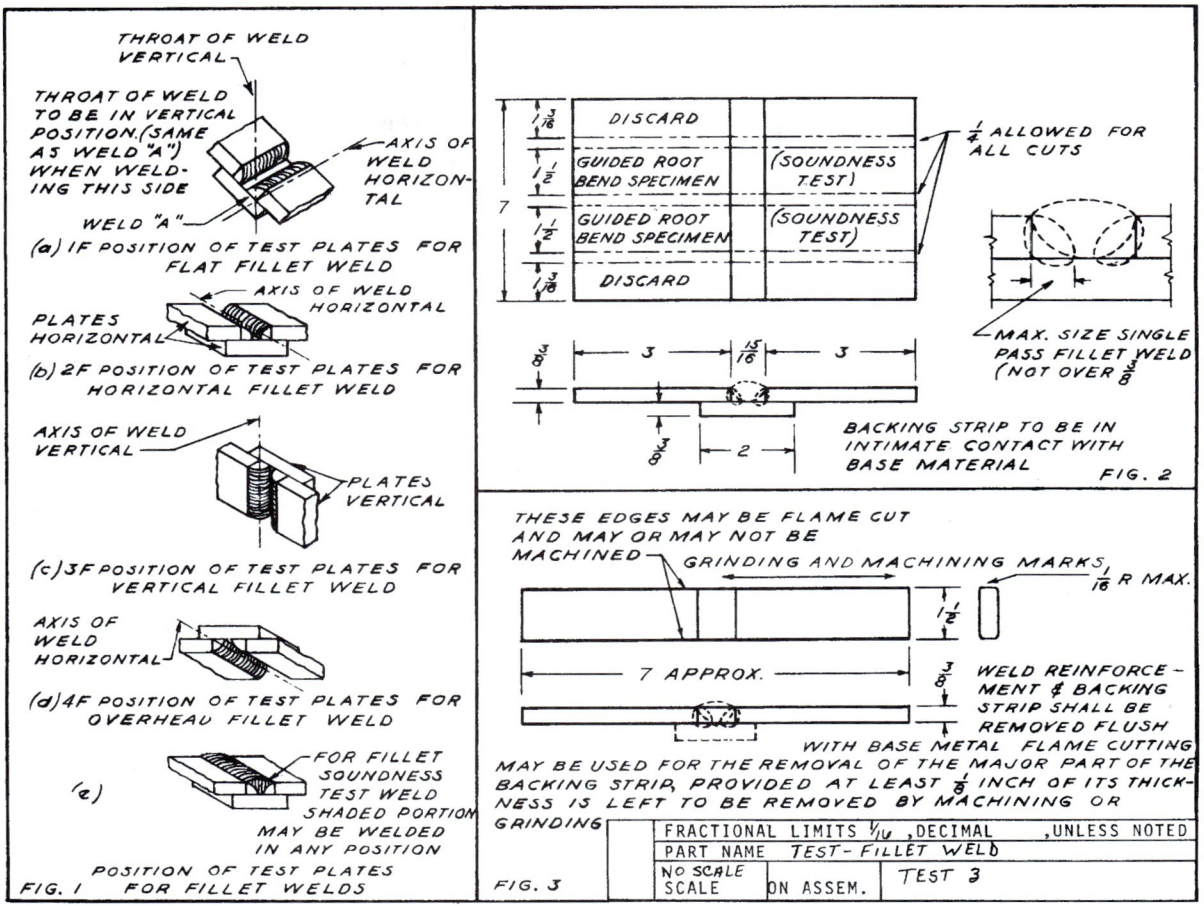

Fig. 15-46 Test 3, Figs. 1 to 3.

- *Fillet welds:* One test weld as shown in Test 3, Fig. 2, is made for each position in which the student is to be tested.
- *Pressure welds:* One vessel that conforms to Test 4 (p. 458) is constructed.

Welding Procedure

The test welds are made in accordance with the following illustrations and jobs.

- *Groove welds* (Test 1: groove weld with backing bar—¼-inch root opening).
  1. Flat position: See Fig. 15-48 (p. 460) and review Job 14-J28 (p. 393).
  2. Horizontal position: See Fig. 15-49 (p. 460) and review Job 14-J30 (p. 397).
  3. Vertical position: See Fig. 15-50 (p. 460) and review Job 14-J38 (p. 414).
  4. Overhead position: See Fig. 15-51 (p. 460) and review Job 15-J54 (p. 450). On occasion, larger root openings may be required. This may require two or three root passes. Note Figs. 15-52 through 15-55, page 461.
- *Groove welds* (Test 2: groove weld without backing bar—standard root gap). Figure 15-56 (p. 462) shows the back side of a joint. It is absolutely necessary to obtain this type of penetration through the back side, or root-bend test specimens will fracture upon bending.
  1. Flat position: See Fig. 15-57 (p. 462) and review Job 14-J33 (p. 403).
  2. Horizontal position: See Fig. 15-58 (p. 462) and review Job 15-J51 (p. 444).
  3. Vertical position: See Fig. 15-59 (p. 462) and review Job 15-J43 (p. 429).
  4. Overhead position: See Fig. 15-60 (p. 463) and review Job 15-J55 (p. 452).

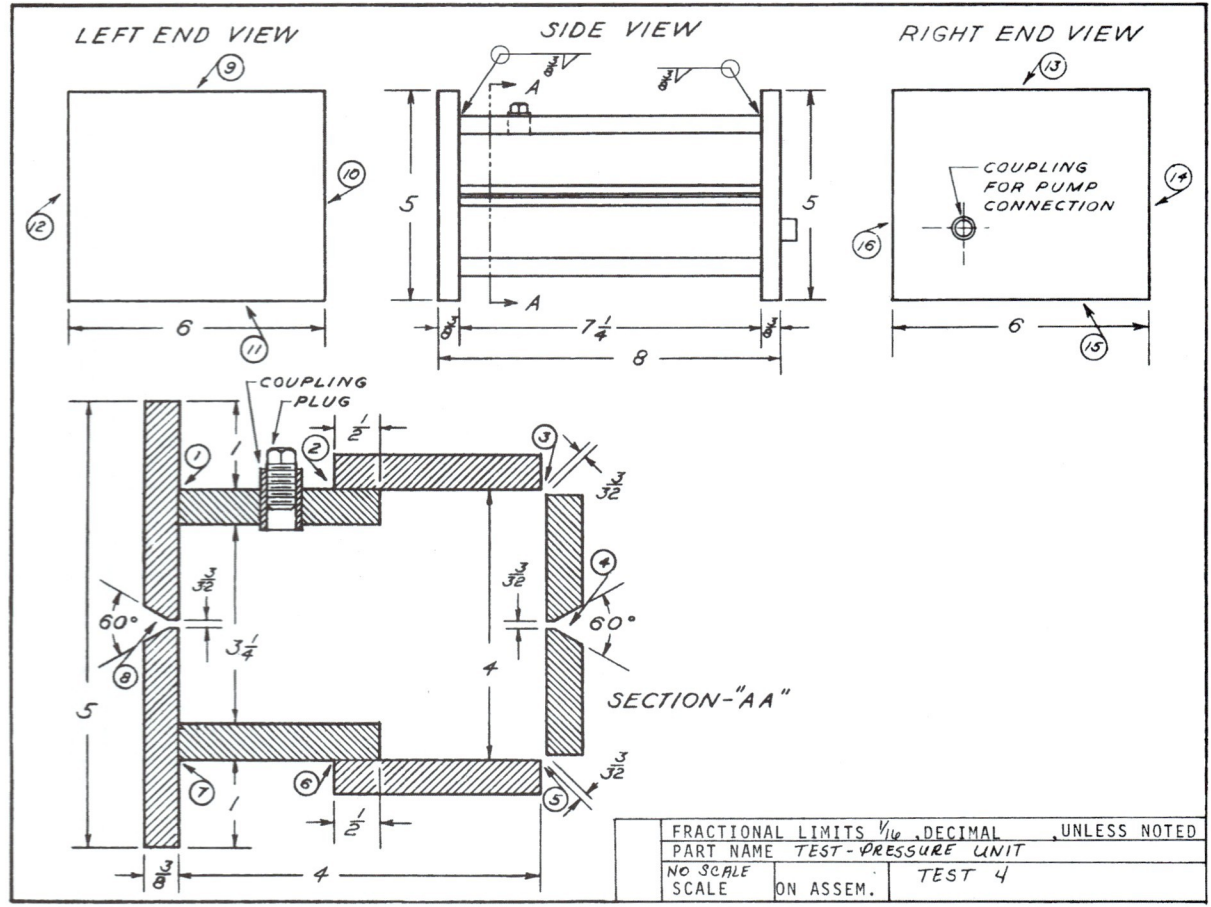

Fig. 15-47 Test 4.

- *Fillet welds* (Test 3).
  1. Flat position: See Fig. 15-61 (p. 463) and review Job 13-J17 (p. 369).
  2. Horizontal position: See Fig. 15-62 (p. 463) and review Job 13-J9 (p. 354) and Job 13-J24 (p. 382).
  3. Vertical position: See Fig. 15-63 (p. 463) and review Jobs 13-J21 (p. 375) and 14-J35 (p. 408).
  4. Overhead position: See Fig. 15-64 (p. 463) and review Job 15-J44 (p. 431) and Job 15-J46 (p. 435).
- *Pressure welds:* [Test 4. See Table 15-1 (p. 459) and Figs. 15-65 and 15-66 (p. 464)].

Test Specimens—Number, Type, and Preparation

- *Groove welds:* One root-bend specimen and one face-bend specimen are prepared from the finished test weld as shown in Test 1, Fig. 3 (p. 455) and Test 2, Fig. 3 (p. 456).
- *Fillet welds:* Two fillet-weld soundness specimens are prepared from the finished test weld as shown in Test 3, Fig. 3 (p. 457).
- *Pressure welds:* The finished vessel, Test 4 (p. 458) is the test specimen.

Method of Testing Specimens

Groove and Fillet Weld Specimens Each specimen is bent in a jig having the contour shown in Fig. 15-67, p. 464. Any convenient means—mechanical, electrical, or hydraulic—may be used for moving the plunger member with relation to the die member. Table 15-3 indicates the number of specimens, and the range of thickness qualified as described in the AWS D1.1 Structural Welding Code—Steel for Welder Qualification.

Table 15-1 Welding Procedures for Test 4

Joints	Operation	Welding Technique	Position	DCEP	DCEN	a.c.	AWS Specification Number
1. T-joint	Fillet welding, single pass, travel up	Job 14-J35	Vertical 3F	X		X	E6010–E6011
2. Lap joint	Fillet welding, single pass, travel up	Job 13-J21	Vertical 3F	X		X	E6010–E6011
3. Outside corner joint	Groove welding, multipass, weaved, travel up	Job 14-J40	Vertical 3F	X		X	E6010–E6011
4. Single-V butt joint	Groove welding, multipass, weaved, travel up	Job 15-J43	Vertical 3G	X		X	E6010–E6011 root E7016–E7018 fill and cap
5. Outside corner joint	Groove welding, multipass, weaved	Job 14-J32	Flat 1F	X		X	E6010–E6011
6. Lap joint	Fillet welding, single pass	Job 13-J8	Horizontal 2F		X	X	E6013
7. T-joint	Fillet welding, single pass	Job 15-J44	Overhead 4F	X		X	E6010–E6011
8. Single-V butt joint	Groove welding, multipass, weaved	Job 15-J55	Overhead 4G	X		X	E6010–E6011 root E7016–E7018 fill and cap
9. Lap joint	Fillet welding, single pass, travel down	Job 13-J12	Vertical 3F		X	X	E6013
10. T-joint	Fillet welding, multipass, travel up	Job 14-J36	Vertical 3F	X		X	E6010–E6011
11. T-joint	Fillet welding, multipass, weaved, travel up	Job 14-J37	Vertical 3F	X		X	E7018
12. T-joint	Fillet welding, multipass	Job 15-J50	Overhead 4F	X		X	E6010–E6011
13. T-joint	Fillet welding, multipass	Job 13-J23	Horizontal 2F		X	X	E6013
14. Fitting	Fillet welding, single pass	Job 15-J52	Horizontal 2F	X		X	E7016–E7018

Table 15-2 Electrode Classification Groups

Group Designation	AWS Electrode Classification[1]
F4	EXX15, EXX16, EXX18, EXX15-X, EXX16-X, EXX18-X
F3	EXX10, EXX11, EXX10-X, EXX11-X
F2	EXX12, EXX13, EXX14, EXX13-X
F1	EXX20, EXX24, EXX27, EXX28, EXX20-X, EXX27-X

[1] The letters XX used in the classification designation stand for the various strength levels 60, 70, 80, 90, 100, 110, and 120 of electrodes.

The specimen is placed on the die member of the jig with the weld at midspan. Face-bend specimens are placed with the face of the weld directed toward the gap. Root-bend and fillet-weld–soundness specimens are placed with the root of the weld directed toward the gap. The two members of the jig are forced together until the curvature of the specimen is such that a wire $1/32$ inch in diameter cannot be passed between the curved portion of the male member and the specimen. Then the specimen is removed from the jig and carefully examined. Table 15-4 (p. 465) indicates the various jig dimensions for various strengths of bend test specimens.

Pressure Welds The finished vessel is subjected to 2,500 p.s.i. internal pressure. The vessel begins to bulge at about 2,800 p.s.i. pressure. *Do not use air.* Use water or oil as the liquid. Water is preferred because it is cleaner to work with. Any hand- or motor-powered liquid pump may be used as the means of producing pressure.

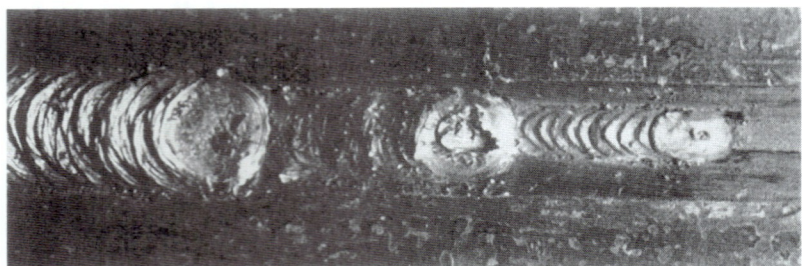

Fig. 15-48 V-groove butt test joint with a backing bar and a ¼-inch root opening welded in the flat position (1G). Note the sequence of weld beads and the appearance. *Edward R. Bohnart*

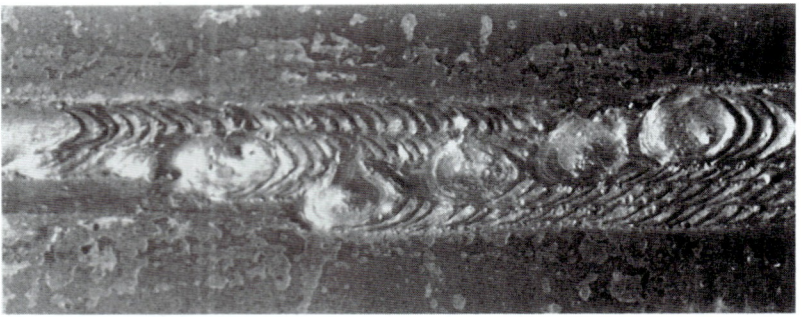

Fig. 15-49 V-groove butt test joint with a backing bar and a ¼-inch root opening welded in the horizontal position (2G). Note the sequence of weld beads and the appearance. *Edward R. Bohnart*

Fig. 15-50 V-groove butt test joint with a backup bar and a ¼-inch root opening welded in the vertical position, travel up (3G). Note the sequence of weld beads and the appearance. *Edward R. Bohnart*

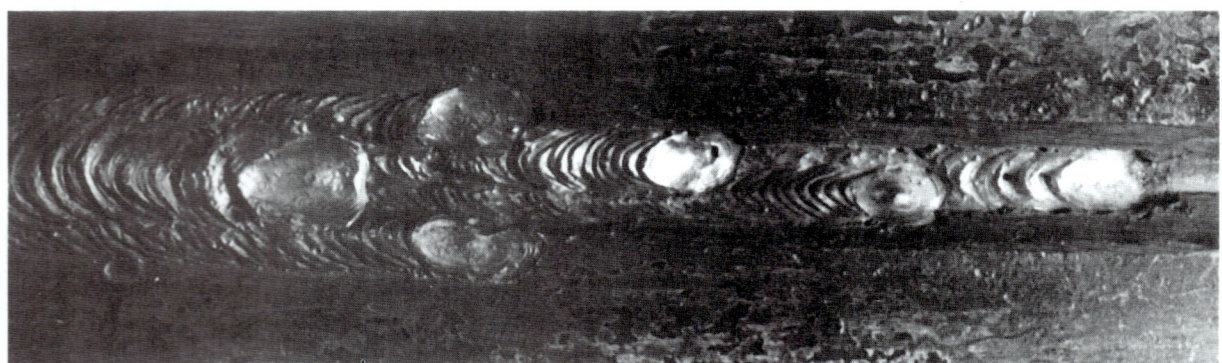

Fig. 15-51 V-groove butt test joint with a backing bar and a ¼-inch root opening welded in the overhead position (4G). Note the sequence of weld beads and the appearance. *Edward R. Bohnart*

Do not use chilled water. Make certain all air is removed (vented) from the vessel. This operation should be carried out only by qualified personnel and in accordance with ASME Section VIII-Division 1, UG-99, which covers hydrostatic testing.

Evaluation of Test Results

Groove and Fillet Weld Specimens The outside surface of the specimens is examined for the appearance of cracks or other open defects. To be acceptable, the convex surface

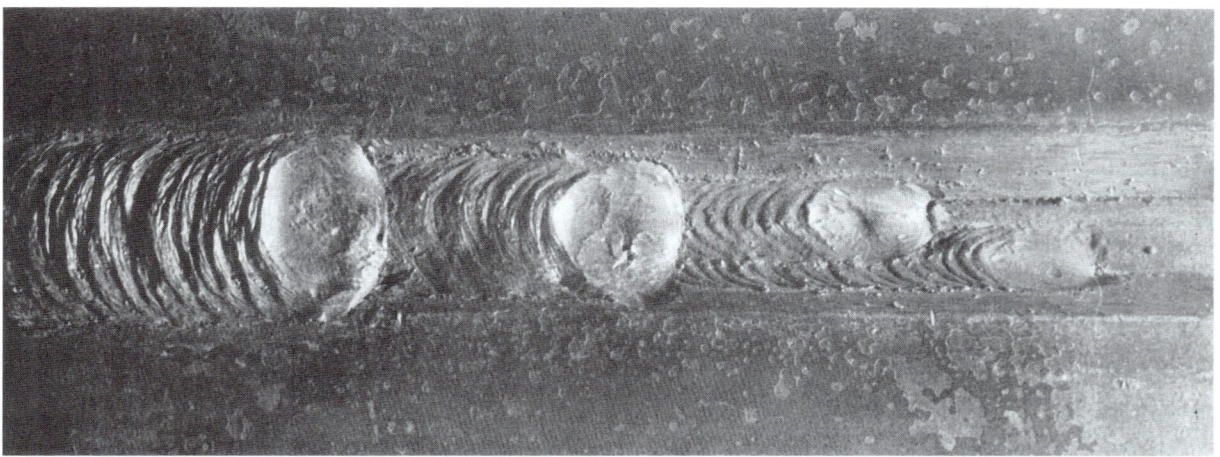

Fig. 15-52 V-groove butt test joint with a backing bar and a ⅜-inch root opening welded in the flat position (1G). Note the sequence of weld beads and the appearance. Edward R. Bohnart

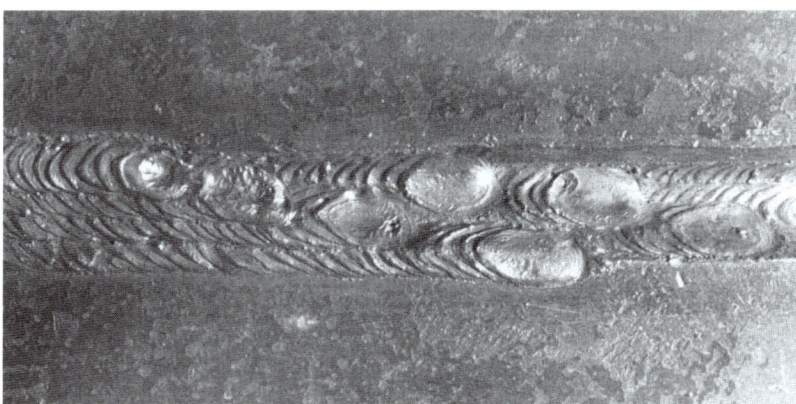

Fig. 15-53 V-groove butt test joint with a backing bar and a ⅜-inch root opening welded in the horizontal position (2G). Note the sequence of weld beads and the appearance. Edward R. Bohnart

Fig. 15-54 V-groove butt test joint with a backing bar and a ⅜-inch root opening welded in the vertical position, travel up. Note the sequence of weld beads and the appearance. Edward R. Bohnart

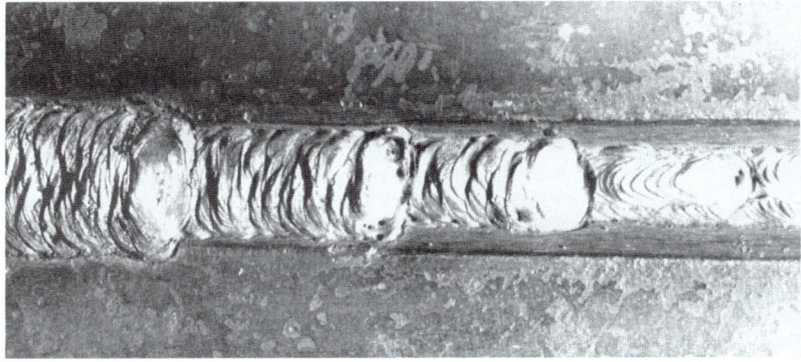

Fig. 15-55 V-groove butt test joint with a backing bar and a ⅜-inch root opening welded in the overhead position (4G). Note the sequence of weld beads and the appearance. Edward R. Bohnart

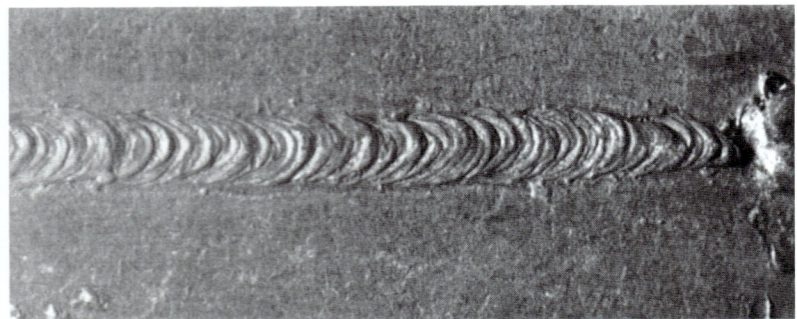

Fig. 15-56 Penetration through the back side of a V-groove butt joint without a backing bar. The formation of a small bead without undercut is absolutely necessary for successful root-bend test specimens. Edward R. Bohnart

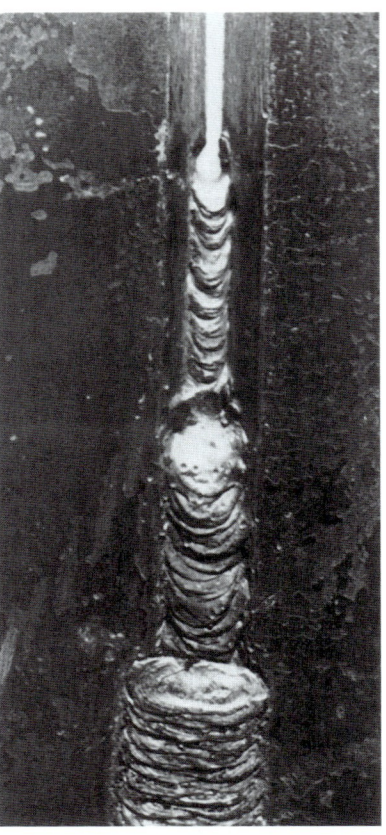

Fig. 15-59 V-groove butt test joint without a backing bar welded in the vertical position, travel up (3G). Note the sequence of weld beads, the keyhole, and the appearance. Edward R. Bohnart

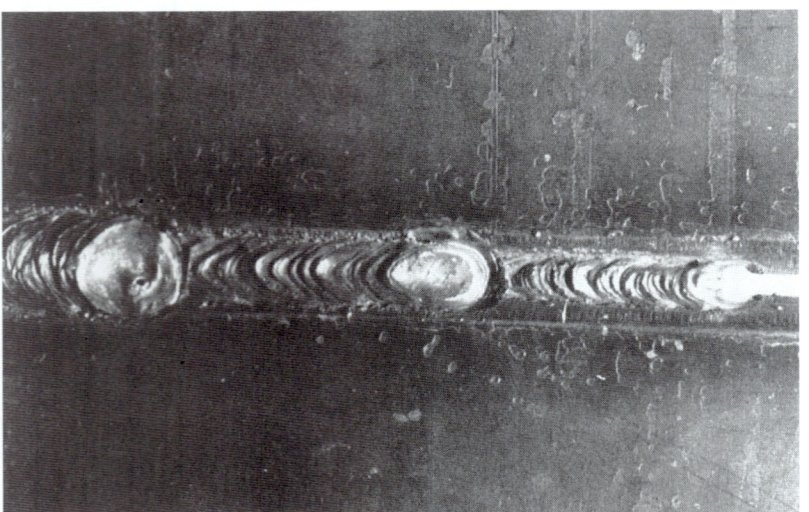

Fig. 15-57 V-groove butt test joint without a backing bar welded in the flat position (1G). Note the sequence of weld beads, the keyhole, and the appearance. Edward R. Bohnart

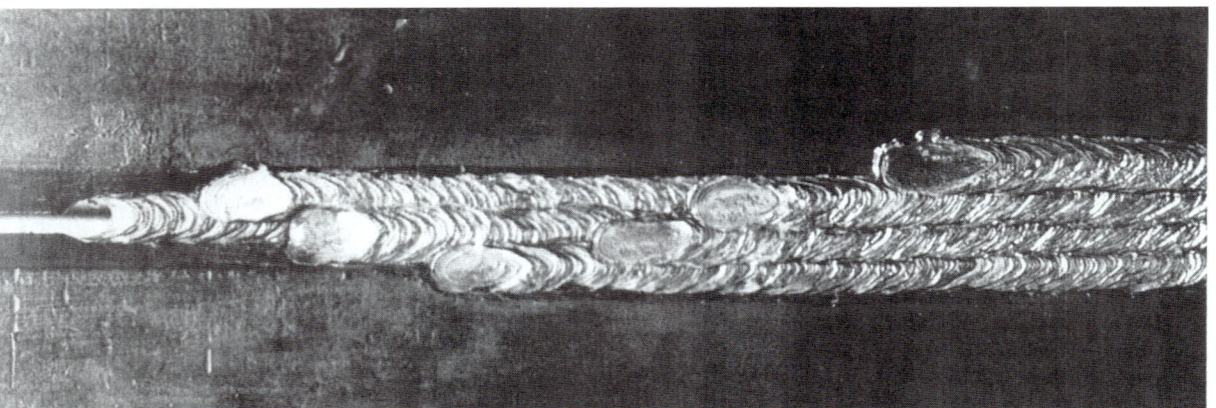

Fig. 15-58 V-groove butt test joint without a backing bar welded in the horizontal position (2G). Note the sequence of weld beads, the keyhole, and the appearance. Edward R. Bohnart

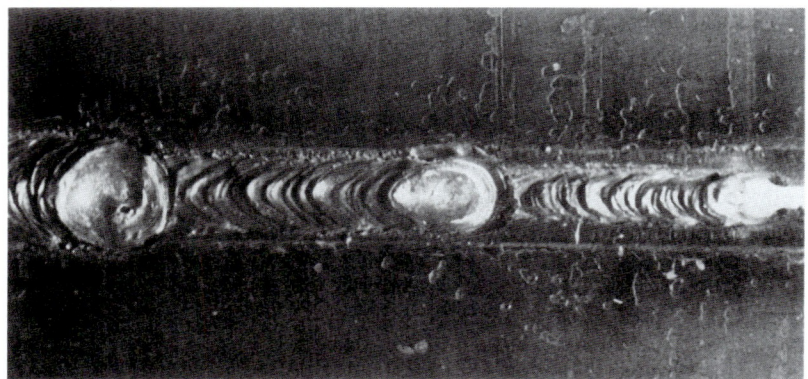

Fig. 15-60 V-groove butt test joint without a backing bar welded in the overhead position (4G). Note the sequence of weld beads, the keyhole, and the appearance. *Edward R. Bohnart*

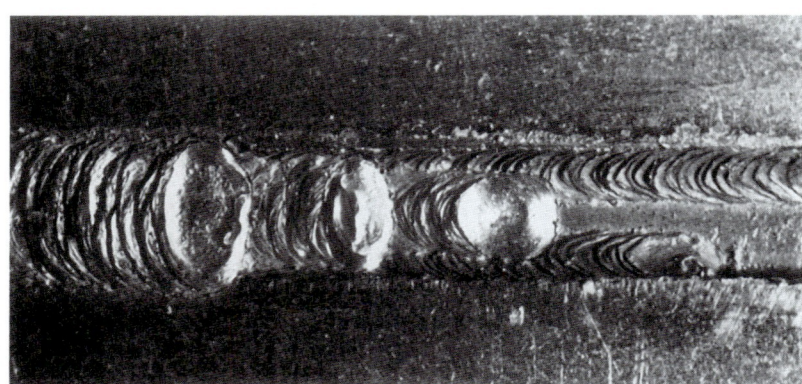

Fig. 15-61 Joint for the fillet weld soundness test welded in the flat position (1F). Note the sequence of weld beads and the appearance. *Edward R. Bohnart*

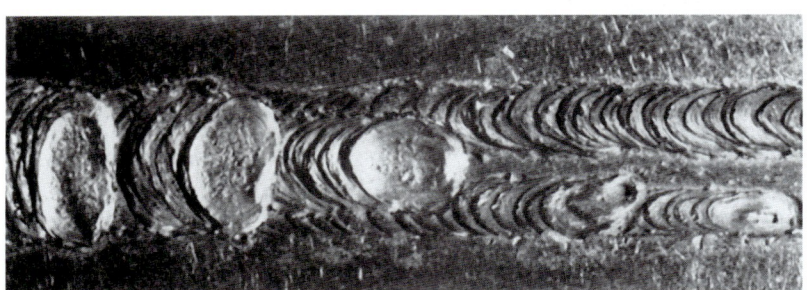

Fig. 15-62 Joint for the fillet weld soundness test welded in the horizontal position (2F). Note the sequence of weld beads and the appearance. *Edward R. Bohnart*

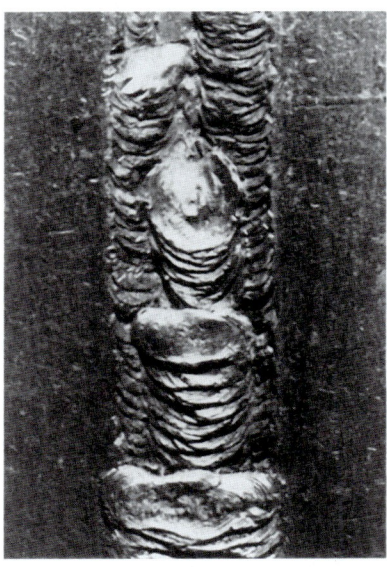

Fig. 15-64 Joint for the fillet weld soundness test welded in the overhead position (4F). Note the sequence of weld beads and the appearance. *Edward R. Bohnart*

Fig. 15-63 Joint for the fillet weld soundness test welded in the vertical position, travel up (3F). Note the sequence of weld beads and the appearance. *Edward R. Bohnart*

Fig. 15-65 One side of the pressure vessel test unit: a corner joint, lap joint, and T-joints are visible. Welds include both fillet and groove welds. Edward R. Bohnart

Fig. 15-66 Another side of the pressure vessel test unit: a corner joint, open V-groove butt joint, and a T-joint are visible. Welds include both fillet and groove welds. Edward R. Bohnart

Fig. 15-67 A shop-built testing machine. The gauge and jack may be purchased, and the rest of the unit can be built in the school shop. Tensile testing is done between the first two levels. Compression testing is carried on between the second and third crossheads. A standard root- and face-bend jig is mounted between these levels. The maximum tensile pull depends on the size of the jack. The unit illustrated has a capacity of 60,000 p.s.i. Edward R. Bohnart

Table 15-3 Number of Specimens and Range of Thickness Qualified

		Test Plates					
Groove Welds		Number of Specimens				Plate Dimensions Qualified (in.)	
Type of Test	Thickness of Test Plates (T) (in.)	Face Bend	Root Bend	Side Bend	Macroetch	Min.	Max.
Groove	3/8	1	1	—	—	1/8	3/4
Groove	3/8 < T < 1	—	—	2	—	1/8	2T max.
Groove	≥1	—	—	2	—	1/8	Unlimited

Table 15-4 Bend Test Jig Dimensions for Various Strengths of Steel

Specified or Actual Base Metal Yield Strength (ksi)	Plunger Dimension		Die Member	
	Width (in.)	Radius of End (in.)	Opening Width (in.)	Radius (in.)
≤50	1 ½	¾	2 ⅜	1 3/16
50–90	2	1	2 ⅞	1 7/16
≥90	2 ½	1 ¼	3 ⅜	1 11/16

of the face and root-bend specimens shall meet both of the following requirements:

1. No single indication shall exceed ⅛ inches, measured in any direction on the surface.
2. The sum of the greatest dimensions of all indications on the surface that exceed 1/32 inches, but are less than or equal to ⅛ inches, shall not exceed ⅜ inches.

Cracks occurring at the corner of the specimens shall not be considered unless there is definite evidence that they result from slag inclusions or other internal discontinuities. The weld and heat-affected zone must be entirely within the bend radius.

Pressure Welds The internal pressure within the vessel is brought to 2,500 p.s.i. Follow all safety precautions. Any leaks or evidence of sweating is considered as a test failure.

Requirements for Passage of Tests

The student is required to pass all the tests in this section in at least one position. The positions in which these tests have been passed will be stated in the student's record.

CHAPTER 15 REVIEW

Multiple Choice

Choose the letter of the correct answer.

1. The V-groove butt joint weld without backing is typically done on which type of application? (Obj. 15.1)
   a. Pipe welding
   b. Critical plate welding
   c. Light gauge sheet metal
   d. Both a and b
2. What code authority normally utilizes open-root V-groove butt type joints? (Obj. 15-1)
   a. AWW
   b. AWS
   c. ASME
   d. None of these
3. When doing an open-root butt joint with an E6010 or E6011 electrode, what manipulation technique is used? (Obj. 15-1)
   a. Stringer bead
   b. Whipping
   c. Weave
   d. Beading
4. The keyhole is the indication that penetration is being achieved on an open-root butt joint with the SMAW process. (Obj. 15-1)
   a. True
   b. False
5. The work angle for a V-groove butt joint with weave beads is _____. (Obj. 15-1)
   a. 0–10°
   b. 20–45°
   c. 60–75°
   d. 90°
6. The travel angle for a vertical-up V-groove butt joint is _____. (Obj. 15-1)
   a. Push of 0–10°
   b. Drag of 0–10°
   c. Push of 80°
   d. Drag of 90°

7. With weave beads, moving the electrode across the face of the weld slowly will help reduce convexity. (Obj. 15-1)
   a. True
   b. False
8. Holding a close arc and hesitating at the toes of the weld will prevent undercut. (Obj. 15-1)
   a. True
   b. False
9. On open-root V-groove butt joints how should the tack welds be prepared prior to the application of the root pass? (Obj. 15-1)
   a. No preparation is required; these are deep penetrating electrodes and will burn out all slag and other defects
   b. Remove slag and wire brush tacks
   c. Remove slag, wire brush, grind, and feather tacks
   d. None of these
10. When doing vertical welding, progression can be from top to bottom (down) or bottom to top (up) and thus must be specified in the welding procedure. (Obj. 15-1)
    a. True
    b. False
11. When welding overhead fillet welds, which type of technique is generally used? (Obj. 15-2)
    a. Beading
    b. Stringers
    c. Weave
    d. Hip-hop
12. The work angle for the first pass of an overhead fillet weld is _____ from the vertical plate. (Obj. 15-2)
    a. 5–15°
    b. 20–30°
    c. 60–80°
    d. 75–85°
13. The travel angle for a fillet weld in the overhead position is _____. (Obj. 15-2)
    a. Push 5–15°
    b. Drag 5–15°
    c. Push 75–85°
    d. Drag 70–90°
14. When doing a multipass fillet weld, the work angle will change with each pass to allow the arc force to deposit the weld metal in the proper location. (Obj. 15-2)
    a. True
    b. False
15. All weld starts and stops should be free of depressions and high spots. (Objs. 15-1 and 15-2)
    a. True
    b. False
16. Overhead lap welds can be found in what type of industries? (Obj. 15-2)
    a. Storage tank
    b. Structural
    c. Shipbuilding
    d. All of these
17. The hydrostatic pressure testing is done to what pressure? (Obj. 15-3)
    a. 2,500 p.s.i.
    b. 2.5 k.s.i.
    c. 250 pounds
    d. Both a and b
18. What is the width of the test specimens for the face and root-bend tests? (Obj. 15-3)
    a. 1 inch
    b. 1⅛ inches
    c. 1½ inches
    d. 2 inches
19. The Test 2 weld is considered easier than the Test 1 weld. (Obj. 15-3)
    a. True
    b. False
20. If welds made in the 3G and 4G positions are passed, what other positions and welds are also qualified? (Obj. 15-3)
    a. 1G and 1F
    b. 2G and 2F
    c. 3F and 4F
    d. All of these
21. Grind and machining marks should run lengthwise with the bend test specimens. (Obj. 15-4)
    a. True
    b. False
22. If a tear occurs at the corner of the bend test specimen, is this a failure? (Obj. 15-4)
    a. Yes.
    b. No.
    c. Yes, if it is due to a discontinuity in the weld.
    d. Sufficient information not provided.
23. What is the corner radius of the bend test specimen? (Obj. 15-4)
    a. 1/16 inch
    b. 1/8 inch
    c. 5/32 inch
    d. 1/4 inch

24. Fillet weld test qualification for welders also qualifies them for _____. (Obj. 15-4)
    a. Groove welds
    b. Pressure welds
    c. Only fillet welds
    d. Groove welds in the flat position only
25. The student welder should pass test plate using both F3 and F4 electrodes. Why? (Obj. 15-4)
    a. Because of the drastic difference in manipulative techniques.
    b. Because the F3 electrodes are more difficult than the F4 group.
    c. Because it is important to produce welds with defects and inclusions.
    d. Because the AWS will not allow students to weld with a lower F number electrode simply by qualifying with a higher F number.

Review Questions

Write the answers in your own words.

26. Describe how the whipping technique controls the keyhole size when welding the root pass on an open-root V-groove butt joint.
27. Describe the arc manipulation technique to be used with low hydrogen electrodes.
28. By now, you are familiar with the various types of welded joints. Name the welds used for the following joints: beading on flat plate, square butt joint, lap joint, corner joint, single-V butt joint, T-joint, parallel joint, and double-bevel butt joint.
29. It is desired to weld a single-V butt joint on a $\frac{3}{8}$-inch plate, without backing bar, in the vertical position, travel down. Explain the entire welding procedure including root condition, number of passes, size of electrode, and welding heats.
30. How are single-V butt joints in the vertical position tested?

INTERNET ACTIVITIES

Internet Activity A

Do a Web search on the following labor unions and make notes on if they are active in making pipe welds: United Association of Plumbers and Pipe Fitters, Iron Workers International, and International Training Institute (Sheet Metal Workers).

Internet Activity B

Using your favorite search engine do a search on the chemicals used to make E7016 and E7018 electrodes. What is the main difference between these two electrodes? *Tip*: Remember who some of the leading manufacturers of welding electrodes are.

Design credits: Cinema and movie icon: Denis Meshkov/123RF; and Illustration of Welding icons: Mr. Nachai Sorasee/123RF

Table 15-5 Job Outline: Shielded Metal Arc Welding Practice: Jobs 15-J43–J55 (Plate)

				Type of Electrode				
				DCEP	DCEN	a.c.	AWS Specif. No. d.c.	and/or	AWS Specif. No. a.c.	Text Reference

Job No.	Joint	Type of Weld	Position	DCEP	DCEN	a.c.	AWS Specif. No. d.c.	and/or	AWS Specif. No. a.c.	Text Reference
15-J43	Single-V butt joint	Groove welding—weaved—multipass—travel up	Ver. (3G)	X		X	E6010 E7016–18		E6011 E7016–18	440
15-J44	T-joint	Fillet welding—single pass	Over. (4F)	X		X	E6010		E6011	443
15-J45	T-joint	Fillet welding—stringer—multipass	Over. (4F)	X		X	E6010		E6011	445
15-J46	Lap joint	Fillet welding—single pass	Over. (4F)	X		X	E6010		E6011	446
15-J47	Lap joint	Fillet welding—single pass	Over. (4F)	X		X	E7016, 7018		E7016, 7018	447
15-J48	Single-V butt joint	Groove welding—weaved—multipass—travel down	Ver. (3G)	X		X	E6010		E6011	450
15-J49	T-joint	Fillet welding—multipass	Over. (4F)	X		X	E6010		E6011	454
15-J50	T-joint	Fillet welding—weaved—multipass	Over. (4F)	X		X	E6010		E6011	455
15-J51	Single-V butt joint	Groove welding—stringer—multipass	Hor. (2G)	X		X	E6010 E7016–18		E6011 E7016–18	458
15-J52	Fittings	Fillet welding—single pass	Hor. (2F)	X		X	E7016–18		E7016–18	458
15-J53	Fittings	Fillet welding—single pass	Hor. (2F)	X		X	E6010		E6011	459
15-J54	Single-V butt joint backing bar	Groove welding—stringer—multipass	Over. (4G)	X		X	E6010 E7016–18		E6011 E7016–18	461
15-J55	Single-V butt joint	Groove welding—weaved—multipass	Over. (4G)	X		X	E6010 E7016–18		E6011 E7016–18	463

NA = not applicable.

16

Pipe Welding and Shielded Metal Arc Welding Practice:

Jobs 16-J1–J17 (Pipe)

Chapter Objectives

After completing this chapter, you will be able to:

16-1 Describe pipe welding and the pipe welding industry.

16-2 Describe pipe welding codes/standards and inspection testing methods.

16-3 Produce groove weld butt joints with and without backing specimens in various positions with various electrodes.

16-4 Produce groove and fillet weld branch connection specimens in various positions with various electrodes.

Introduction

The next several chapters present information and practice jobs in shielded metal arc welding carbon steel pipe and in gas tungsten arc welding and gas metal arc welding steel, stainless steel, and aluminum plate and pipe. You should undertake these jobs only after you have successfully mastered all of the skills practiced in the previous jobs and passed all of the welding tests given in the previous chapter. Many of the welding skills necessary to master these new welding processes are very similar to those you have already learned. You should have very little difficulty learning these new skills.

You are strongly urged to make every effort to master pipe welding skills. There is a need for people with pipe welding skills, and this need will continue in the years ahead.

You will learn that piping systems are a very safe and effective method of transporting products. For these reasons pipelines and piping systems will continue to be built and existing systems will need upgrading, and welders will be required to meet these demands. Table 16-1 gives some indications where piping systems are used. For example, it is forecasted that peak demand will grow at an annual rate as high as 1.5 percent through 2030. Power plants rely heavily on piping systems, pressure vessels, and various other welded structures.

Table 16-1 Pipefitter Welders Needed in the Following Industries

Industry	Piping Systems
Petroleum	Crude oil pipelines, finished product pipelines, offshore drilling platforms, refineries, pumping stations.
Natural gas	Distribution pipelines, compressor stations, local distribution stations.
Water and sewage treatment	Water pipelines for drinking, fire safety and irrigation purposes, sewage and water filtering and treating plants.
Power plants	Reactors, pressure vessels, heating and cooling piping systems. These can be land based or shipboard.

The History of Pipe Welding

The first piping system in recorded history was used to carry water. Early piping systems were made of stone or wood. Welded piping was of considerable importance in the growth of the modern pipeline industry.

The first attempt at welding pipeline involved the oxyacetylene process and took place in 1911 near Philadelphia. In 1922, the first attempt was made to arc weld a pipeline. The job involved 150 joints on a 12-inch line in Mexico. A major advancement came in 1926 when large diameter, seamless steel pipe was made available. The covered electrode was introduced to pipeline welding in 1930 when it was used with great success on a 32-mile, 20-inch line in Kansas.

The early pipelines had small diameters, but in 1942 the "Big Inch," a 30-inch diameter line, was completed from Houston, Texas, to Linden, New Jersey.

Pipe Welding Today

Today piping is made out of many materials and in many sizes to meet the demands of modern industry. The ferrous metals include wrought iron, carbon steels, chromium-molybdenum alloy steels, low temperature steels, stainless steels, and various lined and clad steels. The nonferrous metals include aluminum and aluminum alloys (Fig. 16-1), nickel and nickel alloys, titanium and titanium alloys, and copper and copper alloys. Nonmetallic materials include polyvinyl chloride, fiberglass, and composites. In addition, metal piping may be lined with nonmetallic materials such as glass, plastics, cement, and wood.

The number of materials required for piping systems is expanding constantly due to the need for better materials to meet increasingly severe operating conditions. Industries requiring special piping include power plants, nuclear plants, alternative power generation such as wind turbines, solar, etc.; refineries; chemical and petrochemical plants; paper mills; textile mills; aerospace plants; silicon chip manufacturing; and shipyards.

The most common welded piping in buildings is low pressure steam and hot water systems for heating. A large building or manufacturing plant needs piping for plumbing, sprinkler systems, air conditioning systems, gas and air lines, and lines that carry the various materials used in the manufacturing process. These pipelines must service a combination of low and high pressure and temperature conditions.

Power plant piping systems and nuclear piping systems, Fig. 16-2, are usually under very high temperature and pressure and often contain radioactive fluids. Joint tightness is absolutely essential because of the dangers of radioactive contamination. The temperature in these installations may exceed 1,000°F, and the pressure may exceed 2,500 p.s.i. Highly corrosive materials and conditions

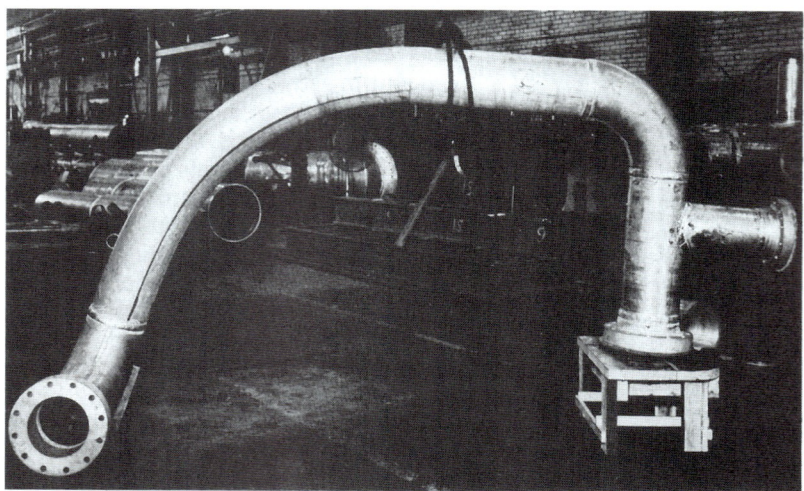

Fig. 16-1 Aluminum piping fabricated in the shop for a large chemical plant. Crane Co/DuPont

Fig. 16-2 Inside a pipe-fabricating shop. Several pipe fabrications for a power plant are shown: two pipe headers and a butt joint groove weld in a long length of pipe. Crane Co/DuPont

make it necessary to use piping of stainless steel, nickel steels, titanium, and various other nonferrous materials when high temperature or pressure is not involved. Welded pipeline, Fig. 16-3, forms an underground network for transporting natural gas, crude oil, refined petroleum products, and many other products including drinking water. The pipeline system is set up to deliver these products to all sections of the United States. Our nation's pipelines supply us with commodities that are fundamental to the American way of life—a system as essential as electric and telephone wires. Pipelines transport the fuel for our cars, trucks, planes, and ships—the energy needed for inexpensive shipment of our factory products and for our mobile lifestyles. Pipelines also deliver, at very low cost, the crude oil that refineries convert into essential materials for core American industries such as plastics, pharmaceuticals, and agriculture. Drinking water and sewage treatment are also provided by a safe, reliable pipeline system. The companies that build and operate interstate pipelines have created the safest mode of transportation today—safer than highway, rail, airborne, and waterborne transport. Yet, few people are aware of the work done by pipelines because this national infrastructure has been built underground for safety and aesthetic reasons.

There are well over a million miles of pipelines. Eighty percent of these lines are welded, and all new pipelines are completely welded. On small diameter and low pressure lines, plastic pipe has become very popular. Even it is welded and this will be covered in Chapter 31. On short pipelines, diameters of 80 and 96 inches have become common. Water pipelines are not as long as the lines for gas and other products, but many of them are much larger. The Hoover Dam penstocks, for example, are 30 feet in diameter and have a wall thickness up to $2\frac{3}{4}$ inches. A line named Capline is a giant hauler. It is an underground steel pipe 630 miles long and 40 inches in diameter that is capable of moving 4,000,000 barrels of oil a day into Illinois from the Louisiana oilfields. If it were gasoline, a day's supply would keep an average family car on the road for 30,000 years. The Maritimes and Northeast Pipeline is a historic infrastructure project. For the first time, a major energy project places the northeastern United States at the beginning of North America's natural gas pipeline network with close proximity to a significant supply basin. This historic project introduces natural gas—a clean, efficient, and cost-competitive energy source—to areas in Maine and the Northeast that do not currently have access to natural gas. Maritimes provides up to 530,000 MBtu/day of incremental transportation capacity, bringing newly developed natural gas reserves from the Sable Island area, offshore Nova Scotia, to markets in Atlantic Canada and the northeastern United States and was placed in service on December 1, 1999.

Another major pipeline is the Alliance Pipeline. The Canadian portion of the system consists of 211 miles of

Fig. 16-3 Overland pipeline being welded in the field. Note that a welder works from each side of the pipe. Nigel Bowles/Alamy Stock Photo

42-inch and 758 miles of 36-inch diameter steel pipe. There are 40 receipt points connecting with lateral pipelines totaling about 434 miles, ranging in length from about 0.2 to 96 miles and in diameter from 4 to 24 inches. There are seven mainline compressor stations of about 31,000 to 40,000 horsepower each, spaced about 120 miles apart. Mainline block valves are spaced about every 20 miles. The U.S. portion of the system consists of 888 miles of 36-inch diameter steel pipe. This high-pressure natural gas transmission system delivers rich natural gas from the Western Canadian Sedimentary Basin to the Chicago market hub. As a side note, Alliance Pipeline contributes $1.8 million annually toward community investment initiatives and programs to support education, environmental stewardship, community development, leadership development, and safety in the communities across the pipeline system.

Details of the Alliance Pipeline:
Initial capacity   1,325,000,000 ft^3/day
Overall mainline length   1,857 mi
    Canada   969 mi
    United States   888 mi
Compressor stations (14)
31,000–40,000 hp
    Canada (7)
    United States (7)
Mainline pipe diameter   36–42 in.
Mainline pipe thickness   0.621 in.
Mainline operating pressure   1,740 p.s.i.
Lateral gathering system (44)   434 mi
Lateral pipe diameter   4.5–24 in.
Rich gas system   Up to 1,188 Btu/ft^3

There are approximately three times as many miles of pipeline in the United States as there are miles of railroad track. The lines start in the gas and oil fields and feed natural gas into towns and cities and crude oil into big refineries. Petroleum products flow out of the refineries into still more miles of pipelines. Gasoline, for example, flows through pipe to terminals where it is picked up by tank trucks and delivered to service stations.

A pipeline system can be compared with the Mississippi River. It begins as a multitude of tiny tributaries and ends in a maze of channels at the delta.

These interstate pipelines deliver over 12.9 billion barrels of petroleum each year. (There are 42 gallons in a barrel.) This amount combined with the natural gas and water delivered through pipelines accounts for approximately 30 percent of the nation's total overland cargo. The cost to transport a barrel of petroleum products from Houston to the New York harbor is about $1, or about 2½ cents per gallon at your local gasoline station. A pipeline network delivers a product that is an integral part of America's economy and does it safely and efficiently, at low cost. Replacing even a modest-sized pipeline, which might transport 150,000 barrels per day, would require 750 tanker truckloads per day, a load delivered every 2 minutes around the clock. Replacing the same pipeline with a railroad train of tank cars carrying 2,000 barrels each would require a 75-car train to arrive and be unloaded every day. As an example, the cost efficiency for petroleum pipelines depends on a relatively small national workforce of about 16,000 skilled employees, yet that workforce transports over 600 billion ton-miles of freight each year. These workers accomplish this job so efficiently that America's oil pipelines transport 17 percent of all U.S. freight, but cost only 2 percent of the nation's freight bill.

Product pipelines transport more than 50 types of refined petroleum products items such as various grades of motor gasoline, home heating oil, diesel fuel, aviation gasoline, jet fuels, and kerosene. For instance, Colonial Pipeline, the major product pipeline that stretches from Texas to New Jersey, transports almost 40 different formulations of gasoline alone—different grades of each mandated type of gasoline, the requirements for which vary seasonally and regionally. Liquefied ethylene, propane, butane, and some petrochemical feedstocks are also transported through oil pipelines. These various products shipped are referred to as *batches*. The physical principles of hydraulics keep the batches of liquid from blending and contaminating one another except where they actually touch. These *interfaces* between different shipments are separated out when they arrive at their destination and are reprocessed. Sometimes batches are separated by metal "pigs" or plugs that keep batches from touching. Pigs are also used for cleaning the interior surfaces of pipelines to help prevent corrosion. Specially developed "smart" pigs containing instrumentation packages are used to double-check pipeline integrity.

The speed at which products move through the pipeline averages from 3 to 8 miles per hour depending upon line size, pressure, and other factors such as the density and viscosity of the liquid being transported. At these rates, it takes from 14 to 22 days to move liquids from Houston, Texas, to New York City.

Building a Pipeline

Running a pipeline across the country has been reduced to a smooth step-by-step production process that gets the job done with a minimum of effort and time. Usually the workers who build pipelines are specialists who travel from country to country, following the jobs wherever they are.

On the job site, the bulldozers and scrapers clear and grade the right-of-way. Then come the trucks, bearing sections of pipe which may be as long as 80 feet. Caterpillar tractors with side booms lay out the pipe along the trail. If there is a curve in the right-of-way, a tractor with huge jaws grabs a section of the pipe and bends it.

The tractor is followed by a ditching machine, which scoops out the trench at the rate of a mile a day. And solid rock that is in the way of the digger is blasted out by the dynamite crew. The pipe is strung out along the trench where the crews of welders join the pipe joints together to make a continuous pipeline, Fig. 16-4. The welding crews may consist of a crew who line up the pipe, a crew of tackers, and a crew of welders who apply the root pass, the intermediate pass, and the finishing pass. Radiograph technicians and/or ultrasonic technicians check the completed welds. If a weld shows any faults, it is cut out and rewelded.

The pipe is ready to be put into the trench after it has been properly wrapped and cathodic protection applied. The line receives its final check and is placed in the trench.

When put into service, the line is continually checked for leaks by metering devices at each end of the line to check input and output. Leaks can also be spotted from the air. The big pipeline companies have fleets of small planes. A pilot can patrol 100 miles of line in an hour and find a spot of oil no bigger than a hat.

The use of pipelines as cargo carriers is in its infancy. If a rubber ball or an electronic sensing device can be sent through a pipeline, someday a capsule loaded with mail, ore, oranges, or even people may be transported this way.

Fig. 16-4 Making a position weld on a pipeline. Travel is downhill, but travel uphill is a welding procedure used on some applications. *Edward R. Bohnart*

Safety Concerns

A pipeline safety bill created standards for oil and natural gas pipelines. This would subject operators who violate the law to higher fines. More than 1.6 million miles of natural gas pipelines and 155,000 miles of hazardous liquid pipelines weave through the country.

Key aspects for the bill are:

- Requires operators to inspect pipelines every five years, although the Transportation Department's inspector general would have some discretion to lift the mandate.
- Requires operators to submit to state or federal oversight agencies a plan to enhance the qualifications of pipeline personnel.
- Requires owners and operators of pipelines to report to the Transportation Department any spill of more than five gallons.
- Increases from $25,000 to $500,000 the civil penalty for each failure by an operator to mark accurately the location of pipelines near construction and demolition operations or violations of other safety standards. The maximum civil penalty was doubled to $1 million.
- Expands state oversight of hazardous liquid and natural gas pipelines.
- Increases outlays for safety efforts by $13 billion over several years. This includes research into technologies to detect pipeline flaws better.
- Improves whistle-blower protections.

Welded-Pipe Application

In the past, pipe was joined together only by mechanical means such as threaded joints, flanges with bolts and gaskets, and lead and hemp joints. At the present time, practically all pipes 2 inches in diameter and larger are joined by welding. The materials to be handled, the operating temperature, and the internal pressure required make any other method of joining out of the question for many piping systems. Many of today's industrial processes would not have been possible with threaded joints in piping installations. The discovery, development, and application of nuclear energy certainly would not have taken place.

The important advantages of welded piping installations include:

- Permanent installation
- Improved overall strength
- Low maintenance costs because of leakproof, care-free joints
- Ease of erection

- Improved flow characteristics
- Reduction in weight due to compactness and lighter weight fittings
- More pleasing appearance
- More economical application of insulation

Cost Welded fittings are available that meet or lower than the installed cost of threaded fittings of comparable size. It is estimated that if all joints on a piping system are welded, a saving of 50 percent can be realized on the cost of material alone. Threading cuts the effective thickness of pipe wall practically in half. When welding fittings are used, pipe with much thinner walls can be used to reduce cost without lowering the efficiency of the piping system. Thus, welded piping systems make sense not only from the standpoint of service durability and safety but also from the standpoint of cost.

Strength Welded fittings have the same thickness as the pipe with which they are used, Fig. 16-5. They are joined to the pipe by an equal thickness of weld. There is no loss of strength caused by thread cutting. The life expectancy of a welded system is double that of threaded pipe.

Joint Security Welded fittings and pipe are permanently joined by a metal bond so that pipe, fittings, and welds all become integral parts of the system. The security of the joints does not depend upon the strength used in wrenching nor the amount of dope applied. There is no chance for leaks due to vibration or stress.

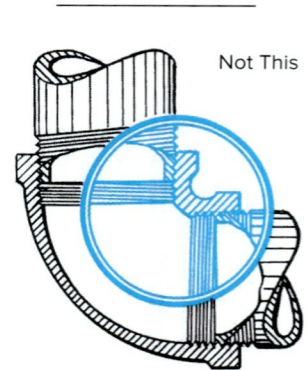

Fig. 16-5 Welded piping improves utility and appearance. It is also leakproof.

Flow The inside diameters of welded fittings and pipe meet exactly. There are no step-ups or step-downs as in a threaded system. Pockets that create turbulence, clogging, and erosion are eliminated so that water or coolant flows freely, Fig. 16-6.

Installation When assembling a welded piping system, there is no need to calculate overlaps for threading nor to allow for differences in the depth of joints. Roughing-in is quicker because the pipe is simply cut to length, placed in position, and welded. Assemblies can be fabricated conveniently in the shop or at the floor level of a construction job, and then they can be quickly tied into the system. Less manual labor is required than for a threaded system because wrenching is not required. Welding equipment is lighter and easier to handle than threading equipment.

Testing and Repair When contractors install a threaded piping system, they know that there will be some leaks which must be tightened before the system goes into operation. They also know that in some cases they will have to repair leaks after the system has been completed. Contractors may have to take out three or more fittings or repair a leaking joint. They even may have to recut some of the pipe. As a rule of thumb, contractors estimate that 5 to 10 percent of their cost for labor will be for this kind of work. Such repair is not necessary on welded piping because no hemp, gaskets, or other sealing materials are used.

Easy Insulation The outside diameters of the pipe and fittings are identical, so that the application of insulation is simple. There is just one smooth surface and no bulky fittings or difficult angles. Sleeve or tubular insulation can be slipped on over the pipe and fittings without difficulty. Tape or wraparound insulation can be applied over joints as smoothly as over pipe lengths, Fig. 16-7. Once applied, insulation never has to be removed to get at loosened and leaking joints.

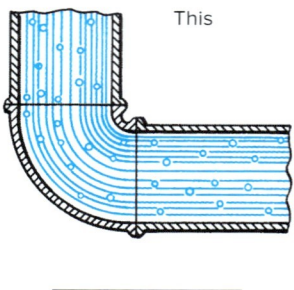

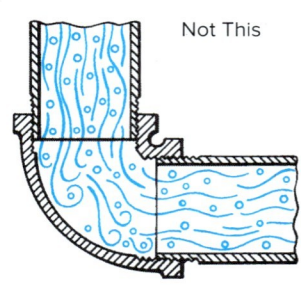

Fig. 16-6 Welded piping improves the flow of materials.

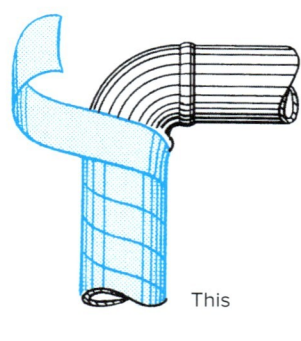

Fig. 16-7 Welded piping ensures the easy application of insulation.

Weight Since the ends of the fittings need not encompass welded pipe, considerable weight is saved through the reduced diameter and shorter length of each fitting. The fact that pipe with thinner walls can be used also cuts down on the weight of the system.

Appearance The smooth lines of a welded piping system present a much more professional appearance than a threaded system. There are no unsightly bulges and exposed threads.

Space Welded-pipe systems are ideal for installations where space is a problem. Fittings are compact, and very little clearance is needed when the fabricator is welding instead of swinging a wrench or turning a fitting. A space shuttle is an example of a limited area in which piping systems are required.

Equipment If a threaded system with a full range of pipe sizes from 2 to 6 inches is being installed, the worker may need up to three different threading outfits, a dozen pipe wrenches, expensive dies, reamers, and pipe cutters. Welding, on the contrary, requires only one unit for pipes of all sizes. The exact equipment will vary, depending on the welding process.

Pipe Specifications and Materials

The American Society for Testing Materials (ASTM) and the American Petroleum Institute (API) issue specifications for ferrous and nonferrous pipe. Study Table 16-2 which lists standard pipe sizes.

Carbon Steel Pipe Carbon steel is by far the most widely used material in piping systems today. The relatively low cost of carbon steel pipe and its excellent forming and

Table 16-2 Pipe Dimensions

Nominal Pipe Size	Outside Diameter	Nominal Wall Thickness for Schedule Numbers									
		Schedule 10	Schedule 20	Schedule 30	Schedule 40	Schedule 60	Schedule 80	Schedule 100	Schedule 120	Schedule 140	Schedule 160
1/8	0.405	—	—	—	0.068	—	0.095				
1/4	0.540	—	—	—	0.088	—	0.119				
3/8	0.675	—	—	—	0.091	—	0.126				
1/2	0.840	—	—	—	0.109	—	0.147	—	—	—	0.187
3/4	1.050	—	—	—	0.113	—	0.154	—	—	—	0.218
1	1.315	—	—	—	0.133	—	0.179	—	—	—	0.250
1 1/4	1.660	—	—	—	0.140	—	0.191	—	—	—	0.250
1 1/2	1.900	—	—	—	0.145	—	0.200	—	—	—	0.281
2	2.375	—	—	—	0.154	—	0.218	—	—	—	0.343
2 1/2	2.875	—	—	—	0.203	—	0.276	—	—	—	0.375
3	3.5	—	—	—	0.216	—	0.300	—	—	—	0.437
3 1/2	4.0	—	—	—	0.226	—	0.318				
4	4.5	—	—	—	0.237	—	0.337	—	0.437	—	0.531
5	5.563	—	—	—	0.258	—	0.375	—	0.500	—	0.625
6	6.625	—	—	—	0.280	—	0.432	—	0.562	—	0.718
8	8.625	—	0.250	0.277	0.322	0.406	0.500	0.593	0.718	0.812	0.906
10	10.75	—	0.250	0.307	0.365	0.500	0.593	0.718	0.843	1.000	1.125
12	12.75	—	0.250	0.330	0.406	0.562	0.687	0.843	1.000	1.125	1.312
14 OD	14.0	0.250	0.312	0.375	0.437	0.593	0.750	0.937	1.062	1.250	1.406
16 OD	16.0	0.250	0.312	0.375	0.500	0.656	0.843	1.031	1.218	1.437	1.562
18 OD	18.0	0.250	0.312	0.437	0.562	0.718	0.937	1.156	1.343	1.562	1.750
20 OD	20.0	0.250	0.375	0.500	0.593	0.812	1.031	1.250	1.500	1.750	1.937
24 OD	24.0	0.250	0.375	0.562	0.687	0.937	1.218	1.500	1.750	2.062	2.312
30 OD	30.0	0.312	0.500	0.625							

welding properties demand its use except for extreme conditions. Carbon steel pipe with a tensile strength up to 70,000 p.s.i. is available for high temperature, high pressure service.

High Yield-Strength Pipe High yield-strength pipe has been developed for high pressure transmission lines because of the cost advantage of using lighter weight pipe for long, cross-country lines. This pipe must be readily weldable in the field. Therefore, carbon steels or high strength, low alloy steels are preferred.

Chrome-Moly Pipe The addition of chromium and molybdenum inhibits graphitization and oxidation and increases high temperature strength. These alloying elements have advanced the useful temperature limits of piping systems, but there are certain drawbacks. Chrome steels have a pronounced tendency to air-harden, particularly when the amount of chromium is 2 percent or more. In such cases, special care in fabrication, special welding procedures, and carefully controlled annealing are necessary.

Chrome-Nickel (Stainless-Steel) Pipe Special consideration must be given to the austenitic stainless steels when they are to be welded, hot formed, or used for corrosion resistance. At temperatures from 800 to 1,500°F, chromium carbides are formed along the grain boundaries and reduce corrosion resistance. This characteristic is minimized by the addition of small amounts of stabilizing elements such as niobium or titanium or by other methods.

Low Temperature Pipe There is an ever-increasing demand for pipe that can provide satisfactory service at low temperatures. There are various grades of steel pipe for low temperature service. Other grades may be furnished to the general requirements of this specification.

Nonferrous Pipe Nonferrous pipe is used for such extreme conditions as very low temperatures, very high temperatures, and corrosive conditions beyond the abilities of ferrous metals. The most common nonferrous pipe materials are nickel, copper, titanium, aluminum, and their many alloyed compositions: Monel, Inconel, Hastelloy, red brass, yellow brass, and cupro-nickel.

Shielded Metal Arc Welding of Pipe

Shielded metal arc welding is one of the principal processes for welding pipe both in the shop and in the field. Welds of X-ray quality are produced on a production basis. This process may be used for nearly all ferrous and nonferrous piping. Standard welding power sources that produce alternating or direct current such as a rectifier, transformer, inverter, or an engine-driven machine may be used. A wide range of electrode types and sizes are available.

Welding may be done in all positions, and the direction of welding may be up or down.

Pipe Welding Electrodes

E6010 Classification This type of shielded arc electrode is generally used in the welding of carbon steel pipe. It is also used to weld the stringer bead on higher strength pipe. The deposits of electrodes in this classification have a tensile strength of 50,000 to 65,000 p.s.i.

Some electrode manufacturers make two types of electrodes under the E6010 classification: one that meets the general classification and another that is especially designed for pipe. The electrode for pipe has a more accurately directed arc and improved wash-in action on the inside of the pipe. These features permit a slightly narrower gap but still produce a flat inside bead. The narrower gap reduces the volume of weld metal in the joint so that overall welding speed is increased. This type also reduces undercut on the back side of the joint and generally improves X-ray quality. Some welders feel they can carry a slightly higher current and larger weld pool on the stringer pass.

E6011 Classification The operating characteristics, mechanical properties, and welding applications of the E6011 resemble those of the E6010, but the E6011 is used with alternating current. Although it may also be used with direct current electrode positive, it loses many of its beneficial characteristics with this current and polarity.

E7010 Classification E7010 electrodes are used for higher strength pipe than E6010 electrodes. Welds made with these electrodes will equal or exceed the minimum yield strength of high strength pipe. They do not necessarily exceed the ultimate tensile strength of these materials since the pipe will frequently sustain 100,000 p.s.i. The deposits of these electrodes have a tensile strength of 70,000 to 78,000 p.s.i.

Here again, manufacturers make several types with special characteristics. One type is specifically designed for vertical-down pipe welding and is used on all passes. It produces less slag interference so that it minimizes the problem of pinholes caused by slag running under the arc in downhill welding.

E7018, E9018G, and E11018G Classifications EXX18 electrodes are low hydrogen electrodes used to weld high tensile and low alloy pipe. In addition to the low hydrogen characteristics, they may also contain small amounts of chromium and molybdenum. They may be used in all positions.

Electrodes in the EXX18 classification produce hydrogen-free weld deposits. The lack of hydrogen reduces the tendency for underbead cracking and microcracking when welding low alloy steel. It also reduces the amount of preheat required. The preheat temperature may be as much as 300°F less than that required for other electrodes. They are typically used for intermediate and cover passes. However, they can be used for open-root beads if joint fitup tolerances are maintained with minimum root face and maximum root opening. The current and polarity should be a.c. or DCEN on root beads only.

Low hydrogen electrodes are used to weld the low alloy pipe used in many inner plant high pressure and high temperature applications. Mild steel pipe with wall thickness over ½ inch is sometimes welded with these electrodes because of their superior cleaning properties and an overall speed advantage over E6010 electrodes. The E7018 electrodes produce weld deposits with a tensile strength of 65,000 to 80,000 p.s.i.; the E9018G electrodes, 90,000 to 105,000 p.s.i.; and the E11018G electrodes, 110,000 to 115,000 p.s.i.

Fig. 16-8 Typical 30-inch headers for a gas conditioning system. The headers are made of pipe with a 30-inch outside diameter and 5/16-inch wall thickness. Nozzles of the same material have outside diameters of 30 inches, 24 inches, and 12¾ inches. Welding saddles were used for all nozzles except the 30-inch size, which were reinforced with welded rings. *Crane Co/DuPont*

Joint Design

The most common types of joint in welded piping systems are the *circumferential butt joint* and the *socket* or *(lap) joint*. A groove weld is used for the butt joint, and a fillet weld is used for the socket joint. Fillet-welded joints often join flanges, valves, and fittings to pipe 2 inches and smaller in diameter. The fabrication shown in Fig. 16-8 is composed of several of the basic types of joints and welds used in the fabrication of piping systems.

Butt Joint

The butt joint, Fig. 16-9, is readily welded by all of the standard welding processes. Butt joints may not have inside support or they may be set up with an inner liner called a *backing ring* or *chill ring*. The butt joint is not difficult to prepare for welding, and it can be welded in all positions without great difficulty. This joint design provides good stress distribution and has maximum strength while permitting the unobstructed flow of materials through the pipe. It is also pleasing in appearance. Its general field of application is pipe to pipe, pipe to flanges, pipe to valves, and pipe to other types of fittings. The butt joint can be used for any size or thickness of pipe and for any type of service.

While there are no universally accepted welding grooves, the bevels specified in Table 16-3 are considered,

Fig. 16-9 This welding student is making a circumferential V-groove weld on a butt joint in pipe in the horizontal fixed position (5G). *Location: Northeast Wisconsin Technical College Mark A. Dierker/McGraw Hill*

for all practical purposes, as standard for industrial piping. Bevels may be made by mechanical means or flame cut.

Socket Joints

Socket joints are fillet welded. They are generally used for joining pipe to pipe, pipe to flanges, pipe to valves, and pipe to socket joints in pipe about 3 inches in diameter and

Table 16-3 Recommended Pipe Bevels

Type of Material	Position of Welding	Direction of Welding	No Backing Ring	With Backing Ring	Bevel Angle[1]	Root Face[2]	Root Opening[3]
Carbon steel	All	Up	×		30 to 37½°	1/16	1/16
Carbon steel	All	Up		×	30 to 37½°	1/8	3/16
Carbon steel	All	Down	×		37½°	1/16	1/16

[1]Bevel angle may be ±2½°.
[2]Root face may be ±1/32.
[3]Root opening may be ±1/16.

under. This type of joint permits unrestricted fluid flow in small pipe. Figure 16-10 illustrates three typical fillet welded socket joints. Adequate penetration of the pieces being joined is an absolute requirement.

Note the 1/16-inch minimum clearance in Fig. 16-10, which is required by many codes. If the pipe is bottomed out in the socket with no clearance, the weld will crack in service. The function of the gap is to permit thermal expansion of the pipe. Without the gap, the heat of welding can cause the pipe to expand at a faster rate than the fitting. If the pipe is resting against the socket fitting, the thermal growth of the pipe stains the weld and will lead to cracking. Failures have also occurred in socket welds subjected to thermal and mechanical cycling during service. The same situation that occurs during welding can occur during service with rapid changes in service temperature causing the pipe to expand against the bottom of the fitting, straining and cracking the weld. A Gap-a-Let is a device inserted into the socket to hold this clearance.

Socket joints are not always acceptable. Butt joints with groove welds with complete penetration to the inside of the piping are required for piping systems containing radioactive solutions, food products, or gases and those solutions subjected to corrosive service with materials that are likely to cause crevice or stress corrosion.

Intersection Joints

Intersection joints such as medium and large Ts, laterals, Ys, and openings in vessels are usually the most difficult to weld. Factory-wrought fittings are preferred for the layout of the piping system since they are the equivalent in strength to the pipe being used. Moreover, their installations involve butt joints only. Cutting and beveling may be done manually or with a machine beveler designed for that purpose.

Figure 16-11 illustrates the three forms of preparation for 90° intersection joints. In type A, the header opening is equal to the inside diameter of the branch, and only

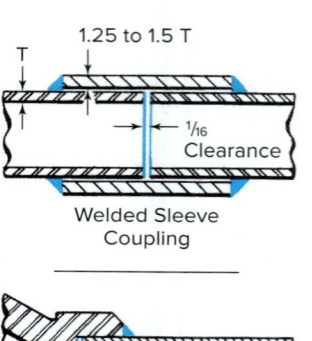
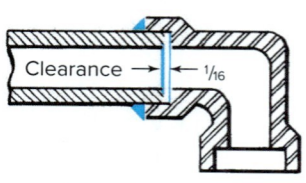

Fig. 16-10 Typical fillet welded lap joints.

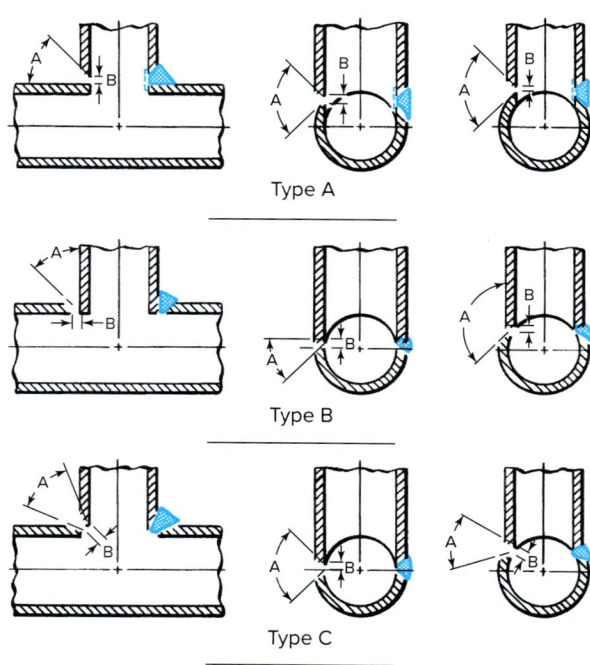

"B" may be increased when backing ring is used.

Type A

Type B

Type C

Angle "A" to be not less than 45° for any wall thickness of pipe.

"B" to be not less than 1/16" nor more than 1/4".

Acceptable types of preparation for 90° full size and 90° reduced size branch connections.

Fig. 16-11 Acceptable types of preparation for 90° full size and 90° reduce size branch connections.

the branch is beveled. This type of branch permits the use of a specially shaped backing ring if necessary. In type B, the header opening is large enough to permit insertion of the branch so that only the header opening must be beveled. Type C is a third form of preparation in which both the header and the branch are beveled. This form of unreinforced branch is adequate only when the pipe is to be used at pressures of 50 to 75 percent of the full pressure that the pipe can withstand. Reinforcement or manufactured welded fittings are required to make 90° intersections equivalent in strength to the pipe from which they are fabricated. Intersection joints at an angle of 45° or less produce a condition at the heel of the intersection that makes it difficult, if not impossible, to secure the degree of penetration and soundness of weld deposit that is necessary for severe service conditions. Wherever possible, intersection joints should be made under shop conditions where the work can be positioned for welding in the flat position.

Welded Fittings

Pipe manufacturers have kept pace with design for welding by providing the industry with a line of seamless and welded fittings especially for welding. They are available in most grades of material suitable for welding and in many combinations of size and thickness. They combine the best characteristics for unimpaired flow conditions with wide availability, ease of welding, and maximum strength.

Insulation can be applied to ready-made fittings without difficulty. The system can be installed in less space than when fittings are fabricated for the job, and system changes can be made without difficulty. The piping requires less maintenance, and new systems are easy to design. The use of these fittings also saves a great deal of time and eliminates a considerable amount of the cutting, tacking, and fitting required in the hand fabrication of various types of joints. They also enhance the appearance of the job. They make the use of hand-mitered construction obsolete.

Manufactured fittings for welding that are available are shown in Fig. 16-12. The application of welding to flanges is illustrated in Fig. 16-13, page 480. The sequence of manufacturing steps from flat plate to the finished fitting is shown in Fig. 16-14, page 480.

Fig. 16-12 Machine-beveled manufactured welding fittings. Sypris Technologies, Inc., Tube Turns Division

Backing Rings The term *backing ring* is applied to a ring-shaped structure, Fig. 16-15, page 480, which is fitted to the inside surface of the pipe before welding. Its functions are to assist the welder in securing complete penetration and

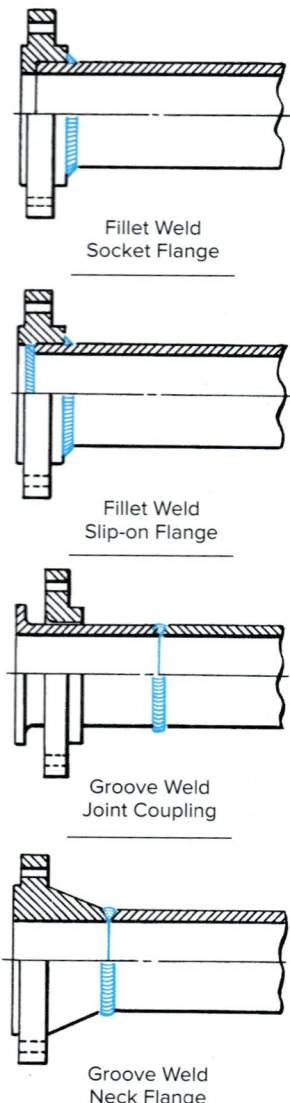

Fig. 16-13 Typical welding flanges.

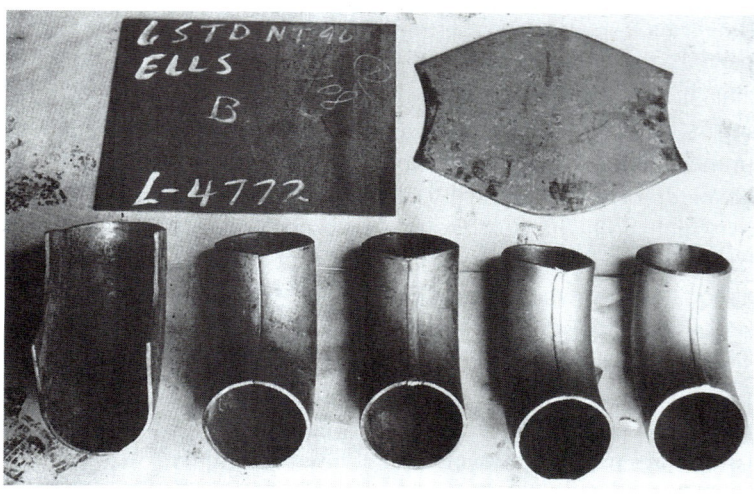

Fig. 16-14 The manufacturing steps necessary to fabricating a welded L with machined ends. Crane Co/DuPont

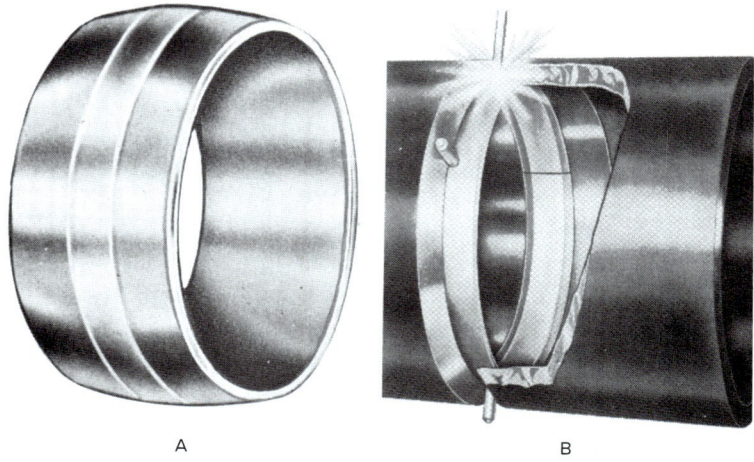

Fig. 16-15 (A) A machined ring. (B) A ring with chamfered nubs for quick "strike off" sets the pipe root opening for the root pass. Nubs melt with the weld metal. Robvon Backing Ring Co.

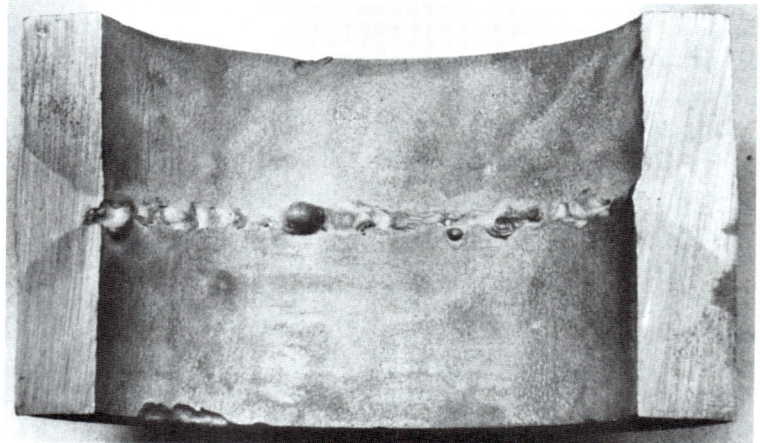

Fig. 16-16 The root side of a V-groove butt joint on pipe welded without a backing ring. Note the excess melt-through (icicles, grapes, etc.) in several places. Edward R. Bohnart

fusion without excessive melt-through (Fig. 16-16), to prevent spatter and slag from entering the pipe at the joint, and to prevent the formation on the inside of the joint of irregularly shaped masses of metal, sometimes called *icicles* or *grapes*. The prevention of icicles is especially important in pipe joints ahead of such units as valves and turbines. Backing rings also help to secure proper alignment of the pipe ends. There are two types of backing rings: split and solid. Split backing rings are designed to fit on the inside of the pipe without machining. They can be expanded or contracted to fit the inside diameter of the pipe. The solid type is machined and requires that the pipe also be machined to receive it. Its use is usually restricted to pipe that is to sustain severe service conditions. The installation cost of solid rings is higher than that of split rings due to the higher cost of the ring itself and the cost of preparing the pipe. A solid ring is shown in Fig. 16-15A, and a split ring, in Fig. 16-15B. These types of devices are also available for square and rectangular tubing.

Correct spacing between the pipe ends is essential to securing sound welds. Most backing rings contain a series of small nubs, ranging from 1/16 to 3/4 inch in length, on the outer surface of the ring. Backing rings for pipe sizes 4 inches and under are 3/32 inch thick by 5/8 inch wide. For pipe sizes 5 inches and over, they are 1/8 inch thick by 1 inch wide. Root-spacer nubs may be 1/8 to 3/16 inch in diameter. The material of the ring should be substantially the same chemical analysis as the pipe or tube that is to be welded.

Consumable insert rings, Fig. 16-17, improve the quality of the weld. They must be of the proper composition and dimensions. In piping for atomic reactors, stainless-steel piping is best installed with consumable insert rings. These rings provide the most favorable weld contour to resist cracking from weld metal shrinkage and *hot-shortness* (brittleness in hot metal), and they eliminate notches at the weld root. They also ensure weld metal composition with the best possible properties of strength, ductility, and toughness. See Fig. 16-18 for end preparation of pipe for consumable insert rings. These are typically welded autogenously with the GTAW process.

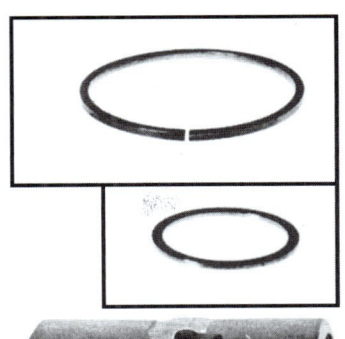

Fig. 16-17 The consumable insert ring becomes part of the root weld bead and ensures complete penetration and smoother flow of materials inside the pipe. Fitup is easy. The GTAW process is used on root pass when consumable insert rings are used. Robvon Backing Ring Co.

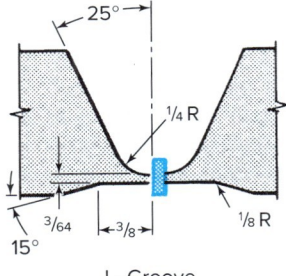

J - Groove

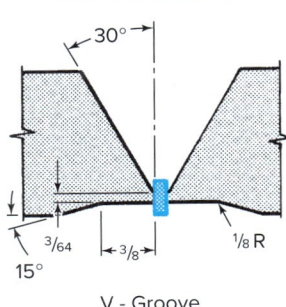

V - Groove

Fig. 16-18 Typical pipe end preparation for consumable insert rings for pipe wall thicknesses up to 1 inch.

Pipe Clamps One difficulty encountered in assembling pipe to be fabricated is the positioning of the pipe before tacking. One method is to clamp the pipe in a fixture in the exact position desired for welding. The joint must be carefully mounted in the proper alignment in the pipe clamp or fixture. Pipe clamps of various types and sizes are available to align and hold pipe joints in preparation for tacking. Figures 16-19 through 16-25 (pp. 481–482) show various pipe joints properly clamped and ready for tacking.

Special Fabrications Sometimes manufactured welded fittings are not available, and the welders must fabricate their own fittings or make special connections on the job. There are a variety of special designs in welded-pipe construction that may be fabricated when necessary. Among these are Ls and laterals, side outlet fittings, mitered joints, elbows, Ys, and expansion joints, Figs. 16-26 through 16-28, page 483. Joints should be carefully laid

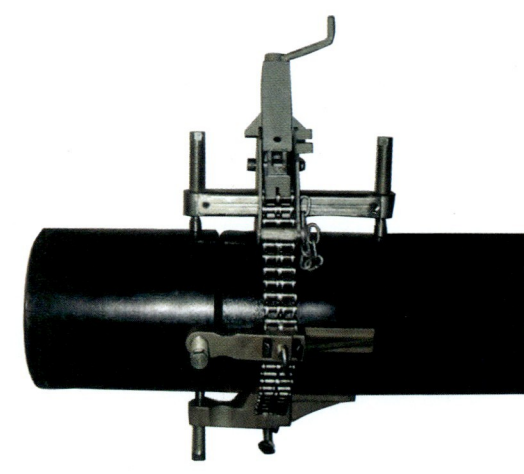

Fig. 16-19 Standard butt joint clamped and ready for tacking. Mathey Dearman/Diana Sullivan, Mathey

Fig. 16-20 An "L" is being set up with a length of pipe for a butt joint. Mathey Dearman/Diana Sullivan, Mathey

Fig. 16-21 A flange is being set up with a length of pipe for a butt joint. Note the internal clamp. Mathey Dearman/Diana Sullivan, Mathey

Fig. 16-23 This fast, easy-to-use clamp can be used to hold small fittings in position for welding. This clamp can adjust "Hi-Lo"™ of the elbow level checks the alignment. Mathey Dearman/Diana Sullivan, Mathey

Fig. 16-24 Clamps are available in several types and models from 2- to 60-inch diameters. This type of cage clamp is fast and efficient when only alignment is required. Mathey Dearman/Diana Sullivan, Mathey

Fig. 16-22 The Spacing Screw is used to obtain a precise root opening between pipes or other fittings. Spacing Screw assemblies are available for steel and stainless-steel applications. Mathey Dearman/Diana Sullivan, Mathey

Fig. 16-25 This hydraulic cage clamp produces the extra strength necessary for aligning large pipe diameters. Pipes can be effortlessly aligned due to the design of the Hydraulic Closure Mechanism. Clamps are available for 16- to 60-inch pipes. Mathey Dearman/Diana Sullivan, Mathey

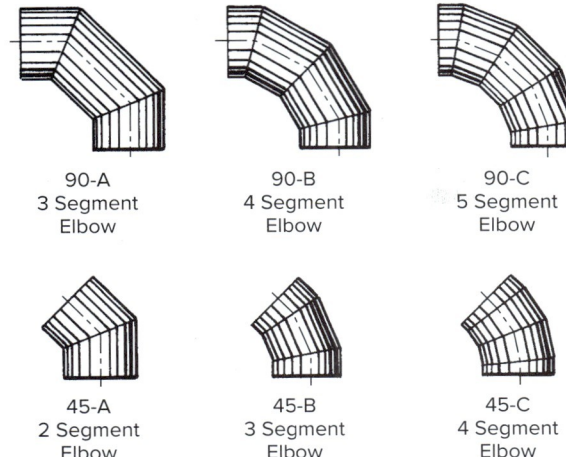

Fig. 16-26 Hand-fabricated segmental 45° and 90° Ls.

Fig. 16-27 A steel 90° three-segment elbow with flange, tacked and ready to be loaded into a welding positioner for final welding.
John Kriesel/Piping Systems, Inc.

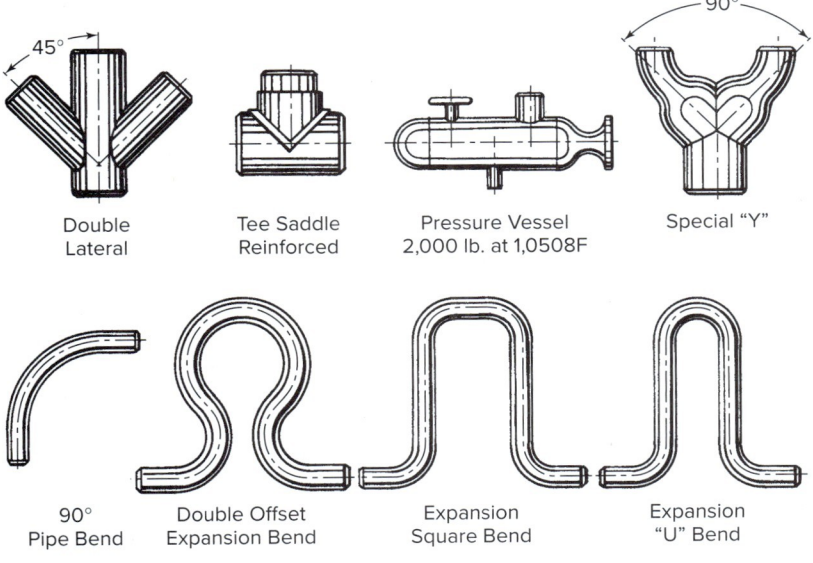

Fig. 16-28 Several types of hand-fabricated branch connections and expansion bends.

out with templates and standard layout curves. These fittings may be formed on the job from straight lengths of pipe by hand or machine flame cutting.

Codes and Standards

You will recall that piping systems are subject to wide variations in pressure and temperature, that corrosion is a problem in many systems, and that many of the materials carried are dangerous to life and property. Piping systems must be insurable against any loss resulting from the failure of the pipe or the welding.

Code-Making Organizations

In order to ensure uniform practices in the design, installation, and testing of piping systems, various codes and standards applicable to welded piping systems have been prepared by committees of leading engineering societies, trade associations, and standardization groups. These are generally written to cover minimum requirements of quality and safety. A few insurance companies, manufacturers, and the military set up their own codes to cover the fabrication of welded piping systems. All of these codes have been set up for the purpose of establishing welding procedures and inspecting and testing methods to provide assurance that the work meets the purchaser's specifications and to give protection against accident. Our prime concern in this chapter will be with the qualification tests for welders rather than with the procedure qualification tests.

Following is a brief description of a few of the principal code-making organizations.

The American Society of Mechanical Engineers (ASME) This society is responsible for the ASME Boiler and Pressure Vessel Code, which covers piping connected to boilers. It is recognized in almost all states and is a prerequisite to acceptance of the installation by the states and insurance companies. It covers the installation of piping systems in connection with power boilers, nuclear vessels, and unfired pressure vessels.

The American Petroleum Institute (API) This group is concerned with the standards for welding pipelines and related facilities. The standard includes regulation of the gas and arc welding of piping used in the compression, pumping, and transmission of crude petroleum, petroleum products, fuel gases, and certain distribution systems.

The American Water Works Association This group has issued standards covering the fabrication of piping for water purification plants.

The Heating, Piping, and Air Conditioning Contractors National Association This piping contractors association has set up standard welding procedures for the installation of piping systems by their member contractors. The National Certified Pipe Welding Bureau is one of several organizations which supervises and certifies welder qualification tests in a uniform manner in accordance with the Association's standard procedures.

The American Welding Society (AWS) This organization is responsible for a great many of the uniform practices in the field of welding. One document that deals with the welding of pipe and tubes is D18.1. It covers the specification for welding austenitic stainless-steel tube and pipe systems in sanitary (hygienic) applications. Not all pipe is used to carry products. Sometimes pipe is used as a structural material, and the D1.1 Structural Welding Code—Steel covers this aspect. The AWS was one of the early pioneers in the development of uniform training and testing practices for welders and welding operators. Today the society is one of the outstanding authorities in all matters pertaining to welding.

These are the principal code bodies concerned with the design, installation, and testing of piping systems. There are a number of other groups concerned with special purpose installations.

It is suggested that you obtain a copy of each of the following publications in order to become thoroughly familiar with all of the information in regard to the common codes and testing situations:

- The American Welding Society, B1.10 Guide for the Nondestructive Inspection of Welds, B1.11 Guide for the Visual Inspection of Welds, B2.1 Standard for Welding Procedure and Performance Qualification, D18.1 Specification for Welding of Austenitic Stainless Steel Tube and Pipe System in Sanitary (Hygienic) Applications, and the D1.1 Structural Welding Code—Steel.
- American Petroleum Institute, Standard for Welding Pipe Lines and Related Facilities (API Std. 1104).
- The American Society of Mechanical Engineers, Section IX—ASME Boiler and Pressure Vessel Code—Welding Qualifications.

Procedure and Welding Qualification

Code welding requires the setting up and acceptance of the qualifications for welding procedures and for welders who are going to fabricate the piping installation following the welding procedure adopted.

Procedure Qualification Tests The welding procedure qualification is set up and approved before the start of production welding. It is recorded on a form called a Welding Procedure Qualification Record (WPQR). This establishes the fact that the procedure can produce welds having suitable mechanical properties and soundness.

The details of each qualified procedure are recorded on a Welding Procedure Specification (WPS), and include a record of the following:

- Material for pipe and fitting
- Diameter and wall thickness of the pipe
- Joint design
- Joint preparation, including root opening and bevel angle if necessary
- Position of welding
- Welding process
- Type and size of electrode or filler rod
- Type of current
- Current setting
- Number of passes
- Welding technique
- Use of a backing strip if necessary
- Preheat, between-pass heat, and postheat temperatures
- Special information peculiar to each job

It must be understood that the purpose of the welding procedure is to make certain that welds made in compliance with it have the potential mechanical properties to meet its application. This is often called "fitness for purpose." Many codes will describe changes that may or may not be allowed. These are welding procedure variables, which are further defined as:

Essential variables: Those in which a change is considered to affect the mechanical properties of the weldments and shall require requalification of the WPS.
Supplementary essential variables: Required for metal for which notch-toughness tests is specified and are in addition to the essential variables and are welding process specific.
Nonessential variables: Those in which a change, as described in the specific code variables, may be made in the WPS without requalification.

The specific code being used should be consulted to determine if these items are considered essential or nonessential variables.

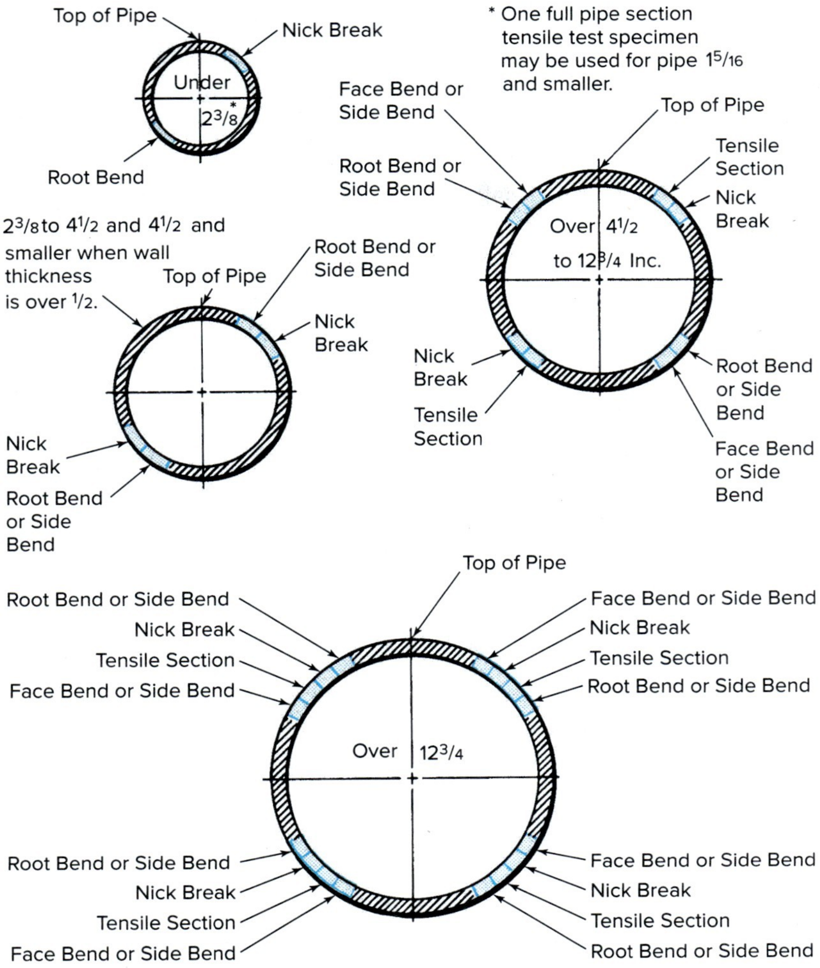

Figure 16-29 shows the location of the test coupons that are to be removed and indicates the type of testing to which they must be subjected in accordance with the API standard. Figures 16-30 and 16-31 give this same information in order to qualify the work under the ASME codes. The preparation for these specimens for the ASME codes are shown in Chapter 28 under Preparation of Test Specimens, pages 922 to 926. The special nature of the test coupons for the API code are shown in Figs. 16-32 through 16-36. Consult Table 16-4 (p. 487) for type and number of test specimens for procedure qualification test requirements for API code and Table 16-5 (p. 487) for the ASME code. A tested procedure can produce quality welding only if the welder has the ability to apply the procedure. Each welder must be properly qualified by demonstrating the ability to make acceptable welds with the tested procedure that has been accepted for a particular job. Welder qualification is usually recognized only on a particular job or for the current employer and only in the range allowed by the procedure, such as pipe diameters, wall thickness range, welding positions, fillet or groove welds, type of pipe materials, and electrode groups. Some piping contractors require the welders in their employ to retest every 6 months.

Fig. 16-29 Location of groove weld butt joints test specimens for procedure qualification.

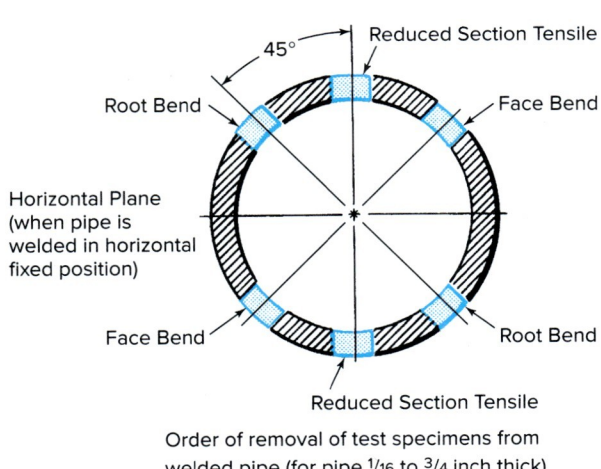

Fig. 16-30 Order of removal of test specimens from welded pipe 1/16 to 3/4 inch thick.

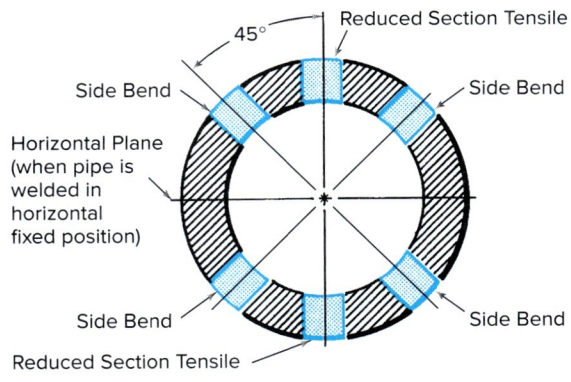

Fig. 16-31 Order of removal of test specimens from welded pipe over 3/4 inch thick. This order may also be used for pipe 3/8 to 3/4 inch thick.

Pipe Welding and Shielded Metal Arc Welding Practice: Jobs 16-J1–J17 (Pipe)   **Chapter 16**   **485**

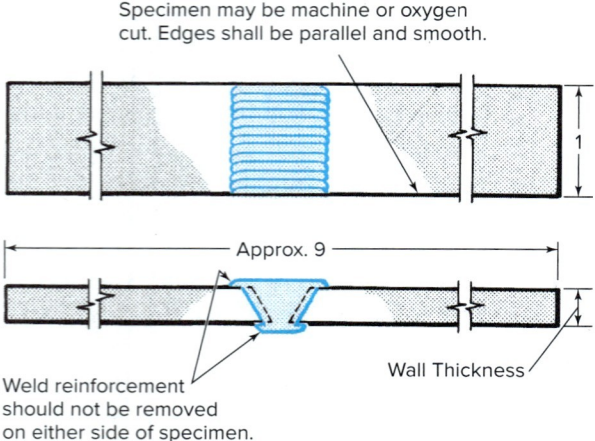

Fig. 16-32 Tensile test specimen.

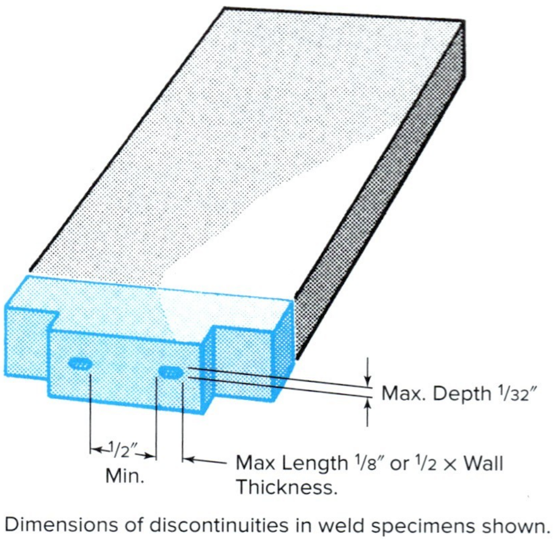

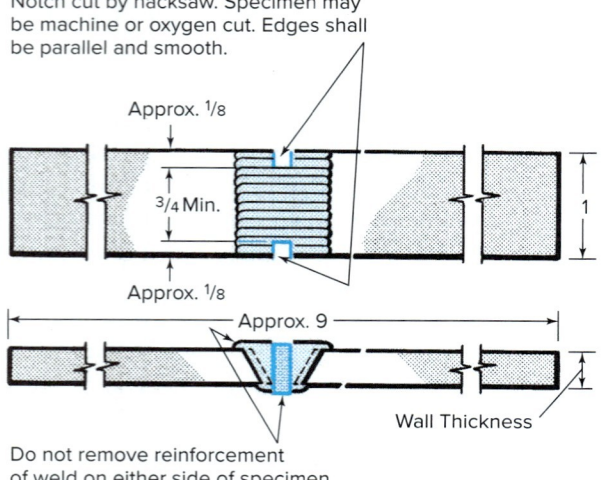

Fig. 16-33 Nick-break test specimen.

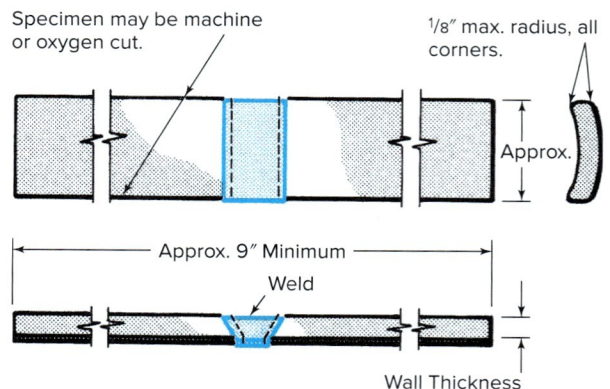

Fig. 16-34 Allowable discontinuities in the specimen shown in Fig. 16-33.

Fig. 16-35 Root- and face-bend test specimen.

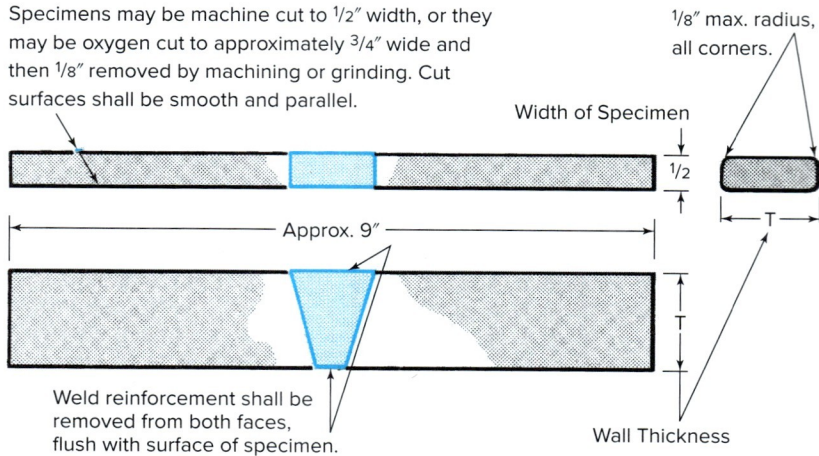

Fig. 16-36 Side-bend test specimen.

The Welder Qualification Tests The welder qualification tests are the ones that the student welder is asked to perform and pass. They are generally referred to as Welder Performance Qualifications (WPQs). The purpose of the welder qualification test is to determine the ability of the welder to make sound welds using a previously qualified welding procedure specification (WPS).

If the welder wishes to qualify under the API code, test coupons should be removed from the pipe as indicated in Fig. 16-37, page 488. The preparation of the test coupons for the ASME codes is shown in Figs. 16-30

Table 16-4 Test Requirements for Procedure Qualification in Pipeline According to the APL Standard 1104

Type and Number of Test Specimens for Procedure Qualification Test

Pipe Size, Outside Diameter (in.)	Tensile	Nick-Break	Root-Bend	Face-Bend	Side-Bend	Total
Wall Thickness (½ in. and under)						
Under 2³⁄₈	0	2	2	0	0	4*
2³⁄₈ to 4½ inclusive	0	2	2	0	0	4
Over 4½ to 12³⁄₈ inclusive	2	2	2	2	0	8
Over 12¾	4	4	4	4	0	16
Wall Thickness (over ½ in.)						
4½ and smaller	0	2	0	0	2	4
Over 4½ to 12¾ inclusive	2	2	0	0	4	8
Over 12¾	4	4	0	0	8	16

*One nick-break and one root-bend specimen from each of two test welds or for pipe ¹⁵⁄₁₆ in. and smaller, one full pipe section tensile specimen.

Source: American Petroleum Institute

Table 16-5 Test Requirement for Procedure Qualification in Pressure Piping According to the ASME Code Section IX

Thickness, T, of Test Coupon Welded (in.)	Range of T of Base Metal Qualified (in.)[1 and 2]		Thickness, t, of Deposited Weld Metal Qualified (in.)[1 and 2]	Type and Number of Tests Required[2]			
	Min.	Max.	Max.	Tension	Side-Bend	Face-Bend	Root-Bend
<¹⁄₁₆	T	2T	2t	2	—	2	2
¹⁄₁₆ to ≤³⁄₈	¹⁄₁₆	2T	2t	2	(3)	2	2
>³⁄₈ to <¾	³⁄₁₆	2T	2t	2	(3)	2	2
¾ < to 1½	³⁄₁₆	2T	2t when t < ¾	2[4]	4	—	—
¾ < to 1½	³⁄₁₆	2T	2T when t ≥ ¾	2[4]	4	—	—
≥1½	³⁄₁₆	8[5]	2t when t < ¾	2[4]	4	—	—
≥1½	³⁄₁₆	8[5]	8[5] when t ≥ ¾	2[4]	4	—	—

[1] See code for further limits on range of thickness qualified and also for allowable exceptions.
[2] For combination of welding procedures, reference the code.
[3] Four side-bend test may be substituted for the required face- and root-bend test, when thickness T is ³⁄₈ in. and over.
[4] See code for details on multiple specimens when coupon thicknesses are over 1 in.
[5] For welding processes specified in code only; otherwise per note 1 use 2T or 2t, whichever is applicable.

Reprinted from ASME 2010 BPVC, Section IX, The American Society of Mechanical Engineers

and 16-31. The preparation for these specimens is described in Chapter 28 under Preparation of Test Specimens, pages 922–926. Consult Table 16-6 (p. 488) for type and number of test specimens for welder qualification test requirements for the API standard and Table 16-7A (p. 488) for the ASME code. The diameter of pipe qualified on is an important consideration. This will determine what range of pipe diameter can be welded on the job. Table 16-7B (p. 488) is representative of the ASME code. The smaller the diameter of the pipe, the more difficult it is to weld. The larger the diameter of the pipe, the easier it is to weld. In some cases, a plate

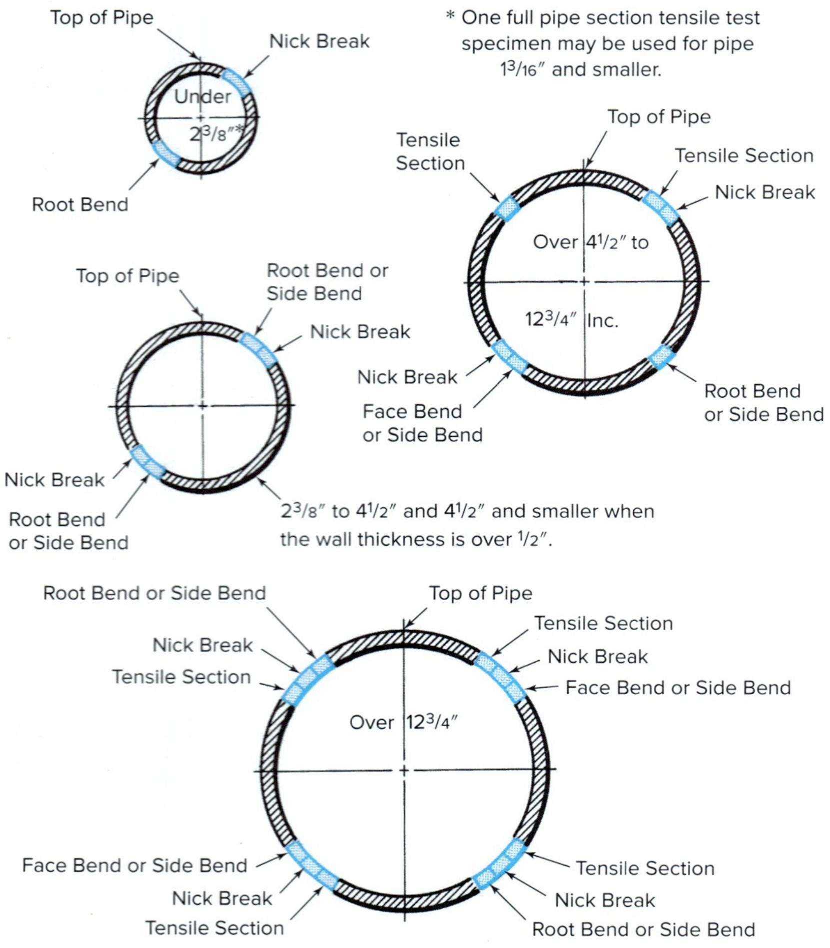

Fig. 16-37 Location of groove weld butt joint test specimens for welder qualification.

test can qualify a welder to weld on pipe 24 inches or greater in diameter.

The above information concerns butt joint groove welds only. The procedure and welder qualifications for fillet welds are as follows. For the API standard, Table 16-8 lists the type and number of test specimens required. Figure 16-38 indicates the location of these test specimens in the pipe joint, and Fig. 16-39, page 490, indicates the nature of the preparation of the test specimens.

Figure 16-40, page 491, shows a sample form that can be used as a record of testing. This form comes from the API standard and can be used for both procedure and welder qualification on butt joints. The record form in Fig. 16-41, page 492, is from the ASME code and is used for welder qualification for groove and fillet welds.

For qualification, the exposed surfaces of the specimen must show complete penetration and no more than 6 gas pockets per square inch of surface area with the greatest dimension not to exceed $1/16$ inch. Slag inclusions must not be greater than $1/32$ inch in depth, or greater than $1/8$ inch or $1/2$ the nominal wall thickness of the thinner member in length, whichever is smaller. Such inclusions must be separated by at least $1/2$ inch of sound metal.

Table 16-6 Test Requirements for Welder Qualification in Pipeline According to the API Standard 1104

Type and Number of Test Specimens for Welder Qualification Test and for Destructive Testing of Production Welds—Butt Joints

Pipe Size, Outside Diameter (in.)	Number of Specimens					
	Tensile	Nick-Break	Root-Bend	Face-Bend	Side-Bend	Total
	Wall Thickness ($1/2$ in. and under)					
Under $2^3/_8$	0	2	2	0	0	4*
$2^3/_8$ to $4^1/_2$ inclusive	0	2	2	0	0	4
Over $4^1/_2$ to $12^3/_4$ inclusive	2	2	2	0	0	6
Over $12^3/_4$	4	4	2	2	0	12
	Wall Thickness (over $1/2$ in.)					
$4^1/_2$ and smaller	0	2	0	0	2	4
Over $4^1/_2$ to $12^3/_4$ inclusive	2	2	0	0	2	6
Over $12^3/_4$	4	4	0	0	4	12

*Obtain from two welds or one full pipe section tensile specimen for pipe $15/16$ in. and smaller.
Source: American Petroleum Institute

Table 16-7A Test Requirements for Welder Qualification in Pressure Piping According to the ASME Code Section IX

		Performance Qualification			
Type of Joint	Thickness of Test Coupon Welded (in.)[1]	Thickness, t, of Deposited Weld Metal Qualified (in.)[2] (max.)	Type and Number of Tests Required[3,4,8]		
			Side-Bend	Face-Bend	Root-Bend[5]
Groove	≤3/8	2t	6	1	1
Groove	>3/8 < 1/2	2t	7	1	1
Groove	≥1/2[9]	Max. to be welded	2		

[1] When using one, two, or more welders, the thickness t of the deposited weld metal for each welder with each process shall be determined and used individually in the Thickness column.

[2] Two or more pipe test coupons of different thicknesses may be used to determine the deposited weld metal thickness qualified, and that thickness may be applied to production welds to the smallest diameter for which the welder is qualified in accordance with the code.

[3] Thickness of test coupon of 1/2 in. or over shall be used for qualifying a combination of three or more welders, each of which may use the same or different welding process.

[4] To qualify for positions 5G and 6G, as prescribed in the code, two root-bend and two face-bend specimens or four side-bend specimens, as applicable to the test coupon thickness, are required.

[5] Face- and root-bend tests may be used to qualify a combination test of (a) one welder using two welding processes, or (b) two welders using the same or a different welding process.

[6] For a 3/8-in. thick coupon, a side-bend test may be substituted for each of the required face- and root-bend tests.

[7] A side-bend test may be substituted for each of the required face- and root-bend tests.

[8] Test coupons shall be visually examined per code and (a) coupons shall be visually examined over the entire circumference, inside and outside, and (b) coupons shall show complete joint penetration with complete fusion of weld metal and base metal.

[9] Test coupon weld deposit shall also consist of a minimum of three layers of weld metal.

Source: Reprinted from ASME 2010 BPVC, Section IX, The American Society of Mechanical Engineers

Table 16-7B Groove-Weld Diameter Limits According to the ASME Code Section IX[1,2]

Outside Diameter of Test Coupon (in.)	Outside Diameter Qualified (in.)	
	Min.	Max.
<1	Size welded	Unlimited
≥1 but <2 7/8	1	Unlimited
≥2 7/8	2 7/8	Unlimited

[1] The type and number of tests required shall be in accordance with Table 16-7A.

[2] The outside diameter of 2 7/8 in. is the equivalent of NPS 2 1/2 in.

Table 16-8 Fillet Weld Test Specimen Requirements for Procedure Qualification According to the API Standard 1104

Pipe Size (OD in.)	Number of Root-Bend Specimens
Under 2 3/8	4 (Obtain from 2 welds)
2 3/8 to 12 3/4 inclusive	4
Over 12 3/4	6

Source: American Petroleum Institute

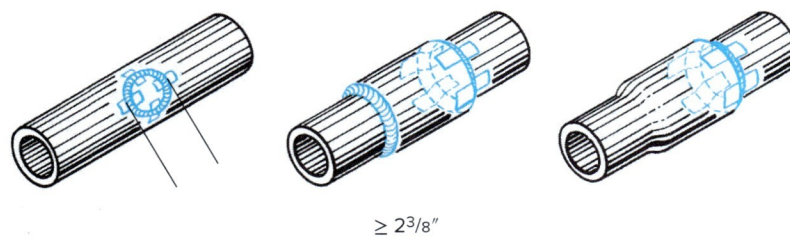

≥ 2 3/8"

For joints under 2 3/8" cut nick-break specimens from same general location but remove two specimens from each of two test welds.

Fig. 16-38 Location of nick-break test specimens in fillet welds for both procedure and welder qualification.

Study Chapter 28 for fillet weld testing for the ASME code.

It should also be noted that a particular specimen qualifies within the limits of the acceptance criteria. If any of the essential variables are changed beyond their limits, the welder must requalify. The following are some examples of typical essential variables that would require the welder to be requalified:

- A change from one welding process to any other welding process or combination of welding processes

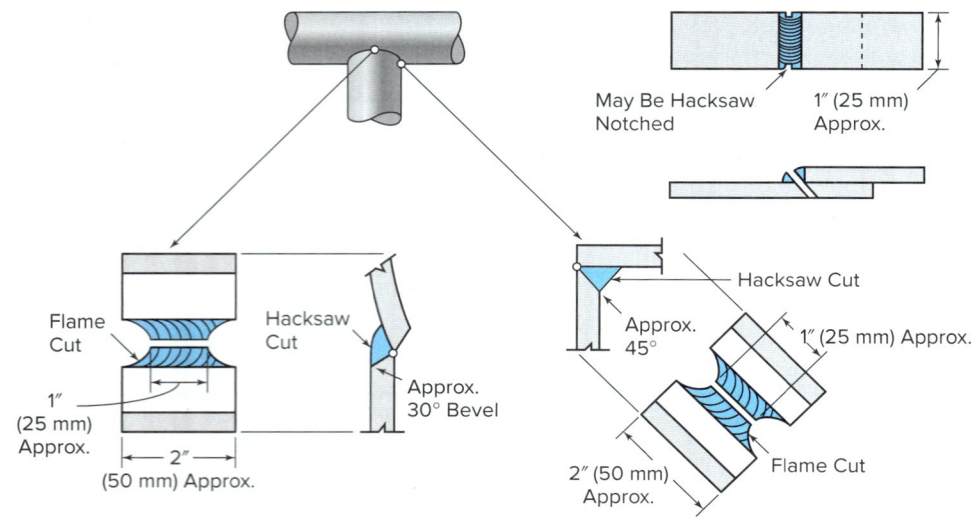

Fig. 16-39 Preparation of fillet weld specimens.

- A change in the direction of welding from vertical-up to vertical-down or vice versa
- A change in filler metal from one classification group to another
- A change in pipe diameter from one group to another
- A change in deposited weld metal thickness beyond the range qualified
- A change in pipe wall thickness and a change in pipe material from one group to another
- A change in position other than that already qualified
- A change in the joint design (from backing strip to no backing strip or from V-bevel to U-bevel)

After you have passed these tests, you will have earned the right to look upon yourself as a skilled welder with better than average knowledge and skill. Code pipe welding is one of the most demanding forms of the welding trade, and it is the highest paid of all the fields of welding.

Making the Test Weld Skilled welders may fail a test because of reasons that have nothing to do with their ability to deposit a sound weld bead. Following are test conditions that must be given careful attention to ensure satisfactory test results:

- The test plate should be of a material similar to that used on the job.
- Proper plate preparation is a must. Correct angle of bevel, root face, root opening, and proper tacking are essential.
- Pipe strength must be similar to that on the job. If it is greater than the weld metal, the weld will break when subjected to the face- and root-bend tests.
- Use an electrode similar to that used on the job.
- Sound root penetration is essential.
- Do not become overly concerned about weld appearance. Good penetration and fusion are more important.
- Welding procedure should be the same as that employed on the job.
- Keep the plates hot during welding and allow them to cool slowly after the completion of the test joint. Never quench the pipe in water nor blow air on them. If test is taken outside, provide a shield from the wind.
- The test specimen must be free of nicks or deep scratches. All grinding or machining must be lengthwise on the specimen.
- Remove all face and root reinforcement and round the edges of the specimen. Do not quench in water after grinding.

Methods of Testing and Inspection

A variety of testing methods are used in determining the quality of pipe joints and piping systems, and fall into two major categories: nondestructive and destructive testing. Following is a listing of the tests used. For additional information you are urged to read Chapter 28.

Visual Inspection Visual inspection (VI) is actually a form of nondestructive testing and is best applied before, during, and after welding is completed. Before welding, make sure the welding procedure requirements are being met. Check the condition of the welding equipment. Check quality and condition of base and filler metals to be used. Verify the joint preparation, fitup, alignment, and cleanliness. If preheat is required, make sure it is appropriate. During welding, frequent observations are made to determine whether or not the prescribed welding procedure is being followed by each welder. Welding current, number of passes, interpass temperature control, and cleaning between passes are particularly important and frequently checked.

Completed welds are inspected visually for general appearance. The amount of reinforcement, the presence or absence of undercutting, the nature of the penetration, and any signs indicating a lack of fusion are noted. These are the same characteristics that you have learned to look for in your previous welding practice.

COUPON TEST REPORT

Date _____ Test No. _____
Location _____
State _____ Weld Position: Roll ❏   Fixed ❏
Welder _____ Mark _____
Welding time _____ Time of day _____
Mean temperature _____ Wind break used _____
Weather conditions _____
Voltage _____ Amperage _____
Welding machine type _____ Welding machine size _____
Filler metal _____
Reinforcement size _____
Pipe type and grade _____
Wall thickness _____ Outside diameter _____

	1	2	3	4	5	6	7
Coupon stenciled							
Original specimen dimensions							
Original specimen area							
Maximum load							
Tensile strength							
Fracture location							

❏ Procedure   ❏ Qualifying test   ❏ Qualified
❏ Welder   ❏ Line test   ❏ Disqualified

Maximum tensile _____ Minimum tensile _____ Average tensile _____
Remarks on tensile-strength tests _____
1. _____
2. _____
3. _____
4. _____
Remarks on bend tests _____
1. _____
2. _____
3. _____
4. _____
Remarks on nick-break tests _____
1. _____
2. _____
3. _____
4. _____

Test made at _____ Date _____
Tested by _____ Supervised by _____

Note: Use back for additional remarks. This form can be used to report either a procedure qualification test or a welder qualification test.

Fig. 16-40   Sample record form for a weld coupon test report.   Source: American Petroleum Institute

QW-484 SUGGESTED FORMAT FOR WELDER/WELDING OPERATOR PERFORMANCE QUALIFICATIONS (WPQ)
(See QW-301, Section IX, ASME Boiler and Pressure Vessel Code)

Welder's name _____ Clock number _____ Stamp no. _____
Welding process(es) used _____ Type _____
(Manual, semiautomatic, machine, automatic)
Identification of WPS followed by welder during welding of test coupon _____
Base material(s) welded _____ Thickness _____
Filler metal specification (SFA) Class (QW-404) _____

Manual or Semiautomatic Variables for Each Process (QW-350)	Actual Values	Range Qualified
Backing (metal, weld metal, welded from both sides, flux, etc.) (QW-402)		
ASME P-No. _____ to ASME P-No. _____ (QW-403)		
( ) Plate ( ) Pipe (enter diameter, if pipe) (QW-403)		
Base metal thickness — OFW (QW-403)		
Filler metal F-No. (QW-404)		
Filler metal product form [solid/cored/flux-cored — GTA/PAW (QW-404)]		
Consumable insert for GTAW or PAW (QW-404)		
Weld deposit thickness for each welding process (QW-404)		
Welding position (1G, 5G, etc.) (QW-405)		
Progression (uphill/downhill) (QW-405)		
Backing gas for GTAW, PAW, or GMAW; fuel gas for OFW (QW-408)		
GMAW transfer mode (QW-409)		
GTAW welding current type/polarity (QW-409)		

Automatic/Machine Welding Variables for the Process Used (QW-860)	Actual Values	Range Qualified
Direct/remote visual control		
Automatic voltage control (GTAW)		
Automatic joint tracking		
Welding position (1G, 5G, etc.)		
Consumable insert		
Backing (metal, weld metal, welded from both sides, flux, etc.)		
Multiple or single pass per side		
Change from automatic to machine		
Filler for EBW or LBW		
Laser type		
Drive type for FRW		
Vacuum type for EBW		

Guided Bend Test Results

Guided Bend Tests Type	( ) QW-462.2 (Side) Results	( ) QW-462.3(a) (Trans. R & F) Type	( ) QW-462.3(b) (Long, R & F) Results

Visual examination results (QW-302.4) _____
Radiographic test results (QW-304 and QW-305) _____
(For alternative qualification of groove welds by radiography)
Fillet Weld — Fracture test _____ Length and percent of defects _____ in.
Macro test fusion _____ Fillet leg size _____ in. × _____ in. Concavity/convexity _____ in.
Welding test conducted by _____
Mechanical tests conducted by _____ Laboratory test no. _____
We certify that the statements in this record are correct and that the test coupons were prepared, welded, and tested in accordance with the requirements of Section IX of the ASME Code.

Organization _____
Date _____ By _____

This form (E00008) may be obtained from the Order Dept., ASME, 22 Law Drive, Box 2300, Fairfield, NJ 07007-2300.

Fig. 16-41 Recommended form for the manufacturer's record of welder performance qualification tests on groove and fillet welds. Source: Reprinted from ASME 2010 BPVC, Section IX, by permission of The American Society of Mechanical Engineers. All rights reserved.

Radiographic Inspection One of the most important and reliable nondestructive testing tools is radiographic inspection. On a commercial basis, this is done with X-ray or gamma ray sources. Digital radiography is now a viable option for onshore, offshore, and refinery use. Generally, X-ray work is done in the shop where equipment can be under better control. Portable radiographic equipment, such as the pipeline crawler, also known as a smart pig, is available for field inspection as well. The pipeline crawler is an automated, self-propelled radiograph machine for inspecting circumferential welds in cross-country pipelines. The crawler is battery operated and requires no external cables or connectors.

The crawler is positioned at each weld by means of a radioactive isotope locator placed at a predetermined position on the outside of the pipeline. The machine then stops at the joint to be radiographed, Fig. 16-42. The radiograph beam produces an image of the weld on a strip of film wrapped around the outside of the joint. Discontinuities are revealed when the film is developed. When the exposure is completed, the crawler moves along on command from the isotope locator, Fig. 16-43, which is placed at the next weld, and the process is repeated.

The crawler can be used in pipelines with diameters ranging from 6 to 72 inches, Fig. 16-44.

Fig. 16-44 Real-time radiographic inspection of pipeline welds. Digital X-ray technology has advanced to becoming truly real time and in high resolution. Utilizing advanced X-ray detector technology and pipeline-specific software, the inspection process has become a replacement for conventional RT. A typical pipeline crawler is used inside the pipeline, but externally the film is replaced by an RTR scanning mechanism that runs on a typical band. Scott James/Shaw Pipeline Services

Fig. 16-42 Smart pigs come in a variety of designs and capabilities. They are typically able to inspect pipe from 6 to 72 inches in diameter with either gamma or X-ray inspection techniques. These crawlers are available to meet quality and rigid production quotas. Operators must be qualified and well-experienced. Scott James/Shaw Pipeline Services

A

B

Fig. 16-43 The smart pig is located in the pipe and is making its way into position to radiograph a girth weld in a cross-country pipeline. Signals are sent to the crawler at each weld being inspected. Scott James/Shaw Pipeline Services

 For video of the Digital RT operation, please visit www.mhhe.com/welding.

Ultrasonic Inspection Ultrasonic inspection for discontinuities in pipe welds is also used. The more common pulse echo or now the more advanced automated phase array methods can be used, Fig. 16-45. Only a highly experienced inspector may be able to interpret the difference between a discontinuity and a defect, noncritical evidence of slag, porosity, and changes in surface contour with the pulse echo method. On the other hand, the automatic phase array system can give a better probability of detection.

 For video of the phase array UT testing, please visit www.mhhe.com/welding.

Liquid Penetrant Inspection Dye penetrant inspection is used extensively where narrow cracks are suspected. This method is extensively used on nonmagnetic materials where magnetic particle inspection processes could not be used.

Magnetic Particle Inspection Often referred to as Magnaflux® testing, this method can be used only on magnetic materials. Fluorescent magnetic particles in a liquid medium (usually water) are applied to the work for the detection of extremely narrow cracks.

Hydrostatic Testing Hydrostatic testing is used extensively for welded-piping components and pipelines. The nondestructive testing of sections between valves is often done. This form of testing is also applied to shop-fabricated pipe assemblies. For most work, clean water is used at about the atmospheric temperature to prevent sweating of the pipe and weld joints. Pressure is applied gradually by means of a pump and registers on a gauge. The test pressure should be that called for by the governing code or contract. It should be at least 1½ times the maximum working pressure to which the piping will be subjected. For power piping, the test pressure should never be less than 50 p.s.i. nor more than twice the service pressure of any valve included in the test.

Fig. 16-45 Equipment used for mechanized ultrasonic testing for the complete inspection of pipeline girth welds. This inspection technique allows for rapid inspection and feedback to the welder, thus allowing tighter controls on the welding process to reduce weld repair rates for the project. This type of "total weld inspection" philosophy includes weld examination using focused probe technology, time of flight diffraction (TOFD), and phased array technology. *Scott James/Shaw Pipeline Services*

The use of compressed air or other gas is not recommended because of its explosive force. A failure could be highly destructive and hazardous to human life. Air testing is used to a limited extent by certain small tank fabricators and at low pressure. Helium tests and soap bubble tests are also used.

Destructive hydrostatic testing, Fig. 16-46, is usually carried out under controlled conditions in a shop or laboratory on a particular component of the piping system.

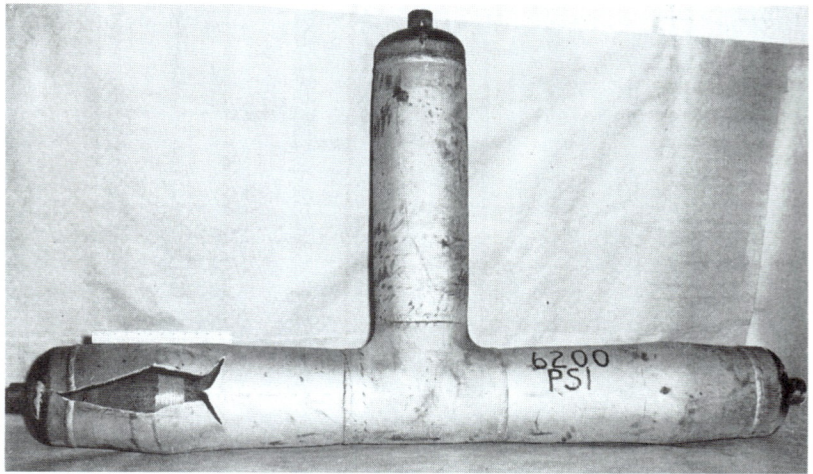

Fig. 16-46 A unit subjected to destructive hydrostatic testing. Note that the bursting point was in the pipe and that the welds are intact. The pipe burst at 6,200 p.s.i., a pressure that is well beyond service requirements. *Edward R. Bohnart*

Oil or water may be used as the internal liquid and subjected to high pressure. The pressure is applied to the yield point in the material or even to the bursting point. This is for the purpose of determining the stress concentrations, strengths, and weaknesses.

Service Tests After the piping installation has been completed, a service test, which is essentially a trial operation of a system under normal service conditions, is performed. Care must be taken to make sure that all valves and regulating devices are set for proper operation and that all hangers and anchor structures are secured.

The customer expects that the system is free from leaks and is able to perform as called for in the specifications. If the welders have done their work well, the system will perform as designed. This is a serious responsibility since the contractor is responsible for any leaks and any damage due to leaks.

Coupon Testing The most common type of destructive testing that you will be concerned with as a pipe welder is done by taking test coupons from pipe joints that you have welded in accordance with the requirements and specifications of whatever code you are working under. Coupons may be cut accurately in the field by means of a test coupon cutter, Fig. 16-47. The cutter uses an oxyacetylene cutting torch. It may be used for both plate and pipe.

Miscellaneous Tests Several other destructive test methods are also employed to maintain control over weld quality.

Trepanning is generally done by removing a cylindrical plug from the weld by means of a power-driven hole saw. This permits checking the inside weld surface for porosity, cracking, penetration, and fusion. Some codes and standard no longer allow use of this method.

Sometimes, narrow strips are removed with a saw or high speed cutting wheels. After the specimen has been removed, it is ground and polished, and the prepared surface is etched. Examination of the macro-etched surface discloses such defects as cracks, lack of fusion, weld profile, porosity, and slag inclusions.

Common Defects in Pipe Welds and Acceptance Criteria

Following is a list of weld discontinuities that occur frequently in pipe welds as described in API (1104). The welding process is dependable and will consistently produce work that is sound, Fig. 16-48. The variable is the welder. Much poor welding is caused not by welders' lack of skill, but by their carelessness or lack of interest in the work. Skilled welders are concerned welders who are careful about everything they do.

Essentially, incomplete joint penetration is defined as the weld metal not extending through the joint thickness. Incomplete fusion is defined as the lack of bond between beads or between the weld metal and base metal. Incomplete penetration and incomplete fusion are separate and distinct conditions that occur in different forms as follows.

- *Incomplete joint penetration (IP).* Inadequate penetration that is not caused by a high or low condition is

Fig. 16-47 A test coupon cutter used in the field for quick, on-the-spot testing. Mathey Dearman/Diana Sullivan, Mathey

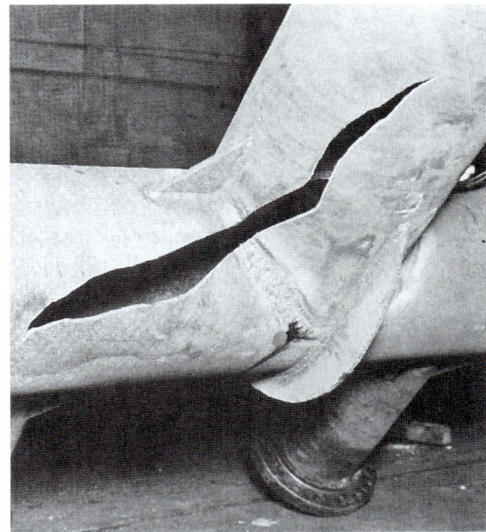

Fig. 16-48 A pipe fabrication that fractured in service. Note that the welds are intact. Crane Co/DuPont

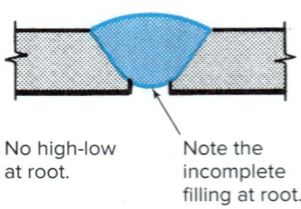

No high-low at root. Note the incomplete filling at root.

One or both bevels may be inadequately filled at inside surface.

Fig. 16-49 Incomplete joint penetration of the weld groove.

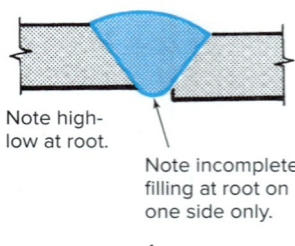

Note high-low at root. Note incomplete filling at root on one side only.

A

Both sides completely tied-in by weld metal.

B

Fig. 16-50 Incomplete joint penetration due to high-low pipe positions.

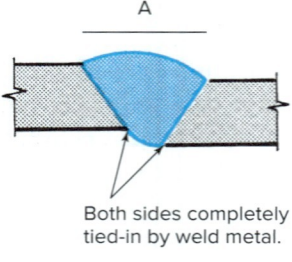

Root bead fused to both inside surfaces but center of root pass is slightly below inside surface of pipe.

Fig. 16-51 Incomplete joint penetration due to internal concavity.

defined as incomplete filling of the weld root, Fig. 16-49.

The acceptance criterion for IP is an individual IP length ≤ 1 inch. If the aggregate IP length in any continuous 12-inch length of weld is ≤1 inch and the aggregate IP length is ≤8 percent of the weld length in any weld less than 12 inches in length, the IP is considered a discontinuity, not a defect.

- **Incomplete joint penetration (IPD) due to a high-low condition.** High-low is defined as a condition in which the pipe and/or fitting surfaces are misaligned, Fig. 16-50A. The weld in a high-low misalignment is acceptable when the length of IPD ≤2 inches or the aggregate length of IPD is any continuous 12-inch length of weld ≤3 inches. This amount of IPD would be considered a discontinuity, not a defect. It is permissible provided that the root of the adjacent pipe and/or fitting joints is completely fused by weld metal, Fig. 16-50B.

- **Internal concavity (IC).** This applies to a bead that is properly fused to and completely penetrates the pipe wall thickness along both sides of the bevel, but the center of the bead lacks buildup and is somewhat below the inside surface of the pipe wall, Fig. 16-51. The magnitude of the IC is measured by the perpendicular distance between an axial extension of the internal pipe wall surface and the lowest point on the weld bead profile. The acceptance criterion for IC is any length as long as the radiographic image density does not exceed that of the thinnest adjacent base metal. The IC is considered a discontinuity, not a defect, if any area exceeds the radiographic image density of the thinnest adjacent base metal, but meets the following criteria: pipe outside diameter ≥2⅜ inch, IC dimension ≤¼ inch, and for any continuous 12-inch length of weld or the total weld length, whichever is less, the aggregate dimension of separate IC is ≤½ inch.

- **Incomplete fusion (IF).** Incomplete fusion is the lack of fusion at the root of the joint or at the top of the joint between the weld metal and the base metal, Fig. 16-52. The acceptance criterion for IF is length ≤1 inch. The IF is considered a discontinuity, not a defect, if its aggregate length in any continuous 12-inch length of weld is ≤1 inch or the aggregate length is ≤8 percent of the weld length in any weld less than 12 inches in length.

- **Excessive melt-through or burn-through (BT).** Excessive BT is a portion of the root bead where excessive penetration has caused the weld pool to be blown into the pipe, Fig. 16-16, page 473. The acceptance criteria for BT on pipe ≥2⅜-inch outside diameter and BT dimension is ≤¼ inch. The BT is considered a discontinuity, not a defect, if the aggregate dimension of separate BT is ≤½ inch for any continuous 12-inch length of weld or the total weld length, whichever is less.

- **Incomplete fusion due to cold lap (IFD).** Incomplete fusion due to cold lap occurs between two adjacent weld beads or between a weld bead and the base metal, Fig. 16-53. The acceptance criterion for IFD is a length ≤2 inches. If the aggregate length of IFD in any continuous 12-inch length of weld is ≤2 inches or the aggregate length of IFD is ≤8 percent of the weld length, the IFD is considered a discontinuity, not a defect.

- **Slag inclusion (SI).** A slag inclusion is a nonmetallic solid entrapped in the weld metal or between the weld metal and the pipe metal. Slag inclusions are further subdivided based on the location and extent of the inclusion. Elongated slag

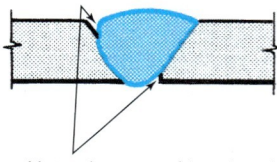

Note absence of bond and that discontinuity is surface connected.

Fig. 16-52 Incomplete fusion at the root of the bead and the face of the weld.

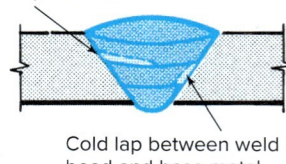

Cold lap between adjacent beads.

Cold lap between weld bead and base metal.

Note: Cold lap is not surface connected.

Fig. 16-53 Incomplete fusion due to cold lap.

inclusions (ESIs) are continuous or broken slag lines or "wagon tracks" that are usually located at the fusion zone. Isolated slag inclusions (ISIs) are irregularly shaped and may be located anywhere in the weld. For evaluation purposes, the size of a radiographic indication of slag is considered its length and is the maximum dimension of the slag. The acceptance criteria for an SI on pipe $\geq 2\frac{3}{8}$ inch is ≤ 2 inches in length. If parallel ESI indications are separated by approximately the width of the root bead, wagon tracks shall be considered as a single indication. If they are $>\frac{1}{32}$ inch wide, they are considered separate indications. All the following are considered discontinuities, not defects: (1) The aggregate length of ESI indications in any continuous 12-inch length of weld is ≤ 2 inches. (2) The width of an ESI is $\leq \frac{1}{16}$ inch. (3) The aggregate length of ISI indications in any continuous 12-inch length of weld exceeds $\frac{1}{2}$ inch. (4) The width of ISI is $\leq \frac{1}{8}$ inch if ≤ 4 ISI indications are present in any continuous 12-inch length of weld. (5) The aggregate length of ESI and ISI indications exceeds 8 percent of the weld length. For pipe with an outside diameter $<2\frac{3}{8}$ inch, the following acceptance criteria apply for SIs. The ESI indication is $\leq$ three times the thinner of the nominal wall thickness joined. If parallel ESI indications are separated by approximately the width of the root bead, wagon tracks shall be considered as a single indication. If an ESI indication is $>\frac{1}{32}$ inch wide, it is considered a separate indication. All the following are considered discontinuities, not defects: (1) The width of an ESI indication is $\leq \frac{1}{16}$ inch. (2) The aggregate length of an ISI indication is $\leq$ two times the thinner of the nominal wall thicknesses joined and the width is $\leq$ one-half the thinner of the nominal wall thicknesses joined. The aggregate length of an ESI or ISI indication is ≤ 8 percent of the weld length.

- **Porosity.** Porosity is gas trapped by solidifying weld metal before the gas has a chance to rise to the surface of the molten pool and escape. Porosity is generally spherical but may be elongated or irregular in shape, such as piping (wormhole) porosity. See Table 16-9 for the acceptance criteria for porosity.
- **Cracks.** Cracks shall be acceptable when the following conditions exist. They are shallow crater cracks or star cracks and are $\leq \frac{5}{32}$ inch. These shallow crater cracks or star cracks are formed in the crater or stopping point of a weld by the weld metal contracting during solidifications.
- **Undercut.** Undercut is a groove melted into the base metal adjacent to the toe or root of the weld and left unfilled by weld metal. Undercut adjacent to the cover-pass EU or root-pass IU is acceptable if the aggregate length of the EU and IV in any combination, in any continuous 12-inch length of weld, is ≤ 2 inches, or if the aggregate length of EV and IV indications, in any combination, is $\leq \frac{1}{6}$ of the weld length. When visual testing (VT) with mechanical measuring devices, the depth shall not exceed that given in Table 16-10 (p. 498).
- **Accumulation of discontinuities.** Excluding incomplete penetration due to high-low and undercutting, any accumulation of discontinuities (AD) is acceptable if the following conditions exist: The aggregate length of indications in any continuous 12-inch length of weld is ≤ 2 inches or if the aggregate length of indications are ≤ 8 percent of the weld length.

It must be understood that these acceptance criteria have been developed over many years of experience and that pipe joints being welded in accordance with API Standard 1104 are "fit for purpose" using this criteria. As a student, you must understand that you are practicing and being trained in ideal conditions, as compared to production pipe welds on a cross-country pipeline or a turn around at a refinery. Your instructor will require you to strive for perfection so that you will have the necessary skills and

Table 16-9 Acceptance Criteria for Porosity According to API Standard 1104

Type of Porosity	Size of Pore	Length	Diameter of Cluster Area	Distribution	Aggregate Length in 12 in. of Weld
Individual scattered (P)	$\leq \frac{1}{8}$ in. or 25% of wall thickness	NA	NA	Less than concentration specified in standard	NA
Cluster porosity (CP) final pass only. All other passes see (P)	$\leq \frac{1}{16}$ in.	NA	$\leq \frac{1}{2}$ in.	NA	$\leq \frac{1}{2}$ in.
Hollow bead (HB)	NA	$\leq \frac{1}{2}$ in.	NA	$\leq \frac{1}{4}$ in. in length separated by ≥ 2 in.	≤ 2 in. or $\leq 8\%$ weld length

NA = not acceptable.

Table 16-10 Maximum Dimensions of Undercutting According to API Standard 1104

Depth	Length
>1/32 in. or >12.5% of pipe wall thickness, whichever is smaller	NA
>1/64 in. or >6% to ≤12.5% of pipe wall thickness, whichever is smaller	2 in. in a continuous 12-in. weld length or one-sixth the weld length, whichever is smaller
>1/64 in. or ≤6% of pipe wall thickness, whichever is smaller	Acceptable regardless of length

NA = not acceptable.

knowledge to function on the job. When welding in ideal conditions, you should be able to eliminate nearly all the discontinuities that have been described.

Practice Jobs

Instructions for Completing Practice Jobs

In practicing the jobs presented in the Job Outline, Table 16-11 (pp. 519–520), follow the same general procedures presented in Chapters 13 through 15. For example, the welding procedures used in the practice Job 16-J2 are essentially the same as those used in practice Job 14-J36. The Job Outline lists the jobs in the previous chapters that give the techniques to be followed in completing the jobs in this chapter. Review the job drawing, the General Job Information, and the Welding Technique in the job to which the Job Outline refers you. Then study the information on the pages in this chapter to which the Job Outline refers you.

The sequence followed in completing the basic operations for the jobs in the Course Outline should be as follows.

Joint Preparation

The preparation of the pipe to be welded should be done so as to conform as nearly as possible to the specifications as shown in Table 16-3, page 478, and Fig. 16-54. Poor beveling and poor preparation and setup of the parts to be welded are responsible for many weld failures. A joint properly prepared is insurance against failure. Narrow groove butt joints are becoming more popular on heavy wall pipe when automatic welding equipment is being used. The groove angle of these narrow groove butt joints can be 10°. This greatly reduces the volume of weld metal required.

1. Refer to Table 16-3, page 478, for the specifications that refer to the joint and position being practiced.

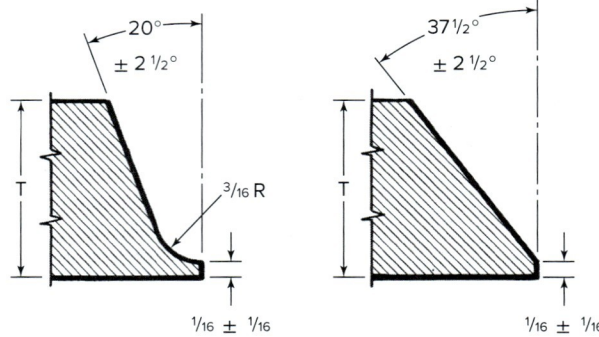

The recommended practice for the detail of welding bevel is as follows:
For wall thicknesses 3/16" to 3/4" inclusive, 37 1/2° ± 2 1/2°, straight bevel; root face 1/16" ± 1/32".
For wall thicknesses greater than 3/4", 20° ± 2 1/2°, J-bevel, 3/16" radius; root face 1/16" ± 1/32".
Welding ends having thicknesses less than 3/16" are prepared with a slight chamfer or square in accordance with manufacturer's practice.

Fig. 16-54 Recommended bevel sizes for pipe joints. (Left) Recommended standard J-bevel of welding ends for thickness (T) greater than 3/4". (Right) Recommended standard straight bevel of welding ends for thickness (T) 3/16" to 3/4" inclusive.

2. Bevel the pipe by hand, machine, or machine oxyacetylene cutting. Be careful to leave the inside edge of the pipe intact so that it serves as a point of contact in preserving proper alignment of the joint when it is set up. Make sure that the bevel angle conforms to the specifications. An insufficient angle of bevel will make it very difficult to secure the proper root penetration and fusion.

3. Clean the bevel face and the pipe surface for at least 1 inch from the edge of the welding groove to remove rust, scale, paint, oil, and grease. Grind flame-cut surfaces smooth enough to remove all traces of scale and any cutting irregularities. Surface finishing may be done with a file, scraper, emery cloth, chipper, power grinder, or sandblasting equipment.

Setting Up and Tacking

Tack welding should be done with great care and precision. The tacks must hold the joint in proper alignment and keep distortion at a minimum. If improperly applied, tack welds contribute to weld failure. There must be enough tack welds of the proper size to hold the pipe in place.

1. Refer to Figs. 16-19 through 16-25 to see how the various joints are clamped and prepared for tacking.
2. Make sure that the proper root opening between the pipe ends has been set up. This is very important because an opening that is too small frequently causes weld failure on the root- and side-bend tests. The root opening should be 3/32 to 1/8 inch. The simplest way to secure and maintain an accurate root opening is to use a 1/8-inch gas welding wire of the proper size. By crossing the wire, you have accurate spacing at four points along the circumference of the joint if you have maintained a square root face around the pipe.
3. Tack the joint. On piping up to and including 4 inches in diameter, make two 1/2-inch long tacks on opposite sides of the joint. On larger pipe sizes, make tack welds at 4- to 6-inch intervals around the circumference of the pipe. In no case, should tack welds be longer than 1 inch. They should penetrate the joint and fuse through the full thickness of the root face. Make sure that the faces of the welds are flat.
4. Clean the tack welds thoroughly and remove any high spots or crater cracks by grinding. Feather the tack weld out.

Welding Technique

It is well at this point to review a few of the important welding techniques that you will need to practice if you are going to master the welding of pipe in all positions.

Direction of Travel Two methods of welding used are vertical-down travel and vertical-up travel. *Vertical-down welding* requires higher welding current and faster travel speeds than vertical-up so that the joint is made with several smaller beads. Root openings are less than those required for vertical-up welding, or there may be no root openings at all. The bevel angle of the joint is also less than in vertical-up. The vertical-down method is faster and more economical on pipe under 1/2 inch in wall thickness. Since most cross-country pipelines have walls less than 1/2 inch thick, they are usually welded by this method. The welds are of excellent quality and able to meet ordinary test requirements.

Vertical-up welding uses lower current and slower travel speed to produce a joint with fewer but heavier beads. Since there are fewer beads to clean, cleaning time is greatly reduced, giving the vertical-up method a speed advantage on heavy wall pipe. The slower travel speed of vertical-up welding and the highly liquid pool melts out gas holes more effectively than vertical-down welding. Welds made by this method are better able to meet radiographic requirements for the high pressure, high temperature piping found in refineries and power plants. Vertical-up welding requires a larger root spacing and bevel angle than vertical-down. Thus, more electrode is required per joint.

Compare the specifications given in Figs. 16-55 and 16-56. The vertical-down method requires 55 to 80 more amperes than the vertical-up method. Although the number of passes is the same, larger electrode sizes are specified for vertical-down. The bevel angle is 15° less for vertical-down, and the root opening is 1/16 inch less. Thus the beads are smaller for welds of equal size, and less filler metal is needed. For the joint design shown, the travel speed for vertical-down is more than twice that for vertical-up.

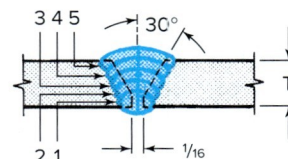

Pass	Electrode	Amps
1	5/32	150–175
2	5/32	170–200
3	3/16	170–205
4	3/16	170–205
5	3/16	170–205

Fig. 16-55 Vertical-down welding technique.

Number of Passes The number of passes required for welding various joints in ferrous piping varies with the following factors:

- Diameter and wall thickness of the pipe
- Joint design
- Position of welding
- Size and type of electrode
- Direction of travel
- Adjustment of the welding current employed

When welding in the rolled (1G) and fixed horizontal positions (5G), a generally accepted technique is to deposit one layer for each 1/8 inch of pipe wall thickness. The size of electrodes may vary from 1/8 to 5/32 inch in diameter for the root pass. The electrode for the intermediate and final passes should have a diameter of 5/32 inch. For pipeline welding using the downhill

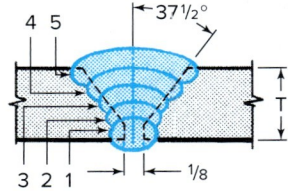

Pass	Electrode	Amps
1	1/8 E6010	85/95
2	5/32 E6010	115/125
	1/8 E7018	90–150
3	5/32 E6010	115/125
	1/8 E7018	90–150
4	5/32 E6010	115/125
	1/8 E7018	90–150
5	5/32 E6010	115/125
	1/8 E7018	90–150

Fig. 16-56 Vertical-up welding technique.

method and occasionally for other piping systems, electrodes with diameters of 3/16 or even 1/4 inch are used for the intermediate and final passes. In welding medium carbon and alloy steels, the number of layers is increased by the use of smaller electrodes and thinner passes. This reduces the heat concentration and ensures complete grain refinement of the weld metal.

When the pipe is in the fixed vertical position and welding is on a horizontal plane (3G), the weld metal is deposited in series of overlapping stringer beads for the intermediate and final passes. Occasionally, a weave bead is used for the final pass. Few welders are able to master this technique, however, and its use is discouraged. The root pass is made with 1/8- or 5/32-inch electrodes. The intermediate and cover passes may be made with 5/32- and even 3/16-inch electrodes.

Control of the Arc Before beginning to weld, it is essential to establish a good work connection and make sure that both cable connections to the machine are tight. The work connection should be attached close to the pipe joint being welded. Current control is extremely important in welding the root pass. A high current setting causes excessive penetration on the inside of the pipe and makes control of the molten weld pool impossible. If the current setting is too low, there is little penetration and fusion at the root.

You will recall that the length of the arc determines the voltage across the arc. The voltage and the amperage provide the heat to melt the electrode and the base metal to form the weld pool. High voltage caused by a long arc gap provides a wide weld bead with little directional control and excessive weld spatter. Low voltage caused by a short arc length causes the bead to pile up in the groove with little fusion and penetration. Low voltage also causes an uneven burnoff rate of the electrode coating so that the slag covers the weld bead inadequately. Poor slag coverage causes porosity in the weld metal.

The speed of travel is another important element in the formation of the weld bead that is deposited. A high speed tends to cause undercutting and a high, narrow bead. Undercutting produces a stress concentration at the point of undercut and can result in joint failure. A low speed causes too much metal to pile up. This causes poor fusion along each edge of the weld and seriously affects the soundness of the weld.

Electrode Angle The angle of the electrode while welding has a significant effect on the final weld. The angle is especially critical in pipe welding since it changes constantly as the weld progresses around the pipe, Fig. 16-57. The welder can maintain control of the weld pool and

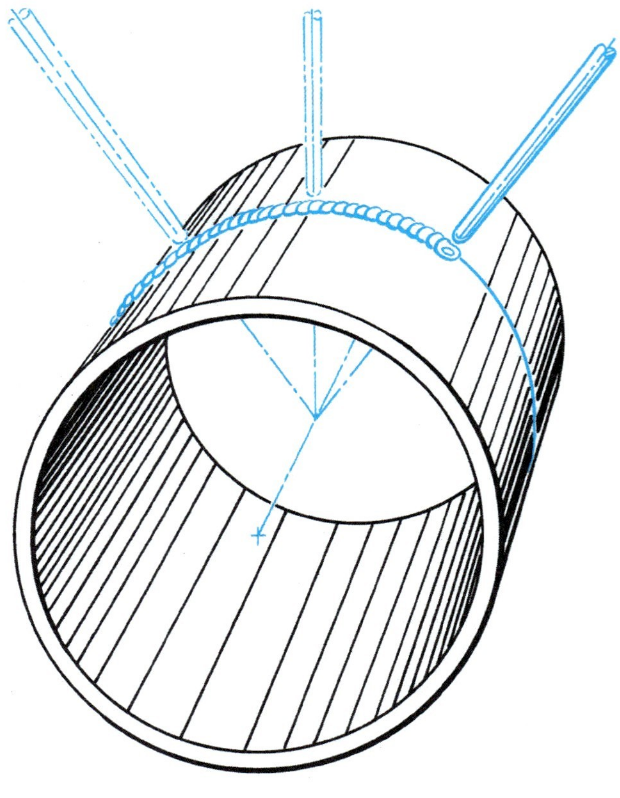

Fig. 16-57 Angle of the electrode as the welder follows the contour of the pipe.

reduce the erratic effects of arc blow by varying the angle of the electrode to meet the conditions of welding.

Refer to Figs. 15-49 through 15-67, pages 460 to 464, which show properly made welds.

Beading and Making a Butt Joint V Groove Weld on Pipe Axis in the Rolled Horizontal Position (1G): Jobs 16-J1 and J2 Before proceeding with jobs 16-J1 and J2, review the welding procedures and photographs for Jobs 13-J4 and J5 (pp. 345–347) and 14-J33 (p. 403). Also study Fig. 16-56.

Beading practice around the outside of the pipe in Job 16-J1 is for the purpose of getting used to the changing position of the electrode holder while following the contour around the pipe. This is an essential technique in pipe welding. The practice should be both downhill and uphill, and both stringer and weaved beads should be deposited.

For the V groove weld, prepare the pipe for welding as shown in Fig. 16-58. Clean, set up, and tack the pieces

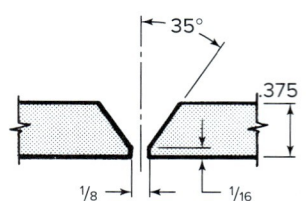

Fig. 16-58 Butt joint preparation.

of pipe as previously directed. The root pass is welded with a ⅛-inch electrode and a current setting of 90 to 125 amperes. The vertical-up and vertical-down directions of welding should be practiced. Restarting the weld bead after a stop may cause you some trouble. Practice the procedure outlined on pages 503–504.

Root Pass: Vertical-Up Method Start just below the 2 o'clock point and weld uphill to the 12 o'clock point. Make sure that you are getting penetration through the back side of the pipe. The back side and face side of the bead should be flat (without undercut or excessive buildup) and show uniform fusion without undercut.

Rotate the pipe clockwise so that the weld crater at the end of the completed weld bead is just below the 2 o'clock point. After making sure that the crater is clean, restart the weld bead and weld uphill to the 12 o'clock point. Figure 16-59 shows the pass sequence on butt joint V groove-welded pipe. There is a hole formation at the tip of the root pass. This is referred to as the "keyhole." If the keyhole is allowed to get too big, it will cause internal undercut or excessive melt-through. The size of the keyhole is affected by the amount of current, electrode angle, and electrode pressure. Excessive current must be avoided. It is necessary to achieve this formation as you weld in order to secure complete penetration on the inside of the pipe, Fig. 16-60. Proper bead formation inside the pipe ensures maximum strength and the inside of the pipe is smooth to permit the free flow of materials. Clean the weld thoroughly.

Root Pass: Vertical-Down Method Start at about the 12 o'clock point and weld downhill to just below the 2 o'clock point. Rotate the pipe counterclockwise so that the crater of the previous weld is at the 12 o'clock point. After making sure that the crater is clean, restart the weld bead and weld downhill to the 2 o'clock point. Make sure that the slag does not run ahead of the weld and prevent fusion. Check the weld for the characteristics described for the vertical-up method. Clean the weld thoroughly.

Filler and Cover Passes The direction of welding and rotation of the pipe are the same as for the root pass. Filler and cover passes, however, may be made with 5/32-inch E6010 electrodes or ⅛-inch E7018 for vertical-up technique and a current setting of 120 to 160 amperes. All passes must be cleaned thoroughly before proceeding with the next pass. Each pass must fuse and penetrate the underneath pass and the side walls of the pipe. Make sure that the pool fills up at each side of the weld and does not leave an undercut. Travel must be rapid enough to avoid an excessively convex contour. Every effort should be made to produce welds with a flat face.

Inspection and Testing After the weld has been completed, clean it by brushing. Inspect the weld for the characteristics listed in Inspection, Job 14-J36, page 410. Compare the weld appearance with Figs. 16-59 and 16-60 and note any defects in your weld.

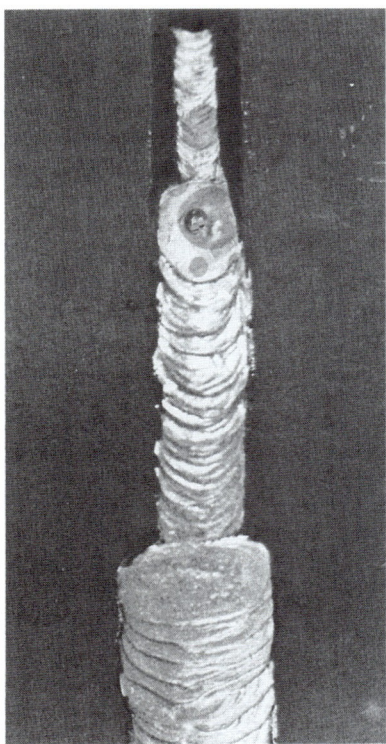

Fig. 16-59 Vertical V-groove butt joint weld in pipe welded in the horizontal rolled position (1G) (travel up). Note the sequence of passes. Edward R. Bohnart

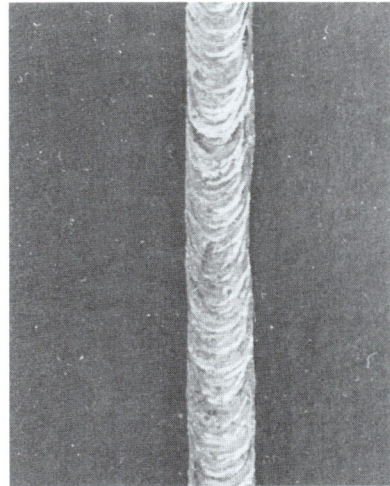

Fig. 16-60 Root penetration on the inside of the pipe. Note that there is complete fusion and penetration without excess melt-through. Edward R. Bohnart

Continue to practice this joint until you can consistently produce sound welds of good appearance with both E6010 and E7018 electrodes. When you feel that you have mastered the welding of this joint in the 1G position, cut two test coupons. Remove the test coupons as instructed in Fig. 16-37, page 488, and subject them to face- and root-bend, nick-break, and tensile tests, as described in Chapter 28, pages 922–926. Examine the bended coupons carefully for signs of lack of fusion and slag inclusions. Apply acceptance criteria for all tests.

Horizontal Butt Joint V-Groove Weld on Pipe Axis in the Fixed Vertical Position (2G): Jobs 16-J3 and J4
Before proceeding with Jobs 16-J3 and J4, review the procedures and photographs for Job 15-J51, page 444. Prepare the two pieces of pipe for welding as shown in Fig. 16-61. Set up the joint and tack it as previously instructed.

Root Pass The root pass is made with a ⅛-inch E6010 electrode and a current setting of 80 to 110 amperes. You may use 5/32-inch electrodes if you can handle this size. Insert the electrode well into the bottom of the groove. Direct the arc toward the top and bottom bevel of the joint at the same time in order to obtain fusion. Manipulate the electrode with a slight circular or back-and-forth motion and play most of the heat on the upper bevel. The work angle of electrode should vary from 5 to 15° from the horizontal and lean from 30 to 45° travel angle in the direction of welding, Fig. 16-62. Control the pool by making small changes in the electrode angle as you proceed around the pipe.

As the weld progresses, inspect the root pass for signs of not enough penetration, too much penetration, undercut, and poor fusion. If you observe any of these faults, check the speed of travel, the angle of the electrode, the heat setting, the arc gap, and the welding technique.

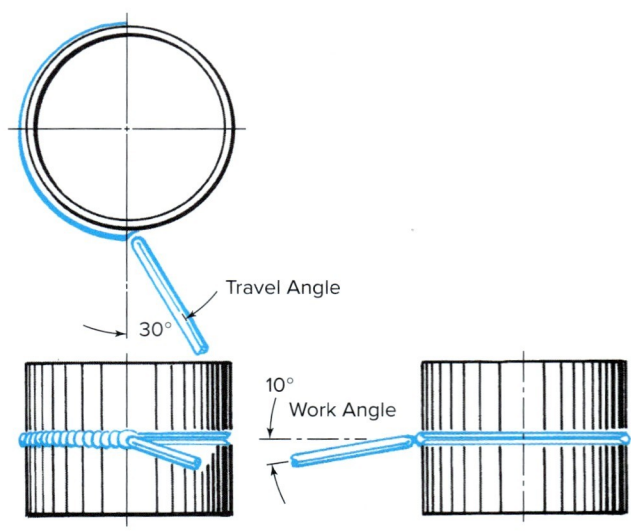

Fig. 16-62 Angle of the electrode for a horizontal groove weld in pipe in the vertical fixed (2G) position.

A procedure may call for a hot pass to be run as the second pass. This is needed if the root pass is small or if it has a number of minor defects in the face of the weld. The hot pass will cause more metal to be deposited in the groove and will eliminate the defects.

Filler Passes Make intermediate passes like those in Job 15-J54, page 450. The second and third passes are usually made with a ⅛-inch E6010 or E7018 electrode, although a 5/32-inch size may be used. The remaining intermediate passes should be made with ⅛-inch E7018 or 5/32-inch electrodes at 120 to 160 amperes. In making these passes, be sure that complete fusion is obtained with the underneath passes and the wall of the pipe. The second pass must be fused with the root pass and the lower pipe bevel. The third pass is placed immediately above the second pass and must be fused to the second pass, the root pass, and the upper pipe bevel. Each layer is started at the lower side of the groove and built upward.

If the travel is too slow or the heat setting too low, poor fusion and excessive buildup will occur. If the travel is too fast or the current is too high, poor bead formation and undercut are likely to occur. Care must be taken to eliminate deep valleys along the edges of each bead as it is applied. Valleys are caused by convexity in the center or poor fillup along the edges of the weld. They are hard to clean and increase the chances of trapping slag. Careful cleaning is essential.

Cover Pass The final layer should be practiced with both the stringer bead technique and the weave bead technique required for Job 13-J24, page 382. Procedure qualifications differ.

When practicing stringer beads, be sure to start each bead in a slightly different area around the pipe so that you do not start and end each bead in the same location. Weld with a ⅛-inch E7018 or 5/32-inch E6010 electrode and a current setting of 120 to 160 amperes. Undercut is a problem along the weld edges, especially along the edges of the first and last stringer bead. This can be overcome by careful attention to the speed of travel and current adjustment.

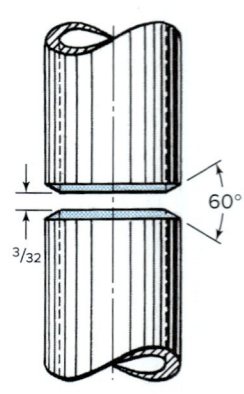

Fig. 16-61 Joint preparation in the vertical fixed (2G) position.

In practicing the weave bead technique, be sure to pause at the top of the weave in order to prevent undercut. Sag is also a problem and can be prevented by weaving at an angle and a rapid advance toward the bottom of the weave.

Inspection and Testing After the weld has been completed, clean and inspect it as instructed for Jobs 16-J1 and J2. Compare it with Figs. 16-63 and 16-64. After you have made a number of these welds with both E6010 and E7018 and they seem satisfactory on visual inspection, cut some test coupons and subject them to applicable testing.

Butt Joint V-Groove Weld on Pipe Axis in the Fixed Horizontal Position (5G) (Bell-Hole) (Travel-Up): Job 16-J5 Two welding procedures are used to weld pipe in the fixed horizontal position (5G). On power piping, the usual method is to start the weld at the bottom and progress upward. For cross-country pipeline work, the weld is started at the top and progress is downhill. You will be required to learn both procedures. Before proceeding with Jobs 16-J5 and J6, review the procedures and photographs for Jobs 15-J43 (p. 429), 15-J48 (p. 439), and 15-J55 (p. 452). Prepare the two pieces of pipe as shown in Fig. 16-58. Set up and tack the joint as previously instructed.

Root Pass Practice welding from the bottom to the top. The root pass is the most important weld that must be made in completing this joint. Fusion must be complete on both pipe sections, and penetration must be through the inside of the pipe, Fig. 16-65. If there is insufficient buildup inside the pipe, the weld will fail in service and also when subjected to the root-bend test. Too much buildup on the inside will restrict the flow of materials inside the pipe and may also cause failure of root-bend specimens under test.

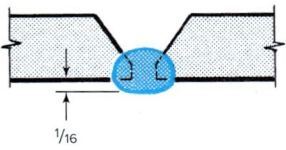

Fig. 16-65 Proper contour of the root pass.

The root pass should be made with a ⅛-inch E6010 electrode and a current setting of 80 to 110 amperes. Generally, best results are obtained by using currents on the low side of the range.

The root pass should never be started at the absolute bottom of the pipe since this is the location from which one of the test coupons is taken, and the possibility of failure is increased. Start the root pass at either the 5 o'clock or 7 o'clock position and proceed across the bottom to the top of the pipe joint, Fig. 16-66. Never stop at the top center of the pipe since this is another test location. Stop either at the 1 or the 11 o'clock position. Take care in welding over the tack welds in the joint. They must be completely fused and become a part of the weld bead. Chip out and grind any unsound or large tacks before welding the root pass. As you carry the pass around the pipe, a small hole (keyhole) of about 3/16 inch in diameter should precede the weld pool, Figs. 16-67 and 16-68, page 504. Make sure the hole is not too large. The hole at the root of the joint is caused by the complete melting down of the root face. It can be maintained by the use of a circular movement of the electrode tip or by moving back and forth in a straight line. The angle of electrode should be about a 10° push travel angle in the direction of travel as shown in Fig. 16-69, page 504.

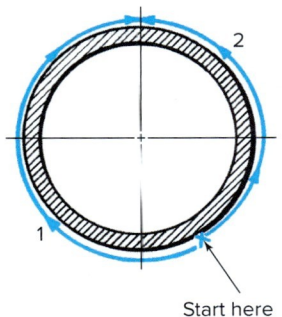

Fig. 16-66 Starting point for the first pass in pipe in the horizontal fixed (2G) position (note the sequence).

Changing Electrodes You will probably have some trouble at first with restarting after a stop to change electrodes. The bead inside the pipe will have a crater or depression at the end which is lower than the inside wall of the pipe and lower than the bead already laid at the point where the new weld must join the weld just completed. This can cause a future failure in the pipeline when it is in service and certainly will cause failure of the root- and side-bend test

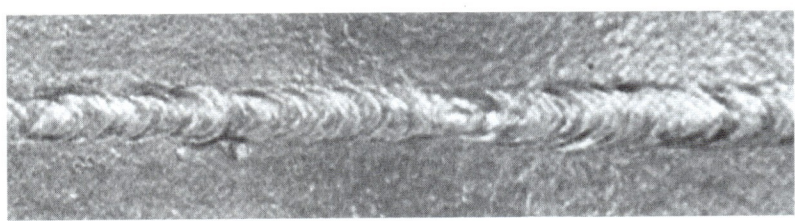

Fig. 16-63 Root penetration on the inside of a pipe welded in the vertical fixed (2G) position. *Edward R. Bohnart*

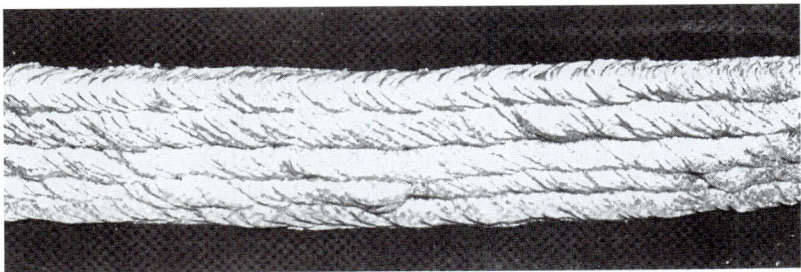

Fig. 16-64 Cover passes on pipe welded in the vertical fixed (2G) position with stringer bead technique. *Edward R. Bohnart*

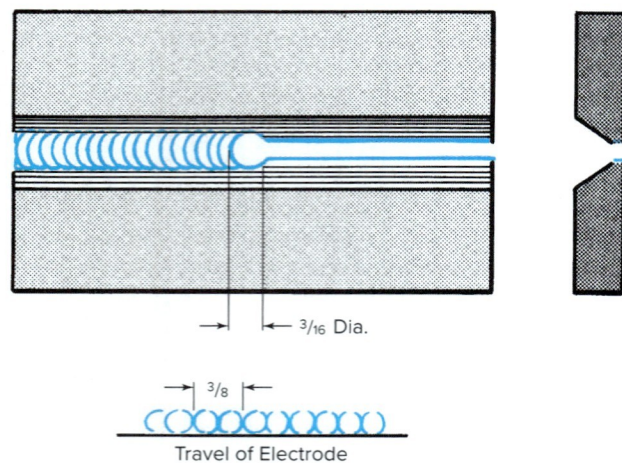

Fig. 16-67 Method of applying the root pass.

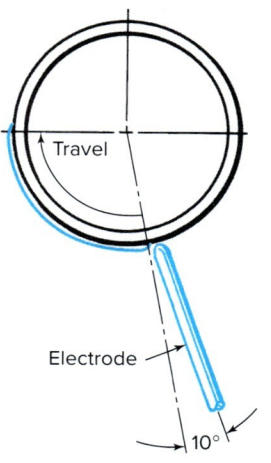

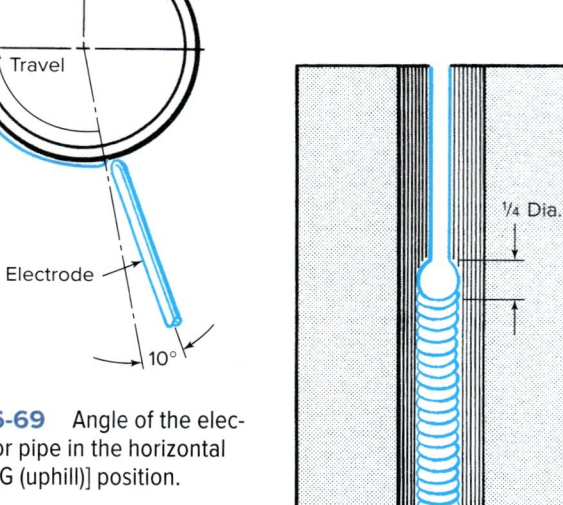

Fig. 16-69 Angle of the electrode for pipe in the horizontal fixed [5G (uphill)] position.

Fig. 16-70 The keyhole at the front edge of the weld pool is enlarged before breaking the arc at the finish of the electrode.

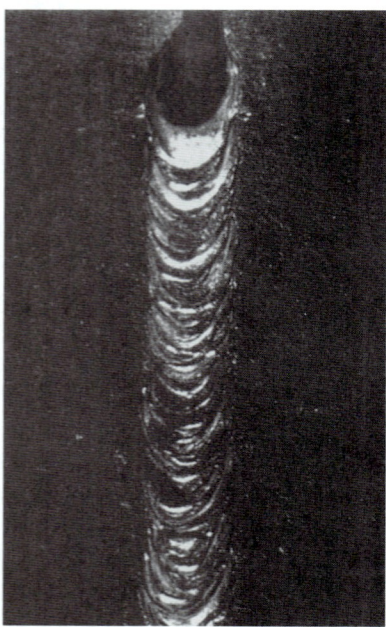

Fig. 16-68 Back side of the root pass welded in the horizontal fixed [5G (uphill)] position. Note the keyhole formation. Edward R. Bohnart

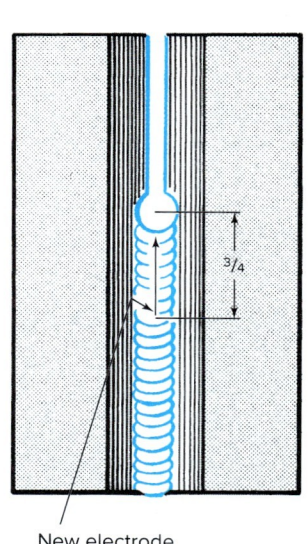

Fig. 16-71 Method for starting the new electrode at the end of the previous weld bead.

specimens. The following precautions are necessary to make a smooth, uninterrupted bead on the inside wall of the pipe at a start-and-stop point:

1. When an electrode is used up, enlarge the hole (keyhole) preceding the weld pool to ¼ inch in diameter before breaking the arc, Fig. 16-70.
2. In restarting the weld bead with a new electrode, strike the arc ½ to 1 inch back on the previous weld and run slowly up to the end of the weld. This heats the weld metal and the base metal enough to ensure a hot and fluid start, Fig. 16-71.
3. As you reach the hole at the end of the previously laid weld bead, insert the tip of the electrode clear through the hole into the pipe and incline the electrode at an angle of 45° in the direction of travel. Maintain this position only long enough to deposit new weld metal on the end of the previous weld bead on the inside of the pipe, Fig. 16-72. Resume the normal electrode position and proceed as before.
4. Inspect the back side of the completed root pass carefully. It should be no more than 1/16 inch higher than the inside wall of the pipe, and it should have good fusion along the edges of the weld. There should be no evidence of icicles, grapes (globular deposits of metal), excess melt-through, or undercut inside the pipe. The face of the pass should be fairly flat with smooth, even ripples, and there should be no undercut along the weld edges.

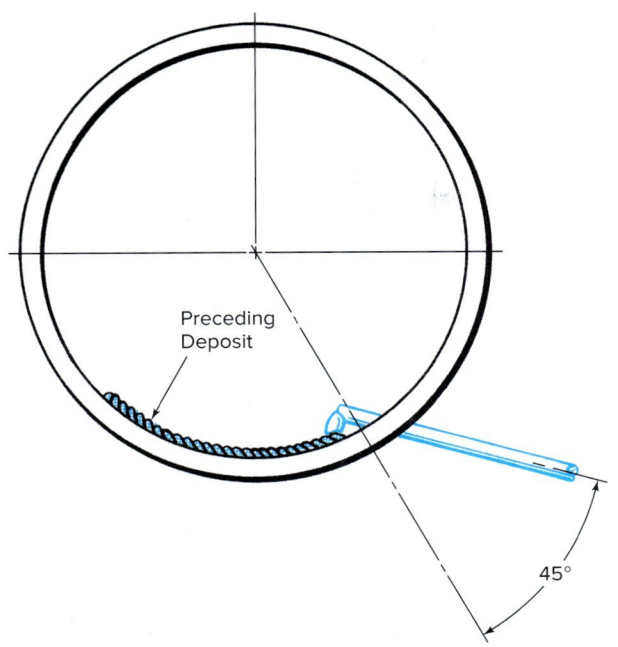

Fig. 16-72 The method used in restarting the weld. Weld metal is deposited at the end of the previous weld on the inside of the pipe.

Filler Pass Before making the next pass, clean the face of the root pass carefully. Chip off any high spots or weld spatter, and chip out any holes or depressions along the edges of the weld. Start the second bead in the same general area as the first bead, but try to avoid the exact starting point. Start on the opposite side of the joint or just ahead of or in back of the previous starting point.

The single filler or intermediate pass is applied with a weaving technique using a 1/8-inch E7018 or a 5/32-inch E6010 electrode and a current setting of 120 to 160 amperes.

The current should be hot enough to ensure complete fusion with the face of the root pass and both beveled faces of the pipe. You should watch for any tendency to undercut or to trap slag in the weld. If either defect is not remedied, it can become the cause of failure of the weld joint when in service. Avoid piling up the weld metal. The face of the weld bead should be flat, and the bead should not be thicker than 1/8 inch. In joining the bead from one side to that deposited from the other side of the pipe, be sure to fill the crater by moving backward over the finished bead for a short distance.

Cover Pass The third or cover pass is made with a 1/8-inch E7018 or a 5/32-inch E6010 electrode and a current setting of 120 to 160 amperes. Be sure to clean the preceding weld and chip off spatter and excess weld metal. The cover pass is also a weave pass, and the welding technique is similar to that used for the second pass. Avoid undercutting by hesitating at the sides of the weld as you weave. There is also the tendency to widen the pass as you proceed with the weld. The weld bead should extend about 1/16 to 1/8 inch beyond the edge of the groove on each side. The face of the weld should have an oval shape with the bead no higher than 1/16 inch above the surface of the pipe at the center of the weld. The edges of the weld should flow smoothly into the surface of the pipe. Note the even width and close ripples of the bead shown in Fig. 16-73.

Inspection and Testing After you have completed this weld, use the visual inspection procedures described previously. The weld should look like that shown in Fig. 16-74. After you have made a number of these joints that are visually satisfactory, cut a few test coupons from the joint and subject them to the usual tests.

Butt Joint V-Groove Weld on Pipe Axis in the Horizontal Fixed Position (Travel-Down) (5G): Job 16-J6 This job involves practice on the same pipe joint as in Job 16-J5, but welding is from the top to the bottom. Vertical-down welding is a cross-country pipeline technique. (Study the photographs of the rolling groove weld made in Job 16-J2, page 500.)

Root Pass Here again you are cautioned to take extra precaution in making the root pass. Always keep in mind that the root pass is the basis for success or failure in making

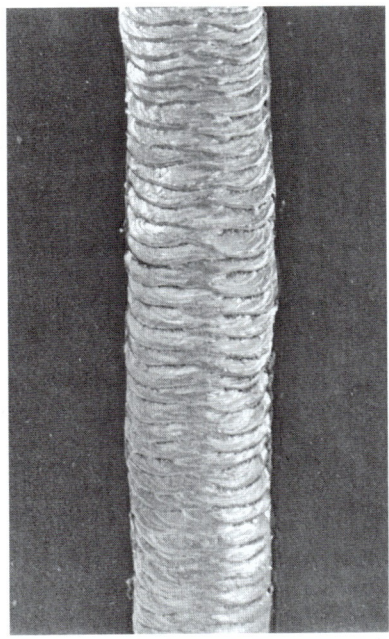

Fig. 16-73 Cover pass on pipe welded in the bell-hole (horizontal fixed) (5G) position. Edward R. Bohnart

Pipe Welding and Shielded Metal Arc Welding Practice: Jobs 16-J1–J17 (Pipe) **Chapter 16 505**

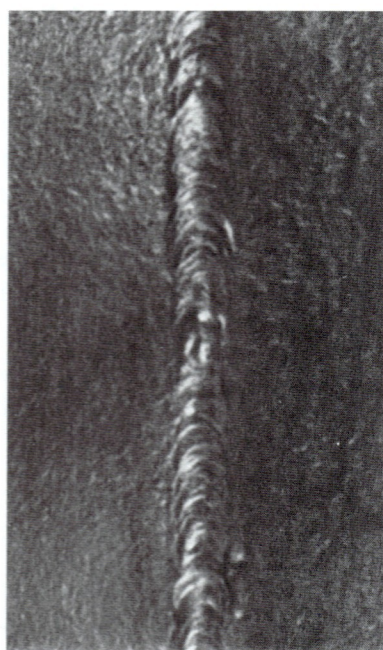

Fig. 16-74 Root penetration on the inside of a pipe welded with the vertical-down technique. *Edward R. Bohnart*

a test weld. It is also the basis of a sound weld. The root pass should be made with a 5/32-inch E6010 electrode and a current setting of not more than 200 amperes nor less than 150 amperes.

Correct joint preparation is important. Internal undercut will occur if the root face is too small, the root opening is too large, or a severe high and low condition of the pipe ends exists. Refer to Table 16-3, page 478. Joint preparation determines the welding heat. A small root face and large root opening require a low heat setting. A large root face and minimum root opening require a high heat setting.

Start the pass at the 11 o'clock or 1 o'clock position. Weld across the top of the pipe and downward past the 6 o'clock position to the 7 o'clock or 5 o'clock position, depending upon the side you are welding from. Make the stringer bead with a drag technique. Rest the electrode coating on the bevel as you drag the electrode downhill around the pipe. Sufficient pressure should be applied to the electrode to force the arc to the inside of the pipe. When observing a welder using this technique, it is common to see the electrode actually bowing as a result of the pressure being applied to it.

The electrode angle should be about 30 to 45° from the horizontal center line (drag travel angle), and it must vary little as you travel around the pipe. Maintain a close arc but be careful that you do not push the electrode through the back side of the pipe.

To secure adequate penetration on the inside of the pipe, it is necessary to maintain a small visible keyhole at all times. If the hole becomes too large, it will cause internal excess melt-through, sag, or undercut. Make sure that enough heat is applied against the cold bevel of the pipe to ensure good fusion.

The speed of travel is faster than that used in welding uphill, and it changes from time to time, depending upon the joint condition and weld formation. You can also maintain control of the weld pool by varying the speed of travel, electrode angle, and electrode movement. After the root pass is completed on one side of the pipe, weld the other side in the same manner. Be sure to overlap the weld on the other side about ½ inch at the top and bottom.

One method of determining whether or not there is complete penetration is to listen to the sound of the arc. A sound somewhat like compressed air being released inside the pipe may be heard when penetration is complete. Insufficient penetration or lack of fusion may be caused by an excessive root face, a root opening that is not wide enough, or too low a welding current. In this event, current settings will have to be increased.

Excess root face, insufficient root opening, or low welding current can result in lack of penetration through the root of the joint. Other defects that can occur are lack of fusion between weld and base metal, excess melt-through, and globular deposits (*grapes*) on the inside of the pipe.

Burning away the side walls of the groove results in undercut. External undercut along the edges of the weld bead is called "wagon tracks." Undercut can also occur on the inside of the pipe. Internal undercut can be repaired only from inside the pipe which is either nearly impossible or a costly procedure.

If a tendency to undercut is observed while welding, it can be overcome by tilting the electrode just a few degrees toward the undercut side. Excess melt-through, metal deposits on the inside of the pipe, and external and internal undercut are usually caused by too high a current, too wide a root opening, or too little root face. Holes and slag pockets are common faults that can be avoided only by the correct heat setting and complete control of electrode manipulation. When inspecting the root pass, look for the same satisfactory characteristics discussed previously and refer to Fig. 16-74.

Hot Pass Be sure to remove all slag before making the hot pass (second pass). Undercut along the edges of the root pass makes slag removal difficult. An old power hacksaw blade, taped on one end to serve as a handle, is a good cleaning tool. The bead should be thoroughly brushed, preferably with a power brush. Any surface defects or excessive spatter should be removed by chipping or grinding.

The second pass not only helps to fill up the groove, but it also reinforces the root pass. The hot pass is welded with a current setting similar to that used for the root pass: about 160 to 190 amperes. Make the second pass with a $5/32$-inch electrode or a $3/16$-inch electrode if you can handle the larger size. The hot pass should be started within 5 minutes of the completion of the root pass.

Welding should start at the top of the joint outside the area of the previous starting point, proceed downhill, and stop at the bottom outside the area of the previous stopping point.

The angle of the electrode should be about 30 to 45° down (drag travel angle) from the horizontal position. A slight electrode movement is necessary, especially in the 4 to 8 o'clock sections of the pipe. A close arc gap, uniform speed of travel, and a steady electrode angle are extremely important. Failure to do so causes holes and permits the slag to run ahead and extinguish the arc.

The hot pass must be applied with enough heat to ensure good penetration into the root pass. This is necessary to burn out wagon tracks and to float any remaining slag to the surface. Too fast a rate of travel results in surface pinholes.

Make sure that the weld is started and finished away from previous starting and stopping points and outside the area from which the test coupons are going to be cut.

Stripper Pass Welds made with the downhill welding technique may be thin at the 2 to 4 o'clock and the 8 to 10 o'clock positions. A weld pass known as a *stripper pass* is used to build these sections up to the same height as the rest of the weld bead. Use a welding technique similar to that used in making the hot pass and a slightly longer arc.

Filler Passes Since the beads applied when welding downhill are thinner than beads welded in the vertical-up direction, additional passes are necessary to complete the pipe joint. The tendency for these beads to be concave in the center may cause problems for the other passes. This can be overcome by reducing the heat used, by slowing the rate of travel, or by a slower weave motion.

The filler passes should be made with a $3/16$-inch electrode and a current setting of 170 to 210 amperes. Pay close attention to the angle of the electrode and the length of the arc gap. Use a slight side-to-side weave and make sure that the weld deposit fills the groove and fuses into the side walls. Improper electrode manipulation or too high a rate of travel can cause porosity, undercut, and a concave face. Travel that is too slow causes the weld deposit to pile up and run ahead of the slag.

Cover Pass This pass is also referred to as a cap pass. Before applying the cover pass, it is sometimes necessary to make the concave portions of the weld flush with the stripper passes. Concavity usually occurs in the 2 to 4 o'clock and 8 to 10 o'clock positions on the pipe.

The cover pass is made with a $5/32$-inch electrode and a current setting of 130 to 160 amperes. A $3/16$-inch electrode may be used with a heat setting of about 160 to 180 amperes. Be sure to start and stop in an area outside of the starting and stopping points of the previous weld bead. A weave motion with some hesitation at each side is used to prevent undercut. Take great care and maintain the same electrode angle, arc gap, and speed of travel practiced previously. The finished weld should be from $1/32$ to $1/16$ inch higher than the pipe wall and should overlap the groove by $1/16$ to $1/8$ inch on each side. Better stress distribution is obtained by keeping the cover pass as flat and narrow as possible. A narrow bead is also less susceptible to surface porosity. If the bead is wider than $3/4$ inch, use two weave beads around the pipe.

Inspection and Testing After you have completed this weld, use the visual inspection procedures described previously. The root pass should look like that shown in Fig. 16-74, and the cover pass should look like that shown in Fig. 16-73. After you have made a number of these joints and they are visually satisfactory, cut a few test coupons from the joint and subject them to the usual face- and root-bend tests.

Butt Joint V-Groove Weld on Pipe Axis with Backing in the 45° from Horizontal Fixed Position (6G) (Arkansas Bell-Hole) (Travel Up): Job Qualification Test 1 This is a popular test position since the welder who can qualify in this position will generally be qualified in all positions and on groove and fillet welds on plate or pipe. This is a versatile test that can qualify the welder for many job applications. The applicable code should be consulted to see what the actual limits are with the 6G test. Fig. 16-75, page 508, detail B. See AWS WPS ANSI/AWS B2.1-1-208 for additional information. The techniques used for the 6G position are a combination of the 2G and 5G techniques. Fit the joint with the appropriate backing ring. Tack in place, and remove root opening spacer nubs.

Root Pass Practice welding from the bottom to the top using a combination of the fill and cover pass techniques used on Jobs 16-J3 and J5. Since this joint has a backing ring in place, the first pass is more like a fill pass than a root pass on an open root butt joint. Use E7018 electrodes for all passes, $3/32$-inch diameter at 70 to 110 amperes or $1/8$-inch diameter at 90 to 150 amperes. The root pass should fuse the edges of both pipes to the backing ring. Clean the crater area before restarting the next bead. Begin the next

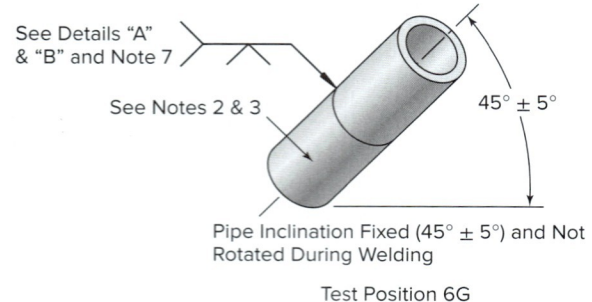

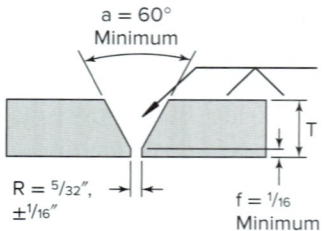

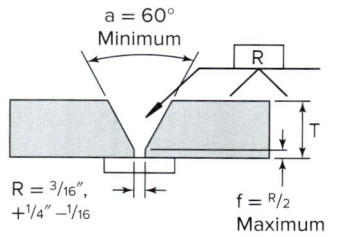

Detail "A" – Joint Geometry Without Backing

Detail "B" – Joint Geometry with Backing

Notes:
1. Administration of this performance qualification test in accordance with AWS QC7, Supplement G, supersedes AWS QC11 and AWS EG3.0 requirements of performance qualification for SMAW of carbon steel pipe.
2. 6" or 8" ϕ Schedule 80 carbon steel pipe (M-1/P-1/S-1, Group 1 or 2).
3. The standard pipe groove test weldment for performance qualification shall consist of two pipe sections, each a minimum of 3 in. (76 mm) long joined by welding to make one test weldment a minimum of 6 in. (150 mm) long.
4. With or without backing. Backing ring to suit diameter and nominal wall thickness of pipe. Refer to Details "A" and "B".
5. All welding done in position according to applicable performance qualification requirements.
6. All parts may be mechanically cut or machine OFC.
7. Use WPS ANSI/AWS B2.1-1-022 for performance qualification without backing. Use WPS ANSI/AWS B2.1-1-016 for performance qualification with backing.

Fig. 16-75 A typical 6G pipe test for welders.

weld approximately ¾ inch back from the crater on top of the prior weld. Let the arc stabilize, and with a short arc gap move into the crater and continue the weld. Never hold a long arc gap with the E7018 low hydrogen electrode, because porosity will occur and the weld will be rejected.

Fill and Cover Passes Before making the next pass, clean the face of the root pass carefully. Chip off or grind any high spots or weld spatter, and chip or grind out any holes or depressions along the toes of the root pass. The E7018 is not a deep penetrating electrode, so the root pass must be smooth with a flat contour as should all subsequent passes. Use E7018 ³⁄₃₂-inch diameter electrodes at 70 to 110 amperes or ⅛-inch diameter electrodes at 90 to 150 amperes.

Inspection and Testing After you have completed this weld, use the visual inspection procedures described previously. The weld should look like that shown in Fig. 16-64. After you have made a number of these joints that are visually satisfactory, cut a few test coupons from the joint and subject them to the usual tests.

Butt Joint V-Groove Weld on Pipe Axis Without Backing in the 45° from Horizontal Fixed Position (6G) (Arkansas Bell-Hole) (Travel Up): Job Qualification Test 2 The techniques used for the 6G position are a combination of the 2G and 5G techniques, Fig. 16-75, detail A. See AWS WPS ANSI/AWS B2.1-1-201 for additional information.

Root Pass Practice welding from the bottom to the top using a combination of techniques learned from Jobs 16-J3 and J5. Because of the incline of the pipe axis, which is 45° from the horizontal, the travel and work angles need to direct the arc and weld pool into the root of the groove. Use ⅛-inch E6010 electrodes at 75 to 125 amperes. Maintain the ³⁄₁₆-inch keyhole, Figs. 16-67 and 16-68. Use proper restarting and stopping techniques as shown in Figs. 16-70 through 16-72.

Fill and Cover Passes Before making the next pass, clean the face of the root pass carefully. Chip off or grind any high spots or weld spatter, and chip or grind out any holes or depressions along the toes of the root pass. The E7018 is not a deep penetrating electrode, so the root pass must be smooth with a flat contour as should all subsequent passes. Use E7018 electrodes, ³⁄₃₂-inch diameter at 70 to 110 amperes or ⅛-inch diameter at 90 to 150 amperes.

Inspection and Testing After you have completed this weld, use the visual inspection procedures described previously. The weld should look like that shown in Figs. 16-64 and 16-74. After you have made a number of these joints that are visually satisfactory, cut a few test coupons from the joint and subject them to the usual tests.

Butt Joint Bevel Groove Weld on Pipe Axis Without Backing in the 45° from Horizontal Fixed Position (6GR) (Arkansas Bell-Hole) with Restricting Ring (Travel Up): Job Qualification Test 3 This qualification covers welding in restricted areas such as T, K, and Y pipe and tubular connections. Check applicable code for the extent of this 6GR qualification, Fig. 16-76. See AWS WPS AWS3-SMAW-1 for additional information.

Root Pass Practice welding from the bottom to the top. The work and travel angles will have to be adjusted as this is a beveled butt joint versus a V-groove butt joint that you have been practicing on. The electrode angle should direct the weld pool to fuse into the square surface of the top pipe and totally consume the root face on the beveled bottom pipe. Care must be taken not to consume or melt the root edge of the top pipe. Do not overpenetrate the joint. Use 1/8-inch diameter E6010 electrodes at 75 to 125 amperes. Maintain the keyhole in the beveled pipe. Use proper restarting and stopping techniques.

Fill and Cover Passes Before making the next pass, clean the face of the root pass carefully. Chip off or grind any high spots or weld spatter, and chip or grind out any holes or depressions along the toes of the root pass. The E7018 is not a deep penetrating electrode, so the root pass must be smooth with a flat contour as should all subsequent passes. Use E7018 electrodes, 3/32-inch diameter at 70 to 110 amperes or 1/8-inch diameter at 90 to 150 amperes.

Inspection and Testing After you have completed this weld, use the visual inspection procedures described previously. The weld should look like that shown in Figs. 16-64 and 16-74. After you have made a number of these joints that are visually satisfactory, cut a few test coupons from the joint and subject them to the usual tests.

Pipe Fabrication Practice: Jobs 16-J7 Through J17 Because of the availability of manufactured fittings for practically every installation purpose, the amount of hand fabrication performed in the field is kept at a minimum. Custom fittings are more costly and do not meet the service requirements of pipe fabricated with manufactured fittings. Since it is still a practice in the field, it is necessary for the student welder to practice this type of construction. It is not the purpose of this book to teach the layout necessary to hand fabricate fittings. It is assumed this will be taught in a related class using a textbook that presents all phases of pipe fabrication. The cost of manufactured welded fittings is very high, and this rules out their use for school practice of T's, Y's, and laterals.

The 90° branch connections and 45° lateral connections (Jobs 16-J7 through J12) are merely a series of lap and fillet welds. The manufacturer's procedure determines whether or not beveling is required and whether the branch connection or the hole in the header is beveled to ensure complete fusion into the inside of the pipe. This would then be considered a groove weld. Welding should be done with 5/32-inch electrodes. The weld is started at a point about 2 inches on either side of the intersection at the heel of the joint. The arc should be in continuous operation when passing over the intersection and the start of the bead. Branch and lateral connections may be made with either the stringer bead or the weave bead technique.

The welding of 45° and 90° L's and other fittings (Jobs 16-J13–J17) are groove welds, and the welding technique is similar to that used in making groove welds in straight pipe sections.

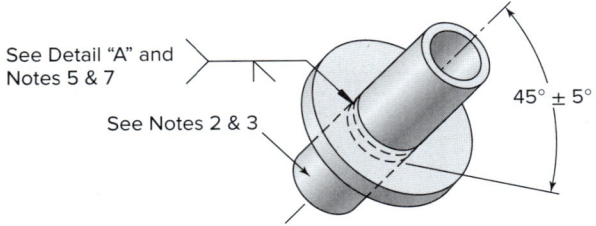
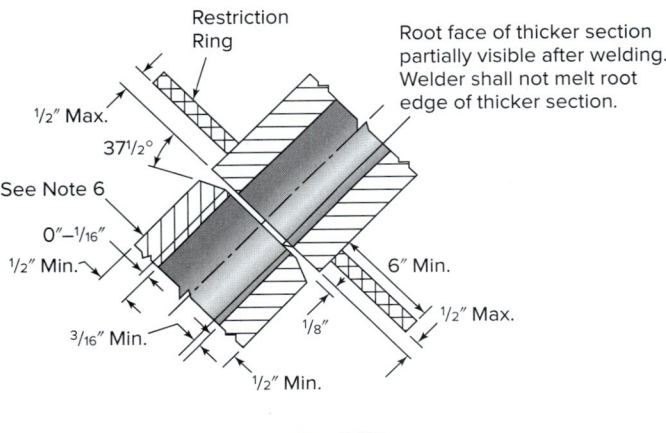

Fig. 16-76 Performance qualification test for SMAW in the 6GR position.

Notes:
1. Duplicate performance qualification tests are not required if welder is tested under AWS QC12 using the AWS QC7 option.
2. 12" φ Schedule 80 M-1/P-1/S-1 carbon steel pipe. Other schedules of pipe 6" φ or greater may be used provided the requirements for minimum offset root and minimum wall thickness can be met. See Detail "A."
3. The standard pipe groove test weldment for performance qualification shall consist of two pipe sections a minimum of 3 in. (76 mm) long joined by welding to make one test weldment a minimum of 6 in. (150 mm) long.
4. Without backing. Restriction ring to suit diameter of pipe.
5. All welding done in position.
6. All parts may be mechanically cut or machine OFC unless otherwise specified. I.D. of chamfered section bored to obtain a 0.500 in. (12.7 mm) minimum wall thickness. Depth of bore 1 1/2 in. minimum. Refer to Detail "A."

Use the visual inspection techniques described previously and strive for welds that meet these standards. You may also cut out test coupons from the fillet lap and butt sections of the joints for macro-etching. The specimens must be ground, polished, and etched with acid. Inspect them thoroughly for such flaws as lack of fusion at the root and to the surface of the pipe, slag inclusions, gas pockets, and undercutting.

Figures 16-77 through 16-82 illustrate some of the various pipe units that can be fabricated in a modern pipe-fabricating shop.

Fig. 16-79 An example of a laydown yard for storage of completed and painted pipe spools can be stored until required on the job site. *John Kriesel/Piping Systems, Inc.*

Fig. 16-77 A 45° lateral in 8 inch diameter pipe schedule 80 pipe with a weld ring for reinforcement plus the attached flange. *John Kriesel/Piping Systems, Inc.*

Fig. 16-80 Tacking up a fabricated L. It is 18 inches in diameter and has a wall thickness of 2 inches. *Crane Co.*

Fig. 16-78 A manifold with two elbows, 3 Tees, and 4 flanges. This type pipe spool must accurately bolt into position as specified on the drawings. While meeting all the weld acceptance criteria per code. *John Kriesel/Piping Systems, Inc.*

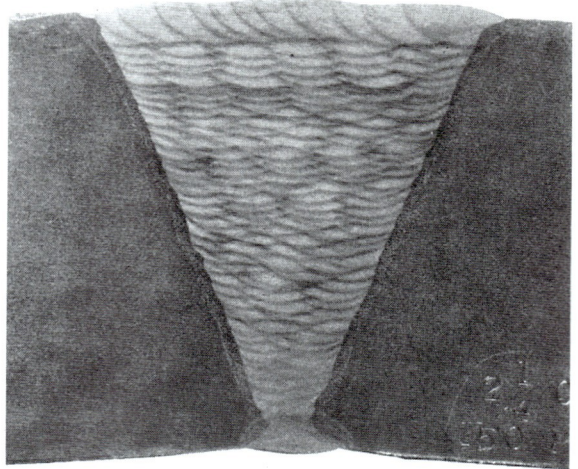

Fig. 16-81 Cross section of an elbow with an outside diameter of 20 inches and a wall thickness of 4 1/16 inches. It was formed from chrome-molybdenum plate, and 150 passes were required for welding the seam. *Crane Co.*

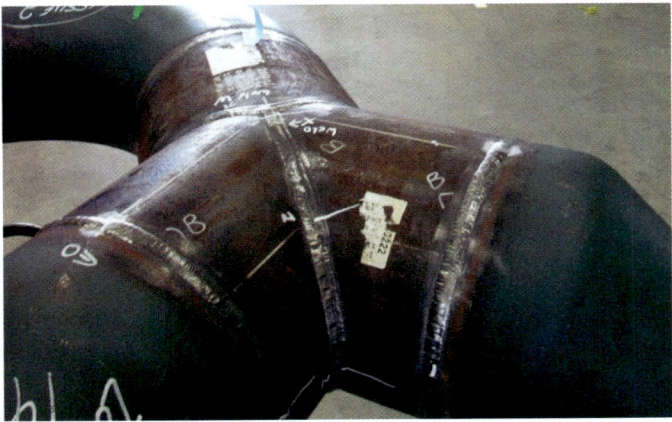

Fig. 16-82 A 10 inch diameter schedule 40 special Y, also know as "pant legs" with an elbow and two reducers welded to it. Piping Systems, Inc.

Fig. 16-83 Cutting a large diameter pipe with the automatic pipe saw. Mathey Dearman

Tools for Pipe Fabrication

Although every effort is made to fabricate as much pipe as possible in the shop, there is still a considerable amount done in the field. The following power and hand tools speed up this work and improve the accuracy of the work.

Power Tools

Pipe Saws Pipe saws consist of a motor driving either a circular saw blade or a milling cutter through a friction clutch with a worm and gear train. The motor and gears of the saw are totally enclosed so that water that usually collects in a trench does not damage the working of the machine.

Pipe saws will cut cast iron and steel pipes used for sewage, gas, water, and other systems. They can be used for cutting pipes to length before laying and also for cutting pipes already laid. The cutting of finished pipe frequently has to be done under very difficult conditions, and it must be done as quickly as possible when fractures occur.

Cuts are smooth and clean. The ends of the pipes are free from flaws and cracks. The cutting tool does not dig in, and the saw requires little space in which to operate, Fig. 16-83.

Pipe-Beveling Machines Pipe ends that are to be welded require a smooth, metallically clean surface that is free of oxide. All types of bevels can be prepared economically so that they are metallically clean with the beveling machine, Fig. 16-84. Pipe-beveling machines are available as electric or pneumatic tools.

Fig. 16-84 A pipe-beveling machine preparing pipe with a heavy wall thickness. Edward R. Bohnart

Figures 16-85 and 16-86, page 512, are photographs of commercial oxyacetylene pipe cutting and beveling machines. These portable machines are built to cut and bevel pipe of any size with speed, economy, and accuracy, either in the field or in the shop. A welding student is using a commercial pipe-cutting machine in Fig. 16-87, page 512.

Pipe Welding and Shielded Metal Arc Welding Practice: Jobs 16-J1–J17 (Pipe) Chapter 16 511

Fig. 16-85 Cutting and beveling machine that can be used in both shop and field. An out-of-round attachment permits the torch carrier to follow the contour of the pipe. It may be operated manually or motor driven. *Mathey Dearman*

Fig. 16-87 Welding apprentice using a commercial pipe-beveling machine. The short pipe practice coupons can easily be beveled on both ends with this machine. Location: UA Local 400 *Mark A. Dierker/McGraw Hill*

Fig. 16-88 A portable power hacksaw has the advantages of being light and highly portable. This one is shown with 3-inch stainless-steel pipe. *Edward R. Bohnart*

Portable Power Hacksaw Power hacksaws, Fig. 16-88, are highly portable tools that make it possible to cut pipe at any location. They are electric powered and pneumatic powered for areas with explosion hazards. Saw blades are made of high speed steel and super high speed steel for stainless steel and tough ferrous metals. The machine can be used for bevel-cutting angles up to 45° on pipe 5 inches and smaller. Straight cuts can be made on pipe 12 inches in diameter. These machines are of particular value when space is limited and when it is difficult to cut materials by hand.

Fig. 16-86 Band-type cutting and beveling machine cuts out-of-round pipe. *Mathey Dearman*

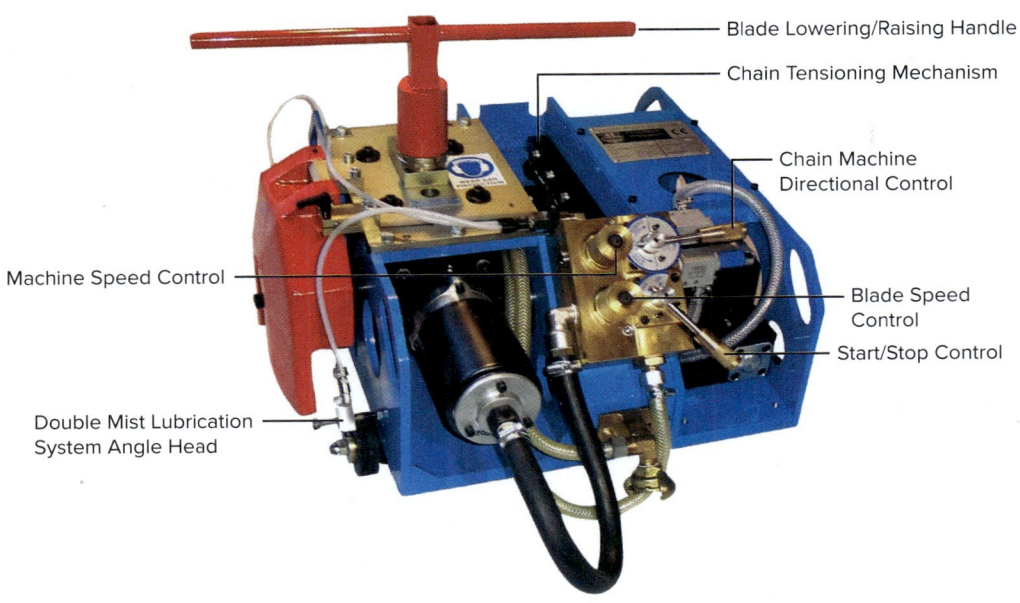

Fig. 16-89 Air-operated cutter and pipe saw. This tool cuts and bevels in one operation. Mathey Dearman

Cutter Pipe Saw The cutter pipe saw, Fig. 16-89, travels around the pipe and is adjustable to all pipe sizes 6 through 72 inches. The machine is held to the pipe by a tensioned timing chain. The chain acts as a flexible ring gear and provides the guide and feed for the machine. It can cut and bevel all grades of steel. The fact that it is powered by air motors makes it possible to use underwater. It can both cut and bevel for weld preparation in only one trip around the pipe.

Power Saw The power saw, Figs. 16-90, 16-91, and 16-92, page 514, can be electric or engine driven. With a diamond abrasive wheel, Fig. 16-91, it is highly useful in cutting cement pipe.

Fig. 16-90 A power saw for cutting various diameters and wall thicknesses of pipe. This is 8 inch schedule 5 stainless-steel pipe. Location: UA Local 400 Mark A. Dierker/McGraw Hill

Fig. 16-91 An abrasive power saw with a diamond cutting wheel. E.H. Wachs Co.

Pipe Welding and Shielded Metal Arc Welding Practice: Jobs 16-J1–J17 (Pipe)   Chapter 16   513

Fig. 16-92 Welding apprentice using a digital control panel to cut pipe practice coupons on a power feed vertical band saw. Location: UA Local 400 Mark A. Dierker/McGraw Hill

Fig. 16-93 A machine cutting tool being used to bevel pipe for both in the shop and in the field. It is capable of machining square, bevel and J's for various types of groove welds. Location: Piping System's Inc. Mark A. Dierker/McGraw Hill

These types of saws can be used to cut steel, cast iron, stainless steel, and alloy steel pipe as well as bar stock, structurals, and rail.

Bev-L Grinder® The Bev-L Grinder®, Fig. 16-93, is a fast, accurate grinding machine that produces clean-faced, uniform weld bevels without causing any change in the physical properties of the pipe. It is possible to get good line-up and fit in the field. The cutting action is fast, accurate, and easy to control. The operator simply rotates the grinder head around the spindle.

The grinder is available with an electric or pneumatic motor. It can be used on all steel, stainless steel, high alloy, and aluminum pipe from 3 to 18 inches in diameter.

Hand Tools

Hand Cutting Torch A great deal of pipe welding is done in maintenance work and in the renewing of piping systems. The standard oxyacetylene hand cutting torch is an ideal tool to take out sections of pipe that must be replaced. You are again urged to practice hand cutting as much as possible in order to develop and perfect your skill, Fig. 16-94.

Pipe Layout Tools A number of tools have been developed to make the job of laying out and fabricating pipe faster and with a greater degree of accuracy. These tools prevent layout mistakes and pipe that is out of alignment.

Contour Marker Figure 16-95 shows the contour marker being used to lay out a section of pipe. The marker prevents errors in layout and takes no longer than conventional layout tools, regardless of the type or size of the pipe joint to be welded. With an adaptor the four sides of the marker can be used on pipe up to 30 inches. With it the fabricator can work compound angles in one setting of the tool, and mark over nipples and old welds without having to break the wrap far back on wrapped pipe. The contour marker eliminates "cut and try" and the need for calculation.

Fig. 16-94 Welding apprentice using an oxyacetylene hand cutting torch to flame cut and bevel pipe. A great deal of field fabrication is cut with this method. Location: UA Local 400 Mark A. Dierker/McGraw Hill

Fig. 16-96 This level is "Quick" to accommodate for the typical two-hole alignment for the top dead center of the pipe. Also for vertical and horizontal positioning of the flange. Location: Piping System's Inc. Mark A. Dierker/McGraw Hill

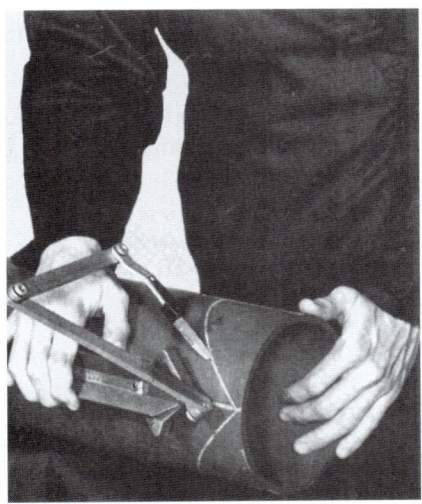

Fig. 16-95 Contour marker for marking off pipe joints and structural angle cuts. Edward R. Bohnart

Figure 16-96 shows the level being used to level a pipe flange. This ensures that flange holes will be level and in alignment with top dead center of pipe. It can be used on any size flange based on the size of the two-step plugs being used.

Circle-Ellipse Projector Figure 16-97 shows the circle-ellipse projector marking out a header to receive the branch. This tool is designed for marking perfect circles, ellipses, and oblongs. The tool can be used for work on pipe, tanks, boilers, cones, and any other irregular, flat, or round surface.

The projector has a strong magnetic base to hold it in position while the marking is being done. It has sliding rods with a protractor and locking lever to adjust it for the desired layout and a chalkholder for round or flat soapstone that will also receive a steel scribe or pencil.

The tool can be used on circles 1½ to 18 inches in diameter, and it can be adjusted from 0 to 70°.

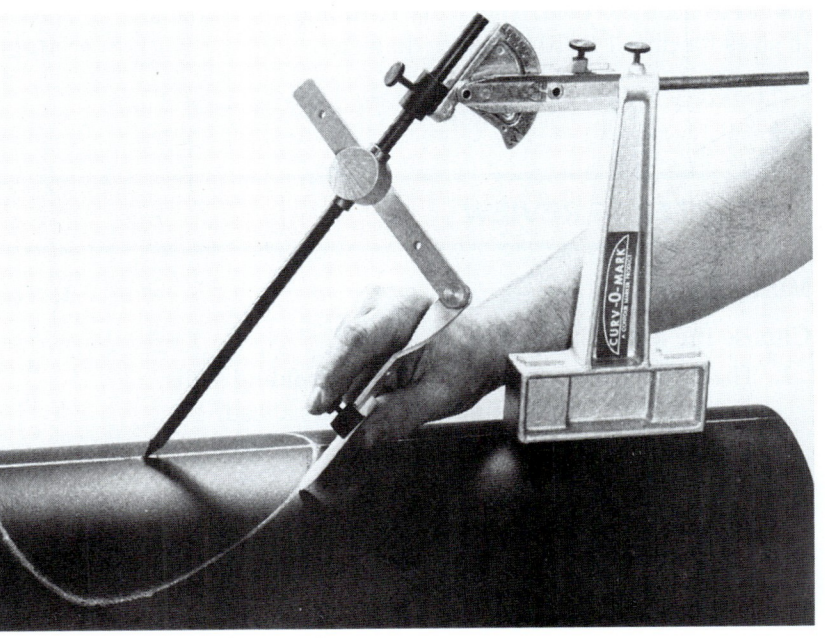

Fig. 16-97 Circle-ellipse protractor for marking circles, ellipses, and oblongs. Edward R. Bohnart

Fig. 16-98 Center head equipped with a dial-set level and center punch for locating the center line of pipe. *Edward R. Bohnart*

Fig. 16-99 Wrap-A-Round® for marking off square pipe ends. *Edward R. Bohnart*

Centering Head Figure 16-98 shows the use of the centering head to locate the center of the pipe. The tool determines the center line at any degree and measures the degree of declivity. It has a dial-set level and center punch.

Wrap-A-Round Figure 16-99 shows the Wrap-A-Round being used for marking around a pipe. It is also used as a straightedge. This device ensures square-cut pipe ends. It is made of flexible gasket material which is reasonably resistant to heat and cold. The Wrap-A-Round® is available in sizes for pipe 1 to 10 inches in diameter.

Miscellaneous Pipe Tools The following tools are also very useful to the pipe welder in the fabrication of pipe in the field:

- *Fitter welder protractors* determine branch angles.
- *Pipe-flange aligners* automatically align with the pipe axis and align pipe flanges quickly.
- *Radius markers* mark circles.
- *Magnetic protractor levels* solve angle and setting problems.
- *Multi-trammel heads* make large circles.

CHAPTER 16 REVIEW

Multiple Choice

Choose the letter of the correct answer.

1. There is a need for people with pipe-welding skills, and this need will continue in the years ahead. (Obj. 16-1)
   a. True
   b. False

2. The "big inch" pipe is generally considered pipe over what diameter? (Obj. 16-1)
   a. 10 in.
   b. 24 in.
   c. 30 in.
   d. All of these

3. Today piping is made out of many materials such as _____. (Obj. 16-1)
   a. Ferrous metals
   b. Nonferrous metals
   c. Nonmetallic materials (plastics, fiberglass, and composites)
   d. All of these

4. A power plant piping system may see pressures in excess of 2,500 p.s.i. and temperatures exceeding 1,000°F. (Obj. 16-1)
   a. True
   b. False

5. There are not many advantages to welding pipe versus using threaded mechanical fasteners. (Obj. 16-1)
   a. True
   b. False
6. A schedule 80 pipe has a thinner wall thickness than a schedule 40 pipe. (Obj. 16-1)
   a. True
   b. False
7. The 1/16 clearance is used on socket joints to _____. (Obj. 16-1)
   a. Get better penetration
   b. Make it easier to fit up
   c. Keep weld from cracking
   d. It is really not needed
8. A typical pipe end prep bevel is 37½°, which would make the groove angle 75°. (Obj. 16-1)
   a. True
   b. False
9. A pipe welding fitting can be _____. (Obj. 16-1)
   a. An elbow
   b. A tee
   c. A flange
   d. All of these
10. Which code-making organization is responsible for boilers and pressure vessels? (Obj. 16-2)
    a. ASME
    b. API
    c. AWWA
    d. AWS
11. Which standard-making organization is responsible for pipelines and related facilities? (Obj. 16-2)
    a. ASME
    b. API
    c. AWWA
    d. AWS
12. Which code-making organization is responsible for sanitary (hygienic) applications and structural pipe and tube? (Obj. 16-2)
    a. ASME
    b. API
    c. AWWA
    d. AWS
13. The purpose of a welding procedure is to make certain that welds made with it will have the appropriate _____ properties. (Obj. 16-2)
    a. Appearance
    b. Size
    c. Length
    d. Mechanical
14. *Essential variables* are those in which a change is considered to affect the properties of the weldments and shall require requalification of the welding procedure specification. (Obj. 16-2)
    a. True
    b. False
15. A welder qualification test is to determine a welder's ability to deposit sound weld metal. (Obj. 16-2)
    a. True
    b. False
16. The larger the diameter of pipe, the more difficult it is to weld because it takes so long to weld. (Obj. 16-2)
    a. True
    b. False
17. Changing the welding direction from vertical-up to vertical-down is not a major change, so it is not an essential variable for welder qualification. (Obj. 16-2)
    a. True
    b. False
18. Test specimens must be free of nicks or deep scratches and all tool marks should run lengthwise on the specimen. (Obj. 16-2)
    a. True
    b. False
19. The location of where the test specimens are taken from the pipe is not important. You can take out the weld specimen where it looks the best. (Obj. 16-2)
    a. True
    b. False
20. Visual inspection is a form of nondestructive testing. (Obj. 16-2)
    a. True
    b. False
21. Which of the following is *not* a destructive test? (Obj. 16-2)
    a. Tensile test coupons
    b. Ultrasonic test coupons
    c. Nick-break coupon
    d. Bend test coupon
22. When does a discontinuity become a defect? (Obj. 16-2)
    a. When the instructor does not like the way it looks
    b. If you drop the weld coupon on the floor and it breaks
    c. When its size or location exceeds the acceptance criteria
    d. When it is transverse to the weld direction
23. A narrow groove butt joint for automatic welding on heavy wall pipe may have a groove angle of _____. (Obj. 16-3)
    a. 5°
    b. 10°
    c. 20°
    d. 40°

24. The normal groove angle for pipe welding is from _____. (Obj. 16-3)
    a. 15 to 30°
    b. 30 to 37½°
    c. 40 to 60°
    d. 60 to 75°
25. How far back from the edges of the joint should the pipe be cleaned? (Obj. 16-3)
    a. 1 in.
    b. 4 in.
    c. Clean the whole pipe
    d. Not necessary to clean since the heat of the arc will burn off all the contaminants
26. Different root opening and root faces are required when doing vertical-up than when doing vertical-down techniques on pipe. (Obj. 16-3)
    a. True
    b. False
27. The electrode angle is *not* critical in pipe welding. (Obj. 16-3)
    a. True
    b. False
28. Stripper passes are used to _____. (Obj. 16-3)
    a. Fill in where you need a bigger root pass
    b. Build up weld areas when welding vertical-down usually in the 2 to 4 o'clock and 8 to 10 o'clock areas
    c. Build up on the cover pass so there will be more reinforcement
    d. Weld 2G position where additional fill is required
29. Branch and lateral connections are best done with manufactured fittings to reduce the amount of hand fitting on the job site. (Obj. 16-4)
    a. True
    b. False
30. Branch or lateral connections can be lap joints with fillet and/or groove welds depending on the design and procedure used. (Obj. 16-4)
    a. True
    b. False

Review Questions

Write the answers in your own words.

31. When welding a 6-inch schedule 40 pipe, how much undercut is allowable along the cover pass per the acceptance criteria from API 1104? (Obj. 16-2)
32. What is the acceptance criteria for melt-through on pipe ≥2⅜-inch outside diameter? (Obj. 16-2)
33. Define and describe porosity; how is it formed? (Obj. 16-2)
34. List the stages in building a pipeline. (Obj. 16-1)
35. Name at least five materials used for pipe. (Obj. 16-1)
36. Name four of the various code authorities that are concerned with pipe welding. (Obj. 16-2)
37. Draw a simple sketch showing the welding procedure used for the following pipe thicknesses and positions of welding: (Obj. 16-2)
    a. Horizontal roll position, butt joint, ¼- and ⅜-inch wall
    b. Horizontal fixed-position, butt joint, ⅜- and ½-inch wall
    c. Vertical fixed-position, butt joint, ⅜- and ½-inch wall
38. What is a hot pass? (Obj. 16-3)
39. Name at least six hand tools used by the pipe welder. (Obj. 16-1)
40. Name at least seven different methods of testing of pipe welds. (Obj. 16-2)

INTERNET ACTIVITIES

Internet Activity A

Search the AWS Web page and list the titles and specification numbers for all the standard welding procedures' specification for welding pipe with the SMAW process.

Internet Activity B

Using your favorite search engines check out how pipes are sized versus how tubes are sized. Note inside or outside diameters, wall thickness, and if there are differences if it is welded or seamless. Also note the weight per foot.

Design credits: Cinema and movie icon: Denis Meshkov/123RF; and Illustration of Welding icons: Mr. Nachai Sorasee/123RF

Table 16-11 Job Outline: Shielded Metal Arc Welding Practice (Pipe)

Job No.	Type of Joint	Type of Weld	Welding Position	Welding Technique	Pipe Specifications Dia. (in.)	Pipe Specifications Weight	Electrode Specifications Type[1]	Electrode Specifications Size (in.)	Polarity	Text References This Unit	Text References Other Jobs
16-J1	Pipe	Surface	Horizontal roll (1)	Surface stringer & weaved.	4–10	Schedule 40	E6010 E7018	3/32, 1/8, & 5/32	EP	513	13-J4 & J5
16-J2	Butt	V-Groove	Horizontal roll (1G)	1st pass stringer; 3 passes weaved.	6–10	Schedule 40	E6010 E7018	1P-1/8-6010 Others 5/32-6010 1/8-7018	EP	513	14-J33
16-J3	Butt	V-Groove	Vertical fixed (2G)	7 passes stringer.	6–10	Schedule 40	E6010 E7018	1 to 3P-1/8-6010 4 to 7P-5/32-6010 2 to 7P-1/8-7018	EP	514	15-J51
16-J4	Butt	V-Groove	Vertical fixed (2G)	7 passes stringer, Cover pass weave.	6–10	Schedule 40	E6010 E7018	1 to 3P-1/8-6010 4 to 7P-5/32-6010 Weave 5/32-6010 2 to 7P-1/8-7018	EP	514	13-J24
16-J5	Butt	V-Groove	Horizontal fixed (5G)	1st pass stringer; 2 passes weaved; travel up.	6–10	Schedule 40	E6010 E7018	1P-1/8-6010 2P-5/32-6010 3P-5/32-6010 2 to 3P-7018	EP	515	15-J43 & J55
16-J6	Butt	V-Groove	Horizontal fixed (5G)	1st pass stringer; 3 passes weaved; travel down.	6–10	Schedule 40	E6010	1P-1/8 2P-5/32 3-4P-3/16	EP	518	15-J48 & J55
16-JQT1[2]	Butt with backing	V-Groove	45° from horizontal fixed (6G)	Stringer passes.	6–8 M-1/P-1 group 1 or 2	Schedule 80	E7018	3/32 or 1/8 all passes	EP	520	NA
16-JQT2[3]	Butt without backing	V-Groove	45° from horizontal fixed (6G)	Stringer passes.	6–8 M-1/P-1 group 1 or 2	Schedule 80	E6010 E7018	1P-1/8-6010 Subsequent passes 3/32-1/8-7018	EP	521	NA
16-JQT3[4]	Butt without backing	Bevel-Groove	45° from horizontal fixed with restricting (6GR)	Stringer passes.	12 M-1/P-1 group 1 or 2	Schedule 80	E6010 E7018	1P-1/8-6010 Subsequent passes 3/32-1/8-7018	EP	521	NA
16-J7	90° branch	Groove fillet	Top (branch)	Multipass stringer.	4–10 Small to large. Size on size.	Schedule 40	E6010	1/8 & 5/32	EP	522	14-J33 13-J25
16-J8	90° branch	Groove fillet	Horizontal (branch)	Multipass: 1st pass stringer; others weaved.	4–10 Small to large. Size on size.	Schedule 40	E6010	1/8 & 5/32	EP	522	15-J43 14-J36
16-J9	90° branch	Groove fillet	Bottom (branch)	Multipass: Last pass weaved.	4–10 Small to large.	Schedule 40	E6010	1/8 & 5/32	EP	522	15-J55 13-J24

(Continued)

Table 16-11 (Concluded)

Job No.	Type of Joint	Type of Weld	Welding Position	Welding Technique	Pipe Specifications Dia. (in.)	Pipe Specifications Weight	Electrode Specifications Type[1]	Electrode Specifications Size (in.)	Polarity	Text References This Unit	Text References Other Jobs
16-J10	45° branch	Groove fillet	Top (branch)	Multipass stringer.	4–10 Small to large. Size on size.	Schedule 40	E6010	1/8 & 5/32	EP	522	14-J33 13-J25
16-J11	45° branch	Groove fillet	Horizontal (branch)	Multipass: 1st pass stringer; others weaved.	4–10 Small to large. Size on size.	Schedule 40 or 80	E6010	1/8 & 5/32	EP	522	15-J43 14-J36
16-J12	45° branch	Groove fillet	Bottom (branch)	Multipass: Last pass laced.	4–10 Small to large. Size on size.	Schedule 40	E6010	1/8 & 5/32	EP	522	15-J55
16-J13	45° L	Groove fillet	Horizontal fixed	Multipass: 1st pass stringer; others weaved.	4–10	Schedule 40	E6010	1/8 & 5/32	EP	522	15-J43-J55
16-J14	90° L	Groove fillet	Horizontal fixed	Multipass. 1st pass stringer; others weaved.	4–10	Schedule 40	E6010	1/8 & 5/32	EP	522	15-J43-J55
16-J15	Blunt pipe head	Groove fillet	Horizontal fixed	Multipass.	4–6	Schedule 40	E6010	1/8 & 5/32	EP	522	15-J43-J55
16-J16	Orange peel head	Groove fillet	Horizontal fixed	Multipass.	4–6	Schedule 40	E6010	1/8 & 5/32	EP	522	14-J33 15-J43-J55
16-J17	90° Y	Groove fillet	Fixed	Multipass.	4–6	Schedule 40	E6010	1/8 & 5/32	EP	522	14-J33 15-J43-J55

Notes:

At this point the student should hand-cut, bevel, and weld several pipe joints in all three positions.

The student should be able to pass bend tests of pipe coupons cut from pipe welded in the horizontal fixed (5G) and vertical fixed positions (2G).

The student should be able to produce acceptable pipe test pieces by using manual and machine oxyacetylene and plasma arc cutting equipment.

At this point the student should repeat those jobs on which additional practice is necessary, using pipe of different diameters and wall thicknesses. The student is now ready to take the official API or ASME Code Certification Test for Carbon Steel Pipe. Welding is to be in the horizontal fixed (5G) and the vertical fixed positions (2G).

Students who qualify on standard carbon steel pipe for SMAW may take additional training in SMAW of heavy, extra heavy, and alloy pipe and in GTAW and GMAW of alloy and aluminum pipe.

[1]E7010 electrodes should also be used for practice.

E6011 electrodes should be used with alternating current.

E7018, E9018, and E10018 electrodes should be used for low alloy steel pipe.

[2]For additional information follow AWS WPS ANSI/AWS B2.1-1-208 for carbon steel pipe.

[3]For additional information follow AWS WPS ANSI/AWS B2.1-1-201 for carbon steel pipe.

[4]For additional information follow AWS WPS AWS3-SMAW-1.

UNIT 3

Arc Cutting and Gas Tungsten Arc Welding

Chapter 17

Arc Cutting Principles and Arc Cutting Practice: Jobs 17-J1–J7

Chapter 18

Gas Tungsten Arc and Plasma Arc Welding Principles

Chapter 19

Gas Tungsten Arc Welding Practice: Jobs 19-J1–J19 (Plate)

Chapter 20

Gas Tungsten Arc Welding Practice: Jobs 20-J1–J17 (Pipe)

17

Arc Cutting Principles and Arc Cutting Practice:

Jobs 17-J1–J7

In the majority of fabricating shops and in the field, the welder must be able to do manual arc cutting, Fig. 17-1. The various arc cutting devices have become universal tools. They are widely used in foundries for the removal of risers from castings, for cleaning castings, and as a means of scrapping obsolete metal structures. Cutting devices are also used in the fabrication of metal structures, Fig. 17-2.

Cutting processes have made it possible to fabricate structures requiring heavy thicknesses of metal from rolled metal. Formerly, these structures had to be cast. The combination of cutting and welding processes created an industry devoted to the fabrication of heavy machinery and equipment from rolled metal. Arc cutting increases the speed of fabrication and eliminates many costly joining, shaping, and finishing operations.

Arc Cutting

Arc cutting processes melt metal along a desired line of cut with the heat generated by an electric arc. A number of the processes also use oxygen, compressed air, or one of the inert gases in addition to the arc. Several of the arc processes compare favorably with oxyfuel gas cuts in quality of cut. The primary advantage of arc cutting is that it can be used on all types of metals. Some of its applications include cast iron, scrap, aluminum, magnesium,

Chapter Objectives

After completing this chapter, you will be able to:

17-1 Describe arc cutting principles.

17-2 Identify plasma arc cutting (PAC) techniques.

17-3 Identify air carbon arc cutting (CAC-A) techniques.

17-4 Perform visual inspection of cuts and gouges.

17-5 Describe equipment setup and use.

17-6 Perform troubleshooting of cut and gouge quality.

17-7 Perform straight line, curve square cuts, bevel cuts, and gouges on a variety of shapes and metals with PAC.

17-8 Perform gouges with CAC-A.

Fig. 17-1 Welding students being instructed in the proper use of a PAC torch in a lab. Prographics

Plasma Arc Cutting (PAC)

Plasma arc cutting, Fig. 17-3, is similar in many respects to the gas tungsten arc (TIG). This process was introduced by the Linde Division of Union Carbide in 1955. Both automated and manual equipment produce economical, high quality, ready-to-weld cuts.

An oxyacetylene cutting torch cannot cut aluminum, magnesium, or stainless steels because these metals form oxides when exposed to oxygen. The oxides resist further oxidation—the basis of oxyacetylene cutting. The plasma arc process, Fig. 17-4, page 524, can cut these metals because the arc stream is much hotter than the melting temperatures of both the metals and their oxides. It is also a high speed process, Fig. 17-5, page 524.

The only requirement for plasma cutting is that the metal being cut must be able to conduct electricity. By blowing out molten metal, the forceful plasma cutting jet forms the kerf. However, the kerf can be 1½ to 2 times the width of a kerf made with oxy-fuel gas cutting (OFC).

Plasma is the fourth state of matter; the others are gas, liquid, and solid. Unlike gas, plasma is ionized so that it can conduct electric current. (An *ion* is an atom or group of atoms that has lost or gained electrons so that it can carry a positive or negative electrical charge. High voltage or high temperature can ionize a gas.) Thus, the purpose of the gas used in the plasma arc process is different from

Fig. 17-2 Plasma arc cutting a thin wall stainless-steel pipe that will be used in the food processing industry. Flame retardant material has been placed inside the pipe to protect the inner surfaces from the molten dross of the cut. Note the gloves protecting from heat and flying sparks. Miller Electric Mfg. Co.

Fig. 17-3 Heavy cut in carbon steel with the plasma arc cutting process. Various sizes of stiffener plates are being cut automatically. Accuracy is to +0.002 inches. Team Industries, Inc.

titanium, copper, carbon steel, and stainless steels. Arc cutting equipment is also used for hole piercing, rivet cutting, gouging, and other special uses.

Arc cutting is done with the plasma arc, the carbon arc, and the metal arc.

Fig. 17-4 Plasma arc cutting makes shape and line cuts on hard-to-work metals such as this 1¾-inch stainless-steel plate. Team Industries, Inc.

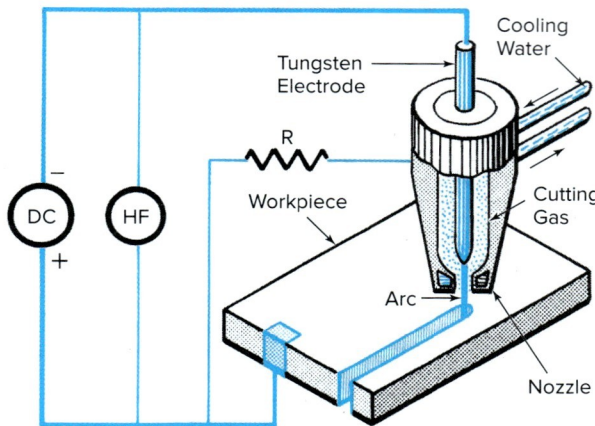

Fig. 17-6 Schematic diagram of plasma arc cutting equipment.

Fig. 17-5 Cutting various branch connections like T, K, and Y in pipe with the plasma process. Pipe diameter 3–48 inches can be accommodated. This machine can use plasma or oxyfuel for cutting. Equipment conveys pipe from outside storage rack to the cutting machine. The cutting head is a CNC controlled with a 5-axis and a touch sensitive screen for ease and accuracy of production requirements. Team Industries, Inc.

that used in the oxyacetylene process and the inert gas processes (TIG and MIG/MAG). Instead of producing a flame for cutting the work or shielding the operation from the atmosphere, the gas is superheated so that it becomes a conductor and can actually maintain an electric arc.

Because maximum transfer of heat to the work is essential, plasma arc torches use a *transferred arc* for cutting in which both the material being cut and the torch act as electrodes in the electric circuit, Fig. 17-6. The work is thus subjected to plasma arc heat.

In the plasma arc torch, the tip of the electrode is located within the nozzle. The nozzle has a relatively small opening that constricts the arc. The high pressure gas must flow through the electric arc where it is heated to the plasma temperature of approximately 25,000°F, a temperature far higher than that of any flame. The expanding gas as a result of the heat is forced through the small opening in the nozzle where it emerges in the form of a sonic jet at high velocity. The hot jet of gas can melt any known metal. The high velocity gas blasts the molten metal through the kerf to produce a high quality cut, which is free of metallic oxides or dross. The cut is done at high travel speed which reduces the heat input as compared with OFC. Argon, nitrogen, air, oxygen, and nitrogen-hydrogen and argon-hydrogen mixtures are used in the plasma arc process.

For video of how PAC works, please visit www.mhhe.com/welding.

Good ventilation should be provided when doing plasma arc cutting to remove all fumes from the area of the cutting action. Because of the cutting speed and large volume of metal removed, PAC produces large quantities of fumes. It is not just the metal fumes but certain coating fumes such as the zinc found on galvanized steel or cadmium coated parts as examples that need to be guarded against.

Mechanized Plasma Arc Cutting The equipment that is needed for the mechanized plasma arc cutting system includes a torch, a cutting control box, gas regulators, power supply, a carriage unit, a supply of cutting gases, and water.

Torch The typical torch can be used for cutting all metals. It is water cooled and may be equipped with a variety of

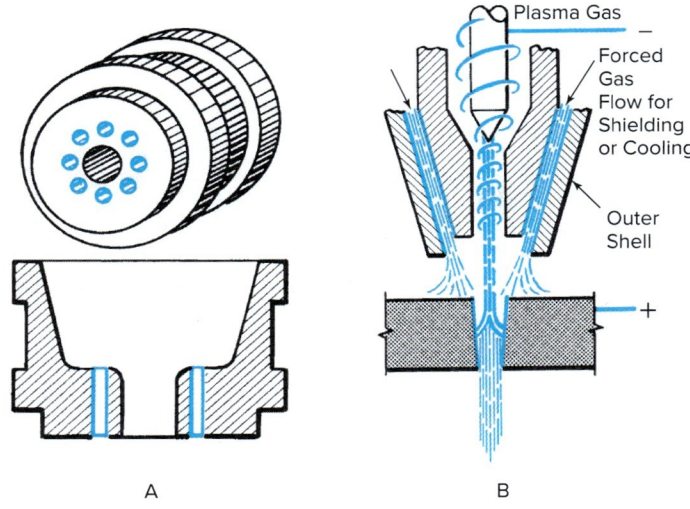

Fig. 17-7 (A) Multiport plasma arc cutting nozzle. (B) Dual flow plasma arc cutting nozzle.

Fig. 17-8 Control unit for the mechanized plasma arc cutting head. ESAB

nozzles to permit the use of different gases. The amount of current also affects nozzle selection. Single-port nozzles, Fig. 17-6, are generally used with argon-hydrogen and nitrogen-hydrogen mixtures, but multiport and dual flow nozzles, Fig. 17-7, produce better results when compressed air or oxygen is the plasma gas, orifice gas, or cutting gas. Torches can be supplied by two gases, one as the plasma gas, orifice gas, or cutting gas and another as the shielding gas, or in the case of air plasma cutting, a gas to help cool the torch head.

PAC torches operate at extremely high temperatures, and various parts of the torch must be considered to be consumable. The electrode and tip (restricting orifice) are vulnerable to wear during cutting. Cutting performance will deteriorate as they wear. Current plasma torches have self-alignment and self-adjusting consumable parts. If they are assembled with the manufacturer's specified parts and following the manufacturer's instructions, no further adjustment should be required. Periodic inspection of the other torch parts such as shielding cups, insulators, seals (O-rings), swirl ring, and shields should be performed per the manufacturer's maintenance recommendations.

Controls The control unit, Fig. 17-8, provides the sequence of operations and control of all functions such as arc starting, varying gas flow, varying power level, carriage travel, and flow of water.

Interlocks are used with PAC systems. An inadequate supply of plasma gas may cause internal arcing and damage to the torch. Pressure interlock switches will shut the system down if the gas systems are not adequate. The same also pertains to coolant systems for water-cooled torches. The PAC system will not function if these interlock switches are not properly supplied.

Environmental Controls A tremendous amount of noise and fumes are generated with high powered mechanized plasma arc cutting. A common approach to overcome noise and fumes is to cut over a water table and surround the arc with a water shroud. This requires a cutting table filled with water up to the work-supporting surface, a water shroud attachment to the torch, and a circulating pump to circulate filtered water from the water table through the shroud.

Another method is to completely submerge the part to be cut under approximately 3 inches of water. Thus, the working end of the torch and cut are completely submerged. While the torch is under water and not cutting, a flow of compressed air prevents water from entering the torch. Coloring agents can be added to water to reduce glare.

Electrodes Special tungsten electrodes, held in place by a collet, are used in the cutting torch. Electrode type, shape, and location are critical for proper operation.

Power Supply Direct current, electrode negative constant current (drooper) type power sources are used. A power source with an open circuit voltage range of 150 to 400 volts is required, Fig. 17-9, page 526. For heavy cutting, a machine capable of producing 400 open-circuit volts and 200 kilowatts or more may be necessary. Both transformer-rectifier and inverter type power units are available. These units may also be connected in series to meet higher voltage requirements.

Regulators Gas pressure regulators and flowmeters are required for controlling the flow of the plasma and shielding gas. See Chapter 5, pages 151–153. All gas goes through the main port of the cutting torch at 60 to 350 cubic feet per hour.

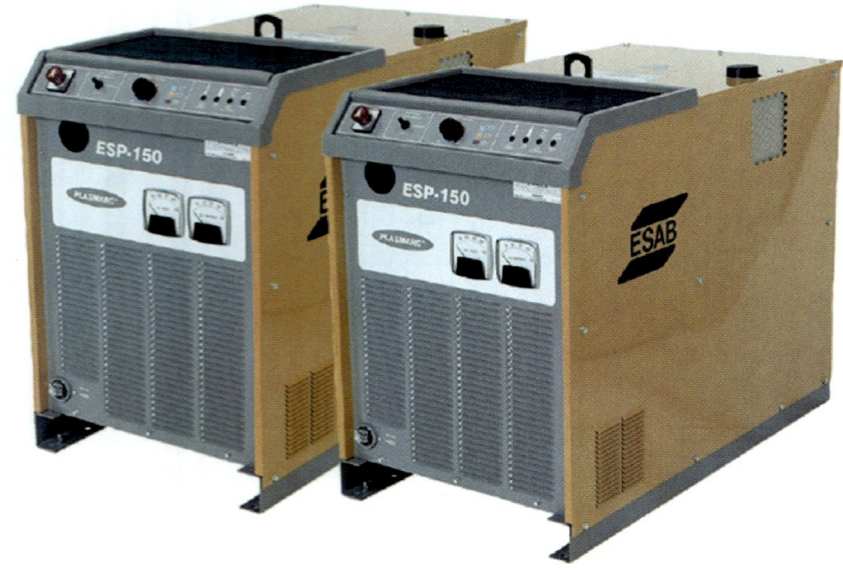

Fig. 17-9 Transformer-rectifier power supply with a total capacity of 4-inch thick cutting (370 volts d.c., open circuit). ESAB

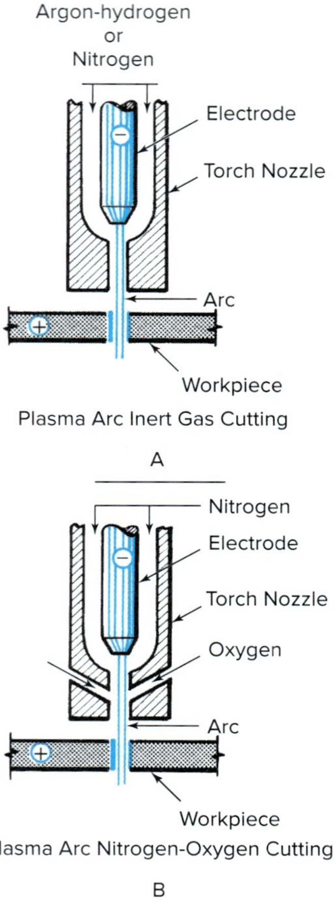

Fig. 17-10 (A) Plasma arc inert gas cutting. (B) Plasma arc nitrogen-oxygen cutting.

Cutting Gases Aluminum, stainless steels, and other nonferrous metals require a nonoxidizing gas for cutting, such as a mixture of argon-hydrogen or nitrogen, Fig. 17-10A. Carbon steel, cast iron, and certain alloy steels require an oxidizing gas that provides additional heat from the iron-oxygen reaction at the cutting point. Separately supplied nitrogen, oxygen, or compressed air may be used for these metals. The life of electrodes operating in oxygen is short. Oxygen is usually supplied as an outer sheath surrounding a nitrogen plasma, Fig. 17-10B. Typical cutting conditions are given in Tables 17-1 through 17-3, pp. 528–529.

Water supply Water at high pressure and a high rate of flow is necessary to dissipate the heat generated in the torch. Usually it is necessary to install a circulatory system in order to provide an adequate flow.

Manual Plasma Arc Cutting Manual plasma arc cutting, Fig. 17-11, is used for workpieces that cannot be adapted to a mechanized setup, for remote locations, and for specialized work on odd-shaped pieces. Typical manual cutting capabilities include $1\frac{1}{4}$-inch thick stainless steel and 1-inch thick aluminum. The equipment needed includes a manual torch, supply of electrodes, and power source with control unit. The electrodes and power source for the mechanized process may also be used for manual cutting.

Torch An air-cooled torch with a 100-ampere capacity is a popular model. It has a right-angle head. The pilot arc is established by pressing the switch on the torch handle. The cutting arc is established when the torch is brought to within $\frac{1}{2}$ inch of the workpiece. The arc is immediately extinguished when the welder releases the torch switch. Some manual torches are water cooled. Because of the high voltage nature of the PAC process, water leaks or moisture on or around the equipment should be prevented prior to operations.

Plasma Arc Starting with High Frequency When the PAC power source is turned on and the trigger or torch switch is closed or turned on, there will be approximately 2 to 3 seconds for preflow of gas or air before the pilot arc starts. The pilot arc is an arc between the electrode and the torch tip. This pilot is a nontransferred, noncutting arc. When the cutting gas or air reaches the pilot arc, it is superheated to over 25,000°F. When this pilot arc is brought into close proximity to the workpiece, the electric circuit is completed. This is referred to as the transferred or cutting arc. A pilot arc relay will open and shut the pilot power off. Figure 17-12 represents the transferred and nontransferred high frequency starting method. The most common pilot arc starting method has been to strike a high frequency spark between the electrode and the torch tip. The pilot arc is created by high frequency, which is a high voltage produced by a transformer and a spark

Fig. 17-11 Cutting a carbon steel pate manually with the plasma arc process and very portable power source. The process has the same flexibility as oxyacetylene without the large cumbersome cylinders. Note the protective clothing and fall restraint.
Miller Electric Mfg. Co.

gap oscillator. This device is referred to as an arc starter (high frequency generator). A major concern is that the high frequency may interfere with telephones, computers, or machine [computer numeric controls (CNC), etc.] controls if the system is not properly installed. See Chapter 18 as this is the same type of high frequency used in the gas tungsten arc welding (GTAW) process.

Plasma Arc Starting with Contact Starts The contact start pilot arc (without using high frequency) is formed between the nozzle and the inner electrode. This is created when the electrode is in contact with the torch tip and retracts when the trigger is pulled. The airflow pulls or forces the electrode to retract while electricity is flowing through the electrode. The narrow opening or orifice of the torch tip accelerates the expanding plasma toward the workpiece. When the torch and workpiece are close enough, the pilot arc crosses or "jumps" the gap, which then allows the main electric plasma current to follow. This then becomes the plasma or columnar transferred cutting arc.

Verify the type of starting system on the PAC equipment you will be using.

Control Unit The control unit generally built into the power source for manual cutting regulates the electrical, gas, and water supplies. Separate flowmeters may be provided for argon and hydrogen. However, most manual cutting is done with compressed air. Solenoid valves provide for the flow of pure argon to the torch to establish the pilot arc at the start of the cutting cycle.

Cutting Gases The manual torch is designed to use compressed air as a plasma gas and secondary cooling gas. A mixture of 80 percent argon and 20 percent hydrogen

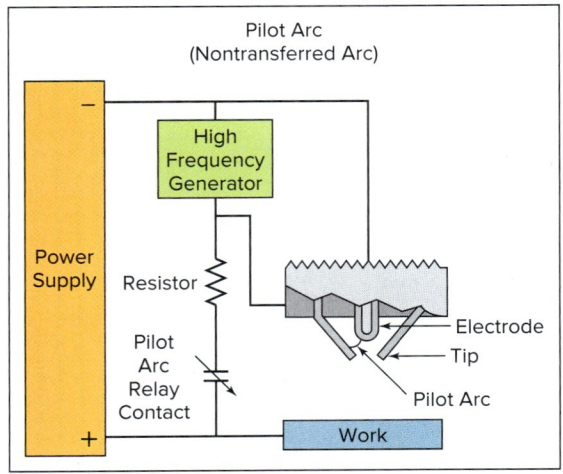
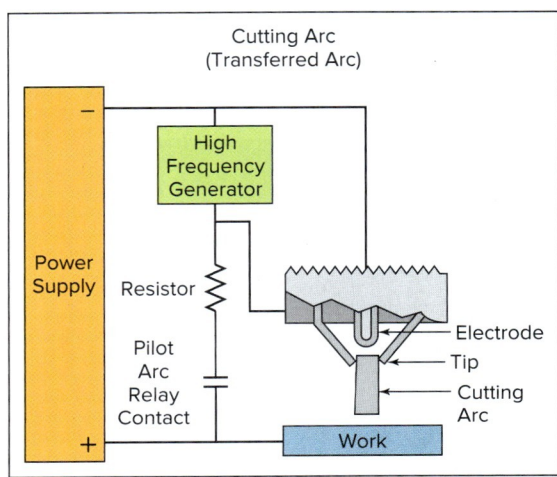

Fig. 17-12 Pilot and cutting arc. Source: Miller Electric Mfg. Co.

Arc Cutting Principles and Arc Cutting Practice: Jobs 17-J1–J7 **Chapter 17**   **527**

may be used based on cut quality, brightness, fume generation, and cost. The secondary or cooling gas, when used, can be argon or nitrogen. Since manual PAC is generally done with air, a compressor external to the power source or special power sources with air compressor built in are available. Other gases can be supplied premixed in cylinders or as individual gases supplied to a gas mixer device.

Advantage of Plasma Cutting Both the mechanized and manual plasma cutting processes produce economical, high speed, ready-to-weld cuts. They are intended to replace less efficient, slower methods such as sawing, powder-cutting, and oxyacetylene cutting on some applications. The plasma arc process has the following advantages:

- Dross-free cuts on carbon and stainless steels, nickel, Monel, Inconel, cast iron, clad steels, aluminum, copper, and magnesium
- Clean cuts on most metals up to 5 inches thick
- Precision cuts with a narrow kerf (especially when high current-density equipment is used)
- Minimum heat-affected zone
- Cutting speeds up to 300 inches per minute (many times faster than oxyacetylene cutting)
- Cuts of such quality that machining or finishing is not needed in many cases
- There is almost no distortion of metals when using the plasma arc cutting processes. There is no bowing or cambering. Except for a microscopically thin layer at the cut surface, magnetic permeability and hardness are little affected.

Plasma arc cutting is finding increased use as a fabricating tool in the transportation, nuclear power, and chemical industries. It is also used in the forging and casting industry for removing risers and gates. The stack cutting of several sheets of 1/16- to 1/4-inch thickness is possible. In addition to cutting, the process can be used for the piercing of holes and gouging, including pad washing and scarfing.

Cutting Carbon Steel The plasma arc processes produce dross-free cuts in carbon steel with smooth surfaces and sharp edges. No preheating is required. Stack cutting of sheets produces cuts comparable to those obtained when cutting one sheet of equal thickness. The edges are not fused when cutting carbon steel. Net cost per foot compares favorably with other processes in thicknesses up to 2 inches.

While it is possible to obtain fairly good results if inert gases are used to cut carbon steel, superior results are obtained when nitrogen and oxygen are used. The cutting speed is higher than that for oxyacetylene cuts, thus resulting in less heat input. See Table 17-1.

Cutting Stainless Steel Completely dross-free plasma arc cuts in stainless steel up to 2 inches thick have eliminated the need for further finishing. Radiographic quality welds can be produced without further cleaning of cut surfaces. Cut quality of high strength alloys, including those with a high nickel or cobalt content, is similar to those of stainless steel.

Stainless steel and nonferrous metals are generally cut with mixtures of argon and hydrogen or with nitrogen mixtures. See Table 17-2.

Cutting Aluminum Plasma arc cutting methods provide equal or better quality at much faster speeds than other cutting methods. Prior to its development, other flame cutting attempts were unsatisfactory on a production basis. Plasma arc cuts in aluminum are dross-free for thicknesses up to 5 inches. Excellent cuts are also obtained on magnesium at higher cutting speeds than aluminum. See Table 17-3.

Table 17-1 Typical Conditions for Plasma Arc Cutting of Carbon Steel

Thickness		Speed		Orifice Diameter[1]		Current (DCEN) (A)	Power (kW)
in.	mm	in./min	mm/s	in.	mm		
1/4	6	200	86	1/8	3.2	275	55
1/2	13	100	42	1/8	3.2	275	55
1	25	50	21	5/32	4.0	425	85
2	51	25	11	3/16	4.8	550	110

[1]Plasma gas flow rates vary with orifice diameter and gas used from about 200 ft³/h (94 L/min.) for a 1/8-in. (3.2 mm) orifice to about 300 ft³/h (104 L/min.) for a 3/16-in. (4.8 mm) orifice. The gases used are usually compressed air, nitrogen with up to 10% hydrogen additions, or nitrogen with oxygen added downstream from the electrode (dual flow). The equipment manufacturer should be consulted for each application.

Table 17-2 Typical Conditions for Plasma Arc Cutting of Stainless Steels

Thickness		Speed		Orifice Diameter[1]		Current (DCEN) (A)	Power (kW)
in.	mm	in./min	mm/s	in.	mm		
1/4	6	200	86	1/8	3.2	300	60
1/2	13	100	42	1/8	3.2	300	60
1	25	50	21	5/32	4.0	400	80
2	51	20	9	3/16	4.8	500	100
3	76	16	7	3/16	4.8	500	100
4	102	8	3	3/16	4.8	500	100

[1]Plasma gas flow rates vary with orifice diameter and gas used from about 100 ft³/h (47 L/min.) for a 1/8-in. (3.2 mm) orifice to about 200 ft³/h (94 L/min.) for a 3/16-in. (4.8 mm) orifice. The gases used are nitrogen and argon with hydrogen additions from 0 to 35%. The equipment manufacturer should be consulted for each application.

Table 17-3 Typical Conditions for Plasma Arc Cutting of Aluminum Alloys

Thickness		Speed		Orifice Diameter[1]		Current (DCEN) (A)	Power (kW)
in.	mm	in./min	mm/s	in.	mm		
1/4	6	300	127	1/8	3.2	300	60
1/2	13	200	86	1/8	3.2	250	50
1	25	90	38	5/32	4.0	400	80
2	51	20	9	5/32	4.0	400	80
3	76	15	6	3/16	4.8	450	90
4	102	12	5	3/16	4.8	450	90
6	152	8	3	1/4	6.4	750	170

[1]Plasma gas flow rates vary with orifice diameter and gas used from about 100 ft³/h (47 L/min.) for a 1/8-in. (3.2 mm) orifice to about 250 ft³/h (120 L/min.) for a 1/4-in. (6.4 mm) orifice. The gases used are nitrogen and argon with hydrogen additions from 0 to 35%. The equipment manufacturer should be consulted for each application.

Air Carbon Arc Cutting

The air carbon arc cutting (CAC-A) process is a method of cutting and gouging by melting the work with an electric arc and blowing away the molten metal with a strong jet of compressed air. Because the process does not depend on oxidation, it can be used on virtually all metals. The metals are not harmed by the process but a small heat affected zone is produced.

For video of CAC-A in operation, please visit www.mhhe.com/welding.

An arc is struck between the carbon electrode and the metal to be cut. The metal melts instantly, and high velocity jets of air blast the molten metal away. The air blast is continuous and directed through and behind the point of arcing. The electrode is pushed forward at a rapid rate. The travel speed depends on the size of electrode, type of material, amperage, and air pressure. The depth and contour of the groove is controlled by the electrode angle, travel speed, and current. The width is determined by the size of the electrode. Two plates may be grooved simultaneously, Fig. 17-13, page 530.

The air carbon arc process is commonly used in steel foundries to remove defects from castings. It is also used a good deal in metal fabrication. The process can remove weld defects, clean out the roots of welds, widen grooves that have pulled together during welding, and make U-grooves in plates. As a maintenance process, it is used by the railroads to gouge out cracks before damaged railroad cars and tracks are welded. In refineries it cuts stainless-steel pipe, removes stainless-steel liners, and patches tank bottoms. Air carbon arc cutting also prepares metal parts for hard-facing and repair welding. The process can be used on all carbon, manganese, stainless steels, copper and nickel alloys, cast iron, and other hard-to-cut metals.

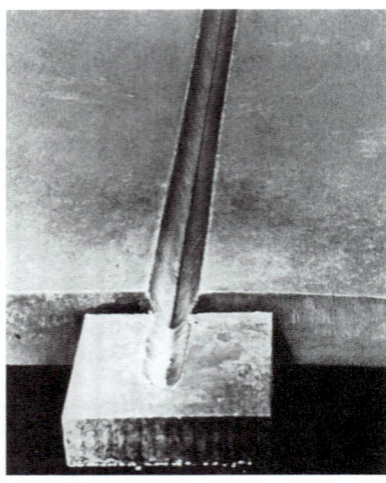

Fig. 17-13 Nature of the gouge that is possible with the air carbon arc torch. Both plates were set up as a square groove butt joint, and a U-groove was cut into the joint. The groove is ready for welding. *Thermadyne Industries, Inc.*

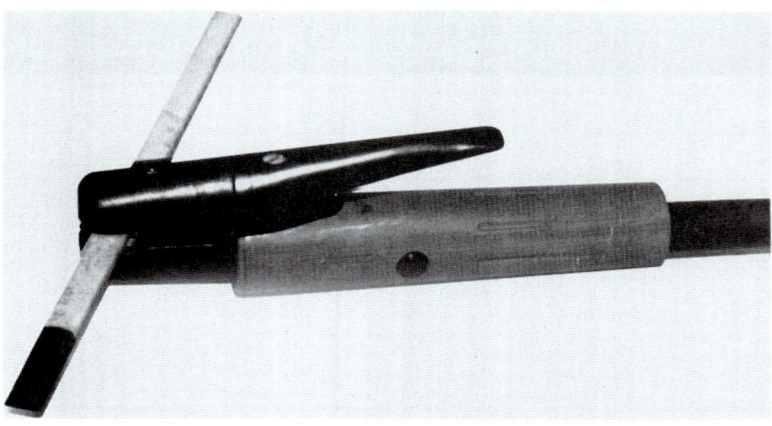

Fig. 17-14 This air carbon arc torch has a flat electrode that makes it possible to cut with more precision. *Thermadyne Industries, Inc.*

The air carbon arc electrode is made of carbon and graphite, and it is coated with copper. Sizes range from $5/32$ to 1 inch. A flat electrode, Fig. 17-14, allows greater flexibility. Table 17-4 lists types of current and polarities for different metals.

The compressed air passes through holes in the electrode holder that directs it parallel to the electrode. The electrode holder contains an air control valve and a cable that carries both the current and air. The holder may be air or water cooled. The cable is connected to a welding machine and a source of compressed air.

The regular constant current welding machine may be used. See Tables 17-5 through 17-7 (pp. 531–532). Direct current electrode positive polarity is used for most applications, but an electrode for alternating current is available. There is equipment for manual, machine, and automatic operation. Figure 17-15, page 532, illustrates mechanized equipment in operation.

Ordinary compressed air is supplied by a compressor. Most applications require from 80 to 100 p.s.i. for heavy duty and light duty manual cutting with as little as 40 p.s.i. at a flow rate over 80 cubic feet per hour, depending upon the thickness of the material. The air pressure must be of such volume that a clean, dross-free surface is ensured.

Oxygen Arc Cutting

Oxygen arc cutting (OAC) is a method of cutting, piercing, and gouging metals with an electric arc and a stream of oxygen.

Table 17-4 Electrode and Current Recommendations for Air Carbon Arc Cutting of Several Alloys

Alloy	Electrode Type	Current Type	Remarks
Carbon, low alloy, and stainless steels	d.c.	dcrp	
	a.c.	ac	Only 50% as efficient as dcrp
Cast irons	a.c.	dcsp	At middle of electrode current range
	a.c.	ac	
	d.c.	dcrp	At maximum current only
Copper alloys:			
copper 60% or less	d.c.	dcrp	At maximum current
copper over 60%	a.c.	ac	
Nickel alloys	a.c.	ac	
	a.c.	dcsp	
Magnesium alloys	d.c.	dcrp	Before welding, surface must be cleaned.
Aluminum alloys	d.c.	dcrp	Electrode extension should not exceed 4 in. (100 mm). Before welding, surface must be cleaned.

Table 17-5 Ampere Settings (a.c. or d.c.) and Oxygen Pressures

Chrome Nickel, Straight Chrome, Monel Nickel			Brass, Bronze, Copper			Cast iron			Aluminum			Low Alloy (Alloy Content Less than 12%)			Carbon Steel and Low Alloy High Tensile		
Thickness	Amps	Oxygen Pounds per Square Inch	Thickness	Amps	Oxygen Pounds per Square Inch	Thickness	Amps	Oxygen Pounds per Square Inch	Thickness	Amps	Oxygen Pounds per Square Inch	Thickness	Amps	Oxygen Pounds per Square Inch	Thickness	Amps	Oxygen Pounds per Square Inch
1/4	175	3–5	1/4	180	10–15	1/4	180	10	1/4	200	30	1/4	175	30–35	1/4	175	75
1/2	185	5–10	1/2	185	10–15	1/2	185	10–15	1/2	200	30	1/2	180	35–40	1/2	175	75
3/4	195	10–15	3/4	190	15–20	3/4	190	15–20	3/4	200	30	3/4	190	40–45	3/4	175	75
1	200	15–20	1	200	20–25	1	200	20–25	1	200	30	1	200	45–50	1	175	75
1 1/4	210	20–25	1 1/4	210	25–30	1 1/4	210	25–30	1 1/4	200	35	1 1/4	205	50–55	1 1/4	200	75
1 1/2	215	25–30	1 1/2	215	30–35	1 1/2	215	30–35	1 1/2	200	35	1 1/2	210	55–60	1 1/2	200	75
1 3/4	220	35–40	1 3/4	220	35–40	1 3/4	220	35–40	1 3/4	200	40	1 3/4	215	60–65	1 3/4	200	75
2	220	40–45	2	225	40–45	2	225	40–45	2	200	40	2	220	65–70	2	200	75
2 1/4	225	45–50	2 1/4	225	45–50	2 1/4	225	45–50	2 1/4	175	45	2 1/4	225	70–75	2 1/4	225	75
2 1/2	225	50–55	2 1/2	230	50–55	2 1/2	230	50–55	2 1/2	175	45	2 1/2	225	75	2 1/2	225	75
2 3/4	230	55–60	2 3/4	230	55	2 3/4	230	55–60	2 3/4	175	45	2 3/4	230	75	2 3/4	225	75
3	230	60	3	235	55	3	235	60	3	175	45	3	230	75	3	225	75

Source: Arcos Corp.

Table 17-6 Power Sources for Air Carbon Arc Cutting and Gouging

Type of Current	Type of Power Source	Remarks
d.c.	Constant current motor-generator, rectifier, or resistor grid unit	Recommended for all electrode sizes.
d.c.	Constant potential motor-generator or rectifier	Recommended for 1/4-in. (6.4-mm) and larger diameter electrodes only. May cause carbon deposit with small electrodes. Not suitable for automatic torches with voltage control.
a.c.	Constant current transformer	Recommended for a.c. electrodes only.
a.c. or d.c.	Constant current	D.C. supplied from three-phase transformer-rectifier supplies is satisfactory, but d.c. from single-phase sources gives unsatisfactory arc characteristics. A.C. output from a.c.-d.c. units is satisfactory provided a.c. electrodes are used.

Table 17-7 Electrode and Current Recommendations for Air Carbon Arc Cutting of Several Alloys

Alloy	Electrode Type	Current Type	Remarks
Carbon, low alloy, and stainless steels	d.c.	DCEP	
	a.c.	a.c.	Only 50% as efficient as DCEP
Cast irons	a.c.	DCEN	At middle of electrode current range
	a.c.	a.c.	
	d.c.	DCEP	At maximum current only
Copper alloys:			
copper 60% or less	d.c.	DCEP	At maximum current
copper over 60%	a.c.	a.c.	
Nickel alloys	a.c.	a.c.	
	a.c.	DCEN	
Magnesium alloys	d.c.	DCEN	Before welding, surface must be cleaned.
Aluminum alloys	d.c.	DCEN	Electrode extension should not exceed 4 in. (100 mm). Before welding, surface must be cleaned.

Fig. 17-15 Air carbon arc torch and carriage for mechanized operation. The current is d.c. electrode positive polarity, variable voltage. Electrode sizes are 5/16 through 5/8 inches. *Thermadyne Industries, Inc.*

In the oxygen arc process, a stream of oxygen is directed into a pool of molten metal. The pool is formed and kept molten by an arc established between the base metal and a tubular coated electrode, which is consumed during the cutting operation, Fig. 17-16. In addition to providing the source of the arc, the coating on the electrode provides insulation, acts as an arc stabilizer, and aids the flow of molten metal from the cut.

The cutting action is fast, and preheating is not required. The cut material is not contaminated in any way. The finished cuts require very little grinding or machining. The cut surfaces, however, are usually rougher than those produced by other arc cutting processes. Experienced welders are able to use the process for the first time with very little practice.

The oxygen arc process can be used to cut those metals that have always been considered nearly impossible by standard methods. Electrodes were first developed

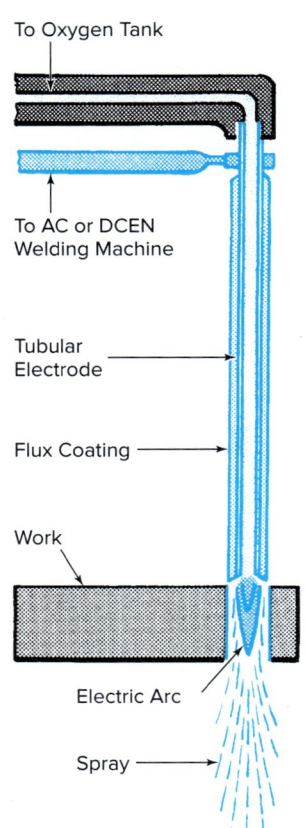

Fig. 17-16 Schematic of the oxygen arc electrode in operation.

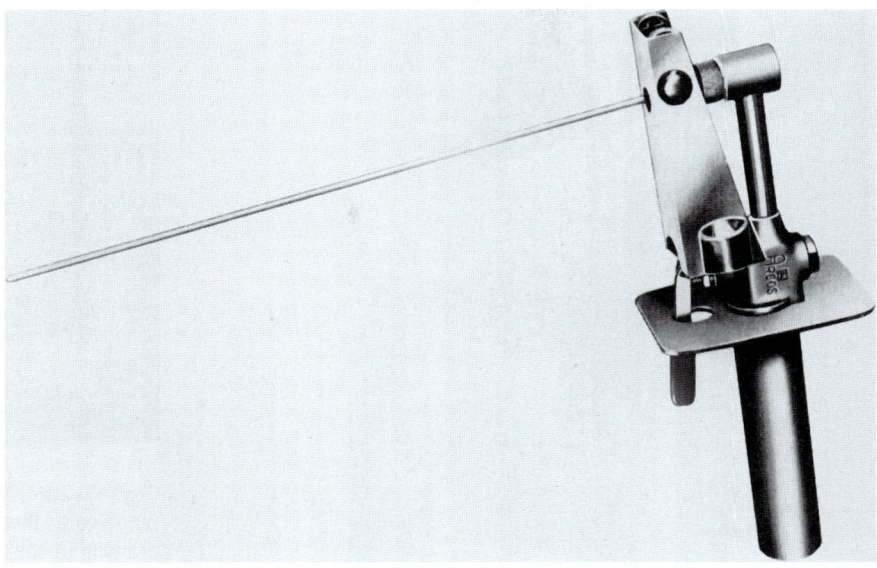

Fig. 17-17 The oxygen arc electrode and holder. Arcos Corp.

primarily for use in underwater cutting. It is possible to cut ferrous and nonferrous metals in any thickness and position. Such metals as stainless steels, nickel-clad steel, bronze, copper, brass, aluminum, and cast iron are cut without difficulty.

Equipment includes an oxygen arc electrode holder, oxygen arc coated tubular electrodes, an a.c. or d.c. welding machine, a tank of oxygen, and oxygen-regulating gauges.

The oxygen arc cutting electrode is a ferrous metal tube covered with a nonconductive coating. The function of the tube is to conduct current for the establishment and maintenance of the arc. The bore of the tube directs the oxygen to the metal being cut. These electrodes are available in 3/16- and 5/16-inch diameters with bores of 1/16 and 1/10 inch, respectively. They are 14 and 18 inches long.

The electrode is held by a special holder, Fig. 17-17. This holder is similar in appearance to a welding electrode holder. When used for cutting under water, a fully insulated holder equipped with a suitable flashback arrester is required. Both current and oxygen are fed to the electrode through the holder. It is easy to insert the electrode into the holder and remove the stub end after cutting. The flow of oxygen is controlled by a valve built into the holder and triggered by the operator with the hand that grasps the holder.

Both the cutting and piercing operations are begun by tapping the tip of the electrode on the work to establish an arc and release the oxygen. For piercing, the electrode is pushed into and through the plate. For cutting, the electrode is dragged along the plate surface at the same rate as the cut. The coating is kept in contact with the material being cut at all times. The coating burns off more slowly than the electrode, thus maintaining the arc. Figure 17-18 illustrates the process at the point of cutting.

The speed of cut varies with the thickness and composition of the plate, the oxygen pressure, the current, and the size of the electrode, Fig. 17-19, page 534. Table 17-3 gives the ampere settings and oxygen pressures for various types of metal that can be cut by the oxygen arc process.

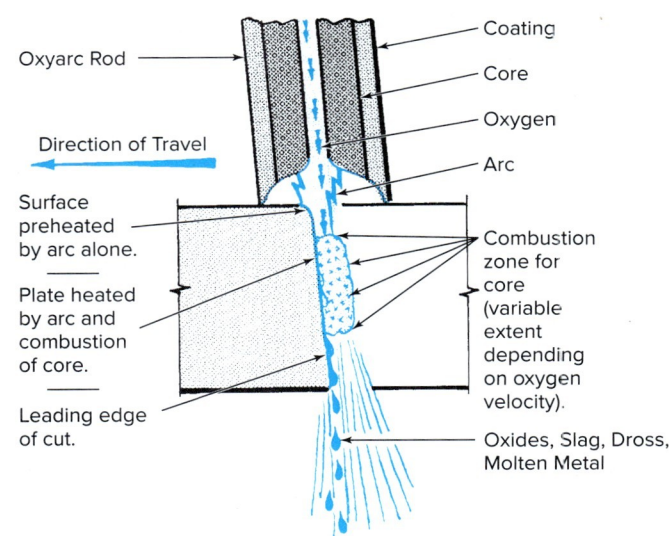

Fig. 17-18 The oxygen arc process at the point of cutting.

Arc Cutting Principles and Arc Cutting Practice: Jobs 17-J1–J7 **Chapter 17** 533

Fig. 17-19 The oxygen arc cutting of cast iron risers in the foundry. Risers are 5 inches square. The operation required 5 minutes cutting time.
Arcos Corp.

Fig. 17-20 This custom control system will closely track all pipe spools throughout the production process, from the isometric drawing to shipment. It can pinpoint materials shortages, allocates material to each pipe spool as required while tracking labor-hours required. Automated systems like this are required to speed production and delivery while ensuring the accuracy of each pipe spool.
Team Industries, Inc.

Gouging is performed by inclining the electrode until it is almost parallel to the plate surface and pointed away from the operator along the line of the gouge.

Goggles Welders must wear protective goggles to prevent harm to their eyes from sparks, hot particles of metal, and glare.

Gloves The heat coming from a cutting job may be very intense. There may also be a shower of sparks and hot material which makes it necessary for welders to protect their hands with gloves. Gloves should be kept free from grease and oil because of the danger involved in contact with oxygen.

Arc Cutting Machines

A very large part of the cutting done today is performed by arc cutting machines, Fig. 17-20. These machines have a device to hold the cutting torch and guide it along the work at a uniform rate of speed. It is possible to produce work of higher quality and at a greater speed than with the hand cutting torch. Machines may be used for cutting straight lines, bevels, circles, and other cuts of varied shape.

Small portable cutting machines are used with only one torch. Large permanent installations can make use of several cutting torches to make a number of similar shapes at the same time. A multiple-torch cutting machine and its automatic controls are shown in Figs. 17-21 and 17-22.

Maximum productive capacity is achieved through the use of stationary cutting machines developed for production cutting of regular and irregular shapes of practically any design. These machines can be particularly adapted to operations in which the same pattern or design is to be cut repeatedly, Fig. 17-23. A number of cutting torches are

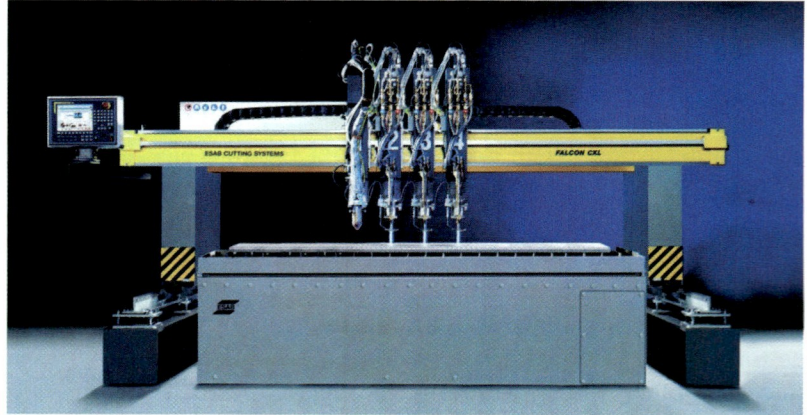

Fig. 17-21 Gantry-type mechanized cutting machine with PAC speeds up to 400 inches per minute. ESAB

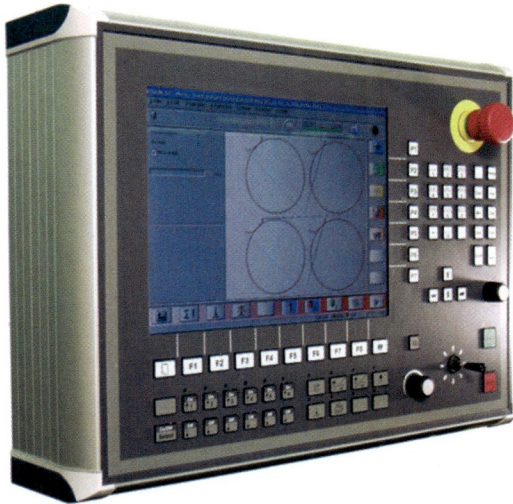

Fig. 17-22 An automatic digital control unit. ESAB

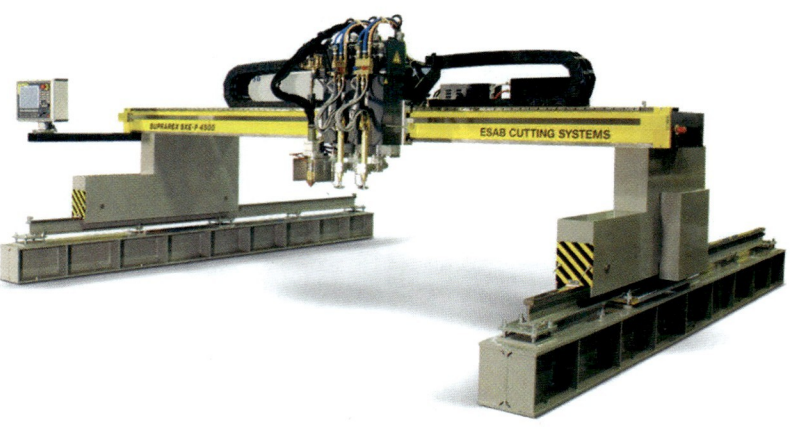

Fig. 17-23 A large gantry system with multiple-head capacity. Travel speed from 2–1,400 inches per minute. Note the vision computer numerical control. ESAB

mounted on the machine so that a number of parts of the same shape can be cut simultaneously. These machines can be used for straight line or circle cutting. They can be guided by hand or a template.

Cutting machines may be guided by various types of tracing devices. One type follows a pattern line of tracer ink and electrically controls the movement of the torch by means of a servomechanism.

Some units make use of a tracer roller, which is magnetized and kept in contact with a steel pattern. The tracer follows the exact outer contour of the pattern and causes the cutting tools to produce a cut in exactly the same shape.

Stack Cutting

In addition to cuts made through a single thickness of material, cuts are made through several thicknesses at the same time. This is a machine cutting process known as stack cutting. The plates in the stack must be clean and flat and have their edges in alignment where the cut is started. The plates must be in tight contact so that there is a minimum of air space between them. It is usually necessary to clamp them together.

Beam Cutter

The beam cutter is a portable structural fabricating tool. From one rail setting the operator can trim, bevel, and cope beams, channels, and angles. The beam rail is positioned across the flanges. Two permanent magnets lock and square the rail in position. Variable speed power units are used on both the horizontal and vertical drives. A squaring gauge enables the operator to adjust the torch from bevel to straight trim cuts.

This all-position arc cutting machine weighs only 60 pounds. It is easily moved and set in position by one operator, so setup time is minimal. The beam cutter provides clean, accurate cuts in a minimum amount of time. It has become a very important tool for use in the construction industry.

Learning Arc Cutting Skills

Your instructor will assign appropriate practice in arc cutting from the jobs listed in the Job Outlines, Tables 17-9, page 543, and 17-11, page 546. Before you begin a job, refer to Chapter 1 on safety and then study the specifications given in the Job Outline. The Text Reference column lists the pages in the chapter on which the job drawings and job description appear. Pay particular attention to the drawing that accompanies the job description.

Practice Jobs

Plasma Arc Cutting

The type and thickness of the metal to be cut will determine the amperage setting and cutting speed. The manual air plasma cutting power source and equipment will generally be rated in amperage and the thickness and type of material it is designed for. In most cases the electrode, tip, and gas pressure need not be varied for doing manual PAC with compressed air. Always follow the specific information provided by the equipment manufacturer. Figure 17-24, page 536, is a manual air plasma cutting system. Figure 17-25, page 536, represents a typical plasma cutting torch, its major parts, and inspection

Fig. 17-24 Typical plasma cutting system. Miller Electric Mfg. Co.

procedures for worn consumables. See Table 17-8 for plasma arc cutting troubleshooting.

Torch positioning and travel speed are the key variables to control for successful cutting. Some torches at low power can be dragged directly on the work for higher power cutting. The torch must be held approximately 1/8 inch from the work. Figure 17-26 indicates the standoff distance. Some torches can be equipped with a drag shield that will allow the torch to be rested on the work. This allows for much steadier operation.

When possible while cutting, use a template or a straightedge, or a piece of angle iron bar stock as a pattern to guide the torch for making the plasma cut. Be careful to line up the center of the top with the layout line. Because the speed of plasma cutting is high and the typical torch head is large, it can be difficult to see and follow a marked line. Free-hand cutting for these reasons should be limited.

When setting up and using PAC equipment, follow the manufacturer's recommendations for proper setup and operation.

1. Connect the power source to the primary power.
2. Check the torch, compressed air (plasma gas), settings, and work clamp connection.
3. Put on personal safety equipment.
4. Check the torch tip, electrode, and shield cup.
5. Check the compressed air pressure and flow requirements.
6. Check the power source amperage and control settings.
7. Check the torch operation.
8. Turn on the power source and begin cutting.

See Fig. 17-27 for the PAC procedure.

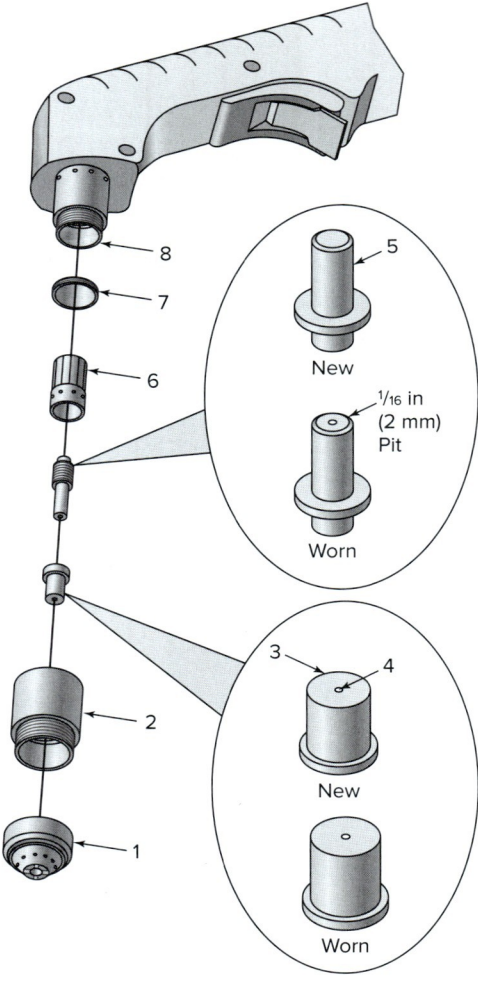

Fig. 17-25 Typical 50- to 80-ampere torch parts and consumable inspection. 1. Drag shield. 2. Retaining cup (remove and inspect for cracks replace if necessary). 3. Tip. 4. Opening (remove and inspect for deformed hole, or if 50 percent oversized, compare to new tip, clean inside of tip if not clean, and brighten with steel wool; remove all steel wool residue). 5. Electrode (inspect; if center has a pit more than 1/16 inch deep, remove and replace electrode). 6. Swirl ring (remove and inspect; replace if side holes are plugged). 7. O-ring (inspect and coat with thin film of silicone lubricant; replace if damaged). 8. Plunger area (inspect for debris or foreign materials, clean out if necessary). Carefully reassemble parts in reverse order. Source: Miller Electric Mfg. Co.

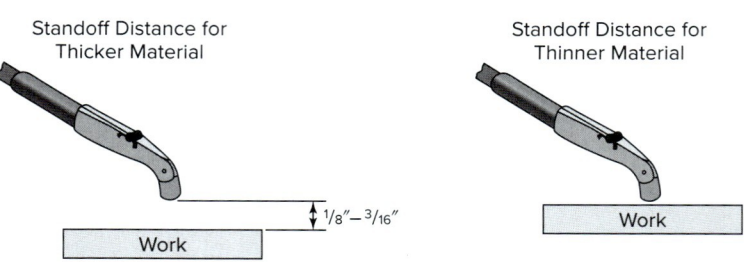

Fig. 17-26 Torch standoff distance. Source: Miller Electric Mfg. Co.

Table 17-8 Plasma Arc Cutting Troubleshooting

Problem	Correction	View
Acceptable cut	• Minimum beveling of edges • Minimum top edge rounding • Smooth surface, minimal or no grinding required • No dross • Kerf approximately 1½–2 times orifice diameter	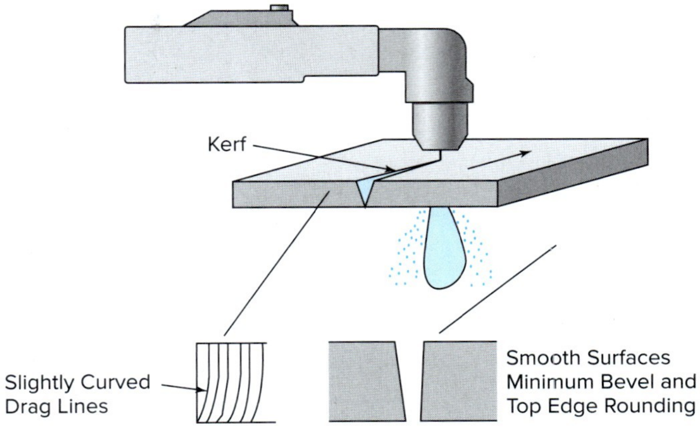 Kerf; Slightly Curved Drag Lines; Smooth Surfaces Minimum Bevel and Top Edge Rounding
Top edge of kerf melted and rounded, heavy dross on bottom of kerf easily removed.	• Increase travel speed. • Reduce power level and maintain current speed.	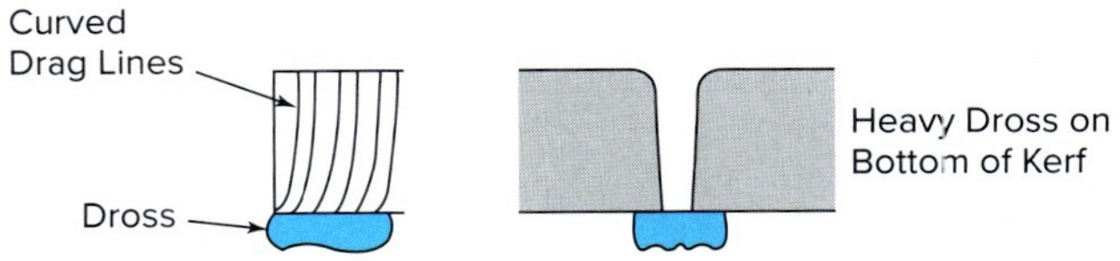 Curved Drag Lines; Dross; Heavy Dross on Bottom of Kerf
Drag lines very curved, dross on bottom of kerf hard to remove. Cut not completely through part.	• Decrease travel speed. • Increase power level.	 Very Curved Drag Lines; Dross; Light Dross on Bottom of Kerf
Wide kerf, irregular drag lines, rough cut and heavy dross on bottom of kerf.	• Check tip and electrode for wear, clean and/or replace. • Check standoff distance (too far away likely).	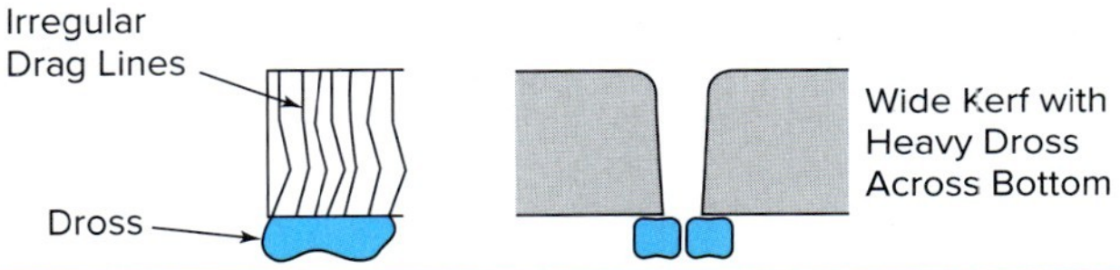 Irregular Drag Lines; Dross; Wide Kerf with Heavy Dross Across Bottom

Source: From Welding Handbook, 9/e.

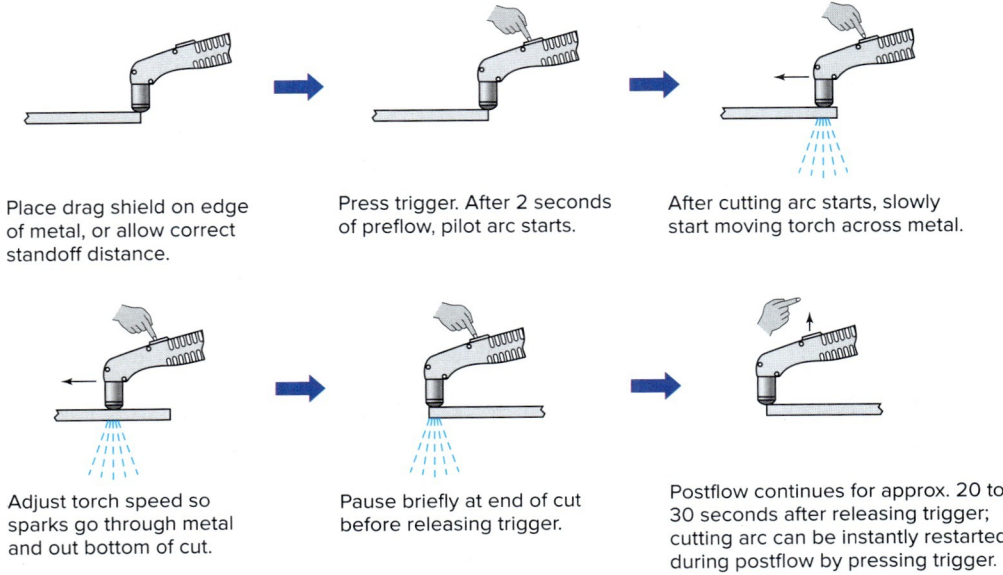

Place drag shield on edge of metal, or allow correct standoff distance.

Press trigger. After 2 seconds of preflow, pilot arc starts.

After cutting arc starts, slowly start moving torch across metal.

Adjust torch speed so sparks go through metal and out bottom of cut.

Pause briefly at end of cut before releasing trigger.

Postflow continues for approx. 20 to 30 seconds after releasing trigger; cutting arc can be instantly restarted during postflow by pressing trigger.

Fig. 17-27 Plasma cutting sequence. Source: Miller Electric Mfg. Co.

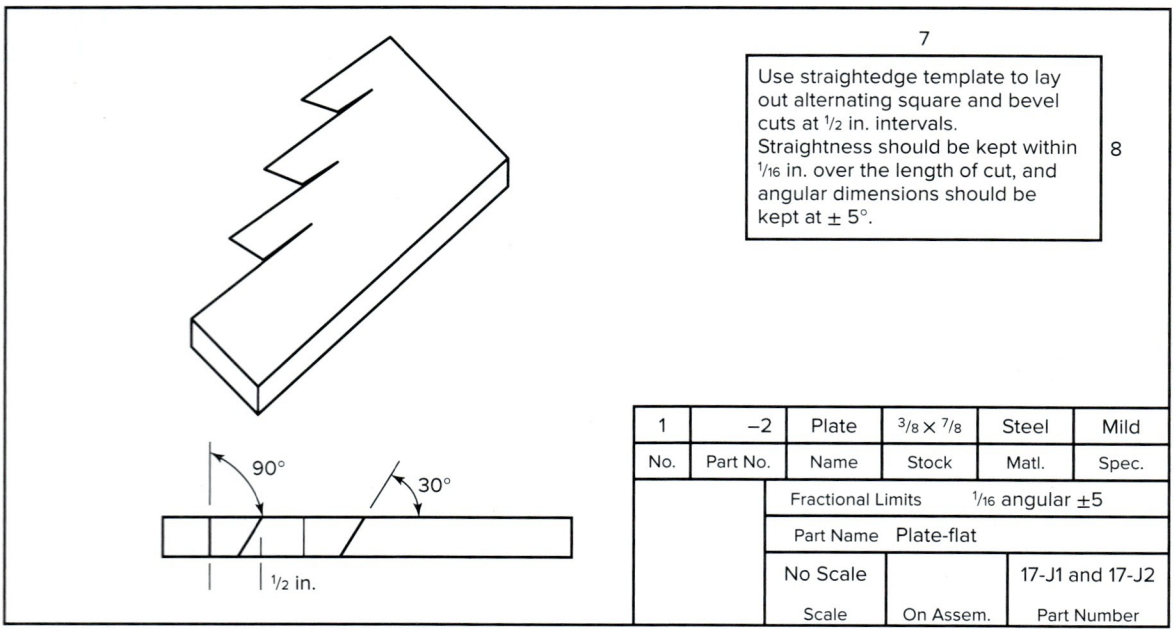

Fig. 17-28 Job drawings J1 and J2.

Job 17-J1 Square Cutting with PAC

Objective

To make a square cut of acceptable quality in the flat position with a manual air plasma cutting system. See the Job drawing in Fig. 17-28.

Cutting Technique

The plasma torch needs to be positioned so the work and travel angles are at a 90° angle to the piece being cut. Review Fig. 17-27. Use a template to keep the cut straight. Compare your cut surfaces to Fig. 17-29. With the PAC process you will note that one side of the cut is square and the other side has a slight bevel of 4 to 6°. This is due to the swirling, cutting gas as it exits the torch tip or nozzle. Since you are attempting to make a square cut, one piece will be considered as the production piece and the other will be considered as scrap. Figure 17-30 explains how to use the direction of travel to ensure a quality production piece.

Job 17-J2 Bevel Cutting with PAC

Objective

To make a bevel cut of acceptable quality in the flat position with a manual air plasma cutting system. The thickness to be cut is determined not by the metal thickness but by the angle of the bevel. A ⅜-inch plate cut with a 30° bevel will require a cut capacity of approximately ½ inch. See the Job drawing in Fig. 17-28.

Cutting Technique

The plasma torch needs to be positioned so the work angle is approximately 30° from the vertical and with a travel angle of 90°. Use a template to keep the cut straight. Review Fig. 17-27.

Job 17-J3 Gouging with PAC

Objective

To make a gouge of acceptable quality in the flat position with a manual air plasma cutting system. See the Job drawing in Fig. 17-31, page 540.

Gouging Technique

Plasma arc gouging is a variation of the PAC process. Gouging utilizes a different torch tip that produces a reduction in the arc constriction, which results in a lower arc stream velocity. Figure 17-32, page 540, shows the larger diameter orifice of the gouging tip. This larger diameter orifice provides the reduction in arc constriction, which results in a lower arc stream velocity. It gives a softer, wider arc and proper stream velocity. Gouging may be used for edge preparation (J- or U-grooves), removal of tack welds, braces that have been welded, or defective welds. It can be

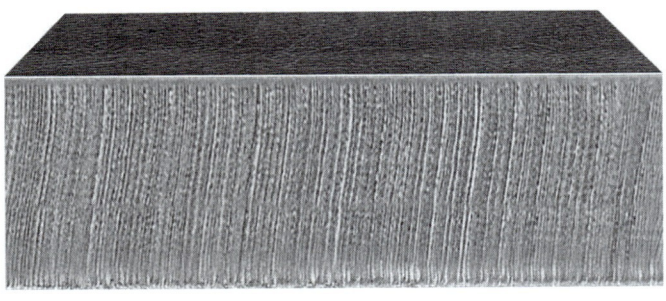

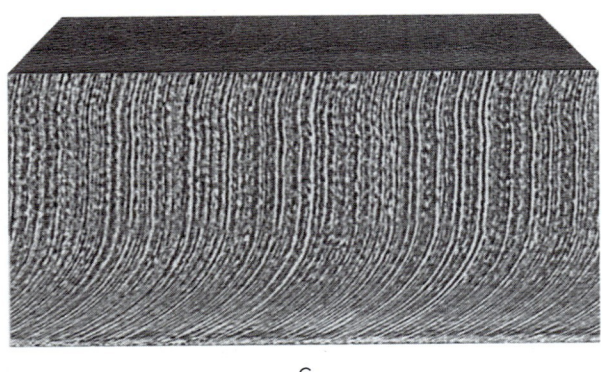

Fig. 17-29 Plasma arc cutting quality of square cut. Miller Electric Mfg. Co.

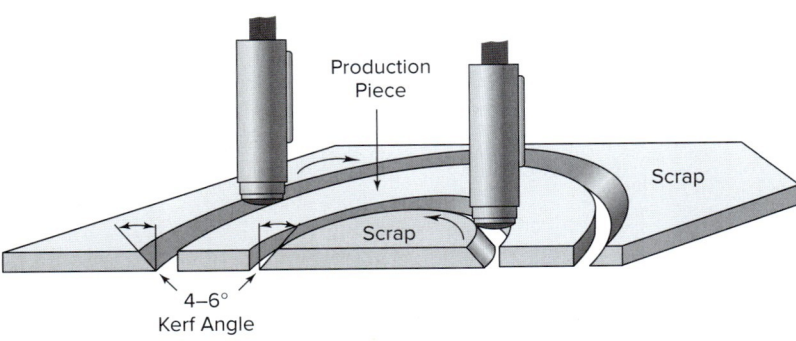

On a torch with a clockwise swirl, the straight side of the kerf will be the right-hand side, in the direction of travel.

Fig. 17-30 PAC direction of travel and getting a square cut.
Source: Miller Electric Mfg. Co.

used in all positions with the proper protection. The gouge surface should be clean and bright. It is particularly useful on aluminum and stainless steel, which should require little cleanup prior to welding.

Position the torch with a travel angle of 90° and a work angle of approximately 30 to 45° from the vertical as shown in Fig. 17-33, page 540. This torch angle and the speed of travel will determine the gouging depth. It is important that not too much material be removed in a single pass. Use multiple passes to achieve the required depth and width.

Job 17-J4 Hole Piercing

Objective

To be able to perform hole piercing to an acceptable quality level. See the Job Drawing in Fig. 17-34, page 541.

Hole Piercing Technique

Make certain you have replaced the gouging tip with a proper cutting tip. Set the torch in the same position that was used for gouging, approximately 30 to 45° from the vertical. As the material is blown out of the hole area, rotate the torch into an approximate 90° position and then move the torch to cut the proper size hole.

Job 17-J5 Shape Cutting with PAC

Objective

To make various shape cuts of acceptable quality in the flat position with a manual air plasma cutting system. See the Job Drawing in Fig. 17-34, page 541.

Shape Cutting Technique

Use the techniques already learned for square, bevel, and hole piercing to accomplish this job. See Fig. 17-35, page 542. Templates can be used instead of laying out the parts and is generally preferable to free-hand cutting with the PAC process. Templates should be made or provided by your instructor for this job.

Air Carbon Arc Cutting

The torch should be held to give a work angle of approximately 90° and with a travel angle of 45°. The arc and molten weld pool will be blown out ahead of the arc, so use a push angle on the electrode. Since the air jet ports are located in the torch where the electrode is clamped, they need to be in fairly close proximity to the work. The maximum extension should not be over

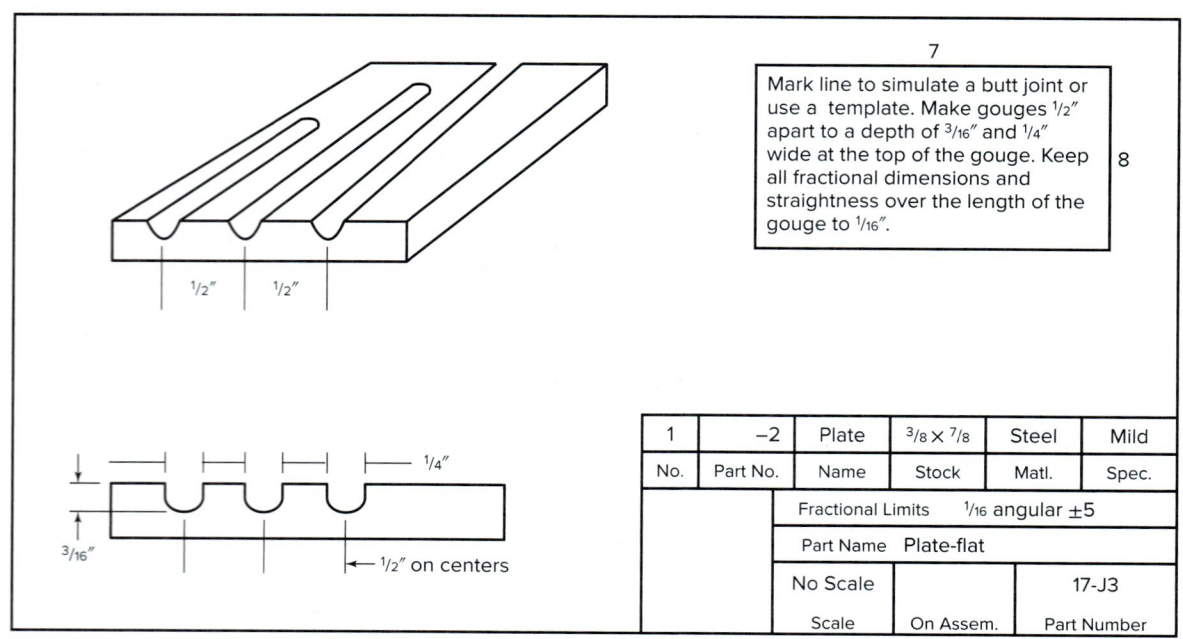

Fig. 17-31 Job 17-J3.

Fig. 17-32 Gouging tip and cutting tip comparison.
Miller Electric Mfg. Co.

7 inches, while the minimum should never be less than 1½ inches or the torch may be damaged. Always locate the air jet between the work and the electrode, Fig. 17-36, page 543.

Keep the arc length as short as possible without touching the work with the carbon electrode. The travel speed will have to be quite fast to keep up with the molten pool that is being blown away. Slow travel speed generates more heat into the base metal and produces a deeper groove. High travel speed decreases metal removal resulting in a

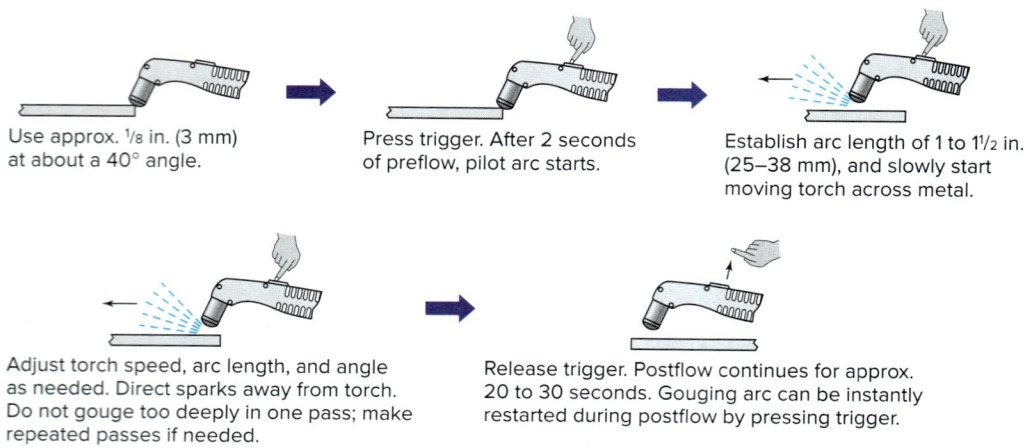

Fig. 17-33 Plasma gouging sequence. Source: Miller Electric Mfg. Co.

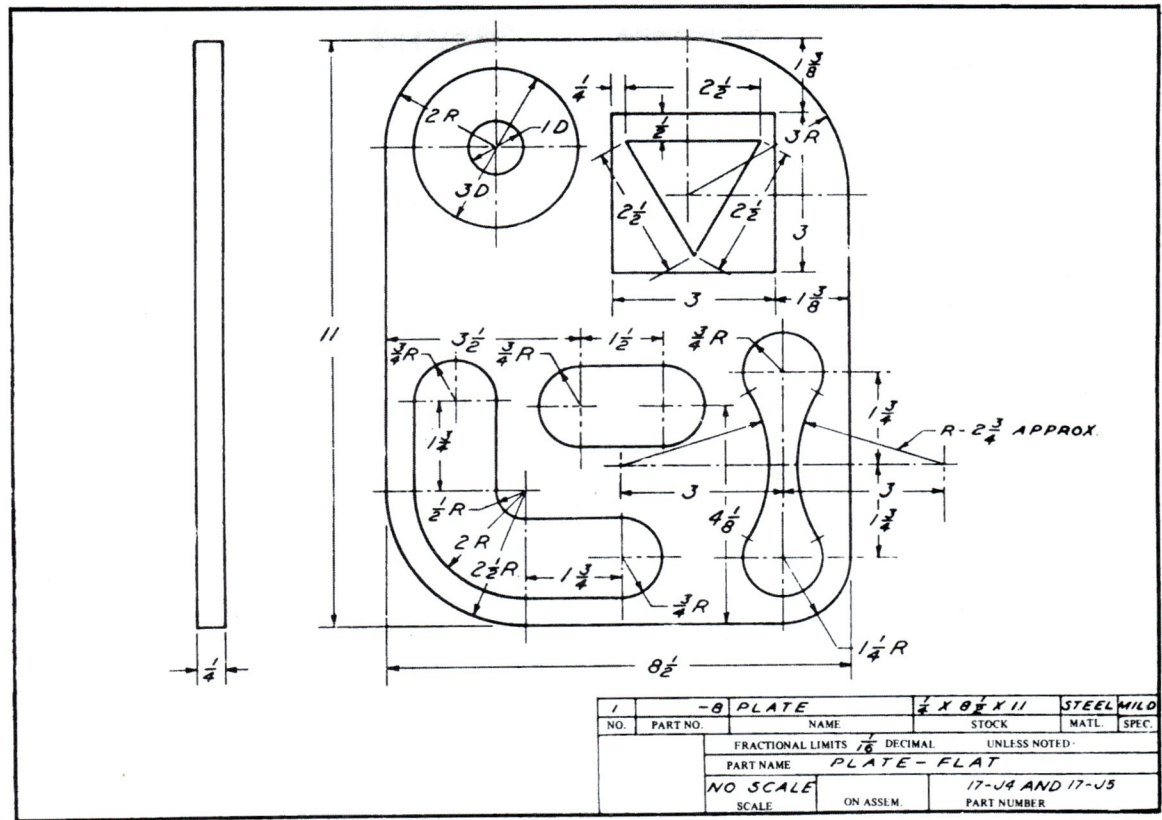

Fig. 17-34 Jobs 17-J4 and 17-J5.

shallow groove. The depth of the groove can also be controlled with the electrode angle. Steeper electrode angles will result in a deeper groove, while flatter electrode angles will result in a shallower groove. The current level can affect the shape of the groove. When the current is lower than recommended, the groove will retain the U-shape but it will be shallow. If it is raised above the recommended amperage by about 15 percent, the groove will change from a U- to V-shape.

Always turn on the air supply at the torch prior to starting and arc.

Setting Up and Using CAC-A Equipment

Follow the manufacturer's recommendations for proper setup and operation, Fig. 17-37, page 543:

1. Connect the power source to the primary power.
2. Make sure the electrical power is turned off.
3. Bolt the electrode cable from the power source to the connector on the carbon arc torch. Use direct current electrode positive (DCEP).
4. Connect the air hose to the carbon arc torch. This is usually done with a quick disconnect system.
5. Make sure both the electrical connection and air hose connection to the torch are properly protected with an appropriate "boot cover."
6. Attach the work clamp and cable to the power source and workpiece. Use DCEP.
7. Turn on the air supply to the torch and check for proper pressure.
8. Select the proper size and shape of carbon electrode.
9. Before starting make sure all safety precautions are taken. Use a hot work permit as required.
10. Put on personal safety equipment (arc rays, sparks, fumes, and noise must be guarded against).
11. Check the power source amperage and control settings.
12. Turn on the power source, turn on the air with the torch on/off valve, and begin cutting.

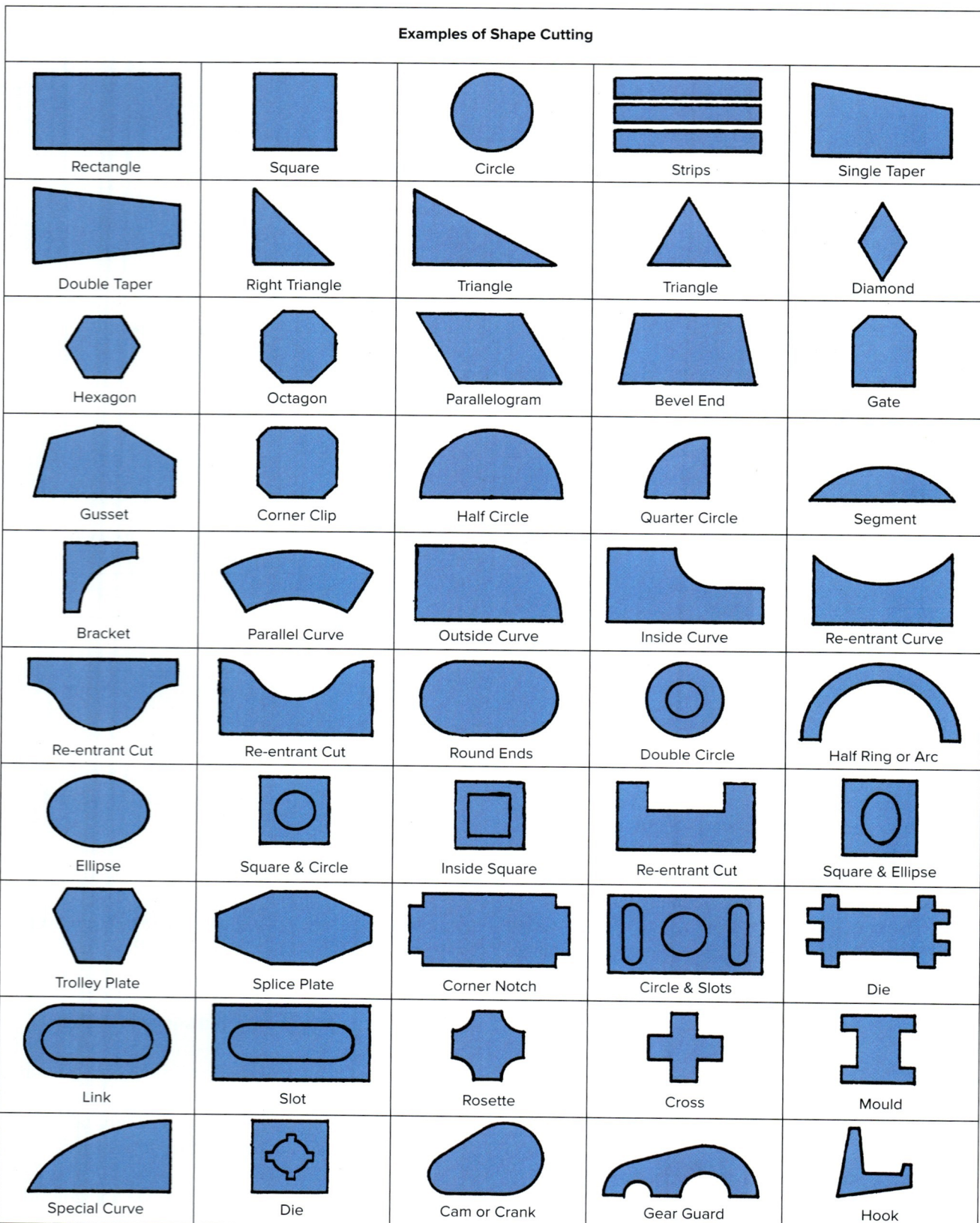

Fig. 17-35 Examples of shape cutting.

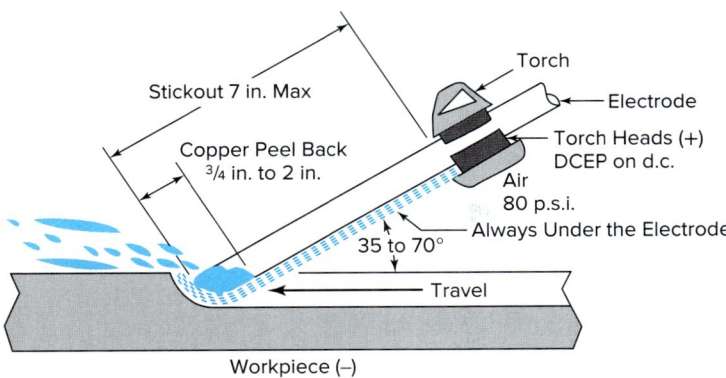

Fig. 17-36 How a standard CAC-A torch works. Source: From Welding Handbook, 9/e.

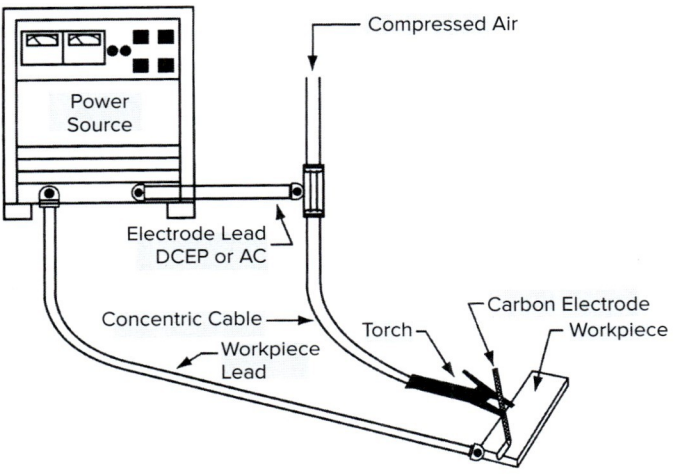

Fig. 17-37 Typical air carbon arc cutting equipment setup. Source: American Welding Society, *Welding Handbook*, Vol. 2, 9th ed., p. 653, Fig. 15.8.

Job 17-J6 Gouging with CAC-A

Position the torch as shown in Fig. 17-38, page 544.

Objective

To make a gouge in a butt joint to form a U-groove, with acceptable quality to make a complete joint penetration (CJP) weld that is capable of passing a radiograph test. See the Job Drawing in Fig. 17-39, page 544.

General Job Information

Make a square groove butt joint with an E7018 electrode or substitute electrode or welding process per your instructor's instructions. This is to simulate the making of a CJP weld. The practice is to gouge to the proper depth to expose sound weld metal. This may take several passes and should be done carefully so that excessive gouging is not done. Additional plates can be welded together to increase practice opportunity. If practice pieces are available in the scrap area, they can be used to simulate this job. If they have been welded on both sides, just gouge out one weld to the dimensions indicated on the Job Drawing. When the proper technique is used, the CAC-A process produces little, if any, dross on the finished cut. If dross remains, it is rich in carbon and must be removed. This can be done with a chipping hammer or grinding. If you attempt to weld through this dross, the resulting weld will be of very low quality. Use Table 17-10 for proper settings.

Table 17-9 Job Outline: Plasma Arc Cutting Practice: Jobs 17-J1–J5

Job Number	Type of Job	Material[1]	Thickness (in.)	Minimum Output Rating	Minimum Air Pressure and Flow Rate	Text Reference
17-J1	Square cut	Carbon steel	3/8	25 A @ 90 V	60 p.s.i. @ 4.5 ft^3/min	556
17-J2	Bevel cut	Carbon steel	3/8	50 A @ 110 V	70 p.s.i. @ 5.3 ft^3/min	556
17-J3	Gouging	Carbon steel	3/8	50 A @ 110 V	70 p.s.i. @ 5.3 ft^3/min	556
17-J4	Hole piercing	Carbon steel	1/4	50 A @ 110 V	70 p.s.i. @ 5.3 ft^3/min	557
17-J5	Shape cutting	Carbon steel	1/4	25 A @ 90 V	60 p.s.i. @ 4.5 ft^3/min	557

[1]Once carbon steel has been mastered, practice on stainless steel and aluminum.
Source: From Welding Handbook, 9/e.

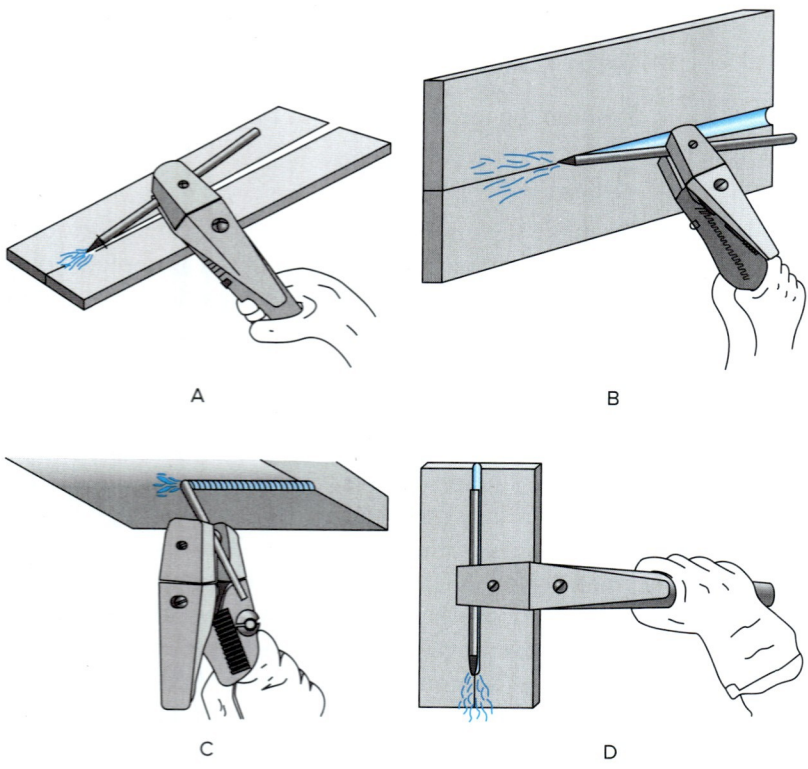

Fig. 17-38 CAC-A torch positions for all-position gouging.
Source: American Welding Society, C5.3-91, *Air Carbon Arc Gouging and Cutting*, pp. 6 and 7, Figs. 6 through 9.

Job 17-J7 Weld Removal with CAC-A

Objective

Remove weld metal with minimal gouging into the base metal. See the Job Drawing in Fig. 17-40.

General Job Information

Select appropriate fillet welds from scrap area. If none are available, set up as per the Job Drawing and weld prior to CAC-A gouging. Use appropriate techniques similar to those used in the Job Drawing.

Reduce the electrode to approximately 30°, and direct the electrode toward the center of the fillet weld. Use sufficient travel speed to maintain the arc while removing the weld with minimal gouging into the base metal. You are looking for the depth of fusion into the joint. When you have gouged deep enough, you will see a very distinct line that is the joint root. Table 17-10 gives some general setting information. Table 17-12, page 546, covers troubleshooting issues.

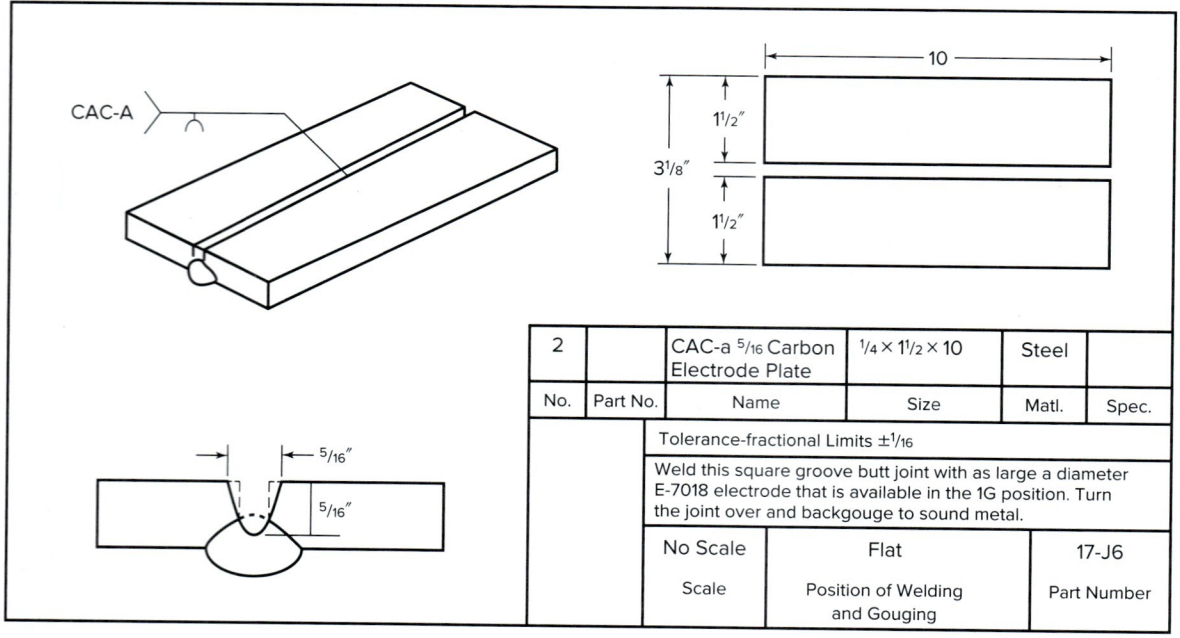

Fig. 17-39 Job 17-J6.

544 Chapter 17 Arc Cutting Principles and Arc Cutting Practice: Jobs 17-J1–J7

Table 17-10 CAC-A Settings for Gouging

Electrode Diameter	Desired Groove Depth	Travel Speed (in./min)	DCEP Polarity and Power Data		Compressed Air	
			Amperes	Volts	ft³/min	p.s.i.
5/16	1/8	65	400	42	30	60
5/16	3/16	45	400	42	30	60
5/16	1/4	36	400	42	30	60
5/16	5/16	33	400	42	30	60
1/4	NA	NA	300–400	42	30	60

Source: American Welding Society

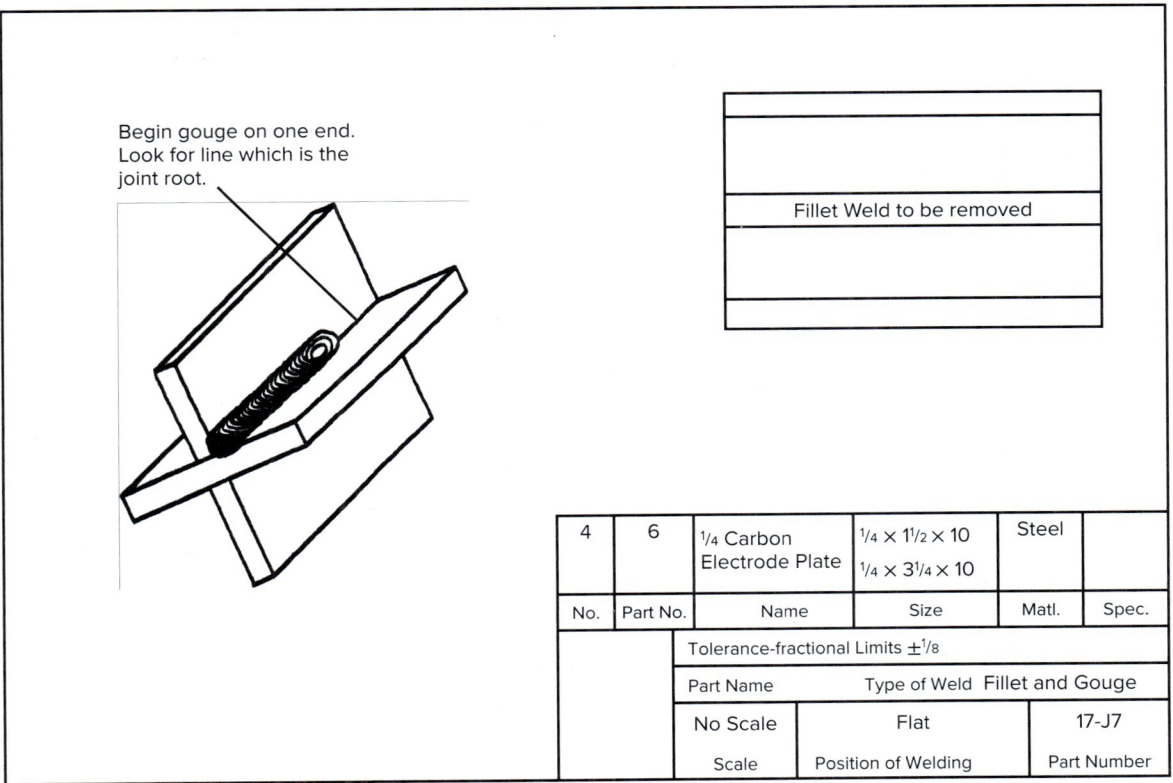

Fig. 17-40 Job 17-J7.

Shutting Down the CAC-A Equipment

1. Turn off the air supply by using the torch valve. Allow enough airflow time to help cool the carbon electrode. Do not let the electrode come into contact with the workpiece until the power source is turned off.
2. Turn off the power source.
3. Turn off the main air supply.
4. Remove the carbon electrode from the holder. Place it in its proper storage area after it has cooled.
5. Disconnect the air supply and power cable to the torch. Disconnect the workpiece connection.
6. Wrap up the work and electrode cables. Wrap up the air hose and store. Place the carbon air arc torch in proper storage.
7. Mark on hot metal with "HOT."
8. Return all unused carbon electrodes to proper storage.
9. Dispose of all dross, used electrodes, and scrap metal.
10. Turn off the main power to both the power source and air compressor.
11. Conduct a fire watch for smoldering combustibles a minimum of one hour after work stoppage.

Table 17-11 Job Outline: Air Carbon Arc Cutting: Jobs 17-J6–J7

Job Number	Type of Job	Material[1]	Thickness (in.)	Carbon Diameter	Electrical Requirements DCEP		Minimum Air Pressure and Flow Rate	Text Reference
					Amperes	Volts		
17-J6	Gouge U-groove	Carbon steel	1/4	5/16	400	42	60 p.s.i. @ 30 ft³/min	561
17-J7	Weld removal	Carbon steel	1/4	1/4	350	42	60 p.s.i. @ 30 ft³/min	562

Table 17-12 Troubleshooting CAC-A

Problem	Correction
Carbon deposits at beginning of groove	Turn on air jet before starting arc; make sure air jet is between electrode and workpiece.
Unsteady arc, requiring slow travel speed even on shallow grooves	There is insufficient amperage for the electrode diameter. Use a larger power source or, if not available, smaller diameter carbon.
Erratic groove with arc wandering and electrode heating rapidly	The current and polarity must be DCEP.
Erratic arc action, with arc going on and off with rough groove surface	Speed up the rate of travel. This is a very fast process and you must keep the arc length short. Keep in mind that the weld pool is constantly being blown away with the high volume of air pressure. Position yourself to be able to freely move to keep up with the metal removal.
Carbon deposits at various locations in the groove	The carbon electrode was shorted out on the workpiece and/or travel speed was too fast.
Irregular groove depth	Keep your travel speed, arc length, and electrode angle steady.
Dross adhering to the edges of the groove	There is not enough air pressure or flow rate. The volume of the air is very important. Make sure at least a minimum of 3/8 in. inside diameter hoses are used. No restriction should exist in the air lines.

CHAPTER 17 REVIEW

Multiple Choice

Choose the letter of the correct answer.

1. Arc cutting can be used on which of the following metals? (Obj. 17-1)
   a. Cast iron
   b. Aluminum
   c. Titanium
   d. All of these

2. Arc cutting is done with the heat of an electric arc and which of the following? (Obj. 17-1)
   a. Oxygen
   b. Compressed air
   c. Inert gas
   d. All of these can be used

3. The basic principle of PAC is that an electric arc constricted through a small nozzle opening

increases the arc temperature to approximately _____°F. (Obj. 17-2)
   a. 250,000
   b. 25,000
   c. 2,500
   d. It is too hot to measure
4. The arc that starts when the trigger on a plasma arc torch is depressed is called the _____. (Obj. 17-2)
   a. Starter arc
   b. Main arc
   c. Cutting arc
   d. Pilot arc
5. Plasma is considered _____. (Obj. 17-2)
   a. The fourth state of matter
   b. A rare earth element
   c. An ionized gas
   d. Both a and c
6. Which of the following electrode angles are used for CAC-A? (Obj. 17-3)
   a. Push
   b. Drag
   c. Work
   d. Backhand
7. Which of the following will control the depth of a CAC-A groove? (Obj. 17-3)
   a. Electrode angle
   b. Travel speed
   c. Current
   d. All of these
8. To change the shape of the groove from a "U" to a "V" with the CAC-A process, which control will have the most effect? (Obj. 17-3)
   a. Air pressure
   b. Current level
   c. Electrode angle
   d. Travel speed
9. The carbon electrode should extend beyond the torch _____ inches. (Obj. 17-3)
   a. ½ to 1½
   b. 1½ to 7
   c. 7 to 10
   d. It is gripped on the very end of the electrode and used down to nothing
10. Where should the air jet ports on the CAC-A torch be located? (Obj. 17-3)
    a. Between the electrode and the work
    b. On top of the electrode away from the work
    c. Alongside the electrode
    d. It does not matter where they are located because the air is forced through the electrode center
11. When CAC-A is being done, when should the air be turned on? (Obj. 17-3)
    a. When the arc is struck and a weld pool has formed
    b. When the cut is complete to cool the plate down
    c. Never turn air on with this process because the heat of the arc melts the plate
    d. When you are ready to start the arc so that carbon deposits are not left on the plate
12. The fine lines left on a cut surface are called _____. (Obj. 17-4)
    a. Drag lines
    b. Kerf lines
    c. Dross lines
    d. Cut lines
13. If one side of the PAC has a slight bevel of 4 to 6°, this is considered _____. (Obj. 17-4)
    a. A poor cut and the torch angle needs to be held at 90°
    b. Normal for this process
    c. A poor cut and faster travel speed is required
    d. A poor cut and the power setting on the machine needs to be reduced
14. The CAC-A should be _____. (Obj. 17-4)
    a. Clean
    b. Smooth
    c. Free of dross
    d. All of these
15. If the bottom of a CAC-A groove is excessively V shaped, what can be done? (Obj. 17-4)
    a. Decrease torch angle
    b. Increase torch angle
    c. Decrease current by 15 percent
    d. Increase current by 15 percent
16. Which of the following is *not* part of setting up and using PAC equipment? (Obj. 17-5)
    a. Connect the power source to primary power
    b. Check the torch, air (plasma gas) settings, and work clamp connections
    c. Turn on and set the fuel gas flow rate
    d. Check the torch tip, electrode, and shield cup
17. Which of the following is *not* part of setting up and using CAC-A equipment? (Obj. 17-5)
    a. Connect the power source to primary power and make sure the power is turned off
    b. Connect the electrical power and air supply to the torch
    c. Attach the work clamp and cable to the power source using DCEP
    d. Select the proper size and shape of metallic electrode

18. PAC with too high a rate of travel speed will cause _____. (Obj. 17-6)
    a. The lines along the edge of the cut to be very curved
    b. Dross on the bottom of the cut to be hard to remove
    c. The part to not be completely cut through
    d. All of these
19. If CAC-A gouging the arc action is erratic and goes on and off with a rough groove surface, which of the following is most likely to be the cause of this problem? (Obj. 17-6)
    a. Air was not turned on prior to starting the arc
    b. There is insufficient amperage
    c. The current and polarity are wrong
    d. The travel speed is too slow
20. If CAC-A gouging the dross adheres to the cut, what corrective actions should you take? (Obj. 17-6)
    a. Slow down; the travel speed is too fast
    b. Increase the amperage; you are cutting too cold
    c. Increase the air pressure and make sure the hose is large enough to get the proper flow rate
    d. Correct the electrode angle; it is too steep or too flat

Review Questions

Write the answers in your own words.

21. What makes plasma arc cutting different from oxyfuel cutting? (Obj. 17-2)
22. List seven advantages of PAC. (Obj. 17-1)
23. A tremendous amount of noise and fumes are generated with high-powered mechanized plasma arc cutting. What can be done to overcome this? (Obj. 17-1)
24. List some inspection points and considerations for proper maintenance of a PAC torch. (Obj. 17-5)
25. List the proper cutting procedure and technique for making an air carbon arc gouge with a $5/16$-inch carbon to a depth of $3/16$ inch. (Obj. 17-3)

INTERNET ACTIVITIES

Internet Activity A

Search any one of the Web pages of major plasma arc cutting equipment manufacturers and make a report to your instructor on what are the most recent equipment innovations.

Internet Activity B

Using your favorite search engines, check out the various types of plasma and shielding gases that are recommended for the PAC process.

Design credits: Cinema and movie icon: Denis Meshkov/123RF; and Illustration of Welding icons: Mr. Nachai Sorasee/123RF

18

Gas Tungsten Arc and Plasma Arc Welding Principles

Chapter Objectives

After completing this chapter, you will be able to:

18-1   Describe the principles of gas tungsten arc welding.
18-2   Describe the principles of plasma arc welding.
18-3   Describe the differences between nonconsumable and consumable electrode processes.
18-4   Name the essential equipment for GTAW.
18-5   Describe the function of the shielding gas.
18-6   List controls and features found on a GTAW power source.
18-7   State the proper use of alternating current, DCEN, and DCEP for GTAW.
18-8   List the proper tungsten to be used for various applications.
18-9   Describe the GTAW hot wire method.
18-10 Show a properly set up GTAW system.
18-11 Explain the differences between transferred and non-transferred PAW arcs.
18-12 Explain the main difference between PAW and GTAW.

Gas Shielded Arc Welding Processes

In the gas shielded arc welding processes, the weld is produced by the arc maintained between the end of a metal electrode and the part to be welded. The electrode may be consumable or nonconsumable. A shield of gas ejected from the torch surrounds the arc and weld region. The shielding gas may or may not be *inert* (chemically inactive), and pressure is not used. Filler metal may or may not be added.

History

You have learned that a vital characteristic of sound welding depends on shielding the electric arc and the weld pool from contamination from the surrounding air. The first experiments with gas shielding were carried out in the early 1920s. The development of coated electrodes in the 1930s eliminated interest in the process at that time. Welds produced by coated electrodes are actually gas shielded welds because the electric arc produces a gas as it burns off the coating.

Coated electrode welding served industry well through World War II (1945). But when the war ended and industry began retooling for consumer products, competition spurred intensive research and experimentation to develop new welding methods.

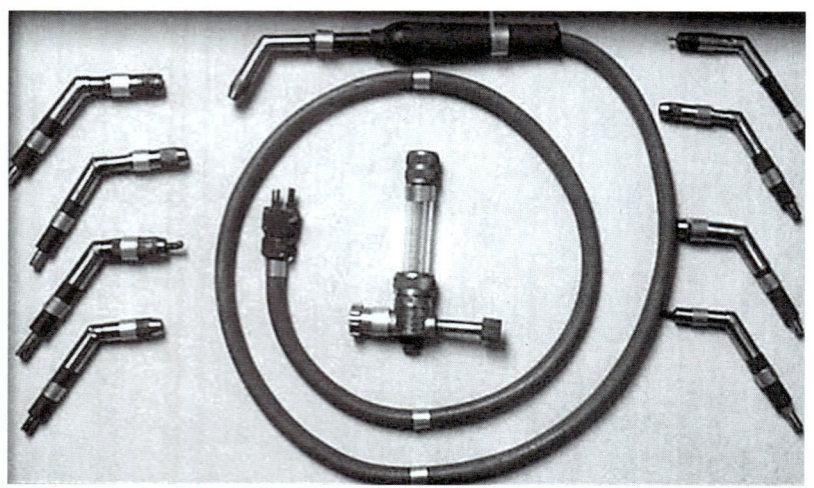

Fig. 18-1 Original torch developed by Meredith in 1944. Miller Electric Mfg. Co.

This effort led to the development of the inert gas shielded arc welding processes.

In 1930, Henry M. Hobart and Phillip K. Devers of the General Electric Company were granted patents that covered the basic principle of gas shielded arc welding. In 1944, Russell Meredith of the Northrop Aircraft Company was issued a patent on the welding of magnesium and magnesium alloys. The process was by an electric arc in helium and argon shielded atmospheres with a tungsten electrode. One of Meredith's original torches is shown in Fig. 18-1. Meredith's process is the gas tungsten arc process, known as GTAW or TIG. The process was originally developed for welding corrosion-resistant and other difficult-to-weld metals such as magnesium, stainless steel, and aluminum. Today it is a standard process (Figs. 18-2 through 18-4), widely used in both factory and field to weld practically all commercial metals either manually, semiautomatically, mechanically, or automatically.

Continued experimentation to make the process faster led to the development of the gas metal arc welding (GMAW or MIG/MAG) process in 1948. It was first applied to the welding of aluminum, and it was found to be several times faster than the TIG method. By 1951, it was discovered that by adding a small amount of oxygen to the argon, the arc action was much improved and could be applied to the welding of carbon and stainless steels. Carbon dioxide was also found to be effective as a shielding gas. Flux cored and metal cored wires are expansions of the basic MIG/MAG process.

Overview of the Processes

There are several gas shielded arc welding processes in which the welding student should develop skill. These processes have several of the characteristics of both oxyacetylene and shielded metal arc welding.

- The **nonconsumable electrode process,** Fig. 18-5, is known as *inert gas shielded, tungsten arc welding*. It is more often referred to as *gas tungsten arc welding (GTAW)* or by the shop term *tungsten inert gas (TIG)*.

Fig. 18-2 Welding an aluminum cylinder with the gas tungsten arc process. Miller Electric Mfg. Co.

Fig. 18-3 A welder putting the finishing touches on a chemical reactor. The lining of 1/16-inch thick pure silver is being welded with the gas tungsten arc process. Nooter Corp.

Fig. 18-4 Mechanized welding on a heavy wall pipe joint with the gas tungsten arc welding process and cold wire feed. Arc Machines, Inc.

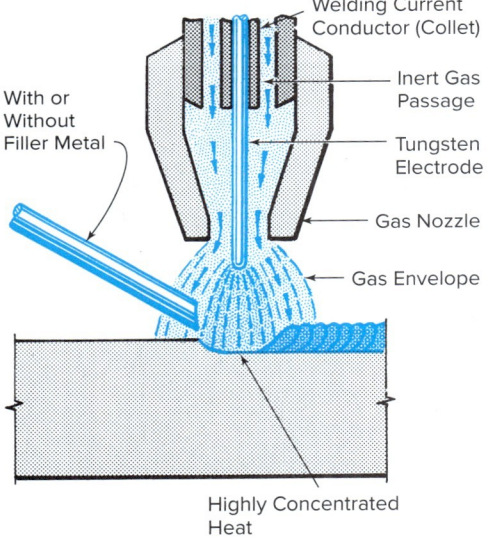

Fig. 18-5 Inert gas shielded, tungsten arc welding (nonconsumable electrode).

- The *consumable electrode process*, shown in Fig. 18-6, is known as the *metal inert gas (MIG) arc welding* process, or, with active gas, as the *metal active gas (MAG) arc welding* process. It is more appropriately referred to as *GMAW* or by the shop term *MIG/MAG*.

A method of applying these processes in a semiautomatic or automatic method is *gas shielded arc spot welding*. This method does not require a lot of practice.

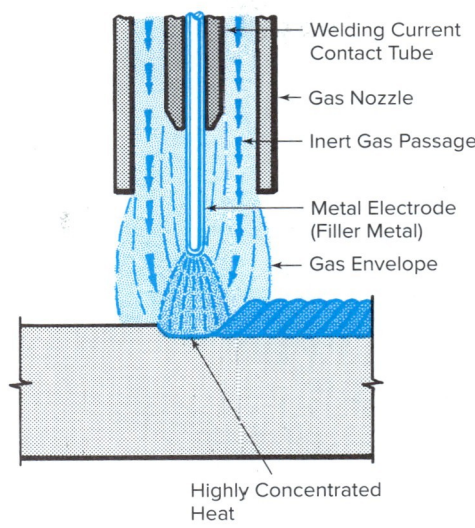

Fig. 18-6 Inert gas shielded, gas metal arc welding (consumable electrode).

Gas Tungsten Arc Process In the gas tungsten arc process, the heat necessary to melt the metal is provided by an intense electric arc that is established between a nonconsumable tungsten electrode and the metal workpiece (Fig. 18-7). The electrode does not melt or become a part of the weld. An edge or corner joint may be fused together without the addition of filler metal, or autogenous welding, using a technique similar to that used with the oxyacetylene flame. On joints where filler metal is required, a welding rod is fed into the weld zone and melted with the base metal in the same manner as that used with oxyacetylene welding. The weld pool is shielded, to protect it, with an inert gas (such as helium or argon, or a mixture of the two). The gas is fed through the welding torch.

Gas tungsten arc welding will be our major focus in this chapter. It is known by such trade names as *Heliarc, Heliwelding,* or, as previously noted, the common shop term of *TIG.* This process can be used to weld such difficult-to-weld metals as aluminum, magnesium, and

ABOUT WELDING

Stick Electrode

The stick electrode is commonly used for field pipe welding, but it is also possible to use a continuous welding process. GTAW (TIG) is the best method of applying the root pass in pipe welding. For the fill and cover passes, GMAW (MIG/MAG) can be used. Also, flux cored arc welding (FCAW) can be used. These semiautomatic processes can be used in the field because it is easy to carry the new portable feeders.

Gas Tungsten Arc and Plasma Arc Welding Principles **Chapter 18** 551

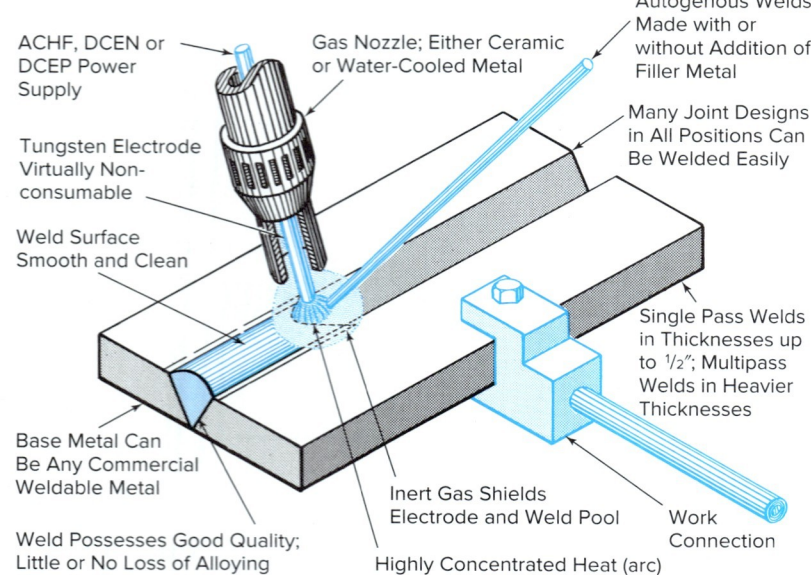

Fig. 18-7 Essentials of the gas tungsten arc welding (GTAW) process.

titanium. It can weld dissimilar metals. Because there is no flux, there is no contaminating residue and no cleaning problem. This fact, in itself, results in considerable cost savings, especially when multipass welding is necessary. Because the shielding gas is transparent, the welder can observe the weld pool clearly as it is formed and carried along during welding. The almost total absence of smoke, fumes, and sparks contributes to neater and sounder welds and to the comfort of the welder.

Welding can be performed in all positions. Because heat concentration and amperage can be more closely controlled in TIG welding than in stick electrode welding, there is less distortion of the base metal near the weld, less weld cracking, and fewer locked-up stresses.

A TIG weld is sound, smooth, strong, ductile, uniform, and bright. These characteristics make it ideal for use in industries such as the food business, chemical business, hospital equipment, and aerospace products.

For video of GTAW applications, please visit www.mhhe.com/welding.

Gas Metal Arc Process Gas metal arc welding is a consumable electrode process. It is also known by such trade names as *Aircomatic, Sigma, Millermatic,* and *Micro Wire* welding, or, as previously noted, the common shop term of MIG/MAG. It generally uses DCEP and a shield gas that can be either inert or active. Typical gases used are argon, helium, oxygen, carbon dioxide, and mixtures of these gases.

A small diameter wire serves both as electrode and filler metal. It is fed into the welding gun automatically, and then into the weld pool at high speed. MIG/MAG welding will be studied in Chapter 21. Compare Figs. 18-5 and 18-6 for a basic understanding of TIG and MIG/MAG welding.

Gas Shielded Arc Spot Welding The gas shielded arc spot welding method is produced with an arc spot pistol grip gun, Fig. 18-8. This gun may be provided with a tungsten electrode or a consumable electrode.

The gun makes a weld in the manner shown in Fig. 18-9. An arc is drawn to the surface of the metal under the protection of the shielding

SHOP TALK

Common Abbreviations

CAC-A Air carbon arc cutting
c.c. Constant current machine
c.v. Constant voltage machine
c.p. Constant potential machine
DCEN (or P) Direct current electrode negative (or positive)
FCAW-G Flux cored arc welding gas shielded
FCAW-S Self-shielded
GMAW Gas metal arc welding
GMAW-P Gas metal arc welding–pulse
GMAW-S Gas metal arc welding–short circuit
GTAW Gas tungsten arc welding
HAZ Heat-affected zone
HVOF High velocity oxyfuel
IR Infrared
ISO International Organization for Standardization
LVOF Low velocity oxyfuel
MIG Metal inert gas arc process
MIG-P Pulsed MIG
MPW Magnetic pulse welding
OSHA Occupational Safety and Health Administration
RSW Resistance spot welding
SMAW Shielded metal arc stick electrode welding
SAW Submerged arc welding
THSP Thermal spraying
TIG Tungsten inert gas
UV Ultraviolet

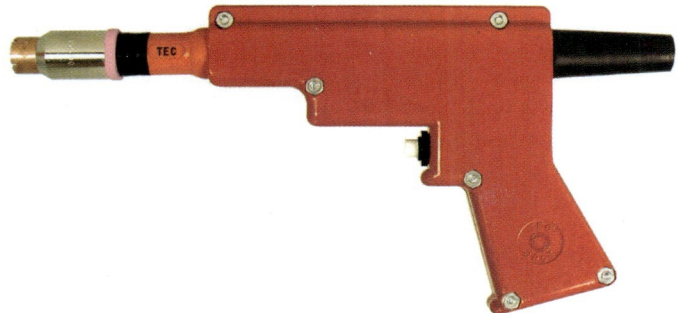

Fig. 18-8 Gun for gas shielded arc spot welding GTAW process. ©TEC Torch Company

gas. The time of the current flow is controlled so that the arc melts through the surface plate to the plate beneath it and forms a weld.

Gas shielded arc spot welding is used mainly in the welding of light gauge sheet metal up to $\frac{1}{8}$ inch thick. It can be used to weld a wide range of metals, such as carbon alloy, stainless steels, and most nonferrous metals. It is used in assembling auto bodies, refrigerators, and other home appliances. Figure 18-10 shows a diagram of a manual gas tungsten arc spot welding system.

Gas shielded arc spot welding has the following advantages over resistance spot welding:

- It is necessary to have access to only the front side of the joint.
- The equipment is portable and commonly available since it is also used for MIG/MAG and TIG welding.
- There is a minimum of spatter, smoke, and sparks.
- There is little distortion of the workpiece.

Comparison of the Gas Shielded and Shielded Metal Arc Welding Processes In any type of welding, the best weld is one that has about the same chemical, metallurgical, and physical properties as the base metal. To obtain such conditions, the molten weld pool must be protected from the atmosphere during the welding operation. Otherwise, atmospheric oxygen and nitrogen combine with the molten weld metal and cause a brittle, porous weld.

In gas shielded arc welding, the weld zone is protected from the atmosphere by an inert gas that is fed through the welding torch. Because of this complete protection from the atmosphere, welds made by a gas shielded arc process are stronger, more ductile, and more resistant to corrosion than those made by the

SHOP TALK

Arc
The space where the electric current jumps between the end of the electrode and the base metal is called an *arc*. It is this gap that causes heat, due to a resistance of current flow.

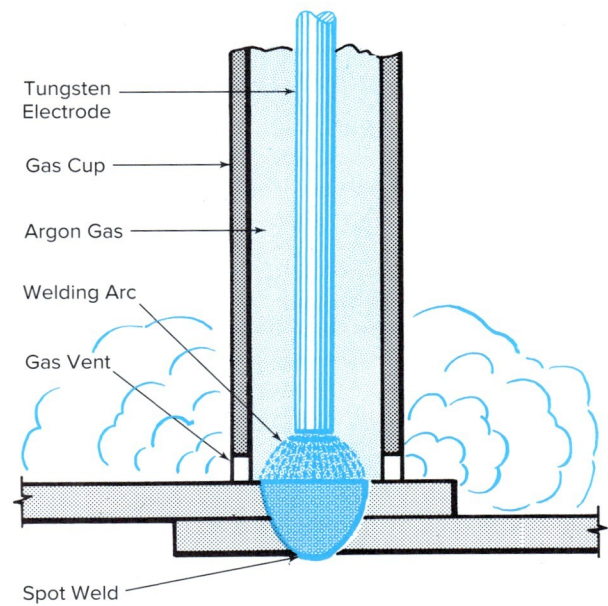

Fig. 18-9 Gas shielded arc spot welding.

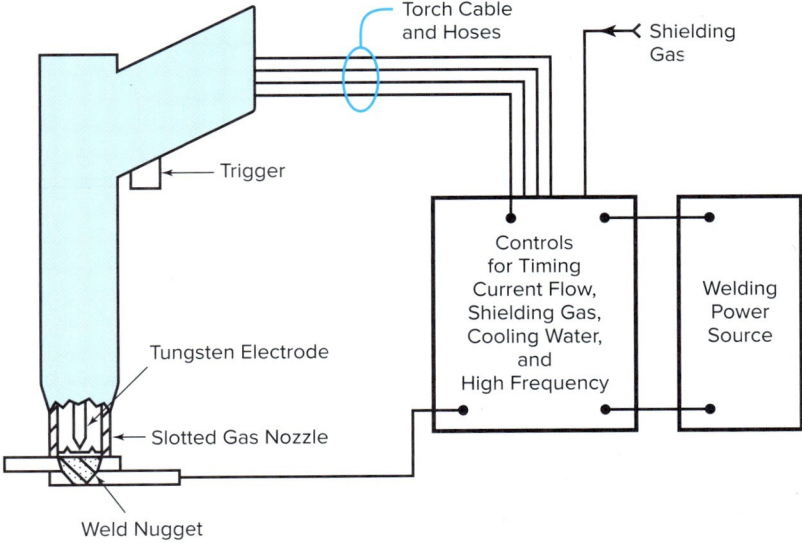

Fig. 18-10 Diagram of manual gas tungsten arc spot welding system.

Gas Tungsten Arc and Plasma Arc Welding Principles **Chapter 18** 553

shielded metal arc (stick electrode) process. The gas shielded arc processes have the following additional advantages:

- The welding heat is confined to a small area. Thus there is a narrow heat-affected zone, and faster welding speeds are possible. Distortion of the welded joint is reduced. The pure heat source can be used on a great variety of metals. And since no metal is transferred across the arc, it can be used effectively in all welding positions. Welds can be made at very high quality levels.
- Welding takes place without spatter, sparks, and fumes. Therefore, weld finishing is kept to a minimum. Welds usually require no finishing at all so that production costs are kept low.
- There is no need for flux. The absence of slag not only reduces the amount of weld finishing, but also produces smoother, nonporous welds. The welder can see (and thus control) the weld pool better.
- The addition of filler metal is by hand and is independent of maintaining the arc. Thus, it may be added only when necessary. Filler metal can often be eliminated when welding thin and medium stock in which good fitup is secured. For fusion welding, the filler metal should match the composition of the base metal. If filler metal is not added, the welder can see the weld pool clearly, thereby improving the appearance and quality of the weld.
- Fusion welds can be made in nearly all common metals. Metals that can be welded by a gas shielded arc process include plain carbon and low alloy steels, cast iron, aluminum and aluminum alloys, stainless steels, brass, bronze, and even silver. Combinations of dissimilar metals can also be welded. Hard-facing and surfacing materials can be applied to steel.

Gas Tungsten Arc Welding

If you have mastered oxyacetylene and shielded metal arc welding, you will have little difficulty learning to weld with the common gas shielded arc processes.

Figure 18-11 illustrates the essential equipment needed for manual TIG welding:

- TIG torch (A)
- Supply of inert gas (B)
- Gas regulator and flowmeter (C)
- Welding transformer (D)
- Water supply (E) and return

A supply of tungsten electrodes, hose for the gas and water, electrical cable, and a fuse assembly or shutoff valve to protect the torch from overheating are also necessary.

Shielding Gases

From your previous studies, you are familiar with the function of the shielding gas. It permits welding to take place in a controlled atmosphere. In shielded metal arc

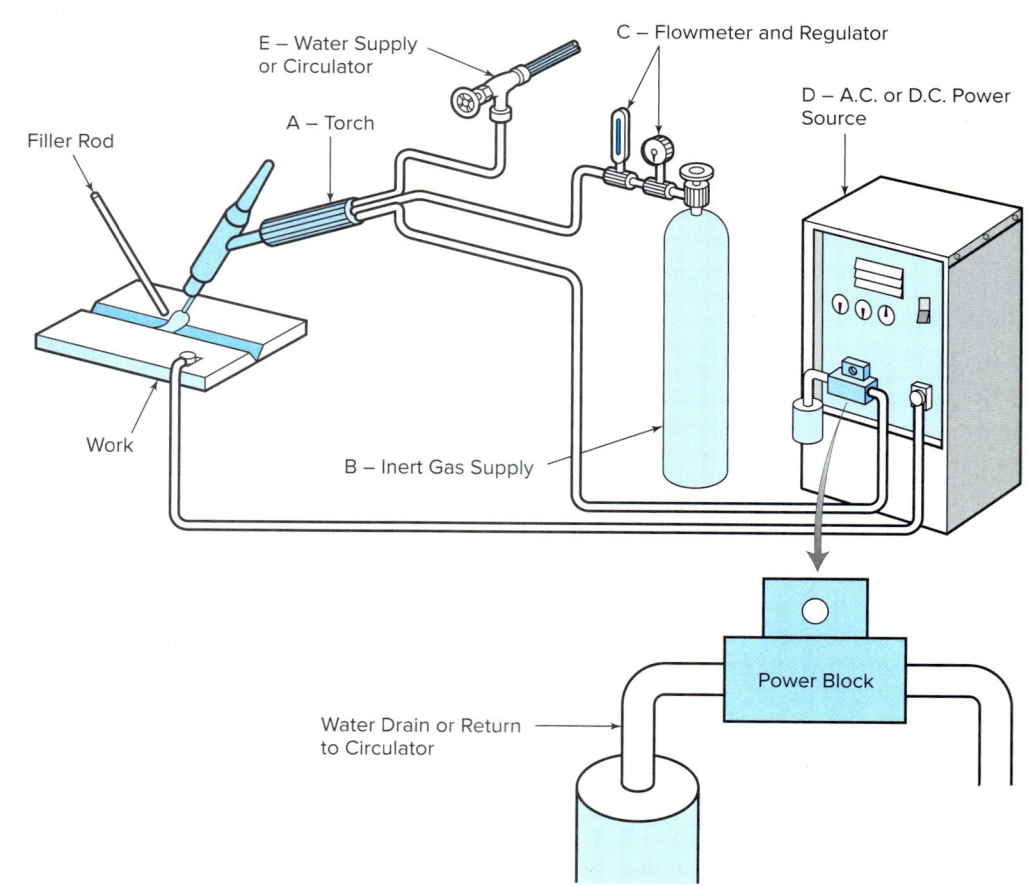

Fig. 18-11 Typical gas tungsten arc installation.

welding, this is accomplished by placing a coating on the electrode that produces a protective atmosphere as it burns in the welding arc. TIG and MIG/MAG welding are similar in that they surround the arc with gases.

In the gas shielded arc welding processes, however, the torch (instead of the electrode) supplies the gas. The gas displaces the air in the arc area before the arc is struck. Because it will displace air in the arc area, it can also displace the breathing air for the welder! These gases can be odorless and tasteless, so there may be no outward indication of a hazard. If welding in confined or enclosed spaces, use an oxygen analyzer to check for hazardous conditions.

Weld Contamination By volume, air is made up of 21 percent oxygen, 78 percent nitrogen, 0.94 percent argon, and 0.06 percent other gases (primarily carbon dioxide). The atmosphere also contains a certain amount of water vapor, depending upon its humidity at any given time. There is hydrogen in the water vapor. Of all the elements that are in the air, the three that cause the most contamination of weld deposits are oxygen, nitrogen, and hydrogen.

Oxygen Oxygen combines readily with other elements in the weld pool to form unwanted oxides and gases. Deoxidizers, such as manganese and silicon, combine with the oxygen to form a light oxide island that floats to the top of the weld pool. In the absence of a deoxidizer, the oxygen combines with iron to form compounds that are trapped in the weld metal and reduce its mechanical properties. Free oxygen also can combine with the carbon in the metal to form carbon monoxide. If carbon monoxide is trapped in the weld metal as it cools, it collects in pockets. This causes porosity in the weld.

Nitrogen Nitrogen is one of the most serious problems in the welding of steel materials because there is so much of it in the air. Nitrogen forms compounds called **nitrides** during the welding of steel that cause hardness, a decrease in ductility, and lower impact resistance. These conditions often lead to cracking in and next to the weld metal. In large amounts, nitrogen also causes serious porosity in the weld metal.

Hydrogen The presence of hydrogen when welding carbon steels produces an erratic arc and affects the soundness of the weld metal. As the weld pool solidifies, the hydrogen is rejected. Part of it, however, is trapped and collects at certain points where it causes stresses. These stresses lead to small cracks in the weld metal that may later enlarge. Other fracture surface discontinuities, such as *fish eyes* and *underbead cracking*, may also result from hydrogen contamination.

Inert Gases and Gas Mixtures Generally, argon or helium serves well for TIG welding. Mixtures containing oxygen may be used on some MIG jobs. Oxygen is not used for TIG welding because tungsten electrodes have a low tolerance for oxygen; however, small amounts of hydrogen (1–7 percent) can be added for special applications.

Argon and helium are both chemically inert. That is, they do not form compounds with other materials. Thus, they do not affect tungsten electrodes or the work material. The welding arc in an atmosphere of either argon or helium is remarkably smooth and quiet. Many other gases and gas mixtures have been experimented with, but they all have some deficiencies such as causing rapid erosion of the tungsten electrode, porosity in the weld metal, arc instability, or high cost. Some noninert gases are active and react chemically with some metals, but not with others. For example, nitrogen can form an effective shield for welding copper. However, it cannot be used for welding mild steel because, at arc temperatures, nitrogen causes porosity in the steel and also forms iron nitrides that embrittle the weld.

If one gas will not provide all the traits desired, a mixture will often improve the action of one gas, or combine the best features of two or more gases. For example, mixing argon with helium gives a balance between penetration and arc stability. Adding 25 percent argon to helium makes penetration deeper than that obtained with argon alone, and the arc stability approaches that of pure argon. Tables 18-1, page 556 and 18-2, page 557, list the performance characteristics of helium and argon.

Helium and Argon Helium is a product of the natural gas industry. It is found in large quantities in oil and natural gas fields. Helium is distributed in steel cylinders similar to oxygen cylinders. At one time it was used extensively for inflating balloons because of its lightness (next to hydrogen, helium is the lightest of all gases) and because it is not flammable.

Argon comprises a little less than 1 percent of the Earth's atmosphere. It is a byproduct of the liquefaction of air during the production of oxygen. Argon is distributed in cylinders like helium. The cylinders are usually yellow or brown.

Although helium and argon are alike in that they do not react chemically with other materials, they have differences that are important in welding.

Argon is used more extensively for TIG welding than helium. It has a smoother, quieter arc and greater cleaning action in the welding of such metals as aluminum and magnesium with alternating current.

Helium is more expensive than argon. Its lightness is also a disadvantage, since it has a tendency to rise from the weld very rapidly. Two or three volumes of helium are required to give the equivalent protection of one volume of argon. Thus, helium is consumed in larger quantities, increasing the cost disadvantage. Furthermore, since argon is denser than helium, the use of argon materially decreases cylinder handling and reduces the amount of space required for cylinder storage.

Table 18-1 Suitability of Argon and Helium for Use as Shielding Gases in Gas Tungsten Arc Welding of Various Metals[1]

Aluminum alloys	Argon (with alternating current) is preferred; offers arc stability and good cleaning action. Argon plus helium (with alternating current) gives less stable arc than argon, but good cleaning action, higher speed, and greater penetration. Helium (with DCEN) gives a stable arc and high welding speed on chemically clean material.
Aluminum bronze	Argon reduces penetration of base metal in surfacing (for which aluminum bronze is used).
Brass	Argon provides stable arc, little fuming.
Cobalt-base alloys	Argon provides good arc stability, is easy to control.
Copper-nickels	Argon provides good arc stability, is easy to control. Used also in welding copper-nickels to steel.
Deoxidized copper	Helium preferred; gives high heat input to counteract thermal conductivity. A mixture of 75% helium and 25% argon provides a stable arc, gives lower heat input than helium alone, and is preferred for thin work metal (1/16 in. or less).
Inconel	Argon provides good arc stability, is easy to control. Helium is preferable for high speed automatic welding.
Low carbon steel	Argon preferred for manual welding; success depends on welder skill. Helium preferred for high speed automatic welding; gives more penetration than argon.
Magnesium alloys	Argon (with alternating current) preferred; offers arc stability and good cleaning action.
Maraging steels	Argon provides good arc stability, is easy to control.
Molybdenum-0.5 titanium alloy	Purified argon or helium equally suitable; welding in chamber preferred, but not necessary if shielding is adequate. For good ductility of the weld, the nitrogen content of the welding atmosphere must be kept below 0.1%, and the oxygen content below 0.005%.
Monel	Argon provides good arc stability, is easy to control.
Nickel alloys	Argon provides good arc stability, is easy to control. Helium is preferred for high speed automatic welding.
PH stainless steels	Helium preferred; provides more uniform root penetration than argon. Argon and argon-helium mixtures also have been used successfully.
Silicon bronze	Argon minimizes hot shortness in the base metal and weld deposit.
Silicon steels	Argon provides good arc stability, is easy to control.
Stainless steel	Helium preferred. Provides greater penetration than argon, with fair arc stability.
Titanium alloys	Argon provides good arc stability, is easy to control. Helium is preferred for high speed automatic welding.

[1]Welding is with direct current electrode negative unless noted otherwise.
Adapted from ASM International.
www.asminternational.org.

More heat is liberated at the arc with helium than argon. Welds of deeper penetration are produced, and welding is faster. Helium is preferred, therefore, for the welding of thick sections of steel and for metals with high thermal conductivity such as aluminum and copper. Argon, on the other hand, is used extensively for welding thin metal and dissimilar metals. Its low thermal conductivity results in a weld deposit with a relatively wide top bead.

In TIG welding, helium produces greater arc voltage (40 percent) per unit of arc length. This results in a hotter arc, deeper penetration, and greater welding speeds. It also reduces the effects of heat on the work. For these reasons, helium is preferred for automatic high production welding.

Table 18-2 Characteristics and Comparative Performance of Argon and Helium as Shielding Gases

Argon	
Arc stability	Arc stability is greater than with helium.
Automatic welding	May cause porosity and undercutting with welding speeds of more than 25 in./min. Problem varies with different metals and thicknesses, and can be corrected by changing to helium or a mixture of argon and helium.
Easy arc starting	Particularly important in welding of thin metal.
Good cleaning action	Preferred for metals with refractory oxide skins, such as aluminum alloys, or ferrous alloys containing a high percentage of aluminum.
Low arc voltage	Results in less heat; thus, argon is used almost exclusively for manual welding of metals less than $1/16$ in. thick.
Low gas volume	Being heavier than air, argon provides good coverage with low gas flows, and it is less affected by air drafts than helium.
Thick work metal	For welding metal thicker than $3/16$ in., a mixture of argon and helium may be beneficial.
Vertical and overhead welding	Sometimes preferred because of better weld pool control, but gives less coverage than helium.
Welding dissimilar metals	Argon is normally superior to helium.
Helium	
Automatic welding	With welding speeds of more than 25 in./min, welds with less porosity and undercutting may be attained (depending on work metal and thickness).
High arc voltage	Results in a hotter arc, which is more favorable for welding thick metal (over $3/16$ in.) and metals with high heat conductivity.
High gas volume	Helium being lighter than air, gas flow is normally 1½ to 3 times greater than with argon. Being lighter, helium is more sensitive to small air drafts, but it gives better coverage for overhead welding, and often for vertical position welding.
Small heat-affected zone	With high heat input and greater speeds, the heat-affected zone can be kept narrow. This results in less distortion and often in better mechanical properties.

Adapted from ASM International. www.asminternational.org.

The lower arc voltage of an argon shielded arc permits changes in the arc length during welding without breaking the arc. With alternating current, helium gives inconsistent starting of the TIG arc. When mixed in the correct proportions with argon, however, helium produces increased speeds, greater current density, and deeper penetration than pure argon. The mixture is best used with square wave or enhanced square wave welding power sources. A mixture of the two gases combines the arc stability of argon shielding and the higher heat input of helium. Refer to Tables 18-1 and 18-2 for additional information.

Hydrogen Hydrogen used with TIG welding has resulted in increased arc potential. Additions of hydrogen increase the heat input, permitting faster travel, increased penetration, better wetting action, and broader weld bead profile. It also produces a cleaner weld bead surface. An argon-hydrogen mixture also improves weld shielding at low gas flow rates. A typical mixture would be 95 percent argon and 5 percent hydrogen. This mixture is most often used on austenitic stainless steels, nickel, and nickel alloys. *Caution:* All hydrogen mixtures are potentially flammable and explosive.

Nitrogen Nitrogen has been used to TIG weld deoxidized copper, duplex stainless steel, and other alloys when mixed with either argon or helium, or both. It is generally in a 1 to 5 percent mixture of nitrogen. Since this gas dissolves in steel at welding temperatures and causes aging effects, it seems unlikely that it will ever be used to weld mild steel. Nitrogen is not a chemically inert gas, but it has been used with success in purging austenitic stainless steel, carbon steel, copper, and low alloy steel.

It should not be used to purge reactive or high nickel alloys.

Refer to Chapter 21 for more detailed information on shielding gases.

Gas Control Equipment

Gas control equipment should provide a uniform flow of the desired quantity of gas around the arc and weld pool during welding. The gas is usually supplied in high pressure cylinders or by a piped system supplied by a cylinder manifold or liquid bulk tank.

Flowmeter Flowmeters indicate the rate of flow of inert gas to the torch. They are calibrated to indicate flows in cubic feet per hour.

Conventional welding regulators reduce the gas pressure, but they do not provide any measure of gas flow. As a result, a gas flowmeter has been incorporated with a regulator, Fig. 18-12. A separate flowmeter may also be used with a standard gas regulator, Fig. 18-13.

Gas Flow Rates It is difficult to predict the exact gas flow rate for a given welding application. A general rule to follow is that the gas flow rate should be high enough to protect the weld pool and the end of the tungsten electrode.

Too much gas is wasteful, and it can cause porosity in the weld deposit because the gas cannot flow from the weld metal fast enough during cooling. High flow rates can produce turbulence exiting the gas nozzle. This can introduce atmospheric contamination into the gas

Fig. 18-13 An individual flowmeter can be used with a standard regulator. Thermadyne Industries, Inc.

shielding. Not enough gas flow permits contamination of the weld pool by the surrounding air and causes oxidation of the weld deposit.

Careful inspection of the end of the tungsten electrode will indicate if the proper quantity of gas is being provided. Proper gas flow leaves the tungsten electrode smooth and rounded on the tip. Improper gas flow causes burning, scale, and deformation of the electrode tip.

> **JOB TIP**
>
> **Job Web Sites**
> Employment can be found on the Web. Search the Web for job Web sites and share with classmates.

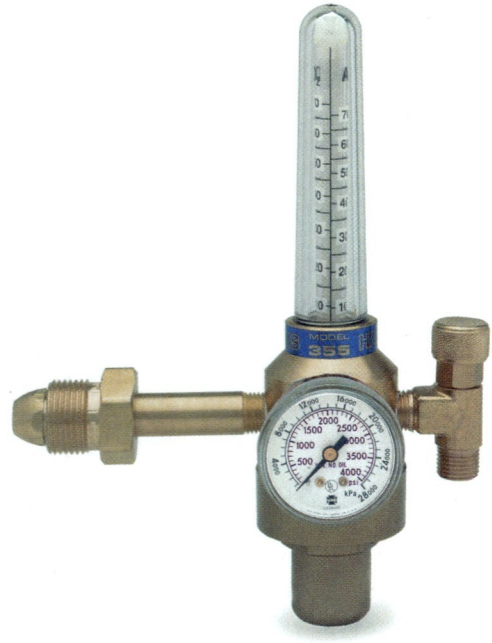

Fig. 18-12 Combination regulator and flowmeter unit.
The Lincoln Electric Co

558   Chapter 18   Gas Tungsten Arc and Plasma Arc Welding Principles

The correct gas flow for TIG welding depends on the following variables:

- Type of shielding gas
- Type and design of the weld joint
- Distance of the nozzle from the work
- Size and shape of the nozzle
- Size of the weld pool
- Amount and type of welding current
- Position of welding
- Position of the torch in relation to the work
- Design and type of jigs
- Speed of welding
- Air currents in the welding area

This last condition is very important since a strong draft across the weld area can dispel most of the shielding gas.

Some standards allow up to 5 miles per hour of drafting air without the use of screening for protection. However, when doing precision GTAW, this use should be limited to 1 to 2 miles per hour.

Shutoff Valves In most welding operations, the arc is maintained for only short periods, and consequently there are many starts and stops during a day's operation. In order to avoid wasting gas, a GTAW power source will have a built-in gas solenoid valve to turn the gas off and on at the proper time. These gas solenoid valves have fittings that allow the gas supplied from the gas cylinder and gas hose to the torch to be connected, Fig. 18-14. The solenoid valve is connected to a timer circuit to control postflow time. If the power source is designed for automatic welding, it will also have a preflow timer. Some electrode holders are equipped with gas shutoff valves in the handle.

Fig. 18-14 Gas solenoid valve used to control gas flow. Note the fitting for the gas hose connection. Miller Electric Mfg. Co.

Power Sources

Alternating current and direct current, either electrode negative or electrode positive, are used in gas tungsten arc welding. With direct current, electrode negative is used more than electrode positive.

Welding power sources are often classified as being either *rotating* or *static*. The rotating types are generators/alternators, driven most often by an internal combustion engine. They put out direct current or both alternating and direct current. The static types are *transformers* with a.c. output and *transformer-rectifiers*, solid-state thyristor controlled transformer-rectifiers, and fast switching high cycle inverters, all with d.c. or a.c.-d.c. output.

The machines used for TIG welding are classified as constant current (variable voltage) machines. A constant current arc welding power source is one that has a drooping output slope. It produces a relatively constant current with a limited change in load voltage. This type of supply can also be used for stick electrode welding. (For a detailed explanation of the output slope, refer to Chapter 11, pp. 279–287.)

In a constant current power source, the welding current remains substantially constant even though the arc length varies and the arc voltage changes. Thus, it is highly suited to most manual operations, including TIG welding, where variations in arc length are most apt to occur. This type of power source is also called *drooping voltage* because output voltage goes down (droops) as the arc length is shortened. It is also said to droop as the current (amperage) goes up. You will note in Fig. 18-15 that when the voltage is 20, amperage is 210. When the voltage drops to 10, the amperage increases only slightly. This volt-amp curve is from an inverter-type power source and maintains the amperage very precisely.

Power sources for precision TIG welding on thin metals may be rated as low as 50 amperes and provide current selections separated by ½-ampere steps, Fig. 18-16, page 560. These machines can make perfect welds on metals as thin

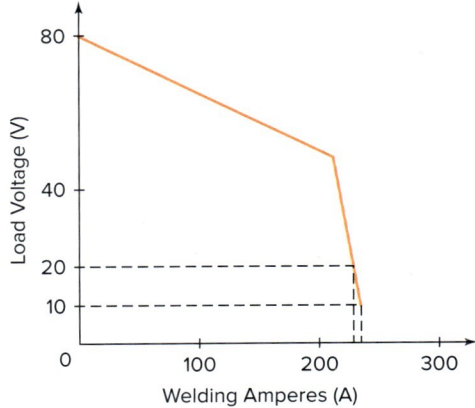

Fig. 18-15 The volt-amp curve of a constant current power source.

Fig. 18-16 This power source provides output from 5 to 300 amperes. It is an a.c.-d.c. inverter GTAW/SMAW power source. It features frequency control and expanded balance control for enhanced square wave output. Its weight is only 90 pounds. Miller Electric Mfg. Co.

as 0.001 inch. For the general run of work, machines of 150 to 300 amperes are generally used. The open circuit (no load) voltage on most constant current machines is usually between 55 and 100 volts. During welding, the closed circuit voltage (load voltage) usually drops to between 15 to 40 volts on constant current welders.

One of the most important ratings for a welding power source, the **duty cycle**, expresses as a percentage, the portion of time that the power supply must deliver its rated output in each 10-minute interval. For manual TIG welding, industrial units rated at a 60 percent duty cycle are generally satisfactory. For automatic processes, a 100 percent duty cycle is required. Limited service and limited input power supplies rated at a 20 percent duty cycle are available for light work, such as in garages, repair shops, and on farms.

Welding power sources designed for TIG will incorporate the following controls and features:

1. Process control TIG or stick.
2. Current control with a remote or the amperage adjustment control on the machine.
3. Output contactor control is On for stick or Remote for TIG.
4. Arc start mode off for scratch start or stick, lift arc for frequency start for TIG; high frequency start is only generally for d.c. welding and high frequency continuous is for a.c. welding.
5. Voltmeter generally an option needed for procedure work.
6. Ammeter generally an option needed for procedure work.
7. Amperage adjustment control.
8. Output selector switch to control a.c. or DCEN or DCEP.
9. Power switch to turn the power source on and off.
10. Postflow timer prevents atmospheric contamination of tungsten, weld pool, and filler metal as it cools 1 second for each 10 welding amperes.
11. Preflow timer for automatic welding or critical work to eliminate the possibility of start contamination.
12. Balance control of a.c. TIG and dig control for stick.

Pulse Control

1. Pulse control to vary from a continuous current to a pulsed current.
2. Low amperage (background current) adjustment of the pulse.
3. Adjust the number of pulses per second.
4. High amperage (peak current) percentage of time at this level versus the low amperage (background current) peak amperage is set on the main amperage adjustment control.

Pulse Current Versus Continuous Current

1. At a given current level less heat input.
2. Less distortion.
3. More uniformity on unequal thickness and thin metals.
4. Set parameters can be used in all positions.
5. Bridging gaps and for open root joints.
6. High frequency pulsing mixes weld pool for greater alloy consistency.
7. Paces inexperienced welders by counting pulses.

For video of GTAW-P function of controls, please visit www.mhhe.com/welding.

See Fig. 18-17 for examples of these controls and features.

Welding Current Characteristics The different types of operating current directly affect weld penetration, profile, heat input, cleaning action, and ripple pattern.

In d.c. welding, the welding current circuit may be hooked up as either electrode negative (the nonstandard term is straight polarity) or electrode positive (the nonstandard term is reverse polarity). The work would be connected to the positive terminal for DCEN. In other words, the electrons flow from the electrode to the work,

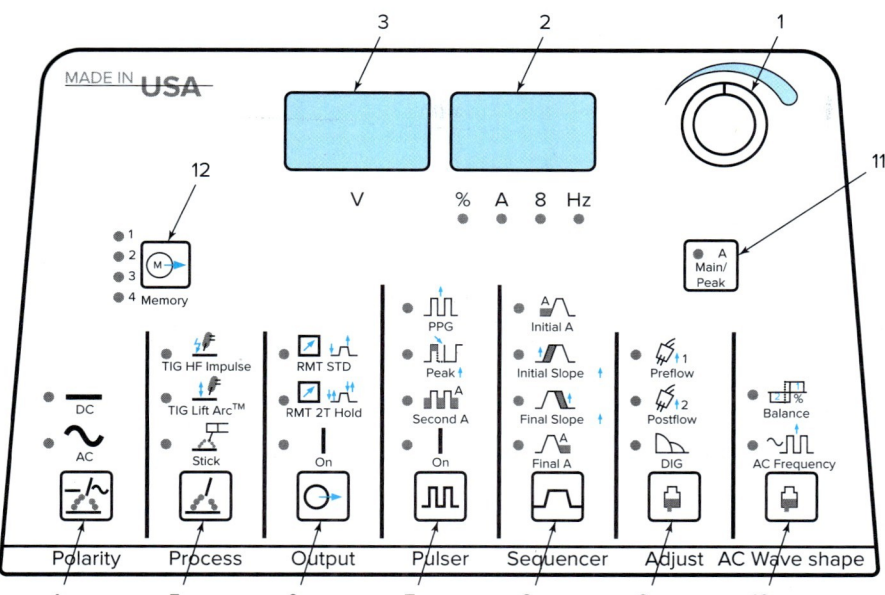

Fig. 18-17 Front panel layout of a typical a.c.-d.c. square wave GTAW power source.

1. Encoder Control: Use this control in conjunction with applicable from panel function switch pad to set values for that function.

2. Meter: For display of actual amperage while welding. Meter also displays preset parameters for the following: % of balance selected, A amperage, S time, and Hz frequency of a.c. arc. The corresponding LED, located directly below the meter will inform you of what function is being displayed.

3. Voltmeter: Displays output or open circuit voltage. If output is off, the voltmeter will display a series of three (– – –). Open circuit voltage is displayed if power is on and output is available.

4. Polarity Control: Press switch pad until desired LED is illuminated. For d.c. the machine will set for DCEN when the TIG process has been selected and DCEP when the stick process has been selected. A.C. can be used for TIG or stick welding.

5. Process Controls: Press switch until desired process LED is illuminated. For TIG HF impulse (for noncontact a.c. or d.c. TIG, for TIG lift arc) electrode must come into contact but no HF will be generated or stick (SMAW) will have adaptive hot start and DIG enabled.

6. Output Controls: Press switch until desired output LED is illuminated. (RMT STD) remote standard when a remote foot, hand, or torch button is used, (RMT 2T hold) remote trigger hold is used for long extended welds to reduce welder fatigue of having to hold switch on, and simply on for stick (SMAW) or use of lift-arc for TIG without use of remote control.

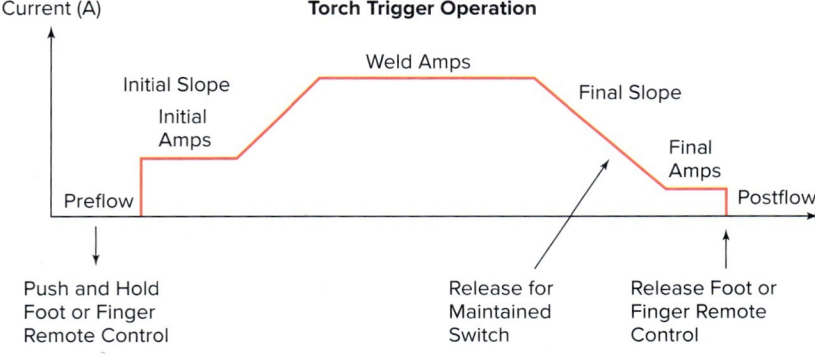

7. Pulser Controls: Press switch until desired pulser LED feature is illuminated. On to indicate pulser is activated, pps for pulse frequency from 1–500 pulses per second, peak time 5–95% of each pulse cycle cannot be spent at the peak amperage level, and background amperage to set the low pulse of the weld amperage which cools the weld pool.

8. Sequencer Controls: For setting initial amperage, initial slope, final slope, and final amperage. These controls are typically used for mechanized welding where a remote foot or hand amperage control is not usable.

9. Adjust Controls: For setting preflow of shielding gas, postflow of shielding gas, and DIG for stick welding.

10. A.C. Waveshape: Balance to control time at the positive and negative half cycles 50–90% electrode negative, and a.c. frequency from 20–250 Hz for more directional control of the welding arc. As a.c. frequency increases, weld bead/pool becomes narrower and the arc becomes more focused.

11. Amperage Control: Press amperage switch pad and turn encoder control to preset amperage from 5–300 amperes.

12. Memory Function Keys: Allow the selection of up to four separate machine settings, similar to a radio being able to select a predetermined station by simply depressing its button.

Gas Tungsten Arc and Plasma Arc Welding Principles **Chapter 18** 561

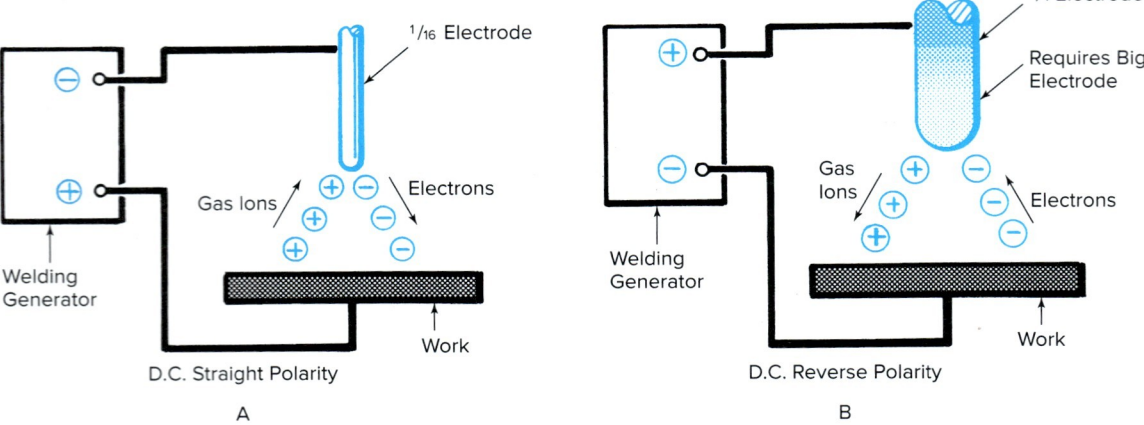

Fig. 18-18 Current characteristics and flow: (A) DCEN. (B) DCEP.

concentrating heat at the joint, as shown in Fig. 18-18A. The work is connected to the negative terminal for DCEP. Positive gas ions flow to the joint, cleaning it, while the electrons flow from the work to the electrode, as shown in Fig. 18-18B.

Direct Current Electrode Negative In DCEN welding, the electrons carry up to 70 percent of the heat energy in the arc to the plate. The work tends to melt rapidly since it receives the greater part of the heat. The positive gas ions are directed toward the negative electrode; the positive electrode experiences only 30 percent of the heat effects. No cleaning action for the removal of oxides from the metal is present with DCEN.

With DCEN, higher currents may be used on tungsten electrodes of the same size. This increased current density will yield deeper penetration, permit higher speeds, and form a narrow, deep bead, Fig. 18-19A. The narrower bead is due to the smaller electrode used, while the increased penetration is due to the heat being carried by the electrons. The narrower bead also results in a narrower heat-affected zone surrounding the weld. Because there are fewer contraction stresses, less trouble with hot cracking of some metals is encountered. The DCEN type of operating current is preferred for welding metals that do not have a refractory oxide such as aluminum or magnesium.

For video of GTAW DCEN heat transfer, please visit www.mhhe.com/welding.

Direct Current Electrode Positive In DCEP welding, the tungsten electrode acquires extra heat because the electrons are flowing toward it instead of the work. The heat tends to melt the tip of the electrode. Because of this excess heating of the electrode, a much larger size electrode is required for DCEP than for DCEN. Electrode size limits welding to relatively low current values. Large diameter electrodes are undesirable because they tend to spread the arc over too large an area, cut down the welder's visibility, and increase the instability of the arc.

With DCEP the work stays comparatively cool, and shallow penetration results, Fig. 18-19B. An important characteristic of DCEP welding is its oxide cleaning ability. This is especially helpful with such metals as aluminum, magnesium, and beryllium-copper. Electrode positive is not generally recommended for TIG because of the excess heat it puts into the tungsten. However, it is present during the electrode positive half cycle of a.c. welding. Although the exact reason for the surface cleaning action is not known, one explanation is that the relatively heavy positive gas ions, in relation to the much smaller and less dense electrons, act much like a miniature sandblaster, chipping away at the dense surface oxides. Then the electrons exiting this surface disperse the oxides from the arc area. This cleaning action is not intended to be used to remove excess oxides, dirt, grease, oil, or other contaminants from the joint prior to welding. Other cleaning methods, mechanical or chemical, should be used prior to arc cleaning.

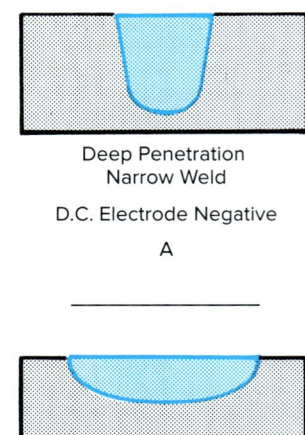

Fig. 18-19 Effect of polarity on weld shape. Comparative weld contours.

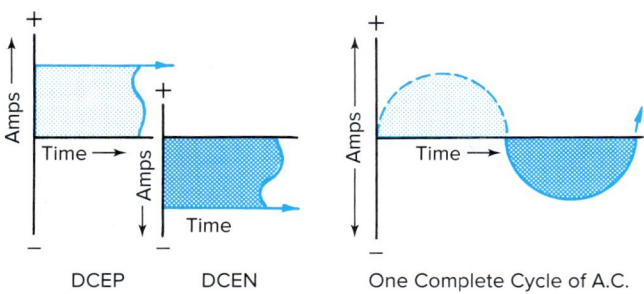

Fig. 18-20 Alternating current cycle.

For video of GTAW DCEP, please visit www.mhhe.com/welding.

Alternating Current Welding Alternating current is widely used for TIG welding. It is recommended principally for light metals that have dense refractory oxides, such as aluminum and magnesium.

For video of GTAW AC, please visit www.mhhe.com/welding.

Theoretically, a.c. welding is a combination of DCEN and DCEP welding. This can be best explained by showing the three current waves visually. As shown in Fig. 18-20, half of each complete a.c. cycle is DCEN; the other half is DCEP.

Since the heat is evenly distributed at the two ends of the arc, the depth of weld penetration is less than that obtained with DCEN, but more than that which would be obtained with DCEP. The profile of the weld lies between that of the deep, narrow type produced by electrode negative and the wide, shallow type profile produced by electrode positive, Fig. 18-21.

This is true with the a.c. sine wave power sources. Current production GTAW welding power sources do not produce sine wave output as shown in Fig. 18-20, but square wave output as shown in Fig. 18-22.

If a.c. TIG welding is going to be done with a.c. sine wave power sources, or an add-on high frequency unit is added to an a.c. sine wave stick machine, then rectification will take place. To protect the power source, derate it as specified by the manufacturer or use Table 18-3.

Square Wave Current Welding This type of welding has many advantages over sine wave welding, by giving the welder better control over the arc, weld pool, bead profile, and etched cleaning action along the toes of the weld. These types of power sources have become the standard of the industry because of their versatility.

Advantages of Square Wave Output

- Reduced arc rectification for a more stable arc.
- No tungsten spitting or inclusions for X-ray quality welding.
- Control over the electrode positive half cycle time for the correct amount of cleaning.
- Control over the electrode negative half cycle time for increased speed and penetration.

Weld Result Summary

D.C. Electrode Negative

D.C. Electrode Positive

A.C. Welding

Fig. 18-21 Comparison of weld contours.

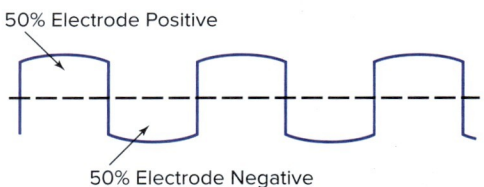

Fig. 18-22 Square wave output waveform.

Table 18-3 Derating A.C. Sine Wave SMAW Power Source for TIG 100% Duty Cycle

Power Source Present Duty Cycle (%)	Power Source Rated Amp Multiple (%)	Non-TIG Application Duty Cycle (%)	Derated Duty Cycle A.C. TIG Welding Derated Amp Multiple (%)
60	75	100	70
50	70	100	70
40	55	100	70
30	50	100	70
20	45	100	70

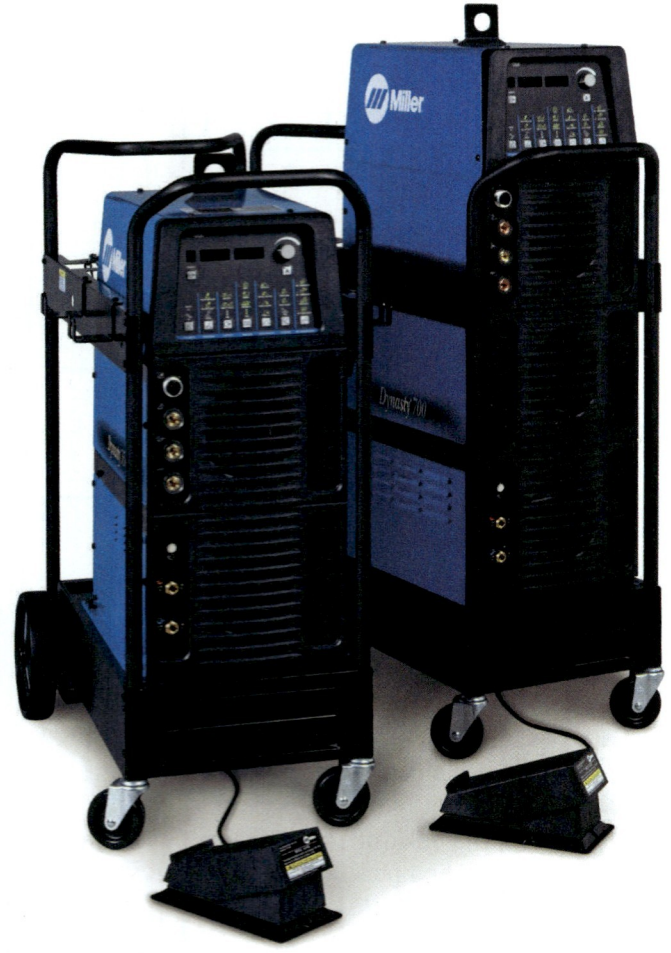

Fig. 18-23 How well the arc can be controlled is important when selecting equipment. The welder must be able to shape the arc and control the weld bead. Miller's Aerowave 300 (an a.c.-d.c. power source) features advanced square wave arc, independent current control, adjustable frequency, a.c. balance control, and lift arc for TIG arc starts. Examples of industrial uses of this machine include aerospace, tube and pipe manufacture, bicycle assembly, fillet welds, and automation. Miller Electric Mfg. Co.

- Control over arc frequency for greater control over arc direction for reduced arc wandering.*
- Control over the electrode positive current level for a further reduction of the cleaning action.*
- Control over the electrode negative current level expands speed and penetration action.*

These controls—required for waveform shaping, which affects weld appearance, efficiency, and quality—are shown in Fig. 18-23.

Arc Starting Methods Since the TIG process uses a nonconsumable electrode, the electrode is not compatible with the metal being welded. Any contact of the hot electrode with

---

*These controls are available only on enhanced square wave power sources.

JOB TIP

Job Hunting
Looking for a job is a job! Make a list of what you plan to do in the next week. Assess what kind of job you want. As you complete items on your list, you not only will be closer to your goal, but also will be in control of the job-hunting process and will be less stressed out.

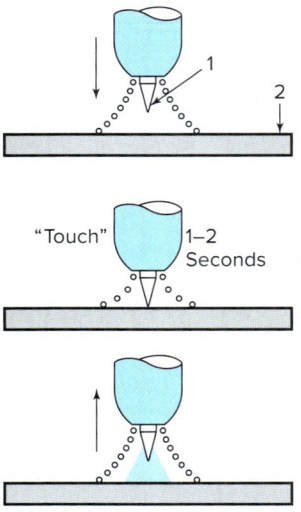

Fig. 18-24 Note technique used with lift arc starting system.

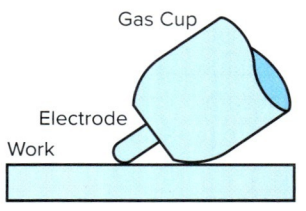

Fig. 18-25 Proper starting technique with high frequency start.

the molten weld pool or filler metal will cause contamination to take place. If the electrode is contaminated with the metal being welded or the filler rod, the arc will become unstable resulting in loss of arc direction, plus a black, sooty smoke will be noticed in the weld area. If the weld bead becomes contaminated with tungsten, this is referred to as an inclusion and may be considered a defect, depending upon the acceptance criteria being applied.

Scratch starting can be used for some applications. This method is very similar to the arc starting method used for stick welding. It is used with DCEN, and on some machines with the proper techniques, X-ray quality welds can be made. On other machines, scratch starting is very difficult and generally leads to tungsten inclusions. Lift arc is another method similar to scratch starting. The power source has a switch to select this mode. This reduces the power level so that there is little chance of tungsten inclusion. Figure 18-24 is an example of this method, which can be used on a.c. or d.c. welding.

High frequency is another method that can be used for arc starting and arc stability, especially when using an older sine wave a.c. power source. Figure 18-25 is an example of a good starting technique with high frequency starting. High frequency presents other problems, it is fundamentally a radio signal generator that can cause interference problems with computers, communication

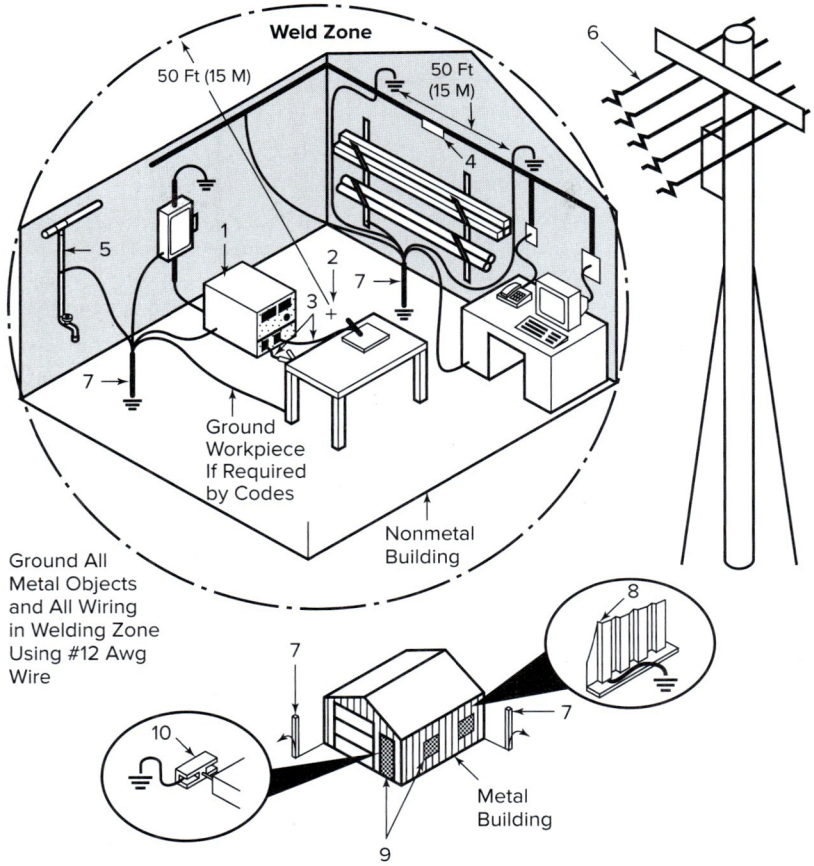

1. HF Source (Welding with Built-in HF or Separate HF Unit): Ground metal machine case, work output terminal, line disconnect device, input supply, and worktable.

2. Welding Zone and Center Point: A circle 50 ft (15 m) from center point between HF source and welding torch in all directions.

3. Weld Output Cables: Keep cables short and close together.

4. Conduit Joint Bonding and Grounding: Electrically join (bond) all conduit sections using copper straps or braided wire. Ground conduit every 50 ft (15 m).

5. Water Pipes and Fixtures: Ground water pipes every 50 ft (15 m).

6. External Power or Telephone Lines: Locate HF source at least 50 ft (15 m) away from power and phone lines.

7. Grounding Rod: Consult the National Electrical Code for specifications.

8. Metal Building Panel Bonding Methods: Bolt or weld building panels together, install copper straps or braided wire across seams, and ground frame.

9. Windows and Doorways: Cover all windows and doorways with grounded copper screen of not more than 1/4-inch (6.4-mm) mesh.

10. Overhead Door Track: Ground the track.

Fig. 18-26 An HF equipped power source, installed in compliance with the FCC.

systems, and certain medical devices. The federal government, through the Federal Communications Commission (FCC), regulates any industrial or medical device that generates radio signals. Figure 18-26 shows how a high frequency equipped power source must be installed.

Advantages of High Frequency (HF)

- The arc may be started without touching the electrode to the work. Thus, less contamination is picked up on the tungsten electrode, and the weld deposit is cleaner.
- Better arc stability is obtained with a.c. sine wave and square wave machines at low current settings.
- A longer arc can be maintained for some applications, such as surfacing and hard surfacing.
- There is an increase in electrode life of almost 100 percent.

- The use of wider current ranges for a specific diameter electrode is possible. Also, less current is required for a given weld.
- It is easier to make all position welds.

Resolving Arc Starting Difficulties If difficulties are encountered especially when doing automatic welding, the following points will prove helpful:

- Use good quality tungsten.
- Use good weld cables that are as short as possible. Long cables should not be coiled up. Insulation must not be cracked or missing. HF is generated at several thousand volts.
- Attach work cables as close to the work as possible; don't pass electric current through gears or bearings; disconnect all printed circuit boards on any vehicle or project being welded.
- Remove any other cables attached to the welding power source such as SMAW electrode leads.
- Keep the torch power cable and connections from touching any metal parts that may be connected to the building ground.
- Make sure all the connections are clean and tight. Aluminum oxide is not a very good conductor.
- Increase the HF intensity knob if available.
- Check the spark gap adjustment of the power source.
- Use pure argon. Helium makes arc starting more difficult.
- Concrete can bleed off the HF through the weld cables. Insulate long runs of cable by placing them on boards or other suitable forms of insulation.
- Use the shortest torch possible. If a long distance from the power source is required, use one of the add-on HF units.
- Use the smallest diameter tungsten for the amperage being used.

Arc Rectification with GTAW Moisture, oxides, and scale on the surface of the plate tend to prevent the flow of current in the electrode positive direction. This is called rectification. For example, if no current at all flowed in the electrode positive direction, the current wave would look like that shown in Fig. 18-27.

The arc would be unstable and sometimes even go out. Figure 18-28 shows partial rectification, which the welder will notice as a change in the arc sound, weld pool agitation, and tungsten spitting due to overcurrent spikes. To prevent this from happening, it is a common practice to introduce into the welding current a high voltage, high frequency, low power, and additional current.

Figure 18-29 shows this high frequency current being superimposed over the normal a.c. sine wave. This

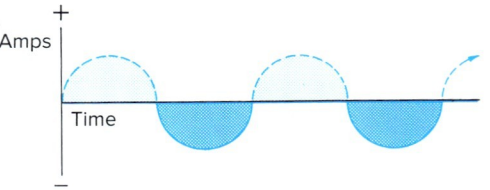

Two Complete Cycles of A.C. with Electrode Positive Completely Rectified.

Fig. 18-27 Rectified or interrupted a.c. cycle.

current jumps the gap between the electrode and the workpiece. It helps break down the oxide film, thereby forming a path for the welding current to follow. Arc stability can also be improved if the arc is not allowed to dwell on the molten and highly cleaned weld pool. To further improve arc stability, keep the arc moving and/or add filler rod.

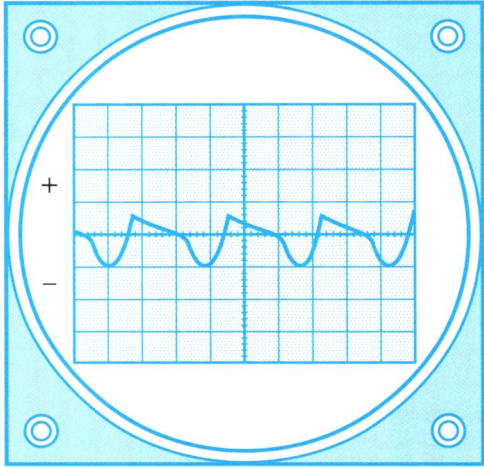

Fig. 18-28 Partial rectification as represented on an oscilloscope. The positive half cycles have been clipped off. This clipping is the result of partial rectification in the arc of a sine wave power source. Source: Miller Electric Mfg. Co.

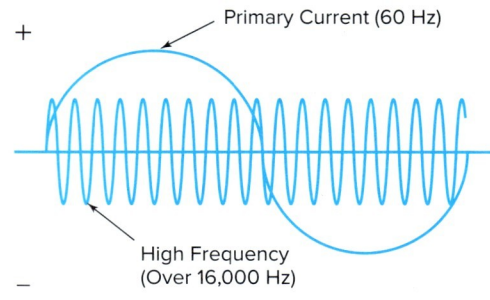

Fig. 18-29 High frequency compared to low frequency a.c. sine wave.

Table 18-4 Current Selection for TIG Welding[1]

Material	Alternating Current[2] With High Frequency Stabilization	Direct Current Electrode Negative	Direct Current Electrode Positive
Magnesium up to 1/8 in. thick	1	N.R.	2
Magnesium above 3/16 in. thick	1	N.R.	N.R.
Magnesium castings	1	N.R.	2
Aluminum up to 3/32 in. thick	1	N.R.	2
Aluminum over 3/32 in. thick	1	N.R.	N.R.
Aluminum castings	1	N.R.	N.R.
Stainless steel	2	1	N.R.
Brass alloys	2	1	N.R.
Silicon copper	N.R.	1	N.R.
Silver	2	1	N.R.
Hastelloy alloys	2	1	N.R.
Silver cladding	1	N.R.	N.R.
Hard-facing	1	1	N.R.
Cast iron	2	1	N.R.
Low carbon steel, 0.015–0.030 in.	2	1	N.R.
Low carbon steel, 0.030–0.125 in.	N.R.[3]	1	N.R.
High carbon steel, 0.015–0.030 in.	2	1	N.R.
High carbon steel, 0.030 in. and up	2	1	N.R.
Deoxidized copper[4]	N.R.	1	N.R.

[1] Key: 1. Excellent operation. 2. Good operation. N.R. Not recommended.
[2] Where a.c. is recommended as a second choice, use about 25% higher current than is recommended for DCEN.
[3] Do not use a.c. on tightly jigged part.
[4] Use brazing flux or silicon bronze flux for 1/4-in. stock and thicker.

Current Selection for TIG Welding Table 18-4 is a handy guide as to the type of current that should be used for a given job. It also indicates the different kinds of metals that can be welded with the TIG process. The training course, as outlined in Chapter 19, Table 19-15, page 614, is limited to welding on mild steel, stainless steel, and aluminum. After these have been mastered, other metals may be welded easily.

DC Power Sources for TIG Welding For d.c. welding power, any standard engine-driven welding generator-alternator or transformer-rectifier with an adequate amperage capacity for the job may be used. It is important that the generator-alternator or transformer-rectifier have good current control at the lower end of its current range. Arc stability is essential, particularly for welding thin-gauge materials. Refer to Chapter 11, pages 279–286, for a detailed discussion of d.c. welding machines used for both TIG and stick electrode welding.

Power Supplies for TIG Welding The common requirement of all a.c. TIG welding machines is that they produce a constant current output and have arc starting and arc stabilization features. Standard a.c. welding transformers cannot be used successfully for TIG welding without the addition of an HF unit. The open circuit voltage required to start current flowing is far above that supplied by the ordinary a.c. transformer for noncontact starts.

There are a.c. transformer power sources designed especially for the TIG welding process. They are equipped with high frequency controls and gas and water solenoids and other current controls such as start and crater fill.

Usually the power source has a switch with four arc starting modes: (1) Off, (2) continuous, (3) start only, and (4) lift arc. When used at the start only mode, the high frequency is on when the foot control is depressed. It helps establish the arc and then shuts off automatically. When continuous, the high frequency is on as long as the foot pedal is depressed. In the lift arc mode, the power level is reduced while the tungsten is shorted to the work. Most TIG power sources have a control that varies the high frequency intensity. Experienced TIG welders set this control for optimum arc starting.

Many of these machines can be used for stick electrode, TIG, TIG spot, and automatic welding. Figures 18-30 through 18-39 represent various types of constant current welding machines used for TIG welding. These machines can be controlled by remote control, Figs. 18-40 through 18-43.

Example of Machine Controls The following outline is presented so that you can better understand the purpose and function of each control on a typical combination gas tungsten arc and shielded metal arc welding machine. Proper setting of the controls is necessary to produce sound welds of good appearance. Figure 18-34, page 570, illustrates the listed controls.

Transformer-Rectifier A.C.-D.C. Arc Welding Solid-State Machine This example of an a.c.-d.c. transformer-rectifier solid-state-controlled machine has square wave output from 5 to 310 amperes and the following functions.

Fig. 18-30 A constant current a.c.-d.c. square wave power source for TIG welding of various metals. It would be used in light industry or in vocational schools. Miller Electric Mfg. Co.

A. Capabilities
   1. Both TIG and stick electrode welding.
   2. Ability to join all ferrous and nonferrous alloys that can be arc welded.

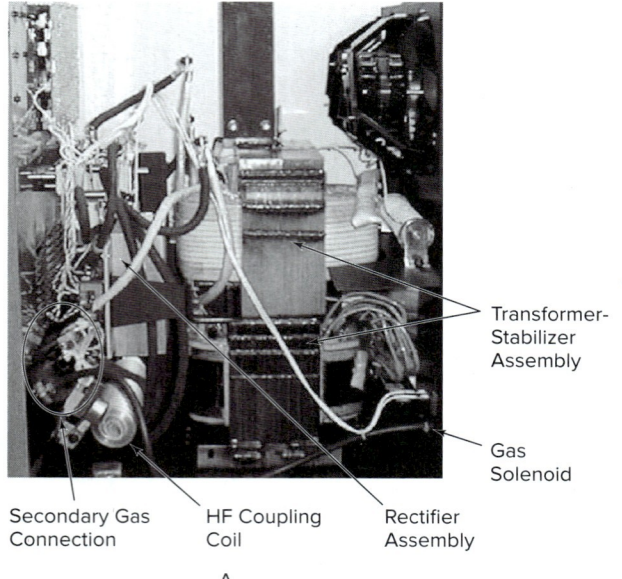
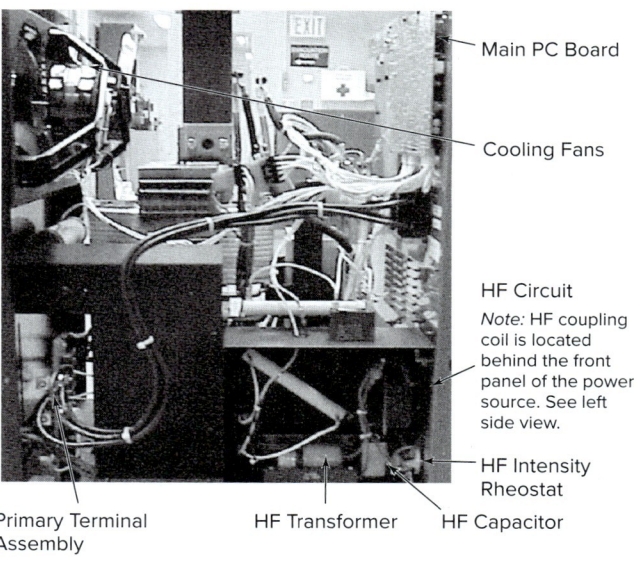

Fig. 18-31 (A) Right side view of the interior construction. (B) Left side view of the interior construction. Miller Electric Mfg. Co.

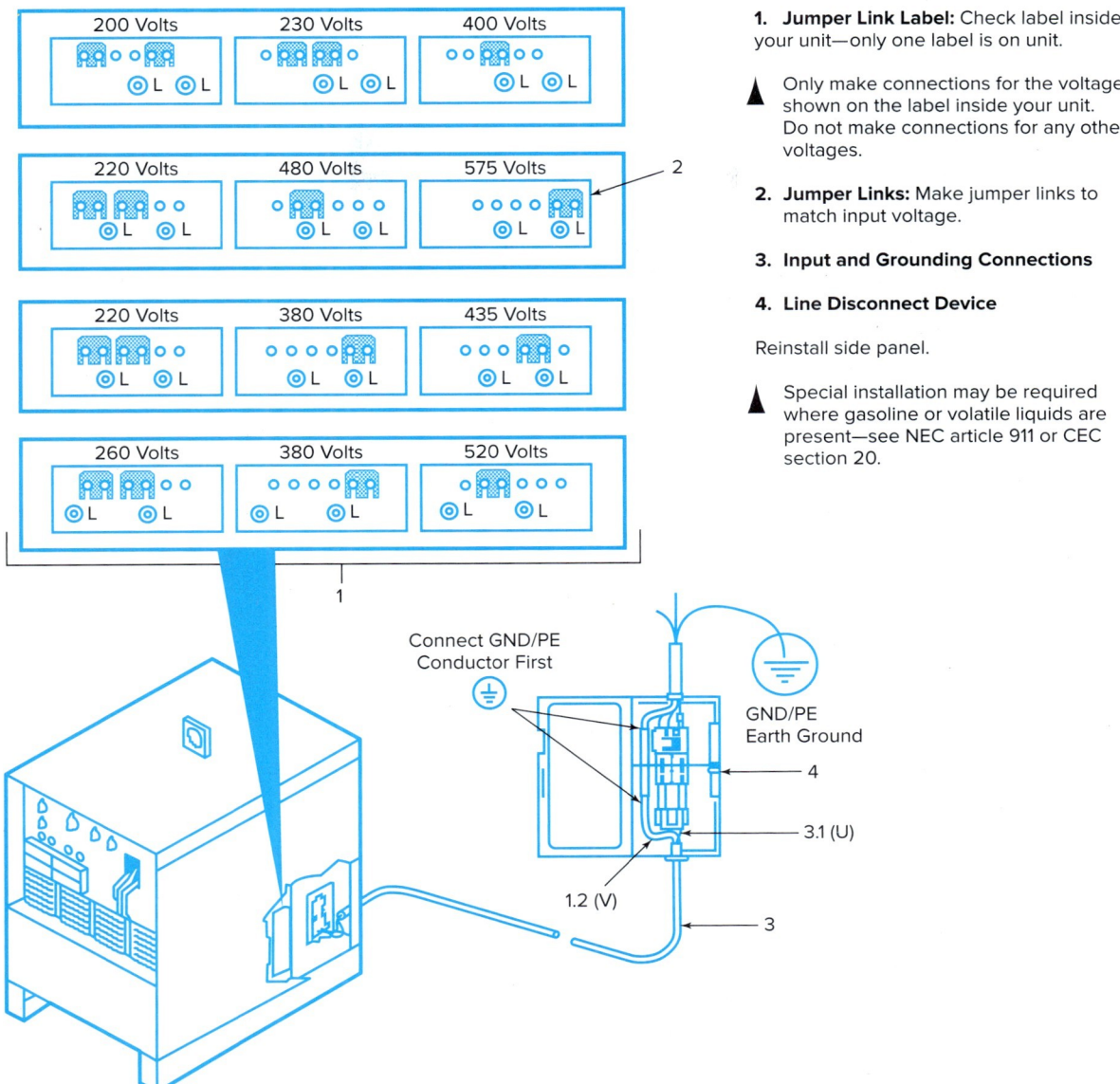

Fig. 18-32 Primary terminal hookup and linkage for a typical welding machine.

3. Current range of 5 to 310 amperes, a.c. and d.c., weld material from gauge metal through heavy plate thickness.
4. Elimination of tungsten spitting and arc rectification (melting of electrode tip).
5. Balance control for adjustable penetration and cleaning action for better TIG arc performance and dig control for stick welding penetration and spatter control.
6. Last procedure recall, which automatically resets operations selected based on polarity switch.
7. Optional remote hand or foot amperage controls for finer resolution and cratering out.
8. Optional pulse control to reduce heat input and distortion.
9. Optional sequencer for mechanized control of initial amperage and time and also final amperage and final slope as well as spot time.
10. Digital ammeters and voltmeters for viewing of actual and preset values for greater accuracy and repeatable welding procedures for alternating or direct current.
11. Fan on demand reduces maintenance by reducing the amount of airborne particles pulled into the machine. It also reduces power consumption.

Fig. 18-33 This power supply is portable, with coolant system and foot remote control. Miller Electric Mfg. Co.

Fig. 18-35 This machine is recommended for pipe welding and other precision-type welding. It can be used for manual or automatic welding and for gas tungsten arc or shielded arc welding. It has a single output from 3 to 400 amperes. Miller Electric Mfg. Co.

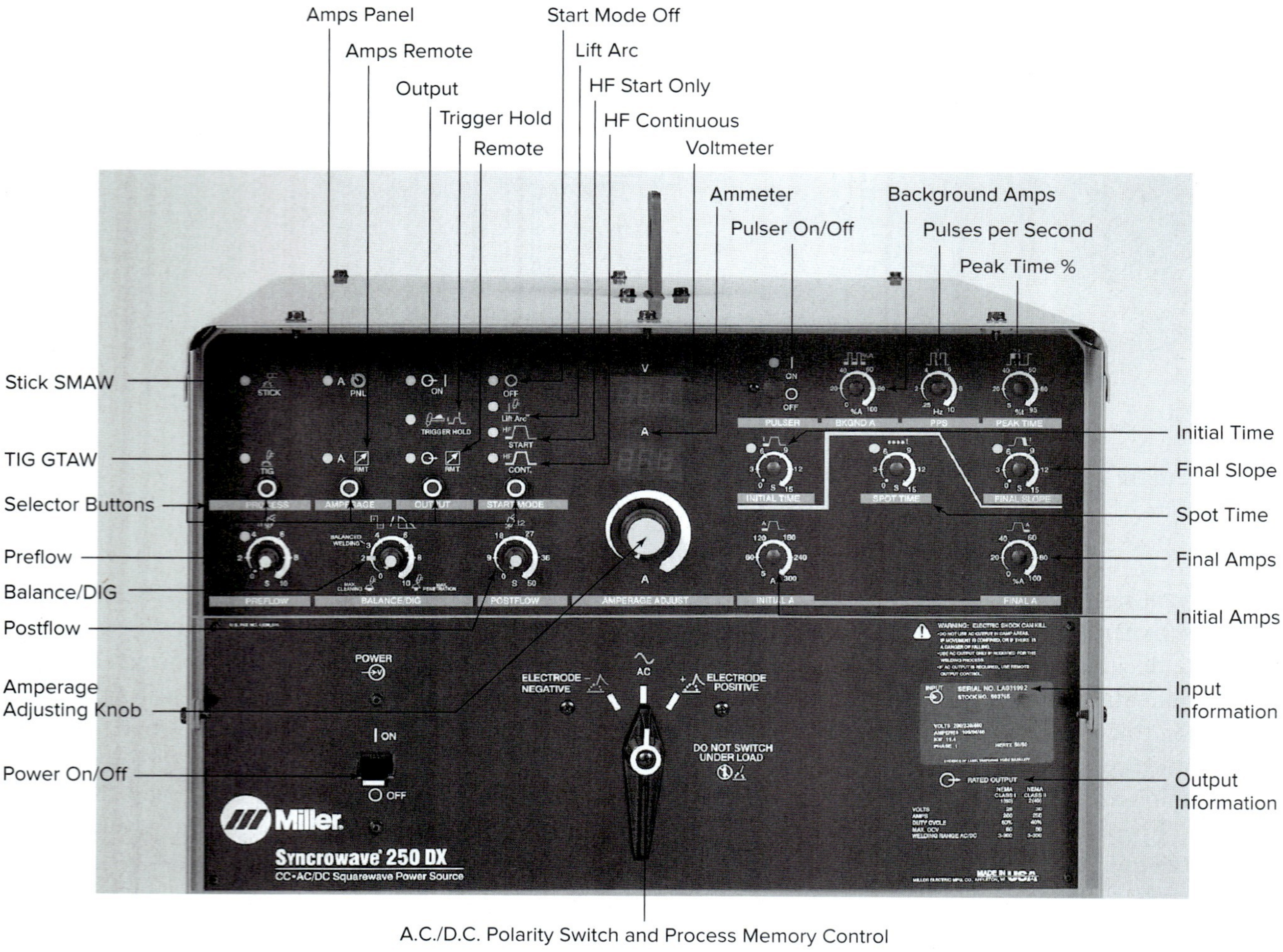

Fig. 18-34 Close-up of the operating panel of the welder shown in Fig. 18-33. Miller Electric Mfg. Co.

570   Chapter 18   Gas Tungsten Arc and Plasma Arc Welding Principles

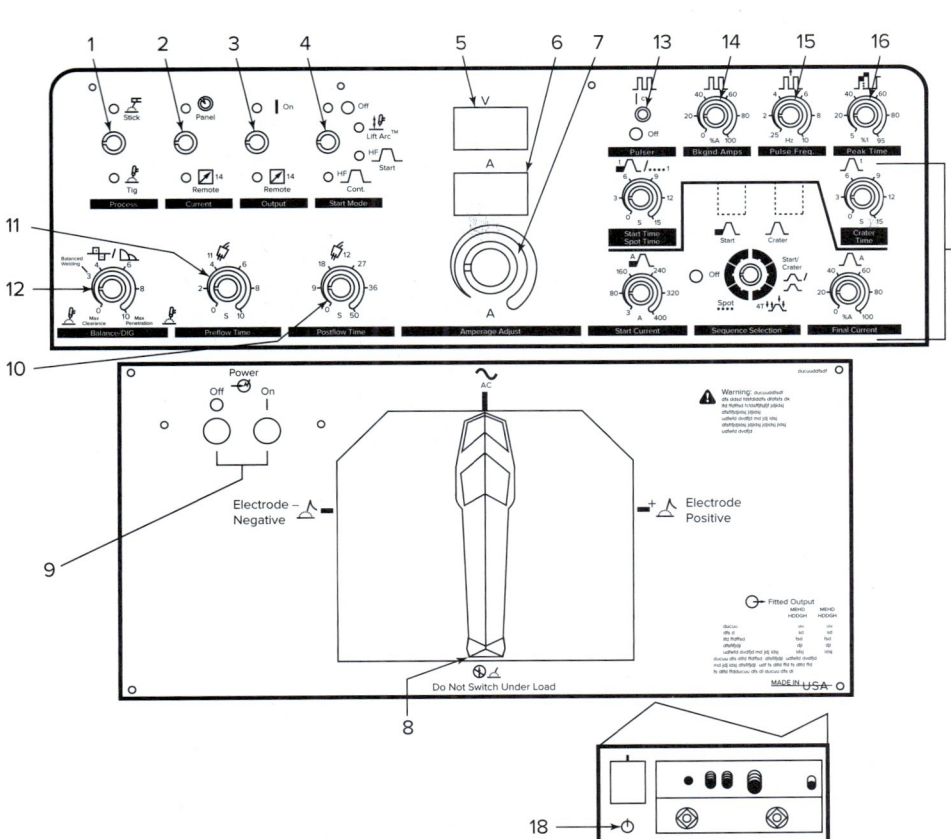

Fig. 18-36 Control panel for machine shown in Fig. 18-34. Note preflow, pulser, and sequencer controls for automatic applications.

Top row of lights in upper left corner are on for SMAW. Bottom row are on for GTAW. 1. Process control, 2. Current control, 3. Output control, 4. Start mode button, 5. Voltmeter, 6. Ammeter, 7. Amperage adjustment control, 8. Output selector switch, 9. Power switch push buttons

Use buttons to turn unit off and on. 10. Postflow time control, 11. Preflow time control, 12. Balance/DIG control

Pulse Controls: 13. Pulser on/off switch, 14. Background amperage control, 15. Pulses frequency control, 16. Peak time control, 17. Sequence controls (optional), 18. High frequency control. Source: Miller Electric Mfg. Co.

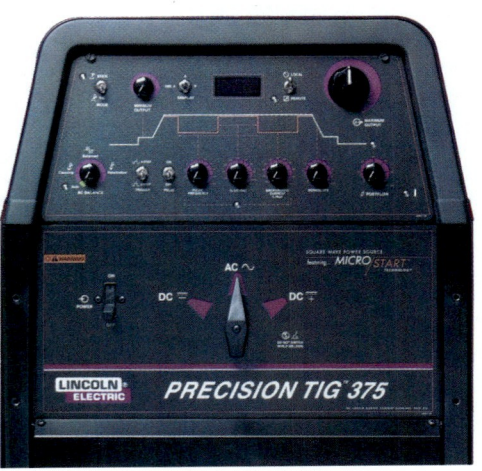

Fig. 18-37 A 375-ampere a.c.-d.c., constant current transformer with solid-state controls and square wave output. This machine can be used for a.c.-d.c. TIG and a.c.-d.c. stick electrode welding. Note the accessories and the cart with built-in water cooler. Lincoln Electric Co.

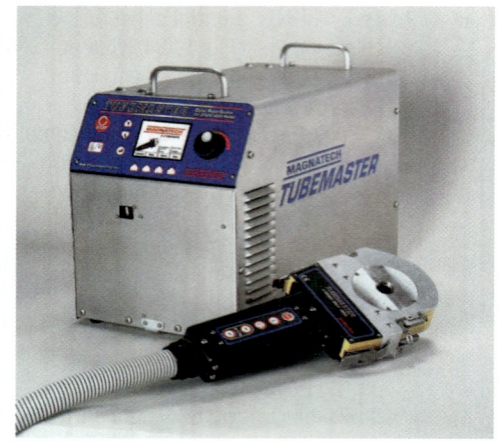

Fig. 18-38 This orbital welding equipment has an all-digital programmable portable power source rated at 200 amps. It has repeatability, precision, and QA/QC verification possibilities. Note the weld head, which makes fusion, welds meeting aerospace, pharmaceutical, food processing, and instrumentation quality requirements. *Magnatech Limited Partnership*

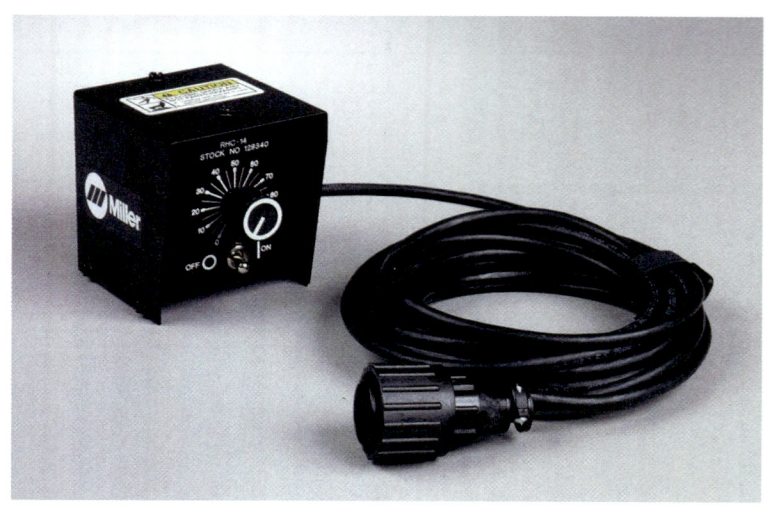

Fig. 18-40 Remote hand control unit with contactor control used with the machine shown in Fig. 18-30. *Miller Electric Mfg. Co.*

Fig. 18-41 Wireless foot controls for the turn table and power source that are so equipped. This means that there are two fewer cords to maneuver around and get damaged in the job. Stainless Steel Pipe purges and ready for GTA welding in the 1G position. *Edward R. Bohnart*

Fig. 18-39 A 350-amp a.c.-d.c. "Heliarc" welding machine. This machine is capable of SMAW and GTAW processes. Note the presentation box with torch and accessories. *ESAB*

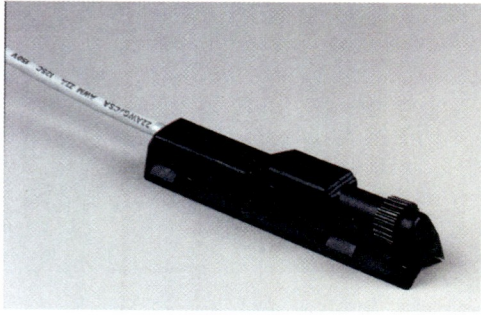

Fig. 18-42 Remote fingertip control, which can be mounted on a torch handle. The control wheel is moved perpendicular to the torch handle. *Miller Electric Mfg. Co.*

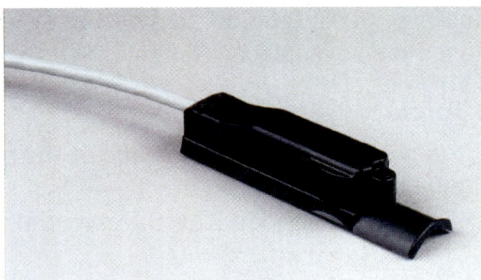

Fig. 18-43 Remote fingertip control, which can be mounted on a torch handle. The control wheel is moved parallel to the torch handle. Miller Electric Mfg. Co.

Table 18-5 Current Range for an A.C.-D.C. Transformer-Rectifier Solid-State Machine

Current	Minimum (A)	Maximum (A)
a.c.	5	310[1]
d.c.	5	310[1]

[1] Finer resolution of the remote amperage control can be done by setting the main amperage control on the machine at the maximum amperage required for the welding being done.

12. Side-by-side stacking and positioning to reduce floor space requirements.

B. Advantages of solid-state electronic phase controls with microprocessor memory
   1. Constant arc heat produced during welding because internal circuits compensate for the following:
      a. Changes of input line voltage
      b. Changes in operating temperature to maintain output current as set by the controls
   2. Instantaneous output response with every control adjustment.
   3. Easy-to-use and accurate current setting because the effect of fine current control adjustments are essentially linear over the entire range.
   4. Selectable TIG start control from hot, normal, or cold (with adaptive hot start for stick).
   5. Process memory control that remembers the setting from the polarity switch, DCEN TIG or DCEP stick. By selecting the polarity, the unit recalls the last procedure used on that setting, which reduces confusion and setup time.

C. Complete welding controls
   1. Current adjustment from 5 to 310 amperes that is accomplished in a single range with digital control on a single amperage control knob in 1-ampere increments. Allows crater out ability down to the minimum output of the machine. Table 18-5 describes the range control.
   2. Fine adjustment current control.
   3. A.C., DCEN, DCEP selector switch with process memory.
   4. Remote control switch and receptacle for connecting current control, arc start switch, or crater fill control.
   5. Balance control for TIG or DIG for stick.
   6. High frequency intensity rheostat to adjust strength of high frequency spark.
   7. Process selector switch for TIG or stick (depending on how it was last set up).
      a. Controls gas solenoid
      b. Provides high frequency start, continuous or lift arc
      c. Proper output volt-amp curve
      d. Remote-control activation or deactivation
   8. Postflow timer that controls gas flow after welding.
   9. Gas inlets and outlets.
   10. 115 volts a.c. power (1,150 watts) available from receptacles in the nameplate.
   11. Power on/off switch.
   12. Pilot light that indicates when machine is on.

D. Current control
   1. Variable current while welding for the following:
      a. Making critical TIG or stick electrode welds
      b. Filling craters
   2. Selectable start for TIG and adaptive hot start for stick that automatically increases the output amperage at the start if needed to prevent the electrode from sticking to the plate.
   3. Output arc control that can be on for stick or remote control for TIG, or set for trigger hold for TIG. This not only stops or starts the arc but also the high frequency, welding power, and gas flow.
   4. Hand controls that fasten to the torch for fingertip control of amperage and output arc power.
   5. Foot control provider that provides maximum freedom for both hands.
   6. Twenty and twenty-five control leads that let the welder take the control to the work, with extensions cords of 25 to 75 feet.

E. Optional features
1. Output arc start switch, which is required for TIG welding. This can be a simple push-button switch or output contact control built into the various foot and hand controls.
2. Sequencer control for mechanized welding that allows arc fade control, which automatically controls crater filling. It adjusts the rate of final slope from 0 to 15 seconds and allows setting of the final current level as well.
3. Sequencer control for mechanized welding to allow initial amperage control (higher than weld amperage for a hot start or lower for a cold start). The start can also be timed from 0 to 15 seconds.
4. Pulser with on/off switch, background current control, pulses per second [0.25–10 pulses per second (p.p.s.)] and percentage of time (5–95 percent) at peak amperage.
5. A condenser for power factor correction that reduces operating costs by improving the supply line power factor.
6. Undercarriage, which consists of a coolant reservoir circulator for a water-cooled torch, a frame for two cylinders to be racked, and four rubber tires for moving the machine and getting the machine off of the floor and raising the controls to a more advantageous height.

The TIG Torch

The TIG torch (electrode holder) is one of the most important items of apparatus. Figure 18-44 is a sectional drawing of a holder that identifies the essential working parts.

The torch feeds both the current and the inert gas to the weld zone. The torch is made up of a torch body, a collet, a ceramic collet cup, a nozzle, and the electrode, Fig. 18-45.

A hose and cable assembly is attached to the holder that connects the torch to the power source, the shielding gas supply, and the source of water it requires for cooling. Some torches have a manually operated switch on the handle to control the gas. All torches are properly insulated to safely handle the maximum current for which they are designed.

The current is fed to the welding zone through the tungsten electrode that is held firmly in place by a steel collet in the electrode holder. The electric arc is the intense source of heat for welding. Most of the heat of the arc is absorbed by the base metal, but a portion of the heat of the arc goes into the electrode. The electrode material is nonconsumable, and only

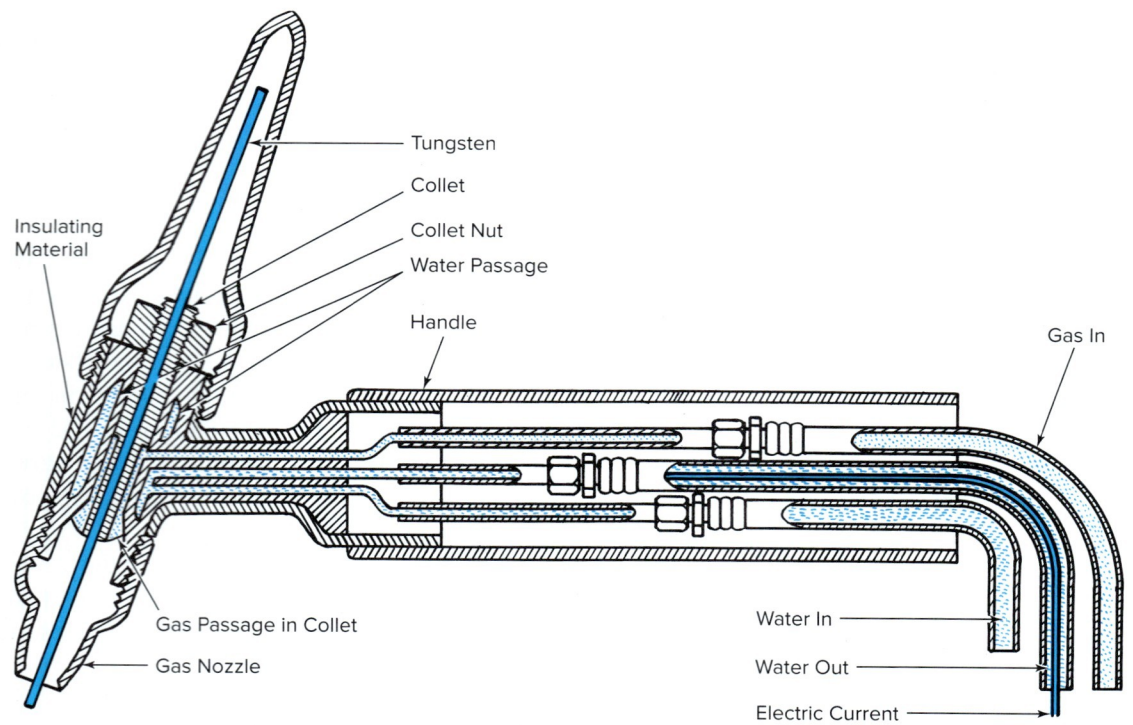

Fig. 18-44 Sectional view of a TIG welding torch.

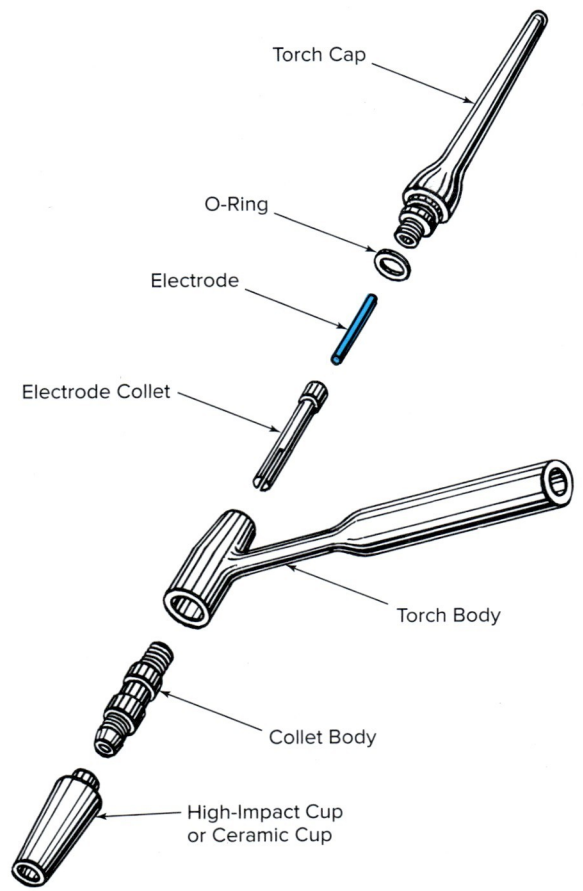

Fig. 18-45 Exploded view of a TIG welding torch.

deposit. Metal cups that become contaminated can be wirebrushed clean and reused.

A number of manufacturers are indicating the size of nozzles by numbers such as 4, 5, 6, 7, and 8. These numbers indicate the nozzle size in sixteenths of an inch. For example, a size no. 4 nozzle indicates $4/16$- ($1/4$-) inch diameter. In general, select a nozzle with a diameter that is from four to six times the diameter of the electrode. Nozzles with too small a hole size tend to overheat and break. A hole that is too large provides poor gas shielding and wastes gas.

Electrode holders are water cooled, air cooled, or air/gas cooled. Air/gas cooled holders are generally confined to very light and intermittent duty. They are particularly desirable, however, for work requiring currents up to 100 amperes because of their low weight, small size, and simplicity of operation. Holders above 100 amperes in size are water cooled. (Electrode holders with capacities up to 500 amperes are available.)

A recent development for TIG welding torches is the **gas lens**, Fig. 18-46. This is a permeable barrier of concentric fine-mesh, stainless-steel screens that produces an unusually stable stream of shielding gas. This device prevents turbulence of the gas stream that tends to pull in air, causing weld contamination. The source of gas turbulence is the torch nozzle where the velocity of the entering gas is much greater than that of the exiting gas. The gas lens reduces the difference in velocities and directs the gas into a coherent stream.

tungsten has been found to be usable. Various sizes of electrodes may be inserted into the holders for a range of welding heats.

The shielding gas is fed to the weld zone through a gas cup, or nozzle, at the head of the torch. A large variety of torch heads are available. Some have straight heads, others have heads at different angles, and a few have movable heads that can be adjusted to any angle. The material of the nozzle must be highly heat resistant and capable of conducting the heat away from the lower edge so that the nozzle does not overheat by radiation from the electrode. Gas nozzles are usually made of a refractory ceramic composition, glass, or metal, for welding up to 250 amperes. Metal nozzles are used for high welding currents over 250 amperes, and the torches are usually water cooled. A glass cup eliminates blind spots by permitting the operator to watch the weld pool.

Ceramic cups that become charred and burned from overheating, or that have a buildup of contaminated material on the inside, should be discarded. If used, charred cups will spread contamination to the weld

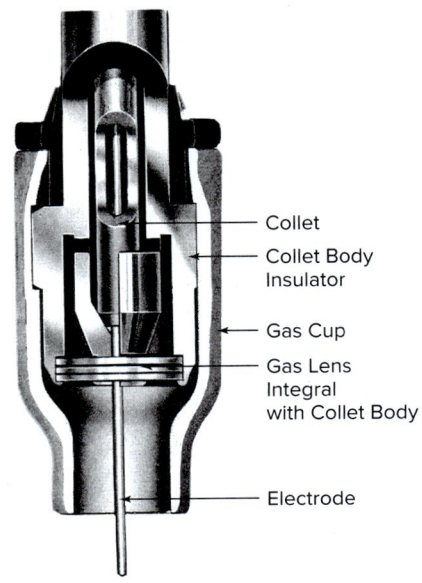

Fig. 18-46 Sectional view of a gas lens. Praxair, Inc.

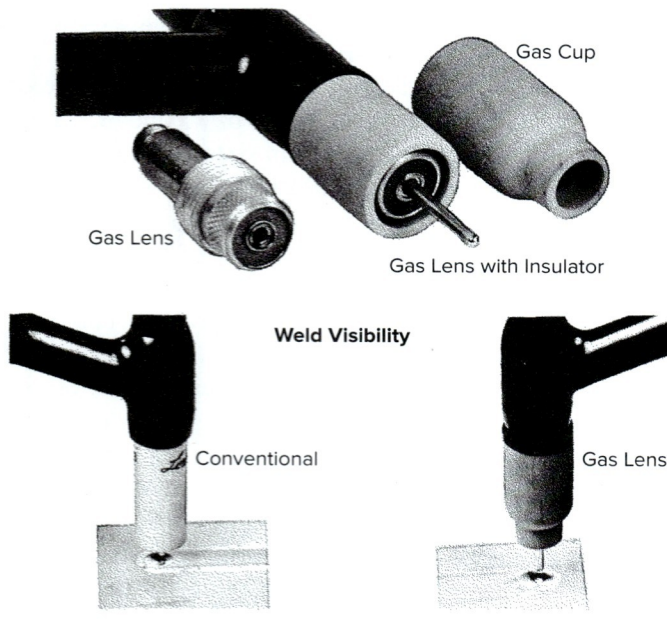

Fig. 18-47 Electrode extension and improved visibility are possible with use of the gas lens. Praxair, Inc.

When a torch is equipped with a gas lens, the shielding gas can be projected to a greater distance. Welding is possible at nozzle elevations up to 1 inch. The tungsten electrode can be extended well beyond the cup. This improves visibility by eliminating the blind spot at the weld pool, Fig. 18-47. Inside corners and other hard-to-reach places can be readily welded.

Electrodes

It is important that the correct type and size of electrode be used for each welding application. The proper electrode is essential to good welding.

Tungsten Electrodes Tungsten is a metal with a melting point higher than any other metal; of all the elements it is second only to carbon. Carbon has a melting point of 6,740°F, and tungsten, 6,170°F. The symbol of the element tungsten is "W." Tungsten also has low electrical resistance, conducts heat well, and is a good emitter of electrons especially when hot. Tungsten electrodes are available as pure tungsten, or they may be alloyed with 1 to 2 percent thorium, or 1 percent zirconium, or 2 percent cerium, or 1 to 2 percent lanthanum added. They have been found to be practically nonconsumable. Pure tungsten is generally used for a.c. welding. The zirconium type is also excellent for a.c. welding. It has a high resistance to contamination, produces a more stable arc than pure tungsten, and has a higher current-carrying capacity.

Thoriated Tungsten Electrodes Thoriated tungsten electrodes are available for DCEN welding. They are generally not recommended for a.c. welding as they do not maintain the spherical balled electrode tip. They have proved to be effective with the enhanced square wave power sources. Although they are more costly than pure tungsten electrodes, their lower rate of consumption makes them more economical to use. Thoriated tungsten electrodes run cooler, and their tips do not become molten so the arcing end is kept intact. When the electrode accidentally touches it, the work does not become as contaminated and may be usable on noncritical work without cleaning. Touch starting is facilitated and the arc stability is improved, especially at low currents. Thoriated tungstens have come under some scrutiny due to their slight radioactivity. The level of radiation has not been found to represent a great health hazard during welding, but rather the grinding dust from electrodes may be a concern. Alternative rare earth doped tungsten electrodes are available such as the ceriated and lanthanated electrodes.

Ceriated Tungsten Electrodes Ceriated tungsten electrodes were developed as possible replacements for thoriated tungsten electrodes because cerium, unlike thorium, is not a radioactive material. Tungsten is alloyed with a nominal 2 percent cerium oxide. Compared with pure tungsten, the ceriated tungsten electrode proved similar in current levels and improved in arc starting and arc stability characteristics to thoriated tungsten electrodes. They are recommended for DCEN welding, but they also do well on a.c. welding, especially with the enhanced square wave power sources.

Lanthanated Tungsten Electrodes The current levels, advantages, and operating characteristics of lanthanated tungsten electrodes are very similar to that of the ceriated tungsten electrodes. These electrodes operate at slightly different arc voltages than thoriated or ceriated tungsten electrodes. They are alloyed with pure tungsten with 1 or 2 percent lanthanum oxide.

Tungsten electrodes have a small colored band to indicate their alloy. Table 18-6 will help in the identification and proper classification of these tungsten electrodes.

Electrode Sizes and Capacities Tungsten electrodes are available in diameters of 0.010, 0.020, 0.040, $\frac{1}{16}$, $\frac{3}{32}$, $\frac{1}{8}$, $\frac{5}{32}$, $\frac{3}{16}$, and $\frac{1}{4}$ inch. They are manufactured in lengths of 3, 6, 7, 18, and, in some instances, 24 inches. Seven inches is the convenient size for most manual operations. They have either a chemically cleaned finish or a ground finish that holds the diameter to a close tolerance. The ground finish costs somewhat more than the cleaned finish, but makes for better electrical pickup at the collet, thus reducing electrical resistance and heat buildup in the torch.

The current-carrying capacity of tungsten electrodes depends on a number of variables such as the following:

- Type of shielding gas
- Length of the electrode extending beyond the collet
- Effectiveness of the cooling afforded by the electrode holder
- Position of welding
- Type of current used

Note the difference in the current capacity of electrodes with DCEN, DCEP, and alternating current indicated in Table 18-7. Slightly higher current values are used with alloyed tungsten electrodes than with pure tungsten electrodes. These are only approximations. (In the welding industry, amperes are often referred to as *heats*.) A good working rule is to select an electrode with the smallest diameter that can be employed without dripping molten tungsten into the weld pool. Too high a current density with helium may result in a black deposit of tungsten on the work. Too low a current density may result in an arc that is hard to direct and may be hard to start.

Electrode Preparation and Maintenance There is some disagreement as to the practice of grinding an electrode to a pencil point. Experience has indicated that grinding should be done only when welding with DCEN for carbon and stainless steels. Grinding should be done on a fine-grit, hard, and abrasive wheel. This wheel should not be used to grind any other material to prevent contaminating the tungsten electrode. Figure 18-48, page 578, shows proper grinding techniques. The vertex angle of the tip controls several factors and is most critical when doing automatic GTA welding, Fig. 18-49, page 579.

Another method of electrode preparation is to ball the electrode so that a rounded end forms on the electrode. Using DCEP and striking an arc on a piece of copper forms the ball end. This ball must not be over 1½ times the diameter of the electrode. (A larger ball

Table 18-6 Tungsten Electrode Identification

Common Name	AWS Classification	Color Code
Pure	EWP	Green
2% cerium	EWCe-2	Orange
1% lanthanum	EWLa-1	Black
1.5% lanthanum	EWLa-1.5	Gold
2% lanthanum	EWLa-2	Blue
1% thorium	EWTh-1	Yellow
2% thorium	EWTh-2	Red
Zirconium	EWZr-1	Brown

Table 18-7 Current Ranges for Tungsten Electrodes

Electrode Diameter (in.)	ACHF Current[1] (A)		DCEN Current[2] (A)				DCEP Current[3] (A)
	Pure Tungsten Argon	Thoriated Argon	Pure Tungsten		Thoriated		Either Gas, Either Electrode
			Argon	Helium	Argon	Helium	
0.010	≤15	≤20	≤15	≤20	≤25	≤30	—
0.020	10–20	10–25	5–30	15–35	15–35	15–45	—
0.040	20–30	20–60	20–70	25–80	15–80	30–90	—
1/16	30–80	60–120	70–135	80–145	50–150	60–160	10–20
3/32	60–130	100–180	150–225	160–235	135–250	140–260	15–30
1/8	100–180	160–250	220–360	230–390	250–400	260–420	25–40
5/32	160–240	200–320	360–450	380–500	400–500	410–525	40–55
3/16	190–300	290–390	440–740	480–780	500–750	510–800	55–80
1/14	250–400	340–525	740–950	750–1,000	750–1,000	780–1,100	80–125

[1]Recommended for welding aluminum, magnesium, and their alloys. With square wave current the range can be increased by 20 percent.
[2]Recommended for welding steels, stainless steels, and other metals.
[3]Recommended only when minimum penetration and maximum surface cleaning are desired. It is seldom used.

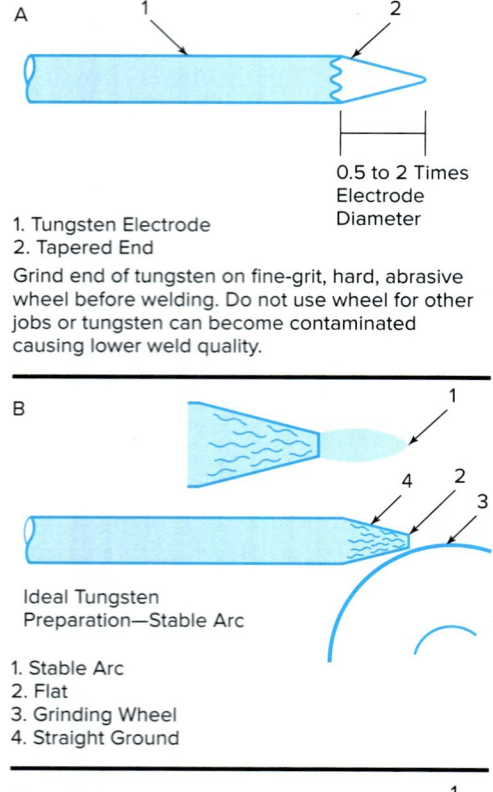

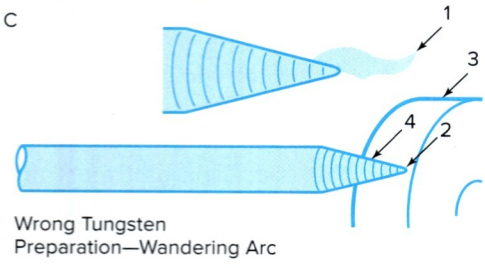

Fig. 18-48 Electrode preparation on a grinder, with a special wheel used only for this operation. (A) A diagram of the tungsten electrode. (B) Correct tungsten preparation produces a stable arc. (C) Incorrect tungsten preparation produces a wandering arc. Source: ESAB Welding and Cutting Products

Maximum electrode life can be obtained if proper operating conditions are employed in the welding process. The following factors reduce tungsten electrode life:

- **Erosion due to low temperature** Caused by operation at too low a current density.
- **Melting and dripping** Caused by operation at too high a current density.
- **Oxidation** Usually caused by improper shielding during the cooling period when the gas postflow has been shut off too soon. When cool, tungsten electrodes should show a mirror-bright arcing end with no discoloration.
- **Contamination by other metal** Caused by the hot electrode touching the filler rod or base metal. This is a particular problem with aluminum. The aluminum apparently vaporizes and disrupts the arc, causing sputtering and resulting in black soot on the work. This continues until the aluminum is removed. During this process, the weld may become contaminated with the tungsten that is thrown off.

Water Supply The tremendous heat of the arc and high current usually necessitate water cooling of the torch and power cable. If the water flow is stopped or reduced, the torch and electrode power cable will overheat and burn. Although adequate protection from heat is important, the equipment must also be lightweight and flexible for easy handling.

There are two methods of preventing overheating:

- A safety switch that opens and shuts off the electrical power source until the flow of water resumes
- A water-cooled, fusible link in the welding current circuit, Fig. 18-51

When the temperature begins to rise, the switch closes or the fuse blows, cutting off all power. When welding in close quarters, however, care must be taken so that the torch does not become overheated due to the heat generated by welding.

If tap water is used to cool the torch head, it should not be allowed to flow continuously, as cold tap water may be below the dew point and could cause moisture buildup inside the torch body. This may lead to weld zone contamination until the torch temperature exceeds the dew point. Tap water is generally not recommended as a coolant because of its inherent mineral content that can build up over a period of time and clog the small cooling orifices in the torch head. Conservation also dictates use of less wasteful methods, such as water reservoir or radiator style water recirculating systems.

indicates that the current is too high.) In practical application on noncritical welding, many welders will take new tungsten or tungsten that has had any contamination removed and ball the tungsten with a higher than normal alternating current, thus speeding up this operation. The balled point is used with sine wave alternating current for aluminum and aluminum-magnesium alloys. A more pointed shape is usable with the square wave and enhanced square wave a.c. waveforms. Figure 18-50 shows a proper balled electrode shape.

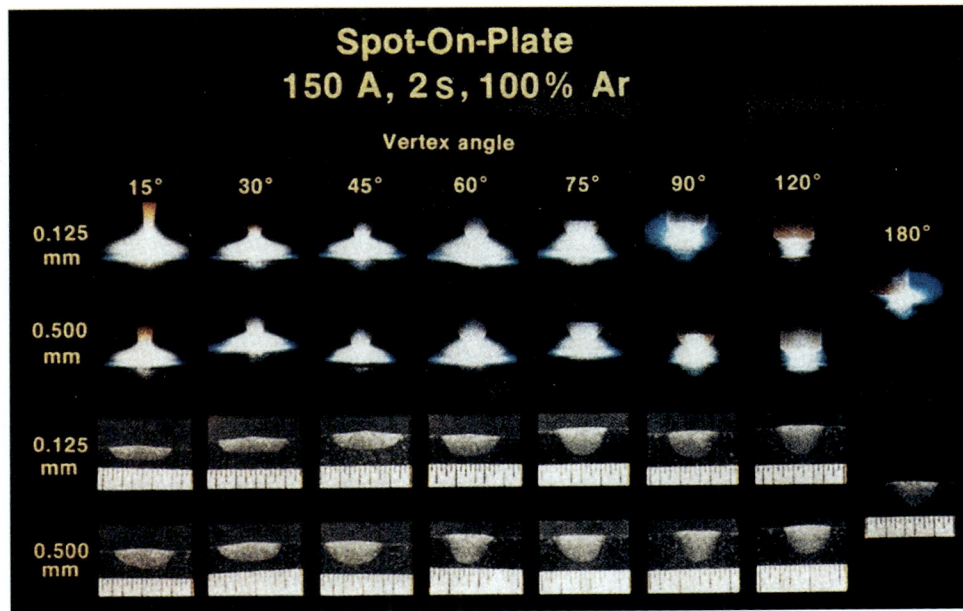

Fig. 18-49 Arc shape and weld bead geometry as a function of electrode tip angle in pure argon shielding. Key, J. F. "Anode/Cathode Geometry and Shielding Gas Interrelationships in GAW." *Welding Journal* 59, no. 12 (1980):364–370/American Welding Society

coolant system manufacturer for a recommendation on proper coolant solution. All coolants must be clean. Otherwise, blocked passages may cause overheating and damage the equipment. It is advisable to use a water strainer or filter on the coolant supply source. This prevents scale, rust, and dirt from entering the hose assembly. If the fuse becomes clogged, it should be disassembled and cleaned.

The rate of coolant flow through the torch is important. Rates that are too low may decrease cooling efficiency. Rates that are too high damage the torch and service line. The direction the coolant flows through the torch is critical. It should flow from the coolant source directly through the water hose to the torch head. The torch head is the hottest spot in the cooling system and should be cooled first with the coolant at its most efficient thermal transfer temperature. This coolant, upon leaving the torch head, should cool the electrode power cable on its return to the recirculating system or drain.

The recirculating coolant must be of the proper type. Since high frequency is being used, it should be deionized to prevent the coolant from bleeding off the high frequency prior to the coolant making it to the arc. If the ambient temperature can drop below freezing, the coolant must also be protected. Do not use antifreeze that contains leak preventers or other additives. Some method of reducing algae growth is advisable as well as including a lubricant for the circulating pump. Consult with the

Cables and Hoses

Tungsten electrode torches must be supplied with electric current and shielding gas, and, if water cooled, they must be provided with a supply of coolant. Electricity, gas, and coolant are conducted to the torch by means of a copper

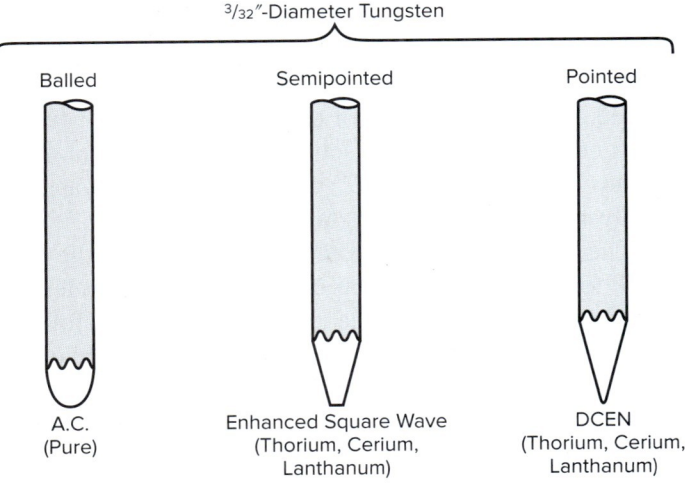

Fig. 18-50 Balled tungsten used for an a.c. sine wave power source.

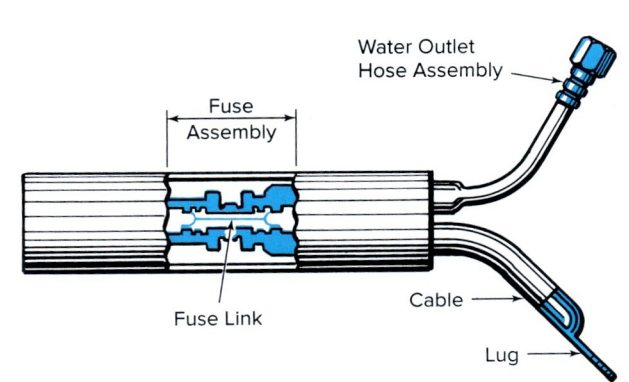

Fig. 18-51 Fuse and hose assembly.

cable and flexible hose. When the torch is a manual one, it is usually desirable to cool the current-conductor cable by enclosing it in one of the water supply hoses, which permits a very good flexible conductor, Fig. 18-44, page 574. Plastic hoses are used for inert gases, since helium will diffuse through ordinary rubber and fabric welding hose. In this case air could enter from the outside, causing contamination.

TIG Hot Wire Welding

TIG hot wire welding is a mechanized gas tungsten arc welding process. The filler wire is preheated so that it is in a molten state as it enters the weld pool. The molten filler metal is added behind the arc to form the weld, Fig. 18-52. The heat of the arc, therefore, is concentrated on the weld—not the wire.

TIG hot wire welding produces TIG quality welds at the higher speed of MIG welding. The hot wire process is up to four times faster than TIG cold wire welding. Welding progresses faster since the heat of the arc is concentrated entirely on the workpiece. Arc power is not used for melting the filler wire. Production time is further reduced by the fewer number of passes required to complete the weld. For a weld comparison, see Table 18-8 and Fig. 18-53.

Refer to Table 18-9 for a comparison of the TIG hot wire, TIG cold wire, and MIG processes.

Table 18-8 Weld Comparison ⅝-in. Thick 18% Maraging Steel

	Cold Wire	Hot Wire
Deposition rate (lb/h)	3	11.5
Travel speed (in./min)	8	12
Number of passes	9	5
Welding time (min/ft)	4.4	5.3

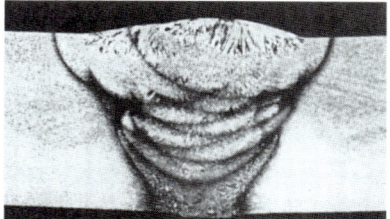

Tig Cold Wire

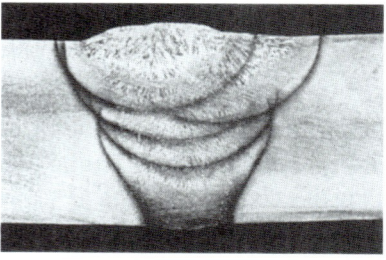

Tig Hot Wire

Fig. 18-53 A cross-sectional etched view of hot wire and cold wire TIG welds. Note that the hot wire process required fewer passes and appears to have a finer grain structure than the cold wire process. *Praxair, Inc.*

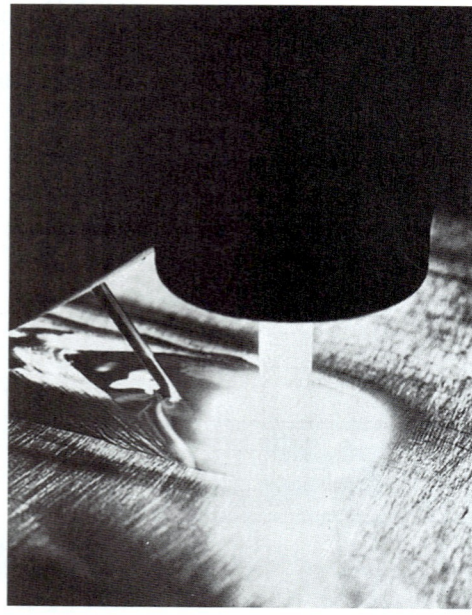

Fig. 18-52 The weld pool and the addition of hot wire at the trailing edge. Note the highly fluid weld pool. *Praxair, Inc.*

When too much cold filler wire is fed under the arc, taking up most of the arc heat, incomplete fusion may result. In hot wire welding, the molten filler wire is introduced at the trailing edge of the weld. Incomplete fusion does not occur for two reasons: (1) the filler metal never passes under the arc; and (2) the arc heat concentrates directly on fusing the base metal.

The view of the weld pool is not blocked by the cold wire feed assembly. Fusion of the workpiece under the arc is clearly visible since the hot wire torch follows the TIG torch. Welding operators can see the weld of the base metals without filler metal from one side and with filler metal from the other side. They also have a clear front view.

The welds are smooth, strong, and ductile. Postweld cleaning is not necessary because oxidation and contamination are eliminated. This type of welding offers two

Table 18-9 Typical Welding Conditions for Low and High Alloy Steels and Heat- and Corrosion-Resistant Materials

	MIG	TIG Hot Wire	TIG Cold Wire
Arc current (A)	350	450	350
Voltage (V)	26–28	12–14	11–13
Welding speed (in./min)	12–15	12–15	5–8
Deposition rate (lb/h)	10–12	11–16	2–4
No. of passes, 1-in. plate	15	12	35
Weld soundness	Usually good; wire quality is a factor.	Excellent; wire preheat eliminates major cause of porosity.	Very good
Filler metals	Pass-through arc; it may be harmful to transfer efficiency.	Commercial alloys satisfactory	High transfer efficiency; commercial alloys are usually satisfactory.
Heat input	High (spray) or low (short arc)	Controllable	High
Shielding gas	Argon-oxygen, helium, carbon dioxide singly or in combination	Inert: argon helium	Inert: argon helium
Equipment	Manual or mechanized	Mechanized	Manual or mechanized
Process application	Limited by transfer efficiency and reactive gases	Same as TIG cold wire plus exceptional weld soundness	Very broad; transfer efficiency and inert gas retain desired properties.

important benefits: (1) porosity is eliminated since preheating tends to drive off contaminants; and (2) transfer efficiency approaches 100 percent because alloys are not burned up by passing under the arc.

The following equipment is needed for the process:

- Torch that transfers the welding current to the wire, guides the filler metal into the pool, and provides inert gas shielding as the wire heats, Figs. 18-54 and 18-55 (p. 582)
- Wire feeding unit that provides instant change of wire feed speed from 50 to 825 inches per minute, Fig. 18-54
- Constant voltage a.c. power source, Fig. 18-54

TIG hot wire welding can be used to weld a variety of steels, titanium, and nickel alloys. It is not recommended for aluminum or copper due to the low resistance of filler wires. It is ideal for the fabrication of heavy-wall missiles, rocket motor casings, pressure vessels, hydrospace equipment, and corrosion-resistant piping.

Plasma arc welding (PAW), Fig. 18-56 (p. 582), was developed from the TIG process and is sometimes referred to as super TIG. **Plasma arc cutting** was introduced first, making it possible to cut aluminum and nonferrous metals with speed and precision. Plasma arc surfacing then became available. It permits the deposition of a wide range of wear-resistant alloys faster and with better control than practically any other surfacing method. Finally, plasma arc welding on a commercial basis was developed. Today the plasma arc process is being used for cutting, plating or coating, weld surfacing, and welding. **Plasma arc welding** is a suitable process for tough-to-weld metals such as stainless steels, titanium, and zirconium.

Plasma arc and gas tungsten arc are much alike electrically. Both protect the electrode and weld pool with inert gas. Both use tungsten electrodes. There is a big difference, however, in the constriction (tightness) of the arc and in the use of the tungsten electrode. The TIG electrode sticks out of the gas cup on the torch. Its arc is conical, and its heat pattern on the work is wide with shallow penetration. In TIG welding, a small change in torch standoff changes the heat pattern at the work a great deal, Fig. 18-57A (p. 583). By comparison, the concentrated heat and the forceful jet of plasma arc produce a deep, but narrow, weld, Fig. 18-57B. Variations in torch standoff change little in the area of the arc spot on the work.

Plasma Arc Characteristics

What occurs in a plasma torch, Fig. 18-58, is that a plasma-forming gas is passed through a d.c. arc maintained between the **cathode** (negative pole), which is the tungsten electrode, and an **anode** (positive pole). The nozzle of the torch is small enough so that the

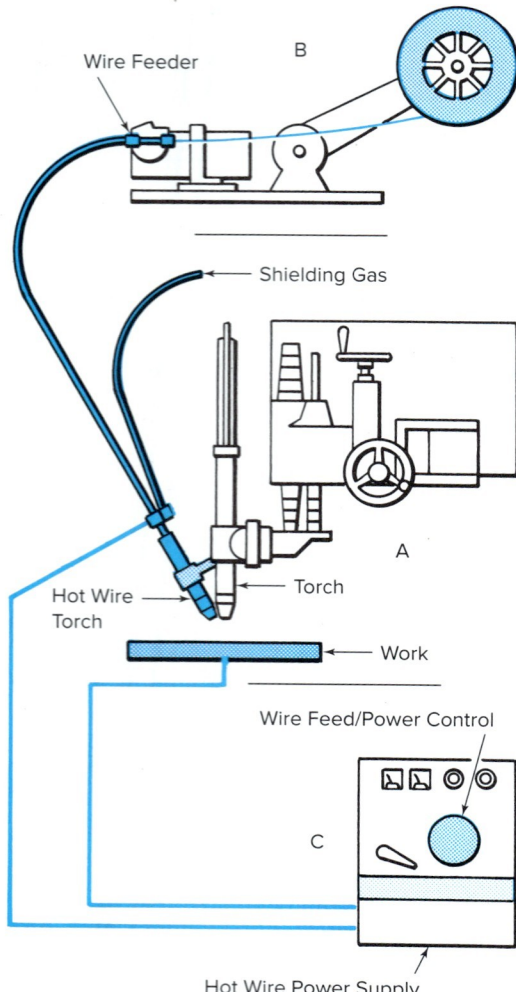

Fig. 18-54 A typical TIG hot wire installation: (A) torch, (B) wire feeder, and (C) power supply.

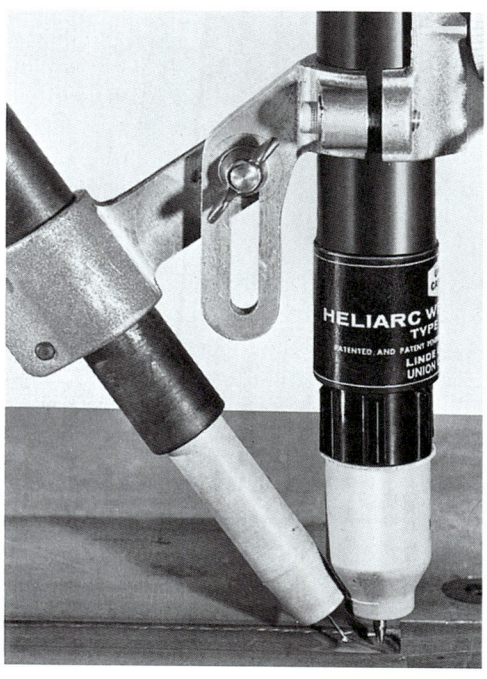

Fig. 18-55 The TIG hot wire welding system employs a special wire feed torch mounted behind the TIG welding torch to supply molten filler metal to the weld pool. Praxair, Inc.

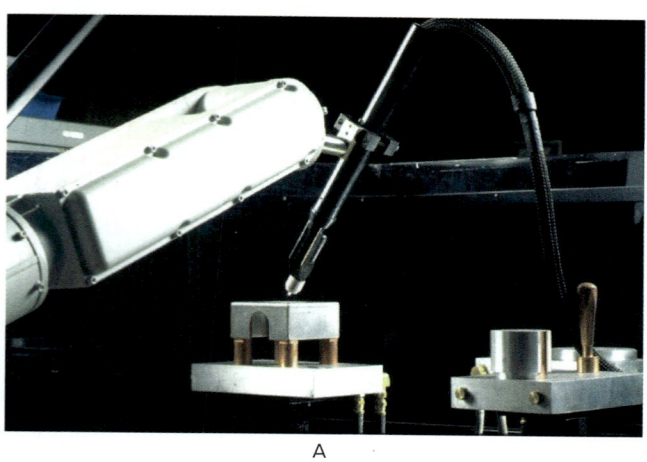

Fig. 18-56 (A) A PAW torch mounted on a robot arm. (B) A close-up view of the torch shows that the pilot arc is on. Note that the arc has not yet transferred to the weldment. Thermadyne Industries, Inc.

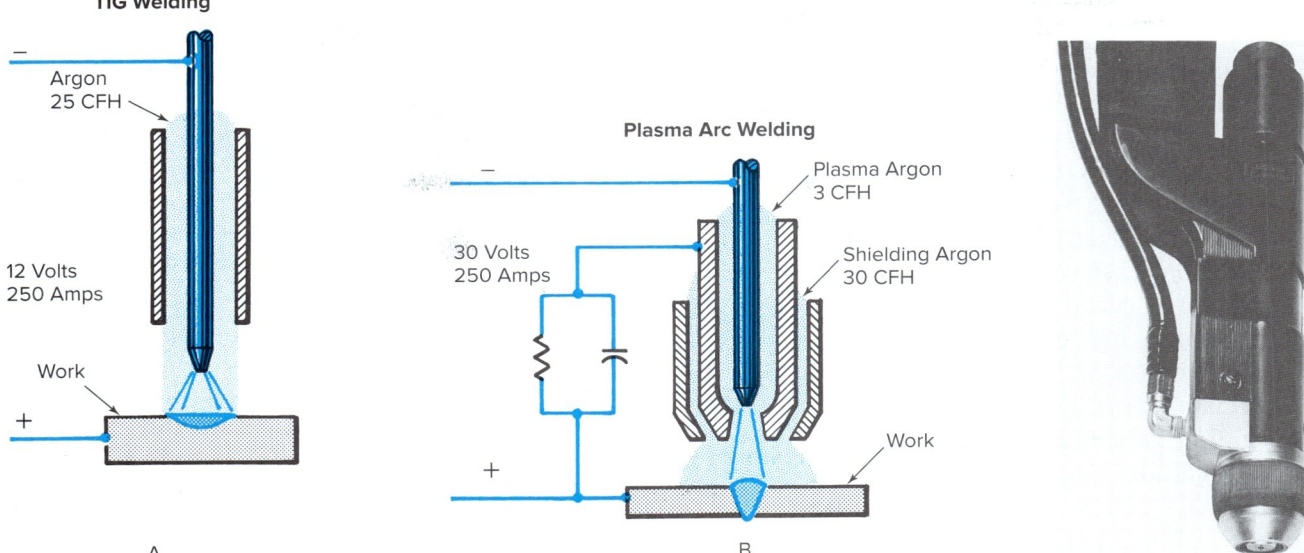

Fig. 18-57 Except for the pilot arc, gas tungsten arc and plasma arc welding torches are nearly alike, electrically. The difference between them is the constricted arc of plasma which forces the arc into a column, increases arc temperature, and concentrates the heat at the work.

Fig. 18-58 Plasma arc welding torch. Praxair, Inc.

gas is forced into intimate contact with the arc. As explained in Chapter 6, the intense heat changes the gas into plasma. Its molecules are broken down into ionized atoms with a high energy content. Present working temperatures for the process fall in the 10,000 to 50,000°F range, depending upon the application. Laboratory processes have created temperatures up to 100,000°F. Compare these flame temperatures with the 5,600 to 6,300°F of the oxyacetylene flame and other welding arcs.

In the plasma torch, the arc never touches the nozzle wall, and a layer of cool nonionized gas insulates the arc. Total power consumption is as important as power concentration. (If 50 kilowatts are delivered through a nozzle of 1/8-inch diameter, a power concentration of 3 megawatts p.s.i. results.)

Transferred and Nontransferred Arcs There are two types of arcs employed by the plasma equipment: transferred and nontransferred arcs. The **transferred arc,** Fig. 18-59, travels between the electrode and the work, which acts as an anode. Thus the arc heats the work with electric energy and hot gas. This arc is used for plasma welding, weld surfacing, and cutting.

The **nontransferred arc** is struck over a short distance and is entirely contained within the torch housing. The arc is struck between a tungsten electrode negative cathode and usually a water-cooled copper anode. The transferred arc puts more heat on the work and is used mostly in plasma welding.

Fusion welding is performed with a transferred arc. The workpiece is connected into the electric circuit by means of a work clamp. If the workpiece is not

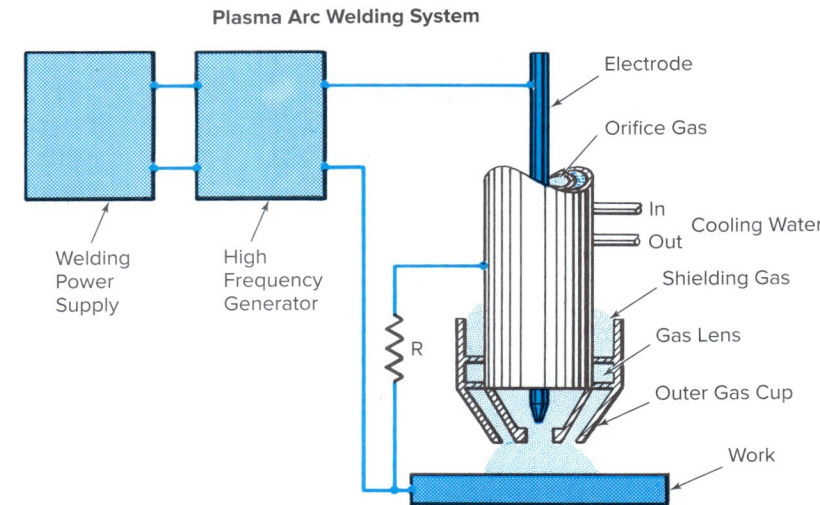

Fig. 18-59 A cutaway view of the plasma torch and plasma arc welding system. Note that the work is part of the electric circuit so that the arc is transferred.

Gas Tungsten Arc and Plasma Arc Welding Principles Chapter 18 583 grounded, however, the arc is not transferred to the work, but remains within the torch nozzle. Only the superheated jet of gas emerges from the torch nozzle and provides an effective heating tool for both metals and nonconductive materials. The nontransferred arc may be used in special welding applications when a lower heat concentration is desirable. Its main use is in joining or cutting nonconductive materials and for plating or coating.

Unique to plasma welding is the **keyhole effect** in which the plasma jet penetrates the workpiece completely. Plasma keyholes the front of the puddle because the jet blows aside the molten metal and lets the arc pass through the seam. As the torch moves, molten metal, supported by surface tension, flows in behind and fills the hole. The keyhole assures the operator there will be full-depth welds of uniform quality.

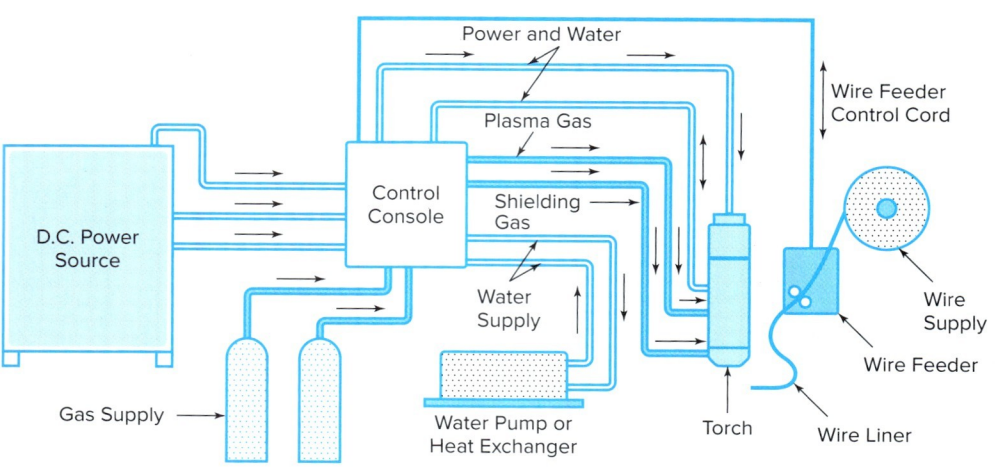

Fig. 18-61 A typical plasma arc welding installation, showing the relationship of each unit.

Fig. 18-60 A 0.5- to 150-amp PAW power source that will operate on 208- to 460-volt a.c. single- or three-phase power. It is equipped with interfaces for manual or automatic controls.
Thermadyne Industries, Inc.

Equipment

In addition to the torch, the plasma arc system includes the power supply, control console (Fig. 18-60), and the welder's remote-control station. The controls provide simplicity of operation. Push-button starting and stopping of the torch is remotely controlled. A water-cooling pump is usually needed to ensure a controlled flow of water to the torch at a steady pressure. This is necessary to cool the torch nozzle and electrode. Study Fig. 18-61 which shows the relationship of the various equipment units needed to complete the plasma system.

The power supply shown in Fig. 18-62 is a heavy-duty d.c. rectifier welder that can be used for high current plasma welding, TIG welding, and stick electrode welding. (Most plasma welding is performed using direct current.) It can be used as a plasma welder for welding steel or stainless steel in thicknesses from 0.001 inch to 0.125 inch in one pass. Variable polarity and enhanced a.c. square wave power sources are available to weld aluminum and magnesium alloy metals. The welder includes water and gas solenoids, a postweld shielding gas timer, and a high frequency generator. Remote and foot controls are available. The controls contain circuitry for plasma needle arc welding (low current) and are mounted on top of the power supply. The water-cooled torch operates at 75 amperes on a 100 percent duty cycle or 100 amperes on a 60 percent duty cycle.

The plasma arc must be mechanized to utilize its advantages of speed and penetration. The torch is mounted on a carriage and is operated with voltage control units or wire feeders. The size of the main port in the nozzle depends on the amount of welding

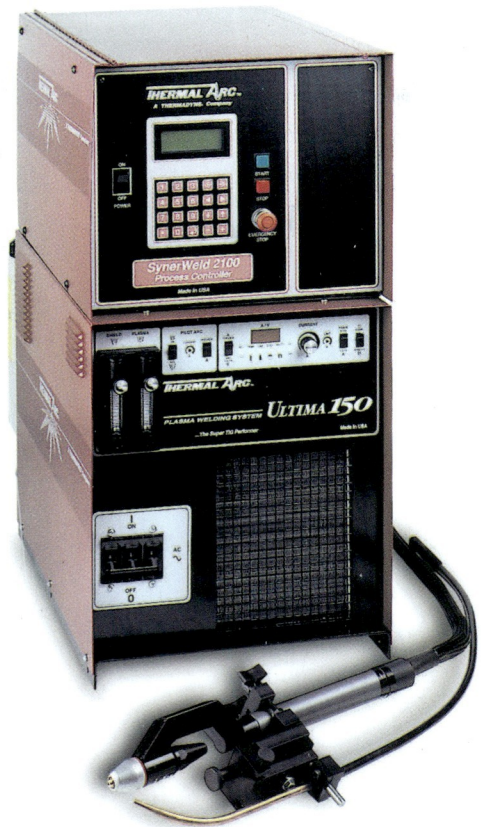

Fig. 18-62 A plasma welding unit. Note control on top of power source and automatic torch with wire delivery system. Thermadyne Industries, Inc.

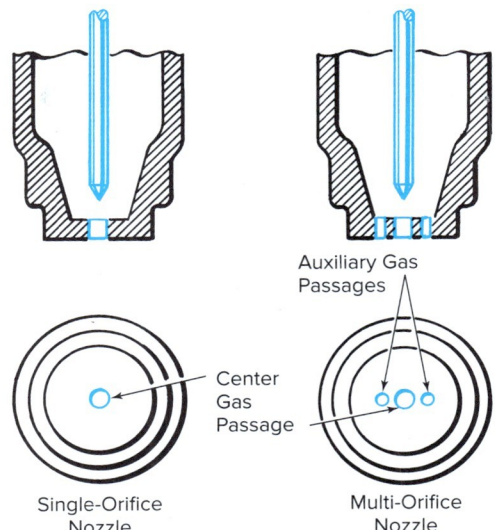

Fig. 18-63 Multi-orifice and single orifice nozzles for plasma arc welding.

current it must carry: the higher the current, the bigger the orifice.

A multi-orifice nozzle, Fig. 18-63, changes the shape of the arc jet and improves certain applications. The multi-orifice nozzle converts the cylindrical plasma arc jet into an oval or elongated shape. The resulting change in heat pattern allows an additional increase in welding speed of 50 to 100 percent with a narrower heat-affected zone for many applications.

Welding Gases

In plasma welding, the shielding gas is usually the same as the plasma gas. Argon can be used with all metals, but it is not always the best gas for all jobs. Hydrogen mixed with argon gives a hotter arc and better transfer of heat, and it is faster. Too much hydrogen, however, causes porosity. Argon-hydrogen is used for plasma and shielding gases for keyhole welding stainless steel, Inconel, nickel, and copper-nickel alloys. The amount of hydrogen in the mixture varies from 5 to 15 percent. In general, the thinner the work, the more hydrogen can be used.

Argon should be used when welding the reactive alloys such as titanium and zirconium and for carbon and high strength steels.

Helium has limited use as a plasma gas. It overheats the nozzle, shortens its life, and cuts its current-carrying capacity. Helium is used only for melt-in welds. In a 50 percent mixture with argon, it gives a hotter flame for a given current. Argon-helium mixtures are generally used for filler and cover passes.

Plasma Needle Arc Welding

The **plasma needle arc welding** process is a manual operation. It uses a small diameter, constricted arc in a water-cooled nozzle. The arc is exceptionally stable, and in the low current range it can be maintained from 15 amperes down to less than $1/10$ amperes. Since the needle arc is cylindrical in shape, variations in torch-to-work distance do not substantially change the area of arc action on the work. This makes it excellent for welding contoured work.

Plasma needle arc welding in the higher current range (15 to 100 amperes) is a manual version of plasma arc welding. The system functions in both the melt-in (conventional fusion welding) and keyhole modes.

Keyholing provides 100 percent penetration on $1/16$- to $1/8$-inch metal. Arc focusing, brought about by arc constriction, makes more efficient use of arc heat. This results in narrower welds, and distortion is reduced because less total heat passes into the work.

Table 18-10 Typical Conditions for Welding Butt Joints with the Plasma Needle Arc Process

Material	Thickness (in.)	Arc Current DCEN (A)	Shielding Gas Mix	Welding Speed (in./min)
Stainless steel	0.031	10	Argon	3
	0.030	10	1% hydrogen-argon	5
	0.010	6	1% hydrogen-argon	8
	0.010	5.6	3% hydrogen-argon	15
	0.005	2	1% hydrogen-argon	5
	0.005	1.6	50% helium-argon	13
	0.003	1.6	1% hydrogen-argon	6
	0.001	0.3	1% hydrogen-argon	5
Titanium	0.022	10	75% helium-argon	7
	0.015	5.8	Argon	5
	0.008	5	Argon	5
	0.003	3	50% helium-argon	6
Inconel 718	0.016	3.5	1% hydrogen-argon	6
	0.012	6	75% helium-argon	15
Hastelloy X	0.020	10	Argon	10
	0.010	5.8	Argon	18
	0.005	4.8	Argon	10
Copper	0.003	10	75% helium-argon	6

With the exception of aluminum and magnesium (these metals can be welded with variable polarity or enhanced square wave power sources), all metals that can be welded with the gas tungsten arc can be welded with the plasma needle arc. Pipe of all kinds can be welded in all positions. Wall thicknesses of 1 inch and over must be beveled.

Table 18-10 gives the welding conditions for butt joints welded with the plasma needle arc process.

Figure 18-64 shows the torch position when welding a butt joint in the flat position with the plasma needle arc process. Figure 18-65 also shows the proper position for an edge joint.

Fig. 18-64 Welding with the manual plasma needle arc welding process. Praxair, Inc.

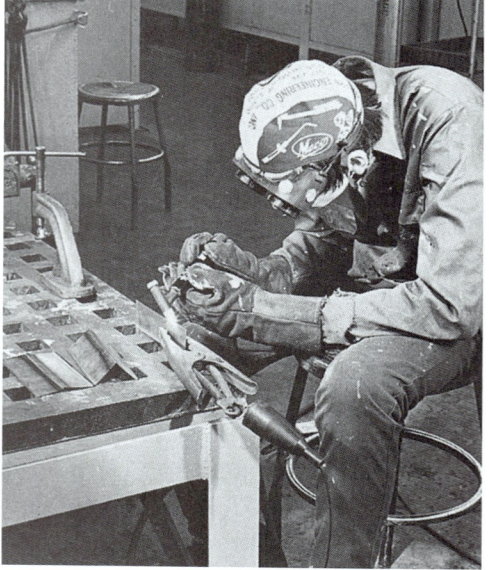

Fig. 18-65 Making an edge joint with the plasma needle arc. Edward R. Bohnart

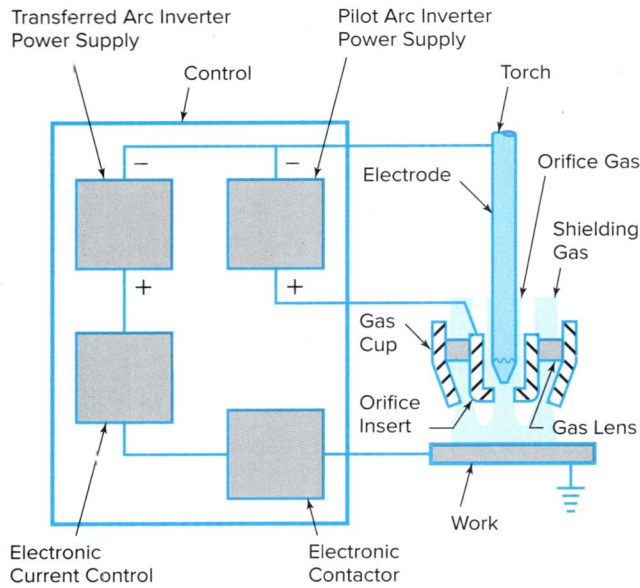

Fig. 18-66 Plasma needle arc operation schematic. Note that the tungsten electrode is sharpened to a point.

The torch is lightweight and water cooled. The torch head is angled so that the welder can inspect the arc and workpiece continuously. Electrode replacement is easy, requiring only the removal of the knurled cap on top of the torch head. The torch can be adapted to mechanized welding operations.

Advantages In general, plasma welding is designed for direct current electrode negative welding of metals other than aluminum. Some variable polarity and enhanced square wave power sources can operate with DCEN and a small amount of DCEP. This small amount of DCEP is required for the arc cleaning action required for welding aluminum.

Plasma needle arc welding has many advantages.

1. A characteristic of plasma welding is high heat concentration that provides the following advantages:
   a. Narrower weld beads restrict the heat-affected zone. This reduces the expansion and contraction of adjacent areas, resulting in less distortion and stress.
   b. The constricted arc provides greater penetration.
   c. Thicknesses from 1.8 inches down to 0.001 inch can be welded. Dissimilar metals of varying thicknesses can be welded.
   d. There are no spatters, sparks, fumes, or flux.
   e. A minimum of current is required.
2. A second characteristic of plasma welding is that greater standoff distance is permissible between

Since the electrode is recessed in the torch nozzle, Fig. 18-66, there is no danger of electrode contamination or tungsten inclusions in the weld pool. These factors, combined with exceptional arc stability and noncritical torch-to-work distance, make the system exceptionally good for many applications.

Figure 18-67 shows a low current needle arc welder and accessories. The control power supply unit includes the power supply and controls in a compact cabinet designed for bench mounting. Connections for water and gas supplies are at the rear of the cabinet. A built-in water pressure switch shuts the unit off automatically if the water supply fails. All control elements are located on the front panel. Three separate flowmeters regulate plasma and shielding gases and permit precise mixing of the shielding gases for best welding performance. A foot switch is plugged into the front panel to provide on/off control of the arc current. Figure 18-68, page 588, is a plasma needle arc interconnection diagram.

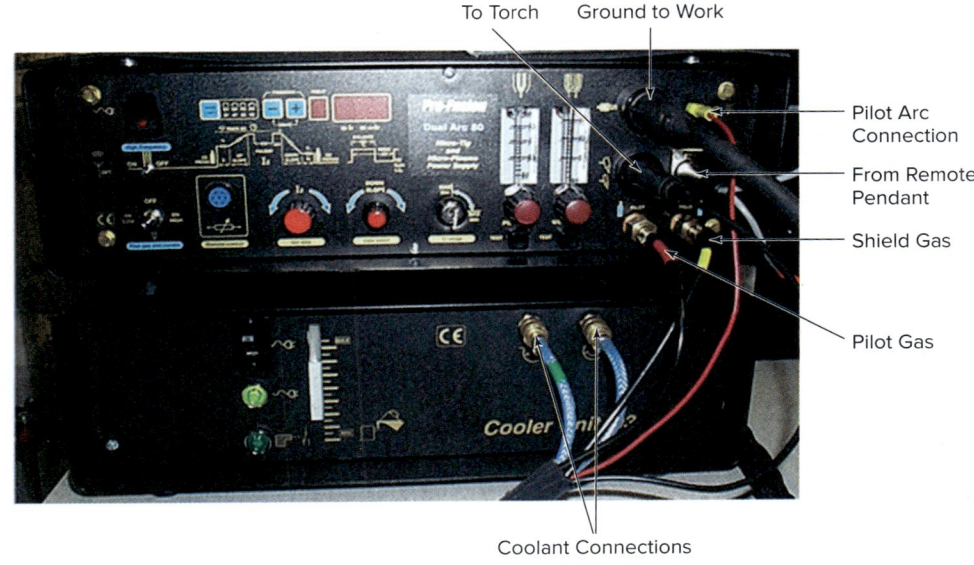

Fig. 18-67 A power source and cooler unit designed for low amperage plasma arc welding.
Elderfield & Hall, Inc.

Gas Tungsten Arc and Plasma Arc Welding Principles **Chapter 18** 587

Fig. 18-68 Connects to the unit shown in Fig. 18-67. Note gas and power supply connections.
Elderfield & Hall, Inc.

torch and work. Standoff distance provides the following advantages:
  a. Pilot arc starting is possible.
  b. Less operator skill is required, and it is easy for a person with welding experience to learn the process.
  c. Torch-to-work distance and torch angle are not critical.
  d. The operator can see the workpiece continuously. The long arc permits visual inspection, and the pilot arc spotlights welds.
  e. Filler wire or overlay material can be added more easily.
  f. The process can be either manual or mechanized.
3. A third characteristic of plasma welding is the protection of the tungsten electrode by the plasma torch. Electrode protection provides the following advantages:
  a. Tungsten contamination of the weld is eliminated. Noncontamination is especially important when welding reactive metals.
  b. Erosion of the electrode tip is retarded, and electrode life is prolonged.
  c. Less downtime is required to change and adjust electrodes.
  d. The lightweight torch is maneuverable and minimizes welder fatigue.
  e. Electrode protection is a factor in reliable arc starting.
4. A fourth characteristic of plasma welding is the soft and stiff arcs that may be obtained. Variable arc characteristics provide the following advantages:
  a. A stiff arc is easy to direct. There is no wandering.
  b. A stiff arc is usable in areas that have minor air drafts and magnetic fields.
  c. The stiff arc is used to obtain maximum penetration.
  d. The soft arc is used for fusion welding.

SHOP TALK

Starting Welds

It is common for welders to start welding prior to having stabilized the electrode force. They start before stabilization in order to speed up production. Even so, electrodes must get up to at least 95 percent of the chosen weld force setting before a weld should be started.

Table 18-11 Surfacing Process Comparisons

Process	Average Deposition Rate (lb/h)	Minimum Weld Dilution (%)	Minimum Deposit Thickness (in.)	Surfacing Material Form	Type of Operation
Plasma arc weld surfacing	7	5	0.010	Powder	Mechanized
Oxyacetylene	4	1	1/32	Rod	Mechanized & manual
TIG	5	10	3/32	Rod, wire	Mechanized & manual
Submerged arc single wire	15	20	1/8	Wire	Mechanized & semiautomatic
Submerged arc series circuit	30	15	3/16	Wire	Mechanized
MIG single wire	12	30	1/8	Wire	Mechanized & semiautomatic
MIG with aux. wire	25	20	3/16	Wire	Mechanized

Plasma Arc Surfacing

Metal surfacing is done to give the surface of a part greater resistance to corrosion, abrasion, or impact. The extent to which the operation is successful depends upon keeping dilution with the base metal at a minimum. Study the process comparison in Table 18-11.

The plasma process is really a weld surfacing process. It permits precision controlled thin overlays of metals resistant to heat, wear, and corrosion. The deposit is fully bonded. Plasma arc surfacing provides close thickness tolerances that require less grinding and machining time to finish than deposits applied by other surfacing processes.

Because plasma arc surfacing uses powdered metals and alloys, it is not limited by wire availability. All metals can be powdered, but not all of them can be made into rod or wire. The plasma arc can be tailored to meet specific requirements and can be used to apply the following materials to carbon steel: cobalt, nickel- and iron-base hard-facing alloys, stainless steels, copper, and tin. Tungsten-carbide particles in a cobalt or nickel-base alloy matrix can also be applied.

The source of heat is the constricted tungsten arc. The powder is carried to the torch from a hopper. The arc melts the powder and fuses it to the work. Dilution is low, and deposition is high. Plasma arc surfacing is a true welding process, not a metal spray process. A power source produces the transferred arc and controls the amount of heat delivered to the work. Argon forms the plasma, transports the metal powder, and shields both the overlay and base metal from oxidation and harmful gas pickup, Fig. 18-69.

Plasma arc surfacing works best on high production jobs. Typical surfacing applications include wear rings and plates, valve parts, wedges, slides, tillage and mower blades, and oil drill tool joints. Plasma arc surfacing has the following advantages:

- The depth of penetration into the base metal is precisely controlled to within 0.005 inch.
- The dilution of the surfacing material by the base metal is precisely controlled. The range of dilution is 5 to 50 percent.
- Layers as thin as 0.010 inch or thick as ¼ inch can be deposited in each pass.
- Strips as narrow as ⅛ inch or wider than ¼ inch can be deposited.
- Speeds over 20 inches per minute can be attained with 95 percent deposition efficiency.
- Flat, smooth deposits require little finishing.
- There is a wide choice of alloy powders for surfacing.

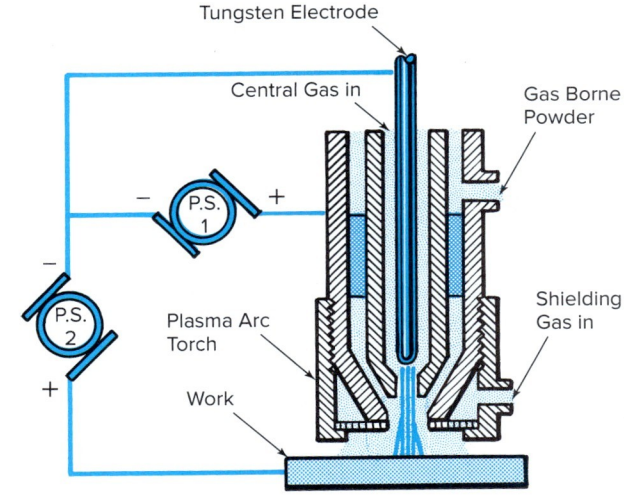

Fig. 18-69 The plasma surfacing system.

Practice Jobs

The practice of gas tungsten arc welding is covered in Chapter 10 for the welding of plate and Chapter 13 for the welding of pipe. Practice in this process is best done when you have mastered the gas welding process and the shielded metal arc welding process and have successfully passed at least the welder qualification tests in the flat, horizontal, and vertical positions. You will find that welding with the gas tungsten arc process will be relatively easy because of the thorough practice you have had in the course of oxyacetylene welding and shielded metal arc welding. Once you have mastered the handling of the flame and the arc, understand the equipment, and are able to tell when fusion and penetration are taking place, you will have much better understanding and less difficulty in mastering any other arc welding process.

CHAPTER 18 REVIEW

Multiple Choice

Choose the letter of the correct answer.

1. The heat necessary to weld with the GTAW process is produced by _____. (Obj. 18-1)
   a. Gas flame
   b. Resistance heating
   c. Gas plasma (arc)
   d. Flux

2. What equipment is required for GTAW? (Obj. 18-4)
   a. Welding generator or transformer
   b. Gas supply
   c. TIG torch
   d. All of these

3. The purpose of the GTAW torch is to _____. (Obj. 18-4)
   a. Hold the tungsten electrode
   b. Direct the electric current to the welding zone
   c. Direct the shielding gas to the welding zone
   d. All of these

4. Argon is an inert gas that shields the arc and weld area from the atmosphere and also has the following characteristics. (Obj. 18-5)
   a. High arc voltage for more heat
   b. High gas volume required because argon is lighter than air
   c. Good arc starting
   d. All of these

5. Helium is an inert gas that shields the arc and weld area from the atmosphere and also has the following characteristics. (Obj. 18-5)
   a. Good cleaning action
   b. High arc voltage for more heat
   c. Low gas volume required because helium is heavier than air
   d. Both a and b

6. Gas flow rates should be high enough to protect the _____. (Obj. 18-5)
   a. Weld pool
   b. Work connection
   c. End of the electrode
   d. Both a and c

7. The power source used for GTAW is said to be of which type? (Obj. 18-6)
   a. Drooper
   b. Constant current
   c. Variable voltage
   d. All of these

8. Square wave refers to the _____. (Obj. 18-6)
   a. Shape of the filler metal
   b. Shape of the electrode
   c. Shape of the output power
   d. Both a and b

9. Which of the following are appropriate arc starting methods for GTAW? (Obj. 18-6)
   a. Poke, scratch, HF
   b. Lift, bang, HF
   c. Lift, scratch, HF
   d. Scratch, slide, HF

10. Direct current electrode negative produces 70 percent of the heat in the _____ and 30 percent of the heat in the _____. (Obj. 18-7)
    a. Shield gas; electrode
    b. Arc; coolant
    c. Plate; electrode
    d. Current; electrode

11. Alternating current is widely used for GTAW because it _____. (Obj. 18-7)
    a. Has greater penetration than DCEP
    b. Cleans the refractory oxide off of metals like aluminum and magnesium better than DCEN
    c. Welds faster than DCEN
    d. Both a and b

12. What type of current should be used for welding stainless steel with the GTAW process? (Obj. 18-7)
    a. Alternating current high frequency
    b. DCEP
    c. DCEN
    d. Both b and c

13. Tungsten is a metal with a melting point higher than any other metal, and it melts at approximately _____. (Obj. 18-8)
    a. 6,740°F
    b. 3,260°F
    c. 5,280°F
    d. 6,170°F

14. A 1.5 percent lanthanum electrode would be color coded _____. (Obj. 18-8)
    a. Gold
    b. Red
    c. Green
    d. Brown

15. What is the proper current range for a 3/32-inch diameter thoriated tungsten electrode when welding carbon steel in an argon shield? (Obj. 18-8)
    a. 100–180 amperes
    b. 135–240 amperes
    c. 160–235 amperes
    d. 220–360 amperes

16. What is the proper electrode shape for welding aluminum with the enhanced square wave power source? (Obj. 18-8)
    a. Pointed similar to welding with DCEN
    b. Square end (right out of the box)
    c. A balled end as used on all a.c. TIG welding
    d. Both b and c

17. The principal advantage of TIG hot wire is _____. (Obj. 18-9)
    a. Faster travel speed similar to MIG
    b. Higher deposition than conventional TIG
    c. Equipment less costly and complicated than conventional TIG
    d. Both a and b

18. Which of the following is *not* a method for resolving arc starting difficulties? (Obj. 18-10)
    a. Use short weld cables with good insulation and kept straight with no coils
    b. Use pure argon; helium makes arc starting more difficult
    c. Use the smallest diameter tungsten for the amperage being used
    d. Attach the work cable as far away from the arc area as possible

19. A gas nozzle size is indicated with a number 8 indicating 8/16 or 1/2 inch. What size tungsten electrode would be appropriate to use with this nozzle? (Obj. 18-10)
    a. 0.010 inch
    b. 1/16 inch
    c. 1/8 inch
    d. 1/4 inch

20. The purpose of a water cooler in a GTAW setup is to _____. (Obj. 18-10)
    a. Cool the electrode power cable and torch
    b. Cool the arc and weld area so it solidifies faster
    c. Cool the work, work connection, and filler metal
    d. Cool the welding operators and make them more productive workers.

21. The first industrial application for plasma arc was _____. (Obj. 18-12)
    a. Surfacing
    b. Cutting
    c. Welding
    d. Metal marking

22. Plasma needle arc welding is done with currents as low as _____. (Obj. 18-12)
    a. 10 amperes
    b. 1 ampere
    c. 0.5 ampere
    d. 0.1 ampere

23. When the plasma jet completely penetrates the workpiece, this is known as _____. (Obj. 18-12)
    a. The keyhole effect
    b. The pigeonhole effect
    c. Complete joint penetration
    d. Partial joint penetration

Review Questions

Write the answers in your own words.

24. Describe in two sentences the GTAW process. (Obj. 18-3)
25. What is plasma arc welding? (Obj. 18-2)
26. List four major characteristics of plasma arc welding. (Obj. 18-2)
27. Describe the differences between nonconsumable and consumable electrode processes. (Obj. 18-3)
28. List the basic equipment needed for gas tungsten arc welding. (Obj. 18-4)
29. List features found on a GTAW power source. (Obj. 18-6)

30. State the proper use of alternating current, DCEN, and DCEP for GTAW. (Obj. 18-7)
31. Explain the difference between transferred and non-transferred arcs. (Obj. 18-11)
32. In two sentences, describe the similarity and differences between GTAW and PAW. (Objs. 18-2, 18-11, 18-12)

INTERNET ACTIVITIES

Internet Activity A

Search on the Web for books about welding at England's Cambridge International Science Publishing. Make a list of books that sound interesting to you and share your findings.

Internet Activity B

Using the Internet, search for safety and health guidelines for welding and write a report on your findings. You may want to try the American Welding Society (AWS) Web site.

Design credits: Cinema and movie icon: Denis Meshkov/123RF; and Illustration of Welding icons: Mr. Nachai Sorasee/123RF

19

Gas Tungsten Arc Welding Practice:

Jobs 19-J1–J19 (Plate)

Chapter Objectives

After completing this chapter, you will be able to:

19-1 List the various types of metal weldable with the GTAW process.

19-2 Describe proper GTAW techniques.

19-3 Explain GTAW joint design and material thickness requirements.

19-4 List procedures for setting up the GTAW equipment.

19-5 Name the major safety precautions of the GTAW process.

19-6 Describe problems, causes, and remedies for GTAW process troubleshooting.

19-7 Produce acceptable weld beading on flat plate.

19-8 Produce acceptable fillet and groove welds on aluminum, mild steel, and stainless steel with the GTAW process.

Gas Tungsten Arc Welding of Various Metals

The gas tungsten arc welding (GTAW) process can be used for nearly all types of metals. The welds are of the highest physical and chemical quality obtainable. Study the following material carefully so that you will be familiar with the GTAW characteristics of each metal.

Table 19-15, on page 614, provides a Job Outline for Chapter 19. It is recommended that students complete the projects in the order shown.

Aluminum

Aluminum is one of the most abundant metals. It has the characteristics of being light and strong, highly corrosion resistant, ductile, and malleable. It has good electrical conductivity and is relatively inexpensive.

Aluminum melts at 1,218°F and its oxide melts at about 3,700°F. (Steel melts at about 2,800°F, and copper at approximately 1,980°F.) Aluminum oxidizes readily at room temperature. The oxide must be removed prior to welding. This can be done mechanically, chemically, or with the cleaning action of the welding arc.

Selection of the *arc welding method* for joining aluminum depends largely on the individual application.

The thickness of the metal, the design of the parts and assemblies, production quantities, and the equipment available must be considered.

The best welding methods for aluminum are the *gas tungsten arc welding* (*GTAW,* also known as *TIG*) and the *gas metal arc welding* (*GMAW,* also known as *MIG*) processes. Because of the shielding gas, each offers good protection for the weld pool. The gas shield is transparent so that the welder can see the fusion zone. This helps in making neater and sounder welds.

One disadvantage in the welding of aluminum is the fact that it does not change color when it approaches the melting point, as most other metals do. The flux required when welding with other processes also causes a glare in the weld pool and emits a considerable amount of smoke. When TIG welding, there is no glare or smoke, and the weld pool is clearly visible to the welder.

The gas tungsten arc welding process is preferred for welding aluminum sections less than ⅛ inch in thickness. This method can also be used on heavier sections, but the gas metal arc welding process, Chapter 21, is usually chosen for its higher welding speed and economy. In this chapter, we are concerned with the TIG welding of aluminum. The welding of aluminum with the MIG process is discussed in Chapters 21 and 22.

All types of aluminum alloys can be welded with the TIG process, including those alloys in the 1000, 1100, 3000, 5000, 6000, and 7000 series. (Refer to Table 3-23, page 94, for the classification system of aluminum alloys.) Alternating current with stabilization is recommended. Welding is possible with direct current electrode negative, but it is not as successful as with stabilized alternating current. The shielding gas is usually argon.

SHOP TALK

Aluminum

Aluminum is harder to resistance weld than steel because, comparatively, it

1. Melts at a lower temperature (1,220°F).
2. Has a shorter weld time due to higher thermal conductivity
3. Has 4 to 5 times lower electrical resistance
4. Has a surface of protective oxide
5. Has different dynamic material resistance

Plate Preparation Aluminum plates may be prepared for welding by mechanical or thermal cutting processes. Plasma arc cutting (PAC) is used a great deal. Mechanical processes include machining, shearing, sawing, chipping, and filing. Any contaminant, such as oil or grease, must be removed from the prepared surface.

In addition to the contamination that may be on the surface of the plate due to joint preparation, there is also an oxide film. It is necessary to remove this material from the plate and filler metal before welding. Contamination may be removed from the surface with caustic soda, acids, and certain solutions. Mechanical methods include stainless-steel wire brushing, scraping, filing, and the use of stainless-steel wool. The interval between cleaning and welding should be as short as possible, since aluminum reoxidizes so rapidly.

When aluminum is welded from one side only, some type of backup may be desirable to control penetration. Backup strips may be made of steel, copper, or stainless steel. For certain welding jobs, especially pipe, the backing may consist of an inert gas shield.

Preheating *Aluminum alloys* are of two types, the *work-hardenable alloys* such as EC, 1100, 3003, 5052, 5083, and 5086; and the *heat-treatable alloys* such as 6061, 6062, 6063, 7005, and 7039. Alloys in the 2000 and 7000 series can also be welded with either the gas metal arc welding process or resistance welding.

Heat-treatable alloys may be preheated to keep cracking to a minimum. These alloys are heated at temperatures above 900°F and then given a low temperature aging treatment above 300°F. Heat treatment is not generally required in the welding of work-hardenable aluminum alloys.

Preheating is necessary if the mass of the base metal is such that heat is conducted away from the joint so fast that the welding arc cannot supply the heat required to produce fusion. Insufficient heat results in poor fusion of the weld bead and inadequate melting of the base metal. Preheating the parts being joined helps produce a satisfactory weld, reduces distortion and cracking in the finished product, and increases welding speed. Mechanical properties of certain aluminum alloys will decline with excessive preheat temperatures. Special consideration should be given to the amount of time heat is applied for each application.

In gas tungsten arc welding, preheating is necessary when welding plate over ⅜ inch thick. With the enhanced square wave a.c. power sources, preheating may be eliminated on thicker sections.

One of the advantages of gas metal arc welding is that preheat is seldom required, regardless of plate thickness.

Welding Technique Because aluminum alloys have relatively high coefficients of thermal expansion compared with most weldable metals, there may be a problem with distortion whenever unequal expansion or contraction occurs. You will recall that control of distortion is a matter of proper design, joint preparation and fitup, choice of welding process, and the use of a proper welding sequence.

In planning the sequence of welding, individual pieces and members must have freedom of movement. Joints that are likely to contract a great deal should be welded first. Welding should be done on both sides of the structure at the same time and alternated from one side to the other to equalize stresses. The weld should be completed as rapidly as possible without interruption.

Because of the high thermal conductivity of aluminum, a very high heat input must be maintained in the weld zone to balance the heat loss to the adjacent metal. Because of the intense temperature of the tungsten arc, little difficulty is encountered.

The molten pool of aluminum has a high degree of surface tension and solidifies rapidly so that it is possible to weld in all positions, Fig. 19-1.

When welding aluminum, the old practice was to **ball** or round the end of the electrode before welding to keep the arc steady. The ball was required with conventional a.c. sine wave power sources. A more tapered tungsten can now be used with the enhanced square wave power source. This type of power source with frequency control and a tapered tungsten allows the arc to be directed into the joint. This allows the aluminum to be welded with a.c. much like stainless steel is welded with DCEN, with its small focused weld pool and no arc wandering.

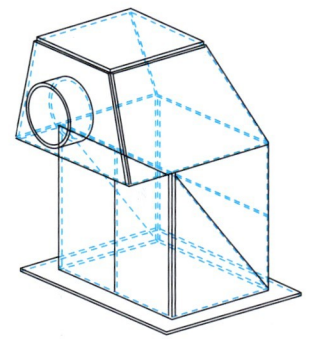

Fig. 19-1 This aluminum weldment was manually welded using the GTAW-P process without a foot control but with the use of a sequence control button or switch mounted on the torch. Note the degree of weld pool control where the three corner joints came together, and consistent weld pool control across the joints. The weldment has been rotated 90° for picture purposes. The pictorial view shows the position it was welded in. The corner and butt joints were all CJP while the lap and T joints could have no melt through to the back side of the joints. Material was approximately ⅛ inch thick, pipe OD was 2¼ inches, and the top piece was approximately 4 inches square.
Edward R. Bohnart

For video of GTAW a.c. square wave arc on aluminum with no arc wander, please visit www.mhhe.com/welding.

If the end is uneven, the arc may move from side to side. The method for balling the end of the electrode is given in Chapter 18, pages 577–579. Consult Tables 19-1 and 19-2 (p. 596) for the common filler metals to be used and the proper operating conditions for welding aluminum with alternating current.

Carbon and Low Alloy Steels

The welding of carbon steels is not taught until you have mastered the welding of aluminum because they are a little easier to weld than the stainless steels. There is a great deal of similarity, however, and the practice gained in welding these steels will make it easier for you to learn to weld the stainless steels. There is also the matter of cost to consider. By practicing on carbon and low alloy steels, you can gain a great deal of experience with the gas tungsten arc process at the lowest possible cost.

TIG welding of carbon steels is usually done on light gauge materials that require relatively less heat than heavy plate. A given size electrode also requires less heat in the welding of carbon steel than the same size used in the welding of aluminum. Lower current results in a smaller arc. The small arc is more difficult to see than the arc for aluminum welding. For this reason, a lighter shade of welding lens is recommended when welding carbon steels.

When welding carbon steels, the best results are obtained with direct current electrode negative. Use argon

Table 19-1 Chemical Compositions

Chemical Composition Requirements for Steel Solid Electrodes and Rods[1]

AWS Classification	UNS Number	C	Mn	Si	P	S	Ni	Cr	Mo	V	Cu	Ti	Zr	Al
ER70S-3	K10726	0.06–0.15	0.090–1.40	0.45–0.75	0.025	0.035	[2]	[2]	[2]	[2]	0.50			
ER70S-6	K11140	0.06–0.15	1.40–1.85	0.80–1.15	0.025	0.035	[2]	[2]	[2]	[2]	0.50			

Chemical Composition Requirements for Aluminum Electrodes and Rods[3]

AWS Classification	UNS Number	Si	Fe	Cu	Mn	Mg	Cr	Ni	Zn	Ti	Other Elements Each	Other Elements Total	Al
ER4043	A94043	4.5–6.0	0.8	0.30	0.05	0.05	—	—	0.10	0.20	0.05[4]	0.15	Remainder
ER5356	A95356	0.25	0.40	0.10	0.05–0.20	4.5–5.5	0.05–0.20	—	0.10	0.06–0.20	0.05[4]	0.15	

Chemical Composition Requirements for Stainless-Steel Electrodes and Rods[5]

AWS Classification	UNS Number	C	Cr	Ni	Mo	Mn	Si	P	S	N	Cu	Other Elements Element	Other Elements Amount
ER308L	S30883	0.03	19.5–22.0	9.0–11.0	0.75	1.0–2.5	0.30–0.65	0.03	0.03	—	0.75		
ER309L	S30983	0.03	23.0–25.0	12.0–14.0	0.75	1.0–2.5	0.30–0.65	0.03	0.03	—	0.75		
ER316L	S31683	0.03	18.0–20.0	11.0–14.0	2.0–3.0	1.0–2.5	0.30–0.65	0.03	0.03	—	0.75		

[1]Weight percent single values are maximums.
[2]These residual elements shall not exceed 0.50% in total.
[3]Weight percent single values are maximums. If other elements are detected, they cannot exceed the limits specified for other elements.
[4]Beryllium shall not exceed 0.0008%.
[5]Weight percent single values are maximums. If other elements are detected, the amount of those elements shall be determined and cannot exceed 0.5%, excluding irons.
Source: Data extracted from AWS filler metal specifications A5.9, A5.10, and A5.18.

Table 19-2 Operating Conditions with Stabilized A.C. Current and Argon Shielding Gas: Aluminum[1]

Work Thickness (in.)	Welding Current (A) Flat	Welding Current (A) Vertical	Welding Current (A) Overhead	Tungsten Diameter (in.)	Filler Rod Size (in.)	Cup Size (in.)	Gas Flow (ft³/h)
1/16	60–90	60–90	60–90	1/16	1/16	1/4, 5/16, 3/8	15
1/8	125–160	115–135	120–160	3/32	3/32	3/8, 7/16	17
3/16	190–240	190–220	180–210	1/8	1/8	7/16, 1/2	21
1/4	260–340	220–260	210–250	3/16	1/8, 3/16	1/2, 5/8, 3/4	25
3/8	330–400	250–300	250–300	3/16, 1/4	3/16, 1/4		29
1/2	400–470	290–350	250–375	3/16, 1/4	3/16, 1/4		31

[1]Optimum conditions for each application should be determined by trial.
Source: Eutectic Corp.

as the shielding gas and cerium, lanthanum, or thoriated electrode with a tapered electrode tip.

The sharpened electrode makes it easier to control the arc and the weld bead size. Because the current is DCEN, it takes a lot of heat to melt the sharpened electrode. Make sure that the electrode does not touch the weld pool. The molten metal will contaminate it and cause an erratic arc. Use a filler rod that matches the base metal.

Set the high frequency switch in the start only position. It is not necessary to use continuous high frequency as was needed with the conventional a.c. sine wave power source. If the power source is equipped with a lift arc or touch start control feature, this can be selected. The scratch start is not recommended, because the weld or tungsten can be contaminated, especially with the inverter-type power source.

Less heat is needed to start the weld pool in carbon steel than aluminum. Since the machine heat setting may be held fairly steady from start to finish, a foot-operated heat control is not absolutely necessary. The welding heat can be controlled by varying travel speed. In order to obtain the highest weld quality, however, a heat control should be used. Consult Tables 19-1 and 19-3 for the common filler metals to be used and the proper operating conditions for the welding of low carbon steel with DCEN.

Stainless Steel Stainless steel is one of the most widely used of all the alloys. It is suitable for all types of welded fabrications in which strength and resistance to high temperatures, pressures, and corrosion are desired. Root passes laid down by gas tungsten arc welding are always specified for X-ray quality welds in large, heavy-walled pipe for nuclear power plants, hygienic applications, and other critical services. Stainless steel is made in all standard steel shapes and forms.

You will recall from Chapter 3 that there are three general types of stainless steel, and all can be welded with the gas tungsten arc process. *Steels* in the 300 series, which are referred to as the **austenitic type,** are chromium-nickel steel alloys. They are highly weldable. These steels are hardenable only by cold working. A second type is the **martensitic steels.** They are straight-chromium stainless steels in the 400 and 500 series. These steels are hardenable by rapid cooling from a high temperature. A third type in the 400 series is **ferritic steels.** They are a straight-chromium content of approximately 11.5 to 18 percent. Ferritic steels are nonhardenable in certain high carbon-chromium combinations. Lower chromium grades may be air hardened.

Stainless steels have a high resistance to corrosion and high temperatures, excellent strength-to-weight ratios, and a high degree of ductility. They can be fabricated by the same methods as carbon steel. Their thermal conductivity, however, is about 50 percent less than that of carbon steel, so heat stays in the weld zone. Because their thermal expansion is about 50 percent greater than that of carbon steel, there is a problem with distortion. Distortion and heat retention can be minimized through the use of proper jigs and fixtures. Reduction of heat input by using GTAW-Pulse is also very helpful to control distortion, Fig. 19-2, page 598.

The gas tungsten arc welding process makes it possible to weld stainless steels without the problems of flux and weld spatter. The dangers involved in multipass welding due to heavy slag are not present.

Stainless steel can be successfully welded with either direct current electrode negative or alternating current with stabilization. Much greater penetration and welding speed can be obtained with direct current electrode negative. When welding with DCEN, high frequency is usually used only to start the weld. Consult Tables 19-1

Table 19-3 Operating Conditions with DCEN and Argon Shielding Gas: Carbon Steel Plate

Plate Thickness (in.)	Tungsten Diameter (in.)	Ceramic Gas Cup Size (in.)	Argon Flow (ft^3/h)	Welding Current DCEN (A)	Filler Rod Size (in.)
1/16	1/16	3/8	10–12	80–120	1/16
3/32	3/32	1/2	12–14	100–140	3/32
1/8	3/32	1/2	12–16	100–140	3/32
3/16	3/32	1/2	12–18	100–140	1/8
1/4	3/32	1/2	14–18	130–175	3/16
3/8	1/8	1/2	16–20	170–200	3/16
1/2	1/8	1/2	18–22	200–250	1/4

and 19-4 for the common filler metals to be used and the proper operating conditions for welding stainless steel with DCEN.

A 1 or 2 percent cerium, lanthanum, or thoriated tungsten electrode is used.

Either *argon* or *helium* (or a mixture of these gases) may be used as a shielding gas. Argon provides smoother arc action and gives good results when welding the thinner gauges. Helium shielding produces a hotter arc than argon and permits higher welding speeds and deeper penetration, particularly on heavier materials. Welding research has determined that argon-hydrogen mixtures of about 1 to 7 percent hydrogen are the equivalent of helium and produce sound welds in austenitic stainless steels. Shielding the back side of the weld may be necessary to prevent oxidation and to promote maximum corrosion resistance. One method is to introduce argon or a backup powder under the weld with some type of backing device to confine the gas to the weld area.

See Fig. 19-3 for the color indications that denote the amount of acceptable oxidation when welding stainless-steel pipe and tubing as used in the food processing industry. The type of welding current and polarity have a large effect on welding penetration. Developments have been made in producing chemical fluxes that affect the surface tension of the weld pool molecules and allow improved penetration on certain metals. The flux is applied prior to welding, and for a given amperage, penetration will be increased. Figure 19-4 shows examples of weld profiles with and without the use of a welding flux.

One of the principal differences between welding stainless steel and aluminum is that aluminum requires more heat and a faster speed of travel. Take care in the selection of filler rod size. If the rod diameter is too large, it will soak up a good deal of heat and make welding more difficult.

It is important that the filler rod match the type of stainless steel being welded. The rod must also deposit metal that has the physical and chemical properties that the job requires.

Fig. 19-2 This stainless-steel weldment was manually welded using the GTAW-P process without a foot control but the use of a sequence control button or switch mounted on the torch. Note the degree of weld pool and heat input control where the three corner joints come together. This was presented for inspection in the as-welded condition. Note the uniform heat color marks along the joints and the lack of buckling and distortion. All corner and butt joints required CJP, while T and lap joints could have no melt-through on the back side of the joint. The small tube extending from one end is for purging. The material was approximately 14-gauge 304 stainless steel. All welding was done with the weldment orientated in the position shown.
Edward R. Bohnart

Table 19-4 Operating Conditions with DCEN and Argon Shielding Gas: Stainless Steel[1]

Work Thickness (in.)	Welding Current (A)			Tungsten Diameter (in.)	Filler Rod Size (in.)	Cup Size (in.)	Gas Flow (ft³/h)
	Flat	Vertical	Overhead				
1/16	80–110	70–100	70–100	1/16	1/16	1/4, 5/16, 3/8	11
3/32	100–130	90–120	90–120	1/16	3/32	1/4, 5/16, 3/8	11
1/8	120–150	110–135	105–140	1/16, 3/32	3/32	1/4, 5/16, 3/8	11
3/16	200–275	150–225	150–225	3/32, 1/8	1/8	3/8, 7/16, 1/2	13
1/4	275–375	200–275	200–275	1/8	3/16		13
1/2	350–375	225–280	225–280	1/8, 3/16	1/4		15

[1]Optimum conditions for each application should be determined by trial.
Source: Eutectic Corp.

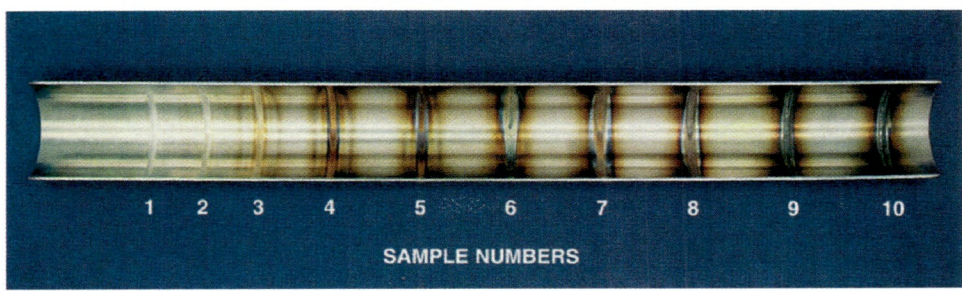

Fig. 19-3 Weld discoloration levels inside of austenitic stainless-steel pipe. Sample 1 has 10 ppm oxygen, sample 5 has 200 ppm, and sample 10 has over 25,000 ppm. Samples 5 and under are generally considered acceptable in the as-welded condition. Proper purging is critical for hygienic quality welds. © American Welding Society. D18.1:1999, Figure 2.

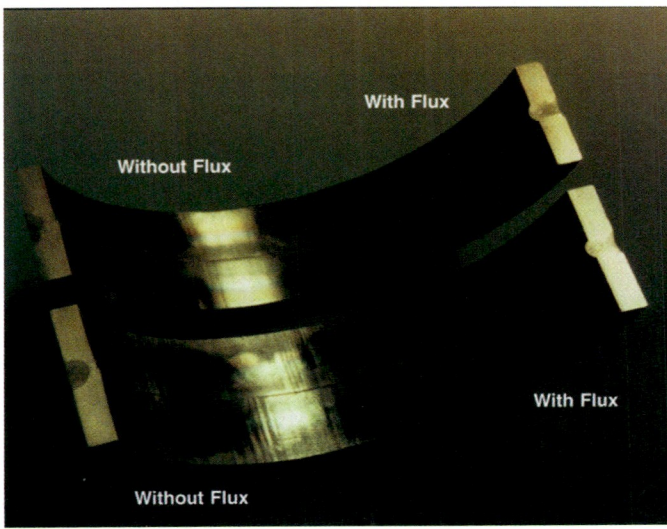

Fig. 19-4 At a given amperage level, note the increased penetration achieved with the flux on this application.
Navy Joining Center

Travel speed is slow with relatively low heat, and it takes considerable skill to keep from burning through thin materials. Because stainless steel distorts much more than carbon and low alloy steels during welding, proper tacking, weld sequencing, and clamping are important.

Magnesium

Many people mistake *magnesium* for aluminum because they are similar in appearance and characteristics. Magnesium is a very light material. It is about two-thirds the weight of aluminum and one-quarter the weight of steel. The melting point of magnesium is 1,204°F, while the melting point of aluminum is 1,218°F. The strength-to-weight ratio of magnesium alloys is good. The thermal conductivity is relatively high—a little less than aluminum.

Welding magnesium is similar to welding aluminum. The same type of joints and method of joint preparation are used. Careful cleaning of the workpiece is always required, since magnesium oxidizes readily. The parts should be degreased by chemical cleaning and/or mechanically cleaned with abrasives.

Rolled or extruded magnesium sections can be joined easily to each other or to castings. All welds are made without flux and usually have more than 90 percent of the tensile strength of the annealed base metal.

Either alternating current with stabilization or DCEN may be used. The shielding gas may be argon or helium.

Consult Table 19-5 (p. 600) for the proper operating conditions for the welding of magnesium with alternating current.

Titanium

Titanium is lightweight, has excellent corrosion resistance, and has a high strength-to-weight ratio that makes it a desirable metal for applications in the chemical, aerospace, marine, and medical fields. Its use in the petrochemical industry and in the manufacture of sports equipment are some more recent applications. Many consider titanium to be very hard to weld. Titanium alloys can be embrittled by failure to follow proper welding techniques, but titanium is much more readily welded than typically believed.

Before welding titanium, it is essential that the weld area and the filler metal be cleaned. All mill scale, oil, grease, dirt, grinding dust, and any other contamination must be removed. If the titanium is scale-free, degreasing is all that is required. If oxide scale is present, it should be degreased prior to descaling. An area at least 1 inch around where the weld is to be made should be cleaned. The joint edges should be brushed with a stainless-steel wire brush and degreased with acetone just prior to welding. Any titanium part handled after cleaning should be done so using a *white glove* procedure to eliminate recontamination of the weld area. The cleaned parts should be welded within a few hours or properly stored by wrapping in lint-free and oil-free materials.

Table 19-5 Operating Conditions with Stabilized A.C. Current and Argon Shielding Gas: Magnesium[1]

Work Thickness (in.)	Welding Current, Flat Position (A) Backup	Welding Current, Flat Position (A) No Backup	Tungsten Diameter (in.)	Filler Rod Size (in.)	Cup Size (in.)	Gas Flow (ft³/h)
1/16	60	35	1/16	3/32	1/4, 5/16, 3/8	13
3/32	90	60	1/16	1/8	1/4, 5/16, 3/8	16
1/3	115	85	1/16	1/8	1/4, 5/16, 3/8	19
	One pass	Two passes				
3/16	120	75	1/16	5/32	1/4, 5/16, 3/8	19
1/4	130	85	3/32	5/32		19
3/8	180	100	3/32	3/16		24
1/2	—	260	5/32	3/16		24
3/4	—	370	3/16	1/4		36

[1]Optimum conditions for each application should be determined by trial.
Source: Eutectic Corp.

Table 19-6 Operating Condition with DCEN and Argon Shielding Gas: Titanium[1]

Work Thickness (in.)	Welding Current, Flat Position (A)	Tungsten Diameter (in.)	Filler Rod Size (in.)	Cup[2] Size	Gas Flow[3] (ft³/h)
1/16	65–105	1/16	1/16	—	15
3/32	75–125	1/16	1/16	—	15
1/8	95–135	1/16	3/32	—	15
1/16	150–225	3/32	3/32	—	20
1/4	225–300	1/8	1/8	—	20
3/8	225–350	1/8	1/8	—	20
1/2	250–350	1/8	1/8	—	20

[1]Optimum conditions for each application should be determined by trial.
[2]Cup size should be as large as possible, that is, 1 in. or larger. A gas lens should be used, and a trailing gas shoe can also be adapted.
[3]This is the torch (primary) shielding gas flow rate; a trailing (secondary) shield gas flow rate should be two to four times this rate. A trailing shielding gas is generally required for welding titanium.

If grinding or sanding is used to clean titanium or prepare a joint, be very cautious of the fine titanium dust particles. Titanium is flammable, and the smaller the dust particles, the more flammable they are. Excess heat input from grinding operations can cause oxidation on the surface of the metal that could lead to weld failure.

Consult Table 19-6 for the proper operating conditions for the welding of titanium with DCEN.

These welding parameters are usable on the three types of titanium alloys. The three types of titanium alloys are:

- **Titanium (CP)** Commercially pure (98 to 99.5 percent Ti) can be strengthened by small additions of oxygen, nitrogen, carbon, and iron.
- **Alpha alloy** Alpha alloys are generally single-phase alloys that contain up to 7 percent aluminum and a small amount (<0.3 percent) of oxygen, nitrogen, and carbon.
- **Alpha-beta alloys** Alpha-beta alloys have a characteristic two-phase microstructure brought about by the addition of up to 6 percent aluminum and varying amounts of beta formers. Beta forming alloys are vanadium, chromium, and molybdenum.

Table 19-7 shows some of the relative weldability of these alloy groups. Also displayed are recommended filler metals. When welding titanium or one of its alloys, the filler

Table 19-7 Titanium Weldability Rating

Alloy	Alloy Designators	Rating[1]	Filler Metal
Commercially pure (CP)	Ti-0.15 O_2	A	ERTi-1
	Ti-0.20 O_2	A	ERTi-2
	Ti-0.35 O_2	A	ERTi-4
Alpha alloys	Ti-0.2 Pd	A	ERTi-7
	Ti-5 Al-2.5 Sn	B	ERTi-6
	Ti-5 Al-2.5 Sn ELI[2]	A	ERTi-6ELI
Alpha-beta alloys	Ti-6 Al-4V ELI	A	ERTi-5ELI
	Ti-7 Al-4Mo	C	ERTi-12
	Ti-8 Mn	D	Welding not recommended

[1] A: Excellent; useful as welded, near 90% joint efficiency if base metal annealed condition. B: Fair to good; useful as welded, near 100% joint efficiency if base metal annealed condition. C: Limited to special applications; cracking can occur under high restraint. D: Welding not recommended; cracking under moderate restraint; use preheat (300–350°F) followed by postweld heat treatment.

[2] ELI: Extralow interstitial impurities are specified. These interstitial impurities are carbon, hydrogen, oxygen, and nitrogen, and both the filler metal and base metal are low in these impurities.

metal should closely match the alloy content of the base metal being welded.

Shielding of the titanium weld and surrounding metal (including the hot end of the filler rod) that reach temperatures of 500°F is required. When doing manual *open air* welding (not in a bubble or totally enclosed chamber), care must be taken to prevent atmospheric contamination of the titanium. Since titanium has a very low thermal conductivity, it stays hot for a long time after the welding arc has moved along the joint. Thus, a trailing gas is essential. This can be accomplished with a large gas lens on the torch or a trailing gas shoe that attaches to the TIG torch. This metal shoe (chamber) has a porous metal diffuser to allow the gas to blanket the titanium until it has cooled below its oxidation temperature. Figure 19-5 is an example of a trailing gas shield. The primary gas shielding is what is flowing through the torch, and the secondary gas shielding is what is flowing through the trailing shield. If the back side of the joint is going to be exposed to oxidation temperatures over 500°F, it must also be protected from the atmosphere by a backing gas shielding or, in the case of pipe or tubing, purging the inside of the pipe or tube. Figures 19-6, 19-7, and 19-8, page 602, show the color indications that denote the amount of acceptable oxidization when welding titanium.

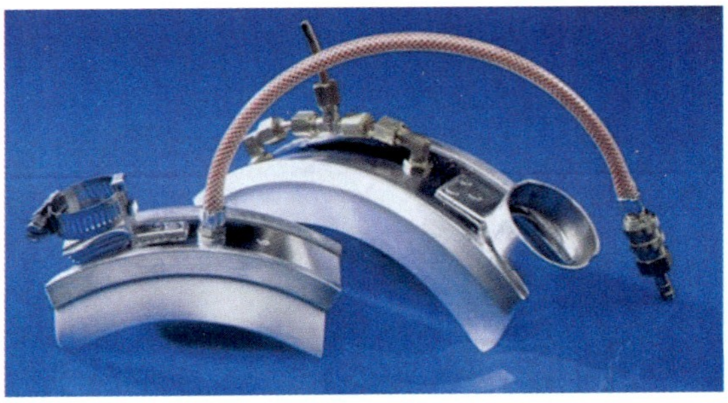

Fig. 19-5 Torch trailing shield for TIG welding of titanium and other reactive metals. © Huntingdon Fusion Techniques

Copper and Its Alloys Copper is one of the oldest metals used. It is not a scarce metal and is readily mined. It has a melting point of 1,981°F. Copper lends itself to all of the modern methods of fabrication. The weldability of each copper alloy group depends largely upon the

ABOUT WELDING

Resistance Weld

Which would you rather try to resistance weld: copper, aluminum, or stainless steel? Copper and aluminum are metals that do not resist electricity well, which makes them difficult to resistance weld. But stainless steel is a good resistor and resistance welds easily.

Fig. 19-6 A glossy silver appearance indicates an acceptable weld. © Navy Joining Center. Images are for representational purposes only and should not be used for weld color inspection because of possible color distortion in the print reproduction process.

Fig. 19-7 A gray appearance with multicolor titanium oxides indicates a rejectable weld. © Navy Joining Center. Images are for representational purposes only and should not be used for weld color inspection because of possible color distortion in the print reproduction process.

Fig. 19-8 A gray appearance with multicolor titanium oxides indicates a rejectable weld. © Navy Joining Center. Images are for representational purposes only and should not be used for weld color inspection because of possible color distortion in the print reproduction process.

alloying elements. Copper is used widely in all industries and has a special application in the fabrication of electrical equipment.

Copper is one of the best conductors of electricity and heat. It is highly resistant to corrosion. When mixed with certain other elements, copper is both ductile and wear resistant, and it can be heat treated.

Copper has the following disadvantages for welding:

- It is brittle at high temperatures and presents problems in jigging.
- It is strongest when cold worked. The high welding temperatures cause it to lose strength.
- Stress reversals cause the metal to become hard and brittle so that it breaks from fatigue.

While pure copper presents no great difficulty in welding, many of its alloys require special treatment. The following copper alloys lend themselves to welding:

- Copper-silicon alloys such as Everdur® and Herculoy®
- The copper-aluminum alloys known as aluminum bronze
- The copper-phosphorus alloys known as phosphor bronze
- Copper-nickel alloys

Those copper alloys containing zinc, tin, or lead are either difficult or impossible to weld. These materials have a low melting point that causes them to volatize under the intense heat of the arc. They can be successfully brazed, however, with the oxyacetylene process.

DCEN is used for welding pure copper and most copper alloys. Alternating current with stabilization is recommended, however, for beryllium copper, aluminum bronze, and copper alloys less than 20 gauge thick. Current settings should be higher for copper and its alloys than for most other metals because of their high thermal conductivity. Always provide good ventilation when welding beryllium copper and when using certain fluxes. Copper has a tendency to form oxides, which must be removed just before welding.

Consult Table 19-8 for the proper operating conditions for welding deoxidized copper with DCEN.

Nickel and Nickel-Base Alloys *Nickel* has wide application in those industries in which corrosion and low and high temperatures are encountered. It is very ductile and can be worked readily. Its tensile strength, elasticity, melting point, and magnetic properties are similar to those of steel. Following is a list of the important high nickel alloy metals:

- **The nearly pure nickels and Duranickel**®, a high strength, low alloy nickel
- **The Monels**®, which are about two-thirds nickel and one-third copper
- **The Inconels**®, which are higher in nickel and iron content than the Monels®. They are outstanding in their ability to resist damage from abrupt changes in temperature.
- **The Nimonics**®, which contain approximately 80 percent nickel and 20 percent chromium. These alloys are used in gas turbine engines.
- **The Hastelloys**® are alloys of nickel, molybdenum, and iron. They have a high resistance to acids.

While nickel can be welded with the shielded metal arc welding process, gas tungsten arc welding has the advantage of eliminating slag entrapment in the weld. Generally, DCEN is recommended. On thin material, however, alternating current with stabilization has the advantage of lower heat input. The current values and electrode sizes are similar to those used to weld carbon steel. Argon, helium, and a mixture of these gases are recommended for most applications, but argon is recommended for thin materials.

Consult Table 19-9 for the proper operating conditions for welding the Hastelloys® with DCEN.

Table 19-8 Operating Conditions with DCEN and Argon Shielding Gas: Deoxidized Copper[1]

Work Thickness (in.)	Welding Current, Flat Position (A)	Preheat Temp. (°F)	Tungsten Diameter (in.)	Filler Rod Size (in.)	Cup Size (in.)	Gas Flow (ft³/h)
1/16	110–150	—	1/16	1/16	3/8	15
1/8	175–250	—	3/32	3/32	3/8, 1/2	15
3/16	250–325	500	1/8	1/8	1/2	15
1/4	300–375	600	1/8	1/8	1/2	15
3/8	375–450	800	3/16	3/16	1/2	17
1/2	500–700	900	3/16	1/4	1/2, 5/8	17

[1]Optimum conditions for each application should be determined by trial.
Source: Eutectic Corp.

Table 19-9 Operating Conditions with DCEN and Argon Shielding Gas: Hastelloy Alloys[1]

Work Thickness (in.)	Welding Current (A)		Tungsten Diameter (in.)	Filler Rod Size (in.)	Gas Flow (ft³/h)
	Flat	Vertical			
1/16	60–90	55–75	1/16	1/16, 3/32	20
1/8	100–125	80–110	3/32	3/32	25
3/16	130–175	100–140	3/32	3/32	30
1/4	130–175	100–140	3/32	1/8	35
3/8	140–200	110–160	3/32	1/8	40
1/2	200–250	160–250	1/8	5/32	40

[1]Optimum conditions for each application should be determined by trial.
Source: Eutectic Corp.

Table 19-10 Welding Dissimilar Metals

Materials to Be Joined	Type of Current	Rod Type
Stainless steel to cast iron	DCEN	Oxweld® 26
	DCEN	Nickel and stainless steel
Stainless to carbon or low alloy steel	—	310 stainless
Copper to stainless steel	DCEN	None required
Copper to Everdur®	—	Oxweld® 26
Cupro-nickel to Everdur®	—	Oxweld® 26
Nickel to steel	DCEN	Nickel
Hastelloy® Alloy C to steel	DCEN	Hastelloy® W; Inconel®; Nickel; or 310 stainless
Aluminum to steel	A.C. with stabilization	No. 25M bronze rod; B.T. silver brazing alloy; Oxweld® 14 Al. Rod
Stainless steel to Inconel®	DCEN	310 stainless
Tungsten to molybdenum	—	Platinum
Copper and Everdur® to steel	DCEN	Copper or 26 Everdur®

Welding Dissimilar Metals

Many industrial fabrications make it necessary to join dissimilar metals. Although both the oxyacetylene and the shielded metal arc welding processes may be used, a greater variety of materials can be joined with gas tungsten arc welding.

Consult Table 19-10 for the listing of dissimilar metals that can be joined and the type of current and filler rod to use. The filler rod should be selected with care. Table 19-10 lists one company's recommendations.

Hard-facing

Hard-facing is a process of applying a hard, wear-resistant layer of metal to the surfaces or edges of parts. Hard-facing rods are deposited for the purpose of improving resistance to impact or abrasion, or both. This may be done to build up worn areas to make them as good as new or to put hard, wear-resistant cutting edges on soft, ductile materials. Hard-facing is used for stone-crushing equipment, power shovel buckets, farm implements, and many other applications. While both the oxyacetylene and the shielded metal arc welding processes may be used, somewhat better results can be obtained with the gas tungsten arc process. Consult Table 19-11 for a listing of materials that may be hard-faced and the proper operating conditions.

Filler Metals

The composition of filler metals for the various metals welded by the gas tungsten arc welding process are selected on the basis of the following criteria:

- The requirements of the specific job
- The type of metal being welded
- Ease of welding
- Characteristics desired, such as strength; ductility; hardness; machinability; and resistance to corrosion, impact, abrasion, and oxidation
- Preheat and postheat requirements
- Color matching, when this is important
- Type of shielding gas being used

You are urged to become familiar with the AWS specifications for filler metals. On most welding jobs, the type of filler metal will be specified by the engineering department. It is necessary, however, for the welder to know the types of filler metals and their uses.

Joint Design and Practices

There is no limit to the types of joints that may be welded with the gas tungsten arc welding process. The basic types of joints used for plate welding are butt, lap, edge, corner, and T-joints. They are similar to those used with other welding processes. Selection of the proper design for a particular application depends primarily on the following factors:

- Physical properties desired in the weld
- Cost of preparing the joint and making the weld
- Type of metal being welded
- Size, shape, and appearance of the assembly to be welded

Table 19-11 Hard-facing

Base Metal	Surfacing Material	Current Type	Current Amperes	Rod Type	Welding Technique	Argon Flow (ft³/h)	Deposit Rockwell (Rc) Hardness	Remarks
Mild & stainless steels	Haynes Stellite alloys	a.c.[1]		Stellite #1	Backhand	25	54	
		a.c.[1]		Stellite #6	Backhand	25	39	
		a.c.[1]		Stellite #12	Backhand	25	47	
		a.c.[1]		Stellite #93	Backhand	25	62	
		a.c.[1]		Hascrome	Backhand	25	23–43	Extruded rod has better weld characteristics than rolled rod
Copper	Stellite #6 alloy	DCEN	180–230 for 3/16-in. material	Stellite #6	Forehand	15	42	Arc directed mainly at welding rod
Steel, copper, & silicon bronze	Aluminum bronze	DCEN		Aluminum bronze rods	Forehand	10	150–300	
Mild steel & cast iron	Bronze & copper	a.c.[1] or DCEN	150 for 1/2-in. material	Aluminum bronze & copper rods	Forehand	10		
Stainless steel	Silver	a.c.[1]	160 for 1/2-in. material		Either	10		Plates pickled prior to surfacing
Mild steel	Stainless steel	a.c.[1] or DCEN			Forehand	10		
Mild steel	Lead	DCEN	75		Forehand	10		Steel ground or pickled and then coated with liquid soldering flux before surfacing
Carbon & alloy tool steels	Tungsten carbide	DCEN	300–375	Tube of 8/15 mesh tungsten particles		30		

[1]A.C. (alternating current with stabilization) develops maximum hardness values; DCEN will permit higher welding speeds.

Other considerations include the following:

- Number and size of tacks
- Purging and shielding gas
- Root face, root opening, and bevel
- Number of passes required
- Size of filler rod
- Whether or not rod is required for the first pass and whether or not a dressing pass is permitted
- Method of striking and breaking the arc
- Whether the direction of travel is moving vertically up or down (aluminum should not be welded vertically down; porosity will be worse)
- Allowable protrusion and reinforcement
- Type of electrode
- Type of welding power
- Tolerances of fitup and alignment

Whereas gas tungsten arc welding is particularly suited to the welding of materials up to 1/8 inch in thickness, it may also be used to weld heavier thicknesses of metals. The MIG/MAG process may be used on heavy stock with better results. The nature and application of the joint is a major consideration. Cost must also be considered in determining the welding process for a particular job.

Usually, filler rod need not be used for thinner materials. Careful consideration should be given to the welding of heavy carbon steel since it is quite possible that the gas metal arc or MAG process is more suitable.

No matter what type of joint is used, proper cleaning of the work before welding is essential if welds of good appearance and sound physical properties are to be obtained. This is of special importance in welding some alloys such as aluminum and magnesium. Welds in these metals will be defective if even minute quantities of foreign material contaminate the inert gas atmosphere.

On small assemblies, manual cleaning with a stainless-steel wire brush, stainless-steel wool, or a chemical solvent is usually sufficient. Do not grind aluminum or magnesium on an emery wheel. Be sure to remove completely all oxide, scale, oil, grease, dirt, and rust from the work surfaces.

Precautions should be taken when using certain chemical solvents for cleaning purposes. The fumes from some chlorinated solvents break down in the heat of the electric arc and form a toxic gas. Ventilating equipment should be provided to remove fumes and vapor from the work area.

Weld Backup

The joint should be backed up on many gas tungsten arc welding applications. **Backing** protects the underside of the weld from atmospheric contamination. Atmospheric contamination causes weld porosity, poor surface appearance, cracking, and burn through.

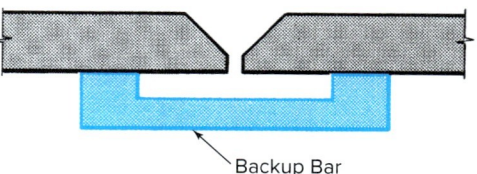

Fig. 19-9 Grooved backup bar for square and beveled-butt joints.

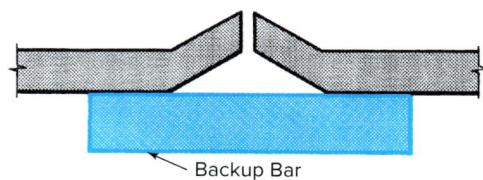

Fig. 19-10 Flat backup bar for upset butt joints.

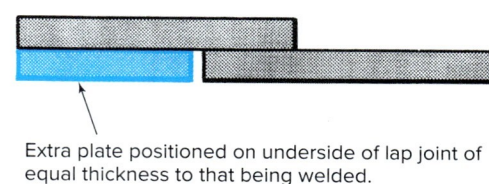

Fig. 19-11 Backing for lap joints.

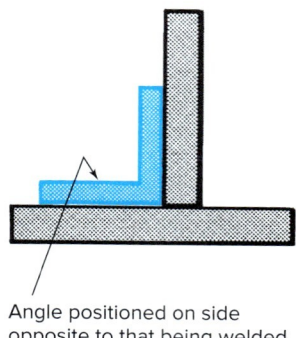

Fig. 19-12 Backup for T-joints.

The weld may be backed up by (1) metal or ceramic backup bars, (2) an inert gas atmosphere on the weld underside, (3) a combination of the first two methods, or (4) flux painted on the weld underside.

Metal backup bars should not actually touch the weld zone. (See Figs. 19-9 through 19-12 for typical backup for butt, lap, and T-joints.) The material used for making a backup bar is determined by the composition of the material being welded. A copper bar may be used to back up welds in stainless steel. For the welding of aluminum or magnesium, the bar should be made of stainless steel or steel. Carbon steel can be used for carbon steel welding. Very often, backup bars are water cooled to carry off the heat of the welding operation.

When the final weld composition must conform to extremely rigid specifications, extra care must be taken to exclude all atmospheric contamination from the weld. This is accomplished by introducing an atmosphere of inert gas on the back side of the weld. Nitrogen may be used for stainless steels. Argon should be used for other metals that oxidize readily or react with nitrogen at high temperatures. Review Chapter 18 for a detailed presentation of shielding gases.

Setting Up the Equipment

Gas tungsten arc welding is a precision technique. Care must be taken to make sure that the equipment is set up in the proper way and that all of the variables are correct for the particular welding job to be performed. The procedure for setting up the equipment before welding requires checking every detail. You should make the checks described in the following paragraphs before starting to weld. Consult Chapter 18 to review many of these points.

1. Make sure that you have a torch of the proper type and size to meet the requirements of the welding job.
2. Check the size, type, appearance, and position of the tungsten electrode in the torch. It should have the diameter recommended for the amount of current and the electrode holder used. See Table 18-7, page 577, and Table 19-12. The end of the electrode should be clean and smooth. A contaminated electrode end indicates that during a previous use the inert gas was shut off before the electrode cooled, that there was a gas leakage in the gas supply system or the torch, or that the electrode tip was contaminated by touching metal. If the tip is not too rough, it may be cleaned with a fine emery cloth. When welding aluminum, use a rounded tip with conventional a.c. sine wave and a tapered tip for a.c. with square or enhanced square wave power sources. When using DCEN for steel, stainless steel, titanium, and so on use a pointed tip.
3. Check the torch for the proper gas cup type and size. Make sure that it is clean and free of spatter. If it is important to see the weld clearly, you may wish to use a glass cup.
4. Check all the connections on the gas supply for leaks. If there is some reason to doubt a connection, check it with the proper leak check solution.
5. Check all work connections. Pay special attention to the location of the work connection to the work and the location of the work connection in relation to the joint design and the direction of welding.
6. Open the main shutoff valve of the cylinder of inert gas and adjust the gas flow to meet the needs of the particular job. See Chapter 18, pages 558–559.
7. Before turning on the coolant supply, make sure that the pressure is not higher than that recommended by the torch manufacturer. Make sure that there are no leaks in the coolant supply. The water should be routed to cool the torch head first, and then the power cable.
8. Adjust the current range for the joint being welded and the size of the tungsten electrode. The position of welding also makes a difference here. As with the shielded metal arc process, vertical and overhead welding generally requires a lower current setting than other positions.
9. It is assumed that you have the proper eye and skin protection for welding and that you will take the necessary safety precautions.

SHOP TALK

Torch Sleeve

A *torch sleeve* is a flexible material that goes over the cables of water-cooled torches. These lightweight sleeves keep the thermal plastic hoses from contacting hot materials and melting.

Table 19-12 General Sizes of TIG Electrode Holders and Diameters of the Electrodes That Fit Them

General Sizes of TIG Electrode Holders	Electrode Diameter (in.)
Small holder (100 A)	0.010, 0.020, 0.040, $1/16$, & $3/32$
Medium holder (250 A)	0.020, 0.040, $1/16$, $3/32$, & $1/8$
Large or heavy-duty holder (500 A)	$1/16$, $3/32$, $1/8$, $5/32$, $3/16$, & $1/4$
	(Some special holders—$5/16$, $3/8$)

Safe Practices

In welding with the gas tungsten arc welding process, observe the same precautions and safe practices that would apply to any other electric welding operation. In any form of electric welding, there are potential shock hazards, burn hazards, and fire hazards. In addition, welding operations on certain metals and alloys may produce unpleasant or dangerous fumes. For a detailed treatment of this subject, you are urged to secure materials from such organizations as the American Welding Society and various welding equipment companies.

The welder should also be properly protected from the rays of the arc. This requires suitable clothing to cover all exposed skin surfaces and a welder's helmet with the proper shade of glass to protect the eyes and face. The shade of the glass lens depends on the intensity of the arc. Table 19-13 lists the recommended lens shade for different current ranges.

Gas Tungsten Arc Welding Safety Rules

Observe the following rules when welding with the TIG process:

- Always use a welding helmet equipped with a shaded number lens suitable to the welding current you are using. The helmet should be mounted on an adjustable headband or, if there are overhead hazards, on a safety hardhat.
- Cover all skin surfaces with leather, heavy clothing, or other adequate protection against burns from sparks, the arc ray, and spatter. (See Fig. 19-13.)

   Recommended personal protective equipment includes the following:
   Helmet
   Safety glasses
   Leather cape (or treated clothing)
   Sleeves
   Apron
   Ear plugs, if welding vertically or overhead
   Gauntlet gloves
   Heavy, flameproof shirt and pants (without cuffs)
   Safety shoes

- Do not weld in or near flammable gases, powders, or liquids, or in or on untreated containers that have held such materials.
- Remove or protect all combustible material in the welding area. If uncertain of the adequacy of the protection, have a worker stand by with a fire extinguisher. Make sure a fully charged, proper type fire extinguisher is available. If you are uncertain, don't weld until the safety issues have been resolved.
- If welding is being done in an area where welding is not normally done, a *hot work permit* will be used. These permits are used when hazards are involved in cutting and welding operations.
- The welding area should be dry and uncluttered to avoid electric shock and falls.
- Be sure you have adequate ventilation. Use special precautions when toxic fumes, whether from cleaning fluids, coatings, or the metal itself, are given off.
- Use nonreflective welding curtains to protect others in the area. Even short exposure to tungsten arc rays can burn the eyes, and skin burns from arc rays or molten metal spatter must be prevented.
- The TIG torch, power supply unit, service leads, work connections, inert gas equipment, and protective equipment should be inspected and in good condition before welding.

Table 19-13 Correct Lenses for Various Current Settings

Shade No.	Welding Current (A)
6	Up to 30
8	30–75
10	75–200
12	200–400
14	Above 400

Fig. 19-13 Student is properly protected and is in the proper position for bench welding. Location: Northeast Wisconsin Technical College Mark A. Dierker/McGraw Hill

- Never change polarity under load, alter current settings under load if prohibited by the instructions, nor in any way overload any of the electrical equipment.
- Know and obey all safety rules pertaining to the particular area in which the welding is to be done, shop practices, and personal protection.

Arc Starting

Arc starting with the gas tungsten arc welding process is somewhat more difficult than with shielded metal arc welding. The following list will aid in getting good arc starts:

- Maintain a short arc during startup procedures.
- Use the smallest diameter tungsten possible.
- Use as high a welding current as possible.
- Use the highest quality tungsten available (of the proper alloy).
- The tungsten must be clean and properly shaped.
- Use the shortest length torch possible.
- Use proper gas cup size.
- Use premium quality cable for torch and work leads.
- Keep torch and work leads as short as possible. Move the power source as close as possible to the work. If the power source cannot be moved closer and a high frequency arc starter is being used, move it closer to the weld.
- Attach the work clamp as close as possible to the weld.
- The work connection must be clean and tight.
- The weld surface must be clean.
- Avoid long cable runs over bare concrete floors, or insulate cables from the floor by laying them on boards.
- If the welding machine is being used for both GTAW and SMAW, make sure the SMAW electrode holder is detached from the machine when using the GTAW process.
- Check and tighten all external connections.
- Keep the torch cable from contacting any grounded metal such as work benches, steel floor plates, and the machine case.
- Use 100 percent argon shielding gas if possible.
- Check the secondary current path and tighten all connections.
- If the machine has adjustable high frequency spark gaps, increase the gap to the manufacturer's recommended maximum.
- Check for mineral deposit buildup in water-cooled torches to avoid high frequency shunting back to ground through deposit material.
- Increase intensity adjustment if available.
- Have faulty connections in the equipment checked.

Four arc starting methods are generally used in TIG welding: high frequency starting, scratch starting, lift arc starting, and high voltage starting.

High Frequency, No-Touch Starting

In welding with alternating current, the electrode does not have to touch the work to start the arc. The superimposed high frequency current establishes a path for the welding current to follow by jumping the gap between the electrode and the work. The torch position shown in Fig. 19-14 illustrates the recommended method of starting the arc with high frequency when the torch is held manually. In this way the welder can position the torch in the joint area and, after lowering the welding hood, close the contractor switch and initiate the arc. By resting the gas cup on the base metal, there is little danger of touching the electrode to the work. After the arc is initiated (Fig. 19-15), the torch can be raised to the proper angle for welding. The arc will then start. For most applications, an arc length of approximately one electrode diameter may be maintained. With argon the arc length is much less critical than with helium.

The high frequency, no-touch method may also be used with direct current electrode negative polarity, if the d.c. welding machine is equipped with a high frequency unit.

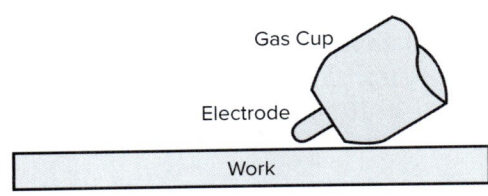

Fig. 19-14 Place the gas cup directly on the work in preparation for a high frequency start. Source: Miller Electric Mfg. Co.

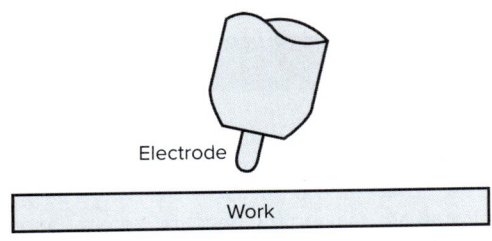

Fig. 19-15 Move from the resting cup position to bring the torch into proper position while maintaining a short arc length. Source: Miller Electric Mfg. Co.

The high frequency current is used only for starting the arc and not for welding. The high frequency is automatically turned off by means of a current relay when the arc is started. With a.c. sine wave welding, the high frequency current is maintained for the welding operation. With a.c. square wave or enhanced square wave, high frequency can be set on start only.

Scratch Starting

In welding with direct current without high frequency, the same motion as SMAW is used for striking the arc. The electrode must touch the work in order for the arc to start. Never use a carbon starting block. The arc can be started on a separate plate of aluminum, copper, or steel and carried to the workpiece. As soon as the arc is struck, withdraw the electrode approximately $\frac{1}{8}$ inch above the work to avoid contaminating the electrode in the molten pool.

Lift Arc Start

This method of arc starting was developed to eliminate the tungsten contamination associated with the scratch start method. With the lift arc, or touch start, the tungsten is brought into contact with the workpiece. When this occurs, the power source senses a short circuit and establishes a low voltage current in the weld circuit. This voltage and current are not great enough to establish an arc, but they do contribute to heating the electrode. When the electrode is lifted from the workpiece, the power source senses the absence of the short circuit condition and automatically switches to the current set on the machine. The fact that the electrode has been preheated assists in arc initiation. This function, if available, is found on the power source along with the high frequency control switch. Figure 19-16 indicates the proper method of use for this function.

High Voltage Starting

Another method of starting is through the use of a high surge of voltage as the electrode is brought close to the base metal. After the arc is established, the high voltage is cut off, and the power returns to its normal voltage value for welding.

It is common practice to warm the electrode and the nozzle on a practice piece to give better starting results on the job.

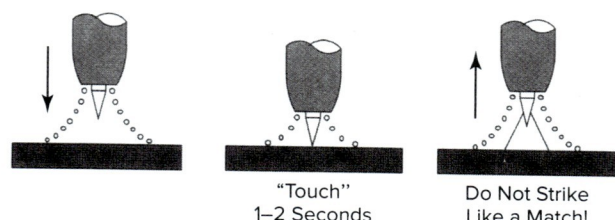

Fig. 19-16 Lift arc starting technique. Source: Miller Electric Mfg. Co.

Arc Wandering

Arc wandering, a condition in which the arc wavers as it leaves the electrode, may occur in gas tungsten arc welding. It is caused by the following problems:

- A current setting that is too low
- Contamination of the electrode
- Magnetic effects
- Drafts in the work area
- Balled tungsten and a.c. sine wave power source
- Weld in narrow joint

A low current setting and contaminated carbon are distinguished by a very rapid movement of the arc from side to side during welding. When the current is correct, the entire end of the electrode is molten and completely covered by the arc.

In DCEN welding, a current setting that is too low generates only enough heat to make part of the electrode end molten. The arc emerges only from this small molten spot which continually shifts, pulling the arc with it.

When welding using sine wave alternating current, a current setting that is too low has a different effect in that the arc wants to jump from the large round ball formed with using a.c. and a sine wave. Electrons want to jump from the hot spot on the electrode and thus wander about the balled end. When these conditions prevail, raise the operating current level.

Carbon contamination is the result of striking the arc with a carbon pencil or on a carbon block. This should never be done. Contamination can also come from dipping the electrode into the weld pool or touching the electrode with the filler metal. Contaminants

SHOP TALK

Using Aluminum

Welders working with aluminum often choose the GTAW method. This process has a very concentrated arc, making the heat input more controlled. Normally these welds are smooth, not requiring removal of spatter.

inside the gas nozzle can disrupt the gas flow causing arc wander.

Magnetic influences usually draw the arc to one side or the other along the entire length of the weld. Magnetic effects great enough to create serious disturbances are not encountered frequently. However, when such influences do occur, shifting the position of the work connection will give some relief.

Air movement causes varying amounts of arc wandering. Great care must be taken to make sure that the welding operation is shielded from drafts.

Electricity wants to take the path of least resistance. A large balled tungsten for use on an a.c. sine wave power source does not have as good of access to the root of a weld joint, so the arc wants to wander to the nearest surface and arc. You will need to try and manipulate the arc into the root. However, this usually leads to a larger weld pool than needed.

For video of GTAW a.c. sine wave T-joint on aluminum, please visit www.mhhe.com/welding.

Welding Technique

You will find that many of the techniques learned in mastering the oxyacetylene and shielded metal arc welding processes will assist you in learning how to weld with the gas tungsten arc welding process. The objective is to complete the weld without disturbing the molecular structure of the surrounding base metal. In order to accomplish this, you should work with the lowest machine setting possible for a given condition. The right amount of heat must be applied in the right place at the right time, without exceeding a safe time limit.

When welding light gauge carbon and stainless steel or when heat must be concentrated on a limited area, the electrode should be sharpened to a point with a smooth taper. This ensures that the arc will jump off the designated place each time you start an arc and will continue arcing in the same direction during the entire operation. If you are welding aluminum or magnesium with sine wave alternating current, ball the end of the tungsten. If the electrode becomes contaminated or deformed, do not increase the machine setting. Reshape the tungsten instead.

A short arc should be maintained. This helps confine the heat to the immediate area. Be careful to direct the arc only in the exact spot where it is desired to liquefy the metal. To apply heat to any other spot only increases distortion and deterioration. Play the arc on the seam to be welded until a molten pool appears. Then proceed as rapidly as possible and make sure that both pieces to be joined become molten under the arc. If the pool is slow in forming on both pieces, it may be necessary to use a slight oscillating motion to distribute the heat evenly. Care must be taken not to leave the molten pool, not to stay in one spot too long, and not to remelt metal that has just been melted.

If the addition of filler rod is necessary, choose a rod one size smaller than the thickness of the material being welded. Do not play the rod into the arc. Instead, move the rod out with the arc and then carry the correct amount back to the weld zone in order to make a uniform distribution. Correct motions to use in different positions are very important for successful results. A different, distinct motion is used in each position; each of these must be learned from an experienced welder by demonstration. In the case of corner welds, apply the heat in the corner first and then work away from the corner toward the outside.

The extension of the tungsten electrode beyond the cup is governed by the shape of the object and the type of joint. The longer the extension, the less effective the shielding. Consequently, some adjustment in rate of gas flow is necessary where shapes require longer extension and loss of shielding occurs.

A gas lens can be used where a long electrode extension is required to aid in getting proper gas coverage. The cup size is governed by the type of joint, the shape of the object, and the amount of current used.

All torch connections must be kept tight to prevent siphoning of air through connections. A small amount of contamination can be serious.

General Recommendations

Study Table 19-14, Troubleshooting Guide (page 612), thoroughly before you begin welding practice.

Look up your defects in the table during practice and correct them as instructed before beginning the next job. You will find the following recommendations helpful:

- Keep the tungsten electrode clean and free of contamination, and keep the tip properly shaped.
- Remove all surface oxides by wire brushing or grinding before welding over any previous weld. Do not remove the hot end of the filler rod from the gas shield as this will cause it to become oxidized.

Table 19-14 Troubleshooting Guide

Problem	Cause	Remedy
Wasteful electrode consumption	1. Improper inert shielding (resulting in oxidation of electrode)	1. Clean nozzle; bring nozzle closer to work; step up gas flow, shorten tungsten extension from electrode holder.
	2. Operating on electrode positive	2. Employ larger electrode or change to electrode negative.
	3. Improper size electrode for current required	3. Use larger electrode.
	4. Excessive heating in holder	4. Use ground finish electrodes; change collet; check for improper collet contact.
	5. Contaminated electrode	5. Remove contaminated portion—erratic results will continue as long as contamination exists.
	6. Electrode oxidation during cooling	6. Keep gas flowing after stoppage of arc for at least 10–15 seconds. Rule: 1 second for each 10 amperes.
Erratic arc	1. Dirty, greasy base material	1. Use appropriate chemical cleaners, wire brush, or abrasives.
	2. Too narrow joint	2. Open joint groove; bring electrode closer to work; decrease voltage.
	3. Contaminated electrode	3. Remove contaminated portion of electrode.
	4. Too large diameter electrode	4. Use smaller electrode—use smallest diameter needed to properly handle current.
	5. Arc too long	5. Bring holder closer to work to shorten arc.
Porosity	1. Entrapped gas impurities (hydrogen, nitrogen, air, water vapor)	1. Purge air from all lines before starting arc; remove condensed moisture from lines; use welding grade (99.995%) inert gas.
	2. Possible use of old acetylene hoses	2. Use only new hoses. Acetylene impregnates hose.
	3. Gas and water hoses interchanged	3. Never interchange water and gas hoses. (Color coding helps.)
	4. Oil film on base material	4. Clean with chemical cleaner not prone to break-up in arc. *Do not weld while wet.*
Tungsten contamination of workpiece (tungsten spitting)	1. Scratch starting with electrode	1. Use high frequency starter; lift arc.
	2. Electrode melting and becoming an inclusion in the weld base plate	2. Use less current or larger electrode; use cerium, lanthanum, thorium, or zirconium tungsten (these run cooler).
	3. Shattering of electrode by thermal shock	3. Make certain electrode ends are not slivered or cracked when using high current values. Use embrittled tungsten to facilitate easy and clean breakage. Properly grind tungsten.

Source: Eutectic Corp.

- Prevent shrinkage holes from forming in the weld craters. Do not finish a weld by breaking the arc abruptly. Taper off the weld bead by carrying it off to the side of the weld and by increasing the travel speed so that the weld pool diminishes in size before breaking the arc. Or, use the remote hand or foot current control to gradually reduce the welding current; or activate the crater control on the power source, if so equipped.
- Electrodes are normally operated at the highest possible current density in order to obtain maximum arc stability. This often results in electrode contamination and excessive electrode consumption. The problem is reduced by using the next size larger electrode and grinding it to a taper.
- Control the length of the arc carefully. A long arc length—$\frac{1}{8}$ to $\frac{3}{16}$ inch—gives poor penetration and wide weld beads. A long arc length is particularly harmful on the first pass of fusion welds. A short arc length—$\frac{1}{32}$ to $\frac{3}{32}$ inch—gives good penetration and a narrow weld bead. This arc length is excellent for the first pass in welding plate or pipe.

> **JOB TIP**
>
> **Job Listing**
> From a typical welding technician job listing: "The job includes witnessing procedure and welder testing in accordance with welding standards; and liaising with current and prospective clients. You must have demonstrated initiative and excellent people skills. Preferably you would have a visual welding inspector qualification, have been a technician, or have teaching background and supervisory experience."

- Whether cover passes are made with stringer beads or weaved beads depends on the following considerations:
  - The kind of material: weave beads cause cracks in some metals.
  - The thickness of the material: excessive penetration occurs in thin metals.
  - The position of welding.
  - The type of current: for example, since DCEN causes deep penetration, weaved beads increase the possibility of excessive penetration.

Practice Jobs

The practice jobs in this chapter require welding the three metals that are most commonly welded by the gas tungsten arc welding process. The major application of TIG welding is in those fields that make use of aluminum, carbon and low alloy steels, and stainless steels. Few welders will have occasion to weld the other metals that can be welded by the process. If you are called upon to weld such materials as titanium, magnesium, or copper, however, the techniques that you have learned in welding the three major metals can be applied with little difficulty.

Welding practice will begin with aluminum because this metal is one of the easiest to weld with the TIG process, it has the widest application in industry, and it creates a high degree of interest for the student. The welding of the basic weld joints in aluminum is followed by practice in the welding of mild steel and stainless steel, in that order.

Instructions for Completing Practice Jobs

Since the gas tungsten arc welding process may be used to weld so many different materials and the type of welding current varies with the material, it is very difficult to work out a series of jobs that gives the exact current range for all situations. Also, since the students are accomplished gas and arc welders, it is unnecessary to go into detailed instructions in regard to welding technique. Only those characteristics that are peculiar to TIG welding are presented.

The jobs listed in the Job Outline in Table 19-15, page 614, are presented for you to perform. They provide experiences in all positions of welding. As in your previous practice, the simplest and easiest jobs are presented first, followed by those of increasing difficulty. Each new weld contains procedures previously learned plus new techniques. It is very important to master each new technique before going on to the next job. Failure to do this will seriously hamper you in your efforts to master the gas tungsten arc welding process.

Consult the Job Outline before you begin welding each joint. The outline presents the type of joint, the sequence of welding, and the proper settings for making each weld. Then turn to the pages listed in the Text Reference column for a discussion of the welding technique recommended for the job.

Weld Beading on Flat Plate: Job 19-J1

After the arc has been started, hold the torch at about a 75° angle to the surface of the work. Preheat the starting point by moving the torch in small circles until a molten pool is formed, Fig. 19-17, page 615. The size of the pool is determined by the electrode size and the welding current. (The metal thickness determines the electrode size and current value.) Pool size can be increased by widening the circular motion of the torch.

The end of the electrode should be held approximately ⅛ inch above the work. When the pool becomes bright and fluid, move the torch slowly and steadily along the plate at a speed that will produce a bead of uniform width. The welding technique is similar to that used with oxyacetylene welding.

Filler rod must be held within the shielding envelope of inert gas while it is hot to prevent oxidation. Hold the filler rod at an angle of about 15° to the work and about 1 inch away from the starting point, Fig. 19-18, page 615. When the pool is fluid, move the arc to the rear of the pool and add filler rod to the leading edges of the pool at one side of the center line, Fig. 19-19, page 615. Move both hands together with a slight backward and forward motion along the joint.

Make sure that fusion takes place between the weld metal and the base metal. By watching the edges of the weld pool, you will learn to judge its fluidity and the extent

Table 19-15 Job Outline: Gas Tungsten Arc Welding Practice (Plate)

Job Number	Type of Joint	Type of Weld	Position[1]	Material Type	Material Thickness (in.)	Tungsten Size (in.)	Current Type[2]	Current Amperes	Shielding Gas Type	Shielding Gas Flow (ft³/h)	Cup Size (in.)	Filler Rod Size (in.)	Text Reference
19-J1	Flat plate	Beading	1	Aluminum	1/8	3/32	a.c.	90–140	Argon	12–20	3/8	3/32	631
19-J2	Edge	Beading	1	Aluminum	1/8	3/32	a.c.	90–140	Argon	12–20	3/8	3/32	634
19-J3	Outside corner	Fillet	1	Aluminum	1/8	3/32	a.c.	100–140	Argon	12–20	3/8	3/32	635
19-J4	Lap	Fillet	2	Aluminum	1/8	3/32–1/8	a.c.	90–140	Argon	12–20	3/8	3/32	636
19-J5	T	Fillet	2	Aluminum	1/8	3/32–1/8	a.c.	90–150	Argon	12–20	5/16	3/32	637
19-J6	Square butt	Groove	1	Aluminum	1/8	3/32	a.c.	80–130	Argon	12–20	3/8	3/32	639
19-J7	Single V butt	Groove	1	Aluminum	1/8	3/16	a.c.	190–280	Argon	22–28	1/2	3/32–1/8	639
19-JQT1[3]	Butt & T	Fillet & groove	1 & 2	Aluminum	18–10 gauge	1/8 max.	a.c.	40–125	Argon	20–40	1/4–5/8	3/32–1/8	Fig. 19-44
19-J8	Flat plate	Beading	1	Mild steel	1/8	1/16–3/32	DCEN	90–120	Argon	10–14	1/4–3/8	3/32	641
19-J9	Edge	Beading	1	Mild steel	1/8	1/16–3/32	DCEN	90–120	Argon	10–14	1/4–3/8	3/32	641
19-J10	Outside corner	Fillet	1	Mild steel	1/8	1/16–3/32	DCEN	90–120	Argon	10–14	1/4–3/8	3/32	641
19-J11	Lap	Fillet	2	Mild steel	1/8	1/16–3/32	DCEN	100–130	Argon	10–14	1/4–3/8	3/32	641
19-J12	T	Fillet	2	Mild steel	1/8	1/16–3/32	DCEN	100–130	Argon	10–14	1/4–3/8	3/32	641
19-JQT2[4]	Butt & T	Fillet & groove	1, 2, 3, & 4	Mild steel	18–10 gauge	1/8 max.	DCEN	45–130	Argon	15–25	1/4–5/8	1/16–3/32	Fig. 19-45
19-J13	Flat plate	Beading	1	Stainless steel	1/8	1/16–3/32	DCEN	80–130	Argon	10–12	1/4–3/8	3/32	641
19-J14	Outside corner	Fillet	2	Stainless steel	1/8	1/16–3/32	DCEN	80–130	Argon	10–12	1/4–3/8	3/32	641
19-J15	Lap	Fillet	2	Stainless steel	1/8	1/16–3/32	DCEN	80–140	Argon	10–12	1/4–3/8	3/32	641
19-J16	T	Fillet	2	Stainless steel	1/8	1/16–3/32	DCEN	80–140	Argon	10–12	1/4–3/8	3/32	641
19-J17	Square butt	Groove	1	Stainless steel	1/8	1/16–3/32	DCEN	70–120	Argon	10–12	1/4–3/8	3/32	641
19-J18	Single V butt	Groove	1	Stainless steel	1/8	3/32	DCEN	90–150	Argon	12–15	5/16–3/8	3/32–1/8	641
19-J19	Single V butt	Groove	1	Stainless steel	1/4	1/8	DCEN	110–180	Argon	12–15	1/2	1/8–5/32	641
19-JQT3[5]	Butt & T	Fillet & groove	1, 2, & 3	Stainless steel	18–10 gauge	1/8 max.	DCEN	35–130	Argon	15–25	1/4–5/8	1/16–3/32	Fig. 19-46

Notes:
- Also practice in the vertical and overhead positions.
- For vertical and overhead welding, reduce the amperage by 10 to 20 amperes.
- When two rod and tungsten sizes are listed, the smaller is for vertical and overhead position welding.
- Ceramic or glass cups should be used for currents to 250 amperes.
- Water-cooled cups should be used for currents above 250 amperes.
- The gas flow should be set at the maximum rate for vertical and overhead welding.
- The type of filler rod must always match the type of base metal being welded.
- This table is typical. Tungsten electrode size, filler rod size, and welding current will vary with the welding situation and the welder's skill.
- Visual examination will be done in accordance with the requirements of AWS QC10: 2008, Table 3.

[1]1 = flat, 2 = horizontal, 3 = vertical, 4 = overhead.
[2]a.c. with stabilization.
[3]For additional information follow AWS Standard Welding Procedure Specification (WPS) B2.1-015 for aluminum.
[4]Also practice in the vertical and overhead positions prior to doing this qualification test. For additional information follow AWS Standard Welding Procedure Specification (WPS) B2.1-008 for carbon steel.
[5]Also practice in the vertical and overhead positions prior to doing this qualification test. For additional information follow AWS Standard Welding Procedure specification (WPS) B2.1-009 for stainless steel.

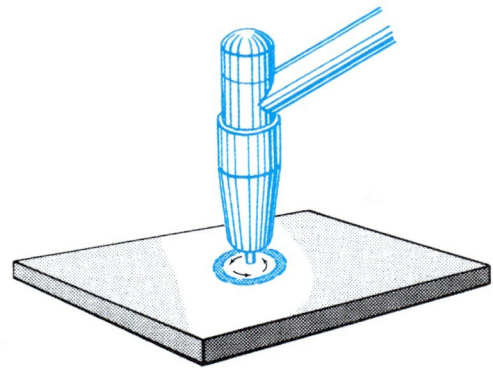

Fig. 19-17 Forming a molten pool with a TIG torch.

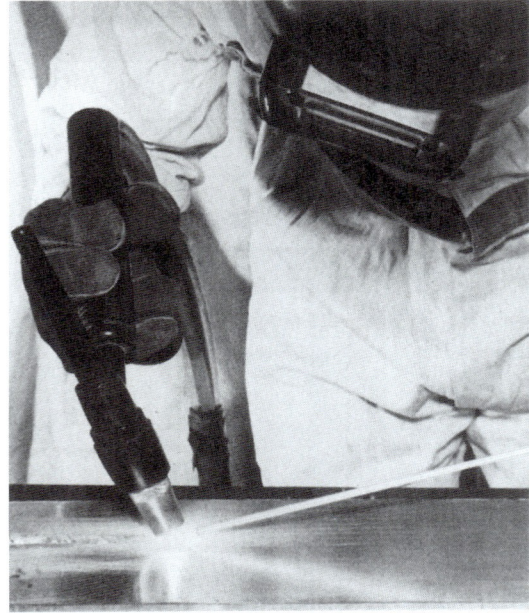

Fig. 19-18 Correct positions of the torch and filler rod at the start of bead welding. © Kaiser Aluminum

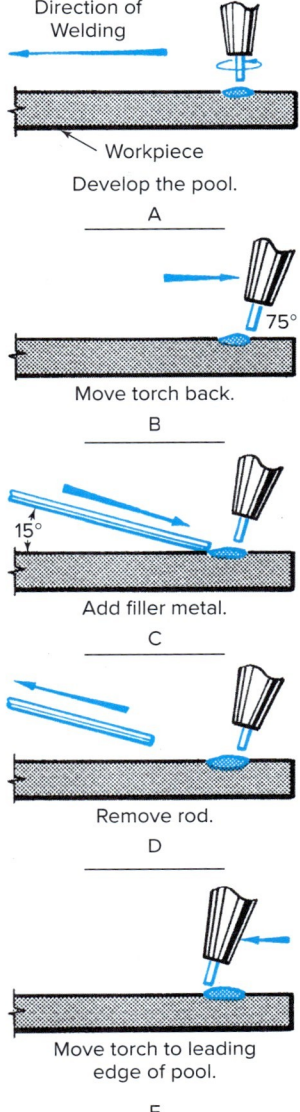

Fig. 19-19 Filler metal is fed by hand in a manner similar to that used when welding with the oxyacetylene process.

of buildup and fusion into the base metal. Incorrect torch angle, improper manipulation, too high a welding current, or too slow a welding speed can cause undercutting in the base plate along one or both edges of the weld bead. Make sure that the tungsten electrode does not touch the filler rod, and do not withdraw the hot end of the filler rod from the gas shield.

A short arc length must be maintained to obtain adequate penetration and to avoid undercutting and excessive width of the weld bead. It is important that you maintain control of penetration and weld contour. One rule is to use an arc length approximately equal to the diameter of the tungsten electrode. It is also important that you see the arc and weld pool. Always use a gas cup or nozzle that is as small as possible for a given weld yet still affords adequate gas shielding.

There are two different techniques for manipulating the torch and filler rod. In the two-step technique, addition of the filler rod to the weld pool is alternated with torch movement. In the uninterrupted technique, the torch is moved forward steadily, and the filler rod is fed intermittently as the weld pool needs it. This latter technique gives a somewhat better appearing weld bead, which needs little or no finishing.

For beading on a vertical surface, the torch is held perpendicular to the work. The weld is usually made from top to bottom. When the filler rod is used, it is added from the bottom (leading edge) of the pool in the same manner as described previously. Figure 19-20, page 616, shows the correct positioning of the rod and torch relative to the workpiece.

The direction of the vertical weld can be changed from bottom to top by reversing the torch and filler rod angle. The bottom-to-top direction is usually used when welding on aluminum and helps reduce porosity-type discontinuities. On thicker sections of steel and stainless steel where the weld pool, penetration, and fusion require more accurate control, the vertical-up technique is preferred.

Various methods can be used to break the arc or stop welding. If the arc is broken abruptly and the torch removed from the weld area, shrinkage cracks will probably occur in the weld crater and cause a defective weld. Cracking can be prevented by one of the following methods:

- Move the torch faster until the metal is no longer molten.

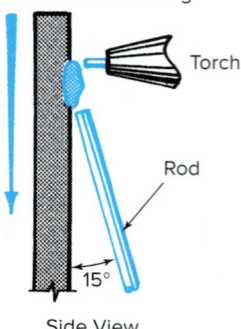

Fig. 19-20 Addition of filler metal when welding in the vertical position.

- Backtrack the bead slightly before breaking the arc.
- Lengthen the arc gradually and break and restart it while adding filler metal to the crater.
- Reduce the current by operating a remote hand or foot control.
- Use sequencer on welding machine and button on torch to reduce the current.

After the arc has been broken, it is necessary to let the gas continue to flow for about 15 seconds or more to permit the end of the electrode to cool. The tungsten electrode requires an inert gas shield until it is cool to prevent oxidation. The weld pool and the end of the filler rod need to be kept in the gas shield as they cool as well. This flow of gas after the arc is broken (i.e., after shutoff) is called *postflow*. The time is set on the postflow timer.

Inspection and Testing Compare the appearance of the beads you have made with those in Fig. 19-5B. Beads should be somewhat convex in contour. The face should be smooth and the ripples close together. There should be good fusion at the weld edges and no surface porosity. Figure 19-21 shows both the surface appearance and etched cross sections of three beads on a flat plate. The welding current selected for each weld determines its quality. The weld bead shown in *A* indicates that the current is too high. In *B* the current is correct, and in *C* the current is too low.

Weld beads made with sufficient and insufficient shielding gas are illustrated in Fig. 19-22. Insufficient shielding gas produces an unsound weld bead having a great deal of porosity and very poor appearance, Fig. 19-22B. Using too much shielding gas is wasteful.

Continue to practice until you can make beads that compare favorably with Fig. 19-21B and that are acceptable to your instructor.

Welding a Parallel Joint: Job 19-J2

The **parallel joint** is the easiest type of weld to make with the gas tungsten arc welding process. The technique is similar to that used with the oxyacetylene and shielded metal arc welding processes. Parallel joints should be used only on light gauge metal, and they usually require no filler rod. This is referred to as **autogenous welding,** which means GTAW without the addition of filler metal. Preparation is simple, and welding is economical and fast. A parallel joint should not be used where direct tension or bending stresses will be applied because it may fail at the root under low stress loads. The joint may be used for small tank bottoms.

Uniform speed of travel will produce beads that are smooth and even. Joints may be welded with or without filler rod. A welding speed that is too slow causes the bead to pile up and roll over the edges of the plates.

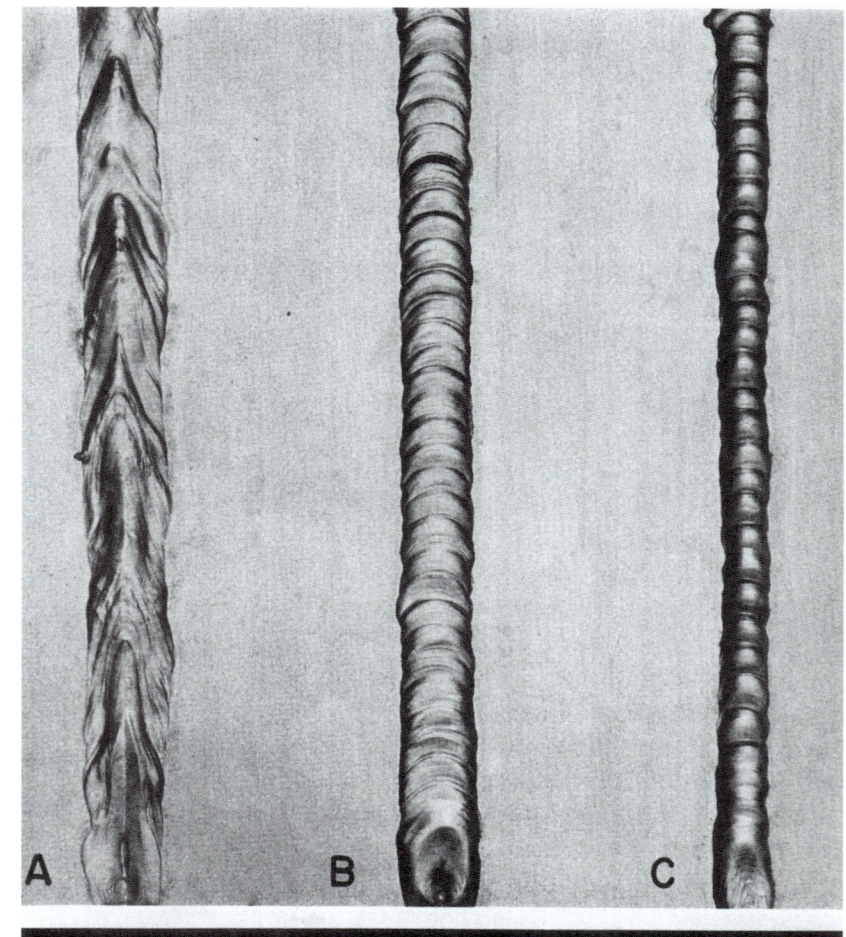

Fig. 19-21 Surface appearance and etched cross section of three aluminum weld beads: (A) Welding current too high. (B) Correct welding current. (C) Welding current too low. © Kaiser Aluminum

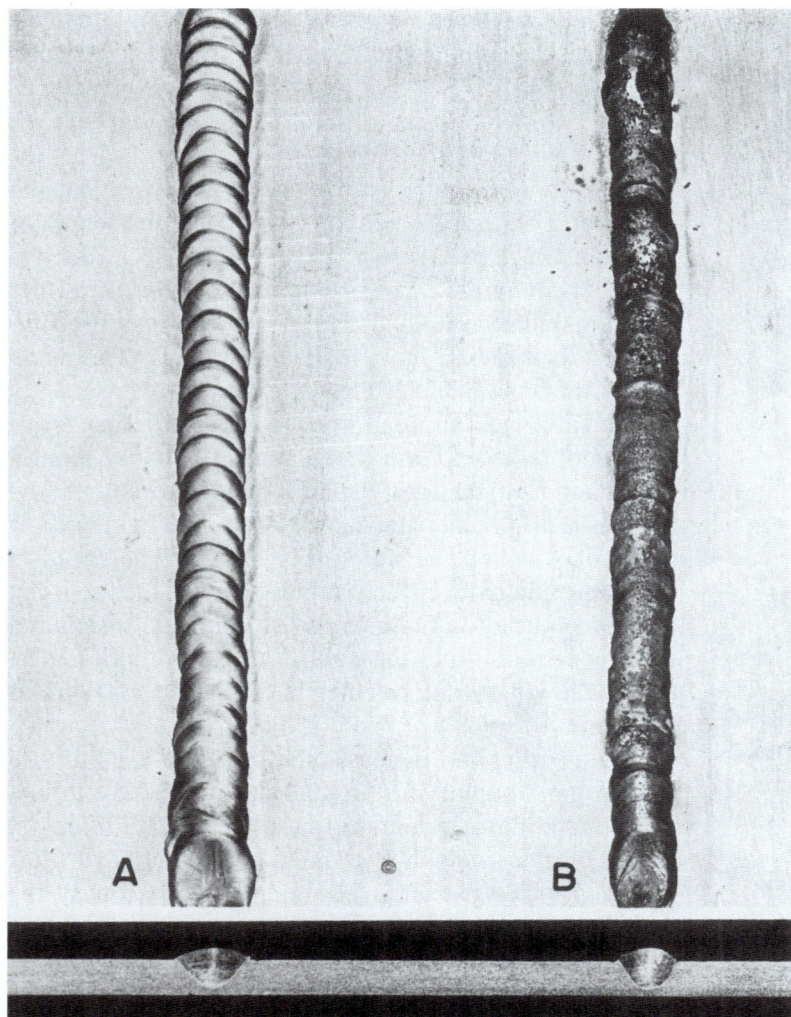

Fig. 19-22 Surface appearance and etched cross section of aluminum weld beads made with gas shielding: (A) Sufficient shielding gas. (B) Insufficient shielding gas. © Kaiser Aluminum

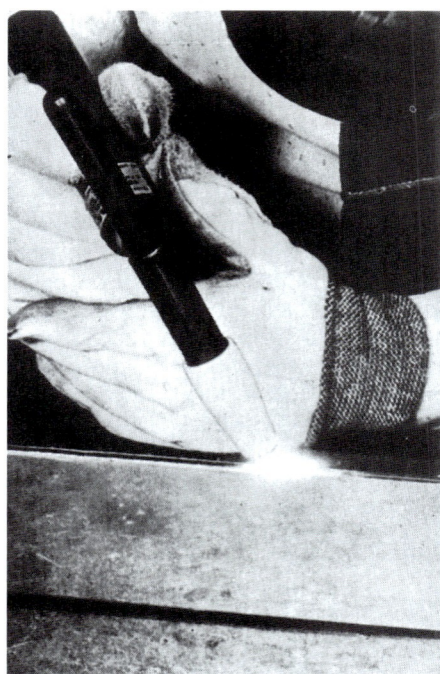

Fig. 19-23 Welding a parallel joint in the flat position. No filler rod added. © Praxair, Inc.

Inspection and Testing Complete a number of parallel joints in all positions and compare them with the bead shown in Fig. 19-24. Pry the plates apart to see if you are getting penetration at the inside corners of the plate and fusion to the surface of the plate edges. Welds should pass the inspection of your instructor.

Welding a Corner Joint: Job 19-J3

Corner joints are used in the manufacture of boxes, pans, guards, and all types of containers. An open corner joint may be used for thicknesses up to $5/32$ inch. Filler rod is

Too high a speed causes skimming of the surface and poor penetration in the joint.

Make sure that the abutting edges of the two plates are fused along the center line and that the entire edge surface of the plates is fused. The position of the torch is shown in Fig. 19-23.

In welding a parallel joint in aluminum, use a balled tip on the tungsten for the a.c. sine wave power sources and a pointed tip tungsten for the a.c. square wave or enhanced square wave a.c. power sources. This will help keep the a.c. arc steady. On carbon steel, stainless steel, and titanium use a tapered tip, since tapered a.c. tip operation helps secure a pinpoint concentration of heat. The welding technique for stainless steel usually requires a slower travel speed than for aluminum or carbon steel.

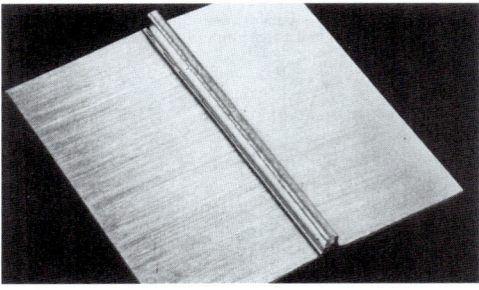

Fig. 19-24 Aluminum parallel joint. Note the smooth appearance of the bead. © Kaiser Aluminum

ABOUT WELDING

Aerowave

The nuclear facility at General Electric uses aluminum chambers from Aero Vac, a fabrication shop. To avoid problems with tungsten spitting, the shop uses the asymmetric technology of an Aerowave, an a.c. TIG machine from Miller Electric. The Aerowave has a low primary current draw. The technology provides a fast travel speed and the ability to weld thick metals at a given amperage. The welder can independently adjust the current in each a.c. half cycle from 1 to 375 amperes. The duration of the electrode negative portion of the cycle can be changed from 30 to 90 percent. The frequency can be adjusted from 40 to 400 hertz. A welder can fine-tune the penetration depth and width ratios of the weld bead.

required. Certain types of corner joints, using heavier materials, require the beveling of one of the plates in order to secure complete penetration. Both pieces must be in good contact all along the seam. The number of passes required depends on the thickness of the material and the bevel angle.

Welding this joint is somewhat similar to making a parallel joint, but it is usually necessary to add filler rod during welding. The amount of filler rod required depends on the thickness of the material and the degree to which it is a full open joint design. It is necessary to get complete fusion through the root of the joint. The inside buildup should not be high, but it must be complete. The position of the welding torch is like that used in the welding of a parallel joint.

Inspection Practice making corner joints until you are satisfied with their general appearance (Fig. 19-25) and you obtain fusion through to the back side. Place a welded joint on an anvil and hammer it flat. You should be able to flatten the joint without breaking the weld through the throat.

Welding a Lap Joint: Job 19-J4

A lap joint does not require edge preparation, but the plates must be in close contact along the entire length of the joint. On materials less than $3/16$ inch thick, filler rod may not be required. The general practice is to use a filler rod in making the weld. This type of joint is not recommended for material more than $1/4$ inch thick. The thickness of the material determines the number of passes.

The fillet weld in a lap joint is started by first forming a pool on the bottom sheet. When the pool becomes bright and fluid, shorten the arc. Play the torch over the upper sheet. The pool will be V-shaped. The center of the pool is called the **notch,** Fig. 19-26. The speed at which the notch travels determines how fast the torch can be moved ahead. In order to secure complete fusion and good penetration into the joint root and on both plates, this notch will have to be filled in for the entire length of the seam.

Adding filler rod increases the speed of welding. Be sure to get complete fusion, and not merely lay in bits of filler rod on cold, unfused base metal. The filler rod should be alternately dipped into the weld pool and withdrawn $1/4$ inch or so. It is very important to control the melting rate of the edge of the top plate. If this edge is melted too rapidly and too far back, a uniform weld is impossible.

There are two conditions necessary to obtain a uniform bead that has the proper proportions and good penetration.

- The top plate must be kept molten along the edge to the joint root, and the bottom plate must be molten on the flat surface to the corner.
- Just the right amount of filler metal must be added where it is needed, Fig. 19-27.

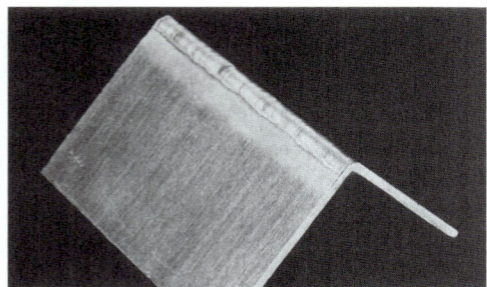

Fig. 19-25 Aluminum corner joint. Note the appearance of the bead. © Kaiser Aluminum

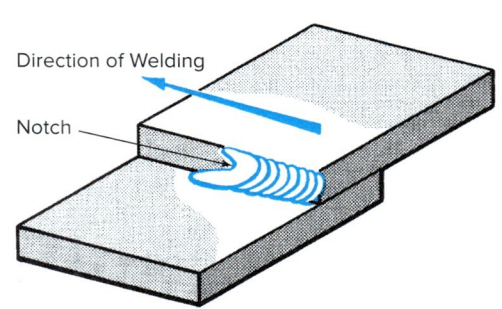

Fig. 19-26 Lap welding technique.

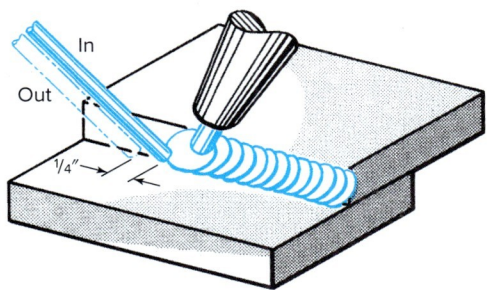

Progress of the weld with filler rod.

Fig. 19-27 Method of adding filler rod to a lap weld. Move filler rod in and out rapidly about ¼ inch.

may require a beveled edge on the web plate. When there is no edge preparation, be sure that current values are high enough for the thickness of the web plate.

The procedure for making a fillet weld in a T-joint is like that described for a lap weld. It is important to proportion the weld properly between the two plates so that each leg of the fillet weld is of equal length.

The design of the joint makes it somewhat difficult to avoid undercutting the vertical plate. If filler rod is used, it can be fed to the weld pool in such a manner that it provides protection for the upper plate. See Figs. 19-29 and 19-30 for the proper positions of the torch and filler rod.

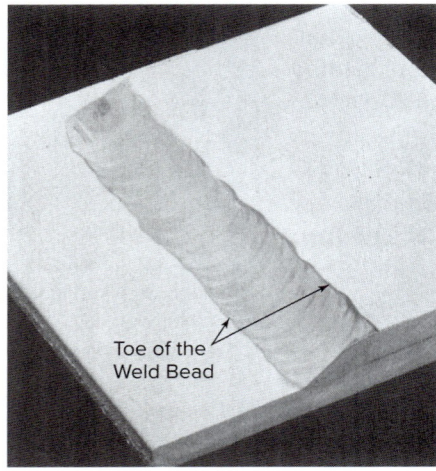

Fig. 19-28 A lap joint in aluminum plate. Note the complete fusion at the toe of the weld bead. © Kaiser Aluminum

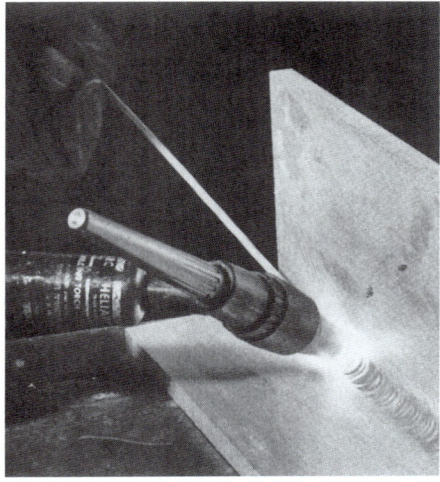

Fig. 19-29 The correct positioning of the torch and filler rod to prevent undercut and obtain fusion in making a horizontal fillet weld. © Kaiser Aluminum

Inspection and Testing Practice making lap joints until you are satisfied with their general appearance (Fig. 19-28) and feel that you are getting good fusion and penetration. Make up a test joint and pry it apart. The top plate should be capable of bending 90° without breaking, and there should be evidence of complete penetration at the joint root.

Welding a T-Joint: Job 19-J5

All T-joints require the addition of filler rod to provide the necessary buildup and to give the strength required. The number of passes depends on the thickness of material and the size of the weld desired. The joint may be welded from one or both sides. Complete penetration

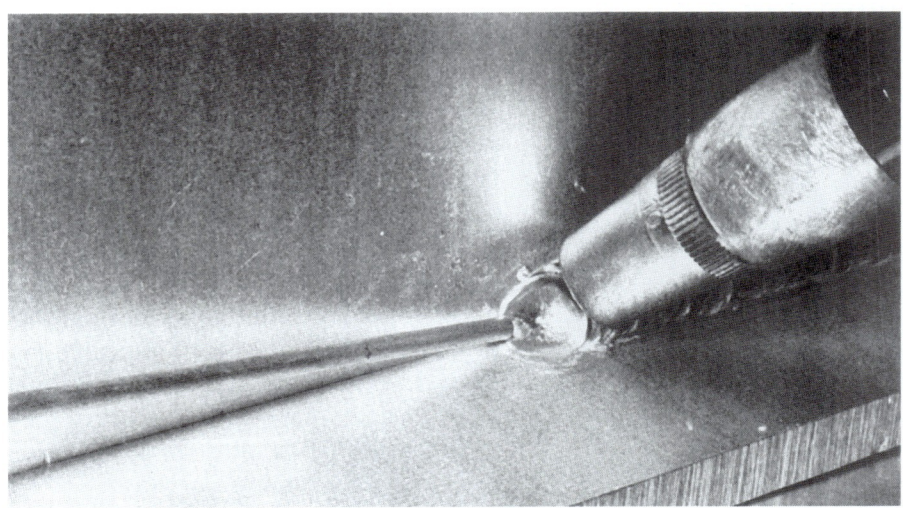

Fig. 19-30 A closeup of the positions shown in Fig. 19-29. Note the relationship of the torch, rod, and crater. © Kaiser Aluminum

Gas Tungsten Arc Welding Practice: Jobs 19-J1–J19 (Plate)   **Chapter 19**   **619**

Undercutting can also be caused by using too high a welding current or by traveling too fast.

Make sure that you obtain complete penetration to the joint root and that complete fusion is taking place on both plate surfaces. Incomplete penetration is caused by not forming the weld pool in the base metal before adding filler rod to the leading edge of the pool. This defect can also be caused by welding with inadequate current or by welding too fast.

When welding a T-joint in aluminum, care should be taken to avoid excessive penetration. In starting the weld, begin about 1 inch away from one edge and weld to the nearest edge. Then return to the start of this weld and begin welding in the opposite direction. Start the second weld pool about ¼ inch into the short bead. This is necessary in order to prevent cracking the plate along the edge of the weld or through the center of the weld.

Do not travel too fast or add filler rod too often. Filler rod absorbs a good deal of heat. If a large amount is added quickly, the weld pool chills and causes a loss of penetration and fusion.

A good fillet weld looks shiny, and every ripple is evenly spaced. (See Fig. 19-29.) A dull-colored weld indicates that the weld was made either too hot or too cold. If there is excessive penetration, the weld was run too hot. If the weld was run too cold, the edges of the weld overlap the plate surface, and the ripples are coarse and rough.

In making a fillet weld in the vertical position, you must make sure that the current setting is not too high and that the weld pool does not become too large. A large weld pool is difficult to handle and spills over. A slight weave assists in smoothing out the bead. Keep the face of the weld as flat as possible. When weaving, a slight hesitation at each side prevents undercut. See Fig. 19-31 for torch and filler rod position.

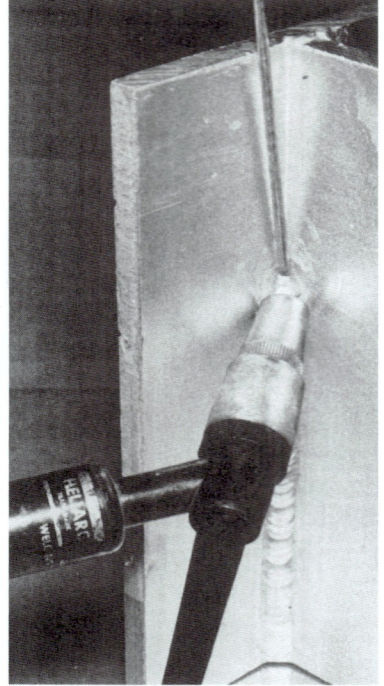

Fig. 19-31 Positions of the torch and filler rod when making a vertical fillet weld in aluminum plate.
© Kaiser Aluminum

When making a fillet weld in the overhead position, a lower current setting and slower travel speed are used than in flat position welding, but the flow of shielding gas is higher. Care must be taken to avoid sagging and poor penetration. These defects result from adding too much filler rod and carrying too large a weld pool. Let the pool wet out enough before adding more filler rod. You may find the overhead position awkward. Therefore, try to get in as comfortable and relaxed a position as possible when welding. This helps in maintaining steady, even manipulation of the torch and filler rod. Figure 19-32 illustrates the correct torch and filler rod positions.

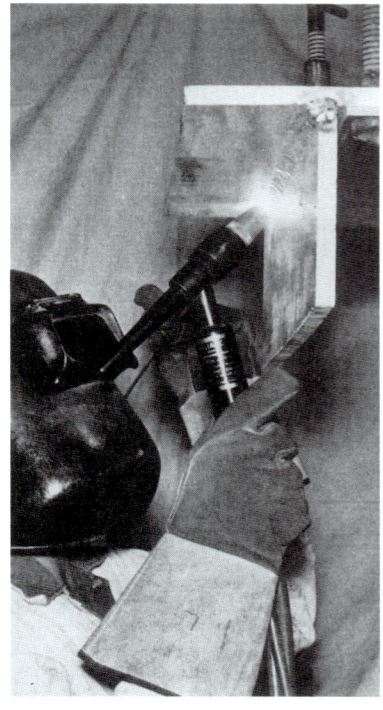

Fig. 19-32 Positions of the torch and filler rod when making an overhead fillet weld in aluminum plate.
© Kaiser Aluminum

Multipass Welding Multipass welding is generally necessary for material over ³⁄₁₆ inch thick. The number of passes required depends on the thickness of the material, the design of the joint, the position of welding, and the nature of the assembly being fabricated. The first pass is always to be considered as the *root weld*. It must provide complete fusion and penetration at the joint root. Subsequent passes can be welded at higher current values, since the root pass acts as a backup. Complete fusion into the root weld and the fusion face of the joint are necessary to prevent areas of incomplete fusion. It is also important to provide for clear flow of the weld pool to prevent inclusions. Both weaved and stringer bead techniques may be used.

Practice making both single- and multiple-pass fillet welds. By making multipass welds, you get the fullest use out of the material. Figure 19-33 gives the sequence of passes for each of the positions of welding.

Inspection and Testing Practice making T-joints until you are satisfied with their general appearance and feel that you are getting good fusion to the plate surfaces

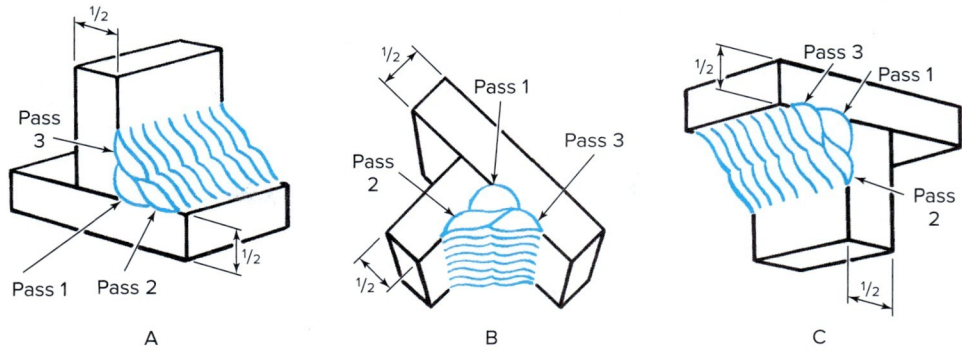

Fig. 19-33 Weld pass-sequence for (A) horizontal, (B) vertical, and (C) overhead positions.

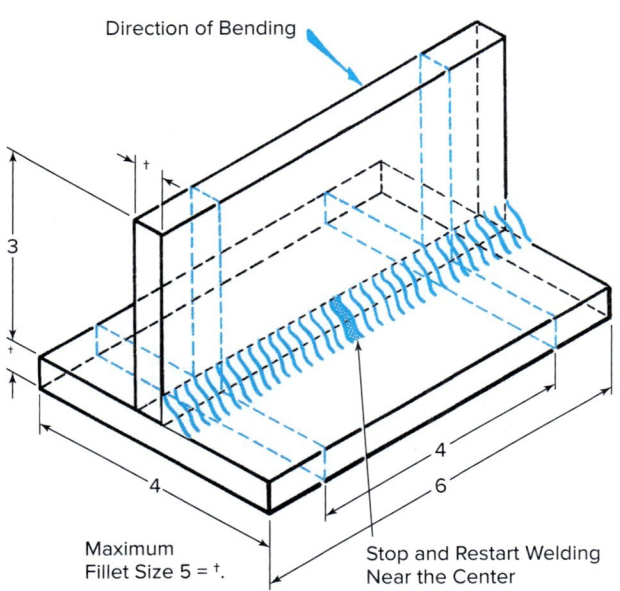

Fig. 19-34 Fillet weld test specimen. Source: Kaiser Aluminum & Chemical Corporation.

and good penetration at the root of the weld. Make a test joint in aluminum like that shown in Fig. 19-34. Bend the specimen back until it breaks or folds over on itself, and etch a cross section of the weld. Examine the cross section for root tie-in, porosity, incomplete fusion, undercut, and inclusions. Fusion must be to the root of the joint. The legs of the fillet must show fusion to the base plates. For fillet welds in plates of different thicknesses, the fillet leg should at least equal ¾ of the thickness of the thinner plate. The weld convexity or concavity must not exceed 1/16 inch beyond a flat face. A flat weld is the most efficient.

Welding a Butt Joint: Jobs 19-J6 and J7

Butt joints are common to all welding. The square butt joint is the easiest to prepare. It can be welded with or without filler rod, depending on the thickness of the material. Complete penetration can be secured. You must try to avoid both incomplete penetration and excess melt-through.

The single-V butt joint is used when complete penetration is required on material thicknesses ranging between ⅜ and 1 inch. Filler rod is necessary for multipass welding. The angle of bevel should be approximately 30°. The root face should measure from ⅛ to 3/16 inch, depending on the type and thickness of the metal being welded.

A double-V butt joint is generally used on material thicker than ½ inch when welding can be done from both sides. The angle of bevel is also 30°, and the root face varies from ⅛ to 3/16 inch. The root opening is best left tight together. With proper welding, complete penetration and fusion are ensured.

The procedure for making a butt joint is similar to that described previously for beading on flat plate. There is the added problem, however, of securing complete penetration through the joint root to the back side. It is, of course, necessary to make sure that the melt-through on the back side is not excessive. Refer to Figs. 19-35 through 19-38, page 622, for the correct position of the torch and rod when welding aluminum plate in the different positions.

In welding aluminum, complete penetration is desirable for plate thicknesses ⅛ inch or under. The back side should look smooth and have close ripples. A good aluminum groove weld on thin plate has a thickness through the throat of the weld equal to the thickness of the material. Excess melt-through is a problem. Hold the

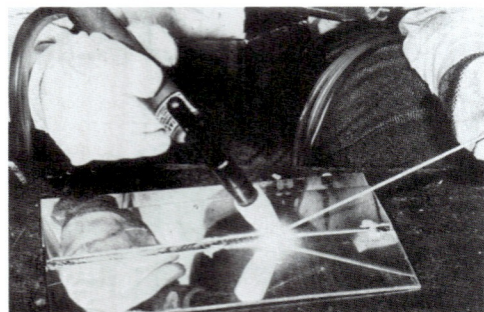

Fig. 19-35 Positions of the torch and filler rod when welding a butt joint in the flat position in aluminum plate.
© Praxair, Inc.

Fig. 19-37 Positions of the torch and filler rod when welding a butt joint in the vertical position in aluminum plate.
© Kaiser Aluminum

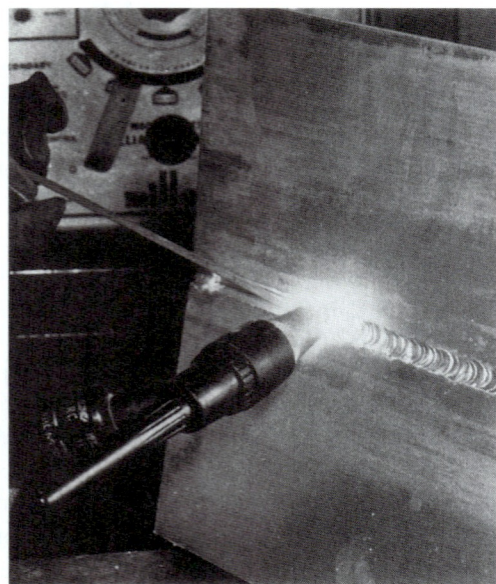

Fig. 19-36 Positions of the torch and filler rod when welding a butt joint in the horizontal position in aluminum plate.
© Kaiser Aluminum

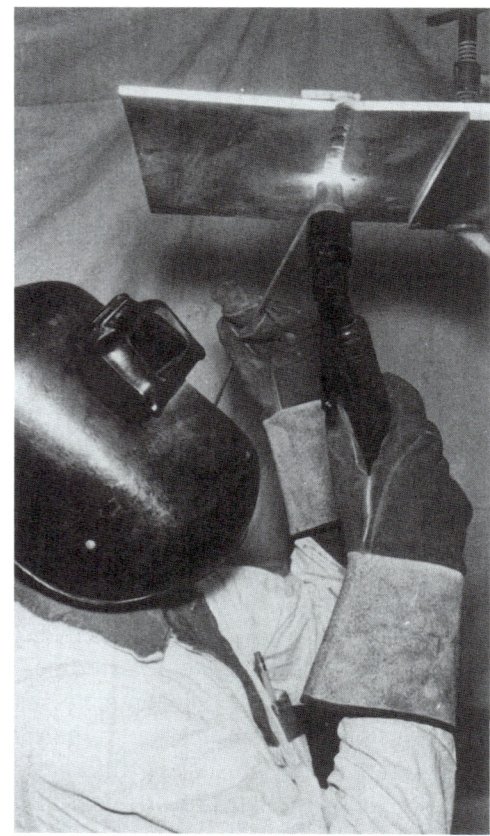

Fig. 19-38 Positions of the torch and filler rod when welding a butt joint in the overhead position in aluminum plate.
© Kaiser Aluminum

JOB TIP

Opportunities

Future opportunities in welding may come from

1. The use of mesh as a reinforcement option
2. The replacement of old concrete bridges with fabricated steel

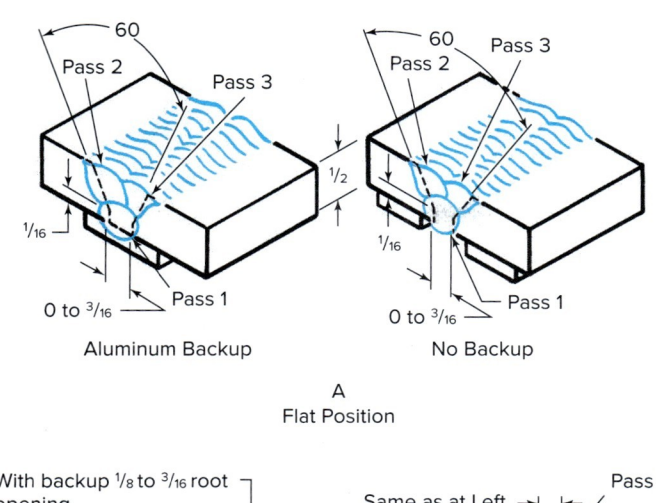
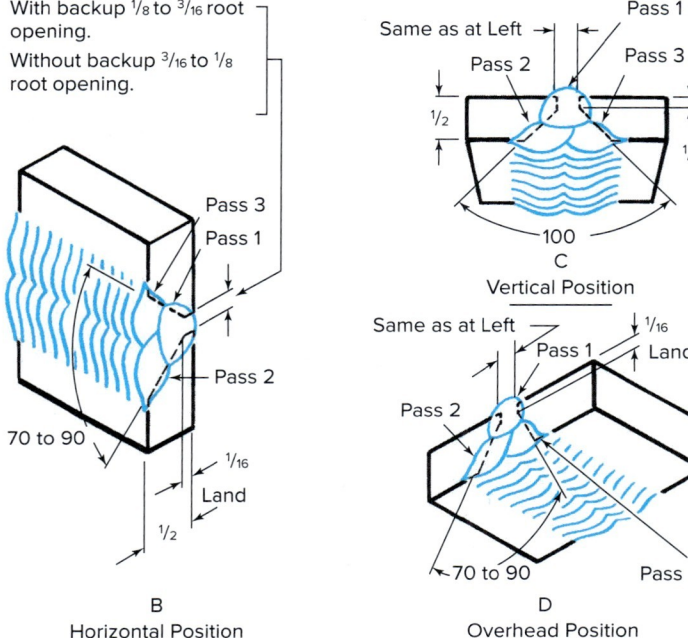

Fig. 19-39 Sequence of weld passes for groove welds in (A) flat, (B) horizontal, (C) vertical, and (D) overhead positions in heavy aluminum plate.

SHOP TALK

Flux Core

There is a wire or filler metal on the market now that is flux core, which eliminates the need for a purge for some applications.

torch forward, and add filler rod when you see that the pool is about to fall through. Do not start the weld at the edge of the plate. Fill the crater hole at the finish of the weld with the technique described earlier. The aluminum oxide needs to be removed from the groove face, groove floor, and the weld area on the front and back sides of the joint.

Practice making both single- and multiple-pass groove welds. Figure 19-39 shows the sequence of passes for multipass welds in heavy plate for each position of welding.

Inspection and Testing Practice making butt joints until you are satisfied with their general appearance and there is evidence that you are getting complete penetration without excessive buildup on the back side. Make up a test weld in $3/16$- to $3/8$-inch aluminum plate, Fig. 19-40, p. 624.

Prepare two specimens sectioned from the test plate for bend testing, as instructed in Fig. 19-41, page 624.

The specimens must bend 180° without cracking or fracturing in a fixture, as specified in Fig. 19-42, page 625.

The hydraulic guided-bend test fixture shown in Fig. 19-43, page 625, can be made in the school. The hydraulic jack and pressure gauge will need to be purchased. Bending the specimens in a fixture of this kind ensures full control of the testing situation and provides a uniform test procedure for all specimens tested. Students have complete confidence in the procedure.

Carbon and Stainless Steel Practice: Jobs 19-J8–J19

In welding carbon and stainless steel, follow much the same procedure that you followed in the welding of aluminum. Excess melt-through is a problem because it is harder to control when welding steel than when welding aluminum. Use a $3/32$- to $3/16$-inch root opening for the steel and stainless steel, as compared to the tight fit used for aluminum. This will allow better control over the amount of melt-through. The larger aluminum pool makes it easier to judge when the weld pool is about to fall through. The steel weld pool is about half as large as the aluminum pool. A copper or ceramic backup strip prevents excessive penetration. It may be necessary to heat the copper backup before beginning the weld. This method is used for welder certification when excess melt-through is a cause for failure.

When welding stainless steel, you will notice that the pool seems to stay in the same place until you move the torch. You will have to push the weld pool with the torch. Other than this, the welding technique is not much different from that used for carbon steel or aluminum.

Stainless steel requires less heat than aluminum. The tungsten electrode should be tapered. The choice of the filler rod size is crucial. If the rod diameter is too large, it will soak up a good deal of heat and make welding more difficult.

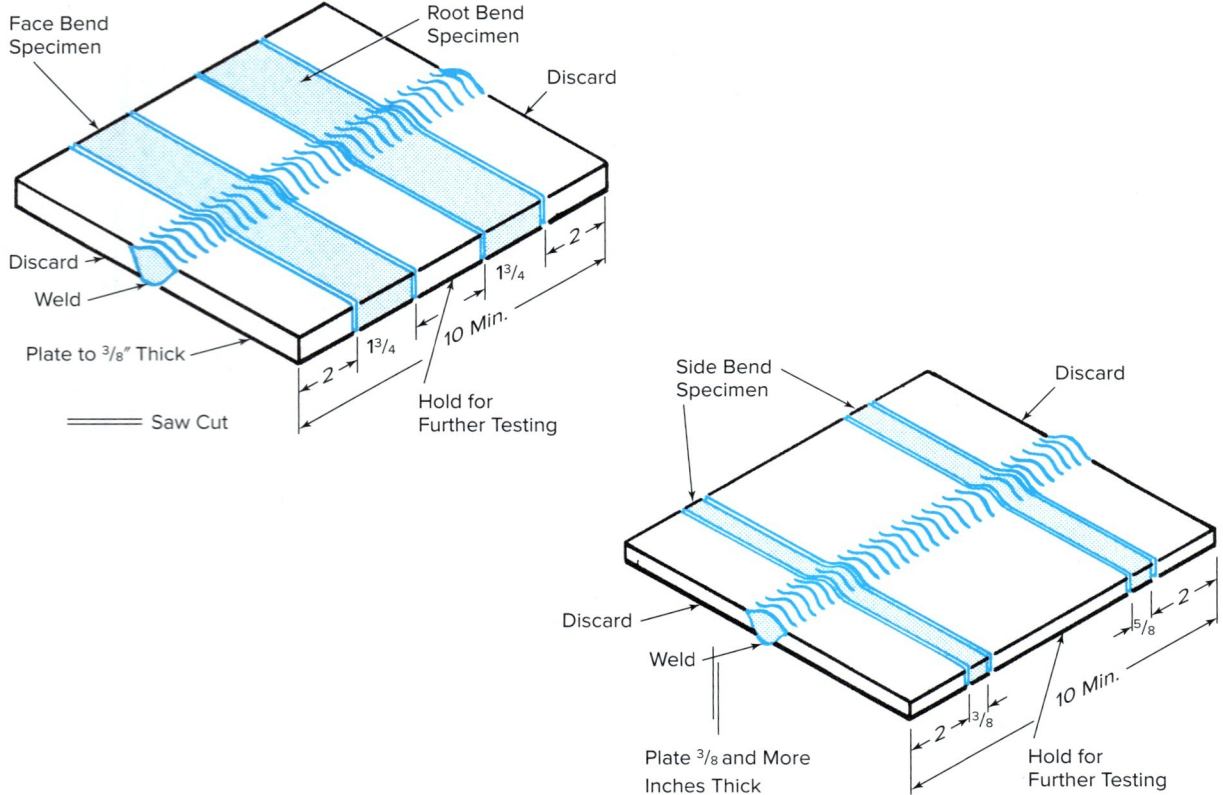

Fig. 19-40 Test plates for groove welds in aluminum.

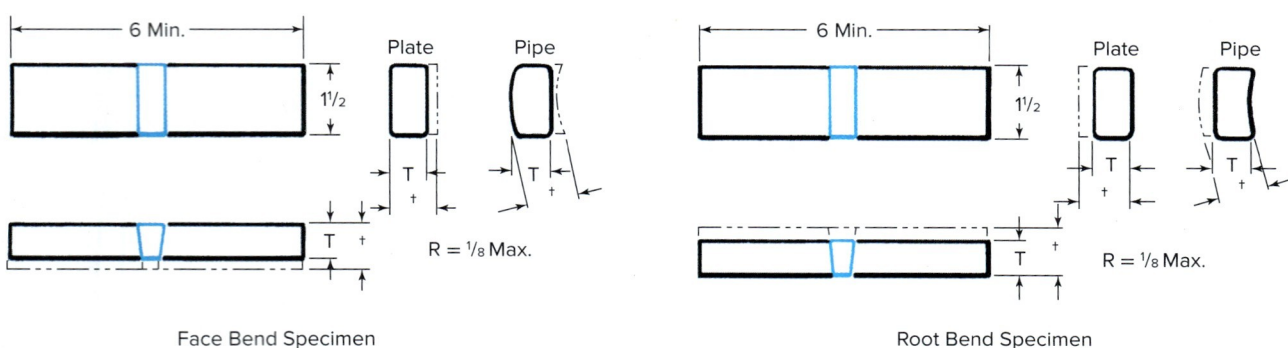

Fig. 19-41 Shapes of face- and root-bend specimens.

Back purging with an inert gas is required when welding stainless steel to prevent the root side of the joint from being contaminated. If the back side is exposed to the atmosphere, it will be severely oxidized; this is often referred to as **sugaring.** Back purging is not generally required on aluminum and steel.

Distortion is greater in stainless steels than in aluminum and carbon steel. Proper clamping and tacking techniques are very important. Numerous tack welds placed close together will help control distortion to a degree. Although travel speed can be increased by increasing the heat setting, high heat requires great skill on the part of the welder. The adding of filler rod must be timed perfectly, or the weld pool will overheat and excess will melt through.

Groove welds in stainless steel are made with exactly the same technique as groove welds in carbon steel, except for the slower travel speed. A dark purplish or purplish-blue bead with hardly noticeable ripples indicates too much heat.

Inspection and Testing Practice the jobs listed in the Job Outline (in the order shown), page 614, for each type of metal. When you are satisfied with their appearance

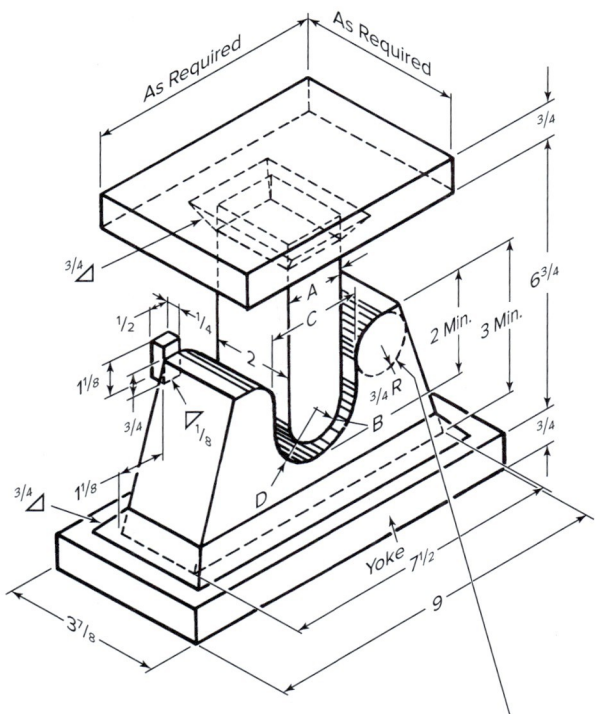

Shoulders hardened and greased, or preferably hardened rollers 1½" dia. May be substituted for jig shoulders.

Fig. 19-42 Guided-bend test jig for aluminum specimens.

Fig. 19-43 Hydraulic guided-bend test fixture. An aluminum test coupon is being bent in the jig shown in Fig. 19-42.   Edward R. Bohnart

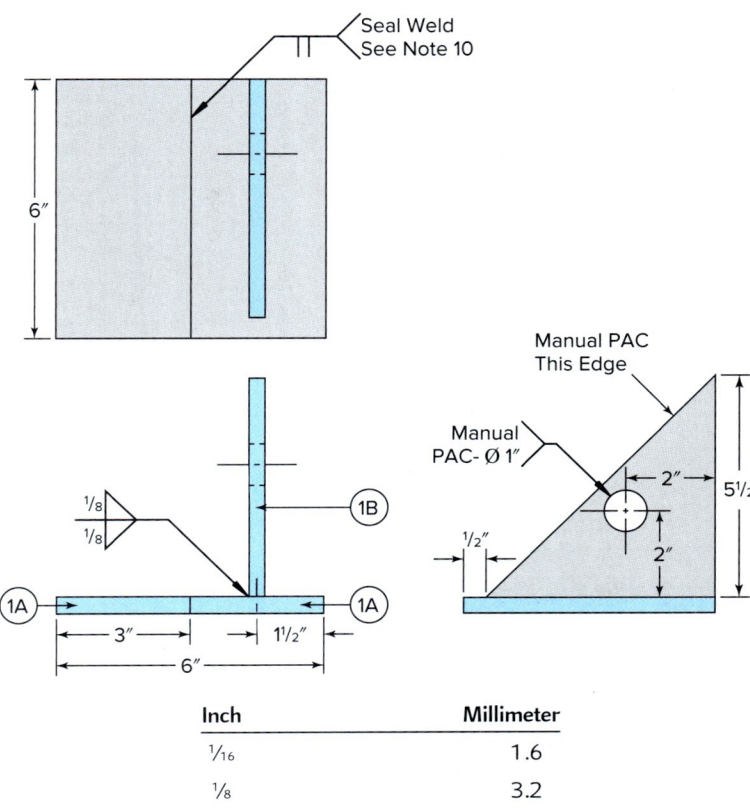

Inch	Millimeter
1/16	1.6
1/8	3.2
1/4	6.4
1/2	12.7
1	25.4

Notes:

1. All dimensions U.S. customary unless otherwise specified.
2. 10 ga.–18 ga. thickness aluminum material. Optional choice of thickness within range specified.
3. The welder shall prepare a bill of materials in U.S. customary units of measure prior to cutting.
4. The welder shall convert the above bill of materials to S.I. metric units of measure.
5. All parts may be mechanically cut or machine PAC unless specified manual PAC.
6. All welds with the GTAW process.
7. Fit and tack entire assembly on bench before welding.
8. All welding done in position according to drawing orientation.
9. Employ boxing technique where applicable.
10. Melt-through not required.
11. Use WPS B2.1.015 for aluminum (M22 or P-22).
12. Use WPS AWS-5-GTAW for aluminum (M-23).
13. Visual examination in accordance with the requirements of AWS QC10, Table 1.

Fig. 19-44 GTAW aluminum work performance qualification.
Source: AWS QC 10

and are getting good penetration and fusion on all types of joints, make up the usual test plates and test the specimen in the usual manner.

Figures 19-44 through 19-46 discuss qualifications in welding work, pages 625–627.

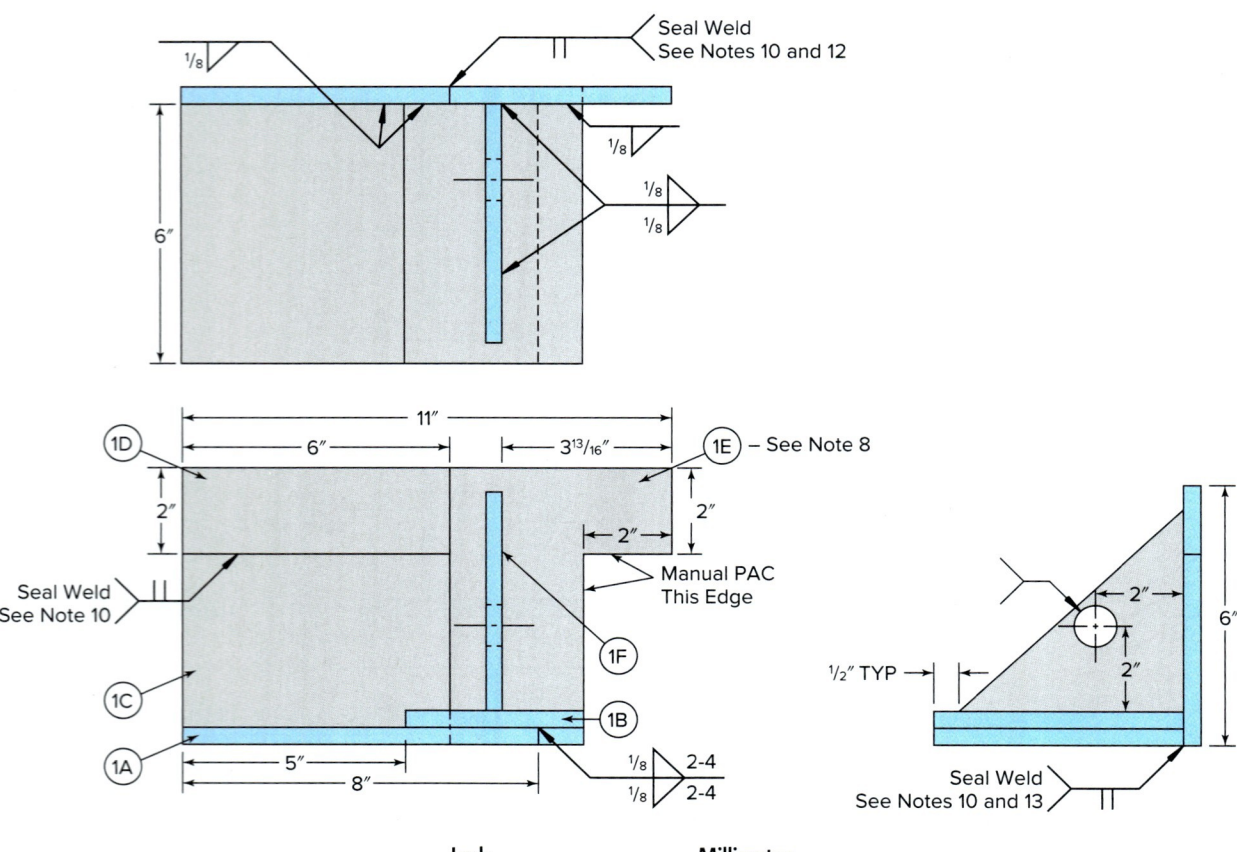

Inch	Millimeter
1/16	1.6
1/8	3.2
1/4	6.4
1/2	12.7
1	25.4

Notes:

1. All dimensions U.S. customary unless otherwise specified.
2. 10 ga.–18 ga. thickness plain carbon steel material. Optional choice of thickness within range specified.
3. The welder shall prepare a bill of materials in U.S. customary units of measure prior to cutting.
4. The welder shall convert the above bill of materials to S.I. metric units of measure.
5. All parts may be mechanically cut or machine PAC unless specified manual PAC.
6. All welds GTAW.
7. Fit and tack entire assembly on bench before attaching to positioning fixture arm.
8. Attach 2 in. × 2 in. extension tab of part 1E to positioning fixture arm. All welding done in position according to drawing orientation.
9. Employ boxing technique where applicable.
10. Melt-through not required.
11. Use WPS B2.1-008.
12. Weld joins Parts 1C and 1D to 1E.
13. Weld joins Parts 1C and 1E to 1A.
14. Visual examination in accordance with the requirements of AWS QC10, Table 1.

Fig. 19-45 GTAW plain carbon steel work quality performance qualification. Source: AWS QC 10

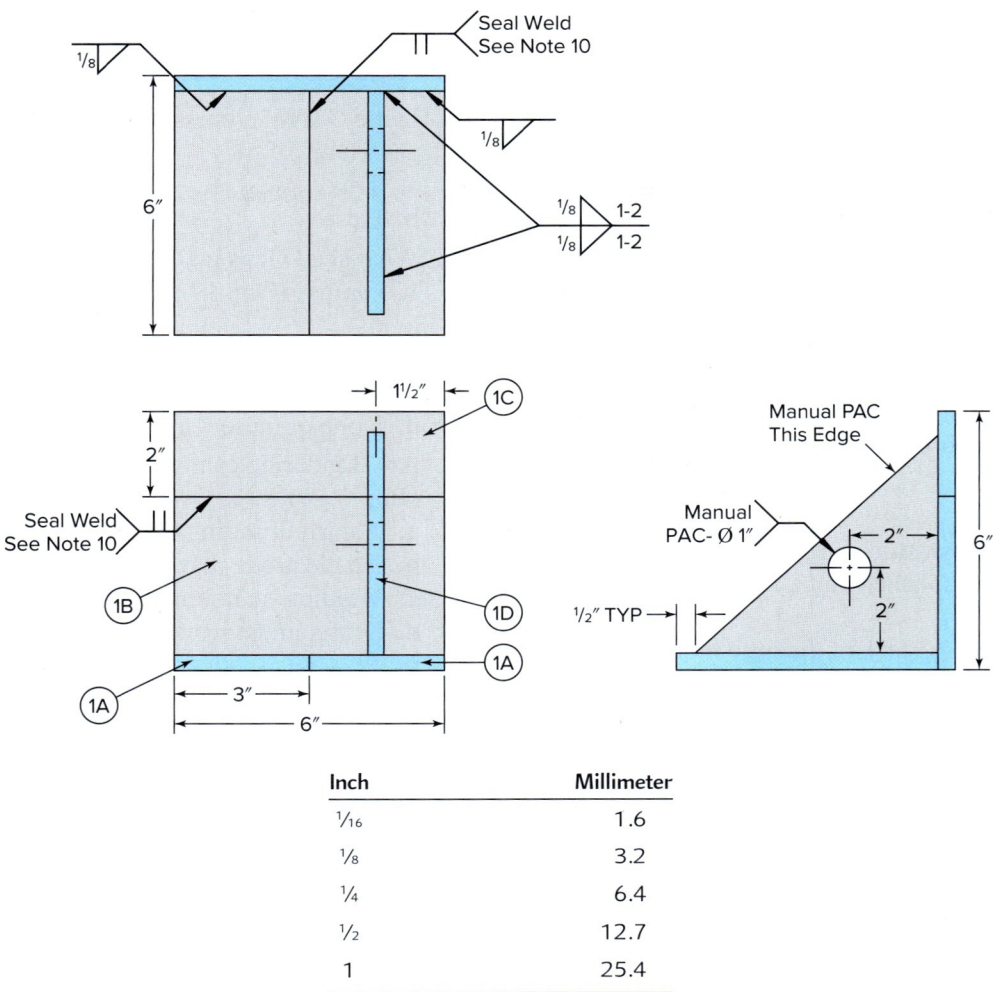

Inch	Millimeter
1/16	1.6
1/8	3.2
1/4	6.4
1/2	12.7
1	25.4

Notes:

1. All dimensions U.S. customary unless otherwise specified.
2. 10 ga.–18 ga. thickness stainless-steel material. Optional choice of thickness within range specified.
3. The welder shall prepare a bill of materials in U.S. customary units of measure prior to cutting.
4. The welder shall convert the above bill of materials to S.I. metric units of measure.
5. All parts may be mechanically cut or machine PAC unless specified manual PAC.
6. All welds GTAW.
7. Fit and tack entire assembly on bench before welding.
8. All welding done in position according to drawing orientation.
9. Employ boxing technique where applicable.
10. Melt-through not required.
11. Use WPS B2.1.009.
12. Visual examination in accordance with the requirements of AWS QC10, Table 1.

Fig. 19-46 GTAW stainless-steel work performance qualification. Source: AWS QC 10

CHAPTER 19 REVIEW

Multiple Choice

Choose the letter of the correct answer.

1. Which of the following metals can be welded with the GTAW process? (Obj. 19-1)
   a. Titanium
   b. Aluminum
   c. Steel
   d. All of these

2. Aluminum should never be welded using alternating current with stabilization. (Obj. 19-1)
   a. True
   b. False

3. What is the proper torch angle for GTAW once the weld pool is established and a weld is being made? (Obj. 19-2)
   a. 90° from the surface of the base metal
   b. 75° from the surface of the base metal
   c. 15° from a line perpendicular to the base metal
   d. Both b and c

4. What is the proper filler rod angle for GTAW? (Obj. 19-2)
   a. 15° from the surface of the base metal
   b. 75° from a line perpendicular to the base metal
   c. 45° from the surface of the base metal
   d. Both a and b

5. Only certain types of joints can be welded with the GTAW process. (Obj. 19-3)
   a. True
   b. False

6. Materials less than what inch thickness are best welded with the GTAW process? (Obj. 19-3)
   a. ½"
   b. ¼"
   c. ⅛"
   d. All of these

7. A setup procedure for the GTAW equipment must be followed, as the process requires precision welding techniques. (Obj. 19-4)
   a. True
   b. False

8. Which of the following GTAW arc starting methods will produce the least chance of tungsten contamination? (Obj. 19-4)
   a. Lift arc
   b. Scratch
   c. High frequency
   d. Both a and c

9. In the GTAW process, hydrogen is the best shielding gas. (Obj. 19-4)
   a. True
   b. False

10. What shade lens is recommended for GTAW with 125 amps? (Obj. 19-5)
    a. 6
    b. 8
    c. 10
    d. None of these

11. Arc wandering can be caused by _____. (Obj. 19-6)
    a. A current setting that is too high
    b. An electrode that is not contaminated
    c. Welding on nonmagnetic metals
    d. Drafts in the work area

12. Problems that will cause the tungsten to contaminate the workpiece are _____. (Obj. 19-6)
    a. Scratch starting the arc
    b. The electrode melting and becoming an inclusion in the base metal
    c. Shattering of the electrode due to thermal shock
    d. All of these

13. Which method should be followed in breaking the arc to stop welding and fill the crater? (Obj. 19-7)
    a. Move the torch slower until the weld pool gets larger
    b. Backtrack the bead slightly before breaking the arc
    c. Shorten the arc quickly and do not add any more filler rod
    d. Both a and b

14. To control distortion when welding stainless steel, use more tack welds closer together, and provide for more jigs and fixtures to hold this metal. (Obj. 19-7)
    a. True
    b. False

15. Why is direct current electrode negative so popular with the GTAW process? (Obj. 19-7)
    a. Cleaning action
    b. Tungsten runs hotter
    c. Faster welding speed
    d. None of these

16. When making fillet welds on lap and T-joints, the weld pool must be observed so that the notch is

filled in the entire length of the joint for joint root penetration to be achieved. (Obj. 19-8)
   a. True
   b. False
17. What test methods can be used to examine T-joints? (Obj. 19-8)
   a. Tensile pull test
   b. Etch test
   c. Hardness test
   d. All of these
18. What test method can be used to examine butt joints? (Obj. 19-8)
   a. Bend test (face, root, and side bends)
   b. Etch test
   c. Hardness test
   d. Both a and c

Review Questions

Write the answers in your own words.

19. What are the proper filler rod and torch angles for gas tungsten arc welding? (Obj. 19-1)
20. List four aluminum welding techniques used in weld sequencing for distortion control. (Obj. 19-2)
21. Explain the method that should be followed in breaking the arc to stop welding. (Obj. 19-2)
22. Name and explain the various starting methods. (Obj. 19-2)
23. List some of the different compositions that make up metal backup bars. Identify typical weld joints that can use backup and how to place the bar in the weld zone. (Obj. 19-3)
24. List five factors that affect welding technique in controlling distortion in aluminum. (Obj. 19-5)
25. List eight safety precautions required for GTAW. (Obj. 19-5)
26. List four causes of arc wandering. (Obj. 19-6)
27. Give two reasons why direct current electrode negative is recommended for welding stainless steel. (Obj. 19-8)
28. Why is it necessary to provide more jigs and fixtures when welding stainless steel? (Obj. 19-8)

INTERNET ACTIVITIES

Internet Activity A
Using your favorite search engine, look up GTAW and report your findings.

Internet Activity B
Find a supply store on the Internet and make a list of items you would need to complete GTAW jobs.

Design credits: Cinema and movie icon: Denis Meshkov/123RF; and Illustration of Welding icons: Mr. Nachai Sorasee/123RF

20

Gas Tungsten Arc Welding Practice:

Jobs 20-J1–J17 (Pipe)

Chapter Objectives

After completing this chapter, you will be able to:

20-1 List the advantages for the GTAW process for pipe welding.

20-2 Describe the joint designs associated with GTAW on pipe.

20-3 Describe proper techniques for making tack, root, filler, and cover pass welds.

20-4 List causes of porosity in GTAW welds.

20-5 Produce acceptable weld beading on pipe.

20-6 Produce acceptable groove welds on steel, stainless steel, and aluminum pipe in all positions.

After you have mastered the gas tungsten arc welding of commercial metals and their alloys in the form of plate, you may gain further skill in the process by practicing on pipe. Gas tungsten arc welding produces welds in pipe that are unusually smooth, fully penetrated, and free from obstructions and crevices on the inside. These welds have maximum strength and are highly resistant to corrosion.

The process may be used to weld both ferrous and nonferrous piping materials. It is a highly desirable process for the welding of nonferrous materials.

Gas tungsten arc welding is used in the fabrication of aluminum, stainless-steel, nickel, nickel alloys, and alloy steel piping. See Figs. 20-1 and 20-2. For temperatures above 750°F and for power plant work, low alloy steel pipe containing small amounts of chromium and molybdenum is used. Stainless-steel piping is used for high and low temperature work, for chemical and process piping, and in those situations in which the pipe must resist corrosive chemicals. Copper, nickel, and their alloys are selected for resistance to chemical attack, especially by saltwater. Aluminum pipe has the advantages of being light in weight, resisting corrosion, and not forming toxic chemicals in the processing of food and drugs. The GTAW process has application in the welding of nuclear piping systems, reactor

Fig. 20-1 An A106B pipe of 2-¾ inch O.D. and a wall thickness of ⅝ inch in the 6G position. For a welder qualification test coupon, this can save 80% of the weld time and material vs. a 6 inch O.D. double extra heavy pipe coupons. This coupon will qualify for unlimited thickness and diameters as small as 1 inch O.D. using two weld processes. Such as GTAW on the root and filling half the joint and SMAW for the remainder of the joint. When welded in accordance to the current ASME Section IX requirements, the purge set up may or may not be required. Always follow the welding procedure. © Renee Bohnart

vessels, and their auxiliary vessels. In these applications, it is necessary to contain radioactive fluids under high pressures at both low and high temperatures. Joint tightness and cleanliness are absolutely essential because of the dangers of radioactive contamination.

Substantial quantities of carbon steel pipe are welded with the gas tungsten arc process for the root pass only.

The thinnest section that can be manually welded is approximately $\frac{1}{32}$ inch. For pipe wall thicknesses from ¼ to ⅜ inch, it is generally more economical to complete the pipe weld, after the gas tungsten arc root pass, with the gas metal arc or submerged arc process. The shielded metal arc welding or flux cored arc welding process may also be used. Some type of gas or metal backing improves the quality of the weld and makes the job of the welder somewhat easier.

Accurate joint preparation and good fitup are essential. Cleaning in preparation for welding must be carefully performed, and in some cases the base metal may require chemical treatment. Some type of gas or metal backing improves the conditions for welding. Welding may be done in all positions.

Joint Design

The following information may be applied to the welding of steel, aluminum, and stainless-steel pipe. Five basic joints are in general use in the gas tungsten

Fig. 20-2 Welding on aluminum pipe, in the 1G pipe rotated position. The welding machine is a 300-ampere a.c.-d.c. enhanced square wave GTAW power source. © Renee Bohnart

arc welding of pipe. Selection of a particular design depends on the type of material, the thickness of the material, and the nature of the installation. It is recommended that all pipe with a wall thickness of more than 1/8 inch be beveled to ensure sound welds.

Standard V-Groove Butt Joint

The V-groove, Fig. 20-3, with the 75° groove angle and a root face of 1/16 inch is considered standard. On noncritical jobs, the joint may be butted together and welded without a filler rod on the root pass. Code quality welding requires that the joint be spaced at the root from 1/16 to 3/32 inch as shown.

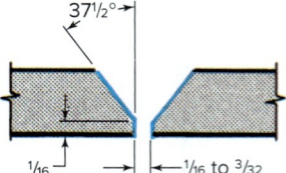

Fig. 20-3 Standard V-groove butt joint.

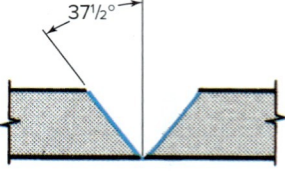

Fig. 20-4 Sharp V-groove butt joint.

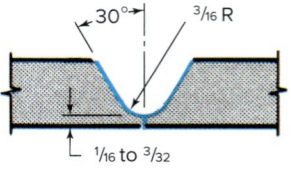

Fig. 20-5 U-groove butt joint.

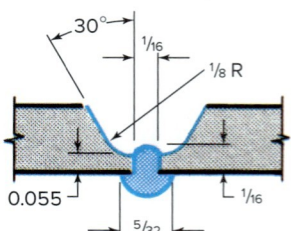

Fig. 20-6 Consumable insert on U-groove butt joint.

Sharp V-Groove Butt Joint

This joint, Fig. 20-4, does not have a root face like the standard V-groove. It can be prepared by cutting with an oxyacetylene cutting torch. The root pass must be welded with filler rod.

U-Groove Butt Joint

The U-groove joint, Fig. 20-5, is used when uniform welds of high quality are required. This joint is particularly adaptable to position welding because it is possible to obtain complete and uniform penetration. Root passes should be made with filler rod, although good penetration can be secured without it.

Consumable Insert Butt Joint

This type of joint, Fig. 20-6, produces welds of the highest quality to specification. The insert may be thought of as a ring of filler metal, which may have a special composition to prevent porosity and to enable the weld metal to meet specific requirements. A disadvantage of the consumable insert joint is the close control of dimensions and fitup needed to produce satisfactory results.

Joint Preparation

When pipe is of such thickness that beveling is necessary, it should conform to the specifications given in Figs. 20-3 through 20-6 and Table 20-1. All beveling should be done by machine. Grinding may deposit impurities on the surface of the metal, which contaminate the completed weld and make the welding operation more difficult. Prevention of surface contamination is especially important when welding aluminum pipe.

Backing or Purging

In order to obtain the smooth interior surface specified in Fig. 20-7 and shown in Figs. 20-8 and 20-9, page 634, it is necessary to protect the back as well as the face of the weld bead from oxidation. This may be done by introducing argon or helium into the pipe. This operation is known as *purging*. The gas shields the molten pool on the inside of the pipe. It should flow at a rate of about 6 cubic feet per minute during the welding operation. Carbon steel and aluminum do not typically require purging; verify by reviewing the welding procedure.

A fixture similar to that shown in Fig. 20-10, page 634, may be used for purging. This fixture can be made with materials usually found in the school welding shop, and it may be fabricated and welded by the students.

The conical pipe inserts D and E may be made with any desired diameter. They will fit all sizes of pipe up to the maximum diameter of the cone at its largest end. The cone D is welded to the tube and is fixed while cone E is adjustable by slipping the clamp F along the tube H in order to adjust to various lengths of pipe.

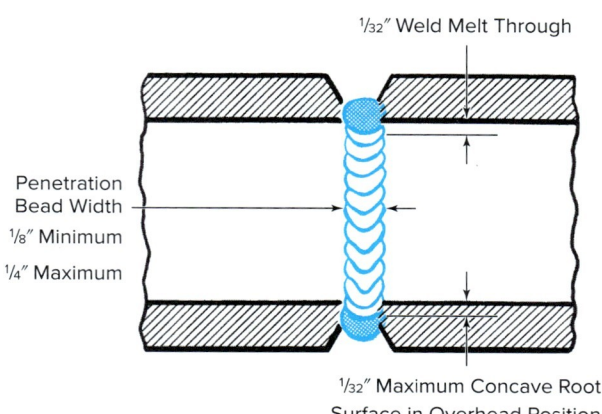

Fig. 20-7 Specification for penetration bead contours in a standard V-groove butt joint.

Table 20-1 Joint Preparation for the Welding of Pipe with the Gas Tungsten Arc Welding Process

Type of Material	Position of Welding	No Backing Ring	With Backing Ring	Type of Gas Shielding	Bevel Bead Angle[1]	Root Face[2]	Root Opening[3]
Aluminum	2 & 5	x		Helium	37½°	1/16	None
Aluminum	All		x	Helium	37½°	1/16	3/16
Aluminum	5 bottom only	x		Helium	60°	1/16	None
Aluminum	1	x		Argon	37½°	1/16	1/16
Aluminum	2 & 5	x		Argon	55°	1/16	1/16
Aluminum	1		x	Argon	37½°	1/16	3/16
Aluminum	2 & 5		x	Argon	55°	1/16	3/16
Steel & stainless steel 1/16 to 5/32 thick	All	x		Argon	None	Total	1/16
	All		x	Argon	None	Total	1/8
5/32 to 1/4 thick	All	x		Argon	37½°	1/16	3/32
	All		x	Argon	37½°	1/16	3/16

[1] Bevel angle may be ±2½°.
[2] Root face may be ±1/32 inch.
[3] Root opening may be ±1/16 inch.

The nut adjustment B permits positioning of the weld at any point around the circle. The pipe support C may be inserted into a larger fixed pipe in a stand, or it may be attached to the welding table so that the height of the unit can be adjusted. The purging gas passes from a flow valve through the welding hose A and into the tube which is plugged at the far end. This forces the gas to flow through the gas outlet holes G in the tubing and inside the pipe joint for purging.

In order to be able to weld under the most favorable of conditions, some kind of pipe-rotating device is necessary for the rolled-position welds. Figure 20-11, page 634, illustrates a manually operated unit that may be made quickly in the school shop. The gas-purging unit for the pipe interior, shown in Fig. 20-10, may also be attached to the rotating unit. This unit may be driven by an electric motor having a control unit with a speed control range up to 10 r.p.m.

The inside of the pipe may also be protected by painting with the proper flux. Welds backed up by flux are high quality, and they are almost equal in physical properties to those made with gas purging. Flux permits more penetration through the back side, whereas gas purging reduces the inside buildup and causes a very smooth inside bead.

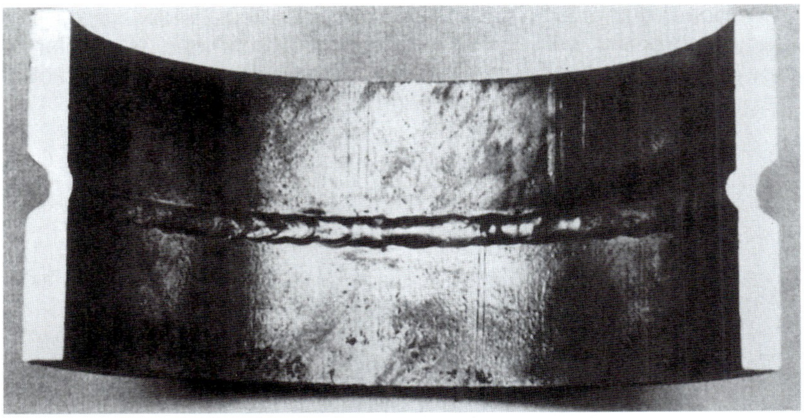

Fig. 20-8 The underside of a root pass in steel pipe welded with the gas tungsten arc welding process. Filler metal has been added. The joint shows complete penetration free of notches, icicles, crevices, and other obstructions to flow. Reinforcement is between 1/32 and 3/32 inch. Edward R. Bohnart

Tack Welds

Tack welds are used to hold the sections of pipe together in preparation for welding. In pipe welding by the gas tungsten arc process, tack

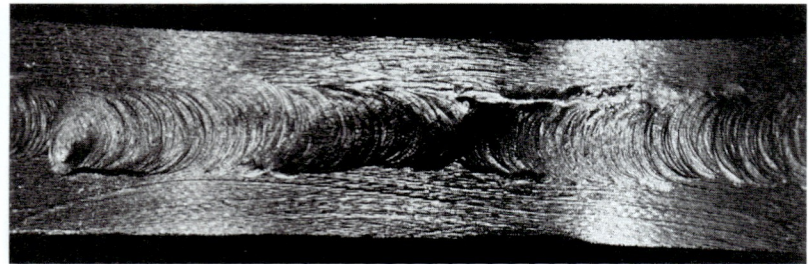

Fig. 20-9 The back side of a root pass in stainless-steel pipe. This is the inside of the pipe. Edward R. Bohnart

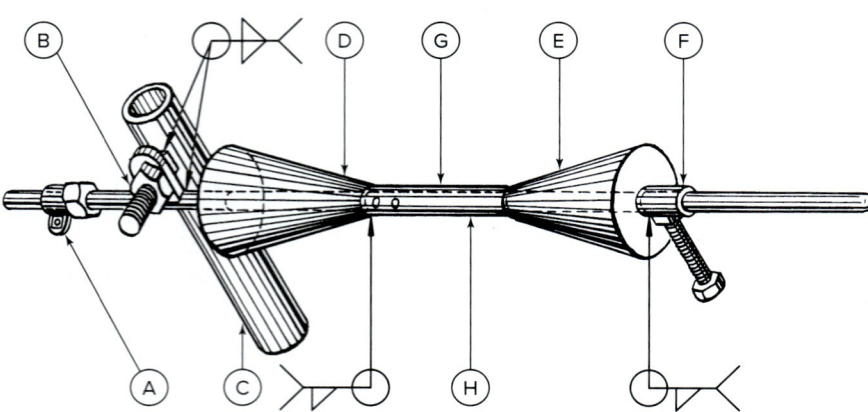

Fig. 20-10 Conical fixture used for purging the inside of piping and tubing. This can be made in the school welding shop.

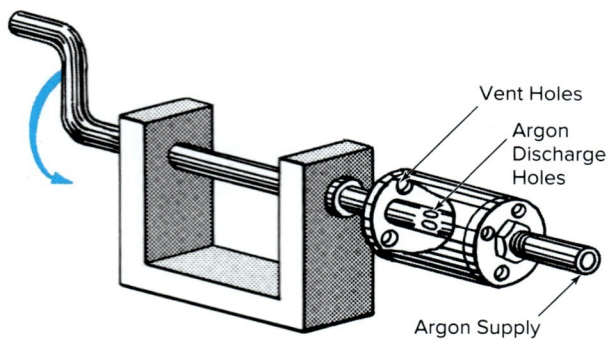

Fig. 20-11 Pipe-rotating and purging device.

If filler rod is to be used for the root pass, you should use filler rod for the tack weld. Tack welds should be about 1 inch long. Four to six tacks per joint, depending on the pipe size, should be spaced evenly around the pipe. Too large a tack weld interferes with the weld that is to follow. If the torch is withdrawn too rapidly, the tack weld will crack.

When the root pass is welded, the tack weld must be re-fused and become a part of the weld. To aid in this, the beginning and end of the tack weld should be filed or ground with a cutting disk, starting approximately $3/8$ inch from the edge and tapering to a feathered edge, as shown in Fig. 20-12. This will allow proper fusion when coming off a tack weld or when approaching the tack weld.

In some critical applications, the tack welds are completely removed. This requires the root pass to be stopped slightly before the tack weld is reached. The tack weld is then removed and the root pass is continued until the next tack weld is encountered, and the procedure is repeated until the root pass is complete and all tack welds have been removed. Care must be taken when removing the tack welds so as not to alter the original joint design. Another technique is the use of "bridge tack" welds. This prevents any issues in the root of the joint from the tack welds. They are removed as the root pass is made around the joint, Figs. 20-13 and 20-14.

welding must be done with care and skill because poor tacks are often the cause of defects in the final weld.

Inert gas purging or flux is applied inside the pipe, and the tack is made as small as possible. The torch is drawn to the side of the bevel until the pool diminishes in size, and then the arc is broken.

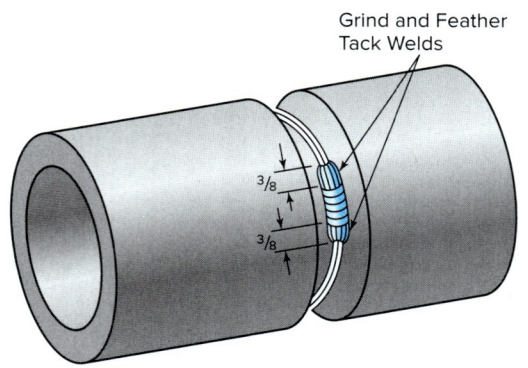

Fig. 20-12 Properly prepared tack welds to allow for proper fusion. Edward R. Bohnart

Fig. 20-13 Bridge tacks being used on a 4-inch Schedule 80, V-groove butt joint on carbon steel pipe. Note the careful joint preparation prior to tacking. *Edward R. Bohnart*

Fig. 20-14 Another form of bridge tacks being used to support large diameter heavy wall pipe and elbow as they are being positioned for welding. These bridge tacks will be evenly spaced around the pipe. *Mark A. Dierker/McGraw Hill*

Porosity in Gas Tungsten Arc Welds

It is important at this time to call your attention again to the matter of porosity in weld metal. There are two main causes: (1) that resulting from the composition of the base metal, and (2) contamination from the surface of the joint or from the surrounding air.

Base Metal Porosity

Some steels contain iron oxides that react with the carbon in the steel during welding to release carbon monoxide and carbon dioxide. These gases cause porosity in the weld. Gas formation can be controlled by the use of deoxidizing filler rod.

Porosity from Contamination

Contamination may be caused by iron oxides on the joint surface or filler rod. Heavy rust or scale on the surface produces a condition like that produced by oxides within the material. Therefore, rust and scale inside and outside the pipe should be removed for a distance of at least 2 inches on each side.

Oil, grease, and moisture on the joint surface or filler rod also contaminate the weld. These materials vaporize when heated and produce gases that may be trapped in the weld and cause porosity. All surfaces must be cleaned with care.

Porosity may also result from atmospheric contamination. Inadequate gas shielding leaves the weld pool open to attack. Inadequate shielding has the following causes:

- Gas flow that is too slow
- An arc that is too long
- Too much electrode extending beyond the end of the gas cup
- Air currents in the work area

Practice Jobs

The equipment needed is the same as that used for the gas tungsten arc welding of plate, Fig. 20-15, page 636.

Instructions for Completing Practice Jobs

Complete the jobs listed in the Job Outline, Table 20-2 (p. 652) as assigned by your instructor. Before you begin a job, study the specifications given in the Job Outline.

After you have demonstrated that you are able to make satisfactory welds in carbon steel pipe of various diameters, you are ready to proceed with the other pipe jobs listed in the Job Outline. These jobs are on aluminum and stainless-steel pipe. The welding technique is similar to that used in your earlier practice on plate.

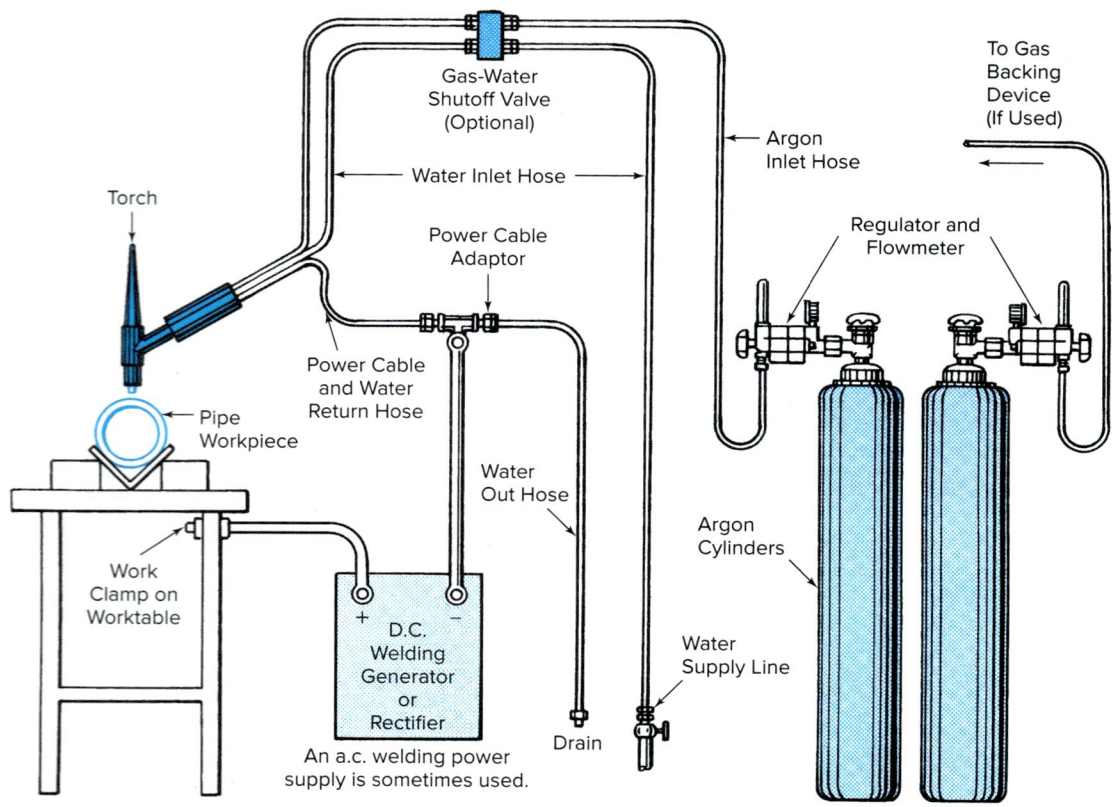

Fig. 20-15 A typical gas tungsten arc pipe welding installation.

Practice on steel, aluminum, and stainless-steel pipe in sizes ranging from 2 to 6 inches in diameter. You are expected to complete the jobs calling for mild steel pipe before practicing with aluminum and stainless-steel pipe. Welding should be done in the rolled, horizontal fixed, and vertical fixed positions. Use the same general welding techniques you learned in Chapter 19 for metal of the same thickness.

The procedures used in welding these practice joints in pipe are those that are practiced on the job. No single point is overlooked in making sure that the procedure will produce welds that are sound. Piping installations are such that a weld failure would not only damage property but also endanger the safety of human beings.

Study the various joint designs and weld procedures carefully. The nature of the material, the size of pipe, and the position of welding are the three important variables that determine the joint design, the pass sequence, and all that this implies. Quality welding requires close control. It is truly a scientific process that demands a great deal of skill in its application.

It is highly important that students who feel they want to specialize in pipe welding become highly proficient in the gas tungsten arc welding of aluminum and stainless pipe, since it is extensively used for piping made of these materials and alloys of steel.

Roll Butt Joint Groove Welds: Jobs 20-J1–J7

Root Pass Practice on pipe 4 to 6 inches in diameter. After you have practiced stringer and weaved beading around the outside of the pipe (Job 20-J1), begin practice on joints with the standard V-joint (Job 20-J2). Prepare the edges as instructed in Fig. 20-3, page 632. Tack the pipe in about three places with small tacks so that they do not interfere with welding. Brush, file or grind, and clean tacks before proceeding with the welding. Set up the pipe so that it can be purged with argon to protect the underside of the weld bead and to assist in making a smooth bead. Flux may be used instead of gas purging. A 2 percent thoriated-tungsten electrode with a diameter of $3/32$ or $1/8$ inch is preferred for welding. It should be ground to a point as shown in Fig. 20-16.

Point the electrode toward the center of the pipe, Fig. 20-17. The arc length should be $1/32$ to $1/16$ inch long. Be careful to manipulate the torch so that the tungsten electrode does not become contaminated.

Start the arc halfway up the groove face approximately centered in the area of the tack weld. Bring the weld pool down the groove face onto the tack weld. The filler rod can be preplaced in the tapered area of the prepared tack weld. The filler rod can be cut on an angle so it lies

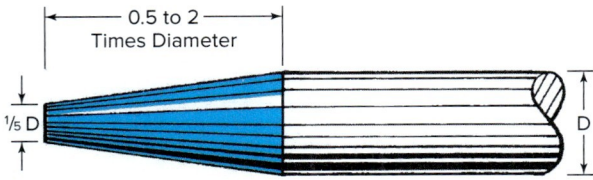

Fig. 20-16 Correct electrode taper.

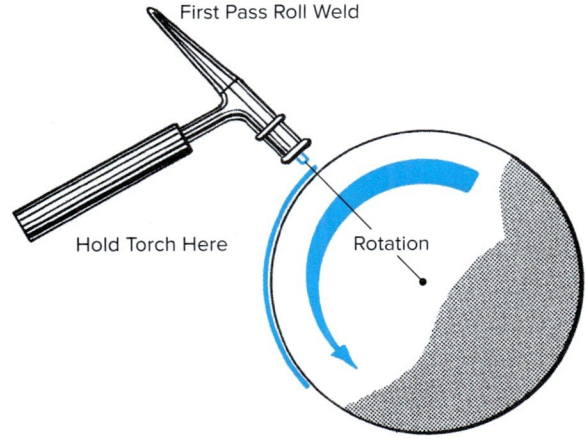

Fig. 20-17 Position of the torch for a rotational weld.

Fig. 20-18 The relative position of the torch and filler rod when making a multi-pass rotational weld on a stainless steel flange to heavy wall pipe butt joint. This is being done on a positioner. Note the completed weld of the pipe to the "T" branch section, also the well-organized tool board in the background. © Team Industries, Inc.

directly into the tapered tack weld. Play the weld pool on the center of the tack weld. This breaks down the tack weld, which is evident by a small "birds-eye" of silica, which will dance rapidly in the center of the weld pool. You have achieved full penetration, and it is time to move. *Don't delay.* Excess heat is detrimental to the weld and may lead to a concave root surface.

Try to weld from tack to tack to reduce the number of starts and stops. Minimizing the number of starts and stops will reduce the number of possible defects and is always a good procedure to follow when welding. Check your fitup, and begin welding the root pass on the tack where the least amount of root opening is present. As you weld the pipe, contraction forces will close up the root opening. So weld the tightest root opening area of the joint first.

Weld the root pass with the stringer bead technique. Start just below the 2 o'clock position and weld upward to the 12 o'clock position. Rotate the pipe clockwise so that the crater of the bead just deposited is just below the 2 o'clock position. After making sure that the weld crater is clean, continue welding as before.

The filler rod should be added almost tangent to the pipe surface, and the torch should be slanted about 15 to 20° toward the rod with an arc length of about 1/16 inch, Fig. 20-18. When the pool increases to about 1/8 inch in thickness, remove the rod and allow the pool to flatten out. When penetration appears to be complete, as shown by the shape of the pool, add more filler rod and advance the weld. Hold the tungsten electrode perpendicular to the surface of the work in all positions. Excessive angulation may result in inadequate shielding and reduce control over the reinforcement contour.

Another technique, rather than inserting and removing the filler rod, is to keep it in place. This is done by laying the filler rod in the joint tangent to the pipe. To facilitate this, bend the filler rod to conform to the shape of the pipe. If a 3/32-inch root opening is being used, a 1/8-inch diameter filler rod would be appropriate for this technique. Approximately one-third of the filler rod diameter should be in the joint and two-thirds of the filler rod diameter is exposed to the arc. A sharp V-groove with no root face works best with this technique. This technique works much like a consumable insert joint.

A fluid weld pool changes shape when certain conditions are present. You should learn how to "read" the weld pool by observing its changing contours as the associated conditions of penetration and bead formation change.

When welding begins, hold the electrode stationary and point it toward the center line of the pipe. Preheating is done by moving the torch in small circles over the two bevels. The pool continues to grow until sharp points are formed on opposite sides of the weld pool at the bottom of the V. When you see these points, advance the electrode in a straight line. The sharp point on the leading edge of the weld pool indicates whether or not the electrode is proceeding in a straight line. Figure 20-19, page 638, shows the point formation after a portion of the pipe has been welded. As long as this point formation is present, penetration is being obtained on the inside of the pipe. If you advance the electrode too rapidly, the point on the weld pool becomes rounded, or it may even form a notch (re-entrant angle),

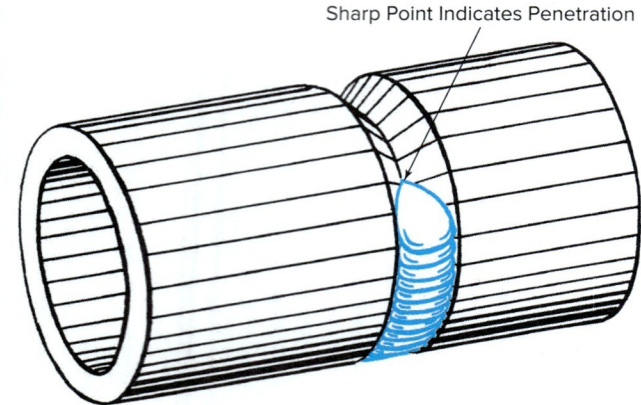

Fig. 20-19 Contour of the weld pool when penetration is obtained.

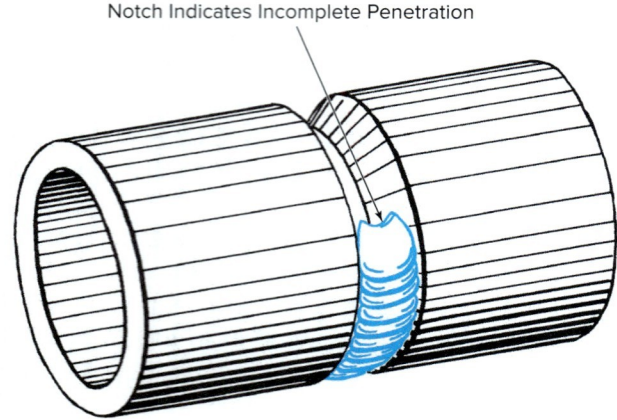

Fig. 20-20 Contour of the weld pool when penetration is not obtained.

Fig. 20-20. The presence of the notch in the weld pool indicates the absence of penetration on the inside of the pipe.

There are other signs that indicate penetration and contour of the weld bead on the inside of the pipe. When the pool is first formed, the surface across the center is raised (convex). When the point formation is obtained on the lead edge of the weld pool, observe the convex surface carefully. Soon after the point formation is obtained, the convex surface of the pool suddenly becomes flat. At this instant, advance the electrode along the line of weld until this surface again appears to be convex. If the pool is allowed to pass through the flat stage to the concave stage, too much penetration will result.

Excessive penetration can be overcome by adding more filler rod to the pool. Slant the torch toward the filler rod so that more rod is melted and increase the speed of welding. This directs more heat to the rod and less to the pipe surface to reduce the amount of heat in the weld pool.

This first pass is critical, and it must be mastered before the welding of the pipe joint may be considered satisfactory. Control of penetration is the most important factor in successful root pass welding. Such control can be obtained only by repeated practice. Satisfactory service of the pipe joint can be obtained only if the root pass is solid. See Fig. 20-9, page 634, for the proper appearance of the back side of the root weld and Fig. 20-21 for the appearance of the face of the weld.

Improper termination and restart of the weld beads increase the probability of defects in the weld. Figures 20-22 and 20-23 show the proper termination of the weld bead. The weld bead is overlapped approximately three times the width of the bead, and the size of the weld pool is reduced by rapidly increasing the weld travel speed. The weld pool is directed off the side of the joint so that it may be terminated in a section that is relatively heavy in wall thickness. This decreases the chill on the weld pool and prevents cracks in the weld crater. In starting the weld after the interruption, begin

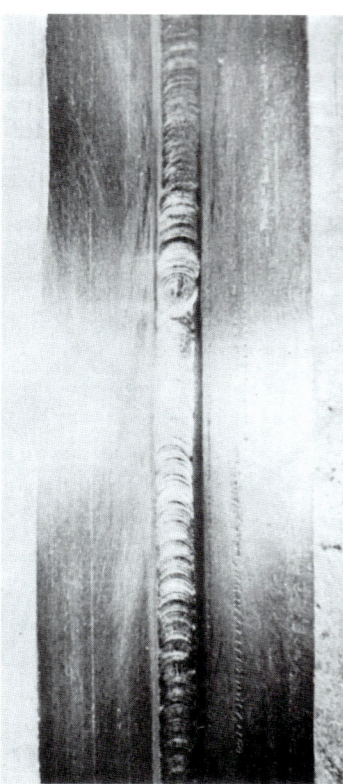

Fig. 20-21 Face contour of a satisfactory root pass in stainless-steel pipe. © Crane Co.

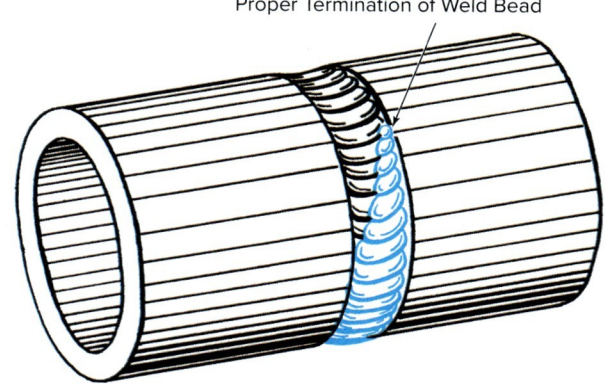

Fig. 20-22 Proper ending of the weld bead prevents crater cracks.

Fig. 20-23 Face of a root pass that shows proper termination where the bevels meet. This is a V-groove butt joint in stainless-steel pipe. *Edward R. Bohnart*

the pool ¼ inch or more back from the end of the previous bead to ensure good fusion of the two beads, Fig. 20-24.

With GTAW and especially the root pass on a pipe joint, the idea is to use finesse, not muscle. Get into the most relaxed and comfortable position possible. Welders develop their own personal techniques. It is generally more difficult to maintain control of the arc length and the weld pool using the "freehand" method. The "walking the cup" method allows for more control on the root pass.

For video of GTAW cup walking technique, please visit www.mhhe.com/welding.

In "walking the cup" the gas cup size is very important. The cup should fit into the groove and not want to slip out. Removing any sharp edges on the gas cup that will be contacting the pipe groove face and at the same time putting a radius on the gas cup will keep it in the groove. Apply light pressure to easily slide or step the cup forward in a uniform side-to-side motion. If you apply too much muscle and not enough finesse, the cup will slip out of the bevel or forward and control is lost. With the gas cup resting directly in the groove, a much more relaxed feel can be developed.

The tungsten should be extended approximately ¼ inch beyond the gas cup. This should provide a torch angle that is 45 to 60°, which is correct for the "walking the cup" method. Note this is much different than the freehand method, which has the electrode pointed at the center of the pipe.

Filler Passes Filler passes are used to fill the pipe joint to within approximately $\frac{1}{16}$ inch of the top surface of the pipe. Heavier passes can be used for the filler pass when making a rolled weld than when welding in a fixed position since the weld pool does not have a tendency to sag out of the joint due to gravity.

Follow the same procedure as for the root pass. Either stringer beads or weave beads can be used for the filler passes in carbon and low alloy steel pipe welded in the rolled 1G or fixed horizontal 5G position.

The stringer bead technique should be used for welding aluminum and stainless-steel pipe in any position when the wall thickness is more than ¼ inch. Stringer beads should also be used for carbon and low alloy steel welded with the pipe in the vertical fixed 2G position.

Each stringer bead should be made completely around the pipe before starting the next bead. Stringer beads on stainless steel should be kept small since the weld bead solidifies quickly. Fast freezing prevents the precipitation of carbides in these steels and reduces the tendency for hot-short cracking.

Care must be taken so that succeeding welding operations do not interfere with, or destroy, the inside surface of the root pass weld. For the root pass, the torch is directed toward the center line of the pipe. When making filler passes on pipe, however, you must incline the torch at an angle of approximately 45° to a point tangent to the surface of the pipe as shown in Fig. 20-25.

Clean the root pass and each succeeding pass carefully with a stainless-steel wire brush.

The filler layer may be made with a single weave bead or with two stringer beads. If a weave bead is used, weld with a $\frac{3}{32}$-inch filler rod. Clean the first pass thoroughly in order to remove surface oxides. The electrode should be held as instructed above. Move the arc from side to side, and add filler rod as rapidly as it melts off. Direct the arc on the side walls of the pipe to avoid burning through the first pass, Fig. 20-26A. If you have difficulty with the weave bead technique, run stringer beads. Make the welds along the sides of the pipe with $\frac{1}{16}$-inch filler rod and overlap at the center, Fig. 20-26B. Advance the torch with a slight oscillating motion in the direction of welding, and

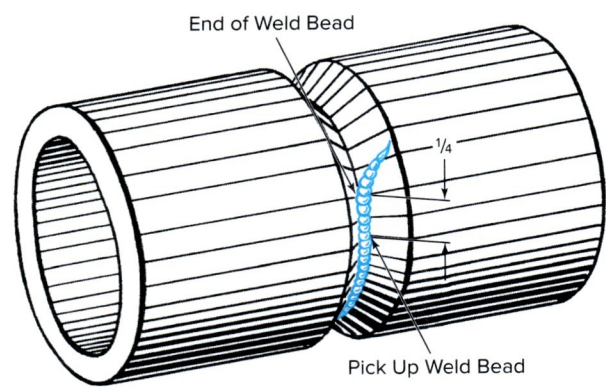

Fig. 20-24 Restarting a weld bead.

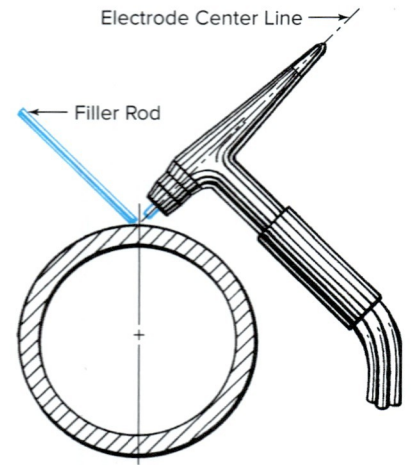

Fig. 20-25 Proper relation of the electrode to the pipe for addition of filler rod on the filler and cover passes.

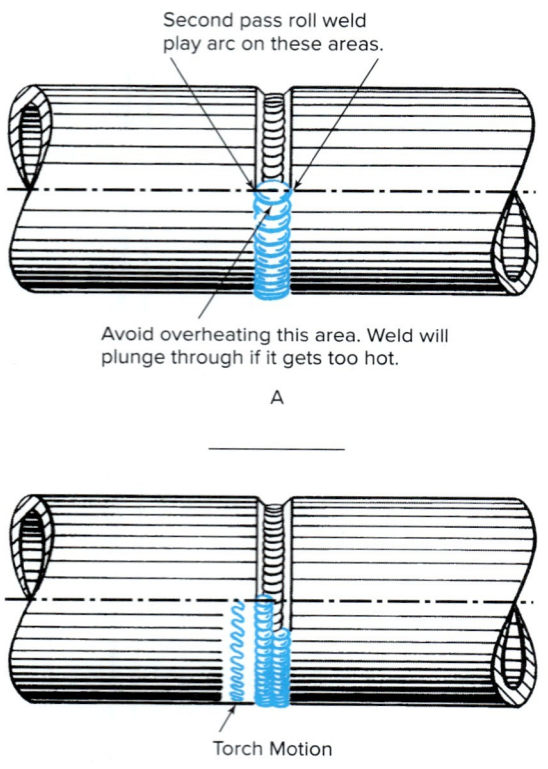

Fig. 20-26 Filler pass welding techniques.

add filler rod as rapidly as possible. Making stringer beads reduces the problem of burning through due to excessive heat in the center of the first pass. Figure 20-27 shows the proper appearance of stringer beads.

Cover Passes Smooth, uniform finish passes can be made only if the filler passes are smooth and free of undercut at the edges, and the pipe joint is filled to a uniform level. Care spent in making smooth filler passes will result in finished welds of excellent appearance as well as quality.

Follow the procedure described for the root and filler passes. Stringer beads or weave beads may be used for carbon and low alloy steel pipe in the rolled or fixed horizontal 5G position. Stringer beads must be used for carbon and low alloy steel pipe in the vertical fixed position, and for stainless or aluminum pipe in any position.

Finished cover passes should be about $3/32$ inch wider than the beveled edges on each side of the joint, and the overlap should be spaced evenly on each side of the joint. The face of the weld should be slightly convex, and the reinforcement should be about $1/16$ inch above the surface of the pipe. Weld edges should be straight and without any undercutting. Figure 20-28 shows the proper appearance of the cover pass. Also see Fig. 20-47, page 646.

When you are able to make welds that have good penetration through to the back side, good appearance, and are satisfactory to your instructor, make up a test weld and remove the usual test coupons for root- and face-bend testing.

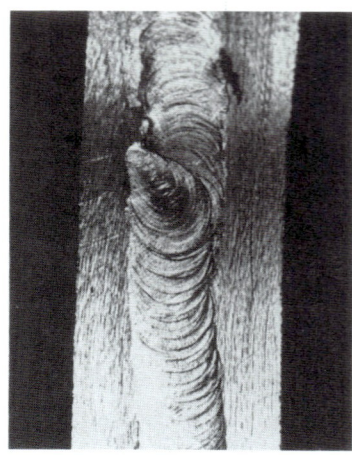

Fig. 20-27 Face appearance of a filler pass in a V-groove butt joint in stainless-steel pipe. *Edward R. Bohnart*

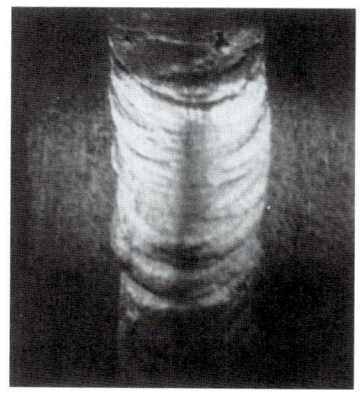

Fig. 20-28 Face appearance of a cover pass in stainless-steel pipe. *Edward R. Bohnart*

Horizontal Fixed Position with 180° Rotation

This weld is another version of Job 20-J2, but welding is done on the top side only. The pipe is in the horizontal 5G position and the weld is vertical. The weld is completed in two halves. Start the weld slightly below

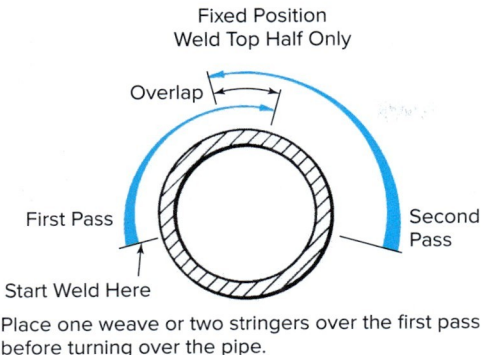

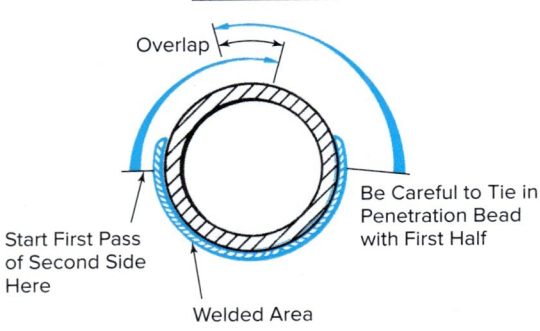

Fig. 20-29 Welding the top side of the pipe in the horizontal fixed (5) position.

Weld on quarters with good overlap of first pass to ensure good penetration bead.

Fig. 20-30 Welding procedure for the root pass in pipe in the vertical fixed (2) position.

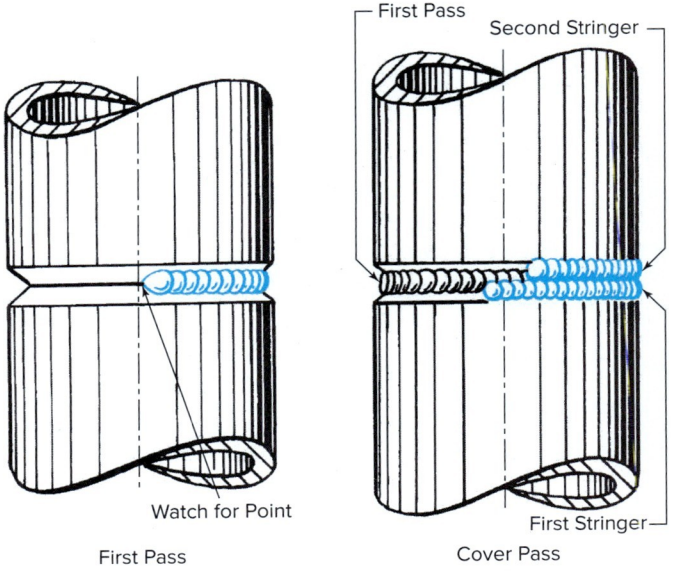

Fig. 20-31 Welding procedure for the first pass and the cover pass on pipe in the vertical fixed (2) position.

the center line of the pipe as shown in Fig. 20-29, and weld toward the top and beyond top center. The other side of the pipe is started in the same way and proceeds to the top. The second weld overlaps the first by about three widths of the weld. The filler and cover passes are welded as previously described with either the weave or stringer bead technique.

When the cover passes are completed, rotate the pipe 180° and weld the second half as described above. Be sure that the welds in the first pass overlap at both their start and finish to avoid incomplete fusion.

Vertical Fixed Position: Jobs 20-J8, J9, J10, and J11

Root Pass The pipe is in the vertical 2G position, and the joint is in the horizontal position. The first pass is made in four quarters. You should change position as each quarter is completed to make sure that you can observe the leading edge of the weld as it progresses. The first pass should overlap at each start and at the finish, Fig. 20-30. The first pass and the cover passes should be made as shown in Fig. 20-31.

Although the shape of the pool and the flow of the molten pool for the vertical fixed position are the same as for rolled pipe, special techniques are required to compensate for the sagging of the pool due to gravity.

Position your torch as shown in Figs. 20-32 and 20-33, page 642. The weld pool is formed on the upper pipe bevel and is kept slightly above the center line of the joint. Move the torch in small circles. Proceed from the top of the pool, around the pool to the bottom, and then up the other side to the top. Do not let the arc dwell too long on the bottom, but let it favor the top of the weld. This

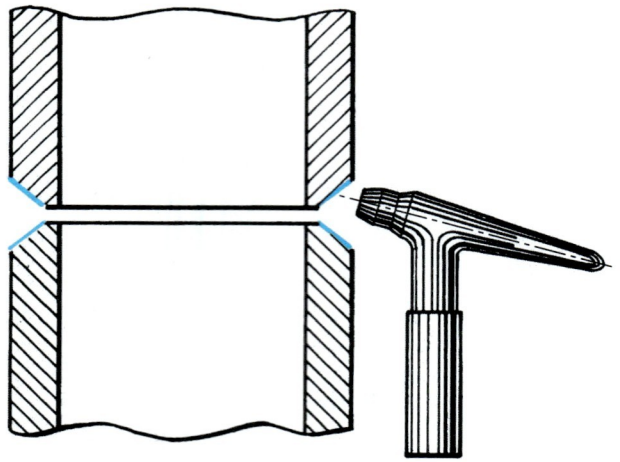

Fig. 20-32 Torch position for welding with pipe axis in the vertical orientation. The weld is being made in the horizontal (2) position.

Fig. 20-34 Welding the root pass without the use of a filler rod. Joint preparation is U-groove, and the pipe is being rolled with its axis in the horizontal direction. The weld is being made in the flat (1) position. © Crane Co.

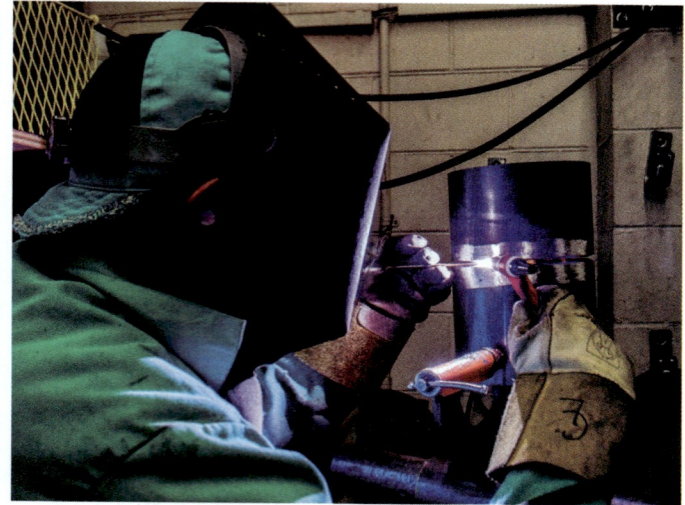

Fig. 20-33 A properly cleaned pipe inside and out being tacked together. The pipe axis in the vertical orientation. The weld will be made in the horizontal (2) position. Filler metal is being used on this V groove weld on a butt joint. Location: Northeast Wisconsin Technical College Mark A. Dierker/McGraw Hill

electrode should be pointed straight in or slightly upward for this pass.

The rest of the passes, including the cover passes, should overlap the preceding passes. The torch should be pointed up or down, depending on the pass sequence. Keep the welds small and make sure that you are securing good fusion with the preceding welds and the pipe surfaces. The pool must be kept fluid, and the forward edge should flow without being forced by the filler rod or torch. Add the filler rod sparingly so that the pool does not get too large and so that cold rod is not deposited on the bead or pipe surfaces. Be sure to clean each pass thoroughly before applying the next pass.

When you are able to make welds that show good penetration through to the back side, have good appearance, and are satisfactory to your instructor, make up a test weld and remove the usual test coupons for root- and face-bend testing.

circular motion ensures fusion of the bottom of the joint with the filler rod, and yet it does not undercut the upper side of the weld bead.

When the pipe is prepared with a U-groove, consumable insert, or rolled edge, the root pass may be welded without filler rod, Fig. 20-34.

Filler and Cover Passes The filler and cover passes should be made as shown in Fig. 20-35. The first filler pass should be a stringer bead made with $1/16$- or $3/32$-inch filler rod and a slight oscillating technique. This pass should cover all or two-thirds of the root pass, depending on the thickness of pipe wall. The tungsten

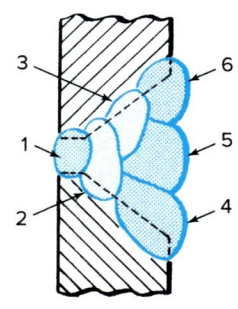

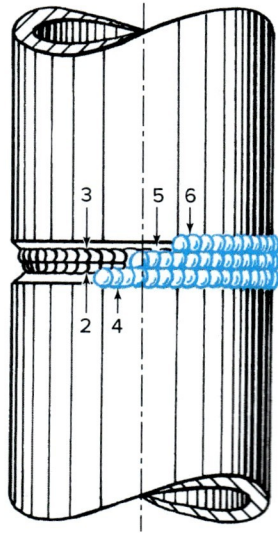

Fig. 20-35 Stringer beads on pipe with axis in the vertical orientation in the horizontal (2) position with weld sequence (top).

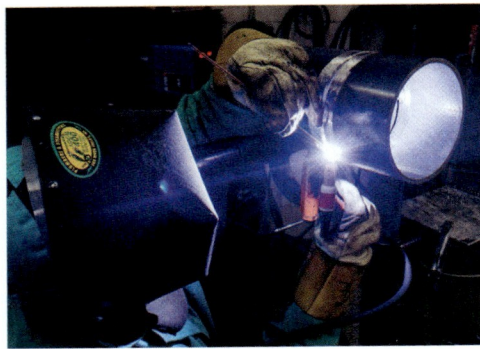

Fig. 20-36 Welding of carbon steel with the pipe axis in the horizontal fixed orientation. The weld is started at the bottom, and travel is up. This is a (5) position weld. Filler metal is being used on this V groove weld on a butt joint. Location: Northeast Wisconsin Technical College Mark A. Dierker/McGraw Hill Education

Fig. 20-37 Welding aluminum with the pipe axis in the horizontal fixed orientation. Note the tight situation for the welder to make this position 5 weld. The travel direction is up. © Sypris Technologies, Inc., Tube Turns Division

Horizontal Fixed Position (Bell Hole Weld): Jobs 20-J12–J17

The pipe is in the horizontal position, and the weld is vertical, Figs. 20-36 and 20-37.

When making a pipe weld in the horizontal fixed 5G position, you will practice all the welding positions: flat, vertical, and overhead. The weld will be made at a constantly changing angle. The ability to make welds of this nature that pass the regular tests marks the welder as a craftsperson. Figures 20-38 through 20-40 show the change of position from bottom to top.

Fig. 20-38 Welding aluminum pipe in the bottom portion of a bell hole. This is an overhead welding, travel up and is a (5) position weld. © Kaiser Aluminum

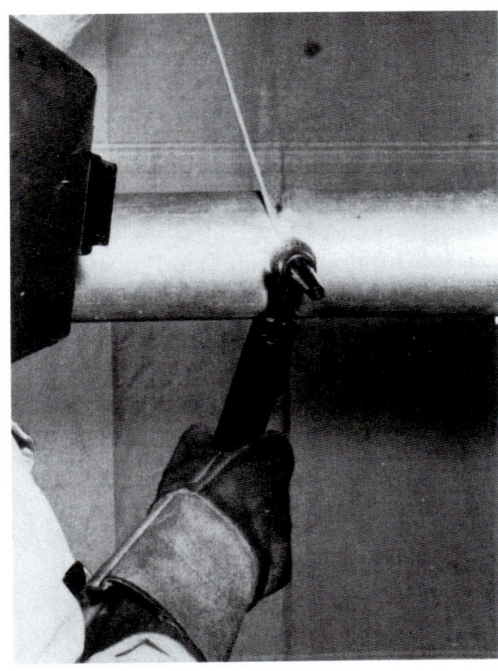

Fig. 20-39 Welding aluminum pipe in the side portion of a bell hole. The welding position changes from the overhead to the vertical position. This is a position 5 weld. © Kaiser Aluminum

Root Pass The same general welding technique followed previously for the welding of rolled pipe may also apply to welding in the horizontal fixed 5G position. You must control the weld pool more closely, however, to get the proper penetration without getting a concave root surface when you are welding overhead. Your response to the demands of the pool must be immediate. You may find it helpful to weld with lower

Gas Tungsten Arc Welding Practice: Jobs 20-J1–J17 (Pipe) **Chapter 20** 643

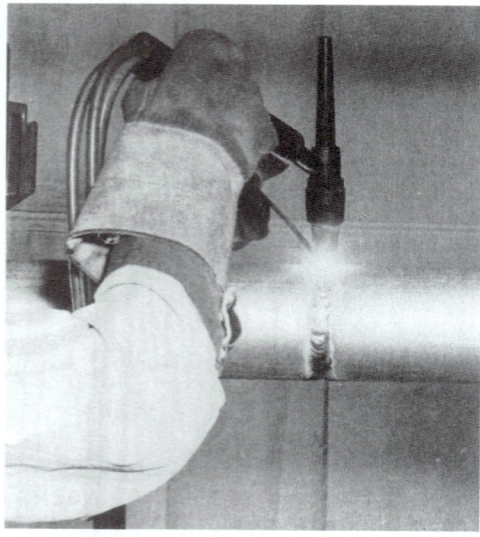

Fig. 20-40 Welding aluminum pipe in the top portion of a bell hole. The welding position changes from the vertical to the flat position. This is a position 5 weld. © Kaiser Aluminum

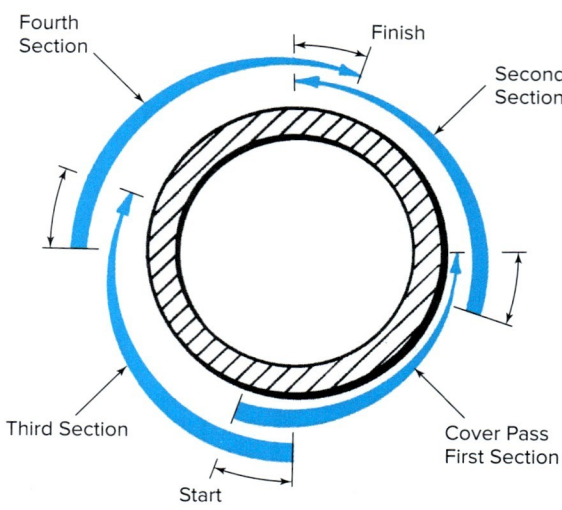

Fig. 20-42 Welding procedure for filler and cover pass on pipe in the bell-hole (5G) position.

current values in order to secure a smaller weld pool. Keep the face of the weld as flat as possible and free of undercut.

The weld bead should be started past bottom center. The welding should progress downhill across the bottom of the pipe and then uphill to the top of the pipe past the 12 o'clock point, Fig. 20-41. If the welding is interrupted in order to change position as you travel around the pipe, be careful to overlap the bead at each start after removing the surface oxides.

The second half of the first pass should overlap the starting point of the first half of the weld. The welding should then proceed to the top and overlap the previous weld at the top, Fig. 20-41. There will be a tendency for the middle of the first pass to be concave and wide. Avoid welding with too little current. Travel as fast as possible.

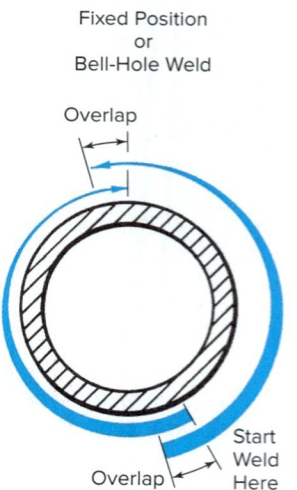

Fig. 20-41 Welding procedure for root pass on pipe in bell-hole position. This is correctly called the 5G position for groove welds.

The Filler and/or Cover Passes The filler and/or cover passes should be made as shown in Fig. 20-42. These passes may be either stringer or weave beads. The first filler pass must be applied with great care because the root pass made with the gas tungsten arc process is likely to be thin, and the danger of burning through is always present. Keep the filler rod in contact with the bottom of the joint, and do not remove it from the weld pool. This limits the penetration into the root pass and helps prevent burn-through. Do not let the arc dwell in the center of the bead. Keep it on one side when making stringer beads, or move it quickly from one side of the joint to the other for weave beads. Make sure that the pool blends into the sides of the joint and that it is fused to the bead underneath. Be sure to clean each pass thoroughly before applying the next pass.

When you are able to make welds that show good penetration through to the back side, have good appearance, and are satisfactory to your instructor, make up a test weld and remove the usual test coupons for root- and face-bend testing.

Inspection and Testing

When completing the remaining jobs assigned from the Job Outline, inspect each pass carefully for weld defects. Compare root passes with the specifications given in Fig. 20-7 and the actual appearance shown

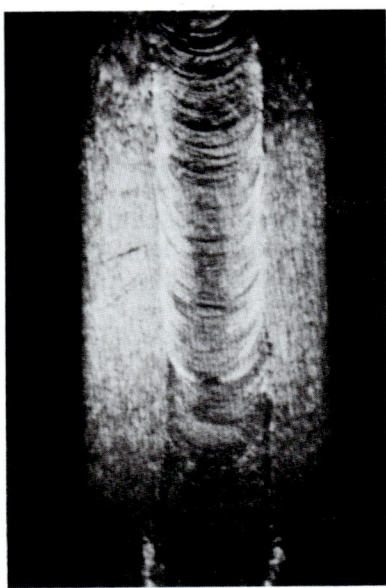

Fig. 20-43 Face appearance of the cover pass in a butt joint in stainless-steel pipe welded with the TIG process. Slight reinforcement. *Edward R. Bohnart*

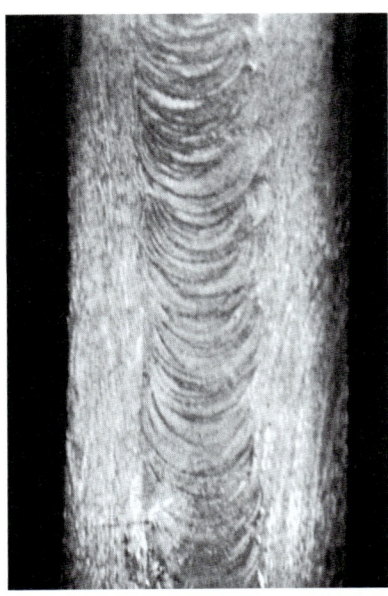

Fig. 20-44 Face appearance of the cover pass in a butt joint in stainless-steel pipe welded with the gas tungsten arc welding process. Relatively flat. *Edward R. Bohnart*

in Figs. 20-8 and 20-9, pages 633–634. Also inspect the outside surface appearance of the root pass and compare it with Figs. 20-21, page 638, and 20-23, page 639.

Finished welds in stainless-steel and aluminum pipe should have the appearance shown in Figs. 20-43 through 20-48. Look for essentially the same good weld characteristics that you have been striving for in shielded metal arc welding practice and gas tungsten arc welding practice on plate. Review the visual acceptance criteria covered in Tables 4-1 through 4-3, pages 117–129.

Testing to determine welder qualification for stainless-steel pipe is usually done by welding together two 6-inch lengths of stainless-steel pipe 1 inch in diameter. One joint is welded in the horizontal fixed 5G position and a second is welded in the vertical fixed 2G position. The specimens are subjected to a tensile pull test. The weld reinforcement is removed. Successful specimens will reach a load limit of more than 75,000 p.s.i. before breaking. Failure may take place in the weld or the pipe wall.

Testing may also be done on larger diameter pipe. To test your practice welds in both stainless steel and aluminum, remove face- and root-bend specimens from the top, bottom, and sides of the pipe and subject them to bending stress. This test is similar to the testing procedures for the shielded metal arc welding of plate and pipe and for the gas tungsten arc welding of plate.

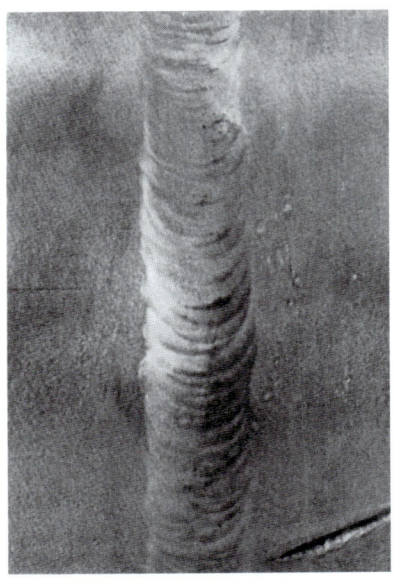

Fig. 20-45 Surface appearance of the cover pass in a butt joint in aluminum pipe welded with the TIG process. *Edward R. Bohnart*

Branch Welds

After you have demonstrated that you are able to make satisfactory groove welds in pipe of various materials, you may wish to practice a number of branch (lateral) joints. Both 90 and 45° branches are suggested. These are similar to those done with the shielded metal arc process, using the stringer bead technique. Completed welds should have the appearance shown in Figs. 20-46 and 20-48.

Fig. 20-46 Butt and branch welds made with the gas tungsten arc welding process. Pipe material is Inconel. Diameters are 2½ and 3½ inches; wall thickness is ¼ inch. Joint gap for groove welds was ⅛ inch, and for the laterals, 3⁄16 inch. Helium was used both for purging and as a shielding gas. (Argon can also be used.) Tungsten electrode size was 3⁄32, filler rod size was 3⁄32, and welding heat was about 55 to 75 amperes. Stringer bead technique was used throughout.
© Pipefitters Union, St. Louis, MO

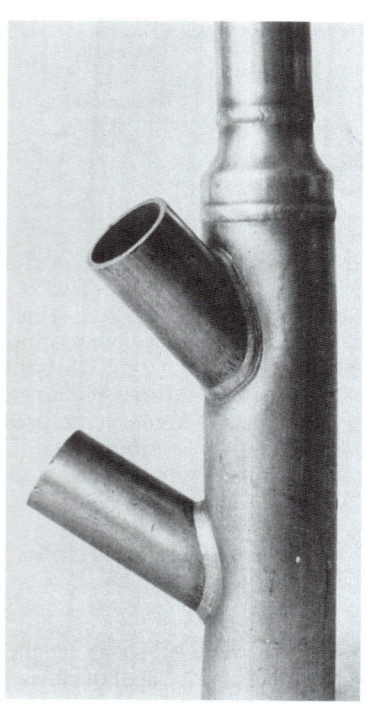

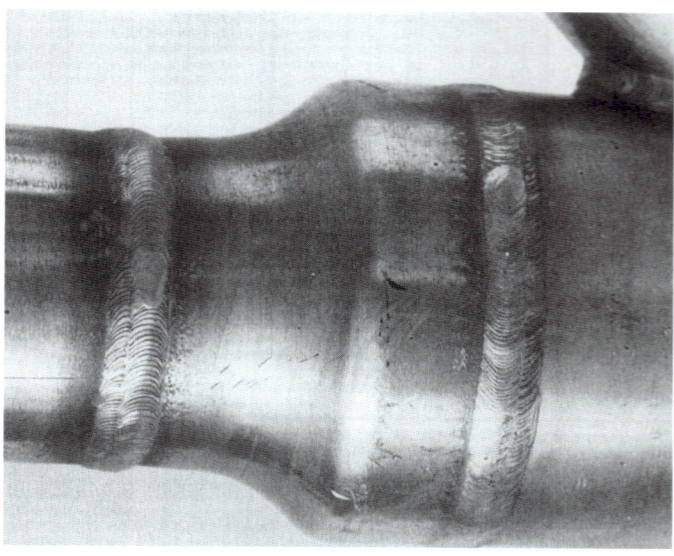

Fig. 20-47 Close-up of groove welds shown in Fig. 20-46.
© Pipefitters Union, St. Louis, MO

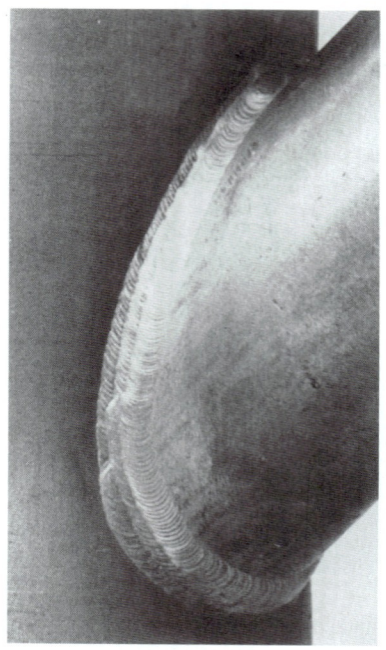

Fig. 20-48 Close-up of branch weld shown in Fig. 20-46. © Pipefitters Union, St. Louis, MO

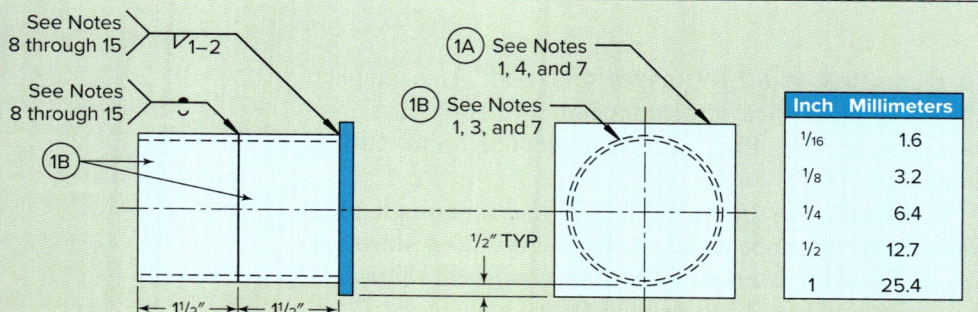

Inch	Millimeters
1⁄16	1.6
1⁄8	3.2
1⁄4	6.4
1⁄2	12.7
1	25.4

Notes:
1. 1 each required, carbon steel, aluminum and stainless steel.
2. All dimensions U.S. customary unless otherwise specified.
3. 3″ φ, 10 ga.–18 ga. thickness round tubing. Optional choice of wall thickness within range specified.
4. 10 ga.–18 ga. material thickness for sheet. Optional choice of thickness within range specified.
5. The welder shall prepare a bill of materials in U.S. customary units of measure prior to cutting.
6. The welder shall convert the above bill of materials to S.I. metric units of measure.
7. Part 1A manual PAC. Part 1B may be mechanically or manual PAC cut. Uniform radius all sharp corners.
8. All Welds GTAW. Fillet weld size = 1½ times the nominal wall thickness of the round tubing. Groove weld root opening = ½ the nominal wall thickness of the round tubing.
9. Groove weld joint geometry 16–18 gauge use square groove. Groove weld joint geometry 10–14 gauge use V-groove.
10. No melt through on fillet welds.
11. Root shielding gas required for aluminum and stainless steel.
12. Fit and tack entire assembly on bench before welding.
13. All welding done in position according to drawing orientation.
14. Use WPS ANSI/AWS B2.1.008 for carbon steel. Use WPS AWS2-1-GTAW for aluminum (M-23/P-23/S-23). Use WPS AWS2-1.1-GTAW for aluminum (M-22/P-22/S-22). Use WPS ANSI/AWS B2.1.009 for stainless steel.
15. Visual examination in accordance with the requirements of AWS QC11, Table 1.

Fig. 20-49 GTAW work qualification test per AWS QC11 for fillet and groove welds. Adapted from AWS QC11

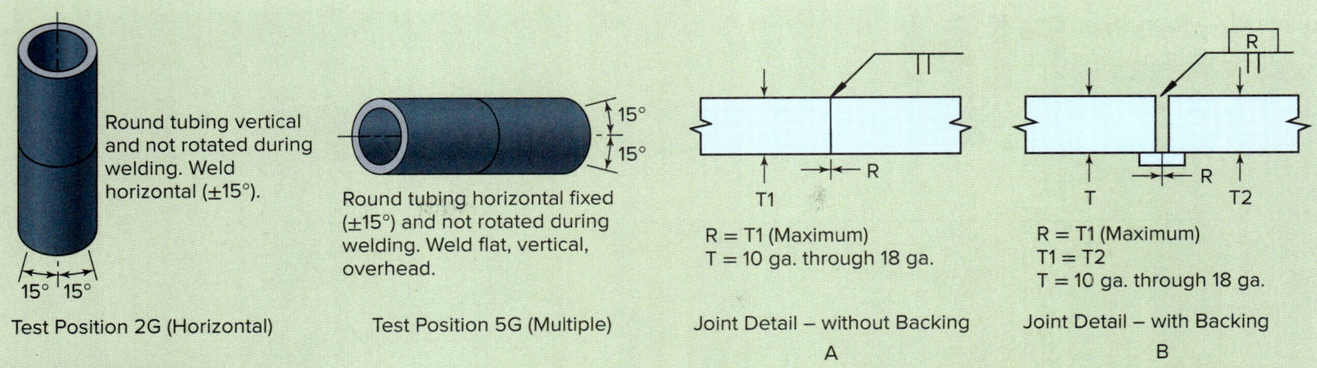

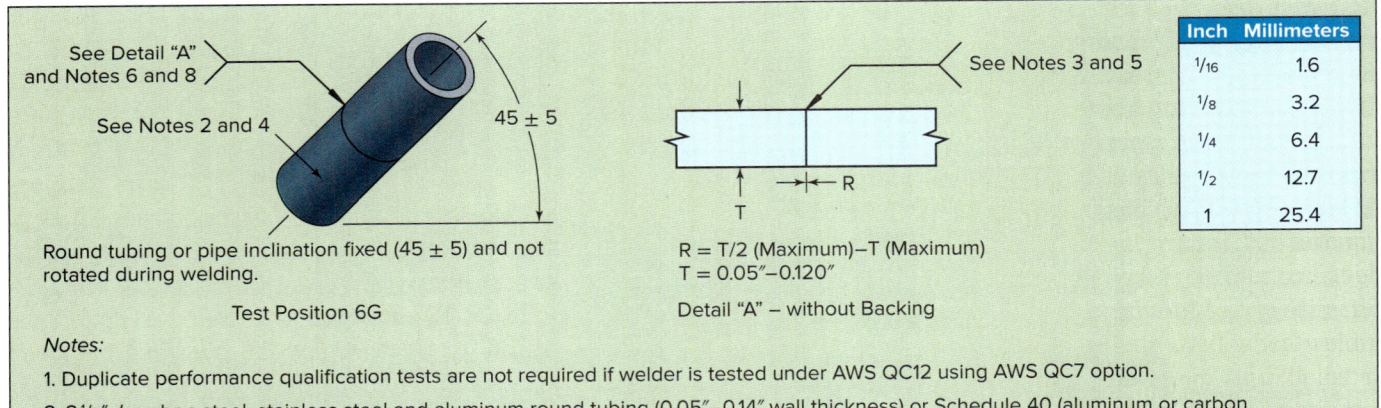

Inch	Millimeters
1/16	1.6
1/8	3.2
1/4	6.4
1/2	12.7
1	25.4

Notes:
1. Administration of this performance qualification test in accordance with AWS QC7, Supplement G, supersedes AWS QC11 and AWS EG3.0 requirements of work qualification for GTAW of carbon steel round tubing.
2. 1″– 2 7/8″ φ, 10 ga.–18 ga. carbon steel round tubing. Optional choice of diameter and wall thickness within range specified.
3. The standard round tubing groove test weldments for performance qualification shall consist of two round tubing sections, each a minimum of 3 in. (76 mm) long joined by welding to make one test weldment a minimum of 6 in. (150 mm) long.
4. Performance Qualification #1 = 2G position. Performance Qualification #2 = 5G position.
5. All welding done in position according to applicable performance qualification requirement.
6. With or without backing. Backing ring to suit diameter and nominal wall thickness of pipe. Refer to Details "A" and "B."
7. Root shielding gas optional.
8. All parts may be mechanically cut or machine OFC.
9. Use WPS ANSI/AWS B2.1.008.
10. Visual examination in accordance with the requirements of AWS B2.1, Sections 3.5.1 and 3.5.3.1. Bend test in accordance with the requirements of AWS B2.1, Sections 3.5.3.1, 3.7.1.2 and Figure 3.7.1C. Bend specimens = 1/2″ (13 mm) wide × 6 in (150 mm) long.

Fig. 20-50 Performance qualification test per AWS QC11 with the GTAW process on carbon steel in the 2G and 5G positions. Adapted from AWS QC11

Inch	Millimeters
1/16	1.6
1/8	3.2
1/4	6.4
1/2	12.7
1	25.4

Notes:
1. Duplicate performance qualification tests are not required if welder is tested under AWS QC12 using AWS QC7 option.
2. 2 1/2″ φ carbon steel, stainless steel and aluminum round tubing (0.05″–0.14″ wall thickness) or Schedule 40 (aluminum or carbon steel) or 10S pipe (0.120″ wall thickness). Round tubing diameter and wall thickness optional within range specified. Pipe Schedule optional according to material requirements.
3. Round tubing or pipe wall thickness greater than 0.0625″ chamfer both pipe sections to form V-groove. Round tubing less than or equal to 0.0625″ wall thickness use Square-groove.
4. The standard pipe groove test weldment for performance qualification shall consist of two pipe sections, each a minimum of 3 in. (76 mm) long joined by welding to make one test weldment a minimum of 6 in. (150 mm) long.
5. Without backing. Refer to Detail "A." Root shielding required. Consumable insert required for one stainless steel test. Consumable insert type optional.
6. All welding done in position.
7. Parts may be mechanically cut, machine OFC or machine PAC.
8. For carbon steel use WPS ANSI/AWS B2.1.008. For stainless steel use WPS ANSI/AWS B2.1.009, ANSI/AWS B2.1.010, AWS3—GTAW—1, or AWS3—GTAW—2. For aluminum use WPS AWS2-1-GTAW or AWS2-1.1-GTAW.
9. Visual examination in accordance with the requirements of QC12, Table 1. Bend test or Radiograph in accordance with the requirements of QC12, Table 2 and Annex D, Figures 7 and 8. Except as specified in AWS EG4.0, Section 4.2 Performance Qualification.

Fig. 20-51 GTAW performance qualification test per AWS QC12, in the 6G position. Adapted from AWS QC12, in the G6 position

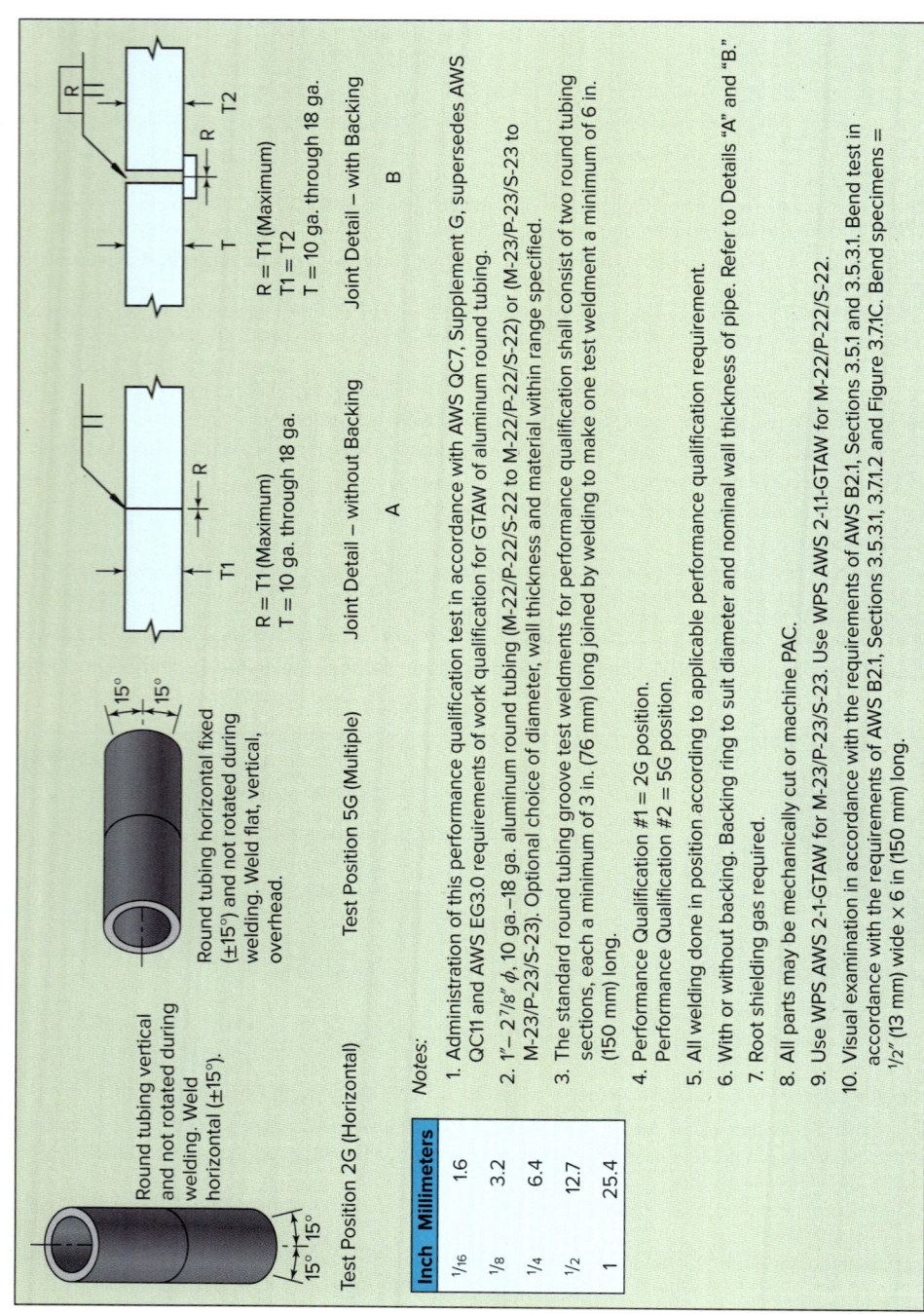

Fig. 20-52 Performance qualification test per AWS QC11 with the GTAW process on aluminum in the 2G and 5G positions. Adapted from AWS QC11

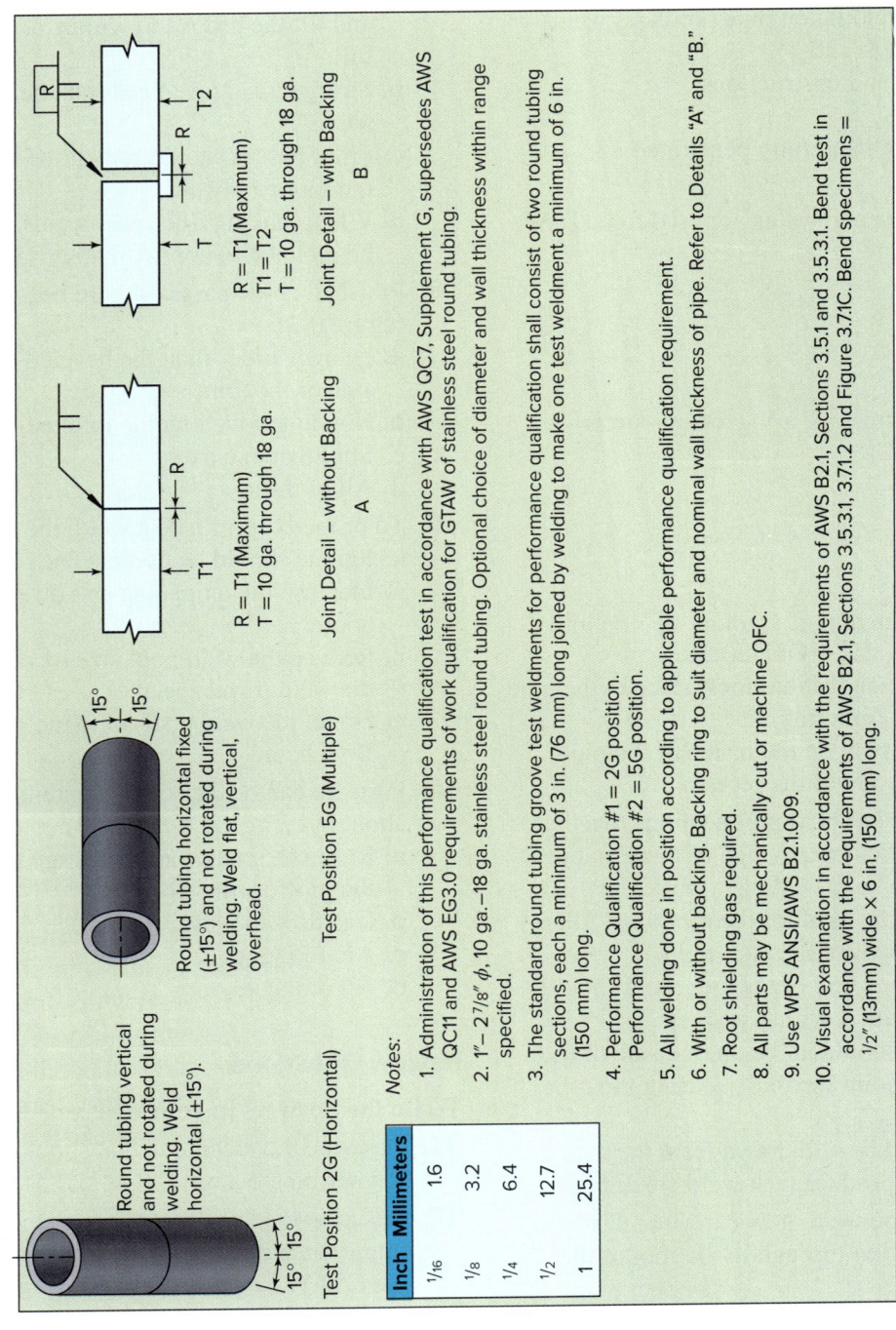

Fig. 20-53 Performance qualification test per AWS QC11 with the GTAW process on stainless steel in the 2G and 5G positions. Adapted from AWS QC11 with the GTAW process on stainless steel

CHAPTER 20 REVIEW

Multiple Choice

Choose the letter of the correct answer.

1. Gas tungsten arc welding on pipe produces welds that are _____. (Obj. 20-1)
   a. Full of crevices and obstructions
   b. Corrosion prone
   c. Unusually smooth and fully penetrated
   d. All of these

2. The following joints and welds are used for GTAW on pipe. (Obj. 20-2)
   a. V-Groove butt
   b. U-groove butt
   c. Consumable insert butt
   d. All of these

3. The proper bevel angle for a V-groove joint is _____. (Obj. 20-2)
   a. 22°
   b. 45°
   c. 55°
   d. None of these

4. Purging, the introduction of argon or helium into a pipe, is done to _____. (Obj. 20-3)
   a. Permit oxidation on the interior surface of the pipe
   b. Protect the joint from flux
   c. Shield the weld pool on the inside of the pipe
   d. Heat-treat the inside of the pipe

5. Tack welds are used to hold pipe sections together in preparation for welding. Which of these statements is true about tack welding? (Obj. 20-3)
   a. If filler rod is to be used for the root pass, filler rod should be used for the tack weld.
   b. The beginning and end of the tack weld should be worn away with a drill or reamer.
   c. The tack welds should all be on one side of the joint to provide joint flexibility during the root weld.
   d. The torch should be withdrawn from the tack weld quickly to produce tack weld cracking.

6. When making a pipe weld, how can you tell if you are getting penetration through to the root of the joint? (Obj. 20-3)
   a. The weld pool will have a sharp point of the leading edge.
   b. The weld pool will flatten out.
   c. The "birds-eye" of silica will begin dancing in the middle of the weld pool.
   d. All of these.

7. Which of the following statements concerning filler passes is *not* true? (Obj. 20-3)
   a. Filler passes are placed on top of the root pass and fill the groove to within $1/16$ inch of being full.
   b. Stringer or weave beads can be used for filler passes.
   c. Filler passes should remelt the inside surface of the root pass weld.
   d. When making filler pass welds, the torch should be inclined to approximately 45°.

8. Finished cover passes should be _____. (Obj. 20-3)
   a. $3/32$ inch wider than the beveled edges on each side of the joint
   b. Not quite touching the joint edges
   c. Slightly concave
   d. All of these

9. To properly terminate a weld, the following technique should be used: (Obj. 20-4)
   a. Increase the amperage and dwell on the weld pool.
   b. Reduce the weld pool size by rapidly increasing the weld travel speed.
   c. Direct the weld pool off to the side of the joint.
   d. Both b and c.

10. Porosity can be caused by contamination brought about by _____. (Obj. 20-4)
    a. Rust, oil, grease, and moisture on the surface of the pipe or filler rod
    b. Gas flow too low
    c. Arc too long
    d. All of these

Review Questions

Write the answers in your own words.

11. Describe the technique that should be used to properly terminate a weld. (Obj. 20-5)

12. Describe the "walking the cup" technique and the importance of the gas cup size for welding pipe with the GTAW process. (Obj. 20-6)

13. Explain the terms *root pass, filler pass,* and *cover pass*. (Obj. 20-3)

14. Explain the proper welding technique when ending a weld bead. (Obj. 20-5)

15. Explain the general rule for determining the width of a cover pass weld bead on pipe. (Obj. 20-6)

INTERNET ACTIVITIES

Internet Activity A

Use your favorite search engine and see what the most recent studies are on the flow of fluids through a pipe. Look for information that would pertain to the weld profile on the inside of the pipe or any other possible restrictions that might impact flow, turbulance, erosion, and so on.

Internet Activity B

Using your favorite search engine, check out how pipe sizes compare between the U.S. Customary measuring system and the Standards International (metric) measuring system. You may want to start your search by finding manufacturers of pipe.

Design credits: Cinema and movie icon: Denis Meshkov/123RF; and Illustration of Welding icons: Mr. Nachai Sorasee/123RF

Table 20-2 Job Outline: Gas Tungsten Arc Welding Practice (Pipe)

Recommended Job Order[1]	Number in Text	Type of Joint	Type of Weld	Position[2]	Welding Technique	Pipe Specifications Material	Diameter (in.)	Wall Thickness
1st	20-J1	NA	Beading	1 (pipe rotated)	Stringer-up, weave-up	Steel	4–6	1/4 in.
2nd	20-J2	Butt	V-groove	1 (pipe rotated)	1 stringer-up, 1 weave-up	Steel	4–6	1/4 in.
3rd	20-J8	Butt	V-groove	2	5 stringers	Steel	4–6	1/4 in.
4th	20-J12	Butt	V-groove	5	1 stringer-up, 1 weave-up	Steel	4–6	1/4 in.
Test	20-JQT1[6]	Butt & T	Groove & fillet	5	Stringer or weave	Steel	3	18–10 gauge
Test	20-JQT2[7]	Butt	Sq. or V-groove	2 & 5	Stringer or weave	Steel	1–2 7/8	18–10 gauge
Test	20-JQT3[8]	Butt	Sq. or V-groove	6	Stringer or weave	Steel	2 1/2	0.05–0.120 in.
5th	20-J3	NA	Beading	1 (pipe rotated)	Stringer-up, weave-up	Al.	4–6	1/4 in.
6th	20-J4	Butt	V-groove	1 (pipe rotated)	1 stringer-up, 1 weave-up	Al.	4–6	1/4 in.
7th	20-J9	Butt	V-groove	2	5 stringers	Al.	4–6	1/4 in.
8th	20-J13	Butt	V-groove	5	1 stringer-up, 2 weave-up	Al.	4–6	1/4 in.
9th	20-J14	Butt	V-groove	5	1 semiweave	Al.	2–3	1/8 in.
10th	20-J15	Butt	V-groove	5	1 semiweave	Al.	2–3	1/8 in.
11th	20-J10	Butt	V-groove	2	1 semiweave	Al.	2–3	1/8 in.
Test	20-JQT4[6]	Butt & T	Groove & fillet	5	Stringer or weave	Al.	3	18–10 gauge
Test	20-JQT5[7]	Butt	Sq. or V-groove	2 & 5	Stringer or weave	Al.	1–2 7/8	18–10 gauge
Test	20-JQT6[8]	Butt	Sq. or V-groove	6	Stringer or weave	Al.	2 1/2	0.05–0.120 in.
12th	20-J5	NA	Beading	1 (pipe rotated)	Stringer-up, weave-up	S. Stl	4–6	1/4 in.
13th	20-J6	Butt	V-groove	1 (pipe rotated)	1 stringer-up, 1 weave-up	S. Stl	4–6	1/4 in.
14th	20-J11	Butt	V-groove	2	5 stringers	S. Stl	4–6	1/4 in.
15th	20-J16	Butt	V-groove	5	1 stringer-up, 2 weave-up	S. Stl	4–6	1/4 in.
16th	20-J7	Butt	Sq.-groove	1	1 semiweave	S. Stl	2–3	1/8 in.
17th	20-J17	Butt	Sq.-groove	5	1 semiweave	S. Stl	2–3	1/8 in.
Test	20-JQT7[6]	Butt & T	Groove & fillet	5	Stringer or weave	S. Stl	3	18–10 gauge
Test	20-JQT8[7]	Butt	Sq. or V-groove	2 & 5	Stringer or weave	S. Stl	1–2 7/8	18–10 gauge
Test	20-JQT9[8]	Butt	Sq. or V-groove	6	Stringer or weave	S. Stl	2 1/2	0.05–0.120 in.

[1] It is recommended that the student do the jobs in this order. In the text, the jobs are grouped according to the type of operation to avoid repetition. The tests will be administered per your instructor's instructions.

[2] 1 = flat, 2 = horizontal, 5 = multiple, 6 = multiple axis 45°.

[3] Ceramic or glass cups should be used for currents up to 250 A. Water-cooled cups should be used for currents above 250 A.

[4] The gas flow should be set for the maximum rate for the vertical and overhead positions. Gas backing is optional for steel and aluminum. It is required for stainless steel.

[5] The type of filler rod must always match the type of base metal being welded.

Tungsten Electrode Size (in.)	Current		Shielding gas		Cup Size (in.)	Filler Rod Size[5] (in.)	Text Reference
	Type	Amperes[3]	Gas Type	Gas Flow (ft³/h)[4]			
3/32	DCEN	140–160	Argon	8–10	3/8	3/32–1/8	654
3/32	DCEN	120–140 140–160	Argon	8–10	3/8	3/32–1/8	654
3/32	DCEN	120–160	Argon	8–10	3/8	3/32–1/8	659
3/32	DCEN	120–140	Argon	8–10	3/8	3/32–1/8	661
1/8 max.	DCEN	45–130	Argon	15–25	1/4–5/8	1/16–3/32	Fig. 20-49
1/8 max.	DCEN	45–130	Argon	15–25	1/4–5/8	1/16–3/32	Fig. 20-50
1/8 max.	DCEN	45–130	Argon	15–25	1/4–5/8	1/16–3/32	Fig. 20-51
1/8	A.C.	150–280	Argon	22–28	1/2	3/32–1/8	654
1/8	A.C.	150–280 160–300	Argon	22–28	1/2	3/32–1/8	654
1/8	A.C.	150–200 160–210	Argon	25–30	1/2	3/32–1/8	659
1/8	A.C.	120–200 140–220	Argon	25–30	1/2	3/32–1/8	661
3/32	A.C.	120–145	Argon	15–20	3/8	3/32–1/8	661
3/32	A.C.	100–120	Argon	12–22	3/8	3/32–1/8	661
3/32	A.C.	100–120	Argon	15–20	3/8	3/32–1/8	659
1/8 max.	A.C.	40–125	Argon	20–40	1/4–5/8	3/32–1/8	Fig. 20-49
1/8 max.	A.C.	40–125	Argon	20–40	1/4–5/8	3/32–1/8	Fig. 20-52
1/8 max.	A.C.	40–125	Argon	20–40	1/4–5/8	3/32–1/8	Fig. 20-51
3/32	DCEN	120–160	Argon	12–15	1/2	3/32–1/8	654
3/32	DCEN	110–150 110–160	Argon	12–15	1/2	3/32–1/8	654
3/32	DCEN	110–150 110–160	Argon	12–15	1/2	3/32–1/8	659
3/32	DCEN	110–150 110–160	Argon	14–18	1/2	3/32–1/8	661
1/16 to 3/32	DCEN	60–120	Argon	12–15	3/32–1/8	3/32–1/8	654
	DCEN	80–140	Argon	12–15	3/32–1/8	3/32–1/8	661
1/8 max.	DCEN	35–130	Argon	15–25	1/4–5/8	1/16–3/32	Fig. 20-49
1/8 max.	DCEN	35–130	Argon	15–25	1/4–5/8	1/16–3/32	Fig. 20-53
1/8 max.	DCEN	35–130	Argon	15–25	1/4–5/8	1/16–3/32	Fig. 20-51

[6]For additional information follow AWS Standard Welding Procedure Specification (WPS) B2.1.008 for carbon steel, AWS2-1-GTAW or AWS2-1.1-GTAW for aluminum, and (WPS) B2.1.009 for stainless steel.

[7]For additional information follow AWS Standard Welding Procedure Specification (WPS) B2.1.008 for carbon steel.

[8]For additional information follow AWS Standard Welding Procedure Specification (WPS) B2.1.008 for carbon steel; AWS3-GTAW-1, AWS3-GTAW-2, AWS2-1-GTAW, or AWS2-1.1-GTAW for aluminum; and AWS B2.1.009 and AWS B2.1.010 for stainless steel.

Note: The conditions given here are basic. They vary with the job situation, the results desired, and the skill of the welder. It is also generally acknowledged if you are capable of welding a more difficult weld and joint like a groove weld on a butt joint that you are capable of making easier fillet welds on lap joints and T-joints.

UNIT 4

Gas Metal Arc, Flux Cored Arc, and Submerged Arc Welding

Chapter 21
Gas Metal Arc and Flux Cored Arc Welding Principles

Chapter 22
Gas Metal Arc Welding Practice with Solid and Metal Core Wire: Jobs 22-J1–J23 (Plate)

Chapter 23
Flux Cored Arc Welding Practice (Plate), Submerged Arc Welding, and Related Processes: FCAW-G Jobs 23-J1–J11, FCAW-S Jobs 23-J1–J12, SAW Job 23-J1

Chapter 24
Gas Metal Arc Welding Practice: Jobs 24-J1–J15 (Pipe)

21

Gas Metal Arc and Flux Cored Arc Welding Principles

Overview

The gas metal arc welding (GMAW) commonly referred to by welders as MIG or MAG, is becoming increasingly popular.

Gas metal arc welding, a continuous wire process, was introduced to industry in the early 1950s. At that time the process was limited by high cost, consumable wire electrodes; relatively crude wire-feeding systems; and welding machines that were not suited for MIG/MAG welding. The process was used principally for the welding of stainless steel and aluminum. The wires used were within the $\frac{1}{16}$- to $\frac{3}{32}$-inch range. Practical minimum stock thickness was ¼ inch.

Today MIG/MAG welding has become one of the more flexible all-round welding tools in the fabricating industry. A great number of types and sizes of wires have been developed, and simplified wire-feeding systems and guns are available. A significant step in power source improvement was the invention of the constant voltage welding machine. Thus, the door was opened to the short-circuiting arc with its greater flexibility. For the first time, all-position MIG/MAG welding was possible, and sheet metal as thin as 22 gauge could be welded. Constant voltage machines have been further improved by the addition of variable voltage control, variable slope control, and variable inductance control. Control of these three factors

Chapter Objectives

After completing this chapter, you will be able to:

21-1 Explain the principles of gas metal arc and flux cored arc welding processes.

21-2 Name the gas metal arc and flux cored arc welding equipment.

21-3 Describe the shielding gases and electrodes used with semiautomatic arc welding processes.

Fig. 21-1 GMAW being performed on a carbon steel trailer frame. Note the positioner, which is able to optimize the welding into flat and horizontal positions. Miller Electric Mfg. Co.

permits the welder to fine-tune the welding characteristics for any desired condition. The inverter-type power sources expand the control over the short-circuiting mode of transfer and allow the pulse-spray mode of transfer as well. The introduction of carbon dioxide as a shielding gas extended the application of the process to a wide variety of mild steels on an economical basis. Figure 21-1 shows shop application of the MIG/MAG process.

The complete name for GMAW is gas metal arc welding. Slang and trade names are often applied, such as MIG for metal inert gas (aluminum and magnesium are typical when only an inert gas is used) or MAG for metal active gas (carbon steel and stainless-steel welding when carbon dioxide [CO_2] and/or oxygen [O_2] is added to the inert gas). Some slang names relate more specifically to the part of the process like CO_2 or wire welding. Some typical trade names of various manufacturers of the weld equipment are *Sigma Welding* (ESAB), *Millermatic Welding* (Miller), *STT* (Lincoln), *Regulated Metal Deposition* (Miller), *Micro-wire Welding* (Hobart), and *Aircomatic Welding* (Airco). The short-circuiting method of metal transfer is known by the trade name of Dipmatic, and buried arc CO_2 welding is derived from the use of CO_2 as a shielding gas.

A process similar to GMAW in that it uses much of the same equipment is the **flux cored arc welding (FCAW)** process. This process started gaining industrial use in the 1950s for the express purpose of improved metallurgical properties derived from the flux and the slag that supports and shapes the weld bead. With certain flux cored electrodes, the need for an external shielding gas is eliminated.

Instead of the solid wire used for GMAW, a tubular wire is used for FCAW. The outside of the wire acts as the electrical conducting sheath or the electrode and provides the bulk of the weld bead forming metal, while the core contains the flux and other ingredients.

Some of these electrodes generate sufficient gas shielding and are thus referred to as self-shielding (*FCAW-S*). They act much like an SMAW coated electrode turned inside out. Since they generate their own shielding, the equipment is less complicated. A great deal of smoke is produced due to their nature and high deposition capability as compared to the SMAW coated electrode. FCAW-S is favored for work where the externally supplied shielding gases may be blown away and where smoke buildup is not an issue. A very common trade name applied to this type of electrode is Innershield (the Lincoln Electric Co.).

Other flux cored electrodes require an external shielding gas. These electrodes are referred to as gas shielded (*FCAW-G*). They typically run on DCEP and have very good penetration and fusion characteristics. They also tend to have greater welder appeal and produce less smoke than the FCAW-S type electrodes. A very common trade name for this type of electrodes is Dual Shield (ESAB Welding and Cutting).

Gas metal arc welding is similar to gas tungsten arc in some respects. For example, gas shields the weld area in both processes. In gas metal arc welding, however, an electrode filler wire or a consumable bare electrode wire is used instead of a nonconsumable tungsten electrode. The wire is fed continuously into the weld by a wire-feed mechanism through a torch or gun. The weld area is surrounded by a gas blanket to protect it from atmospheric contamination in GMAW and FCAW-G. GMAW is shown in Fig. 21-2, page 658, while FCAW-G and FCAW-S are shown in Fig. 21-3, page 658. With the FCAW-S process some shielding is provided by the vaporization of the core ingredients, keeping air from the immediate arc area. In addition, scavengers combine with unwanted elements that would contaminate the weld pool. This along with the slag formers surround the weld from the air, Fig. 21-4, page 658. The basic system is shown in Fig. 21-5, page 659.

Advantages of Gas Metal Arc Welding and Flux Cored Arc Welding

Gas metal arc welding produces high quality welds at high speeds without the use of flux and limited postweld cleaning. It is very desirable for both small jobs and high production metal joining. It frequently replaces another

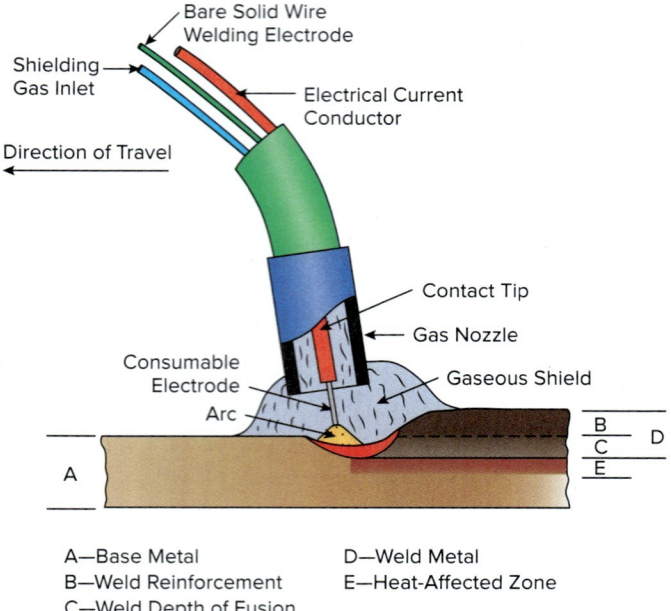

Fig. 21-2 Gas metal arc welding process (GMAW). Source: American Welding Society, *Welding Handbook,* 9th ed., Volume 2, page 150, Fig. 4.2.

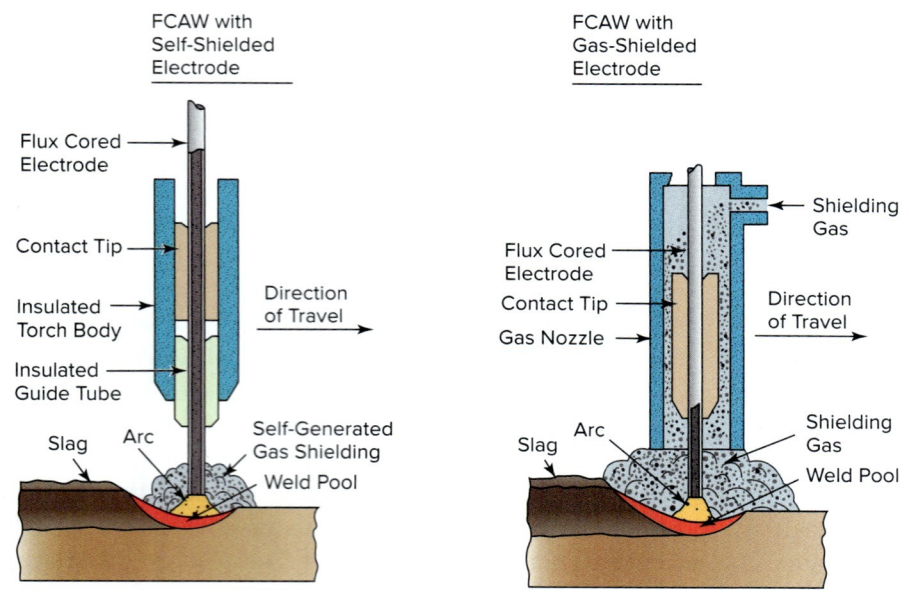

Fig. 21-3 Flux cored arc welding, both self-shielded and gas shielded. Adapted from *Welding Handbook, 9/e*

Fig. 21-4 Welding with the flux cored arc welding process. The welder is working on the longitudinal seam on the inside of a heat exchanger. Miller Electric Mfg. Co.

joining process such as riveting, brazing, silver-soldering, or resistance welding. It may be used instead of the following fusion welding processes: oxyacetylene welding, shielded metal arc welding, submerged arc welding, flux cored arc welding, and gas tungsten arc welding.

When selecting a welding process for a given job, the welding method chosen is that which will do the best job at the lowest cost. The decision is based upon a consideration of the costs of labor, equipment, electrodes and gas, material preparation, actual arc time, and postweld cleaning as well as the importance of weld soundness and appearance. The following are a few of the advantages of GMAW welding:

- Welders who are proficient in the use of other welding processes can be readily trained in the GMAW/FCAW processes. The equipment is simple to set up, and control of the process is incorporated into the welding equipment. The welder must watch the angle of the torch relative to the workpiece, the speed of travel, electrode extension, and the gas-shielding pattern.

- Welders can weld as fast as they are able. One of the principal advantages is the elimination of weld starting and stopping due to the changing of electrodes. This prevents weld failures due to slag inclusions, cold lapping, poor penetration, crater cracking, and poor fusion, which may result from starting and stopping to change electrodes.

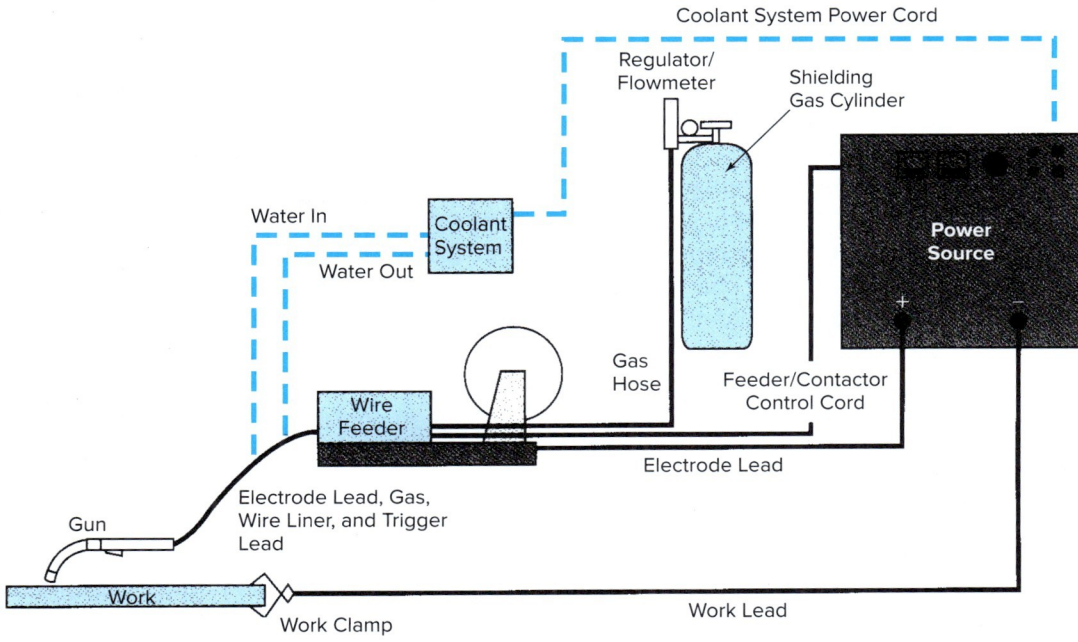

Fig. 21-5 Schematic diagram of a gas metal or flux cored arc installation. Source: Miller Electric Mfg. Co.

- There is no welding slag or flux to remove when solid wire or metal cored wires are used, and weld spatter is at a minimum. The process gives a smooth and good-appearing weld surface. These characteristics provide substantial cost savings since metal finishing is frequently a costly production item. Many manufacturers are able to paint or plate over GMAW welds without additional surface preparation.
- Electrode costs are somewhat less due to the very small amount of electrode loss. It is estimated that when using shielded metal arc welding electrodes, there is an average stub end loss of up to 17 percent and spatter and flux coating losses up to 27 percent. In figuring their costs, many fabricators provide for an electrode use of about 60 percent. In GMAW welding, approximately 95 percent of the consumable electrode wire is deposited in the welded joint. The stub loss is totally eliminated.
- Welding may be done in all positions and on both light and heavy gauge materials. Weldable metal thicknesses range from 24 gauge to unlimited with edge preparation. For heavy materials a narrower groove angle with a thicker root face can be used. A narrow groove for the weld joint reduces preparation time and the amount of filler metal deposit needed. A reduction of welding time and heat input keeps distortion at a minimum.
- The concentration of current (current density) is higher for GMAW/FCAW than for stick electrode welding, Fig 21-6. **Current density** may be defined as the amperage per square inch of the cross-sectional area of the electrode. This provides for higher deposition rates than manual welding. An arc with a high current density concentrates more energy at one point than one with low density. The GMAW arc has a high current density, but the metal arc produced by the SMAW does not. The stream of the GMAW arc is sharp and incisive, but the stream of the SMAW arc is relatively soft and widespread.

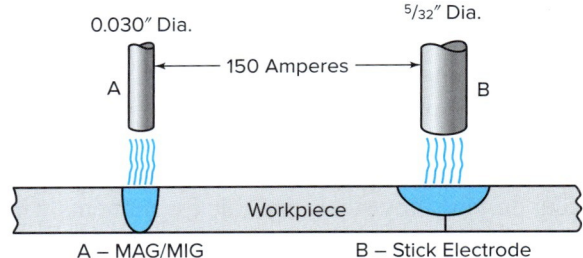

Fig. 21-6 Arc concentration and deposit comparison with the same current: (A) MAG/MIG and (B) stick electrode.

SHOP TALK

MIG Guns
Some MIG guns have trigger wire replacements within the handle, so that you don't have to spend time going for another wire.

Gas Metal Arc and Flux Cored Arc Welding Principles   Chapter 21   659

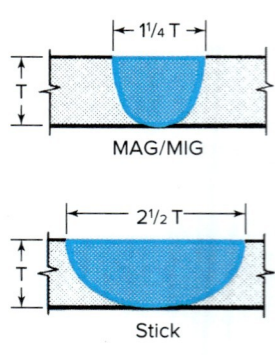

Fig. 21-7 Weld bead comparisons.

The weld deposit made with a stick electrode is wider and more bowl-shaped than a deposit made with the GMAW process, Fig. 21-7. The bead width-to-depth ratio is greater in shielded metal arc welding and the heat input is less. Because of this characteristic, the speed of shielded metal arc welding is slower, and more heat is applied per linear inch of weld. Since there is a greater volume of electrode deposited per linear inch of weld with the shielded metal arc process than the gas metal arc process, the heat input is greater for stick electrodes. With gas metal arc welding, there is greater penetration into the workpiece because there is higher current density at the electrode tip. The width of the normal shielded metal arc weld deposit on ¼-inch plate is about 2½ to 3 times the thickness of the plate, whereas the width of the gas metal arc bead is about 1¼ the thickness of the plate when the weld is shielded with carbon dioxide.

The GMAW process is considered a low hydrogen process. The welding gases used have a very low dew point and the wires (solid or metal cored) offer little opportunity for moisture pickup. The deposited weld metal should be free of hydrogen. This reduces underbead and microcracking on low alloy steels and thick weldments. Underbead cracks usually occur in base metal just below the weld metal and are caused by hydrogen absorption from the arc atmosphere. Low hydrogen processes like MIG/MAG produce sound welds on troublesome steels such as high carbon and low alloy grades.

Forms of Metal Transfer

In gas metal arc welding, an electric arc is established between the metal being welded and a consumable wire electrode that is fed continuously through the gun at a controlled constant speed. At the same time a shielding gas is fed through the gun into the weld zone to protect the molten weld pool. GMAW is a process that can be applied semiautomatically, mechanically, automatically, or robotically. Welding current and wire-feed speed are electrically interlocked so that the welding arc is self-correcting.

The welder first sets the wire-feed speed (WFS) to provide the correct amount of current and weld speed for the job at hand. Once the WFS is set, the welder adjusts the voltage to produce the correct arc length. During welding, if the welder holds the gun too close to the work, the current automatically increases, and the wire is burned off faster than it is fed until the correct arc length is reestablished. If the welder raises the gun too high over the work, the current automatically decreases and the burnoff rate slows down allowing the constant feed rate to shorten the arc to the correct length. Varying the electrode extension as just described is for short term transient situations and should not be done long term. Electrode extension is an important variable, and the length of electrode extension must be known and maintained. Incomplete penetration, incomplete fusion, and porosity are likely to result if the electrode extension is not kept in control.

In GMAW, filler metal is transferred directly through the welding arc. There are two basic types of metal transfer. In the **open arc method** the molten metal is separated from the welding electrode, moved across the arc gap, and deposited as weld metal in the joint. In the **short circuit method** the weld metal is deposited by direct contact of the welding electrode with the base metal.

Open Arc Transfer There are several types of open arc transfer methods: *spray transfer, rational spray and non-rational spray, globular transfer, pulsed-spray transfer,* and *buried arc transfer.* The particular type of metal transfer depends upon the electrode wire size, the shielding gas, the welding current, and the arc voltage.

Spray Transfer Spray transfer is a high heat method with rapid deposition of weld metal. **Spray** describes the way in which the molten metal is transferred to the work. Very fine droplets of electrode metal are transferred rapidly through the arc from the electrode to the workpiece. The droplets are equal to or smaller than the diameter of the filler wire. There is almost a constant spray of metal, Fig. 21-8.

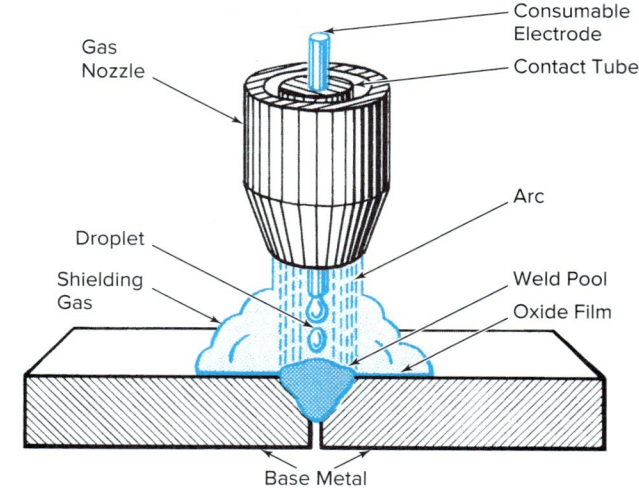

Fig. 21-8 Spray transfer.

Table 21-1 Globular to Spray Transition Currents for Various Electrodes

Electrode Type	Electrode Diameter (in.)	Shielding Gas	Spray Arc Transition Current (Amps)
Low carbon steel	0.023	98% argon + 2% O_2	135
	0.030	98% argon + 2% O_2	150
	0.035	98% argon + 2% O_2	165
	0.045	98% argon + 2% O_2	220
	0.062	98% argon + 2% O_2	275
	0.035	95% argon + 5% O_2	155
	0.045	95% argon + 5% O_2	200
	0.062	95% argon + 5% O_2	265
	0.035	92% argon + 8% CO_2	175
	0.045	92% argon + 8% CO_2	225
	0.062	92% argon + 8% CO_2	290
	0.035	85% argon + 15% CO_2	180
	0.045	85% argon + 15% CO_2	240
	0.062	85% argon + 15% CO_2	295
	0.035	80% argon + 20% CO_2	195
	0.045	80% argon + 20% CO_2	255
	0.062	80% argon + 20% CO_2	345
Stainless steel	0.035	99% argon + 1% O_2	150
	0.045	99% argon + 1% O_2	195
	0.062	99% argon + 1% O_2	265
	0.035	Argon + helium + CO_2	160
	0.045	Argon + helium + CO_2	205
	0.062	Argon + helium + CO_2	280
	0.035	Argon + H_2 + CO_2	145
	0.045	Argon + H_2 + CO_2	185
	0.062	Argon + H_2 + CO_2	255
Aluminum	0.030	Argon	95
	0.047	Argon	130
	0.062	Argon	180
Deoxidized copper	0.035	Argon	180
	0.045	Argon	210
	0.062	Argon	310
Silicon bronze	0.035	Argon	165
	0.045	Argon	205
	0.062	Argon	270

Spray transfer is used with inert gas shields and, mostly with direct current, electrode positive polarity. The spray arc is almost spatter-free, provides deep weld penetration, and has self-regulating characteristics. Filler wire diameters for spray arc transfer are generally between 0.030 and 3/32 inch in diameter. For each wire diameter, there is a certain minimum welding current that must be exceeded to achieve spray transfer, Table 21-1.

Because of its high deposition rate, spray transfer is recommended for materials that are 1/8 inch or thicker, stock requiring heavy, single or multipass welds, and for any filler pass application where speed is advantageous. Because of high arc stability, the high rate of metal transfer, and the axial nature of the spray transfer, it can be directed by the welder. Thus, it is suitable for welding in the vertical and overhead positions. This is true only on certain metal like aluminum and magnesium. On steel and stainless steel, the weld pools are too fluid to be used effectively for anything other than flat or horizontal position welding.

For video of GMAW spray transfer, please visit www.mhhe.com/welding.

Rotational Spray and Nonrotational Spray These high current density spray transfers get their name from specific arc characteristics brought about by a combination of very high wire-feed speed, long electrode extensions, and specific shielding gases. Filler metal deposition rates can be as high as 40 pounds per hour. The rotational spray is created by use of extra-long electrode extensions. This produces resistance heating of the wire electrode and causes the electrode end to become molten. The electromechanical forces generated by the current flow in the wire cause the molten wire end to rotate in a helical path, Fig. 21-9. The shielding gas affects the rotational transition current by changing the surface tension of the molten electrode end. Gases such as argon-carbon dioxide-oxygen or argon-oxygen at wire-feed speeds as high as 1,500 inches per minute, using a shielding gas with increased thermal conductivity and increasing the surface tension on the molten end of the electrode, can suppress the rotational nature of this transfer mode. Using shielding gases higher in carbon dioxide and/or helium does this. This will raise the rotational spray transition currents and will suppress the tendency to rotate. This nonrotational arc will look more like conventional axial spray. The arc is narrower than that produced by the rotational spray. This more concentrated arc can increase the depth of penetration as compared to the rotational mode at the same current level.

Globular Transfer In **globular transfer**, the molten ball at the end of the electrode can grow in size until its diameter reaches two or three times the diameter of the electrode before it separates from the electrode and is transferred across the arc, Fig. 21-10. The physical forces react on the globule of metal and cause it to be highly unstable. Shielding gas composition has the most influence with this transfer mode. Anything less than approximately 80 percent argon will produce a globular transfer. There is a great deal of spatter with the globular method. The globular method of transfer is used with various power levels and carbon dioxide shielding gas. The mechanical properties of the deposited weld metal may be reduced due to this shielding gas and its high arc voltage operation. Gravity imports most of the force acting on the transfer of the globule, unlike the axial spray which uses various electromagnetic forces to nick down the wire and force the small droplets to transfer directionally (axially) along the center line of the electrode wire.

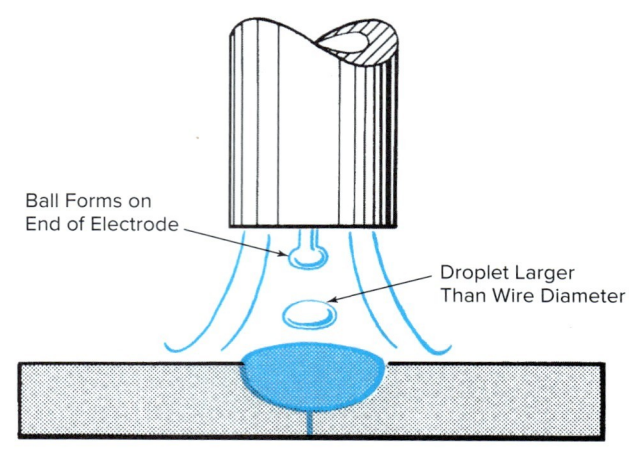

Fig. 21-10 Globular transfer. Source: Miller Electric Mfg. Co.

For video of GMAW globular transfer, please visit www.mhhe.com/welding.

Pulsed-Arc Transfer Pulsed-arc transfer is known as GMAW-P. It is achieved by pulsing the current back and forth between two levels. It permits the use of spray transfer at much lower current levels than usual. It provides for the spray transfer to take place in evenly spaced pulses rather than continuously. The metal transfer from

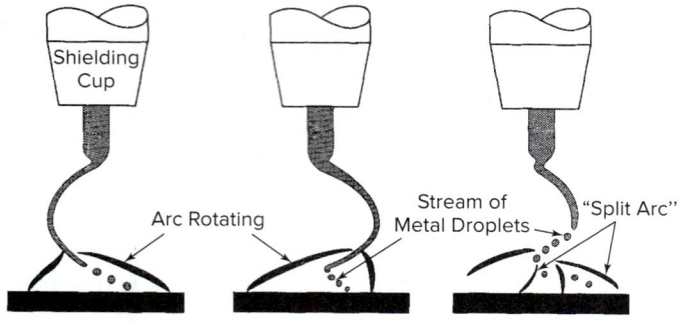

Fig. 21-9 Rotational spray transfer due to high current density operation with electrode extensions from ⅞ to 1½ inches. Deposition rates from 18 to 30 pounds per hour are capable.

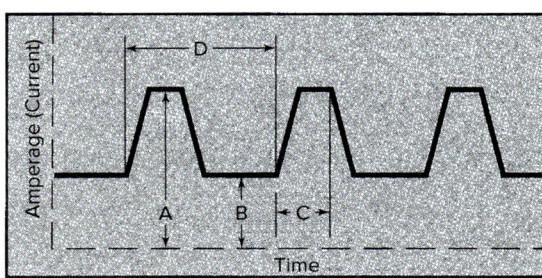

Fig. 21-11 Output current waveform of the gas metal arc welding pulse mode of transfer. (GMAW-P): A = peak current, B = background current, C = pulse width (time at peak), D = pulse frequency (pulses per second). Source: Miller Electric Mfg. Co.

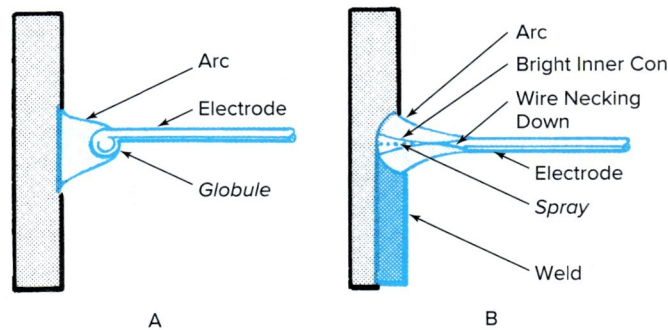

Fig. 21-12 An arc transferring metal by the (A) globular and (B) continuous and pulsed-spray modes is applied in the vertical welding position.

the electrode tip to the base metal occurs at regular intervals. The pulsing current has its peak in the spray transfer current range and its minimum value well below the transition current range.

Two levels of current are employed to achieve the pulsed level of transfer, Fig. 21-11. Both levels of current come from a single power source that is typically of the inverter power source design. A standard single or three-phase power unit (see Fig. 21-30, p. 673) supplies a steady background current that is just enough to maintain the arc between pulses while heating the filler wire and weld joint. This pulsed-current level should be set just above the minimum required for spray transfer. These pulse parameters are no longer set by the welder one parameter at a time. With the software-driven technology today, more accurate pulsing is available. The welder simply sets wire type (aluminum, stainless steel, carbon steel, silicon bronze, etc.), wire diameter, shielding gas, wire-feed speed, and arc length. Once these have been set, the welder has only two knobs to control: heat (WFS) and longer or shorter arc. This new software can sample, monitor, and control the process hundreds of times faster than just a few years ago with the older style synergic controls. This allows more precise control of the arc length (shorter arcs) and weld pool to fill in at weld toes, reducing undercut tendency while improving weld profiles.

During the pulsed-current cycle, the weld metal is sprayed smoothly across the arc and deposited in the weld joint. This type of transfer differs from the normal spray transfer in that the molten drops of filler metal are separated from the tip of the electrode at a regular frequency, corresponding to the frequency of current pulses supplied from the power source.

Pulsed-arc transfer permits use of the spray transfer of filler metal at lower current levels than usual. This extends the advantages of the GMAW process to all-position welding and to lighter gauges of metal than previously possible. It is effective in the welding of thin materials where burn-through is a problem due to high heat. It is also used for out-of-position work where it is difficult to maintain the molten pool of weld metal, Fig. 21-12. Larger diameter wires can be used at lower heat input. This is especially helpful when welding aluminum. All the advantages of open arc transfer are possible at average current levels, from the minimum possible for continuous spray transfer down to current values low in the globular transfer range. The method produces essentially spatter-free metal transfer.

For video of GMAW pulse transfer, please visit www.mhhe.com/welding.

The Buried Arc Transfer The buried arc method of transfer is a process in which the metal transfer occurs below the surface of the base metal, Fig. 21-13. Relatively high current and voltage are necessary. It is not unusual for a filler wire 0.045 inch in diameter to be run at 400 to 425 amperes at 35 to 37 volts. As a result of the high current values, the force of the arc digs a crater into the material being welded. The crater acts as a crucible for the weld metal and reduces spatter. This method of transfer is used with carbon dioxide as the shielding gas to weld

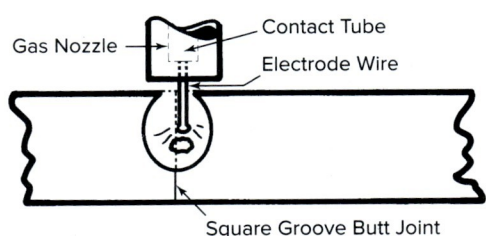

Fig. 21-13 Buried arc mode of transfer.

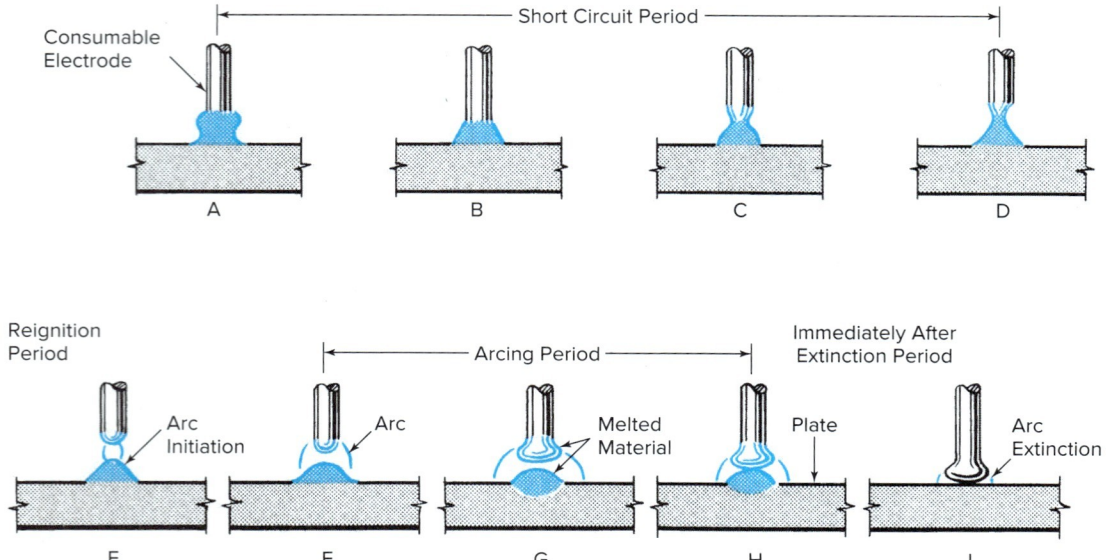

Fig. 21-14 Complete short circuit cycle.

mild steel. The weld profile may be objectionable. Grinding may be required.

Short Circuit Transfer (GMAW-S) In the early development of the gas metal arc process, spray transfer was used almost exclusively. Welding with a short-circuiting arc was developed years later. No metal is transferred across the welding arc. Metal transfer takes place only when the electrode makes contact with the material being welded. Metal deposited in this manner is less fluid and less penetrating than that formed with the spray arc.

At the start of the short arc cycle, the high temperature electric arc melts the filler wire into a droplet of liquid metal. The electrode wire is fed at such a high rate that the molten tip of the filler wire contacts the workpiece before the droplet can separate from the electrode. The contact with the workpiece causes a short circuit, the arc is extinguished, and metal transfer begins due to arc forces and surface tension, Fig. 21-14. The process is actually a series of periodic short circuits that occur as the molten droplet of metal contacts the workpiece and momentarily extinguishes the arc. **Pinch force** is a squeezing action common to all current carriers due to the magnetic field that forms around them, breaks the molten metal bridge at the tip of the electrode, and a small drop of molten metal transfers to the weld pool. Electrical contact is broken, causing the arc to reignite, and the short arc cycle begins again.

Short circuit transfer employs low currents, low wire-feed speed, low voltages, and small diameter filler wires.

Currents range from 50 to 225 amperes; and voltages from 12 to 22 volts. Filler wires with diameters of 0.030, 0.035, and 0.045 inch are used. Shorting occurs at a steady rate of 20 to over 200 times a second according to preset conditions. The faster the wire-feed speed, the more short circuits per second. It can also be used for MIG brazing GMAW-B by changing the electrode to the proper type, Fig. 21-15.

The low heat input of this technique minimizes distortion, burn-through, and spatter. It is particularly useful for welding thin gauge materials in all positions and running open root joints. Short circuit transfer is finding increased use in the welding of heavy thicknesses of

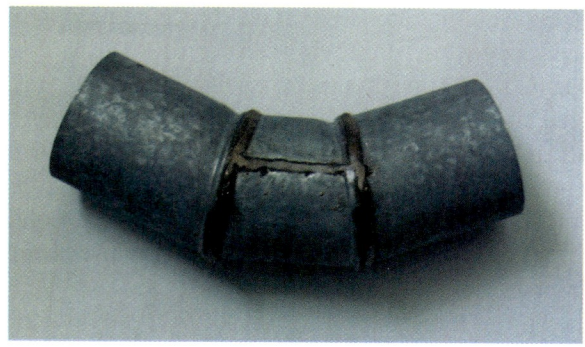

Fig. 21-15 Silicon bronze applied to a 22-gauge galvanized steel pipe elbow using the GMAW-B process. © Wisconsin Wire Work

664 Chapter 21 Gas Metal Arc and Flux Cored Arc Welding Principles quenched and tempered steels. However, other modes of transfer and other processes like FCAW are generally preferred for welding heavy plate. On materials greater than 3/16-inch thick incomplete fusion can be an issue with short circuit transfer. It is used for the welding of carbon steels, low alloy steels, stainless steels, and light gauge metals. When there is relatively poor fitup on a job, the short circuit transfer method permits the welder to bridge the wide gaps. It is very useful on open root joints.

A full range of short circuit transfer cannot be obtained with constant current power source units. Constant voltage machines are available with adjustable slope and appropriate voltage and inductance controls that produce the specific current surges that are needed for short circuit metal transfer over a full range.

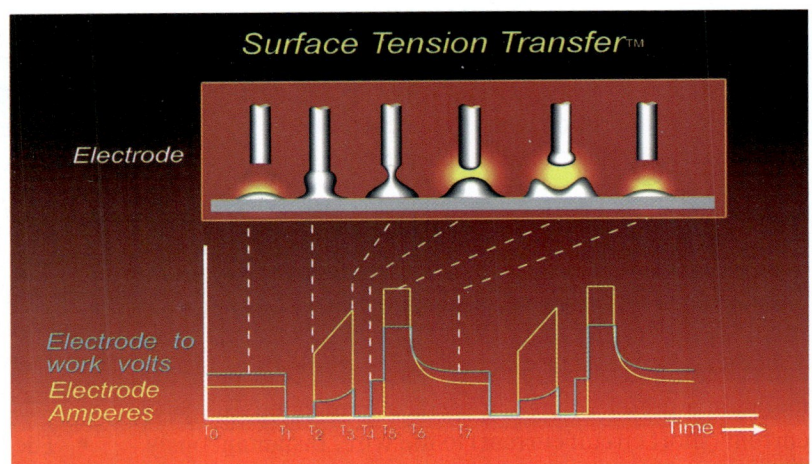

Fig. 21-16 Surface tension mode of transfer with waveform required from unique power The Lincoln Electric Co.

For video of GMAW short-circuiting transfer, please visit www.mhhe.com/welding.

All shielding gases may be used with the short circuit transfer method. Pure argon and helium and their mixtures are used with thin aluminum. Carbon dioxide or a mixture of 25 percent carbon dioxide and 75 percent argon may be used in the welding of carbon steel. There is a growing use of helium-argon-carbon dioxide mixtures for the welding of stainless steels.

Enhanced Short Circuit Transfer A modification of the short circuit mode of transfer has developed over the last decade. Before electronic controls and inverter power sources became available, the short circuit cycle was difficult to control and maintain easily. Use of slope, voltage, inductance, and wire-feed speed were the choices. Now with the capability to monitor and control amps and volts up to 10,000 times per second, control and consistency are available. This enhanced short circuit mode is known by the trade names of the manufacturer. The Lincoln Electric Co. refers to it as Surface Tension Transfer (STT), while Miller Electric Mfg. Co. refers to it as Regulated Metal Deposition (RMD). Other manufacturers of course have their own trade names. They all require a unique power source capable of outputting a very controlled and unique waveform. The power source is neither constant current nor constant voltage but requires amperes and volts to change throughout the entire transfer cycle. The requirements of the arc are constantly being monitored through a sense lead that feeds data to the electronic circuitry, Fig. 21-16.

GMAW/FCAW Welding Equipment

The success of the process depends upon the proper design and the matching of the various components. (Refer to Fig. 21-5, p. 659.) The basic equipment includes the following:

- Welding machine or power source. This may be an engine-driven generator, transformer-rectifier, or inverter with constant voltage output. (Constant voltage is also known as *constant potential*.)
- Consumable electrode wire-feed unit, including the controls.
- Welding gun. A source of cooling water and flow controls are necessary for water-cooled guns.
- Interconnecting hose and cable assemblies leading to the gun. They provide for the gas, water, wire, and cable control wire needed for the process.
- Cylinder of shielding gas, cylinder regulator, and flowmeter.
- Work lead and work clamp.
- A reel of consumable electrode wire of the proper type and size for the particular welding job being performed.

Constant Voltage Characteristics

It is important that the student understand how the constant voltage system functions.

In GMAW, as in all arc welding, the heat required is generated by the arc that is produced between the workpiece and the end of the electrode. To sustain this arc for GMAW/FCAW, control of five items is necessary:

- The proper arc voltage on the wire
- Current going through the wire

- Appropriate wire-feed speed to replace continuously the wire melted by the heat of the arc to form the weld deposit
- Burnoff rate proportional to the current
- Arc action proportional to the arc voltage

Keeping these five factors in mind, let us see what happens when the welding operation is started. The welding power source is turned on, and when the electrode wire touches the properly connected workpiece, current flows in the closed circuit. For an instant, the arc voltage is zero, and because the power source is designed to maintain a preset voltage, it sends a tremendous surge of current (short circuit current) to the wire. Instantaneously, the wire gets white hot. At the same moment, tremendous magnetic fields set up around the wire. The molten wire along with the pinching or stripping action of the magnetic fields deposits a portion of the electrode wire onto the base metal and the arc is established. The nature of the arc is determined by the voltage setting set by the welder on the power source. These flawless, instantaneous arc starts are a characteristic of a constant voltage (potential) machine.

Once the arc is started, it is self-regulating over a wide range. The current is capable of wide variations while the arc voltage remains constant. Thus, the wire is fed at a constant speed. The welding current automatically adjusts itself to maintain the constant physical arc length and arc voltage. The power source supplies enough current to burn the wire off as fast as it is fed and to maintain the proper arc dictated by the voltage adjustment.

The constant voltage machine is flexible in meeting amperage and voltage demands over a wide area. Two students are urged to try this experiment: Using any size welding electrode wire, set the proper wire-feed speed and voltage for welding. One student should weld while the other observes the machine's amperage and voltage meters. When the student welding begins, the other will observe that amperage and voltage readings hold steady as long as the electrode extension remains about the same. The student welder should then lengthen the electrode extension without halting the welding operation and breaking the arc. The observer will note that as the electrode extension is lengthened, the amperage reading is reduced, but the voltage reading remains constant. It should also be noted that the wire-feed speed does not change. Thus it can be seen that voltage stays constant as set and that the machine makes internal changes in amperage.

Thus constant voltage power sources ensure uniform welds due to three inherent characteristics:

- The flow of welding current to the work is automatically adjusted to the rate at which the electrode wire is fed to the work.
- The wire-feed speed is constant.
- Instant arc starting eliminates defective spots at the start of the weld since the wire is practically vaporized by the instant current surge. Instant arc starting prevents the wire from sticking to the work, while the quick recovery to normal current prevents burn-back of the wire.

Output Slope

As stated previously, the power source is referred to as *constant voltage* or *constant potential* because the voltage at the output terminals varies relatively little over wide ranges of current.

The volt-ampere curve of the constant current power source, generally used for stick electrode welding, has a drooping output slope, Fig. 21-17A. When the arc length is constant, the welding current that will be obtained is shown by the point at which the voltage reading intersects the output slope. In shielded metal arc welding, the welder is not able to maintain a constant arc length and, therefore, the voltage may deviate from the normal value. The deviation may be high or low, depending on arc manipulation. Figure 21-18 shows that a rather large increase or

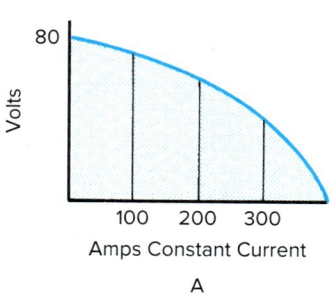

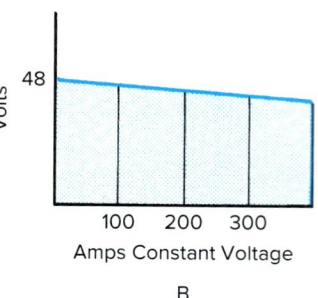

Fig. 21-17 Comparable volt-ampere curves.

JOB TIP

Job Interview

So you're going to a job interview! That's the opportunity to show how professional you will be at work. Leave your pager and cell phone at home. Be on time for the interview. Shake hands firmly. Don't be afraid to ask a question. After you get home, write a thank-you letter or e-mail right away.

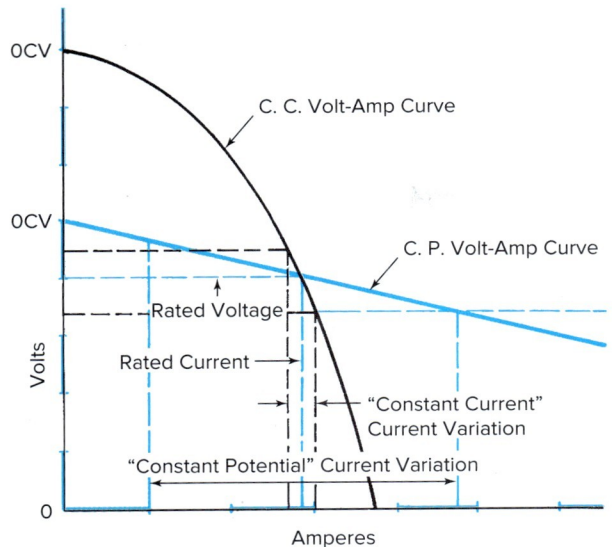

Fig. 21-18 Comparison of constant current and constant voltage characteristic volt-ampere curves.

decrease in welding voltage, due to change in arc length when welding, does not result in a large change in current output. Welders can control the arc length by varying the speed at which they feed the electrode into the welding arc. The constant current power source changes its voltage output in order to maintain a constant welding current level.

The volt-ampere curve of the constant voltage power source used for GMAW has a relatively flat output slope, Fig. 21-17B. Small changes in arc voltage result in relatively large changes in welding current. Figure 21-18 shows that when arc length shortens slightly, a large increase in welding current occurs. This increases the burnoff rate, which brings the arc length back to the desired level.

The constant voltage power source puts out enough current so that the burnoff rate equals the wire-feed rate. If the wire-feed rate is increased, the power source puts out more welding current so that the burnoff rate and the wire-feed rate are again balanced. The arc length is controlled by setting the voltage level on the welding power source, and the welding current is controlled by adjusting the wire-feed speed. Arc voltage is set by the voltage control knob.

Short circuit current can be as high as several thousand amperes, usually six to eight times the rated current capacity.

Types of Controls

All constant voltage machines have voltage control. Open circuit voltage on these machines is lower than on constant current machines. The maximum is usually about 50 volts.

Once the voltage has been set, the machine automatically maintains an arc length while the electrode wire is fed continuously at a set speed. Changes in filler metal, shielding gas, or granular flux may require a voltage adjustment.

Self-regulation of arc length is possible because small changes in voltage cause the machine to produce large changes in welding current. Thus any change in arc length (voltage) results in a large change in current and a corresponding, almost instantaneous change, in the burnoff rate of the electrode wire. This regulates the current at the arc.

Constant voltage machines do not have a heat adjustment like the constant current machines. They do have a voltage control and slope and inductance adjustments for arc stabilization. A small transformer is included to supply single-phase power for operating auxiliary equipment such as the wire drive motor, coolant pump, and gas and water solenoid valves. A built-in contactor controls the flow of the welding current and voltage to the electrode when the trigger on the gun is pulled.

Constant voltage power supplies provide extreme flexibility. With variable slope, voltage, and inductance, each unit can be adjusted, within its current and voltage range, to provide ideal arc characteristics for any specific application. A single machine, typically inverters, can be used for short-circuiting applications, spray arc, TIG, submerged arc, flux cored, stick electrode, and any other welding process within its range of operation. These machines combine the two basic types of power supplies in one single unit. By turning a selector switch on the machine, a steep slope (constant current) or a flat slope (constant voltage) can be produced.

Voltage Controls Voltage controls may be either tapped (stepped) or continuously variable. *Tapped-voltage control* is used for simple types of GMAW. A voltage range tap switch is employed. The control panel shown in Fig. 21-19, page 668, is of this type.

This type of equipment is popular with the home hobbyist and autobody repair type welder. Figure 21-20, page 668, shows a continuous type of control for voltage.

- *Voltmeter and ammeter set* These meters permit accurate, repeatable selection of weld current and voltage.
- *Voltage range switch* This switch offers the operator a choice of three open circuit voltage ranges: high (39–51), medium (30–39), and low (24–31). Wire size and work will determine which voltage range is required.

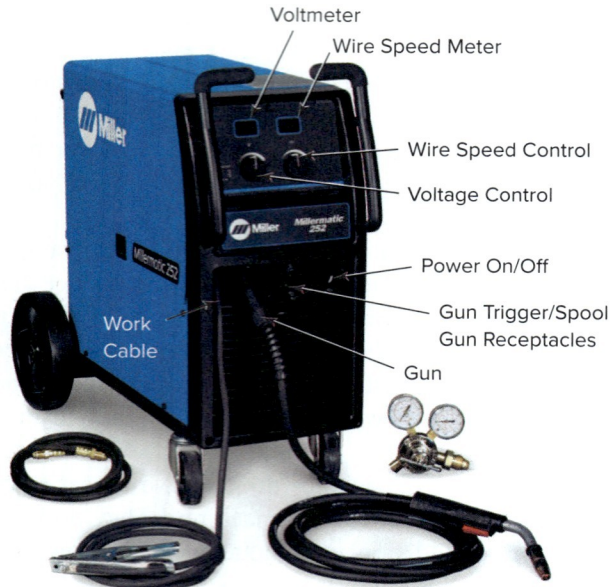

Fig. 21-19 Control panel of a typical 250 ampere constant current power source, wire-feeder combination with adjustable voltage and wire-feed speed control. With digital read out of voltage and wire-feed speed. Gas metal and flux cored arc capable. As well as aluminum welding with push pull or spool gun. Miller Electric Mfg. Co.

Fig. 21-20 Note power source digital meters on this 650-ampere industrial constant voltage machine. Miller Electric Mfg. Co.

- *Fine voltage adjustment* The hand control on the panel of the welder provides infinite fine voltage adjustment within the range selected. Fine voltage changes can be made *while welding*. This permits the operator to obtain exact welding settings under welding load conditions.
- *Voltage indicator* This indicator shows open circuit voltage settings on the welding machine for the particular range selected.

- *115-volt a.c. power receptacle* This receptacle provides 115-volt power for operation of the wire drive equipment. Power for this circuit is supplied from a built-in control transformer.
- *Contactor control receptacle and switch* Provisions are made for a 2-conductor lead from the wire drive control to the contactor control receptacle. This completes the circuitry necessary for remote control of the welding power. The actuating mechanism for closing the circuit is the gun trigger switch. When it is closed, the primary contactor in the power source is energized, and welding power is available.
- *115-volt duplex receptacles* These receptacles provide power for water coolant systems, fixtures, and so on. The power for them comes from the same control transformer that supplies power for the wire drive equipment.
- *On/off power switch* This switch controls the electric power to the fan motor and the 115-volt power outlets.
- *Cooling fan* The fan is driven by a prelubricated, completely sealed ball bearing motor and provides adequate cooling to all components.
- *Current surge protector* This device guards the transformer and rectifier in the event of abnormal operation of welder.

Continuously variable voltage control is commonly used for more critical spray arc MIG welding, and short circuit MIG weldings, Figs. 21-21 and 21-22.

Current Controls There is no current adjustment dial on a constant voltage unit. Current is adjusted by the wire-feed speed setting on the wire drive unit.

Slope Controls Slope controls the amount of current change. It is caused by an impedance to the current flowing through the welding power circuit. (Impedance means the slowing down of a moving object.) Since voltage is the force that causes current to flow but does not flow itself, the impedance is directed toward limiting the flow of current. It is not intended that the current be stopped, only limited. As more impedance is added to the welding circuit, there is a steeper slope to the volt-ampere curve. The steeper slope limits the available short circuit current and slows the machine's rate of response to changing arc conditions. Constant voltage d.c. transformer-rectifiers are available with fixed slope or variable slope.

Fixed-slope machines are generally used for all types of GMAW. A certain amount of fixed slope is designed into the machine, and no adjustment is needed for most jobs, Figs. 21-23 and 21-24.

In the past, the term *slope* was used only in connection with the gas tungsten arc and the shielded metal arc

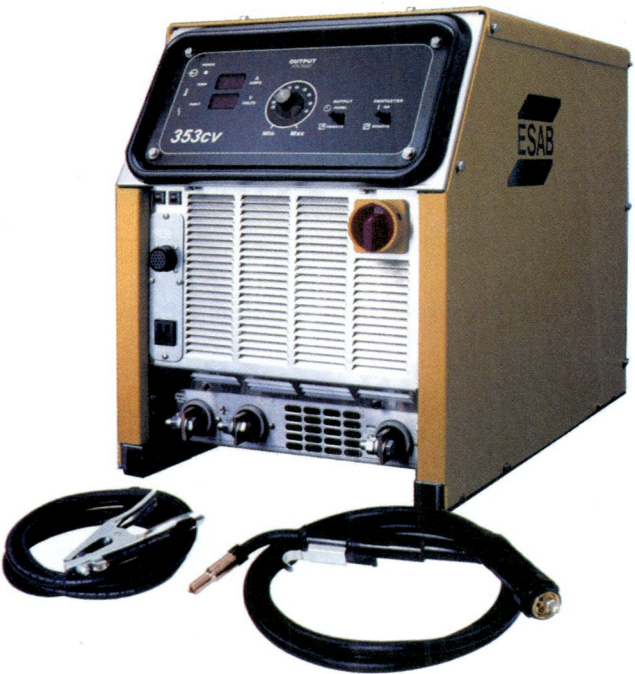

Fig. 21-21 A 350-ampere, d.c. constant voltage transformer rectifier with solid-state control for short circuit transfer MAG/MIG welding. The single knob control changes voltage in infinite units for precise control over the arc. It is capable of running continuous spray and flux cored arcs up to its amperage capability. © ESAB

Fig. 21-23 A 450-ampere, d.c. constant voltage/constant current inverter power source. Capable of running all modes of gas metal arc transfers including pulse. A variable inductance control provides additional arc performance. Miller Electric Mfg. Co.

Fig. 21-22 A 300-ampere d.c. constant voltage transformer-rectifier with solid-state control and built-in predetermined slope control. The Lincoln Electric Co.

Fig. 21-24 A 300-ampere, d.c. constant voltage transformer-rectifier with mechanical brush control for voltage and tapped-inductance control. Miller Electric Mfg. Co.

processes. An increase of welding current was referred to as *upslope;* and a decrease of current, as *downslope.* The operating characteristics and versatility of the constant voltage power sources used for gas metal arc welding have been accomplished through the introduction of variable slope control.

Variable slope machines permit the use of a wider range of wire types and sizes under a variety of GMAW conditions. By changing the slope of the flat volt-ampere curve, the short-circuiting transfer method of welding has been improved. The sudden surge that takes place when the electrode makes contact in starting the weld is decreased, and the weld pool can be kept more fluid. Adjustment of the slope from flat to steep permits precise control of short circuit current. Slope adjustments also reduce weld spatter. Variable slope machines are used for the welding of stainless steel.

Variable slope controls may be tapped or continuously variable, and they are similar to those controlling voltage. Tapped-slope machines are more versatile, but continuously variable slope controls are more precise. Figure 21-25 indicates the output curves possible with a machine having seven tapped-slope settings. Figure 21-26 illustrates the output curves possible with a machine equipped with continuously variable control.

The reactor used for variable slope control consists of one or more current-carrying coils placed around an iron core. In tapped-current control, the welder adjusts the slope by means of the mechanical connections through taps that are connected with the coil. In continuously variable control, contact brushes move over the face of one side of the coil that has been machined or the physical location of the coils is changed.

Inductance Controls Inductance controls the rate of current change. This will affect arc starting. Adjustment of inductance is most common with short-arc GMAW to control weld spatter and how fluid the weld pool is. *Fixed inductance* is common on many constant voltage machines, and no further adjustment is normally required.

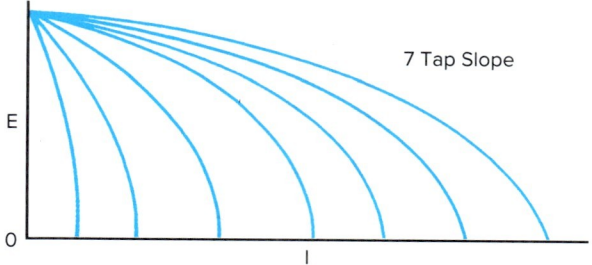

Fig. 21-25 Tapped slope (250-ampere power source).

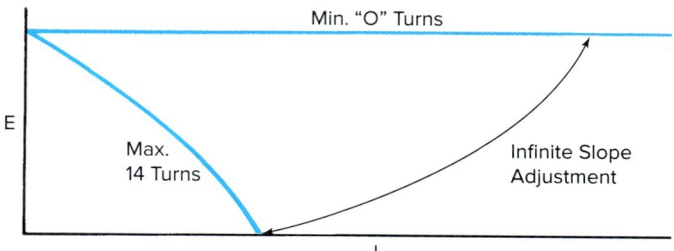

Fig. 21-26 Variable slope.

Variable inductance may be tapped or continuously variable. Controls are similar to those used for voltage and slope.

Variable voltage and inductance controls give precise control of the weld pool, regulate the frequency of short circuiting in short circuit arc transfer, help flatten the weld profile, and reduce spatter. See Table 21-2.

Inductance can best be understood if you consider the constant voltage volt-ampere curve. Visualize what must happen on the volt-ampere curve when the arc starts. The welding machine must be capable of responding to

Table 21-2 Effects of Inductance on GMAW and FCAW

Inductance Setting	Effect	Result
0 or minimum setting	Maximum current rise in both speed and amount of energy available	Blasting high energy starts may be required with large electrode wire and/or high wire-feed speed rates. Small diameter wire and/or low wire-feed speed may cause burn-back to contact tube and excess spatter.
50% or setting in middle of available range	Medium current rise in both speed and amount of energy available	Can be adjusted to give a controlled start. Too much energy too rapidly can cause a blasting start, but too little energy would create a sluggish start. Makes the weld pool more fluid principally with the short circuit transfer. Especially effective when welding stainless steel. Helpful when welding out of position. Helpful to control weld spatter when all other parameters are correct.
100% or maximum setting	Slowest current rise in both speed and amount of energy available	Will cause very sluggish starts. Wire may stub into workpiece without sufficient energy to burn free and start. Provides the most fluid weld pool. Arc will be sluggish and may not respond to manipulation, as for weave beads.

the tremendous demand for current to burn the electrode free, without overcurrenting and creating a blasting start and without undercurrenting and creating a stumbling start. Consideration must be given to the arc length and amperage changes required during the short circuit mode of transfer. Keep in mind the short circuiting mode of transfer can take place at nearly 200 times per second. As the energy moves along the volt-ampere curve, the speed with which it is allowed to move is controlled by the inductor control on the machine. Think about your ability to move through a room if the floor is dry. Then think how you would be slowed down if the room had several feet of water on the floor (50 percent inductance). If the room were full of water and you had to swim, think about how much slower you would be able to move through it (100 percent inductance). This analogy should aid in your understanding of this control.

Other Controls A weld timer most often can be added to the control panel of the standard wire-feeder controller. It permits GMAW spot welding and other timed welding.

Remote control of the wire feed and weld output contactor are generally provided through the wire-feeder controller and power source. However, for additional remote control over the process and parameters a microprocessor, such as that shown in Fig. 21-27, may be used. These type of controls must be matched to specific types of power sources, but they may be used with a variety of wire-feeder types, from standard bench to push-pull or spool-on-gun type feeder systems.

Dual schedule controls can be added to many wire-feeder control panels. Many of the digital microprocessor-based wire feeders have this feature as standard. On more conventional feeders, it is generally considered an option. The power source must be capable of being remote controlled for voltage. The mechanically controlled (brushes on secondary coil) power source, as shown in Fig. 21-24, page 669, is not capable of dual schedule operation. The dual schedule control must be matched with the type of power source being used. Also a special gun trigger or switch must be added to select the schedule of operation that is desired. Figure 21-28, page 672, provides the following dual schedule features:

- Hot starts give extra penetration.
- The control provides a low range for root passes and a high range for hot and cover passes.
- It is ideal for tack welding and can bridge extra-wide gaps.
- The welder can select one condition for overhead or vertical welding and switch to another preselected condition for flat welding.
- A change in the welding condition from one size of material to another can be made without readjustment.

The operator can change from Schedule A to Schedule B or vice versa with the gun trigger at any time during the course of the weld without interrupting the arc. A pilot light on the panel indicates which setting is being used.

Types of Constant Voltage Welding Machines

Constant voltage machines are classified by the manner in which they produce welding current. There are three general types:

- *Transformer-rectifiers* that have an electrical transformer to step down the a.c. line power voltage to a.c. welding voltage and a rectifier to convert the alternating current to direct current.
- *Inverters* that convert the incoming high voltage alternating current to direct current. This high voltage direct current is then switched up to 50 kilohertz and supplied to a transformer-rectifier assembly. These machines are light and portable. Many are capable of running on single- or three-phase primary power. Some are equipped to automatically adjust to whatever primary voltage with which they are supplied. Inverters can run all the various modes of metal transfer, and many are capable of running constant current for SMAW and GTAW processes.
- *Engine-driven generators* that have an internal combustion engine to drive a generator that produces direct current.

D.C. Transformer-Rectifiers Constant voltage transformer-rectifiers may be used for MIG/MAG, submerged arc, FCAW, and multiple-operator systems. They are three-phase power supplies with voltage control, slope, and/or inductance control. Selection depends on the type of welding process, the type and size of filler wire used, the end use of the weldment, cost, and welder preference.

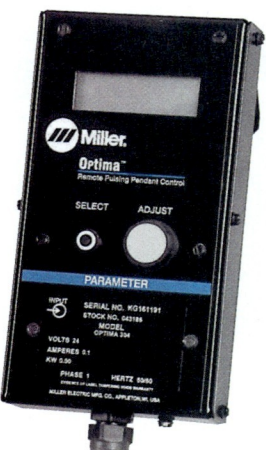

Fig. 21-27 The Optima® remote microprocessor control for standard GMAW, manual GMAW-P, and synergic GMAW-P. Miller Electric Mfg. Co.

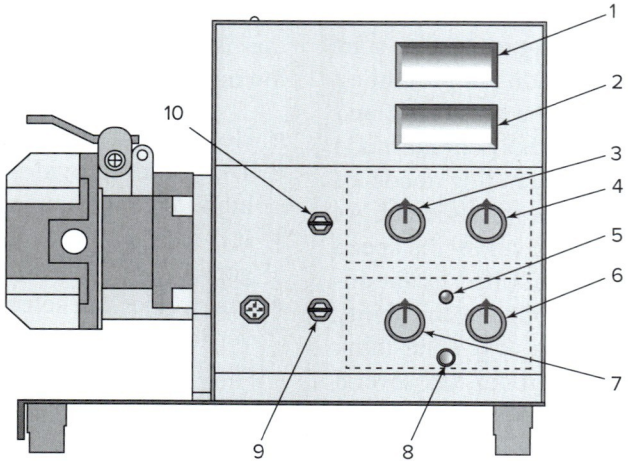

1. Voltmeter (Optional)
2. Wire Speed Meter (Optional)
3. Schedule A Wire Speed Control
The scale is calibrated in inches per minute × 100 and meters per minute.
4. Schedule A Voltage Control (Optional)
When a digital voltage control is used with an inverter-type welding power source, the control functions as a remote digital voltage control to preset arc voltage.

For dual schedule applications, a dual schedule switch is required for the gun. Obtain a proper dual schedule switch and install according to its instructions.
5. Schedule B Indicator Light
6. Schedule B Voltage Control
7. Schedule B Wire Speed Control
8. Press to Set Button
Press and hold button to preset Schedule B wire-feed speed and/or voltage.

9. Jog/Purge Switch
Push up to momentarily feed welding wire at speed set on wire speed control without energizing welding circuit or shielding gas valve.
Push down to momentarily energize gas valve to purge air from gun or adjust gas regulator.
Center position is off.
10. Trigger Hold Switch

Fig. 21-28 Dual schedule control for use with the Miller constant voltage d.c. power sources. It provides two individual preset voltage and amperage (wire-feed speed) welding settings with one machine and one wire feeder.

Figures 21-29 to 21-32 show the transformer-rectifiers for different MIG/MAG welding applications. The welder shown in Fig. 21-32 can be used for stick, MIG/MAG, and TIG welding. This is possible because of a controlled-slope characteristic. Controlled slope permits 15 separate output settings plus fine voltage adjustment control of all positions.

Inverters Advancements in high power electronics have allowed these types of power sources to evolve. The primary advantage of inverters is arc performance, plus their availability to run all the processes requiring either constant voltage (c.v.) or constant current (c.c.). In the constant voltage mode, slope is preset electronically while voltage and inductance are fully adjustable. With the flip of a switch they can be converted to constant current operation with control over amperage and arc force. Since these types of power sources use high speed electronic switching devices, they are capable of pulsing output for both GMAW and GTAW applications. Some units have total control over the output waveform for all modes of metal transfer such as short-circuiting and pulse spray.

Many inverters are capable of running on single- or three-phase input power. They are compact, lightweight,

Fig. 21-29 Shop portable GMAW welding outfit includes 350-ampere c.v. single or three phase input inverter power source. It is equipped with up to 35 feet, of 200 amp push-pull aluminum gun. Note four wheeled cart and single gas cylinder of argon being used to weld an aluminum trailer frame.
Miller Electric Mfg. Co.

and very portable. Some are designed to adjust themselves automatically to whatever primary voltage they are supplied (within a range). Figures 21-31 to 21-32 show various inverter-type machines for MIG/MAG applications.

Fig. 21-30 This 250-ampere, constant voltage solid-state controlled transformer-rectifier has an output rating of 7 to 32 volts and 30 to 300 amperes available in one continuous range. Note the quick twist connections for high and low inductance and the polarity output hook-up. GMAW and FCAW up to its output capability are the appropriate applications. The Lincoln Electric Co

Fig. 21-32 This 450-ampere d.c. inverter power source uses the digital microprocessor-based feeder for standard MIG synergic programming of the following wire sizes: 0.030, 0.035, 0.045, 0.052, and 1/16 in. The following wire types are also preprogrammed: steel, alternate steel, stainless steel, 4000 aluminum, 5000 aluminum, and silicon bronze. This power source is also multiprocess capable as it can output both c.v. and c.c. © ESAB

Fig. 21-31 A 450-ampere d.c. inverter power source. This machine has the capability to output c.v.-c.c., so it is multiprocess capable. It is designed specifically for the GMAW-P process for all position welding on thick and thin materials without spatter. It has a built-in microprocessor that allows both manual as well as synergic (adaptive or nonadaptive) control of the pulse output. It is also capable of running all modes of metal transfer as well as FCAW, GTAW, and SMAW processes. SAW and CAC-A up to its output capability are also recommended. Miller Electric Mfg. Co.

Engine-Driven Generators Constant voltage welding generators usually supply d.c. welding current. They are used for MIG/MAG, FCAW, SMAW, and GTAW.

Engine-driven welding generators are used in the field and for emergency work when electric line power is not available. The engine is usually a gasoline or diesel type. Fuel is gasoline, diesel oil, or propane gas.

These generators differ from constant current generators in their controls. Constant voltage generators have controls for voltage as well as current. Controls for slope and inductance are also available. Controls may be continuous or tapped. Figure 21-33, page 674, is an engine-driven generator capable of c.v.-c.c. output.

Machine Selection

Constant voltage machines are used with the GMAW, MCAW, FCAW, and SAW processes. They are better

Fig. 21-33 Engine-drives of this type generator are available up to 800-amps. For industrial auxiliary strength and weld power in the field, it has c.v.-c.c. versatility. For stick welding, it also provides semiautomatic welding with large diameter, high deposition flux cored wires. The welder is using a suitcase type feed, with FCAW-S electrode wire, and trailer for mobility. *Miller Electric Mfg. Co.*

suited for welding processes in which the consumable electrode wire is fed continuously into the arc at a fixed rate of speed because they permit self-regulation of arc length.

GMAW power source units for both manual and automatic operation are almost always d.c. constant voltage machines. Alternating current has been found to be generally unsuitable because the burnoff rate is erratic. A few a.c. applications are possible but very limited.

Most MIG/MAG welding is done with DCEP which gives better melting control, deeper penetration, and good cleaning action. This is especially important in welding such oxide-forming metals as aluminum and magnesium. DCEN is seldom used. The wire burnoff rate is greater with DCEN, and the arc instability and spatter are problems for most applications.

The selection of the proper machine size for a particular job is always a problem. It should be pointed out, however, that each machine has a wide range of application.

Spray arc welding generally requires from 100 to 400 amperes for semiautomatic applications. Automatic installations may range from 50 to 600 amperes.

Short-circuiting process gas requires only 25 to 250 amperes.

Flux cored wire utilizes from less than 100 up to 500 amperes in applications without shielding gas. With carbon dioxide shielding gas in automatic setups, the current ranges from 100 to 750 amperes.

Automatic submerged arc welding requires from 75 to 1,000 amperes.

As can be seen, the range of current for these constant voltage machines for the continuous electrode process is very wide, from 25 to over 1,000 amperes. However, amperage is not the only issue; the machines must be capable of providing these amperages at specific voltages. Some of the transformer-rectifiers and engine-driven generators that are designed for high output applications (over 450 amperes) may have difficulty running a 0.030-inch diameter wire in the short circuit mode of transfer at 16 volts and 60 amperes. The output volt-amp curve should be consulted to see if both low end and high end performance is available from the machine.

Wire-Feeding Unit and Control

The wire-feeding unit, also called a *wire feeder* or *wire drive*, drives the electrode wire from the coil automatically, Figs. 21-34 and 21-35. It feeds the wire through the cable assembly to the gun, arc, and weld pool. Since a constant rate of wire feed is required, the wire feeder is adjustable to provide for different welding currents. Wire feeders are rated for their minimum and maximum wire-feed speed and the current-carrying capacity of their gun power pickup.

Fig. 21-34 A typical bench style wire feeder that is designed for constant wire-feed speed. This allows it to operate all the GMAW modes of transfer as well as FCAW applications. Input power required from the power source to operate the feeder is 24 V a.c., 7 amperes, 50/60 hertz. The feeder is designed to handle electrode wire diameters of 0.023 to 5/64 inch at wire-feed speeds from 75 to 750 inches per minute with a maximum spool capacity of 60 pounds. Note the simple controls for wire-feed speed, trigger hold, wire jog, and shielding gas purge. *Miller Electric Mfg. Co.*

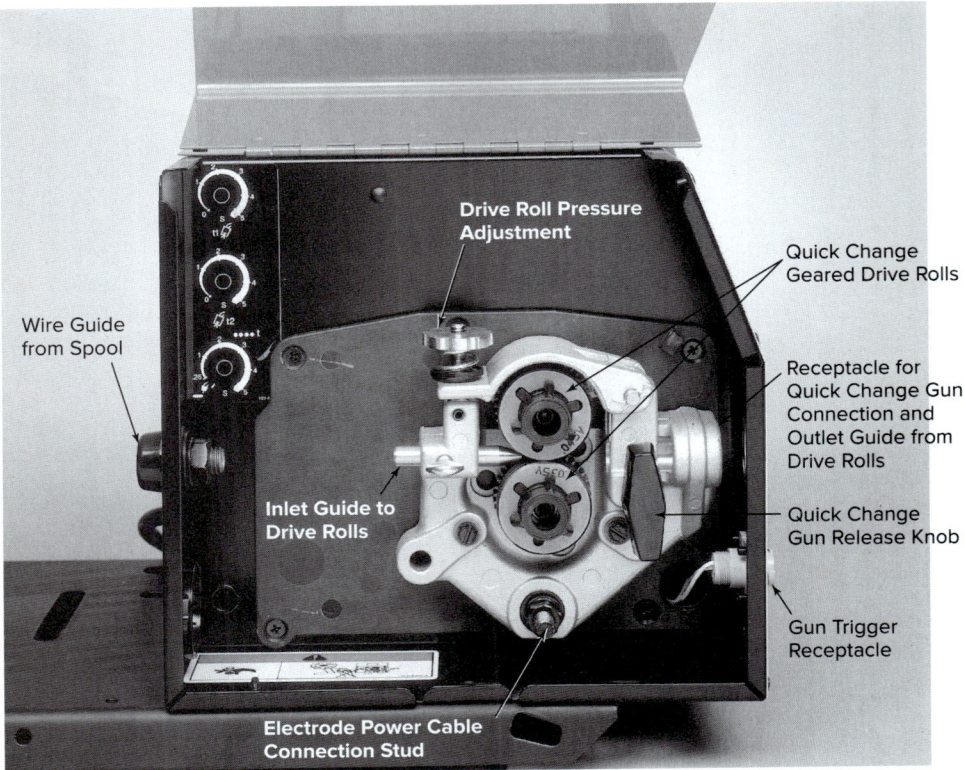

Fig. 21-35 The internal construction of the wire feeder in Fig. 21-34. The relationship of the wire guides to the wire drive rolls is essential for good feeding. There should be as little unsupported wire as possible. The guides should be of the appropriate size and be placed as close to the drive rolls as possible. Miller Electric Mfg. Co.

Usually, the machine is a one-piece unit. A control cabinet wire feeder and wire mount are attached to an all-welded frame. The wire feeder is mounted with the power source in a portable unit, Fig. 21-36. In some cases, the unit is separate from the power source and is mounted on an overhead crane to allow the welder to cover a greater area. For field or maintenance work, the unit is small and portable and may be a great distance from the power source, Fig. 21-37 (p. 676).

The portable feeder shown in Fig. 21-38, page 676, is of the voltage sensing variety. Most wire feeders are of the constant speed variety and work off of a fixed voltage, such as 115, 42, or 24 volts. They have electronic control circuits and drive motors designed to work off of these set voltages to deliver a very accurate constant wire-feed speed rate. Voltage sensing feeders are designed to work off of the open circuit voltage of the power source and while welding the load or arc voltage of the power source. They can be much more simplified and compact in design than constant speed feeders. The basic voltage sensing feeder works directly off of the voltage being supplied to it. A high voltage will cause the wire-feed

Fig. 21-36 This digital wire feeder allows much more rapid setup for a welding procedure and keeps the process in control by accurately controlling and displaying voltage and wire-feed speed. It operates from input power (24 V a.c., 10 amperes, 50/60 hertz) provided by the power source. It is capable of handling electrode wire diameters of 0.023 to ⅛ inch and wire-feed speeds of 50 to 780 inches per minute. Note the digital readout on voltage and wire-feed speed and the simple controls. Miller Electric Mfg. Co.

Fig. 21-37 This highly portable control-feeder weighs less than 25 pounds and fits through a 18x24 inch porthole. The small modular design and tough flame-retardant, nonmetallic case materials will withstand burning, breaking, or loss of rigidity with heat or sun. This is a voltage sensing feeder and so operates on the open circuit arc voltage of the power source and must be supplied 15 to 100 V d.c. The clamp and cable shown must be connected to the work piece. Electrode wire diameter capacity is 0.023 to 0.052 inch. Wire-feed speed is from 50 to 700 inches per minute depending on the arc voltage supplied. The maximum spool size on this feeder is 8 inches and 14 pounds. Digital display of voltage, wire feed speed, and also amperage if desired. Meters can be seen clearly even in direct sunlight. *Miller Electric Mfg. Co.*

Fig. 21-38 This voltage sensing feeder is powered across the arc by the welding current. As with all voltage sensing feeders no control cables are needed. To operate, simply connect the weld cable and attach the work clip, and the feeder is ready to weld. This portable unit is designed for MIG/MAG or flux cored wire diameters of 0.023 to 5/64 inch with a wire-feed speed range from 50 to 700 inches per minute. The totally enclosed case accepts 10- to 44-pound wire packages. The feeder dimensions are 16 × 15 × 31 inches with a net weight of 65 pounds. *The Lincoln Electric Co.*

motor to speed up, causing the wire to accelerate toward the weld pool. This will shorten the arc and lower the voltage. Since this feeder motor works directly off of the arc voltage, the motor will see less voltage and begin to slow down. As less wire is being fed toward the weld pool, the arc will get longer, the voltage will go up, and the wire-feed speed will increase accordingly. This simple form of voltage sensing feeder is being self-regulated off of the arc voltage and will quickly stabilize to a steady arc length. When the feeder is being operated in this voltage sensing mode, it is generally only used for GMAW spray arcs and FCAW, because the short-circuiting mode of transfer and the pulse mode of transfer place too many various arc load voltage conditions on the wire-feed motor to get the constant wire-feed speed that is required for these two modes of metal transfer. Many voltage sensing feeders have a switch that can place them into a constant wire-feed speed mode. They are still working on load (arc) voltage, but electronic circuits are used to filter this power and supply a relatively constant voltage to the drive motor. This allows such equipped voltage sensing feeders to run the short-circuiting and pulse modes of transfer.

Voltage sensing feeders also have the advantage of being usable off of the constant current type of welding machine. This type of feeder can compensate for the reduced short circuit current and slower response time of the constant current machines. Again, only solid and metal cored electrodes with spray arcs should be used. Flux cored electrodes can also be used as long as the electrode selected is not too voltage sensitive. This type of setup works very well with the voltage sensing spool gun and a constant current power source for GMAW of aluminum.

Table 21-3 Arc Length (Voltage) and Amperage Settings with Various Types of Wire Feeders and Power Sources

To Set Arc Length (Voltage), Adjust		To Set Amperage, Adjust
C.V. power source	when used in conjunction with	Constant speed wire feeder
Voltage sensing wire feeder	when used in conjunction with	C.C. power source
C.V. power source	when used in conjunction with	Voltage sensing wire feeder in the constant speed mode

Fig. 21-39 Drive roll types and applications. Miller Electric Mfg. Co.

Adjustment of the wire feeder and power source is just the opposite of that of the constant voltage power source and constant speed wire feeder. See Table 21-3.

The welder has trigger control over gas flow, wire feed, and the welding-current contactor right at the work. When the welder presses the torch trigger, gas and water flow and wire feed start automatically. The arc strikes as the wire touches the workpiece. When the trigger is released, gas, water, and wire feed stop. The wire-feed rate and electrical variables are set on the wire feeder before welding and are automatically controlled during welding.

Wire drives are manufactured so that different sizes of drive rolls can be installed quickly. All sizes of wire can be used, thus making it possible to weld a wide range of metal thicknesses. Different rolls are required for hard wires than for soft wires. Figure 21-39 shows the various types of drive rolls and the application for each. The size and type of drive roll used is critical for good feedability of the continuous electrode.

Most wire-feeding units push the wire through the cable to the torch. The push type of drive is generally used with hard wires such as carbon steel and stainless steel. A number of torches have an internal mechanism that pulls the wire from a wire reel unit through the torch to the arc. Figures 21-40 through 21-42, pages 677–678, show two types of wire-drive, pull-type guns. The pull type of drive is preferable for use with soft wires such as aluminum and magnesium. Other torches have a small spool of wire and a drive motor in the torch which feeds the wire directly

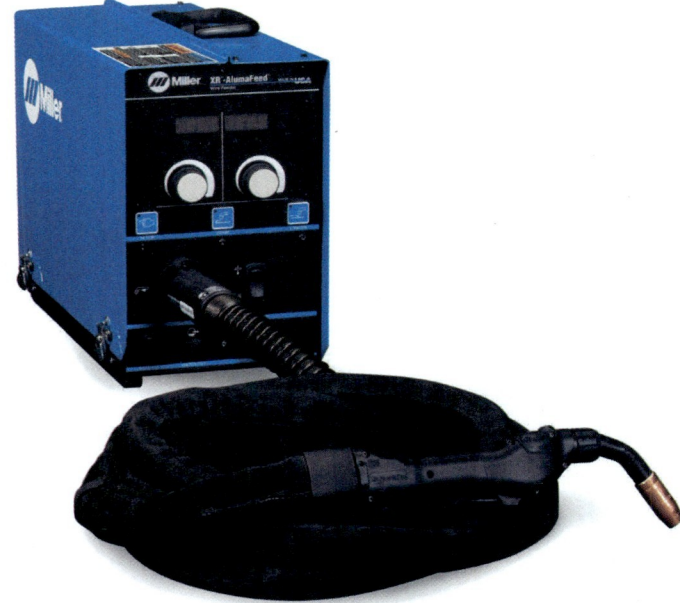

Fig. 21-40 This push-pull wire-feed system is designed specifically for aluminum MIG welding. There is a pull drive motor in the gooseneck gun, while the control and wire spool housing has a push motor. Guns are available from the spool supply housing of 15, 25, and 35 feet for maximum reach. The maximum allowable spool diameter is 12 inches. This water-cooled version is rated at 400 amperes at 100 percent duty cycle. It has an electrode wire diameter capacity of 0.030 to 0.047 inch with a wire-feed speed range of 70 to 875 inches per minute. Note the gun trigger and wire-feed speed control located in the base of the gun handle. Miller Electric Mfg. Co.

XR-Control Panel

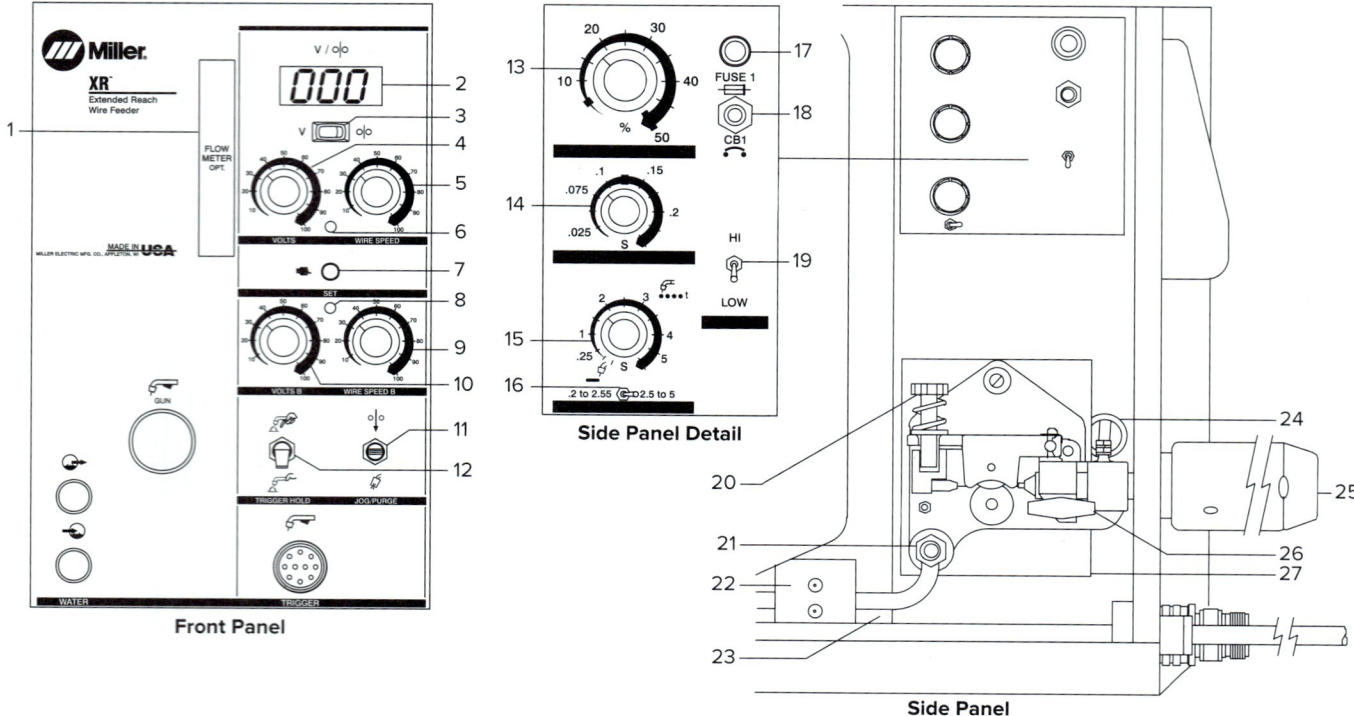

1. Gas Flowmeter*
2. Voltage/Wire Speed Meter
3. Voltage/Wire Speed Switch
4. Voltage Control*
5. Wire Speed Control
6. Schedule A Indicator LED*
7. Press To Set Push Button*
8. Schedule B Indicator LED*
9. Wire Speed B Control*
10. Remote Volts B Control*
11. Wire Jog/Gas Purge Switch
12. Trigger Hold Switch
13. Run-In Speed Control
14. Burnback Time Control*
15. Continuous/Spot Time Control*
16. Spot Time Range Switch*
17. Fuse F1
18. Circuit Breaker CB1
19. Motor Torque Switch
20. Drive Tensioner Adjustment Knob
21. Weld Cable Terminal in Feeder
22. Reed Relay
23. Weld Cable
24. Gas Hose
25. Gun Connector
26. Gun Securing Knob
27. Gun Connector Block

*Optional

Fig. 21-41 Note the control panel callouts. This system can be set up for dual schedule operation. A dual schedule switch will be required. Source: Miller Electric Manufacturing Company. Copyright ©Miller Electric Manufacturing Company. All rights reserved. Used with permission.

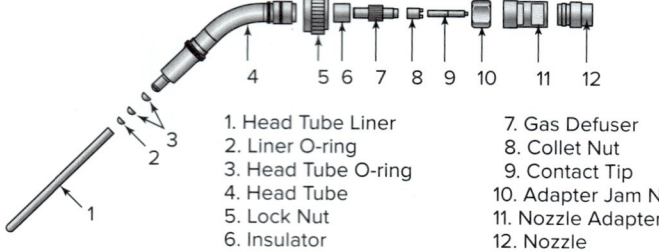

1. Head Tube Liner
2. Liner O-ring
3. Head Tube O-ring
4. Head Tube
5. Lock Nut
6. Insulator
7. Gas Defuser
8. Collet Nut
9. Contact Tip
10. Adapter Jam Nut
11. Nozzle Adapter
12. Nozzle

Fig. 21-42 Note the callouts on the working end of this gun from the push-pull system shown in Figs. 21-40 and 21-41. These are the gun parts that are in the closest proximity to the welding arc and will require the most maintenance. Source: Miller Electric Manufacturing Company. Copyright ©Miller Electric Manufacturing Company. All rights reserved. Used with permission.

through the torch, Figs. 21-43 and 21-44. Still another type of torch acts as a feeder gun and operates in connection with a canister, Fig. 21-45.

Welding Guns and Accessories

The function of the gun, Fig. 21-46 (p. 680), is to deliver the electrode wire and the shielding gas from the wire feeder and the welding current from the power source to the welding area through the contact tube (tip). The MIG/MAG gun is comparable to the electrode holder used for stick electrode welding.

Guns are rated for the minimum and maximum wire diameters on which they are designed to run, current-carrying capacity, and the type of shielding gas this current rating is based on, Table 21-4 (p. 680). A typical 200-ampere air-cooled GMAW gun may have a rating of 0.035- to 0.045-inch diameter, wire, 200 amperes at 100 percent duty cycle with CO_2 gas. The amperage rating should be reduced approximately 50 percent if argon is substituted for the CO_2 shielding gas. Guns are also sized by their reach from the

Fig. 21-43 This spool gun provides reach and accessibility. The gun contains the wire and the drive motor system. Since the wire is pushed only a few inches, smooth feeding of difficult-to-feed wire like aluminum is easily accomplished. Ideal for light industrial applications, the gun is rated for 150 amperes at 60 percent duty cycle and has a 20-foot weld control/cable. The electrode wire diameter capacity is 0.023 to 0.035 inches for aluminum, steel, and stainless steel, with wire-feed speeds of from 115 to 715 inches per minute. The maximum spool size capacity is 4 inches.   Miller Electric Mfg. Co.

Fig. 21-44 For more industrial type applications this spool gun is rated for 200 amperes and 100 percent duty cycle. The electrode wire diameter capacity is 0.023 to 1/16 inch for aluminum wire and stainless steel and steel wire up to 0.045 inch, with wire-feed speeds of from 70 to 875 inches per minute. The maximum spool size is 4 inches. Miller Electric Mfg. Co.

Fig. 21-45 A group of welders MIG welding an aluminum boat hull. The push-pull gun and power source can be up to 35 feet apart, allowing extreme flexibility and reach for projects like this one.   Miller Electric Mfg. Co.

ELECTRIC SHOCK can kill.
- Wear dry, hole-free insulating gloves and body protection.
- Insulate yourself from work and ground using dry insulating mats or covers big enough to prevent any physical contact with the work or ground.
- Additional safety precautions are required when any of the following electrically hazardous conditions are present: in damp locations or while wearing wet clothing; on metal structures such as floors, gratings, or scaffolds; when in cramped positions such as sitting, kneeling, or lying; or when there is a high risk of unavoidable or accidental contact with the workpiece or ground. For these conditions, use the following equipment in order presented: (1) a semiautomatic DC constant voltage (wire) welder, (2) a DC manual (stick) welder, or (3) an AC welder with reduced open-circuit voltage. In most situations, use of a DC, constant voltage wire welder is recommended.

Gas Metal Arc and Flux Cored Arc Welding Principles   **Chapter 21**

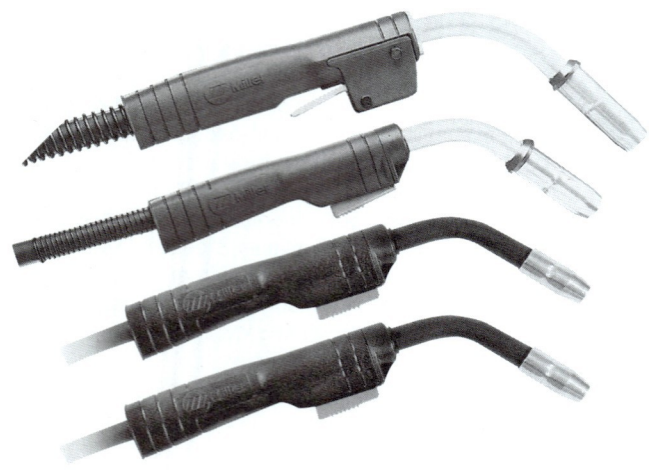

Fig. 21-46 GMAW guns from 400 to 100 amperes. Check how these guns are rated in Table 21-4 Miller Electric Mfg. Co.

wire-feeding mechanism, such as 10 to 15 feet. With a push-type feeder the gun length should be kept as short as possible, yet still give the welder access to the weld joint.

Guns are designed for different types of service. An air-cooled gun may be used for light-duty service and with carbon dioxide as the shielding gas, Figs. 21-47 and 21-48. Carbon dioxide actually assists in the cooling of the gun. For heavy-duty service and inert gas, the gun may be water cooled, Figs. 21-49 and 21-50. Coolant is usually necessary when currents of over 250 amperes are used. Welding continuously with high current generates a great amount of heat in the gun. Coolant equipment, Fig. 21-51 (p. 682), provides a positive system of cooling the gun. These units have an internal cooling system and force coolant through the gun.

Guns for semiautomatic operation may be shaped like a pistol, or have a straight handle or a handle with a gooseneck nozzle end. All types have a trigger switch that turns on the controls for the wire feed, shielding gas, and welding-current contactor.

Nozzles Many different types of materials are used for nozzles: copper and copper alloys are most common. A clean smooth nozzle will allow easy removal of weld spatter. The welder should not roughen or abuse the gas nozzle on the gun. Nozzles are available in various diameters and lengths. They must be large enough to allow proper gas shielding, yet small enough to allow access to the weld joint.

Cable Assemblies The cable assemblies are an integral part of the gun. They generally consist of a single molded conductor that encloses all the necessary components. The electrode wire is guided through a flexible liner contained in the cable assembly. The liner is made of spring steel for strength and protection. Plastic liners are sometimes used for aluminum and magnesium alloys. The liner serves as a guide for the electrode wire and provides uninterrupted travel of the wire without buckling or

> **ABOUT WELDING**
>
> **Gas Leak**
> Gas leaking out the back of a MIG gun can be minimized by a design using a shrink tube wrapping over the 5 feet of liner right after the wire feeder.

Table 21-4 GMAW Gun Ratings

Cable Length (ft)	Rated Output (A)	Duty Cycle	Standard Wire Size (in.)	Weight (lb)
10	100	100% with CO_2 gas, 60% with mixed gas	0.030–0.035	3.2
10	150	100% with CO_2 gas, 60% with mixed gas	0.030–0.035	4.4
12				4.7
15				6.0
10	250	100% with CO_2 gas, 60% with mixed gas	0.030–0.035	5.4
12				6.0
15			0.030–0.035 0.035–0.045	7.2
15	250	100% with CO_2 gas, 60% with mixed gas	0.030–0.035	7.5
10	400	100% with CO_2 gas, 60% with mixed gas	0.035–0.045	8.3
12				9.5
15				10.9

Source: Miller Electric Manufacturing Company. Copyright ©Miller Electric Manufacturing Company. All rights reserved. Used with permission.

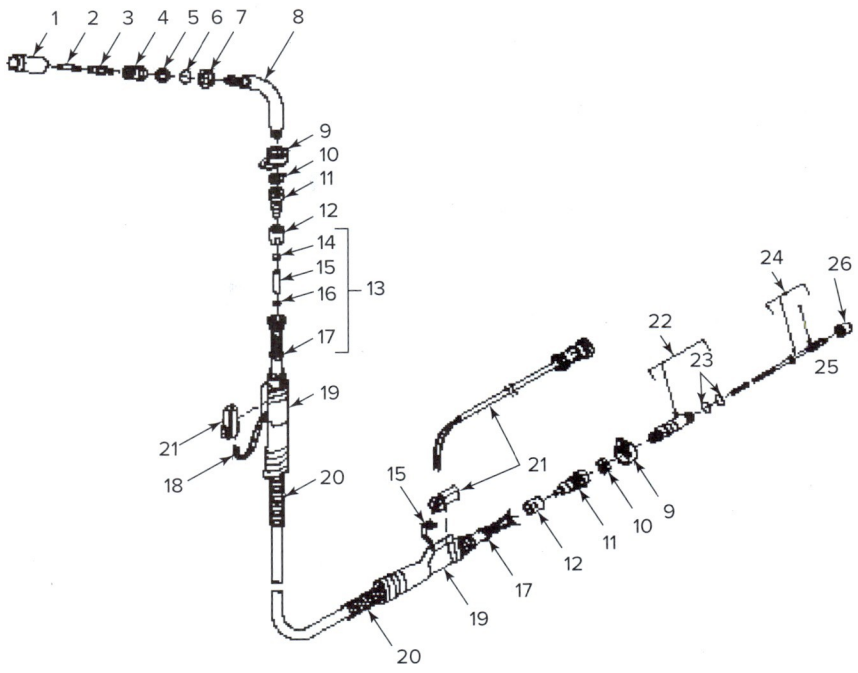

Fig. 21-47 Internal components of GMAW/FCAW gas shielded welding gun. 1. Nozzle. 2. Contact tip. 3. Contact tip adapter. 4. Nozzle adapter. 5. Retaining ring. 6. O-ring. 7. Shock washer. 8. Head tube. 9. Handle locking nut. 10. Jam nut. 11. Cable connector. 12. Connector nut. 13. Unicable clamping kit. 14. Compression clip. 15. Support tube. 16. Inner clamp. 17. Jacket clamp. 18. Switch lead connector. 19. Handle. 20. Cable strain relief. 21. Trigger cord assembly. 22. Feeder connection. 23. O-ring. 24. Monicoil liner kit. 25. O-ring. 26. Outlet guide. 27. Trigger switch. Source: Miller Electric Manufacturing Company. Copyright ©Miller Electric Manufacturing Company. All rights reserved. Used with permission.

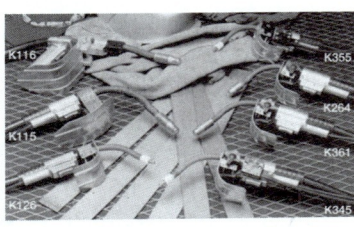

Fig. 21-48 FCAW-S guns rated from 250 to 600 amperes. Note the exposed contact tips on the guns set up for short electrode extension electrodes and the insulated guide tubes on other guns for the long electrode extension electrodes. Use the appropriate gun setup per the manufacturer's specification for the FCAW electrode extension being used. Miller Electric Mfg. Co.

kinking from the drive rolls to the contact tip of the gun. The cable also contains tube passages for shielding gas flow.

Dirt and foreign material gradually build up inside the liner and interfere with the free flow of the electrode wire through the cable. The liner should be blown out each time the electrode wire spool is changed. Short cable assemblies allow for easier feeding and keep the cable as straight as possible.

The Shielding Gas System

The shielding gas system supplies and controls the flow of gas that shields the arc area from contamination by the surrounding atmosphere. The system consists of the following equipment:

- One or more gas storage cylinders (or bulk-manifolded system)
- Hose
- Pressure-reducing regulator
- Flowmeter
- Solenoid control valves

The gas from the cylinder is controlled by means of a pressure-reducing regulator, flowmeter, and the solenoid control valves. It is fed to the torch through an innerconnecting hose system. Some power supplies have certain of these units incorporated into the machine. It is important that the correct regulators and flowmeters be used with the particular shielding gas being employed. The gas cylinders can be replaced with a central gas supply system when there is a sufficient volume of MIG/MAG welding being done.

To reduce waste and misadjustment of flow devices, a fixed orifice in-line system can be used. The manifold system must have the pressure regulated very accurately to

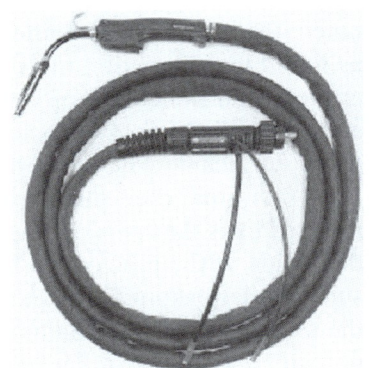

Fig. 21-49 A water-cooled gun rated at 450 amperes with carbon dioxide shielding and 400 amperes with argon or argon mixed gases at 100 percent duty cycle. The Lincoln Electric Co.

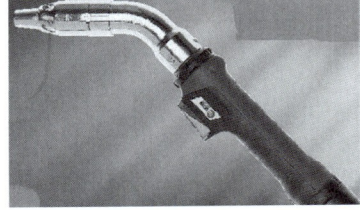

Fig. 21-50 To help reduce fumes some guns are equipped with extractor devices directly above the gas delivering nozzle system. This gun is so equipped, is water cooled, and is rated at 300 amperes at 100 percent duty cycle with carbon dioxide shielding. The Lincoln Electric Co.

Fig. 21-51 Water-circulating units of various profiles for cooling heavy-duty MIG/MAG guns. Miller Electric Mfg. Co.

each feeder system for this to be useful. Another waste in shielding gas is leaks. Leak checks can quickly be done by observing pressure drop. If the system is not in use and the cylinder valve(s) or manifold system is shut down, the pressure should hold. If a pressure drop is noted, the leak must be found. Detection devices as well as liquid leak detection material are available. Not only are leaks an indication of waste and possible safety issues in confined or low spaces (many common shielding gases are heavier than air), but if they are in a volume flow rate area, they may cause contamination. This is referred to as the *venturi effect*, where a high velocity on the inside of the pipe or gas hose will cause a low pressure in relationship to the outside and will draw air into the system. This will contaminate the gas and result in porosity and oxide formation in the welds.

Gas Flow Rates In order to provide a good gas shield, the gas must be fed through the nozzle at such a rate as to ensure *laminar flow* (a straight line of flow), Fig. 21-52. If the flow rate is too high, it causes *turbulence* (swirling). The turbulent gas shield mixes with the surrounding air under the nozzle and causes contamination of the weld deposit. If the gas shield fails to cover the arc or is contaminated, porosity results. Almost all porosity in MIG/MAG welds is caused by poor gas coverage. Too much is just as bad as too little.

The gas flow rate is a function of the shape and cross-sectional area of the gas passages in the nozzle. The recommended gas flow rate for a particular nozzle and a specified gas is suitable for all wire-feed speeds. Some manufacturers have nozzles with a diffuser or lens similar to the gas lens on a GTAW torch to give a more uniform pattern to the gas flow. Always refer to the manufacturer's instructions for recommendations concerning a particular nozzle.

For video of GMAW gun nozzle comparing gas flow patterns, please visit www.mhhe.com/welding.

Specific Gravity The chief factor influencing the effectiveness of a gas for arc shielding is the specific gravity of the gas. This in turn determines the flow rate selected for the welding operation. The *specific gravity* of a gas is a measure of its density relative to the density of air. The density of air is fixed as one. The specific gravity of helium is 0.137, or approximately 1/10 as heavy as air. Thus helium tends to rise quickly from the weld area, and a high flow rate is needed to provide adequate shielding. A gas of low density such as helium is also greatly affected by cross-drafts of air across the arc and weld pool.

Argon and carbon dioxide are much heavier than helium, and lower flow rates are required for equivalent shielding. The specific gravity of argon is 1.38, and the specific gravity of carbon dioxide is 1.52. Both gases,

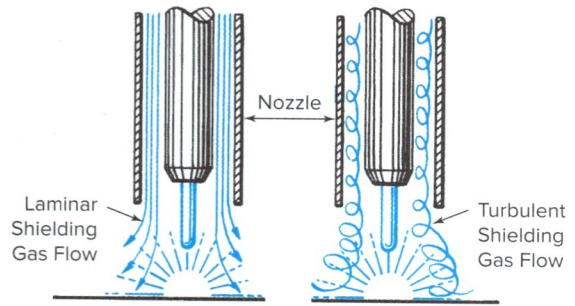

Fig. 21-52 For an effective gas shield, the gas should be fed through the gun nozzle in a laminar flow (left). A swirling gas shield results in contamination of the weld deposit (right).

therefore, are heavier than air. They blanket the weld area. Carbon dioxide has the advantage of being heavier than argon, thus ensuring good protection. It is also more resistant to crossdrafts.

Other Factors Affecting Flow Rates Flow rates may vary from 15 to 50 cubic feet per hour depending upon the type and thickness of the material to be welded, the position of the weld, the type of joint, the shielding gas, and the diameter of the electrode wire. The welder can tell if the amount of shielding gas is correct by the rapid cracking and sizzling arc sound. Inadequate shielding produces a popping, erratic sound and a considerable amount of spatter. The weld is also discolored and has a high degree of porosity.

The following additional factors also have an influence on effective gas shielding:

- Size and shape of the nozzle
- Nozzle-work distance (standoff distance)
- Size of the weld pool
- Amount of welding current
- Air currents
- Design of welding fixture
- Welding speed
- Joint design (fillet weld vs. outside corner)
- Inclination of the torch
- Excessive shielding gas pressure
- Leaking gas fittings

Shielding Gases

Metals are meeting the stringent requirements of the space age with their high corrosion resistance, excellent strength-to-weight ratios, and good physical properties. Add the advantages of welded fabrication, and the metals become an engineer's dream come true.

However, in the molten state, many of the metals absorb oxygen, nitrogen, and hydrogen from the air. These gases cause porosity and brittleness. The prime consideration when welding them is protection of the weld zone from atmospheric elements. In Chapter 18, we have seen that this is accomplished by a shielding gas.

Selecting a shielding gas for MIG/MAG welding can be a problem. Arc atmosphere governs, to a large extent, arc stability, bead shape, depth of penetration, freedom from porosity, and allowable welding speed. Considering these points along with the cost of various gases will lead to the best choice.

The three principal gases for TIG and MIG/MAG welding are argon, helium, and carbon dioxide. These three gases coupled with the two welding processes are producing the porosity-free welds specified for the missiles, rockets, jets, and nuclear reactors that are fabricated from space-age metals.

For MIG welds, gases used singly or in mixtures can be completely inert. If such an atmosphere is desired, argon or helium with a purity greater than 99.9 percent may be best. Other inert gases (xenon, krypton, radon, and neon) are too rare and expensive. MAG welds would use reactive gases like carbon dioxide, oxygen, and hydrogen.

Study Tables 21-5 and 21-6 (p. 684) carefully, and you will have a good understanding of the proper shielding gas to use for a given situation.

Table 21-5 Proper Wire and Gas Combinations

	Mild Steel Solid Wire	Mild Steel and Alloy Metal Cored Wire	Mild Steel and Alloy Flux Cored Wire	Aluminum	Copper	Stainless	Bronze
Carbon dioxide	X	X	X				X
75% argon + 25% carbon dioxide	X	X				X	X
90% argon + 10% carbon dioxide	X	X	X			X	
100% argon				X	X		
95–98% argon + 5–2% oxygen	X	X				X	X
100% helium				X	X		
75% helium + 25% argon				X	X		
90% helium + 7½% argon + 2½% carbon dioxide						X	

Table lists wire and gas combinations for semiautomatic welding.

Table 21-6 Gas Selection for Gas Metal Arc Welding

Metal Type	Thickness	Transfer Mode	Recommended Shielding Gas	Advantages/Description
Carbon steel	Up to 14 gauge	Short circuit	Argon + CO_2 Argon + CO_2 + O_2	Good penetration and distortion control to reduce potential burn-through.
	14 gauge–⅛ in.	Short circuit	Argon + 8 to 25% CO_2 Argon + He + CO_2	Higher deposition rates without burn-through. Minimum distortion and spatter. Good pool control for out-of-position welding.
	Over ⅛ in.	Short circuit	Carbon dioxide Argon + 15–25% CO_2	High welding speeds. Good penetration and pool control. Applicable for out-of-position welds.
		Short circuit globular	Argon + 25% CO_2	Suitable for high current and high speed welding.
		Short circuit	Argon + 50% CO_2	Deep penetration; low spatter; high travel speeds. Good out-of-position welding.
		Short circuit globular (buried arc)	Carbon dioxide	Deep penetration and fastest travel speeds but with higher melt-through potential. High current mechanized welding.
		Spray transfer	Argon + 1–8% O_2	Good arc stability; produces a more fluid pool as O_2 increases; good coalescence and bead contour. Good weld appearance and pool control.
		Spray transfer	Argon + 5–20% CO_2	Fluid pool and oxidizing to weld metal causing higher amounts of oxides and scale, as CO_2 increases. Good arc stability, weld soundness, and increasing width of fusion.
		Short circuit Spray transfer	Argon + CO_2 + O_2 Argon + He + CO_2 Helium + Ar + CO_2	Applicable to both short-circuiting and spray transfer modes. Has wide welding current range and good arc performance. Weld pool has good control which results in improved weld contour.
		High current density rotational	Argon + He + CO_2 + O_2 Argon + CO_2 + O_2	Used for high deposition rate welding where 15–30 lb/h (7 to 14 kg/h) is typical. Special welding equipment and techniques are sometimes required to achieve these deposition levels.
	Over 14 gauge	Pulsed spray	Argon + 2–8% O_2 Argon + 5–20% CO_2 Argon + CO_2 + O_2 Argon + He + CO_2	Used for both light gauge and heavy out-of-position weldments. Achieves good pulse spray stability over a wide range of arc characteristics and deposition ranges.
Low and high alloy steel	Up to 3/32 in.	Short circuit	Argon + 8–20% CO_2 Helium + Ar + CO_2 Argon + CO_2 + O_2	Good coalescence and bead contour. Good mechanical properties.
		Short circuit globular	Argon + 20–50% CO_2	High welding speeds. Good penetration and pool control. Applicable for out-of-position welds. Suitable for high current and high speed welding.
	Over 3/32 in.	Spray transfer (high current density & rotational)	Argon + 2% O_2 Argon + 5–10% CO_2 Argon + CO_2 + O_2 Argon + He + CO_2 + O_2	Reduces undercutting. Higher deposition rates and improved bead wetting. Deep penetration and good mechanical properties.
Low and high alloy steel		Pulsed spray	Argon + 2% O_2 Argon + 5% CO_2 Argon + CO_2 + O_2 Argon + He + CO_2	Used for both light gauge and heavy out-of-position weldments. Achieves good pulse spray stability over a wide range of arc characteristics and deposition ranges.

Table 21-6 (Concluded)

Metal Type	Thickness	Transfer Mode	Recommended Shielding Gas	Advantages/Description
Steel, stainless, nickel, nickel alloys	Up to 14 gauge	Short circuit	Argon + 2 to 5% CO_2	Good control of burn-through and distortion. Used also for spray arc welding. Pool fluidity sometimes sluggish, depending on the base alloy.
	Over 14 gauge	Short circuit	Helium + 7.5 Ar + 2.5 CO_2 Argon + 2–5% CO_2 Argon + He + CO_2 Helium + Ar + CO_2	Low CO_2 percentages in He mix minimizes carbon pickup, which can cause intergranular corrosion with some alloys. Helium improves wetting action and contour. CO_2 percentages over 5% should be used with caution on some alloys. Applicable for all position welding.
		Spray transfer	Argon + 1–2% O_2 Argon + He + CO_2 Helium + Ar + CO_2	Good arc stability. Produces a fluid but controllable weld pool; good coalescence and bead contour. Minimizes undercutting on heavier thicknesses.
Stainless steel	Over 14 gauge	Pulsed spray	Argon + 1–2% O_2 Argon + He + CO_2 Helium + Ar + CO_2 Argon + CO_2 + H_2	Used for both light gauge and heavy out-of-position weldments. Achieves good pulse spray stability over a wide range of arc characteristics and deposition ranges.
Copper, copper-nickel alloys	Up to 1/8 in.	Short circuit	Helium + 10% argon Helium + 25% argon Argon + helium	Good arc stability, weld pool control and wetting.
	Over 1/8 in.	Spray transfer	Helium + argon Argon + 50% helium Argon or helium	Higher heat input of helium mixtures offset high heat conductivity of heavier gauges. Good wetting and bead contour. Can be used for out-of-position welding. Using 100% helium on heavier material thickness improves wetting and penetration.
		Pulsed spray	Argon + helium	Used for both light gauge and heavy out-of-position weldments. Achieves good pulse spray stability over a wide range of arc characteristics and deposition ranges.
Aluminum	Up to 1/2 in.	Spray transfer pulsed spray	Argon	Best metal transfer, arc stability, and plate cleaning. Little or no spatter. Removes oxides when used with DCEP (reverse polarity).
	Over 1/2 in.	Spray transfer pulsed spray	Helium + 20–50% argon Argon + helium	High-heat input. Produces fluid pool, flat bead contour, and deep penetration. Minimizes porosity.
Magnesium, titanium, & other reactive metals	All thicknesses	Spray transfer	Argon	Excellent cleaning action. Provides more stable arc than helium-rich mixtures.
		Spray transfer	Argon + 20–70% helium	Higher heat input and less chance of porosity. More fluid weld pool and improved wetting.

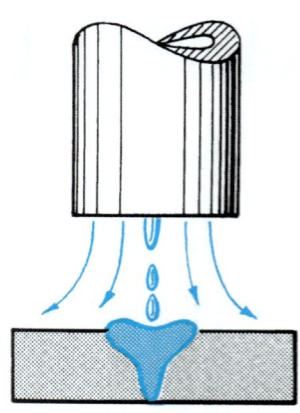

Fig. 21-53 Argon-shielded weld.

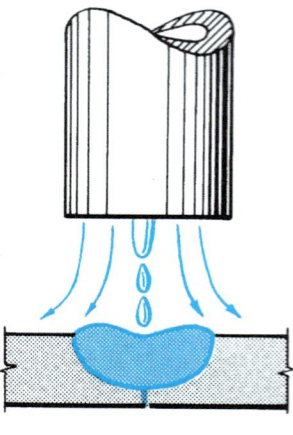

Fig. 21-54 Helium-shielded weld.

Argon Argon is in abundant supply. The amount of air covering one square mile of the earth's surface contains approximately 800,000 pounds of argon. The gas is a by-product of the manufacture of oxygen.

In the argon shield, the welding arc tends to be more stable than in other gas shields. For this reason, argon is often used as a mixture with other gases for arc shielding. The argon gives a quiet arc and thereby reduces spatter.

Argon has low *thermal* (heat) *conductivity*. The arc column is constricted so that high arc densities are produced. A high arc density permits more of the available arc energy to go into the work as heat. The result is a relatively narrow weld bead with deep penetration at the center of the deposit, Fig. 21-53. In some welding applications, however, argon does not provide the penetration characteristics needed for thicker metals.

Argon ionizes more readily than helium, and thus can transmit some electric energy. Therefore, argon requires a lower arc voltage for a given arc length than helium does. Since the result is less heat, it is believed that the argon arc is not as hot as the helium arc. The cooler arc produced with argon makes it preferable for use in welding light gauge metal and materials of low thermal conductivity.

Pure argon is used in MIG welding (however, pure argon is not used for MAG welding) of such materials as carbon steel since it causes poor penetration, undercutting, and poor bead contour. Its primary use is in the welding of nonferrous metals such as copper and alloys containing aluminum and magnesium.

The combination of constricted penetration and reduced spatter makes argon desirable when welding in other than the flat position. Argon is recommended for manual operations because changes in arc length do not produce as great a change in arc voltage and heat input to the work as when helium is used.

Helium Helium is produced from natural gas. Natural gas is cooled and compressed. The hydrocarbons are drawn off first, then nitrogen, and finally helium. The gases are liquefied until helium is produced at $-345°F$.

Helium is a light gas that tends to rise in a turbulent fashion and disperse from the weld region. Therefore, higher flow rates are generally required with helium than with argon shielding. Helium, because of greater arc heat, is usually preferable for welding heavy sections and materials with high thermal conductivity. It produces a wider weld bead with slightly less penetration than argon, Fig. 21-54.

When pure helium is used with MIG and DCEP, there is no cleaning action. Metal transfers from the electrode in large drops and tends to spatter. The beads are broad and flat, and penetration is uniform. Helium lacks the deep central penetration of argon. For a given current and arc length, arc voltage is higher and wire burnoff is greater than with argon.

Helium is used more with automatic and mechanized processes than argon because control of arc length is not a problem. It is used primarily for the nonferrous metals such as aluminum, magnesium, copper, and their alloys.

Carbon Dioxide Carbon dioxide (CO_2) is not an inert gas such as argon or helium. It is a chemical compound composed of one part carbon (C) and two parts oxygen (O_2). It is manufactured from flue gases that are given off by the burning of natural gas, fuel oil, and coke. Carbon dioxide is also a by-product of the manufacture of ammonia and the fermentation of alcohol. The carbon dioxide from these processes is almost 100 percent pure.

The use of carbon dioxide as a shielding gas is increasing. It is cheaper than the inert gases and has several desirable characteristics. It produces a wide weld pattern and deep penetration, thus making it easier to avoid incomplete fusion. The bead contour is good, and there is no tendency toward undercutting.

One disadvantage of carbon dioxide is the tendency for the arc to be unstable and cause weld spatter. This is particularly serious on thin materials when appearance is important. The amount of weld spatter may be reduced by maintaining a very close arc. On heavier materials, the arc may be buried in the workpiece. Spatter can also be reduced by using the FCAW process.

The oxidation effect of the carbon dioxide shield is about equal to a mixture of 91 percent argon and 9 percent oxygen. Under a carbon dioxide shield, about 50 percent of the manganese and 60 percent of the silicon in the welding wire are converted to oxides as it passes through the arc. It is for this reason that the electrode wire contains a balance of deoxidizers and that a short arc is maintained.

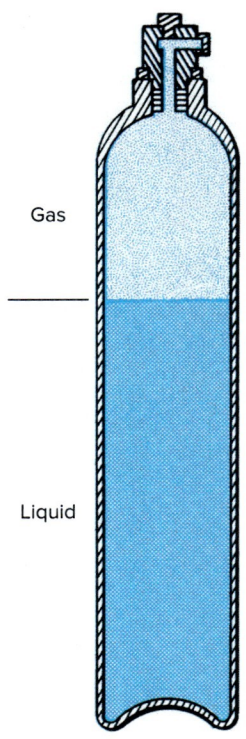

Fig. 21-55 Standard carbon dioxide cylinder.

With the proper care, X-ray quality welds can be consistently produced in carbon steels and some low alloy steels.

In the heat of the arc, carbon dioxide tends to break down into its component parts. At normal arc lengths, about 7 percent of the total volume of the gas shield converts to carbon monoxide (a compound made up of one atom carbon and one atom oxygen). At excessive arc lengths, the quantity reaches 12 percent. Because carbon monoxide is poisonous, welding should never be done in unventilated areas. Welders should be able to see the welding operation clearly without getting so close to the shielding gas that they inhale the carbon monoxide.

Carbon dioxide is available in liquid or gaseous form for welding purposes. Because of the small gas requirement of a single welding arc, liquid carbon dioxide in cylinders is usually the most economical. Bulk systems for distributing carbon dioxide at low pressure to various welding areas are also in use.

Cylinders are filled on the basis of weight. The liquid level on a full cylinder is about two-thirds up from the bottom, Fig. 21-55. Space for gas over the liquid is necessary. The liquid in the cylinder absorbs heat from the atmosphere and boils. Gas is formed until the pressure in the cylinder raises the boiling temperature above the temperature of the liquid. Boiling, and consequently gas production, ceases until either the atmospheric temperature increases or some gas is drawn off, lowering the pressure in the cylinder. The pressure in the cylinder is a function of the atmospheric temperature. Typical pressures for various temperatures are given in Table 21-7.

Table 21-7 Cylinder Pressure of Carbon Dioxide at Various Temperatures

Degrees Fahrenheit	Pounds per Square Inch
100	1,450
70	835
30	476
0	290

At a constant temperature, the pressure in the cylinder remains constant as long as there is liquid remaining. Continued use of the gas after the liquid has completely evaporated noticeably reduces cylinder pressure. Cylinders should be changed when the pressure goes down to 200 p.s.i.

Since even small amounts of moisture in carbon dioxide cause porosity in the weld deposit, it is important that it not become contaminated (See Figs. 21-56 and 21-57.) *Dew point* is the temperature at which water will condense out of the shielding gas. Welding gases have very

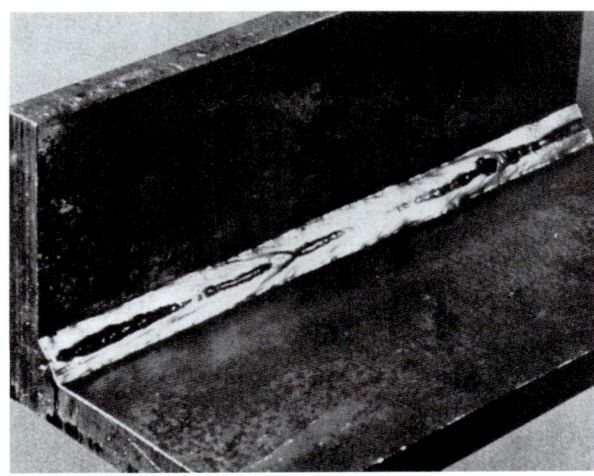

Fig. 21-56 This weld was made using carbon dioxide that contained moisture. Moisture poisoned the gas shield, causing a weak, porous structure. Allegheny Technologies Incorporated

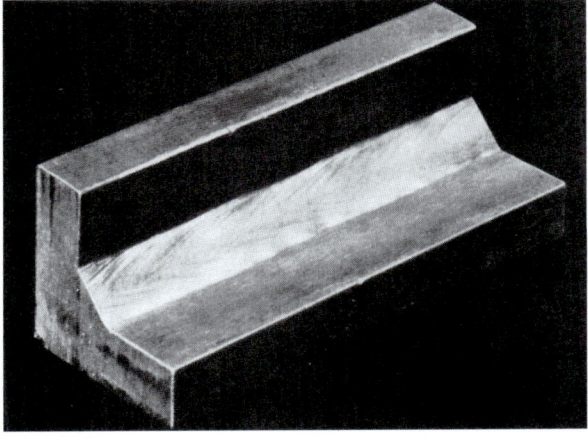

Fig. 21-57 This high quality weld was made by the same welder who made the weld in Fig. 21-56. The same equipment and flux cored electrodes were used, but the carbon dioxide was free of moisture. ©Allegheny Technologies Incorporated

Table 21-8 Dew Point of Carbon Dioxide versus Percentage of Moisture in Carbon Dioxide

Dew Point (°F)	% Moisture by Weight	p.p.m. Moisture in CO_2
−90	0.000353	4
−80	0.00078	8
−70	0.00166	17
−60	0.0034	34
−50	0.0067	67
−40	0.0128	128
−30	0.0235	235
−20	0.0422	422
−10	0.074	740

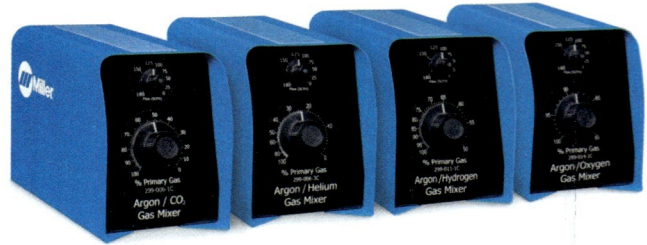

Fig. 21-58 Some jobs and welding procedures require different gas mixes to get the best results. Being able to mix gases on site with a proportional gas mixer eliminates inventorying and handling of costly premixed gases. Individual gases can be mixed in specially designed units such as these for Argon-CO_2, Argon-Helium, Argon-Hydrogen and Argon-Oxygen. *Miller Electric Mfg. Co.*

low dew points, Table 21-8. The parts per million (p.p.m.) of moisture in carbon dioxide at −60°F is approximately 32 with a gas purity of 99.8 percent.

There is a considerable refrigeration effect in the high pressure section of the regulator. Sometimes, on very heavy duty cycles, frost forms on the outside of the regulator. This indicates only that the regulator is getting cold. When frost appears, the small amount of moisture in the gas can form ice in the regulator and either stop the gas flow entirely or break off and be blown into the weld. In either case porosity can, and usually does, occur in the weld metal. If regulator freezing is a problem, use an approved heated regulator. Bulk systems are not prone to freezing because of the small expansion ratio and the large heat absorption area of piping.

The normal gas requirement for MIG/MAG welding is 25 to 30 cubic feet per hour. A single cylinder may not be large enough to support gas production at this rate. It is generally necessary to connect two or more cylinders to a manifold.

Experience has shown the above gas flow rates to be proper for most applications. Higher than normal flow rates are usually necessary on very high speed applications or procedures utilizing high contact tube heights. Too low a flow rate results in a partial or complete lack of gas coverage, which causes a high degree of porosity in the weld. Too much gas flow produces turbulence about the arc so that the carbon dioxide mixes with air. This also causes porosity of the weld deposit.

Gas Mixtures As has been shown in Tables 21-4 through 21-7, gases are mixed in specified percentages to obtain desired characteristics.

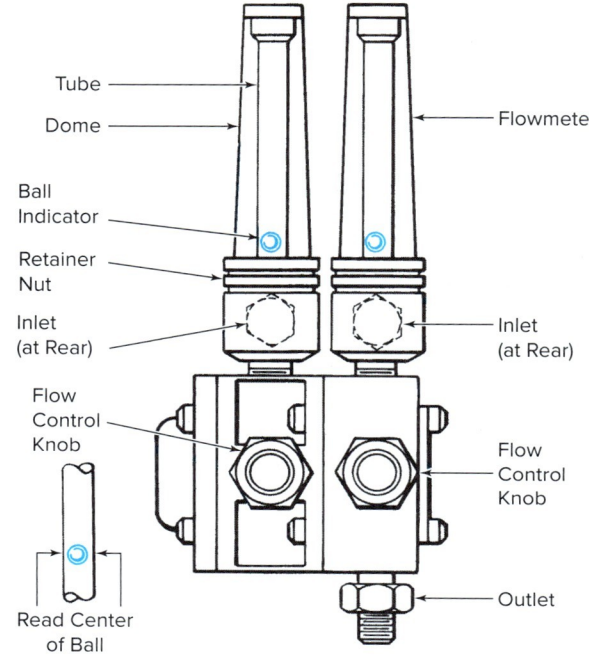

Fig. 21-59 The controls and working parts of a two-module gas mixer similar to that shown in Fig. 21-58.

Figures 21-58 and 21-59 show combination gas mixers and flow controllers. The mixer in Fig. 21-58 will mix three gases or any two of the three gases the model was designed to mix. For example, if a model is designed to mix argon, carbon dioxide, and helium, three mixtures of two gases (argon-carbon dioxide, argon-helium, and helium-carbon dioxide) can also be produced.

The mixer has three flow control knobs, one for each of the three gases to be mixed. By turning a control knob while watching the flowmeter above it, welders can adjust the flow of that gas independently. This gives welders the ability to change mixtures when they change jobs or

Fig. 21-60 Regulator used with gas mixer to reduce the tank pressure to a steady 50 p.s.i. for welding. ©Allegheny Technologies Incorporated

experiment with the mixture on any job until they obtain the best weld.

The shielding gas or gas mixture is an essential variable for most welding procedures. The mixture cannot be indiscriminately changed without consulting the engineer responsible for the welding procedure. For students learning in an experimental environment, trying various gas mixtures is a helpful learning tool. However, in production welding to a welding procedure, the welding procedure must be followed.

It is necessary to use a gas regulator with the mixer in order to reduce the high tank pressure to a steady usable working pressure, Fig. 21-60.

Argon-Helium For many MIG/MAG welding applications the advantages of both argon and helium may be achieved through the use of mixtures of the two gases. The mixture is especially desirable whenever a completely inert gas is essential or desirable for MIG/MAG welding. Mixing argon and helium gives a balance between penetration and arc stability. Adding 25 percent argon to helium makes penetration deeper than that obtained with argon alone, and the arc stability approaches that of pure argon.

ABOUT WELDING

Pulsed MIG/MAG

In pulsed MIG/MAG, a modified spray transfer mode is used. The wire does not touch the weld pool, so no spatter is produced. This method can be used for 14-gauge metal and up. In comparison, pulsed TIG is better for thinner materials.

The percentage of helium to argon may be 20–80 percent to 50–50 percent. The gases are mixed by the user. Separate cylinders of each gas are connected to a control unit and the amount of gas used is regulated through a flowmeter.

When helium is too hot or argon too cool for the desired arc characteristics, they may be mixed to obtain any combination of properties. When argon is added in amounts above 20 percent, the arc becomes more stable. Adding 75 percent helium to the mix practically eliminates any spatter tendency in MIG/MAG welding. Bead contour shows a penetration pattern characteristic of both gases.

Argon-helium mixtures are used to weld aluminum and copper-nickel alloys. The mixtures provide greater heat and produce less porosity than argon, but they have a quieter, more readily controlled arc action than helium. These mixtures are especially useful in the welding of heavy sections of nonferrous metals. The heavier the material, the greater the percentage of helium in the mixture.

Argon-Carbon Dioxide Shielding gas mixtures are not confined to the inert gases themselves. Argon and helium can be mixed with other gases to improve welding characteristics.

The addition of carbon dioxide to argon stabilizes the arc, promotes metal transfer, and reduces spatter. The penetration pattern changes when welding carbon and low alloy steels. The molten weld pool is highly fluid along the fusion edges so that undercutting is prevented. For this reason, argon-carbon dioxide mixtures are used principally in the MAG welding of mild steel, low alloy steel, and some types of stainless steel. The 75 percent argon and 25 percent carbon dioxide is very popular for welding carbon steel with the short circuit mode of transfer. When spray arcs are being used on steel, a minimum of 80 percent argon is required.

Some differences of opinion exist in regard to the amount of carbon dioxide that should be used in the mixture. Some welding engineers believe that the mixture should not exceed 20 to 30 percent carbon dioxide. Others feel that mixtures containing up to 70 percent produce good results. Electrode filler wires used with an oxygen-bearing shielding gas such as carbon dioxide must contain deoxidizers to offset the effects of the oxygen. Up to about 20 percent carbon dioxide permits spray transfer with a solid steel wire.

One of the important reasons for using as much carbon dioxide as possible in the mixture is to reduce the cost of welding. Carbon dioxide costs about 85 percent less than argon. Premixed cylinders of argon and carbon dioxide sell for the same price as pure argon.

Helium-Argon-Carbon Dioxide This mixture of inert gases and a compound gas is used mostly for the welding of austenitic stainless steels with the short circuit method of metal transfer. It is usually purchased premixed from the supplier in cylinders that contain 90 percent helium, 7½ percent argon, and 2½ percent carbon dioxide. You will note that the percentage of helium is high.

This combination of gases produces a weld with very little buildup of the top bead profile. Such a weld is highly desirable when high bead buildup is detrimental to the weldment. A low bead profile also reduces or eliminates postweld grinding. This gas mixture has found considerable use in the welding of stainless steel pipe with the MAG process.

Argon-Oxygen The addition of small amounts of oxygen, usually 1 to 5 percent mixtures, improves and expands the use of argon. Oxygen provides wider penetration, improves bead contour, and eliminates the undercut at the edge of the weld that is obtained with pure argon when welding steel, Fig. 21-61. Argon-oxygen mixtures are used principally in the welding of low alloy steel, carbon steels, and stainless steels.

Best results with shielding gas mixtures, other than argon-helium mixtures, have been obtained with oxygen. Argon-oxygen is only used for MAG welding since oxygen would cause rapid loss of a tungsten electrode for TIG welds. The oxygen oxidizes the base plate slightly but evenly and provides a cathode surface with uniform emission properties. It also eliminates undercutting tendencies. A maximum of 2 percent oxygen delivers these advantages on low alloy steel. Above this percentage, there are no additional benefits, and a more expensive wire with higher deoxidation characteristics is required for welding. A mixture of 5 percent oxygen improves arc stability when welding stainless steels with DCEN.

Research has shown that the droplet rate (metal transfer) can be materially increased with these percentages of oxygen. The increased rate, with no change in current density, permits welding at higher speeds without undercutting. At the increased speed, coalescence of the weld metal is improved so that lower current densities are possible, and larger diameter wires can be used for a given welding current.

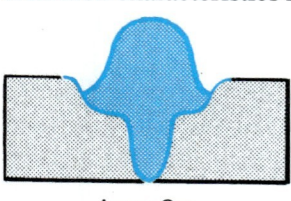
Argon Gas

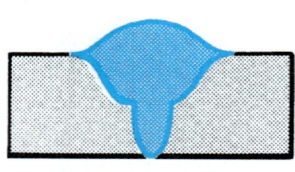
Argon-Oxygen Gas

Fig. 21-61 Typical MIG/MAG welds in stainless steel.

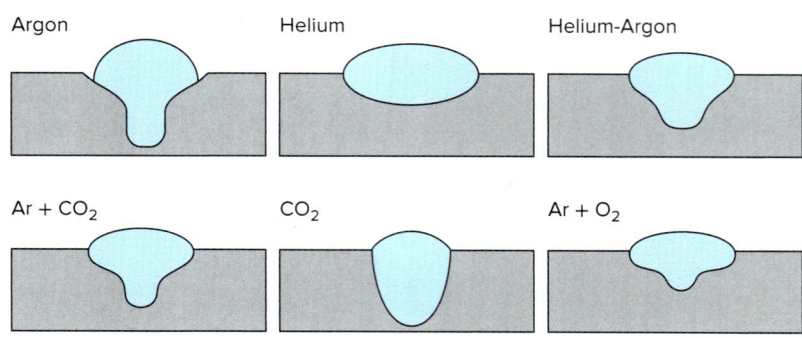

Fig. 21-62 Weld penetration comparisons in cross sections of welds made with several shielding gases.

A comparison of penetration characteristics for several main shielding gases are shown in Fig. 21-62.

Argon-oxygen mixtures can be used on direct current electrode negative jobs, but transfer is not as good as with electrode positive. DCEN applications of the gas mixture are generally limited to overlay work. In such cases, the reduced penetration and dilution of the base metal are advantages.

Other Gases Nitrogen has been used as a shielding gas, either in pure form or in combination with argon, for the welding of copper and copper alloys. When used as a mixture, it is usually in the percentage of 70 percent argon and 30 percent nitrogen. The addition of argon smoothes out the arc and reduces the agitation in the welding pool.

Shielding Gas Designators The American Welding Society has developed a standard to help control the shielding gases used for welding and cutting. This A5.32 standard covers gas types, purity, and dew points. See Fig. 21-63.

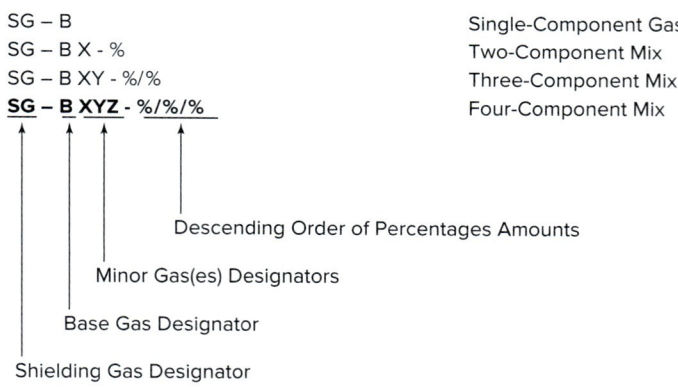

Fig. 21-63 AWS shielding gas designators. Source: AWS A5.32

The designator system is based on volumetric percentages. The gas classification system is based on the following designator and number arrangement:

SG *(shielding gas designator):* An *SG* at the beginning of each classification designates it as a shielding gas. It is always followed by a hyphen.

SG-B *(base gas designator):* The first letter directly to the right of the *SG* indicates the single or major gas in the shielding gas mixture.

SG-B-XYZ *(minor gas component designator):* The letter or letters immediately following the base gas indicate the minor individual gases in descending order of percentage. These letters are followed by a hyphen.

SG-B-XYZ-%/%/% *(percentage designators):* A slash shall be used to separate the individual minor gas percentages for two or more component mixtures.

S-B-G *(special gas mixture):* Shielding gases may be classified as special and carry the *G* designation. The base gas must be identified. However, the percentage of each component shall be as agreed upon between the purchaser and the supplier.

Electrode Filler Wire

Solid Electrode Wire The solid filler wires used for gas metal arc welding are generally quite smaller in diameter than those used for stick electrode welding. Wires range in size from 0.023/0.025 to 1/8 inch in diameter. Thin materials are usually welded with 0.023/0.025, 0.030, or 0.035 inch. Medium thickness materials call for wire sizes 0.045 or 1/16 inch, and thick materials require wire sizes 3/32 or 1/8 inch. The position of welding is important in selecting the proper wire size. Generally, smaller wire sizes are used for out-of-position work such as vertical or overhead welding. Cost is another factor. The smaller the diameter of the electrode wire, the more its cost per pound. Metal deposition rates, however, must also be considered.

You must not apply the SMAW coated electrode thought process to selecting the proper diameter of GMAW solid or metal-cored electrodes. In certain situations that require heavy welding or high deposition, going to a larger diameter electrode is not necessarily the best action. Since a small diameter electrode's cross-sectional area is so much smaller than that of a larger diameter electrode, its current density will be much higher. Table 21-9 compares a few electrode diameters with their corresponding cross-sectional areas. The area of a circle can be easily calculated by the following formula: $A = \pi r^2$, where A = area of circle, π = 3.1416, and r = radius. Current density in this situation is the amperage per square inch of electrode cross-sectional area. To calculate the current density, simply divide the amperage by the cross-sectional area of the electrode. See Table 21-10 for some examples. Higher current density leads to:

- Higher deposition
- Faster travel speed
- Less heat input
- Enhanced weld profile
- Greater penetration depth-to-width ratio, Fig. 21-6, page 659

If too high an amperage is applied to SMAW electrodes, they will overheat. The overheating is due to the resistance of the current flow through the electrode's length to reach the arc. This heating adversely affects the coating and arc performance. But with the GMAW process, this overheating is not a factor due to the short distance from the contact tube to the arc. Special high speed wire feeders are designed to supply wire at over 1,500 inches per minute to take advantage of the higher current densities of using smaller diameter electrodes.

Electrode diameter is an issue when dealing with critical modes of transfer like pulse spray on aluminum. See Table 21-11 (p. 692) for an example as to what is specified versus what is available from various manufacturers. These large variations in weld metal volume can cause various weld inconsistencies due to fluctuations in amperage, voltage, and weld profile.

Table 21-9 Current Density Calculation Chart

Electrode Diameter (in.)	Cross-Sectional Area (in.²)
0.025	0.00031
0.030	0.00071
0.035	0.00096
0.045	0.00160
3/64–0.047	0.00173
3/32–0.094	0.0069
1/8–0.125	0.01227
5/32–0.156	0.0192

Table 21-10 Current Density Comparison

Electrode Diameter (in.)	Amperage	Cross-Sectional Area (in.²)	Current Density
0.030	150	0.00071	211,268
0.045	150	0.00160	93,750
5/32–0.156	150	0.0192	78,125

Table 21-11 Wire Diameter Change versus Cross-Sectional Volume Change

AWS A5.10			Select Manufacturer		
Diameter (in.)	Tolerance	% Volume Change	Diameter (in.)	Tolerance	% Volume Change
0.030	+0.001−0.002	22.6	0.0310	+0.0000−0.0002	1.3
0.035	+0.001−0.002	19.0	0.0347	+0.0000−0.0002	1.1
0.047	+0.001−0.002	13.8	0.0471	+0.0000−0.0003	1.3
0.0625	+0.001−0.002	10.2	0.0616	+0.0000−0.0003	1.0
0.094	+0.001−0.002	6.6	0.0937	+0.0000−0.0003	0.6

Because of the small wire size and the high currents used in MIG welding, the filler metal melts at a rapid rate. The rate ranges from 40 to over 900 inches per minute. Because of this high burnoff rate and the need to be able to weld without interruption, the wires are provided as spools or in coils. Standard spools usually carry 1, 2, 3, 10, or 25 pounds of wire. To take advantage of lower prices and less downtime during spool changes, very large users purchase their filler wire in drums that contain 300, 700, or 1,000 pounds of coiled wire.

The specific type of filler metal is selected for the following reasons:

- To match the composition of the base metal
- To control various weld properties
- To deoxidize the weld deposit
- To promote arc stability and desirable metal transfer characteristics

Wire composition usually matches that of the base metal being welded. In many cases, however, it is necessary to use a filler wire of completely different composition. This is because some alloys lose some of their characteristics as weld deposits. Filler metal alloys that are both favorable for welding and produce the required weld metal properties must be chosen. Among the materials that require filler wires of different composition than the base metal are copper and zinc alloys, high strength aluminum, and high strength steel alloys. Another situation where the filler wire does not match the base metal is when gas metal arc braze welding. A copper-based electrode is used such as aluminum bronze or silicon bronze. It is usually applied in a short circuit mode of transfer with argon gas to thin steel in order to prevent melt-through and reduce distortion. It is also used when welding galvanized steel. The low heat input reduces the amount of coating melted and the copper-based braze weld has better corrosion resistance.

In addition to the alloying elements used in filler wires, deoxidizers and other scavenging agents are nearly always added. This is to prevent porosity or damage to the mechanical properties of the weld metal. When welding steel, some shielding gases such as carbon dioxide cause severe oxidation losses of alloying elements across the arc. The deoxidizers most frequently used in steel filler wires are manganese, silicon, and aluminum. Titanium and silicon are the principal ones used in nickel alloys. Copper alloys may be deoxidized with titanium, silicon, or phosphorus. Deoxidants are not used in titanium, zirconium, aluminum, and magnesium filler wires since these metals are highly reactive. They must be welded with oxygen-free inert gas and with complete shielding, or in closed chambers filled with inert gas.

The chemical compositions of some of the common solid filler metal electrodes are shown in Table 21-12.

Other considerations for the electrode wire are:

- Bare or copper-coated wire, Table 21-13
- Cast of the wire, Fig. 21-64
- Helix of the wire, Fig. 21-65, page 694

Correct cast and helix improves wire feeding, reduces arc wander, and reduces contact tip and liner wear.

Metal Core Electrode Wire Metal core electrode wire is a composite tubular filler metal electrode consisting of a metal sheath and a core of various powdered materials, producing no more than silicon islands on the face of a weld bead. See Table 21-14 (p. 694) for chemical composition. Much of the core material is metallic and adds to the deposition rate of these

> **JOB TIP**
>
> **Interested?**
> Ready to be a welding astronaut?
> That job and many others will be available in the future. Expect jobs in alternative materials such as plastics, composites, and new alloys.

Table 21-12 Chemical Compositions

Chemical Composition Requirements for Steel Solid Electrodes and Rods[1]

AWS Classification	UNS Number	C	Mn	Si	P	S	Ni	Cr	Mo	V	Cu	Ti	Zr	Al
ER70S-3	K10726	0.06–0.15	0.090–1.40	0.45–0.75	0.025	0.035	[2]	[2]	[2]	[2]	0.50			
ER70S-6	K11140	0.06–0.15	1.40–1.85	0.80–1.15	0.025	0.035	[2]	[2]	[2]	[2]	0.50			

Chemical Composition Requirements for Aluminum Electrodes and Rods[3]

AWS Classification	UNS Number	Si	Fe	Cu	Mn	Mg	Cr	Ni	Zn	Ti	Other Elements Each	Other Elements Total	Al
ER4043	A94043	4.5–6.0	0.8	0.30	0.05	0.05	—	—	0.10	0.20	0.05[4]	0.15	Remainder
ER5356	A95356	0.25	0.40	0.10	0.05–0.20	4.5–5.5	0.05–0.20	—	0.10	0.06–0.20	0.05[4]	0.15	

Chemical Composition Requirements for Stainless-Steel Electrodes and Rods[5]

AWS Classification	UNS Number	C	Cr	Ni	Mo	Mn	Si	P	S	N	Cu	Other Elements Element	Other Elements Amount
ER308L	S30883	0.03	19.5–22.0	9.0–11.0	0.75	1.0–2.5	0.30–0.65	0.03	0.03	—	0.75		
ER309L	S30983	0.03	23.0–25.0	12.0–14.0	0.75	1.0–2.5	0.30–0.65	0.03	0.03	—	0.75		
ER316L	S31683	0.03	18.0–20.0	11.0–14.0	2.0–3.0	1.0–2.5	0.30–0.65	0.03	0.03	—	0.75		

[1]Weight percentage single values are maximums.
[2]These residual elements shall not exceed 0.50% in total.
[3]Weight percentage single values are maximums. If other elements are detected, they cannot exceed the limits specified for other elements.
[4]Beryllium shall not exceed 0.0008%.
[5]Weight percentage single values are maximums. If other elements are detected, the amount of those elements shall be determined and cannot exceed 0.5%, excluding irons.
Source: Data extracted from AWS filler metal specifications A5.9, A5.10, and A5.18.

Table 21-13 Copper-Free versus Copper-Coated GMAW Wires

	Copper-Free	Copper-Coated
Arc starts	Best in class	Excellent
Arc stability	Best in class	Excellent
Feedability	Best in class	Excellent
Smoke	Best in class	
Spatter	Best in class Paint ready	Minimal
Copper flakes	None	
Deoxidizers	Alloy specific	Alloy specific
Weld appearance	Excellent	Excellent
Postweld cleaning	Alloy specific	Alloy specific

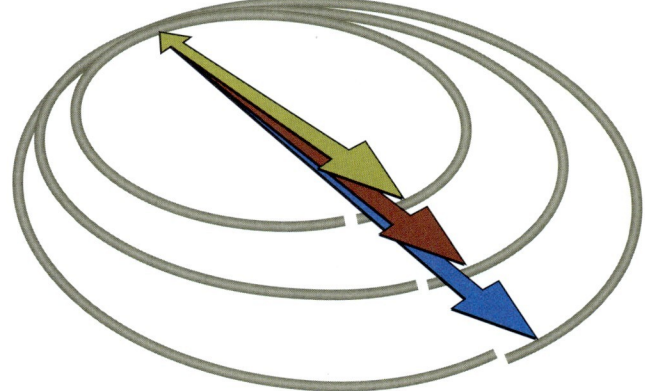

Fig. 21-64 Cast is the diameter of the circle formed by a length of wire thrown loosely on the floor. Cast is normally checked before it enters the wire-feed system. Source: National Standard

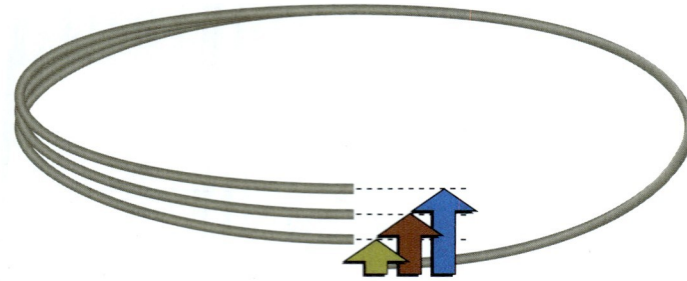

Fig. 21-65 Helix is the "pitch" of a single strand of weld wire measured as the distance one end of a strand of wire lying on a flat surface rises off that surface. Helix is normally checked before it enters the wire-feed system. Source: National Standard

electrodes. Much like adding iron powder to the coating of an E7018 SMAW coated electrode enhances its productivity over the E7016 electrode, the sheathing acts as the conductor. Since this presents much less area to the arc when compared to a solid wire, much higher current densities are achieved, which leads to a faster melt rate and higher deposition at a given amperage or heat input. The metal cored electrodes are intended for single or multipass applications. They are characterized by a spray arc and excellent bead wetting characteristics. The wetting action helps the weld metal to flow into the toes of the weld minimizing undercut. External shielding gas is required and is generally CO_2 or 75 to 80 percent Ar/balance CO_2.

Flux-Cored Electrode Wires Flux protects the weld pool during the process of solidification. This is done in several ways. In submerged arc welding and in most brazing operations, the flux is supplied as a bulk powder and added during welding. Stick electrodes carry their flux on the outside of the wire filler metal. Flux cored filler wires used for FCAW carry their flux as the core of the filler wire.

Flux cored filler wires are available in diameters of 0.030, 0.035, 0.045, 0.052, 0.068, $1/16$, $5/64$, $3/32$, $7/64$, $1/8$, and $5/32$ inch, and are provided in continuous coils. They usually require CO_2 or Ar-CO_2 shielding gas (FCAW-G).

The G is for gas shielded. Certain classifications of flux cored electrodes are designed to be used without shielding gas (FCAW-S). The S is for self-shielded. Flux cored electrodes are unique in their design, formulation, and manufacture. For these reasons the specific manufacturer's information must be consulted for proper operation, Table 21-15. The following information must be followed:

- Voltage range (as measured at the arc)
- Wire-feed speed (WFS) range
- Amperage range
- Deposition characteristics
- Electrode extension (may vary from ½ inch to as much as 4 inches)
- Single or multipass capability

The flux is primarily a method of carrying the alloying and deoxidizing elements to the weld. This is less costly than providing these elements through the sheath of the electrode. Flux cored wire provides the following advantages over stick electrodes:

- Lower cost to manufacture.
- Less chance for handling and shipping damage
- Continuous welding. There is no need to stop and change electrodes.
- Deeper penetration. Fewer passes are necessary when welding thicker sections.
- Greater arc time and no stub loss.
- V-groove butt joints with a 30° groove angle are possible instead of the 60° groove angle required with manual methods. Thus, there is a saving in welding time and filler metal costs.

Flux cored wire welding is a high-deposition process that is fast and economical. It can deposit metal at a higher rate than many other processes.

During the manufacture of most flux cored wire, a seam is formed along the length of the wire, Fig. 21-66, through this seam moisture can be absorbed. Even though flux cored arc welding electrodes are generally considered low hydrogen, this can present problems. Flux cored wires can absorb significant moisture if stored in a humid

Table 21-14 Chemical Compositions Metal Cored Electrode Wires[1]

AWS Classification	UNS Number	C	Mn	Si	S	P	Ni	Cr	Mo	V	Cu
E70C-3	W07703	0.12	1.75	0.90	0.03	0.03	[2]	[2]	[2]	[2]	0.50
E70C-6	W07706	0.12	1.75	0.90	0.03	0.03	[2]	[2]	[2]	[2]	0.50

[1]Weight in percentage. Single values are maximums.
[2]To be reported if intentionally added: the sum of Ni, Cr, Mo, and V shall not exceed 0.50%.

Table 21-15 Chemical Compositions Flux Cored Electrode Wires[1]

AWS Classification	UNS Number	C	Mn	Si	S	P	Cr[2]	Ni[2]	Mo[2]	V[2]	Al[2]	Cu[2]
E7XT-1	W07601											
E7XT-5	W07605	0.18	1.75	0.90	0.03	0.03	0.20	0.50	0.30	0.08	—	0.35
E7XT-9	W07609											
E7XT-4	W07604											
E7XT-6	W07606											
E7XT-7	W07607	Not specified, but determined and reported	1.75	0.60	0.03	0.03	0.20	0.50	0.30	0.08	1.8	0.35
E7XT-8	W07608											
E7XT-11	W07611											
E7XT-G	—	Not specified, but determined and reported	1.75	0.90	0.03	0.03	0.20	0.50	0.30	0.08	1.8	0.35
E7XT-12	W07612	0.15	1.60	0.90	0.03	0.03	0.20	0.50	0.30	0.08	—	0.35
E7XT-13	W06613											
E7XT-2	W07602											
E7XT-31	W07603		3	3	3	3	3	3	3	3	3	3
E7XT-10	W07610											
E7XT-14	W07613											
E7XT-GS	W07614											

[1]Weight in percentage. Single values are maximums.
[2]Reported only if intentionally added. Aluminum only for self-shielded electrodes; gas shielded electrodes need not have significant additions.
[3]Not specified.

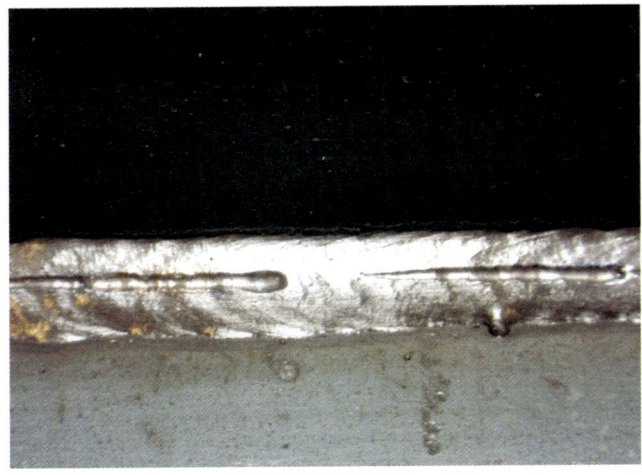

Fig. 21-66 A fillet weld made with the flux core process. Note the elongated surface porosity (worm tracks) due to contaminated wire and or too little electrode extension.
Edward R. Bohnart

environment, in damaged or open packages, or if left unprotected for extended periods. Even overnight, in some situations there is a possibility of hydrogen-induced cracking and/or porosity such as worm tracks. The electrode manufacturer's recommendations should be followed in regard to storage and reconditioning of the wire.

Filler Wire Classifications The American Welding Society worked many years in its attempts to classify continuous wires on a basis similar to those established for stick electrodes. For stick electrodes the fabricator specifies an AWS classification, such as E6010 or E7018. For continuous wires, they can now specify a wire classification such as ER70S-3 or ER70S-4.

The specifications cover mild steel electrodes for MIG/MAG welding. They include gas shielded solid and metal core wires, gas shielded and gasless flux cored wires. Solid wires carry the letter S, as in ER70S-3. All flux cored wires have the letter T for tubular, as in E71T-1. All metal cored wires have C for composite, as in E70C-. Electrode specifications booklets can be ordered from the American Welding Society. The booklet numbers are AWS A5.18 for solid and metal cored wires and AWS A5.20 for flux cored wires. The classification code is demonstrated in Fig. 21-67, page 696. In many respects it is somewhat similar to that used for shielded metal arc electrodes.

The specifications for SAW cover bare mild steel electrodes and fluxes for submerged arc welding. Electrode wires, divided into groups according to manganese content, and a number of companion powder fluxes are listed in the specifications. These specifications are also available from the American Welding Society. The booklet number is AWS A5.17.

1. GMAW Solid Electrode

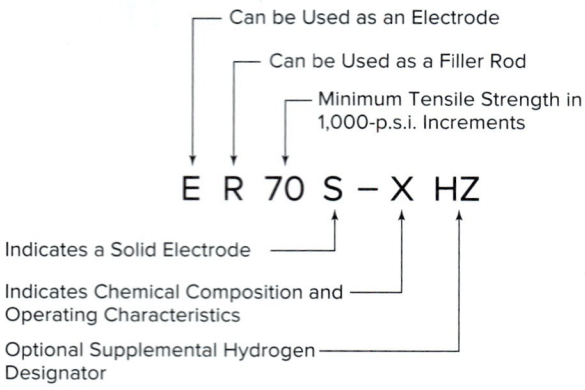

2. GMAW Metal Cored Electrode

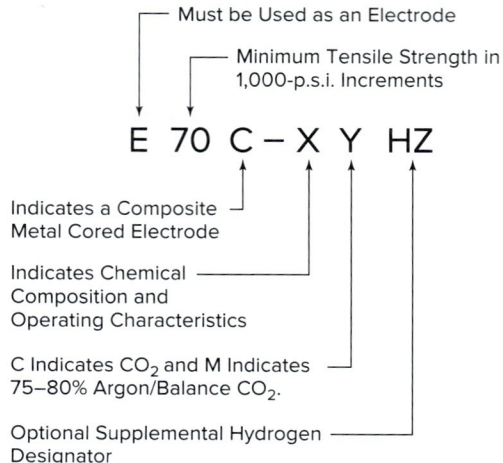

3. FCAW Tubular Flux Cored Electrode

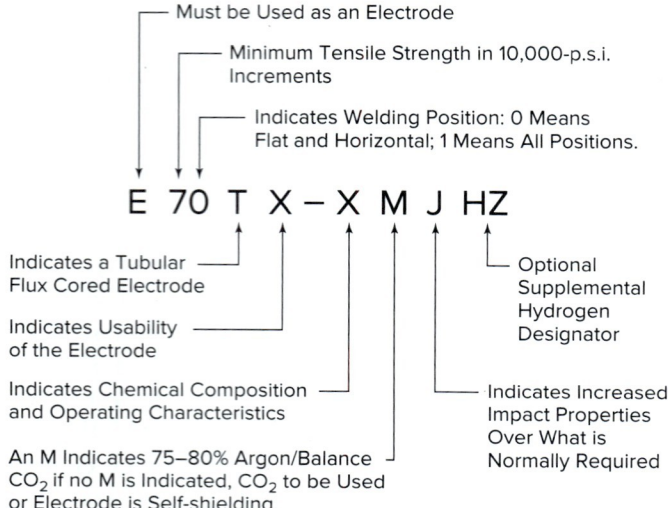

4. SAW Flux and Electrode Combination

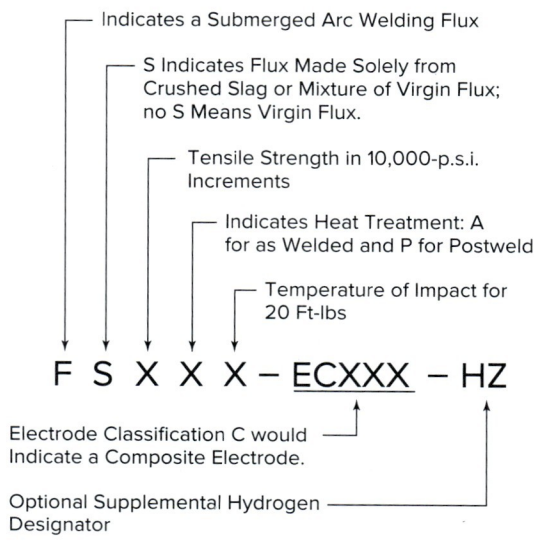

Fig. 21-67 American Welding Society electrode classifications.

In the welding of mild steel, there are enough continuous wires to replace any stick electrode used in shielded metal arc welding. In the flat position, the ER70T-1 electrode, for instance, is a good replacement for the E7010 stick electrode. The gasless cord wire, E70T-4, can replace the E6024. Solid wire, such as ER70S-3, is used for MIG/MAG welding in all positions.

The first classification in each of the three types (ER70S-2, E70T-1, and E70C-3) contains minimum amounts of deoxidizers (manganese and silicon). Study Table 21-16 and the following material carefully for an understanding of the nature of each filler wire.

ER70S-2 The ER70S-2 classification covers a grade of solid wire with considerable amounts of deoxidizers. Consequently, it can be used for welding steels with a rusty or dirty surface. There is some slight sacrifice of weld quality, depending on the degree of contamination. ER70S-2 electrodes can be used to weld semikilled, rimmed, and killed steels. They can be used with argon-oxygen mixtures, carbon dioxide, and argon-carbon dioxide mixtures. They are preferred for out-of-position welding by the short-circuiting type of metal transfer because of ease of operation. Typical steels welded with this electrode are ASTM A36, A285-C, A515-55, and A516-70.

ER70S-3 The wires in the ER70S-3 classification are very similar to those in the ER70S-2 classification except that greater amounts of deoxidizers (manganese and silicon)

Table 21-16 Continuous Wire Data for Carbon Steel

Electrodes	Comments	Shielding Gas	Minimum Tensile Strength (p.s.i.)	Minimum Yield Strength at 0.2% Offset (p.s.i)
Solid Wires				
ER70S-2	Triple-deoxidized wire (aluminum, zirconium, titanium). Premium quality for welding over dirty surfaces.	SG-AC SG-AO	70,000	58,000
ER70S-3	Most popular solid wire. Good substitute for E6012. Single- and multiple-pass welds. Similar to E60S-1, but contains more silicon. Meets AWS requirements with argon-oxygen or carbon dioxide.	SG-AC SG-AO	70,000	58,000
ER70S-4	Higher silicon than E60S-3. Also, higher tensile. Good where longer arcs or more deoxidants are needed.	SG-C SG-AC SG-AO	70,000	58,000
ER70S-5	Another triple-deoxidized wire (aluminum, manganese, silicon). Welds rusty steels; not available in fine wire.	SG-C	70,000	58,000
ER70S-6	Premium wire. Demand is building up for this single- and multiple-pass wire. Mechanical properties should be closely monitored for multipass welding with argon mixtures.	SG-C SG-AC SG-AO	70,000	58,000
ER70S-7	They may permit welding with higher travel speeds with better wetting action.	SG-C SG-AC SG-AO	70,000	60,000
ER70S-G	A catch-all classification.	SG-C SG-AC SG-AO	70,000	58,000
ER80S-D2	Has deoxidizers. Produces radiographic quality welds in low carbon and low alloy steels. Out-of-position welding. Single- and multiple-pass welds. Molybdenum is added for strength.	SG-C SG-AC SG-AO	80,000	68,000
Metal Cored Wires				
ER70C-3	For single- or multiple-pass applications, characterized by a spray arc and excellent bead wash characteristics. This electrode requires impact values at 0°F.			
ER70C-6	For single- or multiple-pass applications, characterized by a spray arc and excellent bead wash characteristics. This electrode requires impact values at −20°F.	SG-C and the "M" wires SG-AC	70,000	58,000
ER70C-G	Only certain mechanical property requirements are specified, for single- and multiple-pass applications. The filler metal supplier should be consulted for the composition, properties, characteristics, and intended use of these classifications.			
ER70C-GS	Only certain mechanical property requirements are specified, for single-pass applications only, tolerance to mill scale, etc. The filler metal supplier should be consulted for the composition, properties, characteristics, and intended use of these classifications.	SG-C and the "M" wires SG-AC	70,000	Not specified

(Continued)

Table 21-16 (Continued)

Electrodes	Comments	Shielding Gas	Minimum Tensile Strength (p.s.i.)	Minimum Yield Strength at 0.2% Offset (p.s.i.)
Flux Core Wires				
EXXT-1	For single- and multiple-pass welding. E70 position 1 and 2. E71 positions 1, 2, 3 up, and 4. Spray transfer, low spatter loss, flat to slightly convex bead contour and produce high deposition rates.	SG-C DCEP	70,000	58,000
EXXT-1M	For single- and multiple-pass welding. E70 position 1 and 2. E71 positions 1, 2, 3 up, and 4. Spray transfer, low spatter loss, flat to slightly convex bead contour and produce high deposition rates. One of the most popular gas shielded flux cored electrodes.	SG-AC DCEP	70,000	58,000
EXXT-2	Higher manganese and/or silicon content and are intended for single-pass applications only. E70 position 1 and 2. E71 positions 1, 2, 3 up, and 4. Welding of dirty metal.	SG-C DCEP	70,000	Not specified
EXXT-2M	Higher manganese and/or silicon content and are intended for single-pass applications only. E70 position 1 and 2. E71 positions 1, 2, 3 up, and 4. Welding of dirty metal.	SG-AC DCEP	70,000	Not specified
EXXT-3	Spray-type transfer, high welding speeds, for single-pass applications in the flat, horizontal, and vertical down on sheet metal. Should not be used on materials thicker than approx. 1/4 in.	No gas required. DCEP	70,000	Not specified
EXXT-4	Globular-type transfer, high deposition rates, low in sulphur welds, resistant to hot cracking, shallow penetration, for poor fit up. Single- and multiple-pass welding.	No gas required. DCEP	70,000	58,000
EXXT-5	Single- and multiple-pass welding. E70 position 1 and 2. E71 positions 1, 2, 3 up, and 4. Globular transfer, slightly convex bead contour, good impact properties and hot and cold crack resistance.	SG-C DCEP May use DCEN for out of position.	70,000	58,000
EXXT-5M	Single- and multiple-pass welding. E70 position 1 and 2. E71 positions 1, 2, 3 up, and 4. Globular transfer, slightly convex bead contour, good impact properties, and hot and cold crack resistance.	SG-AC DCEP May use DCEN for out of position.	70,000	58,000
EXXT-6	Spray-type transfer, good low temperature impact properties, good penetration, easy slag removal. For single- and multiple-pass applications in the flat and horizontal positions.	No gas required. DCEP	70,000	58,000
EXXT-7	Droplet- to spray-type transfer. E70 position 1 and 2. E71 positions 1, 2, 3 up, and 4. Single- and multiple-pass applications. Produce very low sulphur weld metal, resistant to cracking.	No gas required. DCEN	70,000	58,000
EXXT-8	Droplet- to spray-type transfer. E70 position 1 and 2. E71 positions 1, 2, 3 up, and 4. Single- and multiple-pass applications. Good low temperature notch toughness and crack resistance.	No gas required. DCEN	70,000	58,000
EXXT-9	For single- and multiple-pass welding. E70 position 1 and 2. E71 positions 1, 2, 3 up, and 4 with improved impact properties EXXT-1.	SG-C DCEP	70,000	58,000

Table 21-16 (Concluded)

Electrodes	Comments	Shielding Gas	Minimum Tensile Strength (p.s.i.)	Minimum Yield Strength at 0.2% Offset (p.s.i.)
Flux Core Wires *(Continued)*				
EXXT-9M	For single- and multiple-pass welding. E70 position 1 and 2. E71 positions 1, 2, 3 up, and 4 with improved impact properties EXXT-1.	SG-AC DCEP	70,000	58,000
EXXT-10	Small droplet-type transfer, high travel speeds, for single-pass applications in positions 1 and 2.	No gas required. DCEN	70,000	Not specified
EXXT-11	Smooth spray-type transfer, general-purpose electrodes for single- and multiple-pass welding. E70 position 1 and 2. E71 positions 1, 2, 3 down, and 4. Not recommended on thicknesses greater than 3/4 in. unless preheat and interpass temperature control are maintained. One of the most popular self-shielded flux cored electrodes.	No gas required. DCEN	70,000	58,000
EXXT-12	Improved impact properties to meet the lower manganese requirements, with decrease in tensile strength and hardness. Hardness levels must be checked for applicability. For single- and multiple-pass welding. E70 position 1 and 2. E71 positions 1, 2, 3 up, and 4.	SG-C DCEP	70,000–90,000	58,000
EXXT-12M	Improved impact properties to meet the lower manganese requirements, with decrease in tensile strength and hardness. Hardness levels must be checked for applicability. For single- and multiple-pass welding. E70 position 1 and 2. E71 positions 1, 2, 3 up, and 4.	SG-AC DCEP	70,000–90,000	58,000
E61T-13	Usually welded with a short-arc transfer, for positions 1, 2, 3 down, and 4 on single-pass welds only. Cross country line pipe root pass only.	No gas required. DCEN	60,000	Not specified
E71T-13	Usually welded with a short-arc transfer, for positions 1, 2, 3 down, and 4 on single-pass welds only. Cross country line pipe root pass only.	No gas required. DCEN	70,000	Not specified
E71T-14	Smooth spray-type transfer. For positions 1, 2, 3 down, and 4, at high travel speeds. For single-pass welding on metal up to 3/16 in. thick and for coated steels. Electrode manufacturer should be consulted for specific recommendations.	No gas required. DCEN	70,000	Not specified
E6XT-G	Multiple-pass electrodes that are not covered by any presently defined classifications. Except to ensure a carbon steel deposit and tensile strength as specified. EX0 positions 1 and 2, EX1 all positions.	Not specified	60,000	48,000
E7XT-G	Multiple-pass electrodes that are not covered by any presently defined classifications, except to ensure a carbon steel deposit and tensile strength as specified. EX0 positions 1 and 2, EX1 all positions.	Not specified	70,000	58,000
E6XT-GS	Single-pass electrodes that are not covered by any presently defined classifications, except to ensure a carbon steel deposit and tensile strength as specified. EX0 positions 1 and 2, EX1 all positions.	Not specified	60,000	Not specified
E7XT-GS	Single-pass electrodes that are not covered by any presently defined classifications, except to ensure a carbon steel deposit and tensile strength as specified. EX0 positions 1 and 2, EX1 all positions.	Not specified	70,000	Not specified

> **ABOUT WELDING**
>
> **Shielding Gas**
> Shielding gas prevents atmospheric contamination of the weld pool.

are included. They should be used with argon-mixed shielding gas. These wires are widely used in production welding on single-pass weldments although they can be used on multipass welds in killed or semikilled steels. They can be used for out-of-position welding with small diameter electrodes, the short-circuiting type of metal transfer, and argon-carbon dioxide mixture as gas shielding.

ER70S-4 These wires contain slightly greater quantities of silicon and produce a weld deposit of higher tensile strength than the two previous classifications. They are used with carbon dioxide shielding and a somewhat longer arc. This classification does not require impact testing.

ER70S-5 The ER70S-5 classification contains aluminum in addition to manganese and silicon as deoxidizers. It can be used when welding rimmed, killed, or semikilled steels with carbon dioxide shielding gas and high welding currents. ER70S-5 wires can be used on dirty steels. They are not used with the short-circuiting type of metal transfer.

ER70S-6 These wires contain additional manganese and still higher amounts of silicon. Welds have a high resistance to impact when shielded with carbon dioxide. ER70S-6 wires may be used with high currents for welding rimmed steels and on sheet metal where smooth weld beads are desired. They can be used for out-of-position welding with the short-circuiting type of metal transfer. They are suitable for moderately dirty steels. These wires exhibit outstanding mechanical properties and increased hot-short crack resistance, especially on high carbon steels such as rail steel.

ER70S-7 These wires are intended for single- and multiple-pass welding. They may permit welding with higher travel speeds compared with ER70S-3 filler metals. They also provide somewhat better wetting action and bead appearance when compared with those filler metals. Typical base metal specifications are often the same as those for the ER70S-2 classifications.

ER70S-G (general) This classification includes those solid electrodes that are not included in the mild steel classification. They include certain types of alloy electrodes. The filler metal supplier should be consulted for the composition, properties, characteristics, and intended use of these classifications.

ER-80S-D2 These electrodes contain a properly balanced chemical content with adequate deoxidizers to control porosity during welding with carbon dioxide as the shielding gas. They give X-ray quality welds to both standard and difficult-to-weld low carbon and low alloy steels. They can be used in out-of-position welding and for single- and multiple-pass welding. Molybdenum has been added for greater strength.

E70C-3 These wires are metal cored electrodes and intended for single- or multiple-pass applications. They are characterized by a spray arc and excellent bead wash characteristics. The electrodes may be classified with either carbon dioxide or 75 to 80 percent argon with the balance carbon dioxide. This is shown with the *C* or *M* suffix. This electrode requires impact values at 0°F.

E70C-6 These wires are metal cored electrodes and intended for single- or multiple-pass applications. They are characterized by a spray arc and excellent bead wash characteristics. The electrodes may be classified with either carbon dioxide or 75 to 80 percent argon with the balance carbon dioxide. This is shown with the *C* or *M* suffix. This electrode requires impact values at −20°F.

E70C-G These wires are those metal cored filler metals not included in the preceding classes and for which only certain mechanical property requirements are specified. They are intended for single- and multiple-pass applications. The filler metal supplier should be consulted for the composition, properties, characteristics, and intended use of these classifications. The proper shielding gas must be agreed upon by the purchaser and supplier.

E70C-GS These wires are those metal cored filler metals not included in the preceding classes and for which only certain mechanical property requirements are specified. They are intended for single-pass applications only. These electrodes may have higher alloy contents that improve single-pass applications (such as tolerance to mill scale, etc.) but could preclude their use on multiple-pass applications due to higher alloy recovery. The filler metal supplier should be consulted for the composition, properties, characteristics, and intended use of these classifications. The proper shielding gas must be agreed upon by the purchaser and supplier.

EXXT-1 and EXXT-1M These flux-cored tubular wire groups are classified with carbon dioxide shielding gas. However, other gas mixtures such as SG-AC may be used to improve the arc characteristics, especially for out-of-position work, when recommended by the manufacturer. Increasing the argon in the SG-AC mixture will increase the manganese and silicon contents in the weld metal. The increased manganese and silicon will increase the yield and tensile strengths and may affect impact properties.

Electrodes of the EXXT-1M are intended to be used with 75 to 80 percent argon/balance carbon dioxide shielding gas. With greater percentages of carbon dioxide in the shielding gas, arc characteristics and out-of-position characteristics will deteriorate. This will also cause a reduction of the manganese and silicon in the weld metal and will reduce the yield and tensile strengths and may affect impact properties.

Both EXXT-1 and EXXT-1M electrodes can be used for single- and multiple-pass welding using DCEP polarity. The larger diameter wires (usually $5/64$ inch and larger) are used for welding in the flat and horizontal positions (EX0T-1 and EX0T-1M), while the smaller diameter wires (usually $1/16$ inch and smaller) are generally used for welding in all positions (EX1T-1 and EX1T-1M).

These electrodes are characterized by a spray transfer, low spatter loss, flat to slightly convex bead contour, and moderate volume of slag which completely covers the weld bead. Most electrodes of this type have a rutile base slag and produce high deposition rates.

EXXT-2 and EXXT-2M These wires are essentially the same as the EXXT-1 and 1M types. They have higher manganese and/or silicon content and are intended for single-pass applications only. They can be used for welding base metals that have heavier mill scale, rust, or other foreign matter that cannot be tolerated by some electrodes.

EXXT-3 These wires are of the self-shielded type and are used on DCEP polarity with a spray-type transfer. High welding speeds are provided by the slag system. These electrodes are intended for single-pass applications in the flat, horizontal, and vertical (up to 20° incline) positions (vertical down progression) on sheet metal. These electrodes are sensitive to the effects of base-metal quenching and should not be used on T-joints or lap joint on materials thicker than $3/16$ inch, or on butt, edge, or corner joints on materials thicker than $1/4$ inch. Consult with the electrode manufacturer for specific recommendations.

EXXT-4 These wires are of the self-shielded type and are used on DCEP polarity with a globular-type transfer. High deposition rates and welds that are very low in sulphur and thus very resistant to hot cracking are provided by the slag system. Very low penetration is typical, so they can be used on joints that have been poorly fit and for single- and multiple-pass welding.

> **SHOP TALK**
>
> **MIG/MAG Gun Design**
> Some MIG/MAG guns are built so that you can easily separate the body tube from the handle. This design will allow you to quickly fix a jammed wire, for instance, one that has flaking copper. To get rolling again, you simply take out the short worn section of jumper liner from the tube and put in a new one.

EXXT-5 and EXXT-5M These wires are intended to be used with carbon dioxide shielding gas; however, as with the EXXT-1 classification they may be used with SG-AC mixtures to reduce spatter. The *M* classifications are intended to be used with 75 to 80 percent argon/balance carbon dioxide. These electrodes are primarily used for single- and multiple-pass welding in the flat position and for fillet welds in the horizontal position. They have a globular-type transfer, slightly convex bead contour, and a thin slag that may or may not completely cover the weld bead. The base slag is lime-fluoride that produces deposits that have impact properties and hot and cold crack resistance greater than those obtained with the rutile base slags. With DCEN they can be used in all positions; however, the welder appeal of these electrodes is not as good as the rutile base slags.

EXXT-6 These wires are self-shielding and operate on DCEP with a spray-type transfer. They have good low-temperature impact properties, good penetration, and excellent slag removal due to their slag system. These electrodes are used for single- and multiple-pass applications in the flat and horizontal positions.

EXXT-7 These electrodes are self-shielding and operate on DCEN and have a droplet- to spray-type transfer. The larger sizes are to be used in the flat and horizontal position, while the smaller sizes can be used in all positions. These electrodes can be used for single- and multiple-pass applications and produce very low sulphur weld metal that is very resistant to cracking.

EXXT-8 These electrodes are self-shielding and operate on DCEN and have a droplet- to spray-type transfer. These electrodes are suitable for welding in all positions, and the weld metal has very good low temperature notch toughness and crack resistance. The electrodes are used for single- and multiple-pass welds.

EXXT-9 and EXXT-9M These flux cored tubular wire groups are classified with carbon dioxide shielding gas. However, other gas mixtures such as SG-AC may be used to improve the arc characteristics, especially for out-of-position work, when recommended by the manufacturer. Increasing the argon in the SG-AC mixture will increase the manganese and silicon contents in the weld metal. The increased manganese and silicon will increase the yield and tensile strengths and may affect impact properties.

EXXT-9M electrodes are intended to be used with 75 to 80 percent argon/balance carbon dioxide shielding gas. With greater percentages of carbon dioxide in the shielding gas, arc characteristics and out-of-position characteristics will deteriorate. This will also cause a reduction of the manganese and silicon in the weld metal and will reduce the yield and tensile strengths and may affect impact properties.

Both EXXT-9 and EXXT-9M electrodes can be used for single- and multiple-pass welding using DCEP polarity. The larger diameter wires (usually $5/64$ inch and larger) are used for welding in the flat and horizontal positions (EX0T-9 and EX0T-9M), while the smaller diameter wires (usually $1/16$ inch and smaller) are generally used for welding in all positions (EX1T-9 and EX1T-9M). These electrodes are essentially EXXT-1 and 1M types with improved impact properties.

Some electrodes in this classification require that joints be relatively clean and free of oil, excessive oxide, and scale in order that welds of radiographic quality can be obtained.

EXXT-10 These wires are of the self-shielded type and are used on DCEN polarity with a small droplet-type transfer. High welding speeds are provided by the slag system. These electrodes are intended for single-pass applications in the flat, horizontal, and vertical (up to 20° incline) positions.

EXXT-11 These wires are of the self-shielded type and are used on DCEN polarity with a smooth spray-type transfer. They are general-purpose electrodes for single- and multiple-pass welding in all positions. Their use is generally not recommended on thicknesses greater that ¾ inch unless preheat and interpass temperature control are maintained. The electrode manufacturer should be consulted for specific recommendations.

EXXT-12 and EXXT-12M These electrodes are essentially EXXT-1 and 1M types with improved impact properties to meet the lower manganese requirements of the A-1 analysis group in the ASME Boiler and Pressure Vessel Code, Section IX. They, therefore, have an accompanying decrease in tensile strength and hardness. Since the welding procedure will influence all weld metal properties, users are urged to check hardness on any application where there is a required hardness level.

EXXT-13 These wires are of the self-shielded type and are used on DCEN polarity and are usually welded with a short-arc transfer. They are designed for all-position welding on the root pass of pipe joints specifically. They are not recommended for multiple-pass welding.

EXXT-14 These wires are of the self-shielded type and are used on DCEN polarity with a smooth spray-type transfer. They are designed for all-position welding at high travel speeds. They are typically used on sheet metal up to $3/16$ inch thick and are specially formulated for use on galvanized, aluminized, or other coated steels. These electrodes are sensitive to the effects of base metal quenching and should not be used on T-joints or lap joints on materials thicker than $3/16$ inch, or on butt, edge, or corner joints on materials thicker than ¼ inch. Consult the electrode manufacturer for specific recommendations.

EXXT-G This classification is for multiple-pass electrodes that are not covered by any presently defined classifications. Except for chemical requirements to ensure a carbon steel deposit and the specified tensile strength, the requirements for this classification are not specified. The requirements are those that are agreed to by the purchaser and the supplier.

EXXT-GS This classification is for single-pass electrodes that are not covered by any presently defined classifications. Except for the specified tensile strength, the requirements for this classification are not specified. The requirements are those that are agreed to by the purchase and the supplier.

The specifications for mild steel electrodes and fluxes used for submerged arc welding are described in Chapter 23.

Summary

You are now ready to begin practice with the GMAW and FCAW processes. Because of the thorough training you have had up to this point with the other welding processes, you will find that welding with these processes will be relatively easy to learn. The learning situation can be improved by mastering the material covered in this chapter. Any skill is more readily mastered if the technical and related information is understood so that it can be applied while practicing the skill.

Tables 21-17A, 21-17B, and 21-18 (p. 704) summarize the basic information concerning the various processes and the application of the various gases and filler metal. Become so familiar with the information contained in these tables that you will know how to meet each welding problem as it develops in the shop and field. Table 21-19 (p. 705) lists various AWS filler metal specifications. Your school technical library should be able to provide them for review.

Table 21-17A GMAW Filler Metals

Base Metal Type	Recommended Electrode – Material Type	Recommended Electrode – Electrode Classification	AWS Filler Metal Specification (Use Latest Edition)	Current Range – Electrode Diameter (in.)	Current Range – Amperes
Aluminum and aluminum alloys	1100	ER1100 or ER4043	A5.10	0.030	50–175
	3003, 3004	ER1100 or ER5356		3/64	90–250
	5052, 5454	ER5554, ER5356, or ER5183		1/16	160–350
				3/32	225–400
	5083, 5086, 5456	ER5556 or ER5356		1/8	350–475
	6061, 6063	ER4043 or ER5356			
Magnesium alloys	AZ10A	ERAZ61A, ERAZ92A	A5.19	0.040	150–300*
	AZ31B, AZ61A AZ80A	ERAZ61A, ERAZ92A		3/64	160–320*
	ZE10A	ERAZ61A, ERAZ92A		1/16	210–400*
	ZK21A	ERAZ61A, ERAZ92A		3/32	320–510*
	AZ63A, AZ81A AZ91C	ERAZ92A		1/8	400–600*
	AZ92A, AM100A	ERAZ92A			
	HK31A, HM21A, HM31A	EREZ33A			
	LA141A	EREZ33A			
Copper and copper alloys	Deoxidized copper	ECu	A 5.6	0.035	150–300
	Cu-Ni alloys	ECuNi		0.045	200–400
	Manganese bronze	ECuAl-A2		1/16	250–450
	Aluminum bronze	EcuAl-B		3/32	350–550
	TW bronze	ECuSn-A			

*Spray transfer mode.

Table 21-17B GMAW Filler Metals

Base Metal Type	Recommended Electrode – Material Type	Electrode Classification	AWS Filler Metal Specification (Use Latest Edition)	Current Range – Electrode Diameter (in.)	Current Range – Amperes
Nickel and nickel alloys	Monel* Alloy 400	ERNiCu-7	A5.14	0.020	
	Inconel* Alloy 600	ERNiCrFe-5		0.030	
				0.035	100–160
				0.045	150–260
				1/16	100–400
Titanium and titanium alloys	Commercially pure	Use a filler metal one or two grades lower	A5.16	0.030	
	Ti-0.15Pd	ERTi-0.2Pd		0.035	
	Ti-5A1-2.5Sn	ERTi-5A1-2.5Sn or comm. pure		0.045	
Austenitic stainless steels	Type 201	ER308	A5.9	0.020	
	Types 301, 302, 304, & 308	ER308		0.025	
	Type 304L	ER308L		0.030	75–150
	Type 310	ER310		0.035	100–160
	Type 316	ER316		0.045	140–310
	Type 321	ER321		1/16	280–450
	Type 347	ER347		5/64	
				3/32	
				7/64	
				1/8	
Carbon steels	Hot-rolled or cold-drawn plain carbon steels	ER70S-3, or ER70S-1	A5.18	0.020	
		ER70S-2, ER70S-4		0.025	
		ER70S-5, ER70S-6		0.030	
				0.035	40–220
				0.045	60–280
				0.052	125–380
				1/16	160–450
				5/64	275–475
				3/32	
				1/8	

*Trademark of the International Nickel Co.

Table 21-18 GMAW, FCAW, and SAW Process Selection

	Submerged Arc	GMAW Inert Gas Shielded	GMAW Carbon Dioxide Shielded		FCAW	
			Large Wire	Small Wire	–S	–G
Metals to be welded	Low carbon and medium carbon steels, low alloy high strength steels, quenched and tempered steels, many stainless steels	Aluminum and aluminum alloys, stainless steels and phosphorus steels, nickel and nickel alloys, copper alloys, titanium, etc., as well as carbon steels	Low carbon and medium carbon steel, low alloy high strength steels, some stainless steels		Rebuilding and hard-surfacing most steels	Low carbon and medium carbon steels, low alloy high strength steels
Metal thickness	16 gauge (0.062 in.) to ½ in. (no preparation); single pass to 1½ in. and thicker; maximum thickness practically unlimited	12 gauge (0.109 in.) to ⅜ in. (no preparation); single-pass welding up to 1 in.; maximum thickness practically unlimited	10 gauge (0.140 in.) up to ½ in. (no preparation); practical max. 1 in.	20 gauge (0.038 in.) up to ¼ in.; above ¼ in. not economical	Primarily for hard-surfacing	¼ to ½ in. (no preparation); maximum thickness practically unlimited
Welding positions	Flat and horizontal	All positions	Flat and horizontal	All positions	Flat	Flat and horizontal
Major advantages	Highest deposition rate; wide range of materials; wide range of thickness; no visible arc; (no arc shielding necessary)	Welds most nonferrous metals; no flux required; all positions (with small wire); visible arc; negligible clean-up	Low cost; high speed; deep penetration; visible arc	All positions; thin material; will bridge gaps; minimum clean-up; visible arc	Fast deposition; simple; rugged	Smooth surface; deep penetration; sound welds; visible arc
Limitations	Flux removal and handling; visibility obstructed; possible flux entrapment	Minimum thickness limited; cost of gases; spatter removal sometimes required	Uneconomical in heavy thicknesses; gas coverage essential		Minimum thickness limited	Lower efficiency; slag removal
Quality	Good properties, X-ray	Good properties, X-ray	Good properties, X-ray		High hardness	X-ray
Appearance of weld	Very smooth surface; no spatter; flat contour	Fairly smooth convex surface	Relatively smooth; some spatter	Smooth surface; relatively minor spatter	Relatively smooth	Smooth surface; some spatter
Travel speeds	Up to 200 in./min	Up to 100 in./min	Up to 300 in./min	Max. 30 in./min (semiautomatic)	Up to 20 lb/h	Up to 150 in./min
Range of wire sizes (in.)	Diameter—1/16, 5/64, 3/32, 1/8, 5/32, 3/16, 7/32, 1/4	Diameter—0.023/0.025, 0.045, 1/16, 3/32, 1/8	Diameter 0.045, 1/16, 5/64, 3/32, 1/8	Diameter 0.023/0.025, 0.035, 0.045	Diameter 3/32, 7/64, 1/8, 0.030, 0.035, 0.045, 1/16	
Range of welding current	50 min.—1,200 max. amps.	50 min.—600 max. amps.	75 min.—900 max. amps.	25 min.—200 max. amps.	200 min.—500 max. amps.	300 min.—700 max. amps.
Electrode and shielding costs	Least expensive electrode wire; flux relatively inexpensive	Expensive electrode wires; relatively expensive gas	Reasonably priced electrode wires; least expensive gas		Fairly expensive electrode wire	Relatively expensive wire; least expensive gas

Table 21-18 (Concluded)

		Submerged Arc	GMAW Inert Gas Shielded	GMAW Carbon Dioxide Shielded — Large Wire	GMAW Carbon Dioxide Shielded — Small Wire	FCAW –S	FCAW –G
Equipment required	Overall welding costs	Least expensive for heavy metal	Least expensive for nonferrous metals	Least expensive for medium thickness	Least expensive for thin material	Less expensive than manual	Least expensive on low alloy steels
	Welding power	Constant voltage rectifier conventional for semiautomatic 500–1,000 amps	Constant voltage rectifier or engine generator 200, 300, 500, 900 amps	Constant voltage rectifier or engine generator 500, 900 amps.	200, 300 amps (and 350A gas engine).	Conventional or constant voltage or engine generator rectifier 500, 900 amps	
	Wire feeder	Constant speed or arc voltage control (semiautomatic)	Constant speed head with gas valve	Constant speed head with gas valve		Constant speed or arc voltage control head	Constant speed head with gas valve
	Nozzle or gun required — Automatic	Flux nozzle (waterless)	Concentric gas nozzle (water-cooled)	Side-delivery gas nozzle (waterless)		Waterless nozzle	
	Nozzle or gun required — Semi-automatic	Flux flow or flux gun and hopper	Pistol-grip gun (water-cooled)	Water-cooled or waterless guns		Open arc gun (waterless)	Pistol-grip gun (watercooled or waterless)
	Shielding and regulation	Flux hopper or flux flow flux recovery unit	Gas regulator and flowmeter	Gas regulator and flowmeter		None	Gas regulator and flowmeter
Consumables	Wire	Match base metal	Match base metal	Match base metal		Specified usage	Match base metal
	Shielding medium	Pellitized flux	Argon or helium (and inert mixtures)	Carbon dioxide and argon-carbon dioxide mixtures		None	Carbon dioxide

Table 21-19 AWS Filler Metal Specifications

Base Metal Type	GMAW	FCAW	SAW	ESW	EGW
Carbon steel	+5.20	+5.20	A5.17	A5.25	A5.26
Low alloy steel	+5.28	+5.29	A5.23	A5.25	A5.26
Stainless steel	A5.9				
	A5.22	A5.22	A5.9	A5.9	A5.9
Cast iron	A5.15	A5.15			
Nickel alloys	A5.14	A5.34	A5.14	A5.14	
Aluminum alloys	A5.10				
Copper alloys	A5.7				
Titanium alloys	A5.16				
Ziroconium alloys	A5.24				
Magnesium alloys	A5.19				
Surfacing alloys	A5.21	A5.21	A5.21		
Shielding gases	A5.32	A5.32			A5.32

CHAPTER 21 REVIEW

Multiple Choice

Choose the letter of the correct answer.

1. The American Welding Society uses the abbreviation GMAW to refer to _____. (Obj. 21-1)
   a. Gas metal active welding
   b. Gas metal arc welding
   c. Gaseous metallica action welding
   d. Got material anytime welding

2. The American Welding Society uses the abbreviation FCAW to refer to _____. (Obj. 21-1)
   a. Forced capillary action welding
   b. Flux composite arc welding
   c. Flux cored arc welding
   d. Friction composite action welding

3. Which mode of transfer is used for light gauge metal and to minimize spatter? (Obj. 21-1)
   a. Spray
   b. Globular
   c. Short circuit
   d. Pulse

4. Which type of power source is used for GMAW and FCAW primarily? (Obj. 21-2)
   a. C.V.
   b. C.C.
   c. Constant voltage
   d. Both a and c

5. Which controls are found on GMAW and FCAW equipment? (Obj. 21-2)
   a. Voltage, wire-feed speed, slope, and inductance
   b. Voltage, wire-feed speed, arc length, and amperage
   c. Slope, inductance, limited current change, and rate of current change
   d. Slope of the gun angle, reluctance, and power

6. Which type of power source is most effective in providing all the various modes of metal transfer? (Obj. 21-2)
   a. Engine-driven generator
   b. Transformer-rectifier
   c. Inverter
   d. Both a and b

7. When the gun trigger is depressed by the welder, what will take place? (Obj. 21-2)
   a. Nothing unless the work cable is properly connected to the work.
   b. The gas solenoid will activate, the wire feeder will start to feed wire, and the welding current contactor will energize.
   c. The wire feeder control will turn on, the power source will turn on, and the gas cylinder valve will open.
   d. The gun angle will be corrected, electrode extensions will be corrected, and travel speed will be properly set.

8. Shielding gases are used to protect the molten metal from _____. (Obj. 21-3)
   a. Heat and distortion
   b. Being overheated and cooling too fast
   c. Porosity and brittleness
   d. Being too cold and not penetrating

9. List a reactive gas used for MAG welding. (Obj. 21-3)
   a. Carbon dioxide
   b. Oxygen
   c. Argon plus oxygen
   d. All of these

10. Electrodes used for GMAW and FCAW are _____. (Obj. 21-3)
    a. Solid
    b. Metal cored
    c. Flux cored
    d. All of these

Review Questions

Write the answers in your own words.

11. List various slang and trade names for the GMAW process. (Obj. 21-1)

12. List at least six advantages of the GMAW and FCAW processes. (Obj. 21-1)

13. List the modes of metal transfer with the GMAW process and describe each. (Obj. 21-1)

14. Draw and label the arc and gun nozzle area for GMAW, FCAW-G, and FCAW-S. (Obj. 21-2)

15. List five items necessary for the constant voltage power source and constant speed wire feeder to function properly to sustain the welding arc. (Obj. 21-2)

16. Describe the use of the following GMAW and FCAW machine controls: wire-feed speed control, voltage control, slope control, inductance control. (Obj. 21-2)

17. List the factors that go into a GMAW or FCAW gun rating. (Obj. 21-2)
18. List the GMAW, FCAW, and SAW electrodes and in the case of SAW the type of flux combination being used in your welding lab. (Obj. 21-2)
19. List the shielding gases you are using in your welding lab. (Obj. 21-3)
20. Draw the weld profile for each of the shielding gas, electrode wire combinations you are using in your welding lab. (Obj. 21-3)

INTERNET ACTIVITIES

Internet Activity A

Use the Internet to find out about track (railroad) welding. Are most people prepared to do track welding after leaving school? What kinds of equipment do you need?

Internet Activity B

See what you can find out about friction stir welding on the Internet. What are the features on a friction stir welder? What are some advantages of using friction stir welding?

Design credits: Cinema and movie icon: Denis Meshkov/123RF; and Illustration of Welding icons: Mr. Nachai Sorasee/123RF

22

Gas Metal Arc Welding Practice with Solid and Metal Core Wire:
Jobs 22-J1–J23 (Plate)

The practice jobs outlined on the following pages include GMAW practice on a variety of materials in all positions, Fig. 22-1. Since you are already an accomplished welder, having mastered the gas welding, shielded metal arc, and gas tungsten arc processes, it will not be very difficult to master this new process. It is estimated that most students will become competent in 30 to 45 hours of practice. Each job requires some of the skills already learned. This process and the jobs provided also require the learning of new skills and the application of the technical information given in this chapter and Chapter 21. As with all previous practice, each job should be mastered before going on to the next.

Operating Variables That Affect Weld Formation

Welding variables are those factors that affect the operation of the arc and the weld deposit. Sound welding of good appearance results when the variables are in balance. In order for you to develop a feel for the process and understand the arc characteristics and metal formation, it is necessary to become familiar with all the variables and study their effect on the weld deposit through experience. As an advanced student, you will be familiar with some of the variables since they are present in all welding processes to a certain extent. You will find,

Chapter Objectives

After completing this chapter, you will be able to:

22-1 Describe GMAW operating variables.
22-2 Describe GMAW weld defects.
22-3 Describe GMAW safe operation.
22-4 Describe and demonstrate proper care, use, and troubleshooting of equipment.
22-5 Describe and demonstrate welding techniques.
22-6 Make various groove and fillet welds with the various modes of metal transfer with both solid and metal cored electrodes.

Fig. 22-1 This student is practicing gas metal arc welding. Note the dual-digital wire-feeder in the back ground, the wire cutter and practice coupon. Location: Northeast Wisconsin Technical College Mark A. Dierker/McGraw Hill

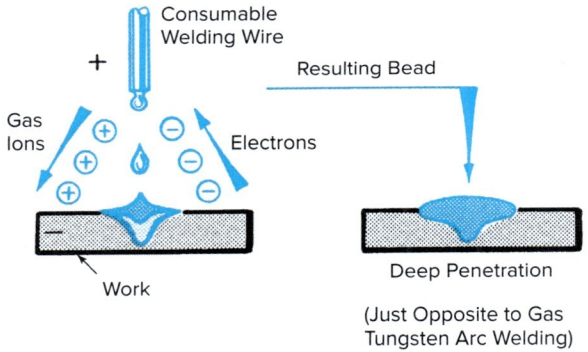

Fig. 22-2 Gas metal arc DCEP welding: wire positive, work negative.

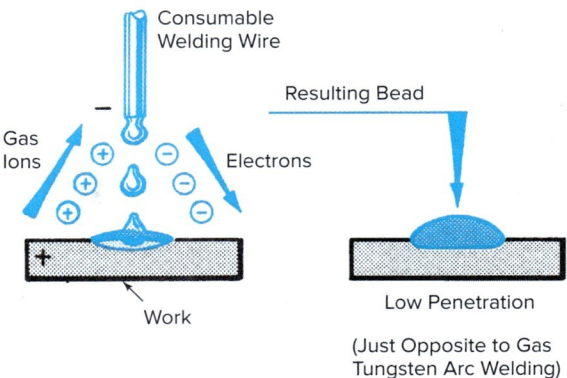

Fig. 22-3 Gas metal arc DCEN welding: wire negative, work positive.

however, that each electric welding process has its own arc characteristics.

Type of Current

Direct Current Electrode Positive (DCEP) Electrode positive is generally used for gas metal arc welding. Because it provides maximum heat input into the work, it allows relatively deep penetration to take place, Fig. 22-2. It also assists in the removal of oxides from the plate, which contributes to a clean weld deposit of high quality. Low current values produce the globular transfer of metal from the electrode. A gradual increase in current increases the electrode melting rate so that at relatively high current values the spray transfer of metal is produced. Although current is the first requirement to achieve a spray transfer, shielding gas and voltage are also very important. On carbon steel, the shielding gas must contain a minimum of 80 percent argon. The voltage level must be high enough to just keep the electrode from dipping into the weld pool.

The welding of ferrous metals with DCEP, solid or metal cored electrode, and pure argon causes the arc to be unstable and introduces porosity into the weld metal. This condition may be corrected by the addition of 2 to 5 percent oxygen to the gas mixture.

Direct Current Electrode Negative (DCEN) Electrode negative has a limited use in the welding of thin gauge materials. The greatest amount of heat occurs at the electrode tip. The wire meltoff rate is a great deal faster than with DCEP. The arc is not stable at the end of the filler wire, and it becomes very difficult to direct the transfer of weld metal where it is desired. The erratic arc results in poor fusion and a considerable amount of spatter. Penetration is also less than with DCEP, Fig. 22-3.

The unstable arc can be corrected to a considerable extent by the use of a shielding gas mixture of approximately 5 percent oxygen added to argon. The normally high electrode meltoff rate, however, is reduced substantially when oxygen is added so that any advantage of DCEN is cancelled out.

Alternating Current Alternating current is seldom used in gas metal arc welding. The arc is unstable because current and voltage both pass through the zero point many times each second as the current reverses. Since alternating current is a combination of DCEN and DCEP polarity, the rate of metal transfer and the depth of penetration fall between those of both polarities. It has found some use for the welding of aluminum.

Shielding Gas

Argon and helium were first used as the shielding gases for the gas metal arc process, and they continue to be the basic gases. Argon is used more than helium on ferrous metals to keep spatter at a minimum. Argon is also heavier than air and therefore gives better weld coverage. Oxygen or carbon dioxide is added to the pure gases to improve arc stability, minimize undercut, reduce porosity, and improve the appearance of the weld.

Helium may be added to argon to increase penetration with little effect on metal transfer characteristics. Hydrogen and nitrogen are used for only a limited number of special applications in which their presence will not cause porosity or embrittlement of the weld metal.

Carbon dioxide is popular as a shielding medium because of the following advantages:

- Low cost
- High density, resulting in low flow rates
- Less burn-back problems because of its shorter arc characteristics.

Review Chapters 18 and 21 for additional information regarding the shielding gases.

The following recommendations concerning specific metals are helpful:

Aluminum alloys: argon. With direct current electrode positive, argon removes surface oxide.
Magnesium and aluminum alloys: 75 percent helium, 25 percent argon. Correct heat input reduces the tendency toward porosity and removes surface oxide.
Stainless steels: argon plus oxygen. When direct current electrode positive is used, 2 percent oxygen improves arc stability.
Magnesium: argon. With direct current, electrode positive, argon removes surface oxide.
Deoxidized copper: 75 percent helium, 25 percent argon preferred. Good wetting and increased heat input counteract high thermal conductivity.
Low alloy steel: argon, plus 2 percent oxygen. Oxygen eliminates undercutting tendencies and removes oxidation.
Mild steel: 15 percent argon, 25 percent carbon dioxide (dip transfer); 100 percent CO_2 may also be used with deoxidized wire. They promote high quality. It is suitable for low current, out-of-position welding. There is little spatter.
Nickel, Monel®, and Inconel®: argon. Good wetting increases the fluidity of the weld metal.
Titanium: argon. Argon reduces the size of the heat-affected zone and improves metal transfer.
Silicon bronze: argon. Argon reduces crack sensitivity on this hot-short material.
Aluminum bronze: argon. Argon reduces penetration of the base metal. It is commonly used as a surfacing material.

Joint Preparation

Joint designs like those used with other arc welding processes may be used with the gas metal arc process. Costs can be reduced, however, by employing somewhat different designs. Any joint design should provide for the most economical use of filler metal. The correct design for a particular job depends on the type of material being welded, the thickness of the material, the position of welding, the welding process, the final results desired, the type and size of filler wire, and welding technique.

The arc in gas metal arc welding is somewhat more penetrating and narrower than the arc in shielded metal arc welding. Therefore, heavier root faces and smaller root openings may be used for groove welds. It is also possible to provide a narrower groove angle for MIG/MAG welding. The angle in a single-V or double-V butt joint is about 75° for shielded metal arc welding. For gas metal arc welding, this angle may be reduced to 30 to 45°, Fig. 22-4. These changes in joint design increase the speed of welding, cut the time necessary for joint preparation, and reduce the amount of weld metal that is required. Thus, they lower the cost of materials and labor. Typically the higher energy spray arcs are used with the narrower groove angle and heavier root face joint designs.

The penetration achieved with shielded arc electrodes is about ⅛ inch maximum in steel. With the MIG/MAG process, 100 percent penetration may be secured in ¼-inch plate in a square butt joint welded from both sides.

For 60° single- or double-V butt joints, no root face is recommended, and the root opening should range from 0 to ³⁄₃₂ inch. Double-V joints may have wider root openings than single-V joints. Poor fitup and root overlap should be avoided. If the root opening is large, a backing bar should be used.

Plates thicker than 1 inch should have U-groove preparation. They require considerably less weld metal. The root face

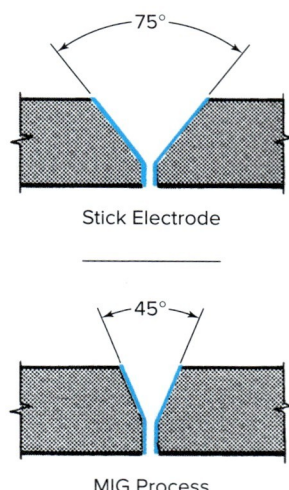

Fig. 22-4 V-groove, butt joint comparison.

thickness should be less than 3/32 inch, and the root spacing should be between 1/32 and 3/32 inch.

In multipass welding, the absence of slag ensures easier cleaning and so reduces the problem of porosity in the weld metal.

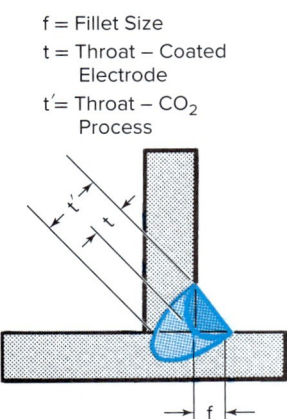

f = Fillet Size
t = Throat – Coated Electrode
t′ = Throat – CO_2 Process

Fig. 22-5 Comparison of penetration in a fillet weld: carbon dioxide shielded MAG weld versus coated electrode weld.

In making fillet welds with the MIG/MAG process, advantage is taken of the deeper penetration by depositing smaller weld beads on the surface of the material. The throat area of the weld is not reduced since a greater part of the weld bead is beneath the surface of the base metal, Fig. 22-5.

Certain types of joints are backed up to prevent the weld from projecting through the back side. The usual materials include blocks, strips, and bars of copper, steel, or ceramics.

Electrode Diameter

The electrode diameter influences the size of the weld bead, the depth of penetration, and the speed of welding. As a general rule, for the same current (wire-feed speed setting needs to be increased), the arc becomes more penetrating as the electrode diameter decreases. At the same time, the speed of welding is also affected because the deposition (burnoff) rate also increases.

When welding with wires below 0.045 inch, the smaller wire operating at a given current density burns off faster than the larger wire. To get the maximum deposition rate at a given current density, use the smallest wire possible that is consistent with an acceptable weld profile.

With wires 0.045 inch and larger for a given welding current, the next size larger wire provides a lower deposition rate. This is because the effects of preheating on the larger wires are less. Large wires deposit wider beads than small wires do under identical conditions.

As in all welding, the selection of the type and size of wire is important. Generally, filler wires should be of the same composition as the materials being welded. The position of welding or other special conditions may affect the size of the electrode. For most purposes, however, filler wires with diameters of 0.023/0.025, 0.030, and 0.035 inch are best for welding thin materials. Diameters of 0.045 inch or 1/16 inch are used for medium thickness, and a diameter of 1/8 inch is best for heavy materials. Filler wires with small diameters are recommended for welding in the vertical and overhead positions. Large diameter wires are desirable for those applications in which penetration is undesirable. Hard-surfacing, overlays, and buildup work are examples.

Electrode Extension

Electrode extension is that length of filler wire that extends past the contact tube, Fig. 22-6. This is the area where preheating of the filler wire occurs. Electrode extension is also called the *stickout*. It controls the dimensions of the weld bead since the length of the extension affects the burnoff rate. Electrode extension exerts an influence on penetration through its effect on the welding current. As the extension length is increased, the preheating of the wire increases and the current is reduced. The current reduction in turn decreases the amount of penetration into the work. Stickout distance may vary from 1/8 to 1 1/4 inches.

Short electrode extensions (1/8–1/2 inch) are used for the short circuit mode of transfer, generally with the smaller diameter electrodes (0.023–0.045 inch). For stainless steel, favor the shorter electrode extension because of its higher resistivity (1/8–1/4 inch). The longer electrode extensions are used for spray arcs (1/2–1 1/4 inches). These are generally done with the larger diameter electrodes. Excessive long arcs with active gases reduce the mechanical properties in the weld, because of various alloys being burned out as the metal is transferred across the longer arc.

The relationship that exists between electrode extension and current and electrode extension and penetration should be understood, Table 22-1, page 712. Tests have indicated that when electrode extension is increased from 3/16 to 5/8 inch, the welding current then drops approximately

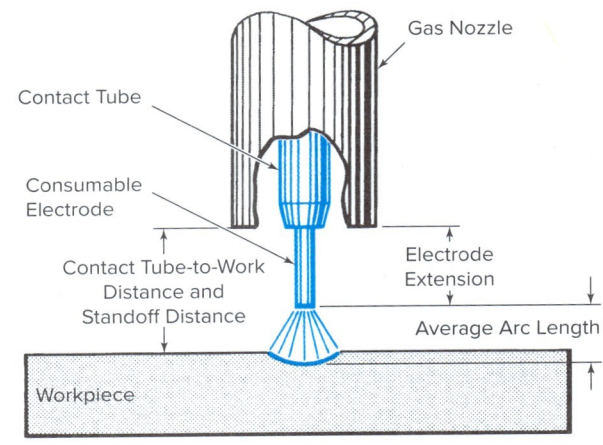

Fig. 22-6 Nomenclature of area between nozzle and workpiece.

Table 22-1 Effect of Electrode Extension on Weld Characteristic

Increased electrode extension	Increase deposition bead height
	Decrease welding current penetration bead width
Decreased electrode extension	Decrease deposition bead height
	Increase welding current penetration bead width

60 amperes. The current is reduced because of the change in the amount of preheating that takes place in the wire. As the electrode extension is increased, the preheating of the wire increases. Thus, less welding current is required from the power source at a given feed rate. Because of the self-regulating characteristics of the constant voltage power source, the welding current is decreased. As the welding current is decreased, the depth of penetration also decreases. (Increased electrode extension also increases the weld deposition rate.) On the other hand, if the electrode extension decreases, the preheating of the filler wire is reduced, and the power source furnishes more current in order to melt the wire at the required rate. This increase in welding current causes an increase in penetration.

Position of the Gun

The position of the welding gun with respect to the joint is expressed by two angles: the travel (gun) angle and the work (gun) angle, Fig. 22-7.

The bead shape, as well as the penetration pattern, can be changed by changing the direction of the wire as it goes into the joint in the line of travel as well as the location of the wire in the joint.

Travel Angle The travel angle can be compared to the angle of the electrode in shielded metal arc welding. The *drag* and *push* nozzle angles are shown in Fig. 22-8. The drag technique results in a high, narrow bead with relatively deeper penetration. The penetration is deeper because the arc tends to run into the pool and create a greater concentration of heat. The force of the arc pushes the molten

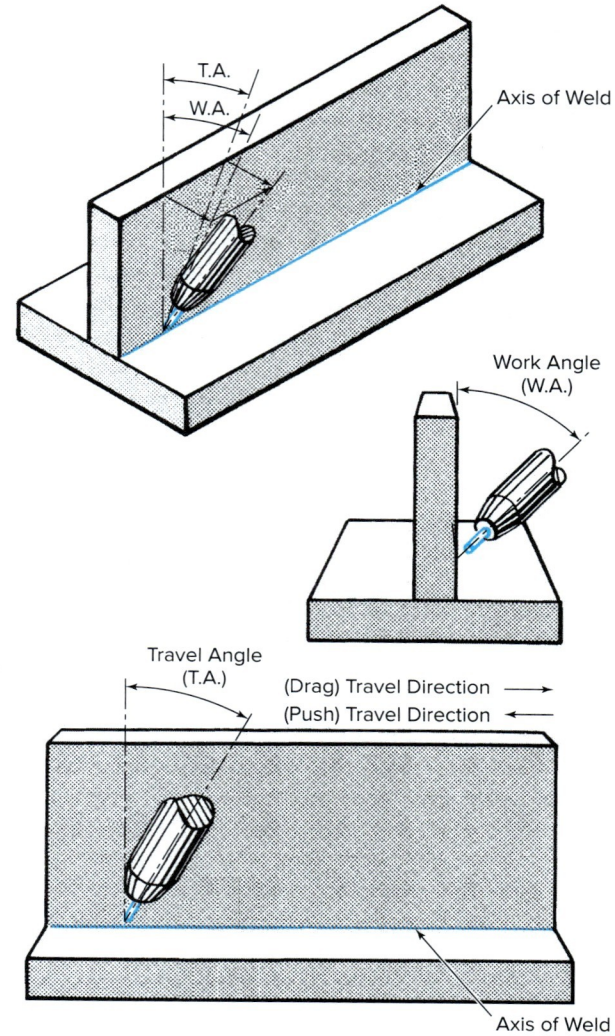

Fig. 22-7 Travel and work gun angles.

metal back for the rounded-bead contour. Maximum penetration is obtained when a drag angle of 10° is used. As the drag angle is reduced, the bead height decreases, and the width increases.

The arc in the push technique strikes cold base metal so that penetration is shallow. The force of the arc pushes the metal ahead of the bead and flattens the contour of the

SHOP TALK

Thebes
A wall painting in an ancient tomb in Thebes, from 1475 B.C., shows how brazing was done.

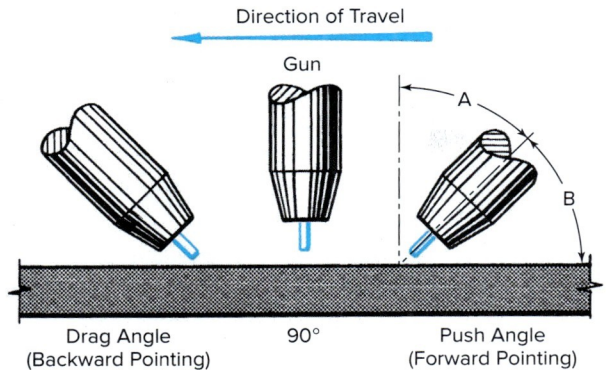

Fig. 22-8 Drag and push travel angles.

bead. Increased travel speeds are a characteristic of the push technique.

Work Angle The work angle refers to the position of the wire to the joint in a plane perpendicular to the line of travel, Fig. 22-7. For fillet weld joints, the work angle is normally half of the joint angle between the plates forming the joint. For butt joint, the work angle is normally 90° to the surface of the plate being joined.

The work angle utilizes the natural arc force to push (wash) the weld metal against a vertical surface to prevent undercut and provide good bead contour. This has particular significance in welding lap and T-joints. High travel speeds usually require greater work angles to ensure the proper washing action.

Wire Location The wire is typically located in the root of the joint. However, on heavy weldments and/or mechanized welding in the horizontal position, offsetting the wire location by 1 to 1.5 diameters onto the bottom plate helps reduce undercut by allowing the weld pool to flow up on the vertical plate. Incorrect wire location can lead to unequal leg fillet welds, Fig. 22-9.

Arc Length

The constant voltage welding machine used for gas metal arc welding provides for the self-adjustment of the arc length. The power source supplies enough current to burn off the filler wire as fast as it is being fed to maintain the arc length appropriate to the voltage setting.

If for any reason the arc length is shortened, the arc voltage will be reduced. This increases the current so that the filler wire melts at an increased rate until the correct arc length is reestablished. But if the arc length is lengthened, the arc voltage will be increased. This reduces the current and slows down the melting rate of the filler wire until the correct arc length is reestablished. No change in the wire-feed speed occurs. The arc length is corrected by the automatic increase or decrease of the burnoff rate of the filler wire. The welder has complete control of the welding current and the arc length by setting the wire-feed speed on the wire feeder and the voltage on the welding machine.

Arc Voltage

In the gas metal arc process, the voltage remains constant as set by the person doing the welding. Furthermore, the burnoff rate of the metal electrode is constant at a given voltage setting.

In the gas metal arc process, the arc voltage has a decided effect upon surface heating, weld profile namely in, bead height, and bead width. The chief function of voltage is to stabilize the welding arc and to provide a smooth, spatter-free weld bead, Fig. 22-10. Thus, for any given welding current there is a particular voltage that will provide the smoothest possible arc and the best weld profile with little or no spatter or undercut.

A higher or lower arc voltage causes the arc to become unstable and affects the surface heating. High arc voltage produces a wider, flatter bead. Excessive voltage increases the possibility of porosity in the weld metal

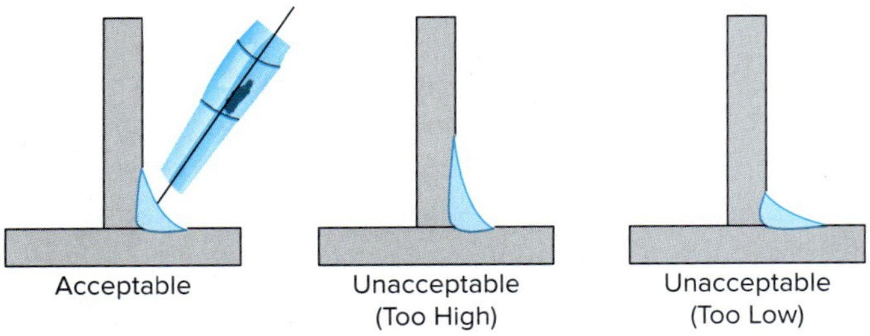

Fig. 22-9 The wire should be pointed at the root of the joint, neither too high nor too low. Unequal leg fillet welds can affect quality and productivity of the welds.

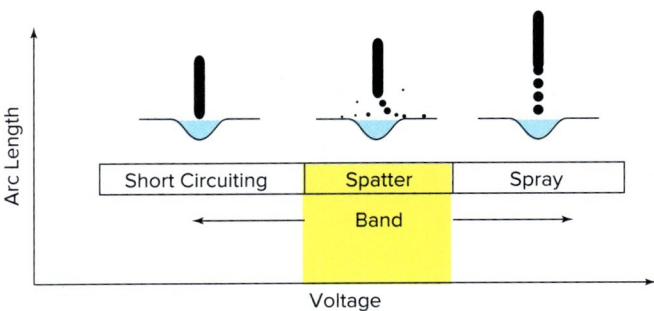

Fig. 22-10 Adjust the arc voltage appropriately for the mode of metal transfer being used.

and also increases the spatter. In fillet welds, it increases undercut and produces a concave fillet subject to cracking. Low voltage causes the bead to be high and narrow. Extremely low voltage causes the wire to stub on the plate. Therefore, changes in voltage have opposite effects on bead height and bead width. As the arc voltage increases, bead height decreases and bead width increases. (See Figs. 22-11 and 22-12 and Table 22-2.) This does not change the overall size of the bead, as in the case of welding current and travel speed, but it changes the profile of the bead.

Generally speaking, high arc voltages result in *globular transfer* of metal from the wire to the weld pool. Globular transfer is spatter-prone and reduces deposition efficiency. High arc voltage also reduces burnoff rates because of greater radiation losses.

For a fixed deposition rate, low arc voltages allow faster travel speeds and downhill operation. High arc voltages necessitate slower travel rates to allow the weld deposit to accumulate properly. A long arc may also cause contamination of the gas field since the shielding gas may not be completely contained, and the contaminating air is permitted to enter the gas shield. The proper arc voltage for the short circuit transfer has a sharp crackling sound, while the spray arc will have a hissing sound like paint being sprayed out of a spray gun. An occasional crackle with the spray arc is a good indication that minimum voltage is being used. The proper voltage has a sharp, crackling sound. The narrow short arc produces good penetration and fusion, and the weld bead formation is excellent.

Arc voltage is generally not set to control penetration. Voltage is a much better control of weld profile and arc stability. However, if the arc voltage is excessive, the energy is spread over a larger weld pool area. This will reduce the focus of the energy and reduce penetration. Conversely, if excessively low voltage is used, the wire may stub into the weld pool and produce a poor weld profile. However, penetration will be increased. A general analogy about voltage is that it heats the plate. The higher the voltage, the more heat goes into surface heating the plate. This is good to remember when trying to control the weld profiles; however, its effect on penetration must also be considered.

Wire-Feed Speed

The special wire feeder and the constant voltage welding machine constitute the heart of the MIG/MAG welding process. There is a fixed relationship between the rate of filler wire burnoff and welding current. The electrode wire-feed speed determines the welding current. Thus, current is set by the wire-feed speed control on the wire feeder. The welding machine supplies the amount of current (amperes) necessary to melt the electrode at the rate required to maintain the preset voltage and resultant arc length.

An increase in the electrode wire-feed speed requires more electrode to be melted to maintain the preset voltage and arc length. Increased wire-feed speed causes higher current to be supplied by the welding machine, and the melting rate and deposition rate increase. More weld metal and more heat applied to the weld joint produce deeper penetration and larger weld beads. If wire-feed speeds are excessive, the welding machine cannot put out enough current to melt the wire fast enough, and *stubbing* or *roping* of the wire occurs, Fig. 22-13. An excessive wire-feed speed causes convex weld beads and poor appearance. A decrease in electrode wire-feed

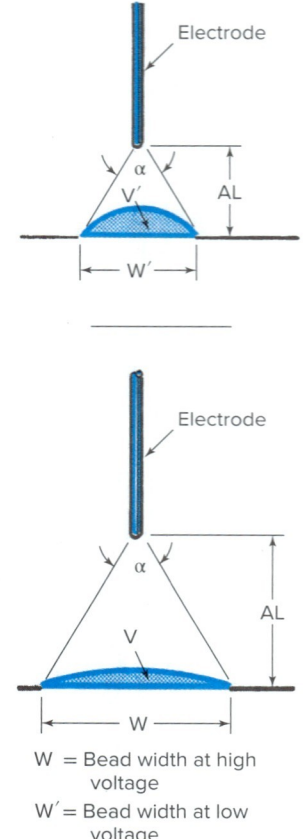

W = Bead width at high voltage
W' = Bead width at low voltage
V' & V = Volume of bead V = V'
AL = Arc length
α = Included angle of arc stream

Fig. 22-11 The relationship of arc length to weld bead width.

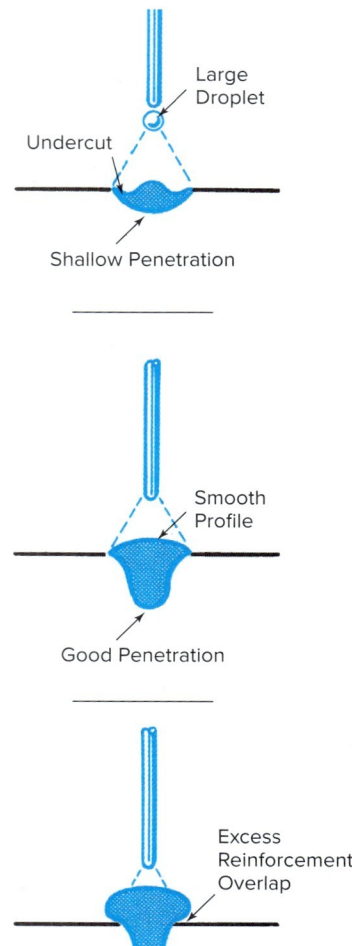

Fig. 22-12 Shallow penetration results when arc voltage is too high for travel speed. Excess reinforcement and overlap occurs when arc voltage is too low for speed. With proper arc voltage for speed, smooth profile in center illustration is obtained.

Table 22-2 Recommended Variable Adjustments for GMAW

Change of Welding Variables			Effect on Weld Deposit			
Variables	Increase	Decrease	Penetration	Deposition	Bead Height	Bead Width
Voltage	X		Increase[1]		Decrease	Increase
Voltage		X	Decrease[1]		Increase	Decrease
Current[3]	X		Increase	Increase	Increase	Increase[2]
Current		X	Decrease	Decrease	Decrease	Decrease
Travel speed	X		Decrease		Decrease	Decrease
Travel speed		X	Increase		Increase	Increase
Stickout	X		Decrease	Increase	Increase	Decrease
Stickout		X	Increase	Decrease	Decrease	Increase
Travel angle (drag—max. 25°)			Maximum		High	Narrow
Travel angle (push)			Decrease		Decrease	Increase

[1] Up to an optimum arc voltage—raising or lowering beyond this point reduces penetration.
[2] Up to an optimum current setting—raising or lowering beyond this point reduces bead width.
[3] Adjusted by wire-feed speed control.

speed results in less electrode being melted. The welding machine supplies less current and so reduces the deposition rate. Less weld metal and less heat are applied to the weld joint so that penetration is shallow and the weld bead is smaller.

Generally, for a given filler wire size, a high setting of the filler wire speed rate results in a short arc. A slow speed setting contributes to a long arc and possible burnback.

Welding Current

The setting at the wire-feed speed control determines the amount of current that will be delivered at the arc. The term current is often related to current density. *Current density* is the amperage per square inch of cross-sectional area of the electrode. Thus, at a given amperage the current density of an electrode 0.035 inch in diameter is higher than that of an electrode 0.045 inch in diameter.

The area of the cross section of a solid wire is easy to calculate. However, the area of the current-carrying sheath of a metal cored electrode is more complex to calculate. Since the sheath of the metal cored electrode is the conductor and the arc jumps and oscillates around the sheath of the electrode, current densities are much higher with the metal cored electrodes than the solid wire. Generally speaking, if you are going to maintain a given amperage and switch from solid wire to metal core, you can either jump one wire diameter size and keep the wire-feed speed the same or you can keep the same wire size and increase the wire-feed speed. This higher current density of the burnoff around the sheath of these metal cored electrodes accounts for their higher deposition rates at a given heat input.

Each type and size of electrode has a minimum and maximum current density. The best working range lies between them.

The depth of penetration, bead formation, filler wire burnoff, speed of travel, and the size and appearance of the weld profile are all affected by the amount of welding current.

There is a direct relationship between the welding current and penetration. In general, for any change in welding current, there is a corresponding change in penetration. As the welding current increases, the penetration increases, and as the welding current decreases, the penetration decreases. Increasing the current also increases the wire meltoff rate and the rate of deposition. If the current is too low for a given electrode size, metal transfer is sluggish, and poor penetration and poor fusion result. The bead is rough, and excessively convex. If the current is too high, penetration may be too deep. Excessive penetration causes burn-through and undercut. The weld bead has a

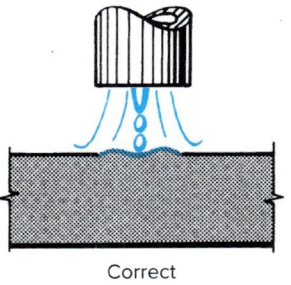

Correct

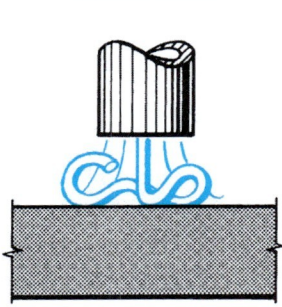

Excessive

Fig. 22-13 Effect of wire-feed speeds.

Table 22-3 GMAW Parameters for Steel

Shielding Gas	Electrode Dia. (in.)	Amperage Range	Wire-Feed Speed Range (in./min)	Arc Voltage Range
\multicolumn{5}{c}{Short Circuit Transfer}				
75% Ar–25% CO_2	0.023	30–90	100–400	14–19
CO_2 (adds 1–2 V)	0.030	40–145	90–340	15–21
	0.035	50–180	80–380	16–22
	0.045	75–250	70–270	17–23
\multicolumn{5}{c}{Spray Transfer}				
Ar + 1–5% O_2*	0.030	135–230	330–650	24–28
	0.035	165–300	340–625	24–30
	0.045	200–375	225–410	24–32
	1/16	275–500	200–300	26–34

*Percent of O_2 needed usually depends upon metal deoxidizers and thickness of material.

poor profile. Undercut is a particular problem in making fillet welds with current settings that are too high.

Each filler wire size and type has a minimum and maximum current range that gives the best results. Study Table 22-3 for comparative current ranges, as well as other important parameters for welding carbon steel, stainless steel, and aluminum. Note how current and wire-feed speed are related. If the current is too high, there is the possibility of electrode burn-back into the contact tube, the arc is unstable, and gas shielding is disturbed. Spatter results, and the deposited weld metal has poor physical characteristics. Increases in current will increase bead height and width. As the current is increased, the voltage must also be increased.

If the current is too low, the filler wire may become short circuited to the work, the electrode may become red hot, and the arc will be extinguished. If the welding can be maintained, the arc is unstable, and poor fusion occurs. The bead formation is high and limited to the surface of the base metal with little or no penetration. This destroys the advantage of gas metal arc welding, which owes its success to the sharp concentration of high current density at the electrode tip.

Too much or too little current affects the physical properties of the weld metal. Tensile strength and ductility are reduced. Porosity may occur. Excessive oxides and other impurities may be present. Bead formation is also affected.

Travel Speed

Travel speed has a decided effect on penetration, bead size, and appearance. It is described as the linear rate at which the arc moves along the weld joint. In most cases, the travel speed should be sufficient to keep the welding arc on the leading edge of the weld pool.

At a given current density, slower travel speeds provide proportionally larger weld beads and more heat input in the base metal per unit length of weld. The longer heating time of the base metal increases the weld deposit per unit length resulting in a higher and wider bead contour. If the speed of travel is too slow, unusual weld buildup occurs. Excessive buildup causes poor fusion, decreases penetration, porosity, inclusions, and a rough, uneven bead.

Progressively increased travel speeds have opposite effects. Less weld metal is deposited with lower heat input per unit length of weld. This produces a narrower weld bead and lower contour. Excessively fast speeds cause undercut. The bead is irregularly shaped because there is too little weld metal deposited per unit length of weld.

Travel speed is a variable that is as important as wire-feed speed and arc voltage. Travel speed is influenced by the thickness of the metal being welded, joint design, cleanliness, joint fitup, and welding position. As travel speed is increased, it is necessary to increase the wire-feed speed, which in turn increases the current and the burnoff rate, to produce the same weld cross section. Excessive travel speed produces undercutting and a low rate of weld metal deposit. A speed of travel that is too slow produces overlap of the base metal, and even burn-through on this material.

Keep in mind the arc is what is melting and penetrating into the base metal. The only place you can be assured of getting good penetration is directly under the arc. You cannot rely on the molten pool to achieve good fusion and penetration. Refer to Figs. 22-14 and 22-15.

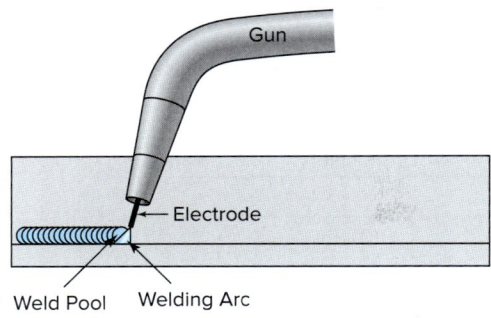

Fig. 22-14 Arc must be on leading edge of weld pool to ensure penetration and fusion.

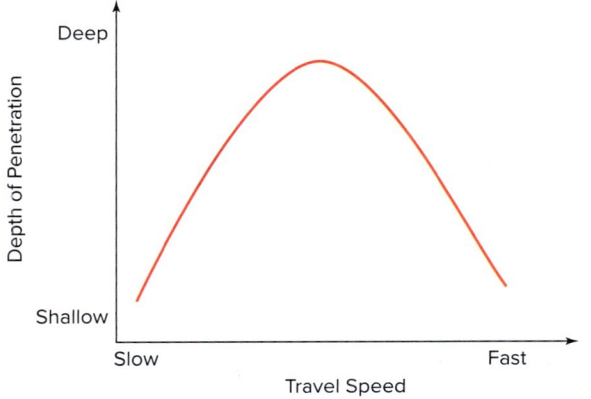

Fig. 22-15 There is an optimum travel speed for each situation. Excessively slow or fast travel speed will reduce penetration.

Summary of Operating Variables

You will recall that the height and width of the bead obtained with the gas metal arc process depend on the adjustment and control of these variables:

- Size and type of filler wire
- Characteristics of the shielding gas
- Joint preparation
- Gas flow rate
- Voltage
- Speed of travel
- Arc length
- Polarity
- Electrode extension
- Work and travel angle
- Wire-feed speed (current)

These variables are adjusted on the basis of the type of material being welded, the thickness of the material, the position of welding, the deposition rate required, and the final weld specifications. Proper adjustment and control determine formation of the weld bead by affecting such things as penetration, bead width, bead height, arc stability, deposition rate, weld soundness, and appearance. An understanding of these variables and their control is essential if you are to master the gas metal arc process. See Tables 22-3 through 22-5, page 718.

Welding current and *travel speed* have a similar effect on both bead height and width. Each variable increases or decreases both bead height and width at the same time. If travel speed is decreased, both bead height and width are increased. Any change in these variables affects the amount of filler metal being deposited per a given length of joint. Therefore, if a given travel speed and welding current (wire-feed speed) are not providing enough weld metal to fill a particular joint, either the travel speed must be decreased or the welding current (wire-feed speed) must be increased.

In multipass welding, excessive bead height interferes with fusion between weld beads, especially along the weld toes. This condition causes voids in the joint, which appear as wagon tracks in radiographic inspection. If the bead width is too small, similar defects can occur due to the undercut at the toe of the weld. This is one of the reasons why this text has stressed multipass beads with a flat face.

Arc voltage has the opposite effect on bead height that it has on bead width. As the arc voltage increases, bead height decreases and bead width increases. While changes in travel speed and welding current (wire-feed speed) change the overall size of the bead, changes in arc voltage affect the shape of the bead. (Refer to Fig. 22-11, p. 714.) For example, if a weld bead has excessive convexity, it can be corrected by increasing the arc voltage and thereby flattening the bead.

Travel angle also affects bead contour. A trailing gun angle tends to produce a high, narrow bead. The bead height decreases and the width increases as the trailing angle is reduced. This effect continues into the leading angle range. If the leading angle is increased too far, the bead starts to become narrow again. Refer to Table 22-2, page 715, for a review of the effect of variable adjustments on the physical characteristics of the weld bead.

Weld Defects

Each completed job requires the close and careful inspection that you have followed with all the other welding processes presented in this text. High quality welds require the application of precise welding procedures. The defects found in welds made by the gas metal arc process are similar to those found in other welding processes. However,

Table 22-4 GMAW Parameters for Stainless Steel

Shielding Gas	Electrode Dia. (in.)	Amperage Range	Wire-Feed Speed Range (in./min)	Arc Voltage Range
Short Circuit Transfer				
90% He + 7½% Ar + 2½% CO_2	0.030	60–125	150–280	16–23
	0.035	75–160	125–280	16–23
	0.045	100–200	110–230	16–24
Spray Transfer				
Ar + 1–2% O_2†	0.035	180–300	290–600	24–33
	0.045	200–450	250–475	24–35
	1/16	220–500	180–300	24–36

*Voltage slightly lower if using AR + CO_2 or AR + O_2.

†Percent of O_2 needed usually depends upon metal thickness.

Table 22-5 GMAW Parameters for Aluminum (ER4043)

Shielding Gas	Wire Size	Arc Voltage	Amperes	WFS (in./min)
Short Circuit Transfer				
Ar	0.030	16–19	50–120	250–550
	0.035	17–20	65–140	240–425
	3/64	17–22	75–170	160–325
Spray Transfer				
Ar	0.030	20–27	95–200	550–1200
	0.035	20–27	110–220	425–850
	3/64	22–31	130–290	250–650
	1/16	22–32	160–360	140–425
	3/32	23–33	190–450	100–210

Note: Parameters may vary with wire series used. Parameters may vary for the same size and series electrode. Check manufacturer's recommendations.

the causes and the corrective action recommended are entirely different.

Incomplete Penetration

Incomplete penetration is the result of too little heat input in the weld area. This can be corrected by increasing the wire-feed speed and reducing the electrode extension distance to obtain maximum current for the particular wire-feed setting. If the wire-feed speed and stickout are correct, check the speed of travel. It may be too fast or too slow.

Incomplete penetration may also be caused by improper welding techniques. It is important that the arc be maintained on the leading edge of the weld pool in order to secure maximum penetration into the base metal. A travel angle of the proper degree is required.

Excessive Penetration

Excessive penetration usually causes excessive melt-through. It is the result of too much heat in the weld area. This can be corrected by reducing the wire-feed speed to obtain lower amperage or increasing the speed of travel.

Improper joint design is another cause of excessive penetration. If the root opening is too wide or if the root face is too small, excessive melt-through is likely to occur. This difficulty can be prevented before welding by checking the groove angle if there is one, the root opening, and the root face to make sure that they are correct for the position of welding. Excessive penetration can be remedied during welding by increasing the electrode extension distance as far as good working practice will allow and weaving the gun.

Whiskers

Whiskers are short lengths of electrode wire sticking through the weld on the root side of the joint. They are caused by pushing the electrode wire past the leading edge of the weld pool. Whiskers can be prevented by reducing the travel speed, reducing the wire-feed speed, increasing the electrode extension distance, and weaving the gun. It is important that the welder does not allow the electrode wire to get ahead of the weld pool.

Voids

Voids are sometimes referred to as *wagon tracks* because of their resemblance in radiographs to ruts in a dirt road. They may be continuous along both sides of the weld deposit. They are found in multipass welding. When the underneath pass has a bead with a large contour or a bead with too much **convexity** or **undercut**, the next bead may not completely fill the void between the previous pass and the plate. Voids may be prevented by making sure that the edges of all passes are filled in so that undercut cannot take place. Excess convexity can be reduced by welding the next pass with a slightly higher arc voltage. Voids can also be corrected by increasing the travel speed on the next pass and making sure that the arc melts the previous bead and fuses into the sides of the joint.

Incomplete Fusion

Incomplete fusion is largely the result of improper gun handling, low heat, and improper speed of travel. It is important that the arc be directed at the base metal and the leading edge of the pool.

To prevent this defect, give careful consideration to the following:

- Direct the arc so that it covers all areas of the joint. The arc, not the pool, should do the fusing.
- Keep the electrode at the leading edge of the pool.
- Reduce the size of the pool as necessary by adjusting the travel speed.
- Check current values carefully. Keep a short electrode extension.

Porosity

The most common defect in welds produced by any welding process is porosity. Porosity that exists on the face of the weld is readily detected, but porosity in the weld metal below the surface must be determined by radiograph ultrasonic or other testing methods. The causes of most porosity are contamination by the atmosphere, a change in the physical qualities of the filler wire, and improper welding technique. Porosity is also caused by entrapment of the gas evolved during weld metal solidification. The intense heat of the arc separates water vapor and other hydrogen-bearing compounds from the metal. Because these compounds are lighter than molten metal, they tend to rise to the surface as gas bubbles before solidification. If they do not reach the surface, they will be entrapped as internal porosity. If they reach the surface, they appear as external porosity. Study the following causes of porosity carefully:

- Travel is so fast that part or all of the shielding gas is lost, and atmospheric contamination occurs.
- The shielding gas flow rate is too low so that the gas does not fully displace all the air in the arc area.
- The shielding gas flow rate is too high. This draws the air into the arc area and causes turbulence, which reduces the effectiveness of the shield.
- The shielding gases must be of the right type for the metal being welded. The gases must be pure and dry.
- The gas shield may be blown away by wind or drafts. The welder should protect the weld area with a wind break or the position of their body. In some cases increasing the gas flow rate may be helpful when welding outdoors.
- There may be defects in the gas system. This could be the result of spatter clogging the nozzle, a broken gas line, defective fittings in the gas system, a defective gas valve, or a frozen regulator. It is important that the filler wire is in the center of the shielding gas flow. If the wire is off center, it can produce an erratic arc that can cause porosity.
- Excessive voltage for the arc required can cause the loss of its deoxidizers and alloying elements. This not only causes porosity, but it also seriously affects the physical characteristics of the weld.

JOB TIP

Job Planning

When getting yourself ready to look for work, ask yourself:

1. What do I expect out of this job?
2. Do I plan on continuing my education?
3. In 5 years, what do I want to be doing?

- Foreign material such as oil, dirt, rust, grease, and paint is on the wire or material to be welded. Electrode wire should be stored in clean, dry areas. The material to be welded may be cleaned with chemicals, sandblasted, brushed, or scraped.
- Improper welding techniques are used. Excessive electrode extension coupled with an improper work and travel gun angles and movement is a serious cause of porosity. Removal of the gun and the shielding gas before the weld pool has solidified causes crater porosity and cracks. The gun should be held at its normal position at the end of the weld until the wire feed and the flow stop. Travel speed should not be so fast that the molten metal does not solidify under the gas shield.

Other Defects

Warpage Warpage occurs when the forces of expansion and contraction are poorly controlled. A thorough understanding of these forces and their relationship to each other will enable you to reduce warpage in a weldment to a minimum.

Weld Cracking Weld cracking is brought about by compositional problems, poor joint design, and poor welding technique. The manganese content of the filler metal may be too low, or the sulfur content may be too high. The weld bead may be too small to withstand expansion and contraction movements in a joint, or the speed of travel may be too fast. Cracking can be prevented by making sure that the filler metal has a composition suitable for the base metal and by providing for expansion and contraction forces during welding.

GMAW electrodes may have additional deoxidizers added to assist when using high CO_2 percentage shielding gases. If high argon percentage shielding gases are substituted with these triple deoxidized electrodes, the deoxidizers will build up in the weld metal on multipass welding. This buildup may reduce ductility, impact properties, and make the weld metal prone to cracking.

Spatter Spatter is made up of very fine particles of metal on the plate surface adjoining the weld area. It is usually caused by high current, a long arc, an irregular and unstable arc, improper shielding gas, improper gun angles, electrode extension, or a clogged nozzle.

Irregular Weld Shape Irregular welds include those that are too wide or too narrow, those that have an excessively convex or concave surface, and those that have coarse, irregular ripples. Such characteristics may be caused by poor gun manipulation as in correct work and travel angles, a speed of travel that is too fast or too slow, current that is too high or too low, improper arc voltage, improper electrode extension, or improper shielding gas.

Undercutting

Undercutting is a cutting away of the base material along the toes of the weld. It may be present in the cover pass weld bead or in multipass welding. This condition is usually the result of high voltage, excessive travel speed, wrong wire location, wrong work angle, wrong travel angle, poor gun technique, improper gas shielding, or the wrong filler wire. To correct undercutting after proper settings are ensured, move the welding gun from side to side in the joint, and hesitate at each side before returning to the opposite side.

Producing High Quality Welds

You are urged to study the material concerning weld defects carefully and to be ready to recognize poor welding when you see it. You must take great care and have pride of work to produce quality welds. The gas metal arc process is able to produce high quality welds in a wide variety of materials. Make sure that you understand the interrelationships of all the variables and that you apply this knowledge with the manual skill of precise gun technique. Study the characteristics of these defects and their correction.

Safe Practices

In any industrial activity, safety is a most important consideration to both the worker and the employer. If workers wish to enjoy long and profitable careers, they must have a concern for their own health and well-being. Accidents and a disregard of the basic safeguards for good health cut short thousands of careers each year. Disability resulting from accidents and poor health costs the employer thousands of dollars in compensation and increased insurance costs. In these times of serious skill shortages, the loss to the workforce is also costly to the nation.

Welding is no more dangerous than other industrial operations. The safety precautions and protective equipment required for the MIG/MAG process are essentially the same as for any other electric welding process. There are a few factors, however, that need attention. You are urged to review the general precautions contained in Chapter 32 in addition to observing the specific precautions listed here.

Eye, Face, and Body Protection

The welding helmets and protective clothing worn when working with the other electric welding processes are necessary. The radiant energy, particularly in the ultraviolet range, that is produced by the gas-shielded process, is

5 to 30 times more intense than that produced by shielded metal arc welding. The lowest intensities are produced by the gas tungsten arc; the highest, by the gas metal arc. Argon produces greater intensities than helium does. Radiant energy is particularly dangerous to the bare skin and unprotected eyes. The greater intensity of ultraviolet radiation also causes rapid disintegration of cotton clothing. There is, however, less spatter from hot metal and slag, and there is less danger of clothing fires than with shielded metal arc welding. The following clothing regulations should be enforced during practice:

- Standard arc welding helmets with lenses ranging in shade from no. 6 for work using up to 30 amperes to no. 14 for work using more than 400 amperes should be worn. A glass that is too dark should be avoided to prevent eyestrain. When welding in a shop or a dark area, a lighter shade lens should be used than is normally used for welding outdoors. The arc should never be viewed with the naked eye when standing closer than 20 feet. The intensity at 2 feet away from the arc is high enough to burn the eyes in a few seconds. Good safety practice also requires that safety glasses be worn under the welding helmet. If these have a light tint, they protect the welder from getting a flash from the work of other welders working in the same area.
- The skin should be covered completely to prevent burns and other damage from ultraviolet light. The skin can be burned if exposed to the arc for a few minutes at a distance of 2 feet from the arc. The back of the head and neck should be protected from radiation reflected from bright surfaces such as aluminum. Gloves should always be worn, and when welding in all positions, special leather jackets are appropriate.
- Shirts should be dark in color to reduce reflections, thereby preventing ultraviolet burns to the face and neck underneath the helmet. The shirt should have a tight collar and long sleeves. Leather, wool, and aluminum-coated cloth withstand the action of radiant energy reasonably well.

Handling of Gas Cylinders

The shielding gases are provided in steel cylinders. The same general precautions should be observed in the handling of these cylinders as those containing oxygen and acetylene. (Refer to Chapters 8 and 32.)

- Stored cylinders should be in a protected area away from fire, cold, and grease and away from the general shop activity.
- Cylinders must be secured to equipment to prevent their being knocked over.
- The proper regulators and flowmeters must be used with each special type of cylinder. High pressure cylinders must be treated with extreme care.
- Cylinders should not be dropped, used as rollers, lifted with magnets, connected into an electric circuit, or handled in any other way that might damage the cylinder or regulator. When cylinders are empty, they should be stored in an upright position with the valve closed.

Ventilation

The ventilation requirements for shielded metal arc welding are ample for gas metal arc welding.

- Ozone is generated in small quantities, generally below the allowable limits of concentration. Adequate ventilation ensures complete safety.
- Nitrogen dioxide is also present around the area of the arc in quantities below allowable limits. The only point where high concentrations may be found is in the fumes 6 inches from the arc. With only natural ventilation, this concentration is reduced quickly to safe levels. Special ventilation should be provided during the MIG/MAG cutting of stainless steel and when using nitrogen as a shielding gas.
- Carbon dioxide shielding may create a hazard from carbon monoxide and carbon dioxide if the welder's head is in the path of the fumes or if welding is done in a confined space. In these conditions, special ventilation to control carbon dioxide fumes should be provided.
- Eye, nose, and throat irritation can be produced when welding near such degreasers as carbon tetrachloride, trichloroethylene, and perchloroethylene. These degreasers break down into phosgene under the action of the powerful rays from the arc. It is necessary to locate degreasing operations far away from the welding activities to remove this hazard.
- During welding, certain metals emit toxic fumes that may cause respiratory irritation and stomach upset. The most common toxic metal vapors are those given off by the welding of lead, cadmium, copper, zinc, and beryllium. These fumes can be controlled by general ventilation, local exhaust ventilation, or respiratory protective equipment.
- Much of the welding smoke and fumes can be engineered out of the GMAW arc by use of a higher argon percent (over 80 percent) and the pulse-spray mode of transfer.
- Welding guns can be purchased with a smoke extractor capability.

Electrical Safety

The electrical hazard associated with gas metal arc welding is less than that with the shielded metal arc process because the open circuit voltage is considerably less. In some types of equipment, however, a 115-volt control circuit may be utilized in the welding gun. This is no different from the voltage present in many other hand tools, and ordinary precaution ensures safety.

Electrical maintenance of the power source and wire feeder should be done only by a qualified person. The equipment should never be worked on in the electrical "hot" condition. Routine maintenance such as gun cleaning, blowing out the cable assembly, and alignment of the cable assembly to the wire feeder can be handled by the welder. Properly maintained welding equipment is safe welding equipment.

Wire-Feeder Safety

When aligning and adjusting drive rolls, make sure the power is turned off. Avoid pinch points when working near drive rolls. When adjusting tension, make sure guards are in place so that accidental contact with moving parts is minimized. When feeding wire, keep in mind the force being applied to the wire is sufficient to push it through your hand or other body parts. Never let the exposed wire come in contact with or be pointed at your body. The wire is sharp and/or may be hot, and you can get punctured or burned by being careless.

Fire Safety

Here again, some of the same hazards are present in gas metal arc welding that are encountered in other welding processes. There is less spatter, and the spatter that does occur is smaller and does not travel as far away from the arc zone as in shielded metal arc welding. Good practice, however, dictates certain precautions.

- Welding should not be done near areas where flammable materials or explosive fumes are present.
- Paint spray or dipping operations should not be located close to any welding operation.
- Combustible material should not be used for floors, walls, welding tables, or in the immediate vicinity of the welding operation.
- When welding on containers that have previously contained combustible materials, special precautions should be taken.
- Use a "hot work permit" as required.

Care and Use of Equipment

The efficient operation of welding equipment depends on its proper application and care. Although gas metal arc equipment is somewhat similar to that used for manual shielded metal arc welding, there are some major differences.

Do not push the gun into the arc like an electrode. If you do, you will melt off the nozzle tip or contact tube. The wire feeder pushes the wire into the arc; sit back and let the machine do the work. All you have to do is lead it down the joint. You may proceed straight down the seam, or you may weave the arc back and forth as you would do with a stick electrode. Fill craters at the end of the weld just as you always have.

Either a push or drag travel angle can be used. Push arc technique, lends itself to tracking joints accurately at fast travel speeds. Under the same set of conditions, push technique produces a flatter bead shape than a drag technique, Fig. 22-16.

A drag technique lends itself to producing high beads and fillets. The arc force holds the molten metal back from the crater.

Whenever possible, welding should be done in the flat welding position to take advantage of the increased penetration and deposition rate that are characteristic of the MIG/MAG process. This is especially true of large fillets and large grooves.

Small fillets and groove welds should be positioned so that the arc can run slightly downhill. Travel speed is faster, and a flat-to-concave bead contour is produced. An angle of 5 to 15° is sufficient. Sheet metal that is 14 gauge and under can be welded vertical down.

The equipment has to be kept clean, in proper adjustment, and in good mechanical condition. The wire-feeding system requires special attention. Maladjustment leads to erratic wire feeding which in turn causes porosity.

Care of Nozzles

Keep the gun nozzle, contact tube, and wire-feeding system clean to eliminate wire-feeding stoppages. The nozzle

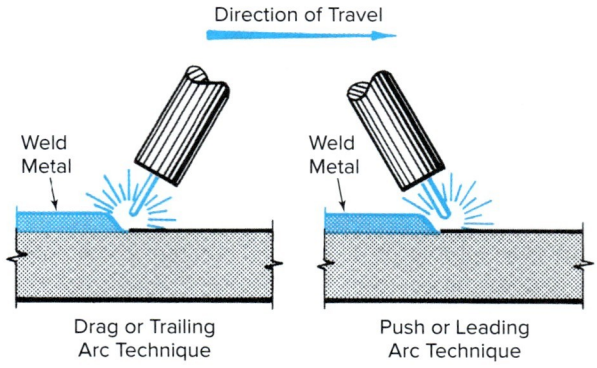

Fig. 22-16 Both the drag and push methods are used in semi-automatic welding. The drag, or trailing arc, produces large wide beads; the push or leading arc, produces a flatter bead shape.

ABOUT WELDING

The First Welded Bridges
A 62-foot long bridge in Pittsburgh was the United States' first all-welded one for trains. In 1928 it was revolutionary in that it used no rivets. To test it out, they drove a locomotive over it! It was built and used by Westinghouse. Its European counterpart, designed in 1927, was over a river in Poland. It lasted 50 years until it was replaced by a wider version.

itself is a natural spatter collector, Fig. 22-17. It surrounds the contact tube and provides a good target area as the spatter flies out of the arc stream. This will cause poor shielding.

If the spatter builds up thick enough, it can actually bridge the gap and electrically connect the insulated nozzle to the contact tube. If you accidentally touch the nozzle to a grounded surface when this happens, there will be a flash. It is quite likely the nozzle will be ruined. If the face of the nozzle is burned back, the shielding gas pattern over the arc may be changed so much that the nozzle must be replaced.

To remove spatter, use a soft, blunt tool such as an ice cream stick for prying. If the nozzle is kept clean, shiny, and smooth, the spatter almost falls out by itself. An antispatter compound may be applied to the gun nozzle and contact tube end. However, the root cause of the spatter should be found and corrected. Antispatter should not be used to cover up a problem.

Do not clean guns or torches by tapping or pounding them on a solid object. Bent gun nozzles distort the gas pattern. Threads are damaged so that it is difficult to replace nozzle tips and contact tubes. The high temperature insulation in the nozzle breaks, and damaged insulation causes arcing in the handles. Trigger switches are broken, and in the case of water-cooled guns, leaks are sure to develop.

Care of Contact Tubes

The contact tube transfers the welding current to the electrode wire. New contact tubes have smooth, round holes of the proper minimum diameter. The hole has to be big enough to allow a wire with a slight cast to pass through easily.

With use, the wire wears the hole to an oval shape, especially on the end closest to the arc. The wire slides more easily, but the transfer of current is not as good. Arcing in the tube results, and movement of the wire can actually be stopped. When this happens, the contact tube has to be replaced. It is not a good idea to salvage these tubes.

When the hole in the contact tube becomes too big through wear, spatter flies up into the bore and wedges against the wire. The wire slows down because of the friction. Spatter in the tube also produces a long arc that burns back up to the contact tube and fuses the wire to the tube. Worn contact tubes also allow the wire to flop around as it emerges from the gun so that it is difficult to track a seam. It can also result in poor arc starts.

The contact tube must be tightly secured in the gun. With the heating and cooling of the gun, it may loosen over time. Periodically check the contact tube for tightness.

Care of Wire-Feed Cables

The wire-feed conduit is not unlike an automobile speedometer cable. It is a flexible steel tube that does not stretch. Because wire-feed cables are the main source of friction in the wire-feed system, they should always be as clean and straight as possible. When welding, do not lay the cables in a loop or bend them around a sharp corner. Bending pinches the wire.

Under normal operating conditions, steel and copper chips loosened by the driving action of the feed rolls rub off the wire and collect in the wire-feed cables. If these chips are allowed to collect for any period of time, they fill up the clearance between the wire and the cable and jam the wire. It is not uncommon for this type of stoppage to completely ruin a feed cable. Jerky wire feeding is an indication of chip buildup.

Wire-feed cables should be cleaned with dry compressed air. The conduit should be removed at the feed roll end, and the air nozzle placed on the tip of the contact tube. This results in a reverse flush. Lubricating a blown-out wire-feed cable with dry powdered graphite reduces

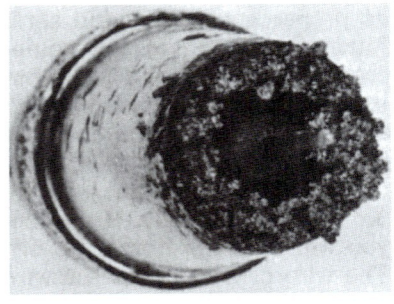

Fig. 22-17 This plugged gun nozzle is an example of improper care of welding equipment, which can damage parts. *Edward R. Bohnart*

the friction in the cable and results in smooth wire feeding. Wear safety glasses when cleaning and lubricating the cable.

Do not put so much lubricant in the cable that there is not room for chips. Pour about ¼ teaspoon of graphite into the feed end of the cable. Blow the graphite through the cable and reassemble the cable and wire driver. Wire-feed cables should be cleaned every time a spool or coil is changed.

Liquid lubricating compounds should be avoided because a paste is formed with the chips that tends to pack very tightly. If a liquid lubricant has been applied, cleaning with compressed air does not work. The cable must be washed with a solvent. This is not only messy, but time-consuming.

Bird Nesting

Occasionally, the wire coils sideways between the wire-feed cable and the drive rolls. This happens with greater frequency when small wires are pushed because the column strength (stiffness) of the small wire is not great enough to withstand the push of the drive motor. Worn contact tubes and other sources of friction previously discussed increase the chances of a "bird nest."

Bird nesting can be prevented by accurate alignment of the wire-feed cable inlet guide. The inlet guide should be aligned exactly with the rollers so that the wire does not have to make a reverse bend, Fig. 22-18A, B, and C. It should be as close to the rollers as possible without touching them, Fig. 22-18D. The notch in the drive rolls must be in perfect alignment to provide smooth passage for the wire, Fig. 22-18E and F.

Cleanliness of the Base Metal

The gas metal arc process produces quality welds free of porosity and cracks when welding on steels that have been cleared of rust, scale, burned edges, and chemical coating. The quantity and type of the contamination that can be tolerated varies from application to application.

These contaminants by and large are gas producers. On high speed applications the weld metal solidifies before the gas has a chance to float to the surface and escape. Porosity is the result. On slow speed applications, porosity is not as common. The gas has a chance to escape from the molten metal, and some of the contaminant actually has time to burn away in front of the crater because of the intense heat of the arc. This reduces the amount of gas formed quite effectively.

A good general rule to follow is to clean the area thoroughly before welding. Good ventilation is essential

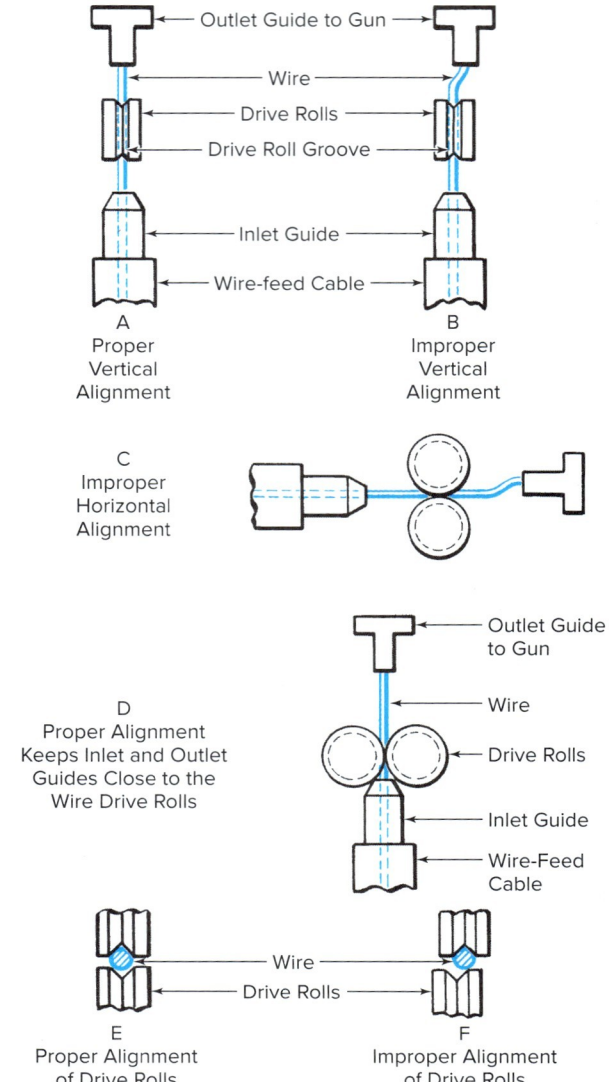

Fig. 22-18 Wire-feed alignment.

when welding galvanized, zinc-coated, aluminum-coated, cadmium-plated, lead-plated, and red-lead-painted surfaces to prevent breathing toxic fumes.

Arc Blow

You will recall from your practice with shielded metal arc welding that arc blow is always a problem when welding with direct current whether it is electrode negative or positive polarity.

To understand what arc blow is, think of the ionized gases carrying the arc from the end of the electrode wire to the work field as a flexible conductor. This conductor has a magnetic field around it. If it is placed in a location such as the corner of a joint or the end of a plate, the magnetic field is distorted and pulled in another direction. The magnetic field reacts to this force by trying to return to a state of equilibrium. The magnetic field actually moves the flexible conductor to another location where the magnetic forces are in balance. Thus, we have a condition of pull and counter-pull: the arc is *blown* to one side or the other as if by a draft. Hence, the term **arc blow.**

Arc blow does not occur with a.c. welding arcs because the forces exerted by the magnetic field on the flexible conductor are reversed 120 times (60 cycles) per second. This tends to keep the magnetic field and the flexible conductor in a constant state of equilibrium. Because of the short arc length in the GMAW process, arc blow is not as prevalent as with other processes like SMAW.

Connecting the Work to Minimize Arc Blow The return path for the welding power from the workpiece to the power source is just as important as the path from the power source to the electrode. The following suggestions will shorten the trial-and-error process of locating the best work connection location:

- Attach the work lead or leads directly on the workpiece if possible. Do not connect through ball or roller bearings, hinges, a hold-down clamp, or swivels. Remove rust or paint on the weldment at the point of the connection.
- Connect both ends of long, narrow weldments.
- Use electrical conductors of the proper length. Extra-long cables that must be coiled up act as a reactor. Stray magnetic fields that affect the action of the arc are also set up.
- Weld away from the work connection.
- On parts that rotate, use a rotating work connection or allow the work cable to wind up no more than one or two turns.
- In making longitudinal welds on cylinders, use two work connections—one on each side of the seam as close as possible to the point of starting.
- If multiple work connections are necessary, make sure that the cables are the same size and length and that they have identical terminals. Both cables should run from the workpiece to the power source.
- On multiple-head installations, all heads should weld in the same direction and away from the work connection.
- Use individual work circuits on multiple-head installations.
- Do not place two or more arcs close to one another on weldments that are prone to magnetic disturbance with one arc such as tubes or tanks requiring longitudinal seams.

Setting Up the Equipment

Examine the photographs of gas metal arc equipment in Chapter 21. Figure 22-19 shows an MIG/MAG welding station in the plant. The following equipment is required:

- Constant voltage d.c. power source. The welding machine may be engine driven or a static transformer-rectifier.
- Wire-feeding mechanism with controls and spooled or reeled filler wire mounted on a fixture.
- Gas-shielding system consisting of one or more cylinders of compressed gas, depending on the gas mixture, pressure-reducing cylinder regulator, flowmeter assembly. (The regulator and flowmeter may be separate or a combination device.)
- Combination gas, water, wire, and cable control assembly and welding gun of the type and size for the particular job.
- Connecting hoses and cables, work lead, and work clamp.
- Face helmet, gloves, sleeves (if necessary), and an assortment of hand tools.

The following safety precautions are assumed:

- The welding equipment has been installed properly.
- The welding machine is in a dry location, and there is no water on the floor of the welding booth.
- The welding booth is lighted and ventilated properly.
- All connections are tight, and all hoses and leads are arranged so that they cannot be burned or damaged.

Fig. 22-19 Typical gas metal arc welding station in industry. Note the inverter power source with microprocessor-based wire-feeder/controller on this steel back-hoe bucket. Miller Electric Mfg. Co.

- Gas cylinders are securely fastened so that they cannot fall over and are not part of the electrical circuit.

Starting Procedure

1. Check the power cable connections. Connect the gun cable to the proper welding terminal on the welding machine. Make sure that the work cable end is connected to the proper terminal on the welding machine. (For DCEP the cable should be connected to the negative terminal.) Connect the work cable clamp to either the work or the work table.
2. Start the welding machine by pressing the *on* button or, in the case of an engine drive, start the engine.
3. Turn on the wire-feed unit. Make sure that the wire and feed rollers are clean and properly adjusted.
4. Check the gas-shielding supply system. Make sure that it is properly connected and that all connections are tight. Open the gas cylinder valve. Open the flowmeter valve slowly, and at the same time squeeze and hold down the gun trigger. Adjust the gas flow rate for the particular welding operation you are about to perform. It is usually 15 to 25 cubic feet per hour. Release the gun trigger.
5. Check the water flow if the gun is water cooled.
6. Set the wire-feed speed control for the type and size of filler wire and for the particular job. Remember that the wire-feed speed determines the welding current in a constant voltage system. If you find that more current is required during welding, increase the wire-feed speed. If less current is required, decrease the wire-feed speed. The current value can be checked on the ammeter only while welding. Wire-feed speed determines deposition and welding speed, so adjust speed accordingly.
7. The voltage rheostat should be set to conform to the type and thickness of material being welded, the diameter of the filler wire, the type of shielding gas, and the type of arc. Voltage may range from 18 to as high as 32 volts. The lower voltages are normally used for thinner metals at low amperages with the short circuit mode of transfer. For heavier materials the higher voltage and amperage values apply with a spray arc. Adjust the voltage rheostat until you get a smooth arc.
8. Adjust for the proper electrode extension beyond the contact tube. This varies slightly to meet different welding conditions. Usually a minimum of ¼ inch and a maximum of ⅝ inch are satisfactory. The tip of the tube is sometimes flush with the nozzle surface to maintain normal extension. An extended tube is used for very low amperages and the short circuit transfer. For spray arcs, the tube is normally recessed for a distance of no more than ⅛ inch from the nozzle opening. The distance from the nozzle to the work may vary from ½ to 1 inch maximum.
9. To start the arc, touch the end of the electrode wire to the proper place on the weld joint, usually just ahead of the weld bead, with the current shut off. Lower your helmet and then press the gun trigger on the torch. Depressing the gun trigger on the torch causes the wire to feed, the current to flow, and the shielding gas to flow. Do not press the trigger on the torch before you are ready to start welding. If this happens, cut off the excessive filler wire extending out of the nozzle with a pair of wire cutters or pliers. A steady crackling or hissing sound of the arc is a good indicator of correct arc length and wire feed.

Shutting Down the Equipment

1. Stop welding and release the gun trigger.
2. Return the feed speed to the zero position.
3. Close the gas outlet valve in the top of the gas cylinder.
4. Squeeze the welding gun trigger, hold it down, and bleed the gas lines.
5. Close the gas flowmeter valve until it is finger-tight. This will prevent the possibility of breaking the glass tube on the flowmeter when the gas cylinder is turned on again.
6. Shut off the welding machine and wire feeder.
7. Hang up the welding gun and cable assembly.

Welding Technique

Make sure that the filler wire is of the correct type and size for the type of metal, type of joint, and thickness of metal being welded. Shielding gas flow, power supply, and wire-feed speed should be adjusted for the welding procedure employed. Voltage and wire-feed speed (current) are interdependent with a constant voltage power supply. Adjustment of one may require readjustment of the other.

Starting the Weld

The arc is started with either a **running start** or a **scratch start.** In the running start, the arc is started at the beginning of the weld. The electrode end is put in contact with the base metal, and the trigger on the torch is pressed. Because this type of start tends to be too cold at the beginning of the weld, it may cause poor bead contour, incomplete penetration, incomplete fusion, and porosity at this point.

The scratch start usually gives somewhat better results. In this type of start, the arc is struck approximately 1 inch ahead of the beginning of the weld. Then the arc is quickly moved back to the starting point of the weld, the direction of travel is reversed, and the weld is started. The arc may also be struck outside of the weld area on a starting tab.

Finishing the Weld

When completing the weld bead, the arc should be manipulated to build up the weld crater. The gun trigger can be turned on and off several times at the end of a weld to fill the crater. Another method is reversing the travel direction at the end of the weld and leaving the crater on top of the weld, not at the end of the weld. The gun should be kept over the weld until the gas stops flowing in order to protect the weld until the metal has solidified.

Travel Angle

A push travel angle of 5 to 15° is generally employed when welding in the flat position. Care should be taken that the push angle is not changed or increased as the end of the weld is approached.

When welding uniform thicknesses, the work angle should be equal on all sides. When welding in the horizontal position, best results are obtained by pointing the gun upward slightly. When welding thick-to-thin joints, it is helpful to direct the arc toward the heavier section. A slight drag angle may help when additional penetration is required.

Control of the Arc

Keep in mind that arc travel speed has a major effect on bead size, penetration, and fusion. Arc voltage is also very important to satisfactory welding. It controls surface heating, bead contour, and to some degree, such defects as undercutting, porosity, and weld discontinuities. Because a long arc is smoother and quieter than a short arc, it is frequently used at the expense of quality and economy. The arc should be occasionally noisy for most applications of spray arcs. The wire location should be such as to keep the weld size and location of the weld appropriate.

Process and Equipment Problems

You will probably experience very little difficulty in welding with the MIG/MAG process. All the equipment in use today is in an advanced stage of development and is dependable over a long service life. All things of a mechanical nature, however, are subject to wear and unpredictable breakdown. As a welder in the shop, it will be to your advantage to understand thoroughly the process and the equipment that you are working with. Study the two charts that are presented here very carefully. Table 22-6, page 728, lists problems with the MIG/MAG short arc process and their correction. Table 22-7, pages 729–730, lists problems with the MIG/MAG process and equipment, their causes, and possible remedies.

Practice Jobs

Instructions for Completing Practice Jobs

You will practice gas metal arc welding on mild steel, aluminum, and stainless steel. Complete the practice jobs according to the specifications given in the Job Outline, pages 746 and 747, in the order assigned by your instructor. Before beginning each job, study the pages listed under Text Reference in the outline.

These jobs should provide approximately 25 hours practice, depending on your skill. In addition, practice other forms of joints in all positions. Use various types and sizes of filler wire and different shielding gases. (Review pp. 693–703 in Chapter 21 for information on filler wire classifications.) Compare the spray arc and short arc transfer. Compare solid wire to metal core. Experiment with a variety of current values. Make sure that the practice plates are properly prepared, are clean, and have been tack welded with great care.

MIG/MAG Welding of Carbon Steel

The bulk of all welding is done on carbon steel. In practically every industry in which welding is a part of the fabrication process, the use of MIG/MAG as the welding process is on the increase. The process produces welds that are of the highest quality, and the cost is lower than with other processes. Welders find it relatively easy to master the process, and they can consistently produce sound welds at a high rate of speed.

Groove Welds: Jobs 22-J1 and J2 Plate up to $\frac{1}{8}$ inch thick may be groove welded with square edges with a root opening of 0 to $\frac{1}{16}$ inch. Heavier plate thicknesses of $\frac{3}{16}$ and $\frac{1}{4}$ inch may be welded without beveling the edges if a $\frac{1}{16}$- to $\frac{3}{32}$-inch opening is provided. The bead should be wider than the root spacing for proper fusion. Two passes, one from each side, are usually necessary with backgouging to sound metal as required.

For code welding and when the requirements are unusually high, plate thicknesses from $\frac{3}{16}$ to 1 inch should

Table 22-6 Short Arc Mode of Transfer Troubleshooting

Observed Fault / Possible Action	Voltage	Current	Slope	Inductance	Polarity[1]	Wire Composition	Gas Type	Torch Manip.	Pool Size	Travel Speed	Fitup	Extension	Wind[2]	Gas Shielding[3]	Conduit[4]	Wire-Feed System	Contact Tip[5]	Oil, Rust, Paint	Type Of Steel[6]	Surface Treatment	Slag, Glass	Arc Blow	Procedure
Porosity, internal						✓								✓				✓	✓	✓			
Porosity, visible													✓	✓				✓		✓			
Incomplete penetration					✓			✓			+	−										✓	
Too much penetration							✓					+											
Incomplete fusion	✓	✓			✓			✓															
Wagon tracks							✓	✓	✓												✓		✓
Cold lapping						✓	✓		✓	✓	+												✓
Cracks																			✓				✓
Whiskers	✓	✓						✓			−	−											
Spatter	✓	✓	✓	+			✓																
Stubbing	+	−																					
Blobbing	−											−											
Bad starts	+	−	−									−									✓		
Unstable arc	✓												✓	✓	✓	✓	✓				✓		
Effervescent puddle													✓	✓					✓				
Too fluid puddle				−		✓	✓																
Too viscous puddle				+		✓	✓																
Crowned bead	+	+		+		✓		✓		✓													
Undercutting	−			−				✓		✓												✓	

[1] Should be DCEP, DCEN looks very similar in short arc welding.
[2] Windshields are recommended for winds over 5 mi/h.
[3] Clogged nozzles, loose connections, and damaged hoses.
[4] Wrong size, clogged.
[5] Wrong size, damaged.
[6] Rimming grades require highly deoxidized wires.

+ Raise or increase.
− Lower or decrease.
√ Check.

be beveled. A 60° single- or double-V without a root face is recommended. A root opening of 0 to 1/16 inch should be maintained. Wider root openings may be provided for double-V joints than for single-V joints. In single-V grooves, a backing pass from the reverse side is generally required unless the fitup is uniform. If possible, the double-V joint should be used. Less filler metal is needed to fill the joint, and less distortion results when welding from both sides of the joint. Uniform penetration can be obtained in joints having no root face if the root opening is held between 0 and 3/32 inch, Fig. 22-20, page 731.

The open root joint should be run using short circuit or pulse spray for the ferrous metals. Practice the 3G using both uphill and downhill techniques on the root pass.

U-grooves should be employed on plate thicker than 1 inch. Root spacing between 1/32 and 3/32 inch should be maintained. A root face of 3/32 inch or less should be provided to ensure adequate penetration. The U-groove butt joint requires less filler metal than the V-groove butt joint.

For spray arc welding of carbon steel, an argon-oxygen mixture containing 1 to 5 percent oxygen is generally recommended. The addition of oxygen produces a more

Table 22-7 Troubleshooting: MIG/MAG Process and Equipment

Introduction

When troubleshooting gas metal arc welding process and equipment problems, it is well to isolate and classify them as soon as possible into one of the following categories:

1. Electrical
2. Mechanical
3. Process

This eliminates much needless lost time and effort. The data collected here for your benefit discusses some of the common problems of gas metal arc welding processes. A little thought will probably enable you to solve your particular problem through the information provided.

The assumption of this data is that a proper welding condition has been achieved and has been used until trouble developed. In all cases of equipment malfunction, the manufacturer's recommendations should be strictly adhered to and followed.

Problem	Probable Cause	Possible Remedy

A. Feeder-control stops feeding electrode wire while welding.	1. Power source.
	  a) Fuse blown in power source primary.
	  b) Control circuit fuse blown.
	2. Primary line fuse blown.
	3. Wire-feeder control.
	  a) Control relay defective.
	  b) Protective fuse blown.
	  c) Wire drive rolls misaligned.
	  d) Drive roll pressure too great.
	  e) Spindle friction too great.
	  f) Excess loading of drive motor.
	  g) Drive rolls worn; slipping.
	  h) Drive motor burned out.
	4. Torch and casing assembly.
	  a) Casing liner dirty, restricted.
	  b) Broken or damaged casing or liner.
	  c) Torch trigger switch defective or wire leads broken.
	  d) Contact tube orifice restricted; burn-back of electrode.
	  e) Friction in torch.
		1. Power source.
		  a) Replace fuse.
		  b) Replace fuse.
		2. Replace fuse.
		3. Wire-feeder control.
		  a) Replace contact relay.
		  b) Replace fuse. Find overload cause.
		  c) Realign drive rolls.
		  d) Loosen and readjust drive rolls.
		  e) Loosen and readjust nut pressure.
		  f) Clear restriction in drive assembly.
		  g) Replace drive rolls.
		  h) Test motor; replace if necessary.
		4. Torch and casing assembly.
		  a) Remove liner, blow out with compressed air.
		  b) Replace faulty part.
		  c) Replace switch; check connections.
		  d) Replace.
		  e) Check wire passage; clean, replace parts as required.

B. Electrode wire feeds but is not energized. No welding arc.	1. Power source.
	  a) Contactor plug not tight in socket.
	  b) Contactor control leads broken.
	  c) Remote-standard switch defective or in wrong position.
	  d) Primary contactor coil defective.
	  e) Primary contactor points defective with electronic solid-state contactor. See owner's manual.
	  f) Welding cables loose on terminals.
	  g) Work connection loose.
	2. Wire-feeder control.
	  a) Contactor plug not properly seated.
	  b) Contact relay defective.
		1. Power source.
		  a) Tighten plug in receptacle.
		  b) Repair or replace.
		  c) Repair or replace; position correctly.
		  d) Replace.
		  e) Replace points or contactor.
		  f) Have serviced by qualified technician.
		  g) Tighten connections.
		  h) Connect to work; tighten connection.
		2. Wire-feeder control.
		  a) Tighten plug in receptacle.
		  b) Repair or replace.

(Continued)

Table 22-7 (Concluded)

Problem	Probable Cause	Possible Remedy

C. Porosity in the weld deposit.	1. Dirty base metal; heavy oxides, mill scale.
	2. Gas cylinder and distribution system.
	   a) Gas cylinder valve off.
	   b) Regulator diaphragm defective.
	   c) Flowmeter cracked or broken.
	   d) Gas supply hose connections loose.
	   e) Gas supply hose leaks.
	   f) Insufficient shielding gas flow.
	   g) Moisture in shielding gas.
	   h) Freezing of carbon dioxide regulator/flowmeter.
	3. Wire-feeder control.
	   a) Gas solenoid defective.
	   b) Gas hose connections loose.
	4. Gun and casing assembly.
	   a) Gun body and/or accessories aspirating atmosphere or air.
	   b) Check O-rings on quick connect type guns.
	   c) Contact tube extended too far.
	   d) Nozzle-to-work distance too great.
	   e) Improper gun angles.
	   f) Welding speed too fast.
	   g) Electrode not centered in nozzle.
	   h) Air turbulence.
	5. Improper electrode wire composition.
		1. Clean base metal before welding.
		2. Gas cylinder and distribution system.
		   a) Turn cylinder valve on.
		   b) Replace diaphragm or regulator.
		   c) Replace and repair.
		   d) Tighten fittings.
		   e) Repair or replace.
		   f) Increase flow rate of gas.
		   g) Replace gas cylinder or supply.
		   h) Thaw unit; install gas line heater.
		3. Wire-feeder control.
		   a) Replace solenoid.
		   b) Tighten connections.
		4. Gun and casing assembly.
		   a) Test; replace or repair faulty units.
		   b) Replace defective O-rings.
		   c) Distance from nozzle end maximum $1/8$ in.
		   d) Should be as recommended by equipment manufacturer.
		   e) Use correct travel angle; approximately $15°$ angle.
		   f) Adjust welding condition for slower speed.
		   g) Adjust contact tube, nozzle, and wire.
		   h) Set up wind screen.
		5. Obtain and use correct electrode wire.

D. Welding electrode wire stubs into workpiece.	1. Power source.
	   a) Excessive slope numerical values set.
	   b) Arc voltage too low.
	   c) Excessive inductance value set.
	2. Wire-feeder control.
	   a) Excess wire-feed speed.
		1. Power source.
		   a) Reduce slope settings as required.
		   b) Increase voltage at power source.
		   c) Reduce inductance setting as required.
		2. Wire-feeder control.
		   a) Reduce wire-feed speed rate.

E. Excessive spatter while welding.	1. Shielding gas system.
	   a) Excessive gas flow rates.
	   b) Insufficient gas flow.
	2. Power source.
	   a) Excessive arc voltage.
	   b) Insufficient slope setting value.
	   c) Insufficient inductance setting value.
	3. Gun contact tube recessed in nozzle too far.
	4. Improper electrode, shielding gas combination.
		1. Shielding gas system.
		   a) Adjust gas flow rate as required.
		   b) Adjust gas flow rate as required.
		2. Power source.
		   a) Reduce voltage at power source.
		   b) Increase slope setting as required.
		   c) Increase inductance setting as required.
		3. Replace contact tube with longer one.
		4. Obtain and use correct electrode wire, shielding gas combination.

F. Weld bead appearance indicates need for more amperage and/or larger bead.	1. Power source.
	   a) Volt-ampere condition too low.
		1. Power source.
		   a) Increase voltage slowly with applicable increase in wire feed.

G. Weld bead appearance indicates need for less amperage and/or smaller bead.	1. Power source.
	   a) Volt-ampere condition too high.
		1. Power source.
		   a) Reduce voltage and wire feed as required.

Source: Miller Electric Manufacturing Co.

Fig. 22-20 GMAW root pass 2G short circuit transfer on ⅜-inch thick plate. Note the excellent penetration and fusion and no whiskers protruding on the back side of this test piece. *Edward R. Bohnart*

stable arc, improves the flow of weld metal, and reduces the tendency to undercut.

Argon with 10 percent CO_2 is sometimes used. With 100 percent carbon dioxide, the arc is not a true spray arc. For short arc welding of carbon steel, a mixture of 75 percent argon and 25 percent carbon dioxide may be used. Straight carbon dioxide is popular for MAG small wire welding.

Code welding requires great care in its application. Good fusion and a minimum of porosity are necessary. The following precautions should be observed when doing the practice jobs, Fig. 22-21:

- Avoid excessive current values. If the current seems to be too high and cannot be reduced without affecting the transfer of metal, switch to a larger size wire.

Fig. 22-21 GMAW final passes short circuit transfer on ⅜-inch thick plate. Groove angle 60°, SG-AC-25, and 0.035 wire diameter. Note the complete fusion, uniform bead width, location, and the ripple pattern created by the slight whipping motion. *Edward R. Bohnart*

- Check your welding speed. Welding speeds that are too high cause porosity, and speeds that are too slow may cause an incomplete fusion.
- Make sure that the gas flow is adequate. The entire weld area must have the protection of the gaseous shield.
- Keep the wire location centered in the gas pattern and in the center of the joint. Make sure that the correct electrode angle is maintained at all times.
- Select the proper filler wire for material being welded and for such special situations as rust, scale, and excessive oxygen.
- When welding from both sides of the plate, be sure that the root pass on the first side is deeply penetrated by the root pass on the second side. Backgouge to sound metal as required.

For video of GMAW troubleshooting, please visit www.mhhe.com/welding.

Fillet Welds: Jobs 22-J3–J10 Fillet welds are used in T-joints, lap joints, and corner joints. Much of the welding done with the gas metal arc process is fillet welding. The deposit rate and rate of travel are high, and there is deep penetration. Since the strength of a fillet weld depends on the throat area, the deep penetration that gas metal arc welding provides permits smaller fillet welds than is possible with stick electrode welding.

The position of the nozzle and the speed of welding are important. You must make sure that penetration is secured at the root of the weld in order to take advantage of the deep penetration characteristics. The welding technique should provide protection for the vertical plate to avoid undercutting. Many of the skills learned in making fillet welds with stick electrodes and the TIG process can be applied to MAG welding.

Welding may be single pass or multipass, depending on the requirements of the job. Multipass welding may be done with weave beads or stringer beads, Fig. 22-22. The sequence of weld passes is the same as that used with stick electrodes. Each pass must be cleaned carefully. In making each pass, fusion must be secured with the underneath pass and the surface of the plate. Review the precautions listed previously for groove welds.

Fig. 22-22 A 3F vertical-up fillet weld using GMAW short circuit transfer. SG-AC-25 with 0.035 wire root pass was stringer. Note the uniform ripple pattern, uniform width, and minimal spatter on this as-welded ⅜-inch weld. *Edward R. Bohnart*

Inspection and Testing After each weld has been completed, use the same inspection and testing methods that you have used in previous welding practice. Look for surface defects. Keep in mind that it is important to have good appearance and uniform weld contour. These characteristics usually indicate that the weld was made properly and that the weld metal is sound throughout. Observe the appearance of the welds shown in Figs. 22-23 through 22-29, page 734.

- Butt joint in the horizontal position Fig. 22-19A and 19B.
- Tee joint in the vertical position Fig. 22-19C

JOB TIP

Salary

When negotiating a salary with a new employer, avoid disclosing what you made in your last job. Many businesses decide what your salary will be by knowing your last pay rate. Instead, say what range you are expecting.

- Corner joint in the flat position: Figs. 22-23 and 22-24
- Lap joint in the horizontal position: Fig. 22-25
- Lap joint in the vertical position; direction of travel, down: Fig. 22-26
- T-joint in the horizontal position: Fig. 22-27
- Beveled-butt joint in the flat position; no backing bar; Figs. 22-28 and 22-29. Note that penetration must be secured through the back side in open root joints like that required when welding with other processes.

To determine the soundness of your welds, test lap, T-, and butt joints in the usual manner.

Fillet and Groove Welding Combination Project: Job Qualification Test 1 This combination test project will allow you to demonstrate your ability to read a drawing, develop a bill of materials (SI conversions are optional), thermally cut, fit components together, tack, and weld a carbon steel project. You will be using the techniques developed in Jobs 22-J1 through J10 using the short-circuiting mode of metal transfer. Follow the instructions found in the notes on Fig. 22-30, page 735.

Inspection and Testing After the project has been tacked, have it inspected for compliance to the drawing. On multiple-pass welds, have your instructor inspect after each weld unless otherwise instructed. Use the following acceptance criteria to judge your welds. Look for surface defects. Keep in mind that it is important to have good appearance and uniform weld contour. These characteristics usually indicate that the weld was made properly and that the weld metal is sound throughout. Only visual inspection will be used on this test project to the following acceptance requirements:

- There shall be no cracks or incomplete fusion.
- There shall be no incomplete joint penetration in groove welds except as permitted for partial joint penetration groove welds.
- Your instructor shall examine the weld for acceptable appearance and shall be satisfied that the welder is skilled in using the process and procedure specified for the test.
- Undercut shall not exceed the lesser of 10 percent of the base metal thickness or ¹⁄₃₂ inch.
- Where visual examination is the only criterion for acceptance, all weld passes are subject to visual examination, at the discretion of your instructor.
- The frequency of porosity shall not exceed one in each 4 inches of weld length, and the maximum diameter shall not exceed ³⁄₃₂ inch.
- Welds shall be free from overlap.
- Only minimal weld spatter shall be accepted, as viewed prior to cleaning.

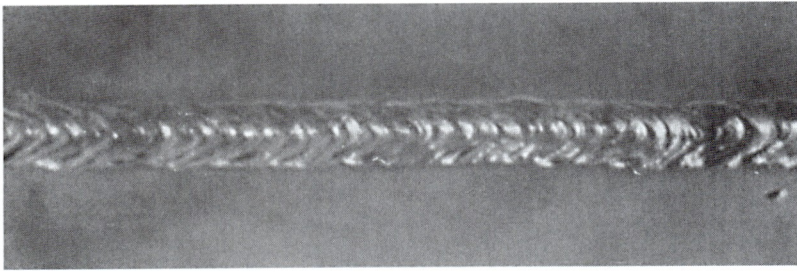

Fig. 22-23 Outside corner joint in steel plate welded with the gas metal arc welding process in the flat position. Edward R. Bohnart

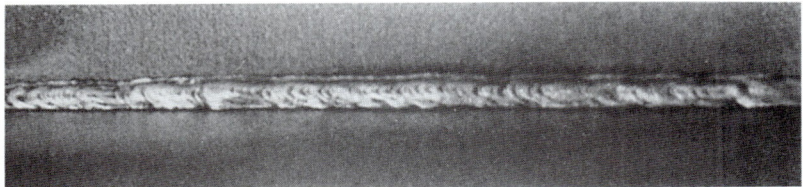

Fig. 22-24 Penetration through the back side of a corner joint welded in the flat position. Edward R. Bohnart

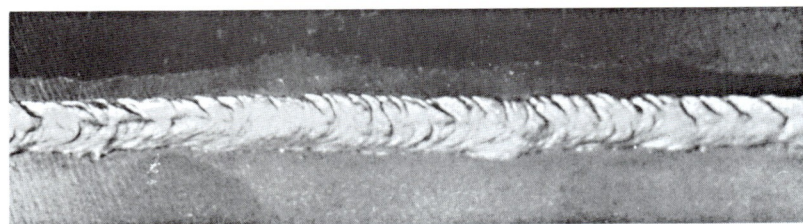

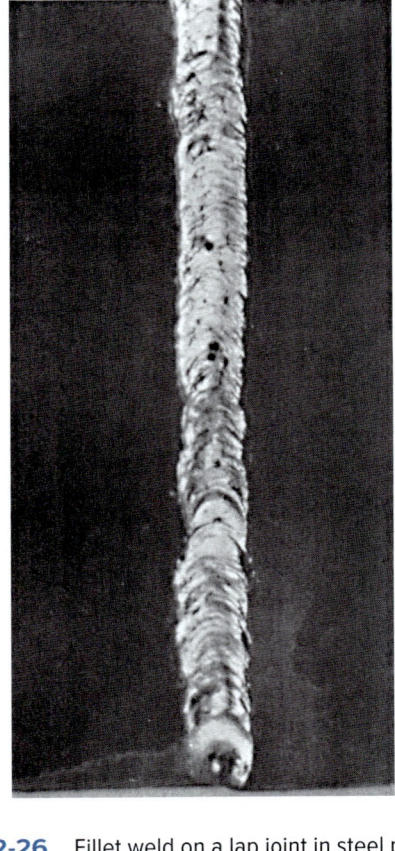

Fig. 22-26 Fillet weld on a lap joint in steel plate welded with the gas metal arc welding process in the 3F position, downhill. Note the porosity caused by poor gas shielding. Edward R. Bohnart

Fig. 22-25 Fillet weld on a lap joint in steel plate welded with the gas metal arc welding process in the 2F position. Edward R. Bohnart

Fillet and Groove Welding Combination Project: Job Qualification Test 2 This combination test project will allow you to demonstrate your ability to read a drawing, develop a bill of materials (SI conversions are optional), thermally cut, fit components together, tack, and weld a carbon steel project. You will be using the techniques developed in Jobs 22-J1 through J10 using the spray arc mode of metal transfer. Follow the instructions found in the notes on Fig. 22-31, page 736.

Inspection and Testing After the project has been tacked, have it inspected for compliance to the drawing. On multiple-pass welds, have your instructor inspect after each weld unless otherwise instructed. Use the following acceptance criteria to judge your welds. Look for surface defects. Keep in mind that it is important to have good appearance and uniform weld contour. These characteristics usually indicate that the weld was made properly and that the weld metal is sound throughout. Only visual inspection will be used on this test project to the following acceptance requirements:

- There shall be no cracks or incomplete fusion.
- There shall be no incomplete joint penetration in groove welds except as permitted for partial joint penetration groove welds.
- Your instructor shall examine the weld for acceptable appearance and shall be satisfied that the welder is skilled in using the process and procedure specified for the test.
- Undercut shall not exceed the lesser of 10 percent of the base metal thickness or $\frac{1}{32}$ inch.
- Where visual examination is the only criterion for acceptance, all weld passes are subject to visual examination, at the discretion of your instructor.
- The frequency of porosity shall not exceed one in each 4 inches of weld length, and the maximum diameter shall not exceed $\frac{3}{32}$ inch.

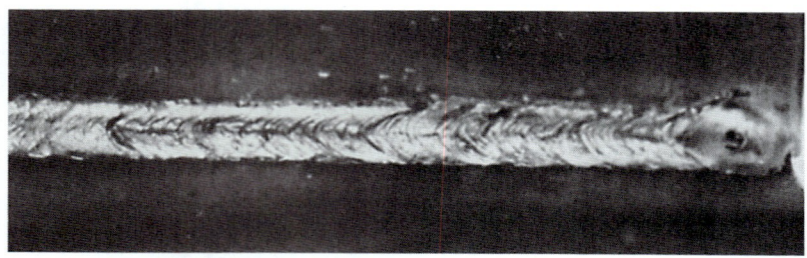

Fig. 22-27 Fillet weld on a T-joint welded in the 2F position with the gas metal arc welding process in steel plate. *Edward R. Bohnart*

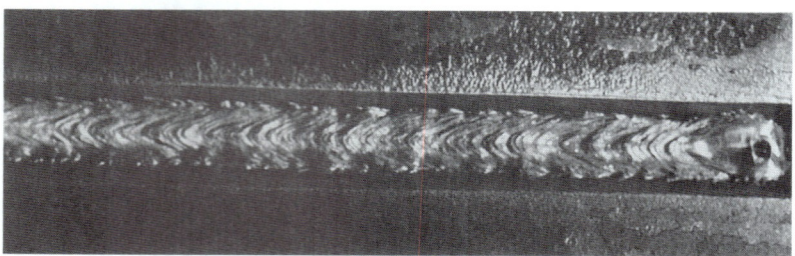

Fig. 22-28 The first (root) pass of a V-groove butt joint welded in the 1G position with the gas metal arc welding process in steel plate. *Edward R. Bohnart*

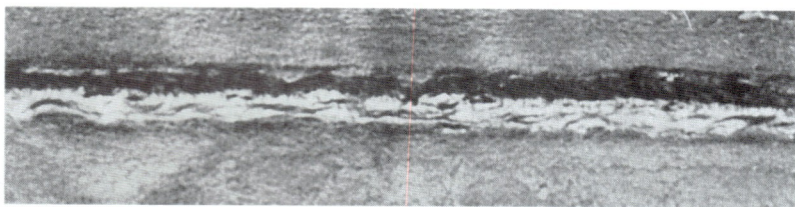

Fig. 22-29 Penetration through the back side of a V-groove butt joint welded in the 1G position. *Edward R. Bohnart*

- Welds shall be free from overlap.
- Only minimal weld spatter shall be accepted, as viewed prior to cleaning.

Groove Weld Project: Job Qualification Test 3 This test project will allow you to demonstrate your ability to read a drawing, fit components together, tack, and weld a carbon steel unlimited thickness test plate. You will be using the techniques developed in Jobs 22-J1 through J10 using the spray arc mode of metal transfer. Follow the instructions found in the notes on Fig. 22-32, page 737.

Inspection and Testing After the project has been tacked, have it inspected for compliance to the drawing. After the project has been completely welded, use visual inspection and cut specimens for bend testing. Use the following acceptance criteria to visually judge your welds. Look for surface defects. Keep in mind that it is important to have good appearance and uniform weld contour. These characteristics usually indicate that the weld was made properly and that the weld metal is sound throughout. Once visual inspection is completed to the following criteria, you will perform side bend tests. Follow side bend test procedures as outlined in Chapter 28.

- There shall be no cracks or incomplete fusion.
- There shall be no incomplete joint penetration in groove welds except as permitted for partial joint penetration groove welds.
- Your instructor shall examine the weld for acceptable appearance and shall be satisfied that the welder is skilled in using the process and procedure specified for the test.
- Undercut shall not exceed the lesser of 10 percent of the base metal thickness or $\frac{1}{32}$ inch.
- Where visual examination is the only criterion for acceptance, all weld passes are subject to visual examination, at the discretion of your instructor.
- The frequency of porosity shall not exceed one in each 4 inches of weld length and the maximum diameter shall not exceed $\frac{3}{32}$ inch.
- Welds shall be free from overlap.
- Only minimal weld spatter shall be accepted, as viewed prior to cleaning.

Side bend acceptance criteria as measured on the convex surface of the bend specimen are the following:

- No single indication shall exceed $\frac{1}{8}$ inch measured in any direction on the surface.
- The sum of the greatest dimensions of all indications on the surface, which exceed $\frac{1}{32}$ inch, but are less than or equal to $\frac{1}{8}$ inch, shall not exceed $\frac{3}{8}$ inch.
- Cracks occurring at the corner of the specimens shall not be considered unless there is definite evidence that they result from inclusions, fusion, or other internal discontinuities.

MIG Welding of Aluminum

You will recall that aluminum is readily joined by welding, brazing, soldering, adhesive bonding, and mechanical fastening. Aluminum is lightweight, and yet some of its alloys have strengths comparable to mild steel. Pure aluminum can be alloyed readily with many other metals to produce

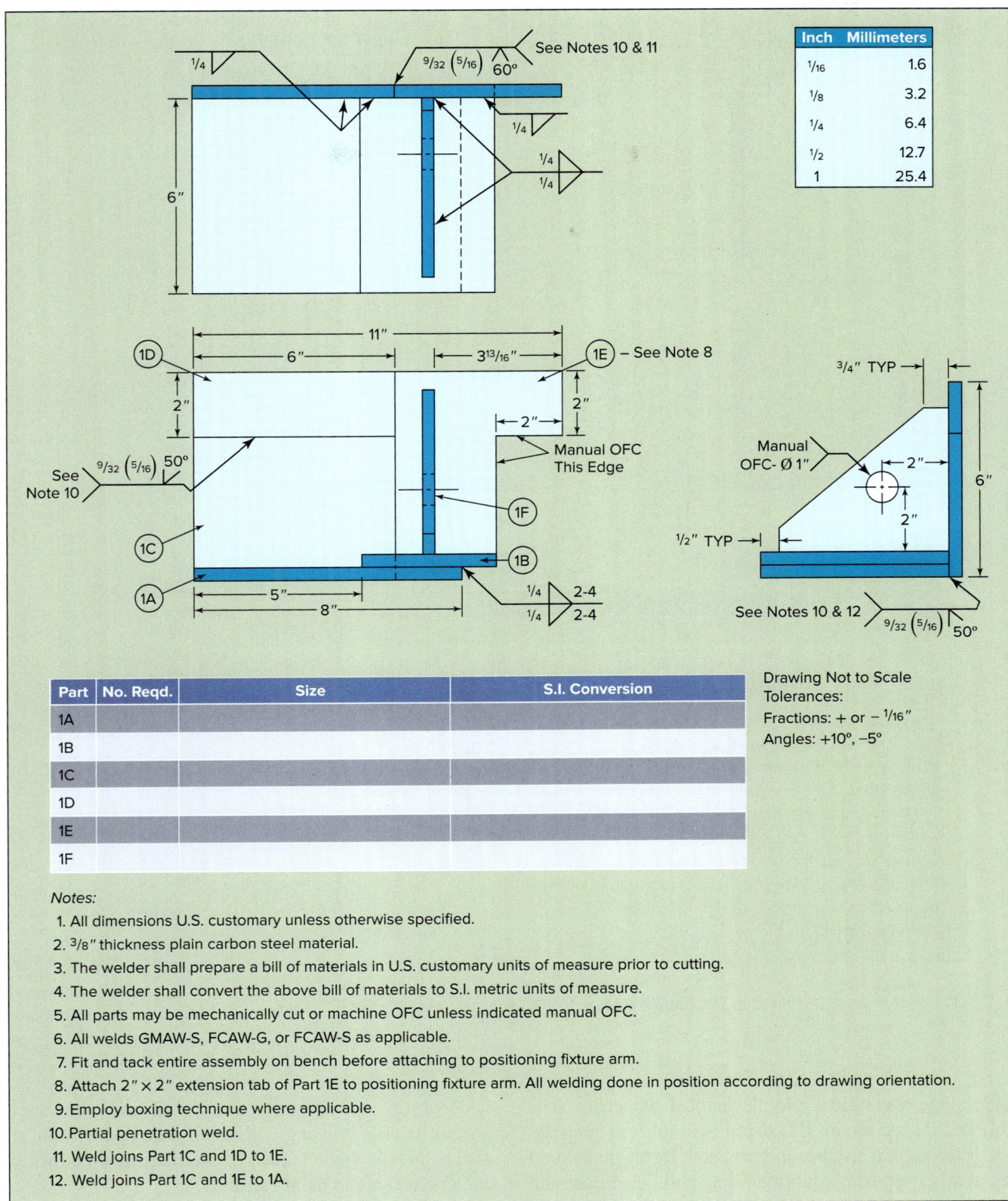

Fig. 22-30 Performance Qualification Test GMAW-S. Adapted from AWS SENSE Program

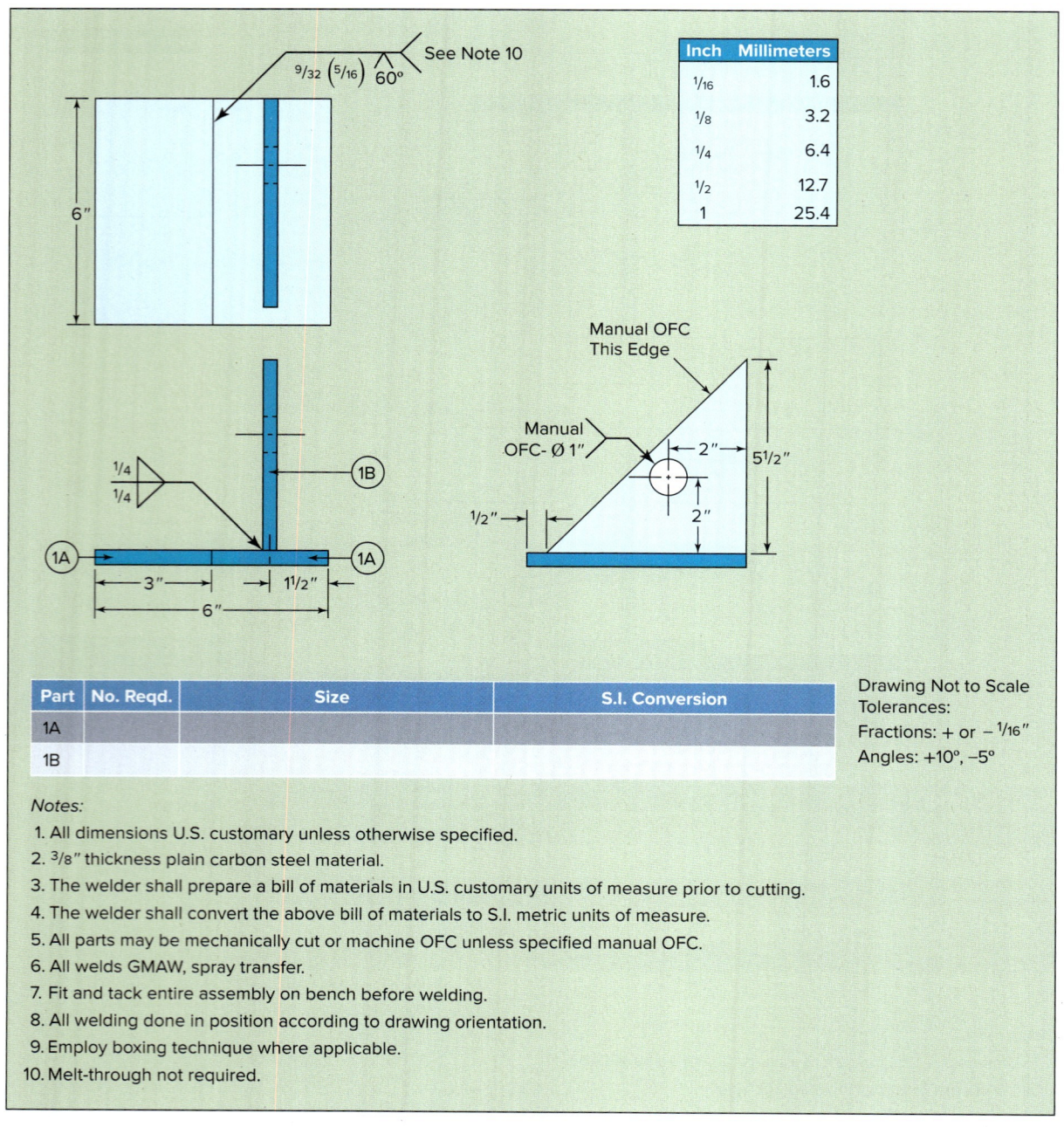

Fig. 22-31 Performance Qualification Test GMAW, Spray Transfer. Adapted from AWS SENSE Program

a wide range of physical and mechanical properties. It is highly ductile and retains that ductility at subzero temperatures. It has high resistance to corrosion, forms no colored salts, and is not toxic. Aluminum has good electrical and thermal conductivity and high reflectivity to both heat and light. It is nonsparking and nonmagnetic.

Aluminum is easy to fabricate. It can be cast, rolled, stamped, drawn, spun, stretched, and roll formed. The metal may also be hammered, forged, and extruded into a wide variety of shapes. Machining ease and speed are important factors in using aluminum parts. Aluminum may also be given a wide variety of mechanical, electrochemical, chemical, and paint finishes.

Pure aluminum melts at 1,220°F, and aluminum alloys have an approximate melting range of 900 to 1,220°F, depending upon the alloying elements. Aluminum does not change color when heated to the welding or brazing range. This makes it difficult to judge when the metal is near the

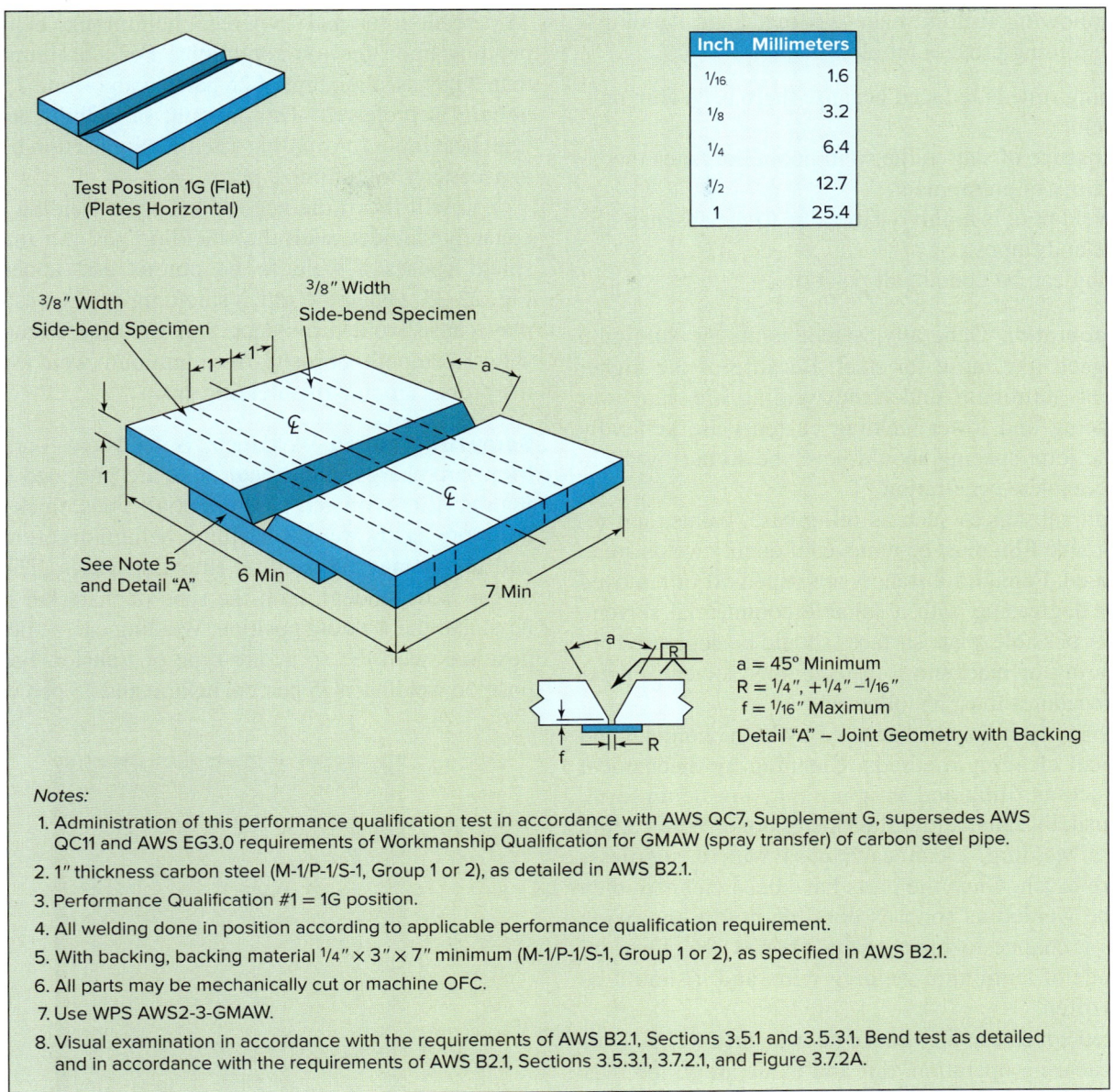

Fig. 22-32 Performance Qualification Test GMAW, Spray Transfer, Carbon Steel, 1G Position. Adapted from AWS SENSE Program

melting point during welding. The high thermal conductivity necessitates high heat input for fusion welding.

Aluminum and its alloys rapidly develop a tenacious, refractory oxide film when exposed to air. The melting point for aluminum oxide is 3,600°F or three times the melting temperature of aluminum. The oxide film must be removed or broken up during welding to permit the base and filler material to flow together properly when fusion welding or to permit flow in brazing or soldering. The oxide may be removed by fluxes, by the action of the welding arc in an inert gas atmosphere, or by mechanical and chemical means.

Aluminum is used by all industries, particularly the automotive, aircraft, electrical, chemical, and food industries. Review Chapter 3 for a more detailed discussion of aluminum and its properties.

MIG and TIG welding have all but replaced stick electrode welding for aluminum and its alloys. The gas tungsten arc and GMAW processes are generally used in welding the lighter gauges of aluminum. Heavier gauges are welded with the gas metal arc process. The type of joint and the position of welding determine to a great extent the process to be used on thicknesses ⅛ inch and under.

The following factors make gas metal arc welding a desirable joining process for aluminum:

- Cleaning time is reduced because there is no flux on the weld.
- The absence of slag in the weld pool eliminates the possibility of entrapment.
- The weld pool is highly visible due to the absence of smoke and fumes.
- Welding can be done in all positions.

Joint Preparation Generally, welded joints for aluminum are designed like those for steel. Because of the higher fluidity of aluminum under the welding arc, narrower joint spacing and lower welding currents are generally used. The joint spacing should never be so narrow as to prevent complete penetration.

Foreign substances such as oil, grease, paints, and refractory oxide film must be removed if quality welds are to be produced. Foreign substances are wiped off or removed by vapor degreasing with a suitable commercial solvent. Whenever possible, plate surfaces should be degreased before shearing or machining. In using any solvent, proper safety procedures must be followed.

The oxide film may be removed by both chemical and mechanical cleaning methods. Cleaning by mechanical means such as filing and scraping may not be uniform, but it is usually satisfactory if properly done. When doing multipass welding, clean each pass with a stainless-steel wire brush. Once the parts have been cleaned, they should be welded as soon as possible before the oxide film has a chance to form again. Oxides also form on the surface of aluminum welding wire, and it should be checked often.

Sheared edges can also cause poor quality welds. During the shearing operation, dirt and oxide are rolled over and trapped in the sheared edge. In many cases, this results in weld inclusions and porosity. To eliminate this problem, sheared edges should be thoroughly degreased and, if possible, mechanically cleaned before welding.

Shielding Gas Argon is preferred for welding aluminum plate thicknesses up to 1 inch. When compared to helium, argon provides better metal transfer and better arc stability, thus reducing spatter.

For the welding of plate thicknesses from 1 to 2 inches, the following shielding may be used: pure argon, a mixture of 50 percent argon and 50 percent helium, or a mixture of 75 percent argon and 25 percent helium. Helium provides a high heat input rate, and argon provides excellent cleaning action.

For the welding of plate thicknesses from 2 to 3 inches, a mixture of 50 percent argon and 50 percent helium or 25 percent argon and 75 percent helium may be used, depending upon the job conditions. For aluminum thicker than 3 inches, a mixture of 25 percent argon and 75 percent helium is preferred. This amount of helium provides a high heat input for welding thick sections. High heat input is necessary to minimize porosity.

You will recall the need to keep the welding area adequately shielded with the shielding gas. An inadequate shield causes a weld to be porous and appear dirty, Fig. 22-33. Using too much shielding gas is wasteful and may cause weld turbulence and porosity. Figure 22-34 shows a smooth, porosity-free aluminum weld made with the proper amount of shielding gas.

Spray Arc Welding Weld metal is deposited continuously. More arc energy and greater heat are provided for melting the filler wire and base material. Thus, thick sections are more easily welded. Helium, helium-argon mixtures, and argon may be used as shielding gases. The choice of gas is dependent upon the type of material, its thickness, and the welding position. Welding can be done in all positions with the spray-arc type of transfer. For out-of-position welding, a 75 percent helium and 25 percent argon

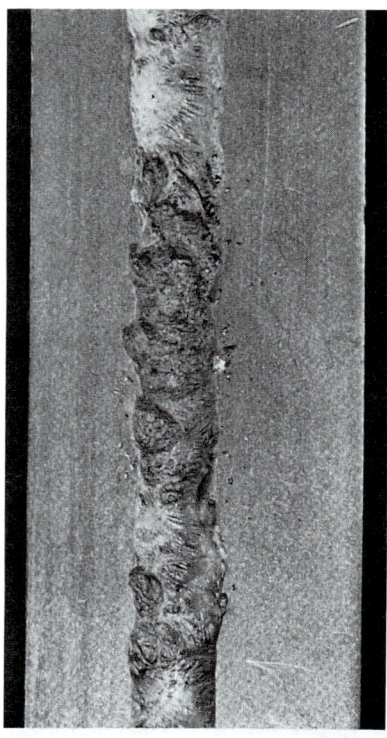

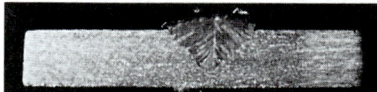

Fig. 22-33 Aluminum weld made with insufficient shielding gas. © Kaiser Aluminum

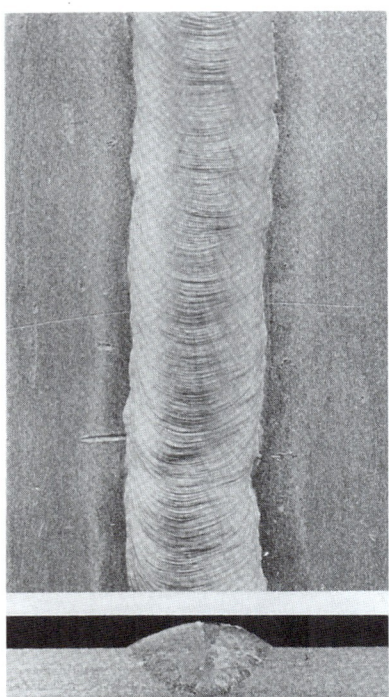

Fig. 22-34 Aluminum weld made with sufficient shielding gas. © Kaiser Aluminum

mixture, straight helium, or straight argon is the shielding gas. The GMAW-P mode of transfer is very effective when welding aluminum. It gives excellent control over the heat input and makes welding thin sections and out-of-position welding much easier.

Out-of-Position Welding Out-of-position welding of aluminum with the gas metal arc welding process is no more difficult than when welding out of position with any one of the other welding processes.

Horizontal Position In welding butt joints and T-joints in the horizontal position, Figs. 22-35 and 22-36, care must be taken to penetrate to the root of the joint. Overheating in any one area causes sagging, undercutting, or melt-through to the back side of the joint. The weld metal should be directed against the upper plate. In multipass welding be sure that there is no incomplete fusion between passes.

Vertical Position Fillet and groove welds in the vertical position must be welded with the travel-up technique, Figs. 22-37 and 22-38, page 740. Do not use too high a welding current nor deposit too large a weld bead. If the molten pool is too large, the effect of gravity makes it difficult to control. A slight side-to-side motion may be helpful. In multipass welding make sure that there is no incomplete fusion between passes.

Overhead Position Fillet and groove welds are made in the overhead position without difficulty. Welding current and travel speed are lower than for the flat position. Because the shielding gas has a tendency to leave the weld area, the gas flow rate is higher. Extreme care must be taken to avoid sagging and poor penetration. Trying to deposit too much metal and carrying too large a weld pool are direct causes of such conditions. You may find overhead welding with the MIG torch somewhat awkward. Assume as

Fig. 22-35 Position of the MIG gun when welding a T-joint in aluminum plate in the 2F position. © Kaiser Aluminum

Fig. 22-36 Position of the MIG gun when welding a V-groove butt joint in aluminum plate in the 2G position. © Kaiser Aluminum

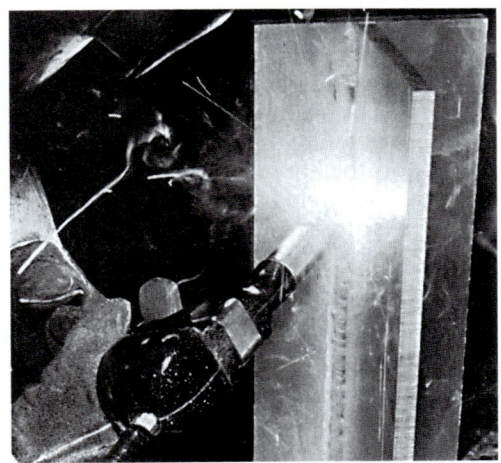

Fig. 22-37 Position of the MIG gun when welding a T-joint in aluminum plate in the 3F position, uphill. © Kaiser Aluminum

Fig. 22-38 Position of the MIG gun when welding a V-groove butt joint in aluminum plate in the 3G position, uphill. © Kaiser Aluminum

comfortable and relaxed a position as possible, Figs. 22-39 and 22-40. This will help to keep the gun steady, which is necessary for quality welding.

Butt Joints: Jobs 22-J11 and J12 Butt joints are easy to design, require a minimum of base material, present good appearance, and perform better under fatigue loading than other types of joints. They require accurate alignment and edge preparation, and it is usually necessary to bevel the edge on thicknesses of ¼ inch or more to permit satisfactory root pass penetration. On heavier plate, chipping the back side and welding the back side with one pass are recommended to ensure complete penetration and fusion. Sections with different thicknesses should be beveled before welding.

Lap Joints: Job 22-J13 Lap joints are more widely used on aluminum alloys than on most other materials. In thicknesses of aluminum up to ½ inch, it is more economical to use double-welded, single-lap joints than double-welded butt joints. Lap joints require no edge preparation, are easy to fit, and require less jigging than butt joints do.

T-Joints: Jobs 22-J14–J16 T-joints have several of the advantages of lap joints. They seldom require edge preparation on material ¼ inch or less in thickness. This is because the fillet welds on T-joints, as on lap joints, are fully penetrated if the weld is fused into the root of the joint. Edge preparation may be used on thick material to reduce welding costs and minimize distortion. T-joints are easily fitted and normally require no back chipping. Any necessary jigging is usually quite simple. Welding a

Fig. 22-39 Position of MIG gun when welding a T-joint in aluminum plate in the 4F position. © Kaiser Aluminum

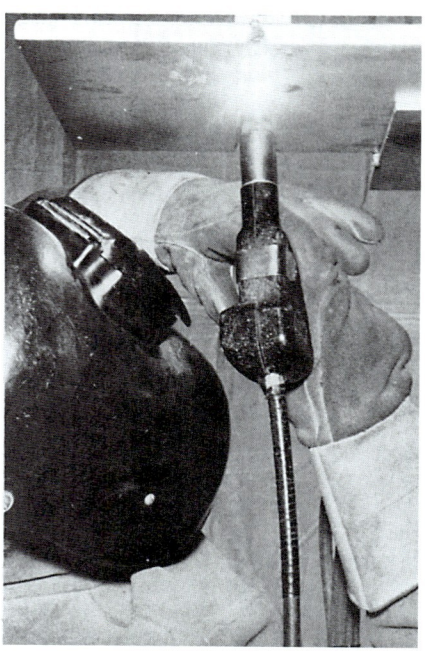

Fig. 22-40 Position of MIG gun when welding a V-groove butt joint in aluminum plate in the 4G position. © Kaiser Aluminum

T-joint on one side only is not ordinarily recommended. Although this type of joint may have adequate shear and tensile strength, a weld on one side acts as a hinge under load so that it is very weak. It is better to put a small, continuous fillet weld on each side of the joint, rather than a large weld on one side of intermittent welds on both sides. Continuous fillet welding is recommended over intermittent welding for longer fatigue life.

Edge and Corner Joints These joints are economical from the standpoint of preparation, base metal used, and welding requirements. However, they are harder to fit up and are prone to fatigue failure. The edges do not require preparation.

Inspection and Testing After the weld has been completed, inspect it carefully for defects. Use the same inspection and testing procedures that you learned in previous practice. Look for surface defects. Keep in mind that it is important to have good appearance and uniform weld contour. These characteristics usually indicate that the weld was made properly and that it is sound throughout. Remember that high quality welds in aluminum can be produced only if proper welding conditions and good cleaning procedures have been established and maintained. Figures 22-41 through 22-43 show some acceptable and unacceptable aluminum weld beads.

The following weld defects are found most often in the welding of aluminum:

- Cracking in the weld metal or in the heat-affected zone. Weld metal cracks are generally in crater or longitudinal form. *Crater cracks* often occur when the arc is broken sharply and leaves a crater. Manipulating the gun properly eliminates this problem. Longitudinal cracks are caused by:
  - Incorrect weld metal composition
  - Improper welding procedure

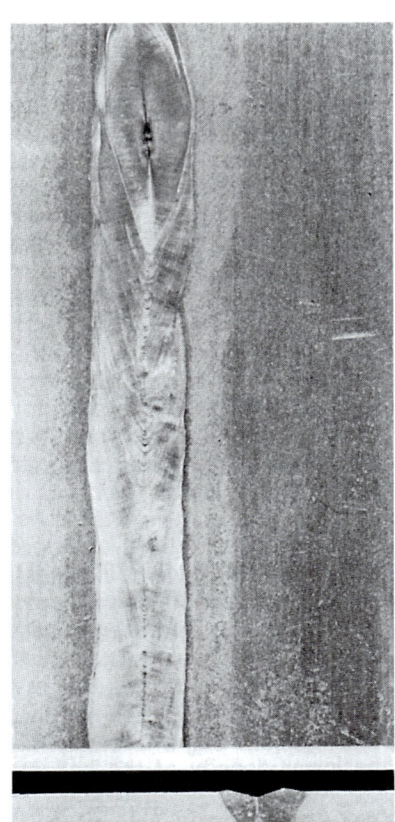

Fig. 22-41 Aluminum weld bead made with current that is too high. Note the flat or concave appearance and excessive penetration. © Kaiser Aluminum

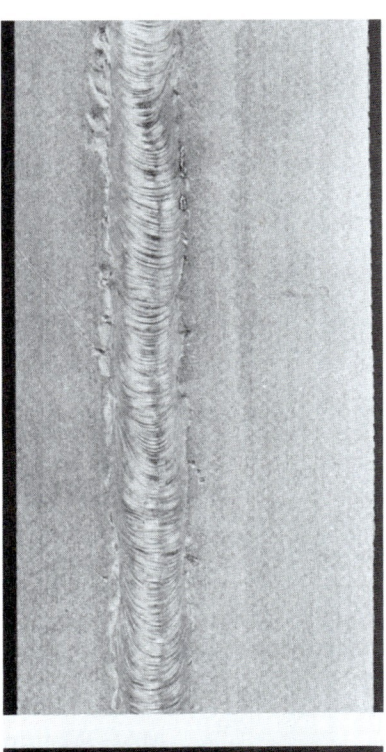

Fig. 22-42 Aluminum weld bead made with current that is too low. Note the incomplete penetration and the excess reinforcement and narrow bead. © Kaiser Aluminum

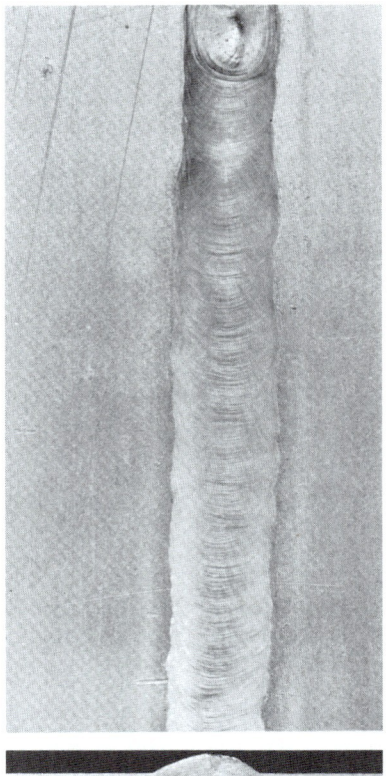

Fig. 22-43 Aluminum weld bead made with correct current. Note the smooth, even ripples; smooth contour; and even penetration. © Kaiser Aluminum

- High stresses imposed during welding by poor joint design or poor jigging
- Porosity is a major concern. A small amount of porosity scattered uniformly throughout the weld has little or no influence on the strength of joints in aluminum. Clusters or gross porosity can adversely affect the weld joint. The main causes of porosity in aluminum welds are:
  - Hydrogen in the weld area
  - Moisture, oil, grease, or heavy oxides in the weld area
  - Improper voltage or arc length
  - Improper or erratic wire feed
  - Contaminated filler wire (Use as large a diameter as possible and GMAW-P if lower heat is needed.)
  - Leaky gun
  - Contaminated or insufficient shielding gas
- Incomplete fusion of the weld metal with the base metal. The major causes are:
  - Incomplete removal of the oxide film before welding
  - Unsatisfactory cleaning between passes
  - Insufficient bevel or back chipping
  - Improper amperage (WFS) or voltage
- Inadequate penetration at the root of the weld and into the side walls of the joint. This is generally caused by:
  - Low welding current (WFS)
  - Improper filler metal size
  - Improper joint preparation
  - Too fast travel speeds for the selected wire-feed speed
- Aluminum welds may have metallic and nonmetallic inclusions. These may be caused by:
  - Copper inclusions caused by burn-back of the electrode to the contact tube
  - Metallic inclusions from cleaning the weld with a wire brush that leaves bristles in the weld
  - The nonmetallic inclusions from poor cleaning of the base metal
  - Always use the push gun travel angle when welding aluminum. (The cleaning action is enhanced.)

Groove Weld Project: Job Qualification Test 4 This test project will allow you to demonstrate your ability to read a drawing, fit components together, tack, and weld aluminum test plates. You will be using the techniques developed in Jobs 22-J11 through J16 using the spray arc mode of metal transfer. Follow the instructions found in the notes on Fig. 22-44.

Inspection and Testing After the project has been tacked, have it inspected for compliance to the drawing. After the project has been completely welded, use visual inspection and cut specimens for bend testing. Use the following acceptance criteria to visually judge your welds. Look for surface defects. Keep in mind that it is important to have good appearance and uniform weld contour. These characteristics usually indicate that the weld was made properly and that the weld metal is sound throughout. Once visual inspection is completed to the following criteria, you will perform side bend tests. Follow side bend test procedures as outlined in Chapter 28.

- There shall be no cracks or incomplete fusion.
- There shall be no incomplete joint penetration in groove welds except as permitted for partial joint penetration groove welds.
- Your instructor shall examine the weld for acceptable appearance and shall be satisfied that the welder is skilled in using the process and procedure specified for the test.
- Undercut shall not exceed the lesser of 10 percent of the base metal thickness or $\frac{1}{32}$ inch.
- Where visual examination is the only criterion for acceptance, all weld passes are subject to visual examination, at the discretion of your instructor.
- The frequency of porosity shall not exceed one in each 4 inches of weld length and the maximum diameter shall not exceed $\frac{3}{32}$ inch.
- Welds shall be free from overlap.
- Only minimal weld spatter shall be accepted, as viewed prior to cleaning.

Side bend acceptance criteria are shown as measured on the convex surface of the bend specimen.

- No single indication shall exceed $\frac{1}{8}$ inch measured in any direction on the surface.
- The sum of the greatest dimensions of all indications on the surface, which exceed $\frac{1}{32}$ inch, but are less than or equal to $\frac{1}{8}$ inch, shall not exceed $\frac{3}{8}$ inch.
- Cracks occurring at the corner of the specimens shall not be considered unless there is definite evidence that they result from slag inclusions or other internal discontinuities.

MAG Welding of Stainless Steel

You will recall that stainless steel is a heat- and corrosion-resistant alloy that is made in a wide variety of compositions. It always contains a high percentage of chromium in addition to nickel and manganese. Stainless steel has excellent strength-to-weight ratios, and many of the alloys possess a high degree of ductility. It is widely used in products such as tubing and piping, kitchen equipment, heating elements, ball bearings, and processing

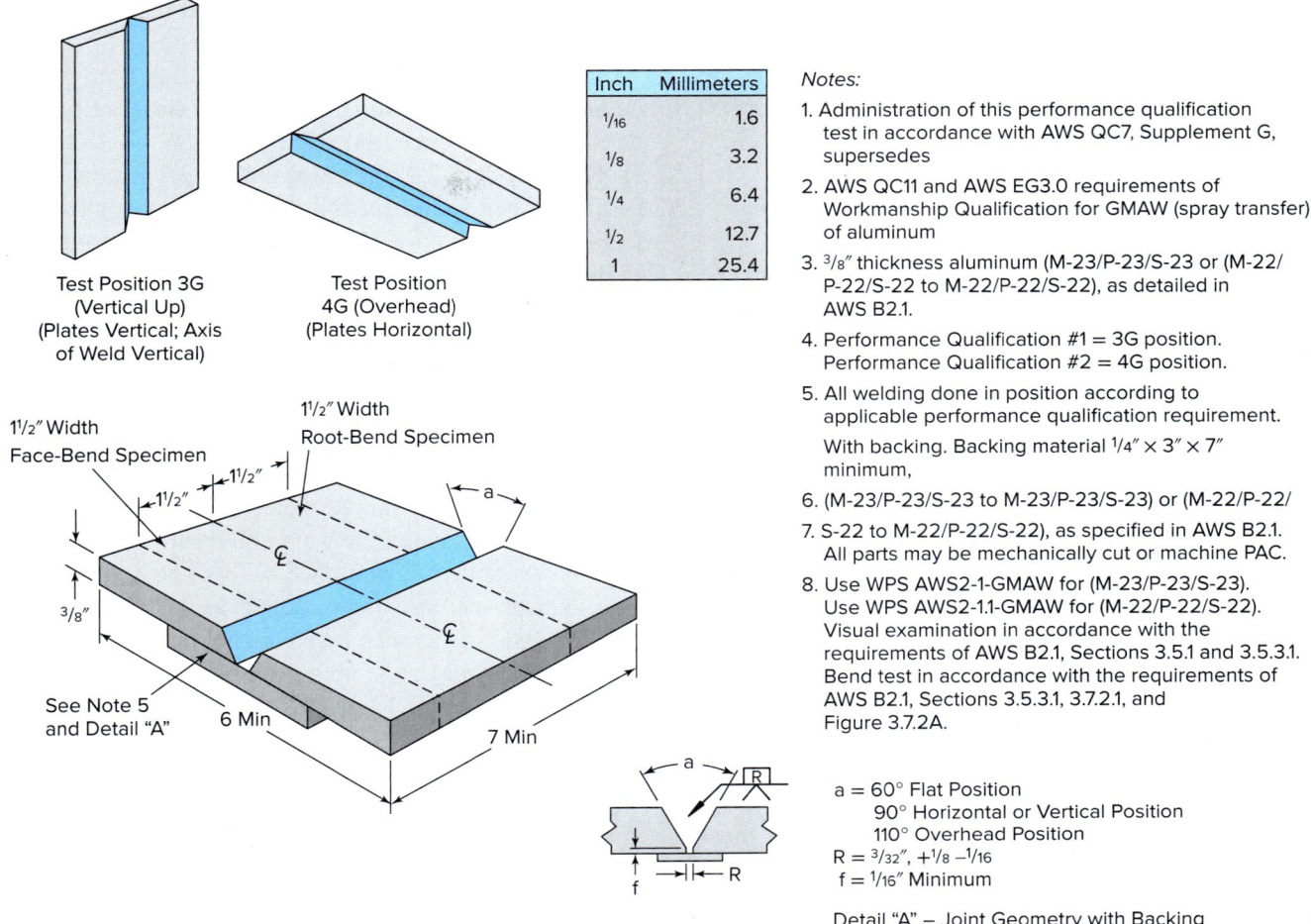

Fig. 22-44 Performance Qualification Test GMAW Spray Transfer, Aluminum, 3G and 4G Positions.

equipment for a wide variety of industries. Stainless steel is supplied in sheets, strip, plate, structural shapes, tubing, pipe, and wire extrusions in a wide variety of alloys and finishes.

Because stainless steel has a lower rate of thermal conductivity than carbon steel, the heat is retained in the weld zone much longer. On the other hand, its thermal expansion is much greater than that of carbon steel, thus causing greater shrinkage stresses and the possibility of warpage. These difficulties can be overcome by the proper use of jigs and fixtures. Stainless steel also has a tendency to undercut, which must be provided for in the welding procedure.

All standard forms of joints are used in stainless-steel fabrication. Sheet up to 3/16 inch can be square-edge groove-welded from one side. Plates 3/16 inch and thicker are beveled to provide access to the root of the joint. Butt joint designs for stainless steel include the single-V groove with 60 and 90° groove angles, the double-V groove, and the single-U groove. Standard lap, corner, and T-joint designs are also employed.

Copper backing bars with fillet welds are necessary for welding stainless-steel sections up to 1/16 inch thick. Backing is also needed when welding plate 1/4 inch and thicker from only one side. No air must be permitted to reach the underside of the weld while the weld pool is solidifying. The oxygen and nitrogen in the air weaken molten stainless steel during cooling. If it is difficult to use a backing bar, argon should be used as a purge gas shield.

Although the shielded metal-arc process is still used for welding stainless steel, the MAG and TIG welding of stainless is increasing. Light gauge materials are welded with TIG, and GMAW-S and P, and heavier materials are welded with the GMAW spray mode. MAG welding stainless steel has the following advantages:

- The absence of slag-forming flux reduces cleaning time and makes it possible to observe the weld pool.

> **SHOP TALK**
>
> **Spraying Safety**
> It's worth being cautious in areas of arc and plasma spraying that chlorinated hydrocarbon solvent vapor is not present. Those activities create ultraviolet radiation; if such radiation contacts vapors, phosgene gas can be produced. This gas is hazardous.

- Continuous wire feed permits uninterrupted welding.
- MAG lends itself to automation.
- Welding may be performed with the short-circuiting, spray, or the pulsed spray modes of transfer.

Spray Arc Welding Electrode diameters as large as $3/32$ inch can be used for stainless steel. Usually $1/16$-inch wire is used with high currents to create the spray arc transfer of metal. Approximately 300 to 350 amperes are required for an electrode $1/16$ inch in diameter, depending on the shielding gas and type of stainless wire. The amount of spatter is determined by the composition and flow rate of the shielding gas, wire-feed speed, and the characteristics of the welding power supply. DCEP is used for most stainless-steel welding. For metal thicknesses up to and including $3/16$ inch, a mixture of argon and 1 to 2 percent oxygen is used. This argon-oxygen mixture improves the rate of transfer and arc stability. It produces a more fluid and controllable weld pool and good coalescence and bead contour while minimizing undercutting. For metal thicknesses greater than $3/16$ inch, a mixture of argon and 2 percent oxygen is used. This mixture provides better arc stability, weld coalescence, and higher welding speed on heavier materials. It is recommended for single-pass welding.

A push travel angle technique should be employed on plate $1/4$ inch thick or more. The gun should be moved back and forth in the direction of travel and, at the same time, moved slightly from side to side. On thinner metal, only the back-and-forth motion along the joint is used.

Short Arc Welding (GMAW-S) Short arc welding requires a low current ranging from 20 to 175 amperes, a low voltage of 12 to 20 volts, and small diameter wires. Metal transfer occurs when the filler wire short circuits with the base metal. The stable arc with low energy and heat input that results is ideally suited for most stainless-steel welding on thicknesses from 16 gauge to $3/16$ inch. The short arc transfer should also be employed for the first pass in those situations in which fitup is poor or copper backing is unsuitable. It is also very desirable in the vertical and overhead positions for the first pass.

For short arc welding of stainless steel in light gauges, the argon-oxygen shielding gas mixtures do not produce the best coalescence. A triple mixture consisting of 90 percent helium, $7\frac{1}{2}$ percent argon, and $2\frac{1}{2}$ percent carbon dioxide gives good arc stability and excellent coalescence. It does not lower corrosion resistance, and it produces a small heat-affected zone that eliminates undercutting and reduces distortion. The flow rates must be increased because of the lower density of the helium gas.

Stainless steel produces a very sluggish welding pool. It does not like to wet out and flow especially with the short circuit mode of transfer. The high heat effect of helium helps make the weld pool more fluid. However, this may not be sufficient to eliminate excess convexity. Increasing voltage does not flatten the weld and creates more spatter. If the power source is equipped with slope and/or inductance control, increase them as much as possible. If excessive slope and inductance is set, the arc may stubble at the start or simply may rope up on the base metal, similar to that shown in Fig. 22-13, page 715.

Pulse Spray Arc (GMAW-P) Pulse spray arc welding can be done with lower current levels similar to short arc welding. It can also be done at higher wire-feed speeds like conventional spray but at lower heat input. Thus it can be used on all thickness ranges. A spray-type gas must be used, one having 1 and 2 percent oxygen with the remainder being argon is most common. Some 90 percent argon and 10 percent CO_2 welding is being done. But corrosion resistance may be jeopardized. Since the arc is on all the time with spray arcs, the weld is more fluid and flows out better. Spatter is also reduced on thin base metals as compared to the short-circuiting mode of transfer.

Hot Cracking Some stainless steels have a tendency toward hot shortness and hot cracking. Type 347 is an example. When these metals are welded, more welding passes are needed. Stringer beads are recommended instead of weave beads. Stringer beads reduce contraction stresses, and cooling is more rapid through the hot-short temperature range.

In welding sections 1 inch or thicker, bead contour and hot cracking can be reduced by preheating to about 500°F. Hot cracking may also be reduced by GMAW-S or P welding. These modes prevent excessive dilution of the weld metal with the base metal, a condition that produces strong cracking characteristics.

Stainless-Steel Sensitization The austenitic types of stainless steel are susceptible to sensitizing the chromium out of the individual grains. This is also known as **carbide precipitation.** This occurs most readily in the 1,200°F

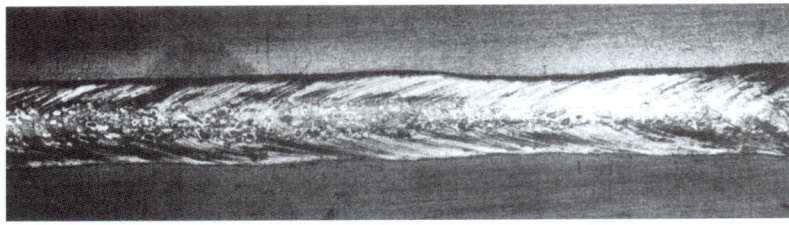

Fig. 22-45 Fillet weld on a lap joint in ⅜-inch stainless-steel plate welded in the 1F position with the gas metal arc welding process. *Edward R. Bohnart*

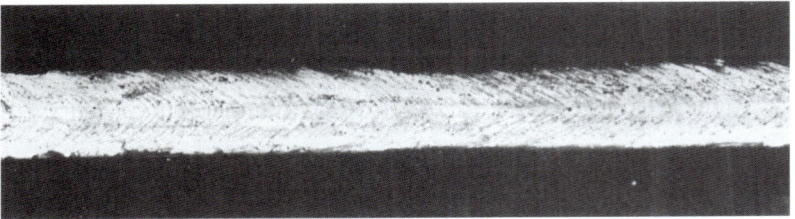

Fig. 22-46 Fillet weld on a T-joint in ⅜-inch stainless-steel plate welded in the 1F position with the gas metal arc welding process. *Edward R. Bohnart*

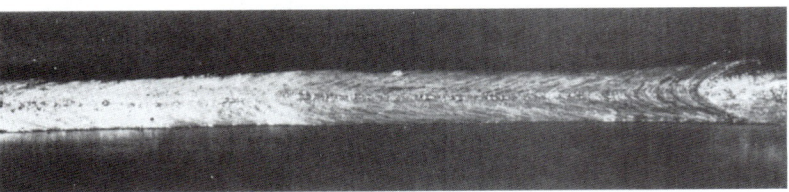

Fig. 22-47 Fillet weld on a T-joint in ⅜-inch stainless-steel plate welded in the 2F position with the gas metal arc welding process. *Edward R. Bohnart*

heat range. The GMAW process with its rapid speed and high deposition rate greatly reduces this situation over the slower GTAW or SMAW processes, minimizing the time at this critical temperature range, using stabilized and low carbon grades of stainless steel, and using proper filler metals such as ER 308L. The L indicates low carbon. Lowering the carbon content also reduces the possibility of carbide precipitation.

Inspection and Testing: Jobs 22-J17–J23 After each weld has been completed, inspect it carefully for defects. Use the inspection and testing procedures that you have learned in previous welding practice. Look for surface defects. Keep in mind that it is important to have good appearance and uniform weld contour. These characteristics usually indicate that the weld was made properly and that it is sound throughout. Be conscious of the tendency of stainless steel to undercut along the edges of the weld and be excessively convex.

Examine the appearance of fillet welds made on heavy stainless-steel plate, Figs. 22-45 through 22-47.

Gas Metal Arc Welding of Other Metals

Earlier in this chapter it was stated that the MIG/MAG process is capable of welding any metal or alloy that can be welded by the other arc and gas welding processes. Thus most aluminum, magnesium, iron, nickel, and copper alloys, as well as titanium and zirconium, can be MIG/MAG welded. Welds of the highest quality are produced at production welding speeds in these metals by the MIG/MAG process.

The Job Outline Table 22-8, pages 746–747, includes practice jobs on three of the major metals that are likely to be encountered by the new welder in industry: carbon steel, aluminum, and stainless steel. A new welder in a plant is not likely to be called upon to weld the other metals listed here. It is the purpose of this text to make sure that the student welder has an opportunity to become familiar with the often used metals. You are urged, however, to secure pieces of metals not included in the course and practice with them. The following information will provide you with the necessary information about these metals, and the instructor can readily demonstrate the welding procedure.

Copper and Its Alloys

Copper can be alloyed with zinc, tin, nickel, aluminum, magnesium, iron, beryllium, lead, and other metals. Copper and many of its alloys, including manganese-bronze, aluminum-bronze, silicon-bronze, phosphor-bronze, cupro-nickel, and some of the tin bronzes may be welded successfully by the gas metal arc process. Electrolytic copper can be joined by using special techniques, but its weldability is not good. The various grades of deoxidized copper are readily weldable with the MIG process. Deoxidized filler wires are necessary for welding deoxidized copper. For welding other copper-base alloys, with the exception of the zinc-bearing type, filler wires of approximately matching

ABOUT WELDING

Breathing Safely

Ionizing radiation comes from electron beam welding. During the grinding process, radioactive dust is made. Welders avoid the dust by using local exhaust and sometimes a respirator.

Table 22-8 Job Outline: Gas Metal Arc Welding Practice with Solid Core Wire (Plate)

Recommended Job Order[1]	Number in Text	Material Type	Thickness	Type of Weld	Type of Joint	Weld Position	No. of Passes	Electrode[2] Type AWS	Size (in.)	Shielding Gas	Gas Flow (ft³/h)	Arc (V)	Welding Current[3] DCEP Amperes	Wire-Feed Speed[4] (WFS) (in./min)	Text Reference
1st	22-J1	Carbon steel	1/8	Groove	Square butt	Flat (1G)	2	E70S-3 E70S-6	0.035	Carbon dioxide	20–25	19–21	110–160	170–340	751
2nd	22-J3	Carbon steel	3/16	Fillet	Lap	Flat (1F)	1	E70S-3 E70S-6	0.035	Carbon dioxide	20–25	19–21	110–170	170–360	752
3rd	22-J4	Carbon steel	3/16	Fillet	Lap	Vertical down-hill (3F)	1	E70S-3 E70S-6	0.035	Carbon dioxide	20–25	19–21	120–160	190–340	752
4th	22-J5	Carbon steel	3/16	Fillet	Lap	Overhead (4F)	1	E70S-3 E70S-6	0.035	Carbon dioxide	20–25	19–21	120–160	190–340	752
5th	22-J6	Carbon steel	1/8	Fillet	T	Horizontal (2F)	1	E70S-3 E70S-6	0.035	Argon 75% carbon dioxide 25%	20–25	19–21	110–170	170–360	752
6th	22-J7	Carbon steel	1/4	Fillet	T	Horizontal (2F)	1	E70S-3 E70S-6	0.045	Argon 98% Oxygen 2%	40–50	24–32	200–375	225–410	752
7th	22-J8	Carbon steel	3/8	Fillet	T	Vertical uphill (3F)	3	ER70S-6	0.035	Carbon dioxide	20–25	21–23	150–160	300–340	752
8th	22-J9	Carbon steel	3/8	Fillet	T	Horizontal (2F)	3	ER70S-3	1/16	Argon 95% Oxygen 5%	40–50	26–33	275–400	200–280	752
9th	22-J10	Carbon steel	3/8	Fillet	T	Overhead (4F)	6	ER70S-6	0.035	Carbon dioxide	20–25	21–23	160–180	340–380	752
10th	22-J2	Carbon steel	1/2	Groove	V-butt 60°	Vertical downhill root (3G) Vertical uphill (3G)	2 2	ER70S-6	0.035	Carbon dioxide	20–25	21–23	160–190 160–180	340–400 340–380	751 751
22-J1-J10	22-JQT1[5]	Carbon steel	3/8	Fillet and groove	Butt, lap, T-, corner	2F 3F 4F 2G 3G 4G	As required to get weld size	ER70S-3	0.035	Argon 75% Carbon dioxide 25%	20–25	19–21	160–180	340–380	753
22-J1-J10	22-JQT2[5]	Carbon steel	3/8	Fillet and groove	Butt, T	2F 1G	As required to get weld size	ER70S-3	0.045	Argon 98% Oxygen 2%	40–50	24–32	200–275	225–410	754
22-J1-J10	22-JQT3[5]	Carbon steel	1	Groove	Butt	1G	As required to get weld size	ER70S-3	0.045	Argon 98% Oxygen 2%	40–50	24–32	200–275	225–410	757
11th	22-J11	Aluminum	1/8	Groove	Square butt	Flat (1G)	2	ER1100	3/64	Argon	30–35	20–24	150–190	280–400	760
12th	22-J13	Aluminum	1/8	Fillet	Lap	Horizontal (2F)	1	ER1100	3/64	Argon	30–35	20–24	150–190	280–400	761
13th	22-J14	Aluminum	3/16	Fillet	T	Horizontal (2F)	3	ER1100	3/64	Argon	30–35	21–25	160–200	290–430	761
14th	22-J15	Aluminum	1/4	Fillet	T	Vertical uphill (3F)	2	ER4043	3/64	Argon	30–35	21–25	160–190	290–400	761
15th	22-J16	Aluminum	1/4	Fillet	T	Overhead (4F)	3	ER4043	3/64	Argon	30–35	21–25	170–190	320–400	761

Job #	Material	Size	Weld type	Joint	Position	Passes	Electrode	Wire dia	Gas					Page
16th	Aluminum	3/8	Groove	V-butt 60% backup	Vertical uphill (3G)	4	ER4043	1/16	Argon 75% Helium 25%	30-40	22-26	200-250	260-330	760
22-J11–J16														
22-JQT4[5]	Aluminum	3/8	Groove	Butt	3G 4G	As required to get weld size	ER4043	3/64	Argon	30-35	21-25	160-190	320-400	763
17th 22-J17	Stainless steel	1/8	Beading	Plate	Flat (1C)	Cover plate	ER308	0.035	Helium 90% Argon 7½% Carbon dioxide 2½%	22-24	24-26	130-160	160-280	766
18th 22-J18	Stainless steel	1/4	Beading	Plate	Flat (1C)	Cover plate	ER308	0.035	Argon 98% Oxygen 2%	25-35	24-27	130-160	160-280	766
19th 22-J19	Stainless steel	1/8	Fillet	Lap	Horizontal	1	ER308	0.035	Helium 90% Argon 7½% Carbon dioxide 2½%	22-24	24-26	140-160	175-280	766
20th 22-J20	Stainless steel	1/8	Fillet	T	Horizontal	1	ER308	0.035	Helium 90% Argon 7½% Carbon dioxide 2½%	22-24	24-26	150-180	190-290	766
21st 22-J21	Stainless steel	1/4	Fillet	T	Horizontal	6	ER308	0.035	Argon 95% Oxygen 5%	25-35	24-27	180-240	290-390	766
22nd 22-J22	Stainless steel	1/4	Fillet	T	Vertical uphill (3F)	6	ER308	0.035	Argon 98% Oxygen 2%	25-35	20-26	140-200	175-320	766
23rd 22-J23	Stainless steel	3/8	Groove	V-butt 60° backup	Vertical uphill (3G)	1 down 5 up	ER308	0.035	Argon 98% Oxygen 2%	25-35	20-28	140-200	175-320	766

Note: The conditions indicated here are basic. They will vary with the job situation, the results desired, and the skill of the welder.

[1] It is recommended that the student do the jobs in this order. In the text, the jobs are grouped according to the type of operation to avoid repetition.

[2] On all carbon steel work, use metal cored wire to practice (E70C-1C or E70C-1M, depending on the shielding gas being used). You will need to increase the wire-feed speed or go to the next sized electrode diameter to compensate for the higher current density.

[3] Pulse spray arcs should be practiced on all jobs. Use the equipment manufacturer's recommended parameter setting. Meet or exceed the wire-feed speed for the other modes of transfer.

[4] WFS = wire-feed speed.

chemistry are generally used. Copper-zinc alloys are not suitable as filler wire because zinc boils at a low temperature (1,663°F) and vaporizes under the intense heat of the electric arc. These alloys can be welded, however, with aluminum-bronze filler wires.

Argon is the preferred shielding gas for welding material 1 inch and thinner. A flow of 50 cubic feet per hour is sufficient. For heavier materials, mixtures of 65 percent helium and 35 percent argon are used.

Joint design is like that for any other metal. Steel backup is usually necessary for sheets ⅛ inch and thinner. Backup is not needed on plates more than ⅛ inch thick. Because of the high heat conductivity of copper, welding currents on the high side are required. Preheat is not required when welding thicknesses of ¼ inch or less. Preheating to 400°F has proved to be helpful when welding copper ⅜ inch or more in thickness.

Always provide good ventilation when welding copper and its alloys. This is of particular importance when welding beryllium-copper. The dust, fumes, and mist produced by beryllium compounds are highly toxic. Precautions should be taken to reduce the dust, fumes, and mist to zero.

A variation of the GMAW process is GMAW-B, where B indicates brazing or just MIG brazing. It uses a silicon-bronze type electrode with inert shielding. Argon at 100 percent is most common. It is generally done with small diameter wire and in the short-circuiting or pulse mode of transfer. The main application is for coated carbon steel sheet metal (light gauge). The coating, such as zinc, is generally applied for corrosion resistance in the automotive and sheet metal industries. The lower melting point of this electrode plus the lower heat input of the short-circuiting and pulse mode does little to disturb the coating in the weld area. The base metal is not melted, and thus it is considered a brazing operation.

Nickel and Nickel-Copper Alloys

Nickel, nickel-copper alloy (Monel®), nickel-chromium-iron (Inconel®), and most other nickel alloys can be welded using the gas metal arc process. Always remove all foreign materials in the vicinity of the weld or heated area. Nickel alloys are susceptible to severe embrittlement and cracking when heated in contact with such foreign materials as lead, phosphorus, and sulfur.

Argon is generally preferable for welding nickel and most nickel alloys up to about ⅜ inch in thickness. Above that thickness, argon-helium mixtures are usually more desirable. The higher heat input of 50 and 75 percent helium mixtures offsets the high heat conductivity of heavier gauges. Oxygen should not be added to the inert shielding gases because it produces oxide films and inclusions in the weld and rough, heavily oxidized weld surfaces.

Joint preparation is like that used with other metals.

Magnesium

Magnesium is a silvery white metal that is two-thirds the weight of aluminum and one-quarter the weight of steel. It has a melting point of 1,204°F, which is near that of aluminum. Its strength-to-weight ratio is high when compared to that of steel.

Welding techniques for magnesium are like those for aluminum. The rate of expansion of magnesium is greater than that of aluminum. This must be taken into consideration when preparing the joint for welding and in choosing the type of restraint to put upon the assembly. Severe warpage will result if proper precautions are not taken. As with aluminum, care must be taken that the surface is clean before welding. The surface may be mechanically cleaned with abrasives or chemically cleaned.

The arc characteristics of helium and argon are somewhat different with magnesium than they are with other metals. The burnoff rates of the wires are equal for both gases at the same current. Penetration is greater with argon-helium mixtures. Argon is recommended in most cases because of the excellent cleaning action obtained. The argon-helium mixtures might be preferred in multipass welding in which the rounded type of penetration pattern is most desirable.

Titanium and Zirconium

Titanium is a bright white metal that burns in air, and it is the only element that burns in nitrogen. It has a melting point of about 3,500°F. Its most important compound is titanium dioxide, which is used extensively in welding electrode coatings. Titanium is also used extensively as a stabilizer in stainless steel.

Zirconium is a bright gray metal with a melting point above 4,500°F. It is very hard and brittle and readily scratches glass. Because of its hardness, it is sometimes used in hard-facing materials. Zirconium is often alloyed with iron and aluminum.

Both titanium and zirconium and many of their alloys may be welded by the gas metal arc process. Special precautions must be taken, however, to protect the welding operation during the period when the metal is hot and susceptible to atmospheric contamination. Welding may be done in an enclosed chamber filled with inert gas, or other special gas shielding methods may be necessary to ensure adequate inert gas coverage. Argon or helium-argon mixtures may be used.

CHAPTER 22 REVIEW

Multiple Choice

Choose the letter of the correct answer.

1. _____ is the length of electrode that extends beyond the contact tube. (Obj. 22-1)
   a. Visible stickout
   b. Setback
   c. Electrode extension
   d. Standoff

2. The _____ angle is the angle the electrode makes between the major workpiece surface or groove face and the axis of the electrode. (Obj. 22-1)
   a. Work
   b. Travel
   c. Perpendicular
   d. Drag

3. If incomplete penetration and fusion are a problem, the length of electrode extending beyond the contact tube should be _____. (Obj. 22-2)
   a. Lengthened
   b. Shortened
   c. Does not matter, has no effect
   d. Both a and b

4. How can metallic fumes be reduced in the GMAW process? (Obj. 22-3)
   a. Use of higher argon content gases
   b. Use of GMAW-P
   c. Lower deposition rate
   d. Both a and b

5. If the contact tube has internal arcing and is roughening the electrode passing through it, it can be salvaged by cleaning the hole of the contact tube with an OFC tip cleaner. (Obj. 22-4)
   a. True
   b. False

6. The gas metal arc welding gun or conduit should be coiled up and kinked in use. (Obj. 22-4)
   a. True
   b. False

7. GMAW can weld directly through rust, scale, burned edges, and chemical coatings due to the dioxidizers in the electrodes and the fluxing action of the shielding gases. (Obj. 22-4)
   a. True
   b. False

8. If the power source energizes (voltage on meter) and the wire feeds but there is no arc, the problem is _____. (Obj. 22-4)
   a. Primary power fuse blown
   b. Wire feeder not on
   c. Loose work connection
   d. Shielding gas not on

9. Arc starting and the beginning of the weld bead is not an issue with the GMAW process because of the constant voltage power source used. (Obj. 22-5)
   a. True
   b. False

10. Travel speed has a major effect on bead size, penetration, and fusion. (Obj. 22-5)
    a. True
    b. False

Review Questions

Write the answers in your own words.

11. List the five operational variables and describe each. (Obj. 22-1)
12. Describe the three main modes of metal transfer used with the GMAW process. (Obj. 22-1)
13. Name and explain a minimum of eight defects associated with the GMAW process. (Obj. 22-2)
14. List the safety concerns for GMAW that are different than those for the SMAW and GTAW processes. (Obj. 22-3)
15. Describe how bird nesting can be eliminated. (Obj. 22-4)
16. List the equipment needed for the GMAW process. (Obj. 22-4)
17. Describe the indicator you will have if the proper arc length (voltage) is set when a spray arc mode of metal transfer is being used. (Obj. 22-5)
18. Describe the crater filling technique. (Obj. 22-5)
19. List the types of metal that can be welded with the GMAW process. (Obj. 22-6)
20. List the four causes of incomplete penetration and fusion when welding aluminum with the GMAW process. (Obj. 22-6)

INTERNET ACTIVITIES

Internet Activity A

Use your favorite search engine to find out if women welders during World War II have received any honors for their wartime efforts. If so, what are they?

Internet Activity B

What does the future hold for welding? Use the Internet to find out what people think about the future of welding.

Design credits: Cinema and movie icon: Denis Meshkov/123RF; and Illustration of Welding icons: Mr. Nachai Sorasee/123RF

23

Flux Cored Arc Welding Practice (Plate), Submerged Arc Welding, and Related Processes:

FCAW-G Jobs 23-J1–J11, FCAW-S Jobs 23-J1–J12; SAW Job 23-J1

Chapter Objectives

After completing this chapter, you will be able to:

23-1 Describe the flux cored arc welding process.

23-2 Describe the welding variables for flux cored arc welding.

23-3 Demonstrate the ability to make various fillet and groove welds with the flux cored arc welding process.

23-4 Define the operational differences between the two main types of flux cored electrodes.

23-5 Explain and demonstrate an understanding of the submerged arc welding process.

23-6 Describe the electroslag and electro gas welding processes.

Flux Cored Wire Welding

Flux cored arc welding (FCAW) has grown very rapidly. This growth has kept pace with the growth of other gas metal arc welding processes. The biggest use is in the fabrication of medium-to-heavy weldments of carbon and alloy steel. **Flux cored wire** increases welding speeds and deposition rates considerably. Figures 23-1, 23-2, and 23-3, page 752, show applications and equipment for FCAW.

All the major producers of welding equipment and suppliers are involved in the process. Hobart refers to its self-shielded flux cored wire as Fabshield® and its gas shielded flux cored wire as FabCO®; ESAB's self-shielded version is CoreShield®, and its gas shielded version is Dual Shield®; Lincoln's self-shielded version is Innershield®, and its gas shielded version is Outershield®.

There are two basic types of flux cored wire welding:

1. Gas shielded, flux cored arc welding (FCAW-G)
2. Self-shielded, flux cored arc welding (FCAW-S)

These two electrodes can produce welds of the highest quality. They combine all the advantages of GMAW, such as:

1. High deposition rates
2. Good performance on fillet and groove welds
3. Scavengers and deoxidizers for less-than-clean metal

Fig. 23-1 Structural steel being welded with the flux cored arc welding process and a self-shielded electrode. The Lincoln Electric Co.

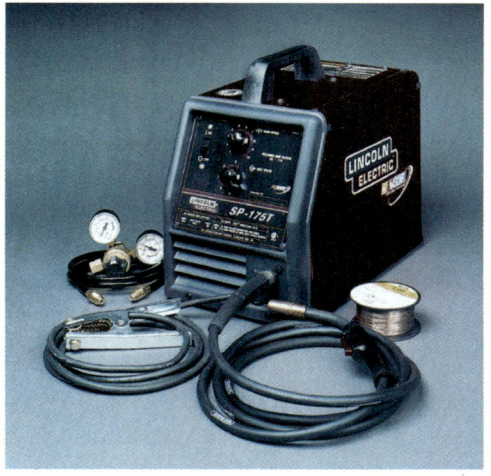

Fig. 23-2 A small wire-feeder power source combination using a 0.035 self-shielded flux cored electrode at approximately 16 volts at 100 amperes being used to make a field repair of a hay wagon rack. Note the convenience of not needing a shielding gas. Flammable materials have been moved from the direct weld area. The Lincoln Electric Co.

4. A slag that retards the cooling rate and supports the molten weld pool
5. The ability to weld in all positions with the correct electrode

These welds have excellent appearance and meet the requirements of many welding codes.

FCAW-G is a process using a continuous consumable electrode to do gas shielded, flux cored arc welding. Its core is filled with flux and alloying agents. These types of electrodes require an external shielding gas source, and are not considered self-shielded (Fig. 23-5A). The solid metal portion of the electrode comprises about 80 to 85 percent of its weight. The core material (shown in Fig. 23-4) makes up the remainder and performs the following functions:

- Acts as a deoxidizer and cleans the weld metal
- Forms a protective slag to cover the deposited weld metal until it has solidified
- Stabilizes the arc so that it is smooth and reduces spatter
- Adds alloying elements to the weld metal to increase strength and provide other desirable weld properties
- Provides shielding gas in addition to that supplied externally

FCAW-S is a process employing a continuous consumable electrode that has its core filled with flux, gasifiers, and alloying agents. These electrodes generate their own shielding gas and require no external shielding source (Fig. 23-5B).

Fig. 23-3 Prior to welding, this is a more than 25-foot-long rack gear for a large capacity crawler crane. There will be approximately 100 feet of ⅝-inch weld size J-groove weld on this project when completed. The plate and weld areas will be preheated to 500°F, and it will take approximately 20 hours to complete the fabrication. With careful execution of the preheat, welding procedure with staggering and skip welding techniques twist and flatness was maintained over its length to less than ½ the electrode diameter. Head and tail stocks were used to position the welds in the flat position. © The Manitowoc Company, Inc.

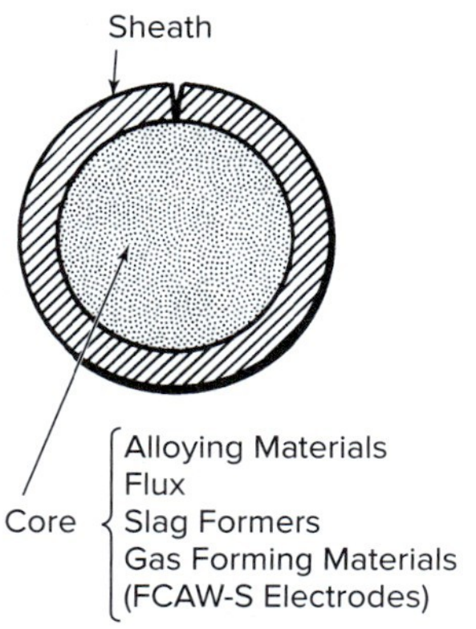

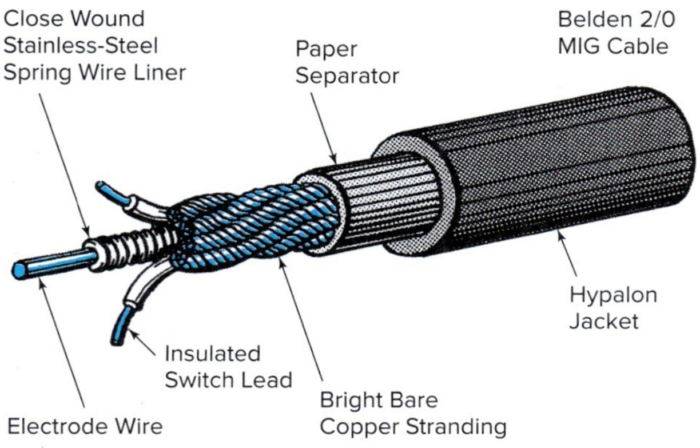

Fig. 23-4 (A) Flux cored electrode cross section. (B) Typical composite flux cored arc welding gun cable construction. Note the wire liner and lack of shielding gas hose. This gun cable is designed for the self-shielded flux cored wire.

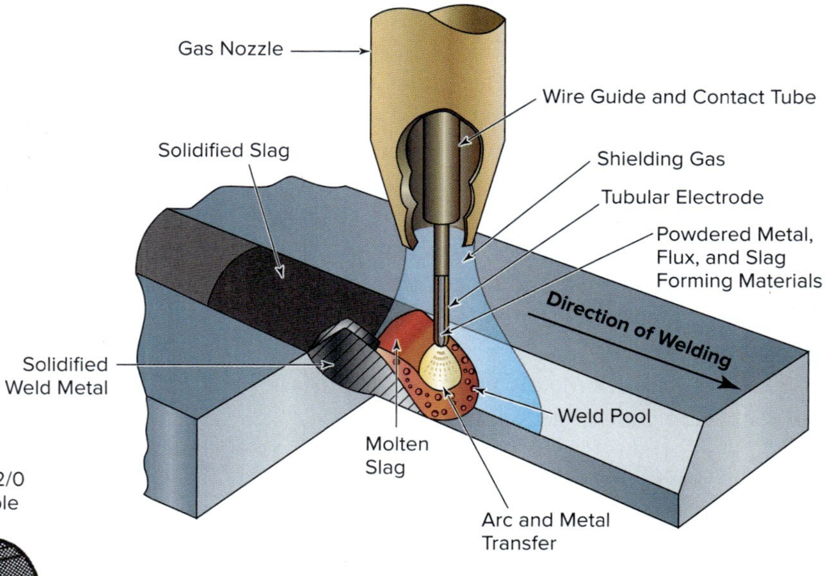

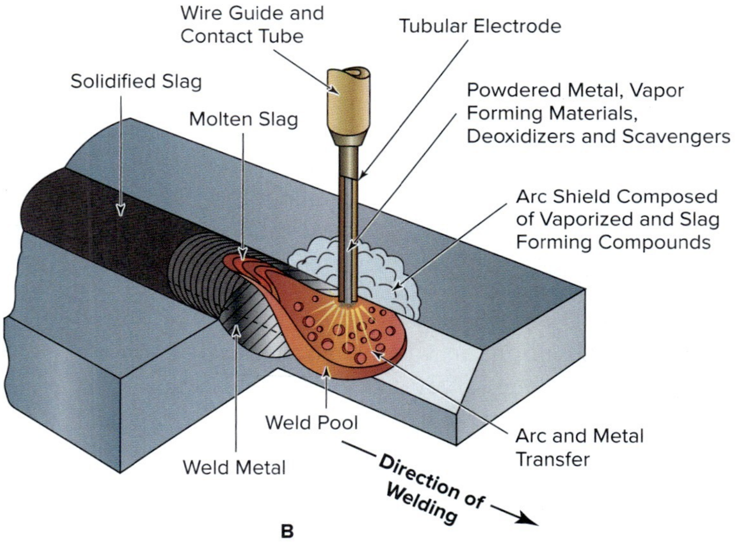

Flux cored wires are made from a continuous tube. Steel strip is formed into a U-shape to hold core flux powders. After flux loading, the U is closed, and the wire is drawn to the required size.

Flux cored filler metal is, in effect, an inside-out covered electrode.

The FCAW-G electrodes have deposition efficiencies up to 90 percent and are able to achieve deposition rates as high as 25 pounds per hour, while FCAW-S electrodes have less deposition efficiencies, up to 87 percent, and are able to achieve deposition rates as high as 40 pounds per hour. These high deposition rates are generally only achievable with automatic/mechanized application of the process.

The metal of the cored electrode wire is transferred through the intense heat (approximately 12,000°F) of the arc column to the work. These electrodes must be used with semiautomatic or automatic/mechanized equipment.

Fig. 23-5 (A) Process schematic for flux cored arc welding with a gas shielded electrode. (B) Process schematic for flux cored arc welding with a self-shielded electrode. Adapted from Welding Handbook, 9/e

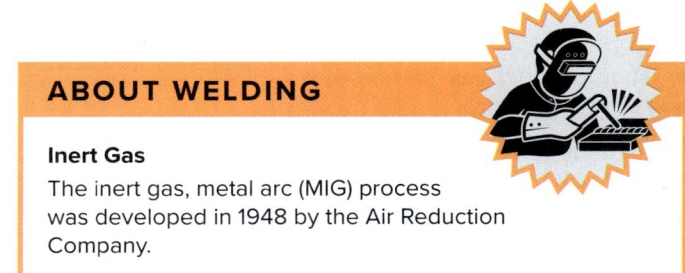

ABOUT WELDING

Inert Gas

The inert gas, metal arc (MIG) process was developed in 1948 by the Air Reduction Company.

FCAW (Plate), SAW, and Related Processes: FCAW-G Jobs 23-J1–J11, FCAW-S Jobs 23-J1–J12; SAW Job 23-J1   Chapter 23   753

The **flux cored electrode** was first developed about 1954 and introduced in its present form in 1957. Flux cored electrodes are available in diameters of 0.030, 0.035, 0.045, 0.052, $\frac{1}{16}$, $\frac{5}{64}$, $\frac{3}{32}$, $\frac{7}{64}$, and $\frac{1}{8}$ inch. Electrodes are available for the welding of mild steel. There are also low alloy grades for high strength structural steel, low nickel wire for low temperature applications, and electrodes for stainless steels. Hard-surfacing electrodes are also available.

Flux cored filler metals are produced on the basis of three general groupings.

1. The so-called *single-pass* filler metals that operate in carbon dioxide shielding gas are intended for welding rusted or mill-scaled plate.
2. The so-called **multipass filler metals** that also operate in carbon dioxide shielding gas provide ductile weld metal with high impact strength at both low and high tensile strengths.
3. **Self-shielding filler metals** are used without auxiliary shielding gas. They offer a convenient simplicity of equipment that is efficient for field construction operations. Also, no shielding gas to be blown away.

The flux cored processes are particularly suited for those applications in which rust and poor fitup are problems, and when larger size fillet welds than those provided by solid core wire are desired. They are also replacing other processes in root pass work in which weld metal drop-through or a lack of visibility is a problem.

In addition, the gas shielded, flux cored arc welding process has the following desirable characteristics:

- It has high deposition rates with little electrode loss.
- It can be adapted to semiautomatic or full automatic operation.
- Welds of high quality that can pass radiographic tests are produced, which are suitable for code work.
- The shielding gas, carbon dioxide, is low in cost.
- Deep penetration reduces weld size.
- The highly stable arc reduces spatter loss.
- Slag can be removed with a minimum of labor.
- Weld appearance is highly desirable.
- Small wires of specific composition can be used in all positions.

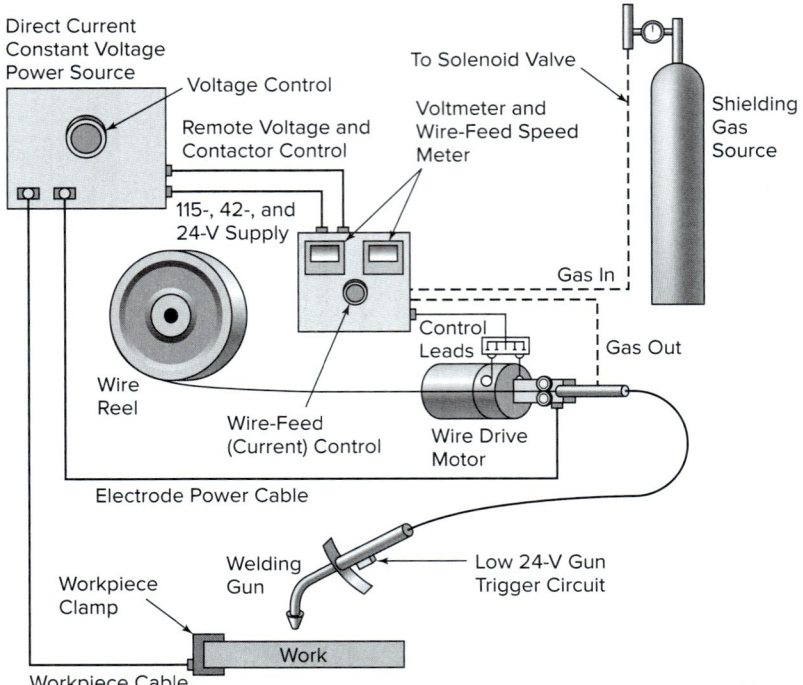

Fig. 23-6 Equipment needed for flux cored arc welding. Note that gas shielding requires only the gas shielded type of electrodes.

- All mild and low alloy steels can be welded.
- Cost is lower than that of other processes.

The major FCAW equipment components, Fig. 23-6, are the following:

- Shielding gas and control unit if carbon dioxide is used (FCAW-G)
- Power source
- Wire-feeding mechanism and controls
- Electrode wire
- Welding gun and cable assembly (The gun is different for FCAW-G and FCAW-S.)

A constant voltage power source with DCEP is generally used with the continuous-feed electrode process. It can be a d.c. rectifier, Figs. 23-7 and 23-8, or an inverter, Fig. 23-9. The generator is generally engine driven. Direct current, electrode negative, is used for most FCAW-S electrodes.

You will recall that in GMAW with a constant voltage power source, current output is determined by the wire feeder adjustment. A specially designed wire-feed and control unit is necessary, Figs. 23-10 and 23-11, page 756. There are a variety of gun styles: Figs. 23-4B, page 753, 23-12, and 23-13, pages 756–757. Guns may be air or water cooled. Review Chapter 21 as necessary concerning the fundamentals of the GMAW process.

Fig. 23-7 Constant voltage/constant current family of d.c. welding machines. These are industrial rated machines at 100 percent duty cycle. Output ratings from 300 through 1,000 amperes are available to meet demanding needs. All the major processes like FCAW, SAW, GMAW, SMAW, and GTAW are capable with this type of equipment. These machines are capable of running ArcReach™ which allows control of the power source with the displayed remote control connected at the wire-feeder. Miller Electric Mfg. Co.

Fig. 23-8 Constant voltage and constant current, 350 ampere at 34 volts, 100 percent duty cycle machine with wire feeder. This power source is designed to operate off of single- or three-phase primary power and is a d.c. inverter machine weighing in at 150 pounds. With its variable inductance and selectable slope, it allows ultrafine tuning and improves arc characteristics for all types of GMAW modes of metal transfer plus FCAW, SMAW, and GTAW. © ESAB

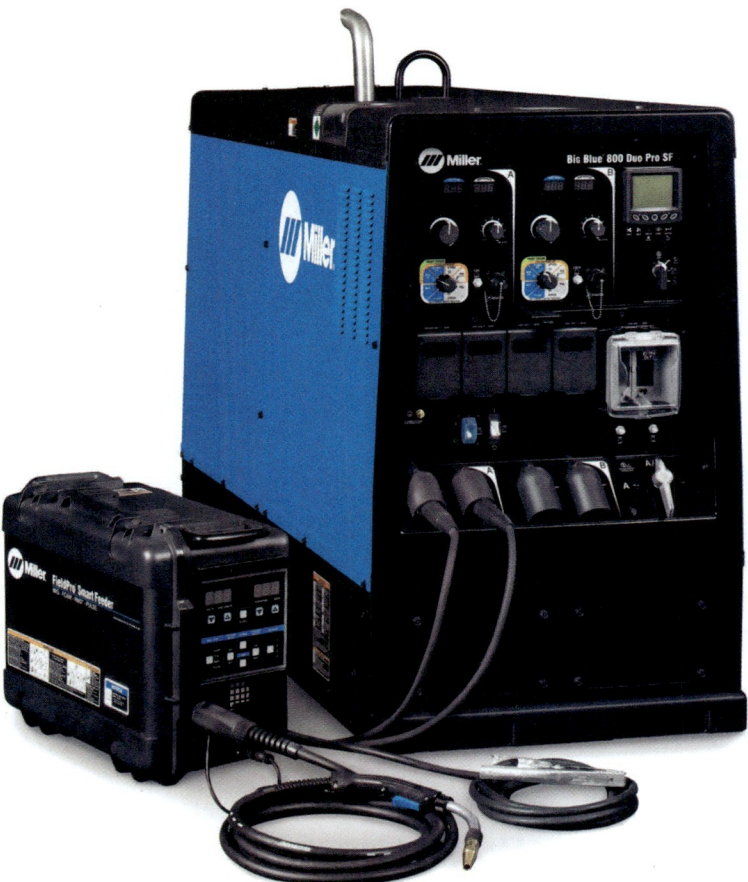

Fig. 23-9 Constant voltage and constant current, d.c., engine-driven generator for heavy-duty field applications. The air-cooled diesel engine develops 65.7 horsepower at 1,850 rpm and can output up to 800 amperes. The totally enclosed voltage sensing feeder operates off of arc voltage. Therefore, the control cords are not required. Two welders can weld at the same time with no enter arc reaction up to 400 amps each at 100 percent duty cycle. Miller Electric Mfg. Co.

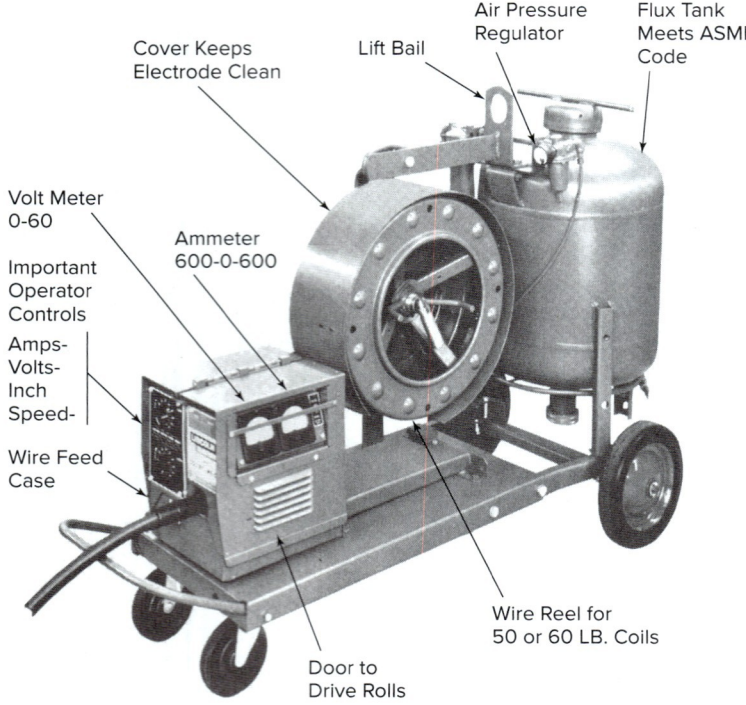

Fig. 23-10 This wire feeder can be used for both solid, metal core, and flux cored wire with gas metal arc, flux cored arc, and submerged arc welding. It feeds a wide range of wire sizes. The pressure tank is used to force the granular submerged arc flux out to the gun. *The Lincoln Electric Co.*

When FCAW-G wires are used with a shielding gas such as carbon dioxide or argon plus carbon dioxide, then ductility, penetration, and toughness of the weld metal are improved. Flux cored wire is also superior on dirty or corroded base metal. See Table 23-1, a comparison of the gas metal arc processes.

Flux cored electrodes require currents in the range of 90 to 960 amperes. Because of their deep penetrating qualities and the highly fluid weld pool, the best welds are obtained by welding in the flat and horizontal positions. Flux cored electrodes are especially intended for large single- or multiple-pass fillet welds in either the flat or horizontal position with DCEP. The wires are also suitable for long groove welds in heavy plate. Flux cored electrodes often replace the E7018 and E7028 shielded metal arc electrodes.

The flux cored wire weld deposit is fully covered by a dense, easily removed slag. The combination of full slag coverage and the gas shielded arc results in radiographic sound welds with excellent mechanical properties. The chemical composition of the weld is constant because alloying elements are built into the cored electrode.

Typical applications include farm equipment, truck bodies, earthmoving equipment, ships, pressure vessels, railroad cars, machine bases, structural steel, storage tanks, and repair of castings.

Fig. 23-11 A gas shielded, flux cored arc weld being made in an awkward-to-reach area. The skill of this welder to weld in any position encountered is very important. It is not always possible to make welds in the most comfortable situations. *The Lincoln Electric Co.*

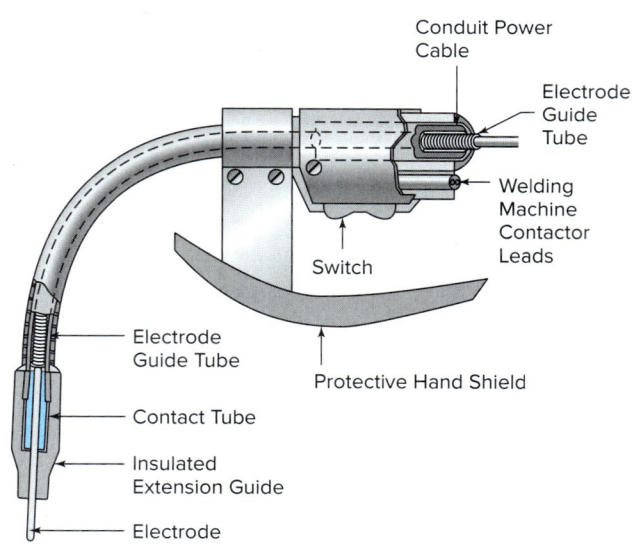

Fig. 23-12 Internal construction of a typical self-shielded flux cored arc welding gun. *Source: From Welding Handbook, 9/e*

Fig. 23-13 Gun for heavy-duty, solid, metal cored or flux cored welding. It is rated at 600 amperes, 100 percent duty cycle with CO_2 shielding gas. It is available in 10- to 15-foot lengths and for wire sizes 0.035 to 1/16 inch. Nozzles are available from 1/2- to 5/8-inch inside diameter. Miller Electric Mfg. Co.

Wire Classifications

Wire classifications for flux cored electrodes of both the self-shielded and gas shielded types are covered in the AWS A5.20—Carbon Steel, A5.22—Stainless Steel, and A5.29—Low Alloy Steel documents, Fig. 23-14, page 758.

Those wires used with gas shielding are the E70T-1, E70T-1M, E71T-1, E71T-1M, E70T-2, E70T-2M, E71T-2, E71T-2M, E70T-5, E70T-5M, E71T-5, E71T-5M, E70T-9, E70T-9M, E71T-9, E71T-9M, E70T-12, E70T-12M, E71T-12, and E71T-12M electrode wires. Those wires requiring no shielding gas are the E70T-3, E70T-4, E70T-6, E70T-7, E71T-7, E70T-8, E71T-8, E70T-10, E70T-11, E71T-11, E61T-13, E71T-13, and E71T-14 electrode wires. See Tables 23-2 through 23-5, pages 759–761.

Even though flux cored electrodes carry an AWS classification number, it does not mean you can necessarily switch between the various manufacturers indiscriminately. Flux cored electrodes have *personality*, which needs to be considered. It is not that one manufacturer's electrodes are better than another's, but they are different. For example, an E71T-11 of 0.045-inch diameter from one manufacturer may require different operation parameters, such as electrode extension, arc voltage, amperage, wire-feed speed, and even mechanical properties. It is best to always consult the manufacturer for the optimum operating conditions. Tables 23-4 and 23-5 (pp. 760–761) compare various manufacturers by name with the AWS classification number.

Hydrogen Control of Flux Cored Electrodes

In the AWS Specification 5.20, the flux cored process is referred to as "a low hydrogen welding process." This specification also makes available optional and supplementary designators for maximum diffusible hydrogen levels of 4, 8, and 16 milliliters per 100 grams of deposited weld metal.

Some of the gas shielded and self-shielded electrodes are designed and manufactured to meet very stringent hydrogen requirements. These electrodes have an H added to their classification. These electrodes will remain fundamentally

Table 23-1 Comparison of Various Semiautomatic Processes

Characteristic	Solid Wire	Metal Core	Flux Core, Gas Shielded	Flux Core, Self-Shielded
Penetration	Shallow to deep	Shallow to medium	Medium to deep	Shallow to medium
Deposition rate	Low to high	Low to high	Medium to high	Low to medium
Deposition efficiency	95–98%	85–99%	85–92%	59–91%
Fluidity of pool	Viscous to fluid	Fluid	Fluid	Fluid
Bead width	Narrow	Wide	Wide	Wide
Bead shape	Convex	Flat	Flat to convex	Flat to convex
Wire size range	0.023–0.125 in.	0.045–3/32 in.	0.035–1/8 in.	0.035–5/32 in.
Shielding gas	Required	Required	Required	None
Variety	Medium	Medium	Wide	Medium
Travel speed	Slow to fast	Slow to fast	Slow to fast	Slow to fast
Fitup	Good to poor	Good to medium	Good to medium	Excellent to medium
Slag forming	Little	Little	Covers weld	Covers weld

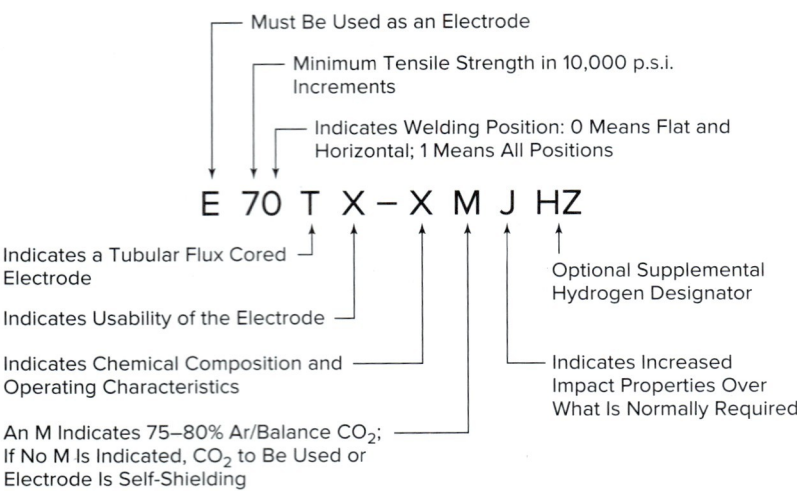

Fig. 23-14 FCAW tubular flux cored electrode.

dry if stored according to their manufacturer in the original unopened package or container.

Sometimes electrodes must be supplied in hermetically sealed packages. These electrodes are used in critical applications where hydrogen must be controlled at 8 milliliters per 100 grams of deposited weld metal or lower. Another good example would be when shipping, handling, and storage conditions are not known or controlled. This system of packaging is much like the way low hydrogen covered electrodes are supplied for shielded metal arc welding.

Once the packaging has been opened, electrodes should not be exposed to conditions exceeding 80 percent relative humidity for a period greater than 16 hours, or any less humid condition for greater than 24 hours.

When electrodes have been exposed to these excess conditions, they need to be conditioned for critical low hydrogen applications. Conditioning requires using temperatures in the 205 to 255°F range for a period of 6 to 12 hours and then cooling. These electrodes can then be stored in sealed 4-millimeter minimum thickness poly bags. Electrodes that are rusty should be discarded. Care must be taken not to cook off the lubricants present on the wire. The best procedure is to follow the specific electrode manufacturer's recommendations.

Joint Design

Flux cored wire can save time and weld metal. Part of the savings results from continuous welding with high deposition rates. Further savings are achieved by designing joints to take full advantage of the deep penetration and sound weld metal.

The volume of weld metal required to complete a butt joint can be effectively reduced by reducing the root opening, increasing the root face, and using smaller groove angles. Deep penetration enables the welder to compensate for poor fitup, and strong fillets are possible with fewer passes and less metal. Smaller joint openings can be used in metals thicker than ½ inch. Welding is usually on material thicknesses greater than ⅛ inch. Thicknesses up to ½ inch are weldable with no edge preparation, and with edge preparation maximum thickness is practically unlimited.

Fillet welds can be reduced in exterior size and retain comparable or greater strength because of the deep penetration of the arc, Fig. 23-15, page 760.

Figures 23-16 to 23-21 show the various joint designs and procedures for welding butt joints.

Shielding Gas

The variety of shielding gases that may be used for flux cored wire welding is limited.

M-type electrodes require 75 to 80 percent argon/balance CO_2. If no M is present, CO_2 is to be used, or with some types of electrodes no shielding gas is required. In addition to carbon dioxide, a mixture consisting of 98 percent argon and 2 percent oxygen or 95 percent argon and 5 percent oxygen may be used. A mixture of 75 percent argon and 25 percent carbon dioxide is also satisfactory. In general, the argon-rich mixtures offer more stable (less penetrating) controllable arcs over 100 percent CO_2. It can be less smoky in operation, but it is a more expensive gas.

When using carbon dioxide, a gas flow rate of 30 to 45 cubic feet per minute must be maintained. It may be necessary to increase flow rates 25 to 50 percent when welding outdoors or in drafty locations. To maintain this flow rate, two or more standard cylinders should be

JOB TIP

Types of Work

Types of work needing welders that may interest you:

Aerospace
Defense
Automotive
Ship building
Bridge building
Skyscraper construction
Sculpting

Table 23-2 Position of Welding, Shielding, Polarity, and Application Requirements for Flux Cored Electrodes

AWS Classification	Welding Position[1]	Shielding[2]	Current[3]	Application[4]	AWS Classification	Welding Position[1]	Shielding[2]	Current[3]	Application[4]
E70T-1	H and F	CO_2	DCEP	M	E70T-9	H and F	CO_2	DCEP	M
E70T-1M	H and F	75–80% Ar/bal CO_2	DCEP	M	E70T-9M	H and F	75–80% Ar/bal CO_2	DCEP	M
E71T-1	H, F, VU, OH	CO_2	DCEP	M	E71T-9	H, F, VU, OH	CO_2	DCEP	M
E71T-1M	H, F, VU, OH	75–80% Ar/bal CO_2	DCEP	M	E71T-9M	H, F, VU, OH	75–80% Ar/bal CO_2	DCEP	M
E70T-2	H and F	CO_2	DCEP	S	E70T-10	H and F	None	DCEN	S
E70T-2M	H and F	75–80% Ar/bal CO_2	DCEP	S	E70T-11	H and F	None	DCEN	M
E71T-2	H, F, VU, OH	CO_2	DCEP	S	E71T-11	H, F, VD, OH	None	DCEN	M
E71T-2M	H, F, VU, OH	75–80% Ar/bal CO_2	DCEP	S	E70T-12	H and F	CO_2	DCEP	M
E70T-3	H and F	None	DCEP	S	E70T-12M	H and F	75–80% Ar/bal CO_2	DCEP	M
E70T-4	H and F	None	DCEP	M	E71T-12	H, F, VU, OH	CO_2	DCEP	M
E70T-5	H and F	CO_2	DCEP	M	E71T-12M	H, F, VU, OH	75–80% Ar/bal CO_2	DCEP	M
E70T-5M	H and F	75–80% Ar/bal CO_2	DCEP	M	E61T-13	H, F, VD, OH	None	DCEN	S
E71T-5	H, F, VU, OH	CO_2	DCEP or DCEN[5]	M	E71T-13	H, F, VD, OH	None	DCEN	S
E71T-5M	H, F, VU, OH	75–80% Ar/bal CO_2	DCEP or DCEN[5]	M	E71T-14	H, F, VD, OH	None	DCEN	S
E70T-6	H and F	None	DCEP	M	EX0T-G	H and F	Not specified	Not specified	M
E70T-7	H and F	None	DCEN	M	EX1T-G	H, F, VD or VU, OH	Not specified	Not specified	M
E71T-7	H, F, VU, OH	None	DCEN	M	EX0T-GS	H and F	Not specified	Not specified	S
E70T-8	H and F	None	DCEN	M	EX1T-GS	H, F, VD or VU, OH	Not specified	Not specified	S
E71T-8	H, F, VU, OH	None	DCEN	M					

[1] H = horizontal position; F = flat position; OH = overhead position; VD = vertical position with downward progression; VU = vertical position with upward progression.

[2] Properties of weld metal from electrodes that are used with external gas shielding (EXXT-1, EXXT-1M, EXXT-2, EXXT-2M, EXXT-5, EXXT-5M, EXXT-9, EXXT-9M, EXXT-12, and EXXT-12M) vary according to the shielding gas employed. Electrodes classified with the specified shielding gas should not be used with other shielding gases without first consulting the manufacturer of the electrode.

[3] The term DCEP refers to direct current electrode positive (d.c., reverse polarity). The term DCEN refers to direct current electrode negative (d.c., straight polarity).

[4] M = single- or multiple-pass; S = single-pass only.

[5] Some E71T-5 and E71T-5M electrodes may be recommended for use on DCEN for improved out-of-position welding.

Source: HOBART® Welding Products—An Illinois Tool Works Company

Table 23-3 Typical Mechanical Properties of Several Flux Cored Electrodes (AW)[1]

AWS Classification	Tensile Strength (p.s.i.)	Yield Strength (p.s.i.)	Elongation % in 2 in.	Reduction of Area	Typical Charpy V-Notch Impact Values (ft-lb)			
					−0°F	−20°F	−40°F	−50°F
E71T-1	85,500	79,400	26.5	67.5	85	75	65	60
E71T-8	81,900	67,500	25.5	69.2	—	80	—	50

[1]AW indicates "as welded," as compared to SR which indicates the welds have been stress relieved.

Table 23-4 Comparative Index of Self-Shielded Flux Cored Wires

AWS Class	HOBART	COREX	ESAB	Lincoln
E70T-4	FABSHIELD 4	Self-Shield 4	Coreshield 40	Innershield NS-3M
E70T-7	FABSHIELD 31, 7027	Self-Shield 7	Coreshield 7	Innershield NR-311
E71T-8	FABSHIELD 7018	—	—	Innershield 202, NR-203M, NR-232, NR-203MP
E70T-10	FABSHIELD HSR	Self-Shield 10	Coreshield 10	Innershield NR-131
E71T-11	FABSHIELD 21B	Self-Shield 11	Coreshield 11	Innershield NR-211-MP
E71T8-K6	FABSHIELD 3Nil	—	—	Innershield NR-207, NR-400, NR-203
E71T-GS	FABSHIELD 23	Self-Shield 11 GS	Coreshield 15	Innershield NR-151, NR-152, NR-157, NR-204-H

Source: HOBART® Welding Products—An Illinois Tool Works Company.

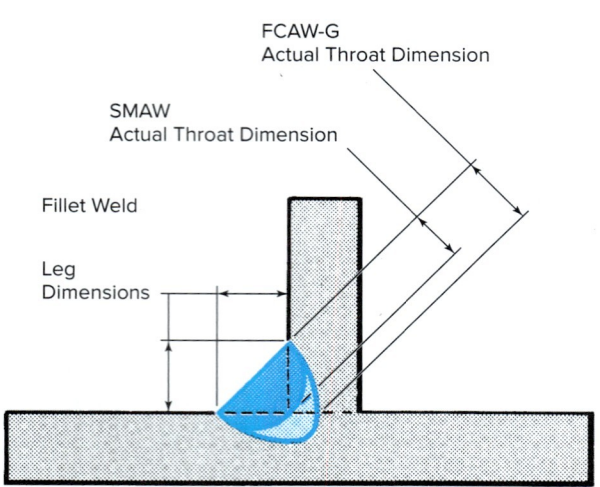

Fig. 23-15 A fillet weld made with SMAW has less penetration at the root than a weld made with FCAW-G. A fillet weld made with flux cored wire may be reduced in leg size and still secure maximum strength because of the greater actual throat dimension.

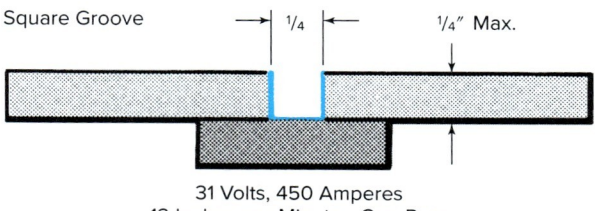

Fig. 23-16 Butt joints with backing up to ⅜ inch thick do not require beveling. Thickness up to ¼ inch can be welded with one pass.

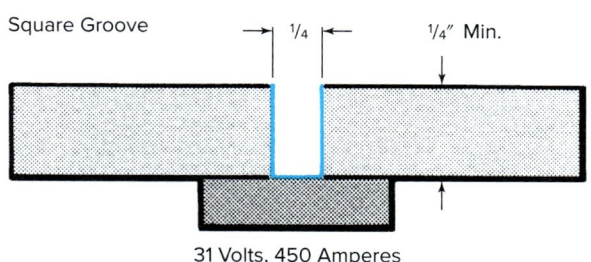

Fig. 23-17 Butt joints with backing in plate more than ⅜ inch thick require beveling. Plates ¼ to ½ inch thick are welded with two passes.

Table 23-5 Comparative Index of Gas Shielded Flux Cored Wires

AWS Class	HOBART	COREX	ESAB	Lincoln	McKay	Tri-Mark
E70T-1	FabCO RXR, FabCO TR-70	Flux-Cor 7	Dual Shield 111 A-C, R-70	Outershield 70	Speed Alloy 71	TM-11, TM-RX7
E70T-5	FabCO 85	Tuf-Cor 5	Dual Shield T-75	Outershield 75-H	Speed Alloy 75	TM-55
E71T-1 (CO_2)	FabCO 802, EXCEL-ARC 71	Verti-Cor 71 Versatile	Dual Shield 7100 Ultra	Outershield 71, 71-H, 71M	Speed Alloy 71V	TM-711M, Triple 7
E71T-1 (75 Ar/25 CO_2)	FabCO 825 EXCEL-ARC 71	Verti-Cor 71 Versatile	Dual Shield 7000 Dual Shield 7100 Ultra	Outershield 71, 71M	Speed Alloy 71V	TM-711M, Triple 7
E71T-1 (CO_2) Low H_2	XL-550	Verti-Cor II	Dual Shield II-71 Ultra	Outershield 71M-H	Speed Alloy 71-VC	TM-771
E71T-1 (75 Ar/25 CO_2) Low H_2	XL-525	Verti-Cor II	Dual Shield II-70 Dual Shield II-70T12 Ultra	Outershield 71C-H	—	TM-770
E81T1-Ni1	XL8Ni1	Verti-Cor II Ni1	Dual Shield II-80-Ni Dual Shield 800003	Outershield 81Ni1-H	Speed Alloy 81Ni1-V	TM-881N1, TM-811N1
E81T1-Ni2	FabCO 803	Verti-Cor 81Ni2	Dual Shield 8000-Ni2	—	Speed Alloy 81Ni2-V	TM-811N2
E110T5-K4	FabCO 115	Tuf-Cor 115	Dual Shield T-115	—	Speed Alloy 115	TM-115

Source: HOBART® Welding Products—An Illinois Tool Works Company.

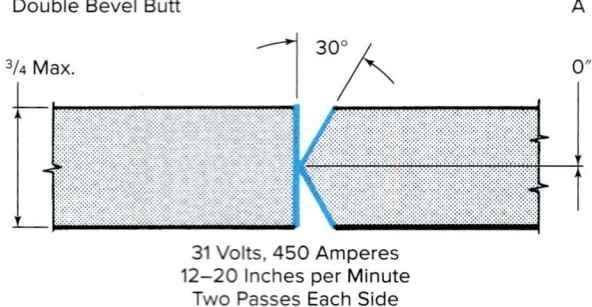

Fig. 23-18 Butt joints without backing in plate up to ¾ inch thick require 30° double bevel groove welds.

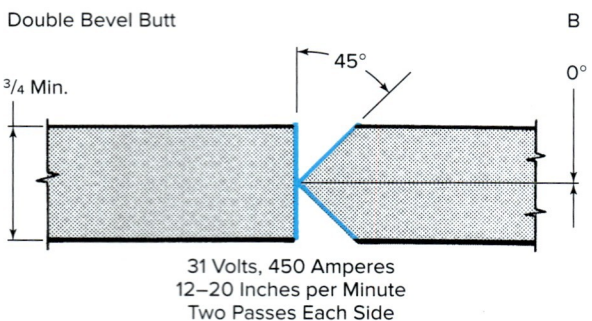

Fig. 23-19 Butt joints without backing in plate over ¾ inch thick require 45° double bevel groove welds.

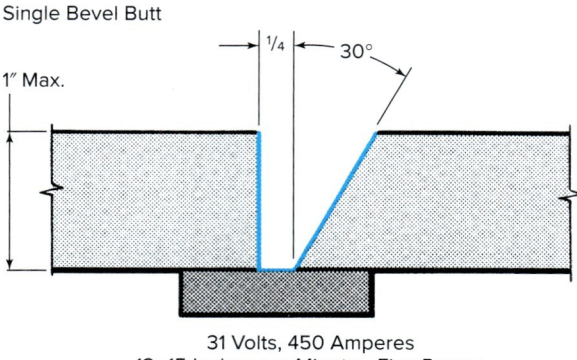

Fig. 23-20 Butt joints with backing in plate up to 1 inch thick require 30° single bevel groove weld and no root face.

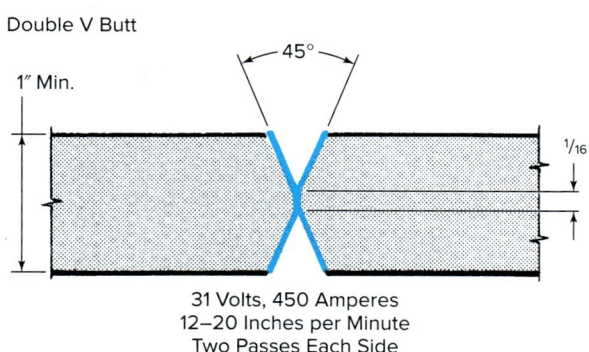

Fig. 23-21 Butt joint with double V-groove welds on plates, 1 inch and thicker, may have a heavy root face with no root opening and a 22½° bevel on each plate.

manifolded together. A pressure-reducing regulator and flowmeter are needed for controlling the shielding gas in the weld zone.

Weld porosity is a result of poor shielding of the arc and molten weld pool when the gas flow is inadequate. Excessive gas flow can cause turbulence of the arc. This causes porosity, weld spatter, and irregular beads.

Welding Variables

You will recall the effect that changes in the various variables had on solid and metal cored wire welding. The effect on flux cored wire is similar. Consistently good welds throughout a wide range of conditions are easily obtained when the variables are understood and controlled. All variables must be in balance for sound welds of good appearance.

Review the material in Chapter 22 concerning the effects changing the welding variables has on the weld bead.

Gun Angles Gun travel and work angle affect the flux cored arc welding process as it does the MCAW and GMAW processes, Fig. 23-22.

A push travel angle causes the gas shield to be directed over the molten pool. A portion of the arc is insulated from the base metal by the molten pool. Less penetration results, and a flatter and wider weld bead is deposited. Slag may be forced ahead of the weld pool to the extent that incomplete fusion and slag inclusions may results. A push angle is seldom used for FCAW. Generally a push angle is used only for the 3F and 3G positions vertical up. A drag angle of 2 to 15° is recommended. Welders usually prefer to use the drag angle because there is a better view of the arc action and the weld being deposited.

The work angle and wire location like MCAW and GMAW must be appropriate for bead placement and ensuring proper penetration and fusion. Undercut can also be affected by the work angle and wire location.

Electrode Extension You will recall from Chapter 22 that electrode extension is the length of the electrode wire extending from the tip of the contact tube. This extended length is subject to resistance heating called *electrode preheat*. Electrical extension has an effect on weld quality, penetration, arc stability, and deposition rate.

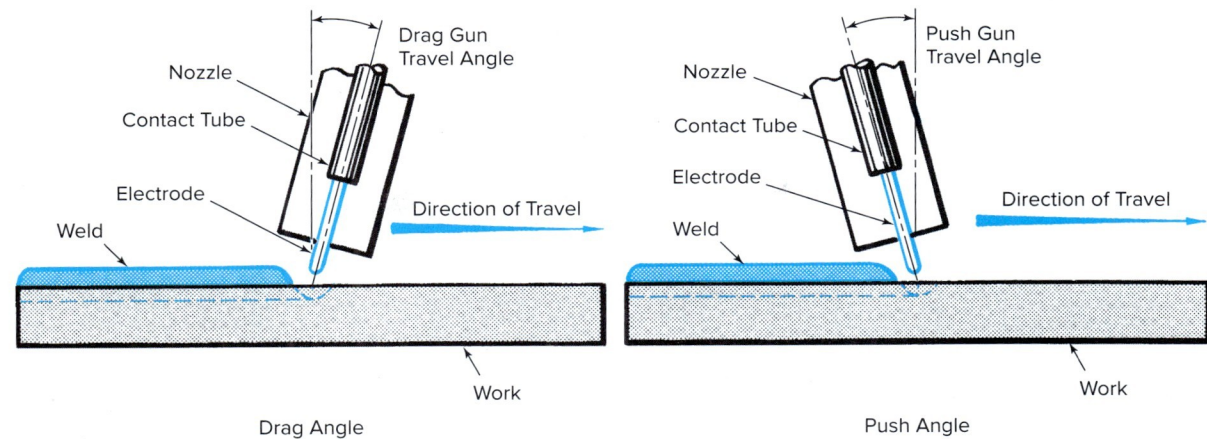

Fig. 23-22 Gun travel angles.

When welding with the FCAW process, electrode extension has an effect on the weld characteristics similar to that produced by solid and metal core wire. There is a difference, however, in the length of electrode extension.

Typical minimum and maximum extensions for FCAW-G electrodes are ¼ to 1½ inches.

In FCAW-S, the minimum extension is ⅛ inch and the maximum extension may go as high as 4½ inches. The manufacturer's specification for the electrodes should be followed to get optimum welding results.

With the FCAW-S electrodes, long extensions are required to properly preheat the electrode before it reaches the welding arc. This preheating activates the vaporizing and slag forming ingredients in the core of the electrode. It also burns off drawing compound (lubricant) residues from the sheath of the electrode. Preheating helps eliminate hydrogen contamination and porosity.

Preheating the FCAW-G electrodes also helps burn off any residual drawing compounds (lubricants) from the sheath of these electrodes as well. It also helps reduce porosity (worm tracking) if the electrode extension is not too short. With the FCAW-G electrodes, the external shielding gas may be lost if the extension is excessive and the gas nozzle contact tube is not properly adjusted. Excessively long electrode extensions cause spatter, irregular arc action, incomplete fusion, incomplete penetration, and slag inclusions. With excessively short electrode extension, spatter will build up on the end of the gun, porosity may increase, and the weld profile may be affected.

It is also recommended when using a long extension with FCAW-G electrodes that the contact tube tip be recessed a distance of ½ to ¾ inch from the end of the gas nozzle to reduce the spatter buildup and the possibility of overheating the contact tube. The correct nozzle-to-work distance ensures complete gas shielding and proper electrode preheating. A correct setting at the nozzle of the welding gun assures maximum weld quality, penetration, deposition, and appearance.

With the FCAW-S electrodes, long extensions (2½ to 4½ inches) require insulated wire guides. These insulated guides allow the welder to more closely control the electrode extension by a visual stickout of, for example, 1 inch. The guides can reduce the effects of helix and cast in these type of electrodes, Fig. 23-23, page 764. On a horizontal fillet weld electrode extension and electrode placement can help control penetration and undercut, Fig. 23-24, page 764.

Flux Cored Arc Welding—Gas Shielded Practice Jobs

Instructions for Completing Practice Jobs

Complete the practice jobs according to the specifications given in the Job Outline, Table 23-10, page 787, in the order assigned by your instructor. Before beginning welding practice, turn to Chapter 22 and review the steps for setting up the equipment and the welding recommendations.

These jobs should provide about 20 to 30 hours of practice, depending upon the skill of the individual student. After you have completed these jobs, you may wish to practice with other forms of joints, other sizes of filler wire, and a wide range of current values. If you have any particular trouble in your practice, consult Tables 23-6 through 23-8 (pp. 765–766).

Welding Technique

Welding current is DCEP. The extension is about ½ to 1 inch. The work angle of the gun is about 60° from the lower plate when welding a lap joint and about 45°

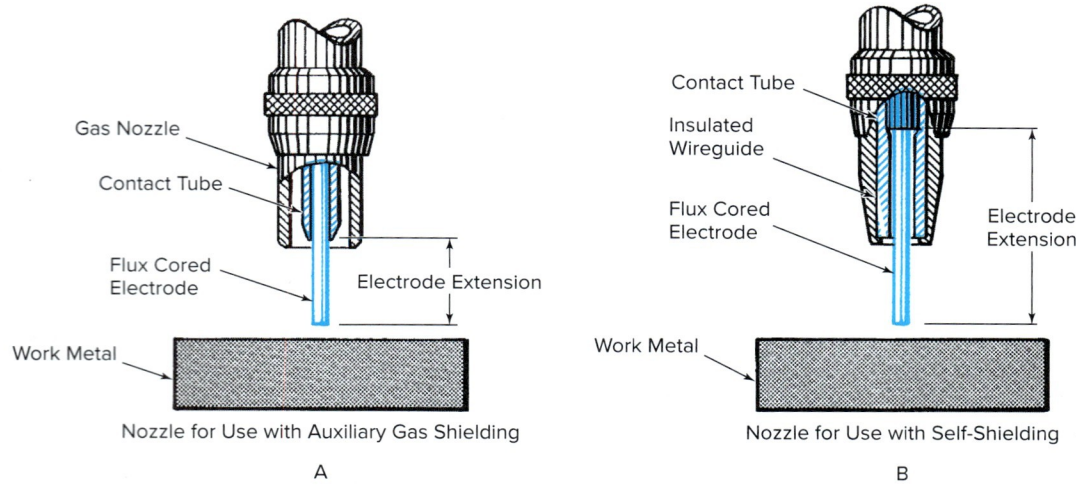

Fig. 23-23 (A) Electrode extension for the gas shielded electrodes. (B) Electrode extension for the self-shielded electrodes.

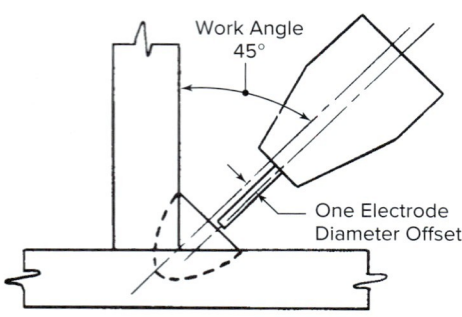

Fig. 23-24 Flux cored arc welding electrode position on a T-joint in the 2F position.

from the lower plate when welding a T-joint. A backing material strip is required when welding a butt joint. The groove angle for butt joints is about 30°.

Check the extension distance carefully. Excessive extension reduces the gas shield and overheats the wire. The arc and the welding area must be properly shielded from drafts that can blow the shielding gas away from the weld area. Do not let the weld metal overheat. If the weld gets too hot, the flux on the bead surface is hard to remove. Take particular care not to overheat the weld metal when making multiple-pass welds. Also be very careful to remove all the flux from the underneath passes when making multiple-pass welds. Make use of all the welding skills that you have learned up to this point in order to ensure sound welds and good appearance.

You will find that the arc is smooth, steady, and essentially spatter-free when operated at the proper wire-speed feed and voltage settings. It is also forceful and penetrating. The arc transfer has a semispray characteristic, and the arc appears to be buried in the pool. The weld deposits are even lower in hydrogen than those made with low hydrogen stick electrodes. The chief source of any hydrogen and porosity in the weld deposit is moisture absorbed from the metal surfaces or from the electrode.

Spatter may be caused by an arc that is too long. In such a case, the voltage setting is too high. Another possible cause of spatter is poor arc stability at low wire-feed speeds. This can be corrected by increasing the wire-feed speed or by shortening the extension. Normal voltage is 29 to 35 volts; but higher voltage may be needed if the welding cables are long.

Penetration is directly related to the travel speed. At a given wire-feed speed, penetration decreases if travel speed is increased or decreased too greatly. Keep the arc toward the leading edge of the weld pool. Penetration is deeper with the drag travel angle welding technique than with the push travel angle technique.

Groove Welding You are reminded to make sure that smooth, even penetration is obtained at the root of the weld. The bead should be equally proportioned and fused to the root face and to the groove face of each beveled workpiece. Pay particular attention to the width of the bead formation so that it is not more than $\frac{1}{16}$ inch on each side beyond the width of the groove. Regulate your speed of travel so that the face of the weld is not more than $\frac{1}{8}$ inch higher than the plate surface.

Fillet Welding The angle of the electrode is most important in making a fillet weld. For 2F the electrode wire is pointed at the bottom plate, close to the corner of the joint,

Table 23-6 Troubleshooting

If a weld should be defective, there is one or more welding conditions that must be changed or corrected in order to obtain a satisfactory weld. This table lists some of the weld discontinuities and the possible causes of these troubles.

Trouble	Possible Causes
Excess melt-through	Current too high Excessive root opening between plates Travel speed too slow Bevel angle too large Root face too small Wire size too large Insufficient metal holddown or clamping
Excess reinforcement or excess concavity	Current too high or low Voltage too high or low Travel speed too slow or fast Improper weld backing Improper spacing in welds with backing Workpiece not level
Penetration too deep or too shallow	Current too high or low Voltage too high or low Improper gap between plates Improper wire size Travel speed too slow or fast
Porosity	Flux too shallow (SAW) Improper cleaning Contaminated weld backing Contaminated electrode (electrode extension) Contaminated gas shield Contaminated plate Improper fitup in welds with manual backing Insufficient penetration in double welds
Reinforcement narrow and steep-sloped (pointed)	Insufficient width of flux (SAW) Voltage too low
"Mountain range" reinforcement	Flux too deep (SAW)
Undercutting	Travel speed too high Improper wire position (fillet welding) Voltage too high Use slight pause at edge of weave beads
Incomplete fusion and cracks	Improper cooling Failure to preheat Improper fitup Concave reinforcement (fillet weld) Verify correct electrode extension
Longitudinal depression in weld face	Verify correct voltage Verify correct amperage and wire-feed speed Contaminated electrode (electrode extension)

Table 23-7 Troubleshooting Adjustments for Flux Cored Arc Welding

Problem	Solution[1]				
	Current	Voltage	Speed	Stickout	Drag Angle
Porosity	5↑	1↓	4↓	2↑	3↑
Spatter	4↓[2]	1↑	5↓	3↓	2↓
Convexity	4↓	1↑	5↓	2↓	3↑
Back arc blow	4↓	3↓	5↓	2↑	1↑
Insufficient penetration	2↑	3↓	4↑	1↓	5↑
Not enough follow	4↑	1↓	5↓	2↑	3↑
Stubbing	4↓	1↑	—	3↓	2↓

[1]Arrows indicate the need to increase or decrease the setting to correct the problem. Numbers indicate order of importance.
[2]With E70T-G electrodes, increasing the current reduces droplet size and decreases spatter.
Source: The Lincoln Electric Co.

Table 23-8 Flux Indicators for Mechanical Properties

Designator Number	Tensile Strength (p.s.i.)	Minimum Yield Strength (p.s.i.)	Percent Elongation in 2 in.
6	60,000–80,000	48,000	22
7	70,000–95,000	58,000	22
8	80,000–100,000	68,000	20
9	90,000–110,000	78,000	17
10	100,000–120,000	88,000	16

so that the weld metal will wash up on the vertical plate and form a weld that does not have undercut along its top edge. Here again, speed of travel is important in bead formation. Travel speed that is too fast for the current forms an undersized bead that has poor penetration at the root, undercut along the edges of the weld, and convexity at the center. Travel speed that is too slow for the current causes excessive pile up of weld metal, slag entrapment, and porosity.

Fillet and Groove Welding Combination Project with FCAW-G: Job Qualification Test 1

This combination test project will allow you to demonstrate your ability to read a drawing, develop a bill of materials (SI conversions are optional), thermally cut, fit components together, tack, and weld a carbon steel project. You will be using the techniques developed in prior jobs in this chapter using the FCAW-G process and electrodes. Follow the instructions found in Table 23-10, Job 23-JQT1, page 787, and the notes in Fig. 23-25.

Inspection and Testing

After the project has been tacked, have it inspected for compliance to the drawing. Examine Figs. 23-26 through 23-28, page 768, carefully for weld formation and appearance. On multiple-pass welds have your instructor inspect after each weld unless otherwise instructed. Use the following acceptance criteria to judge your welds. Look for surface defects. Keep in mind that it is important to have good appearance and uniform weld contour. These characteristics usually indicate that the weld was made properly and that the weld metal is sound throughout. Only visual inspection will be used on this test project to the following acceptance requirements:

- There shall be no cracks or incomplete fusion.
- There shall be no incomplete joint penetration in groove welds except as permitted for partial joint penetration groove welds.
- Your instructor shall examine the weld for acceptable appearance and shall be satisfied that you are skilled in using the process and procedure specified for the test.
- Undercut shall not exceed the lesser of 10 percent of the base metal thickness or $\frac{1}{32}$ inch.
- Where visual examination is the only criterion for acceptance, all weld passes are subject to visual examination, at the discretion of your instructor.
- The frequency of porosity shall not exceed one in each 4 inches of weld length and the maximum diameter shall not exceed $\frac{3}{32}$ inch.
- Welds shall be free from overlap.

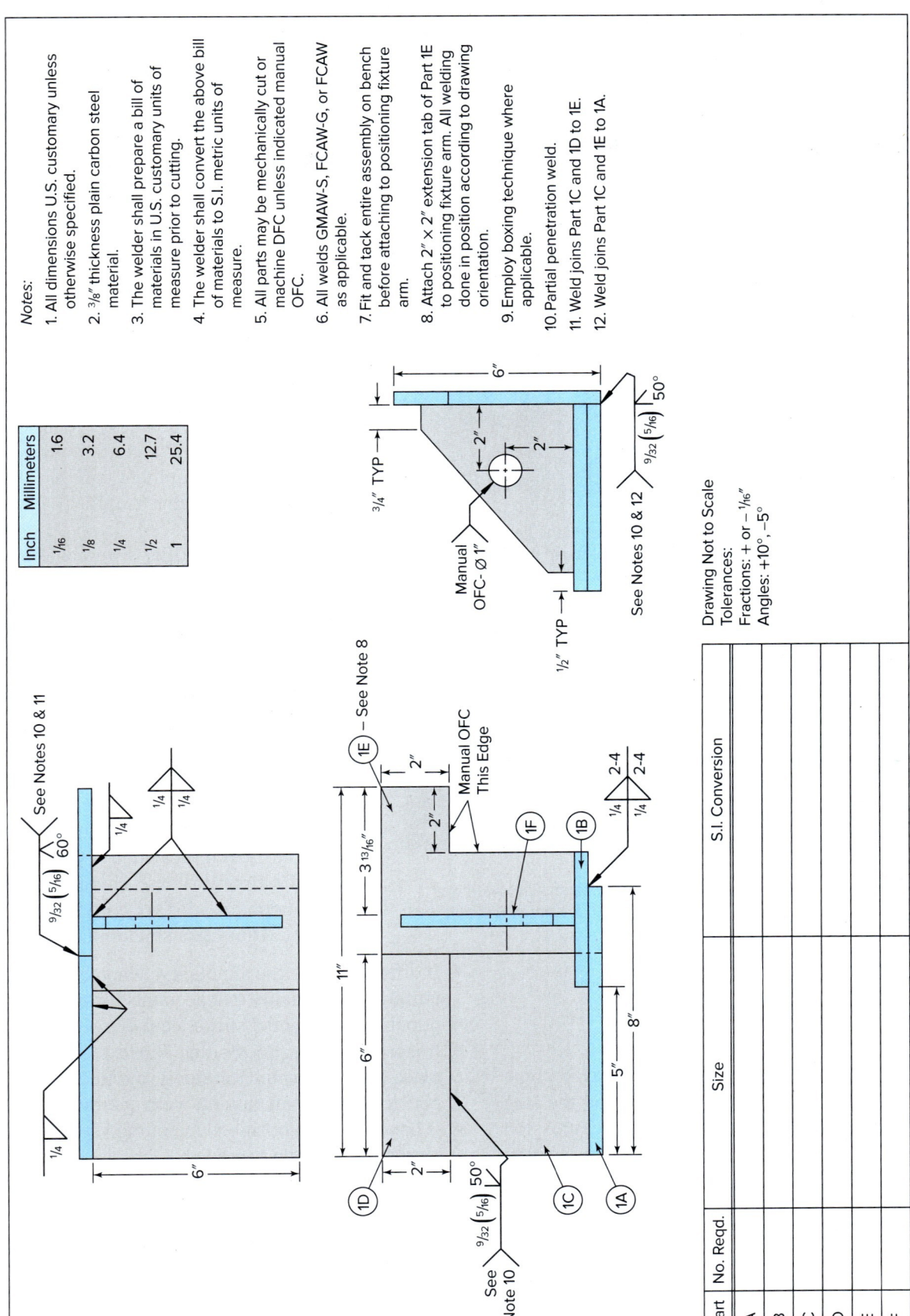

Fig. 23-25 Performance qualification test for FCAW-G or FCAW-S. Source: Modified from AWS S.E.N.S.E. Entry Level E62.0-95, Fig. 2.

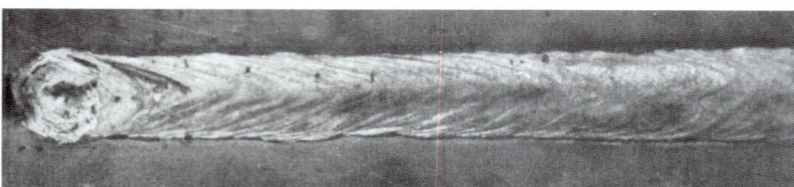

Fig. 23-26 Square butt joint in 3/8-inch mild steel plate with a backup strip. The joint was welded in the flat position, and the weld was made with flux cored filler wire 3/32 inch in diameter. Welding required 450 amperes and 32 volts.
Edward R. Bohnart

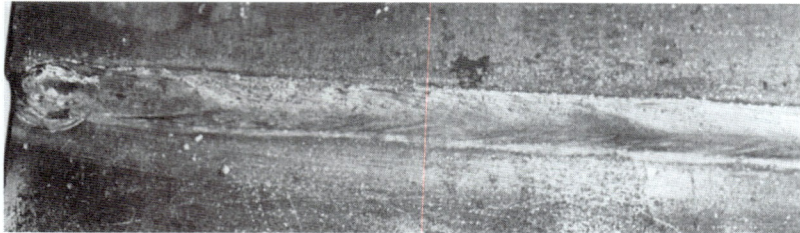

Fig. 23-27 Lap joint in 3/8-inch mild steel plate welded in the flat position. The horizontal fillet weld was made in a single pass with flux cored filler wire 3/32 inch in diameter. Welding required 450 amperes and 32 volts. Edward R. Bohnart

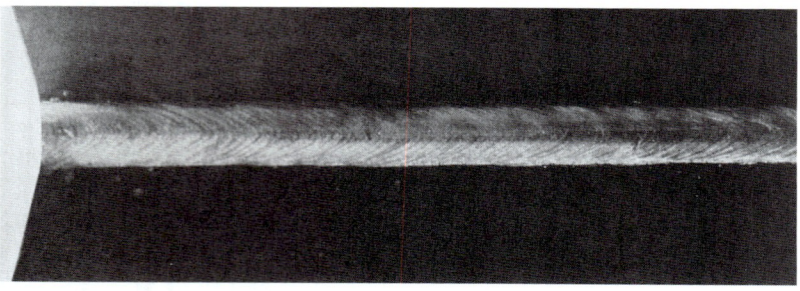

Fig. 23-28 T-joint in 3/8-inch mild steel plate. The horizontal fillet weld was made in a single pass with flux cored filler wire 3/32 inch in diameter. Welding required 450 amperes and 32 volts. Edward R. Bohnart

- Only minimal weld spatter shall be accepted, as viewed prior to cleaning.

When using FCAW-G electrodes, gases may become trapped between the solidifying weld pool and the slag layer. This gas causes a longitudinal surface depression in the face of the weld. This type of discontinuity is often referred to as *worm tracking*. If there is no evidence of a porosity "worm hole" at either end of a track depression, these tracks need not be considered detrimental to the overall joint efficiency. However, tracks do detract from the appearance of the weld and if appearance is important, corrective action may be required. Table 23-4, page 760, lists some overall troubleshooting pointers.

Flux Cored Arc Welding—Self-Shielded

The AWS-accepted term for this welding process is flux cored arc welding—self-shielded (FCAW-S). It is a semiautomatic welding process. One form is known in the trade by the name of Innershield®, a process introduced by the Lincoln Electric Company. In this process, a vapor produced by a special electrode shields the molten weld metal during the welding operation. The flux cored electrode is a continuous wire that also serves as a filler wire, Fig. 23-29.

The equipment needed to carry on this process is essentially the same as that used for other gas metal arc processes: constant voltage power source, and a continuous wire-feed mechanism. The equipment can be fully automatic or semiautomatic. Auxiliary gas shielding is not used.

The tubular steel filler wire contains all the necessary ingredients for shielding, deoxidizing, and fluxing. These materials melt at a lower temperature than the steel electrode metal and form a vapor shield around the arc and molten weld metal. Also included in the electrode are the alloying materials needed to provide high grade weld metal. Electrodes are available for welding in all positions, Fig. 23-30, page 770.

Process Advantages

The following information on the Innershield® process has been provided by the Lincoln Electric Company. These characteristics may also be applied to all flux cored wire welding without auxiliary gas shielding.

- It offers much of the simplicity, adaptability, and uniform weld quality that accounts for the continuing popularity of manual stick electrode welding.
- It is a visible arc process that allows the welder to place the weld metal accurately and to control the pool visually for maximum weld quality.
- It operates in all positions including vertical-up, vertical-down, and overhead.
- Welding can be done outdoors and in drafty locations without wind screens because the shielding does not blow away.
- The job needs only simple fixturing or none at all. Often existing stick electrode fixtures can be used.
- Simple wire-feeding equipment is not encumbered by flux-feeding systems or gas bottles. Installation is

quicker and more flexible. Welding rigs are more portable.
- Welder fatigue is minimized during sustained welding operations because the guns are lightweight, flexible, and easy to handle.
- Compact guns fit into places where other semiautomatic guns or stick electrode holders with the long electrodes cannot go.
- Uninterrupted wire feeding results from special electrode lubrication, freedom from spatter-clogged guns, and the ability to resist crushing of the tubular electrode by high drive roll pressure.
- Arc starts are quick and positive without sticking, skipping, or excessive spatter. The process corrects poor fitup and resists cracking in many crack-sensitive applications.
- Welding costs are cut because higher practical welding currents provide increased deposition rates and travel speeds.
- The process has been job-proven on many applications including repair welding, machinery fabrication, assembly welding, ship and barge building, field welding of storage tanks, and erection of structural steel for small and large buildings.

Flux Cored Arc Welding—Self-Shielded Practice Jobs

Instructions for Completing Practice Jobs

Complete the practice jobs according to the specifications given in the Job Outline, Table 23-11, page 788, in the order assigned by your instructor. Set up for welding as you did when practicing with the FCAW-G process.

For video of an ironworker testing in 3G position with FCAW-S electrode, please visit www.mhhe.com/welding.

These jobs should provide about 20 to 30 hours of practice, depending upon the skill of the individual student. After completing these jobs, you may wish to practice with other forms of joints, various thicknesses of material, and different filler wire sizes. A wide range of current values should be employed. Develop a good understanding of the results of changes in the basic variables. Study Table 23-7, page 766, carefully.

Power Sources

Constant voltage (c.v.) d.c. power sources are used for the FCAW-S process and electrodes. These machines can be of the transformer rectifier type, inverter type, or engine-driven generator type. Many of these self-shielded type of electrodes are very voltage sensitive. As little as 0.5 volt above or below their effective voltage range can create problems such as weld appearance, improper weld profile, and a reduction of mechanical and/or physical properties.

Wire Feeders

Wire feeders are very similar to continuous wire feeders for other gas metal arc processes. They are generally of the push type. In some cases, the pull-push type of feeder can also be used.

For portable work, the voltage sensing feeder is very popular. It operates off of the

Fig. 23-29 The self shielded, flux cored arc welding process.

Fig. 23-30 An ironworker testing in the 3G position on 1-inch plate with backing. Note the head position in relation to the fumes and good body posture. This is a typical test that might be conducted on a job site in a temporary structure. Miller Electric Mfg. Co.

output voltage from the power source and does not require a separate power and power cord to run the feeder motor. Controls are simplified, which reduces weight. Because these electrodes are self-shielding, the feeder does not require a gas solenoid valve. This feature also eliminates the need for gas hoses and a shield gas supply (cylinder).

The voltage sensing feeder may be equipped with a weld contactor to turn power to the arc on and off.

The drive rolls should be of the V knurled type for the hard sheath wires. For the soft sheathed wires, use a V or U cogged-type drive roll. To determine if a hard or soft sheath is provided on your specific wire, take a length of wire and try bending it. If it has a hard sheath, it will break in a few bends. If it has a soft sheath, it will bend many times; in fact, it can be tied in a knot.

Guns

Guns for semiautomatic FCAW-S welding should be light and maneuverable to facilitate high speed work. They should be equipped with a small guide tip for reaching into deep grooves. Guns are made in light-, medium-, and heavy-duty models to provide for different current ranges and electrode diameters. The electrode in the gun is cold until the trigger is pressed. If a voltage sensing feeder is used, then a weld contactor is required. Medium- and heavy-duty guns have a shield to protect the welder's hand from excessive heat.

Electrode Extension Electrode extension may range from ½ to 4½ inches from the contact tip. (See Fig. 23-23, p. 764.) The *visible* extension is the length of electrode extending from the end of the nozzle on the gun. A 2¾-inch electrical extension may be only a 1⅜-inch visible extension. The specified electrode extensions are obtained by using the proper guide tip and visible stickout on the welding gun. When a long guide tip is used to provide a 4½-inch electrode extension, both the voltage and amperage must be increased. When the electrode extension is ½ to 1 inch, a guide tip is not used. Check the electrode manufacturer's requirements for the electrode extension to be used.

Welding Technique

Before starting, check all welding control settings. Direct current electrode negative (DCEN) is very common on self-shielded electrodes. Check the welding procedure or the electrode manufacturer for correct polarity. Drive rolls and wire guide tubes should be correct for the wire size and type being used. Also, the gun, cable, and nozzle contact tip should be correct for the wire size and electrode extension.

Starting the Arc After the proper electrode extension has been set (cut the end of the electrode), the tip of the electrode is positioned just above the work so that it is lightly touching the work. The trigger is pressed to start the arc. The mechanical feed will take care of advancing the electrode. Welding is stopped by releasing the trigger or quickly pulling the gun from the work.

Gun Angles Use a drag angle of about the same angle as for stick electrode welding. The electrode-to-joint (work) angle varies with the type of joint and thickness of material. For horizontal fillets 5⁄16 inch and larger, the electrode points at the bottom plate at about a 45° angle. The wire location for most other positions and joints are to keep the weld size and bead location appropriate. This causes the weld material to be washed up on the vertical plate. For fillets ¼ inch and smaller, the electrode is pointed directly into the corner of the joint at an angle of about 40°.

A stringer bead should be applied with steady travel to avoid excess melt-through. Excess melt-through makes the metal sag on the underside of the joint and causes weld porosity. Joint fitup should be tight.

For out-of-position welding with E71T-G wire, the work is positioned downhill (vertical-down). Stringer beads are applied with settings in the middle to high range. The drag angle is in the direction of travel so that the arc force keeps the molten weld from spilling out of the joint.

In vertical-up and overhead welding, whipping, breaking the arc, moving out of the weld pool, or moving too fast in any direction should be avoided. Currents are in the low range.

Operating Variables

Four major variables affect the welding performance with self-shielded electrodes: arc voltage, current (WFS), travel speed, and electrode extension. These are all interdependent, and if one is changed, one or more of the others will require adjustment. Review the material in Chapter 22 concerning the effects changing these variables has on the weld bead. Also study Table 23-7, page 766, for an understanding of the corrections that you should make when problems arise.

Inspection and Testing

After the weld is completed, use the same inspection and testing procedures that you have learned in previous welding practice. Examine the welds for bead formation and appearance. It is impossible to determine the physical characteristics of a weld by its appearance. However, a weld that shows good fusion along the edges, has normal convexity, is free of undercut and surface defects, and has fine, smooth appearance usually meets the physical requirements.

Fillet and Groove Welding Combination Project with FCAW-S: Job Qualification Test 2

This combination test project will allow you to demonstrate your ability to read a drawing, develop a bill of materials (SI conversions are optional), thermally cut, fit components together, tack, and weld a carbon steel project. You will be using the techniques developed in prior jobs in this chapter using the FCAW-S process and electrodes. Follow the instructions found in Table 23-11, Job 23-JQT2, page 788, and the notes on Fig. 23-25, page 767.

Inspection and Testing

After the project has been tacked, have it inspected for compliance to the drawing. After the project has been completely welded, use the same inspection and testing methods that you have used in previous welding practice. On multiple-pass welds have your instructor inspect after each weld unless otherwise instructed. Use the following acceptance criteria to judge your welds. Look for surface defects. Keep in mind that it is important to have good appearance and uniform weld contour. These characteristics usually indicate that the weld was made properly and that the weld metal is sound throughout. Only visual inspection will be used on this test project to the following acceptance requirements:

- There shall be no cracks or incomplete fusion.
- There shall be no incomplete joint penetration in groove welds except as permitted for partial joint penetration groove welds.
- Your instructor shall examine the weld for acceptable appearance and shall be satisfied that you are skilled in using the process and procedure specified for the test.
- Undercut shall not exceed the lesser of 10 percent of the base metal thickness or $1/32$ inch.
- Where visual examination is the only criterion for acceptance, all weld passes are subject to visual examination, at the discretion of your instructor.
- The frequency of porosity shall not exceed one in each 4 inches of weld length, and the maximum diameter shall not exceed $3/32$ inch.
- Welds shall be free from overlap.
- Large spatter shall be unacceptable, as viewed prior to cleaning.

Automatic or Mechanized Welding Applications

Industry is continually finding new applications for automatic/mechanized welding applications, Fig. 23-31, page 772. These applications have economically replaced many other joining methods such as rivets, bolts, resistance welding, and castings. The difference between automatic and mechanized applications of welding is that in mechanized applications an operator must guide the electrode in the joint. In automatic applications, sensors are used to guide the electrode in the joint.

Automatic or mechanized arc welding has many characteristics that are highly desirable for manufacturing:

- Repeatability
- High quality
- High production
- Low welding costs
- Uniform welds
- Desirable weld appearance

Fig. 23-31 Welds being made on a vessel with the SAW process. The manipulator is equipped with flux delivery and recovery systems. The traveler car and rails allow for making longitudinal welds while the turning rolls allow for making the circumferential welds. Note the operator catwalk and ladder for convenience and safety. © Pandjiris

Fig. 23-32 A side beam carriage is set up with the SAW process. The digital readout control panel, electrode supply coil, wire drive head, and flux hopper are clearly visible. Edward R. Bohnart

- Continuous output
- Ability to conform to various welding conditions
- Almost 100 percent duty cycle

Each company needs to study the individual job and the general plant conditions in making the decision to use automatic/mechanized welding. A few considerations:

- Can the job be redesigned for automatic or mechanized welding?
- How does the cost compare with the costs of other processes?
- Are new work processes, such as handling the material, introduced that will offset the saving in welding time?
- Which automatic or mechanized system is best for the job?
- Are there sufficient floor space and power available for the new equipment?
- Is skilled labor available to operate the automatic/mechanized system?
- Is the work repetitive enough to gain the full advantage of automatic/mechanized welding?

Submerged Arc Welding (SAW)

Submerged arc welding may be applied via automatic, mechanized, or by semiautomatic means. Welding takes place beneath a blanket of granular, fusible flux. A bare filler wire electrode is fed continuously into the arc and is completely hidden by a mound of flux. Equipment includes the power source, an automatic wire-feeding device, a flux-feed system, a gun in semiautomatic welding, and flux pickup, Fig. 23-32.

The filler wire is not in actual contact with the workpiece. The current is carried across the arc gap through the flux. The weld pool is completely covered at all times, and the welding operation is without sparks, spatter, smoke, or flash, Fig. 23-33. Thus protective shields and helmets are not needed. Goggles should be worn, however, for safety. Some smoke and fumes are created that require proper ventilation.

Welds made under the protective layer of flux have unusually good ductility, impact strength, uniformity, density, corrosion resistance, and low hydrogen content. Physical properties are equal or better than the base metal when the base metal is free from rust, scale, moisture, and other surface impurities. In heavy sections, cracking can be minimized by the use of multipass welding and preheating.

Automatic/mechanized submerged arc welding is used extensively in the welding of carbon steel, chromium-molybdenum alloy steel, and stainless-steel piping. Applications include the longitudinal seam welding of rolled-plate piping, circumferential welds on large piping, and straight fillet welds on fabricated plate attachments. The process can be used to make groove welds, fillet welds, and plug welds in both production and repair work, Figs. 23-34 and 23-35, page 774. It is also suitable for hard-surfacing and the welding of castings.

Because the bed of flux must be supported, welding is performed in the flat position. In special situations the

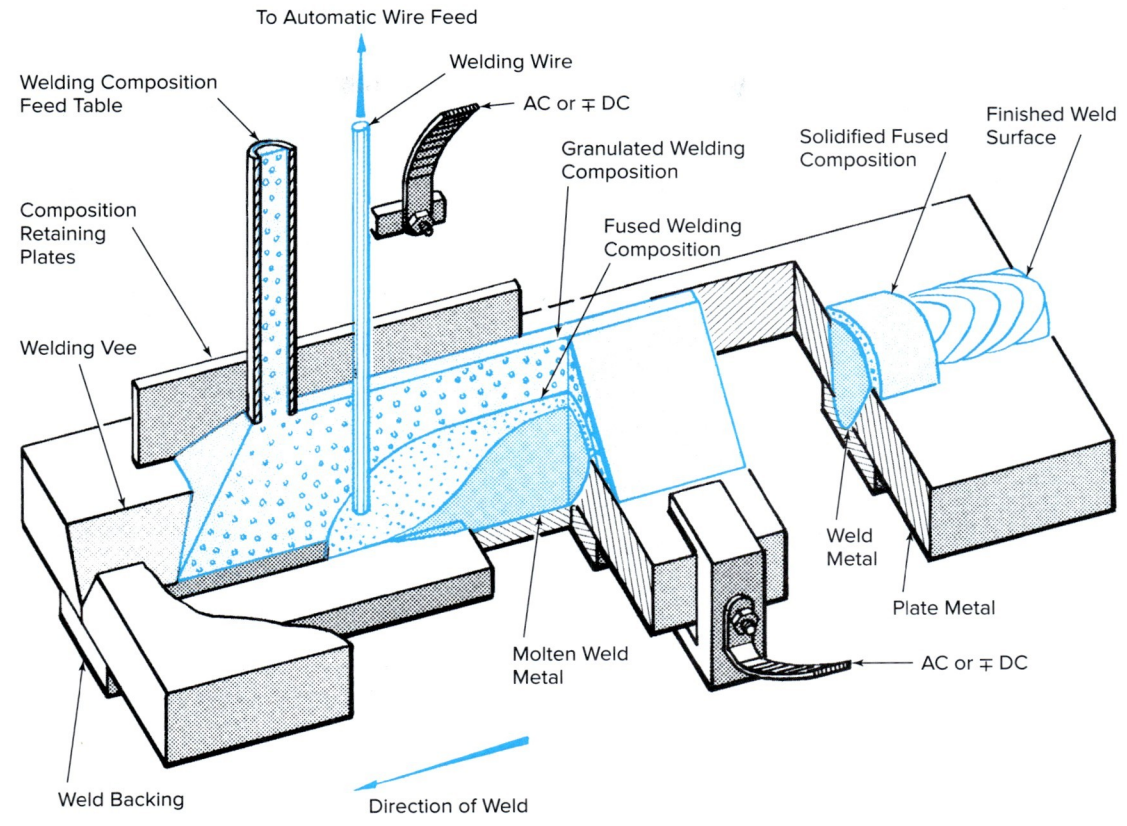

Fig. 23-33 Cutaway view of the welding zone in a single V-groove welded with the submerged arc welding process.

flux can be supported to aid weld positioning (Fig. 23-36, p. 774). Welds can be made in one or two passes in any thickness of steel from 16 gauge to 3 inches or more. High welding speeds are possible, and the current may be as high as 4,000 amperes at 55 volts. Backing rings are often used for groove welds. When fitup is poor, the first pass should be made by manual shielded metal arc welding or semiautomatic GMAW or FCAW processes.

When a backing ring is not used, the first pass may be made by these same processes or the gas tungsten arc welding process. In the welding of alloy pipe, the alloying elements should be contained in the filler wire and not in the flux.

Power Source Welding power may be provided by a d.c. rectifier, or an a.c. transformer. Welding with direct current provides versatile control over the bead shape, penetration, and speed, and arc starting is easier. DCEP provides the best control of bead shape and maximum penetration. DCEN provides the highest deposition rates and minimum penetration. Alternating current provides penetration that is less than DCEP but greater than DCEN. It minimizes arc blow at high amperage, and it is preferred for single-electrode welding with high currents and for multiple-wire, multiple-power welding. The cost of producing current with transformers is less than with generators.

Both constant current (variable voltage) and constant voltage power supplies are used for submerged arc welding. In the **constant current** welding machine, a voltage sensing wire control system maintains a constant welding

SHOP TALK

IIW

The International Institute of Welding (IIW) offers a searchable on-line research database. This group of organizations from 40 countries began in 1947. It presents conferences and exhibitions regarding the science and application of joining technology. Its journal, *Welding in the World,* examines welding, brazing, soldering, thermal cutting, thermal spraying, adhesive bonding, and microjoining.

Fig. 23-34 Mechanized Submerged Arc Weld Made on a Bridge Pile Casing. Note the starting and or run off tab. These tabs assure good weld soundness at the ends of the production joint. © American Welding Society

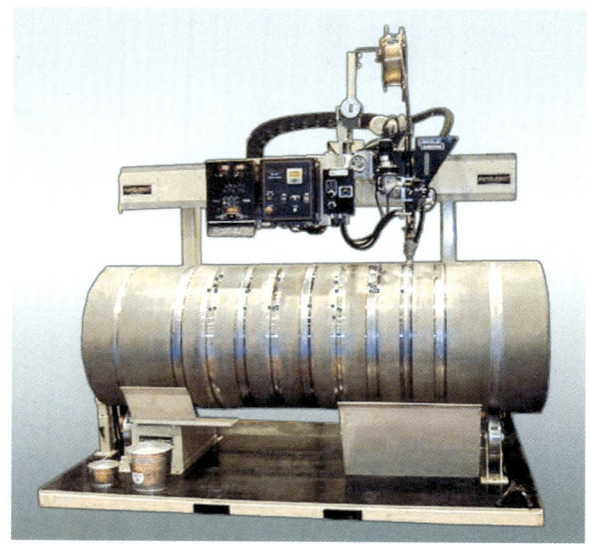

Fig. 23-35 A side beam carriage and turning rolls set up to do both longitudinal and circumferential welds on a vessel. The SAW process is being used on this project. It could have been easily set up for the FCA or GMAW as well. This setup is being used to develop the WPSs for welding on stainless-steel vessels. © Pandjiris

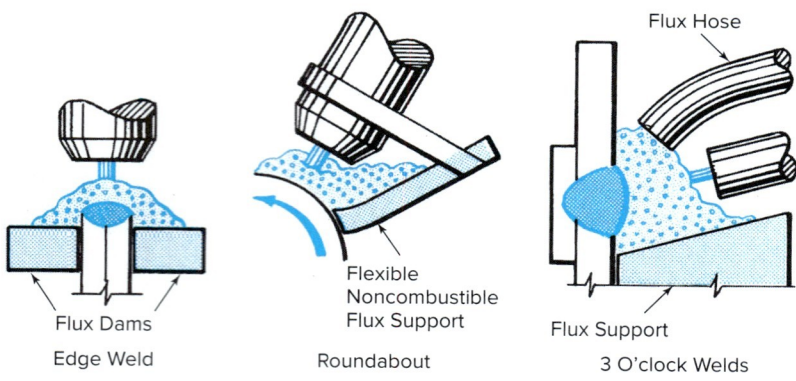

Fig. 23-36 Various types of flux supports for submerged arc welding.

current. The wire feeder varies the wire-feed speed to produce a constant arc voltage and keep the desired arc length. This system is generally used for welding with high currents and large electrodes ($5/32$ inch and over). It is also a good choice for hard-surfacing and welding alloys.

The **constant voltage** welding machine is used with the constant speed wire-feed control system. This system maintains a constant speed wire feed. Voltage is selected at the machine and is held constant regardless of current demand. Constant voltage, constant speed systems are preferred for small diameter filler wires and for high speed welding of thin materials.

Wire and Flux Classifications Wire and flux specifications for the submerged arc welding process are covered in the AWS A5.17 for carbon steel and AWS A5.23 for low alloy steel specifications. These specifications list multiple welding wires and fluxes.

The wires are divided into three groups according to manganese content. The low manganese steel wires are EL8, EL8K, and EL12. In the medium manganese steel category are EM11K, EM12, EM12K, EM13K, and EM15K. The high manganese wires are EH10K, EH11K, EH12K, and EH14.

The classification system follows the standard AWS pattern for filler metal specifications. For example, let us consider the classification EL8K. The prefix E designates an electrode as in other specifications. The letter L indicates that this is an electrode that has comparatively low manganese content (0.60 percent maximum). The letters M and H indicate a medium (1.50 percent maximum) and high (2.20 percent maximum) manganese content, respectively. The number 8 in the designation indicates the nominal carbon content of the electrode. The letter K, which appears in some designations, indicates that the electrode is made from a heat of steel that has been silicon killed.

The companion fluxes for these wires are designated with an F to represent flux. If only the F is designated, it means a virgin flux that has not been mixed with any crushed slag. If it is expressed as FS, the S indicates that it is made solely from crushed slag or is a blend of crushed slag with unused (virgin) flux. A number following the F or FS indicates the mechanical properties achievable. See Table 23-8, page 766.

Fluxes are further described by their reaction with the weld pool and weld metal. There are three general types of fluxes.

1. **Active fluxes** are those that contain controlled amounts of manganese and/or silicon. Improved resistance to porosity and cracking are the results of these alloys. These fluxes are used when the plate has a certain amount of contamination on it. This type of flux is used for making single-pass welds with the fewest defects and the highest quality.
2. **Neutral fluxes** are those that will not produce any significant change in the all-weld metal composition. This type of flux is used for multiple-pass welding on plate that exceeds 1 inch in thickness.
3. **Alloy fluxes** are those that are used with a plain carbon steel electrode to make alloy weld deposits. The alloying ingredients are found in the flux, not the electrode. It is essential that the welding procedure be followed to achieve the proper alloy percentages. This type of flux is primarily used on low alloy steel and for hard-facing.

Fluxes are granular, fusible mineral compounds. They contain various amounts of silicon, manganese, aluminium, titanium, zirconium, and deoxidizers, which are bound together with a binder.

Fluxes are classified on the basis of the mechanical properties of a weld deposit made with the flux in combination with any of the electrodes classified in the specification. For example, let us consider the classification F6A0-EH14. The prefix designates a flux. The number 6, which immediately follows this prefix, designates the mechanical properties. The letter A indicates "as welded." A letter P would have indicated postweld heat treatment (PWHT). The digit after the letter designates the required minimum impact strength, Table 23-9. The zero in the example means that 20 foot-pounds are required at 0°F. The suffix that is placed after the hyphen gives the electrode classification with which the flux will meet the specified mechanical property requirements. The electrode in the example is in the high manganese classification.

The fluxes used with submerged arc welding have the special characteristic of being able to carry the high welding currents required by the process. In all other respects, they produce results like those provided by other fluxes.

Table 23-9 Impact Requirements of the Weld Deposit According to Classification for Submerged Arc Flux

Classification	Required Minimum Impact Strength
0	
1	20 ft-lb at 0°F
2	20 ft-lb at −20°F
3	20 ft-lb at −30°F
4	20 ft-lb at −40°F
5	20 ft-lb at −50°F
6	20 ft-lb at −60°F
8	20 ft-lb at −80°F

- They protect the weld pool from the surrounding air in an envelope of molten flux.
- They act as a cleaning agent for the base metal.
- They may satisfy special metallurgical or chemical needs.
- They may provide minerals or alloys to the weld metal.
- They guard against porosity caused by rusty plate.
- They provide maximum resistance to weld cracking.
- Specific fluxes are designed to work best with certain electrodes, materials, and welding conditions.
- They improve weld appearance.

Wire Sizes Submerged arc electrodes are furnished in continuous lengths wound into coils or drums or on liners. Standard sizes include 1/16, 5/64, 3/32, 1/8, 5/32, 3/16, 7/32, and 1/4 inch. Wire 1/16 inch in diameter is used for making speed welds on steel ranging from 14 gauge to 1/4 inch thick. Wire 5/64 inch in diameter is used for welding material 12 gauge and thicker. Wires 3/32 inch and thicker are used when the gun is carried mechanically. The stiff wire decreases the flexibility of the cable, and the large pool of molten metal is hard to handle when welding is semi-automatic. Large sizes require high current and provide high deposition rates. They can bridge the gaps when fitup is poor. Arc starting, however, becomes more difficult as wire size increases.

Polarity DCEP is recommended for most applications. It produces smoother welds and better bead shape. It has greater penetration and better resistance to porosity. Fillet welds also have deep penetration.

DCEN has meltoff rates that are about one-third greater and provide less penetration than reverse polarity. It is suitable for the following applications:

- Conventional fillets in clean and rust-free plate
- Hard-surfacing
- Hard-to-weld steels in which cracking and porosity, due to the admixture resulting from deep penetration, must be controlled
- Prevention of cracking due to deep penetration and heavy buildup in the root pass of deep groove welds

When changing from DCEP to DCEN at the same current, increase the voltage about 4 volts to maintain a similar bead shape.

Alternating Current Alternating current is recommended for two specific applications:

- Tandem arc welding for increased welding speed
- Single-arc applications in which arc blow cannot be overcome and travel speed is slowed

To maintain good arc stability, a higher current density is needed for alternating current than for direct current.

Joint Design Basic joints common to all welding are used for submerged arc welding. The process is a deep, penetrating one. To avoid excess melt-through, the plates are generally either butted tightly together or a backing bar is used.

Cleaning Rust, scale, and moisture cause porosity. A thoroughly cleaned joint gives the best welding results. All substances on the joint edges must be removed. If not, foreign matter will become entrapped in the weld zone and cause porosity on the surface and beneath the surface.

Use clean, rust-free wire and flux that has been screened to remove large particles and foreign matter.

Position of Welding Practically all welding should be done with the work level or in the flat position. Some sheet metal may be welded slightly downhill. Heavy, deep-groove welding is sometimes done with the plate at an uphill angle of 2 to 5°. This helps keep the molten metal from running ahead of the arc.

Fitup Joint fitup should be uniform and accurate since it affects the appearance and quality of the weld. When establishing welding conditions, the seams should be butted tightly unless a root gap is specified. A gap may be required to secure penetration or to prevent weld cracking or distortion of the plates. An insufficient gap gives a weld an excessive crown or reinforcement. If the gap or bevel angle is excessive, burnthrough or a concave weld will result.

Gaps greater than $1/16$ inch may be filled with SMAW, GMAW, or FCAW processes. Use the appropriate electrode for the base metal and weld quality required.

Starting and Stopping Tabs It is general practice when welding long seams on a tank to tack-weld pieces of steel at each end of the seam. By starting and ending in these tabs, the possibility of cracking and weld craters may be eliminated. They enable the welding conditions to stabilize and maintain uniformity at the beginning and end of the joint.

In general, tabs should be similar in material and design to the weld joint. They should be large enough to support the flux and molten metal properly. Tabs are welded or supported on the base metal in a manner that will prevent molten metal from dropping through any gaps.

Flux Coverage Insufficient flux coverage permits the arc to flash-through and does not provide proper shielding. Excessive flux produces a narrow hump bead. For applications like roundabout, edge, and horizontal welds, a support may be needed to hold the flux around the arc while welding. It can be a piece of fire-resistant material clamped to the nozzle or flux dams tacked or clamped to the work, Fig. 23-36, page 774.

Flux Depth A proper amount of flux is required to establish the best welding conditions. A suitable depth of flux gives a fast, quiet welding action. If the layer of flux is too deep, the gases generated during welding cannot escape. The weld may have a shape somewhat like a mountain range—rough and uneven—and it will be porous. A good indication of proper flux depth is smoke rising out of the flux layer. A shallow layer of flux permits arc flashing and a porous weld.

It is very important that the flux be kept clean and dry. Dirty or damp flux produces an unsound, porous weld. Damp flux should be dried before welding.

Flux Recovery The unfused flux can be recovered during the welding operation. Recovery systems are available that use air power and others that use electrical power, much like a shop vacuum system. There is great savings in reclaiming the unfused flux, Figs. 23-37 and 23-38. The fused flux can also be recrushed and reprocessed. This is usually done off-site by a company specializing in reprocessing fused flux. The reprocessed fused flux is mixed with new flux in certain ratios.

Weld Backing The relatively high heats used in submerged arc welding cause a large pool of molten metal to be formed. Because the weld pool is highly fluid, it will fall through if the joint is not supported on the back

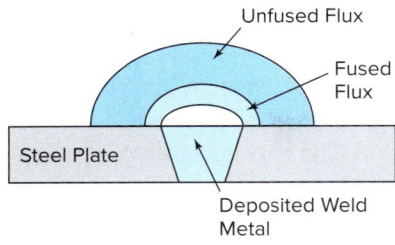

Number of Wires	Deposited Weld Metal (lb)	Amount of Flux Fused (lb)	Amount of Flux Deposited (lb)	Flux Savable (lb)
Single	1	1	2	1
Twin wire	1	1	3	2
Tandem wire	1	1	3–4	2–3

Fig. 23-37 Flux recovery data.

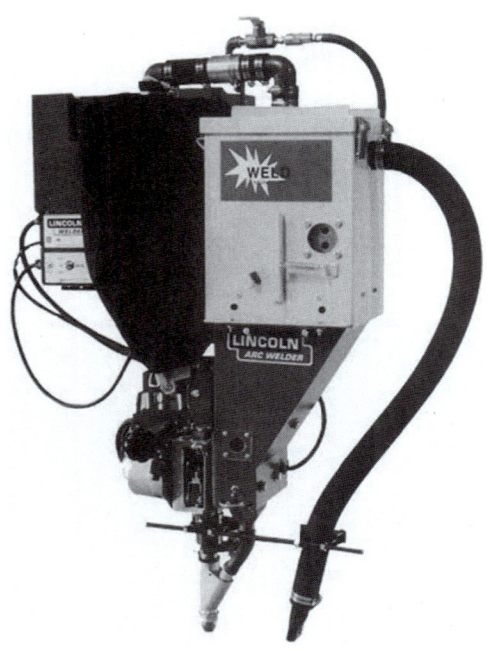

Fig. 23-38 Air powered flux recovery system that mounts easily on submerged arc welding equipment. © Welding Engineering Co. Inc.

Operating Variables The welds produced by the submerged arc welding process and the GMAW or FCAW processes are affected by similar changes and adjustments in the operating variables (voltage, current, wire-feed speed, travel speed, electrode extension, and nozzle angle). Review Chapter 22 as necessary. There is some difference in regard to electrode extension. Normal electrode extension (Fig. 23-39A) may be used for most applications. Some welding engineers, however, recommend long electrode extension. Nozzle attachments are available that increase the electrode extension to either 2¼ or 3¼ inches, Fig. 23-39B. The welding current passing through the longer lengths of exposed filler wire preheats the wire so that it melts more quickly in the arc. This reduces costs by increasing the deposition rate and the speed of welding.

Electrode Size At a fixed current setting, electrode size affects the depth of penetration. Penetration decreases as the size of electrode increases. Smaller filler wires are generally used in semiautomatic equipment. Sizes of 0.035 to 3/32 inch are generally used.

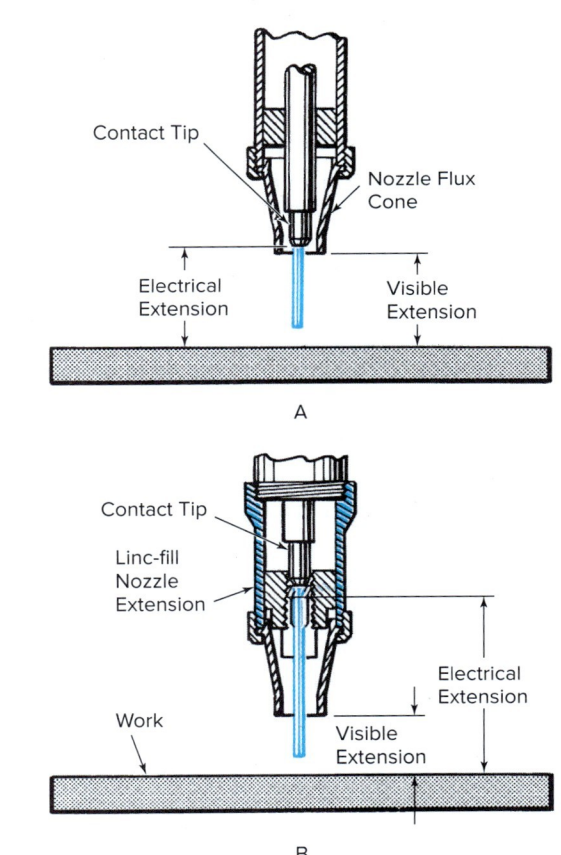

Fig. 23-39 (A) Normal electrode extensions. (B) Use of nozzle extension for long electrode extension. Source: The Lincoln Electric Co.

side. The most common forms of weld backing are steel backing bars, weld metal backing copper, or flux. When steel bars are used, they usually remain as part of the weldment. Weld metal backup in the form of a weld bead on the back side also remains as part of the weldment. Copper is one of the best materials to use. It is nonfusible and is a good conductor of heat. The copper backing bar is often liquid cooled. Flux may be placed at the bottom of the groove to act as a support for the molten weld pool.

Fig. 23-40 A SAW set up for making continuous or intermittent dual fillet welds on bridge girders, stiffeners, or box beams. A side beam carriage is being used in conjunction with head and tail stocks. This allows for the most advantageous welding positions for the highest productivity and quality. This would be considered an automatic application of the SAW process because of the sensing prods being used to track the joint. © Pandjiris

groove welds. The procedure is difficult to set up, Fig. 23-41, and can be justified only on long welding runs or on production work. At least one arc should be a.c.

The *two-wire series* power technique obtains high deposition rates with minimum penetration into the base metal. It is used extensively for hard-surfacing materials. Each filler wire operates independently; it has its own feed motor and voltage control. The power supply cable is connected to one welding head, and the return power cable is connected to the second welding head instead of the workpiece. The two filler wires are in series. The welding current travels from one electrode to the other through the weld pool and surrounding material. The two electrodes are mounted at 45° to each other. Either alternating or direct current can be used, depending upon the application. Alternating current is preferred for mild steel or stainless steel, whereas direct current should be used for nonferrous metals. DCEP gives deeper penetration than DCEN.

For automatic/mechanized submerged arc welding, filler wires are larger to take advantage of higher currents and higher deposition rates. Sizes of 5/64 to 7/32 inch are generally used with currents from 200 to 1,200 amperes. Large electrodes have hard starting characteristics.

With all other conditions (other than WFS) held constant, an increase in wire size reduces the deposition rate and penetration. An increase in current corresponding to an increase in electrode size increases the deposition rate and penetration.

Multiple-Wire Techniques Multiple arcs, Fig. 23-40, increase meltoff rates and direct the arc blow to provide an increase in welding speed. Two electrodes fed through the same jaws from one power source increase the deposition rate by 50 percent on work such as large, flat-position fillets and wide groove welds in which fill-in is a major consideration. The two arcs pull together, causing back blow at the front arc and forward blow at the trailing arc.

When two wires are being run into the same welding pool and power is being supplied by one power source, this is generally referred to as the twin electrode SAW process.

Multiple-wire, multiple-power arcs, in which two or more electrodes each have a separately controlled power source, provide high speed on both fill-in and square

Fig. 23-41 This is a submerged arc tandem set up with two electrodes; there could be three, four or more electrodes. Each electrode has its own head as can be seen. In this case the heads are staggered to weld on the opposite sides of the weldments to help control distortion. Note the side beam carriage that will move the arcs along the fixed weldments. The Lincoln Electric Co.

Thinner materials may be welded with a square edge. A backing bar is often necessary. Heavy materials may be beveled. The material may be only partially beveled so that a relatively heavy root face remains. Welding is often done from both sides. Fillet welds up to ¾ inch may be made without beveling. Fillet weld penetration is deep.

The flux is supplied from a hopper that is either mounted directly on the welding head or connected to the head by tubing. The bare filler rod is fed into the head from a coil or reel.

Submerged Arc Surfacing The term **surfacing** is used to describe the application of welding a layer of metal on a surface to obtain a desired surface dimension or other physical properties such as corrosion resistance. The SAW process is often used to surface carbon steel with stainless steel. This is a very economical way to obtain a corrosion-resistant vessel for the petrochemical or power plant industry. In some cases a typical round electrode is used; in other situations a strip is used. The strip produces a much wider weld with less penetration in the carbon steel. As the arc moves back and forth across the edge of the strip, it moves the heat energy over a much wider area. This reduces the dilution effect and makes the whole process more efficient. In many cases the required chemical analysis can be obtained in one pass, whereas when using a solid round electrode, several layers may be required. It takes a special welding head and drive system to change from feeding a conventional round electrode to feeding a strip. The strip widths commonly used vary from $9/16$ to over 9 inches, with a thickness of 0.024 inch, which enables a band of the strip width and some $3/32$-inch depth to be deposited in a single pass. Deposition rates can be as high as 44 pounds per hour.

Semiautomatic Submerged Arc Welding Semiautomatic submerged arc welding is being replaced by solid wire, metal cored wire, and flux cored wire welding. It is still used to some extent for hard-surfacing. Two examples of this process are Squirt Welding®, a Lincoln process, and Union Melt®, an ESAB process. If the school welding shop has this type of equipment, it would be an added experience for the advanced student to practice a number of butt, lap, and T-joints with the process.

If your school does not have submerged arc welding equipment, the process can still be observed by using existing GMAW or FCAW equipment. Use of the constant voltage power source and wire feeder can be used up to its output and wire diameter capability. A joint can be set up in the flat position and the flux placed directly on the joint to facilitate not having a flux delivery system. Since this will be done semiautomatically, care must be taken not to get the end of the gun in the molten slag or weld pool.

Semiautomatic submerged arc welding can be used on those jobs for which submerged arc welding would be desirable, but the fully automatic process would not be suited for economic reasons or physical limitations.

The semiautomatic application is similar to the automatic/mechanized application in that the welding operation takes place beneath a blanket of flux, but the hand gun is guided manually. The electrode is a continuous wire that is fed through the center of a flexible welding cable and through a gun to the arc. The flux is deposited by gravity on the joint from a welding gun. The flux also can be deposited by force feed from an air supplied system.

Equipment includes the welding machine, the conical flux container and welding nozzle, wire reel, wire-feeding mechanism, and the control unit for the control of wire feed and arc voltage, Fig. 23-42, page 780.

Semiautomatic submerged arc welding is essentially a small wire process. It is possible to weld ¾-inch plate having a square edge with filler wire $5/64$ inch in diameter and 600 amperes of welding current. Each side is welded with one pass. The resultant weld is smooth, penetrating, and spatter-free. Appearance is similar to that obtained with flux cored electrode welding.

Process Characteristics In comparison with automatic/mechanized submerged arc welding, the semiautomatic process has the following characteristics:

- It can follow irregular shapes.
- Welding can be done without fixtures or with only simple fixtures.
- Equipment is easily portable, and the process is highly versatile.
- The cost of equipment is lower than for the automatic/mechanized process.

The semiautomatic process can be used for any of the following broad applications:

- When the gun can be dragged along the joint, providing accurate guiding
- When the work can be rotated and the gun held in position by hand

> **ABOUT WELDING**
>
> **Safety**
> When you're not using a piece of equipment, turn if off. If the equipment is going to be left unattended, or is not in service, just completely disconnect it from its power source.

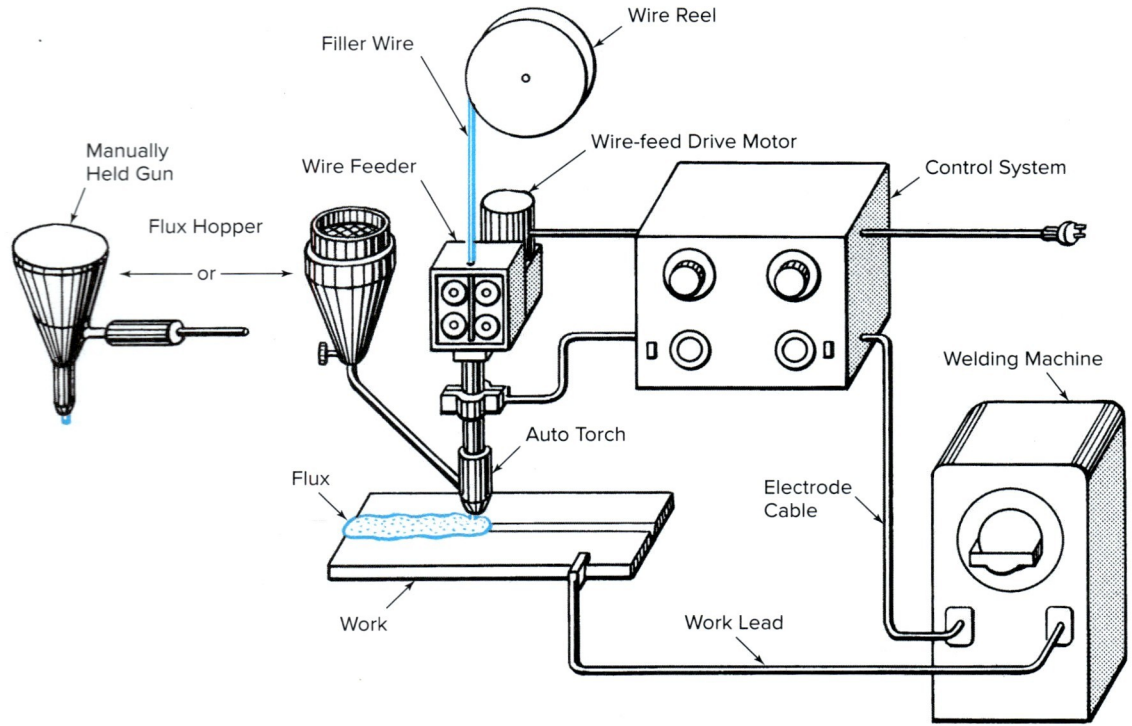

Fig. 23-42 Schematic diagram of component units needed for submerged arc welding. Source: Hobart Brothers Co.

- When the work can be rotated and the gun held in a simple locating fixture
- When both the gun and the work can be moved by special fixtures.

Submerged Arc Welding Semiautomatic Practice Job

Instructions for Completing the Practice Job

Complete the practice job according to the specifications given in the Job Outline, Table 23-12, page 789. Set up for welding as you did when practicing with the FCAW-S process.

If your school has automatic or mechanized SAW equipment, you should practice its operation as well. Set up for welding as you did when practicing with the GMAW process.

This job should provide about 4 to 6 hours of practice depending upon the skill of the individual student and the capability of the equipment. After completing this job, you may wish to practice with other forms of joints and various thicknesses of material. A wide range of current (WFS) and voltage should be employed. Develop a good understanding of the results of changes in the basic variable. Note the bead appearance and penetration characteristics of these welds compared to the other arc welding processes you have experience with.

Power Source Constant voltage d.c. power sources are used for the SAW process. These machines can be of the transformer rectifier type, inverter type, or engine-driven generator type. Make certain the power source has sufficient output voltage, amperage, and duty cycle for the welding you will be doing.

Wire Feeder Use the same feeder as for the FCAW practice setup with as large a diameter as possible for the feeder and power source being used; $1/16$ inch would be best.

Gun Use the same gun as for the GMAW or FCAW-G practice. Since shielding is provided by the flux, you will not be using the gas shielding. Leave the gas nozzle in place and recess the contact tube $3/32$ to $1/4$ inch. This will help you gauge the electrode extension while welding. Use the largest diameter gas nozzle available.

Electrode Extension The electrode extension is set before welding starts. It can range from 1 to 2 inches. You will have to visualize the extension by looking at the joint and the flux depth plus the recess into the gas nozzle, Fig. 23-43.

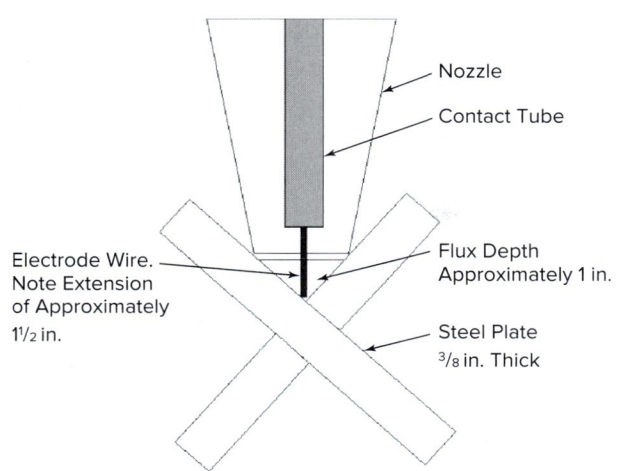

Fig. 23-43 Practice job using semiautomatic application of submerged arc welding process.

Welding Technique Before starting, check all welding control settings. Drive rolls and the wire guide tube should be correct for the wire size and type being used. The gun, cable, and nozzle contact tip should also be correct for the wire size and electrode extension being used.

Starting the Arc After the proper electrode extension has been established, pour a layer of flux into the joint. Flux at both ends of the joint can be supported with sheet metal. Rest the electrode tip in the root of the joint approximately 1 inch from the edge of the joint. The trigger is pressed to start the arc. The mechanical feed will take care of advancing the electrode. Welding is stopped by releasing the trigger. Since you will not be able to see the weld pool (it is submerged below the layer of flux), you will have to judge your travel speed on feel. If you weld too slowly, the fillet weld will be too large; if you weld too quickly, the fillet weld will be too small. The nozzle should be able to be dragged on the joint to help steady and control the electrode extension.

Gun Angle Use a drag travel angle about the same as that used for stick electrode welding. The work angle varies with the type of joint and thickness of material. Drag the nozzle along the joint to deposit a single stringer bead of approximately ½-inch leg size.

Operating Variable Four major variables affect the welding performance with the SAW process: arc voltage, current (WFS), travel speed, and electrode extension. The flux depth must not be too deep or too shallow. Do not allow flash-through to occur, and only a slight amount of smoke should be observed coming out of the layer of flux.

Inspection and Testing After the weld is completed, use the same inspection and testing procedures that you have learned in previous welding practice. Examine the welds for bead formation and fine ripple appearance. It is impossible to determine the physical characteristics of a weld by its appearance. However, a weld that shows good fusion along the edges, has normal convexity, is free from undercut and surface defects, and has fine, smooth ripples usually meets the physical requirements.

Electroslag Welding (ESW)

Electroslag welding (ESW) was developed for the welding of vertical plates, ranging in thickness from 1¼ to 14 inches, with a single pass. The plate edges require no preparation.

Electroslag welding process can be used in the automatic or mechanized applications. The equipment used for electroslag welding consists of a carriage assembly, which moves upward along the joint, and a multiple set of feed-wire guide assemblies that can be made to oscillate horizontally, Fig. 23-44, page 782. Copper shoes are positioned against the joint to act as a dam. The filler wire is fed to the weld zone through a nozzle in a vertically down feed (while the weld progression is vertical-up). The welding operations are controlled from a panel board and may be automatically, mechanically, or semiautomatically controlled.

The fusion of the base metal and the continuously fed filler wires takes place under a heavy layer of high temperature, electrically conductive molten flux. The filler wires may be either solid, metal cored, or flux cored. For welding plates up to 5 inches in thickness, only one electrode may be used. Two electrodes are generally used for plates 4 to 10 inches thick, and three electrodes for 10- to 14-inch thicknesses. When necessary, the electrodes may be oscillated to provide better distribution of weld metal. It is a high heat process. Voltage ranges from 42 to 52; and amperage from 500 to 640, depending on the thickness of metal. The power is alternating current.

The plates are set up in a vertical plane with the square edges spaced from 1 to 1⁵⁄₁₆ inches, according to the thickness of the plate. Water-cooled copper shoes form a mold around the joint gap and give form to the weld. The shoes are mechanized so that they can move vertically upward as the weld proceeds. A prepared block is placed under the plate edges to close the joint cavity. Granular flux is poured into the cavity, and the weld pool is established with the filler wires. At first, only the flux is fused into molten slag.

The extreme heat produced by the resistance heating in the weld pool, molten flux, and electrode cause the base

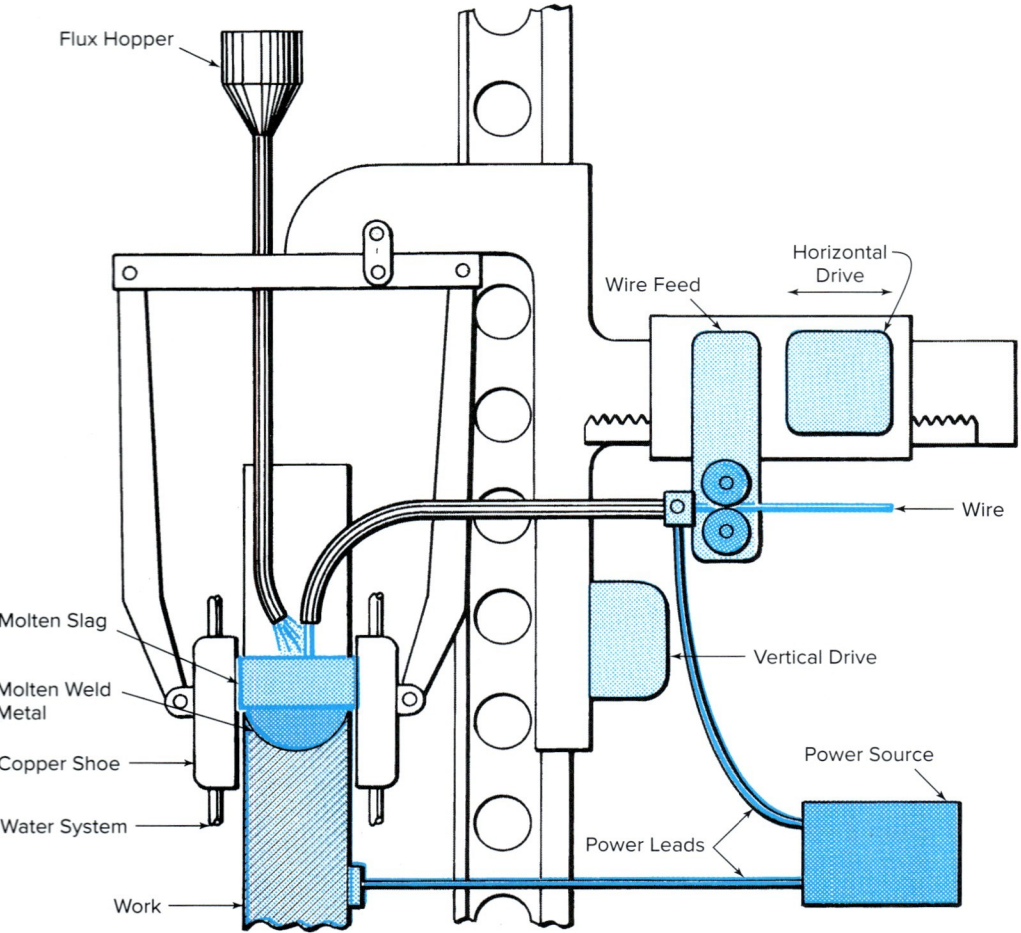

Fig. 23-44 Basic components of an electroslag welding operation. Note the travel rail that provides the vertical up-and-down movement.

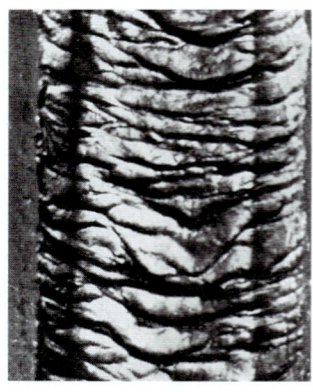

Fig. 23-45 A square groove butt joint made with the electroslag process in steel plate 3 inches thick. It had a 1-inch root opening and was welded in 1 pass. The completed weld is 1¾ inches wide with ⅛-inch face reinforcement. Welding was done in the vertical position, travel up. *Edward R. Bohnart*

metal to melt. The weld is formed by the water-cooled plates. It is homogeneous and has good penetration into the base metal and smooth, clean weld faces, Fig. 23-45.

Electroslag Welding with a Consumable Guide Consumable guide welding, or *CG* as it is referred to in the trade, is a method that is used to weld vertical beam joints.

A tube that is coated with slag-forming and alloying elements guides a filler wire from the wire-feeding unit into a bath formed by the two sides of the joint being welded and two water-cooled copper shoes, Fig. 23-46. This bath contains the molten weld metal and slag. The tube, or guide, is connected to the positive side of a rectifier power source DCEP. The heat necessary to melt the guide, the filler wire, and the joint edges being welded is generated by the passage of the welding current through the electrode, molten flux, and weld pool. The guide tube and filler wire melt at a rate that determines the welding speed.

The consumable guide process operates equally well on direct current or alternating current. The constant voltage power source is recommended.

Consumable guide welding is not a true submerged arc welding process. It uses an arc only at the start of the process to generate heat for the melting of the slag. As soon as the bath of molten slag is established, the slag causes the arc to be extinguished automatically. Another characteristic of the slag is low conductivity. This increases resistance heating from the passage of the electric current.

- An increase in welding speed increases productivity and improves the mechanical properties of the weld metal and the heat-affected zones. The weld metal is free from porosity and slag inclusions. There are fewer problems with residual stresses and plate deformation.

Electrogas Welding (EGW)

Electrogas welding can be done with solid metal cored or flux cored electrodes. It is an *arc welding process* using an approximately vertical welding progression with backing being provided by the molten weld metal. It is used with or without externally supplied shielding gas, which is determined by what type of electrode is supplied.

The *electrogas welding process* is similar to electroslag welding. It is a fully automated method for the welding of butt, corner, and T-joints in the vertical position. It differs only in the welding current and the medium used for protecting the weld pool from atmospheric contamination. Direct current, electrode positive is used instead of alternating current, and shielding gas instead of granular flux may be fed into the weld pool. It is possible to weld metal sections of $\frac{1}{2}$ inch to over 2 inches in thickness with a single pass and without any edge preparation.

Here again, water-cooled copper shoes span the joint cavity and form a dam to contain the molten weld metal. Flux cored, metal cored, or solid filler wire is fed into the cavity by means of a curved guide, Fig. 23-47, page 784. An electric arc is established and continuously maintained between the filler wire and the weld pool. Helium, argon, carbon dioxide, or mixtures of these gases may be fed continuously into the cavity to provide a suitable atmosphere for shielding the arc and the weld pool. The flux core of the filler wire provides deoxidizers and slagging materials for cleaning the weld metal. The base metal is melted and fused as a result of the high temperature from the arc and the molten slag. The molten slag forms a protective coating between the shoes and the gases of the weld and a seal between the shoes and the surfaces of the work to prevent air from entering the weld pocket. As the welding progresses, the copper shoes move upward. Welding is done with DCEP.

The edges are square cut and are spaced from $\frac{11}{16}$ to $\frac{7}{8}$ inch or more. The minimum spacing must be large enough to admit the wire guide and permit it to oscillate without arcing on the plate surfaces. Spacing that is too wide requires excessive filler wire and increases the welding time. Welding is done within a range of 500 to 700 amperes. Filler wire diameters can be used.

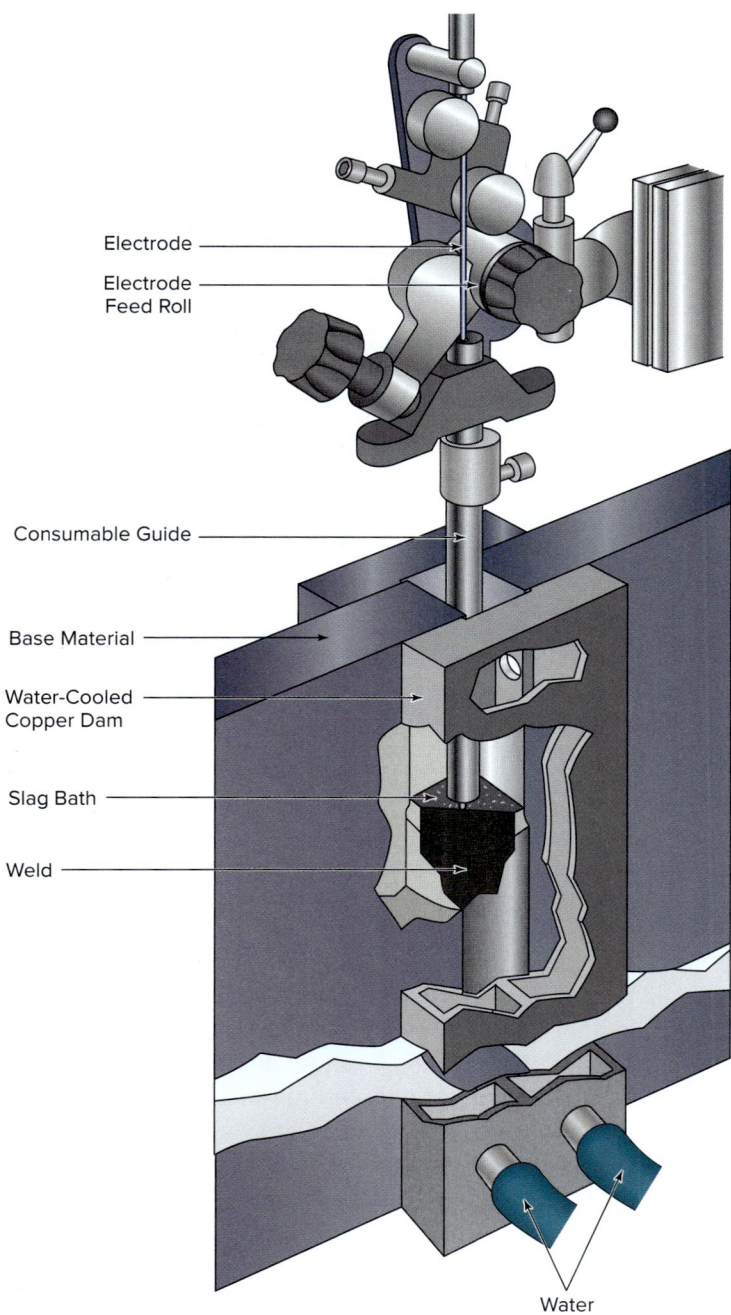

Fig. 23-46 Essentials of consumable guide welding. Adapted from ESAB Welding and Cutting Products

Because of resistance heating, the length of the tube is limited to approximately 40 inches. The tube can be extended, however, if a movable current contact is used.

The consumable guide process has the following advantages over conventional electroslag welding:

- The welding machine is portable, lightweight, and relatively easy to operate.
- Welding of thinner metals is possible.

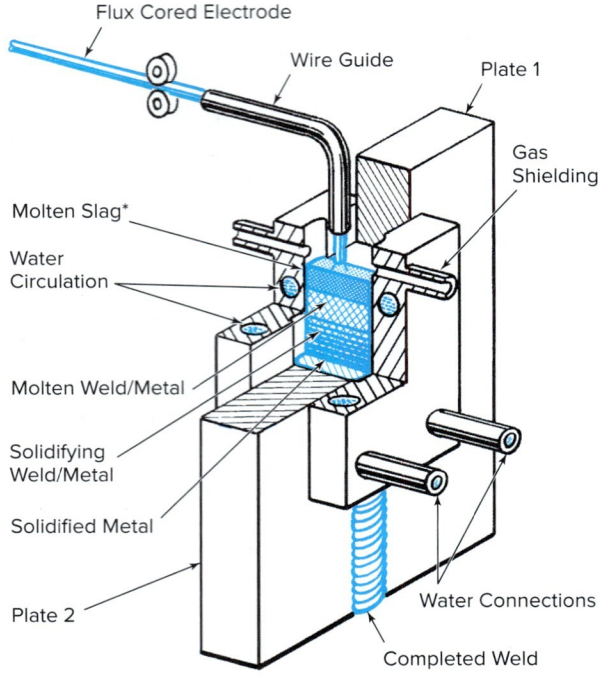

* If welding done with an FCAW electrode.

Fig. 23-47 Basic components for electrogas welding. Note the point of entrance of the shielding gas.

Fig. 23-48 Verti-Shield stationary fixture and controls. The physical specifications are: weight—57 pounds; dimensions—19.25 inches high, 24.25 inches wide, and 18.38 inches deep.
The Lincoln Electric Co.

The Lincoln Electric Company refers to a variation of the EGW as the Verti-Shield process, which is a mechanized vertical-up welding process designed to maximize productivity when joining steel plates. Welding is done vertically in a single pass on either butt joints or T-joints. There are two different methods that can be used:

1. Consumable guide tube process
2. Movable dam process (butt joint with groove welds only)

The Verti-Shield stationary fixture and control consists of a consumable guide fixture, automatic controls, and 10-foot cable for 0.120-inch (3.0 millimeter) electrode, and is for use with ½-inch diameter consumable guide welding, Fig. 23-48.

The electrogas process is used in the field erection of pressure vessels and liquid storage tanks and in the shop fabrication of large pressure vessels and heavy structures. Carbon, low alloy, high tensile, medium alloy, air-hardening, and chrome-nickel stainless steels are successfully welded by this process. In field welding, the work is enclosed in an all-weather shelter to protect it from the elements and air currents that might disturb the gas shield during welding. Alloy steels do not require preheat and may be welded at 32°F, providing there is no frost. Shrinkage, warpage, and distortion are avoided by this process.

Choice of Welding Process

In general, the selection of the welding process (manual, semiautomatic, mechanized, or automatic) depends on the proper evaluation of each job. The three processes have the following applications and advantages.

Manual Welding Applications
- Out-of-position welding in which a large, highly fluid weld pool would spill
- Relatively short welds

JOB TIP

Behavior at Work

When at work, leave your stereotypes behind. They get in the way of good teamwork. With your fellow workers, you have a common goal to keep the company prosperous. With customers, you have a mutually profitable exchange partnership, where you provide the service and the customer provides the marketplace.

- Light and heavy gauge metals
- Nonrepetitive jobs
- Jobs that are costly or difficult to fixture
- Jobs in which fitup cannot be controlled
- Jobs in which it would be difficult to retain flux because of the shape of the work
- Jobs where obstructions of one kind or another make it impossible to make an uninterrupted weld

Manual Welding Advantages

- Can be done indoors or outdoors
- Can be done in any position and in inaccessible locations
- Can weld a wide range of alloys and dissimilar metals
- Low cost, portable equipment

Semiautomatic Applications

- Jobs in which you can take advantage of the additional meltoff provided by currents higher than those possible with manual processes
- Jobs that are repetitive enough so that a high degree of skill can be acquired
- Medium and heavy metals
- When the continuous wire feed increases the welding time (duty cycle)
- When complicated shapes or extremely large weldments make fixturing for the automatic/mechanized application too difficult
- When the penetration, which is deeper than that produced by other manual processes, is an advantage
- When the contour of the work is irregular and fitup is not accurate enough for automatic/mechanized guiding

Semiautomatic Advantages

- Produces welds of desirable appearance and high quality weld metal characteristics
- Higher welding speeds than hand welding
- Less slag and weld spatter
- Reduced costs

Mechanized and Automatic Welding Application

- Highly repetitive jobs that can be fixtured

Mechanized and Automatic Advantages

- Pushbutton arc striking
- High rate of weld metal deposition
- Welds have smooth, even appearance
- Heavier construction than manual and semiautomatic equipment
- Higher current capacity than manual or semi-automatic equipment
- Increased welding speeds
- Self-contained travel mechanism
- Reduced electrode loss
- Machine precision
- Minimum slag removal
- Reduced problem of heat distortion
- Accurate, continuous control and fewer weld rejects due to automation
- High mechanical properties of welds

CHAPTER 23 REVIEW

Multiple Choice

Choose the letter of the correct answer.

1. Which piece of equipment is not required for FCAW-S? (Obj. 23-1)
   a. Shielding gas and control unit
   b. Power source
   c. Wire-feeding mechanism and control
   d. Gun

2. FCAW can be used for code work and produces radiographic quality welds. (Obj. 23-1)
   a. True
   b. False

3. Because of the shallow penetration characteristics of FCAW, groove angles must be increased, the root opening expanded, and a smaller root face must be used to ensure CJP. (Obj. 23-1)
   a. True
   b. False

4. Which gun travel angle is used for FCAW to keep the slag from running ahead of the weld pool? (Obj. 23-2)
   a. Push
   b. Work
   c. Drag
   d. Compound

5. FCAW-G electrodes require an electrode extension of approximately how many inches? (Obj. 23-2)
   a. ¼ to ½
   b. ⅜ to 1½
   c. ¼ to 1½
   d. ¾ to 2
6. FCAW-S electrodes may require electrode extensions as great as _____. (Obj. 23-2)
   a. 1 inch
   b. 2 inches
   c. 3 inches
   d. 4 inches
7. When making a fillet (2F) weld the electrode should be pointed at the bottom plate, with one electrode diameter offset so that the weld metal will wash up on the vertical plate. (Obj. 23-3)
   a. True
   b. False
8. Which type of electrodes are best used in outdoors or in drafty areas? (Obj. 23-4)
   a. FCAW-S
   b. FCAW-G
   c. Innershield®
   d. Both a and c
9. Which flux cored electrode requires a different gun than GMAW? (Obj. 23-4)
   a. FCAW-G
   b. FCAW-S
   c. Dual Shield®
   d. FabCO®
10. Which of the following weld processes is not considered to be an arc welding process? (Objs. 23-5 and 23-6)
    a. FCAW
    b. SAW
    c. ESW
    d. EGW

Review Questions

Write the answers in your own words.

11. What is the function of the flux in the core of an FCAW electrode? (Obj. 23-1)
12. Name and explain the various welding variables used for FCAW. (Obj. 23-2)
13. Describe the gun travel and work angle, offset, polarity, and electrode extension for making a 2F weld on a T-joint with the FCAW-G electrode. (Obj. 23-3)
14. When making a groove weld with FCAW-G electrodes, describe the width and height of the weld in relation to the groove. Also what is required to ensure good root penetration? (Obj. 23-3)
15. What does the M designator mean on an E71T-1M electrode classification? (Obj. 23-3)
16. Sketch the FCAW-G and FCAW-S process schematics and label and highlight the differences. (Obj. 23-4)
17. List six functions of the flux as used for the SAW process. (Obj. 23-5)
18. Describe how the flux depth should be determined for SAW. (Obj. 23-5)
19. With all other conditions held constant with SAW, an increase in wire size will affect penetration and deposition in what way? Fully explain your answer. (Obj. 23-5)
20. Sketch the basic components of the ESW and EGW processes. Label and highlight the differences. (Obj. 23-6)

INTERNET ACTIVITIES

Internet Activity A

Use the Internet to find out about OSHA Regulations (Standards 29CFR) Part 1926 Safety and Health Regulations for Construction, Subpart J Welding and Cutting Part 1926.353 Ventilation and protection in welding, cutting, and heating, note (a) (4).

Internet Activity B

Use your favorite search engine to find out about the U.S. DOT and use of ESW on bridges.

Design credits: Cinema and movie icon: Denis Meshkov/123RF; and Illustration of Welding icons: Mr. Nachai Sorasee/123RF

Table 23-10 Job Outline: Flux Cored Arc Welding Practice with Gas Shielded Electrodes (Plate)

Recommended Job Order	Material	Plate Thickness (in.)	Joint	Weld	Weld Position	No. of Passes	Electrode[1] Type AWS	Electrode[1] Size (in.)	Shielding Gas[2] Type	Shielding Gas[2] Gas Flow (ft³/h)	Welding Current Volts	Welding Current Amperes	Welding Current Wire-Feed Speed (WFS) (in./min)	Electrode Extension (in.)
1st	Steel	1/4	Flat plate	Surfacing	Flat (1C)	Cover plate	E71T-1	1/16	Carbon dioxide	30–40	25–30	310–365	275–375	1
2nd	Steel	1/4	T	Fillet	Flat (1F)	2	E71T-1	1/16	Carbon dioxide	30–40	25–30	310–365	275–375	1
3rd	Steel	1/4	Lap	Fillet	Horizontal (2F)	1	E71T-1	1/16	Carbon dioxide	30–40	23–30	310–365	275–375	1
4th	Steel	3/8	T	Fillet	Horizontal (2F)	6	E71T-1	1/16	Carbon dioxide	35–40	23–30	310–365	275–375	1
5th	Steel	3/16	Butt	Square-groove with backing	Flat (1G)	1	E71T-1	1/16	Carbon dioxide	30–40	23–30	310–365	275–375	1
6th	Steel	3/8	Butt	V-groove with backing	Flat (1G)	2	E71T-1	1/16	Carbon dioxide	35–40	23–30	310–365	275–375	1
7th	Steel	3/8	T	Fillet	Vertical (3F)	2	E71T-1	0.045	Carbon dioxide	35–40	25–27	180–220	275–340	3/4
8th	Steel	1/2	Butt	V-groove with backing	Vertical (3G)	3	E71T-1	0.045	Carbon dioxide	35–40	25–27	180–220	275–340	3/4
9th	Steel	1/2	Butt	V-groove with backing	2G	6	E71T-1	0.045	Carbon dioxide	35–45	26–30	220–275	340–500	3/4
10th	Steel	1/2	T	Fillet	4F	3	E71T-1	0.045	Carbon dioxide	35–45	25–27	180–220	275–340	3/4
11th	Steel	1/2	Butt	V-groove with backing	4G	6	E71T-1	0.045	Carbon dioxide	35–45	27–28	190–220	300–340	3/4
23-JQT1[3]	Steel	3/8	Butt, lap, T	Fillet and groove	2F 3F 4F 2G 3G	As required to attain weld size	E71T-1	0.045	Carbon dioxide	35–45	25–30	180–275	275–500	3/4

Note: The conditions given here are basic and will vary with the job situation, the results desired, and the skill of the welder.

[1] Other FCAW-G electrode types and sizes may be substituted. The specific manufacturer's parameter recommendations should be used. E70T-1 electrodes can be used for the flat (1) and horizontal (2) positions.
[2] Use of 75 to 80% Argon, balance CO_2 can be used with the E71T-1M electrodes, for less smoke, spatter, and a very smooth stable arc. Penetration will be reduced from 100% CO_2. Better weld pool control will be provided for out-of-position welding.

Table 23-11 Job Outline: Flux Cored Arc Welding Practice with Self-Shielded Electrodes

Recommended Job Order	Material	Plate Thickness	Joint	Weld	Weld Position	No. of Passes	Electrode[1] Type	Electrode[1] Size (in.)	Welding Current DCEN Arc Volts	Welding Current DCEN Amperes	Welding Current DCEN Wire-Feed Speed (in./min)	Electrode Extension (in.)
1	Steel	3/16 in.	Flat plate	Surfacing	1C	Cover plate	E71T-11	0.068	18–20	190–270	75–130	3/4
2	Steel	3/16 in.	Butt	Square groove with backing	1G	1	E71T	0.068	15–18	125–190	40–75	5/8
3	Steel	3/16 in.	T	Fillet	2F	1	E71T	0.068	18–20	190–270	75–130	5/8
4	Steel	10 gauge	T	Fillet	2F	1	E71T	0.068	15–18	125–190	40–75	3/4
5	Steel	1/4 in.	T	Fillet	3F	1	E71T	0.045	15–18	140–160	90–110	3/8
6	Steel	3/16 in.	Lap	Fillet	1F	1	E71T	0.068	18–20	190–270	75–130	3/4
7	Steel	10 gauge	Lap	Fillet	1F	1	E71T	0.068	18–18	125–190	40–75	3/4
8	Steel	1/4 in.	T	Fillet	4F	1	E71T	0.045	16–17	140–160	90–110	3/8
9	Steel	3/8 in.	Butt	V-groove with backing	1G	2	E71T	0.068	20–23	270–300	130–175	3/4
10	Steel	1/2 in.	Butt	V-groove with backing	2G	1 2–3 4–6	E71T	0.045	15–16 16–17 17–18	120–140 140–160 160–170	70–90 90–110 110–130	3/8
11	Steel	1/2 in.	Butt	V-groove with backing	3G	1–3	E71T	0.045	17–18	160–170	110–130	3/8
12	Steel	1/2 in.	Butt	V-groove with backing	4G	1 2–3 4–6	E71T	0.045	15–16 16–17 17–18	124–140 140–160 160–170	70–90 90–110 110–130	3/8
23-JQT2[2]	Steel	3/8 in.	Butt, lap, T	Grooves and fillets	2F 3F 4F 2G 3G	As required to attain weld size	E71T	0.045	16–18	140–170	90–130	

[1] Other FCAW-S electrode types and sizes may be substituted. Always use the specific manufacturer's parameter recommendations. Because of the unique nature of the self-shielded electrode the Lincoln Innershield NR-211MP was used for the development of this table.

Table 23-12 Job Outline: Semiautomatic Application of the Submerged Arc Welding Process

Recommended Job Order	Material	Plate Thickness (in.)	Joint	Weld	Weld Position	No. of Passes	Electrode[1]/flux Type	Size (in.)	Welding Current DCEN Arc Volts	Amperes	Electrode Extension (in.)
1	Steel	3/8	T	Fillet	1F	1	FA6-EM13K	1/16	37	400	1–2

24

Gas Metal Arc Welding Practice:
Jobs 24-J1–J15 (Pipe)

This chapter deals only with the techniques for gas metal arc welding standard (schedule 40) and heavy (schedule 80) wall carbon steel pipe.

Industrial Applications of GMAW Pipe Welding

Spurred on by the rapid growth of nuclear power, space and rocket exploration, marine construction, and the chemical, oil, and gas industries, welded-pipe fabrication is increasing at a tremendous rate. Over 90 percent of all steel piping installations are welded. There are over one million miles of pipeline in this country. Over 10 percent of the steel produced in the United States is used in the production of pipe. The Hoover Dam piping installations required pipe 30 feet in diameter with a wall thickness up to 2¾ inches.

There is a considerable increase in the use of the gas metal arc process in the field for power piping, process piping, and construction (Figs. 24-1 and 24-2) and in shop fabrication (Figs. 24-3 and 24-4). Much of this pipe welding is being done with the *GMAW-S* and *GMAW-P processes*. The practice jobs in this chapter specify techniques for carbon steel pipe. Pipe is available in such ferrous metals as low and high strength carbon steels, carbon-molybdenum steels, chromium-molybdenum steels, low temperature steels, stainless steels, and clad and lined

Chapter Objectives

After completing this chapter, you will be able to:

24-1 Explain the use of the gas metal arc welding process on piping structures and systems.

24-2 Describe the gas metal arc equipment used for pipe welding.

24-3 Demonstrate understanding of the gas metal arc welding operations for pipe welding.

24-4 Demonstrate understanding and ability of gas metal arc welding by producing satisfactorily welded joints.

Fig. 24-1 Gas metal arc welding with the short circuit mode of metal transfer on a 5G position V-groove weld on a pipe butt joint for the natural gas industry. This highly skilled journeyman has been trained and skilled in a multitude of welding and cutting processes. The feeder is enclosed to protect it from the elements. The engine driven generator power source is located some distance away. Note the type of conditions that are typically encountered when doing welding in this field. © United Association

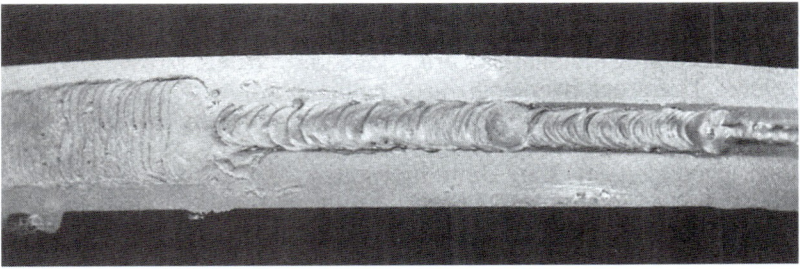

Fig. 24-2 Typical groove welded pipe butt joint used in piping construction. It was welded with the GMAW process—uphill. Downhill procedures are also used. Edward R. Bohnart

Fig. 24-3 Welding carbon steel pipe in the shop with the GMAW short circuit mode of metal transfer on the root pass. Note the arc visible on the inside of the pipe. This ensures that complete penetration is taking place. A pipe gripper is attached to a turning fixture that is rotating the pipe. Miller Electric Mfg. Co.

steels. Pipe is also available in such nonferrous metals as aluminum, nickel, copper, titanium, and their alloys.

Welding may be done in all positions. The direction of welding for vertical welds may be up or down. The usual practice is to weld the first pass down followed by passes that may be welded up or down.

It is not difficult to obtain radiographic quality welds. The MIG/MAG pipe welding process is used extensively on both noncritical and critical piping in which the weld must meet requirements of the codes for such critical applications as nuclear piping, steam power plant piping, and chemical process piping.

It is assumed that you are already skilled in the welding of pipe with the shielded metal arc process and the gas tungsten arc process. You are urged to take the utmost care in the practice of these jobs. Gas metal arc pipe welding is a growing field and offers many opportunities for those who develop a high degree of skill in this welding process.

Advantages

The piping industry is one of the most progressive of all industries. The installation of piping is costly, and for this reason the industry is alert to new developments in the fabrication of pipe. The gas metal arc welding of pipe was first studied in the laboratory and then tested in field experiments. The results led to the use of MIG/MAG as a fabrication tool in the welding of pipe. The tests and the production experiences indicated the following desirable advantages:

- The process is fast and in many instances faster than other welding processes.
- The elimination of flux and slag reduces the cleaning time considerably.
- Heavier passes can be made, thus cutting down the number of passes per joint.
- Since the electrode is continuous, fewer starts and stops are necessary, and there is no electrode stub loss.
- The process can be used on all pipe sizes and pipe wall thicknesses.
- Weld metal is high quality and meets requirements of most codes.
- It is considered a low hydrogen process.
- Weld appearance is good.
- Good penetration, fusion, and a smooth weld bead can be produced inside the pipe on the underside of the root pass.

- Backup rings may not be required.
- Because of the concentration of heat when welding, distortion and warpage are reduced.

The root pass is thicker and stronger than that produced by the TIG and stick electrode process so that the danger of first pass cracking is reduced.

Use of Equipment and Supplies

Equipment

The equipment needed for the gas metal arc welding of pipe does not differ in any way from that required for other forms of MIG/MAG welding. Review the procedures used in setting up and starting the equipment in Chapter 22, pages 725–726. Your procedure will vary depending on whether the welding machine is an engine-driven generator, a d.c. rectifier, or an inverter. Refer to Fig. 24-5 for the equipment in the MIG/MAG system. It includes the following:

- D.C. power source, constant voltage type (engine-driven generator, inverter, or rectifier)
- Wire-feeding mechanism with controls, spooled electrode wire, and spool support

Fig. 24-4 Welding carbon steel pipe in the shop with GMAW-S process. The pipe is being rolled with a pipe positioner so the weld is being done 1G. Note arc and complete penetration weld taking place on this pipe to flange single V-groove weld on this butt joint.
Location: Piping Systems, Inc. Mark A. Dierker/McGraw Hill

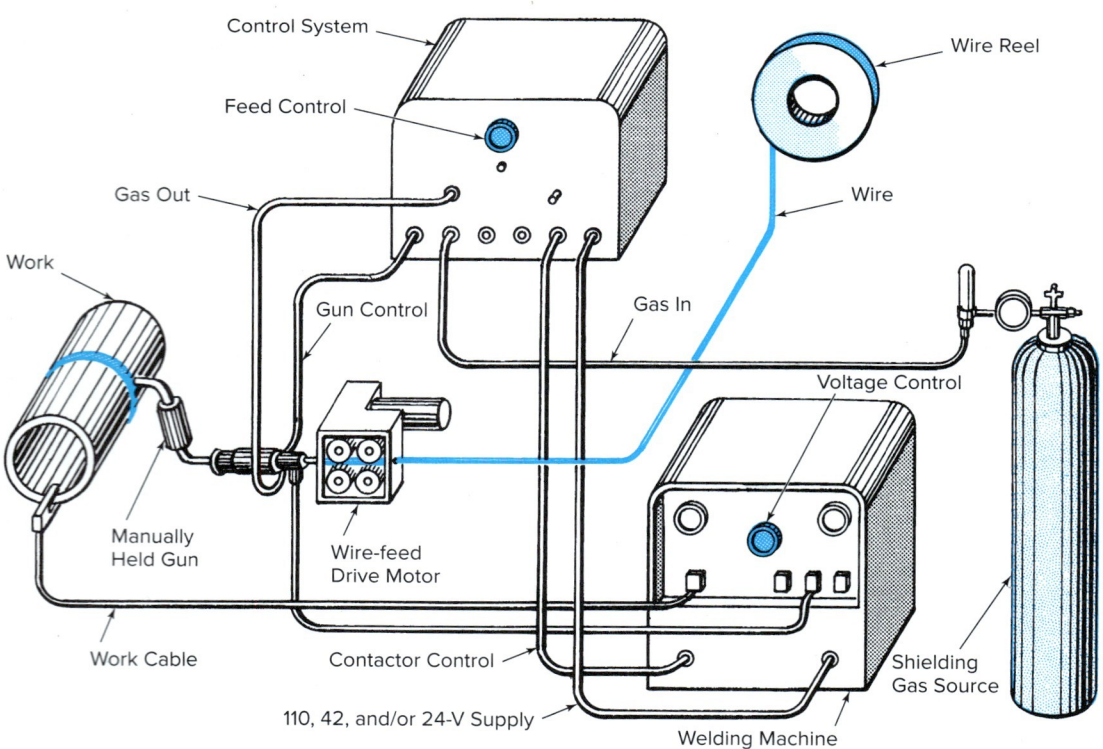

Fig. 24-5 Schematic diagram of the GMAW system for welding pipe.

- Gas shielding system: one or more cylinders of carbon dioxide (CO_2) or a mixture of 75 percent argon and 25 percent carbon dioxide, whichever is used, and a combination pressure-reducing regulator-flowmeter assembly
- Welding gun and cable assembly
- Connecting hoses and cables
- Face helmet, gloves, protective clothing, and hand tools

The engine-driven generator is suitable for outdoor welding applications such as pipelines and bridge and building construction, and in locations where easy access to electric power is not available. These welding generators are capable of generating 120/240 volts auxiliary power for running power tools and lighting.

The inverter power sources are portable and have enhanced arc performance to allow for all modes of transfer including short circuiting, spray, and pulse spray. In many cases, they have inductor controls and with microprocessor technology have the ability to shape their output waveform. They can have stored programs, which ensures that many of the essential variables are being followed per the welding procedure.

The d.c. rectifier, Fig. 24-6, is found in fabricating shops where the input power supply fluctuates very little. It is quieter than the other power sources, and it draws current only during welding.

It is an advantage to use a machine with variable slope control and variable inductance control. It should also be equipped with a "hot start" feature, which enables the welder to select a higher voltage for a timed period at the beginning of a pass or at a new start. A hot start ensures good bead contour and the elimination of discontinuities caused by a cold start. All other equipment is the same as that used for welding plate with the gas metal arc process.

In some cases instead of a hot start in the power source, the wire feeder will be equipped with enhanced starting characteristics, such as a slow start. When the gun trigger is pulled, the wire-feed speed is momentarily slowed to allow for a more positive start, without producing overlap, excess convexity, or a cold start. In other cases the feeder may be equipped with a touch start. This circuit detects when the electric circuit is completed for welding. In this case when the gun trigger is pulled, the power source output energizes, the gas solenoid valve opens, but the wire does not feed until the wire comes in contact with the work. Once the electrode contacts the work, the circuit is completed, the arc is started, and wire begins to feed. This momentary lack of filler metal being added gives the arc time to preheat the metal and make it ready to accept

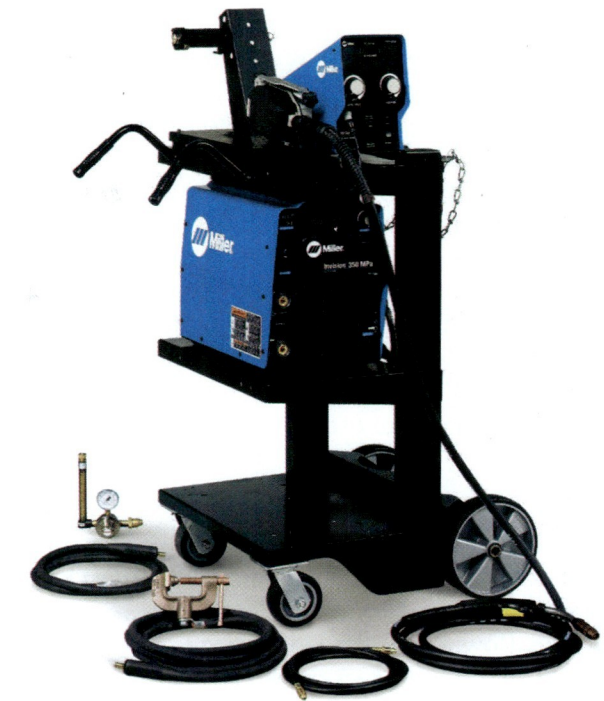

Fig. 24-6 A 300 amp d.c. inverter power source, constant speed wire feeder/controller, and shielding gas cylinder. This setup is lightweight and very portable. It is capable of running off of a variety of primary voltages and with either single or three-phase power. Miller Electric Mfg. Co.

the weld metal. Starts must be crisp and positive for high quality pipe welding. Even with positive starts, any excess convexity must be removed by some mechanical means such as grinding or filing, prior to depositing weld metal on them.

The welding gun may be equipped with special nozzles and contact tubes. Because of the groove welds typically done on piping systems, access to the root of the joint is critical. Special tapered contact tubes and nozzles greatly aid in getting complete fusion. In many cases the contact tube will be flush or slightly protruding beyond the end of the gas nozzle. This ensures that the proper electrode extension can be maintained while allowing the welder a view of the weld pool.

Shielding Gas

For welding carbon steel with the GMAW-S mode of metal transfer, argon with a balance of carbon dioxide is most common. The 75 percent argon and 25 percent CO_2 or 80 percent argon and 20 percent CO_2 are most common. When the GMAW-P mode of metal transfer is used, 90 percent argon and 10 percent CO_2 is popular. Another mix that is also used is 98 percent argon and 2 percent oxygen. Remember it takes at least 80 percent argon in the shielding gas for a spray (conventional or pulse mode) to take place. When welding stainless steel with the GMAW-S mode of metal transfer, a mix of three separate gases is used. This is often referred to as a "tri-mix" and consists of 90 percent helium, 7½ percent argon, and 2½ percent CO_2. While pulse spray arcs on stainless steel can be done with the same mixes as used for carbon steel, high percentages of reactive gases like CO_2 and oxygen should be avoided. For the chrome-moly pipe, mixtures of CO_2 and helium are used. Gas metal arc welding with a silicon bronze filler metal (GMAW-B) on galvanized pipe is done with 100 percent argon. For aluminum, argon or argon-helium mixes are used.

Gas Flow Rate

Be sure that you have the correct regulator and flowmeter. The calibrations of such equipment are designed for specific gases.

The gas flow rate is the key to good gas shielding. A rate of 15 to 25 cubic feet per hour is adequate in most indoor welding situations. When welding on the construction site or inside the shop with the doors open, drafts may disturb the gas shielding. Increasing the gas flow rate by 5 to 10 cubic feet per hour may have a slightly beneficial effect, but generally it is necessary to erect draft shields made from canvas or like material. In some codes, wind speeds over 5 miles per hour require draft shields.

Filler Wire

Pipe is usually welded with E70S-3 or E80S-B2 filler wire. A filler wire diameter of 0.035 inch gives the best results. It enables welders to control the pool better than other sizes do and gives them more leeway in torch manipulation. Adjustment of the power supply is easy. An added advantage of the 0.035-inch diameter wire is that it costs somewhat less than 0.030-inch diameter wire. If welders wish to use 0.030-inch diameter wire, they may use the same welding procedures, but must increase the wire-feed speed to obtain the same level of welding current. Filler wires of ³⁄₆₄-inch diameter are available, but they are not recommended for all-position short-arc pipe welding. The weld pool tends to be very large and fluid, and it is hard to manage so that it is difficult to do code quality work.

Welding Operations

Edge Preparation

The surface of the pipe end must be smooth, and the edge must be square with the pipe length. A poorly prepared pipe end can be the cause of an unsatisfactory weld.

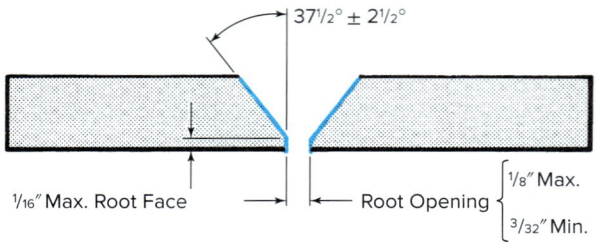

Fig. 24-7 Detail of edge preparation and fitup for ASME code work.

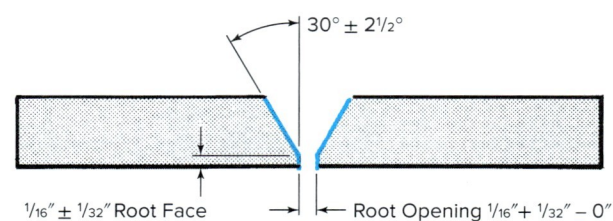

Fig. 24-8 Detail of edge preparation and fitup for API code work.

A bevel of 37½° is widely used in industry for ASME Boiler Code welding, Fig. 24-7. The American Petroleum Institute (API) code recommends a 30° bevel, but a 37½° bevel is sometimes used, Fig. 24-8.

Beveling may be done by any suitable means such as machining, oxyacetylene cutting, abrasive wheel cutting, or cutting on a lathe. Care must be taken with the root face dimension. If the root face is greater than ¹⁄₁₆ inch, adequate penetration cannot be obtained. Some welders like to use a sharp V-joint (thus a root edge) and increase the root opening. You will recall from your past practice that the root bead is the most important bead and that proper edge preparation is necessary to ensure good root beads.

It is important that the beveled edges and the pipe within 1 to 3 inches of the joint be free of oil, paint, scale, rust, and any other foreign material that could cause porosity and other contaminants in the weld.

Fitup and Tacking

In setting up the weld joint for work according to the ASME code, make sure that the root opening is no less than ³⁄₃₂ inch and no more than ⅛ inch to ensure adequate penetration, Fig. 24-7. On pipeline work according to the API code, it is common to use higher currents than in the shop. A root opening of ¹⁄₁₆ to ³⁄₃₂ inch is used in order to limit excessive melt-through on the inside of the pipe, Fig. 24-8.

Tack welds should be a minimum of ¾ inch long and preferably about 1 inch long. For highly critical work welded in the shop, gas tungsten arc welding is often used for tacking and may be used for the root pass.

Both ends for all tacks should be "feathered" by grinding as shown in Fig. 24-9. There are two important reasons for feathering: (1) to remove any possible defects in the ends of the tack, and (2) to reduce the mass of metal at the ends of the tack and, therefore, ensure good fusion of the root bead to the tack. Four tacks are all that are necessary for diameters up to 12 inches.

Some welding procedures require a backing ring to be used. In this case the root opening should be not less than 3/16 inch, and the root edge should be prepared as shown in Fig. 24-10. This provides for excellent fusion of the root bead to the backing ring.

Interpass Cleaning

You will recall from your previous welding practice that it is very important to clean between each pass in multipass welding. Although there will not be a heavy flux deposit to remove, there will be a little dust and a black glassy material in spots on the surface of the bead. This material should be removed with a power wire brush. If the materials are not removed, they will become entrapped in the weld, causing porosity, and the arc will be unstable.

With either vertical uphill or vertical downhill progression techniques, the root pass will generally require a good deal of grinding. When the root pass is being made, little attention is paid to its weld face contour. What is being focused on is that good root penetration is occurring on the inside of the pipe. The root pass must also have good fusion into the groove face as well. Any weld face or weld toe discontinuities should be removed by grinding or filing.

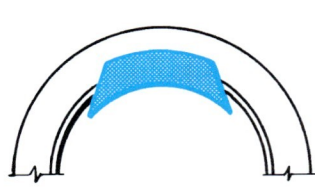

Fig. 24-9 Detail of pipe showing "feathered" tack.

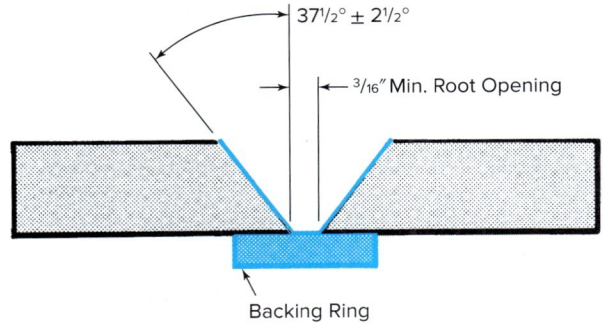

Fig. 24-10 Edge preparation and fitup if backing ring is used.

Practice Jobs

Instructions for Completing Practice Jobs

Your instructor will assign appropriate practice in the gas metal arc welding of pipe from the jobs listed in the Job Outline, Table 24-2, pages 813–814. Before you begin a job, study the specifications given in the Job Outline. Then turn to the pages indicated in the Text Reference column for that job and study the welding technique described.

Note that the specifications are basic. Specific materials and techniques vary with the job situation, the results desired, and the skill of the welder.

Study Table 24-1, (p. 796), carefully before you begin welding practice. Look up your discontinuities in the table during practice and correct them as instructed before beginning the next job.

Beading Practice: Jobs 24-J1 and J2

Beading practice around the outside of the pipe is for the purpose of getting used to changing the position of the gun while following the contour around the pipe. The direction of travel should be both downhill and uphill.

Practice with various electrode extension distances. With an electrode extension of about ¼ inch, you will notice that the penetration is deep and that you may even melt through the wall of the pipe. With an electrode extension distance of ½ to 5/8 inch, you will notice that the penetration is not as deep. A longer extension will also bridge a gap with less melt-through.

Take a piece of pipe of the size specified in the Job Outline. Hold your gun at an angle of 20 to 25° in the direction of travel (drag angle), Fig. 24-11, page 796, and make a stringer bead. Start at the 12 o'clock position and weld downhill to the 3 o'clock position. Reposition the pipe by turning it counterclockwise and continue this until the bead is joined at the starting point. Move over about ¼ to 3/8 inch and make a second bead. Then make a weave pass between the stringer beads, Fig. 24-12, page 796. Weave the gun and pause at the side of each bead to permit fusion to take place. This pass is like the second and third passes in a groove weld or a cap pass.

Repeat this procedure, but weld uphill for both the stringer and weaved passes. When welding uphill, hold

Table 24-1 Troubleshooting for the GMAW Process

Weld Defects	Causes	Remedies
1. Convex bead	Arc voltage too low	Raise voltage
	Electrode extension too long	Shorten electrode extension
	Oscillation too narrow	Widen oscillation
2. Gross porosity	Breeze blowing shielding gas away	Erect wind shields
3. Scattered porosity	Failure to remove large islands of silicon "glass"	Use power wire brush to remove "glass" Clean prior to welding
	Oil or other foreign material on pipe or electrode	
4. Wagon tracks	Failure to remove "glass" from edges of previous pass	Remove "glass"
	Convex root or fill pass	Raise voltage or widen oscillation
5. Undercutting in overhead position	Voltage too high	Lower voltage
	Insufficient dwell at edge of bead	Increase dwell
	Welding current too high	Reduce wire-feed speed
6. Overlapping	Welding current and thus disposition rate too high	Lower wire-feed speed
	Welding speed too low	Increase forward speed
	Arc not on leading edge of pool	Keep arc ahead of pool on base metal
7. Excess melt-through in root	Root opening too wide	Reduce root opening
		Oscillate torch
		Increase drag angle
8. Incomplete fusion in root	Root opening too narrow	Increase root opening
	Root face too thick	Decrease root face thickness
	Arc voltage too high	Decrease voltage
9. Incomplete penetration or "suckback" in overhead position	Travel speed too slow	Speed up travel
	Root opening too wide	Decrease root opening
10. Incomplete fusion to ends of tacks	Root opening too narrow	Increase root opening
	Tacks not adequately feathered	Feather tacks properly
11. Unstable arc	"Glass" left on previous pass	Remove "glass"
	Voltage too high or too low for amperage	Use correct voltage for wire-feed speed
	Contact tip clogged	Renew contact tip
12. Erratic start	Hot start voltage too high	Use a hot start voltage of 2 V over welding voltage for about 1 s

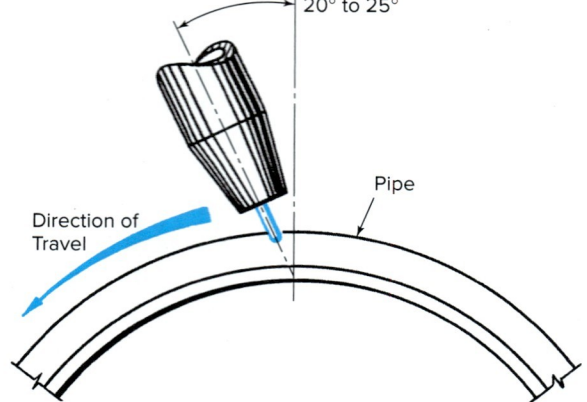

Fig. 24-11 Position of gun for surfacing or butt joint groove welding around the pipe downhill or the pipe can be rolled.

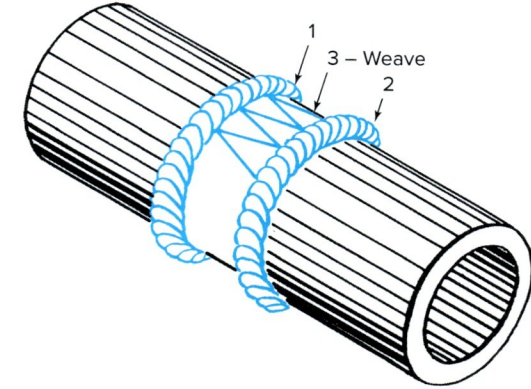

Fig. 24-12 Stringer and weave beads around the outside surface of a pipe welded in position number 1.

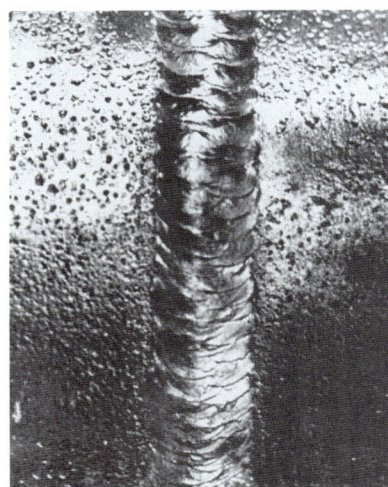

Fig. 24-13 Sample of a weaved finish pass on a butt joint in pipe welded in the roll position. © Plumbers and Pipefitters Union, Alton, IL

the gun nozzle at a 90° angle to the pipe surface. Start at about the 2 o'clock position and weld to just beyond the 12 o'clock position. Here again, experiment with various electrode extensions and with various voltage and amperage settings. Be very careful that the weld metal does not pile up and run off.

Compare the beads with those shown in Fig. 24-13 and inspect them carefully for the usual discontinuities. Pay particular attention to weld contour, incomplete fusion, spatter, undercut, and surface porosity.

Butt Joint

The butt joint is the most commonly used pipe joint in welded pipe systems. In the field the welder is called upon to weld this joint in all positions, such as 2G, 5G, and 6G.

Pressure piping usually requires heavy wall pipe. The bevel angle is 37½°, and the direction of travel is uphill. Cross-country and distribution piping usually has thinner walls. The bevel angle is 30°, and the direction of travel is downhill. As in the previous pipe practice, the beveling may be done by flame-cutting or machine cutting.

Job 24-J3 will be the welding of a butt joint in pipe in the horizontal roll position (1G), travel down, from the 12 o'clock to the 3 o'clock position.

Horizontal Pipe Axis Roll Position: Job 24-J3

Root Pass Select two lengths of pipe of the size and weight specified in the Job Outline. Set up two nipples with the pipe axis vertical as shown in Fig. 24-14 and tack weld at four equally spaced places (the 12, 6, 3, and 9 o'clock positions in that order). Make sure that the pipe remains equally spaced all the way around. Also make sure that tack welds have good penetration and are not too heavy. Feather them by grinding, filing, or chiseling until they are just thick enough to hold the two pipe nipples together as the joint is being welded. See Fig. 24-9, page 795.

Place the pipe axis horizontally and weld the root pass with the stringer bead technique. The gun should always be on the weld and the motion is U-shaped. Start at the 12 o'clock position and weld downhill to the 3 o'clock position. At this point, stop and rotate the pipe in a counterclockwise direction until the weld crater is again in the 12 o'clock position. Be certain to grind the crater and feather it. Clip the end of the wire to remove any ball and to get the proper electrode extension. Start the arc back up on the prior root bead ¼ to ½ inch. Let the arc stabilize, and then move down into and fuse into the feathered crater area and continue with the root pass. Then repeat the welding procedure. The electrode extension should be between ¼ and ⅜ inch. At certain times it may be as long as ⅝ inch because of the 45° drag angle. The stringer bead should be no more than ⅛ inch thick. The root opening often varies. If the opening is wider than 3/32 inch, it may be necessary to weave the gun slightly, Fig. 24-15.

Be careful when welding over a tack. The tack must be completely fused

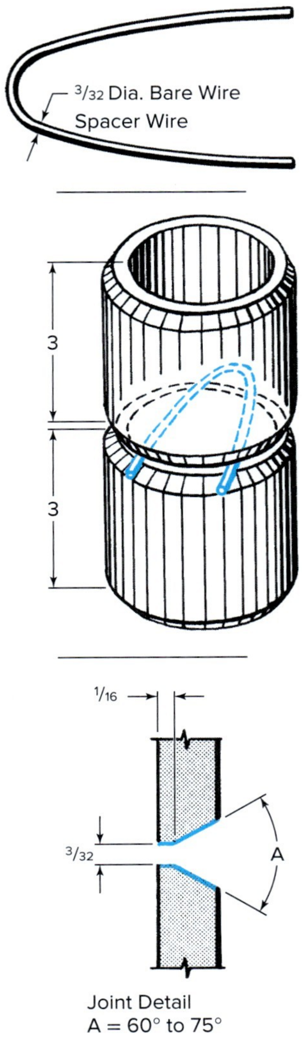

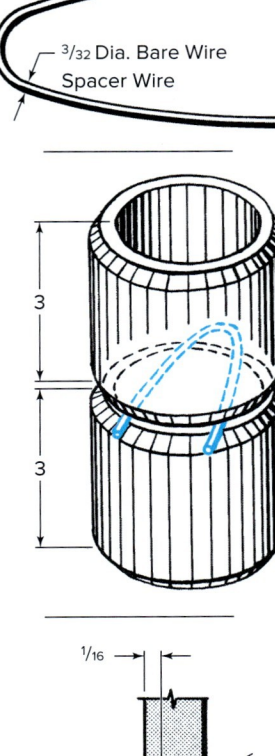

Fig. 24-14 Setting up and spacing the pipe before tack welding. The diameter of the spacing wire determines the root opening.

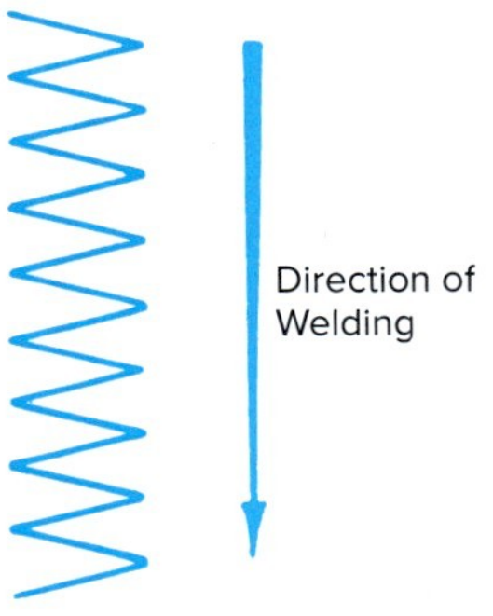

Fig. 24-15 Suggested pattern for torch oscillation for root pass welding downhill.

Gas Metal Arc Welding Practice: Jobs 24-J1–J15 (Pipe)   Chapter 24

and become a part of the weld bead. Many welders have failed qualification tests because of their failure to completely remelt the tack. As you approach the leading edge of the tack, position the gun about 20 to 25° from the perpendicular, Fig. 24-16. As you leave the tack, position the gun about 45 to 55° from the perpendicular, Fig. 24-17.

Each time the weld is restarted, it is necessary to "tie-in" with the previously deposited bead carefully. As you move up into the pool, let the metal wash up against the sides of the groove, and then move downhill. Also be sure to reposition the pipe when stopping the weld. A stop in welding a root pass in an open joint can cause shrinkage, cracks, cavities, and cratering. The weld should be carried slowly ¼ to ⅜ inch onto the pipe bevel wall, Fig. 24-18.

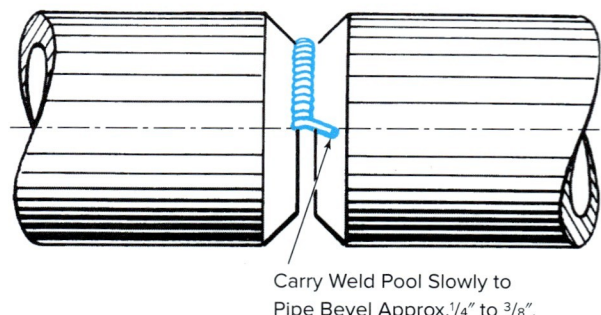

Carry Weld Pool Slowly to Pipe Bevel Approx. ¼" to ⅜".

Fig. 24-18 Recommended technique for eliminating crater cracks when welding is stopped in an open joint.

However, the crater area must be feathered before a "tie-in" is attempted.

Inspect the weld carefully for penetration through the back side (Fig. 24-19) and fusion to the side walls. Grind the weld face to make sure that it is flat. Grind the toes to eliminate any undercut or incomplete fusion. The surface of the weld should be flat, Fig. 24-20. Brush the weld before applying the next pass.

Filler Pass Start the second pass about 2 inches past the original starting point for the root pass. Begin welding by moving your gun from side to side, Fig. 24-21. Pause briefly at each edge of the root pass in order to permit the

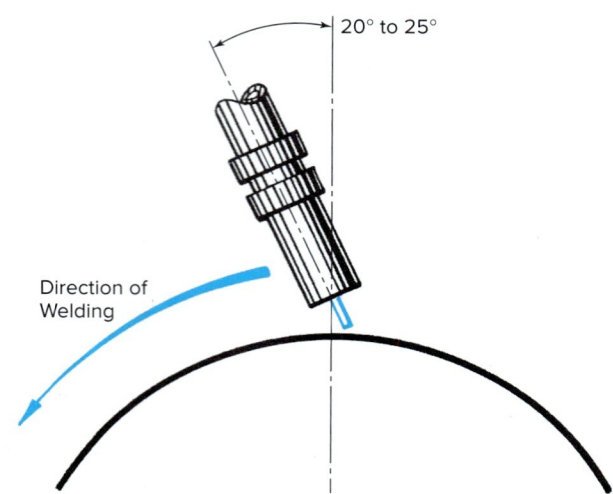

Fig. 24-16 Gun position at the start on the tack.

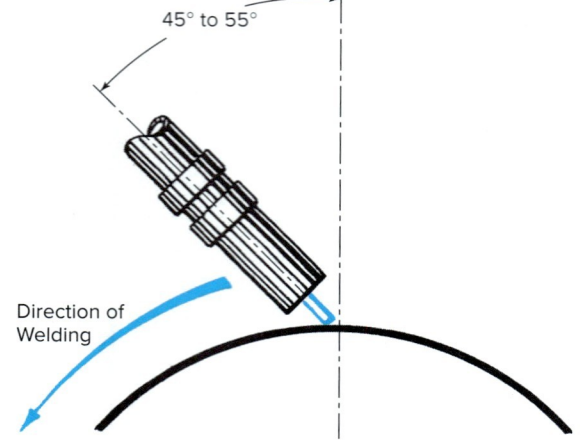

Fig. 24-17 Gun position after leaving the tack at the beginning of the root pass.

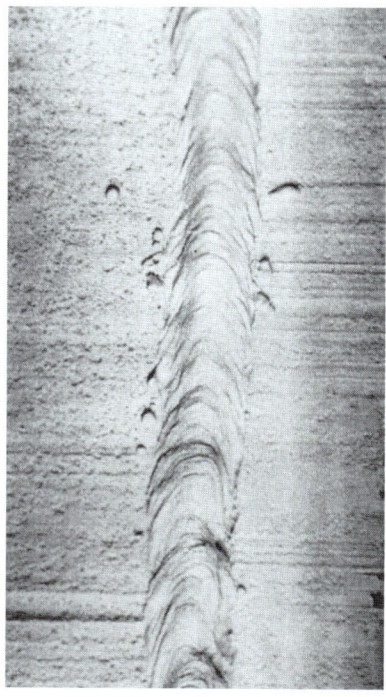

Fig. 24-19 Root penetration through the back side of an open butt joint welded with the downhill technique. © Plumbers and Pipefitters Union, Alton, IL

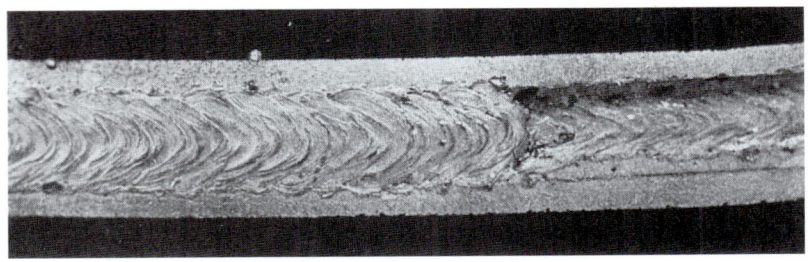

Fig. 24-20 Typical face appearance of root and cover passes in an open butt joint in pipe welded with the downhill technique. Note the flat face. © Plumbers and Pipefitters Union, Alton, IL

weld metal to fill up the groove and obtain proper fusion. Do not depend on the weld metal to wash up on the sides of the groove. This may result in an overlap. You must direct the filler wire where you want complete fusion without undercut. As with the root pass, weld from the 12 to the 3 o'clock position before rotating the pipe. Repeat your procedures for making good tie-ins and stops. Inspect the pass and clean it.

Cover Pass Make the cover pass like the filler pass. Be sure to pause briefly at each edge of the filler pass to obtain proper fusion. The bevel edges on each side should be your guide to determine the final width of the cover pass, Fig. 24-22. The bead should be about 1/16 inch at the crown and should taper out to the edges of the bead.

Inspect the cover pass for smooth tie-in, lack of undercut, good fusion, good appearance, and contour, Fig. 24-20.

Inspection and Testing Weld a butt joint, following the same procedures that you have used in completing your practice. Inspect the weld for visual defects and cut the usual face- and root-bend test coupons from four positions on the pipe: top, bottom, and each side. After preparing the test coupons, bend them on the standard test jig. The welds should show no separations of any kind.

Vertical Pipe Axis Fixed Position (2G): Jobs 24-J4 and J5 The pipe axis is fastened in the vertical fixed position, and the weld is horizontal (2G). This position is found mostly in power piping and construction installations. Very little cross-country transmission line piping is welded in this position.

Select two pieces of pipe 4 to 6 inches long of the size and weight specified in the Job Outline. Bevel the edges to 30 or 37½° as directed by your instructor and leave a 1/16-inch root face on each bevel. Refer to Figs. 24-7 and 24-8, page 794. Tack weld the two nipples together as shown in Fig. 24-23. Place the tack-welded nipples in a welding fixture with the pipe axis vertical.

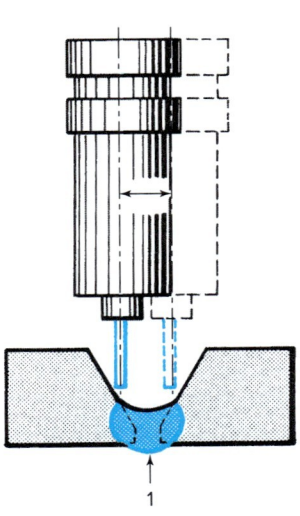

Fig. 24-21 Detail of a joint with the root pass in place, showing the correct width of the gun movement for a filler pass.

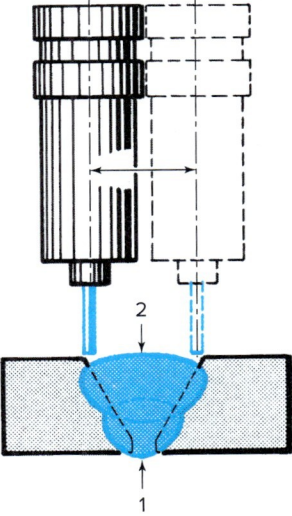

Fig. 24-22 Detail of a joint with the final fill pass in place, showing its correct relationship to the outside diameter of the pipe. The correct width of the torch movement for the cover pass is also shown.

Fig. 24-23 Tacking 2G position joint with appropriate bent filler wire to hold root opening. It is being held in place by a shop build pipe stand. A small surface is used for initial set up like high low alignment. It can then be clamped into place for positions 2, 5, and 6 as required. The height can also be adjusted. Location: UA Local 400 Mark A. Dierker/McGraw Hill

Root Pass Weld the root pass with the stringer bead technique. Start welding at a properly feathered tack. Hold the gun travel angle at a 10 to 15° angle from a point perpendicular to the center of the pipe and use a drag angle, Fig. 24-24A. It is also important to lower your gun 5° from the 90° position, Fig. 24-24B. Carry your wire on the leading edge of the pool to get complete penetration. Make sure that fusion is taking place along the root edges of each pipe bevel, Fig. 24-25. Also make sure that the weld does not sag along the bottom bevel so that it causes overlap and incomplete fusion on the bottom edge. Also be careful not to undercut the top bevel. Penetration through the back side is necessary, Figs. 24-26 and 24-27.

Torch oscillation, if necessary, is best carried out according to the pattern illustrated in Fig. 24-28. When completing a section of the root pass and approaching a feathered tack, change the torch angle gradually over a distance of ¼ inch before the tack from the 15° push angle to a 5° drag angle. This gun position ensures a good tie-in at the tack, Fig. 24-29.

Fig. 24-25 This pipe stand allows the welder to maneuver around it to maintain proper eye contact with the leading edge of the weld pool. The pipe is positioned so that it is a good height for the body mechanics required. Along with contact tip to work distance, travel speed, travel angle, and work angle to assure complete joint penetration and weld profile. Location: UA Local 400 Mark A. Dierker/McGraw Hill

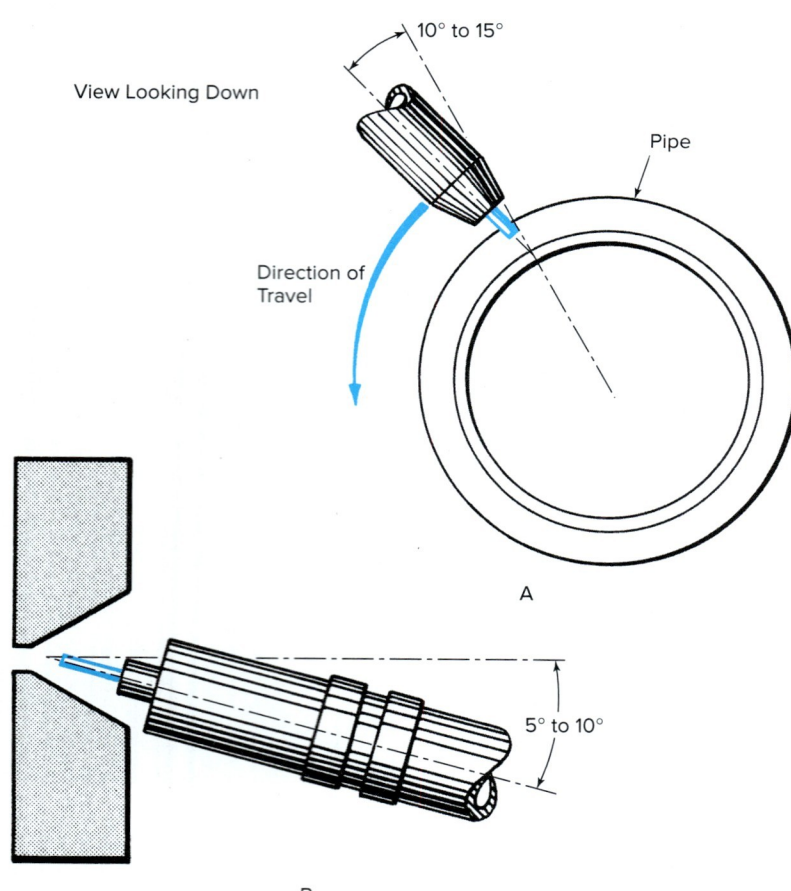

Fig. 24-24 Gun position when making a 2G weld. The pipe axis is fixed vertically. Source: Plumbers and Pipefitters Union, Alton, IL

Filler Passes The second and intermediate passes may be made with the stringer bead technique shown in Fig. 24-30, page 802, or the weave bead technique shown in Fig. 24-31, page 802. The choice of technique depends upon the requirements outlined in the welding procedure. If there is nothing specified, then welders should use a technique that they can handle with good results. Starts and stops should be outside the area of previous starts and stops. The bead sequence that may be used for Schedule 40 pipe is shown in Fig. 24-30B, and the sequence for Schedule 80 pipe is shown in Fig. 24-30C.

In making a circumferential weld around the pipe, you will have to make a number of starts and stops as required by the change of position. Tie-ins require that the electrode be started back in the heel of the crater and that the weld metal be washed up against the sides of the groove until the tie-in is made.

Cover Passes Use the stringer bead technique and about the same electrode position as described for the root pass. Make sure that the arc is directed toward the surface that is being welded. Proceed at a uniform rate of travel so that good fusion and penetration can take place with the bevel surface and the previous welds without undercut and with good bead formation. Each layer of weld metal should not be thicker than ⅛ inch. Brush each

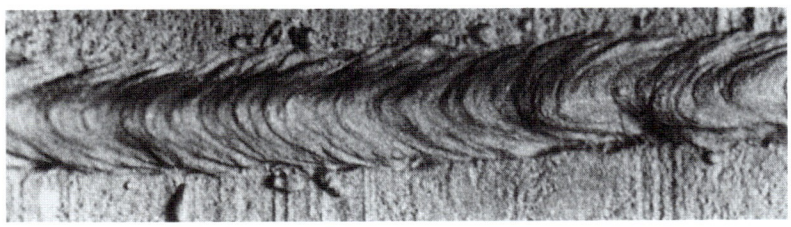

Fig. 24-26 Closeup view of root pass penetration on the inside of a pipe obtained when making a 2G weld on pipe. © Plumbers and Pipefitters Union, Alton, IL

Fig. 24-27 Inside of pipe showing penetration of the root pass in steel pipe welded with the GMAW process and a short circuiting mode of metal transfer in the 2G position. Note the absence of obstruction to flow. Edward R. Bohnart

Select two pieces of pipe 4 to 6 inches long of the size and weight specified in the Job Outline. Bevel the edges 30 or 37½° as directed by your instructor and leave a 1/16-inch root face on each bevel. Refer to Figs. 24-7 and 24-8, page 794. Tack weld the two nipples together as shown in Fig. 24-23, page 799. Place the tack-welded pipe in a welding fixture in the horizontal position.

Special care must be taken when welding in the horizontal fixed position. Before making the tack welds and welding the pipe joint, check the electrode extension carefully (¼ to ⅜ inch) and regulate the heat until the arc is smooth and active. Make sure that the pipe has a root opening of 3/32 inch all around. This root opening is most important and must be maintained. Feather the tack welds so that they do not become an obstruction when you pass over them with the root pass. Tack welds that are too thick cause an incomplete penetration and fusion so that the finished weld is not radiographic quality.

Fig. 24-28 Suggested pattern for gun manipulation in making the root pass for a 2G weld.

pass so that it is clean and free from contamination.

Inspection and Testing Inspect each pass as you complete it. The final layer of stringer beads should show good fusion with each other and be equally spaced, Figs. 24-32, page 802, and 24-33, page 803.

Use the same testing procedure as that used for the roll position (1G).

Horizontal Pipe Axis Fixed Position (5G↓), Downhill Travel: Job 24-J6 In this position, commonly referred to as the bell-hole position, the pipe axis is horizontal and is not turned as the welding progresses. The bell-hole position is encountered in all types of pipe installations. Travel may be uphill or downhill. The downhill welding technique is employed in cross-country transmission line piping, and the uphill welding technique is employed in pressure and power piping and in general construction. Either technique produces sound welds, but the downhill technique is faster. When welding downhill, some welders find it difficult to keep the weld metal from running ahead of the weld pool. You should weld several joints in this position and practice various welding techniques. Do Job 24-J6 with the downhill technique.

Root Pass The root pass should be started at the 11 or 1 o'clock position, continued across the top of the pipe, and carried down past the 6 o'clock position to the 5 or 7 o'clock position, depending on the side being welded.

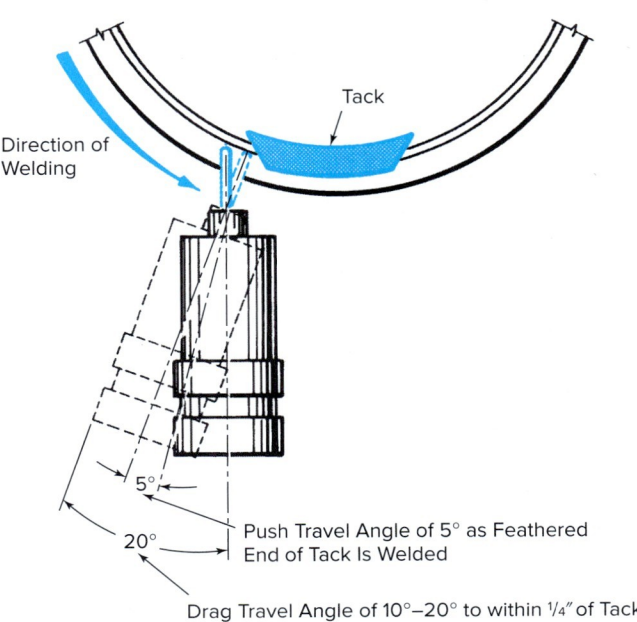

Fig. 24-29 Top view of a pipe in the 2G position, showing rotation of the torch from a drag angle of 20° to a push angle of 5° as a tack weld is approached when welding a root pass.

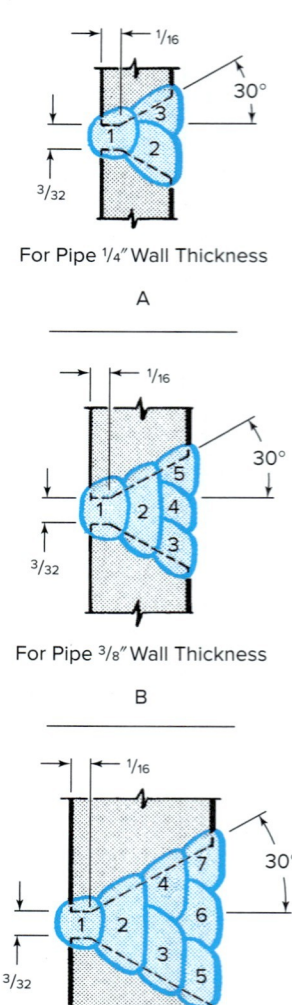

Fig. 24-30 Weld sequence for various pipe wall thicknesses welded with the GMAW-S process. The pipe axis is vertical. The weld is made in the 2G position.

A drag angle of approximately 45 to 55° to the center (axis of the pipe) should be maintained as the weld progresses past the feathered end of the tack, Fig. 24-17, page 798. It may be necessary to oscillate the gun narrowly as required during the weld, Fig. 24-15, page 797. Side movement is necessary to make sure that fusion is taking place on the root face of each pipe section.

When you reach that part of the pipe between the 4 and 6 o'clock positions, you may need to decrease the drag angle of the gun to approximately 20 to 25°. A uniform travel speed is important. If the travel speed is too slow, the base metal becomes highly fluid and sags so that there is a incomplete penetration inside the pipe. You are again cautioned to keep the thickness of each pass to about 1/8 inch.

Use care when stopping a weld and restarting a weld. If for any reason you find it necessary to stop the root pass, there is always the danger that cracks will form in the crater of the weld pool. To eliminate these cracks grind and feather the crater and start the weld just above the crater. This preheats the bead so that the crater is hot when you reach it, Fig. 24-34. This ensures a good tie-in with the weld, and there is no danger of the cracks remaining in the deposit.

After the first half of the root pass is completed, inspect it thoroughly and clean it. Return to the top of the pipe and weld the other half of the pipe to the 6 o'clock position with the same procedure as for the first half. Make the tie-in at the top and at the bottom of the pipe very carefully. These are very critical points and often cause trouble. Compare Figs. 24-35 and 24-36.

The downhill technique of welding is preferred for the root pass, but if the fitup is close, the uphill technique may be employed. Follow the procedure previously described but hold the gun at a 90° angle to the center (axis of pipe), Fig. 24-37, page 804. The welding technique is like that for welding pipe with the shielded metal arc process and plate with the gas metal arc process.

Filler Passes You are now ready to weld the filler passes. Recheck your electrode extension and heat setting. The current may have to be increased slightly.

Start welding at the top, outside of the area of the previous starting point, and travel downhill. Stop at the bottom, outside of the area of the previous stopping point, Fig. 24-38, page 804. The second pass is made with a slight weave by moving the gun from side to side and pausing briefly at each edge of the underneath pass, Fig. 24-39, page 804. Do not depend on the metal to wash up on the sides of the groove. In order to avoid overlap and to secure complete fusion, the electrode wire (arc) must be directed over the surface to be joined. Clean and inspect the weld.

To weld the second pass on the other side of the pipe, start at the top at the end of the previous weld and tie into the end of the previous weld at the bottom. Manipulate the gun as instructed for the first half of the pass. All tie-ins should be staggered so that they are not all made at the same point, one over another. The proper technique

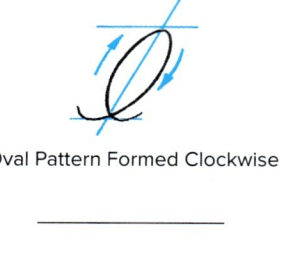

Oval Pattern Formed Clockwise

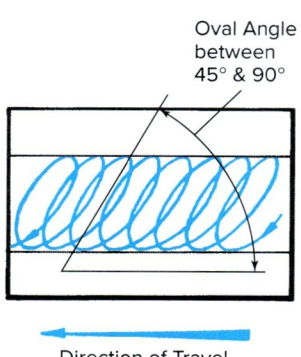

Fig. 24-31 Weaving technique in making a cover pass in the 2G position. The pipe axis is vertical.

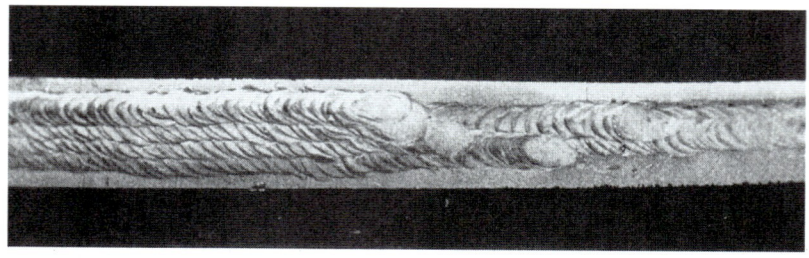

Fig. 24-32 Typical appearance of stringer beads welded in the proper sequence in the 2G position with the pipe axis vertical. The weld was made with the GMAW-S process. © Plumbers and Pipefitters Union, Alton, IL

Fig. 24-33 Closeup view of a completed sequence in a weld like that shown in Fig. 24-32. © Plumbers and Pipefitters Union, Alton, IL

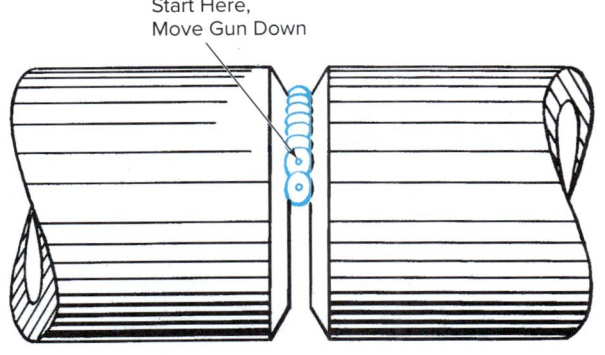

Fig. 24-34 Method of restarting the root pass to eliminate crater cracks.

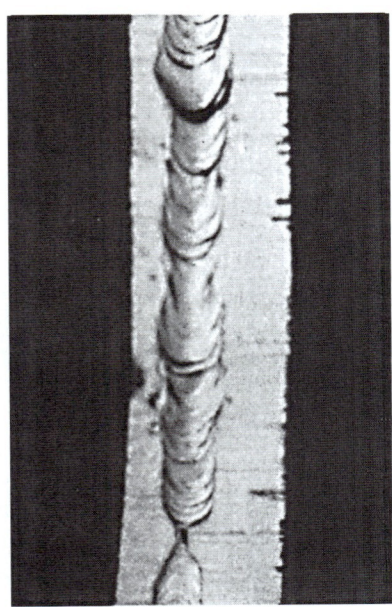

Fig. 24-35 Inside of pipe showing penetration of the root pass in the 5G position. The pipe axis is horizontal. Note that the welder failed to make the proper tie-in at the bottom of the pipe (6 o'clock). This would cause a test or service failure. © Plumbers and Pipefitters Union, Alton, IL

is shown in Fig. 24-40, page 804. Clean and inspect the weld, Fig. 24-41, page 804.

Make a third filler pass like the second pass. Clean and inspect.

Cover Pass The cover pass is welded like the second and third passes. Use the edge of each bevel as a guideline to determine the width of the weld. To prevent undercut, hesitate at the sides to permit the weld pool to fill up. The bead should have about $1/16$ inch convexity, and it should taper out to the edges of the bead.

It is important to keep the arc ahead of the pool when doing filler and cover passes downhill. Any attempt to slow down the travel speed in order to deposit more weld metal allows the molten pool to run ahead of the arc and causes overlapping or incomplete fusion. Keep the metal thin. The maximum pass width that can be handled with any degree of success is $1/2$ inch. Split layer passes are necessary beyond this width.

Practice starts and stops so that you will develop the skill to perform the technique with a minimum of problems. You will find this somewhat difficult when making filler and cover passes. Avoid the tendency to add too much weld deposit, because it results in poor fusion and very poor appearance.

Fig. 24-36 Proper penetration on the inside of a pipe root pass was welded downhill. Bottom of pipe in the 6 o'clock position on this fixed 5G practice nipple. Location: UA Local 400 Mark A. Dierker/McGraw Hill

Fig. 24-37 MIG welding pipe in the 5G position. Care must be taken to follow the contour of the pipe. The welder is just leaving the top of the pipe (12 o'clock) portion of the pipe. The root pass is being done in the downhill direction. Location: UA Local 400 Mark A. Dierker/McGraw Hill

Fig. 24-38 The gun is in the 3 o'clock position of the pipe. Some welders find the 5G position very difficult. It involves a mixing of the flat, overhead, and vertical positions. Location: UA Local 400 Mark A. Dierker/McGraw Hill Education

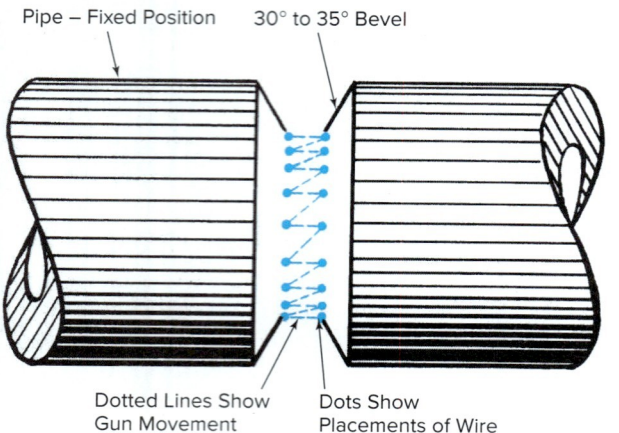

Fig. 24-39 Gun movement when welding filler passes in the 5G downhill position. The pipe axis is in the horizontal fixed position.

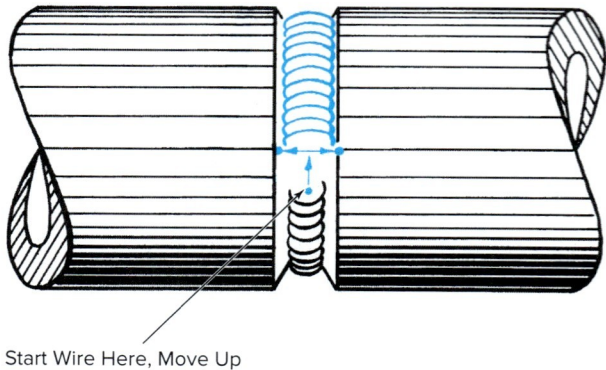

Start Wire Here, Move Up

Fig. 24-40 Welding technique for making tie-ins in the 5G position downhill.

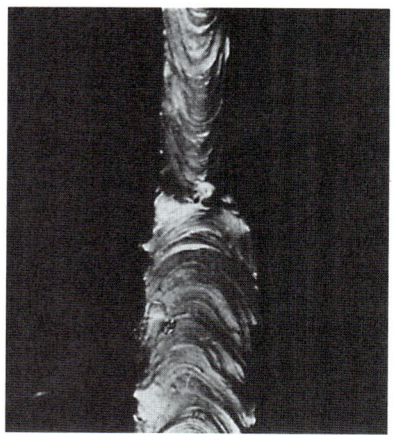

Fig. 24-41 Typical face appearance of a filler pass and a root pass in the 5G position downhill. © Plumbers and Pipefitters Union, Alton, IL

Inspection Inspect your completed welds carefully. Keep in mind the weld characteristics that are indicative of sound welding. Study Figs. 24-42 and 24-43,

For video of GMAW RMD on root pass pipe inside pipe, please visit www.mhhe.com/welding.

which show the penetration inside the pipe. Figure 24-44 shows the desirable face appearance of a root pass in a weldment. Figures 24-45 and 24-46, page 806, illustrate the faces of the root, filler, and cover passes when welding in the vertical downhill direction (5G↓).

Horizontal Pipe Axis Fixed Position (5G↑) Uphill Travel: Jobs 24-J7 and J8 After you have mastered the downhill welding technique, practice Jobs 24-J7 and J8 with the downhill welding technique for the root pass and the uphill welding technique for all other passes.

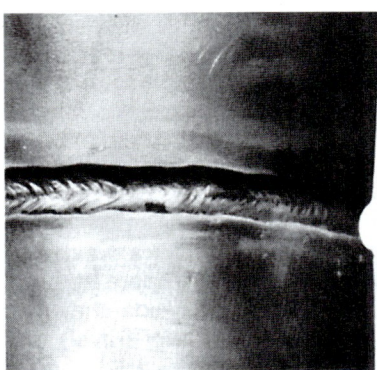

Fig. 24-42 The face of the root pass in steel pipe. The joint shows complete penetration. The inside of the pipe shows root penetration, which is free of notches, icicles, crevices and other obstructions to flow. The size of the root pass is between ⅛ and 5⁄32 inch. Note the concave contour, which will accept the next pass with minimal chance of discontinuities.
Edward R. Bohnart

Fig. 24-44 GMAW-S process was used to weld the root pass in this 1½-inch thick, mild steel pressure vessel. The joint members were spaced ⅛ to 5⁄32 inch apart to give proper penetration and reinforcement on the inside. Tack welds for holding parts in proper alignment were removed by grinding before the root pass was completed. Ends of the root pass increments were tapered with an 8-inch wide disc grinder wheel at the same time tack welds were removed to ensure complete fusion at the tie-in.
Edward R. Bohnart

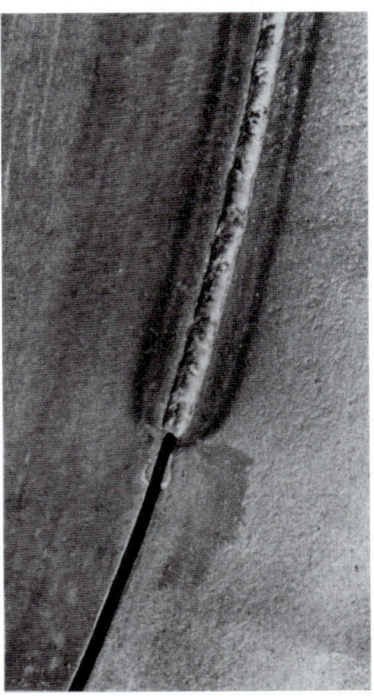

Fig. 24-43 Inside of vessel showing penetration of a root pass deposited in the pressure vessel shown in Fig. 24-44. The joint shows complete penetration free of all obstruction. Root reinforcement is between 1⁄32 and 3⁄32 inch.
Edward R. Bohnart

Select the pipe nipples specified in the Job Outline. Prepare them as previously instructed, tack them, and set them up in the welding fixture.

Root Pass Weld the root pass downhill with the welding technique practiced in Job 24-J6. Clean and inspect the weld for discontinuities.

Filler and Cover Passes Weld the filler and cover passes with the uphill welding technique. Reduce the wire-feed speed to reduce the current. Adjust the voltage until you have a smooth arc. Start at the bottom at the 5 or 7 o'clock position and weld across the 6 o'clock position up toward the 11 or 1 o'clock position at the top of the pipe, depending on the side of the pipe you are welding from. Hold the gun as illustrated in Fig. 24-37, page 804. Weave the gun from side to side and pause at each edge of the previous weld. This gives the weld metal a chance to fuse to the metal surface and allows the edges to fill up so that undercut will not take place. Weld passes should not be more than ⅛ inch thick. The filler passes should have a flat face. The cover pass may have a convexity of reinforcement about 1⁄16 inch high. You are again cautioned to be very careful when making stops and tie-ins. Use the technique shown in Fig. 24-47, page 806. Clean and inspect for welding defects. Compare your welds with those shown in Fig. 24-48, page 806.

Whether the direction of travel is up or down makes some difference in the face appearance of the welds. Generally, uphill travel produces a weld with close ripples.

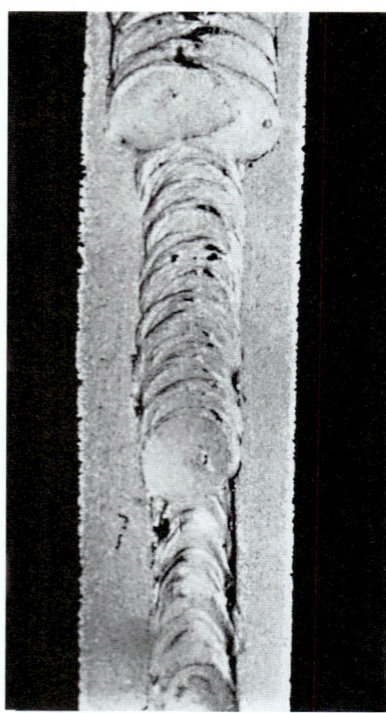

Fig. 24-45 Face appearance of root, filler, and cover passes in pipe welded in the 5G position downhill. © Plumbers and Pipefitters Union, Alton, IL

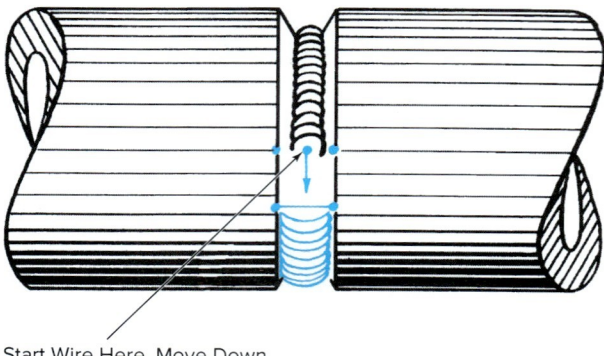

Start Wire Here, Move Down

Fig. 24-47 Welding technique for making tie-ins when welding in the 5G position uphill.

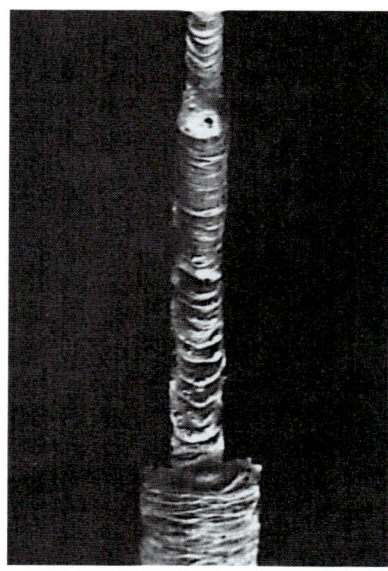

Fig. 24-48 Face appearance of filler passes and cover pass welded with pipe axis horizontal and fixed. Welds were made in the 5G position uphill. Edward R. Bohnart

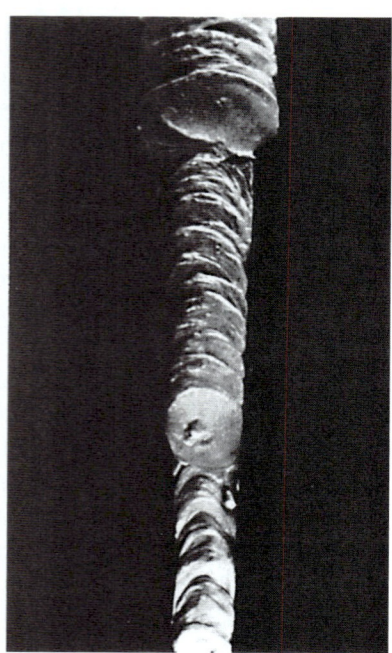

Fig. 24-46 Face appearance of root, filler, and cover passes, welding with pipe axis horizontal and fixed. Weld was made in the 5G position downhill. Travel was somewhat slower than welding shown in Fig. 24-45 so welds are heavier. © Plumbers and Pipefitters Union, Alton, IL

Compare the appearance of the welds shown in Fig. 24-49 (downhill travel) and Fig. 24-50 (uphill travel).

Inspection and Testing Make up two test joints. One should be welded with the travel-down technique, and the other should be welded with the travel-up technique. Cut the usual test coupons from these joints, prepare them carefully, and subject them to the usual face- and root-bend tests. Inspect the surface of the bends carefully for any cracks or voids. None greater than 1/16 inch in any direction should be present.

Intersection Joints: Jobs 24-J9 through J14 Intersection joints such as T's, Y's, K's, and laterals are found in pressure and powerhouse installations and in most general construction work. As you have already discovered when welding these joints with other processes, they

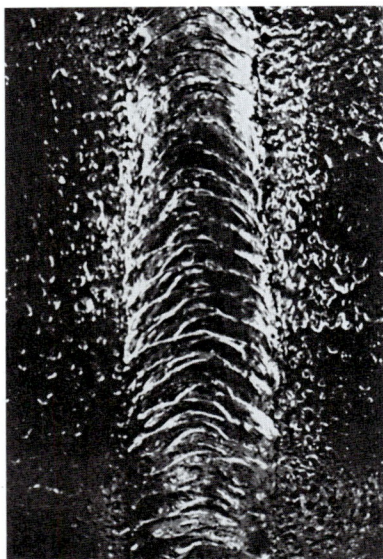

Fig. 24-49 Face appearance of a cover pass in steel pipe welded in the 5G position downhill. The weld has a coarse appearance compared to that shown in Fig. 24-48. © Plumbers and Pipefitters Union, Alton, IL

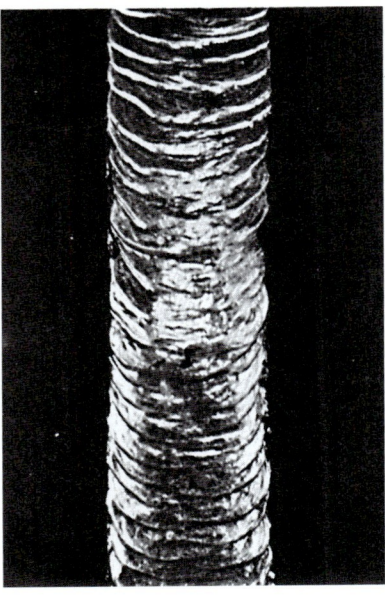

Fig. 24-50 Face appearance of a cover pass in steel pipe welded in the 5G position uphill. Note that the ripples are close together. Compare with the weld shown in Fig. 24-49. Uphill welding will usually produce a smoother weld. Edward R. Bohnart

are difficult to fabricate and even more difficult to weld. Great care must be taken in beveling and fitup.

You will recall that prefabricated fittings are available that require only the application of a butt joint for installation. These fittings are expensive, however, and many companies prefer to have the welder fabricate the joints on the job. It is always a problem to determine which method is less costly.

Practice Jobs 24-J9 through 14 as specified in the Job Outline. The intersection joint requires the welding of T and lap joints with fillet and groove welds. The usual progression is from a fillet weld to a groove weld, back to fillet, then to a groove, and finally back to the original fillet weld that is the starting point.

Follow the general procedure previously recommended for gas metal arc pipe welds. Prepare the pipe properly, tack weld it, and place it in the position called for in the Job Outline. Welding may be performed with both the downhill and uphill techniques. Both stringer and weave beads are used, depending on the position of the joint and the recommended procedures on the job. The root pass is always a stringer bead, and travel may be downhill or uphill. The intermediate and cover passes may be stringer or weave beads, depending on the position of the pipe and requirements of the job. You have already learned the techniques required because the jobs in the Job Outline have provided practice in all of them.

You are again instructed to take extreme care in making the root pass. Be sure that there is penetration to the root of the joint and fusion to the pipe surfaces. All other passes must be fused to the underneath pass and to the pipe wall. When multiple-pass welds are necessary, make sure that all starts and stops are in different areas. The starts and stops must not be at the same point.

The welds should be started on either side of the intersection at the heel of the joint. The heel is the coldest point of the joint. Because it requires the greatest amount of heat to weld, it should be welded after the joint has been heated from the welding of other areas. This procedure permits the welding of the heel intersection in a preheat condition when the arc is in continuous operation.

In welding these joints, there is a tendency to undercut at the heel and the sides. Making sure that the weld pool fills up eliminates this possibility.

Clean and inspect each pass. Make sure that the welds are free of surface discontinuities. After the joint is completed and you have carefully inspected its surface, cut a number of sections from the joint and macro-etch them. Inspect the sections for evidence of incomplete penetration, incomplete fusion, porosity, and undercut.

Fillet and Groove Welding Combination Project with GMAW: Job Qualification Test 1

This combination test project will allow you to demonstrate your ability to read a drawing, develop a bill of materials (SI conversions are optional), thermally cut, fit components together, tack, and weld a carbon steel project. You will be using the techniques developed in prior jobs in this chapter using the GMAW process. Follow the instructions found in Table 24-2 under Job 24-JQT1, and the notes in Fig. 24-51, page 808.

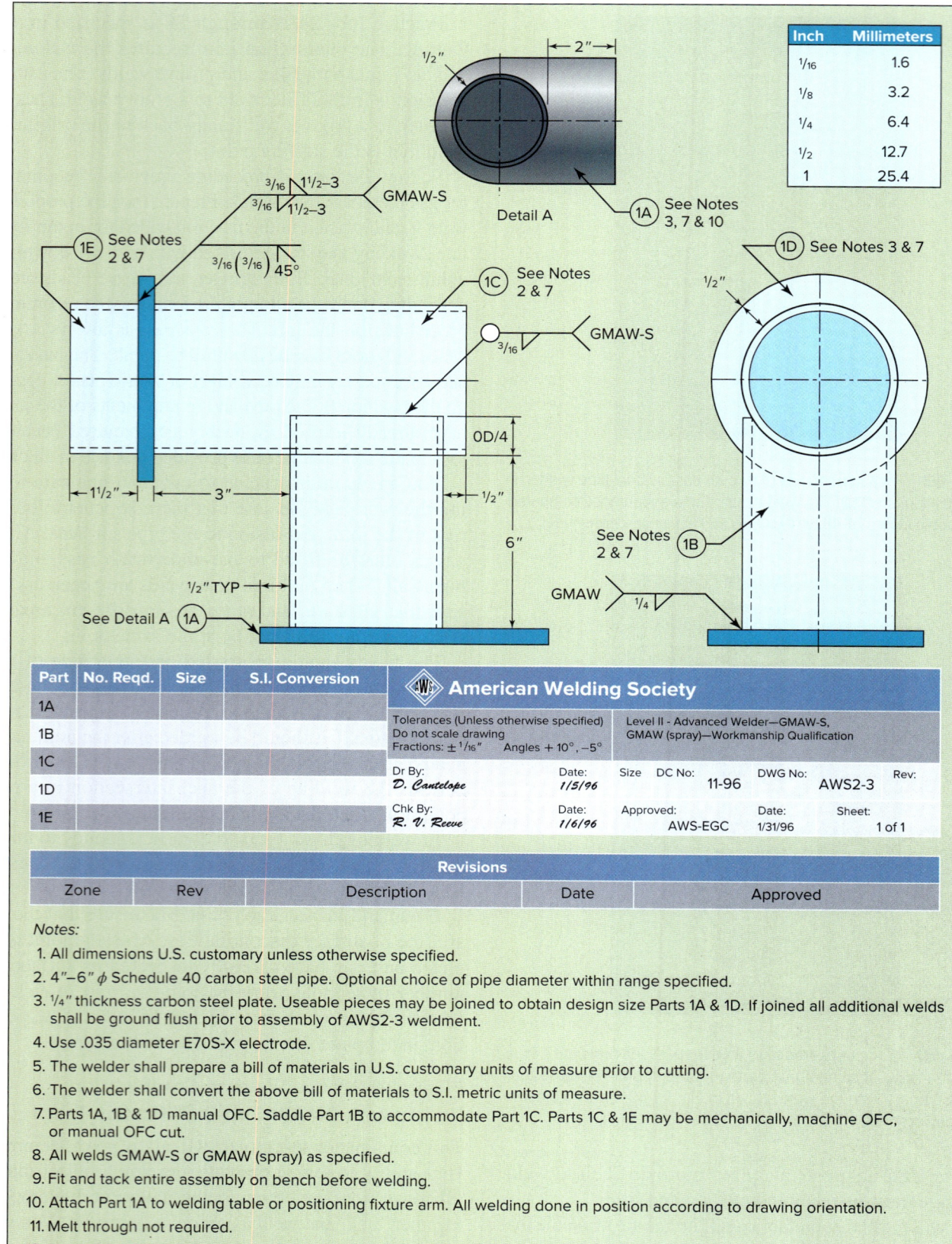

Fig. 24-51 Performance qualification test for GMAW advanced level. Adapted from AWS SENSE Program

Inspection and Testing Use the criteria outlined at the end of this chapter.

Groove Weld Project: Job Qualification Test 2

This test project will allow you to demonstrate your ability to read a drawing, fit components together, tack, and weld a carbon steel pipe test nipple. You will be using the techniques developed in Jobs 24-J1 through J14 using the short circuit mode of metal transfer. Follow the instructions found in the notes in Fig. 24-52 and in Table 24-2, pages 813–814 under Job 24-JQT2.

Inspection and Testing Use the criteria outlined at the end of this chapter.

Practice on Aluminum Pipe

Horizontal Fixed Position: Job 24-J15 The opportunity to weld aluminum pipe with the gas metal arc process is offered as a bonus for those welding students who have mastered the welding of pipe with the stick electrode and TIG processes with a high degree of skill.

Secure two lengths of 5- or 6-inch 6061 aluminum alloy pipe and 4043 or 5356 alloy filler wire $1/16$ inch in diameter. The pipe should be prepared, tacked, and set up as shown in Fig. 24-53. Brush the beveled surfaces thoroughly to remove the protective oxide coating and other contaminants.

Shielding gas should be argon or helium or a mixture of argon and helium. Gas flow should be approximately 60 cubic feet per hour. The high gas flow is necessary because of the tendency for the gas to leave the weld area when welding in the overhead position. The welding current should be set at about 150 to 190 amperes. Follow the welding pass sequence shown in Fig. 24-53.

In making the root pass, a slight motion may be necessary to properly control the metal. Special care must be taken to direct the arc so that overheating is not caused in any one area. This burns away the root face of the pipe. It also causes excessive penetration on the inside of the pipe and excessive buildup and sagging of the weld bead. On aluminum, all welding passes should be uphill.

In welding the filler and cover passes, a slight weave may help you to control the metal. Be careful to keep the molten pool under control. If the pool is too large, the effect of gravity on the molten metal causes it to spill over and sag. Bead size, weld speed, and bead sequence must be such that there is no incomplete fusion between passes.

Clean the welds and inspect the inside of the pipe for the root pass appearance and the outside of the pipe for the face appearance of the

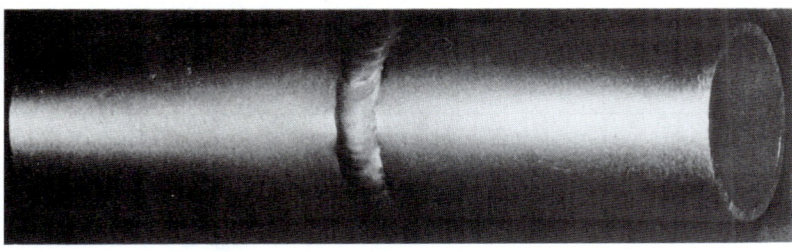

Fig. 24-52 Performance qualification test GMAW-S, carbon steel or aluminum GMAW or GMAW-P, 6G position. © Kaiser Aluminum

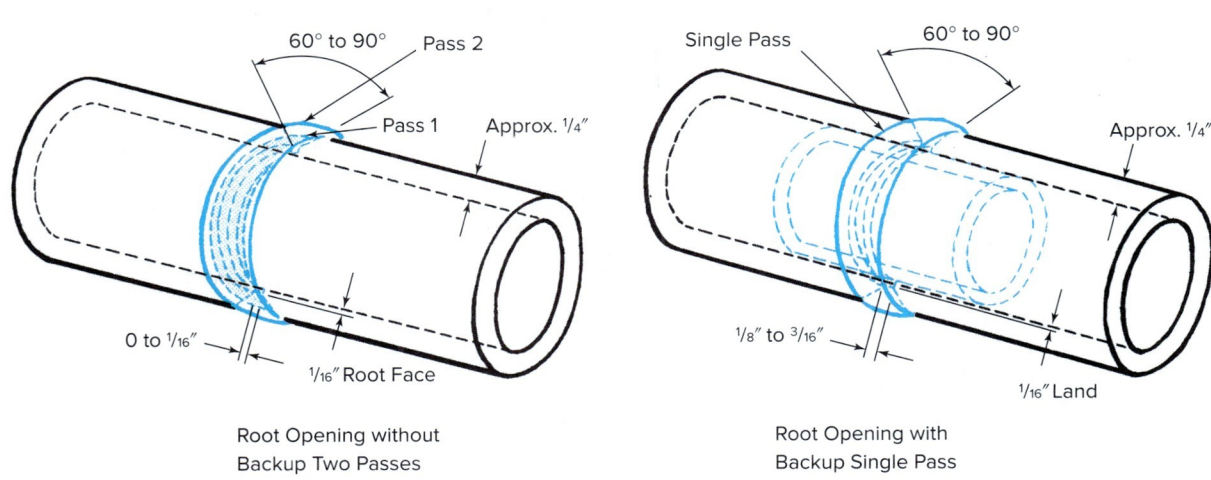

Fig. 24-53 Joint design and weld pass sequence for MIG welding of aluminum pipe in the 5G position uphill.

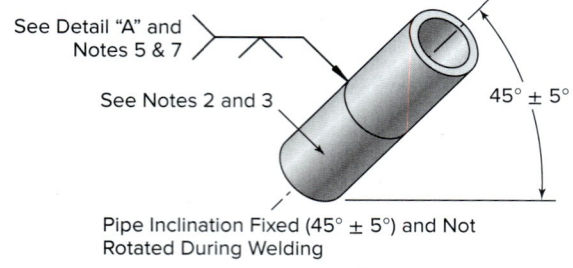
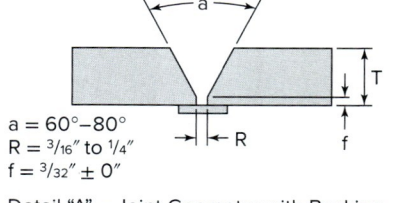
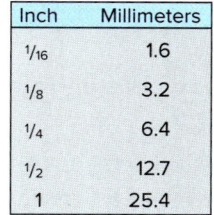

Test Position 6G

Notes: 1. All dimensions U.S. customary unless otherwise specified.

2. 2½″–6″ ⌀ Schedule 40 M-22/P-22/S-22 or M-23/P-23/S-23 aluminum pipe. Pipe diameter and material optional within range specified.
3. The standard pipe groove test weldment for performance qualification shall consist of two pipe sections, each a minimum of 3 in. (76 mm) long joined by welding to make one test weldment a minimum of 6 in. (150 mm) long.
4. With backing. Refer to Detail "A". Backing ring to suit diameter and nominal wall thickness of pipe.
5. All welding done in position.
6. All parts may be mechanically cut or machine PAC.
7. For M-22/P-22/S-22 use WPS AWS3—GMAW-P—1. For M-23/P-23/S-23 use WPS AWS3—GMAW-P—2.
8. Visual examination and bend test per criteria in text.

Tolerance:
Fractions + or −1/16″
Angles +10 degrees, −5 degrees

Fig. 24-54 Performance qualification test GMAW-P, aluminum 6G position.

cover pass. Look for the same weld characteristics that you have learned are indicative of a sound weld.

Groove Weld Project: Job Qualification Test 3

This test project will allow you to demonstrate your ability to read a drawing, fit components together, tack, and weld an aluminum pipe test nipple. You will be using the techniques developed in the various GMAW aluminum welding jobs throughout this text. Follow the instructions found in the notes in Fig. 24-54 and in Table 24-2 under Job 24-JQT3.

Inspection and Testing After the project has been tacked, have it inspected for compliance to the drawing. After the project has been completely welded, use visual inspection and cut specimens for bend testing. Use the following acceptance criteria to visually judge your welds. Look for surface defects. Keep in mind that it is important to have good appearance and uniform weld contour. These characteristics usually indicate that the weld was made properly and that the weld metal is sound throughout. Once visual inspection is completed to the following criteria, you will perform side bend tests. Reference Fig. 24-55 for specimen locations.

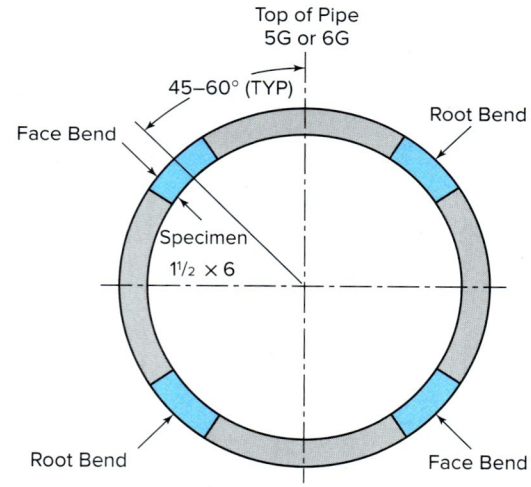

Fig. 24-55 GMAW test pipe specimen location.

Follow face- and root-bend test procedures as outlined in Chapter 28.

- There shall be no cracks or incomplete fusion.
- There shall be no incomplete joint penetration in groove welds except as permitted for partial joint penetration groove welds.
- Your instructor shall examine the weld for acceptable appearance and shall be satisfied that the welder is

skilled in using the process and procedure specified for the test.
- Undercut shall not exceed the lesser of 10 percent of the base metal thickness or 1/32 inch.
- Where visual examination is the only criterion for acceptance, all weld passes are subject to visual examination, at the discretion of your instructor.
- The frequency of porosity shall not exceed one in each 4 inches of weld length, and the maximum diameter shall not exceed 3/32 inch.
- Welds shall be free from overlap.
- Only minimal weld spatter shall be accepted, as viewed prior to cleaning.

Root- and face-bend acceptance criteria as measured on the convex surface of the bend specimen are as follows:
- No single indication shall exceed 1/8 inch measured in any direction on the surface.
- The sum of the greatest dimensions of all indications on the surface, which exceed 1/32 inch, but are less than or equal to 1/8 inch, shall not exceed 3/8 inch.
- Cracks occurring at the corner of the specimens shall not be considered unless there is definite evidence that they result from slag inclusions or other internal discontinuities.

CHAPTER 24 REVIEW

Multiple Choice

Choose the letter of the correct answer.

1. Radiographic quality welds on pipe are not possible with the GMAW process. (Obj. 24-1)
   a. True
   b. False
2. The GMAW-S and GMAW-P mode of metal transfer are too slow to be used for many pipe applications. (Obj. 24-1)
   a. True
   b. False
3. The GMAW process is considered _____. (Obj. 24-1)
   a. Autogenous
   b. A low hydrogen process
   c. A poor quality process
   d. Both a and b
4. For GMAW-S, which type of power source is to be used? (Obj. 24-2)
   a. Constant current
   b. Drooper
   c. Constant voltage
   d. Rising
5. For good arc starting (*hot start*) _____. (Obj. 24-2)
   a. The wire feeder power source must be appropriate
   b. The wire must be clipped
   c. The electrode extension must be short
   d. All of these
6. Some codes require draft shields to protect the shielding gas if the wind speed exceeds _____ miles per hour. (Obj. 24-3)
   a. 2
   b. 5
   c. 10
   d. 15
7. What is the proper groove angle for welding with the boiler code? (Obj. 24-3)
   a. 15°
   b. 30°
   c. 37½°
   d. 75°
8. A root edge has a _____-inch surface. (Obj. 24-3)
   a. 0 (it is sharp)
   b. 1/16
   c. 3/32
   d. 1/8
9. Feathering a tack is not required because the GMAW process penetrates so much. (Obj. 24-3)
   a. True
   b. False
10. The root opening for welding aluminum pipe can be zero. (Obj. 24-4)
    a. True
    b. False

Review Questions

Write the answers in your own words.

11. List several of the metals welded in the piping industry. (Obj. 24-1)

12. List 10 advantages of GMA pipe welding. (Obj. 24-1)
13. List five items included in an MIG/MAG welding system for pipe. (Obj. 24-2)
14. Describe why inverter power sources are becoming so popular for field pipe welding. (Obj. 24-2)
15. Describe the "hot start" features for GMA pipe welding. (Obj. 24-2)
16. Describe the shielding gases used for GMAW-S and GMAW-P on carbon steel and stainless steel. (Obj. 24-2)
17. Sketch and label the V-groove butt joint used for GMA pipe welding and the ASME code. (Obj. 24-3)
18. Why must grinding be done on the root pass? (Obj. 24-3)
19. Sketch and dimension the length of a feathered tack weld on pipe and describe two reasons why it is done. (Obj. 24-3)
20. Describe the proper tie-in procedure for a root pass on a pipe weld. (Obj. 24-4)

INTERNET ACTIVITIES

Internet Activity A
Using your favorite search engines find as many gas metal arc welding procedures as you can for welding pipe. Make a list of where they come from and the procedure number and present to your instructor.

Internet Activity B
Using your favorite search engines list as many organizations as you can find that are involved with gas metal arc welding of pipe training issues in your particular state. These can be unions, trade associations, membership societies, institutions, or schools. Don't list companies that do the welding. Present the list to your instructor.

Design credits: Cinema and movie icon: Denis Meshkov/123RF; and Illustration of Welding icons: Mr. Nachai Sorasee/123RF

Table 24-2 Job Outline: Gas Metal Arc Welding Practice with Solid Wire (Pipe)

Job No.	Type of Joint	Type of Weld	Welding Position	Welding Technique	Pipe Specifications Material	Diameter (in.)	Weight Schedule	Wall Thickness (in.)	Electrode Specifications[1] Type	Size (in.)	Welding Current DCEP Arc Volts	Amperes	Wire-Feed Speed	Shielding Gas[2]	Gas Flow (ft³/h)	Text Reference
24-J1	Flat surface	Surfacing	1C	Stringer downhill Weaved downhill	Carbon steel	4–8	40	1/4–5/16	ER70S-6	0.035	18–21	150–160 130–150	250–290 220–250	Carbon dioxide	15–25	817
24-J2	Flat surface	Surfacing	1C	Stringer uphill Weaved uphill	Carbon steel	4–8	80	5/16–1/2	ER70S-6	0.035	19–23	130–160	220–290	Carbon dioxide	15–20	817
24-J3	Butt	V-groove	1G	1 Stringer—downhill 2 Weaved—downhill	Carbon steel	6	40	5/16	ER70S-6	0.035	18–21	130–140 140–160	220–235 235–290	Carbon dioxide	15–20	819
24-J4	Butt	V-groove	2G	5 Stringer	Carbon steel	6	40	5/16	ER70S-6	0.035	18–21	1-130 to 140 4-140 to 160	220–235 235–290	Carbon dioxide	15–25	821
24-J5	Butt	V-groove	2G	7 Stringer	Carbon steel	8	80	1/2	ER70S-6	0.035	19–23	1-140 to 150 6-150 to 170	235–250 250–330	Carbon dioxide	15–25	821
24-J6	Butt	V-groove	5G	1 Stringer—downhill 3 Weaved—downhill	Carbon steel	8	40	5/16	ER70S-6	0.035	19–21	120–130 130–150	210–220 220–250	Carbon dioxide	15–20	823
24-J7	Butt	V-groove	5G	1 Stringer—downhill 2 Weaved—uphill	Carbon steel	8	40	5/16	ER70S-6	0.035	19–21 19–23	120–130 110–120	210–220 180–210	Carbon dioxide	15–20	827
24-J8	Butt	V-groove	5G	1 Stringer—downhill 2 Weaved—uphill	Carbon steel	8	80	1/2	ER70S-6	0.035	19–21 19–24	120–130 110–125	210–220 180–215	Carbon dioxide	15–20	827
24-J9	90° branch	V-groove fillet	Header horizontal fixed; branch vertical fixed top (2F, 2G)	3 Stringer	Carbon steel	6–8	40	5/16	ER70S-6	0.035	18–21	130–140	220–235	Carbon dioxide	15–20	829

(Continued)

Table 24-2 (Concluded)

Job	Joint	Type	Position	Pass Sequence	Material	Thickness		Size	Electrode	Diameter	Voltage	Amperage		Shielding Gas	Flow	Page
24-J10	90° branch	V-groove, fillet	Header vertical fixed; branch horizontal fixed (5F, 5G)	1 Stringer—downhill 2 Weaved uphill	Carbon steel	6–8	40	5/16	ER70S-6	0.035	18–21	120–130 110–120	210–220 180–210	Carbon dioxide	15–20	829
24-J11	90° branch	V-groove, fillet	Header horizontal fixed; branch vertical fixed bottom (5F, 5G)	3 Stringer	Carbon steel	6–8	40	5/16	ER70S-6	0.035	18–21	120–130	210–220	Carbon dioxide	15–20	829
24-J12	45° branch	V-groove, fillet	Header horizontal fixed; branch top	3 Stringer	Carbon steel	6–8	40	5/16	ER70S-6	0.035	18–21	130–140	220–235	Carbon dioxide	15–20	829
24-J13	45° branch	V-groove, fillet	Header vertical fixed; branch side	1 Stringer—downhill 2 Weaved uphill	Carbon steel	6–8	40	5/16	ER70S-6	0.035	18–21 18–23	120–130 110–120	210–220 180–210	Carbon dioxide	15–20	829
24-J14	45° branch	V-groove, fillet	Headers horizontal fixed; branch bottom	5 Stringer	Carbon steel	6–8	40	5/16–1/2	ER70S-6	0.035	18–23	130–140	220–235	Carbon dioxide	15–20	829
25-J15	Butt	V-groove	Horizontal roll (1G)	1 Stringer 1 Weaved	Aluminum	5–6	80	1/4	ER4043	3/64	20–24	180–190 190–210	365–390 390–426	Argon	30–40	831
24-JQT1	T and lap	Groove, fillet	2F, 5F, 5G	Stringer and weave	Carbon steel	4–6	40	1/4–5/16	ER70S-6	0.035	18–23	110–160	180–290	Carbon dioxide	15–25	829
24-JQT2	Butt	Groove	6G	Stringer and weave	Carbon steel	2½–6	40	0.203–5/16	ER70S-6	0.035	19–21	120–150	210–250	Carbon dioxide	15–25	831
24-JQT3	Butt	Groove	6G	Stringer and weave	Aluminum	2½–6	40	0.203–5/16	ER4043	3/64	(3)	(3)	(3)	Argon	30–40	832

[1] Electrode type ER 80S-D2 may also be used with excellent results.

[2] Argon-rich gas mixtures may also be used. GMAW-P may also be used if equipment is available. Various root pass techniques can also be used, such as GTAW and GMAW-S. GMAW-P is very difficult unless the joint is accurately fit and welding is done in the 1G position.

[3] Use a synergic pulse parameter as specified by your specific equipment manufacturer for 3/64-in. wire.

Note: The conditions indicated here are basic and will vary with the job situation, the results desired, and the skill of the welder. For additional practice the FCAW process can be substituted. Stainless-steel practice can be done using appropriate filler metal and shielding gases. Stainless-steel pipe is expensive and limited practice can be done using carbon steel pipe with stainless electrode and appropriate shielding gas.

UNIT 5

High Energy Beams, Automation, Robotics, and Weld Shop Management

Chapter 25
High Energy Beams and Related Welding and Cutting Process Principles

Chapter 26
General Equipment for Welding Shops

Chapter 27
Automatic and Robotic Arc Welding Equipment

Chapter 28
Joint Design, Testing, and Inspection

Chapter 29
Reading Shop Drawings

Chapter 30
Welding Symbols

Chapter 31
Welding and Bonding of Plastics

Chapter 32
Safety

25

High Energy Beams and Related Welding and Cutting Process Principles

Introduction

This chapter introduces some of the more nonconventional welding and cutting processes. The various processes are evaluated based upon their distinguishing features. Two of the most obvious distinguishing features are the heat source and how the molten pool is shielded from the atmosphere. This text has covered many of the arc welding processes that use the electric arc for a heat source and internal or external gas shielding. None of the processes covered in this chapter utilize the arc for a heat source. They use some very unique sources of energy for welding and cutting. The molten pool can be shielded by normal welding shielding gases or perhaps evacuating the atmosphere in a vacuum chamber.

As a professional in the welding industry (welder, welding technician, welding engineer, and so on), it is important that you understand the various processes that may be encountered. Correct process selection will have a major impact on manufacturing costs. Most of these processes are applied in mechanized or automatic modes. They require extremely accurate joint geometry and positioning. In many cases the welds are autogenous, and so additional filler metal is not required. They may or may not require any shielding gas, so much of the normal consumables required for other welding processes are not necessary. Equipment for the various welding and cutting processes

Chapter Objectives

After completing this chapter, you will be able to:

- **25-1** Explain the use of high energy-density beam processes.
- **25-2** Describe the water jet cutting process.
- **25-3** Describe the friction welding processes.
- **25-4** Explain explosion welding.

can run from a few hundred dollars to millions of dollars. Making the right choices in process and equipment can mean the difference in being competitive in a world market or out of business. This chapter is a survey of various processes, and you are encouraged to investigate these processes in more detail based upon your needs. Additional sources of information are available from the American Welding Society and through the Internet.

High Energy Beam Processes

High energy beams are concentrated heat sources that have been measured as high as 65,000,000 watts per square inch. Compare this to holding your hand a safe distance above a 100-watt lightbulb to feel the heat. Then multiply this by 650,000 to get an idea of the amount of energy that is being released. Compare the weld made with the GTAW process as compared to the electron beam weld in Fig. 25-1.

All the normal type of joints, such as butt, corner, lap, edge, and T, can be welded with the high energy beam processes. These are typically done with groove welds. Fillet welds are difficult and generally not attempted with high energy beam processes. When groove welds are made, they are usually CJP-type welds made using the keyhole technique, Fig. 25-2.

These processes generally produce a very narrow weld with very deep penetration characteristics. These type of welds must be applied to tightly fitting joints with tolerance in the 0.0001-inch range. The heat-affected zones in keyhole welding are normally very narrow and are accompanied by very rapid cooling rates. This may be advantageous on some alloys, but on others where it may be detrimental, preheat must be used to slow the cooling rate.

Certain types of alloy and thickness are more readily welded than others. Table 25-1 (p. 818) covers the capabilities of various welding processes.

Electron Beam Welding (EBW)

As with all high energy-density beam processes, **electronic beam welding (EBW)** is used for high precision and high production applications. Electron beam welding is accomplished by the use of a concentrated stream of high velocity electrons that is formed into a beam. This beam provides a source of intense local heating. Equipment consists of the following:

- Vacuum chamber
- Controls
- Electron beam gun
- Three-phase power source
- Optical viewing system
- Tracking device
- Work-handling equipment

Electron beam welding is shown in Fig. 25-3, page 819. Much like the picture tube in your TV set, electrons are dispersed from the electrically excited, negatively charged cathode. The electron beam is partially shaped by the bias cup grid. The electrons are very small, but they are moving at 30 to 70 percent of the speed of light, so they carry

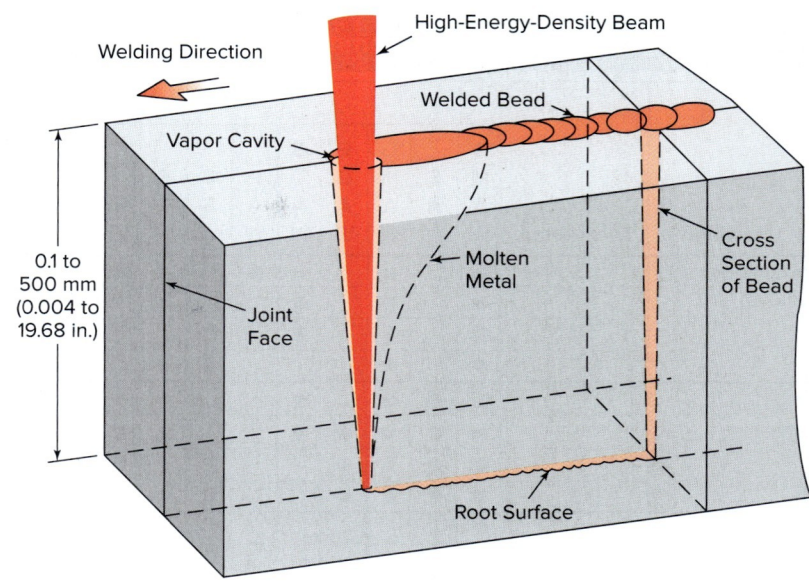

Fig. 25-2 Cross section of a keyhole weld. Adapted from American Welding Society, *Welding Handbook,* Volume 3, Welding Processes Part 2

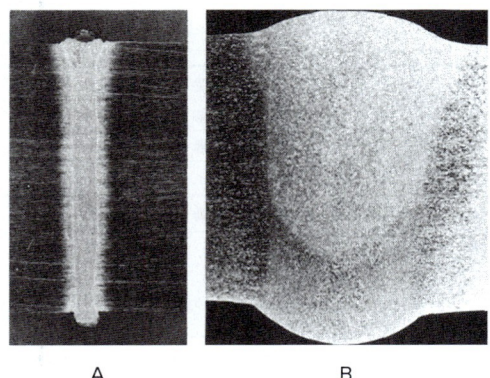

Fig. 25-1 Comparison of electron beam weld on the left (A) and the gas tungsten arc weld on the right (B) in ½-inch thick type 2219 aluminum alloy plate. © American Welding Society. *Welding Handbook,* Vol. 3, Welding Processes Part 2, 9th ed., Chapter 13, p. 465, Fig. 13.16 A, B.

Table 25-1 Capabilities of the Commonly Used Joining Processes

Material	Thickness[2]	SMAW	SAW	GMAW	FCAW	GTAW	PAW	ESW	EGW	RW	FW	OFW	DFW	FRW	EBW	LBW	TB	FB	RB	IB	DB	IRB	DFB	S
Aluminum and alloys	S	●		●		●	●			●	●	●	●	●	●	●	●	●	●	●	●	●	●	●
	I	●		●		●				●	●		●	●	●	●				●		●	●	●
	M	●		●		●					●		●		●	●				●			●	
	T	●		●				●	●		●		●		●								●	
Carbon steel	S	●[3]	●	●		●				●	●	●		●		●	●	●	●	●	●	●	●	●
	I	●	●	●	●	●				●	●	●		●	●	●	●	●	●	●	●	●	●	●
	M	●	●	●	●						●	●		●		●		●					●	
	T	●	●	●	●			●	●		●	●		●		●		●					●	
Cast iron	I	●									●						●	●	●				●	●
	M	●	●	●	●						●						●	●	●				●	●
	T	●	●	●	●						●						●						●	
Copper and alloys	S			●		●	●			●[4]	●				●		●	●	●	●			●	●
	I			●		●				●			●		●		●	●					●	●
	M			●		●				●			●		●		●						●	
	T			●		●							●		●		●						●	
Low alloy steel	S	●	●	●		●				●	●	●		●	●	●	●	●	●	●	●	●	●	●
	I	●	●	●	●	●				●	●	●	●	●	●	●	●	●	●	●			●	●
	M	●	●	●	●						●	●		●	●	●	●						●	
	T	●	●	●	●			●			●	●		●	●	●					●		●	
Magnesium and alloys	S			●		●				●					●	●	●	●		●			●	
	I			●		●				●					●	●	●	●		●			●	
	M			●									●	●	●		●						●	
	T			●											●								●	
Nickel and alloys	S	●		●		●	●			●	●	●			●	●	●	●	●	●	●	●	●	●
	I	●	●	●		●	●			●	●			●	●	●	●	●	●	●			●	●
	M	●	●	●		●				●			●	●	●	●	●	●					●	
	T	●		●		●		●		●			●	●	●		●						●	
Refractory alloys	S			●		●	●			●					●	●	●	●	●		●		●	
	I			●			●			●	●				●	●	●	●					●	
	M									●														
	T																							
Stainless steel	S	●	●	●		●	●			●	●	●	●		●	●	●	●	●	●	●	●	●	●
	I	●	●	●	●	●	●			●	●		●	●	●	●	●	●	●	●			●	●
	M	●	●	●	●	●	●				●		●	●	●	●	●	●					●	
	T	●	●	●	●			●			●		●	●	●		●						●	
Titanium and alloys	S			●		●	●			●			●		●	●	●	●			●		●	
	I			●		●	●			●			●	●	●		●	●					●	
	M			●		●	●			●			●	●	●		●						●	
	T			●			●			●			●	●	●		●						●	

[1] SMAW = shielded metal arc welding; SAW = submerged arc welding; GMAW = gas metal arc welding; FCAW = flux cored arc welding; GTAW = gas tungsten arc welding; PAW = plasma arc welding; ESW = electroslag welding; EGW = electrogas welding; RW = resistance welding; FW = flash welding; OFW = oxyfuel gas welding; DFW = diffusion welding; FRW = friction welding; EBW = electron beam welding; LBW = laser beam welding; TB = torch brazing; FB = furnace brazing; RB = resistance brazing; IB = induction brazing; DB = dip brazing; IRB = infrared brazing; DB = diffusion brazing; and S = soldering.

[2] S = sheet (up to 1/8 in., 3 mm); I = intermediate [1/8 to 1/4 in. (3 to 6 mm)]; M = medium [1/4 to 3/4 in. (6 mm to 19 mm)]; T = thick [3/4 in. (19 mm) and up].

[3] Commercial process.

[4] Copper requires molybdenum-coated tips.

Adapted from *Welding Handbook*, 9/e

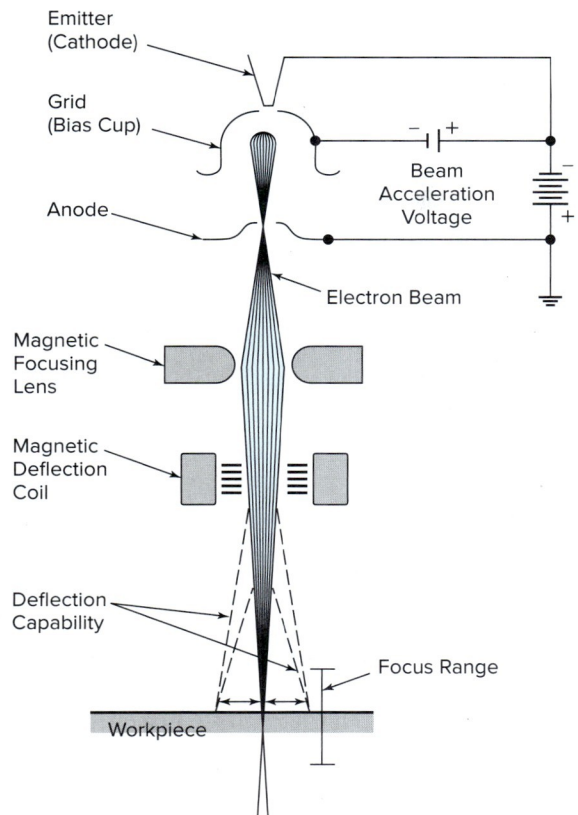

Fig. 25-3 Schematic representation of electron beam welding.
Source: From *Welding Handbook*, 9/e

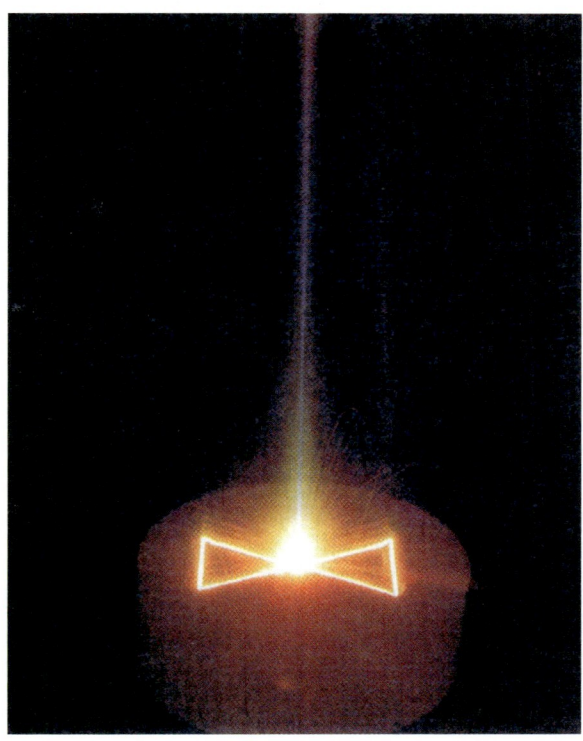

Fig. 25-4 Beam deflection capability of an electron beam column as shown by a "bow tie" pattern on a workpiece.
© American Welding Society. *Welding Handbook,* 8th ed., Vol. 2, p. 675, Fig. 21.2.

a tremendous amount of energy. The beam is further shaped by the positive anode and magnetic focusing lens. The beam can be focused onto a very small spot approximately 0.04 inch in diameter.

The electrons bombarding the metal cause a rapid buildup of heat. A vapor hole is produced that is surrounded by molten metal that will form the weld. The joint geometry must be very accurate. It must be precisely positioned, and travel speed must be very accurate for full penetration to take place. Review Fig. 25-2. The beam can also be defocused and used for vaporization purposes as well as for refining the surfaces of various materials. Figure 25-4 is an example of an electron beam being deflected.

Electron beam welding is done in three variations based on the degree of vacuum used. A high vacuum (EBW-HV) is referred to as a hard vacuum and utilizes a pressure of 1×10^{-3} torr. (A torr is the accepted industry term for a pressure of 1 millimeter of mercury. The standard atmospheric pressure can be expressed as 760 torr or 760 mmHg.) A medium vacuum (EBW-MV) utilizes a pressure of 2×10^{-1} torr. The process can also be done in a nonvacuum condition, which is known as EBW-NV, so the pressure would be 760 torr, Fig. 25-5, page 820.

Electron beam welding has many unique capabilities, Fig. 25-6, page 821. The following advantages should be considered:

- EBW directly converts electric energy into beam energy, so the process is very efficient.
- The depth-to-width ratios for one-pass welding on thick sections is high.
- Heat input is very low.
- There is a narrow heat-affected zone.
- There is minimal distortion.
- The vacuum mode results in high purity welds.
- The beam can be projected over distances in a vacuum (space is a vacuum).
- There are rapid travel speeds.
- The beam can be magnetically deflected to produce various weld shapes.
- The beam has a long focal length, so it can tolerate a broad range of work distances.

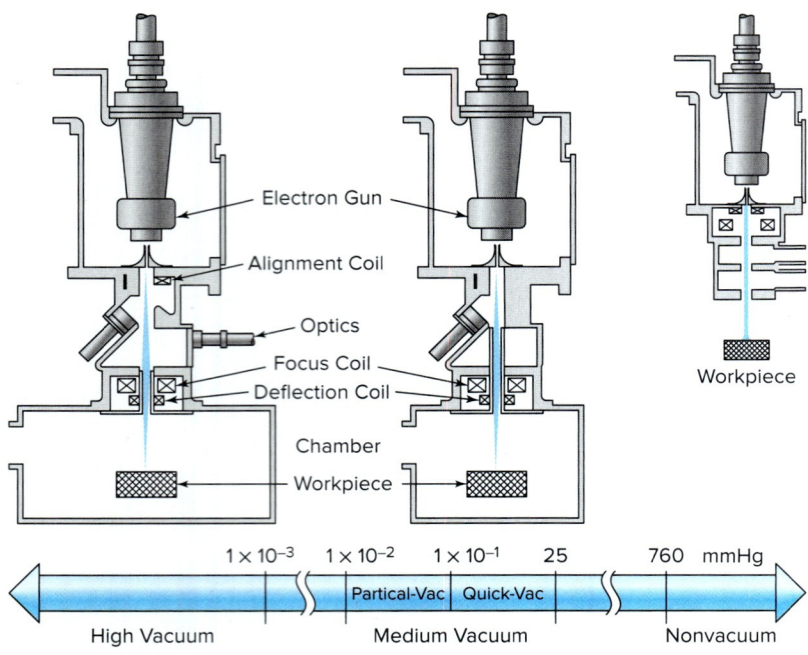

Fig. 25-5 The basic modes of electron beam welding, with corresponding vacuum scale. Source: From *Welding Handbook*, 9/e

- Dissimilar metals can be welded.
- High thermal conductive metals like copper can be welded.

Limitations of the process are as follows:

- Capital cost of equipment is very high.
- Joint preparation is very extensive.
- Rapid solidification can cause cracking.
- For high and medium vacuums, chamber size is a limitation.
- A long time is required to draw a vacuum.
- Partial penetration may have root voids and porosity.
- The beam can be magnetically deflected, so material must be nonmagnetic or demagnetized.
- No-vacuum welding requires the part to be very close to the bottom of the electron beam gun column.
- Radiation shielding must be used, and X-rays are produced.
- Ventilation is required to remove ozone and other noxious gases with the nonvacuum mode.

Laser Beam Welding and Cutting

Laser is an acronym for *l*ight *a*mplification by *s*timulated *e*mission of *r*adiation. You may recall using a magnifying glass to concentrate the rays from the sun on a single spot in order to char or burn paper. The laser energy source is a refinement of this process. The first beam was produced in 1960 using a ruby crystal rod.

The laser is a very helpful tool. It has various uses, such as the high speed bar-code readers at checkout counters or a laser light show at a music concert or other attraction. Presenters use small pocket lasers for enhancing their presentations. On construction sites lasers are used for measuring distances or to aid in the alignment of such things as drop ceilings. Lasers can be used for marking material (scribing) or for the sealing of nonmetallic materials. In automation use, they can direct a robotic welding arm along a weld joint (the laser beam is used for tracking). Medical and military uses of lasers are in the news daily, from no-blood surgery to laser-guided smart bombs. Lasers are very versatile and can be used to process metal, wood, plastics, and composites. For the welding industry lasers are used for welding, cutting, and drilling operations. These are the three applications that will be covered in this section. Lasers can be easily adapted to computer control for doing complex contour work. Figure 25-7 shows an industrial application for laser beam welding.

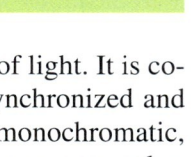

For video of EBW free-form fabrication of a titanium part, please visit www.mhhe.com/welding.

The laser is a very concentrated beam of light. It is coherent light in that the light waves are synchronized and travel parallel to each other. It is also monochromatic, meaning that the light has one frequency, one color. Figure 25-8 shows an example of this light's ability to stay in a tight column, unlike most other light.

Because the laser beam can be used for welding, cutting, and drilling, these will be covered together. A laser light is produced when intense light or electric current excites certain materials. The two different laser types are covered in Table 25-2. Lasers can be operated continuously or pulsed. The ruby or Nd-glass laser can only pulse at low frequencies (1–50 pulses per second), while the Nd-YAG or CO_2 laser can pulse at rates up to 2,000 per second. Pulsing reduces the heat buildup in the laser equipment and is effective for piercing and drilling applications.

Since the laser's light is used to heat the surface it is being focused on, it is considered a noncontact process.

There is no physical contact of equipment with the part other than the beam being focused on it. The material being worked does not have to be a conductor like the base metal in all the arc welding processes.

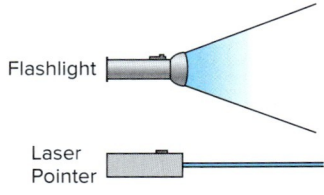

Fig. 25-8 Laser pointer-flashlight beam comparison.

Laser welding is a melting and burning process. The heat from a 10-kilowatt rated laser can weld ½-inch thick material at 60 inches per minute. A 6-kilowatt unit can weld thin material (0.008 inch thick) at 3,000 inches per minute, which is a very fast travel speed. Units are commercially available with power levels up to 25 kilowatts and can do full penetration single-pass welds up to 1¼ inches thick. For cutting, ½ inch is generally considered the limit because other processes are faster than laser on thicknesses greater than ½ inch. A laser can cut materials up to 2 inches thick if speed is not a concern. Drilling can be done up to 1 inch in depth and diameters in a range from 0.0001 to 0.060 inch. Holes smaller than 0.020 inch use the laser to best potential. Material thickness to be processed and travel speed is determined by:

- Part geometry
- Reflectivity of the materials surface
- Heat conductivity
- Vaporization point of the alloy being used
- Type of alloy
- Surface tension of the molten material

Fig. 25-6 Electron beam welding a gear in medium vacuum.
© American Welding Society. "Welding Processes," Vol. 2 of *Welding Handbook*, 8th ed., Fig. 21.5, p. 678.

Laser Beam Welding (LBW) Laser beam welding (LBW) is generally done with an inert shielding gas to shield the weld pool. Since this is a heat source, it can be used without filler metal (autogenous) or with filler metal. Figure 25-2 shows the keyhole method of laser welding. Because of the highly concentrated heat source and fast travel speed, little distortion is created.

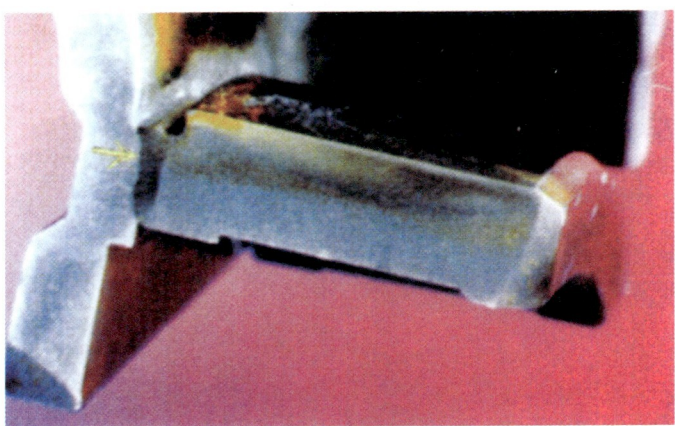

Fig. 25-7 Cross-section of a laser beam weld joining a boss to a ring. A 2.5-kilowatt CO_2 laser produced a travel speed of 60 in./min. Penetration was 0.187 inch. © American Welding Society. *Welding Handbook*, 8th ed., Vol 2, p. 733, Fig. 22.21.

Table 25-2 Laser Types and Uses

Type	Laser Material	Uses
Solid state	Ruby rod, neodymium-doped (Nd), Yttrium-aluminum-garnet (YAG) Neodymium-glass (Nd-glass)	Mostly used for drilling
Gas	Carbon dioxide (CO_2) Nitrogen Helium or mixtures of these	Mostly used for cutting and welding

For video of laser welding in an auto plant, please visit www.mhhe.com/welding.

Laser Beam Cutting (LBC) Laser beam cutting (LBC) is very common. This is accomplished by reducing the spot size of the beam from 0.011 to 0.004 inch, turning the laser beam into a sharp cutting tool. It can be used to cut expensive alloys or traditional metals like stainless steel and copper. It can cut nonmetals like plastic, wood, or cloth. Cut width or kerf can be very small at 0.010 inch. Very small heat-affected zones (HAZs) are produced. Speed drops off dramatically when material approaches ½ inch, where other cutting processes become more efficient and cost effective. As thickness goes up, blowouts may occur. **Blowouts** are unwanted molten metal flying out of the cut, interrupting the cut path.

When cutting, an assist gas is used. This helps improve combustion and physically blows metal from the kerf. Table 25-3 shows assist gases, material they are used on, and various comments. Use of high pressure up to 160 p.s.i. has been used on titanium to reduce cracks in the recast layer.

Pulsing the laser power can help overcome thermal problems. Figure 25-9 shows a laser cutting operation.

The surface of the material must be free of scale, coatings, dirt, and any impurities. On steel it should be cleaned, pickled, and oil-free. The mill scale on hot-rolled steel creates problems. Cold-rolled steel has a much better surface condition for laser cutting. The cutting parameter will be greatly affected by the surface finish, tolerances, HAZ requirements, and flatness.

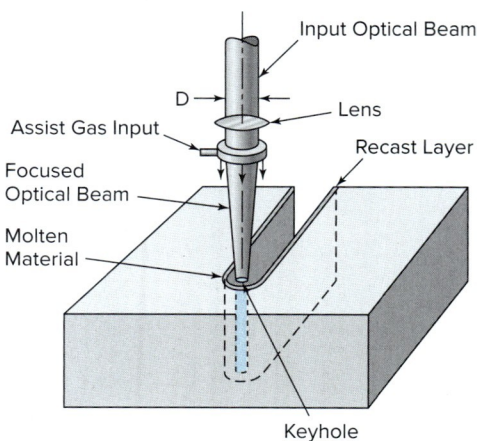

Fig. 25-9 Schematic view of laser cutting operation.
Source: From *Welding Handbook*, 9/e

Table 25-3 Assist Gases Used for Laser Beam Cutting of Various Materials

Assist Gas	Material	Comments
Air	Aluminum	Good result up to 0.060 in. (1.5 mm)
	Plastic	
	Wood	
	Composites	
	Alumina	All gases react similarly; air is the least expensive.
	Glass	
	Quartz	
Argon	Titanium	Inert assist gas required to produce good cutting of various materials
Nitrogen	Stainless steel	
	Aluminum	Clean, oxide-free edges to ⅛ in. (3 mm)
	Nickel alloys	
Oxygen	Carbon steel	Good finish, high speed; oxide layer on surface
	Stainless steel	Heavy oxide on surface
	Copper	Good surface up to ⅛ in. (3 mm)

Adapted from *Welding Handbook*, 9/e

> **SHOP TALK**
>
> **Laser Welding Safety**
> When doing laser welding, keep in mind the six sources of danger.
> 1. Radiation—visible and invisible
> 2. Fire from hitting flammables
> 3. Fumes and mists from vaporized metals
> 4. Mechanical malfunctions sending the beam elsewhere
> 5. Electric shock from the power source
> 6. Neglecting to use eyewear and skin protection

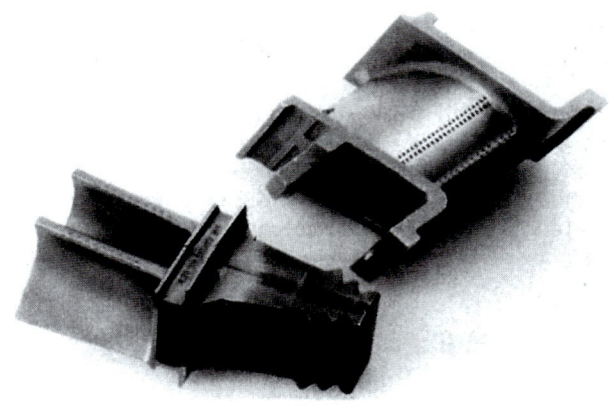

Fig. 25-10 Jet engine blades and a rotor component showing laser drilled holes. © American Welding Society. "Welding Science and Technology," Vol. 1 of *Welding Handbook*, 9th ed., Fig. 16.2, p. 503.

Laser Beam Drilling (LBD) Laser beam drilling (LBD) can be done on very hard materials, like synthetic diamonds, tungsten carbide, quartz, glass, and ceramics. Holes as small as 0.0001 inch can be drilled. LBD is very fast, but the possibility of blowouts can present safety concerns. In this case a blowout is the throwing of molten metal outside the desired hole area. Blowouts occur when the hole diameter is small in relation to the thickness of the material being drilled. Continuous wave lasers have a much higher possibility of blowout when piercing metal or drilling. The pulse wave laser is typically used. Lowering gas pressure to a minimum to allow combustion and checking the focal point can reduce blowouts. The surface must be clean of any oil or dirt. Figure 25-10 shows holes a laser beam drilled in some jet turbine blades.

The equipment required for laser beam applications is shown in Fig. 25-11. This setup is using the solid-state type neodymium-doped, yttrium-aluminum-garnet (YAG) material as a lasing source. The power supply, which operates at high voltage, supplies power to the flash lamps. A cryogenic cooler is used to cool the power supply and the laser rod and flash lamps. The rear mirror is fully reflective and bounces the coherent monochromatic beam back through the laser rod to the partially reflective front mirror. This helps in producing the coherent-synchronized wave light beam. Once the beam exits it can be treated as light and can be reflected, deflected, and focused. An energy monitor is placed in the beam to measure and compare the beam's power level. The beam dump and shutter control the beam. Various optics are placed in the beam path to expand

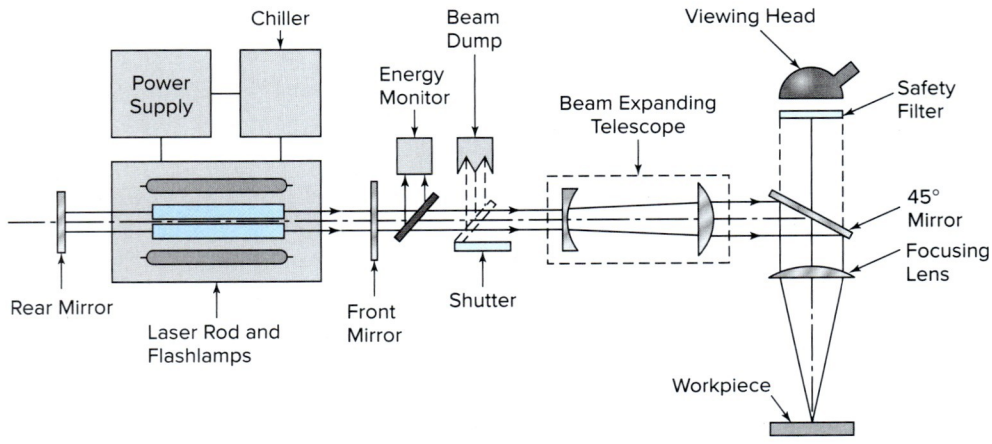

Fig. 25-11 Schematic representation of the elements of an Nd: YAG laser.
Source: From *Welding Handbook*, 9/e

or focus the beam. A 45° angle mirror is used to deflect the yes path 90°. A viewing head with a safety filter is used to monitor the beam's operation as the workpiece is processed. This could be an LBW, LBC, or LBD application depending on the power level, spot size, focus point, whether constant or pulsed power is used, and the amount and type of assist gases used.

Laser beams have the following advantages:

- There is very low heat input, which allows for hermetic sealing near glass-to-metal seals.
- Single-pass welds can be made on material up to 1¼ inch thick.
- It is a noncontact process.
- Beams are readily focused, aligned, and directed by optical elements.
- Very small beams can be used to make very small welds, cuts, or holes.
- Dissimilar metals can be welded.
- A wide variety of materials can be processed.
- They can be readily automated for high speed work.
- They are not affected by magnetic fields.
- Metals with dissimilar physical properties can be processed.
- Nonvacuum or X-ray shielding is required.
- High 10:1 depth-to-width ratios are attainable.
- Beams can be deflected for multiple uses (welding, cutting, marking, drilling).

Laser beams have the following limitations:

- The joint must be very accurate.
- Surfaces must be forced together.
- There is a limitation of beam power for welding, cutting, and drilling.
- Highly reflective materials will deflect the beam.
- High power lasers require a mechanism to deal with the plasma.

SHOP TALK

Heat Input

Some welding power sources have low energy efficiency, such as CO_2 lasers and YAG lasers. Some inverter-based sources, such as MIG/MAG and SMAW, have high efficiencies. When choosing the best joining process, consider, in addition to efficiency, the required and allowed heat input. When these factors are considered, lasers fare as well as MMA.

- Laser machines have a low power conversion efficiency of less than 10 percent.
- Rapid cool rates can lead to porosity and brittleness in the welds.

Laser Assisted Arc Welding Laser assisted arc welding is a hybrid application of several processes. In this case, it is the laser beam and the gas metal arc. With this hybrid welding application, a laser beam (CO_2 or Nd:YAG) is combined in one weld pool with the gas metal arc. As you have learned, laser beam welding is limited due to joint fitup tolerances and being autogenous. The benefit of combining these two processes would be reducing the requirement for precise fitup. It would also allow greater flexibility in which materials can be joined and the types of welds that can be made by the addition of filler metal.

The basic effects of the hybrid application of these two processes are that the arc process has the ability to bridge a gap and not only because of its filler material but also because of its wider flared arc and resultant wider weld pool. The gas metal arc power determines the width of the weld. The laser process is more related to the formation of a keyhole. The laser power determines the depth of penetration. An additional advantage of this hybrid application is that the laser-induced plasma reduces the ignition resistance of the arc, which makes the gas metal arc more stabile, Figs. 25-12 and 25-13.

For video of laser/MIG welding in action, please visit www.mhhe.com/welding.

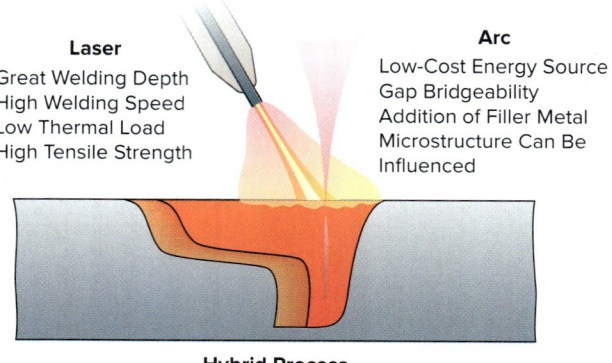

Fig. 25-12 Mechanized gas metal arc welding.
Adapted from Fronius International GmbH. www.fronius.com

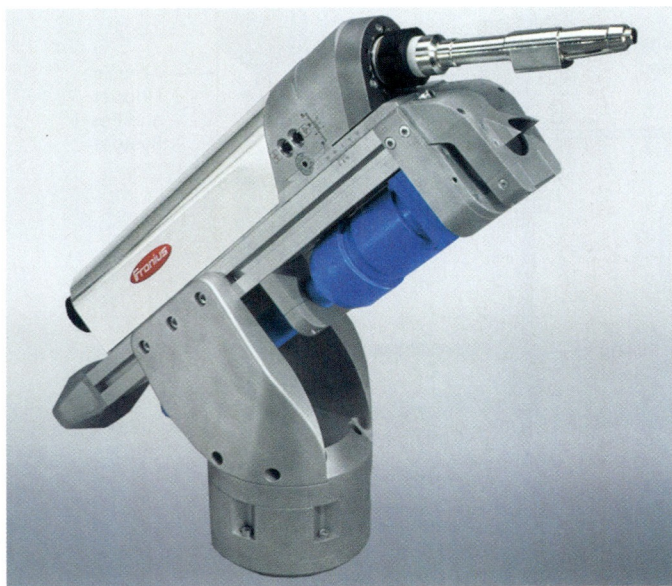

Fig. 25-13 Laser/MIG welding head designed for mounting on a robot arm. Weight is approximately 42 pounds, and size is approximately 30 × 6 × 16 inches. The MIG portion is rated at 100 percent duty cycle at 250 amps and the laser is rated at 4 kilowatts.
© Fronius International GmbH

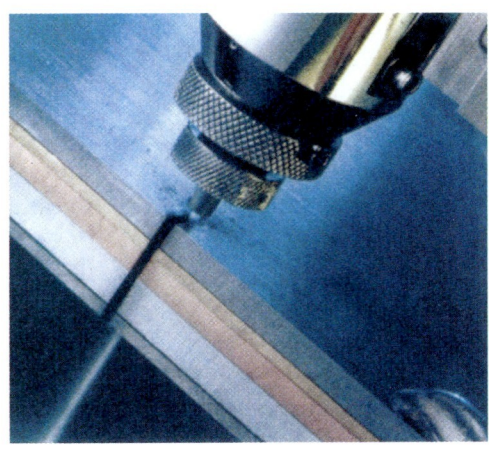

Fig. 25-14 Water jet stack cutting of carbon steel, brass, copper, aluminum, and stainless steel. © American Welding Society. *Welding Handbook,* Vol. 1, p. 47, Fig. 1.57.

The laser-assisted arc welding developments are expected to help in applications like mill coil joining, tailored blanks for the automotive industry, and welding of dissimilar metals. Current research is being done on aluminum alloys, low carbon, and high strength low alloy steels. The use of the laser beam, gas tungsten arc, friction stir, and the plasma arc welding process are also being investigated.

Water Jet Cutting A high velocity jet of water is used to cut a variety of materials including metals and nonmetallics with the **water jet cutting** process. A manufactured sapphire nozzle with a hole from 0.004 to 0.024 inch has water forced through it at high pressures of from 30,000 to 60,000 p.s.i. This is not a thermal cutting process like LBC or PAC, but it does create a kerf from the concentrated water jet erosion. In water jet cutting, water sometimes mixed with abrasive additives is used to erode the material to affect the cut. Since a machine tool is not used in making the cut, this process is more closely related to the thermal cutting techniques than it is to machine tool methods. Dissimilar materials can be easily stack cut, Fig. 25-14.

There are two types of water jet cutting. One type simply uses water, and the other uses water mixed with an abrasive material. Kerf tapering is normally associated with the simple water method because the water has a tendency to spread as it leaves the sapphire nozzle. With abrasive cutting the kerf taper is not a concern unless too high of a travel speed is being used, Fig. 25-15, page 826.

The orifice in the sapphire or ruby nozzle will wear quickly and usually needs replacing every 2 to 4 hours of continuous cutting. Diamonds are also used for nozzles since these orifices wear much better and can give 20 times greater service life. Deburring is generally not required and minimal lateral forces are applied, so fixturing is simple. The process is easily automated and can be used with robotic control.

Materials are cut cleanly, without ragged edges, without heat, and generally faster than with a band saw. Narrow kerfs of 0.030 to 0.100 inch with very smooth edges are typically produced, which is very cost effective since material usage is maximized. This is especially important when cutting expensive materials such as titanium, bronze, Kevlar, and Teflon. Water jet cutting is very versatile because of the many types of materials that it can be used to cut, Table 25-4, page 827.

The following is a list of water jet cutting advantages over conventional cutting methods:

- Cold cutting (no heat-affected zones, no hardening, no cracking).
- Reduces dust and hazardous gases.
- Environmentally friendly.
- Cuts in any direction.
- Perforates most materials without starting holes.
- Cuts virtually any material (including food products).
- Net-shape or near-net-shape parts (no secondary processing required in many applications).
- Minimal fixturing required.
- Saves raw materials (small cutting kerf width, nesting capabilities).

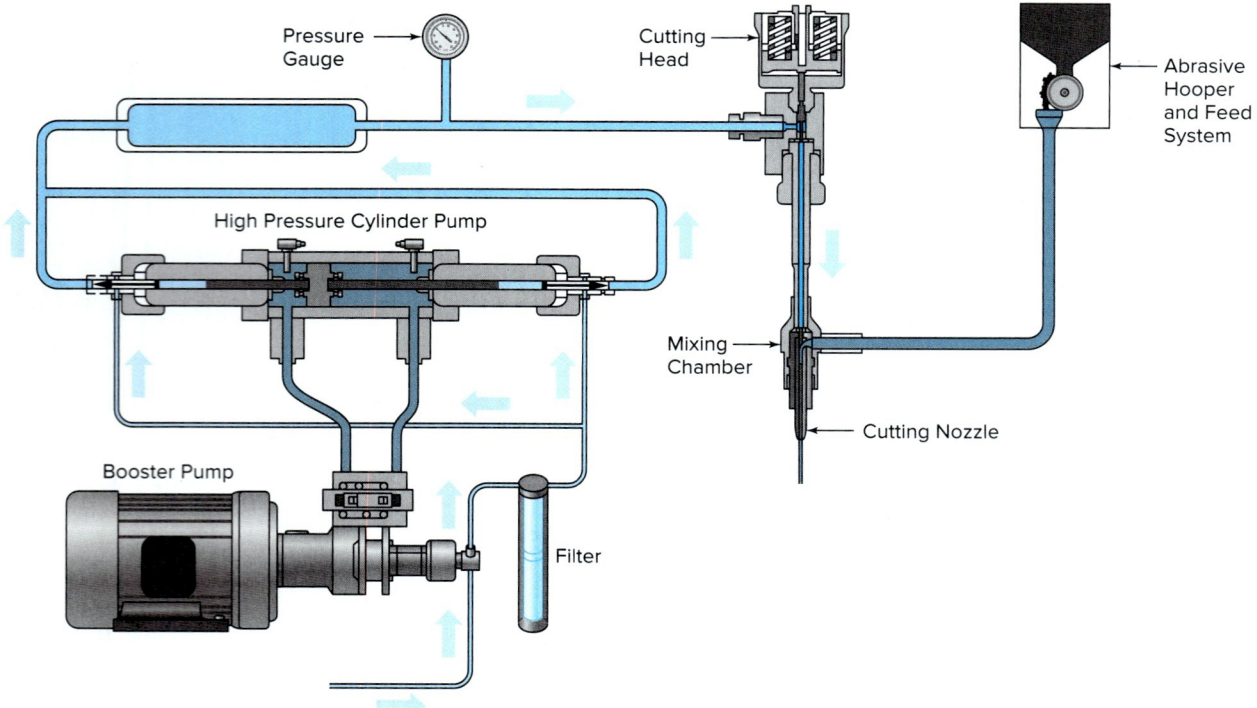

Fig. 25-15 The figure shows a waterjet schematic. The bold arrows indicate water flow direction. Water pressure is boosted from the low tap water pressure to approximately 40,000 pounds per square inch. It is then forced through a tightly constricting orifice (cutting nozzle) and exits at speeds approaching 2.5 times the speed of sound.

- Faster than many conventional cutting tools.
- Does not induce stresses into material while it is being cut.
- Flexible machine integration.

The cold cutting properties of water jet cutting is one of its main advantages. Other cutting methods may burn, melt, or cause cracking in the heat-affected zone. Thermal cutting processes cause surface hardening, warping, and emission of hazardous gases. With water jet cutting the materials undergo no thermal stress, eliminating such undesirable results. Water jet cutting is considered very environmentally friendly as particulate is carried away in the water jet stream and dealt with in a controlled manner.

Friction Welding (FRW) **Friction welding (FRW)** is a solid-state process that uses the heat produced by compressive forces generated by materials rotating together in a friction mode, Fig. 25-16. **Solid-state welding (SSW)** is a group of processes that produces coalescence by the application of pressure at a welding temperature below the melting temperatures of the base metal. Mechanical energy is converted into heat energy. Shielding gases, flux, and filler metal are not required. In the friction welding processes can be used to join a wide variety of dissimilar metals such as aluminum to steel. It is principally used in the oil, defense, aerospace, automotive, electrical, medical, agricultural, and marine industries.

Friction Stir Welding (FSW) A process variation where a probe or tap with a diameter of 0.20 to 0.24 inch is rotated between the square groove faying edges on a butt joint is called **friction stir welding (FSW)**. A shouldered nonconsumable collar has a smoothing effect on the top surface. Temperatures of 840 to 900°F are generated as the lightly tilted rotating probe travels along the length of the joint. The stirring action heats and moves the hot metal from the front of the probe to the rear of the probe creating a weld, Fig. 25-17, page 828.

Welding speeds up to 24 inches per minute on 0.25-inch aluminum sections are typical. Currently the only material being commercially welded with this process is aluminum. With material thicknesses in a range of 0.06 to 0.5 inch, a conventional milling machine can be used. The design of the probe or tap is critical to proper operation. High quality welds with no porosity or cracks are achievable. These welds can be destructively tested with good results even in the heat-affected zone.

The principal advantage for friction stir welding over other friction welding is that parts do not need to be rotated

Table 25-4 Cutting Speeds on Various Materials with Abrasive Water Jet[1]

Material	Thickness (in.)	Travel Speed (in./min)
Aluminum	0.125	40
Aluminum	0.50	18
Aluminum	0.75	5
Armor plate	0.75	10
Brass	0.125	20
Brass	0.425	5
Bronze	1.0	1
Carbon steel	0.75	8
Cast iron	1.5	1
Ceramic (99.6% aluminum)	0.025	6
Copper	0.063	35
Copper	0.625	8
Fiberglass	0.100	200
Fiberglass	0.250	100
Glass	0.250	100
Glass	0.75	40
Graphite/epoxy	0.250	80
Graphite/epoxy	1.0	15
Inconel	0.625	8
Inconel 718	1.25	1
Kevlar	0.375	40
Kevlar	1.0	3
Lead	2.0	8
Lexan	0.5	12
Metal-matrix composition	0.125	30
Pheonolic	0.5	10
Plexiglass	0.175	50
Rubber belting	0.300	200
Stainless steel	0.1	25
Stainless steel (304)	1.0	4
Stainless steel (304)	4.0	1
Titanium	0.025	60
Titanium	0.500	12
Tool steel	0.250	10

[1]Garnet is the abrasive material normally used.
Source: From *Welding Handbook*, 9/e

a very high velocity is called **explosion welding (EXW)**. The subsequent impact and sliding forces create a weld. Explosion welding is typically done at an ambient temperature. The explosion propels the prime component toward the base component at a speed that causes the formation of a metallic bond between them when they collide, Fig. 25-18, page 828.

The explosion deforms the prime component locally, and the detonation quickly progresses. The prime component quickly crosses the stand-off distance and impacts the base component. The welding is accomplished by the plastic flow of the metal pieces across the faying surfaces. Heat may be a by-product of the detonation and collision but is not required for the weld to take place.

EXW is used to join metals that have sufficient strength to withstand the detonation forces. A typical application is to join thin metals to dissimilar thicker metals. An example would be the joining of aluminum and steel. This would be a transition material to allow the carbon steel hull structure of a ship to be welded to the aluminum superstructure. Another example would be to join thin gauge stainless steel (the prime component) to thick

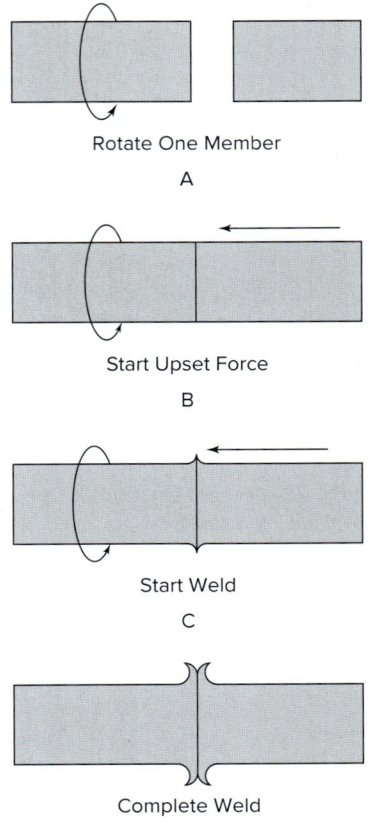

Fig. 25-16 Schematic illustration of friction welding. Note: No shielding or filler metal is required. Source: From *Welding Handbook*, 9/e

against each other. The primary limitation is the limited type of metals that the process is currently usable on.

Explosion Welding (EXW) Another solid-state process that uses a controlled detonation to impact two workpieces at

JOB TIP

Your Pay

Suppose you are in an interview and you're asked what salary you want. Learn in advance what jobs usually pay where you live. Rather than answer, ask the interviewer what the company is planning on paying the best candidate. Evaluate the answer for yourself, based on what you know the field pays.

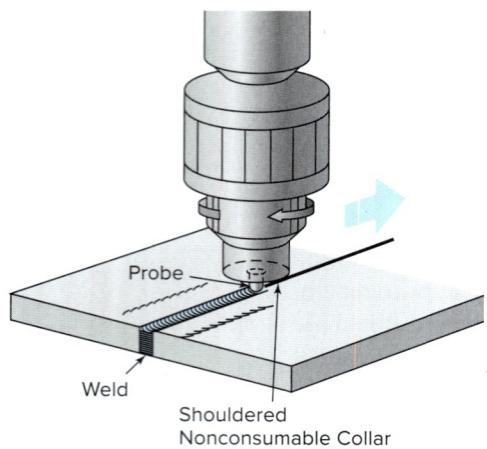

Fig. 25-17 Schematic of friction stir welding. Source: From *Welding Handbook*, 9/e

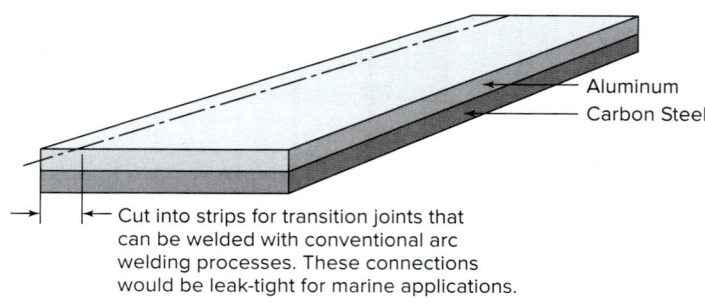

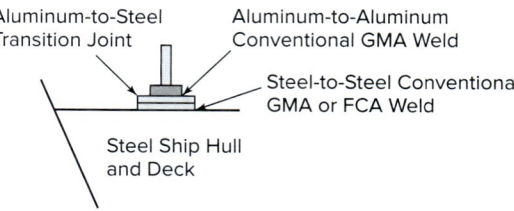

Fig. 25-19 Transition joint created by explosion welding.

They are further broken down to the following.

- Energy source
- Thermal source
- Mechanical loading (that is, pressure normal to the faying surfaces)
- Shielding
- Process description and abbreviation

These processes are applied in the following manner:

- Manual
- Semiautomatic
- Mechanized
- Automatic
- Robotic
- Adaptive control

The processes in this chapter are never applied manually or semiautomatically. They may or may not have adaptive controls. Adaptive controls automatically determine changes in the process conditions and direct the equipment

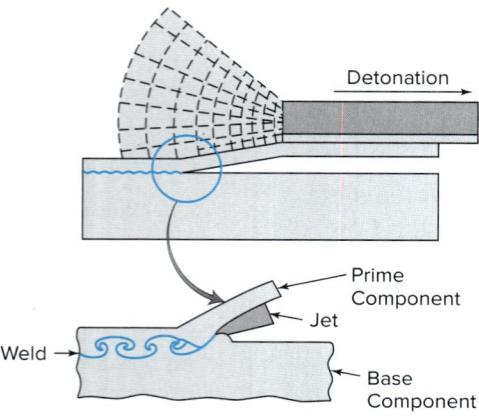

Fig. 25-18 Schematic of explosion welding. Source: From *Welding Handbook*, 9/e

mild steel (the base component) to create a corrosion-resistant surface on a massive object. In this last example a backing may support the base component if it is not large enough to sustain the detonation impact, Fig. 25-19.

Summary

The American Welding Society has defined and described over 110 various joining and cutting processes and process variations. These are contained in a master chart of processes, Fig. 25-20. As a professional in the welding industry you need to understand the advantages and limitations of these processes. These processes are generally grouped into three areas.

- Fusion welding
- Solid-state welding
- Brazing and soldering

ABOUT WELDING

Shipbuilding 1943

In 1943, during World War II, the California Shipbuilding Corp. employed 6,000 welders as well as 160 operators of submerged arc welding equipment. In June, that company made 20 ships and consumed 2,700,000 pounds of welding electrodes. Sixteen other companies were almost just as busy.

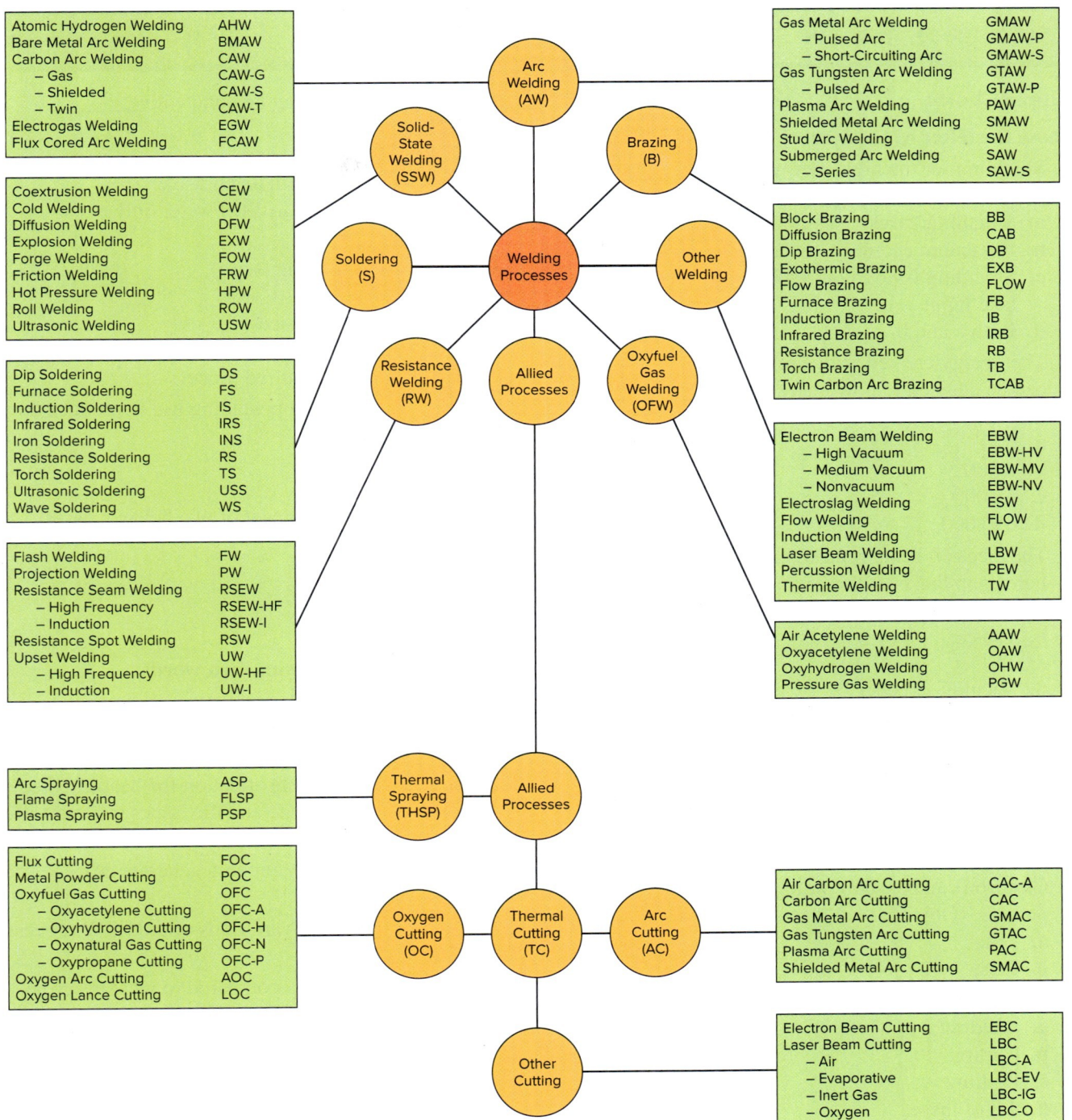

Fig. 25-20 Master chart of welding and allied processes. From AWS Standard Terms and Definitions, AWS3.0:2010, Figures A.1 and A.2

to take appropriate action. More will be covered on adaptive controls and the methods used to apply various processes in Chapters 26 and 27.

The analogy of selecting the best process has been described as the modes of transportation one might use when going on a trip. To get from point A to B you might walk, ride a bike, drive a car, or take a train or plane. But the best method may be a combination of several, such as walking to a car, driving it to the airport, taking a plane, and reversing the order to get to your final destination. Understanding the joining and cutting processes and picking the correct ones for each application is the surest way of producing the highest quality product in the shortest period of time, with the least amount of defects.

High Energy Beams and Related Welding and Cutting Process Principles **Chapter 25** 829

CHAPTER 25 REVIEW

Multiple Choice

Choose the letter of the correct answer.

1. Shielding for the high energy-density beams can be provided by_____. (Obj. 25-1)
   a. Shielding gases
   b. A vacuum chamber
   c. Shielding is not required because of the high heat input
   d. Both a and b

2. The energy created with a high energy-density beam can be equated to how many 100-watt lightbulbs? (Obj. 25-1)
   a. 650
   b. 6,500
   c. 65,000
   d. 650,000

3. The electron beam process can be compared to what home appliance? (Obj. 25-1)
   a. Radar range
   b. Toaster
   c. TV set
   d. Vacuum sweeper

4. The EBW process is usually done on CJP groove welds with what technique? (Obj. 25-1)
   a. Downhill
   b. Uphill
   c. Keyhole
   d. None of these

5. Cast iron can be easily welded with the EBW process. (Obj. 25-1)
   a. True
   b. False

6. Laser beams can be used for_____. (Obj. 25-1)
   a. Welding
   b. Cutting
   c. Drilling
   d. All of these

7. Which of the following is not an issue with LBW? (Obj. 25-1)
   a. Magnetic blow
   b. Reflection of the beam
   c. Porosity
   d. Brittleness in the weld

8. Water jet cutting can be done with_____. (Obj. 25-2)
   a. Plain water
   b. Water mixed with an abrasive
   c. Deionized and distilled water only
   d. Both a and b

9. Friction stir welding is a(n)_____. (Obj. 25-3)
   a. Arc process
   b. Gas shielded process
   c. Solid-state process
   d. Filler metal required process

10. What does explosion welding use to produce the weld? (Obj. 25-4)
    a. Heat
    b. Plastic flow
    c. Low velocity impact of two heavy masses
    d. None of these

Review Questions

Write the answers in your own words.

11. What types of joints and welds are typically done with the high energy beam processes? (Obj. 25-1)

12. Describe the keyhole effect on the heat-affected zone and cooling rate and the issues this may affect. (Obj. 25-1)

13. How does the EBW process generate heat? (Obj. 25-1)

14. Describe the three variations of the EBW process. (Obj. 25-1)

15. Laser beams can be compared to what everyday activities you are aware of? (Obj. 25-1)

16. How does laser light generate so much energy? (Obj. 25-1)

17. Describe the metal surface required for LBC and why it is required. (Obj. 25-1)

18. List 10 advantages of the water jet cutting process. (Obj. 25-2)

19. What is solid-state welding? (Obj. 25-3)

20. Describe how the EXW process works. (Obj. 25-4)

INTERNET ACTIVITIES

Internet Activity A

Connect with others about welding on an online bulletin board. Use the Internet to locate two bulletin boards about welding. Then describe what you found. (Before opening any bulletin board, read the description.)

Internet Activity B

Talk to others in a welding chat room. Look on the Internet to find a welding chat room. One place you will want to look is on the AWS Web site. Ask a question in the chat room. Share the question and response with your class. (Before opening any chat room, read the description.)

Design credits: Cinema and movie icon: Denis Meshkov/123RF; and Illustration of Welding icons: Mr. Nachai Sorasee/123RF

26

General Equipment for Welding Shops

Chapter Objectives

After completing this chapter, you will be able to:

26-1 Describe the function and use of work-holding devices that position the work for the most advantageous welding position.

26-2 Describe the function and use of devices that move the welding process along a fixed workpiece.

26-3 Describe clamping and support operations required in a complete welding shop.

26-4 Describe the function and use of various work cutting and shaping devices.

Companies that do welded fabrication make use of a variety of equipment. Some of this equipment is used in carrying out the welding operation. Other types are used for the related processes such as metal forming, cutting, and finishing that are necessary in fabricating welded structures. It is important for you, as a welding student, to learn how to use the tools and equipment presented in this chapter so that you will be able to perform these processes on the job.

Screens and Booths

Whenever welding is done in a shop where others are doing other jobs, these workers must be protected from the effects of arc rays, the spatter of molten metal, and sparks. In areas dedicated to the welding of small parts, permanent booths are erected. They are made of sheet metal or heavy canvas, and they are painted with a special protective paint. The booths are often equipped with exhaust fans for removing fumes and ducts for introducing fresh air.

A portable screen is used to shield large work when the welding equipment must be taken to the job. A semitransparent curtain constitutes the near-ideal welding curtain, exhibiting good visibility, minimizing the arc glare, and

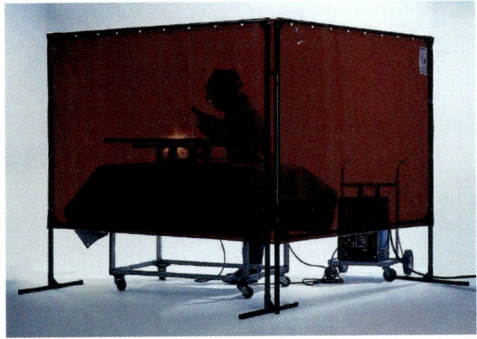

Fig. 26-1 Spectra Orange See-Thru vinyl and movable frame system. © Wilson Industries, Inc.

reflecting usable light back into the work area. There are a variety of colors available:

- Spectra orange
- Yellow
- Green
- Gray
- Blue

One of the best semitransparent curtain colors is Spectra orange. No other transparent welding curtain material manipulates light waves like the Spectra orange curtain. With its special patented polyvinyl chloride (PVC) formula, Spectra orange absorbs, scatters, and filters the light spectrum to create a safer working environment for the welder and surrounding coworkers. This type of curtain is available in 14-mil and 40-mil thicknesses for the manufacture of screen panels or curtain applications, Fig. 26-1. The green color can be had in extra dark to stop all ultraviolet rays and blue light. This dark color is recommended for PAC and other bright welding applications. Where visibility is not an issue, the opaque Rip-Stop vinyl laminated polyester design provides for added durability and strength. Rip-Stop is formulated to withstand abrasion and abuse, while resisting oil, water, and most acids, and is often used in combination with See-Thru vinyl to create a specialty curtain or partition.

Special ventilation systems are required to remove smoke, fumes, and particulate matter from the breathing zone. Welding fumes should not be randomly discharged into the atmosphere and need to be filtered. The cost of cooling or heating a fabrication area dictates the need to recirculate filtered clean air and not discharge it outdoors. A clean, safe environment is critical to protecting people and resources. Various electrostatic and dry cartridge type air-cleaning systems are available. These products are available for in-plant air filtration and pollution control systems. These systems help eliminate dangerous dust, fumes, and

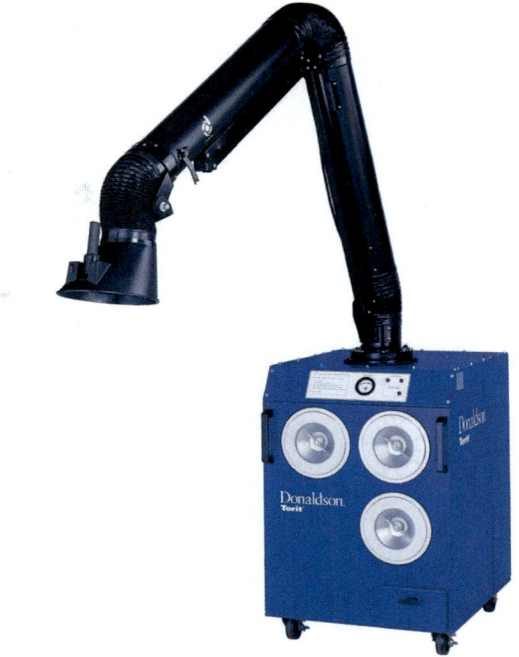

Fig. 26-2 Portable source capture system, which recirculates the air directly back into the shop. Note the flexible hood can be equipped with a light for added convenience. © Donaldson Company, Inc.

mist contaminants. They can be source capture, Fig. 26-2, or large central systems consisting of hoods and ductwork, and are used on heavy production applications, Fig. 26-3, page, 834.

Work-Holding Devices

Weld Fixtures

Once parts for a weldment are accurately produced, they must be held in position for welding. This is done with devices called *fixtures, jigs, or tooling*, each of which has essentially the same meaning. Their function is to hold parts in proper alignment during the assembly of a weldment. Fixtures promote good fitup tolerances, resulting in consistently high-quality weldments. Manual or semiautomatic welding may tolerate fixture inadequacy; however, for automatic and robotic applications, fixtures need to be very accurate and easily accessible by the welding torch or gun. A fixture may be used to hold parts for assembly and tacking, or in some cases they must be built heavy enough to support the weldment during the entire welding operation.

The following advantages of fixtures should be considered. They

1. Improve fitup of parts to achieve tighter tolerances
2. Locate and orient parts for easier loading of parts by the operator

General Equipment for Welding Shops

Fig. 26-3 Central collections systems are highly efficient and recirculate the air directly back into the shop. © Donaldson Company, Inc.

Fig. 26-4 Modular tooling being used to hold an irregular shape for welding. © Bluco Corporation

3. Help identify parts that are out of tolerance
4. Help control weld distortion
5. Reduce labor cost to produce a weldment
6. Provide more consistent quality
7. Reduce production errors by having fixtures accurately identified

Fixture design is crucial for proper use. The fixture must be easy to load and unload. It must be capable of withstanding the tremendous expansion and contraction forces as the weldment goes through its thermal cycle. They need to be as simple and inexpensive as possible yet meet intended purposes. The following sequence should be used for fixture design:

1. The weldment form, fit, and function need to be fully understood to develop initial design ideas.
2. Heat transfer, the work connection, capabilities, direction of part loading, accessibility, part flow, part staging, wear, and maintenance are important operating criteria that need to be considered in the initial design stage.
3. The welding process, weld area accessibility weldment orientation, clamp and holding device orientation, work connection, magnetic properties, and inspection need to be understood and worked into the design concept.
4. Consider the operator from ergonomic issues, safety, and part-handling issues and continue design concepts.
5. The design concept should be reviewed to make the most economical, simplest fixture per weldment as possible.

Modular Tooling

By using modular tooling in the form of platens, locators, bracing, and various clamping devices, much greater flexibility is permitted than when custom-built fixtures intended for a single weldment are used. The platen has a precision hole pattern, which allows the location of various devices. This type of product is accurate, rigid, and easy to set up and use. Modular tooling can be used for the repeatable or one-of-a-kind location of parts to be welded, tacked, or held in position for subsequent operations, Figs. 26-4 and 26-5.

It is obvious that modular tooling must be designed for the intended task. A fixture to hold parts that are only going to be tack welded together need not be designed to

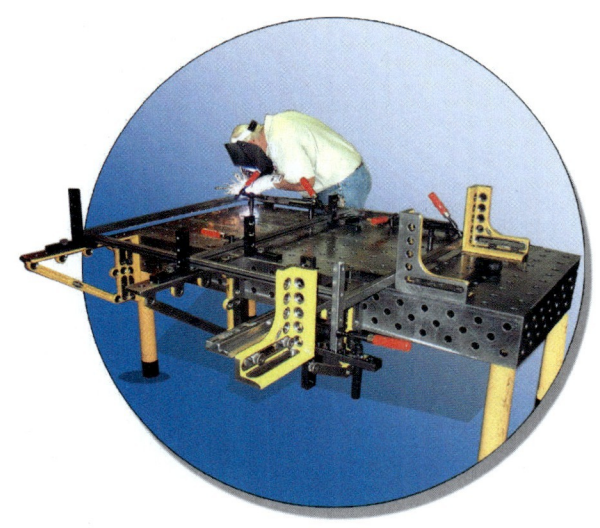

Fig. 26-5 Modular tooling system available for small parts fixturing. Adaptors are available to allow use of the small systems on the larger systems as shown in Fig. 26-4. © Bluco Corporation

take the heat and thermal stresses that will be imparted into a fixture used for welding.

Welding Positioners

Welding positioners, Figs. 26-6 and 26-7, permit the placing of weldment joints in a flat position for optimum welding. They have plane table areas that can be tilted and rotated in any direction. On the smaller positioners, the table is moved by handwheels or gears; and on the larger positioners, by electric gear drives. Parts are to be welded, and in some cases, production jigging is secured to the plane table of the positioner.

The use of positioners to enable welding in the flat position has increased production, reduced costs, improved quality, and promoted safety in both production and repair welding. Flat position welding permits the use of larger electrodes, higher currents, fewer passes, and better control of the weld pool. A vertical-up weld takes almost three times as long to do as the same weld done in the flat position.

Weldments vary in size and weight. For this reason, welding positioners are available in a wide range of sizes. Standard capacities run from 100 to 60,000 pounds. Some positioners have a capacity of 200,000 pounds. The positioning table may be rotated or tilted as desired, Fig. 26-6. It permits welding on the sides, top, and bottom of a job. The table may be tilted through 135° (45° beyond the vertical position). Regardless of the angle of tilt, the whole table may be rotated through a complete circle. Rotation allows all joints to be welded in the flat position without resetting or handling the weldment. Positioners can be used in conjunction with various other devices. The intent of these devices is to make welding as easy as possible. The easier the welding, the more productive and efficient it will be, Fig. 26-7, page 836.

Since welding positioners enable average welders to do quality work, their use is timely with an increase in production with a reduced workforce.

Turning Rolls

Welding fabricators who do a great deal of tank fabrication depend on turning rolls to make *circumferential seams* in the flat position without interruption in the travel. Turning rolls, Fig. 26-8, page 836, support the workpiece on its outside diameter and rotate it.

There are two main types of turning rolls: those with steel or rubber-tired rolls and those with roller chain slings.

With the positioner table at 0°, a horizontal fillet weld could be made around the base, while an overhead fillet weld could be made around the top flange. Neither of these would be in the optimum position.

A

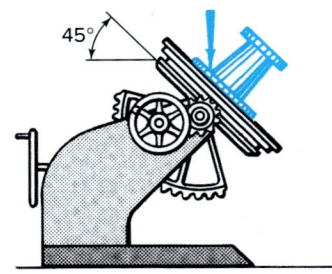

Set the positioner at 135° or 45° from vertical so the flange can be welded in the more advantageous, flat position.

C

The positioner is set at 45° for a flat position fillet weld on the base. This would be a better position for both quality and productivity.

B

Adjustable Roller Stand

A positioner with an adjustable roller stand can support longer cylinder-type objects.

D

Fig. 26-6 This positioner can rotate the part as well as be set at an angle that makes the welding position optimal. Flat position welds are noted by the arrows. The Welding Encyclopedia

General Equipment for Welding Shops **Chapter 26** **835**

Fig. 26-7 A weld positioner can be used in conjunction with other devices like a roller stand and manipulator. © Pandjiris

Fig. 26-8 Car-mounted tandem power rolls. Each roll has a 90-ton turning capacity and a 30-ton weight capacity. Rolls are driven by synchronous motors to give a combined rotating capacity of 180 tons for high eccentric loading with a weight capacity of 60 tons. The fixed center distance handles workpieces 72 through 120 inches in diameter. Manual cars facilitate setup changes for vessels of various lengths. © Pandjiris

The wheel-type turning rolls are more common. They come in two varieties: the separate driver-idler type and the unit frame. In the *driver-idler* type, the driver turns the workpiece. The idler is a matching unit that is not powered. Several idlers can be used with a single driver. They range in capacity from 600 to over 1 million pounds. *Unit-frame* turning rolls have one fixed-location driving axle. The other axle is adjustable for various center distances that handle diameters from 3 inches to 6 feet or more. Weight capacities range up to 30,000 pounds; and frame lengths, up to 12 feet or more. Models that rotate and tilt are available. *Sling-type* turning rolls have a roller chain sling with rubber or brass feet. The chain cradles the workpiece so that large areas of thin-walled cylinders can be supported. The load capacity is as high as 27 tons.

Figure 26-9 shows a larger tank being rotated on a pair of powered rolls. The submerged arc welding process is being used to weld a *circumferential seam*. The welding wire size is $3/32$ inch, and the welding current is 24 volts at 300 amperes. The welding speed is 50 inches per minute on both longitudinal and circumferential seams.

Weld Grippers

A weld gripper is a welding work-holding device that has three movable jaws like the jaws of a lathe chuck. The workpiece—a tank or pipe section—can be gripped on the inside diameter or the outside diameter. The action is

Fig. 26-9 A welding positioner teamed with a manual TIG weld being made on a flange to large bore heavy wall stainless steel pipe. The all welded positioner and the three jaw gripper is designed specifically for pipe welding shops. Note the two wireless foot pedals to control weld output and rotational speed. Also notice the well-organized tool rack. © Team Industries, Inc.

Fig. 26-10 Head and tail stocks with three-jaw grippers. They are variable speed with brakes and drive components are housed within the column structure. © Pandjiris

Fig. 26-11 A manipulator teamed with travel cars to expand its capabilities and reach. The vertical column and horizontal slides are required to accommodate various size weldments. © Pandjiris

fast. The gripper is mounted on a turntable or weld positioner for universal movement, Fig. 26-9.

Headstock-Tailstock Positioners

Like weld grippers and rollers, headstock-tailstock positioners rotate weldments about a horizontal axis and permit welding in the flat position. They support the workpiece at each end. These positioners can be used for angular weldments as well as cylindrical ones.

The headstock is powered and may have a constant or variable speed. The tailstock is not powered. When used in pairs, the headstock and tailstock may have a capacity as high as 160,000 pounds. When used singly, they can support 80,000 pounds. The positioners shown in Fig. 26-10 are available with capacities ranging from 100 to 120,000 pounds.

Manipulators

Manipulators are applicable to both the positioned fillet and the simultaneous double fillet welds. They are also widely used for *circumferential* butt joints and other types. Various automatic processes such as submerged arc welding units are mounted on the horizontal arm, Fig. 26-11. The manipulator provides vertical and horizontal travel and may be rotated through a full 360°. Large units may be mounted on a double-rail track for maximum usage in the shop. On large installations, seating is provided for the operator at the welding head. Operation may be remotely controlled.

Turntable

A turntable provides powerized table rotation in either direction at adjustable speeds. Rotation is controlled by the operator, Fig. 26-12, and may be set for variable speeds. The work is positioned on the turntable and rotates under the welding gun.

Turntables are widely used to move weldments in and out of a robotic work cell. The welding operator can unload and load one side of the turntable while the robot is welding components on the other side. Of course, safety curtains and controls must be in place to protect the

Fig. 26-12 With options added or as part of a system, turntables like this can be used in numerous applications: inspection, painting, flame cutting, and welding. Indexing turntables can meet exacting needs such as robotic applications where up to 40,000 pounds capacity with a 20′ diameter that can rotate 180° in 15 seconds and positioned itself within +/− 0.100″ on the outside perimeter of the table. © Pandjiris

Fig. 26-13 Magnetic welding ground (work) connectors rated for 500 amps and designed to accept 3/0 and 4/0 welding cable. The rotating work connector on the left covers 12 square inches giving good stability on rotating weldments. It has 4 inches of current carrying contact surface. The model shown on the right is only a magnetic work connection not intended for rotating applications. © Lenco dba NLC, Inc.

welding operator from arc flash and the mechanical motion of the robot.

Work Connections to Rotating Weldments

When using positioners, turning rolls, head and tailstock, manipulators, or turntables, getting the electrical work connection to the rotating part can be a concern. Typical work connections will become twisted, resulting in damaged equipment or cabling. This may also result in a less than adequate path for the welding current to flow through. In some cases (vessels, tanks, etc.), there will not be a place to which the normal work clamping device can be attached. Figure 26-13 shows an example of a magnetic work clamp designed for application to rotary-type weldments.

Weld Seamers

Seamers, Fig. 26-14, provide accurate clamping and backup for welding external longitudinal seams of cylinders, cones, boxes, and flat sheets. The system consists of a tabletop mounted on a mainstay that is fixed to the base. The tabletop contains the copper-tipped clamping fingers. A machined track is mounted on the tabletop; the track is fitted with hardened and ground roundways. A carriage rides on bearings on the roundways to provide smooth movement of the welding equipment over the length of the seam welder. The carriage is powered by a variable-speed motor that is controlled by a microprocessor. Travel movement is linear within +0.015 inch per 10-foot length.

Fig. 26-14 Seamers of this type are available for lengths from 24 inches to 10 feet. Some models can clamp materials as thin as 0.005 inch and with linear accuracy of 0.005 inch over 10 feet for high precision capability. © ITW Jetline - Cyclomatic

Clamping of the part being welded is air-operated and effected by two hoses that actuate the movement of two banks of copper-tipped, aluminum clamping fingers. Air regulation is provided to affect a clamping force of up to 5,000 pounds per foot. This clamping force can clamp 0.020- to ⅜-inch thick parts.

This equipment is well-suited to semiautomatic and automatic welding processes. There are two basic types of seamers: one permits welding on the inside so that the weld reinforcement is on the inside of the workpiece, Fig. 26-15; the other permits the welding on the outside so that the weld reinforcement occurs on the outside, Fig. 26-16.

Side Beam Carriage

For precision linear travel, a side beam carriage is an effective tool. The system can be used to carry any welding process, Fig. 26-17, page 840. This manufacturer has travel lengths up to 20 feet long. The side beam track is mounted on a longitudinal seamer. This is only one of the applications for this type of side beam carriage. This type of side beam track is manufactured from a rigid, box-section beam that has been stress-relieved prior to precision machining. Two case-hardened roundways are utilized as the main part of the track; the carriage has cluster bearings that ride on these roundways. Linear accuracy of the track is held to +0.005 inch for every 10 feet of travel length. In addition to the dimensional accuracy of the track, the drive

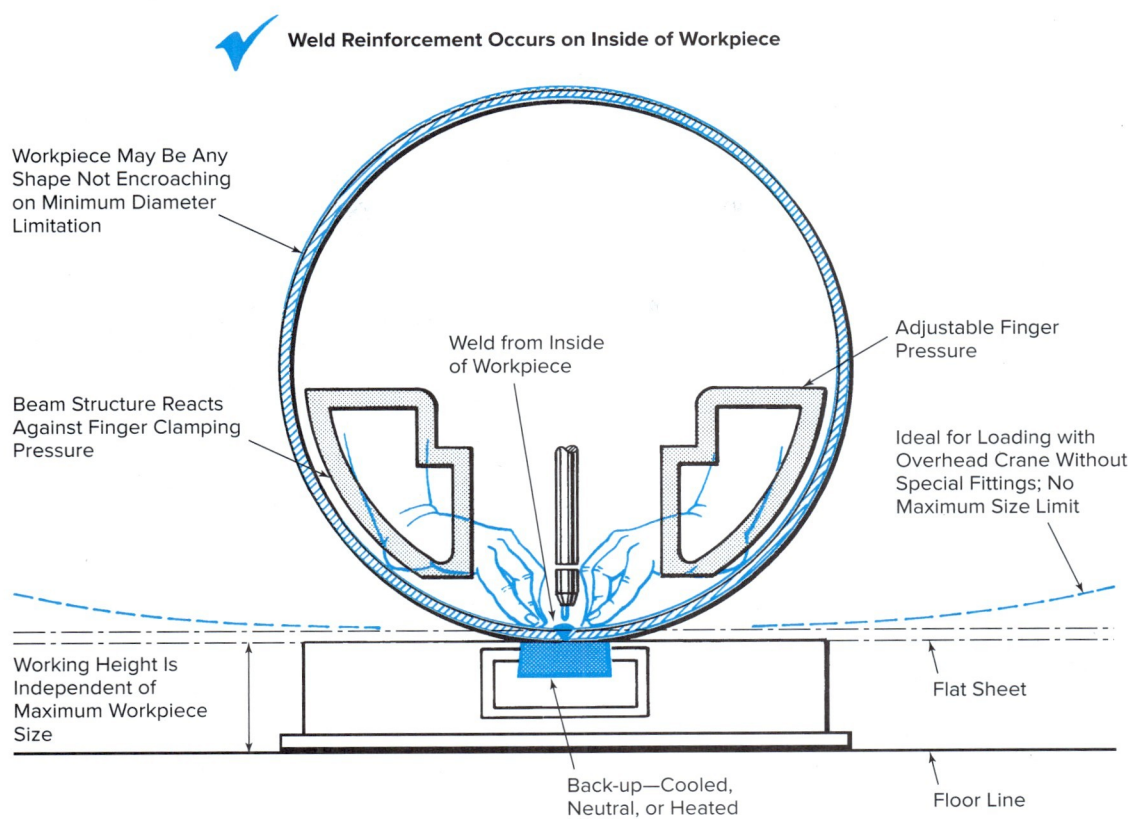

Fig. 26-15 Weld seamer procedure for internal welding. Jetline Engineering Inc.

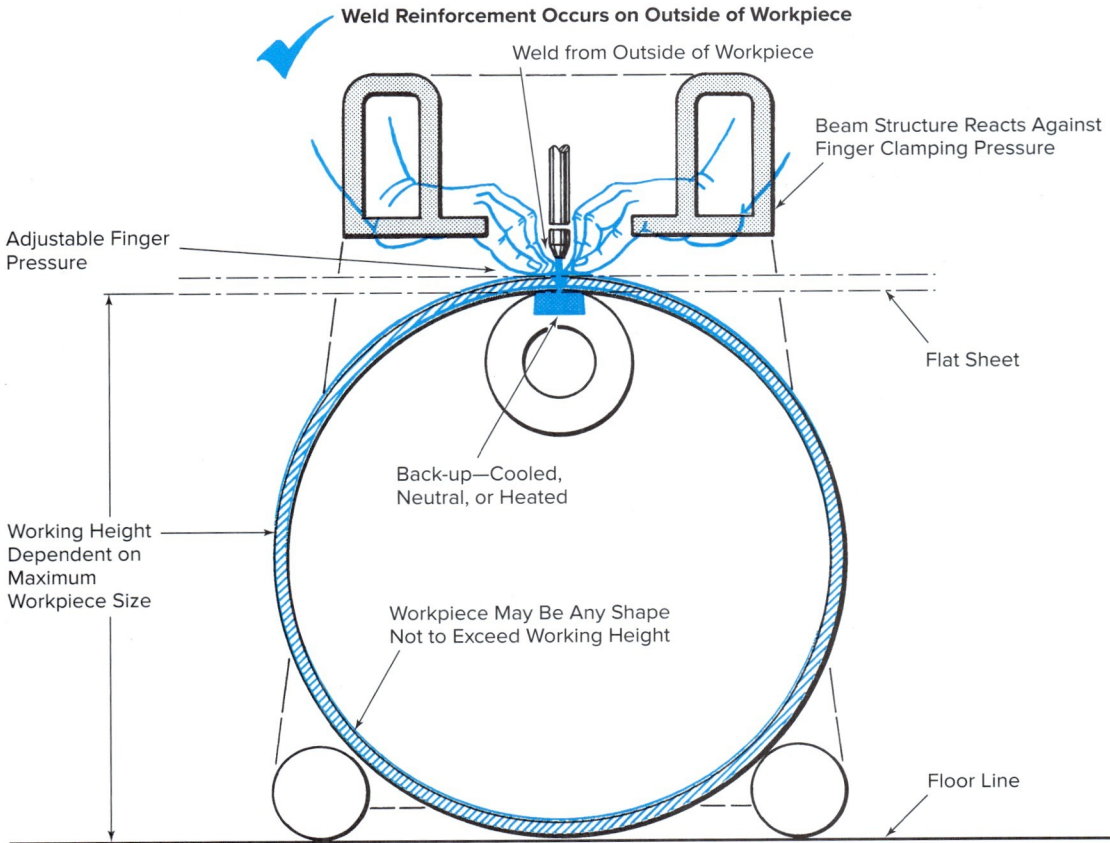

Fig. 26-16 Weld seamer procedure for external welding.

General Equipment for Welding Shops Chapter 26

Fig. 26-17 Side beam carriage for doing precision linear welds with a variety of processes. © ITW Jetline - Cyclomatic

Fig. 26-19 A 3 o'clock welding machine as used to make girth welds on large storage vessel construction in the field or shop. This unit is set up to do submerged arc welding. The unit is hung and supported off the top edge of the structure being welded. The system is motorized and moves around the vessel carrying the welding head and controls as well as the operator and safety cage. © Pandjiris

Weld Elevator

This mechanical unit permits welders to be raised or lowered along a vertical surface so that they are always in the best possible position in the shortest possible time. It is a platform type controlled by the welder, Fig. 26-18. The unit is mounted on rollers for movement in the shop or over rough terrain. This unit has a work platform of 6 by 11 feet and is available with lift heights from 30 to 47 feet with a weight capacity up to 1,750 pounds. This unit is powered by a liquid-cooled dual fuel engine. A joystick is used to control lift and drive controls for smooth and simple operation. Other types of vertical lifts are available that can also move the welding process in a vertical direction. In some instances it is necessary to move the welding process in a horizontal direction around a large vessel, for example. These are sometimes referred to as 3 o'clock welding machines, Fig. 26-19. Figure 26-20 is a view of a pipe fabrication shop using several of the work holding devices discussed previously.

Magnetic Grip Fixtures

There are a number of clamps and holding devices that take advantage of magnetic attraction to speed the work of the welder. Several types are shown in Fig. 26-21. These

Fig. 26-18 A motorized scissor lift. A joystick is used to control lift and steering. © NES Rentals

system is designed to provide smooth, accurate movement of the travel carriage. This is achieved by the use of a linear drive. The linear drive consists of a hardened and ground shaft that is rotated by the drive motor. The carriage is connected to this drive shaft through a special bearing block in which the bearings are set at an angle to the axis of the shaft. As the shaft rotates, it imparts a linear motion to the carriage. This type of drive is not only accurate, but also backlash-free. The track and carriage system is supplied with the latest in the range of motor controls. This is microprocessor-based with closed loop speed control which provides repeatable travel conditions.

types of fixtures are permanent magnets that require no outside power sources or internal batteries. Such fixtures are portable. They provide safe holding strengths from 6 to more than 250 pounds. Whenever holding and positioning are required, this type of fixture can speed the welding operation and increase production.

Magnetic grip devices are equipped with 90 and 180° protractors and V-shaped shoes. They act as fixtures, jigs, and hold-down tools. They provide safe, positive holding and positioning of sheets, parts, rods, and tubes during welding and other fabricating operations. Magnetic fixtures eliminate tacking, clamps, and makeshift holding setups. The position of the work can be changed quickly since it can be held or released instantly.

Care must be taken with magnetic clamps because they may adversely affect the welding arc by inducing *magnetic arc blow*.

Track and Trackless Carriage Systems

Figure 26-22, page 842, shows a track system. This is a mechanized system; note the operator running the various controls and monitoring joint alignment, which utilizes standard manual or semiautomatic welding equipment. The semiautomatic welding gun is visible in the picture. An oxyfuel or plasma cutting torch could be used for joint preparation or metal removal. Metal surfacing can also

Fig. 26-20 A well-organized bay setup for welding stainless steel pipe. It is critical when working with certain alloys like stainless steel to segregate them so cross contamination does not take place. Note each work cell with its own positioner and overhead crane for material handling operations. © Team Industries, Inc.

Fig. 26-21 A permanent magnetic grip welding fixture. © Bunting Magnetic Co.

Fig. 26-22 A vertical V-groove butt joint is being welded uphill with the FCAW-G process. This is a mechanized application using semiautomatic welding equipment. The carriage is performing accurate speed control, side-to-side weave can also be machine controlled. However, the operator is required to monitor and control joint alignment and various other weld variables as required by the weld pool and bead formation. © BUG-O

Fig. 26-23 A BUG-GY® trackless, self-guiding welding carriage. Intermittent (stitch) fillet welds are being placed on a T-joint. Guide wheels and electro magnets are used to track the weldments. Controls are shown in Fig. 26-24. © BUG-O

be done with these types of systems. Very precise coverage is possible, and a minimum of base metal dilution takes place. This is an important benefit because dilution of the base metal seriously affects such characteristics of the surfacing material as resistance to corrosion, abrasion, and impact. The vertical track is magnetically or mechanically held to the weldments in this case. The track can just as easily be set up in any position required. Curved and flexible tracks are also available.

Trackless carriage systems are available that follow the contour of the weldments, Fig. 26-23. These systems are designed for fillet welds on T, lap, and corner joints. They can also be used for groove welds on various joints such as butt with certain weldment modifications. It is a lightweight, portable carriage that uses a powerful magnet and guide wheels to clamp and track on the weldment. It can be programmed for continuous or intermittent (stitch) welding as shown. It has a feature called "weld back" for crater filling. Dynamic braking and closed loop feedback on carriage speed allow this system to make crisp starts and stops and to travel at precise speeds. Controls are listed in Fig. 26-24.

These types of carriage systems can cost much less than custom-designed machines and robots for welding and cutting operation.

Orbital Welding Machine

This type of welding machine is used to make groove welds on pipe and tube butt joints, Figs. 26-25 and 26-26. Certain welding heads are designed to make tube-to-sheet welds for boilers and heat exchangers, Fig. 26-27, page 844. Orbital welding is typically done using the GTAW or GMAW process. When the GTAW process is used and joint thickness and design are appropriate, the welds can be made autogenously. There are two basic types of orbital heads, those with open arc and those with enclosed heads. Because of their design, the enclosed heads are typically used for smaller diameter pipe and tube with thin wall thickness that can be autogenously welded. Orbital systems are used when the pipe is held stationary and the welding head moves around the pipe (2G, 5G, and 6G positions). It takes sophisticated computer controls to deal with the various welding procedures required for the welding positions that will be encountered. These units are highly portable and when combined with the appropriate inverter welding power source and controls produce very high quality and consistent welds.

For video of orbital welding, please visit www.mhhe.com/welding.

Miscellaneous Equipment

In addition to the positioning equipment previously described, cranes, chain hoists, jacks, clamps, and tongs are required for handling and positioning of the work.

Controls:

A. Digital meter, shows speed when traveling or distance when programming.
B. Weld length setting button [0.1 in. (or 1 cm) increments].
C. Skip length setting button [0.1 in. (or 1 cm) increments].
D. Reverse travel distance setting button, for crater fill [0.1 in. (or 1 cm) increments].
E. Programming button.
F. Travel start button, left.
G. Travel stop button.
H. Travel start button, right.
I. Travel speed knob.
J. Mode select switch, continuous or stitch.
K. Arc enable on/off switch.
L. Magnet on/off switch.
M. Power switch.
N. Weld contactor.

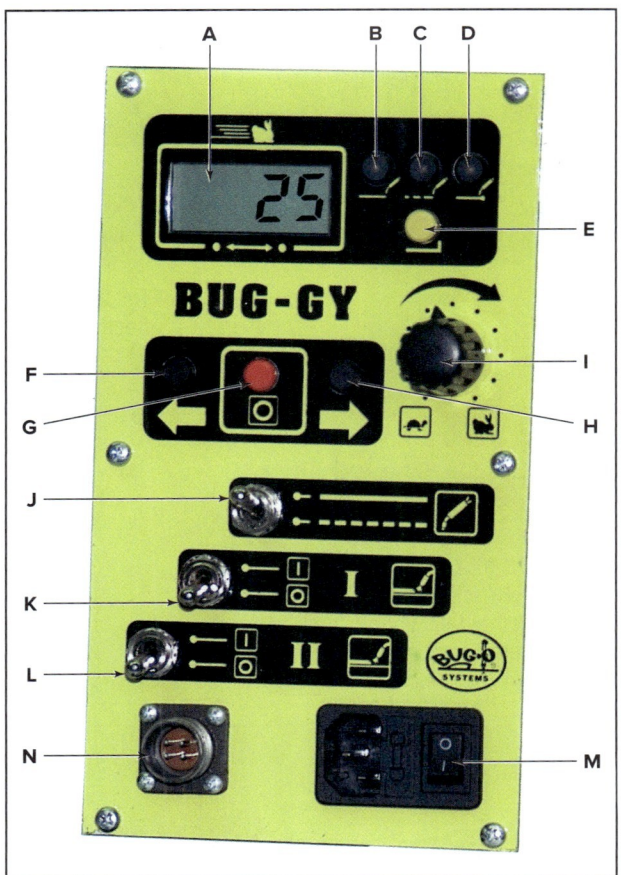

Fig. 26-24 Trackless welding carriage controls. © BUG-O

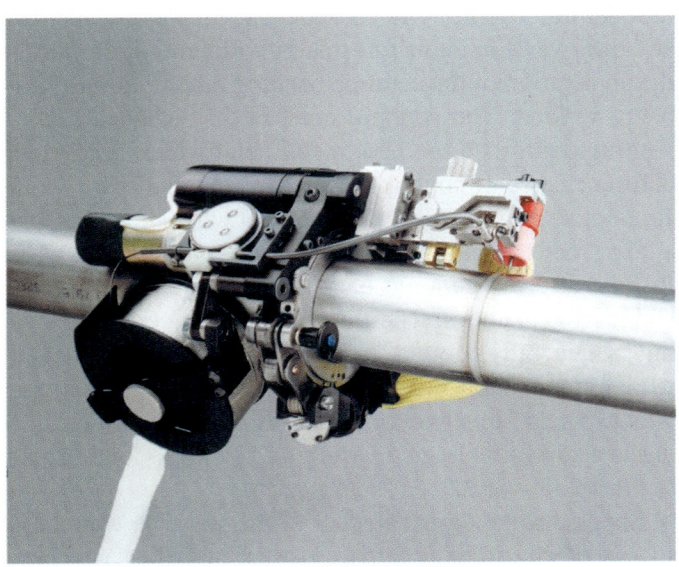

Fig. 26-25 Open arc head that rotates around the pipe. Since the entire head rotates it is mounted on the pipe using guide rings. It covers pipe sizes from 1 to 14 inches outside diameter. It can be used for autogenous welding or equipped as in this case with a filler wire feeder. © Magnatech Limited Partnership

Fig. 26-26 This enclosed head orbital welding machine is used for autogenous tube welding applications. Tubes/fittings are clamped using collets on both sides of the head. The welding arc is totally enclosed within the chamber of the head, which provides 360° inert gas shielding of the weld bead. The torch, which holds the tungsten electrode, rotates around the tube while the body remains stationary. Tube sizes from 0.5 to 4.5 inches outside diameter are typically covered. © Magnatech Limited Partnership

General Equipment for Welding Shops **Chapter 26** 843

Fig. 26-27 This style head is for tube-to-sheet seal welding applications for boilers and heat exchangers. GTA welds can be made autogenously or with filler metals. Various tube to head joint geometries can be accommodated such as the tube and head flush, extended, or recessed. © Magnatech Limited Partnership

Gullco Katbak Nonmetalic Weld Backing

From This...
Katbak's Self-Adhesive Aluminum Foil Tape Makes It Easy to Apply...

To This...
...Enabling Full Penetration from One Side and a Uniform Back Bead Finish Quality

Fig. 26-29 Ceramic backing material being used to produce cost effective radiographic quality welds. © Gullco

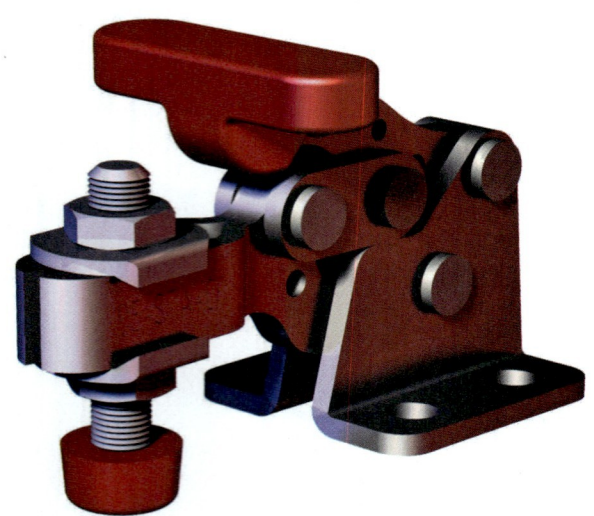

Fig. 26-28 A horizontal hold-down clamp. © Destaco

A generous supply of C-clamps of all types and sizes, hold-down clamps, wedges, bars, and blocks is necessary for the proper spacing and lining up of parts. One type of hold-down clamp is shown in Fig. 26-28.

Backing Materials

The quantity, quality, and appearance of welds are improved by the use of backing materials of all kinds. Copper in bars, strips, and blocks supports the molten metal when welding some types of joints in certain metals. Copper strips laid along the parallel edges of a seam help to keep the welds of uniform width. By carrying away some of the heat, they also reduce distortion. Copper is chosen because of its high heat conductivity and because it resists fusion with the base metal being welded.

In some cases copper may not be acceptable due to possible contamination with the base metal. In these cases there are other materials that can be used for backing materials such as ceramics.

Ceramic weld backing is usually supplied in strip form approximately 24 inches long and 1 inch wide. It can be in the form of a solid tile, or it can be segmented with knuckle-type joints to allow for use on flat or curved surfaces. The ceramic surface the weld is made up against can be flat or have a depression to allow for some root reinforcement. Figure 26-29 shows an example allowing root reinforcement. The ceramic backing is attached to an adhesive aluminum foil that is wide enough to allow the application of the ceramic backing directly to the joint being welded. The ceramic material has a much higher melting point than the base metal being welded, so there is little chance of contamination. Since ceramic is a nonconductor, the welding arc may go out when moved across a large root opening in the case of weaving the root pass. In this case the ceramic backing may be coated with a conductive material. The thickness of the ceramic backing is generally around ¼ inch; however, for high amperage application the backing can be made thicker and wider to help prevent excess melt-through. Ceramic backing is also available on a variety of shapes to lend its use to most CJP welds.

Shops that do a great amount of maintenance work have shapes of all kinds in carbon, plastic, and fireclay. These materials form dams or molds on operations that require the building up of pads and shoulders to certain limits. Round carbon sticks are used to retain the shape of holes in hot metal and to protect threads in tapped holes from the heat of welding.

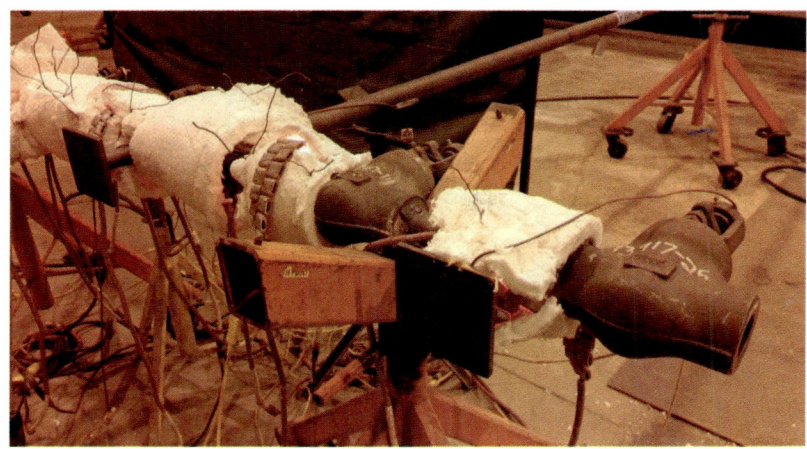

Fig. 26-30 Post Weld Heat Treatment (PWHT) on several groove weld butt joints. Resistance heating sources are being used along with the white insulation to hold heat in and reduce heating in the valves. © Team Industries, Inc.

Preheating and Annealing Equipment

As discussed in Chapter 3, heat treating, annealing, and normalizing are specific metalworking processes required for some types of work. Often the equipment is permanent, and it is found in most big shops. It may be large enough to handle only small parts, or it may be designed to take very large pressure vessels. Sometimes, the preheating is done in temporary ovens built of firebrick. Heat-treating ovens can be fired by electricity, gas or oil burners, coal, or coke fires.

Portable on-site field use of preheating and postweld heat treatment may be required. This can be done with equipment as simple as oxyfuel flames or as sophisticated as a microprocessor-controlled resistance system or an induction heating system.

The resistance system uses fingers that are connected to a power source. The number of fingers used and how they are located determines the heat generated by current flowing through these fingers to heat the weldments. Figure 26-30 shows an application on a piping system.

With induction heating, a coil is placed around or near the weldments. This coil is used to induce a current into the part to be heated. The circulating current in the part causes it to heat up. This heating is the result of the resistance of the current flow in the part. This method can be expanded by the use of high frequency (25–800 kilohertz) induction heating. The principle is the same; however, with the addition of high frequency, heat is also generated by the vibration or molecular friction of the molecules moving against one another. The inductor coil is the key tool that precisely couples the required energy into the weldment, thus creating the desired heating effect. The shape of the inductor is designed to the shape of the weldments to be heated, or in the case of Fig. 26-31 a rolling induction coil.

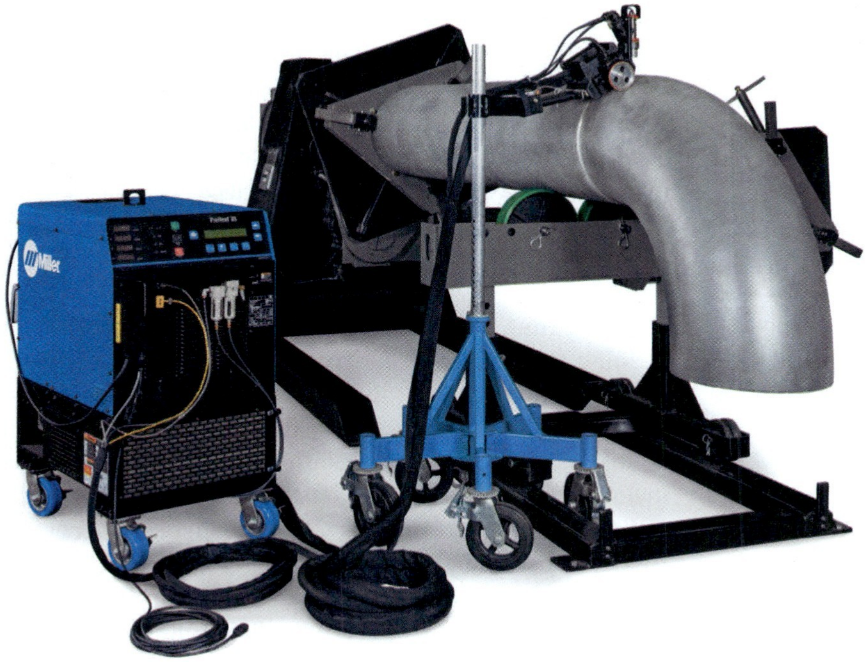

Careful consideration must be taken when designing an induction heating system and heating solutions. The following are typical prerequisites for the system in order to obtain the optimal effect and desired result:

- Material to be heated
- Frequency selection
- Power requirements
- Depth of heat penetration
- Quenching (cooling)

Induction heating has developed in its special application fields, which have surpassed other heating methods. The following are examples of induction heating applications listed with some related workpieces:

Fig. 26-31 A 35-kW high frequency induction heating system setup on a sample pipe and 90° elbow. The rolling induction system allows consistent heating which allows the welder to set target preheat temperature and maintain it as the positioner rolls the pipe and the weld is being performed. Miller Electric Mfg. Co.

- *Annealing* Wire, cable, and stock for drawing and extrusion, etc.
- *Hardening* Cutting tools, stamping dies, gears, chains, and crank and cam shafts, etc.
- *Tempering* Hardening and tempering of inertia welds.
- *Brazing; high temperature, hard and soft* Carbide-tipped tooling (e.g., saws, drills) and electrical apparatuses (e.g., motors, generators and power distribution equipment).
- *Melting*
- *Preheating before welding or weld repair* Steam and gas turbine components, etc.
- *Stress-free annealing after welding or forming processes*
- *Joining and dismantling of shrink fittings* Couplings, turbine bolts, retaining and collector rings, etc.
- *Bonding* Curing of adhesives and glued joints. Automotive and aerospace structural components, etc.
- *Forging* Annealing of hot forms. Metal billets.
- *Curing and removal of coatings* Paints, varnishes, and lacquers.

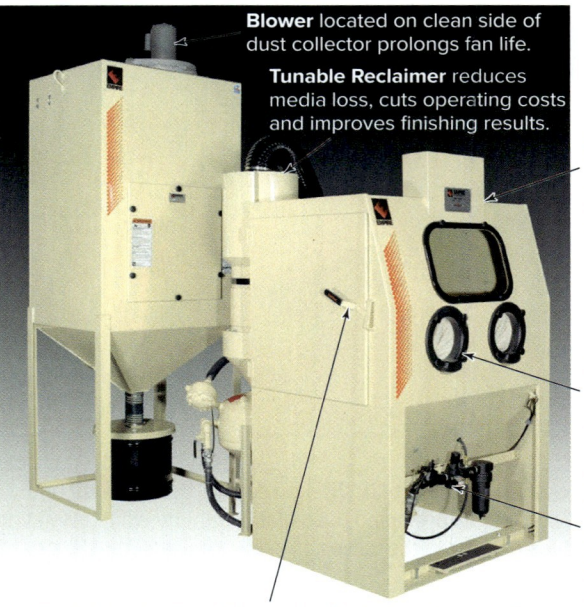

Fig. 26-32 Sandblasting equipment can be very useful in welding shop.
Empire Abrasive Equipment Company

When heat-treating metal, it is necessary to control the rate of cooling. Many of the shops are equipped with sand and lime pits to be used in delayed cooling, and they have containers of water, oil, or pickling solutions for various hardening processes. A generous supply of heat-resistant powder and sheet may also be required.

Sandblasting Equipment

Shops that do work that requires a great deal of preweld and postweld treatment are equipped with sandblasting equipment to clean the surface for welding and remove scale, slag, and rust after welding.

Cleaning the material to be welded is one of the important operations that must be performed in preparation for welding. Clean base metal makes a considerable difference in the physical and chemical qualities of the weld.

The ease and simplicity of abrasive cleaning reduces the physical effort of the worker to a minimum. The operator simply directs the blasting stream on the surface to be cleaned and moves as fast as the foreign material is removed. When combined with the tremendous force of compressed air, the small abrasive particles remove the most stubborn foreign material from areas too confining for other cleaning tools. Abrasive cleaning makes cleaning around screw and bolt heads, in narrow corners, and in deep indentations as easy and simple as doing any flat surface.

Cabinet blasters, Fig. 26-32, eliminate such cleaning processes as scraping, sanding, wire brushing, chipping, and etching. The cleaning particles may be aluminum oxide, metal grit, or sand which produce an etched or frosted finish on the pieces being blasted. The finish may be fine, medium, or coarse depending on the size of the abrasive. Abrasives such as walnut shells, corn cobs, and glass beads remove foreign material without affecting the surface being blasted.

Spot Welder

Most welding shops have a spot-welding and/or seam welding resistance welder in the shop. The spot welder is the most common of the resistance welding machines. It is generally used for the welding of light gauge sheet metal and offers great flexibility in the fabrications of metal parts.

Spot welding is a process where two lapped pieces are welded together by heat and pressure. The two pieces to be joined are pressed together between two metal electrodes. These electrodes carry the low voltage, high density electric current needed, and also provide the pressure needed. The pressure may be applied by compression spring forces, hydraulic forces, pneumatic forces, or magnetic forces. The pressure varies from a few ounces to hundreds of pounds for different size spot welders.

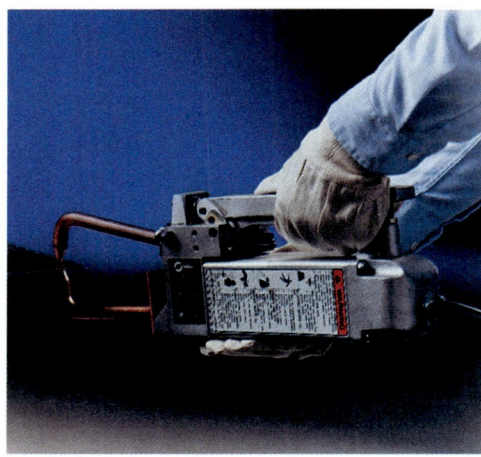

Fig. 26-33 A light weight portable resistance spot welding machine is being used. © Miller Electric Mfg. Co.

The machine may be controlled by a foot pedal, or in this case hand force, Fig. 26-33. Spot welders are available in a great variety of sizes from small bench units to very large machines. The capacity of a spot welder, that is, the thicknesses of metal that can be spot welded together, depends on the KVA (kilovolt-ampere) rating of the machine.

Hydraulic Tools

To a great extent, hydraulic tools have replaced hand tools in today's welding shop. The demands of production, fabrication, testing, maintenance, and setup operations are such that special tools must be used. Hydraulic tools can do anything that the hands can do, but faster, with tons of controlled force. High pressure hydraulic units can package 5 tons of linear force in less than 2 cubic inches of space. One person can carry a hydraulic unit capable of generating over 50 tons with precision control. These tools are used in the shop and on the construction site for the following applications:

- Pressing
- Bending
- Forming
- Clamping
- Pulling
- Straightening
- Lifting
- Materials handling
- Holding
- Spreading
- Pushing
- Positioning
- Testing

Principles of Hydraulics

The basic function of hydraulic tools is simple. Hydraulic tools multiply force and put it to work. The required hydraulic pressure is easily generated by piston-type hand or power pumps that transmit oil from the pump reservoir into a closed system. High pressure flexible hoses with plug-in couplers form the union in the line. The line transmits the oil under pressure from the pump to the main cylinder doing the work. Gauges permit accurate reading of the forces that are generated. Pressure generated by the pump is converted by a hydraulic cylinder into an applied force that is hundreds of times greater than the input force at the pump.

Figure 26-34, page 848, shows the variety of hydraulic tools that are available. In Fig. 26-35, page 848, a hydraulic-powered C-clamp is being used in a trailer body manufacturing plant for pulling and squeezing I-beam rings around trailer tanks before welding. The rings give added strength to the tank shells for handling heavy loads of fluids and chemicals. An air-hydraulic system clamps work instantly in tack-welding operation so that it reduces setup time and labor costs, Fig. 26-36, page 849. As shown in Fig. 26-36A, page 849, clamping cylinders are directly mounted on a shop-built welding fixture that swings down and locks into the closed position. The welder then activates a hand or foot valve to provide instant clamping of the work to be welded, Fig. 26-36B, page 849. Deactivation of the valve, in turn, provides instant release of work after welding is completed.

Hydraulic Bending Machines

The manufacture of tanks and cylinders comprises a large amount of the welding fabrication being done in this country. In order to produce tanks and cylinders, the shop must have the capacity to bend and roll plate. The ability to do quality plate-rolling efficiently and economically often spells the difference between profit and loss in a job.

Figure 26-37, page 849, shows a hydraulic bending machine, also called a roll, rolling ½-inch stainless-steel plate into a cylinder. The steps necessary to carry out the complete process from flat plate to finished cylinder are shown in Fig. 26-38, page 849.

Bending Brakes

Bending and forming machines for sheet metal and plate are natural descendants of the wooden cornice brakes used in fabricating architectural metals as early as the 1830s. Massive carved stone and terracotta building blocks used for topping out buildings in those days were both expensive and unwieldy. The introduction of the cornice brake made it possible to bend lightweight, easy-to-handle sheet metals for cornices and face trim. Modern hand and powered bending machines are the response to the demand of

Fig. 26-34 Assortment of hydraulic tools available to the welder. © Enerpac Inc.

welding fabricators for a line of machines to form both light and heavy gauge metals.

Power Press Brakes

A power press brake like that shown in Fig. 26-39, page 850, has a workpower capacity of 30 to 45 tons and a metal-bending capacity of a 10-gauge sheet 48 inches wide. Press brakes with hydraulic controls have a capacity of from 200 to 2,000 tons. In addition to their capacity to bend materials, power press brakes can punch, blank, form, and notch sheet metal. These brakes are all-welded in their construction. This is another example of the use of tools for welding fabrication in which the tools themselves are also of welded construction.

Hand Bending Brake

A hand brake, Fig. 26-40, page 850, also known as a box and pan brake makes bending and forming easier for the sheet metal fabricator. A great deal of welding fabrication is concerned with boxes and pans of all types that can be formed on these machines. Figure 26-41, page 850, shows examples of metalwork made with the hand bending brake.

Hand Box and Pan Brakes

A box and pan brake, Fig. 26-40, page 850, incorporates all the features of a standard hand brake. Because

Fig. 26-35 A welder with a hydraulic C-clamp. Using proper safety precautions. © Enerpac Inc.

Fig. 26-36 The use of hydraulics in connection with jigs and fixtures makes setting up and tacking easy. © Enerpac Inc.

Fig. 26-37 A hydraulic bending machine with a capacity of 12 ft × ¾-inch thick mild steel. The machine is bending 8 ft × ½-inch stainless-steel plate. Note the use of the hydraulic loading support. © Baileigh Industrial, Inc.

it has removable, sectioned fingers, nose bars, and bending bars, as well as greater clearances than the bending brake, it is more versatile. For example, the brake can form a box or pan having four sides and a bottom from one sheet of metal. Radius bends, such as those in modern metal furniture and cabinets, can be made with the round nose bars. Figure 26-42, page 851, shows examples of the various shapes that can be fabricated. Corners are usually welded.

Universal Bender

The universal bender is an indispensable piece of equipment for a welding fabrication shop. It is a bending and forming tool that can bend radii and angles on a wide variety of shapes ranging from small rod through pipe and tubing to flat stock and angle iron. Its versatility, capacity, and fast, easy setup make it ideal for short run production, custom fabrication, and maintenance work.

No tools are necessary for the assembly of dies that bend pipe, rounds, flats, squares, tubing, and angle iron to specifications. Finger adjustment of mounting pins is all that is required. Figure 26-43 shows tubing being bent to a radius. Figure 26-44 shows the bender being used to bend flat stock edgewise to a radius. Figure 26-45 shows how a round rod can be bent into a U-shape. The machines shown are hand operated. Hard-to-pull bends can be accomplished by adding a

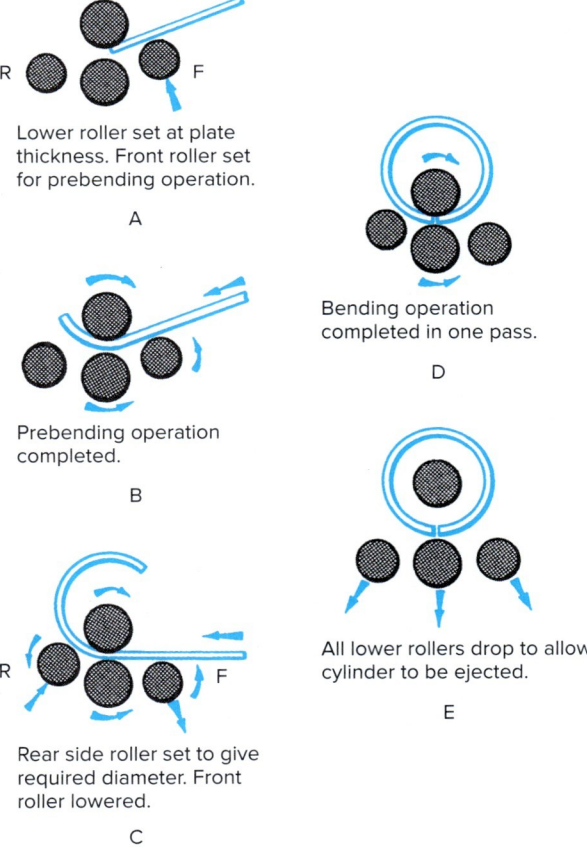

Fig. 26-38 Steps in rolling steel sheet. This is the process performed by the bending machine shown in Fig. 26-37.

General Equipment for Welding Shops Chapter 26 849

Fig. 26-39 An instructor demonstrating the use of a power press brake prior to welding. Note the proper hand position, as the part is bending in the upward direction. © Renee Bohnart

Fig. 26-40 A welding student bending a part in hand bending brake. © Renee Bohnart

hydraulic power attachment. Simple, positive foot control of the hydraulic power makes it possible to use both hands for controlling stock.

The universal bender is found in most job shops, maintenance departments, ornamental iron shops, and on construction jobs that require a large amount of steel fabrication.

Power Squaring Shears

The power squaring shear, Fig. 26-46, page 852, is an indispensable piece of equipment in the welding shop or school. The fabrication of metals requires that they be cut with accuracy, speed, economy, and safety. In the school shop, square edges on the material to be welded are necessary for quality welding and the efficient use of training time.

The power squaring shears shown in Fig. 26-47, page 852, have a capacity that ranges from 16-gauge sheet metal to ½-inch thick plate. Shears with a capacity of 1 inch are available. The machine is of welded construction. Every effort has been made in the interest of safety of operation. Drives are fully enclosed, and the rear of the machine is free from exposed, rotating shafts and eccentrics that pose potential hazards. Flange-mounted motors,

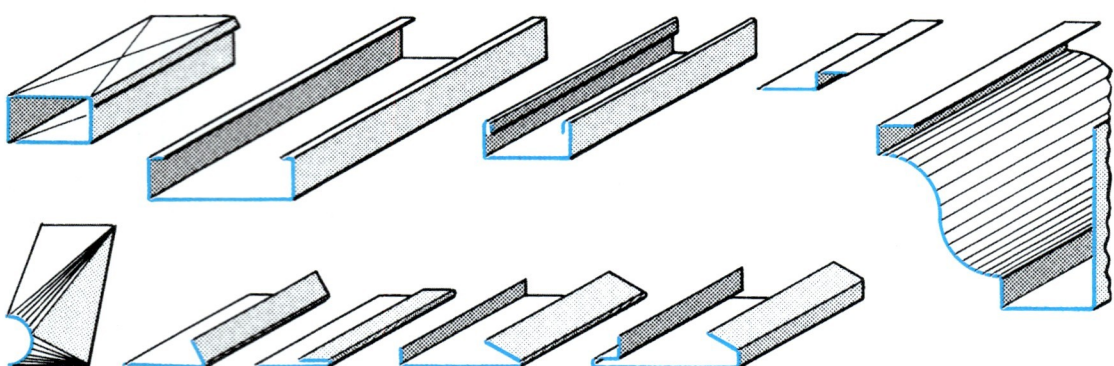

Fig. 26-41 Forms that can be produced in sheet metal on a hand bending brake. These may become a part of a total weldment. Source: Dreis and Krump Manufacturing Company.

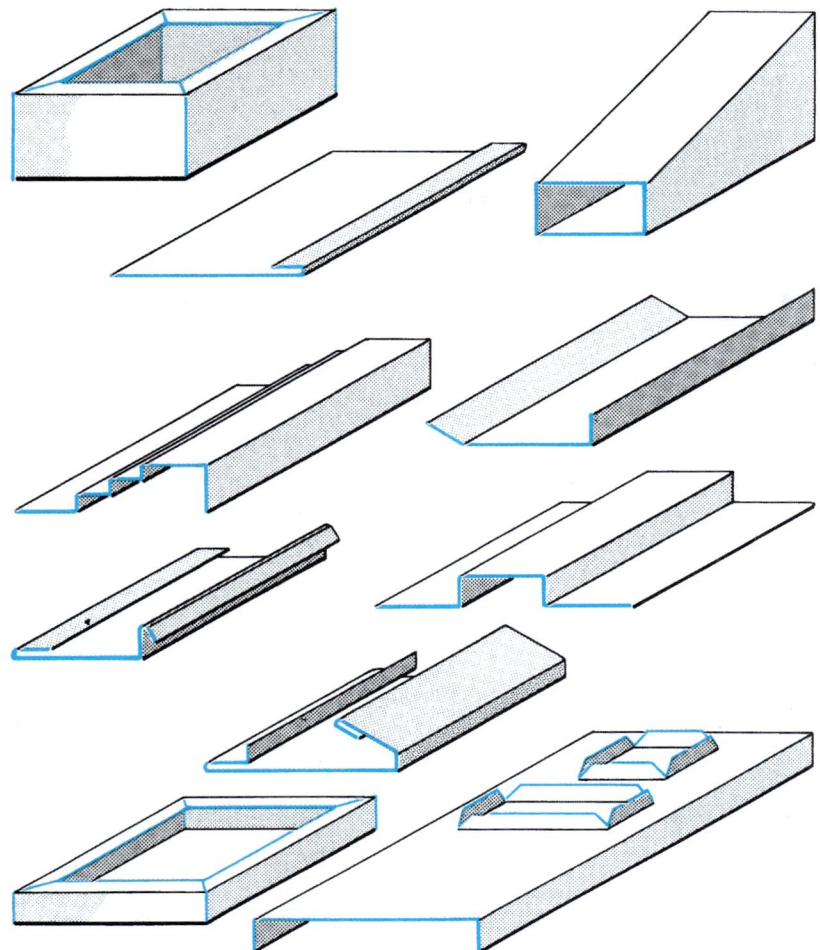

Fig. 26-42 Bending profiles produced in sheet metal on a box and pan brake. Source: Dreis and Krump Manufacturing Company.

welded must be free of rust, oil, grease, and paint. These materials cause brittleness and gas pockets in the welded joint. Multipass welding requires cleaning between each pass. Cleaning after the joint has been welded is also important for the sake of appearance and in preparation for painting. Steel wire brushes, Fig. 26-48, page 852, chisels, and cleaning hammers, Fig. 26-49, page 852, remove slag, rust, and other dry material from metal surfaces.

The chipping hammer has one end shaped like a chisel for general cleaning. The other end is pointed in order to reach the slag in corners, along the edges, and in deep ripples. Some chipping hammers are mounted with a wire brush, Fig. 26-50, page 852. Electric or pneumatic tools may also be used for preweld and postweld cleaning. These include grinders, wire wheels, and powered chipping hammers. Sandblasting is required for a flawless surface. Nonporous metals are chemically cleaned, especially when a high grade paint or enamel finish is desired.

A welder will also have use for the hand tools shown in Fig. 26-51, page 853, which are identified as follows:

A. Identification stamps are necessary in shops where critical welding is done. They are for the purpose of identifying the welder with the job.

direct-connected through splines, eliminate belts, guards, and overhanging flywheels.

Small Hand Tools

The welder must frequently use one of several methods to clean the work. The surfaces of the joint to be

Fig. 26-44 Bending flat strip. © Hossfeld Manufacturing Company

Fig. 26-43 Bending tubing. © Hossfeld Manufacturing Company

Fig. 26-45 Bending round rod. © Hossfeld Manufacturing Company

Fig. 26-46 A welding student is using a squaring shear on an ironworker to create some practice plates for welding. Plate up to 10 inches wide and ⅜ inch thick can be sheared on this 50 ton capacity machine. These type machines are capable of doing much more than just square shearing. © Renee Bohnart

Fig. 26-47 A welding student is using a squaring shear with digital control over width of cut. Steel plate up to 10 feet wide and ½ inch thick can be sheared on this machine. © Renee Bohnart

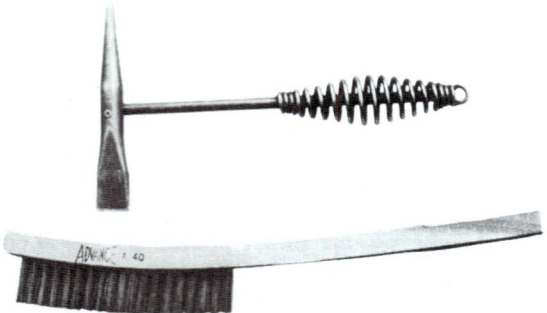

Fig. 26-48 Wire brush for cleaning welds. Edward R. Bohnart

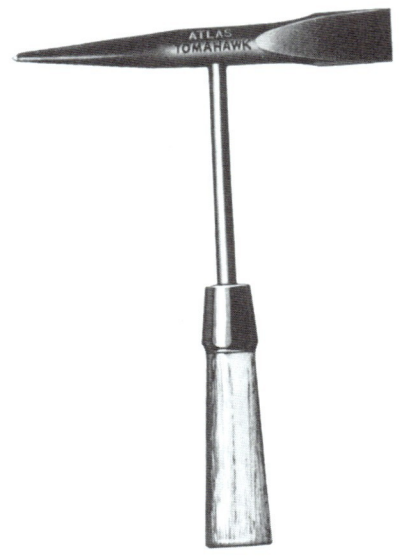

Fig. 26-49 Weld cleaning hammer. © Atlas Welding Accessories

B. Weld gauges make it possible for the welder to measure the size of welds and determine their correctness according to specifications.
C. The combination square and scale may be useful in squaring stock, squaring angles, marking off and measuring dimensions, and doing general layout work.
D. The bevel protractor is essential in work requiring the tacking up and welding of fittings at an angle.
E. Thickness gauges and micrometers determine the thickness of material. On some jobs only a thickness gauge is necessary. For jobs that involve a great deal of light gauge material and when accuracy is necessary as in aircraft construction, the welder will use the micrometer for checking thicknesses.
F. The flexible scale is useful for all types of measuring, especially around radii and other hard-to-secure measurements.

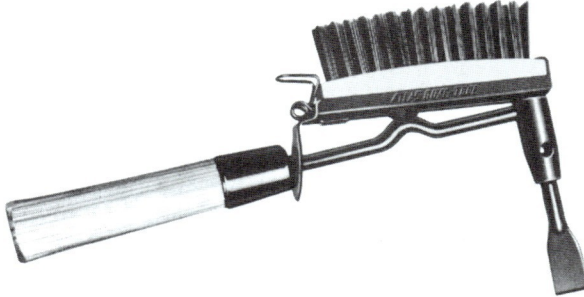

Fig. 26-50 Combination wire brush and cleaning hammer. © Atlas Welding Accessories

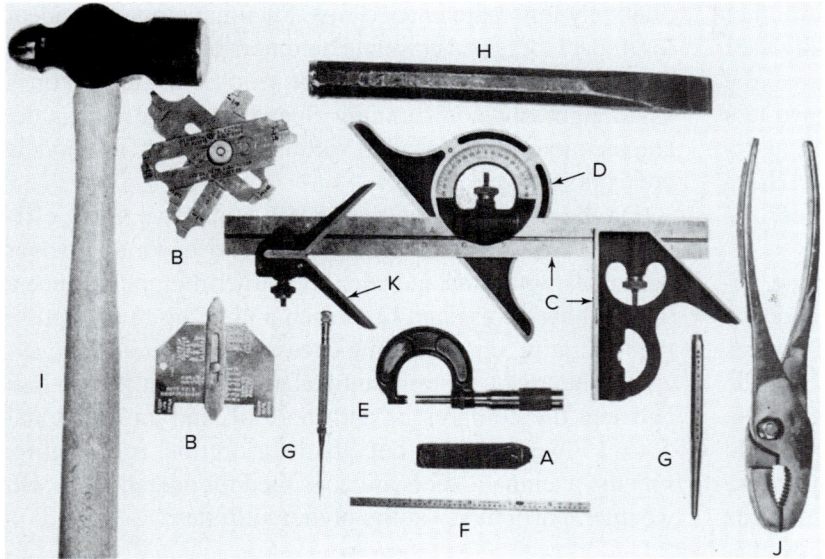

Fig. 26-51 Set of hand tools needed by an experienced welder for layout, fitting up, material checking, and weld checking. *Edward R. Bohnart*

drills, hand and floor grinders, edge preparation tools, and weld-shaving devices.

Efficient and safe use of portable power tools requires a basic understanding of their construction and their power source as well as an appreciation of the advantages, disadvantages, and limitations of each type.

Power tools use a variety of energy sources such as electrical, pneumatic (compressed air), and hydraulic (pressurized fluid) for power. All three types do the same work. The productivity of each tool, however, varies with the application. The operator has constant control over speed and load. Tool size has an effect on operator fatigue and, hence, efficiency.

Types of Power Source

Electric Motors Electric motors are available in various types. The most commonly used are the universal and a.c. induction motors for power tools.

Universal Electric Universal electric tools are the most common portable tools. They operate on standard 110- or 220-volt a.c. or d.c. single-phase current, the cheapest power available. These tools are best suited for intermittent operation, typified by maintenance, installation, and field work.

Since universal motors operate on a.c. or d.c. power and are intended for field use, they can operate on all types of engine-driven welding generators. In some cases the welding generators only output d.c. auxiliary power, which will damage an a.c. motor. Universal motors are of fixed speed.

Universal electric tools have some drawbacks. One of the most significant is that the units have commutators and carbon brushes as well as armature windings, all of which require attention. The maintenance of universal electric tools is costly and they should, under no circumstances, be used continuously as in assembly line operations.

A variation of these types of motors are those with speed control. For example, normally in the trigger mechanism of a drill motor the further you pull the trigger, the faster the buildup to the maximum speed of the tool. These types of tools generally have a reverse capability as well, which is activated with a slide switch. These types of speed-regulated tools are to be used only with a.c. power. Always check the motor nameplate for the correct operation power. If used on d.c. power, they may go to maximum speed causing an injury or tool damage.

These types of motors can be double insulated for additional electrical safety purposes. They can be used in a

G. The scribe and center punch mark lines and other layout information accurately on steel. The scribe, also called a *scratch awl*, has a hardened steel point in order to make an impression in the metal. After the line has been made, it is often found that the finely scribed line must be further marked to be seen clearly. This is especially true of lines that serve as a guide for cutting torch operations. The additional marking is made with the center punch. A center punch is also useful in spotting a point that is to be drilled. The impression made by the punch prevents the drill from wandering all over the surface to be drilled.

H. The cold chisel is handy for preparing the joint for welding and for removing slag, burrs, spatter, high places in the weld, and evidence of poor fusion after welding.

I. The hand hammer is an indispensable tool. It provides the force for using the center punch in layout work and the cold chisel in cleaning.

J. Pliers are used for carrying and handling hot metal and for holding parts that are to be tack welded to the main structure.

K. The center head is used to find the center of round bars and pipe.

A hand metal-cutting hacksaw (not illustrated) is useful for cutting a small piece of round or flat steel.

Portable Power Tools

The fabrication of metal and the preparation of plate for welding require a variety of hand power tools such as chipping hammers, peening hammers, hand and bench

situation where a ground wire is not warranted. Another variation is the battery power tool that fundamentally uses a d.c. motor. Typically, the higher the battery voltage, the more robust the tool will be. A 6-volt tool will have less capability than an 18-volt tool. Battery power is very desirable for increased portability and convenience. However, batteries will need recharging and maintenance.

A.C. Induction Motor The stator windings of an a.c. induction motor are distributed around the stator to produce a roughly sinusoidal distribution. These types of motors can run on single- or three-phase power, but it must be alternating current. When three-phase a.c. voltages are applied to the stator windings, a rotating magnetic field is produced. These types of motors must have the phase connected properly in order to have the motor turn in the proper direction. Old-style motor-generator welding machines had these types of motors. They had an arrow located close by the armature on the generator so that the direction of the armature could be determined. If it was rotating in the incorrect direction, any two of the three phases would need to be switched around. Single-phase a.c. induction motors always rotate in the proper direction.

The rotor of an induction motor also consists of windings or more often a copper squirrel cage embedded within iron laminates. An electric current is induced in the rotor bars, which also produce a magnetic field. The rotating magnetic field of the stator drags the rotor around. The rotor does not quite keep up with the rotating magnetic field of the stator. It falls behind or slips as the field rotates.

These types of motors are very simple and reliable. The power sources are well suited for production and assembly line work, especially in operations requiring high torque under constant load. Slowdown under load is less than with the universal motor, which can be as high as 25 percent, and the pneumatic motor, which can be as high as 36 percent.

Pneumatic Pneumatic power tools require compressed air to operate. The most common air compressors are electrically driven by a three-phase, 220-volt or 440-volt motor. On construction sites, these units are often powered by either diesel or gasoline engines.

There are two types of pneumatic motors: the turbine and the piston. Piston motors are rarely used except on certain types of reciprocating tools. The turbine type is better suited for short stroke applications of 3 inches or less. The turbine motor is used in about 95 percent of pneumatic tools.

The pneumatic tool has several advantages over the universal electric tool. Most important are its suitability for continuous operation such as assembly line work and its relatively low maintenance cost. Pneumatic power is ideal for impact tools and chipping hammers because pneumatic motors have a high tolerance for vibration. The motor is small, light, and cool-running. Speed is infinitely variable, and the motor can be stalled without damage. Most models are explosion-proof.

On the negative side, pneumatic tools have a low efficiency. Torque under constant load is low, and motor power decreases with time and use. The larger motors produce a loud exhaust noise and have such a high air consumption that operating costs become excessive. Moreover, the average pneumatic installation has a 20 percent power loss between the compressor and the tool, and an additional 5 to 15 percent loss due to deterioration from aging. Finally, pneumatic tools are not suited for operation in cold weather, particularly at the higher altitudes.

Hydraulic The hydraulic power tool is employed most in applications requiring extremely high torque at very low speeds, such as tapping over ¾ inch, drilling over 1¼ inch, reaming, and tube expanding. Operation is somewhat similar to that of the pneumatic tool except that a hydraulic fluid is used instead of air. A pump takes the place of the compressor. For the most part, hydraulic tools are used for continuous use and special applications.

Summary When operation is intermittent and the tool is used for different jobs, the universal electric tool is the best choice. High frequency and pneumatic tools are more desirable for production line work when operation is sustained. Of the latter two, the high frequency tool is the more versatile and efficient. The pneumatic tool excels as a small tool for continuous, light-duty applications.

Portable Electric Hand Drills

The electric hand drill is a tool that the welder uses frequently in fabrication and maintenance work. As indicated previously, drill motors may be electric or pneumatic. They can take drill sizes from 1/32 inch to 1¼ inch in diameter. Figure 26-52 shows the internal construction of a ¾-inch, heavy-duty electric drill. Study the photograph carefully to understand the electrical and mechanical functions involved. Figure 26-53 shows a ½-inch battery-powered drill motor and related battery-powered tools.

Electric Hammers

Figure 26-54, page 856, shows the internal construction of a heavy-duty portable electric hammer. This type of hammer is used by the welder for chipping, peening, channeling, and masonry drilling. It is used with punches, chisels, seaming tools, scaling tools, and many sizes and types of masonry drills.

A small riveting hammer is one of the hand tools with which the welding student should have some experience. Many fabricated units are joined with a combination of welding and riveting. Riveting is commonly used for those items in which light and medium gauge sheet metal is being fabricated. Figure 26-55, page 856, shows two students riveting two aluminum plates together with a pneumatic riveting hammer.

Sanders and Grinders

Sanders and grinders are abrasive tools that perform almost every kind of surfacing job for the welder. Equipped with abrasive disks, they handle all sanding, from fast material removal to satin-smooth finishing. Equipped with flaring cup wheels and depressed center wheels, they smooth welds and casting ridges and cut off studs, bolts, and rivets. Equipped with wire cup brushes, they remove paint, rust, and scale from welds and clean castings, tanks, sheet metal, and soldered joints.

Angle sanders and grinders are usually rated to take abrasives with a diameter of 6 to 9 inches. They have a speed of 4,000 to 8,000 r.p.m. Straight grinders may be used with wheels of ½ to 6 inches in diameter. Wheels from 6 inches to 2½ inches are run at speeds of from 3,750 to 14,500 r.p.m. The larger the wheel, the slower the running speed. Wheels from 2 inches to ½ inch are run at speeds of from 14,500 to 30,000 r.p.m.

The internal construction of a portable electric angle sander and grinder is shown in Fig. 26-56, page 856. Figure 26-57, page 857, shows an a.c. electric angle grinder grinding a weld. The pneumatic belt sander in Fig. 26-58, page 857, is using a belt running at 20,000 r.p.m to debur the cut edges on some square tubing. The weld profile is being modified in Fig. 26-59, page 857, with a pneumatic die grinder.

Shears and Nibblers

The welder who works in fabrication or repair frequently cuts metal into various shapes and sizes. Sometimes flame and arc cutting methods cannot be used because they may cause color change or warpage in the work or they may create a fire hazard. In such cases, metal must be cut by a shearing method. Most portable electric shears and nibblers can cut steel as thick as 8 gauge and aluminum up to ¼ inch. These machines are fast cutting and produce a smooth edge without distorting the body of the metal. The nibbler uses a straight up-and-down punching action, and the shears cut with a powerful scissor-like action.

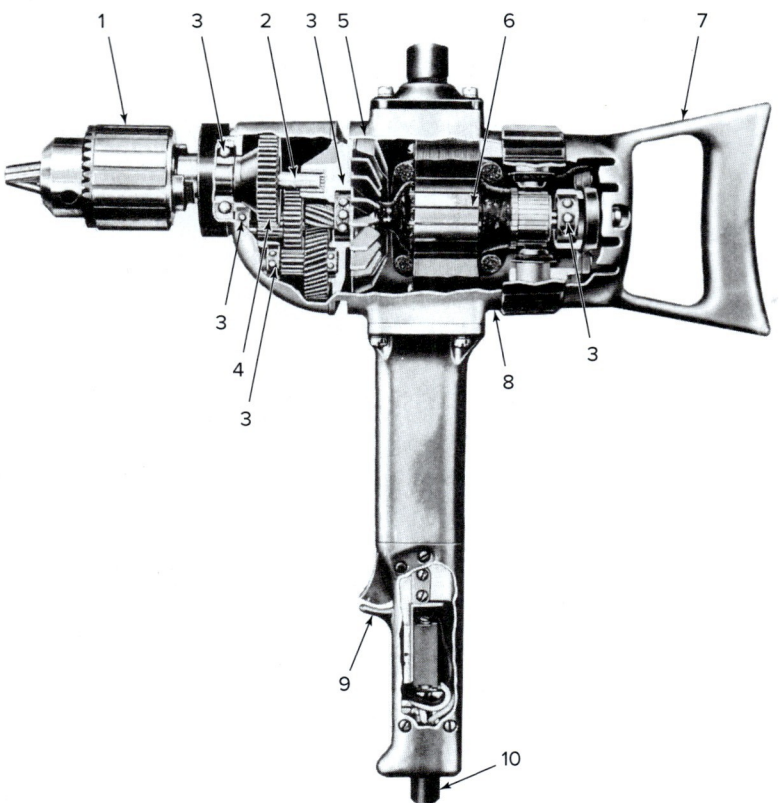

Fig. 26-52 Internal construction of a heavy-duty portable electric drill. 1. Half-inch hardened-steel key-operated drill chuck. 2. Spindle mounted on ball bearings for long life. 3. Heavy-duty ball bearings. 4. Heat-treated gears mounted with ball and roller bearings. 5. Powerful ventilating fan maintains cool operating temperatures. 6. Universal motor mounted with ball bearings. 7. Handle for one-handed operation of tool. 8. Lightweight, extra-rugged aluminum housing. 9. Heavy-duty trigger switch. 10. Heavy-duty three-wire cable. © Black & Decker, Inc.

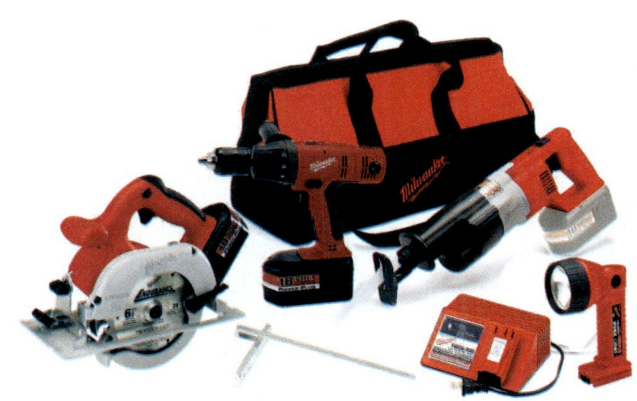

Fig. 26-53 Industrial grade 18-volt tool package. Includes a circular saw, sawzall, drill-driver, and worklight. The contractor bag allows storage for all the tools and accessories the professional needs for the job. © Milwaukee Electric Tool Corporation

General Equipment for Welding Shops **Chapter 26** **855**

Figure 26-60, page 858, shows the internal construction of a portable electric nibbler. The nibbler in Fig. 26-61, page 858, is cutting ⅛-inch steel plate at a speed of 4 feet per minute. Figure 26-62, page 858, shows portable electric shears cutting 16 gauge galvanized sheet metal. On stainless steel it is rated for 18 gauge. Its minimum radius cut is approximately 1 inch. This fast light weight tool (5.0 pounds) is very portable.

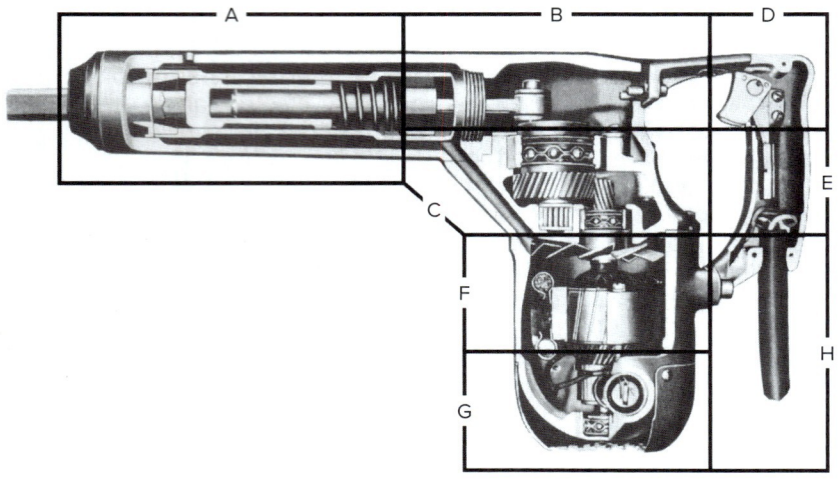

Fig. 26-54 Internal construction of a heavy-duty portable electric hammer. A. Heat-treated piston and ram. B. Heat-treated crank and connecting rod. C. Heat-treated helical gears, ball-and-roller-bearing mounted. D. Heavy-duty trigger switch. E. Pistol-grip handle. F. Universal motor, ball-bearing mounted. G. Aluminum housings. H. Three-conductor cable. Source: Reproduced with permission of Black & Decker. © Black & Decker, Inc.

Fig. 26-55 Welding students using a pneumatic riveting hammer. Edward R. Bohnart

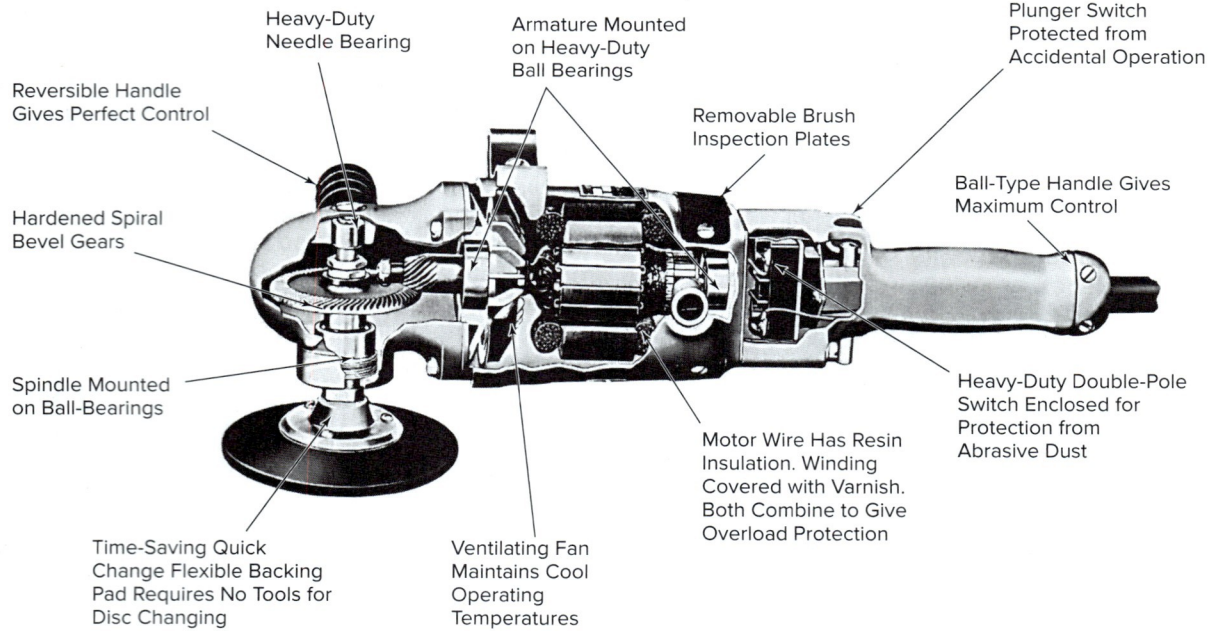

Fig. 26-56 Internal construction of a heavy-duty portable electric angle sander and grinder.
Source: Reproduced with permission of Black & Decker. © Black & Decker, Inc.

Fig. 26-57 A welding student using an electric side grinder. This type tool is very helpful for dressing weld profiles, removing weld defects or in this case joint preparation. Note the ear plugs, gloves, and double eye protection. © Renee Bohnart

Fig. 26-59 A welding student using a pneumatic die grinder with a cutting disk to profile a weld. Because of the high speed 20,000 rpm and guarding issues this type work should be done with double eye protection. That is safety glasses along with a full face shield. Grinding should be done as an exception for poor weld profile at starts, stops, root passes and not as a rule because of poor welding technique which results in the entire weld having to be ground. © Renee Bohnart

Magnetic-Base Drill Press

On many construction and maintenance jobs, it is necessary to drill a hole in a weldment that is too big to be taken to a drill press for drilling. Because of the size of the hole or the need for accuracy, hand drilling with a portable hand drill may be too slow or inaccurate. The magnetic-base drill press, Fig. 26-63, page 859, is a powerful heavy-duty drill mounted on a base with great magnetic holding force. This unit attaches to steel surfaces in the horizontal, vertical, and overhead positions. Thus, the machine provides both the speed and the accuracy of a regular bench- or floor-type drill press. The unit is available with either electric or pneumatic power.

Beveling Machine

The edge preparation of plate, tubing, and pipe before welding has always been a problem because quality cuts are usually costly. Industry has used the traditional methods of sawing, shearing, grinding, planning, and flame and arc cutting with various degrees of success. Each method works well within its limitations. Beveling by machine has the advantage of low initial purchase and operating cost. Beveling machines are available as electric or pneumatic tools. They are capable of beveling mild steel, alloy steel, and stainless steel

Fig. 26-58 A welding student using a pneumatic belt sander to remove burs on some square tubing. The belt is approximately ½ inch wide and 24 inches long and is moving at 20,000 r.p.m. © Renee Bohnart

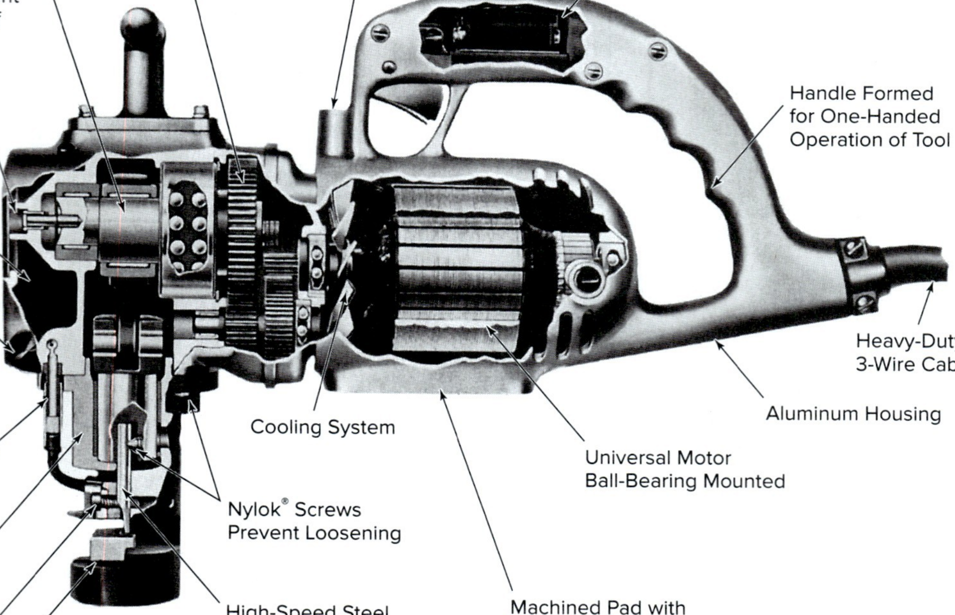

Fig. 26-60 Internal construction of a heavy-duty portable electric nibbler. Source: Reproduced with permission of Black & Decker. © Black & Decker, Inc.

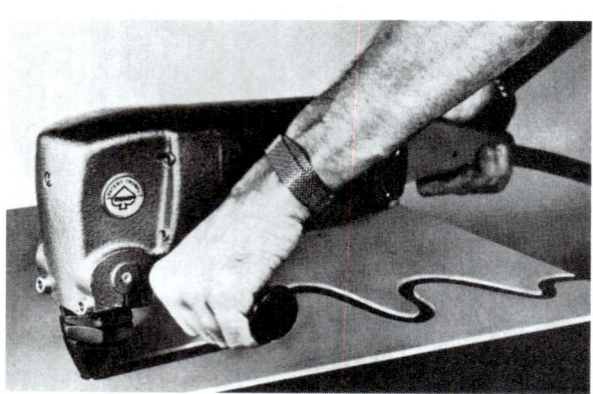

Fig. 26-61 Using a portable electric nibbler. Its cutting capacity is 1/8-inch mild steel plate at a cutting speed of 4 feet per minute. © Black & Decker, Inc.

Fig. 26-62 Using a portable electric shear on a 16 gauge galvanized sheet metal. It can be used for straight lines, curved lines, or circle cutting. The cutting speed is 10 feet per minute. © Renee Bohnart

858   Chapter 26   General Equipment for Welding Shops

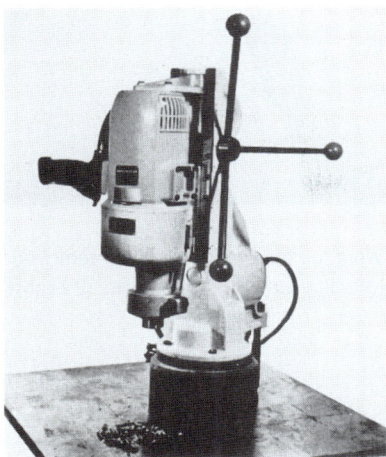

Fig. 26-63 A magnetic base drill press with a four-speed drill. Edward R. Bohnart

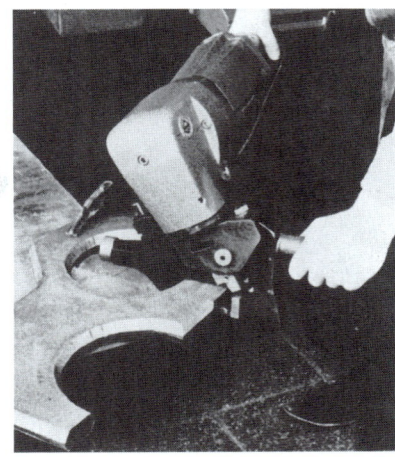

Fig. 26-64 A portable electric beveling tool. It can cut along straight edges, convex and concave curves, and circles. The machine's capacity is ⅝ inch for single-V bevels and 1¼ inch for K- or double-K bevels. © Renee Bohnart

as well as aluminum and soft metals. Beveled angles ranging from 15 to 55° are possible. These machines have the following features:

- They are portable and can be hand held or mounted in a stationary stand or vise.
- Curves can be beveled. Any angle can be followed on convex curves. On concave curves the machines can handle diameters as small as 1½ inch.
- All types of pipes can be beveled.
- The angle of cut can be varied from 10 to 55°.
- Any thickness of metal can be worked on.
- It is possible to maintain a constant and smooth feed rate and to work forward and backward.
- The machine can start and stop at any desired point.
- The machine can operate upside down so that large pieces do not have to be turned when preparing weld edges.
- The cutting tool is easily sharpened and inexpensive to replace. Figure 26-64 indicates the extreme flexibility of the tool.

A wall mounted, pipe-beveling machine, Fig. 26-65, can square up ends of pipe or various other grooves like "J" or bevels. Angles and radius are dependent upon settings and the cutting tool shape. This unit is set up to prepare test pipe nipples.

Weld Shavers

The weld shaver was designed specifically for aircraft, missile, and industrial operations that require the precision shaving, milling, and grooving of high strength materials. The tool is suitable for stainless steel, titanium, aluminum, Monel®, magnesium, and copper. Shavers replace other grinding, buffing, and finishing tools. They cut medium

Fig. 26-65 This 80 lb. BevelMaster by Tri Tool is designed to handle pipe from 4–12 inches in size with wall thicknesses up to 1.32 inches. It is pneumatic and requires 85 cfm at 90 psi. At this school it is mounted and the short pipe nipples are brought to the machine. Lighter units are available for application where the bevel machine must be brought to the pipe. © Renee Bohnart

size weldments up to ¾ inch wide down to surface flushness or to a preselected height controlled to within 0.0005 inch. The depth of cut per pass depends on the hardness of the material being cut. Also available are grooving cutters capable of grooving to a maximum of

Fig. 26-66 Removing weld reinforcement with a pneumatic portable weld shaver. © Zephyr Manufacturing Company

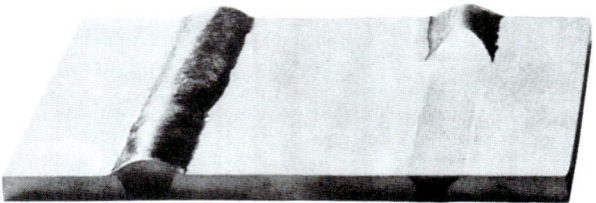

Fig. 26-67 Weld reinforcement has been removed flush with the plate surface. Note how smooth the finish is. © Zephyr Manufacturing Company

Lathe

The lathe, Fig. 26-68, is a turning tool that shapes metal by revolving the workpiece against the cutting edge of the tool. The lathe performs many kinds of external and internal machining operations. Turning, Fig. 26-69, can produce straight, curved, and irregular cylindrical shapes. Other operations include knurling, thread cutting, drilling, boring, and reaming. The lathe can be used to cut and bevel pipe for welding.

Milling Machine

A milling machine is a machine tool that cuts metal with a multiple-tooth cutting tool called a *milling cutter*. The workpiece is mounted on the milling machine table and is fed against the revolving milling cutter.

There are two basic types: (1) the horizontal milling machine, Fig. 26-70, and (2) the vertical milling machine, Figs. 26-71 and 26-72, page 862. These machines can be used to machine flat surfaces, shoulders, T-slots, dovetails,

0.290 inch. Shavers and grooving cutters are pneumatic tools.

The shaver is fitted with carbide and high speed steel cutters which are precision ground and fluted for hard usage and long life. Cutters are easily removed and replaced, and they may be reground many times. Figure 26-66 shows the tool removing a weld bead reinforcement. Figure 26-67 is an example of the weld finish after removal of the reinforcement.

Machine Tools

Many large fabricating plants have completely equipped machine shops where metal machining of all kinds can be done. Those welders who wish to work in the fields of fabrication, maintenance, and repair will find it an advantage to be able to operate such standard machine tools as the lathe, the shaper, and the milling machine.

Basically, a *machine tool* is a machine that cuts metal. It holds both the material being worked on and the cutting tool. Its purpose is to produce metal parts by changing the shape, size, and finish of metal pieces.

In addition to maintenance work, these machines can be used to cut and bevel pipe, bevel plate, and machine-test weld specimens.

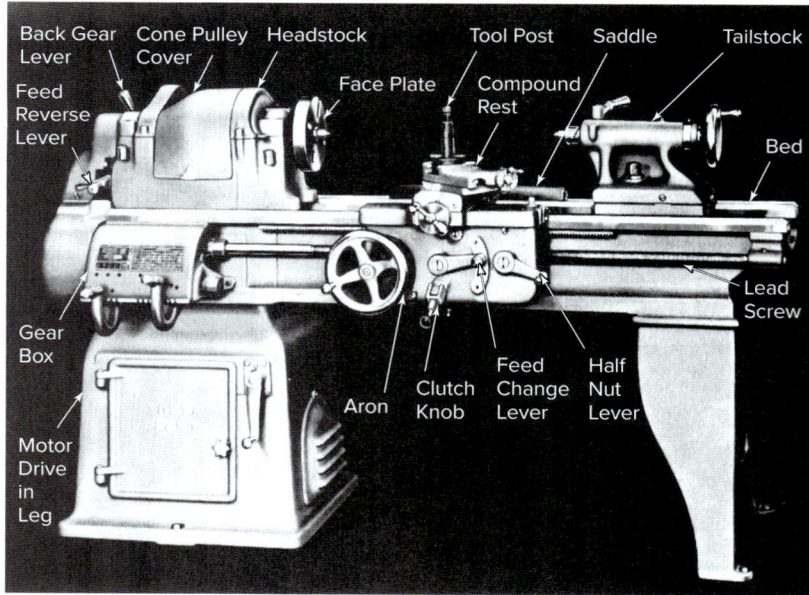

Fig. 26-68 The parts of a standard lathe. © South Bend Lathe Co

Fig. 26-69 Setting up to bevel a 3-inch diameter section of steel pipe on the lathe in preparation for welding. © Renee Bohnart

There are three types of wheels which are classified according to the means of bonding:

- *Vitrified wheels* in which the bonding is a kind of earth or clay. The wheels are baked at about 3,000°F. These wheels are as large as 36 inches in diameter. They are usually used for rough grinding.
- *Silicate wheels* in which the bond is silicate or water glass. They are as large as 60 inches in diameter. They may be used for fine grinding.
- *Elastic wheels* in which the bond is rubber, shellac, or bakelite. Very thin, strong wheels are made with this bond so that they can be run at very high speed.

Wheels are made in many different shapes such as cylindrical, straight, tapered, recessed, cup, and dish. Generally, the straight, tapered, and cup wheels are used in the shop. They may range from 1 to 3 inches in thickness and up to 14 inches in diameter.

Surface Grinder

The preparation of such weld test coupons as the tensile, face, and root bend coupons requires that the surface of the coupons be ground parallel to their length. A smooth surface is necessary. Coarse grinding or surface cuts may cause a weld failure when the coupon is tested. After the coupon is cut from the original test specimen, the surface grinder is the ideal tool to produce the smooth surface, Figs. 26-74, page 862, and 26-75, page 863. These grinders can handle tough materials to close tolerances.

and keyways. Irregular or curved surfaces can also be formed with specially designed cutters. One of the important uses is in the cutting of gear teeth. The vertical miller can also be used for drilling, reaming, countersinking, boring, counterboring, and friction stir welding.

A milling machine is an excellent machine to prepare weld test specimens and remove excess weld metal where the work is portable.

Pedestal Grinder

The pedestal grinder, Fig. 26-73, page 862, is one of the most important pieces of equipment in the welding shop. It polishes or cuts metal with an abrasive wheel. It prepares metal parts for welding, grinds weld metal, removes rust and scale, and sharpens tools. The floor-type grinder usually has a grinding wheel on each end of a shaft that extends through an electric motor.

A grinding wheel is made of abrasive grains cemented or bonded together to form a wheel.

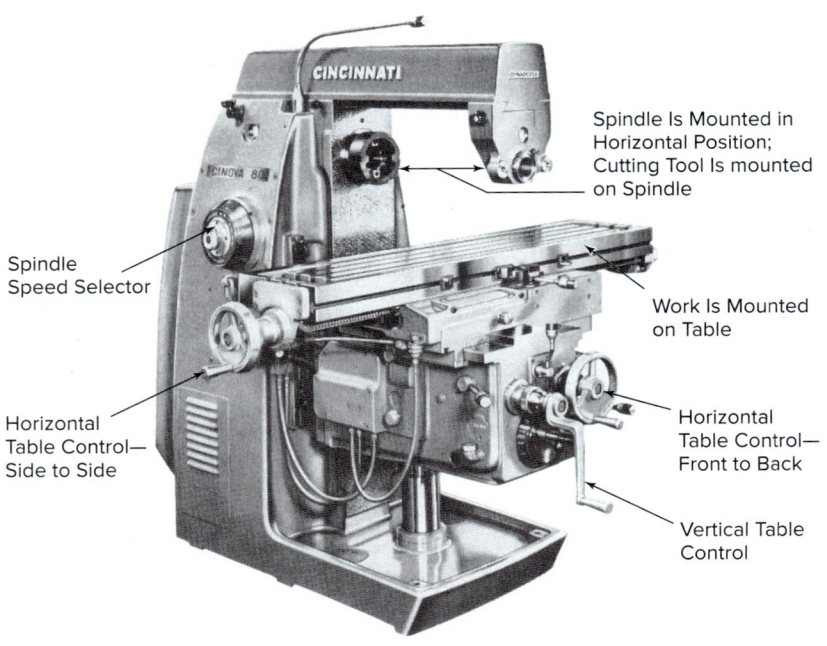

Fig. 26-70 Horizontal milling machine. The cutting tool is mounted on the horizontal head. © MAG IAS, LLC

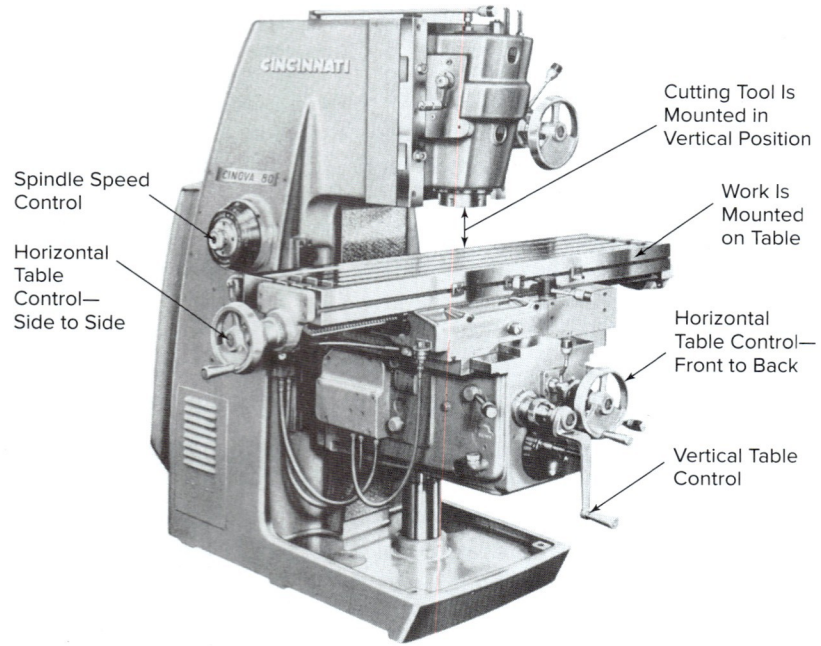

Fig. 26-71 Vertical milling machine. The cutting tool is mounted on the vertical head. © MAG IAS, LLC

Fig. 26-73 A pedestal grinder (2 horsepower, 1,725 r.p.m.) takes a wheel 12 inches in diameter with a 2¼-inch face. It should be equipped with eyeshields and lights. © Baldor Electric Company

Fig. 26-72 Instructor observing student using a vertical milling machine. Gloves are not used around machine tool operations where they could get caught up in the moving equipment. © Renee Bohnart

Fig. 26-74 A precision surface grinder can be used to machine weld test specimens. © DoALL Company

Drill Press

In a shop that does a great deal of fabrication and repair, a drill press is an indispensable machine tool. Many of the holes that need to be drilled can be done with a portable power drill. In heavy plate thicknesses and when accuracy is required, however, a bench or pedestal drill press is necessary. The pedestal press shown in Fig. 26-76 has a 20-inch throat and can drill at speeds from 150 to 2,000 r.p.m. Figure 26-77 shows a pedestal-type drill press in operation.

Fig. 26-75 Surface grinders can be used for various weld preparation such as tensile test specimens. © Renee Bohnart

Power Punch

Another method of making holes in metal is with a power punch. It is important for the shop to provide a machine that can punch holes in different shapes and sizes with a minimum of tool change time.

There are combination machines available that will punch and shear, such as the power iron worker shown in Figs. 26-78 and 26-79, page 864, which is capable of exerting 18 tons of force to punch a ⅞-inch hole through ¼-inch thick material. It can also shear angle iron, round bar, square bar, flat bar, and channels, and be used as a conventional press for forming. It has a blade guard, foot lever guard, and gear guard for operator protection.

Hydroforming

Using high pressure water to change the shape of a piece has been done for many years. You can experience the tremendous forces that can be exerted by water by simply observing a garden hose. When the water nozzle is open, the tap pressure expels the water from the end of the hose nozzle. However, if the nozzle is adjusted to stop the flow of water, the outside diameter of the hose will increase because of the pressure buildup. The same principle is used for hydroforming. Figure 26-80, page 864, shows some typical parts that can be made with this process.

The process is quite simple as illustrated in Fig. 26-81, page 864. It is similar in operation to a punch press; however, it uses half the tooling and

Fig. 26-76 A pedestal model of an electric drill press with dial-controlled variable speeds. © Clausing Industrial, Inc.

Fig. 26-77 Pedestal drill press in operation. © Renee Bohnart

Fig. 26-78 Iron Worker used to punch a hole in a steel plate. Note the punch and die. The safety door is open to show interior components. © Renee Bohnart

a computer program adjusts the water pressure to shape the desired form, unlike the traditional system of top and bottom dies used in a press operation. Figure 26-82 is a small hydroforming press capable of production and use for development work.

There are many advantages to hydroforming. It

- Can process complex components in a single operation
- Saves materials and money through part elimination and simplified assembly
- Reduces tooling costs
- Has dimensional precision
- Reduces part weight
- Reduces the number of weld joints
- Increases part stiffness

Tooling costs are one of the major concerns when working on weldments. In simple terms, hydroforming is not a tool-intensive technology, unlike pressure punch or stamping

Fig. 26-79 An 18-ton power iron worker capable of sheering, punching, and forming. It is capable of approximately 48 strokes per minute. Note the safety guards. © Rogers Manufacturing Inc.

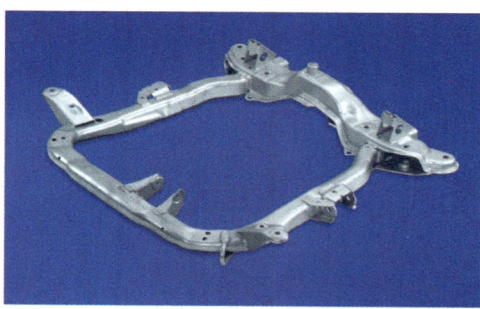

Fig. 26-80 Hydroformed tubes and structures that may become part of a weldment. © Schuler AG

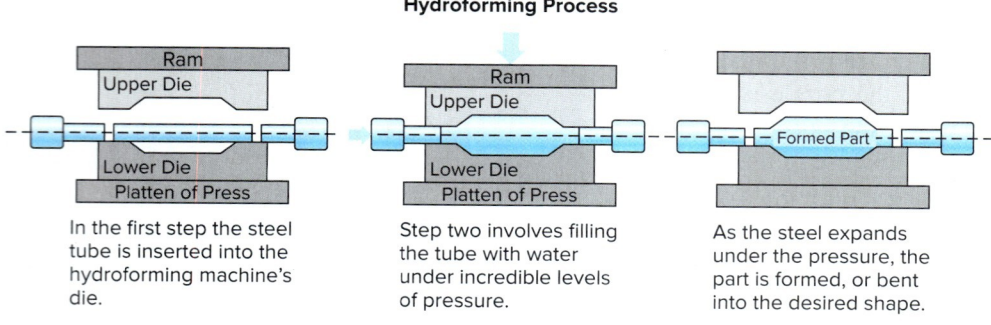

Fig. 26-81 Schematic of hydroforming operation. Advanced Manufacturing

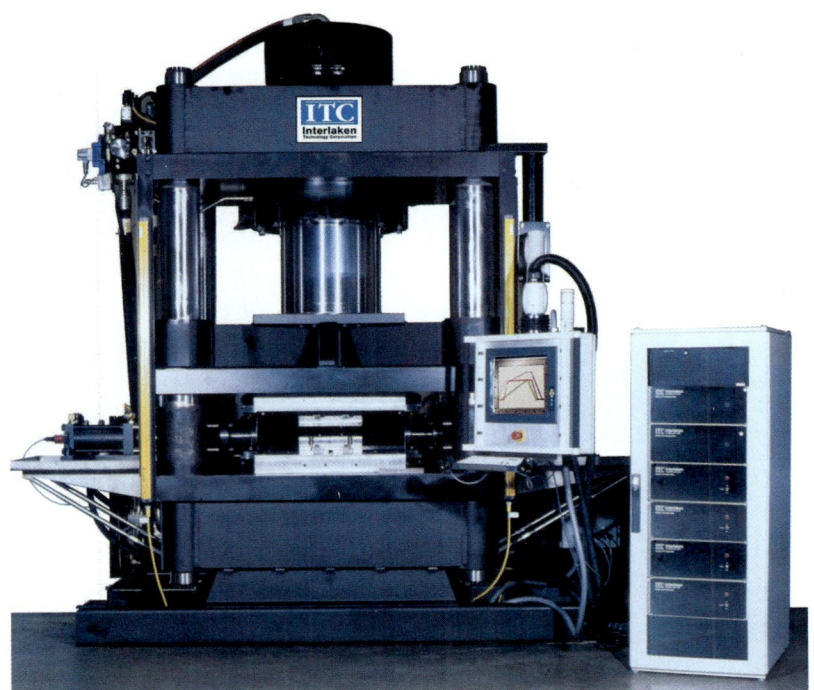

Fig. 26-82 A hydroforming press and computer control capable of doing production and development operations. It is capable of exerting 75 tons of clamping force with a pressure intensifier capable of water pressures of 20,000 pounds per square inch.
© Interlaken Technology Corp

Metal-Cutting Band Saws

A power saw is used more than any other tool in the welding shop. These saws cut rectangular and round bars, angle iron, and a variety of other structural shapes. The metal-cutting band saw shown in Fig. 26-83 is a portable machine that is actually two machines in one. As a horizontal cut-off saw, it features a quick-action vise, which swivels to 45°; an adjustable blade guide; and an automatic shutoff switch. Only seconds are required to convert the tool for vertical use. The head is swung into an upright position, a work table is attached, and the saw is ready for cutting angles, slots, notches, and bevels, Fig. 26-84, page 866. Figures 26-85 and 26-86, page 866, show in use the vertical and horizontal band saw, respectively.

A more sophisticated model of the vertical band saw is shown in Fig. 26-87, page 866.

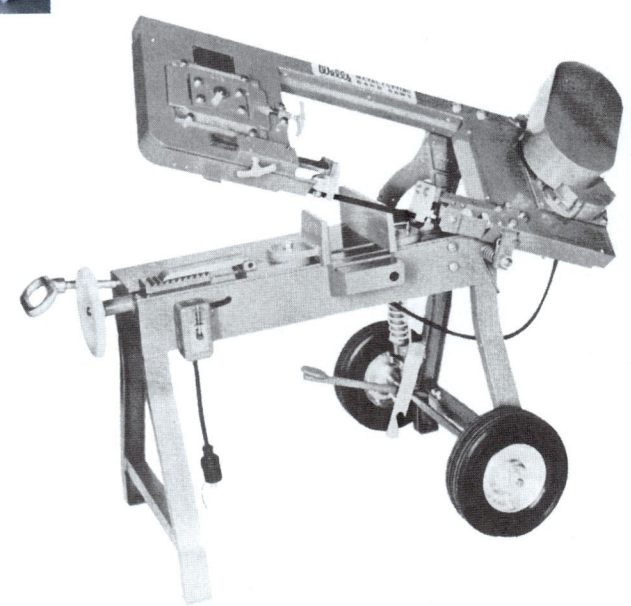

operations. The hydroform tooling is simple, low cost, and capable of being used for short or long runs. Table 26-1 represents some cost considerations.

Limitations are that hydroforming must be done on newly designed parts, because the existing part design may not lend itself to hydroforming. The raw material tube will be more expensive than a sheet of steel that may be used for a punch pressed part. Hydroforming cycle times are longer than those of a punch press. There is wall thinning in areas where maximum stretch is required. As more work is placed on designing materials for hydroforming, tailored tubes with differing properties and wall thicknesses will be produced.

Fig. 26-83 A horizontal metal-cutting band saw for cutting bars, angles, and pipe. Edward R. Bohnart

Table 26-1 Cost Comparison of Hydroforming and Conventional Punch Press Operation for an Engine Cradle

Type	Number of Parts	Weight	Tool Costs (U.S. $)	Part Costs[1] (U.S. $)
Conventional manufactured	34	24.56	5,359.090	51.00
Hydroforming steel	30	20.50	3,712.636	42.83
Hydroforming aluminum[1]	30	14.41	3,891.727	73.17

[1]Aluminum; 25% less weight, 30% higher part costs.
Advanced Manufacturing

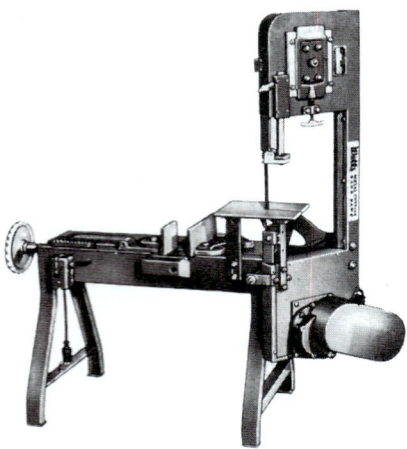

Fig. 26-84 A metal-cutting band saw in the vertical position.
Edward R. Bohnart

Fig. 26-86 Horizontal band saw in use. Once the angle iron length to be cut is measured and securely clamped in place, the saw blade guides will be repositioned as close as possible to the part being cut. © Renee Bohnart

Fig. 26-85 Vertical band saw in use. Note the material handing tool to keep hands away from the blade and also the well-lighted work surface. © Renee Bohnart

Fig. 26-87 A metal-cutting band saw and the many forms that can be cut with this machine. © Baileigh Industrial, Inc.

This type of machine is often called a *band machine* because of its continuous sawblade. Many people in industry feel that it offers many advantages over other types of cutting tools. Unlike other machine tools, it cuts directly to a layout line and removes material in sections instead of chips. Chipless cutting saves both time and material.

The cutting tool on a band saw is a continuous band in which each single-point tooth is a precision cutting tool. It cuts continuously and fast. Wear spreads evenly over all the teeth to extend tool life. Because reciprocating machines waste motion on the return stroke, the band machine's continuous action allows it to do more work in the same amount of time.

The thin band tool, a fraction of an inch thick, saves material. Because of its size, it takes less horsepower to overcome material resistance than do the bigger, wider cutters on other machine tools.

A significant advantage of band machining is its unrestricted versatility. There is no limit to the length, angle,

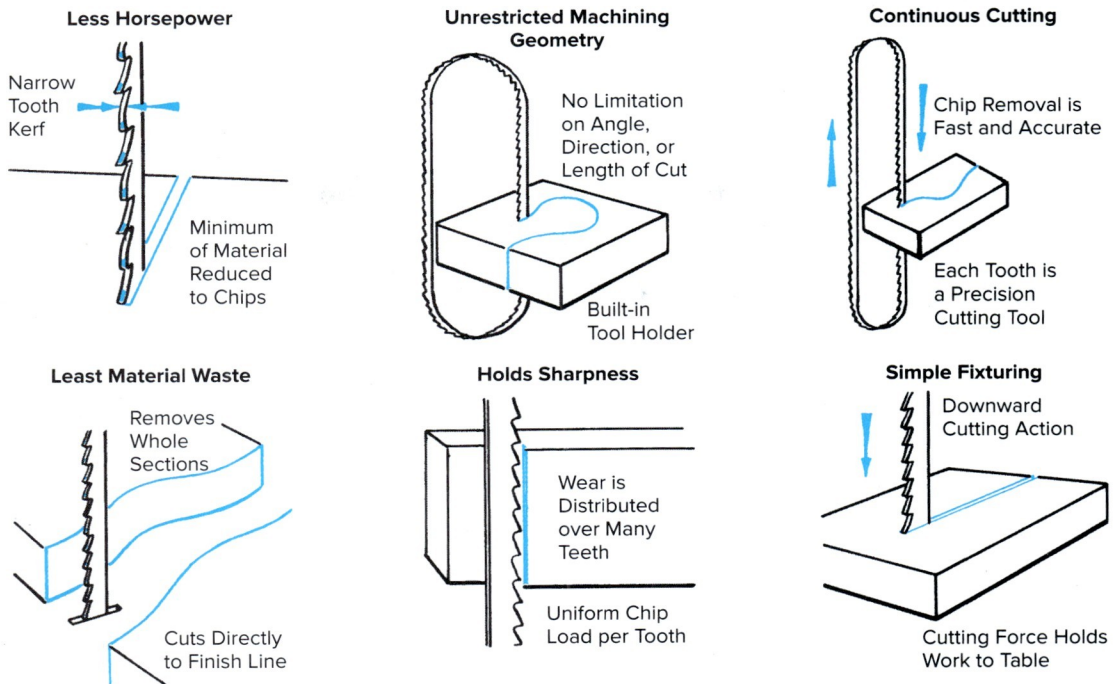

Fig. 26-88 Advantages of the band machine.

contour, or radius that can be cut. The constant downward force of the band holds the work to the table so that it is easy to hold or fixture work for production runs. Making a setup or changeover is fast, operation is easy, and cost per cut is low. Figure 26-88 summarizes the advantages of band machining.

If the band machine's saw blade could be replaced with a high pressure water jet, even more versatility could be achieved while maintaining many of the same advantages. Water jet cutting technology is unique in that it can cut almost all materials cost effectively from the hardest metals to the softest food products. The energy required for cutting materials is obtained by creating ultrahigh water pressures and forming an intense cutting stream by focusing this high speed water through a small, precious-stone orifice. There are two main steps involved in the water jet cutting process. First, the ultrahigh pressure pump or intensifier generally pressurizes normal tap water to levels above 40,000 p.s.i. (2,760 bar), to produce the energy required for cutting. Second, water is then focused through a small precious-stone orifice to form an intense cutting stream. The stream moves at a velocity of up to 2.5 times the speed of sound, depending on how the water pressure is exerted. The process is applicable to both water only and water with abrasives. The cutting nozzle can be stationary or integrated into motion equipment, which allows for intricate shapes and designs

Fig. 26-89 Water jet being used to cut a gear out of a plate.
© ESAB

to be cut much like with the band machine, Fig. 26-89. Motion equipment can be more sophisticated than that of the band machine and may range from a simple cross-cutter to two-dimensional systems and three-dimensional machines as well as multiple-axis robots. Computer aided design/computer aided manufacturing (CAD/CAM) software combined with computer numerical control (CNC) controllers translate drawings or commands into a digitally programmed path for the cutting head to follow. Chapter 25 has additional information on water jet cutting.

Since much of the work that can be done on band saws, band machines, or water jets can also be done with arc or flame cutting, the welder or fabricator will have to determine which method is better for a particular job to be done.

For video of water jet cutting technology, please visit www.mhhe.com/welding.

CHAPTER 26 REVIEW

Multiple Choice

Choose the letter of the correct answer.

1. Ventilation systems are used to remove what in the welding booth? (Obj. 26-1)
   a. Smoke
   b. Fumes
   c. Particulate
   d. All of these

2. Transparent welding curtains should_____. (Obj. 26-1)
   a. Exhibit good visibility
   b. Minimize arc glare
   c. Reflect usable light back into the work area
   d. All of these

3. Welding positioners are used to_____. (Obj. 26-1)
   a. Put the weldments in the most advantageous welding position
   b. Move the arc along a fixed workpiece
   c. Allow the use of manual welding processes
   d. Allow the use of high energy-density beams

4. Turning rolls are devices used to make_____. (Obj. 26-1)
   a. Longitudinal welds on tanks
   b. Circumferential welds on tanks
   c. Girth welds on tanks
   d. Both b and c

5. Manipulators are used to provide travel in which direction? (Obj. 26-1)
   a. Vertical
   b. Horizontal
   c. Circular
   d. Both a and b

6. Seamers move the arc along a fixed weldment. (Obj. 26-2)
   a. True
   b. False

7. Side beam carriages move the arc along a fixed weldment. (Obj. 26-2)
   a. True
   b. False

8. Magnetic clamps are useful work holding devices; however, they may cause_____. (Obj. 26-3)
   a. Marking of the metal
   b. Deflection of the shielding gas
   c. Magnetic arc blow
   d. Muscle strain when trying to remove them

9. Modular tooling is_____. (Obj. 26-3)
   a. Seldom used and not very flexible
   b. Only used for locating of parts but never used for welding
   c. Very flexible and used for a variety of applications
   d. Both b and c

10. Portable power tools can be powered from_____. (Obj. 26-4)
    a. Electric motors
    b. Pneumatics
    c. Hydraulics
    d. All of these

Review Questions

Write the answers in your own words.

11. Explain why positioners are used with other devices. (Obj. 26-1)

12. How are turntables used with robotic applications? (Obj. 26-1)

13. Explain why the side beam carriage can have such close linear tolerances and what a typical linear tolerance would be over 10 feet of carriage movement. (Obj. 26-2)

14. Describe how a trackless carriage system works and what type of joints they can be used on. (Obj. 26-2)

15. Describe the applications and typical welding positions encountered with the enclosed head orbital welding machines. (Obj. 26-2)

16. Why is ceramic backing material in some cases preferred over copper backing? (Obj. 26-3)

17. List eight induction heating applications and list related workpieces. (Obj. 26-3)

18. Which type of hand power tools should be used with engine-driven welding generators that only have d.c. auxiliary power and why? (Obj. 26-4)
19. List some typical applications the power iron worker can be used for. (Obj. 26-4)
20. List five advantages of both the band machine and the water jet cutting machine. (Obj. 26-4)

INTERNET ACTIVITIES

Internet Activity A

Using your favorite search engines find the largest weld positioner that is currently available on the market. Record the date, manufacturer, and the pertinent specifications.

Internet Activity B

Using your favorite search engines find information about the advantages and limitations of pneumatic-powered hand tools as compared to electric-powered hand tools.

Design credits: Cinema and movie icon: Denis Meshkov/123RF; and Illustration of Welding icons: Mr. Nachai Sorasee/123RF

27

Automatic and Robotic Arc Welding Equipment

Companies wishing to do automatic and robotic arc welding must understand the various issues that will be encountered when moving from manual and semiautomatic welding application methods. It is helpful to understand just what must be controlled and by whom to make a weld, Table 27-1.

The methods of application are defined in AWS A3.0 Terms and Definitions as[*]:

Adaptive control, adj. Pertaining to process control that automatically determines changes in process conditions and directs the equipment to take appropriate action.

Automatic, adj. Pertaining to the control of a process with equipment that requires only occasional or no observation of the welding, and no manual adjustment of the equipment controls.

Manual, adj. Pertaining to the control of a process with the torch, gun, or electrode holder held and manipulated by hand.

Mechanized, adj. Pertaining to the control of a process with equipment that requires manual adjustment of the equipment controls in response to visual observation of the operation, with the torch, gun, wire guide assembly, or electrode holder held by a mechanical device.

Chapter Objectives

After completing this chapter, you will be able to:

- **27-1** Explain automatic and robotic welding requirements.
- **27-2** Describe arc control devices.
- **27-3** Describe seam trackers.
- **27-4** Describe arc monitoring equipment.
- **27-5** Describe various weld controls including those that are microprocessor based.
- **27-6** Explain robotic arc welding systems.

---

[*]From AWS Standard Terms and Definitions, AWS3.0:2010, pages 3, 5, 27, 35, and 37.

Table 27-1 Methods of Process Application and Key Factors for Arc Welding

Key Factors for Arc Welding Methods of Application	Starting the Arc	Controls Electrode Feeding into the Arc	Controls Arc Length	Controls Heat for Correct Fusion	Controls Travel Speed	Tracks the Weld Pool Along the Joint	Controls Electrode and Gun Torch Angle	Makes Corrections to Overcome Deviations
Manual	Welder	Welder	Welder	Welder	Welder	Welder	Welder	Welder
Semiautomatic	PS, WF	WF	PS	Welder	Welder	Welder	Welder	Welder
Mechanized	PS, WF, CP	WF	PS	MCD	MCD	Operator	Operator	Operator
Automatic	PS, WF, CP	WF	PS	PS, WF, CP, MCD, SR	MCD	MCD-SR or along set path	MCD	MCD
Robotic	PS, WF, CP	WF	PS	PS, WF, CP, MCD, SR	Robotic control arm	Robotic control arm—SR or along set path	Robotic control arm	PS, WF, CP, MCD, SR
Adaptive control	PS, WF, CP	WF	PS	PS, WF, CP, MCD, SR	MCD (robot)	MCD (robot)—SR	MCD (robot)—SR	PS, WF, CP, MCD (robot)—SR

Note: PS = power source, WF = wire feeder, CP = control panel, MCD = motion control device, SR = sensor required.

Robotic, adj. Pertaining to process controlled by robotic equipment.

Semiautomatic, adj. Pertaining to the manual control of a process with equipment that automatically controls one or more of the process conditions.

The term *adaptive control* is unusual and requires further explanation. Adaptive controls can be applied to virtually all the process application methods. In its purest form, adaptive controls have been around for many decades. One very common use was in small light-duty gas metal arc welding systems. These all-in-one power source and wire feeder systems were designed for sheet metal and autobody repair facilities. The voltage control was directly coupled to the wire-feed speed motor, so any variation in voltage would impact wire-feed speed in an attempt to keep the process in control. These types of all-in-one GMAW systems are still in use today. Adaptive control can be applied to manual welding applications by the welder's ability to shape and control the output volt-amp curve on certain constant current power sources. As Table 27-1, page 871, shows, the welder is capable of controlling all the key factors required for depositing quality weld metal.

Adaptive controls can be applied to semiautomatic and mechanized application methods. An example of this would be when using the GMAW-P mode of metal transfer. Most equipment can be selected for adaptive versus nonadaptive. The adaptive mode generally has sensors, which monitor arc length. If anything occurs that affects the arc length, feedback circuits will attempt to correct the situation to get the arc length back within proper parameters. As can be seen in Table 27-1, welders or operators still have some key factors under their control. In some cases, especially in semiautomatic application situations, the welder may want to control the heat, fusion, and penetration by variations in electrode extension. In this case, the adaptive control will react in the opposite manner than that desired by the welder. In order to prevent these types of situations, the adaptive control can be set for nonadaptive allowing welders to use their skill and understanding of the weld pool to correct deviations as they occur.

In the automatic and robotic methods of applications, welders or operators are not required to control the key factors influencing the weld. This will require the highest level of adaptive control. Multiple sensors may be required to monitor various facets of the weld pool, weld profile, penetration, and fusion patterns. Sampling can occur much faster than welders or operators can possibly see and react to. Data can be collected and reacted to thousands of times per second. These data are sent to a computer controller (which will direct the power source), wire feeder, or motion control device to take appropriate action. However, if the parts can be accurately cut, bent, formed, and fixtured, adaptive controls in the form of sensors, computers, and electromechanical servo systems may not be required. These devices are expensive and add greatly to the complexity of the welding system. What is being attempted with these high level automatic and robotic applications with adaptive control is to replicate the eyes, hands, skill, and knowledge of a welder or machine operator. This is being done in order to make instantaneous corrections to eliminate weld defects. Some typical joint fitup tolerances that will generally be acceptable for automation and robotic welding are the material thickness or 0.060 inch, whichever is less, for gas metal arc welding and 10 percent of the thinnest material thickness for laser beam welding. Another way of judging fitup tolerance is to use ±½ the electrode diameter or in the case of lasers, the focused laser beam diameter. If a 0.035-inch diameter GMAW electrode is being used, ±0.018 inch is a very tight tolerance.

Primary issues that need to be dealt with when doing mechanized, automatic, or robotic welding applications are:

- **Overall dimensions of parts** These must be very accurate. (See the previous paragraph for fitup tolerances.)
- **Parts that are formed or pressed** These must be consistent in shape. [Spring-back from heat number to heat number on parts must be controlled, mechanical properties must be understood and controlled, material test reports (MTRs) must be checked.]
- **Weldment design** This must be appropriate for intended welding operations and weld locations.
- **Fixturing and weld tooling** These must be appropriate for the application.
- **Welding process and mode of metal transfer** These must be appropriate in the case of GMAW.

Cost considerations must be reviewed since the higher the degree of accuracy and the more sophisticated the fixtures, the more expensive the parts and the welding fixture system becomes.

Arc Control Devices

Magnetic Arc Control

In certain situations, it is more advantageous to move the arc and thus the weld pool than it is to require the motion control device to change the torch or gun work or travel

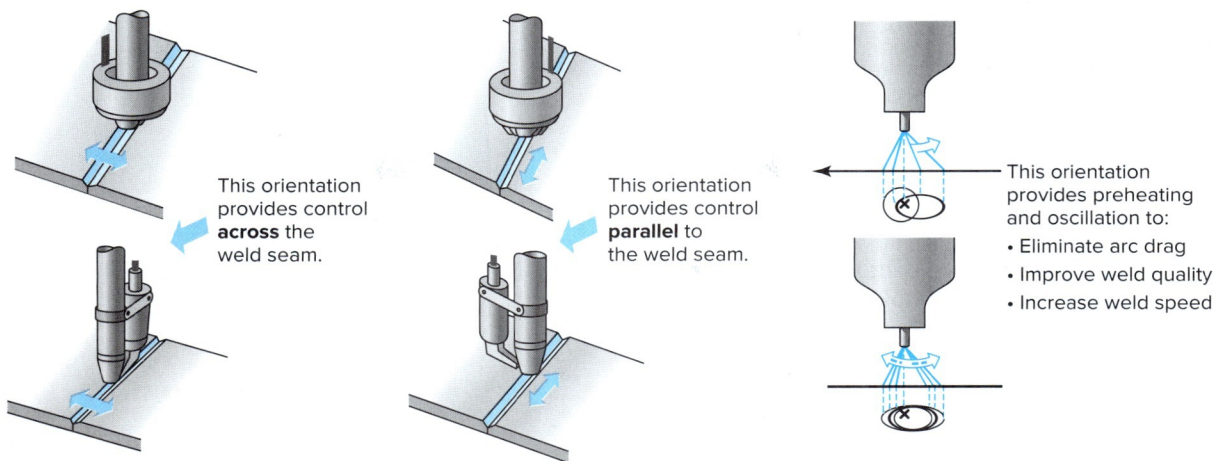

Fig. 27-1 Probe orientation for weld pool control. ITW Jetline-Cyclomatic

angle. These motions may be required in a very rapid fashion to correct deviations as they are forming. These devices may also be used to change from stringer beads to weave beads.

Magnetic arc blow is generally considered a negative action that prevents proper welding conditions and reduces the ability of positioning the weld where it belongs. These same forces can be used to your advantage if properly controlled with a magnetic arc control and probe. These devices control welding at the point of its greatest impact—the welding arc itself. By adding stability and control over arc oscillation and positioning, magnetic arc control ensures a quality weld, even on exotic metals. They are typically used with the GMAW, GTAW, and PAW processes. These magnetic arc control systems can be attached to present welding torches and guns that are being used for mechanized and automatic applications. The operating principles are simple in that magnetic probes are positioned around the arc to position, oscillate, and stabilize it. Magnetic probe orientation is shown in Fig. 27-1.

Magnetic arc control systems require a control unit, magnetic probe, and interconnecting control cable, Figs. 27-2 and 27-3. Oscillation of the welding arc can be carried out at frequencies up to 50 hertz. These devices provide control over:

- Heat distribution
- Excessive undercut
- Excessive porosity
- Incomplete penetration
- Incomplete fusion
- Unwanted magnetic arc blow

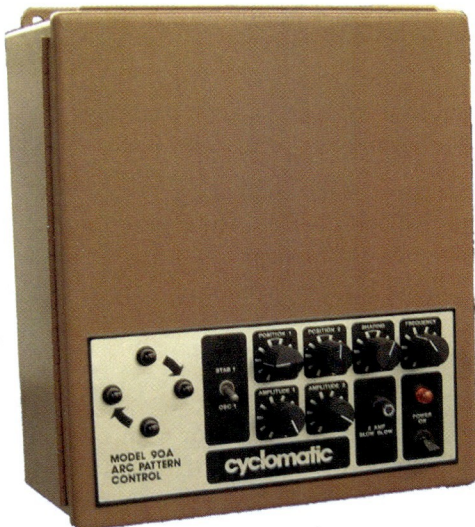

Fig. 27-2 This unit controls the sweep frequency, sweep amplitude, arc position, dwell time (on each toe), and shaping via the magnetic probe. © ITW Jetline-Cyclomatic

Fig. 27-3 Magnetic probe used in conjunction with control shown in Fig. 27-2. The probe mounts on the torch or gun and creates a magnetic field that precisely positions, oscillates, and stabilizes the arc. © ITW Jetline-Cyclomatic

Automatic and Robotic Arc Welding Equipment   **Chapter 27**   **873**

Mechanical Oscillators

The arc can be moved by other means than magnetics, but this necessitates a mechanical oscillator. This mechanical device grips the welding head, torch, and/or gun and physically moves them in much the same manner as a welder doing manual or semiautomatic applications would. These mechanical devices have less dexterity than the welder, as the motion is usually simply side to side, pendulum in shape, or triangular. It is set up in a predetermined fashion and does not receive any direct feedback or sensing from the arc or weld pool.

Mechanical oscillators consist of a control unit and slide- or pendulum-type oscillator mechanism. They are generally rated in load-carrying capacity of the oscillator mechanism, such as 120 pounds. The capacity must be capable of carrying the welding head, gun, and/or torch. They are generally powered by a stepper motor driving the pendulum or slide mechanism. The control provides full remote control of the movement of the oscillator.

A control unit for a cross slide oscillator would generally have the following specifications:

- Leftward speed (0–100 inches per minute)
- Rightward speed (0–100 inches per minute)
- Stroke width (0–4 inches)
- Center adjustment (0–4 inches)
- Left dwell time (0.03–3 seconds)
- Right dwell time (0.03–3 seconds)

More sophisticated mechanical oscillators are available such as the microprocessor-controlled pendulum. This system provides independent control of all torch movements in the horizontal plane and provides automatic torch centering using patented Thru-Arc™ sensing technology. The Thru-Arc™ sensor provides four modes of operation. The first mode provides a constant width oscillation with automatic torch centering. The second mode provides variable width and center line control using a depth-of-penetration parameter to control the width. The third mode provides for single right-side tracking with constant width, and the final mode provides for single left-side tracking with constant width. The entire seam tracking information is derived from the welding arc and oscillator position. The arc voltage and welding current are measured with an external voltage probe and Hall-effect sensing devices provided as part of this system. This particular manufacturer's control unit provides a user definable 50 sequence programmable logic controller (PLC) with four 24-volt d.c. inputs and two normally open (N.O.) relay contacts. Using the PLC you can provide external program control and

Fig. 27-4 A pendulum-type weld oscillator and control system. The ⅝-inch diameter shaft is for attachment of gun or torch mounting assembly. This WOC-1000™ system features Thru-Arc™ tracking technology that allows automatic torch centering and width control. © Computer Weld Technology, Inc.

simple power source interface. An RS-232 serial port is provided for off-line programming and system configuration. This port can also be used to remotely control this particular unit, Fig. 27-4.

Seam Trackers and Arc Length Controls

In some situations, multiple powered cross slides can be incorporated. This allows the operator to simply use a joystick to control the torch position rather than having to manually turn all the hand wheels on the various cross slides, Fig. 27-5. Once the cross slides are powered by the stepper motor, they can be signaled with sensors through a monitor and controller for adaptive control of the torch or gun positions. Figure 27-6 shows a seam tracker set up on a side beam carriage. The T-joint fillet weld has been purposely set up out of alignment to demonstrate the seam tracker's capability. As the probe starts to lose contact with one surface of the joint, it sends a signal to the controller which in turn activates the proper combination of motorized cross slide movements to get the gun back into the proper position.

A similar system would be able to control the arc length when using the GTAW or PAW processes. Arc length control with these processes is done by moving the torch up and down by use of a motorized slide to maintain the arc voltage (arc length). These arc voltage control systems are capable of maintaining the arc within ±1 percent or 0.1 volt, whichever is greater. Most codes only require voltage tolerances of ±7 percent, which may not be accurate enough for high precision GTAW and PAW. They are able to compensate for arc length variation up to travel speeds of 24 inches per

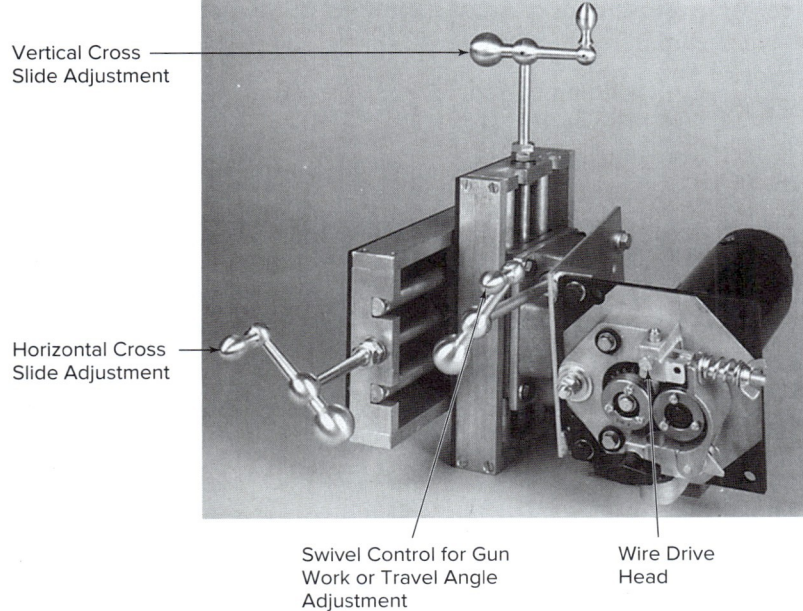

Fig. 27-5 Manually adjusted cross slides. This setup utilizes two cross slides and a swivel for accurate manual alignment and proper gun work or travel angle in the joint. This type of system is used for mechanized welding.
© Computer Weld Technology, Inc.

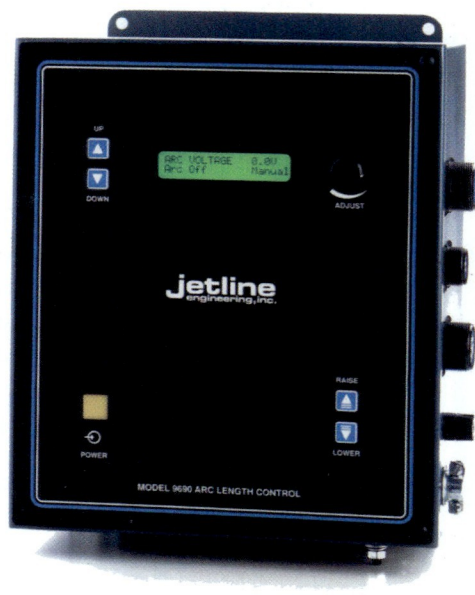

Fig. 27-7 Arc voltage control helps maintain the proper arc length at high speed on material with uneven surfaces, or when total adaptive control of this important variable is required for automatic welding.
© ITW Jetline-Cyclomatic

Fig. 27-6 Side beam carriage setup for GMAW with scam tracker and microprocessor weld control. It is using a tactile probe scam tracking system. © Miller Electric Mfg. Co.

minute. A control along with a drive unit are required, Figs. 27-7 and 27-8, page 876. The torch and cable carrying capacity of this unit totals 20 pounds. If excessive melt-through is encountered, the system will automatically shut off power and retract the welding torch to protect the operator and equipment. Excessive melt-through would cause the arc voltage to rapidly go up. If the control cannot recognize this, the reaction of the drive unit would be to rapidly direct the torch toward the weldment in an attempt to reduce the overvoltage condition. In this case the weldment has not moved; only a burn-through hole has been created, so the control must be programmed to recognize these types of situations and not overreact.

Arc Monitoring

When doing manual, semiautomatic, and mechanized welding applications, welders or operators do the monitoring. They must be highly skilled and trained to monitor and control all the key factors influencing the weld in order to maintain weld quality. Refer to Table 27-1, page 871, for a review of these key factors. On routine high production work, there is a desire to

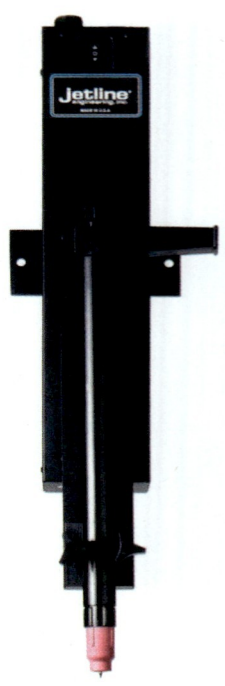

Fig. 27-8 Automatic voltage control drive unit. It features a stepper motor driving a precision lead screw to provide the motion. It is filtered against high frequency and has minimal play and backlash. Manual up and down jog buttons are provided on the side of the unit.
© ITW Jetline-Cyclomatic

relieve the human welder or operator from this tedious type of work. The safety of the welder or operator if the work has to be done on materials that are extremely hot, radioactive, or explosive in nature would necessitate automation or robotics with adaptive controls. Issues dealing with productivity, quality, and a projected smaller workforce must be factored into how a welding process is going to be applied.

As you are aware, welding processes are complex. There are many variables that must be dealt with that may adversely affect the completed weldment. These variables are in a constant state of change, and they have to be observed even in a very harsh environment. It is analogous to a situation where you want to get from point A to B on a trip. While you can read the road numbers on your map, you are having trouble seeing the road signs along the side of the road due to a blizzard going on outside your car. This is a very uncomfortable position to be in. Yet in welding, the work must be dealt with at the required quality level even in the harsh environment of high temperatures, intense electromagnetic radiation, high electric and magnetic fields, molten metal spatter, and fumes as found during a high production weld.

There are four basic control elements that must be dealt with. They are the welding process itself, manipulation of the input variables, what is happening with the disturbing input variables, and what process response variables need to be manipulated to get it back in control, Fig. 27-9. The manipulated input variables are those that directly affect the process response variables. For example, increasing the energy will result in an increase in the penetration level. The variable affecting this energy may be preset and not changed during the welding operation; however, it is very advantageous to be able to change these as required during the welding operation. Examples of the variables that may require changing are:

- Welding current
- Wire-feed speed
- Voltage
- Arc length
- Time and force (for resistance welding)
- Travel speed
- Electrode extension (arc welding processes)
- Electrode position (arc welding processes)
- Electrode angle (arc welding processes)

The disturbing input variables are, of course, undesirable and generally are unavoidable. They will have an effect on the process but are not controlled. Some examples of disturbing input variables are:

- Weld current fluctuations (due to variation in input line voltage to the power source)
- Irregular material thickness (especially for resistance welding)
- Variation in joint geometry and location (primarily when arc welding)
- Variation in joint gap (in brazing and soldering)

The following process response variables are the result of the welding process being used:

- Current
- Voltage
- Travel speed
- Cooling rate (time and temperature)

The above process response variables will produce the following required weld properties:

- Weld size
- Weld profile
- Microstructure
- Soundness

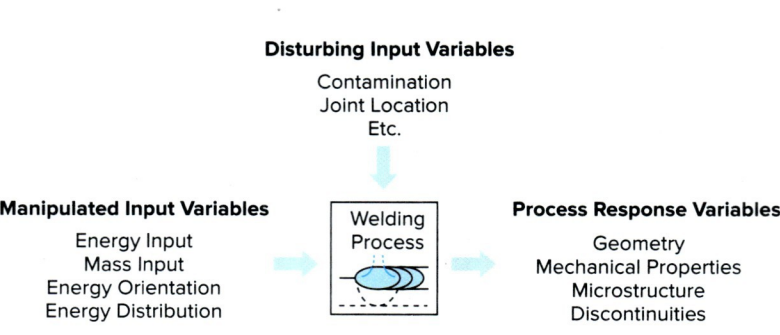

Fig. 27-9 Flow diagram of the welding process as it relates to monitoring and control. American Welding Society (AWS) Welding Handbook

Table 27-2 Common Sensors and Units of Measure

Physical Property	Sensor(s)	Units
Time	Timer, counter	Second, cycle[1]
Temperature	Thermocouple, resistive-temperature device, thermistor, pyrometer	Fahrenheit (Celsius)
Force	Load cell, piezoelectric, linear variable differential transformer, and capacitive force sensors	Newton, pound (kilogram)
Pressure	Displacement- and diaphragm-type pressure sensors	Pounds force per square inch (kilopascal)
Flow rate	Differential pressure flowmeter, mechanical flowmeter, mass flowmeter	Cubic feet per minute (liters per minute)
Electric current	Current shunt, Hall-effect sensor, Rogowski coil	Amperes
Electric potential	Voltmeter	Volt
Displacement	Potentiometer, voltage differential transformer, inductive and capacitive sensors; synchro; resolver; encoder; ultrasonic sensor; machine vision sensor	Inch (millimeter)
Velocity	Potentiometer; voltage differential transformer; inductive and capacitive sensors; synchro; resolver; encoder; ultrasonic sensor; machine vision sensor	Inches per minute (millimeters per second)
Acceleration	Potentiometer; voltage differential transformer; inductive and capacitive sensors; synchro; resolver; encoder; ultrasonic sensor; machine vision sensor	Inches per minute squared (millimeters per second squared)
Radiation (visible light, infrared, ultraviolet)	Photodiode, phototransistor, solar cell sensor	Lumen (candela)
Acoustic (sound) energy	Microphone	Decibel

[1]Cycle = Interval of time based on frequency.
American Welding Society

It is very difficult to sense many of the process response variables since no sensor exists that can measure such things as mechanical properties and microstructures. However, sensors do exist that can measure things such as:

- Temperature
- Weld profile
- Weld size
- Penetration
- Radiation

Sensing Devices

Sensing devices are generally transducers that convert energy from one form to another. They can be as simple as a thermocouple or as complex as a three-dimensional laser camera. Thermocouples detect heat and convert it into an electrical voltage that can be used to regulate a multitude of devices, for example, the motion control device, which can increase or decrease the travel speed and impacts the heat input and resultant temperature, and the power source, wire-feed controllers, which varies weld parameters to affect the heat input. Three-dimensional laser cameras can be used to control the weld volume and profile through a robot control system. Some common sensors and the units they can measure are shown in Table 27-2. Some typical sensors are shown in Figs. 27-10 through 27-12, page 878.

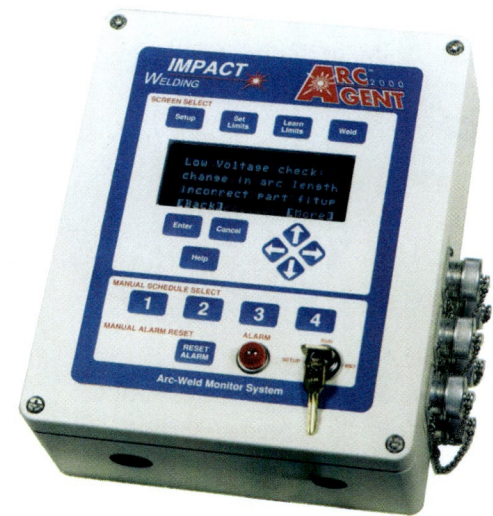

Fig. 27-10 Welding current sensor capable of measuring up to 1,000 amps d.c. This production unit is for permanent attachment in a welding system. © IMPACT Engineering

A less typical sensor, a high speed laser profiling camera, is shown in Figs. 27-13 and 27-14.

> For video of an electrode submerged arc welding setup with vision control, please visit www.mhhe.com/welding.

The laser vision camera has been designed for factory environments. However, more severe environments will require more intensive maintenance. Wherever possible,

Fig. 27-11 Gas turbine flow monitor precisely measures the flow rate of welding gases. It can be used for argon, helium, CO_2, or any mixed inert welding gas. Flow rates can be digitally displayed from 5 to 255 SCFH in 1 SCFH increments. The accuracy of the 1 percent of full scale reading is much better than a conventional ball-type flowmeter. © Computer Weld Technology, Inc.

Fig. 27-13 A precision laser sensor and laser vision camera for applications in the shipbuilding industry. AUTO-TRAC is for seam finding, seam tracking, and adaptive control for hard automation (special-purpose CNC machines) welding processes like GMAW being shown here. It is capable of detecting weld slope, mismatch, undercut, bead width, convexity, open porosity, pinholes, and metal drops. The laser profiler acquisition rate is 16,000 points per second while the sensor acquisition rate is 2,000 Hz allowing it to easily keep up with the hybrid laser/GMAW process. Advantages of laser vision-based seam tracking systems over mechanical probes, human adjustments, or other means include optimized adaptive welding parameters, extremely accurate wire positioning, no material damage due to contactless sensing, preweld joint variation information, and a perfect overlap for cylindrical parts required to be leakproof. It is designed for precise torch positioning, the modular components allow for Ethernet connectivity, and the adaptive software module can be interfaced directly to the welding equipment. All this makes for faster welding speeds, decreases the amount of operator monitoring, lessens intervention requirements, reduces tooling cost, and increases quality and productivity with no mechanical probe parts to replace. © Servo-Robot Corp.

Fig. 27-12 This wire-feed speed sensor uses optical encoder technology to transmit information to the weld monitor. This, coupled with the amperage sensor, can be used to detect electrode extension. These data can be supplied to the motion control device and electrode extension can be accurately maintained. Weld volume, profile, deposition rates, as well as other issues may require accurate control and monitoring of wire-feed speed. © Computer Weld Technology, Inc.

Fig. 27-14 Four electrode submerged arc welding setup with vision control. Electrode wire diameters are 3/16 inch and travel speed is varied around 20 inches per minute based upon joint gap being measured by the laser camera. © Servo-Robot Corp.

it is important to ensure that the camera is properly isolated to avoid any electrical contact with the tool (robot or other motion control device) on which it is installed. The camera head has been specifically designed for arc welding environments. Circulating compressed air in the cooling channel is sufficient to maintain the camera at its proper operating temperature. For heavy welding applications, shop air should be switched over to an air refrigerating system. The compressed air supplied to the camera for cooling must be oil-free. The compressed air also acts as an air knife preventing weld spatter and smoke from accumulating on the camera window. This air supply must be oil-free as well. Oily air will progressively cover the permanent camera window with a thin film of oil and reduce the camera's sensitivity. Vision systems like this consist of three major components:

- Control unit
- Vision head (camera)
- Cables to link them

The welding arc itself can be a sensor in that the voltage or current being supplied to it can be measured and these data supplied to a controller. Then, the controller can be used to signal motorized cross slides or the robot arm to keep the weld located in the joint, Fig. 27-15.

Sensors of the various types are of no value unless their data can be monitored and used to control the deviation that is taking place. Monitors and controls will be covered separately.

Process Monitoring Systems

Data collection and storage functions are done with the welding process monitoring system's computer. The data processing capability will include:

- Decision making
- Statistical analysis
- Signal filtering

These systems can compare what has been set per the welding procedure and what is actually occurring, and if it exceeds the plus or minus tolerance, it can sound an alarm or shut the welding operation down. These monitors can store information from a number of welds to allow for off-line interpretation and statistical process control methods.

Process Control Systems

In process monitoring systems, the monitor has limited control. If tolerance limits are exceeded, an alarm sounds or the system can be shut down. In process control systems, actual weld controllers are used for the following operations:

- Sequence of operation
- Process variables
- Multiple schedules
- Input and output signaling capability
- Motion

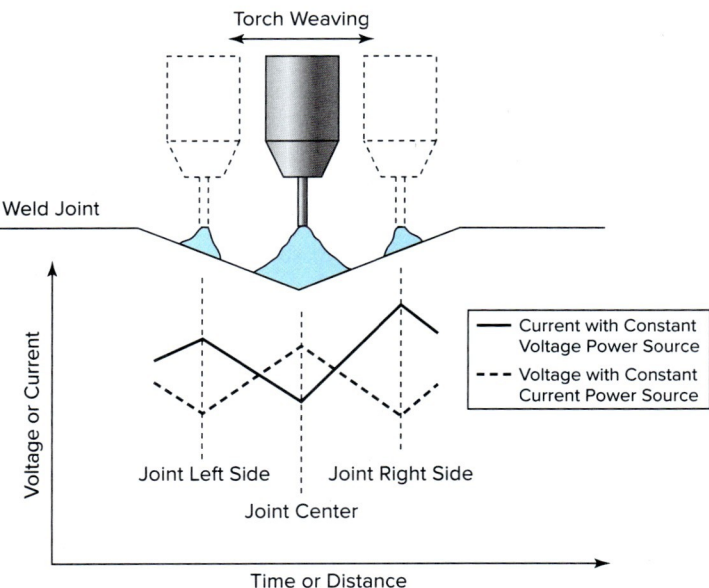

Fig. 27-15 Through-arc sensing for joint tracking. As the weld moves away from the joint center the edge of the joint is sensed. These data are fed back through the controller to motion control devices, which will keep the weld tracking the joint. American Welding Society (AWS) Welding Handbook Committee. Welding Science and Technology, *Welding Handbook*, Vol. 1, 9th ed. Miami: American Welding Society, 2001, Fig 10.8, p. 435.

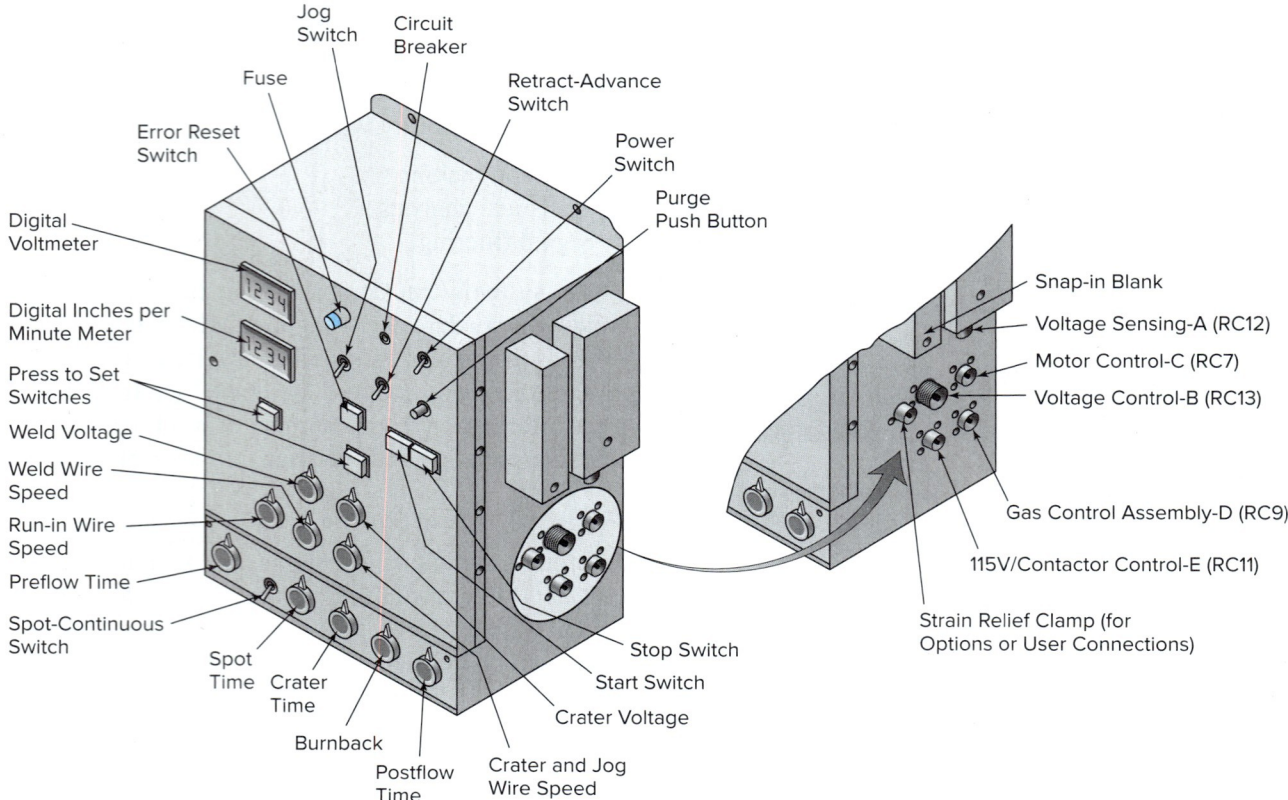

Fig. 27-16 A self-contained weld controller. It is capable of controlling the entire welding sequence for such processes as GMAW, FCAW, and SAW. It has various input and output signaling capability for controlling such activities as part clamping, motion control devices, and so on. Miller Electric Mfg. Co.

These controls may be incorporated into the monitoring cabinetry, or the robot controller, or they may be a free-standing controller, Fig. 27-16. This weld controller displays the voltage and wire-feed speed digitally. It is not microprocessor based, so it can only remember the last welding procedure it was set up to run.

Control over a welding operation for automatic or robotic applications is very involved, but the fundamental goal is to deposit a satisfactory weld. Having good control over the sequence of operation of the weld or "weld sequence" is essential. A typical weld sequence requires control over the following:

- Start button activation (this may be remote controlled when the fixture is loaded and the operator is in a safe position)
- Preflow of the shielding gas timer (flux in the case of SAW)
- Run-in speed (improves starting characteristics by controlling wire-feed speed prior to arc starting)
- Arc initiation (can sense when an arc is started)
- Weld voltage
- Weld wire-feed speed
- Stop button activation (this may be remote controlled by a timer or a limit switch when the torch or gun has reached a specific position)
- Crater voltage
- Crater wire-feed speed
- Crater time
- Burnback time (keeps the wire from freezing in the solidifying weld pool)
- Postflow timer

Auxiliary equipment can be controlled by the weld controller and must be properly sequenced with the weld. A control relay is an electromechanical device that uses a magnetic coil to open and close small switches, Fig. 27-17. For example, when the start button is pushed, some control relays activate (CR54 and CR52). The CR54 relay will stay energized until the postflow timer times out, while the CR52 relay will deenergize when the stop button is activated. If part clamping is required prior to arc initiation and is required through the weld cooling down, CR54 would be the correct relay to use. If the motion

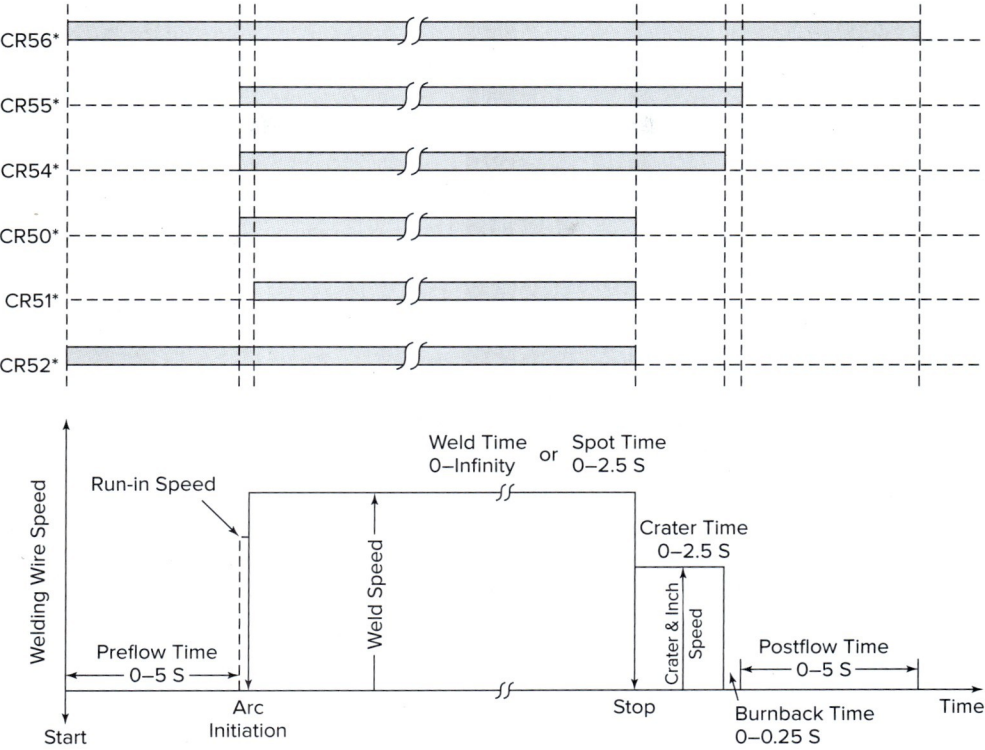

* The dotted line condition indicates the time during the weld cycle that the control relay is deenergized. The solid line condition indicates the time during the weld cycle that the control relay is energized.

Fig. 27-17 A typical weld sequence for a gas metal arc weld. There are 6 control relays that can be used for various input and output signaling capability. These signals can be used for controlling such activities as part clamping, motion control devices, and so on. Miller Electric Mfg. Co.

control device were only to move when the welding arc is on, CR51 would be selected as it energizes upon arc initiation and deenergizes when the stop button is activated. In other cases, you may want the arc to start when the motion control device is in motion. Take a "running start" and keep moving during the crater fill operation. The control relay in this case would be CR54.

Connections to the control relays are done on a terminal strip. This allows for easy connection of auxiliary equipment. In some cases connections are done through receptacles, Fig. 27-18, page 882. These examples are from one equipment manufacturer's owner's manual. Always consult the specific equipment manufacturer's owner's manual on the equipment that is being used.

Microprocessor-Based Controllers

These types of controllers use computer power to store and control various welding conditions, Fig. 27-19, page 882. These systems allow digital readouts and accurate setting of the weld sequence. A tack-feedback sensor located on the wire drive motor maintains preset wire-feed speeds very accurately. Voltage sense leads can be located as close to the arc as possible to maintain the set arc voltage. Voltage and wire-feed speed (amperage) are essential variables, and this system uses closed-loop feedback controls to maintain them, Fig. 27-20, page 883. This system is primarily used with continuous electrode processes like GMAW, FCAW, and SAW. Other microprocessor controls are available for other processes like GTAW and PAW. This control is capable of running all the various modes of metal transfers including pulse. It can be set up for adaptive or nonadaptive pulsing. The eight programs can be modified to meet specific welding needs. Additional programs can be stored on data cards. See Table 27-3, page 873, for some sample gas metal arc pulse programs.

Controller Communication

Automated factories and workplaces depend on accurate control systems to make them work smoothly and efficiently. The control system is like the nervous system of

Remote 14	Socket/Pin Letter	Information
24 V a.c. for output contactor	A	24 V a.c. circuit breaker protected.
	B	Contact closure to A completes 24 V a.c. contactor control circuit.
115 V a.c. output contactor	I	115 V a.c. circuit breaker protected.
	J	Contact closure to I completes 115 V a.c. contactor control circuit.
Remote output control	C	0 to +10 V d.c. output to remote control.
	D	Remote-control circuit common.
	E	0 to +10 V d.c. input command signal from remote control.
	M	Mode select.
A/V amperage voltage	F	Current feedback; +1 V d.c. per 100 A.
	H	Voltage feedback; +1 V d.c. per 10 output receptacle volts.
Electrical ground	G	Circuit common for 24 and 115 V a.c. circuits.
	K	Chassis common.

Fig. 27-18 When receptacle connections are used they are generally hard wired from the manufacturer. In some instances they may need to be field connected by qualified technicians. Understanding what each socket/pin is used for is essential for proper equipment operation. Always consult the specific equipment manufacturer's owner's manual for specific details.

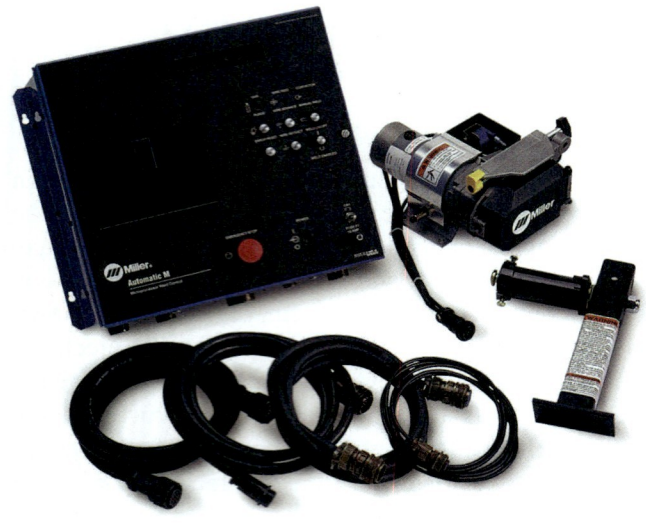

Fig. 27-19 Microprocessor-based weld control for automatic systems and robots lacking weld control capability. This model is ready to operate, with eight preprogrammed synergic pulse (adaptive/nonadaptive) programs that can be easily customized. Additional memory space and the ability to transfer data from one control to another is available on data cards. The unit features 17 dedicated inputs, 14 dedicated outputs, and two user-defined outputs. A quick disconnect allows easy interface with PLC or other control systems. This controller comes with interconnecting cables, wire drive head, and wire supply support assembly.
Miller Electric Mfg. Co.

the body, making everything work in harmony. Communication systems for on-site programming and testing of the controller traditionally have been wireless infrared. While it is inexpensive and robust and has done the job well, it has shortcomings. For instance, infrared requires a direct, uninterrupted line of sight and this is not always easy to achieve on a busy weld shop floor or workspace. The search for a better communication system is vital for automation applications. Power sources, wire feeders, and related equipment need to be prepared for communication using standard protocols including TCP/IP (LAN), ArcLink, Anybus, Profibus, CAN, or even straight communication with a PLC. Other communication strategies are being reviewed. This will allow those manufacturers' controllers to link a range of industrial systems into a single network, which will allow the end user to move closer to the fully integrated manufacturing plant.

Robotic Arc Welding Systems

In Chapter 26, various motion control devices were covered that can move the part under a fixed arc. These devices are head and tail stocks, turntables, turning rolls, and so on. Motion control devices can also move the welding process along a fixed weldment. These devices

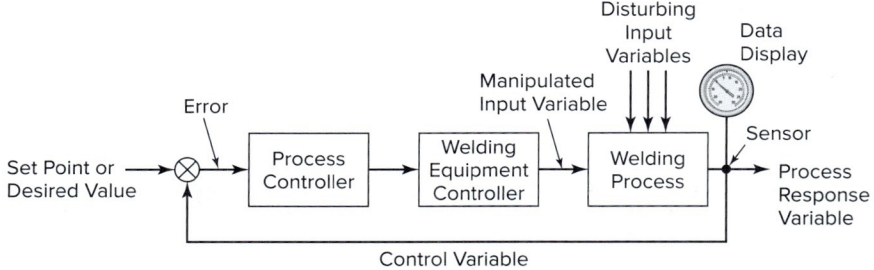

Fig. 27-20 Closed-loop feedback control of the welding process. American Welding Society (AWS) Welding Handbook

Table 27-3 Sample Pulse Welding Programs Electrode Wire Type and Shielding Gas

Program	Hardwire Selections	Softwire Selection (Aluminum)
A	0.035-in. steel, argon-oxygen	0.045-in. 4043, argon
B	0.045-in. steel, argon-oxygen	0.045-in. 5356, argon
C	0.052-in. steel, argon-oxygen	0.035-in. 4043, argon
D	0.035-in. 309 stainless steel, argon-carbon dioxide	0.035-in. 5356, argon
E	0.045-in. 309 stainless steel, argon-carbon dioxide	$1/16$-in. 4043, argon
F	0.052-in. 309 stainless steel, argon-carbon dioxide	$1/16$-in. 5356, argon
G	0.045-in. steel, argon-carbon dioxide	$3/64$-in. 4043, argon
H	0.035-in. nickel alloy, argon	$3/64$-in. 5356, argon

would be seamers, side beam carriages, orbital welding machines, and so on. These systems work fine for the job they were intended to do. However, they are not very flexible. Systems designed for flat plate generally cannot weld pipe, and vice versa. This section will deal with robotic arms, which can move the arc along a fixed weldment. The weldment can also be mounted on manipulating devices (perhaps another robot arm) to move the weld joint into the most advantageous position. The robotic work cell can be set up to articulate the movement of the robot arm and the weldment-manipulating device for the best welding conditions. This also allows for the greatest flexibility in the type of weldments that can be worked on. Figure 27-21, page 884, shows an example of a dual robot work cell. The parts in this example had been welded by a welder using semiautomatic applications and the flux cored arc welding process with 0.052-inch diameter electrodes. This requires a 30-minute cycle time per part to make the 12 to 16 welds needed. The aggressive application of an integrated robot work cell and the use of 0.045 solid wire reduced the cycle time per part to 6 minutes. Some welds were as wide as 1 inch, and material thickness in some cases was $5/8$ inch. Fitup was an issue in that the parts would vary from 0.020 to 0.100 inch in and out of tolerance.

As defined by the Robotic Industries Association (RIA), a robot is "an automatically controlled, reprogrammable multipurpose manipulator programmable in three axes or more which may be either fixed in place or mobile for use in industrial automation applications." The two main types of robot arms are the *articulated* and the *rectilinear*. The articulated arm typically has six axes of motion, Fig. 27-22, page 884, while the rectilinear arm has less than six axes of motion and cannot orient the welding torch or gun in the best fashion. Since the articulated arms are the most versatile and have a good cost-to-performance ratio, they make up 90 percent of the welding robots.

In the last three decades, over 1.1 million industrial robots have been sold. These have typically been of the fixed type. It is not easy to determine just how many of these are welding robots. Sales of mobile robots are expected to increase dramatically. Space exploration was one of the first areas where the public became very aware of mobile robot units. In 1997 NASA's robot lander, Pioneer, crawled across the surface of Mars. Mobile robots are called upon to do jobs located in places where people cannot safely go, such as inside pipelines and in areas of high radiation or high heat. RIA indicates the current price of a robot is one-fifth what it was a decade ago. Return on investment is down to approximately 2 years. Robots have also increased in capabilities and functions. They are becoming more robust and have life expectancies from 12 to 16 years.

The pool of professional welders is dwindling; in fact, 50 percent of the current welding workforce will be retiring in the next decade. This will leave a shortfall of skilled, knowledgeable welders. Statistical information indicates there are fewer people engaged in the manufacturing sector of the industry, yet production is up. This is due to improved tooling and manufacturing methods and the use of automatic and robotic welding machines. There are many industry segments that do not lend themselves to the degree of machine use, as does the manufacturing sector. Industry and work sectors such as custom fabrication shops, construction, maintenance, and repair industry segments will continue to rely heavily on the professional welder. Highly skilled and knowledgeable professional welders will always be in demand. If you want to pursue a career in the manufacturing industry, you should look more into the inspection, technician, or engineering level. With all the technology being placed on sensors, controls, and communication systems, the welding operator will be relegated to loading and unloading weld fixtures and simply pushing start and stop buttons. High manipulative skill levels will not be required, but much greater technical knowledge will be. Technician-level personnel will need to deal with the complexities of the welding operation and the sophisticated equipment required.

Fig. 27-21 Dual robot work cell. Note the robots are mounted overhead for easier access to welding joints. © Reis Group Holding GmbH & Co. KG

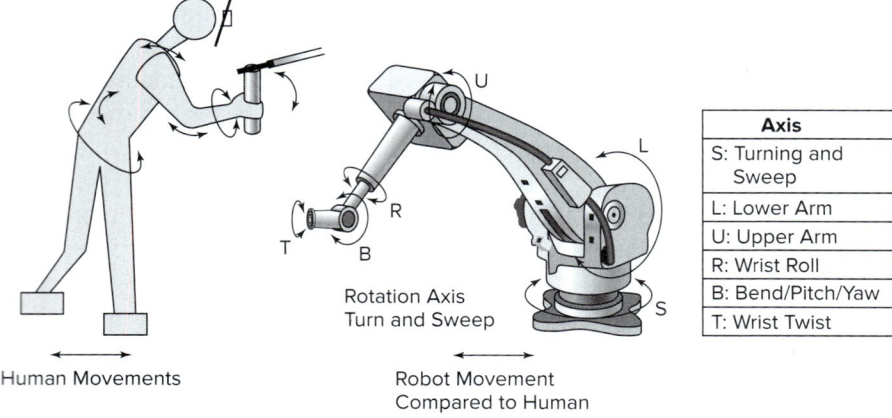

Fig. 27-22 Schematic illustration comparing a human's motion to that of an industrial robot. Servo-Robot Corp.

A typical robotic arc welding system requires some very specifically designed equipment, Figs. 27-23 and 27-24.

Fig. 27-23 Typical arc welding cell including: 1. Robot with arc welding equipment, 2. Welding power source, 3. Positioner, 4. Guarding, 5. Fume extraction. © Motoman, Inc.

Fig. 27-24 Articulated robot arm fitted with laser camera for seam tracking and seam finding, and capable of doing weld inspection. This robot cell, along with a laser camera, provides real-time joint tracking, adaptive control, and visual inspection system integrated with Servo-Robot advanced 3-D laser vision techniques and advanced sensing devices. One benefit of the unique hybrid sensor technology is that it can dramatically speed up and simplify applications traditionally done with 2-D sensors. Sensors of this type can be applied to various industrial sectors to ensure process quality and efficiency. © Servo-Robot Corp.

Robot Ratings

Load Capacity

The end-of-arm load-carrying capacity of a robot is generally in the 6- to 35-pound range. Robots have various load-carrying capacities at other points that allow the mounting of electrode wire drive motors and wire spools, Fig. 27-25.

Repeatability and Accuracy

Robots have the ability to return to the exact same position each time. This makes their movements very repeatable. A welding robot should be able to return the welding gun or torch to within ±0.004 inch of the same point after each program is executed. Accuracy is best described as the robot's ability to move a predetermined distance and direction and the ability to follow a path precisely between programmed points. Accuracy will vary within the robot's work envelope. For off-line programming, accuracy is a very important rating. Offline programming is done to reduce interruption of the robot from production work. The points being programmed are not oriented to the actual weldment but are entered as numeric locations in the robot's work envelope. When these off-line programs are downloaded into

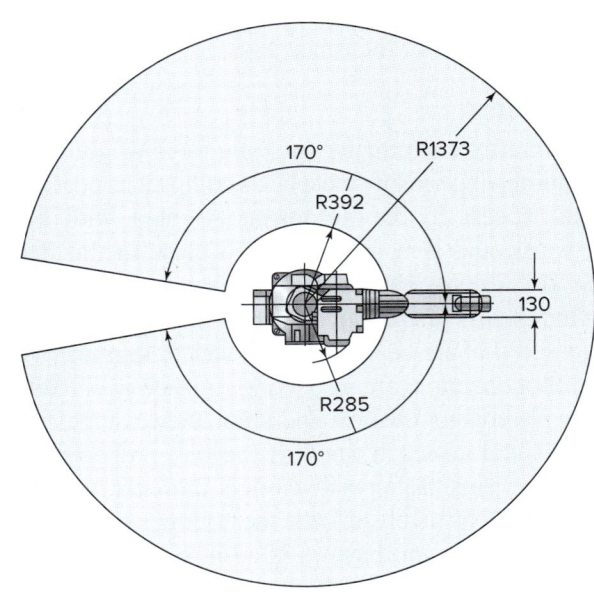

Fig. 27-25 Typical robot load-carrying capacity distribution. American Welding Society (AWS) Welding Handbook Committee. *Welding Science and Technology. Welding Handbook,* Vol. 1, 9th ed. Miami: American Welding Society, 2001, Fig 11.18, p. 471.

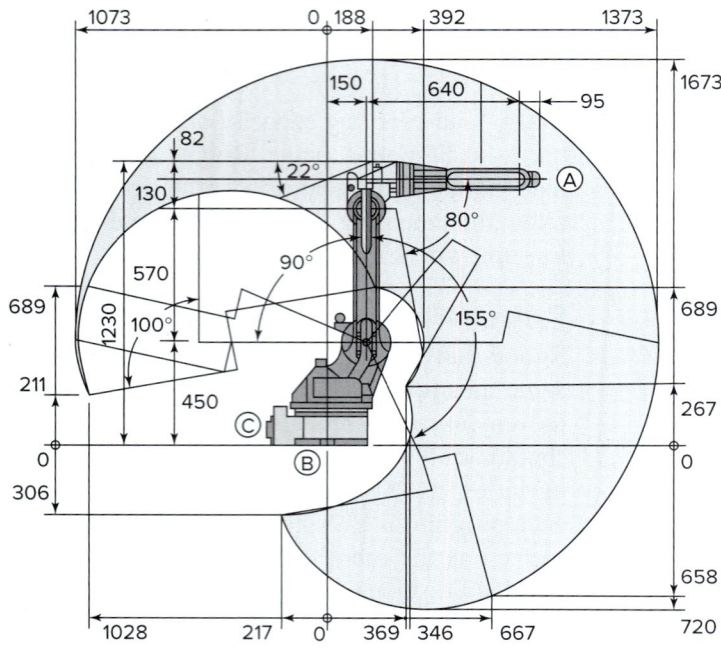

Fig. 27-26 Typical robotic work envelope. All dimensions are metric (mm) and for reference only. Motoman, Inc.

the production machine, they will require less touch-up the more accurate the robot is.

Work Envelope

There are some very large robots with very large work envelopes, and there are bench-top robots with very limited work envelopes. The weldment should be easily located in the work envelope of the robot, Fig. 27-26.

Compactness

The robot arm, end effecter (welding torch or gun mount), gun or torch, and sensors must be as compact as possible for the best access to the weldment. A robot with a slim base, waist, and arm can be placed close to the fixture, which also improves accessibility to the weld joint. You may find that there are welds situated in small, difficult-to-reach areas that are best left for the welder to make. While these robot operators are inspecting finished weldment and loading fixtures for the next operation, they can make these small, difficult-to-reach welds, if they are properly trained as a welder. This has a positive effect in that it keeps these workers actively involved with the welding operation. As with any activity you have to keep in practice if you are going to build quality into the finished product.

Speed

Robots are rated by how many degrees an axis can move in a second. A small wrist twist axis may be able to move 500° per second. This would mean it could twist a welding gun around this axis nearly 1½ revolutions in 1 second. A larger axis such as the base doing a turn or sweep may be rated at 140° per second. But since half a circle is 180°, it can move from one side of the work envelope to the other in a little over 1 second. This may be a distance of 8 feet. Robots are very fast machines, so generally speed is not a rating issue. Because they are so fast, operator safety is paramount when the operator is in the robot's work envelope.

Reliability

This rating is based on mean time between failures (MTBF). A typical rating might be 52,000 hours. Comparing this to a person working 2,000 hours per year, this would equate to nearly 26 years. Maintenance of a robot generally involves just periodic lubrication. Items that are subject to failure are designed for easy replacement. The welding equipment used in the robot cell is also very reliable if given preventive maintenance.

Robot Programming

The robot must be programmed to make the welds in the appropriate place on the weldment and according to the proper welding procedure. Programming is the creation of a detailed sequence of steps that will safely and efficiently take the robot through its work motions. Once the program has been written and proven to work, it can be stored for future use. Some robot controllers can be used to program a number of robot arms and peripheral equipment. On occasion you may have to program up to 27 axes of motion.

Prior to programming:

A. Determine the tool center point. Some robots come with a special tool for conducting this calibration operation.
B. Locate the weldment in the work envelope. Verify process and welding positions that will be encountered for any limitations that might occur. Check accessibility and torch or gun work and travel angle management. It may be necessary to adjust the tool center point.

Programming should include the following steps:

1. Set the path that will be followed as the weld is made.
2. The welding procedure will need to be developed. This will set the weld parameters based

in conjunction with the robot's work-motion programming.
3. Touch up the program while checking and verifying how well it works. Correction of the torch or gun path may be required to get the proper weld bead profile and size.

Training, Qualification, and Certification

The American Welding Society has identified a number of specifications relating to robotics. They are:

- D16.1 Specification for Robotic Arc Welding Safety
- D16.2 Guide for Components of Robotic and Automatic Arc Welding Installations
- D16.3 Risk Assessment Guide for Robotic Arc Welding

It has also identified the qualification of various levels of robotic operations personnel in D16.4 Specification for the Qualification of Robotic Arc Welding Personnel. There are three different levels with the following training recommendations:

Level 1

Skills and Ability Requirements

1. Have the ability to power up the robot and peripheral equipment such as all power sources, coolant pumps, and torch cleaners.
2. Be capable of servicing the robotic welding torch and wire feeding system. This includes servicing the torch, contact tips, gas diffusers, insulators, nozzles, and drive rolls, and changing welding wire.
3. Have a basic understanding of the robot as it is outlined in the company's routine maintenance procedures.
4. Have a basic understanding of the robot control panel so that the robot can be brought back to operation after work has been performed inside the work cell. This includes resetting any safety circuits and making sure that the robot is in the home position.
5. Have knowledge of general safety requirements.
6. Have a working knowledge of all the robotic peripheral equipment.
7. Have the ability to perform routine and preventive maintenance on such items as the torch cleaner, wire feeder, torch mount, and torch cable support hardware.

Level 2

Skills and Ability Requirements + Level 1

1. Have the ability to visually inspect the welds on the component to the applicable standard and make changes as allowed by the welding procedure to bring the welds within specifications. The individual should have a strong welding background and a thorough understanding of the robotic program and its function.
2. Have the ability to document information on any robot-related problems and communicate them to the welding engineer or technician. Have good written and oral communication skills.
3. Be capable of evaluating weld cross sections.

Level 3

Skills and Ability Requirements + Levels 1 and 2

1. Have the ability to make changes to the weld data, torch angles, electrode stickout, starting techniques, and other welding variables. Have an extensive welding background and a thorough understanding of the robotic interfacing system.
2. Demonstrate a thorough understanding of all aspects of the robotic work cell. Demonstrate programming, robotic arc welding, seam tracking, fixturing, and any other welding- or robotic-related functions. Have the capability to enter the work cell and make changes to the weld program, main program, torch clean program, or any other related programs. Be capable of fixture changes to improve part fitup and locating.
3. Be capable of performing file management tasks such as saving, copying, and deleting program files.
4. Demonstrate expertise in the welding operations including all of the arc welding robots, automated welding equipment, and all manual welding operations.
5. Be responsible for the initial weld inspection and be familiar with the tools that measure the weldment quality.
6. Have the ability to perform weld cross sectioning by cutting, polishing, and etching appropriate samples when necessary.
7. Keep accurate and up-to-date records, including issuing revised weld procedures as needed.

The AWS QC19 Standard for Certification of robotic arc welding operators and technicians has two levels of certification. The operator certification covers levels 1 and 2 out of the D16.4 standard, and the technician certification includes mastery of levels 1, 2, and 3 out of D16.4 plus the individual must be a current AWS certified welding inspector. Testing would consist of taking a closed-book test and a performance test. The closed-book test would cover all the knowledge requirements of the D16.4. The performance test information can be found on the AWS Web site and is titled Certification of Robotic Arc Welding Operators and Technicians Exam and may consist of the following:

Craw-O/T Performance Test

Part 1: Required Safety Tasks

Step 1
Before starting the exam, inspect the robot welding system and identify any potential safety hazards.

Step 2
Identify the pinch points of the robotic welding system.

Step 3
Show the basic robot cell operation to the Test Proctor.

Step 4
Show how to safely enter into the welding cell for service.

Step 5
Indicate the emergency stops and use one of them.

Step 6
Recover the robot system from this e-stop condition.

Part 2: Equipment Familiarization Tasks

Step 7
Identify each of the following welding cell components to the Test Proctor using the robotic arc welding cell:
  Emergency stops
  Identify each axis of the robot
  Operator start button
  Positioner "if applicable"
  Robot arm
  Robot Breakaway
  Robot controller
  Teach pendant
  Welding cell safety switches
  Welding drive rolls
  Welding gas supply system
  Welding power supply
  Welding torch
  Wire-feed unit

Step 8
Inspect the diffuser, contact tip, welding torch, and drive roll. Then show how to replace them.

Step 9
Show how to turn on the welding power source and robot controller.

Step 10
Show how to route welding wire from the wire-feeding mechanism through the wire-feeding system to the contact tip.

Step 11
Inspect the operation of the wire-feeding system by using the teach pendant or wire jog on the feeder.

Step 12
Make sure the shielding gas supply system operates correctly by purging the system.

Step 13
If a water-cooled torch is used, make sure the welding torch water circulator system is on and functioning.

Before starting Part 3, you will be allowed to practice actual welding on scrap pieces so you can fine tune the welding parameters and achieve a satisfactory weld.

Part 3: Test Piece Preparation and Programming

Step 14
Move the robot by using the teach pendant as instructed by the Test Proctor.

Step 15
Make sure the correct Tool Center Point has been chosen for the torch being used.

Step 16
Program the following points with your robot system, relative to the Test Piece.
  Home Position
  Pounce Position

Step 17
Put the sample part in a position so the robot can access all welds required for the Test Piece.

Step 18
Write a basic welding program for the Test Piece specified. This program will include these basic types of points.
  1. Joint move over the part to a pounce position
  2. Linear move to the start of the first weld
  3. Weld starting point
  4. Weld end point
  5. Linear move between each weld end and next weld start

Repeat steps 3 through 5 for all of the welds shown on the Test Piece drawing provided.

Linear move from the stop of the last weld to a retract position. Joint move from the retract position to a safe position.

NOTE: Use two welding schedules while programming this sample part and be sure to save your program.

Step 19
Without welding the Test Piece, show the safe operation of the welding program.

Step 20
The Test Piece will now be rotated in any direction approximately 2" by the Test Proctor.

Step 21
Secure the Test Piece to the positioner, table, or fixture with clamps, making sure the clamps will not interfere with your welding program.

Step 22
Edit the welding program points so these points are now in the correct position to weld the Test Piece. At the same time, input a delay or wait command into the welding program.

Part 4: Welding the Test Piece and Evaluation

Step 23
Weld the Test Piece once Test Proctor has verified your work up to this point.

Step 24
Record the welding parameters that were input into the welding program used to weld the Test Piece on the *Performance Test Record* Form.

Step 25
Visually inspect the welds on the Test Piece using the acceptance criteria contained on the *Weld Quality Assessment Form*. Before sectioning the welds for etching, talk about your findings with the Test Proctor prior to sectioning the welds for etching. Be sure to clearly mark the location of the welds to be sectioned if you are not doing the sectioning and macroetch of the welds yourself. Complete the *Weld Quality Assessment Form* based on the visual inspection and macroetch results. The Test Proctor will initial the *Assessment Form* when the weld inspection is acceptable.

Conclusion

Today, automation can mean solutions from a single robot through to a full production line. The trend has been to focus on individual pieces of robot equipment. That idea is changing, and now end users are looking for systems that make a whole production process work together. This would be full manufacturing solutions, including software and hardware. The basic integration platform for flexible automation is the hub of an automated manufacturing system. End users can plug in their specific requirements into this platform, and it will perform the tasks of control, network communication, and information management. The integration platform is the integral part of the move toward industrial information technology. The advantage to end users is that they no longer have to order isolated bits of high-tech equipment and try to integrate them. Now end users will be able to order fully integrated, seamless manufacturing plants from select manufacturers. Arc welding is one of the fastest growing segments in automation and robotics. It has been determined that only about 10 percent of the small- to medium-sized manufacturing companies have installed robots, so the growth potential is very great in these areas. This will result in the need for additional qualified personnel.

CHAPTER 27 REVIEW

Multiple Choice

Choose the letter of the correct answer.

1. Companies moving from manual and semiautomatic welding to automatic and robotic welding have few issues to consider. (Obj. 27-1)
   a. True
   b. False

2. Automatic welding requires the operator to control _____. (Obj. 27-1)
   a. Torch movement along the joint
   b. Gun work and travel angle
   c. Electrode extension
   d. None of these

3. Adaptive control is most used with which process application method? (Obj. 27-1)
   a. Manual and semiautomatic
   b. Semiautomatic and mechanized
   c. Automatic and robotic
   d. Both a and b

4. Magnetic arc controls are used to_____. (Obj. 27-2)
   a. Position the arc
   b. Oscillate the arc
   c. Stabilize the arc
   d. All of these
5. Mechanical oscillators move the arc in which manner? (Obj. 27-2)
   a. Side to side
   b. Pendulum shape
   c. Triangular shape
   d. All of these
6. Seam trackers are used to_____. (Obj. 27-3)
   a. Make travel speed adjustments to compensate for poor fitup
   b. Position the gun in the joint and maintain this position while welding
   c. Make voltage and wire-feed speed adjustments to maintain weld size
   d. Both b and c
7. Arc monitoring is used for_____. (Obj. 27-4)
   a. Relieving the welder from tedious work
   b. Safety reasons
   c. No particular reason; it is just easy and inexpensive
   d. Both a and b
8. Arc monitors and sensors cannot easily check which of the following? (Obj. 27-4)
   a. Temperature and heat input
   b. Weld size and profile
   c. Mechanical properties and microstructures
   d. Penetration and radiation
9. The weld controller may be_____. (Obj. 27-5)
   a. A stand-alone controller
   b. Built into the monitor cabinetry
   c. Built into the robot controller cabinetry
   d. All of these
10. A typical robotic arc welding cell would include_____. (Obj. 27-6)
    a. The robot with arc welding equipment
    b. Equipment guarding
    c. Fume extraction
    d. All of these

Review Questions

Write the answers in your own words.

11. List five issues that must be considered before doing automatic or robotic welding applications. (Obj. 27-1)
12. Magnetic oscillators are used to control what six things? (Obj. 27-2)
13. How could a sophisticated microprocessor-controlled oscillator provide external program control and power source interface? (Obj. 27-2)
14. Describe how an arc length control would work for the gas tungsten arc or plasma arc welding processes. (Obj. 27-3)
15. List the four basic control elements that must be dealt with when doing automatic and robotic welding. (Obj. 27-4)
16. List some various types of sensors that are used in the welding industry. (Obj. 27-4)
17. List a complete welding sequence that a weld controller would need to be capable of dealing with. (Obj. 27-5)
18. What is the function of the control relay, how does it work, and how is it connected into the welding system? (Obj. 27-5)
19. Describe the two most common types of welding robots. (Obj. 27-6)
20. Describe robot preprogramming and programming steps that should be followed. (Obj. 27-6)

INTERNET ACTIVITIES

Internet Activity A

Using the American Welding Society's Web site, review and make a report to your instructor on the certification of robotic personnel.

Internet Activity B

Using your favorite search engines, find information about the robot arm you are using in your shop. If it has become outdated, list the specification on the current model available. If your shop does not have a robot arm, check with your instructor to get the name of a company in your area that is using a robot. Find out what robot arm the company is using and make a report on it.

Design credits: Cinema and movie icon: Denis Meshkov/123RF; and Illustration of Welding icons: Mr. Nachai Sorasee/123RF

28

Joint Design, Testing, and Inspection

Chapter Objectives

After completing this chapter, you will be able to:

28-1 Describe the various types of weld joint designs.

28-2 Understand the implications of doing code welding.

28-3 Describe various nondestructive weld test methods.

28-4 Describe various destructive weld test methods.

28-5 Demonstrate the ability to do groove and fillet weld soundness tests.

28-6 Describe and conduct visual weld inspection.

28-7 Explain the various gauges used for weld inspection.

Welding was first used as a means of patching and repairing. It was rarely employed as a means of fabricating pressure vessels, pipelines, and other structures such as buildings and bridges that would be a hazard to life and property if they failed. As welding began to be used as a fabrication process, it became essential for welded joints to be strong enough to meet the service requirements for which they were designed (fitness for purpose). Methods for testing the quality of the weld, the ability of the welder, as well as the ability of the inspector had to be devised.

For a long time, it had been considered necessary only to look at a completed weld in order to judge its quality and the welder's ability. If carried out by a competent inspector and/or welder, visual inspection may be satisfactory for welds that are designed primarily to hold parts together and that are not subject to high stress in service. This kind of inspection is limited since there is no way of knowing if the weld metal has internal defects. The outer appearance of the weld may be entirely satisfactory, and yet it may be porous, contain cracks, and lack both complete fusion and penetration. The weld metal may have serious defects due to poor welding technique.

Critical welding demands that the weld metal and joint be tested for strength, soundness, and other physical qualities required in the design. The reliability of the welded joint can be determined by the degree to which the weld metal is kept free of foreign materials such as slag, porosity, cracks, and by the degree to which it is fused to the base material.

Joint Design

You are familiar with the five basic joints (butt, corner, edge, lap, parallel, and T) and the type of weld applied to these types of joints, namely, fillet and groove welds. Variations of these basic joints are shown in Table 28-1 on page 893. In order to do code quality work, it is necessary to understand and apply good joint design to get the best weld quality along with the most economic performance.

Open and Closed Roots

Open roots are spaces between the edges of the members to be welded. They are used to secure complete root penetration in butt joints and to secure attachment to a backing member, Fig. 28-1.

The term *penetration* refers to the depth to which the base metal is melted and fused with the metal of the filler rod or electrode. In those cases in which there is no root opening, some of the weld metal from the first pass is partly removed by chipping or machining from the reverse side. Whether a certain type of joint should be set up as an open or closed root depends upon the following factors:

- The thickness of the base metal
- The kind of joint
- The nature of the job
- The position of welding
- The type and size of electrode
- The structural importance of the joint in the fabrication (whether it is a prime load-carrying joint)
- The physical properties required of the weld

Parallel Joints and Edge Welds

Joints can use many different types of welds. Over the years, five types of welds have been taken on for clarification. They are butt, corner, edge, lap, and T-joint. They are easily detectable except for the "parallel joint," so parallel is a recent term and structure being used to replace the parallel joint. To take on the parallel joint with parallel surfaces or more workpieces with ends that are flush. The parallel joint can locate weld areas; it does not get the orientation of the members. A parallel joint is designed by parallel surfaces of two or more workpieces, Fig. 28-2, which means the ends need to be flushed.

Butt Joints

Closed Square-Groove Butt Joint The square-groove butt joint can be welded in several different ways. It is satisfactory for all usual load conditions. Preparation of the joint is simple and inexpensive since it requires only the butting together of the plate edges.

Complete joint penetration (CJP) of the base metal is necessary if the closed butt joint is to be used for code work. Welding from one side, Fig. 28-3A, cannot secure complete joint penetration in most stock. Because of the unwelded root, the joint is weak at this point. Welding from both sides, Fig. 28-3B, materially increases its strength. Constant and severe loading, however, causes failure of the joint because of the unwelded areas at the root.

On metal ⅛ inch or thinner, complete penetration can be obtained by welding from one side, Fig. 28-3C. On metal ³⁄₁₆ inch or thinner, complete penetration is possible by welding from both sides, Fig. 28-3D. Shielded metal arc welding

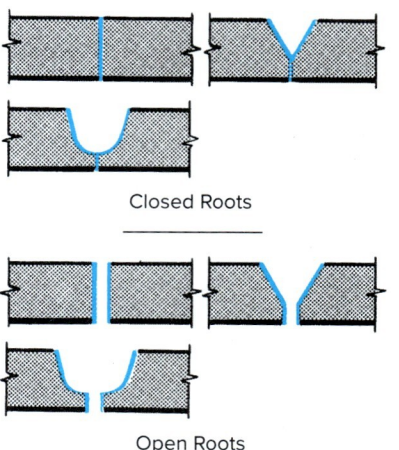

Fig. 28-1 Closed and open roots.

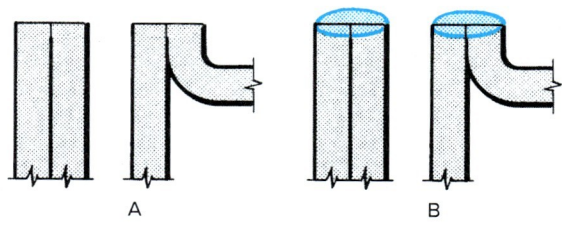

Fig. 28-2 Parallel joints and Flanged Parallel joints as shown in Fig. 28-2 A are economical for noncode work since the cost of penetration is low. They are not suitable, however, for severe load conditions. This parallel joint and the edge weld shown in Fig. 28-2 B should not be used if either member is subjected to direct tension or bending of the weld root. McGraw Hill

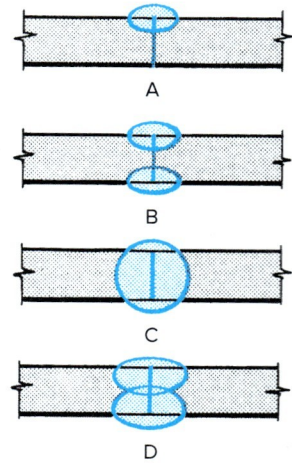

Fig. 28-3 Closed square-groove butt joints.

Table 28-1 Forms of Weld Joints

Parallel Joints	Fig. No.
Parallel joint (the joint between the surfaces of two or more parallel or nearly parallel members) Flanged corner joint	1

Butt Joints	Fig. No.
Closed single-flanged butt joint	2
Open single-flanged butt joint	3
Closed double-flanged butt joint	4
Open double-flanged butt joint	5
Closed upset butt joint	6
Open upset butt joint	7
Closed square-groove butt joint	8
Open square-groove butt joint	9
Closed single V-groove butt joint	10
Open single V-groove butt joint	11
Closed double V-groove butt joint	12
Open double V-groove butt joint	13
Closed single bevel-groove butt joint	14
Open single bevel-groove butt joint	15
Closed double bevel-groove butt joint	16
Open double bevel-groove butt joint	17
Closed single U-groove butt joint	18
Open single U-groove butt joint	19
Closed double U-groove butt joint	20
Open double U-groove butt joint	21
Strapped closed square-groove butt joint	22
Strapped open square-groove butt joint	23
Strapped closed single V-groove butt joint	24
Strapped open single V-groove butt joint	25
Strapped closed single U-groove butt joint	26
Strapped open single U-groove butt joint	27

Lap Joints	Fig. No.
Single lap joint	28
Double lap joint	29
Single-strap lap joint	30
Double-strap lap joint	31
Closed joggled single lap joint	32
Open joggled single lap joint	33
Flanged single lap joint	34
Flanged closed joggled single lap joint	35
Flanged open joggled single lap joint	36
Linear slotted lap joint	37
Circular slotted lap joint	38

Corner Joints	Fig. No.
Closed lapped corner joint	39
Open lapped corner joint	40
Closed corner joint	41
Open corner joint	42

T-Joints	Fig. No.
Closed square T-joint	43
Open square T-joint	44
Closed single bevel J-groove T-joint	45
Open single bevel J-groove T-joint	46
Closed double bevel J-groove T-joint	47
Open double bevel J-groove T-joint	48
Closed single J-groove T-joint	49
Open single J-groove T-joint	50
Closed double J-groove T-joint	51
Open double J-groove T-joint	52

may be used on metal ¼ inch thick and the submerged arc welding process on metal ⅝ inch thick. For welds in which complete joint penetration is necessary on metal more than 3/16 inch thick, it is recommended that the root of the first pass be chipped or gouged out from the reverse side to sound metal before depositing the second weld.

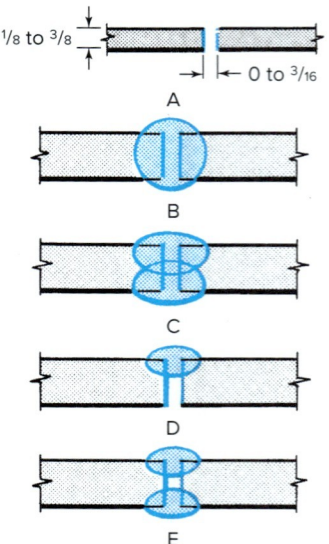

Fig. 28-4 Open square-groove butt joints.

Open Square-Groove Butt Joint Securing penetration on open square-groove butt joints, Fig. 28-4A, is easier than on closed square-groove butt joints. Because of this fact, heavier sections can be welded. It is possible to weld 3/16-inch material or less from one side, Fig. 28-4B, and up to ¼ inch from both sides, Fig. 28-4C, with complete joint penetration. If complete joint penetration is not achieved, however (Figs. 28-4D and E), the open square-groove butt joint will not be any stronger than the closed type, and it will have the same possibility of failure at the root of the weld under load.

Metal ⅜ inch thick may be welded with the shielded metal arc process, and metal ¾ inch thick, with the submerged arc welding process if the root of the first pass is chipped out from the reverse side to sound metal before depositing the second weld. The cost of joint preparation is the same as for closed square-groove butt joints. Oftentimes, it is a little more difficult to line up the work with the proper gapping all along the entire length of the joint.

Single V-Groove Butt Joint Single V-groove butt joints, Fig. 28-5, are superior to square-groove butt joints and are used a great deal for important work. They provide for 100 percent penetration and offer a better plate edge preparation for welding than square-groove butt joints. Metal preparation, however, is more costly than for the square-groove butt joint, and a greater amount of electrode deposit is used in welding. The single V-groove type is ordinarily used on plate thicknesses ranging from ¼ inch to ⅝ inch. If welding is to be from one side only, Fig. 28-6A, full penetration to the root of the weld must be obtained. Failure to do so will cause a fracture if the joint is subjected to severe loading.

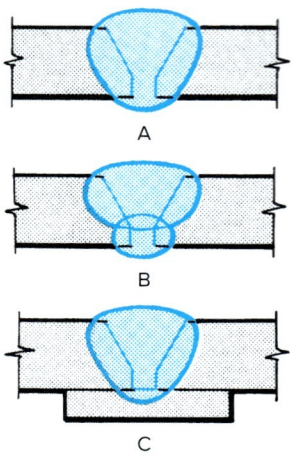

Fig. 28-6 Single V-groove butt joints.

Joints welded from both sides with complete joint penetration provide full strength and meet the requirements of code welding. Welding from both sides, Fig. 28-6B, can be accomplished only where the work will permit the operator to weld from both sides of the plate. It is easier to obtain complete penetration through the entire thickness in this way. If a backing strip is used, Fig. 28-6C, it is possible to weld faster and use larger electrodes, especially on the first or root pass. A removable backing is also used when welding from one side with the submerged arc process. Metal thickness up to 1½ inch can be welded in this manner.

Double V-Groove Butt Joint The double V-groove butt joint, Fig. 28-7, is suitable for most severe load conditions. It is used on heavier plate than single V-groove butt joints, usually ¾ to 1½ inch thick. For metal thicknesses greater than 1½ inch, the double U-Groove butt joint is recommended because less electrode metal is needed. The cost of joint preparation is greater than for the single V-groove butt

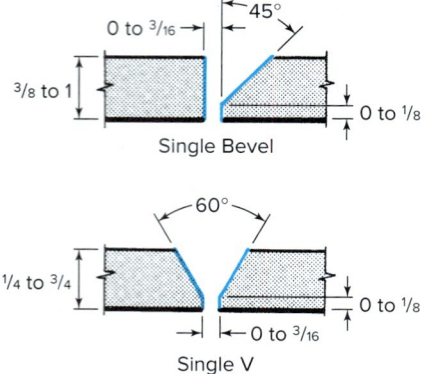

Fig. 28-5 Proportions for single V-groove butt joints.

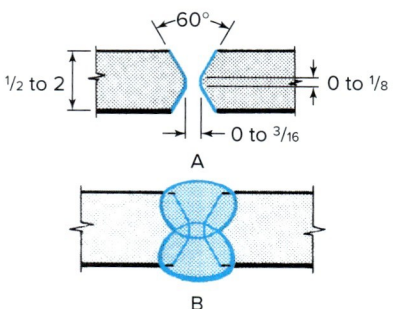

Fig. 28-7 Double V-groove butt joint.

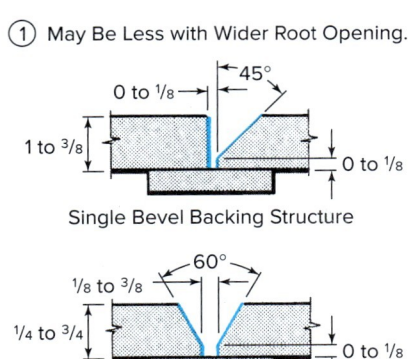

Fig. 28-8 Proportions for single bevel-groove butt joints.

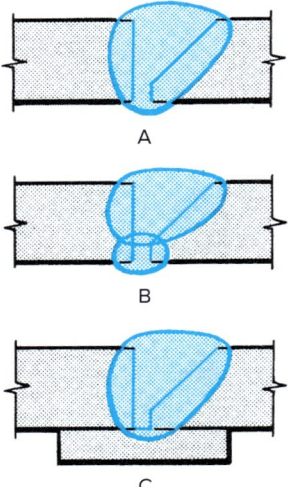

Fig. 28-9 Single bevel-groove butt joints.

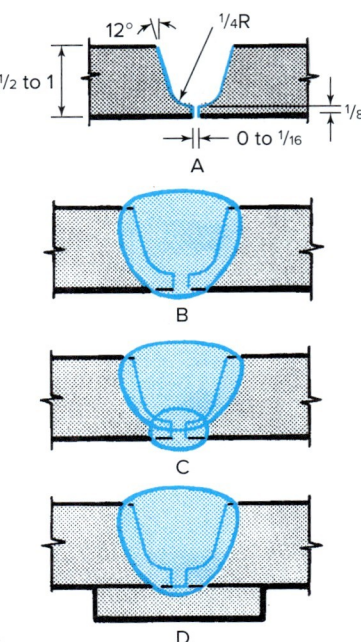

Fig. 28-11 Single U-groove butt joints.

joint, but the amount of filler metal needed in welding is less.

It is essential that complete root penetration be achieved. The work must permit welding from both sides, and the back side of the first pass must be chipped before applying the second pass from the other side. Welding from both sides permits an even distribution of heat through the joint, thus reducing the concentration of stress at the joint and the amount of warpage and distortion.

Beveled-Groove Butt Joints Single bevel-groove butt joints (Figs. 28-8 and 28-9) and double bevel-groove butt joints (Fig. 28-10) are used in some areas. They are suggested for work where load demands are greater than can be met by square butt joints and less than values requiring V-groove butt joints. They join metal up to ¾ inch thick,

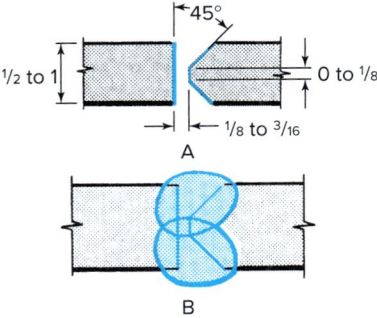

Fig. 28-10 Double bevel-groove butt joint.

and less filler metal is required than for a V-groove butt joint, thus reducing the number of electrodes needed. The cost of preparation is less than for V-groove butt joints since it is necessary to bevel only one plate edge. For full strength the root of the first pass should be chipped from the reverse side to sound metal before depositing the second pass. The welder will find it difficult to obtain good fusion to the perpendicular wall of the square plate and to secure complete joint penetration.

Single U-Groove Butt Joint Single U-groove butt joints, Fig. 28-11, are used for very important work, such as fired and nonfired pressure vessels. The cost of preparation is greater than for bevel and V-groove butt joints, but fewer electrodes are required in welding. The single U-groove butt joint is used on plate thicknesses ranging from ½ inch to ¾ inch. Heavier metal may be welded with the submerged arc process. It is often the practice to make the first pass with the shielded metal arc or the MIG process. Complete penetration is necessary for the single U-groove butt joint to give satisfactory service. It is easier to obtain complete penetration on single U-groove butt joints welded from both sides (Fig. 28-11C), and on joints with a backup strip (Fig. 28-11D), than on joints welded from one side only (Fig. 28-11B). The joint is usually welded with free-flowing electrodes.

Double U-Groove Butt Joint Double U-groove butt joints, Fig. 28-12, page 896, are used on work of the same nature as single U-groove butt joints but when plate thicknesses

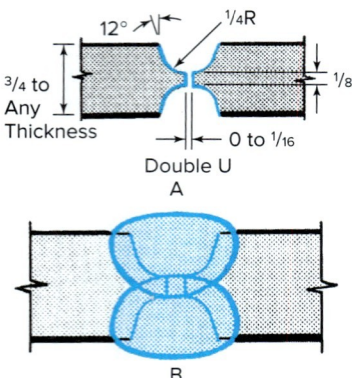

Fig. 28-12 Double U-groove butt joint.

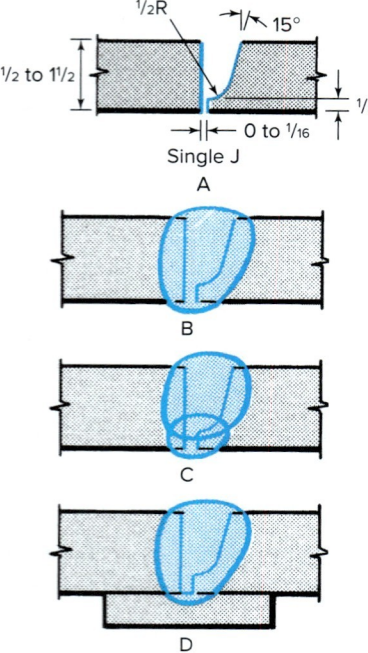

Fig. 28-13 Single J-groove butt joints.

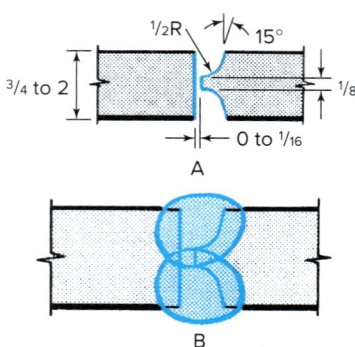

Fig. 28-14 Double J-groove butt joint.

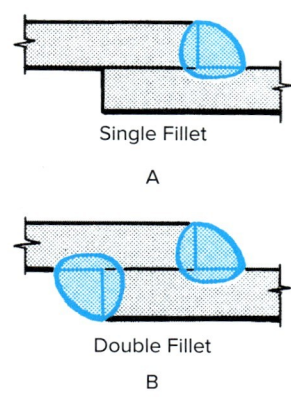

Fig. 28-15 Lap joints with fillet welds.

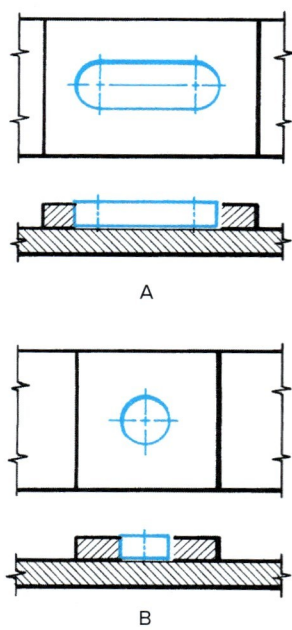

Fig. 28-16 Slot and plug welds on lap joints.

are greater and welding can be done from both sides. Plate thicknesses range up to ¾ inch. Although the cost of preparation is greater than for single U-groove butt joints, double joints may be welded with fewer electrodes. Welding from both sides permits a more even distribution of stress and reduces distortion.

The choice between double-U and double V-groove butt joints should be made on the basis of the relative costs of metal preparation and welding.

J-Groove Butt Joints Single J-groove butt joints, Fig. 28-13, and double J-groove butt joints, Fig. 28-14, are used on work similar to that requiring U-groove butt joints, but when load conditions are not as demanding. The cost of preparation is less since only one plate edge must be prepared. Less filler metal is required to fill the groove. It is difficult to secure good fusion and thorough penetration because of the perpendicular wall of the square member.

Lap Joints

Lap joints are used frequently on all kinds of work. There is no plate preparation involved. The single-fillet lap joint (Fig. 28-15A), while not as strong as the double-fillet lap joint (Fig. 28-15B), is more often used on noncode work. Single-fillet lap joints should not be used if the root of the joint is to be subjected to bending. In both cases fusion to the root of the joint is necessary. The welder must make sure that the edge of the upper plate is not burned away. A lap joint should never replace the butt joint on work under severe load.

Slot and Plug Welds on a Lap Joint The slot and plug weld on a lap joint, Fig. 28-16,

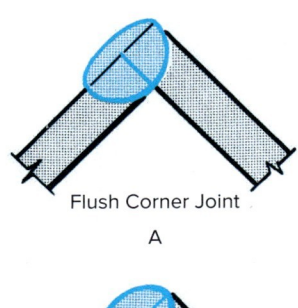

Flush Corner Joint
A

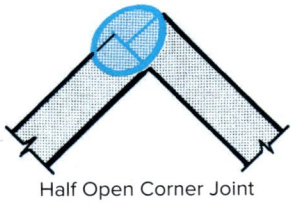

Half Open Corner Joint
B

Full Open Corner Joint
C

Fig. 28-17 Corner joints.

are used infrequently. They join one plate or bracket to another when it is desirable to conceal the weld or when there is a lack of an edge to weld on. In order to withstand a heavy load, the unit to be welded requires a series of these welds, and the cost of preparation is high. If the slots are small, it is difficult for the welder to make welds that are free of porosity and slag inclusions.

Corner Joints

Flush corner joints, Fig. 28-17A, can be used on light gauge sheet metal, usually under 12 gauge. Heavier plates can be welded if load is not severe and if there is no bending action at the root of the weld. No edge preparation is needed, and fitup is usually simple.

Half-open corner joints, Fig. 28-17B, may be used on 12-gauge to 7-gauge plate. This type of joint forms a groove and permits weld penetration to the root and good appearance. No edge preparation is required, and fitup is usually simple.

Full-open corner joints, Fig. 28-17C, can be used on any plate thickness. If welding is to be from one side, penetration must be secured through the root of the weld. If welded from both sides, the joint is suitable for severe loads. It has a good stress distribution, and no edge preparation is required.

More filler metal is required than for the half-open joint, and fitup is likely to be difficult. Plates must be cut absolutely square, and suitable clamping and holding devices are often needed to facilitate fitup. This type of joint is used in production welding.

T-Joints

Square-Groove T-Joint The square-groove T-joint, Fig. 28-18, may be used on ordinarily plate thicknesses up to ½ inch. Preparation of the plate is not necessary, and fitup can be fast and economical. Electrode costs are high.

The single-fillet T-joint will not withstand bending action at the root of the weld and should be used with caution. If it is possible to weld from both sides, Fig. 28-18B, the joint will withstand high load conditions.

Single Bevel-Groove T-Joint
The single-bevel groove T-joint, Fig. 28-19, is able to withstand more severe loads than the square-groove T-joint. It can be used on plate thicknesses ranging from ⅜ to ⅝ inch. Plate of greater thickness can be welded with the submerged arc process. Cost of preparation is greater than for the square-groove T-joint, and fitup is likely to take longer. Electrode costs are less because these are groove welds not fillet welds.

If it is possible to weld from one side only, Fig. 28-19A, complete joint penetration (CJP) must be obtained so that bending does not cause failure. If welding can be done from both sides, Fig. 28-19B, the load resistance of the joint is materially increased.

Double Bevel-Groove T-Joint
The double bevel-groove T-joint, Fig. 28-20, page 898, is used for heavy plate thicknesses up to 1 inch. Welding is done from both sides of the plate. This joint may be used for severe loads. The welder must make sure that fusion is obtained with both the flat and vertical plates. Complete joint penetration is necessary. Joint preparation is more expensive than for the square-groove T- or single bevel-groove joint, but weld time and electrode costs are less.

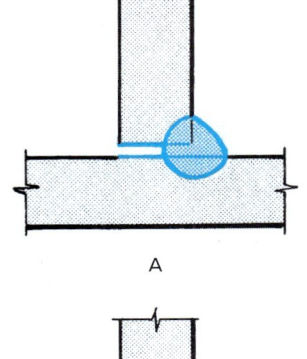

A

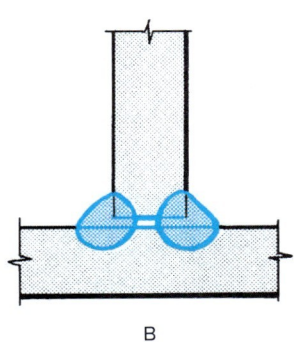

B

Fig. 28-18 Square-groove T-joints.

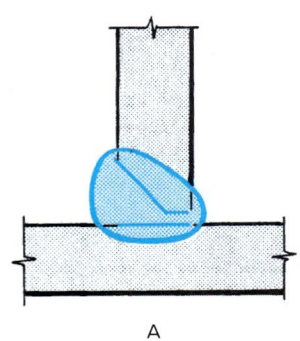

A

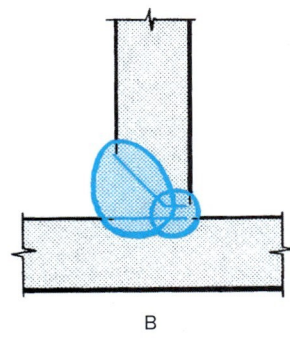

B

Fig. 28-19 Single bevel-groove T-joints.

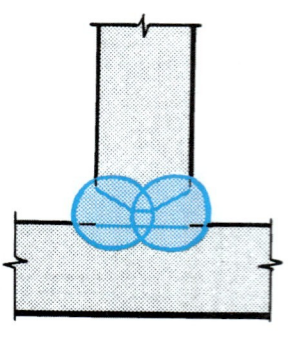

Fig. 28-20 Double bevel-groove T-joint.

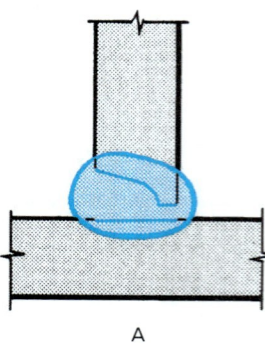

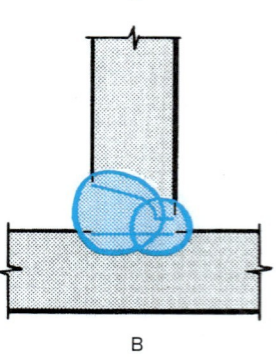

Fig. 28-21 Single J-groove T-joint.

Single J-Groove T-Joint Single J-groove T-joints, Fig. 28-21, can be used for most severe load conditions. They are generally used on plates 1 inch or heavier. If welding is to be done from one side, Fig. 28-21A, great care should be taken to secure complete joint penetration. If welding from both sides is possible, efficiency of the joint can be increased materially by putting the bead on the side opposite J-groove T-joint, Fig. 28-21B. This reduces the tendency of failure at the root as a result of load at this point. The cost of plate edge preparation is higher than for the bevel-groove T-joint, but there is a saving in electrode costs.

Double J-Groove T-Joint The double J-groove T-joint, Fig. 28-22, will withstand the most severe load conditions. It is used on plates 1¼ inch or heavier. The welder must be able to weld from both sides of the plate. Complete joint penetration and surface fusion are essential to prevent failure under severe load. Plate edge preparation is higher than for V-groove T-joints and single J-groove T-joints; however, electrode costs are lower. Frequently both J's are not of equal dimensions, Fig. 28-22B.

Summary

The material you have just studied indicates the importance of joint design and the weld requirements. In general, the soundness and service life of the product will depend upon the proper joint design and flawless welding. We have seen that proper plate preparation and setup depends upon a number of variables such as the thickness of plate, the plate edge preparation, and the root preparation. (See Figs. 28-4A through 28-14A, pages 894–896.) These variables are determined by the nature of the components to be welded and the service that the weldment will be expected to give. Correct joint design and proper welding procedure keep distortions to a minimum, reduce cracking due to shrinkage, and make it easier for the welder to produce sound welds with good appearance at the lowest possible cost.

Figures 28-23 through 28-28 (pp. 899–904) show the proper penetration of plate edges as recommended by the American Welding Society. Study these elements of joint design carefully because they form the basis for understanding welded construction.

Code Welding

It is usually the practice of a particular industry to develop a code. A *code* is a set of regulations governing all the elements of welded construction in a certain industry. Codes provide for human safety and protect property against failure of the weldment. A universal testing procedure for all types of welding in all locations does not exist. Thus in the field of piping, we find that welding must conform to a number of standards, depending on the nature of the installation. The welding of pressure piping conforms to the Code for Pressure Piping of the American Standards Association. The welding of piping connected to boilers and pressure vessels conforms to the Code for Boilers and Pressure Vessels, Section IX, developed by the American Society of Mechanical Engineers (ASME). The welding of pipelines conforms to the Standard for Welding Pipelines and Related Facilities (API Standard 1104) developed by the American Petroleum Institute. The welding of buildings, bridges, and aircraft must conform to the codes and standards set for these structures. Generally, these standards are set by the federal, state, and local governments, insurance companies, and various professional organizations. Many state and local bridge and building codes accept the procedure and operator qualification provisions of the AWS Structural Welding Code—Steel D1.1. A code of growing national importance in the food and hygienic welding industry is AWS D18.1, which is for welding of austenitic stainless-steel tube and pipe systems in sanitary (hygienic) applications. In aerospace and ground support systems,

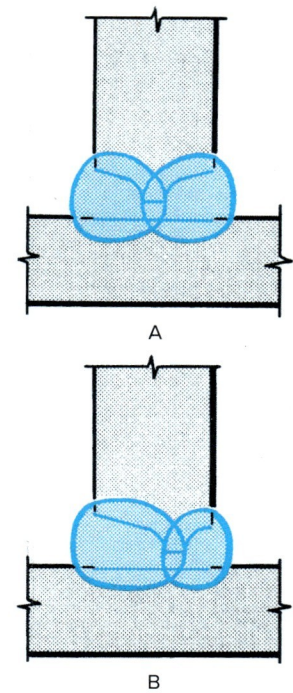

Fig. 28-22 Double J-groove T-joint.

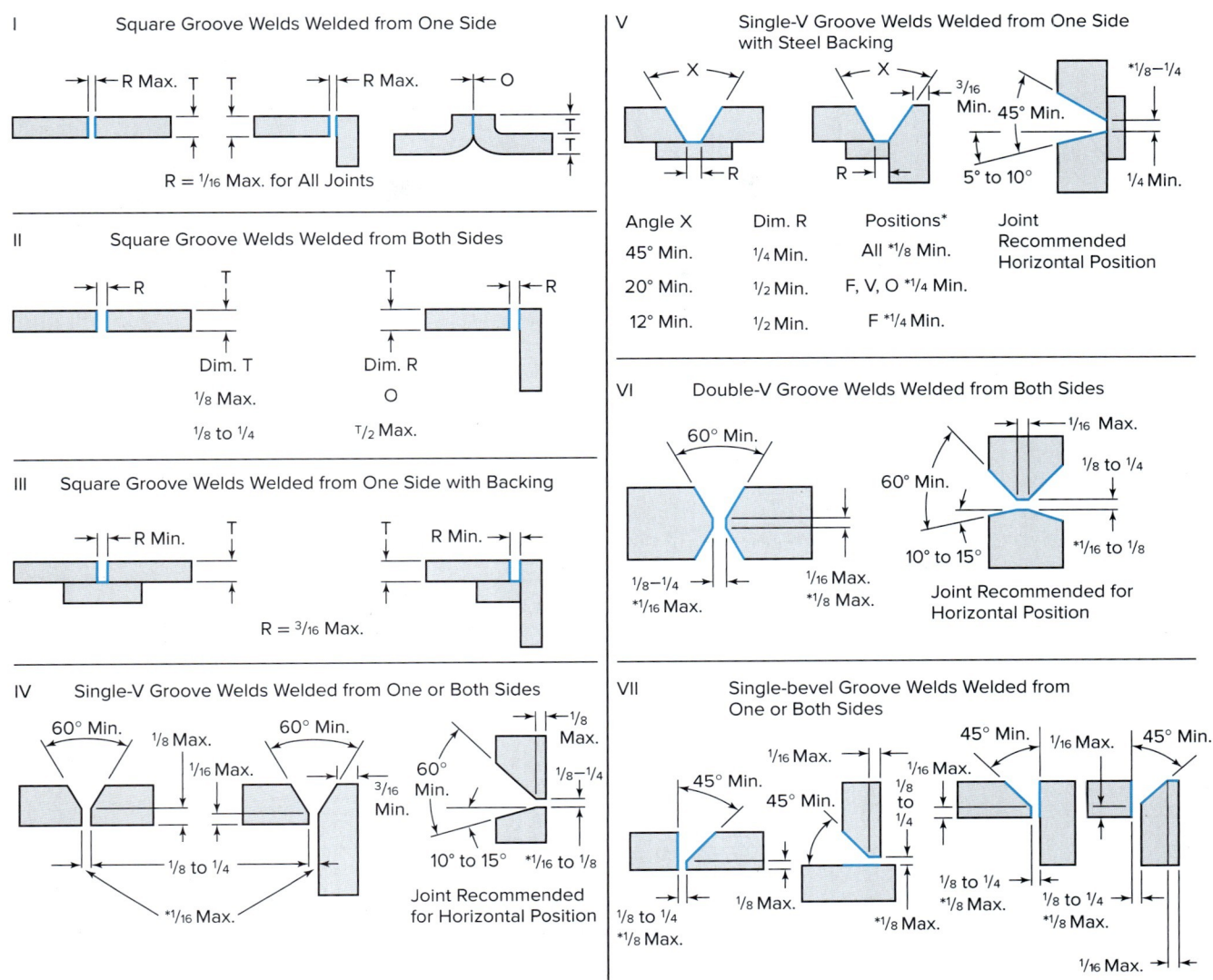

Fig. 28-23 Recommended dimensions of grooves for shielded metal arc welding, gas metal arc welding, and gas welding (except pressure gas welding). *Note:* Dimensions marked * are exceptions that apply specifically to designs for gas metal arc welding.

AWS D17.1 covers fusion welding for aerospace applications. This specification represents the most significant change to aviation welding standards in more than three decades and is being adopted by many aerospace companies. It is being looked at to replace government-sponsored standards. The structural codes for steel, aluminum, stainless steel, and sheet metal, and specifications for the welding of hygienic piping and aerospace systems can be obtained through the American Welding Society.

The welder does not have to be thoroughly informed about the details of all the existing codes. The employer, through the engineering and production departments, makes sure that the work meets the standards required for it. The welder should, however, have a good understanding of the different weld tests and know what to look for in any visual inspection.

There are two broad categories of welding tests. A *procedure qualification test* is a test conducted for the purpose of determining the correctness of the method of welding for a specific welding project. The American Welding Society and various code authorities have established standard procedures for welding. The welding procedure meets specifications for base metal filler metal, joint preparation, position of welding, the welding process, and welding techniques. Such requirements as current setting, electrode size, electrode manipulation and preheat, interpass, and postheat temperatures are

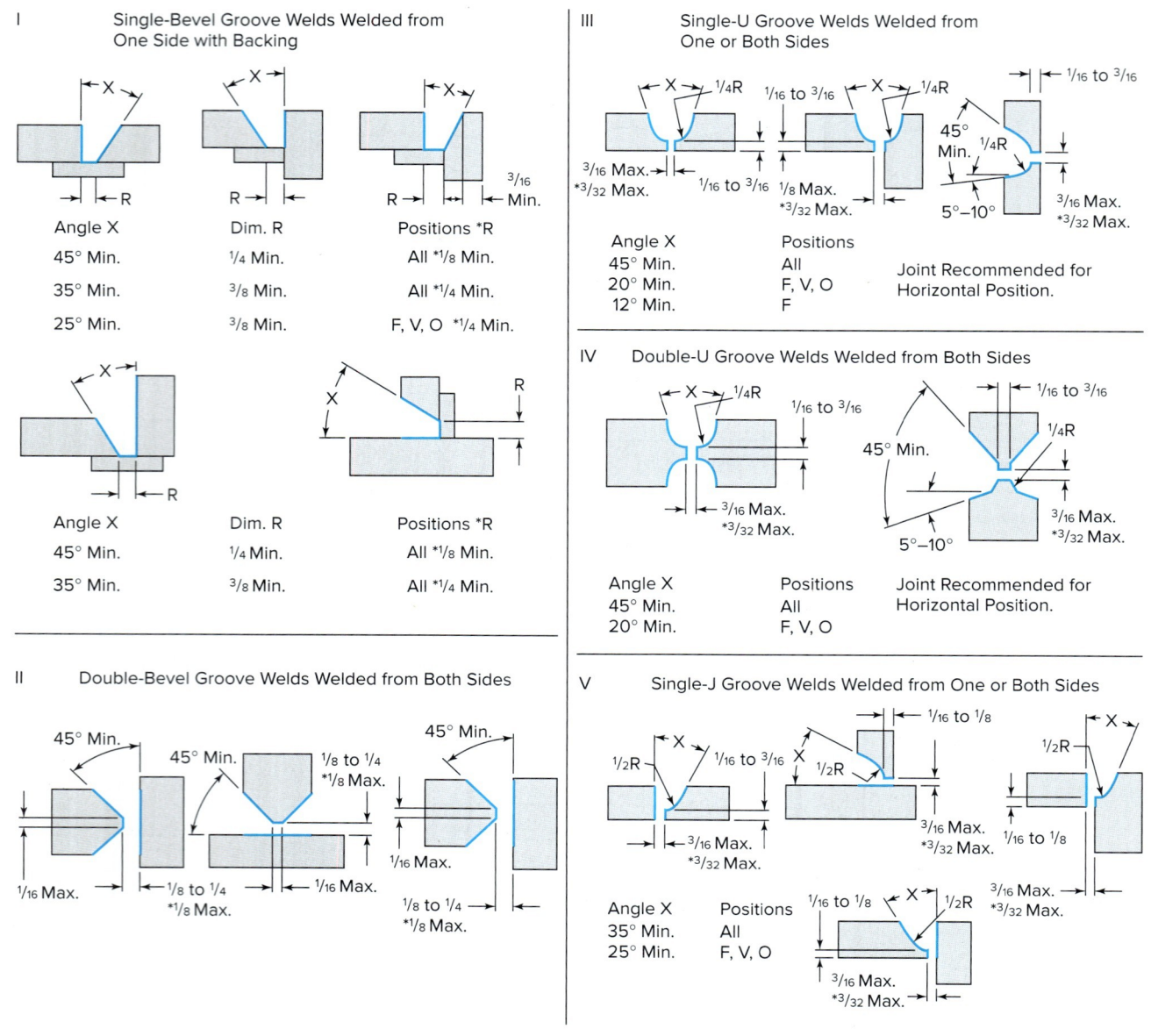

Fig. 28-24 Recommended dimensions of grooves for shielded metal arc welding, gas metal arc welding, and gas welding (except pressure gas welding). *Note:* Dimensions marked * are exceptions that apply specifically to designs for MIG welding.

also specified. Following a particular procedure ensures uniform results. Tests that certify welders for codework may be known as *welder qualification tests* or *performance qualification tests*. They are given to find out whether or not the welder has the knowledge and the skill to follow and apply a procedure of welding as developed for the class of work at hand. Testing may be with handheld welding equipment (the stick-electrode holder, the gas welding torch, or the semiautomatic welding gun) or with fully automatic welding equipment. Every effort has been made to simplify these tests so that they can be administered at low cost and still establish the soundness of the weld.

Reliability of welding is based on the use of appropriate inspection controls. The methods of testing that determine the quality of a weld are divided into three very broad classifications: (1) nondestructive testing, (2) destructive testing, and (3) visual inspection.

A *nondestructive* test, as the name implies, is any test that does not damage the weld or the finished product. Modern inspection equipment and techniques have made nondestructive testing an effective inspection tool.

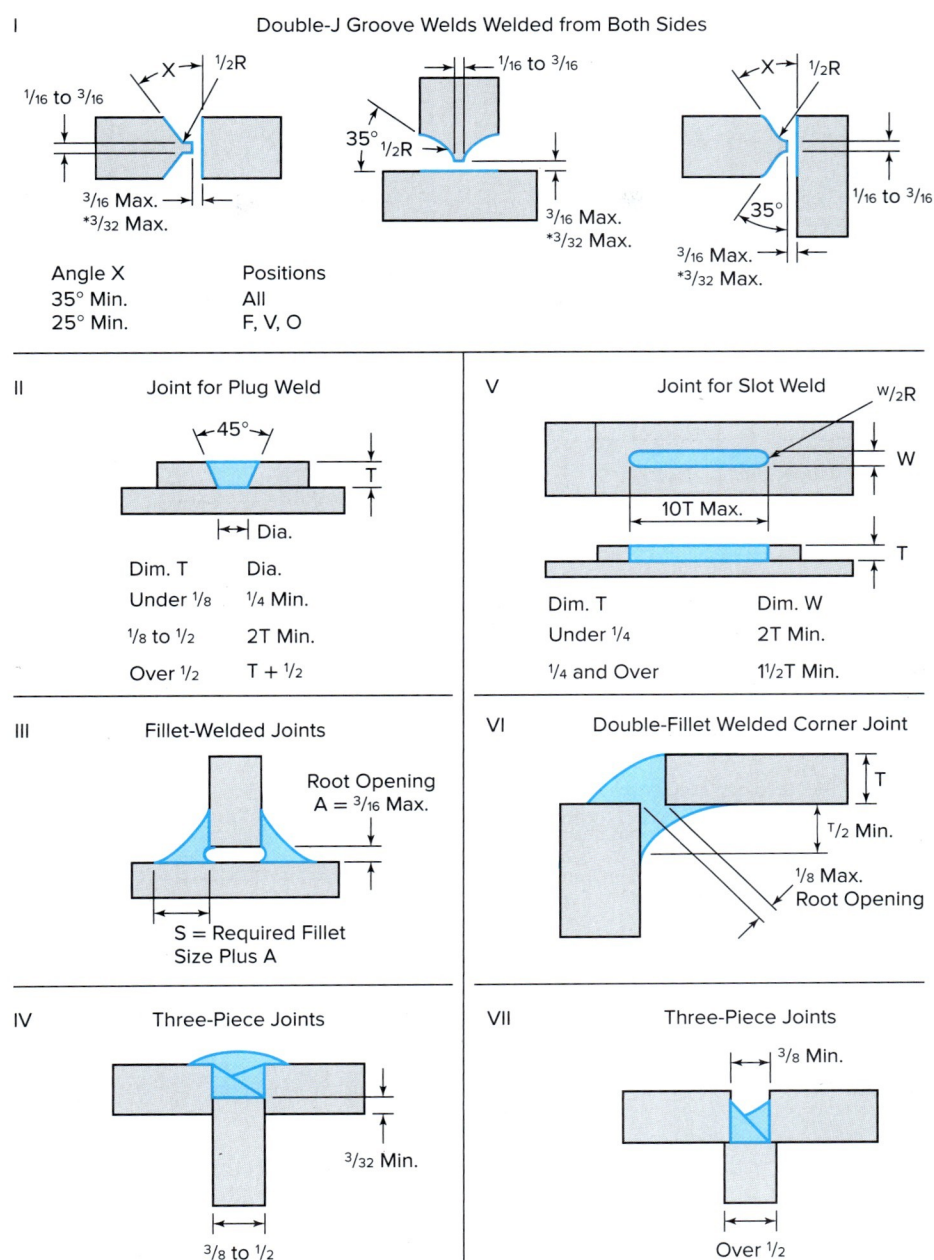

Fig. 28-25 Recommended dimensions of grooves for shielded metal arc welding, gas metal arc welding, and gas welding (except pressure gas welding). *Note:* Dimensions marked * are exceptions that apply specifically to gas metal arc welding.

Nondestructive testing is usually done by specialists who have been trained in the use of the equipment and the interpretation of test results. The original cost of the equipment may be high, nondestructive testing can be the fastest and least expensive of the inspection methods.

Destructive testing, also called *mechanical testing,* usually requires that a test specimen be taken from the fabrication or that sample plates be welded from which the test specimens are cut. The weld is damaged beyond use. Destructive testing is normally used to determine the mechanical and metallurgical qualities of a proposed welding procedure and to determine the ability of a welder before the actual production work is started.

In *visual inspection,* which should be the first inspection method used, the surface of the weld and the base metal are observed for visual imperfections. Certain inspection tools and gauges may be used with the observation procedure. Visual inspection is the testing method most commonly used

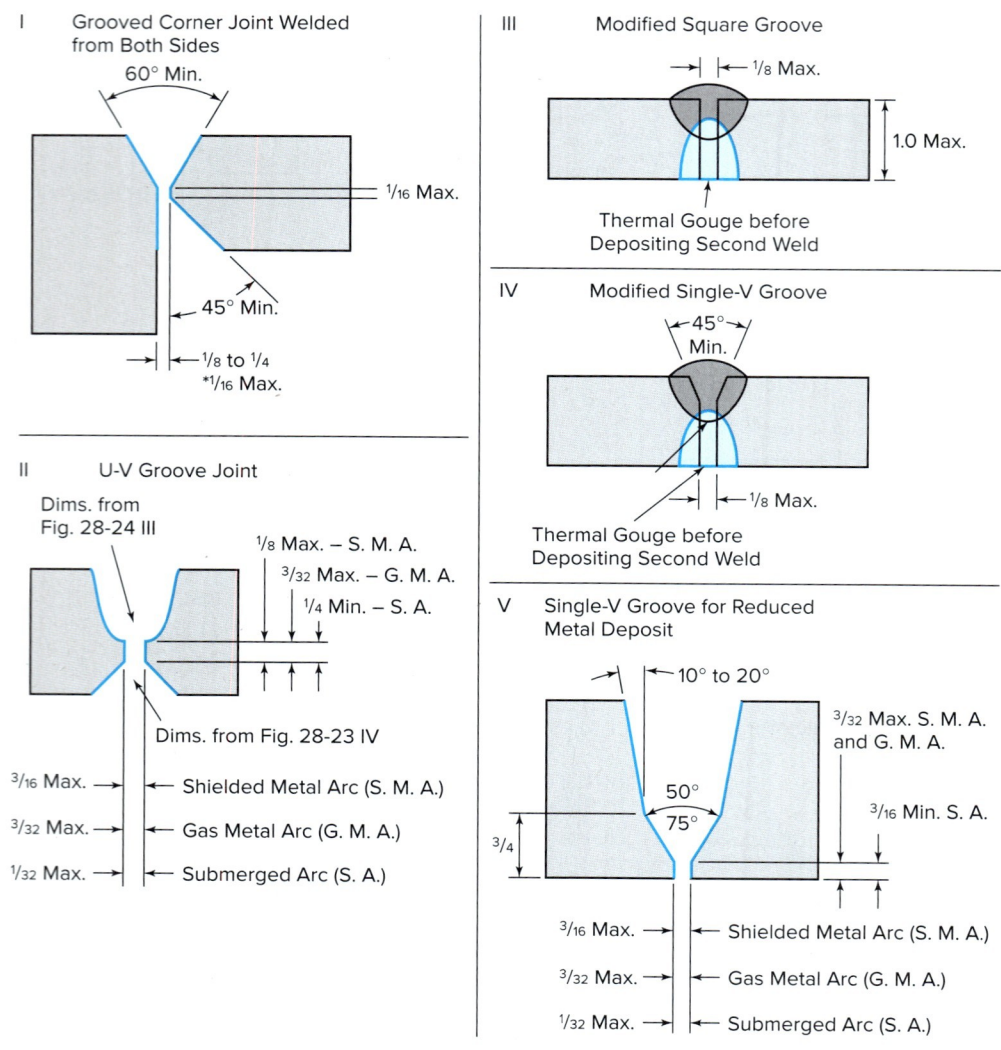

Fig. 28-26 Recommended dimensions of mixed grooves for arc welding.

by welders themselves as well as by the welding inspector and welding supervisor.

This method is highly effective when applied before, during, and after the welding operation by properly trained and skilled welders and inspectors. See Table 28-2, page 904, for an example of a well laid out inspection program. The American Welding Society has a certification program QC-1 to certify welding inspectors at three different levels. This certification is for visual inspection only, but additional add-ons are available for other nondestructive and destructive methods.

Nondestructive Testing (NDT)

Magnetic Particle Testing

Magnetic particle testing is one of the most easily used nondestructive tests. It is used to inspect plate edges before welding for surface imperfections. It tests welds for such defects as surface cracks, lack of fusion, porosity, undercut, incomplete root penetration, and slag inclusions. This method is limited to magnetic materials such as steels and cast iron. It cannot be used with such nonmagnetic materials as the stainless steels, aluminum, and copper. Very often the method is referred to as the *Magnaflux®* method. Magnaflux® is the name of a particular brand of testing equipment, Fig. 28-29, page 905.

Magnetic particle testing, Fig. 28-30, page 905, detects the presence of internal and surface cracks too fine to be seen by the naked eye. Defects can be detected to a depth of ¼ to ⅜ inch below the surface of the weld. Defects much deeper than this are not likely to be found.

The part to be examined must be smooth, clean, dry, and free from oil, water, and excess slag from the welding operation. Wire brushing or sandblasting is usually satisfactory preparation for most welds. The part is magnetized by using an electric current to set up a magnetic

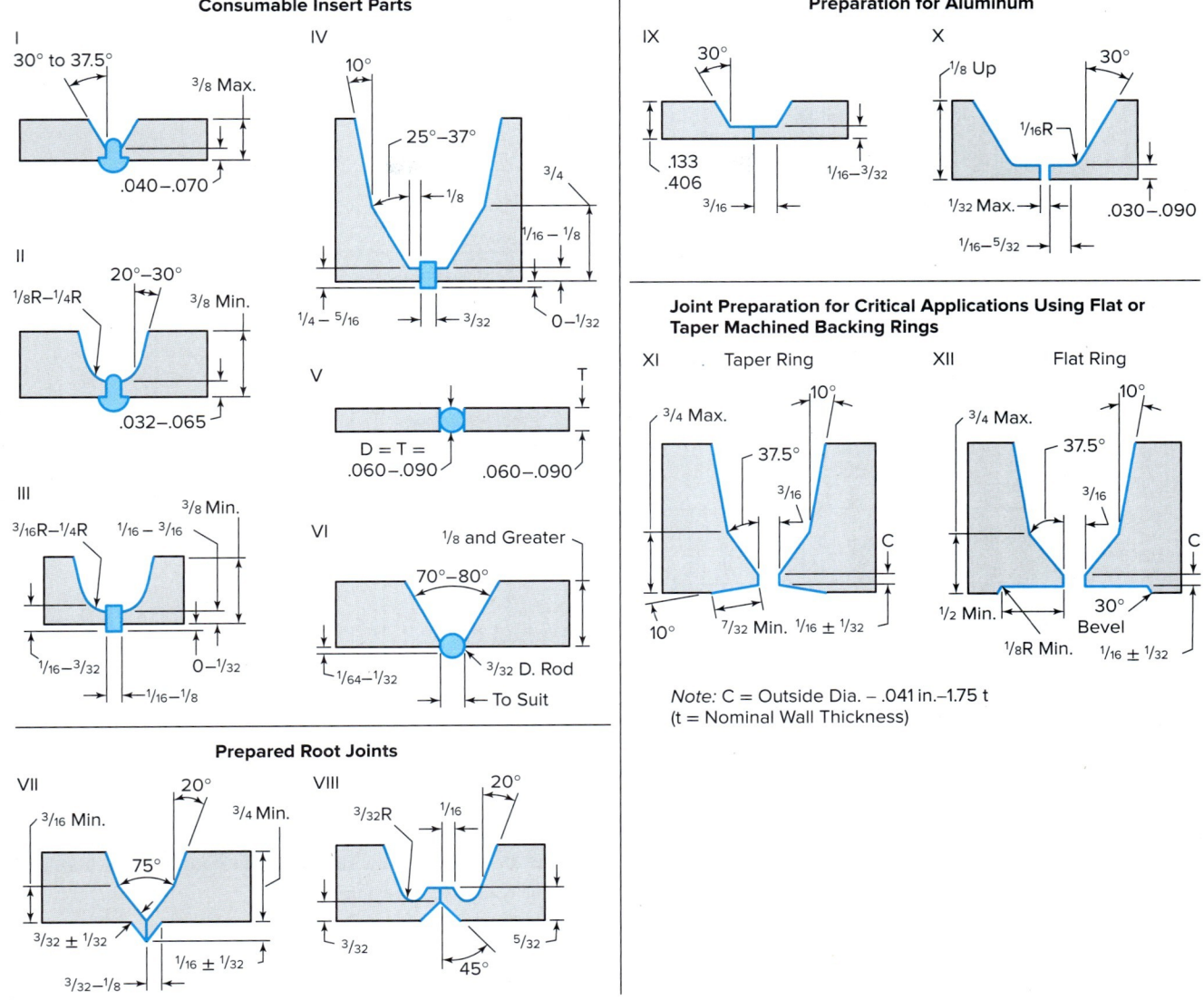

Fig. 28-27 Recommended dimensions of grooves for gas tungsten arc welding processes to obtain controlled and complete penetration. *Note:* Dimensions are for steel except as noted.

field within the material or by putting the piece in an electric coil. The magnetized surface is covered with a thin layer of magnetic powder such as blast resin or red iron oxide, Fig. 28-31, page 905. Another method uses a fluorescent powder that glows in black light. These powders can be applied dry or they may flow over the surface if they are in a suspension of oil, water, or any other low viscosity liquid. The layer of powder can be blown off the surface when there are no defects. If there is a defect, the powder is held to the surface at the defect because the powerful magnetic field in the workpiece sets up a north magnetic pole at one end of the defect and a south magnetic pole at the other. This is referred to as "flux leakage." These poles have a stronger attraction for the magnetic particles than the surrounding surface of the material.

Figure 28-32, page 906, gives a simple explanation of the principles of magnetic particle testing. As shown in Part 1, an open magnet has two poles: north and south. The magnetic field between the two poles attracts and holds a nail. If the magnet is bent until the poles almost touch, the magnetic field between the two poles attracts iron powder (Part 2). If the magnet is completely closed as in Part 3, it will not attract or hold iron powder because the magnetic field is in a circle inside the ring and there can be no polarity. This is the reason that the powder can

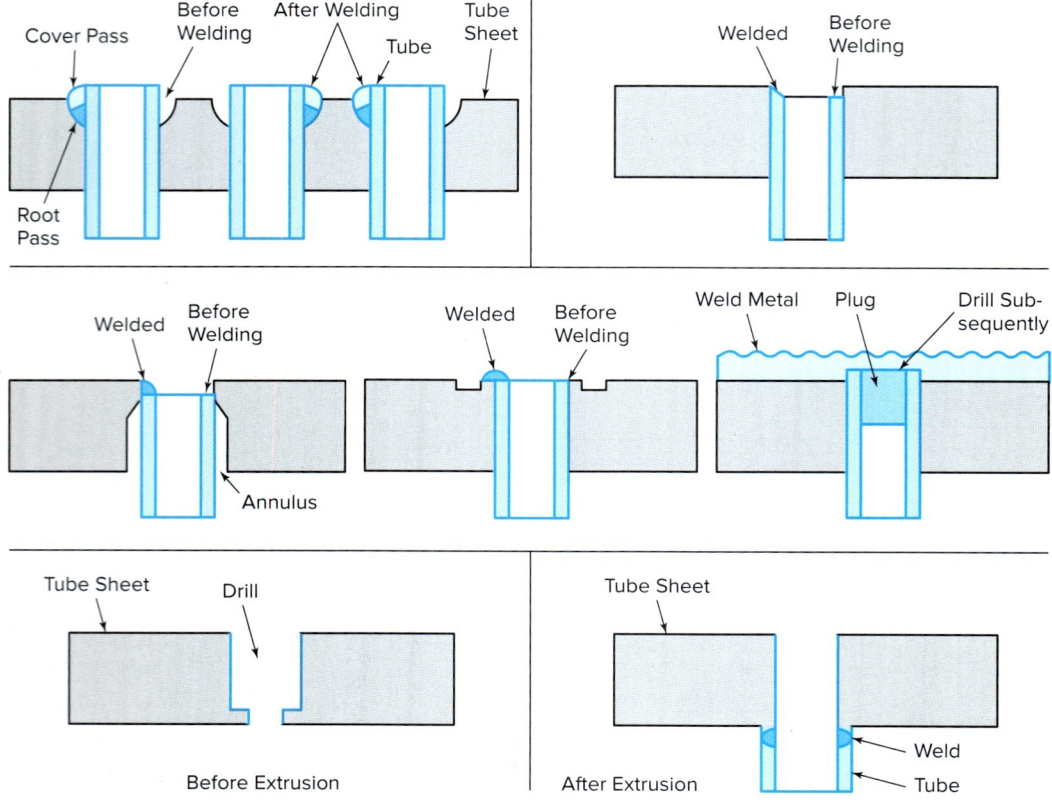

Fig. 28-28 Recommended dimensions of grooves for the welding of tubes to tube sheets.

Table 28-2 Welding Inspection Program

Phase 1	Initial review	1. Review purchase order, all codes, and drawings. 2. Develop all necessary inspection plans. 3. Check welding procedures. 4. Check welder status (determine whether they qualified for the work to be done). 5. Establish inspection documentation system. 6. Publish nonconforming product identification system. 7. Create a corrective action program.
Phase 2	Prewelding checks	1. Check suitability and condition of welding equipment. 2. Check conformance of base and filler materials. 3. Check the positioning of weldments and specific joints. 4. Check joint preparation, fitup, cleanliness. 5. Check fixturing for adequacy of alignment maintenance. 6. Check preheat (or initial) temperature.
Phase 3	During welding inspection	1. Check compliance with the WPS provisions. 2. Check quality and placement of key weld passes. 3. Check weld bead sequencing and placement. 4. Check interpass temperature and cleaning. 5. Check adequacy of back-gouging. 6. Monitor any specified in-process nondestructive examination.
Phase 4	Postwelding activities	1. Check finished weld appearance and soundness. 2. Check weld sizes and dimensions. 3. Check dimensional accuracy of weldment. 4. Carry out or monitor and evaluate specified nondestructive examination. 5. Monitor any PWHT or other postweld work. 6. Finalize and collate inspection documentation.

Fig. 28-29 A portable magnetic particle testing unit. This method speeds inspection of welds and stressed areas during fabrication and repair. The unit can be used both in shop and field. © Magnaflux

Fig. 28-31 A portable magnetic particle test unit checking critical welds during the construction of a Detroit bank building. Note the use of magnetic powder as the unit is applied. © American Welding Society. B1.10M/B1.10:2009, Figure 17, p. 19.

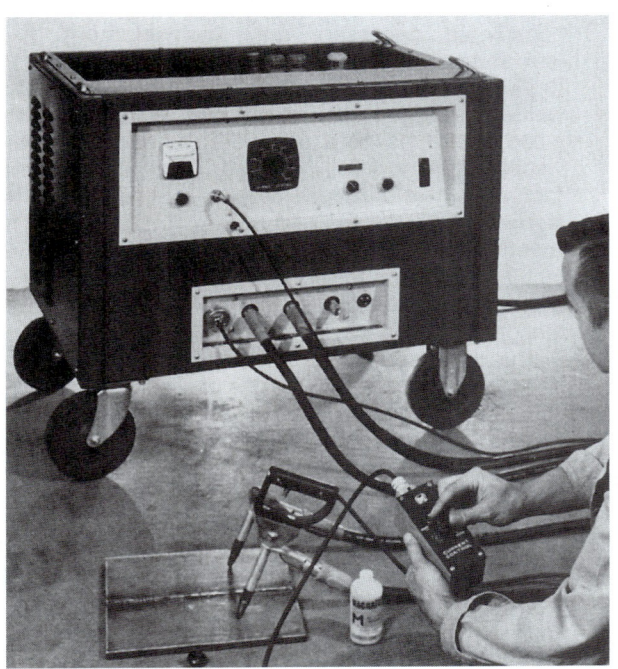

Fig. 28-30 Remote control used with a magnetic particle test unit. Welds can be inspected for incomplete penetration and fusion, slag inclusions, crater cracks, and porosity. © Magnaflux

be blown off the surface of a weld without defects. If there is a crack in the weld, a flux leakage will occur at the crack which will hold the powder (Part 4).

Cracks must be at an angle to the magnetic lines of force in order to show. The method illustrated by Fig. 28-32, page 906 (Part 3) applies electrical current directly to both ends of the piece being inspected. This is often referred to as a "Headshot." A magnetic field is produced at right angles to the flow of current, which may be represented by circular lines of force within the workpiece (Part 3). When these lines of force encounter a longitudinal (lengthwise) crack, they leak through and become points of surface attraction for the magnetic powder dusted on the surface (Part 4). A transverse (crosswise) crack would not show because the lines of force would be parallel with the crack, Fig. 28-33, page 906. This is known as circular magnetism. Another method is to magnetize a workpiece by putting it inside a coil. In this method, the magnetic lines of force are longitudinal and parallel with the workpiece so that transverse cracks show up, Fig. 28-34, page 906. This is known as longitudinal magnetism.

Joint Design, Testing, and Inspection **Chapter 28** **905**

Circular Magnetization

1. Open Magnet

2. Partially Closed Magnet

3. Completely Closed Magnet

4. Cracked Magnet

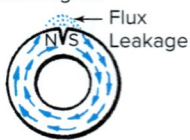

Fig. 28-32 Circular magnetization produces the lines of force shown in Parts 3 and 4. Compare these patterns with those evident in the common horseshoe and closed magnets.

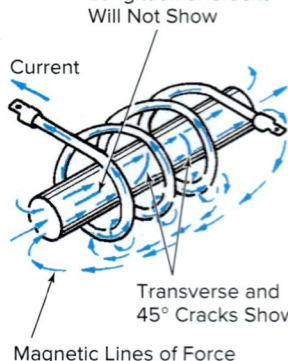

Fig. 28-34 Longitudinal magnetism is when the magnetic field is produced with a coil, the lines of force are parallel and longitudinal. A longitudinal crack will not show, but a crack angled against the lines of force is indicated.

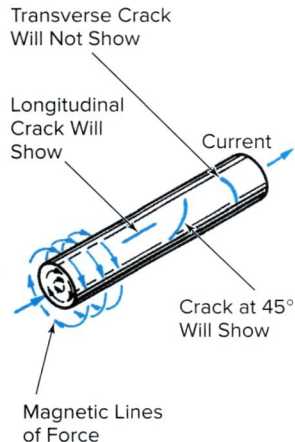

Fig. 28-33 Circular magnetism is when a current is passed through the workpiece, the magnetic lines of force are at right angles to the current, and discontinuities that are angled against the lines of force will create the flux leakage needed to produce magnetic poles on the surface. Under the illustrated arrangement, a transverse crack would not give an indication, but by changing the position of the probes 90°, it would be at right angles to the lines of force and would show.

A direct magnetization method is normally used with direct current, half-wave direct current, or full-wave direct current, Fig. 28-35. These types of currents have penetrating abilities that generally allow slight subsurface discontinuities to be detected. Direct magnetization may also be used with alternating current, which is limited to the detection of surface discontinuities only. An indirect magnetization method may also be used, Fig. 28-36. This method uses an electrically supplied coil wrapped around a soft iron core to produce an electromagnet.

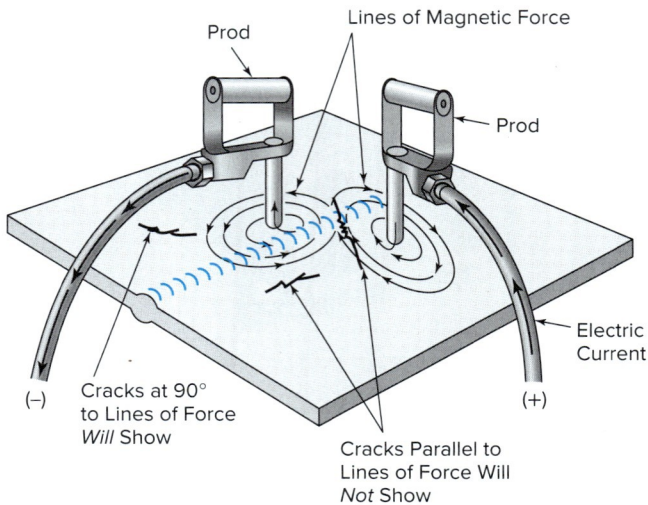

Fig. 28-35 Direct magnetization using d.c. prods. American Welding Society (AWS) B1.10M/B1.10:2009, Guide for Nondestructive Examinations of Welds, Fig 13, p. 15.

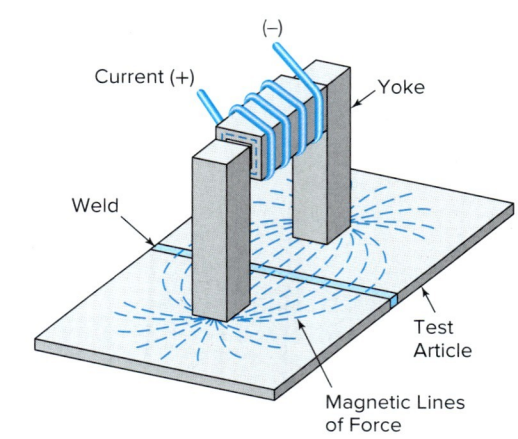

Fig. 28-36 Indirect magnetization using a yoke. American Welding Society (AWS) B1.10M/B1.10:2009, Guide for Nondestructive Examinations of Welds, Fig 14, p. 15.

In some cases where portability is an issue and no source of electric power is readily available, permanent bottle magnets can be used.

The volume and variety of products tested by the magnetic particle method is enormous. In addition to weldments of all types, finished steel products, castings, pipe and tubing, racing cars, aircraft, and missiles can be checked for structural defects.

Radiographic Inspection

Radiography is a nondestructive test method that shows the presence and type of microscopic defects in the

Table 28-3 Gamma Ray Sources, Strength, and Usage

Gamma Source	Half Life	Energy Level	Strength (Ci)	Usage
Cobalt-60 (Co-60)	5.3 years	1.33 and 1.17 MeV	20	For examination of thicker sections, typically 4–8 in. thick and of medium density such as steel or copper. This source requires heavy shielding, the weight of which makes for difficult transportation and setup.
Iridium-192 (Ir-192)	74 days	0.31, 0.47, and 0.60 MeV	50	For examining steels up to 3 in. thick. It is used extensively for weld inspection. Its relative low activity makes for easy shielding, and thus simplifies handling.
Thulium-170 (Tm-170)	130 days	52 and 84 keV	—	Produces a soft disintegration of the gamma rays. Has an energy level similar to that of a 110-kV X-ray machine. The best application is on 3/8- to 1/2-in. thick metal.

Note: eV = electron volt, Ci = curie.

interior of welds. The method utilizes either the X-ray or gamma ray. The source of X-rays is the X-ray tube. Gamma rays are similar to X-rays except for their shorter wavelengths. They are produced by the atomic disintegration of radium or one of the several commercially available radioisotopes. While gamma rays, because of their short wavelengths, can penetrate a considerable thickness of material, exposure time is much longer than for X-rays. The film produced by X-rays or gamma rays are referred to as *radiographs*.

The size of X-ray equipment is rated on the basis of its electric energy which in turn determines the intensity of the X-ray produced. The voltage across the tube controls the X-ray wavelength and penetrating power. Voltage is measured in kilovolts: 1 kilovolt equals 1,000 volts. Generally, industrial applications run from 50 kilovolts (50,000 volts), which is used for microradiography, to 2,000 kilovolts (2,000,000 volts) which can penetrate 9-inch thick steel. These thickness limits increase with softer metals. One hundred fifty kilovolts (150,000 volts) have a thickness limit of 1 inch in steel and 4½ inches in aluminum. Special units have a capacity of 24,000 kilovolts (24,000,000 volts).

Gamma rays can also be used as a source of radiation. These types of sources are typically used for field radiographic inspection. See Table 28-3. These radioisotopes are constantly emitting radiation and are typically provided in small cylinders. They must be properly stored and exposure requires special equipment.

In radiographic testing, a photograph is taken of the internal condition of the weld metal. The photographic film is placed on the side opposite the source of radiation, Fig. 28-37, page 908. The distance between the film and the surface of the workpiece must not be greater than 1 inch. The radiation rays penetrate the metal and produce an image on the film. Different materials absorb radiation at different rates. Since slag absorbs less radiation than steel, the presence of slag permits more radiation to reach the film. Thus, the area of the slag inclusion shows up darker than the steel on the film, and this indicates a discontinuity in the weld metal. A radiograph can establish the presence of a variety of defects and record their size, shape, and relative location. As can be seen in Fig. 28-38, page 908, the orientation of the discontinuity with the source may or may not make it detectable.

Figure 28-39, page 909, shows common defects as they appear in radiographs. Porosity, which is caused by gas being entrapped in the weld deposit, shows up as small dark spots (Fig. 28-39A). A nonmetallic inclusion, such as slag, usually shows up as an irregular shape (Fig. 28-39B). These images may be fine or coarse, and they may be widely scattered or closely grouped. Cracks show up as a line darker than the film background (Fig. 28-39C). Both longitudinal and transverse cracks are detectable. Incomplete fusion gives dark shadows (Fig. 28-39D). Incomplete root penetration is indicated as a straight dark line (Fig. 28-39E). Undercutting shows up as a dark, linear shadow at the edge of the weld (Fig. 28-39F).

Penetrant Inspection

Penetrant inspection is a nondestructive method for locating defects open to the surface. Like radiographs, it can be used on nonmagnetic materials such as stainless steel, aluminum, magnesium, tungsten, and plastics. The penetrant method cannot detect interior defects.

Red Dye Penetrant The surface to be inspected must be clean and free of grease, oil, and other foreign materials. It is sprayed with the dye penetrant, which penetrates into

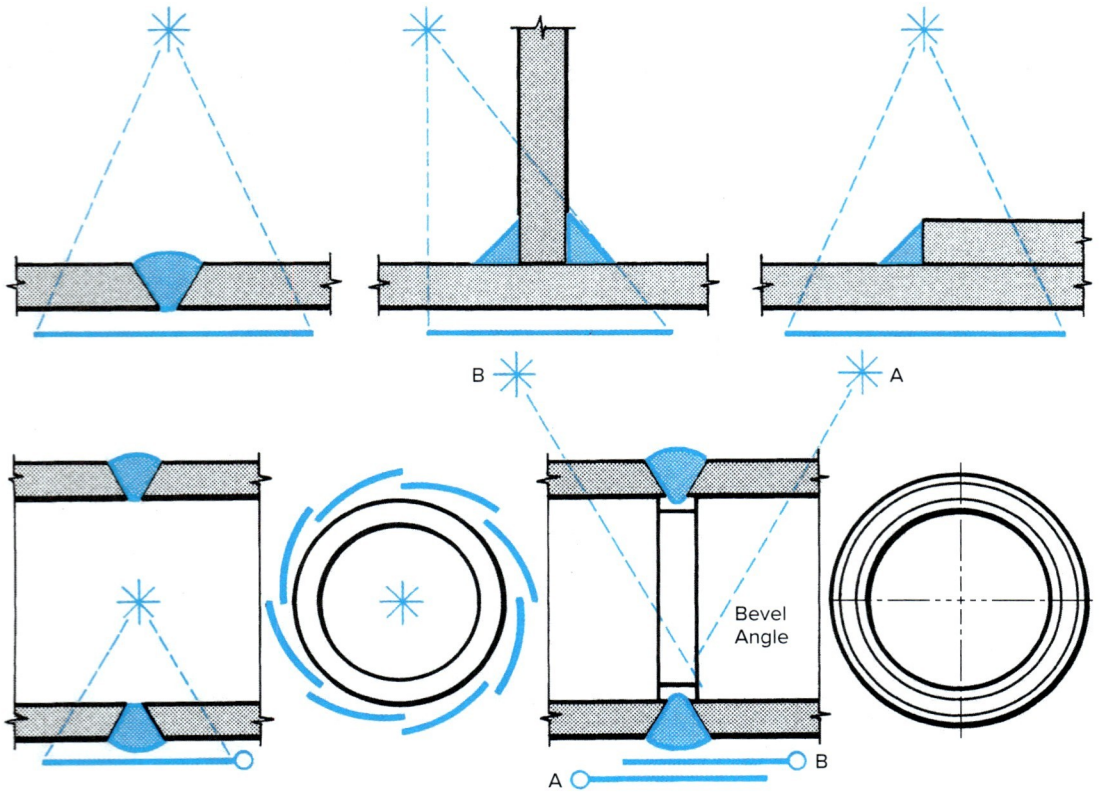

Fig. 28-37 Typical arrangements of the radiation source and film in weld radiography. The angle of exposure and the geometry of the weld influence the interpretation of the negative. Note that multiple exposures may be necessary for pipe welds. American Welding Society (AWS) B1.10M/B1.10:2009, Guide for Nondestructive Examinations of Welds, Fig 17, p. 19.

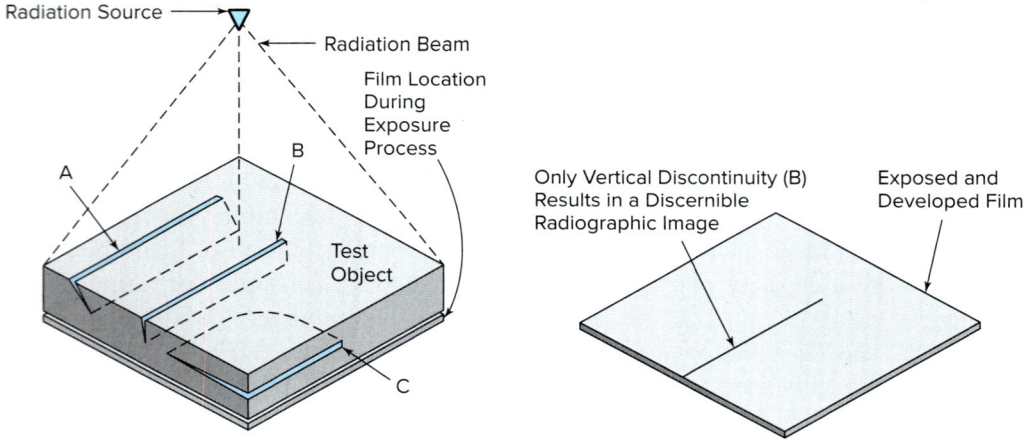

Fig. 28-38 Orientation of discontinuities with radiographic inspection. American Welding Society (AWS) B1.10M/B1.10:2009, Guide for Nondestructive Examinations of Welds, Fig 13, p. 15

cracks and other irregularities. The excess penetrant is wiped clean with a solvent. Then the part is sprayed with a highly volatile liquid that contains a fine white powder. This is known as a *developer*.

Evaporation of the liquid will leave the dry white powder, which has a blotting paper action on the red dye left in the cracks, drawing it out by capillary action so that the defects are marked clearly in red.

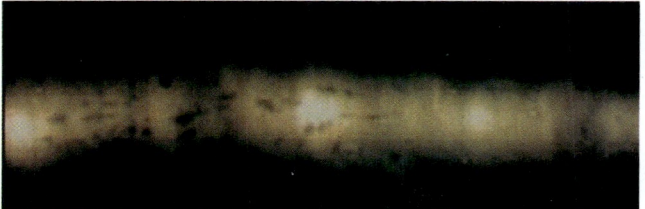
A. Porosity as indicated by the dark areas in the lighter denser weld metal. Some are spherical and some are elongated.

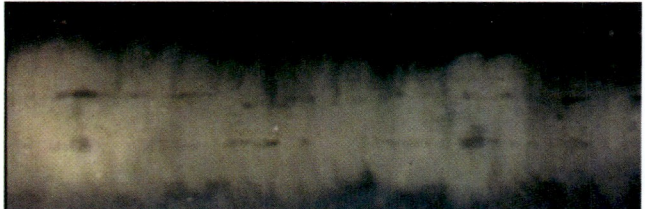
B. Slag inclusion as indicated by the darker less dense areas. These may be located along the toes of the root pass.

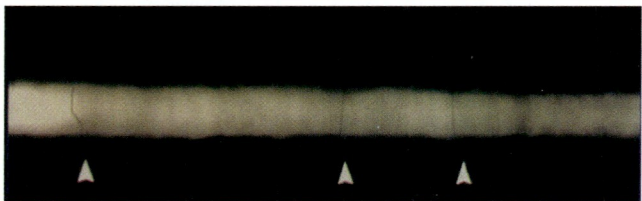
C. Transverse cracks.

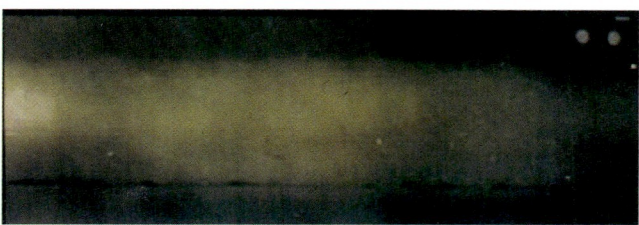
D. Incomplete fusion, less dense area along edge of weld.

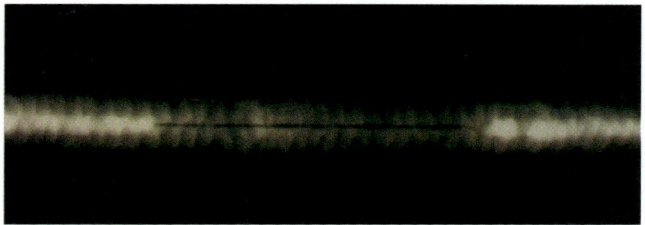
E. Incomplete penetration in root pass. Note higher density created by melt-through on the CJP weld without backing.

F. Undercut as shown by less dense areas along toe of cap pass. The vertical lines just to the right of center are the wire type image quality indicators. This film has many other indications that may or may not need further investigation. The lighter spots are greater density areas and if viewing the weld shows these have no reference they could be scratches on the film and are sometimes referred to as "relics."

Fig. 28-39A-F Weld discontinuities as indicated on radiographic film. © American Welding Society

For video of an RT inspection of a weld, please visit www.mhhe.com/welding.

Red dye penetrant is used on pressure and storage vessels and in critical piping applications to check the initial weld pass made by the TIG welding process for hairline cracks. It is often applied to the inspection of aircraft jet engines and weldments made of aluminum, magnesium, and stainless steels.

Spotcheck® Spotcheck®, Fig. 28-40, is a dye penetrant test for defects open to the surface. Like other dye penetrants, it relies on penetration of the defect by a dye, removal of the excess dye, and development of the indication.

Spotcheck® dye penetrant is a highly sensitive process. Small cracks show up against the white developer background. It locates cracks, pores, leaks, and seams invisible to the unaided eye. It marks them clearly and distinctly in red, right on the part of material.

Fig. 28-40 Spotcheck® visible penetrant kit which includes penetrant, developer, and cleaner. © Magnaflux

Spotcheck® is used on almost all materials. Examples include steel, aluminum, brass, carbides, glass, and plastics. Figure 28-41A and B, page 910, shows inspections with Spotcheck®. Spotcheck® readily locates such defects

Joint Design, Testing, and Inspection **Chapter 28** 909

Fig. 28-41A Applying Spotcheck®. © Magnaflux

Fig. 28-41B Spotcheck® of a welded joint. © Magnaflux

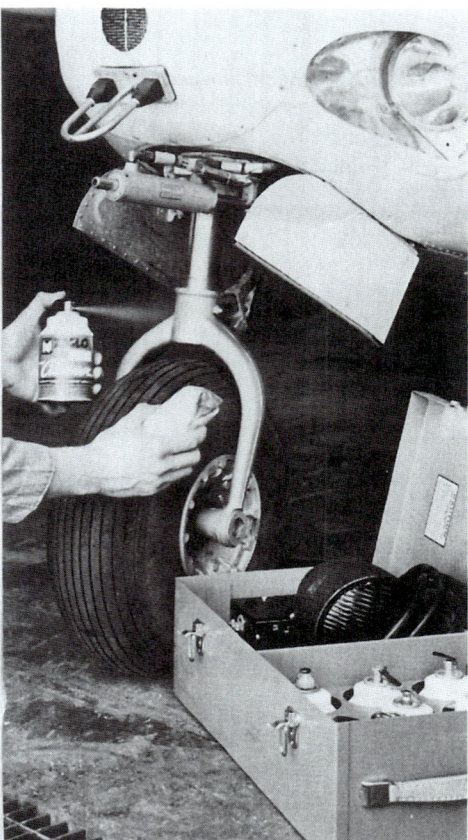

Fig. 28-42 Preparing the nose wheel fork of an aircraft for fluorescent penetrant. Note the black light in the inspection kit. © Magnaflux

as cracks from shrinkage, fatigue, grinding, heat treating, porosity, and cold shuts. It also points out seam defects and forging laps and bursts, as well as lack of bond between joined metals and through-leaks in welds. Spotcheck® is particularly recommended for the testing of moderate numbers of medium-to-small parts.

Spotcheck® offers the following advantages:

- Complete portability for critical inspection at remote shop or field locations
- Fast inspection of small, critical sections suspected of being defective
- Ease of application and dependable interpretation of results
- Low initial investment and low per-part cost in moderate volume uses

Fluorescent Penetrant Fluorescent penetrants may be used instead of the red dye penetrant, and the technique is similar to that used in the dye method. The treated metal surface is examined under ultraviolet or black light in semidarkness. This method is referred to as *fluorescent penetrant inspection*. The sharp contrast between the fluorescent material and the base or weld metal background clearly indicates cracks or other defects in the metal. Fluorescent penetrant is most useful for leak detection in lined or clad vessels. It is also used for brazed joints. The joint in Fig. 28-42 is being cleaned in preparation for fluorescent inspection.

Ultrasonic Inspection

Ultrasonic inspection is a nondestructive test method. Ultrasonic inspection is rapid and has the ability to probe deeply without damaging the weldment (200 inches). Because it can be closely controlled, it is able to supply precise information without elaborate test setups. It can detect, locate, and measure both surface and subsurface defects in the weld or base metal.

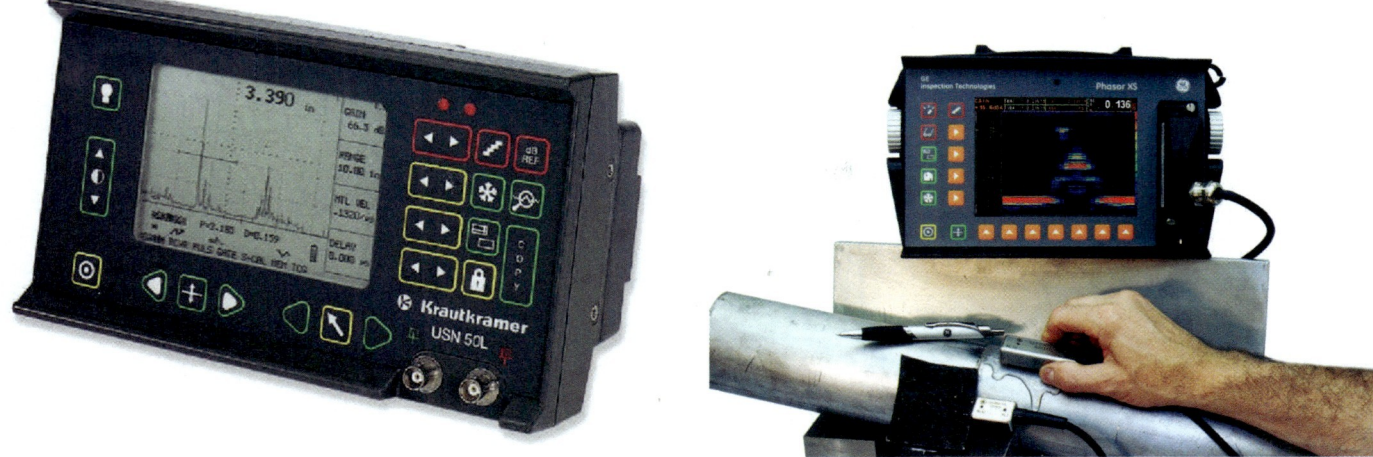

Fig. 28-43 Portable ultrasonic weld flaw detector with built-in trigonometric flaw location calculations with curvature correction and AWS D1.1 weld rating calculations. It combines a phased array imaging device and a conventional flaw detector into one instrument, with a phased array probe that allows significantly increased confidence in inspection data. Compared with traditional thickness gauge and flaw detector inspection, it offers greater probability of detection as well as faster and more reliable scanning. (left) © Agfa Corporation; (right) © General Electric Company

Ultrasonic inspection requires considerable skill and experience both in its application and in the interpretation of the echoes that appear on the screen. For example, a properly welded backing ring may produce the same pattern as a weld with unacceptable lack of penetration or root cracking. Only a highly experienced inspector may be able to tell the difference between serious defects and normal weld conditions. This disadvantage limits its application. Many companies use this method to locate defects and then apply the radiographic method to those areas that are doubtful.

Ultrasonic testing is done by means of an electrically timed wave that is similar to a sound wave, but of higher pitch and frequency. The term *ultrasonic* comes from the fact that these frequencies are above those heard by the human ear. The ultrasonic waves are passed through the material being tested and are reflected back by any density change. Three basic types of waves are used: shear (angle) beams, longitudinal (straight) beams for surface and subsurface flaws, and surface waves for surface breaks and cracks. The waves are generated by a unit similar to a high fidelity amplifier, to which a search unit is attached. The reflected signals appear on a screen as vertical reflections of the horizontal baseline, Fig. 28-43. Figure 28-44 shows an example of a butt joint.

The search unit is called a transducer. The transducer contains a piezoelectric device that converts electric energy into mechanical energy (sound) and then converts the mechanical energy (sound) back into an electric signal. This electric signal can be displayed on the older cathode ray tube (CRT) or the newer liquid crystal display (LCD).

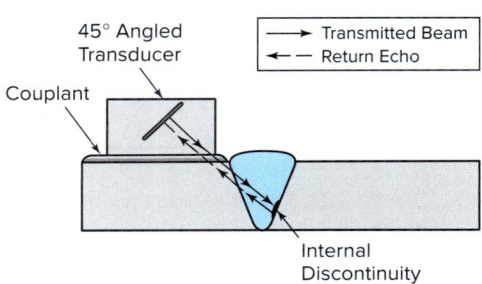

Fig. 28-44 A CJP weld on a V-groove butt joint being inspected with an angle transducer.

The search unit must be closely coupled to the part to be inspected. This is done with a couplant material.

When the search unit is applied to the material, two reference pips appear on the screen. The first pip is the echo from the surface contacted (referred to as the *main bang*), and the second pip is the echo from the bottom or opposite surface of the material. The distance between these pips is carefully calibrated, and this pattern indicates that the material is in satisfactory condition. When a defect is picked up by the search unit, it produces a third pip, which registers on the screen between the first and second pip, Fig. 28-45, page 912, since the flaw must be located between the top and bottom surfaces of the material. The distance between the pips and the relative height indicate the location and the severity of the discontinuity. An angle sound beam can also be used as shown in Fig. 28-44. Great advancements have been made in automated UT inspection using technology referred to as time of flight defraction (TOFD) and phase array.

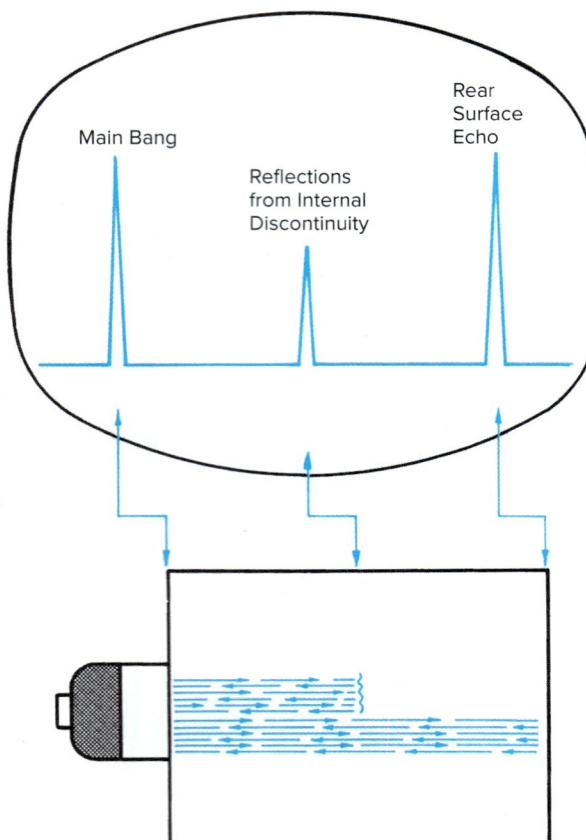

Fig. 28-45 Short pulses appear as pips and register on the ultrasonic testing screen.

This method of testing is finding increased use by the piping industry not only for detecting defects, but also for checking on the progress of corrosion and wear by taking periodic measurements of metal thickness. As the pipe material is reduced in thickness by corrosion and wear, the two pips move closer together, and the rate of deterioration can be measured. Applications of ultrasonic testing in both field and shop are shown in Figs. 28-46 through 28-48.

 For video of automated UT testing on a pipeline, please visit www.mhhe.com/welding.

The flaw (indication) must be compared to the acceptance criteria. It will remain a discontinuity unless it exceeds the acceptance criteria, which would cause it to be termed a defect. The defect would need to be repaired and reinspected or the weldment replaced. In some instances even though the discontinuity exceeds the acceptance criteria, many codes allow the engineer of record to accept the weldment. This would only be done after the engineer conducted appropriate review of the criticality of the

Fig. 28-46 This code-compliant conventional ultrasonic flaw detector can also perform weld or phased array inspections utilizing simple menu-driven interfaces. Training can be minimal and these systems can solve very demanding inspection applications in less time at affordable pricing. This UT technician is performing critical inspection on a piping system in a petrochemical plant. David Mack

Fig. 28-47 Ultrasonic testing of a weld in a large steel beam in the field. © Agfa Corporation

discontinuity. This may require further testing by fracture mechanics analysis to determine if the weldment is fit for purpose.

Eddy Current Testing

Like magnetic particle testing, the eddy current method makes use of electromagnetic energy to detect defects in

Fig. 28-48 Using portable ultrasonic instrument to check a structural weld on the seventy-sixth floor of the John Hancock Building in Chicago. © Magnaflux

the material. Testing equipment is shown in Figs. 28-49 and 28-50.

When a coil that has been energized with alternating current at high frequency is brought close to a conductive material, it will produce eddy currents in the material. *Eddy currents* are secondary currents induced in a conductor (in this case the metal being tested). Eddy currents are caused by a variation in the magnetic field. A search coil is used in addition to the energizing coil. The search coil may be connected to meters, recorders, liquid crystal displays, or oscilloscopes, which pick up the signals from the weldment. A level of individual current or voltage is noted for the material. Any defect in the material distorts the magnetic field and is indicated on the recording instrument. The size of the defect is shown by the amount of this change.

For example, in the inspection of butt-welded tubing, the tubing is passed through the energizing coil to induce an a.c. eddy current in the tube. The search coil also circles the tube. This coil is connected to a sensitive instrument. The eddy currents induced in the tubing induce eddy currents in the search coil, and the meter records the current or voltage in the coil. If the tube is sound throughout, a steady level of current or voltage is induced in the search coil and the meter reading holds steady. A flaw in the tubing distorts the magnetic field induced in the tubing by the energizing coil. When the flow passes within the range of the search coil, the needle on the search meter moves.

The eddy current method is suitable for both ferrous and nonferrous materials, and it is being used extensively in testing welded tubing, pipe, and rails,

Fig. 28-49 Eddy current testing for industrial applications requiring surface and subsurface crack detection and metal sorting. This method and equipment can deliver productivity, quality, and safety in a very portable unit. Randy Montoya/U.S. Coast Guard

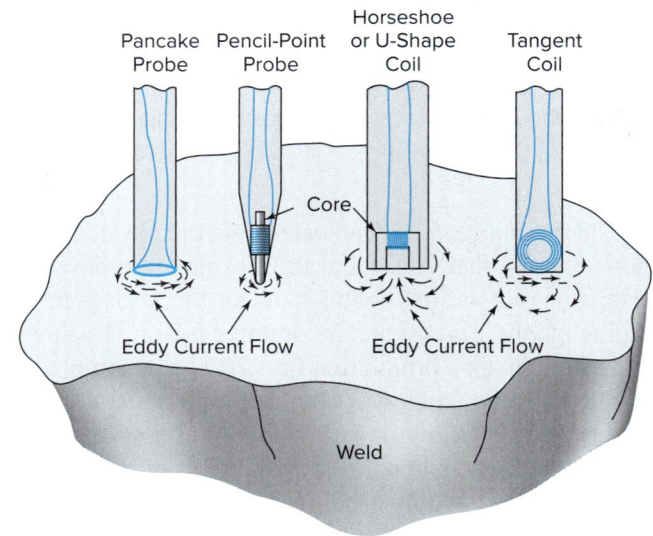

Fig. 28-50 Typical eddy current surface prods for the examination of welds. American Welding Society (AWS) B1.10M/B1.10:2009, Guide for Nondestructive Examinations of Welds, Fig 28, p. 25.

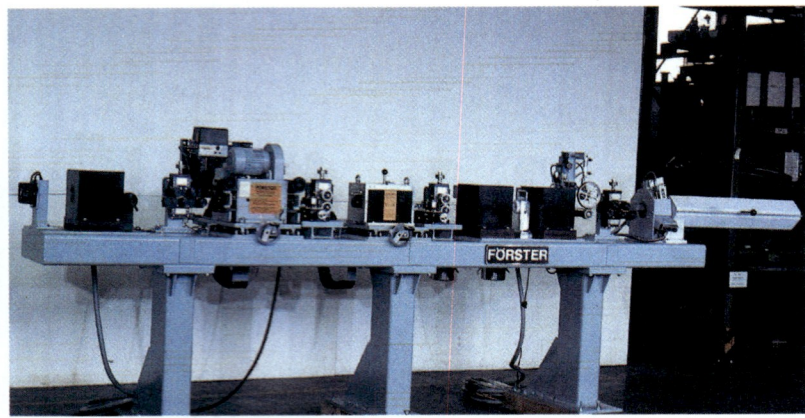

Fig. 28-51 Valve spring wire is tested on a compact eddy current testing line with an encircling, through-type coil and rotating scanning probes. The test line also contains a demagnetization system and a flaw marking system. © Foerster Instruments

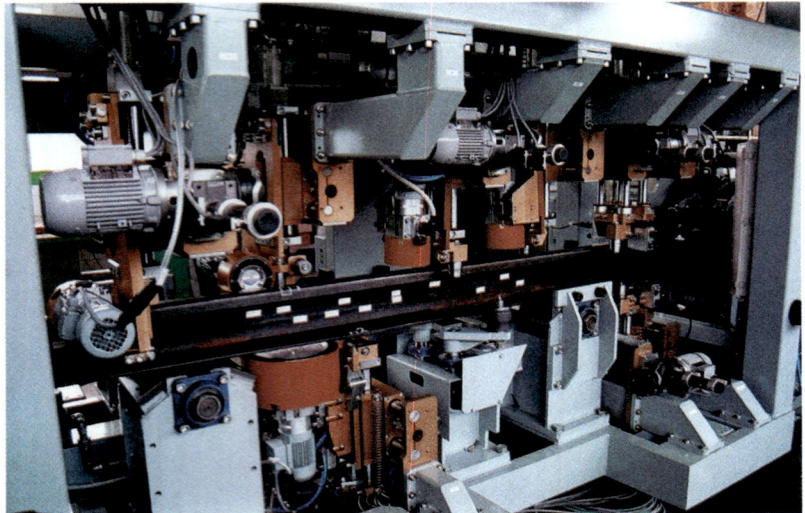

Fig. 28-52 Eddy current system for testing rails for surface flaws with a rotating disk and shape-adapted segment coils. © Foerster Instruments

Figs. 28-51 and 28-52. It can determine the physical characteristics of a material, the wall thickness in tubing, and the thickness of various coatings. It can check for porosity, slag inclusions, cracks, and incomplete fusion. Testing of tubing on a factory production line may be accomplished at a rate of approximately 3,000 feet per minute.

Eddy current testing is only good up to approximately $3/16$-inch thickness, and calibration blocks are required for all the types of weld material to be inspected. These two areas limit its use.

Leak Tests

Leak tests are made by means of pneumatic or hydraulic pressure. A load is applied that is equal to or greater than that expected in service. If the test is to determine leakage, it is not necessary to test beyond the working load of the weldment. If, however, failure of the weldment might cause great property loss or personal injury, the test pressure applied usually exceeds the working pressure. This method of testing is usually used to test pressure vessels and pipelines. If used as a destructive method, pressure is applied until the unit bursts, Fig. 28-53.

Water is usually used to test for leaks, but very small leaks are not always detected. When air or oil of low viscosity is used, it indicates leaks that water cannot. Hydrogen will leak where oil will not, but hydrogen is a fire hazard and must be used with great care.

In cases where a leak would be critical, helium can be used. Helium can detect very small leaks safely. Helium, like hydrogen, is a very small atom.

When the expansive force of air or gas is high, adequate safety measures must be taken to guard against explosion. When air or gas is used, the weld seam may be painted with liquid soap or another chemical solution to cause the formation of bubbles at the point of leakage. Small tanks may be immersed in a water or liquid bath. On work that is of a critical nature, the pressure should be held for a period of time and the pressure noted on a gauge that is connected to the vessel.

NDT Process Comparison

The welding student is urged to study Tables 28-4 through 28-7, pages 915–917. In Chapter 4, you

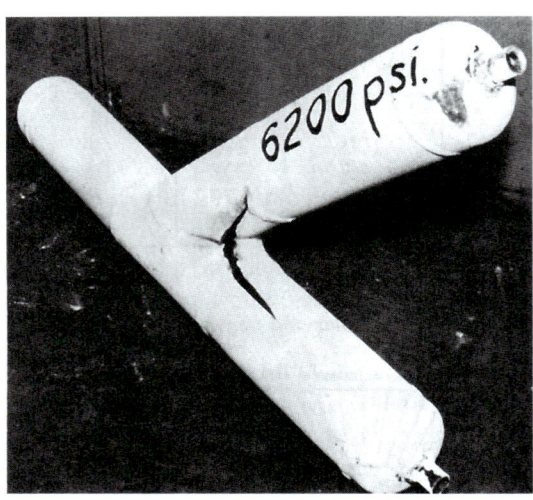

Fig. 28-53 A piping unit that has been subjected to hydraulic testing. The weld failed at 6,200 p.s.i. © Crane Co.

Table 28-4 Discontinuities Commonly Encountered with Welding Processes

Welding Process	Cracks	Incomplete Fusion	Incomplete Joint Penetration	Overlap	Porosity	Slag	Undercut
Arc							
EGW—Electrogas welding	●	●	●	●	●		●
GTAW—Gas tungsten arc welding	●	●	●		●		●
PAW—Plasma arc welding	●	●	●		●		●
SAW—Submerged arc welding	●	●	●	●	●	●	●
SW—Stud welding	●	●			●		●
CAW—Carbon arc welding	●	●	●	●	●	●	●
FCAW—Flux cored arc welding	●	●	●	●	●	●	●
GMAW—Gas metal arc welding	●	●	●	●	●		●
SMAW—Shielded metal arc welding	●	●	●	●	●	●	●
Oxyfuel Gas							
OAW—Oxyacetylene welding	●	●	●	●	●		●
OHW—Oxyhydrogen welding	●	●	●		●		
PGW—Pressure gas welding	●	●			●		
Resistance							
PW—Projection welding	●	●	●				
RSEW—Resistance seam welding	●	●	●		●[1]		
RSW—Resistance spot welding	●	●	●		●[1]		
FW—Flash welding	●	●	●				
UW—Upset welding	●	●	●				
Solid-State[2]							
CW—Cold welding	●	●					
DFW—Diffusion welding	●	●					
EXW—Explosion welding		●					
FOW—Forge welding		●					
FRW—Friction welding		●					
USW—Ultrasonic welding		●					
Other							
EBW—Electron beam welding	●	●	●		●		
ESW—Electroslag welding	●	●	●	●	●	●	●
IW—Induction welding	●	●					
LBW—Laser beam welding	●	●	●		●		
PEW—Percussion welding	●	●					
TW—Thermite welding	●	●			●	●	

[1] Porosity in resistance welds is more properly called voids.
[2] Solid-state is not a fusion process, so incomplete joining is incomplete welding rather than incomplete fusion.

Adapted from American Welding Society (AWS) B1.10M/B1.10:2009, Guide for Nondestructive Examinations of Welds, Table 2, p. 12.

learned about various weld joints, types of welds, and discontinuities found in welds. You are now learning about the various nondestructive testing (NDT) methods. It is critical to your understanding of testing and inspection to tie all this information together. By understanding which discontinuities are prone to be found with which welding process, which inspection method is best used to find a particular type of discontinuity, which type of joint is best inspected with which test method, and the mechanics of each inspection method, you are well on your way to gaining the skills and knowledge required for weld testing. It should be noted that proof of the existence of a discontinuity does not indicate that the discontinuity will seriously affect the fitness for purpose of the weldment. The size, location, and criticality of the discontinuity must be judged against the acceptance criteria to determine whether it is a defect and must be repaired or replaced.

Hardness Tests

It is often important to test the hardness of the weld deposit or the base metal in the area of the weld. It is important to know the hardness of the weld deposit if the weld

Table 28-5 Common Weld Inspection Methods Versus Discontinuities

Discontinuities	Inspection Methods						
	RT	UT	PT[1]	MT[2,3]	VT[1]	ET[4]	LT[5]
Cracks	O	A	A	A	A	A	A
Incomplete fusion	O	A	U	O	O	O	U
Incomplete joint penetration	A	A	U	O	O	O	U
Laminations	U	A	A	A	A	U	U
Overlap	U	O	A	A	O	O	U
Porosity	A	O	A	O	A	O	A
Slag inclusions	A	O	A	O	A	O	O
Undercut	A	O	A	O	A	O	U

Notes:
[1]Surface.
[2]Surface and slightly subsurface.
[3]Magnetic particle examination is applicable only to ferromagnetic materials.
[4]Weld preparation or edge of base metal.
[5]Leak testing is applicable only to enclosed structures that may be sealed and pressurized during testing.

Legend: RT—radiographic testing; UT—ultrasonic testing; PT—penetrant testing, including both DPT (dye penetrant testing) and FPT (fluorescent penetrant testing); MT—magnetic particle testing; VT—visual testing; ET—eddy current testing; LT—leak testing; A—applicable method; O—marginal applicability (depending on other factors such as material thickness, discontinuity size, orientation, and location); U—usually not used.

Adapted from American Welding Society (AWS) B1.10M/B1.10:2009, Guide for Nondestructive Examinations of Welds, Table 3, p. 27.

Table 28-6 Applicable Examination Methods—Five Weld Joint Types

Joints	Inspection Methods						
	RT	UT	PT	MT	VT	ET	LT
Butt	A	A	A	A	A	A	A
Corner	O	A	A	A	A	O	A
Parallel	O	O	A	A	A	O	A
Lap	O	O	A	A	A	O	A
T	O	A	A	A	A	O	A

Notes: RT—radiographic examination; UT—ultrasonic testing; PT—penetrant examination, including both DPT (dye penetrant testing) and FPT (fluorescent penetrant testing); MT—magnetic particle examination; VT—visual testing; ET— electromagnetic examination; A—applicable method; O—marginal applicability (depending on other factors such as material thickness, discontinuity size, orientation, and location).

Adapted from American Welding Society (AWS) B1.10M/B1.10:2009, Guide for Nondestructive Examinations of Welds, Table 4, p. 27.

Table 28-7 Examination Method Selection Guide

	Visual	Penetrant	Magnetic Particle	Radiography (Gamma)
Equipment Needs	Magnifiers, color enhancement, projectors, other measurement equipment (i.e., rulers, micrometers, optical comparators, light source).	Fluorescent or dye penetrant, developers, cleansers (solvents, emulsifiers, etc.). Suitable cleaning gear. Ultraviolet light source if fluorescent dye is used.	Prods, yokes, coils suitable for inducing magnetism into the weld. Power source (electric). Magnetic powders—some applications require special facilities and ultraviolet lights.	Gamma ray sources, gamma ray camera projectors, film holders, film, lead screens, film processing equipment, film viewers, exposure facilities, radiation monitoring equipment.
Applications	Weldments that have discontinuities only on the surface.	Weldments that have discontinuities only on the surface.	Weldments that have discontinuities on or near the surface.	Weldments that have voluminous discontinuities such as porosity, incomplete joint penetration, and slag. Lamellar-type discontinuities such as cracks and incomplete fusion can be detected with a lesser degree of reliability. It may also be used in certain applications to evaluate dimensional requirements such as fitup, root conditions, and wall thickness.
Advantages	The method is economical and expedient, and requires relatively little training and relatively little equipment for many applications.	The equipment is portable and relatively inexpensive. The inspection results are expedient. Results are easily interpreted. Requires no electric energy except for light sources.	The method is relatively economical and expedient. Inspection equipment is considered portable. Unlike dye penetrants, magnetic particle can detect some discontinuities slightly below the surface.	The method is generally not restricted by type of material or grain structure. The method detects surface and subsurface discontinuities. Radiographic images aid in characterizing discontinuities. The film provides a permanent record for future review.
Limitations	The method is limited to external or surface conditions only and by the visual acuity of the observer or inspector.	Surface films such as coatings, scale, and smeared metal may mask or hide discontinuities. Bleed out from porous surfaces can also mask indications. Parts must be cleaned before and after inspection.	The method is applicable only to ferromagnetic materials. Parts must be cleaned before and after inspection. Thick coatings may mask rejectable discontinuities. Some applications require the part to be demagnetized after inspection. Magnetic particle inspection requires use of electric energy for most applications.	Planar discontinuities must be favorably aligned with radiation beam to be reliably detected. Radiation poses a potential hazard to personnel. Cost of radiographic equipment, facilities, safety programs, and related licensing is relatively high. A relatively long time between exposure process and availability of results. Accessibility to both sides of the weld required.

Adapted from American Welding Society (AWS) B1.10M/B1.10:2009, Guide for Nondestructive Examinations of Welds, Annex A, pp. 29–30.

Fig. 28-54 A bench model Brinell hardness tester. This unit is capable of applying eight standard Brinell loads. © Newage Testing Instruments, Inc.

is going to be machined or if it will be subject to surface wear of one kind or another.

There are a number of nondestructive hardness tests, and the choice depends upon the material being tested. The most common tests are the Brinell, Rockwell, Vickers, and Knoop.

The preparation of the specimen is very important. The surface should be flat and free of scratches.

Brinell The Brinell hardness test consists of impressing a hardened steel ball into the metal to be tested at a given pressure for a predetermined time, Fig. 28-54. The diameter of the impression is measured, and this indicates a Brinell number on a chart.

The ball is 10 ± 0.0025 millimeters. It is forced into the specimen by hydraulic pressure of about 3,000 kilograms for a period of 15 seconds. The Brinell hardness number (BHN) can be related to the actual tensile strength of carbon steel. Simply multiply the BHN by 500, and this will equal its approximate tensile strength. The load for soft metals such as brasses and bronzes is 500 kilograms applied for the same amount of time.

Because the impression the Brinell ball makes is large, it can be used only for obtaining values over a large area and when the impression on the surface is not objectionable.

Rockwell The Rockwell system is similar to the Brinell system, but it differs in that the readings can be obtained from the dial. The Rockwell tester, Fig. 28-55, measures the depth of residual penetration made by a small hardened steel ball or diamond cone. A minor load of 10 kilograms is applied. This seats the penetrator (ball or cone) in the surface of the specimen and holds it in position. The

Fig. 28-55 A bench model Rockwell hardness tester. This unit meets ASME E-18 requirements. © Newage Testing Instruments, Inc.

full load of 150 kilograms is then applied. After the major load is removed, the hardness number is indicated on the dial gauge. The hardness numbers are based on the difference of penetration between the major and minor loads.

The two Rockwell scales are known as the B-scale and the C-scale. See Table 28-8. The C-scale is used for the harder metals. A cone-shaped diamond penetrator is used instead of a ball, and it is applied at a load of 150 kg, Fig. 28-56. The B-scale is used for the softer metals, and the penetrator is a hardened steel ball $\frac{1}{8}$ or $\frac{1}{16}$ inch in diameter applied at a lesser load of 100 kilograms. A lesser load is applied to the ball than to the cone.

Microhardness Testing The microhardness test uses a range of loads and diamond indenters to make an indentation, which is measured and converted to a hardness value, Fig. 28-57. It is very useful for testing on a wide variety of materials as long as test samples are carefully prepared. For welding the macroetched specimen can have its hardness checked from the unaffected base metal through the heat-affected zone (HAZ) and into the weld metal. There are two types of indenters, a square base pyramid-shaped diamond for testing in a Vickers tester and a narrow rhombus shaped indenter for a Knoop tester. Typically loads are very light, ranging from a few grams to one or several kilograms, although "macro" Vickers loads can range up to 30 kg or more. The microhardness methods are used to test on metals, ceramics, and composites—almost any type of material. The Vickers method is more commonly used for:

- Testing of very thin materials like foils
- Measuring the surface of a part

Table 28-8 Typical Readings for Standard Hardness Tests

Material	Rockwell C	Rockwell B	Brinell	Scleroscope	Vicker	Tensile Load Strength PSI
Hard	69		755	98	1050	368
	60		631	84	820	311
	51		510	71	553	261
	39		370	52	369	182
	24	100	245	34	248	116
	16	95	207	29	207	97
	10	89	179	25	187	83
Soft	0	79	143	21	143	72

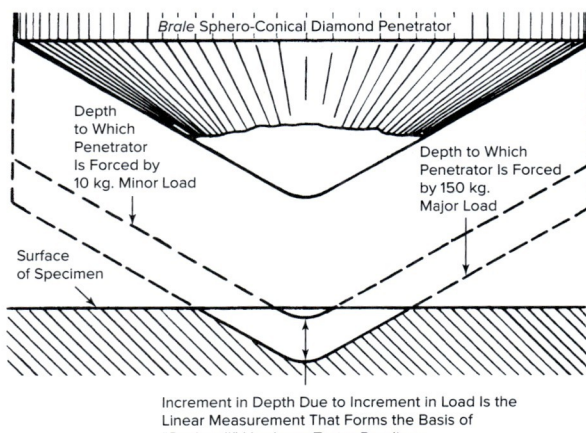

Fig. 28-56 Spheroconical diamond penetrator.
Source: Wilson Instrument Division, ACCO

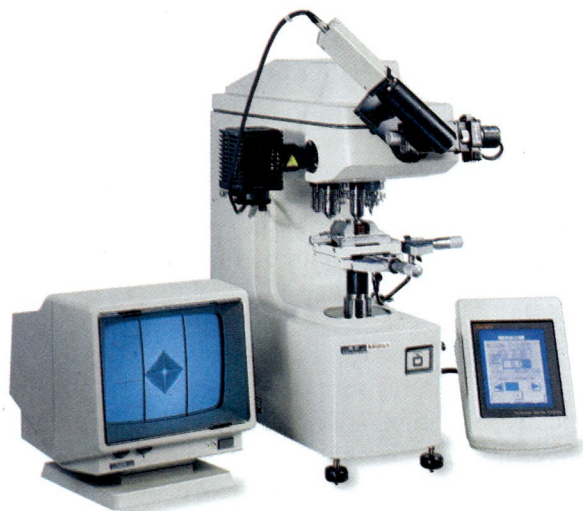

Fig. 28-57 A bench model microhardness testing system. Designed to test both Vickers and Knoop hardness scales.
© Mitutoyo

- Small parts or small areas
- Measuring individual microstructures
- Measuring the depth of case hardening by sectioning a part and making a series of indentations to describe a profile of the change in hardness

The Knoop method is commonly used for:

- Closely spaced requirements
- Testing close to an edge due to the narrow shape of the indentation
- More resolution due to the width of the Knoop indentation
- Thinner materials because indentation is less deep

There are a number of considerations in microhardness testing. Sample preparation is usually necessary in order to provide a specimen that can fit into the tester, make a sufficiently smooth surface to permit a regular indentation shape and good measurement, and be held perpendicular to the indenter. Test specimens are usually mounted in a plastic medium for easier preparation and testing. Better resolution is achieved by using the largest indentation possible. (Error is magnified as indentation size decreases.) The test procedure is subject to problems of operator influence on the test results.

Impact Hardness Tester The Equotip hardness tester shown in Fig. 28-58, page 920, is a portable machine that is easy to operate and extremely accurate. Each hardness test takes only 1 second to perform. An integral loading mechanism is part of the testing probe that allows the operator to position, load, and test in one motion. Hardnesses are displayed digitally and a simple electronic adjustment permits testing in any direction, automatically adjusted for gravity. This unit converts electronically to Brinell, Rockwell B and C, Shore, and Vikers. The electronic module allows for downloading data. This system uses the comparative method of evaluating hardness; the loading

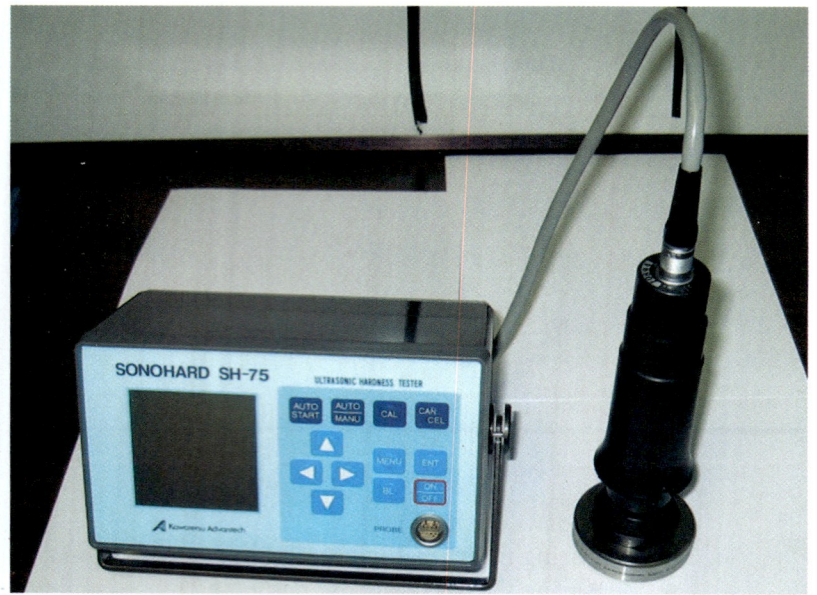

Fig. 28-58 Compact portable sonic hardness tester designed for massive workpieces. It can measure in all directions and in poorly accessible places. The RS 232 C port allows for download from memory storage. © Micro Photonics Inc.

spring is not of a calibrated tension. By testing on a calibrated test block, the equipment does not need to be zeroed or adjusted. The harder the material, the greater the rebound of the impact body; and the softer the material, the less the rebound. Typical metals to be tested are:

- Cast steel
- Irons
- Tool steel
- Aluminum
- Yellow metals (brass, copper)
- Stainless steel

The tester is designed for use on all metallic materials from 80 Brinell to about 68 Rockwell C. Accuracy is in the ±0.5 percent.

Destructive Testing

Destructive testing is the mechanical testing of weld samples to determine their strength and other properties. Testing methods are relatively inexpensive and highly reliable. Therefore, they are widely used for welder and procedure qualification. Destructive testing is usually performed on test specimens that are taken from a welded plate that duplicates the material and the weld procedures used on the job. Certain general test procedures have been developed by the American Welding Society that are the standard for the industry and various code-making bodies. While tests may differ somewhat in detail and application from industry to industry, they all make use of the basic procedures and types of test specimens recommended by the American Welding Society. The following list summarizes the more common mechanical tests that will be presented in this chapter.

Groove Welds

- **Reduced-section tension test** Determines tensile strength, yield strength, and ductility; used for procedure qualification.
- **Free-bend test** Determines ductility; used for procedure qualification.
- **Root-bend test** Determines soundness; used widely for welder qualification; also used for procedure qualification.
- **Face-bend test** Determines soundness; used widely for welder qualification; also used for procedure qualification.
- **Side-bend test** Determines soundness; used widely for welder qualification; also used for procedure qualification.
- **Nick-break test** Determines soundness; at one time used widely for welder qualification; used infrequently today.

Fillet Welds

- **Longitudinal and transverse shear tests** Determines shear strength; used for procedure qualification.
- **Fillet weld soundness test** Determines soundness; used extensively for welder qualification.
- **Fillet weld break test** Determines soundness; used infrequently today.
- **Fillet weld fracture and macro test** Determines soundness; ASME test for procedure qualification and welder qualification.

General Requirements

The material that follows will give the student the information needed to understand the specifications for the various test specimens, their preparation, and method of testing.

In general, all codes require essentially the same qualifying procedures for plate or pipe. Each position of welding, type of joint, and weld has a designated number for identification. Generally face-, root-, and side-bend test specimens are required for groove welds in plate or pipe, and T-joint break or macroetch test specimens are required for fillet welds in plate. The required number and type of test specimens vary with the thickness of material.

Under most welding codes, the tests remain in effect indefinitely unless:

- The welder does not work with the qualified process for a period of more than six months. Requalification is required only on 3/8-inch thick plate.
- There is reason to be dissatisfied with the work of the welder. If a welder fails their test, they may retest as follows:
  An immediate retest consists of two test welds of each type failed. All test specimens must pass.
  A complete retest may be made if the welder has had further training or practice since the last test.

Procedure Qualification Tests These tests are conducted for the purpose of determining the correctness of the welding method. The American Welding Society and various code authorities have established standard procedures for welding various types of structures. The welding procedure test should cover base metal specifications, filler metals, joint preparation, position of welding, and welding process, techniques and characteristics, including current setting, electrode size and electrode manipulations, preheat, interpass and postheat temperatures. The close adherence to the procedures established will ensure uniform results.

The welding procedure can be thought of as analogous to a chef and the use of a recipe. Once a good recipe is produced, a skilled chef should be able to follow it and produce a consistent dish. If a skilled welder follows the welding procedure specification (WPS) (the recipe), a consistent high quality weldment should be produced.

Following are the procedure qualification requirements for different types of welds.

Groove weld The specimens required for groove welds are given in Table 28-9. See Fig. 28-59. In most cases groove weld tests qualify the welding procedure for use with both groove and fillet welds within the range tests.

Fillet welds Longitudinal shear test specimens are welded as shown in Fig. 28-60, page 922, and prepared for testing as shown in Fig. 28-61, page 922. Transverse shear test specimens are made as shown in Fig. 28-62, page 922.

The test weld for the bend and soundness tests is made as shown in Fig. 28-63, page 923. Two root-bend test specimens are taken from the test weld which are prepared for testing as shown in Fig. 28-63. Two specimens for the fillet weld soundness test are also taken from the test weld and prepared as shown in Fig. 28-64, page 924. Table 28-10 (p. 924) covers the requirement for a fillet weld soundness test for WPS qualification.

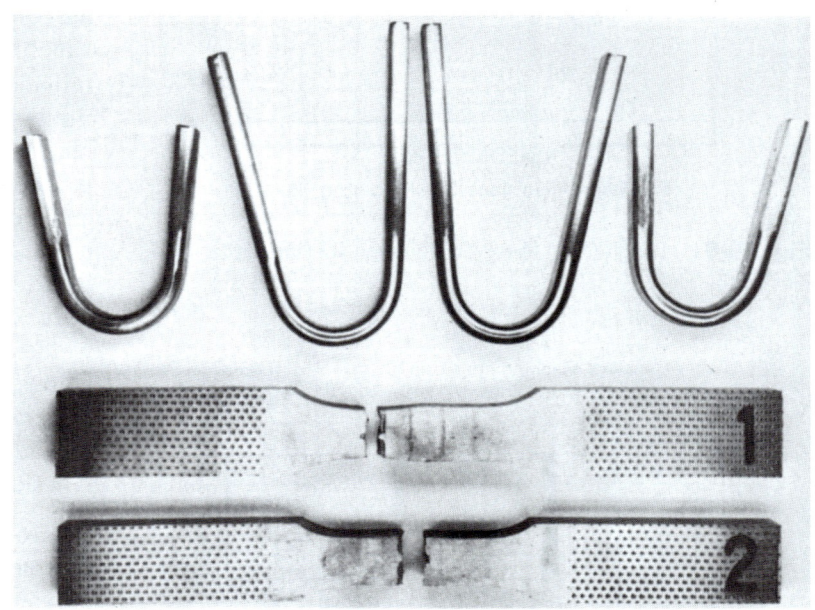

Fig. 28-59 Test specimens for ASME procedure qualification: 2 tensile, 2 face-, and 2 root-bend specimens. Note that the tensile specimens are pulled until broken. © Nooter Corp.

Table 28-9 Typical WPS Qualification Requirement for Complete Joint Penetration Groove Welds—Number and Type of Test Specimens and Range of Thickness Qualified

Nominal Plate Thickness (T) Tested (in.)	Number of Specimens				Nominal Plate, Pipe, or Tube Thickness Qualified (in.)	
	Reduced Section Tensile	Root Bend	Face Bend	Side Bend	Minimum	Maximum
≤ 1/8 ≤ T ≤ 3/8	2	2	2	—	1/8	2T
3/8 < T < 1	2	—	—	4	1/8	2T
1 and over	2	—	—	4	1/8	Unlimited

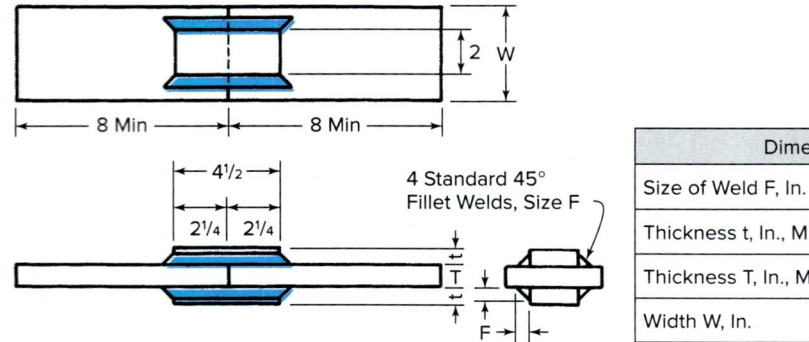

Fig. 28-60 Longitudinal fillet weld shearing specimen after welding. Source: CRC-Evans Pipeline International, Inc.

Dimensions				
Size of Weld F, In.	1/8	1/4	3/8	1/2
Thickness t, In., Min.	3/8	1/2	3/4	1
Thickness T, In., Min.	3/8	3/4	1	1 1/4
Width W, In.	3	3	3	3 1/2

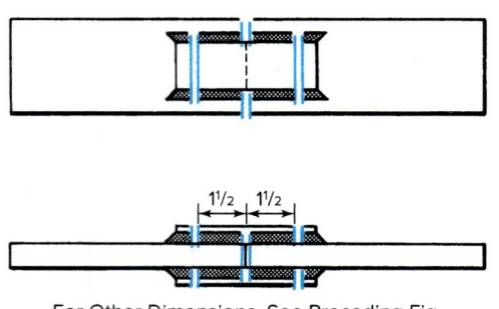

For Other Dimensions, See Preceding Fig.

Fig. 28-61 Longitudinal fillet weld shearing specimen after machining.

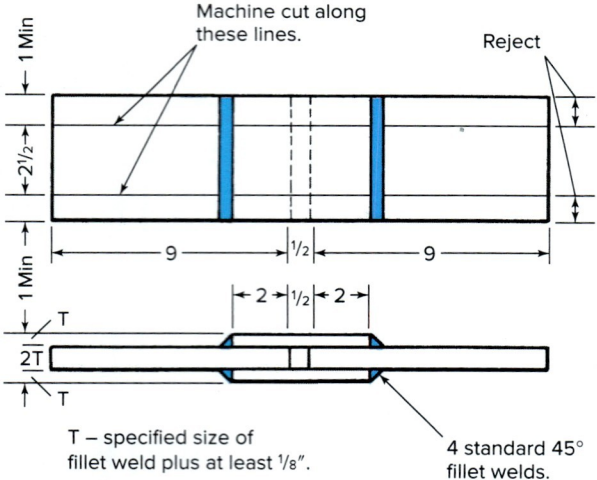

Fig. 28-62 Transverse fillet weld shearing specimen.

Welder Qualification Tests (Performance)

This is also referred to as a *performance qualification test*. The test is conducted for the purpose of determining whether the welder has the knowledge and skill to make sound welds and to follow and apply a procedure of welding as determined for the class of work at hand. Most of the test specimens are similar to those used for the procedure qualification test. However, not as many are required, and the preparation of the test specimen is simplified. The cost of testing is reduced and the determination of weld soundness is achieved. There are time limits imposed by specific codes on how long a welder remains qualified. With some codes like the AWS D1.1, it is indefinitely if the welder is using the welding process and procedures continuously. If six months passed between use or if the welder's ability is being questioned they will have to take a requalification test.

It is the general practice for code welding to qualify welders on the groove weld tests in the 3G and 4G positions, Table 28-11 (p. 925). Passing of these tests also permits the welder to weld plate in all positions. Following are the welder qualification requirements for different types of welds.

Groove Welds Table 28-12 (p. 925) covers the various complete joint penetration groove weld tests. This is based on various plate thicknesses. This will directly relate to the type and number of specimens to be tested and the thickness of materials that can be welded in production.

Fillet Welds There are two different kinds of fillet weld test specimens to serve a variety of purposes.

Many companies use the fillet weld test procedure as represented by Fig. 28-63. Welds are made in each position for which the welder or welding operator is to be qualified. Two root-bend tests are made.

Another form of fillet weld test that may be used for welders or welding operators is shown in Fig. 28-65 (p. 926). Test specimens must be subjected to a fracture test, Fig. 28-66, page 926, and etched. Table 28-13 (p. 926) covers the fillet bend and break tests. The number, type of test, and plate thickness that can be typically welded in production are also covered.

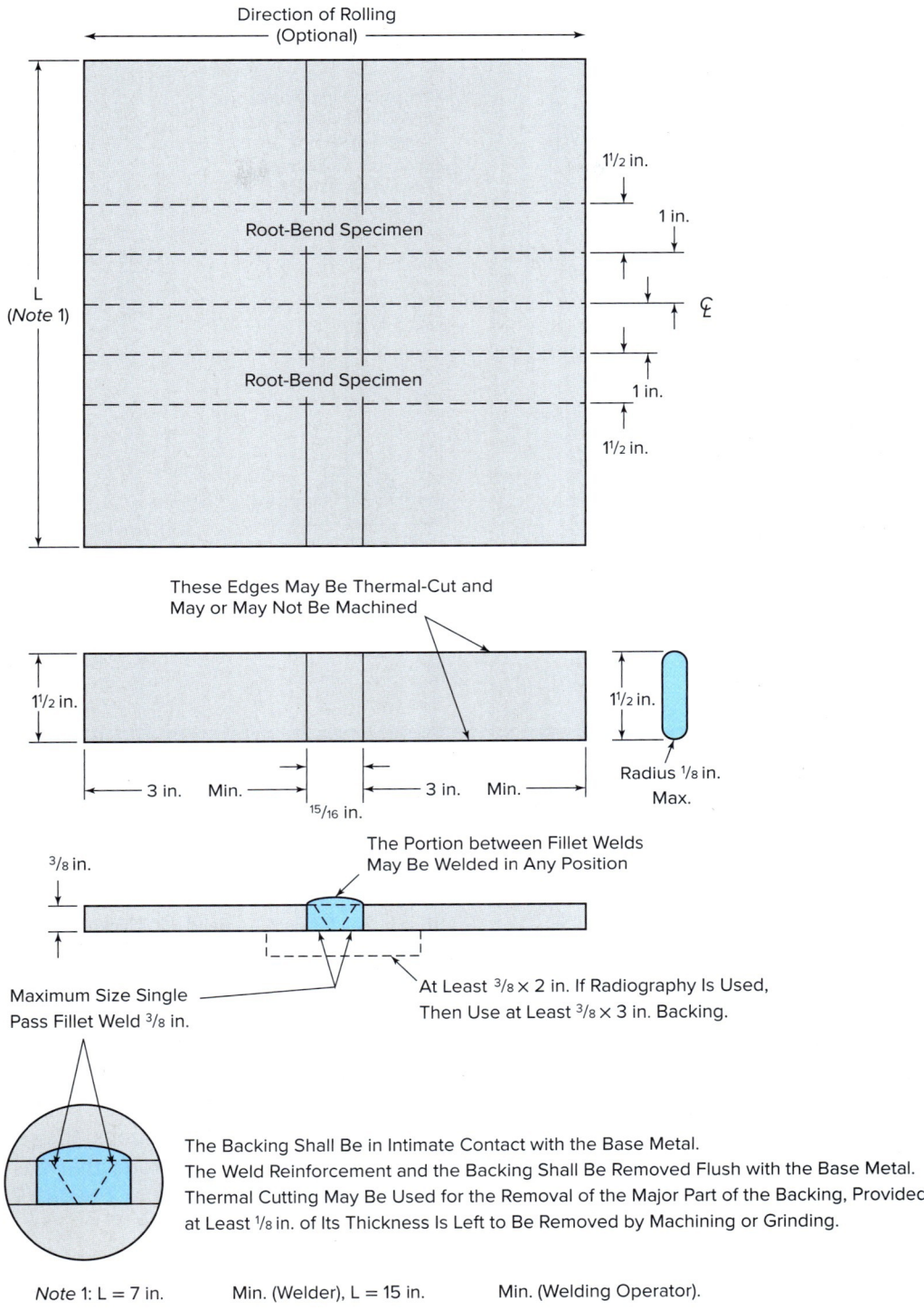

Fig. 28-63 Fillet weld root-bend test plate—welder or welding operator qualification.
American Welding Society (AWS) Structural Welding Code Steel D1.1/D1.1M:2010, Fig 4.32, p. 145.

Preparation of Test Specimens The material that follows will give the student the information needed to understand the nature of the various test specimens, their preparation, use, method of testing, and results required for passing. A word of caution to the student regarding the welding of a test specimen: While these tests are designed to determine the capability of welders, many welders have failed them for reasons not related

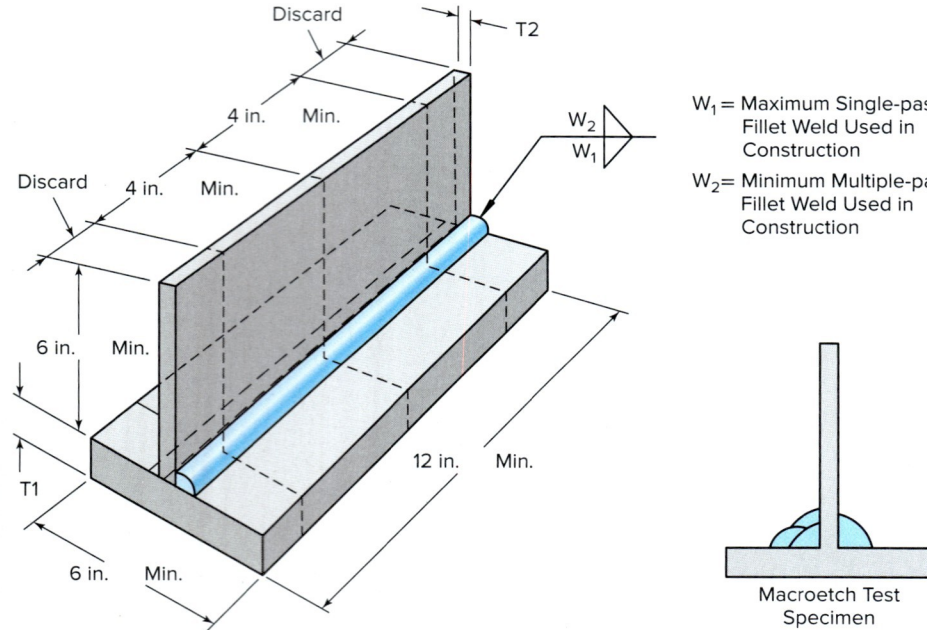

Fig. 28-64 Fillet weld soundness test for WPS qualification. American Welding Society (AWS) Structural Welding Code Steel D1.1/D1.1M:2010, Fig 4.19, p. 130.

Inches		
Weld Size	T1 Min*	T2 Min*
3/16	1/2	3/16
1/4	3/4	1/4
5/16	1	5/16
3/8	1	3/8
1/2	1	1/2
5/8	1	5/8
3/4	1	3/4
>3/4	1	1

*Where the maximum plate thickness used in production is less than the value shown in the table, the maximum thickness of the production pieces may be substituted for T1 and T2.

Table 28-10 Typical Number and Type of Test Specimens and Range of Thickness Qualified for WPS Qualification: Fillet Welds

		Test Specimens Required				Sizes Qualified	
Test Specimen	Fillet Size	Number of Welds per WPS	Macroetch	All-Weld Metal Tension	Side Bend	Plate Thickness	Fillet Size
Plate T-test	Single pass, max. size to be used in construction	1 in each position to be used	3 faces	—	—	Unlimited	Max. tested single pass and smaller
	Multiple pass, min. size to be used in construction	1 in each position to be used	3 faces	—	—	Unlimited	Min. tested multiple pass and larger

to their welding ability. This is due principally to carelessness in the application of the weld and in the preparation of the test unit and the test specimen.

Selecting and Preparing Plates It is important that both the test plate and backup strip (where it is used) are weldable, ductile low carbon steel. The test is designed so that both the plate and the weld bend and stretch during the test. If the plate has tensile strength much above that of the weld metal, it will not stretch at all and will force the weld metal to stretch far beyond its yield point and fail. The test specimens must be removed so that they are bent with the internal grain structure, not across it.

Welding the Plates Proper electrode selection is the first step in producing a sound weld. Since plates are generally welded in all positions, they should be welded with an all-position electrode that has good ductility. Electrodes should be in the E6010, E6011, or E7018 classifications.

The most important part of the entire job is the first pass in the groove welded butt joint, and the root pass in the fillet test. Every effort should be made to get especially good penetration, fusion, and sound weld metal on these root beads. Welders should not let their desire to have a bead of good appearance interfere with the objective of securing good penetration and fusion. These weld characteristics are the difference between a satisfactory weld and failure.

No preheat or postheat treatment is permissible to pass the test. It is helpful to keep the plates fairly hot and allow

Table 28-11 Typical Welder Qualification—Production Welding Positions Qualified by Plate or Pipe Tests

Structural Member Tested on	Qualification Test Weld Type	Qualification Test Positions	Production Plate Capable of Being Welded Groove CJP	Production Plate Capable of Being Welded Groove PJP	Production Plate Capable of Being Welded Fillet	Production Pipe Butt Groove CJP	Production Pipe Butt Groove PJP	Production Pipe T—K—Y Groove CJP	Production Pipe T—K—Y Groove PJP	Production Pipe Fillet
Plate	Groove	1G 2G 3G 4G 3G and 4G	F F, H F, H, V F, OH All	F F, H F, H, V F, OH All	F, H F, H F, H, V F, OH All	F F, H F, H, V F, OH All (1)	F F, H F, H, V F, OH All (1)	Does not qualify	F F, H F, H, V F, OH All (1, 2)	F, H F, H F, H, V F, OH All
Plate	Fillet	1F 2F 3F 4F 3F and 4F	Does not qualify	Does not qualify	F F, H F, H, V F, OH All	Does not qualify	Does not qualify	Does not qualify	Does not qualify	F F, H F, H, V F, OH All
Pipe or tubular	Groove (Pipe or box)	1G 2G 5G 6G 2G and 5G	F F, H F, V, OH All All	F F, H F, V, OH All All	F, H F, H F, V, OH All All	F F, H F, V, OH All All (3)	F F, H F, V, OH All All (3)		F F, H F, V, OH All All (2, 3)	F F, H F, V, OH All All
Pipe or tubular	Groove (Pipe or box)	6GR Circular	All	All	All	All (3, 4)	All (3)	All (2, 3)	All (2, 3)	All
Pipe or tubular	Groove (Pipe or box)	6GR Circular and square	All	All	All	All (3, 4)	All (3)	All (2, 3)	All (2, 3)	All
Pipe or tubular	Pipe fillet	1F rotated 2F 2F rotated 4F 5F	Does not qualify	Does not qualify	F F, H F, H F, H, OH All	Does not qualify	Does not qualify	Does not qualify	Does not qualify	F F, H F, H F, H, OH All

[1]Only qualified for pipe over 24 in. in diameter welded with backing, back-gouging, or both.
[2]Not qualified for welds having a groove angle less than 30°.
[3]Qualification using box tubing also qualifies for pipe over 24 in. in diameter.
[4]Not qualified for joints welded from one side without backing, or welded from two sides without back-gouging.

Table 28-12 Typical Welder and Welding Operator Qualification—Number of Specimens, Type of Specimen, and Range of Thickness Qualified for CJP Groove Welds

Plate Thickness of Production Groove Weld Tests	Number of Specimens[1] Face Bend[2]	Number of Specimens[1] Root Bend[2]	Number of Specimens[1] Side Bend[2]	Number of Specimens[1] Macroetch	Plate Thickness Qualified on for Production Welding Minimum	Plate Thickness Qualified on for Production Welding Maximum
3/8 in. (T)	1	1	—	—	1/8	3/4[3]
3/8 ≤ T < 1	—	—	2/8	—	1/8	2T[3]
1 in. or over	—	—	2	—	1/8	Unlimited[3]

[1]All welds will be visually inspected prior to conducting any subsequent tests.
[2]Radiographic inspection of the test plate, pipe, or tubing may be made instead of conducting bend test for welder or welding operator qualification.
[3]Also qualifies for welding any fillet or PJP weld size on any thickness of plate, pipe, or tubing.

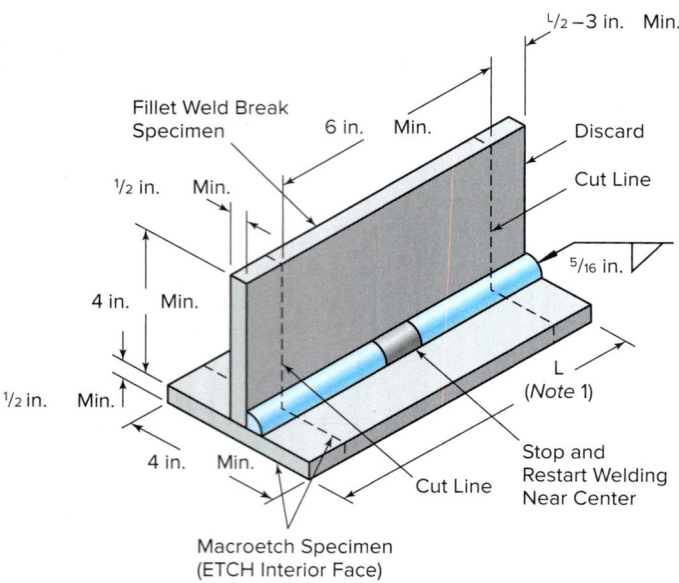

Fig. 28-66 Method of rupturing fillet weld break specimen.

Fig. 28-65 Fillet weld break and macroetch test plate—welder and welding operator qualification. American Welding Society (AWS) Structural Welding Code Steel D1.1/D1.1M:2010, Fig 4.37, p. 189.

Notes:
1. L = 8 in. Min. (Welder), 15 in. Min. (Welding Operator).
2. Either end may be used for the required macroetch specimen. The other end may be discarded.

them to cool slowly after the welding is completed. The welder should not, under any conditions, quench the plates in cold water or in any other way accelerate the cooling rate.

Finishing the Specimen It is important that care be taken in finishing the specimen because poor finishing can cause a sound weld to fail the test. As shown in Fig. 28-67, all grinding and machining marks must be lengthwise on the sample. Otherwise they produce a notch effect that may cause failure. The smoother the finish, the better the chance of passing the test. Even a slight nick across the sample may open up under the severe stress of the test. Unless there is a distinct lack of penetration, the surface should be ground or machined until the entire bend area cleans up, leaving no low or irregular spots.

Any weld reinforcement either on the face or on the root side must be removed, Fig. 28-68. The edges of the specimen should have a smooth $\frac{1}{8}$-inch radius, which can be done with a file. When grinding specimens, do not quench them in water when hot. This may create small surface cracks that become larger during the bend test.

After the test specimens have been bent, the outside surface of the bend-test specimen shall be visually examined for surface discontinuities. This would be the root side for root bends, and the face side for face bends. To be acceptable, the surface shall contain no discontinuities exceeding the following dimensions:

- $\frac{1}{8}$ inch as measured in any direction on the surface.
- $\frac{3}{8}$ inch as the sum of all the discontinuities exceeding $\frac{1}{32}$ inch but less than $\frac{1}{8}$ inch.
- $\frac{1}{4}$-inch maximum corner crack, except when the corner crack results from visible slag inclusion or other fusion-type discontinuities, in which case the $\frac{1}{8}$-inch maximum size shall apply.
- If a corner crack exceeds $\frac{1}{4}$ inch and there was no evidence of slag or fusion-type discontinuities, then it shall be discarded and a replacement taken from the original test plate.

Table 28-13 Typical Welder and Welding Operator Qualification—Number of Specimens, Type of Specimen, and Range of Thickness Qualified for Macroetch, Bend, and Break Tests on Fillet Welds

Type of Test Weld[1]	Nominal Test Plate Thickness T (in.)	Fillet Weld Break Test	Macroetch	Side Bend[2]	Root Bend[2]	Face Bend[2]	Plate Thickness Qualified on for Production Welding	
							Minimum	Maximum
Fillet (Fig. 28-35)	$\frac{1}{2}$	1	1	—	—	—	$\frac{1}{8}$ in.	Unlimited
Fillet (Fig. 28-29)	$\frac{3}{8}$	—	—	—	2	—	$\frac{1}{8}$ in.	Unlimited

[1]All welds will be visually inspected prior to conducting any subsequent tests.
[2]Radiographic inspection of the test plate, pipe, or tubing may be made instead of conducting bend tests for welder or welding operator qualification.

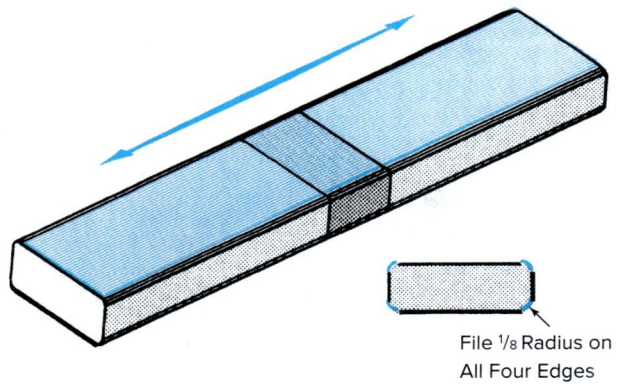

Grinding marks must parallel specimen edges. Otherwise they may have a notch effect which could cause failure.

Fig. 28-67 Longitudinal grinding and rounded edges give the welders a fair chance and reduce failures due to causes beyond their control.

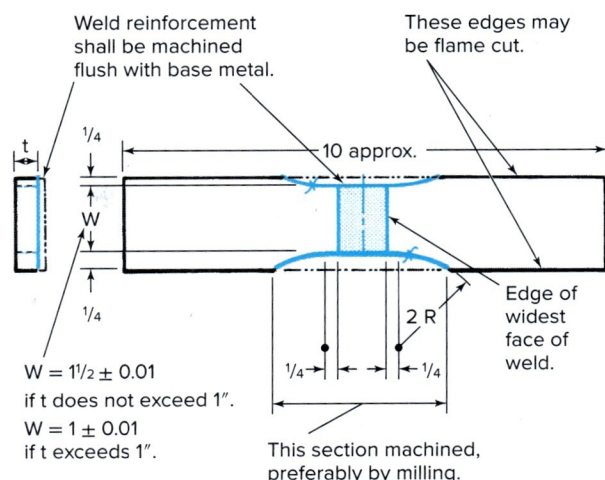

Fig. 28-69 Tension specimen (plate).

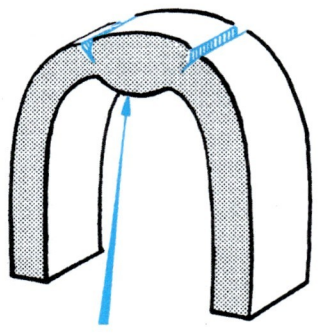

If weld reinforcement is not removed before root bend test, stretching is concentrated in two places and failure results.

Fig. 28-68 AWS specifies that bead reinforcement be removed by grinding or machining. The drawing shows why.

Groove Weld Soundness Tests

Reduced-Section Tension Test

Purpose The reduced-section tension test determines the tensile strength of the weld metal.

It is used only for procedure qualification tests, but the test procedure gives spectacular results and can be used to test welders in school. The pulling and breaking of the specimen is of great interest to students.

The test is suitable for butt joints in plate or in pipe.

Usual Size and Shape of Specimens Refer to Figs. 28-69 through 28-71, page 928.

Method of Testing The test is made by subjecting the specimen to a longitudinal load great enough to break it or pull it apart, Fig. 28-59, page 921. This is usually accomplished by means of a tensile-testing machine designed for this purpose, Fig. 28-72, page 928. Before testing, the least width and corresponding thickness of the reduced section is measured in inches. The specimen is ruptured under tensile load and the maximum load in pounds is determined. The cross-sectional area is obtained as follows:

Cross-sectional area = width × thickness

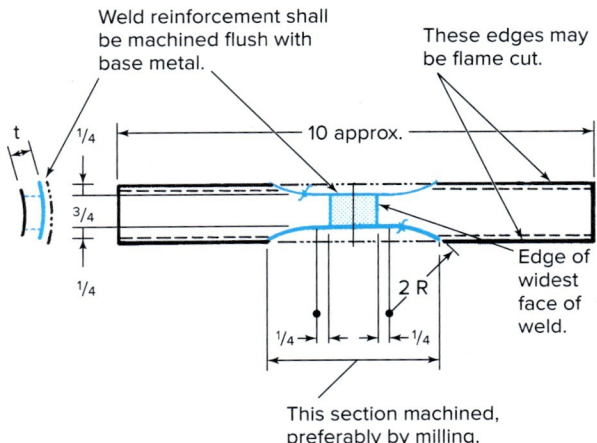

Fig. 28-70 Tension specimen (pipe) exceeding 2 inches in diameter.

The tensile strength in pounds per square inch is obtained by dividing the maximum load by the cross-sectional area.

On jobs for which a quick test is desired and for which the testing agent is interested only in determining whether the joint is stronger than the plate, the weld reinforcement is not removed and the specimen is pulled until failure occurs. No calculations are made when this type of test is done.

Usual Test Results Required The specimen shall have a tensile strength equal to or greater than:

- The minimum specified tensile strength of the base material.
- The lower of minimum specified tensile strengths of dissimilar materials.

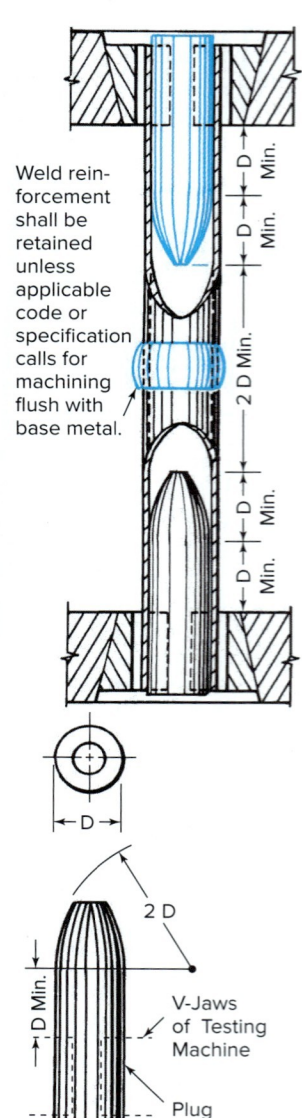

Fig. 28-71 Tension specimen (pipe) not exceeding 2 inches in diameter.

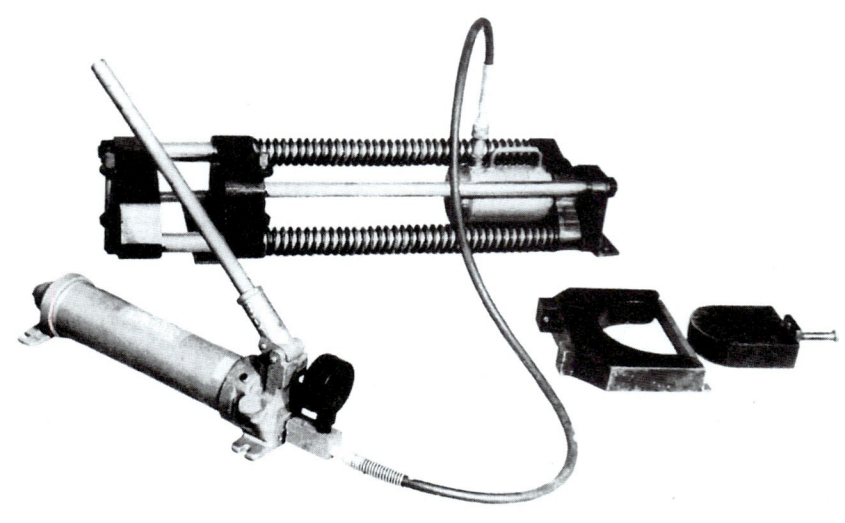

Fig. 28-72 Tensile and guided-bend testing machine. It is portable and has a range of 60,000 to 225,000 p.s.i. A large, direct-reading dial gauge shows the amount of pressure applied. © CRC-Evans Pipeline International, Inc.

- The specified tensile strength of the weld metal if the weld metal is of a lower strength than the base metal.
- Five percent below the specified minimum tensile strength of the base metal if the specimen breaks in the base metal outside the weld.

Root-, Face-, and Side-Bend Soundness Tests

Purpose These tests are very common and are for the purpose of revealing lack of soundness, penetration, and fusion in the weld metal. They are procedure and welder qualification tests that are applied to groove welds in both plate and pipe. The face-bend test checks the quality of fusion to the side walls and the face of the weld joint, porosity, slag inclusion, and other defects. It also measures the ductility of the weld. The root-bend test checks the penetration and fusion throughout the root of the joint. The side-bend test also checks for soundness and fusion. Location of where these bend test specimens are to be taken can be found in the code being used.

Usual Size and Shape of Specimens Refer to Figs. 28-73 through 28-75, page 930.

Method of Testing The tests are rather severe, and the ability to make welds that can meet these test requirements is a mark of good welding ability. These tests lend themselves to school and shop applications. The testing equipment needed is inexpensive and can be built in the school shop. See Figs. 28-76 through 28-78 (pp. 930–931) for three designs that students find easy to work with in testing their weld specimens.

Each specimen is bent in a jig having the contour and other features shown in Fig. 28-76. You will note that there is a size for plate and pipe. Any convenient means—manual, mechanical, electrical, or hydraulic—may be used for moving the male member in relation to the female member.

The specimen is placed on the female member of the jig with the weld at midspan. Face-bend specimens are placed with the face of the weld directed toward the gap. Root-bend specimens are placed with the root of the weld directed toward the gap. Side-bend specimens are placed with the side showing the greater discontinuity, if any, directed toward the gap. The two members of the jig are forced together until the curvature of the specimen is such that a $1/32$-inch diameter wire cannot be passed between the curved portion of the male member and the specimen. Then the specimen is removed from the jig.

Usual Test Results Required The convex surface of the specimen is visually examined for the appearance of cracks or other discontinuities. This would be the root side for root bends, the face side for face bends, and the side for

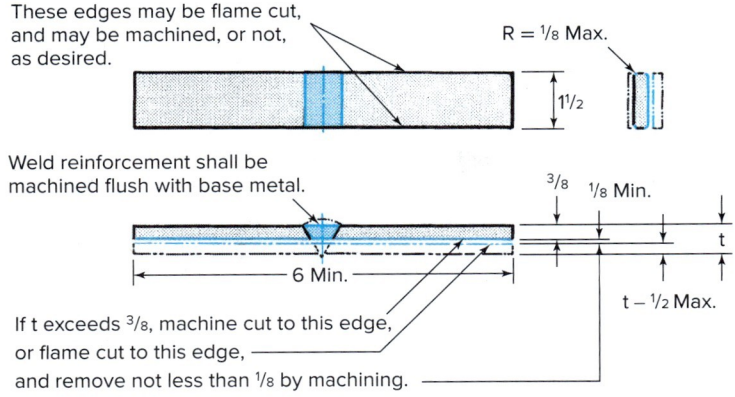

Fig. 28-73 Face- and root-bend specimens (plate).

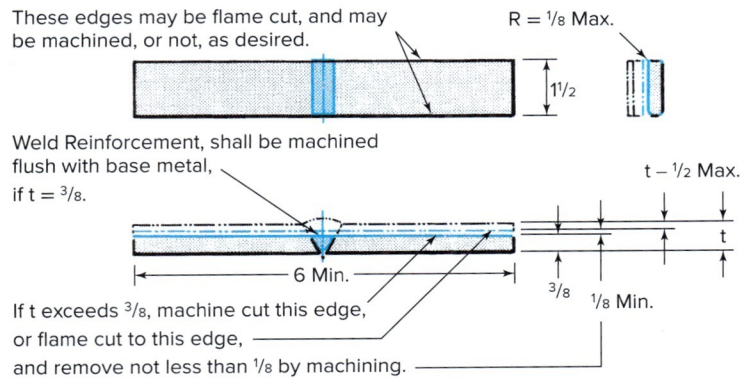

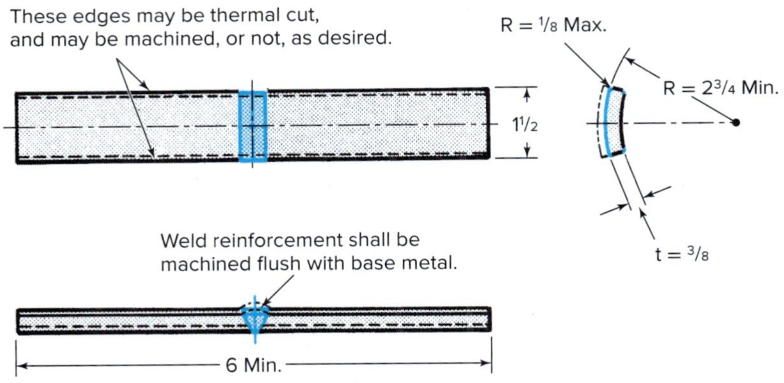

Fig. 28-74 Face- and root-bend specimens (pipe).

side bends. To be acceptable, the surface shall contain no discontinuities exceeding these dimensions:

- ⅛ inch as measured in any direction on the surface.
- ⅜ inch as the sum of all the discontinuities exceeding ¹⁄₃₂ inch but less than ⅛ inch.

- ¼-inch maximum corner crack, except when the corner crack results from visible slag inclusion or other fusion-type discontinuities, in which case the ⅛-inch maximum size shall apply.
- If a corner crack exceeds ¼ inch and there was no evidence of slag or fusion-type discontinuities, then it shall be discarded and a replacement taken from the original test plate.

Consult Figs. 28-79 and 28-80, page 932, for a comparison of satisfactory and defective specimens.

Nick-Break Test

Purpose The nick-break test is for the purpose of determining the soundness of the weld.

Usual Size and Shape of Specimens Refer to Figs. 28-81 and 28-82, page 932.

Method of Testing The weld reinforcement is not removed from the specimen to be tested. The specimen is notched in the sides by cutting it with a saw. Then it is supported and struck with quick, sharp blows by a hammer or heavy weight, Fig. 28-83 (p. 932). This causes a failure to occur between the saw cuts. The weld metal is examined for defects such as slag and oxide inclusions, porosity, and incomplete fusion, Fig. 28-84 (p. 933).

Base Metal Yield Strength	A	R_A	B	R_B
≤50 ksi	1½	¾	2⅜	1³⁄₁₆
>50 ≤ 90 ksi	2	1	2⅞	1⁷⁄₁₆
>90 ksi	2½	1¼	3⅜	1¹¹⁄₁₆

Usual Test Results Required The requirements for passing this test are that the fractured surface does not have any discontinuities exceeding these limits. The greatest dimension of any porosity pore shall not exceed ¹⁄₁₆ inch, and the combined area of all porosity shall not exceed 2 percent of the exposed surface area. Slag inclusions shall not be more than ¹⁄₃₂ inch in depth and shall not be more than ⅛ inch or one-half the nominal thickness in length, whichever is smaller. No incomplete fusion is allowed. There shall be at least ½-inch separation between adjacent discontinuities. See Fig. 28-85, page 933, for where these measurements are made. A

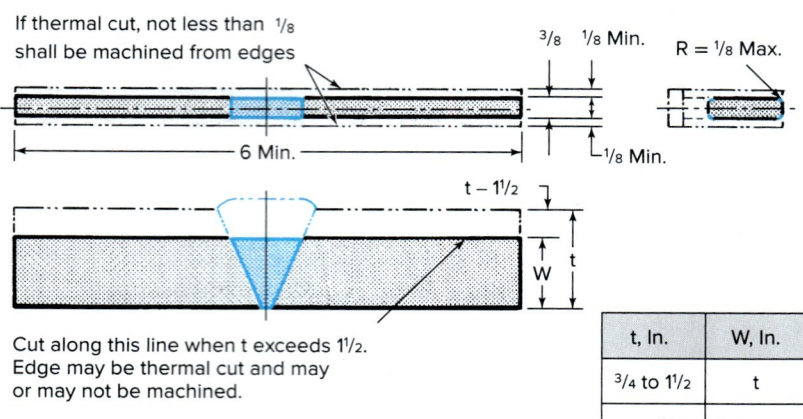

Fig. 28-75 Side-bend specimen.

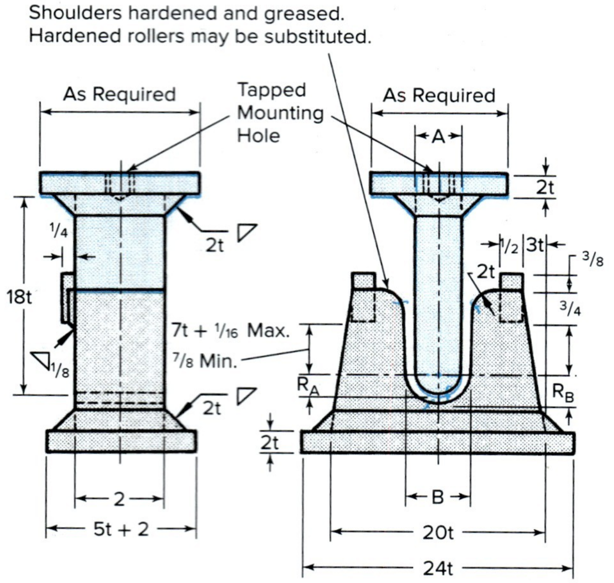

Fig. 28-76 Jig for guided-bend test used to qualify welders for work done under AWS and various other specifications. It can be made in the school shop.

fisheye per AWS 3.0 Terms and Definitions is "A discontinuity found on the fracture surface of a weld in steel that consists of a small pore or inclusion surrounded by an approximately round, bright area." Fisheyes are not to be considered a cause for rejection.

Fillet Weld Soundness Tests

Longitudinal and Transverse Shear Tests

Purpose These tests determine the shearing strength of fillet welds. They are ordinarily used for procedure qualifications.

Usual Size and Shape of Specimens Refer to Figs. 28-80 through 28-82, page 932, for standard specifications. Figures 28-86 and 28-87, page 933, show the prepared longitudinal weld specimens.

Method of Testing The specimen is ruptured by pulling in a tensile testing machine, and the maximum load in pounds is determined, Fig. 28-88, page 933.

a. *Transverse Welds.* The shearing strength of transverse fillet welds is determined in the following manner. Before pulling the specimen, the width of the specimen is measured in inches, and the size of the fillet welds is recorded. The shearing strength of the welds in pounds per *linear* inch is obtained by dividing the maximum force by twice the width of the specimen.

$$\text{Shearing strength (lb/in.)} = \frac{\text{maximum load}}{2 \times \text{width of specimen}}$$

The shearing strength of the welds in pounds per *square* inch is obtained by dividing the shearing strength in pounds per linear inch by the average theoretical throat dimension of the welds in inches. The theoretical throat dimension may be obtained by multiplying the size of the fillet weld by 0.707.

$$\text{Shearing strength (p.s.i.)} = \frac{\text{shearing strength (lb/in.)}}{\text{throat dimension of weld}}$$

b. *Longitudinal Welds.* The shearing strength of longitudinal fillet welds, Figs. 28-86 and 28-87, page 933, is determined in the following manner. Before pulling the specimens, the length of each weld is measured in inches. The shearing strength of the welds in pounds per *linear* inch is obtained by dividing the maximum force by the sum of the lengths of the welds that ruptured.

$$\text{Shearing strength (lb/in.)} = \frac{\text{maximum load}}{\text{sum of lengths}}$$

Usual Test Results Required For the longitudinal shear test specimen, the shearing strength of the welds in pounds per square inch cannot be less than two-thirds of the minimum specified tensile range of the base material. For the

930   Chapter 28   Joint Design, Testing, and Inspection

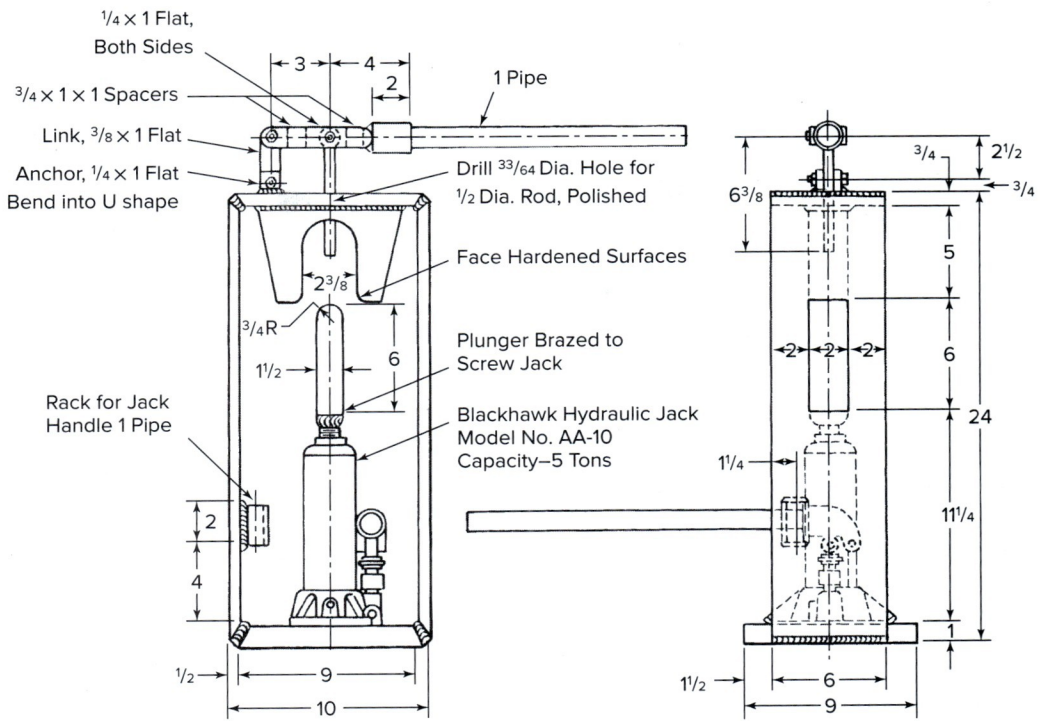

Fig. 28-77 A school-built guided-bend tester.

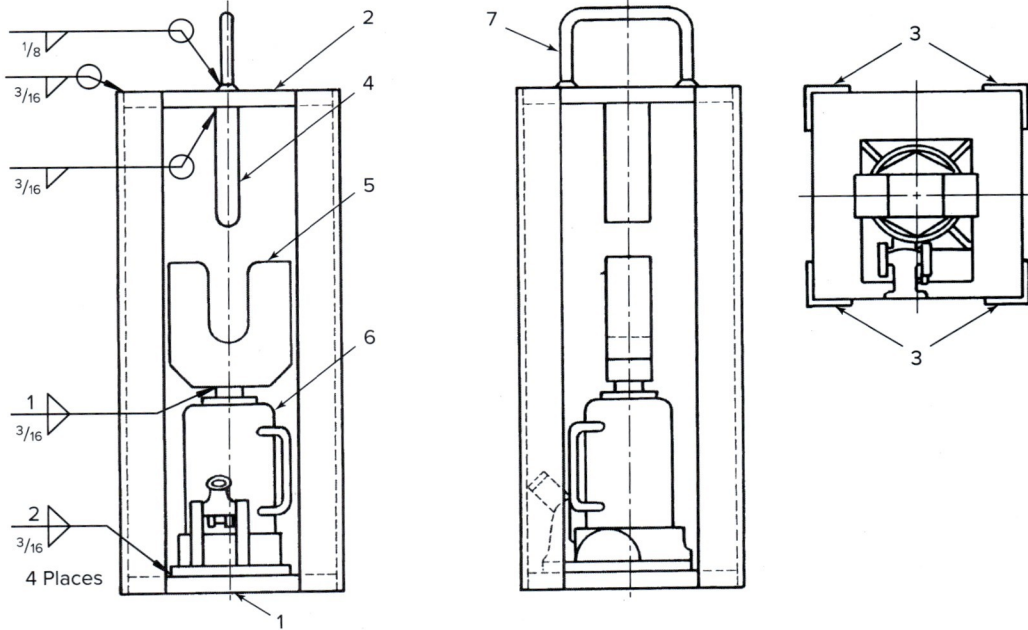

Fig. 28-78 A school-built guided-bend tester.

transverse shear test specimens, the shearing strength of the welds in pounds per square inch cannot be less than seven-eighths of the minimum specified tensile range of the base material.

Fillet-Weld Break Test

Purpose This test is for the purpose of determining the soundness of a fillet weld.

Fig. 28-79 Face-bend specimens. The specimen on the right withstood the test satisfactorily whereas the specimen on the left cracked and failed before the bending could be completed. Edward R. Bohnart

Fig. 28-80 Root-bend specimens. The specimen on the right withstood the test satisfactorily whereas the specimen on the left cracked and failed before the bending could be completed. Note the full penetration in the specimen that passed. Edward R. Bohnart

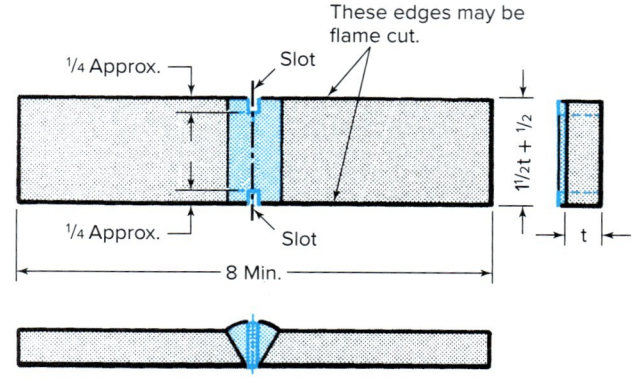

Fig. 28-81 Nick-break specimen (plate).

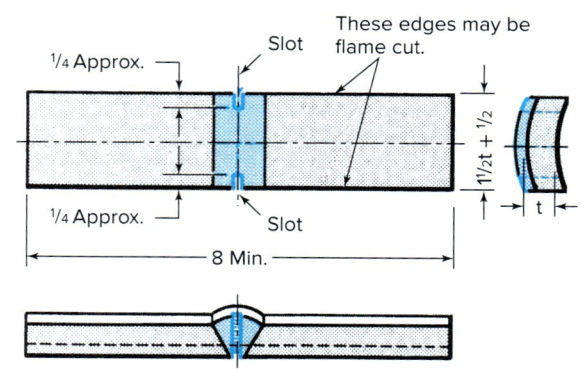

Fig. 28-82 Nick-break specimen (pipe).

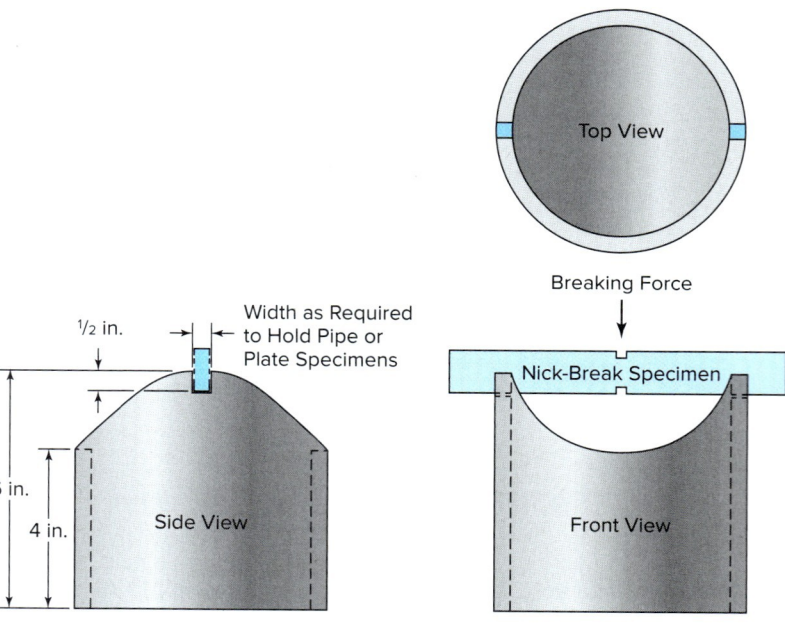

Fig. 28-83 A school-built stand for the rupturing nick-break specimens. It is made out of a piece of 6-inch schedule 40 carbon steel pipe.

There are several different types of fillet weld tests. One is the form recommended by the American Welding Society. It is a quick, easy method of testing for welder qualification and is widely used, especially by schools and industries that are doing fillet welds to code work.

Method of Testing (AWS Form, Fig. 28-65) The specimen is formed by fillet welding one flat plate or bar at right angles to another in the form of a T-joint. The entire length of the fillet weld shall be visually examined prior to conducting the fillet weld break test. A 6-inch long specimen is removed with at least one start and stop located within the test specimen. The specimen is then fractured by application of pressure from a press, testing machine, or hammer. The fillet weld is fractured at the root, Fig. 28-66, page 927. The fractured weld metal is then examined for defects such as slag and oxide inclusions, porosity, fusion, incomplete root penetration, and uneven distribution of weld metal. If the specimen does not break, increase the load and/or repeat loading. However, if it still does not break, it shall be considered acceptable if it bends flat upon itself.

Usual Test Results Required (AWS) The weld must fracture through the throat of the weld, Fig. 28-89. It should have complete penetration to the root of

Fig. 28-84 A nick-break specimen after it has been broken for inspection of the weld deposit. A sound weld is free of slag and oxide inclusions and gas porosity and shows complete fusion. Edward R. Bohnart

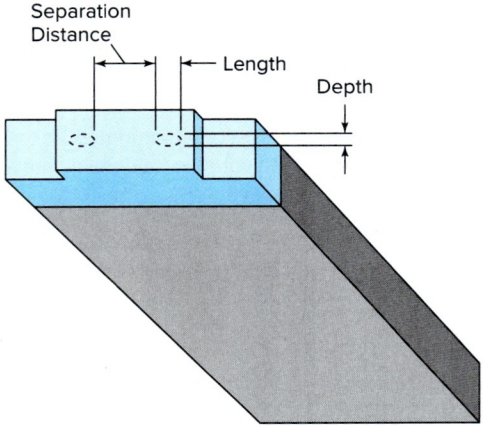

Fig. 28-85 How to measure discontinuities on the broken surface of a nick-break specimen.

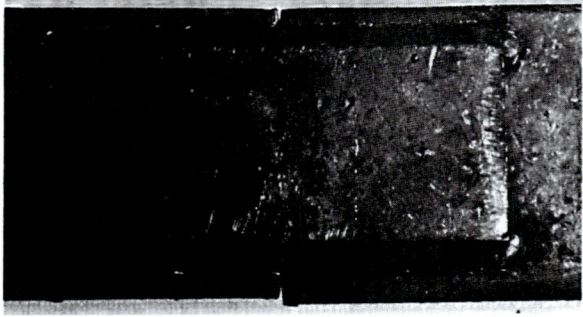

Fig. 28-86 Top side of a longitudinal fillet weld specimen. Edward R. Bohnart

Fig. 28-87 Side view of a longitudinal fillet weld specimen. Edward R. Bohnart

Fig. 28-88 The bottom photograph shows the construction of a transverse fillet weld specimen. The top photograph shows a specimen after it has been ruptured by pulling. Note that the specimen withstood a tensile pull of 81,660 pounds. Edward R. Bohnart

Fig. 28-89 A ruptured fillet weld break-test specimen. Note that the weld broke through the center. This indicates a weld of even proportion and penetration. Edward R. Bohnart

the joint and no inclusions. The visual examination prior to conducting the break test shall show:
- Uniform appearance
- Uniform weld size, with fillet weld size not varying over ⅛ inch
- No overlap
- No cracks
- No undercut greater than 0.010 inch
- No visible surface porosity

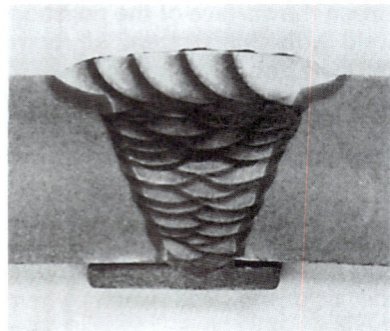

Fig. 28-90 Etched chromium steel specimen showing multiple-pass groove welds in a single-V butt joint. Note the weld metal, HAZ, and unaffected base metal. © Nooter Corp.

Upon passing the visual examination of the fillet weld, the fracture test can be conducted. The fillet break test specimen shall pass if:

- The specimen bends flat upon itself.
- The fractured surface shows complete fusion to the start of the joint root.
- No inclusions or porosity larger than ³⁄₃₂ inch are measured in any direction on the fractured surface.
- The sum of the acceptable inclusions and porosity do not exceed ³⁄₈ inch in the 6-inch long specimen.

Other Tests

In addition to the destructive test outlined in the foregoing section, the welding industry makes use of such other methods of testing as etching, impact, fatigue, corrosion, and specific gravity testing.

Etching Often a weld must be sectioned and etched to be examined for defects. Etching reveals the penetration and soundness of a weld cross section, Fig. 28-90. The process has the following objectives:

- To determine the soundness of a weld
- To make visible the boundaries between the weld metal and the base metal and between the layers of weld metal
- To determine the location and the depth of penetration of the weld
- To determine the location and number of weld passes
- To examine the metallurgical structure of the heat-affected zone

A deep etch greatly exaggerates defects such as normally harmless small cracks or porosity. Therefore, the surface examination must be made as soon as the weld is

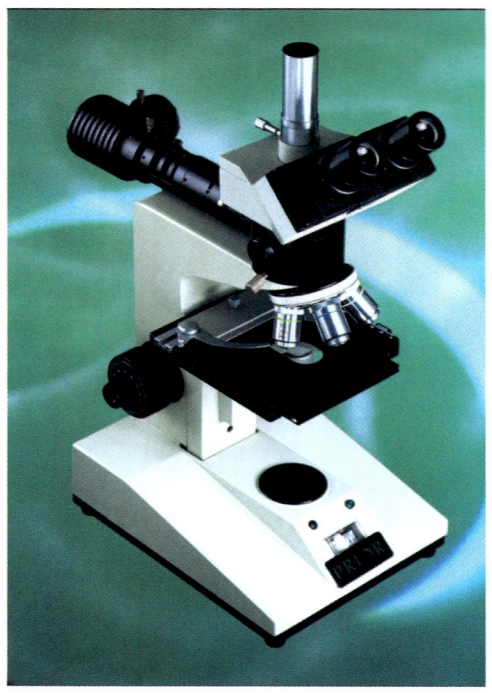

Fig. 28-91 Metallurgical microscope featuring 50× to 1500× magnification range and 20-watt variable halogen illumination for examination of polished weld specimens. © Prior Scientific

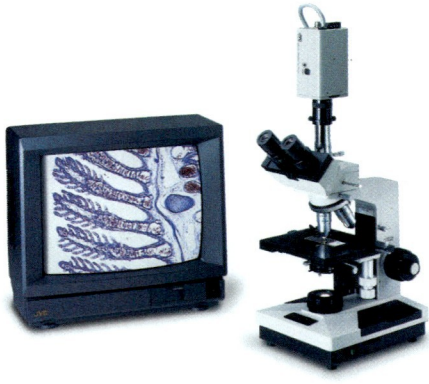

Fig. 28-92 This system consists of color cameras, monitors, power supply, and interconnection cables. Systems such as this greatly aid accuracy and efficiency by presenting larger images that reduce eye fatigue. The images can be captured and stored on film, video tape, and digitally. © Prior Scientific

clearly defined so that over-etching will not destroy the value of the sample. The surface may be preserved with a thin, clear lacquer. It may be inspected with a polarizing microscope, Fig. 28-91, or photographed with a metallograph, Fig. 28-92.

A transverse section is cut from the welded joint to be etched. Cutting with a fine tooth saw is preferred. The face of the weld and the base material should be filed to a smooth surface and then polished with fine emery cloth. The surface is exposed to one of the following etching solutions:

Iodine and Potassium Iodide This solution is made by mixing 1 part of powdered iodine (solid) with 12 parts of a solution of potassium iodide by weight. The resulting solution should consist of 1 part of potassium iodide to 5 parts of water by weight. Brush the surface of the polished weld with this reagent at room temperature.

Nitric Acid This solution is made by mixing 1 part of concentrated nitric acid to 3 parts of water by volume. You must be cautious in the use of nitric acid because it causes bad stains and severe burns. *Always pour the acid into the water when mixing.*

Nitric acid solution may be applied to the surface of the polished weld with a glass stirring rod at room temperature, or the specimen may be immersed in the boiling solution. Make sure that the room is well-ventilated. Nitric acid etches rapidly.

After etching, the specimen should be immediately rinsed in clear hot water. Remove the excess water and immerse the etched surface in ethyl alcohol. Then remove and dry the specimen in a warm air blast. The appearance may be preserved by coating with a thin clear lacquer.

Hydrochloric Acid This etching solution consists of equal parts by volume of concentrated hydrochloric acid, also called *muriatic acid,* and water. The weld specimen is immersed in the solution at a temperature near boiling. Both polished and unpolished surfaces will react to the acid treatment. The solution usually dissolves slag inclusions, thus causing cavities, and enlarges porosity.

You must be cautious in the use of any acid because it causes bad stains and severe burns. *Always pour the acid into the water when mixing.*

Ammonium Persulphate A solution is made by mixing 1 part of solid ammonium persulphate with 9 parts of water by weight. The surface of the polished weld specimen is rubbed vigorously with cotton that has been saturated with this reagent at room temperature.

Safety Concerns The macroetch test (less than 10× magnification) must be done only by authorized personnel in an appropriate, safe environment due to the dangerous chemicals being used. This is a very simple and visual test that greatly aids in your understanding of a sound weld; what a proper weld profile looks like; and the observation of grain structure as found in the weld metal, heat-affected zone, and the unaffected base metal.

Usual Results Required on a Macroetch Fillet Weld Test (AWS) The fillet weld will have complete fusion to the start of the root of the T-joint but not necessarily beyond it. The minimum fillet weld size shall be what is specified. The fillet weld macroetch test shall have:

- No cracks
- No incomplete fusion
- Weld profiles that blend smoothly into base metal
- No convexity or concavity exceeding $1/16$ inch
- No undercut exceeding 0.010 inch
- No porosity or inclusions greater than $1/32$ inch
- No accumulated acceptable porosity or inclusions exceeding $1/4$ inch

Impact Testing The purpose of impact testing is to determine the impact strength of welds and base metal in welded products. The impact strength is the ability of a metal to withstand a sharp, high velocity blow. The test provides the information for comparing the toughness of the weld metal with the base metal. A weld or metal may show high tensile strength and high ductility in a tension stress and yet break if subjected to a sharp, high velocity blow.

There are two standard methods of testing: the *Izod* and the *Charpy* tests. The specimen is broken by a single blow. The impact strength is measured in foot-pounds, which is a unit of work. (*Work* in physics is defined as the product of a force and the distance through which it moves.) The two types of specimens used for these tests and the method of applying the load are shown in Fig. 28-93, page 936. Both tests are made in an impact-testing machine.

The Izod specimen is notched about 1 inch from the end. It is supported between two vise jaws in a vertical position, and the free end above the notch is subject to a swinging blow.

The Charpy specimen is notched in the center. The specimen is supported horizontally between two anvils, and a weight (pendulum) swings against the specimen opposite the notch. The amount of energy in the falling pendulum is known. The distance through which the pendulum swings after breaking the specimen indicates how much of the total energy was used in breaking it. The shorter the distance traveled by the pendulum, the higher the reading on the indicating scale and the tougher the metal. The weaker the metal specimen, the longer the distance traveled by the pendulum and

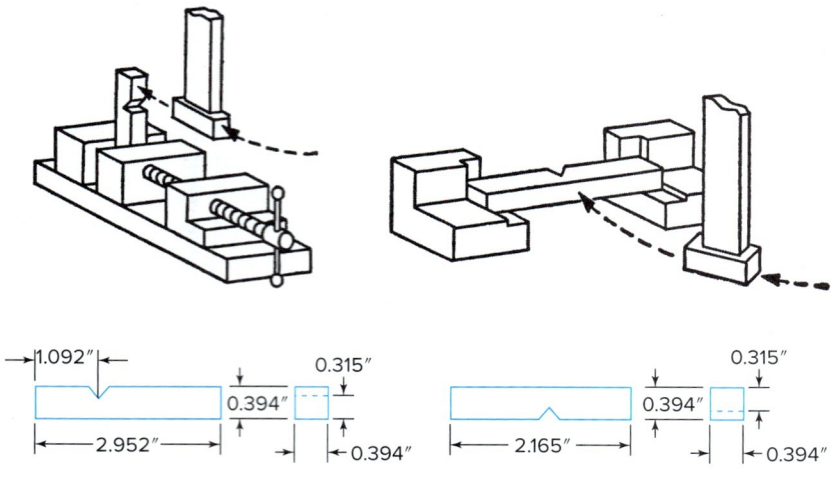

Fig. 28-93 Typical Izod (left) and Charpy (right) impact test specimens and methods of holding and applying the test load. The V-notch specimens (shown) have an included angle of 45° and a bottom radius of 0.010 inch in the notch. Source: The Lincoln Electric Co.

corrosion. Any defect in the weld material shows as a difference in the rate of corrosion when compared to the base metal.

Specific Gravity The specific gravity test is carried out to make sure that no fine porosity exists. It is performed in a laboratory. The test specimen is a cylinder of all-weld metal ⅝ inch in diameter and 2 inches long taken from the weld bead. The weight and volume of the specimen are determined in metric units. The specific gravity is obtained by dividing the weight in grams, by the volume in cubic centimeters. A weld specimen of high quality will have a specific gravity of 7.80 grams per cubic centimeter. The smallest defects materially affect the results. The density of the metal being tested must be considered.

the lower the reading on the indicating scale. The reading on the scale is in foot-pounds of energy absorbed by the metal.

Fatigue Testing Fatigue testing is for the purpose of finding out how well a weld can resist repetitive stress as compared with the base metal. Improperly made welds that contain such defects as porosity, slag inclusions, lack of penetration, cracks, and rough and uneven welds that don't blend smoothly into the base metal, are very low in fatigue strength as compared with the base metal. The improper use of heat in welding also reduces the fatigue resistance of the base metal next to the weld.

Two principal methods of testing are used. In one, the specimen is bent back and forth in a regular fatigue-testing machine. In the second method, the specimen is rotated under load in a testing machine. In each case, the specimen is subject to tension and compression stresses. The number of cycles imposed before failure is recorded. In both methods careful attention is given to duplicating service conditions.

Corrosion Testing The environmental conditions to which many weldments are subject require the use of corrosion-resisting materials such as brass, copper, stainless steels, and other alloy steels. In welding a corrosion-resistant material, great care must be taken to ensure that the weld metal is equal to or better than the base material.

In testing for corrosion resistance, the weld metal and base metal are subjected to similar corrosive conditions, and the materials are compared for their resistance to

Visual Inspection

Visual inspection is probably the most widely used of all inspection methods. It is quick and does not require expensive equipment. A good magnifying glass (10× or less) is recommended. A great deal can be learned from the surface condition of a weld, and a careful evaluation of the appearance can determine the suitability for a given job. The service conditions to which a weldment may be subject must also be considered.

Visual inspection of the weldment is required before more expensive NDE methods are applied. This method of inspection should be employed by the welder, the welding inspector, and supervisor from the beginning to the end of the welding job.

Principal Defects

Every welder should be familiar with the principal defects that contribute to the failure of welded joints. This will aid the welder in the effort to produce work of the highest quality. It takes little additional time to inspect a weld, and it often prevents embarrassment when the work is finally inspected by the shop supervisor or inspector. When taking important tests, the knowledge of these defects and an awareness of the disastrous results that usually follow when defects are present will increase the chances of the welder to meet the required specifications.

The material to be welded should be inspected carefully for surface defects and the presence of contaminating materials. The faces and edges of the material should be free of laminations, blisters, nicks, and seams.

These types of defects are caused by the steel plate making process. Heavy scale, oxide films, grease, paint, and oil should be removed. Pieces of material that are warped or bent should be corrected or rejected. The material should be checked for size, edge preparation, and angle of bevel. Make sure that the material is a type suitable for welding. In setting up or assembling the job, all parts should fit and be in alignment. It is important to understand the purpose of the design, the use of the weldment, and the welding procedure to be followed.

It will be helpful for the student to review the weld characteristics described in Chapter 4, pages 129 to 138, at this time.

The following defects are commonly found in welded steel joints:

- Incomplete penetration
- Incomplete fusion
- Undercutting
- Inclusions
- Porosity
- Cracking
- Brittle welds
- Dimensional defects

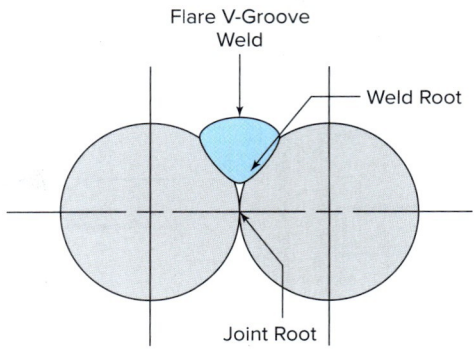

Fig. 28-95 Two solid shafts lying side by side with a flare V-groove weld joining them. It is very difficult to get complete joint penetration with this type of design. The weld pool will bridge the two shafts and effectively block the arc from reaching the root of the joint.

Incomplete Penetration This term is used to describe the failure of the filler and base metal to fuse together at the root of the joint, Fig. 28-94. The root-face sections of a welding groove may fail to reach the melting temperature for their entire depth, or the weld metal may not reach the root of the fillet joint, leaving a void caused by the bridging of the weld metal from one plate to the other, Figs. 28-95 and 28-96. Bridging occurs in groove welds when the deposited metal and base metal are not fused at the root of the joint because the root face has not reached fusion temperature along its entire depth.

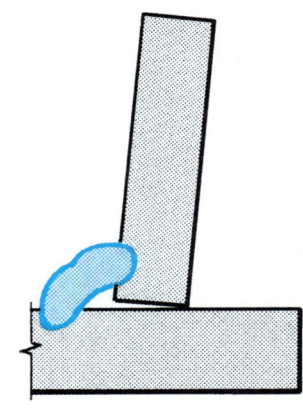

Fig. 28-96 Bridging in a fillet weld.

Although in a few cases incomplete penetration may be due to unclean surfaces, the heat transfer conditions at the root of the joint are a more frequent cause. If the metal being joined first reaches the melting point at the surfaces above the root of the joint, molten metal may bridge the gap between these surfaces and screen off the heat source before the metal at the root melts. In arc welding, the arc establishes itself between the electrode and the closest part of the base metal. All other areas of the base metal receive heat mainly by conduction. If the portion of the base metal closest to the electrode is far from the root of the joint, the conduction of heat may be insufficient to raise the temperature of the metal at the root to the melting point.

Incomplete penetration will cause weld failure if the weld is subjected to tension or bending stresses such as those produced in tensile and bend testing. Even though the service stresses in the completed structure may not

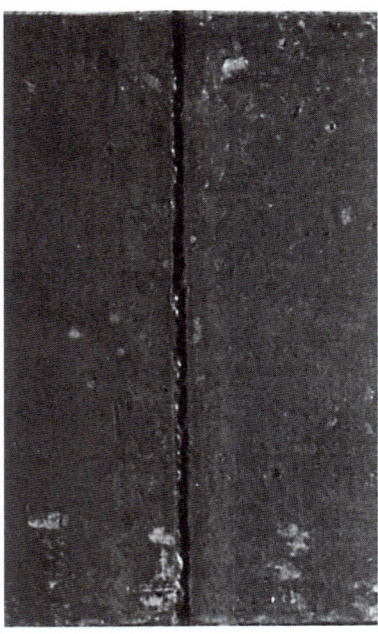

Fig. 28-94 Incomplete penetration and incomplete fusion through the back side of a bevel butt joint. This will cause failure when the joint is subjected to load. *Edward R. Bohnart*

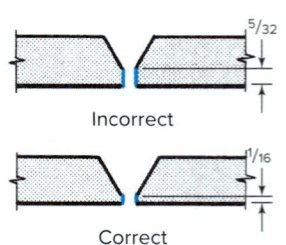

Fig. 28-97 Root face dimensions.

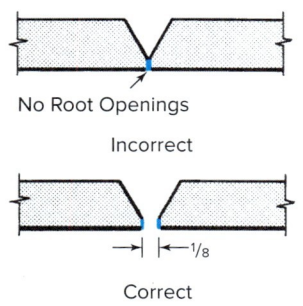

Fig. 28-98 Root opening.

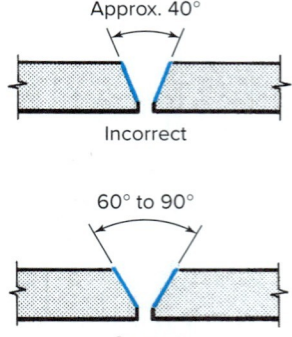

Fig. 28-99 Groove angle.

require tension or bending at the point of incomplete penetration, distortion and shrinkage stresses in the parts during welding frequently cause a crack at the unfused section. Such cracks may grow as successive beads are deposited until they extend through the entire thickness of the weld.

The most frequent cause of incomplete penetration is a joint design that is not suitable for the welding process or the conditions of construction. When the groove is welded from one side only, incomplete penetration is likely to result under the following conditions:

- The root-face dimension is too big even though the root opening is adequate, Fig. 28-97. For example, the proper root-face dimension for V-groove joints in 3/8-inch plate is about 1/16 to 3/32 inch. Root-face dimensions of 3/16 or 1/4 inch are too big and keep the root surfaces from melting, thus preventing fusion and penetration from taking place.
- The root opening is too small, Fig. 28-98. For example, the opening at the root of a V-groove joint in 3/8-inch plate should not be less than 3/32 to 1/8 inch. A root opening of 1/16 inch or less would make it difficult to melt through to the other side, causing incomplete penetration.
- The groove angle of a V-groove is too small, Fig. 28-99. For example, the groove angle of V-groove joints in 3/8-inch plate should be

about 60°. An angle of 40° does not allow for enough freedom of movement for the electrode to be manipulated at the root of the joint, causing the root face to be burned away. Fusion also takes place on the surfaces of the groove faces bridging the root gap as shown in Fig. 28-95.

Even if the joint design is adequate, incomplete fusion will result from the following errors in technique:

- The electrode is too large. For example, running the root pass in a V-groove joint in 3/8-inch plate in the overhead position should be done with a 1/8- or 5/32-inch electrode; a 3/16- or 1/4-inch electrode is too large.
- The rate of travel is too high. Traveling too fast causes the metal to be deposited only on the surface above the root.
- The welding current is too low. If there is not enough current or if the current setting is incorrect, the weld metal cannot be forced from the electrode to the root of the joint and the arc is not strong enough to melt the metal at the root.

Incomplete Fusion Many welders confuse incomplete fusion with lack of penetration. Incomplete fusion is the failure of a welding process to fuse together layers of weld metal or weld metal and base metal. (See Figs. 28-100 through 28-102.) In Fig. 28-100, note that the weld metal just rolls over the plate surfaces. This is generally referred to as *overlap*. Failure to obtain fusion may occur at any point in the welding groove. Overlap at the toe of the weld, Fig. 28-101. Very often the weld has good fusion at the root of the joint and the plate surface but because of poor welding technique and heat conduction, the toe of the weld does not fuse.

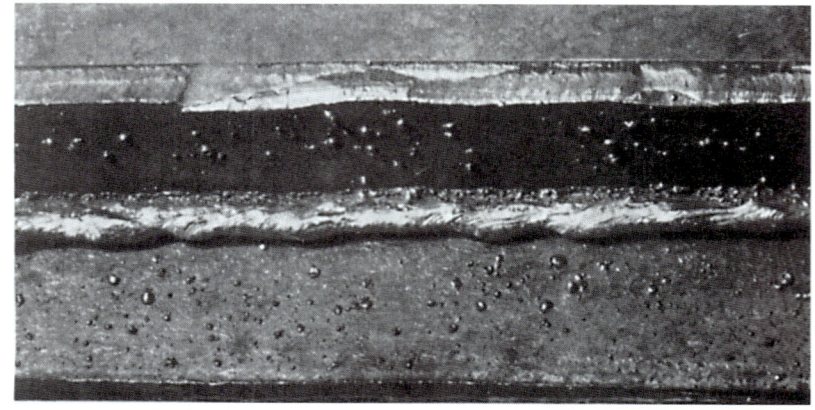

Fig. 28-100 Fillet weld with incomplete fusion. Note that the weld metal just rolls over the plate surfaces. This is generally referred to as overlap. Edward R. Bohnart

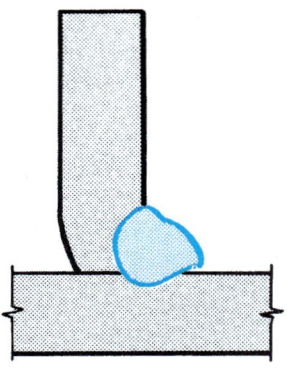

Fig. 28-101 Overlap, which may be attributed generally to improper welding technique or improper welding heat.

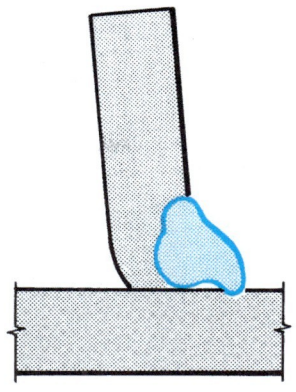

Fig. 28-102 Incomplete fusion.

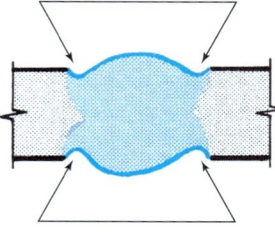

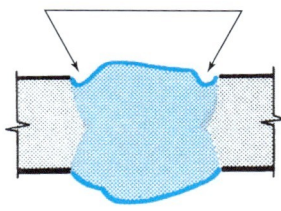

Fig. 28-103 Undercutting.

Incomplete fusion is caused by the following conditions, Fig. 28-102:

- Failure to raise the temperature of the base metal or the previously deposited weld metal to the melting point. Reasons for this failure include (1) an electrode that is too small, (2) a rate of travel that is too fast, (3) an arc length that is too close, and (4) welding current that is too low.
- Improper fluxing, which fails to dissolve the oxide and other foreign material from the surfaces to which the deposited metal must fuse. Incomplete fusion is not common with the SMAW process unless the surfaces being welded are covered with a material that prevents the molten weld metal from fusing to them.

With the GMAW-S and GMAW-P modes of metal transfer, the arc must be played out to the toes of the weld to prevent incomplete fusion and overlap. The molten weld pool will not ensure complete fusion. The heat directly under the arc is required.

Incomplete fusion is avoided by:

- Making sure that the surfaces to be welded are free of foreign material
- Selecting the proper type and size of electrodes
- Selecting the correct current adjustment wire-feed speed and voltage
- Using good welding technique

The welder does not have to melt away large portions of the side walls of the groove in order to be sure of obtaining fusion. It is only necessary to bring the surface of the base metal to the melting temperature to obtain fusion between the base metal and the weld metal.

Undercutting Undercutting is the burning away of the base metal at the toe of the weld, Fig. 28-103 (top). On multilayer welds, it may also occur at the juncture of a layer with the wall of a groove, Fig. 28-103 (bottom). Figure 28-104 shows severe undercutting on the vertical plate of a T-joint.

Undercutting of both types is usually due to improper electrode manipulation. Different types of electrodes have varying tendencies to undercut. For example, reverse polarity electrodes have a greater tendency to undercut than straight polarity electrodes, and a different technique of welding must be employed if undercut is to be avoided. With some electrodes, even the most skilled welder may be unable to avoid undercutting under certain conditions. In addition to poor welding technique and the type of electrode required, undercutting may be caused by:

- Current adjustment that is too high.
- Arc length that is too long.
- Failure to fill up the crater completely with weld metal. This permits the arc to range over surfaces that are not to be covered with weld metal.

Undercutting may be a very serious defect. To prevent any serious effect upon the completed joint, it must be corrected before depositing the next bead. A well-rounded

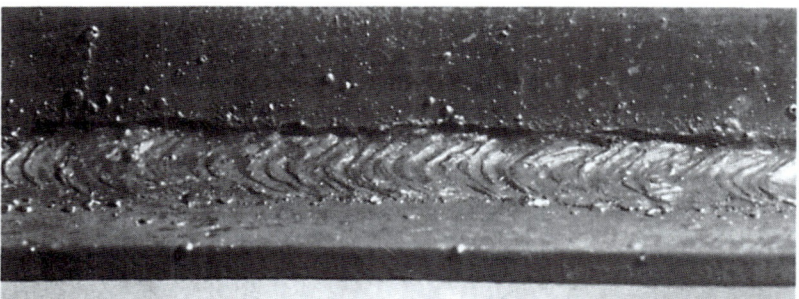

Fig. 28-104 Fillet weld that is undercut severely along the upper edge of the weld (toe). Edward R. Bohnart

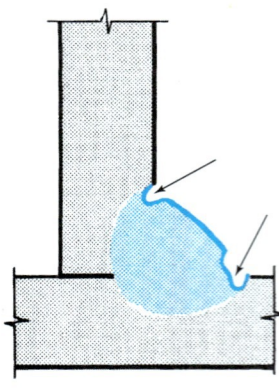

Fig. 28-105 Undercutting in a fillet weld reduces the cross section of the members and acts as a stress raiser.

chipping tool is used to remove the sharp recess that might otherwise trap slag. If the undercutting is slight and the welder is careful in applying the next bead, it may not be necessary to chip.

Undercutting at the surface of a joint should not be permitted since it materially reduces the strength of the joint, Fig. 28-105. For example, on material less than 1 inch thick, undercut shall not exceed $\frac{1}{32}$ inch deep. There is an exception if the undercut is $\frac{1}{16}$ inch deep and is less than 2 inches long; these 2 inches can be cumulative as well; in any 12-inch length of weld it will be acceptable. If the weld being inspected is part of a primary member and the weld is transverse to the tensile stress, then undercut shall be no more than 0.010 inch in depth. Special undercut gauges are made for making this very precise measurement. All other undercut on primary members should not exceed $\frac{1}{32}$ inch in depth no matter what their length is. Fortunately, welders can always see this type of undercutting when they examine the weld, and they can usually correct it by the deposition of additional metal. Good craftspeople develop such skill in welding that they rarely cause undercutting of the base metal.

Inclusions Inclusions are entrapped foreign solid materials such as slag, flux, tungsten, or oxides. They are usually elongated or globular in shape. Inclusions may be caused by contamination of the weld metal by foreign bodies. In arc welding, slag inclusions are generally made up of electrode coating materials or fluxes. In multilayer welding operations, failure to remove the slag between the layers causes slag inclusions (Figs. 28-106 and 28-107).

During the deposition and solidification of the weld metal, chemical reactions occur between the metal, the air, and the electrode coating materials or the gases produced by the arc flame. Some of the products of these reactions are metallic compounds that are only slightly soluble in the molten weld metal. The oxide may be forced below the surface by the stirring action of the arc, or it may flow ahead of the arc, causing the metal to be deposited over the oxide.

When oxide inclusions are present in the molten metal from any cause, they tend to rise to the surface because of their lower density. Any factors such as high viscosity of the weld metal, rapid chilling, or too low a temperature may prevent inclusions from being released. Slag is frequently trapped in the weld when weld metal is deposited by any of the slag forming processes over a sharp V-shaped recess. Under such conditions, the arc may fail to raise the temperature of the bottom of the recess high enough for the metal to fill it and allow the slag, which is ahead of the arc, to float out. The same problem arises if a sharp recess is present due to undercutting by the previous bead. Slag inclusions of this type are usually elongated, and if individual inclusions are of considerable size or are closely spaced, they may reduce the strength of the joint. It is usually not desirable to remove small or isolated inclusions.

Entrapped foreign material forced into the metal by the arc or formed there by chemical reactions usually appears as finely divided or globular inclusions. Inclusions of this type are likely to be a particular problem in overhead welding.

Fig. 28-106 Cross section of a fillet weld, exposing internal defects caused by slag inclusions in the weld metal deposit. These voids reduce the mechanical properties of the weld metal. *Edward R. Bohnart*

Fig. 28-107 Slag inclusions.

Most inclusions can be prevented by:

- Preparing the groove and weld properly before each bead is deposited
- Taking care to avoid leaving any contours that will be difficult to penetrate fully with the arc
- Making sure that all slag has been cleaned from the surface of the previous bead

The release of slag from the molten weld metal is aided by all factors that make the metal less viscous (thick) and that delay its solidification, such as preheating and high heat input per inch per unit of time.

Porosity Porosity, Fig. 28-108, is the presence of pockets that do not contain any solid material. Porosity differs from slag inclusions in that the pockets contain gas rather than a solid. The gases forming the voids are derived from:

- Gas released by the cooling weld metal because of its reduced solubility as the temperature drops
- Gases formed by chemical reactions in the weld

Excessive porosity in welds has a serious effect on the mechanical properties of the joint. Certain codes permit a specified maximum amount of porosity. Pockets may be found scattered uniformly throughout the entire weld, isolated in small groups, or concentrated at the root.

Porosity is best prevented by avoiding:

- Overheating and underheating of the weld metal
- Excess moisture in the covered electrode
- Contaminated base metal or consumables
- Too high a current setting
- Too long an arc

A metal temperature that is too high increases unnecessarily the amount of gas dissolved in the molten metal. This excess gas is available for release from solution upon cooling. If the welding current and/or the arc length is excessive, the deoxidizing elements of the electrode coating are used up during welding so that there are not enough of them left to combine with the gases in the molten metal during cooling. Underheating does not permit the weld pool to be molten long enough to allow the trapped gases to escape.

Reducing all sources of contamination to a minimum will greatly reduce possible gasifiers and help eliminate hydrogen pick-up. With the gas shielded processes the shielding gas must be pure, delivered at the proper flow rate, and protected from being blown away.

For example, porosity on CJP groove welds that are transverse to the tensile loading (are most critical) shall have no visible piping porosity. For all other groove and fillet welds, the piping porosity larger than $\frac{1}{32}$ inch in diameter (or the largest dimension if not round) should not cumulatively measure more than $\frac{3}{8}$ inch in any 1-inch length of weld or $\frac{3}{4}$ inch in any 12-inch length of weld. Any porosity larger than $\frac{3}{32}$ inch in diameter (or the largest dimension if not round) is cause for rejection.

Cracking Cracks are linear ruptures of metal under stress. When they are large, they can be seen easily, but they are often very narrow separations. Cracks may occur in the weld metal, in the plate next to the weld, or in the heat-affected zone. Common weld metal and base metal cracks are shown in Fig. 28-109, page 942.

Cracking results from localized stress that exceeds the ultimate strength of the material. Little deformation is apparent because the cracks relieve stress when they occur during or as a result of welding.

There are three major classes of cracking: hot cracking, cold cracking, and microfissuring.

Hot cracking occurs at elevated temperatures during cooling shortly after the weld has been deposited and has started to solidify. Stress must be present to induce cracking. Slight stress causes very small cracks that can be detected only with some of the nondestructive testing techniques such as radiographic and liquid penetrant inspection. Most welding cracks are hot cracks.

Cold cracking refers to cracking at or near room temperature. These cracks may occur hours or days after cooling. These cracks usually start in the base metal in the heat-affected zone. They may appear as underbead cracks parallel to the weld or as toe cracks at the edge of the weld. Cold cracking occurs more often in steels than in other metals.

Microfissures (microcracks) may be either hot or cold cracks. They are too small to be seen with the naked eye and are not detectable at magnifications below 10 power. These cracks usually do not reduce the service life of the fabrication.

Weld Metal Cracking Three different types of cracks occur in weld metal: transverse cracks, longitudinal cracks, and crater cracks.

Transverse cracks run across the face of the weld and may extend into the base metal. They are usually caused by excessive restraint during welding.

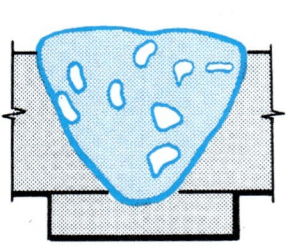

Fig. 28-108 Porosity.

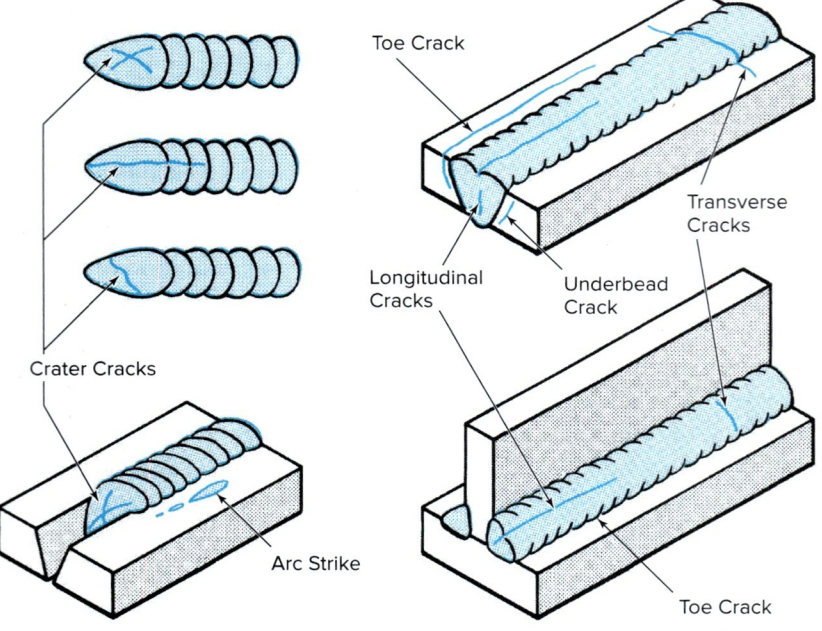

Fig. 28-109 Cracks in welded joints.

Longitudinal cracks are usually confined to the center of the weld deposit. They may be the continuation of crater cracks or cracks in the first layer of welding. Cracking of the first pass is likely to occur if the bead is thin. If this cracking is not eliminated before the following layers are deposited, the crack will progress through the entire weld deposit. Longitudinal cracking can be corrected by:

- Increasing the thickness of the root pass deposit
- Controlling the heat input
- Decreasing the speed of travel to allow more weld metal to build up
- Correcting electrode manipulation
- Preheating and postheating

Whenever the welding operation is interrupted or improperly terminated, there is a tendency for the formation of a crack in the crater. *Crater cracks* usually proceed to the edge of the crater and may be the starting point for longitudinal weld cracks.

Base Metal Cracking Base metal cracking usually occurs within the heat-affected zone of the metal being welded. The possibility of cracking increases when working with hardenable materials. These cracks usually occur along the edges of the weld and through the heat-affected zone into the base metal. The *underbead crack,* limited mainly to steel, is a base metal crack usually associated with hydrogen. Reducing contaminants and excess moisture will help eliminate this type of cracking. Toe cracks in steel can be of similar origin. They are caused by hot cracking in or near the fusion line.

Incorrect welding procedure causes cracking in the base metal. Arc strikes—accidental touching of the electrode to the work outside the weld joint—may cause small cracks. If the weld is started at the edge of the plate and proceeds into the plate, a crack will occur along the edge of the weld at the toe. Cracks can also be started as a result of undercutting. Root cracks, which start from the root of the weld and progress through the weld metal, often produce cracking in the plate on the side opposite the weld.

A nonfused area at the root of a weld may also crack without apparent deformation if this area is subject to tensile stress. In welding two plates together, the root of the weld is subject to such tensile stress as successive layers are deposited that the nonfused root will permit a crack to start that may progress through practically the entire thickness of the weld.

Care must be taken in the type of steel selected for a particular job and in the electrode chosen for welding. Improper design with little regard for expansion and contraction also contributes to cracking.

Brittle Welds A brittle weld has poor elongation, a very low yield point, very poor ductility, and poor resistance to

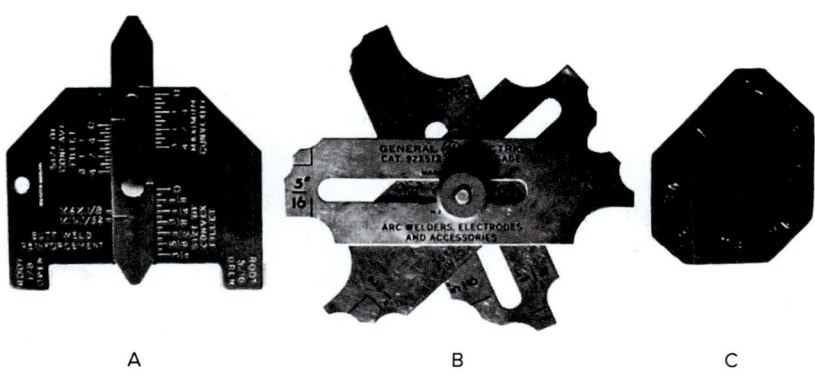

Fig. 28-110 Three types of weld gauges commonly used by welders and inspectors for determining weld size. *Edward R. Bohnart*

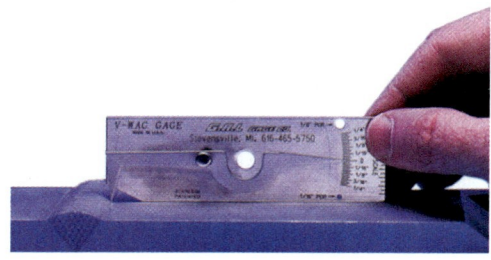

Fig. 28-111 To measure depth of undercut, set the gauge on base metal with the pointer in the bottom of the undercut. Directly read the depth on the scale as indicated. The locking screw can be tightened to hold the reading for future reference. © G.A.L. Gage Co.

stress and strain. Brittle welds are highly subject to failure and may fail without warning any time during the life of the weldment.

One of the principal causes of brittle welds is the use of excessive heat which burns the metal. Multilayer welds are recommended to avoid brittleness. Such welds have a tendency to anneal the previously deposited weld beads and the base metal weld area. Careful selection of the material and the electrode also ensure against brittle weld deposits.

Dimensional Defects Dimensional defects include longitudinal contraction, transverse contraction, warping, and angular distortion. They are caused by improper welding procedure and/or technique. The use of such controls as welding jigs, proper welding sequences, correct welding procedure, suitable joint design, and preheat and postheat processes prevent distortion or keep it at an acceptable minimum. Refer to Chapter 3, pages 100–105, for a detailed discussion of the causes and control of distortion.

Weld Gauges

Incorrect weld size and contour are defects that can be detected by visual inspection with the aid of weld gauges. Weld gauges are tools that the welder can use to make sure that the completed weld is within the limits specified by the engineering design and weld procedure.

Figure 28-110 shows three types of weld gauges: (A) combination butt and fillet weld gauge, (B) fillet weld gauge, and (C) a second type of fillet weld gauge made in the shop. Figure 28-111 shows an additional gauge.

Figure 28-112, page 944, illustrates the method of using the combination butt and fillet weld gauge shown in Fig. 28-110A. Figures 28-113 and 28-114, page 944, illustrate the method of using the fillet weld gauge shown in Fig. 28-110B. Figure 28-115, page 945, illustrates the method of using the shop-made fillet weld gauge shown in Fig. 28-110C. Note that there are two designs: one for measuring concave fillets and another for measuring convex fillets. The gauge for convex fillets must have a concave face. A shop-made gauge cannot measure concave fillet welds accurately.

Summary

With the advent of laser scanners, the ability to inspect welds has reached a new stage. These handheld devices are being used to inspect welds in the fabrication shop as well as for field inspection. They are being used to inspect welds for the steel construction, shipbuilding, automotive, military, and various other industries, Fig. 28-116 (p. 945). These laser scanners can be used for preweld inspection to determine if the joint design meets the specification. See Tables 28-14 and 28-15 (p. 946). The interface allows the operator to select the type of inspection to perform, and to view measurement and inspection results (accept or reject) on a bright color display. The data can be stored in memory or downloaded to a separate computer.

Welding demands constant visual examination during the entire operation. In order to make corrections during the welding operation, all welding requires careful attention to such considerations as the welding process, cleaning, preheating, joint preparation, the proper electrode, chipping, peening and postheating, proper weld size, and surface defects. Special attention should be given to the root pass because it tends to solidify quickly and may trap slag or gas. The root pass is also susceptible to cracking which may continue through all layers.

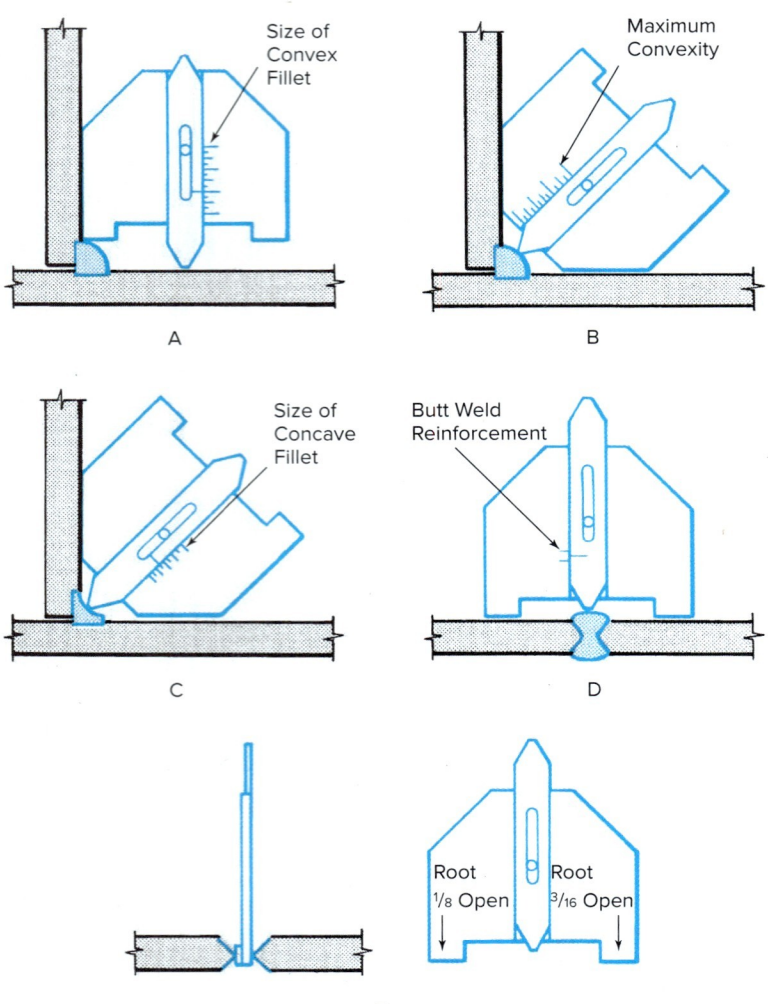

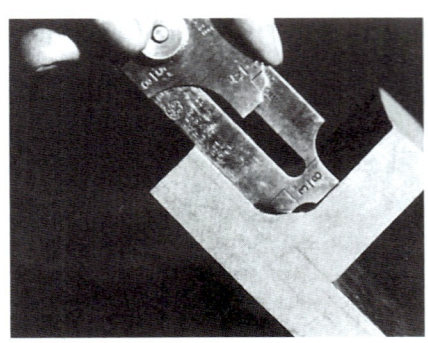

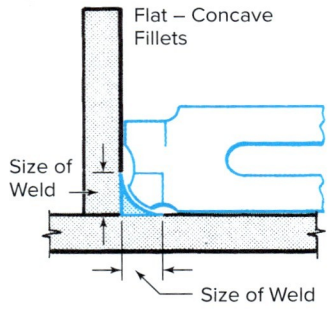

Fig. 28-113 Method of using fillet weld gauge to determine the size of flat and concave fillet welds. © General Electric Company

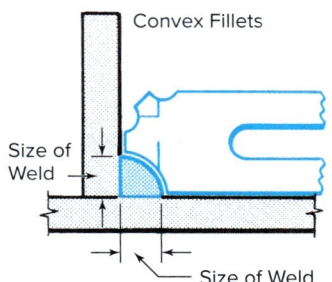

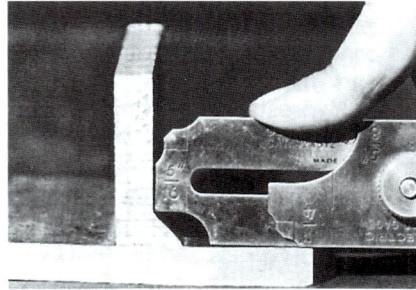

Fig. 28-114 Method of using fillet weld gauge to determine the size of a convex fillet weld. © General Electric Company

Fig. 28-112 Method of using butt and fillet weld gauge. (A) To determine the size of a convex fillet weld, place the gauge against the toe of the shortest leg of the fillet. Slide the pointer out until it touches the structure. Read "size of convex fillet" on the face of the gauge. (B) To determine the convexity of a convex fillet weld after its size has been determined, place the gauge against the structure, and slide the pointer out until it touches the face of the fillet weld. The maximum convexity should not be greater than indicated by "maximum convexity" for the size of fillet being checked. (C) To determine the size of a concave fillet weld, place the gauge against the structure, and slide the pointer out until it touches the face of the fillet weld. Read "size of concave fillet" on the face of the gauge. (D) To determine the reinforcement of a butt weld, place the gauge so that the reinforcement will come between the legs of the gauge and slide the pointer out until it touches the face of the weld. The permissible reinforcement is that indicated on the face of the gauge. (E) To determine the root opening (⅛ or 3/16 inch) of a butt joint, place one leg of the gauge in the space separating the parts. If the gauge fits snugly, the root opening is that indicated on the face of the gauge. Source: General Electric Co.

The completed weld should be examined carefully for proper size and contour, surface defects, and warping and distortion of the weldment. Special attention should be given to unfilled craters, poor starts and stops, crater cracks, and cracking at the edge of the welds. Tables 28-5 through 28-7 and 28-16, page 946, summarize the various weld and base metal defects that may be encountered, and the various methods of testing that will establish the

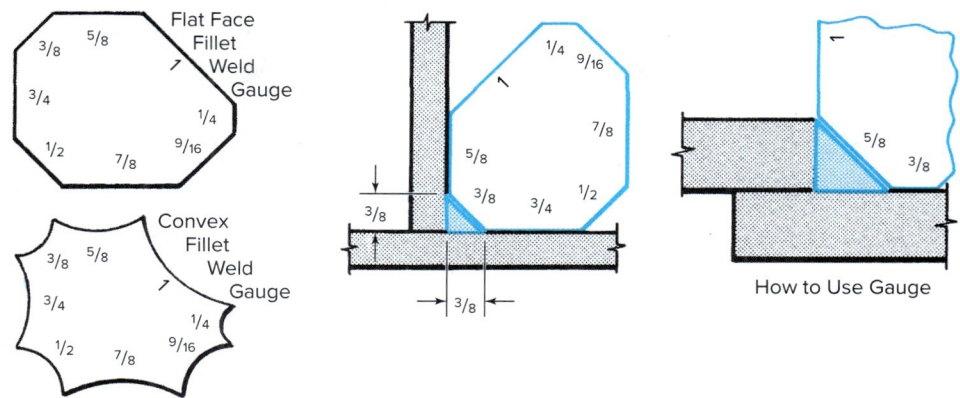

Fig. 28-115 How to use a shop-made fillet weld gauge.

Fig. 28-116 Handheld laser scanner welding inspection system. Requires control and the gun that is shown. © Servo-Robot Corp.

presence of these defects. In Table 28-7 are the recommended inspection methods to be used for evaluating fillet and butt joints. An understanding of these tables will give you adequate knowledge of testing and its application to weld and base metal. The following points should be remembered:

- Visual inspection is the most convenient method. Everyone can participate in it, Fig. 28-117.
- Radiographic inspection permits looking into the weld for defects that fall within the sensitivity range of the process. It provides a permanent record of the results.

Fig. 28-117 A welding instructor makes a visual inspection of a student's practice weld. © Prographics

- Magnetic particle inspection is outstanding for detecting surface cracks and is used to advantage on heavy weldments and assemblies.

Joint Design, Testing, and Inspection **Chapter 28** 945

Table 28-14 Preweld Inspection (Joint Preparation) with Handheld Laser Scanner

Features \ Joint Type	Groove Joint[1]	T-Joint	Lap Joint	Square Butt Joint	Corner Joint	Joint
Angle between plates	•	•	•	•	•	•
Bevel angle	•	•	—	—	—	—
Groove angle	•	—	—	—	—	—
Joint cross-sectional area	•	•	—	•	•	—
Joint gap	•	•	•	•	•	•
Top opening	•	•	—	—	—	—
Root opening	•	•	—	—	—	—

[1] Groove joints include V-, U-, J-, bevel grooves.

Source: Servo-Robot

Table 28-15 Postweld Inspection (Welds) with Handheld Laser Scanner

Features \ Weld Type	Fillet Weld	Groove Weld	Lap Weld	Plug & Slot
Leg size	•	—	•	—
Mismatch	—	•	•	—
Concavity/Convexity	•	•	•	•
Theoretical throat	•	—	•	—
Toe angle	•	•	•	—
Undercut	•	•	•	—
Weld face	•	•	•	—
Weld size	•	—	•	—
Cross section area	•	•	•	•
Reinforcement	—	•	—	—

Source: Servo-Robot

Table 28-16 Tests for Weld and Base Metal Defects

Defects	Methods of Testing
Dimensional defects:	
Warpage	Visual inspection with proper mechanical gauges and fixtures
Incorrect joint separation	Visual inspection with proper mechanical gauges and fixtures
Incorrect weld size	Visual inspection with approved weld gauge
Incorrect weld profile	Visual inspection with approved weld gauge
Defective properties:	
Low tensile strength	All-weld-metal tension test, transverse tension test, fillet-weld shear test, base metal tension test
Low yield strength	All-weld-metal tension test, transverse tension test, base metal tension test
Low ductility	All-weld-metal tension test, guided-bend test, base metal tension test
Improper hardness	Hardness tests
Impact failure	Impact tests
Incorrect composition	Chemical analysis
Improper corrosion resistance	Corrosion tests

- Dye penetrant is easy to use for detecting surface cracks. Its indications are readily interpreted.
- Ultrasonic inspection is excellent for detecting subsurface discontinuities, but it requires expert interpretation.
- Hydrostatic testing determines the tightness of welds in fabricated vessels. Various liquids and gases may be used which, when escaping in minute amounts, are easily detected. The method is simple to apply.
- Hardness tests indicate the approximate tensile strengths of metals and show whether or not the base metal and weld metal strengths are matched. The method of testing is fairly easy to apply.
- Destructive testing gives an absolute measure of the strength of the sample tested. Samples may not always be representative of all the welding on a component. They do, however, have the potential of predicting the serviceability of a weldment. Destructive testing is a good means of determining a welder's ability. Results are permanent. The method is fairly easy to apply in shop and field.

CHAPTER 28 REVIEW

Multiple Choice

Choose the letter of the correct answer.

1. Which of the following is both a joint and a weld? (Obj. 28-1)
   a. V-groove
   b. Double-fillet
   c. Edge
   d. Flange

2. Which type of butt joint design will you have the best chance of getting complete joint penetration (CJP) on? (Obj. 28-1)
   a. Closed square groove
   b. Open square groove
   c. Single bevel groove without root opening
   d. V-groove with root opening

3. Which of the following is not a nondestructive test method? (Obj. 28-2)
   a. Radiographic
   b. Ultrasonic
   c. Tensile
   d. Magnetic particle

4. Liquid penetrant inspection can detect surface and interior defects. (Obj. 28-2)
   a. True
   b. False

5. Only a red dye penetrant can be used with the liquid penetrant method. (Obj. 28-2)
   a. True
   b. False

6. Destructive test specimens should be ground in the direction they are to be stressed. (Obj. 28-3)
   a. True
   b. False

7. Which of the following are used to destructively test groove welds? (Obj. 28-3)
   a. Tension test
   b. Bend test
   c. Nick-break test
   d. All of these

8. Which of the following are used to destructively test fillet welds? (Obj. 28-3)
   a. Shear test
   b. Break test
   c. Macroetch test
   d. All of these

9. A welding procedure can be thought of as a _____. (Obj. 28-4)
   a. Code for welding
   b. Standard for welding
   c. Recipe for welding
   d. All of these

10. A welder qualification test is _____. (Obj. 28-4)
    a. A performance test
    b. For groove welds only
    c. For fillet welds only
    d. None of these

11. Which is the most important weld bead and is generally the most difficult to make? (Obj. 28-5)
    a. The final bead
    b. The filler beads
    c. The root bead
    d. All of these

12. Which of the following tests should be conducted first? (Obj. 28-6)
    a. Radiographic
    b. Ultrasonic
    c. Visual
    d. Bend

13. One of the most frequent causes of incomplete penetration is _____. (Obj. 28-6)
    a. The welding process used
    b. The electrode being used
    c. The shielding gas used
    d. The joint design
14. Separate gauges must be used for measuring concave and convex fillet weld sizes. (Obj. 28-7)
    a. True
    b. False

Review Questions

Write the answers in your own words.

15. Sketch the joint design and label the appropriate dimensions for a ⅜-inch wall thickness carbon steel butt joint to be welded with the gas tungsten arc welding process and a consumable insert. (Obj. 28-1)
16. Describe the radiographic test method and how it is used to inspect welds. (Obj. 28-2)
17. Explain the transducer as used for ultrasonic testing and why it is called a *piezoelectric device*. (Obj. 28-2)
18. Describe the difference between face bends, root bends, and side bends and list the acceptance criteria for bend tests. (Obj. 28-3)
19. When doing a break test on a fillet weld, what are the acceptance criteria? (Obj. 28-3)
    The fillet break test specimen shall pass if: (list criteria).
20. Explain the difference between a procedure qualification test and a welder qualification test. (Obj. 28-4)
21. Sketch a bend-test specimen and label the important features it should contain to allow it the greatest chance of passing the bend test. (Obj. 28-5)
22. List six defects that are commonly found in welded steel joints. (Obj. 28-6)
23. Describe how undercut is measured. (Obj. 28-7)
24. Describe how reinforcement on a groove weld is measured. (Obj. 28-7)
25. Sketch the measurement of fillet weld sizes on concave and convex fillet welds. (Obj. 28-7)

INTERNET ACTIVITIES

Internet Activity A

Using the American Welding Society's Web site, review and make a report to your instructor on the certification of welding inspection personnel. List the names of each level of certification and how many years of experience you must have to sit for the certification examination.

Internet Activity B

Using your favorite search engines, look up and make a list of companies that provide weld inspection tools.

Design credits: Cinema and movie icon: Denis Meshkov/123RF; and Illustration of Welding icons: Mr. Nachai Sorasee/123RF

29 Reading Shop Drawings

Chapter Objectives

After completing this chapter, you will be able to:

- 29-1 Describe the reasons for drawings.
- 29-2 Identify and describe uses of various drawings.
- 29-3 Describe various dimensioning techniques.
- 29-4 Describe various geometric shapes.
- 29-5 Identify and explain various views found on drawings.
- 29-6 Demonstrate the ability to read various drawings.

Introduction

Drawings and shop sketches are the sign language of industry. They transmit ideas from design to finished product.

Drawings are the universal language of craftspeople the world over. It is possible for workers who can read *drawings* to produce identical parts even though they work in different lands. Tanks, pipelines, ships, buildings, airplanes, and machines are made from the information given on drawings.

The Purpose of Shop Drawings

Pictures and notes on drawings are only a method of passing information from one person to another in a condensed form. Information is passed from the engineer who designs the job to the worker who fabricates it. The engineer under whose direction the drawings are prepared must have a thorough knowledge of design, mathematics, mechanics, physics, chemistry, and drafting. Workers who interpret the finished drawings do not need such theoretical knowledge, but they must understand the drawing well enough to carry out all instructions shown.

Skilled welders are not only competent in welding, but they are also able to read the drawings that describe the work that they must do. Welders are much more valuable to the employer if they can read drawings. With this knowledge, they can learn from a drawing just what is to be built, and they can picture the object in their minds as it will be after it is completed. They must also be able to interpret the notations on the drawing. The drawing serves as an instruction sheet to which they can refer from time to time without having to ask the supervisor question after question.

Welders will need to rely on the ability to read drawings more and more. In the past, production welders might have worked on the same weldment day after day and week after week. The weldment would have become so familiar to them that they would seldom have needed to refer to the drawing. Mass production has now changed; long production runs are not used. This is being done in order to supply customers with product in a more timely fashion. In the past, manufacturers could back-order a product and several weeks were usually required before that product would come back up on a production schedule. Nowadays many manufacturers are striving for a single-day turnaround on out-of-stock product. Instead of welding parts for only one product day after day, welders may be called upon to weld up parts or products changing their work every few hours. Since they will be exposed to many more parts and products in a day or work week, welders must be able to read and interpret drawings quickly and efficiently.

The ability to read drawings is a common reason for promotion to a position of more responsibility. When welding supervisors are chosen, those who can interpret drawings have a better chance than other qualified welders. These people understand the design requirements because they understand the designer's language and can translate that language into welding that meets the requirements.

Welders should learn to read and interpret plans for the following reasons:

- A skilled welder who can interpret plans is always in demand on any job and is likely to be given responsibility.
- The welder has an accurate idea of what the weldment should look like when completed. An understanding of the total job makes assembly easier and guides the choice of welding procedure.
- The welder is able to fabricate a weldment that duplicates the engineering design as closely as possible.
- The welder can check on the equipment and supplies that are needed for the job.
- In field work, welders can avoid interference with other trades by knowing the extent of the work

even before they start. They can begin preparations for some of their work while carpenters are building their forms or masons are getting ready to pour concrete.

Since sketches and drawings are used repeatedly in this text, you must be able to read drawings properly and understand them, Fig. 29-1. As part of your welding training, you will also be called upon to do production jobs in order to experience real job situations. Very often these jobs come to the school welding shop as a drawing. Such drawings are usually very simple, and you will have to learn only the fundamentals of reading drawings in order to interpret them. After you have completed this chapter, you are urged to study a book on advanced drawing reading so that you will be able to read the more complicated drawings with which you will work in industry.

Types of Drawings

Engineers and/or architects must communicate their concepts of a structure's shape and function in drawings. These engineering and architectural drawings give

Fig. 29-1 This young welder, who has just completed his training course, is reading a drawing of the job he is working on.
© Prographics

overall dimensions, shapes, etc., but they do not give sufficient detail to fabricate it. A person called a *detailer* must take these initial drawings, redraw, and expand them with enough information so that they can be used to fabricate the structure or weldment. These drawings are generally called *shop* or *working drawings*. These are the types of drawings welders will be most familiar with. These shop or working drawings are further broken down into detailed and assembly drawings.

Detailed drawings show individual parts in detail, with dimensions, joint designs, welding symbols, and any other information that may be required to make the part. Figure 29-8, page 955, is an example of a detailed drawing.

Assembly drawings represent how all the parts come together to create the final assembly of the product. Figure D-1, page 977, is an example of an assembly drawing. *Erection drawings* are a type of assembly drawing used by welders to do field erection of a structure. Such drawings would be used for transmission towers, buildings, bridges, and so on.

As a welder, whatever type of industry you are interested in entering will have drawings that meet that specific industry's needs. By developing a fundamental understanding of how to read a drawing, you should be able to read any type of drawing you might encounter with appropriate study.

Methods of Printing Drawings

Original drawings are seldom found on the job site. Copies of the original drawing are used on the job site. Copies come in various sizes. Familiarize yourself with Figs. 29-2 and 29-3. Before a copy can be made, the original drawing must be done. This can still be done manually with a draftsperson at a drawing board. However, computer-aided drafting (CAD) is more common. Another common name applied to CAD is computer-aided drafting and design (CADD), which introduces the design element into the process. A computer with the proper software and a printer is a very powerful tool. However, it is just a tool and still must be put to use by a person knowledgeable in drawing construction who knows how to interpret them, and of course knows the welding symbols. If the person is also responsible for design, they must be very familiar with all the various aspects of welding along with the welding capability of the shop doing the fabrication in order to get the proper detail on the drawing.

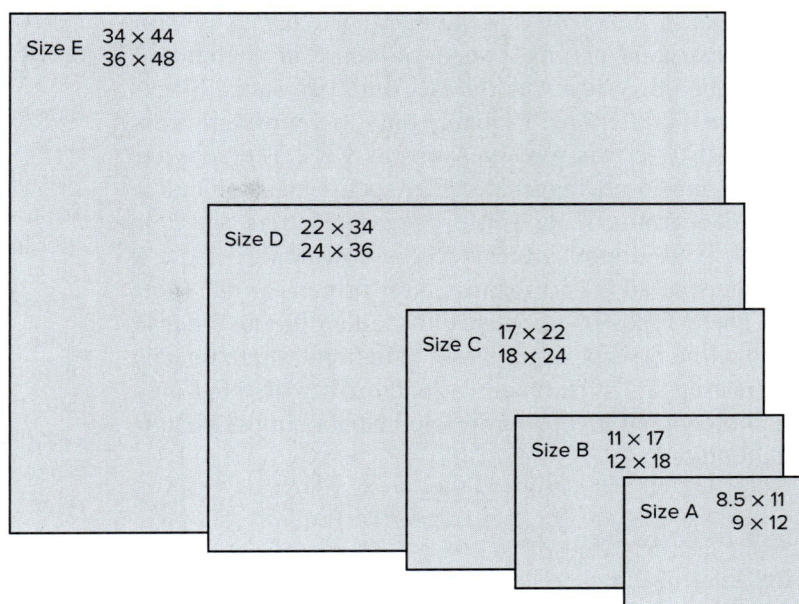

Fig. 29-2 Standard drawing sizes in inches. An "E" size would be the largest while the "A" size would be the smallest. These letter codes may appear on a specification so you must have an idea how large the print will be.

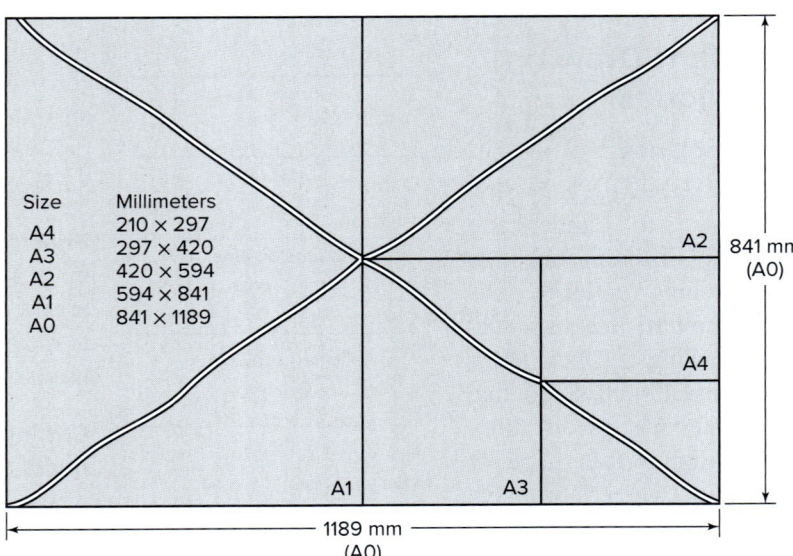

Fig. 29-3 Standard drawing sizes in SI units. These sizes will be encountered when working on prints coming from outside the United States. The "A0" size would be the largest while the "A4" size would be the smallest. These letter-number combination codes may appear on a specification so you must have an idea how large the print will be.

The term *blueprint* is seldom used today. It is akin to an instructor calling a write-on board or chalkboard a blackboard. When was the last time you saw a blackboard in a classroom? Probably only in a museum. This would also be true for blueprints as well. The copying of drawing is no longer done in such a manner that a blue background with white lines and letters are produced. Now digital information is downloaded from a computer to an ink-jet printer, laser printer, or pen plotter. These types of printers and plotters produce dense lines with very good quality. In situations where dozens of drawings must be made, a good quality drawing may be photocopied in an appropriate machine in the quantities required.

Care of Drawings

Any kind of drawing should be handled carefully in the field. Avoid thumbprints by handling by the "thumb room" space left along the edge. Do not put any marks on the drawing. They may lead to costly mistakes or delays. Any notations added to a drawing on the job should be made only by a person in authority, and only if the information will aid in doing better work.

Standard Drawing Techniques

Types of Lines

To read *drawings,* it is first necessary to understand what the different lines on the drawings represent. The line is the basis for all industrial drawings. Lines show an object in such detail that the worker has no difficulty in visualizing the shape of the object when fabricated. Figure 29-4 illustrates the several types of commonly used lines. Note the different weights (thicknesses) of the lines.

Object Lines The shape of an object is shown on a drawing by unbroken heavy lines known as object lines or *visible edge lines.* The lines may vary in thickness depending on the size of the object. A heavy line is also used as a border line around the edge of drawings.

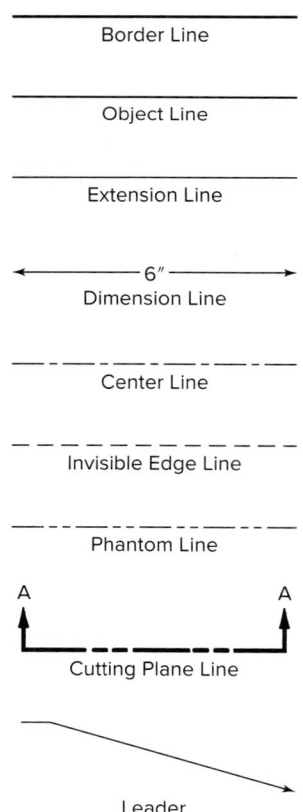

Fig. 29-4 Standard lines.

Extension Lines Extension lines extend from the object lines of an object to a dimension. Thus, they express the size of the object. They extend away from the object and are not a part of the object. An extension line is medium in thickness and unbroken.

Dimension Lines Dimension lines are medium, solid lines, which are unbroken except where the dimension is placed. They have arrowheads at their ends. The distance determined by the dimension line is the distance from the end of one arrowhead to the end of the other or the distance between the extension lines to which the dimension refers.

Center Lines Center lines are often referred to as dot-and-dash lines, but they are actually light broken lines made up of long and short dashes spaced alternately. Center lines are not a part of the drawing of the object. They indicate the center of a whole circle, a part of a circle, or certain dimensions.

Hidden Lines A drawing must include lines that represent all the edges and intersections of the object. In some views, it is not possible to see all the surfaces. These surfaces are represented by lines of medium thickness made up of a series of short dashes. These lines are referred to as hidden or *invisible object lines* since they represent edges or surfaces that cannot be seen.

Phantom Lines Phantom lines are light, broken lines made up of a series of one long and two short dashes. These lines are used to show alternate locations of a part that is designed to move through a range of motion. These lines may also be used to show a machined surface.

Cutting Plane Lines A cutting plane, also called a *sectional line,* is a line of medium thickness that indicates where an imaginary cut is made through the object. The line is a series of one long and two short dashes alternately spaced. Arrowheads are placed at the ends of the line pointing toward the direction in which the section will be shown. The cutting plane line also has letters to identify it with its respective section on the same or on a different drawing.

Section Lines Section lines are a series of lines that can be solid or solid and broken. They can be arranged in specific patterns to represent various materials, Fig. 29-5. They are used to represent imaginary cut surfaces referenced by the cutting plane line or the type of material it is made out of.

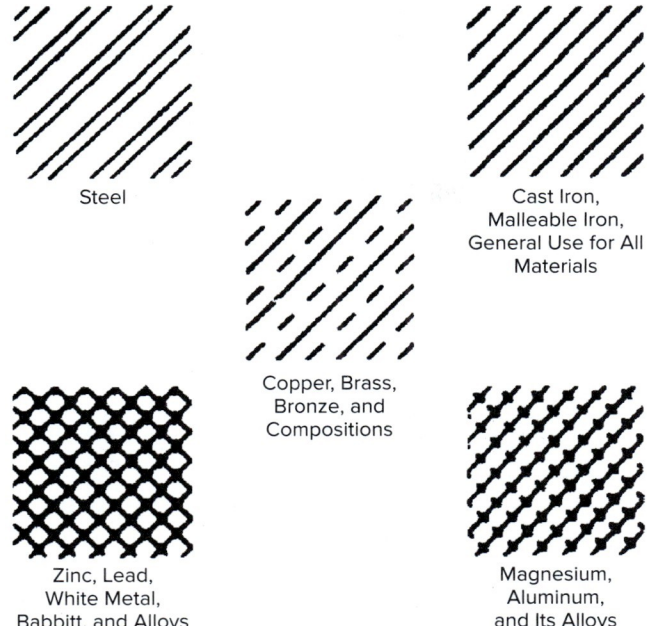

Fig. 29-5 Section lines as used with cutting plane lines or for material identification.

inch-pound system) and the metric system which is based on the meter, liter, and gram. The metric system is a decimal system expressed in multiples of 10. The base unit is 10, and all other units are 10 times the basic unit.

The U.S. system is currently being used in the United States and a few other countries. The International System of units, abbreviated SI, is being used by 92 percent of the world's population, and 65 percent of all world production and trade is based on this system. It is a

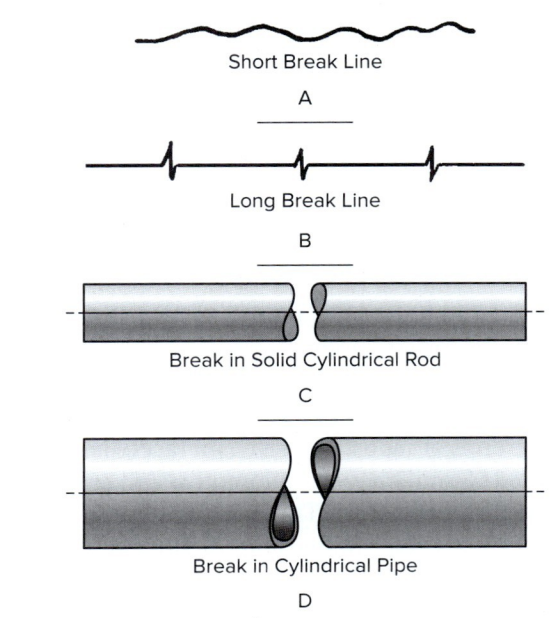

Fig. 29-6 Break lines.

Leaders Leaders usually point to a particular surface to show a dimension or a note. The leader is usually drawn at an angle, and it is a light line with an arrowhead at the pointing end.

Break Lines Break lines are used to show a break in an object or to indicate that only a part of an object is shown. A short break is shown as a heavy irregular line drawn freehand, Fig. 29-6A. A long break is shown as a ruled light line with zigzags in it, Fig. 29-6B. A break in a solid bar or pipe is shown as a curved line, Fig. 29-6C and D.

Projection Lines Projection lines show the relationship of surfaces in one view with the same surfaces in other views. Projection lines on shop drawings are helpful to welders when laying out pipe joints. They do not appear on drawings unless a part is so complicated that it is necessary to show how certain details on the drawing are obtained. Projection lines are fine unbroken lines running from a point in one view to the same point in another view, Fig. 29-7.

Dimensioning

The U.S. Customary and SI Metric Measurement Systems The two most commonly used systems of measurement throughout the world are the U.S. Customary and is abbreviated U.S. (sometimes called the

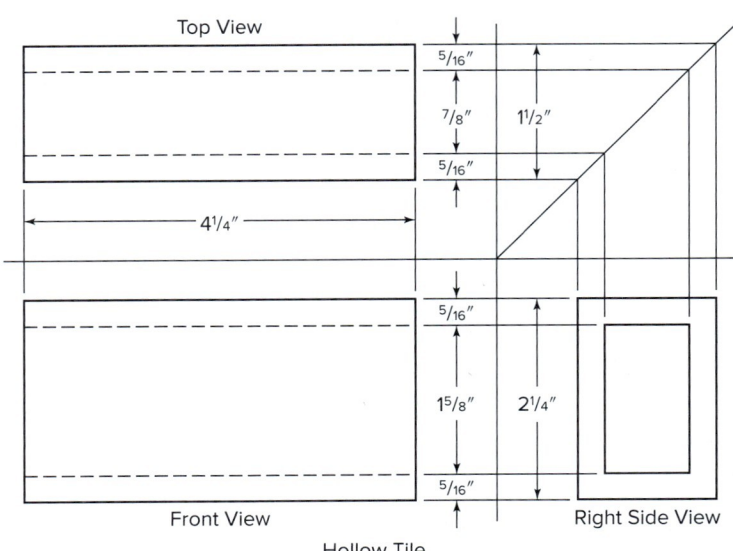

Fig. 29-7 Projection lines.

Table 29-1 Common SI Metric Measurements

Quantity	Unit of Measurement	Symbol
Length	millimeter (1/1000th of a meter)	mm
	meter	m
	kilometer (one thousand meters)	km
Area	square meter	m^2
	hectare (ten thousand square meters)	ha
Volume	cubic centimeter	cm^3
	cubic meter	m^3
	milliliter (1/1000th of a liter)	ml
	liter (1/1000th of a cubic meter)	l
Mass	gram (1/1000th of a kilogram)	g
	kilogram	kg
	ton (1,000 kilograms)	t

Table 29-2 SI Conversions

Property	To Convert From:	To:	Multiply By:
Area dimensions	$in.^2$	mm^2	645.2
	mm^2	$in.^2$	0.001550
Current density	$A/in.^2$	A/mm^2	0.001550
	A/mm^2	$A/in.^2$	645.2
Deposition rate	lb/h	kg/h	0.454
	kg/h	lb/h	2.205
Flow rate	ft^3/h	L/min	0.4719
	L/min	ft^3/h	2.119
Heat input	J/in.	J/m	39.37
	J/m	J/in.	0.0254
Linear measure	in.	mm	25.4
	mm	in.	0.03937
	ft	mm	304.8
	mm	ft	0.003281
Mass	lb	kg	0.454
	kg	lb	2.205
Pressure	p.s.i.	kPa	6.895
	p.s.i.	MPa	0.006895
	kPa	p.s.i.	0.145
	MPa	p.s.i.	145.0
	bar	p.s.i.	14.50
	p.s.i.	bar	0.069
Temperature	°F	°C	(°F − 32)/1.8
	°C	°F	(°C × 1.8) + 32
Tensile strength	p.s.i.	MPa	0.006895
	MPa	p.s.i.	145.0
Travel speed	in./min	mm/s	0.4233
	mm/s	in./min	2.362
Vacuum	Pa	torr	0.007501
Wire-feed speed	in./min	mm/s	0.423
	mm/s	in./min	2.362

modernized version of the metric system that has been established by international agreement. Many attempts have been made to introduce the SI metric system into the United States.

Many groups or industries have embraced the SI system such as the U.S. military, the automotive and bridge building industries, and so on. This will have special significance for welders since dimensions on drawings, electrode sizes, and metal shape sizes will all have to be identified by SI metric measurements. The teaching of this system will be required in our schools.

Understanding the SI metric system of measurement involves chiefly the mastery of a new vocabulary. The most commonly used units for length, for example, are meters, centimeters, millimeters, and kilometers; for mass they are grams, milligrams, and kilograms; for volume, liters and milliliters. Converting from the traditional U.S. units to SI metric units, or vice versa, involves complicated computations. However, as the SI metric system comes into full use, converting back and forth will not be necessary. Metric concepts will be introduced in the schools at an early grade level, and students will learn them as they traditionally have learned inches and ounces.

The only quantities that must be learned as equivalents in metric calculation are 10, 100, and 1,000. On the other hand the student is forced to contend with 29 combinations in the U.S. sytem. Many of them are multidigit: for example, 5,280 feet = 1 mile. Table 29-1 lists the common SI metric units of measure, and Table 29-2 supplies approximate conversion factors for welding related issues.

The common metric ruler is divided into small increments using the millimeter (one-thousandth of a meter) as a fine unit of measurement. Major divisions or bold figures are marked in units of 10 millimeters.

Multiples and Submultiples SI metric measurements are worked from a base of 10 and increase by powers of 10. The powers of 10 are referred to as *multiples* and *submultiples*. They are expressed as shown in Table 29-3. Table 29-4 demonstrates the multiples and submultiples of the meter.

Table 29-3 Multiples and Submultiples of Base 10

Multiplication Factors	Prefix	Symbol
$1\,000\,000\,000\,000 = 10^{12}$	tera	T
$1\,000\,000\,000 = 10^{9}$	giga	G
$1\,000\,000 = 10^{6}$	mega	M
$1\,000 = 10^{3}$	kilo	k
$100 = 10^{2}$	hecto	h
$10 = 10^{1}$	deka	da
$1 = 10^{0}$		
$0.1 = 10^{-1}$	deci	d
$0.01 = 10^{-2}$	centi	c
$0.001 = 10^{-3}$	milli	m
$0.000\,001 = 10^{-6}$	micro	µ
$0.000\,000\,001 = 10^{-9}$	nano	n
$0.000\,000\,000\,001 = 10^{-12}$	pico	p
$0.000\,000\,000\,000\,001 = 10^{-15}$	femto	f
$0.000\,000\,000\,000\,000\,001 = 10^{-18}$	atto	a

Table 29-4 Multiples and Submultiples of the Meter

Multifactor		Quantity		Symbol
10^{3} meters	=	kilometer	=	km
10^{2} meters	=	hectometer	=	hm
10^{1} meters	=	dekameter	=	dam
10^{0} meter	=	meter	=	m
10^{-1} meter	=	decimeter	=	dm
10^{-2} meter	=	centimeter	=	cm
10^{-3} meter	=	millimeter	=	mm
10^{-6} meter	=	micrometer	=	µm

Application to Drawings Figure 29-8 shows a simple drawing in the standard U.S. Customary linear measurement system. Figure 29-9, page 956, shows the same drawing in the SI metric linear measurement system. You will note that both drawings are very similar. The only difference is in the use of millimeters instead of inches for the

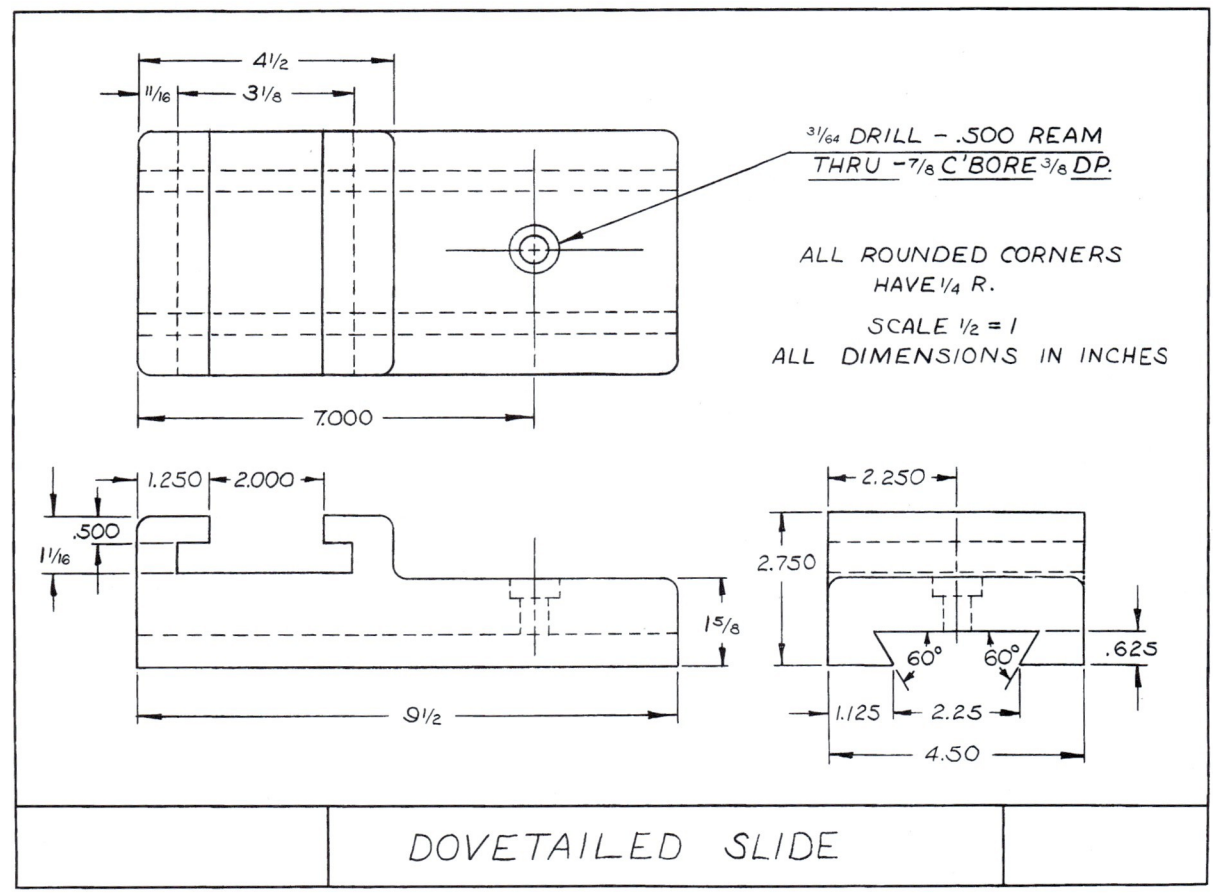

Fig. 29-8 Three-view drawing of a dovetailed slide with the dimensions given in the U.S. Customary measurement system. (Third-angle projections)

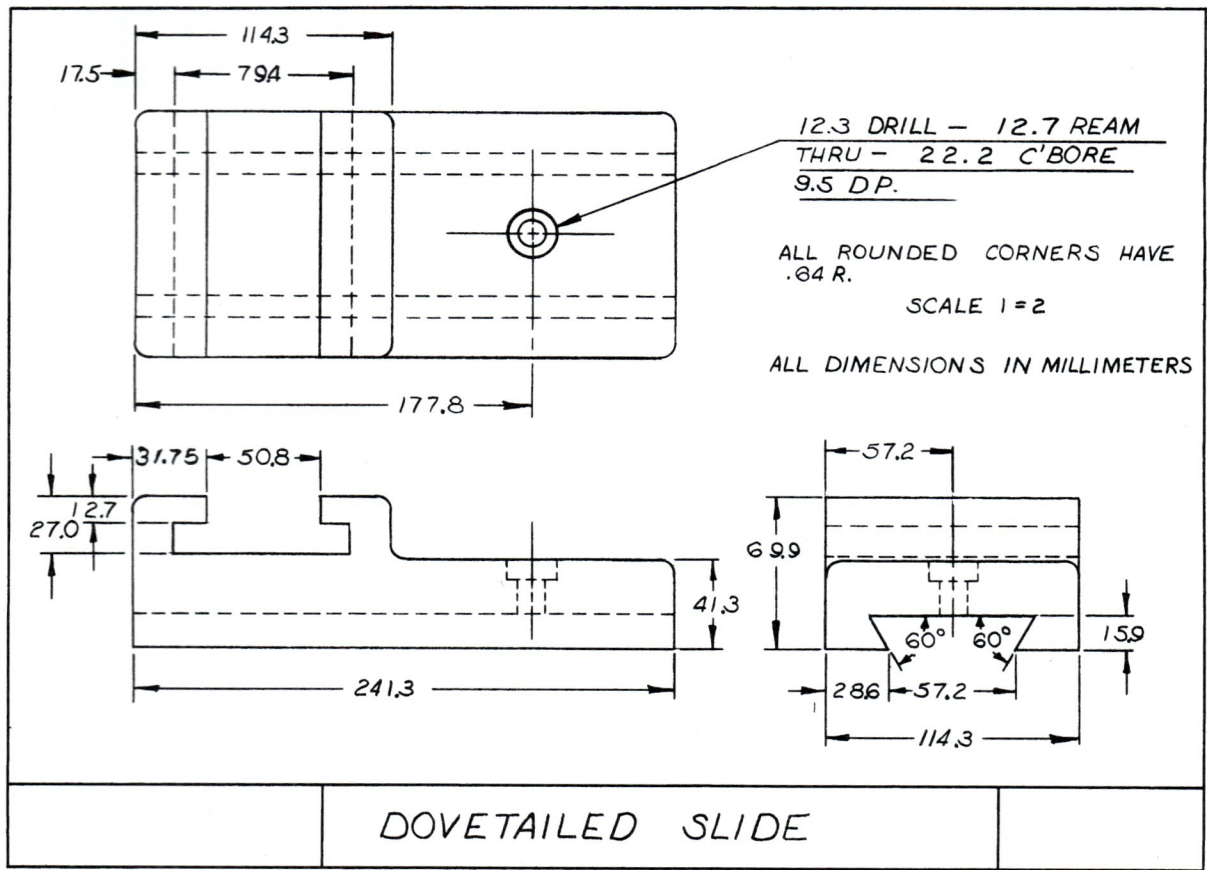

Fig. 29-9 Three-view drawing of the same dovetailed slide with the dimensions given in the SI metric linear measurement system. (Third-angle projection)

dimensions. This is what is considered a hard metric conversion to the SI system from a U.S. Customary drawing. In actual industry practice, hard conversions are done only where required. For example, in this situation the drilled, reamed, and counterbored hole is designed for a ½-inch diameter bolt of some type. The SI system does not have ½-inch bolts or a hard conversion to a 12.7-millimeter diameter bolt. In order to make these conversions realistic, common sense must be used, and in this case a 12- or perhaps a 13-millimeter diameter bolt would be available and the hole should be designed and shown on the drawing to represent what is available in the particular country where it is being fabricated and/or used. If this part were designed outside of the United States and the manufacturer did not use soft metrics conversion, we would have a problem going to a hardware store and getting a 12- or 13-millimeter bolt.

Another issue in dealing with drawings coming from other countries is that they sometimes use a different view system. We use the third-angle projection view, while most other countries use the first-angle projection view. The fundamental difference in these two view methods is the way the views are positioned from the front view. More will be covered on this later in this chapter on drawing views.

It is anticipated that the system of showing all drawings in metric, that is, with only SI metric measurements shown on the drawing itself will become standard practice. Some companies also place a small chart in the upper left-hand corner of the drawing which shows their U.S. Customary equivalents. This helps the worker begin to "think metric."

At the present time many companies that do business with foreign countries have a dual dimensioning system on their drawings. This permits the part to be manufactured with either the U.S. Customary or SI measurements. Drawing practice is to put the U.S. Customary measurement first, and the SI metric measurement in parentheses directly behind or slightly below it. The order may be reversed with the SI measurement first. Because of the extra cost involved and the complex nature of the drawing, it is not likely that this system will continue to be used once

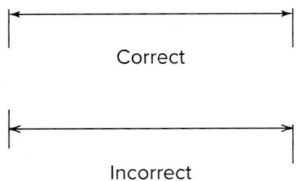

Fig. 29-10 Arrowheads.

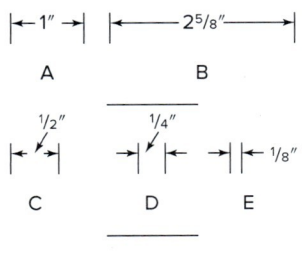

Fig. 29-11 Common methods of dimensioning.

the changeover to SI metric begins in earnest.

Methods of Dimensioning There are several ways in which dimensions are shown on drawings. The size and shape of the object being drawn determine which method is chosen for each dimension. The location of a dimension is indicated by an arrowhead. See Fig. 29-10 for correct and incorrect forms of arrowheads. The placement of the dimension itself depends on the amount of space there is.

- Parts A, B, and C of Fig. 29-11 show the method used when there is enough space at the location for the dimension line, arrowhead, and number.
- Figure 29-11D shows the method used when there is space for the number but not the dimension line and arrowhead.
- Parts E and F of Fig. 29-11 show two methods used when there is no space for the dimension line, arrowhead, and number.
- Parts A and B of Fig. 29-12 show two methods used for dimensioning large circles.
- Parts C, D, and E of Fig. 29-12 show two methods used for small circles when there is no space for the dimension within the circle nor projected from its diameter. Note that diameter is abbreviated as *DIA*.

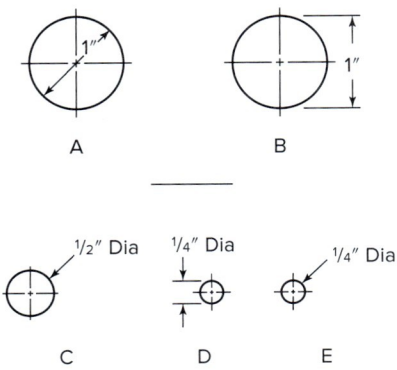

Fig. 29-12 Dimensioning circles.

- Parts A and B of Fig. 29-13 show two methods used for radii (rounded corners) when there is space for the dimension line, arrowhead, and number.
- Parts C and D of Fig. 29-13 show two methods used for radii when there is space for the arrowhead but not the dimension line and number. Note that radius is abbreviated as *R*.
- Figure 29-14 shows the methods for dimensioning a bevel (A); a chamfer (B); an included angle (C); a 45° bevel in a butt-joint (D); and a 45° bevel in a T-joint (E).

Angular Dimensions Angles are measured in degrees, not in feet and inches. Each degree is one three hundred sixtieth (1/360) of a circle. The degree may be divided into smaller units called *minutes*. There are 60 minutes in 1 degree. The minute may be divided into smaller units called *seconds*. There are 60 seconds in each minute.

The symbol for degrees is a small circle (°), for minutes a single apostrophe (') mark, and for seconds a double apostrophe ("). For example, 10 degrees, 15 minutes, 8 seconds would be written thus: 10°15′8″.

Figure 29-15 illustrates the measurement of angles in degrees. As shown in Fig. 29-15A, if a wheel makes one complete turn or revolution, a point on the wheel

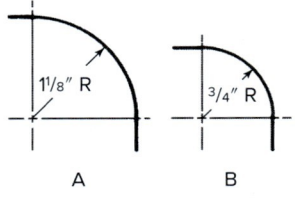
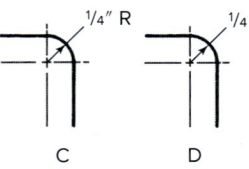

Fig. 29-13 Dimensioning radii.

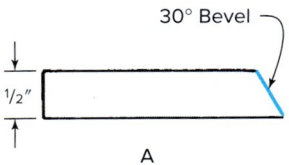
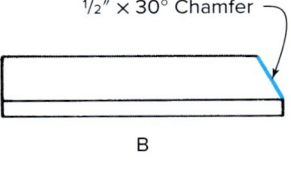
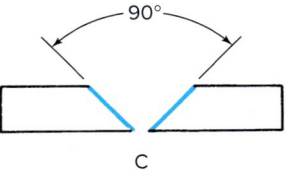
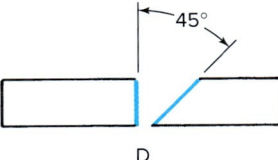
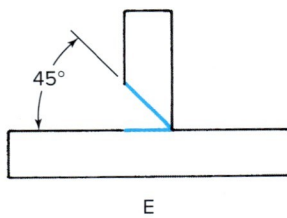

Fig. 29-14 Dimensioning angles.

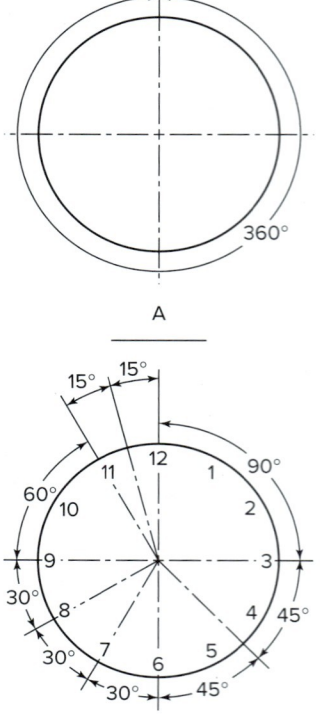

Fig. 29-15 Angular dimensions.

would travel 360°. Measurement in degrees can also be thought of in terms of the movement of the hands on a clock, Fig. 29-15B. If the minute hand moves from 12 around to 12, it travels 360°. The hour hand moves 90° from 12 to 3; 45° from 3 to half past 4, 30° from 6 to 7; 60° from 9 to 11; and 15° from 11 to half past 11.

Dimensions of Holes

Drilled Holes The size of drilled holes is often given in notes such as the one found in Fig. A-2, page 991. The note "Drill ¼ hole" means that the holes are made with a ¼-inch drill and will be ¼ inch in diameter.

Counterbored Holes Counterbored holes are usually designated by giving (1) the diameter of the drilled hole, (2) the diameter of the counterbore, (3) the depth of the counterbore, and (4) the number of holes to be drilled. For example, the note in Fig. 29-16 reads "½ drill, ¾ c'bore, ⅜ deep, 3 holes."

Countersunk Holes Countersunk holes are designated by giving (1) the diameter of the drilled hole, (2) the angle at which the hole is to be countersunk, (3) the diameter at the large end of the hole, and (4) the number of holes to be countersunk. See Fig. 29-17.

Screw Threads and Threaded Holes The two main types of screw threads are called the *National Coarse Thread Series* and the *National Fine Thread Series*. Figure 29-18 shows two ways to represent screw threads on a bolt. The note under part A, "1½″—6 NC" means that the diameter is 1½ inches, there are 6 threads for each inch of the threaded rod, and the thread is in the National Coarse Thread Series. The note under part B, "⅝″—18 NF" means that the diameter is ⅝ inch, there are 18 threads per inch of threaded length of rod, and the thread is in the National Fine Thread Series.

The total thread does not need to be shown on the drawing, so the conventional or simplified drawing of the thread can be found on the drawing, Fig. 29-19.

Figure 29-20 shows the cross section of blocks of metal in which the threaded holes do not go through the blocks. Part A shows the depth of the drilled hole (dimension X) and the depth of the threads (dimension Y).

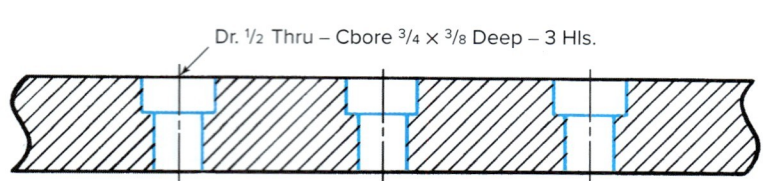

Fig. 29-16 Counterbored holes.

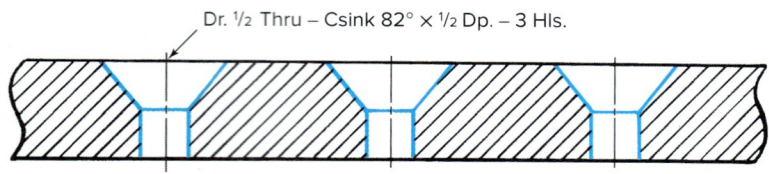

Fig. 29-17 Countersunk holes.

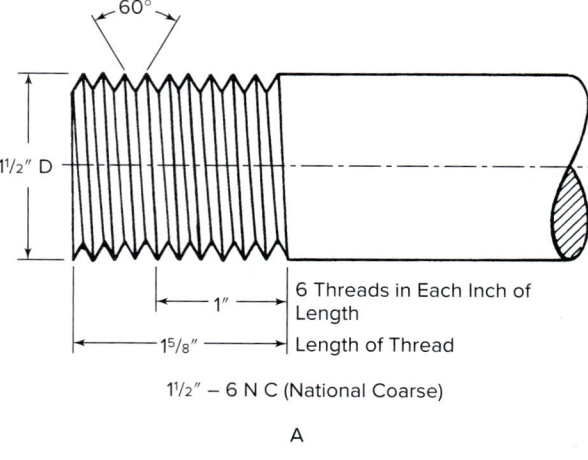

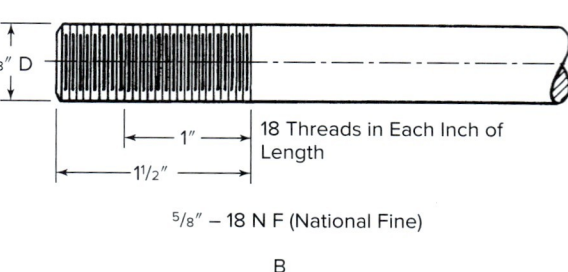

Fig. 29-18 Coarse and fine screw threads.

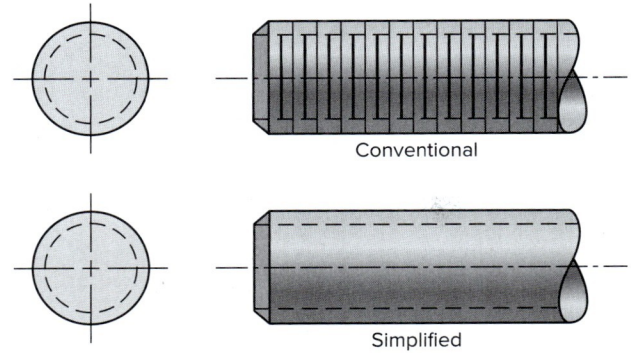

Fig. 29-19 Drawing symbols for external threads.

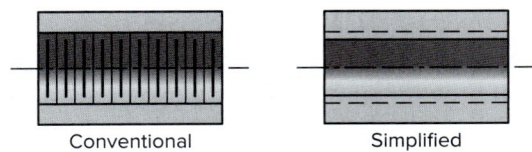

Fig. 29-22 Drawing symbols for internal threads that are used with sections.

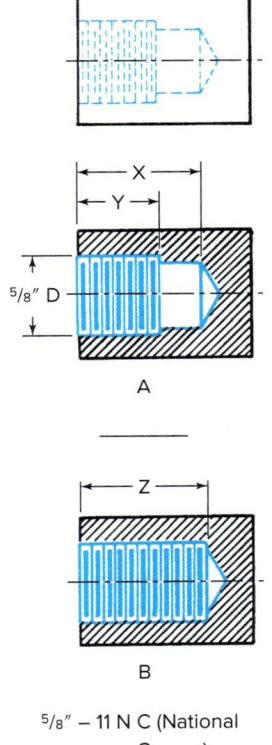

Fig. 29-20 Threaded holes.

In this case, the hole has not been threaded the full length of the drilled hole. In part B, the threads have been tapped the full length of the drilled hole, dimension Z. The thread is ⅝″—11 NC (National Coarse). Internal threads are shown in Figs. 29-21 and 29-22.

Coordinate Dimensions

The system of coordinate dimensions uses a baseline or datum surface off of which all dimensions are based. This baseline or datum surface acts as the starting point for all measurements. This system allows for more accurate location, as measurements are always based off of the starting point and you will not have the chance for cumulative errors when taking measurement at various points, Fig. 29-23.

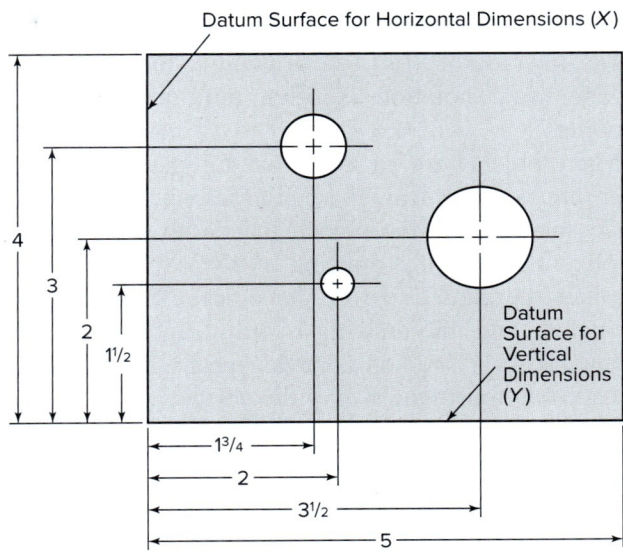

Fig. 29-23 A part dimensioned using coordinate dimensions. Note the baseline or datum surface, which acts as the starting point for all dimensions.

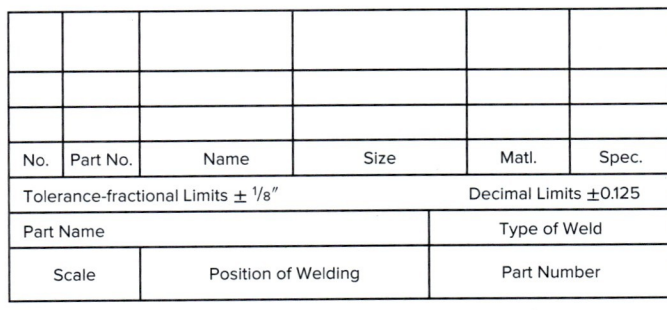

No.	Part No.	Name	Size	Matl.	Spec.
Tolerance-fractional Limits ± ⅛″				Decimal Limits ±0.125	
Part Name				Type of Weld	
Scale		Position of Welding		Part Number	

Fig. 29-24 Title block. Note ±⅛-inch tolerance and ±0.125-inch tolerance.

Tolerances Most drawings indicate the *limits of tolerance*, which is the permissible range of variation in the dimensions of the completed job, Fig. 29-24. A drawing may specify limits of tolerance as ±1/32 inch. This indicates that the dimensions of the object must be held to any size between one thirty-second of an inch larger or one thirty-second of an inch smaller than the size specified on the

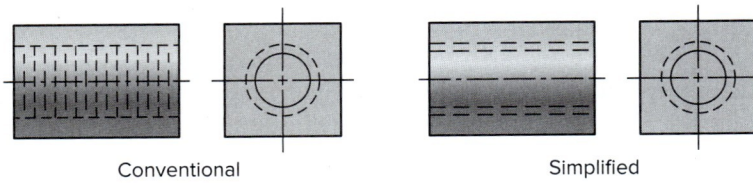

Fig. 29-21 Drawing symbols for internal threads.

drawing. Tolerances may be expressed in fractional, decimal, or SI metric dimensions.

Scale

When practical, drawings are made full size; that is, the drawing is made the same size as the object. The scale of a drawing is the ratio between the actual size of the drawing and the actual size of the object that the drawing represents. The scale of a drawing made full size is indicated as follows: *full size* or *scale 12" = 1'*. In other words, the ratio of a full-size drawing to the object is 1 to 1. The scale notation is often omitted for full-size drawings.

Many objects are so large that full-size drawings of them are not practical. In such cases, the drawings are made smaller than the object, that is, to scale. For example, if the scale of a drawing is ½" = 1", the drawing is one-half the actual size of the object. If the object is 12 inches long, the drawing is only 6 inches long. The notation scale ¼" = 1" on a drawing means that the drawing is made one-fourth the size of the object.

Your attention is called to the fact that the dimensions given on a drawing must be the actual dimensions of the object and are to be read as such. Never measure a drawing to determine the dimension to use. Paper shrinkage and dimension revisions may have occurred. Always use the dimension called for in the drawing.

Symbols and Abbreviations

Plans are specifically designed for the purpose of passing on information in the clearest and briefest manner possible. Sometimes the drawing must be supplemented with additional information. To avoid lengthy notes, symbols and abbreviations have been devised.

A *symbol* is a simple picture that represents an object or an idea. Its purpose is to condense information. The addition sign (+) is a common symbol. Except in a few instances, a drawing symbol usually shows the outline of an object, and it can be easily recognized as the object it represents. A drawing symbol also represents the same object to all trades. The weld symbols given in Chapter 30 are an example of these two features of drawing symbols.

Abbreviations are letters that are used for brevity. Like symbols, abbreviations are standardized so that they have the same meaning for all trades. The drawing symbol for inch is ". The abbreviation is *in*. Except for a few unusual cases, the letters in an abbreviation are contained in the words they stand for.

No matter how many standards are developed, there are always some engineers and architects who, through years of service, have developed symbols of their own.

They usually issue an index sheet with a set of drawings, and this sheet defines the meaning of their own symbols and abbreviations. It is always good practice to check these index sheets on a job before interpreting the drawing.

Table 29-5 lists symbols and abbreviations generally used in industry. Some occur quite frequently, while others are so seldom seen that they can be easily forgotten.

Geometric Shapes

The following geometric shapes are combined in different ways to indicate the shape of the object in a drawing:

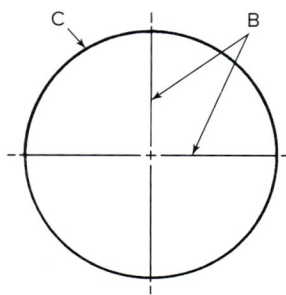
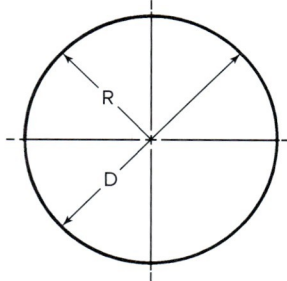

Fig. 29-25 Parts of a circle.

- *Circle* Parts of a circle are drawn as shown in Fig. 29-25. The lines *B* running through the center of the circle are center lines. The line *C* is the *circumference* of the circle (the distance around the circle). The line *D* is the *diameter* of the circle (the distance from one edge of the circle to the other). The diameter may be thought of as the width of the circle. The line *R* is the *radius* of the circle (the distance from the center of the circle to the circumference). The radius is one-half the length of the diameter. A *right angle* is one-fourth (90°) of a circle. Lines *B* form a right angle.
- *Rectangle* A rectangle, Fig. 29-26A (p. 962), is a figure with four right angle (square) corners. The opposite sides are parallel. Two of the sides are of unequal length.
- *Square* A square, Fig. 29-26B, is a rectangle with four sides of equal length. The corners of a square are right angles.
- *Diagonal* A straight line connecting the opposite corners of a rectangle or square is called a diagonal, Fig. 29-26C.
- *Triangle* A triangle, Fig. 29-26D and E, is a flat figure having three straight sides.
- *Equilateral triangle* An equilateral triangle, Fig. 29-26D, is a triangle whose three sides are equal and whose three angles are equal.
- *Right triangle* A right triangle, Fig. 29-26E, is a triangle with one right (square) angle.

Table 29-5 Drawing Symbols and Abbreviations

Alternating current . . . . AC
American Society for
  Testing Materials. . . . . ASTM
American Standard
  Association . . . . ASA
American Welding Society . . . . AWS
American Wire Gauge . . . . AWG
Ampere . . . . AMP
And . . . . &
Angle. . . . . ∠
Approximate. . . . . APPROX
Asbestos . . . . ASB
Average . . . . AVG
Barrel . . . . BBL
Basement . . . . BASMT
Bevel . . . . BEV
Birmingham Wire Gauge. . . . . BWG
Brass . . . . BRS
Brass, SAE # . . . . BRS #
British Thermal Unit . . . . BTU
Bronze, S.A.E. BRZ #
Brown & Sharpe . . . . B&S
Building. . . . . BLDG
Butt welded . . . . BW
Cast iron CI
Center to center . . . . C to C
Centerline . . . . ℄
Chamfer . . . . CHAM
Channel. L or CHAN
Circular. . . . . CIR
Clean out . . . . CO
Cold-rolled steel. . . . . CRS
Column . . . . COL
Composition. COMP
Concrete . . . . CONC
Construction . . . . CONST
Counterbore . . . . CBORE
Countersink . . . . CSK
Copper . . . . COP
Cubic feet . . . . CU FT
Cubic feet per minute . . . . CFM
Cylinder, cylindrical . . . . CYL
Degree . . . . DEG or °
  (Example). . . . . 30 DEG. or 30°
Detail . . . . DET
Diagonal . . . . DIAG
Diameter . . . . DIA
Dimension . . . . DIM
Direct current . . . . DC
Ditto . . . . DO
Down . . . . DN
Drawing . . . . DWG
Drill . . . . DR
Electric . . . . ELEC
Elevation. . . . . EL or ELEV
Extension . . . . EXTEN
Exterior . . . . EXT
Extra heavy . . . . XHVY
Fabricate . . . . FAB
Feet . . . . FT or '
  (Example) . . . . 10 FT or 10'
Feet and inches (Example)
  . . . . 5 FT-11 IN or 5'-11"
Field weld . . . . FW
Figure . . . . FIG
Finish symbol . . . . ∨ or ✱
Flange . . . . FLG
Floor . . . . FL
Floor drain . . . . FD or FLDR
Forging dies . . . . FOD
Front view . . . . FV
Gallon . . . . GAL
Galvanized iron . . . . GI
Gauge . . . . GA
Glass . . . . GL
Hexagonal . . . . HEX
Holes . . . . HLS
Horsepower . . . . HP
Inches . . . . IN. or "
  (Example). . . . . 10 IN. or 10"
Inside diameter . . . . ID
Insulation . . . . INS
Interior . . . . INT
I-beam I
I-beam column . . . . I COL
  (Example) 8" I-beam, 30
  pounds per ft. 8" 30 # I COL
Kilowatt . . . . KW
Lap-Weld . . . . LW
Left hand . . . . LH
Left side view . . . . LSV
Linear feet . . . . LIN FT
Malleable iron . . . . MI
Maximum . . . . MAX
Mechanical . . . . MECH
Mild steel . . . . MSM STL
Minimum . . . . MIN
National coarse (screw
  threads) . . . . NC
National fine (screw
  threads) . . . . NF
National form . . . . N
Number . . . . NO. or #
Object . . . . OBJ
Opening . . . . OPNG
Ounce . . . . OZ
Outside diameter . . . . OD
Overall . . . . OA
Oxidize . . . . OXD
Perpendicular . . . . PERP or ⊥
Piece(s) . . . . PC, PCS
Pitch . . . . P
Plate . . . . ℞
Point . . . . PT
Pound . . . . LB or #
Pounds per square inch . . . . PSI
Projection . . . . PROJ
Radius . . . . R or RAD
Reference . . . . REF
Required . . . . REQD
Revision . . . . REV
Revolutions per minute . . . . RPM
Right hand . . . . RH
Right side view . . . . RSV
Room . . . . RM
Round . . . . RD or ⌀
Screwed . . . . SCRD
Seamless . . . . SMLS
Section . . . . SECT
Sheet steel . . . . SHT STL
Side view . . . . SV
Society of Automotive
  Engineers . . . . SAE
Specification . . . . SPEC
Square . . . . SQ
Square feet . . . . SQ FT or ⬜ FT
Square inch(es) . . . . SQ IN
Steel . . . . STL
Steel casting . . . . STL CSTG
Standard . . . . STD
Technical . . . . TECH
Thread(s) . . . . THD, THDS
Threaded . . . . THRD
Through . . . . THRU
Tongue and groove . . . . T&G
Top view . . . . TV
Typical . . . . TYP
United States Standard . . . . USS
Volt V
Volume . . . . VOL
Watts . . . . W
Weight . . . . WT
Wrought iron . . . . WI
Yard . . . . YD

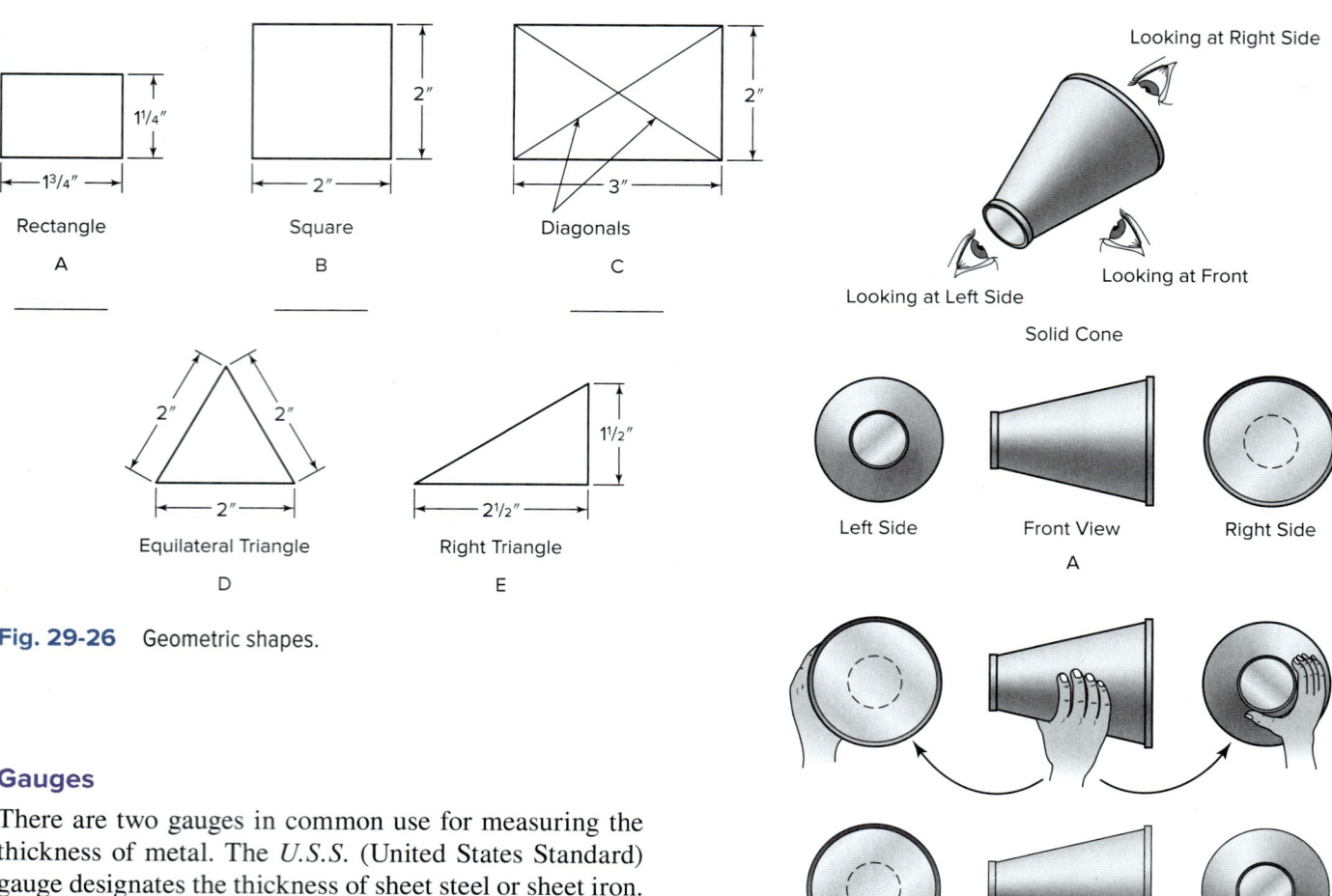

Fig. 29-26 Geometric shapes.

Gauges

There are two gauges in common use for measuring the thickness of metal. The *U.S.S.* (United States Standard) gauge designates the thickness of sheet steel or sheet iron. The *B. & S.* (Brown and Sharpe) gauge designates the thickness of sheet brass and copper wire. In both cases, the larger the gauge number, the thinner the material. For example, 16 gauge is thinner than 9 gauge. Gauge sizes are given in the Appendix.

Types of Views

Drawings are made so that they describe the object in enough detail to be fabricated. Workers must learn to visualize what the object looks like from studying a drawing. It is important for workers to determine immediately what their position is in regard to the picture. Are they looking down at the object, at its side, or at the inside? Workers can obtain a mistaken impression by failing to orient themselves.

This is where third-angle and first-angle projection come into play. In the third-angle projection method, it is as if you are moving your eye around a fixed object. With first-angle projection, it is as if you are stationary and moving the object with your hand while holding it up against the drawing surface. Since you are not moving and the object is being rolled on the drawing surface by your hand, you get a completely different presentation of how the object is drawn. Figure 29-27 shows a simple cone-shaped object in both first-angle and third-angle projection. In most cases the view of the object that gives the best presentation of what the object looks like is drawn and considered the front view. This does not mean it is necessarily the front of the object. For example, a TIG torch front view on a drawing would most likely be actually the side of the torch because it most clearly shows its shape. You must learn to orient yourself to the drawing and how the views are laid out from this front view. If the object were simple, perhaps one view, notes, and the material list would be all that are required. If, however, the object is more complex, perhaps two or three views may be required to convey all the needed visual information. The maximum regular views that can be shown are six.

Fig. 29-27 (A) Third-angle projection (ISO-A), very common in the United States. Much like moving your eye around the object. (B) First-angle projection (ISO-E), very common in Europe and other areas around the world. Much like rolling the object on the drawing surface or on a computer screen.

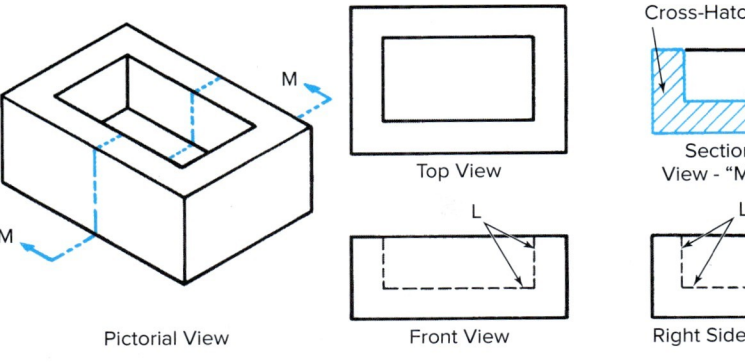

Fig. 29-28 Pictorial view of a core box and three-view drawing.

These would be the front, top, bottom, left and right sides, and the backside views. As trade continues to expand between nations, your ability to understand a drawing in both ISO-A third-angle or ISO-E first-angle projection will become more important.

Three-View Drawings

In the United States, the general practice has been to use third-angle projection to show first the front view, then the top view directly above it, and a side view; see the core box in Fig. 29-28B. These three basic views are enough to show all the details of an object. The inside or hidden surfaces that cannot be seen in the front and side views are shown by the use of invisible edge lines (L).

It is also possible to show the inside construction of an object by resorting to what are called sectional views, Figs. 29-28 and 29-32.

The pictorial drawing is rather easy for the untrained worker to understand. It is useful in showing the general appearance of simple objects. As most manufactured articles are rather complicated, it is necessary to use a system of three-view drawings. In addition to the three views, a pictorial drawing is used in this text to help make clear to you the approximate appearance of the object when seen from a corner. In general, isometric pictorial drawings are used only as an aid in learning to read working drawings. Pictorial drawings are seldom found on industrial drawings.

You should try to visualize the completed object as you study each view shown on the three-view drawing along with the pictorial drawing. Thus, you will understand better the relationship between the object as you are used to seeing it and the way this object is represented on a working drawing. The ability to look at a three-view drawing and visualize the object as it actually appears when looked at from any angle is the basis for reading drawings.

The combination of front, top, and right side views shown in Fig. 29-29 represents the three views most commonly used by drafters to describe simple fabrications. In the pictorial view we can see the top surface, the front surface, and the right side surface of the camera. If you looked directly down on the camera with the front of it toward you, you would see the top as shown in the *top view* of Fig. 29-29. If you looked directly at the front of the camera, you would see the front as shown in the *front view*. If you looked directly at the right side, you would see it as shown in the *right side view*.

In Fig. 29-30, page 964, a steel plate is shown in a pictorial view and the normal three views. The top surface, the front surface, and the right side or end surface are shown in the pictorial view. If you were above the plate looking

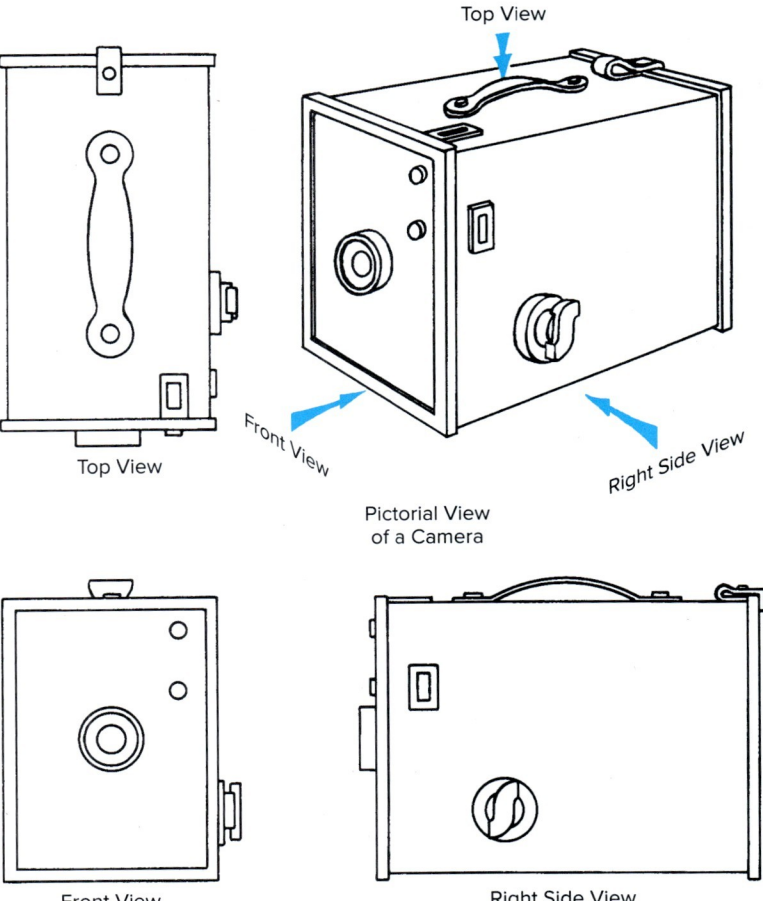

Fig. 29-29 Three-view drawing of a box camera.

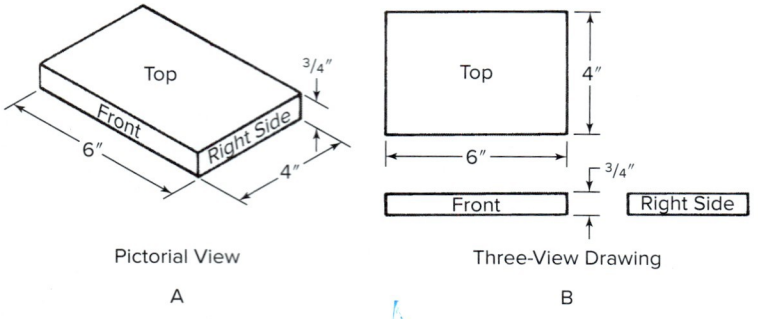

Fig. 29-30 Pictorial view of a steel plate and typical three-view drawing.

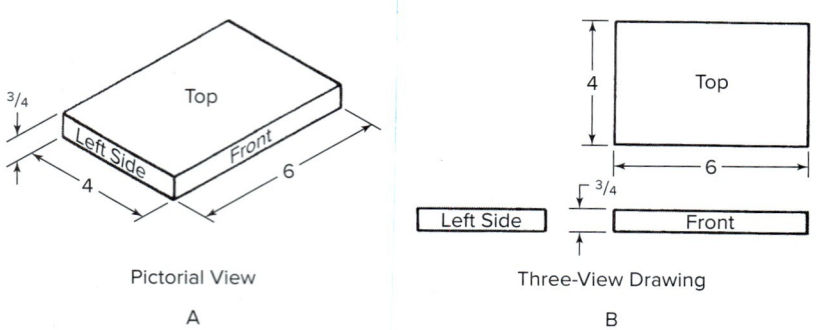

Fig. 29-31 Pictorial view of steel plate and three-view drawing using left side.

down on it, you would see the top surfaces as shown in the top view. If you were to see the plate from the front, you would see the front surface as it is shown in the front view. If you were to the right of the plate looking at the right side, you would see the right side surface as it is shown in right side view.

Sometimes, the left side view is shown instead of or in addition to the right side view. Figure 29-31A is a pictorial view of the same steel plate showing the top, front, and left surfaces. These three faces of the plate are shown in the three-view drawing, Fig. 29-31B.

If the object is very complicated, it may be necessary to use more than three views. Sometimes the top, front, right and left side views, as well as the back and bottom views and special detail views are needed to present the shape, dimensions, and shop operations involved in making the object.

When all the required details can be shown in two views, the third view can be omitted. There are times when only one view shows all the necessary details. In this case the best view is shown on a drawing. Although it is good practice to show an object in the most convenient manner, an attempt is usually made to show it in its natural position.

Sectional and Detail Views

Sometimes in order to give the worker the necessary information to make the article, other views such as details and sections which show the inside of the object are given.

It is possible to show the inside construction of an object with *sectional views*. If we assume an imaginary cut on sectional line *M*, Fig. 29-32, we would have sectional view M-M instead of the usual side view. Line *M*, with arrows at its ends, shows not only the location of the imaginary cut in the top view, but also the direction in which the observer would look in order to see the sectional view M-M. Line *M* is called a *section line*. The diagonal lines indicate that part of the solid section of the box that would be cut through for the purpose of showing the sectional view. The diagonal lines are called *cross-hatching*.

Detail views must be used when the scale of a drawing is so small as to make it impossible to show a clear picture of an object. When this happens, the object is drawn in a simplified form on the main drawing, and then the reader is referred to a drawing of the object in detail by a notation such as *SEE DETAIL A*. By using a much larger scale, the detail drawing can show the object more clearly. This method is also used when large numbers of an object are shown and when a certain assembly is used in several places. The main drawing then gives the location of the assembly and a typical detail is drawn somewhere else to show it more clearly.

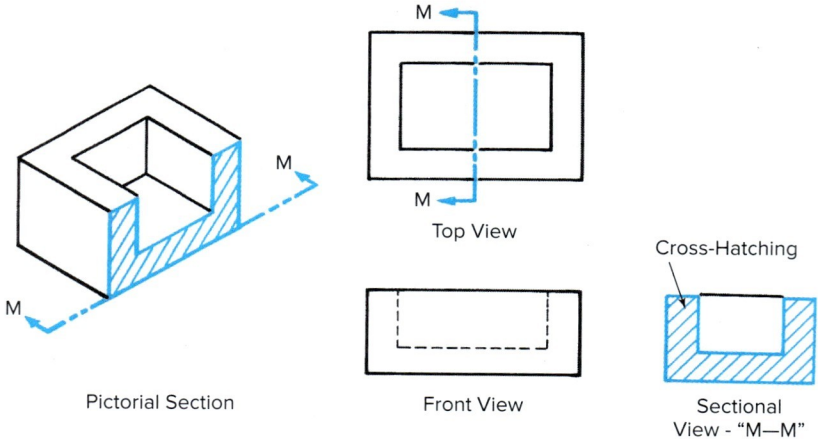

Fig. 29-32 Pictorial view of a core box and method of sectioning.

The welder should study each view of a drawing carefully because each of them shows structural features necessary in the completed weldment.

Piping Drawings

Piping drawings can be very specialized drawings, and they require the use of various symbols to aid in drawing and interpretation. Table 29-6 shows the American National Standards Institute's Symbols for pipe fittings. Pipe drawings are used to show how the assembly of a pipe system goes together. Pipe spools are segments of piping systems that can be fabricated in a shop and then sent on to the field site for final installation. Pipe spools allow for faster, higher quality welding and more accurate assembly in a more controlled environment. Pipe drawings are done with both the two-line and single-line methods. Since the single line is much easier to draw and a skilled pipe fitter/welder can quickly read and interpret its meaning, this is the most popular method used in industry. The drawings are usually done in pictorial view, as the typical three-view drawing would be very difficult to read and interpret, Fig. 29-33.

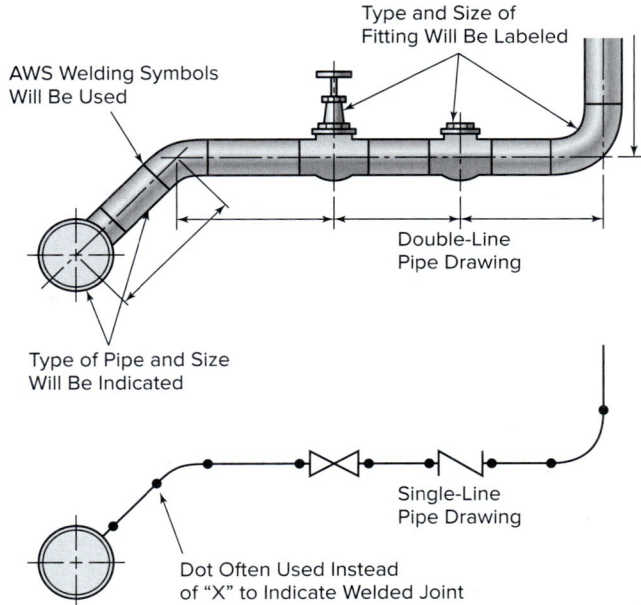

Fig. 29-33 Double and single-line pipe drawings. On double-line drawing dimension points are based on the pipes centerline. As pipe's inside and outside diameters vary based upon diameter and wall thickness (schedule number).

Table 29-6 ANSI Pipe Symbols[1]

Fitting	Flanged	Screwed	Welded[2]
Bushing	N/A	(symbol)	(symbol)
Cap	N/A	(symbol)	N/A
Cross, Straight	(symbol)	(symbol)	(symbol)
Joint, Connecting Pipe	(symbol)	(symbol)	(symbol)
Joint, Expansion	(symbol)	(symbol)	(symbol)
Lateral	(symbol)	(symbol)	N/A
Pipe Plug	N/A	(symbol)	N/A
Reducer, Concentric	(symbol)	(symbol)	(symbol)

Table 29-6 (Concluded)

Fitting	Flanged	Screwed	Welded[2]
Sleeve			
Union			
Elbow			
45°			
90°			
Turned Down			
Turned Up			
Tee			
Straight			
Outlet Up			
Outlet Down			
Valve			
Check			
Gate			
Globe			

[1]These are a few of the many ANSI pipe symbols that are used.
[2]A • can be used instead of the X to represent a welded butt joint.

Adapted from American Welding Society, 2007

CHAPTER 29 REVIEW

Part I. Review

Multiple Choice

A. Choose the letter of the correct answer.

1. Is it important that a welder be able to read drawings? (Obj. 29-1)
   a. No, paper and ink will no longer be used; everyone will have a portable computer to view the object.
   b. No, welders make long production runs of parts that may last for weeks. They can then memorize how the parts go together, so they do not need to read drawings.
   c. Yes, pictures and notes on drawings are only a method of passing information from person to person. Whether the drawings are on paper or the screen, they must be read and interpreted.
   d. Yes, computers will never be used in the harsh environment of a weld shop, so only paper drawings will be used.
2. What is the purpose of a drawing? (Obj. 29-2)
   a. Copies of the original drawing are used on the job site.
   b. Original drawings must be handled very carefully in the shop environment.
   c. Prints are used for law enforcement for identification purposes, as in fingerprints.
   d. Blueprints have no purpose in modern production shops. It is all being done with computers.
3. Which method(s) can be used for copying a drawing? (Obj. 29-2)
   a. Ink-jet printer
   b. Laser printer
   c. Photocopier
   d. All of these
4. What is the meaning of *scale* as applied to drawings? (Obj. 29-2)
   a. Deals with how much the object weighs.
   b. Is not required. All drawings are made the full size of the object they are showing.
   c. The scale of a drawing is the ratio between the actual size of the drawing and the actual size of the object.
   d. All of these.
5. Which projection is used for drawings in the United States? (Obj. 29-2)
   a. First-angle
   b. Second-angle
   c. Third-angle
   d. Fourth-angle
6. What is considered a very accurate dimensioning method? (Obj. 29-3)
   a. Coordinate
   b. Self-imposed
   c. Trigonometric
   d. Trinomial
7. Which of the following is not considered a geometric shape? (Obj. 29-4)
   a. Circle and rectangles
   b. Squares and triangles
   c. Equilateral and right
   d. Both a and b
8. A circle contains how many degrees? (Obj. 29-4)
   a. 180
   b. 270
   c. 360
   d. Circles do not contain degrees; they are not hot.
9. What view of an object is usually drawn? (Obj. 29-5)
   a. Front
   b. Side
   c. Top
   d. All of these
10. A single-view drawing is never used in industry. (Obj. 29-5)
    a. True
    b. False

Review Questions

Write the answers in your own words.

11. Drawings and shop sketches are considered to be what in industry? (Obj. 29-1)
12. What are detailed drawings used for? (Obj. 29-2)
13. Describe the coordinate dimensioning method. (Obj. 29-3)
14. Sketch an equilateral triangle with a 2-inch side. (Obj. 29-4)
15. Explain the two systems used on pipe drawings and the type of view typically used. (Obj. 29-5)
16. Explain how a section view is used on a drawing and what types of lines are used to communicate this information. (Obj. 29-5)
17. Draw a simple three-view sketch of a rectangular block 4 inches long, 3 inches wide, and 2 inches thick. Mark dimensions on the views. (Obj. 29-6)

18. Describe the difference between a hard conversion from U.S. Customary to SI versus a soft conversion and give an example. (Obj. 29-6)
19. What are the two common gauges used to measure the thickness of metals? (Obj. 29-6)
20. What are the two main types of screw threads used in the United States, and should these be hard converted to SI units or should they be soft converted? Why? (Obj. 29-6)

B. Draw the appropriate line for each of the following terms:

1. Border line
2. Visible edge line
3. Invisible edge line
4. Extension line
5. Center line
6. Section line
7. Pointer
8. Dimension line
9. Break line
10. Phantom line

C. Draw the appropriate figure for the following:

1. Rectangle
2. Square
3. Diagonal
4. Triangle
5. Right triangle

Part II. Interpreting Drawings

Following are a number of illustrations that present the views common to shop drawings. Study them carefully and answer the questions that follow each drawing.

A. One-View Drawings. Figures A-1 through A-4 show four different objects that are presented as single-view drawings. Flat objects of this kind can be shown with all dimensions given, and the thickness and type of material are given by a note.

A.1 Washer

- *Center lines:* Your attention is called to lines *B* and *C* which are drawn through the center of the circles. These are center lines and are always used on views of drawings representing circles and generally on drawings representing parts of circles. They are called dot-and-dash lines, but their construction actually consists of short and long dashes. Center lines are not a part of the drawing of the object.

- *Diameter and radius:* Line *G* is a dimension line indicating the outside diameter of the washer. Diameter is often abbreviated *DIA*. One-half of the diameter is the radius, abbreviated *R*.

  1. Of what material is the washer made?
  2. What is the abbreviation for wrought iron?
  3. What is the thickness of the washer in decimals? This is approximately 5/32 inch.
  4. What is the outside diameter of the washer? Do not include the word *diameter* or its abbreviation in your answer.
  5. What is the diameter of the hole?
  6. What is the radius of the outside of the washer? Do not include the word *radius* or its abbreviation in your answer.
  7. What is the distance from the edge of the hole to the outside of the washer?
  8. What kind of lines are lines *F* and *G*?
  9. What kind of lines are lines *H* and *E*?
  10. What kind of lines are lines *B* and *C*?
  11. What do we call lines that refer to or point to some part of a drawing?
  12. For what purpose is a dot-and-dash line used?

A.2 Lock Keeper Figure A-2 is a drawing of a lock keeper that is fastened to a doorjamb or doorframe to receive the bolt from the lock.

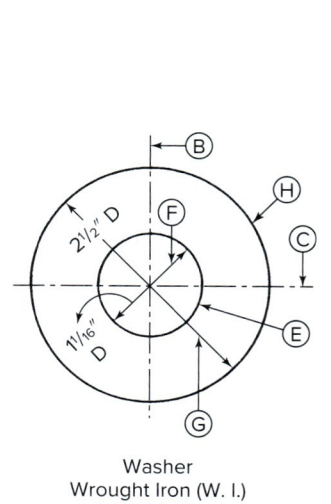

Washer
Wrought Iron (W. I.)
0.1563" Thick

Fig. A-1 Washer.

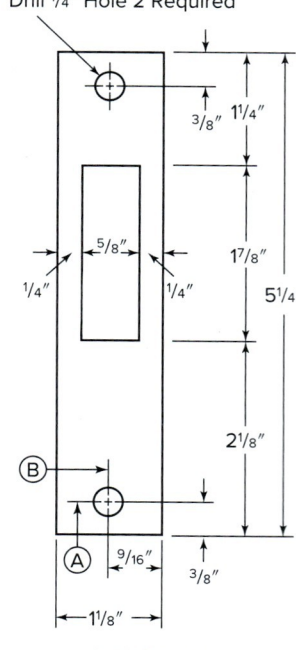

Lock Keeper
Sheet Brass
No. 14 (0.064") B. & S. GA.

Fig. A-2 Lock keeper.

- *Brown and Sharpe Gauge:* The notation "No. 14 (0.064″) B. & S. Ga." means under 14 gauge, Brown and Sharpe. The brass in this size is approximately 1/16 inch thick. This is another one of the several gauges, such as U.S.S., for designating thickness of metals. The Brown and Sharpe gauge is the most common one for sheet brass and copper wire.
- *Dimensioning:* Where a space is too small for the dimension line and the dimension, the dimension lines are placed outside the extension lines. Note the 3/8-inch dimensions at the bottom and the top of the drawing. In this case, the distance dimensioned is from the center line to the extension line, or from arrowhead to arrowhead. A similar example is that of the 1/4-inch dimension which locates the rectangular hole from the sides of the object.
- *Drilled Holes:* The size of drilled holes is often given in notes such as the one found on this drawing. The note "DRILL 1/4″ HOLE 2 REQ'D." means that the holes are made with a 1/4-inch drill and will be 1/4 inch in diameter. This note refers not only to the hole to which it points but also to the hole near the bottom of the lock keeper. Note that lines A and B are center lines that are also used as extension lines for dimensions.

  1. Of what material is the lock keeper made?
  2. What is the overall length?
  3. What is the overall width?
  4. What is the thickness in decimals of an inch?
  5. What number gauge metal is used?
  6. What kind of lines are lines A and B?
  7. What is the diameter of the drilled holes?
  8. How many drilled holes are there?
  9. What is the distance from the center of the drilled hole, near the bottom, to the bottom of the keeper?
  10. What is the distance from the center of the drilled holes to the left side of the keeper?
  11. What is the length of the rectangular hole?
  12. What is the width of the rectangular hole?
  13. What is the distance from the right side of the keeper to the right side of the rectangular hole?
  14. What is the distance from the bottom of the rectangular hole to the bottom of the keeper?

A.3 Dial Pointer Figure A-3 is a dial pointer used on an oil-burning engine. The left end of the pointer is partly circular in shape with a circular hole. The other end tapers to a blunt point. The pointer is made of 16-gauge metal which is 0.0625 inch thick.

1. What kind of line is line K?
2. What kind of line is line Y? (This is not a dimension line.)
3. What is the distance from the center line of the drilled hole to the right end of the pointer?
4. What is the radius of the circular end of the pointer?
5. What is the overall length of the pointer?
6. What is the overall width of the pointer?
7. How wide is the pointer at the right end?
8. What is the diameter of the drilled hole?

A.4 Spacing Washer The spacing washer shown in Fig. A-4 is a part of an electric motor. The washer is made of 28-gauge steel that is 0.0156 inch thick. Its general shape is somewhat like the letter U.

1. What is the outside diameter for the washer?
2. What kind of lines are lines E and F?
3. What kind of lines are lines A and B?
4. What kind of lines are lines G and H?

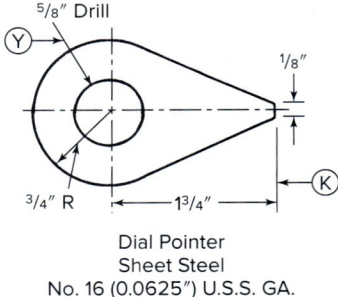

Fig. A-3 Dial pointer.

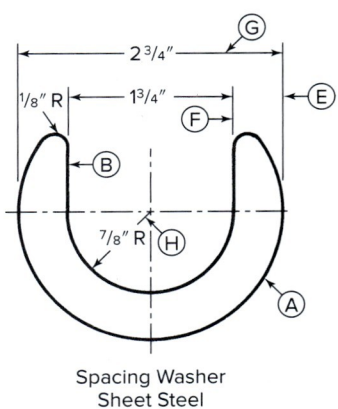

Fig. A-4 Spacing washer.

5. What is the radius of the small ends at the back of the washer (at the top of the U)?
6. What is the distance between the straight edges forming the back part of the opening?

B. Two-View Drawings. Only two views, such as the top and front views or the front and side views may be adequate for some simple objects. The drawing of the spring cap, Fig. B-1, consists of only two views—the front view and the side view. When two views of an object are the same, or if the third view does not give additional information, which an experienced shop worker could not get from the other two views, only two views are given. Since the drawing of the top view of the cap would be exactly the same as the drawing of the front view, the top view is omitted.

B.1 Spring Cap Figure B-1 is a two-view drawing of a spring cap used on an engine.

1. Which lines in the front view represent the cylindrical hole in the body?
2. Which circle in the side view represents the hole in the body?
3. Which circle in the side view represents the short cylinder between the flange and the body?
4. Which circle in the side view represents the outside cylindrical surface of the flange?
5. Which circle in the side view represents the outside surface of the body?
6. Which lines in the front view represent the cylindrical surface of the larger hole at the right end of the spring cap?
7. Which line in the side view represents the larger cylindrical hole?
8. Which line in the front view represents the shoulder (offset) that is formed where the larger hole and the smaller hole meet?
9. Between which two circles in the side view does the shoulder lie?
10. Which line in the front view represents the shoulder formed where the flange and the short cylinder meet (the left side of the flange)?
11. Between which two lines in the side view should the shoulder lie?
12. How many surfaces are visible in the side view?
13. If you were to look at the left end of the spring cap, how many flat surfaces would be visible?
14. What is the diameter of the hole in the body?
15. What is the diameter of the larger hole in the right end?
16. What is the length of the body?

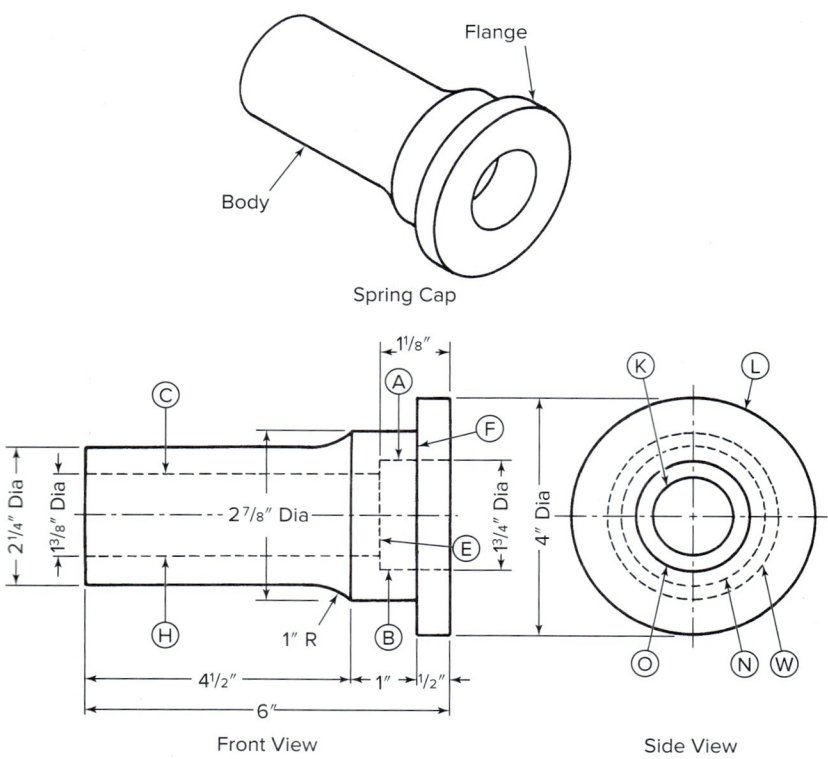

Fig. B-1 Sparing cap.

17. What is the radius of the profile (the outline of the widened part at the right end of the body)?
18. What is the length of the cylinder between the flange and the body?
19. What is the thickness of the flange (the distance from the left to right)?
20. How long is the large cylindrical hole at the right end?
21. What is the diameter of the body?
22. What is the thickness of the wall at the left end of the body?
23. What is the width of the shoulder (offset) between the two cylindrical holes? Note that the diameter of the large hole includes the diameter of the small hole plus two shoulders.
24. What is the diameter of the short cylinder next to the flange?
25. If the left side view of the spring cap were drawn, how many cylindrical surfaces would be represented by dotted lines?

B.2 Porcelain Tube The porcelain tube, Fig. B-2, is made of clay and baked in a kiln. Tubes were formerly used in electrical work when wires were run through joists, studs, partitions, or any wooden member of a building. The building laws of almost all large cities prohibit this type of wiring and require metal pipes called conduits. Knob and tube wiring is still permitted in some districts. The porcelain tube, however, is an excellent object to be studied as a two-view drawing. Center lines are used on drawings of cylindrical or cone-shaped objects.

The outside diameter of the top face of the tube and the outside diameter of the cylindrical part are the same, $^{11}/_{16}$ inch. If these diameters were different, the surface of the cylinder would have to be shown in the top view by a dotted circle. Since they are the same, the visible (solid) circle covers the dotted circle. When looking at the top view, you must understand that circle C represents both the outside edge of the top face of the tube and the outside surface of the cylindrical part.

1. What is the largest diameter?
2. What is the length of the cylindrical part?
3. What is the length of each of the conical parts? Length in this case is taken along the axis or center of the tube.
4. What is the diameter of the base of the conical parts?
5. What is the outside diameter of the cylindrical part?
6. What is the diameter of the hole?
7. What is the thickness of the wall of the cylinder? (Note that the outside diameter of this cylinder includes the diameter of the hole and the thickness of the two walls.)
8. Which circle in the top view represents the hole in the tube?
9. Between which two circles in the top view does the conical surface F in the front view lie?
10. Which lines in the front view represent the hole?
11. Which circle in the top view represents the outside edge of the top surface of the tube?
12. Which circle in the top view represents the outside surface of the cylinder?
13. Between which two circles in the top view does the conical surface G in the front view lie?
14. Which surface in the top view is the top of the tube?
15. How many circles in the top view represent cylindrical surfaces?

C. Three-View Drawings. It has been previously shown that simple objects can be presented as one-view or two-view drawings. Complicated objects must be presented by three-view drawings for the welder to be able to fabricate

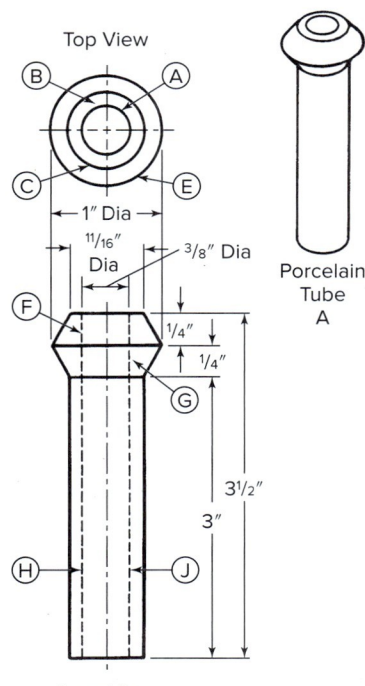

Fig. B-2   Porcelain tube.

the final component. The views are arranged in a group to show the shape and construction of the object.

C.1 Channel The channel shown in Fig. C-1 is a steel structural member used in building construction, shipbuilding, railroad work, and the machine metal trades. Channels of various shapes are used.

The front and back legs of the channel are called *flanges*, and the other member is called the *web*.

In the top view, surface B is the top of the web. This surface is represented by line V in the side view. In the front view, the top surface of the web is not visible. Therefore, line N, representing this surface, is a dotted line to indicate a hidden surface. Line G in the top view and U in the side view represent the inside flat surface of the front flange. Lines E in the top view and W in the side view represent the inside flat surface of the back flange. Neither of these surfaces is visible in the front view.

1. Which line in the top view represents the front of the channel?
2. Which line in the side view represents the front of the channel?
3. Which surface in the front view is the front?
4. Which line in the top view represents the back of the channel?
5. Which line in the side view represents the back of the channel?
6. Which surface in the side view is the right end?
7. Which line in the side view represents the inside flat surface of the back flange?
8. Which line in the top view represents the inside flat surface of the back flange?
9. Which line in the side view represents the curved surface at the top of the back flange?
10. Which surface in the top view is the curved surface at the top of the back flange?
11. Which line in the side view represents the upper flat surface of the web?
12. Which surface in the top view is the upper flat surface of the web?
13. Which line in the front view represents the upper flat surface of the web?
14. Which line in the side view represents the rounded corner where the back flange and the web meet?
15. Which surface in the top view is the rounded corner where the back flange and the web meet?
16. What is the distance from the top of the flat surface of the web to the top of the channel?
17. What is the thickness of the web?
18. What is the thickness of the flanges?
19. What is the distance between the inside flat surfaces of the flanges?
20. What is the radius of the curved surfaces at the top of the flanges?
21. What is the radius of the curved surfaces where the web and the flanges meet?
22. What is the height (width) of the inside flat surface of the front flange?

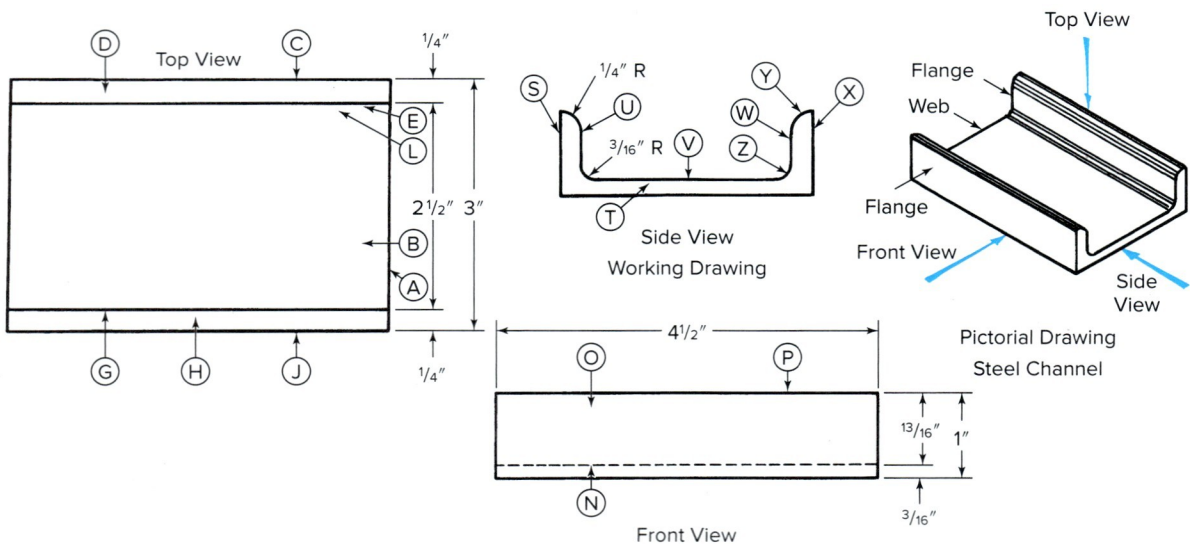

Fig. C-1 Steel channel.

23. What is the width of the upper flat surface of the web?
24. Line *E* in the top view represents a surface. Which line in the side view also represents this surface?
25. Is the surface referred to in question 24 visible in the front view?
26. Which line in the top view represents surface *T* in the side view?
27. Which surface does line *N* in the front view represent in the top view?
28. Which line in the side view represents surface *B* in the top view?
29. Which line in the front view represents line *V* in the side view?
30. Which surface in the top view represents line *Z* in the side view?
31. Which line in the top view represents line *U* in the side view?
32. Is the surface referred to in question 31 visible in the front view?
33. Which line in the top view represents surface *O* in the front view?
34. Which line in the side view represents surface *D* in the top view?
35. Which line in the front view represents the tops of the front and back flanges?
36. How many flat surfaces are visible in the top view?
37. How many curved surfaces are visible in top view?

C.2 Standard Steel Tee Figure C-2 is a drawing of a standard steel tee, so called because its end view resembles the letter *T*. Tees are used in structural members of buildings and steel bridges. T-shaped parts are often used in machinery.

The upright member of the tee is known as the *stem*. The bottom member is known as the *flange*.

1. What is the scale of the drawing?
2. Of what material is the tee made?
3. What is the overall length?
4. What is the overall width?
5. What is the overall height?
6. Which line in the top view represents the back of the flange?
7. Which line in the top view represents the back of the stem?
8. Which line in the front view represents the top of the front part of the flange?
9. Which line in the side view represents the front of the stem?
10. Which flat surfaces (note there are two) in the top view are the top of the flange?
11. Which line in the front view represents the right side of the tee?
12. Which surface in the top view is the top of the stem?
13. Which surface in the front view is the front of the stem?
14. Which surface in the front view is the front of the flange?
15. What is the thickness of the flange?
16. What is the thickness of the stem?
17. What is the distance from the front of the flange to the front flat surface of the stem?
18. What is the distance from the top of the stem to the top flat surface of the flange?
19. Note the rounded or curved corner where the flange and the stem meet. This makes the connection between the flange and the stem much stronger. What is the radius of this curve?
20. The distance from the front of the flange to the front surface of the stem is $11/16$ inch. Since $3/8$ inch of this distance is taken up by the curved surface, the flat surface of the top of the flange at the front is $11/16$ inch wide. This is also true of the flat surface at the back of the flange. What is the width (height) of the flat surface of the front of the stem?
21. Which surface in the front view does line *Q* in the side view represent?
22. Which line in the front view does surface *J* in the top view represent?
23. Which line in the top view does surface *B* in the front view represent?
24. Which line in the side view represents line *P* in the top view?
25. Which surface in the top view does line *X* in the side view represent?
26. Which line in the side view represents surface *L* in the top view?
27. Which surface in the top view does line *T* in the side view represent?
28. Which surface in the front view does line *H* in the top view represent?

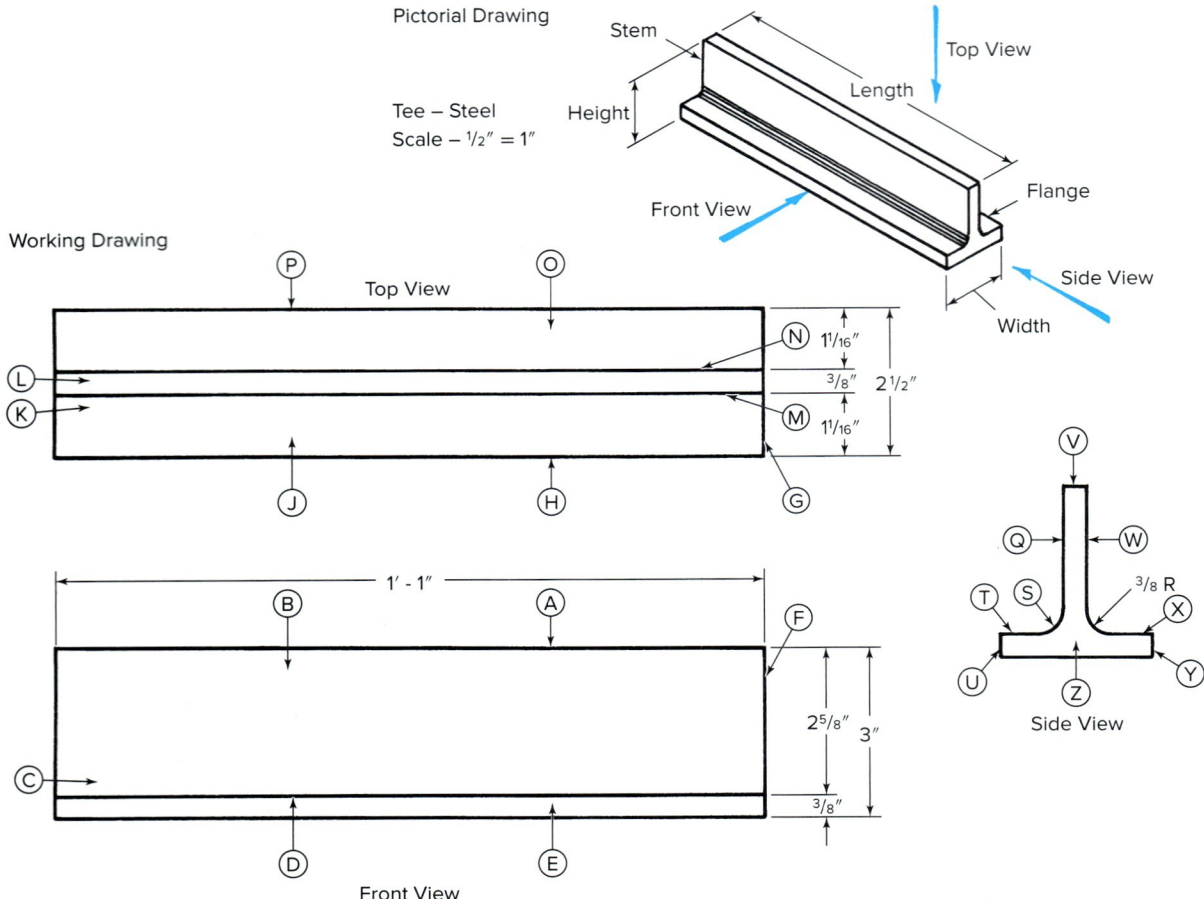

Fig. C-2 Standard steel tee.

29. Which surface in the top view does line *A* in the front view represent?
30. Which line in the top view represents line *W* in the side view?
31. Which line in the top view represents surface *Z* in the side view?
32. Which line in the side view represents surface *E* in the front view?
33. The rounded corner (curved surface) where the front of the stem meets the top of the flange is represented in the side view by line *S*. Which letter indicates this surface in the top view?
34. Which letter in the front view indicates the rounded corner where the front of the stem meets the top of the flange?
35. If the scale of a drawing were 1" = 1", and the length of the drawing of the object were 5", what would be the actual length of the object?
36. If the length of an object were 20 inches, and the scale of the drawing were ½" = 1", the length of the drawing would be one-half the length of the object, or 10 inches. If the scale of a drawing were 1/12" = 1", and the length of the drawing were 7 inches, what would be the actual length of the object?
37. If the actual length of this tee were 13 inches, and the scale of the drawing were ½" = 1", what would be the length of the drawing of the tee?
38. What is the actual length of the tee represented on this drawing?
39. What is the actual height of the tee represented on this drawing?
40. What is the width of the drawing of the tee represented on this drawing? Do not use a rule to measure the drawing, but get your information from the dimensions on the drawing and the scale of the drawing.

C.3 Separator Figure C-3 on the next page is a drawing of a machine part called a separator. It is often used as a shelf for bearings.

In previous problems, center lines were used only for circles or parts of circles. In the top and front views of this drawing, the center line of the separator itself is used for the purpose of dimensioning.

The note "6 Req'd." (required) means that 6 pieces are required of the kind represented on the drawing.

1. What is the scale of the drawing?
2. What is the overall height?
3. What is the overall depth, front to back?
4. What is the overall width?
5. What kind of lines are lines W and R in the side view?
6. What kind of lines are lines B and C in the top view: extension lines, projection lines, object lines, or invisible lines?
7. What kind of object lines are lines J in the front view?
8. What kind of line is line H in the top view?
9. What kind of line is line L in the front view?
10. Which two lines in the top view represent the front surfaces of the back web?
11. Which two lines in the front view represent the top surfaces of the bottom flange?
12. Which surface in the side view is the right surface of the center web?
13. How many holes are represented by line D in the top view?
14. How many holes are represented by lines S in the side view?
15. What is the thickness of the back web?
16. What is the thickness of the center web?
17. What is the thickness of the top flange?
18. What is the distance, top to bottom, of the inside surfaces of the flanges?
19. What is the distance from the centers of the holes in the bottom flange to the back of the separator?
20. What is the distance from the center of the right hole in the upper flange to the right side of the separator?

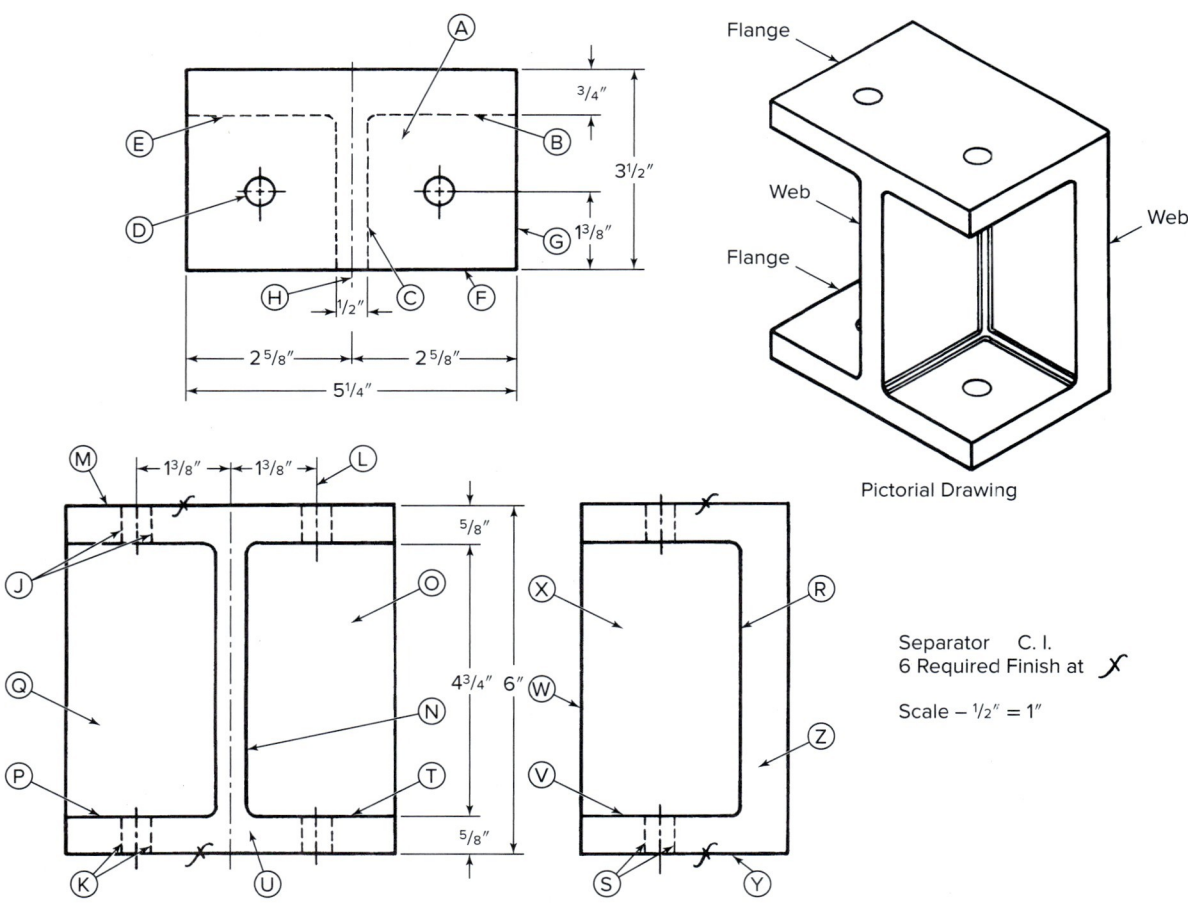

Fig. C-3 Separator.

21. What is the center-to-center distance of the holes in the lower flange?
22. What is the depth of the drilled holes?
23. How many drilled holes are there?
24. Which surface in the front view is represented by line E in the top view?
25. Line V in the side view represents two surfaces. Which lines in the front view represent these surfaces?
26. Which line in the top view represents line W in the side view?
27. Which surface in the side view represents line G in the top view?
28. Which surface in the front view represents line R in the side view?
29. The left hole in the bottom flange is represented by line D in the top view. Which lines in the side view represent this hole?
30. Which surface in the front view represents line F in the top view?
31. Which line in the top view represents surface X in the side view?
32. Which line in the side view represents the surfaces indicated by lines E and B in the top view?
33. Which surface in the side view does line N in the front view represent?
34. Which surface in the top view does line M in the front view represent?
35. Which line in the top view represents surface O in the front view?
36. Which line in the side view represents surface U in the front view?
37. Which line in the front view represents the surface indicated by line C in the top view?
38. Lines K in the front view represent a hole in the bottom flange. Which line in the top view represents this hole?
39. Which of the following surfaces are to be finished: O and U in the front view, A in the top view, or X in the side view?
40. Which of the following lines in the side view represent surfaces to be finished: R, V, or Y?

D.1 Assembly Drawing Figure D-1 on the next page (ISO-A and ISO-E) represents a welded assembly made out of three pieces of ¼-inch steel plate. A V-groove weld on a butt joint and two fillet welds will join the pieces. Welding symbols will be covered in Chapter 30. Tolerances must be carefully considered as they may be cumulative on a drawing.

1. If part 1A is cut $5^{15}/_{16}$ inches long, is it acceptable?
2. If part 1A is beveled at 25°, is it acceptable?
3. How many different size pieces make up this assembly?
4. What type of line is L pointing to?
5. What geometric shape is part 1B?
6. The center of the 1 inch hole is located how far from the bottom?
7. Dimension parts 1A and 1B.
8. How far from the right edge of the assembly is the surface of the vertical member located?
9. What is the maximum and minimum distances the edge of the hole can be from the 5½-inch sides of the vertical member?
10. What is the maximum and minimum allowable dimensions on the top view at the M location?

INTERNET ACTIVITIES

Internet Activity A

Using your favorite search engines, find and list as many companies as possible that are making computer software for doing drawings but that also incorporate the AWS 2.4 Standard on Welding Symbols into their programming.

Internet Activity B

Using the American Welding Society's Web site, review the most current information on the AWS Skills Competition Committee drawings for the U.S. Open Weld Trials or the World Skills Competition. Determine if the drawings are provided in ISO-A or ISO-E or both. Which drawing type would you prefer to work with?

Design credits: Cinema and movie icon: Denis Meshkov/123RF; and Illustration of Welding icons: Mr. Nachai Sorasee/123RF

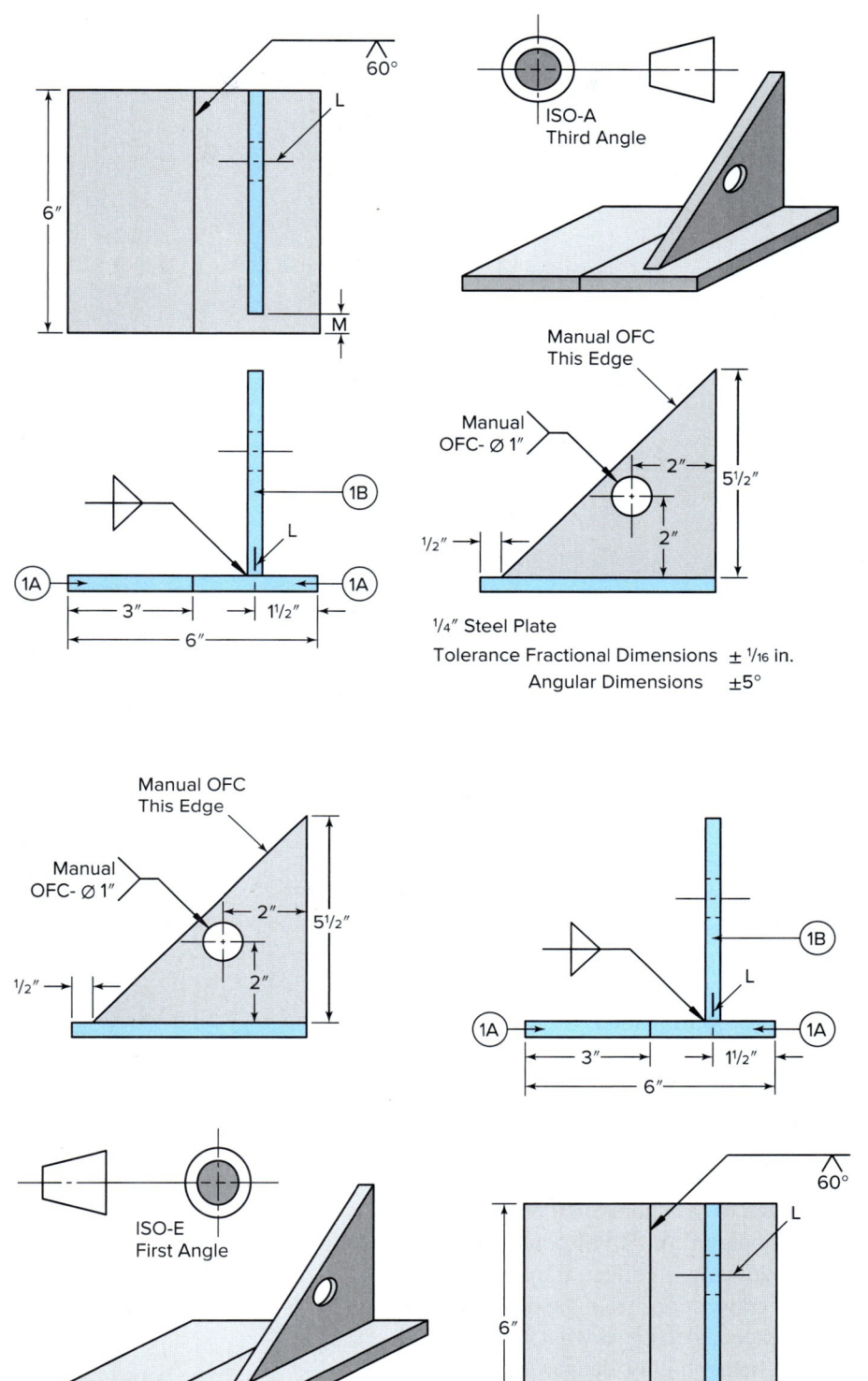

Fig. D-1 Assembly drawing.

30
Welding Symbols

The following material has been developed by the American Welding Society and is being reprinted here through the courtesy of that organization. The symbols presented here include the latest revisions. For those students desiring a more detailed reference, the booklet *Standard Symbols for Welding, Brazing, and Nondestructive Examination* (AWS A2.4-2012) published by the American Welding Society is recommended.

Welding cannot take its proper place as a production tool unless designers can put across their ideas to the workers. Noncode work may not require precise or detailed information on drawings. When the failure of welded structures would endanger life and property, however, simple and specific instructions must be given to the shop. Such practices as writing "To be welded throughout" or "To be completely welded" on the bottom of a drawing, in effect, transfer the design of all attachments and connections from the designer to the welder who cannot be expected to know what strength is necessary. In addition to being highly dangerous, this practice is also costly because certain shops, in their desire to be safe, use much more welding than is necessary.

Welding symbols give complete welding information. They indicate quickly to the designer, drafters, production supervisor, and the welder, which welding technique is needed for each joint to satisfy the requirements for material strength and service conditions. Figure 30-1 shows the basic welding symbol components.

Chapter Objectives

After completing this chapter, you will be able to:

- **30-1** Know the name of the AWS standard for welding symbols.
- **30-2** List the eight elements that may be found on a welding symbol.
- **30-3** List the basic weld symbols.
- **30-4** List the supplementary symbols.
- **30-5** Name the five basic joints.
- **30-6** Identify the applications of the different weld symbols.

Many companies develop their own system of symbols, references, and legends with the AWS symbols as a base. Very often they will need only a few of the symbols presented in this chapter.

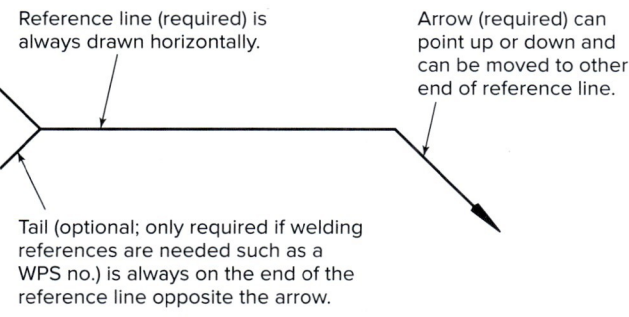

Fig. 30-1 Welding symbol components.

In the AWS American National Standard symbol system, the *joint* is the basis of reference. The joint types are referred to as butt, tee, lap, corner, and parallel. Any welded joint indicated by a symbol will always have an *arrow side* and *other side*. Accordingly, the terms arrow side, other side, and both sides are used to locate the weld with respect to the joint, Fig. 30-2.

Another feature is the use of the *tail* of the reference line for giving the specifications for welding. The size and type of the weld is only part of the information needed to make the weld. The welding process; type and brand of filler metal; whether or not peening, chipping, or preheating are required; and other information must also be known. The specifications placed in the *tail* of the reference line usually conform to the practices of each company. If notations are not used, the tail of the symbol may be omitted.

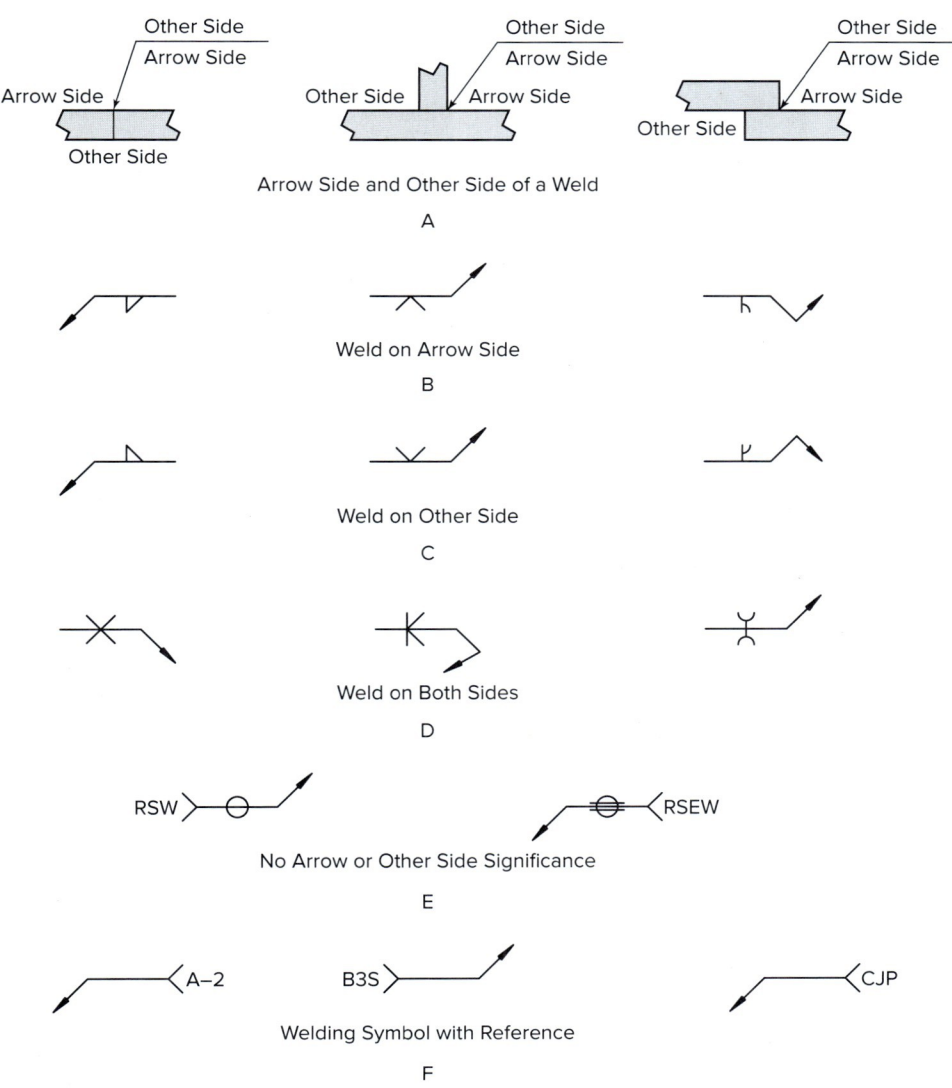

Fig. 30-2 Arrow and other side significance with references in tail.

Welding Symbols **Chapter 30** 979

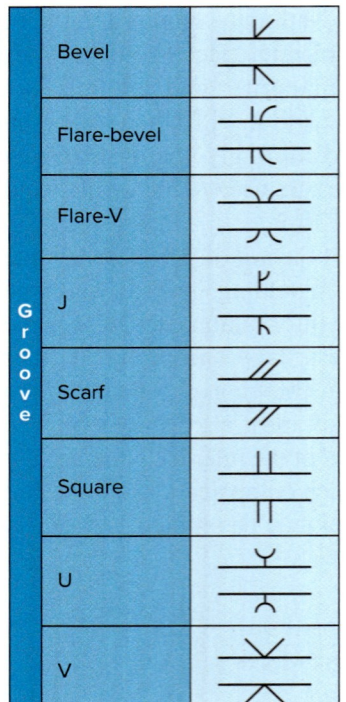

Fig. 30-3 Basic weld symbols. The reference line is shown dashed for illustrative purposes. Adapted from A2.4-2007, Figures 1, 2. American Welding Society, 2007.

can be combined to describe any set of conditions governing a welded joint.

The symbols, together with the specification references, provide for a shorthand system whereby a large volume of information may be communicated accurately with only a few lines, abbreviations, and numbers.

In order to understand welding symbols in their fullest meanings you must review the basic joints shown in Table 30-1 and then apply the basic welds to them. This chapter will deal with only the most common types of welds such as fillets and grooves. These are by far the most common types of welds and weld symbols you will encounter in industry.

Fillet Welds and Symbol

The fillet weld is a right triangle shaped weld placed *on* the joint, not *in* the joint like a groove weld. The fillet weld symbol is also a right triangle in shape, so it is easily recognized. The vertical line of the fillet weld symbol is always drawn to the left of the symbol. The T and lap joints are most commonly welded with fillet welds. Figure 30-6 (p. 984) shows how the location, size, and length of a fillet weld are indicated on a welding symbol and drawing. A sketch is shown to let you know how it would appear on a weldment. The size of the fillet weld is always shown to the left of the symbol and the length always to the right of the symbol. If no length is indicated, the weld should run the full length of the joint.

The AWS standard makes a distinction between the terms weld symbol and welding symbol. The **weld symbol**, Figs. 30-3 and 30-4, is the graphic symbol that indicates the weld required for the job. The assembled **welding symbol**, Fig. 30-5 (p. 983), consists of the following eight elements or as many of these elements as are necessary; however, only the reference line and arrow are required elements:

- Reference line (required)
- Arrow (required)
- Basic weld symbols
- Dimensions and other data
- Supplementary symbols
- Finish symbols
- Tail
- Other specifications

The standard welding symbols are given in Table 30-1. Although it may appear at first that there are a very large number of different symbols, the system of symbols is broken down into a few basic elements. These elements

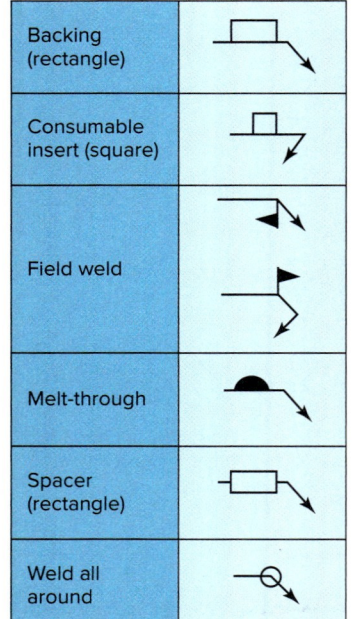

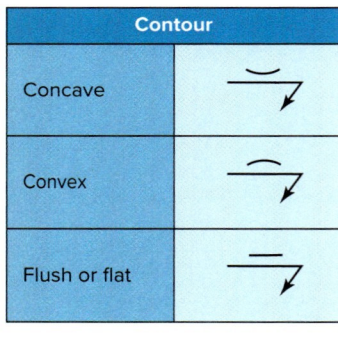

Fig. 30-4 Supplementary symbols. Adapted from A2.4-2007, Figures 1, 2. American Welding Society, 2007.

980 Chapter 30 Welding Symbols

Table 30-1 Welding Symbol Chart

AMERICAN WELDING SOCIETY WELDING SYMBOL CHART

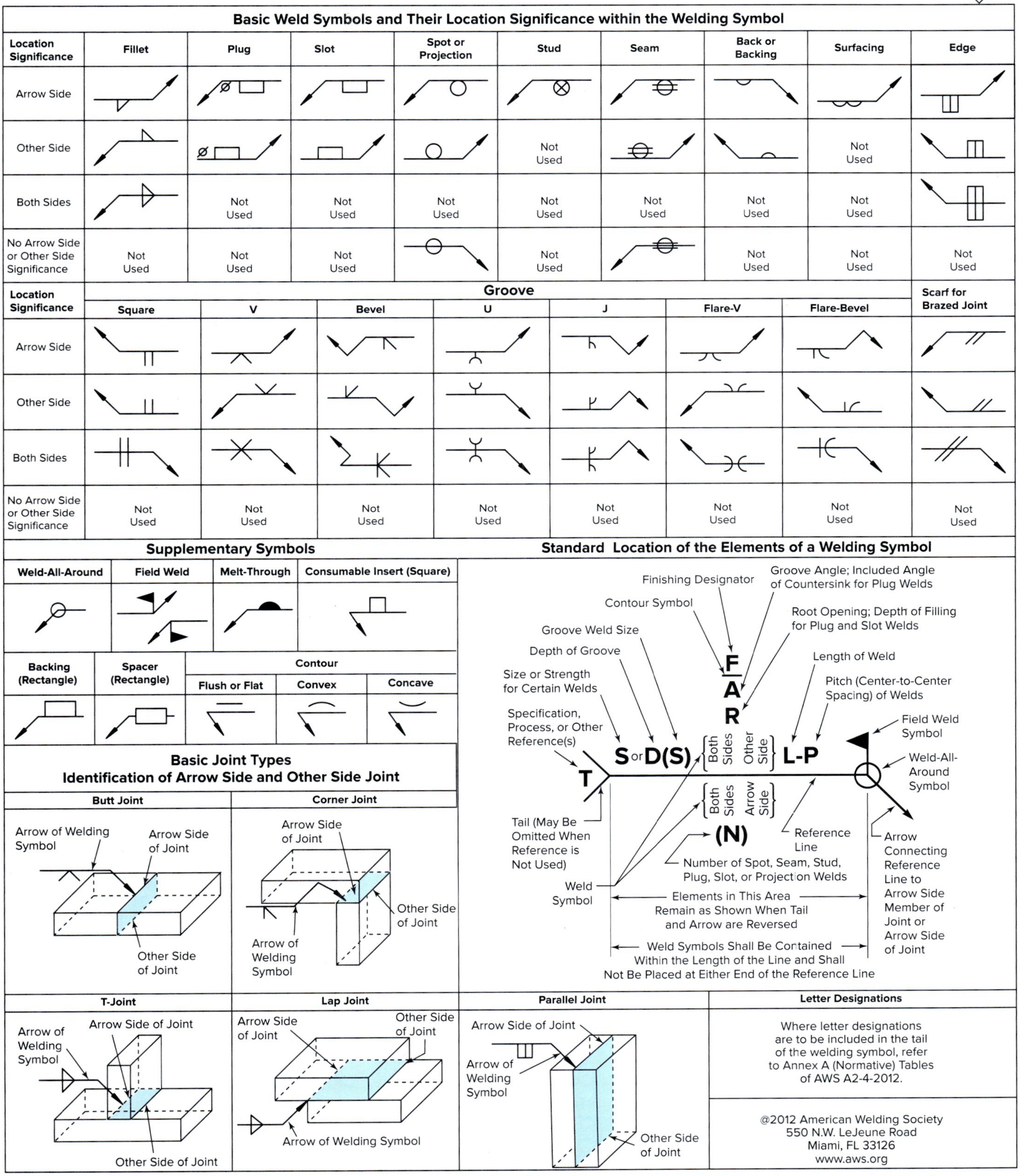

Table 30-1 (Concluded)

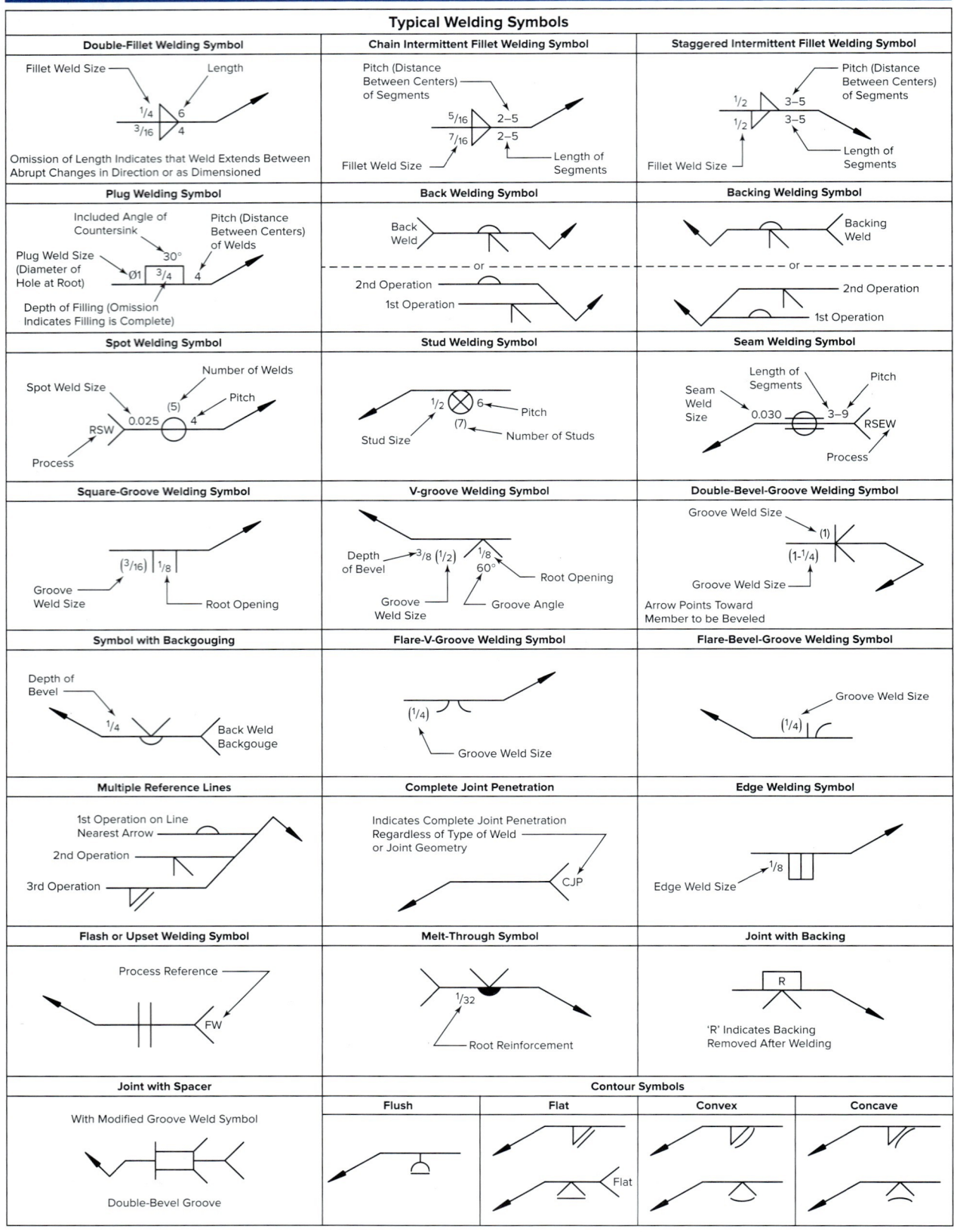

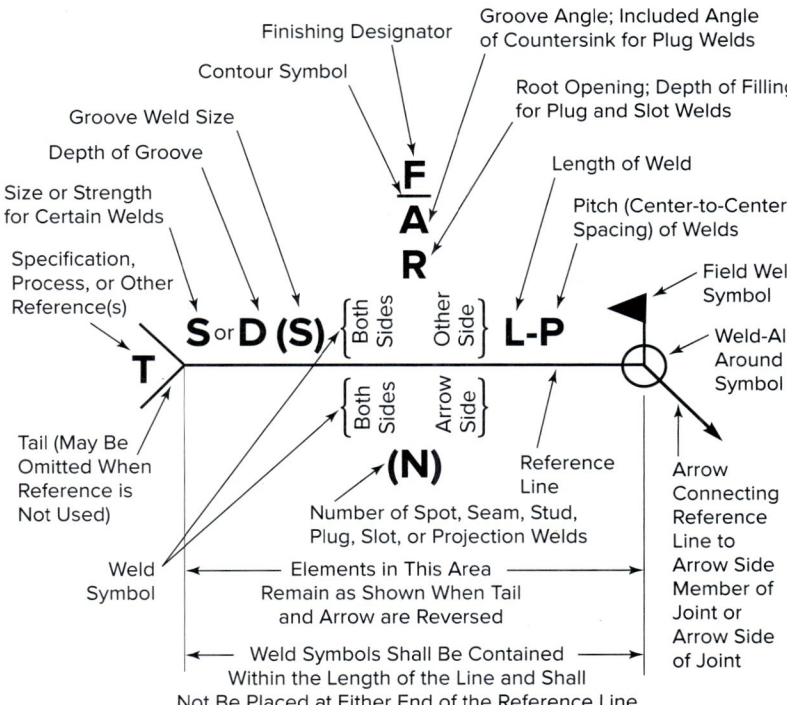

Fig. 30-5 Standard location of elements of a welding symbol.
Source: AWS A2.4-2007, Figure 3, reproduced with permission of the American Welding Society, Miami, FL

Figure 30-7 (p. 985) shows intermittent fillet welds as well as staggered fillet welds. When properly applied, fillet welds do not need to be large in size or necessarily run the full length of the joint. Always follow the welding symbol specification on the drawing and place the proper sized fillet weld in the proper location. Do not overweld or underweld beyond the tolerances indicated on the drawing. To do so is wasteful and may lead to joint failure, excessive distortion, and other heat-related issues.

Weld-All-Around Symbol

The weld-all-around symbol is used when the weld is to extend all the way around the joint or a series of connected joints. The symbol is a circle viewable at the junction of the reference line and arrow. Figure 30-8 (p. 985) shows various applications of the weld-all-around symbol.

Groove Welds

There are seven types of groove welds and symbols to represent them:

- Square
- V
- Bevel
- U
- J
- Flare-V
- Flare-bevel

As in the fillet weld, these symbols take the shape of the groove the weld is to be placed in. Groove welds are placed *in* a joint, while fillet welds are placed *on* a joint. Typically groove welds are used to make butt joints, but they can in some cases be used to make other types of joints. One misunderstanding that comes up is the term *double bevel*. It does not mean that both edges of the butt joint members are beveled. It means that one of the joint members has a bevel on both sides and the joint is to be welded from both sides. If both members of a butt joint are beveled and brought together, the shape of the joint groove is now a V and that is how it must be described. If welding symbols are to aid in communication, we must all understand the same references. Figure 30-9 (p. 985) shows the symbol for a V-groove weld in a butt joint. To properly read and interpret a groove symbol, start with the information found in the groove symbol and then work from the reference line out. In many cases the metal will be prepared for the welder, and the welder will set the joint up and tack it for subsequent welding. The first thing the welder must do in setting the joint up is to set the root opening. The second thing would be to verify the groove angle. This groove angle is 60°, so each plate must have a bevel angle of 30°. The dimensions to the left of the groove weld symbol give the groove depth (D) and the weld size. They are read from left to right. The second thing the welder must do when setting this joint up is to verify the depth of the groove. Once this and the root opening have been set, the weld can be made. In this case, the depth of penetration beyond the groove is 1/8 inch. Note the weld reinforcement is not considered when measuring the size of the groove weld. This is true for all the various types of groove welds. This penetration level will be achievable if the welding procedure is followed. If the welding is being done manually or semiautomatically, the welder will need to use their skill and knowledge to observe the penetration as it is occurring and to manipulate the arc accordingly to achieve what is being specified. Figure 30-10 is a symbol for a double V-groove weld. If a groove weld is not dimensioned for depth and size, the joint shall be prepared and the weld shall be made to ensure complete joint penetration (CJP). The design engineer should specify exactly how the joint should be prepared and the size of groove weld required.

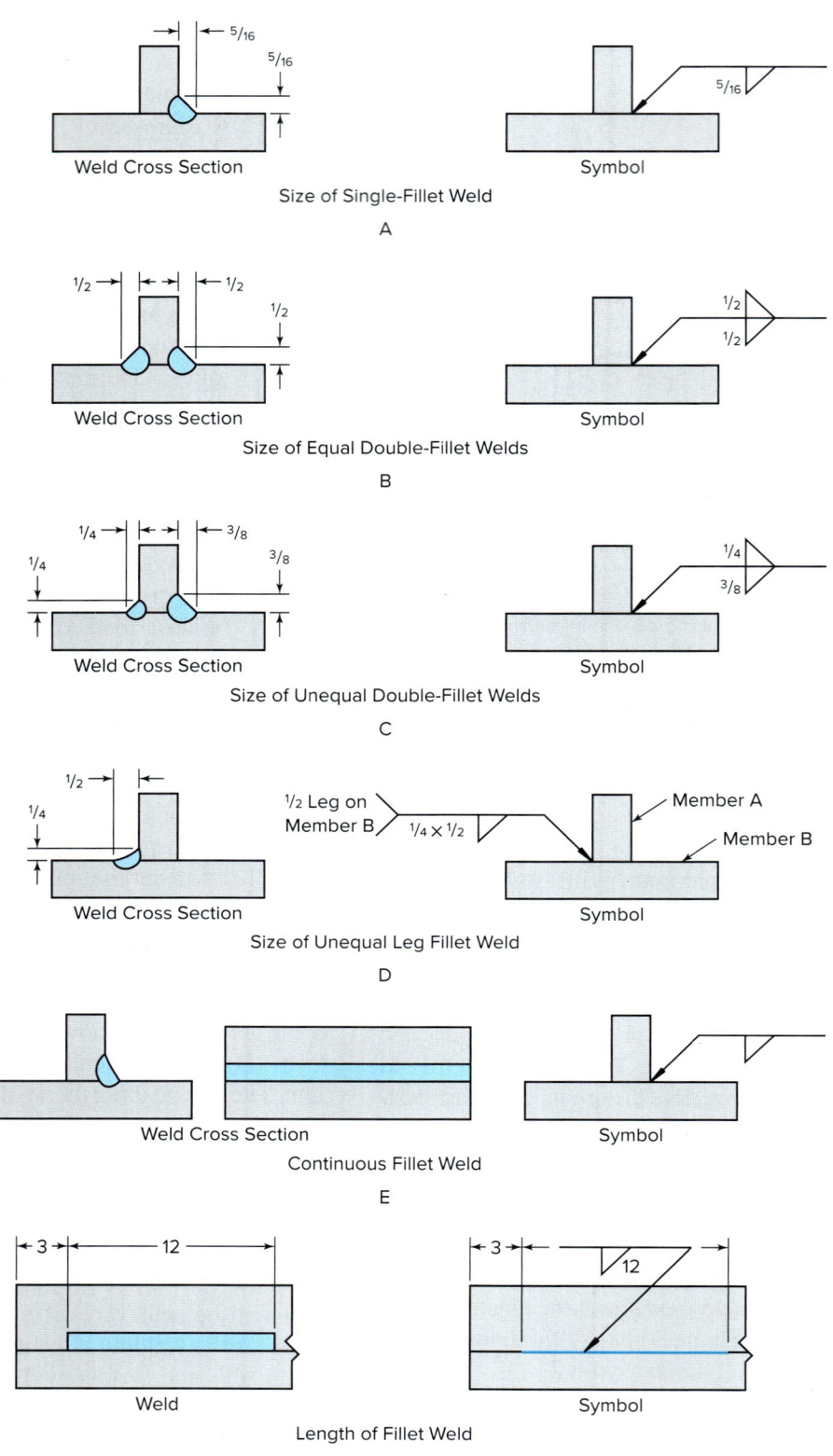

Fig. 30-6 Fillet welds on T-joints.

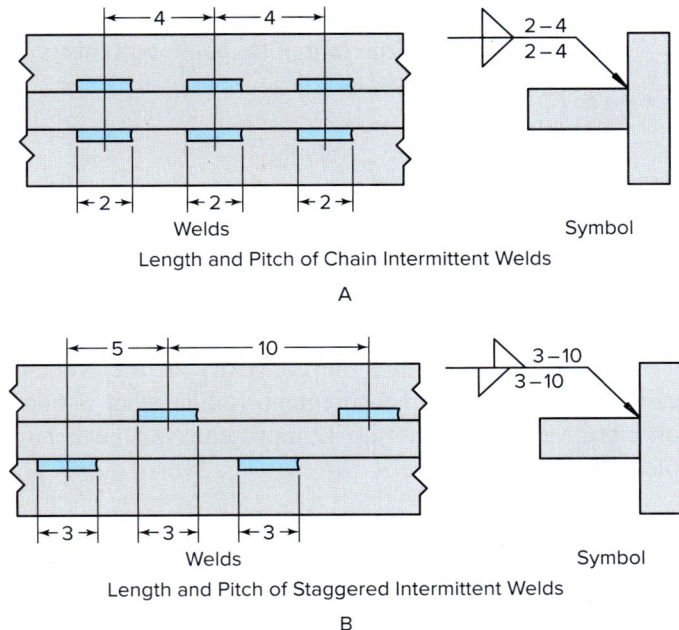

Fig. 30-7 Fillet weld length and pitch as indicated on the welding symbol. Note the chain and staggered intermittent fillet welds.

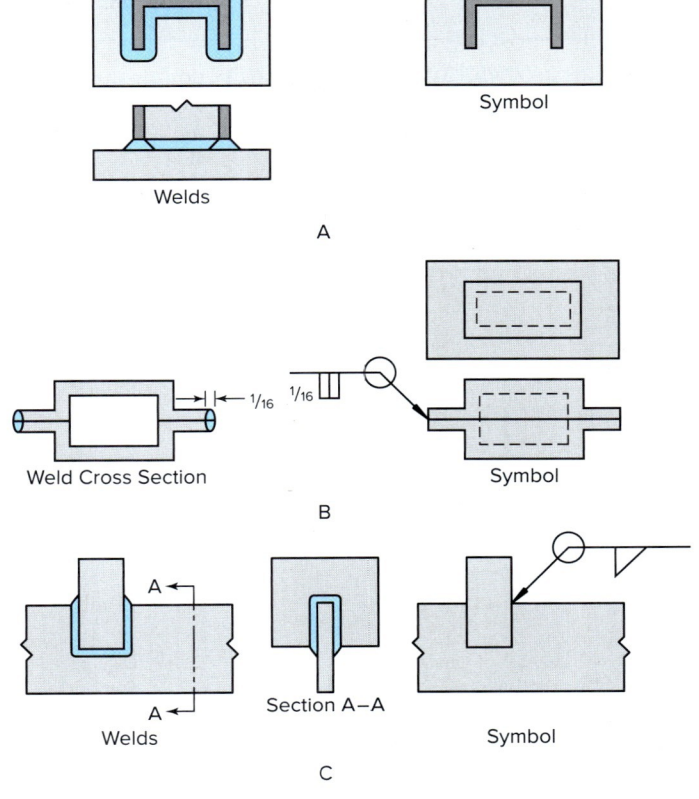

Fig. 30-8 Weld-all-around symbol usage.

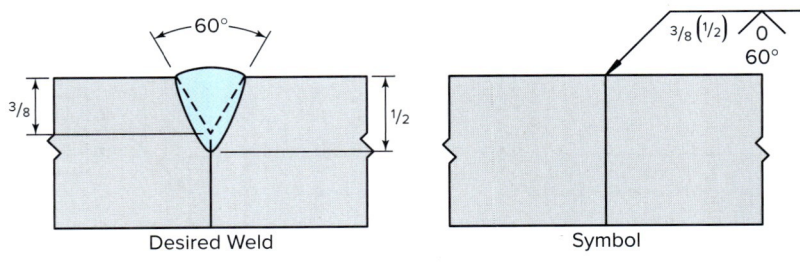

Depth of Bevel—3/8 in.
Groove Weld Size (Always Shown in Parentheses)—1/2 in.
Root Opening—0
Groove Angle—60°

Fig. 30-9 V-groove weld on a butt joint. (PJP)

This should not be left up to personnel on the shop floor. Since groove welds and butt joints are typically used in highly critical applications, they generally will be required to run the full length of the joint. These types of joints are typically found in tensile loading conditions in primary members. If a failure occurs, it may be catastrophic, leading to loss of life and/or property damage. If a groove weld length is required, it will be shown to the right of the groove weld symbol. If no length is shown, the weld shall be made the full length of the joint.

The flare bevel or flare V-grooves are unusual in that they take their shape from the structural member that is making up the joint. Like the square-groove weld, metal is not removed to make the flare groove. In this case the round bar makes for a flare in the bevel, which will make it difficult for the welder to get the heat of the process into the root of the joint. The arc will jump to one side of the joint or the other, and the weld pool will bridge the joint and prevent effective penetration into the root of this type of joint. If the design engineer understands the welding process being used, the joint should be properly designed to be achievable by the welder on the shop floor. Figure 30-11, page 986, shows a properly designed and specified flare-groove weld. Note the groove weld size is less than the depth of the groove, indicating that this designer understands the limits of the welding processes being used. If no groove size or weld size is specified, a CJP should be made, which in the case of a flare V or flare bevel will be virtually impossible with conventional

Welding Symbols **Chapter 30** 985

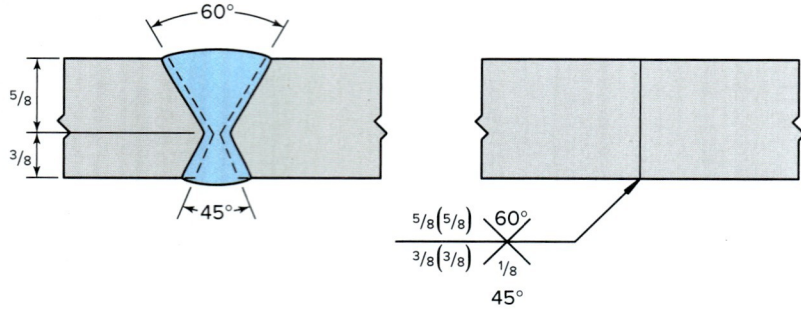

Fig. 30-10 Double V-groove weld on a butt joint. (CJP)

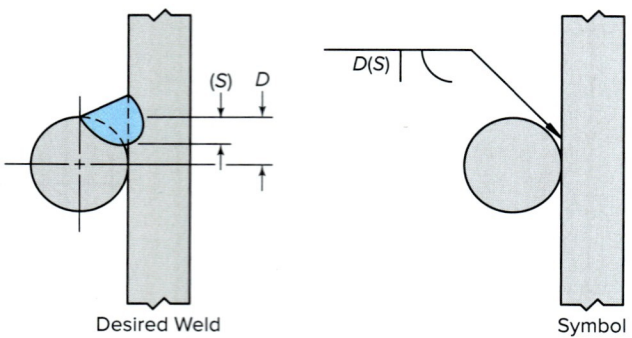

Notes: Variable (S) denotes groove weld size.
The depth of the bevel D equals the radius of the round member.

Fig. 30-11 Flare-bevel groove weld. (PJP)

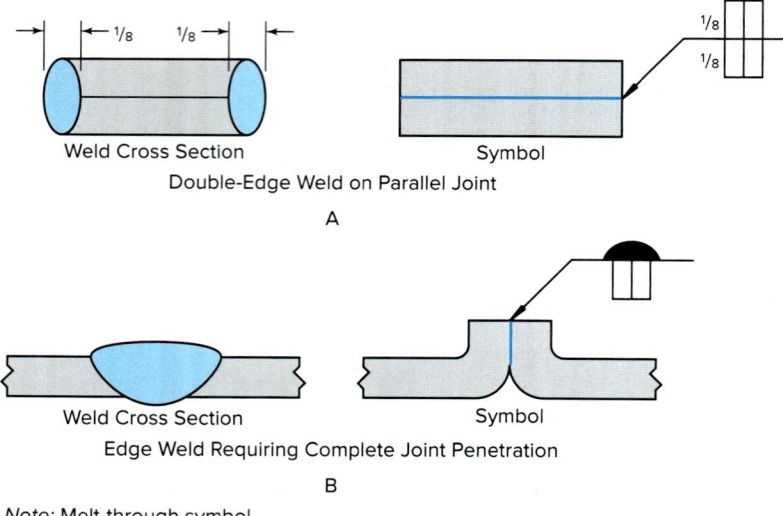

Note: Melt-through symbol.

Fig. 30-12 Edge welds on various joints.

arc welding processes. There just is not sufficient access to the root of the joint with this type of flared groove.

Edge Welds

The edge weld is a weld that totally consumes the full thickness of the edge of the joint it is being applied to. It is typically used on parallel joints, flanged butt joints, or flanged corner joints. Note that the convexity of the weld is considered in its size measurement, unlike what is done on a groove weld. Figure 30-12 represents various examples of edge welds.

Combination Symbols

Weld symbols can be combined into one welding symbol. The amount of information that can be expressed by a welding symbol of this type is illustrated in Fig. 30-13. Written specifications for the same weld would read as follows:

> Arrow side j groove, 1¼-inch depth, 20° groove angle with ½-inch fillet welds. Other side j groove, ¾-inch depth, 20° groove angle with a ⅜-inch fillet weld. All welding to be done in the field in accordance with specification A2.

The specification would be located on the drawing. It could include the welding process, filler wire, current, etc. When a sequence of operations is necessary, multiple reference lines should be used (sketch 3). The reference line nearest the arrow is the first operation.

Contour Symbol

The desired shape of the weld is shown on the welding symbol by use of a contour symbol. It is located above the imaginary face of the weld for groove weld symbols and above the side of the right triangle that most closely represents the face of the weld for fillet weld symbols. The contour symbol is easily understood since it takes the shape of the contour desired. If the weld is to be flat, it is a straight line. If it is to be concave, it takes on that shape. If the weld is to have some convexity, it will have that shape. In some cases, it will not be possible for the weld to attain the particular shape being specified without external assistance. If some postweld treatment is required, this is shown with an abbreviation. Figure 30-14 shows various groove,

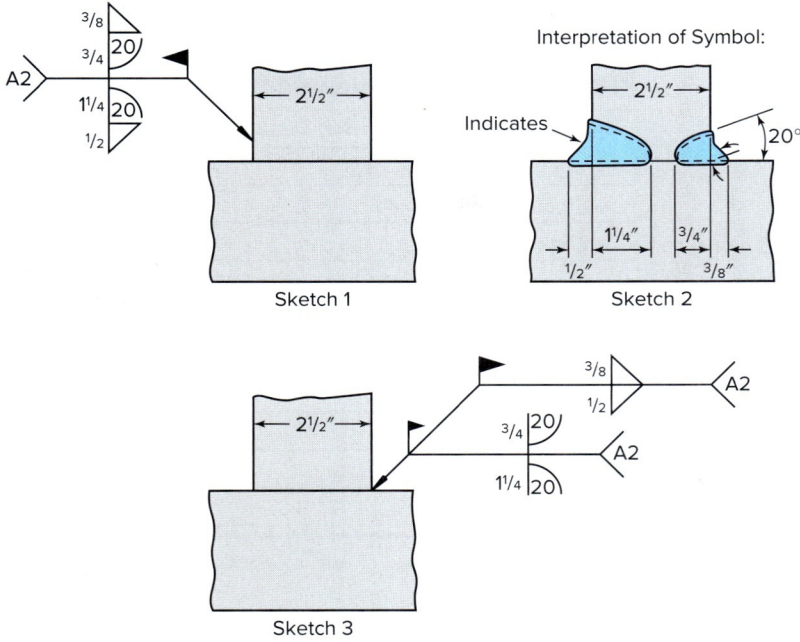

Fig. 30-13 Symbolic method of expressing welding information.

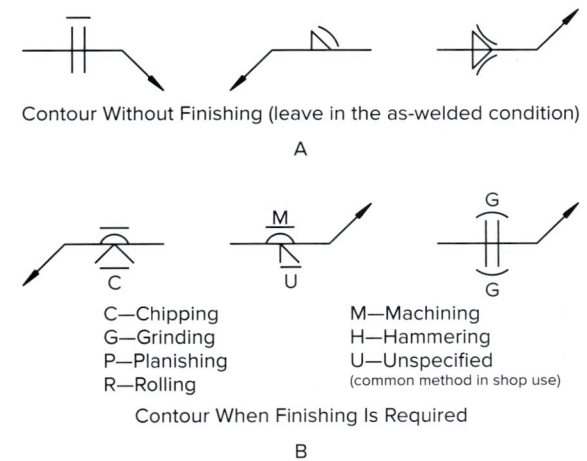

Fig. 30-14 Contour symbol and letters which indicate the finishing method.

fillet, back, and backing welds with various contour symbols. The finishing methods and abbreviations are also covered.

Applications of Welding Symbols

Table 30-1 and application illustrations in this chapter are taken from the American Welding Society standards. Many students find that they can learn to read and understand these symbols more readily through the use of the illustrations than through the use of written instructions. You are urged to study the table and the illustrations carefully in order to be able to recognize the symbols on sight. For any detailed reference required you should consult the AWS Standard A2.4 symbols of welding and brazing.

A nondestructive examination symbol has several elements. Take a look at Fig. 30-15, which shows elements and how they are located with respect to each other. There you will see a reference line to the left and an arrow to the right. In the middle, there are examination method letter designations. In the top middle is extent and number of examinations. You can see supplementary symbols, such as *examine-all-around* and *examine in field*. Finally, at the left you can see the tail (specifications, codes, or other references).

You can see in Fig. 30-15 that the arrow connects the reference line to the part to be examined. The *arrow side of the part* is the side of the part to which the arrow points. Obviously enough, the side without the arrow point is referred to as the *other side*.

As mentioned, there can be supplementary symbols. These are illustrated in Fig. 30-16, page 988: examine-all-around, field examination, and radiation direction.

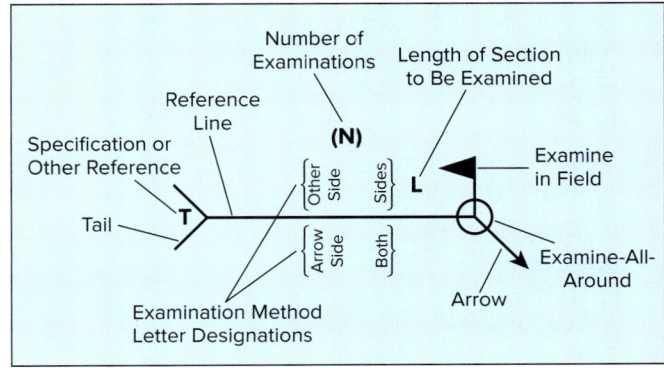

Fig. 30-15 Nondestructive examination symbol elements and how they are located with respect to each other. Adapted from A2.4-2007, Figure 56. American Welding Society, 2007.

Welding Symbols Chapter 30 987

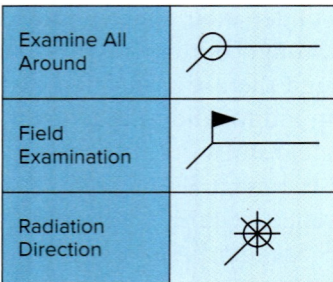

Fig. 30-16 Supplementary symbols. Adapted from A2.4-2007, Figure 50. American Welding Society, 2007.

Each nondestructive examination method has a letter designation. For example, the radiographic method is designated with the letters *RT*. Table 30-2 shows how other methods are designated.

Table 30-2 Nondestructive Examination Method Letter Designations
Examination Method - *Letter Designation*
• Acoustic Emission - *AET*
• Electromagnetic - *ET*
• Leak - *LT*
• Magnetic Particle - *MP*
• Neutron Radiographic - *NRT*
• Penetrant - *PT*
• Proof - *PRT*
• Radiographic - *RT*
• Ultrasonic - *UT*
• Visual - *VT*

Adapted from A2.4-2007, Table A.6. American Welding Society, 2007.

CHAPTER 30 REVIEW

Multiple Choice

Choose the letter of the correct answer.

1. The welding symbols were developed by _____. (Obj. 30-1)
   a. The American Society for Testing
   b. The American Welding Society
   c. The American Petroleum Institute
   d. The Nuclear Industry

2. The tail of the welding symbol should never be used to make a reference on the drawing. (Obj. 30-1)
   a. True
   b. False

3. Symbols have not been developed for use with the gas tungsten arc and gas metal arc processes. (Obj. 30-1)
   a. True
   b. False

4. Which of the following elements are on a welding symbol? (Obj. 30-2)
   a. Reference line
   b. Arrow
   c. Tail
   d. All of these

5. A weld symbol on the bottom of a reference line signifies _____. (Obj. 30-3)
   a. Other side
   b. Both sides
   c. Arrow side
   d. None of these

6. The vertical line of a fillet weld symbol is always drawn on which side? (Obj. 30-3)
   a. Left
   b. Right
   c. No significance
   d. Only a

7. _____ is not a supplementary weld symbol. (Obj. 30-4)
   a. Field weld
   b. Melt-through
   c. Fillet weld
   d. Weld-all-around

8. Which of the following is a basic joint? (Obj. 30-5)
   a. Butt
   b. Corner
   c. T
   d. All of the above

9. In a multiple reference line, the line closest to the arrow signifies _____. (Obj. 30-6)
   a. The last operation
   b. The first operation
   c. It has no significance
   d. Both a and b

10. Companies are never permitted to deviate in any manner from the official AWS symbol system. (Obj. 30-6)
    a. True
    b. False
11. The difference between a back and backing weld symbol is that _____. (Obj. 30-6)
    a. A backing weld is done first
    b. A backing weld is done last
    c. A back weld is done last
    d. Both a and c
12. The first number located inside a groove weld indicates _____. (Obj. 30-6)
    a. The degree of bevel
    b. The contour
    c. The root opening
    d. None of these
13. The number immediately to the left of a fillet weld indicates _____. (Obj. 30-6)
    a. Length
    b. Height
    c. Width
    d. Size
14. Table 30-1, taken from the American Welding Society standards, is concerned with _____. (Obj. 30-6)
    a. Manual welding processes
    b. Semiautomatic welding processes
    c. Both a and b
    d. None of these

Review Questions

Write the answers in your own words.

15. Why are symbols used on drawings of welded fabrications? (Obj. 30-1)
16. Draw the three basic welding symbol components and label accordingly. (Obj. 30-1)
17. Draw a welding symbol showing the standard eight elements. (Objs. 30-1, 30-4, 30-5)
18. List eight elements that may be contained in a welding symbol. (Obj. 30-2)
19. How are specification references shown on the symbol? (Obj. 30-2)
20. Draw the symbol for the following types of welds: (Obj. 30-3)
    a. Bead
    b. Fillet
    c. Plug and slot
21. Draw the symbol for the following types of grooves: (Obj. 30-3)
    a. Square
    b. V
    c. Bevel
    d. J
22. Draw the symbol for the following weld specifications: (Obj. 30-3)
    a. Fillet weld
    b. Weld-all-around
    c. Flush weld
23. Name and draw the supplementary NDT symbols. (Obj. 30-4)
24. Draw the proper symbol for the following specifications: (Obj. 30-6)
    a. ³⁄₈-inch fillet weld on T-joint, welding from one side only (arrow side), single pass
    b. ½-inch fillet weld on T-joint, welding on both sides, 3 passes
    c. V-groove weld, 60° groove angle, ³⁄₈-inch plate, ⅛-inch root opening, 3 passes, to be welded all-around from one side of plate
    d. ¼-inch intermittent fillet welds on T-joint, length of welds 2 inches, welds spaced 4 inches center-to-center (pitch), welding on both sides of joint
    e. A pad is built up on both sides with stringer beads to a height of ½ inch
    f. Draw a multiple reference line, first operation—V-groove arrow side, second operation—back weld other side, third operation—other side flush by grinding

Pressure Vessel

The drawing shown in Fig. 30-17 on the next page will review much of what you have learned in Chapters 29 and 30. Figure 30-18, page 992, is a cut list for the pressure vessel. This drawing uses the SI measurement system since it is designed for the WorldSkills International Competition. This competition is to the skilled tradesperson what the athletic Olympics is to the athlete. This simulated pressure vessel (attempts to get as many different types of welds and processes in a very compact weldment) is only one of a number of tests that must be performed in the 3-day contest (22 hours total). The United States has been very fortunate to have gotten two gold, two silver, and three bronze medals out of the 10 most recent competitions. For additional information on this competition, go to the following Web site: www.worldskills.org.

The typical electrodes and filler metals that would be used on this type of project are:

- EWTh-2
- ER70S-6
- E7018 or E7016
- E71T-1
- ER70S-1B for GTAW

Answer the following questions, and if you are sufficiently skilled with the processes covered in the drawing, review this project with your instructor. If materials and time permit, construct the pressure vessel. This will certainly allow you to demonstrate the skill and knowledge that you have attained through the study of this text and the assistance from your instructor(s). The number 6 weld will be very difficult to make with a low hydrogen electrode in the overhead position. Make sure the root opening and root face are appropriate and use either a.c. or DCEN for at least the root pass. The melt-through and weld tie-ins must look like this weld was done with the GTAW process with a consumable insert in order to achieve a medal at the world competition. The pressure vessel must be properly assembled, and all welds must pass visual inspection. No leaks should be found up to the full pressure of the hydrostatic test. Hydrostatic testing must be done by qualified personnel. The vessel must be completely filled with water, and there should not be air bubbles or air pockets anywhere in the vessel or in the pumping system.

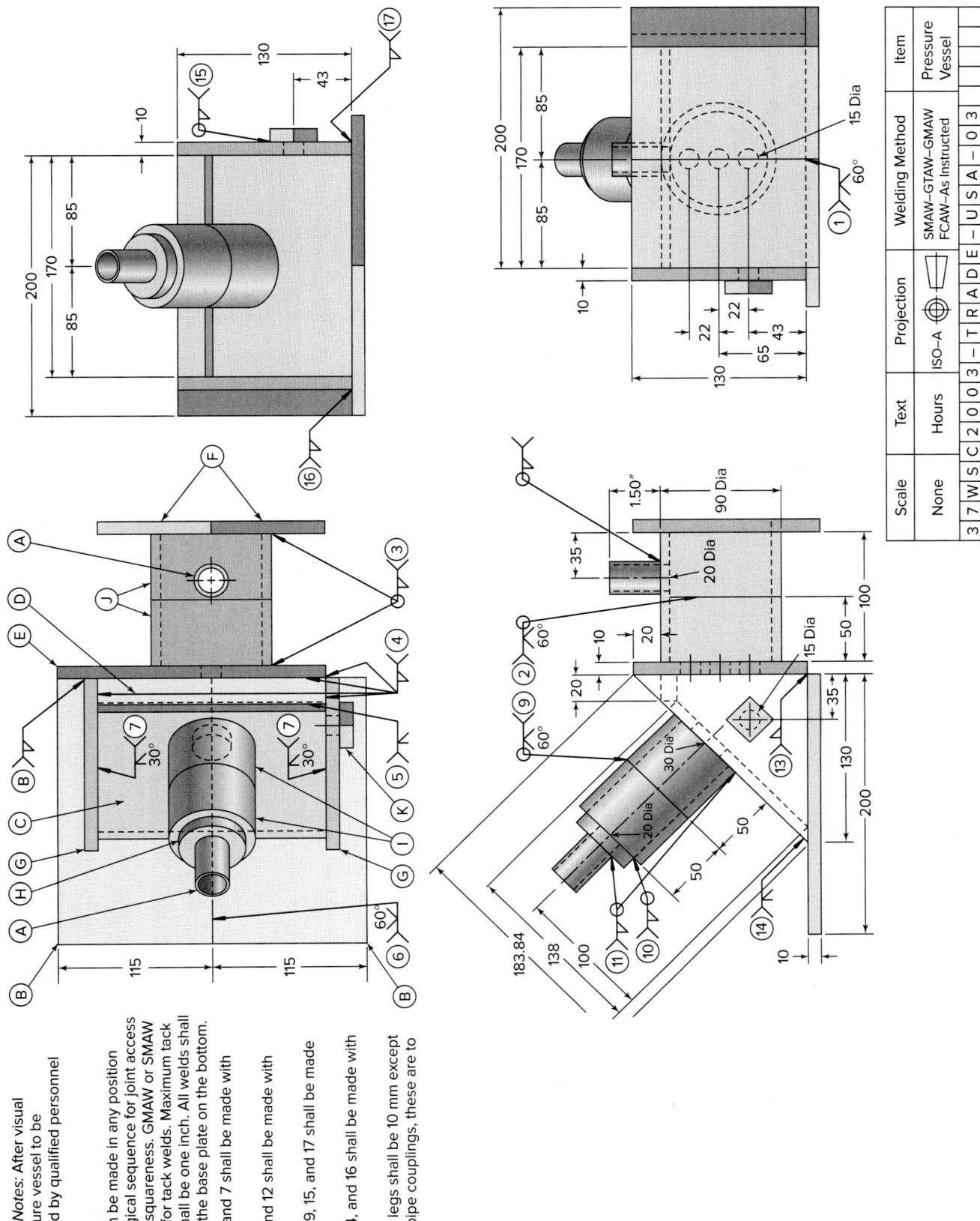

Fig. 30-17 Pressure vessel for additional practice on drawing reading, reading, and interpreting welding symbols, and demonstrating welding skills. Assembly and welding time should be approximately 6 hours.

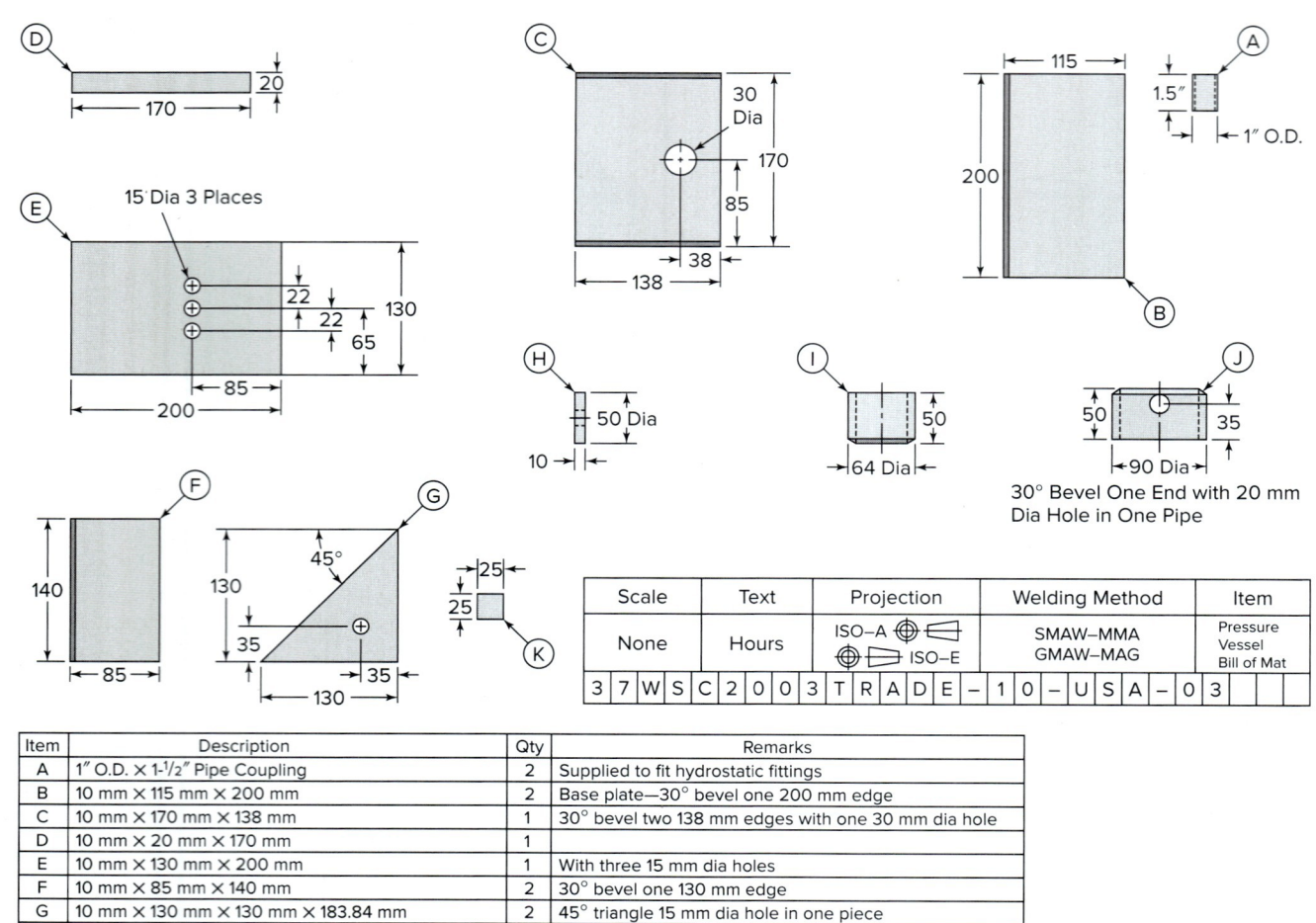

Fig. 30-18 Pressure vessel bill of materials. All materials are from low carbon steel.

Item	Description	Qty	Remarks
A	1" O.D. × 1-1/2" Pipe Coupling	2	Supplied to fit hydrostatic fittings
B	10 mm × 115 mm × 200 mm	2	Base plate—30° bevel one 200 mm edge
C	10 mm × 170 mm × 138 mm	1	30° bevel two 138 mm edges with one 30 mm dia hole
D	10 mm × 20 mm × 170 mm	1	
E	10 mm × 130 mm × 200 mm	1	With three 15 mm dia holes
F	10 mm × 85 mm × 140 mm	2	30° bevel one 130 mm edge
G	10 mm × 130 mm × 130 mm × 183.84 mm	2	45° triangle 15 mm dia hole in one piece
H	10 mm × 50 mm Diameter	1	With 20 mm dia hole centred
I	50 mm I.D. × 50 mm × 7 mm Thick	2	30° bevel one end
J	76 mm I.D. × 50 mm × 7 mm Thick	2	30° bevel one end with 20 mm dia hole in one pipe
K	10 mm × 16 mm × 16 mm	1	

Pressure Vessel Review

Choose the letter of the correct answer.

1. Which projection angle is used to make this drawing?
   a. First
   b. Second
   c. Third
   d. Fourth
2. What type of weld is number 15?
   a. Groove
   b. Fillet
   c. Edge
   d. None of these
3. What is the length of weld 15?
   a. Approximately 4 inches
   b. Approximately 100 millimeters
   c. Approximately 1/3 foot
   d. All of these
4. What type of weld is number 6?
   a. Groove
   b. Fillet
   c. Edge
   d. None of these

5. What level of penetration is required on weld number 6?
   a. PJP
   b. ALP
   c. CJP
   d. Not enough information is known
6. What position will weld number 6 be made in?
   a. 1G
   b. 2G
   c. 3G
   d. 4G
7. What process is weld number 13 to be made with?
   a. SMAW
   b. GTAW
   c. GMAW
   d. FCAW
8. What is the bevel angle required on the plate for weld number 1?
   a. 13°
   b. 30°
   c. 45°
   d. 60°
9. What position will weld number 2 be made in?
   a. 2G
   b. 5G
   c. 6G
   d. Not enough information provided
10. What thickness of plate is required?
    a. Approximately 10 millimeters
    b. Approximately 3/8 inch
    c. Approximately 7 millimeters
    d. Both a and b
11. What is the approximate inside diameter of pipe J?
    a. 76 millimeters
    b. 3 inches
    c. 90 millimeters
    d. Both a and b

Review Questions

Write the answers in your own words.

12. What is the purpose of part A, and how is it attached to the pressure vessel?
13. List the views shown on the pressure vessel drawing.
14. How are the pipe coupling part A's attached to the weldment?
15. What type of weld is required at weld number 7, and does part C or G get the joint preparation?
16. Describe part F and state the dimensions.
17. Describe the welding, positions, and processes that will be used to join part F to the weldment.
18. Describe how weld number 9 will be made. Be as descriptive as possible.
19. Describe how weld number 13 will be made.
20. What angle from the horizontal is the part I pipe located, and how did you make this assumption?
21. What are the geometric shapes for parts G and K?

INTERNET ACTIVITIES

Internet Activity A

Use the Internet to find welding symbols for the International Organization for Standardization (ISO) ISO 2553. How do they compare to the AWS welding symbol standard?

Internet Activity B

Check out the WorldSkills Web site and see what the requirements are for the Trade #10 electric arc welding. Also see who is hosting the next competition and when.

Design credits: Cinema and movie icon: Denis Meshkov/123RF; and Illustration of Welding icons: Mr. Nachai Sorasee/123RF

Welding and Bonding of Plastics

31

The very popular use of plastics (polymeric materials) is generally considered a recent occurrence. However, some of the best polymers like silk, skin, and wood have been around for a very long time. The synthetic polymers (thermoplastics) have been used in the world economy for over 100 years. The two major forms of plastics are the thermoplastics and thermosetting plastics, because thermosetting plastics do not melt but decompose they are not weldable. Thermoplastic welding and fabrication has been a part of the manufacturing industry since the mid-1930s. This was the birth of the *hot-gas* welding technique. If you do a search on plastics using your favorite computer search engine, you will find many products made of plastics that have been welded. Unfortunately many of the articles are about welds that have failed. Like many other manufacturing mediums, plastic welding and fabricating has required a long process of careful and skillful development to produce industry standards. Standards and processes developed by individual manufacturers of equipment and by plastic product producers are in wide practical use in the United States; however, there is no public documentation regarding the requirements for their use, design criteria, and application. Unlike the United States, European countries have traditionally treated plastic welding and fabrication as a skilled trade and have set standards. For example, the German Welding

Chapter Objectives

After completing this chapter, you will be able to:

31-1 Describe uses of plastics.
31-2 Identify types of plastics.
31-3 Describe plastic welding processes.
31-4 Identify common plastic weld faults.
31-5 Make various plastic groove, fillet, and edge welds on sheet, plate, and pipe.

Society (DVS) has programs in place to govern design, training, certification, and inspection of plastic welds. The American Welding Society (AWS) has been working closely with the DVS by conducting conferences on the welding of plastics and other means of promoting standardization. The AWS has plastics welding covered in volume 3 of its *Welding Handbook*. The AWS has a technical committee, G1, made up of volunteers from the industry and is specifically focused on the welding of plastics. The purpose of the G1 committee is to define and publish the information required for all aspects of plastic welding, design, training, qualification, and inspection of welded plastic joints. Without such documentation, the processes of plastic fabrication and welding will continue to vary in quality and reliability from one source to another. In addition, well-documented standards will open many new markets for fabricated plastic products and will ensure the highest possible quality and reliability of product to the end user.

Plastics are useful because:

- They have excellent properties. They are corrosion resistant, lightweight, and fatigue resistant, and when composite structures are used, great strength-to-weight ratios are achievable.
- Manufacturing is simplified. Very complex products and parts can be made in one step, which greatly speeds up production.
- They are abundant and recyclable. Five percent of crude oil is converted to polymers, but polymers are considered more valuable than fuel.

Since the use of plastics continues to grow, the need to join them also expands. One of the preferred methods, and in some cases the only method, used to join certain plastics is fusion welding. Key points for welding being a popular joining method are:

- Welding produces very strong joints (as strong as the bulk polymer used to make it). As in the welding of metals, it does have a heat-affected zone (HAZ) that must be considered.
- Some plastics can only be joined by welding.
- The fusion line is usually the same as for the bulk polymer, so it is easily recyclable.
- There is relative insensitivity to surface preparation as the pressure used to make the weld forces the surface layers from the fusion line.
- The process time is very fast. Cycle time can be less than a second for very high productivity.

An analogy of what plastics are and how they behave is taken from volume 3 of the AWS *Welding Handbook* which references the preparation of spaghetti. When spaghetti is properly prepared, each piece is separate and has a definite shape and flexibility. This is much like the plastic (polymer) molecule consisting of a chain of atoms. Thermoplastic polymers consist of many strands that are intertwined but still capable of moving past each other. Thermoplastics soften and/or melt when heated and can be welded. Putting the heated polymers in a cooled mold of the shape required forms these types of plastics into useful parts and products. With thermosetting polymers the analogy is that the spaghetti was not properly prepared and is all clumped together. The strands are not capable of much independent movement, and the mass behaves as a whole. Thermosetting plastics are usually formed by polymerization of polymers in a heated mold of the shape required for a part or a product. Since thermosetting plastics do not melt (if you try to heat them up, they only decompose), they are thus not weldable by themselves. Thermoset-type plastics are typically joined by mechanical fasteners or by adhesives.

Consider the following requirements before you start any plastic welding:

1. Know the application. This is primarily the design engineer's concern about the original product. Consideration must be given to such factors as temperature usage, strength, corrosion resistance, and resistance to UV radiation.
2. Know the material. This is critical when selecting the filler metal and which joining process to use.
3. Know the welding process. Since welding processes can be applied either manually or automatically, it is important to develop an understanding of the advantages and limitations of each.

Know Your Plastics

In order to select the proper process and filler materials you must know what type of plastic, you are working on. When you are attempting to recognize a plastic, it is not the color that you are trying to match. Color should never be used to identify a type of plastic, since many various plastics can have any one of a number of colors added to them. One of the easiest, surest methods to identify the type of plastic you are working on is to look the part over and see if it has an identification number or symbol on it. The number can be referenced back through the producer of the part to determine the plastic used. If no number system can be found, perhaps the symbol can be identified, Table 31-1. These symbols are used a great deal in

Table 31-1 Symbols Used to Identify Plastics

Plastic Symbols	Name	Abbreviation	Description	Uses
1	Polyethylene terephthalte	PET	A type of polyester. Can be made very thin. Tough and has high tensile strength.	Carbonated bottles, videotape, sails, carpets and textiles (Terylene), and ovenproof dishes
2	High density polyethylene	HDPE	Low cost and tough. It is part of the polyethylene family. Also see LDPE below.	Shampoo bottles, toys, dustpan, garbage cans, water tanks
3	Polyvinyl chloride	PVC	Often called vinyl. Can be made very soft or hard. Good wear resistance. Good for food handling.	Noncarbonated bottles, clingfilm, credit cards, imitation leather, cable covering, drainpipes, and music records
4	Low density polyethylene	LDPE	The most common plastic. Very good electrical and abrasion resistance. It is part of the polythene family.	Detergent bottles, carrier bags, cable covering, squeezy containers
5	Polypropylene	PP	Can flex without breaking. Has a waxy feel. Can be used at high and low temperatures. Hard to glue to, but plastic welds easily.	Car bumpers, garden chairs, microwave ware, hinged briefcases, fluted plastic sheet such as Corex
6	Polystyrene	PS	Most people think of the expanded polystyrene packaging, which is made up of about 98% air. General-purpose polystyrene is crystal clear. Can be toughened by adding rubber to produce high impact polystyrene. Glues very well using the construction kit polystyrene glue. Very good for school vacuum forming.	Plastic jewelry toys, aircraft model kits, vending cups, and TV cabinets. Expanded type used for packaging, ceiling tiles, and boxes to keep food warm.
7	Other	—	Used on other plastics. It should indicate the material(s) being used.	

the recycling of plastics. The following plastics may not have symbols:

Polyurethane (rubbery plastic). A flexible, foamy-type plastic. *Uses:* Roller skate wheels, training shoes, car bumpers, and stretchy Lycra fabrics.

Polycarbonates. A very tough transparent plastic. *Uses:* CDs, glass substitutes, and riot shields.

Acrylic (polymethylmethacrylate). A hard, rigid plastic that can be polished on a buffing machine. Can be cut and polished using ordinary workshop tools. It can be shaped by heating to about 320°F. *Uses:* Shop signs, baths, lenses, and decorative finishes as it colors well.

Nylon. A very tough plastic with high abrasion resistance. As a solid plastic, it can be turned and shaped well on a lathe. A fairly stable plastic. *Uses:* Zippers, catches, rope, gear wheels, socks, carpets, and combs.

ABS (acrylonitrile butadiene styrene). Good finish, heat and impact resistant. *Uses:* Briefcases, computer keyboards, telephone handsets.

PTFE. This plastic has a very low friction value and is expensive. *Uses:* To make nonstick pans and bearings.

The following are **thermoset** plastics and are considered nonweldable. But you still should be familiar with them.

MF (melamine formaldehyde). Like all thermosets it is heatproof and chemical resistant but a bit brittle.

Uses: Sheets that cover wood such as Formica. Can also be molded as a resin into ashtrays.
- **Urea formaldehyde.** Like all thermosets it is heatproof and chemical resistant but a bit brittle. *Uses:* Toilet seats, electric sockets, and plugs.
- **Polyester resin.** Like all thermosets it is heatproof and chemical resistant but a bit brittle. Good for clear casting to embed objects and GRP (glass reinforced plastic). Supplied as a resin with hardener. Colors can be added as required. *Uses:* Canoes, GRP car bodies, clear cast embed objects, and artificial stone.
- **Epoxy resin.** Similar to polyester resin but more dimensionally stable and more expensive. *Uses:* Car repair kits and two-part glues for mending china.
- **Rubbers** (elastomers). These are considered by some people not to be plastics, but they are part of the polymer family. Rubber is very flexible and stretchy. They can be made more rigid by the addition of chalk and carbon. *Uses:* Tires, elastic (rubber) bands, oil seals, and shoe soles.

If you are unable to identify a plastic by use of an identification code or symbol, a four-step method can be used. There are a countless number of plastic materials and blends on the market today, so if you are unable to apply this four-step method, most testing labs offer a service of plastic identification.

Scratching Test This is not a very accurate test. However, it will give you some idea of the type of plastic you are working with. The surface hardness of the plastic can be determined by scratching with your fingernail. If it scratches, you know it is not one of the ABS, PVC, or any other harder plastics. If it does scratch, it is one of the softer-type plastics such as PE, PP, or PTFE.

Sound Test Just like when dropping a 4043 aluminum filler rod and a 5356 filler rod on a hard surface, the harder 5356 alloy filler rod has a very definite ring to it, while the softer 4043 alloy filler rod sounds flat. Plastics have similar properties in that they have different specific weights and surface hardnesses that will cause them to not sound the same. An exact identification is hard to determine, but by taking a solid piece of the material and dropping it on a hard and even surface from a height of approximately 5 to 10 inches, you will hear specific tones. As you work with plastics, you will develop a trained ear. While learning to train your ear, you should have some plastic samples of known types to compare and learn the slight differences in a plastic's tone. This is generally reliable enough to determine the plastic's family.

Floating Test As with the sound test, this will only help you identify the plastic's family. Since all plastics have a specific weight that is higher or lower than the specific weight of water, if you take a glass of clean water that is roughly room temperature and insert a small piece of plastic that you want to test, you can determine if it will float or not. There are only two plastics that will float: PE and PP. All other plastics will sink.

Burning Test This is the most accurate test and can be done easily in any shop, Table 31-2. Since every plastic reacts differently when it is burned and if you know the differences, you can quickly and easily determine the type of plastic being worked with. Always keep samples of plastics of known types to use as a reference. You must have good ventilation as these fumes can be toxic. Your first concern is breathing. Second, remove a thin sample and put it on a surface resistant to heat. Never hold the sample in your hand. Third, light a torch or some other flame source and attempt to ignite the sample. Never use matches as the sulfur smell will mask the smell of the plastic. You want to now observe the reaction. Is it burning with a flame or is no flame present? What is the color of the flame? Is there carbon soot above the flame? Blow out the flame and carefully fan

Table 31-2 Identification of Plastics by Burning

Plastic Material	Observation
PE	Produces a blue/yellow flame, smokes, and smells like paraffin (candle).
PP	Produces a blue/yellow flame, drips, and smells like diesel.
ABS	Smells sweet, lacks sooty flame, and does not extinguish.
Polyamide	Smells like burnt horn, stringy, and does not extinguish.
Polycarbonate	Black sooty smoke, may extinguish.
PVC	Acrid smell, black smoke, and does not extinguish.

the fumes toward your nose so you can determine the smell or odor.

You can use these plastic identification tests separately or together to determine the type of plastic you are working on. As previously stated, you must know the material before you attempt to weld it.

Characteristics of Plastics

There are two basic types of plastics: (1) thermosetting plastics and (2) thermoplastics. Thermosetting plastics harden under heat. Through chemical reaction they are formed into permanent shapes that cannot be changed either by reapplying heat or chemical reaction. These plastics are used for such products as wall panels, countertops, and fiberglass boats. They cannot be welded. Thermoplastics, however, soften when heated and solidify when cooled with no chemical change. These plastics can be machined, formed, and welded. Heating and cooling can take place many times without changing their appearance or chemical makeup.

Types of Thermoplastics

Polyvinyl Chloride (PVC) Polyvinyl chloride, usually known by the abbreviation PVC, is one of the most popular materials for construction because of its excellent physical properties, ease of fabrication, relatively low cost, and the ability to be formed into a wide range of products such as sheeting and roll goods.

PVC has wide forming-temperature ranges and self-extinguishing properties. This is important when flammability is a major consideration in the selection of materials. The primary limitation of PVC is its recommended working temperature range of 140 to 150°F. This range limits its structural strength and its corrosion resistance, which decrease rapidly as the temperature increases.

Rigid Polyvinyl Chloride There are two broad classifications for rigid PVC products. Type I has normal resistance to impact and high resistance to corrosion within its working temperature range. Type II PVC is modified with rubber to increase the impact resistance.

Both type I and type II PVC plastics are the best materials for general corrosion protection because of their physical properties, chemical resistance, and low cost. They can be hot-air welded, cemented, or assembled by mechanical processes. Type I is slightly higher in corrosion resistance and is considered the best to use for this purpose.

Both types may be found in such shapes as sheets, pipe, valves, fittings, structural shapes, round and hex bars, and slab stock. Flexible lining materials are available for tanks, large ducts, and troughs. They are applied with adhesives. Flexible joining strips are used for welded seams and corners on the inside of line tanks.

Modified High Impact Rigid Polyvinyl Chloride This class of rigid polyvinyl chloride (PVC) was developed for intermediate corrosion service, primarily in fume exhaust systems. It is readily formed in press and vacuum operations and can be worked and welded at the same temperature as regular polyvinyl chloride. Its availability is limited to 4 × 8 foot sheets up to ½ inch thick. Corrosion resistance is somewhat less than for type II PVC, but it is superior to ABS formulations. It oxidizes but does not burn. It cannot be exposed to very strong acids and the usual solvents that attack rigid polyvinyl chloride. It has slightly better impact strength than unmodified PVC.

Neither pipe nor structural shapes are made from high impact PVC since the availability of unmodified rigid PVC structural shapes is sufficient to meet the limited need. It can be welded to type I or type II PVC, and the same welding rod as that used to weld type I PVC can be used. It is joined in this manner for the lower corrosion areas in an overall exhaust system. Modified high impact PVC is not recommended generally for tanks, either lined or self-supporting. The weld strength is slightly less than for unmodified rigid polyvinyl chloride, and welding speeds are somewhat slower.

Polyethylene (PE) Polyethylene, abbreviated PE, is available in three classes of material. All three classes are the same chemically. Density and physical properties differ, however, according to the packing of the resins. The three types are referred to as low density, medium density, and high density. In general, the main differences in going from low to high density materials are in corrosion resistance, working temperature, and tensile strength. These characteristics increase from low density to high density. Various forms such as sheet, structural shapes, and pipe are available.

Low Density (Branched) Polyethylene Polyethylene is lighter than metal and floats in water. It also burns. This plastic is produced as film, sheet, rod, tubing, and block stock. It is furnished in natural and black.

Low density polyethylene offers reasonably good corrosion resistance. Its main drawback is that its working temperatures and tensile strengths are lower than those of the other classes of polyethylene. It is generally utilized when stiffness, structural strength, and high working temperatures are not required. It can be readily formed by vacuum- or press-forming operations. Although it cannot be joined by cement, it can be welded using the same class of polyethylene rod. Dry nitrogen is usually recommended as a source for hot-gas welding units.

Welding rod is available in diameters of $\frac{1}{8}$, $\frac{5}{32}$, and $\frac{3}{16}$ inch. It is supplied in spools of 1 to 15 pounds. Straight rod, approximately 4 to 5 feet long, is also supplied.

Medium Density Polyethylene This plastic is produced as film, sheet, rod, tubing, and block. Corrosion resistance is somewhat better than that of low density (conventional) polyethylene. It is not cementable, and it will burn. The tensile strength and maximum working temperature are not as great as for high density (linear) polyethylene. It is used for both pressure and conduit tubing and pipe. Applications for structural fabrication from sheet are limited. Impact strength is good, and it can be both vacuum and press formed.

Hot-gas welding is done with dry nitrogen and medium density welding rod.

High Density (Linear) Polyethylene This class of materials may also be referred to as low pressure polyethylene. It is much lighter than metal and is combustible. Although it cannot be cemented, it can be welded. Because of its strength and resistance to high temperature, this material has replaced some of the other polyethylene classes. It has the highest working stress factor and best corrosion resistance of all three classes of polyethylene. The material has a reasonably high working temperature under low load conditions. At high heats, low load conditions must be maintained for satisfactory usage.

This material is used more than the other two classes of polyethylene for structural fabrication. Applications include tanks, fume exhaust hoods, ductwork, valve bodies, and pipe. It can be vacuum formed. Film, sheet, rod, tubing, pipe, and block are available. Welding is done with dry nitrogen and welding rod that matches the material.

Polypropylene (PP) This plastic has lower impact strength than polyethylene, but its tensile strength is much higher and its working temperatures are superior. It also offers more resistance to organic solvents and degreasing agents than polyethylene. While the material can be joined by welding, it cannot be cemented. Such forms as film, rod, sheet, tubing, and blocks are available. This material is used in the manufacture of hoods, plating barrels, ductwork, and other industrial corrosion-resistant applications.

Welding rod is available in $\frac{1}{8}$-, $\frac{5}{32}$-, and $\frac{3}{16}$-inch coil and flat stock. Dry nitrogen is recommended for welding. Some welding is being done with dry air.

Acrylonitrile Butadiene Styrene (ABS) There are two classifications of rigid ABS plastics. Type I is designed for normal temperatures, and there are various mixtures for desired physical properties. Type II is for use in higher temperatures. Different compounds are available for specific manufacturing operations such as calendaring, press laminating, extrusion, and injection molding. There is also a military specification for both type I and type II. Pipe materials are classified as type I with grades 1, 2, and 3, which express increasing working stress values. A type II plastic is intermediate in working stress between grades 2 and 3 of type I.

ABS is much lighter than rigid polyvinyl chloride and only slightly heavier than water. While cementing is the main joining method, the materials can be hot-gas welded with nitrogen. This plastic offers good corrosion resistance except with oxidizing and very strong acids and organic solvents that normally attack thermoplastic materials such as PVC. The material supports combustion, and the flame is characterized by a smoky soot. The type II material is used mostly for higher temperature applications, and pipe is available in both types. The strength of the weld is limited to 60 to 80 percent of the original strength of the plastic.

A major use of ABS is in heat-formed structural parts in which reasonable corrosion resistance is desirable, along with high impact strength. Semiflexible and rigid sheet, extruded pipe, and many injection-molded items are available in a wide range of colors.

Acrylics Because acrylic plastics are transparent, they are widely used as a substitute for glass. The sheet comes both plain and masked for protection in handling. Premium grades are preshrunk before shipment. Certain grades have built-in properties such as ultraviolet absorption, resistance to scratching and/or shattering, and other specialty requirements. Because of its clarity, acrylic plastic is often used in connection with regular thermoplastic structural fabrications. When ordering, it is necessary to specify corrosion resistance, crazing characteristics, and other specifications desired. The material can be cemented and welded.

Welding as a Method of Joining Plastics

Thermoplastics, particularly polyvinyl chloride, polyethylene, and polypropylene, are enjoying increased use today because of their many advantages. Such advantages include light weight, resistance to chemicals and moisture, strength, pleasant appearance, ease of forming, and low cost. The development of hot-gas welding has also helped to spur their use. In manufacturing, fabricating, construction, and plant maintenance, the possibilities of utilizing thermoplastics are as unlimited as the imagination.

Welding is the ideal fabrication process for making such plastic industrial products as chemical-resistant tanks, hoods, vents, ductwork, special containers, and piping systems, Figs. 31-1 and 31-2. The welding of plastic pipe is increasing in oil refineries and chemical plants. Because there is no need to take the line apart, downtime and replacement costs are saved. Welding seals leaks instantly in new or old installations. Plastic tank linings are installed and repaired by welding. The welding of thermoplastics is somewhat similar to the gas welding of metals. All the basic joint designs are used, and welding is possible in all positions, Fig. 31-3. As with all other welding applications, the joint must be prepared properly, and the welding procedure must be carefully applied. The weld is not difficult to make. It can be done quickly, and it is permanent.

The Preparation of Plastics

The preparation of thermoplastics for assembly by welding is similar to the procedures for metal fabrications. The

Fig. 31-2 Large welded plastic duct. Virginia Industrial Plastics

pieces are laid out, cut, machined, and joined with the same tools, equipment, and skills employed in the metalworking trades. There are, however, some special working requirements.

Layout All layout work should be done directly on the plastic sheet in pencil, soapstone, or china marker. Do not use wax pencil since the wax cannot be removed after the plastic is heated. Because of the high notch-sensitivity of plastics, a scribe or other sharp pointed tool causes fracture and so should not be used for marking.

Shrinkage Where accurate dimensional control is required, the sheet should be preshrunk for approximately 20 minutes at 250°F, depending on the gauge of material. The cooling of the sheet must be controlled so that it does not buckle and deform.

Forming Plastic sheet can be heated with a heat gun and formed around metal forms to make round duct and other curved shapes, Figs. 31-4 and 31-5.

Cutting Thermoplastics can be cut with the same hand or power tools used to cut wood or metal. Most cutting jobs are shop operations requiring power tools.

Fig. 31-1 Laboratory with plastic installations. In this case, resistance to corrosion is the principal advantage of plastic fabrication. Virginia Industrial Plastics

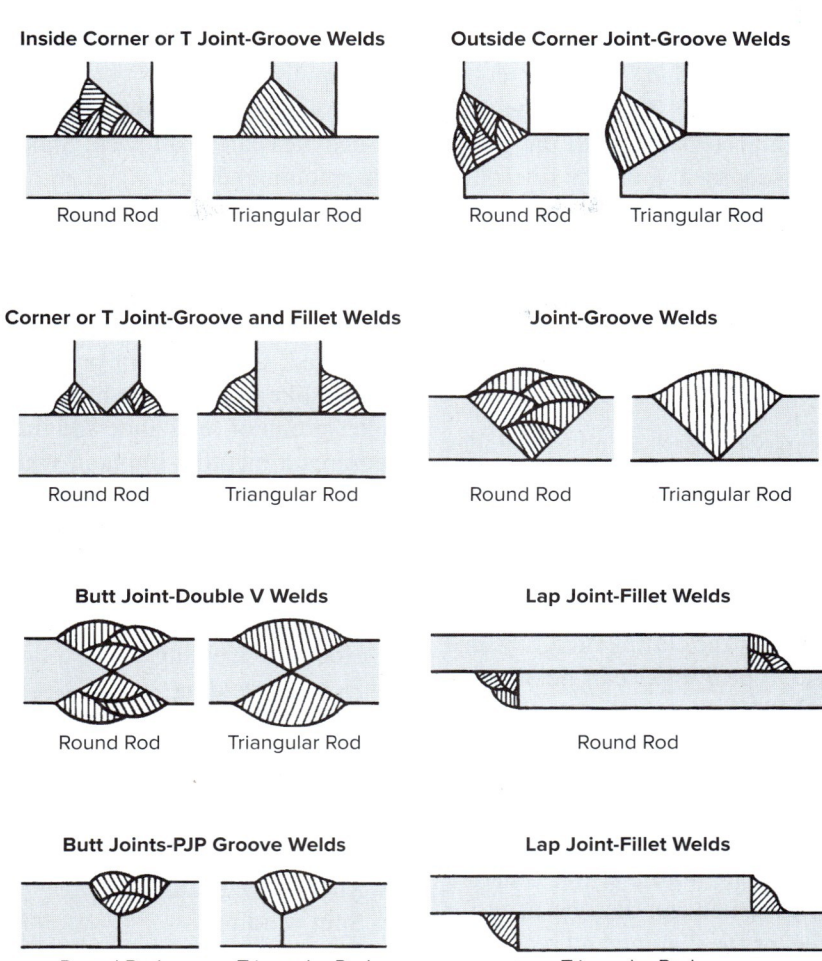

Fig. 31-3 Basic joints that may be formed and welded with round rod and triangular welding rods. Source: Kamweld Products Co.

Fig. 31-4 Heat gun used for forming plastics and other plastic work. Weighs only 2¾ pounds, has adjustable air intake, with a 1³⁄₁₆-inch diameter nozzle. Is equipped with a 6-foot power cord and a toggle on-off-cold switch. © Malcom

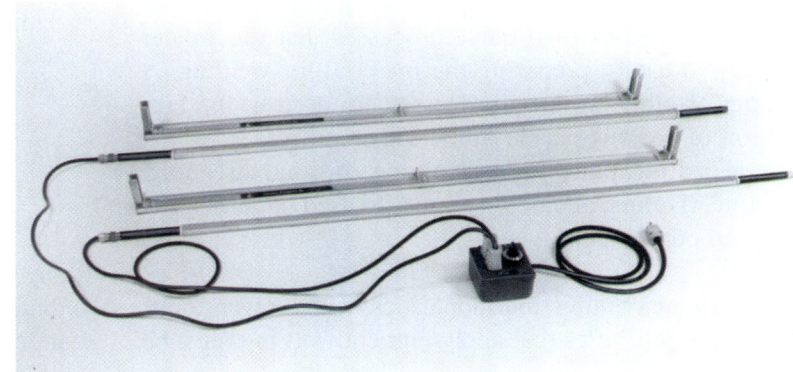

Fig. 31-5 Heat bars for bending and forming plastic. Handles materials up to ½ inch thick. Combination of variable temperature plus two heating surfaces results in adaptability to a wide range of materials, sizes, and thicknesses. © Laramy Products Co., Inc.

Sawing In sawing plastic sheet, there is likely to be concentrated heat buildup in the saw blade because of the poor heat conductivity of the plastic. To allow for this, the blade must be selected in accordance with the gauge of material. Thin materials require the use of fine pitch blades, and blades for thicker materials should be heavier and hollow ground. The number of teeth per inch may vary from 8 to 22 teeth. In order to make a slicing cut in the material, the teeth should have negative rake with little or no set. The saw blade should just show through the material to keep friction at a minimum, and the feed should be slow.

Circular and band saws may be used. Band saw blades should have a slight set and finer pitch than circular saw blades. There is less heating with band saw blades because the length of the blade allows the teeth to cool. Special blades can be obtained for the cutting of plastics.

Shearing Shears can be used for the cutting of light gauge sheets. An ⅛-inch sheet of type I PVC can be sheared easily. Heavier gauge type I PVC shows stress marks. The other plastic types shear better and to a higher gauge. All shearing should be done at room temperature. A cold sheet will crack or shatter.

Routing Beveling is quickly done with a power router, a file, or a rotary sander. The router can also be used for rimming the edges of sheets or for shaping and recessing. The feed must be slow and continuous, and the swarf must be removed by compressed air.

A woodworking jointer may also be used to square or bevel the edge of a sheet. The table and fence must be clean and smooth so that the plastic will not be scratched. The jointer knives should be sharp and finely honed.

Other Working Processes The work processes previously described involve those processes that the welder is likely to use in preparing the material for welding. A number of other working processes are used in the fabrication of various plastic units. These processes, which include drilling, punching, machining, milling, threading, knurling, riveting, and bolting, are also used in fabricating steel and other metals.

Safety In machining thermoplastics with power tools, the same safety rules that apply in metalworking must be followed.

The use of guards on all equipment is vital. Hold-down clamps are used to secure the work to the worktable. Hands must be kept as far away as possible from the cutting tool. The safety switch for shutting off power tools must be in a conspicuous place, and its location must not be changed.

Proper clothing should be worn. No loose shirt sleeves or neckties can be allowed. Shop coats, properly fastened, are recommended.

Plastic Welding Processes

As with metals, plastics are welded by applying enough localized heat to produce fusion of the areas that are to be joined. One group of processes uses a typical heating source like hot gas or a hot plate to melt the plastic. Another group of resistance and induction heating processes requires a metallic implant, which remains in the fusion line. A final group of processes like spin, vibration, and ultrasonic welding uses friction or hysteresis (molecular friction) to heat the plastic. Several different processes have been developed. They include:

- Hot-plate welding
- Hot-gas welding
- Injection welding
- Resistive implant
- High frequency welding
- Induction welding
- Dielectric welding
- Microwave heating
- Spin welding
- Vibration welding
- Ultrasonic welding
- Solvent welding

Hot-gas welding and injection welding are typical manual welding processes when used for repair welding.

Whichever process is selected, the following three parameters must be considered:

- Time required to reach the weld temperature and time to cool down.
- Temperature, since all thermoplastics have a certain weld temperature range that you must stay in.
- Pressure, if required to make the weld, must not be too high or too low or mixing of the molecules will not be possible.

Many of these processes are only applied in machine or automatic modes. Little skill is required on the part of the equipment operator with many of these processes. For additional information on any of these processes review the various American Welding Society publications or search the Internet for up-to-date information and technology developments. The following is an overview of some of the common plastic processes.

Ultrasonic Welding This is basically a high frequency vibration that is directed through a plastic joint. The vibration causes friction, and then heat, often causing a solid fusion in less than a second, Fig. 31-6. Generally frequencies above 20 kilohertz are used. The distance the vibration travels has a great deal to do with determining the classification. Ultrasonic welding is very well suited for rigid thermoplastic parts. If the joint is designed properly, direct application of the ultrasonic shoe is possible. With a smaller contact area, the energy concentration is increased. This results in the use of V-notches, tongues, pins, and other special joints. For remote sealing, thicker walls should be incorporated into the part design. The epoxy molds can be used to reinforce weaker parts when using ultrasonic welding on certain parts.

Advantages
- Fast
- Clean
- Filler materials not needed

Disadvantages
- Many tool designs required
- Design rules not always available nor easily applied

Linear Vibration Welding Very similar to ultrasonic welding, except that frequencies are in the hundreds of hertz and amplitudes are in fractions of an inch. Vibration welding is best used with plastics that have a high coefficient of friction and low viscosity.

Spin/Friction Welding Here two parts are spun and the contact area builds up heat through friction and pressure. The pressure forces a good fusion between the parts and forces out many of the discontinuities. Some flashing may occur with this method.

Advantages
- Produces a good weld.
- Air does not enter during welding.
- Inexpensive machines, such as drill presses may be used.

Disadvantages
- Circular weld joints are required.

Hot-Plate Welding The plastic is brought into contact with a heated plate to soften or melt the plastic. The parts are then removed from the plate and pressed together. This causes the polymer chains to intermingle and a weld is produced. Figure 31-7 is a shop machine; field machines are also available.

Fig. 31-6 Ultrasonic plastics welding machine available in 20, 30, 40, 50, and 70 kHz frequencies. They can be purchased as single units, stand-alone assembly systems, or incorporated into automated machinery. © Dukane

Fig. 31-7 A shop machine hot plate welding machine. It is suitable for PP, PE, PVC, CPVC, and PVDF. Ideal for fabrication of elbows, tees, laterals, and crosses. Sizes range up to approximately 24-inch pipe. Also capable of welding IPS, DIPS, PIP, and sewer materials. © Wegener

Fig. 31-8 Welder using an injection welding machine to attach an access cover on a plastic tank. Drader Manufacturing

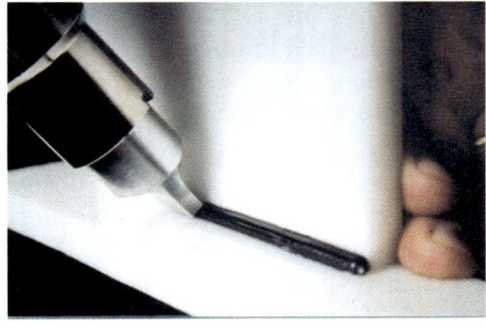

Fig. 31-9 Weld being made on inside of a corner joint with the injector welding machine. The one-hand operation frees up the hand for holding the plastic part in position. Drader Manufacturing

Advantages
- Simple
- Easy to perform

Disadvantages
- Slow speed on heavy sections
- Typically butt joints
- Requires a variety of platens

Injection Welding Unit This type of welding equipment is used in the manual mode, Fig. 31-8. Injection welding of the thermoplastics such as polyethylene, polypropylene, polycarbonate, polyamides, ABS, and polystyrene are readily done. The process injects molten welding rod below the surface of the plastic to create the weld. The injection tip forms the weld zone of molten welding rod. Physical mixing of the plastic substrate and the welding rod makes a strong, high quality weld, Fig. 31-9. Less stress is produced because the entire welding rod is melted and the weld is located closer to the neutral axis of the joint. The automatic feed system lets the welder work the gun with one hand. The welding gun's tip works below the surface layer of plastic, so there is no need to scrape off the oxidation to get a quality weld. This manufacturer provides 12 interchangeable tips. Sheet and pipe can be fabricated obtaining welds from $\frac{1}{16}$ to $\frac{5}{16}$ inch in a single pass. Injection welding can be used for spot and tack welding operations. It can be used to weld products such as high molecular weight polyethylene (HMWPE), polycarbonate (PC), Santoprene, and EPDM, which are considered hard-to-weld plastics.

Hot-Gas Welding Hot-gas welding is one of the principal methods of welding plastics. The two basic requirements for hot-gas welding are a heat source and a welding rod that aids in fusion of the weld to the base material. The joints used in thermoplastic welding are identical to those in metal welding, namely, butt, lap, edge, corner, and T-joints. The same material preparation, fitup, root gap, and beveling are required in plastic welding as in metal welding. Beveling of plastic edges is essential to obtaining a satisfactory weld. Bevel angles of 30 to 35° are used to create the appropriate V-groove angle for hot-gas welding. For injection welding, the groove angles are opened up to 80 to 90°. Flux is not required in the welding of plastics. Some plastics, however, are welded more satisfactorily in an atmosphere of dry nitrogen. Procedures for determining the quality of the completed weld are like the inspection and testing methods in metal welding.

Because of the differences in the physical characteristics of thermoplastics and metals, there are corresponding differences between welding techniques for metals and plastics. In the welding of metal, the welding rod and the base metal become molten and fuse into the required bond to form the welded joint. There is a sharply defined melting point in metal welding. This is not the case in thermoplastic welding.

Because they are poor heat conductors, plastic materials are difficult to heat uniformly. When heat is incorrectly applied, the surface of the plastic welding rod and the base material can char, melt, or decompose before the material immediately below the surface becomes fully softened. You will, therefore, have to develop skills in working within temperature ranges that are narrower than those in metal welding. You will also have to become accustomed to the welding rod not

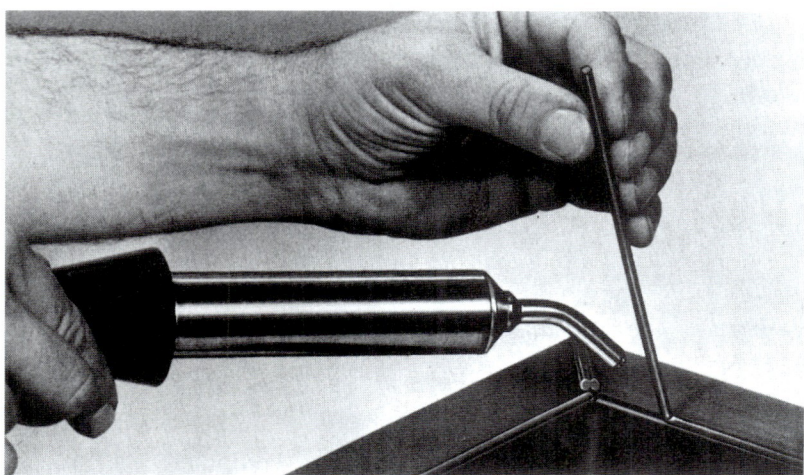

Fig. 31-10 Basic welding procedure for welding plastics. The use of the torch and filler rod is similar to gas welding. Note, however, the distinctive appearance of the weld deposit. The joint is an outside corner joint, and the weld is a fillet weld. © Seelye, Inc.

becoming molten throughout. In fact, the exposed surface of the rod will seem unchanged except for flow lines on either side. The familiar molten crater in metal welding is not found in plastic welding, Fig. 31-10. Only the lower surface of the welding rod becomes fusible while the inner core of the rod merely becomes flexible. Hence, the materials do not flow together as a liquid. You must apply pressure on the welding rod to force the fusible portion into the joint and make the permanent bond.

Plastic Filler Rod Both hot-gas and injection welding require filler rod. The rod needs to match the properties and in some cases the color of the plastic to be welded. Many manufacturers have welding rod available in a variety of colors, sizes (diameters $\frac{1}{8}$–$\frac{3}{16}$ inch and in various lengths), types (PVC, Lexan, PPR, Styre, etc.), and various profiles (round or triangular), Fig. 31-11.

Fig. 31-11 Various types, sizes, and colors of plastic filler rod for hot-gas and injection welding. Drader Manufacturing

With the injection process you are totally melting the filler rod before it is applied, so you do not need to have all the various diameters or shapes on hand as you would for the hot-gas process. Some plastics use filler to enhance the plastic's properties. The amount and quality of filler material inside the plastic parent material and the welding rod can greatly affect the quality of your weld, such as when plastic contains filler materials like wood chips or denim to provide volume and make plastic less expensive. These fillers might be all right in some situations but not in others. Consider what will happen inside the weld when the molecules of the parent plastic material and the welding rod meet; they will not connect, so they cannot mix and good fusion will not be possible. Special welding rods are manufactured for these types of applications. If you have a situation where you are not certain of the base material and/or the filler material, you may want to perform a rod fusion test. This will at least give you some assurance that the welds you will be making are sound. This test is accomplished by the following steps:

1. Scrape an area clean on a spot on the back or some other out-of-the-way location.
2. Melt the rod and fuse it into the base material; cool it with cold water.
3. Pull the rod away from the part to determine the strength at the fusion point.
4. This can be repeated with various rods until the best match is found.

The idea is to use the rod that provides the best fusion. If little fusion is achieved, you should try it again on a similar part. If fusion is again the problem, you may have to use an adhesive instead of fusion welding.

Stretching and Distortion Regardless of the skill of the welder, some stretching of the welding rod will always occur. Stretching should not exceed 15 percent. The welding rod stretches too much when it leans away from the direction of welding. In speed welding, stretching is caused by too much pressure on the rod or by plastic residue on the shoe and in the preheating tube. When a thermoplastic rod is heated enough to form a weld, it becomes soft and tends to stretch. Contraction during cooling causes stresses that produce cracks and checks across the face of the weld. The amount of stretch in a completed weld can be determined by measuring the length of the rod before and after welding.

In multilayer welds, deposited beads are reheated in the process of laying new beads one on another. Stretching in multilayer welds must be held to a minimum since checks and cracks caused by stretching show up as voids in the finished weld and cannot be detected by visual inspection. When making multilayer welds, allow ample time for each weld pass to cool before proceeding with additional welds. To save time and give added strength to multilayer welds, alternate welds from one side of the groove to the other. Thermoplastics are notably poor conductors of heat, and stresses in plastic are confined to a much smaller area than those in metal. Shrinkage of a weld upon cooling is greater near the crown than at the root. As with steel, when welding single-V and fillet welds, the work should be offset before welding to compensate for distortion. Preheating the general area to be welded also helps. Distortion can also be reduced by using speed welding and triangular welding rod instead of making multilayer welds with round welding rod.

Welding PVC The material must be kept clean at all times. It can be cleaned by wiping with methylethylketone (MEK) or a similar solvent. Freshly beveled edges do not need to be cleaned. Welding edges are beveled or offset to provide areas for the welding rod and to permit better adhesion by removing polished outer surfaces. Cut bevels with a jointer, sander, router, or plane. Clamp the sections firmly in place or tack weld. Avoid backup materials that are good heat conductors. Allow a root gap in most assembly procedures except when tack welding. Thickness, shape, size, good construction techniques, and the strength required dictates the type of weld to use.

Welding Polyethylene and Polypropylene Procedures are the same for these materials except for the welding gas and welding temperatures. These should be varied according to Table 31-3 (p. 1007).

In welding polyethylene and polypropylene, a number of precautions should be followed:

- The base material should be freshly cut or scraped and clean.
- The welding rod and the material must be of the same density.
- Polyethylene and polypropylene are subject to stress cracking. The base material and welding rod should not be overheated nor should the welding rod be stretched. Use only 1 foot of rod for 1 foot of weld.
- If the welded joint will be under stress in service, the weld will be subject to chemical attack that would not occur under normal circumstances. This is called "environmental stress cracking." It causes welds to fail and cracks to radiate into the sheet from the weld.
- Polyethylene and polypropylene welding rods tend to loop in the direction of the weld. Do not let this cause you to force the rod and add undue strain on the weld.

Plastic Welding Equipment

The use of the correct welding equipment and process can make a big difference in the quality and appearance of the plastic weld. The most common equipment used when performing plastic repairs are hot-gas welding units and injection welding units. Hot-gas units are usually less expensive; however, their use requires the skill of the welder to make a strong weld. The correct joint preparation, correct heat, and correct pressure applied to the welding rod are extremely important. When attempting to weld thermoformed (vacuum formed, press formed) parts, problems might occur when the heated air hits the surface and starts spreading out in all directions. With the hot-gas process, it is difficult to control the airflow. Since thermoformed pieces have high amounts of stress in the formed areas, a hole or deformation may occur when the hot air reaches one of these areas. The injection welding equipment may be better to use in this application. These units use a heated aluminum tip instead of hot air to melt the parent material. The heat is very localized in the area where the tip touches the plastic. Melting of holes or excess deformation is not as likely to occur. The injection welds will generally show a higher strength in these applications.

The hot-gas torches are divided into two basic types: electrically heated and gas heated. In both types, welding gas (compressed air or nitrogen) passes over a heat source that raises its temperature to 450 to 800°F.

Electric torches, Fig. 31-12 (p. 1008), are used in manufacturing plants and wherever current is available. Many welders prefer this type of torch because it is more compact and easier to handle. Gas-heated torches are used primarily in field operations where electricity is not available.

Figure 31-13, page 1008, shows the internal construction of the typical electrically heated torch. The electricity heats a stainless-steel barrel. A gentle flow of compressed air or nitrogen passes over the heating element. This heats the air rapidly to the desired temperature. The hot air leaves the nozzle from the lower end of

Table 31-3 Thermoplastic Welding Chart

	PVC Type I	PVC Type II	PVC Plasticized	Polyethylene Regular	Polyethylene Linear	Polypropylene	Chlorinated Polyether	FEP Fluorocarbon	Acrylic
Welding temp. (°F)[1]	500–550	475–525	500–800	500–550	550–600	550–600	600–650	550–650	600–650
Welding gas	Air	Air	Air	Inert	Inert	Inert	Air	Air	Air
Groove-weld strength (%)	75–90	75–90	75–90	80–95	50–80	65–90	65–90	80–95	75–85
Maximum continuous service temp. (°F)	160	145	150	140	210	230	250	250	140
Bending & forming temperature (°F)	250	250	100	245	270	300	350	550	280
Cementable	Yes	Yes	Yes	No	No	No	No	No	Yes
Specific gravity	1.35	1.35	1.35	0.91	0.95	0.90	1.4	2.15	1.19
Support combustion	No	No	No	Yes	Yes	Yes	No	No	Yes
Odor under flame	HCL	HCL	HCL	Wax	Wax	Wax	Sweet chlorine	Pungent	Sweet
Color[2]	Gray	Light gray	Black	Translucent or black	White or black	Cream to amber	Olive drab	Bluish translucent	Transparent

[1] Measured 1/4 in. from welding tip.
[2] These are the most commonly used colors and are subject to change. For complete details, refer to the specifications of individual plastic manufacturers.

Source: Laramy Products Co. Inc.

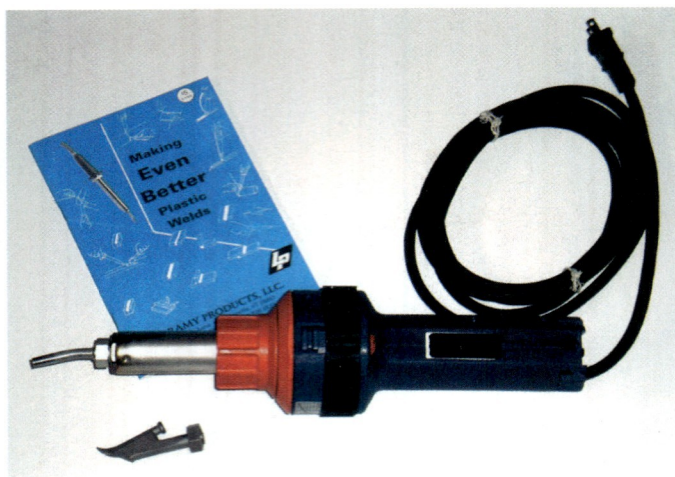

Fig. 31-12 High speed electric welding torch. © Laramy Products Co., Inc.

Fig. 31-14 Electrically heated torch, air regulator/pressure gauge, and neoprene air hose with electrical cable inside. The torch has low, medium, and high heat ranges. © Kamweld

the barrel. From there it goes through a tip or high speed tool to heat both the base material and the welding rod simultaneously. The heat softens the rod and base material so that they become one.

Figure 31-14 shows the simple outfit that is needed for plastic welding.

Figure 31-15 shows a complete portable plastic welding outfit with its own built-in air compressor. It requires only electric power for operation.

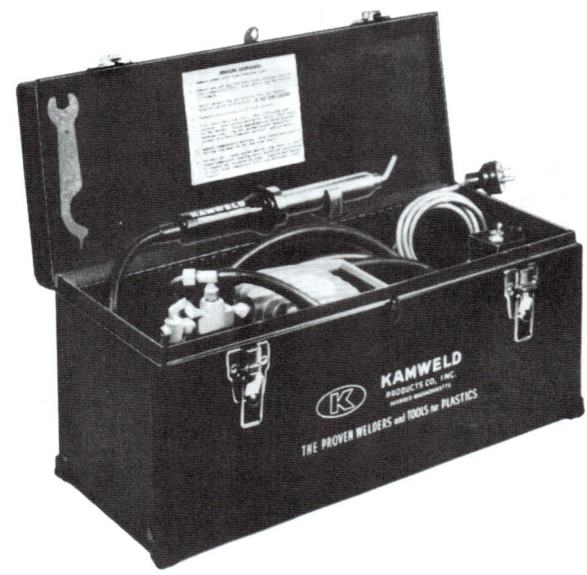

Fig. 31-15 Plastic welding outfit includes an air compressor, and it can be used in both shop and field. © Seelye, Inc.

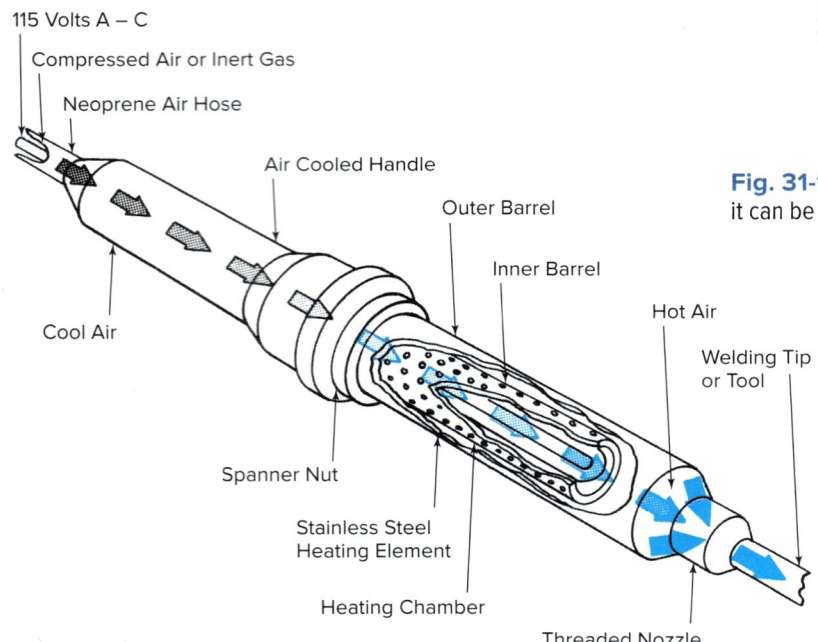

Fig. 31-13 Internal construction of a typical electric welding torch.

The welder of plastics should have a number of miscellaneous tools such as a wire brush, sharp knife, rotary sander, rasp, bending spring, files, saws, and C-clamps. The welder also uses a heat gun, which is an electrically heated blower, for softening plastics for bending and forming before welding. Unlike metal welding, plastic welding requires no gloves, aprons, or goggles since welding temperatures are low and visibility is not obstructed by a flame or arc.

Setting up the Equipment

1. Make sure that you have the proper type of torch for the work at hand.
2. Select the proper heating element.
3. Relieve the regulator-adjusting screw to prevent damage to the regulator due to sudden excessive air pressure.
4. Connect the welding unit to the air or nitrogen supply and adjust the regulator for about 3 pounds pressure.
5. Connect the torch with a 115-volt electric outlet.
6. Let the torch warm up for 3 or 4 minutes. Make sure that the compressed air or nitrogen is flowing continuously through the barrel of the torch to prevent burning out the heating element.
7. Select the proper tip or high speed welding tool for the type of work to be welded, Fig. 31-16. As with other types of gas welding, tips can be changed while the torch is hot. Avoid touching the hot barrel of the torch.
8. Select the proper air pressure for the size of the heating element (in watts) and for the temperature desired at the tip end.

Rules for Welding with Electric Welding Torches

- Be sure the welding gas supply is clean. Moisture or oil in the welding gas may prevent a satisfactory bond in welding and cause a short circuit in the heating element of the torch.
- Never leave electricity on when the welding gas is turned off. Always turn gas on first and turn it off last.
- Always ground the torch to prevent a short circuit, possible electric shock, and damage to the heating element.
- The volume of welding gas passing over the heating element determines the welding temperature. To increase the temperature, reduce the gas volume. To decrease the temperature, increase the gas volume. To determine the temperature of the heated air, hold a thermometer ¼ inch from the end of the welding tip.
- Never touch the end of the torch barrel or welding tip when the torch is turned on.
- To obtain maximum life from the heating element, always use the recommended welding temperature.
- Read the manufacturer's operating instructions before using a torch for the first time.

Rules for Welding with Gas Welding Torches

- Be sure the torch is equipped with the proper jet for the heating gas being used.
- Be sure the welding gas supply is clean. Moisture or oil in the welding gas may prevent a satisfactory bond in the welding.
- When regulating welding temperatures, reduce the volume of welding gas or increase the pressure of the heating gas to raise the temperature. To lower the temperature, increase the volume of welding gas or reduce the pressure of heating gas.
- Never touch the end of the torch barrel or welding tip when the torch is turned on.
- Always turn the welding gas on before lighting the torch.
- Never leave the torch lighted when the welding gas is off. Always turn off the flame before shutting off the welding gas.
- Always read the manufacturer's instructions before using a torch for the first time.

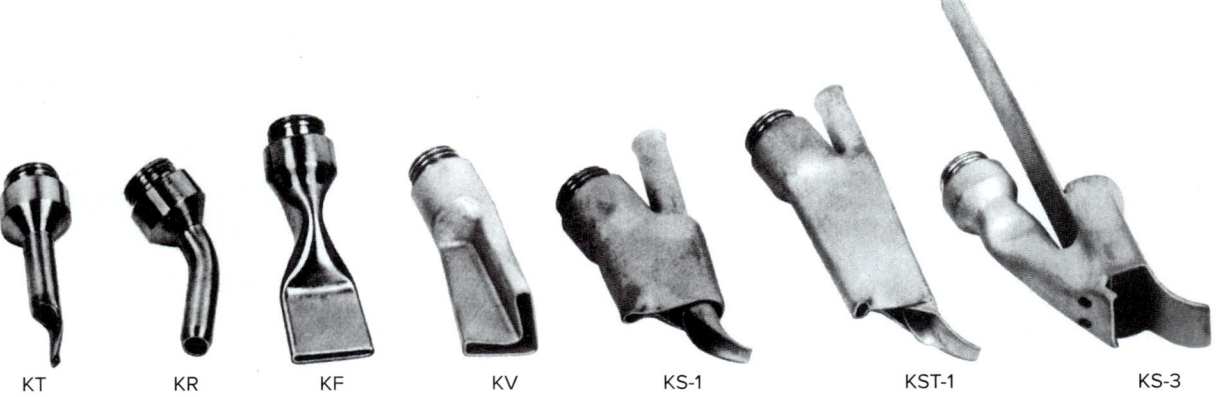

Fig. 31-16 Welding tips and high speed tools. KT—tacker tip; KR—round tip; KF—flat strip tip; KV—flat corner tip; KS-1—high speed tip for round rod; KST-1—high speed tip for triangular rod; KS-3—high speed tip for flat and corner strips. © Kamweld

Inspection and Testing

Good plastic welding can be achieved only if there is an adequate program of inspection and testing. You will recall the importance of testing in metal welding procedures in order to determine fusion, penetration, porosity, and appearance. These same weld characteristics are important to plastic welds, Fig. 31-17.

In general, the strength of a plastic weld is dependent on a combination of six interrelated factors:

- Strength of the base material
- Temperature and type of welding gas
- Pressure on the welding rod during welding
- Proper weld and joint selection
- Proper material preparation before welding
- Skill of the welder

Unlike metal welds, dressing plastic welds decreases the strength of the completed welds by approximately 25 percent. Welds that are equivalent to less than 75 percent of the material strength are considered to be unsatisfactory.

Faulty Welds

Faulty welds are the result of the following errors that sometimes occur in plastic welding:

- Overheating the base material or the plastic filler rod, Fig. 31-18
- Underheating the base material or the plastic filler rod, Fig. 31-19
- Improper penetration through entire root of weld, Fig. 31-20

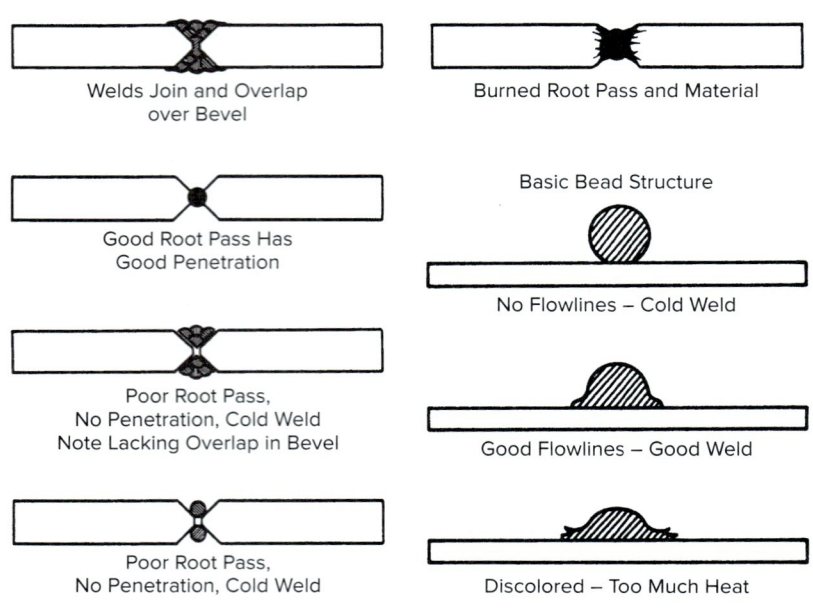

Fig. 31-17 Good and faulty welds. Source: Kamweld Products Co.

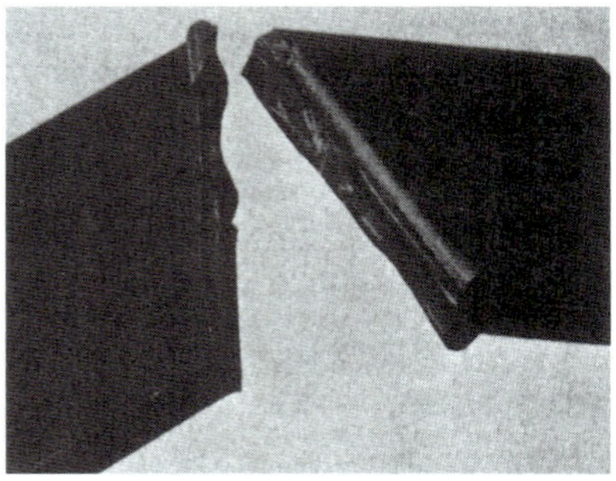

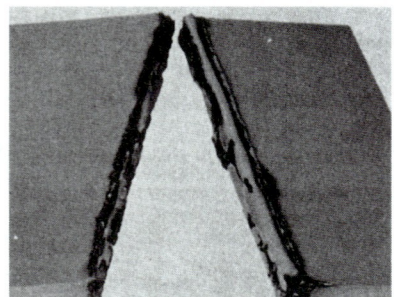

Fig. 31-18 Overheated weld. © Kamweld

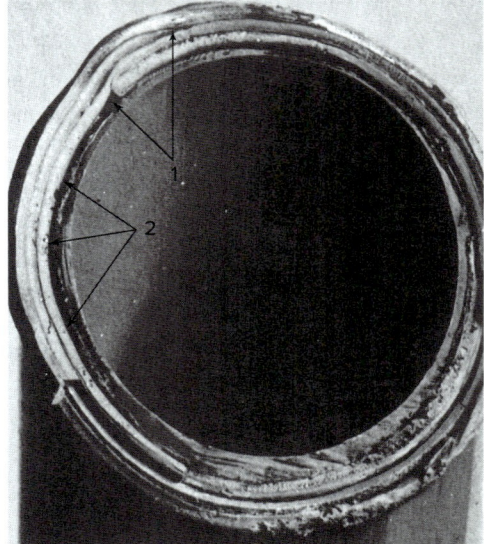

Fig. 31-19 Cold weld. © Kamweld

Fig. 31-20 Voids (1) and scorching (2) can be seen in this broken weld. There is incomplete penetration throughout. © Laramy Products Co., Inc.

- Porosity caused by air inclusions or dirt, Fig. 31-20
- Stretching the filler rod
- Incorrect handling of the welding torch
  - Wrong torch or tip work or travel angle
  - Too slow or too fast travel
  - Lack of or faulty fanning motion of the torch
  - Heat at the torch tip too close or too far away from work
  - Heat at the torch tip not centered on the weld bead

A good plastic weld requires the following:

- Thorough root penetration
- Proper balance between the heat used on the weld and the pressure exerted on the welding rod
- Correct handling of the welding torch
- Correct preparation of the joint to be welded

Study Table 31-4 (p. 1012) which presents in greater detail the causes of common plastic welding troubles and how to correct them.

Visual Inspection

All welding should be inspected before the welded part is put into service. Visual inspection permits only partial evaluation of the weld bead. Such internal defects as incomplete fusion and penetration, air inclusions, and cracks cannot be determined by visual inspection.

A great deal, however, can be learned through visual inspection. A good weld has flowlines (little wavelike lines) along each side of the deposited weld rod and shows no signs of decomposition. If the flowlines are present, continuous, and uniform, it is visual evidence of a good weld. The continuity of these flowlines indicates that there was enough heat on the filler rod to create flow. The continuity also shows whether or not the welder applied the correct pressure to the rod, so that the hot, viscous material was forced out of the weld bed and bonded the plastic parts together.

Visual examination of multilayer welds can be accomplished by cutting across the axis of the weld and polishing the cross section. Close inspection will reveal faults such as voids, scorching, and notching.

Testing of Welds

Welded joints are the sites of potential weakness in a plastic structure. It is, therefore, necessary to show that a welded joint is fit for its intended purpose. This can be achieved by appropriate destructive, nondestructive, and chemical testing techniques. These methods were developed by the plastic pipe industry as quality control techniques for welding processes. In addition, some standard plastic test methods can be applied to plastic welds. Organizations such as the American Society of Testing Materials and the American Welding Society, as well as other plastics organizations, have established procedures for testing plastics and plastic welds. Some of the more common test methods will now be discussed.

Destructive Testing

Tensile Test This test is used primarily to evaluate butt joint-groove welds on rigid sheet. Cut two 4 × 6-inch pieces from $3/16$-inch sheet and make a 30° double bevel with a $1/64$-inch root face. Clamp each piece to the bench, leaving a root gap of $1/64$ inch. Apply two or three beads on each side of the specimen. Use small diameter rod for root welding and larger rod to complete the weld. Cut the test section into $1/2$-inch wide specimens and pull each in a tensile tester at the rate of 0.025 inch per minute. The welding value may be calculated as follows:

$$\frac{\text{Breaking strength of weld} \times 100}{\text{Original tensile strength of material}}$$

$$= \text{welding value}$$

A tensile strength value of 80 to 100 percent of the base material is considered acceptable.

Creep Rupture Test Creep rupture tests can be used to compare the long term performance of plastic welds with that of the parent material. This test is normally carried out using tensile specimens under a constant load and an elevated temperature, and the time to failure is measured. Tests are normally carried out in water, but a surface-active medium can be used to accelerate failure. This is a much more expensive test to perform than short-term static tests. However, the creep tests provide more useful information when designing components that will be under constant load.

Bending Test Weld a test specimen like that used for tensile testing. While it is still hot, bend it double along the axis of the weld. The weld should be fused to the beveled surface of the plastic sheet. Another bend test may be conducted 24 hours after welding. In this test, bend the specimen 90° by hand. It should resist breaking, and the base material should break outside of the weld.

Burst Test This is the most effective way of testing pipe groove welds and fillet welds on fabricated fittings and

Table 31-4 Causes of Faulty Welds

Distortion

Cause:
1. Overheating at joint
2. Welding too slow
3. Rod too small
4. Improper sequence

Solution:
1. Allow each bead to cool
2. Weld at constant speed—use speed tip
3. Use larger-sized or triangular-shaped rod
4. Offset pieces before welding
5. Use double V or backup weld
6. Backup weld with metal

Poor Appearance

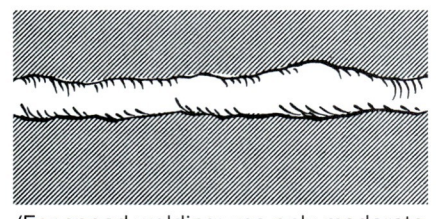

(For speed welding: use only moderate pressure, constant speed, keep shoe free of residue)

Cause:
1. Uneven pressure
2. Excessive stretching
3. Uneven heating

Solution:
1. Practice starting, stopping, and finger manipulation on rod
2. Hold rod at proper angle
3. Use slow uniform fanning motion, heat both rod and material

Poor Fusion

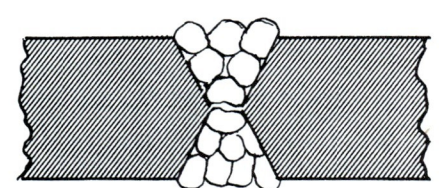

Cause:
1. Faulty preparation
2. Improper welding techniques
3. Wrong speed
4. Improper choice of rod size
5. Wrong temperature

Solution:
1. Clean materials before welding
2. Keep pressure and fanning motion constant
3. Take more time by welding at lower temperatures
4. Use small rod at root and large rods at top—practice proper sequence
5. Preheat materials when necessary
6. Clamp parts securely

Poor Penetration

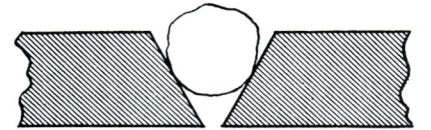

Cause:
1. Faulty preparation
2. Rod too large
3. Welding too fast
4. Not enough root gap

Solution:
1. Use 60° bevel
2. Use small rod at root
3. Check for flowlines while welding
4. Use tacking tip or leave 1/32″ root gap and clamp pieces

Porous Weld

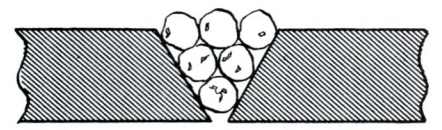

Cause:
1. Porous weld rod
2. Balance of heat on rod
3. Welding too fast
4. Rod too large
5. Improper starts or stops
6. Improper crossing of beads
7. Stretching rod

Solution:
1. Inspect rod
2. Use proper fanning motion
3. Check welding temperature
4. Weld beads in proper sequence
5. Cut rod at angle, but cool before releasing
6. Stagger starts and overlap splices ½″

Table 31-4 (Concluded)

Scorching

Cause:
1. Temperature too high
2. Welding too slow
3. Uneven heating
4. Material too cold

Solution:
1. Increase airflow
2. Hold constant speed
3. Use correct fanning motion
4. Preheat material in cold weather

Stress Cracking

Cause:
1. Improper welding temperature
2. Undue stress on weld
3. Chemical attack
4. Rod and base material not same composition
5. Oxidation or degradation of weld

Solution:
1. Use recommended welding temperature
2. Allow for expansion and contraction
3. Stay within known chemical resistance and working temperatures of material
4. Use similar materials and inert gas for welding
5. Refer to recommended application

Warping

Cause:
1. Shrinkage of material
2. Overheating
3. Faulty preparation
4. Faulty clamping of parts

Solution:
1. Preheat material to relieve stress
2. Weld rapidly—use backup weld
3. Too much root gap
4. Clamp parts properly—backup to cool
5. For multilayer welds—allow time for each bead to cool

Adapted from Laramy Products Co.

---

couplings. The open ends of the pipe are capped, and the weld is subjected to hydrostatic pressure with water as the liquid.

Impact Test The weld is subjected to a sudden impact by hitting it with a hammer. The broken joint is examined for faulty bonding, voids, and scorching.

Fracture Mechanics Tests If more rigorous testing and information is required, fracture mechanics analysis may be applied. The fracture mechanics test methods can be used to quantitatively qualify the characteristics of plastic welds. These processes require extremely accurate notch tip positioning and test procedures. Testing can be conducted using either the three-point bend loading or a single edge notch bend (SENB) specimen. Bend test methods can be the same as those used for other plastic and metallic parent materials. If the plastic being tested is very brittle, the use of linear elastic fracture mechanics (LEFM) can be used. Since many plastics show a great deal of crack tip plasticity, they may require a more complex elastic-plastic fracture mechanics (EPFM) analysis test.

Nondestructive Testing It should be noted here that visual inspection is considered by many to be a nondestructive form of testing.

Spark-Coil Test A high frequency, high voltage spark-coil tester, Fig. 31-21, page 1014, detects pores and cracks in a plastic weld that are not visible by any other inspection method. In using a spark-coil tester, the plastic weld is grounded with a metal backing, and the pencil-like tip of the tester is run along the weld area. If there is porosity or cracking, a straight line of sparks passes through the weld to the metal ground. Sparks are generated with voltages up to 55 kilovolts and frequencies around 200 kilohertz.

Radiography This is perhaps the most efficient method of plastic weld inspection because it shows defects in a plastic weld that are not readily discernible by any other inspection method. It gives a complete, detailed picture of

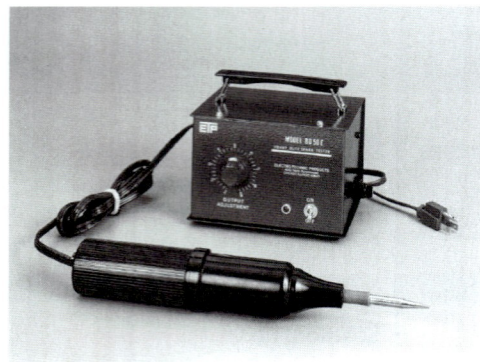

Fig. 31-21 High frequency spark tester. © Electro-Technic Products, Inc.

the internal characteristics of the weld joint and a permanent record of the weld. The one disadvantage of this type of testing is the high cost.

Chemical Tests A PVC weld test specimen is immersed in acetone for 2 to 4 hours. Faulty welds will separate from the base material, and strains and stresses in the weld will be indicated by the swelling of the material. The dye penetrant method is another chemical inspection method suitable for plastic welds. A dye penetrant is painted or sprayed on the weld. A poor weld will be disclosed by the penetrant seeping through and discoloring the weld.

Practice Jobs

Instructions for Completing Practice Jobs with the Hot-Gas Process

Welding plastics is very similar to welding other materials. You must learn, however, a definite knack, or feel, that is somewhat different from that experienced when welding metal. Most mechanics get the feel for plastic welding in a few hours, but it can take as much as 40 hours of diligent practice before you make consistently good thermoplastic welds.

Learn the types of joints and what to look for in good and poor welds. Practice starts, stops, and tie-ins. Be careful that the base material and the rod are not exposed to overheating and underheating.

There are five important "musts" concerned with plastic welding:

- Small beads should form along each side of the weld where the rod meets the base material.
- The rod should hold its basic round shape.
- Neither the rod nor the base material should char or discolor.
- It is very important that the rod does not stretch over the weld. The length of rod used should be no more nor less than the length of the weld. If the rod is forced ahead in the direction of welding, instead of straight down into the weld point, fewer feet of the rod will be used than the length of the weld. This stretches the hot filler rod so that it breaks open at the top of the weld bead. The broken bead is a special problem when the weld is reheated to add additional beads to a multipass weld.
- Do not use oxygen or other flammable gases.

Plastics must be clean and dry prior to welding and during the welding operation. Surface preparation is necessary. Every particle that remains in between the weld will affect the strength of the actual weld. The best way to clean the weld area is to scrape off the first layer of the material surface. Do not use sanding paper because the dust will stay on the surface. Solvents should be used carefully. They are harmful and a solvent's residues will migrate into the weld. The best cleaning tool is a scraping blade. Plastics can also pick up moisture and must be protected and dried. Some plastics, like ABS or polycarbonate, are hygroscopic (this means they absorb the moisture from the surrounding environment). These hygroscopic materials need to be dried before using them for welding purposes. Estimate the material you need to weld and dry only this amount. This will save time as the drying time depends on the quantity of the material and the rod.

The plastic filler rod must be of the same composition as the type of plastic being welded. The type of plastic, the joint design, the thickness of the material, the position of welding, and the type of equipment are considerations in choosing a filler rod.

Heat for plastic welding is supplied by a heated gas instead of a flame. Compressed air, nitrogen, or an inert gas is used. (Never use oxygen or flammable gases.) The gas passes through the torch where it is heated by the heating element and is then directed through the torch tip to the surface of the joint.

The torch is held in one hand, and the filler rod is held and fed with the other hand, Fig. 31-10, page 1005. This is similar to the gas welding position. The rod can also be fed automatically with the use of a high speed welding tip, Fig. 31-22. This tip increases the speed of welding. It feeds the welding rod automatically in the right position and produces a uniform weld bead. One hand is left free to steady the work and insert new rods. Flexible PVC plastics are welded with a special strip feeder and both flat and corner tips. Welding filler rod for flexible plastics is in strip form.

The temperature of the welding gas is regulated by increasing or decreasing the volume of gas supplied to the torch. A slow movement of the gas over the heating element produces more heat. Therefore, a decrease in the welding gas volume increases the temperature of the gas. Conversely, a rapid movement of the gas over the heating element produces less heat, and so an increase in the welding gas volume decreases the temperature of the gas. The tip and the heating element may be changed to alter the heating capacity of the torch further.

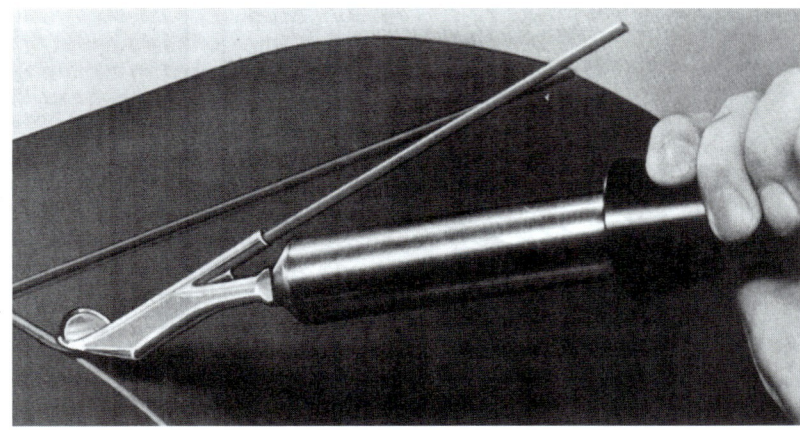

Fig. 31-22 Using the high speed welding tip. A plastic weld is being run on a girth seam of a tank. © Seelye, Inc.

 For video of hot-gas welding, please visit www.mhhe.com/welding.

Tack Welding

As with metal welding, tacking is used to assemble the parts to be welded quickly. Tacking ensures the proper alignment of all pieces. It holds the sections together during welding so that clamps, jigs, or additional workers are not needed for this purpose. Tack-welded sections can be broken apart easily for reassembly and retacking.

Procedure for Tack Welding

1. Attach the tack welding tip to the torch.
2. Wait for 1 or 2 minutes so that the tip can reach its proper temperature.
3. Hold the tip at a work and travel angle of approximately 90° and place it directly on the joint to be tacked.
4. Draw the tacker tip along the joint for the desired length, Fig. 31-23. Tacks should be about ½ to 1 inch long. Pieces that are large or unwieldy should be tacked at regular intervals as when welding metal.
5. The unit is now ready for continuous welding. The tacks should not show any brown or burned spots. This is an indication that too much heat was applied, that the rate of travel was too slow, or a combination of both. Tacks that do not hold the workpieces together are caused by incomplete fusion due to insufficient heat, incorrect travel speed, or a combination of both.
6. Practice welds until you are able to make satisfactory tacks. The material must not be burned and should show good fusion.

Hand Welding (Beading)

The purpose of hand welding is to join two or more pieces permanently together with a rod or strip as a filler. This is similar to other forms of metal welding.

The fusion that takes place is the result of the proper combination of heat and pressure. The welder applies pressure on the filler rod with one hand while applying heat to the rod and base material with hot gas from the welding torch, Fig. 31-24, page 1016. Both pressure and heat must be kept constant and in proper balance. Too much pressure on the rod tends to stretch the bead and cause poor fusion and irregular bead buildup. Too much heat chars, melts, and distorts the material.

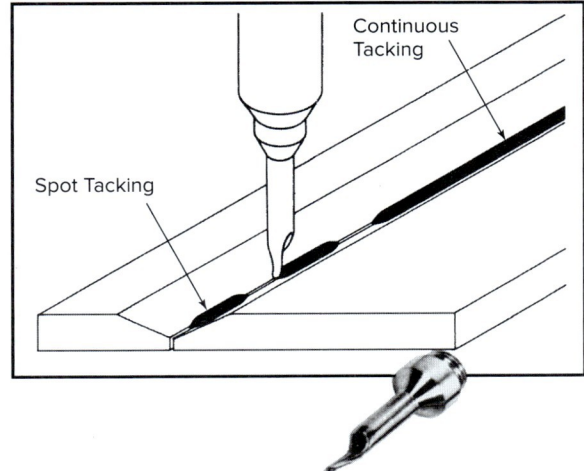

Fig. 31-23 Tack welding.
Source: Kamweld Products Co.

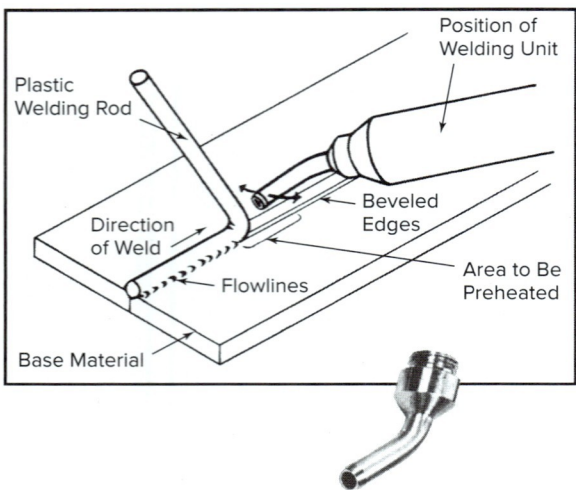

Fig. 31-24 Hand welding with a round tip. Source: Kamweld Products Co.

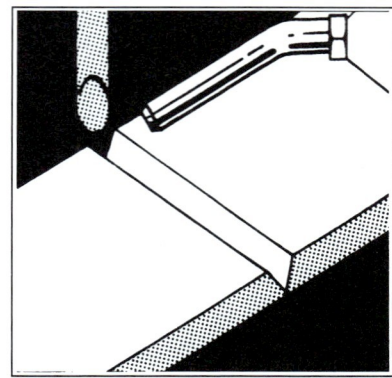

Fig. 31-25 Starting the welding operation. Source: Seelye Plastics

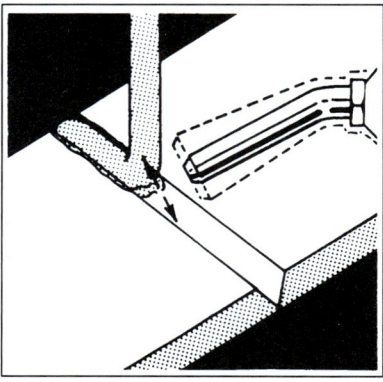

Fig. 31-26 The torch motion during welding. Source: Seelye Plastics

Procedure for Hand Welding PVC Plastics with a Round Tip

1. Install a heating element that produces from 450 to 500°F.
2. Attach a round tip to the torch.
3. Set the air pressure according to the recommendations by the manufacturer of the equipment.
4. Obtain a flat piece of PVC about 6 inches long, 4 inches wide, and at least $3/32$ inch thick. Make sure that the surface is clean. Clamp the piece to the workbench.
5. Secure a PVC filler rod $1/8$ inch in diameter and cut the end at a 60° angle with cutting pliers.
6. Check for correct temperature by holding the tip $1/4$ inch from the material and counting off 4 seconds. At the count of four, the material should show a faint yellowish tinge.
7. Hold the torch $1/4$ to $3/4$ inch from the material to be welded and preheat the starting area on the base material and rod until it appears shiny and becomes tacky. The rod is held at a work angle of 90° to each side of the base material, Fig. 31-25.
8. Move the torch up and down with a fanning or weaving motion in order to heat both the filler rod and base material equally, Fig. 31-26. About 60 percent of the heat should be directed toward the base material and 40 percent to the filler rod.
9. A good start is essential. It should take only a few seconds for the material and rod to be hot enough. Once the weld has been started, continue to fan the torch from the rod to the base material at the rate of two full oscillations per second. Exert only as much pressure on the rod as is necessary to cause fusion to take place.

- Too much forward pressure causes stretching which will lead to cracking as you weld.
- You should notice a small bead forming along both edges of the welding bead and a small roll forming under the welding rod, Fig. 31-27.
- Too much heat in the rod softens it so that pressure bends the rod rather than forcing it into the base material. Too little heat causes it to lay on the surface of the material without being fused to it.
- A slight yellowing of the rod and base material is caused by slight overheat. Run a number of welds that are alternately too hot and too cold in order to be able to recognize this condition.
- When a weld is to be ended, stop all forward motion and direct a quick heat directly at the intersection of the rod and base material. Remove the heat and maintain downward pressure for

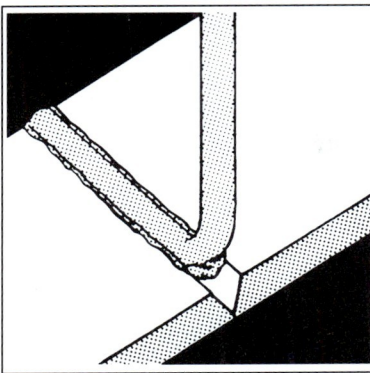

Fig. 31-27 Note the bead being formed along both edges of the weld. Source: Seelye Plastics

Fig. 31-28 Double square-groove weld for butt joints, welded from both sides. Source: Kamweld Products Co.

several seconds until the rod is cool. Then release the downward pressure. Twist the rod with your fingers until it breaks. If a continuation weld is to be made, the deposited bead should be cut at an angle of 30° with a sharp knife or cutting pliers.

Bead Test Test the strength of your bead by trying to pick the rod off the base material with a pair of pliers while the bead is still hot. If fusion has taken place, the rod will stretch and may break, but it will not part from the base material. If the rod lifts readily from the base material, there may have been not enough heat or too much heat.

After you have run a series of satisfactory beads on a plastic pad with 1/8-inch rod, repeat the procedure with different size welding rods. If the beads are laid so that they overlap the preceding bead, change the work angle of the welding rod to 45°. Always allow cooling time between beads. Welds may be cooled with a wet sponge or a jet of cool air.

Hand Welding of Joints

When you have mastered making good bead welds, you are ready to practice welding plastic sheet together.

There are three basic types of butt joints generally used in plastic construction: (1) the square-groove butt joint, (2) the single V-groove butt joint, and (3) the double V-groove butt joint.

Square-Groove Butt Joints This type of joint is generally made in light gauge sheets as thick as 3/32 inch. No preparation of the edge is required. A root gap of approximately 1/64 inch is necessary to permit full penetration through the back side. Welding is from both sides when possible, Fig. 31-28.

The square-groove butt joint is acceptable when the work is not of a critical nature and when cost is a consideration.

1. Obtain two pieces of plastic sheet 6 inches long, 3 inches wide, and 3/32 inch thick.
2. Set up the pieces with a root gap of 1/64 inch to allow the semimolten plastic to flow through to the back side of the joint. Make sure the back side is suspended above the table.
3. Use the same welding technique described for beading. Weld one pass on each side of the plate.
4. Inspect the weld carefully for faults. Test the weld according to the procedure given under Butt Joint Test, below.

Single V-Groove Butt Joints Generally, single V-groove butt joints are used when only one side of the plastic sheet is accessible. If the back side is accessible, a single bead on the back side provides additional strength.

1. Obtain two pieces of plastic sheet 6 inches long, 3 inches wide, and 1/8 inch thick.
2. Prepare the pieces with a 30° bevel and a 1/32-inch flat face at the root. This is like the preparation necessary for steel welds.
3. Set up the pieces with a root gap of about 1/64 inch to allow the semimolten plastic to flow through to the back side of the work. Make sure that the back side is suspended above the table. Use the same welding technique described for beading.
4. Weld the first pass along the root of the weld. Make sure that you obtain penetration through the back side, Fig. 31-29.

Fig. 31-29 Single V-groove weld for butt joints welded from one side only. Source: Kamweld Products Co.

Fig. 31-30 Double V-groove weld for butt joint welded from both sides. Source: Kamweld Products Co.

5. Weld two additional passes along the edge of each sheet. Make sure that the weld is built up evenly. The joint must be completely filled with overlaps on the top beveled edges, Fig. 31-29, page 1017.
6. For the sake of additional practice and after you have checked the back side for faults, weld a bead on the back side. A single V-groove weld, reinforced in this manner, is somewhat stronger than a double V-groove butt joint.
7. Inspect the weld carefully for faults. Test the weld according to the procedure given under Butt Joint Test.

Double V-Groove Butt Joints The double V-groove butt joint has a high strength factor only slightly less than the single V-groove joint with reinforcement on the back side. Both sides of the joint must be accessible for welding.

1. Obtain two pieces of plastic sheet 6 inches long, 3 inches wide, and 1/8 inch thick.
2. Prepare the edges of the sheets to be welded with a 30° bevel. Allow a root gap of 1/32 to 1/64 inch.
3. Weld the first pass along the root of the weld on one side. Make sure that you obtain penetration through to the back side, Fig. 31-30.
4. Weld two additional passes along the edge of each sheet. Make sure that the weld is built up evenly, Fig. 31-30.
5. Inspect your welds carefully for good fusion and appearance.
6. Turn the plate over and repeat the above procedure on the other side.
7. Inspect the welds carefully for faults. Test the weld according to the procedure given under Butt Joint Test.

Butt Joint Test A square butt joint may be tested by fracturing it. Place the joint in the jaws of a vise with the weld bead facing away from you and about 3/16 inch above and parallel to the top of the vise jaws. Cover it with a cloth to prevent injury due to flying chips and pieces. A blow with a hammer on the weld side usually breaks off the top pieces. If a break occurs through the weld bead with

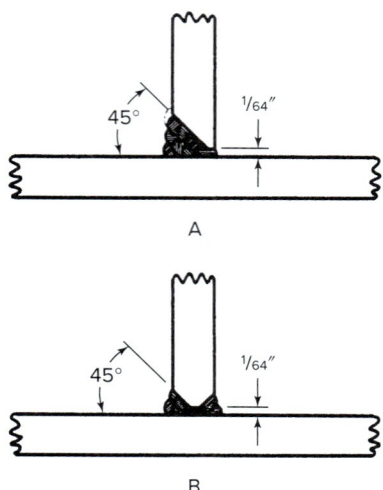

Fig. 31-31 Fillet welds for T-joints. Source: Kamweld Products Co.

some portion of the weld on each piece, the weld may be considered good. A 100 percent weld breaks through the base material. If there is overheating or underheating, the material usually separates at that point. For most plastic fabrication, a weld strength of 75 percent is acceptable, but every effort should be made to achieve 100 percent strength.

Fillet Welds Fillets and corner welds are used to attach two sheets of plastic at 90° to each other in T-joints and corner joints. The vertical plate should be beveled. If welding is to be from one side only, the vertical plate must be beveled from one side only as shown in Fig. 31-31A. If, on the other hand, welding can take place from both sides, the vertical plate must be beveled from both sides as shown in Fig. 31-31B. Welding from both sides produces the stronger of the two welds.

Lap Welds Very often two sheets are joined by forming a lap joint, Fig. 31-32. Welding a plastic angle to a sheet also forms a lap joint. The welds may be made by fusing the overlapping areas with a flat tip, or they may be made with the round tip and filler rod.

Edge Welds Edge welds are used to weld heads and bottoms into tanks or boxes and to weld an angle to the

Fig. 31-32 Method of welding lap joints. Source: Kamweld Products Co.

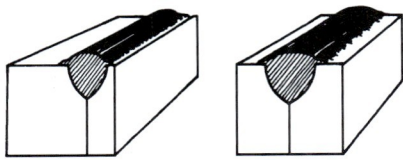

Fig. 31-33 Method of welding edge joints. Source: Kamweld Products Co.

edge of a sheet. Two flat surfaces are set up side by side, and the two edges are welded together. If the material is heavy, the edge of each plate is chamfered and the weld is a groove, Fig. 31-33. The procedure for welding is similar to that used in making V-groove butt joints. *Note:* If you are having trouble making satisfactory welds, review Table 31-4, pages 1012–1013, that lists contributing factors to faulty welding.

Repair of Cracks

The two most common types of repairs are fixing cracks and replacing broken or missing parts. The most effective method of crack repair can be determined by understanding what caused the plastic to crack in the first place. It could have been caused by internal stresses brought about by the manufacturing process, or from improper storage or handling, or even from incorrect use of the piece. In order to prevent that crack from traveling further when you begin your repair weld, drill a hole approximately $3/32$ inch in diameter at each end of the crack. The round shape of the hole will distribute the stress more evenly, reduce the notch effect, and should stop the crack from growing. When the fracture runs the whole length of the plastic part, a bigger gap appears. This also indicates that a lot of stress has been set up inside the plastic part. When you attempt to force the part back together, new stress problems might occur. Care must be taken in these situations. Prepare the crack by opening it up like a V-groove. This will ensure better weld strength because of the increased penetration. Use a stick scraper to get the 60 to 70° groove angle required for hot-gas welding. Once it is prepared, tack weld or clamp the parts in position. The welding rod and weld area must be clean. Any oxidation, dust, grease, or any other contamination reduces the weld strength. Use a blade scraper to do this type of preparation properly. Figure 31-34 shows some examples of tools used when making plastic welds.

One of the most critical parts of making a quality weld is maintaining the proper temperature. Don't think that by turning the heat up you will be able to weld much faster. You have to remember what is going on inside the material, not just on the surface. Keep in mind that there is a

Fig. 31-34 Typical tools used for making plastic welds: (1) scraper, (2) moon knife, (3) roller, (4) temperature gauge, and (5) wire brush. © Wegener

fairly narrow temperature range that you must keep plastics in, to make a quality weld. Too hot and you scorch them, too cold and they won't fuse.

Pressure on the welding rod must be carefully applied to fill the crack. The molecules cannot mix if the pressure is too high or too low. Practice is required to get the proper feel of just what the correct pressure is (3–6 pounds of pressure would be typical). Once you have finished the weld, smooth down the weld bead. This will help avoid a new crack starting. Weld the crack from both sides if at all possible; this will ensure complete joint penetration. It will take longer, but the results will be worth the effort.

Replacing Missing Pieces

Use the same procedures outlined in the previous section on the repair of cracks when repairing a plastic part that has pieces missing. The areas where pieces are missing normally show irregular jagged shapes. The best method is to cut out the entire damaged area in a shape that is round, square, or rectangular. If a square or rectangular shape is used, radius the corners so there are no stress risers. Cracks will start on these high stressed corners if they do not have the proper radius. Take time to get a good fit of the replacement piece. This will make the repair process simpler and quicker and the repair stronger. Match the material thickness to avoid weak spots in the repair. The replacement material can be cut out of sheets or pipes or other damaged parts you may have available. The important issue is that the material, the thickness, and the shape will have to fit the missing area.

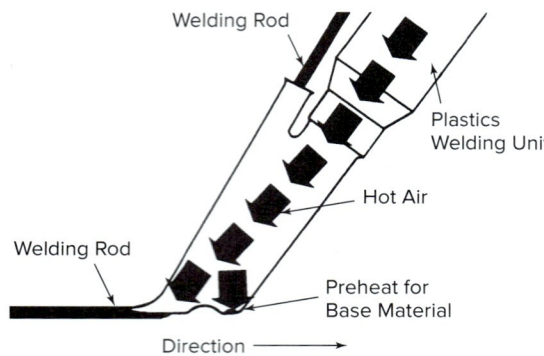

Fig. 31-35 Operation of the high speed welding tip.
Source: Kamweld Products Co.

High Speed Welding

The high speed welding tip increases the average welding speed to over 4 feet per minute on flat or curved surfaces. It feeds the welding rod automatically in the right position and produces a uniform weld head, Fig. 31-35. One hand is left free to steady or turn the work and insert new rods. A handy cutting blade is attached to the tip. The 500-watt heating element is recommended for high speed welding.

Procedure for High Speed Welding with Round Rod

1. Secure two pieces of PVC about 18 inches long, 3 inches wide, and 3/32 inch thick, and clamp them to the workbench.
2. Select the high speed tool designed for the diameter of filler rod to be used. Prepare the rod by cutting one end at a 60° angle.
3. Set up the equipment and allow the unit to warm up.
4. Hold the welding unit straight down at a 90° work and travel angle.
5. Hold the shoe of the high speed tool about ½ to ¾ inch above the surface of the workpiece, and hold it at the starting point, Fig. 31-36.
6. Insert the beveled filler rod into the preheated tube and push it into the softened base material until the rod bends slightly backwards, Fig. 31-36.
7. Change the travel angle of the tip to about 60° in the direction of welding. Apply pressure on the top surface of the rod sticking out until it starts to fuse to the surface of the material, Fig. 31-37.
8. Continue to exert pressure with the shoe and start pulling the torch in the direction of welding. At the same time, exert light pressure on the rod in the preheating tube.
9. Continue to press on the top surface of the rod with the shoe as you proceed with the weld. Do not move forward faster than the fusion of the rod to the base material will permit.
10. Once the weld is started, there can be no hesitation. As you continue with the weld, the rod will feed through, being pulled by its adhesion to the base material. The following conditions must be given careful attention:
    - The speed of the weld can be increased by lowering the travel angle of the welding unit to about 45°.
    - Note the flowlines which are similar to those visible in hand welding. The absence of flowlines

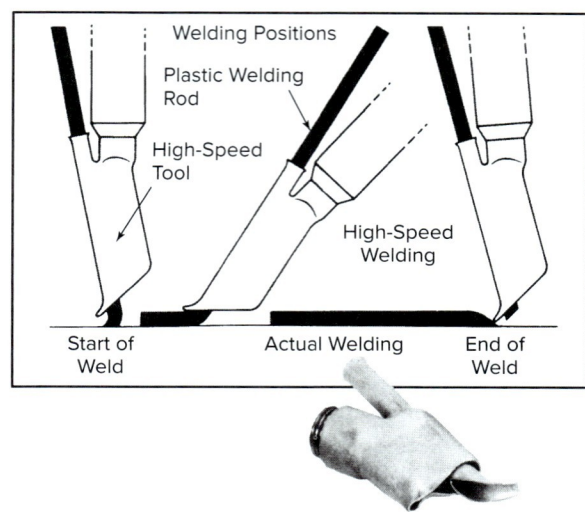

Fig. 31-36 Welding positions for the high speed tip.
Source: Kamweld Products Co.

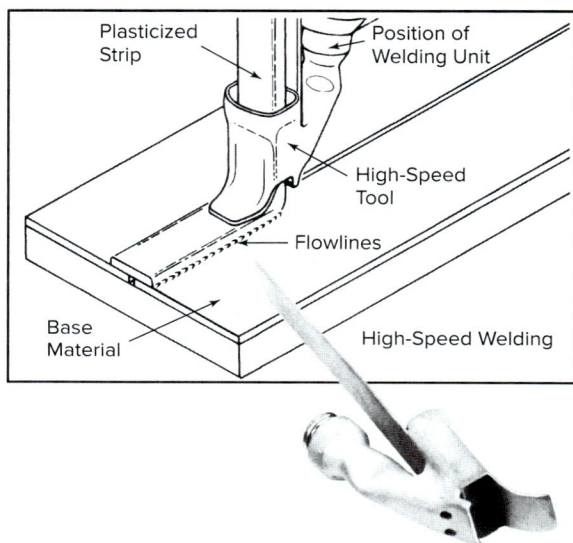

Fig. 31-37 Adhesion of flexible strip to base material during high speed welding. Source: Kamweld Products Co.

indicates a "cold" and unsatisfactory weld. The crown is higher than in hand welding.
- If the tip is not moved forward fast enough, the rod softens and bunches up in the preheating tube. Sometimes it chars or burns. The emerging end of the rod softens, flattens out, stretches, and breaks. When this happens, withdraw the tip and cut the rod. Any residue in the preheat tube can be removed by pushing a cold rod back and forth through the tube until it is cleared.
- Observe the emerging rod constantly so that any corrective action that may be necessary can be taken immediately.

11. If the rod is stretching, the weld is going too slowly and the rod is overheating. When stretching occurs, withdraw the tip, cut off the rod, and make a new start before the point where the rod started to stretch.
12. If flowlines do not appear, the weld is going too fast, and adequate bonding is not taking place. Bring back the tip to the 90° travel angle temporarily in order to slow down the welding rate. Then move it to the desired travel angle for proper welding speed. The rate at which the weld proceeds is governed by the temperature, the consistency of the rod, and the travel angle of the welding unit.
13. Make sure that the preheater hole and the shoe are always in line with the direction of the weld so that only the material in front of the shoe is preheated.
14. To stop the welding process, (a) withdraw the tip quickly until the rod is out of the tube, and (b) bring the tip quickly to the 90° travel angle and cut off the rod with the end of the shoe. Remove the rod from the preheating tube immediately so that it does not clog up the tube. The preheating tube must be kept clean, and the shoe should be cleaned with a soft wire brush to remove residue.
15. A good speed weld in a V-joint has a slightly higher crown than the normal hand weld, and it is more uniform. It should appear smooth and shiny, with a slight bead on each side. Study the weld characteristics shown in Table 31-4, pages 1012–1013, and compare with your weld.

High Speed Weld Tests When you and your instructor feel that you have acquired the skill of making this type of plastic weld, make up test specimens and subject them to the following test procedures:

- Cut through the joint and inspect for complete bonding. Strips should also be cut from the work and subjected to tensile and bending stress tests.
- If a pressure test is desired, make up a small box and subject it to a water-pressure test. The box can be similar to that used for metal arc welding.

High Speed Welding with a Plastic Strip Flexible PVC plastic is welded with a flexible plastic strip instead of a round rod. Plastic strips come in different shapes such as the flat strip and the corner V-strip and are generally supplied in roll form. Strip welding is generally used for tank linings and similar applications.

Only one pass is necessary with a strip. The technique of welding is like that of welding with round rod with the following exceptions:

- The strip must be precut in length, and 1 to 2 inches must be allowed for trimming.
- Start the weld by tamping with the broad shoe of the high speed tool on the top of the first inch of strip in the direction of the weld. Do not drag the tip until the first inch of the strip adheres firmly to the base material. The tip should be held at a travel angle of about 80°, and some pressure should be applied by hand to hold the top of the strip down.
- Guide the strip by hand and continue to weld at sufficient speed. The usual flowlines should appear on both sides of the strip, Fig. 31-37. If travel speed is too slow, excessive heat causes the strip to soften and stretch. This can be corrected by a quick tamping motion with the shoe like that used to start the weld. Then proceed in the direction of the weld.
- To stop the weld, simply remove the tip and allow the remaining strip to pull through the high speed tool.

If a strip weld is to be made from corner to corner, such as on the bottom of a tank, the weld should be started in one corner and proceed only halfway across. Then another weld is started in the other corner and overlaps the scarfed end of the first weld. Where each strip butts to the other or in corners, additional flat strip overlays are necessary. A flat tip is used to preheat the area while the flat strip, held in a gloved hand, is pressed firmly into the corner. All high corners can be heated and pushed down with the blunt end of a knife.

Welding Plastic Pipe

After you have mastered welding plastic sheet, you should practice making butt joint-groove welds in plastic pipe. The preparation and welding of pipe is similar to that used for flat material. There is one major difference. In welding sheet, the welding proceeds in a forward direction only. In welding pipe, the torch and filler rod must follow the direction of the round shape, Fig. 31-38, page 1022.

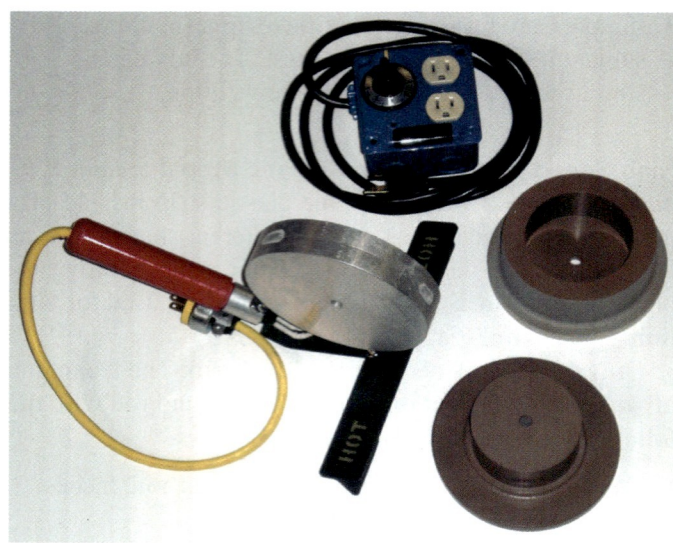

Fig. 31-38 This fusion pipe welding machine uses an electric tubular heater and is mounted on the clampable base. A stepless rheostat is used to control the heat. It can be used in shop or field for welding of PE, PP, and PVDF pipes and couplings. Two sizes are available from this manufacturer; one for ½–2 inch and another for 3–4 inch pipe. © Laramy Products Co., Inc.

Solvent Weld Process (Bonding)

Plastic pipe may be joined by a method referred to as a *solvent weld* or *bonding* process. Pipe and fittings are solvent welded with MEK. The solvent chemically etches the surface of both the pipe and fittings so that when joined, the two surfaces are fused into each other like brazed copper or welded steel. Cutting through a solvent-welded joint shows that the original line of division no longer exists. This joint is now stronger than either the pipe or fitting.

Procedure for Bonding Pipe

1. Use Schedule 40 ABS-DWV pipe and MEK solvent. Cut the pipe and remove all burrs. Be sure both pipe and fitting are clean. Wipe clean with a cloth if necessary. Grease or oil may be removed with paint thinner.
2. In applying solvent, use a brush large enough to pass around the pipe end or fitting socket quickly. Apply solvent to the fitting socket first and then to the pipe end. Use solvent moderately, but spread it thoroughly and evenly. Excess solvent serves no purpose.
3. Insert the pipe into the fitting and position it with a quick rotating motion of a quarter turn or so. This ensures a full, even spread of the solvent. Do not attempt to realign the pipe and fitting after the set has begun. If you notice a mistake, remove the fitting quickly and reapply the solvent as before. Do not attempt to turn the fitting after the set has taken place. This will destroy the joint. In dry weather, joints will dry in 1½ to 2 minutes. Cold, wet weather slows the drying period, but in no event will it exceed 5 minutes. Three minutes is the typical setting period.
4. After full set, water tests may be applied immediately. Pressure systems require a longer drying period because of higher pressure tests.
5. Not all plastics nor solvents react the same. Some plastic piping requires solvent which has a considerably longer setting and drying period. A two-stage process for cleaning and cementing, which increases the time required, is necessary with some solvents. Solvent can be removed from the hands with thinner. Wash hands afterwards with soap and water.
6. Test bonded joints by cutting through the joint and inspect for thorough bonding. Strips should also be cut from the work and subjected to tensile and bending stress tests.

Gluing of Plastics

Plastics like PP and PE show a very low surface energy. The surface energy can be raised by using a primer or an electrical or flame treatment. This will help produce a decent glued joint, but it is only temporary. These types of plastics are much better welded. Adhesives have been used for quite some time to join plastics. Currently there are approximately 250,000 different adhesives for all kinds of applications to choose from. This makes selecting the correct one very challenging. Proper research is required before using a new plastic or adhesive substance. You must understand how the plastic will react to the adhesive. How the joint will be applied and what kind of environment it will be located in are also important considerations. Because of the differences in the chemical structures of adhesives, they all have certain technological properties. A large number of adhesives or glues can be found in each of the following groups:

Melting glue (hot glue). A one-component, thermoplastic adhesive that requires heating up to the melting stage. It has a fast curing time, easy storage and handling, and is solvent-free and efficient. It has low temperature resistance and is not capable of handling large splicing areas.

Polycondensate (phenolics). Typically a two-component, thermoset adhesive. The components react to each other when combined. The advantage is a very strong splicing, a wide range of use, and enough time to work with it as well as good heat resistance. Mixing is not always exact. Storage conditions are critical and can be quite expensive.

Polymerisate (cyanacrylates). A one-component, thermoplastic adhesive that needs a catalyst to start the reaction. Moisture from the surrounding environment will do in this case. It has a fast curing time and a strong bond. Typically a brittle splicing results. Parts need to have good matching surfaces.

Adhesive dispersion (wood glue). Not commonly used in plastic joining. This belongs to the group of adhesives, but is used for woodworking.

Polyaddition adhesive (epoxies). Typically a two-component, thermoset adhesive. The two components react to each other. It has a very strong splicing, a wide range of use, and provides enough time to work with it. It also has good heat resistance. It is difficult to get mixing exact. Storage is critical and can be expensive.

The design of the joint depends on the application. If a lap or T-connection or any other typical joint is chosen, the following will apply when doing splicing. Make sure that the adhesive is able to handle the medium it will come in contact with later. It is always exciting and dangerous when you put something in a Styrofoam cup and it winds up dissolving the cup and running all over the countertop.

The preparation of the splicing surface is a crucial step in ensuring a successful joint. The surfaces never start out clean, so take the time to clean them properly. This is done by removing any heavy dirt, dust, or grease. Follow this initial cleaning with the use of sanding or grinding paper (grit 100–150) to roughen the splicing area. The grinding dust can be removed with oil-free compressed air. If compressed air of proper quality is not available, use a degreaseable solvent to take care of any residue. Remember some plastics are not resistant to certain solvents. To verify, try the cleaner first on a sample piece or in an out-of-the-way area. Such items as alcohol and gasoline are not recommended. They will leave a residue, which will weaken the joint.

Once the cleaning operation is done, avoid touching the prepared surface with bare hands. Fingerprints and any soil left on the surface will weaken the joint. In any case, the splicing should be done immediately after the areas have been prepared.

Before starting the process, the following should be done:

- Clean the area where the joining will occur.
- Make sure you have enough room for the pieces, tools, and personal protection, etc.
- Prepare the equipment you will need.

Some other safety considerations should be observed as well.

- Eating, drinking, and especially smoking in the splicing area should be prohibited at all times. There are fumes and dust particles floating in the surrounding air, and you may swallow or inhale these particles with your food and cigarettes.
- Open flames are not allowed in close proximity because of the flammability of fumes from most adhesives.
- Avoid skin contact because some solvents will find their way directly into the bloodstream; contact could also cause allergic reactions.
- Good ventilation is important whenever adhesives or solvents are used.
- Read and strictly follow the MSDS of the adhesive used.
- Lastly, all adhesives are basically special waste and should be disposed of in the proper way.

Care of Plastic Pipe

As with all welding, there are certain material characteristics that must be provided for.

- Be careful in storing pipe and fittings. They should never be stored in the sun. These materials heat very slowly. Therefore, the side exposed to the sun's rays absorbs the heat faster than the unexposed side, and uneven expansion causes the pipe to "bow." Both pipe and fittings should be kept as free of dirt as possible.
- Care must be taken in cutting pipe. Plastic pipe may be cut with pipe cutters or any crosscut saw. Special pipe cutter wheels for plastic are available to fit all standard pipe cutters. Lightweight cutters designed exclusively for plastic piping are also available. Do not use dull cutters; they may cause too many burrs. All burrs must be removed from the pipe. Care must be taken to ensure a square cut. All socket fittings require a good square pipe to ensure a good fit. Keep in mind that the better the fit, the better the joint.
- Pipe must be laid out and cut with a high degree of accuracy because errors cannot be rectified with stress, heat, or a hammer. A fitting socket must be measured to the full depth of the socket.
- Plastic pipe must be supported properly. Avoid tight supports, which tend to compress or cut the pipe or fitting. Horizontal piping should be supported every 4 or 5 feet and on both sides at vital turns and large fittings to relieve strain. Vertical stacks should be supported at the base and between floor and

ceilings. Provide for expansion on runs over 50 feet in length.
- Never attempt to heat and bend plastic pipe.
- Plastic pipe may be placed underground with safety since it is impervious to the soil acids in this country. It is the only material that can be safely installed in a cinder fill. It requires no painting or coating. Make sure that pipe is fully supported all around before completing the fill in order to protect pipe alignment. The pipe is impervious to corrosion by cement, so no coating or wrapping is required.
- Plastic piping lends itself to prefabrication. Its light weight coupled with its strong welded joints permits preassembly of large assemblies and easy movement from shop to field. Assemblies can be moved 5 minutes after the last joint is made. Always test the assembly before moving it to the field.
- Plastic pipe can be joined to pipe made of other materials only with the appropriate adapter fittings. A complete line of adapter fittings is available. Lead can be poured and caulked directly into a cast iron hub without damage to the plastic pipe.

CHAPTER 31 REVIEW

Multiple Choice

Choose the letter of the correct answer.

1. Approximately how long ago were thermoplastics developed? (Obj. 31-1)
   a. 25 years
   b. 50 years
   c. 75 years
   d. 100 years

2. Another name for *plastic* is _____. (Obj. 31-1)
   a. Perchloralethlene
   b. Polysulfide
   c. Polymer
   d. Both b and c

3. What are the two main types of plastics? (Obj. 31-2)
   a. Thermoplastics and thermoshere plastics
   b. Thermoplastics and thermosetting plastics
   c. Thermoplastics and polyvinyl chloride
   d. Thermoplastics and polyethylene

4. What form of plastic is transparent? (Obj. 31-2)
   a. Polyvinyl chloride
   b. Polyethylene
   c. Acrylics
   d. Acrylonitrile butadiene styrene

5. Plastic welding is done on which joints? (Obj. 31-3)
   a. All the basic joints can be welded
   b. Lap and T-joints
   c. Butt and corner joints
   d. Edge and notch joints

6. Welding processes for joining plastics are broken down into how many groups? (Obj. 31-3)
   a. Two
   b. Three
   c. Four
   d. Five

7. What two welding processes are very common for repair welding? (Obj. 31-3)
   a. Hot-plate welding, sonic welding
   b. Hot-gas welding, injection welding
   c. Resistive implant, induction welding
   d. Spin welding, vibration welding

8. Which is not considered a plastic weld fault? (Obj. 31-4)
   a. Porous weld and scorching
   b. Poor penetration and fusion
   c. Slag and spatter
   d. Stress cracking and warping

9. Plastic welding must have a strength of 100 percent of the base material. (Obj. 31-4)
   a. True
   b. False

10. What is an appropriate V-groove angle for an injection weld? (Obj. 31-5)
    a. 30°
    b. 60°
    c. 45°
    d. 90°

Review Questions

Write the answers in your own words.

11. Why are plastics so useful in making products? (Obj. 31-1)

12. Describe six ways a plastic's type can be identified. (Obj. 31-2)
13. List the type of plastic represented by each abbreviation. (Obj. 31-2)
    a. PET _____
    b. HDPE _____
    c. PVC _____
    d. LDPE _____
    e. PP _____
    f. PS _____
    g. ABS _____
    h. GRP _____
14. List the key points of why welding is such a popular joining method for plastic. (Obj. 31-3)
15. What are the three key parameters that must be controlled in making a plastic weld and why? (Obj. 31-3)
16. Name the two basic types of hot-gas welding torches. (Obj. 31-3)
17. List the six interrelated factors that offset the strength of a plastic weld. (Obj. 31-4)
18. Faulty welds are the result of errors. List six of these errors. (Obj. 31-4)
19. Name the three testing groups used to test plastic welds, and briefly explain each. (Obj. 31-4)
20. Describe how a crack should be repair welded with a hot-gas welding machine. (Obj. 31-5)

INTERNET ACTIVITIES

Internet Activity A

On the AWS Web site, locate the titles of all publications that may cover the topic of plastic joining. Also check AWS news releases over the last 5 years for any information on plastic joining. Make a list of titles of publications and dates of news releases and present to your instructor.

Internet Activity B

Using your favorite search engines list the names of recently (in the last 5 years) developed plastics and note the joining method(s) to be used. Present the list to your instructor.

Design credits: Cinema and movie icon: Denis Meshkov/123RF; and Illustration of Welding icons: Mr. Nachai Sorasee/123RF

32 Safety

Chapter Objectives

After completing this chapter, you will be able to:

32-1 Describe electrical and oxyacetylene safety issues.
32-2 List personal protection equipment (PPE).
32-3 Describe burn protection.
32-4 Identify respiratory safety issues.
32-5 Identify confined space welding issues.

The first attempt at establishing standards for the operation of electric and gas welding and cutting equipment was organized in March 1946 under the sponsorship of the American Welding Society. The committee that carried out this assignment was made up of users and suppliers of welding equipment and welded products, insurance companies, government agencies, and other organizations interested in welding and cutting.

Although the hazards associated with the welding processes are not greater than the hazards connected with any other form of industrial work, it is important that welders recognize them. Welders should know and practice the safety procedures that prevent injury to themselves and to others. The safety recommendations that follow are for the protection of persons from injury and illness and the protection of property and equipment from damage by fire and other hazards caused by the installation, operation, and maintenance of welding and cutting equipment. All the welding processes included in this book are given consideration.

The welding student is urged to secure a copy of the complete welding safety standards published by the American Welding Society, titled *Safety in Welding and Cutting* (Standard Z49.1).

Safety Practices: Electric Welding Processes

Machines

Electric welding machines are subjected to all kinds of abuse. They receive little expert attention as compared with electric equipment for the usual power purposes. Welders and/or operators should make sure that such machines are well-protected against accidental contact by themselves or other workers. The following precautions should be routine:

- The welder should never attempt to install or repair welding equipment. A qualified electrician should be in charge.
- In many cases, the welding machines are moved from one operation to another. Consequently, convenient primary power receptacles should be provided about the shop or construction operation. Welding transformers should not be attached to lighting circuits.
- Welding machines should be furnished with a grounding connection for grounding the frame and case. This should not be confused with the work lead. The ground protects against electric shock if there is a short circuit in the machine.
- Welding machines should not be operated above the current ratings and duty cycle specified by the manufacturer.
- When several welders and/or operators are working on one structure, they should avoid touching two electrode holders at the same time. If this cannot be done, all d.c. machines should be connected with the same polarity and all a.c. machines should be set at the same phase of the supply circuit and with the same instantaneous polarity.
- Before starting operations, all connections to the machine shall be checked to make certain they are secure.
- Equipment should be checked for leaks of water, shielding gas, and engine fuels.
- When the welder stops work for an extended period or when the machine is to be moved, the power supply switch should be disconnected from the source of power.
- If gasoline-powered generators are used inside buildings and confined areas, arrangements must be made for the engine exhaust to be vented outside. Otherwise, enough carbon monoxide and other toxic gases may accumulate to harm the workers.
- A.C. transformers are usually air cooled. If transformers are cooled with a liquid, the liquid should be nonflammable. The transformers should be covered on the top and sides with enclosures of reasonably heavy sheet steel or other suitable material. This prevents injury to transformer windings and other parts and also prevents accidental contact with uninsulated live parts.

Cables and Electrical Connections

Welding cable is subjected to severe abuse since it may be dragged over the work and across sharp corners and even run over by shop trucks. Special welding cables of high quality insulation should be used and kept in good condition. The fact that welding circuit voltages are low may lead to carelessness in keeping the welding cable in good repair. In large shops where the cable is tended by an electrical department, welders and/or operators should observe and report the condition of their cable. In smaller shops, welders and/or operators should learn how to make their own repairs. In all shops, workers should observe the following precautions:

- When several lengths of cable are coupled together, the cable connectors should be insulated on both the work line and the electrode line.
- On large jobs there is apt to be a considerable amount of loose cable lying around. Welders and/or operators must keep this cable arranged neatly and not put it where it could cause a stumbling hazard or become damaged. When possible, it should be strung overhead high enough to permit free passage of persons and vehicles.
- Special care should be taken to see that welding supply cables are not near or across the power supply cables and other high-tension wires.
- Welding cables should be the extra flexible type designed for welding service and be the correct size for current and duty, Table 32-1, page 1028.
- Steel conduits containing electric wiring, chains, wire ropes, cranes, hoists, and elevators should not carry welding current. The work connection should be made as close to the workpiece as practical to avoid these situations.
- If the metal structure of a building carries welding current, make sure that proper electrical contact exists at all joints.
- All electrical grounds should be intact, and all welding work connections should be checked to make sure that they are mechanically sound and electrically adequate for the required current.
- Welding cables must be kept dry and free from grease and oil. They should be arranged so that they do not lie in water or oil or in ditches or on tank bottoms. Rooms in which electric welding is done regularly

Table 32-1 Cable Sizes for Arc Welding Machines Based on Safe Operation Temperatures, Which Are Dependent on Amperage, Distance, and Duty Cycle[1]

	Total Cable (Copper) Length in Weld Circuit Not Exceeding							
	100 ft (30 m) or Less		150 ft (45 m)	200 ft (60 m)	250 ft (70 m)	300 ft (90 m)	350 ft (105 m)	400 ft (120 m)
Welding Amperes	10–60% Duty Cycle	60–100% Duty Cycle	10–100% Duty Cycle					
100	4	4	4	3	2	1	1/0	1/0
150	3	3	2	1	1/0	2/0	3/0	3/0
200	3	2	1	1/0	2/0	3/0	4/0	4/0
250	2	1	1/0	2/0	3/0	4/0	2-2/0	2-2/0
300	1	1/0	2/0	3/0	4/0	2-2/0	2-3/0	2-3/0
350	1/0	2/0	3/0	4/0	2-2/0	2-3/0	2-3/0	2-4/0
400	1/0	2/0	3/0	4/0	2-2/0	2-3/0	2-4/0	2-4/0
500	2/0	3/0	4/0	2-2/0	2-3/0	2-4/0	3-3/0	3-3/0
600	3/0	4/0	2-2/0	2-3/0	2-4/0	3-3/0	3-4/0	3-4/0
700	4/0	2-2/0	2-3/0	2-4/0	3-3/0	3-4/0	3-4/0	4-4/0

[1] Weld cable size is based on American Wire Gauge (AWG) and for a 4-volt drop.

should be wired with enough outlets so that it is not necessary to have extension cables strewn about the workplace, Fig. 32-1.

Electrode Holders, TIG Torches, and MIG/MAG Guns

The two hazards presented by this equipment are overheating and electric shocks. They frequently become hot during welding operations. This is usually caused by using equipment for heavy welding that were designed for lighter work. Loose connections between the cable and the holders may also overheat the holder.

- Care should be taken to use equipment of the proper design for the work being performed.
- If the equipment becomes hot, it should be allowed to cool. The practice of dipping hot electrode holders, TIG torches, or MIG/MAG guns in water must be prohibited.
- All equipment and cables must be well-insulated. If they are not, welders and/or operators may get an electric shock if they remove their gloves or if their clothing is damp.
- Metal and carbon electrodes should be removed from holders when not in use to prevent injuring workers, igniting tanks of fuel gas, or defacing the work. Tungsten electrodes should be removed or retracted within holders. Wire electrodes in semiautomatic guns should be retracted or cut off. When not in use, electrode holders, torches, and guns should not be placed in contact with grounded metal surfaces.
- When welders are working in a sitting or prone position, they should make sure that their body is protected by a mat made of dry insulating material.

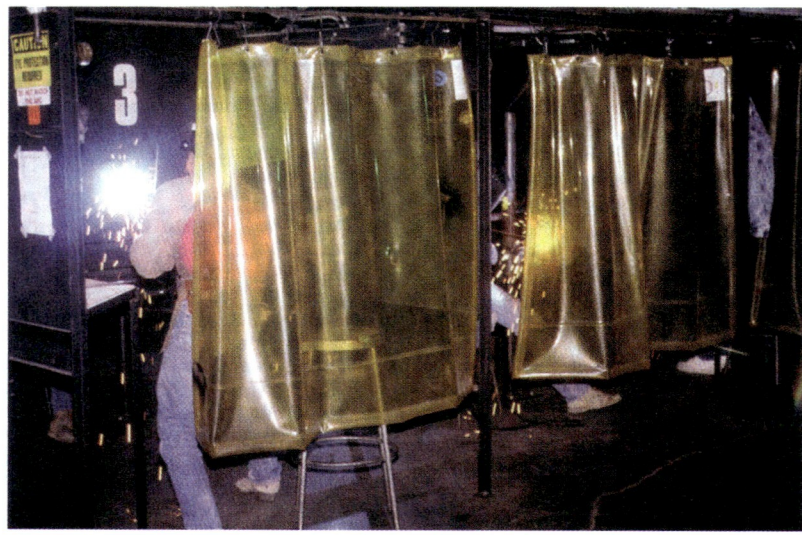

Fig. 32-1 A welding shop with individual stations. Note that the welding cables are suspended so that they are free of the floor and weld curtains protect between booths. © Wilson Industries, Inc.

- Water-cooled holders for TIG, PAC, and MIG welding must not be used if they leak.
- The power supply output must be turned off when changing electrodes in TIG electrode holders and when threading coiled electrodes into MIG/MAG electrode equipment.

Maintenance of Equipment

All arc welding equipment should be maintained in safe working order at all times. Defective equipment should not be used. The following routine care is essential to keeping a welding shop safe:

- Commutators should be kept clean to prevent excessive flashing. Fine sandpaper should be used instead of flammable liquids for cleaning commutators.
- After welding, the entire welding machine should be blown out with clean, dry compressed air.
- Fuel systems on engine-driven machines should be checked regularly for possible leaks.
- Rotating and moving components should be kept properly lubricated and guards kept in place.
- Air filters are not recommended. The reduction of airflow caused by even a clean filter may cause overheating, and a dirty filter reduces airflow dangerously.
- Outdoor welding equipment should be protected from bad weather. The protective cover, however, must not obstruct the ventilation system which prevents overheating of the machine. When not in use, the equipment should be stored in a clean, dry place. Machines that have become wet must be thoroughly dried and tested before being used.

Personal Protective Equipment (PPE)

Personal protective equipment includes items such as goggles, face shields, safety glasses, hard hats, safety shoes, gloves, vests, earplugs, and earmuffs. Breathing protection devices like respirators and electrical insulation materials (rubber mats, gloves, sleeves, and blankets) are also considered PPE. As a welder or operator when you approach your work, you must be properly protected. It is best to use a systematic approach to make sure you have not forgotten any of your PPE. If you begin mentally thinking of your feet and then work up your body to your head, you can quickly review all the PPE you will need to have to perform your work. Eye and fume hazards present some of the more recognizable safety issues.

Your PPE is only one issue; you must also be very aware of your surroundings. Always keep your clothing and work area dry. Never work in confined spaces without proper supervision. Never weld or cut on containers that have held toxic or flammable material without proper supervision. Never weld or cut unless proper ventilation is provided. Always be aware of pinch points. If doing maintenance and repair work, understand the work you are doing and be certain the work is in such a condition that it will be prevented from moving and causing personal injury. Never overlook any safety concerns you might have and bring them to the proper authority, Fig. 32-2. The Occupational Safety and Health Administration (OSHA) Act requires the use of PPE when other measures are not suitable in reducing the risks. Title 29 CFR 1910.132 states what is required of your future employer in regard to PPE. For your PPE use the systematic approach as described, and you will be much better prepared for the work at hand.

Fig. 32-2 A welder properly protected for the MIG/MAG process. A helmet, a leather sleeve on the left arm, and leather gloves are worn for safety. Note that the cuffs of the pants are rolled down so they cannot catch particles of hot metal. © ESAB

Boots Boots should be high topped to protect the lower legs as well as the foot and ankle area. For welding, generally a slip-on steel-toed leather boot is preferred. This allows for easy removal of the boot if a hot spark or metal finds its way inside. Eyelets and laces are places where hot sparks can collect, so the smoother the exposed surface of the boot is, the less chance of collecting sparks. If lace-up boots are preferred, make sure they are fitted with bellows tongues to prevent hot metal and scale from getting into the boot and causing burns. If low-cut boots are worn, spats are available to keep hot particles from entering the top of the boot. Some specific hazards you might encounter that require proper footwear are:

- Hot surfaces
- Wet surfaces
- Slippery, oily surfaces
- Heavy objects (barrels, tools, steel pieces that might roll or fall on your feet)
- Sharp objects (nails, spikes, or other sharp objects that might pierce the soles or uppers on your boots)
- Molten metal or sparks that might be directed toward your feet

Safety footwear must meet minimum compression and impact performance ratings as set by the American National Standards Institute (ANSI). In some industries, metatarsal protection is also required on your footwear.

Slacks Slacks should be of wool or cotton construction. They should be clean and free from frayed edges that will collect sparks and make them less flame resistant. None of the synthetic fabrics should be used for any clothing subject to hot metals in a welding environment. Slacks should not have cuffs that will collect sparks, spatter, and hot slag. Keep pant legs outside of boots to prevent sparks from entering the tops of the boots.

Shirt Shirts should be of wool or cotton construction and long-sleeved. It is better to have snaps instead of buttons, as the arc radiation will degrade the thread holding the buttons on. You should be able to close the shirt collar around your neck. The small V created by the shirt when it is not fully closed and the bottom of the welding helmet often leave exposed skin that can be severely arc burned. If the shirt has breast pockets, they should be equipped with flaps to keep any sparks from entering the pockets. Never carry matches or other flammable materials on your person while welding.

Hands and Forearms Most welding gloves have gauntlets on them to protect the forearms from sparks and hot metal. The gauntlet also protects the exposed skin below your shirtsleeve and the top of a typical glove from arc burn. Use gloves that allow you to feel the work you are doing but yet protect you from the heat source you are using. Lightweight leather and cotton gloves are in most cases very adequate for GTAW; however, for heavy GMAW or FCAW a double insulated leather glove may be more preferable. Rubber gloves with leather protectors are desirable when working in damp locations and when perspiration causes leather or cotton gloves to become wet. Dry conditions are required to maintain electrical safety (by keeping the electrical resistance level as high as practical). Use tongs or pliers to pick up hot weld coupons for examination. Gloves are intended for protection, not for picking up hot metal. The hand protection you are wearing in a welding environment should be able to protect you from:

- Burns
- Punctures
- Cuts
- Bruises
- Abrasions
- Electric shock

Shoulders and Upper Arms Shoulders and upper arms can in most cases be protected with what is referred to as *cape sleeves*. These PPE are generally made out of a green cotton fabric that has been treated with flame retardant, or they can be made of leather. The intent is to protect the covered areas from sparks, spatter, and hot slag. In some cases, the cape sleeves may have snaps for attachment of a bib to give additional protection to the front of the body.

Welding Jackets Welding jackets are used when weather permits for coverage of the total upper body. These are generally made out of green cotton fabric that has been treated with flame retardant, or they can also be made out of leather. Leather is a very durable material and possesses good properties for protecting from heat, sparks, spatter, and hot slag. However, leather is heavier than the cotton material and tends to be warmer in hot environments.

Coveralls In some environments, coveralls are preferred for protecting the legs and upper body. They will need to have additional reinforcement in areas where sparks, spatter, and hot slag may accumulate. This reinforcement must be made out of fire-resistant materials. Undergarments must also be flame resistant in case the first layer of clothing is penetrated by a hot particle.

Apron An apron can be used for protecting the front part of the body. It must be made from nonflammable-type material. Never wear an apron when climbing ladders. It presents a very serious tripping hazard.

Eye Protection The electric arc produces a high intensity of ultraviolet and infrared rays which have a harmful effect on the eyes and skin under continued and repeated exposure. It is necessary, therefore, to have full protection at all times both while engaged in actual welding and while observing welding operations. Ultraviolet rays do not usually cause permanent injury to the eyes unless there is continued and repeated exposure, but temporary effects may be quite painful. This is referred to as "flash burn." Even short exposures have caused painful results and disability. Infrared rays are the heat rays of the spectrum. They do not cause permanent injury to the eyes except from excessive exposure. The intensity of ultraviolet light

Table 32-2 Distance from Arc at Which Ultraviolet Radiation Is Reduced to U.S. Daily Threshold Limit Values for Various Exposure Times and Distances

Process	Parent Metal	Shielding Gas	Current (A)	Distance (m) for 1 min	Distance (m) for 10 min	Distance (m) for 8 h
Shielded metal arc	Mild steel	—	100–200	3	10	70
Gas metal arc	Mild steel	CO_2	80	0.8	9	20
			200	2.1	7	50
			350	4	13	90
(flux-cored wire)	Mild steel	CO_2	175	1.2	3.5	24
			350	2.2	7	50
	Mild steel	95% argon + 5% oxygen	150	3	8	65
			350	6.5	20	140
	Aluminum	Argon	150	3	10	70
			300	5	17	110
	Aluminum	Helium	150	1.3	5	36
			300	3	10	70
GTAW	Mild steel	Argon	50	0.3	1	7
			150	0.9	3	20
			300	1.6	5	40
	Mild steel	Helium	250	3	10	70
	Aluminum	Argon	50 a.c.	0.3	1	7
			150 a.c.	0.8	2.7	18
			250 a.c.	1.3	4	30
	Aluminum	Helium	150 a.c.	0.7	3	20
Plasma arc welding	Mild steel	Argon	200–260	1.5	5	33
		85% argon + 15% hydrogen	200–275	1.7	5.5	40
		Helium	200	2.9	9	65
Plasma arc cutting (dry)	Mild steel	65% argon + 35% hydrogen	400	1.3	4	30
			1,000	2.5	8	55
Plasma arc cutting with water injection	Mild steel	Nitrogen	300	3.2	11	75
			750	1.8	5.5	40

Source: Lyon, T. L., et al. *Evaluation of the Potential Hazards for Actinic Ultraviolet Radiation Generated by Electric Welding and Cutting Arcs.* U.S. Army Environmental Hygiene Agency.

varies inversely with the square of the distance between the source of light and the eyes. While the effects are lessened as the distance is increased, Table 32-2, workers have been affected when arcs have been reflected from walls, ceilings, and even screens having highly reflective surfaces. The direct rays of the electric arc affect the skin like sunburn. The skin may become uncomfortable and even *painful,* but the rays do not cause permanent injury.

Because of the intensity of ultraviolet and infrared rays from the electric arc, welders and/or operators and their assistants must use a hand-shield or helmet. The device protects the skin of the face and neck and is also equipped with a filter glass to provide adequate eye protection. The selection of filter glass depends upon laboratory tests since the transmission of ultraviolet and infrared radiation cannot be determined by visual inspection. Depth of color does not necessarily indicate removal of the invisible radiation that may injure the eyes. Reliable dealers are able to supply filter glasses that have been shown by tests to conform to requirements. Table 32-3, page 1032, is a guide for the selection of

Table 32-3 Guide for Lens Shade Number Selection

Process	Electrode Size in.	Electrode Size mm	Arc Current (A)	Minimum Protective Shade	Suggested Shade No. (Comfort)[1]
Shielded metal arc welding (SMAW)	Less than 3/32	2.5	Less than 60	7	
	3/32–5/32	2.5–4	60–160	8	10
	5/32–1/4	4–6.4	160–250	10	12
	More than 1/4	6.4	250–550	11	14
Gas metal arc and flux cored arc welding (GMAW and FCAW)			Less than 60	7	
			60–160	10	11
			160–250	10	12
			250–500	10	14
Gas tungsten arc welding (GTAW)			Less than 50	8	10
			50–150	8	12
			150–500	10	14
Air carbon arc cutting (CAC-A)					
Light			Less than 500	10	12
Medium			500–1000	11	14
Plasma arc welding (PAW)			Less than 20	6	6–8
			20–100	8	10
			100–400	10	12
			400–800	11	14
Plasma arc cutting (PAC)					
Light[2]			Less than 300	8	9
Medium[2]			300–400	9	12
Heavy[2]			400–800	10	14
Torch brazing (TB)			—	—	3 or 4
Torch soldering (TS)			—	—	2
Carbon arc welding (CAW)			—	—	14

	Plate Thickness in.	Plate Thickness mm			Suggested Shade No. (Comfort)[1]
Oxyfuel gas welding (OFW)					
Light	Under 1/8	Under 3.2			4 or 5
Medium	1/8–1/2	3.2–12.7			5 or 6
Heavy	Over 1/2	Over 12.7			6 or 8
Oxygen cutting (OC)					
Light	Under 1	Under 25			3 or 4
Medium	1–6	25–150			4 or 5
Heavy	Over 6	Over 150			5 or 6

[1] The rule of thumb is that the user should start with a protective shade that is too dark to see the weld zone. Then, a lighter shade that provides sufficient visibility in the weld zone without going below the minimum number can be selected. In oxyfuel gas welding or cutting, in which a torch produces a high yellow light it is desirable to use a filter lens that absorbs the yellow or sodium line in the visible light of the (spectrum) operation.
[2] The suggested filters are for applications where the arc is clearly visible. Lighter shades may be used where the arc is hidden by the work or submerged in water.

Source: American Welding Society *Welding Handbook*, Vol. 1, 9/e

the proper shade numbers. These recommendations may be varied to suit the individual's needs.

The following precautions regarding helmets and hand shields should be observed:

- All hand shields and helmets should have a clear cover glass to protect the filter lenses from spatter. The cover glass should be free from defects that cause eye strain. Ordinary window glass is not usually suitable. The cover glass should be discarded when enough spatter has accumulated to interfere with vision.
- Helmets, hand shields, face shields, safety glasses, and goggles must be made of a material that is not readily flammable and that is an insulator for heat and electricity. It must not corrode readily nor discolor the skin and be capable of sterilization, Figs. 32-3, 32-4, and 32-5.
- Helmets and goggles should not be transferred from one person to another without being sterilized.
- Although some shops employ helpers to chip and clean welds, the welders and/or welding operators frequently do their own chipping and cleaning. Because of the colored filter glass in the face shields and helmets, it is necessary to raise them in order to see the welds properly during cleaning. Additional protection is needed to protect the eyes from flying particles. Arc welders and/or welding operators wear safety glasses with side shields or goggles with clear lenses under the shield or helmet. In other cases, a clear lens is a part of the helmet with a flip front. Auto-darkening lenses that allow the helmet to stay in the down position are also available. This technology will be covered in more detail in a later section of this chapter, Fig. 32-6.

Fig. 32-5 Facial contouring safety glasses with side shields. Features a graphite frame and 56-millimeter hard-coated replaceable polycarbonate lenses available in various shades. © Sellstrom Manufacturing Company

Fig. 32-3 Face shields provide protection from sparks, spatter, grinding, and hot slag. These units are equipped with a shade 5 lens for protection against infrared and ultraviolet rays. Clear acetate lenses are also available. © Sellstrom Manufacturing Company

Fig. 32-4 A welding flip-front goggle made out of PVC and designed for good ventilation. It meets ANSI Z87.1. © Sellstrom Manufacturing Company

Fig. 32-6 A helmet equipped with an auto-darkening lens and a forced air respirator. Note the welder's good head position in relation to the fume plume and the other personal protection equipment (PPE), such as gloves and welder's jacket being effectively used. The helmet can remain in the down position for weld cleaning, inspection, and grinding. This protects the face and keeps the respirator in a proper position. © American Welding Society. *Welding Science and Technology*, Vol 1 of Welding Handbook, 9th ed., Fig. 17.5, p. 723.

- Welds are frequently cleaned by using a power-driven wire brush. Such a brush should be guarded with a hood guard, and the welder and/or welding operator should wear eye protection approved for such operations.
- When more than one welder is working on the job at the same time, all welders and/or welding operators should wear goggles under their helmets to give added protection. These goggles should have shaded lenses with the same shade of side shield attached to them.
- Helpers working with arc welding operators should be equipped with goggles and gloves. If they are exposed to the arc, they should wear welders' helmets or hand shields.
- Special precautions must be taken to protect other workers from the harmful rays given off from arc welding operations and from flying chips during cleaning operations. It is preferable to locate welding jobs in special rooms or booths. If this is impractical, the operations should be screened or enclosed, not only to prevent workers from looking directly at the arc, but also to protect them from reflected rays as much as possible. Further protection against reflected rays is frequently provided by applying paint of low reflective qualities to the screen or enclosure and to other nearby surfaces.
- Laser welding and cutting present some specific eye hazards that must be dealt with. Figure 32-7 shows laser safety eyewear.

Fig. 32-7 Laser eyewear for visitors or peripheral personnel (diffuse viewing only) or for the more critical application by the laser technicians themselves (laminated glass or clear technology). © Uvex Safety

Auto-Darkening Welding Lens Now welders can see their work, at comfortable light levels, with constant eye and face protection. These auto-darkening lenses turn dark the moment an arc is struck and become transparent again when welding stops. At all times, whether dark or transparent, they offer full ultraviolet (UV) and infrared (IR) protection. Figure 32-8 shows a typical helmet designed for use with the auto-darkening lens. The use of an auto-darkening lens in the helmet eliminates the neck strain of helmet flipping while greatly increasing the accuracy of electrode placement. This can greatly eliminate inadvertent arc strikes and other discontinuities that may occur when you attempt to strike the arc and don't know exactly where the electrode is located. This, in turn, reduces the need for grinding and rework. In addition, welders can get into tight, cramped spaces with full protection and have a clear view of their work. This makes welding in extremely awkward places much easier. In high production type work areas, where fixtures are used to hold parts and you are expected to move from one weld quickly and efficiently to the next without damaging the fixture with inadvertent welds, the auto-darkening lens is very effective.

Fig. 32-8 This Speedglas® helmet equipped with an auto-darkening lens is lightweight and very flexible and so is appropriate for the welder to use when getting into hard-to-reach areas. It also has side windows to provide a peripheral view since the helmet can be used in the down mode for most work like grinding, cleaning, and inspection of welds. It has good neck protection and is designed to deflect the welding fumes around the helmet to keep the welder's breathing zone cleaner. © Hornell, Inc. Speedglas®

Figure 32-9 shows the welder's perspective on how this works.

The typical auto-darkening lens can be made of a laminate of many different layers: for example, a UV/IR filter, three polarizers, two liquid crystal elements, and a cover glass, Fig. 32-10. The UV/IR filter continually blocks harmful radiation, whether the lens is on, off, light, or dark.

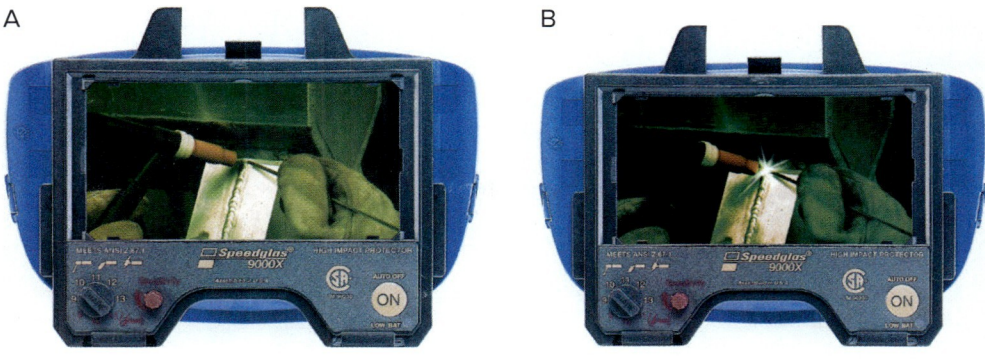

Fig. 32-9 The view on the left is in the clear mode while the right view has auto-darkened so rapidly the welder will not see a flash. This particular manufacturer's Speedglas® provides for lens shade selection as well. © Hornell, Inc. Speedglas®

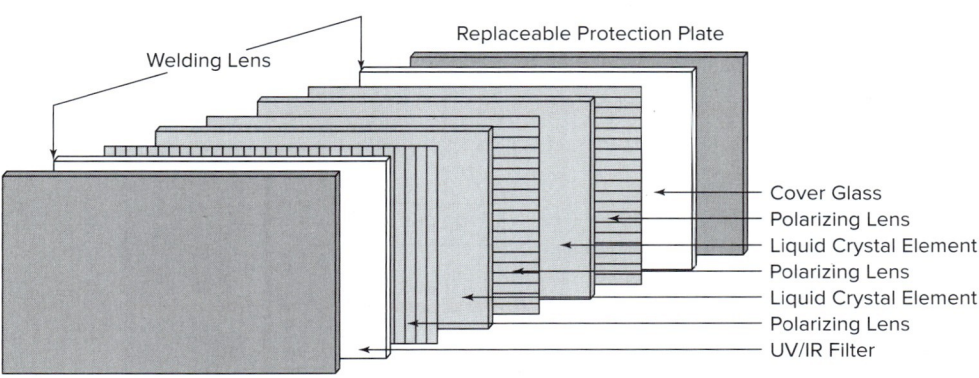

Fig. 32-10 The design of the Speedglas auto-darkening lens. Always use replaceable protective plates on the inside and outside of the helmet to protect the lens from weld spatter. Source: Hornell, Inc. Speedglas®

Aided by surface-mounted electronics, the liquid crystal elements act as shutters that detect and react to the welding arc by instantly shading the lens. Constant UV/IR protection is always provided. This filter provides the same UV/IR filtration protection as a traditional, non-auto-darkening welding lens. Generally when you pick up an unenergized auto-darkening lens, it is in an intermediate shade; you must turn it on to make it go into the light state. If anything catastrophic happens to the lens, it simply defaults back to the darker off shade. As mentioned above, UV/IR protection is always present. These types of helmets meet or exceed all applicable safety standards and approvals such as ANSI Z87.1, American National Standard for Eye and Face Protection, and CSA Z94.3, National Standard of Canada for Industrial Eye & Face Protection.

Ear Protection Hearing loss is one of the leading occupational illnesses in the United States. The level of noise you will be exposed to must first be determined in order to select the best hearing protection devices. Excess noise depends on the following factors:

- Loudness of the noise [as measured in decibels (dBA)]
- Length of exposure to these noises
- Constant or intermittent noise
- Type of noise (Is it an impulse sound like a sharp noise from a needle scaler, hammering, or banging?)
- Whether you are moving between work areas that are noisy and then quiet
- Whether the noise is being produced by one source or multiple sources

The louder the noise you are exposed to, the shorter the duration you can safely be exposed to it. At 90 dBA an 8-hour shift should pose no problems. However, if 115 dBA is encountered for more than 15 minutes, ear protection is required. Table 32-4, pages 1036–1037, is a comparison of various common sounds. Impulse noise is more hazardous than noise at a steady level. The ear has some natural protection to tone sound down, but it cannot react quickly enough to deal with impulse noise. This can be an issue when dealing with GTAW-P and GMAW-P at certain power levels and frequencies.

The simplest form of ear protection is the small device that fits directly in your ear canal, Fig. 32-11, page 1038. Where more harsh noise conditions exist, ear canal protection as well as earmuffs should be used. Earmuffs that completely cover the ear as well as the surrounding bone act together to reduce sounds that can be transmitted to the inner ear, Fig. 32-12, page 1038. Some sound suppression equipment is available that allows normal sound levels to be heard but stops high-energy sound waves, even impulse noise, from causing damage. It is important that emergency

Table 32-4 Comparison of Nonoccupational Noise and Occupational Noise

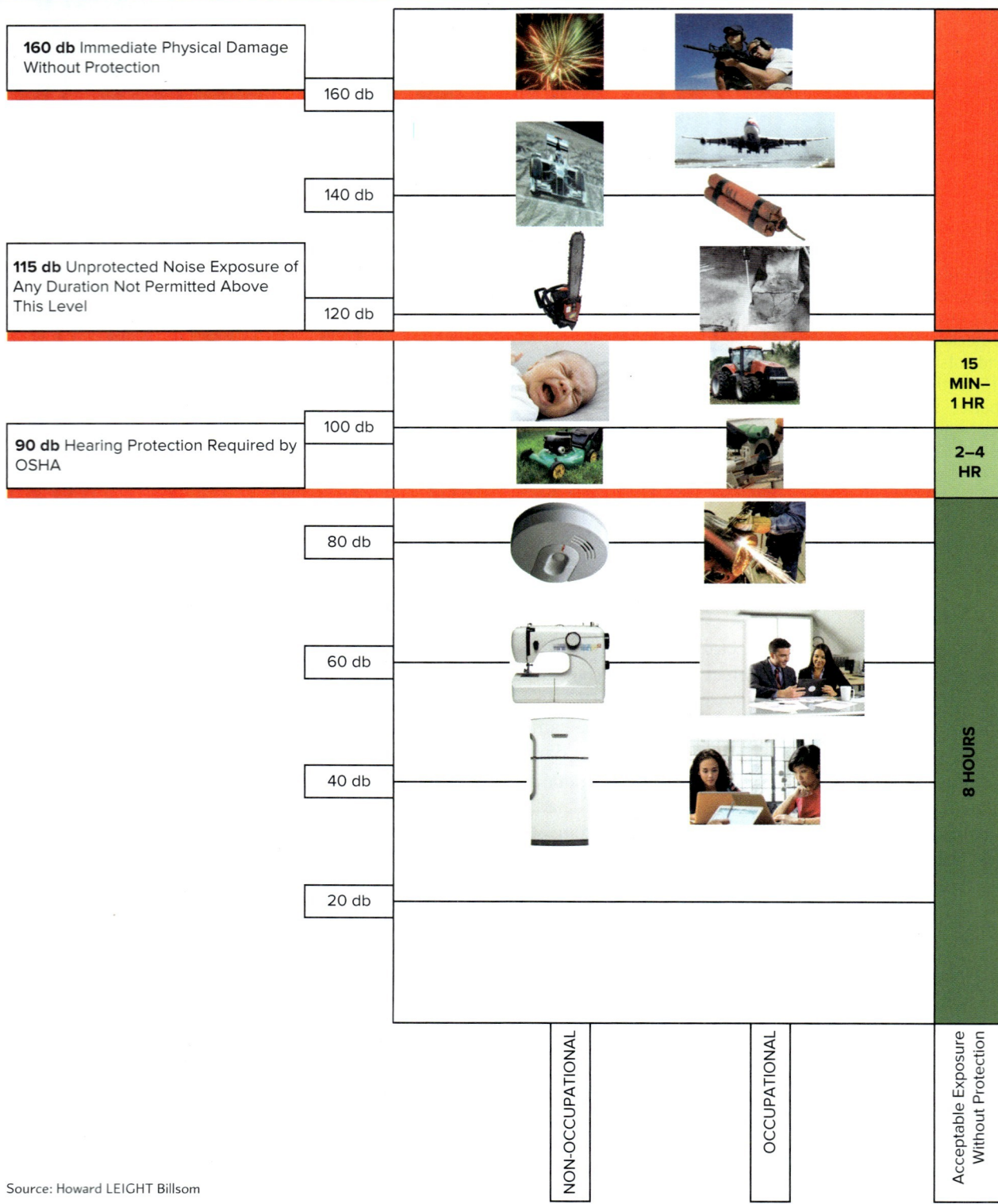

Source: Howard LEIGHT Billsom

Photos: (fireworks) © Lena Kofoed; (shooting) moodboard/Image Source; (jet) © Medioimages/Superstock; (racecar) © Robert Daly/Caia Image/Glow Images; (dynamite) Ingram Publishing/Fotosearch; (chainsaw) Ingram Publishing/Alamy Stock Photo; (cutting torch) © Arcos Corp.; (baby crying) © JGI/Blend Images LLC; (tractor) © tanger.Shutterstock; (lawnmower) © Ingram Publishing; (saw) © John Lund/Getty Images; (fire alarm) D. Hurst/Alamy Stock Photo; (bevel cut on a pipe) Mark A. Dierker/Mc Graw Hill; (sewing machine) graficart.net/Alamy Stock Photo; (conversation) Hurst Photo/Shutterstock; (refrigerator) s-cphoto/Getty Images; (quiet office) © Jose Luis Pelaez Inc/Blend Images LLC

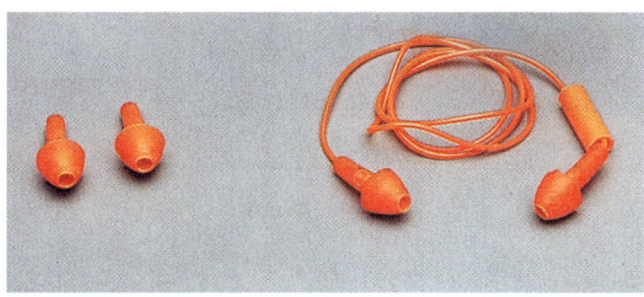

Fig. 32-11 Earplugs with a collapsible air cavity design. They have a noise reduction rating of 24.0 dB and are nontoxic, hypoallergenic, and reusable and can be washed in warm water with mild soap. They meet ANSI S3.19 specifications. © Sellstrom Manufacturing Company

Fig. 32-12 Earmuffs that feature polyfoam ear cushions for extra comfort. They have a noise reduction rating of 21.0 dB and when used with earplugs, as shown in Fig. 32-11, provide for very effective hearing protection. They meet ANSI S3.19 specifications. © Sellstrom Manufacturing Company

and/or warning signals can be seen or heard by the welders or operators when working in the industrial environment.

If sound levels do not dictate ear protection, it may still be advisable for welders to wear small nonflammable earplugs. These devices can be used to prevent foreign material from entering the ear canal. In some cases when welding out of position or in confined spaces, sparks, spatter, and hot slag may find their way to your exposed ear. These earplugs would act as a last stage deterrent from a burn to the inner ear area. Flaps secured to the helmet or bill on the welder's cap should be used to deflect foreign materials away from the ear, especially when welding overhead or bent over in a confined space.

Respiratory Protection One of the most effective ways to protect you from the hazardous effects of welding fumes is quite simple. Keep your head out of the fume plume! This is considered the most important factor. The simple positioning of your head in relation to the source of the fume generation is effective in reducing fumes in your breathing zone. When your head is in the fume plume, this causes the fumes to surround your face and helmet and you will be breathing a much higher level of airborne particles of base metal, welding consumables, or coatings that may be present on the workpiece. Keep your head to one side of the fume plume. In some situations, the work can be positioned to force the fume plume to rise to one side or the other. The shape of the welding helmet and how well it comes under your chin can also help reduce exposure. It must be understood that the welding helmet alone is not adequate respiratory protection.

Certain materials have been determined to be very toxic. You need to guard against breathing them in or ingesting them by any other means. These materials can be present in the consumables, base metals, metal coatings, or in the general atmosphere near welding and cutting operations. The current exposure limit is 1.0 milligram per cubic millimeter or less. Check the (SDS) Safety Data Sheets which are required by the U.S. OSHA Hazard Communication Standard on the materials you are using to see if any of these materials are present. Table 32-5 lists some toxic materials, and Table 32-6 covers possible hazardous materials emitted during welding or thermal cutting.

When welding is carried on outdoors or in large, well-ventilated shops and nontoxic materials are involved,

Table 32-5 Toxic Materials

Element	Chemical Symbol
Antimony	Sb
Arsenic	As
Barium	Ba
Beryllium	Be
Cadmium	Cd
Chromium	Cr
Cobalt	Co
Copper	Cu
Lead	Pb
Manganese	Mn
Mercury	Hg
Nickel	Ni
Selenium	Se
Silver	Ag
Vanadium	V

Table 32-6 Possible Hazardous Materials Emitted During Welding and Thermal Cutting

Base or Filler Metal	Emitted Metals or Their Compounds
Carbon and low alloy steels	Chromium, manganese, vanadium
Stainless steels	Chromium, manganese, nickel
Manganese steels and hard-facing materials	Chromium, cobalt, manganese, nickel, vanadium
High copper alloys	Beryllium, chromium, copper, lead, nickel
Coated or plated steel or copper	Cadmium[1], chromium, copper, lead, nickel, silver

[1]When cadmium is a constituent in a filler metal, a precautionary label must be affixed to the container or coil. Refer to Section 9 of American National Standards Institute (ANSI) Accredited Standards Committee Z49, 2012, Safety in Welding, Cutting, and Allied Processes, ANSI Z49.1:2012, Miami: American Welding Society, pp. 21–23.

Source: American Welding Society Welding Handbook, Vol. 1, 9/e.

welders and/or welding operators suffer no harmful effects. When toxic concentrations of gases, fumes, and dust are generated, however, protection must be provided.

The amount of contamination to which welders may be exposed depends on the following factors:

- The dimensions of the space in which welding is to be done. The height of the ceiling is of major importance (10,000 cubic feet per welder and 16 ft ceiling height).
- The number of welders working in a unit of space.
- The nature of the hazardous fumes, gases, or dust resulting from the metals being welded.
- The nature of the filler metals and shielding gases being used.
- Atmospheric conditions such as humidity.
- Heat generated by the welding operation.
- The presence of volatile solvents. (A volatile substance evaporates rapidly, thus polluting the air.)

The following precautions are taken to ensure a supply of fresh air in the work area when natural ventilation is not effective:

- Local exhaust systems, Fig. 32-13, remove toxic substances. When welding operations are permanently located, the entire booth may be ventilated with a system like those provided for spray-coating booths. Portable exhaust systems are also available for this purpose and should be located as close to the fume source as practical. When a large quantity of air is removed by exhaust systems, it is replaced by fresh air. The exhaust air should be cleaned and returned to the environment.
- Supplied-air respirators, such as air line respirators, hose masks with or without blowers, and self-contained oxygen-breathing apparatuses, are recommended for confined areas and other locations where there are high concentrations of toxic substances, Figs. 32-14 and 32-15, page 1040.

OSHA regulates the use of respirators, and all laws must be followed. An on-site safety specialist or industrial hygienist must determine respirator applicability and filter selection. The safety specialist must fully

Fig. 32-13 In confined areas, where ventilation is restricted and/or ceiling height is not sufficient, fumes can be removed with extractor equipment. The advantage of this type of equipment is that the cleaned air is returned to the environment, thus reducing heating and cooling expenses. Note the position of the pickup arm. Do not position the pickup arm to draw the fumes around your head. Extractor equipment can be equipped with an in-hood light and in-hood remote power switch. © Donaldson Company, Inc.

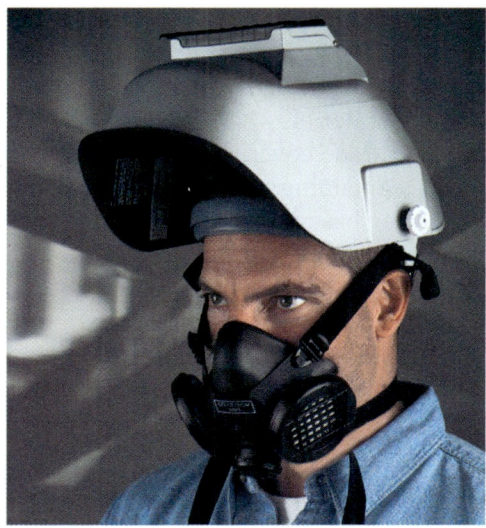

Fig. 32-14 A neoprene half-mask respirator facepiece. The soft nature of neoprene provides a comfortable fit and seal. These facepieces are available in various sizes and must be properly fitted. They comply with NIOSH 42 CFR Part 84 standard. © Sellstrom Manufacturing Company

Fig. 32-15 Particle fume protection for welders. 1. Auto-darkening lens. Different models are available. 2. Narrow helmet design with extended throat protection, heat reflective silver front, and made from lightweight materials. 3. Speedglas® side window for peripheral vision which increases the welder's field of view by 100 percent. 4. Battery, turbo fan, and filter all in one compact belt pack unit. © Hornell, Inc. Speedglas®

review the entire welding application, including the type of base metals, possible coatings on the base metals, filler metals, shielding gases, welding process, and numerous other variables. The safety specialist will collect air samples from the welder's worksite to determine the level of exposure to various contaminants. The welder's breathing zone inside the helmet with the head out of the fume plume should also be monitored. Only then can the safety specialist determine the appropriate protection for the welder, Table 32-7.

For updated information, check current publications of the National Institute for Occupational Safety and Health (NIOSH) which was established by the OSHA Act of 1970. NIOSH is part of the Centers for Disease Control and Prevention (CDC) and is the only federal institute responsible for conducting research and making recommendations for the prevention of work-related illnesses and injuries. The Institute's responsibilities include:

- Investigating potentially hazardous working conditions as requested by an employer or employees
- Evaluating hazards in the workplace, ranging from chemicals to machinery

Table 32-7 The Correct Respirator Filter Is Critical for Each Application

N100 particulate filter (99.97% filter efficiency level): Effective against particulate aerosols free of oils; time use restrictions may apply.	R100 particulate filter (99.97% filter efficiency level): Effective against all particulate aerosols; time use restrictions may apply.	P100 particulate filter (99.97% filter efficiency level): Effective against all particulate aerosols.
N99 particulate filter (99% filter efficiency level): Effective against particulate aerosols free of oil; time use restrictions may apply.	R99 particulate filter (99% filter efficiency level): Effective against all particulate aerosols; time use restrictions may apply.	P99 particulate filter (99% filter efficiency level): Effective against all particulate aerosols.
N95 particulate filter (95% filter efficiency level): Effective against particulate aerosols free of oil; time use restrictions may apply.	R95 particulate filter (95% filter efficiency level): Effective against all particulate aerosols; time use restrictions may apply.	P95 particulate filter (95% filter efficiency level): Effective against all particulate aerosols.

HE high efficiency particulate air filter for powered, air-purifying respirators.

Source: NIOSH, Standard Application Procedures for the Certification of Respirators

Note: The N series respirators are not resistant to oils, which can degrade the filter media. The R series respirators are more resistant to oils, while the P series are significantly more resistant to oils.

- Creating and disseminating methods for preventing disease, injury, and disability
- Conducting research and providing scientifically valid recommendations for protecting workers
- Providing education and training to individuals preparing for or actively working in the field of occupational safety and health

Figure 32-15 is available with the following different filters to meet specific needs:

- **Speedglas Fresh-air II P 9000 system** P stands for particle fume protection; this NIOSH-approved system provides welders with high efficiency (HE) particulate filtration. It is not to be used in atmospheres classified as immediately dangerous to life or health (IDLH) or in atmospheres of less than 19.5 percent oxygen.
- **Speedglas Fresh-air II G 9000 system** G stands for gas and fume protection; this NIOSH-approved system provides welders with protection against organic vapors, sulfur dioxide, chlorine, and hydrogen chloride, as well as high efficiency (HE) particulate filtration. It is not to be used in atmospheres classified as IDLH or in atmospheres of less than 19.5 percent oxygen.
- **Speedglas Fresh-air II SA 9000 system** SA stands for supplied-air; this NIOSH-approved system provides welders with protection against high levels of particle fumes and gases.

  To repeat once more, the final determination of respirator applicability and filter selection must be made by an on-site safety specialist or industrial hygienist.
- If screens are used, they should be arranged so that ventilation is not seriously restricted. Screens should have a space at the bottom of about 2 feet above the floor. The operator should make sure, however, that the screen protects from the arc glare.
- Oxygen from a cylinder or torch is never used instead of air for ventilation.
- Extreme care must be taken when welding or cutting the following materials:
  - Fluxes, electrode coverings, and fusible granular materials containing fluorine compounds.
  - Zinc-bearing base and filler metals and metals coated with zinc-bearing materials
  - Lead-base metals and metals coated with lead-bearing materials such as paint
  - Base and filler metals containing beryllium
  - Cadmium-coated base metals and filler metals
  - Metals coated with mercury-bearing materials, including paint
  - Fluxes, coverings, and other materials containing antimony, arsenic, bismuth, chromium, cobalt, copper, nickel, manganese, magnesium, molybdenum, thorium, vanadium, and their compounds

Skullcaps Where hard hats are not required, some means must be provided to keep sparks, spatter, and hot slag from burning your scalp. Skullcaps must all be made from fire-retardant materials and be heavy enough to stop any hot particles, Fig. 32-16. Lightweight handkerchief-type materials will not be adequate to stop large sparks, spatter, or slag. The pipeliner style cap is very functional in that it has a small bill that does not interfere with the headband system of the welding helmet but allows for additional protection, Fig. 32-17.

Fig. 32-16 Cotton shop beanie fits snugly to protect from dust and shop dirt and is flame retardant for the welding environment, yet it retains a comfortable fit. © Kromer Cap Co., Inc.

Fig. 32-17 Cotton "pipeliner" style cap with balloon top and rugged double-stitch construction. It has an absorbent, cotton sweatband which makes for a very durable and washable welder's cap. © Kromer Cap Co., Inc.

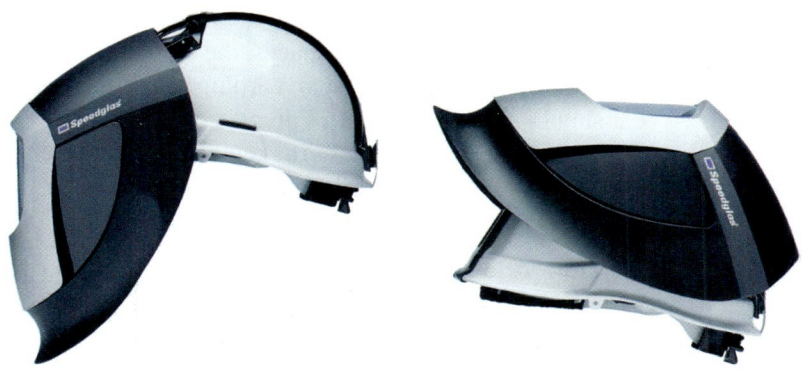

Fig. 32-18 Hard hat equipped with a ratcheting headband for fast, secure sizing. It has a six-point webbing system for shock absorption and comfort. The center rail helmet mount and slide allow for a comfortable low profile in the helmet-up mode. © Hornell, Inc. Speedglas®

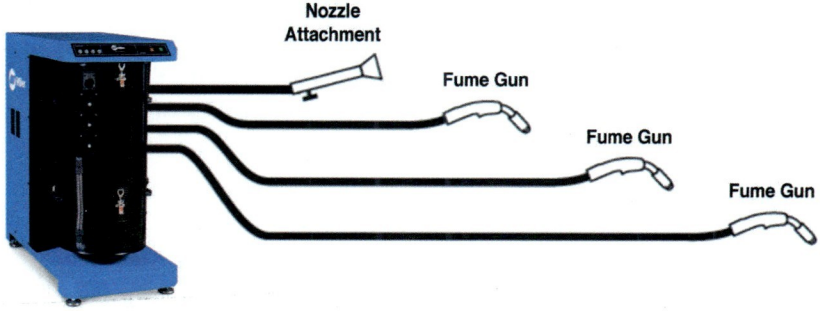

Fig. 32-19 Fume extractor with multiple GMAW-FCAW fume gun installation capable. With reach up to 70 feet per fume gun used, for a total of 200 feet of collection hose. Miller Electric Mfg. Co.

The bill can be rotated around to protect the side of the face and ear from injury.

Hard hats Where the possibility exists for head injuries from falling objects or the nature of the work, hard hats are required. When set up for use with a welding helmet, they become very effective. The helmet can be quickly released from the hard hat when not required. The hard hat should be flame resistant and a nonconductor of electricity. They have an internal mechanism for adjusting to various head sizes, Fig. 32-18.

GMAW and FCAW Smoke Exhaust System The gun smoke extractor system, Fig. 32-19, increases the operator's visibility, thereby improving efficiency and weld quality. Integral tubing permits the operator to weld wherever they can reach without repositioning a separate exhaust duct.

Smoke swirls up and away from the arc and then is directed through the gun to the filter.

Visible smoke particles are captured within the filter and kept from reentering the work area, thus leading to a cleaner work area and less maintenance to other equipment, Fig. 32-20. The GMAW-FCAW welding gun also runs cooler due to airflow through internal ducting, Fig. 32-21.

Wall Reflectivity Where arc welding and cutting are generally done, the walls and any other reflective surface must be properly covered. Of most concern is the reflection of the UV radiation. It is best blocked by using paint made with titanium dioxide or zinc oxide. Any pigment of choice can be added as long as it does not promote reflectivity. Pigment that is based on powdered or flaked metals should not be used, as they are very reflective of UV radiation. Welding screens and curtains can be effective in reducing reflections.

Protection Against Burns Your PPE should protect you in most cases from burns from sparks, spatter, hot slag, and molten metal. Burns are some of the most common and painful injuries in a weld shop and the severity should be understood. In a first-degree burn the surface of the skin is damaged, but the outermost layer of skin is still intact and therefore able to perform its functions. Sunburn is often this type of burn. A second-degree

Fig. 32-20 (Left) Volume of smoke with the smoke exhaust holder and system; (right) volume of smoke without the smoke exhaust holder and system. © NovaTech

Fig. 32-21 GMAW-FCAW gun equipped with a smoke exhaust device. © NovaTech

burn has extended through the surface of the skin and into the second layer of skin. Blisters are the first sign of a second-degree burn. Third-degree burns destroy both the outermost and second layer of skin to a greater depth. Fluid loss, heat loss, and infection that come with second-degree burns also occur with third-degree burns, along with nerve death. So loss of feeling in the area of the burn is possible. The following precautions should be taken:

- When entering the work environment, always wear approved safety glasses with side shields. They should be clean. Even small weld spatter on the lenses can cause eyestrain and eye fatigue. They should be lightweight, well fitted, and designed to reduce fogging. Antifogging cleaning solutions are available. You should have no reason to remove your safety glasses while in the work environment. Always protect your eyes from hot particle burns. If necessary, use lightly tinted lenses to protect from inadvertent stray arc rays.
- On account of the intense heat of the welding arc and flying particles of hot metal, woolen clothing is preferable to cotton because it is not so easily ignited. Fireproof outer clothing may be needed in some instances. Welders and welding operators should not wear oil-soaked clothing or worn clothing with frayed exposed fibers because such clothing ignites easily. Loose-fitting clothing, long hair, and anything that can be caught in rotating equipment should not be worn or unprotected.
- In production work, a sheet-metal screen in front of the worker's legs provides further protection against sparks and molten metal in cutting and welding operations.
- Never carry matches, butane, or propane cigarette lighters in your pockets or other unprotected areas.
- For overhead welding or welding in awkward positions in extremely confined spaces, the ears should be shielded by use of fire-resistant earplugs and the outer ear should be protected with suitable coverings.
- The main key in reducing burns in a welding environment is to consider all metal as hot until you know otherwise.

Protection Against Shock Voltages required for arc welding are low and normally do not cause injury or severe shock. Nevertheless, under some circumstances, voltages may be dangerous to life. The severity of the shock is determined largely by the amount of current flowing through the body, and the current, in turn, is determined by both the voltage and the contact resistance of the skin. Clothing damp from perspiration or wet working conditions reduces contact resistance. Low contact resistance increases a low current to a value high enough to cause such violent muscular contraction that the welder cannot let go of the live part. The following safety precautions reduce the possibility of severe electric shocks:

- Welders must make sure their machines are grounded. While welding, they must protect themselves from electrical contact with the work and the ground.
- Welders should never permit the live metal parts of an electrode holder to touch their bare skin or wet clothing.
- Electrode holders must not be cooled by immersion in water.
- Water-cooled holders for TIG and MIG/MAG welding must not be used if there is a water leak.
- The welding machine supplying output power to the arc must always be turned off when changing electrodes in TIG electrode holders and when threading coiled electrodes into MIG/MAG equipment.
- Special precautions should be taken to prevent shock-induced falls when the welder and/or operator is working above ground level. When working above ground, welders should not coil or loop welding electrode cable around parts of their bodies. They should not use cables with splices or repaired insulation within 10 feet of the holder.

  OSHA-approved fall protection devices should be used where required.
- Although insulated holders are used and the electrode coatings provide insulation, the welder and/or operator is nevertheless exposed to the open circuit voltage when changing electrodes and at any other time when the arc is extinguished. The worker should avoid standing on wet floors or coming into contact with a grounded surface.

Use wooden pallets and/or rubber mats to insulate yourself from the work.
- The danger of electric shock is increased during periods of high temperature and high humidity. Under these conditions the physical reserve of the individual worker is lowered. The electrical resistance of the skin and clothing is reduced by perspiration. Thus, an environment is set up in which electric shock is much more of a potential hazard.
- If the power source is not working properly, carefully turn it off with its insulated switch or at the primary circuit breaker. Never touch electrical equipment that is not working properly and at the same time any electrically grounded structure, such as water pipes; electrical conduit; or portable electrical equipment such as fans, grinders, or fume extractors. This may put your body in-between the electrical fault in the equipment and a path to ground. Some welding machines are equipped with ground fault circuit interrupters. These devices detect if an electrical fault exists and will shut the equipment down. A qualified electrician should always be called upon to correct these types of electrical faults.
- Rings and other metal jewelry should be removed before welding to eliminate this possibility for electric shock.

Protection Against Heat Stress The heat generated by the welding process itself, ambient temperature plus humidity, and if required preheat/interpass temperature control on the base metal can take a toll on the welder or operator. The body will attempt to correct the effect of this higher heat. It does this by increasing blood flow just below the skin which leads to perspiration. If the body cannot keep up with the increasing temperature, heat stress will set in. The symptom can consist of weakness, shivering, irritability, and disorientation. Heat stroke, a very dangerous condition, can occur if safety actions are not followed. The results can be convulsion and loss of consciousness.

This can be a main concern when doing high intensity welding (300+ amps) at high duty cycle welding. At lower intensity and duty cycle, the welder and operator may not show symptoms but productivity and quality may suffer. Heat stress can reduce the ability to focus on the job, so less speed and accuracy will result. The following safety precautions reduce the possibility of heat stress:

- Use a welding helmet with a reflective color on the front, like silver. This will reduce the radiant heat by reflecting it away, making it a few degrees cooler inside the helmet.
- Welding helmet headgear that moves air upward and downward through air vents inside the helmet. This can reduce the heat effect several more degrees. These are typically powered by a belt or back/shoulder battery pack.
- Welding jacket and coats that are lighter and more breathable should be considered for heat stress issues.
- Cooling bandana, towel, and hats can be soaked to give relief from heat.
- Cooling vests are available in a variety of styles for the type and intensity of welding being done.
- Hydration packs will help keep the welder or operator hydrated. Being able to have immediate access to fluids and adequate fluid intake can be very helpful in reducing the effect of heat stress and cramping.

In a weld training environment, heat stress should be readily controllable. On a job, a Safety Director will be aware of this issue and should have available all prior to PPE that may be required for prevention of heat stroke.

Safety Practices: Oxyacetylene Welding and Cutting

Care of Cylinders

Oxyacetylene welding and cutting require the use of a mixture of flammable gases and air that is highly explosive. Generally, compressed gas cylinders are safe for the purpose for which they are intended. Few accidents have occurred, but those that have occurred have been the result of abuse and mishandling in the use and/or storing of flammable gases. The following precautions should be taken routinely:

Handling Cylinders

- Cylinders may be lifted by a crane or derrick only if a cradle or platform is used. Never use a sling or attach a hook to the valve protection caps. Never drop cylinders nor permit them to strike each other violently.
- Cylinders may be moved by tilting and rolling them on their bottom edges. Never drag or slide them. If transferred in a carrier, they should be fastened securely.
- Cylinders should be fastened securely while in use.
- Never move unattached cylinders with the regulators on. Cylinder valves must be closed and capped before being moved. The valves of empty cylinders must be closed, the valve protection caps secured in place, and the cylinder marked MT (empty).
- Cylinder valves must be closed when work is finished.
- Keep cylinders away from welding and cutting operations so that sparks, hot slag, and flame cannot reach them.
- Do not place cylinders near an electric circuit. They must not be used as a ground in arc welding, nor to strike an arc.

- Tampering with the numbers and markings stamped on the cylinders is illegal. Never tamper with the valves or safety devices on valves or cylinders.
- Never attempt to refill a cylinder or to mix gases in a cylinder.

Storage of Cylinders

- Cylinders must be secured and stored in a dry, well-ventilated room away from open flame.
- Cylinders should be protected against the weather. Ice and snow should not accumulate on them. They should be screened against the continuous direct rays of the sun to prevent excessive rises in temperature.
- Cylinders should not be exposed to temperatures above 130°F or below −20°F.
- Cylinders must not be stored near highly flammable substances such as oil and gasoline. Fuel gases should not be stored with oxygen cylinders.
- Cylinders should be stored away from locations where heavy moving objects may strike or fall on them.

Acetylene Cylinders

- Acetylene is a fuel gas and should be referred to by its proper name and not by the word *gas*.
- Always store and use acetylene cylinders valve end up. Never operate them while laying on their side. Acetone will be forced into the valve, regulator, and hose.
- Keep sparks and open flame away from cylinders.
- Handle carefully. A damaged cylinder may leak and cause a fire.
- Do not use acetylene directly from the cylinder. Install a regulating device between the cylinder and the torch. The maximum safe working pressure for acetylene is 15 p.s.i. Before connecting a regulator to a cylinder valve, open the valve slightly and close it immediately, Fig. 32-22. This is called *cracking*. Cracking removes dust and dirt from the valve. Never crack a cylinder near other welding work or near sparks, flame, and other sources of combustion.
- Before removing a regulator from a cylinder valve, close the valve and release the gas from the regulator.
- Never connect two or more cylinders together with a manifold or other connecting device that has not been approved for this purpose.
- The cylinder valve should be opened about a ¾ turn. The wrench used for opening the cylinder should always be kept on the valve when the cylinder is in use.
- Never test for acetylene leaks with an open flame. Use an appropriate solution.
- If the fuse plug or other overpressure or overtemperature safety device is leaking, immediately move the cylinder outdoors. The cylinder's location should be blocked off for some distance and the area kept from any source of ignition. The wind direction should be noted, and the cylinder gas supplier contacted for proper disposal.

Fig. 32-22 Before attaching cylinders, crack the valve slowly to blow any accumulated dust clear. Location: Northeast Wisconsin Technical College Mark A. Dierker/McGraw Hill

Oxygen Cylinders

- Always refer to oxygen by its proper name, *oxygen*, and not as *air*.
- Although oxygen itself does not burn, it supports and speeds up combustion. Never permit oil and grease to come in contact with oxygen cylinders or any of the equipment being used in the welding or cutting operation. A jet of oxygen should never be permitted to strike a substance that has oil or grease on it. Bulk storage and use of liquid oxygen can present other hazards. Liquid oxygen is very cold (nearly 300° Fahrenheit). If permitted to contact skin or non-protective clothing, cold surfaces present on liquid oxygen systems such as valves, lines, or couplings can cause severe frostbite or cryogenic burns. Skin will stick to cold surfaces at cryogenic temperatures, causing additional injury.
- Never use oxygen as a substitute for compressed air.
- Do not store oxygen cylinders near combustible materials or fuel gases.
- Do not use oxygen directly from the cylinder; the pressure must be reduced through regulation.
- Never use a hammer or wrench on the oxygen cylinder valve. If the valve cannot be opened with the hands, return the cylinder to the supplier.
- Before connecting the regulator, open the cylinder valve for an instant to remove dirt and dust. After attaching

the regulator and before the cylinder valve is opened, be sure that the adjusting screw of the regulator is released. Do not open the cylinder valve suddenly. Open the valve completely when the cylinder is in use. High-pressure cylinders (2,200 p.s.i.) have double seating valves that will leak if not fully opened. Face away from the front of the gauge when opening them. If the regulator malfunctions, you want to protect yourself from the possibility that the glass might blow out.
- When connecting two or more cylinders, always use a manifold designed and approved for the purpose, Fig. 32-23.
- Never try to mix gases in an oxygen cylinder.
- Never interchange equipment made for use with oxygen with equipment intended for use with other gases. Only use approved regulators.

Operator Protection

Fire is the greatest hazard to the welder when using the oxyacetylene process. Accordingly, welders should wear protective clothing and be very careful not to introduce fire hazards in the welding area. The following precautions should be observed:

- Never perform a cutting or welding operation without goggles fitted with lenses of the proper shade, see Table 32-3, page 1032.
- The head and hair should be protected by a cap.
- The hands should be protected by gloves.
- The arms should be protected by long sleeves.
- The feet should be protected by high-top boots, pants should be cuffless.
- Keep clothing free from oil and grease.

Torch

The torch and hose assembly must be put together carefully and correctly to ensure safe operation. Figure 32-24 shows the complete assembly. The following precautions are routine:

- Connect the oxygen hose from the oxygen regulator to the hose connection on the torch marked *oxygen*.
- Connect the acetylene hose from the acetylene regulator to the hose connection on the torch marked *acetylene*.
- Select the proper welding tip or cutting nozzle and screw it carefully, but not too tightly, on the torch.

- When changing torches, shut off the gases at the pressure regulators and not by crimping the hose.
- Use a friction lighter (striker) instead of a match, cigarette lighter, or welding arc for lighting the torch. Make sure that the torch is not directed toward another person while you are lighting it.
- Never attempt to light or relight a torch from hot metal in a small hole where gas may accumulate. Light the torch with a friction lighter and adjust the flame before inserting it into the hole.
- When the welding or cutting is finished, extinguish the flame by closing the oxygen valve first and the acetylene valve next. Do not shut off the fuel valve first; this may create a popping sound. This popping

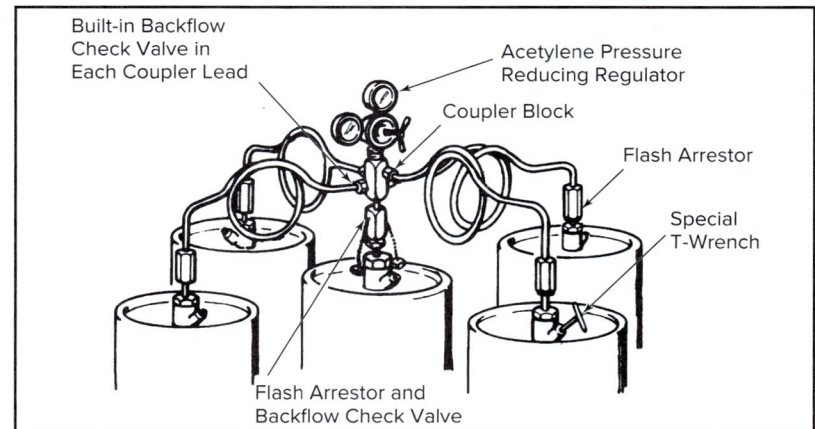

Fig. 32-23 Typical acetylene cylinder manifold for a single-hose line.
Source: ESAB Welding and Cutting Products

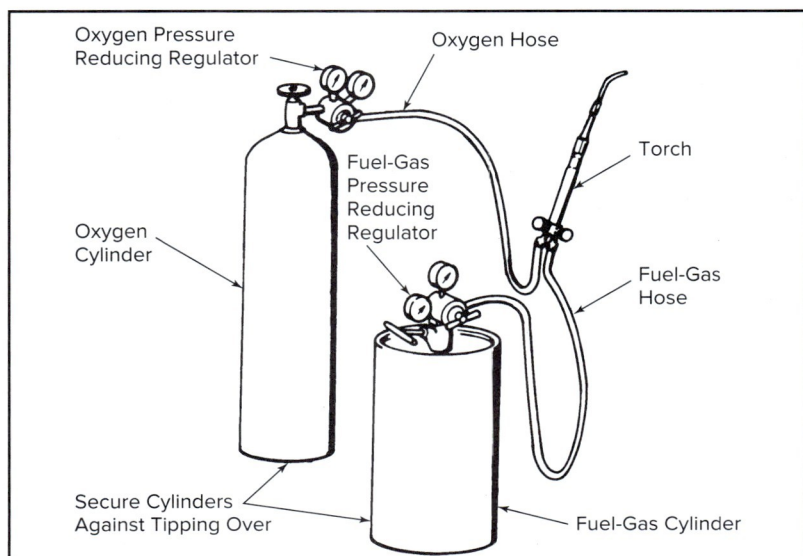

Fig. 32-24 The basic system for oxyacetylene welding. Source: ESAB Welding and Cutting Products

throws carbon soot back into the torch which might clog gas passages and the flashback arrestors.
- When stopping the operation for a few minutes, it is permissible to close only the torch valves. Always extinguish a torch when it is not in your hands.
- When stopping for a longer period (during the lunch hour or overnight), close the cylinder valves. Then release all gas pressure from the regulators by opening the torch valves momentarily. Close the torch valves and release the pressure-adjusting screws. If the equipment is to be taken down, make sure that all pressure-adjusting screws are turned to the left until free.
- Do not leave unlighted torches connected for use in boiler tubes, tanks, and other confined spaces during the lunch hour or when leaving the job for other reasons.
- If leakage develops around the torch valve stems, tighten the packing nuts and repack them if necessary. Use only packing supplied by or recommended by the manufacturer of the torch. Never use oil on the packing or elsewhere.
- If a torch valve does not shut off completely, shut off the gas supply and remove the valve assembly. Wipe the seating portion of the valve stem and the body or replaceable seat with a clean cloth. If the valve still leaks, new parts should be used or the valve body should be reseated.
- If the holes in the torch tip or nozzle become clogged, clean them with the proper size drill or a copper or brass wire. A sharp hard tool, which would enlarge or bellmouth the holes, must not be used. Clean the holes from the inner end wherever possible.

Hose

It is very important to have the correct hose for the kind of gas used. The generally recognized colors are red for acetylene and other fuel gases, green for oxygen, and black for inert gas and air. Hose must be kept in good repair to prevent fires and explosions. The following precautions should be taken for proper use and maintenance:

- Protect the hose from sharp edges, flying sparks, hot slag, and hot objects.
- Do not allow hose to come in contact with oil and grease. These materials cause the rubber to deteriorate and increase the danger of fire when in contact with oxygen.
- Hose should be stored in cool locations. Be careful not to put it on greasy floors or shelves because the rubber absorbs grease and oil.
- New hose is dusted on the inside with fine talc. Blow this dust out with compressed air before using.
- All hose should be examined carefully at frequent intervals for leaks, worn places, and loose connections. This can be done by immersing the hose in water under normal working pressure. Hose and hose connections may also be tested with oil-free air or an oil-free inert gas at twice the normal pressure to which it is subjected in service. In no case should this be less than 200 p.s.i. for hose and 300 p.s.i. for hose connections.
- Leaks must be repaired at once by cutting the hose and inserting a splice. Acetylene escaping from defective hose may ignite and start a serious fire. It may also set fire to the operator's clothing and cause severe burns.
- When hose shows wear at a connection, cut off the worn portion and reinsert connections securely.
- Do not repair hose with tape.
- If a flashback occurs and burns inside the hose, discard that length of hose. A flashback of this sort makes a piece of hose unsafe because it burns the inner walls. Sooner or later this part of the hose will disintegrate and cause trouble by clogging or otherwise interfering with the hose and operation of the torch.
- Hose-line safety devices, if properly installed and operating, should prevent any reverse flow and flashback into the hoses. Check valves with flashback arrestors must be installed, used, inspected, and maintained strictly in accordance with the manufacturer's instructions.
- Hose connections should be of the regulation type conforming to the standards of the Compressed Gas Association (CGA), Fig. 32-25. The nuts on fuel gas fittings have a groove cut around them. This signifies left-hand threads.

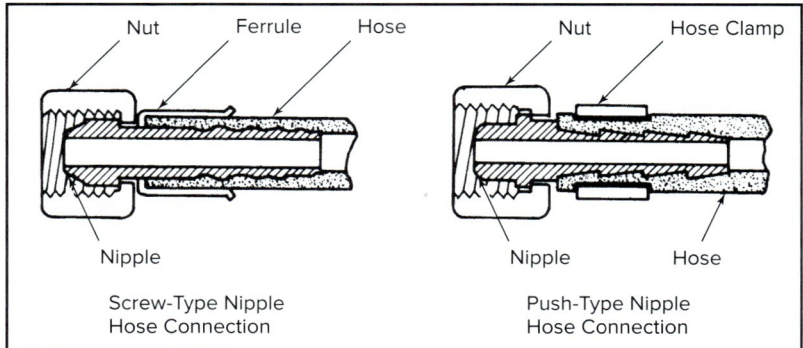

Fig. 32-25 Two types of hose nipples are commonly used for oxyacetylene welding: the push type has a serrated nipple and requires a hose clamp, and the screw type has a ferrule placed over the hose end and a coarsely threaded nipple turned into the hose end. Source: ESAB Welding and Cutting Products

Pressure-Reducing Regulators

Pressure-reducing regulators, both adjustable and nonadjustable, must be used only for the gas and at pressures for which they are intended. The following safety precautions are observed in their use:

- Always use the right regulator for the gas being used.
- Clear all passages before applying pressure.
- Never force a regulator onto or into a cylinder valve.
- Never use a creeping regulator.
- Never use oil on the regulator for any purpose.
- Pressure-adjusting screws on regulators should always be fully released and the regulator drained of gas before the regulator is attached to a cylinder and before the cylinder valve is opened.
- Only skilled mechanics properly instructed in the work should repair regulators and parts of regulators such as gauges.
- The working or low-pressure gauges attached to regulators should be periodically tested by the supplier to ensure accuracy.
- Union nuts and connections on regulators should be inspected before use to detect faulty seats that may cause leakage of gas when the regulators are attached to the cylinder valves. They should be removed from service if damaged.
- Never use adapters to connect a regulator to a valve that was not designed to fit on without approval of the cylinder gas supplier.

Welding and Cutting Operations

The welder must take the following precautions when welding or cutting with the oxyacetylene process:

- Do not use the welding or cutting flame where an open flame of any kind would be dangerous, such as in or near rooms containing flammable vapors or liquids, dust, and loose combustible stock.
- Do not use welding or cutting equipment near dipping or spraying rooms.
- Be careful when welding or cutting around a sprinkler system.
- If the work can be moved, it is better to take it to a safe place rather than perform the work in a hazardous location.
- When the work requires that torches be used near wooden construction and in locations where the combustible material cannot be removed or protected, station extra workers with small hoses, chemical extinguishers, or fire pails nearby. It is advisable to carry a fire extinguisher as regular equipment. Make certain it is of the proper type for the combustibles in the area, Fig. 32-26.
- Never do any hot work such as welding or cutting used drums, barrels, tanks, or other containers until they have been cleaned so thoroughly that it is absolutely certain that no flammable materials are present. Inspection with an analyzer is required. The inside can be filled with water, sand, or an inert gas to displace volatile gases.

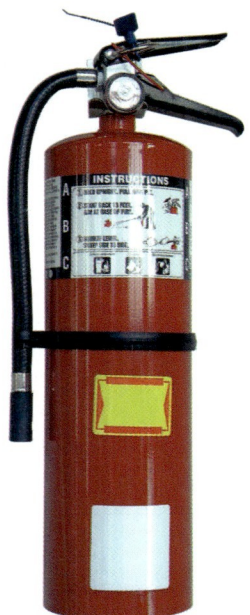

Portable Fire Extinguisher in Preparation for Those Critical First 2 Minutes of a Fire

- Class A fire extinguishers are for ordinary combustible materials such as paper, wood, cardboard, and most plastics. The numerical rating on these types of extinguishers indicates the amount of water it holds and the amount of fire it can extinguish.

- Class B fires involve flammable or combustible liquids such as gasoline, kerosene, grease, and oil. The numerical rating for class B extinguishers indicates the approximate number of square feet of fire it can extinguish.

- Class C fires involve electrical equipment, such as appliances, wiring, circuit breakers, and outlets. Never use water to extinguish class C fires—the risk of electrical shock is far too great! Class C extinguishers do not have a numerical rating. The C classification means the extinguishing agent is nonconductive.

- Class D fire extinguishers are commonly found in a chemical laboratory. They are for fires that involve combustible metals such as magnesium, titanium, potassium, and sodium. These types of extinguishers also have no numerical rating, nor are they given a multipurpose rating—they are designed for class D fires only.

Fig. 32-26 Your fire extinguishers should have ABC ratings on them. Gino Santa Maria/Shutterstock

- Never put down a torch unless the oxygen and acetylene have been completely shut off. Never hang torches from regulators or other equipment so that the flame can come in contact with the oxygen or acetylene cylinders even though you think the valves have been shut off.
- Mount cylinders on portable equipment securely.
- Never support work on compressed gas cylinders.
- Do not cut materials in such a position as to permit the severed section to fall on your legs or feet. Protect your legs and feet from sparks and hot slag.
- Do not allow showers of sparks from a welding or cutting operation to fall upon persons who may be working below. Make sure that sparks do not fall on equipment, flammable material, or into the unprotected head of an acetylene cylinder. The sparks may melt the fusible plugs of the cylinder head and ignite the escaping contents. Such carelessness is both dangerous and wasteful.
- When welding or cutting is being performed in a confined space, such as the interior of a boiler, always leave the cylinders on the outside with an attendant and lead the gas in through the hose to the point where the work is being done.
- Be careful when beginning a cut on a closed container. The air pressure inside will cause it to blow out as soon as a hole is made. Keep your face to one side.
- Never work directly on a concrete floor because when it is heated, the concrete may spall (chip) and fly about, possibly injuring you and other workers.
- Make sure that the room is well-ventilated.
- Do not set the oxygen pressure higher than that required to do the job, Fig. 32-27.
- Only use equipment within its rating limits and approved PPE.

Backfire and Preignition Sometimes the welding or cutting operation will be interrupted by a series of popping sounds at the torch. This is caused by a momentary retrogression (backfire) of the flame into the torch tip. Backfire results from the preignition of the gases. Preignition is caused by a number of conditions that can be avoided.

- It may be caused by touching the torch tip to the work. If this happens, the torch can be relighted instantly if the metal being welded or cut is hot enough to ignite the gases. Otherwise, a lighter should be used. Never relight the torch from hot metal in a small hole.
- The pressure at the regulators may be too low. Adjust the regulator to a higher pressure before relighting.
- The flame may be too small for the tip size. Relight the torch and increase the size of the flame.
- The tip may have carbon deposits or metal particles inside the hole. They become overheated and act as ignitors of the gas before it passes through the hole.
- If welding or cutting in a confined area such as a corner, the tip may become overheated and preignition will take place inside the tip. Correct this condition by cooling the tip.

Flashback (Sustained Backfire) Sustained backfire, known as *flashback,* occurs when the retrogression of the flame back into the mixing chamber is accompanied by a hissing or squealing sound and a characteristic smoky, sharp-pointed flame of small volume. The gas supply should be cut off immediately to prevent overheating and the possible destruction of the torch head. Generally, the torch head must be cooled before it can be reignited.

If the flame flashes back (burns back inside the torch), immediately shut off the torch oxygen valve that controls the flame. Then close the acetylene valve. After a moment, relight the torch in the usual manner. Even with improper

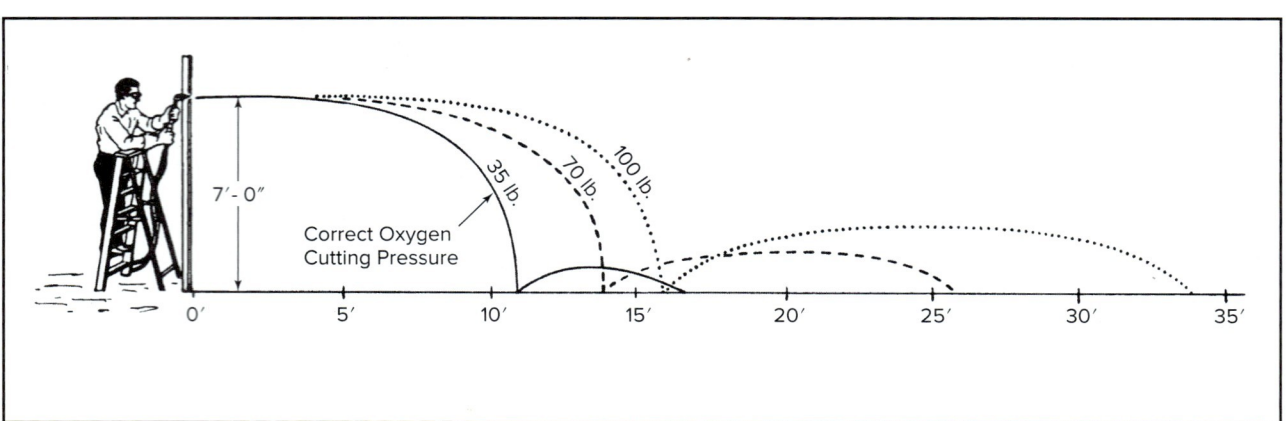

Fig. 32-27 Oxygen pressure that is too high can blow sparks twice as far as need be. They can easily start fires, and oxygen is wasted. Source: Nooter Corp.

handling of a torch, flashback rarely occurs. When one happens, it indicates that something is seriously wrong with the torch or the manner of operating it. In order to prevent flashbacks, proper delivery pressure of both gases should be maintained.

Work in Confined Spaces

By definition, a confined space:

- Is large enough for an employee to enter fully and perform assigned work;
- Is not designed for continuous occupancy by the employee; and
- Has a limited or restricted means of entry or exit.

These spaces may include underground vaults, tanks, storage, bins, pits and diked areas, vessels, silos, and other similar areas.

By definition, a permit-required confined space has one or more of these characteristics:

- Contains or has the potential to contain a hazardous atmosphere;
- Contains a material with the potential to engulf someone who enters the space;
- Has an internal configuration that might cause an entrant to be trapped or asphyxiated by inwardly converging walls or by a floor that slopes downward and tapers to a smaller cross section; and/or
- Contains any other recognized serious safety or health hazards.

Figure 32-28 represents one of these situations. The terms *confined space* and *permit-required confined space* are employed by U.S. Department of Labor Occupational Safety and Health Administration OSHA 3138-01R and OSHA 29 CFR 1910.146. They are used to describe situations and requirements that are applied when welders' activities are hindered by their surroundings and this poses health and safety hazards. The American Welding Society covers confined spaces in Safety in Welding, Cutting and Allied Processes, ANSI Z49.1. For the most up-to-date standards on welding and cutting in confined spaces, reference these two organizations' documents. The following protection should be provided for welders:

- Ventilation is necessary when working in confined spaces.
- Oxygen levels must be measured. They should be in the 19.5 to 23.5 percent range. Normal atmospheric air contains approximately 21 percent oxygen by volume. Oxygen levels should not be allowed to get over 25 percent. Materials like paper and rags that burn normally will violently flare up in an oxygen-rich atmosphere.
- Toxic or flammable gases and vapors must be monitored. The test for these types of materials, as well as for oxygen, must be done with instruments approved by an appropriate agency such as the U.S. Mine Safety and Health Administration (MSHA). Continuous monitoring and audible alarms should be used.
- Gases that are heavier than air will accumulate in low areas such as pits, tank bottoms, and near floors. Heavy gases include argon, methyl-acetylene-propadiene (MPS), propane, and carbon dioxide.
- Gases that are lighter than air will accumulate in high areas such as tank tops and near ceilings or roofs. Light gases include helium and hydrogen.
  - In confined spaces or where the welding space contains partitions, balconies, or other structural barriers to the extent that they significantly obstruct cross ventilation, a minimum ventilation rate shall be 2,000 cubic feet per minute per welder, unless exhaust hoods or respirators are used.
  - Entry permits, if required, should be filled out and posted. A qualified person in a position of responsibility should sign them, and the attendant's responsibilities shall be clearly defined. A preplanned rescue procedure for quickly removing or protecting those working in the confined space must be in place.
  - An emergency signal must be made available for the welder to notify outside personnel for help if the welder is required to enter the confined space through a manhole or other small opening where observation may be restricted.

Fig. 32-28 This welder is working inside of a boiler shell that represents one of these conditions. Miller Electric Mfg. Co.

- If the work in the confined space is considered immediately dangerous to life or health (IDLH), an attendant shall be stationed on the outside of the confined space.
- Where it is impossible to provide local ventilation, air-supplied respirators or hose masks are provided. A worker is stationed outside of the confined space to service the power and ventilation lines, ensuring the safety of those who are working within.
- Gas cylinders and welding machines are left outside of confined spaces.
- Portable equipment mounted on wheels must be securely blocked to prevent accidental movement. Any movable components must be secured from movement. If electrically driven, lock-out–tag-out procedures should be employed.
- Where a welder must enter a confined space through a movable or other small opening, means must be provided for removing the welder quickly in case of emergency. When safety belts and lifelines are used for this purpose, they are attached to the welder's body in such a way that it prevents the body from being jammed in the small exit opening.
- In order to prevent the accumulation of gas because of leaks or improperly closed valves, the gas supply to the torch should be positively shut off outside the confined area whenever the torch is not to be used for a period of time such as during the lunch hour or overnight. If possible, remove the torch and hose from the confined space.
- After welding operations are completed, the welder marks the hot metal or provides some other means of warning other workers.

Fire Protection

Many safety practices for electric and oxyacetylene welding and cutting are required for fire prevention. In addition to these special practices, the following general precautions should be taken wherever any welding or cutting is being done:

- Welding operations should be done in permanent locations free from fire hazards. Booths should be constructed of nonflammable materials such as sheet metal. If it is possible to move the job, it is preferable to take it to a safe location for cutting and welding rather than to perform the work in a hazardous location.
- Fire is a particular hazard when portable welding equipment is used. Before welding operations in which portable equipment is used are started, the location should be thoroughly inspected by a competent employee to determine what fire protection equipment is necessary. A hot-work permit is required and issued by the welding supervisor, a member of the plant fire department, or some other qualified person before welding operations are started. This is particularly important in a hazardous location.
- Welding operations should not be permitted in or near rooms containing flammable vapors, liquids, or dust until all fire and explosion hazards have been eliminated. If welding is necessary in such locations, the area should be thoroughly ventilated. Sufficient draft should be maintained during welding and cutting operations to prevent the accumulation of explosive concentrations of such substances.
- Tanks, drums, and pipelines that have contained flammable liquids should be cleansed of all solid or liquid flammable material, purged of all flammable gases and vapors, and tested for the presence of flammable gases before welding operations are started.
- Welding operations should not be performed on spray booths or ducts that may contain combustible deposits without first making sure that such places are free of flammable materials.
- Where welding has to be done in the vicinity of combustible material, special precautions should be taken to make certain that sparks or hot slag do not reach such material and thus start a fire. If the work cannot be moved, exposed combustible materials should, if possible, be moved a safe distance away. Otherwise, they should be covered with fireproof blankets or sheet metal during welding operations, Fig. 32-29.

Fig. 32-29 Welding blankets are made to size from heat-resistant materials to provide protection from sparks, spatter, and slag generated by welding or thermal cutting applications.
© Wilson Industries, Inc.

Pull the pin at the top of the extinguisher. The pin releases a locking mechanism and will allow you to discharge the extinguisher.

Aim at the base of the fire, not the flames. This is important. In order to put out the fire, you must extinguish the fuel.

Squeeze the lever slowly. This will release the extinguishing agent in the extinguisher. If the handle is released, the discharge will stop.

Sweep from side to side. Using a sweeping motion, move the fire extinguisher back and forth until the fire is completely out. Operate the extinguisher from a safe distance, several feet away, and then move toward the fire once it starts to diminish. Be sure to read the instructions on your fire extinguisher—distances differ with various fire extinguishers. Remember: Aim at the base of the fire, not at the flames!

Fig. 32-30 This acronym is a good quick reference. Print and place this next to each fire extinguisher.

- Wood floors should be swept clean before welding operations are started. Wood floors should be covered with metal or other noncombustible material where sparks of hot metal are likely to fall. In some cases, it is advisable to wet down the floors. Hot metal or slag should not be allowed to fall through cracks in the floor or other openings on combustible materials on the floor below. Particular attention should be taken to see that hot slag or sparks do not fall into machine tool pits.
- Similarly, sheet metal guards or weld blankets should be used to guard cracks and holes in walls, open doorways, and open or broken windows. Make certain that there is no opening where the curtain meets the floor.
- A worker properly trained and equipped with fire-extinguishing equipment should be stationed at or near welding operations in hazardous locations, Fig. 32-30. The worker makes sure that sparks do not lodge in floor cracks or pass through floor or wall openings. This worker should be kept at the job site for as long as 30 minutes after the job is completed to make sure that smoldering fires have not been started.
- A fire watch is appropriate in the following situations: (1) When there are combustibles on walls, floors, or openings that may expose combustibles within a 35-foot radius. (2) When there are metal walls, ceilings, roofs, or pipes adjacent to materials that are likely to ignite by means of radiation or conduction. (3) When performing ship work or during similar situations. The firewatcher is allowed to do other duties as long as those duties do not interfere with performing the fire watch.
- The usual precautions in handling electric power should be observed. The insulation of the welding cable may be burned off if leads are too small to carry the necessary current. The insulation may be cut through by dragging the leads across sharp objects. Electrode holders may be carelessly dropped. Welding sets may be shortened due to mishandling. These hazards can largely be eliminated by good maintenance of equipment.

Other Precautions for Safe Working Conditions

A large number of potentially hazardous miscellaneous situations must also be of concern to the welder who is conscious of the need to work in a safe environment. Very often accidents are caused by a relatively unimportant condition that has not been taken care of. Extremely dangerous hazards usually draw the attention of the welder and are therefore rarely a cause of accidents.

Welders and/or welding operators should be thoroughly instructed in the performance of their work regarding the protection of themselves and others working nearby. They should also realize that good work quality in making sound welds is essential so that others may not be injured because of the failure of the welded part. The American Welding Society has prepared standard codes for welding procedures and operator qualifications that are generally accepted by industry.

The following suggestions reduce unexpected hazards:

- If it is necessary for a welder to work at an elevation of more than 6 feet, adequate provision should be made to prevent falling in case of electric shock or other injury. This can be accomplished by the use of railings, full-body harnesses and lifelines, or some other equally effective safeguard. Lifebelts, full-body harnesses, and similar devices should be of a type that will permit quick escape.
- If welding is performed in confined spaces, such as in tanks and the hulls of ships, some means should be provided for quickly removing the welder in case of emergency. An attendant should always be stationed in a position to readily give assistance to welders working in such places.
- When welders and/or welding operators have occasion to leave their work or to stop work for any length of time, they should always open the main switch in the equipment.
- Before operations are started, heavy portable equipment mounted on wheels should be securely blocked to prevent accidental movement.
- Welding equipment should be maintained in good mechanical and electrical condition to avoid unnecessary electrical hazards.
- Welding equipment used in the open should be protected from inclement weather conditions. When not in use, the equipment should be stored in a clean, dry place. Cables should be neatly coiled and stored

where they will not be damaged or create stumbling hazards to employees.
- After welding operations are completed, the welder and/or welding operator should mark the hot metal or provide some warning sign to prevent other employees from touching them and getting burned.
- Report all injuries at once.
- Good housekeeping should be maintained on all welding jobs. Welders and/or welding operators should not discard electrode stubs on the floor or leave tools or other objects where they will constitute hazards. Welders and/or welding operators should have a receptacle in which to keep their daily supply of electrodes. The same receptacle can be used for disposing of electrode stubs. Structural welders and/or welding operators should be provided with a receptacle with a strap for attachment to a belt to facilitate carrying electrodes.
- Material-handling equipment, such as cranes and hoists, should be maintained in a safe operating condition. Cables, chains, and slings used in moving heavy parts to be welded should receive special attention.
- All tools such as hammers, chisels, brushes, bars, and other hand tools should be maintained in safe condition. A tool box should be used by each welder in which all tools should be kept when not in use.
- When working with engine-driven equipment, be careful not to mishandle fuel. The fuel system must be kept in good working order. Leaks must quickly be repaired, since the ignition system, electrical controls, sparks, spatter, or engine heat may start a fire. The engine-driven equipment should be shut off during refueling operations. Any fuel spills that do occur should be quickly cleaned up and fumes allowed to dissipate before the engine-driven equipment is restarted.
- Be aware of back injuries and always use good body mechanics. Use appropriate lifting techniques when lifting equipment or get assistance when moving heavy or awkward work.
- Use the same care for your *safety procedures* as you do for your welding procedures. Get it right first, and keep safety first.

Safety Checklist
☐ Explosive environments
☐ Fire hazards
☐ Moving machinery
☐ Items that may fall
☐ Trip hazards
☐ Pinch points
☐ Trapped spaces
☐ Confined spaces
☐ Location of fire extinguishers
☐ Exit routes
☐ Fall protection (harnesses and safety belts)
☐ PPE (for noise, light, radiation, hot metal, grinding operations fumes)

CHAPTER 32 REVIEW

A. Safety Precautions: Electric Welding Processes

Multiple Choice

Choose the letter of the correct answer.

1. Who is responsible for installing or repairing arc welding equipment? (Obj. 32-1)
   a. The welder
   b. The operator
   c. The welder's helper
   d. A qualified electrician or service technician
2. What size welding cable is required to weld at 300 amperes, 100 feet away from the power source with the SMAW process? (Obj. 32-1)
   a. 1
   b. ⅔
   c. ⅗
   d. Not enough information known
3. If the electrode holder becomes hot, you should _____. (Obj. 32-1)
   a. Drop it in a bucket of water
   b. Go to a larger capacity holder if it is being overworked
   c. Loosen the connections
   d. Both b and c
4. Water leaks coming from water-cooled TIG torches, PAC torches, or GMAW guns should _____. (Obj. 32-1)
   a. Be ignored
   b. Not be used while they are leaking
   c. Be repaired or replaced by qualified personnel
   d. Both b and c

5. Arc welding current is never strong enough to injure the welder. (Obj. 32-1)
   a. True
   b. False

Review Questions

Write the answers in your own words.

6. Name four safety precautions that have a bearing on the safe handling of the electric arc welding machines. (Obj. 32-1)
7. List six precautions that should be taken to protect yourself against electric shock. (Obj. 32-1)
8. List two things that can be used to insulate yourself from the work. (Obj. 32-1)
9. If an electric arc welding machine is not operating properly, what should be done? (Obj. 32-1)
10. Why should metal jewelry not be worn when doing electric arc welding? (Obj. 32-1)

B. Safety Precautions: Oxyacetylene Welding and Cutting

Multiple Choice

Choose the letter of the correct answer.

1. Both the oxygen and acetylene cylinders can be laid on the ground during use. (Obj. 32-1)
   a. True
   b. False
2. How may the welding and cutting hose be protected from damage? (Obj. 32-1)
   a. Protect hose from sharp edges, flying sparks, hot slag, hot object, and grease and oil, and store in cool area.
   b. Hoses are armored and so are not affected by sharp edges, heat, or contaminants like grease and oil.
   c. Protect hose from sharp edges, flying sparks, hot slag, hot objects, and oil and grease them so they are easier to store in high heat areas.
   d. Expose hose to sharp edges, flying sparks, hot slag, hot objects, and grease and oil, and never store in cool areas.
3. Which are some of the safety precautions that should be taken when hooking up the cutting torch? (Obj. 32-1)
   a. Interchanging hoses is OK; use whatever size tip is in the torch; tighten tip with at least a 12-inch adjustable wrench; fold over hoses and clamp with a vise grip; no need to shut off oxygen or acetylene at the cylinders when changing torches; use match to light torch; and if torch tip holes are clogged, drill out with one size larger drill bit.
   b. Connect hoses to the proper fitting; select the proper size tip; do not overtighten tip in torch; fold over hoses and clamp with a vise grip; no need to shut off oxygen or acetylene at the cylinders when changing torches; use match to light torch; and if torch tip holes are clogged, drill out with one size larger drill bit.
   c. Connect hoses to the proper fitting; select the proper size tip; do not overtighten tip in torch; shut off the oxygen and acetylene at the cylinder prior to changing torches; use friction lighter to light torch; and if torch tip holes are clogged, clean them with proper drill or a copper or brass wire.
   d. Interchanging hoses is OK; use whatever size tip is in the torch; tighten tip with at least a 12-inch adjustable wrench; shut off the oxygen and acetylene at the cylinder prior to changing torches; use friction lighter to light torch; and if torch tip holes are clogged, clean them with proper drill or a copper or brass wire.
4. When the valves on the cutting torch freeze or become hard to operate, what should be done and why? (Obj. 32-1)
   a. Check if the packing nut is too tight. Lubricate with "N O" oil as this is indicated on the regulator as the type of oil to be used.
   b. Check if the packing nut is too loose. Lubricate with "N O" oil as this is indicated on the regulator as the type of oil to be used.
   c. Check if the packing nut is too tight. If this does not free the valve up, send the torch to an authorized repair technician. Never lubricate with any type of oil or grease.
   d. Check if the packing nut is too loose. If this does not free the valve up, send the torch to an authorized repair technician. Never lubricate with any type of oil or grease.
5. What conditions may cause the torch to backfire or cause preignition in the gas torch? (Obj. 32-1)
   a. Tip is too far from work; pressure at regulator is too high; flame is too large for tip size; tip may be too clean—tips work best when dirty; and/or tip is operating too cold.
   b. Tip is too far from work; pressure at regulator is too high; flame is too small for tip size; tip may be dirty with carbon deposits; and/or tip is too hot.
   c. Tip is too close to work; pressure at regulator is too low; flame is too large for tip size; tip may be too clean—tips work best when dirty; and/or tip is operating too cold.

d. Tip is too close to work; pressure at regulator is too low; flame is too small for tip size; tip may be dirty with carbon deposits; and/or tip is too hot.

Review Questions

Write the answers in your own words.

6. How may thermal cutters or cutting operators protect themselves from harm while carrying out gas welding and cutting operations? (Obj. 32-1)
7. List at least six precautions that should be observed when handling gas cylinders. (Obj. 32-1)
8. List at least six precautions that should be observed in the use of acetylene cylinders. (Obj. 32-1)
9. List at least six precautions that should be observed in the use of oxygen cylinders. (Obj. 32-1)
10. List at least six precautions that should be observed in the use of pressure-reducing regulators. (Obj. 32-1)

C. Safety Precautions: Personal Protective Equipment

Multiple Choice

Choose the letter of the correct answer.

1. Which of the following is not PPE? (Obj. 32-2)
   a. Gloves
   b. Safety glasses
   c. Welding helmet
   d. Welding machine
2. A systematic approach to selecting PPE should be used and the following is the best practice. (Obj. 32-2)
   a. Think about the possibility of burns and get all the burn protection equipment together.
   b. Consider the arc rays and get protective equipment together.
   c. Start at your feet and work to the top of your head considering the work at hand and select the appropriate PPE.
   d. Start at your waist and work both down and up to select the appropriate PPE.
3. Foot protection should consist of_____. (Obj. 32-2)
   a. High-top boots
   b. Metatarsal protection
   c. Steel toes
   d. All of these
4. Which shade lens should be used for air carbon arc cutting and gouging at 400 amperes? (Obj. 32-2)
   a. 5
   b. 8
   c. 12
   d. So dark that you can't see through the lens.
5. When you are wearing a welding helmet, you are still required to wear eye protection underneath it. (Obj. 32-2)
   a. True
   b. False

Review Questions

Write the answers in your own words.

6. Explain how the auto-darkening lens works and what its advantages are. (Obj. 32-2)
7. What is the maximum sound level that a person can be exposed to for an 8-hour shift, and what must be done if exposure will exceed this time or sound level? (Obj. 32-2)
8. What agency should be contacted for the most up-to-date information regarding the type of filter to be used in a respirator and why? (Obj. 32-2)
9. Why should lightweight handkerchiefs not be used to protect your head from the welding environment? (Obj. 32-2)
10. A hard hat should have what properties? (Obj. 32-2)

D. Safety Precautions: Burn Protection

Multiple Choice

Choose the letter of the correct answer.

1. PPE is not an issue when dealing with burn protection. (Obj. 32-3)
   a. True
   b. False
2. Burns are generally caused by_____. (Obj. 32-3)
   a. Sparks, spatter
   b. Hot slag, hot metal
   c. Arc rays
   d. All of these
3. Safety glasses are required to protect your eyes from burns when_____. (Obj. 32-3)
   a. You enter the work (shop) area
   b. You are welding and raise your helmet to clean and inspect the weld
   c. Your instructor or supervisor tells you to put them on
   d. Both b and c
4. What type of fabric is most commonly worn in a welding environment? (Obj. 32-3)
   a. Wool or cotton
   b. Fire-resistant fabric
   c. Synthetics like polyester
   d. Both a and b

Review Questions

Write the answers in your own words.

5. For overhead welding or welding in awkward positions how should the ears be protected? (Obj. 32-3)
6. What is the main key to reducing burns in a welding environment? (Obj. 32-3)
7. In order to keep safety glasses on at all times, what must be considered? (Obj. 32-3)
8. Clothing must be in what condition to be worn in a welding environment and why? (Obj. 32-3)

E. Safety Precautions: Respiratory Protection

Multiple Choice

Choose the letter of the correct answer.

1. What is the current exposure limits on toxic materials? (Obj. 32-4)
    a. 1.0 milligram per cubic meter or less
    b. 1.0 milligram per square meter
    c. 1.0 milligram per cubic feet or less
    d. Both a and c
2. Where can the information be found about the toxic level of the material you are working with? (Obj. 32-4)
    a. You must look it up in a catalog
    b. Safety Data Sheet
    c. SDS
    d. Both b and c
3. If you are welding on oily metal from machining or cutting operation and a respirator is required, which series filter would be the best? (Obj. 32-4)
    a. N
    b. R
    c. P
    d. Z
4. What does the abbreviation NIOSH mean? (Obj. 32-4)
    a. Not In Or on Shoes or Head
    b. Northern Institute of Safety and Health
    c. National Institute for Occupational Safety and Health
    d. National Institute for Occasional Safety and Happiness
5. The abbreviation IDLH indicates a condition that is_____. (Obj. 32-4)
    a. Not dangerous if dealing with respirators
    b. Not dangerous if dealing with PPEs
    c. Immediately dangerous to life or health and must seriously be dealt with
    d. Both a and b

Review Questions

Write the answers in your own words.

6. When welding on stainless steel or using stainless-steel filler metals, what three elements are you exposed to that may be potentially hazardous? List the chemical symbols for these elements. (Obj. 32-4)
7. Ventilation should not be an issue if the welding area occupies enough space. How much area and ceiling height is required for each welder? (Obj. 32-4)
8. A respirator filter marked with an HE indicates what? (Obj. 32-4)
9. Who is responsible for reviewing the entire welding application, including the type of base metals, possible coatings on the base metals, filler metals, shielding gases, welding process, and numerous other variables? Those responsible will also collect air samples from the welder's worksite to determine the level of exposure to various contaminants for the purpose of determining if respirators are required and what filter type will be used. (Obj. 32-4)
10. List six factors that must be considered in determining the amount of hazardous fumes that may be encountered in a welding environment. (Obj. 32-4)

F. Safety Precautions: Confined Space Welding Protection

Multiple Choice

Choose the letter of the correct answer.

1. A confined space is_____. (Obj. 32-5)
    a. A relatively small space, such as a tank, boiler, small compartment, small room, or where your activities as a welder are hindered by the surroundings
    b. A wide open expanse where you are free to move around
    c. A room larger than 25 feet × 25 feet × 16 feet in ceiling height
    d. An area where you feel claustrophobic
2. What AWS standard has information about welding in confined spaces? (Obj. 32-5)
    a. A3.0
    b. A2.4
    c. Z49.1
    d. D1.1

3. Ventilation is necessary when working in confined spaces. (Obj. 32-5)
   a. True
   b. False
4. Gas cylinders and welding machines are left outside confined spaces. (Obj. 32-5)
   a. True
   b. False
5. Who should sign the entry permits before work is performed in a confined space? (Obj. 32-5)
   a. A qualified person in a position of responsibility
   b. The senior welder on the job
   c. The AWS certified welding inspector
   d. Whoever happens to be walking by

Review Questions

Write the answers in your own words.

6. What government regulation number and organization covers confined spaces? (Obj. 32-5)
7. What is the range of oxygen that supports a good work environment, and what level of oxygen is considered hazardous because of flammability? (Obj. 32-5)
8. Toxic or flammable gases, vapors, and oxygen levels must be detected before entering a confined space. The inspection test for these types of materials must be done with instruments approved by which agency? (Obj. 32-5)
9. In confined spaces or where the welding space contains partitions, balconies, or other structural barriers to the extent that they significantly obstruct cross ventilation, a minimum ventilation of what rate is required? This rate is only applicable when exhaust hoods or respirators are not being used. (Obj. 32-5)
10. Portable equipment mounted on wheels or any other movable components must be handled in what manner when they involve you working in a confined space? (Obj. 32-5)

INTERNET ACTIVITIES

Internet Activity A

Using your favorite search engines, get on the OSHA Web site. List the date and number of pages devoted to oxyfuel safety issues and report to your instructor.

Internet Activity B

Using your favorite search engines, investigate respiratory issues dealing with welding. Note what materials were involved and what authority the source of your information was. Report your findings to your instructor.

Design credits: Cinema and movie icon: Denis Meshkov/123RF; and Illustration of Welding icons: Mr. Nachai Sorasee/123RF

Appendices

Appendix A
Conversion Tables

Appendix B
Illustrated Guide to Welding Terminology

Appendix C
Welding Abbreviation List

Appendix D
Major Agencies Issuing Codes, Specifications, and Associations

Appendix E
Sources of Welding Information

Appendix F
Metric Conversion Information for the Welding Industry

Appendix A

Conversion Tables

Table A-1. U.S. Standard Gauges and Their Fractional, Decimal, and Metric Equivalents   1059

Table A-2. S.I. Metric and U.S. (Customary) Equivalent Linear Measurements   1060

Table A-3. S.I. Metric and U.S. (Customary) Equivalent Measurements of Area   1060

Table A-4. S.I. Metric and U.S. (Customary) Equivalent Measurements of Mass (Weight)   1061

Table A-5. S.I. Metric and U.S. (Customary) Equivalent Measurements of Volume (Capacity)   1061

Table A-6. Conversion of Decimal Inches to Millimeters and Fractional Inches to Decimal Inches and Millimeters   1062

Table A-7. Conversion of Millimeters to Decimal Inches   1063

Table A-8. Fractional and Number Drill Sizes, Decimal Equivalents, and Tap Sizes   1064

Table A-9. Twist Drill Number Sizes in Inches, Millimeters, and Centimeters   1065

Table A-10. Standard Pipe Data   1066

Table A-11. Commercial Pipe Sizes and Wall Thicknesses   1067

Table A-12. Melting Points of Metals and Alloys of Practical Importance   1068

Table A-13. Properties of Elements and Metal Compositions   1069

Table A-14. Identifying Metals by Spark Testing   1070

Table A-1 U.S. Standard Gauges and Their Fractional, Decimal, and Metric Equivalents

U.S. Standard Gauge	Fractional Equivalent (Inches)			Decimal Equivalent (Inches)	Metric Equivalent (mm)
	Major Fraction	32nds	64ths		
30	...	...	...	0.0125	0.3175
29	...	...	...	0.0140625	0.3572
28	...	...	1	0.015625	0.3969
27	...	...	...	0.0171875	0.4366
26	...	...	...	0.01875	0.4762
25	...	...	...	0.021875	0.5556
24	...	...	...	0.025	0.6350
23	...	...	...	0.028125	0.7144
22	...	1	2	0.03125	0.7938
21	...	...	...	0.034375	0.8731
20	...	...	...	0.0375	0.9525
19	...	...	...	0.04375	1.1112
...	...	...	3	0.046875	1.1906
18	...	...	...	0.05	1.2700
17	...	...	...	0.05625	1.4288
16	1/16	2	4	0.0625	1.5875
15	...	...	...	0.0703125	1.7859
14	...	...	5	0.078125	1.9844
13	...	3	6	0.09375	2.3812
12	...	...	7	0.109375	2.7781
11	1/8	4	8	0.125	3.1750
10	...	...	9	0.140625	3.5719
9	...	5	10	0.15625	3.9688
8	...	...	11	0.171875	4.3656
7	3/16	6	12	0.1875	4.7625
6	...	...	13	0.203125	5.1594
5	...	7	14	0.21875	5.5562
4	...	...	15	0.234375	5.9531
3	1/4	8	16	0.25	6.3500
2	...	...	17	0.265625	6.7469
1	...	9	18	0.28125	7.1438
...	...	...	19	0.296875	7.5406
0	5/16	10	20	0.3125	7.9375
...	...	...	21	0.328125	8.3344
2/0	...	11	22	0.34375	8.7312
...	...	...	23	0.359375	9.1281
3/0	3/8	12	24	0.375	9.5250
...	...	...	25	0.390625	9.9219
4/0	...	13	26	0.40625	10.3188
...	...	...	27	0.421875	10.7156
5/0	7/16	14	28	0.4375	11.1125
...	...	...	29	0.453125	11.5094
6/0	...	15	30	0.46875	11.9062
...	...	...	31	0.484375	12.3031
7/0	1/2	16	32	0.5	12.7000

Table A-2 S.I. Metric and U.S. (Customary) Equivalent Linear Measurements

U.S. (Customary) to Metric	Metric to U.S. (Customary)
1 inch = 25.40 millimeters	1 millimeter = 0.03937 inch
1 inch = 2.540 centimeters	1 centimeter = 0.3937 inch
1 inch = 0.0254000 meters	1 meter = 39.37 inches
1 foot = 30.480 centimeters	1 meter = 3.2808 feet
1 foot = 0.3048 meter	1 meter = 1.0936 yards
1 yard = 91.440 centimeters	1 kilometer = 0.62137 mile
1 yard = 0.9144 meter	
1 mile = 1609.35 meters	
1 mile = 1.609 kilometers	

Table A-3 S.I. Metric and U.S. (Customary) Equivalent Measurements of Area

U.S. (Customary) to Metric	Metric to U.S. (Customary)
1 sq. inch = 645.16 sq. millimeters	1 sq. millimeter = 0.00155 sq. inch
1 sq. inch = 6.4516 sq. centimeters	1 sq. centimeter = 0.1550 sq. inch
1 cu. inch = 16.387 cu. centimeters	1 cu. centimeter = 0.0610 cu. inch
1 sq. foot = 929.03 sq. centimeters	1 sq. meter = 10.7640 sq. feet
1 sq. foot = 9.290 sq. decimeters	1 sq. meter = 1.196 sq. yards
1 sq. foot = 0.0929 sq. meter	1 cu. meter = 35.314 cu. feet
1 cu. foot = 0.02832 cu. meter	1 cu. meter = 1.308 cu. yards
1 sq. yard = 0.836 sq. meter	1 sq. hectometer = 2.471 acres
1 cu. yard = 0.765 cu. meter	1 hectare = 2.471 acres
1 acre = 0.4047 sq. hectometer	1 sq. kilometer = 0.386 sq. mile
1 acre = 0.4047 hectare	
1 sq. mile = 2.59 sq. kilometers	

Table A-4 S.I. Metric and U.S. (Customary) Equivalent Measurements of Mass (Weight)

U.S. (Customary) to Metric	Metric to U.S. (Customary)
1 ounce (dry) = 28.35 grams	1 gram = 0.03527 ounce
1 pound = 0.4536 kilogram	1 kilogram = 2.2046 pounds
1 short ton (2000 lb.) = 907.2 kilograms	1 metric ton = 2204.6 pounds
1 short ton (2000 lb.) = 0.9072 metric ton	1 metric ton = 1.102 tons (short)

Table A-5 S.I. Metric and U.S. (Customary) Equivalent Measurements of Volume (Capacity)

U.S. (Customary) to Metric	Metric to U.S. (Customary)
1 fluid ounce = 2.957 centiliters	1 centiliter = 10 cubic centimeters
1 fluid ounce = 29.57 cubic centimeters	1 centiliter = 0.338 fluid ounce
1 pint (liq.) = 4.732 deciliters	1 deciliter = 100 cubic centimeters
1 pint (liq.) = 473.2 cubic centimeters	1 deciliter = 0.0528 pint (liq.)
1 quart (liq.) = 0.9463 liter	1 liter = 1 cubic decimeter
1 quart (liq.) = 0.9463 cubic decimeter	1 liter = 1.0567 quarts (liq.)
1 gallon (liq.) = 3.7853 liters	1 liter = 0.26417 gallon (liq.)
1 gallon (liq.) = 3.7853 cubic decimeters	1 hectoliter = 26.417 gallons (liq.)

Table A-6 Conversion of Decimal Inches to Millimeters and Fractional Inches to Decimal Inches and Millimeters

Inches dec.	mm.	Inches dec.	mm.	Inches frac.	dec.	mm.	Inches frac.	dec.	mm.
0.01	0.2540	0.51	12.9540	1/64	0.015625	0.3969	33/64	0.515625	13.0969
0.02	0.5080	0.52	13.2080	1/32	0.031250	0.7938	17/32	0.531250	13.4938
0.03	0.7620	0.53	13.4620	3/64	0.046875	1.1906	35/64	0.546875	13.8906
0.04	1.0160	0.54	13.7160	1/16	0.062500	1.5875	9/16	0.562500	14.2875
0.05	1.2700	0.55	13.9700						
0.06	1.5240	0.56	14.2240	5/64	0.078125	1.9844	37/64	0.578125	14.6844
0.07	1.7780	0.57	14.4780	3/32	0.093750	2.3812	19/32	0.593750	15.0812
0.08	2.0320	0.58	14.7320						
0.09	2.2860	0.59	14.9860	7/64	0.109375	2.7781	39/64	0.609375	15.4781
0.10	2.5400	0.60	15.2400	1/8	0.125000	3.1750	5/8	0.625000	15.8750
0.11	2.7940	0.61	15.4940						
0.12	3.0480	0.62	15.7480	9/64	0.140625	3.5719	41/64	0.640625	16.2719
0.13	3.3020	0.63	16.0020	5/32	0.156250	3.9688	21/32	0.656250	16.6688
0.14	3.5560	0.64	16.2560						
0.15	3.8100	0.65	16.5100	11/64	0.171875	4.3656	43/64	0.671875	17.0656
0.16	4.0640	0.66	16.7640	3/16	0.187500	4.7625	11/16	0.687500	17.4625
0.17	4.3180	0.67	17.0180						
0.18	4.5720	0.68	17.2720	13/64	0.203125	5.1594	45/64	0.703125	17.8594
0.19	4.8260	0.69	17.5260	7/32	0.218750	5.5562	23/32	0.718750	18.2562
0.20	5.0800	0.70	17.7800						
0.21	5.3340	0.71	18.0340	15/64	0.234375	5.9531	47/64	0.734375	18.6531
0.22	5.5880	0.72	18.2880	1/4	0.250000	6.3500	3/4	0.750000	19.0500
0.23	5.8420	0.73	18.5420						
0.24	6.0960	0.74	18.7960	17/64	0.265625	6.7469	49/64	0.765625	19.4469
0.25	6.3500	0.75	19.0500	9/32	0.281250	7.1438	25/32	0.781250	19.8437
0.26	6.6040	0.76	19.3040						
0.27	6.8580	0.77	19.5580	19/64	0.296875	7.5406	51/64	0.796875	20.2406
0.28	7.1120	0.78	19.8120	5/16	0.312500	7.9375	13/16	0.812500	20.6375
0.29	7.3660	0.79	20.0660						
0.30	7.6200	0.80	20.3200	21/64	0.328125	8.3344	53/64	0.828125	21.0344
0.31	7.8740	0.81	20.5740	11/32	0.343750	8.7312	27/32	0.843750	21.4312
0.32	8.1280	0.82	20.8280						
0.33	8.3820	0.83	21.0820	23/64	0.359375	9.1281	55/64	0.859375	21.8281
0.34	8.6360	0.84	21.3360	3/8	0.375000	9.5250	7/8	0.875000	22.2250
0.35	8.8900	0.85	21.5900						
0.36	9.1440	0.86	21.8440	25/64	0.390625	9.9219	57/64	0.890625	22.6219
0.37	9.3980	0.87	22.0980	13/32	0.406250	10.3188	29/32	0.906250	23.0188
0.38	9.6520	0.88	22.3520						
0.39	9.9060	0.89	22.6060	27/64	0.421875	10.7156	59/64	0.921875	23.4156
0.40	10.1600	0.90	22.8600	7/16	0.437500	11.1125	15/16	0.937500	23.8125
0.41	10.4140	0.91	23.1140						
0.42	10.6680	0.92	23.3680	29/64	0.453125	11.5094	61/64	0.953125	24.2094
0.43	10.9220	0.93	23.6220	15/32	0.468750	11.9062	31/32	0.968750	24.6062
0.44	11.1760	0.94	23.8760						
0.45	11.4300	0.95	24.1300	31/64	0.484375	12.3031	63/64	0.984375	25.0031
0.46	11.6840	0.96	24.3840	1/2	0.500000	12.7000	1	1.000000	25.4000
0.47	11.9380	0.97	24.6380						
0.48	12.1920	0.98	24.8920						
0.49	12.4460	0.99	25.1460						
0.50	12.7000	1.00	25.4000						

For converting decimal-inches in "thousandths," move decimal point in both columns to left.

Source: L. S. Starrett Co.

Table A-7 Conversion of Millimeters to Decimal Inches

mm	Inches	mm	Inches	mm	Inches	mm	Inches	mm	Inches
0.01	0.00039	0.41	0.01614	0.81	0.03189	21	0.82677	61	2.40157
0.02	0.00079	0.42	0.01654	0.82	0.03228	22	0.86614	62	2.44094
0.03	0.00118	0.43	0.01693	0.83	0.03268	23	0.90551	63	2.48031
0.04	0.00157	0.44	0.01732	0.84	0.03307	24	0.94488	64	2.51968
0.05	0.00197	0.45	0.01772	0.85	0.03346	25	0.98425	65	2.55905
0.06	0.00236	0.46	0.01811	0.86	0.03386	26	1.02362	66	2.59842
0.07	0.00276	0.47	0.01850	0.87	0.03425	27	1.06299	67	2.63779
0.08	0.00315	0.48	0.01890	0.88	0.03465	28	1.10236	68	2.67716
0.09	0.00354	0.49	0.01929	0.89	0.03504	29	1.14173	69	2.71653
0.10	0.00394	0.50	0.01969	0.90	0.03543	30	1.18110	70	2.75590
0.11	0.00433	0.51	0.02008	0.91	0.03583	31	1.22047	71	2.79527
0.12	0.00472	0.52	0.02047	0.92	0.03622	32	1.25984	72	2.83464
0.13	0.00512	0.53	0.02087	0.93	0.03661	33	1.29921	73	2.87401
0.14	0.00551	0.54	0.02126	0.94	0.03701	34	1.33858	74	2.91338
0.15	0.00591	0.55	0.02165	0.95	0.03740	35	1.37795	75	2.95275
0.16	0.00630	0.56	0.02205	0.96	0.03780	36	1.41732	76	2.99212
0.17	0.00669	0.57	0.02244	0.97	0.03819	37	1.45669	77	3.03149
0.18	0.00709	0.58	0.02283	0.98	0.03858	38	1.49606	78	3.07086
0.19	0.00748	0.59	0.02323	0.99	0.03898	39	1.53543	79	3.11023
0.20	0.00787	0.60	0.02362	1.00	0.03937	40	1.57480	80	3.14960
0.21	0.00827	0.61	0.02402	1	0.03937	41	1.61417	81	3.18897
0.22	0.00866	0.62	0.02441	2	0.07874	42	1.65354	82	3.22834
0.23	0.00906	0.63	0.02480	3	0.11811	43	1.69291	83	3.26771
0.24	0.00945	0.64	0.02520	4	0.15748	44	1.73228	84	3.30708
0.25	0.00984	0.65	0.02559	5	0.19685	45	1.77165	85	3.34645
0.26	0.01024	0.66	0.02598	6	0.23622	46	1.81102	86	3.38582
0.27	0.01063	0.67	0.02638	7	0.27559	47	1.85039	87	3.42519
0.28	0.01102	0.68	0.02677	8	0.31496	48	1.88976	88	3.46456
0.29	0.01142	0.69	0.02717	9	0.35433	49	1.92913	89	3.50393
0.30	0.01181	0.70	0.02756	10	0.39370	50	1.96850	90	3.54330
0.31	0.01220	0.71	0.02795	11	0.43307	51	2.00787	91	3.58267
0.32	0.01260	0.72	0.02835	12	0.47244	52	2.04724	92	3.62204
0.33	0.01299	0.73	0.02874	13	0.51181	53	2.08661	93	3.66141
0.34	0.01339	0.74	0.02913	14	0.55118	54	2.12598	94	3.70078
0.35	0.01378	0.75	0.02953	15	0.59055	55	2.16535	95	3.74015
0.36	0.01417	0.76	0.02992	16	0.62992	56	2.20472	96	3.77952
0.37	0.01457	0.77	0.03032	17	0.66929	57	2.24409	97	3.81889
0.38	0.01496	0.78	0.03071	18	0.70866	58	2.28346	98	3.85826
0.39	0.01535	0.79	0.03110	19	0.74803	59	2.32283	99	3.89763
0.40	0.01575	0.80	0.03150	20	0.78740	60	2.36220	100	3.93700

For converting millimeters in "thousandths," move decimal point in both columns to left.

Source: L. S. Starrett Co.

Table A-8 Fractional and Number Drill Sizes, Decimal Equivalents, and Tap Sizes

Fraction or Drill Size	Decimal Equivalent	Tap Size	Fraction or Drill Size	Decimal Equivalent	Tap Size	Fraction or Drill Size	Decimal Equivalent	Tap Size
Number size drills 80	0.0135		26	0.1470		25/64	0.3906	7/16–20
79	0.0145		25	0.1495	10–24	X	0.3970	
1/64	0.0156		24	0.1520		Y	0.4040	
78	0.0160		23	0.1540		13/32	0.4062	
77	0.0180		5/32	0.1562		Z	0.4130	
76	0.0200		22	0.1570	10–30	27/64	0.4219	1/2–13
75	0.0210		21	0.1590	10–32	7/16	0.4375	
74	0.0225		20	0.1610		29/64	0.4531	1/2–20
73	0.0240		19	0.1660		15/32	0.4687	
72	0.0250		18	0.1695		31/64	0.4844	9/16–12
71	0.0260		11/64	0.1719		1/2	0.5000	
70	0.0280		17	0.1730		33/64	0.5156	9/16–18
69	0.0292		16	0.1770	12–24	17/32	0.5312	5/8–11
68	0.0310		15	0.1800		35/64	0.5469	
1/32	0.0312		14	0.1820	12–28	9/16	0.5625	
67	0.0320		13	0.1850	12–32	37/64	0.5781	5/8–18
66	0.0330		3/16	0.1875		19/32	0.5937	11/16–11
65	0.0350		12	0.1890		39/64	0.6094	
64	0.0360		11	0.1910		5/8	0.6250	11/16–16
63	0.0370		10	0.1935		41/64	0.6406	
62	0.0380		9	0.1960		21/32	0.6562	3/4–10
61	0.0390		8	0.1990		43/64	0.6719	
60	0.0400		7	0.2010	1/4–20	11/16	0.6875	3/4–16
59	0.0410		13/64	0.2031		45/64	0.7031	
58	0.0420		6	0.2040		23/32	0.7187	
57	0.0430		5	0.2055		47/64	0.7344	
56	0.0465		4	0.2090		3/4	0.7500	
3/64	0.0469	0–80	3	0.2130	1/4–28	49/64	0.7656	7/8–9
55	0.0520		7/32	0.2187		25/32	0.7812	
54	0.0550	1–56	Letter size drills 2	0.2210		51/64	0.7969	
53	0.0595	1–64, 72	1	0.2280		13/16	0.8125	7/8–14
1/16	0.0625		A	0.2340		53/64	0.8281	
52	0.0635		15/64	0.2344		27/32	0.8437	
51	0.0670		B	0.2380		55/64	0.8594	
50	0.0700	2–56, 64	C	0.2420		7/8	0.8750	1–8
49	0.0730		D	0.2460		57/64	0.8906	
48	0.0760		1/4 — E	0.2500		29/32	0.9062	
5/64	0.0781		F	0.2570	5/16–18	59/64	0.9219	
47	0.0785	3–48	G	0.2610		15/16	0.9375	1–12, 14
46	0.0810		17/64	0.2656		61/64	0.9531	
45	0.0820	3–56, 4–32	H	0.2660		31/32	0.9687	
44	0.0860	4–36	I	0.2720	5/16–24	63/64	0.9844	1 1/8–7
43	0.0890	4–40	J	0.2770		1	1.0000	
42	0.0935	4–48	K	0.2810		1 3/64	1.0469	1 1/8–12
3/32	0.0937		9/32	0.2812		1 7/64	1.1093	1 1/4–7
41	0.0960		L	0.2900		1 1/8	1.1250	
40	0.0980		M	0.2950		1 11/64	1.1719	1 1/4–12
39	0.0995		19/64	0.2968		1 7/32	1.2187	1 3/8–6
38	0.1015	5–40	N	0.3020		1 1/4	1.2500	
37	0.1040	5–44	5/16	0.3125	3/8–16	1 19/64	1.2968	1 3/8–12
36	0.1065	6–32	O	0.3160		1 11/32	1.3437	1 1/2–6
7/64	0.1093		P	0.3230		1 3/8	1.3750	
35	0.1100		21/64	0.3281		1 27/64	1.4219	1 1/2–12
34	0.1110	6–36	Q	0.3320	3/8–24	1 1/2	1.5000	
33	0.1130	6–40	R	0.3390				
32	0.1160		11/32	0.3437				
31	0.1200		S	0.3480				
1/8	0.1250		T	0.3580				
30	0.1285		23/64	0.3594				
29	0.1360	8–32, 36	U	0.3680	7/16–14			
28	0.1405	8–40	3/8	0.3750				
9/64	0.1406		V	0.3770				
27	0.1440		W	0.3860				

Pipe Thread Sizes

Thread	Drill	Thread	Drill
1/8–27	R	1 1/2–11 1/2	1 47/64
1/4–18	7/16	2–11 1/2	2 7/32
3/8–18	37/64	2 1/2–8	2 5/8
1/2–14	23/32	3–8	3 1/4
3/4–14	59/64	3 1/2–8	3 3/4
1–11 1/2	1 5/32	4–8	4 1/4
1 1/4–11 1/2	1 1/2		

Source: L. S. Starrett Co.

Table A-9 Twist Drill Number Sizes in Inches, Millimeters, and Centimeters

Number	Inches	Millimeters	Centimeters	Number	Inches	Millimeters	Centimeters
1	0.2280	5.7912	0.5791	41	0.0960	2.4384	0.2438
2	0.2210	5.6134	0.5613	42	0.0935	2.3622	0.2362
3	0.2130	5.4102	0.5410	43	0.0890	2.2606	0.2261
4	0.2090	5.3086	0.5309	44	0.0860	2.1844	0.2184
5	0.2055	5.2070	0.5207	45	0.0820	2.0828	0.2083
6	0.2040	5.1816	0.5182	46	0.0810	2.0574	0.2057
7	0.2010	5.1054	0.5105	47	0.0785	1.9812	0.1981
8	0.1990	5.0800	0.5080	48	0.0760	1.9304	0.1930
9	0.1960	4.9784	0.4978	49	0.0730	1.8542	0.1854
10	0.1935	4.9022	0.4902	50	0.0700	1.7780	0.1778
11	0.1910	4.8514	0.4851	51	0.0670	1.7018	0.1702
12	0.1890	4.8006	0.4801	52	0.0635	1.6256	0.1626
13	0.1850	4.6990	0.4699	53	0.0595	1.4986	0.1499
14	0.1820	4.6228	0.4623	54	0.0550	1.3970	0.1397
15	0.1800	4.5720	0.4572	55	0.0520	1.3208	0.1321
16	0.1770	4.4958	0.4496	56	0.0465	1.1684	0.1168
17	0.1730	4.3942	0.4394	57	0.0430	1.0922	0.1092
18	0.1695	4.2926	0.4292	58	0.0420	1.0668	0.1067
19	0.1660	4.2164	0.4216	59	0.0410	1.0414	0.1041
20	0.1610	4.0894	0.4089	60	0.0400	1.0160	0.1016
21	0.1590	4.0386	0.4039	61	0.0390	0.9906	0.0991
22	0.1570	3.9878	0.3988	62	0.0380	0.9652	0.0965
23	0.1540	3.9116	0.3912	63	0.0370	0.9398	0.0940
24	0.1520	3.8608	0.3861	64	0.0360	0.9144	0.0914
25	0.1495	3.7846	0.3785	65	0.0350	0.8890	0.0889
26	0.1470	3.7338	0.3734	66	0.0330	0.8382	0.0838
27	0.1440	3.6576	0.3658	67	0.0320	0.8128	0.0813
28	0.1405	3.5560	0.3556	68	0.0310	0.7874	0.0787
29	0.1360	3.4544	0.3454	69	0.0292	0.7366	0.0737
30	0.1285	3.2512	0.3251	70	0.0280	0.7112	0.0711
31	0.1200	3.0480	0.3048	71	0.0260	0.6604	0.0660
32	0.1160	2.9464	0.2946	72	0.0250	0.6350	0.0635
33	0.1130	2.8702	0.2870	73	0.0240	0.6096	0.0610
34	0.1110	2.8194	0.2819	74	0.0225	0.5588	0.0559
35	0.1100	2.7940	0.2794	75	0.0210	0.5334	0.0533
36	0.1065	2.6924	0.2692	76	0.0200	0.5080	0.0508
37	0.1040	2.6416	0.2642	77	0.0180	0.4572	0.0457
38	0.1015	2.5654	0.2565	78	0.0160	0.4064	0.0406
39	0.0995	2.5146	0.2515	79	0.0145	0.3556	0.0356
40	0.0980	2.4892	0.2489	80	0.0135	0.3302	0.0330

Table A-10 Standard Pipe Data

Nominal Pipe Diameter in Inches	Actual Inside Diameter in Inches	Actual Outside Diameter in Inches	Weight per Foot-Pounds	Nominal Pipe Diameter in Inches	Actual Inside Diameter in Inches	Actual Outside Diameter in Inches	Weight per Foot-Pounds
1/8	0.269	0.405	0.244	2½	2.469	2.875	5.793
1/4	0.364	0.540	0.424	3	3.068	3.500	7.575
3/8	0.493	0.675	0.567	3½	3.548	4.000	9.109
1/2	0.622	0.840	0.850	4	4.026	4.500	10.790
3/4	0.824	1.050	1.130	4½	4.560	5.000	12.538
1	1.049	1.315	1.678	5	5.047	5.563	14.617
1¼	1.380	1.660	2.272	6	6.065	6.625	18.974
1½	1.610	1.900	2.717	8	7.981	8.625	28.554
2	2.067	2.375	3.652	10	10.020	10.750	40.483

Table A-11 Commercial Pipe Sizes and Wall Thicknesses

The following table lists the pipe sizes and wall thicknesses currently established as standard, or specifically:

1. The traditional standard weight, extra strong, and double extra strong pipe.
2. The pipe wall thickness schedules listed in American Standard B36.10, which are applicable to carbon steel and alloys *other than* stainless steels.
3. The pipe wall thickness schedules listed in American Standard B36.19, and ASTM Specification A409, which are applicable *only* to corrosion-resistant materials. (NOTE: Schedule 10S is also available in carbon steel in sizes 12″ and smaller.)

ASA-B36.10 and B36.19

Nominal Wall Thickness for change

Nominal Pipe Size	Outside Diameter	Schedule 5S*	Schedule 10S*	Schedule 10	Schedule 20	Schedule 30	Standard†	Schedule 40	Schedule 60	Extra Strong†	Schedule 80	Schedule 100	Schedule 120	Schedule 140	Schedule 160	XX Strong
⅛	0.405	—	0.049	—	—	—	0.068	0.068	—	0.095	0.095	—	—	—	—	—
¼	0.540	—	0.065	—	—	—	0.088	0.088	—	0.119	0.119	—	—	—	—	—
⅜	0.675	—	0.065	—	—	—	0.091	0.091	—	0.126	0.126	—	—	—	—	—
½	0.840	0.065	0.083	—	—	—	0.109	0.109	—	0.147	0.147	—	—	—	0.188	0.294
¾	1.050	0.065	0.083	—	—	—	0.113	0.113	—	0.154	0.154	—	—	—	0.219	0.308
1	1.315	0.065	0.109	—	—	—	0.133	0.133	—	0.179	0.179	—	—	—	0.250	0.358
1¼	1.660	0.065	0.109	—	—	—	0.140	0.140	—	0.191	0.191	—	—	—	0.250	0.382
1½	1.900	0.065	0.109	—	—	—	0.145	0.145	—	0.200	0.200	—	—	—	0.281	0.400
2	2.375	0.065	0.109	—	—	—	0.154	0.154	—	0.218	0.218	—	—	—	0.344	0.436
2½	2.875	0.083	0.120	—	—	—	0.203	0.203	—	0.276	0.276	—	—	—	0.375	0.552
3	3.5	0.083	0.120	—	—	—	0.216	0.216	—	0.300	0.300	—	—	—	0.438	0.600
3½	4.0	0.083	0.120	—	—	—	0.226	0.226	—	0.318	0.318	—	—	—	—	—
4	4.5	0.083	0.120	—	—	—	0.237	0.237	—	0.337	0.337	—	0.438	—	0.531	0.674
5	5.563	0.109	0.134	—	—	—	0.258	0.258	—	0.375	0.375	—	0.500	—	0.625	0.750
6	6.625	0.109	0.134	—	—	—	0.280	0.280	—	0.432	0.432	—	0.562	—	0.719	0.864
8	8.625	0.109	0.148	—	0.250	0.277	0.322	0.322	0.406	0.500	0.500	0.594	0.719	0.812	0.906	0.875
10	10.75	0.134	0.165	—	0.250	0.307	0.365	0.365	0.500	0.500	0.594	0.719	0.844	1.000	1.125	1.000
12	12.75	0.156	0.180	—	0.250	0.330	0.375	0.406	0.562	0.500	0.688	0.844	1.000	1.125	1.312	1.000
14 O.D.	14.0	0.156	0.188	0.250	0.312	0.375	0.375	0.438	0.594	0.500	0.750	0.938	1.094	1.250	1.406	—
16 O.D.	16.0	0.165	0.188	0.250	0.312	0.375	0.375	0.500	0.656	0.500	0.844	1.031	1.219	1.438	1.594	—
18 O.D.	18.0	0.165	0.188	0.250	0.312	0.438	0.375	0.562	0.750	0.500	0.938	1.156	1.375	1.562	1.781	—
20 O.D.	20.0	0.188	0.218	0.250	0.375	0.500	0.375	0.594	0.812	0.500	1.031	1.281	1.500	1.750	1.969	—
22 O.D.	22.0	0.188	0.218	0.250	0.375	0.500	0.375	—	0.875	0.500	1.125	1.375	1.625	1.875	2.125	—
24 O.D.	24.0	0.218	0.250	0.250	0.375	0.562	0.375	0.688	0.969	0.500	1.218	1.531	1.812	2.062	2.344	—
26 O.D.	26.0	—	—	0.312	0.500	—	0.375	—	—	0.500	—	—	—	—	—	—
28 O.D.	28.0	—	—	0.312	0.500	0.625	0.375	—	—	0.500	—	—	—	—	—	—
30 O.D.	30.0	0.250	0.312	0.312	0.500	0.625	0.375	—	—	0.500	—	—	—	—	—	—
32 O.D.	32.0	—	—	0.312	0.500	0.625	0.375	0.688	—	0.500	0.688	—	—	—	—	—
34 O.D.	34.0	—	—	0.312	0.500	0.625	0.375	0.688	—	0.500	0.688	—	—	—	—	—
36 O.D.	36.0	—	—	0.312	0.500	0.625	0.375	0.750	—	0.500	0.750	—	—	—	—	—
42 O.D.	42.0	—	—	—	—	—	0.375	—	—	0.500	—	—	—	—	—	—

All dimensions are given in inches.

The decimal thicknesses listed for the respective pipe sizes represent their nominal or average wall dimensions. The actual thicknesses may be as much as 12.5% under the nominal thickness because of mill tolerance. Thicknesses shown in lightface for Schedule 60 and heavier pipe are not currently supplied by the mills, unless a certain minimum tonnage is ordered.

*Schedules 5S and 10S are available in corrosion resistant materials and Schedule 10S is also available in carbon steel.   †Thicknesses shown in italics are available also in stainless steel, under the designation Schedule 40S.

‡Thicknesses shown in italics are available also in stainless steel, under the designation Schedule 80S.

Source: American Standards Association

Table A-12 Melting Points of Metals and Alloys of Practical Importance

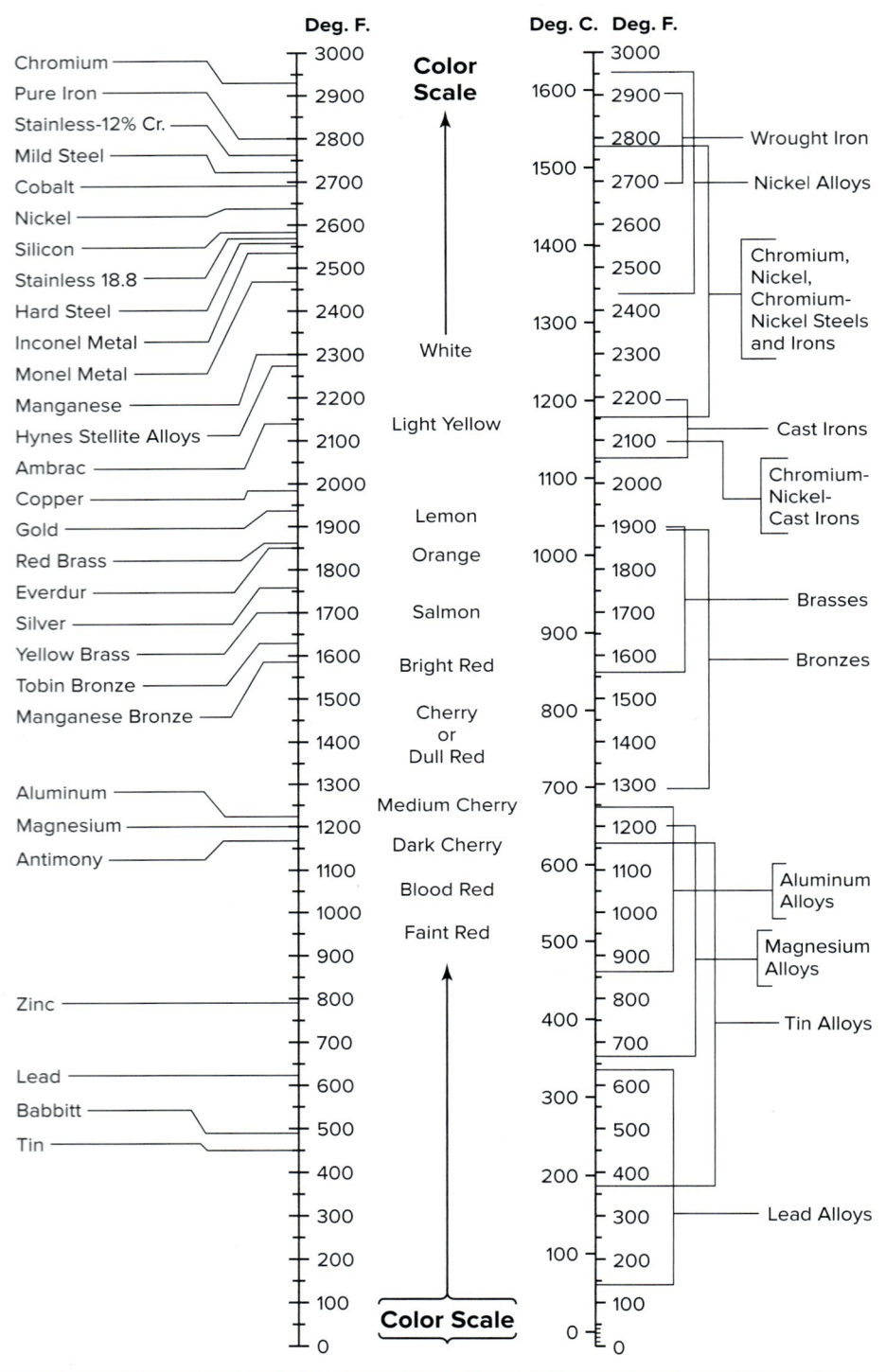

Source: Linde Division, Union Carbide Corp.

Table A-13 Properties of Elements and Metal Compositions

Elements	Symbol	Density (Specific Gravity)	Weight per cu. ft.	Specific Heat	Melting Point °C	Melting Point °F
Aluminum	Al	2.7	166.7	0.212	658.7	1217.7
Antimony	Sb	6.69	418.3	0.049	630	1166
Armco iron	...	7.9	490.0	0.115	1535	2795
Carbon	C	2.34	219.1	0.113	3600	6512
Chromium	Cr	6.92	431.9	0.104	1615	3034
Copper	Cu	8.89	555.6	0.092	1083	1981.4
Gold	Au	19.33	1205.0	0.032	1063	1946
Hydrogen	H	0.070*	0.00533	...	−259	−434.2
Iridium	Ir	22.42	1400.0	0.032	2300	4172
Iron	Fe	7.865	490.9	0.115	1530	2786
Lead	Pb	11.37	708.5	0.030	327	621
Manganese	Mn	7.4	463.2	0.111	1260	2300
Mercury	Hg	13.55	848.84	0.033	−38.7	−37.6
Nickel	Ni	8.80	555.6	0.109	1452	2645.6
Niobium	Nb	7.06	452.54	...	1700	3124
Nitrogen	N	0.97*	0.063	...	−210	−346
Oxygen	O	1.10*	0.0866	...	−218	−360
Phosphorus	P	1.83	146.1	0.19	44	111.2
Platinum	Pt	21.45	1336.0	0.032	1755	3191
Potassium	K	0.87	54.3	0.170	62.3	144.1
Silicon	Si	2.49	131.1	0.175	1420	2588
Silver	Ag	10.5	655.5	0.055	960.5	1761
Sodium	Na	0.971	60.6	0.253	97.5	207.5
Sulfur	S	1.95	128.0	0.173	119.2	246
Tin	Sn	7.30	455.7	0.054	231.9	449.5
Titanium	Ti	5.3	218.5	0.110	1795	3263
Tungsten	W	17.5	1186.0	0.034	3000	5432
Uranium	U	18.7	1167.0	0.028		
Vanadium	V	6.0	343.3	0.115	1720	3128
Zinc	Zn	7.19	443.2	0.093	419	786.2
Bronze (90 percent Cu 10 percent Sn)	...	8.78	548.0	...	850–1000	1562–1832
Brass (90 percent Cu 10 percent Zn)	...	8.60	540.0	...	1020–1030	1868–1886
Brass (70 percent Cu 30 percent Zn)	...	8.44	527.0	...	900–940	1652–1724
Cast pig iron	...	7.1	443.2	...	1100–1250	2012–2282
Open-hearth steel	...	7.8	486.9	...	1350–1530	2462–2786
Wrought-iron bars	...	7.8	486.9	...	1530	2786

*Density compared with air.

Source: Linde Division, Union Carbide Corp.

Table A-14 Identifying Metals by Spark Testing

Spark tests should be made on a high speed power grinder, and the specimen should be held so that the sparks will be given off horizontally. For most accurate results, the sparks should be examined against a dark background, preferably in a dark corner of the shop.

The color, shape, average length, and activity of the sparks are details that are characteristics of the material tested. Spark testing can be a very accurate method of identifying metals but it requires considerable practice and experience to become an expert. Several common sparks are given in the table. If the operators learn the technique for identifying these metals readily, they will soon be able to expand their experience to include others by observation and comparison with the sparks from known samples.

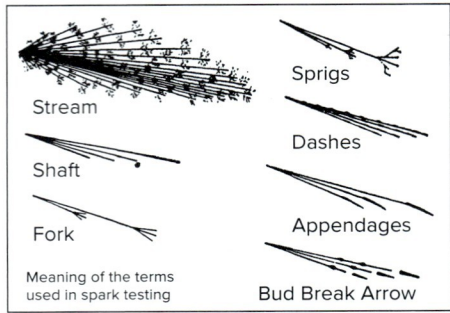

Meaning of the terms used in spark testing: Stream, Shaft, Fork, Sprigs, Dashes, Appendages, Bud Break Arrow

Wrought Iron	Low Carbon Steel*	High Carbon Steel	Alloy Steel**
Color—Straw Yellow. Average stream length with power grinder—65 in. Volume—Large. Long shafts ending in forks and arrowlike appendages. Color—White	Color—White. Average length of stream with power grinder—70 in. Volume—Moderately large. Shafts shorter than wrought iron and in forks and appendages. Forks become more numerous and sprigs appear as carbon content increases	Color—White. Average stream length with power grinder—55 in. Volume—Large. Numerous small and repeating sprigs	Color—Straw Yellow. Stream length varies with type and amount of alloy content. Color—White. Shafts may end in forks, buds or arrows, frequently with break between shaft and arrow; few, if any, sprigs

White Cast Iron	Gray Cast Iron	Malleable Iron	Nickel***
Color—Red. Color—Straw Yellow. Average stream length with power grinder—20 in. Volume—Very small. Sprigs—finer than gray iron, small and repeating	Color—Red. Color—Straw Yellow. Average stream length with power grinder—25 in. Volume—Small. Many sprigs, small and repeating	Color—Straw Yellow. Average stream length with power grinder—30 in. Volume—Moderate. Longer shafts than gray iron ending in numerous small, repeating sprigs	Color—Orange. Average stream length with power grinder—10 in. Short shafts with no forks or sprigs

* These data apply also to cast steel.
** Spark shown is for stainless steel.
*** Monel metal spark is very similar to nickel.

Appendix B

Illustrated Guide to Welding Terminology

Arc seam weld, Fig. B-59 1081	**Double-J groove weld,** Fig. B-17 1075
Arc spot weld, Fig. B-60 1081	**Double-U groove weld,** Fig. B-16............................ 1075
Backhand welding, Fig. B-70 1083	**Double-V groove weld,** Fig. B-14............................ 1075
Backing welds, Fig. B-72 1083	**Double-welded joint,** Fig. B-74 1083
Backstep sequence, Fig. B-53 1080	**Drag and kerf,** Fig. B-57 1080
Back welds, Fig. B-73 1083	**Dross,** Fig. B-57 1080
Base metal zone, heat-affected zone, and weld metal zone, Fig. B-78 1084	**Edge weld,** Fig. B-4 1074
Bead, Fig. B-21 1076	**Edge preparation,** Fig. B-31 1077
Bevel angle, Fig. B-37 1077	**Fillet weld,** Fig. B-20 1075
Block sequence, Fig. B-54 1080	**Fillet weld profile,** Fig. B-81 1085
Brazed or solder joint, Fig. B-83 1085	**Fillet weld size,** Fig. B-46 1079
Butt joint, Fig. B-2 1074	**Flanged butt joint edge weld,** Fig. B-62 1081
Cascade sequence, Fig. B-56 1080	**Flanged corner joint edge weld,** Fig. B-61 1081
Chain intermittent fillet welds, Fig. B-24 1076	**Flash weld,** Fig. B-29 1077
Concavity, Fig. B-46 1079	**Forehand welding,** Fig. B-71 1083
Convexity, Fig. B-42 1078	**Fusion,** Fig. B-48 1079
Corner joint, Fig. B-3 1074	**Groove,** Fig. B-32 1077
Direct Current Electrode Positive (DCEP) reverse polarity is an outdated term, Fig. B-76 1084	**Groove angle,** Fig. B-38 1077
Direct Current Electrode Negative (DCEN) straight polarity is an outdated term, Fig. B-77 ... 1084	**Groove radius,** Fig. B-35 1077
Depth of fusion, Fig. B-48 1079	**Groove weld profiles,** Fig. B-82 1085
Double-bevel groove weld, Fig. B-15...................... 1075	**Heat Affected Zone (HAZ),** Fig. B-78..................... 1084
Double-flare bevel groove weld, Fig. B-18............. 1075	**Incomplete fusion and incomplete penetration,** Fig. B-75 1083
Double-flare V-groove weld, Fig. B-19 1075	**Kerf,** Fig. B-57 1080
	Lap joint, Fig. B-5 1074

Layers, Fig. B-50 1079
Leg of fillet weld, Fig. B-44 1078
Melt-through welds, Fig. B-64 1081
Multipass sequence, Fig. B-55 1080
Nugget size, Fig. B-65 1081
Parallel joint, Fig. B-4 1074
Passes, Fig. B-49 1079
Plug weld, Fig. B-22 1076
Positions for welding for groove welds, Fig. B-1A 1073
Positions of pipe during welding, Fig. B-69 1082
Positions of welding for fillet welds, Fig. B-1B 1073
Projection welds, Fig. B-27 1076
Reinforcement, Fig. B-42 1078
Root edge, Fig. B-33 1077
Root face, Fig. B-34 1077
Root opening, Fig. B-36 1077
Root penetration and joint penetration, Fig. B-47 1079
Seam weld, Fig. B-28 1077
Single-bevel groove weld, Fig. B-9 1075
Single-flare bevel groove weld, Fig. B-12 1075
Single-flare V-groove weld, Fig. B-13 1075
Single-J groove weld, Fig. B-11 1075
Single-U groove weld, Fig. B-10 1075
Single-V groove weld, Fig. B-8 1075
Single welded joint, Fig. B-67 1082
Slot weld, Fig. B-23 1076
Socket joint for fillet welding or brazing, Fig. B-79 1084
Spot weld, Fig. B-26 1076
Square groove weld, Fig. B-7 1075
Staggered intermittent fillet welds, Fig. B-25 1076
Stringers beads, Fig. B-52 1080
Surface or bead weld, Fig. B-21 1076
Surfacing welds, Fig. B-68 1082
T-joint, Fig. B-6 1074
Temper bead welding (TBW), Fig. B-80 1084
Types of groove welds, Fig. B-63 1081
Types of resistance welds, Fig. B-66 1082
Under cut, Fig. B-43 1078
Undercut and overlap, Fig. B-58 1080
Upset weld, Fig. B-30 1077
Weaving, Fig. B-51 1080
Weld face, Fig. B-40 1078
Weld root, Fig. B-39 1078
Weld throat, Fig. B-45 1078
Weld toe, Fig. B-41 1078

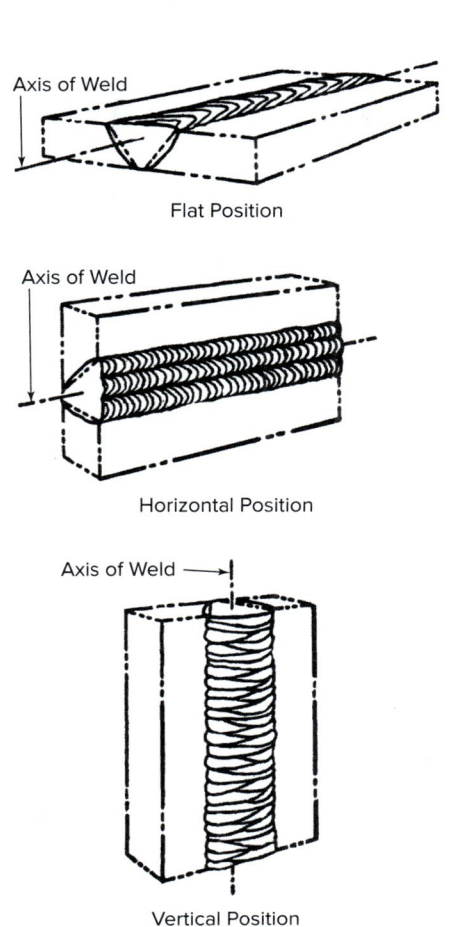
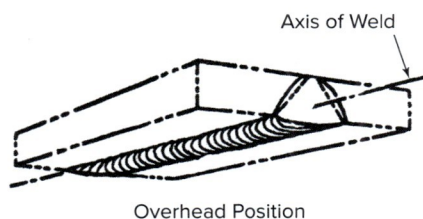

Fig. B-1A Positions for welding for groove welds.

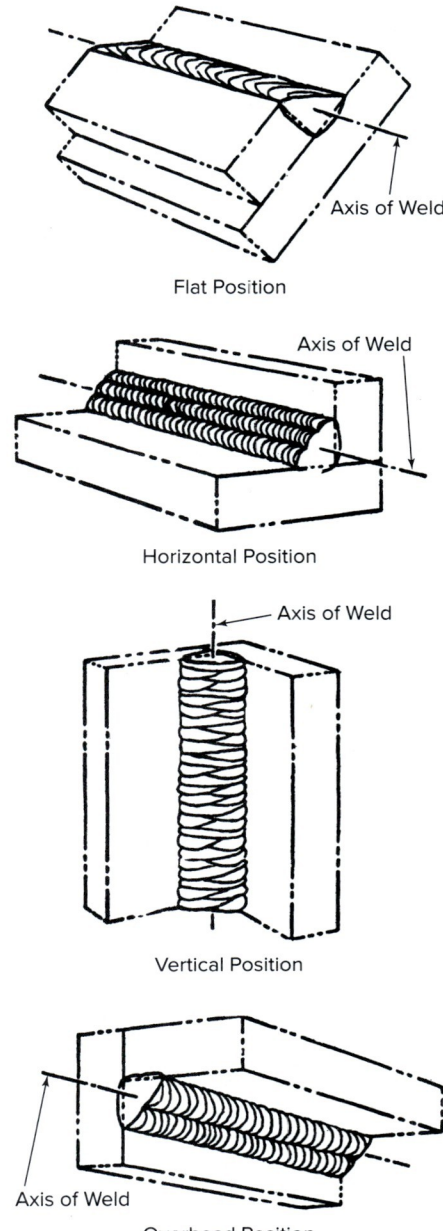

Fig. B-1B Positions of welding for fillet welds.

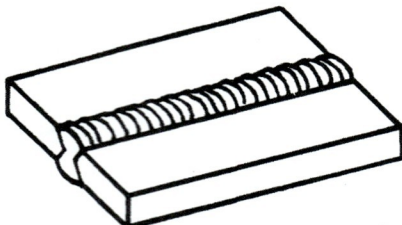

Types of Welds Applicable to Butt Joints

Single Square Groove
Double Square Groove
Single-V Groove
Double-V Groove (Illustrated)
Single-Bevel Groove
Double-Bevel Groove
Single-U Groove
Double-U Groove
Single-J Groove
Double-J Groove
Single-Flare Bevel Groove
Double-Flare Bevel Groove
Braze

Fig. B-2 Butt joint.

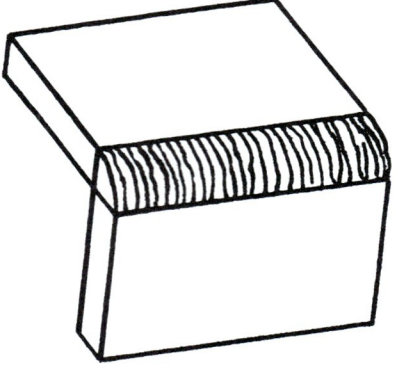

Types of Welds Applicable to Corner Joints

Fillet (Illustrated)
Square Groove
Single-V Groove
Single-Bevel Groove
Double-Bevel Groove
Single-U Groove
Single-J Groove
Double-J Groove
Projection (Resistance)
Single-Flare Bevel Groove
Double-Flare Bevel Groove
Slot
Spot
Seam
Plug
Braze

Fig. B-3 Corner joint.

Types of Welds Applicable to Parallel Joints

Edge (Illustrated)
V-Groove
U-Groove
Flare Bevel Groove
J-Groove
Square Groove
Seam
Spot
Projection
Braze

Fig. B-4 Parallel joint.

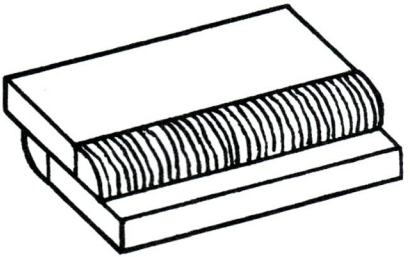

Types of Welds Applicable to Lap Joints

Fillet (Illustrated)
Plug
Slot
Spot (Resistance)
Seam (Resistance)
Projection (Resistance)
Bevel Groove
Flare Bevel Groove
J-Groove
Braze

Fig. B-5 Lap joint.

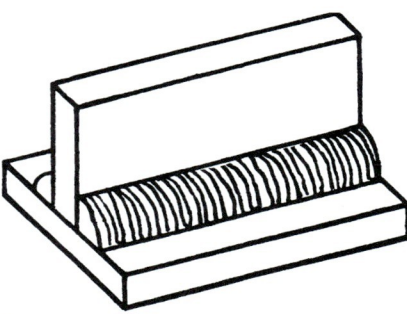

Types of Welds Applicable to T Joints

Fillet (Illustrated)
Single-Bevel Groove
Double-Bevel Groove
Single-J Groove
Double-J Groove
Single-Flare Bevel Groove
Double-Flare Bevel Groove
Single Square Groove
Double Square Groove
Plug
Slot
Spot Seam
Projection
Braze

Fig. B-6 T-joint.

Fig. B-7 Square groove weld.

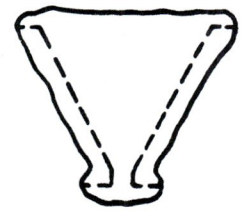

Fig. B-8 Single-V groove weld.

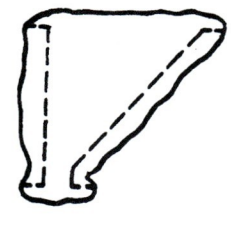

Fig. B-9 Single-bevel groove weld.

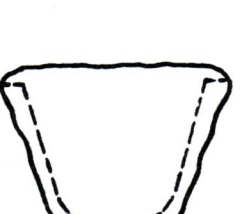

Fig. B-10 Single-U groove weld.

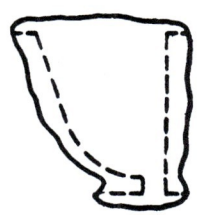

Fig. B-11 Single-J groove weld.

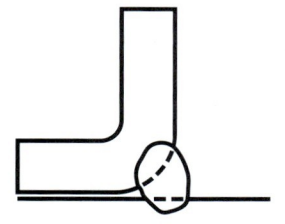

Fig. B-12 Single-flare bevel groove weld.

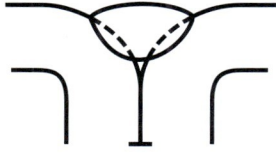

Fig. B-13 Single-flare V-groove weld.

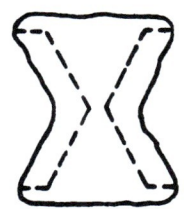

Fig. B-14 Double-V groove weld.

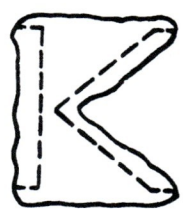

Fig. B-15 Double-bevel groove weld.

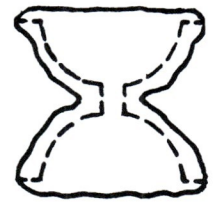

Fig. B-16 Double-U groove weld.

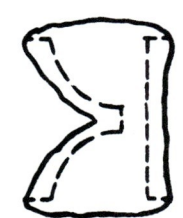

Fig. B-17 Double-J groove weld.

Fig. B-18 Double-flare bevel groove weld.

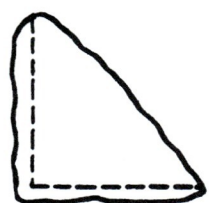

Fig. B-19 Double-flare V-groove weld.

Fig. B-20 Fillet weld.

Illustrated Guide to Welding Terminology **Appendix B**

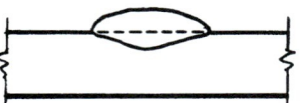

Fig. B-21 Surface or bead weld.

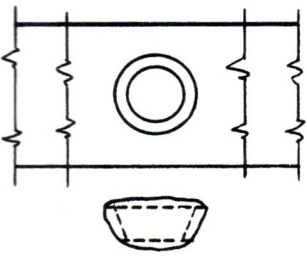

Fig. B-22 Plug weld.

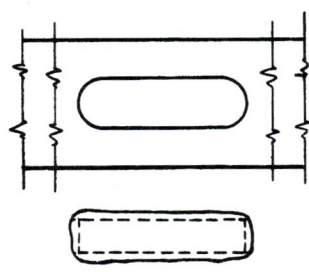

Fig. B-23 Slot weld.

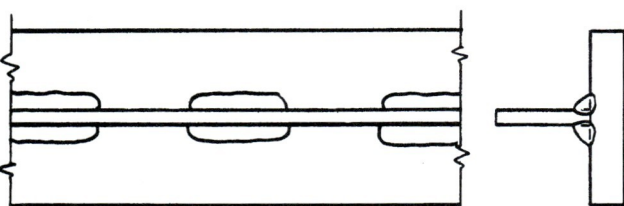

Fig. B-24 Chain intermittent fillet welds.

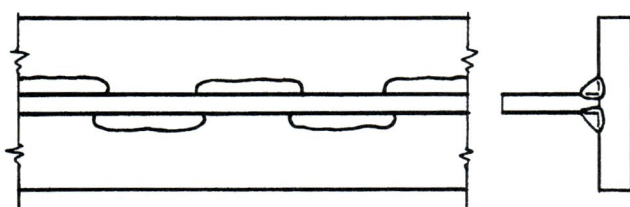

Fig. B-25 Staggered intermittent fillet welds.

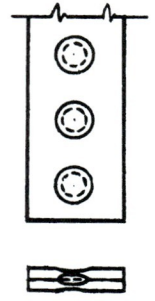

Fig. B-26 Spot weld.

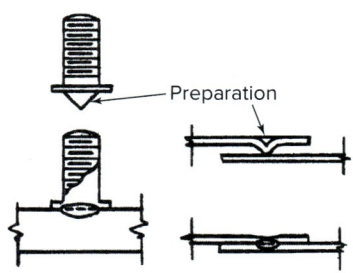

Fig. B-27 Projection welds.

1076 **Appendix B** Illustrated Guide to Welding Terminology

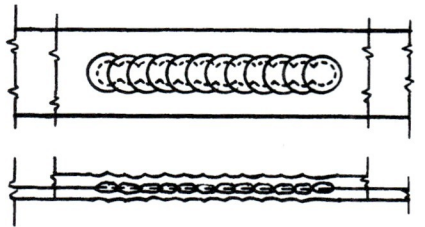
Fig. B-28 Seam weld.

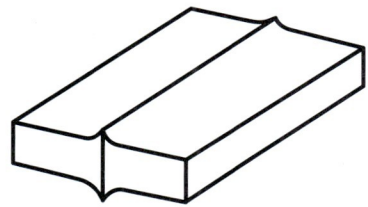
Fig. B-29 Flash weld.

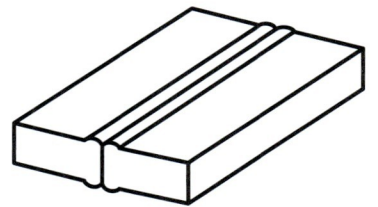
Fig. B-30 Upset weld.

Fig. B-31 Edge preparation.

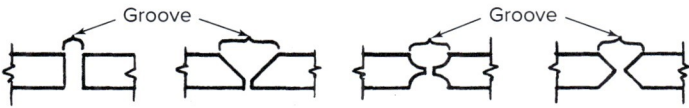
Fig. B-32 Groove.

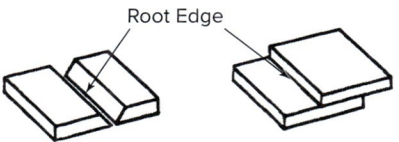
Fig. B-33 Root edge.

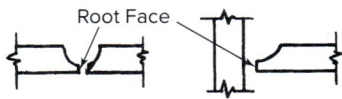
Fig. B-34 Root face.

Fig. B-35 Groove radius.

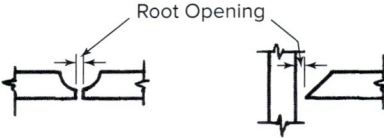
Fig. B-36 Root opening.

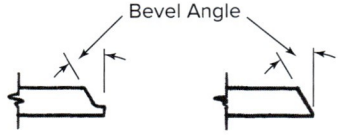
Fig. B-37 Bevel angle.

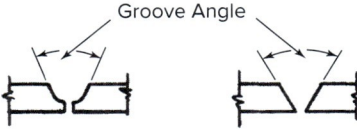
Fig. B-38 Groove angle.

Illustrated Guide to Welding Terminology **Appendix B** 1077

Fig. B-39 Weld root.

Fig. B-40 Weld face.

Fig. B-41 Weld toe.

Fig. B-42 Convexity and reinforcement.

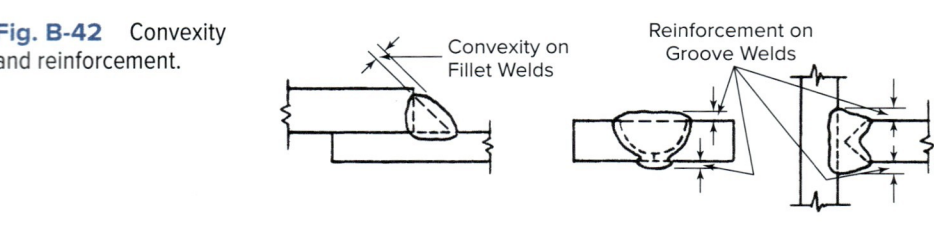

Fig. B-43 Under cut.

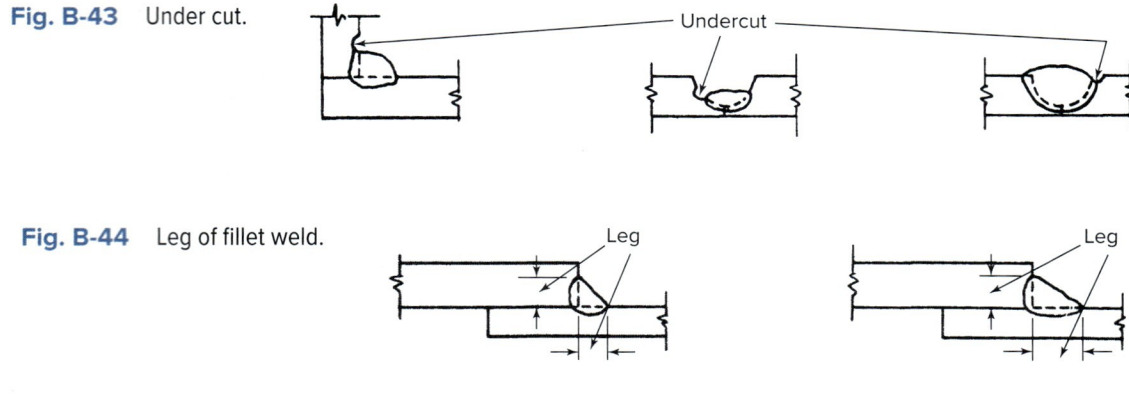

Fig. B-44 Leg of fillet weld.

Fig. B-45 Weld throat.

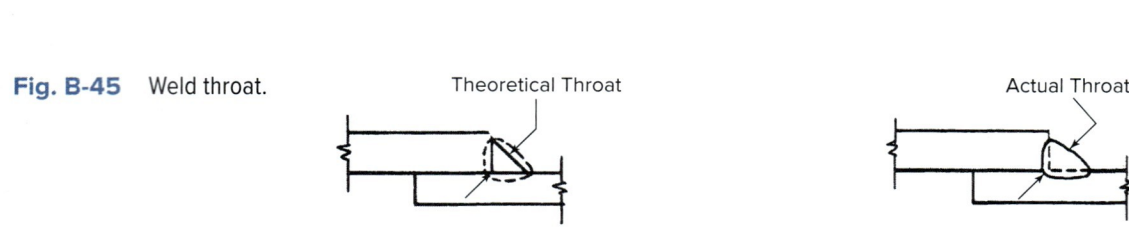

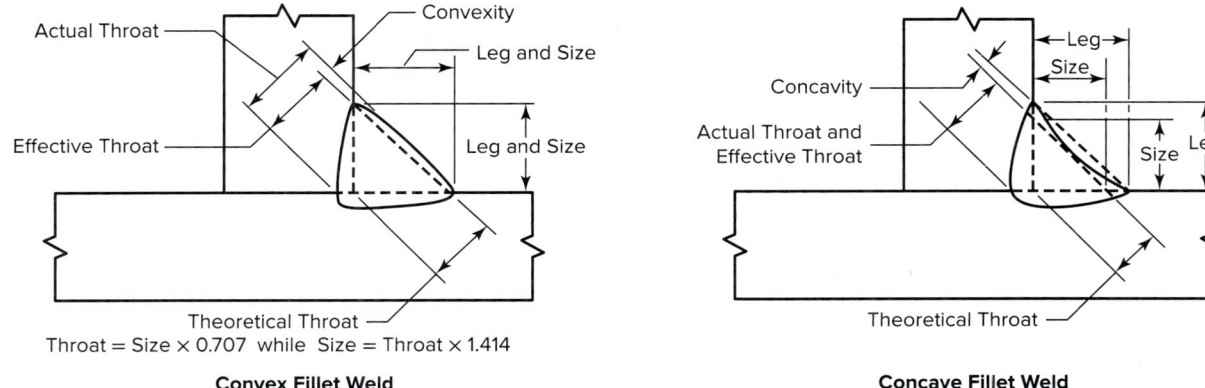

Note: The size of a fillet weld is the length of the largest inscribed right isosceles triangle. While the fillet weld leg is the distance from the joint root to the toe of the fillet weld.

Fig. B-46 Fillet weld size.

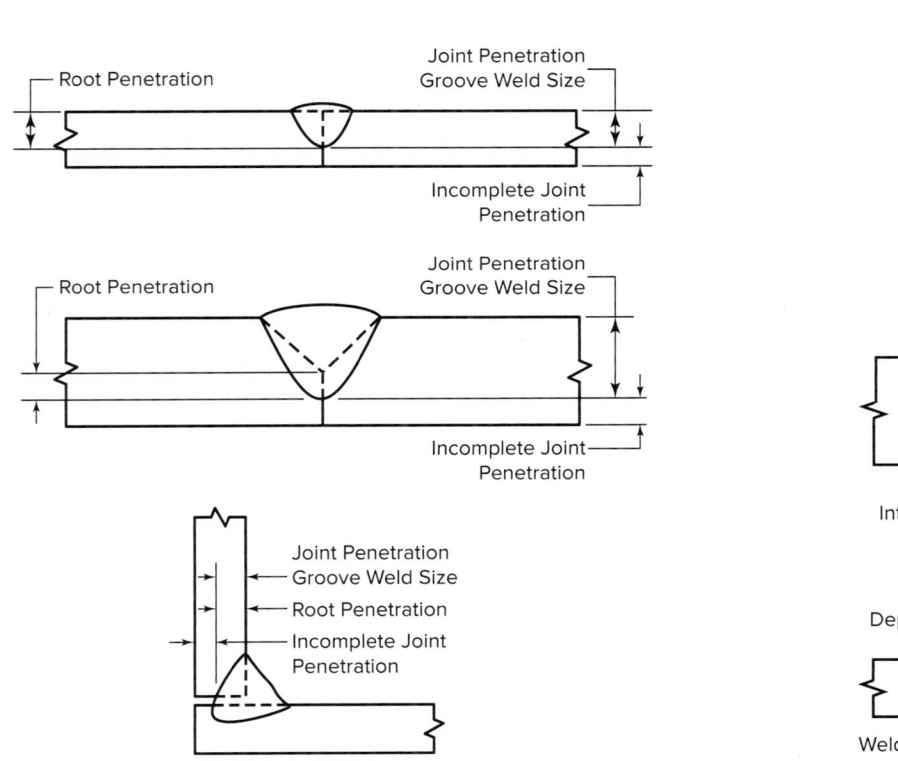

Fig. B-47 Root penetration and joint penetration.

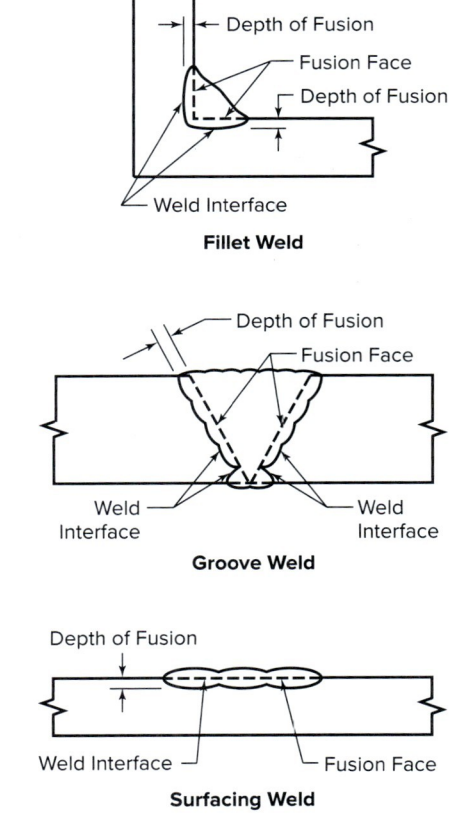

Fig. B-48 Depth of fusion.

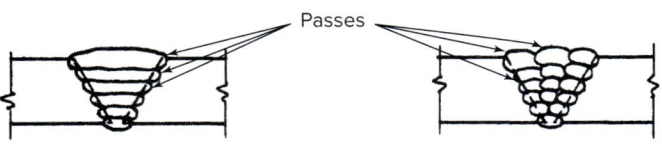

Fig. B-49 Passes.

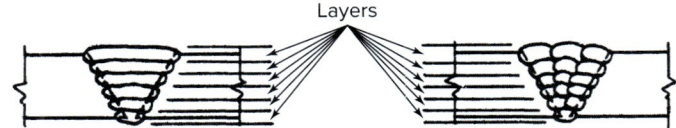

Fig. B-50 Layers.

Illustrated Guide to Welding Terminology   **Appendix B**   1079

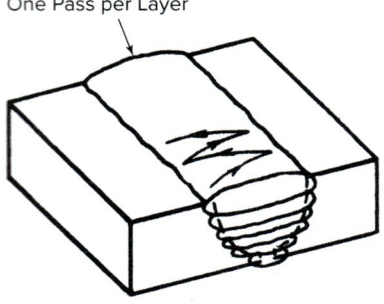

Fig. B-51 Weaving.

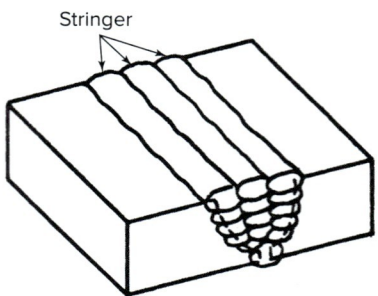

Fig. B-52 Stringer beads.

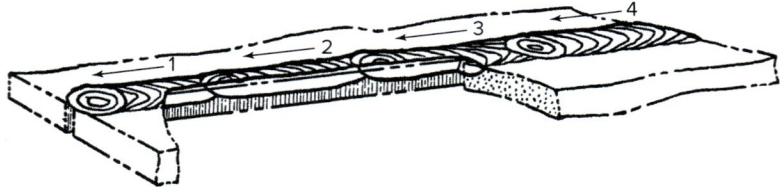

Fig. B-53 Backstep sequence.

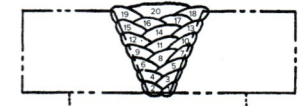

Fig. B-55 Multipass sequence.

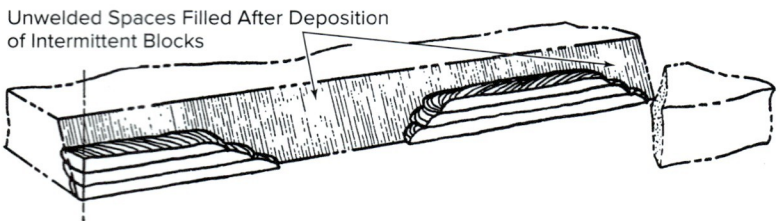

Fig. B-54 Block sequence.

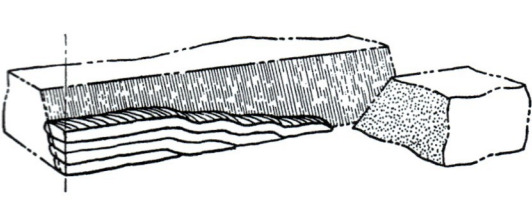

Fig. B-56 Cascade sequence.

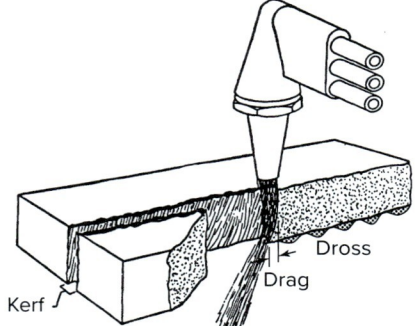

Fig. B-57 Drag, kerf, and dross.

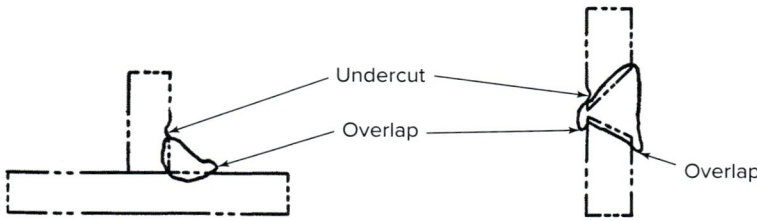

Fig. B-58 Undercut and overlap.

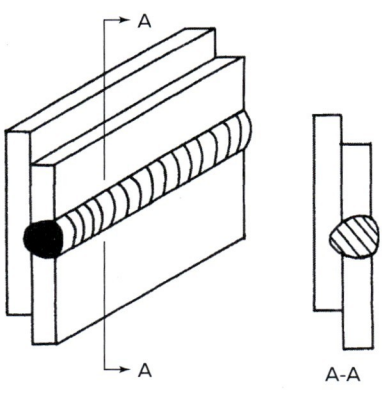

Fig. B-59 Arc seam weld.

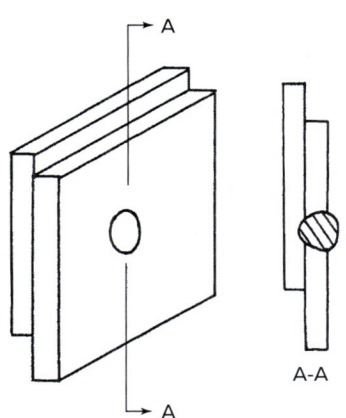

Fig. B-60 Arc spot weld.

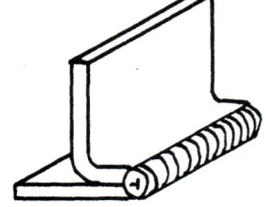

Fig. B-61 Flanged corner joint edge weld.

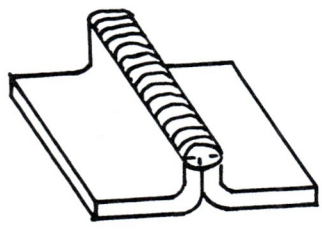

Fig. B-62 Flanged butt joint edge weld.

Flare-Bevel Groove Weld

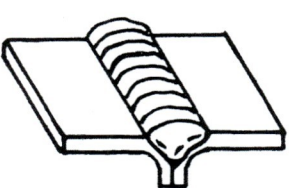

Flare-Vee Groove Weld

Fig. B-63 Types of groove welds.

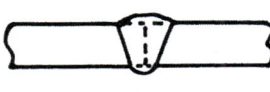

A

C

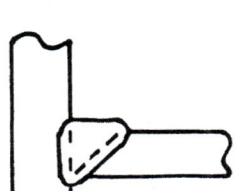

B

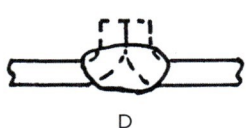

D

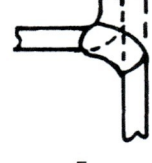

E

Fig. B-64 Melt-through welds.

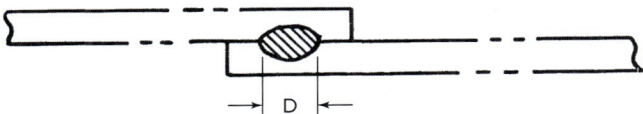

Fig. B-65 Nugget size.

Illustrated Guide to Welding Terminology **Appendix B** 1081

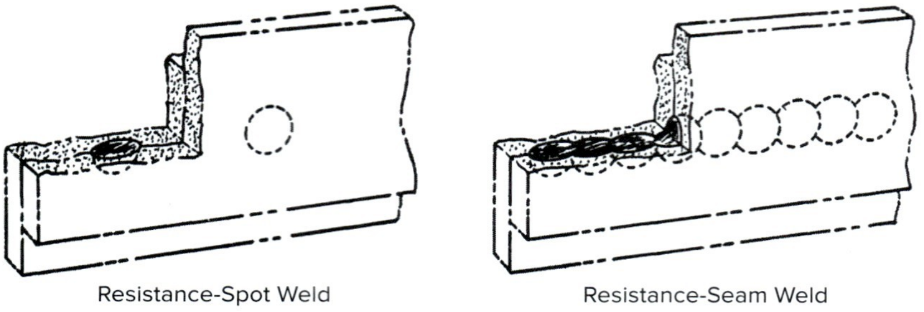

Fig. B-66 Types of resistance welds.

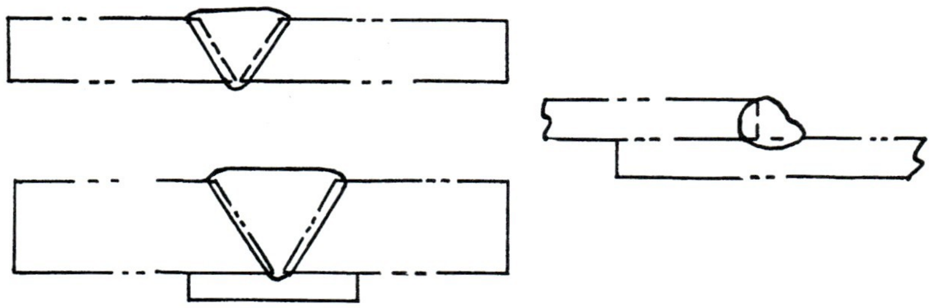

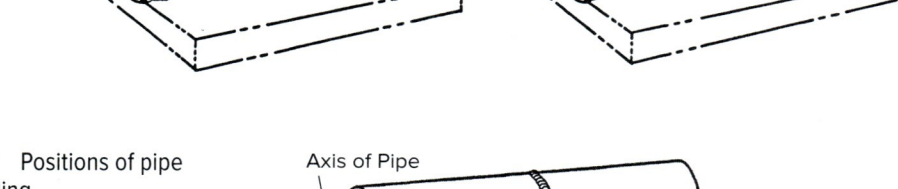

Fig. B-67 Single welded joints.

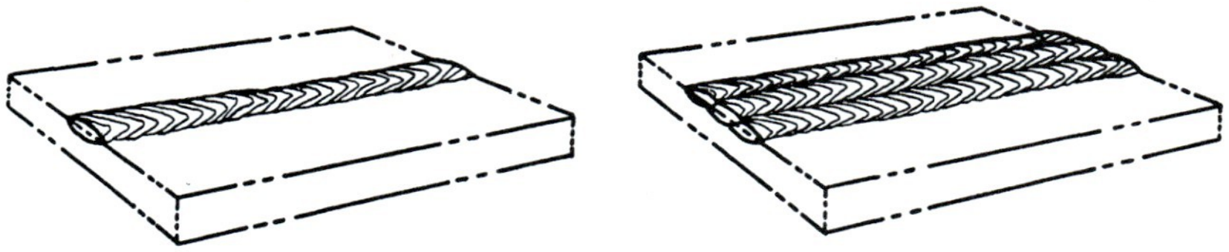

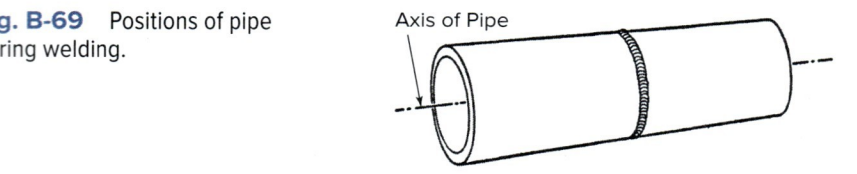

Fig. B-68 Surfacing welds.

Fig. B-69 Positions of pipe during welding.

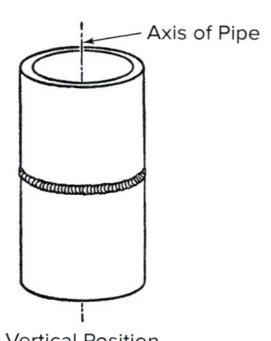

Horizontal Fixed Position (Pipe Stationary During Welding)
Horizontal Rolled Position (Pipe Rotated During Welding)

Vertical Position

1082 Appendix B Illustrated Guide to Welding Terminology

Fig. B-70 Backhand welding.

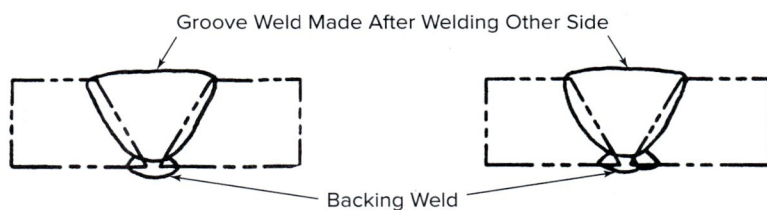

Fig. B-72 Backing welds.

Fig. B-71 Forehand welding.

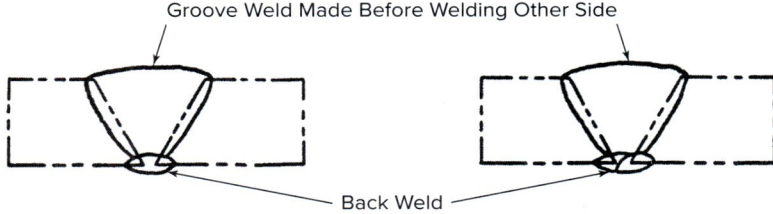

Fig. B-73 Back welds.

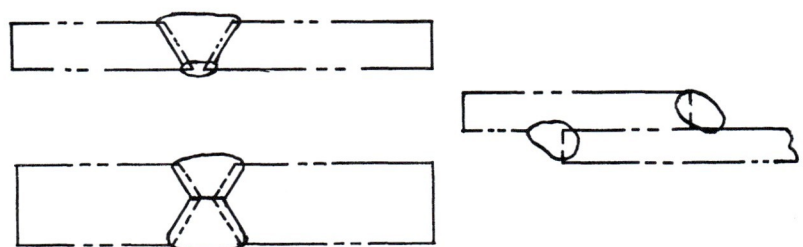

Fig. B-74 Double-welded joint.

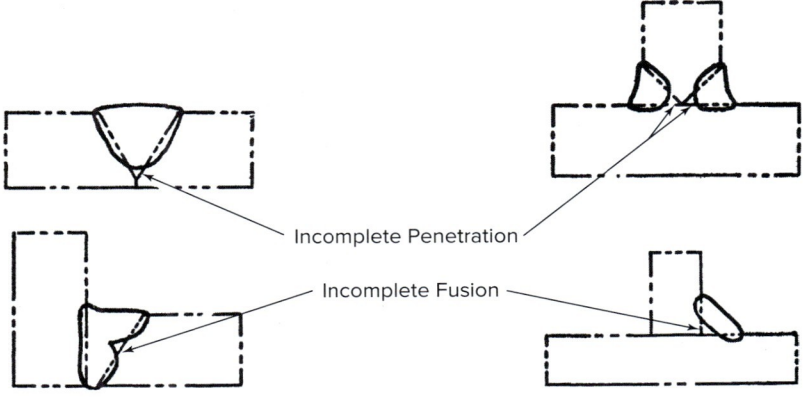

Fig. B-75 Incomplete fusion and incomplete penetration.

Illustrated Guide to Welding Terminology **Appendix B**

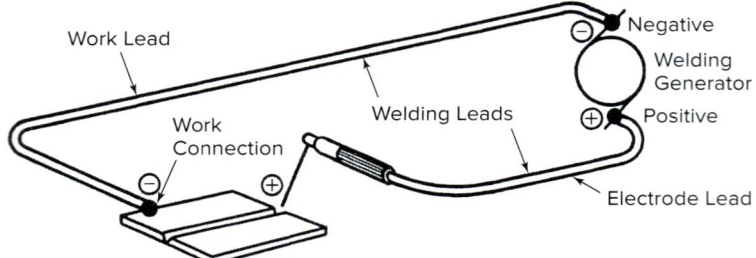

Fig. B-76 Direct Current Electrode Positive (DCEP) reverse polarity is an outdated term.

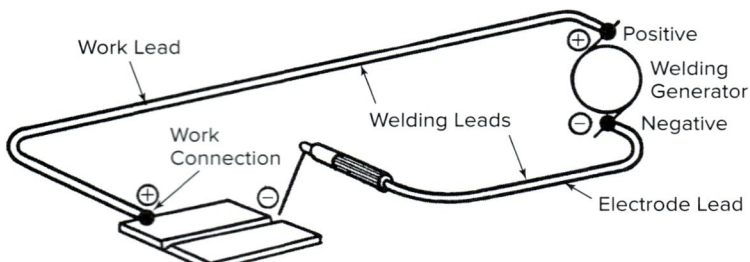

Fig. B-77 Direct Current Electrode Negative (DCEN) straight polarity is an outdated term.

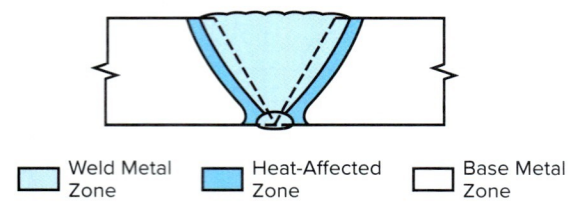

Fig. B-78 Base metal zone, heat-affected zone, and weld metal zone.

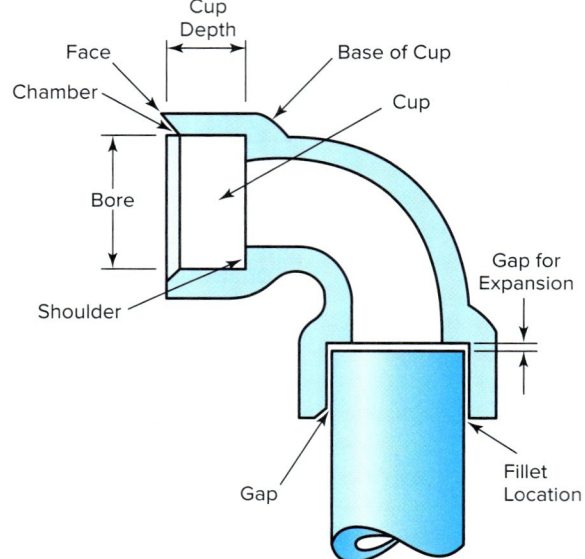

Fig. B-79 Socket joint for fillet welding or brazing.

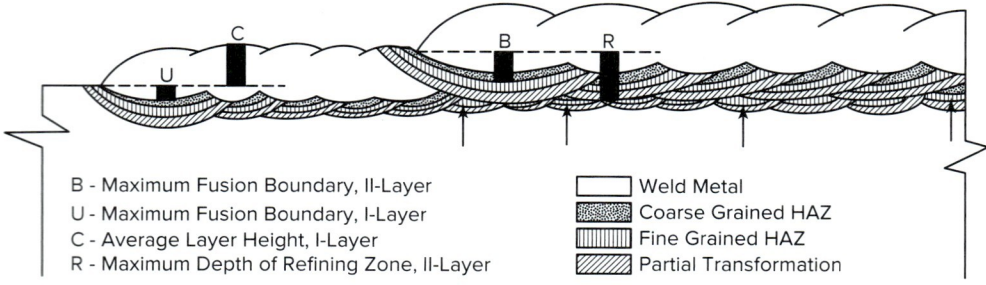

B - Maximum Fusion Boundary, II-Layer
U - Maximum Fusion Boundary, I-Layer
C - Average Layer Height, I-Layer
R - Maximum Depth of Refining Zone, II-Layer

Weld Metal
Coarse Grained HAZ
Fine Grained HAZ
Partial Transformation

Fig. B-80 Temper bead welding (TBW).

1084 Appendix B Illustrated Guide to Welding Terminology

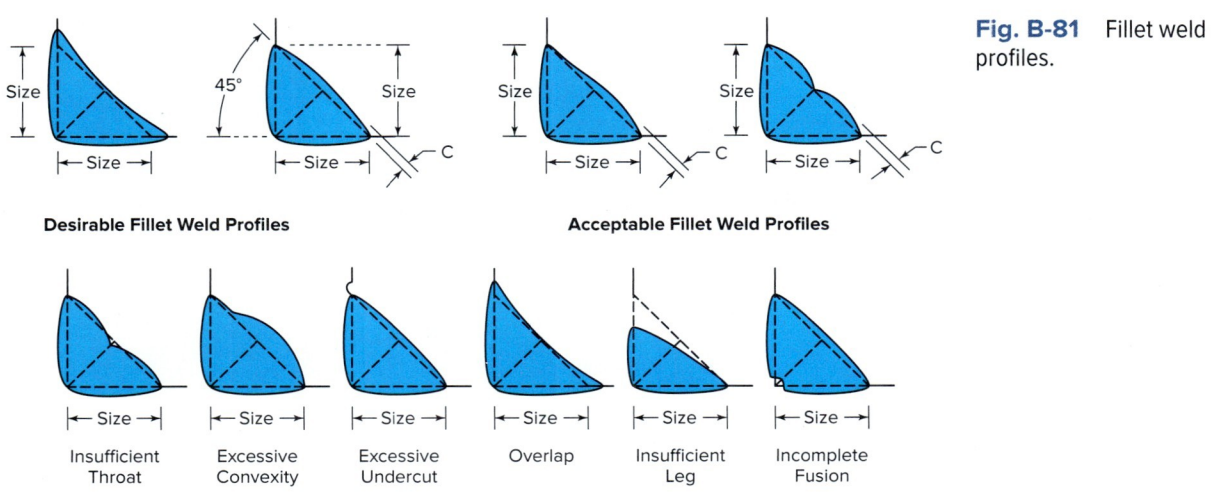

Fig. B-81 Fillet weld profiles.

Note: Convexity, C, of a weld or individual surface bead shall not exceed 0.07 times the actual face width of the weld or individual bead, respectively, plus 0.06 in. (1.5 mm).

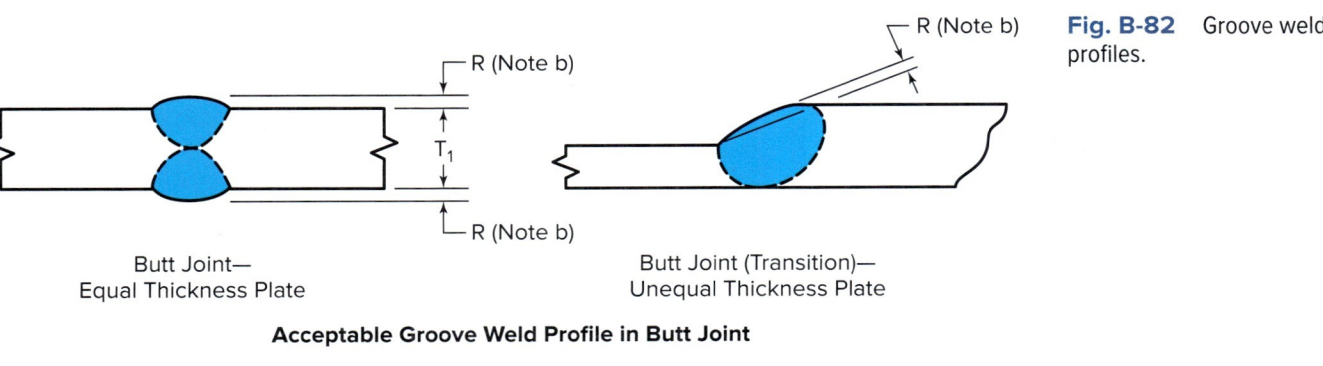

Fig. B-82 Groove weld profiles.

bReinforcement R shall not exceed 1/8 in [3 mm]

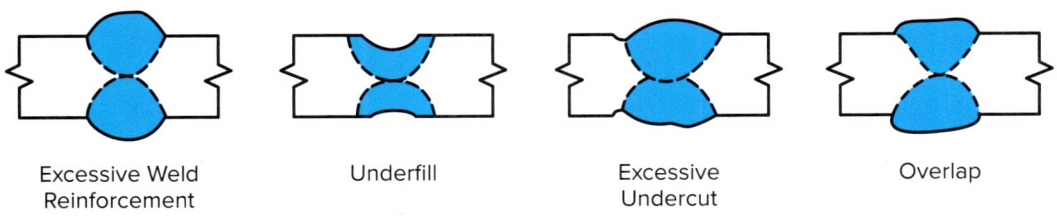

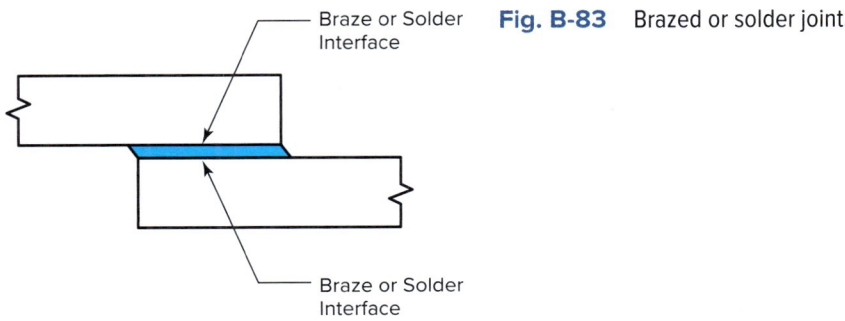

Fig. B-83 Brazed or solder joint.

Appendix C

Welding Abbreviation List

These terms are American Welding Society Standards

Welding Processes

Oxyfuel Gas Welding (OFW)

air acetylene welding	AAW
oxyacetylene welding	OAW
oxyhydrogen welding	OHW
pressure gas welding	PGW

Arc Welding (AW)

arc stud welding	SW
atomic hydrogen welding	AHW
bare metal arc welding	BMAW
carbon arc welding	CAW
—gas	CAW-G
—shielded	CAW-S
—twin	CAW-T
electrogas welding	EGW
flux cored arc welding	FCAW
—gas-shielded	FCAW-G
—self-shielded	FCAW-S
gas metal arc welding	GMAW
—pulsed arc	GMAW-P
—short circuiting arc	GMAW-S
gas tungsten arc welding	GTAW
—pulsed arc	GTAW-P
magnetically impelled arc welding	MIAW
plasma arc welding	PAW
shielded metal arc welding	SMAW
submerged arc welding	SAW
—series	SAW-S

Brazing (B)

block brazing	BB
Carbon arc brazing	CAB
diffusion brazing	DFB
dip brazing	DB
Electron beam brazing	EBB
exothermic brazing	EXB
furnace brazing	FB
induction brazing	IB
infrared brazing	IRB
Laser beam brazing	LBW
resistance brazing	RB
torch brazing	TB
twin carbon arc brazing	TCAB

Soldering (S)

dip soldering	DS
furnace soldering	FS
induction soldering	IS
infrared soldering	IRS

iron soldering. .... INS
resistance soldering .... RS
torch soldering. .... TS
ultrasonic soldering. .... USS
pressure gas soldering .... PGS
wave soldering .... WS

Solid State Welding (SSW)

coextrusion welding .... CEW
cold welding .... CW
diffusion welding. .... DFW
hot isolated pressure welding .... HIPW
explosion welding. .... EXW
forge welding.. FOW
friction welding... FRW
    —direct drive friction welding. .... FRW-DD
    —friction stir welding. .... FSW
    —inertia friction welding .... FRW-I
hot pressure welding .... HPW
roll welding. .... ROW
ultrasonic welding .... USW

Resistance Welding (RW)

flash welding... FW
pressure-controlled resistance welding .... RW-PC
projection welding.. PW
resistance seam welding.. RSEW
    —high-frequency seam welding.................. RSEW-HF
    —induction seam welding. .... RSEW-I
    —mash seam welding. .... RSEW-MS
resistance spot welding. .... RSW
upset welding .... UW
    —high-frequency. .... UW-HF
    —induction. .... UW-I

Other Welding and Joining

adhesive bonding.. AB
braze welding .... BW
electron beam welding .... EBW
electroslag welding. .... ESW
flow welding .... FLOW
induction welding. .... IW
laser beam welding... LBW
percussion welding. .... PEW
thermit welding. .... TW

Allied Processes

Thermal Cutting (TC)
Oxygen Cutting (OC)

flux cutting.. OC-F
metal powder cutting.. OC-P
oxygen arc cutting .... OAC
oxyfuel gas cutting .... OFC
    —oxyacetylene cutting. .... OFC-A
    —oxyhydrogen cutting. .... OFC-H
    —oxynatural gas cutting. .... OFC-N
    —oxypropane cutting. .... OFC-P
oxygen gouging. .... OG
oxygen lance cutting. .... OLC

Arc Cutting (AC)

carbon arc cutting... CAC
air carbon arc cutting. .... CAC-A
gas metal arc cutting. .... GMAC
gas tungsten arc cutting.. GTAC
plasma arc cutting. .... PAC
shielded metal arc cutting .... SMAC

High Energy Beam Cutting

electron beam cutting .... EBC
laser beam cutting. .... LBC
    —air.. LBC-A
    —evaporative... LBC-EV
    —inert gas. .... LBC-IG
    —oxygen... LBC-O

Thermal Spraying (THSP)

arc spraying.. ASP
flame spraying... FLSP
    —wire flame spraying .... FLSP-W
high velocity oxyfuel spraying. .... HVOF
plasma spraying. .... PSP
vacuum plasma spraying. .... VPSP

Applying Process

Adaptive Control.. AD
Automatic Welding .... AU
Mechanized Welding.. ME
Manual Welding... MA
Robotic .... RO
Semi-Automatic Welding... SA

Appendix D

Major Agencies Issuing Codes, Specifications, and Associations

Amer. Assn. of State Highway & Transportation Officials (AASHTO) www.transportation.org
Washington, DC

American Bureau of Shipping (ABS)
www.eagle.org
Spring, TX

American Institute of Steel Construction (AISC)
www.aisc.org
Chicago, IL

American Iron & Steel Inst. (AISI)
www.steel.org
Washington, DC

American Petroleum Institute (API)
www.api.org
Washington, DC

American Society of Civil Engineers (ASCE)
www.asce.org
Reston, VA

American Gas Association (AGA)
www.aga.org
Washington, DC

The Aluminum Association (AA)
www.aluminum.org
Washington, DC

American Society for Quality (ASQ)
www.asq.org
Milwaukee, WI

American National Standards Institute (ANSI)
www.ansi.org
Washington, DC

Fabricators and Manufacturers' Association (FMA)
www.fmanet.org
Elgin, IL

American Society of Mechanical Engineers (ASME)
www.asme.org
New York, NY

American Society for Testing & Materials (ASTM)
www.astm.org
West Conshehocken, PA

American Water Works Assn. (AWWA)
www.awwa.org
Denver, CO

American Welding Society (AWS)
www.aws.org
Miami, FL.

National Association of Corrosion Engineers (NACE)
www.nace.org
Houston, TX

Society of Automotive Engineers (SAE)
www.sae.org
Warrendale, PA

Association of American Railroads (AAR)
www.aar.org
Washington, DC

Copper Development Association (CDA)
  www.copper.org
New York, NY

National Electrical Mfg. Association (NEMA)
  www.nema.org
Rosslyn, VA

Resistance Welding Mfg. Alliance (RWMA)
  www.rwma.org
Miami, FL 33166
(305) 443-9353 Ext. 295
Contact: Adrian Bustillo or email RWMA

American Society for Nondestructive Testing (ASNT)
  www.asnt.org
Columbus, OH

International Titanium Association (ITA)
  www.titanium.org
Northglenn, CO

Appendix E
Sources of Welding Information

Airco Plating Inc.
www.aircoplating.com
3650 N.W. 46th St.
Miami, FL 33142

American Foundry Society
www.afsnew.org
1695 N Penny LN
Schaumburg, IL 60173

American Welding Society
www.aws.org
8669 NW 36 St., 130
Miami, FL 33166-6672

Ductile Iron Society
www.ductile.org
W175 N 11117
Stonewood Dr. Suite 280
Germantown, WI 53022

Edison Welding Institute (EWI)
www.ewi.org
1250 Arthur E. Adams Dr.
Columbus, OH 43221-3585

ESAB Welding and Cutting Products
www.esabna.com
2800 Airport Rd.
Denton, TX 76207

Hobart Brothers
www.hobartbrothers.com
101 Trade Square East
Troy, OH 45373

Lincoln Electric Co.
www.lincolnelectric.com
22801 St. Clair Ave.
Cleveland, OH 44117

Miller Electric Manufacturing Co.
www.millerwelds.com
1635 W. Spencer St.
Appleton, WI 54914

National Certified Pipe Welding Bureau
www.mcaa.org/ncpwb
1385 Piccard Dr.
Rockville, MD 20850

The Harris Products Group
www.jwharris.com
2345 Murphy Blvd.
Gainesville, GA 30504

Thermadyne Industries
www.thermadyne.com
101 S Hanley Rd
St. Louis, MO 63015

Tescom Corporation
www.tescom.com
12616 Industrial Blvd NW
Elk River, MN 55330

Welding Research Council (WRC)
www.forengineers.org/wrc
20600 Chargrin Blvd
Shake Heights, OH 44122

Appendix F

Metric Conversion Information for the Welding Industry

The American Welding Society's *Metric Practice Guide for the Welding Industry,* AWS A1.1., is based on the International System of Units (SI), as defined in the U.S. Federal Register/Vol. 73, No. 96/Friday, May 16, 2008/ Notices *Interpretation of the International System of Units for the United States.*

American Welding Society
   www.aws.org
8669 NW 36 ST., Suite 130
Miami, FL 33136

Glossary

If students expect to make progress in a new field of training, it is vitally important that they understand the terms that describe the processes and equipment. Very often different terms may have the same meaning. Many terms developed out of expressions used in the field by welders and others working in the occupation.

The following list of terms and their definitions are those standardized and approved by the American Welding Society. Also included are a number of terms from the fields of electricity and metallurgy which are applicable to welding and weld testing. Only those terms that are concerned with the manual and semiautomatic processes are included because they are generally used in the field by welders and must be understood if the student expects to become competent in the craft.

Although these terms are presented in the Glossary for easy reference, students are urged to begin the study of the definitions at the beginning of the training program and to continue the effort for the duration of the training period. Constant reference after entrance into the occupation is also recommended.

Those students who wish a more complete listing of terms, including those for automatic and resistance welding processes, should secure a copy of the AWS publication *Terms and Definitions* (AWS A3.0).

The figure numbers in the Glossary refer to Appendix B, the Illustrated Guide to Welding Terminology, pages 1095–1109.

A

Abrasion: associated with surfaces that are subjected to continuous grinding, rubbing, or gouging action.

Acetone: a liquid chemical having the property of dissolving or absorbing many times its own volume of acetylene.

Acetylene: the most widely used of all the fuel gases, both for welding and cutting. It is generated as the result of the chemical reaction that takes place when calcium carbide comes in contact with water.

Acetylene feather: one of the three zones of the flame that contains white hot carbon particles, some of which are introduced into the weld pool during welding.

Active fluxes: fluxes formulated to produce a weld metal composition that is dependent on the welding parameters, especially arc voltage. Typically used with the submerged arc welding process.

Actual throat: the part of a fillet weld that is the same as the effective throat on a concave fillet weld.

Actual throat (throat of fillet weld): the distance from the root to the face of the weld. See Figs. B-45 and B-46.

Added stiffening: this welding distortion correction method can be used only on plate. It consists of pulling the plate into line with strong backs and welding additional stiffeners to the plate to make it retain its plane.

Admixture: the interchange of filler metal and base metal during welding resulting in weld metal of composition borrowed from both.

Air-hardening: characteristic of a steel that becomes partially or fully hardened (martensitic) when cooled in air from above its critical point. May not apply when the object to be hardened has considerable thickness.

Alloy: a material having metallic characteristics and made up of two or more elements, one of which is a metal.

Alloy fluxes: a flux containing ingredients that react with the filler metal and base metal to establish a desired alloy content in the weld metal. The weld parameters must be closely controlled. Typically used with the submerged arc welding process.

All-weld metal test specimen: a test specimen wherein the portion being tested is composed wholly of weld metal.

Alternating current (a.c.): an electrical current that has positive and negative values alternately. The current flows in one direction during any half cycle, and reverses direction during the next half cycle, causing the alternation of direction of current flow.

Alternating current arc welding: an arc welding process wherein the power supply at the arc is alternating current.

Aluminum-magnesium-chromium alloy: this alloy is strong and highly resistant to corrosion. Good ductility permits the material to be worked.

Aluminum-magnesium-silicon alloy: this alloy is readily welded and can be heat treated.

Aluminum-silicon-magnesium-chromium alloy: this alloy has silicon and magnesium as its main alloys. The welds using this alloy are not as strong as the material being welded, but weld strength can be improved by heat treatment.

Ampere: ampere is another name for current and is the unit of electrical rate measurement.

Angle of bevel: see *Bevel angle*.

Annealing: the process of softening a metal by heating it, usually above the upper critical temperature and then cooling it at a slow rate.

Anode: the positive terminal of a power source; the electrode when using direct current electrode positive.

Arc blow: a condition created by the magnetic field causing a pull or push of the arc resulting in the arc being blown to one side or the other as if by a draft.

Arc cutting: a group of cutting processes wherein the severing of metals is affected by melting with the heat of an arc between an electrode and the base metal.

Arc gouging: an application of arc cutting wherein a chamfer or groove is formed.

Arc seam welding: an arc welding process wherein coalescence at the faying surfaces is produced continuously by heating with an electric arc between an electrode and the work. The weld is made without preparing a hole in either member. Filler metal or a shielding gas or flux may or may not be used. See Fig. B-59.

Arc spot welding: an arc welding process wherein coalescence at the faying surfaces is produced in one spot by heating with an electric arc between an electrode and the work. The weld is made without preparing a hole in either member. Filler metal or a shielding gas or flux may or may not be used. See Fig. B-60.

Arc strike: an indication resulting from an arc, consisting of any localized remelted metal, heat-affected metal, or change in the surface profile of any metal object.

Arc voltage: the voltage generated between the electrode and the work during welding.

Arc welding: a group of welding processes wherein coalescence is produced by heating with an arc or arcs, with or without the application of pressure and with or without the use of filler metal.

Arrow side: the side of the joint the arrow is pointing to on a welding symbol.

Atom: the smallest unit of an element that can take part in a chemical reaction.

Austenite: a solid solution of carbon or iron-carbide in face-centered cubic iron.

Austenitic chromium-nickel steels: includes the basic type [18-8 (types 302 and 304), 18 percent chromium and 8 percent nickel] and should be heat treated if the part is to be put into severe chemical service.

Austenitic stainless steel: a chromium-nickel steel in which the nickel content usually ranges from 3.5 to 22 percent, and the chromium content, from 16 to 26 percent.

Autogenous welding: welding without the addition of filler metal.

Automatic gas cutting: see *Machine oxygen cutting*.

Automatic welding: welding with equipment that automatically controls the entire welding operation (including feed, speed, etc.).

Axis of a weld: a line through the length of a weld, perpendicular to the cross section at its center of gravity, parallel to the root. See Figs. B-1A and B-1B.

B

Backfire: the loud snap or pop that occurs when the flame goes out due to the improper operation of the welding torch or defective equipment.

Backhand welding: a gas welding technique wherein the torch or gun is directed opposite to the progress of the welding. See Fig. B-70.

Backing: material (metal, ceramic, carbon, granular flux, etc.) backing up the joint during welding to facilitate obtaining a sound weld at the root.

Backing bead: see *Backing weld*.

Backing ring: backing in the form of a ring generally used in the welding of pipe.

Backing weld: backing in the form of a weld. See Fig. B-72.

Backstep sequence: a longitudinal sequence wherein the weld bead increments are deposited in the direction opposite to the progress of welding the joint. See Fig. B-53.

Back weld: a weld deposited at the back of a single-groove weld. See Fig. B-73.

Balanced forces: one shrinkage force can be balanced with another by prebending and presetting in a direction opposite to the move.

Balanced welding sequence: an equal number of welders weld on opposite sides of a structure at the same time, thus introducing balanced stresses.

Ball: the process of rounding the end of the electrode before welding aluminum to keep the arc steady.

Base metal (parent metal): the metal to be welded or cut. Often called "parent metal."

Beading (parallel beads): a technique of depositing weld metal without oscillation of the electrode, which generally results in a relatively narrow pass. See Fig. B-52.

Bead welds: welds made by single-pass deposits of weld metal that are used to build up a pad of metal and to replace metal on worn surfaces. See Fig. B-21.

Bend test: see *Free-bend test specimen* and *Guided-bend test*.

Bevel: an angular type of edge preparation.

Bevel angle: the angle formed between the prepared edge of a member and a plane perpendicular to the surface of the member. See Fig. B-37.

Bevel cutting: one of the common operations used in beveling the edges of plate and pipe for welding.

Beveling: a type of chamfering.

Billet: a type of steel ingot that is square or oblong, but considerably smaller than a bloom.

Black iron: untreated flat steel that is hot rolled.

Block sequence: a combined longitudinal and buildup sequence for a continuous multiple-pass weld wherein separated lengths are completely or partially built up in cross section before intervening lengths are deposited. See Fig. B-54.

Bloom: a type of steel ingot that is square or oblong with a minimum cross-sectional area of 36 square inches.

Blowhole: a defect in metal caused by hot metal cooling too rapidly when excessive gaseous content is present. Specifically, in welding, a gas pocket in the weld metal resulting from the hot metal solidifying without all of the gases having escaped to the surface.

Blowouts: unwanted molten metal flying out of a laser beam cut, interrupting the cut path.

Blowpipe: see *Welding torch* or *Cutting torch*.

Bond: a weld that compares favorably with fusion welds where a molecular union is formed between the bronze welding rod and the prepared surface of the work.

Bond line: the junction of the weld metal and the base metal or the junction of the base metal parts when weld metal is not present.

Bore (brazing): the inside diameter of the cup on a fitting. See Fig. B-79.

Bottle: see *Cylinder*.

Brazed joint: a union of two or more members produced by the application of a brazing process.

Braze welding: a method of welding whereby a groove, fillet, plug, or slot weld is made using a nonferrous filler metal having a melting point below that of the base metals but above 800°F. The filler metal is not distributed in the joint by capillary attraction. (*Bronze welding*, formerly used, is a misnomer for this term.)

Brazing: a group of joining processes wherein coalescence is produced by heating to suitable temperatures above 800°F and by using a nonferrous filler metal having a melting point below that of the base metals. The filler metal is distributed between the closely fitted surfaces of the joint by capillary attraction.

Brazing filler metal: a nonferrous metal or alloy to be added in making a braze. Its melting temperature is above 800°F, but below the melting temperatures of the base metals being joined.

Brittleness: the tendency of a material to fail suddenly by breaking without any permanent deformation of the material before failure.

Bronze surfacing: a process in which worn surfaces are built up before they are braze welded.

Bronze welding: a term erroneously used to denote braze welding. See also the preferred term, *Braze welding*.

Buildup sequence: the order in which the weld beads of a multiple-pass weld are deposited with respect to the cross section of the joint. See Fig. B-55.

Burner: see *Oxygen cutter*.

Butt joint: a welded joint between two abutting parts lying in approximately the same plane. See Fig. B-2.

C

Capillary attraction: the phenomenon of surface tension setting up adhesion and cohesion forces which cause molten metals to flow between closely fitted solid surfaces. The term is usually used in connection with brazing processes.

Carbide: the chemical combination of carbon with some other element. A metallic carbide takes the form of very hard crystals.

Carbide precipitation: this precipitation occurs most readily in the 1,200°F heat range. The gas metal arc welding process with its rapid speed and high deposition rate greatly reduces this situation over the slower gas tungsten arc welding or shielded metal arc welding processes.

Carbon arc cutting (CAC): the process of severing metals by melting with the heat of an arc between a carbon electrode and the base metal.

Carbon arc welding (CAW): an arc welding process wherein coalescence is produced by heating with an arc between a carbon electrode and the work, and no shielding is used. Pressure may or may not be used, and filler metal may or may not be used.

Carbon electrode: a non-filler-metal electrode used in arc welding consisting of a carbon or graphite rod.

Carbon-electrode arc welding: a group of arc welding processes wherein carbon electrodes are used. See also *Inert-gas carbon arc welding* and *Carbon arc welding*.

Carburizing flame: the excess fuel flame. It is also called the reducing flame.

Cascade sequence: a combined longitudinal and buildup sequence wherein weld beads are deposited in overlapping layers. See Fig. B-56.

Cathode: the negative terminal of a power source; the electrode when using direct current electrode negative. In an electrolytic cell, the cathode is the source of electrons.

Caulk weld: see preferred term, *Seal weld*.

Cementite: iron carbide (Fe_3C), constituent of steel and cast iron. It is hard and brittle.

Certification: written verification of a welder's or procedure's ability to produce sound welds, usually conducted by an independent third party. Testing must be done to a prescribed standard.

Chain intermittent fillet welds: two lines of intermittent fillet welding on a joint wherein the fillet increments welded in one line are approximately opposite to those in the other line. See Fig. B-24.

Chamfer: see *Edge preparation*.

Chamfering: the preparation of a contour, other than for a square groove weld, on the edge of a member for welding.

Chemical dip brazing: a dip brazing process wherein the filler metal is added to the joint before immersion in a bath of molten chemicals.

Chill ring: see *Backing ring*.

Coalescence: the growing together, or growth into one body, of the base metal parts.

Coarse adjustment dial: in constant current generators with dual continuous control this dial continuously adjusts current. It is also referred to as the job selector or electrode selector by some manufacturers.

Coated electrode: see *Lightly coated electrode*.

Cold shortness: the characteristic tendency of a metal toward brittleness at room temperature or lower.

Commercially pure wrought aluminum: this aluminum is 99 percent pure with just a little iron and silicon added.

Complete fusion: fusion that has occurred over the entire base metal surfaces exposed for welding and between all layers and passes.

Complete joint penetration: see *Joint penetration*.

Composite electrode: a filler metal electrode used in arc welding consisting of more than one metal component combined mechanically. It may or may not include materials that protect the molten metal from the atmosphere, improve the properties of the weld metal, or stabilize the arc.

Composite joint: a joint wherein welding is used in conjunction with a mechanical joining process. (Welding and riveting.)

Composition: the contents of an alloy, in terms of what elements are present and in what amount (by percentage of weight).

Compressive strength: the resistance of a material to a force that is tending to deform or fail it by crushing.

Concave fillet weld: a fillet weld having a concave face. See Fig. B-46.

Concavity: the maximum distance from the face of a concave fillet weld perpendicular to a line joining the toes.

Conduction: the transfer of heat energy through direct contact of the material's solid structure. The process used for the transfer of heat from the product through the shell and through the thickness of the rolls by their direct contact.

Conductor: a conductor is an electrical path. Metals like copper and aluminum that offer little resistance to current flow are considered good conductors.

Cone: the conical part of a gas flame next to the orifice of the tip.

Constant current: a power source with a volt-ampere relationship giving a small current change from a large arc voltage change.

Constant current welding machines: machines that are primarily used for shielded metal arc welding and gas tungsten arc welding because current remains fairly constant regardless of changes in arc length. They are also called *drooping voltage* or *droopers* because the load voltage decreases, or droops, as the welding current increases.

Constant voltage: constant voltage and constant potential are terms applied to welding machines. They indicate the ability of the machine to produce welding current of stable voltage regardless of the amperage output.

Consumable electrode: a metal electrode (filler wire) that establishes the arc and gradually melts away, being carried across the arc (deposited) to provide filler metal into the joint.

Continuous casting: the process by which molten steel is solidified into a semifinished billet, bloom, or slab for subsequent finishing.

Continuous weld: a weld that extends without interruption for its entire length.

Contour: the shape of the face of the weld.

Convection: heat transfer by moving airflow. It is the heat transfer mechanism in a cooling system that occurs by quickly moving sprayed water droplets or mist from spray nozzles.

Convex fillet weld: a fillet weld having a convex face. See Fig. B-46.

Convexity: the maximum distance from the face of a convex fillet weld perpendicular to a line joining the toes. See Fig. B-46.

Corner-flange weld: a flange weld with only one member flanged at the location of welding. See Fig. B-61.

Corner joint: a welded joint between two parts located approximately at right angles to each other in the form of an *L*. See Fig. B-3.

Corrosion: the destruction of a surface from atmospheric chemical contamination and from oxidation or scaling at elevated temperatures.

Covered electrode: a composite filler metal electrode made up of a core wire to which a covering sufficient to provide a slag layer on the weld metal has been applied.

Cover glass (lens): a clear, transparent material used in goggles, hand shields, and helmets to protect the filter glass from spattering material.

CO_2 welding: see preferred term, *Gas metal arc welding*.

Crack propagation: the development, growth, or progress of a crack through a solid.

Crater: in arc welding a depression at the termination of a weld bead, or in the weld pool beneath the electrode.

Crater filling: using a technique like "backstepping," "triggering," or lowering power levels so that the crater meets weld size requirements.

Creep: the slow deformation (for example, elongation) of a metal under prolonged stress. Not to be confused with the deformation that results immediately upon application of a stress.

Cupping process: a process in which heated plate is formed around cup-shaped dies.

Current density: the amperage per square inch of the cross-sectional area of the electrode.

Cutting attachment: a device that is attached to a gas welding torch to convert it into an oxygen cutting torch.

Cutting process: a process wherein the severing or removing of metals is affected.

Cutting tip: a tip in a cutting torch that has a central orifice through which the high pressure oxygen flows.

Cutting torch: this torch mixes oxygen and acetylene or other fuel gases in the proportions necessary for cutting.

Cycle: applied to a.c. current, it refers to the alternating flow of current during a unit of time, $\frac{1}{60}$ second. The cycle is repeated as long as current is flowing in the circuit.

Cylinder (bottle): a portable steel container for storage of a compressed gas.

Cylinder manifold: see *Manifold*.

D

D.C. electrode negative (DCEN): the electrode is connected to the negative terminal of the power source, and the work is connected to the positive terminal.

D.C. electrode positive (DCEP): the electrode is connected to the positive terminal of the power source, and the work is connected to the negative terminal.

Decarburization: the loss of carbon from a ferrous alloy in a reactive atmosphere at high temperatures.

Deoxidizer (degasifier): an element or compound added to the electrode flux, or core wire, to remove oxygen and its derivatives from the weld.

Deoxidizing: the removal of oxygen from the molten weld metal, usually by chemical combination with other elements forming inorganic compounds that float to the surface of the molten metal to form slag when cooled.

Deposited metal: filler metal that has been added during a welding operation.

Deposition rate: the speed with which filler metal is added to a weld joint, usually stated in terms of volume of metal deposited per minute.

Deposition sequence: the order in which the increments of weld metal are deposited.

Deposit sequence: see *Deposition sequence*.

Depth of fusion: the distance that fusion extends into the base metal from the surface melted during welding. See Fig. B-48.

Differential hardening: the characteristic of certain steels to develop high surface or case hardness and to retain a tough, soft core. Any hardening procedure designed to produce only localized or superficial hardness.

Diffusion brazing: a special process that employs a filler metal that diffuses into the base metal under a specific set of conditions of time, temperature, and pressure.

Direct current (d.c.): an electrical current that flows in one direction only and has either a positive or a negative polarity. It differs from a.c. current in that there is no change in the flow of current.

Direct current arc welding: an arc welding process wherein the power source at the arc is direct current.

Distortion: also called shrinkage. This term usually means the overall motion of parts being welded from the position occupied before welding to that occupied after welding.

Distortion effects: the nonuniform expansion and contraction of the weld metal and adjacent base metal during the heating and cooling cycle of the welding process. The kind of welding process used has an influence on distortion.

Double-bevel groove weld: a type of groove weld. See Fig. B-15.

Double-groove weld: see *Groove weld*.

Double-J groove weld: a type of groove weld. See Fig. B-17.

Double-U groove weld: a type of groove weld. See Fig. B-16.

Double-V groove weld: a type of groove weld. See Fig. B-14.

Double-welded butt joint: a butt joint welded from both sides.

Double-welded joint: in arc and gas welding, any joint welded from both sides. See Fig. B-74.

Double-welded lap joint: a lap joint in which the overlapped edges of the members to be joined are welded along the edges of both members.

Downhand: see *Flat position (downhand)*.

Drag: the distance between the point of exit of the cutting oxygen stream and the projection, on the exit surface, of the point of entrance. See Fig. B-57.

Drop forging: a forging process in which a piece of roughly shaped metal is placed between die-shaped faces of the exact form of the finished piece. Many automobile parts are made in this way.

Dross: when doing thermal cutting, the remaining solidified oxidized metallic material adhering to the workpiece adjacent to the cut surface. Often mistakenly called slag.

Ductile: descriptive of material that is easily hammered, bent, or drawn into a new shape or form.

Ductile iron: also referred to as nodular iron. Iron to which amounts of magnesium and/or cerium are added when it is produced which lowers the tensile strength, toughness, and ductility of the iron.

Ductility: the ability of a material to become permanently deformed without failure.

Duplex stainless steel (DDS): a steel that is characterized by a low-carbon body-centered-cubic ferrite, face-centered-cubic austenite microstructure.

Duranickel: a high strength, low alloy nickel.

Duty cycle: expresses as a percentage the portion of time that the power supply must deliver its rated output in each 10-minute interval.

E

Edge-flange weld: a flange weld with two members flanged at the location of welding. See Fig. B-62.

Edge preparation: a prepared contour on the edge of a part to be joined by a groove weld. See Fig. B-31.

Effective throat: the part of a fillet weld that is measured from the depth of the joint root penetration.

Elasticity: the ability of a material to return to original shape and dimensions after a deforming load has been removed.

Elastic limit: the maximum stress to which a material can be subjected without permanent deformation or failure by breaking.

Electrode (arc welding): see *Carbon electrode, Composite electrode, Lightly coated electrode, Flux cored electrode,* and *Tungsten electrode*.

Electrode extension: the length of electrode wire that extends past the contact tube.

Electrode holder: a device used for mechanically holding the electrode and conducting current to it.

Electrode lead: the electrical conductor between the source of arc welding current and the electrode holder.

Electrode negative (EN): see *D.C. electrode negative (DCEN)*.

Electrode positive (EP): see *D.C. electrode positive (DCEP)*.

Electron beam welding (EBW): a welding process that uses a high velocity stream of electrons hitting the joint to produce the heat required for welding.

Electroslag welding (ESW): a welding process that uses the resistance heating effect of the molten conductive slag. A continuous filler metal or electrode is fed into the molten slag which melts the filler metal and surfaces of the workpieces.

Element: a substance that can't be broken down into two other substances. Everything on earth is a combination of such elements, of which there are only 103.

Elongation: the stretching of a material by which any straight line dimension increases.

Endurance limit: the maximum stress that a material will support indefinitely under variable and repetitive load conditions.

Endurance limit of the material: the maximum level of load that can be applied to a material at which no failure will occur, no matter how many cycles the load is applied.

Explosion welding (EXW): a solid-state welding process that makes a weld with a high velocity impact of the workpieces together. This is done by a controlled detonation of explosives.

F

Face of weld: the exposed surface of a weld on the side from which welding was done. See Fig. B-40.

Face reinforcement: reinforcement of the groove weld at the side of the joint from which welding was done.

Face shield: see *Hand (face) shield*.

Fast-freeze electrodes: electrodes that have the ability to deposit weld metal that solidifies, or freezes, rapidly.

Fatigue failure: the cracking, breaking, or other failure of a material as the result of repeated or alternating stressing below the material's ultimate tensile strength.

Fatigue limit: that load, usually expressed in pounds per square inch, that may be applied to a material for an indefinite number of cycles without causing failure of the material.

Fatigue strength: the resistance of a material to repeated or alternating stressing without failure.

Ferrite: pure iron of body-centered cubic crystal structure. It is soft and ductile.

Ferritic stainless steel: a magnetic, straight-chromium steel containing 14 to 26 percent chromium.

Ferritic steels: straight high chromium steels (15 to 27 percent chromium, no nickel, basic type 430), which are not hardenable, should be preheated to 300 to 500°F, especially in cold weather.

Ferrous: descriptive of a metallic material that is dominated by iron in its chemical composition.

Ferrous metals: metals that have a high iron content. They include the many types of steel and its alloys, cast iron, and wrought iron.

Filler metals: the metal or alloy to be added in making a braze, solder, or welded joint.

Fillet weld: a weld that is approximately triangular in cross-sectional shape. These types of welds join two surfaces that are at approximately right angles to one another. These welds are typically used on lap, T, and corner joints. See Figs. B-3, B-5, B-6, and B-20.

Fillet weld size: see *Size of weld*.

Filter glass (lens): a glass, usually colored, used in goggles, helmets, and hand shields to exclude harmful light rays.

Fixture: a device designed to hold parts to be joined in proper relation to each other.

Flame gouging: a means for quickly and accurately removing a narrow strip of surface metal from steel plate, forgings, and castings without penetrating through the entire thickness of metal.

Flame neutral: the flame that has no chemical effect on the molten weld metal during welding.

Flame scarfing: a process used mainly in the steel mills to remove cracks, surface seams, scabs, breaks, decarburized surfaces, and other defects on the surfaces of unfinished steel shapes.

Flange joints: these joints are formed by turning over the edge of the sheet to a height equal to the thickness of the plate. They are sometimes referred to as edge joints.

Flare-bevel groove weld: a type of groove weld. See Fig. B-63.

Flare-V-groove weld: a type of groove weld. See Fig. B-63.

Flash arrestor: a sintered metal alloy that prevents the flashback from traveling upstream into the hose, regulator, and gas supply systems.

Flashback: this occurs when the flame burns back inside the torch and causes a shrill hissing or squealing.

Flash welding: a resistance welding process wherein coalescence is produced simultaneously over the entire area of abutting surfaces by the heat obtained from flashing due to the flow of electric current between the two surfaces and by the application of pressure after heating is substantially completed. Flashing and upsetting are accompanied by expulsion of metal from the joint. See Fig. B-29.

Flat contour fillet weld: a fillet weld with a flat contour face. The design engineer specifies on the weld symbol that the weld size and the leg are the same.

Flat position (downhand): the position of welding wherein welding is performed from the upper side of the joint, and the face of the weld is approximately horizontal. See Figs. B-1A and B-1B.

Flat position (number 1): the welding position used to weld from the upper side of the joint at a point where the weld axis is approximately horizontal and the weld face lies in an approximately horizontal plane.

Flush weld: a groove weld made with a minimum reinforcement.

Flux: a cleaning agent of fusible material or gases used to dissolve oxides, release trapped gases, nitrides, slag, or other undesirable inclusions.

Flux cored arc welding (FCAW): an arc welding process wherein coalescence is produced by heating with an arc between a continuous filler metal (consumable) electrode and the work. Shielding is obtained from a flux contained within the electrode. Additional shielding may or may not be obtained from an externally supplied gas or gas mixture.

Flux cored electrode: a continuous filler metal electrode consisting of a metal tube containing flux and various alloying ingredients.

Forehand welding: a welding technique wherein the welding torch or gun is directed toward the progress of welding. See Fig. B-71.

Forge welding (FOW) (blacksmith, roll, hammer): a group of pressure welding processes wherein the parts to be welded are brought to suitable temperature by means of external heating, and the weld is consummated by pressure or blow. Temperature is below that required for fusion.

Free-bend test specimen: a specimen that is tested by bending without constraint of a jig.

Free carbon: this is the percentage of carbon that is not combined and usually exists as free atoms at grain boundaries or in the form of pure carbon flakes of graphite.

Freezing: solidification of a hot, liquid metal. In the case of pure iron, this starts and ends at the same temperature. With the addition of carbon, freezing starts at one temperature and ends at a lower temperature.

Friction stir welding (FSW): this variation of friction welding produces the weld by friction heating and plastic material displacement caused by a high speed rotating tool that moves along the weld joint.

Friction welding (FRW): a solid-state process that produces heat due to compressive forces and workpieces that are rotating or moving relative to each other. The faying surfaces will then displace material in a plastic manner.

Fusion: the melting together of filler metal and base metal or of base metal only, which results in coalescence. See also *Depth of fusion*.

Fusion line: the junction between the metal that has been melted and the unmelted base metal.

Fusion welding: a group of welding processes in which metals are welded together without the application of mechanical pressure or blows. The metals are welded by bringing them to the molten state at the surfaces to be joined. Fusion welding processes can be carried out with or without the addition of filler metal.

Fusion welds: welds in which the two metal members to be joined are fused directly to each other without the addition of filler metal from a consumable electrode or welding rod.

Fusion zone: the area of base metal melted as determined on the cross section of a weld.

G

Galvanized sheets: sheets of steel that are coated with zinc.

Gap (brazing): the space between the outside of the pipe and the inside of the cup. (If both are perfectly round, the gap will be one-half the difference in diameters.) For lap seams and other types of joints, the gap is the clearance space to be filled with alloy.

Gas brazing: a brazing process wherein the heat is obtained from a gas flame.

Gas cutter: see *Oxygen cutter*.

Gas cutting: the process of severing ferrous metals by means of the chemical action of oxygen on elements in the base metal.

Gas gouging: see *Oxygen gouging*.

Gas lens: a permeable barrier of concentric fine-mesh, stainless-steel screens that produce an unusually stable stream of shielding gas.

Gas metal arc cutting: an arc cutting process wherein the severing of metals is affected by melting with an electric arc between a metal (consumable) electrode and the work. Shielding is obtained from a gas, a gas mixture (which may contain an inert gas), or a mixture of a gas and a flux.

Gas metal arc welding (GMAW): also known as metal inert gas (MIG) or metal active gas (MAG) welding. A process that concentrates high heat at a focal point, producing deep penetration, narrow bead width, a small heat-affected zone, and faster welding speeds, resulting in less warpage and distortion of the welded joint and minimum postweld cleaning.

Gas pocket porosity: a weld cavity caused by entrapped gas.

Gas regulator: see *Regulator*.

Gas-shielded arc welding: a general term used to describe gas metal arc welding and gas tungsten arc welding.

Gas torch: see *Welding torch* and *Cutting torch*.

Gas tungsten arc cutting (GTAC): an arc cutting process wherein the severing of metals is affected by melting with an electric arc between a single tungsten (nonconsumable) electrode and the work. Shielding is obtained from a gas or gas mixture (which may contain an inert gas).

Gas tungsten arc welding (GTAW): an arc welding process wherein coalescence is produced by heating with an electric arc between a single tungsten (nonconsumable) electrode and the work. Shielding is obtained from a gas or gas mixture (which may contain an inert gas). Pressure may or may not be used and filler metal may or may not be used. This process is frequently called TIG welding.

Gas welding: a group of welding processes wherein coalescence is produced by heating with a gas flame or flames, with or without the application of pressure.

General rule about warping: all other things being equal, a decrease in speed and an increase in the number of passes increases warping.

Globular transfer: a mode of metal transfer in gas metal arc welding in which the consumable electrode is transferred across the arc in large droplets.

Goggles: see *Welding goggles*.

Gouging: the forming of a bevel or groove by material removal.

Grain: the crystalline body that may be viewed under a microscope as having definable limits.

Graphitic carbon: free, uncombined carbon existing in a metallic material in the form of flakes.

Groove: the opening provided between two members to be joined by a groove weld. See Fig. B-32.

Groove angle: the total included angle of the groove between parts to be joined by a groove weld. See Fig. B-38.

Groove face: that surface of a member included in the groove. See Fig. B-32.

Groove radius: the radius of a J- or U-groove. See Figs. B-16, B-17, and B-35.

Groove weld: a weld made in the groove between two members to be joined. The standard types of groove welds are as follows:

Square groove weld, Fig. B-7

Single-V groove weld, Fig. B-8

Single-bevel groove weld, Fig. B-9

Single-U groove weld, Fig. B-10

Single-J groove weld, Fig. B-11

Single-flare bevel groove weld, B-12

Single-flare V-groove weld, B-13

Double-V groove weld, Fig. B-14

Double-bevel groove weld, Fig. B-15

Double-U groove weld, Fig. B-16

Double-J groove weld, Fig. B-17

Ground connection: the connection of the work lead to the work correctly called work connection.

Ground lead: see *Work lead*.

Guided-bend test: a bending test wherein the specimen is bent to a definite shape by means of a jig.

Gun (*arc welding*): in semiautomatic, machine, and automatic welding, a manipulating device to transfer current and guide the electrode into the arc. It may also include provisions for shielding and arc initiation.

H

Hand (face) shield: a protective device used in arc welding for shielding the face and neck; equipped with suitable filter glass lens and designed to be held by handle.

Hard crack: see *Underbead cracks*.

Hard-facing: a process of applying a hard, wear-resistant layer of metal to the surfaces or edges of parts.

Hard surfacing: see *Surfacing*.

Hastelloys: a nickel alloy group that is an alloy of nickel, molybdenum, and iron.

Headers: pipes or groups of pipes used as inlet or outlet systems to control flow through a variety of openings.

Heat-affected zone (HAZ): that portion of the base metal which has not been melted, but whose mechanical properties or microstructures have been altered by the heat of welding or cutting. See Fig. B-78.

Heat distribution: the welding heat on structures should be distributed as evenly as possible through a planned welding sequence and planned weld positions.

Helmet: a protective device used in arc welding for shielding the face and neck; equipped with suitable filter glass lens and designed to be worn on the head.

Hematite (Fe_2O_3): known as red iron, this ore contains about 70 percent iron. It is widely mined in the United States.

High energy beams: an energy beam produced by a group of thermal processes for cutting, welding, and surfacing. Energy density has been measured at 650,000,000 watts per square inch.

Horizontal fixed position: the position of a pipe joint wherein the axis of the pipe is approximately horizontal and the pipe is not rotated during welding.

Horizontal position (*fillet weld*): the position of welding wherein welding is performed on the upper side of an approximately horizontal surface and against an approximately vertical surface. See Fig. B-1B. (*Groove weld*) the position of welding wherein the axis of the weld lies in an approximately horizontal plane and the face of the weld lies in an approximately vertical plane. See Fig. B-1A.

Horizontal position (number 2): for fillet welds, the welding position in which the weld is on the upper side of an approximately horizontal surface and against an approximately vertical surface. For groove welds, the welding position in which the weld face lies in an approximately vertical plane and the weld axis at the point of welding is approximately horizontal.

Horizontal rolled positions (pipe welding): the position of a pipe joint wherein welding is performed in the flat position by rotating the pipe. See Fig. B-69.

Hot-shortness: the characteristic tendency of a material toward brittleness at elevated temperatures.

Hot starting: to get a good weld with consumable electrode processes and prevent arc strikes, start the arc ahead of where you want to begin welding, then rapidly move to the starting point of the weld. With nonconsumable electrodes use a higher power level creating the weld pool prior to adding filler metal. Both techniques will preheat the starting point and help stabilize the arc and provide good fusion/penetration at the start of the weld and in the arc strike area.

Hydrogen embrittlement: this condition is caused by the loss of ductility in weld metal as a result of the metal absorbing hydrogen. Loss of ductility causes cracking in some metals.

Hydrogen pickup: an increase in moisture that causes porosity and slag inclusions in the weld deposit.

I

Impact: a striking action perpendicular to a member that occurs as a result of chipping, upsetting, cracking, or crushing forces and causes metal to be lost or deformed.

Impedance: the combination of resistance and reactance, which opposes the flow of current, in an alternating current circuit.

Included angle: see *Groove angle*.

Inclusions: these are usually nonmetallic particles such as slag that are trapped in the weld as a result of fast freezing or poor manipulation.

Incomplete fusion (IF): fusion that is less than complete. See Fig. B-75.

Incomplete joint penetration: joint penetration that is less than that specified.

Inconels: a nickel alloy group that is higher in nickel and iron content than the Monels.

Increase speed with heat: as the heat input is increased, the welding speed should be increased.

Inert gas: a gas such as helium or argon which does not chemically combine with other elements. Such a gas serves as an effective shield of the welding arc and protects the molten weld metal against contamination from the atmosphere until it freezes.

Inert-gas carbon arc welding: see *Gas carbon arc welding*.

Inert-gas metal arc welding: see preferred term, *Gas metal arc welding (GMAW)*.

Inert-gas tungsten arc welding: see preferred term, *Gas tungsten arc welding (GTAW)*.

Inner cone: the pale blue core of the flame.

Insert (brazing): a ring of silver-brazing alloy set into a machined groove in the cup of a fitting.

Intermittent welds: a series of short welds spaced at intervals. These welds cannot be used where maximum strength is required or where it is necessary that the work be watertight or airtight. See Figs. B-24 and B-25.

Interpass temperature: in a multiple-pass weld, the lowest temperature of the deposited weld metal before the next pass is started.

Irons: these iron-base alloys with high carbon content have the characteristics of cast iron and are used for facing heavy cast iron machinery parts.

Izod: one method used to conduct an impact test. In notched Izod testing, a sample is mounted in a vise fixture with the notch facing the pendulum. The pendulum is released and swings through a path breaking the sample. The energy absorbed by the sample is in joules (foot-pounds).

J

Jasper: an iron-bearing rock. The ore is predominantly magnetite or hematite.

J-groove welds: see *Groove weld*.

Jigs: devices designed to hold parts to be joined in proper relationship to each other. Essentially the same as fixtures.

Joint: the junction of members or the edges of members that are to be joined or have been joined.

Joint design: the joint geometry together with the required dimensions of the welded joint.

Joint fitup: the resulting condition of the workpiece or workpieces that have been brought together for welding.

Joint penetration: the minimum depth a groove weld extends from its face into a joint, exclusive of reinforcement. See Fig. B-47.

Joint welding procedure: the materials, detailed methods, and practices employed in the welding of a particular joint.

Joint welding sequence: see *Buildup sequence*.

K

Kerf: the gap created as material is removed by cutting. See Fig. B-57.

Keyhole effect: the plasma jet penetrates the workpiece completely. Plasma keyholes the front of the puddle because the jet blows aside the molten metal and lets the arc pass through the seam. As the torch moves, molten metal, supported by surface tension, flows in behind and fills the hole. The keyhole assures the operator there will be full-depth welds of uniform quality.

Keyholing: the technique of forming a hole (keyhole) on the leading edge of the weld pool. It completely penetrates the joint. As the heat source progresses, molten metal fills in behind the hole to form the weld bead.

K-Monel: a wrought alloy of nickel, copper, aluminum, and titanium that possesses the corrosion resistance that is characteristic of Monel, but has greater strength and hardness.

L

Lack of fusion: see *Incomplete fusion (IF)*.

Lack of joint penetration: see *Incomplete joint penetration*.

Land: see preferred term, *root face (shoulder)*.

Lap joint: a welded joint in which two overlapping parts are connected by means of fillet, plug, slot, spot projection, or seam welds. See Fig. B-5.

Laser beam cutting (LBC): a welding process that uses a laser beam to cut materials.

Laser beam drilling (LBD): a welding process that uses a laser beam to drill materials.

Laser beam welding (LBW): a welding process that uses a laser beam to fuse the joint.

Layer: a stratum of weld metal consisting of one or more weld beads. See Fig. B-50.

Lead burning: a misnomer for the welding of lead.

Leg of a fillet weld: the distance from the root of the joint to the toe of the fillet weld. See Figs. B-44 and B-46.

Lens: see filter glass.

Lightly coated electrode: a filler metal electrode used in arc welding consisting of a metal wire with a light coating applied subsequent to the drawing operation primarily for stabilizing the arc.

Limonite ($2Fe_2O_3 \bullet H_2O$): an ore that contains from 52 to 66 percent iron.

Linear coefficient of thermal expansion: the increase in the length of a bar 1 inch long when its temperature is raised 1°C.

Liquidus temperature: the temperature at which the brazing material is completely melted.

Load: the amount of a force applied to a material or structure.

Load voltage: the voltage at the output terminals of the welding machine when an arc is going. Load voltage is the combination of the arc voltage plus the voltage drop in the welding circuit.

Local preheating: preheating a specific portion of a structure.

Local stress-relief heat treatment: stress-relief heat treatment of a specific portion of a structure.

Locked-up stress: see *Residual stresses*.

Longitudinal sequence: the order in which the increments of a continuous weld are deposited with respect to its length.

Lower critical temperature: the temperature at which an alloy completes its transformation from one type of solid structure to another as it is cooling.

Low hydrogen electrodes: electrodes that have coverings containing practically no hydrogen.

M

Machine oxygen cutting: oxygen cutting with equipment that performs the cutting operation under the constant observation and control of an operator. The equipment may or may not perform the loading and unloading of the work.

Magnesium: this alloy is used extensively by the aircraft industry. It is two-thirds as heavy as aluminum and less than one-quarter as heavy as steel.

Magnetic burning square: a tool that makes it possible to cut straight lines with a high degree of accuracy.

Magnetite (Fe_3O_4): a brownish ore that contains about 65 to 70 percent iron. This is the richest iron ore and the least common.

Malleability: the pliability of a material in forming operations. It is the characteristic of a metal to be deformed without rupturing.

Manifold: a multiple header for interconnecting several cylinders to one or more torch supply lines.

Manual oxygen cutting: oxygen cutting wherein the entire cutting operation is performed and controlled by hand.

Manual welding: welding wherein the entire welding operation is performed and controlled by hand.

Mapp gas: a liquified acetylene compound that is fuel gas for oxyfuel heating and cutting.

Martensite: a structure resulting from transformation of austenite at temperature considerably below the usual range; achieved by rapid cooling. It is made up of ultra-hard, needlelike crystals that are a supersaturated solid solution of carbon in iron.

Martensitic stainless steel: an air-hardening steel containing chromium as its principal alloying element in amounts ranging from 4 to 12 percent.

Martensitic steels: straight chromium steels (12 to 18 percent chromium, basic type 410) that have the characteristic of air hardening. These steels should be preheated locally or completely to 300 to 500°F, which prevents excessive air-hardening and possible cracking.

Matrix: the principal, physically continuous metallic constituent in which crystals or free atoms of other constituents are embedded. It serves as a binder, holding the entire mass together.

Melting range: the difference between the solidus and liquidus temperatures.

Melting rate: the weight or length of electrode melted in a unit of time.

Melting ratio: the ratio of the volume of weld metal below the original surface of the base metal to the total volume of the weld metal.

Melt point (brazing): the temperature below which an alloy is substantially solid.

Melt-thru: complete joint penetration of weld metal in a joint welded from one side with visible root reinforcement. See Fig. B-64.

Metal expansion: metals with a high coefficient of expansion distort more than those with a lower coefficient.

Metallurgy: the science and technology of extracting metals from their ores, refining them, and preparing them for use.

Metal surfacing: the process of welding on wearing metal surfaces a coating, edge, or point of metal that is highly capable of resisting abrasion, corrosion, erosion, high temperature, or impact. This process is also called hard-facing.

Micro-cracks: cracks or fissures in a metallic structure which cannot be seen except with the aid of a microscope.

Microstructure: the detailed structure of a metal or alloy, as revealed by microscopic examination, showing the various continuous phases as well as any nonmetallic inclusions.

MIG welding: see preferred term, *Gas metal arc welding (GMAW)*.

Mixing chamber: that part of a gas welding or cutting torch wherein the gases are mixed for combustion.

Modulus of elasticity: the ratio of tensile stress to the strain it causes within that range of elasticity where there is a straight-line relationship between stress and strain. The higher the modulus, the lower the degree of elasticity.

Monels: a nickel alloy group that is about two-thirds nickel and one-third copper.

Multipass filler metals: a filler metal that is alloyed and deoxidized such that it can be used to weld multiple passes into a joint without negative impact on physical or mechanical properties.

N

Neutral flame: a gas welding flame wherein the portion used is neither oxidizing nor reducing.

Neutral fluxes: fluxes formulated to produce a weld metal composition that is not dependent on the welding parameters, especially arc voltage. Typically used with the submerged arc welding process.

Nimonics: a nickel alloy group that contains approximately 80 percent nickel and 20 percent chromium.

Nitrides: compounds formed by nitrogen during the welding of steel that cause hardness, a decrease in ductility, and lower impact resistance.

Nonconsumable electrode process: known as inert gas shielded tungsten arc welding. It is more often referred to as gas tungsten arc welding (GTAW) or by the shop term TIG.

Nonferrous: lacking iron in sufficient percentage to have any dominating influence on properties of the material.

Nonferrous metals: metals that are almost free of iron. The group includes copper, lead, zinc, titanium, aluminum, nickel, tungsten, manganese, brass, and bronze. The precious metals (gold, platinum, and silver) and radioactive metals such as uranium and radium are also nonferrous.

Nonpressure welding: a group of welding processes wherein the weld is made without pressure (an obsolete term).

Nontransferred arc: an arc employed by plasma equipment that is struck over a short distance and entirely contained within the torch housing. The arc is struck between a tungsten cathode and usually a water-cooled copper anode.

Nozzle: a device at the end of a gun that directs gas-shielding media.

Nucleus: the central core of the atom, which carries all of the positive charge.

Nugget: the weld metal joining the parts in resistance spot, resistance seam, or projection welds.

Nugget size: the diameter of width of the nugget measured in the plane of the interface between the pieces joined. See Fig. B-65.

O

Ohm: a unit of electrical resistance. There is one ohm resistance when a pressure of one volt causes a current of one ampere to flow in a conductor.

Open arc: occurs when the weld arc is exposed and the welder requires a welding helmet to observe the operation. Unlike submerged arc or electroslag welding.

Open circuit voltage: the voltage reading at the terminals of a welding machine when it is turned on but is not engaged in a welding operation. The welding circuit is not complete, and no current is flowing.

Open joints: see *Root opening*. See Fig. B-36.

Ore: the rock or earth in which we find metals in their natural form.

Oscillation: a back-and-forth motion across the face of the weld relative to the direction of travel. It is used for welding, brazing, soldering, cutting, or spraying process devices. It can be considered weaving.

Other side: the side of the joint opposite where the arrow is pointing to on a welding symbol.

Output slope: the relationship between the output voltage and output current (amperage) of the machine as the current or welding workload is increased or decreased. It is also called the *volt-ampere characteristic* or *curve*.

Overhead position: the position of welding wherein welding is performed from the underside of the joint. See Figs. B-1A and B-1B.

Overhead position (number 4): the welding position in which welding is performed from the underside of the joint.

Overheating: sufficient exposure of a metal to an extremely high temperature for an undesirable coarse grain structure to develop. The structure often can be corrected by suitable heat treatment, cold working, or a combination of these.

Overlap: protrusion of weld metal beyond the bond at the toe or root of the weld. See Fig. B-58.

Overlap fusion: this lack of fusion is largely the result of improper landing, low heat, and improper speed of travel.

Oxide: the scale that forms on metal surfaces when they are exposed to air and especially when they are heated.

Oxidize (of a metal): to combine chemically with oxygen, forming another composition that is called an oxide.

Oxidizing flame: a gas welding flame wherein the portion used has an oxidizing effect.

Oxyacetylene cutting (OFC-A): an oxygen cutting process wherein the severing of metals is affected by means of the chemical reaction of oxygen with the base metal at elevated temperatures, the necessary temperature being maintained by means of gas flames obtained from the combustion of acetylene with oxygen.

Oxyacetylene welding (OAW): a gas welding process wherein coalescence is produced by heating with a gas flame or flames obtained from the combustion of acetylene with oxygen with or without the application of pressure and with or without the use of filler metal.

Oxygen: the gaseous chemical element in the air that is necessary for life.

Oxygen cutter: one who is capable of performing a manual oxygen cutting operation.

Oxygen cutting operator: one who operates machine or automatic oxygen cutting equipment.

Oxygen gouging: an application of oxygen cutting wherein a chamfer or groove is formed.

Oxygen lance cutting (OLC): a method of cutting heavy sections of steel that would be very difficult by any other means.

Oxygen machining: a process of shaping ferrous metals by oxygen cutting or oxygen grooving.

Oxyhydrogen welding (OHW): a gas welding process wherein coalescence is produced by heating with a gas flame or flames obtained from the combustion of hydrogen with oxygen without the application of pressure and with or without the use of filler metal.

P

Parallel beads: see *Beading (parallel beads)*.

Parallel joint: a welded joint connecting the edges of two or more parallel or nearly parallel parts. See Fig. B-4.

Parent metal: the metal to be welded or otherwise worked upon; also called the base metal.

Partial joint penetration: joint penetration that is less than complete.

Pass: the weld metal deposited by one general progression along the axis of a weld. See Fig. B-49.

Pass sequence: the method of depositing a weld with respect to its length.

Pasty range: the difference between the melting and solid temperatures.

Patterns: devices used to replicate mass quantities of a product.

Pearlite: a continuous granular mixture composed of alternate plates or layers of cementite (pure iron-ferrite and iron-carbide).

Peel test: a destructive method of inspection wherein a lap joint is mechanically separated by peeling.

Peening: mechanical working of metal by means of impact blows.

Penetration: the penetration, or depth of fusion, of a weld is the distance from the original surface of the base metal to that point at which fusion ceases. See Fig. B-47.

Percent of elongation: the percentage a material can be stretched before it breaks. Metals such as gold, silver, copper, and iron may be stretched, formed, or drawn without tearing or cracking.

Phosphor bronzes: copper-tin alloy (ECuSn-A) bronzes.

Physical property: an inherent physical characteristic of a material that is not directly an ability to withstand a physical force of any kind.

Pickling: acid cleaning that removes rust, scale, oxides, and sulfides.

Pickup: the absorption of base metal by the weld metal as the result of admixture. Depending upon the materials involved, this can be an asset and not a liability.

Piercing: a process in which a heated steel bar is pierced by a mandrel and rolled to the desired diameter and wall thickness.

Pig iron: iron that is hard and brittle. It contains considerable amounts of dissolved carbon, manganese, silicon, phosphorus, and sulfur. The name "pig iron" comes from the shape of the crude cast iron oblong blocks.

Pinch force: this squeezing action, common to all current carriers, breaks the molten metal bridge at the tip of the electrode, and a small drop of molten metal transfers to the weld pool.

Pitch: the center-to-center distance of the welds which can be used to determine the space between the welds.

Plasma arc cutting (PAC): an arc cutting process wherein severing of the metal is obtained by melting a localized area with a constricted arc and removing the molten material with a high velocity jet of hot, ionized gas issuing from the orifice.

Plasma arc welding (PAW): an arc welding process wherein coalescence is reduced by heating with a constricted arc between an electrode and the workpiece (transferred arc) or the electrode and the constricting nozzle (nontransferred arc). Shielding is obtained from the hot, ionized gas issuing from the orifice which may be supplemented by an auxiliary source of shielding gas. Shielding gas may be an inert gas or a mixture of gases. Pressure may or may not be used, and filler metal may or may not be supplied.

Plasma needle arc welding: this process is a manual operation that uses a small diameter constricted arc in a water-cooled nozzle. It is excellent for welding contoured work.

Plug welds: welds made in a circular hole, usually on a lap joint. See Fig. B-22.

Porosity: cavity-type discontinuities formed by gas entrapment during solidification.

Positioned weld: a weld made in a joint that has been so placed as to facilitate making the weld.

Position of welding: see flat, horizontal, vertical, and overhead positions and horizontal rolled, horizontal fixed, and vertical pipe positions.

Postheating: heat applied to the base metal after welding or cutting for the purpose of tempering, stress-relieving, or annealing.

Powder brazing: a form of brazing that is used in mass production industries.

Precipitation-hardening (PH) stainless steel: a steel that has the ability to develop high strength with a reasonable simple heat treatment.

Preheating: the application of heat to the base metal prior to a welding or cutting operation.

Preinserted-ring type (*brazing*): describes the kind of fittings that have inserts to supply silver-brazing alloy to their joints.

Pressure welding: any welding process or method wherein pressure is used to complete the weld.

Procedure qualification: the demonstration that welds made by a specific procedure can meet prescribed standards.

Projection welding (RPW): a resistance welding process wherein localization of heat between two surfaces or between the end of one member and surface of another is affected by projections. See Fig. B-27.

Propane: a hydrocarbon present in petroleum and natural gas. It is used primarily for oxyfuel heating, cutting, soldering, and brazing.

Properties: those features or characteristics of a metal that make it useful and distinctive from all others.

Proportional elastic limit: the stress point beyond which an increase in stress results in permanent deformation.

Protective atmosphere: a gas envelope surrounding the part to be brazed or welded in which the gas composition is controlled with respect to chemical composition, dew point, pressure, flow rate, and so on. Examples are inert gases, combusted fuel gases, hydrogen, and a vacuum

Q

Qualification: the ability of a welder or procedure to produce sound welds.

R

Radial crack: a crack originating in the fusion zone and extending into the base metal, usually at right angles to the line of fusion. This type of crack is due to the high stresses involved in the cooling of a rigid structure.

Radiation: the transfer of heat energy from the surface to the atmosphere. It is the predominant form of heat transfer in the upper regions of the secondary cooling chamber.

Radiographic quality: soundness of a weld that shows no internal or underbead cracks, voids, or inclusions when inspected by X-ray or gamma ray techniques.

Rectifier: an electrical assembly used in rectifier welders to change alternating current to direct current.

Red hardness: the property of metallic material to retain its hardness at high temperatures.

Reducing flame: an oxyacetylene flame that has an excess of acetylene gas in the mixture (contrasted with an oxidizing flame, which has an excess of oxygen in the mixture).

Refractory: quality of resistance to melting. Popular usage is any metal melting above 3,632°F.

Regulator: a device for controlling the delivery of gas at some substantially constant pressure regardless of variation in the higher pressure at the source.

Reinforcement of weld: weld metal in excess of the metal necessary for the specified weld size. See Fig. B-42.

Residual stresses: the stresses that remain after the welded members have cooled to a normal temperature and affect the weldment. These stresses must be relieved or they will cause cracking or fracture in the weld and/or the plate.

Resistance: an electrical unit that opposes the passage of current. The ohm is the unit of measure for resistance.

Resistance welding (RW): one of the three major types of welding. It includes spot welding, seam welding, projection welding, flash welding, and other similar processes that are performed on machines.

Reverse polarity (electrode positive): the arrangement of direct arc welding leads wherein the work is the negative pole and the electrode is the positive pole of the welding arc.

Ripples: the usually smooth, fine, consistent lines that appear in the face of a weld. These result in the uniform solidification of the weld pool in stages.

Root: see *Root of joint* and *Root of weld*.

Root crack: a crack in the weld or base metal occurring at the root of a weld.

Root edge: the edge of the part to be welded that is adjacent to the root. See Fig. B-33.

Root face (shoulder or land): the portion of the prepared edge of a part to be joined by a groove weld that may or may not have been beveled or grooved. See Fig. B-34.

Root of joint: that portion of a joint to be welded where the members approach closest to each other. In cross section the root of the joint may be a point, a line, or an area. See Figs. B-33 and B-34.

Root of weld: the point at the bottom of the weld. See Fig. B-39.

Root opening: the separation at the root between parts to be joined by a groove weld. See Fig. B-36.

Root penetration: the depth a groove weld extends into the root of a joint measured on the centerline of the root cross section. See Fig. B-39.

Root radius: the radius near the root portion of the prepared edge of a part to be joined by a U- or J-groove weld.

Root reinforcement: reinforcement of groove weld at the side other than that from which welding was done.

Root surface: the exposed surface of a weld on the side other than that from which welding was done.

Running start: in this type of start the electrode end is put in contact with the base metal and the trigger is pulled.

S

Scarf: see *Groove (scarf)* or *Edge preparation*.

Scarfing: see *Chamfering*.

Scratch start: in this type of start (optional feeder control) the trigger is pulled which energizes the wire but not the drive motor. When the wire is touched to the workpiece, the drive motor will start. This is sometimes preferred for a better weld profile and complete fusion at the weld start.

Seal weld: any weld used primarily to obtain tightness against leakage.

Seam welding: a resistance welding process wherein coalescence is produced by the heat obtained from resistance to the flow of electric current through the work parts held together under pressure by circular electrodes. The resulting weld is a series of overlapping spot welds made progressively along a joint by rotating the electrodes. See Fig. B-28.

Self-shielding filler metals: typically used to describe a flux cored arc welding electrode that requires no external shielding gas.

Semiautomatic arc weld: a weld made with equipment that automatically controls only the feed of the electrode, the manipulation of the electrode being controlled by hand.

Shielded metal arc cutting (SMAC): a method of metal arc cutting wherein the severing of metals is affected by melting with the heat of an arc between a covered metal electrode and the base metal.

Shielded metal arc welding (SMAW): an arc welding process wherein coalescence is produced by heating with an electric arc between a covered metal electrode and the work. Shielding is obtained from decomposition of the electrode covering. Pressure is not used, and filler metal is obtained from the electrode.

Shielding: primarily the protection of molten metal by the arc from oxidizing or otherwise reacting with elements in the surrounding air. Usually, the shielding also stabilizes the arc.

Short circuit method: in this metal transfer method the weld metal is deposited by direct contact of the welding electrode with the base metal. GMAW

Shoulder (*brazing*): the machined ridge at the bottom or inner end of the cup where the end of the pipe seats. See Fig. B-79.

Shoulder (*welding*): see *Root face (shoulder)*.

Shrinkage: this welding distortion correction method consists of alternate heating and cooling, frequently accompanied by hammering or mechanical working, thus shrinking excess material in a wrinkle or buckle.

Shrinkage stress: see *Residual stresses*.

Shrink welding: this welding distortion correction method is a variation of shrinkage in which the heat is applied by running beads of weld metal on the convex side of a buckled area.

Siderite ($FeCO_3$): an ore that contains about 48 percent iron.

Silicon bronzes: copper-silicon alloy (ECuS) bronzes.

Silver alloy brazing (*sliver soldering*): a brazing process wherein a silver alloy is used as a filler metal.

Silver-brazing alloy: metal that contains silver in its composition and which is designed especially for relatively low-temperature joining of metal parts such as pipe, tubing, and fittings.

Single-groove welds: see *Groove weld*.

Single-welded butt joint: a butt joint welded from one side only.

Single-welded joint: in arc and gas welding, any joint welded from one side only. See Fig. B-67.

Single-welded lap joint: a lap joint in which the overlapped edges of the members to be joined are welded along the edge of one member.

Size of weld:

Groove weld—the joint penetration (depth of chamfering plus the root penetration when specified).

Fillet weld—for equal-leg fillet welds, the leg length of the largest isosceles right triangle that can be inscribed within the fillet weld cross section. See Fig. B-46.

Flange weld—the weld metal thickness measured at the root of the weld. For unequal leg fillet welds, the leg length of the largest right triangle that can be described within the fillet weld cross section.

Skip sequence: see *Wandering sequence.*

Skip-stop, backstep method: this method is a combination of skip and backstep welding.

Slab: a type of steel ingot that is oblong. It varies in thickness from 2 to 6 inches and in width from 5 to 6 feet.

Slag: the nonmetallic layer that forms on top of molten metal. It is usually a complex of chemicals (oxides, silicates, etc.) that float to the top of the hot molten metal. When a bead of weld metal cools, the slag "cap" on the bead can be readily chipped or ground away.

Slag inclusion: nonmetallic solid material entrapped in weld metal or between weld metal and base metal.

Slope: slope is of importance when setting the current for the gas metal arc welding process. It indicates the shape of the volt-ampere output curve.

Slot weld: a weld made in an elongated hole in one member of a lap joint joining that member to that portion of the surface of the other member that is exposed through the hole. The hole may be open at one end (i.e., may extend to the edge of the member), and may or may not be filled completely with weld metal. See Fig. B-23.

Soaking: prolonged heating of a metal at a selected temperature.

Soldering: a group of joining processes wherein coalescence of materials is produced by heating them to a maximum of 840°F, which melts the soldering alloy. The base metal is not melted. Capillary action draws the solder into the tight-fitting joints.

Soldering flux: a liquid, solid, or gaseous material which, when heated, improves the wetting of metals with solder.

Solid-state welding (SSW): a group of welding processes producing coalescence with the application of pressure without melting any of the joint material.

Solidus temperature: the highest temperature at which a brazing material is completely solid.

Spacer strip: a metal strip or bar inserted in the root of a joint prepared for a groove weld to serve as a backing and to maintain root opening during welding.

Spatter: in arc and gas welding, the metal particles expelled during welding and which do not form a part of the weld.

Spatter loss: the difference in weight between the weight of the electrode deposited and the weight of the electrode consumed (melted); metal lost due to spatter.

Spelter: copper-zinc alloys that were developed for joining iron and steel at the beginning of the Iron Age.

Spool: a filler metal package type consisting of a continuous length of electrode wound on a cylinder (called the barrel) which is flanged at both ends. The flange extends below the inside diameter of the barrel and contains a spindle hole. Used for MIG welding.

Spot gouging: the process for removing defects in the weld.

Spot welding: a resistance welding process wherein coalescence is produced by the heat obtained from resistance to the flow of electric current through the work parts held together under pressure by electrodes. The size and shape of the individually formed welds are limited primarily by the size and contour of the electrodes. See Fig. B-26.

Spray transfer: a mode of metal transfer in gas metal arc welding in which the consumable electrode is propelled axially across the arc in small droplets.

Square groove weld: see *Groove weld.*

Stack cutting: oxygen cutting of stacked metal plates arranged so that all the plates are severed by a single cut.

Staggered intermittent fillet welds: two lines of intermittent fillet welding in a T- or lap joint in which the increments of welding in one line are staggered with respect to those in the other line. See Fig. B-25.

Stainless steel: the popular term for the chromium and chromium-nickel steels. It is a tough, strong material that is highly resistant to corrosion, high temperatures, oxidation, and scaling.

Static electricity: a term applied to stationary electricity which is nonactive; may be produced by friction between two or more bodies.

Steel: a metal that is a combination of iron and carbon.

Stepback sequence: see *Backstep sequence.*

Stick electrode: see preferred term, *Covered electrode.*

Stick electrode welding: another term for shielded metal arc welding.

Stitch welding: the use of intermittent welds to join two or more parts.

Straight polarity (electrode negative): the arrangement of direct current arc welding leads wherein the work is the positive pole and the electrode is the negative pole of the welding arc. See Fig. B-77.

Strain: the physical effect of stress, usually evidenced by stretching or other deformation of the material.

Stress: the load, or amount of a force, applied to a material tending to deform or break it.

Stress cracking: cracking of a weld or base metal containing residual stresses.

Stress-relief heat treatment: uniform heating of a structure or portion thereof to a sufficient temperature, below the critical range, to relieve the major portion of the residual stresses, followed by uniform cooling. (*Note:* Terms such as *normalizing, annealing*, etc., are misnomers for this application.)

Stringer bead: a weld made by moving the weld pool along the intended path in a straight line. See Fig. B-52.

Subcritical annealing: heating of a metal to a point below the critical or transformation range, in order to relieve internal stresses.

Sublime: to cause to pass directly from the solid to the vapor state and condense back to solid form. As in the CAC-A process using carbon (graphite) electrode.

Submerged arc welding: an arc welding process wherein coalescence is produced by heating with an electric arc or arcs between a bare metal electrode or electrodes and the work. The arc is shielded by a blanket of granular, fusible material on the work. Pressure is not used, and filler metal is obtained from the electrode and sometimes from a supplementary welding rod.

Sugaring: the severe oxidation of the back side of stainless steel caused by its exposure to the atmosphere.

Surfacing: the deposition of filler metal on a metal surface to obtain desired properties or dimensions. Also referred to as buildup, buttering, cladding, and hard-facing. See Fig. B-68.

T

Tack weld: a weld that holds parts together prior to the final welds being made.

Taconite: a low grade ore that contains from 11 to 40 percent iron and a large amount of silica. It is green in color.

Tail: that part of the welding symbol that is on the opposite end of the reference line from the arrow end. It is where special instructions can be placed.

Temper:
1. The amount of carbon present in the steel: 10 temper is 1.00 percent carbon.
2. The degree of hardness that an alloy has after heat treatment or coldworking, for example, the aluminum alloys.

Temper bead welding (TBW): Each layer of beads provides heat for the thermal treatment of the microstructure of the previous weld bead or the layer, thus removing coarse grain structure and improving mechanical properties. See Fig. B-80.

Temporary weld: a weld made to attach a piece or pieces to a weldment for temporary use in handling, shipping, or working on the weldment.

Tensile: involving a pulling stress applied to a material. A compressive stress is the opposite of a tensile stress. See also *Ultimate tensile strength*.

Tensile strength: the resistance of a material to a force which is acting to pull it apart.

Terne plate: a sheet of steel that is coated with an alloy of lead and tin.

Theoretical throat: the part of a weld that extends from the point where two base metal members join (the beginning of the joint root), to the top of the weld, minus any convexity on the convex fillet weld and concavity on the concave fillet weld, to the top of the largest triangle that can be inscribed in the weld. See Figs. B-45 and B-46.

Thermal conductor: a material such as aluminum that conducts heat about three times faster than iron and requires higher heat input than that used in welding steel.

Thermal stress: the stresses produced in a structure or member caused by differences in temperature or coefficients of expansion.

Thermit welding (TW): a group of welding processes wherein coalescence is produced by heating with superheated liquid metal and slag resulting from a chemical reaction between a metal oxide and aluminum with or without the application of pressure. Filler metal, when used, is obtained from the liquid metal.

Throat (of a groove weld): see preferred term, *Size of weld*.

TIG welding: see preferred term, *Gas tungsten arc welding (GTAW)*.

Tinning: a term used for melting a small amount of the rod and letting it spread over the joint.

Tin plate: sheets of steel that are coated with tin.

T-joint: a welded joint at the junction of two parts located approximately at right angles to each other in the form of a *T*. See Fig. B-6.

Toe crack: a crack originating at the junction between the face of the weld and the base metal. It may be any one of three types: (1) radial or stress crack; (2) underbead crack extending through the hardened zone below the fusion line; or (3) the result of poor fusion between the deposited filler metal and the base metal.

Toe of weld: the junction between the face of the weld and the base metal. See Fig. B-41.

Torch: see *Welding torch* or *Cutting torch*.

Torch brazing: a welding process where a bronze filler rod supplies the weld metal, and the oxyacetylene flame furnishes the heat.

Torch tip: see *Welding tip* or *Cutting tip*.

Transferred arc: an arc that travels between the electrode and the work, which acts as an anode. This arc heats the work with electric energy and hot gas and is used for plasma welding, weld surfacing, and cutting.

Transformation range: unless otherwise specified, the temperature range during which a metal, when cooled, is changing from one type of crystal structure to the structure it will have permanently at room temperature.

Transverse contraction: occurs when two butt-welded plates that are free to move draw together at the opposite end due to contraction of the weld metal upon cooling. This contraction can be controlled.

Travel angle: the angle between the electrode axis and a line perpendicular to the weld axis. It can be referenced as the drag or push angle. It can be used to define positions of torches, guns, rods, and beams.

Trepanning: is a method of testing that can be used on a production part. A hole is cut with a hole saw through the base metal with the weld centered in the cut, in order to keep the core sample. Which can be polished and etched. The plug then can be welded back into hole it came from.

Tungsten electrode: a non-filler-metal electrode used in arc welding consisting of a tungsten wire.

U

Ultimate tensile strength: a material's resistance to being pulled apart (test to fracture), usually specified in pounds per square inch.

Underbead cracks: cracks usually found in the heat-affected zone that may not open to the surface of the base metal initially.

Undercut: a groove melted into the base metal adjacent to the toe of the weld and left unfilled. See Figs. B-43 and B-58.

Upper critical temperature: the temperature at which an alloy begins to transform from one solid structure to another as it cools.

Upset welding: a resistance welding process wherein coalescence is produced simultaneously over the entire area of abutting surfaces by the heat obtained from resistance to the flow of electric current between the two surfaces and by the application of pressure after heating is substantially completed. There is no flashing, only resistance heating leading to upsetting of metal from the joint. See Fig. B-30.

V

Vertical position (number 3): the welding position in which the weld axis, at the point of welding, is approximately vertical and the weld face lies in an approximately vertical plane. See Figs. B-1A and B-1B.

Vertical position (pipe welding): the position of a pipe joint wherein the axis of the pipe is vertical, welding is performed in the horizontal position, and the pipe may or may not be rotated. See Fig. B-69.

Voids: are sometimes referred to as wagon tracks because of their resemblance to ruts in a dirt road. They are found in multipass welding and may be continuous along both sides of the weld deposits.

Volt: the force that causes current to flow.

Voltage regulator: an automatic electrical control device for maintaining a constant voltage supply to the primary of a welding transformer.

Volt-ampere curve: curve that indicates the current output of a welding machine and is formed through the use of voltage and amperage values.

W

Wandering block sequence: a block sequence wherein successive blocks are completed at random after several starting blocks have been completed.

Wandering sequence: a method along with balanced welding that contributes to the simultaneous completion of welded connections in large fabrications with minimal distortions.

Washing: a procedure used in the steel mills to remove unwanted metal such as fins on castings and to blend in rise pads and sand washouts in castings.

Water jet cutting: also referred to as hydrodynamic machining. This technique severs materials with the use of a high velocity water jet. Abrasives may or may not be added to the stream of water.

Watt: a unit of electrical power measurement.

Weave bead: a weld made by moving the weld pool along the intended path but with a side-to-side oscillation. See Fig. B-51.

Weld: a localized coalescence of metal wherein coalescence is produced by heating to suitable temperatures with or without the application of pressure and with or without the use of filler metal. The filler metal either has a melting point approximately the same as the base metals or has a melting point below that of the base metals but above 800°F.

Weldability: the capacity of a metal to be welded under the fabrication conditions imposed into a specific, suitably designed structure and to perform satisfactorily in the intended service.

Weld bead: a weld deposit resulting from a pass.

Weld crack: a crack in weld metal.

Welded joint: a union of two or more members produced by the application of a welding process.

Welder: one who is capable of performing a manual or semiautomatic welding operation.

Welder certification: certification in writing that a welder has produced welds meeting prescribed standards.

Welder qualification: the demonstration of a welder's ability to produce welds meeting prescribed standards.

Weld gauge: a device designed for checking the shape and size of welds.

Welding: the joining together of two pieces of metal by heating to a temperature high enough to cause softening or melting, with or without the application of pressure and with or without the use of filler metal.

Welding current: the current flowing through the welding circuit during the making of a weld.

Welding direction: the direction of welding should be away from the point of restraint and toward the point of maximum freedom.

Welding electrode: see *Electrode (arc welding)*.

Welding from both sides: a welding process that reduces distortion.

Welding generator: a generator used for supplying current for welding.

Welding goggles: goggles with tinted lenses used during welding, brazing, or oxygen cutting, which protect the eyes from harmful radiation and flying particles.

Welding ground: see *Work lead*.

Welding leads: the work lead and electrode lead of an arc-welding circuit.

Welding machine: equipment used to perform the welding operation. For example, spot welding machine, arc welding machine, seam welding machine, and so on.

Welding operator: one who operates machine or automatic welding equipment.

Welding procedure: the detailed methods and practices including joint welding procedures involved in the production of a weldment.

Welding process: a metal joining process wherein coalescence is produced by heating to suitable temperature with or without the application of pressure and with or without the use of filler metal. See also *Forge welding (FOW) (blacksmith, roll, hammer); Thermit welding (TW); Gas welding; Arc welding; Resistance welding (RW);* and *Brazing*.

Welding rod: a form of filler metal, used for welding or brazing wherein the filler metal does not conduct the electrical current.

Welding sequence: the order of making the welds in a weldment.

Welding symbol: all the parts that make up the symbol, such as arrow, reference line, tail, weld type, supplemental symbols. Only the arrow and reference line are considered absolute requirements for a welding symbol. All other items are optional as required.

Welding technique: the details of a manual, machine, or semiautomatic welding operation which, within the limitations of the prescribed joint welding procedure, are controlled by the welder or welding operator.

Welding tip: a welding torch tip designed for welding.

Welding torch: an apparatus for mixing oxygen and acetylene in the proportions necessary to carry out the welding operation.

Welding torch (blowpipe): a device used in gas welding or torch brazing for mixing and controlling the flow of gases.

Welding transformer: a transformer used for supplying current for welding.

Weld line: see *Bond line*.

Weldment: an assembly whose component parts are joined by welding.

Weld metal: that portion of a weld which has been melted during welding.

Weld metal area: the area of the weld metal as measured on the cross section of a weld.

Weldor: see *Welder*.

Weld penetration: see *Joint penetration* and *Root penetration*.

Weld symbol: a graphic character connected to the reference line of a welding symbol specifying the weld type.

Wetting: the bonding or spreading of a liquid filler metal or flux on a solid base metal.

Whipping: a manual welding technique in which the arc or flame is moved back and forth in the direction of the weld path (with or without lengthening the arc or flame).

Whipping the electrode: see *Oscillation*.

Whiskers: these are short lengths of electrode wire sticking through the weld on the root side of the joint.

Work angle: the angle less than 90° between a line perpendicular to the major workpiece surface and a plane determined by the electrode axis and the weld axis. On T- and corner joints it is perpendicular to the nonbutting member. It can also be used to define the position of guns, torches, rods, and beams.

Work connection: the connection of the work lead to the work.

Work hardening: the capacity of a material to harden as the result of cold rolling or other cold working involving deformation of the metal.

Work lead: the electric conductor between the source of arc welding current and the work. See Figs. B-76 and B-77.

Wrought aluminum-manganese alloy: this alloy contains about 1.2 percent manganese and a minimum of 97 percent aluminum. It is stronger and less ductile than commercially pure wrought aluminum.

Y

Yield point: the point in a load test where a definite increase in the length of the specimen occurs with no increase in the load.

Yield strength: the stress point at which permanent deformation results.

Z

Zero drag: the condition referred to when the line cut by the torch is vertical from top to bottom.

Index

A

AAW. *See* air-acetylene welding
abrasion, 321
abrasive water jets, 827 (table)
ABS. *See* acrylonitrile butadiene styrene
acceptance criteria, 497 (table)
accumulation of discontinuities (AD), 497
a.c.-d.c. transformer-rectifier, 286–287, 573 (table)
acetone, 141
acetylene, 4, 141–142
acetylene cylinders, 142, 1044
acetylene feather, 206
acid cleaning, 257
a.c. induction motors, 854
acrylics (polymethylmethacrylate), 996, 999
acrylonitrile butadiene styrene (ABS), 996, 999
active fluxes, 775
a.c. transformer welding machines, 286–287
a.c. transformers, 281
a.c. welding machines, 5
adaptive controls, 870, 872
adhesive dispersion (wood glue), 1023
air-acetylene welding (AAW), 160
air carbon arc cutting, 529–530, 531, 532
    electrodes and current recommendations, 532 (table)
    gouging with, 543
    job outline for, 546 (table)
    settings for gauging, 545 (table)
    shutting down equipment for, 545
    weld removal with, 544
air carbon arc cutting and gouging power sources, 532 (table)
air filters, 285
Aircomatic, 552
aircraft welding, 14–16
AISI numbering system, 87–92
alkaline degreasing, 257
alloy flux, 775
alloy steels. *See also* stainless steels
    carbon equivalency, 87
    carburization, 83
    chromium, 83
    classifications, 83–86
    hard facing (surfacing), 247
    high strength low alloy steels, 83
    low alloy steels, 83
    postweld heat treatment, 86
    sulfidation, 83
    tool steels, 87
    typical welding conditions, 581 (table)
alternating current welding, 709–710
aluminum and aluminum-base alloys, 81
    aluminum-magnesium-chromium alloy, 96
    aluminum-magnesium-silicon alloy, 96
    aluminum-silicon-magnesium-chromium alloy, 96
    brazeable metals, 268–269
    casting alloys, 238
    categories of, 238
    characteristics of, 238–239
    commercially pure wrought aluminum, 96
    dissimilar metals, 604
    filler metals, 604
    filler rod for, 239
    flame adjustment, 239–240
    flux, 239
    gas metal arc welding, 594, 718 (table)
    gas tungsten arc welding, 594
    group designations for, 96 (table)
    hard facing (surfacing), 604
    heat-treatable alloys, 594
    hot-shortness, 238
    magnesium, 599
    plasma arc cutting, 528 (table)
    preheating of, 594
    pure aluminum, 238
    rod sizes, tip sizes, and gas pressures for welding of, 240 (table)
    sheet aluminum welding, 241–242
    stabilized a.c. current and shielding gas conditions, 596 (table)
    thermal conditions, 238
    types of, 594
    welding methods for, 594
    welding rods and electrodes, 324 (table)
    welding technique, 240–241
    work-hardenable alloys, 594
    wrought aluminum alloys, 238
    wrought aluminum-manganese alloy, 96

aluminum bronze, 602
aluminum-making, 95–96
aluminum pipe, 809–810
aluminum-silicon filler metals (BAlSi), 261
American Petroleum Institute (API), 475, 483, 794
American Society of Mechanical Engineers (ASME), 475, 483, 794, 898
American Water Works Association, 484
American Welding Society (AWS), 484
amperage settings
   and oxygen pressures, 531 (table)
   wire feeders (wire drives) and power sources, 677 (table)
angular distortion, 102–103
angularity, 197
anisotropy property, 66
annealing, 108
   equipment, 845–846
ANSI pipe symbols, 965 (table)
antimony, 252
arc blow, 287, 332, 873
arc control devices
   arc length controls, 874–875
   magnetic arc control, 872–873
   mechanical oscillators, 874
   seam trackers, 874–875
arc cutting. *See also* air carbon arc cutting; plasma arc cutting
   bevel cutting, 538
   air carbon arc cutting, 529–530
   gas tungsten arc cutting, 523
   oxygen arc cutting, 530–534
   practice job, 187–193
   shape cutting, 539–542
   square cutting with, 538
   stack cutting, 535
air carbon arc cutting (CAC-A), 529–530
arc cutting machines, 534–535
arc length and arc voltage, 332, 677 (table)
arc monitoring
   control elements, 875–877
   microprocessor-based controllers, 881
   process control systems, 879–881
   process monitoring systems, 879
   sensing devices, 877–879
arc power sources, 288–289
arc seam weld, 1081
arc spot weld, 1081
arc strike, 942
arc voltage, 281
arc wandering, 610–611
arc welding
   machines
      cable sizes for, 291 (table), 1028 (table)
      types and uses of, 282–283 (table)
   process, 4
   process applications and key factors for, 871 (table)

argon, 686
   argon and helium in GTAW, 556 (table)
   argon-carbon dioxide, 689
   argon-helium, 689
   argon-oxygen, 690
   gas selection for gas metal arc welding, 684–685 (table)
   helium-argon-carbon dioxide, 690
   and helium in GTAW, 556 (table)
   wire and gas combinations, 683 (table)
Arkansas bell-hole, 507–508
ASME. *See* American Society of Mechanical Engineers
assembly drawings, 951, 977
assist gases, laser beam cutting, 822 (table)
ASTM numbering system, 92
atomic structure, 70
austenitic stainless steels, 319
auto-darkening welding lens, 1034–1035
autogenous welding
   flat position, 213–214
   horizontal position, 217–218
   overhead position, 218–219
   welding technique, 616
automatic welding
   processes, 771–780
   terminology, 870
automotive welding, 16
AWS electrode classification system, 305–306, 306 (table)
AWS filler metal specifications, 705 (table)

B

backhand welding, 224, 1083
backing, 632–633, 1083
backing rings, 477
backstep sequence, 1080
back welds, 1083
balanced forces, 106
base metal
   cracking, 942
   testing, 310, 946 (table)
   zone, 1084
bauxite ore, 95
   beading
      horizontal position, 218
      right and left handed, 214
      vertical position, 216–217
bead welds, 113, 1076, 1086
beam cutter, 535
bell-hole job, 503–505
bending, 76
bending test, 465 (table), 621, 623
bevel angle, 1077
bevel cutting, 182–183, 538
bevel-groove butt joints, 895
BHN. *See* Brinell hardness number

billet, 65
bismuth (Bi), 251
black hose, 158
black iron, 66
blast furnace, 46–49
block sequence, 1080
bloom, 65
blowouts, 822
body-centered cubic (BCC), 71
body-centered tetragonal (BCT), 71
boron, 77–80
Bourbonville, Eugene, 139
boxing (end return), 106, 125
branch welds, 645–646
brass, 81
brazed or solder joint, 1085
braze welding
    brazing *vs.*, 251
    bronze filler rod, 233
    bronze surfacing, 233
    capillary action, 232–233
    flux for, 233–234
    industrial uses of, 233
    job outline, 249 (table)
    malleable cast iron, 233
    practice, 234–235
    powder brazing, 235–236
    tinning, 235
    torch brazing, 233
brazing
    braze welding *vs.*, 251
    capillary action, 273
    copper tubing, 250–274
    definition, 251
    of fittings, 271–272
    fluxing, 272–273
    of pipe, 271
    practice jobs, 269–274
    soldering *vs.*, 251
    timing of, 273
brazing filler metals, 262–263 (table)
    classifications of, 261
brazing fluxes
    application, 265 (table)
    materials in, 264
Brinell hardness test, 76, 918
brittleness, 76
brittle welds, 942–943
bronze, 81
burning test, for plastic, 997–998, 997 (table)
burn protection, 1041–1042
butt joint-groove welds, 500–502
    roll, 636–640
    symbols, 983–985
    in welding qualification tests, 486–490

butt joints. *See also* single V-butt joints
    bend testing, 623
    bevel-groove, 895
    closed square-groove, 892, 894
    consumable inserts, 632
    double J-groove, 896
    double U-groove, 895–896
    double V-groove, 894–895
    joint design, 893–896
    joint design of pipe, 478
    open square-groove, 894
    pipe welding, 797–807
    with plasma needle arc process, 586 (table)
    shrinkage effects on, 103
    single J-groove, 896
    single U-groove, 895
    single V-groove, 894, 895
    square, in flat position, 399–400
    square, in vertical position, 416–418
    square, practice jobs, 220–221
    U-groove, 632
    V-groove, 632
buttonhead rivet heads removal, 185

C

cables and fasteners, 290–293
cable sizes
    for arc welding machines, 1028 (table)
    size selection, 291 (table)
CAD. *See* computer-aided drafting
CADD. *See* computer-aided drafting and design
cadmium-silver solders, 253
cadmium-zinc solders, 253
calcium carbonate, 43
capillary action, 251
capped steels, 63
carbide, 141
carbon, 46, 77
carbon and low alloy steels, 595–599.
    See also stainless steels
carbon content in steels
    uses for steel by, 92 (table)
    of various steels, 82
carbon dioxide
    cylinder pressure of, at various temperatures, 687 (table)
    percent of moisture *vs.*, 688 (table)
carbon equivalency, 87
carbon steel electrodes, AWS classification, 305–306
carbon steel pipe, 475–476
carbon steel plate, 597 (table)
carbon steels
    application, mechanical properties and chemical compositions of, 93 (table)
    continuous wire data, 697 (table)
    pig iron, 82

carbon steels—cont.
   plasma arc cutting of, 528 (table)
   welding with MIG/MAG, 727–734
carburizing, 67–68
cascade sequence, 1080
case hardening, 67–69
casting vs. welding, 14, 20
cast iron
   brazeable metals, 268–269
   cast iron welding practice, 238
   compositions of, 94 (table)
   filler rod, 237
   flux, 237
   hard facing (surfacing), 247
   preheating of, 236–237
   techniques for cutting, 197–198
   types of, 95
   welding applications of, 236
   welding of, 236–238
   welding procedures, 237 (table)
cast steel, 62
cementation process, 49
Centers for Disease Control and Prevention
      (CDC), 1039
ceriated tungsten electrodes, 576
chain intermittent fillet welds, 104, 1076
changing electrodes, 503–505
Charpy tests, 935–936
chemical cleaning, 257
chill rings, 477
chrome-moly pipe, 476
chrome-nickel (stainless-steel) pipe, 476
chromium, 80
chromium steels, 90 (table)
chromium-vanadium steels, 90 (table)
circumferential seams, 836
clamping parts, 106
cleaning
   acid, 257
   for aluminum welding, 240–241
   chemical, 257
   interpass, 795
   mechanical, 257
   postweld, 241
   precleaning and surface preparation, 257
closed square-groove butt joint, 892, 894
coal, 37
coarse-grained metals, 72
cobalt, 81
cobalt-base rods, 245
code-making organizations, 483–484
codes and standards (piping), 483–498
code welding
   destructive testing, 901
   mechanical testing, 901
   nondestructive testing, 900
   performance qualification tests, 900
   procedure qualification tests, 899–900
   visual inspection, 901–902
   welder qualification tests, 900
coefficient of thermal expansion, 75, 100
Coffin, C. L., 4
coke, 39–40
cold cracking, 941
cold peening, 108
cold working, 84, 108, 120
columbium (niobium), 81
columnar transferred cutting arc, 527
combination symbols, 986
commercial pipe sizes and wall thicknesses,
      1067 (table)
common defects and acceptable criteria, 495–498
Compressed Gas Association (CGA), 1046
compression, 75
computer-aided drafting (CAD), 951
computer-aided drafting and design (CADD), 951
conduction, 62
constant current welding machines, 280, 281, 773–774
constant potential welding machines, 280
constant voltage welding machines, 671–672
construction machinery welding, 16–18
consumable electrode process, 551
consumable guide welding (CG), 782–783
consumable insert butt joint, 632
contact electrodes, 318
continuous casting, 58–62
continuous current, pulse vs., 560
continuous welds, 119
continuous wire data, 697–699 (table)
contour symbol, 986–987
controller communication, 881–882
convection, 62
conversion factors, metric system (SI), 954 (table)
conversion (of units)
   inches to millimeters, 1062 (table)
   millimeters to decimal inches, 1063 (table)
convexity on fillet welds, 1078
copper and copper-base alloys, 81
   brazeable metals, 268–269
   copper-free GMAW wires vs., 693 (table)
   DCEN and argon shielding gas conditions, 597 (table)
   gas metal arc welding, 745–748
   hard facing (surfacing), 247
   MIG brazing, 748
   soldering, 251
   surfacing materials, 245
   weldability, 601–603
copper filler metals (BCu), 261
copper-free GMAW wires, copper vs., 693 (table)
copper-phosphorus filler metals (BCuP), 261
copper-silicon alloys (ECuSi), 325
copper-tin alloys (ECuSn-A), 325

copper tubing
    braze welding, 251
    capillary action, 251
    fluxes, 253–254
    joint design in soldering, 254–255
    soldering and brazing, 251–274
copper-zinc filler metals (RBCuZn), 261, 262
corner joint, 1074
corrosion, 321
    and heat-resisting alloys, 90 (table)
    testing, 936
corrosion-resistant materials, 581 (table)
coupling to flat plate joint, in the horizontal position, 447–449
covered electrodes
    shielded arc properties of weld metal, 337 (table)
    weld characteristics in mild rolled steel, 338 (table)
cracks and cracking
    base metal cracking, 942
    crater cracks, 497, 741, 942
    hot cracking, 744
    hydrogen cracking, 130–131
    longitudinal cracks, 942
    plastic welding crack repair, 1019
    star cracks, 497
    transverse cracks, 941
    underbead, 131, 555
    weld cracks, 720
crater cracks, 942
crater filling, 342
cross-hatching, 964
cross section volume change *vs.* wire diameter change, 692 (table)
crosswise shrinkage (transverse contraction), 101–102
crucible process, 49
cryogenic stress relieving, 108
crystalline structures, 71 (table)
current controls, 285
current density calculation chart, 691 (table)
current density comparison, 691 (table)
current output, 290
current ranges
    for a.c.-d.c. transformer-rectifier machine, 573 (table)
    for tungsten electrodes, 577 (table)
current selection for tungsten inert gas welding (TIG), 567 (table)
current values, 332–333
current variations with electrode classifications and sizes, 333 (table)
cutting. *See also* air carbon arc cutting; arc cutting; oxyfuel gas cutting; plasma arc cutting
    AWS C4.1 Surface Roughness Guide for Oxygen Cutting, 197
    cast iron, 198–201
    of different metals, 181–182
    draglines, 197
    dross, 197
    equipment, 187–193
    flatness, 197
    flux cutting, 198
    inspection of, 196
    oscillatory motion, 182
    preheating, 181
    roughness, 197
    speeds with abrasive water jets, 827 (table)
    techniques, 182–185
    waster plate, 181
    wire feed, 181–182
cutting tip size, speed, pressure and gas flow rates, 172 (table)
cutting torch function, 179
cyaniding, 67–69
cyclic loading, 77
cylinders
    acetylene cylinders, 142, 1044
    capacity of, 143
    carbon dioxide cylinder pressure, 687 (table)
    care and handling, 1043–1044
    gas cylinders handling, 721
    handling, storage, and operation safety
        acetylene generators, 147–148
        OSHA regulations, 143–144
    Mapp gas, 142–143
    oxyacetylene welding cylinders, 143
    oxygen cylinders, 141, 1044–1045
    storage, 1044

D

Davey, Edmund, 138
DCEN. *See* direct current electrode negative
DCEP. *See* direct current electrode positive
defects. *See also* cracks and cracking; porosity; testing and inspection methods
    brittle welds, 942–943
    convexity, 719
    dimensional defects, 943
    distortion, 943
    fish eyes, 555
    flux inclusions in soldering, 259
    inclusions, 940–941
    incomplete fusion, 719, 938–939
    incomplete penetration, 937–938
    nature of, 132
    in pipe welding, 495–498
    in plastic welding, 1010–1011
    spatter, 720
    troubleshooting guide, 334–336 (table)
    undercutting, 720, 939–940
    wagon tracks, 719
    warpage, 720
    weld and base metal tests for, 946 (table)
    weld contamination, 555

degasifier, 316
degassing equipment. *See also* vacuum furnaces
    consumable electrode vacuum arc melting, 55
    vacuum degassing, 55, 57
    vacuum induction melting, 55, 57–59
density, 75
deoxidizer, 316
depth of fusion, 1079
destructive testing
    corrosion testing, 936
    etching, 934–935
    fatigue testing, 936
    fillet welds, 920
        break test, 931–934
        soundness tests, 930–934
    groove welds, 920
        soundness tests, 927–930
    hydrostatic, 494–495
    impact testing, 935–936
    longitudinal shear tests, 930–931
    nick-break test, 929–930
    performance qualification tests, 922–923
    plastic welds, 1011, 1013, 1014, 1019
    procedure qualification tests, 921–922
    reduced-section tension test, 927–928
    side-bend test, 928–929
    soundness tests, 928–929
    specific gravity, 936
    test specimens, preparation of, 923–927
    transverse shear tests, 930–931
detailed drawings, 951
dimensional defects, 943
dimensioning
    angular dimensions, 957–958
    coordinate dimensions, 959
    dimensions of holes, 958–959
    methods of, 957
    metric system, 953
    tolerances, 959–960
    U.S. customary systems, 953–957
dip brazing (DB), 267–268
direct current and alternating current-direct current inverter, 288
direct current electrode negative (DCEN), 285, 331
    carbon steel plate, 597 (table)
    copper and copper-base alloys, 603 (table)
    gas metal arc welding, 709
    Hastelloy, 603 (table)
    reverse polarity, 1084
    straight polarity, 1084
    stainless steels, 598 (table)
    titanium and titanium-base alloys, 600 (table)
direct current electrode positive (DCEP), 285, 331, 709
direct current transformer-rectifier welding machines, 286, 670–672
direct current welding machines, 6, 280

direction of travel
    vertical-down welding, 499
    vertical-up welding, 499
discontinuities. *See also* cracks and cracking; defects; porosity
    fillet welds, profiles of, 126–128
    heat-affected zone, 131
    inclusions, 131–132
    incomplete fusion, 131
    incomplete joint penetration, 131
    inspection methods *vs.*, 916 (table)
    insufficient throat, 126–127
    nature of, 132–134
    underfill, 132
    weld discontinuity, 125
    with welding processes, 915 (table)
dissimilar metals, 604 (table)
distortion, 100, 105, 943
    prevention before welding, 103–104
    types of, 101–103
distortion control, 104–105
double-bevel groove weld, 1075
double-flare bevel groove weld, 1075
double-flare V-groove weld, 1075
double J-groove butt joints, 896
double-J groove weld, 1075
double U-groove butt joints, 895–896
double-U groove weld, 1075
double V-groove butt joint, 894–895
double-V groove weld, 1075
double-welded joint, 1083
downslope controls, 669
drag cutting, 182–185
drag, kerf, and dross, 1080
draglines, 197
drag technique, 420
drawings. *See also* dimensioning; views
    assembly, 977, 951
    blueprint, 952
    computer-aided drafting, 951
    geometric shapes, 960–962
    lines, types of, 952–953
    NDT letter designations, 988 (table)
    pipe symbols, 965–966 (table)
    purpose of, 949–950
    scale of drawings, 960
    steel drawing, 67
    symbols and abbreviations, 960, 961 (table)
    welding symbol chart, 981–982 (table)
driver-idler rolls, 836
dross, cutting, 197
DSS. *See* duplex stainless steels
dual control, 285
dual shield electrodes, 657
dual tapped-current control, 285
ductility, 3, 77

duplex stainless steels (DSS), 86 (table), 319
Duranickel, 603
duty cycle, 284, 290, 560
dye penetrant, 907–909
dye penetration inspection, 494

E

early developments in welding, 4–5
edge preparation, 240, 794, 1077
edge weld, 349–352, 986, 1074
efficiency, 290
elasticity, 76
elastic limit, 76
elements, properties of, 1069 (table)
electric air processes, 294
electric arc furnace, 49, 50
electrical conductivity, 75
electrical resistance, 75
electric arc furnace, 50
electric arc processes, 6
electric induction furnace, 52
electric shock, 1042
electrode angle, 333
electrode cables, 290–291, 1028 (table)
electrode classification
    alloy steel, 320
    carbon steel
        AWS classification, 305–306, 306 (table)
        mild steel, welding characteristics of,
            304 (table)
    general stainless steel classifications, 320
    groups, 306 (table)
    hard facing
        abrasion, 321
        corrosion, 321
        ferrous base alloys, 321–323
        impact, 321
        nonferrous base alloys, 323
        properties of, 322 (table)
    sizes and current variations, 333 (table)
electrode classifications, specialized
    aluminum electrodes, 323–325
    high nickel electrodes, 325
    phosphor bronzes, 325
    silicon bronzes, 325
    welding rods and electrodes weld metal
        composition, 324 (table)
    specific types, 315–325
    E6010
        all-position, 315–316
        cellulosic type of electrode coating, 315
        DCEP (fast-freeze type), 315–316
        degasifier, 316
        deoxidizer, 316
        pipe welding, 476

E6011
    high cellulose potassium type electrode coating, 316
    pipe welding, 476
E6013, 316
E7010 pipe welding, 476
E7014, 316
E7015 low hydrogen electrodes, 317
E7016 low hydrogen electrodes, 317
E7018
    low hydrogen electrodes, 317
    pipe welding, 476–477
E7024 low hydrogen electrodes, 318
E7028 low hydrogen electrodes, 317
E7048 low hydrogen electrodes, 317–318
E9018G pipe welding, 476–477
E11018G pipe welding, 476–477
electrode classifications, stainless steels
    duplex, 319
    identification of, 320 (table)
    precipitation-hardening, 319
    stainless steel
        austenitic, 319
        ferritic, 319
        martensitic, 319
electrode coverage
    and base metal, 302
    coating, 302
    iron powder, 302
    low hydrogen electrodes, 302
electrode coverings
    arc characteristics and electrode coating, 302
    composition of, 302–303
    functions of, 302
    materials for, 303
    polarity interchangeability, 303
    protective gaseous atmosphere, 302
    shielded arc properties of weld metal, 337 (table)
    slag covering, 302
    weld characteristics in mild rolled steel, 338 (table)
electrode extension, 712 (table)
electrode filler wire
    filler wire classifications, 695–702
    flux-cored electrode wires, 694–695
    metal core electrode wire, 692–694
    solid electrode wire, 691–692
electrode holders
    metal, 293–294
    TIG, sizes of, 607 (table)
electrode numbering system, 305–306
electrode ovens, 326–328
electrodes
    alloy steels, current range for, 305 (table)
    chemical compositions of, 596 (table), 693 (table)
    and current recommendations, air carbon arc
        cutting, 530 (table)
    hard facing, properties of, 322 (table)

electrodes—*cont.*
  markings of, 326
  mild steel
    current range for, 305 (table)
    welding characteristics of, 304 (table)
  moisture control, 326–328
  packing and protection of, 325–328
  shielded metal arc welding, AWS filler metal specifications, 325 (table)
  stainless steel, identification of, 320 (table)
  standard sizes and lengths of, 325–326
electrode selection
  base metal, 310
  electrode selection and welding position, 307, 312, 314
  factors affecting, 308 (table), 315
  joint design and fitup, 310
  operating characteristics of electrodes, 308–310
  production efficiency, 314–315
  size of electrodes, 307
  welding current, 310
  welding position, 312, 314
electrode selector dial, 285
electrode sizes
  and classifications with current variations, 333 (table)
  and heat ranges, 333 (table)
  in submerged arc welding, 777–778
electrode wire types, pulse-welding programs, 883 (table)
electroslag welding (ESW), 4, 781–783
elements, 77
embrittlement, 108
employment as pipefitter, 470 (table)
endurance limit, 77
engine-driven generators, 280, 671, 673
environmental protection, 65
epoxy resin, 997
erection drawings, 951
essential variables, 484
etching, 934–935
Everdur, 602
excess acetylene flame, 206
excessive convexity, 127
excessive melt-through on burning through (BT), 496
excessive penetration, 718–719
excess undercut, 127
expansion and contraction, 102
explosion welding (EXW), 827–828
extrusion, 66–67

F

fabrication, 14
face-centered cubic (FCC), 75
fall protection, 1052
fast-fill electrodes, 308–309

fast-follow electrodes, 309
fast-freeze electrodes, 309
fatigue, 76
  failure, 77
  testing, 77, 936
FCAW. *See* flux cored arc welding
FCAW equipment, 754
  AWS filler metal specifications, 705 (table)
  constant voltage characteristics, 665–666
  continuously variable voltage control, 668
  current controls, 668
  dual schedule controls, 671
  electrode filler wire, 691–702
  inductance controls, 670–671
  output slope, 666–667
  remote control, 671
  shielding gases, 683–691
  shielding gas system, 681–683
  slope controls, 668–670
  tapped-voltage control used for, 667–668
  voltage controls, 667–668
  welding guns and accessories, 678–681
  welding machine selection, 673–674
  wire feeders (wire drives), 674
ferritic stainless steels, 84–85, 84 (table), 319, 597
ferrous metals, 34, 207 (table)
ferrous steels, 91 (table)
filler metals
  aluminum and aluminum-base alloys, 554
  aluminum-silicon filler metals, 261
  brazing filler metals, 261, 262–263 (table)
  copper filler metals, 261
  copper-phosphorus filler metals, 261
  copper-zinc filler metals, 261, 263
  forms of, 261–264
  for gas metal arc welding, 703 (table)
  gold filler metals, 261
  magnesium filler metals, 261
  multipass filler metals, 754
  nickel filler metals, 261–263
  self-shielding filler metals, 754
  silver filler metals, 263
  single-pass filler metals, 754
  SMAW electrode specifications, 325 (table)
  soldering, 251–257
  specifications, 705 (table)
  torch brazing, 261
filler rods, 159
filler wire for gas metal arc welding (GMAW), 794
fillet and groove welding combination project
  with FCAW-G, 732–733
  with FCAW-S, 771
  job qualification, 807–809
fillet welds, 113, 1075
  break test, 931–934
  with excessive convexity, 127

with excess undercut, 127
with incomplete fusion, 127
with insufficient leg, 127
with insufficient throat, 126–127
with other discontinuities, 127, 128
with overlap, 127
and porosity, 127–128
profiles, 126–128, 1085
shrinkage effects on, 103
size, 1079
soundness tests, 920, 930–934
specimens, 458–459
and symbol, 980–983
welder qualifications for, 924 (table)
welding operator qualification, 926 (table)
fill-freeze electrodes, 309
filter lenses, 161
fine-grained metals, 72
fitness, for purpose, 484
fixed inductance controls, 670
fixed-slope machines, 668
flame
   adjustment, 209–211
   recommended for metals and alloys, 206 (table)
flame cutting. *See also* oxyfuel gas cutting
   data for mild steel, 184 (table), 190 (table)
   principles, 179–181
   surface appearance of cuts, 186–187
flame gouging, 184–185
flame hardening, 68
flame piercing, 183–184
flame test, metals identification by, 313 (table)
flame washing, 184
flanged butt joint edge weld, 1081
flanged corner joint edge weld, 1081
flare groove weld, 981–982 (table), 983–986
flash arresters, 145
flashback, 211
flashback arrestors, 211
flash welding, 915 (table), 1077
flat position
   aluminum, 241
   autogenous welding, 214
   beading, 214–216
   keyhole welding, 401
   open root weld procedures, 400
   outside corner joints, 400–403
   parallel joint, 349–352
   single V-butt joints, 102, 393–395
   square butt joints, 399–400
   striking arc short stringer beading, 339–341
   stringer beading, 341–343, 345–347, 389–391
   T-joints, 363–365
   traveling up, 370
   vertical-down welding, 499
   weaved beading, 343–345, 347–349

floating test, for plastic, 997
flowmeters, 558–559
fluorescent penetrant inspection, 910
flux cored arc welding (FCAW)
   advantages of GMAW, 658–659
   on bridge, 27
   current densities, 659
   and gas metal arc welding, 657–660
   inductance effects, 670 (table)
   job outline, 787 (table), 788 (table)
   joint designs for, 758
   process selection, 704–705 (table)
   shielding gas for, 758–762
   technique for, 763–766
   troubleshooting guide, 765–766 (table)
   welding variables for, 762–763
   wire classifications for electrodes in, 757
flux cored arc welding (self-shielded)
   commercial names of, 768
   equipment for, 768
   practice jobs, 769–771
   process advantages, 768–769
   welding technique, 770–771
flux cored electrodes, 754
   chemical compositions of wires, 695 (table)
   hydrogen control of, 757
   mechanical properties of, 760 (table)
   position of welding, polarity and applications, 759 (table)
flux cored wire, 751
flux cored wire welding, 751
fluxes, 159–160
   for brazing, 264
   classifications, 253, 774
   corrosive, 254
   functions of, 253
   inorganic, 253
   intermediate, 254
   noncorrosive, 254
   organic, 253
   paste, 254
   rosin, 253
   soldering, 253
flux indicators, for mechanical properties, 766 (table)
flux selector guide, 254 (table)
forcible restraints, 106
forehand welding, 1083
Fouche, Edmond, 138
fractional and number drill sizes, 1064 (table)
free-cutting steels, 88 (table)
friction stir welding (FSW), 827–828
friction welding (FRW), 826
front view, 963
F-35 Lightning II Joint Strike Stealth Fighter (5th generation), 15

fuel gases
    bushy flame, 181
    comparison of, 180–181
    cutting, special applications, 181
    cutting gases, cost of, 180
    efficiencies, 139 (table)
    flame temperature, 180
    heat concentration, 180
    properties of, 168 (table)
fuels, 37–40
    coal, 37
    coke, 39–40
    natural gas, 38–39
    oil, 38
full annealing, 108
fume extractor, with multiple GMAW-FCAW fume gun installation capable, 1041
furnace brazing (FB), 267
fusibility, 75
fusion, 204
    incomplete, 1083
    welding, 4

G

galvanized sheets, 66
gamma ray sources, strength and usage, 907 (table)
gas control equipment, 558–559
gases, 139–143. See also fuel gases; inert gases and gas mixtures; oxygen; shielding gases
gas flow rates, 558, 794
gas lens, 575
gas metal arc welding (GMAW), 551. See also MIG/MAG welding
    advantages of, 791–792
    aluminum, parameters for, 718 (table)
    arc length, 713
    arc voltage, 713–714, 717
    of certain metals, 745–748
    current, types of, 709–710
    drag nozzle angle, 712
    electrode diameter, 711
    electrode extension, 711–712
    equipment for, 792–793
    filler metals for, 703 (tables)
    filler wire for, 794
    and flux cored arc welding, 657–660
    gas flow rates for, 794
    gas selection for, 684–685 (table)
    globular transfer, 714
    gun ratings, 680 (table)
    invention of, 6
    job outline, 746–747 (table), 813–814 (table)
    joint preparation, 710–711
    operating variables of, 708–720
    pipe welding applications, 790–792
    process of, 552, 656
    process selection, 704–705 (table)
    push nozzle angle, 712
    roping, 714
    shielding gas for, 710, 793–794
    stainless steel, parameters for, 718 (table)
    steel, parameters for, 716 (table)
    stickout, 711
    stubbing, 714–715
    trade names for, 657
    travel angle, 712–713, 717
    travel speed, 716–717
    troubleshooting guide, 728 (table), 796 (table)
    variable adjustments for, 715 (table)
    welding current, 715–716
    welding technique, 726–727
    wire and gas combinations, 683 (table)
    wire feed speed, 714–715
    wire location, 713
    work angle, 713
gas metal arc welding (GMAW) equipment
    arc blow, 724–725
    AWS filler metal specifications, 705 (table)
    bird nesting, 724
    care and use of, 722–726
    cleanliness of base metal, 724
    constant voltage characteristics, 665–666
    contact tubes, care of, 723
    continuously variable voltage control, 668
    current controls, 668
    dual schedule controls, 672
    electrode filler wire, 691–702
    equipment setup, 725–726
    inductance controls, 670–671
    nozzles, care of, 722–723
    output slope, 666–667
    remote control, 671
    shielding gases, 683–691
    shielding gas system, 681–683
    slope controls, 668–670
    starting procedure, 726
    tapped-voltage control used for, 667–668
    voltage controls, 667–668
    welding guns and accessories, 678–681
    welding machine selection, 673–674
    wire-feed cables, care of, 723–724
    wire feeders (wire drives), 674–678
gas shielded arc welding
    compared with shielded metal arc welding, 553–554
    flux cored, 752
    inert gases and gas mixtures, 555–558
    shielding gases for, 555–558
    spot, 553
    weld contamination, 555
gas shielded flux cored wires, 761 (table)
gas shielding history, 549–550

gas tungsten arc cutting (GTAC), 551–552
gas tungsten arc welding (GTAW), 550–551.
    See also tungsten inert gas welding
  argon and helium shielding gases, 556 (table)
  backing, 606
  current ranges for tungsten electrodes, 577 (table)
  current selection for, 567 (table)
  electrodes, 576–579
  equipment setup, 607
  gas control equipment, 558–559
  high frequency, no-touch starting with, 609–610
  inductance effects, 670 (table)
  invention of, 5–6
  job outline, 614 (table), 652–653 (table)
  joint design and practice for, 604–607
  joint preparation for welding of pipe, 633 (table)
  machine controls, 568–574
  piping welding, 630–631
  porosity in, 635
  power sources, 559–561, 567–568
  safety rules for, 608–609
  torches, 574–576
  troubleshooting guide, 612 (table)
  weld contamination, 555
  welding current characteristics, 560–566
gas welding
  of ferrous metals, 207 (table)
  job outline, 249 (table)
  processes, 160
  rods, 159
gas welding equipment
  equal-pressure (balanced-pressure) torch, 153–154
  equal welding torch, 153–154
  friction spark lighter, 158–159
  for GMAW/FCAW, 665–705
  injector welding torch, 153
  master regulator, 153
  for oxyacetylene welding, 148–160
  oxygen and acetylene hose, 157–158
  pressure regulators, 149–152
  protective equipment, 148
  single-stage cylinder regulator, 152
  single-stage station regulator, 152
  torch tips, 154–158
  welding torches, 152–153
gauges, 962
German silver, 81
globular to spray transition welding, 661 (table)
gloves, 297
GMAW. *See* gas metal arc welding
goggles, 161–162, 534
gold filler metals (BAu), 261
gouging
  with air carbon arc cutting, 543
  with plasma arc cutting, 538–539

grain (crystal)
  orientation, 66
  size, 72
grapes, 480, 506
groove, 1077
  angle, 938, 981–982 (table), 1077
  radius, 1077
groove weld project, 809–811
groove weld project beam processes
  electron beam welding, 817–820
  joining and cutting processes, 828–829
  laser beam welding and cutting, 820–828
groove welds, 113–114, 1081
  with discontinuities, 127
  double bevel, 983
  with excessive convexity, 128–129
  flare bevel, 985–986
  with insufficient size, 128
  with overlap, 129
  profiles, 128–129, 1085
  soundness tests, 920, 927–930
  specimens, 458
  symbols for, 983–986
  with undercut, 129
  V-groove weld, 983–986
  welder qualifications for, 921 (table)
  welding operator qualification, 925 (table)
GTAC. *See* gas tungsten arc cutting
GTAW. *See* gas tungsten arc welding
gun angle
  gas metal arc welding, 717
  submerged arc welding, 781
  welding technique, 726–727
  welding variables, 762–763

H

hand and head shields, 294–296
hand flame cutting data for mild steel, 190 (table)
handheld laser scanner
  postweld inspection with, 946 (table)
  preweld inspection with, 946 (table)
hand tools, 851–853
  centering head, 516
  circle-ellipse projector, 515–516
  contour marker, 514–515
  hand cutting torch, 514
  pipe layout tools, 514
  wrap-a-round, 516
hand welding (beading) of plastics, 1015–1017
hard facing (surfacing)
  application of, 246–247
  cobalt-base rods, 245
  hardness, 76
  lead, 244–247
  materials, 245, 605 (table)

hard facing—*cont.*
  metal surfacing, 244–247
  submerged arc surfacing, 779
  surfacing material, 245
  techniques, 247, 605 (table)
hardness tests
  Brinell hardness test, 918
  impact hardness tester, 919–920
  Knoop tester, 919
  microhardness testing, 918–919
  Rockwell hardness test, 918
  typical readings, 917 (table)
  Vickers tester, 919
Hare, Robert, 138
Hastelloy, 603, 603 (table)
HAZ. *See* heat-affected zone
hazardous materials, 1038 (table)
headers, 21
headband system, 1040–1041
head shields, 294–296
hearing protection, 1035–1037
heat-affected zone (HAZ), 995, 1084
heat distribution, 106
Heating, Piping, and Air Conditioning Contractors National Association, 484
heating methods, 267–268
heat ranges and electrode sizes, 333 (table)
heat-resistant materials, 581 (table)
heat transfer, 62
heat treatment, 67
  annealing, 70
  case hardening, 67–69
  normalizing, 70
  tempering, 70
heavy steel plate and pipe
  backhand method, 224
  forehand method, 224
  heavy plate welding practice, 223–224
  of pipe, 224–227
  of plate, 224
  vertical welds in pipe, 228–229
HELIARC. *See* gas tungsten arc welding
helium, 557 (table)
  argon-helium, 689
  helium, 686
  helium-argon-carbon dioxide, 689–690
helmets, 295–296
hematite, 36
Herculoy, 602
Heroult, Paul, 49–50
hexagonal close-package (HCP), 71–72
high carbon and high speed tool steels, 268
high density (linear) polyethylene, 999
high density polyethylene (HDPE), 996 (table)
high frequency, no-touch starting, 609–610
high speed steel, 247

high speed welding of plastics, 1020–1021
high voltage starting, 610
high yield-strength pipe, 476
hole piercing, 539
hoods, 295–296
hopper cars, 23–24
horizontal position, 241
hose and hose connectors, 157–158
hot cracking, 744, 941
hot-gas welding, 994, 1014–1015
hot pass, 502, 506–507
hot shortness, 75, 323
hot starting, 340
Hunter process, 97
hydraulics principles, 847
hydraulic tools
  bending breaks, 847–848
  hand bending brakes, 848
  hand box and pan brakes, 848–849
  hydraulic been machines, 847
  power press brakes, 848
  universal bender, 849–850
hydroforming, 865 (table)
hydrogen pickup, 317
hydrogen weld contamination, 555

I

icicles, 479–480
identifying electrodes, 305–306
IF. *See* incomplete fusion
IFD. *See* incomplete fusion due to cold lap
ilmenite ore, 97
impact, 321
  hardness tester, 919–920
  resistance, 76
  testing, 935–936
inclusions, 940–941
incomplete fusion (IF), 204, 496, 719, 938–939
incomplete fusion due to cold lap (IFD), 496
incomplete joint penetration (IP), 204, 495–496, 718, 937–938
Inconel, 603
indium (In), 252
indium soldering, 252
inductance effects, on GTAW and FCAW, 670 (table)
induction brazing (IB), 267
industrial welding applications, 7–10
inert gases and gas mixtures
  argon and helium, 555–557
  hydrogen, 557
  nitrogen, 557–558
inert gas welding, development of, 5–6
infrared brazing (IRB), 268
infrared radiation, 294
ingots, 62

Innershield electrode, 657
inspection program, welding, 904 (table). *See also* testing and inspection methods
interface
    product pipeline, 472
    weld, 114
intermittent welds, 119
internal concavity (IC), 496
intersection joints, 478–479, 806–807
inverters, 281, 672
IRB. *See* infrared brazing
iron-base rods, 245
iron blast furnace slag, 46
iron carbon phase, 71–72
iron ore, 36
iron powder, 309
iron smelting, 46–49
irregular weld shape, 720
Izod tests, 935–936

J

jasper, 36
jigs and fixtures, 19, 104, 241, 833
job outline
    advanced gas welding and braze welding practice, 249 (table)
    air carbon arc cutting, 546 (table)
    braze welding, 249 (table)
    flux cored arc welding, 787 (table), 788 (table)
    gas metal arc welding, 813–814 (table)
    gas tungsten arc welding, 614 (table), 652–653 (table)
    plasma arc cutting, 543 (table)
    shielded metal arc welding, 387 (table), 427 (table), 468 (table), 519–520 (table)
    submerged arc welding, 789 (table)
job selector dial, 285
joining processes comparison chart, 818 (table)
joint design, 892–898
    butt joints, 892–896
    closed roots, 892
    corner joints, 897
    distortion, 103–104
    for flux cored arc welding, 757–758
    for gas tungsten arc welding, 604–607
    for pipe welding, 498–511, 631–634 (table)
    lap joints, 896–897
    lateral, 806–807
    open roots, 892
    parallel joints, 892
    penetration, 892
    T-joints, 897–898
    tack welding, 499
    y-joints, 806–807
joint design, for pipe
    backing ring, 477, 481
    butt joint of, 477
    chill ring, 477
    circumferential butt-joint, 477
    consumable insert rings, 481
    intersection joints of, 478–479
    pipe clamps, 481
    socket (lap) joint, 477–483
    solid backing ring, 480
    special fabrications, 481, 483
    split backing rings, 480
    thermal expansion, 478
    welded fittings in pipe, 479–483
joint penetration, 1079
joints
    inspection methods, 916 (table)
    preparation, 256, 498, 632
    types of, 113, 893 (table)

K

kerf cutting, 182
keyhole, 501
keyhole effect, 584
keyhole welding, 401
    with lasers, 822
killed steels, 63
k-joints, 806–807
Kroll process, 97

L

lanthanated tungsten electrodes, 576
lap joints, 1074
    back-and-forth whipping motion, 435–436
    in horizontal position, 352–356, 361–363
    multipass procedures, 378
    in overhead position, 435–439
    in vertical position, 359–361, 375–378
laser beam welding and cutting
    cutting, 822
    drilling, 823–824
    laser assisted arc welding, 824–825
    welding, 821
lasers, 820
    types and uses, 821 (table)
laser vision camera, 878–879
lateral joints, 806–807
laydown yard, for storage of pipe spools, 510
layers, 1079
laying out and cutting odd shapes, 197–198
LDPE. *See* low density polyethylene
lead, 81–82, 243–244
    burning, 243
LeChatelier, 138
leg of fillet weld, 1078
lengthwise shrinkage (longitudinal contraction), 101

lenses
   for current settings, 608 (table)
   shade number selection guide, 1032 (table)
leuxocene ores, 97
lift arc start, 610
limestone, 43–44, 45–48
limonite, 36
linear coefficient of thermal expansion, 75
liquid vessel systems, 147
load voltage, 281
long arc technique, 392
longitudinal cracks, 942
low carbon and low alloy steel, 268
low density (branched) polyethylene, 998–999
low density polyethylene (LDPE), 996 (table)
low-hydrogen electrodes, 302, 309, 316–318, 430
low temperature pipe, 476

M

machine flame cutting data, 184 (table)
machine tool
   drill presses, 862
   elastic wheels, 861
   hydroforming, 863–865
   lathes, 860
   metal-cutting band saws, 865–868
   milling machines, 860–861
   pedestal grinder, 861
   power punches, 863
   silicate wheels, 861
   steel *vs*. cast iron construction, 19–20
   surface grinders, 861
   vitrified wheels, 861
   water jet cutting, 867
magnesium and magnesium-base alloys
   brazeable metals, 268
   gas metal arc welding, 748
   oxyacetylene process, 242–243
   stabilized A.C. current and shielding gas
      conditions, 600 (table)
magnesium filler metals (BMg), 261
magnetic arc blow, 873
magnetic arc control, 872–873
magnetic particle testing
magnetite, 36
MAG welding, 551. *See also* gas metal arc welding;
      MIG/MAG welding
   carbide precipitation, 744–745
   hot cracking, 744
   hot shortness, 744
   process, 551
   pulse arc spray, 744
   short arc welding, 744
   spray arc welding, 744
   of stainless steel, 742–745

maintenance, 14, 284
malleability, 76
manganese, 80
manganese steels, 88 (table)
manifold distribution, 145–147
manifold system, 145, 510
manipulators, 837
manual flame cutting (oxyfuel cutting), 178
Mapp gas, 142–143
maraging steel, 556 (table), 580 (table)
markings of electrodes, 326
martensitic stainless steel, 83–84, 84 (table), 319, 597
mean time between failures (MTBF), 886
mechanical cleaning, 257
mechanical loading, 108
mechanical properties, flux indicators for, 766 (table)
medium density polyethylene, 999
melamine formaldehyde (MF), 996–997
melting glue (hot glue), 1022
melting point, 72–73, 1068 (table)
melting ranges
   solders containing other metals, 252 (table)
   of tin/lead solders, 252 (table)
melt-through welds, 1081
Meritens, August de, 4
metal active gas (MAG) welding. *See* gas metal arc welding;
      MAG welding
metal compositions properties, 1069 (table)
metal core electrode wires, chemical compositions
      of, 694 (table)
metal expansion, 105
metal identification
   by appearance, 312 (table)
   by chips, 314 (table)
   by flame test, 313 (table)
   by spark testing, 1070 (table)
metal inert gas (MIG). *See* gas metal arc welding; MIG welding
metal internal structures, 70–72
metallurgy, 70
metal products, unique applications and markets, 99 (table)
metals
   alloying elements, 77
   expansion of, 105
   phases (crystalline structures), 71 (table)
   physical and chemical properties, 72–77, 78–79 (table)
   properties related to energy, 72–75
   stress resistance, 75–77
metal solderability chart and flux selector guide, 254 (table)
metal surfacing, 244
metal transfer
   buried arc transfer, 663–664
   enhanced short circuit transfer, 665
   globular transfer, 662
   nonrotational spray, 662
   open arc method, 660
   open arc transfer, 660–664

pinch force, 664
pulsed-arc transfer, 662–663
Regulated Metal Deposition, 665
rotational spray, 662
short circuit method, 660, 664–665
spray transfer, 660–662
Surface Tension Transfer, 665
metalworking
  cupping, 66
  flash welding, 66
  forging, 65
  history, 3–7
  piercing, 66
  rolling, 65–66
Metal-X, 98
metric system (SI)
  common measurements, 954 (table)
  conversion factors, 954 (table)
  multiples and submultiples of base 10, 955 (table)
microfissures (microcracks), 941
microhardness testing, 918–919
Micro Wire, 552
MIG/MAG welding, 551. *See also* gas metal arc welding; MAG welding
  of carbon steel, 727–734
  fillet and groove welding combination projects, 766
  fillet welds, 731
  groove welds, 727–731
  overhead position, 739–740
  overview, 656–657
  process, 552
  troubleshooting guide, 728–730 (table)
MIG welding, 551. *See also* gas metal arc welding
  of aluminum, 734–742
  butt joints, 740
  corner joints, 741
  crater cracks, 741
  groove weld project, 742
  inclusions, 742
  joint preparation, 738
  lap joints, 740
  parallel joints, 741
  process, 551
  shielding gases, 738
  spray arc welding, 738–739
  T-joints, 740–741
  welding positions, 738
mild steel
  carbon content, 82
  current range for, 305 (table)
  flame cutting data for, 184 (table)
  hand flame cutting data for, 190 (table)
  machine flame cutting data for, 184 (table)
  welding characteristics of, 304 (table)
Millermatic, 552
modified high impact rigid polyvinyl chloride, 998

modular tooling, 834–835
modulus of elasticity, 76–77, 100
moisture control, 326–328
mold, 62
molybdenum, 80
molybdenum steels, 89
Monel, 603
MTBF. *See* mean time between failures
multipass filler metals, 754
multipass sequence, 1080
multiple-operator welding systems, 289
multiples and submultiples
  of base 10, 955 (table)
  of the meter, 955 (table)
multi-purpose face shield, 162

N

National Institute for Occupational Safety and Health (NIOSH), 1039
natural gas, 38–39
neutral fluxes, 775
nick-break test, 929–930
nickel and nickel-base alloys, 80–81
  brazeable metals, 268
  filler metals, 261–263
  gas metal arc welding, 748
  gas tungsten arc welding, 603
  nickel-chromium steels, 89
  nickel steels, 89 (table)
  rods, 245
nickel-base rods, 245
Nimonics, 603
niobium (columbium), 81
NIOSH. *See* National Institute for Occupational Safety and Health
nitrides, 555
nitride weld contamination, 555
nitriding, 68
nitrogen weld contamination, 555
noise protection, 1035–1037
nonconsumable electrode process, 550–551
nondestructive testing (NDT). *See also* testing and inspection methods; visual inspection
  dye penetrant, 907–909
  eddy current testing, 912–914
  fluorescent penetrant, 910
  handheld weld scanners, 132–134
  hardness tests, 916–918
  leak tests, 914–916
  letter designations, 988 (table)
  magnetic particle testing, 902–907
  palm organizer, 132–134
  penetrant inspection, 907–910
  for plastic welds, 1013–1014
  radiographic inspection, 906–907

nondestructive testing—*cont.*
   surface inspection for soldering, 259
   ultrasonic inspection, 910–912
   water pressure test for soldering, 259–260
nonferrous metals, 34
nonferrous pipe, 476
nonmetals, 77–80
nonoccupational noise and occupational noise comparison, 1036 (table)
notch, 197
nuclear power, 20–21
nugget size, 1081
nylon, 996

O

OAC. *See* oxygen arc cutting
OAW. *See* oxyacetylene welding
Occupational Safety and Health Administration (OSHA) Act, 1029
OFC. *See* oxyfuel gas cutting
OHW. *See* oxyhydrogen welding
oil, 38
OLC. *See* oxygen lance cutting
open circuit and arc voltage open circuit voltage, 281–283
open root weld procedures, 400
open square-groove butt joint, 894
OSHA Hazard Communication Standard, 1037
output slope, 280, 666–667
   constant potential, 666
   constant voltage, 665
outside corner joints
   in flat position, 400–403
   in vertical position, 418–420
overhead position, 121
   aluminum, 241
   autogenous welding, 218–219
   beading, 219
   4G test, 452
   lap joints, 435–439
   lead, 243
   MIG welding, 739–740
   single V-butt joints, 452–465
   single V-butt joints (backing bar construction), 450–452
   stringer beading, 389–391
   T-joints, 431–435, 441–444
   undercutting, 393
   weaved beading, 391–393
overheating, 75
overlap, 1080
   fillet welds with, 127
   groove welds with, 128 (table)
overpeening, 108
overwelding, 106
oxidizing flame, 239

oxyacetylene cutting (OFC-A), 166–167, 1043–1052
oxyacetylene flames, 206
oxyacetylene welding (OAW), 137
   acetylene, 141–142
   and cutting, 1043–1052
   cylinder, 143
   development of, 4
   gases, 139–141
   history of, 138–139
   lead, 243–244
   magnesium alloys, 242–243
   Mapp gas, 142–143
   oxygen, 139–141
   portable equipment setup, 242–243
   propane gas, 142
   safe handling, 143
oxyacetylene welding, practice jobs, 212–223
   autogenous welding
      flat position, 213–214
      overhead position, 218–219
   beading
      flat position, 214–216
      overhead position, 219
   edge and corner joints, 219–220
   lap joint, 222, 223
   pipe, 224–229
   plate, 223–224
   square butt joint, 220–221
   T-joint, 222–223
oxyacetylene welding equipment
   backflow protection, 148
   flashback protection, 148
   hoses, 157–158
   pressure regulators, 149–152
   torches, 152–157
   torch tips, 154–157
oxyacetylene welding flame
   acetylene feather, 206
   carburizing flame, 206
   excess acetylene flame, 206
   excess oxygen flame, 206–207
   inner core, 206
   neutral flame, 206
   outer envelope, 206
   oxyacetylene flames, 206
   reducing flame, 206
oxyfuel gas cutting (OFC), 179
   beam cutting, 174
   cutting tips, 169, 172
   equipment for, 168–171
   friction lighter, 169
   gloves, 171
   goggles, 169, 171
   high speed tip, 172
   job outline, 193 (table)
   lighters, 169

oxyfuel gas cutting machines, 171–173
stack cutting, 173–174
standard tip, 172
oxygen, 36–37, 41–42
production and distribution, 140–141
safety, in handling oxygen gases, 141
steel's appetite for, 41
valve mechanisms, 140–141
weld contamination, 555
oxygen and acetylene hose, 157–158
oxygen arc cutting (OAC), 530–534
oxygen cylinders, 141, 1044–1045
oxygen lance cutting (OLC), 175
oxygen process, 53–55
oxyhydrogen welding (OHW), 160

P

packing and protection of electrodes, 325–328
parallel joint, 349–352, 616–618, 986, 1074
passes, number of, 499–500, 1079
paste solders, 253
pasty range, 252
peening, 103, 108
penetrant inspection
fluorescent penetrant, 910
red dye penetrant, 907–909
Spotcheck, 909–910
penetration, 204
incomplete, 1083
percent of elongation, 77
perpendicular shrinkage, 103
personal protective equipment (PPE), 1029–1043
apron, 1030
auto-darkening welding lens, 1034–1035
boots, 1029–1030
burn protection, 1041
cape sleeves, 1030
clothing, 1029–1030
coveralls, 1030
ear protection, 298, 1035–1037
electric shock, 1042
eye protection, 720–721, 1030–1035
face and body protection, 720–721
fall protection, 1042–1043
filter lenses, 161
flash guard check valves, 162–163
gloves, 162, 1030
hand and head shields, 294–296, 1031–1033
hazardous fumes, 1038–1040
headgear, 1040–1041
helmets, 1031–1033
protective clothing, 162, 297–298, 1029–1030
respirator filter selection, 1039 (table)
respiratory protection, 1037–1040
shirts, 1030

safety glasses, 297
shock protection, 1042
slacks, 1030
sound sources (noise) comparison, 1036 (table)
toxic materials, 1037 (table)
welding and cutting goggles, 161–162
welding jackets
phosphor bronzes, 325
phosphorus, 80
pickling, 257
pig iron, 47
pilot arc, 526
pipe
carbon steel pipe, 475–476
chrome-moly pipe, 476
chrome-nickel (stainless-steel) pipe, 476
dimensions, 475 (table)
high yield-strength, 476
low temperature, 476
nonferrous, 476
plastic. See plastic pipe
sizes, 273
specifications and materials, 475–476
welding. See pipe welding
pipe fabrication
hand tools, 514–516
power tools, 511–514
practice, 509–511
pipeline
construction, 472–473
crawlers, 493
safety bill, 473
welding, procedure qualification tests, 487 (table), 488 (table)
pipe positions during welding, 1082
pipe welding
backhand, 227–229
and branch welds, 645–649
cover pass, filler pass, 664
edge preparation, 794
fitup and tacking, 794–795
gas tungsten arc welding, 360–361
history of, 470
in the horizontal fixed position (bell hole weld), 643–644
horizontal fixed position with 180° rotation, 640–641
inspection and testing of pipe welds, 644–645
interpass cleaning for, 795
joint design for, 631–634
practice
horizontal fixed position (forehand technique), 226–227
roll positions, 226
and roll butt joint groove welds, 636–640
shielded metal arc welding, 475–483
in vertical fixed position, 641–643
vertical welds in horizontal fixed position, 228–229

pipe welding, butt joints, 797–807
    with backing in 45 degree from horizontal fixed position (6G), 507–508
    fixed horizontal position (5G), 503–507
    fixed vertical position, 502–503
    horizontal pipe axis
        bell-hole position, 801–804
        fixed horizontal position (5G uphill travel), 804–806
        fixed position (5G) downhill, 801–804
        roll positions, 797–799
    intersection joints, 806–807
    rolled horizontal position (1G), 500–502
    vertical pipe axis, fixed position (2G), 799–801
    without backing in 45 degree horizontal fixed position (6GR), 508–509
pipe welding electrodes. *See* electrode classification
piping, 21–22. *See also* joint design for piping; pipe
    bevels, 478 (table)
    chill rings, 477
    code-making organizations, 483–484
    codes and standards, 483–498
    dimensions, 475 (table)
    pipe weld defects and acceptance criteria, 495–498
    and pipelines, 470–472
    procedure for welding qualifications, 484–490
    symbols, 965–966 (table)
    testing and inspection methods, 490–495
pitch, 119
plasma arc, starting, 526–527
plasma arc cutting (PAC)
    advantages of, 528–529
    of aluminum alloys, 529 (table)
    bevel cutting with, 538
    of carbon steel, 528 (table)
    gouging with, 538–539
    ion, 523
    job outline, 543 (table)
    manual plasma arc cutting, 526–528
    mechanized plasma arc cutting, 524–526
    plasma, 523
    shape cutting with, 539–542
    square cutting with, 538
    of stainless steels, 529 (table)
    transferred arc, 524
    troubleshooting guide for, 537 (table)
plasma arc welding (PAW), 581
    equipment for, 584–585
    welding gases, 585
plasma needle arc process, 586 (table)
plasticity, 76
plastic pipe
    care of, 1023–1024
    welding, 1021–1022
plastics
    burning test, 996–997, 997 (table)
    characteristics of, 998–999
    gluing of, 1022–1023
    overview, 994–995
    preparation of, 1000–1002
    symbols used to identify, 996 (table)
    tests to identify, 997–998
    thermoplastics, types of, 998–999
    welding of, 999–1009. *See also* plastic welding
    without symbols, 996–997
plastic strip welding, 1021
plastic welding
    causes of faulty welds, 1012–1013 (table)
    crack repair, 1019
    hand welding (beading) of, 1015–1017
    high speed welding, 1020–1021
    hot gas welding, 1004–1005
    linear vibration welding, 1003
    pipe, 1021
    procedure, 1009
    tack welding of, 1015
    thermoplastic welding chart, 1007 (table)
    ultrasonic welding, 1003
plastic welds
    defects in, 1010–1011
    destructive testing for, 1011–1014
    joint types, 1017–1019
    nondestructive testing for, 1013
plug welds, 114, 1076
pneumatic motors, 854
polarity, 331–332
polarity switch, 285
polyaddition adhesive (epoxies), 1023
polycarbonates, 996
polycondensate (phenolics), 1022–1023
polyester resin, 997
polyethylene (PE), 998–999
polyethylene terephthalte (PET), 996 (table)
polymerisate (cyanacrylates), 1023
polypropylene (PP), 999
polystyrene (PS), 996 (table)
polyurethane (rubbery plastic), 996
polyvinyl chloride (PVC), 998
pores, 127
porosity, 75, 127–128
    acceptance criteria for, 497 (table)
    base metal porosity, 635
    causes of, 719–720, 941
    contamination, 635
    limits, 128 (table)
    in pipe welds, 497
portable power tools, 853–854
    beveling machines, 857–859
    electric hammers, 854–855
    electric hand drill, 854
    grinders, 855
    magnetic-base drill press, 857
    nibblers, 855–857

sanders, 855
shears, 855–857
weld shavers, 859–860
postheating, to control residual stress, 107
postweld inspection, 946 (table)
postweld heat treatment (PWHT), 845
power cable, 290
power factor, 290
power saw for cutting various diameters, 513
power sources
    air carbon arc cutting and gouging, 532 (table)
    arc length and amperage settings, 677 (table)
    drooping voltage (droopers), 559
    duty cycle, 560
    gas tungsten arc welding, 559–562, 567–568
    pulse control, 560
    pulse *vs.* continuous current, 560
    rotating welding power source, 559
    shielded metal arc welding, 279–280, 288–289
    SMAW derating for TIG welding, 563 (table)
    static welding power source, 559
    submerged arc welding, 773–774
    transformer-rectifiers, 559
    tungsten inert gas welding, 559–560, 567
    types, 287–288
    welding machine controls, 568–574
power squaring shears, 850–851
power supplies, 280
    ratings, 290
power tools
    abrasive power saw, 513
    Bev-L Grinder, 514
    cutter pipe saw, 513
    pipe-beveling machines, 511–512
    pipe saw, 511
    portable power hacksaw, 512
prebending, 103–104
precipitation-hardening (PH) stainless steel, 319
precleaning and surface preparation, 257
preheating, 107, 236–237
    equipment, 845–846
    temperatures for, 108 (table)
pressure piping welding, procedure qualification tests, 487 (table)
pressure regulators
    design, 149
    fire safety precautions, 149–151
    regulator construction, 151–152
    regulator design, 149
    single-stage regulator, 152
    two-stage regulator, 152
pressure vessel, 990
    bill of materials, 992
    construction, 29–31
    drawing, 991
pressure welds, test weld evaluations, 459–460

preweld inspection, 946 (table)
printing drawings, 951–952
procedure qualification tests
    pipeline welding, 487 (table), 488–489 (table)
    pressure piping welding, 487 (table)
    procedures for, 484–486, 490
production efficiency, electrode, 314–315
projection welds, 1076
propane gas, 142
protection against heat stress, 1043
protective clothing. *See* personal protective equipment
protective lenses. *See* eye protection
PTFE, 996
pulse current, continuous *vs.*, 560
pulse-welding programs, shielding gases and electrode wire type, 883 (table)
purging, 632–633
PWHT. *See* postweld heat treatment

Q

qualifications for welding, 8–10
    code pipe welding, 490
    fillet welds, 924 (table), 926 (table)
    groove welds, 921 (table), 925 (table)
    pipeline welding, 487 (table), 488 (table)
    pressure piping welding, 487 (table)
    procedures for, 484–490
    production welding qualified by plate or pipe welds, 925 (table)
    robotic arc welding personnel, 887–889
    Welder Performance Qualification Test, 486
    Welding Procedure Qualification Record, 484
    Welding Procedure Specification, 484

R

radiation, 62
radiographs, 907
railroad equipment, 22–25
    hopper cars, 23–24
    tank cars, 23
rate of heating and distortion, 103
red dye penetrant, 907–909
refractory materials, 44, 46
reinforcement, 205, 1078
remote control, 287, 286
residual stress, 100, 107–109
resilience, 77
resistance brazing (RB), 267
resistance welding, 7
resistance welds, 1082
respirator filter selection, 1039 (table)
reverse drag cutting, 182
RG-45 gas welding rods, 159
RG-60 gas welding rods, 159

RG-65 gas welding rods, 159
RIA. *See* Robotic Industries Association
right side view, 963
rimmed steels, 63–64
riveting *vs.* welding, 27, 29–30
robotic arc welding systems
    articulated robot arms, 883, 885
    personnel qualifications, 887–889
    rectilinear robot arms, 883
    robot, 883
Robotic Industries Association (RIA), 883
robots
    programming, 886–887
    ratings of, 885–886
    robotic welding applications, 872
Rockwell hardness test, 76, 918
rods
    chemical composition, 596 (table)
    sizes for welding of aluminum, 240 (table)
roll butt joint groove welds
    freehand method, 639
    and pipe welding, 636–640
    "walking the cup" method, 639
root edge, 1077
root face, 1077
root opening, 1077
rubbers (elastomers), 997
rutile, 97

S

SAE/AISI classification system
    American Iron and Steel Institute, 87
    chromium steels, 90 (table)
    chromium-vanadium steels, 90 (table)
    corrosion and heat-resisting alloys, 90 (table)
    manganese steels, 88 (table)
    molybdenum steels, 89
    nickel-chromium steels, 89 (table)
    nickel steels, 89 (table)
safety equipment. *See* personal protective equipment
safety practices. *See also* personal protective equipment
    acetylene cylinders, 1044
    backfire, 1048
    cables and electrical connection safety, 1027–1028
    cylinders
        care and handling of, 1043–1044
        storage of, 1044
    electrical safety, 722
    electric welding process, 1027–1043
    electrode holder safety, 1028–1029
    fire protection, 1050–1051
    fire safety, 722
    flashback, 1048–1049
    gas cylinders, 721, 1043–1045
    in gas welding and gas cutting, 212
    hazardous materials, 1037 (table)
    hose safety, 1046
    maintenance equipment and safety, 1029
    MIG/MAG gun safety, 1028–1029
    operating procedures, 163
    for oxyacetylene welding and cutting, 1043–1052
    oxygen cylinders, 1044–1045
    pacemakers, 9
    pipeline safety bill, 473
    for plastic welding, 1022–1023
    preignition, 1048
    pressure-reducing regulators, 1047
    TIG torch safety, 1028–1029
    torch safety, 1045–1046
    toxic materials, 1037 (table)
    ultraviolet radiation threshold limits, 1031 (table)
    ventilation, 721
    welding and cutting operations, 1047–1049
    welding machine safety, 1027
    wire-feeder safety, 722
    work in confined spaces, 1049–1050
    working conditions, 10
scratching test, for plastic, 997
scratch starting, 610
screens and booths, 832–833
section lines, 964
seam weld, 1077
selenium, 80
self-shielded flux cored arc welding (FCAW-S), 657, 752–753
self-shielded flux cored wires, 760 (table)
self-shielding filler metals, 754
semikilled steels, 63
sensors, units of measure and, 877 (table)
shape cutting, 539–542
shear, 75
shielded metal arc welding (SMAW), 278
    a.c.-d.c. transformer-rectifier welding machines, 286–287
    a.c. power sources, 287
    a.c. transformer welding machines, 287–288
    arc power sources, 288–289
    constant current welding machines, 281
    controls of the engine-driven generator, 285
    cost comparisons, 288–289
    d.c. transformer-rectifier welding machines, 286
    development of, 5
    drooping voltage (droopers), 281
    electrodes AWS filler metal specifications, 325 (table)
    electrodes for, 301, 476–477
    engine-driven generator welding machines, 283–285
    vs. gas shielded arc welding process, 553–554
    multiple-operator welding systems, 289
    operating principles of, 279
    output slope, 280, 281, 283
    of pipe, 476
    power source derating for TIG welding, 563 (table)

power source types, 280–281
practice, job outline, 387 (table), 427 (table), 468 (table), 519–520 (table)
welding machines, 286–287, 288
welding power sources, 279–283
welding systems, 289

shielding gases
argon, 686
argon and helium in GTAW, 556 (table)
argon-carbon dioxide, 689
argon-helium, 689
argon-oxygen, 690
carbon dioxide, 686–688
dew point, 687
droplet rate, 690
for flux cored arc welding, 764
for gas metal arc welding, 793–794
gas mixtures, 688–691
helium, 686
helium-argon-carbon dioxide, 690
nitrogen, 690
pulse-welding programs, 883 (table)
shielding gas designators, 690–691
thermal conductivity, 686

shielding gas system
flow rates, factors affecting, 683
gas flow rates, 682–683
laminar flow, 682
specific gravity, 682
turbulence, 682

shipbuilding, 25–26
shock protection, 1042–1043
shop drawings, 949–950
shrinkage, 100
shrink welding, 105
shutoff valves, 559
siderite, 36
silicon, 80
silicon bronzes, 325
silicon-manganese steels, 90 (table)

silver
filler metals, 263
soldering, 251

S.I. metric and U.S. (customary) equivalent linear measurements, 1060 (table)
S.I. metric and U.S. (customary) equivalent measurements of area, 1060 (table)
S.I. metric and U.S. (customary) equivalent measurements of mass (weight), 1061 (table)
S.I. metric and U.S. (customary) equivalent measurements of volume (capacity), 1061 (table)
SI metric measurement systems, 953–957
single-bevel groove weld, 1075
single-flare bevel groove weld, 1075
single-flare V-groove weld, 1075
single J-groove butt joints, 896

single-J groove weld, 1075
single-pass filler metals, 754
single U-groove butt joints, 895
single-U groove weld, 10
single V-butt joints, 452
in position, 393–395, 403–406
in horizontal position, 406, 444–447
in horizontal position (2F), 406–408
in horizontal position (2G), 397–399, 444–447
in overhead position, 450–454
undercutting, 453
in vertical position, 414–416, 422–424, 429–431, 439–441
single V-groove butt joints, 894
single-V groove weld, 1075
single welded joints, 1082
size of welding machines, 284
skip-stop step-back method, 105
slab, 65
slag, 46
slag inclusions (SI), 496–497
slang and trade names, 657
Slavianoff, N. G., 4
sling-type rolls, 836
slope controls, 668–670
slot welds, 114, 1076
smart pig, 493
SMAW. *See* shielded metal arc welding
smelting, of iron, 46–49
soaking pit, 62
Society of Automotive Engineers (SAE), 87
socket joint for fillet welding or brazing, 1084
solderability chart, 254 (table)
soldering
brazing *vs.*, 251
cadmium-silver solders, 253
cadmium-zinc solders, 253
clearance, 254
copper tubing, 250–251
definition, 251
filler metals, 251–253
fluxes, 253–254, 254 (table)
flux inclusions, 259
heating methods, 255–256
joint design, 254–255
paste solders, 253
practice jobs, 257–260
preparation, 256–257
solders, forms of, 253
surface inspection, 259
tin-antimony-lead solders, 253
tin-antimony solders, 253
tin-lead solders, 252
tin-zinc solders, 253
water pressure test, for soldering, 259–260
zinc-aluminum solder, 253
solders, melting ranges of, 252 (table)

solidification, 62–64
solid-state welding, 826
solvent
   degreasing, 257
   welding, 1022
sound sources comparison (noise levels), 1036 (table)
sound test, for plastic, 997
sound weld characteristics, 204–205
spacing of parts, 104
spark testing, metal identification by, 1070 (table)
spatter, 720
specific gravity, 75
specific gravity test, 936
spot gouging, 184
spot weld, 1076
spot welder, 846–847
Spotcheck, 909–910
square butt joints
   in flat position, 399–400
   in vertical position, 416–418
square cutting, with plasma arc, 538
square groove, 1075
stabilized a.c. current and shielding gas conditions
   aluminum, 596 (table)
   magnesium, 600 (table)
stack cutting, 535
staggered intermittent fillet welds, 104, 1076
stainless steels
   austenitic, 86, 85, 85 (table), 319, 597
   brazeable metals, 268–269
   classifications of, 83–87, 319–320, 597
   DCEN and argon shielding gas, 598 (table)
   duplex stainless steel alloys, 86, 86 (table), 319
   electrodes, 320 (table)
   ferritic, 84–85, 84 (table), 319, 597
   GMAW parameters, 718 (table)
   heat-resisting steels, 83–87
   helium, 598
   MAG welding, 743–744
   martensitic, 83–84, 84 (table), 319, 597
   oxidation, 83–87
   pipe specifications and materials, 475–476
   plasma arc cutting, 529 (table)
   precipitation-hardening, 319
   stainless steel sensitization, 744–745
   uses for, 83
   welding technique for, 623
standard pipe data, 1066 (table)
star cracks, 497
start and stop controls, 285
states of matter, 70
steel
   alloy steels, 82–87
   carbon content of, 82
   carbon steels, 82
   effects of common elements on, 77–82
   GMAW parameters, 716 (table)
   history of, 35–36
   production, 35
   raw materials for making of, 36–46
   scrap, 41–43
   stainless. *See* stainless steels
   stainless heat-resisting steels, 83–87
   uses by carbon content, 92 (table)
steelmaking, 35–36, 51–69
steel weldments, 19–20
stick electrode welding, 278, 300
stick welding process, 5
stiffening, 105
straight line cutting, 182
   and bevel cutting, 193–197
strain hardening, 108
strength, 76
   of welds, 124–125
stress, 100
   concentrations of, 125
   and distortion, 100–103
   relieving, 107–108
stringer beading, 120
   arc establishment, 339–340
   in flat position, 341–343, 345–347, 370–372
   in horizontal position, 356–357
   in overhead position, 389–391
   travel angle, 341
   in vertical position, 357–359
stringer beads, 1080
stripper pass, 507
strong backs, 104
structural steel construction
   bridges, 27
   industrial and commercial buildings, 27–29
submerged arc welding (SAW)
   alternating current, 776
   automatic submerged arc welding, 772
   electrode size, 777–778
   flux coverage, 776
   flux depth, 776
   flux recovery, 776
   gun angle for, 781
   impact requirements, 775 (table)
   job outline, 789 (table)
   joint fitup, 776
   multiple-wire technique, 778–779
   operating variables, 777
   polarity, 775–776
   power sources, 773–774
   process selection, 704–705 (table)
   semiautomatic, 779–780
      electrogas welding, 783–784
      electroslag welding, 781–783
   starting and stopping tabs, 776
   submerged arc surfacing, 779

weld backing, 776–777
welding technique for, 781
wire and flux classifications, 774–775
wire sizes, 775
sulphur, 80
supplementary essential variables, 484
Surface Tension Transfer (STT), 665
surface welds. *See* bead welds
surfacing. *See* beading; hard facing (surfacing)
sweating, 246
symbols. *See also* drawings and drawing information
    and abbreviations, 960, 961 (table)
    ANSI pipe symbols, 965–966 (table)
    combination symbols, 986
    groove welds, 983–986
    for nondestructive examination, 987
    used to identify plastics, 996 (table)
    weld all around, 983
    welding applications, 987–988
    welding symbols, 981–982 (table)

T

tack welds, 107, 119–120, 633–634, 1015–1024
tail, 979
TBW. *See* temper bead welding
temper bead welding (TBW), 1084
tank cars, 23
tank and pressure vessel construction, 29–31
tension, 76
terne plate, 66
testing and inspection methods. *See also* destructive testing; nondestructive testing; visual inspection
    coupon testing, 495
    fillet weld specimens, 458–459
    groove weld specimens, 458–459
    hydrostatic testing, 494
    inspection methods
        vs. discontinuities, 916 (table)
        by joint types, 916 (table)
        selection guide, 917 (table)
    liquid penetrant inspection, 494
    magnetic particle inspection, 494
    pressure welds, 459–460
    pressure welds, test weld evaluations, 465
    radiographic inspection, 493
    service tests, 495
    test specimens, 458, 464, 465
    trepanning, 495
    ultrasonic inspection, 494
    weld and base metal defects, 946 (table)
tests. *See also* qualifications for welding
    base metal and its preparation, 455
    electrode classifications, 459
    electrodes, types of used for testing, 456
    fillet welds, 457

    groove welds
        with backing bar, 457
        without backing bar, 457
    group F-numbers, 456
    number of test welds required, 456–457
    requirements for passage of, 465
    results, evaluation of, 460, 465
    specimens, 458–460, 464–465, 464 (table)
    test welds, 455–456, 490
    types of, 454–455
    welding procedure, 457–458, 460–464
thermal conductivity, 75, 100
thermoplastic polymers, 995
    types of, 998–999
    welding, 994, 1007 (table)
thermoset plastics, 996
thermosetting polymers, 995
thoriated tungsten electrodes, 576
Thru-Arc™ sensor, 874
TIG welding. *See* tungsten inert gas welding
tin (Sn)
    plate, 66
    soldering, 252
tin-antimony solders, 253
tin-antimony-lead solders, 253
tin-lead solders, 252
tin-zinc solders, 253
tip sizes
    for steel plates of different thicknesses, 209 (table)
    for welding of aluminum, 240 (table)
titanium and titanium-base alloys
    alpha-beta titanium alloy, 600
    alpha titanium alloy, 600
    chemical compositions of, 99 (table)
    DCEN and argon shielding gas, 600 (table)
    gas metal arc welding, 748
    mechanical properties of, 98 (table)
    open air welding, 601
    production of, 97–98
    properties and uses of, 97–98
    titanium (CP), 600
    trailing gas shield, 601
    types of, 600
    weldability ratings, 601 (table)
    welding, effects of on metal, 99–109
    and zirconium, 81
T-joints, 1074
    drag technique, 420
    in flat position, 363–369
    in horizontal position, 378–385, 395–397, 420–422
    long arc technique, 408, 411
    in overhead position, 431–435, 441–444
    practice jobs, 807
    in vertical position, 410–414
    vertical-up welding, 412

T-joints—*cont.*
    weld symbol, 420
    whipping the electrode, 408, 411
tolerance limits, 959–960
tooling, 833–835. *See also* jigs and fixtures
tools
    hand, 514–516
    hydraulic, 847
    portable power, 853–854
    power, 511–514
top edge rounding, 197
top view, 963
torch and filler metal angles, 150
torch brazing (TB)
    brazeable metals, 268–269
    filler metals, 261–264
    fluxes, 264
    hard solders, 260
    heating methods, 267–268
    industrial applications, 260–261
    joint design for, 265–267
    liquidus temperature, 260
    melting range, 260
    solidus temperature, 260
torsion, 76
toughness, 76
toxic materials, 1037 (table)
transformer-rectifiers, 280
transverse contraction (crosswise shrinkage), 101–102
transverse cracks, 941
transverse shrinkage, 103
travel angle. *See* gun angle
travel-down job, 505
traveling up, 370
travel-up job, 503–505
troubleshooting guide
    flux cored arc welding, 765–766 (table)
    gas metal arc welding, 728 (table), 796 (table)
    gas tungsten arc welding, 612 (table)
    MIG/MAG process and equipment, 729–730 (table)
    plasma arc cutting, 537 (table)
    shielded metal arc welding, 334–336 (table)
tundish, 62
tungsten, 82
tungsten carbide rods, 245
tungsten electrodes, 576
    ceriated, 576
    current ranges for, 577 (table)
    identification, 577 (table)
    lanthanated, 576
    thoriated, 576
tungsten inert gas welding (TIG), 551
    balled electrode shape, 578, 579
    cables and hoses, 579–580
    ceriated tungsten electrodes, 576
    current selection for, 567 (table)
    D.C. power sources for, 567
    development of, 5
    electrode holder sizes and electrode diameters, 607 (table)
    electrodes, 576–579
    equipment for, 554
    gas lens, 575
    lanthanated tungsten electrodes, 576
    power sources, 559–560, 567–568
    thoriated tungsten electrodes, 576
    TIG torches, 1028–1029
    TIG welding water supply, 578–579
    tungsten electrodes, 576–578
    tungsten preparation and maintenance, 577–578
    welding current characteristics, 560, 562
tungsten inert gas welding, hot wire, 580–590
    equipment for plasma arc welding, 584–585
    keyhole effect, 584
    nontransferred arc, 583–584
    plasma arc
        characteristics, 582–584
        cutting, 581
        surfacing, 589
        welding, 581, 583–584
    plasma needle arc welding, 585–588
    super TIG, 581
    transferred arc, 583
    welding gases, 585
tungsten steels, 90 (table)
turning rolls, 835–836
twist drill number sizes (in inches, millimeters, and centimeters), 1065 (table)
two-wire series power technique, 778

U

U-groove butt joint, 632
U-grooves, 185
ultraviolet radiation, 294
    threshold limits, 1031 (table)
undercutting, 497, 720, 939–940, 1078, 1080
Unified Numbering System (UNS), 92, 94 (table)
unit-frame rolls, 836
units of measure and sensors, 877 (table)
universal electric motors, 853–854
UNS. *See* Unified Numbering System
upset welding, 915 (table), 982 (table), 1077
upslope controls, 670
urea formaldehyde, 997
U.S. standard gauges, 1059 (table)

V

vacuum arc remelt (VAR) process, 97
vacuum furnaces, 55, 57–58
vanadium, 82
vapor condensation type solvents, 257

variable inductance controls, 670–671
variable slope machines, 670
variable voltage welding machine, 280
vertical cutting progressions, 179
vertical-down welding, 501
vertical-up welding, 501
vertical welding, 241
V-groove butt joint, 632. *See also* single V-butt joints
vibratory stress relieving, 108
Vickers tester, 918
views. *See also* drawings and drawing information
    detail views, 964–965
    piping drawings, 965
    sectional views, 964–965
    three-view drawings, 963–964
    types of, 964–965
visual inspection, 132–134, 205
    defects, 936–943
    methods of, 490
    palm organizer, 132
    of straight line and bevel cutting, 196
    weld gauges, 943
voids, 719
voltage sensing feeders, 675–676
volt-ampere characteristic (curve), 280
volt-ampere meters, 285
volts, 280

W

wandering sequences, 104, 106
warpage, 720
warping (contraction of weld deposit), 102
water jet cutting, 825–826
weaved beading, 120
    in flat position, 343–345, 347–349
    in overhead position, 391–393
    in vertical position, 372–374
weaving, 1080
weldability, 75
weld-all-around symbol, 983
weld appearance, satisfactory, 205
weld backing, 776–777
weld contamination, 555
weld cracking, 720. *See also* cracks and cracking
welded-pipe applications and advantages, 473–475
Welder Performance Qualification Test (WPQ), 486
welder qualification tests, 486–490
    code pipe welding, 490
    for fillet welds, 924 (table), 926 (table)
    for groove welds, 921 (table), 925 (table)
    production welding qualified by plate or pipe welds, 925 (table)
welding
    definition, 2
    market segments of, 279
    with shielded metal arc electrodes, 334–338

welding associations, 7
welding cable, size selection, 291 (table), 1028 (table)
welding current, 310
    alternating current welding, 563
    arc rectification with gas tungsten arc cutting, 566
    arc starting difficulties, 566
    arc starting methods, TIG welding, 564
    direct current electrode, 562
    high frequency arc starting, 565–566
    rectification, 566
    reverse polarity, 560
    square wave current welding, 563–564
    straight polarity, 560
welding direction, 106
welding fixtures, 833–834
welding gases, 139–147. *See also* fuel gases; inert gases and gas mixtures; oxygen; shielding gases
welding guns and accessories cable assemblies, 680–681
    nozzles, 680
    push-pull wire-feed system, 677
welding lenses. *See* eye protection; lenses
welding machines
    automatic submerged arc welding, 674
    basic types of, 289 (table)
    constant voltage welding machines, 673–674
    flux cored wire, 674
    maintenance of, 284–285
    shielded metal arc welding, 280
    short-circuiting process gas, 674
    size of, 284
    spray-arc welding, 674
welding occupations, 9–10
welding operator qualification
    fillet welds, 926 (table)
    groove welds, 925 (table)
welding positioners, 106, 835
welding positions
    for groove welds, 1073
    for fillet welds, 1073
welding procedure for Test 4, 459 (table)
Welding Procedure Qualification Record (WPQR), 484
Welding Procedure Specification (WPS), 484
welding symbols, 981–982 (table), 987–988. *See also* symbols
welding technique. *See also* specific joints
    aluminum, 240–241
    autogenous welding, 213–214, 217–219, 616
    bend testing, 620–621, 623
    butt joint, 621–623
    carbon and stainless steel, 623–627
    corner joints, 617–618
    fillet welding, 771
    flux cored arc welding, 770–771
    gas metal arc welding, 726–727
    gas tungsten arc welding, 611–627
    groove welding, 764
    lap joints, 618–619

welding technique.—*cont.*
　　parallel joints, 616–618
　　submerged arc welding, 780–781
　　T-joints, 619–621
　　weld beading on flat plate, 613–616
welding tips, gas pressures for, 156 (table)
welding variables
　　electrode extension, 762–763
　　flux cored arc welding, 762–763
　　gun angle, 762
weld face, 1078
weld gauges, 943
weld length, 119–120
weld metal cracking, 941–942
　　crater cracks, 942
　　longitudinal cracks, 942
　　transverse cracks, 941
weld metal zone, 1084
weld positions
　　for aluminum, 241
　　and electrode selection, 312
　　flat position (number 1), 120
　　horizontal position (number 2), 120–121
　　overhead position (number 4), 121
　　pipe positions, 122
　　pipe weld designations, 121
　　plate positions, 122
　　plate weld designations, 121
　　test positions, 123
　　vertical position (number 3), 121
welds
　　profiles, 126–129
　　reinforcement, 205
　　removal, 544–546
　　root, 1078
　　throat, 1078
　　toe, 1078
　　types, 113–114
weld size
　　actual throat size, 117–118
　　complete joint penetration welds, 114
　　concave fillet welding, 116–117
　　contour, 116
　　convex fillet weld, 116
　　effective throat size, 117
　　equal leg fillet, 119
　　fillet welds, 116–119
　　flat contour fillet, 116
　　groove welds, 114
　　partial joint penetration welds, 114
　　seal welds, 114–115
　　theoretical throat size, 117
　　unequal leg fillet weld, 118–119
whipping technique, 392
whiskers, 719
Willson, Thomas, L., 138
wire diameter *vs.* cross section volume
　　　　change, 692 (table)
　　and flux classifications, 774–775
wire feeders (wire drives) arc length and amperage
　　　　settings, 677 (table)
　　voltage sensing feeders, 675–676
wireless foot controls, 572
work cables, 291–292
work clamps, 292–293
work-holding devices
　　backing materials, 844
　　headstock-tailstock positioners, 837
　　magnetic grip fixtures, 840–841
　　manipulators, 837
　　orbital welding machine, 842, 843
　　seamers, 838
　　side beam carriage, 838, 840
　　track and trackless carriage systems, 841–842
　　turning rolls, 835–836
　　turntables, 837–838
　　weld elevator, 840
　　weld fixtures, 833–834
　　weld grippers, 836
working drawings, 951
WPS (weld qualifications). *See* qualifications for welding
Wrought steel, 62

Y

yield point, 77, 100
Y-joints, 806

Z

zero drag cutting, 182
zinc, 252
zinc-aluminum solders, 253
zirconium, 748